Get the most out of each worked example by using all of its features.

EXAMPLE 1   Here, we state the given problem.

Strategy   Then, we explain what will be done to solve the problem.

Why   Next, we explain why it will be done this way.

Solution   The steps that follow show how the problem is solved by using the given strategy.

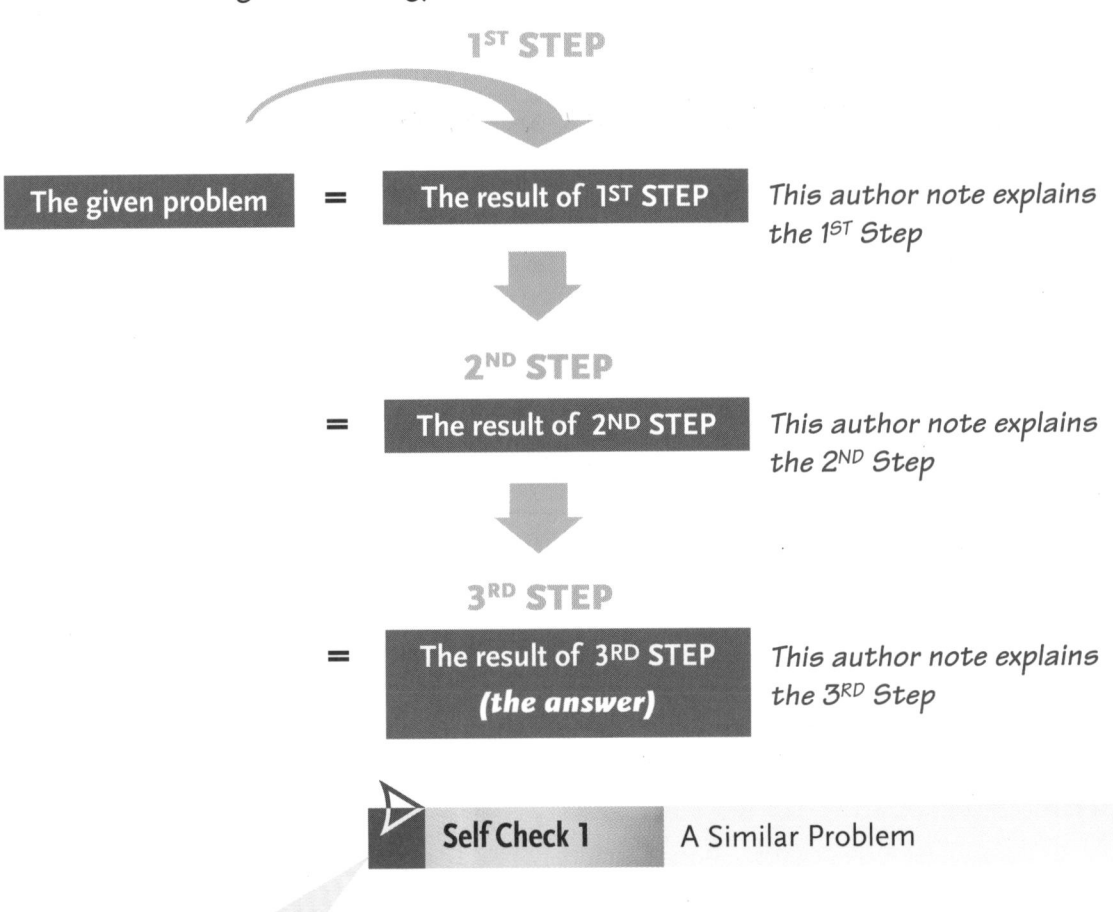

1ST STEP

| The given problem | = | The result of 1ST STEP | This author note explains the 1ST Step

2ND STEP

= | The result of 2ND STEP | This author note explains the 2ND Step

3RD STEP

= | The result of 3RD STEP **(the answer)** | This author note explains the 3RD Step

Self Check 1   A Similar Problem

After reading the example, try the Self Check problem to test your understanding. The answer is given at the end of the section, right before the Study Set.

Now Try   Problem 45

After you work the Self Check, you are ready to try a similar problem in the Guided Practice section of the Study Set.

Elementary and Intermediate Algebra

Alan S. Tussy | R. David Gustafson

CENGAGE
Learning™

Australia • Brazil • Japan • Korea • Mexico • Singapore • Spain • United Kingdom • United States

CENGAGE
Learning™

Elementary and Intermediate Algebra

Alan S. Tussy | R. David Gustafson

Executive Editors:
Michele Baird
Maureen Staudt
Michael Stranz

Project Development Manager:
Linda deStefano

Senior Marketing Coordinators:
Sara Mercurio
Lindsay Shapiro

Senior Production / Manufacturing Manager:
Donna M. Brown

PreMedia Services Supervisor:
Rebecca A. Walker

Rights & Permissions Specialist:
Kalina Hintz

Cover Image:
Getty Images*

* Unless otherwise noted, all cover images used by Custom Solutions, a part of Cengage Learning, have been supplied courtesy of Getty Images with the exception of the Earthview cover image, which has been supplied by the National Aeronautics and Space Administration (NASA).

For product information and technology assistance, contact us at
Cengage Learning Customer & Sales Support, 1-800-354-9706

For permission to use material from this text or product, submit all requests online at **cengage.com/permissions**
Further permissions questions can be emailed to
permissionrequest@cengage.com

ISBN-13: 978-0-495-73975-3

ISBN-10: 0-495-73975-8

Cengage Learning
5191 Natorp Boulevard
Mason, Ohio 45040
USA

Cengage Learning is a leading provider of customized learning solutions with office locations around the globe, including Singapore, the United Kingdom, Australia, Mexico, Brazil, and Japan. Locate your local office at:
international.cengage.com/region

Cengage Learning products are represented in Canada by Nelson Education, Ltd.

For your lifelong learning solutions, visit **custom.cengage.com**

Visit our corporate website at **cengage.com**

Printed in Canada
1 2 3 4 5 6 7 12 11 10 09 08

In memory of my mother, Jeanene,
and in honor of my dad, Bill.

—AST

In memory of my teacher and mentor,
Professor John Finch.

—RDG

CONTENTS

PREFACE

Elementary and Intermediate Algebra, Fourth Edition, is more than a simple upgrade of the third edition. Substantial changes have been made to the example structure, the Study Sets, and the pedagogy. Throughout the process, the objective has been to ease teaching challenges and meet students' educational needs.

Algebra, for many of today's developmental math students, is like a foreign language. They have difficulty translating the words, their meanings, and how they apply to problem solving. With these needs in mind (and as educational research suggests), the fundamental goal is to have students read, write, think, and speak using the *language of algebra*. Instructional approaches that include vocabulary, practice, and well-defined pedagogy, along with an emphasis on reasoning, modeling, communication, and technology skills have been blended to address this need.

The most common student question as they watch their instructors solve problems and as they read the textbook is . . . *Why?* The new fourth edition addresses this question in a unique way. Experience teaches us that it's not enough to know *how* a problem is solved. Students gain a deeper understanding of algebraic concepts if they know *why* a particular approach is taken. This instructional truth was the motivation for adding a **Strategy** and **Why** explanation to the solution of each worked example. The fourth edition now provides, on a consistent basis, a concise answer to that all-important question: *Why?*

This is just one of several changes in this revision, and we trust that all of them will make the course a better experience for both instructor and student.

NEW TO THIS EDITION

- New Example Structure
- New Chapter Opening Applications
- New *Study Skills Workshops*
- New Chapter Objectives
- New *Guided Practice* and *Try It Yourself* sections in the *Study Sets*
- New End-of-Chapter Organization

Chapter Openers Answering The Question: When Will I Use This?

Have you heard this question before? Instructors are asked this question time and again by students. In response, we have written chapter openers called *From Campus to Careers*. This feature highlights vocations that require various algebraic skills. Designed to inspire career exploration, each includes job outlook, educational requirements, and annual earnings information. Careers presented in the openers are tied to an exercise found later in the *Study Sets*.

Examples That Tell Students Not Just How, But WHY

Why? That question is often asked by students as they watch their instructor solve problems in class and as they are working on problems at home. It's not enough to know how a problem is solved. Students gain a deeper understanding of the algebraic concepts if they know why a particular approach was taken. This instructional truth was the motivation for adding a *Strategy* and *Why* explanation to each worked example.

Examples That Offer Immediate Feedback

Each example includes a *Self Check*. These can be completed by students on their own or as classroom lecture examples, which is how Alan Tussy uses them. Alan asks selected students to read aloud the *Self Check* problems as he writes what the student says on the board. The other students, with their books open to that page, can quickly copy the *Self Check* problem to their notes. This speeds up the note-taking process and encourages student participation in his lectures. It also teaches students how to read mathematical symbols. Each *Self Check* answer is printed adjacent to the corresponding problem in the Annotated Instructor's Edition for easy reference. *Self Check* solutions can be found at the end of each section in the student edition before the *Study Sets* begin.

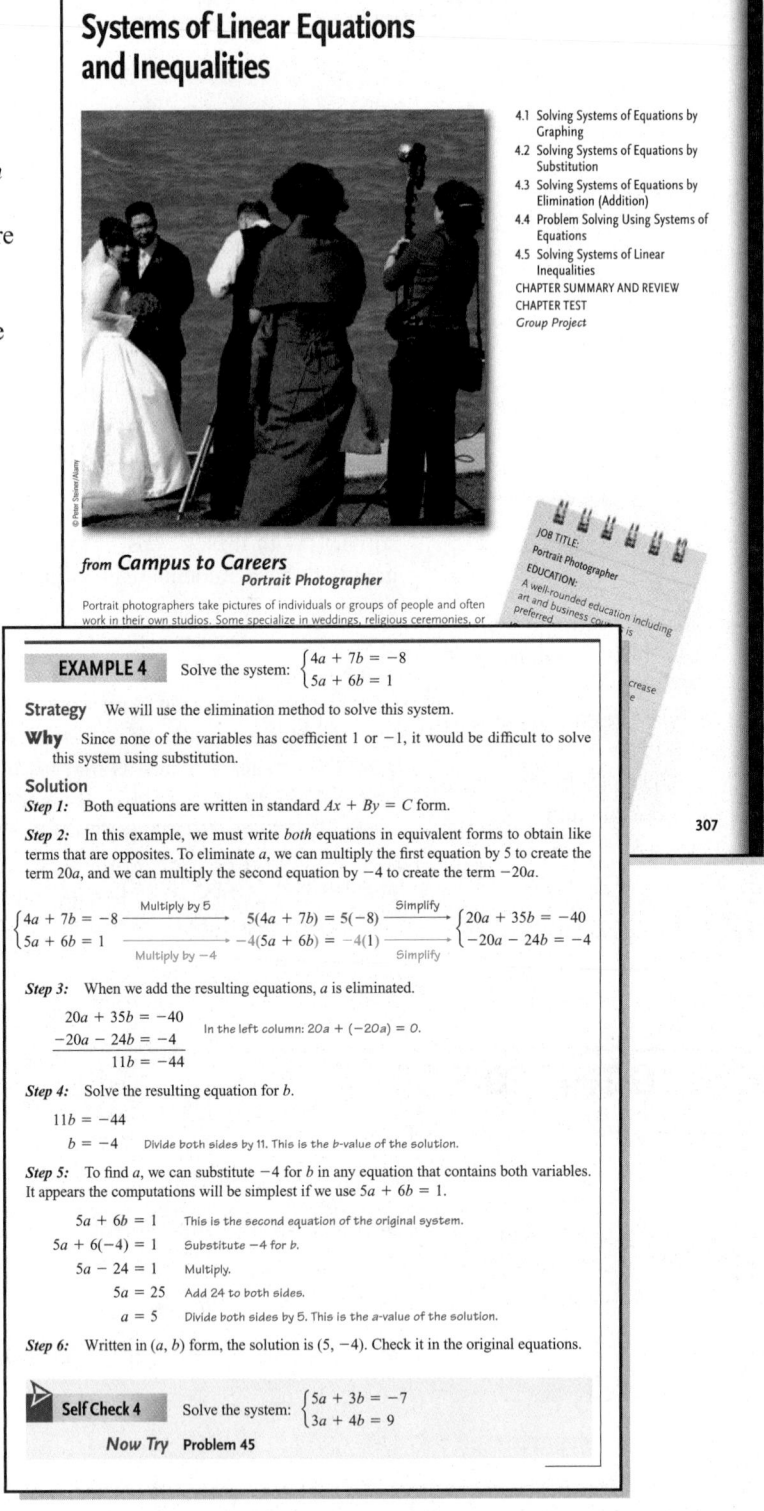

Systems of Linear Equations and Inequalities

4.1 Solving Systems of Equations by Graphing
4.2 Solving Systems of Equations by Substitution
4.3 Solving Systems of Equations by Elimination (Addition)
4.4 Problem Solving Using Systems of Equations
4.5 Solving Systems of Linear Inequalities
CHAPTER SUMMARY AND REVIEW
CHAPTER TEST
Group Project

from **Campus to Careers**
Portrait Photographer

Portrait photographers take pictures of individuals or groups of people and often work in their own studios. Some specialize in weddings, religious ceremonies, or

JOB TITLE:
Portrait Photographer
EDUCATION:
A well-rounded education including art and business co...
preferred...

307

EXAMPLE 4   Solve the system: $\begin{cases} 4a + 7b = -8 \\ 5a + 6b = 1 \end{cases}$

Strategy   We will use the elimination method to solve this system.

Why   Since none of the variables has coefficient 1 or -1, it would be difficult to solve this system using substitution.

Solution

Step 1:   Both equations are written in standard $Ax + By = C$ form.

Step 2:   In this example, we must write *both* equations in equivalent forms to obtain like terms that are opposites. To eliminate a, we can multiply the first equation by 5 to create the term $20a$, and we can multiply the second equation by -4 to create the term $-20a$.

$\begin{cases} 4a + 7b = -8 \\ 5a + 6b = 1 \end{cases}$ $\xrightarrow[\text{Multiply by} -4]{\text{Multiply by 5}}$ $\begin{array}{l} 5(4a + 7b) = 5(-8) \\ -4(5a + 6b) = -4(1) \end{array}$ $\xrightarrow{\text{Simplify}}$ $\begin{cases} 20a + 35b = -40 \\ -20a - 24b = -4 \end{cases}$

Step 3:   When we add the resulting equations, a is eliminated.

$\begin{array}{r} 20a + 35b = -40 \\ -20a - 24b = -4 \\ \hline 11b = -44 \end{array}$ In the left column: $20a + (-20a) = 0$.

Step 4:   Solve the resulting equation for b.

$11b = -44$

$b = -4$ Divide both sides by 11. This is the b-value of the solution.

Step 5:   To find a, we can substitute -4 for b in any equation that contains both variables. It appears the computations will be simplest if we use $5a + 6b = 1$.

$5a + 6b = 1$ This is the second equation of the original system.

$5a + 6(-4) = 1$ Substitute -4 for b.

$5a - 24 = 1$ Multiply.

$5a = 25$ Add 24 to both sides.

$a = 5$ Divide both sides by 5. This is the a-value of the solution.

Step 6:   Written in (a, b) form, the solution is $(5, -4)$. Check it in the original equations.

Self Check 4   Solve the system: $\begin{cases} 5a + 3b = -7 \\ 3a + 4b = 9 \end{cases}$

Now Try   **Problem 45**

Examples That Ask Students To Try

Each example ends with a *Now Try* problem. These are the final step in the learning process. Each one is linked to similar problems found within the *Guided Practice* section of the *Study Sets*.

Emphasis on Study Skills

Each chapter begins with a *Study Skills Workshop*. Instead of simple suggestions printed in the margins, each workshop contains a *Now Try This* section offering students actionable skills, assignments, and projects that will impact their study habits throughout the course.

Useful Objectives Help Keep Students Focused

Objectives are now numbered at the start of each section to focus students' attention on the skills that they will learn as they work through the section. When each objective is introduced, the number and heading will appear again to remind them of the objective at hand.

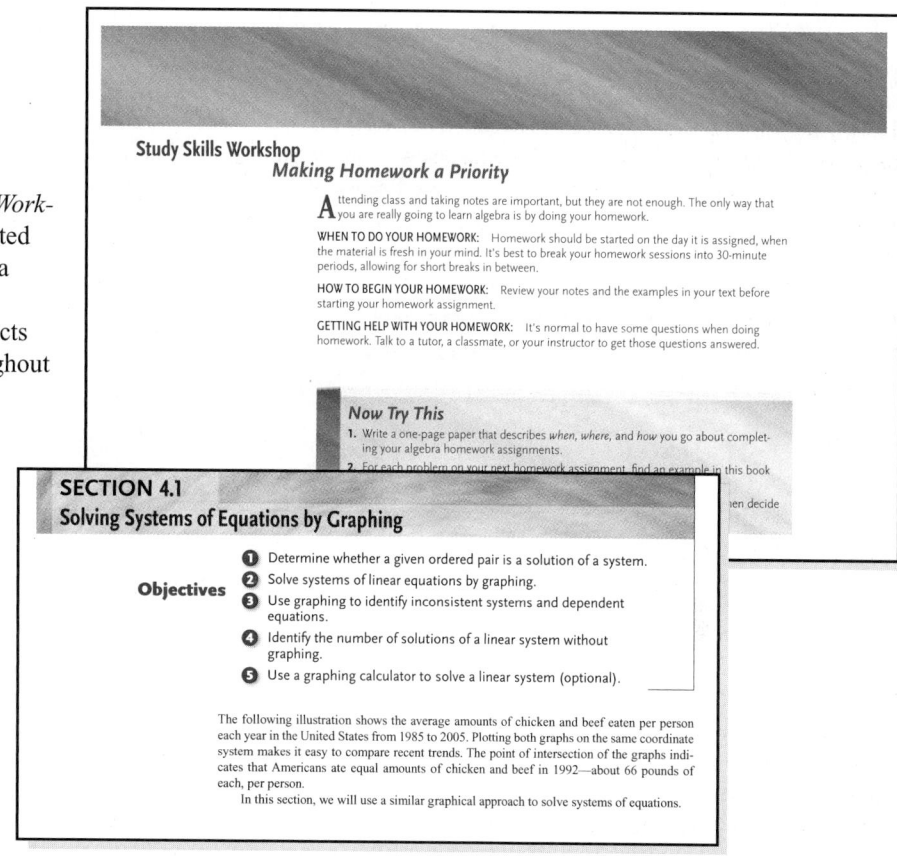

Heavily Revised Study Sets

The *Study Sets* have been thoroughly revised to ensure every concept is covered even if the instructor traditionally assigns every other problem. Particular attention was paid to developing a gradual level of progression.

Guided Practice

All of the problems in the *Guided Practice* portion of the *Study Sets* are linked to an associated worked example from that section. This feature will promote student success by referring them to the proper example(s) if they encounter difficulties solving homework problems.

Try It Yourself

To promote problem recognition, some *Study Sets* now include a collection of *Try It Yourself* problems that do not have the example linking. The problem types are thoroughly mixed and are not linked, giving students an opportunity to practice decision making and strategy selection as they would when taking a test or quiz.

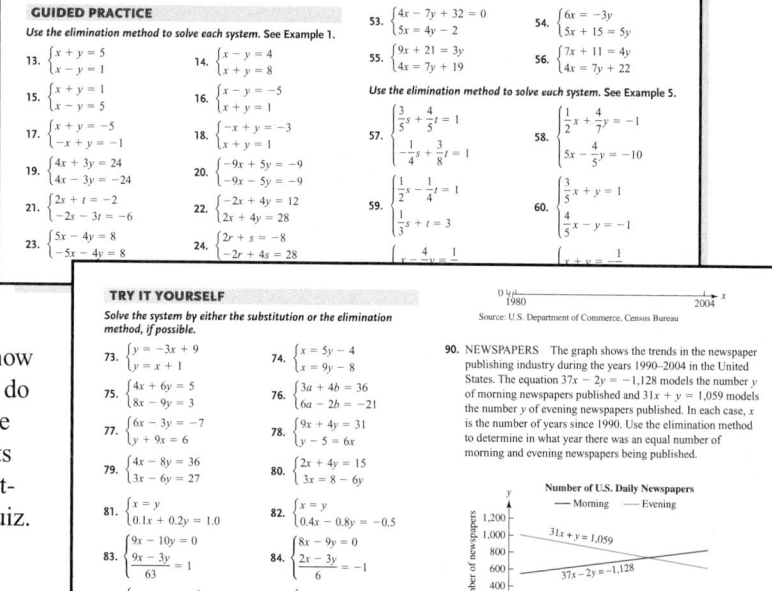

Comprehensive End-of-Chapter Summary with Integrated Chapter Review

The end-of-chapter material has been redesigned to function as a complete study guide for students. New Chapter Summaries that include definitions, concepts, and examples, by section, have been written. Review problems for each section have been placed after each section summary.

CHAPTER 4
Summary & Review

SECTION 4.1   Solving Systems of Equations by Graphing

DEFINITIONS AND CONCEPTS	EXAMPLES
When two equations are considered at the same time, we say that they form a **system of equations**. A **solution of a system** of equations in two variables is an ordered pair that satisfies both equations of the system.	Is $(4, 3)$ a solution of the system $\begin{cases} x + y = 7 \\ x - y = 5 \end{cases}$? To answer this question, we substitute 4 for x and 3 for y in each equation. $x + y = 7 \qquad\qquad x - y = 5$ $4 + 3 \stackrel{?}{=} 7 \qquad\qquad 4 - 3 \stackrel{?}{=} 5$ $\qquad 7 = 7 \quad$ True $\qquad\quad 1 = 5 \quad$ False Although $(4, 3)$ satisfies the first equation, it does not satisfy the second. Because it does not satisfy both equations, it is not a solution of the system.

To **solve a system graphically:**

1. Graph each equation on the sa[...] system.

2. Determine the coordinates of [...] section of the graphs. That or[...] solution.

3. Check the solution in each eq[...] nal system.

SECTION 4.3   Solving Systems of Equations by Elimination (Addition)

DEFINITIONS AND CONCEPTS	EXAMPLES
To solve a system of equations in x and y using **elimination (addition):** 1. Write each equation in the standard $Ax + By = C$ form. 2. Multiply one (or both) equations by nonzero quantities to make the coefficients of x (or y) opposites. 3. Add the equations to eliminate the terms involving x (or y). 4. Solve the equation obtained in step 3. 5. Find the value of the other variable by substituting the value of the variable found in step 4 into any equation containing both variables. 6. Check the solution in the equations of the original system. With the elimination method, the basic objective is to obtain two equations whose sum will be one equation in one variable. If in step 3 both variables drop out and a false statement results, the system has **no solution.** If a true statement results, the system has **infinitely many solutions.**	Use elimination to solve: $\begin{cases} 2x - 3y = 4 \\ 3x + y = -5 \end{cases}$ **Step 1:** Both equations are written in $Ax + By = C$ form. **Step 2:** Multiply the second equation by 3 so that the coefficients of y are opposites. **Step 3:** $\quad 2x - 3y = 4$ $\quad \underline{9x + 3y = -15}$ $\quad 11x \quad\;\; = -11 \qquad$ Add the like terms, column by column. **Step 4:** Solve for x. $\quad 11x = -11$ $\quad\;\; x = -1 \qquad$ Divide both sides by 11. **Step 5:** Find y. $\quad 3x + y = -5 \qquad$ This is the second equation. $\quad 3(-1) + y = -5 \qquad$ Substitute -1 for x. $\quad\qquad\quad y = -2$ The solution is $(-1, -2)$. **Step 6:** Check $\quad 2x - 3y = 4 \qquad\qquad\qquad 3x + y = -5$ $\quad 2(-1) - 3(-2) \stackrel{?}{=} 4 \qquad\quad 3(-1) + (-2) \stackrel{?}{=} -5$ $\quad\quad -2 + 6 \stackrel{?}{=} 4 \qquad\qquad\quad -3 - 2 \stackrel{?}{=} -5$ $\qquad\qquad 4 = 4 \;$ True $\qquad\qquad -5 = -5 \;$ True

TRUSTED FEATURES

- **The Study Sets** found in each section offer a multifaceted approach to practicing and reinforcing the concepts taught in each section. They are designed for students to methodically build their knowledge of the section concepts, from basic recall to increasingly complex problem solving, through reading, writing, and thinking mathematically.

  Vocabulary—Each Study Set begins with the important Vocabulary discussed in that section. The fill-in-the-blank vocabulary problems emphasize the main concepts taught in the chapter and provide the foundation for learning and communicating the language of algebra.

  Concepts—In Concepts, students are asked about the specific subskills and procedures necessary to successfully complete the practice problems that follow.

  Notation—In Notation, the students review the new symbols introduced in a section. Often, they are asked to fill in steps of a sample solution. This helps to strengthen their ability to read and write mathematics and prepares them for the practice problems by modeling solution formats.

  Guided Practice—The problems in Guided Practice are linked to an associated worked example from that section. This feature will promote student success by referring them to the proper examples if they encounter difficulties solving homework problems.

  Try It Yourself—To promote problem recognition, the Try It Yourself problems are thoroughly mixed and are not linked, giving students an opportunity to practice decision-making and strategy selection as they would when taking a test or quiz.

  Applications—The Applications provide students the opportunity to apply their newly acquired algebraic skills to relevant and interesting real-life situations.

  Writing—The Writing problems help students build mathematical communication skills.

  Review—The Review problems consist of randomly selected problems from previous chapters. These problems are designed to keep students' successfully mastered skills fresh and at the forefront of their minds before moving on to the next section.

  Challenge Problems—The Challenge Problems provide students with an opportunity to stretch themselves and develop their skills beyond the basics. Instructors often find these to be useful as extra-credit problems.

- **Detailed Author Notes** that guide students along in a step-by-step process continue to be found in the solutions to every example.

- **The Language of Algebra** boxes draw connections between mathematical terms and everyday references to reinforce the language of algebra thread that runs throughout the text.

- **The Notation, Success Tips, Caution,** and **Calculators** boxes offer helpful tips to reinforce correct mathematical notation, improve students' problem-solving abilities, warn students of potential pitfalls and increase clarity, and offer tips on using scientific calculators.

- **Using Your Calculator** (formerly called Accent on Technology) sections are designed for instructors who wish to use calculators as part of the instruction in this course. These sections introduce keystrokes and show how scientific and graphing calculators can be used to solve problems. In the Study Sets, icons are used to denote problems that require a graphing calculator.

- **Strategic use of color** has been implemented within the new design to help the visual learner.

- **Chapter Tests** are available at the end of every chapter as preparation for the class exam.

- **The Cumulative Review** following the end-of-chapter material keeps students' skills sharpened before moving on to the next chapter. Each problem is now linked to the associated section from which the problem came for ease of reference. The final Cumulative Review, found at the end of the last chapter, is often used by instructors as a Final Exam Review.

CHANGES TO THE TABLE OF CONTENTS

Based on feedback from colleagues and users of the third edition, the following changes have been made to the table of contents in an effort to further streamline the text and make it even easier to use.

- Chapter 2 topics have been reorganized and the section *Simplifying Algebraic Expressions Using Properties of Real Numbers* has been moved from Chapter 2 to Section 1.9.

  2.1 *Solving Equations Using Properties of Equality*
  2.2 *More about Solving Equations*
  2.3 *Applications of Percent* (Commission and discount problems were added)
  2.4 *Formulas*
  2.5 *Problem Solving* (Consecutive integer, commission, and set-up fee/cost per item problems were added)
  2.6 *More about Problem Solving*
  2.7 *Solving Inequalities*

- Parallel and perpendicular lines are now introduced in Section 3.4 *Slope and Rate of Change*.

- For those instructors wishing to discuss functions in the first half of the course, Chapter 3 now includes Section 3.8, Introduction to Functions. This topic is a natural fit after studying linear equations in two variables. This section can, however, be omitted without consequence because the topic of function is reintroduced in Section 8.8 of the transition chapter.

- To give more attention to applications, the material at the end of Chapter 6: Factoring and Quadratic Equations has been separated into two sections. Section 6.7 now focuses solely on solving quadratic equations by factoring while the newly written Section 6.8 is exclusively devoted to applications of quadratic equations.

- Chapter 8, Transition to Intermediate Algebra, has been reorganized slightly. Compound inequalities, formerly introduced in Section 8.1, are now discussed in Section 8.2. The review of rational expressions, Section 8.6, now includes a review of rational equations. The introduction to functions, formerly found in Sections 8.7 and 8.8, has been incorporated into one section, Section 8.8.

- Section 11.1: *Algebra and Composition of Functions* now includes examples and problems where sum, difference, product, and quotient functions are evaluated graphically.

- There is greater emphasis on *f(x)* function notation in Chapter 11: *Exponential and Logarithmic Functions*.

- To give more attention to applications, Section 12.2 of the third edition has been separated into two sections. Section 12.2 now focuses solely on solving systems of equations in three variables while the newly written Section 12.3 is exclusively devoted to problem solving using systems of three equations.

- Section 14.4: Permutations and Combinations and Section 14.5: Probability have been deleted and are now available online.

GENERAL REVISIONS AND OVERALL DESIGN

- We have edited the prose so that it is even more clear and concise.

- Strategic use of color has been implemented within the new design to help the visual learner.

- Added color in the solutions highlight strategic steps and improve readability.

- We have updated all data and graphs and have added scaling to all axes in all graphs.

- We have added more real-world applications and deleted some of the more "contrived" problems.

- We have included more problem-specific photographs.

INSTRUCTOR RESOURCES

Print Ancillaries

INSTRUCTOR'S RESOURCE BINDER (0-495-38982-X)
Maria H. Andersen, *Muskegon Community College*

NEW! Offered exclusively with Tussy/Gustafson. Each section of the main text is discussed in uniquely designed Teaching Guides containing instruction tips, examples, activities, worksheets, overheads, assessments, and solutions to all worksheets and activities.

COMPLETE SOLUTIONS MANUAL (0-495-38977-3)
Kristy Hill, *Hinds Community College*

The Complete Solutions Manual provides worked-out solutions to all of the problems in the text.

TEST BANK (0-495-38978-1)
Carol M. Walker & David J. Walker, *Hinds Community College*

Drawing from hundreds of text-specific questions, an instructor can easily create tests that target specific course objectives. The Test Bank includes multiple tests per chapter, as well as final exams. The tests are made up of a combination of multiple-choice, free-response, true/false, and fill-in-the-blank questions.

ANNOTATED INSTRUCTOR'S EDITION (0-495-38974-9)
The Instructor's Edition provides the complete student text with answers next to each respective exercise.

Electronic Ancillaries

WebAssign ENHANCED WEBASSIGN (0-495-38984-6)

Instant feedback and ease of use are just two reasons why WebAssign is the most widely used homework system in higher education. WebAssign's homework delivery system allows you to assign, collect, grade, and record homework assignments via the web. And now, this proven system has been enhanced to include links to textbook sections, video examples, and problem-specific tutorials. Enhanced WebAssign is more than a homework system—it is a complete learning system for math students.

CENGAGENOW™ (0-495-39455-6)

CengageNOW™ is an online teaching and learning resource that gives you more control in less time and delivers the results you want—NOW.

POWERLECTURE™: A 1-STOP MICROSOFT® POWERPOINT® TOOL (0-495-55701-3)

NEW! The ultimate multimedia manager for your course needs. The PowerLecture CD-ROM includes the Complete Solutions Manual, ExamView®, JoinIn™, and custom PowerPoint® lecture slides.

TEXT SPECIFIC DVDs (0-495-38979-X)

These text specific DVDs provide additional guidance and support to students when they are preparing for an upcoming quiz or exam.

STUDENT RESOURCES

Print Ancillaries

STUDENT WORKBOOK (0-495-55478-2)
Maria H. Andersen, *Muskegon Community College*

NEW! Get a head start. The Student Workbook contains all of the Assessments, Activities, and Worksheets from the Instructor's Resource Binder for classroom discussions, in-class activities, and group work.

STUDENT SOLUTIONS MANUAL (0-495-38976-5)
Alexander H. Lee, *Hinds Community College*

The Student Solutions Manual provides worked-out solutions to the odd-numbered problems in the text.

Electronic Ancillaries

WebAssign ENHANCED WEBASSIGN (0-495-38984-6)

Get instant feedback on your homework assignments with Enhanced WebAssign (assigned by your instructor). This online homework system is easy to use and includes helpful links to textbook sections, video examples, and problem-specific tutorials.

INSTANT ACCESS CODE, CENGAGENOW™ (0-495-39460-2)

Instant Access gives students without a new copy of Tussy/Gustafson's *Elementary Algebra, Fourth Edition,* one access code to all available technology associated with this textbook. CengageNOW, a powerful and fully integrated teaching and learning system, provides instructors and students with unsurpassed control, variety, and all-in-one utility. CengageNOW ties together the fundamental learning activities: diagnostics, tutorials, homework, personalized study, quizzing, and testing. Personalized Study is a learning companion that helps students gauge their unique study needs and makes the most of their study time by building focused personalized learning plans that reinforce key concepts. Pre-Tests give students an initial assessment of their knowledge. Personalized study plans, based on the students' answers to the Pre-Test questions, outline key elements for review. Post-Tests assess student mastery of core chapter concepts. Results can even be e-mailed to the instructor!

PRINTED ACCESS CARD, CENGAGENOW™ (0-495-39459-9)

This printed access card provides entrance to all the content that accompanies Tussy/Gustafson's *Elementary Algebra, Fourth Edition*, within CengageNOW.

WEBSITE *academic.cengage.com/math/tussy*
Visit us on the web for access to a wealth of free learning resources, including tutorials, final exams, chapter outlines, chapter reviews, web links, videos, flashcards, and more!

ACKNOWLEDGMENTS

We want to express our gratitude to Steve Odrich, Maria H. Andersen, Diane Koenig, Alexander Lee, Ed Kavanaugh, Karl Hunsicker, George Carlson, Jim Cope, Arnold Kondo, John McKeown, Kent Miller, Donna Neff, Eric Robitoy, Maryann Rachford, Chris Scott, Rob Everest, Cathy Gong, Dave Ryba, Terry Damron, Marion Hammond, Lin Humphrey, Doug Keebaugh, Robin Carter, Tanja Rinkel, Bob Billups, Jeff Cleveland, Jo Morrison, Sheila White, Jim McClain, Paul Swatzel, Bill Tussy, Liz Tussy, and the Citrus College Library staff (including Barbara Rugeley) for their help with this project. Your encouragement, suggestions, and insight have been invaluable to us.

We would also like to express our thanks to the Brooks/Cole editorial, marketing, production and design staff for helping us craft this new edition: Charlie Van Wagner, Danielle Derbenti, Greta Kleinert, Laura Localio, Lynh Pham, Cassandra Cummings, Donna Kelley, Sam Subity, Cheryll Linthicum, Vernon Boes, and Graphic World.

Additionally, we would like to say that authoring a textbook is a tremendous undertaking. A revision of this scale would not have been possible without the thoughtful feedback and support from the following colleagues listed below. Their contributions to this edition have shaped this revision in countless ways.

Alan S. Tussy
R. David Gustafson

Advisory Board

Kim Caldwell, Volunteer State Community College
Peter Embalabala, Lincoln Land Community College
John Garlow, Tarrant Community College–Southeast Campus
Becki Huffman, Tyler Junior College
Mary Legner, Riverside Community College
Ann Loving, J. Sargeant Reynolds Community College

Trudy Meyer, El Camino College
Carol Ann Poore, Hinds Community College
Jill Rafael, Sierra College
Pamelyn Reed, Cy-Fair College
Patty Sheeran, McHenry Community College
Valerie Wright, Central Piedmont Community College
Loris Zucca, Kingwood College

Reviewers

Maria Andersen, Muskegon Community College

Scott Barnett, Henry Ford Community College

David Behrman, Somerset Community College

Jeanne Bowman, University of Cincinnati

Carol Cheshire, Macon State College

Suzanne Doviak, Old Dominion University

Peter Embalabala, Lincoln Land Community College

Joan Evans, Texas Southern University

Rita Fielder, University of Central Arkansas

Anissa Florence, Jefferson Community and Technical College

Pat Foard, South Plains College

Tom Fox, Cleveland State Community College

Heng Fu, Thomas Nelson Community College

Kim Gregor, Delaware Technical Community College–Wilmington

Haile Kebede Haile, Minneapolis Community and Technical College

Jennifer Hastings, Northeast Mississippi Community College

Kristy Hill, Hinds Community College

Laura Hoye, Trident Technical College

Becki Huffman, Tyler Junior College

Angela Jahns, North Idaho College

Cynthia Johnson, Heartland Community College

Ann Loving, J. Sargeant Reynolds Community College

Lynette King, Gadsden State Community College

Mike Kirby, Tidewater Community College

Mary Legner, Riverside Community College

Wayne (Paul) Lee, Saint Philip's College

Yixia Lu, South Suburban College

Keith Luoma, Augusta State University

Susan Meshulam, Indiana University/ Purdue University Indianapolis

Trudy Meyer, El Camino College

Molly Misko, Gadsden State Community College

Elsie Newman, Owens Community College

Charlotte Newsom, Tidewater Community College

Randy Nichols, Delta College

Stephen Nicoloff, Paradise Valley Community College

Charles Odion, Houston Community College

Jason Pallett, Longview Community College

Mary Beth Pattengale, Sierra College

Naeemah Payne, Los Angeles Community College

Carol Ann Poore, Hinds Community College

Jill Rafael, Sierra College

Pamela Reed, North Harris Montgomery Community College

Nancy Ressler, Oakton Community College

Emma Sargent, Tennessee State University

Ned Schillow, Lehigh Carbon Community College

Debra Shafer, University of North Carolina

Hazel Shedd, Hinds Community College

Donald Solomon, University of Wisconsin

John Squires, Cleveland State Community College

Robin Steinberg, Pima Community College

Eden Thompson, Utah Valley State College

Carol Walker, Hinds Community College

Diane Williams, Northern Kentucky University

Loris Zucca, Kingwood College

Class Testers

Candace Blazek, Anoka Ramsey Community College

Jennifer Bluth, Anoka Ramsey Community College

Vicki Gearhart, San Antonio College

Megan Goodwin, Anoka Ramsey Community College

Haile Haile, Minneapolis Community and Technical College

Vera Hu-Hyneman, SUNY–Suffolk Community College

Marlene Kutesky, Virginia Commonwealth University

Richard Leedy, Polk Community College

Wendiann Sethi, Seton Hall University

Eleanor Storey, Frontrange Community College

Cindy Thore, Central Piedmont Community College

Gowribalan "Ana" Vamadeva, University of Cincinnati

Cynthia Wallin, Central Virginia Community College

John Ward, Jefferson Community and Technical College

Focus Groups

Khadija Ahmed, Monroe Community College

Maria Andersen, Muskegon Community College

Chad Bemis, Riverside Community College

A. Elena Bogardus, Camden Community College

Carilynn Bouie, Cuyahoga Community College

Kim Brown, Tarrant Community College

Carole Carney, Brookdale Community College

Joe Castillo, Broward Community College

John Close, Salt Lake Community College

Chris Copple, Northwest State Community College

Mary Deas, Johnson County Community College

Maggie Flint, Northeast State

Douglas Furman, SUNY Ulster Community College

Abel Gage, Skagit Valley College

Amy Hoherz, Johnson County Community College

Pete Johnson, Eastern Connecticut State University

Ed Kavanaugh, Schoolcraft College

Leonid Khazanov, Borough of Manhattan Community College

MC Kim, Suffolk County Community College

Fred Lang, Art Institute of Washington

Hoat Le, San Diego Community College

Richard Leedy, Polk Community College

Daniel Lopez, Brookdale Community College

Ann Loving, J. Sargeant Reynolds Community College

Charles Odion, Houston Community College

Maggie Pasqua Viz, Brookdale Community College

Fred Peskoff, Borough of Manhattan Community College

Sheila Pisa, Riverside Community College–Moreno Valley

Jill Rafael, Sierra College

Christa Solheid, Santa Ana College

Jim Spencer, Santa Rosa Junior College

Teresa Sutcliffe, Los Angeles Valley College

Rose Toering, Kilian Community College

Judith Wood, Central Florida Community College

Mary Young, Brookdale Community College

Workshops

Andrea Adlman, Ventura College

Rodney Alford, Calhoun Community College

Maria Andersen, Muskegon Community College

Hamid Attarzadeh, Jefferson Community and Technical College

Victoria Baker, University of Houston–Downtown

Betty Barks, Lansing Community College

Susan Beane, University of Houston–Downtown

Barbara Blass, Oakland Community College

Charles A. Bower, St. Philip's College

Tony Craig, Paradise Valley Community College

Patrick Cross, University of Oklahoma

Archie Earl, Norfolk State University

Melody Eldred, State University of New York at Cobleskill

Joan Evans, Texas Southern University

Mike Everett, Santa Ana College

Betsy Farber, Bucks County Community College

Nancy Forrest, Grand Rapids Community College

Radu Georgescu, Prince George's Community College

Rebecca Giles, Jefferson State Community College

Thomas Grogan, Cincinnati State

Paula Jean Haigis, Calhoun Community College

Haile Haile, Minneapolis Community and Technical College

Kelli Jade Hammer, Broward Community College

Julia Hassett, Oakton Community College

Alan Hayashi, Oxnard College

Joel Helms, University of Cincinnati

Jim Hodge, Mountain State University

Jeffrey Hughes, Hinds Community College

Leslie Johnson, John C. Calhoun State Community College

Cassandra Johnson, Robeson Community College

Ed Kavanaugh, Schoolcraft College

Alex Kolesnik, Ventura College

Marlene Kustesky, Virginia Commonwealth University

Lider-Manuel Lamar, Seminole Community College

Roger Larson, Anoka Ramsey Community College

Alexander Lee, Hinds Community College, Rankin Campus

Richard Leedy, Polk Community College

Marcus McGuff, Austin Community College

Owen Mertens, Missouri State University

James Metz, Kapi'olani Community College

Pam Miller, Phoenix College

Tania Munding, Ohlone College

Charlie Naffziger, Central Oregon Community College

Oscar Neal, Grand Rapids Community College

Doug Nelson, Central Oregon Community College

Katrina Nichols, Delta College

Megan Nielsen, St. Cloud State University

Nancy Ressler, Oakton Community College

Elaine Richards, Eastern Michigan University

Harriette Roadman, New River Community College

Lilia Ruvalcaba, Oxnard College

Wendiann Sethi, Seton Hall University

Karen Smith, Nicholls State University

Donald Solomon, University of Wisconsin–Milwaukee

Frankie Solomon, University of Houston–Downtown

Michael Stack, South Suburban College

Kristen Starkey, Rose State College

Kristin Stoley, Blinn College

Eleanor Storey, Front Range Community College–Westminster Campus

Fariheh Towfiq, Palomar College

Gowribalan Vamadeva, University of Cincinnati

Beverly Vredevelt, Spokane Falls Community College

Andreana Walker, Calhoun Community College

Cynthia Wallin, Central Virginia Community College

John Ward, Kentucky Community and Technical College–Jefferson Community College

Richard Watkins, Tidewater Comunity College

Antoinette Willis, St. Philip's College

Nazar Wright, Guilford Technical Community College

Shishen Xie, University of Houston–Downtown

Catalina Yang, Oxnard College

Heidi Young, Bryant and Stratton College

Ghidei Zedingle, Normandale Community College

APPLICATIONS INDEX

Examples that are applications are shown with **boldface page numbers**.
Exercises that are applications are shown with lightface page numbers.

CHAPTER 1

An Introduction to Algebra

© AP/Wide World Photo

from *Campus to Careers*

Lead Transportation Security Officer

Since 9/11, Homeland Security is one of the fastest-growing career choices in the United States. A lead transportation security officer works in an airport where he or she searches passengers, screens baggage, reviews tickets, and determines staffing requirements. The job description calls for the ability to perform arithmetic computations correctly and solve practical problems by choosing from a variety of mathematical techniques such as formulas and percentages.

In **Problem 93** of **Study Set 1.5,** we will make some arithmetic computations using data from two of the busiest airports in the United States, Orlando International and New York La Guardia.

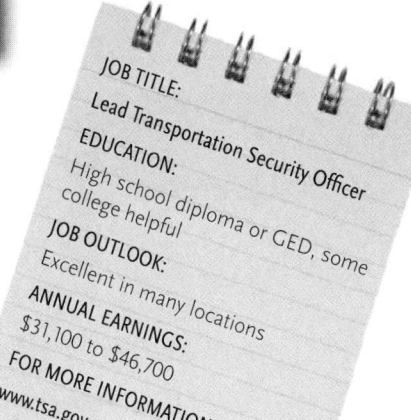

JOB TITLE:
Lead Transportation Security Officer
EDUCATION:
High school diploma or GED, some college helpful
JOB OUTLOOK:
Excellent in many locations
ANNUAL EARNINGS:
$31,100 to $46,700
FOR MORE INFORMATION:
www.tsa.gov

Study Skills Workshop
Committing to the Course

Starting a new course is exciting, but it might also make you a bit nervous. In order to be successful in your algebra class, you need a plan.

MAKE TIME FOR THE COURSE: As a general guideline, 2 hours of independent study time is recommended for every hour in the classroom.

KNOW WHAT IS EXPECTED: Read your instructor's syllabus thoroughly. It lists class policies about attendance, homework, tests, calculators, grading, and so on.

BUILD A SUPPORT SYSTEM: Know where to go for help. Take advantage of your instructor's office hours, your school's tutorial services, the resources that accompany this textbook, and the assistance that you can get from classmates.

Now Try This

Each of the forms referred to below can be found online at:
http://www.thomsonedu.com/math/tussy.

1. To help organize your schedule, fill out the *Weekly Planner Form*.
2. Review the class policies by completing the *Course Information Sheet*.
3. Use the *Support System Worksheet* to build your course support system.

SECTION 1.1
Introducing the Language of Algebra

Objectives

1. Read tables and graphs.
2. Use the basic vocabulary and notation of algebra.
3. Identify expressions and equations.
4. Use equations to construct tables of data.

Algebra is the result of contributions from many cultures over thousands of years. The word *algebra* comes from the title of the book *Ihm Al-jabr wa'l muqābalah,* written by an Arabian mathematician around A.D. 800. Using the vocabulary and notation of algebra, we can mathematically **model** many situations in the real world. In this section, we begin to explore the language of algebra by introducing some of its basic components.

1 Read Tables and Graphs.

In algebra, we use tables to show relationships between quantities. For example, the following table lists the number of bicycle tires a production planner must order when a given number of bicycles is to be manufactured. For a production run of, say, 300 bikes, we locate 300 in the left column and then scan across the table to see that the company must order 600 tires.

Bicycles to be manufactured	Tires to order
100	200
200	400
300	600
400	800

The Language of Algebra
Horizontal is a form of the word *horizon*. Think of the sun setting over the *horizon*. *Vertical* means in an upright position. Pro basketball player LeBron James' *vertical* leap measures more than 40 inches.

The information in the table can also be presented in a **bar graph,** as shown below. The **horizontal axis,** labeled "Number of bicycles to be manufactured," is scaled in units of 100 bicycles. The **vertical axis,** labeled "Number of tires to order," is scaled in units of 100 tires. The height of a bar indicates the number of tires to order. For example, if 200 bikes are to be manufactured, we see that the bar extends to 400, meaning 400 tires are needed.

Another way to present this information is with a **line graph.** Instead of using a bar to represent the number of tires to order, we use a dot drawn at the correct height. After drawing the data points for 100, 200, 300, and 400 bicycles, we connect them with line segments to create the following graph, on the right.

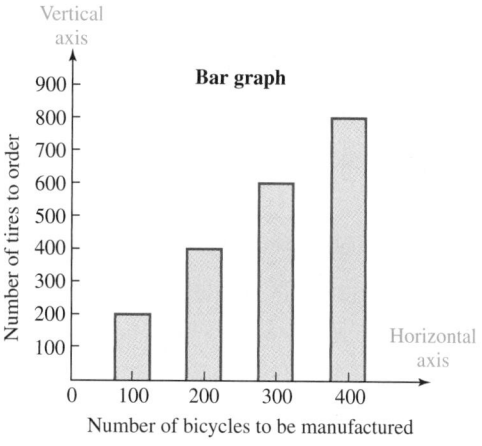

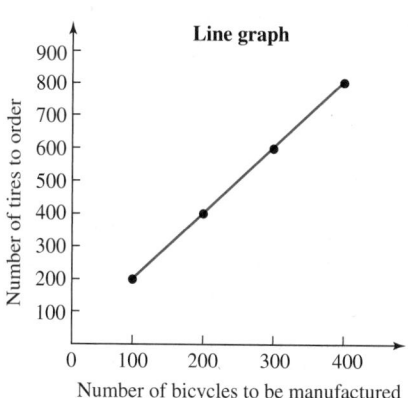

EXAMPLE 1 Use the line graph to find the number of tires needed if 250 bicycles are to be manufactured.

Strategy Since we know the number of bicycles to be manufactured, we will begin on the horizontal axis of the graph and scan up and over to read the answer on the vertical axis.

Why We scan up and over because the number of tires is given by the scale on the vertical axis.

Solution   We locate 250 between 200 and 300 on the horizontal axis and draw a dashed line upward to intersect the graph. From the point of intersection, we draw a dashed horizontal line to the left that intersects the vertical axis at 500. This means that 500 tires should be ordered if 250 bicycles are to be manufactured.

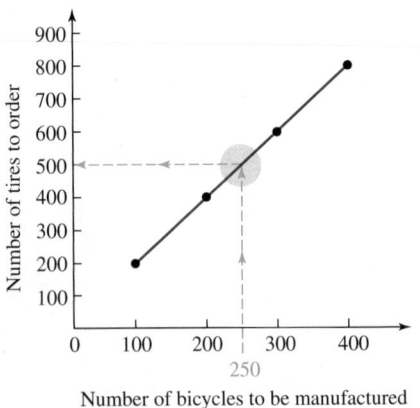

Number of bicycles to be manufactured

Self Check 1   Use the graph to find the number of tires needed if 350 bicycles are to be manufactured.

Now Try   **Problem 31**

2   **Use the Basic Vocabulary and Notation of Algebra.**

From the table and graphs, we see that there is a relationship between the number of tires to order and the number of bicycles to be manufactured. Using words, we can express this relationship as a **verbal model:**

"The number of tires to order is two times the number of bicycles to be manufactured."

Since the word **product** indicates the result of a multiplication, we can write:

"The number of tires to order is the *product* of 2 and the number of bicycles to be manufactured."

To indicate other arithmetic operations, we will use the following words.

- A **sum** is the result of an addition: the sum of 5 and 6 is 11.
- A **difference** is the result of a subtraction: the difference of 3 and 2 is 1.
- A **quotient** is the result of a division: the quotient of 6 and 3 is 2.

Many symbols used in arithmetic are also used in algebra. For example, a + symbol is used to indicate addition, a − symbol is used to indicate subtraction, and an = symbol means *is equal to.*

Since the letter x is often used in algebra and could be confused with the multiplication symbol $\times$, we usually write multiplication using a **raised dot** or **parentheses.**

Symbols for Multiplication	$\times$	Times symbol	$6 \times 4 = 24$
	$\cdot$	Raised dot	$6 \cdot 4 = 24$
	( )	Parentheses	$(6)4 = 24$ or $6(4) = 24$ or $(6)(4) = 24$

In algebra, the symbol most often used to indicate division is the *fraction bar*.

Symbols for Division			
	÷	Division symbol	$24 \div 4 = 6$
	$\overline{)}$	Long division	$\dfrac{6}{4\overline{)24}}$
	—	Fraction bar	$\dfrac{24}{4} = 6$

EXAMPLE 2   Write each statement in words, using one of the words *sum, product, difference,* or *quotient:*   **a.** $\dfrac{22}{11} = 2$ **b.** $22 + 11 = 33$

Strategy   We will examine each statement to determine whether addition, subtraction, multiplication, or division is being performed.

Why   The word that we should use (*sum, product, difference,* or *quotient*) depends on the arithmetic operation that we have to describe.

Solution

a. Since the fraction bar indicates division, we have: The quotient of 22 and 11 equals 2.

b. The symbol indicates addition: The sum of 22 and 11 equals 33.

Self Check 2   Write the following statement in words: $22 - 10 = 12$

Now Try   **Problems 33 and 35**

 Identify Expressions and Equations.

Another way to describe the tires–to–bicycles relationship uses *variables*. **Variables** are letters (or symbols) that stand for numbers. If we let the letter *b* represent the number of bicycles to be manufactured, then the number of tires to order is two times *b*, written 2*b*. In the notation, the number 2 is an example of a **constant** because it does not change value.

The Language of Algebra
Since the number of bicycles to be manufactured can *vary*, or change, it is represented using a *variable*.

When multiplying a variable by a number, or a variable by another variable, we can omit the symbol for multiplication. For example,

$2b$ means $2 \cdot b$ xy means $x \cdot y$ $8abc$ means $8 \cdot a \cdot b \cdot c$

We call $2b$, xy, and $8abc$ *algebraic expressions*.

Algebraic Expressions	Variables and/or numbers can be combined with the operations of addition, subtraction, multiplication, and division to create **algebraic expressions**.

Here are some other examples of algebraic expressions.

The Language of Algebra
We often refer to *algebraic expressions* as simply *expressions*.

$4a + 7$ — This expression is a combination of the numbers 4 and 7, the variable *a*, and the operations of multiplication and addition.

$\dfrac{10 - y}{3}$ — This expression is a combination of the numbers 10 and 3, the variable *y*, and the operations of subtraction and division.

$15mn(2m)$ — This expression is a combination of the numbers 15 and 2, the variables *m* and *n*, and the operation of multiplication.

In the bicycle manufacturing example, if we let the letter t stand for the number of tires to order, we can translate the **verbal model** to mathematical symbols.

The number of tires to order	is	two	times	the number of bicycles to be manufactured.
t	$=$	2	$\cdot$	b

The statement $t = 2 \cdot b$, or more simply, $t = 2b$, is called an *equation*. An **equation** is a mathematical sentence that contains an $=$ symbol. The $=$ symbol indicates that the expressions on either side of it have the same value. Other examples of equations are

$$3 + 5 = 8 \qquad x + 5 = 20 \qquad 17 - 2r = 14 + 3r \qquad p = 100 - d$$

The Language of Algebra
The equal symbol = can be represented by verbs such as:

is   are   gives   yields

The symbol ≠ is read as *"is not equal to."*

EXAMPLE 3   Translate the verbal model into an equation.

The number of decades	is	the number of years	divided by	10.

Strategy   We will represent the unknown quantities using variables and we will use symbols to represent the words *is* and *divided by*.

Why   To translate a verbal (word) model into an equation means to write it using mathematical symbols.

Solution   We can represent the two unknown quantities using variables: Let $d =$ the number of decades and $y =$ the number of years. Then we have:

The number of decades	is	the number of years	divided by	10.
d	$=$	y	$\div$	10

If we write the division using a fraction bar, then the verbal model translates to the equation $d = \frac{y}{10}$.

▷ **Self Check 3**   Translate into an equation: The number of unsold tickets is the difference of 500 and the number of tickets that have been purchased.

Now Try   **Problems 41 and 45**

In the bicycle manufacturing example, using the equation $t = 2b$ to describe the relationship has one major advantage over the other methods. It can be used to determine the number of tires needed for a production run of any size.

EXAMPLE 4   Use the equation $t = 2b$ to find the number of tires needed for a production run of 178 bicycles.

Strategy   In $t = 2b$, we will replace b with 178. Then we will multiply 178 by 2 to obtain the value of t.

Why   The equation $t = 2b$ indicates that the number of tires is found by multiplying the number of bicycles by 2.

Solution

$t = 2b$ This is the describing equation.

$t = 2(178)$ Replace b, which stands for the number of bicycles, with 178. Use parentheses to show the multiplication. We could also write $2 \cdot 178$.

$t = 356$ Multiply.

If 178 bicycles are manufactured, 356 tires will be needed.

Self Check 4   Use the equation $t = 2b$ to find the number of tires needed if 604 bicycles are to be manufactured.

Now Try   **Problem 53**

4   **Use Equations to Construct Tables of Data.**

Equations such as $t = 2b$, which express a relationship between two or more variables, are called **formulas.** Some applications require the repeated use of a formula.

EXAMPLE 5   Find the number of tires to order for production runs of 233 and 852 bicycles. Present the results in a table.

Strategy   We will use the equation $t = 2b$ twice.

Why   There are two different-sized production runs: one of 233 bikes and another of 852 bikes.

Solution

Step 1:   We construct a two-column table. Since b represents the number of bicycles to be manufactured, we use it as the heading of the first column. Since t represents the number of tires needed, we use it as the heading of the second column. Then we enter the size of each production run in the first column, as shown.

The Language of Algebra
To *substitute* means to put or use in place of another, as with a *substitute* teacher. Here, we *substitute* 233 and 852 for b.

Bicycles to be manufactured b	Tires to order t
233	466
852	1,704

Step 2:   We **substitute** 233 and 852 for b in $t = 2b$ and find each corresponding value of t. The results are entered in the second column.

$t = 2b$ $t = 2b$

$t = 2(233)$ $t = 2(852)$

$t = 466$ $t = 1,704$

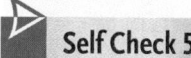

Self Check 5   Find the number of tires needed for production runs of 87 and 487 bicycles. Present the results in a table.

Now Try   **Problem 55**

ANSWERS TO SELF CHECKS   **1.** 700   **2.** The difference of 22 and 10 equals 12.   **3.** $u = 500 - p$
4. 1,208   **5.**

b	t
87	174
487	974

STUDY SET
1.1

VOCABULARY

Fill in the blanks.

1. A _____ is the result of an addition. A _____ is the result of a subtraction. A _____ is the result of a multiplication. A _____ is the result of a division.

2. _____ are letters (or symbols) that stand for numbers.

3. A number, such as 8, is called a _____ because it does not change.

4. Variables and numbers can be combined with the operations of addition, subtraction, multiplication, and division to create algebraic _____.

5. An _____ is a mathematical sentence that contains an = symbol.

6. An equation such as $t = 2b$, which expresses a relationship between two or more variables, is called a _____.

7. The _____ axis of a graph extends left and right and the vertical axis extends up and down.

8. The word _____ comes from the title of a book written by an Arabian mathematician around A.D. 800.

CONCEPTS

Classify each item as an algebraic expression or an equation.

9. **a.** $m + 18 = 23$ **b.** $m + 18$

10. **a.** $30x$ **b.** $30x = 600$

11. **a.** $\dfrac{c - 7}{5}$ **b.** $\dfrac{c - 7}{5} = 7c$

12. **a.** $r = \dfrac{2}{3}$ **b.** $\dfrac{2}{3}r$

13. What arithmetic operations does the expression $\dfrac{12 + 9t}{25}$ contain? What variable does it contain?

14. What arithmetic operations does the equation $4y - 14 = 5(6)$ contain? What variable does it contain?

15. Construct a line graph using the data in the following table.

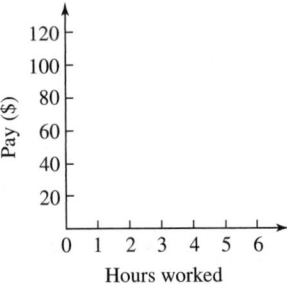

Hours worked	Pay (dollars)
1	20
2	40
3	60
4	80
5	100

16. Use the data in the graph to complete the table.

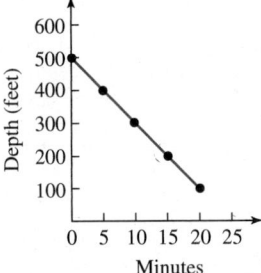

Minutes	Depth (feet)

NOTATION

Fill in the blanks.

17. The symbol $\neq$ means _____ _____ _____ _____.

18. The symbols (   ) are called _____.

19. Write the multiplication 5×6 using a raised dot and then using parentheses.

20. Give four verbs that can be represented by an equal symbol =.

Write each expression without using a multiplication symbol or parentheses.

21. $4 \cdot x$ 22. $P \cdot r \cdot t$

23. $2(w)$ 24. $(x)(y)$

Write each division using a fraction bar.

25. $32 \div x$ 26. $30\overline{)90}$

27. $5\overline{)55}$ 28. $h \div 15$

GUIDED PRACTICE

Use the line graph in Example 1 to find the number of tires needed to build the following number of bicycles. See Example 1.

29. 150 **30.** 400

31. Explain what the dashed lines help us find in the graph.

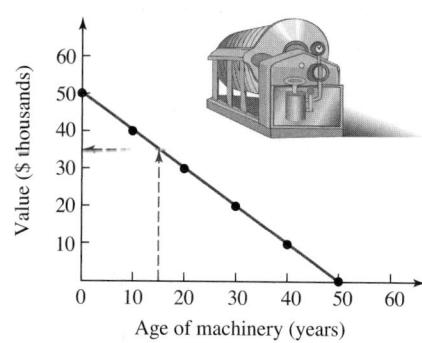

Value ($ thousands)

Age of machinery (years)

32. Use the line graph to find the income received from 30, 50, and 70 customers.

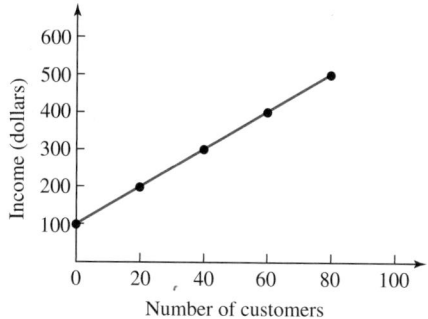

Income (dollars)

Number of customers

Express each statement using one of the words sum, product, difference, *or* quotient. *See Example 2.*

33. 8(2) = 16   **34.** 45 · 12 = 540

35. 11 − 9 = 2   **36.** 65 + 89 = 154

37. $x + 2 = 10$ **38.** $16 − t = 4$

39. $\dfrac{66}{11} = 6$ **40.** 12 ÷ 3 = 4

Translate each verbal model into an equation. (Answers may vary, depending on the variables chosen.) See Example 3.

41.

| The sale price | is | $100 | minus | the discount. |

42.

| The cost of dining out | equals | the cost of the meal | plus | $7 for parking. |

43.

| 7 | times | the age of a dog in years | gives | the dog's equivalent human age. |

44.

| The number of centuries | is | the number of years | divided by | 100. |

45. The amount of sand that should be used is the product of 3 and the amount of cement used.

46. The number of waiters needed is the quotient of the number of customers and 10.

47. The weight of the truck is the sum of the weight of the engine and 1,200.

48. The number of classes still open is the difference of 150 and the number of classes that are closed.

49. The profit is the difference of the revenue and 600.

50. The distance is the product of the rate and 3.

51. The quotient of the number of laps run and 4 gives the number of miles run.

52. The sum of the tax and 35 gives the total cost.

Use the formula to complete each table. See Examples 4 and 5.

53. $d = 360 + L$

Lunch time (minutes) L	School day (minutes) d
30	
40	
45	

54. $b = 1{,}024k$

Kilobytes k	Bytes b
1	
5	
10	

55. $t = 1{,}500 − d$

Deductions d	Take-home pay t
200	
300	
400	

56. $w = \dfrac{s}{12}$

Inches of snow s	Inches of water w
12	
24	
72	

Use the data in the table to complete the formula.

57. $d = \dfrac{e}{\boxed{}}$

Eggs e	Dozens d
24	2
36	3
48	4

58. $p = \boxed{}\, c$

Canoes c	Paddles p
6	12
7	14
8	16

59. $I = \boxed{}\, c$

Couples c	Individuals I
20	40
100	200
200	400

60. $t = \dfrac{p}{\boxed{}}$

Players p	Teams t
5	1
10	2
15	3

APPLICATIONS

61. TRAFFIC SAFETY   As the railroad crossing guard drops, the measure of angle 1 (written $\angle 1$) increases while the measure of $\angle 2$ decreases. At any instant the *sum* of the measures of the two angles is 90°. Complete the table. Then use the data to construct a line graph. Scale each axis in units of 15°.

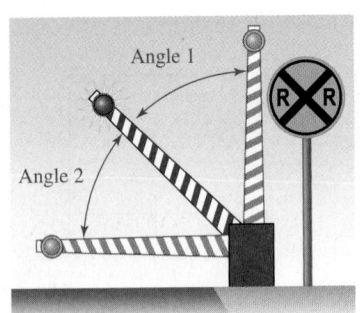

Angle 1 (degrees)	Angle 2 (degrees)
0	
15	
30	
45	
60	
75	
90	

62. U.S. CRIME STATISTICS   Property crimes include burglary, theft, and motor vehicle theft. Graph the following property crime rate data using a bar graph. Scale the vertical axis in units of 50.

Year	Crimes per 1,000 households
1992	325
1994	310
1996	266
1998	217

Year	Crimes per 1,000 households
2000	178
2002	159
2004	161

Source: Bureau of Justice Statistics.

WRITING

63. Many students misuse the word *equation* when discussing mathematics. What is an equation? Give an example.

64. Explain the difference between an algebraic expression and an equation. Give an example of each.

65. In this section, four methods for describing numerical relationships were discussed: tables, words, graphs, and equations. Which method do you think is the most useful? Explain why.

66. In your own words, define *horizontal* and *vertical*.

CHALLENGE PROBLEMS

67. Complete the table and the formula.

$t = \boxed{}$

s	t
10	
18	19
	34
47	48

68. Suppose $h = 4n$ and $n = 2g$. Complete the following formula: $h = \boxed{}\, g$.

SECTION 1.2
Fractions

Objectives

1. Factor and prime factor natural numbers.
2. Recognize special fraction forms.
3. Multiply and divide fractions.
4. Build equivalent fractions.
5. Simplify fractions.
6. Add and subtract fractions.
7. Simplify answers.
8. Compute with mixed numbers.

In arithmetic, we add, subtract, multiply, and divide **natural numbers:** 1, 2, 3, 4, 5, and so on. Assuming that you have mastered those skills, we will now review the arithmetic of fractions.

 Factor and Prime Factor Natural Numbers.

To compute with fractions, we need to know how to *factor* natural numbers. To **factor** a number means to express it as a product of two or more numbers. For example, some ways to factor 8 are

$$1 \cdot 8, \qquad 4 \cdot 2, \qquad \text{and} \qquad 2 \cdot 2 \cdot 2$$

The numbers 1, 2, 4, and 8 that were used to write the products are called *factors* of 8. In general, a **factor** is a number being multiplied.

Sometimes a number has only two factors, itself and 1. We call such numbers *prime numbers.*

> **The Language of Algebra**
> When we say "factor 8," we are using the word *factor* as a verb. When we say "2 is a *factor* of 8," we are using the word *factor* as a noun.

Prime Numbers and Composite Numbers

A **prime number** is a natural number greater than 1 that has only itself and 1 as factors. The first ten prime numbers are 2, 3, 5, 7, 11, 13, 17, 19, 23, and 29.

A **composite number** is a natural number, greater than 1, that is not prime. The first ten composite numbers are 4, 6, 8, 9, 10, 12, 14, 15, 16, and 18.

Every composite number can be factored into the product of two or more prime numbers. This product of these prime numbers is called its **prime factorization.**

EXAMPLE 1   Find the prime factorization of 210.

Strategy   We will use a series of steps to express 210 as a product of only prime numbers.

Why   To *prime factor* a number means to write it as a product of prime numbers.

Solution   First, write 210 as the product of two natural numbers other than 1.

$$210 = 10 \cdot 21$$ The resulting prime factorization will be the same no matter which two factors of 210 you begin with.

> **The Language of Algebra**
> Prime factors are often written in *ascending* order. To *ascend* means to move upward.

Neither 10 nor 21 are prime numbers, so we factor each of them.

$$210 = 2 \cdot 5 \cdot 3 \cdot 7 \quad \text{Factor 10 as } 2 \cdot 5 \text{ and factor 21 as } 3 \cdot 7.$$

Writing the factors in ascending order, the **prime-factored form** of 210 is $2 \cdot 3 \cdot 5 \cdot 7$. Two other methods for prime factoring 210 are shown below.

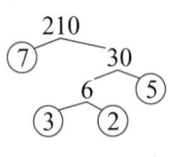

Factor tree   *Division ladder*

Work downward. Factor each number as a product of two numbers (other than 1 and itself) until all factors are prime. Circle prime numbers as they appear at the end of a branch.

↑ Work upward. Perform repeated division until the final quotient is a prime number. It is helpful to start with the smallest prime, 2, as a trial divisor. Then, in order, try larger primes as divisors: 3, 5, 7, 11, and so on.

Either way, the factorization is $2 \cdot 3 \cdot 5 \cdot 7$. To check it, multiply the prime factors. The product should be 210.

Self Check 1   Find the prime factorization of 189.

Now Try **Problem 15**

2 Recognize Special Fraction Forms.

In a fraction, the number above the **fraction bar** is called the **numerator,** and the number below is called the **denominator.**

Fraction bar $\longrightarrow \dfrac{5 \leftarrow \text{numerator}}{6 \leftarrow \text{denominator}}$

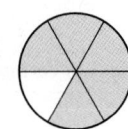

Fractions can describe the number of equal parts of a whole. For example, consider the circle with 5 of 6 equal parts colored red. We say that $\frac{5}{6}$ (five-sixths) of the circle is shaded.

Fractions are also used to indicate division. For example, $\frac{8}{2}$ indicates that the numerator, 8, is to be divided by the denominator, 2:

$$\frac{8}{2} = 8 \div 2 = 4 \quad \text{We know that } \frac{8}{2} = 4 \text{ because of its related multiplication statement:} \quad 2 \cdot 4 = 8.$$

If the numerator and denominator of a fraction are the same nonzero number, the fraction indicates division of a number by itself, and the result is 1. Each of the following fractions is, therefore, a **form of 1.**

$$1 = \frac{1}{1} = \frac{2}{2} = \frac{3}{3} = \frac{4}{4} = \frac{5}{5} = \frac{6}{6} = \frac{7}{7} = \frac{8}{8} = \frac{9}{9} = \cdots$$

If a denominator is 1, the fraction indicates division by 1, and the result is simply the numerator. For example, $\frac{5}{1} = 5$ and $\frac{24}{1} = 24$.

Special Fraction Forms	For any nonzero number a,

$$\frac{a}{a} = 1 \quad \text{and} \quad \frac{a}{1} = a$$

3 **Multiply and Divide Fractions.**

The rule for multiplying fractions can be expressed in words and in symbols as follows.

Multiplying Fractions	To multiply two fractions, multiply the numerators and multiply the denominators. For any two fractions $\frac{a}{b}$ and $\frac{c}{d}$,

$$\frac{a}{b} \cdot \frac{c}{d} = \frac{a \cdot c}{b \cdot d}$$

EXAMPLE 2   Multiply: $\dfrac{7}{8} \cdot \dfrac{3}{5}$

Strategy   To find the product, we will multiply the numerators, 7 and 3, and multiply the denominators, 8 and 5.

Why   This is the rule for multiplying two fractions.

Solution

$$\frac{7}{8} \cdot \frac{3}{5} = \frac{7 \cdot 3}{8 \cdot 5} \qquad \text{Multiply the numerators.}$$
$$\text{Multiply the denominators.}$$
$$= \frac{21}{40}$$

Self Check 2   Multiply: $\frac{5}{9} \cdot \frac{2}{3}$

Now Try   **Problem 27**

One number is called the **reciprocal** of another if their product is 1. To find the reciprocal of a fraction, we invert its numerator and denominator.

$\frac{3}{4}$ is the reciprocal of $\frac{4}{3}$, because $\frac{3}{4} \cdot \frac{4}{3} = \frac{12}{12} = 1$.

$\frac{1}{10}$ is the reciprocal of 10, because $\frac{1}{10} \cdot 10 = \frac{10}{10} = 1$.

We use reciprocals to divide fractions.

Dividing Fractions	To divide two fractions, multiply the first fraction by the reciprocal of the second. For any two fractions $\frac{a}{b}$ and $\frac{c}{d}$, where $c \neq 0$,

$$\frac{a}{b} \div \frac{c}{d} = \frac{a}{b} \cdot \frac{d}{c}$$

EXAMPLE 3   Divide: $\frac{1}{3} \div \frac{4}{5}$

Strategy   We will multiply the first fraction, $\frac{1}{3}$, by the reciprocal of the second fraction, $\frac{4}{5}$.

Why   This is the rule for dividing two fractions.

Solution

$$\frac{1}{3} \div \frac{4}{5} = \frac{1}{3} \cdot \frac{5}{4} \qquad \text{Multiply } \tfrac{1}{3} \text{ by the reciprocal of } \tfrac{4}{5}. \text{ The reciprocal of } \tfrac{4}{5} \text{ is } \tfrac{5}{4}.$$

$$= \frac{1 \cdot 5}{3 \cdot 4} \qquad \begin{array}{l}\text{Multiply the numerators.}\\ \text{Multiply the denominators.}\end{array}$$

$$= \frac{5}{12}$$

Self Check 3   Divide: $\frac{6}{25} \div \frac{1}{2}$

Now Try   **Problem 31**

4 **Build Equivalent Fractions.**

The two rectangles on the right are the same size. The first rectangle is divided into 10 equal parts. Since 6 of those parts are red, $\frac{6}{10}$ of the figure is shaded.

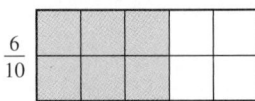

The second rectangle is divided into 5 equal parts. Since 3 of those parts are red, $\frac{3}{5}$ of the figure is shaded. We can conclude that $\frac{6}{10} = \frac{3}{5}$ because $\frac{6}{10}$ and $\frac{3}{5}$ represent the same shaded part of the rectangle. We say that $\frac{6}{10}$ and $\frac{3}{5}$ are *equivalent fractions*.

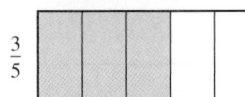

Equivalent Fractions	Two fractions are **equivalent** if they represent the same number.

Writing a fraction as an equivalent fraction with a larger denominator is called **building** the fraction. To build a fraction, we multiply it by a form of 1. Since any number multiplied by 1 remains the same (identical), 1 is called the **multiplicative identity element.**

Multiplication Property of 1	The product of 1 and any number is that number. For any number a, $$1 \cdot a = a \qquad \text{and} \qquad a \cdot 1 = a$$

EXAMPLE 4   Write $\frac{3}{5}$ as an equivalent fraction with a denominator of 35.

Strategy   We will compare the given denominator to the required denominator and ask, "By what must we multiply 5 to get 35?"

Success Tip
Multiplying $\frac{3}{5}$ by $\frac{7}{7}$ changes its appearance, but does not change its value, because we are multiplying it by a form of 1.

Why The answer to that question helps us determine the form of 1 to be used to build an equivalent fraction.

Solution We need to multiply the denominator of $\frac{3}{5}$ by 7 to obtain a denominator of 35. It follows that $\frac{7}{7}$ should be the form of 1 that is used to build $\frac{3}{5}$. Multiplying $\frac{3}{5}$ by $\frac{7}{7}$ changes its appearance but does not change its value, because we are multiplying it by 1.

$$\frac{3}{5} = \frac{3}{5} \cdot \frac{7}{7} \qquad \frac{7}{7} = 1$$

$$= \frac{3 \cdot 7}{5 \cdot 7} \qquad \text{Multiply the numerators.}$$
$$\phantom{= \frac{3 \cdot 7}{5 \cdot 7}} \qquad \text{Multiply the denominators.}$$

$$= \frac{21}{35}$$

 Self Check 4 Write $\frac{5}{8}$ as an equivalent fraction with a denominator of 24.

Now Try **Problem 35**

Building Fractions

To build a fraction, multiply it by 1 in the form of $\frac{c}{c}$, where c is any nonzero number.

 Simplify Fractions.

Every fraction can be written in infinitely many equivalent forms. For example, some equivalent forms of $\frac{10}{15}$ are:

The Language of Algebra
The word *infinitely* is a form of the word *infinite*, which means endless.

$$\frac{2}{3} = \frac{4}{6} = \frac{6}{9} = \frac{8}{12} = \mathbf{\frac{10}{15}} = \frac{12}{18} = \frac{14}{21} = \frac{16}{24} = \frac{18}{27} = \frac{20}{30} = \cdots$$

Of all of the equivalent forms in which we can write a fraction, we often need to determine the one that is in *simplest form*.

Simplest Form of a Fraction

A fraction is in **simplest form,** or **lowest terms,** when the numerator and denominator have no common factors other than 1.

To **simplify a fraction,** we write it in simplest form by removing a factor equal to 1. For example, to simplify $\frac{10}{15}$, we note that the greatest factor common to the numerator and denominator is 5 and proceed as follows:

$$\frac{10}{15} = \frac{2 \cdot \mathbf{5}}{3 \cdot \mathbf{5}} \qquad \text{Factor 10 and 15.}$$

$$= \frac{2}{3} \cdot \frac{\mathbf{5}}{\mathbf{5}} \qquad \text{Use the rule for multiplying fractions in reverse: write } \frac{2 \cdot 5}{3 \cdot 5} \text{ as the product of two fractions, } \frac{2}{3} \text{ and } \frac{5}{5}.$$

$$= \frac{2}{3} \cdot 1 \qquad \text{A nonzero number divided by itself is equal to 1: } \frac{5}{5} = 1.$$

$$= \frac{2}{3} \qquad \text{Use the multiplication property of 1: any number multiplied by 1 remains the same.}$$

To simplify $\frac{10}{15}$, we removed a factor equal to 1 in the form of $\frac{5}{5}$. The result, $\frac{2}{3}$, is equivalent to $\frac{10}{15}$.

We can easily identify the greatest common factor of the numerator and the denominator of a fraction if we write them in prime-factored form.

EXAMPLE 5   Simplify each fraction, if possible: **a.** $\frac{63}{42}$ **b.** $\frac{33}{40}$

Strategy   We will begin by prime factoring the numerator and denominator of the fraction. Then, to simplify it, we will remove a factor equal to 1.

Why   We need to make sure that the numerator and denominator have no common factors other than 1. If that is the case, then the fraction is in *simplest form*.

Solution

a. After prime factoring 63 and 42, we see that the greatest common factor of the numerator and the denominator is $3 \cdot 7 = 21$.

$$\frac{63}{42} = \frac{3 \cdot 3 \cdot 7}{2 \cdot 3 \cdot 7} \qquad \text{Write 63 and 42 in prime-factored form.}$$

$$= \frac{3}{2} \cdot \frac{3 \cdot 7}{3 \cdot 7} \qquad \text{Write } \tfrac{3 \cdot 3 \cdot 7}{2 \cdot 3 \cdot 7} \text{ as the product of two fractions, } \tfrac{3}{2} \text{ and } \tfrac{3 \cdot 7}{3 \cdot 7}.$$

$$= \frac{3}{2} \cdot 1 \qquad \text{A nonzero number divided by itself is equal to 1: } \tfrac{3 \cdot 7}{3 \cdot 7} = 1.$$

$$= \frac{3}{2} \qquad \text{Any number multiplied by 1 remains the same.}$$

b. Prime factor 33 and 40.

$$\frac{33}{40} = \frac{3 \cdot 11}{2 \cdot 2 \cdot 2 \cdot 5}$$

Since the numerator and the denominator have no common factors other than 1, the fraction $\frac{33}{40}$ is in simplest form (lowest terms).

Self Check 5   Simplify each fraction, if possible:
a. $\frac{24}{56}$ **b.** $\frac{16}{125}$

Now Try   **Problem 45**

> **The Language of Algebra**
> What do Calvin Klein, Queen Latifah, and Tom Hanks have in common? They all attended a community college. The word *common* means shared by two or more. In this section, we will work with *common* factors and *common* denominators.

To streamline the simplifying process, we can replace pairs of factors common to the numerator and denominator with the equivalent fraction $\frac{1}{1}$.

EXAMPLE 6   Simplify: $\frac{90}{105}$

Strategy   We will begin by prime factoring the numerator, 90, and denominator, 105. Then we will look for any factors common to the numerator and denominator and remove them.

Why When the numerator and/or denominator of a fraction are large numbers, s[...]
90 and 105, writing their prime factorizations is helpful in identifying any common
factors.

Solution

$$\frac{90}{105} = \frac{2 \cdot 3 \cdot 3 \cdot 5}{3 \cdot 5 \cdot 7}$$ Write 90 and 105 in prime-factored form.

$$= \frac{2 \cdot \overset{1}{\cancel{3}} \cdot 3 \cdot \overset{1}{\cancel{5}}}{\underset{1}{\cancel{3}} \cdot \underset{1}{\cancel{5}} \cdot 7}$$ Slashes and 1's are used to show that $\frac{3}{3}$ and $\frac{5}{5}$ are replaced by the equivalent fraction $\frac{1}{1}$. A factor equal to 1 in the form of $\frac{3 \cdot 5}{3 \cdot 5} = \frac{15}{15}$ was removed.

$$= \frac{6}{7}$$ Multiply the remaining factors in the numerator: $2 \cdot 1 \cdot 3 \cdot 1 = 6$. Multiply the remaining factors in the denominator: $1 \cdot 1 \cdot 7 = 7$.

Self Check 6   Simplify: $\frac{126}{70}$

Now Try   **Problem 53**

We can use the following steps to simplify a fraction.

Simplifying Fractions

1. Factor (or prime factor) the numerator and denominator to determine their common factors.
2. Remove factors equal to 1 by replacing each pair of factors common to the numerator and denominator with the equivalent fraction $\frac{1}{1}$.
3. Multiply the remaining factors in the numerator and in the denominator.

The procedure for simplifying fractions is based on the following property.

The Fundamental Property of Fractions

If $\frac{a}{b}$ is a fraction and c is a nonzero real number,

$$\frac{ac}{bc} = \frac{a}{b}$$

Caution   When all common factors of the numerator and/or the denominator of a fraction are removed, forgetting to write 1's above the slashes can lead to a common mistake.

Correct	*Incorrect*
$\dfrac{15}{45} = \dfrac{\overset{1}{\cancel{3}} \cdot \overset{1}{\cancel{5}}}{\underset{1}{\cancel{3}} \cdot 3 \cdot \underset{1}{\cancel{5}}} = \dfrac{1}{3}$	$\dfrac{15}{45} = \dfrac{\cancel{3} \cdot \cancel{5}}{\cancel{3} \cdot 3 \cdot \cancel{5}} = \dfrac{0}{3} = 0$

6   Add and Subtract Fractions.

In algebra as in everyday life, we can only add or subtract objects that are similar. For example, we can add dollars to dollars, but we cannot add dollars to oranges. This concept is important when adding fractions.

Consider the problem $\frac{2}{5} + \frac{1}{5}$. When we write it in words, it is apparent we are adding similar objects.

two-**fifths**   +   one-**fifth**

└── Similar objects ──┘

Because the denominators of $\frac{2}{5}$ and $\frac{1}{5}$ are the same, we say that they have a **common denominator.**

Adding and Subtracting Fractions that Have the Same Denominator	To add (or subtract) fractions that have the same denominator, add (or subtract) their numerators and write the sum (or difference) over the common denominator. For any fractions $\frac{a}{d}$ and $\frac{b}{d}$, $$\frac{a}{d}+\frac{b}{d}=\frac{a+b}{d} \quad \text{and} \quad \frac{a}{d}-\frac{b}{d}=\frac{a-b}{d}$$

Caution
We do **not** add fractions by adding the numerators and adding the denominators!

$$\frac{2}{5}+\frac{1}{5}\neq\frac{2+1}{5+5}=\frac{3}{10}$$

The same caution applies when subtracting fractions.

For example,

$$\frac{2}{5}+\frac{1}{5}=\frac{2+1}{5}=\frac{3}{5} \quad \text{and} \quad \frac{18}{23}-\frac{9}{23}=\frac{18-9}{23}=\frac{9}{23}$$

Caution Be careful when adding (or subtracting) numerators and writing the result over the common denominator. Only *factors* common to the numerator and the denominator of a fraction can be removed. For example, it is incorrect to remove the 5's in $\frac{5+8}{5}$ because 5 is not used as a factor in the expression $5+8$. This error leads to an incorrect answer of 9.

Correct

$$\frac{5+8}{5}=\frac{13}{5}$$

Incorrect

$$\frac{5+8}{5}=\frac{\overset{1}{5}+8}{\underset{1}{5}}=\frac{9}{1}=9$$

Success Tip
To determine the LCD of two fractions, list the multiples of one of the denominators. The first number in the list that is exactly divisible by the other denominator is their LCD. For $\frac{2}{5}$ and $\frac{1}{3}$, the multiples of the first denominator, 5, are

5, 10, ⑮, 20, 25, . . .

Since 15 is the first number in the list that is exactly divisible by the second denominator, 3, the LCD is 15.

Now we consider the problem $\frac{2}{5}+\frac{1}{3}$. Since the denominators are not the same, we cannot add these fractions in their present form.

two-**fifths**   +   one-**third**

└── Not similar objects ──┘

To add (or subtract) fractions with different denominators, we express them as equivalent fractions that have a common denominator. The smallest common denominator, called the **least** or **lowest common denominator,** is usually the easiest common denominator to use.

Least Common Denominator (LCD)	The **least** or **lowest common denominator (LCD)** for a set of fractions is the smallest number each denominator will divide exactly (divide with no remainder).

The denominators of $\frac{2}{5}$ and $\frac{1}{3}$ are 5 and 3. The numbers 5 and 3 divide many numbers exactly (30, 45, and 60, to name a few), but the smallest number that they divide exactly is 15. Thus, 15 is the LCD for $\frac{2}{5}$ and $\frac{1}{3}$.

To find $\frac{2}{5}+\frac{1}{3}$, we find equivalent fractions that have denominators of 15 and we use the rule for adding fractions.

$$\frac{2}{5}+\frac{1}{3}=\frac{2}{5}\cdot\frac{3}{3}+\frac{1}{3}\cdot\frac{5}{5}$$ Multiply $\frac{2}{5}$ by 1 in the form of $\frac{3}{3}$. Multiply $\frac{1}{3}$ by 1 in the form of $\frac{5}{5}$.

$$=\frac{6}{15}+\frac{5}{15}$$ Multiply the numerators and multiply the denominators. Note that the denominators are now the same.

$$= \frac{6 + 5}{15}$$

Add the numerators.
Write the sum over the common denominator.

$$= \frac{11}{15}$$

When adding (or subtracting) fractions with unlike denominators, the least common denominator is not always obvious. Prime factorization is helpful in determining the LCD.

Finding the LCD Using Prime Factorization

1. Prime factor each denominator.
2. The LCD is a product of prime factors, where each factor is used the greatest number of times it appears in any one factorization found in step 1.

EXAMPLE 7   Subtract: $\dfrac{3}{10} - \dfrac{5}{28}$

Strategy   We will begin by expressing each fraction as an equivalent fraction that has the LCD for its denominator. Then we will use the rule for subtracting fractions with *like* denominators.

Why   To add or subtract fractions, the fractions must have like denominators.

Solution   To find the LCD, we find the prime factorization of both denominators and use each prime factor the *greatest* number of times it appears in any one factorization:

$$\left.\begin{array}{l} 10 = 2 \cdot 5 \\ 28 = 2 \cdot 2 \cdot 7 \end{array}\right\} \text{LCD} = \mathbf{2 \cdot 2 \cdot 5 \cdot 7} = 140$$

2 appears twice in the factorization of 28.
5 appears once in the factorization of 10.
7 appears once in the factorization of 28.

Since 140 is the smallest number that 10 and 28 divide exactly, we write $\frac{3}{10}$ and $\frac{5}{28}$ as fractions with the LCD 140.

$$\frac{3}{10} - \frac{5}{28} = \frac{3}{10} \cdot \frac{14}{14} - \frac{5}{28} \cdot \frac{5}{5}$$

We must multiply 10 by 14 to obtain 140.
We must multiply 28 by 5 to obtain 140.

$$= \frac{42}{140} - \frac{25}{140}$$

Multiply the numerators and multiply the denominators.
Note that the denominators are now the same.

$$= \frac{42 - 25}{140}$$

Subtract the numerators.
Write the difference over the common denominator.

$$= \frac{17}{140}$$

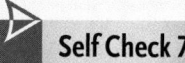

Self Check 7   Subtract: $\frac{11}{48} - \frac{7}{40}$

Now Try   Problem 65

We can use the following steps to add or subtract fractions with different denominators.

Adding and Subtracting Fractions that Have Different Denominators	1. Find the LCD. 2. Rewrite each fraction as an equivalent fraction with the LCD as the denominator. To do so, build each fraction using a form of 1 that involves any factors needed to obtain the LCD. 3. Add or subtract the numerators and write the sum or difference over the LCD. 4. Simplify the result, if possible.

7 **Simplify Answers.**

When adding, subtracting, multiplying, or dividing fractions, remember to express the answer in simplest form.

EXAMPLE 8 Perform the operations and simplify:

a. $45\left(\dfrac{4}{9}\right)$ **b.** $\dfrac{5}{12} + \dfrac{3}{2} - \dfrac{1}{4}$

Strategy We will perform the indicated operations and then make sure that the answer is in simplest form (lowest terms).

Why Fractional answers should always be given in simplest form.

Solution

a. $45\left(\dfrac{4}{9}\right) = \dfrac{45}{1}\left(\dfrac{4}{9}\right)$ Write 45 as a fraction: $45 = \dfrac{45}{1}$.

$= \dfrac{45 \cdot 4}{1 \cdot 9}$ Multiply the numerators.
  Multiply the denominators.

$= \dfrac{5 \cdot \overset{1}{\cancel{9}} \cdot 4}{1 \cdot \underset{1}{\cancel{9}}}$ To simplify the result, factor 45 as $5 \cdot 9$. Then remove the common factor 9 of the numerator and denominator.

$= 20$ Multiply the remaining factors in the numerator. Multiply the remaining factors in the denominator. $\dfrac{20}{1} = 20$.

> **Caution**
> Remember that an LCD is **not** needed when multiplying or dividing fractions.

b. Since the smallest number that 12, 2, and 4 divide exactly is 12, the LCD is 12.

$\dfrac{5}{12} + \dfrac{3}{2} - \dfrac{1}{4} = \dfrac{5}{12} + \dfrac{3}{2} \cdot \dfrac{6}{6} - \dfrac{1}{4} \cdot \dfrac{3}{3}$ $\dfrac{5}{12}$ already has a denominator of 12. Build $\dfrac{3}{2}$ and $\dfrac{1}{4}$ so that their denominators are 12.

$= \dfrac{5}{12} + \dfrac{18}{12} - \dfrac{3}{12}$ Multiply the numerators and multiply the denominators. The denominators are now the same.

$= \dfrac{20}{12}$ Add the numerators, 5 and 18, to get 23. From that sum, subtract 3. Write that result, 20, over the common denominator.

$= \dfrac{\overset{1}{\cancel{4}} \cdot 5}{3 \cdot \underset{1}{\cancel{4}}}$ To simplify $\dfrac{20}{12}$, factor 20 and 12, using their greatest common factor, 4. Then remove $\dfrac{4}{4} = 1$.

$= \dfrac{5}{3}$

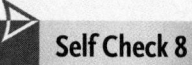 **Self Check 8**   Perform the operations and simplify: **a.** $24\left(\frac{7}{6}\right)$
b. $\frac{1}{15} + \frac{31}{30} - \frac{3}{10}$

Now Try   **Problems 67 and 71**

8   **Compute with Mixed Numbers.**

A **mixed number** represents the sum of a whole number and a fraction. For example, $5\frac{3}{4}$ means $5 + \frac{3}{4}$.

EXAMPLE 9   Divide: $5\frac{3}{4} \div 2$

Strategy   We begin by writing the mixed number $5\frac{3}{4}$ and the whole number 2 as fractions. Then we use the rule for dividing two fractions.

Why   To multiply (or divide) with mixed numbers, we first write them as fractions, and then multiply (or divide) as usual.

Solution

> **The Language of Algebra**
> Fractions such as $\frac{23}{4}$, with a numerator greater than or equal to the denominator, are called **improper fractions**. In algebra, such fractions are often preferable to their equivalent mixed number form.

$$5\frac{3}{4} \div 2 = \frac{23}{4} \div \frac{2}{1}$$

Write $5\frac{3}{4}$ as an improper fraction by multiplying its whole-number part by the denominator: $5 \cdot 4 = 20$. Then add the numerator to that product: $3 + 20 = 23$. Finally, write the result, 23, over the denominator 4. Write 2 as a fraction: $2 = \frac{2}{1}$.

$$= \frac{23}{4} \cdot \frac{1}{2}$$

Multiply by the reciprocal of $\frac{2}{1}$, which is $\frac{1}{2}$.

$$= \frac{23}{8}$$

Multiply the numerators.
Multiply the denominators.

$$= 2\frac{7}{8}$$

Write $\frac{23}{8}$ as a mixed number by dividing the numerator, 23, by the denominator, 8. The quotient, 2, is the whole-number part; the remainder, 7, over the divisor, 8, is the fractional part.

  Self Check 9   Multiply: $1\frac{1}{8} \cdot 9$

Now Try   **Problem 77**

EXAMPLE 10   *Freeway Signs.*   How far apart are the Downtown San Diego and Sea World Drive exits?

Strategy   We can find the distance between exits by finding the difference in the mileages on the freeway sign: $6\frac{1}{2} - 1\frac{3}{4}$.

Why   The word *difference* indicates subtraction.

INTERSTATE
CALIFORNIA
5

Downtown San Diego 1 3/4
Sea World Dr   6 1/2

Success Tip

This problem could also be solved by writing the mixed numbers $6\frac{1}{2}$ and $1\frac{3}{4}$, as improper fractions and subtracting them.

Solution

$$6\frac{1}{2} = \quad 6\frac{2}{4} = \quad 5\frac{2}{4} + \frac{4}{4} = \quad 5\frac{6}{4}$$

$$-1\frac{3}{4} = -1\frac{3}{4} = -1\frac{3}{4} \quad = -1\frac{3}{4}$$

$$\overline{\qquad\qquad\qquad\qquad\qquad\qquad} \quad 4\frac{3}{4}$$

Using vertical form, express $\frac{1}{2}$ as an equivalent fraction with denominator 4. Then, borrow 1 in the form of $\frac{4}{4}$ from 6 to subtract the fractional parts of the mixed numbers.

The Downtown San Diego and Sea World Drive exits are $4\frac{3}{4}$ miles apart.

 Now Try Problem 107

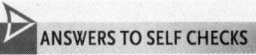 **ANSWERS TO SELF CHECKS**   **1.** $189 = 3 \cdot 3 \cdot 3 \cdot 7$ **2.** $\frac{10}{27}$ **3.** $\frac{12}{25}$ **4.** $\frac{15}{24}$ **5. a.** $\frac{3}{7}$ **b.** In simplest form **6.** $\frac{9}{5}$ **7.** $\frac{13}{240}$ **8. a.** 28 **b.** $\frac{4}{5}$ **9.** $\frac{81}{8} = 10\frac{1}{8}$

STUDY SET
1.2

VOCABULARY

Fill in the blanks.

1. A factor is a number being _____.

2. Numbers that have only 1 and themselves as factors, such as 23, 37, and 41, are called _____ numbers.

3. When we write 60 as $2 \cdot 2 \cdot 3 \cdot 5$, we say that we have written 60 in _____ form.

4. The _____ of the fraction $\frac{3}{4}$ is 3, and the _____ is 4.

5. Two fractions that represent the same number, such as $\frac{1}{2}$ and $\frac{2}{4}$, are called _____ fractions.

6. $\frac{2}{3}$ is the _____ of $\frac{3}{2}$, because their product is 1.

7. The _____ common denominator for a set of fractions is the smallest number each denominator will divide exactly.

8. The _____ number $7\frac{1}{3}$ represents the sum of a whole number and a fraction: $7 + \frac{1}{3}$.

CONCEPTS

Complete each fact about fractions.

9. **a.** $\dfrac{a}{a} =$ **b.** $\dfrac{a}{1} =$

    c. $\dfrac{a}{b} \cdot \dfrac{c}{d} =$ **d.** $\dfrac{a}{b} \div \dfrac{c}{d} =$

    e. $\dfrac{a}{d} + \dfrac{b}{d} =$ **f.** $\dfrac{a}{d} - \dfrac{b}{d} =$

10. What two equivalent fractions are shown?

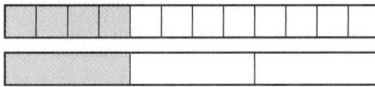

11. Complete each statement.

    a. To simplify a fraction, we remove factors equal to ___ in the form of $\frac{2}{2}$, $\frac{3}{3}$, or $\frac{4}{4}$, and so on.

    b. To build a fraction, we multiply it by ___ in the form of $\frac{2}{2}$, $\frac{3}{3}$, or $\frac{4}{4}$, and so on.

12. What is the LCD for fractions having denominators of 24 and 36?

NOTATION

Fill in the blanks.

13. **a.** Multiply $\frac{5}{6}$ by a form of 1 to build an equivalent fraction with denominator 30.

$$\frac{5}{6} \cdot \quad =$$

    b. Remove common factors to simplify $\frac{12}{42}$.

$$\frac{12}{42} = \frac{2 \cdot \quad \cdot 3}{2 \cdot 3 \cdot \quad} =$$

14. **a.** Write $2\frac{15}{16}$ as an improper fraction.

    b. Write $\frac{49}{12}$ as a mixed number.

GUIDED PRACTICE

Find the prime factorization of each number. **See Example 1.**

15. 75   **16.** 20
17. 28   **18.** 54
19. 81   **20.** 125
21. 117   **22.** 147
23. 220   **24.** 270
25. 1,254   **26.** 1,144

Perform each operation. **See Examples 2 and 3.**

27. $\frac{5}{6} \cdot \frac{1}{8}$ **28.** $\frac{2}{3} \cdot \frac{1}{5}$

29. $\frac{7}{11} \cdot \frac{3}{5}$ **30.** $\frac{13}{9} \cdot \frac{2}{3}$

31. $\frac{3}{4} \div \frac{2}{5}$ **32.** $\frac{7}{8} \div \frac{6}{13}$

33. $\frac{6}{5} \div \frac{5}{7}$ **34.** $\frac{4}{3} \div \frac{3}{2}$

Build each fraction or whole number to an equivalent fraction with the indicated denominator. **See Example 4.**

35. $\frac{1}{3}$, denominator 9   **36.** $\frac{3}{8}$, denominator 24

37. $\frac{4}{9}$, denominator 54   **38.** $\frac{9}{16}$, denominator 64

39. 7, denominator 5   **40.** 12, denominator 3

41. 5, denominator 7   **42.** 6, denominator 8

Simplify each fraction, if possible. **See Examples 5 and 6.**

43. $\frac{6}{18}$ **44.** $\frac{6}{9}$

45. $\frac{24}{28}$ **46.** $\frac{35}{14}$

47. $\frac{15}{40}$ **48.** $\frac{22}{77}$

49. $\frac{33}{56}$ **50.** $\frac{26}{21}$

51. $\frac{26}{39}$ **52.** $\frac{72}{64}$

53. $\frac{36}{225}$ **54.** $\frac{175}{490}$

Perform the operations and, if possible, simplify. **See Objective 6 and Example 7.**

55. $\frac{3}{5} + \frac{3}{5}$ **56.** $\frac{4}{9} - \frac{1}{9}$

57. $\frac{6}{7} - \frac{2}{7}$ **58.** $\frac{5}{13} + \frac{6}{13}$

59. $\frac{1}{6} + \frac{1}{24}$ **60.** $\frac{17}{25} - \frac{2}{5}$

61. $\frac{7}{10} - \frac{1}{14}$ **62.** $\frac{9}{8} - \frac{5}{6}$

63. $\frac{2}{15} + \frac{7}{9}$ **64.** $\frac{7}{25} + \frac{3}{10}$

65. $\frac{21}{56} - \frac{9}{40}$ **66.** $\frac{13}{24} - \frac{3}{40}$

Perform the operations and, if possible, simplify. **See Example 8.**

67. $16\left(\frac{3}{2}\right)$ **68.** $30\left(\frac{5}{6}\right)$

69. $18\left(\frac{2}{9}\right)$ **70.** $14\left(\frac{3}{7}\right)$

71. $\frac{2}{3} - \frac{1}{6} + \frac{5}{18}$ **72.** $\frac{3}{5} - \frac{7}{10} + \frac{7}{20}$

73. $\frac{5}{12} + \frac{1}{3} - \frac{2}{5}$ **74.** $\frac{7}{15} + \frac{1}{5} - \frac{4}{9}$

Perform the operations and, if possible, simplify. **See Examples 9 and 10.**

75. $4\frac{2}{3} \cdot 7$ **76.** $7 \cdot 1\frac{3}{28}$

77. $8 \div 3\frac{1}{5}$ **78.** $15 \div 3\frac{1}{3}$

79. $8\frac{2}{9} - 7\frac{2}{3}$ **80.** $3\frac{4}{5} - 3\frac{1}{10}$

81. $3\frac{3}{16} + 2\frac{5}{24}$ **82.** $15\frac{5}{6} + 11\frac{5}{8}$

TRY IT YOURSELF

Perform the operations and, if possible, simplify.

83. $\frac{3}{5} + \frac{2}{3}$ **84.** $\frac{4}{3} + \frac{7}{2}$

85. $21\left(\frac{10}{3}\right)$ **86.** $28\left(\frac{4}{7}\right)$

87. $6 \cdot 2\frac{7}{24}$ **88.** $3\frac{1}{2} \cdot \frac{1}{5}$

89. $\frac{2}{3} - \frac{1}{4} + \frac{1}{12}$ **90.** $\frac{3}{7} - \frac{2}{5} + \frac{2}{35}$

91. $\frac{21}{35} \div \frac{3}{14}$ **92.** $\frac{23}{25} \div \frac{46}{5}$

93. $\frac{4}{3}\left(\frac{6}{5}\right)$ **94.** $\frac{21}{8}\left(\frac{2}{15}\right)$

95. $\frac{4}{63} + \frac{1}{45}$ **96.** $\frac{5}{18} + \frac{1}{99}$

97. $3 - \frac{3}{4}$ **98.** $4 - \frac{7}{3}$

99. $\frac{1}{2} \cdot \frac{3}{5}$ **100.** $\frac{3}{4} \cdot \frac{5}{7}$

101. $3\frac{1}{3} \div 1\frac{5}{6}$ **102.** $2\frac{1}{2} \div 1\frac{5}{8}$

103. $\frac{11}{21} - \frac{8}{21}$ **104.** $\frac{19}{35} - \frac{12}{35}$

APPLICATIONS

105. FORESTRY   A ranger cut down a pine tree and measured the widths of the outer two growth rings.

   a. What was the growth over this 2-year period?

   b. What is the difference in the widths of the rings?

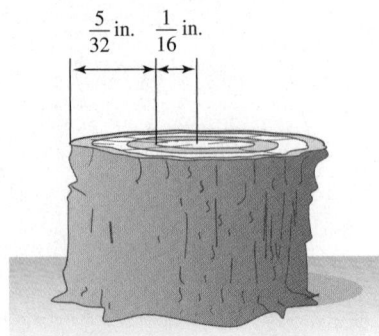

$\frac{5}{32}$ in. $\frac{1}{16}$ in.

106. HARDWARE   To secure the bracket to the stock, a bolt and a nut are used. How long should the threaded part of the bolt be?

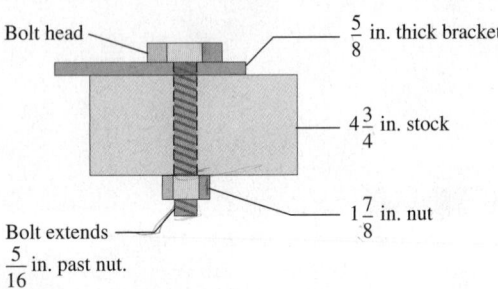

Bolt head

$\frac{5}{8}$ in. thick bracket

$4\frac{3}{4}$ in. stock

$1\frac{7}{8}$ in. nut

Bolt extends $\frac{5}{16}$ in. past nut.

107. COOKING   How much butter is left in a $10\frac{1}{2}$-pound tub of butter if $4\frac{3}{4}$ pounds are used to make a wedding cake?

108. CALORIES   A company advertises that its mints contain only 3 calories a piece. What is the calorie intake if you eat an entire package of 20 mints?

109. FRAMES   How many inches of molding are needed to make the square picture frame?

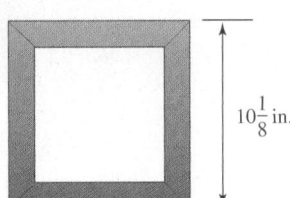

$10\frac{1}{8}$ in.

110. DECORATING   The materials used to make a pillow are shown in the next column. Examine the inventory list to decide how many pillows can be manufactured in one production run with the materials in stock.

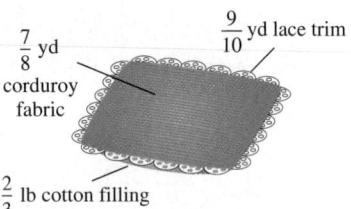

$\frac{7}{8}$ yd corduroy fabric

$\frac{9}{10}$ yd lace trim

$\frac{2}{3}$ lb cotton filling

Factory Inventory List

Materials	*Amount in stock*
Lace trim	135 yd
Corduroy fabric	154 yd
Cotton filling	98 lb

WRITING

111. Explain how to add two fractions having unlike denominators.

112. To multiply two fractions, must they have like denominators? Explain.

113. What are equivalent fractions?

114. Explain the error in the following addition.

$$\frac{4}{3} + \frac{3}{2} = \frac{4+3}{3+2} = \frac{7}{5}$$

REVIEW

Use the formula to complete each table.

115. $T = 15g$

Number of gears g	Number of teeth T
10	
12	

116. $p = r - 200$

Revenue r	Profit p
1,000	
5,000	

CHALLENGE PROBLEMS

117. Which is larger: $\frac{11}{12}$ or $\frac{8}{9}$?

118. If the circle represents a whole, find the missing value.

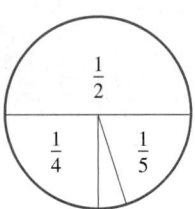

$\frac{1}{2}$

$\frac{1}{4}$

$\frac{1}{5}$

SECTION 1.3
The Real Numbers

Objectives

 Define the set of integers.

2 Define the set of rational numbers.

3 Define the set of irrational numbers.

4 Classify real numbers.

5 Graph sets of numbers on the number line.

6 Find the absolute value of a real number.

A **set** is a collection of objects, such as a set of golf clubs or a set of dishes. In this section, we will define some important sets of numbers that are used in algebra.

1   **Define the Set of Integers.**

Natural numbers are the numbers that we use for counting. To write this set, we list its **members** (or **elements**) within **braces** { }.

Natural Numbers

The set of **natural numbers** is {1, 2, 3, 4, 5, . . . }.   Read as "the set containing one, two, three, four, five, and so on."

The natural numbers, together with 0, form the set of **whole numbers.**

Whole Numbers

The set of **whole numbers** is {0, 1, 2, 3, 4, 5, . . . }.

Notation

The symbol . . . used in the previous definitions is called an **ellipsis** and it indicates that the established pattern continues forever.

Whole numbers are not adequate for describing many real-life situations. For example, if you write a check for more than what's in your account, the account balance will be less than zero.

We can use the **number line** below to visualize numbers less than zero. A number line is straight and has uniform markings. The arrowheads indicate that it extends forever in both directions. For each natural number on the number line, there is a corresponding number, called its *opposite,* to the left of 0. In the diagram, we see that 3 and −3 (negative three) are opposites, as are −5 (negative five) and 5. Note that 0 is its own opposite.

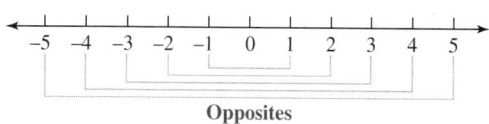

Opposites

Opposites

Two numbers that are the same distance from 0 on the number line, but on opposite sides of it, are called **opposites.**

The whole numbers, together with their opposites, form the set of **integers.**

Integers	The set of **integers** is $\{\ldots, -4, -3, -2, -1, 0, 1, 2, 3, 4, \ldots\}$.

The Language of Algebra
The *positive integers* are:
1, 2, 3, 4, 5, . . .
The *negative integers* are:
−1, −2, −3, −4, −5, . . .

On the number line, numbers greater than 0 are to the right of 0. They are called **positive numbers.** Positive numbers can be written with or without a **positive sign** +. For example, 2 = +2 (positive two). They are used to describe such quantities as an elevation above sea level (+3000 ft) or a stock market gain (25 points).

Numbers less than 0 are to the left of 0 on the number line. They are called **negative numbers.** Negative numbers are always written with a **negative sign** −. They are used to describe such quantities as an overdrawn checking account (−$75) or a below-zero temperature (−12°).

Positive and negative numbers are called **signed numbers.**

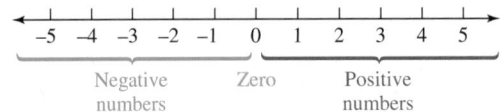

2 **Define the Set of Rational Numbers.**

We use fractions to describe many situations in daily life. For example, a morning commute might take $\frac{1}{4}$ hour or a recipe might call for $\frac{2}{3}$ cup of sugar. Fractions such as $\frac{1}{4}$ and $\frac{2}{3}$, that are quotients of two integers, are called *rational numbers.*

Rational Numbers	A **rational number** is any number that can be expressed as a fraction with an integer numerator and a nonzero integer denominator.

The Language of Algebra
Rational numbers are so named because they can be expressed as the *ratio* (quotient) of two integers.

Some other examples of rational numbers are

$$\frac{3}{8}, \qquad \frac{41}{100}, \qquad \frac{25}{25}, \quad \text{and} \quad \frac{19}{12}$$

To show that negative fractions are rational numbers, we use the following fact.

Negative Fractions	For any numbers a and b where b is not 0, $$-\frac{a}{b} = \frac{-a}{b} = \frac{a}{-b}$$

To illustrate this rule, we consider $-\frac{11}{16}$. It is a rational number because it can be written as $\frac{-11}{16}$ or as $\frac{11}{-16}$.

Positive and negative mixed numbers are also rational numbers because they can be expressed as fractions. For example,

$$7\frac{5}{8} = \frac{61}{8} \qquad \text{and} \qquad -6\frac{1}{2} = -\frac{13}{2} = \frac{-13}{2}$$

Any natural number, whole number, or integer can be expressed as a fraction with a denominator of 1. For example, $5 = \frac{5}{1}$, $0 = \frac{0}{1}$, and $-3 = \frac{-3}{1}$. Therefore, every natural number, whole number, and integer is also a rational number.

> **The Language of Algebra**
> To *terminate* means to bring to an end. In the movie *The Terminator,* actor Arnold Schwarzenegger plays a heartless machine sent to Earth to bring an end to his enemies.

Many numerical quantities are written in decimal notation. For instance, a candy bar might cost \$0.89, a dragster might travel at 203.156 mph, or a business loss might be −\$4.7 million. These decimals are called **terminating decimals** because their representations terminate (stop). As shown below, terminating decimals can be expressed as fractions. Therefore, terminating decimals are rational numbers.

$$0.89 = \frac{89}{100} \qquad 203.156 = 203\frac{156}{1000} = \frac{203{,}156}{1{,}000} \qquad -4.7 = -4\frac{7}{10} = \frac{-47}{10}$$

Decimals such as 0.3333 . . . and 2.8167167167 . . . , which have a digit (or block of digits) that repeats, are called **repeating decimals.** Since any repeating decimal can be expressed as a fraction, repeating decimals are rational numbers.

The set of rational numbers cannot be listed as we listed other sets in this section. Instead, we use **set-builder** notation.

Rational Numbers

The set of rational numbers is

$$\left\{ \frac{a}{b} \,\middle|\, a \text{ and } b \text{ are integers, with } b \neq 0. \right\}$$

Read as "the set of all numbers of the form $\frac{a}{b}$, such that a and b are integers, with $b \neq 0$."

> **The Language of Algebra**
> A fraction and its *decimal equivalent* are different notations that represent the same value.

To find the *decimal equivalent* for a fraction, we divide its numerator by its denominator. For example, to write $\frac{1}{4}$ and $\frac{5}{22}$ as decimals, we proceed as follows:

```
    0.25
 4)1.00     Write a decimal point and additional zeros to the right of 1.
    8
    20
    20
     0      The remainder is 0.
```

```
     0.22727 . . .
 22)5.00000      Write a decimal point and additional zeros to the right of 5.
    4 4
      60
      44
      160        60 and 160 continually appear as remainders. Therefore, 2
      154        and 7 will continually appear In the quotient.
        60
        44
        160
```

> **The Language of Algebra**
> Since every natural number belongs to the set of whole numbers, we say the set of natural numbers is a *subset* of the set of whole numbers. Similarly, the set of whole numbers is a subset of the set of integers, and the set of integers is a subset of the set of rational numbers.

The decimal equivalent of $\frac{1}{4}$ is 0.25 and the decimal equivalent of $\frac{5}{22}$ is 0.2272727 . . . . We can use an **overbar** to write repeating decimals in more compact form: $0.2272727\ldots = 0.2\overline{27}$. Here are more fractions and their decimal equivalents.

Terminating decimals	***Repeating decimals***
$\frac{1}{2} = 0.5$	$\frac{1}{6} = 0.166666\ldots$ or $0.1\overline{6}$
$\frac{5}{8} = 0.625$	$\frac{1}{3} = 0.333333\ldots$ or $0.\overline{3}$
$\frac{3}{4} = 0.75$	$\frac{5}{11} = 0.454545\ldots$ or $0.\overline{45}$

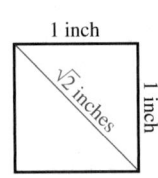

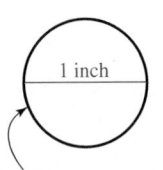

The distance around the circle is π inches.

3 Define the Set of Irrational Numbers.

Not all numbers are rational numbers. One example is the square root of 2, written $\sqrt{2}$. It is the number that, when multiplied by itself, gives 2: $\sqrt{2} \cdot \sqrt{2} = 2$. It can be shown that $\sqrt{2}$ *cannot* be written as a fraction with an integer numerator and an integer denominator. Therefore, it is not rational; it is an irrational number. It is interesting to note that a square with sides of length 1 inch has a diagonal that is $\sqrt{2}$ inches long.

The number represented by the Greek letter π (pi) is another example of an irrational number. A circle, with a 1-inch diameter, has a circumference of π inches.

Expressed in decimal form,

$$\sqrt{2} = 1.414213562\ldots \quad \text{and} \quad \pi = 3.141592654\ldots$$

These decimals neither terminate nor repeat.

| Irrational Numbers | An **irrational number** is a nonterminating, nonrepeating decimal. An irrational number cannot be expressed as a fraction with an integer numerator and an integer denominator. |

The Language of Algebra
Since π is irrational, its decimal representation has an infinite number of *decimal places*. In 2002, a University of Tokyo mathematician used a super computer to calculate π to over one trillion *decimal places*.

Other examples of irrational numbers are:

$$\sqrt{3} = 1.732050808\ldots \qquad -\sqrt{5} = -2.236067977\ldots$$
$$-\pi = -3.141592654\ldots \qquad 3\pi = 9.424777961\ldots \quad \text{3π means 3 · π.}$$

We can use a calculator to approximate the decimal value of an irrational number. To approximate $\sqrt{2}$ using a scientific calculator, we use the square root key $\sqrt{}$. To approximate π, we use the *pi* key π.

$$\sqrt{2} \approx 1.414213562 \quad \text{and} \quad \pi \approx 3.141592654$$

Rounded to the nearest thousandth, $\sqrt{2} \approx 1.414$ and $\pi \approx 3.142$.

4 Classify Real Numbers.

The set of **real numbers** is formed by combining the set of rational numbers and the set of irrational numbers. Every real number has a decimal representation. If it is rational, its corresponding decimal terminates or repeats. If it is irrational, its decimal representation is nonterminating and nonrepeating.

| The Real Numbers | A **real number** is any number that is a rational number or an irrational number. |

The following diagram shows how various sets of numbers are related. Note that a number can belong to more than one set. For example, -6 is an integer, a rational number, and a real number.

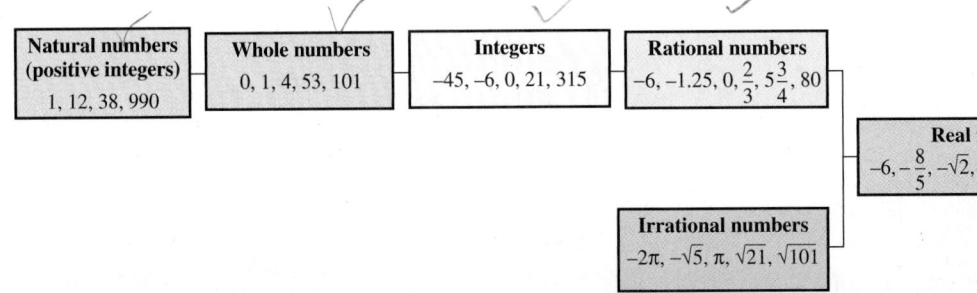

EXAMPLE 1 Which numbers in the following set are natural numbers, whole numbers, integers, rational numbers, irrational numbers, real numbers? $\left\{-3.4, \frac{2}{5}, 0, -6, 1\frac{3}{4}, \pi, 16\right\}$

Strategy We begin by scanning the given set, looking for any natural numbers. Then we scan it five more times, looking for whole numbers, for integers, for rational numbers, for irrational numbers, and finally, for real numbers.

Why We need to scan the given set of numbers six times, because numbers in that set can belong to more than one classification.

Solution

Natural numbers: 16   16 is a member of $\{1, 2, 3, 4, 5, \dots\}$.

Whole numbers: 0, 16   0 and 16 are members of $\{0, 1, 2, 3, 4, 5, \dots\}$.

Integers: 0, −6, 16   0, −6, and 16 are members of $\{\dots, -3, -2, -1, 0, 1, 2, 3, \dots\}$.

Rational numbers:

$-3.4, \frac{2}{5}, 0, -6, 1\frac{3}{4}, 16$ A rational number can be expressed as a fraction of two integers: $-3.4 = \frac{-34}{10}, 0 = \frac{0}{1}, -6 = \frac{-6}{1}, 1\frac{3}{4} = \frac{7}{4}$, and $16 = \frac{16}{1}$.

Irrational numbers: π $\pi = 3.1415\dots$ is a nonterminating, nonrepeating decimal.

Real numbers:

$-3.4, \frac{2}{5}, 0, -6, 1\frac{3}{4}, \pi, 16$ Every natural number, whole number, integer, rational number, and irrational number is a real number.

Self Check 1 Use the instructions for Example 1 with: $\left\{0.1, \sqrt{2}, -\frac{2}{7}, 45, -2, \frac{13}{4}, -6\frac{7}{8}\right\}$

Now Try **Problem 27**

5 **Graph Sets of Numbers on the Number Line.**

Every real number corresponds to a point on the number line, and every point on the number line corresponds to exactly one real number. As we move right on the number line, the values of the numbers increase. As we move left, the values decrease. On the number line below, we see that 5 is greater than −3, because 5 lies to the right of −3. Similarly, −3 is less than 5, because it lies to the left of 5.

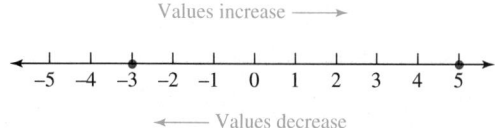

Values increase ⟶

⟵ Values decrease

The **inequality symbol** $>$ means "is greater than." It is used to show that one number is greater than another. The inequality symbol $<$ means "is less than." It is used to show that one number is less than another. For example,

$5 > -3$ Read as "5 is greater than −3."

$-3 < 5$ Read as "−3 is less than 5."

To distinguish between these inequality symbols, remember that each one points to the smaller of the two numbers involved.

$$5 > -3 \qquad\qquad\qquad -3 < 5$$

Points to the smaller number.

EXAMPLE 2   Use one of the symbols $>$ or $<$ to make each statement true:

a. $-4 \quad 4$ **b.** $-2 \quad -3$ **c.** $4.47 \quad 12.5$ **d.** $\dfrac{3}{4} \quad \dfrac{5}{8}$

Strategy   To pick the correct inequality symbol to place between a given pair of numbers, we need to determine the position of each number on a number line.

Why   For any two numbers on a number line, the number to the *left* is the smaller number and the number to the *right* is the larger number.

Solution

a. Since -4 is to the left of 4 on the number line, we have $-4 < 4$.

b. Since -2 is to the right of -3 on the number line, we have $-2 > -3$.

c. Since 4.47 is to the left of 12.5 on the number line, we have $4.47 < 12.5$.

d. To compare fractions, express them in terms of the same denominator, preferably the LCD. If we write $\frac{3}{4}$ as an equivalent fraction with denominator 8, we see that $\frac{3}{4} = \frac{3 \cdot 2}{4 \cdot 2} = \frac{6}{8}$. Therefore, $\frac{3}{4} > \frac{5}{8}$.

To compare the fractions, we could also convert each to its decimal equivalent. Since $\frac{3}{4} = 0.75$ and $\frac{5}{8} = 0.625$, we know that $\frac{3}{4} > \frac{5}{8}$.

Self Check 2   Use one of the symbols $<$ or $>$ to make each statement true:
a. $1 \_ -1$ **b.** $-5 \_ -4$ **c.** $6.7 \_ 4.999$ **d.** $\frac{3}{5} \_ \frac{2}{3}$

Now Try   Problems 37 and 44

To **graph a number** means to mark its position on the number line.

EXAMPLE 3   Graph each number in the set: $\left\{ -2.43, \ \sqrt{2}, \ 1, \ -0.\overline{3}, \ 2\dfrac{5}{6}, \ -\dfrac{3}{2} \right\}$

Strategy   We locate the position of each number on the number line, draw a bold dot, and label it.

Why   To *graph a number* means to make a drawing that represents the number.

Solution

• To locate -2.43, we round it to the nearest tenth: $-2.43 \approx -2.4$.

• To locate $\sqrt{2}$, we use a calculator: $\sqrt{2} \approx 1.4$.

• To locate $-0.\overline{3}$, we recall that $0.\overline{3} = 0.333 \ldots = \frac{1}{3}$. Therefore, $-0.\overline{3} = -\frac{1}{3}$.

• In mixed-number form, $-\frac{3}{2} = -1\frac{1}{2}$. This is midway between -1 and -2.

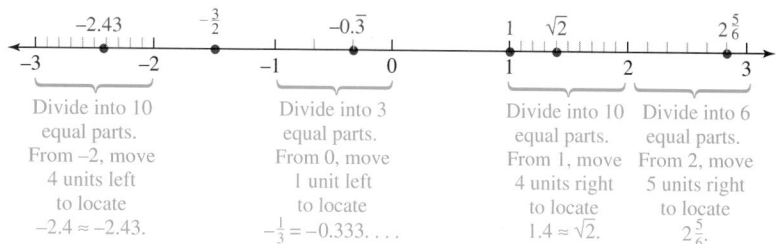

Self Check 3   Graph each number in the set: $\left\{1.7,\ \pi,\ -1\frac{3}{4},\ 0.\overline{6},\ \frac{5}{2},\ -3\right\}$

Now Try   **Problem 57**

6   **Find the Absolute Value of a Real Number.**

A number line can be used to measure the distance from one number to another. For example, in the following figure we see that the distance from 0 to -4 is 4 units and the distance from 0 to 3 is 3 units.

To express the distance that a number is from 0 on a number line, we can use absolute values.

Absolute Value	The **absolute value** of a number is its distance from 0 on the number line.

Success Tip
Since absolute value expresses distance, the absolute value of a number is always positive or zero, but never negative.

To indicate the absolute value of a number, we write the number between two vertical bars. From the figure above, we see that $|-4| = 4$. This is read as "the absolute value of negative 4 is 4" and it tells us that the distance from 0 to -4 is 4 units. It also follows from the figure that $|3| = 3$.

EXAMPLE 4   Find each absolute value: **a.** $|18|$ **b.** $\left|-\frac{7}{8}\right|$ **c.** $|98.6|$ **d.** $|0|$

Strategy   We need to determine the distance that the number within the vertical absolute value bars is from 0.

Why   The absolute value of a number is the distance between 0 and the number on a number line.

Solution

a. Since 18 is a distance of 18 from 0 on the number line, $|18| = 18$.

b. Since $-\frac{7}{8}$ is a distance of $\frac{7}{8}$ from 0 on the number line, $\left|-\frac{7}{8}\right| = \frac{7}{8}$.

c. Since 98.6 is a distance of 98.6 from 0 on the number line, $|98.6| = 98.6$.

d. Since 0 is a distance of 0 from 0 on the number line, $|0| = 0$.

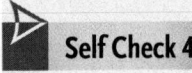

Self Check 4 Find each absolute value:

a. $|100|$ b. $|-4.7|$ c. $\left|\sqrt{2}\right|$

Now Try **Problems 61 and 67**

ANSWERS TO SELF CHECKS   **1.** Natural numbers: 45; whole numbers: 45; integers: 45, -2; rational numbers: $0.1, -\frac{2}{7}, 45, -2, \frac{13}{4}, -6\frac{7}{8}$; irrational numbers: $\sqrt{2}$; real numbers: all   **2. a.** $>$ **b.** $<$ **c.** $>$ **d.** $<$ **3.**   **4. a.** 100   **b.** 4.7   **c.** $\sqrt{2}$

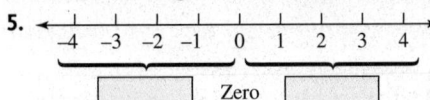

STUDY SET
1.3

VOCABULARY

Fill in the blanks.

1. The set of _____ numbers is $\{0, 1, 2, 3, 4, 5, \ldots\}$.
2. The set of _____ numbers is $\{1, 2, 3, 4, 5, \ldots\}$.
3. The set of _____ is $\{\ldots, -2, -1, 0, 1, 2, \ldots\}$.
4. Positive and negative numbers are called _____ numbers.
5.

```
  ┌──┬──┬──┬──┬──┬──┬──┬──┬──┐
 -4 -3 -2 -1  0  1  2  3  4
```
   [_____] Zero [_____]

6. The symbols $<$ and $>$ are _____ symbols.
7. A _____ number is any number that can be expressed as a fraction with an integer numerator and a nonzero integer denominator.
8. 0.25 is called a _____ decimal and 0.333 . . . is called a _____ decimal.
9. An _____ number cannot be expressed as a quotient of two integers.
10. An irrational number is a nonterminating, nonrepeating _____.
11. Every point on the number line corresponds to exactly one _____ number.
12. The _____ of a number is the distance on the number line between the number and 0.

CONCEPTS

13. Represent each situation using a signed number.
    a. A loss of $15 million
    b. A building foundation $\frac{5}{16}$ inch above grade

14. Show that each of the following numbers is a rational number by expressing it as a fraction with an integer numerator and a nonzero integer denominator: $6, -9, -\frac{7}{8}, 3\frac{1}{2}, -0.3, 2.83$.

15. Give the opposite of each number.
    a. 20   b. $-\frac{2}{3}$

16. What two numbers are a distance of 8 away from 5 on the number line?

17. What two numbers are a distance of 5 away from -9 on the number line?

18. Refer to the graph below. Use an inequality symbol, $<$ or $>$, to make each statement true.
    a. a ▢ b b. b ▢ a
    c. b ▢ 0 and a ▢ 0   d. $|a|$ ▢ $|b|$

```
  ←──●──────┼────●──→
     a      0    b
```

NOTATION

Fill in the blanks.

19. $\sqrt{2}$ is read "the _____ _____ of 2."
20. $|-15|$ is read "the _____ _____ of -15."
21. The symbol $\approx$ means ___ _____ _____ ___.
22. The symbols $\{\ \}$ are called _____.
23. The symbol π is a letter from the _____ alphabet.
24. To find the decimal equivalent for the fraction $\frac{2}{3}$ we divide: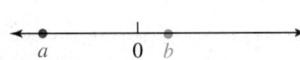
25. Fill in the blanks: $-\frac{4}{5} = \frac{\ \ }{5} = \frac{4}{\ \ }$

26. Write each repeating decimal using an overbar.

    a. 0.666 . . .   **b.** 0.2444 . . .

    c. 0.717171 . . .   **d.** 0.456456456 . . .

GUIDED PRACTICE

Place check marks in the table to show the set or sets to which each number belongs. For example, the check shows that $\sqrt{2}$ *is irrational. See Example 1.*

27.

	5	0	−3	$\frac{7}{8}$	0.17	$-9\frac{1}{4}$	$\sqrt{2}$	π
Real	✗	✗	✓	✗	✗	✗	✗	✗
Irrational							✓	✗
Rational	✗	✗	✗	✗	✗	✗		
Integer	✗	✗	✗			✗		
Whole	✗	✗			✗			
Natural	✗							

28. Which numbers in the following set are natural numbers, whole numbers, integers, rational numbers, irrational numbers, real numbers? $\left\{67, \frac{4}{13}, -5.9, 11\frac{2}{3}, \sqrt{2}, 0, -3, \pi\right\}$

Determine whether each statement is true or false. See Example 1.

29. Every whole number is an integer.

30. Every integer is a natural number.

31. Every integer is a whole number.

32. Irrational numbers are nonterminating, nonrepeating decimals.

33. Irrational numbers are real numbers.

34. Every whole number is a rational number.

35. Every rational number can be written as a fraction of two integers.

36. Every rational number is a whole number.

Use one of the symbols < or > to make each statement true. See Example 2.

37. 0   −4   **38.** 27   115

39. 5   4   **40.** 0   32

41. 917   971   **42.** 898   889

43. −2   −3   **44.** −5   −4

45. $-\frac{5}{8}$ $-\frac{3}{8}$ **46.** $-19\frac{2}{3}$ $-19\frac{1}{3}$

47. $\frac{2}{3}$ $\frac{3}{5}$ **48.** −2.27   −5.25

Write each fraction as a decimal. If the result is a repeating decimal, use an overbar. See Objective 2.

49. $\frac{5}{8}$ **50.** $\frac{3}{32}$

51. $\frac{1}{30}$ **52.** $\frac{7}{9}$

53. $\frac{1}{60}$ **54.** $\frac{5}{11}$

55. $\frac{21}{50}$ **56.** $\frac{2}{125}$

Graph each set of numbers on a number line. See Example 3.

57. $\left\{-\pi, 4.25, -1\frac{1}{2}, -0.333 . . . , \sqrt{2}, -\frac{35}{8}, 3\right\}$

58. $\left\{-2\frac{1}{8}, \pi, 2.75, -\sqrt{2}, \frac{17}{4}, 0.666 . . . , -3\right\}$

59. The integers between −5 and 2

60. The whole numbers less than 4

Find each absolute value. See Example 4.

61. $|83|$ **62.** $|29|$

63. $\left|\frac{4}{3}\right|$ **64.** $\left|\frac{9}{16}\right|$

65. $|-11|$ **66.** $|-14|$

67. $|-6.1|$ **68.** $|-25.3|$

Insert one of the symbols >, <, or = in the blank. See Examples 2 and 4.

69. $|3.4|$ −3   **70.** 0.08   0.079

71. $|-1.1|$ 1.2   **72.** −5.5 $-5\frac{1}{2}$

73. $\left|-\frac{15}{2}\right|$ 7.5   **74.** $\sqrt{2}$ π

75. $\frac{99}{100}$ 0.99   **76.** $|2|$ $|-2|$

77. 0.3   0.333 . . .   **78.** $\left|-2\frac{2}{3}\right|$ $\frac{7}{3}$

79. 1 $\left|-\frac{15}{16}\right|$ **80.** −0.666 . . .   −0.6

APPLICATIONS

81. DRAFTING   Which dimensions of the aluminum bracket shown below are natural numbers, whole numbers, integers, rational numbers, irrational numbers, and real numbers?

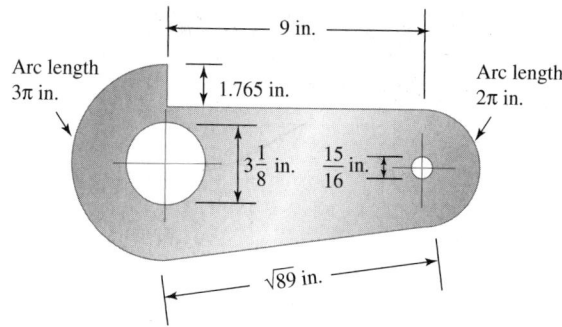

82. HISTORY   Refer to the time line shown on the next page.

    a. What basic unit was used to scale the time line?

    b. What symbolism is used to represent zero?

    c. Which numbers could be thought of as positive and which as negative?

    d. Express the dates for the Maya civilization using positive and negative numbers.

MAYA CIVILIZATION

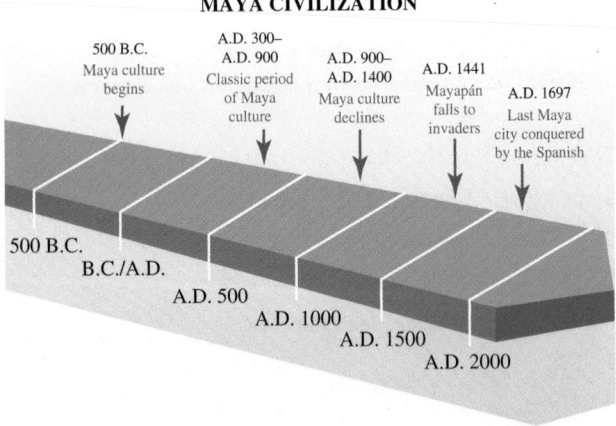

Based on data from *People in Time and Place, Western Hemisphere* (Silver Burdett & Ginn, 1991), p. 129.

83. ARCHERY   Which arrow landed farther from the target? How does the concept of absolute value apply here?

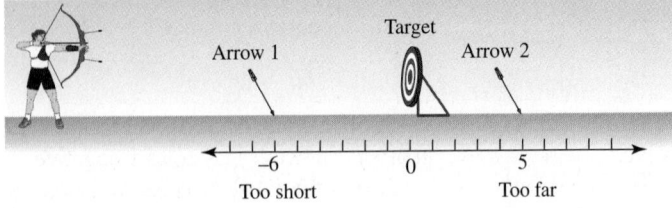

84. DRAFTING   On an architect's scale, the edge marked 16 divides each inch into 16 equal parts. Find the decimal form for each fractional part of one inch that is highlighted on the scale.

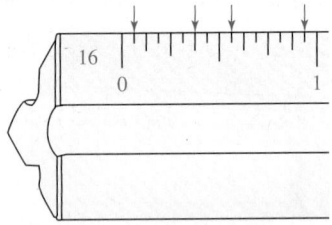

85. TRADE   Each year from 1990 through 2005, the United States imported more goods and services from Japan than it exported to Japan. This caused trade deficits, which are represented by negative numbers on the following graph.

  a. In which year was the deficit the worst? Express that deficit using a signed number.

  b. In which year was the deficit the smallest? Express that deficit using a signed number.

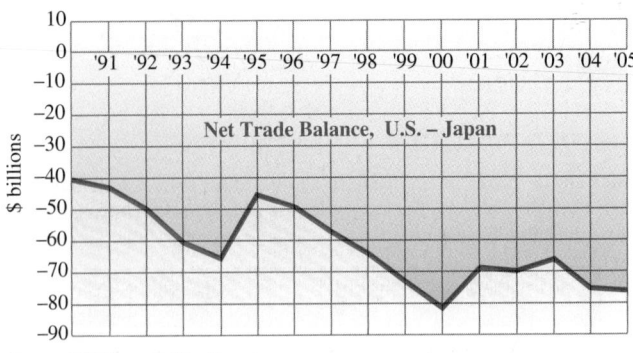

Source: U.S. Bureau of the Census

86. U.S. BUDGET   A budget *deficit* is a negative number that indicates the government spent more money than it took in that year. A budget *surplus* is a positive number that indicates the government took in more money than it spent that year. Refer to the graph.

  a. In which year was the federal budget deficit the worst? Express that deficit using a signed number.

  b. In which year was the federal budget surplus the greatest? Estimate that surplus.

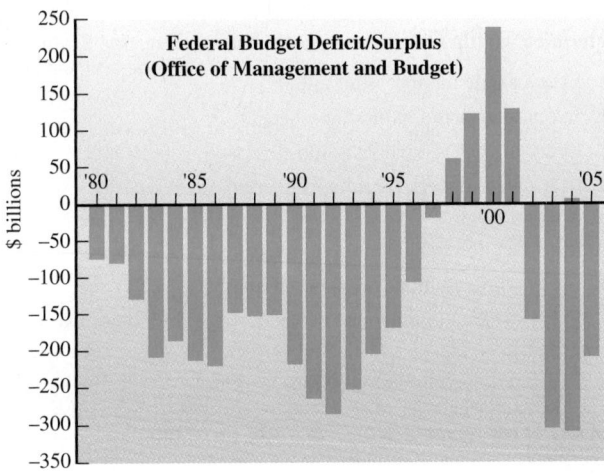

Source: U.S. Bureau of the Census

WRITING

87. Explain the difference between a rational and an irrational number.

88. Can two different numbers have the same absolute value? Explain.

89. Explain how to find the decimal equivalent of a fraction.

90. What is a real number?

91. *Pi Day* (or *Pi Approximation Day*) is an unofficial holiday held to celebrate π. Why do you think Pi Day is observed each year on March 14?

92. Explain why $0.1\overline{333}$ is not the simplest way to represent $0.1333\ldots$.

REVIEW

93. Simplify: $\frac{24}{54}$

94. Find: $\frac{3}{4}\left(\frac{8}{5}\right)$

95. Find: $5\frac{2}{3} \div 2\frac{5}{9}$

96. Find: $\frac{3}{10} + \frac{2}{15}$

CHALLENGE PROBLEMS

97. What is the set of nonnegative integers?

98. Is 0.10100100010000 . . . a repeating decimal? Explain.

Find a rational number between each pair of numbers.

99. $\frac{1}{8}$ and $\frac{1}{9}$

100. $1.7\overline{1}$ and $1.7\overline{2}$

SECTION 1.4
Adding Real Numbers; Properties of Addition

Objectives

1 Add two numbers that have the same sign.

2 Add two numbers that have different signs.

3 Use properties of addition.

4 Identify opposites (additive inverses).

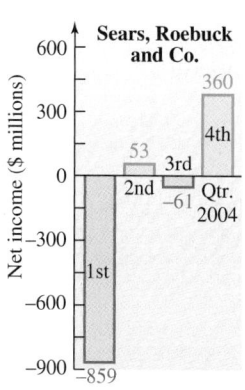

Source: 2004 Sears Annual Report

In the graph on the left, signed numbers are used to show the financial performance of Sears, Roebuck and Company for the year 2004. Positive numbers indicate *profits* and negative numbers indicate *losses*. To find Sears' net income (in millions of dollars), we need to calculate the following sum:

Net income $= -859 + 53 + (-61) + 360$

In this section, we discuss how to perform this addition and others involving signed numbers.

1 **Add Two Numbers That Have the Same Sign.**

A number line can be used to explain the addition of signed numbers. For example, to compute $5 + 2$, we begin at 0 and draw an arrow five units long that points right. It represents 5. From the tip of that arrow, we draw a second arrow two units long that points right. It represents 2. Since we end up at 7, it follows that $5 + 2 = 7$.

The Language of Algebra
The names of the parts of an addition fact are:

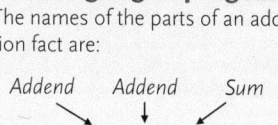

$$\overset{\text{Addend}}{5} + \overset{\text{Addend}}{2} = \overset{\text{Sum}}{7}$$

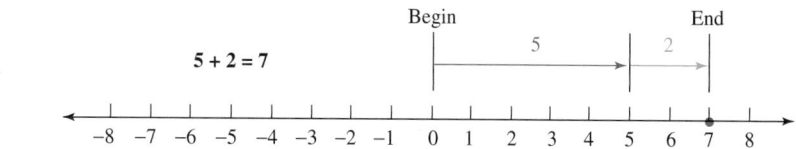

$5 + 2 = 7$

To compute $-5 + (-2)$, we begin at 0 and draw an arrow five units long that points left. It represents -5. From the tip of that arrow, we draw a second arrow two units long that points left. It represents -2. Since we end up at -7, it follows that $-5 + (-2) = -7$.

Notation
To avoid confusion, we write negative numbers within parentheses to separate the negative sign $-$ from the addition symbol $+$.

$$-5 + (-2)$$

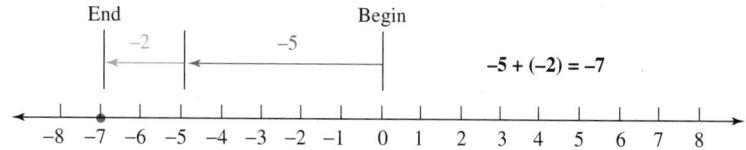

$-5 + (-2) = -7$

To check this result, think of the problem in terms of money. If you lost $5 (−5) and then lost another $2 (−2), you would have lost a total of $7 (−7).

When we use a number line to add numbers with the same sign, the arrows point in the same direction and they build upon each other. Furthermore, the answer has the same sign as the numbers that we added. These observations suggest the following rules.

Adding Two Numbers That Have the Same (Like) Signs	1. To add two positive numbers, add them as usual. The final answer is positive.
	2. To add two negative numbers, add their absolute values and make the final answer negative.

EXAMPLE 1   Add: **a.** $-20 + (-15)$ **b.** $-7.89 + (-0.6)$

c. $-\dfrac{1}{3} + \left(-\dfrac{1}{2}\right)$

Strategy   We will use the rule for adding two numbers that have the same sign.

Why   In each case, we are asked to add two negative numbers.

Solution

> *The Language of Algebra*
> Two negative numbers, as well as two positive numbers, are said to have *like* signs.

a. $-20 + (-15) = -35$ Add their absolute values, 20 and 15, to get 35. Then make the final answer negative.

b. Add their absolute values, 7.89 and 0.6.

$$\begin{array}{r} 7.89 \\ +0.6 \\ \hline 8.49 \end{array}$$ Remember to align the decimal points when adding decimals.

Then make the final answer negative: $-7.89 + (-0.6) = -8.49$.

> *Success Tip*
> The sum of two positive numbers is *always* positive. The sum of two negative numbers is *always* negative.

c. Add their absolute values, $\dfrac{1}{3}$ and $\dfrac{1}{2}$.

$$\dfrac{1}{3} + \dfrac{1}{2} = \dfrac{2}{6} + \dfrac{3}{6}$$ The LCD is 6. Build each fraction: $\dfrac{1}{3} \cdot \dfrac{2}{2} = \dfrac{2}{6}$ and $\dfrac{1}{2} \cdot \dfrac{3}{3} = \dfrac{3}{6}$.

$$= \dfrac{5}{6}$$ Add the numerators and write the sum over the LCD.

Then make the final answer negative: $-\dfrac{1}{3} + \left(-\dfrac{1}{2}\right) = -\dfrac{5}{6}$.

 Self Check 1   Add: **a.** $-51 + (-9)$ **b.** $-12.3 + (-0.88)$

c. $-\dfrac{1}{4} + \left(-\dfrac{2}{3}\right)$

Now Try   **Problems 15, 21, and 25**

---

2   Add Two Numbers That Have Different Signs.

To compute $5 + (-2)$, we begin at 0 and draw an arrow five units long that points right. From the tip of that arrow, we draw a second arrow two units long that points left. Since we end up at 3, it follows that $5 + (-2) = 3$. In terms of money, if you won $5 and then lost $2, you would have $3 left.

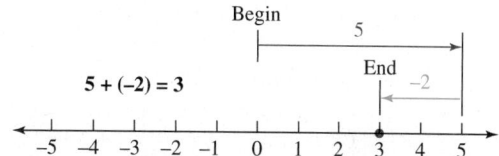

To compute $-5 + 2$, we begin at 0 and draw an arrow five units long that points left. From the tip of that arrow, we draw a second arrow two units long that points right. Since we end up at -3, it follows that $-5 + 2 = -3$. In terms of money, if you lost $5 and then won $2, you have lost $3.

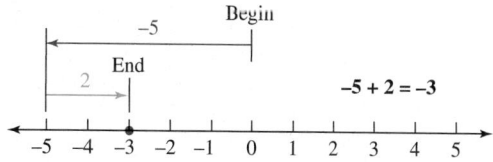

When we use a number line to add numbers with different signs, the arrows point in opposite directions and the longer arrow determines the sign of the answer. If the longer arrow represents a positive number, the sum is positive. If it represents a negative number, the sum is negative. These observations suggest the following rules.

The Language of Algebra
A positive number and a negative number are said to have *unlike* signs.

Adding Two Numbers That Have Different (Unlike) Signs

To add a positive number and a negative number, subtract the smaller absolute value from the larger.

1. If the positive number has the larger absolute value, the final answer is positive.

2. If the negative number has the larger absolute value, make the final answer negative.

EXAMPLE 2   Add: **a.** $-20 + 32$ **b.** $5.7 + (-7.4)$ **c.** $-\dfrac{19}{25} + \dfrac{2}{5}$

Strategy   We will use the rule for adding two numbers that have different (unlike) signs.

Why   In each case, we are asked to add a positive number and a negative number.

Solution

a. $-20 + 32 = 12$ *Subtract the smaller absolute value from the larger: $32 - 20 = 12$. The positive number, 32, has the larger absolute value, so the final answer is positive.*

b. Subtract the smaller absolute value, 5.7, from the larger, 7.4.

$$\begin{array}{r} 7.4 \\ -5.7 \\ \hline 1.7 \end{array}$$ *Remember to align the decimal points when subtracting decimals.*

Since the negative decimal, -7.4, has the larger absolute value, make the final answer negative: $5.7 + (-7.4) = -1.7$.

c. Since $\dfrac{2}{5} = \dfrac{10}{25}$, the fraction $-\dfrac{19}{25}$ has the larger absolute value. We subtract the smaller absolute value from the larger:

$$\dfrac{19}{25} - \dfrac{2}{5} = \dfrac{19}{25} - \dfrac{10}{25}$$ *Replace $\dfrac{2}{5}$ with the equivalent fraction $\dfrac{10}{25}$.*

$$= \dfrac{9}{25}$$ *Subtract the numerators and write the difference over the LCD.*

Success Tip
The sum of two numbers with different signs may be positive or negative. The sign of the sum is the sign of the number with the greater absolute value.

Calculators

Entering negative numbers
We don't do anything special to enter positive numbers on a calculator. To enter a negative number, say -7.4, some calculators require the $-$ sign to be entered before entering 7.4 while others require the $-$ sign to be entered after entering 7.4. Consult your owner's manual to determine the proper keystrokes.

Since the negative fraction $-\frac{19}{25}$ has the larger absolute value, make the final answer negative: $-\frac{19}{25} + \frac{10}{25} = -\frac{9}{25}$.

Self Check 2   Add: **a.** $63 + (-87)$ **b.** $-6.27 + 8$
c. $-\frac{1}{10} + \frac{1}{2}$

Now Try   **Problems 29, 33, and 35**

EXAMPLE 3   ***Accounting.***   Find the net earnings of Sears, Roebuck and Company for the year 2004 using the data in the graph on page 35.

Strategy   To find the net income, we will add the quarterly profits and losses (in millions of dollars), performing the additions as they occur from left to right.

Why   The phrase *net income* means that we should combine (add) the quarterly profits and losses to determine whether there was an overall profit or loss that year.

Solution

$$-859 + 53 + (-61) + 360 = -806 + (-61) + 360 \qquad \text{Add: } -859 + 53 = -806.$$
$$= -867 + 360 \qquad \text{Add: } -806 + (-61) = -867.$$
$$= -507$$

In 2004, Sears' net income was $-\$507$ million.

> **The Language of Algebra**
> *Net* refers to what remains after all the deductions (losses) have been accounted for. *Net income* is a term used in business that often is referred to as the *bottom line*. Net income indicates what a company has earned (or lost) in a given period of time (usually one year).

Self Check 3   Add: $650 + (-13) + 87 + (-155)$
Now Try   **Problem 43**

---

③   **Use Properties of Addition.**

The addition of two numbers can be done in any order and the result is the same. For example, $8 + (-1) = 7$ and $-1 + 8 = 7$. This example illustrates that addition is **commutative.**

The Commutative Property of Addition

Changing the order when adding does not affect the answer.
For any real numbers a and b,
$$a + b = b + a$$

> **The Language of Algebra**
> *Commutative* is a form of the word *commute*, meaning to go back and forth. *Commuter* trains take people to and from work.

In the following example, we add $-3 + 7 + 5$ in two ways. We will use grouping symbols ( ), called **parentheses,** to show this. Standard practice requires that the operation within the parentheses be performed first.

Method 1: Group -3 and 7   *Method 2: Group 7 and 5*
$$(-3 + 7) + 5 = 4 + 5 \qquad\qquad -3 + (7 + 5) = -3 + 12$$
$$= 9 \qquad\qquad\qquad\qquad = 9$$

It doesn't matter how we group the numbers in this addition; the result is 9. This example illustrates that addition is **associative.**

| The Associative Property of Addition |  | Changing the grouping when adding does not affect the answer. For any real numbers a, b, and c, $$(a + b) + c = a + (b + c)$$ |

Sometimes, an application of the associative property can simplify a computation.

EXAMPLE 4   Find the sum: $98 + (2 + 17)$

Strategy   We will use the associative property to group 2 with 98. Then, we evaluate the expression by following the rules for the order of operations.

The Language of Algebra
Associative is a form of the word *associate*, meaning to join a group. The NBA (National Basketball Association) is a group of professional basketball teams.

Why   It is helpful to regroup because 98 and 2 are a pair of numbers that are easily added.

Solution

$$98 + (2 + 17) = (98 + 2) + 17 \quad \text{Use the associative property of addition to regroup.}$$
$$= 100 + 17 \quad \text{Do the addition within the parentheses first.}$$
$$= 117$$

▷ **Self Check 4**   Find the sum: $(39 + 25) + 75$
Now Try   **Problem 53**

EXAMPLE 5   ***Game Shows.***   A contestant on *Jeopardy!* correctly answered the first question to win $100, missed the second to lose $200, correctly answered the third to win $300, and missed the fourth to lose $400. What is her score after answering four questions?

Strategy   We can represent money won by a positive number and money lost by a negative number. Her score is the sum of 100, -200, 300, and -400. Instead of doing the additions from left to right, we will use another approach. Applying the commutative and associative properties, we will add the positives, add the negatives, and then add those results.

Why   It is easier to add numbers that have the same sign than numbers that have different signs. This method minimizes the possibility of an error, because we only have to add numbers that have different signs once.

Solution

$$100 + (-200) + 300 + (-400)$$
$$= (100 + 300) + [(-200) + (-400)] \quad \text{Reorder the numbers. Group the positives together. Group the negatives together using brackets [ ].}$$
$$= 400 + (-600) \quad \text{Add the positives. Add the negatives.}$$
$$= -200 \quad \text{Add the results.}$$

After four questions, her score was $-\$200$, which represents a loss of $200.

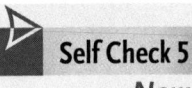 **Self Check 5**   Add: $-6 + 1 + (-4) + (-5) + 9$

Now Try   **Problem 49**

The Language of Algebra
Identity is a form of the word *identical*, meaning the same. You have probably seen *identical* twins.

Whenever we add 0 to a number, the result is the number. Therefore, $8 + 0 = 8$, $2.3 + 0 = 2.3$, and $0 + (-16) = -16$. These examples illustrate the **addition property of 0**. Since any number added to 0 remains the same, 0 is called the **identity element** for addition.

Addition Property of 0 (Identity Property of Addition)	When 0 is added to any real number, the result is the same real number. For any real number a, $$a + 0 = a \quad \text{and} \quad 0 + a = a$$

4 Identify Opposites (Additive Inverses).

Recall that two numbers that are the same distance from 0 on a number line, but on opposite sides of it, are called **opposites**. To develop a property for adding opposites, we will find $-4 + 4$ using a number line. We begin at 0 and draw an arrow four units long that points left, to represent -4. From the tip of that arrow, we draw a second arrow, four units long that points right, to represent 4. We end up at 0; therefore, $-4 + 4 = 0$.

The Language of Algebra
Don't confuse the words *opposite* and *reciprocal*. The opposite of 4 is -4. The reciprocal of 4 is $\frac{1}{4}$.

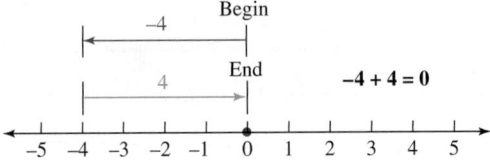

This example illustrates that when we add opposites, the result is 0. Therefore, $1.6 + (-1.6) = 0$ and $-\frac{3}{4} + \frac{3}{4} = 0$. Also, whenever the sum of two numbers is 0, those numbers are opposites. For these reasons, opposites are also called **additive inverses**.

Addition Property of Opposites (Inverse Property of Addition)	The sum of a number and its opposite (additive inverse) is 0. For any real number a and its opposite or additive inverse $-a$, $$a + (-a) = 0 \quad \text{Read } -a \text{ as "the opposite of } a\text{."}$$

EXAMPLE 6   Add: $12 + (-5) + 6 + 5 + (-12)$

Strategy   Instead of working from left to right, we will use the commutative and associative properties of addition to add pairs of opposites.

Why   Since the sum of a number and its opposite is 0, it is helpful to identify such pairs in an addition.

Solution

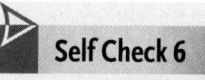

$$12 + (-5) + 6 + 5 + (-12) = 0 + 0 + 6$$

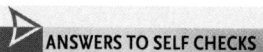

$$= 6$$

Self Check 6 Add: $8 + (-1) + 6 + 5 + (-8) + 1$
Now Try **Problem 61**

ANSWERS TO SELF CHECKS **1. a.** -60 **b.** -13.18 **c.** $-\frac{11}{12}$ **2. a.** -24 **b.** 1.73 **c.** $\frac{2}{5}$ **3.** 569
4. 139 **5.** -5 **6.** 11

STUDY SET
1.4

VOCABULARY

Fill in the blanks.

1. In the addition statement $-2 + 5 = 3$, the result, 3, is called the _____.

2. Two numbers that are the same distance from 0 on a number line, but on opposite sides of it, are called _____ or additive _____.

3. The _____ property of addition states that changing the order when adding does not affect the answer. The _____ property of addition states that changing the grouping when adding does not affect the answer.

4. Since any number added to 0 remains the same (is identical), the number 0 is called the _____ element for addition.

CONCEPTS

5. For each pair of numbers, which one has the larger absolute value?
   a. 6 or 5 **b.** 8.9 or -9.2

6. Determine whether each statement is true or false.
   a. The sum of a number and its opposite is always 0.
   b. The sum of two negative numbers is always negative.
   c. The sum of two numbers with different signs is always negative.

7. For each addition, just determine the sign of the answer.
   a. $39 + (-64)$ **b.** $-189 + 198$

8. Complete each property of addition. Then give its name.
   a. $a + (-a) =$
   b. $a + 0 =$
   c. $a + b = b +$
   d. $(a + b) + c = a +$

9. Use the commutative property of addition to complete each statement.
   a. $-5 + 1 =$ _____
   b. $15 + (-80.5) =$ _____
   c. $-20 + (4 + 20) = -20 + ($ _____ $)$
   d. $(2.1 + 3) + 6 = ($ _____ $) + 6$

10. Use the associative property of addition to complete each statement.
   a. $(-6 + 2) + 8 =$ _____
   b. $-7 + (7 + 3) =$ _____

11. What properties were used in Step 1 and Step 2 of the solution?

$$(99 + 4) + 1 = (4 + 99) + 1 \quad \text{Step 1}$$
$$= 4 + (99 + 1) \quad \text{Step 2}$$
$$= 4 + 100$$
$$= 104$$

12. Consider: $-3 + 6 + (-9) + 8 + (-4)$
   a. Add all the positives in the expression.
   b. Add all of the negatives.
   c. Add the results from parts **a** and **b**.

NOTATION

13. a. Express the commutative property of addition using the variables x and y.

b. Express the associative property of addition using the variables x, y, and z.

14. Fill in the blank: We read $-a$ as "the _____ of a."

GUIDED PRACTICE

Add. See Example 1.

15. $-8 + (-1)$

16. $-3 + (-2)$

17. $-5 + (-12)$

18. $-4 + (-14)$

19. $-29 + (-45)$

20. $-23 + (-31)$

21. $-4.2 + (-6.1)$

22. $-5.1 + (-5.1)$

23. $-\dfrac{3}{4} + \left(-\dfrac{2}{3}\right)$

24. $-\dfrac{1}{5} + \left(-\dfrac{3}{4}\right)$

25. $-\dfrac{1}{4} + \left(-\dfrac{1}{10}\right)$

26. $-\dfrac{3}{8} + \left(-\dfrac{1}{3}\right)$

Add. See Example 2.

27. $-7 + 4$

28. $-9 + 7$

29. $50 + (-11)$

30. $27 + (-30)$

31. $15.84 + (-15.84)$

32. $9.19 + (-9.19)$

33. $-6.25 + 8.5$

34. $21.37 + (-12.1)$

35. $-\dfrac{7}{15} + \dfrac{3}{15}$

36. $-\dfrac{8}{11} + \dfrac{3}{11}$

37. $\dfrac{1}{2} + \left(-\dfrac{1}{8}\right)$

38. $\dfrac{5}{6} + \left(-\dfrac{1}{4}\right)$

Add. See Examples 3 and 5.

39. $8 + (-5) + 13$

40. $17 + (-12) + (-23)$

41. $21 + (-27) + (-9)$

42. $-32 + 12 + 17$

43. $-27 + (-3) + (-13) + 22$

44. $53 + (-27) + (-32) + (-7)$

45. $-20 + (-16) + 10$

46. $-13 + (-16) + 4$

47. $19.35 + (-20.21) + 1.53$

48. $33.12 + (-35.7) + 2.98$

49. $-60 + 70 + (-10) + (-10) + 205$

50. $-100 + 200 + (-300) + (-100) + 200$

Apply the associative property of addition to find the sum. See Example 4.

51. $-99 + (99 + 215)$

52. $67 + (-67 + 127)$

53. $(-112 + 56) + (-56)$

54. $(-67 + 5) + (-5)$

55. $\dfrac{1}{8} + \left(\dfrac{7}{8} + \dfrac{2}{3}\right)$

56. $\left(\dfrac{1}{2} + \dfrac{9}{16}\right) + \dfrac{7}{16}$

57. $(12.4 + 1.9) + 1.1$

58. $87.6 + (2.4 + 1.7)$

Add. See Example 6.

59. $-1 + 9 + 1$

60. $5 + 8 + (-5)$

61. $-7 + 5 + (-10) + 7$

62. $-3 + 6 + (-9) + (-6)$

63. $-8 + 11 + (-11) + 8 + 1$

64. $2 + 15 + (-15) + 8 + (-2)$

65. $-2.1 + 6.5 + (-8.2) + 2.1$

66. $0.9 + 0.5 + (-0.2) + (-0.9)$

TRY IT YOURSELF

Add.

67. $-9 + 81 + (-2)$

68. $11 + (-21) + (-13)$

69. $0 + (-6.6)$

70. $0 + (-2.14)$

71. $-\dfrac{9}{16} + \dfrac{7}{16}$

72. $-\dfrac{3}{4} + \dfrac{1}{4}$

73. $-6 + (-8)$

74. $-4 + (-3)$

75. $-167 + 167$

76. $-25 + 25$

77. $19.2 + (-41.3)$

78. $57.93 + (-93.27)$

79. $2,345 + (-178)$

80. $-4,061 + 5,000$

81. $3 + (-6) + (-3) + 74$

82. $4 + (-3) + (-4) + 5$

83. $-\dfrac{1}{4} + \left(-\dfrac{2}{7}\right)$

84. $-\dfrac{3}{32} + \left(-\dfrac{1}{2}\right)$

85. $-0.2 + (-0.3) + (-0.4)$

86. $-0.9 + (-1.9) + (-2.9)$

APPLICATIONS

87. MILITARY SCIENCE   During a battle, an army retreated 1,500 meters, regrouped, and advanced 2,400 meters. The next day, it advanced another 1,250 meters. Find the army's net gain.

88. HEALTH   Find the point total for the six risk factors (in blue) on the medical questionnaire. Then use the table to determine the patient's risk of contracting heart disease in the next 10 years.

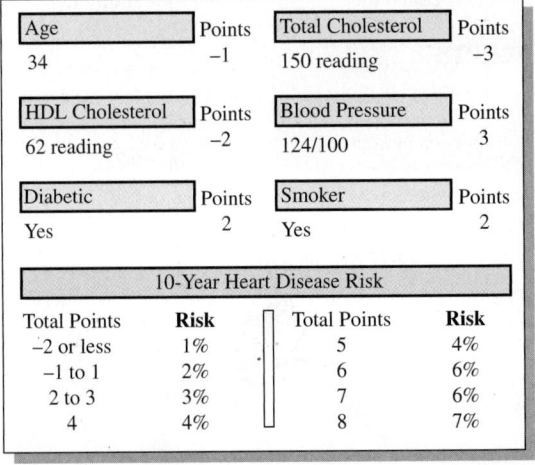

Age	Points	Total Cholesterol	Points
34	−1	150 reading	−3

HDL Cholesterol	Points	Blood Pressure	Points
62 reading	−2	124/100	3

Diabetic	Points	Smoker	Points
Yes	2	Yes	2

10-Year Heart Disease Risk			
Total Points	**Risk**	**Total Points**	**Risk**
−2 or less	1%	5	4%
−1 to 1	2%	6	6%
2 to 3	3%	7	6%
4	4%	8	7%

89. GOLF The leaderboard below shows the top finishers from the 1997 Masters Golf Tournament. Scores for each round are compared to *par,* the standard number of strokes necessary to complete the course. A score of −2, for example, indicates that the golfer used two strokes less than par to complete the course. A score of 5 indicates five strokes more than par. Determine the tournament total for each golfer.

Leaderboard

	Round				
	1	**2**	**3**	**4**	**Total**
Tiger Woods	−2	−6	−7	−3	
Tom Kite	+5	−3	−6	−2	
Tommy Tolles	0	0	0	−5	
Tom Watson	+3	−4	−3	0	

90. SUBMARINES A submarine was cruising at a depth of 1,250 feet. The captain gave the order to climb 550 feet. Compared to sea level, find the new depth of the sub.

91. CREDIT CARDS Refer to the monthly statement. What is the new balance?

Previous Balance	New Purchases, Fees, Advances & Debits	Payments & Credits	New Balance
3,660.66	1,408.78	3,826.58	
04/21/08 Billing Date	05/16/08 Date Payment Due	9,100 Credit Line	

92. POLITICS The following proposal to limit campaign contributions was on the ballot in a state election, and it passed. What will be the net fiscal impact on the state government?

Proposition

212 **Campaign Spending Limits**   YES ☐   NO ☐

Limits contributions to $200 in state campaigns. Financial impact: Cost of $4.5 million for enforcement. Increases state revenue by $6.7 million by eliminating tax deductions for lobbying.

93. MOVIE LOSSES According to the Numbers Box Office Data website, the movie *Stealth,* released in 2005 by Sony Pictures, cost about $176,350,000 to produce, promote, and distribute. It reportedly earned back just $76,700,000 worldwide. Express the dollar loss suffered by Sony as a signed number.

94. STOCKS The last entry on the line for June 12 indicates that one share of Walt Disney Co. stock lost $0.81 in value that day. How much did the value of a share of Disney stock rise or fall over the 5-day period?

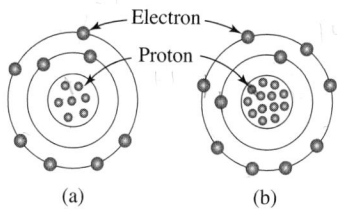

June 12	43.88	23.38	Disney	.21	0.5	87	−43	40.75	−.81
June 13	43.88	23.38	Disney	.21	0.5	86	−15	40.19	−.56
June 14	43.88	23.38	Disney	.21	0.5	87	−50	41.00	+.81
June 15	43.88	23.38	Disney	.21	0.5	89	−28	41.81	+.81
June 16	43.88	23.38	Disney				−15	41.19	−.63

Based on data from the *Los Angeles Times*

95. CHEMISTRY An atom is composed of protons (with a charge of +1), neutrons (with no charge), and electrons (with a charge of −1). Two simple models of atoms are shown. What is the overall charge of each atom?

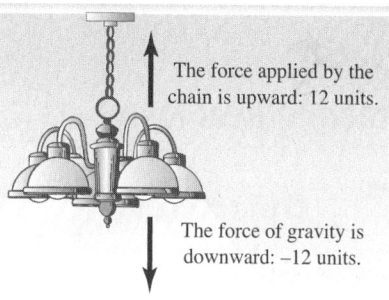

(a)   (b)

96. PHYSICS In the illustration, arrows show the two forces acting on a lamp hanging from a ceiling. What is the sum of the forces?

The force applied by the chain is upward: 12 units.

The force of gravity is downward: −12 units.

97. THE BIG EASY The city of New Orleans lies, on average, 6 feet below sea level. What is the elevation of the top of an 85-foot tall building in New Orleans?

98. ELECTRONICS A closed circuit contains two batteries and three resistors. The sum of the voltages in the loop must be 0. Is it?

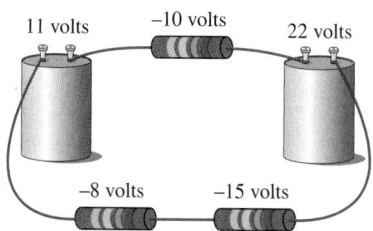

11 volts   −10 volts   22 volts

−8 volts   −15 volts

99. ACCOUNTING   The 2004 quarterly profits and losses of Greyhound Bus Lines are shown in the table. Losses are denoted using parentheses. Calculate the company's total net income for 2004.

Quarter	Net income ($ million)
1st	(21.1)
2nd	(4.3)
3rd	2.6
4th	(0.4)

Source: www.greyhound.com

100. POLITICS   Six months before an election, the incumbent trailed the challenger by 18 points. To overtake her opponent, the incumbent decided to use a four-part strategy. Each part of the plan is shown below, with the expected point gain. With these gains, will the incumbent overtake the challenger on election day?

- TV ads   +10
- Voter mailing   +3
- Union endorsement   +2
- Telephone calls   +1

WRITING

101. Explain why the sum of two positive numbers is always positive and the sum of two negative numbers is always negative.

102. Explain why the sum of a negative number and a positive number is sometimes positive, sometimes negative, and sometimes zero.

REVIEW

103. True or false: Every real number can be expressed as a decimal.

104. Multiply: $\dfrac{1}{3} \cdot \dfrac{1}{3}$

105. What two numbers are a distance of 6 away from -3 on the number line?

106. Graph: $\left\{ -2.5, \ \sqrt{2}, \ \frac{11}{3}, \ -0.333 \ldots, \ 0.75 \right\}$

CHALLENGE PROBLEMS

107. A set is said to be *closed under addition* if the sum of any two of its members is also a member of the set. Is the set $\{-1, 0, 1\}$ a closed set under addition? Explain.

108. Think of two numbers. First, add the absolute value of the two numbers, and write your answer. Second, add the two numbers, take the absolute value of that sum, and write that answer. Do the two answers agree? Can you find two numbers that produce different answers? When do you get answers that agree, and when don't you?

SECTION 1.5
Subtracting Real Numbers

Objectives ❶ Use the definition of subtraction.
❷ Solve application problems using subtraction.

In this section, we discuss a rule to use when subtracting signed numbers.

❶ **Use the Definition of Subtraction.**

A minus symbol $-$ is used to indicate subtraction. However, this symbol is also used in two other ways, depending on where it appears in an expression.

$5 - 18$ This is read as "five minus eighteen."

-5 This is usually read as "negative five." It could also be read as "the additive inverse of five" or "the opposite of five."

$-(-5)$ This is usually read as "the opposite of negative five." It could also be read as "the additive inverse of negative five."

In $-(-5)$, parentheses are used to write the opposite of a negative number. When such expressions are encountered in computations, we simplify them by finding the opposite of the number within the parentheses.

$-(-5) = 5$ Read as "the opposite of negative five is five."

This observation illustrates the following rule.

Opposite of an Opposite	The opposite of the opposite of a number is that number. For any real number a,
	$$-(-a) = a$$ Read as "the opposite of the opposite of a is a."

EXAMPLE 1   Simplify each expression: **a.** $-(-45)$ **b.** $-(-h)$ **c.** $-|-10|$

Strategy   To simplify each expression, we will use the concept of opposite.

Why   In each case, the outermost $-$ symbol is read as "the opposite."

Solution

a. The number within the parentheses is -45. Its opposite is 45. Therefore, $-(-45) = 45$.

b. The opposite of the opposite of h is h. Therefore, $-(-h) = h$.

c. The notation $-|-10|$ means "the opposite of the absolute value of negative ten." Since $|-10| = 10$, we have:

$$-|-10| = -10$$ The absolute value bars is do not affect the $-$ symbol outside them. Therefore, the result is negative.

Self Check 1   Simplify each expression: **a.** $-(-1)$ **b.** $-(-y)$
c. $-|-500|$

Now Try   **Problems 13, 15, and 17**

To develop a rule for subtraction, we consider the following illustration. It represents the subtraction $5 - 2 = 3$.

The Language of Algebra
The names of the parts of a subtraction fact are:

Minuend   Subtrahend
$5 - 2 = 3$
Difference

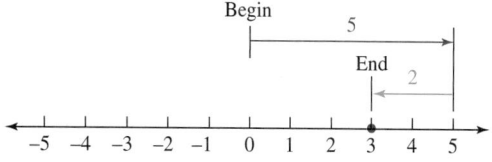

The illustration above also represents the addition $5 + (-2) = 3$. We see that

Subtracting 2 from 5   is the same as   adding the opposite of 2 to 5.

$$5 - 2 = 3 \qquad\qquad 5 + (-2) = 3$$

The results are the same.

This observation suggests the following definition.

Subtraction of Real Numbers	To subtract two real numbers, add the first number to the opposite (additive inverse) of the number to be subtracted.
	For any real numbers a and b,
	$$a - b = a + (-b)$$

EXAMPLE 2   Subtract and check the result.

$$\textbf{a.}\ -13 - 8 \quad \textbf{b.}\ -7 - (-45) \quad \textbf{c.}\ \frac{1}{4} - \left(-\frac{1}{8}\right)$$

Strategy   To find each difference, we will apply the rule for subtraction: Add the first number to the opposite of the number to be subtracted.

Why   It is easy to make an error when subtracting signed numbers. We will probably be more accurate if we write each subtraction as addition of the opposite.

Solution

a. We read $-13 - 8$ as "negative thirteen *minus* eight." Subtracting 8 is the same as adding -8.

Change the subtraction to addition.

$$-13 - 8 \quad = \quad -13 + (-8) \quad = \quad -21$$

Change the number being subtracted to its opposite.

To check, we add the *difference*, -21, and the *subtrahend*, 8, to obtain the *minuend*, -13.

Check: $-21 + 8 = -13$

b. We read $-7 - (-45)$ as "negative seven *minus* negative forty-five." Subtracting -45 is the same as adding 45.

Add

$$-7 - (-45) \quad = \quad -7 + 45 = 38$$

the opposite

Check: $38 + (-45) = -7$

c. $\dfrac{1}{4} - \left(-\dfrac{1}{8}\right) = \dfrac{2}{8} - \left(-\dfrac{1}{8}\right)$ Express $\frac{1}{4}$ in terms of the LCD 8: $\frac{1}{4} \cdot \frac{2}{2} = \frac{2}{8}$.

$$= \frac{2}{8} + \frac{1}{8} \qquad \text{To subtract, add the opposite.}$$

$$= \frac{3}{8}$$

Check: $\frac{3}{8} + \left(-\frac{1}{8}\right) = \frac{2}{8} = \frac{1}{4}$

> **Self Check 2**   Subtract: **a.** $-32 - 25$ **b.** $17 - (-12)$
> **c.** $-\frac{1}{3} - \left(-\frac{3}{4}\right)$
>
> **Now Try**   **Problems 25, 37, and 53**

The Language of Algebra
When we change a number to its opposite, we say we have *changed* (or *reversed*) its sign.

The Language of Algebra
The rule for subtracting real numbers is often summarized as: *Subtracting a number is the same as adding its opposite.*

Calculators
The subtraction key
When using a calculator to subtract signed numbers, be careful to distinguish between the *subtraction* key $-$ and the keys that are used to enter negative values: $+/-$ on a scientific calculator and $(-)$ on a graphing calculator.

EXAMPLE 3   **a.** Subtract 0.5 from 4.6   **b.** Subtract 4.6 from 0.5

Strategy   We will translate each phrase to mathematical symbols and then perform the subtraction. We must be careful when translating the instruction to subtract one number *from* another number.

Why   The order of the numbers in each word phrase must be reversed when we translate it to mathematical symbols.

Solution

a. The number to be subtracted is 0.5.

Subtract 0.5 from 4.6

$4.6 - 0.5 = 4.1$ To translate, reverse the order in which 0.5 and 4.6 appear in the sentence.

b. The number to be subtracted is 4.6.

Subtract 4.6 from 0.5

$0.5 - 4.6 = 0.5 + (-4.6)$ To translate, reverse the order in which 4.6 and 0.5 appear in the sentence. Add the opposite of 4.6.

$= -4.1$

Caution   Notice from parts **a** and **b** that $4.6 - 0.5 \neq 0.5 - 4.6$. This result illustrates an important fact: subtraction is *not* commutative. When subtracting two numbers, it is important that we write them in the correct order, because, in general, $a - b \neq b - a$.

Self Check 3   **a.** Subtract 2.2 from 4.9   **b.** Subtract 4.9 from 2.2
Now Try   **Problem 57**

EXAMPLE 4   Perform the operations: $-9 - 15 + 20 - (-6)$

Strategy   This expression contains addition and subtraction. We will write each subtraction as addition of the opposite and then evaluate the expression.

Why   It is easy to make an error when subtracting signed numbers. We will probably be more accurate if we write each subtraction as addition of the opposite.

Solution

$$-9 - 15 + 20 - (-6) = -9 + (-15) + 20 + 6$$
$$= -24 + 26 \qquad \text{Add the negatives. Add the positives. Add the results.}$$
$$= 2$$

Self Check 4   Perform the operations: $-40 - (-10) + 7 - (-15)$
Now Try   **Problem 63**

2   Solve Application Problems Using Subtraction.

Subtraction finds the *difference* between two numbers. When we find the difference between the maximum value and the minimum value of a collection of measurements, we are finding the **range** of the values.

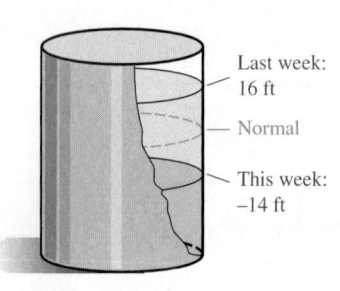

134° —

0° —

−80° —

Difference in temperature extremes

EXAMPLE 5   *U.S. Temperatures.*   The record high temperature in the United States was 134°F in Death Valley, California, on July 10, 1913. The record low was −80°F at Prospect Creek, Alaska, on January 23, 1971. Find the temperature range for these extremes.

Strategy   We will subtract the lowest temperature from the highest temperature.

Why   The *range* of a collection of data indicates the spread of the data. It is the difference between the largest and smallest values.

Solution

$$134 - (-80) = 134 + 80 \qquad \text{134° is the higher temperature and −80° is the lower.}$$
$$= 214$$

The temperature range for these extremes is 214°F.

 Now Try   **Problem 89**

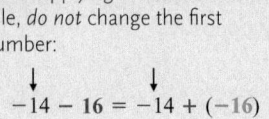

Last week: 16 ft

Normal

This week: −14 ft

EXAMPLE 6   *Water Levels.*   In one week, the water level in a storage tank went from 16 feet above normal to 14 feet below normal. Find the change in the water level.

Strategy   We can represent a water level above normal using a positive number and a water level below normal using a negative number. To find the change in the water level, we will subtract.

Why   In general, *to find the change in a quantity, we subtract the earlier value from the later value.*

Solution

$$-14 - 16 = -14 + (-16) \qquad \text{The earlier water level, 16, is subtracted from the later water level, −14.}$$
$$= -30$$

The negative result indicates that the water level fell 30 feet that week.

Caution

When applying the subtraction rule, *do not* change the first number:

$$\overset{\downarrow}{-14} - \overset{\downarrow}{16} = -14 + (\mathbf{-16})$$

 Now Try   **Problem 91**

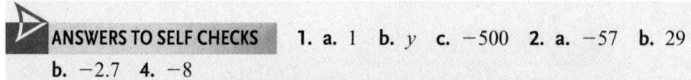 **ANSWERS TO SELF CHECKS**   **1. a.** 1   **b.** y **c.** −500   **2. a.** −57   **b.** 29   **c.** $\frac{5}{12}$ **3. a.** 2.7   **b.** −2.7   **4.** −8

STUDY SET
1.5

VOCABULARY

Fill in the blanks.

1. _____ finds the difference between two numbers.
2. In the subtraction $-2 - 5 = -7$, the result of -7 is called the _____.
3. The difference between the maximum and the minimum value of a collection of measurements is called the _____ of the values.
4. To find the _____ in a quantity, subtract the earlier value from the later value.

CONCEPTS

5. Find the opposite, or additive inverse, of each number.
   a. 12
   b. $-\dfrac{1}{5}$
   c. 2.71
   d. 0
6. Complete each statement.
   a. $a - b = a +$
      To subtract two numbers, add the first number to the _____ of the number to be subtracted.
   b. $-(-a) =$
      The opposite of the opposite of a number is that _____.
7. In each case, determine what number is being subtracted.
   a. $5 - 8$
   b. $-5 - (-8)$
8. Apply the rule for subtraction and fill in the blanks.
   $$1 - (-9) = 1 \quad\boxed{}\quad = \boxed{}$$
9. Use addition to check this subtraction: $15 - (-8) = 7$. Is the result correct?
10. Write each subtraction in the following expression as addition of the opposite.
    $$-10 - 8 + (-23) + 5 - (-34)$$

NOTATION

11. Write each phrase using symbols. Then find its value.
    a. One minus negative seven
    b. The opposite of negative two
    c. The opposite of the absolute value of negative three
    d. Subtract 6 from 2
12. Write each expression in words.
    a. $-(-m)$
    b. $-2 - (-3)$
    c. $x - (-y)$

GUIDED PRACTICE

Simply each expression. **See Example 1.**

13. $-(-55)$
14. $-(-27.2)$
15. $-(-x)$
16. $-(-t)$
17. $-|-25|$
18. $-|-100|$
19. $-\left|-\dfrac{3}{16}\right|$
20. $-\left|-\dfrac{4}{3}\right|$

Subtract. **See Example 2.**

21. $4 - 7$
22. $1 - 6$
23. $2 - 15$
24. $3 - 14$
25. $-6 - 4$
26. $-3 - 4$
27. $8 - (-3)$
28. $17 - (-21)$
29. $0 - 6$
30. $0 - 9$
31. $0 - (-1)$
32. $0 - (-8)$
33. $-1 - (-3)$
34. $-1 - (-7)$
35. $20 - (-20)$
36. $30 - (-30)$
37. $-2 - (-7)$
38. $-9 - (-1)$
39. $-14 - 55$
40. $-13 - 47$
41. $-44 - 44$
42. $-33 - 33$
43. $0 - (-12)$
44. $0 - 12$
45. $-0.9 - 0.2$
46. $-0.3 - 0.2$
47. $6.3 - 9.8$
48. $2.1 - 9.4$
49. $-1.5 - 0.81$
50. $-1.57 - (-0.8)$
51. $-\dfrac{1}{8} - \dfrac{3}{8}$
52. $-\dfrac{3}{4} - \dfrac{1}{4}$
53. $-\dfrac{9}{16} - \left(-\dfrac{1}{4}\right)$
54. $-\dfrac{1}{2} - \left(-\dfrac{1}{4}\right)$
55. $\dfrac{1}{3} - \dfrac{3}{4}$
56. $\dfrac{1}{6} - \dfrac{5}{8}$

Perform the indicated operation. **See Example 3.**

57. Subtract -5 from 17.
58. Subtract 45 from -50.
59. Subtract 12 from -13.
60. Subtract -11 from -20.

Perform the operations. **See Example 4.**

61. $8 - 9 - 10$
62. $1 - 2 - 3$
63. $-25 - (-50) - 75$
64. $-33 - (-22) - 44$
65. $-6 + 8 - (-1) - 10$
66. $-4 + 5 - (-3) - 13$
67. $61 - (-62) + (-64) - 60$
68. $93 - (-92) + (-94) - 95$

TRY IT YOURSELF

Perform the operations.

69. $244 - (-12)$

70. $354 - (-29)$

71. $-20 - (-30) - 50 + 40$

72. $-24 - (-28) - 48 - 44$

73. $-1.2 - 0.9$

74. $-2.52 - 1.72$

75. $\dfrac{1}{8} - \left(-\dfrac{5}{7}\right)$

76. $\dfrac{5}{8} - \left(-\dfrac{2}{9}\right)$

77. $-62 - 71 - (-37) + 99$

78. $-17 - 32 - (-85) - 51$

79. Subtract 47.5 from 0.

80. Subtract 30.3 from 0.

81. Subtract -137 from 12.

82. Subtract 512 from -47.

83. $-1,903 - (-1,732)$

84. $-300 - (-11)$

85. $2.83 - (-1.8)$

86. $4.75 - (-1.9)$

87. $-\dfrac{5}{6} - \dfrac{3}{4}$

88. $-\dfrac{3}{7} - \dfrac{2}{5}$

APPLICATIONS

89. THE EMPIRE STATE   New York state's record high temperature of 108°F was set in 1926, and the record low of −52°F was set in 1979. What is the range of these temperature extremes?

90. EYESIGHT   Nearsightedness, the condition where near objects are clear and far objects are blurry, is measured using negative numbers. Farsightedness, the condition where far objects are clear and near objects are blurry, is measured using positive numbers. Find the range in the measurements shown.

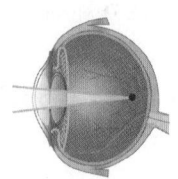

Nearsighted	Farsighted
−2.5	+4.35

91. LAW ENFORCEMENT   A burglar scored −18 on a lie detector test, a score that indicates deception. However, on a second test, he scored +3, a score that is inconclusive. Find the change in the scores.

92. RACING   To improve handling, drivers often adjust the angle of the wheels of their car. When the wheel leans out, the degree measure is considered positive. When the wheel leans in, the degree measure is considered negative. Find the change in the position of the wheel shown in the next column.

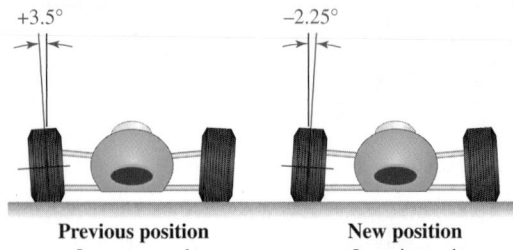

Previous position	New position
+3.5°	−2.25°
Lean outward	Lean inward

93. *from Campus to Careers*
Lead Transportation Security Officer

Determine the change in the number of passengers using each airport in 2006 compared with 2005.

Top 2 Destination Airports in the U.S.
(Number of passengers)

Orlando Int'l Airport Florida	2006*	1,269,000
	2005*	1,309,000
La Guardia Airport New York	2006*	1,149,000
	2005*	1,112,000

*12 months ending August of each year
Source: Bureau of Transportation Statistics

94. U.S. JOBS   The table lists the three occupations that are predicted to have the largest job declines from 2004–2014. Complete the column labeled "Change."

Occupation	Number of jobs		
	2004	2014	Change
Farmers/ranchers	1,065,000	910,000	
Stock clerks	1,566,000	1,451,000	
Sewing machine operators	256,000	163,000	

Source: Bureau of Labor Statistics

95. GEOGRAPHY   The elevation of Death Valley, California, is 282 feet below sea level. The elevation of the Dead Sea in Israel is 1,312 feet below sea level. Find the difference in their elevations.

96. CARD GAMES   Gonzalo won the second round of a card game and earned 50 points. Matt and Hydecki had to deduct the value of each of the cards left in their hands from their score on the first round. Use this information to update the score sheet on the next page. (Face cards are counted as 10 points, aces as 1 point, and all others have the value of the number printed on the card.)

Matt   Hydecki

Running point total	Round 1	Round 2
Matt	+50	
Gonzalo	−15	
Hydecki	−2	

97. FOREIGN POLICY In 2004, Congress forgave $4.1 billion of Iraqi debt owed to the United States. Before that, Iraq's total debt was estimated to be $120.2 billion.

  a. Which expression below can be used to find Iraq's total debt after getting debt relief from the United States?

    i. 120.2 + 4.1   **ii.** 120.2 − (−4.1)

    iii. −120.2 − (−4.1)   **iv.** −120.2 − 4.1

  b. Find Iraq's total debt after getting the debt relief.

98. HISTORY Plato, a famous Greek philosopher, died in 347 B.C. at the age of 81. When was he born?

99. NASCAR Complete the table below to determine how many points the third, fourth, and fifth place finishers were behind the leader.

2006 Final Driver Standings			
Rank	Driver	Points	Points behind leader
1	Jimmie Johnson	6,475	. . .
2	Matt Kenseth	6,419	−56
3	Denny Hamlin	6,407	
4	Kevin Harvick	6,397	
5	Dale Earnhardt, Jr	6,328	

100. GAUGES With the engine off, the ammeter on a car reads 0. If the headlights, which draw a current of 7 amps, and the radio, which draws a current of 6 amps, are both turned on, what will be the new reading?

WRITING

101. Explain what it means when we say that subtraction is *not commutative.*

102. Why is addition of signed numbers taught before subtraction of signed numbers?

103. Explain why we know that the answer to 4 − 10 is negative without having to do any computation.

104. Is the following statement true or false? Explain.

Having a debt of $100 forgiven is equivalent to gaining $100.

REVIEW

105. Find the prime factorization of 30.

106. Write the set of integers.

107. True or false: −4 > −5?

108. Use the associative property of addition to simplify the calculation: −18 + (18 + 89).

CHALLENGE PROBLEMS

109. Suppose x is positive and y is negative. Determine whether each statement is true or false.

  a. $x - y > 0$ **b.** $y - x < 0$

  c. $-x < 0$ **d.** $-y < 0$

110. Find:

$$1 - 2 + 3 - 4 + 5 - 6 + \ldots + 99 - 100$$

SECTION 1.6
Multiplying and Dividing Real Numbers; Multiplication and Division Properties

Objectives

1 Multiply signed numbers.
2 Use properties of multiplication.
3 Divide signed numbers.
4 Use properties of division.

In this section, we will develop rules for multiplying and dividing positive and negative numbers.

① **Multiply Signed Numbers.**

Multiplication represents repeated addition. For example, 4(3) is equal to the sum of four 3's.

$$4(3) = 3 + 3 + 3 + 3$$
$$= 12$$

The Language of Algebra
The names of the parts of a multiplication fact are:

Factor, Factor, Product
4(3) = 12

This example illustrates that *the product of two positive numbers is positive.*
To develop a rule for multiplying a positive number and a negative number, we will find $4(-3)$, which is equal to the sum of four -3's.

$$4(-3) = -3 + (-3) + (-3) + (-3)$$
$$= -12$$

We see that the result is negative. As a check, think in terms of money. If you lose $3 four times, you have lost a total of $12, which is written $-$12. This example illustrates that *the product of a positive number and a negative number is negative.*

Multiplying Two Numbers That Have Different (Unlike) Signs

To multiply a positive number and a negative number, multiply their absolute values. Then make the final answer negative.

EXAMPLE 1   Multiply: **a.** $8(-12)$ **b.** $-151 \cdot 5$ **c.** $(-0.6)(1.2)$ **d.** $\frac{3}{4}\left(-\frac{4}{15}\right)$

Strategy   We will use the rule for multiplying two numbers that have different signs.

Why   In each case, we are asked to multiply a positive number and a negative number.

Success Tip
The product of two numbers with unlike signs is *always* negative.

Solution

a. $8(-12) = -96$ Multiply the absolute values, 8 and 12, to get 96. Since the signs are unlike, make the final answer negative.

b. $-151 \cdot 5 = -755$ Multiply the absolute values, 151 and 5, to get 755. Since the signs are unlike, make the final answer negative.

c. To find the product of these two decimals with unlike signs, first multiply their absolute values, 0.6 and 1.2.

$$\begin{array}{r} 1.2 \\ \times 0.6 \\ \hline 0.72 \end{array}$$ Place the decimal point in the result so that the answer has the same number of decimal places as the sum of the number of decimal places in the factors.

Then make the final answer negative: $(-0.6)(1.2) = -0.72$.

d. $\frac{3}{4}\left(-\frac{4}{15}\right) = -\frac{\overset{1}{\cancel{3}} \cdot \overset{1}{\cancel{4}}}{\cancel{4} \cdot \cancel{3} \cdot 5}$ Multiply the absolute values $\frac{3}{4}$ and $\frac{4}{15}$. Since the signs are unlike, make the final answer negative.

$= -\frac{1}{5}$ To simplify the fraction, factor 15 as $3 \cdot 5$. Remove the common factors 3 and 4 in the numerator and denominator.

Self Check 1   Multiply: **a.** $20(-3)$ **b.** $-3 \cdot 5$
  c. $4.3(-2.6)$ **d.** $-\frac{5}{8} \cdot \frac{16}{25}$
Now Try   **Problems 19, 25, and 27**

To develop a rule for multiplying two negative numbers, consider the following list, where we multiply -4 by factors that decrease by 1. We know how to find the first four products. Graphing those results on a number line is helpful in determining the last three products.

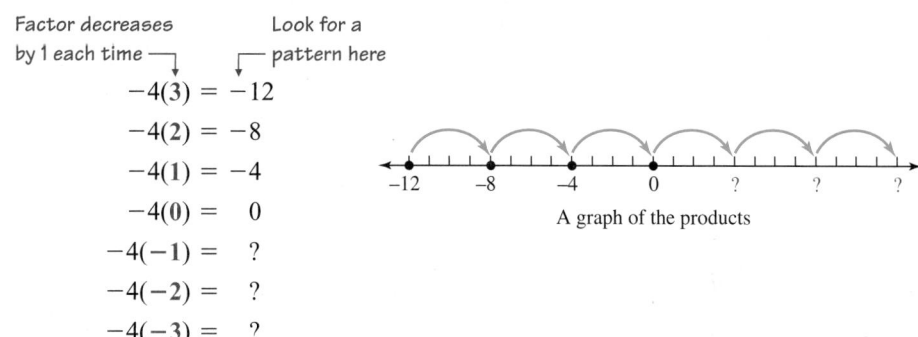

$$-4(3) = -12$$
$$-4(2) = -8$$
$$-4(1) = -4$$
$$-4(0) = 0$$
$$-4(-1) = \ ?$$
$$-4(-2) = \ ?$$
$$-4(-3) = \ ?$$

Factor decreases by 1 each time

Look for a pattern here

A graph of the products

From the pattern, we see that the product increases by 4 each time. Thus,

$$-4(-1) = 4, \qquad -4(-2) = 8, \qquad \text{and} \qquad -4(-3) = 12$$

These results illustrate that *the product of two negative numbers is positive.* As a check, think of losing four debts of \$3. This is equivalent to gaining \$12. Therefore, $-4(-\$3) = \$12$.

Since the product of two positive numbers is positive, and the product of two negative numbers is also positive, we can summarize the multiplication rule as follows.

Multiplying Two Numbers That Have the Same (Like) Signs	To multiply two real numbers that have the same sign, multiply their absolute values. The final answer is positive.

EXAMPLE 2   Multiply: **a.** $-5(-6)$ **b.** $\left(-\frac{1}{2}\right)\left(-\frac{5}{8}\right)$

Strategy   We will use the rule for multiplying two numbers that have the same sign.

Why   In each case, we are asked to multiply two negative numbers.

Solution

a. $-5(-6) = 30$ Multiply the absolute values, 5 and 6, to get 30. Since both factors are negative, the final answer is positive.

b. $\left(-\frac{1}{2}\right)\left(-\frac{5}{8}\right) = \frac{5}{16}$ Multiply the absolute values, $\frac{1}{2}$ and $\frac{5}{8}$, to get $\frac{5}{16}$. Since the two factors have the same sign, the final answer is positive.

Success Tip
The product of two numbers with like signs is *always* positive.

Self Check 2   Multiply: **a.** $-15(-8)$ **b.** $-\frac{1}{4}\left(-\frac{1}{3}\right)$
Now Try   **Problems 31 and 39**

Use Properties of Multiplication.

The multiplication of two numbers can be done in any order; the result is the same. For example, $-9(4) = -36$ and $4(-9) = -36$. This illustrates that multiplication is **commutative.**

The Commutative Property of Multiplication	Changing the order when multiplying does not affect the answer. For any real numbers a and b, $$ab = ba$$

In the following example, we multiply $-3 \cdot 7 \cdot 5$ in two ways. Recall that the operation within the parentheses should be performed first.

Method 1: Group −3 and 7	*Method 2: Group 7 and 5*
$(-3 \cdot 7)5 = (-21)5$	$-3(7 \cdot 5) = -3(35)$
$\quad\quad = -105$	$\quad\quad = -105$

It doesn't matter how we group the numbers in this multiplication; the result is -105. This example illustrates that multiplication is **associative.**

The Associative Property of Multiplication	Changing the grouping when multiplying does not affect the answer. For any real numbers a, b, and c, $$(ab)c = a(bc)$$

EXAMPLE 3   Multiply: **a.** $-5(-37)(-2)$ **b.** $-4(-3)(-2)(-1)$

Strategy   First, we will use the commutative and associative properties of multiplication to reorder and regroup the factors. Then we will perform the multiplications.

Why   Applying of one or both of these properties before multiplying can simplify the computations and lessen the chance of a sign error.

Solution   Using the commutative and associative properties of multiplication, we can reorder and regroup the factors to simplify computations.

a. Since it is easy to multiply by 10, we will find $-5(-2)$ first.

$$-5(-37)(-2) = -5(-2)(-37) \quad \text{Use the commutative property of multiplication.}$$
$$= 10(-37)$$
$$= -370$$

b. $-4(-3)(-2)(-1) = 12(2) \quad \text{Multiply the first two factors and multiply the last two factors.}$
$$= 24$$

 Self Check 3   Multiply: **a.** $-25(-3)(-4)$ **b.** $-1(-2)(-3)(-3)$
Now Try   **Problems 43 and 47**

In Example 3a, we multiplied three negative numbers. In Example 3b, we multiplied four negative numbers. The results illustrate the following fact.

Multiplying Negative Numbers	The product of an even number of negative numbers is positive. The product of an odd number of negative numbers is negative.

Recall that the product of 0 and any whole number is 0. The same is true for any real number. Therefore, $-6 \cdot 0 = 0$, $\frac{7}{16} \cdot 0 = 0$, and $0(4.51) = 0$.

Multiplication Property of 0	The product of 0 and any real number is 0. For any real number a, $$0 \cdot a = 0 \quad \text{and} \quad a \cdot 0 = 0$$

Whenever we multiply a number by 1, the number remains the same. Therefore, $1 \cdot 6 = 6$, $4.57 \cdot 1 = 4.57$, and $1(-9) = -9$. Since any number multiplied by 1 remains the same (is identical), the number 1 is called the **identity element** for multiplication.

Multiplication Property of 1 (Identity Property of Multiplication)	The product of 1 and any number is that number. For any real number a, $$1 \cdot a = a \quad \text{and} \quad a \cdot 1 = a$$

Two numbers whose product is 1 are **reciprocals** or **multiplicative inverses** of each other. For example, 8 is the multiplicative inverse of $\frac{1}{8}$, and $\frac{1}{8}$ is the multiplicative inverse of 8, because $8 \cdot \frac{1}{8} = 1$. Likewise, $-\frac{3}{4}$ and $-\frac{4}{3}$ are multiplicative inverses because $-\frac{3}{4}\left(-\frac{4}{3}\right) = 1$. All real numbers, except 0, have a multiplicative inverse.

Multiplicative Inverses (Inverse Property of Multiplication)	The product of any number and its multiplicative inverse (reciprocal) is 1. For any nonzero real number a, $$a\left(\frac{1}{a}\right) = 1$$

EXAMPLE 4   Find the reciprocal of each number: **a.** $\frac{2}{3}$ **b.** $-\frac{2}{3}$ **c.** -11

Strategy   To find the reciprocal of a fraction, we invert the numerator and the denominator.

Why   We want the product of the given number and its reciprocal to be 1.

Solution

a. The reciprocal of $\frac{2}{3}$ is $\frac{3}{2}$ because $\frac{2}{3}\left(\frac{3}{2}\right) = 1$.   To find the reciprocal of a fraction, invert the numerator and denominator.

b. The reciprocal of $-\frac{2}{3}$ is $-\frac{3}{2}$ because $-\frac{2}{3}\left(-\frac{3}{2}\right) = 1$.

c. The reciprocal of -11 is $-\frac{1}{11}$ because $-11\left(-\frac{1}{11}\right) = 1$.   Think of -11 as $\frac{-11}{1}$ to find its reciprocal.

Caution
Do not change the sign of a number when finding its reciprocal.

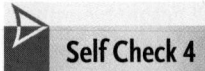

Self Check 4   Find the reciprocal of each number: **a.** $-\frac{15}{16}$

b. $\frac{15}{16}$ **c.** -27

Now Try   **Problems 53 and 55**

3   **Divide Signed Numbers.**

Every division fact can be written as an equivalent multiplication fact.

Division

For any real numbers a, b, and c, where $b \neq 0$,

$$\frac{a}{b} = c \quad \text{provided that} \quad c \cdot b = a$$

The Language of Algebra
The names of the parts of a division fact are:

Dividend ⟍   ⟋ Quotient
$$\frac{15}{5} = 3$$
Divisor ⟋

We can use this relationship between multiplication and division to develop rules for dividing signed numbers. For example,

$$\frac{15}{5} = 3 \quad \text{because} \quad 3(5) = 15$$

From this example, we see that *the quotient of two positive numbers is positive.*
To determine the quotient of two negative numbers, we consider $\frac{-15}{-5}$.

$$\frac{-15}{-5} = 3 \quad \text{because} \quad 3(-5) = -15$$

From this example, we see that the *quotient of two negative numbers is positive.*
To determine the quotient of a positive number and a negative number, we consider $\frac{15}{-5}$.

$$\frac{15}{-5} = -3 \quad \text{because} \quad -3(-5) = 15$$

From this example, we see that *the quotient of a positive number and a negative number is negative.*
To determine the quotient of a negative number and a positive number, we consider $\frac{-15}{5}$.

$$\frac{-15}{5} = -3 \quad \text{because} \quad -3(5) = -15$$

From this example, we see that *the quotient of a negative number and a positive number is negative.*
We summarize the rules from the previous examples and note that they are similar to the rules for multiplication.

Dividing Two Real Numbers

To divide two real numbers, divide their absolute values.

1. The quotient of two numbers that have the same *(like)* signs is positive.
2. The quotient of two numbers that have different *(unlike)* signs is negative.

EXAMPLE 5   Divide and check the result: **a.** $\dfrac{-81}{-9}$ **b.** $\dfrac{45}{-9}$

c. $-2.87 \div 0.7$ **d.** $-\dfrac{5}{16} \div \left(-\dfrac{1}{2}\right)$

Strategy   We will use the rules for dividing signed numbers. In each case, we need to ask, "Is it a quotient of two numbers with the same sign or different signs?"

Why   The signs of the numbers that we are dividing determine the sign of the result.

Solution

a. $\dfrac{-81}{-9} = 9$ Divide the absolute values, 81 by 9, to get 9. Since the signs are like, the final answer is positive.

Multiply to check the result: $9(-9) = -81$.

b. $\dfrac{45}{-9} = -5$ Divide the absolute values, 45 by 9, to get 5. Since the signs are unlike, make the final answer negative.

Check: $-5(-9) = 45$

c. $-2.87 \div 0.7 = -4.1$ Since the signs are unlike, make the final answer negative.

Check: $-4.1(0.7) = -2.87$

d. $-\dfrac{5}{16} \div \left(-\dfrac{1}{2}\right) = -\dfrac{5}{16}\left(-\dfrac{2}{1}\right)$ Multiply the first fraction by the reciprocal of the second fraction. The reciprocal of $-\frac{1}{2}$ is $-\frac{2}{1}$.

$= \dfrac{5 \cdot 2}{16 \cdot 1}$ Multiply the absolute values $\frac{5}{16}$ and $\frac{2}{1}$. Since the signs are like, the final answer is positive.

$= \dfrac{5 \cdot \overset{1}{\cancel{2}}}{\underset{1}{\cancel{2}} \cdot 8 \cdot 1}$ To simplify the fraction, factor 16 as $2 \cdot 8$. Then remove the common factor 2.

$= \dfrac{5}{8}$

Check: $\dfrac{5}{8}\left(-\dfrac{1}{2}\right) = -\dfrac{5}{16}$

Self Check 5   Find each quotient: **a.** $\dfrac{-28}{-4}$ **b.** $\dfrac{75}{-25}$

c. $0.32 \div (-1.6)$ **d.** $\dfrac{3}{4} \div \left(-\dfrac{5}{8}\right)$

Now Try   **Problems 57, 61, 69, and 77**

Success Tip

To perform this decimal division, move each decimal point one place to the right.

$0.7 \overline{)2.87}$

EXAMPLE 6   ***Depreciation.***   Over an 8-year period, the value of a $150,000 house fell at a uniform rate to $110,000. Find the amount of depreciation per year.

Strategy   The phrase *uniform rate* means that the value of the house fell the same amount each year, for 8 straight years. We can determine the amount it depreciated in one year (per year) by dividing the total change in value of the house by 8.

Why   The process of separating a quantity into equal parts (in this case, the change in the value of the house) indicates division.

Solution   First, we find the change in the value of the house.

$$110,000 - 150,000 = -40,000 \quad \text{Subtract the previous value from the current value.}$$

The result represents a drop in value of $40,000. Since the depreciation occurred over 8 years, we divide $-40,000$ by 8.

$$\frac{-40,000}{8} = -5,000 \quad \text{Divide the absolute values, 40,000 by 8, to get 5,000, and make the quotient negative.}$$

The house depreciated $5,000 per year.

 Now Try   Problem 105

> **The Language of Algebra**
> Depreciation is a form of the word depreciate, meaning to lose value. You've probably heard that the minute you drive a new car off the lot, it has depreciated.

4 Use Properties of Division.

Whenever we divide a number by 1, the quotient is that number. Therefore, $\frac{12}{1} = 12$, $\frac{-80}{1} = -80$, and $7.75 \div 1 = 7.75$. Furthermore, whenever we divide a nonzero number by itself, the quotient is 1. Therefore, $\frac{35}{35} = 1$, $\frac{-4}{-4} = 1$, and $0.9 \div 0.9 = 1$. These observations suggest the following properties of division.

> **Division Properties**
>
> Any number divided by 1 is the number itself. Any number (except 0) divided by itself is 1. For any real number a,
>
> $$\frac{a}{1} = a \quad \text{and} \quad \frac{a}{a} = 1 \quad (\text{where } a \neq 0).$$

Caution   Division is *not commutative*. For example, $\frac{6}{3} \neq \frac{3}{6}$ and $\frac{-12}{4} \neq \frac{4}{-12}$. In general, $\frac{a}{b} \neq \frac{b}{a}$.

We will now consider division that involves zero. First, we examine division of zero. Let's look at two examples. We know that

$$\frac{0}{2} = 0 \quad \text{because} \quad 0 \cdot 2 = 0 \quad \text{and} \quad \frac{0}{-5} = 0 \quad \text{because} \quad 0(-5) = 0$$

> **The Language of Algebra**
> When we say a division by 0, such as $\frac{2}{0}$, is *undefined*, we mean that $\frac{2}{0}$ does not represent a real number.

These examples illustrate that *0 divided by a nonzero number is 0*.

To examine division by zero, let's look at $\frac{2}{0}$ and its related multiplication statement.

$$\frac{2}{0} = ? \quad \text{because} \quad ? \cdot 0 = 2$$

> **The Language of Algebra**
> Division of 0 by 0, written $\frac{0}{0}$, is called *indeterminate*. This form is studied in advanced mathematics classes.

There is no number that can make $0 \cdot ? = 2$ true because any number multiplied by 0 is equal to 0, not 2. Therefore, $\frac{2}{0}$ does not have an answer. We say that such a division is **undefined.**

These results suggest the following division facts.

> **Division Involving 0**
>
> For any nonzero real number a,
>
> $$\frac{0}{a} = 0 \quad \text{and} \quad \frac{a}{0} \text{ is undefined.}$$

EXAMPLE 7 Find each quotient, if possible: **a.** $\dfrac{0}{8}$ **b.** $\dfrac{-24}{0}$

Strategy In each case, we need to determine if we have division *of* 0 or division *by* 0.

Why *Division of 0* by a nonzero number is defined, and the result is 0. However, *division by 0* is undefined; there is no result.

Solution

a. $\dfrac{0}{8} = 0$ because $0 \cdot 8 = 0$.   This is division of 0 by 8.

b. $\dfrac{-24}{0}$ is undefined.   This is division of −24 by 0.

Self Check 7 Find each quotient, if possible: **a.** $\dfrac{4}{0}$ **b.** $\dfrac{0}{17}$

Now Try **Problems 73 and 75**

ANSWERS TO SELF CHECKS **1. a.** −60 **b.** −15 **c.** −11.18 **d.** $-\frac{2}{5}$ **2. a.** 120 **b.** $\frac{1}{12}$ **3. a.** −300 **b.** 18 **4. a.** $-\frac{16}{15}$ **b.** $\frac{16}{15}$ **c.** $-\frac{1}{27}$ **5. a.** 7 **b.** −3 **c.** −0.2 **d.** $-\frac{6}{5}$ **7. a.** Undefined **b.** 0

STUDY SET
1.6

VOCABULARY

Fill in the blanks.

1. The answer to a multiplication problem is called a _____. The answer to a division problem is called a _____.

2. The _____ property of multiplication states that changing the order when multiplying does not affect the answer.

3. The _____ property of multiplication states that changing the grouping when multiplying does not affect the answer.

4. Division of a nonzero number by 0 is _____.

CONCEPTS

Fill in the blanks.

5. **a.** The product or quotient of two numbers with like signs is _____.

   b. The product or quotient of two numbers with unlike signs is _____.

6. **a.** The product of an even number of negative numbers is _____.

   b. The product of an odd number of negative numbers is _____.

7. **a.** $\dfrac{-9}{3} = -3$ because ____ · ____ = ____

   b. $\dfrac{0}{8} = 0$ because ____ · ____ = ____

8. Complete each property of multiplication.

   a. $a \cdot b = b \cdot$ ____

   b. $(ab)c =$ ____

   c. $0 \cdot a =$ ____

   d. $1 \cdot a =$ ____

   e. $a\left(\dfrac{1}{a}\right) =$ ____

9. Complete each property of division.

   a. $\dfrac{a}{1} =$ ____

   b. $\dfrac{a}{a} =$ ____

   c. $\dfrac{0}{a} =$ ____

   d. $\dfrac{a}{0} =$ ____

10. Which property justifies each statement?

    a. $-5(2 \cdot 17) = (-5 \cdot 2)17$

    b. $-5\left(-\dfrac{1}{5}\right) = 1$

    c. $-5 \cdot 2 = 2(-5)$

    d. $-5(1) = -5$

    e. $-5 \cdot 0 = 0$

11. Complete each statement using the given property.

  a. Commutative property of multiplication

   $5 \cdot 8 = $

  b. Associative property of multiplication

   $-2(6 \cdot 9) = $

  c. Inverse property of multiplication

   $5\left(\quad\right) = 1$

  d. Multiplication property of 1

   $\quad(-20) = -20$

12. Complete the table.

Number	Opposite (additive inverse)	Reciprocal (multiplicative inverse)
2		
$-\frac{4}{5}$		
1.75		

Let POS stand for a positive number and NEG stand for a negative number. Determine the sign of each result, if possible.

13. a. POS · NEG   **b.** POS + NEG

  c. POS − NEG   **d.** $\dfrac{POS}{NEG}$

14. a. NEG · NEG   **b.** NEG + NEG

  c. NEG − NEG   **d.** $\dfrac{NEG}{NEG}$

NOTATION

Write each sentence using symbols.

15. The product of negative four and negative five is twenty.

16. The quotient of sixteen and negative eight is negative two.

GUIDED PRACTICE

Multiply. See Example 1.

17. $4(-1)$ **18.** $6(-1)$
19. $-2 \cdot 8$ **20.** $-3 \cdot 4$
21. $12(-5)$ **22.** $(-9)(11)$
23. $3(-22)$ **24.** $-8 \cdot 9$
25. $1.2(-0.4)$ **26.** $(-3.6)(0.9)$

27. $\dfrac{1}{3}\left(-\dfrac{3}{4}\right)$ **28.** $\left(-\dfrac{3}{4}\right)\left(\dfrac{4}{5}\right)$

Multiply. See Example 2.

29. $(-1)(-7)$ **30.** $(-2)(-5)$
31. $(-6)(-9)$ **32.** $(-8)(-7)$
33. $-3(-3)$ **34.** $-1(-1)$
35. $63(-7)$ **36.** $43(-6)$
37. $-0.6(-4)$ **38.** $-0.7(-8)$
39. $\left(-\dfrac{7}{8}\right)\left(-\dfrac{2}{21}\right)$ **40.** $\left(-\dfrac{5}{6}\right)\left(-\dfrac{2}{15}\right)$

Multiply. See Example 3.

41. $-3(-4)(0)$ **42.** $15(0)(-22)$
43. $3.3(-4)(-5)$ **44.** $(-2.2)(-4)(-5)$
45. $-2(-3)(-4)(-5)(-6)$ **46.** $-9(-7)(-5)(-3)(-1)$
47. $(-41)(3)(-7)(-1)$ **48.** $56(-3)(-4)(-1)$
49. $(-6)(-6)(-6)$ **50.** $(-5)(-5)(-5)$
51. $(-2)(-2)(-2)(-2)$ **52.** $(-3)(-3)(-3)(-3)$

Find the reciprocal of each number. Then find the product of the given number and its reciprocal. See Example 4.

53. $\dfrac{7}{9}$ **54.** $-\dfrac{8}{9}$

55. -13 **56.** $\dfrac{1}{8}$

Divide. See Example 5.

57. $-30 \div (-3)$ **58.** $-12 \div (-2)$
59. $-6 \div (-2)$ **60.** $-36 \div (-9)$
61. $\dfrac{24}{-6}$ **62.** $\dfrac{-78}{6}$
63. $\dfrac{85}{-5}$ **64.** $\dfrac{-84}{7}$
65. $\dfrac{17}{-17}$ **66.** $\dfrac{-24}{24}$
67. $\dfrac{-110}{-110}$ **68.** $\dfrac{-200}{-200}$
69. $\dfrac{-10.8}{1.2}$ **70.** $\dfrac{-13.5}{-1.5}$
71. $\dfrac{0.5}{-100}$ **72.** $\dfrac{-1.7}{10}$
73. $\dfrac{0}{150}$ **74.** $\dfrac{0}{-12}$
75. $\dfrac{-17}{0}$ **76.** $\dfrac{225}{0}$
77. $-\dfrac{1}{3} \div \dfrac{4}{5}$ **78.** $-\dfrac{2}{3} \div \dfrac{7}{8}$
79. $-\dfrac{9}{16} \div \left(-\dfrac{3}{20}\right)$ **80.** $-\dfrac{4}{5} \div \left(-\dfrac{8}{25}\right)$

TRY IT YOURSELF

Perform the operations.

81. $\dfrac{-23.5}{5}$

82. $\dfrac{-337.8}{6}$

83. $-5.2 \cdot 100$

84. $-1.17 \cdot 1{,}000$

85. $\dfrac{1}{2}\left(-\dfrac{1}{3}\right)\left(-\dfrac{1}{4}\right)$

86. $\dfrac{1}{3}\left(-\dfrac{1}{5}\right)\left(-\dfrac{1}{7}\right)$

87. $\dfrac{550}{-50}$

88. $\dfrac{440}{-20}$

89. $-3\dfrac{3}{8} \div \left(-2\dfrac{1}{4}\right)$

90. $-3\dfrac{4}{15} \div \left(-2\dfrac{1}{10}\right)$

91. $7.2(-2.1)(-2)$

92. $4.6(-5.4)(-2)$

93. $\dfrac{1}{2}\left(-\dfrac{3}{4}\right)$

94. $\dfrac{1}{3}\left(-\dfrac{5}{16}\right)$

95. $\dfrac{16}{25} \div \dfrac{64}{15}$

96. $\dfrac{15}{16} \div \dfrac{25}{8}$

97. $\dfrac{-24.24}{-0.8}$

98. $\dfrac{-55.02}{-0.7}$

99. $-1\dfrac{1}{4}\left(-\dfrac{3}{4}\right)$

100. $-1\dfrac{1}{8}\left(-\dfrac{3}{8}\right)$

Use the associative property of multiplication to find each product.

101. $-\dfrac{1}{2}(2 \cdot 67)$

102. $\left(-\dfrac{5}{16} \cdot \dfrac{1}{7}\right)7$

103. $-0.2(-10 \cdot 3)$

104. $-1.5(-100 \cdot 4)$

APPLICATIONS

105. REAL ESTATE   Over a 5-year period, the value of a $200,000 lot fell at a uniform rate to $160,000. Find the amount of depreciation per year.

106. TOURISM   The ocean liner Queen Mary cost $22,500,000 to build in 1936. The ship was purchased by the city of Long Beach, California, in 1967 for $3,450,000. It now serves as a convention center. What signed number indicates the annual average depreciation of the ship over the 31-year period from 1936 to 1967? Round to the nearest dollar.

107. FLUID FLOW   In a lab, the temperature of a fluid was decreased 6° per hour for 12 hours. What signed number indicates the drop in temperature?

108. STRESS ON THE JOB   A health care provider for a company estimates that 75 hours per week are lost by employees suffering from stress-related illness. In one year, how many hours are lost? Use a signed number to answer.

109. WEIGHT LOSS   As a result of a diet, Tom has been steadily losing $4\dfrac{1}{2}$ pounds per month.

   a. Which expression below can be used to determine how much heavier Tom was 8 months ago?

   i. $-4\dfrac{1}{2} \cdot 8$ **ii.** $-4\dfrac{1}{2}(-8)$

   iii. $4\dfrac{1}{2}(-8)$ **iv.** $-4\dfrac{1}{2} - 8$

   b. How much heavier was Tom 8 months ago?

110. PLANETS   The temperature on Pluto gets as low as $-386°$F. This is twice as low as the lowest temperature reached on Jupiter. What is the lowest temperature on Jupiter?

111. CAR RADIATORS   The instructions on a container of antifreeze state, "A 50/50 mixture of antifreeze and water protects against freeze-ups down to $-34°$F, while a 60/40 mix protects against freeze-ups down to one and one-half times that temperature." To what temperature does the 60/40 mixture protect?

112. ACCOUNTING   For 2004, the net income for Martha Stewart Living Omnimedia, Inc., was about $-$60,000,000$. The company's losses for 2005 were even worse, by a factor of about 1.25. What signed number indicates the company's net income that year?

113. AIRLINES   In the 2005 income statement for Delta Air Lines, numbers within parentheses represent a loss. Complete the statement given these facts. The second and fourth quarter *losses* were approximately the same and totaled $2,200 million. The third quarter loss was about $\dfrac{1}{3}$ of the first quarter loss.

DELTA INCOME STATEMENT				2005
All amounts in millions of dollars	1st Qtr (1,200)	2nd Qtr (?)	3rd Qtr (?)	4th Qtr (?)

Source: Yahoo! Finance

114. COMPUTERS   The formula = A1*B1*C1 in cell D1 of the spreadsheet instructs the computer to multiply the values in cells A1, B1, and C1 and to print the result *in place of the formula* in cell D1. (The symbol * represents multiplication.) What value will be printed in the cell D1? What values will be printed in cells D2 and D3?

		Microsoft Excel - Book 1		
: ⅀ File Edit View Insert Format Tools Data Window				
	A	B	C	D
1	4	−5	−17	= A1*B1*C1
2	22	−30	14	= A2*B2*C2
3	−60	−20	−34	= A3*B3*C3
4				
5				

115. PHYSICS   An oscilloscope displays electrical signals as wavy lines on a screen. See the next page. By switching the magnification dial to ×2, for example, the height of the "peak" and the depth of the "valley" of a graph will be doubled. Use signed numbers to indicate the height and depth of the display for each setting of the dial.

   a. normal   **b.** ×0.5

   c. ×1.5   **d.** ×2

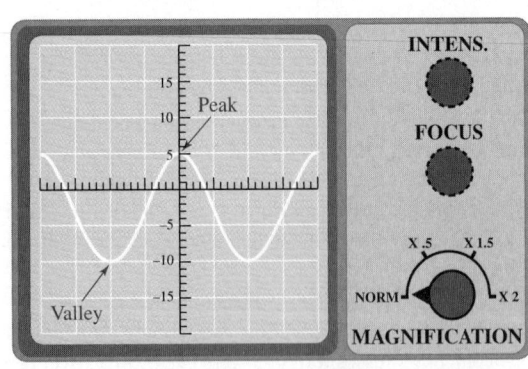

116. LIGHT   Water acts as a selective filter of light. In the illustration, we see that red light waves penetrate water only to a depth of about 5 meters. How many times deeper does

a. yellow light penetrate than red light?

b. green light penetrate than orange light?

c. blue light penetrate than yellow light?

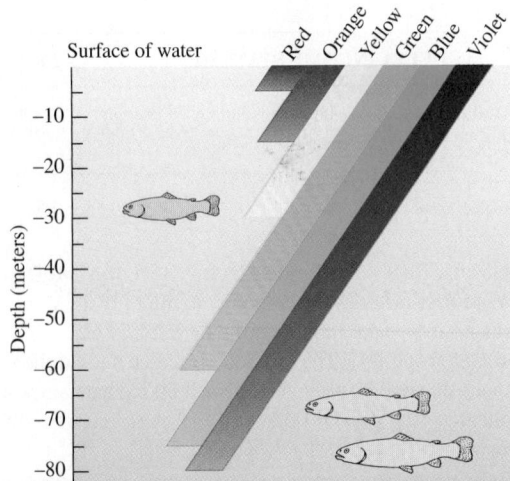

WRITING

117. Explain why $\frac{16}{0}$ is undefined.

118. The commutative property states that changing the order when multiplying does not change the answer. Are the following activities commutative? Explain.

   a. Washing a load of clothes; drying a load of clothes

   b. Putting on your left sock; putting on your right sock

119. What is wrong with the following statement?
A negative and a positive is a negative.

120. If we multiply two different numbers and the answer is 0, what must be true about one of the numbers? Explain your answer.

REVIEW

121. Add: $-3 + (-4) + (-5) + 4 + 3$

122. Write $-3 - (-5)$ as addition of the opposite.

123. Find $\frac{1}{2} + \frac{1}{4} + \frac{1}{3}$. Answer in decimal form.

124. Which integers have an absolute value equal to 45?

CHALLENGE PROBLEMS

125. If the product of five numbers is negative, how many of them could be negative? Explain.

126. Suppose a is a positive number and b is a negative number. Determine whether the given expression is positive or negative.

   a. $-a(-b)$ b. $\dfrac{-a}{b}$

   c. $\dfrac{-a}{a}$ d. $\dfrac{1}{b}$

SECTION 1.7
Exponents and Order of Operations

Objectives

❶ Evaluate exponential expressions.

❷ Use the order of operations rules.

❸ Evaluate expressions containing grouping symbols.

❹ Find the mean (average).

In algebra, we often have to find the value of expressions that involve more than one operation. In this section, we introduce an order-of-operations rule to follow in such cases. But first, we discuss a way to write repeated multiplication using *exponents*.

Evaluate Exponential Expressions.

In the expression $3 \cdot 3 \cdot 3 \cdot 3 \cdot 3$, the number 3 repeats as a factor five times. We can use **exponential notation** to write this product in a more compact form.

Exponent and Base	An **exponent** is used to indicate repeated multiplication. It is how many times the **base** is used as a factor.

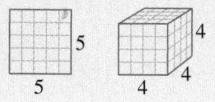
The Language of Algebra
5^2 represents the area of a square with sides 5 units long. 4^3 represents the volume of a cube with sides 4 units long.

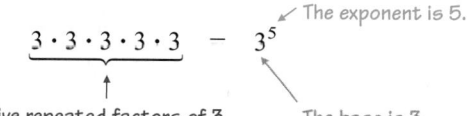

In the **exponential expression** 3^5, 3 is the base, and 5 is the exponent. The expression is called a power of 3. Some other examples of exponential expressions are:

5^2 Read as "5 to the second power" or "5 squared."

4^3 Read as "4 to the third power" or "4 cubed."

$(-2)^5$ Read as "−2 to the fifth power."

EXAMPLE 1   Write each product using exponents: **a.** $7 \cdot 7 \cdot 7$
b. $(-5)(-5)(-5)(-5)(-5)$ **c.** $8 \cdot 8 \cdot 15 \cdot 15 \cdot 15 \cdot 15$
d. $a \cdot a \cdot a \cdot a \cdot a \cdot a$ **e.** $4 \cdot \pi \cdot r \cdot r$

Strategy   We need to determine the number of repeated factors in the expression.

Why   An exponent can be used to represent repeated multiplication.

Solution

a. The factor 7 is repeated 3 times. We can represent this repeated multiplication with an exponential expression having a base of 7 and an exponent of 3: $7 \cdot 7 \cdot 7 = 7^3$.

b. The factor −5 is repeated five times: $(-5)(-5)(-5)(-5)(-5) = (-5)^5$.

c. $8 \cdot 8 \cdot 15 \cdot 15 \cdot 15 \cdot 15 = 8^2 \cdot 15^4$

d. $a \cdot a \cdot a \cdot a \cdot a \cdot a = a^6$

e. $4 \cdot \pi \cdot r \cdot r = 4\pi r^2$

 Self Check 1   Write each product using exponents:
a. $(12)(12)(12)(12)(12)(12)$ **b.** $2 \cdot 9 \cdot 9 \cdot 9$
c. $(-30)(-30)(-30)$ **d.** $y \cdot y \cdot y \cdot y$
e. $12 \cdot b \cdot b \cdot b \cdot c$

Now Try   **Problems 15 and 21**

To **evaluate** (find the value of) an exponential expression, we write the base as a factor the number of times indicated by the exponent. Then we multiply the factors, working left to right.

EXAMPLE 2   Evaluate each expression: **a.** 5^3 **b.** 10^1 **c.** $\left(-\dfrac{2}{3}\right)^3$

d. $(0.6)^2$ **e.** $(-3)^4$ **f.** $(-3)^5$

Strategy   We will rewrite each exponential expression as a product of repeated factors, and then perform the multiplication. This requires that we identify the base and the exponent.

Why   The exponent tells the number of times the base is to be written as a factor.

Solution

a. $5^3 = 5 \cdot 5 \cdot 5$ Write the base, 5, as a factor 3 times.

$\quad\quad = 125$ Multiply, working left to right. We say 125 is the *cube* of 5.

b. $10^1 = 10$ The base is 10. Since the exponent is 1, we write the base once.

c. $\left(-\dfrac{2}{3}\right)^3 = \left(-\dfrac{2}{3}\right)\left(-\dfrac{2}{3}\right)\left(-\dfrac{2}{3}\right)$ Since $-\frac{2}{3}$ is the base and 3 is the exponent, we write $-\frac{2}{3}$ as a factor three times.

$\quad\quad = \dfrac{4}{9}\left(-\dfrac{2}{3}\right)$ Work from left to right. $\left(-\frac{2}{3}\right)\left(-\frac{2}{3}\right) = \frac{4}{9}$

$\quad\quad = -\dfrac{8}{27}$

d. $(0.6)^2 = (0.6)(0.6)$ Since 0.6 is the base and 2 is the exponent, we write 0.6 as a factor two times.

$\quad\quad = 0.36$ We say 0.36 is the *square* of 0.6.

e. $(-3)^4 = (-3)(-3)(-3)(-3)$ Write the base, -3, as a factor 4 times.

$\quad\quad = 9(-3)(-3)$ Work from left to right.

$\quad\quad = -27(-3)$

$\quad\quad = 81$

f. $(-3)^5 = (-3)(-3)(-3)(-3)(-3)$ Write the base, -3, as a factor 5 times.

$\quad\quad = 9(-3)(-3)(-3)$ Work from left to right.

$\quad\quad = -27(-3)(-3)$

$\quad\quad = 81(-3)$

$\quad\quad = -243$

Self Check 2   Evaluate: **a.** 2^5 **b.** 9^1 **c.** $\left(-\dfrac{3}{4}\right)^3$

d. $(-0.3)^2$ **e.** $(-6)^2$ **f.** $(-5)^3$

Now Try   Problems 23, 29, and 33

In Example 2e, we raised -3 to an even power; the result was positive. In part f, we raised -3 to an odd power; the result was negative. These results illustrate the following rule.

Even and Odd Powers of a Negative Number

When a negative number is raised to an even power, the result is positive.

When a negative number is raised to an odd power, the result is negative.

Although the expressions $(-4)^2$ and -4^2 look alike, they are not. When we find the value of each expression, it becomes clear that they are not equivalent.

$$(-4)^2 = (-4)(-4) \quad \text{The base is } -4, \text{ the exponent is 2.} \qquad -4^2 = -(4 \cdot 4) \quad \text{The base is 4, the exponent is 2.}$$

$$= 16 \qquad\qquad\qquad\qquad\qquad\qquad = -16$$

Different results

EXAMPLE 3   Evaluate: -2^4

Strategy   We will rewrite the expression as a product of repeated factors and then perform the multiplication. We must be careful when identifying the base. It is 2, not -2.

Why   Since there are no parentheses around -2, the base is 2.

Solution

$$-2^4 = -(2 \cdot 2 \cdot 2 \cdot 2) \quad \text{Read as "the opposite of the fourth power of two."}$$

$$= -16 \qquad\qquad\qquad \text{Do the multiplication within the parentheses to get 16. Then write the opposite of that result.}$$

Self Check 3   Evaluate: -5^4

Now Try   **Problem 35**

---

② **Use the Order of Operations Rules.**

Suppose you have been asked to contact a friend if you see a Rolex watch for sale when you are traveling in Europe. While in Switzerland, you find the watch and send the text message shown on the left. The next day, you get the response shown on the right.

Something is wrong. The first part of the response (No price too high!) says to buy the watch at any price. The second part (No! Price too high.) says not to buy it, because it's too expensive. The placement of the exclamation point makes us read the two parts of the response differently, resulting in different meanings. When reading a mathematical statement, the same kind of confusion is possible. For example, consider the expression

$$2 + 3 \cdot 6$$

We can evaluate this expression in two ways. We can add first, and then multiply. Or we can multiply first, and then add. However, the results are different.

$$2 + 3 \cdot 6 = 5 \cdot 6 \quad \text{Add 2 and 3 first.} \qquad 2 + 3 \cdot 6 = 2 + 18 \quad \text{Multiply 3 and 6 first.}$$
$$= 30 \quad \text{Multiply 5 and 6.} \qquad\qquad\qquad = 20 \quad \text{Add 2 and 18.}$$

Different answers

If we don't establish a uniform order of operations, the expression has two different values. To avoid this possibility, we will always use the following set of priority rules.

Order of Operations

1. Perform all calculations within parentheses and other grouping symbols following the order listed in Steps 2–4 below, working from the innermost pair of grouping symbols to the outermost pair.
2. Evaluate all exponential expressions.
3. Perform all multiplications and divisions as they occur from left to right.
4. Perform all additions and subtractions as they occur from left to right.

When grouping symbols have been removed, repeat Steps 2–4 to complete the calculation.

If a fraction is present, evaluate the expression above and the expression below the bar separately. Then simplify the fraction, if possible.

It isn't necessary to apply all of these steps in every problem. For example, the expression $2 + 3 \cdot 6$ does not contain any parentheses, and there are no exponential expressions. So we look for multiplications and divisions to perform and proceed as follows:

$$2 + 3 \cdot 6 = 2 + 18 \quad \text{Do the multiplication first.}$$
$$= 20 \quad \text{Do the addition.}$$

EXAMPLE 4   Evaluate: **a.** $3 \cdot 2^3 - 4$ **b.** $-30 - 4 \cdot 5 + 9$
c. $24 \div 6 \cdot 2$ **d.** $160 - 4 + 6(-2)(-3)$

The Language of Algebra
Sometimes, for problems like these, the instruction *simplify* is used instead of *evaluate*.

Strategy   We will scan the expression to determine what operations need to be performed. Then we will perform those operations, one-at-a-time, following the order of operations rules.

Why   If we don't follow the correct order of operations, the expression can have more than one value.

Solution

a. Three operations need to be performed to evaluate this expression: multiplication, raising to a power, and subtraction. By the order of operations rules, we evaluate 2^3 first.

$$3 \cdot 2^3 - 4 = 3 \cdot 8 - 4 \quad \text{Evaluate the exponential expression: } 2^3 = 8.$$
$$= 24 - 4 \quad \text{Do the multiplication: } 3 \cdot 8 = 24.$$
$$= 20 \quad \text{Do the subtraction.}$$

b. This expression involves subtraction, multiplication, and addition. The order of operations rules tell us to multiply first.

$$-30 - 4 \cdot 5 + 9 = -30 - 20 + 9 \qquad \text{Do the multiplication: } 4 \cdot 5 = 20.$$
$$= -50 + 9 \qquad \text{Working from left to right, do the subtraction:}$$
$$\qquad -30 - 20 = -30 + (-20) = -50.$$
$$= -41 \qquad \text{Do the addition.}$$

c. Since there are no calculations within parentheses nor are there exponents, we perform the multiplications and divisions as they occur from left to right. The division occurs before the multiplication, so it must be performed first.

$$24 \div 6 \cdot 2 = 4 \cdot 2 \qquad \text{Working left to right, do the division: } 24 \div 6 = 4.$$
$$= 8 \qquad \text{Do the multiplication.}$$

d. Although this expression contains parentheses, there are no operations to perform within them. Since there are no exponents, we will perform the multiplications as they occur from left to right.

$$160 - 4 + 6(-2)(-3) = 160 - 4 + (-12)(-3) \qquad \text{Do the multiplication, working left to right: } 6(-2) = -12.$$
$$= 160 - 4 + 36 \qquad \text{Complete the multiplication: } (-12)(-3) = 36.$$
$$= 156 + 36 \qquad \text{Working left to right, the subtraction occurs before the addition. The subtraction must be performed first: } 160 - 4 = 156.$$
$$= 192$$

> **Caution**
> A common mistake is to forget to work from left to right and incorrectly perform the multiplication before the division. This produces the wrong answer, 2.
> $$\cancel{24 \div 6 \cdot 2 = 24 \div 12}$$
> $$\cancel{= 2}$$

Self Check 4   Evaluate: **a.** $2 \cdot 3^2 + 17$ **b.** $-40 - 9 \cdot 4 + 10$
c. $18 \div 2 \cdot 3$ **d.** $240 - 8 + 3(-2)(-4)$

Now Try   **Problems 39, 45, 49, and 51**

3 **Evaluate Expressions Containing Grouping Symbols.**

Grouping symbols serve as mathematical punctuation marks. They help determine the order in which an expression is to be evaluated. Examples of grouping symbols are parentheses ( ), brackets [ ], braces { }, absolute value symbols | |, and the fraction bar —.

EXAMPLE 5   Evaluate each expression: **a.** $(6 - 3)^2$ **b.** $5^3 + 2(-8 - 3 \cdot 2)$

Strategy   We will perform the operation(s) within the parentheses first. When there is more than one operation to perform within the parentheses, we follow the order of operations rules.

Why   This is the first step of the order of operations rule.

Solution
a. $(6 - 3)^2 = 3^2$ Do the subtraction within the parentheses: $6 - 3 = 3$.
$$= 9 \qquad \text{Evaluate the exponential expression.}$$

Notation

Multiplication is indicated when a number is next to a parenthesis or bracket.

$$5^3 + 2(-8 - 3 \cdot 2)$$

b. We begin by performing the operations within the parentheses in the proper order: multiplication first, and then subtraction.

$$5^3 + 2(-8 - 3 \cdot 2) = 5^3 + 2(-8 - 6)$$ Do the multiplication: $3 \cdot 2 = 6$.

$$= 5^3 + 2(-14)$$ Do the subtraction: $-8 - 6 = -14$.

$$= 125 + 2(-14)$$ Evaluate 5^3.

$$= 125 + (-28)$$ Do the multiplication: $2(-14) = -28$.

$$= 97$$ Do the addition.

 Self Check 5   Evaluate: **a.** $(12 - 6)^3$ **b.** $1^3 + 6(-6 - 3 \cdot 0)$

Now Try   **Problem 73**

Expressions can contain two or more pairs of grouping symbols. To evaluate the following expression, we begin within the innermost pair of grouping symbols, the parentheses. Then we work within the outermost pair, the brackets.

Innermost pair
↓   ↓
$$-4[2 + 3(4 - 8^2)] - 2$$
↑   ↑
Outermost pair

EXAMPLE 6   Evaluate: $-4[2 + 3(4 - 8^2)] - 2$

Strategy   We will work within the parentheses first and then within the brackets. At each stage, we follow the order of operations rules.

Why   By the order of operations, we must work from the *innermost* pair of grouping symbols to the *outermost*.

The Language of Algebra

When one pair of grouping symbols is inside another pair, we say that those grouping symbols are *nested*, or *embedded*.

Solution

$$-4[2 + 3(4 - 8^2)] - 2$$

$$= -4[2 + 3(4 - 64)] - 2$$ Evaluate the exponential expression within the parentheses: $8^2 = 64$.

$$= -4[2 + 3(-60)] - 2$$ Do the subtraction within the parentheses: $4 - 64 = 4 + (-64) = -60$.

$$= -4[2 + (-180)] - 2$$ Do the multiplication within the brackets: $3(-60) = -180$.

$$= -4[-178] - 2$$ Do the addition within the brackets: $2 + (-180) = -178$.

$$= 712 - 2$$ Do the multiplication: $-4[-178] = 712$.

$$= 710$$ Do the subtraction.

 Self Check 6   Evaluate: $-5[4 + 2(5^2 - 15)] - 10$

Now Try   **Problem 67**

EXAMPLE 7   Evaluate: $\dfrac{-3(3 + 2) + 5}{17 - 3(-4)}$

Strategy   We will evaluate the expression above and the expression below the fraction bar separately. Then we will simplify the fraction, if possible.

Why   Fraction bars are grouping symbols. They group the numerator and denominator. The expression could be written $[-3(3 + 2) + 5] \div [17 - 3(-4)]$.

Solution

$$\frac{-3(3 + 2) + 5}{17 - 3(-4)} = \frac{-3(5) + 5}{17 - (-12)}$$ In the numerator, do the addition within the parentheses. In the denominator, do the multiplication.

$$= \frac{-15 + 5}{17 + 12}$$ In the numerator, do the multiplication. In the denominator, write the subtraction as the addition of the opposite of -12, which is 12.

$$= \frac{-10}{29}$$ Do the additions.

$$= -\frac{10}{29}$$ Write the $-$ sign in front of the fraction: $\frac{-10}{29} = -\frac{10}{29}$. The fraction does not simplify.

 Self Check 7   Evaluate: $\dfrac{-4(-2 + 8) + 6}{8 - 5(-2)}$

Now Try   **Problem 81**

EXAMPLE 8   Evaluate: $10|9 - 15| - 2^5$

Strategy   The absolute value bars are grouping symbols. We will perform the calculation within them first.

Why   By the order of operations, we must perform all calculations within parentheses and other grouping symbols (such as absolute value bars) first.

Solution

$$10|9 - 15| - 2^5 = 10|-6| - 2^5$$ Subtract: $9 - 15 = 9 + (-15) = -6$.

$$= 10(6) - 2^5$$ Find the absolute value: $|-6| = 6$.

$$= 10(6) - 32$$ Evaluate the exponential expression: $2^5 = 32$.

$$= 60 - 32$$ Do the multiplication: $10(6) = 60$.

$$= 28$$ Do the subtraction.

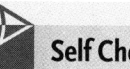 **Self Check 8**   Evaluate: $10^3 + 3|24 - 25|$

Now Try   **Problem 85**

---

 4   **Find the Mean (Average).**

The **arithmetic mean** (or **average**) of a set of numbers is a value around which the values of the numbers are grouped.

<table>
<tr><td colspan="2">Finding an
Arithmetic Mean</td><td>To find the **mean** of a set of values, divide the sum of the values by the number of values.</td></tr>
</table>

Number of rings	Number of calls
1	11
2	46
3	45
4	28
5	20

EXAMPLE 9   *Hotel Reservations.* In an effort to improve customer service, a hotel electronically recorded the number of times the reservation desk telephone rang before it was answered by a receptionist. The results of the week-long survey are shown in the table. Find the average number of times the phone rang before a receptionist answered.

Strategy First, we will determine the total number of times the reservation desk telephone rang during the week. Then we will divide that result by the total number of calls received.

Why To find the *average* value of a set of values, we divide the sum of the values by the number of values.

Solution To find the total number of rings, we multiply each *number of rings* (1, 2, 3, 4, and 5 rings) by the respective number of occurrences and add those subtotals.

Total number of rings = 11(1) + 46(2) + 45(3) + 28(4) + 20(5)

The total number of calls received was 11 + 46 + 45 + 28 + 20. To find the average, we divide the total number of rings by the total number of calls.

$$\text{Average} = \frac{11(1) + 46(2) + 45(3) + 28(4) + 20(5)}{11 + 46 + 45 + 28 + 20}$$

In the numerator, do the multiplications. In the denominator, do the additions.

$$= \frac{11 + 92 + 135 + 112 + 100}{150}$$

$$= \frac{450}{150}$$ Do the addition.

$$= 3$$ Simplify the fraction.

The average number of times the phone rang before it was answered was 3.

 Now Try **Problem 115**

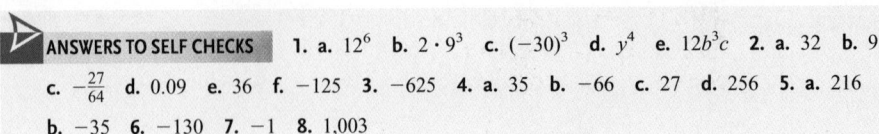
ANSWERS TO SELF CHECKS   1. a. 12^6 b. $2 \cdot 9^3$ c. $(-30)^3$ d. y^4 e. $12b^3c$ 2. a. 32 b. 9 c. $-\frac{27}{64}$ d. 0.09 e. 36 f. -125 3. -625 4. a. 35 b. -66 c. 27 d. 256 5. a. 216 b. -35 6. -130 7. -1 8. 1,003

STUDY SET
1.7

VOCABULARY

Fill in the blanks.

1. In the exponential expression 7^5, 7 is the _____, and 5 is the _____. 7^5 is the fifth _____ of seven.
2. 10^2 can be read as ten _____, and 10^3 can be read as ten _____.
3. An _____ is used to represent repeated multiplication.
4. To _____ the expression $2(-1 + 4^2)$ means to find its value.
5. The rules for the _____ of operations guarantee that an evaluation of a numerical expression will result in a single answer.
6. To find the arithmetic _____ or average of a set of values, divide the sum of the values by the number of values.

CONCEPTS

7. To evaluate each expression, what operation should be performed first?
   a. $24 - 4 + 2$
   b. $32 \div 8 \cdot 4$
   c. $8 - (3 + 5)^2$
   d. $65 \cdot 3^3$
8. To evaluate $\frac{36 - 4(7)}{2(10 - 8)}$, what operation should be performed first in the numerator? In the denominator?

NOTATION

9. a. Give the name of each grouping symbol: ( ), [ ], { }, | |, and —.

   b. In the expression $-8 + 2[15 - (-6 + 1)]$, which grouping symbols are innermost, and which are outermost?

10. What operation is indicated?
$$\downarrow$$
$$2 + 9|5 - (2 + 4)|$$

11. a. In the expression $(-5)^2$, what is the base?
    b. In the expression -5^2, what is the base?
12. Write each expression using symbols. Then evaluate it.
    a. Negative two squared
    b. The opposite of the square of two

Complete the evaluation of each expression.

13. $-19 - 2[(1 + 2)^2 \cdot 3] = -19 - 2[\quad^2 \cdot 3]$
    $= -19 - 2[\quad \cdot 3]$
    $= -19 - 2[\quad]$
    $= -19 - \quad$
    $= \quad$

14. $\dfrac{46 - 2^3}{-3(5) - 4} = \dfrac{46 - \quad}{\quad}$
    $= \dfrac{\quad}{\quad}$
    $= -2$

GUIDED PRACTICE

Write each product using exponents. See Example 1.

15. $8 \cdot 8 \cdot 8$
16. $(-4)(-4)(-4)(-4)$
17. $7 \cdot 7 \cdot 7 \cdot 12 \cdot 12$
18. $5 \cdot 5 \cdot 5 \cdot 5 \cdot 5 \cdot 7 \cdot 7 \cdot 7$
19. $x \cdot x \cdot x$
20. $b \cdot b \cdot b \cdot b$
21. $r \cdot r \cdot r \cdot r \cdot s \cdot s$
22. $m \cdot m \cdot m \cdot n \cdot n \cdot n \cdot n$

Evaluate each expression. See Example 2.

23. 7^2
24. 9^2
25. 6^3
26. 6^4
27. $(-5)^4$
28. $(-5)^3$
29. $(-0.1)^2$
30. $(-0.8)^2$
31. $\left(-\dfrac{1}{4}\right)^3$
32. $\left(-\dfrac{1}{3}\right)^4$
33. $\left(\dfrac{2}{3}\right)^3$
34. $\left(\dfrac{3}{4}\right)^3$

Evaluate each expression. See Example 3.

35. $(-6)^2$ and -6^2
36. $(-4)^2$ and -4^2
37. $(-8)^2$ and -8^2
38. $(-9)^2$ and -9^2

Evaluate each expression. See Example 4.

39. $3 - 5 \cdot 4$
40. $-4 \cdot 6 + 5$
41. $32 - 16 \div 4 + 2$
42. $60 - 20 \div 10 + 5$
43. $9 \cdot 5 - 6 \div 3$
44. $8 \cdot 5 - 4 \div 2$
45. $12 \div 3 \cdot 2$
46. $18 \div 6 \cdot 3$
47. $-22 - 15 + 3$
48. $-33 - 8 + 10$
49. $-2(9) - 2(5)(10)$
50. $-6(7) - 3(-4)(-2)$
51. $2 \cdot 5^2 + 4 \cdot 3^2$
52. $5 \cdot 3^3 - 4 \cdot 2^3$
53. $-2(-1)^2 + 3(-1) - 3$
54. $-4(-3)^2 + 3(-3) - 1$

Evaluate each expression. See Examples 5 and 6.

55. $-4(6 + 5)$
56. $-3(5 - 4)$
57. $(9 - 3)(9 - 9)^2$
58. $-(-8 - 6)(6 - 6)^2$

59. $(-1 - 3^2 \cdot 4)2^2$

60. $-1(28 - 5^2 \cdot 2)3^2$

61. $1 + 5(10 + 2 \cdot 5) - 1$

62. $14 + 3(7 - 5 \cdot 3)$

63. $-(2 \cdot 3 - 2^2)^5$

64. $-(3 \cdot 5 - 2 \cdot 6)^4$

Evaluate each expression. See Example 6.

65. $(-1)^9[-7^2 - (-2)^2]$

66. $[-9^2 - (-8)^2](-1)^{10}$

67. $64 - 6[15 + 2(-3 + 8)]$

68. $4 - 2[26 + 2(5 - 3)]$

69. $-2[2 + 4^2(8 - 9)]^2$

70. $-3[5 + 3^2(4 - 5)]^2$

71. $3 + 2[-1 - (4 - 5)]$

72. $4 + 2[-7 - (3 - 9)]$

73. $8 - 3[5^2 - (7 - 3)^2]$

74. $3 - [3^3 + (3 - 1)^3]$

Evaluate each expression. See Example 7.

75. $\dfrac{-2 - 5}{-7 + (-7)}$

76. $\dfrac{-3 - (-1)}{-2 + (-2)}$

77. $\dfrac{4 \cdot 2^4 - 60 + (-4)}{5^4 - (-4)(-5)}$

78. $\dfrac{(6 - 5)^8 - 1}{(-9)(-3) - 4}$

79. $\dfrac{2(-4 - 2 \cdot 2)}{3(-3)(-2)}$

80. $\dfrac{3(-3^2 + 2 \cdot 2^2)}{(5 - 8)(7 - 9)}$

81. $\dfrac{72 - (2 - 2 \cdot 4)}{10^2 - (9 \cdot 10 + 2^2)}$

82. $\dfrac{13^2 - 5^2}{-3(5 - 3^2)}$

Evaluate each expression. See Example 8.

83. $-2|4 - 8|$

84. $-5|1 - 8|$

85. $-|7 - 2^3(4 - 7)|$

86. $-|9 - 5(1 - 2^3)|$

87. $\dfrac{(3 + 5)^2 + |-2|}{-2(5 - 8)}$

88. $\dfrac{|-25| - 8(-5)}{2^4 - 29}$

89. $\dfrac{|6 - 4| + 2|-4|}{26 - 2^4}$

90. $\dfrac{4|9 - 7| + |-7|}{3^2 - 2^2}$

TRY IT YOURSELF

Evaluate each expression.

91. $[6(5) - 5(5)]^3(-4)$

92. $5 - 2 \cdot 3^4 - (-6 + 5)^3$

93. $8 - 6[(130 - 4^3) - 2]$

94. $91 - 5[(150 - 3^3) - 1]$

95. $-2\left(\dfrac{15}{-5}\right) - \dfrac{6}{2} + 9$

96. $-6\left(\dfrac{25}{-5}\right) - \dfrac{36}{9} + 1$

97. $-5(-2)^3 - |-2 + 1|$

98. $-6(-3)^3 - |-6 + 5|$

99. $\dfrac{18 - [2 + (1 - 6)]}{16 - (-4)^2}$

100. $\dfrac{6 - [6(-1) - 88]}{4 - 2^2}$

101. $-|-5 \cdot 2^4| - 30$

102. $2 + |-3 \cdot 2^2 \cdot 2^2 \cdot 1^2|$

103. $(-3)^3\left(\dfrac{-4}{2}\right)(-1)$

104. $(-2)^3\left(\dfrac{-6}{2}\right)(-1)$

105. $\dfrac{1}{2}\left(\dfrac{1}{8}\right) + \left(-\dfrac{1}{4}\right)^2$

106. $-\dfrac{1}{9}\left(\dfrac{1}{4}\right) + \left(-\dfrac{1}{6}\right)^2$

107. $\dfrac{-5^2 \cdot 10 + 10 \cdot 2^4}{-5 - 3 - 1}$

108. $\dfrac{(-6^2 - 2^4 \cdot 2) + 5}{-4 - 3}$

109. $-\left(\dfrac{40 - 1^3 - 2^4}{3(2 + 5) + 2}\right)$

110. $-\left(\dfrac{8^2 - 10}{2(3)(4) - 5(3)}\right)$

APPLICATIONS

111. LIGHT As light energy passes through the first unit of area, 1 yard away from the bulb, it spreads out. How much area does that light energy cover 2 yards, 3 yards, and 4 yards from the bulb? Express each answer using exponents.

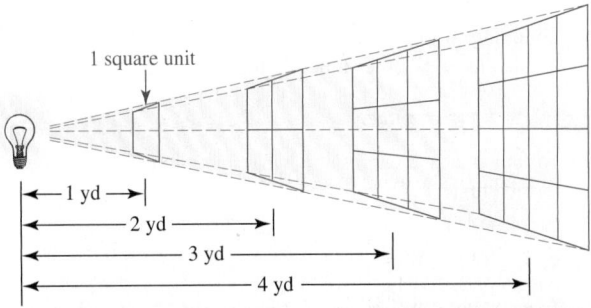

1 square unit

|← 1 yd →|
|← 2 yd →|
|← 3 yd →|
|← 4 yd →|

112. CHAIN LETTERS A woman sent two friends a letter with the following request: "Please send a copy of this letter to two of your friends." Assume that all those receiving letters responded and that everyone in the chain received just one letter. Complete the table and then determine how many letters will be circulated in the 10th level?

Level	Number of letters circulated
1st	$2 = 2^1$
2nd	$= 2$
3rd	$= 2$
4th	$= 2$

113. HURRICANES The table lists the number of major hurricanes to strike the mainland United States by decade. Find the average number per decade.

Decade	Number	Decade	Number
1901–1910	4	1951–1960	8
1911–1920	7	1961–1970	6
1921–1930	5	1971–1980	4
1931–1940	8	1981–1990	5
1941–1950	10	1991–2000	5

Source: National Hurricane Center

114. ENERGY USAGE   Find the average number of therms of natural gas used per month.

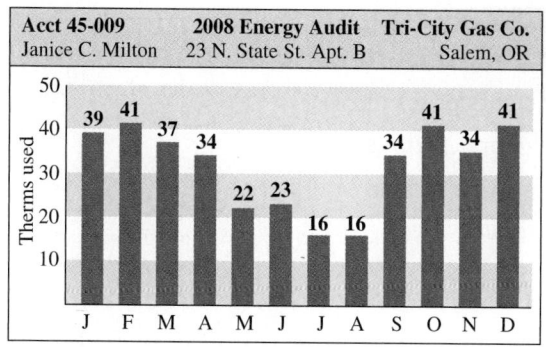

115. CASH AWARDS   A contest is to be part of a promotional kickoff for a new children's cereal. The prizes to be awarded are shown.

   a. How much money will be awarded in the promotion?

   b. What is the average cash prize?

> ### Coloring Contest
> **Grand prize: Disney World vacation plus $2,500**
> Four 1st place prizes of $500
> Thirty-five 2nd place prizes of $150
> Eighty-five 3rd place prizes of $25

116. SURVEYS   Some students were asked to rate their college cafeteria food on a scale from 1 to 5. The responses are shown on the tally sheet. Find the average rating.

Poor		Fair		Excellent
1	2	3	4	5
	|||	|||	ℍℍ	ℍℍ ||||

117. WRAPPING GIFTS   How much ribbon is needed to wrap the package if 15 inches of ribbon are needed to make the bow?

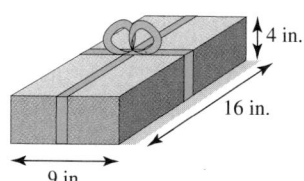

118. SCRABBLE   Write an expression to determine the number of points received for playing the word QUARTZY and then evaluate it. (The number on each tile gives the point value of the letter.)

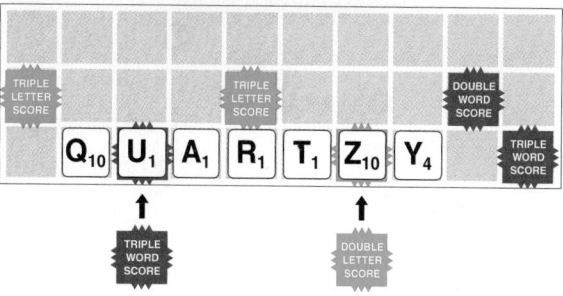

WRITING

119. Explain the difference between 2^3 and 3^2.

120. Why are the order of operations rules necessary?

121. Explain the error. What is the correct answer?

$$40 \div 4 \cdot 2 = 40 \div 8$$
$$= 5$$

122. Explain the error. What is the correct answer?

$$5 + 3(2 - 6) = 5 + 3(-4)$$
$$= 8(-4)$$
$$= -32$$

REVIEW

123. What numbers are a distance of 6 away from -11 on a number line?

124. Fill in the blank with > or <: $0.3 \quad \frac{1}{3}$

CHALLENGE PROBLEMS

125. Using each of the numbers 2, 3, and 4 only once, what is the greatest value that the following expression can have?

$$\left(\square^{\square}\right)^{\square}$$

126. Insert a pair of parentheses into $4 \cdot 3^2 - 4 \cdot 2$ so that it has a value of 40.

Translate the set of instructions to an expression and then evaluate it.

127. Subtract the sum of -9 and 8 from the product of the cube of -3 and the opposite of 4.

128. Increase the square the reciprocal of -2 by the difference of -0.25 and -1.

SECTION 1.8
Algebraic Expressions

Objectives

1. Identify terms and coefficients of terms.
2. Write word phrases as algebraic expressions.
3. Analyze problems to determine hidden operations.
4. Evaluate algebraic expressions.

Since problems in algebra are often presented in words, the ability to interpret what you read is important. In this section, we will introduce several strategies that will help you translate words into algebraic expressions.

 Identify Terms and Coefficients of Terms.

Recall that variables and/or numbers can be combined with the operations of arithmetic to create **algebraic expressions.** Addition symbols separate expressions into parts called *terms*. For example, the expression $x + 8$ has two terms.

$$\underset{\text{First term}}{x} \quad + \quad \underset{\text{Second term}}{8}$$

Since subtraction can be written as addition of the opposite, the expression $a^2 - 3a - 9$ has three terms.

$$a^2 - 3a - 9 = \underset{\text{First term}}{a^2} \quad + \quad \underset{\text{Second term}}{(-3a)} \quad + \quad \underset{\text{Third term}}{(-9)}$$

In general, a **term** is a product or quotient of numbers and/or variables. A single number or variable is also a term. Examples of terms are:

$$4, \quad y, \quad 6r, \quad -w^3, \quad 3.7x^5, \quad \frac{3}{n}, \quad -15ab^2$$

The numerical factor of a term is called the **coefficient** of the term. For instance, the term $6r$ has a coefficient of 6 because $6r = 6 \cdot r$. The coefficient of $-15ab^2$ is -15 because $-15ab^2 = -15 \cdot ab^2$. More examples are shown below.

A term such as 4, that consists of a single number, is called a **constant term.**

> ### Notation
> By the commutative property of multiplication, $r6 = 6r$ and $-15b^2a = -15ab^2$. However, we usually write the numerical factor first and the variable factors in alphabetical order.

> ### The Language of Algebra
> Terms such as x and y have *implied* coefficients of 1. *Implied* means suggested without being precisely expressed.

Term	Coefficient	
$8y^2$	8	
$-0.9pq$	-0.9	
$\frac{3}{4}b$	$\frac{3}{4}$	This term could be written $\frac{3b}{4}$.
$-\frac{x}{6}$	$-\frac{1}{6}$	Because $-\frac{x}{6} = -\frac{1x}{6} = -\frac{1}{6} \cdot x$
x	1	Because $x = 1x$
$-t$	-1	Because $-t = -1t$
27	27	

EXAMPLE 1   Identify the coefficient of each term in the expression: $7x^2 - x + 6$

Strategy   We will begin by writing the subtraction as addition of the opposite. Then we will determine the numerical factor of each term.

Why   Addition symbols separate expressions into terms.

Solution   If we write $7x^2 - x + 6$ as $7x^2 + (-x) + 6$, we see that it has three terms: $7x^2$, $-x$, and 6. The numerical factor of each term is its coefficient.

The coefficient of $7x^2$ is 7 because $7x^2$ means $7 \cdot x^2$.

The coefficient of $-x$ is -1 because $-x$ means $-1 \cdot x$.

The coefficient of the constant 6 is 6.

Self Check 1   Identify the coefficient of each term in the expression: $p^3 - 12p^2 + 3p - 4$

Now Try   **Problem 19**

It is important to be able to distinguish between the *terms* of an expression and the *factors* of a term.

EXAMPLE 2   Is m used as a *factor* or a *term* in each expression?
a. $m + 6$ **b.** $8m$

Strategy   We will begin by determining whether m is involved in an addition or a multiplication.

Why   Addition symbols separate expressions into *terms*. A *factor* is a number being multiplied.

Solution
a. Since m is added to 6, m is a term of $m + 6$.
b. Since m is multiplied by 8, m is a factor of $8m$.

Self Check 2   Is b used as a *factor* or a *term* in each expression?
a. $-27b$ **b.** $5a + b$

Now Try   **Problems 21 and 23**

2   **Write Word Phrases as Algebraic Expressions.**

The tables on the next page show how key phrases can be translated into algebraic expressions.

Caution

Be careful when translating subtraction. Order is important. For example, when a translation involves the phrase *less than*, note how the terms are reversed.

18 less than w

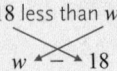
$w \quad - \quad 18$

Addition	
the sum of a and 8	$a + 8$
4 plus c	$4 + c$
16 added to m	$m + 16$
4 more than t	$t + 4$
20 greater than F	$F + 20$
T increased by r	$T + r$
exceeds y by 35	$y + 35$

Subtraction	
the difference of 23 and P	$23 - P$
550 minus h	$550 - h$
18 less than w	$w - 18$
7 decreased by j	$7 - j$
M reduced by x	$M - x$
12 subtracted from L	$L - 12$
5 less f	$5 - f$

Caution

Be careful when translating division. As with subtraction, order is important. For example, s divided by d is *not* written $\frac{d}{s}$.

Multiplication	
the product of 4 and x	$4x$
20 times B	$20B$
twice r	$2r$
double the amount a	$2a$
triple the profit P	$3P$
three-fourths of m	$\frac{3}{4}m$

Division	
the quotient of R and 19	$\frac{R}{19}$
s divided by d	$\frac{s}{d}$
the ratio of c to d	$\frac{c}{d}$
k split into 4 equal parts	$\frac{k}{4}$

Caution

$5 < c$ is the translation of the statement 5 *is less than the capacity c* not 5 less than the capacity c.

EXAMPLE 3   Write each phrase as an algebraic expression:
a. one-half of the profit P **b.** 5 less than the capacity c
c. the product of the weight w and 2,000, increased by 300

Strategy   We will begin by identifying any key phrases.

Why   Key phrases can be translated to mathematical symbols.

Solution

a. **Key phrase:** *One-half of*   **Translation:** multiplication by $\frac{1}{2}$

The algebraic expression is: $\frac{1}{2}P$.

b. **Key phrase:** *less than*   **Translation:** subtraction

Sometimes thinking in terms of specific numbers makes translating easier. Suppose the capacity was 100. Then 5 *less than* 100 would be $100 - 5$. If the capacity is c, then we need to make it 5 less. The algebraic expression is: $c - 5$.

c. **Key phrase:** *product of*   **Translation:** multiplication
Key phrase: *increased by*   **Translation:** addition

In the given wording, the comma after 2,000 means w is first multiplied by 2,000; then 300 is added to that product. The algebraic expression is: $2{,}000w + 300$.

 Self Check 3   Write each phrase as an algebraic expression:
a. 80 less than the total t **b.** $\frac{2}{3}$ of the time T
c. the difference of twice a and 15, squared
Now Try   **Problems 25, 31, and 35**

To solve application problems, we often let a variable stand for an unknown quantity.

EXAMPLE 4   ***Swimming.*** A pool is to be sectioned into 8 equally wide swimming lanes. Write an algebraic expression that represents the width of each lane.

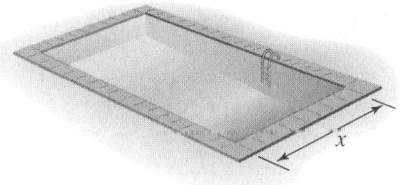

Strategy   We will begin by letting $x =$ the width of the swimming pool in feet. Then we will identify any key phrases.

Why   The width of the pool is unknown.

Solution   The key phrase, *sectioned into 8 equally wide lanes,* indicates division. Therefore, the width of each lane is $\frac{x}{8}$ feet.

Self Check 4   It takes Val m minutes to get to work by bus. If she drives her car, her travel time exceeds this by 15 minutes. How long does it take her to get to work by car?

Now Try   **Problem 61**

EXAMPLE 5   ***Painting.*** A 10-inch-long paintbrush has two parts: a handle and bristles. Choose a variable to represent the length of one of the parts. Then write an expression to represent the length of the other part.

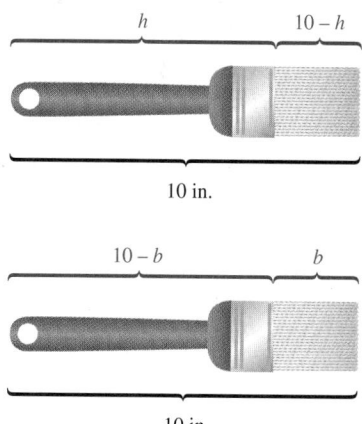

Strategy   There are two approaches. We can let $h =$ the length of the handle or we can let $b =$ the length of the bristles.

Why   Both the length of the handle and the length of the bristles are unknown.

Solution   Refer to the drawing on the top. If we let $h =$ the length of the handle (in inches), then the length of the bristles is $10 - h$.

   Now refer to the drawing on the bottom. If we let $b =$ the length of the bristles (in inches), then the length of the handle is $10 - b$.

Self Check 5   Part of a $900 donation to a college went to the scholarship fund, the rest to the building fund. Choose a variable to represent the amount donated to one of the funds. Then write an expression that represents the amount donated to the other fund.

Now Try   **Problem 13**

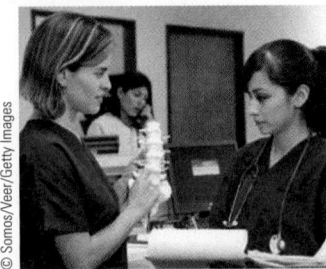

EXAMPLE 6   ***Enrollments.***   Second semester enrollment in a nursing program was 32 more than twice that of the first semester. Let x represent the enrollment for one of the semesters. Write an expression that represents the enrollment for the other semester.

Strategy   We will begin by letting $x =$ the enrollment for the first semester.

Why   Because the second-semester enrollment is related to the first-semester enrollment.

Solution

Key phrase: *more than*   **Translation:** addition

Key phrase: *twice that*   **Translation:** multiplication by 2

The second semester enrollment was $2x + 32$.

Self Check 6   In an election, the incumbent received 55 fewer votes than three times the challenger's votes. Let x represent the number of votes received by one candidate. Write an expression that represents the number of votes received by the other.

Now Try   **Problem 105**

❸   **Analyze Problems to Determine Hidden Operations.**

Many applied problems require insight and analysis to determine which mathematical operations to use.

EXAMPLE 7   ***Vacations.***   Disneyland, in California, was in operation 16 years before the opening of Disney World in Florida. Euro Disney, in France, was constructed 21 years after Disney World. Write algebraic expressions to represent the ages (in years) of each Disney attraction.

Strategy   We will begin by letting $x =$ the age of Disney World.

Why   The ages of Disneyland and Euro Disney are both related to the age of Disney World.

Solution   In carefully reading the problem, we see that Disneyland was built 16 years before Disney World. That makes its age 16 years more than that of Disney World. The key phrase *more than* indicates addition.

$x + 16 =$ the age of Disneyland

Euro Disney was built 21 years *after* Disney World. That makes its age 21 years less than that of Disney World. The key phrase *less than* indicates subtraction.

$x - 21 =$ the age of Euro Disney

Attraction	Age
Disneyland	$x + 16$
Disney World	x
Euro Disney	$x - 21$

    Now Try   **Problem 107**

Number of years	Number of months
1	12
2	24
3	36
x	12*x*

We multiply the number of years by 12 to find the number of months.

EXAMPLE 8   How many months are in *x* years?

Strategy   There are no key phrases so we must carefully analyze the problem. We will begin by considering some specific cases.

Why   It's often easier to work with specifics first to get a better understanding of the relationship between the two quantities. Then we can generalize using a variable.

Solution   Let's calculate the number of months in 1 year, 2 years, and 3 years. When we write the results in a table, a pattern is apparent.
The number of months in *x* years is $12 \cdot x$ or $12x$.

Self Check 8   How many days is *h* hours?

Now Try   **Problems 7 and 67**

In some problems, we must distinguish between *the number of* and *the value of* the unknown quantity. For example, to find the value of 3 quarters, we multiply the number of quarters by the value (in cents) of one quarter. Therefore, the value of 3 quarters is $3 \cdot 25$ cents = 75 cents.

The same distinction must be made if the number is unknown. For example, the value of *n* nickels is not *n* cents. The value of *n* nickels is $n \cdot 5$ cents = $5n$ cents. For problems of this type, we will use the relationship

Number $\cdot$ value = total value

EXAMPLE 9   Find the total value of: **a.** five dimes   **b.** *q* quarters   **c.** $x + 1$ half-dollars

Strategy   To find the total value (in cents) of each collection of coins, we multiply the number of coins by the value (in cents) of one coin, as shown in the table.

Why   Number $\cdot$ value = total value

Solution

Type of coin	Number	Value	Total value
Dime	5	10	50
Quarter	*q*	25	25*q*
Half-dollar	*x* + 1	50	50(*x* + 1)

Multiply: $5 \cdot 10 = 50$.

Multiply: $q \cdot 25$ can be written $25q$.

Multiply: $(x + 1) \cdot 50$ can be written $50(x + 1)$.

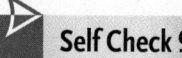

Self Check 9   Find the value of: **a.** six $50 savings bonds   **b.** *t* $100 savings bonds   **c.** $x - 4$ $1,000 savings bonds

Now Try   **Problems 14 and 69**

Evaluate Algebraic Expressions.

To evaluate an algebraic expression, we substitute given numbers for each variable and perform the necessary calculations in the proper order.

EXAMPLE 10   Evaluate each expression for $x = 3$ and $y = -4$:   **a.** $y^3 + y^2$
b. $-y - x$ **c.** $|5xy - 7|$ **d.** $\dfrac{y - 0}{x - (-1)}$

Strategy   We will replace each x and y in the expression with the given value of the variable, and evaluate the expression using the order of operation rules.

Why   To *evaluate an expression* means to find its numerical value, once we know the value of its variable(s).

Solution

a. $y^3 + y^2 = (-4)^3 + (-4)^2$ Substitute −4 for each y. We must write −4 within parentheses so that it is the base of each exponential expression.

$\qquad\qquad = -64 + 16$ Evaluate each exponential expression.

$\qquad\qquad = -48$

> **Caution**
> When replacing a variable with its numerical value, we must often write the replacement number within parentheses to convey the proper meaning.

b. $-y - x = -(-4) - 3$ Substitute −4 for y and 3 for x. Don't forget to write the − sign in front of (−4).

$\qquad\qquad = 4 - 3$ Simplify: −(−4) = 4.

$\qquad\qquad = 1$

c. $|5xy - 7| = |5(3)(-4) - 7|$ Substitute 3 for x and −4 for y.

$\qquad\qquad = |-60 - 7|$ Do the multiplication: 5(3)(−4) = −60.

$\qquad\qquad = |-67|$ Do the subtraction: −60 − 7 = −60 + (−7) = −67.

$\qquad\qquad = 67$ Find the absolute value of −67.

d. $\dfrac{y - 0}{x - (-1)} = \dfrac{-4 - 0}{3 - (-1)}$ Substitute 3 for x and −4 for y.

$\qquad\qquad = \dfrac{-4}{4}$ In the denominator, do the subtraction: 3 − (−1) = 3 + 1 = 4.

$\qquad\qquad = -1$ Simplify the fraction.

 Self Check 10   Evaluate each expression for $a = -2$ and $b = 5$:
a. $|a^3 + b^2|$ **b.** $-a + 2ab$ **c.** $\dfrac{a + 2}{b - 3}$

Now Try   **Problems 79 and 91**

EXAMPLE 11   *Rocketry.*   If a toy rocket is shot into the air with an initial velocity of 80 feet per second, its height (in feet) after t seconds in flight is given by $-16t^2 + 80t$. How many seconds after the launch will it hit the ground?

Strategy   We can substitute positive values for t, the time in flight, until we find the one that gives a height of 0.

Why   When the height of the rocket is 0, it is on the ground.

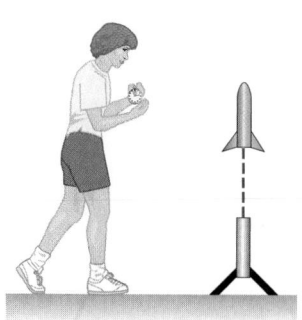

Solution   We begin by finding the height after the rocket has been in flight for 1 second ($t = 1$).

$$-16t^2 + 80t = -16(1)^2 + 80(1) \quad \text{Substitute 1 for } t.$$
$$= 64$$

As we evaluate $-16t^2 + 80t$ for several more values of t, we record each result in a table. The columns of the table can also be headed with the terms **input** and **output.** The values of t are the inputs into the expression $-16t^2 + 80t$, and the resulting values are the outputs.

t	$-16t^2 + 80t$
1	64
2	96
3	96
4	64
5	0

Evaluate for $t = 2$: $-16t^2 + 80t = -16(2)^2 + 80(2) = 96$
Evaluate for $t = 3$: $-16t^2 + 80t = -16(3)^2 + 80(3) = 96$
Evaluate for $t = 4$: $-16t^2 + 80t = -16(4)^2 + 80(4) = 64$
Evaluate for $t = 5$: $-16t^2 + 80t = -16(5)^2 + 80(5) = 0$

Input	Output
1	64
2	96
3	96
4	64
5	0

The height of the rocket is 0 when $t = 5$. The rocket will hit the ground 5 seconds after being launched.

 Self Check 11   In Example 11, suppose the height of the rocket is given by $-16t^2 + 112t$. What will be the height of the rocket 6 seconds after launch?

Now Try   **Problem 97**

 ANSWERS TO SELF CHECKS   **1.** $1, -12, 3, -4$ **2. a.** Factor   **b.** Term   **3. a.** $t - 80$ **b.** $\frac{2}{3}T$
c. $(2a - 15)^2$ **4.** $(m + 15)$ minutes   **5.** $s =$ amount donated to scholarship fund in dollars; $900 - s =$ amount donated to building fund   **6.** $x =$ number of votes received by the challenger; $3x - 55 =$ number of votes received by the incumbent   **8.** $\frac{h}{24}$ **9. a.** \$300   **b.** \$100t
c. \$1,000$(x - 4)$ **10. a.** 17   **b.** -18 **c.** 0   **11.** 96 ft

STUDY SET
1.8

VOCABULARY

Fill in the blanks.

1. Variables and/or numbers can be combined with the operations of arithmetic to create algebraic _____.

2. A _____ is a product or quotient of numbers and/or variables. Examples are: $8x$, $\frac{t}{2}$, and $-cd^3$.

3. Addition symbols separate algebraic expressions into parts called _____.

4. A term, such as 27, that consists of a single number is called a _____ term.

5. The _____ of the term $10x$ is 10.

6. To _____ $4x - 3$ for $x = 5$, we substitute 5 for x and perform the necessary calculations.

CONCEPTS

7. Complete the table below on the left to determine the number of days in w weeks.

8. Complete the table below on the right to determine the number of minutes in s seconds.

Number of weeks	Number of days
1	
2	
3	
w	

Number of seconds	Number of minutes
60	
120	
180	
s	

9. The knife shown below is 12 inches long. Write an expression that represents the length of the blade.

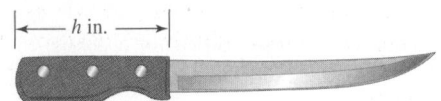

10. A student inherited $5,000 and deposits x dollars in American Savings. Write an expression that represents the amount of money left to deposit in a City Mutual account.

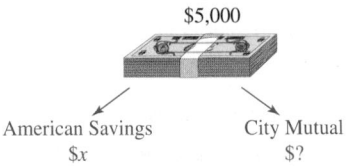

$5,000

American Savings   City Mutual
$\$x$ $\$?$

11. Solution 1 is poured into solution 2. Write an expression that represents the number of ounces in the mixture.

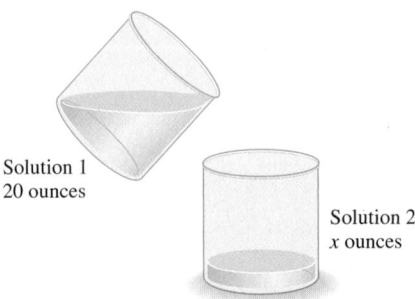

Solution 1
20 ounces

Solution 2
x ounces

12. Peanuts were mixed with p pounds of cashews to make 100 pounds of a mixture. Write an expression that represents the number of pounds of peanuts that were used.

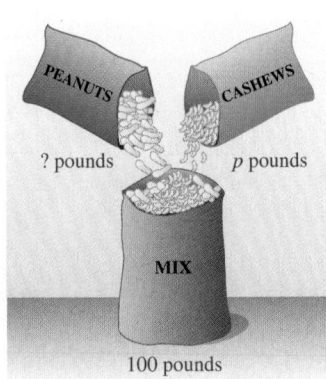

PEANUTS   CASHEWS

? pounds p pounds

MIX

100 pounds

13. **a.** Let b = the length of the beam shown below (in feet). Write an expression that represents the length of the pipe.

   b. Let p = the length of the pipe (in feet). Write an expression that represents the length of the beam.

15 ft

14. Complete the table. Give each value in cents.

Coin	Number	Value	Total value
Nickel	6		
Dime	d		
Half-dollar	$x + 5$		

NOTATION

Complete each solution. Evaluate each expression for a = 5, x = −2, and y = 4.

15. $9a - a^2 = 9(\quad) - (5)^2$
$$= 9(5) - \qquad$$
$$= \qquad - 25$$
$$= 20$$

16. $-x + 6y = -(\quad) + 6(\quad)$
$$= \qquad + 24$$
$$= 26$$

17. Write each term in standard form.
  a. $y8$ **b.** $d2c$
  c. What property of multiplication did you use?

18. Fill in the blanks.
  a. $\dfrac{w}{2} = \qquad w$ **b.** $\dfrac{2}{3}m = \dfrac{\qquad}{3}$

GUIDED PRACTICE

19. Consider the expression $3x^3 + 11x^2 - x + 9$. See Example 1.
  a. How many terms does the expression have?
  b. What is the coefficient of each term?

20. Complete the following table.

Term	$6m$	$-75t$	w	$\frac{1}{2}bh$	$\frac{x}{5}$	t
Coefficient						

Determine whether the variable c is used as a factor or as a term. See Example 2.

21. $c + 32$ **22.** $-24c + 6$

23. $5c$ **24.** $a + b + c$

Translate each phrase to an algebraic expression. If no variable is given, use x as the variable. See Example 3.

25. The sum of the length l and 15

26. The difference of a number and 10

27. The product of a number and 50

28. Three-fourths of the population p

29. The ratio of the amount won w and lost l

30. The tax t added to c

31. P increased by two-thirds of p

32. 21 less than the total height h

33. The square of k minus 2,005

34. s subtracted from S

35. 1 less than twice the attendance a

36. J reduced by 500

37. 1,000 split n equal ways

38. Exceeds the cost c by 25,000

39. 90 more than twice the current price p

40. 64 divided by the cube of y

41. 3 times the total of 35, h, and 300

42. Decrease x by -17

43. 680 fewer than the entire population p

44. Triple the number of expected participants

45. The product of d and 4, decreased by 15

46. The quotient of y and 6, cubed

47. Twice the sum of 200 and t

48. The square of the quantity 14 less than x

49. The absolute value of the difference of a and 2

50. The absolute value of a, decreased by 2

51. One-tenth of the distance d

52. Double the difference of x and 18

Translate each algebraic expression into an English phrase. (Answers may vary.) See Example 3.

53. $\dfrac{3}{4}r$

54. $\dfrac{2}{3}d$

55. $t - 50$

56. $c + 19$

57. xyz

58. $10ab$

59. $2m + 5$

60. $2s - 8$

Answer with an algebraic expression. See Example 4.

61. A model's skirt is x inches long. The designer then lets the hem down 2 inches. What is the length of the altered skirt?

62. A soft drink manufacturer produced c cans of cola during the morning shift. Write an expression for how many six-packs of cola can be assembled from the morning shift's production.

63. The tag on a new pair of 36-inch-long jeans warns that after washing, they will shrink x inches in length. What is the length of the jeans after they are washed?

64. A caravan of b cars, each carrying 5 people, traveled to the state capital for a political rally. How many people were in the caravan?

Answer with an algebraic expression. See Example 8.

65. How many minutes are there in h hours?

66. How many feet are in y yards?

67. How many feet are in i inches?

68. How many centuries in y years?

Answer with an algebraic expression. See Example 9.

69. A sales clerk earns $\$x$ an hour; how much does he earn in an 8-hour day?

70. A cashier earns $d an hour; how much does she earn in a 40-hour week?

71. If a car rental agency charges 49¢ a mile, what is the rental fee if a car is driven x miles?

72. If one egg is worth e cents, find the value (in cents) of one dozen eggs.

73. A ticket to a concert costs $t. What would a pair of concert tickets cost?

74. If one apple is worth a cents, find the value (in cents) of 20 apples.

75. Tickets to a circus cost $25 each. What will tickets cost for a family of x people if they also pay for two of their neighbors?

76. A certain type of office desk that used to sell for $x is now on sale for $50 off. What will a company pay if it purchases 80 of the desks?

Evaluate each expression, for x = 3, y = −2, and z = −4. See Example 10.

77. $-y$

78. $-z$

79. $-z + 3x$

80. $-y - 5x$

81. $3y^2 - 6y - 4$

82. $-z^2 - z - 12$

83. $(3 + x)y$

84. $(4 + z)y$

85. $(x + y)^2 - |z + y|$

86. $[(z - 1)(z + 1)]^2$

87. $-\dfrac{2x + y^3}{y + 2z}$

88. $-\dfrac{2z^2 - x}{2x - y^2}$

Evaluate each expression. See Example 10.

89. $b^2 - 4ac$ for $a = -1$, $b = 5$, and $c = -2$

90. $(x - a)^2 + (y - b)^2$ for $x = -2$, $y = 1$, $a = 5$, and $b = -3$

91. $a^2 + 2ab + b^2$ for $a = -5$ and $b = -1$

92. $\dfrac{a - x}{y - b}$ for $x = -2$, $y = 1$, $a = 5$, and $b = 2$

93. $\dfrac{n}{2}[2a + (n - 1)d]$ for $n = 10$, $a = -4.2$, and $d = 6.6$

94. $\dfrac{a(1 - r^n)}{1 - r}$ for $a = -5$, $r = 2$, and $n = 3$

95. $(27c^2 - 4d^2)^3$ for $c = \frac{1}{3}$ and $d = \frac{1}{2}$

96. $\dfrac{-b^2 + 16a^2 + 1}{2}$ for $a = \frac{1}{4}$ and $b = -10$

Complete each table. See Example 11.

97.

x	$x^3 - 1$
0	
−1	
−3	

98.

g	$g^2 - 7g + 1$
0	
7	
−10	

99.

s	$\dfrac{5s + 36}{s}$
1	
6	
−12	

100.

a	$2{,}500a + a^3$
2	
4	
−5	

101.

Input x	Output $2x - \dfrac{x}{2}$
100	
−300	

102.

Input x	Output $\dfrac{x}{3} + \dfrac{x}{4}$
12	
−36	

103.

x	$(x + 1)(x + 5)$
−1	
−5	
−6	

104.

x	$\dfrac{1}{x + 8}$
−7	
−9	
−8	

APPLICATIONS

105. VEHICLE WEIGHTS   A Hummer H2 weighs 340 pounds less than twice a Honda Element.

   a. Let x represent the weight of one of the vehicles. Write an expression for the weight of the other vehicle.

   b. If the weight of the Element is 3,370 pounds, what is the weight of the Hummer?

106. SOD FARMS   The expression $20{,}000 - 3s$ gives the number of square feet of sod that are left in a field after s strips have been removed. Suppose a city orders 7,000 strips of sod. Evaluate the expression and explain the result.

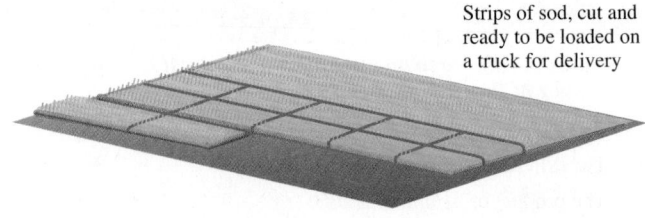

Strips of sod, cut and ready to be loaded on a truck for delivery

107. COMPUTER COMPANIES   IBM was founded 80 years before Apple Computer. Dell Computer Corporation was founded 9 years after Apple.

   a. Let x represent the age (in years) of one of the companies. Write expressions to represent the ages (in years) of the other two companies.

   b. On April 1, 2008, Apple Computer Company was 32 years old. How old were the other two computer companies then?

108. THRILL RIDES   The distance in feet that an object will fall in t seconds is given by the expression $16t^2$. Find the distance that riders on "Drop Zone" will fall during the times listed in the table.

Time (seconds)	Distance (feet)
1	
2	
3	
4	

WRITING

109. What is an algebraic expression? Give some examples.

110. Explain why 2 *less than* x does not translate to $2 < x$.

111. In this section, we substituted a number for a variable. List some other uses of the word *substitute* that you encounter in everyday life.

112. Explain why d dimes are not worth $d\not{c}$.

REVIEW

113. Find the LCD for $\frac{5}{12}$ and $\frac{1}{15}$.

114. Simplify: $\frac{3 \cdot 3 \cdot 5}{3 \cdot 5 \cdot 5 \cdot 11}$

115. Evaluate: $\left(\frac{2}{3}\right)^3$

116. Find the result when $\frac{7}{8}$ is multiplied by its reciprocal.

CHALLENGE PROBLEMS

117. Evaluate: $(8 - 1)(8 - 2)(8 - 3) \ldots (8 - 49)(8 - 50)$

118. Translate to an expression: The sum of a number decreased by six, and seven more than the quotient of triple the number and five.

SECTION 1.9
Simplifying Algebraic Expressions Using Properties of Real Numbers

Objectives

1. Simplify products.
2. Use the distributive property.
3. Identify like terms.
4. Combine like terms.

In algebra, we frequently replace one algebraic expression with another that is equivalent and simpler in form. That process, called *simplifying an algebraic expression,* often involves the use of one or more properties of real numbers.

1 **Simplify Products.**

The commutative and associative properties of multiplication can be used to simplify certain products. For example, let's simplify $8(4x)$.

$$8(4x) = 8 \cdot (4 \cdot x) \quad \text{Rewrite } 4x \text{ as } 4 \cdot x.$$
$$= (8 \cdot 4) \cdot x \quad \text{Use the associative property of multiplication to group 4 with 8.}$$
$$= 32x \quad \text{Do the multiplication within the parentheses.}$$

We have found that $8(4x) = 32x$. We say that $8(4x)$ and $32x$ are **equivalent expressions** because for each value of x, they represent the same number.

Success Tip

By the commutative property of multiplication, we can *change* the order of factors. By the associative property of multiplication, we can change the *grouping* of factors.

© Joel Rogers, www.coastergallery.com

$$\text{If } x = 10$$

$$8(4x) = 8[4(10)] \qquad 32x = 32(10)$$
$$= 8(40) \qquad\qquad = 320$$
$$= 320$$

$$\text{If } x = -3$$

$$8(4x) = 8[4(-3)] \qquad 32x = 32(-3)$$
$$= 8(-12) \qquad\qquad = -96$$
$$= -96$$

EXAMPLE 1   Simplify: **a.** $-9(3b)$ **b.** $15a(6)$ **c.** $3(7p)(-5)$

d. $\dfrac{8}{3} \cdot \dfrac{3}{8}r$ **e.** $35\left(\dfrac{4}{5}x\right)$

Strategy   We will use the commutative and associative properties of multiplication to reorder and regroup the factors in each expression.

Why   We want to group all of the numerical factors of an expression together so that we can find their product.

Solution

a. $-9(3b) = (-9 \cdot 3)b$ Use the associative property of multiplication to regroup the factors.

$\qquad\quad\; = -27b$ Do the multiplication within the parentheses.

b. $15a(6) = 15(6)a$ Use the commutative property of multiplication to reorder the factors.

$\qquad\quad\; = 90a$ Do the multiplication, working from left to right: $15(6) = 90$.

c. $3(7p)(-5) = [3(7)(-5)]p$ Use the commutative and associative properties of multiplication to reorder and regroup the factors.

$\qquad\qquad\;\; = -105p$ Do the multiplication within the brackets.

d. $\dfrac{8}{3} \cdot \dfrac{3}{8}r = \left(\dfrac{8}{3} \cdot \dfrac{3}{8}\right)r$ Use the associative property of multiplication to regroup the factors.

$\qquad\quad\; = 1r$ Multiply within the parentheses. The product of a number and its reciprocal is 1.

$\qquad\quad\; = r$ The coefficient 1 need not be written.

e. $35\left(\dfrac{4}{5}x\right) = \left(35 \cdot \dfrac{4}{5}\right)x$ Use the associative property of multiplication to regroup the factors.

$\qquad\qquad\;\; = \left(\dfrac{\overset{1}{\cancel{5}} \cdot 7 \cdot 4}{\underset{1}{\cancel{5}}}\right)x$ Factor 35 as $5 \cdot 7$ and then remove the common factor 5.

$\qquad\qquad\;\; = 28x$

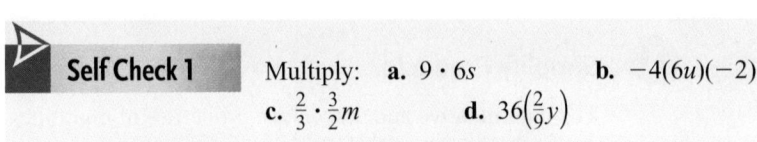

Self Check 1   Multiply: **a.** $9 \cdot 6s$ **b.** $-4(6u)(-2)$

c. $\dfrac{2}{3} \cdot \dfrac{3}{2}m$ **d.** $36\left(\dfrac{2}{9}y\right)$

Now Try   **Problems 15, 25, 29, and 31**

---

2   **Use the Distributive Property.**

Another property that is often used to simplify algebraic expressions is the **distributive property.** To introduce it, we will evaluate $4(5 + 3)$ in two ways.

Method 1
Use the order of operations:

$$4(5 + 3) = 4(8)$$
$$= 32$$

Method 2
Distribute the multiplication:

$$4(5 + 3) = 4(5) + 4(3)$$
$$= 20 + 12$$
$$= 32$$

Each method gives a result of 32. This observation suggests the following property.

The Distributive Property	For any real numbers a, b, and c, $a(b + c) = ab + ac$

To illustrate one use of the distributive property, let's consider the expression $5(x + 3)$. Since we are not given the value of x, we cannot add x and 3 within the parentheses. However, we can distribute the multiplication by the factor of 5 that is outside the parentheses to x and to 3 and add those products.

$$5(x + 3) = 5(x) + 5(3) \quad \text{Distribute the multiplication by 5.}$$
$$= 5x + 15 \quad \text{Do the multiplications.}$$

EXAMPLE 2 Multiply: **a.** $8(m + 9)$ **b.** $-12(4t + 1)$ **c.** $6\left(\dfrac{x}{3} + \dfrac{9}{2}\right)$

Strategy   In each case, we will distribute the multiplication by the factor *outside* the parentheses over each term *within* the parentheses.

Why   In each case, we cannot simplify the expression within the parentheses. To multiply, we must use the distributive property.

Solution

a. $8(m + 9) = 8 \cdot m + 8 \cdot 9 \quad$ Distribute the multiplication by 8.
$$= 8m + 72 \quad \text{Do the multiplications.}$$

b. $-12(4t + 1) = -12(4t) + (-12)(1) \quad$ Distribute the multiplication by -12.
$$= -48t + (-12) \quad \text{Do the multiplications.}$$
$$= -48t - 12 \quad \begin{array}{l}\text{Write the result in simpler form. Recall that adding} \\ -12 \text{ is the same as subtracting 12.}\end{array}$$

c. $6\left(\dfrac{x}{3} + \dfrac{9}{2}\right) = 6 \cdot \dfrac{x}{3} + 6 \cdot \dfrac{9}{2} \quad$ Distribute the multiplication by 6.

$$= \dfrac{2 \cdot \overset{1}{\cancel{3}} \cdot x}{\underset{1}{\cancel{3}}} + \dfrac{\overset{1}{\cancel{2}} \cdot 3 \cdot 9}{\underset{1}{\cancel{2}}} \quad \begin{array}{l}\text{Factor 6 as } 2 \cdot 3 \text{ and then remove the common} \\ \text{factors 3 and 2.}\end{array}$$

$$= 2x + 27$$

Since subtraction is the same as adding the opposite, the distributive property also holds for subtraction.

$$a(b - c) = ab - ac$$

EXAMPLE 3   Multiply: **a.** $3(3b - 4)$ **b.** $-6(-3y - 8)$ **c.** $-1(t - 9)$

Strategy   In each case, we will distribute the multiplication by the factor *outside* the parentheses over each term *within* the parentheses.

Why   In each case, we cannot simplify the expression within the parentheses. To multiply, we must use the distributive property.

Solution

a. $3(3b - 4) = 3(3b) - 3(4)$ Distribute the multiplication by 3.
$= 9b - 12$ Do the multiplications.

b. $-6(-3y - 8) = -6(-3y) - (-6)(8)$ Distribute the multiplication by -6.
$= 18y - (-48)$ Do the multiplications.
$= 18y + 48$ Write the result in simpler form. Add the opposite of -48.

Another approach is to write the subtraction within the parentheses as addition of the opposite. Then we distribute the multiplication by -6 over the addition.

$-6(-3y - 8) = -6[-3y + (-8)]$ Add the opposite of 8.
$= -6(-3y) + (-6)(-8)$ Distribute the multiplication by -6.
$= 18y + 48$ Do the multiplications.

c. $-1(t - 9) = -1(t) - (-1)(9)$ Distribute the multiplication by -1.
$= -t - (-9)$ Do the multiplications.
$= -t + 9$ Write the result in simpler form. Add the opposite of -9.

Caution
A common mistake is to forget to distribute the multiplication over each of the terms within the parentheses.
$3(3b - 4) = 9b - 4$

Success Tip
Notice that distributing the multiplication by -1 *changes the sign* of each term within the parentheses.

Caution   The distributive property does not apply to every expression that contains parentheses—only those where multiplication is distributed over addition (or subtraction). For example, to simplify $6(5x)$, we do not use the distributive property.

Correct	*Incorrect*

$$6(5x) = (6 \cdot 5)x = 30x \qquad 6(5x) = 30 \cdot 6x = 180x$$

The distributive property can be extended to several other useful forms. Since multiplication is commutative, we have:

$$(b + c)a = ba + ca \qquad (b - c)a = ba - ca$$

For situations in which there are more than two terms within parentheses, we have:

$$a(b + c + d) = ab + ac + ad \qquad a(b - c - d) = ab - ac - ad$$

EXAMPLE 4 Multiply: **a.** $(6x + 4)\frac{1}{2}$ **b.** $2(a - 3b)8$
c. $-0.3(3a - 4b + 7)$

Strategy We will multiply each term within the parentheses by the factor (or factors) outside the parentheses.

Why In each case, we cannot simplify the expression within the parentheses. To multiply, we must use the distributive property.

Solution

a. $(6x + 4)\dfrac{1}{2} = (6x)\dfrac{1}{2} + (4)\dfrac{1}{2}$ Distribute the multiplication by $\frac{1}{2}$.

$\qquad\qquad = 3x + 2$ Do the multiplications.

b. $2(a - 3b)8 = 2 \cdot 8(a - 3b)$

$\qquad\qquad = 16(a - 3b)$ Multiply 2 and 8 to get 16.

$\qquad\qquad = 16a - 48b$ Distribute the multiplication by 16.

c. $-0.3(3a - 4b + 7) = -0.3(3a) - (-0.3)(4b) + (-0.3)(7)$

$\qquad\qquad\qquad\qquad = -0.9a + 1.2b - 2.1$ Do each multiplication.

Self Check 4   Multiply: **a.** $(-6x - 24)\frac{1}{3}$
b. $6(c - 2d)9$
c. $-0.7(2r + 5s - 8)$
Now Try **Problems 53, 55, and 57**

We can use the distributive property to find the opposite of a sum. For example, to find $-(x + 10)$, we interpret the $-$ symbol as a factor of -1, and proceed as follows:

$$-(x + 10) = -1(x + 10) \qquad \text{Replace the } - \text{ symbol with } -1.$$

$$\qquad\qquad = -1(x) + (-1)(10) \qquad \text{Distribute the multiplication by } -1.$$

$$\qquad\qquad = -x - 10$$

In general, we have the following property of real numbers.

The Opposite of a Sum	The opposite of a sum is the sum of the opposites. For any real numbers a and b, $$-(a + b) = -a + (-b).$$

EXAMPLE 5   Simplify: $-(-9s - 3)$

Strategy   We will multiply each term within the parentheses by -1.

Why   The $-$ outside the parentheses represents a factor of -1 that is to be distributed.

Solution

$$-(-9s - 3) = -1(-9s - 3)$$ Replace the $-$ symbol in front of the parentheses with -1.

$$= -1(-9s) - (-1)(3)$$ Distribute the multiplication by -1.

$$= 9s + 3$$

 Self Check 5   Simplify: $-(-5x + 18)$

Now Try   **Problem 59** $5x - 18$

---

3   **Identify Like Terms.**

Before we can discuss methods for simplifying algebraic expressions involving addition and subtraction, we need to introduce some new vocabulary.

Like Terms	**Like terms** are terms containing exactly the same variables raised to exactly the same powers. Any constant terms in an expression are considered to be like terms. Terms that are not like terms are called **unlike terms**.

Success Tip
When looking for like terms, don't look at the coefficients of the terms. Consider only the variable factors of each term. If two terms are like terms, only their coefficients may differ.

Here are several examples.

Like terms	*Unlike terms*	
$4x$ and $7x$	$4x$ and $7y$	The variables are not the same.
$-10p^2$ and $25p^2$	$-10p$ and $25p^2$	Same variable, but different powers.
$\frac{1}{3}c^3d$ and c^3d	$\frac{1}{3}c^3d$ and c^3	The variables are not the same.

EXAMPLE 6   List the like terms in each expression: **a.** $7r + 5 + 3r$
b. $6x^4 - 6x^2 - 6x$ **c.** $-17m^3 + 3 - 2 + m^3$

Strategy   First, we will identify the terms of the expression. Then we will look for terms that contain the same variables raised to exactly the same powers.

Why   If two terms contain the same variables raised to the same powers, they are like terms.

Solution

a. $7r + 5 + 3r$ contains the like terms $7r$ and $3r$.

b. Since the exponents on x are different, $6x^4 - 6x^2 - 6x$ contains no like terms.

c. $-17m^3 + 3 - 2 + m^3$ contains two pairs of like terms: $-17m^3$ and m^3 are like terms, and the constant terms, 3 and -2, are like terms.

Self Check 6   List the like terms: **a.** $2x - 2y + 7y$
b. $5p^2 - 12 + 17p^2 + 2$

Now Try   **Problem 63**

---

④   **Combine Like Terms.**

To add or subtract objects, they must be similar. For example, fractions that are to be added must have a common denominator. When adding decimals, we align columns to be sure to add tenths to tenths, hundredths to hundredths, and so on. The same is true when working with terms of an algebraic expression. They can be added or subtracted only if they are like terms.

This expression can be simplified   This expression cannot be simplified
because it contains like terms.   because its terms are not like terms.

$3x + 4x$ $3x + 4y$

Recall that the distributive property can be written in the following forms:

$$(b + c)a = ba + ca \qquad (b - c)a = ba - ca$$

We can use these forms of the distributive property in reverse to simplify a sum or difference of like terms. For example, we can simplify $3x + 4x$ as follows:

$$3x + 4x = (3 + 4)x \qquad \text{Use } ba + ca = (b + c)a.$$
$$= 7x$$

The Language of Algebra
Simplifying a sum or difference of like terms is called *combining like terms.*

We can simplify $15m^2 - 9m^2$ in a similar way:

$$15m^2 - 9m^2 = (15 - 9)m^2 \qquad \text{Use } ba - ca = (b - c)a.$$
$$= 6m^2$$

In each case, we say that we *combined like terms.* These examples suggest the following general rule.

Combining Like Terms

Like terms can be combined by adding or subtracting the coefficients of the terms and keeping the same variables with the same exponents.

EXAMPLE 7   Simplify by combining like terms, if possible: **a.** $2x + 9x$
b. $-8p + (-2p) + 4p$ **c.** $0.5s^3 - 0.3s^3$

d. $4w + 6$ **e.** $\dfrac{4}{9}b + \dfrac{7}{9}b$

Strategy   We will use the distributive property in reverse to add (or subtract) the coefficients of the like terms. We will keep the same variables raised to the same powers.

Why   To *combine like terms* means to add or subtract the like terms in an expression.

Solution

a. Since $2x$ and $9x$ are like terms with the common variable x, we can combine them.

$$2x + 9x = 11x \quad \text{Think: } (2 + 9)x = 11x.$$

b. $-8p + (-2p) + 4p = -6p \quad \text{Think: } [-8 + (-2) + 4]p = -6p.$

c. $0.5s^3 - 0.3s^3 = 0.2s^3 \qquad \text{Think: } (0.5 - 0.3)s^3 = 0.2s^3.$

d. Since $4w$ and 6 are not like terms, they cannot be combined. $4w + 6$ doesn't simplify.

e. $\dfrac{4}{9}b + \dfrac{7}{9}b = \dfrac{11}{9}b \qquad\qquad \text{Think: } \left(\frac{4}{9} + \frac{7}{9}\right)b = \frac{11}{9}b.$

Success Tip

Just as 2 apples plus 9 apples is 11 apples, $2x + 9x = 11x$.

Self Check 7   Simplify, if possible: **a.** $3x + 5x$
b. $-6y + (-6y) + 9y$ **c.** $4.4s^4 - 3.9s^4$
d. $4a - 2$ **e.** $\frac{10}{7}c - \frac{4}{7}c$

Now Try   Problems 67, 71, 79, and 83

EXAMPLE 8   Simplify by combining like terms: **a.** $16t - 15t$
b. $16t - t$ **c.** $15t - 16t$ **d.** $16t + t$

Strategy   As we combine like terms, we must be careful when working with the terms such as t and $-t$.

Why   Coefficients of 1 and -1 are usually not written.

Solution

a. $16t - 15t = t \qquad \text{Think: } (16 - 15)t = 1t = t.$

b. $16t - t = 15t \qquad \text{Think: } 16t - 1t = (16 - 1)t = 15t.$

c. $15t - 16t = -t \qquad \text{Think: } (15 - 16)t = -1t = -t.$

d. $16t + t = 17t \qquad \text{Think: } 16t + 1t = (16 + 1)t = 17t.$

Self Check 8   Simplify: **a.** $9h - h$ **b.** $9h + h$ **c.** $9h - 8h$
d. $8h - 9h$

Now Try   Problems 73 and 77

EXAMPLE 9   Simplify: $6a^2 + 54a - 4a - 36$

Strategy   First, we will identify any like terms in the expression. Then we will use the distributive property in reverse to combine them.

Why   To *simplify* an expression we use properties of real numbers to write an equivalent expression in simpler form.

Solution We can combine the like terms that involve the variable *a*.

$$6a^2 + 54a - 4a - 36 = 6a^2 + 50a - 36 \quad \text{Think: } (54-4)a = 50a.$$

[handwritten: $6a^2 + 50a - 36$]

Self Check 9 Simplify: $7y^2 + 21y - 2y - 6$
Now Try Problem 93   *[handwritten: $7y^2 + 19y - 6$]*

EXAMPLE 10 Simplify: $4(x+5) - 5 - (2x-4)$

Strategy First, we will remove the parentheses. Then we will identify any like terms and combine them.

Why To *simplify* an expression we use properties of real numbers, such as the distributive property, to write an equivalent expression in simpler form.

Solution

> **Success Tip**
> Here, the distributive property is used both *forward* (to remove parentheses) and in *reverse* (to combine like terms).

$$4(x+5) - 5 - (2x-4) = 4(x+5) - 5 - 1(2x-4)$$

Replace the − symbol in front of (2x − 4) with −1.

$$= 4x + 20 - 5 - 2x + 4$$

Distribute the multiplication by 4 and −1.

$$= 2x + 19$$

Think: (4 − 2)x = 2x.
Think: (20 − 5 + 4) = 19.

Self Check 10 Simplify: $6(3y-1) + 2 - (-3y+4)$
Now Try Problem 99

ANSWERS TO SELF CHECKS **1. a.** $54s$ **b.** $48u$ **c.** m **d.** $8y$ **2. a.** $7m+14$ **b.** $-640x-240$
c. $4y+9$ **3. a.** $10x-5$ **b.** $9y+36$ **c.** $-c+22$ **4. a.** $-2x-8$ **b.** $54c-108d$
c. $-1.4r-3.5s+5.6$ **5.** $5x-18$ **6. a.** $-2y$ and $7y$ **b.** $5p^2$ and $17p^2$; -12 and 2 **7. a.** $8x$
b. $-3y$ **c.** $0.5s^4$ **d.** Does not simplify **e.** $\frac{6}{7}c$ **8. a.** $8h$ **b.** $10h$ **c.** h **d.** $-h$
9. $7y^2+19y-6$ **10.** $21y-8$

STUDY SET
1.9

VOCABULARY

Fill in the blanks.

1. To _____ the expression $5(6x)$ means to write it in simpler form: $5(6x)=30x$.

2. $5(6x)$ and $30x$ are _____ expressions because for each value of *x*, they represent the same number.

3. To perform the multiplication $2(x+8)$, we use the _____ property.

4. We call $-(c+9)$ the _____ of a sum.

5. Terms such as $7x^2$ and $5x^2$, which have the same variables raised to exactly the same power, are called _____ terms.

6. When we write $9x + x$ as $10x$, we say we have _____ like terms.

CONCEPTS

7. a. Fill in the blanks to simplify the expression.

$$4(9t) = (\quad \cdot \quad)t = \quad t$$

b. What property did you use in part a?

8. a. Fill in the blanks to simplify the expression.

$$-6y \cdot 2 = \quad \cdot \quad \cdot y = \quad y$$

b. What property did you use in part a?

9. Fill in the blanks.

a. $2(x + 4) = 2x \quad 8$ **b.** $2(x - 4) = 2x \quad 8$

c. $-2(x + 4) = -2x \quad 8$ **d.** $-2(-x - 4) = 2x \quad 8$

10. Fill in the blanks to combine like terms.

a. $4m + 6m = (\quad)m = \quad m$

b. $30n^2 - 50n^2 = (\quad)n^2 = \quad n^2$

c. $12 + 32d + 15 = 32d + \quad$

d. Like terms can be combined by adding or subtracting the _____ of the terms and keeping the same _____ with the same exponents.

11. Simplify each expression, if possible.

a. $5(2x)$ **b.** $5 + 2x$

c. $6(-7x)$ **d.** $6 - 7x$

e. $2(3x)(3)$ **f.** $2 + 3x + 3$

12. Fill in the blanks: Distributing multiplication by -1 changes the _____ of each term within the parentheses.

$$-(x + 10) = \quad (x + 10) = -x \quad 10$$

NOTATION

13. Translate to symbols.

a. Six times the quantity of h minus four.

b. The opposite of the sum of z and sixteen.

14. Write an equivalent expression for the given expression using fewer symbols.

a. $1x$ **b.** $-1d$ **c.** $0m$

d. $5x - (-1)$ **e.** $16t + (-6)$

GUIDED PRACTICE

Simplify each expression. **See Example 1.**

15. $3 \cdot 4t$ **16.** $9 \cdot 3s$

17. $9(7m)$ **18.** $12n(8)$

19. $5(-7q)$ **20.** $-7(5t)$

21. $5t \cdot 60$ **22.** $70a \cdot 10$

23. $(-5.6x)(-2)$ **24.** $(-4.4x)(-3)$

25. $5(4c)(3)$ **26.** $9(2h)(2)$

27. $-4(-6)(-4m)$ **28.** $-5(-9)(-4n)$

29. $\dfrac{5}{3} \cdot \dfrac{3}{5}g$ **30.** $\dfrac{9}{7} \cdot \dfrac{7}{9}k$

31. $12\left(\dfrac{5}{12}x\right)$ **32.** $15\left(\dfrac{4}{15}w\right)$

33. $8\left(\dfrac{3}{4}y\right)$ **34.** $27\left(\dfrac{2}{3}x\right)$

Multiply. **See Examples 2–5.**

35. $5(x + 3)$ **36.** $4(x + 2)$

37. $-3(4x + 9)$ **38.** $-5(8x + 9)$

39. $45\left(\dfrac{x}{5} + \dfrac{2}{9}\right)$ **40.** $35\left(\dfrac{y}{5} + \dfrac{8}{7}\right)$

41. $0.4(x - 4)$ **42.** $2.2(2q - 1)$

43. $6(6c - 7)$ **44.** $9(9d - 3)$

45. $-6(13c - 3)$ **46.** $-2(10s - 11)$

47. $-15(-2t - 6)$ **48.** $-20(-4z - 5)$

49. $-1(-4a + 1)$ **50.** $-1(-2x + 3)$

51. $(3t + 2)8$ **52.** $(2q + 1)9$

53. $(3w - 6)\dfrac{2}{3}$ **54.** $(2y - 8)\dfrac{1}{2}$

55. $4(7y + 4)2$ **56.** $8(2a - 3)4$

57. $25(2a - 3b + 1)$ **58.** $5(9s - 12t - 3)$

59. $-(x - 7)$ **60.** $-(y + 1)$

61. $-(-5.6y + 7)$ **62.** $-(-4.8a - 3)$

List the like terms in each expression, if any. **See Example 6.**

63. $3x + 2 - 2x$

64. $3y + 4 - 11y + 6$

65. $-12m^4 - 3m^3 + 2m^2 - m^3$

66. $6x^3 + 3x^2 + 6x$

Simplify by combining like terms. **See Examples 7 and 8.**

67. $3x + 7x$ **68.** $12y - 15y$

69. $-4x + 4x$ **70.** $-16y + 16y$

71. $-7b^2 + 27b^2$ **72.** $-2c^3 + 12c^3$

73. $13r - 12r$ **74.** $25s + s$

75. $36y + y - 9y$ **76.** $32a - a + 5a$

77. $43s^3 - 44s^3$ **78.** $8j^3 - 9j^3$

79. $-9.8c + 6.2c$ **80.** $-5.7m + 4.3m$

81. $-0.2r - (-0.6r)$ **82.** $-1.1m - (-2.4m)$

83. $\dfrac{3}{5}t + \dfrac{1}{5}t$ **84.** $\dfrac{3}{16}x - \dfrac{5}{16}x$

85. $-\dfrac{7}{16}x - \dfrac{3}{16}x$ **86.** $-\dfrac{5}{18}x - \dfrac{7}{18}x$

Simplify by combining like terms. **See Example 9.**

87. $15y - 10 - y - 20y$ **88.** $9z - 7 - z - 19z$

89. $3x + 4 - 5x + 1$ **90.** $4b + 9 - 9b + 9$

91. $9m^2 - 6m + 12m - 4$ **92.** $6a^2 + 18a - 9a + 5$

93. $4x^2 + 5x - 8x + 9$ **94.** $10y^2 - 8y + y - 7$

Simplify. **See Example 10.**

95. $2z + 5(z - 3)$ **96.** $12(m + 11) - 11$

97. $2(s^2 - 7) - (s^2 - 2)$ **98.** $4(d^2 - 3) - (d^2 - 1)$

99. $-9(3r - 9) - 7(2r - 7)$ **100.** $-6(3t - 6) - 3(11t - 3)$

101. $36\left(\dfrac{2}{9}x - \dfrac{3}{4}\right) + 36\left(\dfrac{1}{2}\right)$ **102.** $40\left(\dfrac{3}{8}y - \dfrac{1}{4}\right) + 40\left(\dfrac{4}{5}\right)$

TRY IT YOURSELF

Simplify each expression.

103. $6 - 4(-3c - 7)$ **104.** $10 - 5(-5g - 1)$

105. $-4r - 7r + 2r - r$ **106.** $-v - 3v + 6v + 2v$

107. $24\left(-\dfrac{5}{6}r\right)$ **108.** $\dfrac{3}{4} \cdot \dfrac{1}{2}g$

109. $a + a + a$ **110.** $t - t - t - t$

111. $60\left(\dfrac{3}{20}r - \dfrac{4}{15}\right)$ **112.** $72\left(\dfrac{7}{8}f - \dfrac{8}{9}\right)$

113. $5(-1.2x)$ **114.** $5(-6.4c)$

115. $-(c + 7) + 2(c - 3)$ **116.** $-(z + 2) + 5(3 - z)$

117. $a^3 + 2a^2 + 4a - 2a^2 - 4a - 8$

118. $c^3 - 3c^2 + 9c + 3c^2 - 9c + 27$

APPLICATIONS

In Exercises 119–122, recall that the **perimeter** *of a figure is equal to the sum of the lengths of its sides.*

119. THE RED CROSS   In 1891, Clara Barton founded the Red Cross. Its symbol is a white flag bearing a red cross. If each side of the cross has length x, write an expression that represents the perimeter of the cross.

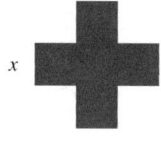

120. BILLIARDS   Billiard tables vary in size, but all tables are twice as long as they are wide.

    a. If the billiard table is x feet wide, write an expression that represents its length.

    b. Write an expression that represents the perimeter of the table.

x ft

121. PING-PONG   Write an expression that represents the perimeter of the Ping-Pong table.

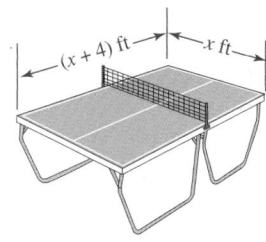

$(x + 4)$ ft x ft

122. SEWING   Write an expression that represents the length of the yellow trim needed to outline a pennant with the given side lengths.

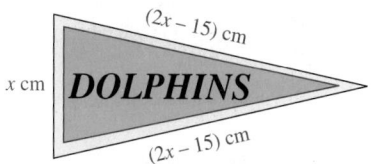

$(2x - 15)$ cm

x cm   *DOLPHINS*

$(2x - 15)$ cm

WRITING

123. Explain why the distributive property applies to $2(3 + x)$ but not to $2(3x)$.

124. Tell how to combine like terms.

REVIEW

Evaluate each expression for $x = -3$, $y = -5$, and $z = 0$.

125. $\dfrac{x - y^2}{2y - 1 + x}$ **126.** $\dfrac{2y + 1}{x} - x$

CHALLENGE PROBLEMS

127. Fill in the blanks: $(\quad - \quad) = -75x + 40$

Simplify.

128. $-2[x + 4(2x + 1)] - 5[x + 2(3x + 4)]$

CHAPTER 1
Summary & Review

SECTION 1.1   Introducing the Language of Algebra

DEFINITIONS AND CONCEPTS	EXAMPLES
Tables, bar graphs, and **line graphs** are used to describe numerical relationships.	See page 3 for examples of tables and graphs.
A **sum** is the result of an addition. A **difference** is the result of a subtraction. A **product** is the result of a multiplication. A **quotient** is the result of a division.	$3 + 15 = 18$ $16 - 1 = 15$ 　sum　　　　　difference $7 \cdot 8 = 56$　　$\dfrac{63}{9} = 7$ 　product　　　　quotient
A **variable** is a letter (or symbol) that stands for a number. **Algebraic expressions** contain variables and numbers combined with the operations of addition, subtraction, multiplication, and division. An **equation** is a statement that two expressions are equal. Equations that express a relationship between two or more variables are called **formulas**.	Variables: x,　a,　and　y Expressions: $5y + 7$,　$\dfrac{12 - x}{5}$,　and　$8a(b-3)$ Equations: $3x-4 = 12$　and　$\dfrac{t}{9} = 12$ $A = lw$　(The formula for the area of a rectangle)

REVIEW EXERCISES

The line graph shows the number of cars in a parking structure from 6 P.M. to 12 midnight on a Saturday.

1. What units are used to scale the horizontal and vertical axes?

2. How many cars were in the parking structure at 11 P.M.?

3. At what time did the parking structure have 500 cars in it?

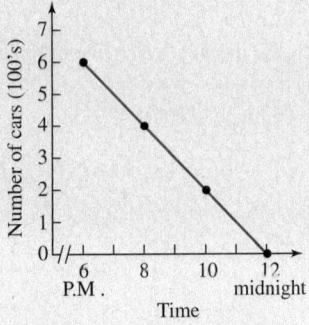

4. When was the structure empty of cars?

Express each statement in words, using one of the words sum, difference, product, or quotient.

5. $15 - 3 = 12$

6. $15 + 3 = 18$

7. $15 \div 3 = 5$

8. $15 \cdot 3 = 45$

9. a. Write the multiplication 4×9 with a raised dot and then with parentheses.

　　b. Write the division $9 \div 3$ using a fraction bar.

10. Write each multiplication without a multiplication symbol.

　　a. $8 \cdot b$　　　　　　**b.** $P \cdot r \cdot t$

11. Classify each item as either an expression or an equation.

　　a. $5 = 2x + 3$　　　　**b.** $2x + 3$

12. Use the formula $n = b + 5$ to complete the table.

Brackets (b)	Nails (n)
5	
10	
20	

SECTION 1.2   Fractions

DEFINITIONS AND CONCEPTS	EXAMPLES
A **factor** is a number being multiplied.	$8 \cdot 9 = 72$ Factor Factor
A **prime number** is a natural number that is greater than 1 that has only itself and 1 as factors. A **composite number** is a natural number, greater than 1, that is not prime.	Primes: $\{2, 3, 5, 7, 11, 13, 17, 19, 23, \ldots\}$ Composites: $\{4, 6, 8, 9, 10, 12, 14, 15, \ldots\}$
Any composite numbers can be factored into the product of two or more prime factors.	Find the prime factorization of 98. $98 = 2 \cdot 49 = 2 \cdot 7 \cdot 7$
In a fraction, the number above the **fraction bar** is the **numerator** and the number below the faction bar is called the **denominator.** Two fractions are **equivalent** if they represent the same number.	$\dfrac{11}{15}$ ← Numerator / ← Denominator Equivalent fractions: $\dfrac{1}{2} = \dfrac{2}{4} = \dfrac{3}{6} = \dfrac{4}{8} = \ldots$
To **multiply two fractions,** multiply their numerators and multiply their denominators.	Multiply: $\dfrac{5}{8} \cdot \dfrac{3}{4} = \dfrac{15}{32}$
One number is the **reciprocal** of another if their product is 1. To **divide two fractions,** multiply the first fraction by the reciprocal of the second fraction.	The reciprocal of $\dfrac{4}{5}$ is $\dfrac{5}{4}$ because $\dfrac{4}{5} \cdot \dfrac{5}{4} = 1$. Divide: $\dfrac{4}{7} \div \dfrac{5}{8} = \dfrac{4}{7} \cdot \dfrac{8}{5} = \dfrac{32}{35}$
Multiplication property of 1: the product of 1 and any number is that number.	$1 \cdot 5 = 5$ and $\dfrac{7}{8} \cdot 1 = \dfrac{7}{8}$
To **build a fraction,** multiply it by a form of 1 such as $\frac{2}{2}, \frac{3}{3}, \frac{4}{4}, \ldots$	Write $\frac{3}{4}$ as an equivalent fraction with a denominator of 20. $\dfrac{3}{4} = \dfrac{3}{4} \cdot \dfrac{5}{5} = \dfrac{15}{20}$
To **simplify a fraction,** remove pairs of factors common to the numerator and the denominator. A fraction is in **simplest form,** or **lowest terms,** when the numerator and denominator have no common factors other than 1.	Simplify: $\dfrac{12}{18} = \dfrac{2 \cdot \cancel{6}}{3 \cdot \cancel{6}} = \dfrac{2}{3}$ $\dfrac{6}{6} = 1$
To find the **LCD** of two fractions, prime factor each denominator and find the product of the prime factors, using each factor the greatest number of times it appears in any one factorization.	Find the LCD of $\frac{5}{12}$ and $\frac{7}{8}$. $\left.\begin{array}{l} 12 = 2 \cdot 2 \cdot 3 \\ 8 = 2 \cdot 2 \cdot 2 \end{array}\right\}$ LCD $= 2 \cdot 2 \cdot 2 \cdot 3 = 24$
To **add (or subtract) fractions that have the same denominator,** add (or subtract) the numerators and keep the common denominator. Simplify, if possible. To **add (or subtract) fractions that have different denominators,** rewrite each fraction as an equivalent fraction with the LCD as the denominator. Then add (or subtract) as usual. Simplify, if possible.	Add: $\dfrac{5}{12} + \dfrac{7}{8} = \dfrac{5}{12} \cdot \dfrac{2}{2} + \dfrac{7}{8} \cdot \dfrac{3}{3}$ The LCD is 24. Build each fraction. $= \dfrac{10}{24} + \dfrac{21}{24}$ The denominators are now the same. $= \dfrac{31}{24}$ This result does not simplify.

SECTION 1.2   Fractions—*continued*

DEFINITIONS AND CONCEPTS	EXAMPLES
A **mixed number** represents the sum of a whole number and a fraction. In some computations, it is necessary to write mixed numbers as fractions.	Mixed numbers: $6\frac{1}{3} = 6 + \frac{1}{3}$ and $1\frac{3}{4} = \frac{7}{4}$

REVIEW EXERCISES

13. **a.** Write 24 as the product of two factors.

   b. Write 24 as the product of three factors.

   c. List the factors of 24.

14. What do we call fractions, such as $\frac{1}{8}$ and $\frac{2}{16}$, that represent the same number?

Give the prime factorization of each number, if possible.

15. 54

16. 147

17. 385

18. 41

Simplify each fraction.

19. $\frac{20}{35}$

20. $\frac{24}{18}$

Build each number to an equivalent fraction with the indicated denominator.

21. $\frac{5}{8}$, denominator 64

22. 12, denominator 3

What is the LCD for fractions having the following denominators?

23. 10 and 18

24. 21 and 70

Perform each operation and simplify, if possible.

25. $\frac{1}{8} \cdot \frac{7}{8}$

26. $\frac{16}{35} \cdot \frac{25}{48}$

27. $\frac{1}{3} \div \frac{15}{16}$

28. $16\frac{1}{4} \div 5$

29. $\frac{17}{25} - \frac{7}{25}$

30. $\frac{8}{11} - \frac{1}{2}$

31. $\frac{17}{24} + \frac{11}{40}$

32. $4\frac{1}{9} - 3\frac{5}{6}$

33. THE INTERNET   A popular website averaged $1\frac{3}{4}$ million hits per day during a 30-day period. How many hits did it receive during that time?

34. MACHINE SHOPS   How much must be milled off the $\frac{17}{24}$-inch-thick steel rod so that the collar will slip over it?

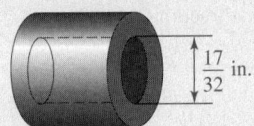   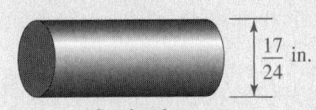

Steel rod

SECTION 1.3   The Real Numbers

DEFINITIONS AND CONCEPTS	EXAMPLES
To write a set, we list its **elements** within **braces** { }.	In the English alphabet, the set of vowels is {a, e, i, o, u}.
The **natural numbers** are the numbers we count with.	Natural numbers: {1, 2, 3, 4, 5, 6, . . .}
The **whole numbers** are the natural numbers together with 0.	Whole numbers: {0, 1, 2, 3, 4, 5, 6, . . .}
Two numbers are called **opposites** if they are the same distance from 0 on the number line but are on opposite sides of it.	Opposites: 3 and -3
The **integers** include the whole numbers and their opposites.	Integers: {. . . , $-3, -2, -1, 0, 1, 2, 3,$ . . .}
The **rational numbers** are numbers that can be expressed as fractions with an integer numerator and a nonzero integer denominator.	Rational numbers: $$-6, \quad -3.1, \quad -\frac{1}{2}, \quad 0, \quad \frac{11}{12}, \quad 9\frac{4}{5}, \quad \text{and} \quad 87$$
Terminating and **repeating decimals** can be expressed as fractions and are, therefore, rational numbers.	Rational numbers: $$-0.25 = -\frac{1}{4} \quad \text{and} \quad 0.\overline{6} = \frac{2}{3}$$

SECTION 1.3   The Real Numbers–*continued*

DEFINITIONS AND CONCEPTS	EXAMPLES
An **irrational number** is a nonterminating, nonrepeating decimal. An irrational number cannot be expressed as a fraction with an integer numerator and a nonzero integer denominator.	Irrational numbers: $\sqrt{5}$, π, and $-\sqrt{7}$
A **real number** is any number that is either a rational or an irrational number. Every real number corresponds to a point on the **number line,** and every point on the number line corresponds to exactly one real number	Graph the set $\left\{ -2, -0.75, 1\frac{3}{4}, \pi \right\}$ on a number line.
Inequality symbols: $>$ is greater than $<$ is less than	$25 > 15$ and $-2 > -7$ $3.3 < 9.7$ and $-10 < -9$
The **absolute value** of a number is the distance on the number line between the number and 0.	$\|5\| = 5$, $\|-7\| = 7$, and $-\left\|-\frac{5}{9}\right\| = -\frac{5}{9}$

REVIEW EXERCISES

35. a. Which number is a whole number but not a natural number?

    b. Write the set of integers.

36. Represent 206 feet below sea level with a signed number.

37. Use one of the symbols $>$ or $<$ to make each statement true.

    a. 0   5   **b.** -12 -13

38. Show that each of the following numbers is a rational number by expressing it as a ratio (quotient) of two integers.

    a. 0.7   **b.** $4\frac{2}{3}$

Write each fraction as a decimal. Use an overbar if the result is a repeating decimal.

39. $\dfrac{1}{250}$ **40.** $\dfrac{17}{22}$

41. Graph each number on a number line:
$\left\{ \pi, 0.33\overline{3} \ldots, 3.75, \sqrt{2}, -\frac{17}{4}, \frac{7}{8}, -2 \right\}$

42. Determine which numbers in the given set are natural numbers, whole numbers, integers, rational numbers, irrational numbers, and real numbers. $\left\{ -\frac{4}{5}, 99.99, 0, \sqrt{2}, -12, 4\frac{1}{2}, 0.66\overline{6}, \ldots, 8 \right\}$

Determine whether each statement is true or false.

43. All integers are whole numbers.

44. π is a rational number.

45. The set of real numbers corresponds to all points on the number line.

46. A real number is either rational or irrational.

Insert one of the symbols >, <, or = in the blank to make each statement true.

47. $\|-6\|$ $\|5\|$ **48.** -9 $\|-10\|$

SECTION 1.4   Adding Real Numbers; Properties of Addition

DEFINITIONS AND CONCEPTS	EXAMPLES
To **add two real numbers with like signs:**	
1. To add two positive numbers, add them as usual. The final answer is positive.	Add: $3 + 5 = 8$
2. To add two negative numbers, add their absolute values and make the final answer negative.	Add: $-5 + (-11) = -16$
To **add two real numbers with unlike signs:**	
1. Subtract their absolute values (the smaller from the larger).	Add: $-8 + 6 = -2$
2. To that result, attach the sign of the number with the larger absolute value.	Add: $12 + (-5) = 7$

SECTION 1.4   Adding Real Numbers; Properties of Addition–*continued*

DEFINITIONS AND CONCEPTS	EXAMPLES
Properties of Addition	
Commutative property: $a + b = b + a$ *Changing the order when adding does not affect the answer.*	$5 + (-9) = -9 + 5$ Reorder.
Associative property: $(a + b) + c = a + (b + c)$ *Changing the grouping when adding does not affect the answer.*	$(3 + 7) + 5 = 3 + (7 + 5)$ Regroup.
Addition property of 0: $a + 0 = a$ and $0 + a = a$	$-6 + 0 = -6$ *0 is the additive identity element.*
Addition property of opposites: $a + (-a) = 0$ and $(-a) + a = 0$	$11 + (-11) = 0$ *11 and −11 are additive inverses.*

REVIEW EXERCISES

Add.

49. $-45 + (-37)$

50. $25 + (-13)$

51. $0 + (-7)$

52. $-7 + 7$

53. $12 + (-8) + (-15)$

54. $-9.9 + (-2.4)$

55. $\dfrac{5}{16} + \left(-\dfrac{1}{2}\right)$

56. $35 + (-13) + (-17) + 6$

57. Determine what property of addition is shown.

   a. $-2 + 5 = 5 + (-2)$

   b. $(-2 + 5) + 1 = -2 + (5 + 1)$

   c. $80 + (-80) = 0$

   d. $-5.75 + 0 = -5.75$

58. TEMPERATURES   Determine Washington State's record high temperature if it is 166° greater than the state's record low temperature of $-48°$F.

SECTION 1.5   Subtracting Real Numbers

DEFINITIONS AND CONCEPTS	EXAMPLES
The opposite of the opposite of a number is that number. For any real number a, $-(-a) = a$.	$-(-13) = 13$
To **subtract two real numbers,** add the first to the opposite (additive inverse) of the number to be subtracted. For any real numbers a and b, $\quad a - b = a + (-b)$	Subtract: $4 - 7 = 4 + (-7) = -3$ $\qquad 6 - (-8) = 6 + 8 = 14$ $\qquad -1 - (-2) = -1 + 2 = 1$
To **check** a subtraction, the difference plus the subtrahend should equal the minuend.	To check $-6 - 2 = -8$, verify that $-8 + 2 = -6$.

REVIEW EXERCISES

Write the expression in simpler form.

59. a. The opposite of 10

   b. The additive inverse of -3

60. a. $-\left(-\dfrac{9}{16}\right)$ **b.** $-|-4|$

Perform the operations.

61. $45 - 64$

62. Subtract $\dfrac{1}{3}$ from $-\dfrac{3}{5}$

63. $-7 - (-12)$

64. $3.6 - (-2.1)$

65. $0 - 10$

66. $-33 + 7 - 5 - (-2)$

67. GEOGRAPHY   The tallest peak on Earth is Mount Everest, at 29,028 feet, and the greatest ocean depth is the Mariana Trench, at $-36,205$ feet. Find the difference in these elevations. Check the result.

68. HISTORY   Plato, a famous Greek philosopher, died in 347 B.C. (-347) at the age of 81. When was he born? Check the result.

SECTION 1.6 Multiplying and Dividing Real Numbers; Multiplication and Division Properties

DEFINITIONS AND CONCEPTS	EXAMPLES
To **multiply two real numbers,** multiply their absolute values.	
1. If the numbers have **like signs,** the final answer is positive.	Multiply: $-5(-7) = 35$ and $14(3) = 42$
2. If the numbers have **unlike signs,** the final answer is negative.	Multiply: $6(-6) = -36$ and $-11(5) = -55$

Properties of multiplication

Commutative property: $ab = ba$ *Changing the order when multiplying does not affect the answer.*	$-8(12) = 12(-8)$ Reorder.
Associative property: $(ab)c = a(bc)$ *Changing the grouping when multiplying does not affect the answer.*	$(-4 \cdot 9) \cdot 7 = -4(9 \cdot 7)$ Regroup.
Multiplication property of 0: $0 \cdot a = 0$ and $a \cdot 0 = 0$	$0 \cdot (-7) = 0$
Multiplication property of 1: $1 \cdot a = a$ and $a \cdot 1 = a$	$1 \cdot 32 = 32$ 1 is the multiplicative identity.
Multiplicative inverse property: $a\left(\frac{1}{a}\right) = 1$ and $\frac{1}{a}(a) = 1$	$4\left(\frac{1}{4}\right) = 1$ 4 and $\frac{1}{4}$ are multiplicative inverses.

To **divide two real numbers,** divide their absolute values.	
1. If the numbers have **like signs,** the final answer is positive.	Divide: $\dfrac{16}{8} = 2$ and $\dfrac{-25}{-5} = 5$
2. If the numbers have **unlike signs,** the final answer is negative.	Divide: $\dfrac{-36}{9} = -4$ and $\dfrac{56}{-7} = -8$

For any real number, $\frac{a}{1} = a$ and $\frac{a}{a} = 1$, where $a \neq 0$.	$\dfrac{25}{1} = 25$ and $\dfrac{-32}{-32} = 1$
Division of zero by a nonzero number is 0. **Division by zero** is undefined.	$\dfrac{0}{17} = 0$ but $\dfrac{17}{0}$ is undefined.
To **check** the division $\frac{a}{b} = c$, verify that $c \cdot b = a$.	To check $\dfrac{6}{-2} = -3$, verify that $-3(-2) = 6$.

REVIEW EXERCISES

Multiply.

69. $-8 \cdot 7$

70. $-9\left(-\dfrac{1}{9}\right)$

71. $2(-3)(-2)$

72. $(-4)(-1)(-3)$

73. $-1.2(-5.3)$

74. $0.002(-1,000)$

75. $-\dfrac{2}{3}\left(\dfrac{1}{5}\right)$

76. $-6(-3)(0)(-1)$

77. ELECTRONICS   The picture on the screen can be magnified by switching a setting on the monitor. What would be the new high and low if every value changed by a factor of 1.5?

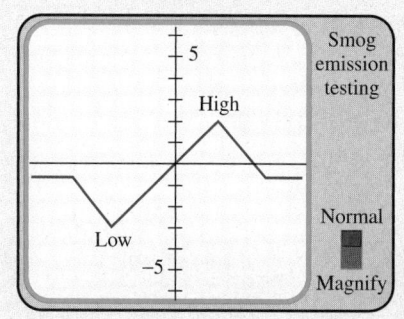

78. Determine what property of multiplication is shown.

  a. $(2 \cdot 3)5 = 2(3 \cdot 5)$

  b. $(-5)(-6) = (-6)(-5)$

  c. $-6 \cdot 1 = -6$

  d. $\frac{1}{2}(2) = 1$

Perform each division, if possible.

79. $\dfrac{44}{-44}$

80. $\dfrac{-272}{16}$

81. $\dfrac{-81}{-27}$

82. $-\dfrac{3}{5} \div \dfrac{1}{2}$

83. $\dfrac{-60}{0}$

84. $\dfrac{-4.5}{1}$

85. Fill in the blanks: $\frac{0}{18} = 0$ because __ · __ = __ .

86. GEMSTONES   A 3-carat yellow sapphire stone valued at \$3,000 five years ago is now worth \$1,200. What signed number indicates the average annual depreciation of the sapphire?

SECTION 1.7   Exponents and Order of Operations

DEFINITIONS AND CONCEPTS	EXAMPLES
An **exponent** represents repeated multiplication.	$8^5 = 8 \cdot 8 \cdot 8 \cdot 8 \cdot 8$ *The exponent of 5 indicates that 8 is to be used as a factor 5 times.*
In a^n, a is the **base** and n is the **exponent.**	In 7^4, the base is 7 and 4 is the exponent.
Order of Operations 1. Perform all calculations within grouping symbols, working from the innermost to the outermost in the following order. 2. Evaluate all exponential expressions 3. Perform all multiplications and divisions as they occur from left to right. 4. Perform all additions and subtractions as they occur from left to right. In fractions, evaluate the numerator and denominator separately. Then simplify the fraction.	Evaluate: $\dfrac{3(6-4^3)-2^4+4}{8 \div 4 \cdot 3} = \dfrac{3(6-64)-2^4+4}{2 \cdot 3}$ *Evaluate: $4^3 = 64$.* $= \dfrac{3(-58)-2^4+4}{6}$ *Subtract within the parentheses.* $= \dfrac{3(-58)-16+4}{6}$ *Evaluate: $2^4 = 16$.* $= \dfrac{-174-16+4}{6}$ *Multiply.* $= \dfrac{-190+4}{6}$ *Subtract.* $= \dfrac{-186}{6}$ *Add.* $= -31$ *Divide.*
$\text{Mean} = \dfrac{\text{sum of values}}{\text{number of values}}$	Find the mean of the test scores of 74, 83, 79, 91, and 73. $\text{Mean} = \dfrac{74+83+79+91+73}{5} = 80$

REVIEW EXERCISES

87. Write each expression using exponents.

  a. $8 \cdot 8 \cdot 8 \cdot 8 \cdot 8$

  b. $9 \cdot \pi \cdot r \cdot r$

88. Evaluate each expression.

  a. 9^2 **b.** $\left(-\dfrac{2}{3}\right)^3$

  c. 2^5 **d.** 50^1

Evaluate each expression.

89. $2 + 5 \cdot 3$

90. $-24 \div 2 \cdot 3$

91. $-(16-3)^2$

92. $43 + 2(-6 - 2 \cdot 2)$

93. $10 - 5[-3 - 2(5 - 7^2)] - 5$

94. $\dfrac{-4(4+2)-4}{2|-18-4(5)|}$

95. $(-3)^3\left(\dfrac{-8}{2}\right) + 5$

96. $\dfrac{2^4 - (4-6)(3-6)}{12 + 4[(-1)^8 - 2^2]}$

97. Write each expression in symbols and then evaluate it.

    a. Negative nine squared

    b. The opposite of the square of nine

98. WALK-A-THONS   Use the data in the table to find the average (mean) donation to a charity walk-a-thon.

Donation	$5	$10	$20	$50	$100
Number received	20	65	25	5	10

SECTION 1.8 Algebraic Expressions

DEFINITIONS AND CONCEPTS	EXAMPLES
Addition symbols separate algebraic expressions into **terms.** In a term, the numerical factor is called the **coefficient.**	Since $a^2 + 3a - 5$ can be written as $a^2 + 3a + (-5)$, it has three terms. The coefficient of a^2 is 1, the coefficient of $3a$ is 3, and the coefficient of -5 is -5.
Key phrases can be translated to algebraic expressions.	*5 more than x* can be expressed as $x + 5$. *25 less than twice y* can be expressed as $2y - 25$. One-half of c can be expressed as $\frac{1}{2}c$.
Number $\cdot$ value = total value	The total value (in cents) of n nickels is $n \cdot 5 = 5n$ cents.
To **evaluate algebraic expressions,** we substitute the values of its variables and use the rules for the order of operations rule.	Evaluate $\frac{x^2 - y^2}{x + y}$ for $x = 2$ and $y = -3$. $\dfrac{x^2 - y^2}{x + y} = \dfrac{2^2 - (-3)^2}{2 + (-3)}$ Substitute 2 for x and -3 for y. $= \dfrac{4 - 9}{-1}$ $= \dfrac{-5}{-1}$ $= 5$

REVIEW EXERCISES

99. How many terms does each expression have?

    a. $3x^2 + 2x - 5$ **b.** $-12xyz$

100. Identify the coefficient of each term of the given expression.

    a. $16x^2 - 5x + 25$ **b.** $\dfrac{x}{2} + y$

Write each phrase as an algebraic expression.

101. 25 more than the height h

102. 15 less than triple the cutoff score s

103. 6 less than one-half of the time

104. The absolute value of the difference of 2 and the square of a

105. HARDWARE   Let n represent the length of the nail. Write an algebraic expression that represents the length of the bolt (in inches).

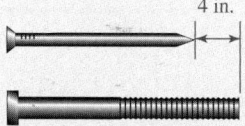
4 in.

106. HARDWARE   Let b represent the length of the bolt. Write an algebraic expression that represents the length of the nail (in inches).

107. How many years are in d decades?

108. Five years after a house was constructed, a patio was added. How old, in years, is the patio if the house is x years old?

109. Complete the table below. The units are cents.

Coin	Number	Value	Total value
Nickel	6	5	
Dime	d	10	

110. Complete the table below.

x	$20x - x^3$
0	
1	
-4	

Evaluate each algebraic expression for the given values of the variables.

111. $b^2 - 4ac$ for $b = -10$, $a = 3$, and $c = 5$

112. $\dfrac{x + y}{-x - z}$ for $x = 19$, and $y = 17$, and $z = -18$

SECTION 1.9   Simplifying Algebraic Expressions Using Properties of Real Numbers

DEFINITIONS AND CONCEPTS	EXAMPLES
We often use the *commutative property of multiplication* to reorder factors and the *associative property of multiplication* to regroup factors when **simplifying expressions**.	Simplify: $-5(3y) = (-5 \cdot 3)y = -15y$ $-45b\left(\dfrac{5}{9}\right) = -45\left(\dfrac{5}{9}b\right) = \left(-45 \cdot \dfrac{5}{9}\right)b = -25b$
The **distributive property** can be used to remove parentheses: $a(b + c) = ab + ac \qquad a(b - c) = ab - ac$ $a(b + c + d) = ab + ac + ad$	Multiply: $7(x + 3) = 7 \cdot x + 7 \cdot 3 = 7x + 21$ $-0.2(4m - 5n - 7) = -0.2(4m) - (-0.2)(5n) - (-0.2)(7)$ $\qquad\qquad\qquad\quad = -0.8m + n + 1.4$
Like terms are terms with exactly the same variables raised to exactly the same powers.	$3x$ and $-5x$ are like terms. $-4t^3$ and $3t^2$ are unlike terms because the variable t has different exponents. $0.5xyz$ and $3.7xy$ are unlike terms because they have different variables.
Simplifying the sum or difference of like terms is called **combining like terms**. Like terms can be combined by adding or subtracting the coefficients of the terms and keeping the same variables with the same exponents.	Simplify: $4a + 2a = 6a$ Think: $(4 + 2)a = 6a$. $5p^2 + p - p^2 - 9p = 4p^2 - 8p$ Think: $(5 - 1)p^2 = 4p^2$ and $\qquad\qquad\qquad\qquad\qquad\qquad\qquad (1 - 9)p = -8p$. $2(k - 1) - 3(k + 2) = 2k - 2 - 3k - 6 = -k - 8$

REVIEW EXERCISES

Simplify each expression.

113. $-4(7w)$ **114.** $3(-2x)(-4)$

115. $0.4(5.2f)$ **116.** $\dfrac{7}{2} \cdot \dfrac{2}{7}r$

Use the distribution property to remove parentheses.

117. $5(x + 3)$ **118.** $-(2x + 3 - y)$

119. $\dfrac{3}{4}(4c - 8)$ **120.** $-2(-3c - 7)(2.1)$

Simplify each expression by combining like terms.

121. $8p + 5p - 4p$ **122.** $-5m + 2 - 2m - 2$

123. $n + n + n + n$ **124.** $5(p - 2) - 2(3p + 4)$

125. $55.7k^2 - 55.6k^2$

126. $8a^3 + 4a^3 + 2a - 4a^3 - 2a - 1$

127. $\dfrac{3}{5}w - \left(-\dfrac{2}{5}w\right)$ **128.** $36\left(\dfrac{1}{9}h - \dfrac{3}{4}\right) + 36\left(\dfrac{1}{3}\right)$

129. GEOMETRY   Write an algebraic expression in simplified form that represents the perimeter of the triangle.

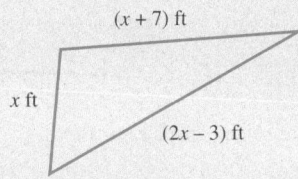

$(x + 7)$ ft

x ft

$(2x - 3)$ ft

130. Write an equivalent expression for the given expression using fewer symbols.
  a. $1x$ **b.** $-1x$
  c. $4x - (-1)$ **d.** $4x + (-1)$

CHAPTER 1
Test

1. Fill in the blanks.

   a. Two fractions, such as $\frac{1}{2}$ and $\frac{5}{10}$, that represent the same number are called _____ fractions.

   b. The result of a multiplication is called a _____.

   c. $\frac{8}{7}$ is the _____ of $\frac{7}{8}$ because $\frac{8}{7} \cdot \frac{7}{8} = 1$.

   d. $9x^2$ and $7x^2$ are _____ _____ because they have the same variable raised to exactly the same power.

   e. For any nonzero real number a, $\frac{a}{0}$ is _____.

2. SECURITY GUARDS
   The graph shows the cost to hire a security guard.

   a. What will it cost to hire a security guard for 3 hours?

   b. If a school was billed $40 for hiring a security guard for a dance, for how long did the guard work?

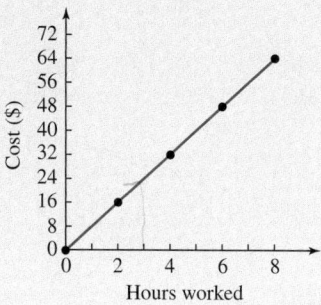

3. Use the formula $f = \frac{a}{5}$ to complete the table.

Square miles (a)	Fire stations (f)
15	3
100	20
350	40

4. Give the prime factorization of 180.

5. Simplify: $\dfrac{42}{105}$

6. Divide: $\dfrac{15}{16} \div \dfrac{5}{8}$

7. Add: $\dfrac{7}{10} + \dfrac{1}{14}$

8. Subtract: $8\dfrac{2}{5} - 1\dfrac{2}{3}$

9. SHOPPING   Find the cost of the fruit on the scale.

Oranges
84 cents a pound

10. Write $\frac{5}{6}$ as a decimal.

11. Graph each member of the set on a number line.

$$\left\{-1\tfrac{1}{4}, \ \sqrt{2}, \ -3.75, \ \tfrac{7}{2}, \ 0.5, \ -3\right\}$$

12. Determine whether each statement is true or false.

    a. Every integer is a rational number.

    b. Every rational number is an integer.

    c. π is an irrational number.

    d. 0 is a whole number.

13. Describe the set of real numbers.

14. Insert the proper symbol, $>$ or $<$, in the blank.

    a. -2 ___ -3 b. $-|-9|$ ___ 8

    c. $|-4|$ ___ $-(-5)$ d. $\left|-\frac{7}{8}\right|$ ___ 0.5

15. TELEVISION   During "sweeps week," networks try to gain viewers by showing flashy programs. Use the data to determine the average daily gain (or loss) of ratings points by a network for the 7-day "sweeps period."

Day	M	T	W	Thr	F	Sa	Su
Point loss/gain	0.6	-0.3	1.7	1.5	-0.2	1.1	-0.2

Perform the operations.

16. $(-6) + 8 + (-4)$

17. $-\dfrac{1}{2} + \dfrac{7}{8}$

18. a. $-10 - (-4)$

    b. Show a check of the result.

19. a. $\dfrac{-126}{-9}$

    b. Show a check of the result.

20. $(-2)(-3)(-5)$

21. $-6.1(0.4)$

22. $\dfrac{0}{-3}$

23. $\left(-\dfrac{3}{5}\right)^3$

24. $3 + (-3)$

25. $0 - 3$

26. $-30 + 50 - 10 - (-40)$

27. ASTRONOMY   *Magnitude* is a term used in astronomy to describe the brightness of planets and stars. Negative magnitudes are associated with brighter objects. By how many magnitudes do a full moon and the sun differ?

Object	Magnitude
Sun	-26.5
Full moon	-12.5

28. What property of real numbers is illustrated?

  a. $(-12 + 97) + 3 = -12 + (97 + 3)$

  b. $2(x + 7) = 2x + 14$

  c. $-2(m)5 = -2(5)m$

  d. $\frac{1}{8}(8) = 1$

  e. $0 + 15 = 15$

29. Write each product using exponents:

  a. $9(9)(9)(9)(9)$ **b.** $3 \cdot x \cdot x \cdot z \cdot z \cdot z$

Evaluate each expression.

30. $8 + 2 \cdot 3^4$ **31.** $\dfrac{3(40 - 2^3)}{-2(6 - 4)^2}$

32. -10^2 **33.** $9 - 3[45 - 5^2(1^5 - 4)]$

34. Evaluate $3(x - y) - 5(x + y)$ for $x = 2$ and $y = -5$.

35. Complete the table.

x	$2x - \dfrac{30}{x}$
5	
10	
-30	

36. Translate to an algebraic expression: seven less than twice the width w.

37. a. MUSIC   A band recorded x songs for a CD. However, two of the songs were not included in the album because of poor sound quality. Write an expression that represents the number of songs on the CD.

  b. MONEY   Find the value of q quarters in cents.

38. How many terms are in the expression $4x^2 + 5x - 7$? What is the coefficient of the second term?

Simply each expression.

39. $5(-4x)$ **40.** $-8(-7t)(4)$

41. $\dfrac{4}{5}(15a + 5) - 16a$ **42.** $-1.1d^3 - 3.8d^3 - d^3$

43. $9x^2 + 2(7x - 3) - 9(x^2 - 1)$

44. Write an expression that represents the perimeter of the rectangle.

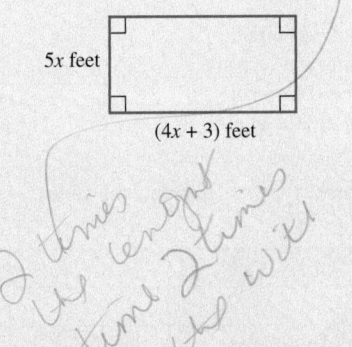

5x feet

$(4x + 3)$ feet

GROUP PROJECT

WRITING FRACTIONS AS DECIMALS

Overview: This is a good activity to try at the beginning of the course. You can become acquainted with other students in your class while you review the process for finding decimal equivalents of fractions.

Instructions: Form groups of 6 students. Select one person from your group to record the group's responses on the questionnaire. Express the results in fraction form and in decimal form.

What fraction (decimal) of the students in your group . . .	Fraction	Decimal
have the letter a in their first names?		
have a birthday in January or February?		
work full-time or part-time?		
have ever been on television?		
live more than 10 miles from the campus?		
say that summer is their favorite season of the year?		

CHAPTER 2

Equations, Inequalities, and Problem Solving

© Jeremy Hardie/Getty Images

from **Campus to Careers**
Automotive Service Technician

Anyone whose car has ever broken down appreciates the talents of automotive service technicians. To work on today's high-tech cars and trucks, a person needs strong diagnostic and problem-solving skills. Courses in automotive repair, electronics, physics, chemistry, English, computers, and mathematics provide a good educational background for a career as a service technician.

Service technicians must be knowledgeable about the repair and maintenance of automobiles and the fuels that power them. In **Problem 75** of **Study Set 2.4,** you will see how the octane ratings of three familiar grades of gasoline, unleaded, unleaded plus, and premium, are calculated using a formula.

JOB TITLE:
Automotive Service Technician

EDUCATION:
Strongly recommended formal training at a vocational school or community college.

JOB OUTLOOK:
Demand for technicians will grow as the number of vehicles in operation increases.

ANNUAL EARNINGS:
$37,000 to $47,000

FOR MORE INFORMATION:
www.bls.gov/oco/home.htm

Study Skills Workshop
Preparing to Learn

Many students feel that there are two types of people—those who are good at math and those who are not—and that this cannot be changed. This isn't true! Here are some suggestions that can increase your chances for success in algebra.

DISCOVER YOUR LEARNING STYLE: Are you a visual, verbal, or audio learner? Knowing this will help you determine how best to study.

GET THE MOST OUT OF THE TEXTBOOK: This book and the software that comes with it contain many student support features. Are you taking advantage of them?

TAKE GOOD NOTES: Are your class notes complete so that they are helpful when doing your homework and studying for tests?

Now Try This

1. To determine what type of learner you are, take the *Learning Style Survey* found online at http://www.metamath.com/multiple/multiple_choice_questions.html. Then, write a one-page paper explaining what you learned from the survey results and how you will use the information to help you succeed in the class.

2. To learn more about the student support features of this book, take the *Textbook Tour* found online at http://www.thomsonedu.com/math/tussy.

3. Rewrite a set of your class notes to make them more readable and to clarify the concepts and examples covered. If they are not already, write them in outline form. Fill in any information you didn't have time to copy down in class and complete any phrases or sentence fragments.

SECTION 2.1
Solving Equations Using Properties of Equality

Objectives

1. Determine whether a number is a solution.
2. Use the addition property of equality.
3. Use the subtraction property of equality.
4. Use the multiplication property of equality.
5. Use the division property of equality.

In this section, we introduce four fundamental properties of equality that are used to solve equations.

1. Determine Whether a Number is a Solution.

An **equation** is a statement indicating that two expressions are equal. An example is $x + 5 = 15$. The equal symbol = separates the equation into two parts: The expression $x + 5$

is the **left side** and 15 is the **right side.** The letter x is the **variable** (or the **unknown**). The sides of an equation can be reversed, so we can write $x + 5 = 15$ or $15 = x + 5$

- An equation can be true: $6 + 3 = 9$
- An equation can be false: $2 + 4 = 7$
- An equation can be neither true nor false. For example, $x + 5 = 15$ is neither true nor false because we don't know what number x represents.

An equation that contains a variable is made true or false by substituting a number for the variable. If we substitute 10 for x in $x + 5 = 15$, the resulting equation is true: $\mathbf{10} + 5 = 15$. If we substitute 1 for x, the resulting equation is false: $1 + 5 = 15$. A number that makes an equation true when substituted for the variable is called a **solution** and it is said to **satisfy** the equation. Therefore, 10 is a solution of $x + 5 = 15$, and 1 is not. The **solution set** of an equation is the set of all numbers that make the equation true.

EXAMPLE 1   Is 9 a solution of $3y - 1 = 2y + 7$?

Strategy   We will substitute 9 for each y in the equation and evaluate the expression on the left side and the expression on the right side separately.

Why   If a true statement results, 9 is a solution of the equation. If we obtain a false statement, 9 is not a solution.

Solution

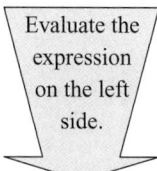   Evaluate the expression on the left side.

$$3y - 1 = 2y + 7$$
$$3(9) - 1 \stackrel{?}{=} 2(9) + 7$$
$$27 - 1 \stackrel{?}{=} 18 + 7$$
$$26 = 25$$

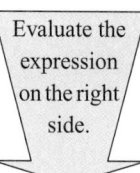   Evaluate the expression on the right side.

Since $26 = 25$ is false, 9 is not a solution of $3y - 1 = 2y + 7$.

 Self Check 1   Is 25 a solution of $10 - x = 35 - 2x$?

Now Try   **Problem 19**

---

❷ Use the Addition Property of Equality.

To **solve an equation** means to find all values of the variable that make the equation true. We can develop an understanding of how to solve equations by referring to the scales shown on the right.

The first scale represents the equation $x - 2 = 3$. The scale is in balance because the weights on the left side and right side are equal. To find x, we must add 2 to the left side. To keep the scale in balance, we must also add 2 to the right side. After doing this, we see that x grams is balanced by 5 grams. Therefore, x must be 5. We say that we have solved the equation $x - 2 = 3$ and that the solution is 5.

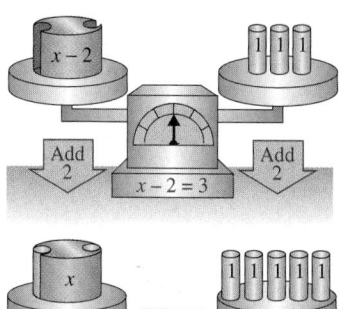

In this example, we solved $x - 2 = 3$ by transforming it to a simpler *equivalent equation*, $x = 5$.

Equivalent Equations	Equations with the same solutions are called **equivalent equations.**

The procedure that we used suggests the following property of equality.

Addition Property of Equality	Adding the same number to both sides of an equation does not change its solution. For any real numbers a, b, and c, if $a = b$, then $a + c = b + c$

When we use this property, the resulting equation is *equivalent to the original one.* We will now show how it is used to solve $x - 2 = 3$ algebraically.

EXAMPLE 2   Solve: $x - 2 = 3$

Strategy   We will use a property of equality to isolate the variable on one side of the equation.

Why   To solve the original equation, we want to find a simpler equivalent equation of the form $x = \textbf{a number}$, whose solution is obvious.

Solution   We will use the addition property of equality to isolate x on the left side of the equation. We can undo the subtraction of 2 by adding 2 to both sides.

$$x - 2 = 3 \qquad \text{This is the equation to solve.}$$
$$x - 2 + 2 = 3 + 2 \qquad \text{Add 2 to both sides.}$$
$$x + 0 = 5 \qquad \text{The sum of a number and its opposite is zero: } -2 + 2 = 0.$$
$$x = 5 \qquad \text{When 0 is added to a number, the result is the same number.}$$

Since 5 is obviously the solution of the equivalent equation $x = 5$, the solution of the original equation, $x - 2 = 3$, is also 5. To check this result, we substitute 5 for x in the original equation and simplify.

$$x - 2 = 3$$
$$5 - 2 \overset{?}{=} 3 \qquad \text{Substitute 5 for x.}$$
$$3 = 3 \qquad \text{True}$$

Since the statement is true, 5 is the solution. A more formal way to present this result is to write the solution within braces as a solution set: $\{5\}$.

Self Check 2   Solve: $n - 16 = 33$
Now Try   Problem 37

The Language of Algebra
We solve equations by writing a series of steps that result in an equivalent equation of the form

$$x = a \ number$$

or

$$a \ number = x$$

We say the variable is *isolated* on one side of the equation. *Isolated* means alone or by itself.

EXAMPLE 3   Solve: **a.** $-19 = y - 7$ **b.** $-27 + y = -3$

Strategy   We will use a property of equality to isolate the variable on one side of the equation.

Why   To solve the original equation, we want to find a simpler equivalent equation of the form $y =$ **a number** or **a number** $= y$, whose solution is obvious.

Solution

a. To isolate y on the right side, we use the addition property of equality. We can undo the subtraction of 7 by adding 7 to both sides.

$$-19 = y - 7 \qquad \text{This is the equation to solve.}$$
$$-19 + 7 = y - 7 + 7 \qquad \text{Add 7 to both sides.}$$
$$-12 = y \qquad \text{The sum of a number and its opposite is zero: } -7 + 7 = 0.$$

Check: $\quad -19 = y - 7 \qquad$ This is the original equation.
$$-19 \overset{?}{=} -12 - 7 \qquad \text{Substitute } -12 \text{ for } y.$$
$$-19 = -19 \qquad \text{True}$$

> **Notation**
> We may solve an equation so that the variable is isolated on either side of the equation. Note that $-12 = y$ is equivalent to $y = -12$.

Since the statement is true, the solution is -12. The solution set is $\{-12\}$.

b. To isolate y, we use the addition property of equality. We can eliminate -27 on the left side by adding its opposite (additive inverse) to both sides.

$$-27 + y = -3 \qquad \text{The equation to solve.}$$
$$-27 + y + 27 = -3 + 27 \qquad \text{Add 27 to both sides.}$$
$$y = 24 \qquad \text{The sum of a number and its opposite is zero: } -27 + 27 = 0.$$

Check: $\quad -27 + y = -3 \qquad$ This is the original equation.
$$-27 + 24 \overset{?}{=} -3 \qquad \text{Substitute 24 for } y.$$
$$-3 = -3 \qquad \text{True}$$

> **Caution**
> After checking a result, be careful when stating your conclusion. Here, it would be incorrect to say:
>
> The solution is -3.
>
> The number we were checking was 24, not -3.

The solution is 24. The solution set is $\{24\}$.

Self Check 3   Solve: **a.** $-5 = b - 38$ **b.** $-20 + n = 29$

Now Try   Problems 39 and 43

③ Use the Subtraction Property of Equality.

Since any subtraction can be written as an addition by adding the opposite of the number to be subtracted, the following property is an extension of the addition property of equality.

Subtraction Property of Equality	Subtracting the same number from both sides of an equation does not change its solution. For any real numbers a, b, and c, $$\text{if } a = b, \text{ then } a - c = b - c$$

When we use this property, the resulting equation is equivalent to the original one.

| **EXAMPLE 4** | Solve: **a.** $x + \dfrac{1}{8} = \dfrac{7}{4}$ **b.** $54.9 + x = 45.2$ |

Strategy   We will use a property of equality to isolate the variable on one side of the equation.

Why   To solve the original equation, we want to find a simpler equivalent equation of the form $x = $ **a number**, whose solution is obvious.

Solution

a. To isolate x, we use the subtraction property of equality. We can undo the addition of $\frac{1}{8}$ by subtracting $\frac{1}{8}$ from both sides.

> **The Language of Algebra**
> We could also isolate x by adding the additive inverse of $\frac{1}{8}$, which is $-\frac{1}{8}$, to both sides:
>
> $x + \frac{1}{8} + \left(-\frac{1}{8}\right) = \frac{7}{4} + \left(-\frac{1}{8}\right)$

$$x + \frac{1}{8} = \frac{7}{4} \qquad \text{This is the equation to solve.}$$

$$x + \frac{1}{8} - \frac{1}{8} = \frac{7}{4} - \frac{1}{8} \qquad \text{Subtract } \tfrac{1}{8} \text{ from both sides.}$$

$$x = \frac{7}{4} - \frac{1}{8} \qquad \text{On the left side, } \tfrac{1}{8} - \tfrac{1}{8} = 0.$$

$$x = \frac{7}{4} \cdot \frac{2}{2} - \frac{1}{8} \qquad \text{Build } \tfrac{7}{4} \text{ so that it has a denominator of 8.}$$

$$x = \frac{14}{8} - \frac{1}{8} \qquad \text{Multiply the numerators and multiply the denominators.}$$

$$x = \frac{13}{8} \qquad \text{Subtract the numerators. Write the result over the common denominator 8.}$$

Verify that $\frac{13}{8}$ is the solution by substituting it for x in the original equation and simplifying.

b. To isolate x, we use the subtraction property of equality. We can undo the addition of 54.9 by subtracting 54.9 from both sides.

$$54.9 + x = 45.2 \qquad \text{This is the equation to solve.}$$

$$54.9 + x - \mathbf{54.9} = 45.2 - \mathbf{54.9} \qquad \text{Subtract 54.9 from both sides.}$$

$$x = -9.7 \qquad \text{On the left side, } 54.9 - 54.9 = 0.$$

Check:

$$54.9 + x = 45.2 \qquad \text{This is the original equation.}$$

$$54.9 + (-9.7) \overset{?}{=} 45.2 \qquad \text{Substitute } -9.7 \text{ for } x.$$

$$45.2 = 45.2 \qquad \text{True}$$

The solution is -9.7. The solution set is $\{-9.7\}$.

 Self Check 4   Solve: **a.** $x + \frac{4}{15} = \frac{11}{5}$ **b.** $0.7 + a = 0.2$

Now Try   **Problems 49 and 51**

4   **Use the Multiplication Property of Equality.**

The first scale shown below represents the equation $\frac{x}{3} = 25$. The scale is in balance because the weights on the left side and right side are equal. To find x, we must triple (multiply by 3) the weight on the left side. To keep the scale in balance, we must also triple the weight on the right side. After doing this, we see that x is balanced by 75. Therefore, x must be 75.

The procedure that we used suggests the following property of equality.

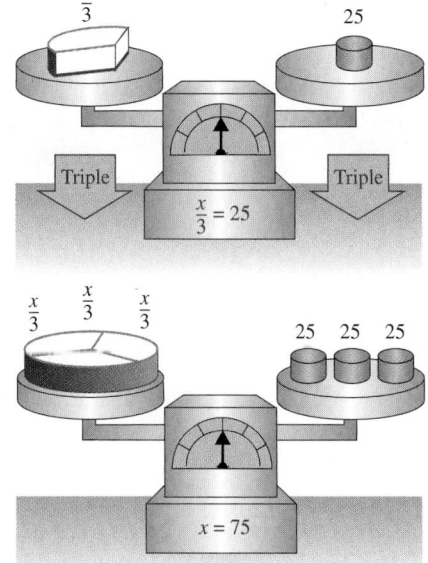

Multiplication Property of Equality	Multiplying both sides of an equation by the same nonzero number does not change its solution.
	For any real numbers a, b, and c, where c is not 0,
	if $a = b$, then $ca = cb$

(handwritten: $a = b$ $c \neq 0$ $ac = bc$)

When we use this property, the resulting equation is equivalent to the original one. We will now show how it is used to solve $\frac{x}{3} = 25$ algebraically.

(handwritten in left margin: Division $a = b$ $\frac{a}{c} = \frac{b}{c}$ $c \neq 0$)

EXAMPLE 5   Solve: $\dfrac{x}{3} = 25$

Strategy   We will use a property of equality to isolate the variable on one side of the equation.

Why   To solve the original equation, we want to find a simpler equivalent equation of the form $x = $ **a number**, whose solution is obvious.

Solution   To isolate x, we use the multiplication property of equality. We can undo the division by 3 by multiplying both sides by 3.

$$\frac{x}{3} = 25 \qquad \text{This is the equation to solve.}$$

$$3 \cdot \frac{x}{3} = 3 \cdot 25 \qquad \text{Multiply both sides by 3.}$$

$$\frac{3x}{3} = 75 \qquad \text{Do the multiplications.}$$

$$1x = 75 \qquad \text{Simplify } \tfrac{3x}{3} \text{ by removing the common factor of 3 in the numerator and denominator: } \tfrac{3}{3} = 1.$$

$$x = 75 \qquad \text{The coefficient 1 need not be written since } 1x = x.$$

If we substitute 75 for x in $\frac{x}{3} = 25$, we obtain the true statement $25 = 25$. This verifies that 75 is the solution. The solution set is $\{75\}$.

> ▷ **Self Check 5**   Solve: $\frac{b}{24} = 3$
>
> **Now Try   Problem 53**

Since the product of a number and its reciprocal (or multiplicative inverse) is 1, we can solve equations such as $\frac{2}{3}x = 6$, where the coefficient of the variable term is a fraction, as follows.

EXAMPLE 6   Solve: **a.** $\frac{2}{3}x = 6$ **b.** $-\frac{5}{4}x = 3$

Strategy   We will use a property of equality to isolate the variable on one side of the equation.

Why   To solve the original equation, we want to find a simpler equivalent equation of the form $x = $ **a number**, whose solution is obvious.

Solution

a. Since the coefficient of x is $\frac{2}{3}$, we can isolate x by multiplying both sides of the equation by the reciprocal of $\frac{2}{3}$, which is $\frac{3}{2}$.

$$\frac{2}{3}x = 6 \qquad \text{This is the equation to solve.}$$

$$\frac{3}{2} \cdot \frac{2}{3}x = \frac{3}{2} \cdot 6 \qquad \text{To undo the multiplication by } \frac{2}{3}, \text{ multiply both sides by the reciprocal of } \frac{2}{3}.$$

$$\left(\frac{3}{2} \cdot \frac{2}{3}\right)x = \frac{3}{2} \cdot 6 \qquad \text{Use the associative property of multiplication to group } \frac{3}{2} \text{ and } \frac{2}{3}.$$

$$1x = 9 \qquad \text{On the left, } \frac{3}{2} \cdot \frac{2}{3} = 1. \text{ On the right, } \frac{3}{2} \cdot 6 = \frac{18}{2} = 9.$$

$$x = 9 \qquad \text{The coefficient 1 need not be written since } 1x = x.$$

Check: $\frac{2}{3}x = 6$ This is the original equation.

$$\frac{2}{3}(9) \overset{?}{=} 6 \qquad \text{Substitute 9 for } x \text{ in the original equation.}$$

$$6 = 6 \qquad \text{On the left side, } \frac{2}{3}(9) = \frac{18}{3} = 6.$$

Since the statement is true, 9 is the solution. The solution set is $\{9\}$.

b. To isolate x, we multiply both sides by the reciprocal of $-\frac{5}{4}$, which is $-\frac{4}{5}$.

$$-\frac{5}{4}x = 3 \qquad \text{This is the equation to solve.}$$

$$-\frac{4}{5}\left(-\frac{5}{4}x\right) = -\frac{4}{5}(3) \qquad \text{To undo the multiplication by } -\frac{5}{4}, \text{ multiply both sides by the reciprocal of } -\frac{5}{4}.$$

Notation

Variable terms with fractional coefficients can be written in two ways. For example:

$$\frac{2}{3}x = \frac{2x}{3} \quad \text{and} \quad -\frac{5}{4}a = -\frac{5a}{4}$$

$$1x = -\frac{12}{5}$$ On the left side, $-\frac{4}{5}\left(-\frac{5}{4}\right) = 1$.

$$x = -\frac{12}{5}$$ The coefficient 1 need not be written since 1x = x.

The solution is $-\frac{12}{5}$. Verify that this is correct by checking.

Self Check 6 Solve: **a.** $\frac{7}{2}x = 21$ **b.** $-\frac{3}{8}b = 2$

Now Try **Problems 61 and 67**

5 **Use the Division Property of Equality.**

Since any division can be rewritten as a multiplication by multiplying by the reciprocal, the following property is a natural extension of the multiplication property.

Division Property of Equality	Dividing both sides of an equation by the same nonzero number does not change its solution. For any real numbers a, b, and c, where c is not 0, if $a = b$, then $\dfrac{a}{c} = \dfrac{b}{c}$

When we use this property, the resulting equation is equivalent to the original one.

EXAMPLE 7 Solve: **a.** $2t = 80$ **b.** $-6.02 = -8.6t$

Strategy We will use a property of equality to isolate the variable on one side of the equation.

Why To solve the original equation, we want to find a simpler equivalent equation of the form $t = $ **a number** or **a number** $ = t$, whose solution is obvious.

Solution

a. To isolate t on the left side, we use the division property of equality. We can undo the multiplication by 2 by dividing both sides of the equation by 2.

The Language of Algebra
Since division by 2 is the same as multiplication by $\frac{1}{2}$, we can also solve $2t = 80$ using the multiplication property of equality. We could also isolate t by multiplying both sides by the *multiplicative inverse* of 2, which is $\frac{1}{2}$:

$$\frac{1}{2} \cdot 2t = \frac{1}{2} \cdot 80$$

$2t = 80$ This is the equation to solve.

$\dfrac{2t}{2} = \dfrac{80}{2}$ Use the division property of equality: Divide both sides by 2.

$1t = 40$ Simplify $\frac{2t}{2}$ by removing the common factor of 2 in the numerator and denominator: $\frac{2}{2} = 1$.

$t = 40$ The product of 1 and any number is that number: 1t = t.

If we substitute 40 for t in $2t = 80$, we obtain the true statement $80 = 80$. This verifies that 40 is the solution. The solution set is $\{40\}$.

b. To isolate t on the right side, we use the division property of equality. We can undo the multiplication by -8.6 by dividing both sides by -8.6.

$$-6.02 = -8.6t \qquad \textit{This is the equation to solve.}$$

$$\frac{-6.02}{-8.6} = \frac{-8.6t}{-8.6} \qquad \textit{Use the division property of equality: Divide both sides by } -8.6.$$

$$0.7 = t \qquad \textit{Do the division: } 8.6\overline{)6.02}. \textit{ The quotient of two negative numbers is positive.}$$

The solution is 0.7. Verify that this is correct by checking.

Self Check 7   Solve: **a.** $16x = 176$ **b.** $10.04 = -0.4r$

Now Try   **Problems 69 and 79**

EXAMPLE 8   Solve: $-x = 3$

Strategy   The variable x is not isolated, because there is a $-$ sign in front of it. Since the term $-x$ has an understood coefficient of -1, the equation can be written as $-1x = 3$. We need to select a property of equality and use it to isolate the variable on one side of the equation.

Why   To find the solution of the original equation, we want to find a simpler equivalent equation of the form $x = \textbf{a number}$, whose solution is obvious.

Solution   To isolate x, we can either multiply or divide both sides by -1.

Multiply both sides by -1:

$$-x = 3 \qquad \textit{The equation to solve}$$
$$-1x = 3 \qquad \textit{Write: } -x = -1x$$
$$(-1)(-1x) = (-1)3$$
$$1x = -3$$
$$x = -3 \qquad \textit{1x = x}$$

Divide both sides by -1:

$$-x = 3 \qquad \textit{The equation to solve}$$
$$-1x = 3 \qquad \textit{Write: } -x = -1x$$
$$\frac{-1x}{-1} = \frac{3}{-1}$$
$$1x = -3 \qquad \textit{On the left side, } \tfrac{-1}{-1} = 1.$$
$$x = -3 \qquad \textit{1x = x}$$

Check: $-x = 3$ *This is the original equation.*

$$-(-3) \stackrel{?}{=} 3 \qquad \textit{Substitute } -3 \textit{ for x.}$$

$$3 = 3 \qquad \textit{On the left side, the opposite of } -3 \textit{ is 3.}$$

Since the statement is true, -3 is the solution. The solution set is $\{-3\}$.

Self Check 8   Solve: $-h = -12$

Now Try   **Problem 81**

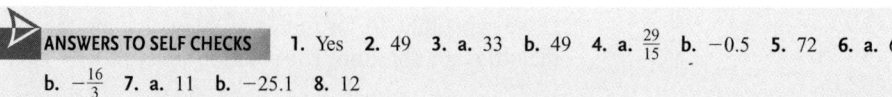

ANSWERS TO SELF CHECKS   **1.** Yes   **2.** 49   **3. a.** 33 **b.** 49   **4. a.** $\frac{29}{15}$ **b.** -0.5 **5.** 72   **6. a.** 6   **b.** $-\frac{16}{3}$ **7. a.** 11   **b.** -25.1 **8.** 12

STUDY SET
2.1

VOCABULARY

Fill in the blanks.

1. An _____, such as $x + 1 = 7$, is a statement indicating that two expressions are equal.

2. Any number that makes an equation true when substituted for the variable is said to _____ the equation. Such numbers are called _____.

3. To _____ an equation means to find all values of the variable that make the equation true.

4. To solve an equation, we _____ the variable on one side of the equal symbol.

5. Equations with the same solutions are called _____ equations.

6. To _____ the solution of an equation, we substitute the value for the variable in the original equation and determine whether the result is a true statement.

CONCEPTS

7. Given $x + 6 = 12$,
   a. What is the left side of the equation?
   b. Is this equation true or false?
   c. Is 5 the solution?
   d. Does 6 satisfy the equation?

8. For each equation, determine what operation is performed on the variable. Then explain how to undo that operation to isolate the variable.
   a. $x - 8 = 24$
   b. $x + 8 = 24$
   c. $\dfrac{x}{8} = 24$
   d. $8x = 24$

9. Complete the following properties of equality.
   a. If $a = b$, then
   $$a + c = b + \quad \text{and} \quad a - c = b - \quad$$
   b. If $a = b$, then $ca = \quad b$ and $\dfrac{a}{c} = \dfrac{b}{\quad}$ $(c \neq 0)$

10. a. To solve $\dfrac{h}{10} = 20$, do we multiply both sides of the equation by 10 or 20?
    b. To solve $4k = 16$, do we subtract 4 from both sides of the equation or divide both sides by 4?

11. Simplify each expression.
    a. $x + 7 - 7$
    b. $y - 2 + 2$
    c. $\dfrac{5t}{5}$
    d. $6 \cdot \dfrac{h}{6}$

12. a. To solve $-\dfrac{4}{5}x = 8$, we can multiply both sides by the reciprocal of $-\dfrac{4}{5}$. What is the reciprocal of $-\dfrac{4}{5}$?
    b. What is $-\dfrac{5}{4}\left(-\dfrac{4}{5}\right)$?

NOTATION

Complete each solution to solve the equation.

13.
$$x - 5 = 45$$
$$x - 5 + \boxed{} = 45 + \boxed{}$$
$$x = \boxed{}$$

Check:
$$x - 5 = 45$$
$$\boxed{} - 5 \quad 45$$
$$\boxed{} = 45 \quad \text{True}$$

$\boxed{}$ is the solution.

14. $8x = 40$
$$\dfrac{8x}{\boxed{}} = \dfrac{40}{\boxed{}}$$
$$x = \boxed{}$$

Check:
$$8x = 40$$
$$8(\boxed{}) \overset{?}{=} 40$$
$$\boxed{} = 40 \quad \text{True}$$

$\boxed{}$ is the solution.

15. a. What does the symbol $\overset{?}{=}$ mean?
    b. If you solve an equation and obtain $50 = x$, can you write $x = 50$?

16. Fill in the blank: $-x = \boxed{} x$

GUIDED PRACTICE

Check to determine whether the given number is a solution of the equation. See Example 1.

17. $6, \ x + 12 = 28$
18. $110, \ x - 50 = 60$
19. $-8, \ 2b + 3 = -15$
20. $-2, \ 5t - 4 = -16$
21. $5, \ 0.5x = 2.9$
22. $3.5, \ 1.2 + x = 4.7$
23. $-6, \ 33 - \dfrac{x}{2} = 30$
24. $-8, \ \dfrac{x}{4} + 98 = 100$
25. $-2, \ |c - 8| = 10$
26. $-45, \ |30 - r| = 15$
27. $12, \ 3x - 2 = 4x - 5$
28. $5, \ 5y + 8 = 3y - 2$
29. $-3, \ x^2 - x - 6 = 0$
30. $-2, \ y^2 + 5y - 3 = 0$
31. $1, \ \dfrac{2}{a + 1} + 5 = \dfrac{12}{a + 1}$
32. $4, \ \dfrac{2t}{t - 2} - \dfrac{4}{t - 2} = 1$
33. $\dfrac{3}{4}, \ x - \dfrac{1}{8} = \dfrac{5}{8}$
34. $\dfrac{7}{3}, \ -4 = a + \dfrac{5}{3}$
35. $-3, \ (x - 4)(x + 3) = 0$
36. $5, \ (2x + 1)(x - 5) = 0$

Use a property of equality to solve each equation. Then check the result. See Examples 2–4.

37. $a - 5 = 66$ **38.** $x - 34 = 19$

39. $9 = p - 9$ **40.** $3 = j - 88$

41. $x - 1.6 = -2.5$ **42.** $y - 1.2 = -1.3$

43. $-3 + a = 0$ **44.** $-1 + m = 0$

45. $d - \dfrac{1}{9} = \dfrac{7}{9}$ **46.** $\dfrac{7}{15} = b - \dfrac{1}{15}$

47. $x + 7 = 10$ **48.** $y + 15 = 24$

49. $s + \dfrac{1}{5} = \dfrac{4}{25}$ **50.** $\dfrac{1}{6} = h + \dfrac{4}{3}$

51. $3.5 + f = 1.2$ **52.** $9.4 + h = 8.1$

Use a property of equality to solve each equation. Then check the result. See Examples 5–8.

53. $\dfrac{x}{15} = 3$ **54.** $\dfrac{y}{7} = 12$

55. $0 = \dfrac{v}{11}$ **56.** $\dfrac{d}{49} = 0$

57. $\dfrac{d}{-7} = -3$ **58.** $\dfrac{c}{-2} = -11$

59. $\dfrac{y}{0.6} = -4.4$ **60.** $\dfrac{y}{0.8} = -2.9$

61. $\dfrac{4}{5}t = 16$ **62.** $\dfrac{11}{15}y = 22$

63. $\dfrac{2}{3}c = 10$ **64.** $\dfrac{9}{7}d = 81$

65. $-\dfrac{7}{2}r = 21$ **66.** $-\dfrac{4}{5}s = 36$

67. $-\dfrac{5}{4}h = -5$ **68.** $-\dfrac{3}{8}t = -3$

69. $4x = 16$ **70.** $5y = 45$

71. $63 = 9c$ **72.** $40 = 5t$

73. $23b = 23$ **74.** $16 = 16h$

75. $-8h = 48$ **76.** $-9a = 72$

77. $-100 = -5g$ **78.** $-80 = -5w$

79. $-3.4y = -1.7$ **80.** $-2.1x = -1.26$

81. $-x = 18$ **82.** $-y = 50$

83. $-n = \dfrac{4}{21}$ **84.** $-w = \dfrac{11}{16}$

TRY IT YOURSELF

Solve each equation. Then check the result.

85. $8.9 = -4.1 + t$ **86.** $7.7 = -3.2 + s$

87. $-2.5 = -m$ **88.** $-1.8 = -b$

89. $-\dfrac{9}{8}x = 3$ **90.** $-\dfrac{14}{3}c = 7$

91. $\dfrac{3}{4} = d + \dfrac{1}{10}$ **92.** $\dfrac{5}{9} = r + \dfrac{1}{6}$

93. $-15x = -60$ **94.** $-14x = -84$

95. $-10 = n - 5$ **96.** $-8 = t - 2$

97. $\dfrac{h}{-40} = 5$ **98.** $\dfrac{x}{-7} = 12$

99. $a - 93 = 2$ **100.** $18 = x - 3$

APPLICATIONS

101. SYNTHESIZERS To find the unknown angle measure, which is represented by x, solve the equation $x + 115 = 180$.

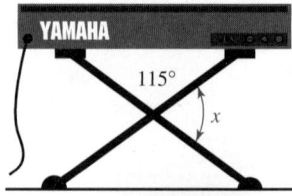

102. STOP SIGNS To find the measure of one angle of the stop sign, which is represented by x, solve the equation $8x = 1{,}080$.

103. SHARING THE WINNING TICKET When a 2006 Florida Lotto Jackpot was won by a group of 16 nurses employed at a Southwest Florida Medical Center, each received $375,000. To find the amount of the jackpot, which is represented by x, solve the equation $\dfrac{x}{16} = 375{,}000$.

104. TENNIS Billie Jean King won 40 Grand Slam tennis titles in her career. This is 14 less than the all-time leader, Martina Navratilova. To find the number of titles won by Navratilova, which is represented by x, solve the equation $40 = x - 14$.

WRITING

105. What does it mean to solve an equation?

106. When solving an equation, we *isolate* the variable on one side of the equation. Write a sentence in which the word *isolate* is used in a different context.

107. Explain the error in the following work.

$$\text{Solve:} \quad x + 2 = 40$$
$$x + 2 - 2 = 40$$
$$x = 40$$

108. After solving an equation, how do we check the result?

REVIEW

109. Evaluate $-9 - 3x$ for $x = -3$.

110. Evaluate: $-5^2 + (-5)^2$

111. Translate to symbols: Subtract x from 45

112. Evaluate: $\dfrac{2^3 + 3(5 - 3)}{15 - 4 \cdot 2}$

CHALLENGE PROBLEMS

113. If $a + 80 = 50$, what is $a - 80$?

114. Find two solutions of $|x + 1| = 100$.

SECTION 2.2
More about Solving Equations

Objectives

① Use more than one property of equality to solve equations.
② Simplify expressions to solve equations.
③ Clear equations of fractions and decimals.
④ Identify identities and contradictions.

We have solved simple equations by using properties of equality. We will now expand our equation-solving skills by considering more complicated equations. We want to develop a general strategy that can be used to solve any kind of *linear equation in one variable*.

Linear Equation in One Variable

A **linear equation in one variable** can be written in the form

$$ax + b = c$$

where a, b and c are real numbers and $a \neq 0$.

① **Use More Than One Property of Equality to Solve Equations.**

Sometimes we must use several properties of equality to solve an equation. For example, on the left side of $2x + 6 = 10$, the variable x is multiplied by 2, and then 6 is added to that product. To isolate x, we use the order of operations rules in reverse. First, we undo the addition of 6, and then we undo the multiplication by 2.

$2x + 6 = 10$	This is the equation to solve.
$2x + 6 - 6 = 10 - 6$	To undo the addition of 6, subtract 6 from both sides.
$2x = 4$	Do the subtractions.
$\dfrac{2x}{2} = \dfrac{4}{2}$	To undo the multiplication by 2, divide both sides by 2.
$x = 2$	Do the divisions.

The solution is 2.

EXAMPLE 1   Solve: $-12x + 5 = 17$

Strategy   First we will use a property of equality to isolate the *variable term* on one side of the equation. Then we will use a second property of equality to isolate the *variable* itself.

Why   To solve the original equation, we want to find a simpler equivalent equation of the form $x = $ **a number**, whose solution is obvious.

Solution   On the left side of the equation, x is multiplied by -12, and then 5 is added to that product. To isolate x, we undo the operations in the opposite order.

- To isolate the variable term, $-12x$, we subtract 5 from both sides to undo the addition of 5.

- To isolate the variable, x, we divide both sides by -12 to undo the multiplication by -12.

The Language of Algebra
We subtract 5 from both sides to isolate the *variable term*, $-12x$. Then we divide both sides by -12 to isolate the *variable*, x.

$$-12x + 5 = 17$$ This is the equation to solve.

$$-12x + 5 - 5 = 17 - 5$$ Use the subtraction property of equality: Subtract 5 from both sides to isolate the variable term −12x.

$$-12x = 12$$ Do the subtractions: 5 − 5 = 0 and 17 − 5 = 12.

$$\frac{-12x}{-12} = \frac{12}{-12}$$ Use the division property of equality: Divide both sides by −12 to isolate x.

$$x = -1$$ Do the divisions.

Caution
When checking solutions, always use the original equation.

Check: $$-12x + 5 = 17$$ This is the original equation.

$$-12(-1) + 5 \stackrel{?}{=} 17$$ Substitute −1 for x.

$$12 + 5 \stackrel{?}{=} 17$$ Do the multiplication on the left side.

$$17 = 17$$ True

The solution is −1. The solution set is $\{-1\}$.

Self Check 1   Solve: $8x - 13 = 43$
Now Try **Problem 15**

EXAMPLE 2   Solve: $\dfrac{5}{8}m - 2 = -12$

Strategy   We will use properties of equality to isolate the variable on one side of the equation.

Why   To solve the original equation, we want to find a simpler equivalent equation of the form $m = \textbf{a number}$, whose solution is obvious.

Solution   We note that the coefficient of m is $\frac{5}{8}$ and proceed as follows.

- To isolate the variable term $\frac{5}{8}m$, we add 2 to both sides to undo the subtraction of 2.
- To isolate the variable, m, we multiply both sides by $\frac{8}{5}$ to undo the multiplication by $\frac{5}{8}$.

$$\frac{5}{8}m - 2 = -12$$ This is the equation to solve.

$$\frac{5}{8}m - 2 + 2 = -12 + 2$$ Use the addition property of equality: Add 2 to both sides to isolate the variable term $\frac{5}{8}m$.

$$\frac{5}{8}m = -10$$ Do the additions: −2 + 2 = 0 and −12 + 2 = −10.

$$\frac{8}{5}\left(\frac{5}{8}m\right) = \frac{8}{5}(-10)$$ Use the multiplication property of equality: Multiply both sides by $\frac{8}{5}$ (which is the reciprocal of $\frac{5}{8}$) to isolate m.

$$m = -16$$ On the left side: $\frac{8}{5}\left(\frac{5}{8}\right) = 1$ and $1m = m$. On the right side: $\frac{8}{5}(-10) = -\dfrac{8 \cdot 2 \cdot \overset{1}{\cancel{5}}}{\underset{1}{\cancel{5}}} = -16.$

The solution is −16. Verify this by substituting −16 into the original equation. The solution set is $\{-16\}$.

Self Check 2   Solve: $\frac{7}{12}a - 6 = -27$

Now Try   **Problem 21**

EXAMPLE 3   Solve: $-0.2 = -0.8 - y$

Strategy   First, we will use a property of equality to isolate the variable term on one side of the equation. Then we will use a second property of equality to isolate the variable itself.

Why   To solve the original equation, we want to find a simpler equivalent equation of the form **a number** $= y$, whose solution is obvious.

Solution   To isolate the variable term $-y$ on the right side, we eliminate -0.8 by adding 0.8 to both sides.

$$-0.2 = -0.8 - y \qquad \text{This is the equation to solve.}$$
$$-0.2 + \mathbf{0.8} = -0.8 - y + \mathbf{0.8} \qquad \text{Add 0.8 to both sides to isolate } -y.$$
$$0.6 = -y \qquad \text{Do the additions.}$$

Since the term $-y$ has an understood coefficient of -1, the equation can be written as $0.6 = -1y$. To isolate y, we can either multiply both sides or divide both sides by -1. If we choose to divide both sides by -1, we proceed as follows.

$$0.6 = -1y$$
$$\frac{0.6}{-1} = \frac{-1y}{-1} \qquad \text{To undo the multiplication by } -1, \text{divide both sides by } -1.$$
$$-0.6 = y$$

The solution is -0.6. Verify this by substituting -0.6 into the original equation.

Self Check 3   Solve: $-6.6 - m = -2.7$

Now Try   **Problem 35**

---

② Simplify Expressions to Solve Equations.

When solving equations, we should simplify the expressions that make up the left and right sides before applying any properties of equality. Often, that involves removing parentheses and/or combining like terms.

EXAMPLE 4   Solve: **a.** $3(k + 1) - 5k = 0$ **b.** $8a - 2(a - 7) = 68$

Strategy   We will use the distributive property along with the process of combining like terms to simplify the left side of each equation.

Why   It's best to simplify each side of an equation before using a property of equality.

Solution

a. $3(k + 1) - 5k = 0$ This is the equation to solve.

$3k + 3 - 5k = 0$ Distribute the multiplication by 3.

$-2k + 3 = 0$ Combine like terms: $3k - 5k = -2k$.

$-2k + 3 - 3 = 0 - 3$ To undo the addition of 3, subtract 3 from both sides. This isolates the variable term $-2k$.

$-2k = -3$ Do the subtractions: $3 - 3 = 0$ and $0 - 3 = -3$.

$\dfrac{-2k}{-2} = \dfrac{-3}{-2}$ To undo the multiplication by -2, divide both sides by -2. This isolates the variable k.

$k = \dfrac{3}{2}$ Simplify: $\dfrac{-3}{-2} = \dfrac{3}{2}$.

Check: $3(k + 1) - 5k = 0$ This is the original equation.

$3\left(\dfrac{3}{2} + 1\right) - 5\left(\dfrac{3}{2}\right) \overset{?}{=} 0$ Substitute $\dfrac{3}{2}$ for k.

$3\left(\dfrac{5}{2}\right) - 5\left(\dfrac{3}{2}\right) \overset{?}{=} 0$ Do the addition within the parentheses. Think of 1 as $\dfrac{2}{2}$ and then add: $\dfrac{3}{2} + \dfrac{2}{2} = \dfrac{5}{2}$.

$\dfrac{15}{2} - \dfrac{15}{2} \overset{?}{=} 0$ Do the multiplications.

$0 = 0$ True

The solution is $\dfrac{3}{2}$ and the solution set is $\left\{\dfrac{3}{2}\right\}$.

b. $8a - 2(a - 7) = 68$ This is the equation to solve.

$8a - 2a + 14 = 68$ Distribute the multiplication by -2.

$6a + 14 = 68$ Combine like terms: $8a - 2a = 6a$.

$6a + 14 - 14 = 68 - 14$ To undo the addition of 14, subtract 14 from both sides. This isolates the variable term $6a$.

$6a = 54$ Do the subtractions.

$\dfrac{6a}{6} = \dfrac{54}{6}$ To undo the multiplication by 6, divide both sides by 6. This isolates the variable a.

$a = 9$ Do the divisions.

> **Caution**
> To check a result, we evaluate each side of the equation following the order of operations rules. On the left side, perform the addition within parentheses first. *Don't distribute the multiplication by 3.*
>
> $3\left(\dfrac{3}{2} + 1\right)$
>
> $\uparrow$
> Add first

Self Check 4   Solve: **a.** $4(a + 2) - a = 11$ **b.** $9x - 5(x - 9) = 1$

Now Try   Problems 45 and 47

When solving an equation, if variables appear on both sides, we can use the addition (or subtraction) property of equality to get all variable terms on one side and all constant terms on the other.

EXAMPLE 5   Solve: $3x - 15 = 4x + 36$

Strategy   There are variable terms ($3x$ and $4x$) on both sides of the equation. We will eliminate $3x$ from the left side of the equation by subtracting $3x$ from both sides.

Why To solve for x, all the terms containing x must be on the same side of the equation.

Solution

$3x - 15 = 4x + 36$	This is the equation to solve.
$3x - 15 - 3x = 4x + 36 - 3x$	Subtract $3x$ from both sides to isolate the variable term on the right side.
$-15 = x + 36$	Combine like terms: $3x - 3x = 0$ and $4x - 3x = x$.
$-15 - 36 = x + 36 - 36$	To undo the addition of 36, subtract 36 from both sides.
$-51 = x$	Do the subtractions.

Check:

$3x - 15 = 4x + 36$	The original equation.
$3(-51) - 15 \stackrel{?}{=} 4(-51) + 36$	Substitute -51 for x.
$-153 - 15 \stackrel{?}{=} -204 + 36$	Do the multiplications.
$-168 = -168$	True

The solution is -51 and the solution set is $\{-51\}$.

> **Self Check 5** Solve: $30 + 6n = 4n - 2$
>
> **Now Try** **Problem 57**

③ Clear Equations of Fractions and Decimals.

Equations are usually easier to solve if they don't involve fractions. We can use the multiplication property of equality to clear an equation of fractions by multiplying both sides of the equation by the least common denominator.

EXAMPLE 6 Solve: $\dfrac{1}{6}x + \dfrac{5}{2} = \dfrac{1}{3}$

Strategy To clear the equations of fractions, we will multiply both sides by their LCD.

Why It's easier to solve an equation that involves only integers.

Solution

$\dfrac{1}{6}x + \dfrac{5}{2} = \dfrac{1}{3}$	This is the equation to solve.
$6\left(\dfrac{1}{6}x + \dfrac{5}{2}\right) = 6\left(\dfrac{1}{3}\right)$	Multiply both sides by the LCD of $\frac{1}{6}$, $\frac{5}{2}$, and $\frac{1}{3}$, which is 6. Don't forget the parentheses.
$6\left(\dfrac{1}{6}x\right) + 6\left(\dfrac{5}{2}\right) = 6\left(\dfrac{1}{3}\right)$	On the left side, distribute the multiplication by 6.
$x + 15 = 2$	Do each multiplication: $6\left(\frac{1}{6}\right) = 1$, $6\left(\frac{5}{2}\right) = \frac{30}{2} = 15$, and $6\left(\frac{1}{3}\right) = \frac{6}{3} = 2$.
$x + 15 - 15 = 2 - 15$	To undo the addition of 15, subtract 15 from both sides.
$x = -13$	

Check the solution by substituting -13 for x in $\frac{1}{6}x + \frac{5}{2} = \frac{1}{3}$.

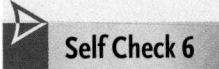

Self Check 6   Solve: $\frac{1}{4}x + \frac{1}{2} = -\frac{1}{8}$

Now Try   **Problem 63**

If an equation contains decimals, it is often convenient to multiply both sides by a power of 10 to change the decimals in the equation to integers.

EXAMPLE 7   Solve: $0.04(12) + 0.01x = 0.02(12 + x)$

Strategy   To clear the equations of decimals, we will multiply both sides by a carefully chosen power of 10.

Why   It's easier to solve an equation that involves only integers.

Solution   The equation contains the decimals 0.04, 0.01, and 0.02. Since the greatest number of decimal places in any one of these numbers is two, we multiply both sides of the equation by 10^2 or 100. This changes 0.04 to 4, and 0.01 to 1, and 0.02 to 2.

$$0.04(12) + 0.01x = 0.02(12 + x)$$

$$100[0.04(12) + 0.01x] = 100[0.02(12 + x)]$$ Multiply both sides by 100. Don't forget the brackets.

$$100 \cdot 0.04(12) + 100 \cdot 0.01x = 100 \cdot 0.02(12 + x)$$ Distribute the multiplication by 100.

$$4(12) + 1x = 2(12 + x)$$ Multiply each decimal by 100 by moving its decimal point 2 places to the right.

$$48 + x = 24 + 2x$$ Distribute the multiplication by 2.

$$48 + x - 24 - x = 24 + 2x - 24 - x$$ Subtract 24 and x from both sides.

$$24 = x$$ Simplify each side.

$$x = 24$$

The solution is 24. Check by substituting 24 for x in the original equation.

> **Success Tip**
> Recall that multiplying a decimal by 10 moves the decimal point 1 place to the right, multiplying it by 100 moves it 2 places to the right, and so on.

> **Success Tip**
> When we write the decimals in the equation as fractions, it becomes more apparent why it is helpful to multiply both sides by the LCD, 100.
>
> $$\frac{4}{100}(12) + \frac{1}{100}x = \frac{2}{100}(12 + x)$$

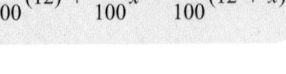

Self Check 7   Solve: $0.08x + 0.07(15,000 - x) = 1,110$

Now Try   **Problem 71**

The previous examples suggest the following strategy for solving equations. It is important to note that not every step is needed to solve every equation.

Strategy for Solving Linear Equations in One Variable	1. **Clear the equation of fractions or decimals:** Multiply both sides by the LCD to clear fractions or multiply both sides by a power of 10 to clear decimals.

1. **Clear the equation of fractions or decimals:** Multiply both sides by the LCD to clear fractions or multiply both sides by a power of 10 to clear decimals.
2. **Simplify each side of the equation:** Use the distributive property to remove parentheses, and then combine like terms on each side.
3. **Isolate the variable term on one side:** Add (or subtract) to get the variable term on one side of the equation and a number on the other using the addition (or subtraction) property of equality.
4. **Isolate the variable:** Multiply (or divide) to isolate the variable using the multiplication (or division) property of equality.
5. **Check the result:** Substitute the possible solution for the variable in the *original* equation to see if a true statement results.

EXAMPLE 8   Solve: $\dfrac{7m + 5}{5} = -4m + 1$

Strategy   We will follow the steps of the equation solving strategy to solve the equation.

Why   This is the most efficient way to solve a linear equation in one variable.

Solution

Success Tip
We remove the common factor 5 in this way:

$$\dfrac{\overset{1}{\cancel{5}}}{1}\left(\dfrac{7m + 5}{\cancel{5}}\right)$$
$$\phantom{\dfrac{5}{1}\left(\dfrac{7m}{}\right)}_{\,1}$$

$$\dfrac{7m + 5}{5} = -4m + 1 \qquad \text{This is the equation to solve.}$$

Step 1 $5\left(\dfrac{7m + 5}{5}\right) = 5(-4m + 1)$ Clear the equation of the fraction by multiplying both sides by 5.

Step 2 $7m + 5 = -20m + 5$ On the left side, remove the common factor 5 in the numerator and denominator. On the right side, distribute the multiplication by 5.

Step 3 $7m + 5 + 20m = -20m + 5 + 20m$ To eliminate the term $-20m$ on the right side, add $20m$ to both sides.

$$27m + 5 = 5 \qquad \text{Combine like terms: } 7m + 20m = 27m \text{ and } -20m + 20m = 0.$$

$$27m + 5 - 5 = 5 - 5 \qquad \text{To isolate the term } 27m, \text{ undo the addition of } 5 \text{ by subtracting } 5 \text{ from both sides.}$$

$$27m = 0 \qquad \text{Do the subtractions.}$$

Step 4 $\dfrac{27m}{27} = \dfrac{0}{27}$ To isolate m, undo the multiplication by 27 by dividing both sides by 27.

$$m = 0 \qquad \text{0 divided by any nonzero number is 0.}$$

Step 5   Substitute 0 for m in $\dfrac{7m + 5}{5} = -4m + 1$ to check that the solution is 0.

Caution
Remember that when you multiply one side of an equation by a nonzero number, you must multiply the other side of the equation by the same number.

 Self Check 8   Solve: $6c + 2 = \dfrac{18 - c}{9}$

Now Try   **Problem 79**

---

4   **Identify Identities and Contradictions.**

Each of the equations in Examples 1 through 8 had exactly one solution. However, some equations have no solutions while others have infinitely many solutions.

An equation that is true for all values of its variable is called an **identity.** One example is

$$x + x = 2x$$ If we substitute −10 for x, we get the true statement −20 = −20. If we substitute 7 for x, we get 14 = 14, and so on.

Since we can replace x with any number and the equation will be true, all real numbers are solutions of $x + x = 2x$. This equation has infinitely many solutions. Its solution set is written as {all real numbers}.

An equation that is not true for any values of its variable is called a **contradiction.** One example is

$$x = x + 1$$ No number is 1 greater than itself.

Since $x = x + 1$ has no solutions, its solution set is the **empty set,** or **null set,** and is written as $\varnothing$.

EXAMPLE 9   Solve: $3(x + 8) + 5x = 2(12 + 4x)$

Strategy   We will follow the steps of the equation solving strategy to solve the equation.

Why   This is the most efficient way to solve a linear equation in one variable.

Solution

Success Tip
Note that at the step
$8x + 24 = 24 + 8x$ we know
that the equation is an identity.

$3(x + 8) + 5x = 2(12 + 4x)$	This is the equation to solve.
$3x + 24 + 5x = 24 + 8x$	Distribute the multiplication by 3 and by 2.
$8x + 24 = 24 + 8x$	Combine like terms: 3x + 5x = 8x. Note that the sides of the equation are identical.
$8x - 8x + 24 = 24 + 8x - 8x$	To eliminate the term 8x on the right side, subtract 8x from both sides.
$24 = 24$	Combine like terms on both sides: 8x − 8x = 0.

In this case, the terms involving x drop out and the result is true. This means that any number substituted for x in the original equation will give a true statement. Therefore, *all real numbers* are solutions and this equation is an identity.

    Self Check 9   Solve: $3(x + 5) - 4(x + 4) = -x - 1$

Now Try   **Problem 87**

EXAMPLE 10   Solve: $3(d + 7) - d = 2(d + 10)$

Strategy   We will follow the steps of the equation solving strategy to solve the equation.

Why   This is the most efficient way to solve a linear equation in one variable.

Solution

The Language of Algebra
Contradiction is a form of the word
contradict, meaning conflicting
ideas. During a trial, evidence
might be introduced that *contra-
dicts* the testimony of a witness.

$3(d + 7) - d = 2(d + 10)$	This is the equation to solve.
$3d + 21 - d = 2d + 20$	Distribute the multiplication by 3 and by 2.
$2d + 21 = 2d + 20$	Combine like terms: 3d − d = 2d.
$2d + 21 - 2d = 2d + 20 - 2d$	To eliminate the term 2d on the right side, subtract 2d from both sides.
$21 = 20$	Combine like terms on both sides: 2d − 2d = 0.

In this case, the terms involving d drop out and the result is false. This means that any number that is substituted for d in the original equation will give a false statement. Since this equation has *no solution,* it is a contradiction.

Self Check 10	Solve: $-4(c - 3) + 2c = 2(10 - c)$
Now Try	**Problem 89**

ANSWERS TO SELF CHECKS   **1.** 7 **2.** -36 **3.** -3.9 **4. a.** 1 **b.** -11 **5.** 16 **6.** $-\frac{5}{2}$ **7.** 6,000
8. 0 **9.** All real numbers; the equation is an identity **10.** No solution; the equation is a contradiction

STUDY SET
2.2

VOCABULARY

Fill in the blanks.

1. $3x + 8 = 10$ is an example of a linear _____ in one variable.

2. To solve $\frac{s}{3} + \frac{1}{4} = -\frac{1}{2}$, we can _____ the equation of the fractions by multiplying both sides by 12.

3. An equation that is true for all values of its variable is called an _____.

4. An equation that is not true for any values of its variable is called a _____.

CONCEPTS

Fill in the blanks.

5. To solve $3x - 5 = 1$, we first undo the _____ of 5 by adding 5 to both sides. Then we undo the _____ by 3 by dividing both sides by 3.

6. To solve $\frac{x}{2} + 3 = 5$, we can undo the _____ of 3 by subtracting 3 from both sides. Then we can undo the _____ by 2 by multiplying both sides by 2.

7. **a.** Combine like terms on the left side of $6x - 8 - 8x = -24$.

   b. Distribute and then combine like terms on the right side of $-20 = 4(3x - 4) - 9x$.

8. Is -2 a solution of the equation?
   a. $6x + 5 = 7$ **b.** $8(x + 3) = 8$

9. Multiply.
   a. $20\left(\frac{3}{5}x\right)$ **b.** $100 \cdot 0.02x$

10. By what must you multiply both sides of $\frac{2}{3} - \frac{1}{2}b = -\frac{4}{3}$ to clear it of fractions?

11. By what must you multiply both sides of $0.7x + 0.3(x - 1) = 0.5x$ to clear it of decimals?

12. **a.** Simplify: $3x + 5 - x$
    b. Solve: $3x + 5 = 9$
    c. Evaluate $3x + 5 - x$ for $x = 9$
    d. Check: Is -1 a solution of $3x + 5 - x = 9$?

NOTATION

Complete the solution.

13. Solve: $2x - 7 = 21$

$$2x - 7 \quad = 21$$
$$2x = 28$$
$$\frac{2x}{\quad} = \frac{28}{\quad}$$
$$x = 14$$

Check: $2x - 7 = 21$
$$2(\quad) - 7 \quad 21$$
$$\quad - 7 \stackrel{?}{=} 21$$
$$\quad = 21$$

_____ is the solution.

14. A student multiplied both sides of $\frac{3}{4}t + \frac{5}{8} = \frac{1}{2}t$ by 8 to clear it of fractions, as shown below. Explain his error in showing this step.

$$8 \cdot \frac{3}{4}t + \frac{5}{8} = 8 \cdot \frac{1}{2}t$$

GUIDED PRACTICE

Solve each equation and check the result. See Examples 1 and 2.

15. $2x + 5 = 17$

16. $3x - 5 = 13$

17. $5q - 2 = 23$

18. $4p + 3 = 43$

19. $-33 = 5t + 2$

20. $-55 = 3w + 5$

21. $\frac{5}{6}k - 5 = 10$

22. $\frac{2}{5}c - 12 = 2$

23. $-\frac{7}{16}h + 28 = 21$

24. $-\frac{5}{8}h + 25 = 15$

25. $\frac{t}{3} + 2 = 6$

26. $\frac{x}{5} - 5 = -12$

27. $-3p + 7 = -3$

28. $-2r + 8 = -1$

29. $-5 - 2d = 0$

30. $-8 - 3c = 0$

31. $2(-3) + 4y = 14$

32. $4(-1) + 3y = 8$

33. $0.7 - 4y = 1.7$

34. $0.3 - 2x = -0.9$

Solve each equation and check the result. See Example 3.

35. $1.2 - x = -1.7$

36. $0.6 = 4.1 - x$

37. $-6 - y = -2$

38. $-1 - h = -9$

Solve each equation and check the result. See Example 4.

39. $3(2y - 2) - y = 5$

40. $2(-3a + 2) + a = 2$

41. $4(5b) + 2(6b - 1) = -34$

42. $9(x + 11) + 5(13 - x) = 0$

43. $-(4 - m) = -10$

44. $-(6 - t) = -12$

45. $10.08 = 4(0.5x + 2.5)$

46. $-3.28 = 8(1.5y - 0.5)$

47. $6a - 3(3a - 4) = 30$

48. $16y - 8(3y - 2) = -24$

49. $-(19 - 3s) - (8s + 1) = 35$ **50.** $2(3x) - 5(3x + 1) = 58$

Solve each equation and check the result. See Example 5.

51. $5x = 4x + 7$

52. $3x = 2x + 2$

53. $8y + 44 = 4y$

54. $9y + 36 = 6y$

55. $60r - 50 = 15r - 5$

56. $100f - 75 = 50f + 75$

57. $8y - 2 = 4y + 16$

58. $7 + 3w = 4 + 9w$

59. $2 - 3(x - 5) = 4(x - 1)$

60. $2 - (4x + 7) = 3 + 2(x + 2)$

61. $3(A + 2) = 2(A - 7)$

62. $9(T - 1) = 6(T + 2) - T$

Solve each equation and check the result. See Example 6.

63. $\frac{1}{8}y - \frac{1}{2} = \frac{1}{4}$

64. $\frac{1}{15}x - \frac{4}{5} = \frac{2}{3}$

65. $\frac{1}{3} = \frac{5}{6}x + \frac{2}{9}$

66. $\frac{2}{3} = -\frac{2}{3}x + \frac{3}{4}$

67. $\frac{1}{6}y + \frac{1}{4}y = -1$

68. $\frac{1}{3}x + \frac{1}{4}x = -2$

69. $\frac{2}{3}y + 2 = \frac{1}{5} + y$

70. $\frac{2}{5}x + 1 = \frac{1}{3} + x$

Solve each equation and check the result. See Example 7.

71. $0.06(s + 9) - 1.24 = -0.08s$

72. $0.08(x + 50) - 0.16x = 0.04(50)$

73. $0.09(t + 50) + 0.15t = 52.5$

74. $0.08(x - 100) = 44.5 - 0.07x$

75. $0.06(a + 200) + 0.1a = 172$

76. $0.03x + 0.05(6,000 - x) = 280$

77. $0.4b - 0.1(b - 100) = 70$

78. $0.105x + 0.06(20,000 - x) = 1,740$

Solve each equation and check the result. See Example 8.

79. $\frac{10 - 5s}{3} = s$

80. $\frac{40 - 8s}{5} = -2s$

81. $\frac{7t - 9}{16} = t$

82. $\frac{11r + 68}{3} = -3$

83. $\frac{5(1 - x)}{6} = -x + 1$

84. $\frac{3(14 - u)}{8} = -3u + 6$

85. $\frac{3(d - 8)}{4} = \frac{2(d + 1)}{3}$

86. $\frac{3(c - 2)}{2} = \frac{2(2c + 3)}{5}$

Solve each equation, if possible. See Examples 9–10.

87. $8x + 3(2 - x) = 5x + 6$

88. $5(x + 2) = 5x - 2$

89. $-3(s + 2) = -2(s + 4) - s$

90. $21(b - 1) + 3 = 3(7b - 6)$

91. $2(3z + 4) = 2(3z - 2) + 13$

92. $x + 7 = \frac{2x + 6}{2} + 4$

93. $4(y - 3) - y = 3(y - 4)$

94. $5(x + 3) - 3x = 2(x + 8)$

TRY IT YOURSELF

Solve each equation, if possible. Check the result.

95. $3x - 8 - 4x - 7x = -2 - 8$

96. $-6t - 7t - 5t - 1 = 12 - 3$

97. $0.05a + 0.01(90) = 0.02(a + 90)$

98. $0.03x + 0.05(2,000 - x) = 99.5$

99. $\frac{3(b + 2)}{2} = \frac{4b - 10}{4}$

100. $\frac{2(5a - 7)}{4} = \frac{9(a - 1)}{3}$

101. $4(a - 3) = -2(a - 6) + 6a$

102. $9(t + 2) = -6(t - 3) + 15t$

103. $10 - 2y = 8$

104. $7 - 7x = -21$

105. $2n - \frac{3}{4}n = \frac{1}{2}n + \frac{13}{3}$

106. $\frac{5}{6}n + 3n = -\frac{1}{3}n - \frac{11}{9}$

107. $-\frac{2}{3}z + 4 = 8$

108. $-\frac{7}{5}x + 9 = -5$

109. $-2(9 - 3s) - (5s + 2) = -25$

110. $4(x - 5) - 3(12 - x) = 7$

WRITING

111. To solve $3x - 4 = 5x + 1$, one student began by subtracting $3x$ from both sides. Another student solved the same equation by first subtracting $5x$ from both sides. Will the students get the same solution? Explain why or why not.

112. What does it mean to clear an equation such as $\frac{1}{4} + \frac{1}{2}x = \frac{3}{8}$ of the fractions?

113. Explain the error in the following solution.

Solve: $2x + 4 = 30$.

$$\frac{2x}{2} + 4 = \frac{30}{2}$$
$$x + 4 = 15$$
$$x + 4 - 4 = 15 - 4$$
$$x = 11$$

114. Write an equation that is an identity. Explain why every real number is a solution.

REVIEW

Name the property that is used.

115. $x \cdot 9 = 9x$

116. $4 \cdot \frac{1}{4} = 1$

117. $(x + 1) + 2 = x + (1 + 2)$

118. $2(30y) = (2 \cdot 30)y$

CHALLENGE PROBLEMS

119. In this section, we discussed equations that have no solution, one solution, and an infinite number of solutions. Do you think an equation could have exactly two solutions? If so, give an example.

120. The equation $4x - 3y = 5$ contains two different variables. Solve the equation by determining a value of x and a value for y that make the equation true.

SECTION 2.3
Applications of Percent

Objectives
1. Change percents to decimals and decimals to percents.
2. Solve percent problems by direct translation.
3. Solve applied percent problems.
4. Find percent of increase and decrease.
5. Solve discount and commission problems.

In this section, we will use translation skills from Chapter 1 and equation-solving skills from Chapter 2 to solve problems involving percents.

1 Change Percents to Decimals and Decimals to Percents.

The word **percent** means parts per one hundred. We can think of the percent symbol % as representing a denominator of 100. Thus, $93\% = \frac{93}{100}$. Since the fraction $\frac{93}{100}$ is equal to the decimal 0.93, it is also true that $93\% = 0.93$.

When solving percent problems, we must often convert percents to decimals and decimals to percents. To change a percent to a decimal, we *divide by 100 by moving the decimal point 2 places to the left and dropping the % symbol.* For example,

$$31\% = 31.0\% = 0.31$$

To change a decimal to a percent, we *multiply the decimal by 100 by moving the decimal point 2 places to the right, and inserting a % symbol.* For example,

$$0.678 = 67.8\%$$

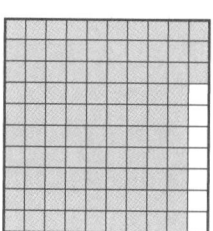
93% or $\frac{93}{100}$ or 0.93 of the figure is shaded.

2   **Solve Percent Problems by Direct Translation.**

There are three basic types of percent problems. Examples of these are:

Type 1   What number is 8% of 215?
Type 2   102 is 21.3% of what number?
Type 3   31 is what percent of 500?

Every percent problem has three parts: the *amount,* the *percent,* and the *base.* For example, in the question *What number is 8% of 215?*, the words "what number" represent the **amount,** 8% represents the **percent,** and 215 represents the **base.** In these problems, the word "is" means "is equal to," and the word "of" means "multiplication."

What number	is	8%	of	215?
↓	↓	↓	↓	↓
Amount	**=**	**Percent**	**·**	**base**

EXAMPLE 1   What number is 8% of 215?

Strategy   We will translate the words of this problem into an equation and then solve the equation.

Why   The variable in the translation equation represents the unknown number that we are asked to find.

The Language of Algebra
Translate the word
• *is* to an equal symbol =
• *of* to multiplication
• *what* to a variable

Solution   In this problem, the phrase "what number" represents the amount, 8% is the percent, and 215 is the base.

What number	is	8%	of	215?
↓	↓	↓	↓	↓
x	=	0.08	·	215 *Change the percent to a decimal: 8% = 0.08.*
x	=	17.2		*Do the multiplication.*

Thus, 8% of 215 is 17.2.

To check, we note that 17.2 out of 215 is $\frac{17.2}{215} = 0.08 = 8\%$.

 Self Check 1   What number is 5.6% of 40?
Now Try   **Problem 13**

We will illustrate the other two types of percent problems with application problems.

3   **Solve Applied Percent Problems.**

One method for solving applied percent problems is to use the given facts to write a **percent sentence** of the form

	is		%	of		?

We enter the appropriate numbers in two of the blanks and the words "what" or "what number" in the remaining blank. As before, we translate the words into an equation and solve it.

EXAMPLE 2   *Aging Populations.*   By the year 2075, the U.S. Bureau of the Census predicts that about 102 million residents will be age 65 or older. The **circle graph** (or **pie chart**) indicates that age group will make up 21.3% of the population. If this prediction is correct, find the population of the United States in 2075. (Round to the nearest million.)

Projection of the 2075 U.S. Population by Age

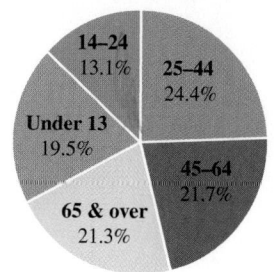

Source: U.S. Bureau of the Census (2000).

Strategy   To find the predicted U.S. population in 2075, we will translate the words of the problem into an equation and then solve the equation.

Why   The variable in the translation equation represents the unknown population in 2075 that we are asked to find.

Solution   In this problem, 102 is the amount, 21.3% is the percent, and the words "what number" represent the base. The units are in millions.

102	is	21.3%	of	what number?
↓	↓	↓	↓	↓
102	=	0.213	·	x

$$\frac{102}{0.213} = \frac{0.213x}{0.213}$$ To undo the multiplication by 0.213, divide both sides by 0.213.

$478.9 \approx x$ Do the divisions.

$479 \approx x$ Round 478.9 to the nearest million.

The U.S. population is predicted to be about 479 million in the year 2075. We can check using estimation: 102 million out of a population of 479 million is approximately $\frac{100 \text{ million}}{500 \text{ million}}$, or $\frac{1}{5}$, which is 20%. Since this is close to 21.3%, the answer 479 seems reasonable.

Self Check 2   By the year 2100, it is predicted that 131 million, or 23%, of the U.S. residents will be age 65 or older. If the prediction is correct, find the population in 2100.

Now Try   **Problem 17**

We pay many types of taxes in our daily lives, such as sales tax, gasoline tax, income tax, and Social Security tax. **Tax rates** are usually expressed as percents.

EXAMPLE 3   *Taxes.*   A maid makes $500 a week. One of the deductions from her weekly paycheck is a Social Security tax of $31. Find her Social Security tax rate.

Strategy   To find the tax rate, we will translate the words of the problem into an equation and then solve the equation.

Why   The variable in the translation equation represents the unknown tax rate that we are asked to find.

Solution

31	is	what percent	of	500?
↓	↓	↓	↓	↓
31	=	x	·	500

31 is the amount, x is the percent, and 500 is the base.

$$\frac{31}{500} = \frac{500x}{500}$$ To undo the multiplication by 500, divide both sides by 500.

$0.062 = x$ Do the divisions.

$6.2\% = x$ Change the decimal 0.062 to a percent.

The Social Security tax rate is 6.2%

We can use estimation to check: $31 out of $500 is about $\frac{30}{500}$ or $\frac{6}{100}$, which is 6%. Since this is close to 6.2%, the answer seems reasonable.

 Self Check 3   The maid mentioned in Example 3 also has $7.25 of Medicare tax deducted from her weekly paycheck. Find her Medicare tax rate.

Now Try   **Problem 21**

 Find Percent of Increase and Decrease.

Percents are often used to describe how a quantity has changed. For example, a health care provider might increase the cost of medical insurance by 3%, or a police department might decrease the number of officers assigned to street patrols by 10%. To describe such changes, we use **percent of increase** or **percent of decrease.**

EXAMPLE 4   *Identity Theft.*   The Federal Trade Commission receives complaints involving the theft of someone's identity information, such as a credit card, Social Security number, or cell phone account. Refer to the data in the table. What was the percent of increase in the number of complaints from 2001 to 2005? (Round to the nearest percent.)

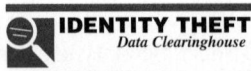

Year	2001	2005
Number of Complaints	86,000	256,000

Strategy   First, we will subtract to find the *amount of increase* in the number of complaints. Then we will translate the words of the problem into an equation and solve it.

Why   A percent of increase problem involves finding the *percent of change,* and the change in a quantity is found using subtraction.

Solution To find the *amount of increase,* we subtract the earlier value, 86,000, from the later value, 256,000.

$$256{,}000 - 86{,}000 = 170{,}000$$

We know that an increase of 170,000 is some unknown percent of the number of complaints in 2001, which was 86,000.

170,000	is	what percent	of	86,000?
↓	↓	↓	↓	↓
170,000	=	x	·	86,000

170,000 is the amount, x is the percent, and 86,000 is the base.

$$\frac{170{,}000}{86{,}000} = \frac{86{,}000x}{86{,}000}$$

To undo the multiplication by 86,000, divide both sides by 86,000.

$$1.977 \approx x \quad \text{Do the divisions.}$$
$$197.7\% \approx x \quad \text{Change 1.977 to a percent.}$$

Rounding 197.7% to the nearest percent, we find that the number of identity theft complaints increased by about 198% from 2001 to 2005.

A 200% increase would be double the number of complaints: 2(86,000) = 172,000. It seems reasonable that 170,000 complaints is a 198% increase.

Self Check 4 In 2004, there were 247,000 complaints of identity theft. Find the percent increase from 2004 to 2005. (Round to the nearest percent.)

Now Try **Problem 43**

Caution
The percent of increase (or decrease) is a percent of the *original* number, that is, the number before the change occurred.

5 **Solve Discount and Commission Problems.**

When the price of an item is reduced, we call the amount of the reduction a **discount.** If a discount is expressed as a percent, it is called the **rate of discount.**

EXAMPLE 5 *Health Club Discounts.* A 30% discount on a 1-year membership for a fitness center amounted to a $90 savings. Find the cost of a 1-year membership before the discount.

Strategy We will translate the words of the problem into an equation and then solve the equation.

Why The variable in the translation equation represents the unknown cost of a 1-year membership before the discount that we are asked to find.

Solution We are told that $90 is 30% of some unknown membership cost.

90	is	30%	of	what number?
↓	↓	↓	↓	↓
90	=	0.30	·	x

90 is the amount, 30% is the percent, and x is the base.

$$\frac{90}{0.30} = \frac{0.30x}{0.30}$$

To undo the multiplication by 0.30, divide both sides by 0.30.

$$300 = x \quad \text{Do the divisions.}$$

A one-year membership cost $300 before the discount.

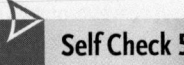

Self Check 5 A shopper saved $6 on a pen that was discounted 5%. Find the original cost.

Now Try **Problem 51**

Instead of working for a salary or at an hourly rate, many salespeople are paid on **commission.** An employee who is paid a commission is paid a percent of the goods or services that he or she sells. We call that percent the **rate of commission.**

EXAMPLE 6 **Commissions.** A real estate agent earned $14,025 for selling a house. If she received a $5\frac{1}{2}\%$ commission, what was the selling price?

Strategy We will translate the words of the problem into an equation and then solve the equation.

Why The variable in the translation equation represents the unknown selling price of the house that we are asked to find.

Solution We are told that $14,025 is $5\frac{1}{2}\%$ of some unknown selling price of a house.

$14,025	is	5.5%	of	what number?	Write $5\frac{1}{2}\%$ as 5.5%.
↓	↓	↓	↓	↓	
$14,025	=	0.055	·	x	14,025 is the amount, 5.5% is the percent, and x is the base.

$$\frac{14{,}025}{0.055} = \frac{0.055x}{0.055}$$ To undo the multiplication by 0.055, divide both sides by 0.055.

$$255{,}000 = x$$ Do the divisions.

The selling price of the house was $255,000.

Self Check 6 A jewelry store clerk receives a 4.5% commission on all sales. What was the price of a gold necklace sold by the clerk if his commission was $15.75?

Now Try **Problem 53**

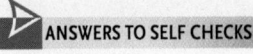

ANSWERS TO SELF CHECKS **1.** 2.24 **2.** 570 million **3.** 1.45% **4.** 4% **5.** $120 **6.** $350

STUDY SET
2.3

VOCABULARY

Fill in the blanks.

1. _____ means parts per one hundred.
2. In the statement "10 is 50% of 20," 10 is the _____, 50% is the percent, and 20 is the _____.
3. In percent questions, the word *of* means _____, and ___ means equals.
4. An employee who is paid a _____ is paid a percent of the goods or services that he or she sells.

CONCEPTS

5. Represent the amount of the figure that is shaded using a fraction, a decimal, and a percent.

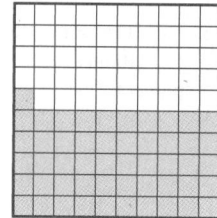

6. Fill in the blank: To solve a percent problem, we translate the words of the problem into an _____ and solve it.

7.

High School Sports Programs Girl's Water Polo—Number of Participants	
2001	**2005**
14,792	17,241

Source: National Federation of State High School Associations

   a. Find the *amount* of increase in participation.
   b. Fill in blanks to find the percent of increase in participation: _____ is _____ % of _____?

8. Fill in the blanks using the words *percent, amount,* and *base.*

   $$\boxed{} = \boxed{} \cdot \boxed{}$$

9. Translate each sentence into an equation. **Do not solve.**
   a. 12 is 40% of what number?
   b. 99 is what percent of 200?
   c. What is 66% of 3?

10. Use estimation to determine if each statement is reasonable.
    a. 18 is 48% of 93.
    b. 47 is 6% of 206.

NOTATION

11. Change each percent to a decimal.
    a. 35%   b. 8.5%
    c. 150%   d. $2\frac{3}{4}\%$

12. Change each decimal to a percent.
    a. 0.9   b. 9
    c. 0.999

GUIDED PRACTICE

See Examples 1–3.

13. What number is 48% of 650?
14. What number is 60% of 200?
15. 78 is what percent of 300?
16. 143 is what percent of 325?
17. 75 is 25% of what number?
18. 78 is 6% of what number?
19. What number is 92.4% of 50?
20. What number is 2.8% of 220?
21. 0.42 is what percent of 16.8?
22. 199.92 is what percent of 2,352?
23. 128.1 is 8.75% of what number?
24. 1.12 is 140% of what number?

APPLICATIONS

25. ANTISEPTICS   Use the facts on the label to determine the amount of pure hydrogen peroxide in the bottle.

HYDROGEN PEROXIDE
3% solution
inactive ingredient: purified water
16 Fluid ounces

26. DINING OUT   Refer to the sales receipt. Compute the 15% tip (*rounded up* to the nearest dollar). Then find the total cost of the meal.

Corner Pub
Nashville, TN

VISA   078392762
Amount: $75.18

+ Tip: _____

= Total: _____

X _____

27. U.S. FEDERAL BUDGET   The circle graph shows how the government spent $2,500 billion in 2005. How much was spent on

a. Social Security/Medicare?

b. Defense/Veterans?

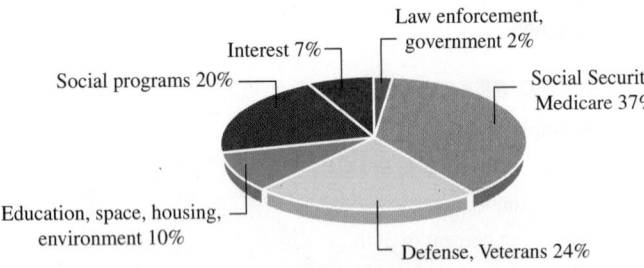

Based on 2006 Federal Income Tax Form 1040

28. TAX TABLES   Use the table to compute the amount of federal income tax to be paid on an income of $39,909.

If your income is over—	But not over—	Your income tax is—	of the amount over—
$0	$7,300	.... 10%	$0
7,300	29,700	$730.00 + 15%	7,300
29,700	71,950	4,090.00 + 25%	29,700

29. PAYPAL   Many e-commerce businesses use PayPal to perform payment processing for them. For certain transactions, merchants are charged a fee of 2.9% of the selling price of the item plus $0.30. What would PayPal charge an online art store to collect payment on a painting selling for $350?

30. eBAY   When a student sold an Xbox on eBay for $153, she was charged a two-part final value fee: 5.25% of the first $25 of the selling price plus 3.25% of the remainder of the selling price over $25. Find the fee to sell the Xbox on eBay.

31. PRICE GUARANTEES   Home Club offers a "10% Plus" guarantee: If the customer finds the same item selling for less somewhere else, he or she receives the difference in price plus 10% of the difference. A woman bought miniblinds at the Home Club for $120 but later saw the same blinds on sale for $98 at another store. How much can she expect to be reimbursed?

32. ROOM TAXES   A guest at the San Antonio Hilton Airport Hotel paid $180 for a room plus a 9% city room tax, a $1\frac{3}{4}$% county room tax, and a 6% state room tax. Find the total amount of tax that the guest paid on the room.

33. COMPUTER MEMORY   The *My Computer* screen on a student's computer is shown in the next column. What percent of the memory on the hard drive Local Disk (C:) of his computer is used? What percent is free? (GB stands for gigabytes.)

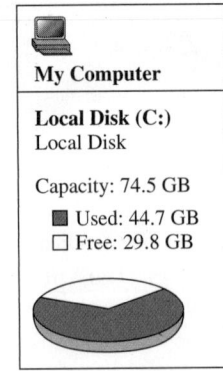

My Computer

Local Disk (C:)
Local Disk

Capacity: 74.5 GB

■ Used: 44.7 GB
□ Free: 29.8 GB

34. GENEALOGY   Through an extensive computer search, a genealogist determined that worldwide, 180 out of every 10 million people had his last name. What percent is this?

35. DENTISTRY   Refer to the dental record. What percent of the patient's teeth have fillings? Round to the nearest percent.

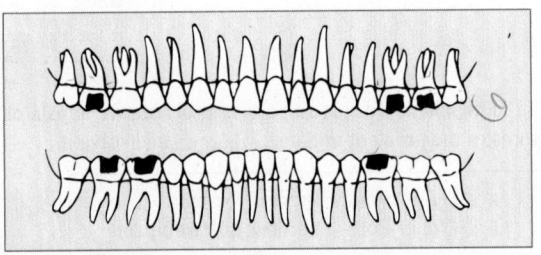

36. TEST SCORES   The score 175/200 was written by an algebra instructor at the top of a student's test paper. Write the test score as a percent.

37. DMV WRITTEN TEST   To obtain a learner's permit to drive in Nevada, a score of 80% (or better) on a 50-question multiple-choice test is required. If a teenager answered 33 questions correctly, did he pass the test?

38. iPODS   The settings menu screen of an Apple iPod is shown. What percent of the memory capacity is still available? Round to the nearest percent. (GB stands for gigabytes.)

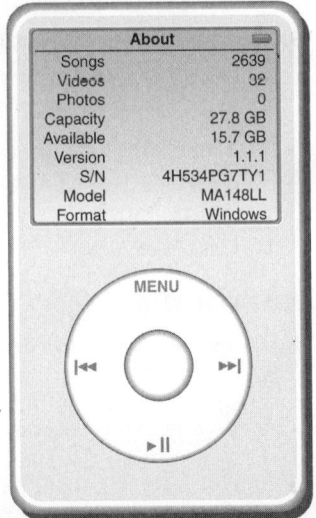

About	
Songs	2639
Videos	32
Photos	0
Capacity	27.8 GB
Available	15.7 GB
Version	1.1.1
S/N	4H534PG7TY1
Model	MA148LL
Format	Windows

39. CHILD CARE   After the first day of registration, 84 children had been enrolled in a day care center. That represented 70% of the available slots. Find the maximum number of children the center could enroll.

40. RACING PROGRAMS   One month before a stock car race, the sale of ads for the official race program was slow. Only 12 pages, or just 30% of the available pages, had been sold. Find the total number of pages devoted to advertising in the program.

41. NUTRITION   The Nutrition Facts label from a can of clam chowder is shown.

   a. Find the number of grams of saturated fat in one serving. What percent of a person's recommended daily intake is this?

   b. Determine the recommended number of grams of saturated fat that a person should consume daily.

Nutrition Facts

Serving Size 1 cup (240mL)
Servings Per Container about 2

Amount per serving

Calories 240   Calories from Fat 140

	% Daily Value*
Total Fat 15 g	23%
Saturated Fat 5 g	25%
Cholesterol 10 mg	3%
Sodium 980 mg	41%
Total Carbohydrate 21 g	7%
Dietary Fiber 2 g	8%
Sugars 1 g	
Protein 7 g	
---

42. COMMERCIALS   Jared Fogle credits his tremendous weight loss to exercise and a diet of low-fat Subway sandwiches. His current weight (about 187 pounds) is 44% of his maximum weight (reached in March of 1998). What did he weigh then?

43. EXPORTS   According to the graph, between what two years was there the greatest percent decrease in U.S. exports to Mexico? Find the percent of decrease.

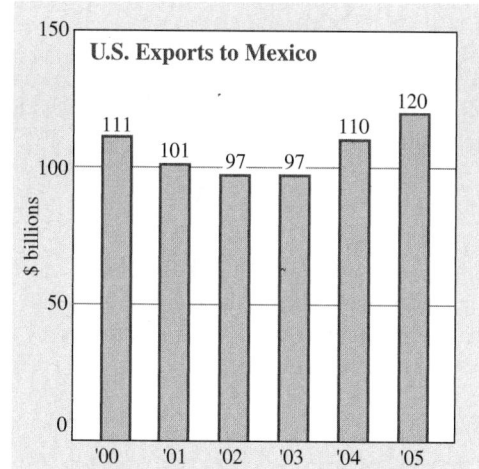

Based on data from www.census.gov/foreign-trade

44. AUCTIONS   A pearl necklace of former First Lady Jacqueline Kennedy Onassis, originally valued at $700, was sold at auction in 1996 for $211,500. Find the percent of increase in the value of the necklace. (Round to the nearest percent.)

45. INSURANCE COSTS   A college student's good grades earned her a student discount on her car insurance premium. Find the percent of decrease to the nearest percent if her annual premium was lowered from $1,050 to $925.

46. U.S. LIFE EXPECTANCY   Use the following life expectancy data for 1900 and 2004 to find the percent of increase for males and for females. Round to the nearest percent.

Years of life expected at birth		
	Male	Female
1900	46.3	48.3
2004	75.2	80.4

Source: National Vital Statistics Reports

47. TALK RADIO   Refer to the table and find the percent increase in the number of news/talk radio stations from 2004 to 2005. Round to the nearest percent.

Number of U.S. news/talk radio stations	
2004: 1,282	**2005:** 1,324

Source: *The World Almanac, 2006*

48. FOOD LABELS   To be labeled "Reduced Fat," foods must contain at least 25% less fat per serving than the regular product. One serving of the original Jif peanut butter has 16 grams of fat per serving. The new Jif Reduced Fat product contains 12 grams of fat per serving. Does it meet the labeling requirement?

49. TV SHOPPING   Jan bought a toy from the QVC home shopping network that was discounted 20%. If she saved $15, what was the original price of the toy?

50. DISCOUNTS   A 12% discount on a watch saved a shopper $48. Find the price of the watch before the discount.

51. SALES   The price of a certain model patio set was reduced 35% because it was being discontinued. A shopper purchased two of them and saved a total of $210. Find the price of a patio set before the discount.

52. TV SALES   The price of a plasma screen television was reduced $800 because it was used as a floor model. If this was a 40% savings, find the original price of the TV.

53. REAL ESTATE   The $3\frac{1}{2}$% commission paid to a real estate agent on the sale of a condominium earned her $3,325. Find the selling price of the condo.

54. CONSIGNMENT   An art gallery agreed to sell an artist's sculpture for a commission of 45%. What must be the selling price of the sculpture if the gallery would like to make $13,500?

55. STOCKBROKERS   A stockbroker charges a 2.5% commission to sell shares of a stock for a client. Find the value of stock sold by a broker if the commission was $640.

56. AGENTS An agent made one million dollars by charging a 12.5% commission to negotiate a long-term contract for a professional athlete. Find the amount of the contract.

WRITING

57. Explain the error:

What is 5% of 8?

$x = 5 \cdot 8$
~~$x = 5 \cdot 8$~~
$x = 40$

40 is 5% of 8.

58. Write a real-life situation that could be described by "9 is what percent of 20?"

59. Explain why 150% of a number is more than the number.

60. Why is the problem "What is 9% of 100?" easy to solve?

REVIEW

61. Divide: $-\frac{16}{25} \div \left(-\frac{4}{15}\right)$

62. What two numbers are a distance of 8 away from 4 on the number line?

63. Is -34 a solution of $x + 15 = -49$?

64. Evaluate: $2 + 3[24 - 2(2 - 5)]$

CHALLENGE PROBLEMS

65. SOAPS A soap advertises itself as $99\frac{44}{100}\%$ pure. First, determine what percent of the soap is impurities. Then express your answer as a decimal.

66. Express $\frac{1}{20}$ of 1% as a percent using decimal notation.

SECTION 2.4
Formulas

Objectives

1 Use formulas from business.
2 Use formulas from science.
3 Use formulas from geometry.
4 Solve for a specified variable.

A **formula** is an equation that states a relationship between two or more variables. Formulas are used in fields such as business, science, and geometry.

Use Formulas from Business.

A formula for retail price: To make a profit, a merchant must sell an item for more than he or she paid for it. The price at which the merchant sells the product, called the **retail price,** is the *sum* of what the item cost the merchant plus the **markup.** Using r to represent the retail price, c the cost, and m the markup, we can write this formula as

$r = c + m$ Retail price = cost + markup

A formula for profit: The **profit** a business makes is the *difference* between the **revenue** (the money it takes in) and the cost. Using p to represent the profit, r the revenue, and c the cost, we can write this formula as

$p = r - c$ Profit = revenue − cost

If we are given the values of all but one of the variables in a formula, we can use our equation solving skills to find the value of the remaining variable.

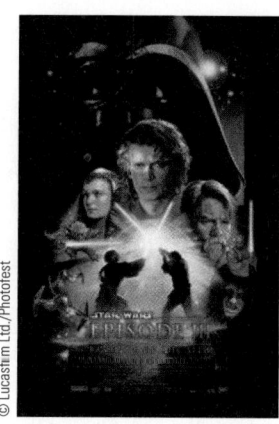

© Lucasfilm Ltd./Photofest

EXAMPLE 1 *Films.* Estimates are that 20th Century Fox made a $309 million profit on the movie *Star Wars: Revenge of the Sith.* If the studio received $424 million in worldwide box office revenue, find the cost to make and distribute the film. (Source: www.the-numbers.com, August 2006)

Strategy To find the cost to make and distribute the film, we will substitute the given values in the formula $p = r - c$ and solve for c.

Why The variable c represents the unknown cost.

Solution The movie made $309 million (the profit p) and the studio took in $424 million (the revenue r). To find the cost c, we proceed as follows.

$$p = r - c$$ This is the formula for profit.

$$309 = 424 - c$$ Substitute 309 for p and 424 for r.

$$309 - 424 = 424 - c - 424$$ To eliminate 424 on the right side, subtract 424 from both sides.

$$-115 = -c$$ Do the subtractions.

$$\frac{-115}{-1} = \frac{-c}{-1}$$ To solve for c, divide (or multiply) both sides by -1.

$$115 = c$$ The units are millions of dollars.

It cost $115 million to make and distribute the film.

Self Check 1 A PTA spaghetti dinner made a profit of $275.50. If the cost to host the dinner was $1,235, how much revenue did it generate?

Now Try Problem 11

A formula for simple interest: When money is borrowed, the lender expects to be paid back the amount of the loan plus an additional charge for the use of the money, called **interest.** When money is deposited in a bank, the depositor is paid for the use of the money. The money the deposit earns is also called interest.

Interest is computed in two ways: either as **simple interest** or as **compound interest.** Simple interest is the *product* of the principal (the amount of money that is invested, deposited, or borrowed), the annual interest rate, and the length of time in years. Using I to represent the simple interest, P the principal, r the annual interest rate, and t the time in years, we can write this formula as

$$I = Prt \quad \text{Interest = principal} \cdot \text{rate} \cdot \text{time}$$

EXAMPLE 2 *Retirement Income.* One year after investing $15,000, a retired couple received a check for $1,125 in interest. Find the interest rate their money earned that year.

Strategy To find the interest rate, we will substitute the given values in the formula $I = Prt$ and solve for r.

Why The variable r represents the unknown interest rate.

Solution   The couple invested $15,000 (the principal P) for 1 year (the time t) and made $1,125 (the interest I). To find the annual interest rate r, we proceed as follows.

$I = Prt$	This is the formula for simple interest.
$1{,}125 = 15{,}000r(1)$	Substitute 1,125 for I, 15,000 for P, and 1 for t.
$1{,}125 = 15{,}000r$	Simplify the right side.
$\dfrac{1{,}125}{15{,}000} = \dfrac{15{,}000r}{15{,}000}$	To solve for r, undo the multiplication by 15,000 by dividing both sides by 15,000.
$0.075 = r$	Do the divisions.
$7.5\% = r$	To write 0.075 as a percent, multiply 0.075 by 100 by moving the decimal point two places to the right and inserting a % symbol.

The couple received an annual rate of 7.5% that year on their investment. We can display the facts of the problem in a table.

	P	$\cdot$ r	$\cdot t =$	I
Investment	15,000	0.075	1	1,125

Self Check 2   A father loaned his daughter $12,200 at a 2% annual simple interest rate for a down payment on a house. If the interest on the loan amounted to $610, for how long was the loan?

Now Try   **Problem 15**

❷ Use Formulas from Science.

A formula for distance traveled: If we know the average rate (of speed) at which we will be traveling and the time we will be traveling at that rate, we can find the distance traveled. Using d to represent the distance, r the average rate, and t the time, we can write this formula as

$d = rt$ Distance = rate · time

EXAMPLE 3   **Whales.**   As they migrate from the Bering Sea to Baja California, gray whales swim for about 20 hours each day, covering a distance of approximately 70 miles. Estimate their average swimming rate in miles per hour (mph).

Strategy   To find the swimming rate, we will substitute the given values in the formula $d = rt$ and solve for r.

Why   The variable r represents the unknown average swimming rate.

Solution   The whales swam 70 miles (the distance d) in 20 hours (the time t). To find their average swimming rate r, we proceed as follows.

$d = rt$	This is the formula for distance traveled.
$70 = r(20)$	Substitute 70 for d and 20 for t.
$\dfrac{70}{20} = \dfrac{20r}{20}$	To solve for r, undo the multiplication by 20 by dividing both sides by 20.
$3.5 = r$	Do the divisions.

The whales' average swimming rate is 3.5 mph. The facts of the problem can be shown in a table.

	r	$\cdot$ t	$= d$
Gray whale	3.5	20	70

Self Check 3   An elevator travels at an average rate of 288 feet per minute. How long will it take the elevator to climb 30 stories, a distance of 360 feet?

Now Try   **Problem 19**

A formula for converting temperatures: In the American system, temperature is measured on the Fahrenheit scale. The Celsius scale is used to measure temperature in the metric system. The formula that relates a Fahrenheit temperature F to a Celsius temperature C is:

$$C = \frac{5}{9}(F - 32)$$

EXAMPLE 4   Convert the temperature shown on the City Savings sign to degrees Fahrenheit.

■ CITY SAVINGS
TEMP 30°C

Strategy   To find the temperature in degrees Fahrenheit,   we will substitute the given Celsius temperature in the formula $C = \frac{5}{9}(F - 32)$ and solve for F.

Why   The variable F represents the unknown temperature in degrees Fahrenheit.

Solution   The temperature in degrees Celsius is 30°. To find the temperature in degrees Fahrenheit F, we proceed as follows.

$$C = \frac{5}{9}(F - 32)$$ This is the formula for temperature conversion.

$$30 = \frac{5}{9}(F - 32)$$ Substitute 30 for C, the Celsius temperature.

$$\frac{9}{5} \cdot 30 = \frac{9}{5} \cdot \frac{5}{9}(F - 32)$$ To undo the multiplication by $\frac{5}{9}$, multiply both sides by the reciprocal of $\frac{5}{9}$.

$$54 = F - 32$$ Do the multiplications.

$$54 + 32 = F - 32 + 32$$ To isolate F, undo the subtraction of 32 by adding 32 to both sides.

$$86 = F$$

30°C is equivalent to 86°F.

The Language of Algebra
In 1724, Daniel Gabriel *Fahrenheit*, a German scientist, introduced the temperature scale that bears his name. The Celsius scale was invented in 1742 by Swedish astronomer Anders *Celsius*.

Self Check 4   Change $-175°C$, the temperature on Saturn, to degrees Fahrenheit.

Now Try   **Problem 25**

3 Use Formulas from Geometry.

To find the **perimeter** of a plane (two-dimensional, flat) geometric figure, such as a rectangle or triangle, we find the distance around the figure by computing the sum of the lengths of its sides. Perimeter is measured in American units, such as inches, feet, yards, and in metric units such as millimeters, meters, and kilometers.

> **The Language of Algebra**
> When you hear the word *perimeter*, think of the distance around the "rim" of a flat figure.

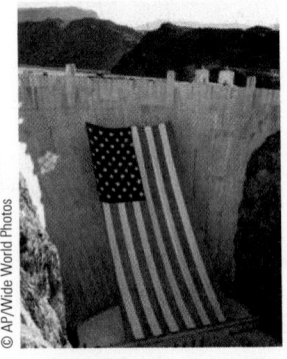

Perimeter formulas

$P = 2l + 2w$ (rectangle)
$P = 4s$ (square)
$P = a + b + c$ (triangle)

EXAMPLE 5   *Flags.*   The largest flag ever flown was an American flag that had a perimeter of 1,520 feet and a length of 505 feet. It was hoisted on cables across Hoover Dam to celebrate the 1996 Olympic Torch Relay. Find the width of the flag.

505 ft

Strategy   To find the width of the flag, we will substitute the given values in the formula $P = 2l + 2w$ and solve for w.

Why   The variable w represents the unknown width of the flag.

Solution   The perimeter P of the rectangular-shaped flag is 1,520 ft and the length l is 505 ft. To find the width w, we proceed as follows.

$P = 2l + 2w$	This is the formula for the perimeter of a rectangle.
$1,520 = 2(505) + 2w$	Substitute 1,520 for P and 505 for l.
$1,520 = 1,010 + 2w$	Do the multiplication.
$510 = 2w$	To undo the addition of 1,010, subtract 1,010 from both sides.
$255 = w$	To isolate w, undo the multiplication by 2 by dividing both sides by 2.

The width of the flag is 255 feet. If its length is 505 feet and its width is 255 feet, its perimeter is $2(505) + 2(255) = 1,010 + 510 = 1,520$ feet, as given.

> **Self Check 5**   The largest flag that consistently flies is the flag of Brazil in Brasilia, the country's capital. It has perimeter 1,116 feet and length 328 feet. Find its width.
>
> **Now Try   Problem 27**

Area formulas

$A = lw$ (rectangle)
$A = s^2$ (square)
$A = \frac{1}{2}bh$ (triangle)
$A = \frac{1}{2}h(B + b)$ (trapezoid)

The **area** of a plane (two-dimensional, flat) geometric figure is the amount of surface that it encloses. Area is measured in square units, such as square inches, square feet, square yards, and square meters (written as in.^{2}, ft^2, yd^2, and m^2, respectively).

EXAMPLE 6   **a.** What is the circumference of a circle with diameter 14 feet? Round to the nearest tenth of a foot.   **b.** What is the area of the circle? Round to the nearest tenth of a square foot.

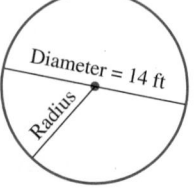

Diameter = 14 ft

Radius

Strategy   To find the circumference and area of the circle, we will substitute the proper values into the formulas $C = \pi D$ and $A = \pi r^2$ and find C and A.

Why   The variable C represents the unknown circumference of the circle and A represents the unknown area.

Solution

a. Recall that the circumference of a circle is the distance around it. To find the circumference C of a circle with diameter D equal to 14 ft, we proceed as follows.

$$C = \pi D$$ This is the formula for the circumference of a circle. πD means $\pi \cdot D$.

$$C = \pi(14)$$ Substitute 14 for D, the diameter of the circle.

$$= 14\pi$$ The exact circumference of the circle is 14π.

$$\approx 43.98229715$$ To use a scientific calculator to approximate the circumference, enter $\boxed{\pi}$ $\boxed{\times}$ $\boxed{14}$ $\boxed{=}$. If you do not have a calculator, use 3.14 as an approximation of π. (Answers may vary slightly depending on which approximation of π is used.)

The circumference is exactly 14π ft. Rounded to the nearest tenth, this is 44.0 ft.

b. The radius r of the circle is one-half the diameter, or 7 feet. To find the area A of the circle, we proceed as follows.

$$A = \pi r^2$$ This is the formula for the area of a circle. πr^2 means $\pi \cdot r^2$.

$$A = \pi(7)^2$$ Substitute 7 for r, the radius of the circle.

$$= 49\pi$$ Evaluate the exponential expression: $7^2 = 49$. The exact area is 49π ft^2.

$$\approx 153.93804$$ To use a calculator to approximate the area, enter 49 $\boxed{\times}$ $\boxed{\pi}$ $\boxed{=}$.

The area is exactly 49π ft^2. To the nearest tenth, the area is 153.9 ft^2.

> **Self Check 6**   Find the circumference of a circle with radius 10 inches. Round to the nearest hundredth of an inch.
>
> *Now Try*   **Problem 28**

The **volume** of a three-dimensional geometric solid is the amount of space it encloses. Volume is measured in cubic units, such as cubic inches, cubic feet, and cubic meters (written as in.^{3}, ft^3, and m^3, respectively).

EXAMPLE 7   Find the volume of the cylinder. Round to the nearest tenth of a cubic centimeter.

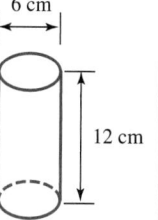

6 cm

12 cm

Strategy   To find the volume of the cylinder, we will substitute the proper values into the formula $V = \pi r^2 h$ and find V.

Why   The variable V represents the unknown volume.

Solution   Since the radius of a circle is one-half its diameter, the radius r of the circular base of the cylinder is $\frac{1}{2}(6 \text{ cm}) = 3$ cm. The height h of the cylinder is 12 cm. To find volume V of the cylinder, we proceed as follows.

Circle formulas

$D = 2r$ (diameter)

$r = \dfrac{1}{2}D$ (radius)

$C = 2\pi r = \pi D$ (circumference)

$A = \pi r^2$ (area)

Notation

When an approximation of π is used in a calculation, it produces an approximate answer. Remember to use an *is approximately equal to* symbol $\approx$ in your solution to show that.

Volume formulas

$V = lwh$ (rectangular solid)

$V = s^3$ (cube)

$V = \dfrac{4}{3}\pi r^3$ (sphere)

$V = \pi r^2 h$ (cylinder)

$V = \dfrac{1}{3}\pi r^2 h$ (cone)

$V = \pi r^2 h$	This is the formula for the volume of a cylinder. $\pi r^2 h$ means $\pi \cdot r^2 \cdot h$.
$V = \pi(3)^2(12)$	Substitute 3 for r and 12 for h.
$= \pi(9)(12)$	Evaluate the exponential expression.
$= 108\pi$	Multiply. The exact volume is 108π cm^3.
≈ 339.2920066	Use a calculator to approximate the volume.

To the nearest tenth, the volume is 339.3 cubic centimeters. This can be written as 339.3 cm^3.

Self Check 7   Find the volume of a cone whose base has radius 12 meters and whose height is 9 meters. Round to the nearest tenth of a cubic meter. Use the formula $V = \frac{1}{3}\pi r^2 h$.

Now Try   **Problem 29**

4 **Solve for a Specified Variable.**

The Language of Algebra
The word *specified* is a form of the word *specify*, which means to select something for a purpose. Here, we select a variable for the purpose of solving for it.

Suppose a shopper wishes to calculate the markup m on several items, knowing their retail price r and their cost c to the merchant. It would take a lot of time to substitute values for r and c into the formula for retail price $r = c + m$ and then repeatedly solve for m. A better way is to solve the formula for m first, substitute values for r and c, and then compute m directly.

To **solve a formula for a specified variable** means to isolate that variable on one side of the equation, with all other variables and constants on the opposite side.

EXAMPLE 8   Solve the formula for retail price, $r = c + m$ for m.

Strategy   To solve for m, we will focus on it as if it is the only variable in the equation. We will use a strategy similar to that used to solve linear equations in one variable to isolate m on one side. (See page 125 if you need to review the strategy.)

Why   We can solve the formula as if it were an equation in one variable because all the other variables are treated as if they were numbers (constants).

Solution

The Language of Algebra
We say that the formula is *solved for m* because m is alone on one side of the equation and the other side does not contain m.

	To solve for m, we will isolate m on this side of the equation.
$r = c + m$	
$r - c = c + m - c$	To isolate m, undo the addition of c by subtracting c from both sides.
$r - c = m$	Simplify the right side: $c - c = 0$.
$m = r - c$	Reverse the sides of the equation so that m is on the left.

Self Check 8   Solve the formula for profit, $p = r - c$, for r.

Now Try   **Problem 31**

EXAMPLE 9   Solve $A = \frac{1}{2}bh$ for b.

Strategy   To solve for b, we will treat b as the only variable in the equation and use properties of equality to isolate it on one side. We will treat the other variables as if they were numbers (constants).

Why   To solve for a specified variable means to isolate it on one side of the equation.

Solution   We use the same steps to solve an equation for a specified variable that we use to solve equations with only one variable.

$$A = \frac{1}{2}bh$$ To solve for b, we will isolate b on this side of the equation.

$$2 \cdot A = 2 \cdot \frac{1}{2}bh$$ To clear the equation of the fraction, multiply both sides by 2.

$$2A = bh$$ Simplify.

$$\frac{2A}{h} = \frac{bh}{h}$$ To isolate b, undo the multiplication by h by dividing both sides by h.

$$\frac{2A}{h} = b$$ On the right side, remove the common factor of h: $\frac{b\overset{1}{\cancel{h}}}{\underset{1}{\cancel{h}}} = b$.

$$b = \frac{2A}{h}$$ Reverse the sides of the equation so that b is on the left.

Self Check 9   Solve $A = \frac{1}{2}r^2a$ for a.

Now Try   **Problem 37**

EXAMPLE 10   Solve $P = 2l + 2w$ for l.

Strategy   To solve for l, we will treat l as the only variable in the equation and use properties of equality to isolate it on one side. We will treat the other variables as if they were numbers (constants).

Why   To solve for a specified variable means to isolate it on one side of the equation.

Solution

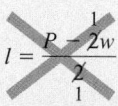
$$P = 2l + 2w$$ To solve for l, we will isolate l on this side of the equation.

$$P - 2w = 2l + 2w - 2w$$ To undo the addition of $2w$, subtract $2w$ from both sides.

$$P - 2w = 2l$$ Combine like terms: $2w - 2w = 0$.

$$\frac{P - 2w}{2} = \frac{2l}{2}$$ To isolate l, undo the multiplication by 2 by dividing both sides by 2.

$$\frac{P - 2w}{2} = l$$ Simplify the right side.

We can write the result as $l = \frac{P - 2w}{2}$.

Self Check 10   Solve $P = 2l + 2w$ for w.

Now Try   **Problem 45**

EXAMPLE 11   In Chapter 3, we will work with equations that involve the variables x and y, such as $3x + 2y = 4$. Solve this equation for y.

Strategy   To solve for y, we will treat y as the only variable in the equation and use properties of equality to isolate it on one side.

Why   To solve for a specified variable means to isolate it on one side of the equation.

Solution

To solve for y, we will isolate y on this side of the equation.

$$3x + 2y = 4$$

$$3x + 2y - 3x = 4 - 3x \qquad \text{To eliminate } 3x \text{ on the left side, subtract } 3x \text{ from both sides.}$$

$$2y = 4 - 3x \qquad \text{Combine like terms: } 3x - 3x = 0.$$

$$\frac{2y}{2} = \frac{4 - 3x}{2} \qquad \text{To isolate } y, \text{ undo the multiplication by 2 by dividing both sides by 2.}$$

$$y = \frac{4}{2} - \frac{3x}{2} \qquad \text{Write } \frac{4 - 3x}{2} \text{ as the difference of two fractions with like denominators, } \frac{4}{2} \text{ and } \frac{3x}{2}.$$

$$y = 2 - \frac{3}{2}x \qquad \text{Simplify: } \frac{4}{2} = 2. \text{ Write } \frac{3x}{2} \text{ as } \frac{3}{2}x.$$

$$y = -\frac{3}{2}x + 2 \qquad \text{On the right side, write the } x \text{ term first.}$$

Success Tip

When solving for a specified variable, there is often more than one way to express the result. Keep this in mind when you are comparing your answers with those in the back of the text.

Self Check 11   Solve $x + 3y = 12$ for y.

Now Try   **Problem 47**

EXAMPLE 12   Solve $V = \pi r^2 h$ for r^2.

Strategy   To solve for r^2, we will treat it as the only variable expression in the equation and isolate it on one side.

Why   To solve for a specified variable means to isolate it on one side of the equation.

Solution

To solve for r^2, we will isolate r^2 on this side of the equation.

$$V = \pi r^2 h$$

$$\frac{V}{\pi h} = \frac{\pi r^2 h}{\pi h} \qquad \pi r^2 h \text{ means } \pi \cdot r^2 \cdot h. \text{ To isolate } r^2, \text{ undo the multiplication by } \pi \text{ and } h \text{ on the right side by dividing both sides by } \pi h.$$

$$\frac{V}{\pi h} = r^2 \qquad \text{On the right side, remove the common factors of } \pi \text{ and } h: \dfrac{\overset{1}{\cancel{\pi}} r^2 \overset{1}{\cancel{h}}}{\underset{1}{\cancel{\pi}} \underset{1}{\cancel{h}}} = r^2.$$

$$r^2 = \frac{V}{\pi h} \qquad \text{Reverse the sides of the equation so that } r^2 \text{ is on the left.}$$

Self Check 12 Solve $V = lwh$ for w.
Now Try Problem 55

ANSWERS TO SELF CHECKS **1.** $1,510.50 **2.** 2.5 yr **3.** 1.25 min **4.** $-283°F$ **5.** 230 ft
6. 62.83 in. **7.** 1,357.2 m^3 **8.** $r = p + c$ **9.** $a = \frac{24}{r^2}$ **10.** $w = \frac{P - 2l}{2}$ **11.** $y = 4 - \frac{1}{3}x$ or
$y = -\frac{1}{3}x + 4$ **12.** $w = \frac{V}{lh}$

STUDY SET
2.4

VOCABULARY

Fill in the blanks.

1. A _____ is an equation that is used to state a known relationship between two or more variables.

2. The distance around a plane geometric figure is called its _____, and the amount of surface that it encloses is called its ____.

3. The _____ of a three-dimensional geometric solid is the amount of space it encloses.

4. The formula $a = P - b - c$ is _____ for a because a is isolated on one side of the equation and the other side does not contain a.

CONCEPTS

5. Use variables to write the formula relating:
   a. Time, distance, rate
   b. Markup, retail price, cost
   c. Costs, revenue, profit
   d. Interest rate, time, interest, principal

6. Complete the table.

	Principal ·	rate ·	time =	interest
Account 1	$2,500	5%	2 yr	
Account 2	$15,000	4.8%	1 yr	

7. Complete the table to find how far light and sound travel in 60 seconds. (*Hint:* mi/sec means miles per second.)

	Rate	· time =	distance
Light	186,282 mi/sec	60 sec	
Sound	1,088 ft/sec	60 sec	

8. Determine which concept (perimeter, area, or volume) should be used to find each of the following. Then determine which unit of measurement, ft, ft^2, or ft^3, would be appropriate.
   a. The amount of storage in a freezer
   b. The amount of ground covered by a sleeping bag lying on the floor
   c. The distance around a dance floor

NOTATION

Complete the solution.

9. Solve $Ax + By = C$ for y.

$$Ax + By = C$$
$$Ax + By - \quad = C - $$
$$By = C - Ax$$
$$\frac{By}{\quad} = \frac{C - Ax}{\quad}$$
$$y = \frac{C - Ax}{\quad}$$

10. a. Approximate 98π to the nearest hundredth.
   b. In the formula $V = \pi r^2 h$, what does r represent? What does h represent?
   c. What does 45°C mean?
   d. What does 15°F mean?

GUIDED PRACTICE

Use a formula to solve each problem. See Example 1.

11. HOLLYWOOD As of 2006, the movie *Titanic* had brought in $1,835 million worldwide and made a gross profit of $1,595 million. What did it cost to make the movie?

12. VALENTINE'S DAY   Find the markup on a dozen roses if a florist buys them wholesale for $12.95 and sells them for $47.50.

13. SERVICE CLUBS   After expenses of $55.15 were paid, a Rotary Club donated $875.85 in proceeds from a pancake breakfast to a local health clinic. How much did the pancake breakfast gross?

14. NEW CARS   The factory invoice for a minivan shows that the dealer paid $16,264.55 for the vehicle. If the sticker price of the van is $18,202, how much over factory invoice is the sticker price?

See Example 2.

15. ENTREPRENEURS   To start a mobile dog-grooming service, a woman borrowed $2,500. If the loan was for 2 years and the amount of interest was $175, what simple interest rate was she charged?

16. SAVINGS   A man deposited $5,000 in a credit union paying 6% simple interest. How long will the money have to be left on deposit to earn $6,000 in interest?

17. LOANS   A student borrowed some money from his father at 2% simple interest to buy a car. If he paid his father $360 in interest after 3 years, how much did he borrow?

18. BANKING   Three years after opening an account that paid simple interest of 6.45% annually, a depositor withdrew the $3,483 in interest earned. How much money was left in the account?

See Example 3.

19. SWIMMING   In 1930, a man swam down the Mississippi River from Minneapolis to New Orleans, a total of 1,826 miles. He was in the water for 742 hours. To the nearest tenth, what was his average swimming rate?

20. PARADES   Rose Parade floats travel down the 5.5-mile-long parade route at a rate of 2.5 mph. How long will it take a float to complete the route if there are no delays?

21. HOT-AIR BALLOONS   If a hot-air balloon travels at an average of 37 mph, how long will it take to fly 166.5 miles?

22. AIR TRAVEL   An airplane flew from Chicago to San Francisco in 3.75 hours. If the cities are 1,950 miles apart, what was the average speed of the plane?

See Example 4.

23. FRYING FOODS   One of the most popular cookbooks in U.S. history, *The Joy of Cooking,* recommends frying foods at 365°F for best results. Convert this to degrees Celsius.

24. FREEZING POINTS   Saltwater has a much lower freezing point than freshwater does. For saltwater that is as saturated as much it can possibly get (23.3% salt by weight), the freezing point is −5.8°F. Convert this to degrees Celsius.

25. BIOLOGY   Cryobiologists freeze living matter to preserve it for future use. They can work with temperatures as low as −270°C. Change this to degrees Fahrenheit.

26. METALLURGY   Change 2,212°C, the temperature at which silver boils, to degrees Fahrenheit. Round to the nearest degree.

See Examples 5–7. If you do not have a calculator, use 3.14 as an approximation of π. Answers may vary slightly depending on which approximation of π is used.

27. ENERGY SAVINGS   One hundred inches of foam weather stripping tape was placed around the perimeter of a rectangular-shaped window. If the length of the window is 30 inches, what is its width?

28. RUGS   Find the amount of floor area covered by a circular throw rug that has a radius of 15 inches. Round to the nearest square inch.

29. STRAWS   Find the volume of a 150 millimeter-long drinking straw that has an inside diameter of 4 millimeters. Round to the nearest cubic millimeter.

30. RUBBER BANDS   The world's largest rubber band ball is $5\frac{1}{2}$ ft tall and was made in 2006 by Steve Milton of Eugene, Oregon. Find the volume of the ball. Round to the nearest cubic foot.   (*Hint:* The formula for the volume of a sphere is $V = \frac{4}{3}\pi r^3$.)

Solve each formula for the specified variable (or expression). See Examples 8–12.

31. $r = c + m$ for c

32. $p = r - c$ for c

33. $P = a + b + c$ for b

34. $a + b + c = 180$ for a

35. $E = IR$ for R

36. $d = rt$ for t

37. $V = lwh$ for l

38. $I = Prt$ for r

39. $C = 2\pi r$ for r

40. $V = \pi r^2 h$ for h

41. $V = \frac{1}{3}Bh$ for h

42. $C = \frac{1}{7}Rt$ for R

43. $w = \frac{s}{f}$ for f

44. $P = \frac{ab}{c}$ for c

45. $T = 2r + 2t$ for r

46. $y = mx + b$ for x

47. $Ax + By = C$ for x

48. $A = P + Prt$ for t

49. $K = \frac{1}{2}mv^2$ for m

50. $V = \frac{1}{3}\pi r^2 h$ for h

51. $A = \dfrac{a + b + c}{3}$ for c

52. $x = \dfrac{a + b}{2}$ for b

53. $2E = \dfrac{T - t}{9}$ for t

54. $D = \dfrac{C - s}{n}$ for s

55. $s = 4\pi r^2$ for r^2

56. $E = mc^2$ for c^2

57. $Kg = \dfrac{wv^2}{2}$ for v^2

58. $c^2 = a^2 + b^2$ for a^2

59. $V = \dfrac{4}{3}\pi r^3$ for r^3

60. $A = \dfrac{\pi r^2 S}{360}$ for r^2

61. $\dfrac{M}{2} - 9.9 = 2.1B$ for M

62. $\dfrac{G}{0.5} + 16r = -8t$ for G

63. $S = 2\pi rh + 2\pi r^2$ for h

64. $c = bn + 16t^2$ for t^2

65. $3x + y = 9$ for y

66. $-5x + y = 4$ for y

67. $-x + 3y = 9$ for y

68. $5y - x = 25$ for y

69. $4y + 16 = -3x$ for y

70. $6y + 12 = -5x$ for y

71. $A = \dfrac{1}{2}h(b + d)$ for b

72. $C = \dfrac{1}{4}s(t - d)$ for t

73. $\dfrac{7}{8}c + w = 9$ for c

74. $\dfrac{3}{4}m - t = 5b$ for m

APPLICATIONS

75. from Campus to Careers
Automotive Service Technician

If your automobile engine is making a knocking sound, a service technician will probably tell you that the octane rating of the gasoline that you are using is too low. Octane rating numbers are printed on the yellow decals on gas pumps. The formula used to calculate them is

© Jeremy Hardie/Getty Images

$$\text{Pump octane number} = \dfrac{(R + M)}{2}$$

where R is the *research octane number*, which is determined with a test engine running at a low speed and M is the *motor octane number*, which is determined with a test engine running at a higher speed. Calculate the octane rating for the following three grades of gasoline.

Gasoline grade	R	M	Octane rating
Unleaded	92	82	
Unleaded plus	95	83	
Premium	97	85	

76. PROPERTIES OF WATER The boiling point and the freezing point of water are to be given in both degrees Celsius and degrees Fahrenheit on the thermometer. Find the missing degree measures.

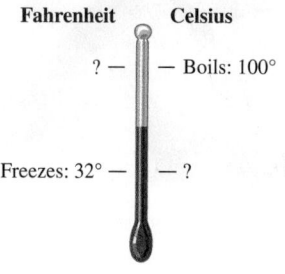

Fahrenheit   Celsius
? —   — Boils: 100°
Freezes: 32° —   — ?

77. AVON PRODUCTS Complete the financial statement.

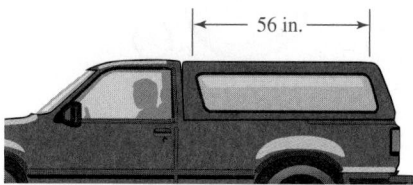

Income statement (dollar amounts in millions)	Quarter ending Sep 04	Quarter ending Sep 05
Revenue	1,806.2	1,886.0
Cost of goods sold	1,543.4	1,638.9
Operating profit		

Source: Avon Products, Inc.

78. CREDIT CARDS The finance charge that a student pays on his credit card is 19.8% APR (annual percentage rate). Determine the finance charges (interest) the student would have to pay if the account's average balance for the year was $2,500.

79. CAMPERS The perimeter of the window of the camper shell is 140 in. Find the length of one of the shorter sides of the window.

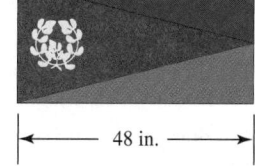

← 56 in. →

80. FLAGS The flag of Eritrea, a country in east Africa, is shown. The perimeter of the flag is 160 inches.
a. What is the width of the flag?
b. What is the area of the red triangular region of the flag?

← 48 in. →

81. KITES 650 in.² of nylon cloth were used to make the kite shown. If its height is 26 inches, what is the wingspan?

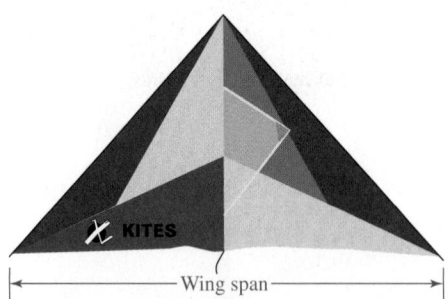

KITES

Wing span

82. MEMORIALS The Vietnam Veterans Memorial is a black granite wall recognizing the more than 58,000 Americans who lost their lives or remain missing. Find the total area of the two triangular-shaped surfaces on which the names are inscribed.

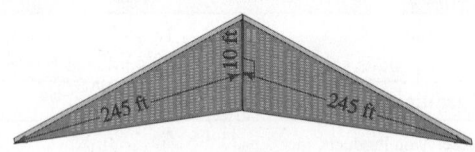

10 ft

245 ft   245 ft

83. WHEELCHAIRS Find the diameter of the rear wheel and the radius of the front wheel.

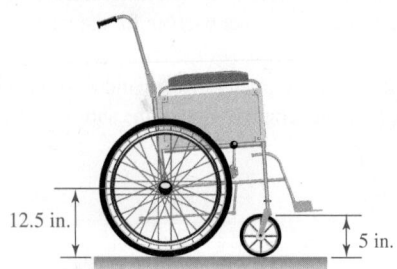

12.5 in.

5 in.

84. ARCHERY The diameter of a standard archery target used in the Olympics is 48.8 inches. Find the area of the target. Round to the nearest square inch.

85. BULLS-EYE See Exercise 84. The diameter of the center yellow ring of a standard archery target is 4.8 inches. What is the area of the bulls-eye? Round to the nearest tenth of a square inch.

86. GEOGRAPHY The circumference of the Earth is about 25,000 miles. Find its diameter to the nearest mile.

87. HORSES A horse trots in a circle around its trainer at the end of a 28-foot-long rope. Find the area of the circle that is swept out. Round to the nearest square foot.

88. YO-YOS How far does a yo-yo travel during one revolution of the "around the world" trick if the length of the string is 21 inches?

89. WORLD HISTORY The Inca Empire (1438–1533) was centered in what is now called Peru. A special feature of Inca architecture was the trapezoid-shaped windows and doorways. A standard Inca window was 70 cm (centimeters) high, 50 cm at the base and 40 cm at the top. Find the area of a window opening. [*Hint:* The formula for the area of a trapezoid is $A = \frac{1}{2}(\text{height})(\text{upperbase} + \text{lowerbase})$].

© DIOMEDIA/Almay

90. HAMSTER HABITATS Find the amount of space in the tube.

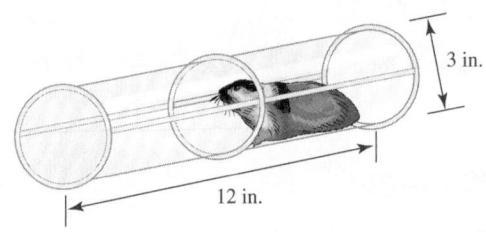

3 in.

12 in.

91. TIRES The road surface footprint of a sport truck tire is approximately rectangular. If the area of the footprint is 45 in.², about how wide is the tire?

$7\frac{1}{2}$ in.

92. SOFTBALL The strike zone in fast-pitch softball is between the batter's armpit and the top of her knees, as shown. If the area of the strike zone for this batter is 442 in.², what is the width of home plate?

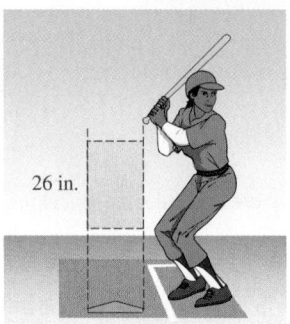

26 in.

93. FIREWOOD The cord of wood shown occupies a volume of 128 ft³. How long is the stack?

4 ft

4 ft

94. TEEPEES The teepees constructed by the Blackfoot Indians were cone-shaped tents about 10 feet high and about 15 feet across at the ground. Estimate the volume of a teepee with these dimensions, to the nearest cubic foot.

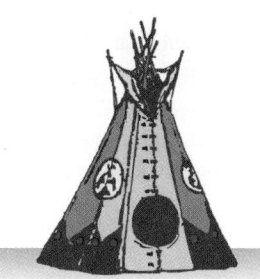

95. IGLOOS During long journeys, some Canadian Eskimos built winter houses of snow blocks stacked in the dome shape shown. Estimate the volume of an igloo having an interior height of 5.5 feet to the nearest cubic foot.

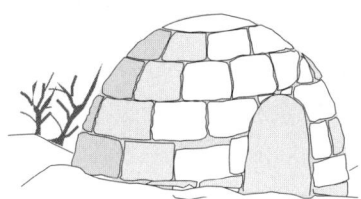

96. PYRAMIDS The Great Pyramid at Giza in northern Egypt is one of the most famous works of architecture in the world. Find its volume to the nearest cubic foot.

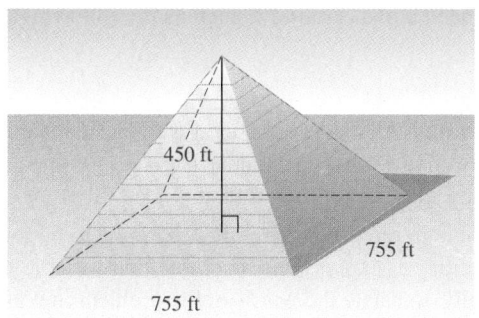

450 ft

755 ft

755 ft

97. COOKING If the fish shown in the illustration in the next column is 18 inches long, what is the area of the grill? Round to the nearest square inch.

98. SKATEBOARDING A half-pipe ramp is in the shape of a semicircle with a radius of 8 feet. To the nearest tenth of a foot, what is the length of the arc that the rider travels on the ramp?

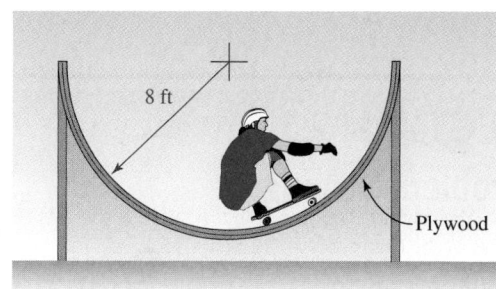

8 ft

Plywood

99. PULLEYS The approximate length L of a belt joining two pulleys of radii r and R feet with centers D feet apart is given by the formula $L = 2D + 3.25(r + R)$. Solve the formula for D.

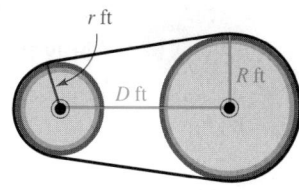

r ft

D ft

R ft

100. THERMODYNAMICS The Gibbs free-energy function is given by $G = U - TS + pV$. Solve this formula for the pressure p.

WRITING

101. After solving $A = B + C + D$ for B, a student compared her answer with that at the back of the textbook. Could this problem have two different-looking answers? Explain why or why not.

Student's answer: $B = A - C - D$

Book's answer: $B = A - D - C$

102. A student solved $x + 5c = 3c + a$ for c. His answer was
$c = \frac{3c + a - x}{5}$ for c. Explain why the equation is not solved
for c.

103. Explain the difference between what perimeter measures and what area measures.

104. Explain the error made below.

$$y = \frac{\overset{1}{\cancel{3x + 2}}}{\underset{1}{\cancel{2}}}$$

REVIEW

105. Find 82% of 168.

106. 29.05 is what percent of 415?

107. What percent of 200 is 30?

108. A woman bought a coat for $98.95 and some gloves for $7.95. If the sales tax was 6%, how much did the purchase cost her?

CHALLENGE PROBLEMS

109. In mathematics, letters from the Greek alphabet are often used as variables. Solve the following equation for α (read as "alpha"), the first letter of the Greek alphabet.

$$-7(\alpha - \beta) - (4\alpha - \theta) = \frac{\alpha}{2}$$

110. When a car of mass collides with a wall, the energy of the collision is given by the formula $E = \frac{1}{2}mv^2$. Compare the energy of two collisions: a car striking a wall at 30 mph, and at 60 mph. Then complete this sentence: Although the speed is only twice as fast, the energy is ___ times greater.

SECTION 2.5
Problem Solving

Objectives

1 Apply the steps of a problem-solving strategy.

2 Solve consecutive integer problems.

3 Solve geometry problems.

In this section, you will see that algebra is a powerful tool that can be used to solve a wide variety of real-world problems.

1 **Apply the Steps of a Problem-Solving Strategy.**

To become a good problem solver, you need a plan to follow, such as the following five-step strategy.

Strategy for Problem Solving

1. **Analyze the problem** by reading it carefully to understand the given facts. What information is given? What are you asked to find? What vocabulary is given? Often, a diagram or table will help you visualize the facts of the problem.

2. **Form an equation** by picking a variable to represent the numerical value to be found. Then express all other unknown quantities as expressions involving that variable. Key words or phrases can be helpful. Finally, translate the words of the problem into an equation.

3. **Solve the equation.**

4. **State the conclusion.**

5. **Check the result** using the original wording of the problem, not the equation that was formed in step 2.

EXAMPLE 1 *California Coastline.* The first part of California's magnificent 17-Mile Drive begins at the Pacific Grove entrance and continues to Seal Rock. It is 1 mile longer than the second part of the drive, which extends from Seal Rock to the Lone Cypress. The final part of the tour winds through the Monterey Peninsula, eventually returning to the entrance. This part of the drive is 1 mile longer than four times the length of the second part. How long is each part of 17-Mile Drive?

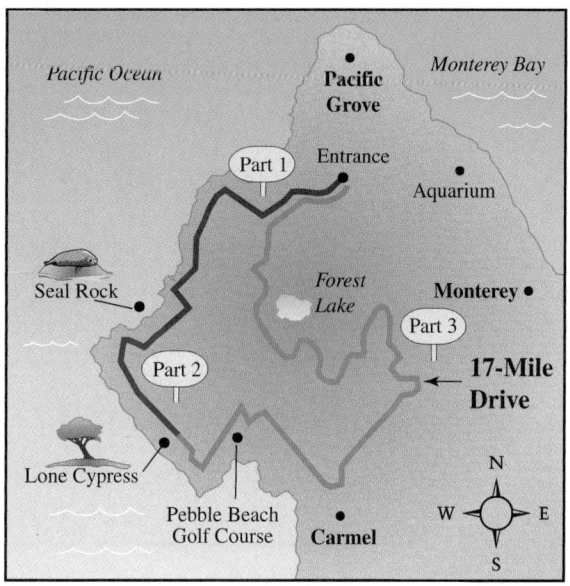

Analyze the Problem The drive is composed of three parts. We need to find the length of each part. We can straighten out the winding 17-Mile Drive and model it with a line segment.

Form an Equation Since the lengths of the first part and of the third part of the drive are related to the length of the second part, we will let x represent the length of that part. We then express the other lengths in terms of x. Let

x = the length of the second part of the drive

$x + 1$ = the length of the first part of the drive

$4x + 1$ = the length of the third part of the drive

> **Caution**
> For this problem, one common mistake is to let
>
> x = the length of each part of the drive
>
> The three parts of the drive have different lengths; x cannot represent three different distances.

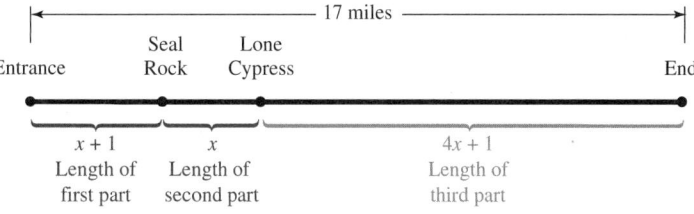

The sum of the lengths of the three parts must be 17 miles.

The length of part 1	plus	the length of part 2	plus	the length of part 3	equals	the total length.
$x + 1$	+	x	+	$4x + 1$	=	17

Solve the Equation

$$x + 1 + x + 4x + 1 = 17$$

$6x + 2 = 17$	Combine like terms: $x + x + 4x = 6x$ and $1 + 1 = 2$.
$6x = 15$	To undo the addition of 2, subtract 2 from both sides.
$\dfrac{6x}{6} = \dfrac{15}{6}$	To isolate x, undo the multiplication by 6 by dividing both sides by 6.
$x = 2.5$	Do the divisions.

Recall that x represents the length of the second part of the drive. To find the lengths of the first and third parts, we evaluate $x + 1$ and $4x + 1$ for $x = 2.5$.

First part of drive	*Third part of drive*	
$x + 1 = 2.5 + 1$	$4x + 1 = 4(2.5) + 1$	Substitute 2.5 for x.
$= 3.5$	$= 11$	

State the Conclusion   The first part of the drive is 3.5 miles long, the second part is 2.5 miles long, and the third part is 11 miles long.

Check the Result   Since 3.5 mi + 2.5 mi + 11 mi = 17 mi, the answers check.

Now Try   **Problem 13**

EXAMPLE 2   *Computer Logos.* A trucking company had their logo embroidered on the front of baseball caps. They were charged $8.90 per hat plus a one time set up fee of $25. If the project cost $559, how many hats were embroidered?

Analyze the Problem

- It cost $8.90 to have a logo embroidered on a hat.
- The set up charge was $25.
- The project cost $559.
- We are to find the number of hats that were embroidered.

Form an Equation   Let $x =$ the number of hats that were embroidered. If x hats are embroidered, at a cost of $8.90 per hat, the cost to embroider all of the hats is $x \cdot \$8.90$ or $\$8.90x$. Now we translate the words of the problem into an equation.

The cost to embroider one hat	times	the number of hats	plus	the set up charge	equals	the total cost.
8.90	$\cdot$	x	$+$	25	$=$	559

Solve the Equation

$8.90x + 25 = 559$	
$8.90x = 534$	To undo the addition of 25, subtract 25 from both sides.
$\dfrac{8.90x}{8.90} = \dfrac{534}{8.90}$	To isolate x, undo the multiplication by 8.90 by dividing both sides by 8.90.
$x = 60$	Do the divisions.

State the Conclusion   The company had 60 hats embroidered.

Check the Result   The cost to embroider 60 hats is 60($8.90) = $534. When the $25 set up charge is added, we get $534 + $25 = $559. The answer checks.

 Now Try   **Problem 21**

EXAMPLE 3   ***Auctions.***   A classic car owner is going to sell his 1960 Chevy Impala at an auction. He wants to make $46,000 after paying an 8% commission to the auctioneer. For what selling price (called the "hammer price") will the car owner make this amount of money?

Analyze the Problem   When the commission is subtracted from the selling price of the car, the owner wants to have $46,000 left.

Form an Equation   Let x = the selling price of the car. The amount of the commission is 8% of x, or $0.08x$. Now we translate the words of the problem to an equation.

The Language of Algebra
Phrases such as *should be* or *will be* translate to an equal symbol =.

The selling price of the car	minus	the auctioneer's commission	should be	$46,000.
x	$-$	$0.08x$	$=$	46,000

Solve the Equation

$$x - 0.08x = 46,000$$

$$0.92x = 46,000 \qquad \text{Combine like terms: } 1.00x - 0.08x = 0.92x.$$

$$\frac{0.92x}{0.92} = \frac{46,000}{0.92} \qquad \text{To isolate } x \text{, undo the multiplication by 0.92 by dividing both sides by 0.92.}$$

$$x = 50,000 \qquad \text{Do the divisions.}$$

State the Conclusion   The owner will make $46,000 if the car sells for $50,000.

Check the Result   An 8% commission on $50,000 is 0.08($50,000) = $4,000. The owner will keep $50,000 − $4,000 = $46,000. The answer checks.

 Now Try   **Problem 27**

---

❷   **Solve Consecutive Integer Problems.**

Integers that follow one another, such as 15 and 16, are called **consecutive integers.** They are 1 unit apart. **Consecutive even integers** are even integers that differ by 2 units, such as 12 and 14. Similarly, **consecutive odd integers** differ by 2 units, such as 9 and 11. When solving consecutive integer problems, if we let x = the first integer, then

- two consecutive integers are x and $x + 1$
- two consecutive even integers are x and $x + 2$
- two consecutive odd integers are x and $x + 2$

EXAMPLE 4

U.S. History.   The year George Washington was chosen president and the year the Bill of Rights went into effect are consecutive odd integers whose sum is 3,580. Find the years.

Analyze the Problem   We need to find two consecutive odd integers whose sum is 3,580. From history, we know that Washington was elected president first and the Bill of Rights went into effect later.

Form an Equation   Let x = the first odd integer (the date when Washington was chosen president). The next odd integer is 2 *greater than* x, therefore $x + 2$ = the next larger odd integer (the date when the Bill of Rights went into effect).

The first odd integer	plus	the second odd integer	is	3,580.
x	$+$	$x + 2$	$=$	3,580

Solve the Equation

$$x + x + 2 = 3,580$$

$2x + 2 = 3,580$ Combine like terms: $x + x = 2x$.

$2x = 3,578$ To undo the addition of 2, subtract 2 from both sides.

$x = 1,789$ To isolate x, undo the multiplication by 2 by dividing both sides by 2.

State the Conclusion   George Washington was chosen president in the year 1789. The Bill of Rights went into effect in $1789 + 2 = 1791$.

Check the Result   1789 and 1791 are consecutive odd integers whose sum is $1789 + 1791 = 3,580$. The answers check.

 Now Try   Problem 33

---

3   Solve Geometry Problems.

EXAMPLE 5

Crime Scenes.   Police used 400 feet of yellow tape to fence off a rectangular-shaped lot for an investigation. Fifty less feet of tape was used for each width as for each length. Find the dimensions of the lot.

Analyze the Problem Since the yellow tape surrounded the lot, the concept of perimeter applies. Recall that the formula for the perimeter of a rectangle is $P = 2l + 2w$. We also know that the width of the lot is 50 feet less than the length.

Form an Equation Since the width of the lot is given in terms of the length, we let $l =$ the length of the lot. Then $l - 50 =$ the width. Using the perimeter formula, we have:

2	times	the length	plus	2	times	the width	is	the perimeter.
2	·	l	+	2	·	$(l - 50)$	=	400

Success Tip
When solving geometry problems, a sketch is often helpful.

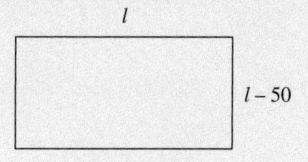

l

$l - 50$

Perimeter = 400 ft

Solve the Equation

$2l + 2(l - 50) = 400$ Write the parentheses so that the entire expression $l - 50$ is multiplied by 2.

$2l + 2l - 100 = 400$ Distribute the multiplication by 2.

$4l - 100 = 400$ Combine like terms: $2l + 2l = 4l$.

$4l = 500$ To undo the subtraction of 100, add 100 to both sides.

$l = 125$ To isolate l, undo the multiplication by 4 by dividing both sides by 4.

State the Conclusion The length of the lot is 125 feet and width is $125 - 50 = 75$ feet.

Check the Result The width (75 feet) is 50 less than the length (125 feet). The perimeter of the lot is $2(125) + 2(75) = 250 + 150 = 400$ feet. The answers check.

 Now Try **Problem 39**

EXAMPLE 6   *Isosceles Triangles.* If the vertex angle of an isosceles triangle is 56°, find the measure of each base angle.

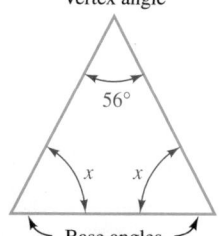

Vertex angle

56°

x x

Base angles

Analyze the Problem An **isosceles triangle** has two sides of equal length, which meet to form the **vertex angle.** In this case, the measurement of the vertex angle is 56°. We can sketch the triangle as shown. The **base angles** opposite the equal sides are also equal. We need to find their measure.

Form an Equation If we let $x =$ the measure of one base angle, the measure of the other base angle is also x. Since the sum of the angles of any triangle is 180°, the sum of the base angles and the vertex angle is 180°. We can use this fact to form the equation.

One base angle	plus	the other base angle	plus	the vertex angle	is	180°.
x	+	x	+	56	=	180

Solve the Equation

$x + x + 56 = 180$

$2x + 56 = 180$ Combine like terms: $x + x = 2x$.

$2x = 124$ To undo the addition of 56, subtract 56 from both sides.

$x = 62$ To isolate x, undo the multiplication by 2 by dividing both sides by 2.

State the Conclusion The measure of each base angle is 62°.

Check the Result Since $62° + 62° + 56° = 180°$, the answer checks.

 Now Try **Problem 43**

STUDY SET
2.5

VOCABULARY

Fill in the blanks.

1. Integers that follow one another, such as 7 and 8, are called
_____ integers.
2. An _____ triangle is a triangle with two sides of the same
length.
3. The equal sides of an isosceles triangle meet to form the
_____ angle. The angles opposite the equal sides are called
_____ angles, and they have equal measures.
4. When asked to find the dimensions of a rectangle, we are to find
its _____ and _____.

CONCEPTS

5. A 17-foot pipe is cut into three sections. The longest section is
three times as long as the shortest, and the middle-sized section
is 2 feet longer than the shortest. Complete the diagram.

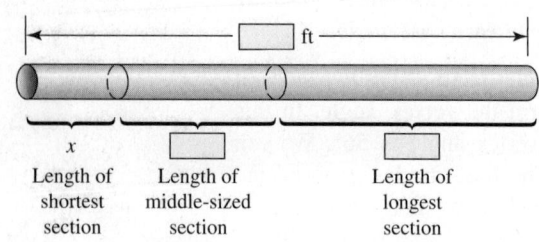

6. It costs $28 per hour to rent a trailer. Write an expression that
represents the cost to rent the trailer for x hours.
7. A realtor is paid a 3% commission on the sale of a house. Write
an expression that represents the amount of the commission if a
house sells for x.
8. The perimeter of the rectangle is 15 feet. Fill in the blanks:
$2(\quad) + 2x = $

x
$5x - 1$

9. What is the sum of the measures of the angles of any triangle?

10. What is x?

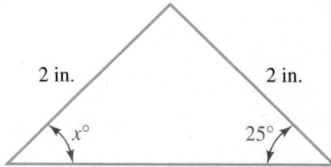

NOTATION

11. **a.** If x represents an integer, write an expression for the next
largest integer.
   b. If x represents an odd integer, write an expression for the
next largest odd integer.
12. What does 45° mean?

GUIDED PRACTICE

See Example 1.

13. A 12-foot board has been cut into two sections, one twice as
long as the other. How long is each section?

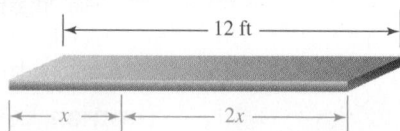

14. The robotic arm will extend a total distance of 18 feet. Find the
length of each section.

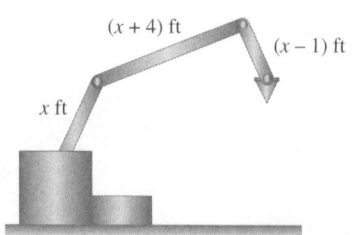

APPLICATIONS

15. NATIONAL PARKS   The Natchez Trace Parkway is a historical 444-mile route from Natchez, Mississippi, to Nashville, Tennessee. A couple drove the Trace in four days. Each day they drove 6 miles more than the previous day. How many miles did they drive each day?

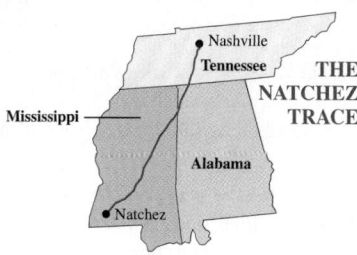

16. TOURING   A rock group plans to travel for a total of 38 weeks, making three concert stops. They will be in Japan for 4 more weeks than they will be in Australia. Their stay in Sweden will be 2 weeks shorter than that in Australia. How many weeks will they be in each country?

17. SOLAR HEATING   One solar panel is 3.4 feet wider than the other. Find the width of each panel.

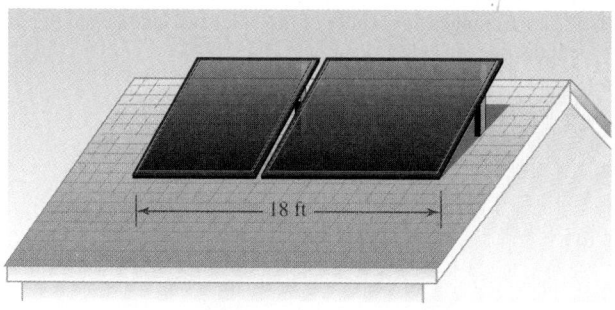

18. ACCOUNTING   Determine the 2005 income of Abercrombie & Fitch Company for each quarter from the data in the graph.

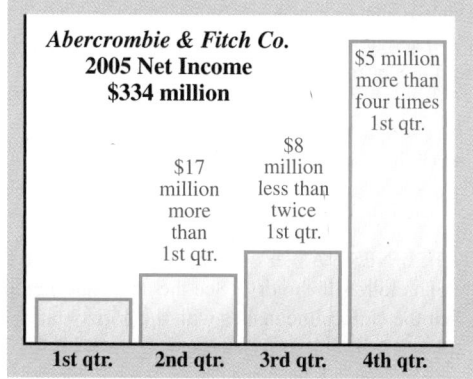

19. COUNTING CALORIES   A slice of pie with a scoop of ice cream has 850 calories. The calories in the pie alone are 100 more than twice the calories in the ice cream alone. How many calories are in each food?

20. WASTE DISPOSAL   Two tanks hold a total of 45 gallons of a toxic solvent. One tank holds 6 gallons more than twice the amount in the other. How many gallons does each tank hold?

21. CONCERTS   The fee to rent a concert hall is $2,250 plus $150 per hour to pay for the support staff. For how many hours can an orchestra rent the hall and stay within a budget of $3,300?

22. TRUCK MECHANICS   An engine repair cost a truck owner $1,185 in parts and labor. If the parts were $690 and the mechanic charged $45 per hour, how many hours did the repair take?

23. FIELD TRIPS   It costs a school $65 a day plus $0.25 per mile to rent a 15-passenger van. If the van is rented for two days, how many miles can be driven on a $275 budget?

24. DECORATIONS   A party supply store charges a set-up fee of $80 plus 35¢ per balloon to make a balloon arch. A business has $150 to spend on decorations for their grand opening. How many balloons can they have in the arch?   (*Hint:* 35¢ = $0.35.)

25. TUTORING   High school students enrolling in a private tutoring program must first take a placement test (cost $25) before receiving tutoring (cost $18.75 per hour). If a family has set aside $400 to get their child extra help, how many hours of tutoring can they afford?

26. DATA CONVERSION   The *Books2Bytes* service converts old print books to Microsoft Word electronic files for $20 per book plus $2.25 per page. If it cost $1,201.25 to convert a novel, how many pages did the novel have?

27. CATTLE AUCTIONS   A cattle rancher is going to sell one of his prize bulls at an auction and would like to make $45,500 after paying a 9% commission to the auctioneer. For what selling price will the rancher make this amount of money?

28. LISTING PRICE   At what price should a home be listed if the owner wants to make $567,000 on its sale after paying a 5.5% real estate commission?

29. SAVINGS ACCOUNTS   The balance in a savings account grew by 5% in one year, to $5,512.50. What was the balance at the beginning of the year?

30. AUTO INSURANCE   Between the years 2000 and 2006, the average cost for auto insurance nationwide grew 27%, to $867. What was the average cost in 2000? Round to the nearest dollar.

Consecutive integer problems

31. SOCCER   Ronaldo of Brazil and Gerd Mueller of Germany rank 1 and 2, respectively, with the most goals scored in World Cup play. The number of goals Ronaldo and Mueller have scored are consecutive integers that total 29. Find the number of goals scored by each man.

32. DICTIONARIES The definitions of the words *job* and *join* are on back-to-back pages in a dictionary. If the sum of those page numbers is 1,411, on what page can the definition of *job* be found?

33. TV HISTORY *Friends* and *Leave It to Beaver* are two of the most popular television shows of all time. The number of episodes of each show are consecutive even integers whose sum is 470. If there are more episodes of *Friends,* how many episodes of each were there?

34. VACATIONS The table shows the average number of vacation days an employed adult receives for selected countries. Complete the table. (The numbers of days are listed in descending order.)

Average Number of Vacation Days per Year	
Country	Days
Italy	42
France	
Germany	
U.S.	13

Consecutive odd integers whose sum is 72.

Source: The World Almanac, 2006.

35. CELEBRITY BIRTHDAYS Elvis Presley, George Foreman, and Kirstie Alley have birthdays (in that order) on consecutive even-numbered days in January. The sum of the calendar dates of their birthdays is 30. Find each birthday.

36. LOCKS The three numbers of the combination for a lock are consecutive integers, and their sum is 81. Find the combination.

Geometry problems

37. TENNIS The perimeter of a regulation singles tennis court is 210 feet and the length is 3 feet less than three times the width. What are the dimensions of the court?

38. SWIMMING POOLS The seawater Orthlieb Pool in Casablanca, Morocco, is the largest swimming pool in the world. With a perimeter of 1,110 meters, this rectangular-shaped pool is 30 meters longer than 6 times its width. Find its dimensions.

39. ART The *Mona Lisa* was completed by Leonardo da Vinci in 1506. The length of the picture is 11.75 inches shorter than twice the width. If the perimeter of the picture is 102.5 inches, find its dimensions.

40. NEW YORK CITY Central Park, which lies in the middle of Manhattan, is rectangular-shaped and has a 6-mile perimeter. The length is 5 times the width. What are the dimensions of the park?

41. ENGINEERING A truss is in the form of an isosceles triangle. Each of the two equal sides is 4 feet shorter than the third side. If the perimeter is 25 feet, find the lengths of the sides.

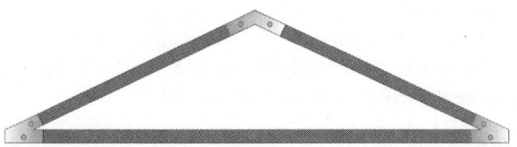

42. FIRST AID A sling is in the shape of an isosceles triangle with a perimeter of 144 inches. The longest side of the sling is 18 inches longer than either of the other two sides. Find the lengths of each side.

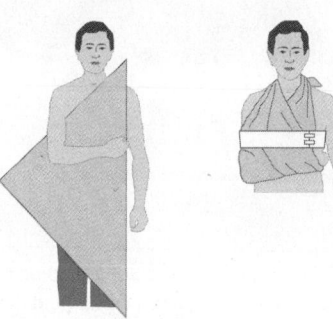

43. TV TOWERS The two guy wires supporting a tower form an isosceles triangle with the ground. Each of the base angles of the triangle is 4 times the third angle (the vertex angle). Find the measure of the vertex angle.

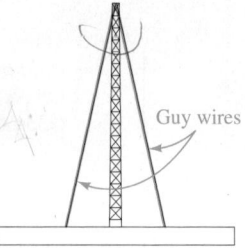

Guy wires

44. CLOTHESLINES A pair of damp jeans are hung in the middle of a clothesline to dry. (See the next page.) Find $x°$, the angle that the clothesline makes with the horizontal.

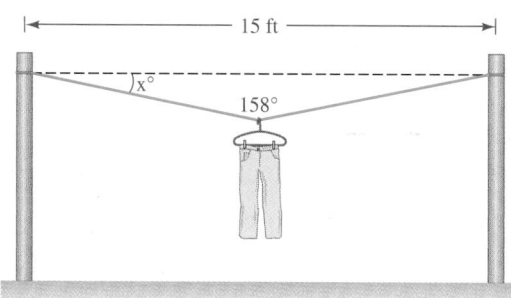

45. MOUNTAIN BICYCLES   For the bicycle frame shown, the angle that the horizontal crossbar makes with the seat support is 15° less than twice the angle at the steering column. The angle at the pedal gear is 25° more than the angle at the steering column. Find these three angle measures.

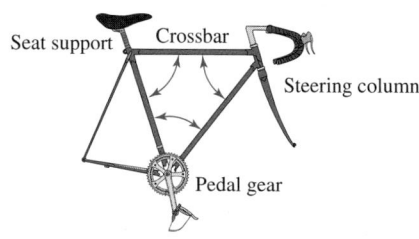

46. TRIANGLES   The measure of ∠1 (read as angle 1) of a triangle is one-half that of ∠2. The measure of ∠3 is equal to the sum of the measures of ∠1 and ∠2. Find each angle measure.

47. COMPLEMENTARY ANGLES   Two angles are called ***complementary angles*** when the sum of their measures is 90°. Find the measures of the complementary angles shown in the illustration.

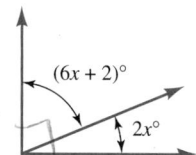

48. SUPPLEMENTARY ANGLES   Two angles are called ***supplementary angles*** when the sum of their measures is 180°. Find the measures of the supplementary angles shown in the illustration.

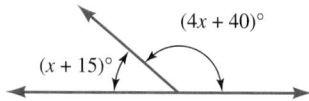

WRITING

49. Create a geometry problem that could be answered by solving the equation $2w + 2(w + 5) = 26$.

50. What information do you need to know to answer the following question?

A business rented a copy machine for $85 per month plus 4¢ for every copy made. How many copies can be made each month?

51. Make a list of words and phrases that translate to an equal symbol $=$.

52. Define the word *strategy*.

REVIEW

Solve.

53. $\dfrac{5}{8}x = -15$

54. $\dfrac{12x + 24}{13} = 36$

55. $\dfrac{3}{4}y = \dfrac{2}{5}y - \dfrac{3}{2}y - 2$

56. $6 + 4(1 - x) = 3(x + 1)$

57. $4.2(y - 4) - 0.6y = -13.2$

58. $16 - 8(b + 4) = 24b + 64$

CHALLENGE PROBLEMS

59. What concept discussed in this section is illustrated by the following day and time?

Two minutes and three seconds past 1 A.M. on the 5th day of April, 2006

60. MANUFACTURING   A company has two machines that make widgets. The production costs are listed below.

Machine 1: Setup cost $400 and $1.70 per widget

Machine 2: Setup cost $500 and $1.20 per widget

Find the number of widgets for which the cost to manufacture them on either machine is the same.

SECTION 2.6
More about Problem Solving

Objectives

① Solve investment problems.
② Solve uniform motion problems.
③ Solve liquid mixture problems.
④ Solve dry mixture problems.
⑤ Solve number-value problems.

In this section, we will solve problems that involve money, motion, and mixtures. Tables are a helpful way to organize the information given in these problems.

① **Solve Investment Problems.**

To find the amount of *simple interest I* an investment earns, we use the formula $I = Prt$, where P is the principal (the amount invested), r is the annual interest rate, and t is the time in years.

EXAMPLE 1   *Paying Tuition.*   A college student invested the $12,000 inheritance he received and decided to use the annual interest earned to pay his tuition cost of $945. The highest rate offered by a bank at that time was 6% annual simple interest. At this rate, he could not earn the needed $945, so he invested some of the money in a riskier, but more profitable, investment offering a 9% return. How much did he invest at each rate?

Analyze the Problem   We know that $12,000 was invested for 1 year at two rates: 6% and 9%. We are asked to find the amount invested at each rate so that the total return would be $945.

Form an Equation   Let $x =$ the amount invested at 6%. Then $12,000 - x =$ the amount invested at 9%. To organize the facts of the problem, we enter the principal, rate, time, and interest earned in a table.

Step 1: List each investment in a row of the table.

Bank			
Riskier Investment			

Step 2: Label the columns using $I = Prt$ reversed and also write Total:.

	P	$\cdot\ r\ \cdot t =$	I
Bank			
Riskier Investment			
		Total:	

Step 3: Enter the rates, times, and total interest.

	P	$\cdot\ r$	$\cdot t =$	I
Bank		0.06	1	
Riskier Investment		0.09	1	
			Total: **945**	

Step 4: Enter each unknown principal.

	P	$\cdot\ r$	$\cdot t =$	I
Bank	x	0.06	1	
Riskier Investment	$12{,}000 - x$	0.09	1	
			Total: 945	

Step 5: In the last column, multiply P, r, and t to obtain expressions for the interest earned.

	P	· r	· t =	I
Bank	x	0.06	1	**0.06x**
Riskier Investment	12,000 − x	0.09	1	**0.09(12,000 − x)**
				Total: 945

← This is x · 0.06 · 1.

← This is (12,000 − x) · 0.09 · 1.

Use the information in this column to form an equation.

The interest earned at 6%	plus	the interest earned at 9%	equals	the total interest.
0.06x	+	[0.09(12,000 − x)]	=	945

Solve the Equation

$$0.06x + 0.09(12,000 − x) = 945$$

$$100[0.06x + 0.09(12,000 − x)] = 100(945)$$ Multiply both sides by 100 to clear the equation of decimals.

$$100(0.06x) + 100(0.09)(12,000 − x) = 100(945)$$ Distribute the multiplication by 100.

$$6x + 9(12,000 − x) = 94,500$$ Do the multiplications by 100.

$$6x + 108,000 − 9x = 94,500$$ Use the distributive property.

$$−3x + 108,000 = 94,500$$ Combine like terms.

$$−3x = −13,500$$ Subtract 108,000 from both sides.

$$x = 4,500$$ To isolate x, divide both sides by −3.

<blockquote>
Success Tip
We can *clear an equation of decimals* by multiplying both sides by a power of 10. Here, we multiply 0.06 and 0.09 by 100 to move each decimal point two places to the right:

$$100(0.06) = 6 \qquad 100(0.09) = 9$$
</blockquote>

<blockquote>
Caution
On the left side of the equation, do not incorrectly distribute the multiplication by 100 over addition **and** multiplication.

$$100[0.06x + 0.09(12,000 − x)]$$
</blockquote>

State the Conclusion   The student invested $4,500 at 6% and $12,000 − $4,500 = $7,500 at 9%.

Check the Result   The first investment earned 0.06($4,500), or $270. The second earned 0.09($7,500), or $675. Since the total return was $270 + $675 = $945, the answers check.

 Now Try   **Problem 17**

2 Solve Uniform Motion Problems.

If we know the rate r at which we will be traveling and the time t we will be traveling at that rate, we can find the distance d traveled by using the formula $d = rt$.

EXAMPLE 2   ***Rescues at Sea.***   A cargo ship, heading into port, radios the Coast Guard that it is experiencing engine trouble and that its speed has dropped to 3 knots (this is 3 sea miles per hour). Immediately, a Coast Guard cutter leaves port and speeds at a rate of 25 knots directly toward the disabled ship, which is 56 sea miles away. How long will it take the Coast Guard to reach the ship? (Sea miles are also called nautical miles.)

Analyze the Problem   We know the *rate* of each ship (25 knots and 3 knots), and we know that they must close a *distance* of 56 sea miles between them. We don't know the *time* it will take to do this.

Form an Equation   Let t = the time it takes the Coast Guard to reach the cargo ship. During the rescue, the ships don't travel at the same rate, but they do travel for the same amount of time. Therefore, t also represents the travel time for the cargo ship.

We enter the rates, the variable t for each time, and the total distance traveled by the ships (56 sea miles) in the table. To fill in the last column, we use the formula $r \cdot t = d$ twice to find an expression for each distance traveled: $25 \cdot t = 25t$ and $3 \cdot t = 3t$.

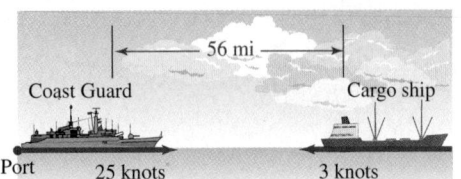

	$r \cdot t = d$		
Coast Guard cutter	25	t	$25t$
Cargo ship	3	t	$3t$
		Total: 56	

Multiply $r \cdot t$ to obtain an expression for each distance traveled.

Use the information in this column to form an equation.

The distance the cutter travels	plus	the distance the ship travels	equals	the original distance between the ships.
$25t$	$+$	$3t$	$=$	56

Solve the Equation

$$25t + 3t = 56$$
$$28t = 56 \qquad \text{Combine like terms: } 25t + 3t = 28t.$$
$$t = \frac{56}{28} \qquad \text{To isolate } t, \text{ divide both sides by 28.}$$
$$t = 2 \qquad \text{Do the division.}$$

State the Conclusion   The ships will meet in 2 hours.

Check the Result   In 2 hours, the Coast Guard cutter travels $25 \cdot 2 = 50$ sea miles, and the cargo ship travels $3 \cdot 2 = 6$ sea miles. Together, they travel $50 + 6 = 56$ sea miles. Since this is the original distance between the ships, the answer checks.

 Now Try   **Problem 27**

EXAMPLE 3   ***Concert Tours.***   While on tour, a country music star travels by bus. Her musical equipment is carried in a truck. How long will it take her bus, traveling 60 mph, to overtake the truck, traveling at 45 mph, if the truck had a $1\frac{1}{2}$-hour head start to her next concert location?

Analyze the Problem   We know the rate of each vehicle (60 mph and 45 mph) and that the truck began the trip $1\frac{1}{2}$ or 1.5 hours earlier than the bus. We need to determine how long it will take the bus to catch up to the truck.

Form an Equation   Let t = the time it takes the bus to overtake the truck. With a 1.5-hour head start, the truck is on the road longer than the bus. Therefore, $t + 1.5$ = the truck's travel time.

We enter each rate and time in the table, and use the formula $r \cdot t = d$ twice to fill in the distance column.

	$r \cdot$	t	=	d
Bus	60	t		$60t$
Truck	45	$t + 1.5$		$45(t + 1.5)$

Enter this information first.

Multiply $r \cdot t$ to obtain an expression for each distance traveled.

Use the information in this column to form an equation.

When the bus overtakes the truck, they will have traveled the same distance.

The distance traveled by the bus	is the same as	the distance traveled by the truck.
$60t$	$=$	$45(t + 1.5)$

Solve the Equation

$60t = 45(t + 1.5)$

$60t = 45t + 67.5$ Distribute the multiplication by 45: 45(1.5) = 67.5.

$15t = 67.5$ Subtract 45t from both sides: 60t − 45t = 15t.

$t = 4.5$ To isolate t, divide both sides by 15: $\frac{67.5}{15} = 4.5$.

State the Conclusion   The bus will overtake the truck in 4.5 or $4\frac{1}{2}$ hours.

Check the Result   In 4.5 hours, the bus travels $60(4.5) = 270$ miles. The truck travels for $1.5 + 4.5 = 6$ hours at 45 mph, which is $45(6) = 270$ miles. Since the distance traveled are the same, the answer checks.

 Now Try **Problem 31**

Success Tip

We used 1.5 hrs for the head start because it is easier to solve $60t = 4.5(t + 1.5)$ than $60t = 45\left(t + 1\frac{1}{2}\right)$.

③ Solve Liquid Mixture Problems.

We now discuss how to solve mixture problems. In the first type, a liquid mixture of a desired strength is made from two solutions with different concentrations.

EXAMPLE 4   *Mixing Solutions.* A chemistry experiment calls for a 30% sulfuric acid solution. If the lab supply room has only 50% and 20% sulfuric acid solutions, how much of each should be mixed to obtain 12 liters of a 30% acid solution?

Analyze the Problem   The 50% solution is too strong and the 20% solution is too weak. We must find how much of each should be combined to obtain 12 liters of a 30% solution.

Success Tip
The strength *(concentration)* of a mixture is always between the strengths of the two solutions used to make it.

Form an Equation   If $x =$ the number of liters of the 50% solution used in the mixture, the remaining $(12 - x)$ liters must be the 20% solution. The amount of pure sulfuric acid in each solution is given by

Amount of solution · strength of the solution = amount of pure sulfuric acid

A table and sketch are helpful in organizing the facts of the problem.

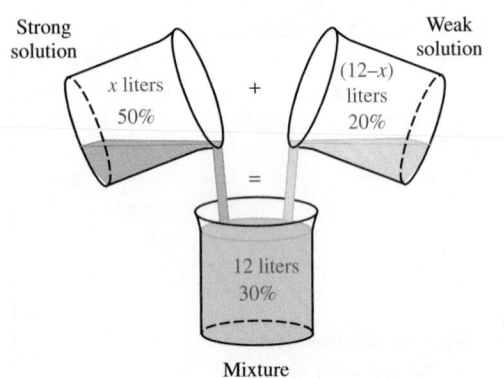

	Amount ·	Strength =	Amount of pure sulfuric acid
Strong	x	0.50	$0.50x$
Weak	$12 - x$	0.20	$0.20(12 - x)$
Mixture	12	0.30	$12(0.30)$

Multiply amount · strength three times to fill in this column.

Enter this information first.

Use the information in this column to form an equation.

The sulfuric acid in the 50% solution	plus	the sulfuric acid in the 20% solution	equals	the sulfuric acid in the mixture.
$0.50x$	$+$	$0.20(12 - x)$	$=$	$12(0.30)$

Solve the Equation

$$0.50x + 0.20(12 - x) = 12(0.30)$$

$0.5x + 2.4 - 0.2x = 3.6$ Distribute the multiplication by 0.20.

$0.3x + 2.4 = 3.6$ Combine like terms: $0.5x - 0.2x = 0.3x$.

$0.3x = 1.2$ Subtract 2.4 from both sides.

$x = 4$ To isolate x, undo the multiplication by 0.3 by dividing both sides by 0.3: $\frac{1.2}{0.3} = 4$.

Success Tip
We could begin by multiplying both sides of the equation by 10 to clear it of the decimals.

State the Conclusion   4 liters of 50% solution and $12 - 4 = 8$ liters of 20% solution should be used.

Check the Result   The amount of acid in 4 liters of the 50% solution is $0.50(4) = 2.0$ liters and the amount of acid in 8 liters of the 20% solution is $0.20(8) = 1.6$ liters. Thus, the amount of acid in these two solutions is $2.0 + 1.6 = 3.6$ liters. The amount of acid in 12 liters of the 30% mixture is also $0.30(12) = 3.6$ liters. Since the amounts of acid are equal, the answers check.

 Now Try   **Problem 39**

 Solve Dry Mixture Problems.

In another type of mixture problem, a dry mixture of a specified value is created from two differently priced ingredients.

EXAMPLE 5   ***Snack Foods.***   Because cashews priced at $9 per pound were not selling, a produce clerk decided to combine them with less expensive peanuts and sell the mixture for $7 per pound. How many pounds of peanuts, selling at $6 per pound, should be mixed with 50 pounds of cashews to obtain such a mixture?

Analyze the Problem   We need to determine how many pounds of peanuts to mix with 50 pounds of cashews to obtain a mixture worth $7 per pound.

Form an Equation   Let x = the number of pounds of peanuts to use in the mixture. Since 50 pounds of cashews will be combined with the peanuts, the mixture will weigh $50 + x$ pounds. The value of the mixture and of each of its ingredients is given by

$$\text{Amount} \cdot \text{the price} = \text{the total value}$$

We can organize the facts of the problem in a table.

	Amount	· Price	= Total value	
Peanuts	x	6	$6x$	Multiply amount · price
Cashews	50	9	450	three times to fill in this
Mixture	$50 + x$	7	$7(50 + x)$	column.

Enter this information first.

Use the information in this column to form an equation.

The value of the peanuts	plus	the value of the cashews	equals	the value of the mixture.
$6x$	$+$	450	$=$	$7(50 + x)$

Solve the Equation

$$6x + 450 = 7(50 + x)$$

$$6x + 450 = 350 + 7x \qquad \text{Distribute the multiplication by 7.}$$

$$450 = 350 + x \qquad \text{To eliminate the term 6x on the left side, subtract 6x from both}$$
$$\text{sides: } 7x - 6x = x.$$

$$100 = x \qquad \text{To isolate x, subtract 350 from both sides.}$$

State the Conclusion   100 pounds of peanuts should be used in the mixture.

Check the Result   The value of 100 pounds of peanuts, at $6 per pound, is $100(6) = \$600$ and the value of 50 pounds of cashews, at $9 per pound, $50(9) = \$450$. Thus, the value of these two amounts is $1,050. Since the value of 150 pounds of the mixture, at $7 per pound, is also $150(7) = \$1,050$, the answer checks.

 Now Try   **Problem 45**

5   **Solve Number–Value Problems.**

When problems deal with collections of different items having different values, we must distinguish between the *number of* and the *value of* the items. For these problems, we will use the fact that

$$\text{Number} \cdot \text{value} = \text{total value}$$

EXAMPLE 6   *Dining Area Improvements.*   A restaurant owner needs to purchase some tables, chairs, and dinner plates for the dining area of her establishment. She plans to buy four chairs and four plates for each new table. She also plans to buy 20 additional plates in case of breakage. If a table costs $100, a chair $50, and a plate $5, how many of each can she buy if she takes out a loan for $6,500 to pay for the new items?

Analyze the Problem   We know the *value* of each item: Tables cost $100, chairs cost $50, and plates cost $5 each. We need to find the *number* of tables, chairs, and plates she can purchase for $6,500.

Form an Equation   The number of chairs and plates she needs depends on the number of tables she buys. So we let t = the number of tables to be purchased. Since every table requires four chairs and four plates, she needs to order $4t$ chairs. Because 20 additional plates are needed, she should order $(4t + 20)$ plates. We can organize the facts of the problem in a table.

	Number	· Value	= Total value
Tables	t	100	$100t$
Chairs	$4t$	50	$50(4t)$
Plates	$4t + 20$	5	$5(4t + 20)$
			Total: 6,500

Multiply number · value three times to fill in this column.

Enter this information first.

Use the information in this column to form an equation.

The value of the tables	plus	the value of the chairs	plus	the value of the plates	equals	the value of the purchase.
$100t$	+	$50(4t)$	+	$5(4t + 20)$	=	6,500

Solve the Equation

$$100t + 50(4t) + 5(4t + 20) = 6,500$$

$$100t + 200t + 20t + 100 = 6,500 \quad \text{Do the multiplications and distribute.}$$

$$320t + 100 = 6,500 \quad \text{Combine like terms: } 100t + 200t + 20t = 320t.$$

$$320t = 6,400 \quad \text{Subtract 100 from both sides.}$$

$$t = 20 \quad \text{To isolate } t, \text{ divide both sides by 320.}$$

To find the number of chairs and plates to buy, we evaluate $4t$ and $4t + 20$ for $t = 20$.

Chairs: $4t = 4(20)$ *Plates:* $4t + 20 = 4(20) + 20$ Substitute 20 for t.

$= 80$ $= 100$

State the Conclusion   The owner needs to buy 20 tables, 80 chairs, and 100 plates.

Check the Result   The total value of 20 tables is $20(\$100) = \$2,000$, the total value of 80 chairs is $80(\$50) = \$4,000$, and the total value of 100 plates is $100(\$5) = \$500$. Because the total purchase is $\$2,000 + \$4,000 + \$500 = \$6,500$, the answers check.

  Now Try   Problem 53

STUDY SET
2.6

VOCABULARY

Fill in the blanks.

1. Problems that involve depositing money are called _____ problems, and problems that involve moving vehicles are called uniform _____ problems.

2. Problems that involve combining ingredients are called _____ problems, and problems that involve collections of different items having different values are called _____ problems.

CONCEPTS

3. Complete the *principal column* given that part of $30,000 is invested in stocks and the rest in art.

	$P \cdot r \cdot t = I$			
Stocks	x			
Art	?			

4. A man made two investments that earned a combined annual interest of $280. Complete the table and then form an equation for this investment problem.

	P	$\cdot$	r	$\cdot t =$	I
Bank	x		0.04	1	
Stocks	$6,000 - x$		0.06	1	
				Total:	

5. Complete the *rate column* given that the east-bound plane flew 150 mph slower than the west-bound plane.

	$r \cdot t = d$		
West	r		
East	?		

6. **a.** Complete the *time column* given that a runner wants to overtake a walker and the walker had a $\frac{1}{2}$-hour head start.

	$r \cdot t = d$		
Runner		t	
Walker		?	

   b. Complete the *time column* given that part of a 6-hour drive was in fog and the other part was in clear conditions

	$r \cdot t = d$		
Foggy		t	
Clear		?	

7. A husband and wife drive in opposite directions to work. Their drives last the same amount of time and their workplaces are 80 miles apart. Complete the table and then form an equation for this distance problem.

	$r \cdot t = d$		
Husband	35	t	
Wife	45		
	Total:		

8. **a.** How many gallons of acetic acid are there in barrel 2?

   b. Suppose the contents of the two barrels are poured into an empty third barrel. How many gallons of liquid will the third barrel contain?

   c. Estimate the strength of the solution in the third barrel: 15%, 35%, or 60% acid?

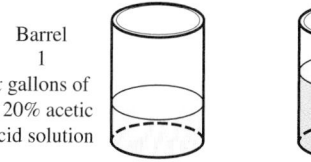

Barrel 1
x gallons of a 20% acetic acid solution

Barrel 2
42 gallons of a 40% acetic acid solution

9. **a.** Two antifreeze solutions are combined to form a mixture. Complete the table and then form an equation for this mixture problem.

	Amount $\cdot$ Strength = Pure antifreeze		
Strong	6	0.50	
Weak	x	0.25	
Mixture		0.30	

   b. Two oil-and-vinegar salad dressings are combined to make a new mixture. Complete the table and then form an equation for this mixture problem.

	Amount $\cdot$ Strength = Pure vinegar		
Strong	x	0.06	
Weak		0.03	
Mixture	10	0.05	

10. The value of all the nylon brushes that a paint store carries is $670. Complete the table and then form an equation for this number-value problem.

	Number $\cdot$ Value = Total value		
1-inch	$2x$	4	
2-inch	x	5	
3-inch	$x + 10$	7	
		Total:	

NOTATION

11. Write 6% and 15.2% in decimal form.

12. By what power of 10 should each decimal be multiplied to make it a whole number?

    a. 0.08   **b.** 0.162

GUIDED PRACTICE

Solve each equation. See Example 1.

13. $0.18x + 0.45(12 - x) = 0.36(12)$

14. $0.12x + 0.20(4 - x) = 0.6$

15. $0.08x + 0.07(15,000 - x) = 1,110$

16. $0.108x + 0.07(16,000 - x) = 1,500$

APPLICATIONS

Investment problems

17. CORPORATE INVESTMENTS   The financial board of a corporation invested $25,000 overseas, part at 4% and part at 7% annual interest. Find the amount invested at each rate if the first-year combined income from the two investments was $1,300.

18. LOANS   A credit union loaned out $50,000, part at an annual rate of 5% and the rest at an annual rate of 8%. They collected combined simple interest of $3,400 from the loans that year. How much was loaned out at each rate?

19. OLD COINS   A salesperson used her $3,500 year-end bonus to purchase some old coins, with hopes of earning 15% annual interest on the gold coins and 12% annual interest on the silver coins. If she saw return on her investment of $480 the first year, how much did she invest in each type of coin?

20. HIGH-RISK COMPANIES   An investment club used funds totaling $200,000 to invest in a bio-tech company and in an ethanol plant, with hopes of earning 11% and 14% annual interest, respectively. Their hunch paid off. The club made a total of $24,250 interest the first year. How much was invested at each rate?

21. RETIREMENT   A professor wants to supplement her pension with investment interest. If she invests $28,000 at 6% interest, how much would she have to invest at 7% to achieve a goal of $3,500 per year in supplemental income?

22. EXTRA INCOME   An investor wants to receive $1,000 annually from two investments. He has put $4,500 in a money market account paying 4% annual interest. How much should he invest in a stock fund that pays 10% annual interest to achieve his goal?

23. 1099 FORMS   The form shows the interest income Terrell Washington earned in 2008 from two savings accounts. He deposited a total of $15,000 at the first of that year, and made no further deposits or withdrawals. How much money did he deposit in account 822 and in account 721?

USA HOME SAVINGS

This is important tax information and is being furnished to the Internal Revenue Service.

2008

RECIPIENT'S name

TERRELL WASHINGTON

Account Number	Annual Percent Yield	Interest earned
822	5%	?
721	4.5%	?

FORM 1099   **Total Interest Income $720.00**

24. INVESTMENT PLANS   A financial planner recommends a plan for a client who has $65,000 to invest. (See the chart.) At the end of the presentation, the client asks, "How much will be invested at each rate?" Answer this question using the given information.

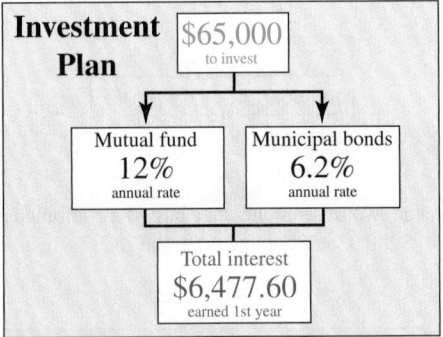

25. INVESTMENTS   Equal amounts are invested in each of three accounts paying 7%, 8%, and 10.5% annually. If one year's combined interest income is $1,249.50, how much is invested in each account?

26. PERSONAL LOANS   Maggy lent her brother some money at 2% annual interest. She lent her sister twice as much money at half of the interest rate. In one year, Maggy collected combined interest of $200 from her brother and sister. How much did she lend each of them?

Uniform motion problems

27. TORNADOES   During a storm, two teams of scientists leave a university at the same time in vans to search for tornadoes. The first team travels east at 20 mph and the second travels west at 25 mph. If their radios have a range of up to 90 miles, how long will it be before they lose radio contact?

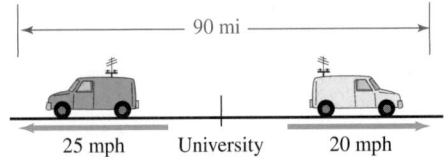

28. UNMANNED AIRCRAFT   Two remotely controlled unmanned aircraft are launched in opposite directions. One flies east at 78 mph and the other west at 82 mph. How long will it take the aircraft to fly a combined distance of 560 miles?

29. HELLO/GOODBYE   A husband and wife work different shifts at the same plant. When the husband leaves from work to make the 20-mile trip home, the wife leaves their home and drives to work. They travel on the same road. The husband's driving rate is 45 mph and the wife's is 35 mph. How long into their drives can they wave at each when passing on the road?

30. AIR TRAFFIC CONTROL   An airliner leaves Berlin, Germany, headed for Montreal, Canada, flying at an average speed of 450 mph. At the same time, an airliner leaves Montreal headed for Berlin, averaging 500 mph. If the airports are 3,800 miles apart, when will the air traffic controllers have to make the pilots aware that the planes are passing each other?

31. CYCLING   A cyclist leaves his training base for a morning workout, riding at the rate of 18 mph. One and one-half hours later, his support staff leaves the base in a car going 45 mph in the same direction. How long will it take the support staff to catch up with the cyclist?

32. PARENTING   How long will it take a mother, running at 4 feet per second, to catch up with her toddler, running down the sidewalk at 2 feet per second, if the child had a 5-second head start?

33. ROAD TRIPS   A car averaged 40 mph for part of a trip and 50 mph for the remainder. If the 5-hour trip covered 210 miles, for how long did the car average 40 mph?

34. CROSS-TRAINING   An athlete runs up a set of stadium stairs at a rate of 2 stairs per second, immediately turns around, and then descends the same stairs at a rate of 3 stairs per second. If the workout takes 90 seconds, how long does it take him to run up the stairs?

35. WINTER DRIVING   A trucker drove for 4 hours before he encountered icy road conditions. He reduced his speed by 20 mph and continued driving for 3 more hours. Find his average speed during the first part of the trip if the entire trip was 325 miles.

36. SPEED OF TRAINS   Two trains are 330 miles apart, and their speeds differ by 20 mph. Find the speed of each train if they are traveling toward each other and will meet in 3 hours.

Liquid mixture problems

37. SALT SOLUTIONS   How many gallons of a 3% salt solution must be mixed with 50 gallons of a 7% solution to obtain a 5% solution?

38. PHOTOGRAPHY   A photographer wishes to mix 2 liters of a 5% acetic acid solution with a 10% solution to get a 7% solution. How many liters of 10% solution must be added?

39. MAKING CHEESE   To make low-fat cottage cheese, milk containing 4% butterfat is mixed with milk containing 1% butterfat to obtain 15 gallons of a mixture containing 2% butterfat. How many gallons of each milk must be used?

40. ANTIFREEZE   How many quarts of a 10% antifreeze solution must be mixed with 16 quarts of a 40% antifreeze solution to make a 30% solution?

41. PRINTING   A printer has ink that is 8% cobalt blue color and ink that is 22% cobalt blue color. How many ounces of each ink are needed to make 1 gallon (64 ounces) of ink that is 15% cobalt blue color?

42. FLOOD DAMAGE   One website recommends a 6% chlorine bleach-water solution to remove mildew. A chemical lab has 3% and 15% chlorine bleach-water solutions in stock. How many gallons of each should be mixed to obtain 100 gallons of the mildew spray?

43. INTERIOR DECORATING   The colors on the paint chip card below are created by adding different amounts of orange tint to a white latex base. How many gallons of Desert Sunrise should be mixed with 1 gallon of Bright Pumpkin to obtain Cool Cantaloupe?

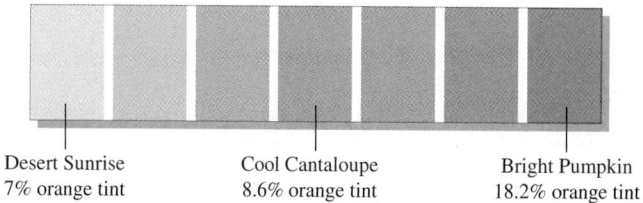

Desert Sunrise
7% orange tint

Cool Cantaloupe
8.6% orange tint

Bright Pumpkin
18.2% orange tint

44. ANTISEPTICS   A nurse wants to add water to 30 ounces of a 10% solution of benzalkonium chloride to dilute it to an 8% solution. How much water must she add?   (*Hint:* Water is 0% benzalkonium chloride.)

Dry mixture problems

45. LAWN SEED   A store sells bluegrass seed for $6 per pound and ryegrass seed for $3 per pound. How much ryegrass must be mixed with 100 pounds of bluegrass to obtain a blend that will sell for $5 per pound?

46. COFFEE BLENDS   A store sells regular coffee for $8 a pound and gourmet coffee for $14 a pound. To get rid of 40 pounds of the gourmet coffee, a shopkeeper makes a blend to put on sale for $10 a pound. How many pounds of regular coffee should he use?

47. RAISINS   How many scoops of natural seedless raisins costing $3.45 per scoop must be mixed with 20 scoops of golden seedless raisins costing $2.55 per scoop to obtain a mixture costing $3 per scoop?

48. FERTILIZER   Fertilizer with weed control costing $38 per 50-pound bag is to be mixed with a less expensive fertilizer costing $6 per 50-pound bag to make 16 bags of fertilizer that can be sold for $28 per bag. How many bags of cheaper fertilizer should be used?

49. PACKAGED SALAD   How many 10-ounce bags of Romaine lettuce must be mixed with fifty 10-ounce bags of Iceberg lettuce to obtain a blend that sells for $2.50 per ten-ounce bag?

Price: $2.20   Price: $3.50

50. MIXING CANDY   Lemon drops worth $3.80 per pound are to be mixed with jelly beans that cost $2.40 per pound to make 100 pounds of a mixture worth $2.96 per pound. How many pounds of each candy should be used?

51. BRONZE   A pound of tin is worth $1 more than a pound of copper. Four pounds of tin are mixed with 6 pounds of copper to make bronze that sells for $3.65 per pound. How much is a pound of tin worth?

52. SNACK FOODS   A bag of peanuts is worth $.30 less than a bag of cashews. Equal amounts of peanuts and cashews are used to make 40 bags of a mixture that sells for $1.05 per bag. How much is a bag of cashews worth?

Number-value problems

53. RENTALS   The owners of an apartment building rent equal numbers of 1-, 2-, and 3-bedroom units. The monthly rent for a 1-bedroom is $550, a 2-bedroom is $700, and a 3-bedroom is $900. If the total monthly income is $36,550, how many of each type of unit are there?

54. WAREHOUSING   A store warehouses 40 more portables than big-screen TV sets, and 15 more consoles than big-screen sets. The monthly storage cost for a portable is $1.50, a console is $4.00, and a big-screen is $7.50. If storage for all the televisions costs $276 per month, how many big-screen sets are in stock?

55. SOFTWARE   Three software applications are priced as shown. Spreadsheet and database programs sold in equal numbers, but 15 more word processing applications were sold than the other two combined. If the three applications generated sales of $72,000, how many spreadsheets were sold?

Software	Price
Spreadsheet	**$150**
Database	**$195**
Word processing	**$210**

56. INVENTORIES   With summer approaching, the number of air conditioners sold is expected to be double that of stoves and refrigerators combined. Stoves sell for $350, refrigerators for $450, and air conditioners for $500, and sales of $56,000 are expected. If stoves and refrigerators sell in equal numbers, how many of each appliance should be stocked?

57. PIGGY BANKS   When a child emptied his coin bank, he had a collection of pennies, nickels, and dimes. There were 20 more pennies than dimes and the number of nickels was triple the number of dimes. If the coins had a value of $5.40, how many of each type coin were in the bank?

58. WISHING WELLS   A scuba diver, hired by an amusement park, collected $121 in nickels, dimes, and quarters at the bottom of a wishing well. There were 500 nickels, and 90 more quarters than dimes. How many quarters and dimes were thrown into the wishing well?

59. BASKETBALL   Epiphanny Prince, of New York, scores 113 points in a high school game on February 1, 2006, breaking a national prep record that was held by Cheryl Miller. Prince made 46 more 2-point baskets than 3-point baskets, and only 1 free throw. How many 2-point and 3-point baskets did she make?

60. MUSEUM TOURS   The admission prices for the Coca-Cola Museum in Atlanta are shown. A family purchased 3 more children's tickets than adult tickets, and 1 less senior ticket than adult tickets. The total cost of the tickets was $73. How many of each type of ticket did they purchase?

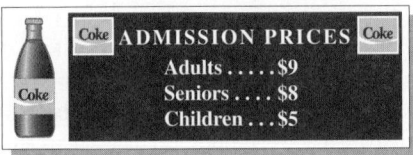

ADMISSION PRICES
Adults $9
Seniors $8
Children . . . $5

WRITING

61. Create a mixture problem of your own, and solve it.

62. Is it possible to mix a 10% sugar solution with a 20% sugar solution to get a 30% sugar solution? Explain.

REVIEW

Multiply.

63. $-25(2x - 5)$

64. $-12(3a + 4b - 32)$

65. $-(3x - 3)$

66. $\frac{1}{2}(4b - 8)$

67. $(4y - 4)4$

68. $3(5t + 1)2$

CHALLENGE PROBLEMS

69. EVAPORATION   How much water must be boiled away to increase the concentration of 300 milliliters of a 2% salt solution to a 3% salt solution?

70. TESTING   A teacher awarded 4 points for each correct answer and deducted 2 points for each incorrect answer when grading a 50-question true-false test. A student scored 56 points on the test and did not leave any questions unanswered. How many questions did the student answer correctly?

71. FINANCIAL PLANNING   A plumber has a choice of two investment plans:

- An insured fund that pays 11% interest
- A risky investment that pays a 13% return

If the same amount invested at the higher rate would generate an extra $150 per year, how much does the plumber have to invest?

72. INVESTMENTS   The amount of annual interest earned by $8,000 invested at a certain rate is $200 less than $12,000 would earn at a rate 1% lower. At what rate is the $8,000 invested?

SECTION 2.7
Solving Inequalities

Objectives

 Determine whether a number is a solution of an inequality.

2 Graph solution sets and use interval notation.

3 Solve linear inequalities.

4 Solve compound inequalities.

5 Solve inequality applications.

In our daily lives, we often speak of one value being *greater than* or *less than* another. For example, a sick child might have a temperature *greater than* 98.6°F or a granola bar might contain *less than* 2 grams of fat.

In mathematics, we use *inequalities* to show that one expression is greater than or is less than another expression.

1 **Determine Whether a Number is a Solution of an Inequality.**

An **inequality** is a statement that contains one or more of the following symbols.

Inequality Symbols			
$<$ is less than	$>$ is greater than	$\neq$ is not equal to
$\leq$ is less than or equal to	$\geq$ is greater than or equal to	

An inequality can be true, false, or neither true nor false. For example,

- $9 \geq 9$ is true because $9 = 9$.
- $37 < 24$ is false.
- $x + 1 > 5$ is neither true nor false because we don't know what number x represents.

The Language of Algebra
Because $<$ requires one number to be strictly less than another number and $>$ requires one number to be strictly greater than another number, $<$ and $>$ are called *strict inequalities*.

An inequality that contains a variable can be made true or false depending on the number that is substituted for the variable. If we substitute 10 for x in $x + 1 > 5$, the resulting inequality is true: $10 + 1 > 5$. If we substitute 1 for x, the resulting inequality is false: $1 + 1 > 5$. A number that makes an inequality true is called a **solution** of the inequality, and we say that the number *satisfies* the inequality. Thus, 10 is a solution of $x + 1 > 5$ and 1 is not.

In this section, we will find the solutions of *linear inequalities in one variable*.

Linear Inequality in One Variable	A **linear inequality in one variable** can be written in one of the following forms where a, b, and c are real numbers and $a \neq 0$.

$$ax + b > c \qquad ax + b \geq c \qquad ax + b < c \qquad ax + b \leq c$$

EXAMPLE 1   Is 9 a solution of $2x + 4 \leq 21$?

Strategy   We will substitute 9 for x and evaluate the expression on the left side.

Why   If a true statement results, 9 is a solution of the inequality. If we obtain a false statement, 9 is not a solution.

Solution

$$2x + 4 \leq 21$$
$$2(9) + 4 \overset{?}{\leq} 21 \quad \text{Substitute 9 for x. Read} \overset{?}{\leq} \text{ as "is possibly less than or equal to."}$$
$$18 + 4 \overset{?}{\leq} 21$$
$$22 \leq 21 \quad \text{This inequality is false.}$$

The statement $22 \leq 21$ is false because neither $22 < 21$ nor $22 = 21$ is true. Therefore, 9 is not a solution.

    Self Check 1   Is 2 a solution of $3x - 1 \geq 0$?

Now Try   **Problem 13**

---

2   **Graph Solution Sets and Use Interval Notation.**

The **solution set** of an inequality is the set of all numbers that make the inequality true. Some solution sets are easy to determine. For example, if we replace the variable in $x > -3$ with a number greater than -3, the resulting inequality will be true. Because there are infinitely many real numbers greater than -3, it follows that $x > -3$ has infinitely many solutions. Since there are too many solutions to list, we use **set-builder notation** to describe the solutions set.

$$\{x \mid x > -3\} \quad \text{Read as "the set of all x such that x is greater than } -3\text{."}$$

We can illustrate the solution set by **graphing the inequality** on a number line. To graph $x > -3$, a **parenthesis** or **open circle** is drawn on the endpoint -3 to indicate that -3 is not part of the graph. Then we shade all of the points on the number line to the right of -3. The right arrowhead is also shaded to show that the solutions continue forever to the right.

> **Notation**
> The parenthesis (opens in the direction of the shading and indicates that an endpoint is not included in the shaded interval.

Method 1: parenthesis

−5 −4 −3 −2 −1 0 1 2 3 4 5

Method 2: open circle

−5 −4 −3 −2 −1 0 1 2 3 4 5

All real numbers greater than −3

The graph of $x > -3$ is an example of an **interval** on the number line. We can write intervals in a compact form called **interval notation.**

Notation
The *infinity* symbol ∞ does not represent a number. It indicates that an interval extends to the right without end.

The interval notation that represents the graph of $x > -3$ is $(-3, \infty)$. As on the number line, a left parenthesis is written next to -3 to indicate that -3 is not included in the interval. The **positive infinity symbol** ∞ that follows indicates that the interval continues without end to the right. With this notation, *a parenthesis is always used next to an infinity symbol.*

The illustration below shows the relationship between the symbols used to graph an interval and the corresponding interval notation. If we begin at -3 and move to the right, the shaded arrowhead on the graph indicates that the interval approaches positive infinity ∞.

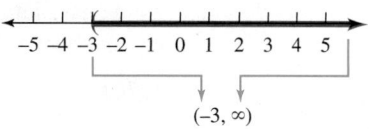

We now have three ways to describe the solution set of an inequality.

Set-builder notation	*Number line graph*	*Interval notation*
$\{x \mid x > -3\}$	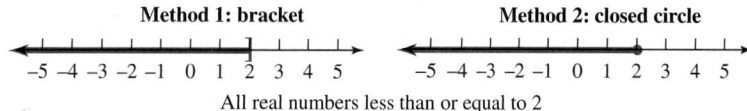	$(-3, \infty)$

EXAMPLE 2   Graph: $x \leq 2$

Strategy   We need to determine which real numbers, when substituted for x, would make $x \leq 2$ a true statement.

Why   To graph $x \leq 2$ means to draw a "picture" of all of the values of x that make the inequality true.

Notation
The *bracket*] opens in the direction of the shading and indicates that an endpoint is included in the shaded interval.

Solution   If we replace x with a number less than or equal to 2, the resulting inequality will be true. To graph the solution set, a **bracket** or a **closed circle** is drawn at the endpoint 2 to indicate that 2 is part of the graph. Then we shade all of the points on the number line to the left of 2 and the left arrowhead.

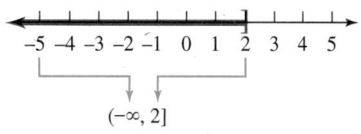

The interval is written as $(-\infty, 2]$. The right bracket indicates that 2 is included in the interval. The **negative infinity symbol** $-\infty$ shows that the interval continues forever to the left. The illustration below shows the relationship between the symbols used to graph the interval and the corresponding interval notation.

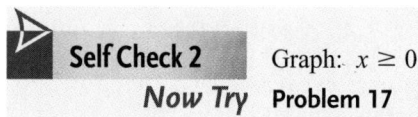

▷ **Self Check 2**   Graph: $x \geq 0$
    Now Try   **Problem 17**

Solve Linear Inequalities.

To **solve an inequality** means to find all values of the variable that make the inequality true. As with equations, there are properties that we can use to solve inequalities.

| Addition and Subtraction Properties of Inequality | Adding the same number to, or subtracting the same number from, both sides of an inequality does not change its solutions. |

Adding the same number to, or subtracting the same number from, both sides of an inequality does not change its solutions.

For any real numbers a, b, and c,

If $a < b$, then $a + c < b + c$.

If $a < b$, then $a - c < b - c$.

Similar statements can be made for the symbols $\leq$, $>$, and $\geq$.

After applying one of these properties, the resulting inequality is equivalent to the original one. **Equivalent inequalities** have the same solution set.

Like equations, inequalities are solved by isolating the variable on one side.

EXAMPLE 3   Solve $x + 3 > 2$. Write the solution set in interval notation and graph it.

Strategy   We will use a property of inequality to isolate the variable on one side.

Why   To solve the original inequality, we want to find a simpler equivalent inequality of the form $x >$ **a number** or $x <$ **a number**, whose solution is obvious.

Solution   We will use the subtraction property of inequality to isolate x on the left side of the inequality. We can undo the addition of 3 by subtracting 3 from both sides.

$$x + 3 > 2 \qquad \text{This is the inequality to solve.}$$
$$x + 3 - 3 > 2 - 3 \qquad \text{Subtract 3 from both sides.}$$
$$x > -1$$

> **Success Tip**
> We solve linear inequalities by writing a series of steps that result in an equivalent inequality of the form
>
> $x >$ *a number*
>
> or
>
> $x <$ *a number*
>
> Similar statements apply to linear inequalities containing $\leq$ and $\geq$.

All real numbers greater than -1 are solutions of $x + 3 > 2$. The solution set can be written in set-builder notation as $\{x \mid x > -1\}$ and in interval notation as $(-1, \infty)$. The graph of the solution set is shown below.

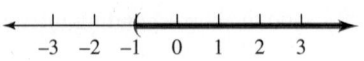

> **Notation**
> Since we use parentheses and brackets in interval notation, we will use them to graph inequalities. Note that parentheses, not brackets, are written next to ∞ and $-\infty$ because there is no endpoint.
>
> $(-3, \infty)$ $(-\infty, 2]$

Since there are infinitely many solutions, we cannot check all of them.

As an informal check, we can pick some numbers in the graph, say 0 and 30, substitute each number for x in the original inequality, and see whether true statements result.

Check:
$x + 3 > 2$
$0 + 3 \overset{?}{>} 2$ Substitute 0 for x.
$3 > 2$ True.

$x + 3 > 2$
$30 + 3 \overset{?}{>} 2$ Substitute 30 for x.
$33 > 2$ True.

The solution set appears to be correct.

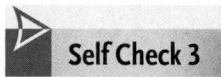

Self Check 3   Solve $x - 3 < -2$. Write the solution set in interval notation and graph it.

Now Try   **Problem 25**

As with equations, there are properties for multiplying and dividing both sides of an inequality by the same number. To develop what is called *the multiplication property of inequality,* we consider the true statement $2 < 5$. If both sides are multiplied by a positive number, such as 3, another true inequality results.

$2 < 5$ This inequality is true.
$3 \cdot 2 < 3 \cdot 5$ Multiply both sides by 3.
$6 < 15$ This inequality is true.

However, if we multiply both sides of $2 < 5$ by a negative number, such as -3, the direction of the inequality symbol is reversed to produce another true inequality.

$2 < 5$ This inequality is true.
$-3 \cdot 2 > -3 \cdot 5$ Multiply both sides by -3 and reverse the direction of the inequality.
$-6 > -15$ This inequality is true.

The inequality $-6 > -15$ is true because -6 is to the right of -15 on the number line.

Dividing both sides of an inequality by the same negative number also requires that the direction of the inequality symbol be reversed.

$-4 < 6$ This inequality is true.
$\dfrac{-4}{-2} > \dfrac{6}{-2}$ Divide both sides by -2 and change $<$ to $>$.
$2 > -3$ This inequality is true.

These examples illustrate the multiplication and division properties of inequality.

> **Caution**
>
> If the inequality symbol is not reversed when both sides of a true inequality are multiplied by a negative number, the result is a false inequality. For example,
>
> $3 < 6$ True
> $-3 \cdot 3 < -3 \cdot 6$
> $-9 < -18$ False
>
> The same is true if we divide both sides of a true inequality by a negative number. Divide both sides of $3 < 6$ by -3 to see for yourself.

> **Multiplication and Division Properties of Inequality**
>
> Multiplying or dividing both sides of an inequality by the same positive number does not change its solutions.
> For any real numbers a, b, and c, where c is positive,
>
> If $a < b$,   then $ac < bc$.   If $a < b$,   then $\dfrac{a}{c} < \dfrac{b}{c}$.
>
> If we multiply or divide both sides of an inequality by a negative number, the direction of the inequality symbol must be reversed for the inequalities to have the same solutions.
> For any real numbers a, b, and c, where c is negative,
>
> If $a < b$,   then $ac > bc$.   If $a < b$,   then $\dfrac{a}{c} > \dfrac{b}{c}$.
>
> Similar statements can be made for the symbols $\leq$, $>$, and $\geq$.

EXAMPLE 4   Solve each inequality. Write the solution set in interval notation and graph it.   **a.** $-\dfrac{3}{2}t \geq -12$ **b.** $-5t < 55$

Strategy   We will use a property of inequality to isolate the variable on one side.

Why   To solve the original inequality, we want to find a simpler equivalent inequality, whose solution is obvious.

Solution

a. To undo the multiplication by $-\frac{3}{2}$, we multiply both sides by the reciprocal, which is $-\frac{2}{3}$.

$$-\frac{3}{2}t \geq -12 \qquad \text{This is the inequality to solve.}$$

$$-\frac{2}{3}\left(-\frac{3}{2}t\right) \leq -\frac{2}{3}(-12) \qquad \begin{array}{l}\text{Multiply both sides by } -\frac{2}{3}. \text{ Since we are multiplying both sides}\\ \text{by a negative number, reverse the direction of the inequality.}\end{array}$$

$$t \leq 8 \qquad \text{Do the multiplications.}$$

The solution set is $(-\infty, 8]$ and it is graphed as shown.

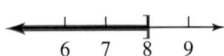

b. To undo the multiplication by -5, we divide both sides by -5.

$$-5t < 55 \qquad \text{This is the inequality to solve.}$$

$$\frac{-5t}{-5} > \frac{55}{-5} \qquad \begin{array}{l}\text{To isolate } t, \text{ undo the multiplication by } -5 \text{ by dividing both sides by } -5.\\ \text{Since we are dividing both sides by a negative number, reverse the direction}\\ \text{of the inequality.}\end{array}$$

$$t > -11$$

The solution set is $(-11, \infty)$ and it is graphed as shown.

> **Self Check 4**   Solve each inequality. Write the solution set in interval notation and graph it.   **a.** $-\frac{h}{20} \leq 10$ **b.** $-12a > -144$
>
> ***Now Try***   **Problems 31 and 33**

EXAMPLE 5   Solve $-5 > 3x + 7$. Write the solution set in interval notation and graph it.

Strategy   First we will use a property of inequality to isolate the *variable term* on one side. Then we will use a second property of inequality to isolate the *variable* itself.

Why   To solve the original inequality, we want to find a simpler equivalent inequality of the form $x > \textbf{a number}$ or $x < \textbf{a number}$, whose solution is obvious.

Solution

$$-5 > 3x + 7 \qquad \text{This is the inequality to solve.}$$

$$-5 - 7 > 3x + 7 - 7 \qquad \begin{array}{l}\text{To isolate the variable term, } 3x, \text{ undo the addition of } 7 \text{ by}\\ \text{subtracting } 7 \text{ from both sides.}\end{array}$$

$$-12 > 3x \qquad \text{Do the subtractions.}$$

$$\frac{-12}{3} > \frac{3x}{3} \qquad \text{To isolate } x, \text{ undo the multiplication by } 3 \text{ by dividing both sides by } 3.$$

$$-4 > x \qquad \text{Do the divisions.}$$

> **Caution**
> Don't be confused by the negative number on the left side. We didn't reverse the > symbol because we divided both sides by *positive* 3.
> $$\frac{-12}{3} > \frac{3x}{3}$$

To determine the solution set, it is useful to rewrite the inequality $-4 > x$ in an equivalent form with the variable on the left side. If -4 is greater than x, it follows that x must be less than -4.

$$x < -4$$

The solution set is $(-\infty, -4)$ whose graph is shown below.

Self Check 5   Solve $-13 < 2r - 7$. Write the solution set in interval notation and graph it.

Now Try   **Problem 39**

EXAMPLE 6   Solve $5.1 - 3k < 19.5$. Write the solution set in interval notation and graph it.

Strategy   We will use properties of inequality to isolate the variable on one side.

Why   To solve the original inequality, we want to find a simpler equivalent inequality of the form $k >$ **a number** or $k <$ **a number**, whose solution is obvious.

Solution

$5.1 - 3k < 19.5$	This is the inequality to solve.
$5.1 - 3k - \mathbf{5.1} < 19.5 - \mathbf{5.1}$	To isolate $-3k$ on the left side, subtract 5.1 from both sides.
$-3k < 14.4$	Do the subtractions.
$\dfrac{-3k}{-3} > \dfrac{14.4}{-3}$	To isolate k, undo the multiplication by -3 by dividing both sides by -3 and reverse the direction of the $<$ symbol.
$k > -4.8$	Do the divisions.

The solution set is $(-4.8, \infty)$, whose graph is shown below.

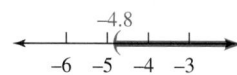

Self Check 6   Solve $-9n + 1.8 > -17.1$. Write the solution set in interval notation and graph it.

Now Try   **Problem 47**

The equation solving strategy on page 125 can be applied to inequalities. However, when solving inequalities, we must remember to *change the direction of the inequality symbol when multiplying or dividing both sides by a negative number.*

EXAMPLE 7   Solve $8(y + 1) \geq 2(y - 4) + y$. Write the solution set in interval notation and graph it.

Strategy   We will follow the steps of the equation solving strategy (adapted to inequalities) to solve the inequality.

Why   This is the most efficient way to solve a linear inequality in one variable.

Solution

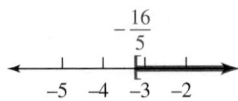

$8(y + 1) \geq 2(y - 4) + y$	This is the inequality to solve.
$8y + 8 \geq 2y - 8 + y$	Distribute the multiplication by 8 and by 2.
$8y + 8 \geq 3y - 8$	Combine like terms: $2y + y = 3y$.
$8y + 8 - 3y \geq 3y - 8 - 3y$	To eliminate $3y$ from the right side, subtract $3y$ from both sides.
$5y + 8 \geq -8$	Combine like terms on both sides.
$5y + 8 - 8 \geq -8 - 8$	To isolate $5y$, undo the addition of 8 by subtracting 8 from both sides.
$5y \geq -16$	Do the subtractions.
$\dfrac{5y}{5} \geq \dfrac{-16}{5}$	To isolate y, undo the multiplication by 5 by dividing both sides by 5. Do not reverse the direction of the $\geq$ symbol.
$y \geq -\dfrac{16}{5}$	

Success Tip

As an informal check, substitute a number on the graph that is shaded, such as 0, into $8(y + 1) \geq 2(y - 4) + y$. A true statement should result. Then substitute a number on the graph that is not shaded, such as -4, into the inequality. A false statement should result.

The solution set is $\left[-\frac{16}{5}, \infty\right)$. To graph it, we note that $-\frac{16}{5} = -3\frac{1}{5}$.

```
                    -16/5
      |    |    [====●====>
     -5   -4   -3   -2
```

Self Check 7   Solve $5(b - 2) \geq -(b - 3) + 2b$. Write the solution set in interval notation and graph it.

Now Try   Problem 53

4   **Solve Compound Inequalities.**

The Language of Algebra

The word *compound* means made up of two or more parts. For example, a *compound* inequality has three parts. Other examples are: a *compound* sentence, a *compound* fracture, and a chemical *compound*.

Two inequalities can be combined into a **compound inequality** to show that an expression lies between two fixed values. For example, $-2 < x < 3$ is a combination of

$$-2 < x \quad \text{and} \quad x < 3$$

It indicates that x is greater than -2 and that x is also less than 3. The solution set of $-2 < x < 3$ consists of all numbers that lie between -2 and 3, and we write it as $(-2, 3)$. The graph of the compound inequality is shown below.

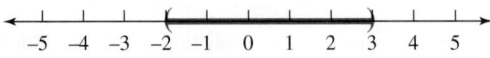

```
      |    |    |    (    |    |    |    |    )    |    |
     -5   -4   -3   -2   -1    0    1    2    3    4    5
```

EXAMPLE 8   Graph: $-4 \leq x < 0$

Strategy   We need to determine which real numbers, when substituted for x, would make $-4 \leq x < 0$ a true statement.

Why   To graph $-4 \leq x < 0$ means to draw a "picture" of all of the values of x that make the compound inequality true.

Solution   If we replace the variable in $-4 \leq x < 0$ with a number between -4 and 0, including -4, the resulting compound inequality will be true. Therefore, the solution set is the interval $[-4, 0)$. To graph the interval, we draw a bracket at -4, a parenthesis at 0, and shade in between.

To check, we pick a number in the graph, such as -2, and see whether it satisfies the inequality. Since $-4 \leq -2 < 0$ is true, the answer appears to be correct.

Self Check 8   Graph $-2 \leq x < 1$ and write the solution set in interval notation.

Now Try   **Problem 61**

To solve compound inequalities, we isolate the variable in the middle part of the inequality. To do this, we apply the properties of inequality to all *three* parts of the inequality.

EXAMPLE 9   Solve $-4 < 2(x - 1) \leq 4$. Write the solution set in interval notation and graph it.

Strategy   We will use properties of inequality to isolate the variable by itself as the middle part of the inequality.

Why   To solve the original inequality, we want to find a simpler equivalent inequality of the form **a number** $< x \leq$ **a number**, whose solution is obvious.

Solution

$$-4 < 2(x - 1) \leq 4 \qquad \text{This is the compound inequality to solve.}$$
$$-4 < 2x - 2 \leq 4 \qquad \text{Distribute the multiplication by 2.}$$
$$-4 + 2 < 2x - 2 + 2 \leq 4 + 2 \qquad \text{To isolate } 2x, \text{ undo the subtraction of 2 by adding 2 to all three parts.}$$
$$-2 < 2x \leq 6 \qquad \text{Do the additions.}$$
$$\frac{-2}{2} < \frac{2x}{2} \leq \frac{6}{2} \qquad \text{To isolate } x, \text{ we undo the multiplication by 2 by dividing all three parts by 2.}$$
$$-1 < x \leq 3 \qquad \text{Do the divisions.}$$

The solution set is $(-1, 3]$ and its graph is shown.

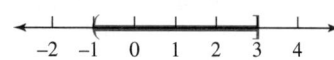

Notation
Note that the two inequality symbols in $-4 \leq x < 0$ point in the same direction and point to the smaller number.

Success Tip
Think of interval notation as a way to tell someone how to draw the graph, from left to right, giving them only a "start" and a "stop" instruction.

> **Self Check 9**   Solve $-6 \leq 3(t + 2) \leq 6$. Write the solution set in interval notation and graph it.
>
> *Now Try*   **Problem 69**

5 **Solve Inequality Applications.**

When solving problems, phrases such as "not more than," or "should exceed" suggest that the problem involves an inequality rather than an equation.

EXAMPLE 10   ***Grades.***   A student has scores of 72%, 74%, and 78% on three exams. What percent score does he need on the last exam to earn a grade of no less than B (80%)?

Analyze the Problem   We know three scores. We are to find what the student must score on the last exam to earn a grade of B or higher.

Form an Inequality   We can let $x =$ the score on the fourth (and last) exam. To find the average grade, we add the four scores and divide by 4. To earn a grade of *no less than* B, the student's average must be *greater than or equal to* 80%.

> **The Language of Algebra**
> Some phrases that suggest an inequality are:
>
> surpass: $>$ at least: $\geq$
> not exceed: $\leq$ at most: $\leq$
> between: $<$ $<$

The average of the four grades	must be no less than	80.
$\dfrac{72 + 74 + 78 + x}{4}$	$\geq$	80

Solve the Inequality

$$\frac{224 + x}{4} \geq 80 \qquad \text{Combine like terms in the numerator: } 72 + 74 + 78 = 224.$$

$$4\left(\frac{224 + x}{4}\right) \geq 4(80) \qquad \text{To clear the inequality of the fraction, multiply both sides by 4.}$$

$$224 + x \geq 320 \qquad \text{Simplify each side.}$$

$$x \geq 96 \qquad \text{To isolate } x, \text{ undo the addition of 224 by subtracting 224 from both sides.}$$

State the Conclusion   To earn a B, the student must score 96% or better on the last exam. Assuming the student cannot score higher than 100% on the exam, the solution set is written as [96, 100]. The graph is shown below.

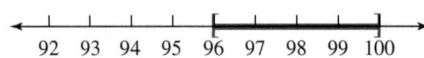

Check the Result   Pick some numbers in the interval, and verify that the average of the four scores will be 80% or greater.

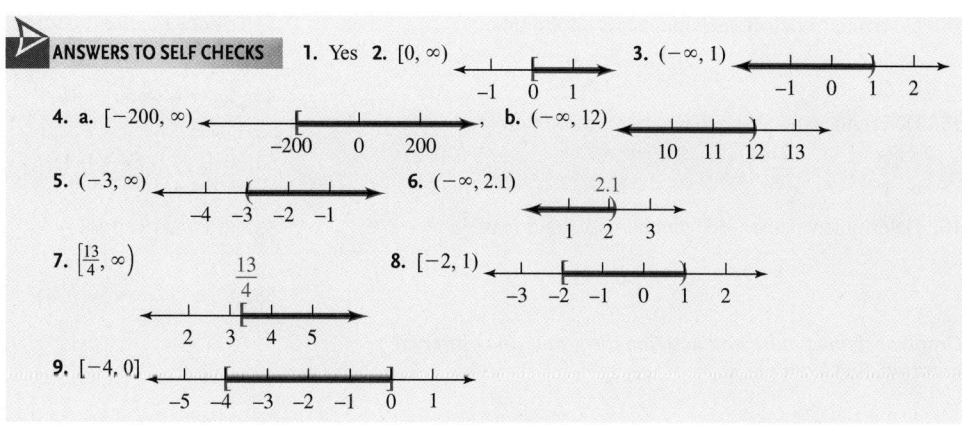

ANSWERS TO SELF CHECKS **1.** Yes **2.** $[0, \infty)$

3. $(-\infty, 1)$

4. a. $[-200, \infty)$, **b.** $(-\infty, 12)$

5. $(-3, \infty)$ **6.** $(-\infty, 2.1)$

7. $\left[\frac{13}{4}, \infty\right)$ **8.** $[-2, 1)$

9. $[-4, 0]$

STUDY SET
2.7

VOCABULARY

Fill in the blanks.

1. An _____ is a statement that contains one of the symbols: $>, \geq, <,$ or $\leq$.

2. To _____ an inequality means to find all the values of the variable that make the inequality true.

3. The solution set of $x > 2$ can be expressed in _____ notation as $(2, \infty)$.

4. The inequality $-4 < x \leq 10$ is an example of a _____ inequality.

CONCEPTS

Fill in the blanks.

5. a. Adding the _____ number to both sides of an inequality does not change the solutions.

b. Multiplying or dividing both sides of an inequality by the same _____ number does not change the solutions.

c. If we multiply or divide both sides of an inequality by a _____ number, the direction of the inequality symbol must be reversed for the inequalities to have the same solutions.

6. To solve $-4 \leq 2x + 1 < 3$, properties of inequality are applied to all _____ parts of the inequality.

7. Rewrite the inequality $32 < x$ in an equivalent form with the variable on the left side.

8. The solution set of an inequality is graphed below. Which of the four numbers, 3, -3, 2, and 4.5, when substituted for the variable in that inequality, would make it true?

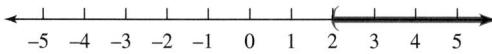

NOTATION

9. Write each symbol.

a. is less than or equal to   **b.** infinity

c. bracket   **d.** is greater than

10. Consider the graph of the interval $[4, 8)$.

a. Is the endpoint 4 included or not included in the graph?

b. Is the endpoint 8 included or not included in the graph?

Complete the solution to solve each inequality.

11. $4x - 5 \geq 7$

$4x - 5 + \quad \geq 7 + \quad$

$4x \geq \quad$

$\dfrac{4x}{\quad} \geq \dfrac{12}{\quad}$

$x \geq 3$ Solution set: $[\quad, \infty)$

12. $-6x > 12$

$\dfrac{-6x}{\quad} \quad \dfrac{12}{-6}$

$x < \quad$ Solution set: $(\quad, -2)$

GUIDED PRACTICE

See Example 1.

13. Determine whether each number is a solution of $3x - 2 > 5$.

a. 5   **b.** -4

14. Determine whether each number is a solution of
$3x + 7 < 4x - 2$.

    a. 12   **b.** 9

15. Determine whether each number is a solution of
$-5(x - 1) \geq 2x + 12$.

    a. 1   **b.** -1

16. Determine whether each number is a solution of $\frac{4}{5}a \geq -2$.

    a. $-\dfrac{5}{4}$ **b.** -15

Graph each inequality and describe the graph using interval notation. See Example 2.

17. $x < 5$ **18.** $x \geq -2$

19. $-3 < x \leq 1$ **20.** $-4 \leq x \leq 2$

Write the inequality that is represented by each graph. Then describe the graph using interval notation.

21.
        -1

22.
        2

23.
    -7 2

24.
    4   6

Solve each inequality. Write the solution set in interval notation and graph it. See Examples 3–4.

25. $x + 2 > 5$ **26.** $x + 5 \geq 2$

27. $g - 30 \geq -20$ **28.** $h - 18 \leq -3$

29. $8h < 48$ **30.** $2t > 22$

31. $-\dfrac{3}{16}x \geq -9$ **32.** $-\dfrac{7}{8}x \leq 21$

33. $-3y \leq -6$ **34.** $-6y \geq -6$

35. $\dfrac{2}{3}x \geq 2$ **36.** $\dfrac{3}{4}x < 3$

Solve each inequality. Write the solution set in interval notation and graph it. See Examples 5–6.

37. $9x + 1 > 64$ **38.** $4x + 8 < 32$

39. $0.5 \geq 2x - 0.3$ **40.** $0.8 > 7x - 0.04$

41. $\dfrac{x}{8} - (-9) \geq 11$ **42.** $\dfrac{x}{6} - (-12) > 14$

43. $\dfrac{m}{-42} - 1 > -1$ **44.** $\dfrac{a}{-25} + 3 < 3$

45. $-x - 3 \leq 7$ **46.** $-x - 9 > 3$

47. $-3x - 7 > -1$ **48.** $-5x + 7 \leq 12$

Solve each inequality. Write the solution set in interval notation and graph it. See Example 7.

49. $9a + 4 > 5a - 16$ **50.** $8t + 1 < 4t - 19$

51. $0.4x \leq 0.1x + 0.45$ **52.** $0.9s \leq 0.3s + 0.54$

53. $8(5 - x) \leq 10(8 - x)$ **54.** $17(3 - x) \geq 3 - 13x$

55. $8x + 4 > -(3x - 4)$ **56.** $7x + 6 \geq -(x - 6)$

57. $\dfrac{1}{2} + \dfrac{n}{5} > \dfrac{3}{4}$ **58.** $\dfrac{1}{3} + \dfrac{c}{5} > -\dfrac{3}{2}$

59. $\dfrac{6x + 1}{4} \leq x + 1$ **60.** $\dfrac{3x - 10}{5} \leq x + 4$

Solve each compound inequality. Write the solution set in interval notation and graph it. See Examples 8–9.

61. $2 < x - 5 < 5$ **62.** $-8 < t - 8 < 8$

63. $0 \leq x + 10 \leq 10$ **64.** $-9 \leq x + 8 < 1$

65. $-3 \leq \dfrac{c}{2} \leq 5$ **66.** $-12 < \dfrac{b}{3} < 0$

67. $3 \leq 2x - 1 < 5$ **68.** $4 < 3x - 5 \leq 7$

69. $-9 < 6x + 9 \leq 45$ **70.** $-30 \leq 10d + 20 < 90$

71. $6 < -2(x - 1) < 12$ **72.** $4 \leq -4(x - 2) < 20$

TRY IT YOURSELF

Solve each inequality or compound inequality. Write the solution set in interval notation and graph it.

73. $6 - x \leq 3(x - 1)$ **74.** $3(3 - x) \geq 6 + x$

75. $\dfrac{y}{4} + 1 \leq -9$ **76.** $\dfrac{r}{8} - 7 \geq -8$

77. $0 < 5(x + 2) \leq 15$ **78.** $-18 \leq 9(x - 5) < 27$

79. $-1 \leq -\dfrac{1}{2}n$ **80.** $-3 \geq -\dfrac{1}{3}t$

81. $-m - 12 > 15$ **82.** $-t + 5 < 10$

83. $-\dfrac{2}{3} \ge \dfrac{2y}{3} - \dfrac{3}{4}$

84. $-\dfrac{2}{9} \ge \dfrac{5x}{6} - \dfrac{1}{3}$

85. $9x + 13 \ge 2x + 6x$

86. $7x - 16 < 2x + 4x$

87. $7 < \dfrac{5}{3}a + (-3)$

88. $5 < \dfrac{7}{2}a + (-9)$

89. $-8 \le \dfrac{y}{8} - 4 \le 2$

90. $6 < \dfrac{m}{16} + 7 < 8$

91. $-2(2x - 3) > 17$

92. $-3(x + 0.2) < 0.3$

93. $\dfrac{5}{3}(x + 1) \ge -x + \dfrac{2}{3}$

94. $\dfrac{5}{2}(7x - 15) \ge \dfrac{11}{2}x - \dfrac{3}{2}$

95. $2x + 9 \le x + 8$

96. $3x + 7 \le 4x - 2$

97. $-7x + 1 < -5$

98. $-3x - 10 \ge -5$

APPLICATIONS

99. GRADES   A student has test scores of 68%, 75%, and 79% in a government class. What must she score on the last exam to earn a B (80% or better) in the course?

100. OCCUPATIONAL TESTING   An employment agency requires applicants average at least 70% on a battery of four job skills tests. If an applicant scored 70%, 74%, and 84% on the first three exams, what must he score on the fourth test to maintain a 70% or better average?

101. GAS MILEAGE   A car manufacturer produces three models in equal quantities. One model has an economy rating of 17 miles per gallon, and the second model is rated for 19 mpg. If government regulations require the manufacturer to have a fleet average that exceeds 21 mpg, what economy rating is required for the third model?

102. SERVICE CHARGES   When the average daily balance of a customer's checking account falls below $500 in any week, the bank assesses a $5 service charge. The table shows the daily balances of one customer. What must Friday's balance be to avoid the service charge?

Day	Balance
Monday	$540.00
Tuesday	$435.50
Wednesday	$345.30
Thursday	$310.00

103. GEOMETRY   The perimeter of an equilateral triangle is at most 57 feet. What could the length of a side be?   (*Hint:* All three sides of an equilateral triangle are equal.)

104. GEOMETRY   The perimeter of a square is no less than 68 centimeters. How long can a side be?

105. COUNTER SPACE   A rectangular counter is being built for the customer service department of a store. Designers have determined that the outside perimeter of the counter (shown in red) needs to exceed 30 feet. Determine the acceptable values for x.

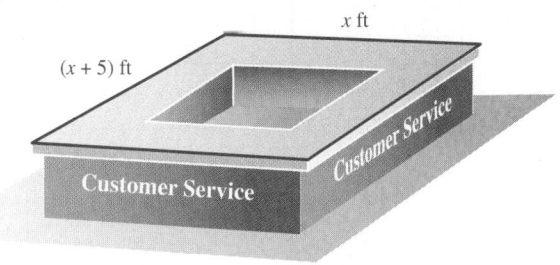

106. NUMBER PUZZLES   What numbers satisfy the condition: Four more than three times the number is at most 10?

107. GRADUATIONS   It costs a student $18 to rent a cap and gown and 80 cents for each graduation announcement that she orders. If she doesn't want her spending on these graduation costs to exceed $50, how many announcements can she order?

108. TELEPHONES   A cellular telephone company has currently enrolled 36,000 customers in a new calling plan. If an average of 1,200 people are signing up for the plan each day, in how many days will the company surpass their goal of having 150,000 customers enrolled?

109. WINDOWS   An architect needs to design a triangular-shaped bathroom window that has an area no greater than 100 in.² If the base of the window must be 16 inches long, what window heights will meet this condition?

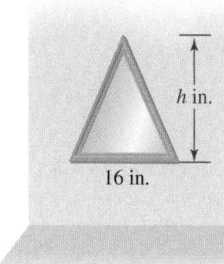

110. ROOM TEMPERATURES   To hold the temperature of a room between 19° and 22° Celsius, what Fahrenheit temperatures must be maintained?   *Hint:* Use the formula $C = \dfrac{5}{9}(F - 32)$.

111. INFANTS   The graph is used to classify the weight of a baby boy from birth to 1 year. Estimate the weight range for boys in the following classifications, using a compound inequality:

   a. 10 months old, "heavy"

   b. 5 months old, "light"

   c. 8 months old, "average"

   d. 3 months old, "moderately light"

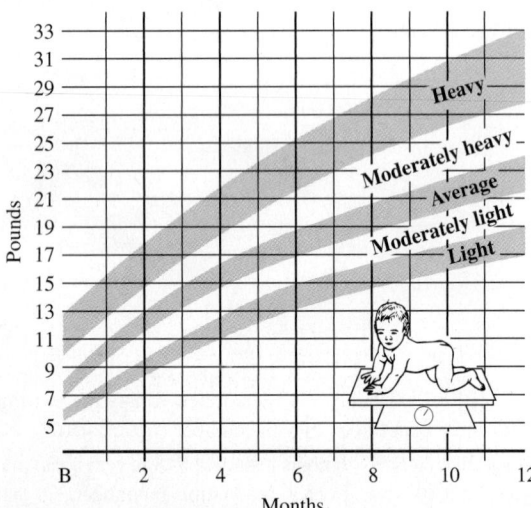

Based on data from *Better Homes and Gardens Baby Book* (Meredith Corp., 1969).

112. NUMBER PUZZLES   What *whole* numbers satisfy the condition: Twice the number decreased by 1 is between 50 and 60?

WRITING

113. Explain why multiplying both sides of an inequality by a negative number reverses the direction of the inequality.

114. Explain the use of parentheses and brackets for graphing intervals.

REVIEW

Complete each table.

115.

x	$x^2 - 3$
-2	
0	
3	

116.

x	$\frac{x}{3} + 2$
-6	
0	
12	

CHALLENGE PROBLEMS

117. Solve the inequality. Write the solution set in interval notation and graph it.

$$3 - x < 5 < 7 - x$$

118. Use a guess-and-check approach to solve $\frac{1}{x} > 1$. Write the solution set in interval notation and graph it.

CHAPTER 2
Summary & Review

SECTION 2.1 Solving Equations Using Properties of Equality

DEFINITIONS AND CONCEPTS	EXAMPLES
An **equation** is a statement indicating that two expressions are equal. The equal symbol = separates an equation into two parts: the left side and the right side.	$2x + 4 = 10 \qquad -5(a + 4) = -11a \qquad \frac{3}{2}t + 6 = t - \frac{1}{3}$
A number that makes an equation a true statement when substituted for the variable is called a **solution** of the equation.	Determine whether 2 is a solution of $x + 4 = 3x$. **Check:** $x + 4 = 3x$ $2 + 4 \overset{?}{=} 3(2)$ Substitute 2 for each x. $6 = 6$ True Since the resulting statement is true, 2 is a solution.
Equivalent equations have the same solutions.	$x - 2 = 6$ and $x = 8$ are equivalent equations because they have the same solution, 8.
To **solve an equation** isolate the variable on one side of the equation by undoing the operations performed on it using properties of equality. **Addition (Subtraction) property of equality:** If the same number is added to (or subtracted from) both sides of an equation, the result is an equivalent equation.	Solve: $x - 5 = 7$ Solve: $c + 9 = 16$ $x - 5 + 5 = 7 + 5$ $c + 9 - 9 = 16 - 9$ $x = 12$ $c = 7$
Multiplication (Division) property of equality: If both sides of an equation are multiplied (or divided) by the same nonzero number, the result is an equivalent equation.	Solve: $\frac{1}{3}m = 2$ Solve: $10y = 50$ $3\left(\frac{1}{3}m\right) = 3(2)$ $\frac{10y}{10} = \frac{50}{10}$ $m = 6$ $y = 5$

REVIEW EXERCISES

Determine whether the given number is a solution of the equation.

1. 84, $x - 34 = 50$

2. 3, $5y + 2 = 12$

3. −30, $\frac{x}{5} = 6$

4. 2, $a^2 - a - 1 = 0$

5. −3, $5b - 2 = 3b - 8$

6. 1, $\frac{2}{y+1} = \frac{12}{y+1} - 5$

Fill in the blanks.

7. An _____ is a statement indicating that two expressions are equal.

8. To solve $x - 8 = 10$ means to find all the values of the variable that make the equation a _____ statement.

Solve each equation and check the result.

9. $x - 9 = 12$

10. $-y = -32$

11. $a + 3.7 = -16.9$

12. $100 = -7 + r$

13. $120 = 5c$

14. $t - \frac{1}{2} = \frac{3}{2}$

15. $\frac{4}{3}t = -12$

16. $3 = \frac{q}{-2.6}$

17. $6b = 0$

18. $\frac{15}{16}s = -3$

SECTION 2.2   More about Solving Equations

DEFINITIONS AND CONCEPTS	EXAMPLES
A five-step **strategy for solving linear equations:**	Solve: $2(y + 2) + 4y = 11 - y$
1. *Clear* the equation of fractions or decimals.	$2y + 4 + 4y = 11 - y$ Distribute the multiplication by 2.
2. *Simplify* each side. Use the distributive property and combine like terms when necessary.	$6y + 4 = 11 - y$ Combine like terms: $2y + 4y = 6y$.
3. *Isolate the variable term.* Use the addition and subtraction properties of equality.	$6y + 4 + y = 11 - y + y$ To eliminate $-y$ on the right, add y to both sides.
4. *Isolate the variable.* Use the multiplication and division properties of equality.	$7y + 4 = 11$ Combine like terms.
5. *Check* the result in the original equation.	$7y + 4 - 4 = 11 - 4$ To isolate the variable term $7y$, subtract 4 from both sides.
	$7y = 7$ Simplify each side of the equation.
	$\dfrac{7y}{7} = \dfrac{7}{7}$ To isolate y, divide both sides by 7.
	$y = 1$
To clear an equation of fractions, multiply both sides of an equation by the LCD.	To solve $\dfrac{1}{2} + \dfrac{x}{3} = \dfrac{3}{4}$, first multiply both sides by 12: $$12\left(\dfrac{1}{2} + \dfrac{x}{3}\right) = 12\left(\dfrac{3}{4}\right)$$
To clear an equation of decimals, multiply both sides by a power of 10 to change the decimals in the equation to integers.	To solve $0.5(x - 4) = 0.1x + 0.2$, first multiply both sides by 10: $$10[0.5(x - 4)] = 10(0.1x + 0.2)$$
An equation that is true for all values of its variable is called an **identity.**	When we solve $x + 5 + x = 2x + 5$, the variables drop out and we obtain a true statement $5 = 5$. All real numbers are solutions.
An equation that is not true for any value of its variable is called a **contradiction.**	When we solve $y + 2 = y$, the variables drop out and we obtain a false statement $2 = 0$. The equation has no solutions.

REVIEW EXERCISES

Solve each equation. Check the result.

19. $5x + 4 = 14$

20. $98.6 - t = 129.2$

21. $\dfrac{n}{5} + (-2) = 4$

22. $\dfrac{b - 5}{4} = -6$

23. $5(2x - 4) - 5x = 0$

24. $-2(x - 5) = 5(-3x + 4) + 3$

25. $\dfrac{3}{4} = \dfrac{1}{2} + \dfrac{d}{5}$

26. $\dfrac{5(7 - x)}{4} = 2x - 3$

27. $\dfrac{3(2 - c)}{2} = \dfrac{-2(2c + 3)}{5}$

28. $\dfrac{b}{3} + \dfrac{11}{9} + 3b = -\dfrac{5}{6}b$

29. $0.15(x + 2) + 0.3 = 0.35x - 0.4$

30. $0.5 - 0.02(y - 2) = 0.16 + 0.36y$

31. $3(a + 8) = 6(a + 4) - 3a$ **32.** $2(y + 10) + y = 3(y + 8)$

SECTION 2.3   Applications of Percent

DEFINITIONS AND CONCEPTS	EXAMPLES

To solve **percent problems,** use the facts of the problem to write a sentence of the form:

☐ is ☐ % of ☐ ?

Translate the sentence to mathematical symbols: *is* translates to an = symbol and *of* means multiply. Then solve the equation.

648   is   30%   of   what number?
↓   ↓   ↓   ↓   ↓
648 =   30%   · x Translate.

$648 = 0.30x$ Change 30% to a decimal: 30% = 0.30.

$\dfrac{648}{0.30} = x$ To isolate x, divide both sides by 0.30.

$2{,}160 = x$ Do the division.

Thus, 648 is 30% of 2,160.

To find the **percent of increase** or **the percent of decrease,** find what percent the increase or decrease is of the original amount.

SALE PRICES   To find the percent of decrease when ground beef prices are reduced from $4.89 to $4.59 per pound, we first find the amount of decrease: $4.89 - 4.59 = 0.30$. Then we determine what percent 0.30 is of 4.89 (the original price).

0.30   is   what%   of   4.89?
↓   ↓   ↓   ↓   ↓
0.30 = x ·   4.89   Translate.

$0.30 = 4.89x$

$\dfrac{0.30}{4.89} = x$ To isolate x, divide both sides by 4.89.

$0.061349693 \approx x$ Do the division.

$0\,0\,6.1349693\% \approx x$ Write the decimal as a percent.

To the nearest tenth of a percent, the percent of decrease is 6.1%.

REVIEW EXERCISES

33. Fill in the blanks.

   a. _____ means parts per one hundred.

   b. When the price of an item is reduced, we call the amount of the reduction a _____.

   c. An employee who is paid a _____ is paid a percent of the goods or services that he or she sells.

34. 4.81 is 2.5% of what number?

35. What number is 15% of 950?

36. What percent of 410 is 49.2?

37. U.S. ONLINE DATA   The circle graph to the right shows Internet usage in the United States by the approximately 288.5 million people, ages 3 and over, in 2007. Determine the number of broadband users and the number of dial-up users. Round to the nearest tenth of one million.

38. COST OF LIVING   A retired trucker receives a monthly Social Security check of $764. If she is to receive a 3.5% cost-of-living increase soon, how much larger will her check be?

39. FAMILY BUDGETS   It is recommended that a family pay no more than 30% of its monthly income (after taxes) on housing. If a family has an after-tax income of $1,890 per month and pays $625 in housing costs each month, are they within the recommended range?

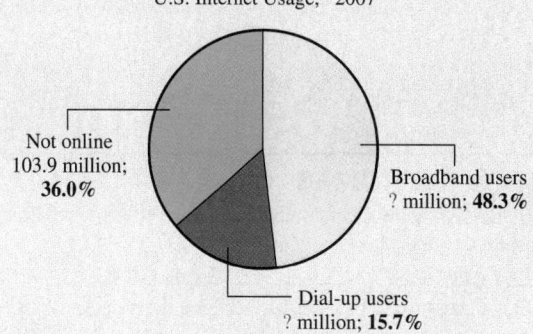

U.S. Internet Usage,* 2007

Not online
103.9 million;
36.0%

Broadband users
? million; **48.3%**

Dial-up users
? million; **15.7%**

Source: *Advertising Age 2007 Fact Pack*

*An Internet user is defined as someone who uses the Internet at least once per month.

40. DISCOUNTS   A shopper saved $148.50 on a food processor that was discounted 33%. What did it originally cost?

41. TUPPERWARE   The hostess of a Tupperware party is paid a 25% commission on her in-home party's sales. What would the hostess earn if sales totaled $600?

42. COLLECTIBLES   A collector of football trading cards paid $6 for a 1984 Dan Marino rookie card several years ago. If the card is now worth $100, what is the percent of increase in the card's value? (Round to the nearest percent.)

SECTION 2.4 Formulas

DEFINITIONS AND CONCEPTS	EXAMPLES
A **formula** is an equation that states a relationship between two or more variables.	Retail price: $r = c + m$ Profit: $p = r - c$ Simple Interest: $I = Prt$ Distance: $d = rt$ Temperature: $C = \dfrac{5}{9}(F - 32)$
The **perimeter** of a plane geometric figure is the distance around it. The **area** of a plane geometric figure is the amount of surface that it encloses. The **volume** of a three-dimensional geometric solid is the amount of space it encloses.	Rectangle: $P = 2l + 2w$ Circle: $C = \pi D = 2\pi r$ $\qquad\quad A = lw$ $\qquad\qquad A = \pi r^2$ Rectangular solid: $V = lwh$ Cylinder: $V = \pi r^2 h$ *See inside the back cover of the text for more geometric formulas.
If we are given the values of all but one of the variables in a formula, we can use our equation solving skills to find the value of the remaining variable.	BEDDING The area of a standard queen-size bed sheet is 9,180 in.² If the width is 102 inches, what is the length? $A = lw$ This is the formula for the area of a rectangle. $9{,}180 = 102w$ Substitute 9,180 for the area A and 102 for the width w. $\dfrac{9{,}180}{102} = w$ To isolate w, divide both sides by 102. $90 = w$ Do the division. The length of a standard queen-size bed sheet is 90 inches.
To solve a formula for a specific variable means to isolate that variable on one side of the equation, with all other variables and constants on the opposite side. Treat the specified variable as if it is the only variable in the equation. Treat the other variables as if they were numbers (constants).	Solve the formula for the volume of a cone for h. $V = \dfrac{1}{3}\pi r^2 h$ This is the formula for the volume of a cone. $3(V) = 3\left(\dfrac{1}{3}\pi r^2 h\right)$ To clear the equation of the fraction, multiply both sides by 3. $3V = \pi r^2 h$ Simplify. $\dfrac{3V}{\pi r^2} = \dfrac{\pi r^2 h}{\pi r^2}$ To isolate h, divide both sides by πr^2. $\dfrac{3V}{\pi r^2} = h$ or $h = \dfrac{3V}{\pi r^2}$

REVIEW EXERCISES

43. SHOPPING   Find the markup on a CD player whose wholesale cost is $219 and whose retail price is $395.

44. RESTAURANTS   One month, a restaurant had sales of $13,500 and made a profit of $1,700. Find the expenses for the month.

45. SNAILS   A typical garden snail travels at an average rate of 2.5 feet per minute. How long would it take a snail to cross a 20-foot long flower bed?

46. CERTIFICATES OF DEPOSIT   A $26,000 investment in a CD earned $1,170 in interest the first year. What was the annual interest rate?

47. JEWELRY   Gold melts at about 1,065°C. Change this to degrees Fahrenheit.

48. CAMPING
 a. Find the perimeter of the air mattress.
 b. Find the amount of sleeping area on the top surface of the air mattress.
 c. Find the approximate volume of the air mattress if it is 3 inches thick.

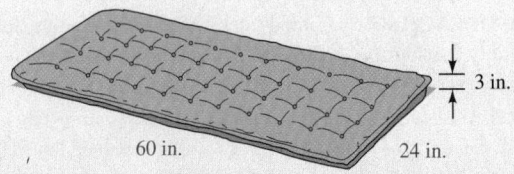

60 in.   24 in.   3 in.

49. Find the area of a triangle with a base 17 meters long and a height of 9 meters.

50. Find the area of a trapezoid with bases 11 inches and 13 inches long and a height of 12 inches.

51. a. Find the circumference of a circle with a radius of 8 centimeters. Round to the nearest hundredth of one centimeter.

    b. Find the area of the circle. Round to the nearest square centimeter.

52. Find the volume of a 12-foot cylinder whose circular base has a radius of 0.5 feet. Give the result to the nearest tenth.

53. Find the volume of a pyramid that has a square base, measuring 6 feet on a side, and a height of 10 feet.

54. HALLOWEEN After being cleaned out, a spherical-shaped pumpkin has an inside diameter of 9 inches. To the nearest hundredth, what is its volume?

Solve each formula for the specified variable.

55. $A = 2\pi rh$ for h

56. $A - BC = \dfrac{G - K}{3}$ for G

57. $C = \dfrac{1}{4}s(t - d)$ for t

58. $4y - 3x = 16$ for y

SECTION 2.5 Problem Solving

DEFINITIONS AND CONCEPTS

To solve application problems, use the five-step problem-solving strategy.

1. Analyze the problem.
2. Form an equation.
3. Solve the equation.
4. State the conclusion.
5. Check the result.

EXAMPLES

INCOME TAXES After taxes, an author kept $85,340 of her total annual earnings. If her earnings were taxed at a 15% rate, how much did she earn that year?

Analyze the Problem The author earned some unknown amount of money. On that amount, she paid 15% in taxes. The difference between her total earnings and the taxes paid was $85,340.

Form an Equation If we let x = the author's total earnings, the amount of taxes that she paid was 15% of x or $0.15x$. We can use the words of the problem to form an equation.

Her total earnings	minus	the taxes that she paid	equals	the money that she kept.
x	$-$	$0.15x$	$=$	85,340

Solve the Equation

$$x - 0.15x = 85,340$$
$$0.85x = 85,340 \quad \text{Combine like terms.}$$
$$x = 100,400 \quad \text{To isolate } x, \text{ divide both sides by 0.85.}$$

State the Conclusion The author earned $100,400 that year.

Check the Result The taxes were 15% of $100,400 or $15,060. If we subtract the taxes from her total earnings, we get $100,400 − $15,060 = $85,340. The answer checks.

REVIEW EXERCISES

59. SOUND SYSTEMS A 45-foot-long speaker wire is to be cut into three pieces. One piece is to be 15 feet long. Of the remaining pieces, one must be 2 feet less than 3 times the length of the other. Find the length of the shorter piece.

60. SIGNING PETITIONS A professional signature collector is paid $50 a day plus $2.25 for each verified signature he gets from a registered voter. How many signatures are needed to earn $500 a day?

61. LOTTERY WINNINGS After taxes, a lottery winner was left with a lump sum of $1,800,000. If 28% of the original prize was withheld to pay federal income taxes, what was the original cash prize?

62. NASCAR The car numbers of drivers Bobby Labonte and Kyle Petty are consecutive odd integers whose sum is 88. If Labonte's number is the smaller, find the numbers of each car.

63. ART HISTORY *American Gothic* was painted in 1930 by Grant Wood. The length of the rectangular painting is 5 inches more than the width. Find the dimensions of the painting if it has a perimeter of $109\frac{1}{2}$ inches.

© SuperStock, Inc./SuperStock

64. GEOMETRY Find the missing angle measures of the triangle.

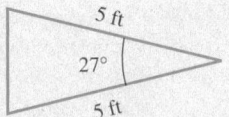

5 ft

27°

5 ft

SECTION 2.6 More on Problem Solving

DEFINITIONS AND CONCEPTS	EXAMPLES

To solve application problems, use the five-step problem solving strategy.

1. Analyze the problem.

2. Form an equation.

3. Solve the equation.

4. State the conclusion.

5. Check the result.

Tables are a helpful way to organize the facts of a problem.

TRUCKING   Two trucks leave from the same place at the same time traveling in opposite directions. One travels at a rate of 60 mph and the other at 50 mph. How long will it take them to be 165 miles apart?

Analyze the Problem   We know that one truck travels at 60 mph and the other at 50 mph. Together, the trucks will travel a distance of 165 miles.

Form an Equation   We enter each rate in the table under the heading *r*. Since the trucks travel for the same length of time, say *t* hours, we enter *t* for each truck under the heading *t*. Since $d = r \cdot t$, the first truck will travel $60t$ miles and the second will travel $50t$ miles. We enter the distances traveled under the heading *d* in the table.

	r	$\cdot$	t	$= d$
Truck 1	60		t	$60t$
Truck 2	50		t	$50t$
			Total:	165

└— Use the information in this column to form an equation.

The distance the first truck travels	plus	the distance the second truck travels	is	165 miles.
$60t$	$+$	$50t$	$=$	165

Solve the Equation $60t + 50t = 165$

$110t = 165$ Combine like terms.

$\dfrac{110t}{110} = \dfrac{165}{110}$ To isolate *t*, divide both sides by 110.

$t = 1.5$

State the Conclusion   The trucks will be 165 miles apart in 1.5 hours.

Check the Result   If the first truck travels 60 mph for 1.5 hours, it will go $60(1.5) = 90$ miles. If the second truck travels 50 mph for 1.5 hours, it will go $50(1.5) = 75$ miles. Since 90 miles + 75 miles = 165 miles, the result checks.

REVIEW EXERCISES

65. INVESTMENT INCOME   A woman has $27,000. Part is invested for 1 year in a certificate of deposit paying 7% interest, and the remaining amount in a cash management fund paying 9%. After 1 year, the total interest on the two investments is $2,110. How much is invested at each rate?

66. WALKING AND BICYCLING   A bicycle path is 5 miles long. A man walks from one end at the rate of 3 mph. At the same time, a friend bicycles from the other end, traveling at 12 mph. In how many minutes will they meet?

67. AIRPLANES   How long will it take a jet plane, flying at 450 mph, to overtake a propeller plane, flying at 180 mph, if the propeller plane had a $2\frac{1}{2}$-hour head start?

68. AUTOGRAPHS   Kesha collected the autographs of 8 more television celebrities than she has of movie stars. Each TV celebrity autograph is worth $75 and each movie star autograph is worth $250. If her collection is valued at $1,900, how many of each type of autograph does she have?

69. MIXTURES   A store manager mixes candy worth 90¢ per pound with gumdrops worth $1.50 per pound to make 20 pounds of a mixture worth $1.20 per pound. How many pounds of each kind of candy does he use?

70. MILK   Cream is about 22% butterfat and low-fat milk is about 2% butterfat. How many gallons of cream must be mixed with 18 gallons of low-fat milk to make whole milk that contains 4% butterfat?

SECTION 2.7   Solving Inequalities

DEFINITIONS AND CONCEPTS	EXAMPLES
An **inequality** is a mathematical statement that contains an $>$, $<$, $\geq$, or $\leq$ symbol.	$3x < 8$ $\frac{1}{2}y - 4 \geq 12$ $2z + 4 \leq z - 5$
A **solution of an inequality** is any number that makes the inequality true.	Determine whether 3 is a solution of $2x - 7 < 5$. **Check:** $2x - 7 < 5$ $2(3) - 7 \overset{?}{<} 5$ Substitute 3 for *x*. $-1 < 5$ True Since the resulting statement is true, 3 is a solution.
We **solve inequalities** as we solve equations. However, if we multiply or divide both sides by a negative number, we must reverse the inequality symbol.	Solve: $-3(z - 1) \geq -6$ $-3z + 3 \geq -6$ Distribute the multiplication by -3. $-3z \geq -9$ To isolate the variable term $-3z$, subtract 3 from both sides. $\dfrac{-3z}{-3} \leq \dfrac{-9}{-3}$ To isolate *z*, divide both sides by -3. Reverse the inequality symbol. $z \leq 3$ Do the divisions.
Interval notation can be used to describe the solution set of an inequality. A **parenthesis** indicates that a number is not in the solution set of an inequality. A **bracket** indicates that a number is included in the solution set.	In interval notation, the solution set is $(-\infty, 3]$, whose graph is shown.

REVIEW EXERCISES

Solve each inequality. Write the solution set in interval notation and graph it.

71. $3x + 2 < 5$

72. $-\dfrac{3}{4}x \geq -9$

73. $\dfrac{3}{4} < \dfrac{d}{5} + \dfrac{1}{2}$

74. $5(3 - x) \leq 3(x - 3)$

75. $\dfrac{t}{-5} - (-1.8) \geq -6.2$

76. $63 < 7a$

77. $8 < x + 2 < 13$

78. $0 \leq 3 - 2x < 10$

79. SPORTS EQUIPMENT   The acceptable weight w of Ping-Pong balls used in competition can range from 2.40 to 2.53 grams. Express this range using a compound inequality.

80. SIGNS   A large office complex has a strict policy about signs. Any sign to be posted in the building must be rectangular in shape, its width must be 18 inches, and its perimeter is not to exceed 132 inches. What possible sign lengths meet these specifications?

CHAPTER 2
Test

1. Fill in the blanks.

  a. To _____ an equation means to find all of the values of the variable that make the equation true.

  b. _____ means parts per one hundred.

  c. The distance around a circle is called its _____.

  d. An _____ is a statement that contains one of the symbols $>$, $\geq$, $<$, or $\leq$.

  e. The _____ property of _____ says that multiplying both sides of an equation by the same nonzero number does not change its solution.

2. Is 3 a solution of $5y + 2 = 12$?

Solve each equation.

3. $3h + 2 = 8$

4. $\dfrac{4}{5}t = -4$

5. $-22 = -x$

6. $\dfrac{11b - 11}{5} = 3b - 2$

7. $0.8(x - 1{,}000) + 1.3 = 2.9 + 0.2x$

8. $2(y - 7) - 3y = -(y - 3) - 17$

9. $\dfrac{m}{2} - \dfrac{1}{3} = \dfrac{1}{4} + \dfrac{m}{6}$

10. $9 - 5(2x + 10) = -1$

11. $24t = -6(8 - 4t)$

12. $6a + (-7) = 3a - 7 + 2a$

13. What is 15.2% of 80?

14. DOWN PAYMENTS   To buy a house, a woman was required to make a down payment of $11,400. What did the house sell for if this was 15% of the purchase price?

15. BODY TEMPERATURES   Suppose a person's body temperature rises from 98.6°F to a dangerous 105°F. What is the percent increase? Round to the nearest percent.

16. COMMISSIONS   An appliance store salesperson receives a commission of 5% of the price of every item that she sells. What will she make if she sells a $599.99 refrigerator?

17. GRAND OPENINGS   On its first night of business, a pizza parlor brought in $445. The owner estimated his profits that night to be $150. What were the costs?

18. Find the Celsius temperature reading if the Fahrenheit reading is 14°.

19. PETS   The spherical fishbowl is three-quarters full of water. To the nearest cubic inch, find the volume of water in the bowl. $\left(\text{*Hint:* The volume of a sphere is given by } V = \tfrac{4}{3}\pi r^3.\right)$

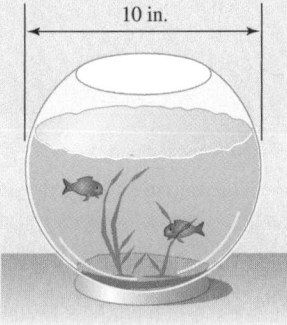

10 in.

20. Solve $A = P + Prt$ for r.

21. IRONS   Estimate the area of the soleplate of the iron.

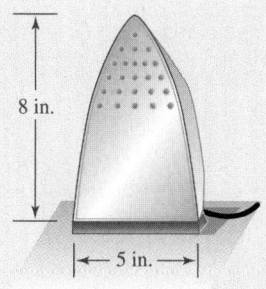

8 in.

←— 5 in. —→

22. TELEVISION   In a typical 30-minute block of time on TV, the number of programming minutes are 2 less than three times the number of minutes of commercials. How many minutes of programming and commercials are there?

23. HOME SALES   A condominium owner cleared $114,600 on the sale of his condo, after paying a 4.5% real estate commission. What was the selling price?

24. COLORADO   The state of Colorado is approximately rectangular-shaped with perimeter 1,320 miles. Find the length (east to west) and width (north to south), if the length is 100 miles longer than the width.

25. TEA   How many pounds of green tea, worth $40 a pound, should be mixed with herbal tea, worth $50 a pound, to produce 20 pounds of a blend worth $42 a pound?

26. READING   A bookmark is inserted between two page numbers whose sum is 825. What are the page numbers?

27. TRAVEL TIMES   A car leaves Rockford, Illinois, at the rate of 65 mph, bound for Madison, Wisconsin. At the same time, a truck leaves Madison at the rate of 55 mph, bound for Rockford. If the cities are 72 miles apart, how long will it take for the car and the truck to meet?

28. PICKLES   To make pickles, fresh cucumbers are soaked in a salt water solution called *brine*. How many liters of a 2% brine solution must be added to 30 liters of a 10% brine solution to dilute it to an 8% solution?

29. GEOMETRY   If the vertex angle of an isosceles triangle is 44°, find the measure of each base angle.

30. INVESTMENTS   Part of $13,750 is invested at 9% annual interest, and the rest is invested at 8%. After one year, the accounts paid $1,185 in interest. How much was invested at the lower rate?

Solve each inequality. Write the solution set in interval notation and graph it.

31. $-8x - 20 \le 4$

32. $-8.1 > \dfrac{t}{2} + (-11.3)$

33. $-12 \le 2(x + 1) < 10$

34. AWARDS   A city honors its citizen of the year with a framed certificate. An artist charges $15 for the frame and 75 cents per word for writing out the proclamation. If a city regulation does not allow gifts in excess of $150, what is the maximum number of words that can be written on the certificate?

GROUP PROJECT

TRANSLATING KEY WORDS AND PHRASES

Overview: Students often say that the most challenging step of the five-step problem-solving strategy is forming an equation. This activity is designed to make that step easier by improving your translating skills.

Instructions: Form groups of 3 or 4 students. Select one person from your group to record the group's responses. Determine whether addition, subtraction, multiplication, or division is suggested by each of the following words or phrases. Then use the word or phrase in a sentence to illustrate its meaning.

deflate	recede	partition	evaporate	amplify
bisect	augment	hike	erode	boost
annexed	diminish	plummet	upsurge	wane
quadruple	corrode	taper off	trisect	broaden

COMPUTER SPREADSHEETS

Overview: In this activity, you will get some experience working with a spreadsheet.

Instructions: Form groups of 3 or 4 students. Examine the following spreadsheet, which consists of cells named by column and row. For example, 7 is entered in cell B3. In any cell you may enter data or a formula. For each formula in cells D1–D4 and E1–E4, the computer performs a calculation using values entered in other cells and prints the result in place of the formula. Find the value that will be printed in each formula cell. The symbol $*$ means multiply, / means divide, and $\wedge$ means raise to a power.

	A	B	C	D	E
1	-8	20	-6	$= 2*B1 - 3*C1 + 4$	$= B1 - 3*A1\wedge2$
2	39	2	-1	$= A2/(B2 - C2)$	$= B3*B2*C2*2$
3	50	7	3	$= A3/5 + C3\wedge3$	$= 65 - 2*(B3 - 5)\wedge5$
4	6.8	-2.8	-0.5	$= 100*A4 + B4*C4$	$= A4/10 + A3/2*5$

CHAPTER 3

Graphing Linear Equations and Inequalities in Two Variables; Functions

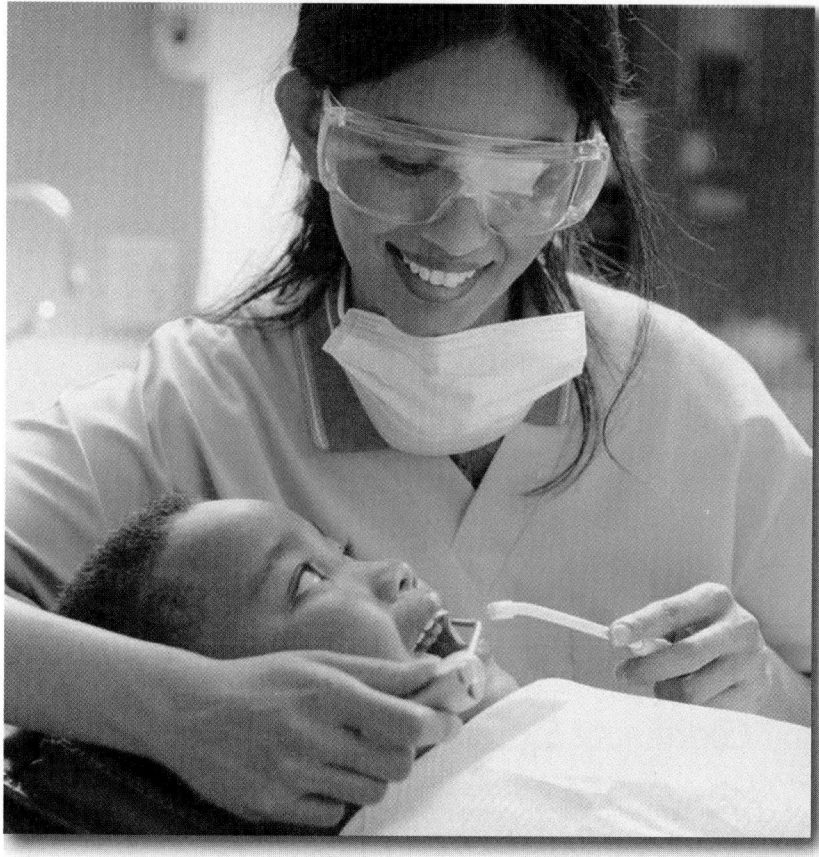

© Karin Dreyer/Getty Images

from *Campus to Careers*
Dental Assistant

A dental assistant is a valuable member of the dental health team who prepares patients for treatment, takes x-rays, sterilizes instruments, and keeps records. Part of the training of a dental assistant includes learning about a *coordinate system* that is used to identify the location of teeth in the mouth. This coordinate system is much like one used in algebra to graph points, lines, and curves. In **Problem 33** of **Study Set 3.1** you will see how the coordinate system used by dentists divides the mouth into four parts called *quadrants*.

JOB TITLE:
Dental Assistant

EDUCATION:
An associate's degree from an accredited program is required.

JOB OUTLOOK:
Excellent—One of the fastest-growing occupations in the health care industry.

ANNUAL EARNINGS:
Median salary $31,739

FOR MORE INFORMATION:
www.bls.gov/oco/ocos163.htm

Taking a math test doesn't have to be an unpleasant experience. Here are some suggestions that can make it more enjoyable and also improve your score.

PREPARING FOR THE TEST: Begin studying several days before the test rather than cramming your studying into one marathon session the night before.

TAKING THE TEST: Follow a test-taking strategy so you can maximize your score by using the testing time wisely.

EVALUATING YOUR PERFORMANCE: After your graded test is returned, classify the type of errors that you made on the test so that you do not make them again.

Now Try This

1. Write a study session plan that explains how you will prepare on each of the 4 days before the test, as well as on test day. For some suggestions, see *Preparing for a Test.**

2. Develop your own test-taking strategy by answering the survey questions found in *How to Take a Math Test.**

3. Use the outline found in *Analyzing Your Test Results** to classify the errors that you made on your most recent test.

*Found online at: http://www.thomsonedu.com/math/tussy

SECTION 3.1
Graphing Using the Rectangular Coordinate System

Objectives

1. Construct a rectangular coordinate system.
2. Plot ordered pairs and determine the coordinates of a point.
3. Graph paired data.
4. Read line graphs.

It is often said, "A picture is worth a thousand words." This is certainly true in algebra, where we often use mathematical pictures called *rectangular coordinate graphs* to illustrate numerical relationships.

1 Construct a Rectangular Coordinate System.

When designing the Gateway Arch in St. Louis, architects created a mathematical model called a **rectangular coordinate graph.** This graph, shown on the next page, is drawn on a grid called a **rectangular coordinate system.** This coordinate system is also called a **Cartesian coordinate system,** after the 17th-century French mathematician René Descartes.

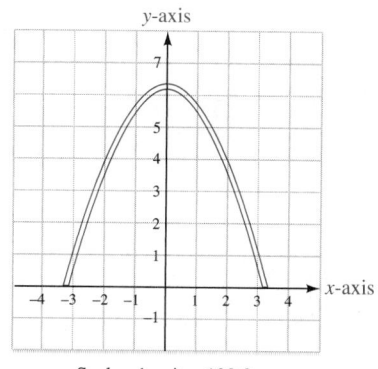

Scale: 1 unit = 100 ft

The Language of Algebra
The word *axis* is used in mathematics and science. For example, Earth rotates on its *axis* once every 24 hours. The plural of *axis* is **axes**.

A rectangular coordinate system is formed by two perpendicular number lines. The horizontal number line is usually called the **x-axis,** and the vertical number line is usually called the **y-axis.** On the x-axis, the positive direction is to the right. On the y-axis, the positive direction is upward. Each axis should be scaled to fit the data. For example, the axes of the graph of the arch are scaled in units of 100 feet.

The point where the axes intersect is called the **origin.** This is the zero point on each axis. The axes form a **coordinate plane,** and they divide it into four regions called **quadrants,** which are numbered counterclockwise using Roman numerals.

The Language of Algebra
A *coordinate plane* can be thought of as a perfectly flat surface extending infinitely far in every direction.

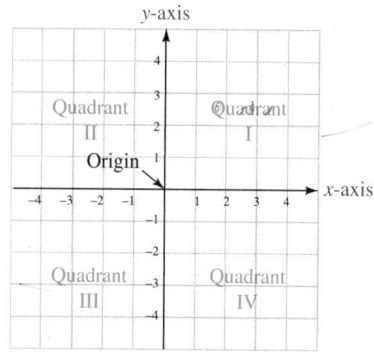

Notation
Don't be confused by this new use of parentheses. (3, −4) represents a point on the coordinate plane, whereas 3(−4) indicates multiplication. Also, don't confuse the ordered pair with interval notation.

Each point in a coordinate plane can be identified by an **ordered pair** of real numbers x and y written in the form (x, y). The first number, x, in the pair is called the **x-coordinate,** and the second number, y, is called the **y-coordinate.** Some examples of such pairs are $(3, -4)$, $\left(-1, -\frac{3}{2}\right)$, and $(0, 2.5)$.

$$(3, -4)$$

The x-coordinate   The y-coordinate

2 **Plot Ordered Pairs and Determine the Coordinates of a Point.**

The process of locating a point in the coordinate plane is called **graphing** or **plotting** the point. On the next page, we use blue arrows to show how to graph the point with coordinates $(3, -4)$. Since the x-coordinate, 3, is positive, we start at the origin and move 3 units to the *right* along the x-axis. Since the y-coordinate, −4, is negative, we then move *down* 4 units and draw a dot. This locates the point $(3, -4)$.

In the figure, red arrows are used to show how to plot the point $(-4, 3)$. We start at the origin, move 4 units to the *left* along the x-axis, then move *up* 3 units and draw a dot. This locates the point $(-4, 3)$.

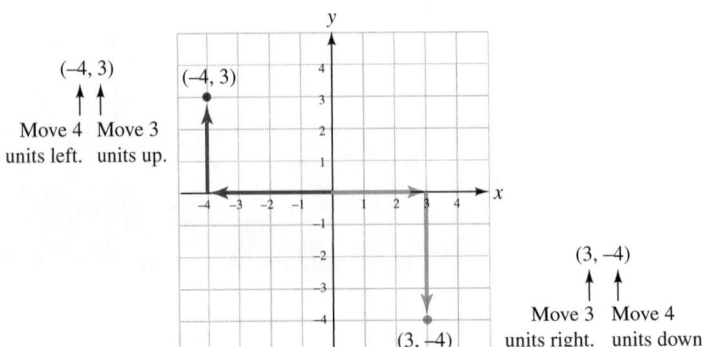

The Language of Algebra
Note that the points $(3, -4)$ and $(-4, 3)$ have different locations. Since the order of the coordinates of a point is important, we call them **ordered pairs**.

EXAMPLE 1   Plot each point. Then state the quadrant in which it lies or the axis on which it lies.   **a.** $(4, 4)$ **b.** $\left(-1, -\dfrac{7}{2}\right)$
c. $(0, 2.5)$ **d.** $(-3, 0)$ **e.** $(0, 0)$

Strategy   After identifying the x- and y-coordinates of the ordered pair, we will move the corresponding number of units left, right, up, or down to locate the point.

Why   The coordinates of a point determine its location on the coordinate plane.

Solution

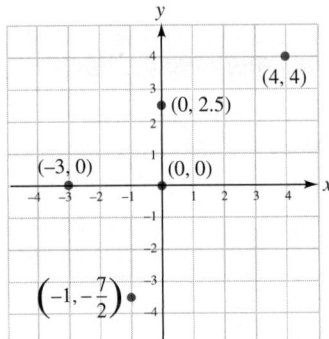

a. Since the x-coordinate, 4, is positive, we start at the origin and move 4 units to the *right* along the x-axis. Since the y-coordinate, 4, is positive, we then move *up* 4 units and draw a dot. This locates the point $(4, 4)$. The point lies in quadrant I.

b. To plot $\left(-1, -\dfrac{7}{2}\right)$, we begin at the origin and move 1 unit to the *left*, because the x-coordinate is -1. Then, since the y-coordinate is negative, we move $\dfrac{7}{2}$ units, or $3\dfrac{1}{2}$ units, *down*. The point lies in quadrant III.

Success Tip
Points with an x-coordinate that is 0 lie on the y-axis. Points with a y-coordinate that is 0 lie on the x-axis. Points that lie on an axis are not considered to be in any quadrant.

c. To plot $(0, 2.5)$, we begin at the origin and do not move right or left, because the x-coordinate is 0. Since the y-coordinate is positive, we move 2.5 units *up*. The point lies on the y-axis.

d. To plot $(-3, 0)$, we begin at the origin and move 3 units to the *left*, because the x-coordinate is -3. Since the y-coordinate is 0, we do not move up or down. The point lies on the x-axis.

e. To plot $(0, 0)$, we begin at the origin, and we remain there because both coordinates are 0. The point with coordinates $(0, 0)$ is the origin.

Self Check 1   Plot the points $(2, -2)$, $(-4, 0)$, $\left(1.5, \dfrac{5}{2}\right)$, and $(0, 5)$.

Now Try   **Problem 17**

Notation
Points are often labeled with capital letters. For example, the notation $A(2, 3)$ indicates that point A has coordinates $(2, 3)$.

EXAMPLE 2 Find the coordinates of points A, B, C, D, E, and F plotted in figure (a) below.

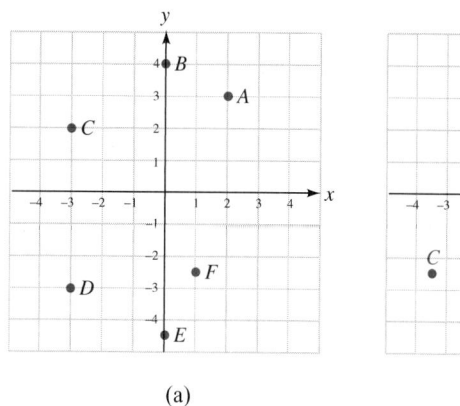

(a)   (b)

Strategy   We will start at the origin and count to the left or right on the *x*-axis, and then up or down to reach each point.

Why   The movement left or right gives the *x*-coordinate of the ordered pair and the movement up or down gives the *y*-coordinate.

Solution   To locate point A, we start at the origin, move 2 units to the right on the *x*-axis, and then 3 units up. Its coordinates are $(2, 3)$. The coordinates of the other points are found in the same manner.

$$B(0, 4) \qquad C(-3, 2) \qquad D(-3, -3) \qquad E(0, -4.5) \qquad F(1, -2.5)$$

Self Check 2   Find the coordinates of each point in Figure (b) above.

Now Try   **Problem 20**

③   Graph Paired Data.

Every day, we deal with quantities that are related:

- The time it takes to cook a roast depends on the weight of the roast.
- The money we earn depends on the number of hours we work.
- The sales tax that we pay depends on the price of the item purchased.

We can use graphs to visualize such relationships. For example, suppose a tub is filling with water, as shown on the next page. Obviously, the amount of water in the tub depends on how long the water has been running. To graph this relationship, we can use the measurements that were taken as the tub began to fill.

Time (mins)	Water in tub (gal)	
0	0	→ (0, 0)
1	8	→ (1, 8)
3	24	→ (3, 24)
4	32	→ (4, 32)

The data in the table can be expressed as ordered pairs (x, y).

↑ x-coordinate ↑ y-coordinate

The data in each row of the table can be written as an ordered pair and plotted on a rectangular coordinate system. Since the first coordinate of each ordered pair is a time, we label the *x*-axis *Time (min)*. The second coordinate is an amount of water, so we label the *y*-axis *Amount of water (gal)*. The *y*-axis is scaled in larger units (multiples of 4 gallons) because the size of the data ranges from 0 to 32 gallons.

After plotting the ordered pairs, we use a straightedge to draw a line through the points. As expected, the completed graph shows that the amount of water in the tub increases steadily as the water is allowed to run.

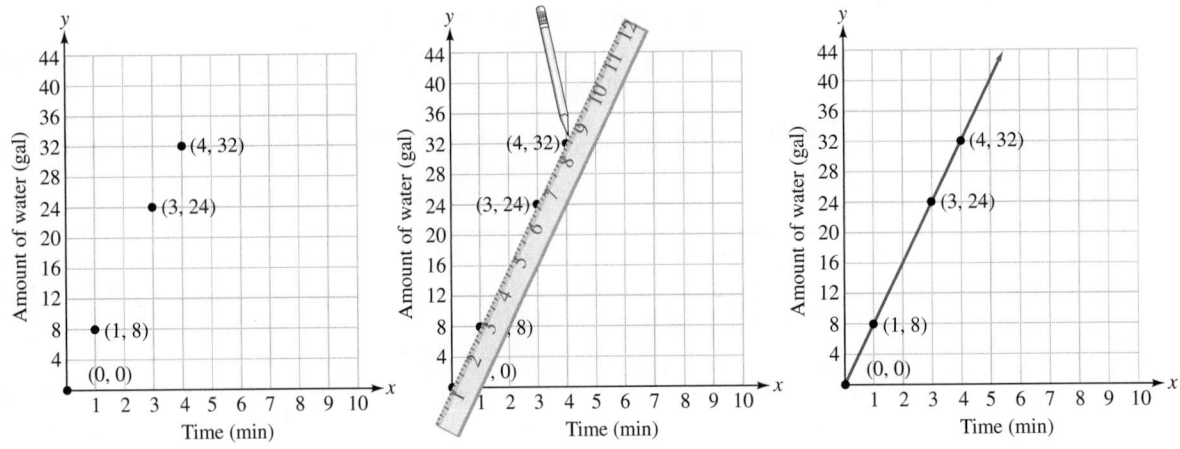

We can use the graph to determine the amount of water in the tub at various times. For example, the green dashed line on the graph shows that in 2 minutes, the tub will contain 16 gallons of water. This process, called **interpolation,** uses known information to predict values that are not known but are *within* the range of the data. The blue dashed line on the graph shows that in 5 minutes, the tub will contain 40 gallons of water. This process, called **extrapolation,** uses known information to predict values that are not known and are *outside* the range of the data.

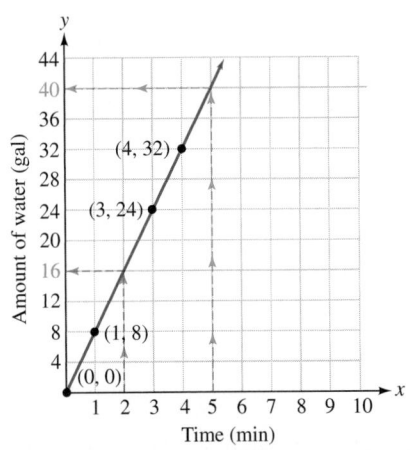

④   Read Line Graphs.

Since graphs are a popular way to present information, the ability to read and interpret them is very important.

The Language of Algebra

A rectangular coordinate system is a *grid*—a network of uniformly spaced perpendicular lines. At times, some large U.S. cities have such horrible traffic congestion that vehicles can barely move, if at all. The condition is called *gridlock*.

EXAMPLE 3   ***TV Shows.***   The following graph shows the number of people in an audience before, during, and after the taping of a television show. Use the graph to answer the following questions.

a. How many people were in the audience when the taping began?

b. At what times were there exactly 100 people in the audience?

c. How long did it take the audience to leave after the taping ended?

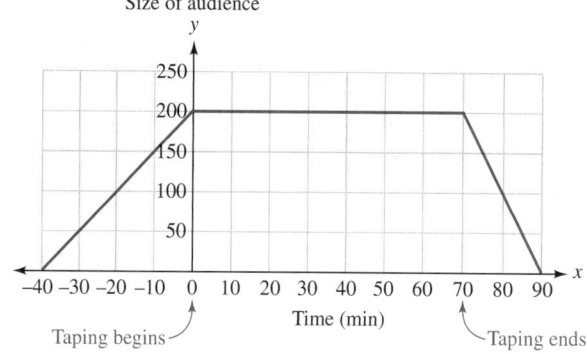

Strategy   We will use an ordered pair of the form *(time, size of audience)* to describe each situation mentioned in parts (a), (b), and (c).

Why   The coordinates of specific points on the graph can be used to answer each of these questions.

Solution

a. The time when the taping began is represented by 0 on the x-axis. The point on the graph directly above 0 is (0, **200**). The y-coordinate indicates that 200 people were in the audience when the taping began.

b. We can draw a horizontal line passing through 100 on the y-axis. Since the line intersects the graph twice, at (**−20**, 100) and at (**80**, 100), there are two times when 100 people were in the audience. The x-coordinates of the points tell us those times: 20 minutes before the taping began, and 80 minutes after.

c. The x-coordinate of the point (**70**, 200) tells us when the audience began to leave. The x-coordinate of (**90**, 0) tells when the exiting was completed. Subtracting the x-coordinates, we see that it took $90 - 70 = 20$ minutes for the audience to leave.

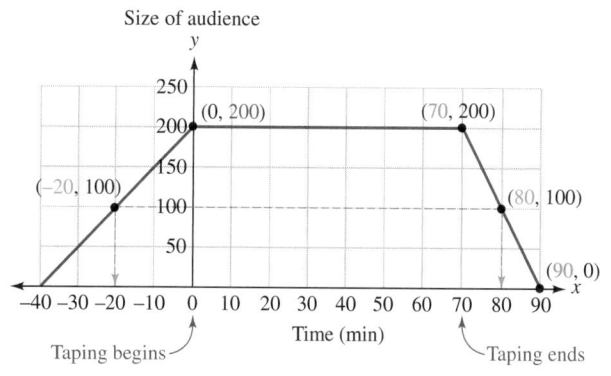

Self Check 3   Use the graph in Example 3 to answer the following questions.

a. At what times were there exactly 50 people in the audience?

b. How many people were in the audience when the taping took place?

c. When were the first audience members allowed into the taping session?

Now Try   **Problems 21 and 23**

ANSWERS TO SELF CHECKS   **1.**

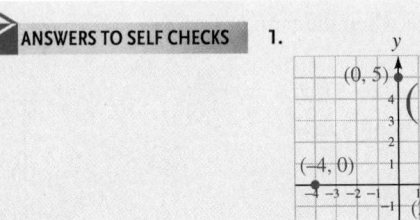

2. $A(4, 0)$; $B(0, 1)$; $C(-3.5, -2.5)$; $D(2, -4)$

3. a. 30 min before and 85 min after taping began   **b.** 200   **c.** 40 min before taping began

---

STUDY SET
3.1

VOCABULARY

Fill in the blanks.

1. (7, 1) is called an _____ pair.

2. In the ordered pair $(2, -5)$, the y-_____ is -5.

3. A rectangular coordinate system is formed by two perpendicular number lines called the x-_____ and the y-_____. The point where the axes cross is called the _____.

4. The x- and y-axes divide the coordinate plane into four regions called _____.

5. The point with coordinates (4, 2) can be graphed on a _____ coordinate system.

6. The process of locating the position of a point on a coordinate plane is called _____ the point.

CONCEPTS

7. Fill in the blanks.

a. To plot $(-5, 4)$, we start at the _____ and move 5 units to the _____ and then move 4 units ____.

b. To plot $\left(6, -\frac{3}{2}\right)$, we start at the _____ and move 6 units to the _____ and then move $\frac{3}{2}$ units _____.

8. In which quadrant is each point located?

a. $(-2, 7)$ **b.** $\left(\frac{1}{2}, \frac{15}{16}\right)$

c. $(-1, -2.75)$ **d.** $(50, -16)$

9. a. In which quadrants are the second coordinates of points positive?

b. In which quadrants are the first coordinates of points negative?

c. In which quadrant do points with a positive x-coordinate and a negative y-coordinate lie?

10. FARMING   Write each row of data in the table as an ordered pair. Then plot the ordered pairs on the following graph and draw a straight line through the points. Use the graph to determine how many bushels will be produced if

a. 6 inches of rain fall.   **b.** 10 inches of rain fall.

Rain (inches)	Bushels produced	
2	10	(,)
4	15	(,)
8	25	(,)

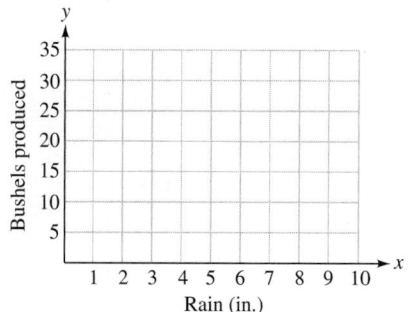

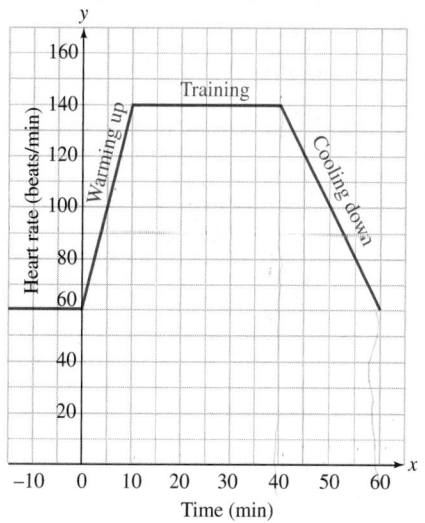

NOTATION

11. Explain the difference between (3, 5) and 3(5).

12. In the table, which column contains values associated with the vertical axis of a graph?

x	y
2	0
5	3
−1	−3

13. Do these ordered pairs name the same point?

$$\left(2.5, -\tfrac{7}{2}\right), \left(2\tfrac{1}{2}, -3.5\right), \left(2.5, -3\tfrac{1}{2}\right)$$

14. Do (3, 2) and (2, 3) represent the same point?

15. In the ordered pair (4, 5), is the number 4 associated with the horizontal or the vertical axis?

16. Fill in the blank: In the notation $P(4, 5)$, the capital letter P is used to name a _____.

GUIDED PRACTICE

See Examples 1 and 2.

17. Graph each point:
$$(-3, 4), (4, 3.5), \left(-2, -\tfrac{5}{2}\right), (0, -4), \left(\tfrac{3}{2}, 0\right), (3, -4)$$

18. Graph each point:
(4, 4), (0.5, −3), (−4, −4), (0, −1), (0, 0), (0, 3), (−2, 0)

19. Complete the coordinates for each point in Figure (a) below.

The following graph gives the heart rate of a woman before, during, and after an aerobic workout. Use it to answer Problems 21–24. See Example 3.

21. a. What was her heart rate before beginning the workout?

  b. After beginning her workout, how long did it take the woman to reach her training-zone heart rate?

22. a. What was the woman's heart rate half an hour after beginning the workout?

  b. For how long did the woman work out at her training zone?

23. a. At what time was her heart rate 100 beats per minute?

  b. How long was her cool-down period?

24. a. What was the difference in the woman's heart rate before the workout and after the cool-down period?

  b. What was her approximate heart rate 8 minutes after beginning?

The following graph shows the depths of a submarine at certain times after it leaves port. Use the graph to answer Problems 25–28. See Example 3.

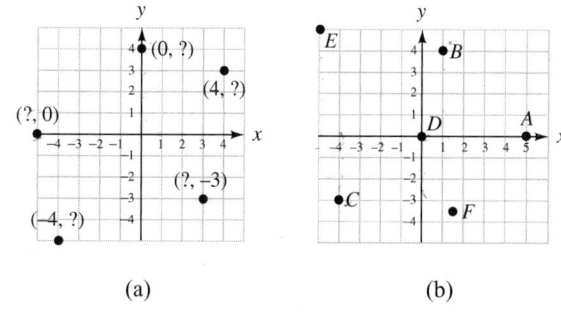

(a)   (b)

20. Find the coordinates of points A, B, C, D, E, and F in Figure (b) above.

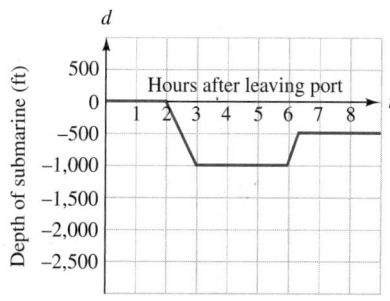

25. a. For how long does the sub travel at sea level?

   b. What is the depth of the sub 5 hours after leaving port?

26. a. Once the sub begins to dive, how long does it take to reach −1,000 feet in depth?

   b. For how long does the sub travel at a depth of 1,000 feet?

27. a. Explain what happens 6 hours after the sub leaves port.

   b. What is the depth of the sub 8 hours after leaving port?

28. a. How long does it take the sub to first reach −500 feet in depth?

   b. Approximate the time when the sub reaches −500 feet in depth for the second time.

APPLICATIONS

29. BRIDGE CONSTRUCTION Find the coordinates of each rivet, weld, and anchor.

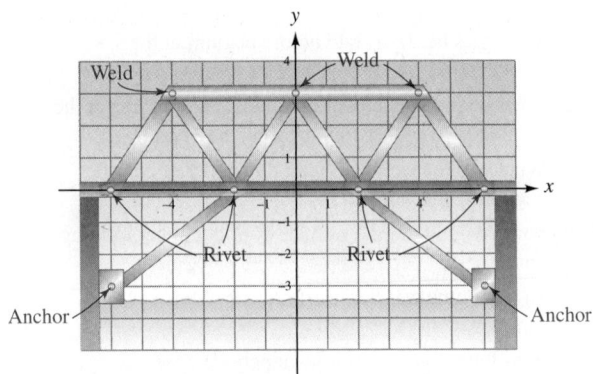

Scale: 1 unit = 8 ft

30. GOLF A golfer is videotaped and then has her swing displayed on a computer monitor so that it can be analyzed. Give the coordinates of the highlighted points.

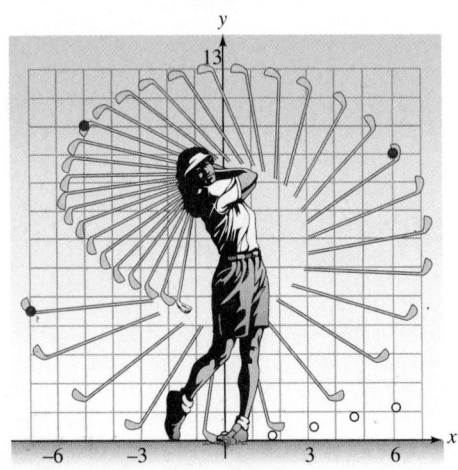

31. GAMES In the game Battleship, coordinates are used to locate ships. What are the coordinates of the ship shown? Express each answer in the form (letter, number).

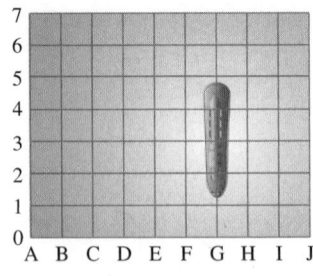

32. MAPS Use coordinates of the form (number, letter) to locate each of the following on the map: Rockford, Mount Carroll, Harvard, and the intersection of state Highway 251 and U.S. Highway 30.

33. from Campus to Careers
   Dental Assistant

Dentists describe teeth as being located in one of four *quadrants* as shown below.

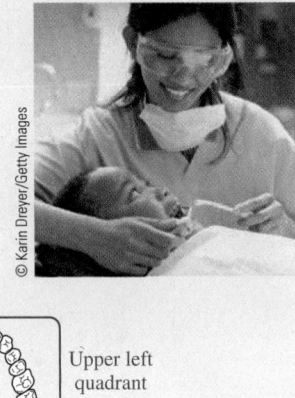

   a. How many teeth are in the *upper left quadrant*?

   b. Why would the upper left quadrant appear on the right in the illustration?

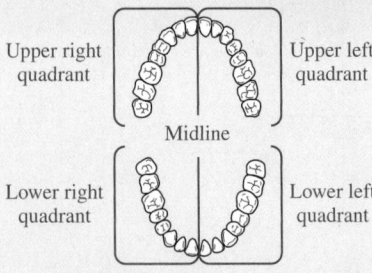

34. WATER PRESSURE The graphs show how the path of a stream of water changes when the hose is held at two different angles.

   a. At which angle does the stream of water shoot up higher? How much higher?

   b. At which angle does the stream of water shoot out farther? How much farther?

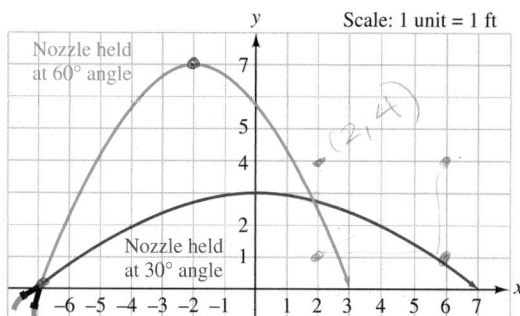

38. BOATING The table below shows the cost to rent a sailboat for a given number of hours. Plot the data in the table as ordered pairs. Then draw a straight line through the points.

   a. What does it cost to rent the boat for 3 hours?

   b. For how long can the boat be rented for $60?

   c. What does it cost to rent the boat for 9 hours?

Rental time (hr)	Cost ($)
2	20
4	30
6	40

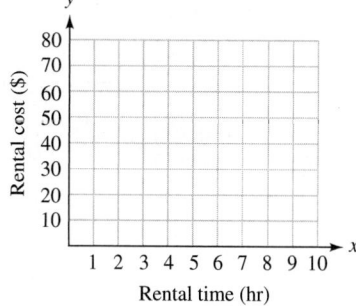

35. GEOMETRY Three vertices (corners) of a rectangle are (2, 1), (6, 1), and (6, 4). Find the coordinates of the fourth vertex. Then find the area of the rectangle.

36. GEOMETRY Three vertices (corners) of a right triangle are $(-1, -7)$, $(-5, -7)$, and $(-5, -2)$. Find the area of the triangle.

39. DEPRECIATION The table below shows the value (in thousands of dollars) of a color copier at various lengths of time after its purchase. Plot the data in the table as ordered pairs. Then draw a straight line passing through the points.

   a. What does the point (3, 7) on the graph tell you?

   b. Find the value of the copier when it is 7 years old.

   c. After how many years will the copier be worth $2,500?

37. TRUCKS The table below shows the number of miles that an 18-wheel truck can be driven on a given number of gallons of diesel fuel. Plot the data in the table as ordered pairs. Then draw a straight line through the points.

   a. How far can the truck go on 4 gallons of fuel?

   b. How many gallons of fuel are needed to travel a distance of 30 miles?

   c. How far can the truck go on 7 gallons of fuel?

Age (yr)	Value ($1,000)
3	7
4	5.5
5	4

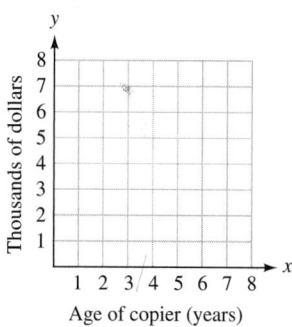

Fuel (gal)	Distance (mi)
2	10
3	15
5	25

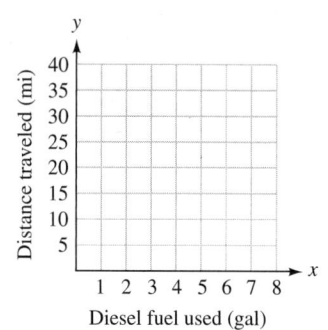

40. SWIMMING   The table below shows the number of people at a public swimming pool at various times during the day. Plot the data in the table as ordered pairs. Then draw a straight line passing through the points.

a. How many people will be at the pool at 6 P.M.?

b. At what time will there be 250 people at the pool?

c. At what time will the number of people at the pool be half of what it was at noon?

Time	Number of people
0	350
3	200
5	100

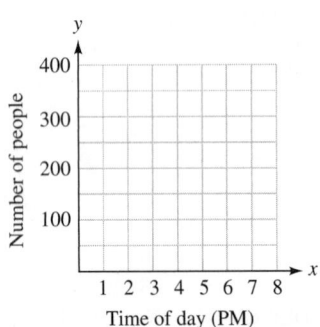

WRITING

41. Explain why the point $(-3, 3)$ is not the same as the point $(3, -3)$.

42. Explain how to plot the point $(-2, 5)$.

43. Explain why the coordinates of the origin are $(0, 0)$.

44. Explain this diagram.

II $(-, +)$	I $(+, +)$
III $(-, -)$	IV $(+, -)$

REVIEW

45. Solve $AC = \frac{2}{3}h - T$ for h.

46. Solve $5(x + 1) \le 2(x - 3)$. Write the solution set in interval notation and graph it.

47. Evaluate: $\dfrac{-4(4 + 2) - 2^3}{|-12 - 4(5)|}$

48. Simplify: $\dfrac{24}{54}$

CHALLENGE PROBLEMS

49. In what quadrant does a point lie if the *sum* of its coordinates is negative and the *product* of its coordinates is positive?

50. Draw line segment $\overline{AB}$ with endpoints $A(6, 5)$ and $B(-4, 5)$. Suppose that the x-coordinate of a point C is the average of the x-coordinates of points A and B, and the y-coordinate of point C is the average of the y-coordinates of points A and B. Find the coordinates of point C. Why is C called the midpoint of $\overline{AB}$?

SECTION 3.2
Graphing Linear Equations

Objectives

❶ Determine whether an ordered pair is a solution of an equation.

❷ Complete ordered-pair solutions of equations.

❸ Construct a table of solutions.

❹ Graph linear equations by plotting points.

❺ Use graphs of linear equations to solve applied problems.

In this section, we will discuss equations that contain two variables. Such equations are often used to describe algebraic relationships between two quantities. To see a mathematical picture of these relationships, we will construct graphs of their equations.

1 Determine Whether an Ordered Pair Is a Solution of an Equation.

We have previously solved **equations in one variable.** For example, $x + 3 = 9$ is an equation in x. If we subtract 3 from both sides, we see that 6 is the solution. To verify this, we replace x with 6 and note that the result is a true statement: $9 = 9$.

In this chapter, we extend our equation-solving skills to find solutions of **equations in two variables.** To begin, let's consider $y = x - 1$, an equation in x and y.

A solution of $y = x - 1$ is a pair of values, one for x and one for y, that make the equation true. To illustrate, suppose x is 5 and y is 4. Then we have:

$$y = x - 1 \quad \text{This is the given equation.}$$
$$4 \overset{?}{=} 5 - 1 \quad \text{Substitute 5 for x and 4 for y.}$$
$$4 = 4 \quad \text{True}$$

Since the result is a true statement, $x = 5$ and $y = 4$ is a solution of $y = x - 1$. We write the solution as the ordered pair (5, 4), with the value of x listed first. We say that (5, 4) *satisfies* the equation.

In general, a **solution of an equation in two variables** is an ordered pair of numbers that makes the equation a true statement.

> **Notation**
> Equations in two variables often involve the variables x and y. However, other letters can be used. For example, $a - 3b = 5$ and $n = 4m + 6$ are equations in two variables.

EXAMPLE 1   Is $(-1, -3)$ a solution of $y = x - 1$?

Strategy   We will substitute -1 for x and -3 for y and see whether the resulting equation is true.

Why   An ordered pair is a *solution* of $y = x - 1$ if replacing the variables with the values of the ordered pair results in a true statement.

Solution
$$y = x - 1 \quad \text{This is the given equation.}$$
$$-3 \overset{?}{=} -1 - 1 \quad \text{Substitute } -1 \text{ for x and } -3 \text{ for y.}$$
$$-3 = -2 \quad \text{False}$$

Since $-3 = -2$ is false, $(-1, -3)$ is not a solution of $y = x - 1$.

 Self Check 1   Is (9, 8) a solution of $y = x - 1$?
Now Try   **Problem 17**

---

2 Complete Ordered-Pair Solutions of Equations.

If only one of the values of an ordered-pair solution is known, we can substitute it into the equation to determine the other value.

EXAMPLE 2   Complete the solution $(-5, \quad)$ of the equation $y = -2x + 3$.

Strategy   We will substitute the known x-coordinate of the solution into the given equation.

Why   We can use the resulting equation in one variable to find the unknown y-coordinate of the solution.

Solution   In the ordered pair $(-5, \quad)$, the x-value is -5; the y-value is not known. To find y, we substitute -5 for x in the equation and evaluate the right side.

$y = -2x + 3$	This is the given equation.
$y = -2(-5) + 3$	Substitute -5 for x.
$y = 10 + 3$	Do the multiplication.
$y = 13$	This is the missing y-coordinate of the solution.

The completed ordered pair is $(-5, 13)$.

Self Check 2   Complete the solution $(-2, \underline{\quad})$ of the equation $y = 4x - 2$.

Now Try   **Problems 29 and 31**

Solutions of equations in two variables are often listed in a **table of solutions** (or **table of values**).

EXAMPLE 3   Complete the table of solutions for $3x + 2y = 5$.

x	y	(x, y)
7		$(7, \quad)$
	4	$(\quad, 4)$

Strategy   In each case we will substitute the known coordinate of the solution into the given equation.

Why   We can solve the resulting equation in one variable to find the unknown coordinate of the solution.

Solution   In the first row, we are given an x-value of 7. To find the corresponding y-value, we substitute 7 for x and solve for y.

x	y	(x, y)
7	-8	$(7, -8)$

$3x + 2y = 5$	This is the given equation.
$3(7) + 2y = 5$	Substitute 7 for x.
$21 + 2y = 5$	Do the multiplication.
$2y = -16$	To isolate the variable term, 2y, subtract 21 from both sides.
$y = -8$	To isolate y, divide both sides by 2. This is the missing y-coordinate of the solution.

A solution of $3x + 2y = 5$ is $(7, -8)$. It is entered in the table on the left.

In the second row, we are given a y-value of 4. To find the corresponding x-value, we substitute 4 for y and solve for x.

x	y	(x, y)
7	-8	$(7, -8)$
-1	4	$(-1, 4)$

$3x + 2y = 5$	This is the given equation.
$3x + 2(4) = 5$	Substitute 4 for y.
$3x + 8 = 5$	Do the multiplication.
$3x = -3$	To isolate the variable term, 3x, subtract 8 from both sides.
$x = -1$	To isolate x, divide both sides by 3. This is the missing x-coordinate of the solution.

Another solution is $(-1, 4)$. It is entered in the table on the left.

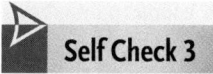

Self Check 3	Complete the table of solutions for $3x + 2y = 5$.

x	y	(x, y)
___	-2	$(\_\_, -2)$
5	___	$(5, \_\_)$

Now Try **Problem 37**

3 **Construct a Table of Solutions.**

To find a solution of an equation in two variables, we can select a number, substitute it for one of the variables, and find the corresponding value of the other variable. For example, to find a solution of $y = x - 1$, we can select a value for x, say, -4, substitute -4 for x in the equation, and find y.

x	y	(x, y)
-4	-5	$(-4, -5)$

$$y = x - 1$$
$$y = -4 - 1 \quad \text{Substitute } -4 \text{ for } x.$$
$$y = -5$$

The ordered pair $(-4, -5)$ is a solution. We list it in the table on the left.

To find another solution of $y = x - 1$, we select another value for x, say, -2, and find the corresponding y-value.

x	y	(x, y)
-4	-5	$(-4, -5)$
-2	-3	$(-2, -3)$

$$y = x - 1$$
$$y = -2 - 1 \quad \text{Substitute } -2 \text{ for } x.$$
$$y = -3$$

A second solution is $(-2, -3)$, and we list it in the table of solutions.

If we let $x = 0$, we can find a third ordered pair that satisfies $y = x - 1$.

x	y	(x, y)
-4	-5	$(-4, -5)$
-2	-3	$(-2, -3)$
0	-1	$(0, -1)$

$$y = x - 1$$
$$y = 0 - 1 \quad \text{Substitute } 0 \text{ for } x.$$
$$y = -1$$

A third solution is $(0, -1)$, which we also add to the table of solutions.

We can find a fourth solution by letting $x = 2$, and a fifth solution by letting $x = 4$.

x	y	(x, y)
-4	-5	$(-4, -5)$
-2	-3	$(-2, -3)$
0	-1	$(0, -1)$
2	1	$(2, 1)$
4	3	$(4, 3)$

$$y = x - 1 \qquad\qquad y = x - 1$$
$$y = 2 - 1 \quad \text{Substitute } 2 \text{ for } x. \qquad y = 4 - 1 \quad \text{Substitute } 4 \text{ for } x.$$
$$y = 1 \qquad\qquad\qquad y = 3$$

A fourth solution is $(2, 1)$ and a fifth solution is $(4, 3)$. We add them to the table.

Since we can choose any real number for x, and since any choice of x will give a corresponding value of y, it is apparent that the equation $y = x - 1$ has *infinitely many solutions*. We have found five of them: $(-4, -5)$, $(-2, -3)$, $(0, -1)$, $(2, 1)$, and $(4, 3)$.

4 **Graph Linear Equations by Plotting Points.**

It is impossible to list the infinitely many solutions of the equation $y = x - 1$. However, to show all of its solutions, we can draw a mathematical "picture" of them. We call this picture the *graph of the equation*.

Notation
The graph only shows a part of the line. The arrowheads indicate that it extends indefinitely in both directions.

To graph $y = x - 1$, we plot the ordered pairs shown in the table on a rectangular coordinate system. Then we draw a straight line through the points, because the graph of any solution of $y = x - 1$ will lie on this line. Furthermore, every point on this line represents a solution. We call the line the **graph of the equation.** It represents all of the solutions of $y = x - 1$.

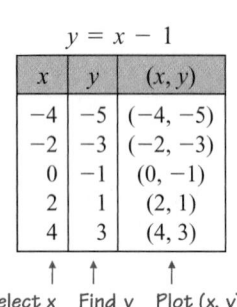

$y = x - 1$

x	y	(x, y)
-4	-5	$(-4, -5)$
-2	-3	$(-2, -3)$
0	-1	$(0, -1)$
2	1	$(2, 1)$
4	3	$(4, 3)$

↑   ↑   ↑
Select x   Find y   Plot (x, y)

Construct a table of solutions.

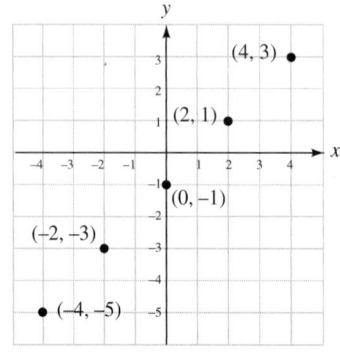

Plot the ordered pairs.

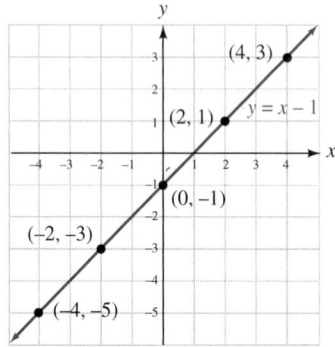

Draw a straight line through the points.
This is the *graph of the equation.*

The equation $y = x - 1$ is said to be *linear* and its graph is a line. By definition, a linear equation in two variables is any equation that can be written in the following **standard** or **general form,** where the variable terms appears on one side of an equal symbol and a constant appears on the other.

Linear Equations

A **linear equation in two variables** is an equation that can be written in the form

$$Ax + By = C$$

where A, B, and C are real numbers and A and B are not both 0.

Success Tip
The exponent on each variable of a linear equation is an understood 1. For example, $y = 2x + 4$ can be thought of as $y^1 = 2x^1 + 4$.

Some more examples of linear equations are

$$y = 2x + 4, \qquad 2x + 3y = 12, \qquad \text{and} \qquad 3x = 5y$$

Linear equations can be graphed in several ways. Generally, the form in which an equation is written determines the method that we use to graph it. To graph linear equations solved for y, such as $y = 2x + 4$, we can use the following method.

Graphing Linear Equations Solved for y by Plotting Points

1. Find three ordered pairs that are solutions of the equation by selecting three values for x and calculating the corresponding values of y.
2. Plot the solutions on a rectangular coordinate system.
3. Draw a straight line passing through the points. If the points do not lie on a line, check your computations.

EXAMPLE 4   Graph: $y = 2x + 4$

Strategy   We will find three solutions of the equation, plot them on a rectangular coordinate system, and then draw a straight line passing through the points.

Why   To *graph* a linear equation in two variables means to make a drawing that represents all of its solutions.

Solution   To find three solutions of this linear equation, we select three values for *x* that will make the computations easy. Then we find each corresponding value of *y*.

Success Tip

When selecting *x*-values for a table of solutions, a rule of thumb is to choose a negative number, a positive number, and 0. When $x = 0$, the computations to find *y* are usually quite simple.

If x = −2	*If x = 0*	*If x = 2*
$y = 2x + 4$	$y = 2x + 4$	$y = 2x + 4$
$y = 2(-2) + 4$	$y = 2(0) + 4$	$y = 2(2) + 4$
$y = -4 + 4$	$y = 0 + 4$	$y = 4 + 4$
$y = 0$	$y = 4$	$y = 8$
$(-2, 0)$ is a solution.	$(0, 4)$ is a solution.	$(2, 8)$ is a solution.

We enter the results in a table of solutions and plot the points. Then we draw a straight line through the points and label it $y = 2x + 4$.

Success Tip

Since two points determine a line, only two points are needed to graph a linear equation. However, we should plot a third point as a check. If the three points do not lie on a straight line, then at least one of them is in error.

$y = 2x + 4$

x	*y*	(x, y)
−2	0	$(-2, 0)$
0	4	$(0, 4)$
2	8	$(2, 8)$

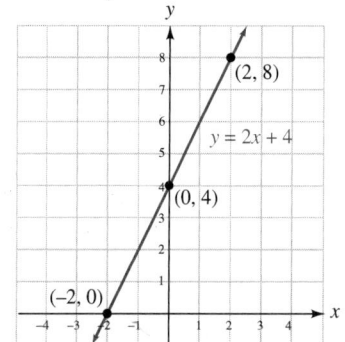

As a check, we can pick two points that the line appears to pass through, such as $(1, 6)$ and $(-1, 2)$. When we substitute their coordinates into the given equation, the two true statements that result indicate that $(1, 6)$ and $(-1, 2)$ are solutions and that the graph of the line is correctly drawn.

Check:

$y = 2x + 4$	$y = 2x + 4$
$6 \stackrel{?}{=} 2(1) + 4$	$2 \stackrel{?}{=} 2(-1) + 4$
$6 \stackrel{?}{=} 2 + 4$	$2 \stackrel{?}{=} -2 + 4$
$6 = 6$ True	$2 = 2$ True

 Self Check 4   Graph: $y = 2x - 2$

Now Try   **Problem 41**

EXAMPLE 5   Graph: $y = -3x$

Strategy   We will find three solutions of the equation, plot them on a rectangular coordinate system, and then draw a straight line passing through the points.

Why   To *graph* a linear equation in two variables means to make a drawing that represents all of its solutions.

Solution   To find three solutions, we begin by selecting three *x*-values: -1, 0, and 1. Then we find the corresponding values of *y*. If $x = -1$, we have

$y = -3x$ This is the equation to graph.

$y = -3(-1)$ Substitute -1 for x.

$y = 3$

$(-1, 3)$ is a solution.

In a similar manner, we find the *y*-values for *x*-values of 0 and 1, and record the results in a table of solutions. After plotting the ordered pairs, we draw a straight line through the points and label it $y = -3x$.

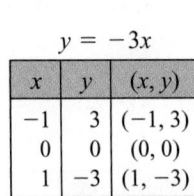 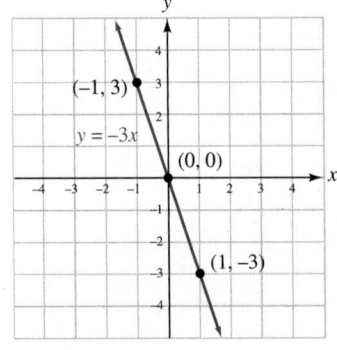

$y = -3x$

x	y	(x, y)
-1	3	$(-1, 3)$
0	0	$(0, 0)$
1	-3	$(1, -3)$

Self Check 5   Graph: $y = -4x$
Now Try   **Problem 45**

To graph linear equations in *x* and *y* using the method discussed in this section, the variable *y* must be isolated on one side of the equation.

EXAMPLE 6   Graph $2x + 3y = -12$ by first solving for *y*.

Strategy   We will use properties of equality to solve the given equation for *y*. Then we will use the point-plotting method of this section to graph the resulting equivalent equation.

Why   The calculations to find several solutions of a linear equation in two variables are usually easier when the equation is solved for *y*.

Solution   To solve for *y*, we proceed as follows.

$2x + 3y = -12$ This is the given equation.

$2x + 3y - 2x = -2x - 12$ To isolate the variable term 3y on the left side, subtract 2x from both sides. When solving for y, it is common practice to write the subtraction (or addition) of a variable term *before* the constant term.

$3y = -2x - 12$ On the left side, combine like terms, 2x − 2x = 0.

Notation

Note that the division by 3 on the right side of the equation is done term-by-term instead of with a single fraction bar. It is common practice to write

$$\frac{-2x}{3} - \frac{12}{3} \quad \text{not} \quad \frac{-2x - 12}{3}$$

Success Tip

When we chose x-values that are multiples of the denominator 3, the corresponding y-values are integers, and not difficult to plot fractions.

$$\frac{3y}{3} = \frac{-2x}{3} - \frac{12}{3} \qquad \text{To isolate } y, \text{ undo the multiplication by 3 by dividing both sides by 3.}$$

$$y = -\frac{2}{3}x - 4 \qquad \text{Write } \frac{-2x}{3} \text{ as } -\frac{2}{3}x. \text{ Simplify: } \frac{12}{3} = 4.$$

Since $y = -\frac{2}{3}x - 4$ is equivalent to $2x + 3y = -12$, we can use it to draw the graph of $2x + 3y = -12$.

To find solutions of $y = -\frac{2}{3}x - 4$, each value of x must be multiplied by $-\frac{2}{3}$. This computation is made easier if we select x-values that are *multiples of the denominator 3,* such as -3, 0, and 6. For example, if $x = -3$, we have

$$y = -\frac{2}{3}x - 4 \qquad \text{This is the equation to graph.}$$

$$y = -\frac{2}{3}(-3) - 4 \qquad \text{Substitute } -3 \text{ for } x.$$

$$y = 2 - 4 \qquad \text{Multiply: } -\frac{2}{3}(-3) = 2. \text{ This step is simpler if we select } x\text{-values that are multiples of 3.}$$

$$y = -2$$

Thus, $(-3, -2)$ is a solution.

Two more solutions, one for $x = 0$ and one for $x = 6$, can be found in a similar way, and entered in a table. We plot the ordered pairs, draw a straight line through the points, and label the line as $y = -\frac{2}{3}x - 4$ or as $2x + 3y = -12$.

$$2x + 3y = -12$$
$$\text{or}$$
$$y = -\frac{2}{3}x - 4$$

x	y	(x, y)
-3	-2	$(-3, -2)$
0	-4	$(0, -4)$
6	-8	$(6, -8)$

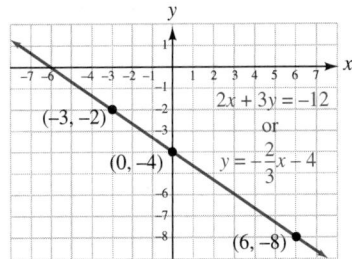

 Self Check 6   Graph $5x - 2y = -2$ by first solving for y.

Now Try   **Problem 67**

⑤ Use Graphs of Linear Equations to Solve Applied Problems.

When linear equations are used to model real-life situations, they are often written in variables other than x and y. In such cases, we must make the appropriate changes when labeling the table of solutions and the graph of the equation.

EXAMPLE 7   *Cleaning Windows.* The linear equation $A = -0.03n + 32$ estimates the amount A of glass cleaning solution (in ounces) that is left in the bottle after the sprayer trigger has been pulled a total of n times. Graph the equation and use the graph to estimate the amount of solution that is left after 500 sprays.

Strategy   We will find three solutions of the equation, plot them on a rectangular coordinate system, and then draw a straight line passing through the points.

Why   We can use the graph to estimate the amount of solution left after any number of sprays.

Solution   Since A depends on n in the equation $A = -0.03n + 32$, solutions will have the form (n, A). To find three solutions, we begin by selecting three values of n. Because the number of trials cannot be negative, and the computations to find A involve decimal multiplication, we select 0, 100, and 1,000. For example, if $n = 100$, we have

$A = -0.03n + 32$ *This is the equation to graph.*

$A = -0.03(\mathbf{100}) + 32$

$A = -3 + 32$ *Multiply: $-0.03(100) = -3$.*

$A = 29$

Thus, (100, 29) is a solution. It indicates that after 100 sprays, 29 ounces of cleaner will be left in the bottle.

In the same way, solutions are found for $n = 0$ and $n = 1,000$ and listed in the table. Then the ordered pairs are plotted and a straight line is drawn through the points.

To graphically estimate the amount of solution that is left after 500 sprays, we draw the dashed blue lines, as shown. Reading on the vertical A-axis, we see that after 500 sprays, about 17 ounces of glass cleaning solution would be left.

Now Try **Problem 77**

Success Tip

Since we selected large n-values, such as 100 and 1,000, the horizontal n-axis was scaled in units of 100. Since the corresponding A-values range from 2 to 32, the vertical A-axis was scaled in units of 4.

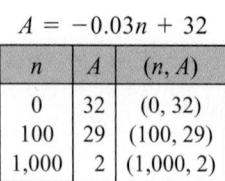

$A = -0.03n + 32$

n	A	(n, A)
0	32	(0, 32)
100	29	(100, 29)
1,000	2	(1,000, 2)

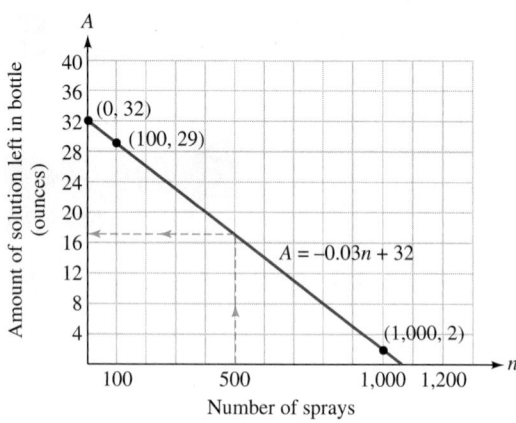

1. Yes   2. $(-2, -10)$ 3.

x	y	(x, y)
3	-2	$(3, -2)$
5	-5	$(5, -5)$

4.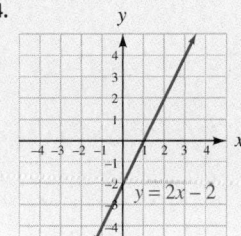

5.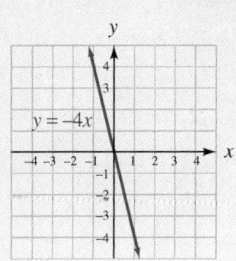

6.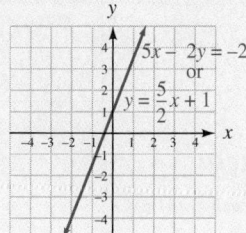

STUDY SET
3.2

VOCABULARY

Fill in the blanks.

1. $y = 9x + 5$ is an equation in _____ variables, x and y.

2. A _____ of an equation in two variables is an ordered pair of numbers that makes the equation a true statement.

3. Solutions of equations in two variables are often listed in a _____ of solutions.

4. The line that represents all of the solutions of a linear equation is called the _____ of the equation.

5. $y = 3x + 8$ is called a _____ equation because its graph is a line.

6. The _____ form of a linear equation in two variables is $Ax + By = C$.

CONCEPTS

7. Consider: $y = -3x + 6$
   a. How many variables does the equation contain?
   b. Does $(4, -6)$ satisfy the equation?
   c. Is $(-2, 0)$ a solution?
   d. How many solutions does this equation have?

8. To graph a linear equation, three solutions were found, they were plotted (in black), and a straight line was drawn through them, as shown in the next column.
   a. Looking at the graph, complete the table of solutions.
   b. From the graph, determine three other solutions of the equation.

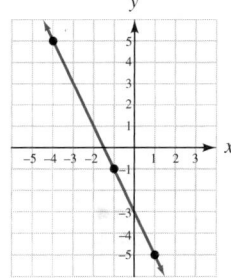

x	y	(x, y)
-4		(,)
-1		(,)
1		(,)

9. The graph of $y = -2x - 3$ is shown in Problem 8. Fill in the blanks: Every point on the graph represents an ordered-pair _____ of $y = -2x - 3$ and every ordered-pair solution is a _____ on the graph.

10. The graph of a linear equation is shown.
    a. If the coordinates of point M are substituted into the equation, will the result be true or false?
    b. If the coordinates of point N are substituted into the equation, will the result be true or false?

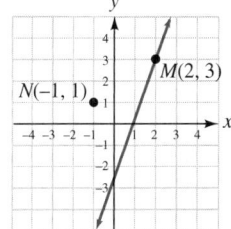

11. Suppose you are making a table of solutions for each given equation. What three x-values would you select to make the calculations for finding the corresponding y-values the easiest?
    a. $y = \frac{4}{5}x + 2$
    b. $y = 0.6x + 500$

12. A table of solutions for a linear equation is shown below. When constructing the graph of the equation, how would you scale the x-axis and the y-axis?

x	y	(x, y)
-20	600	$(-20, 600)$
5	100	$(5, 100)$
35	-500	$(35, -500)$

NOTATION

Complete each solution.

13. Verify that $(-2, 6)$ is a solution of $y = -x + 4$.

$$y = -x + 4$$
$$\underset{=}{\overset{2}{}} = -(\quad) + 4$$
$$6 \overset{2}{=} \quad + 4$$
$$6 = \quad$$

14. Solve $5x + 3y = 15$ for y.

$$5x + 3y - 5x = \quad + 15$$
$$\quad = -5x + 15$$
$$\frac{3y}{\quad} = \frac{-5x}{\quad} + \frac{15}{\quad}$$
$$y = \quad x + \quad$$

15. a. Rewrite the linear equation $y = \frac{1}{2}x + 7$ showing the understood exponents on the variables.

$$y^{\quad} = \frac{1}{2}x^{\quad} + 7$$

b. Explain why $y = x^2 + 2$ and $y = x^3 - 4$ are not linear equations.

16. Complete the labeling of the table of solutions and graph of $c = -a + 4$.

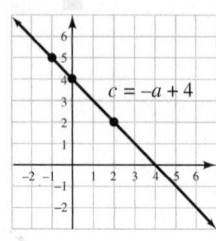

		$(\quad , \quad)$
-1	5	$(-1, 5)$
0	4	$(0, 4)$
2	2	$(2, 2)$

GUIDED PRACTICE

Determine whether each equation has the given ordered pair as a solution. See Example 1.

17. $y = 5x - 4$; $(1, 1)$

18. $y = -2x + 3$; $(2, -1)$

19. $7x - 2y = 3$; $(2, 6)$

20. $10x - y = 10$; $(0, 0)$

21. $x + 12y = -12$; $(0, -1)$

22. $-2x + 3y = 0$; $(-3, -2)$

23. $3x - 6y = 12$; $(-3.6, -3.8)$

24. $8x + 4y = 10$; $(-0.5, 3.5)$

25. $y - 6x = 12$; $\left(\frac{5}{6}, 7\right)$

26. $y + 8x = 4$; $\left(\frac{3}{4}, 2\right)$

27. $y = -\frac{3}{4}x + 8$; $(-8, 12)$

28. $y = \frac{1}{6}x - 2$; $(-12, 4)$

For each equation, complete the solution. See Example 2.

29. $y = -5x - 4$; $(-3, \quad)$

30. $y = 8x + 30$; $(-6, \quad)$

31. $4x - 5y = -4$; $(\quad, 4)$

32. $7x + y = -12$; $(\quad, 2)$

33. $y = \frac{x}{4} + 9$; $(16, \quad)$

34. $y = \frac{x}{6} - 8$; $(48, \quad)$

35. $7x = 4y$; $(\quad, -2)$

36. $11x = 16y$; $(\quad, -3)$

Complete each table of solutions. See Example 3.

37. $y = 2x - 4$

x	y	(x, y)
8		
	8	

38. $y = 3x + 1$

x	y	(x, y)
-3		
	-2	

39. $3x - y = -2$

x	y	(x, y)
-5		
	-1	

40. $5x - 2y = -15$

x	y	(x, y)
5		
	0	

Construct a table of solutions and then graph each equation. See Examples 4 and 5.

41. $y = 2x - 3$

42. $y = 3x + 1$

43. $y = 5x - 4$

44. $y = 6x - 3$

45. $y = x$

46. $y = 4x$

47. $y = -x - 1$

48. $y = -x + 2$

49. $y = -2x + 1$

50. $y = -3x + 2$

51. $y = \frac{x}{3}$

52. $y = -\frac{x}{3} - 1$

53. $y = -\frac{1}{2}x$

54. $y = \frac{3}{4}x$

55. $y = \frac{3}{8}x - 6$

56. $y = -\frac{3}{2}x + 2$

57. $y = \frac{2}{3}x - 2$

58. $y = \frac{5}{6}x - 5$

59. $y = 1.5x - 4$

60. $y = 0.5x + 3$

Solve each equation for y and then graph it. See Example 6.

61. $3y = 12x + 15$

62. $5y = 20x - 30$

63. $-6y = 30x + 12$

64. $-3y = 9x - 15$

65. $8x + 4y = 16$

66. $14x + 7y = 28$

67. $5y - x = 20$

68. $4y - x = 8$

69. $7x - y = 1$

70. $2x - y = -3$

71. $7y = -2x$

72. $6y = -4x$

APPLICATIONS

73. BILLIARDS   The path traveled by the black 8-ball is described by the equations $y = 2x - 4$ and $y = -2x + 12$. Construct a table of solutions for $y = 2x - 4$ using the x-values 1, 2, and 4. Do the same for $y = -2x + 12$, using the x-values 4, 6, and 8. Then graph the path of the 8-ball.

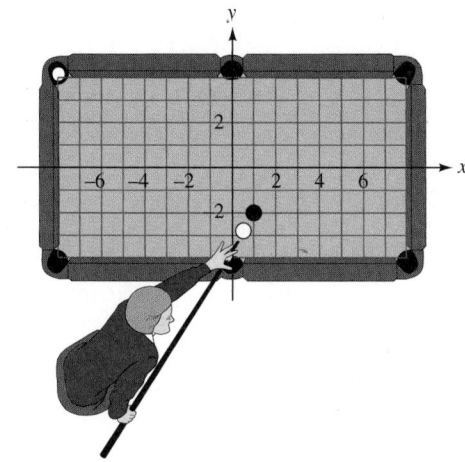

74. PING-PONG   The path traveled by a Ping-Pong ball is described by the equations $y = \frac{1}{2}x + \frac{3}{2}$ and $y = -\frac{1}{2}x - \frac{3}{2}$. Construct a table of solutions for $y = \frac{1}{2}x + \frac{3}{2}$ using the x-values 7, 3, and -3. Do the same for $y = -\frac{1}{2}x - \frac{3}{2}$, using the x-values -3, -5, and -7. Then graph the path of the ball.

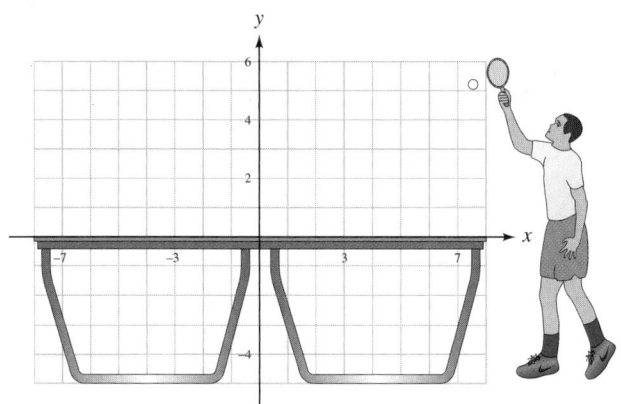

75. DEFROSTING POULTRY   The number of hours h needed to defrost a turkey weighing p pounds in the refrigerator can be estimated by $h = 5p$. Graph the equation and use the graph to estimate the time needed to defrost a 25-pound turkey. (Source: helpwithcooking.com.)

76. OWNING A CAR   In 2006, the average cost c (in dollars) to own and operate a car was estimated by $c = 0.52m$, where m represents the number of miles driven. Graph the equation and use the graph to estimate the cost of operating a car that is driven 25,000 miles. (Source: Automobile Association of America.)

77. HOUSEKEEPING   The linear equation $A = -0.02n + 16$ estimates the amount A of furniture polish (in ounces) that is left in the bottle after the sprayer trigger has been pulled a total of n times. Graph the equation and use the graph to estimate the amount of polish that is left after 650 sprays.

78. SHARPENING PENCILS   The linear equation $L = -0.04t + 8$ estimates the length L (in inches) of a pencil after it has been inserted into a sharpener and the handle turned a total of t times. Graph the equation and use the graph to estimate the length of the pencil after 75 turns of the handle.

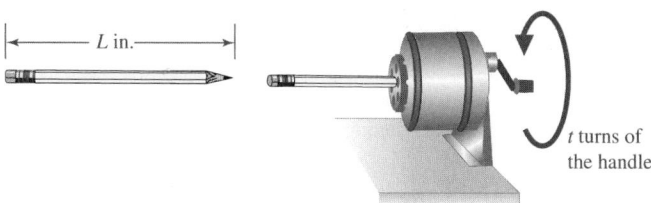

79. NFL TICKETS   The average ticket price p to a National Football League game during the years 1990–2005 is approximated by $p = \frac{12}{5}t + 22$, where t is the number of years after 1990. Graph this equation and use the graph to predict the average ticket price in 2020. (Source: Team Marketing Report, NFL.)

80. U.S. AUTOMOBILE ACCIDENTS   The number n of lives saved by seat belts during the years 1995–2004 is estimated by $n = 615t + 9,900$, where t is the number of years after 1995. Graph this equation and use the graph to predict the number of lives that will be saved by seat belts in 2015. (Source: Bureau of Transportation Statistics.)

81. RAFFLES   A private school is going to sell raffle tickets as a fund raiser. Suppose the number n of raffle tickets that will be sold is predicted by the equation $n = -20p + 300$, where p is the price of a raffle ticket in dollars. Graph the equation and use the graph to predict the number of raffle tickets that will be sold at a price of $6.

82. CATS   The number n of cat owners (in millions) in the United States during the years 1995–2004 is estimated by $n = \frac{13}{20}t + 32$, where t is the number of years after 1995. Graph this equation and use the graph to predict the number of cat owners in the United States in 2015. (Source: Pet Food Institute.)

WRITING

83. When we say that $(-2, -6)$ is a solution of $y = x - 4$, what do we mean?

84. What is a table of solutions?

85. What does it mean when we say that a linear equation in two variables has infinitely many solutions?

86. A linear equation and a graph are two ways of describing a relationship between two quantities. Which do you think is more informative and why?

87. From geometry, we know that two points determine a line. Why is it a good practice when graphing linear equations to find and plot three solutions instead of just two?

88. A student found three solutions of a linear equation in two variables and plotted them as shown. What conclusion can be made about the location of the points?

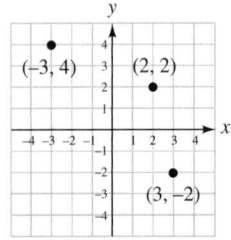

89. Two students were asked to graph $y = 3x - 1$. One made the table of solutions on the left. The other made the table on the right. The tables are completely different. Could they both be correct? Explain.

x	y	(x, y)
0	-1	$(0, -1)$
2	5	$(2, 5)$
3	8	$(3, 8)$

x	y	(x, y)
-2	-7	$(-2, -7)$
-1	-4	$(-1, -4)$
1	2	$(1, 2)$

90. Both graphs below are of the same linear equation $y = 10x$. Why do the graphs have a different appearance?

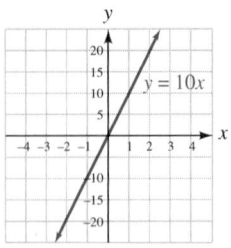 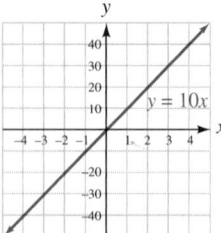

REVIEW

91. Simplify: $-(-5 - 4c)$

92. Write the set of integers.

93. Find the volume, to the nearest tenth of a cubic foot, of a sphere with radius 6 feet.

94. Solve: $-2(a + 3) = 3(a - 5)$

CHALLENGE PROBLEMS

*Graph each of the following **nonlinear** equations in two variables by constructing a table of solutions consisting of seven ordered pairs. These equations are called nonlinear, because their graphs are not straight lines.*

95. $y = x^2 + 1$

96. $y = x^3 - 2$

97. $y = |x| - 2$

98. $y = (x + 2)^2$

SECTION 3.3
Intercepts

Objectives

1 Identify intercepts of a graph.

2 Graph linear equations by finding intercepts.

3 Identify and graph horizontal and vertical lines.

4 Obtain information from intercepts.

5 (Optional) Use a calculator to graph linear equations.

In this section, we will graph linear equations by determining the points where their graphs intersect the *x*-axis and the *y*-axis. These points are called the *intercepts* of the graph.

Identify Intercepts of a Graph.

The graph of $y = 2x - 4$ is shown below. We see that the graph crosses the y-axis at the point $(0, -4)$; this point is called the **y-intercept** of the graph. The graph crosses the x-axis at the point $(2, 0)$; this point is called the **x-intercept** of the graph.

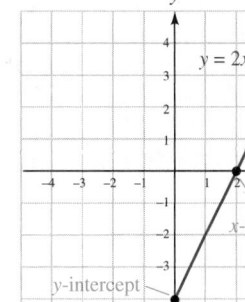

> **The Language of Algebra**
> The point where a line *intersects* the x- or y-axis is called an *intercept*.

 EXAMPLE 1   For the graphs in figures (a) and (b), give the coordinates of the x- and y-intercepts.

a. 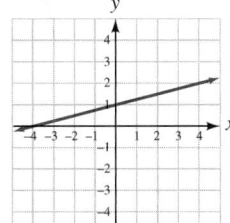   **b.** 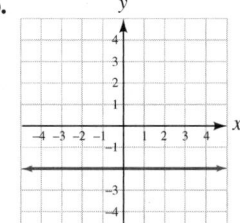   **c.**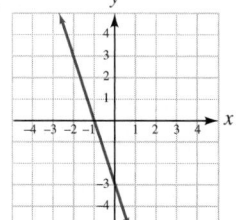

Strategy   We will determine where each graph (shown in red) crosses the x-axis and the y-axis.

Why   The point at which a graph crosses the x-axis is the x-intercept and the point at which a graph crosses the y-axis is the y-intercept.

Solution

a. In figure (a), the graph crosses the x-axis at $(-4, 0)$. This is the x-intercept. The graph crosses the y-axis at $(0, 1)$. This is the y-intercept.

b. In figure (b), the horizontal line does not cross the x-axis; there is no x-intercept. The graph crosses the y-axis at $(0, -2)$. This is the y-intercept.

 Self Check 1   Give the coordinates of the x- and y-intercept of the graph in Figure (c).

Now Try   **Problem 11**

From the previous examples, we see that a y-intercept has an x-coordinate of 0, and an x-intercept has a y-coordinate of 0. These observations suggest the following procedures for finding the intercepts of a graph from its equation.

Finding Intercepts	To find the *y*-intercept, substitute 0 for *x* in the given equation and solve for *y*.
	To find the *x*-intercept, substitute 0 for *y* in the given equation and solve for *x*.

2 **Graph Linear Equations by Finding Intercepts.**

Plotting the *x*- and *y*-intercepts of a graph and drawing a line through them is called the **intercept method of graphing a line.** This method is useful when graphing linear equations written in the standard (general) form $Ax + By = C$.

EXAMPLE 2   Graph $x - 3y = 6$ by finding the *y*- and *x*-intercepts.

Strategy   We will let $x = 0$ to find the *y*-intercept of the graph. We will then let $y = 0$ to find the *x*-intercept.

Why   Since two points determine a line, the *y*-intercept and *x*-intercept are enough information to graph this linear equation.

Solution

y-intercept: $x = 0$

$x - 3y = 6$

$0 - 3y = 6$ Substitute 0 for x.

$-3y = 6$

$y = -2$ To isolate y, divide both sides by −3.

The *y*-intercept is $(0, -2)$.

x-intercept: $y = 0$

$x - 3y = 6$

$x - 3(0) = 6$ Substitute 0 for y.

$x - 0 = 6$

$x = 6$

The *x*-intercept is $(6, 0)$.

Since each intercept of the graph is a solution of the equation, we enter the intercepts in the table of solutions below.

As a check, we find one more point on the line. We select a convenient value for *x*, say, 3, and find the corresponding value of *y*.

$x - 3y = 6$

$3 - 3y = 6$ Substitute 3 for x.

$-3y = 3$ To isolate the variable term, −3y, subtract 3 from both sides.

$y = -1$ To isolate y, divide both sides by −3.

Therefore, $(3, -1)$ is a solution. It is also entered in the table.

We plot the intercepts and the check point, draw a straight line through them, and label the line as $x - 3y = 6$.

> **Success Tip**
> The check point should lie on the same line as the *x*- and *y*-intercepts. If it does not, check your work to find the incorrect coordinate or coordinates.

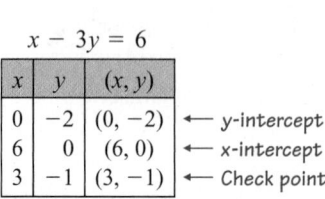

$x - 3y = 6$

x	y	(x, y)	
0	−2	(0, −2)	← y-intercept
6	0	(6, 0)	← x-intercept
3	−1	(3, −1)	← Check point

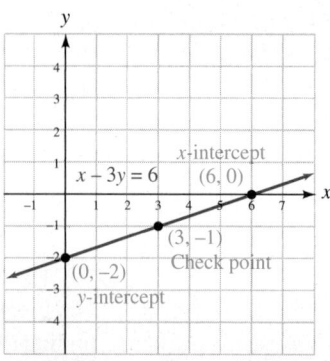

> ▷ **Self Check 2**   Graph $x - 2y = 2$ by finding the intercepts.
> *Now Try* **Problem 27**

The computations for finding intercepts can be simplified if we realize what occurs when we substitute 0 for y or 0 for x in an equation written in the form $Ax + By = C$.

EXAMPLE 3   Graph $40x + 3y = -120$ by finding the y- and x-intercepts.

Strategy   We will let $x = 0$ to find the y-intercept of the graph. We will then let $y = 0$ to find the x-intercept.

Why   Since two points determine a line, the y-intercept and x-intercept are enough information to graph this linear equation.

Solution   When we substitute 0 for x, it follows that the term $40x$ will be equal to 0. Therefore, to find the y-intercept, we can cover the $40x$ and solve the remaining equation for y.

$$40x + 3y = -120 \qquad \text{If } x = 0, \text{ then } 40x = 40(0) = 0. \text{ Cover the } 40x \text{ term.}$$
$$y = -40 \qquad \text{To solve } 3y = -120, \text{ divide both sides by 3.}$$

The y-intercept is $(0, -40)$.

When we substitute 0 for y, it follows that the term $3y$ will be equal to 0. Therefore, to find the x-intercept, we can cover the $3y$ and solve the remaining equation for x.

$$40x + 3y = -120 \qquad \text{If } y = 0, \text{ then } 3y = 3(0) = 0. \text{ Cover the } 3y \text{ term.}$$
$$x = -3 \qquad \text{To solve } 40x = -120, \text{ divide both sides by 40.}$$

The x-intercept is $(-3, 0)$.

We can find a third solution by selecting a convenient value for x and finding the corresponding value for y. If we choose $x = -6$, we find that $y = 40$. The solution $(-6, 40)$ is entered in the table, and the equation is graphed as shown.

The Language of Algebra
This method to find the intercepts of the graph of a linear equation is commonly referred to as the *cover-over method.*

Caution
When using the cover-over method to find the y-intercept, be careful not to cover the sign in front of the y-term.

Success Tip
To fit y-values of 40 and -40 on the graph, the y-axis was scaled in units of 10.

$40x + 3y = -120$

x	y	(x, y)	
0	-40	$(0, -40)$	← y-intercept
-3	0	$(-3, 0)$	← x-intercept
-6	40	$(-6, 40)$	← Check point

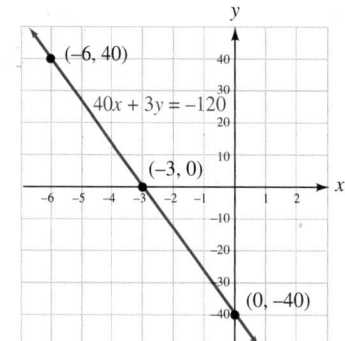

> ▷ **Self Check 3**   Graph $32x + 5y = -160$ by finding the intercepts.
> *Now Try* **Problem 35**

EXAMPLE 4   Graph $3x = -5y + 8$ by finding the intercepts.

Strategy   We will let $x = 0$ to find the y-intercept of the graph. We will then let $y = 0$ to find the x-intercept.

Why   Since two points determine a line, the y-intercept and x-intercept are enough information to graph this linear equation.

Solution   We find the intercepts and select $x = 1$ to find a check point.

y-intercept: x = 0	*x-intercept: y = 0*	*Check point: x = 1*
$3x = -5y + 8$	$3x = -5y + 8$	$3x = -5y + 8$
$3(0) = -5y + 8$	$3x = -5(0) + 8$	$3(1) = -5y + 8$
$0 = -5y + 8$	$3x = 8$	$3 = -5y + 8$
$-8 = -5y$	$x = \dfrac{8}{3}$	$-5 = -5y$
$\dfrac{8}{5} = y$	$x = 2\frac{2}{3}$	$1 = y$
$1\frac{3}{5} = y$	The x-intercept is $\left(2\frac{2}{3}, 0\right)$.	A check point is $(1, 1)$.

The y-intercept is $\left(0, 1\frac{3}{5}\right)$.

The ordered pairs are plotted as shown, and a straight line is then drawn through them.

Success Tip
When graphing, it is often helpful to write any coordinates that are improper fractions as mixed numbers. For example:
$$\left(\tfrac{8}{3}, 0\right) = \left(2\tfrac{2}{3}, 0\right)$$

$3x = -5y + 8$

x	y	(x, y)	
0	$\frac{8}{5} = 1\frac{3}{5}$	$\left(0, 1\frac{3}{5}\right)$	← y-intercept
$\frac{8}{3} = 2\frac{2}{3}$	0	$\left(2\frac{2}{3}, 0\right)$	← x-intercept
1	1	$(1, 1)$	← Check point

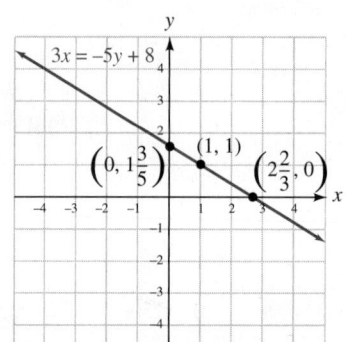

 Self Check 4   Graph $8x = -4y + 15$ by finding the intercepts.
Now Try   Problem 39

EXAMPLE 5   Graph $2x + 3y = 0$ by finding the intercepts.

Strategy   We will let $x = 0$ to find the y-intercept of the graph. We will then let $y = 0$ to find the x-intercept.

Why   Since two points determine a line, the y-intercept and x-intercept are enough information to graph this linear equation.

Solution   When we find the y- and x-intercepts (shown on the next page), we see that they are both $(0, 0)$. In this case, the line passes through the origin. Since we are using two points and a check point to graph lines, we need to find two more ordered-pair solutions.

If $x = 3$, we see that $(3, -2)$ is a solution. And if $x = -3$, we see that $(-3, 2)$ is also a solution. These two solutions and the origin are plotted and a straight line is drawn through them to give the graph of $2x + 3y = 0$.

y-intercept: x = 0	*x-intercept: y = 0*	*Let x = 3*	*Let x = −3*
$2x + 3y = 0$	$2x + 3y = 0$	$2x + 3y = 0$	$2x + 3y = 0$
$2(0) + 3y = 0$	$2x + 3(0) = 0$	$2(3) + 3y = 0$	$2(-3) + 3y = 0$
$3y = 0$	$2x = 0$	$6 + 3y = 0$	$-6 + 3y = 0$
$y = 0$	$x = 0$	$3y = -6$	$3y = 6$
		$y = -2$	$y = 2$
The *y*-intercept is $(0, 0)$.	The *x*-intercept is $(0, 0)$.	$(3, -2)$ is a solution.	$(-3, 2)$ is a solution.

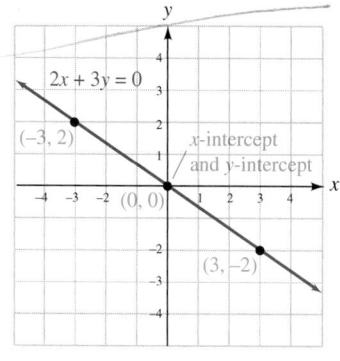

$2x + 3y = 0$

x	y	(x, y)	
0	0	$(0, 0)$	← The x-intercept and y-intercept.
3	−2	$(3, -2)$	← A solution.
−3	2	$(-3, 2)$	← This solution serves as a check point.

▶ **Self Check 5**   Graph $5x - 2y = 0$ by finding the intercepts.

Now Try   **Problem 47**

③   **Identify and Graph Horizontal and Vertical Lines.**

Equations such as $y = 4$ and $x = -3$ are linear equations, because they can be written in the general form $Ax + By = C$. For example, $y = 4$ is equivalent to $0x + 1y = 4$ and $x = -3$ is equivalent to $1x + 0y = -3$. We now discuss how to graph these types of linear equations.

EXAMPLE 6   Graph: $y = 4$

Strategy   To find three ordered-pair solutions of this equation to plot, we will select three values for x and use 4 for y each time.

Why   The given equation requires that $y = 4$.

Solution   We can write the equation in general form as $0x + y = 4$. Since the coefficient of x is 0, the numbers chosen for x have no effect on y. The value of y is always 4. For example, if $x = 2$, we have

$$0x + y = 4 \quad \text{This is the given equation, } y = 4, \text{ written in standard (general) form.}$$
$$0(2) + y = 4 \quad \text{Substitute 2 for x.}$$
$$y = 4 \quad \text{Simplify the left side.}$$

One solution is (2, 4). To find two more solutions, we choose $x = 0$ and $x = -3$. For any x-value, the y-value is always 4, so we enter (0, 4) and (−3, 4) in the table. If we plot the ordered pairs and draw a straight line through the points, the result is a horizontal line. The y-intercept is (0, 4) and there is no x-intercept.

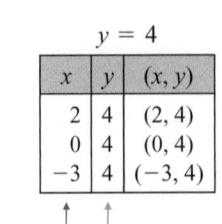

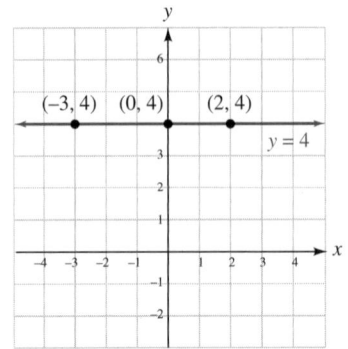

x	y	(x, y)
2	4	(2, 4)
0	4	(0, 4)
−3	4	(−3, 4)

Choose any number for x.   Each value of y must be 4.

Self Check 6   Graph: $y = -2$

Now Try   Problem 51

EXAMPLE 7   Graph: $x = -3$

Strategy   To find three ordered-pair solutions of this equation to plot, we must select −3 for x each time.

Why   The given equation requires that $x = -3$.

Solution   We can write the equation in general form as $x + 0y = -3$. Since the coefficient of y is 0, the numbers chosen for y have no effect on x. The value of x is always −3. For example, if $y = -2$, we have

$x + 0y = -3$ This is the given equation, x = −3, written in standard (general) form.

$x + 0(-2) = -3$ Substitute −2 for y.

$x = -3$ Simplify the left side.

One solution is (−3, −2). To find two more solutions, we choose $y = 0$ and $y = 3$. For any y-value, the x-value is always −3, so we enter (−3, 0) and (−3, 3) in the table. If we plot the ordered pairs and draw a straight line through the points, the result is a vertical line. The x-intercept is (−3, 0) and there is no y-intercept.

$x = -3$

x	y	(x, y)
−3	−2	(−3, −2)
−3	0	(−3, 0)
−3	3	(−3, 3)

Each value of x must be −3.   Choose any number for y.

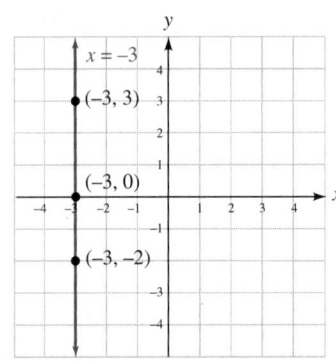

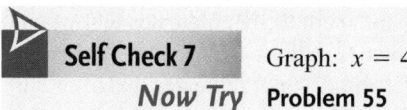

Self Check 7   Graph: $x = 4$
   Now Try   **Problem 55**

From the results of Examples 6 and 7, we have the following facts.

Equations of Horizontal and Vertical Lines	The equation $y = b$ represents the horizontal line that intersects the y-axis at $(0, b)$. The equation $x = a$ represents the vertical line that intersects the x-axis at $(a, 0)$.

The graph of the equation $y = 0$ has special importance; it is the x-axis. Similarly, the graph of the equation $x = 0$ is the y-axis.

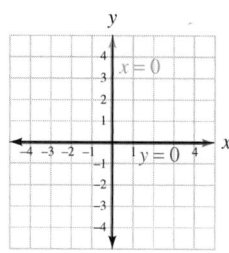

4 **Obtain Information from Intercepts.**

The ability to read and interpret graphs is a valuable skill. When analyzing a graph, we should locate and examine the intercepts. As the following example illustrates, the coordinates of the intercepts can yield useful information.

© Motoring Picture Library/Alamy

EXAMPLE 8   *Hybrid Mileage.*   The following graph shows city mileage data for a 2006 Toyota Prius Hybrid. What information do the intecepts give about the car?

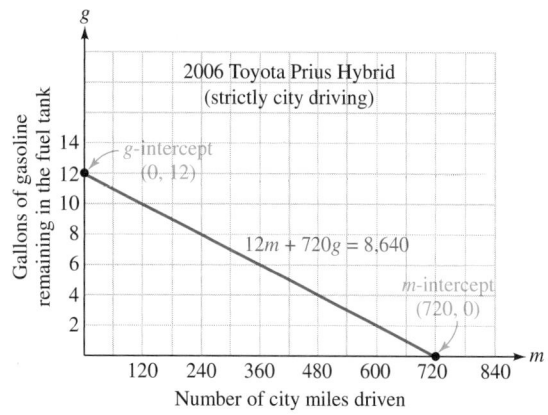

Strategy   We will determine where the graph (the line in red) intersects the g-axis and where it intersects the m-axis.

Why   Once we know the intercepts, we can interpret their meaning.

Solution   The g-intercept $(0, 12)$ indicates that when the car has been driven 0 miles, the fuel tank contains 12 gallons of gasoline. That is, the Prius has a 12-gallon fuel tank.

The m-intercept $(720, 0)$ indicates that after 720 miles of city driving, the fuel tank contains 0 gallons of gasoline. Thus, 720 miles of city driving can be done on 1 tank of gas in a Prius.

 Now Try   **Problem 81**

5 (Optional) Use a Calculator to Graph Linear Equations.

So far, we have graphed linear equations by making tables of solutions and plotting points. A graphing calculator can make the task of graphing much easier. However, a graphing calculator does not take the place of a working knowledge of the topics discussed in this chapter. It should serve as an aid to enhance your study of algebra.

The Viewing Window   The screen on which a graph is displayed is called the **viewing window.** The **standard window** has settings of

$$\text{Xmin} = -10, \qquad \text{Xmax} = 10, \qquad \text{Ymin} = -10, \qquad \text{and} \qquad \text{Ymax} = 10$$

which indicate that the minimum x- and y-coordinates used in the graph will be -10, and that the maximum x- and y-coordinates will be 10.

Graphing an Equation   To graph $y = x - 1$ using a graphing calculator, we press the **Y =** key and enter $x - 1$ after the symbol Y_1. Then we press the **GRAPH** key to see the graph.

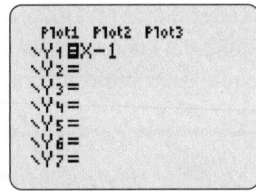 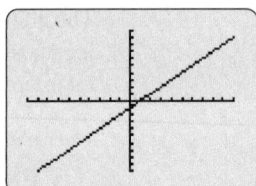

Change the Viewing Window   We can change the viewing window by pressing the **WINDOW** key and entering -4 for the minimum x- and y-coordinates and 4 for the maximum x- and y-coordinates. Then we press the **GRAPH** key to see the graph of $y = x - 1$ in more detail.

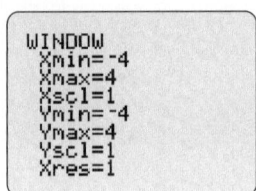 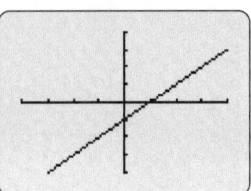

Solving an Equation for y   To graph $3x + 2y = 12$, we must first solve the equation for y.

$$3x + 2y = 12$$
$$2y = -3x + 12 \qquad \text{Subtract 3x from both sides.}$$
$$y = -\frac{3}{2}x + 6 \qquad \text{Divide both sides by 2.}$$

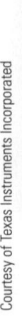 Courtesy of Texas Instruments Incorporated

Next, we press the **WINDOW** key to reenter the standard window settings, press **Y =** and enter $y = -\frac{3}{2}x + 6$, and press **GRAPH** to see the graph.

 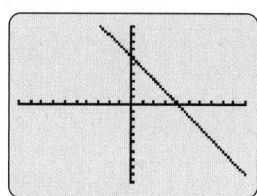

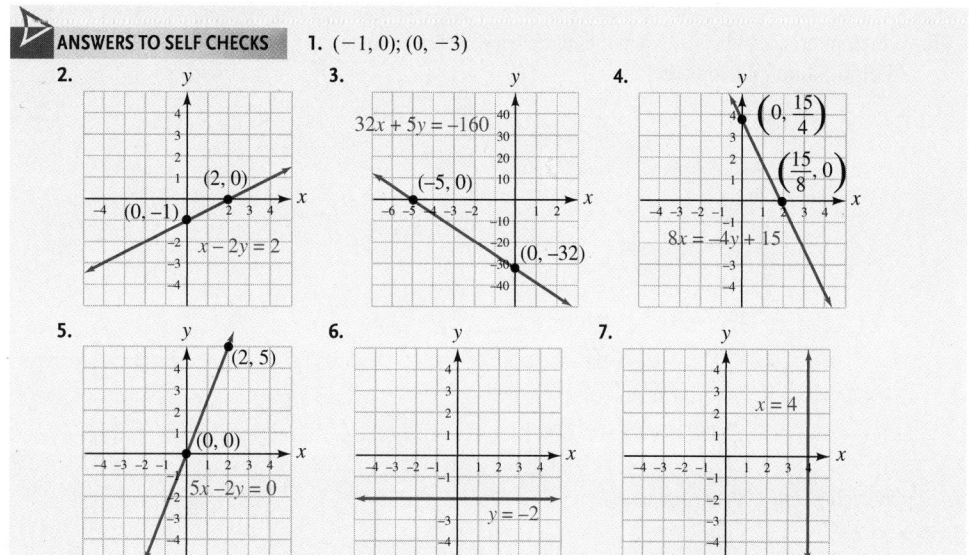

▷ **ANSWERS TO SELF CHECKS**   **1.** $(-1, 0); (0, -3)$

STUDY SET
3.3

VOCABULARY

Fill in the blanks.

1. The _____ of a line is the point where the line intersects the *x*-axis.

2. The *y*-intercept of a line is the point where the line _____ the *y*-axis.

3. The graph of $y = 4$ is a _____ line and the graph of $x = 6$ is a _____ line.

4. The intercept method is useful when graphing linear equations written in the _____ form $Ax + By = C$.

CONCEPTS

5. Fill in the blanks.

   a. To find the *y*-intercept of the graph of a line, substitute ___ for *x* in the equation and solve for ___.

   b. To find the *x*-intercept of the graph of a line, substitute ___ for *y* in the equation and solve for ___.

6. Complete the table of solutions and fill in the blanks.

$$3x + 2y = 6$$

x	y	(x, y)	
0			← -intercept
	0		← -intercept
−2			← point

7. a. Refer to the graph. Which intercept tells the purchase price of the machinery? What was that price?

b. Which intercept indicates when the machinery will have lost all of its value? When is that?

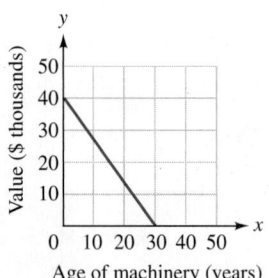

8. Match each graph with its equation.

a. $x = 2$ **b.** $y = 2$ **c.** $y = 2x$
d. $2x - y = 2$ **e.** $y = 2x + 2$ **f.** $y = -2x$

i. 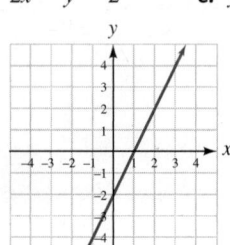   **ii.**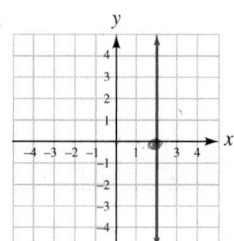

iii. 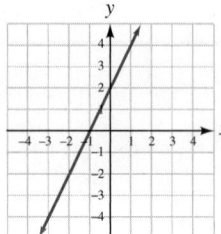   **iv.**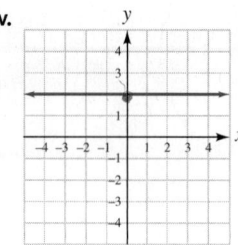

v. 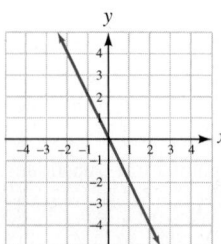   **vi.**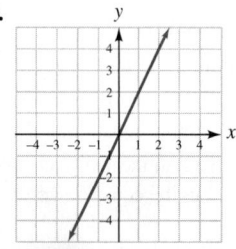

9. What is the equation of the *x*-axis? What is the equation of the *y*-axis?

10. Write the coordinates that are improper fractions as mixed numbers.

a. $\left(\frac{7}{2}, 0\right)$ **b.** $\left(0, -\frac{17}{3}\right)$

GUIDED PRACTICE

Give the coordinates of the intercepts of each graph. See Example 1.

11. 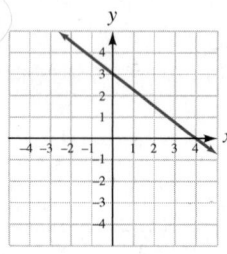   **12.**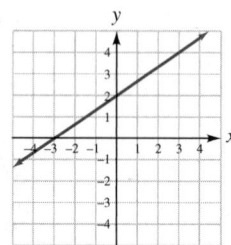

13. 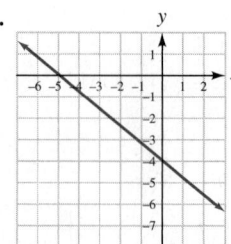   **14.**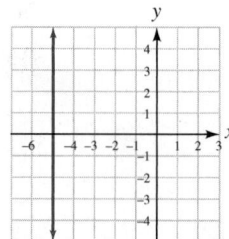

15. 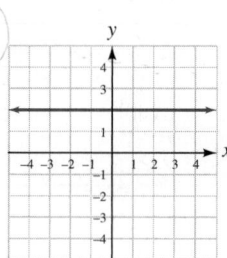   **16.**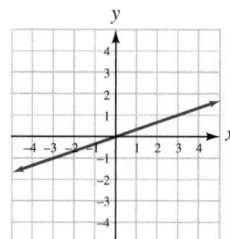

Estimate the coordinates of the intercepts of each graph. See Example 1.

17. 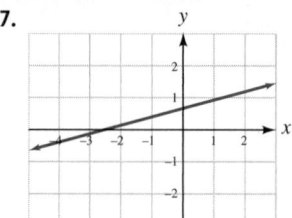   **18.**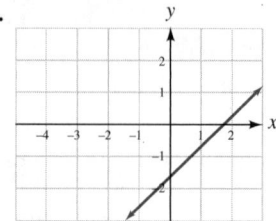

Find the x- and y-intercepts of the graph of each equation. Do not graph the line. See Example 2.

19. $8x + 3y = 24$

20. $5x + 6y = 30$

21. $7x - 2y = 28$

22. $2x - 9y = 36$

23. $-5x - 3y = 10$

24. $-9x - 5y = 25$

25. $6x + y = 9$

26. $x + 8y = 14$

Use the intercept method to graph each equation. See Example 2.

27. $4x + 5y = 20$

28. $3x + 4y = 12$

29. $5x + 15y = -15$

30. $8x + 4y = -24$

31. $x - y = -3$

32. $x - y = 3$

33. $x + 2y = -2$

34. $x + 2y = -4$

Use the intercept method to graph each equation. See Example 3.

35. $30x + y = -30$

36. $20x - y = -20$

37. $4x - 20y = 60$

38. $6x - 30y = 30$

Use the intercept method to graph each equation. See Example 4.

39. $3x + 4y = 8$

40. $2x + 3y = 9$

41. $-9x + 4y = 9$

42. $-5x + 4y = 15$

43. $3x - 4y = 11$

44. $5x - 4y = 13$

45. $9x + 3y = 10$

46. $4x + 4y = 5$

Use the intercept method to graph each equation. See Example 5.

47. $3x + 5y = 0$

48. $4x + 3y = 0$

49. $2x - 7y = 0$

50. $6x - 5y = 0$

Graph each equation. See Examples 6 and 7.

51. $y = 5$

52. $y = -3$

53. $y = 0$

54. $x = 0$

55. $x = -2$

56. $x = 5$

57. $x = \dfrac{4}{3}$

58. $y = -\dfrac{1}{2}$

59. $y - 2 = 0$ (*Hint:* Solve for y first.)

60. $x + 1 = 0$ (*Hint:* Solve for x first.)

61. $5x = 7.5$ (*Hint:* Solve for x first.)

62. $3y = 4.5$ (*Hint:* Solve for y first.)

TRY IT YOURSELF

Graph each equation.

63. $7x = 4y - 12$

64. $7x = 5y - 15$

65. $4x - 3y = 12$

66. $5x - 10y = 20$

67. $x = -\dfrac{5}{3}$

68. $y = \dfrac{5}{2}$

69. $y - 3x = -\dfrac{4}{3}$

70. $y - 2x = -\dfrac{9}{8}$

71. $7x + 3y = 0$

72. $4x - 5y = 0$

73. $-4x = 8 - 2y$

74. $-5x = 10 + 5y$

75. $3x = -150 - 5y$

76. $x = 50 - 5y$

77. $-3y = 3$

78. $-2x = 8$

APPLICATIONS

79. CHEMISTRY The relationship between the temperature T and volume V of a gas at a constant pressure is graphed below. The T-intercept of this graph is a very important scientific fact. It represents the lowest possible temperature, called **absolute zero**.

a. Estimate absolute zero.

b. What is the volume of the gas when the temperature is absolute zero?

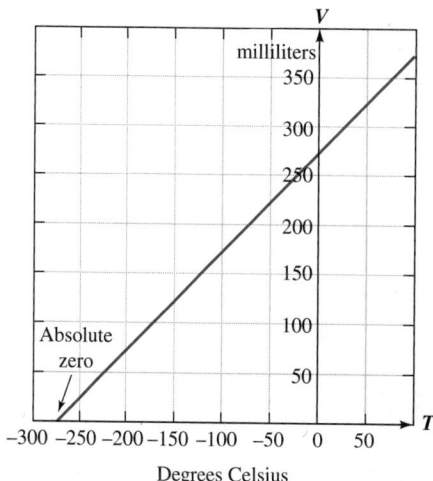

80. PHYSICS The graph shows the length L of a stretched spring (in inches) as different weights w (in pounds) are attached to it. What information about the spring does the L-intercept give us?

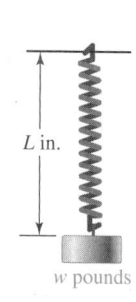

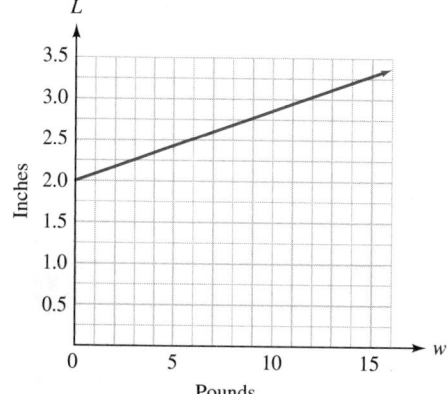

81. LANDSCAPING A developer is going to purchase x trees and y shrubs to landscape a new office complex. The trees cost $50 each and the shrubs cost $25 each. His budget is $5,000. This situation is modeled by the equation $50x + 25y = 5,000$. Use the intercept method to graph it.

  a. What information is given by the y-intercept?

  b. What information is given by the x-intercept?

82. EGGS The number of eggs eaten by an average American in one year has remained almost constant since the year 2000. See the graph below. Draw a straight line that passes through, or near, the data points. What is the equation of the line?

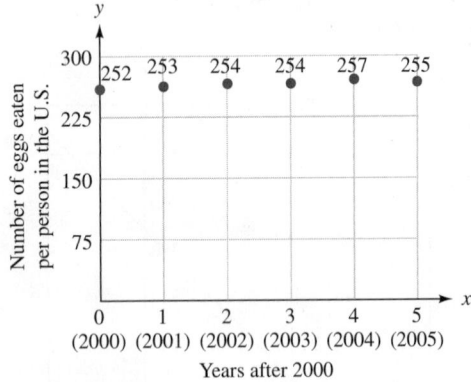

Number of eggs eaten per person in the U.S.

Years after 2000

Source: United Egg

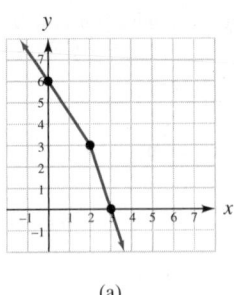

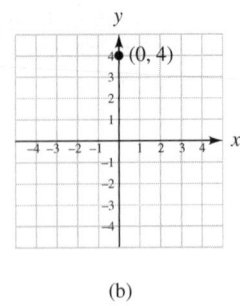

(a)   (b)

84. A student graphed the linear equation $y = 4$, as shown above in figure (b). Explain her error.

85. How do we find the intercepts of the graph of an equation without having to graph the equation?

86. In Section 3.2, we discussed a method to graph $y = 2x - 3$. In Section 3.3, we discussed a method to graph $2x + 3y = 6$. Briefly explain the steps involved in each method.

REVIEW

87. Simplify: $\dfrac{3 \cdot 5 \cdot 5}{3 \cdot 5 \cdot 5 \cdot 5}$

88. Simplify: $4\left(\dfrac{d}{2} - 3\right) - 5\left(\dfrac{2}{5}d - 1\right)$

89. Translate: Six less than twice x

90. Is -5 a solution of $2(3x + 10) = 5x + 6$?

CHALLENGE PROBLEMS

91. Where will the line $y = b$ intersect the line $x = a$?

92. Write an equation of the line that has an x-intercept of $(4, 0)$ and a y-intercept of $(0, 3)$.

93. What is the least number of intercepts a line can have? What is the greatest number a line can have?

94. On a rectangular coordinate system, draw a circle that has exactly two intercepts.

WRITING

83. To graph $3x + 2y = 12$, a student found the intercepts and a check point, and graphed them, as shown in figure (a). Instead of drawing a crooked line through the points, what should he have done?

SECTION 3.4
Slope and Rate of Change

Objectives

① Find the slope of a line from its graph.

② Find the slope of a line given two points.

③ Find slopes of horizontal and vertical lines.

④ Solve applications of slope.

⑤ Calculate rates of change.

⑥ Determine whether lines are parallel or perpendicular using slope.

In this section, we introduce a means of measuring the steepness of a line. We call this measure the *slope of the line,* and it can be found in several ways.

1 **Find the Slope of a Line from Its Graph.**

The **slope of a line** is a ratio that compares the vertical change with the corresponding horizontal change as we move along the line from one point to another.

As an example, let's find the slope of the line graphed below. To begin, we select two points on the line, *P* and *Q*. One way to move from *P* to *Q* is to start at point *P* and count upward 5 grid squares. Then, moving to the right, we count 6 grid squares to reach point *Q*. The vertical change in this movement is called the **rise.** The horizontal change is called the **run.** Notice that a right triangle, called a **slope triangle,** is created by this process.

The slope of a line is defined to be *the ratio of the vertical change to the horizontal change.* So we have

$$\text{slope} = \frac{\text{vertical change}}{\text{horizontal change}} = \frac{\text{rise}}{\text{run}} = \frac{5}{6}$$ This ratio is a comparison of the rise and the run using a quotient.

The slope of the line is $\frac{5}{6}$. This indicates that there is a rise (vertical change) of 5 units for each run (horizontal change) of 6 units.

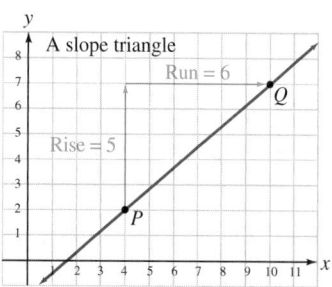

EXAMPLE 1   Find the slope of the line graphed in figure (a) below.

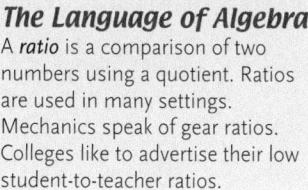 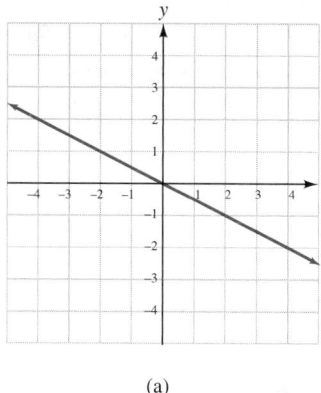 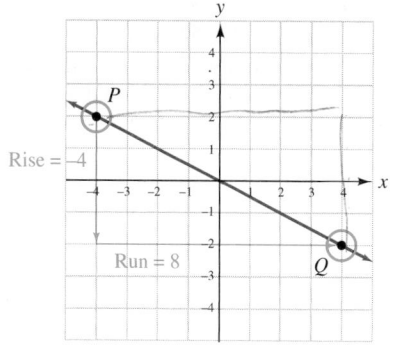 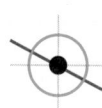

(a)   (b)

Pick two points on the line that also lie on the intersection of two grid lines.

Strategy   We will pick two points on the line, construct a slope triangle, and find the rise and run. Then we will write the ratio of the rise to the run.

Why   The slope of a line is the ratio of the rise to the run.

The Language of Algebra
Many historians believe that *m* was chosen to represent the slope of a line because it is the first letter of the French word *monter*, meaning to ascend or climb.

Solution   We begin by choosing two points on the line, *P* and *Q*, as shown in figure (b), on the previous page. One way to move from *P* to *Q* is to start at point *P* and count *downward* 4 grid squares. Because this movement is downward, the rise is -4. Then, moving right, we count 8 grid squares to reach *Q*. This indicates that the run is 8.

To find the slope of the line, we write a ratio of the rise to the run in simplified form. Usually the letter *m* is used to denote slope, so we have

$$m = \frac{\text{rise}}{\text{run}} = \frac{-4}{8} = -\frac{1}{2}$$

The slope of the line is $-\frac{1}{2}$.

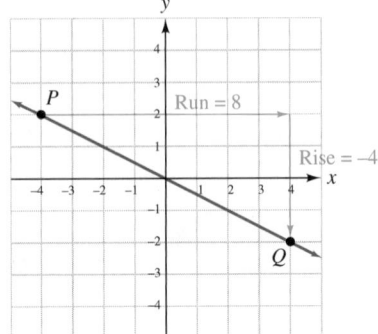

The movement from *P* to *Q* can be reversed. Starting at *P*, we can move to the right, a run of 8; and then downward, a rise of -4, to reach *Q*. With this approach, the slope triangle is above the line. When we form the ratio to find the slope, we get the same result as before:

$$m = \frac{\text{rise}}{\text{run}} = \frac{-4}{8} = -\frac{1}{2}$$

Success Tip
When drawing a slope triangle, movement upward or to the right is positive. Movement downward or to the left is negative.

Self Check 1   Find the slope of the line shown above using two points different from those used in the solution of Example 1.

Now Try   Problem 21

The identical answers from Example 1 and its Self Check illustrate an important fact: *For any line, the same value will be obtained no matter which two points on the line are used to find the slope.*

 2   Find the Slope of a Line Given Two Points.

The Language of Algebra
The prefix *sub* means below or beneath, as in submarine or subway. In x_1, x_2, y_1, and y_2, the subscripts 1 and 2 are written lower than the variable. They are not exponents.

We can generalize the graphic method for finding slope to develop a slope formula. To begin, we select points *P* and *Q* on the line shown in the figure below. To distinguish between the coordinates of these points, we use **subscript notation**. Point *P* has coordinates (x_1, y_1), which are read as "*x* sub 1 and *y* sub 1." Point *Q* has coordinates (x_2, y_2), which are read as "*x* sub 2 and *y* sub 2."

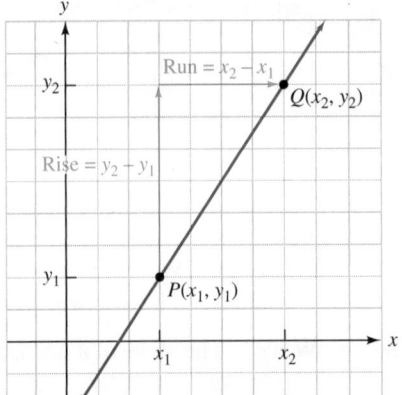

Success Tip
Recall that subtraction is used to measure change.

As we move from point P to point Q, the rise is the difference of the y-coordinates: $y_2 - y_1$. We call this difference the **change in y.** The run is the difference of the x-coordinates: $x_2 - x_1$. This difference is called the **change in x.** Since the slope is the ratio $\frac{\text{rise}}{\text{run}}$, we have the following formula for calculating slope.

Slope of a Line

The **slope** of a line passing through points (x_1, y_1) and (x_2, y_2) is

$$m = \frac{\text{vertical change}}{\text{horizontal change}} = \frac{\text{rise}}{\text{run}} = \frac{\text{change in } y}{\text{change in } x} = \frac{y_2 - y_1}{x_2 - x_1} \quad \text{if } x_2 \neq x_1$$

EXAMPLE 2   Find the slope of the line passing through $(1, 2)$ and $(3, 8)$.

Strategy   We will use the slope formula to find the slope of the line.

Why   We know the coordinates of two points on the line.

Solution   When using the slope formula, it makes no difference which point you call (x_1, y_1) and which point you call (x_2, y_2). If we let (x_1, y_1) be $(1, 2)$ and (x_2, y_2) be $(3, 8)$, then

Success Tip
The slope formula is a valuable tool because it allows us to calculate the slope of a line without having to view its graph.

$m = \dfrac{y_2 - y_1}{x_2 - x_1}$ This is the slope formula.

$m = \dfrac{8 - 2}{3 - 1}$ Substitute 8 for y_2, 2 for y_1, 3 for x_2, and 1 for x_1.

$m = \dfrac{6}{2}$ Do the subtractions.

$m = 3$ Simplify. Think of this as a $\frac{3}{1}$ rise-to-run ratio.

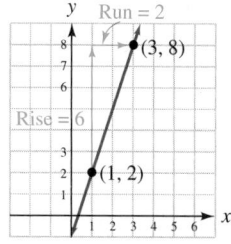

The slope of the line is 3. The graph of the line, including the slope triangle, is shown here. Note that we obtain the same value for the slope if we let $(x_1, y_1) = (3, 8)$ and $(x_2, y_2) = (1, 2)$.

$$m = \frac{y_2 - y_1}{x_2 - x_1} = \frac{2 - 8}{1 - 3} = \frac{-6}{-2} = 3$$

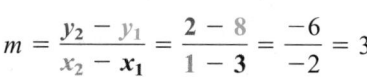

Self Check 2   Find the slope of the line passing through $(2, 1)$ and $(4, 11)$.

Now Try **Problem 33**

Caution   When using the slope formula, always subtract the y-coordinates and their corresponding x-coordinates in the same order. Otherwise, your answer will have the wrong sign.

$$m \neq \frac{y_2 - y_1}{x_1 - x_2} \qquad \text{and} \qquad m \neq \frac{y_1 - y_2}{x_2 - x_1}$$

EXAMPLE 3 Find the slope of the line that passes through $(-2, 4)$ and $(5, -6)$.

Strategy We will use the slope formula to find the slope of the line.

Why We know the coordinates of two points on the line.

Solution Since we know the coordinates of two points on the line, we can find its slope. If we let (x_1, y_1) be $(-2, 4)$ and (x_2, y_2) be $(5, -6)$, then

$$m = \frac{y_2 - y_1}{x_2 - x_1}$$ This is the slope formula.

$$m = \frac{-6 - 4}{5 - (-2)}$$ Substitute -6 for y_2, 4 for y_1, 5 for x_2, and -2 for x_1.

$$m = -\frac{10}{7}$$ Do the subtractions. We can write the result as $\frac{-10}{7}$ or $-\frac{10}{7}$.

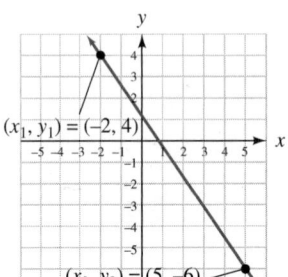

The slope of the line is $-\frac{10}{7}$.

If we graph the line by plotting the two points, we see that the line falls from left to right—a fact indicated by its negative slope.

> **Caution**
> Slopes are normally written as fractions, sometimes as decimals, but never as mixed numbers.
>   As with any fractional answer, always express slope in simplified form (lowest terms).

▷ **Self Check 3** Find the slope of the line that passes through $(-1, -2)$ and $(1, -7)$.

Now Try **Problem 39**

In Example 2, the slope of the line was positive. In Examples 1 and 3, the slopes of the lines were negative. In general, lines that rise from left to right have a positive slope. Lines that fall from left to right have a negative slope.

> **Success Tip**
> To classify the slope of a line as positive or negative, follow it from left to right, as you would read a sentence in a book.

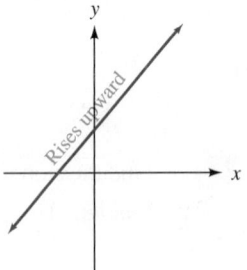

Positive slope

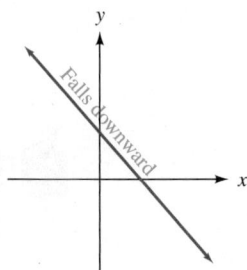

Negative slope

In the following illustration, we see a line with slope 3 is steeper than a line with slope of $\frac{5}{6}$, and a line with slope of $\frac{5}{6}$ is steeper than a line with slope of $\frac{1}{4}$. In general, *the larger the absolute value of the slope, the steeper the line.*

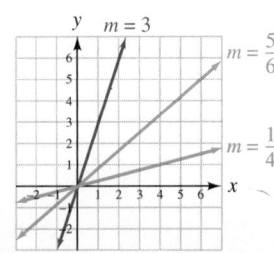

③ Find Slopes of Horizontal and Vertical Lines.

In the next two examples, we calculate the slope of a horizontal line and we show that a vertical line has no defined slope.

EXAMPLE 4   Find the slope of the line $y = 3$.

Strategy   We will find the coordinates of two points on the line.

Why   We can then use the slope formula to find the slope of the line.

Solution   The graph of $y = 3$ is a horizontal line. To find its slope, we select two points on the line: $(-2, 3)$ and $(3, 3)$. If (x_1, y_1) is $(-2, 3)$ and (x_2, y_2) is $(3, 3)$, we have

$$m = \frac{y_2 - y_1}{x_2 - x_1}$$ This is the slope formula.

$$m = \frac{3 - 3}{3 - (-2)}$$ Substitute 3 for y_2, 3 for y_1, 3 for x_2, and -2 for x_1.

$$m = \frac{0}{5}$$ Simplify the numerator and the denominator.

$$m = 0$$

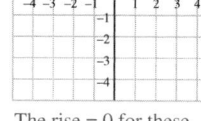

The rise $= 0$ for these two points.

The slope of the line $y = 3$ is 0.

Self Check 4   Find the slope of the line $y = 10$.

Now Try   **Problem 61**

The y-coordinates of any two points on a horizontal line will be the same, and the x-coordinates will be different. Thus, the numerator of $\frac{y_2 - y_1}{x_2 - x_1}$ will always be zero, and the denominator will always be nonzero. Therefore, the slope of a horizontal line is 0.

EXAMPLE 5   If possible, find the slope of the line $x = -2$.

Strategy   We will find the coordinate of two points on the line.

Why   We can then use the slope formula to find the slope of the line, if it exits.

Solution   The graph of $x = -2$ is a vertical line. To find its slope, we select two points on the line: $(-2, 3)$ and $(-2, -1)$. If (x_2, y_2) is $(-2, 3)$ and (x_1, y_1) is $(-2, -1)$, we have

Notation
This example explains why the definition of slope includes the restriction that $x_1 \neq x_2$.

$$m = \frac{y_2 - y_1}{x_2 - x_1}$$ This is the slope formula.

$$m = \frac{3 - (-1)}{-2 - (-2)}$$ Substitute 3 for y_2, -1 for y_1, -2 for x_2, and -2 for x_1.

$$m = \frac{4}{0}$$ Note that $x_1 = x_2$.
Simplify the numerator and the denominator.

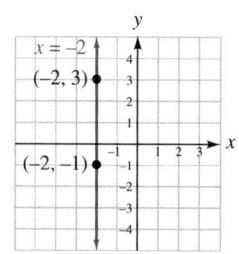

The run $= 0$ for these two points.

Since division by zero is undefined, $\frac{4}{0}$ has no meaning. The slope of the line $x = -2$ is undefined.

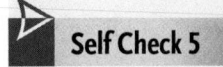

Self Check 5   If possible, find the slope of the line $x = 12$.

Now Try   Problem 67

The y-coordinates of any two points on a vertical line will be different, and the x-coordinates will be the same. Thus, the numerator of $\frac{y_2 - y_1}{x_2 - x_1}$ will always be nonzero, and the denominator will always be 0. Therefore, the slope of a vertical line is undefined.

We now summarize the results from Examples 4 and 5.

Slopes of Horizontal and Vertical Lines	Horizontal lines (lines with equations of the form $y = b$) have slope 0. Vertical lines (lines with equations of the form $x = a$) have undefined slope.

The Language of Algebra
Undefined and *0* do not mean the same thing. A horizontal line has a defined slope; it is **0**. A vertical line does not have a defined slope; we say its slope is *undefined*.

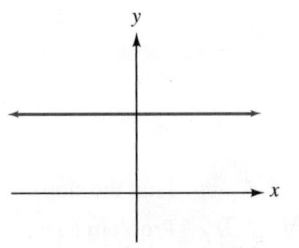

Horizontal line: 0 slope

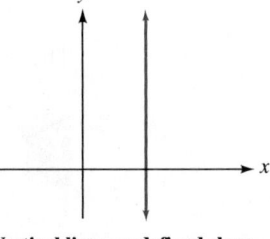

Vertical line: undefined slope

 4   **Solve Applications of Slope.**

The concept of slope has many applications. For example, architects use slope when designing ramps and roofs. Truckers must be aware of the slope, or *grade,* of the roads they travel. Mountain bikers ride up rocky trails and snow skiers speed down steep slopes.

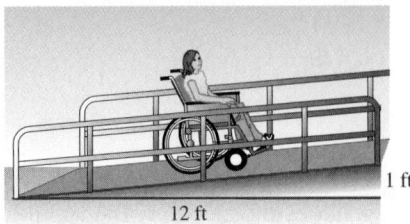

12 ft   1 ft

15 ft

100 ft

The *Americans with Disabilities Act* provides a guideline for the steepness of a ramp. The maximum slope for a wheelchair ramp is 1 foot of rise for every 12 feet of run: $m = \frac{1}{12}$.

The grade of an incline is its slope expressed as a percent: A 15% grade means a rise of 15 feet for every run of 100 feet: $m = \frac{15}{100}$, which simplifies to $\frac{3}{20}$.

EXAMPLE 6   ***Architecture.***   **Pitch** is the incline of a roof expressed as a ratio of the vertical rise to the horizontal run. Find the pitch of the roof shown in the illustration.

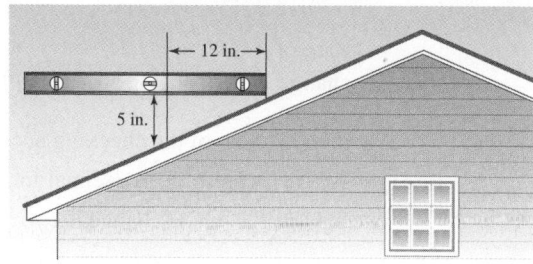

Strategy   We will determine the rise and the run of the roof from the illustration. Then we will write the ratio of the rise to the run.

Why   The pitch of a roof is its slope, and the slope of a line is the ratio of the rise to the run.

Solution   In the illustration, a level is used to create a slope triangle. From the triangle, we see that the rise is 5 and the run is 12.

$$m = \frac{\text{rise}}{\text{run}} = \frac{5}{12} \qquad \text{The roof has a } \tfrac{5}{12} \text{ pitch.}$$

 Now Try   **Problem 99**

5   **Calculate Rates of Change.**

We have seen that the slope of a line is a ratio of two numbers. If units are attached to a slope calculation, the result is called a **rate of change.** In general, a rate of change describes how much one quantity changes with respect to another. For example, we might speak of snow melting at the rate of 6 inches per day or a tourist exchanging money at the rate of 12 pesos per dollar.

EXAMPLE 7   ***Banking.***   A bank offers a business account with a fixed monthly fee, plus a service charge for each check written. The relationship between the monthly cost y and the number x of checks written is graphed below. At what rate does the monthly cost change?

> **Notation**
> In the graph, the symbol $\doteq$ indicates a break in the labeling of the vertical axis. The break enables us to omit a large portion of the grid that would not be used.

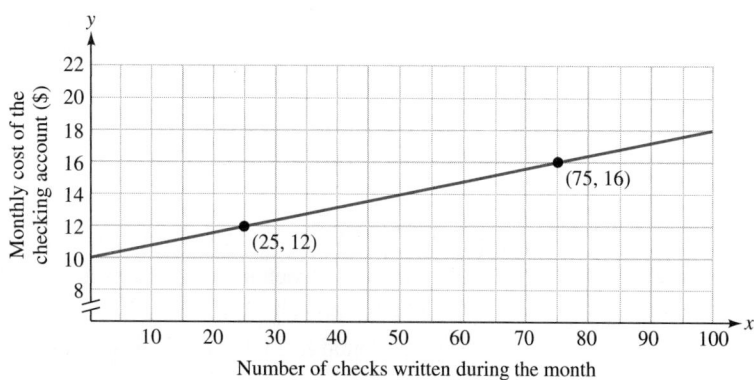

Strategy We will find the slope of the line and attach the proper units.

Why If units are attached to a slope calculation, the result is a rate of change.

Solution From the graph, we see that two points on the line are (25, 12) and (75, 16). If we let $(x_1, y_1) = (25, 12)$ and $(x_2, y_2) = (75, 16)$, we have

$$\frac{\text{Rate of}}{\text{change}} = \frac{(y_2 - y_1)\text{ dollars}}{(x_2 - x_1)\text{ checks}} = \frac{(16 - 12)\text{ dollars}}{(75 - 25)\text{ checks}} = \frac{4\text{ dollars}}{50\text{ checks}} = \frac{2\text{ dollars}}{25\text{ checks}}$$

> **The Language of Algebra**
> The preposition *per* means for each, or for every. When we say the rate of change is 8¢ *per* check, we mean 8¢ for each check.

The monthly cost of the checking account increases $2 for every 25 checks written.

We can express $\frac{2}{25}$ in decimal form by dividing the numerator by the denominator. Then we can write the rate of change in two other ways, using the word *per*, which indicates division.

Rate of change = $0.08 per check   or   Rate of change = 8¢ per check

 Now Try **Problem 103**

6   Determine Whether Lines Are Parallel or Perpendicular Using Slope.

Two lines that lie in the same plane but do not intersect are called **parallel lines.** Parallel lines have the same slope and different *y*-intercepts. For example, the lines graphed in figure (a) are parallel because they both have slope $-\frac{2}{3}$.

> **The Language of Algebra**
> The words *parallel* and *perpendicular* are used in many settings. For example, the gymnast on the *parallel* bars is in a position that is *perpendicular* to the floor.

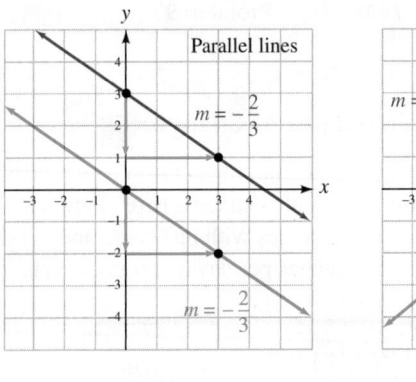

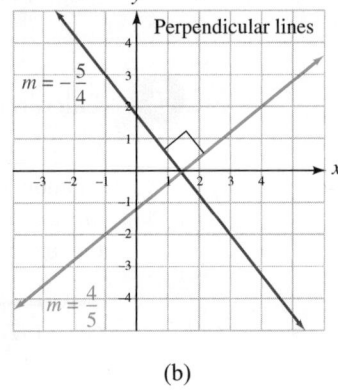

(a)   (b)

Lines that intersect to form four right angles (angles with measure 90°) are called **perpendicular lines.** If the product of the slopes of two lines is -1, the lines are perpendicular. This means that the slopes are **negative (or opposite) reciprocals.** In figure (b), we know that the lines with slopes $\frac{4}{5}$ and $-\frac{5}{4}$ are perpendicular because

$$\frac{4}{5}\left(-\frac{5}{4}\right) = -\frac{20}{20} = -1 \quad \text{\small $\frac{4}{5}$ and $-\frac{5}{4}$ are negative reciprocals.}$$

> **Slopes of Parallel and Perpendicular Lines**
> 1. Two lines with the same slope are parallel.
> 2. Two lines are perpendicular if the product of the slopes is -1; that is, if their slopes are negative reciprocals.
> 3. Any horizontal line and any vertical line are perpendicular.

| EXAMPLE 8 | Determine whether the line that passes through $(7, -9)$ and $(10, 2)$ and the line that passes through $(0, 1)$ and $(3, 12)$ are parallel, perpendicular, or neither. |

Strategy   We will use the slope formula to find the slope of each line.

Why   If the slopes are equal, the lines are parallel. If the slopes are negative reciprocals, the lines are perpendicular. Otherwise, the lines are neither parallel nor perpendicular.

Solution   To calculate the slope of each line, we use the slope formula.

The line through $(7, -9)$ and $(10, 2)$:

$$m = \frac{y_2 - y_1}{x_2 - x_1} = \frac{2 - (-9)}{10 - 7} = \frac{11}{3}$$

The line through $(0, 1)$ and $(3, 12)$:

$$m = \frac{y_2 - y_1}{x_2 - x_1} = \frac{12 - 1}{3 - 0} = \frac{11}{3}$$

Since the slopes are the same, the lines are parallel.

| Self Check 8 | Determine whether the line that passes through $(2, 1)$ and $(6, 8)$ and the line that passes through $(-1, 0)$ and $(4, 7)$ are parallel, perpendicular, or neither. |

Now Try   **Problems 73 and 75**

| EXAMPLE 9 | Find the slope of a line perpendicular to the line passing through $(1, -4)$ and $(8, 4)$. |

Strategy   We will use the slope formula to find the slope of the line passing through $(1, -4)$ and $(8, 4)$.

Why   We can then form the negative reciprocal of the result to produce the slope of a line perpendicular to the given line.

Solution   The slope of the line that passes through $(1, -4)$ and $(8, 4)$ is

$$m = \frac{y_2 - y_1}{x_2 - x_1} = \frac{4 - (-4)}{8 - 1} = \frac{8}{7}$$

The slope of a line perpendicular to the given line has slope that is the negative (or opposite) reciprocal of $\frac{8}{7}$, which is $-\frac{7}{8}$.

| Self Check 9 | Find the slope of a line perpendicular to the line passing through $(-4, 1)$ and $(9, 5)$. |

Now Try   **Problem 85**

ANSWERS TO SELF CHECKS   **1.** $-\frac{1}{2}$ **2.** 5   **3.** $-\frac{5}{2}$ **4.** 0   **5.** Undefined slope   **8.** Neither   **9.** $-\frac{13}{4}$

STUDY SET
3.4

VOCABULARY

Fill in the blanks.

1. The _____ of a line is a measure of the line's steepness. It is the _____ of the vertical change to the horizontal change.

2. $m = \dfrac{}{\text{horizontal change}} = \dfrac{\text{rise}}{} = \dfrac{\text{change in } y}{}$

3. The rate of _____ of a linear relationship can be found by finding the slope of the graph of the line and attaching the proper units.

4. _____ lines do not intersect. _____ lines intersect to form four right angles.

CONCEPTS

5. Which line graphed has
   a. a positive slope?   b. a negative slope?
   c. zero slope?   d. undefined slope?

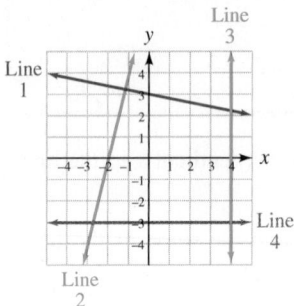

6. Consider each graph of a line and the slope triangle. What is the rise? What is the run? What is the slope of the line?

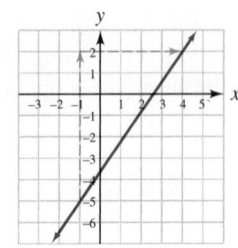   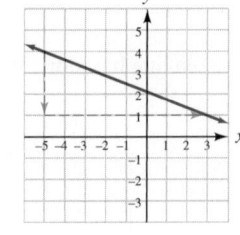

(a)   (b)

7. For each graph, determine which line has the greater slope.

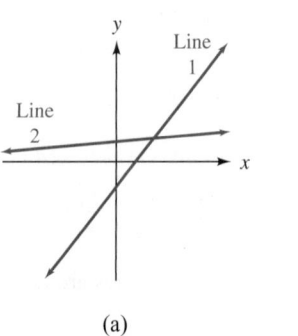   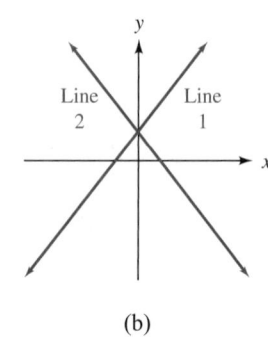

(a)   (b)

8. Which two labeled points should be used to find the slope of the line?

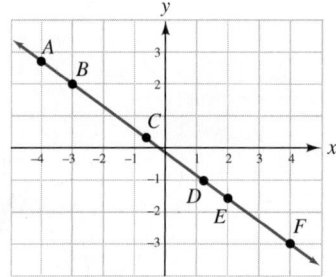

9. Fill in the blank: When calculating the slope of a line, the _____ value will be obtained no matter which two points on the line are used.

10. Evaluate each expression.
    a. $\dfrac{10 - 4}{6 - 5}$ b. $\dfrac{-1 - 1}{-2 - (-7)}$

11. Write each slope in a better way.
    a. $m = \dfrac{0}{6}$ b. $m = \dfrac{8}{0}$

    c. $m = \dfrac{3}{12}$ d. $m = \dfrac{-10}{-5}$

12. Fill in the blanks: _____ lines have a slope of 0. Vertical lines have _____ slope.

13. The *grade* of an incline is its slope expressed as a percent. Express the slope $\frac{2}{5}$ as a grade.

14. GROWTH RATES   The graph on the next page shows how a child's height increased from ages 2 through 5. Fill in the correct units to find the rate of change in the child's height.

$$\dfrac{\text{Rate of}}{\text{change}} = \dfrac{(40 - 31)}{(5 - 2)}$$

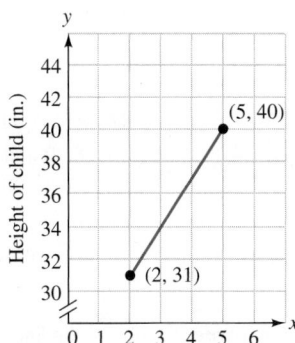

15. Find the negative reciprocal of each number.

   a. 6   **b.** $-\dfrac{7}{8}$ **c.** -1

16. Fill in the blanks.

   a. Two different lines with the same slope are _____.

   b. If the slopes of two lines are negative reciprocals, the lines are _____.

   c. The product of the slopes of perpendicular lines is _____.

NOTATION

17. a. What is the formula used to find the slope of a line passing through (x_1, y_1) and (x_2, y_2)?

   b. Fill in the blanks to state the slope formula in words: m equals y _____ two minus y _____ one _____ x sub _____ minus x sub _____.

18. Explain the difference between y^2 and y_2.

19. Consider the points $(7, 2)$ and $(-4, 1)$. If we let $x_1 = 7$, then what is y_2?

20. The symbol $\doteq$ is used when graphing to indicate a _____ in the labeling of an axis.

GUIDED PRACTICE

Find the slope of each line, if possible. **See Example 1.**

21.   **22.**

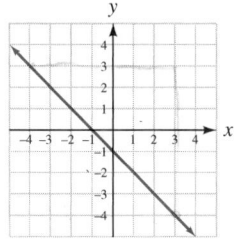

23.   **24.**

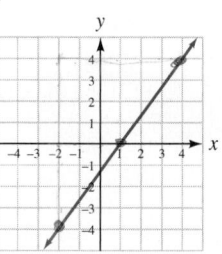 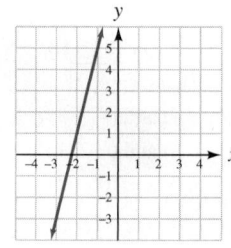

25.   **26.**

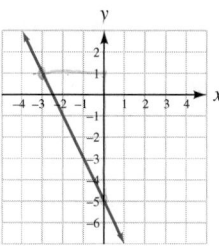 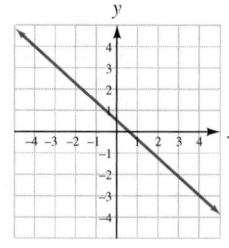

27.   **28.**

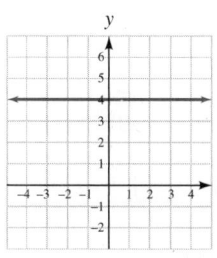 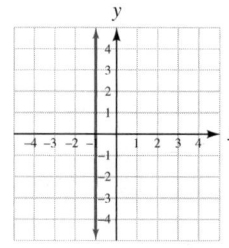

29.   **30.**

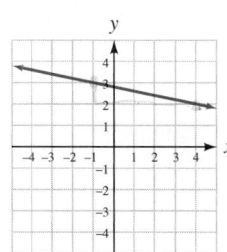 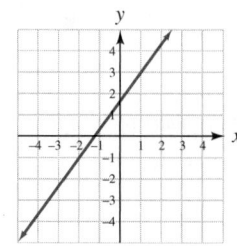

31.   **32.**

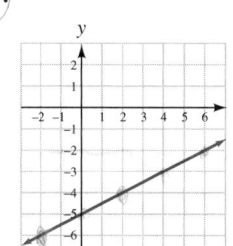 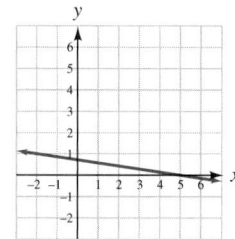

Find the slope of the line passing through the given points, when possible. **See Examples 2 and 3.**

33. $(1, 3)$ and $(2, 4)$ **34.** $(1, 3)$ and $(2, 5)$

35. $(3, 4)$ and $(2, 7)$ **36.** $(3, 6)$ and $(5, 2)$

37. $(0, 0)$ and $(4, 5)$

38. $(4, 3)$ and $(7, 8)$

39. $(-3, 5)$ and $(-5, 6)$

40. $(6, -2)$ and $(-3, 2)$

41. $(-2, -2)$ and $(-12, -8)$

42. $(-1, -2)$ and $(-10, -5)$

43. $(5, 7)$ and $(-4, 7)$

44. $(-1, -12)$ and $(6, -12)$

45. $(8, -4)$ and $(8, -3)$

46. $(-2, 8)$ and $(-2, 15)$

47. $(-6, 0)$ and $(0, -4)$

48. $(0, -9)$ and $(-6, 0)$

49. $(-2.5, 1.75)$ and $(-0.5, -7.75)$

50. $(6.4, -7.2)$ and $(-8.8, 4.2)$

51. $(-2.2, 18.6)$ and $(-1.7, 18.6)$

52. $(4.6, 3.2)$ and $(4.6, -4.8)$

53. $\left(-\frac{4}{7}, -\frac{1}{5}\right)$ and $\left(\frac{3}{7}, \frac{6}{5}\right)$

54. $\left(-\frac{4}{9}, -\frac{1}{8}\right)$ and $\left(\frac{5}{9}, \frac{3}{8}\right)$

55. $\left(-\frac{3}{4}, \frac{2}{3}\right)$ and $\left(\frac{4}{3}, -\frac{1}{6}\right)$

56. $\left(\frac{1}{2}, \frac{3}{4}\right)$ and $\left(-\frac{11}{16}, -\frac{1}{2}\right)$

Determine the slope of the graph of the line that has the given table of solutions. See Examples 2 and 3.

57.

x	y	(x, y)
-3	-1	$(-3, -1)$
1	2	$(1, 2)$

58.

x	y	(x, y)
-3	6	$(-3, 6)$
0	2	$(0, 2)$

59.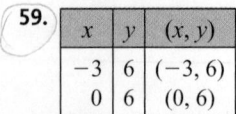

x	y	(x, y)
-3	6	$(-3, 6)$
0	6	$(0, 6)$

60.

x	y	(x, y)
4	-5	$(4, -5)$
4	0	$(4, 0)$

Find the slope of each line, if possible. See Examples 4 and 5.

61. $y = -11$

62. $y = -2$

63. $y = 0$

64. $x = 0$

65. $x = 6$

66. $x = 6$

67. $x = -10$

68. $y = 8$

69. $y - 9 = 0$

70. $x + 14 = 0$

71. $3x = -12$

72. $2y + 2 = -6$

Determine whether the lines through each pair of points are parallel, perpendicular, or neither. See Example 8.

73. $(5, 3)$ and $(1, 4)$
$(-3, -4)$ and $(1, -5)$

74. $(2, 4)$ and $(-1, -1)$
$(8, 0)$ and $(11, 5)$

75. $(-4, -2)$ and $(2, -3)$
$(7, 1)$ and $(8, 7)$

76. $(-2, 4)$ and $(6, -7)$
$(-6, 4)$ and $(5, 12)$

77. $(2, 2)$ and $(4, -3)$
$(-3, 4)$ and $(-1, 9)$

78. $(-1, -3)$ and $(2, 4)$
$(5, 2)$ and $(8, -5)$

79. $(-1, 8)$ and $(-6, 8)$
$(3, 3)$ and $(3, 7)$

80. $(11, 0)$ and $(11, -5)$
$(14, 6)$ and $(25, 6)$

81. $(6, 4)$ and $(2, 5)$
$(-2, -3)$ and $(2, -4)$

82. $(-3, -1)$ and $(3, -2)$
$(8, 2)$ and $(9, 8)$

83. $(4, 2)$ and $(5, -3)$
$(-5, 3)$ and $(-2, 9)$

84. $(8, -3)$ and $(8, -8)$
$(11, 3)$ and $(22, 3)$

Find the slope of a line perpendicular to the line passing through the given two points. See Example 9.

85. $(0, 0)$ and $(5, -9)$

86. $(0, 0)$ and $(5, 12)$

87. $(-1, 7)$ and $(1, 10)$

88. $(-7, 6)$ and $(0, 4)$

89. $\left(-2, \frac{1}{2}\right)$ and $\left(-1, \frac{3}{2}\right)$

90. $\left(\frac{1}{3}, -1\right)$ and $\left(\frac{4}{3}, -2\right)$

91. $(-1, 2)$ and $(-3, 6)$

92. $(5, -4)$ and $(-1, -7)$

APPLICATIONS

93. POOLS　Find the slope of the bottom of the swimming pool as it drops off from the shallow end to the deep end.

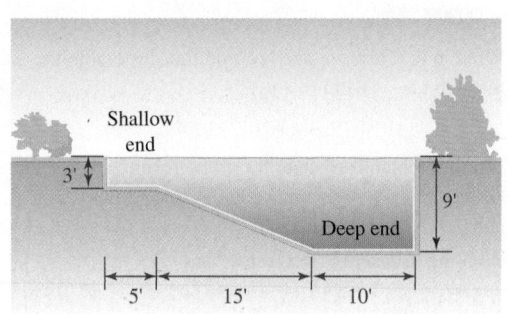

94. DRAINAGE　Find the slope of the concrete patio slab using the 1-foot ruler, level, and 10-foot-long board shown in the illustration.　(*Hint:* 10 feet = 120 in.)

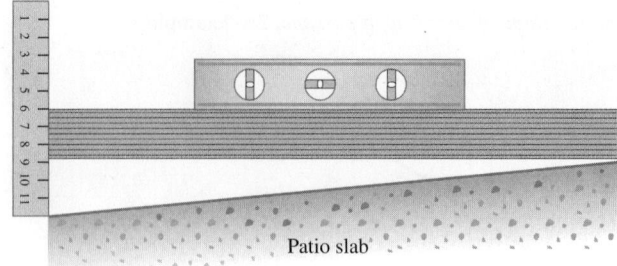

Patio slab

95. GRADE OF A ROAD　Refer to the illustration on the next page. Find the slope of the decline and use that information to find the grade of the road.

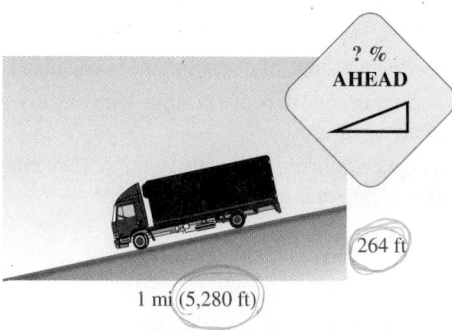

1 mi (5,280 ft)

96. STREETS One of the steepest streets in the United States is Eldred Street in Highland Park, California (near Los Angeles). It rises approximately 220 feet over a horizontal distance of 665 feet. What is the grade of the street?

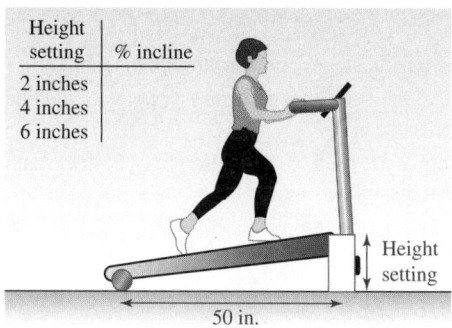

97. TREADMILLS For each height setting listed in the table, find the resulting slope of the jogging surface of the treadmill. Then express each incline as a percent.

Height setting	% incline
2 inches	
4 inches	
6 inches	

98. ARCHITECTURE Locate the coordinates of the peak of the roof if it is to have a pitch of $\frac{2}{5}$ and the roof line is to pass through the two given points in black.

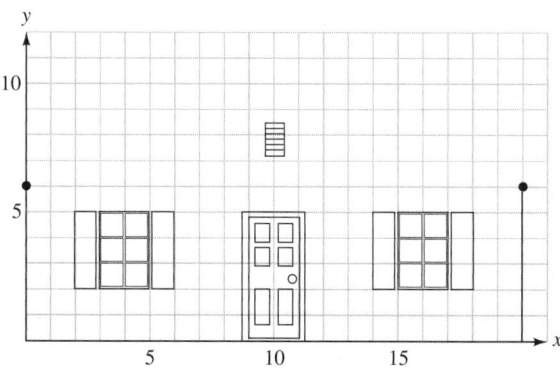

99. CARPENTRY Find the pitch of each roof.

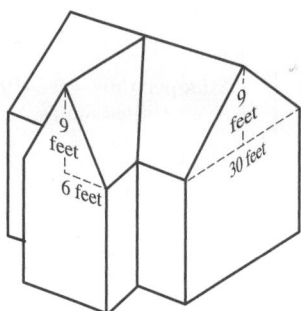

100. DOLL HOUSES Find x so that the pitch of the roof of the doll house is $\frac{4}{3}$.

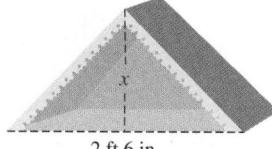

2 ft 6 in.

101. IRRIGATION The graph shows the number of gallons of water remaining in a reservoir as water is used from it to irrigate a field. Find the rate of change in the number of gallons of water in the reservoir.

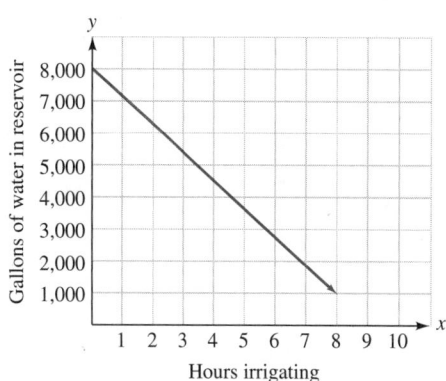

102. COMMERCIAL JETS Examine the graph and consider trips of more than 7,000 miles by a Boeing 777. Use a rate of change to estimate how the maximum payload decreases as the distance traveled increases.

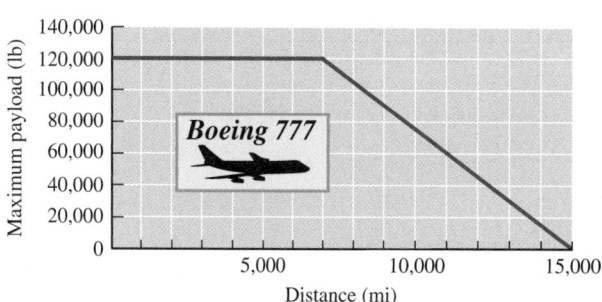

Based on data from Lawrence Livermore National Laboratory and *Los Angeles Times* (October 22, 1998).

103. MILK PRODUCTION The following graph approximates the amount of milk produced per cow in the United States for the years 1996-2005. Find the rate of change.

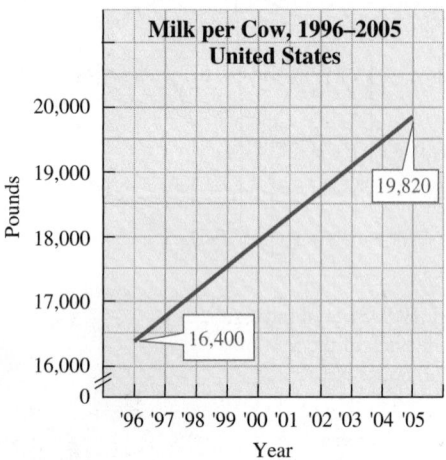

Source: United States Department of Agriculture

104. WAL-MART The graph below approximates the net sales of Wal-Mart for the years 1991–2006. Find the rate of change in sales for the years
   a. 1991–1998
   b. 1998–2006

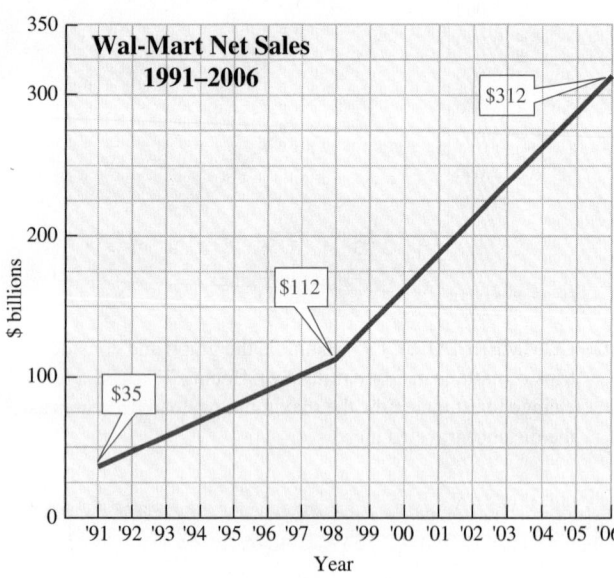

Based on data from the Wal-Mart 2006 Financial Summary

WRITING

105. Explain why the slope of a vertical line is undefined.

106. How do we distinguish between a line with positive slope and a line with negative slope?

107. Explain the error in the following solution: *Find the slope of the line that passes through* (6, 4) *and* (3, 1).

$$m = \frac{1 - 4}{6 - 3} = \frac{-3}{3} = -1$$

108. Explain the difference between a rate of change that is positive and one that is negative. Give an example of each.

REVIEW

109. HALLOWEEN CANDY A candy maker wants to make a 60-pound mixture of two candies to sell for $2 per pound. If black licorice bits sell for $1.90 per pound and orange gumdrops sell for $2.20 per pound, how many pounds of each should be used?

110. MEDICATIONS A doctor prescribes an ointment that is 2% hydrocortisone. A pharmacist has 1% and 5% concentrations in stock. How many ounces of each should the pharmacist use to make a 1-ounce tube?

CHALLENGE PROBLEMS

111. Use the concept of slope to determine whether $A(-50, -10)$, $B(20, 0)$, and $C(34, 2)$ all lie on the same straight line.

112. A line having slope $\frac{2}{3}$ passes through the point $(10, -12)$. What is the y-coordinate of another point on the line whose x-coordinate is 16?

113. Subscripts are used in other disciplines besides mathematics. In what disciplines are the following symbols used?
   a. H_2O and CO_2
   b. C_7 and G_7
   c. B_6 and B_{12}

114. Evaluate $2a_2^2 + 3a_3^3 + 4a_4^4$ for $a_2 = 2$, $a_3 = 3$, and $a_4 = 4$.

SECTION 3.5
Slope–Intercept Form

Objectives

① Use slope–intercept form to identify the slope and y-intercept of a line.

② Write a linear equation in slope–intercept form.

③ Write an equation of a line given its slope and y-intercept.

④ Use the slope and y-intercept to graph a linear equation.

⑤ Recognize parallel and perpendicular lines.

⑥ Use slope–intercept form to write an equation to model data.

Of all of the ways in which a linear equation can be written, one form, called *slope–intercept form,* is probably the most useful. When an equation is written in this form, two important features of its graph are evident.

① **Use Slope–Intercept Form to Identify the Slope and y-Intercept of a Line.**

To explore the relationship between a linear equation and its graph, let's consider $y = 2x + 1$. We can graph this equation using the point-plotting method discussed in Section 3.2.

$y = 2x + 1$

x	y	(x, y)
-1	-1	$(-1, -1)$
0	1	$(0, 1)$
1	3	$(1, 3)$

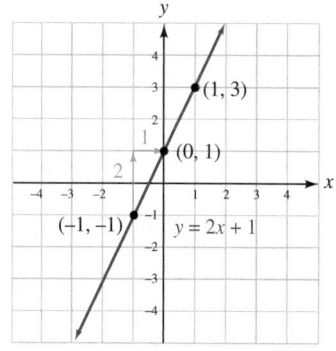

To find the slope of the line, we pick two points on the line, $(-1, -1)$ and $(0, 1)$, and draw a slope triangle and count grid squares:

$$\text{Slope} = \frac{\text{rise}}{\text{run}} = \frac{2}{1} = 2$$

From the equation and the graph, we can make two observations:

- The graph crosses the y-axis at 1. This is the same as the constant term in $y = 2x + 1$.
- The slope of the line is 2. This is the same as the coefficient of x in $y = 2x + 1$.

This illustrates that the slope and y-intercept of the graph of $y = 2x + 1$ can be determined from the equation.

$$y = 2x + 1$$

The slope of the line is 2.   The y-intercept is (0, 1).

These observations suggest the following form of an equation of a line.

Slope–Intercept Form of the Equation of a Line	If a linear equation is written in the form $$y = mx + b$$ the graph of the equation is a line with slope m and y-intercept $(0, b)$.

When an equation of a line is written in slope–intercept form, the coefficient of the x-term is the line's slope and the constant term gives the y-coordinate of y-intercept.

$$y = mx + b$$
$$\uparrow \quad \uparrow$$
Slope y-intercept: $(0, b)$

Linear equation	Equation written in slope–intercept form	Slope	y-intercept
$y = 6x - 2$	$y = 6x + (-2)$	6	$(0, -2)$
$y = -\dfrac{5}{4}x$	$y = -\dfrac{5}{4}x + 0$	$-\dfrac{5}{4}$	$(0, 0)$
$y = \dfrac{x}{2} + 3$	$y = \dfrac{1}{2}x + 3$	$\dfrac{1}{2}$	$(0, 3)$
$y = -\dfrac{7}{8} - x$	$y = -x + \left(-\dfrac{7}{8}\right)$	-1	$\left(0, -\dfrac{7}{8}\right)$

❷   Write a Linear Equation in Slope–Intercept Form.

The equation of any nonvertical line can be written in slope–intercept form. To do so, we apply the properties of equality to solve the equation for y.

EXAMPLE 1   Find the slope and y-intercept of the line with the given equation.
a. $8x + y = 9$ **b.** $x + 4y = 16$ **c.** $-9x - 3y = 11$

Strategy   We will write each equation in slope–intercept form, $y = mx + b$.

Why   When the equations are written in slope–intercept form, the slope and y-intercept of their graphs become apparent.

Solution

a. The slope and y-intercept of the graph of $8x + y = 9$ are not obvious because the equation is not in slope–intercept form. To write it in $y = mx + b$ form, we isolate y on the left side.

$$8x + y = 9$$
$$8x + y - 8x = -8x + 9 \qquad \text{To eliminate the term } 8x \text{ on the left side, subtract } 8x \text{ from both sides.}$$
$$y = -8x + 9 \qquad \text{On the left side, combine like terms: } 8x - 8x = 0.$$

$$y = -8x + 9$$
$$m = -8 \uparrow \qquad \uparrow \quad b = 9$$
The slope is -8.   The y-intercept is $(0, 9)$.

b. To write the equation in slope–intercept form, we solve for y.

$$x + 4y = 16$$

$$x + 4y - x = -x + 16$$ To eliminate the x term on the left side, subtract x from both sides.

$$4y = -x + 16$$ Simplify the left side.

$$\frac{4y}{4} = \frac{-x + 16}{4}$$ To isolate y, undo the multiplication by 4 by dividing both sides by 4.

$$y = \frac{-x}{4} + \frac{16}{4}$$ On the right side, write $\frac{-x + 16}{4}$ as the sum of two fractions with like denominators.

$$y = -\frac{1}{4}x + 4$$ Write $\frac{-x}{4}$ as $-\frac{1}{4}x$. Simplify: $\frac{16}{4} = 4$.

Since $m = -\frac{1}{4}$ and $b = 4$, the slope is $-\frac{1}{4}$ and the y-intercept is $(0, 4)$.

c. To write the equation in $y = mx + b$ form, we isolate y on the left side.

$$-9x - 3y = 11$$

$$-3y = 9x + 11$$ To eliminate the term $-9x$ on the left side, add 9x to both sides: $-9x + 9x = 0$.

$$\frac{-3y}{-3} = \frac{9x}{-3} + \frac{11}{-3}$$ To isolate y, undo the multiplication by -3 by dividing both sides by -3.

$$y = -3x - \frac{11}{3}$$ Simplify.

Since $m = -3$ and $b = -\frac{11}{3}$, the slope is -3 and the y-intercept is $\left(0, -\frac{11}{3}\right)$.

 Self Check 1   Find the slope and y-intercept of the line with the given equation.
a. $9x + y = -4$ **b.** $x + 11y = -22$ **c.** $-10x - 2y = 7$

Now Try   **Problems 11, 35, and 45**

---

3 **Write an Equation of a Line Given Its Slope and y-Intercept.**

If we are given the slope and y-intercept of a line, we can write an equation of the line by substituting for m and b in the slope–intercept form.

EXAMPLE 2   Write an equation of the line with slope -1 and y-intercept $(0, 9)$.

Strategy   We will use the slope–intercept form, $y = mx + b$, to write an equation of the line.

Why   We know the slope of the line and its y-intercept.

Solution   If the slope is -1 and the y-intercept is $(0, 9)$, then $m = -1$ and $b = 9$.

$$y = mx + b$$ This is the slope–intercept form.

$$y = -1x + 9$$ Substitute −1 for m and 9 for b.

$$y = -x + 9$$ Simplify: −1x = −x.

The equation of the line with slope -1 and y-intercept $(0, 9)$ is $y = -x + 9$.

▷ **Self Check 2**   Write an equation of the line with slope 1 and *y*-intercept
(0, −12).

   Now Try   **Problem 51**

EXAMPLE 3   Write an equation of the line graphed in figure (a).

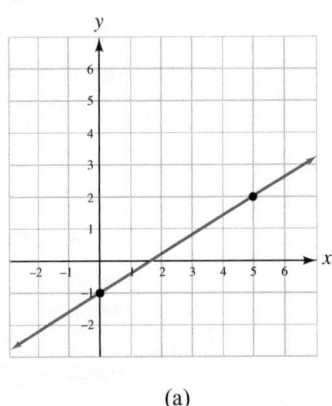

(a)

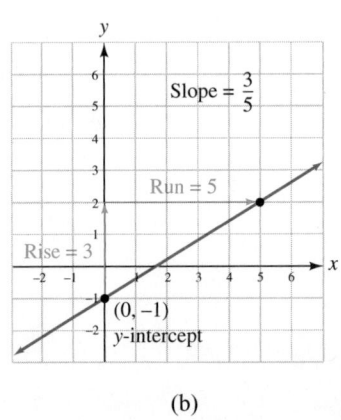

(b)

Strategy   We will use the slope–intercept form, $y = mx + b$, to write an equation of the
line.

Why   We can determine the slope and *y*-intercept of the line from the given graph.

Solution   In figure (b), we see that the *y*-intercept of the line is (0, −1). Using the
y-intercept and a second point on the line, we draw a slope triangle to find that the slope
of the line is $\frac{3}{5}$. When we substitute $\frac{3}{5}$ for *m* and −1 for *b* into the slope–intercept form
$y = mx + b$, we obtain an equation of the line: $y = \frac{3}{5}x - 1$.

▷ **Self Check 3**   Write an equation of the line graphed here.

   Now Try   **Problem 59**

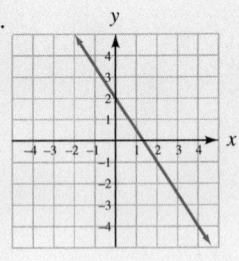

---

4   **Use the Slope and *y*-Intercept to Graph a Linear Equation.**

If we know the slope and *y*-intercept of a line, we can graph the line.

EXAMPLE 4   Use the slope and y-intercept to graph $y = 5x - 4$.

Strategy   We will examine the equation to identify the slope and the y-intercept of the line to be graphed. Then we will plot the y-intercept and use the slope to determine a second point on the line.

Why   Once we locate two points on the line, we can draw the graph of the line.

Solution   Since $y = 5x - 4$ is written in $y = mx + b$ form, we know that its graph is a line with a slope of 5 and a y-intercept of $(0, -4)$. To draw the graph, we begin by plotting the y-intercept. The slope can be used to find another point on the line.

If we write the slope as the fraction $\frac{5}{1}$, the rise is 5 and the run is 1. From $(0, -4)$, we move 5 units *upward* (because the numerator, 5, is positive) and 1 unit to the right (because the denominator, 1, is positive). This locates a second point on the line, $(1, 1)$. The line through $(0, -4)$ and $(1, 1)$ is the graph of $y = 5x - 4$.

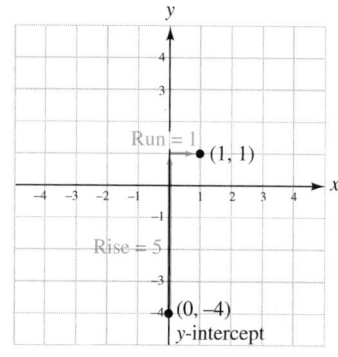

Plot the y-intercept. From $(0, -4)$, draw the rise and run parts of the slope triangle for $m = \dfrac{5}{1}$ to find another point on the line.

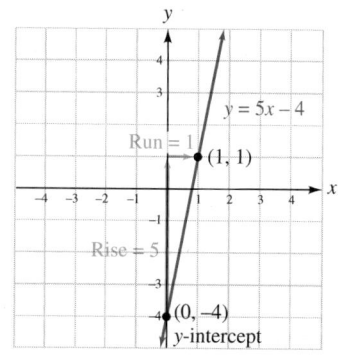

Use a straightedge to draw a line through the two points.

An alternate way to find another point on the line is to write the slope in the form $\frac{-5}{-1}$. As before, we begin at the y-intercept $(0, -4)$. Since the rise is negative, we move 5 units *downward,* and since the run is negative, we then move 1 unit to the *left.* We arrive at $(-1, -9)$, another point on the graph of $y = 5x - 4$.

▷ **Self Check 4**   Use the slope and y-intercept to graph $y = 2x - 3$.
      Now Try   **Problem 67**

EXAMPLE 5   Use the slope and y-intercept to graph $4x + 3y = 6$.

Strategy   We will write the equation of the line in slope–intercept form, $y = mx + b$. Then we will identify the slope and y-intercept of its graph.

Why   We can use that information to plot two points that the line passes through.

Solution   To write $4x + 3y = 6$ in slope–intercept form, we isolate y on the left side.

$$4x + 3y = 6$$

$$3y = -4x + 6 \qquad \text{To eliminate } 4x \text{ from the left side, subtract } 4x \text{ from both sides.}$$

$$\frac{3y}{3} = \frac{-4x}{3} + \frac{6}{3} \qquad \text{To isolate } y, \text{ undo the multiplication by 3 by dividing both sides by 3.}$$

$$y = -\frac{4}{3}x + 2 \qquad m = -\frac{4}{3} \text{ and } b = 2.$$

The slope of the line is $-\frac{4}{3}$ and the y-intercept is $(0, 2)$. To draw the graph, we begin by plotting the y-intercept. If we write the slope as $\frac{-4}{3}$, the rise is -4 and the run is 3. From $(0, 2)$, we then move 4 units *downward* (because the numerator is negative) and 3 units to the *right* (because the denominator is positive). This locates a second point on the line, $(3, -2)$.

We can find another point on the graph by writing the slope as $\frac{4}{-3}$. In this case, the rise is 4 and the run is -3. Again, we begin at the y-intercept $(0, 2)$, but this time, we move 4 units *upward* because the rise is positive. Then we move 3 units to the *left,* because the run is negative, and arrive at the point $(-3, 6)$. The line that passes through $(0, 2)$, $(3, -2)$, and $(-3, 6)$ is the graph of $4x + 3y = 6$.

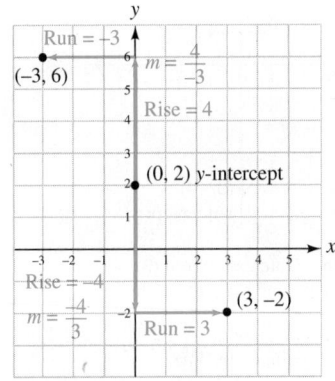

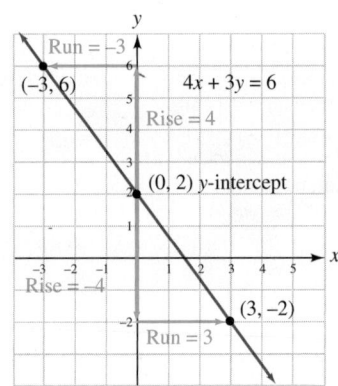

Plot the y-intercept. From $(0, 2)$, draw the rise and run parts of the slope triangle for $m = \frac{-4}{3}$ $\left(\text{or } m = \frac{4}{-3}\right)$ to find another point on the line.

Use a straightedge to draw a line through the points.

Self Check 5   Use the slope and y-intercept to graph $5x + 6y = 12$.

Now Try   **Problem 75**

⑤ Recognize Parallel and Perpendicular Lines.

The slope–intercept form enables us to quickly identify parallel and perpendicular lines.

EXAMPLE 6   Are the graphs of $y = -5x + 6$ and $x - 5y = -10$ parallel, perpendicular, or neither?

Strategy   We will find the slope of each line and then compare the slopes.

Why   If the slopes are equal, the lines are parallel. If the slopes are negative reciprocals, the lines are perpendicular. Otherwise, the lines are neither parallel nor perpendicular.

Solution   The graph of $y = -5x + 6$ is a line with slope -5. To find the slope of the graph of $x - 5y = -10$, we will write the equation in slope–intercept form.

$$x - 5y = -10$$

$$-5y = -x - 10 \qquad \text{To eliminate } x \text{ from the left side, subtract } x \text{ from both sides.}$$

$$\frac{-5y}{-5} = \frac{-x}{-5} - \frac{10}{-5} \qquad \text{To isolate } y, \text{ undo the multiplication by } -5 \text{ by dividing both sides by } -5.$$

$$y = \frac{x}{5} + 2 \qquad m = \tfrac{1}{5} \text{ because } \tfrac{x}{5} = \tfrac{1}{5}x.$$

The graph of $y = \frac{x}{5} + 2$ is a line with slope $\frac{1}{5}$. Since the slopes -5 and $\frac{1}{5}$ are negative reciprocals, the lines are perpendicular. This is verified by the fact that the product of their slopes is -1.

$$-5\left(\frac{1}{5}\right) = -\frac{5}{5} = -1$$

> ### Success Tip
> Graphs are not necessary to determine if two lines are parallel, perpendicular, or neither. We simply examine the slopes of the lines.

Self Check 6   Determine whether the graphs of $y = 4x + 6$ and $x - 4y = -8$ are parallel, perpendicular, or neither.

Now Try   **Problem 83**

6   Use Slope–Intercept Form to Write an Equation to Model Data.

In the following example, to make the equation more descriptive, we replace x and y in $y = mx + b$ with two other variables.

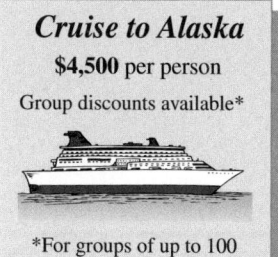

Cruise to Alaska
$4,500 per person
Group discounts available*

*For groups of up to 100

EXAMPLE 7   ***Group Discounts.***   To promote group sales for an Alaskan cruise, a travel agency reduces the regular ticket price of $4,500 by $5 for each person traveling in the group.

a. Write a linear equation that determines the per-person cost c of the cruise, if p people travel together.

b. Use the equation to determine the per-person cost if 55 teachers travel together.

Strategy   We will determine the slope and the y-intercept of the graph of the equation from the given facts about the cruise.

Why   If we know the slope and y-intercept, we can use the slope–intercept form, $y = mx + b$, to write an equation to model the situation.

Solution

a. Since the per-person cost of the cruise steadily decreases as the number of people in the group increases, the rate of change of $-\$5$ per person is the slope of the graph of the equation. Thus, m is -5.

If 0 people take the cruise, there will be no discount and the per-person cost of the cruise will be $4,500. Written as an ordered pair of the form (p, c), we have $(0, 4,500)$. When graphed, this point would be the c-intercept. Thus, b is 4,500.

Substituting for m and b in the slope–intercept form of the equation, we obtain the linear equation that models the pricing arrangement.

$$c = -5p + 4,500$$

b. To find the per-person cost of the cruise for a group of 55 people, we substitute 55 for p and solve for c.

$$c = -5p + 4,500$$
$$c = -5(55) + 4,500 \quad \text{Substitute 55 for } p.$$
$$c = -275 + 4,500$$
$$= 4,225$$

If a group of 55 people travel together, the cruise will cost each person $4,225.

Self Check 7   Write a linear equation in slope–intercept form that finds the cost c of the cruise if a $10-per-person discount is offered for groups.

Now Try   **Problem 91**

ANSWERS TO SELF CHECKS   **1. a.** $m = -9$; $(0, -4)$ **b.** $m = -\frac{1}{11}$; $(0, -2)$ **c.** $m = -5$; $\left(0, -\frac{7}{2}\right)$

2. $y = x - 12$ **3.** $y = -\frac{3}{2}x + 2$

4. 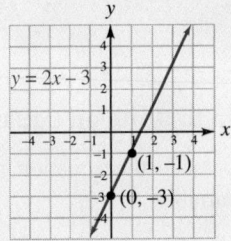   **5.** 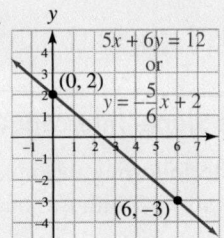   **6.** Neither   **7.** $c = -10p + 4,500$

STUDY SET
3.5

VOCABULARY

Fill in the blanks.

1. The equation $y = mx + b$ is called the _____ form of the equation of a line.

2. The graph of the linear equation $y = mx + b$ has _____ $(0, b)$ and _____ m.

CONCEPTS

3. Determine whether each equation is in slope–intercept form.
   a. $7x + 4y = 2$ b. $5y = 2x - 3$
   c. $y = 6x + 1$ d. $x = 4y - 8$
4. a. How do we solve $4x + y = 9$ for y?

   b. How do we solve $-2x + y = 9$ for y?

5. Simplify the right side of each equation.
   a. $y = \dfrac{4x}{2} + \dfrac{16}{2}$ b. $y = \dfrac{15x}{-3} + \dfrac{9}{-3}$

   c. $y = \dfrac{2x}{6} - \dfrac{6}{6}$ d. $y = \dfrac{-9x}{-5} - \dfrac{20}{-5}$

6. Find the slope and y-intercept of each line graphed below. Then use that information to write an equation for that line.
   a. 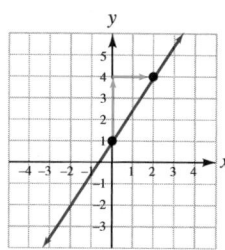 b.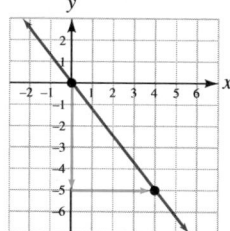

NOTATION

Complete the solution by solving the equation for y. Then find the slope and the y-intercept of its graph.

7. $2x + 5y = 15$
   $2x + 5y - 2x = \boxed{} + 15$
   $ = -2x + 15$
   $\dfrac{5y}{\boxed{}} = \dfrac{-2x}{\boxed{}} + \dfrac{15}{\boxed{}}$

   $y = \boxed{}\, x + \boxed{}$

   The slope is $\boxed{}$ and the y-intercept is $\boxed{}$.
8. What is the slope–intercept form of the equation of a line?

9. Fill in the blanks:
   $-\dfrac{3}{2} = \dfrac{3}{\boxed{}} = \dfrac{\boxed{}}{2}$

10. Determine whether each statement is true or false.
    a. $\dfrac{x}{6} = \dfrac{1}{6}x$ b. $\dfrac{5}{3}x = \dfrac{5x}{3}$

GUIDED PRACTICE

Find the slope and the y-intercept of the line with the given equation. See Example 1.

11. $y = 4x + 2$ 12. $y = 7x + 3$
13. $y = -5x - 8$ 14. $y = -4x - 2$
15. $y = 4x - 9$ 16. $y = 6x - 1$
17. $y = 11 - x$ 18. $y = 12 - 4x$
19. $y = 1 - 20x$ 20. $y = 8 - 15x$
21. $y = \dfrac{1}{2}x + 6$ 22. $y = \dfrac{4}{5}x - 9$
23. $y = \dfrac{x}{4} - \dfrac{1}{2}$ 24. $y = \dfrac{x}{15} - \dfrac{3}{4}$
25. $y = -5x$ 26. $y = 14x$
27. $y = \dfrac{2}{3}x$ 28. $y = \dfrac{3}{4}x$
29. $y = x$ 30. $y = -x$
31. $y = -2$ 32. $y = 30$
33. $-5y - 2 = 0$ 34. $3y - 13 = 0$
35. $x + y = 8$ 36. $x - y = -30$
37. $6y = x - 6$ 38. $2y = x + 20$
39. $7y = -14x + 49$ 40. $9y = -27x + 36$
41. $-4y = 6x - 4$ 42. $-6y = 8x + 6$
43. $2x + 3y = 6$ 44. $4x + 5y = 25$
45. $3x - 5y = 15$ 46. $x - 6y = 6$
47. $-4x + 3y = -12$ 48. $-5x + 2y = -8$
49. $-6x + 6y = -11$ 50. $-4x + 4y = -9$

Write an equation of the line with the given slope and y-intercept and graph it. See Example 2.

51. Slope 5, y-intercept $(0, -3)$ 52. Slope -2, y-intercept $(0, 1)$
53. Slope -3, y-intercept $(0, 6)$ 54. Slope 4, y-intercept $(0, -1)$
55. Slope $\dfrac{1}{4}$, y-intercept $(0, -2)$ 56. Slope $\dfrac{1}{3}$, y-intercept $(0, -5)$
57. Slope $-\dfrac{8}{3}$, y-intercept $(0, 5)$ 58. Slope $-\dfrac{7}{6}$, y-intercept $(0, 2)$

Write an equation for each line shown. See Example 3.

59.

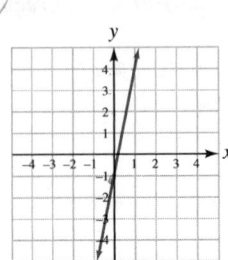

60.

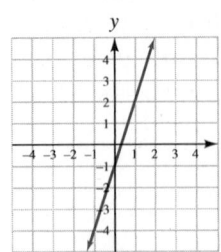

61.

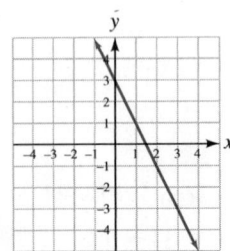

62.

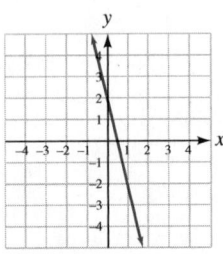

63.

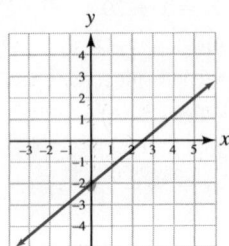

64.

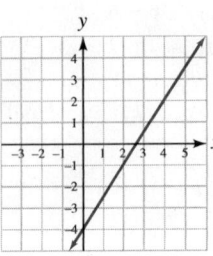

65.

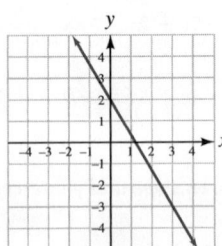

66.

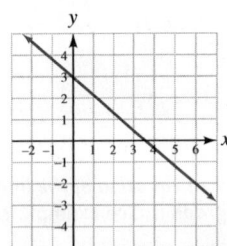

Find the slope and the y-intercept of the graph of each equation and graph it. See Examples 4 and 5.

67. $y = 3x + 3$

68. $y = -3x + 5$

69. $y = -\dfrac{1}{2}x + 2$

70. $y = \dfrac{x}{3}$

71. $y = -3x$

72. $y = -4x$

73. $4x + y = -4$

74. $2x + y = -6$

75. $3x + 4y = 16$

76. $2x + 3y = 9$

77. $10x - 5y = 5$

78. $4x - 2y = 6$

For each pair of equations, determine whether their graphs are parallel, perpendicular, or neither. See Example 6.

79. $y = 6x + 8$
$y = 6x$

80. $y = 3x - 15$
$y = -\dfrac{1}{3}x + 4$

81. $y = x$
$y = -x$

82. $y = \dfrac{1}{2}x - \dfrac{4}{5}$
$y = 0.5x + 3$

83. $y = -2x - 9$
$2x - y = 9$

84. $y = \dfrac{3}{4}x + 1$
$4x - 3y = 15$

85. $3x = 5y - 10$
$5x = 1 - 3y$

86. $-2y = 2 - x$
$2x - 3 = 4y$

87. $x - y = 12$
$-2x + 2y = -23$

88. $y = -3x + 1$
$3y = x - 5$

89. $x = 9$
$y = 8$

90. $-x + 4y = 10$
$2y + 16 = -8x$

APPLICATIONS

91. PRODUCTION COSTS   A television production company charges a basic fee of $5,000 and then $2,000 an hour when filming a commercial.

   a. Write a linear equation that describes the relationship between the total production costs c and the hours h of filming.

   b. Use your answer to part a to find the production costs if a commercial required 8 hours of filming.

92. COLLEGE FEES   Each semester, students enrolling at a community college must pay tuition costs of $20 per unit as well as a $40 student services fee.

   a. Write a linear equation that gives the total fees t to be paid by a student enrolling at the college and taking x units.

   b. Use your answer to part a to find the enrollment cost for a student taking 12 units.

93. CHEMISTRY A portion of a student's chemistry lab manual is shown below. Use the information to write a linear equation relating the temperature F (in degrees Fahrenheit) of the compound to the time t (in minutes) elapsed during the lab procedure.

> Chem. Lab #1   Aug. 13
> **Step 1:** Removed compound from freezer @ $-10°F$.
>
> **Step 2:** Used heating unit to raise temperature of compound $5°\ F$ every minute.

94. RENTALS Use the information in the newspaper advertisement to write a linear equation that gives the amount of income A (in dollars) the apartment owner will receive when the unit is rented for m months.

> **APARTMENT FOR RENT**
> 1 bedroom/1 bath, with garage
> $500 per month + $250 nonrefundable security fee.

95. EMPLOYMENT SERVICE A policy statement of LIZCO, Inc., is shown below. Suppose a secretary had to pay an employment service $500 to get placed in a new job at LIZCO. Write a linear equation that tells the secretary the actual cost c of the employment service to her m months after being hired.

> **Policy no. 23452**—A new hire will be reimbursed by LIZCO for any employment service fees paid by the employee at the rate of $20 per month.

96. VIDEOTAPES A VHS videocassette contains 800 feet of tape. In the long-play mode (LP), it plays 10 feet of tape every 3 minutes. Write a linear equation that relates the number of feet f of tape yet to be played and the number of minutes m the tape has been playing.

97. SEWING COSTS A tailor charges a basic fee of $20 plus $5 per letter to sew an athlete's name on the back of a jacket.
  a. Write a linear equation that will find the cost c to have a name containing x letters sewn on the back of a jacket.
  b. Graph the equation.
  c. Suppose the tailor raises the basic fee to $30. On your graph from part b, draw the new graph showing the increased cost.

98. SALAD BARS For lunch, a delicatessen offers a "Salad and Soda" special where customers serve themselves at a well-stocked salad bar. The cost is $1.00 for the drink and 20¢ an ounce for the salad.
  a. Write a linear equation that will find the cost c of a "Salad and Soda" lunch when a salad weighing x ounces is purchased.
  b. Graph the equation.
  c. How would the graph from part b change if the delicatessen began charging $2.00 for the drink?
  d. How would the graph from part b change if the cost of the salad changed to 30¢ an ounce?

99. BASEBALL Use the following facts to write a linear equation in slope–intercept form that approximates the average price of a Major League Baseball ticket for the years 2000–2006.
  • Let t represent the number of years since 2000 and c the average cost of a ticket in dollars.
  • In 2000, the average ticket price was $16.63.
  • From 2000 to 2006, the average ticket price increased 89¢ per year.
  (Source: Team Marketing Report, MLB)

100. NAVIGATION The graph shows the recommended speed at which a ship should proceed into head waves of various heights.
  a. What information does the y-intercept of the line give?
  b. What is the rate of change in the recommended speed of the ship as the wave height increases?
  c. Write the equation of the line.

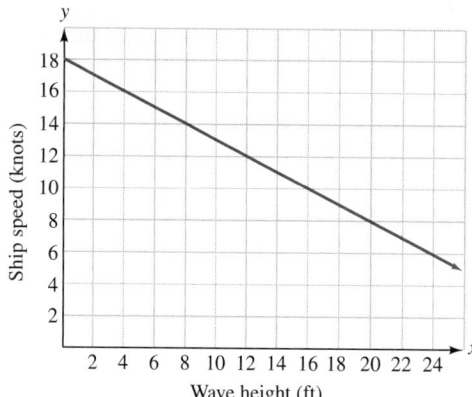

WRITING

101. Why is $y = mx + b$ called the slope–intercept form of the equation of a line?

102. On a quiz, a student was asked to find the slope of the graph of $y = 2x + 3$. She answered: $m = 2x$. Her instructor marked it wrong. Explain why the answer is incorrect.

REVIEW

103. CABLE TV   A 186-foot television cable is to be cut into four pieces. Find the length of each piece if each successive piece is 3 feet longer than the previous one.

104. INVESTMENTS   Joni received $25,000 as part of a settlement in a class action lawsuit. She invested some money at 10% and the rest at 9% simple interest rates. If her total annual income from these two investments was $2,430, how much did she invest at each rate?

CHALLENGE PROBLEMS

105. If the graph of $y = mx + b$ passes through quadrants I, II, and IV, what can be known about the constants m and b?

106. The equation $y = \frac{3}{4}x - \frac{5}{2}$ is in slope–intercept form. Write it in standard (general) form, $Ax + By = C$, where $A > 0$.

SECTION 3.6
Point–Slope Form

Objectives

1 Use point–slope form to write an equation of a line.

2 Write an equation of a line given two points on the line.

3 Write equations of horizontal and vertical lines.

4 Use a point and the slope to graph a line.

5 Write linear equations that model data.

If we know the slope of a line and its y-intercept, we can use the slope–intercept form to write the equation of the line. The question that now arises is, can *any* point on the line be used in combination with its slope to write its equation? In this section, we answer this question.

1   **Use Point–Slope Form to Write an Equation of a Line.**

Refer to the line graphed on the left, with slope 3 and passing through the point $(2,1)$. To develop a new form for the equation of a line, we will find the slope of this line in another way.

If we pick another point on the line with coordinates (x, y), we can find the slope of the line by substituting the coordinates of the points (x, y) and $(2, 1)$ into the slope formula.

$$\frac{y_2 - y_1}{x_2 - x_1} = m$$

$$\frac{y - 1}{x - 2} = m \qquad \text{Let } (x_1, y_1) \text{ be } (2, 1) \text{ and } (x_2, y_2) \text{ be } (x, y).$$
Substitute y for y_2, 1 for y_1, x for x_2, and 2 for x_1.

Since the slope of the line is 3, we can substitute 3 for m in the previous equation.

$$\frac{y - 1}{x - 2} = 3$$

We then multiply both sides by $x - 2$ to clear the equation of the fraction.

$$\frac{y - 1}{x - 2}(x - 2) = 3(x - 2)$$

$$y - 1 = 3(x - 2) \qquad \text{Simplify the left side. Remove the common factor } x - 2 \text{ in the}$$

numerator and denominator: $\dfrac{y-1}{x-2} \cdot \dfrac{x-2}{1}$.

The resulting equation displays the slope of the line and the coordinates of one point on the line:

Slope of the line
$$y - 1 = 3(x - 2)$$
y-coordinate of the point x-coordinate of the point

In general, suppose we know that the slope of a line is m and that the line passes through the point (x_1, y_1). Then if (x, y) is any other point on the line, we can use the definition of slope to write

$$\frac{y - y_1}{x - x_1} = m$$

If we multiply both sides by $x - x_1$ to clear the equation of the fraction, we have

$$y - y_1 = m(x - x_1)$$

This form of a linear equation is called **point–slope form.** It can be used to write the equation of a line when the slope and one point on the line are known.

Point–Slope Form of the Equation of a Line

If a line with slope m passes through the point (x_1, y_1), the equation of the line is

$$y - y_1 = m(x - x_1)$$

EXAMPLE 1   Find an equation of a line that has slope -8 and passes through $(-1, 5)$. Write the answer in slope–intercept form.

Strategy   We will use the point–slope form, $y - y_1 = m(x - x_1)$, to write an equation of the line.

Why   We are given the slope of the line and the coordinates of a point that it passes through.

Solution   Because we are given the coordinates of a point on the line and the slope of the line, we begin by writing the equation of the line in the point–slope form. Since the slope is -8 and the given point is $(-1, 5)$, we have $m = -8$, $x_1 = -1$, and $y_1 = 5$.

$$y - y_1 = m(x - x_1) \qquad \text{This is the point–slope form.}$$
$$y - 5 = -8[x - (-1)] \qquad \text{Substitute } -8 \text{ for } m, -1 \text{ for } x_1, \text{ and 5 for } y_1.$$
$$y - 5 = -8(x + 1) \qquad \text{Simplify within the brackets.}$$

Notation

After writing an equation in point–slope form, it is common practice to solve it for y and write an equivalent equation in slope–intercept form.

To write this equation in slope–intercept form, we solve for y.

$$y - 5 = -8(x + 1)$$

$$y - 5 = -8x - 8 \qquad \text{Distribute the multiplication by } -8.$$

$$y - 5 + 5 = -8x - 8 + 5 \qquad \text{To isolate } y, \text{ undo the subtraction of 5 by adding 5 to both sides.}$$

$$y = -8x - 3$$

In slope–intercept form, the equation is $y = -8x - 3$.

To verify this result, we note that $m = -8$. Therefore, the slope of the line is -8, as required. To see whether the line passes through $(-1, 5)$, we substitute -1 for x and 5 for y in the equation. If this point is on the line, a true statement should result.

$$y = -8x - 3$$

$$5 \overset{?}{=} -8(-1) - 3$$

$$5 \overset{?}{=} 8 - 3$$

$$5 = 5 \qquad \text{True}$$

Self Check 1   Find an equation of the line that has slope -2 and passes through $(4, -3)$. Write the answer in slope–intercept form.

Now Try   **Problems 13 and 19**

② **Write an Equation of a Line Given Two Points on the Line.**

In the next example, we show that it is possible to write the equation of a line when we know the coordinates of two points on the line.

EXAMPLE 2   Find an equation of the line that passes through $(-2, 6)$ and $(4, 7)$. Write the equation in slope–intercept form.

Strategy   We will use the point–slope form, $y - y_1 = m(x - x_1)$, to write an equation of the line.

Why   We know the coordinates of a point that the line passes through and we can calculate the slope of the line using the slope formula.

Solution   To find the slope of the line, we use the slope formula.

$$m = \frac{y_2 - y_1}{x_2 - x_1} = \frac{7 - 6}{4 - (-2)} = \frac{1}{6} \qquad \text{Substitute 7 for } y_2, \text{ 6 for } y_1, \text{ 4 for } x_2, \text{ and } -2 \text{ for } x_1.$$

Either point on the line can serve as (x_1, y_1). If we choose $(4, 7)$, we have

$$y - y_1 = m(x - x_1) \qquad \text{This is the point–slope form.}$$

$$y - 7 = \frac{1}{6}(x - 4) \qquad \text{Substitute } \tfrac{1}{6} \text{ for } m, \text{ 7 for } y_1, \text{ and 4 for } x_1.$$

Success Tip

In Example 2, either of the given points can be used as (x_1, y_1) when writing the point–slope equation. The results will be the same.

Looking ahead, we usually choose the point whose coordinates will make the computations the easiest.

To write this equation in slope–intercept form, we solve for y.

$$y - 7 = \frac{1}{6}x - \frac{2}{3}$$ Distribute the multiplication by $\frac{1}{6}$.

$$y - 7 + 7 = \frac{1}{6}x - \frac{2}{3} + 7$$ To isolate y, add 7 to both sides.

$$y = \frac{1}{6}x - \frac{4}{6} + \frac{42}{6}$$ Simplify the left side. Write 7 as $\frac{42}{6}$ to prepare to add the fractions.

$$y = \frac{1}{6}x + \frac{19}{3}$$ Simplify: $\frac{38}{6} = \frac{\overset{1}{\cancel{2}} \cdot 19}{\cancel{2} \cdot 3} = \frac{19}{3}$. This is slope–intercept form.

The equation of the line that passes through $(-2, 6)$ and $(4, 7)$ is $y = \frac{1}{6}x + \frac{19}{3}$.

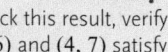

Success Tip
To check this result, verify that $(-2, 6)$ and $(4, 7)$ satisfy $y = \frac{1}{6}x + \frac{19}{3}$.

 Self Check 2   Find an equation of the line that passes through $(-5, 4)$ and $(8, -6)$. Write the equation in slope–intercept form.

Now Try   Problem 29

3 **Write Equations of Horizontal and Vertical Lines.**

We have previously graphed horizontal and vertical lines. We will now discuss how to write their equations.

EXAMPLE 3   Write an equation of each line and graph it.   **a.** A horizontal line passing through $(-2, -4)$ **b.** A vertical line passing through $(1, 3)$

Strategy   We will use the appropriate form, either $y = b$ or $x = a$, to write an equation of each line.

Why   These are the standard forms for the equations of a horizontal and a vertical line.

Solution

a. The equation of a horizontal line can be written in the form $y = b$. Since the y-coordinate of $(-2, -4)$ is -4, the equation of the line is $y = -4$. The graph is shown in the figure.

b. The equation of a vertical line can be written in the form $x = a$. Since the x-coordinate of $(1, 3)$ is 1, the equation of the line is $x = 1$. The graph is shown in the figure.

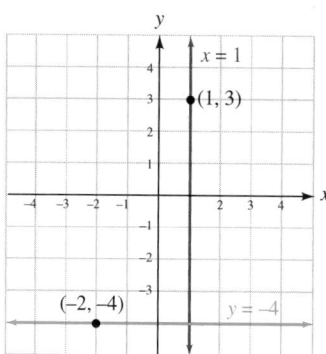

 Self Check 3   Write an equation of each line.   **a.** A horizontal line passing through $(3, 2)$ **b.** A vertical line passing through $(-1, -3)$

Now Try   Problems 41 and 43

Use a Point and the Slope to Graph a Line.

If we know the coordinates of a point on a line, and if we know the slope of the line, we can use the slope to determine a second point on the line.

EXAMPLE 4   Graph the line with slope $\dfrac{2}{5}$ that passes through $(-1, -3)$.

Strategy   First, we will plot the given point $(-1, -3)$. Then we will use the slope to find a second point that the line passes through.

Why   Once we determine two points that the line passes through, we can draw the graph of the line.

Solution   We begin by plotting the point $(-1, -3)$. From there, we move 2 units up and then 5 units to the right, since the slope is $\dfrac{2}{5}$. This puts us at a second point on the line, $(4, -1)$. We then draw a line through the two points.

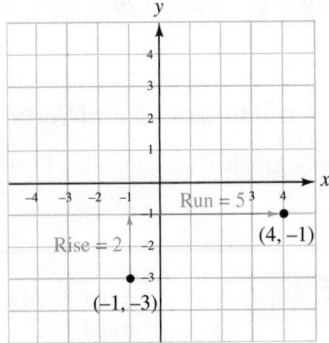   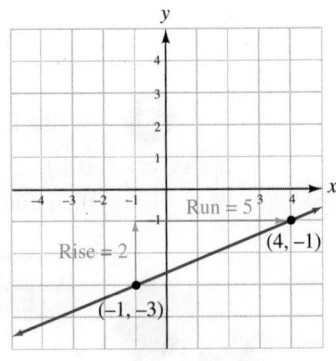

    Self Check 4   Graph the line with slope -4 that passes through $(-4, 2)$.

Now Try   Problem 45

---

⑤   **Write Linear Equations That Model Data.**

Many situations can be described by a linear equation. Quite often, these equations are written using variables other than x and y. In such cases, it is helpful to determine what an ordered-pair solution of the equation would look like.

EXAMPLE 5   *Men's Shoe Sizes.*   The length (in inches) of a man's foot is not his shoe size. For example, the smallest adult men's shoe size is 5, and it fits a 9-inch-long foot. There is, however, a linear relationship between the two. It can be stated this way: Shoe size increases by 3 sizes for each 1-inch increase in foot length.

a.   Write a linear equation that relates shoe size s to foot length L.

b.   Shaquille O'Neal, a famous basketball player, has a foot that is about 14.6 inches long. Find his shoe size.

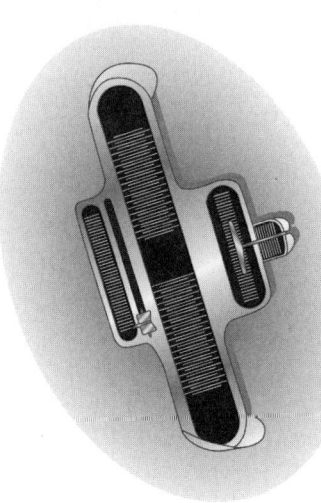

Strategy We will first find the slope of the of the line that describes the linear relationship between shoe size and the length of a foot. Then we will determine the coordinates of a point on that line.

Why Once we know the slope and the coordinates of one point on the line, we can use the point–slope form to write the equation of the line.

Solution

a. Since shoe size s depends on the length L of the foot, ordered pairs have the form (L, s). Because the relationship is linear, the graph of the desired equation is a line.

- The line's slope is the rate of change: $\frac{3 \text{ sizes}}{1 \text{ inch}}$. Therefore, $m = 3$.
- A 9-inch-long foot wears size 5, so the line passes through $(9, 5)$.

We substitute 3 for m and the coordinates of the point into the point–slope form and solve for s.

$s - s_1 = m(L - L_1)$	This is the point–slope form using the variables L and s.
$s - 5 = 3(L - 9)$	Substitute 3 for m, 9 for L_1, and 5 for s_1.
$s - 5 = 3L - 27$	Distribute the multiplication by 3.
$s = 3L - 22$	To isolate s, add 5 to both sides.

The equation relating men's shoe size and foot length is $s = 3L - 22$.

b. To find Shaquille's shoe size, we substitute 14.6 inches for L in the equation.

$$s = 3L - 22$$
$$s = 3(\mathbf{14.6}) - 22$$
$$s = 43.8 - 22$$
$$s = 21.8$$

Since men's shoes only come in full- and half-sizes, we round 21.8 up to 22. Shaquille O'Neal wears size 22 shoes.

 Now Try **Problem 73**

EXAMPLE 6 *Studying Learning.* In a series of trials, a rat was released in a maze to search for food. Researchers recorded the time that it took the rat to complete the maze on a **scatter diagram**. After the 40th trial, they drew a line through the data to obtain a model of the rat's performance. Write an equation of the line in slope–intercept form.

The Language of Algebra
The term *scatter diagram* is somewhat misleading. Often, the data points are not scattered loosely about. In this case, they fall, more or less, along an imaginary straight line, indicating a linear relationship.

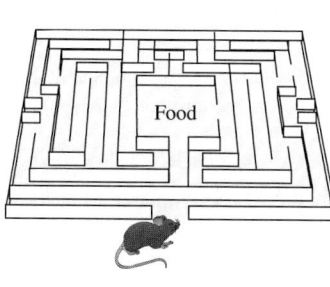

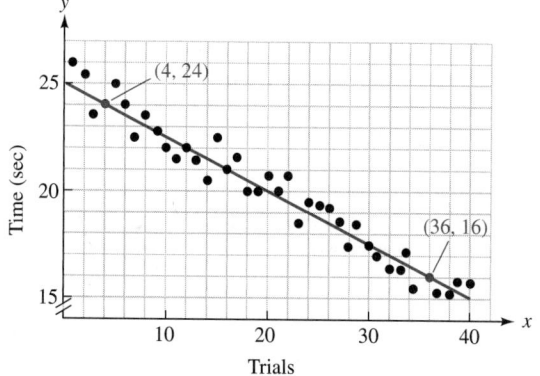

Strategy   From the graph, we will determine the coordinates of two points on the line.

Why   We can write an equation of a line when we know the coordinates of two points on the line. (See Example 2.)

Solution   We begin by writing a point–slope equation. The line passes through several points; we will use (4, 24) and (36, 16) to find the slope.

$$m = \frac{y_2 - y_1}{x_2 - x_1} = \frac{16 - 24}{36 - 4} = \frac{-8}{32} = -\frac{1}{4}$$

Any point on the line can serve as (x_1, y_1). We will use (4, 24).

$y - y_1 = m(x - x_1)$ This is the point–slope form.

$y - 24 = -\dfrac{1}{4}(x - 4)$ Substitute $-\frac{1}{4}$ for m, 4 for x_1, and 24 for y_1.

To write this equation in slope–intercept form, solve for y.

$y - 24 = -\dfrac{1}{4}x + 1$ Distribute the multiplication by $-\frac{1}{4}$: $-\frac{1}{4}(-4) = 1$.

$y = -\dfrac{1}{4}x + 25$ To isolate y, add 24 to both sides.

A linear equation that models the rat's performance on the maze is $y = -\frac{1}{4}x + 25$, where x is the number of the trial and y is the time it took, in seconds.

 Now Try   **Problem 81**

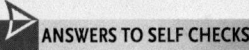 ANSWERS TO SELF CHECKS

1. $y = -2x + 5$
2. $y = -\frac{10}{13}x + \frac{2}{13}$
3. a. $y = 2$ **b.** $x = -1$

4.

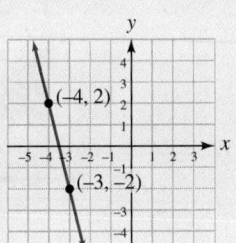

STUDY SET
3.6

VOCABULARY

Fill in the blanks.

1. $y - y_1 = m(x - x_1)$ is called the _____ form of the equation of a line. In words, we read this as *y* minus *y* _____ one equals *m* _____ the quantity of *x* _____ *x* sub _____.

2. $y = mx + b$ is called the _____ form of the equation of a line.

CONCEPTS

3. Determine in what form each equation is written.

    a. $y - 4 = 2(x - 5)$

    b. $y = 2x + 15$

4. What point does the graph of the equation pass through, and what is the line's slope?
   a. $y - 2 = 6(x - 7)$
   b. $y + 3 = -8(x + 1)$

5. Refer to the following graph of a line.
   a. What highlighted point does the line pass through?
   b. What is the slope of the line?
   c. Write an equation of the line in point–slope form.

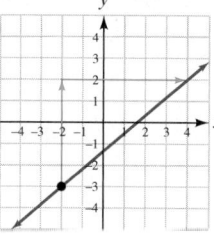

6. On a quiz, a student was asked to write the equation of a line with slope 4 that passes through $(-1, 3)$. Explain how the student can check her answer, $y = 4x + 7$.

7. Suppose you are asked to write an equation of the line in the scatter diagram below. What two points would you use to write the point–slope equation?

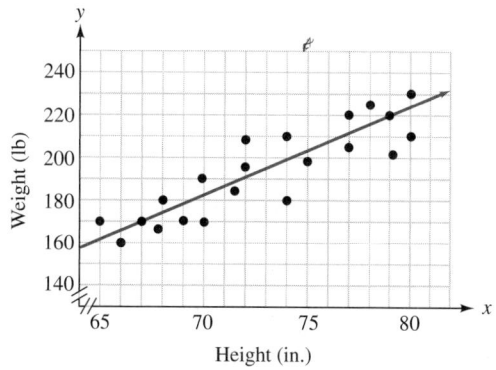

Height (in.)

8. In each case, a linear relationship between two quantities is described. If the relationship were graphed, what would be the slope of the line?
   a. The sales of new cars increased by 15 every 2 months.
   b. There were 35 fewer robberies for each dozen police officers added to the force.
   c. One acre of forest is being destroyed every 30 seconds.

NOTATION

Complete the solution.

9. Write an equation of the line with slope -2 that passes through the point $(-1, 5)$. Write the answer in slope–intercept form.

$$y - y_1 = m(x - x_1)$$
$$y - \boxed{} = -2[x - (\boxed{})]$$
$$y - 5 = -2[x \boxed{} 1]$$
$$y - 5 = -2x - \boxed{}$$
$$y = -2x + \boxed{}$$

10. What is the point–slope form of the equation of a line?

11. Consider the steps below and then fill in the blanks:

$$y - 3 = 2(x + 1)$$
$$y - 3 = 2x + 2$$
$$y = 2x + 5$$

The original equation was in _____ form. After solving for y, we obtain an equation in _____ form.

12. Fill in the blanks: The equation of a horizontal line has the form $\boxed{} = b$ and the equation of a vertical line has the form $\boxed{} = a$.

GUIDED PRACTICE

Use the point–slope form to write an equation of the line with the given slope and point. Leave the equation in that form. See Example 1.

13. Slope 3, passes through $(2, 1)$
14. Slope 2, passes through $(4, 3)$
15. Slope $\dfrac{4}{5}$, passes through $(-5, -1)$
16. Slope $\dfrac{7}{8}$, passes through $(-2, -9)$

Use the point–slope form to write an equation of the line with the given slope and point. Then write the equation in slope–intercept form. See Example 1.

17. Slope 2, passes through $(3, 5)$
18. Slope 8, passes through $(2, 6)$
19. Slope -5, passes through $(-9, 8)$
20. Slope -4, passes through $(-2, 10)$
21. Slope -3, passes through the origin
22. Slope -1, passes through the origin
23. Slope $\dfrac{1}{5}$, passes through $(10, 1)$
24. Slope $\dfrac{1}{4}$, passes through $(8, 1)$
25. Slope $-\dfrac{4}{3}$,

x	y
6	-4

26. Slope $-\dfrac{3}{2}$,

x	y
-2	1

27. Slope $-\dfrac{11}{6}$, passes through $(2, -6)$
28. Slope $-\dfrac{5}{4}$, passes through $(2, 0)$

Find an equation of the line that passes through the two given points. Write the equation in slope–intercept form, if possible. See Example 2.

29. Passes through $(1, 7)$ and $(-2, 1)$
30. Passes through $(-2, 2)$ and $(2, -8)$

31.

x	y
−4	3
2	0

32.

x	y
−1	−4
1	−2

33. Passes through (5, 5) and (7, 5)

34. Passes through (−2, 1) and (−2, 15)

35. Passes through (5, 1) and (−5, 0)

36. Passes through (−3, 0) and (3, 1)

37. Passes through (−8, 2) and (−8, 17)

38. Passes through $\left(\frac{2}{3}, 2\right)$ and (0, 2)

39. Passes through $\left(\frac{2}{3}, \frac{1}{3}\right)$ and (0, 0)

40. Passes through $\left(\frac{1}{2}, \frac{3}{4}\right)$ and (0, 0)

Write an equation of the line with the given characteristics. See Example 3.

41. Vertical, passes through (4, 5)

42. Vertical, passes through (−2, −5)

43. Horizontal, passes through (4, 5)

44. Horizontal, passes through (−2, −5)

Graph the line that passes through the given point and has the given slope. See Example 4.

45. (1, −2), slope −1

46. (−4, 1), slope −3

47. (5, −3), $m = \frac{3}{4}$

48. (2, −4), $m = \frac{2}{3}$

49. (−2, −3), slope 2

50. (−3, −3), slope 4

51. (4, −3), slope $-\frac{7}{8}$

52. (4, 2), slope $-\frac{1}{5}$

TRY IT YOURSELF

Find an equation of the line with the following characteristics. Write the equation in slope–intercept form, if possible.

53. Passes through (5, 0) and (−11, −4)

54. Passes through (7, −3) and (−5, 1)

55. Horizontal, passes through (−8, 12)

56. Horizontal, passes through (9, −32)

57. Slope $-\frac{2}{3}$, passes through (3, 0)

58. Slope $-\frac{2}{5}$, passes through (15, 0)

59. Slope 8, passes through (2, 20)

60. Slope 6, passes through (1, −2)

61. Vertical, passes through (−3, 7)

62. Vertical, passes through (12, −23)

63. Slope 7 and y-intercept (0, 0)

64. Slope 3 and y-intercept (0, 4)

65. Passes through (−2, −1) and (−1, −5)

66. Passes through (−3, 6) and (−1, −4)

67. x-intercept (7, 0) and y-intercept (0, −2)

68. x-intercept (−3, 0) and y-intercept (0, 7)

69. Slope $\frac{1}{10}$, passes through the origin

70. Slope $\frac{9}{8}$, passes through the origin

71. Undefined slope, passes through $\left(-\frac{1}{8}, 12\right)$

72. Undefined slope, passes through $\left(\frac{2}{5}, -\frac{5}{6}\right)$

APPLICATIONS

73. ANATOMY   There is a linear relationship between a woman's height and the length of her radius bone. It can be stated this way: Height increases by 3.9 inches for each 1-inch increase in the length of the radius. Suppose a 64-inch-tall woman has a 9-inch-long radius bone. Use this information to find a linear equation that relates height h to the length r of the radius. Write the equation in slope–intercept form.

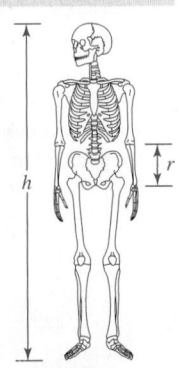

74. AUTOMATION   An automated production line uses distilled water at a rate of 300 gallons every 2 hours to make shampoo. After the line had run for 7 hours, planners noted that 2,500 gallons of distilled water remained in the storage tank. Find a linear equation relating the time t in hours since the production line began and the number g of gallons of distilled water in the storage tank. Write the equation in slope–intercept form.

75. POLE VAULTING   Find the equations of the lines that describe the positions of the pole for parts 1, 3, and 4 of the jump. Write the equations in slope–intercept form, if possible.

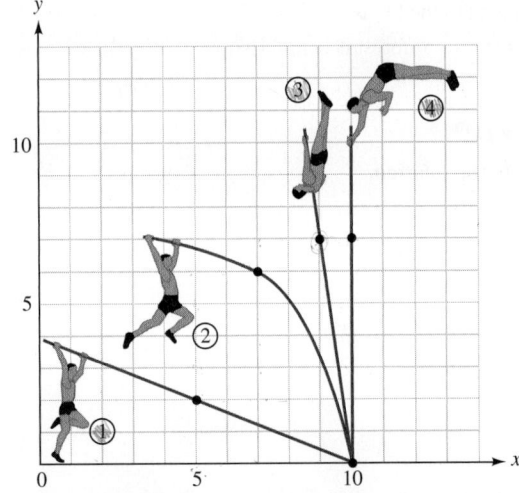

76. FREEWAY DESIGN   The graph below shows the route of a proposed freeway.

   a. Give the coordinates of the points where the proposed freeway will join Interstate 25 and Highway 40.

   b. Write the equation of the line that describes the route of the proposed freeway. Give the answer in slope–intercept form.

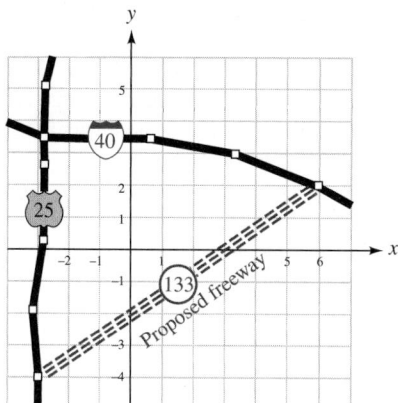

77. TOXIC CLEANUP   Three months after cleanup began at a dump site, 800 cubic yards of toxic waste had yet to be removed. Two months later, that number had been lowered to 720 cubic yards.

   a. Find an equation that describes the linear relationship between the length of time m (in months) the cleanup crew has been working and the number of cubic yards y of toxic waste remaining. Write the equation in slope–intercept form.

   b. Use your answer to part (a) to predict the number of cubic yards of waste that will still be on the site one year after the cleanup project began.

78. DEPRECIATION   To lower its corporate income tax, accountants of a company depreciated a word processing system over several years using a linear model, as shown in the worksheet.

   a. Find a linear equation relating the years since the system was purchased, x, and its value, y, in dollars. Write the equation in slope–intercept form.

   b. Find the purchase price of the system.

Tax Worksheet

Method of depreciation: *Linear*

Property	Years after purchase	Value
Word processing system	2	$60,000
"	4	$30,000

79. TRAMPOLINES   There is a linear relationship between the length of the protective pad that wraps around a trampoline and the radius of the trampoline. Use the data in the table to find an equation that gives the length l of pad needed for any trampoline with radius r. Write the equation in slope–intercept form. Use units of feet for both l and r.

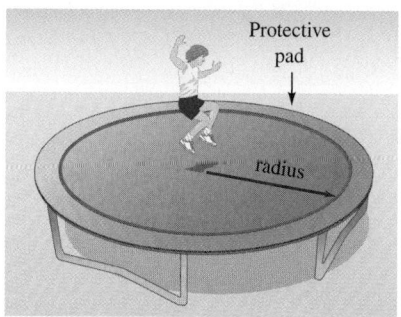

Radius	Pad length
3 ft	19 ft
7 ft	44 ft

80. CONVERTING TEMPERATURES   The relationship between Fahrenheit temperature, F, and Celsius temperature, C, is linear.

   a. Use the data in the illustration to write two ordered pairs of the form (C, F).

   b. Use your answer to part (a) to find a linear equation relating the Fahrenheit and Celsius scales. Write the equation in slope–intercept form.

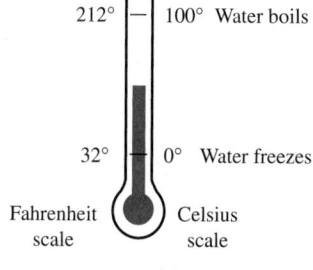

81. GOT MILK   The scatter diagram shows the amount of milk that an average American drank in one year for the years 1980–2004. A straight line can be used to model the data.

   a. Use two points on the line to find its equation. Write the equation in slope–intercept form.

   b. Use your answer to part (a) to predict the amount of milk that an average American will drink in 2020.

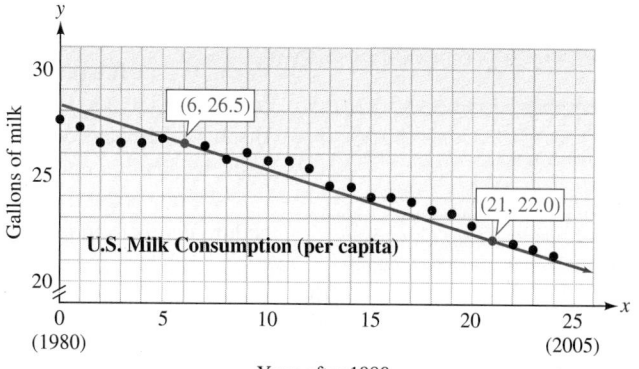

Source: United States Department of Agriculture

82. ENGINE OUTPUT   The horsepower produced by an automobile engine was recorded for various engine speeds in the range of 2,400 to 4,800 revolutions per minute (rpm). The data were recorded on the following scatter diagram. Find an equation of the line that models the relationship between engine speed s and horsepower h. Write the equation in slope–intercept form.

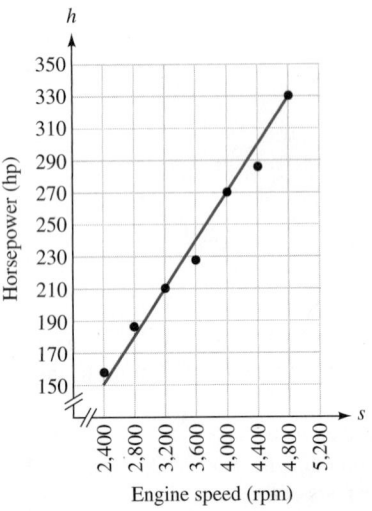

Engine speed (rpm)

WRITING

83. Why is $y - y_1 = m(x - x_1)$ called the point–slope form of the equation of a line?

84. If we know two points that a line passes through, we can write its equation. Explain how this is done.

85. Explain the steps involved in writing $y - 6 = 4(x - 1)$ in slope–intercept form.

86. Think of the points on the graph of the horizontal line $y = 4$. What do the points have in common? How do they differ?

REVIEW

87. FRAMES   The length of a rectangular picture is 5 inches greater than twice the width. If the perimeter is 112 inches, find the dimensions of the frame.

88. SPEED OF AN AIRPLANE   Two planes are 6,000 miles apart, and their speeds differ by 200 mph. They travel toward each other and meet in 5 hours. Find the speed of the slower plane.

CHALLENGE PROBLEMS

89. Find an equation of the line that passes through (2, 5) and is parallel to the line $y = 4x - 7$. Write the equation in slope–intercept form.

90. Find an equation of the line that passes through $(-6, 3)$ and is perpendicular to the line $y = -3x - 12$. Write the equation in slope–intercept form.

SECTION 3.7
Graphing Linear Inequalities

Objectives

1 Determine whether an ordered pair is a solution of an inequality.

2 Graph a linear inequality in two variables.

3 Graph inequalities with a boundary through the origin.

4 Solve applied problems involving linear inequalities in two variables.

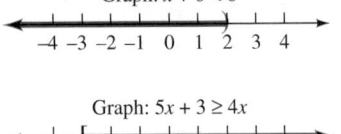

Graph: $x + 6 < 8$

Graph: $5x + 3 \geq 4x$

Recall that an **inequality** is a statement that contains one of the symbols $<$, $\leq$, $>$, or $\geq$. Inequalities in one variable, such as $x + 6 < 8$ and $5x + 3 \geq 4x$, were solved in Section 2.7. Because they have an infinite number of solutions, we represented their solution sets graphically, by shading intervals on a number line.

We now extend that concept to linear inequalities *in two variables,* as we introduce a procedure that is used to graph their solution sets.

1   Determine Whether an Ordered Pair Is a Solution of an Inequality.

If the $=$ symbol in a linear equation in two variables is replaced with an inequality symbol, we have a **linear inequality in two variables.** Some examples are

$$x - y \leq 5, \qquad 4x + 3y < -6, \qquad \text{and} \qquad y > 2x$$

As with linear equations, a **solution of a linear inequality** in two variables is an ordered pair of numbers that makes the inequality true.

EXAMPLE 1 Determine whether each ordered pair is a solution of $x - y \leq 5$. Then graph each solution: **a.** $(4, 2)$ **b.** $(0, -6)$ **c.** $(1, -4)$

Strategy We will substitute each ordered pair of coordinates into the inequality.

Why If the resulting statement is true, the ordered pair is a solution.

Solution

a. For $(4, 2)$:

$$x - y \leq 5 \qquad \text{This is the given inequality.}$$
$$4 - 2 \overset{?}{\leq} 5 \qquad \text{Substitute 4 for } x \text{ and 2 for } y.$$
$$2 \leq 5 \qquad \text{True}$$

Because $2 \leq 5$ is true, $(4, 2)$ is a solution of $x - y \leq 5$. We say that $(4, 2)$ *satisfies* the inequality. This solution is graphed as shown, on the right.

b. For $(0, -6)$:

$$x - y \leq 5 \qquad \text{This is the given inequality.}$$
$$0 - (-6) \overset{?}{\leq} 5 \qquad \text{Substitute 0 for } x \text{ and } -6 \text{ for } y.$$
$$6 \leq 5 \qquad \text{False}$$

Because $6 \leq 5$ is false, $(0, -6)$ is not a solution.

c. For $(1, -4)$:

$$x - y \leq 5 \qquad \text{This is the given inequality.}$$
$$1 - (-4) \overset{?}{\leq} 5 \qquad \text{Substitute 1 for } x \text{ and } -4 \text{ for } y.$$
$$5 \leq 5 \qquad \text{True}$$

Because $5 \leq 5$ is true, $(1, -4)$ is a solution, and we graph it as shown.

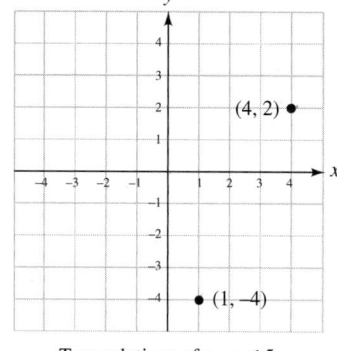

Two solutions of $x - y \leq 5$.

> ### *Notation*
> The symbol $\overset{?}{\leq}$ is read as "is possibly less than or equal to."

Self Check 1 Using the inequality in Example 1, determine whether each ordered pair is a solution: **a.** $(8, 2)$
b. $(4, -1)$ **c.** $(-2, 4)$ **d.** $(-3, -5)$

Now Try **Problem 19**

In Example 1, we graphed two of the solutions of $x - y \leq 5$. Since there are infinitely more ordered pairs (x, y) that make the inequality true, it would not be reasonable to plot them. Fortunately, there is an easier way to show all of the solutions.

2 **Graph a Linear Inequality in Two Variables.**

The graph of a linear inequality is a picture that represents the set of all points whose coordinates satisfy the inequality. In general, such graphs are regions bounded by a line. We call those regions **half-planes,** and we use a two-step procedure to find them.

> | **EXAMPLE 2** | Graph: $x - y \leq 5$ |

Strategy   We will graph the related *equation* $x - y = 5$ to establish a boundary line between two regions of the coordinate plane. Then we will determine which region contains points whose coordinates satisfy the given inequality.

Why   The graph of a linear inequality in two variables is a region of the coordinate plane on one side of a boundary line.

Solution   Since the inequality symbol $\leq$ includes an equal symbol, the graph of $x - y \leq 5$ includes the graph of $x - y = 5$.

Step 1:   To graph $x - y = 5$, we use the intercept method, as shown in part (a) of the illustration. The resulting line, called a **boundary line,** divides the coordinate plane into two half-planes. To show that the points on the boundary line are solutions of $x - y \leq 5$, we draw it as a solid line.

$x - y = 5$

x	y	(x, y)
0	-5	$(0, -5)$
5	0	$(5, 0)$
6	1	$(6, 1)$

Let $x = 0$ and find y. Let $y = 0$ and find x.
As a check, let $x = 6$ and find y.

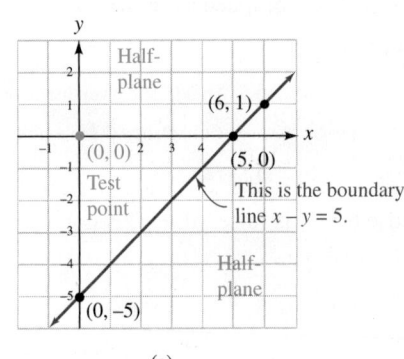

(a)

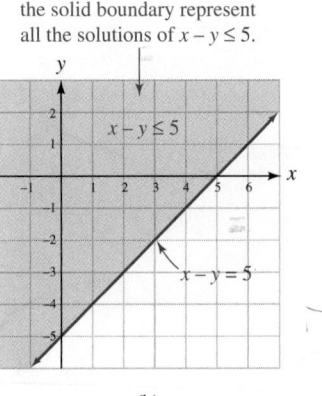

This shaded half-plane and the solid boundary represent all the solutions of $x - y \leq 5$.

(b)

Step 2:   Since the inequality $x - y \leq 5$ also allows $x - y$ to be less than 5, other ordered pairs, besides those on the boundary, satisfy the inequality. For example, consider the origin, with coordinates $(0, 0)$. If we substitute 0 for x and 0 for y in the given inequality, we have

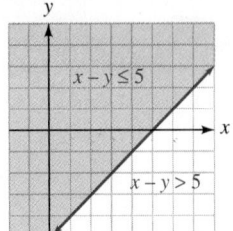
$$x - y \leq 5$$
$$0 - 0 \overset{?}{\leq} 5$$
$$0 \leq 5 \quad \text{True}$$

Because $0 \leq 5$, the coordinates of the origin satisfy $x - y \leq 5$. In fact, the coordinates of every point on the same side of the line as the origin satisfy the inequality. To indicate this, we shade the half-plane that contains the test point $(0, 0)$, as shown in part (b). Every point in the shaded half-plane and every point on the boundary line satisfies $x - y \leq 5$.

As an informal check, we can pick an ordered pair that lies in the shaded region and one that does not lie in the shaded region. When we substitute their coordinates into the inequality, we should obtain a true statement and then a false statement.

For (3, 1), in the shaded region:	***For (5, −4), not in the shaded region:***
$x - y \leq 5$	$x - y \leq 5$
$3 - 1 \overset{?}{\leq} 5$	$5 - (-4) \overset{?}{\leq} 5$
$2 \leq 5 \quad$ True	$9 \leq 5 \quad$ False

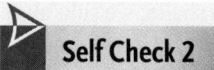

Self Check 2   Graph: $x - y \leq 2$

Now Try   **Problem 35**

The previous example suggests the following procedure to graph linear inequalities in two variables.

Graphing Linear Inequalities in Two Variables

1. Replace the inequality symbol with an equal symbol = and graph the boundary line of the region. If the original inequality allows the possibility of equality (the symbol is either $\leq$ or $\geq$), draw the boundary line as a solid line. If equality is not allowed ($<$ or $>$), draw the boundary line as a dashed line.

2. Pick a test point that is on one side of the boundary line. (Use the origin if possible.) Replace x and y in the inequality with the coordinates of that point. If a true statement results, shade the side that contains that point. If a false statement results, shade the other side of the boundary.

EXAMPLE 3   Graph: $4x + 3y < -6$

Strategy   We will graph the related equation $4x + 3y = -6$ to establish the boundary line between two regions of the coordinate plane. Then we will determine which region contains points that satisfy the given inequality.

Why   The graph of a linear inequality in two variables is a region of the coordinate plane on one side of a boundary line.

Solution   To find the boundary line, we replace the inequality symbol with an equal symbol = and graph $4x + 3y = -6$. Since the inequality symbol $<$ does not include an equal symbol, the points on the graph of $4x + 3y = -6$ will not be part of the graph of $4x + 3y < -6$. To show this, we draw the boundary line as a dashed line. See part (a) of the illustration.

$4x + 3y = -6$

x	y	(x, y)
0	-2	$(0, -2)$
$-\frac{3}{2}$	0	$\left(-\frac{3}{2}, 0\right)$
-3	2	$(-3, 2)$

Let $x = 0$ and find y. Let $y = 0$ and find x.
As a check, let $x = -3$ and find y.

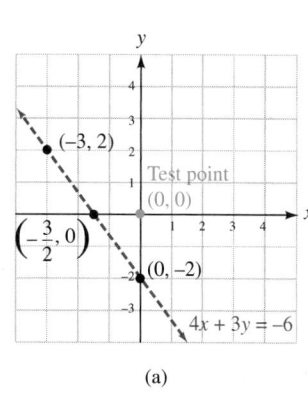

(a)

This shaded half-plane represents all the solutions of $4x + 3y < -6$.

The coordinates of any point on the dashed boundary do not satisfy the inequality.

(b)

To determine which half-plane to shade, we substitute the coordinates of a point that lies on one side of the boundary line into $4x + 3y < -6$. We choose the origin $(0, 0)$ as the test point because the computations are easy when they involve 0. We substitute 0 for x and 0 for y in the inequality.

$$4x + 3y < -6$$
$$4(0) + 3(0) \overset{?}{<} -6 \qquad \text{The symbol } \overset{?}{<} \text{ is read as "is possibly less than."}$$
$$0 + 0 \overset{?}{<} -6$$
$$0 < -6 \qquad \text{False}$$

Since $0 < -6$ is a false statement, the point $(0, 0)$ does not satisfy the inequality. This indicates that it is not on the side of the dashed line we wish to shade. Instead, we shade the other side of the boundary line. The graph of the solution set of $4x + 3y < -6$ is the half-plane below the dashed line, as shown in part (b).

Self Check 3   Graph: $5x + 6y < -15$

Now Try   **Problem 37**

> **Caution**
> When using a test point to determine which half-plane to shade, remember to substitute the coordinates into the given inequality, not the equation for the boundary.

3 **Graph Inequalities with a Boundary through the Origin.**

In the next example, the boundary line passes through the origin.

EXAMPLE 4   Graph: $y > 2x$

Strategy   We will graph the related equation $y = 2x$ to establish the boundary line between two regions of the coordinate plane. Then we will determine which region contains points that satisfy the given inequality.

Why   The graph of a linear inequality in two variables is a region of the coordinate plane on one side of a boundary line.

Solution   To find the boundary line, we graph $y = 2x$. Since the symbol $>$ does *not* include an equal symbol, the points on the graph of $y = 2x$ are not part of the graph of $y > 2x$. Therefore, the boundary line should be dashed, as shown in part (a) of the illustration.

> **Success Tip**
> Draw a solid boundary line if the inequality has $\leq$ or $\geq$. Draw a dashed line if the inequality has $<$ or $>$.

$y = 2x$

x	y	(x, y)
0	0	$(0, 0)$
-1	-2	$(-1, -2)$
1	2	$(1, 2)$

Select three values for x and find the corresponding values of y.

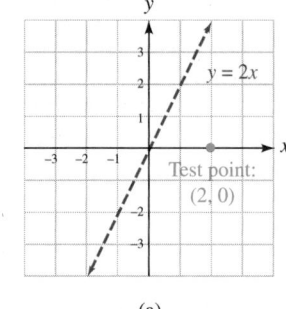

(a)

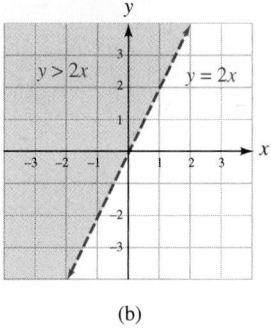

(b)

To determine which half-plane to shade, we substitute the coordinates of a point that lies on one side of the boundary line into $y > 2x$. Since the origin is on the boundary, it cannot serve as a test point. One of the many possible choices for a test point is $(2, 0)$, because it does not lie on the boundary line. To see whether it satisfies $y > 2x$, we substitute 2 for x and 0 for y in the inequality.

$$y > 2x$$
$$0 \overset{?}{>} 2(2) \quad \text{The symbol } \overset{?}{>} \text{ is read as "is possibly greater than."}$$
$$0 > 4 \quad \text{False}$$

Since $0 > 4$ is a false statement, the point $(2, 0)$ does not satisfy the inequality. We shade the half-plane that does not contain $(2, 0)$, as shown in part (b).

Self Check 4   Graph: $y < 3x$

Now Try **Problem 55**

EXAMPLE 5   Graph each linear inequality: **a.** $x < -3$ **b.** $y \geq 0$

Strategy   We will use the procedure for graphing linear inequalities in two variables.

Why   Since the inequalities can be written as $x + 0y < -3$ and $0x + y \geq 0$, they are linear inequalities in two variables.

Solution

a. Because $x < -3$ contains an $<$ symbol, we draw the boundary, $x = -3$, as a dashed vertical line. We can use $(0, 0)$ as the test point.

$$x < -3 \quad \text{This is the given inequality.}$$
$$0 < -3 \quad \text{Substitute 0 for x. The y-coordinate of the test point (0, 0) is not used.}$$

Since the result is false, we shade the half-plane that does not contain $(0, 0)$, as shown in figure (a) below. Note that the solution consists of all points that have an x-coordinate that is less than -3.

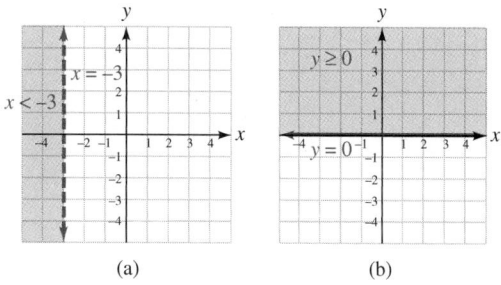

(a)   (b)

b. Because $y \geq 0$ contains an $\geq$ symbol, we draw the boundary, $y = 0$, as a solid horizontal line. (Recall that the graph of $y = 0$ is the x-axis.) Next, we choose a test point not on the boundary. The point $(0, 1)$ is a convenient choice.

$$y \geq 0 \quad \text{This is the given inequality.}$$
$$1 \geq 0 \quad \text{Substitute 1 for y. The x-coordinate of the test point (0, 1) is not used.}$$

Since the result is true, we shade the half-plane that contains $(0, 1)$, as shown in part (b) above. Note that the solution consists of all points that have a y-coordinate that is greater than or equal to 0.

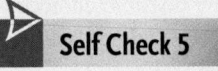 **Self Check 5**   Graph each linear inequality: **a.** $x \geq 2$ **b.** $y < 4$

Now Try **Problems 63 and 65**

4 Solve Applied Problems Involving Linear Inequalities in Two Variables.

When solving applied problems, phrases such as *at least, at most,* and *should not exceed* indicate that an inequality should be used.

EXAMPLE 6   ***Working Two Jobs.***   Carlos has two part-time jobs, one paying $10 per hour and another paying $12 per hour. If x represents the number of hours he works on the first job, and y represents the number of hours he works on the second, the graph of $10x + 12y \geq 240$ shows the possible ways he can schedule his time to earn at least $240 per week to pay his college expenses. Find three possible combinations of hours he can work to achieve his financial goal.

Strategy   We will graph the inequality and find three points whose coordinates satisfy the inequality.

Why   The coordinates of these points will give three possible combinations.

Solution   The graph of the inequality is shown below in part (a) of the illustration. Any point in the shaded region represents a possible way Carlos can schedule his time and earn $240 or more per week. If each shift is a whole number of hours long, the highlighted points in part (b) represent the acceptable combinations. Three such combinations are

(6, 24):   6 hours on the first job, 24 hours on the second job

(12, 12):   12 hours on the first job, 12 hours on the second job

(22, 4):   22 hours on the first job, 4 hours on the second job

To verify one combination, suppose Carlos works 22 hours on the first job and 4 hours on the second job. He will earn

$$\$10(22) + \$12(4) = \$220 + \$48$$
$$= \$268$$

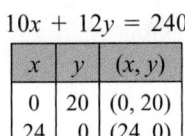

$10x + 12y = 240$

x	y	(x, y)
0	20	(0, 20)
24	0	(24, 0)

Let $x = 0$ and find y.
Let $y = 0$ and find x.

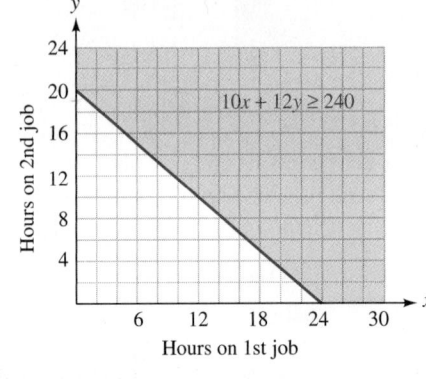

(a)

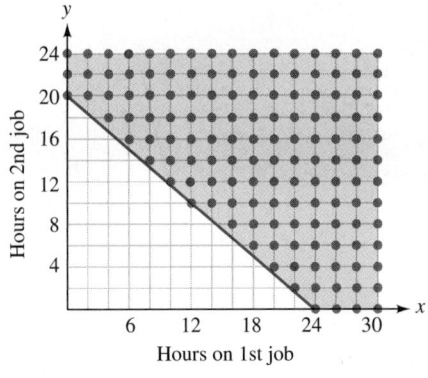

(b)

▷ *Now Try* **Problem 75**

▷ **ANSWERS TO SELF CHECKS**   **1. a.** Not a solution   **b.** Solution   **c.** Solution   **d.** Solution

2. 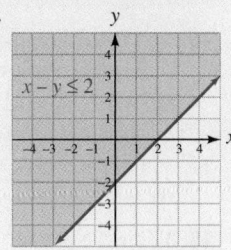   **3.** 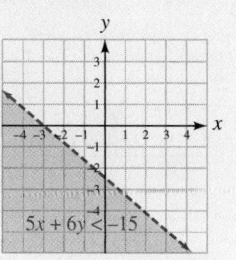   **4.**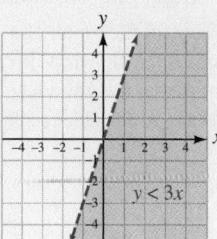

5.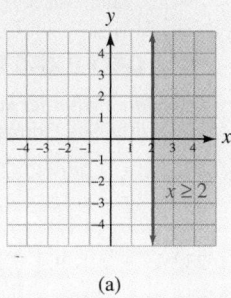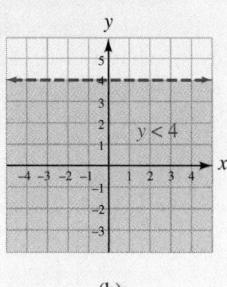

(a)   (b)

STUDY SET
3.7

VOCABULARY

Fill in the blanks.

1. $2x - y \le 4$ is a linear _____ in two variables.

2. A _____ of a linear inequality is an ordered pair of numbers that makes the inequality true.

3. $(7, 2)$ is a solution of $x - y > 1$. We say that $(7, 2)$ _____ the inequality.

4. In the graph, the line $2x - y = 4$ is the _____ line.

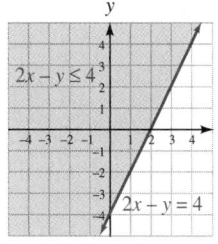

5. In the graph above, the line $2x - y = 4$ divides the coordinate plane into two _____.

6. When graphing a linear inequality, we determine which half-plane to shade by substituting the coordinates of a test _____ into the inequality.

CONCEPTS

7. Determine whether $(-3, -5)$ is a solution of $5x - 3y \geq 0$.

8. Determine whether $(3, -1)$ is a solution of $x + 4y < -1$.

9. Fill in the blanks: A _____ line indicates that points on the boundary are not solutions and a _____ line indicates that points on the boundary are solutions.

10. The boundary for the graph of a linear inequality is shown. Why can't the origin be used as a test point to decide which side to shade?

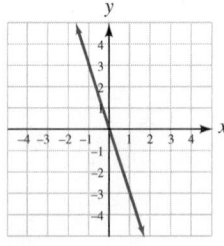

11. If a false statement results when the coordinates of a test point are substituted into a linear inequality, which half-plane should be shaded to represent the solution of the inequality?

12. A linear inequality has been graphed. Determine whether each point satisfies the inequality.

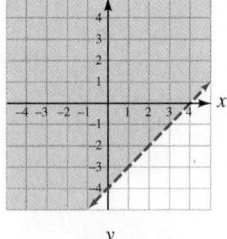

   a. $(1, -3)$
   b. $(-2, -1)$
   c. $(2, 3)$
   d. $(3, -4)$

13. A linear inequality has been graphed. Determine whether each point satisfies the inequality.

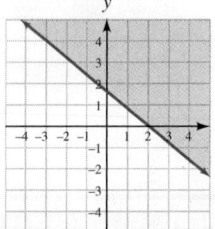

   a. $(2, 1)$
   b. $(-2, -4)$
   c. $(4, -2)$
   d. $(-3, 4)$

14. To graph linear inequalities, we must be able to graph boundary lines. Complete the table of solutions for each given boundary line.

   a. $5x - 3y = 15$

x	y	(x, y)
0		
	0	
1		

   b. $y = 3x - 2$

x	y	(x, y)
-1		
0		
2		

NOTATION

15. Write the meaning of each symbol in words.

   a. $<$ **b.** $\geq$

   c. $\leq$ **d.** $\overset{?}{>}$

16. a. When graphing linear inequalities, which inequality symbols are associated with a dashed boundary line?

   b. When graphing linear inequalities, which inequality symbols are associated with a solid boundary line?

17. Fill in the blanks: The inequality $4x + 2y \leq 9$ means $4x + 2y \quad 9$ or $4x + 2y \quad 9$.

18. Fill in the blanks: The inequality $-x + 8y \geq 1$ means $-x + 8y \quad 1$ or $-x + 8y \quad 1$.

GUIDED PRACTICE

Determine whether each ordered pair is a solution of the given inequality. See Example 1.

19. $2x + y > 6; (3, 2)$ **20.** $4x - 2y \geq -6; (-2, 1)$

21. $-5x - 8y < 8; (-8, 4)$ **22.** $x + 3y > 14; (-3, 8)$

23. $4x - y \leq 0; \left(\frac{1}{2}, 1\right)$ **24.** $9x - y \leq 2; \left(\frac{1}{3}, 1\right)$

25. $-5x + 2y > -4; (0.8, 0.6)$

26. $6x - 2y < -7; (-0.2, 1.5)$

Complete the graph by shading the correct side of the boundary. See Example 2.

27. $x - y \geq -2$ **28.** $x - y < 3$

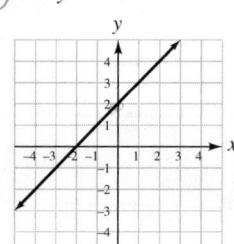   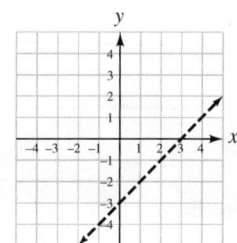

29. $y > 2x - 4$ **30.** $y \leq -x + 1$

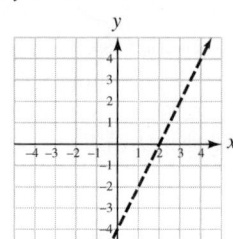   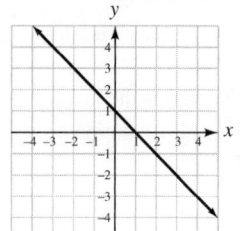

31. $x - 2y \geq 4$

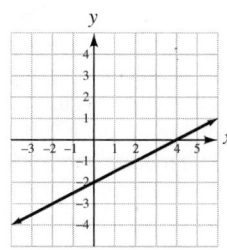

32. $3x + 2y > 12$

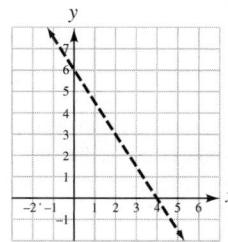

33. $y \leq 4x$

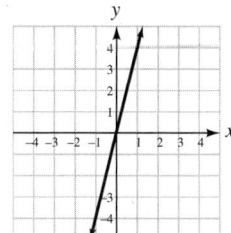

34. $y + 2x < 0$

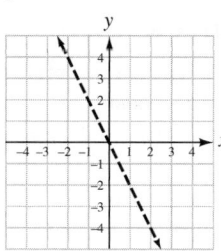

Graph each inequality. See Examples 2 and 3.

35. $x + y \geq 3$ **36.** $x + y < 2$

37. $3x - 4y > 12$ **38.** $5x + 4y \geq 20$

39. $2x + 3y \leq -12$ **40.** $3x - 2y > 6$

41. $y < 2x - 1$ **42.** $y > x + 1$

43. $y < -3x + 2$ **44.** $y \geq -2x + 5$

45. $y \geq -\dfrac{3}{2}x + 1$ **46.** $y < \dfrac{x}{3} - 1$

47. $x - 2y \geq 4$ **48.** $4x + y \geq -4$

49. $2y - x < 8$ **50.** $y + 9x \geq 3$

51. $7x - 2y < 21$ **52.** $3x - 3y \geq -10$

53. $2x - 3y \geq 4$ **54.** $4x + 3y < 6$

Graph each inequality. See Example 4.

55. $y \geq 2x$ **56.** $y < 3x$

57. $y < -\dfrac{x}{2}$ **58.** $y \geq x$

59. $y + x < 0$ **60.** $y - x < 0$

61. $5x + 3y < 0$ **62.** $2x + 5y > 0$

Graph each inequality. See Example 5.

63. $x < 2$ **64.** $y > -3$

65. $y \leq 1$ **66.** $x \geq -4$

67. $y + 2.5 > 0$ **68.** $x - 1.5 \leq 0$

69. $x \leq 0$ **70.** $y < 0$

APPLICATIONS

71. DELIVERIES   To decide the number x of pallets and the number y of barrels that a truck can hold, a driver refers to the graph below. Can a truck make a delivery of 4 pallets and 10 barrels in one trip?

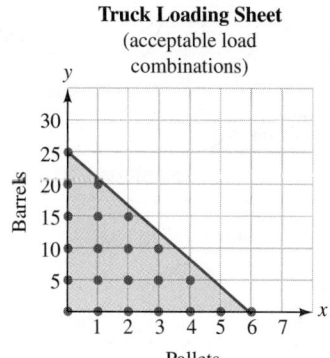

Truck Loading Sheet
(acceptable load combinations)

72. ZOOS   To determine the allowable number of juvenile chimpanzees x and adult chimpanzees y that can live in an enclosure, a zookeeper refers to the graph. Can 6 juvenile and 4 adult chimps be kept in the enclosure?

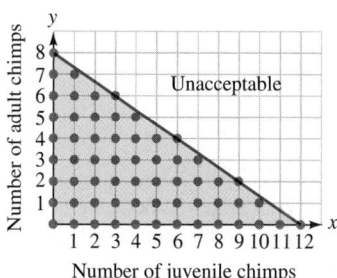

73. ROLLING DICE   The points on the graph represent all of the possible outcomes when two fair dice are rolled a single time. For example, (5, 2), shown in red, represents a 5 on the first die and a 2 on the second. Which of the following sentences best describes the outcomes that lie in the shaded area?

  (i) Their sum is at most 6.

 (ii) Their sum exceeds 6.

(iii) Their sum does not exceed 6.

(iv) Their sum is at least 6.

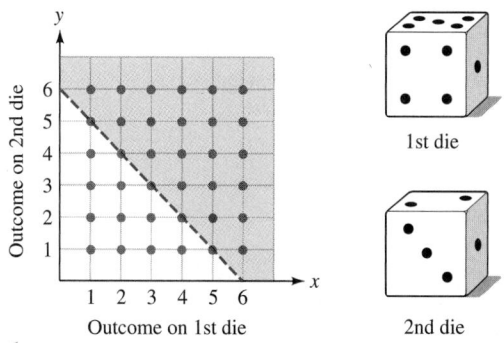

74. NATO In March 1999, NATO aircraft and cruise missiles targeted Serbian military forces that were south of the 44th parallel in Yugoslavia, Montenegro, and Kosovo. Shade the geographic area that NATO was trying to rid of Serbian forces.

Based on data from *Los Angeles Times* (March 24, 1999)

75. PRODUCTION PLANNING It costs a bakery $3 to make a cake and $4 to make a pie. If x represents the number of cakes made, and y represents the number of pies made, the graph of $3x + 4y \leq 120$ shows the possible combinations of cakes and pies that can be produced so that costs do not exceed $120 per day. Graph the inequality. Then find three possible combinations of pies and cakes that can be made so that the daily costs are not exceeded.

76. HIRING BABYSITTERS Mrs. Cansino has a choice of two babysitters. Sitter 1 charges $6 per hour, and Sitter 2 charges $7 per hour. If x represents the number of hours she uses Sitter 1 and y represents the number of hours she uses Sitter 2, the graph of $6x + 7y \leq 42$ shows the possible ways she can hire the sitters and not spend more than $42 per week. Graph the inequality. Then find three possible ways she can hire the babysitters so that her weekly budget for babysitting is not exceeded.

77. INVENTORIES A clothing store advertises that it maintains an inventory of at least $4,400 worth of men's jackets at all times. At the store, leather jackets cost $100 and nylon jackets cost $88. If x represents the number of leather jackets in stock and y represents the number of nylon jackets in stock, the graph of $100x + 88y \geq 4,400$ shows the possible ways the jackets can be stocked. Graph the inequality. Then find three possible combinations of leather and nylon jackets so that the store lives up to its advertising claim.

78. MAKING SPORTING GOODS A sporting goods manufacturer allocates at least 2,400 units of production time per day to make baseballs and footballs. It takes 20 units of time to make a baseball and 30 units of time to make a football. If x represents the number of baseballs made and y represents the number of footballs made, the graph of $20x + 30y \geq 2,400$ shows the possible ways to schedule the production time. Graph the inequality. Then find three possible combinations of production time for the company to make baseballs and footballs.

WRITING

79. Explain how to decide which side of the boundary line to shade when graphing a linear inequality in two variables.

80. Why is the origin usually a good test point to choose when graphing a linear inequality?

81. Why is (0, 0) not an acceptable choice for a test point when graphing a linear inequality whose boundary passes through the origin?

82. Explain the difference between the graph of the solution set of $x + 1 > 8$, an inequality in one variable, and the graph of $x + y > 8$, an inequality in two variables.

REVIEW

83. Solve $A = P + Prt$ for t.

84. What is the sum of the measures of the three angles of any triangle?

85. Simplify: $40\left(\dfrac{3}{8}x - \dfrac{1}{4}\right) + 40\left(\dfrac{4}{5}\right)$

86. Evaluate: $-4 + 5 - (-3) - 13$

CHALLENGE PROBLEMS

87. Find a linear inequality that has the graph shown.

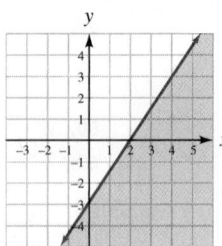

88. Graph the inequality: $4x - 3(x + 2y) \geq -6$

SECTION 3.8
An Introduction to Functions

Objectives

1. Find the domain and range of a relation.
2. Identify functions and their domains and ranges.
3. Use function notation.
4. Graph functions.
5. Use the vertical line test.
6. Solve applications involving functions.

In this section, we will discuss *relations* and *functions*. These two concepts are included in our study of graphing because they involve ordered pairs.

1 Find the Domain and Range of a Relation.

The following table shows the number of medals won by American athletes at seven recent Winter Olympics.

USA Winter Olympic Medal Count						
Year	1984	1988	1992	1994*	1998	2002	2006
Medals	8	6	11	13	13	34	25
	Sarajevo YUG	Calgary CAN	Albertville FRA	Lillehammer NOR	Nagano JPN	Salt Lake City USA	Turin ITA

* The Winter Olympics were moved ahead two years so that the
winter and summer games would alternate every two years.

We can display the data in the table as a set of ordered pairs, where the **first component** represents the year and the **second component** represents the number of medals won by American athletes:

{(1984, 8), (1988, 6), (1992, 11), (1994, 13), (1998, 13), (2002, 34), (2006, 25)}

A set of ordered pairs, such as this, is called a **relation.** The set of all first components is called the **domain** of the relation and the set of all second components is called the **range** of a relation.

EXAMPLE 1   Find the domain and range of the relation {(1, 7), (4, −6), (−3, 1), (2, 7)}.

Strategy   We will examine the first and second components of the ordered pairs.

Why   The set of first components is the domain and the set of second components is the range.

Solution   The relation {(1, 7), (4, −6), (−3, 1), (2, 7)} has the domain {−3, 1, 2, 4} and the range is {−6, 1, 7}. The elements of the domain and range are usually listed in increasing order, and if a value is repeated, it is listed only once.

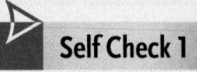

> **Self Check 1**   Find the domain and range of the relation $\{(8, 2), (-1, 10),$
> $(6, 2), (-5, -5)\}$.
>
> ***Now Try***   **Problem 15**

② Identify Functions and their Domains and Ranges.

An **arrow** or **mapping diagram** can be used to illustrate a relation. The data from the Winter Olympics example is shown below in that form.

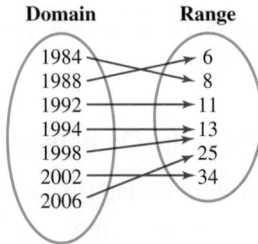

    Notice that for each year, there corresponds exactly one medal count. That is, this relation assigns to each member of the domain exactly one member of the range. Relations that have this characteristic are called *functions*.

Function	A **function** is a set of ordered pairs (a relation) in which to each first component there corresponds exactly one second component.

    We may also think of a function as a rule that assigns to each value of one variable exactly one value of another variable. Since we often worked with sets of ordered pairs of the form (x, y), it is helpful to define a function in an alternate way using the variables x and y.

y is a Function of x	If to each value of x in the domain there is assigned exactly one value of y in the range, then y is said to be a function of x.

> **EXAMPLE 2**   Determine whether the arrow diagram and the tables define y to be a function of x. If a function is defined, give its domain and range.

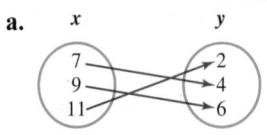

Strategy   We will check to see whether each value of x is assigned exactly one value of y.

Why   If this is true, the arrow diagram or table defines y to be a function of x.

Solution

a. The arrow diagram defines a function because to each value of x there is assigned exactly one value of y: $7 \rightarrow 4$, $9 \rightarrow 6$, and $11 \rightarrow 2$.

The domain of the function is $\{7, 9, 11\}$ and the range is $\{2, 4, 6\}$.

b. The table does not define a function, because to the x value 2 there is assigned more than one value of y: $2 \rightarrow 3$ and $2 \rightarrow 1$.

c. Since to each number x exactly one value y is assigned, the table defines y to be a function of x. It also illustrates an important fact about functions: *The same value of y can be assigned to different values of x.* In this case, each number x is assigned a y-value of 8.

The domain of the function is $\{0, 3, 4, 9\}$ and the range is $\{8\}$.

Self Check 2   Determine whether the arrow diagram and the table define y to be a function of x. If a function is defined, give its domain and range.

a. 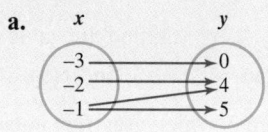   **b.**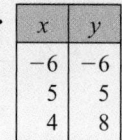

Now Try   **Problems 19 and 25**

---

3   **Use Function Notation.**

A function can be defined by an equation. For example, $y = 2x - 3$ is a rule that assigns to each value of x exactly one value of y. To find the y-value that is assigned to the x-value 4, we substitute 4 for x and evaluate the right side of the equation.

$$y = 2x - 3$$
$$y = 2(4) - 3 \quad \text{Substitute 4 for x.}$$
$$= 8 - 3 \quad \text{Evaluate the right side.}$$
$$= 5$$

The function $y = 2x - 3$ assigns the y-value 5 to an x-value of 4. When making such calculations, the value of x is called an **input** and its corresponding value of y is called an **output.**

A special notation is used to name functions that are defined by equations.

Function Notation	The notation $y = f(x)$ denotes that the variable y is a function of x.

Since $y = f(x)$, the equations $y = 2x - 3$ and $f(x) = 2x - 3$ are equivalent. We read $f(x) = 2x - 3$ as "f of x is equal to $2x$ minus 3."

This is the variable used to
represent the input value.
↓

$$f(x) = 2x - 3$$
↑   ↑

This is the name   This expression shows how to obtain
of the function.   an output from a given input.

Function notation provides a compact way of representing the value that is assigned to some number x. For example, if $f(x) = 2x - 3$, the value that is assigned to an x-value 5 is represented by $f(5)$.

$$f(x) = 2x - 3$$
$$f(5) = 2(5) - 3 \qquad \text{Substitute the input 5 for each } x.$$
$$= 10 - 3 \qquad \text{Evaluate the right side.}$$
$$= 7 \qquad \text{The output is 7.}$$

Thus, $f(5) = 7$. We read this as "f of 5 is 7." The output 7 is called a **function value.**

To see why function notation is helpful, consider these equivalent sentences:

> **Caution**
> The symbol $f(x)$ denotes a function. It does not mean $f \cdot x$ (f times x). Read $f(x)$ as "f of x."

1. If $y = 2x - 3$, find the value of y when x is 5.
2. If $f(x) = 2x - 3$, find $f(5)$.

Sentence 2, which uses $f(x)$ notation is much more compact.

EXAMPLE 3   For $f(x) = 5x + 7$, find each of the following function values:
a. $f(2)$ **b.** $f(-4)$ **c.** $f(0)$

Strategy   We will substitute 2, -4, and 0 for x in the expression $5x + 7$ and then evaluate it.

Why   The notation $f(x) = 5x + 7$ indicates that we are to multiply each input (each number written within the parentheses) by 5 and then add 7 to that product.

Solution

a. To find $f(2)$, we substitute the number within the parentheses, 2, for each x in $f(x) = 5x + 7$, and evaluate the right side of the equation.

> **The Language of Algebra**
> Another way to read $f(2) = 17$ is to say "the value of the function is 17 at 2."

$$f(x) = 5x + 7$$
$$f(2) = 5(2) + 7 \qquad \text{Substitute the input 2 for each } x.$$
$$= 10 + 7 \qquad \text{Evaluate the right side.}$$
$$= 17 \qquad \text{The output is 17.}$$

Thus, $f(2) = 17$.

b.
$$f(x) = 5x + 7$$
$$f(-4) = 5(-4) + 7 \qquad \text{Substitute the input } -4 \text{ for each } x.$$
$$= -20 + 7 \qquad \text{Evaluate the right side.}$$
$$= -13 \qquad \text{The output is } -13.$$

Thus, $f(-4) = -13$.

c. $f(x) = 5x + 7$

$f(0) = 5(0) + 7$ *Substitute the input 0 for each x.*

$\quad\ = 0 + 7$ *Evaluate the right side.*

$\quad\ = 7$ *The output is 7.*

Thus, $f(0) = 7$.

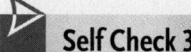

Self Check 3   For $f(x) = -2x + 3$, find each of the following function values: **a.** $f(4)$ **b.** $f(-1)$ **c.** $f(0)$

Now Try   **Problem 31**

We can think of a function as a machine that takes some input x and turns it into some output $f(x)$, as shown in part (a) of the figure. In part (b), the function machine for $f(x) = x^2 + 2x$ turns the input 4 into the output $4^2 + 2(4) = 24$, and we have $f(4) = 24$.

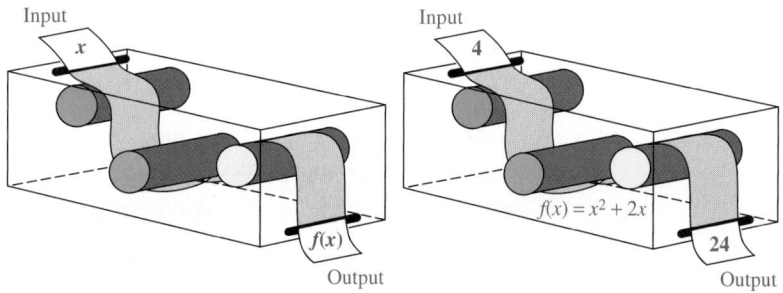

The letter f used in the notation $y = f(x)$ represents the word *function*. However, other letters, such as g and h can be used to name functions.

EXAMPLE 4   For $g(x) = 3 - 2x$ and $h(x) = x^3 - 1$, find   **a.** $g(3)$ **b.** $h(-2)$

Strategy   We will substitute 3 for x in $3 - 2x$ and substitute -2 for x in $x^3 - 1$, and then evaluate each expression.

Why   The numbers 3 and -2, that are within the parentheses, are inputs that should be substituted for the variable x.

Solution

a. To find $g(3)$, we use the function rule $g(x) = 3 - 2x$ and replace x with 3.

$g(x) = 3 - 2x$ *Read g(x) as "g of x."*

$g(3) = 3 - 2(3)$ *Substitute 3 for each x.*

$\quad\ = 3 - 6$ *Evaluate the right side.*

$\quad\ = -3$

Thus, $g(3) = -3$.

b. To find $h(-2)$, we use the function rule $h(x) = x^3 - 1$ and replace x with -2.

$$h(x) = x^3 - 1 \qquad \text{Read } h(x) \text{ as "}h \text{ of } x\text{."}$$
$$h(-2) = (-2)^3 - 1 \qquad \text{Substitute } -2 \text{ for each } x.$$
$$= -8 - 1 \qquad \text{Evaluate the right side.}$$
$$= -9$$

Thus, $h(-2) = -9$.

Self Check 4   Find $g(0)$ and $h(4)$ for the functions in Example 3.

Now Try   **Problem 37**

4 **Graph Functions.**

We have seen that a function such as $f(x) = 4x + 1$ assigns to each value of x a single value $f(x)$. The input-output pairs generated by a function can be written in the form $(x, f(x))$. These ordered pairs can be plotted on a rectangular coordinate system to give the graph of the function.

EXAMPLE 5   Graph: $f(x) = 4x + 1$

Strategy   We can graph the function by creating a table of function values and plotting the corresponding ordered pairs.

Why   After drawing a line though the plotted points, we will have the graph of the function.

Solution   To make a table, we choose several values for x and find the corresponding values of $f(x)$. If x is -1, we have

$$f(x) = 4x + 1 \qquad \text{This is the function to graph.}$$
$$f(-1) = 4(-1) + 1 \qquad \text{Substitute } -1 \text{ for each } x.$$
$$= -4 + 1 \qquad \text{Evaluate the right side.}$$
$$= -3$$

Thus, $f(-1) = -3$. This means that, when x is -1, $f(x)$ is -3, and it indicates that the ordered pair $(-1, -3)$ lies on the graph of $f(x)$.

Similarly, we find the corresponding values of $f(x)$ for x-values of 0 and 1. Then we plot the resulting ordered pairs and draw a straight line through them to get the graph of $f(x) = 4x + 1$. Since $y = f(x)$, the graph of $f(x) = 4x + 1$ is the same as the graph of the equation $y = 4x + 1$.

Notation

A table of function values is similar to a table of solutions, except that the second column is usually labeled $f(x)$ instead of y.

x	$f(x)$

x	y

$f(x) = 4x + 1$

x	$f(x)$	
-1	-3	$\rightarrow (-1, -3)$
0	1	$\rightarrow (0, 1)$
1	5	$\rightarrow (1, 5)$

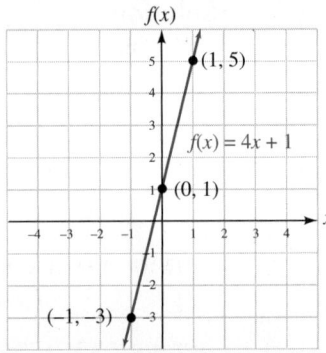

The vertical axis can be labeled y or $f(x)$.

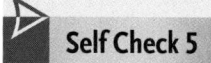

We call $f(x) = 4x + 1$ from Example 5 a **linear function** because its graph is a nonvertical line. Any linear equation, except those of the form $x = a$, can be written using function notation by writing it in slope–intercept form ($y = mx + b$) and then replacing y with $f(x)$.

EXAMPLE 6   Graph: $f(x) = |x|$

Strategy We can graph the function by creating a table of function values and plotting the corresponding ordered pairs.

Why After drawing a "V" shape though the plotted points, we will have the graph of the function.

Solution To create a table of function values, we choose values for x and find the corresponding values of $f(x)$. For $x = -4$ and $x = 3$, we have

$$f(x) = |x| \qquad\qquad f(x) = |x|$$
$$f(-4) = |-4| \qquad\qquad f(3) = |3|$$
$$= 4 \qquad\qquad\qquad = 3$$

Thus, $f(-4) = 4$ and $f(3) = 3$.

Similarly, we find the corresponding values of $f(x)$ for several other x-values. When we plot the resulting ordered pairs, we see that they lie in a "V" shape. We join the points to complete the graph as shown. We call $f(x) = |x|$ an **absolute value function.**

$f(x) = |x|$

x	$f(x)$	
-4	4	$\longrightarrow (-4, 4)$
-3	3	$\longrightarrow (-3, 3)$
-2	2	$\longrightarrow (-2, 2)$
-1	1	$\longrightarrow (-1, 1)$
0	0	$\longrightarrow (0, 0)$
1	1	$\longrightarrow (1, 1)$
2	2	$\longrightarrow (2, 2)$
3	3	$\longrightarrow (3, 3)$
4	4	$\longrightarrow (4, 4)$

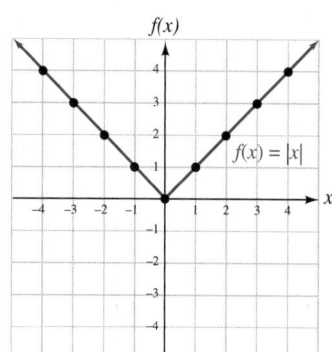

5 **Use the Vertical Line Test.**

If any vertical line intersects a graph more than once, the graph cannot represent a function, because to one value of x there would correspond more than one value of y.

| The Vertical Line Test | If a vertical line intersects a graph in more than one point, the graph is not the graph of a function. |

The graph shown in red does not represent a function, because a vertical line intersects the graph at more than one point. The points of intersection indicate that the *x*-value −1 corresponds to two different *y*-values, 3 and −1.

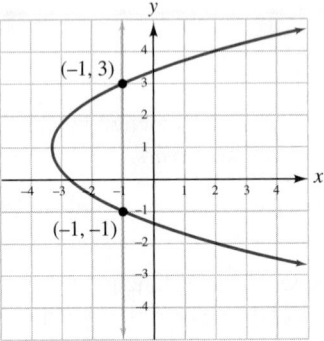

x	y
−1	3
−1	−1

When the coordinates of the two points of intersection are listed in a table, it is easy to see that the x-value of −1 is assigned two different y-values. Thus, this is not the graph of a function.

EXAMPLE 7   Determine whether each of the following is the graph of a function.

a. 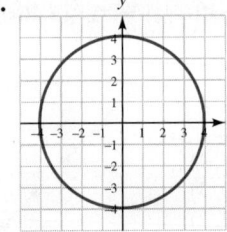   **b.**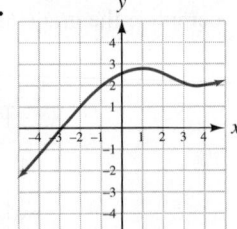

Strategy   We will check to see whether any vertical line intersects the graph more than once.

Why   If any vertical line does intersect the graph more than once, the graph is not a function.

Solution

a. Refer to figure (a) on the right. This is not the graph of a function because the vertical line shown in blue intersects the graph at more than one point. The points of intersection indicate that the *x*-value 3 corresponds to assigned two different *y*-values, 2.5 and −2.5.

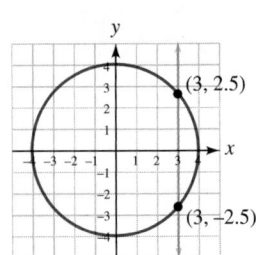

This is not the graph of a function.

(a)

b. Refer to figure (b) on the right. This is a graph of a function because no vertical line intersects the graph at more than one point. Several vertical lines are drawn in blue to illustrate this.

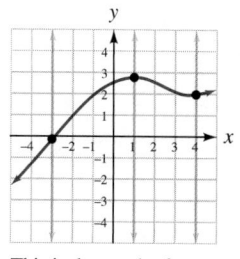

This is the graph of a function.

(b)

Self Check 7

Determine whether each of the following is the graph of a function.

a.

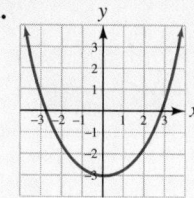

b.

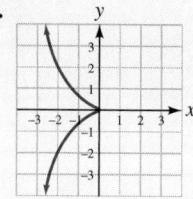

Now Try **Problem 51**

6 **Solve Applications Involving Functions.**

Functions are used to describe certain relationships where one quantity depends upon another. Letters other than f and x are often chosen to more clearly describe these situations.

EXAMPLE 8 ***Party Rentals.*** The function $C(h) = 40 + 5(h - 4)$ gives the cost in dollars to rent an inflatable jumper for h hours. Find the cost of renting the jumper for 10 hours.

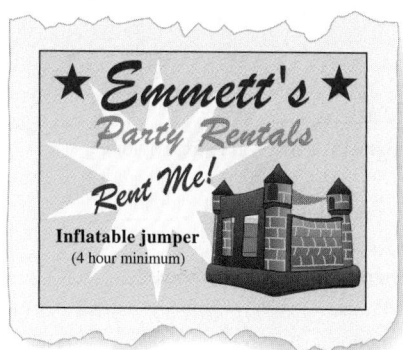

★ *Emmett's* ★
Party Rentals
Rent Me!
Inflatable jumper
(4 hour minimum)

Strategy To find the cost to rent the jumper for 10 hours, we will substitute 10 for each h in $C(h) = 40 + 5(h - 4)$ and evaluate the right side.

Why In $C(h) = 40 + 5(h - 4)$, the variable h represents the number of hours that the jumper is rented. We need to find $C(10)$.

Solution   For this application involving hours and cost, the notation $C(h)$ is used. The independent variable is h and the name of the function is C. If the jumper is rented for 10 hours, then h is 10 and we must find $C(10)$.

$$C(h) = 40 + 5(h - 4) \qquad \text{Read } C(h) \text{ as "}C \text{ of } h\text{."}$$

$$C(10) = 40 + 5(10 - 4) \qquad \text{Substitute 10 for each } h.$$

$$= 40 + 5(6) \qquad \text{Evaluate the right side.}$$

$$= 40 + 30$$

$$= 70$$

It costs $70 to rent the jumper for 10 hours.

Self Check 8   Find the cost of renting the jumper for 8 hours.

Now Try   **Problem 61**

ANSWERS TO SELF CHECKS   **1.** Domain: $\{-5, -1, 6, 8\}$; range: $\{-5, 2, 10\}$
2. a. No   **b.** Yes; domain: $\{-6, 4, 5\}$; range: $\{-6, 5, 8\}$ **3. a.** -5 **b.** 5   **c.** 3   **4.** 3, 63
5.   **6.**   **7. a.** Function   **b.** Not a function
8. $60

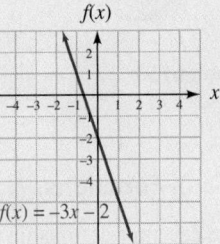

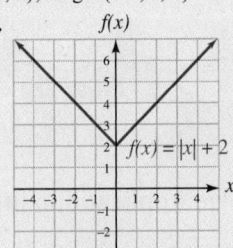

STUDY SET
3.8

VOCABULARY

Fill in the blanks.

1. A set of ordered pairs is called a _____.

2. A _____ is a rule that assigns to each x-value exactly one y-value.

3. The set of all input values for a function is called the _____, and the set of all output values is called the _____.

4. We can think of a function as a machine that takes some _____ x and turns it into some output _____.

5. If $f(2) = -3$, we call -3 a function _____.

6. The graph of a _____ function is a straight line and the graph of an _____ value function is V-shaped.

CONCEPTS

7. FEDERAL MINIMUM HOURLY WAGE   The following table is an example of a function. Use an arrow diagram to show how members of the range are assigned to members of the domain.

Year	1990	1992	1994	1996	1998	2000	2002	2004	2006	2008
Minimum wage ($)	3.80	4.25	4.25	4.75	5.15	5.15	5.15	5.15	5.15	6.55

Source: *Time Almanac 2006 and aflcio.org*

8. The arrow diagram describes a function. What is the domain and what is the range of the function?

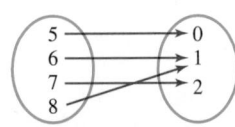

9. For the given input, what value will the function machine output?

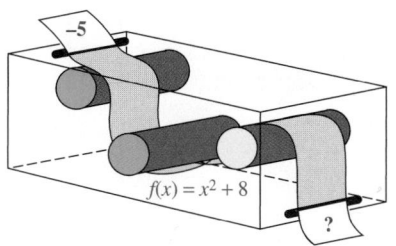

$f(x) = x^2 + 8$

10. a. Fill in the blank: If a _____ line intersects a graph in more than one point, the graph is not the graph of a function.

b. Give the coordinates of the points where the given vertical line inter-sects the graph.

c. Is this the graph of a function?

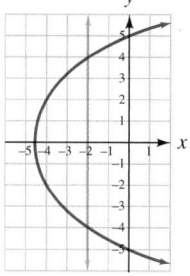

NOTATION

Fill in the blanks.

11. We read $f(x) = 5x - 6$ as "f ____ x is $5x$ minus 6."

12. Since $y =$ _____, the following two equations are equivalent:

$$y = 3x + 2 \quad \text{and} \quad f(x) = 3x + 2$$

13. The notation $f(4) = 5$ indicates that when the x-value ☐ is input into a function rule, the output is ☐. This fact can be shown graphically by plotting the ordered pair (☐ , ☐).

14. When graphing the function $f(x) = -x + 5$, the vertical axis of the coordinate system can be labeled ☐ or ☐.

GUIDED PRACTICE

Find the domain and range of each relation. **See Example 1.**

15. $\{(6, -1), (-1, -10), (-6, 2), (8, -5)\}$

16. $\{(11, -3), (0, 0), (4, 5), (-3, -7)\}$

17. $\{(0, 9), (-8, 50), (6, 9)\}$

18. $\{(1, -12), (-6, 8), (5, 8)\}$

Determine whether each arrow diagram or table defines y as a function of x. If a function is defined, give its domain and range. If it does not define a function, find two ordered pairs that show a value of x that is assigned more than one value of y. **See Example 2.**

19.
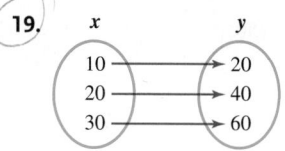

20.
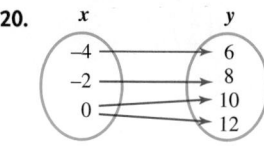

21.
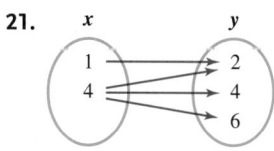

22.
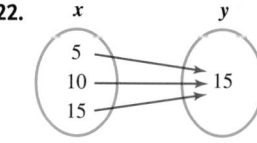

23.

x	y
1	7
2	15
3	23
4	16
5	8

24.

x	y
30	2
30	4
30	6
30	8
30	10

25.

x	y
-4	6
-1	0
0	-3
2	4
-1	2

26.

x	y
1	1
2	2
3	3
4	4

27.

x	y
3	4
3	-4
4	3
4	-3

28.

x	y
-1	1
-3	1
-5	1
-7	1
-9	1

29.

x	y
6	0
-3	-8
1	9
5	4

30.

x	y
1.6	0
-3	-1
2.5	20
-7	0.1
1.6	19

Find each function value. See Examples 3 and 4.

31. $f(x) = 4x - 1$

   a. $f(1)$ **b.** $f(-2)$

   c. $f\left(\dfrac{1}{4}\right)$ **d.** $f(50)$

32. $f(x) = 1 - 5x$

   a. $f(0)$ **b.** $f(-75)$

   c. $f(0.2)$ **d.** $f\left(-\dfrac{4}{5}\right)$

33. $f(x) = 2x^2$

   a. $f(0.4)$ **b.** $f(-3)$

   c. $f(1,000)$ **d.** $f\left(\dfrac{1}{8}\right)$

34. $g(x) = 6 - x^2$

   a. $g(30)$ **b.** $g(6)$

   c. $g(-1)$ **d.** $g(0.5)$

35. $h(x) = |x - 7|$

   a. $h(0)$ **b.** $h(-7)$

   c. $h(7)$ **d.** $h(8)$

36. $f(x) = |2 + x|$

   a. $f(0)$ **b.** $f(2)$

   c. $f(-2)$ **d.** $f(-99)$

37. $g(x) = x^3 - x$

   a. $g(1)$ **b.** $g(10)$

   c. $g(-3)$ **d.** $g(6)$

38. $g(x) = x^4 + x$

   a. $g(1)$ **b.** $g(-2)$

   c. $g(0)$ **d.** $g(10)$

39. $s(x) = (x + 3)^2$

   a. $s(3)$ **b.** $s(-3)$

   c. $s(0)$ **d.** $s(-5)$

40. $s(x) = (x - 8)^2$

   a. $s(8)$ **b.** $s(-8)$

   c. $s(1)$ **d.** $s(12)$

41. If $f(x) = 3.4x^2 - 1.2x + 0.5$, find $f(-0.3)$.

42. If $g(x) = x^4 - x^3 + x^2 - x$, find $g(-12)$.

Complete each table of function values and then graph each function. See Examples 5 and 6.

43. $f(x) = -3x - 2$

x	$f(x)$
-2	
-1	
0	
1	

44. $f(x) = -2x + 8$

x	$f(x)$
-1	
0	
1	
2	

45. $h(x) = |1 - x|$

x	$h(x)$
-2	
-1	
0	
1	
2	
3	
4	

46. $h(x) = |x + 2|$

x	$h(x)$
-5	
-4	
-3	
-2	
-1	
0	
1	

Graph each function. See Examples 5 and 6.

47. $f(x) = \dfrac{1}{2}x - 2$ **48.** $f(x) = -\dfrac{2}{3}x + 3$

49. $h(x) = -|x|$ **50.** $g(x) = |x| - 2$

Determine whether each graph is the graph of a function. If it is not, find ordered pairs that show a value of x that is assigned more than one value of y. See Example 7.

51.

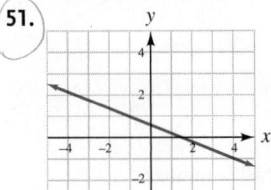

52.

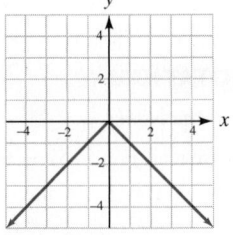

53.

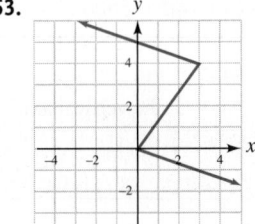

54.

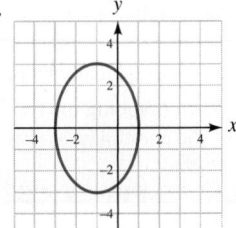

55.

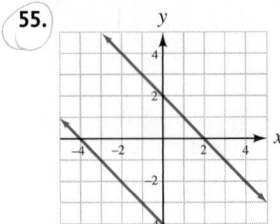

56.

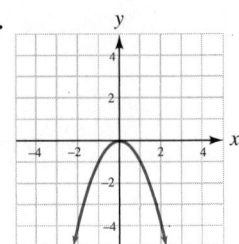

57.

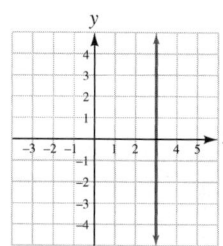

58.

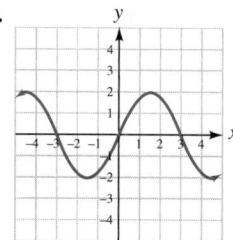

APPLICATIONS

59. REFLECTIONS When a beam of light hits a mirror, it is reflected off the mirror at the same angle that the incoming beam struck the mirror. What type of function could serve as a mathematical model for the path of the light beam shown here?

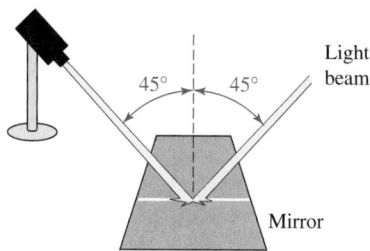

Light beam

45° 45°

Mirror

60. LIGHTNING The function $D(t) = \frac{t}{5}$ gives the approximate distance in miles that you are from a lightning strike, where t is the number of seconds between seeing the lightning and hearing the thunder. Find $D(5)$ and explain what it means.

61. VACATIONING The function $C(d) = 500 + 100(d - 3)$ gives the cost in dollars to rent an RV motor home for d days. Find the cost of renting the RV for a vacation that will last 7 days.

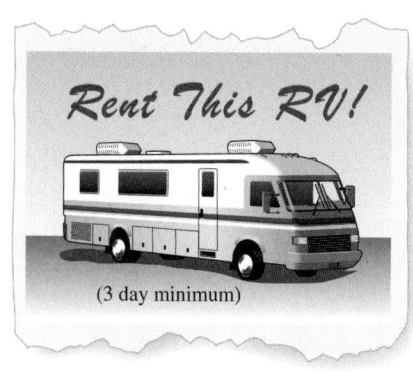

Rent This RV!

(3 day minimum)

62. STRUCTURAL ENGINEERING The maximum safe load in pounds of the rectangular beam shown in the figure is given by the function $S(t) = \frac{1,875t^2}{8}$, where t is the thickness of the beam, in inches. Find the maximum safe load if the beam is 4 inches thick.

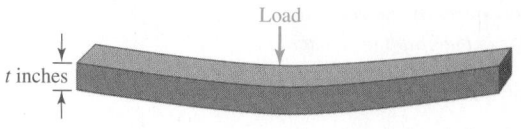

Load

t inches

63. LAWN SPRINKLERS The function $A(r) = \pi r^2$ can be used to determine the area that will be watered by a rotating sprinkler that sprays out a stream of water. Find $A(5)$ and $A(20)$. Round to the nearest tenth.

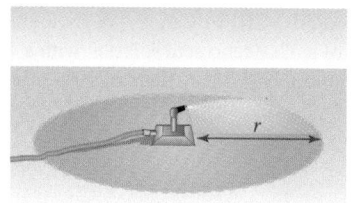

r

64. PARTS LISTS The function $f(r) = 2.30 + 3.25(r + 0.40)$ approximates the length (in feet) of the belt that joins the two pulleys, where r is the radius (in feet) of the smaller pulley. Find the belt length needed for each pulley in the parts list.

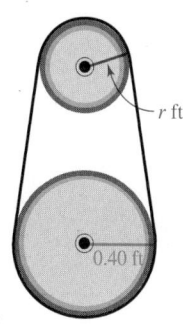

r ft

0.40 ft

Parts list		
Pulley	r	Belt length
P-45M	0.32	
P-08D	0.24	

WRITING

65. In the function $y = -5x + 2$, why do you think x is called the *independent* variable and y the *dependent* variable?

66. Explain what a politician meant when she said, "The speed at which the downtown area will be redeveloped is a function of the number of low-interest loans made available to the property owners."

67. A student was asked to determine whether the graph on the right is the graph of a function. What is wrong with the following reasoning?

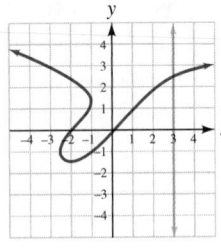

When I draw a vertical line through the graph, it intersects the graph only once. By the vertical line test, this is the graph of a function.

68. In your own words, what is a function?

REVIEW PROBLEMS

69. COFFEE BLENDS A store sells regular coffee for $4 a pound and gourmet coffee for $7 a pound. To get rid of 40 pounds of the gourmet coffee, the shopkeeper plans to make a gourmet blend that he will put on sale for $5 a pound. How many pounds of regular coffee should be used?

70. PHOTOGRAPHIC CHEMICALS A photographer wishes to mix 2 liters of a 5% acetic acid solution with a 10% solution to get a 7% solution. How many liters of 10% solution must be added?

CHALLENGE PROBLEMS

71. Is the graph of $y \geq 3 - x$ a function? Explain.

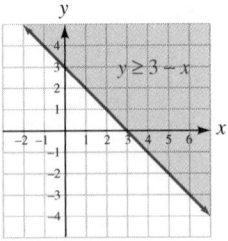

72. If $f(x) = x^2 + x$, find: $f\left(\frac{4}{5}r\right)$

73. Let $f(x) = -2x + 5$. For what value of x is $f(x) = -7$?

74. Let $f(x) = x - 2$ and $g(x) = 3x$. Find $f(g(6))$.

CHAPTER 3
Summary & Review

SECTION 3.1   Graphing Using the Rectangular Coordinate System

DEFINITIONS AND CONCEPTS	EXAMPLES
A **rectangular coordinate system** is composed of a horizontal number line called the **x-axis** and a vertical number line, called the **y-axis.** The two axes intersect at the **origin.** To **plot** or **graph** ordered pairs means to locate their position on a rectangular coordinate system. The x- and y-axes divide the coordinate plane into four regions called **quadrants.**	Plot the points: $(2, 3), (-4, 2), (-3, -1), (0, -2.5), (4, -2)$ To graph each point, start at the origin and count the appropriate number of units in the x-direction and then the appropriate number of units in the y-direction. 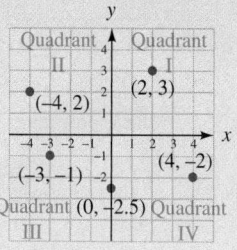

REVIEW EXERCISES

1. Graph the points with coordinates $(-1, 3), (0, 1.5), (-4, -4),$ $\left(2, \frac{7}{2}\right),$ and $(4, 0)$.

2. HAWAII   Estimate the coordinates of Oahu using an ordered pair of the form (longitude, latitude).

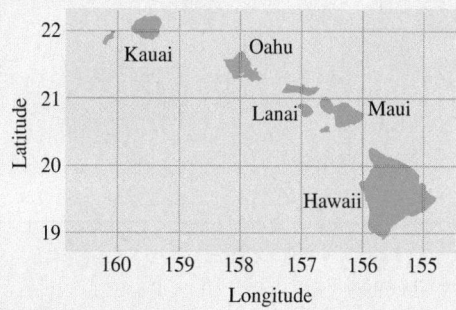

3. In what quadrant does the point $(-3, -4)$ lie?

4. What are the coordinates of the origin?

5. GEOMETRY   Three vertices (corners) of a square are $(-5, 4), (-5, -2),$ and $(1, -2)$. Find the coordinates of the fourth vertex and then find the area of the square.

6. COLLEGE ENROLLMENTS   The graph gives the number of students enrolled at a college for the period from 4 weeks before to 5 weeks after the semester began.

   a. What was the maximum enrollment and when did it occur?

   b. How many students had enrolled 2 weeks before the semester began?

   c. When was the enrollment 2,250?

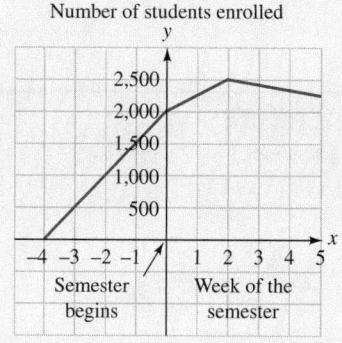

SECTION 3.2   Graphing Linear Equations

DEFINITIONS AND CONCEPTS	EXAMPLES
A **solution of an equation in two variables** is an ordered pair of numbers that makes the equation a true statement when the numbers are substituted for the variables. The **standard** or **general form** of a linear equation is $Ax + By = C$, where A, B, and C are real numbers and A and B are not both zero.	Determine whether $(2, -3)$ a solution of $2x - y = 7$. We substitute the coordinates into the equation. $2x - y = 7$ $2(2) - (-3) \overset{?}{=} 7$ Substitute 2 for x and −3 for y. $4 + 3 \overset{?}{=} 7$ Evaluate the left side. $7 = 7$ True Since the result is true, $(2, -3)$ is a solution of the equation.
If only one coordinate of an ordered-pair solution is known: 1. Substitute it into the equation for the appropriate variable. 2. Solve the resulting equation to find the unknown coordinate.	To complete the solution $(, 8)$ for $3x + y = -1$, we substitute 8 for y and solve the resulting equation for x. $3x + y = -1$ $3x + 8 = -1$ Substitute 8 for y. $3x = -9$ Subtract 8 from both sides. $x = -3$ To isolate x, divide both sides by 3. The solution is $(-3, 8)$.
To **graph a linear equation** solved for y: 1. Find three solutions by selecting three values of x and finding the corresponding values of y. 2. Plot each ordered-pair solution. 3. Draw a line through the points.	Graph: $y = -2x + 1$ We construct a table of solutions, plot the points, and draw the line. $y = -2x + 1$ 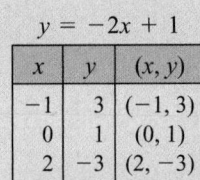 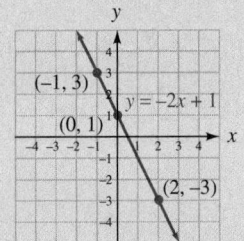

$y = -2x + 1$

x	y	(x, y)
-1	3	$(-1, 3)$
0	1	$(0, 1)$
2	-3	$(2, -3)$

REVIEW EXERCISES

7. Is $(-3, -2)$ a solution of $y = 2x + 4$?

8. Complete the table of solutions.

$$3x + 2y = -18$$

x	y	(x, y)
-2		
	3	

9. Which of the following equations are not linear equations?

$8x - 2y = 6 \quad y = x^2 + 1 \quad y = x \quad 3y = -x + 4 \quad y - x^3 = 0$

10. The graph of a linear equation is shown.

   a. When the coordinates of point A are substituted into the equation, will a true or false statement result?

   b. When the coordinates of point B are substituted into the equation, will a true or false statement result?

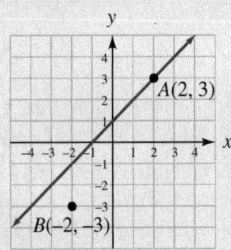

Graph each equation by constructing a table of solutions.

11. $y = 4x - 2$

12. $y = \dfrac{3}{4}x$

13. $5y = -5x + 15$ (*Hint:* Solve for y first.)

14. $6y = -4x$ (*Hint:* Solve for y first.)

15. BIRTHDAY PARTIES   A restaurant offers a party package for children that includes everything: food, drinks, cake, and favors. The cost c, in dollars, is given by the equation $c = 8n + 50$, where n is the number of children attending the party. Graph the equation and use the graph to estimate the cost of a party if 18 children attend.

16. Determine whether each statement is true or false.

    a. It takes three or more points to determine a line.

    b. A linear equation in two variables has infinitely many solutions.

SECTION 3.3   Intercepts

DEFINITIONS AND CONCEPTS	EXAMPLES
The point where a line intersects the x-axis is called the **x-intercept.** The point where a line intersects the y-axis is called the **y-intercept.** To **find the y-intercept,** substitute 0 for x in the given equation and solve for y. To **find the x-intercept,** substitute 0 for y and solve for x. Plotting the x- and y-intercepts of a graph and drawing a line through them is called the **intercept method for graphing a line.**	Use the y- and x-intercepts to graph $3x + 4y = -6$. 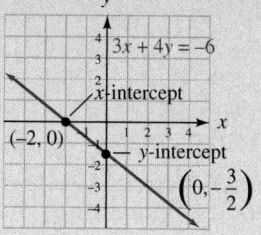 **y-intercept: $x = 0$** $3x + 4y = -6$ $3(0) + 4y = -6$ $4y = -6$ $y = -\dfrac{3}{2}$ **x-intercept: $y = 0$** $3x + 4y = -6$ $3x + 4(0) = -6$ $3x = -6$ $x = -2$ The y-intercept is $\left(0, -\dfrac{3}{2}\right)$ and the x-intercept is $(-2, 0)$.
The equation $y = b$ represents the **horizontal line** that intersects the y-axis at $(0, b)$. The equation $x = a$ represents the **vertical line** that intersects the x-axis at $(a, 0)$.	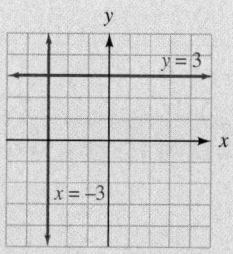

REVIEW EXERCISES

17. Identify the x- and y-intercepts of the graph shown on the right.

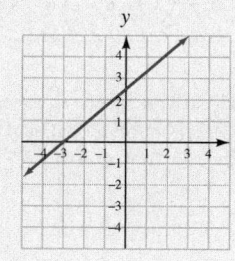

Use the intercept method to graph each equation.

19. $-4x + 2y = 8$

20. $5x - 4y = 13$

21. Graph: $y = 4$

22. Graph: $x = -1$

18. DEPRECIATION   The graph shows how the value of some sound equipment decreased over the years. Find the intercepts of the graph. What information do the intercepts give about the equipment?

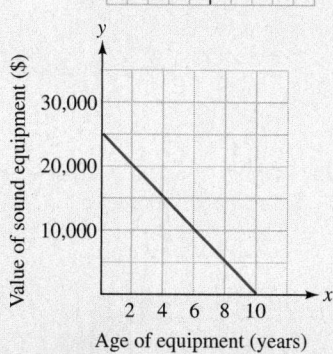

SECTION 3.4   Slope and Rate of Change

DEFINITIONS AND CONCEPTS	EXAMPLES
The **slope** m of a line is a ratio that compares the vertical and horizontal change as we move along the line from one point to another. We can find the slope of a line graphically using the ratio $m = \frac{\text{rise}}{\text{run}}$. We can also find the slope of a line using the **slope formula**: $$m = \frac{y_2 - y_1}{x_2 - x_1} \quad \text{if } x_1 \neq x_2$$ Lines that rise from left to right have a **positive slope,** and lines that fall from left to right have a **negative slope.** Horizontal lines have **zero slope** and vertical lines have **undefined slope.**	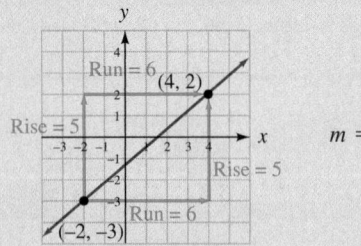 $m = \frac{\text{rise}}{\text{run}} = \frac{5}{6}$ To find the slope of the line that passes through the points $(-2, -3)$ and $(4, 2)$, we substitute into the slope formula: $$m = \frac{y_2 - y_1}{x_2 - x_1} = \frac{2 - (-3)}{4 - (-2)} = \frac{5}{6}$$
When units are attached to a slope, the slope is called a **rate of change.**	An example of a rate of change is: $\dfrac{300 \text{ pounds}}{1 \text{ year}}$ Read as "300 pounds per year."
Parallel lines have the same slope. The slopes of **perpendicular lines** are negative reciprocals. The product of their slopes is -1.	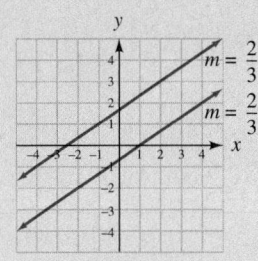 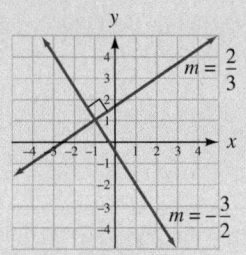 Parallel lines Perpendicular lines

REVIEW EXERCISES

In each case, find the slope of the line.

23.
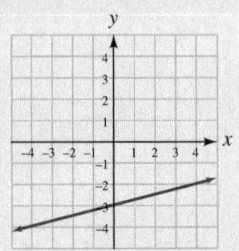

24.
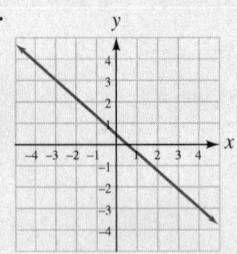

25. The line with this table of solutions

x	y	(x, y)
2	-3	$(2, -3)$
4	-17	$(4, -17)$

26. The line passing through the points $(1, -4)$ and $(3, -7)$

27. Draw a line having a slope that is
    a. Positive   **b.** Negative   **c.** 0   **d.** Undefined

28. CARPENTRY   If a truss like the one shown below is used to build the roof of a shed. Find the slope (pitch) of the roof.

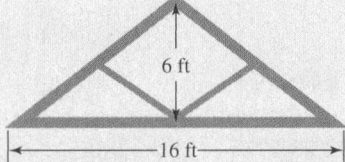

29. RAMPS   Find the grade of the ramp shown below. Round to the nearest tenth of a percent.

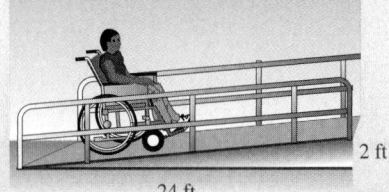

30. TOURISM   The graph shows the number of international travelers to the United States from 1986 to 2004, in two-year increments.

    a. Between 2000 and 2002 the largest decline in the number of visitors occurred. Find the rate of change.

    b. Between 1986 and 1988 the largest increase in the number of visitors occurred. Find the rate of change?

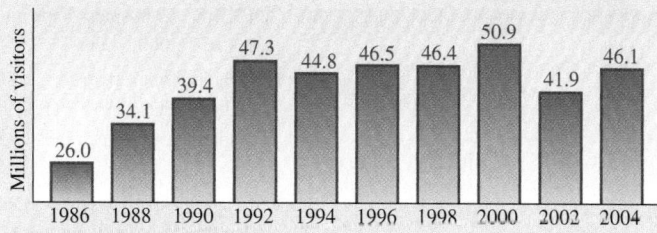

Based on data from *World Almanac* 2006.

31. Without graphing, determine whether the line that passes through (6, 6) and (4, 2) and the line that passes through (2, −10) and (−2, −2) are parallel, perpendicular, or neither.

32. Find the slope of a line perpendicular to the line passing through (−1, 9) and (−8, 4).

SECTION 3.5   Slope–Intercept Form

DEFINITIONS AND CONCEPTS	EXAMPLES
If a linear equation is written in **slope–intercept form** $$y = mx + b$$ the graph of the equation is a line with slope m and y-intercept $(0, b)$.	Find the slope and y-intercept of the line whose equation is $5x + 3y = 3$. To find the slope and y-intercept, we solve the equation for y. $5x + 3y = 3$ $3y = -5x + 3$ Subtract 5x from both sides. $y = -\dfrac{5}{3}x + 1$ To isolate y, divide both sides by 3. $m = -\dfrac{5}{3}$ and $b = 1$. The slope of the line is $-\dfrac{5}{3}$ and the y-intercept is $(0, 1)$.
To **graph a line in slope–intercept form,** plot the y-intercept and use the slope to determine a second point on the line.	Graph: $y = -\dfrac{5}{3}x + 1$ $y = \dfrac{-5}{3}x + 1$ $m = \dfrac{\text{rise}}{\text{run}} = \dfrac{-5}{3}$ $b = 1$ 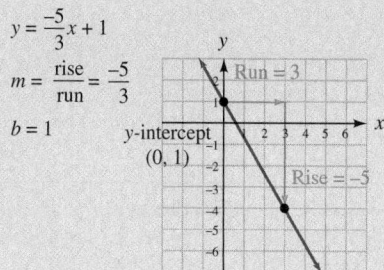
If we know the slope of a line and its y-intercept, we can write its equation.	The equation of a line with slope $\dfrac{1}{8}$ and y-intercept $(0, -5)$ is $y = \dfrac{1}{8}x - 5$.
Two different lines with the same slope are **parallel.**	Lines with equations $y = 3x + 4$ and $y = 3x - 12$ are parallel because each line has slope 3.
If the slopes of two lines are negative reciprocals, the product of their slopes is −1 and the lines are **perpendicular.**	Lines with equations $y = 3x + 4$ and $y = -\dfrac{1}{3}x - 12$ are perpendicular because their slopes, 3 and $-\dfrac{1}{3}$, are negative reciprocals.

REVIEW EXERCISES

Find the slope and the y-intercept of each line.

33. $y = \dfrac{3}{4}x - 2$

34. $y = -4x$

35. $y = \dfrac{x}{8} + 10$

36. $7x + 5y = -21$

37. Graph the line with slope 4 and y-intercept $(0, -1)$. Write an equation of the line.

38. Write an equation for the line shown here.

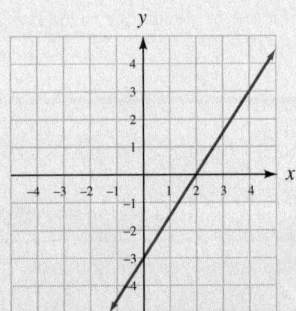

39. Find the slope and the y-intercept of the line whose equation is $9x - 3y = 15$. Then graph it.

40. COPIERS   A business buys a used copy machine that has already produced 75,000 copies.

    a. If the business plans to run 300 copies a week, write a linear equation that would find the number of copies c the machine has made in its lifetime after the business has used it for w weeks.

    b. Use your result in part a to predict the total number of copies that will have been made on the machine 1 year, or 52 weeks, after being purchased by the business.

Without graphing, determine whether graphs of the given pairs of lines are parallel, perpendicular, or neither.

41. $y = -\dfrac{2}{3}x + 6$

    $y = -\dfrac{2}{3}x - 6$

42. $x + 5y = -10$

    $y - 5x = 0$

SECTION 3.6   Point–Slope Form

DEFINITIONS AND CONCEPTS	EXAMPLES
If a line with slope m passes through the point with coordinates (x_1, y_1), the equation of the line in **point–slope form** is $$y - y_1 = m(x - x_1)$$	Find an equation of the line with slope -3 that passes through $(-2, 4)$. Write the equation in slope–intercept form. We substitute the slope and the coordinates of the point into the point–slope form.

$$
\begin{aligned}
y - y_1 &= m(x - x_1) && \text{This is point–slope form.} \\
y - 4 &= -3[x - (-2)] && \text{Substitute.} \\
y - 4 &= -3(x + 2) && \text{Simplify within the brackets.} \\
y - 4 &= -3x - 6 && \text{Distribute.} \\
y &= -3x - 2 && \text{To isolate y, add 4 to both sides. This is}\\
& && \text{slope–intercept form.}
\end{aligned}
$$

| If we know **two points that a line passes through,** we can write its equation. | Find an equation of the line that passes through $(2, 5)$ and $(3, 7)$. Write the equation in slope–intercept form. The slope of the line is |

$$ m = \frac{y_2 - y_1}{x_2 - x_1} = \frac{7 - 5}{3 - 2} = 2 $$

Either point on the line can serve as (x_1, y_1). If we choose $(2, 5)$, we have:

$$
\begin{aligned}
y - y_1 &= m(x - x_1) && \text{This is point–slope form.} \\
y - 5 &= 2(x - 2) && \text{Substitute: } x_1 = 2,\ y_1 = 5,\ \text{and } m = 2. \\
y - 5 &= 2x - 4 && \text{Distribute.} \\
y &= 2x + 1 && \text{To isolate y, add 5 to both sides. This is the}\\
& && \text{slope–intercept form.}
\end{aligned}
$$

REVIEW EXERCISES

Find an equation of the line with the given slope that passes through the given point. Write the equation in slope–intercept form and graph the equation.

43. $m = 3$, $(1, 5)$

44. $m = -\dfrac{1}{2}$, $(-4, -1)$

Find an equation of the line with the following characteristics. Write the equation in slope–intercept form.

45. passing through $(3, 7)$ and $(-6, 1)$

46. horizontal, passing through $(6, -8)$

47. CAR REGISTRATION   When it was 2 years old, the annual registration fee for a Dodge Caravan was $380. When it was 4 years old, the registration fee dropped to $310. If the relationship is linear, write an equation that gives the registration fee f in dollars for the van when it is x years old.

48. U.S. WEDDINGS   The scatter diagram shows the estimated average cost of a wedding for the years 2000–2006. A straight line can be used to model the data.

   a. Use the two highlighted points in red to write the equation of the line. Write the answer in slope–intercept form.

   b. Use your answer to part a to predict the average cost of a wedding in 2020.

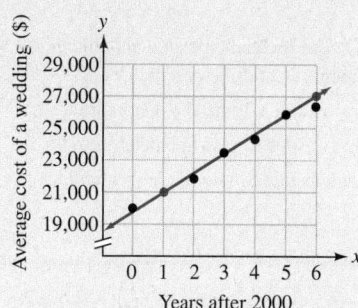

Source: U.S. Wedding Statistics and Marketing website

SECTION 3.7   Graphing Linear Inequalities

DEFINITIONS AND CONCEPTS	EXAMPLES
An ordered pair (x, y) is a **solution of an inequality** in x and y if a true statement results when the variables are replaced by the coordinates of the ordered pair.	Determine whether $(-2, 5)$ is a solution of $x + 3y > -6$. We substitute the coordinates into the equation and see if a true statement results. $x + 3y > -6$ $-2 + 3(5) \overset{?}{>} -6$ $13 > -6$ True Since the result is true, $(-2, 5)$ is a solution.
To graph a linear inequality: **1.** Replace the inequality symbol with an $=$ symbol and graph the boundary line. Draw a solid line if the inequality contains $\leq$ or $\geq$ and a dashed line if it contains $<$ or $>$. **2.** Pick a test point not on the boundary. Substitute its coordinates into the inequality. If the inequality is satisfied, shade the side that contains the test point. If the inequality is not satisfied, shade the other side.	Graph: $2x - y \leq 4$ **1.** Graph the boundary line $2x - y = 4$ and draw it as a solid line because the inequality symbol is $\leq$. **2.** Test the point $(0, 0)$: $2x - y \leq 4$ $2(0) - 0 \overset{?}{\leq} 4$ $0 \leq 4$ True Since the coordinates of the test point satisfy the inequality, we shade the side of the boundary line that contains $(0, 0)$. 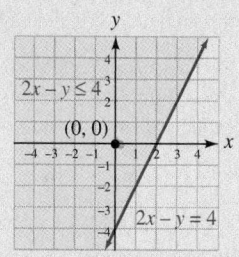

REVIEW EXERCISES

49. Determine whether each ordered pair is a solution of $2x - y \leq -4$.

   a. $(0, 5)$ **b.** $(2, 8)$

   c. $(-3, -2)$ **d.** $\left(\frac{1}{2}, -5\right)$

50. Fill in the blanks: $2x - 3y \geq 6$ means $2x - 3y$ _____ 6 or
   $2x - 3y$ _____ 6.

Graph each inequality.

51. $x - y < 5$ 52. $2x - 3y \geq 6$

53. $y \leq -2x$ 54. $y < -4$

55. The graph of a linear inequality is shown on the right. Would a true
   or a false statement result if the coordinates of

   a. point A were substituted into the inequality?

   b. point B were substituted into the inequality?

   c. point C were substituted into the inequality?

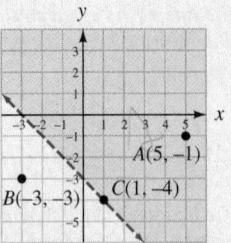

56. **WORK SCHEDULES** A student told her employer that during
   the school year, she would be available for up to 30 hours a week,
   working either 3- or 5-hour shifts. If x represents the number of
   3-hour shifts she works and y represents the number of 5-hour
   shifts she works, the inequality $3x + 5y \leq 30$ shows the possible
   combinations of shifts she can work. Graph the inequality and find
   three possible combinations.

SECTION 3.8   An Introduction to Functions

DEFINITIONS AND CONCEPTS	EXAMPLES
A **relation** is a set of ordered pairs. The set of all **first components** is called the **domain** of the relation and the set of all **second components** is called the **range** of a relation.	The relation $\{(4, 7), (0, -3), (-3, 8), (1, 7)\}$ has the domain $\{-3, 0, 1, 4\}$ and the range is $\{-3, 7, 8\}$.
A **function** is a set of ordered pairs (a relation) in which to each first component there corresponds exactly one second component. If to each value of x in the domain there is assigned exactly one value of y in the range, then **y is a function of x.**	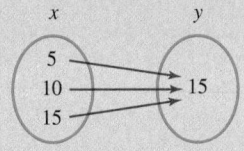 y is not a function of x: $5 \rightarrow 3$ and $5 \rightarrow 9$ y is a function of x
A function can be defined by an equation. The notation $y = f(x)$ indicates that the variable y is a function of x. It is read as "f of x." We can think of a function as a machine that takes some **input** x and turns it into some **output** $f(x)$, called a **function value.**	For the function $f(x) = 8x + 5$, $f(-2)$ is the value of $f(x)$ when $x = -2$. $\quad f(x) = 8x + 5$ $f(-2) = 8(-2) + 5 \quad$ Substitute -2 for each x. $\qquad = -16 + 5 \quad$ Evaluate the right side. $\qquad = -11$ Thus, $f(-2) = -11$
The **vertical line test:** If a vertical line intersects a graph in more than one point, the graph is not the graph of a function.	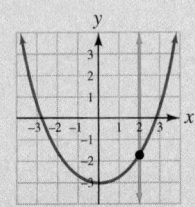 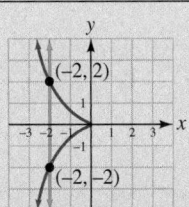 A function Not a function

SECTION 3.8 An Introduction to Functions—*continued*

DEFINITIONS AND CONCEPTS	EXAMPLES
The input-output pairs that a function generates can be written as ordered pairs and plotted on a rectangular coordinate system to give the **graph of a function.**	Graph the function: $f(x) = -\frac{2}{3}x + 3$ We make a table of solutions, plot the points, and draw the graph. 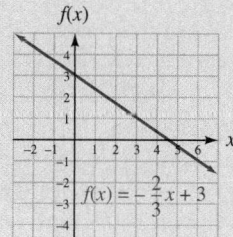

REVIEW EXERCISES

Find the domain and range of each relation.

57. $\{(7, -3), (-5, 9), (4, 4), (0, -11)\}$

58. $\{(2, -2), (15, -8), (-6, 9), (1, -8)\}$

Determine whether each arrow diagram or table defines y to be function of x. If a function is defined, give its domain and range. If it does not define a function, find ordered pairs that show a value of x that corresponds to more than one value of y.

59. 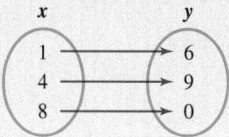   **60.**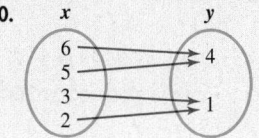

61.

x	y
9	81
7	49
5	25
3	9

62.

x	y
-1	2
0	3
-1	4
1	5

Fill in the blanks.

63. The set of all input values for a function is called the _____, and the set of all output values is called the _____.

64. Fill in the blank: Since $y =$ _____ , the equations $y = 2x - 8$ and $f(x) = 2x - 8$ are equivalent.

For f(x) = x² − 4x, find each of the following function values.

65. $f(1)$ **66.** $f(0)$

67. $f(-3)$ **68.** $f\left(\frac{1}{2}\right)$

For g(x) = 1 − 6x, find each of the following function values.

69. $g(1)$ **70.** $g(-6)$

71. $g(0.5)$ **72.** $g\left(\frac{3}{2}\right)$

Determine whether each graph is the graph of a function. If it is not, find two ordered pairs that show a value of x that corresponds to more than one value of y.

73. 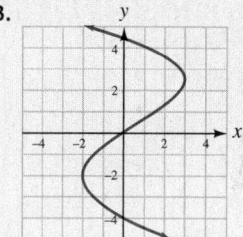   **74.**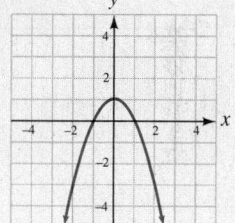

75. Complete the table of function values. Then graph the function.

$f(x) = 1 - |x|$

x	f(x)
0	
1	
2	
-1	
-2	
-3	

76. ALUMINUM CANS The function $V(r) = 15.7r^2$ estimates the volume in cubic inches of a can 5 inches tall with a radius of r inches. Find the volume of the can shown in the illustration. Round to the nearest tenth.

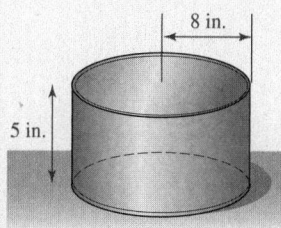

CHAPTER 3
Test

1. Fill in the blanks.

   a. A rectangular coordinate system is formed by two perpendicular number lines called the x-_____ and the y-_____.

   b. A _____ of an equation in two variables is an ordered pair of numbers that makes the equation a true statement.

   c. $3x + y = 10$ is called a _____ equation in two variables because its graph is a line.

   d. The _____ of a line is a measure of steepness.

   e. A _____ is a set of ordered pairs in which to each first component there corresponds exactly one second component.

The graph shows the number of dogs being boarded in a kennel over a 3-day holiday weekend.

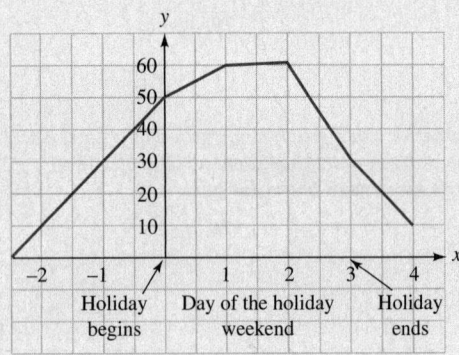

2. How many dogs were in the kennel 2 days before the holiday?

3. What is the maximum number of dogs that were boarded on the holiday weekend?

4. When were there 30 dogs in the kennel?

5. What information does the y-intercept of the graph give?

6. Plot each point on a rectangular coordinate system: $(1, 3)$, $(-2, 4)$, $(-3, -2)$, and $(3, -2)$.

7. Find the coordinates of each point shown in the graph.

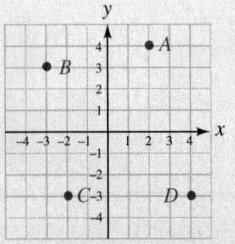

8. In which quadrant is each point located?

   a. $(-1, -5)$ b. $\left(6, -2\frac{3}{4}\right)$

9. Is $(-3, -4)$ a solution of $3x - 4y = 7$?

10. Complete the table of solutions for the linear equation.

$$x + 4y = 6$$

x	y	(x, y)
2		
	3	

11. The graph of a linear equation is shown.

   a. If the coordinates of point C are substituted into the equation, will the result be true or false?

   b. If the coordinates of point D are substituted into the equation, will the result be true or false?

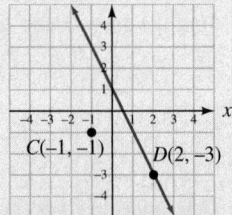

12. Graph: $y = \dfrac{x}{3}$

13. What are the x- and y-intercepts of the graph of $2x - 3y = 6$?

14. Graph: $8x + 4y = -24$

15. Find the slope of the line.

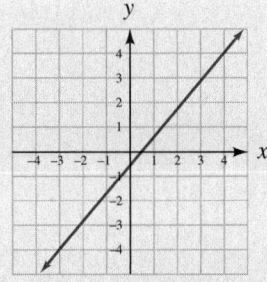

16. Find the slope of the line passing through $(-1, 3)$ and $(3, -1)$.

17. What is the slope of a horizontal line?

18. RAMPS   Find the grade of a ramp that rises 2 feet over a horizontal distance of 20 feet.

19. One line passes through $(9, 2)$ and $(6, 4)$. Another line passes through $(0, 7)$ and $(2, 10)$. Without graphing, determine whether the lines are parallel, perpendicular, or neither.

20. When graphed, are the lines $y = 2x + 6$ and $2x - y = 0$ parallel, perpendicular, or neither?

In Problems 21 and 22, refer to the illustration that shows the elevation changes in a 26-mile marathon course.

21. Find the rate of change of the decline on which the woman is running.

22. Find the rate of change of the incline on which the man is running.

Distance (mi)

23. Graph: $x = -4$

24. Graph the line passing through $(-2, -4)$ having slope $\frac{2}{3}$.

25. Find the slope and the y-intercept of the graph of $x + 2y = 8$.

26. Find an equation of the line passing through $(-2, 5)$ with slope 7. Write the equation in slope–intercept form.

27. Find an equation for the line shown. Write the equation in slope–intercept form.

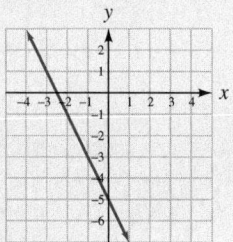

28. DEPRECIATION   After it is purchased, a \$15,000 computer loses \$1,500 in resale value every year.

   a. Write a linear equation that gives the resale value v of the computer x years after being purchased.

   b. Use your answer to part (a) to predict the value of the computer 8 years after it is purchased.

29. Determine whether $(6, 1)$ is a solution of $2x - 4y \geq 8$.

30. WATER HEATERS   The scatter diagram shows how excessively high temperatures affect the life of a water heater. Write an equation of the line that models the data for water temperatures between 140° and 180°. Let T represent the temperature of the water in degrees Fahrenheit and y represent the expected life of the heater in years. Give the answer in slope–intercept form.

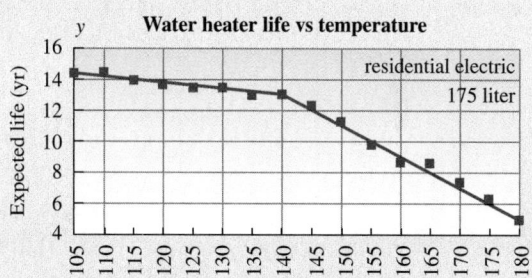

Water heater life vs temperature

residential electric
175 liter

Water: stored temperature (Fahrenheit)

Source: www.uniongas.com/WaterHeating

31. Graph the inequality: $x - y > -2$

32. Find the domain and range of the relation: $\{(5, 3), (1, 12), (-4, 3), (0, -8)\}$

Determine whether the table, arrow diagram, or graph define y to be function of x. If a function is defined, give its domain and range. If it does not define a function, find ordered pairs that show a value of x that corresponds to more than one value of y.

33.

x	y
1	4
2	3
3	2
4	1

34.

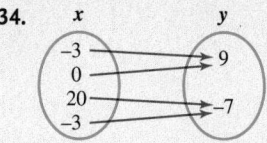

35.

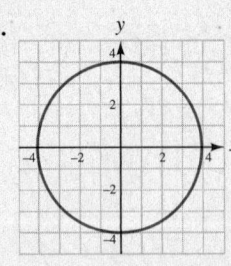

36.

x	y
5	12
10	12
15	12
20	12
25	12

37. If $f(x) = 2x - 7$, find: $f(-3)$

38. If $g(x) = 3.5x^3$ find: $g(6)$

39. TELEPHONE CALLS   The function $C(n) = 0.30n + 15$ gives the cost C per month in dollars for making n phone calls. Find $C(45)$ and explain what it means.

40. Graph: $f(x) = |x| - 1$

GROUP PROJECT

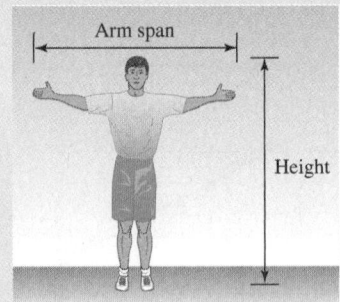

Arm span

Height

Overview: In this activity, you will explore the relationship between a person's height and arm span. Arm span is defined to be the distance between the tips of a person's fingers when their arms are held out to the side.

Instructions: Form groups of 5 or 6 students. Measure the height and arm span of each person in your group, and record the results in a table like the one shown below.

Name	Height (in.)	Arm span (in.)
1.		
2.		
3.		
4.		
5.		
6.		

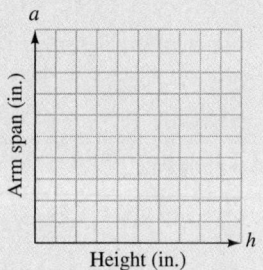

Arm span (in.)

Height (in.)

Plot the data in the table as ordered pairs of the form (height, arm span) on a graph like the one shown above. Then draw a straight-line model that best fits the data points.

Pick two convenient points on the line and find its slope. Use the point–slope form $a - a_1 = m(h - h_1)$ to find an equation of the line. Then, write the equation in slope–intercept form.

Ask a person from another group for his or her height measurement. Substitute that value into your linear model to predict that person's arm span. How close is your prediction to the person's actual arm span?

(From *Activities for Beginning and Intermediate Algebra* by Debbie Garrison, Judy Jones, and Jolene Rhodes)

CUMULATIVE REVIEW
Chapters 1–3

1. Find the prime factorization of 108. [Section 1.2]
2. Write $\frac{1}{250}$ as a decimal. [Section 1.3]
3. Determine whether each statement is true or false. [Section 1.3]
   a. Every whole number is an integer.
   b. Every integer is a real number.
   c. 0 is a whole number, an integer, and a rational number.

Perform the operations.

4. $-27 + 21 + (-9)$ [Section 1.4]
5. $-1.57 - (-0.8)$ [Section 1.5]
6. $-9(-7)(5)(-3)$ [Section 1.6]
7. $\dfrac{-180}{-6}$ [Section 1.6]
8. Evaluate: $\left| \dfrac{(6-5)^4 - (-21)}{-27 + 4^2} \right|$ [Section 1.7]
9. Evaluate $b^2 - 4ac$ for $a = 2$, $b = -8$, and $c = 4$. [Section 1.8]
10. Suppose x sheets from a 500-sheet ream of paper have been used. How many sheets are left? [Section 1.8]
11. How many terms does the algebraic expression $3x^2 - 2x + 1$ have? What is the coefficient of the second term? [Section 1.8]
12. Use the distributive property to remove parentheses. [Section 1.9]
   a. $2(x + 4)$ b. $-2(x - 4)$

Simplify each expression. [Section 1.9]

13. $5a + 10 - a$
14. $-7(9t)$
15. $-2b^2 + 6b^2$
16. $5(-17)(0)(2)$
17. $(a + 2) - (a - 2)$
18. $-4(-5)(-8a)$
19. $-y - y - y$
20. $\dfrac{3}{2}(4x - 8) + x$

Solve each equation. [Sections 2.1 and 2.2]

21. $3x - 5 = 13$
22. $1.2 - x = -1.7$
23. $\dfrac{2x}{3} - 2 = 4$
24. $\dfrac{y - 2}{7} = -3$
25. $-3(2y - 2) - y = 5$
26. $9y - 3 = 6y$
27. $\dfrac{1}{3} + \dfrac{c}{5} = -\dfrac{3}{2}$
28. $5(x + 2) = 5x - 2$
29. $-x = 99$
30. $3c - 2 = \dfrac{11(c - 1)}{5}$

31. PENNIES   A 2006 telephone survey of adults asked whether the penny should be discontinued from the national currency. The results are shown in the circle graph. If 869 people favored keeping the penny, how many took part in the survey? [Section 2.3]

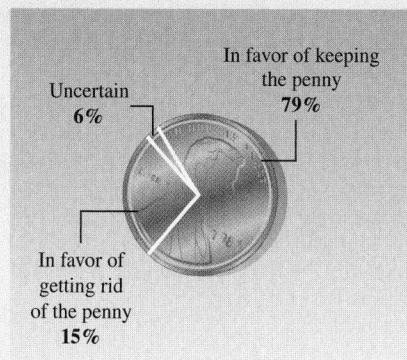

In favor of keeping the penny
79%

Uncertain
6%

In favor of getting rid of the penny
15%

Based on data from Coinstar

32. Solve for h: $S = 2\pi rh + 2\pi r^2$ [Section 2.4]
33. BAND AIDS   Find the perimeter and the area of the gauze pad of the bandage. [Section 2.4]

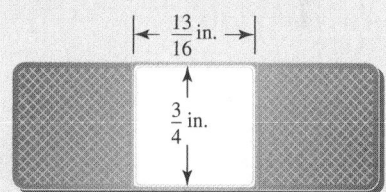

$\frac{13}{16}$ in.

$\frac{3}{4}$ in.

34. HIGH HEELS   Find x. [Section 2.5]

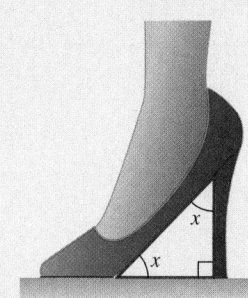

x

x

35. Complete the table. [Section 2.6]

	% acid	Liters	Amount of acid
50% solution	0.50	x	
25% solution	0.25	$13 - x$	
30% mixture	0.30	13	

36. ROAD TRIPS   A bus, carrying the members of a marching band, and a truck, carrying their instruments, leave a high school at the same time. The bus travels at 60 mph and the truck at 50 mph. In how many hours will they be 75 miles apart? [Section 2.6]

37. MIXING CANDY   Candy corn worth $2.85 per pound is to be mixed with black gumdrops that cost $1.80 per pound to make 200 pounds of a mixture worth $2.22 per pound. How many pounds of each candy should be used? [Section 2.6]

Solve each inequality. Write the solution set in interval notation and graph it. [Section 2.7]

38. $-\dfrac{3}{16}x \geq -9$

39. $8x + 4 > 3x + 4$

40. In which quadrants are the second coordinates of ordered pairs positive? [Section 3.1]

41. Is $(-2, 4)$ a solution of $y = 2x - 8$? [Section 3.2]

Graph each equation.

42. $y = x$ [Section 3.2]

43. $4y + 2x = -8$ [Section 3.3]

44. What is the slope of the graph of the line $y = 5$? [Section 3.4]

45. What is the slope of the line passing through $(-2, 4)$ and $(5, -6)$? [Section 3.4]

46. ROOFING   Find the pitch of the roof. [Section 3.4]

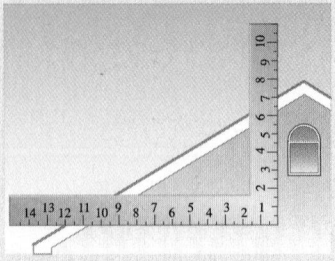

47. Find the slope and the y-intercept of the graph of the line described by $4x - 6y = -12$. [Section 3.5]

48. Write an equation of the line that has slope -2 and y-intercept $(0, 1)$. [Section 3.5]

49. Find an equation of the line that has slope $-\dfrac{7}{8}$ and passes through $(2, -9)$. Write the equation in point–slope form and in slope–intercept form. [Section 3.6]

50. Is $(-2, -4)$ a solution of $x + y \leq -6$? [Section 3.7]

51. Graph: $y \geq x + 1$ [Section 3.7]

52. Graph $x < 4$ on a rectangular coordinate system. [Section 3.7]

53. If $f(x) = x^4 + x$, find: $f(-3)$ [Section 3.8]

54. Is this the graph of a function? [Section 3.8]

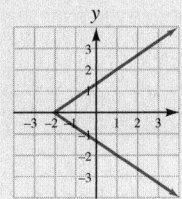

CHAPTER 4

Systems of Linear Equations and Inequalities

© Peter Steiner/Alamy

from **Campus to Careers**
Portrait Photographer

Portrait photographers take pictures of individuals or groups of people and often work in their own studios. Some specialize in weddings, religious ceremonies, or school photographs, and many work on location. Their job responsibilities require a variety of mathematical skills such as scheduling appointments, keeping financial records, pricing photographs, purchasing supplies, billing customers, and operating digital equipment.

Photographers often make packets of pictures available to their customers. In **Problem 29** of **Study Set 4.4,** we will find the costs of two sizes of photographs that are part of a wedding picture packet.

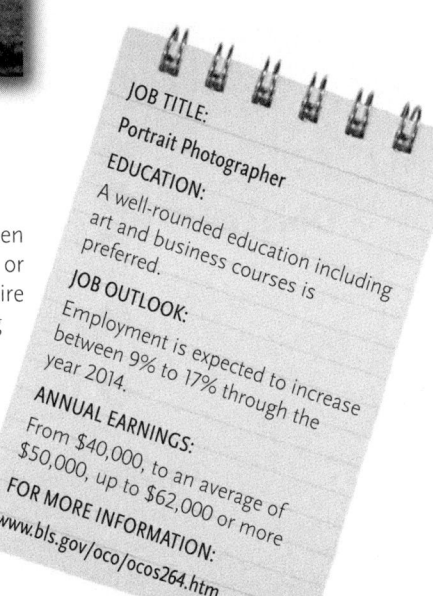

JOB TITLE:
Portrait Photographer

EDUCATION:
A well-rounded education including art and business courses is preferred.

JOB OUTLOOK:
Employment is expected to increase between 9% to 17% through the year 2014.

ANNUAL EARNINGS:
From $40,000, to an average of $50,000, up to $62,000 or more

FOR MORE INFORMATION:
www.bls.gov/oco/ocos264.htm

307

Study Skills Workshop
Making Homework a Priority

Attending class and taking notes are important, but they are not enough. The only way that you are really going to learn algebra is by doing your homework.

WHEN TO DO YOUR HOMEWORK: Homework should be started on the day it is assigned, when the material is fresh in your mind. It's best to break your homework sessions into 30-minute periods, allowing for short breaks in between.

HOW TO BEGIN YOUR HOMEWORK: Review your notes and the examples in your text before starting your homework assignment.

GETTING HELP WITH YOUR HOMEWORK: It's normal to have some questions when doing homework. Talk to a tutor, a classmate, or your instructor to get those questions answered.

Now Try This

1. Write a one-page paper that describes *when, where,* and *how* you go about completing your algebra homework assignments.

2. For each problem on your next homework assignment, find an example in this book that is similar. Write the example number next to the problem.

3. Make a list of questions that you have while doing your next assignment. Then decide whom you are going to ask to get those questions answered.

SECTION 4.1
Solving Systems of Equations by Graphing

Objectives

1. Determine whether a given ordered pair is a solution of a system.
2. Solve systems of linear equations by graphing.
3. Use graphing to identify inconsistent systems and dependent equations.
4. Identify the number of solutions of a linear system without graphing.
5. Use a graphing calculator to solve a linear system (optional).

The following illustration shows the average amounts of chicken and beef eaten per person each year in the United States from 1985 to 2005. Plotting both graphs on the same coordinate system makes it easy to compare recent trends. The point of intersection of the graphs indicates that Americans ate equal amounts of chicken and beef in 1992—about 66 pounds of each, per person.

In this section, we will use a similar graphical approach to solve systems of equations.

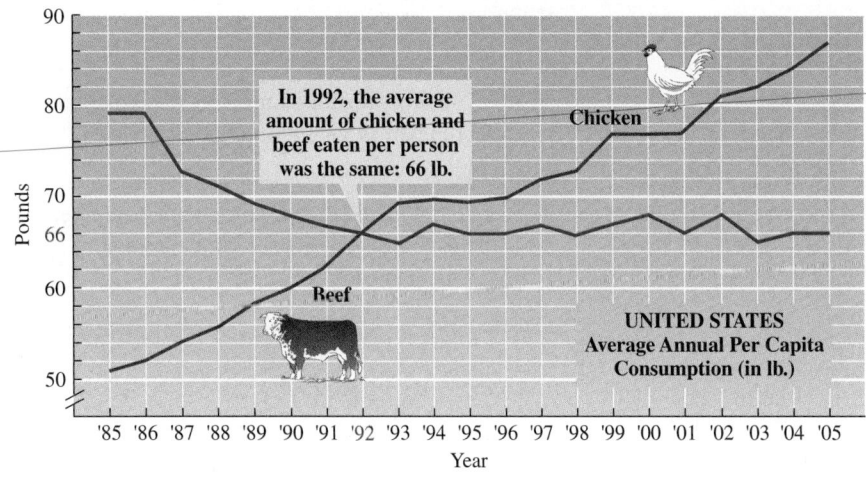

Source: U.S. Department of Agriculture

① Determine Whether a Given Ordered Pair Is a Solution of a System.

We have previously discussed equations in two variables, such as $x + y = 3$. Because there are infinitely many pairs of numbers whose sum is 3, there are infinitely many pairs (x, y) that satisfy this equation. Some of these pairs are listed in Table (a).

Now consider the equation $x - y = 1$. Because there are infinitely many pairs of numbers whose difference is 1, there are infinitely many pairs (x, y) that satisfy $x - y = 1$. Some of these pairs are listed in the Table (b).

> **The Language of Algebra**
> We say that $(2, 1)$ *satisfies* $x + y = 3$, because the x-coordinate, 2, and the y-coordinate, 1, make the equation true when substituted for x and y: $2 + 1 = 3$. To *satisfy* means to make content, as in *satisfy* your thirst or a *satisfied* customer.

$x + y = 3$

x	y	(x, y)
0	3	$(0, 3)$
1	2	$(1, 2)$
2	1	$(2, 1)$
3	0	$(3, 0)$

(a)

$x - y = 1$

x	y	(x, y)
0	-1	$(0, -1)$
1	0	$(1, 0)$
2	1	$(2, 1)$
3	2	$(3, 2)$

(b)

From the two tables, we see that $(2, 1)$ satisfies both equations.

When two equations with the same variables are considered simultaneously (at the same time), we say that they form a **system of equations.** Using a left brace { , we can write the equations from the previous example as a system:

$$\begin{cases} x + y = 3 \\ x - y = 1 \end{cases}$$ Read as "the system of equations $x + y = 3$ and $x - y = 1$."

Because the ordered pair $(2, 1)$ satisfies both of these equations, it is called a **solution of the system.** In general, a system of linear equations can have exactly one solution, no solution, or infinitely many solutions.

> **EXAMPLE 1** Determine whether $(-2, 5)$ is a solution of each system of equations.
>
> **a.** $\begin{cases} 3x + 2y = 4 \\ x - y = -7 \end{cases}$
>
> **b.** $\begin{cases} 4y = 18 - x \\ y = 2x \end{cases}$

Strategy   We will substitute the x- and y-coordinates of $(-2, 5)$ for the corresponding variables in both equations of the system.

Why   If both equations are satisfied (made true) by the x- and y-coordinates, then the ordered pair is a solution of the system.

Solution

a. Recall that in an ordered pair, the first number is the x-coordinate and the second number is the y-coordinate. To determine whether $(-2, 5)$ is a solution, we substitute -2 for x and 5 for y in each equation.

Check:

$$3x + 2y = 4 \quad \text{The first equation.} \qquad x - y = -7 \quad \text{The second equation.}$$
$$3(-2) + 2(5) \stackrel{?}{=} 4 \qquad\qquad -2 - 5 \stackrel{?}{=} -7$$
$$-6 + 10 \stackrel{?}{=} 4 \qquad\qquad\qquad -7 = -7 \quad \text{True}$$
$$4 = 4 \quad \text{True}$$

> **The Language of Algebra**
> A system of equations is two (or more) equations that we consider *simultaneously*—at the same time. Some professional sports teams *simulcast* their games. That is, the announcer's play-by-play description is broadcast on radio and television at the same time.

Since $(-2, 5)$ satisfies both equations, it is a solution of the system.

b. Again, we substitute -2 for x and 5 for y in each equation.

Check:

$$4y = 18 - x \quad \text{The first equation.} \qquad y = 2x \quad \text{The second equation.}$$
$$4(5) \stackrel{?}{=} 18 - (-2) \qquad\qquad 5 \stackrel{?}{=} 2(-2)$$
$$20 \stackrel{?}{=} 18 + 2 \qquad\qquad\qquad 5 = -4 \quad \text{False}$$
$$20 = 20 \qquad \text{True}$$

Although $(-2, 5)$ satisfies the first equation, it does not satisfy the second. Because it does not satisfy both equations, $(-2, 5)$ is not a solution of the system.

Self Check 1   Determine whether $(4, -1)$ is a solution of: $\begin{cases} x - 2y = 6 \\ y = 3x - 11 \end{cases}$

Now Try   **Problem 15**

---

2 **Solve Systems of Linear Equations by Graphing.**

To **solve a system of equations** means to find all of the solutions of the system. One way to solve a system of linear equations is to graph the equations on the same set of axes.

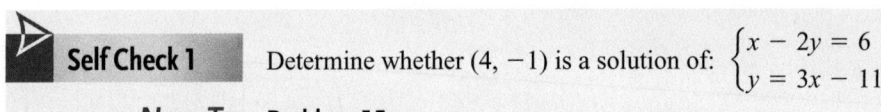

EXAMPLE 2   Solve the system of equations by graphing: $\begin{cases} 2x + 3y = 2 \\ 3x - 2y = 16 \end{cases}$

Strategy   We will graph both equations on the same coordinate system.

Why   Recall that the graph of a linear equation is a "picture" of its solutions. If both equations are graphed on the same coordinate system, we can see whether they have any common solutions.

Solution   The intercept-method is a convenient way to graph equations such as $2x + 3y = 2$ and $3x - 2y = 16$, because they are in standard $Ax + By = C$ form.

	$2x + 3y = 2$			$3x - 2y = 16$	
x	y	(x, y)	x	y	(x, y)
0	$\frac{2}{3}$	$(0, \frac{2}{3})$	0	-8	$(0, -8)$
1	0	$(1, 0)$	$\frac{16}{3}$	0	$(\frac{16}{3}, 0)$
-2	2	$(-2, 2)$	2	-5	$(2, -5)$

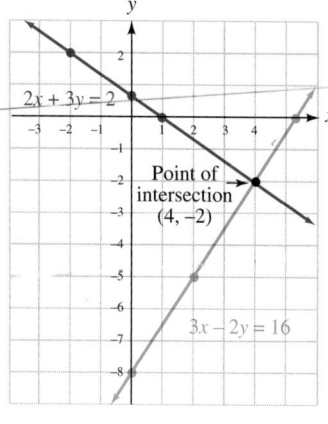

To find the y-intercept, let x = 0 and solve for y.
To find the x-intercept, let y = 0 and solve for x.
As a check, pick another x-value, such as −2 or 2, and find y.

The coordinates of each point on the line graphed in red satisfy $2x + 3y = 2$ and the coordinates of each point on the line graphed in blue satisfy $3x - 2y = 16$. Because the point of intersection is on both graphs, its coordinates satisfy both equations.

It appears that the graphs intersect at the point $(4, -2)$. To verify that it is the solution of the system, we substitute 4 for x and -2 for y in each equation.

Check:

$$2x + 3y = 2 \quad \text{The first equation.} \qquad 3x - 2y = 16 \quad \text{The second equation.}$$
$$2(4) + 3(-2) \stackrel{?}{=} 2 \qquad\qquad 3(4) - 2(-2) \stackrel{?}{=} 16$$
$$8 + (-6) \stackrel{?}{=} 2 \qquad\qquad 12 - (-4) \stackrel{?}{=} 16$$
$$2 = 2 \quad \text{True} \qquad\qquad 16 = 16 \quad \text{True}$$

Since $(4, -2)$ makes both equations true, it is the solution of the system. The solution set is written as $\{(4, -2)\}$.

 Self Check 2 Solve the system of equations by graphing: $\begin{cases} 2x - y = -5 \\ x + y = -1 \end{cases}$

Now Try **Problem 25**

To solve a system of linear equations in two variables by graphing, follow these steps.

The Graphing Method

1. Carefully graph each equation on the same rectangular coordinate system.
2. If the lines intersect, determine the coordinates of the point of intersection of the graphs. That ordered pair is the solution of the system.
3. Check the proposed solution in each equation of the original system.

3 **Use Graphing to Identify Inconsistent Systems and Dependent Equations.**

A system of equations that has at least one solution, like that in Example 2, is called a **consistent system.** A system with no solution is called an **inconsistent system.**

EXAMPLE 3 Solve the system of equations by graphing: $\begin{cases} y = -2x - 6 \\ 4x + 2y = 8 \end{cases}$

Strategy We will graph both equations on the same coordinate system.

Why If both equations are graphed on the same coordinate system, we can see whether they have any common solutions.

Solution Since $y = -2x - 6$ is written in slope–intercept form, we can graph it by plotting the y-intercept $(0, -6)$ and then drawing a slope triangle whose rise is -2 and whose run is 1. We can graph $4x + 2y = 8$ using the intercept method.

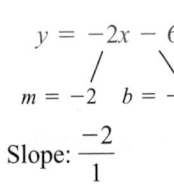

$$y = -2x - 6$$

$m = -2 \qquad b = -6$

Slope: $\dfrac{-2}{1}$

y-intercept: $(0, -6)$

$4x + 2y = 8$

x	y	(x, y)
0	4	$(0, 4)$
2	0	$(2, 0)$
1	2	$(1, 2)$

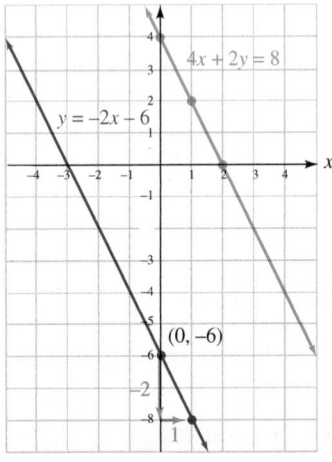

The lines in the graph appear to be parallel. We can verify this by writing the second equation in slope–intercept form and observing that the lines have the same slope, -2, and different y-intercepts, $(0, -6)$ and $(0, 4)$.

$$y = -2x - 6 \qquad 4x + 2y = 8$$
$$2y = -4x + 8 \qquad \text{Subtract 4x from both sides.}$$
$$y = -2x + 4 \qquad \text{To isolate y, divide both sides by 2.}$$

Different y-intercepts

Same slope

Caution
A common error is to graph the parallel lines, but forget to answer with the words *no solution*.

Because the lines are parallel, there is no point of intersection. Such a system has *no solution*. The solution set is the empty set, which is written $\varnothing$.

Self Check 3 Solve the system of equations by graphing: $\begin{cases} y = \dfrac{3}{2}x \\ 3x - 2y = 6 \end{cases}$

Now Try **Problem 33**

Some systems of equations have infinitely many solutions.

EXAMPLE 4 Solve the system of equations by graphing: $\begin{cases} y = 2x + 4 \\ 4x + 8 = 2y \end{cases}$

Strategy We will graph both equations on the same coordinate system.

Why If both equations are graphed on the same coordinate system, we will be able to see if they have any solutions in common.

Solution To graph $y = 2x + 4$, we use the slope and y-intercept, and to graph $4x + 8 = 2y$, we use the intercept method.

$$y = 2x + 4$$

$$m = 2 \quad b = 4$$

Slope: $\dfrac{2}{1}$

y-intercept: $(0, 4)$

$4x + 8 = 2y$

x	y	(x, y)
0	4	$(0, 4)$
-2	0	$(-2, 0)$
-3	-2	$(-3, -2)$

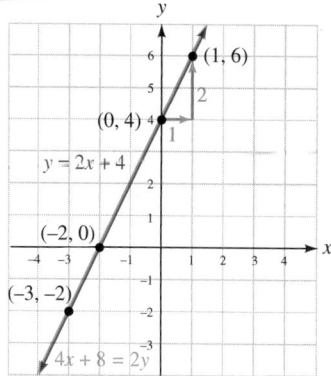

The graphs appear to be identical. We can verify this by writing the second equation in slope–intercept form and observing that it is the same as the first equation.

$y = 2x + 4$ The first equation. $\qquad$ $4x + 8 = 2y$ The second equation.

$$2y = 4x + 8$$

$$\frac{2y}{2} = \frac{4x}{2} + \frac{8}{2} \qquad \text{Divide both sides by 2.}$$

$$y = 2x + 4$$

This confirms that $4x + 8 = 2y$ and $y = 2x + 4$ are different forms of the same equation. Thus, their graphs are identical.

Since the graphs are the same line, they have infinitely many points in common. The coordinates of each of those points satisfy both equations of the system. In cases like this, we say that there are *infinitely many solutions*.

From the graph, it appears that four of the infinitely many solutions are $(-3, -2)$, $(-2, 0)$, $(0, 4)$, and $(1, 6)$. Checks for two of these ordered pairs follow.

Check $(-3, -2)$:

$4x + 8 = 2y$	$y = 2x + 4$
$4(-3) + 8 \stackrel{?}{=} 2(-2)$	$-2 \stackrel{?}{=} 2(-3) + 4$
$-12 + 8 \stackrel{?}{=} -4$	$-2 \stackrel{?}{=} -6 + 4$
$-4 = -4$	$-2 = -2$

Check $(0, 4)$:

$4x + 8 = 2y$	$y = 2x + 4$
$4(0) + 8 \stackrel{?}{=} 2(4)$	$4 \stackrel{?}{=} 2(0) + 4$
$0 + 8 \stackrel{?}{=} 8$	$4 \stackrel{?}{=} 0 + 4$
$8 = 8$	$4 = 4$

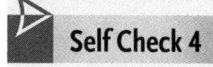 **Self Check 4** Solve the system of equations by graphing: $\begin{cases} 6x - 4 = 2y \\ y = 3x - 2 \end{cases}$

Now Try **Problem 31**

In Examples 2 and 3, the graphs of the equations of the system were different lines. We call equations with different graphs **independent equations**. The equations in Example 4 have the same graph and are equivalent. Because they are different forms of the same equation, they are called **dependent equations**.

There are three possible outcomes when we solve a system of two linear equations using the graphing method:

The two lines intersect at one point.

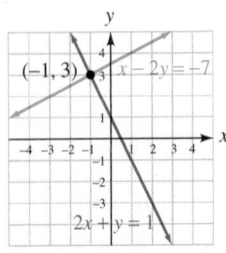

Exactly one solution
(the point of intersection)

Consistent system
Independent equations

The two lines are parallel.

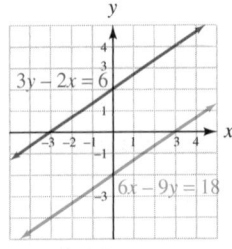

No solution

Inconsistent system
Independent equations

The two lines are identical.

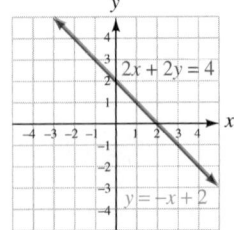

Infinitely many solutions
(any point on the line is a solution)

Consistent system
Dependent equations

 Identify the Number of Solutions of a Linear System Without Graphing.

We can determine the number of solutions that a system of two linear equations has by writing each equation in slope–intercept form.

- If the lines have different slopes, they intersect, and the system has one solution. (See Example 2.)
- If the lines have the same slope and different y-intercepts, they are parallel, and the system has no solution. (See Example 3.)
- If the lines have the same slope and same y-intercept, they are the same line, and the system has infinitely many solutions. (See Example 4.)

EXAMPLE 5   Without graphing, determine the number of solutions of:
$$\begin{cases} 5x + y = 5 \\ 3x + 2y = 8 \end{cases}$$

Strategy   We will write both equations in slope–intercept form.

Why   We can determine the number of solutions of a linear system by comparing the slopes and y-intercepts of the graphs of the equations.

Solution   To write each equation in slope–intercept form, we solve for y.

$5x + y = 5$ The first equation. $3x + 2y = 8$ The second equation.

$y = -5x + 5$ $2y = -3x + 8$

$$y = -\frac{3}{2}x + 4$$

Different slopes

Since the slopes are different, the lines are neither parallel nor identical. Therefore, they will intersect at one point and the system has one solution.

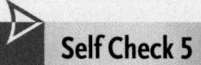

Self Check 5 Without graphing, determine the number of solutions of:
$$\begin{cases} 3x + 6y = 1 \\ 2x + 4y = 0 \end{cases}$$

Now Try **Problem 61**

5 **Use a Graphing Calculator to Solve a Linear System (Optional).**

A graphing calculator can be used to solve systems of equations, such as

$$\begin{cases} 2x + y = 12 \\ 2x - y = -2 \end{cases}.$$

Before we can enter the equations into the calculator, we must solve them for y.

$2x + y = 12$ The first equation. $2x - y = -2$ The second equation.

$\quad y = -2x + 12$ $-y = -2x - 2$

$\quad\quad\quad\quad\quad\quad\quad\quad\quad\quad\quad\quad\quad\quad\quad\quad\quad\quad\quad\quad\quad y = 2x + 2$

We enter the resulting equations as Y_1 and Y_2 and graph them on the same axes. If we use the standard window setting, their graphs will look like figure (a).

To find the solution of the system, we can use the INTERSECT feature found on most graphing calculators. With this feature, after pushing enter three times to identify each graph and a guess for the point, the cursor automatically moves to the point of intersection of the graphs and displays the coordinates of that point. In figure (b), we see that the solution is $(2.5, 7)$.

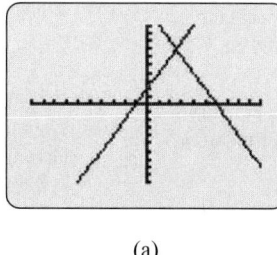 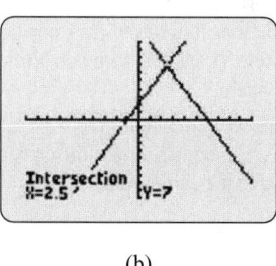

(a)   (b)

ANSWERS TO SELF CHECKS   **1.** No
  2. $(-2, 1)$ **3.** No solution   **4.** Infinitely many solutions

 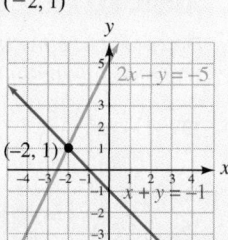 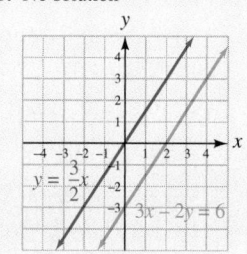 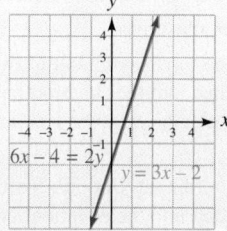

5. No solution

STUDY SET
4.1

VOCABULARY

Fill in the blanks.

1. The pair of equations $\begin{cases} x - y = -1 \\ 2x - y = 1 \end{cases}$ is called a _____ of linear equations.

2. Because the ordered pair (2, 3) satisfies both equations in Problem 1, it is called a _____ of the system of equations.

3. The point of _____ of the lines graphed in part (a) below is (1, 2).

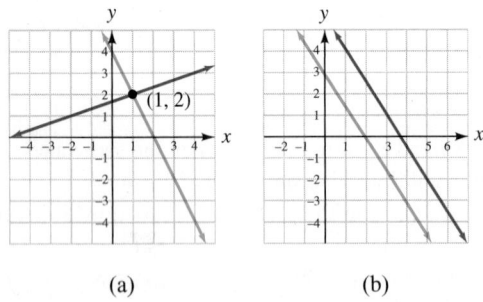

(a)   (b)

4. The lines graphed in part (b) above do not intersect. They are _____ lines.

5. A system of equations that has at least one solution is called a _____ system. A system with no solution is called an _____ system.

6. We call equations with different graphs _____ equations. Because _____ equations are different forms of the same equation, they have the same graph.

CONCEPTS

7. Refer to the illustration.
   a. If the coordinates of point *A* are substituted into the equation for Line 1, will the result be true or false?
   b. If the coordinates of point *C* are substituted into the equation for Line 1, will the result be true or false?

8. Refer to the illustration.
   a. If the coordinates of point *C* are substituted into the equation for Line 2, will the result be true or false?
   b. If the coordinates of point *B* are substituted into the equation for Line 1, will the result be true or false?

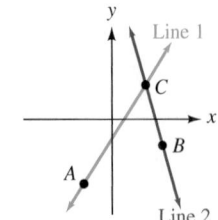

9. a. To graph $5x - 2y = 10$, we can use the intercept method. Complete the table.

x	y
0	
	0

   b. To graph $y = 3x - 2$, we can use the slope and *y*-intercept. Fill in the blanks.

   Slope: $\dfrac{}{} = \dfrac{}{1}$ *y*-intercept: _____

10. What is the apparent solution of the system graphed on the right? Is the system consistent or inconsistent?

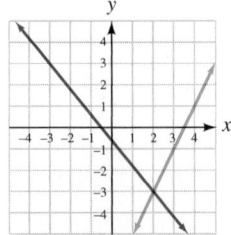

11. How many solutions does the system graphed on the right have? Are the equations dependent or independent?

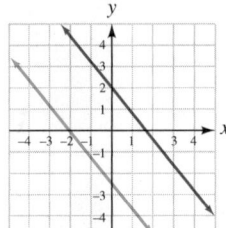

12. How many solutions does the system graphed on the right have? Give three of the solutions. Is the system consistent or inconsistent?

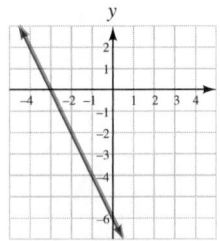

GUIDED PRACTICE

Determine whether the ordered pair is a solution of the given system of equations. See Example 1.

13. (1, 1), $\begin{cases} x + y = 2 \\ 2x - y = 1 \end{cases}$

14. (1, 3), $\begin{cases} 2x + y = 5 \\ 3x - y = 0 \end{cases}$

15. (3, −2), $\begin{cases} 2x + y = 4 \\ y = 1 - x \end{cases}$

16. (−2, 4), $\begin{cases} 2x + 2y = 4 \\ 3y = 10 - x \end{cases}$

17. (12, 0), $\begin{cases} x - 9y = 12 \\ y = 10 - x \end{cases}$

18. (15, 0), $\begin{cases} x - 2y = 15 \\ y = 16 - x \end{cases}$

19. $(-2, -4)$, $\begin{cases} 4x + 5y = -23 \\ -3x + 2y = 0 \end{cases}$ **20.** $(-5, 2)$, $\begin{cases} -2x + 7y = 17 \\ 3x - 4y = -19 \end{cases}$

21. $\left(\dfrac{1}{2}, 3\right)$, $\begin{cases} 2x + y = 4 \\ 4x - 11 = 3y \end{cases}$ **22.** $\left(2, \dfrac{1}{3}\right)$, $\begin{cases} x - 3y = 1 \\ -2x + 6 = -6y \end{cases}$

23. $(2.5, 3.5)$, $\begin{cases} 4x - 3 = 2y \\ 4y + 1 = 6x \end{cases}$ **24.** $(0.2, 0.3)$, $\begin{cases} 20x + 10y = 7 \\ 20y = 15x + 3 \end{cases}$

Solve each system of equations by graphing. If a system has no solution or infinitely many, so state. See Examples 2–4.

25. $\begin{cases} 2x + 3y = 12 \\ 2x - y = 4 \end{cases}$ **26.** $\begin{cases} 5x + y = 5 \\ 5x + 3y = 15 \end{cases}$

27. $\begin{cases} x + y = 4 \\ x - y = -6 \end{cases}$ **28.** $\begin{cases} x + y = 4 \\ x - y = -2 \end{cases}$

29. $\begin{cases} y = 3x + 6 \\ y = -2x - 4 \end{cases}$ **30.** $\begin{cases} y = x + 3 \\ y = -2x - 3 \end{cases}$

31. $\begin{cases} y = x - 1 \\ 3x - 3y = 3 \end{cases}$ **32.** $\begin{cases} y = -x + 1 \\ 4x + 4y = 4 \end{cases}$

33. $\begin{cases} y = -\dfrac{1}{3}x - 4 \\ x + 3y = 6 \end{cases}$ **34.** $\begin{cases} y = -\dfrac{1}{2}x - 3 \\ x + 2y = 2 \end{cases}$

35. $\begin{cases} y = -x - 2 \\ y = -3x + 6 \end{cases}$ **36.** $\begin{cases} y = 2x - 4 \\ y = -5x + 3 \end{cases}$

37. $\begin{cases} -x + 3y = -11 \\ 3x - y = 17 \end{cases}$ **38.** $\begin{cases} 2x - 3y = -18 \\ 3x + 2y = -1 \end{cases}$

39. $\begin{cases} x + y = 2 \\ y = x \end{cases}$ **40.** $\begin{cases} x + y = 4 \\ y = x \end{cases}$

41. $\begin{cases} y = \dfrac{3}{4}x + 3 \\ y = -\dfrac{x}{4} - 1 \end{cases}$ **42.** $\begin{cases} y = \dfrac{2}{3}x + 4 \\ y = -\dfrac{x}{3} + 7 \end{cases}$

43. $\begin{cases} 2y = 3x + 2 \\ 3x - 2y = 6 \end{cases}$ **44.** $\begin{cases} 3x - 6y = 18 \\ x = 2y + 3 \end{cases}$

45. $\begin{cases} 4x - 2y = 8 \\ y = 2x - 4 \end{cases}$ **46.** $\begin{cases} 2y = -6x - 12 \\ 3x + y = -6 \end{cases}$

47. $\begin{cases} x + y = 2 \\ y = x - 4 \end{cases}$ **48.** $\begin{cases} x + y = 1 \\ y = x + 5 \end{cases}$

49. $\begin{cases} x + 4y = -2 \\ y = -x - 5 \end{cases}$ **50.** $\begin{cases} 3x + 2y = -8 \\ 2x - 3y = -1 \end{cases}$

51. $\begin{cases} x = 3 \\ 3y = 6 - 2x \end{cases}$ **52.** $\begin{cases} x = 4 \\ 2y = 12 - 4x \end{cases}$

53. $\begin{cases} y = -3 \\ -x + 2y = -4 \end{cases}$ **54.** $\begin{cases} y = -4 \\ -2x - y = 8 \end{cases}$

55. $\begin{cases} x + 2y = -4 \\ x - \dfrac{1}{2}y = 6 \end{cases}$ **56.** $\begin{cases} \dfrac{2}{3}x - y = -3 \\ 3x + y = 3 \end{cases}$

Find the slope and the y-intercept of the graph of each line in the system of equations. Then, use that information to determine the number of solutions of the system. See Example 5.

57. $\begin{cases} y = 6x - 7 \\ y = -2x + 1 \end{cases}$ **58.** $\begin{cases} y = \dfrac{1}{2}x + 8 \\ y = 4x - 10 \end{cases}$

59. $\begin{cases} 3x - y = -3 \\ y - 3x = 3 \end{cases}$ **60.** $\begin{cases} x + 4y = 4 \\ 12y = 12 - 3x \end{cases}$

61. $\begin{cases} x + y = 6 \\ x + y = 8 \end{cases}$ **62.** $\begin{cases} 5x + y = 0 \\ 5x + y = 6 \end{cases}$

63. $\begin{cases} 6x + y = 0 \\ 2x + 2y = 0 \end{cases}$ **64.** $\begin{cases} x + y = 1 \\ 2x - 2y = 5 \end{cases}$

Use a graphing calculator to solve each system, if possible. See Objective 5.

65. $\begin{cases} y = 4 - x \\ y = 2 + x \end{cases}$ **66.** $\begin{cases} 3x - 6y = 4 \\ 2x + y = 1 \end{cases}$

67. $\begin{cases} 6x - 2y = 5 \\ 3x = y + 10 \end{cases}$ **68.** $\begin{cases} x - 3y = -2 \\ 5x + y = 10 \end{cases}$

APPLICATIONS

69. TRANSPLANTS   Refer to the graph. In what year were the number of donors and the number waiting for a liver transplant the same? Estimate the number.

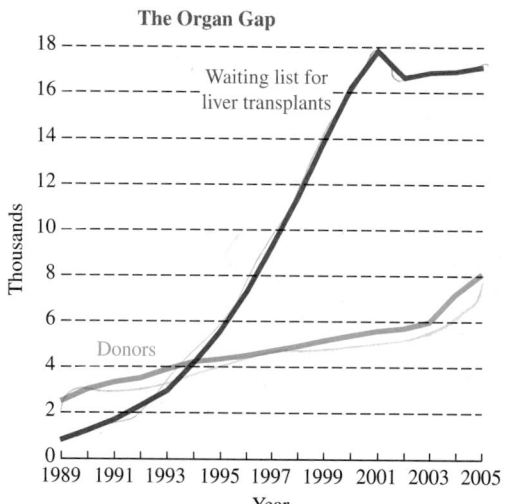

The Organ Gap

Source: Organ Procurement and Transportation Network

70. BEVERAGES Refer to the graph. In what year was average number of gallons of milk and carbonated soft drinks consumed per person the same? Estimate the number of gallons.

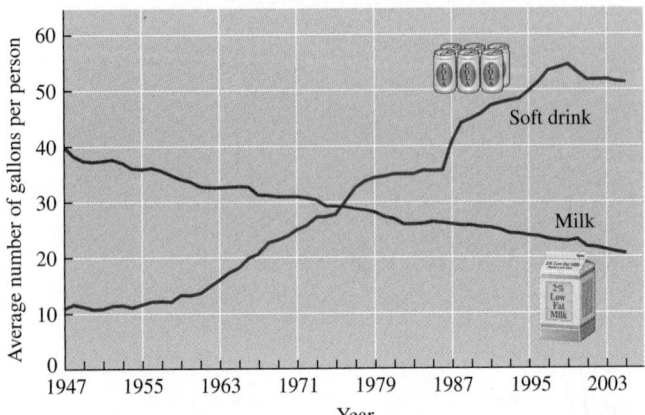

U.S. Milk Consumption vs. Soft Drink Consumption

Source: USDA, Economic Research Service

71. LATITUDE AND LONGITUDE Refer to the following map.
a. Name three American cities that lie on a latitude line of 30° north.
b. Name three American cities that lie on a longitude line of 90° west.
c. What city lies on both lines?

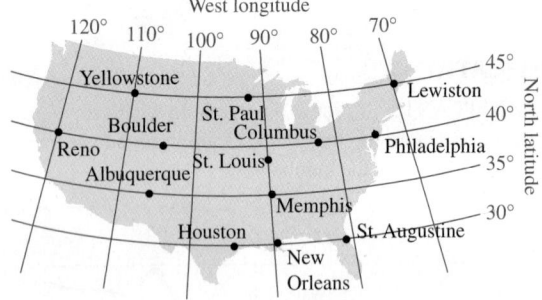

72. ECONOMICS The following graph illustrates the law of supply and demand.
a. Complete each sentence with the word *increases* or *decreases*. As the price of an item increases, the supply of the item _____. As the price of an item increases, the demand for the item _____.
b. For what price will the supply equal the demand? How many items will be supplied for this price?

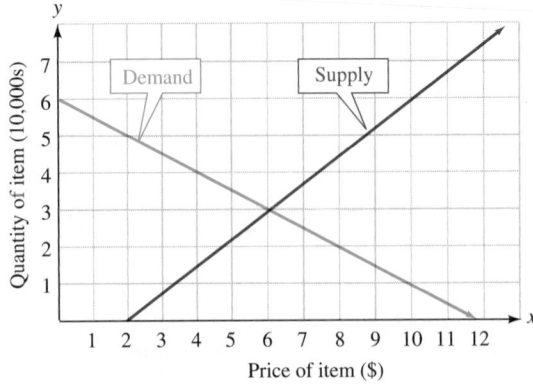

73. DAILY TRACKING POLLS Use the graph to answer the following.
a. Which political candidate was ahead on October 28 and by how much?
b. On what day did the challenger pull even with the incumbent?
c. If the election was held November 4, who did the poll predict would win, and by how many percentage points?

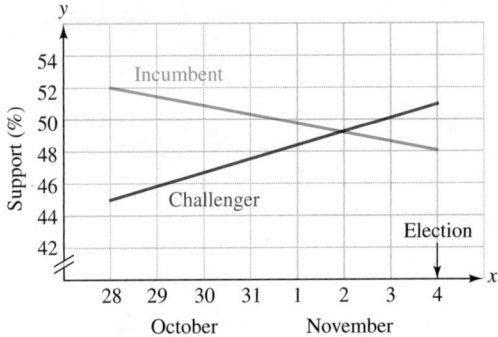

74. AIR TRAFFIC CONTROL The equations describing the paths of two airplanes are $y = -\frac{1}{2}x + 3$ and $3y = 2x + 2$. Graph each equation on the radar screen shown. Is there a possibility of a midair collision? If so, where?

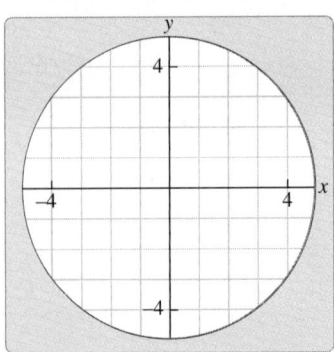

75. TV COVERAGE
A television camera is located at $(-2, 0)$ and will follow the launch of a space shuttle, as shown here. (Each unit in the illustration is 1 mile.) As the shuttle rises vertically on a path described by $x = 2$, the farthest the camera can tilt back is a line of sight given by $y = \frac{5}{2}x + 5$. For how many miles of the shuttle's flight will it be in view of the camera?

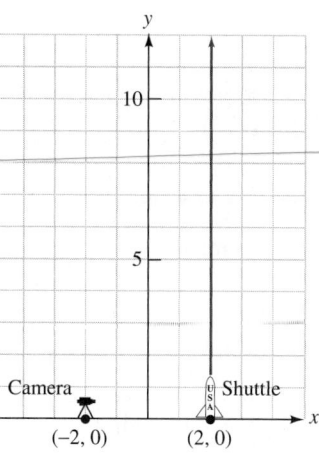

81. Suppose the graphs of the two linear equations of a system are the same line. What is wrong with the following statement? *The system has infinitely many solutions. Any ordered pair is a solution of the system.*

82. Write a definition of the word *parallel.*

REVIEW

Solve each inequality. Write the solution set in interval notation and graph it.

83. $-4(3y + 2) \leq 28$

84. $-5 < 3t + 4 \leq 13$

85. $\frac{r}{8} - 7 \geq -8$

86. $-1 \leq -\frac{1}{2}n$

87. $7x - 16 < 6x$

88. $\frac{1}{3} + \frac{c}{5} > -\frac{3}{2}$

WRITING

76. Explain why it is difficult to determine the solution of the system in the graph.

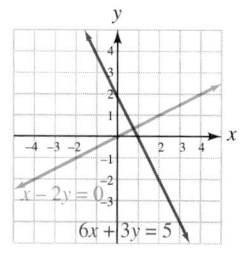

77. Without graphing, how can you tell that the graphs of $y = 2x + 1$ and $y = 3x + 2$ intersect?

78. Could a system of two linear equations have exactly two solutions? Explain why or why not.

79. What is an inconsistent system?

80. What are dependent equations?

CHALLENGE PROBLEMS

89. Can a system of two linear equations in two variables be inconsistent but have dependent equations? Explain.

90. Construct a system of two linear equations that has a solution of $(-2, 6)$.

91. Write a system of two linear equations such that $(2, 3)$ is a solution of the first equation but is not a solution of the second equation.

92. Solve by graphing: $\begin{cases} \dfrac{1}{3}x - \dfrac{1}{2}y = \dfrac{1}{6} \\ \dfrac{2x}{5} + \dfrac{y}{2} = \dfrac{13}{10} \end{cases}$

SECTION 4.2
Solving Systems of Equations by Substitution

Objectives

1 Solve systems of linear equations by substitution.

2 Find a substitution equation.

3 Solve systems of linear equations that contain fractions.

4 Use substitution to identify inconsistent systems and dependent equations.

When solving a system of equations by graphing, it is often difficult to determine the coordinates of the intersection point. For example, a solution of $\left(\frac{7}{8}, \frac{3}{5}\right)$ would be almost impossible to identify. In this section, we will discuss a second, more precise method for solving systems that does not involve graphing.

Solve Systems of Linear Equations by Substitution.

One algebraic method for solving a system of equations is the **substitution method.** It is introduced in the following example.

EXAMPLE 1   Solve the system: $\begin{cases} y = 3x - 2 \\ 2x + y = 8 \end{cases}$

Strategy   Note that the first equation is solved for y. Because y and $3x - 2$ are equal (represent the same value), we will substitute $3x - 2$ for y in the second equation.

Why   The objective is to obtain one equation containing only one unknown. When $3x - 2$ is substituted for y in the second equation, the result will be just that—an equation in one variable, x.

Solution   Since the right side of $y = 3x - 2$ is used to make a substitution, $y = 3x - 2$ is called the **substitution equation.**

$$\begin{cases} y = \boxed{3x - 2} \\ 2x + y = 8 \end{cases}$$

To find the solution of the system, we proceed as follows:

$$2x + y = 8 \quad \text{This is the second equation of the system.}$$
$$2x + 3x - 2 = 8 \quad \text{Substitute } 3x - 2 \text{ for } y.$$

The resulting equation has only one variable and can be solved for x.

$$2x + 3x - 2 = 8$$
$$5x - 2 = 8 \quad \text{Combine like terms: } 2x + 3x = 5x.$$
$$5x = 10 \quad \text{To isolate the variable term, } 5x, \text{ add 2 to both sides.}$$
$$x = 2 \quad \text{Divide both sides by 5. This is the x-value of the solution.}$$

> **Caution**
> When using the substitution method, a common error is to find the value of one of the variables, say x, and forget to find the value of the other. Remember that a solution of a linear system of two equations is an ordered pair (x, y).

We can find the y-value of the solution by substituting 2 for x in either equation of the original system. We will use the substitution equation because it is already solved for y.

$$y = 3x - 2 \quad \text{This is the substitution equation.}$$
$$y = 3(2) - 2 \quad \text{Substitute 2 for x.}$$
$$y = 6 - 2$$
$$y = 4 \quad \text{This is the y-value of the solution. We would have obtained the same result if we had substituted 2 for x in } 2x + y = 8 \text{ and solved for y.}$$

The ordered pair $(2, 4)$ appears to be the solution of the system. To check, we substitute 2 for x and 4 for y in each equation.

Check: $y = 3x - 2$ The first equation. $\qquad$ $2x + y = 8$ The second equation.

$\quad 4 \stackrel{?}{=} 3(2) - 2 \qquad\qquad\qquad\qquad\quad 2(2) + 4 \stackrel{?}{=} 8$

$\quad 4 \stackrel{?}{=} 6 - 2 \qquad\qquad\qquad\qquad\qquad\quad 4 + 4 \stackrel{?}{=} 8$

$\quad 4 = 4 \qquad \text{True} \qquad\qquad\qquad\qquad\quad 8 = 8 \quad \text{True}$

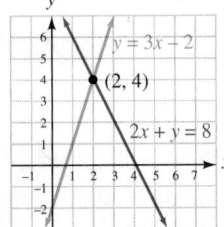

Since $(2, 4)$ satisfies both equations, it is the solution. The solution set is written as $\{(2, 4)\}$. A graph of the equations of the system shows an intersection point of $(2, 4)$. This illustrates an important fact: *The solution found using the substitution method will be the same as the solution found using the graphing method.*

▷ **Self Check 1**   Solve the system: $\begin{cases} x + 4y = 7 \\ x = 6y - 3 \end{cases}$

Now Try **Problem 15**

The substitution method works well for solving systems where one equation is solved, or can be easily solved, for one of the variables. To solve a system of equations in x and y by the substitution method, follow these steps.

The Substitution Method

1. Solve one of the equations for either x or y. If this is already done, go to step 2. (We call this equation the **substitution equation**.)
2. Substitute the expression for x or for y obtained in step 1 into the other equation and solve that equation.
3. Substitute the value of the variable found in step 2 into the substitution equation to find the value of the remaining variable.
4. Check the proposed solution in each equation of the original system. Write the solution as an ordered pair.

EXAMPLE 2   Solve the system: $\begin{cases} 4x + 27 = 7y \\ x = -5y \end{cases}$

Strategy   We will use the substitution method to solve this system.

Why   The substitution method works well when one of the equations of the system (in this case, $x = -5y$) is solved for a variable.

Solution

Step 1:   Because x and $-5y$ represent the same value, we can substitute $-5y$ for x in the first equation.

$\begin{cases} 4x + 27 = 7y \\ x = -5y \end{cases}$ This is the substitution equation.

Success Tip
The basic objective of this method is to use an appropriate substitution to obtain one equation in one variable.

Step 2:   When we substitute $-5y$ for x in the first equation, the resulting equation contains only one variable, and it can be solved for y.

$4x + 27 = 7y$	This is the first equation of the system.
$4(-5y) + 27 = 7y$	Substitute $-5y$ for x. Don't forget the parentheses.
$-20y + 27 = 7y$	Do the multiplication.
$27 = 27y$	To eliminate $-20y$ on the left side, add 20y to both sides.
$1 = y$	Divide both sides by 27. This is the y-value of the solution.

Step 3:   To find x, substitute 1 for y in the equation $x = -5y$.

Caution
We don't have to find the values of the variables in alphabetical order. Here, we found y first.

$x = -5y$	This is the substitution equation.
$x = -5(1)$	Substitute 1 for y.
$x = -5$	This is the x-value of the solution.

Step 4:   The following check verifies that the solution is $(-5, 1)$.

Check:

$4x + 27 = 7y$	The first equation.	$x = -5y$	The second equation.
$4(-5) + 27 \stackrel{?}{=} 7(1)$		$-5 \stackrel{?}{=} -5(1)$	
$-20 + 27 \stackrel{?}{=} 7$		$-5 = -5$	True
$7 = 7$	True		

▷ Self Check 2   Solve the system: $\begin{cases} 3x + 40 = 8y \\ x = -4y \end{cases}$

Now Try   **Problem 19**

② Find a Substitution Equation.

Sometimes neither equation of a system is solved for a variable. In such cases, we can find a substitution equation by solving one of the equations for one of its variables.

EXAMPLE 3   Solve the system: $\begin{cases} 4x + y = 3 \\ 3x + 5y = 15 \end{cases}$

Strategy   Since the system does not contain an equation solved for x or y, we must choose an equation and solve it for x or y. We will solve for y in the first equation, because y has a coefficient of 1. Then we will use the substitution method to solve the system.

Why   Solving $4x + y = 3$ for x or solving $3x + 5y = 15$ for x or y would involve working with cumbersome fractions.

Solution

Step 1:   To find a substitution equation, we proceed as follows:

$4x + y = 3$	This is the first equation of the system.
$y = 3 - 4x$	To isolate y, subtract $4x$ from both sides. This is the substitution equation.

> **Success Tip**
>
> To find a substitution equation, solve one of the equations of the system for one of its variables. If possible, solve for a variable whose coefficient is 1 or −1 to avoid working with fractions.

Because y and $3 - 4x$ are equal, we can substitute $3 - 4x$ for y in the second equation of the system.

$$\begin{cases} 4x + y = 3 & \rightarrow y = \boxed{3 - 4x} \\ 3x + 5y = 15 \end{cases}$$

Step 2:   When we substitute for y in the second equation, the resulting equation contains only one variable and can be solved for x.

$3x + 5y = 15$	This is the second equation of the system.
$3x + 5(3 - 4x) = 15$	Substitute $3 - 4x$ for y. Don't forget the parentheses.
$3x + 15 - 20x = 15$	Distribute the multiplication by 5.
$15 - 17x = 15$	Combine like terms.
$-17x = 0$	To isolate the variable term, $-17x$, subtract 15 from both sides.
$x = 0$	Divide both sides by −17. This is the x-value of the solution.

> **Caution**
>
> Here, use parentheses when substituting $3 - 4x$ for y so that the multiplication by 5 is distributed over both terms of $3 - 4x$.
>
> $3x + 5(3 - 4x) = 15$

Step 3:   To find y, substitute 0 for x in the equation $y = 3 - 4x$.

$y = 3 - 4x$	This is the substitution equation.
$y = 3 - 4(0)$	Substitute 0 for x.

$$y = 3 - 0$$
$$y = 3 \qquad \text{This is the y-value of the solution.}$$

Step 4: The solution appears to be (0, 3). Check it in the original equations.

Check:

$4x + y = 3$ The first equation.	$3x + 5y = 15$ The second equation.
$4(0) + 3 \stackrel{?}{=} 3$	$3(0) + 5(3) \stackrel{?}{=} 15$
$0 + 3 \stackrel{?}{=} 3$	$0 + 15 \stackrel{?}{=} 15$
$3 = 3$ True	$15 = 15$ True

Self Check 3   Solve the system: $\begin{cases} 2x - 3y = 10 \\ 3x + y = 15 \end{cases}$

Now Try   **Problem 31**

EXAMPLE 4   Solve the system: $\begin{cases} 3a - 3b = 5 \\ 3 - a = -2b \end{cases}$

Strategy   Since the coefficient of a in the second equation is -1, we will solve that equation for a. Then we will use the substitution method to solve the system.

Why   If we solve for the variable with a numerical coefficient of -1, we can avoid having to work with fractions.

Solution

Step 1:   To find a substitution equation, we proceed as follows:

$$3 - a = -2b \qquad \text{This is the second equation of the system.}$$
$$-a = -2b - 3 \qquad \text{To isolate the variable term, } -a, \text{ subtract 3 from both sides.}$$

To obtain a on the left side, multiply both sides of the equation by -1.

$$-1(-a) = -1(-2b - 3) \qquad \text{Multiply both sides by } -1. \text{ Don't forget the parentheses.}$$
$$a = 2b + 3 \qquad \text{Do the multiplications. This is the substitution equation.}$$

Because a and $2b + 3$ represent the same value, we can substitute $2b + 3$ for a in the first equation.

$$\begin{cases} 3a - 3b = 5 \\ 3 - a = -2b \rightarrow a = \boxed{2b + 3} \end{cases}$$

Step 2:   Substitute $2b + 3$ for a in the first equation and solve for b.

$$3a - 3b = 5 \qquad \text{This is the first equation of the system.}$$
$$3(2b + 3) - 3b = 5 \qquad \text{Substitute } 2b + 3 \text{ for } a. \text{ Don't forget the parentheses.}$$
$$6b + 9 - 3b = 5 \qquad \text{Distribute the multiplication by 3.}$$
$$3b + 9 = 5 \qquad \text{Combine like terms: } 6b - 3b = 3b.$$
$$3b = -4 \qquad \text{To isolate the variable term, } 3b, \text{ subtract 9 from both sides.}$$
$$b = -\frac{4}{3} \qquad \text{Divide both sides by 3. This is the b-value of the solution.}$$

Step 3:   To find a, substitute $-\frac{4}{3}$ for b in the equation $a = 2b + 3$.

$a = 2b + 3$	This is the substitution equation.
$a = 2\left(-\dfrac{4}{3}\right) + 3$	Substitute $-\dfrac{4}{3}$ for b.
$a = -\dfrac{8}{3} + \dfrac{9}{3}$	Do the multiplication and write 3 as a fraction with a denominator of 3.
$a = \dfrac{1}{3}$	Add. This is the a-value of the solution.

Step 4: The solution is $\left(\dfrac{1}{3}, -\dfrac{4}{3}\right)$. Check it in the original equations.

Self Check 4   Solve the system: $\begin{cases} 2s - t = 4 \\ 3s - 5t = 2 \end{cases}$

Now Try   **Problem 43**

 **Solve Systems of Linear Equations that Contain Fractions.**

It is usually helpful to clear any equations of fractions and combine any like terms before performing a substitution.

EXAMPLE 5   Solve the system: $\begin{cases} \dfrac{y}{4} = -\dfrac{x}{2} - \dfrac{3}{4} \\ 2x - y = -1 + y - x \end{cases}$

Strategy   We will use properties of algebra to write each equation of the system in simpler form. Then we will use the substitution method to solve the resulting equivalent system.

Why   The first equation will be easier to work with if we clear it of fractions. The second equation will be easier to work with if we eliminate the variable terms on the right side.

Solution   We can clear the first equation of fractions by multiplying both sides by the LCD.

$$\frac{y}{4} = -\frac{x}{2} - \frac{3}{4}$$

$$4\left(\frac{y}{4}\right) = 4\left(-\frac{x}{2} - \frac{3}{4}\right) \qquad \text{Multiply both sides by the LCD, 4.}$$
$$\text{Don't forget the parentheses.}$$

$$4\left(\frac{y}{4}\right) = 4\left(-\frac{x}{2}\right) - 4\left(\frac{3}{4}\right) \qquad \text{Distribute the multiplication by 4.}$$

(1) $y = -2x - 3$ Simplify. Call this equation 1.

We can write the second equation of the system in standard $Ax + By = C$ form by adding x and subtracting y from both sides.

$$2x - y = -1 + y - x$$
$$2x - y + x - y = -1 + y - x + x - y$$

(2) $3x - 2y = -1$ Combine like terms. Call this equation 2.

Step 1:   Equations 1 and 2 form an equivalent system, which has the same solution as the original one. To find x, we proceed as follows:

(1) $\begin{cases} y = -2x - 3 \\ 3x - 2y = -1 \end{cases}$ This is the substitution equation.
(2)

Step 2:   To find x, substitute $-2x - 3$ for y in equation 2 and proceed as follows:

$$3x - 2y = -1$$
$$3x - 2(-2x - 3) = -1 \qquad \text{Substitute } -2x - 3 \text{ for y. Don't forget the parentheses.}$$
$$3x + 4x + 6 = -1 \qquad \text{Distribute the multiplication by } -2.$$
$$7x + 6 = -1 \qquad \text{Combine like terms: } 3x + 4x = 7x.$$
$$7x = -7 \qquad \text{To isolate the variable term, 7x, subtract 6 from both sides.}$$
$$x = -1 \qquad \text{Divide both sides by 7. This is the x-value of the solution.}$$

Step 3:   To find y, we substitute -1 for x in equation 1.

$$y = -2x - 3$$
$$y = -2(-1) - 3 \qquad \text{Substitute } -1 \text{ for x.}$$
$$y = 2 - 3 \qquad \text{Do the multiplication.}$$
$$y = -1 \qquad \text{Subtract. This is the y-value of the solution.}$$

Step 4:   The solution is $(-1, -1)$. Check it in the original system.

> ### Caution
> Always use the original equations when checking a solution. Do not use a substitution equation or an equivalent equation that you found algebraically. If an error was made, a proposed solution that would not satisfy the original system might appear to be correct.

Self Check 5   Solve the system: $\begin{cases} \dfrac{y}{6} = \dfrac{x}{3} + \dfrac{1}{2} \\ 2x - y = -3 + y - x \end{cases}$

Now Try   **Problem 45**

4   Use Substitution to Identify Inconsistent Systems and Dependent Equations.

In the previous section, we solved inconsistent systems and systems of dependent equations graphically. We can also solve these systems using the substitution method.

EXAMPLE 6   Solve the system $\begin{cases} 4y - 12 = x \\ y = \dfrac{1}{4}x \end{cases}$ if possible.

Strategy   We will use the substitution method to solve this system.

Why   The substitution method works well when one of the equations of the system $\left(\text{in this case, } y = \frac{1}{4}x\right)$ is solved for a variable.

Solution   To try to solve this system, substitute $\frac{1}{4}x$ for y in the first equation and solve for x.

$$4y - 12 = x$$
$$4\left(\frac{1}{4}x\right) - 12 = x \qquad \text{Substitute } \tfrac{1}{4}x \text{ for y.}$$
$$x - 12 = x \qquad \text{Do the multiplication.}$$
$$x - 12 - x = x - x \qquad \text{To eliminate x on the right side, subtract x from both sides.}$$
$$-12 = 0 \qquad \text{False}$$

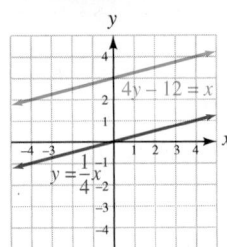

Here, the terms involving x drop out, and we get $-12 = 0$. This false statement indicates that the system has no solution and is inconsistent. The solution set is the empty set, $\varnothing$. The graphs of the equations of the system help to verify this; they are parallel lines.

Self Check 6 Solve the system $\begin{cases} x - 4 = y \\ -2y = 4 - 2x \end{cases}$ if possible.

Now Try **Problem 57**

EXAMPLE 7 Solve the system: $\begin{cases} x = -3y + 6 \\ 2x + 6y = 12 \end{cases}$

Strategy We will use the substitution method to solve this system.

Why The substitution method works well when one of the equations of the system (in this case, $x = -3y + 6$) is solved for a variable.

Solution To solve this system, substitute $-3y + 6$ for x in the second equation and solve for y.

$$2x + 6y = 12$$
$$2(-3y + 6) + 6y = 12 \quad \text{Substitute } -3y + 6 \text{ for } x.$$
$$\qquad\qquad\qquad\qquad \text{Don't forget the parentheses.}$$
$$-6y + 12 + 6y = 12 \quad \text{Distribute the multiplication by 2.}$$
$$12 = 12 \quad \text{True}$$

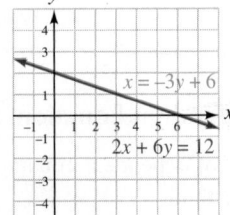

Here, the terms involving y drop out, and we get $12 = 12$. This true statement indicates that the two equations of the system are equivalent. Therefore, they are dependent equations and the system has infinitely many solutions. The graphs of the equations help to verify this; they are the same line.

Any ordered pair that satisfies one equation of this system also satisfies the other. To find several of the infinitely many solutions, we can substitute some values of x, say 0, 3, and 6, in either equation and solve for y. The results are: $(0, 2)$, $(3, 1)$, and $(6, 0)$.

Self Check 7 Solve the system: $\begin{cases} y = 2 - x \\ 3x + 3y = 6 \end{cases}$

Now Try **Problem 59**

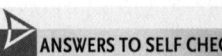**ANSWERS TO SELF CHECKS** **1.** $(3, 1)$ **2.** $(-8, 2)$ **3.** $(5, 0)$ **4.** $\left(\frac{18}{7}, \frac{8}{7}\right)$ **5.** $(-3, -3)$ **6.** No solution **7.** Infinitely many solutions

STUDY SET
4.2

VOCABULARY

Fill in the blanks.

1. To solve the system $\begin{cases} x = y + 1 \\ 3x + 2y = 8 \end{cases}$ using the method discussed in this section, we begin by _____ $y + 1$ for x in the second equation.

2. We say that the equation $y = 2x + 4$ is solved for ___.

CONCEPTS

3. If the substitution method is used to solve $\begin{cases} 5x + y = 2 \\ y = -3x \end{cases}$, which equation should be used as the substitution equation?

4. Suppose the substitution method will be used to solve $\begin{cases} x - 2y = 2 \\ 2x + 3y = 11 \end{cases}$. Find a substitution equation by solving one of the equations for one of the variables.

5. Suppose $x - 4$ is substituted for y in the equation $x + 3y = 8$. Insert parentheses in $x + 3x - 4 = 8$ to show the substitution.

6. Fill in the blank. With the substitution method, the objective is to use an appropriate substitution to obtain one equation in _____ variable.

7. A student uses the substitution method to solve the system $\begin{cases} 4a + 5b = 2 \\ b = 3a - 11 \end{cases}$ and finds that $a = 3$. What is the easiest way for her to determine the value of b?

8. **a.** Clear $\frac{x}{5} + \frac{2y}{3} = 1$ of fractions.
   b. Write $2x + y = x - 5y + 3$ in the form $Ax + By = C$.

9. Suppose $-2 = 1$ is obtained when a system is solved by the substitution method.
   a. Does the system have a solution?
   b. Which of the following is a possible graph of the system?

   i. 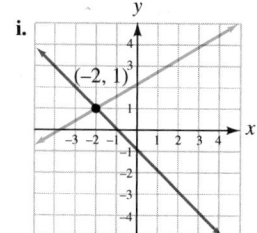 ii.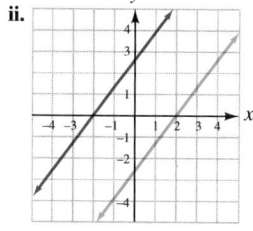

10. Suppose $2 = 2$ is obtained when a system is solved by the substitution method.
    a. Does the system have a solution?
    b. Which graph is a possible graph of the system?

    i. 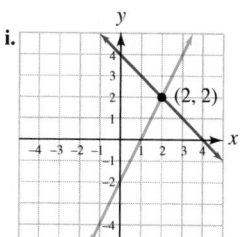 ii.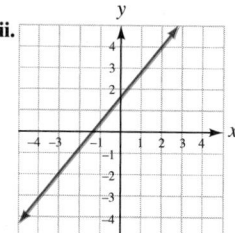

NOTATION

Complete the solution to solve the system.

11. Solve: $\begin{cases} y = 3x \\ x - y = 4 \end{cases}$

   $x - y = 4$ This is the second equation.

   $x - (\quad) = 4$

   $-2x = $

   $x = $

   $y = 3x$ This is the first equation.

   $y = 3(\quad)$

   $y = $

   The solution is $(\quad, \quad)$.

12. The system $\begin{cases} a = 3b + 2 \\ a + 3b = 8 \end{cases}$ was solved, and it was found that $b = 1$ and $a = 5$. Write the solution as an ordered pair.

GUIDED PRACTICE

Solve each system by substitution. **See Examples 1 and 2.**

13. $\begin{cases} y = 2x \\ x + y = 6 \end{cases}$

14. $\begin{cases} y = 3x \\ x + y = 4 \end{cases}$

15. $\begin{cases} y = 2x - 6 \\ 2x + y = 6 \end{cases}$

16. $\begin{cases} y = 2x - 9 \\ x + 3y = 8 \end{cases}$

17. $\begin{cases} 3x + y = -4 \\ x = y \end{cases}$

18. $\begin{cases} x + 2y = -6 \\ x = y \end{cases}$

19. $\begin{cases} x + 3y = -4 \\ x = -5y \end{cases}$

20. $\begin{cases} x + 5y = -3 \\ x = -4y \end{cases}$

21. $\begin{cases} 2x - y = -5 \\ x = -2y - 5 \end{cases}$

22. $\begin{cases} y = -2x \\ 3x + 2y = -1 \end{cases}$

23. $\begin{cases} b = \dfrac{2}{3}a \\ 8a - 3b = 3 \end{cases}$

24. $\begin{cases} a = \dfrac{2}{3}b \\ 9a + 4b = 5 \end{cases}$

25. $\begin{cases} 2x + 5y = -2 \\ y = -\dfrac{x}{2} \end{cases}$

26. $\begin{cases} y = -\dfrac{x}{2} \\ 2x - 3y = -7 \end{cases}$

27. $\begin{cases} x = \dfrac{1}{3}y - 1 \\ x = y + 5 \end{cases}$

28. $\begin{cases} x = \dfrac{1}{2}y + 2 \\ x = y - 6 \end{cases}$

Solve each system by substitution. See Examples 3 and 4.

29. $\begin{cases} r + 3s = 9 \\ 3r + 2s = 13 \end{cases}$

30. $\begin{cases} x - 2y = 2 \\ 2x + 3y = 11 \end{cases}$

31. $\begin{cases} 4x + y = -15 \\ 2x + 3y = 5 \end{cases}$

32. $\begin{cases} 4x + y = -5 \\ 2x - 3y = -13 \end{cases}$

33. $\begin{cases} 6x - 3y = 5 \\ x + 2y = 0 \end{cases}$

34. $\begin{cases} 5s + 10t = 3 \\ 2s + t = 0 \end{cases}$

35. $\begin{cases} 2x + 3 = -4y \\ x - 6 = -8y \end{cases}$

36. $\begin{cases} 5y + 2 = -4x \\ x + 2y = -2 \end{cases}$

37. $\begin{cases} 2a - 3b = -13 \\ -b = -2a - 7 \end{cases}$

38. $\begin{cases} a - 3b = -1 \\ -b = -2a - 2 \end{cases}$

39. $\begin{cases} 8x - 6y = 4 \\ 2x - y = -2 \end{cases}$

40. $\begin{cases} 5x + 4y = 0 \\ 2x - y = 0 \end{cases}$

41. $\begin{cases} 4x + 5y = 2 \\ 3x - y = 11 \end{cases}$

42. $\begin{cases} 5u + 3v = 5 \\ 4u - v = 4 \end{cases}$

43. $\begin{cases} 3x + 4y = -19 \\ 2y - x = 3 \end{cases}$

44. $\begin{cases} 5x - 2y = -7 \\ 5 - y = -3x \end{cases}$

Solve each system by substitution. See Example 5.

45. $\begin{cases} \dfrac{x}{2} + \dfrac{y}{2} = -1 \\ \dfrac{x}{3} - \dfrac{y}{2} = -4 \end{cases}$

46. $\begin{cases} \dfrac{2}{3}a + \dfrac{b}{5} = 1 \\ \dfrac{a}{3} - \dfrac{2}{3}b = \dfrac{13}{3} \end{cases}$

47. $\begin{cases} 5x = \dfrac{1}{2}y - 1 \\ \dfrac{1}{4}y = 10x - 1 \end{cases}$

48. $\begin{cases} \dfrac{x}{4} + y = \dfrac{1}{4} \\ \dfrac{y}{2} + \dfrac{11}{20} = \dfrac{x}{10} \end{cases}$

49. $\begin{cases} x - \dfrac{4}{5}y = 4 \\ \dfrac{y}{3} = \dfrac{x}{2} - \dfrac{5}{2} \end{cases}$

50. $\begin{cases} 3x - 2y = \dfrac{9}{2} \\ \dfrac{x}{2} - \dfrac{3}{4} = 2y \end{cases}$

51. $\begin{cases} y + x = 2x + 2 \\ 6x - 4y = 21 - y \end{cases}$

52. $\begin{cases} y - x = 3x \\ 2x + 2y = 14 - y \end{cases}$

53. $\begin{cases} 4x + 5y + 1 = -12 + 2x \\ x - 3y + 2 = -3 - x \end{cases}$

54. $\begin{cases} 6x + y = -8 + 3x - y \\ 3x - y = 2y + x - 1 \end{cases}$

55. $\begin{cases} 3(x - 1) + 3 = 8 + 2y \\ 2(x + 1) = 8 + y \end{cases}$

56. $\begin{cases} 4(x - 2) = 19 - 5y \\ 3(x - 2) - 2y = -y \end{cases}$

Solve each system by substitution, if possible. See Examples 6 and 7.

57. $\begin{cases} 2a + 4b = -24 \\ a = 20 - 2b \end{cases}$

58. $\begin{cases} 3a + 6b = -15 \\ a = -2b - 5 \end{cases}$

59. $\begin{cases} y - 3x = -5 \\ 21x = 7y + 35 \end{cases}$

60. $\begin{cases} 8y = 15 - 4x \\ x + 2y = 4 \end{cases}$

61. $\begin{cases} 6 - y = 4x \\ 2y = -8x - 20 \end{cases}$

62. $\begin{cases} 9x = 3y + 12 \\ 4 = 3x - y \end{cases}$

63. $\begin{cases} x = -3y + 6 \\ 2x + 4y = 6 + x + y \end{cases}$

64. $\begin{cases} 2x - y = x + y \\ -2x + 4y = 6 \end{cases}$

TRY IT YOURSELF

Solve each system by substitution, if possible. If a system has no solution or infinitely many solutions, so state.

65. $\begin{cases} -y = 11 - 3x \\ 2x + 5y = -4 \end{cases}$

66. $\begin{cases} -x = 10 - 3y \\ 2x + 8y = -6 \end{cases}$

67. $\begin{cases} \dfrac{x}{2} + \dfrac{y}{6} = \dfrac{2}{3} \\ \dfrac{x}{3} - \dfrac{y}{4} = \dfrac{1}{12} \end{cases}$

68. $\begin{cases} \dfrac{c}{2} + \dfrac{d}{14} = 1 \\ \dfrac{c}{5} - \dfrac{d}{2} = -\dfrac{33}{10} \end{cases}$

69. $\begin{cases} y - 4 = 2x \\ y = 2x + 2 \end{cases}$

70. $\begin{cases} x + 3y = 6 \\ x = -3y + 6 \end{cases}$

71. $\begin{cases} a + b = 1 \\ a - 2b = -1 \end{cases}$

72. $\begin{cases} 2b - a = -1 \\ 3a + 10b = -1 \end{cases}$

73. $\begin{cases} x = 7y - 10 \\ 2x - 14y + 20 = 0 \end{cases}$

74. $\begin{cases} y - 1 = 5x \\ 10x - 2y = 2 \end{cases}$

75. $\begin{cases} 4x + 1 = 2x + 5 + y \\ 2x + 2y = 5x + y + 6 \end{cases}$

76. $\begin{cases} 6x = 2(y + 20) + 5x \\ 5(x - 1) = 3y + 4(x + 10) \end{cases}$

77. $\begin{cases} 2a + 3b = 7 \\ 6a - b = 1 \end{cases}$

78. $\begin{cases} 3a + 5b = -6 \\ 5b - a = -3 \end{cases}$

79. $\begin{cases} 2x - 3y = -4 \\ x = -\dfrac{3}{2}y \end{cases}$

80. $\begin{cases} x = -\dfrac{3}{8}y \\ 8x - 3y = 4 \end{cases}$

APPLICATIONS

81. OFFROADING   The *angle of approach* indicates how steep of an incline a vehicle can drive up without damaging the front bumper. The *angle of departure* indicates a vehicle's ability to exit an incline without damaging the rear bumper. The angle of approach a and the departure angle d for an H3 Hummer are described by the system $\begin{cases} a + d = 77 \\ a = d + 3 \end{cases}$. Use substitution to solve the system. (Each angle is measured in degrees.)

Angle of approach $a°$ Angle of departure $d°$

82. HIGH SCHOOL SPORTS   The equations shown model the number of boys and girls taking part in high school soccer programs. In both models, x is the number of years after 2000, and y is the number of participants. If the trends continue, the graphs will intersect. Use the substitution method to predict the year when the number of boys and girls participating in high school soccer will be the same.

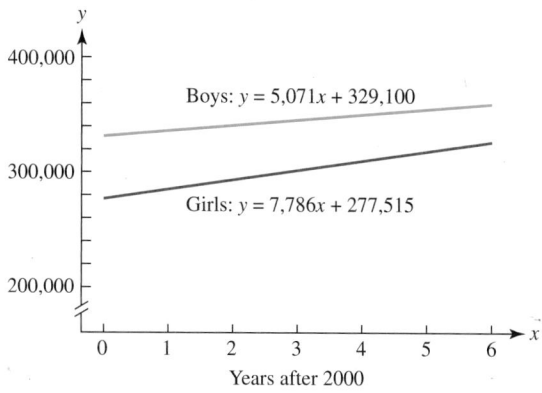

Boys: $y = 5{,}071x + 329{,}100$

Girls: $y = 7{,}786x + 277{,}515$

Years after 2000

Source: National Federation of State High School Associations

WRITING

83. What concept does this diagram illustrate?

$$\begin{cases} 6x + 5y = 11 \\ y = 3x - 2 \end{cases}$$

84. When using the substitution method, how can you tell whether
   a. a system of linear equations has no solution?
   b. a system of linear equations has infinitely many solutions?

85. When solving a system, what advantages are there with the substitution method compared with the graphing method?

86. Consider the equation $5x + y = 12$. Explain why it is easier to solve for y than it is for x.

REVIEW

87. Find the prime factorization of 189.

88. Complete each statement. For any nonzero number a,

   a. $\dfrac{a}{a} =$ **b.** $\dfrac{a}{1} =$

   c. $\dfrac{0}{a} =$ **d.** $\dfrac{a}{0} =$

89. Simplify: $\dfrac{30}{36}$ **90.** Add: $\dfrac{5}{12} + \dfrac{1}{4}$

91. Multiply: $\dfrac{7}{8} \cdot \dfrac{3}{5}$ **92.** Divide: $\dfrac{1}{3} \div \dfrac{4}{5}$

CHALLENGE PROBLEMS

Use the substitution method to solve each system.

93. $\begin{cases} \dfrac{6x - 1}{3} - \dfrac{5}{3} = \dfrac{3y + 1}{2} \\ \dfrac{1 + 5y}{4} + \dfrac{x + 3}{4} = \dfrac{17}{2} \end{cases}$

94. $\begin{cases} 0.5x + 0.5y = 6 \\ 0.001x - 0.001y = -0.004 \end{cases}$

95. The system $\begin{cases} \dfrac{1}{2}x = y + 3 \\ x - 2y = 6 \end{cases}$ has infinitely many solutions. Find three of them.

96. Could the substitution method be used to solve the following system? Explain why or why not. If not, what method could be used?

$$\begin{cases} y = -2 \\ x = 5 \end{cases}$$

SECTION 4.3
Solving Systems of Equations by Elimination (Addition)

Objectives

① Solve systems of linear equations by the elimination method.

② Use multiplication to eliminate a variable.

③ Use the elimination method twice to solve a system.

④ Use elimination to identify inconsistent systems and dependent equations.

⑤ Determine the most efficient method to use to solve a linear system.

In the first step of the substitution method for solving a system of equations, we solve one of the equations for one of the variables. At times, this can be difficult, especially if none of the variables has a coefficient of 1 or -1. This is the case for the system

$$\begin{cases} 2x + 5y = 11 \\ 7x - 5y = 16 \end{cases}$$

Solving either equation for x or y involves working with cumbersome fractions. Fortunately, we can solve systems like this one using an easier algebraic method called the **elimination** or the **addition method.**

 Solve Systems of Linear Equations by the Elimination Method.

The elimination method for solving a system is based on the **addition property of equality:** *When equal quantities are added to both sides of an equation, the results are equal.* In symbols, if $A = B$ and $C = D$, then adding the left sides and the right sides of these equations, we have $A + C = B + D$. This procedure is called *adding the equations.*

Add the terms on the left sides.

$$\begin{aligned} A &= B \\ C &= D \\ \hline A + C &= B + D \end{aligned}$$

Add the terms on the right sides.

EXAMPLE 1   Solve the system: $\begin{cases} 2x + 5y = 11 \\ 6x - 5y = 13 \end{cases}$

Strategy   Since the coefficients of the y-terms are opposites, we will add the left sides and the right sides of the given equations.

Why   When we add the equations in this way, the result will be an equation that contains only one variable, x.

Solution   Since $6x - 5y$ and 13 are equal quantities, we can add $6x - 5y$ to the left side and 13 to the right side of the first equation, $2x + 5y = 11$.

The Language of Algebra
The *elimination* method, or *addition* method as it is also known, is so named because one of the variables is eliminated using addition.

$$2x + 5y = 11$$
$$6x - 5y = 13$$

To add the equations, add the like terms, column by column.

$$\overline{8x \qquad = 24}$$

L 11 + 13 = 24

5y + (-5y) = 0

2x + 6x = 8x

Because the sum of the terms $5y$ and $-5y$ is 0, we say that the variable y has been eliminated. Since the resulting equation has only one variable, we can solve it for x.

$$8x = 24$$

$x = 3$ Divide both sides by 8. This is the x-value of the solution.

To find the y-value of the solution, substitute 3 for x in either equation of the original system.

$2x + 5y = 11$ This is the first equation of the system.

$2(3) + 5y = 11$ Substitute 3 for x.

$6 + 5y = 11$ Multiply.

$5y = 5$ Subtract 6 from both sides.

$y = 1$ Divide both sides by 5. This is the y-value of the solution.

Now we check the proposed solution (3, 1) in the equations of the original system.

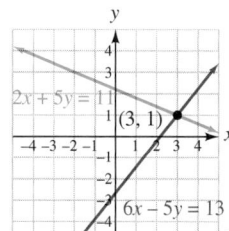

Check: $2x + 5y = 11$ The first equation. $6x - 5y = 13$ The second equation.

$2(3) + 5(1) \overset{?}{=} 11$ $\qquad$ $6(3) - 5(1) \overset{?}{=} 13$

$6 + 5 \overset{?}{=} 11$ $\qquad$ $18 - 5 \overset{?}{=} 13$

$11 = 11$ True $\qquad$ $13 = 13$ True

Since (3, 1) satisfies both equations, it is the solution. The graph on the left helps to verify this. The solution set is written $\{(3, 1)\}$.

Self Check 1 Solve the system: $\begin{cases} -4x + 3y = 4 \\ 4x + 5y = 28 \end{cases}$

Now Try **Problem 19**

To solve a system of equations in x and y by the elimination method, follow these steps.

The Elimination (Addition) Method

1. Write both equations of the system in standard $Ax + By = C$ form.

2. If necessary, multiply one or both of the equations by a nonzero number chosen to make the coefficients of x (or the coefficients of y) opposites.

3. Add the equations to eliminate the terms involving x (or y).

4. Solve the equation resulting from step 3.

5. Find the value of the remaining variable by substituting the solution found in step 4 into any equation containing both variables. Or, repeat steps 2–4 to eliminate the other variable.

6. Check the proposed solution in each equation of the original system. Write the solution as an ordered pair.

Use Multiplication to Eliminate a Variable.

In Example 1, the coefficients of the terms $5y$ in the first equation and $-5y$ in the second equation were opposites. When we added the equations, the variable y was eliminated. For many systems, however, we are not able to immediately eliminate a variable by adding. In such cases, we use the multiplication property of equality to create coefficients of x or y that are opposites.

EXAMPLE 2   Solve the system: $\begin{cases} 2x + 7y = -18 \\ 2x + 3y = -10 \end{cases}$

Strategy   We will use the elimination method to solve this system.

Why   Since none of the variables has a coefficient of 1 or -1, it would be difficult to solve this system using substitution.

Solution

Step 1:   Both equations are in standard $Ax + By = C$ form. We see that neither the coefficients of x nor the coefficients of y are opposites. Adding these equations as written does not eliminate a variable.

Step 2:   To eliminate x, we can multiply both sides of the second equation by -1. This creates the term $-2x$, whose coefficient is opposite that of the $2x$ term in the first equation.

$$\begin{cases} 2x + 7y = -18 \\ 2x + 3y = -10 \end{cases} \xrightarrow[\text{Multiply by } -1]{\text{Unchanged}} \begin{array}{l} 2x + 7y = -18 \\ -1(2x + 3y) = -1(-10) \end{array} \xrightarrow[\text{Simplify}]{\text{Unchanged}} \begin{cases} 2x + 7y = -18 \\ -2x - 3y = 10 \end{cases}$$

Step 3:   When the equations are added, x is eliminated.

$$\begin{array}{r} 2x + 7y = -18 \\ -2x - 3y = 10 \\ \hline 4y = -8 \end{array}$$

In the left column: $2x + (-2x) = 0$.

Success Tip

It doesn't matter which variable is eliminated first. We don't have to find the values of the variables in alphabetical order. Choose the one that is the easier to eliminate. The basic objective of this method is to obtain two equations whose sum will be *one* equation in *one* variable.

Step 4:   Solve the resulting equation to find y.

$4y = -8$

$y = -2$ Divide both sides by 4. This is the y-value of the solution.

Step 5:   To find x, we can substitute -2 for y in either of the equations of the original system, or in $-2x - 3y = 10$. It appears the computations will be simplest if we use $2x + 3y = -10$.

$2x + 3y = -10$ This is the second equation of the original system.

$2x + 3(-2) = -10$ Substitute -2 for y.

$2x - 6 = -10$ Multiply.

$2x = -4$ Add 6 to both sides.

$x = -2$ Divide both sides by 2. This is the x-value of the solution.

Step 6:   The solution is $(-2, -2)$. Check this result in the original equations.

> **Self Check 2**   Solve the system: $\begin{cases} x + 7y = -24 \\ 3x + 7y = -30 \end{cases}$
>
> *Now Try* **Problem 27**

EXAMPLE 3   Solve the system: $\begin{cases} 7x + 2y - 14 = 0 \\ 9x = 4y - 28 \end{cases}$

Strategy   We will use the elimination method to solve this system.

Why   Since none of the variables has coefficient 1 or -1, it would be difficult to solve this system using substitution.

Solution

Step 1:   To compare coefficients, write each equation in the standard $Ax + By = C$ form. Since each of the original equations will be written in an equivalent form, the resulting system will have the same solution as the original system.

$$\begin{cases} 7x + 2y = 14 \\ 9x - 4y = -28 \end{cases}$$ Add 14 to both sides of $7x + 2y - 14 = 0$.
    Subtract 4y from both sides of $9x = 4y - 28$.

Success Tip

We choose to eliminate y because the coefficient -4 is a *multiple* of the coefficient 2. The same cannot be said for 7 and 9, the coefficients of x.

Step 2:   Neither the coefficients of x nor the coefficients of y are opposites. To eliminate y, we can multiply both sides of the first equation by 2. This creates the term $4y$, whose coefficient is opposite that of the $-4y$ term in the second equation.

$$\begin{cases} 7x + 2y = 14 \\ 9x - 4y = -28 \end{cases} \xrightarrow{\text{Multiply by 2}} \begin{matrix} 2(7x + 2y) = 2(14) \\ 9x - 4y = -28 \end{matrix} \xrightarrow{\text{Simplify}} \begin{cases} 14x + 4y = 28 \\ 9x - 4y = -28 \end{cases}$$

Step 3:   When the equations are added, y is eliminated.

$$\begin{array}{r} 14x + 4y = 28 \\ 9x - 4y = -28 \\ \hline 23x \quad\quad = 0 \end{array}$$ In the middle column: $4y + (-4y) = 0$.

Caution

When using the elimination method, don't forget to multiply both sides of an equation by the appropriate number.

Multiply both sides by 2
$2(7x + 2y) = 2(14)$

Step 4:   Since the result of the addition is an equation in one variable, we can solve for x.

$$23x = 0$$
$$x = 0$$ Divide both sides by 23. This is the x-value of the solution.

Step 5:   To find y, we can substitute 0 for x in any equation that contains both variables. It appears the computations will be simplest if we use $7x + 2y = 14$.

$$7x + 2y = 14$$ This is the first equation of the original system.
$$7(0) + 2y = 14$$ Substitute 0 for x.
$$0 + 2y = 14$$ Multiply.
$$2y = 14$$
$$y = 7$$ Divide both sides by 2. This is the y-value of the solution.

Step 6:   The solution is $(0, 7)$. Check this result in the original equations.

▷ **Self Check 3** Solve the system: $\begin{cases} 3x = 10 - 2y \\ 5x - 6y + 30 = 0 \end{cases}$

Now Try **Problem 41**

EXAMPLE 4 Solve the system: $\begin{cases} 4a + 7b = -8 \\ 5a + 6b = 1 \end{cases}$

Strategy We will use the elimination method to solve this system.

Why Since none of the variables has coefficient 1 or -1, it would be difficult to solve this system using substitution.

Solution

Step 1: Both equations are written in standard $Ax + By = C$ form.

Step 2: In this example, we must write *both* equations in equivalent forms to obtain like terms that are opposites. To eliminate a, we can multiply the first equation by 5 to create the term $20a$, and we can multiply the second equation by -4 to create the term $-20a$.

$$\begin{cases} 4a + 7b = -8 \\ 5a + 6b = 1 \end{cases} \xrightarrow[\text{Multiply by } -4]{\text{Multiply by 5}} \begin{array}{l} 5(4a + 7b) = 5(-8) \\ -4(5a + 6b) = -4(1) \end{array} \xrightarrow[\text{Simplify}]{\text{Simplify}} \begin{cases} 20a + 35b = -40 \\ -20a - 24b = -4 \end{cases}$$

Step 3: When we add the resulting equations, a is eliminated.

$$\begin{array}{r} 20a + 35b = -40 \\ -20a - 24b = -4 \\ \hline 11b = -44 \end{array}$$ In the left column: $20a + (-20a) = 0$.

Step 4: Solve the resulting equation for b.

$$11b = -44$$
$$b = -4 \quad \text{Divide both sides by 11. This is the } b\text{-value of the solution.}$$

Step 5: To find a, we can substitute -4 for b in any equation that contains both variables. It appears the computations will be simplest if we use $5a + 6b = 1$.

$$5a + 6b = 1 \quad \text{This is the second equation of the original system.}$$
$$5a + 6(-4) = 1 \quad \text{Substitute } -4 \text{ for } b.$$
$$5a - 24 = 1 \quad \text{Multiply.}$$
$$5a = 25 \quad \text{Add 24 to both sides.}$$
$$a = 5 \quad \text{Divide both sides by 5. This is the } a\text{-value of the solution.}$$

Step 6: Written in (a, b) form, the solution is $(5, -4)$. Check it in the original equations.

▷ **Self Check 4** Solve the system: $\begin{cases} 5a + 3b = -7 \\ 3a + 4b = 9 \end{cases}$

Now Try **Problem 45**

Success Tip

We create the term $20a$ from $4a$ and the term $-20a$ from $5a$. Note that the *least common multiple* of 4 and 5 is 20:

$4, 8, 12, 16, \mathbf{20}, 24, 28, \ldots$
$5, 10, 15, \mathbf{20}, 25, 30, \ldots$

Success Tip

With this method, it doesn't matter which variable is eliminated. We could have created terms of $42b$ and $-42b$ to eliminate b. We will get the same solution, $(5, -4)$.

3 Use the Elimination Method Twice to Solve a System.

Sometimes it is easier to find the value of the second variable of a solution by using elimination a second time.

EXAMPLE 5   Solve the system:
$$\begin{cases} \dfrac{1}{6}x + \dfrac{1}{2}y = \dfrac{1}{3} \\[2mm] -\dfrac{x}{9} + y = \dfrac{5}{9} \end{cases}$$

Strategy   We will begin by clearing each equation of fractions. Then we will use the elimination method to solve the resulting equivalent system.

Why   It is easier to create a pair of terms that are opposites if their coefficients are integers rather than fractions.

Solution

Step 1:   To clear the equations of the fractions, multiply both sides of the first equation by 6 and both sides of the second equation by 9.

$$\begin{cases} \dfrac{1}{6}x + \dfrac{1}{2}y = \dfrac{1}{3} \xrightarrow{\text{Multiply by 6}} 6\left(\dfrac{1}{6}x + \dfrac{1}{2}y\right) = 6\left(\dfrac{1}{3}\right) \xrightarrow{\text{Simplify}} \\[2mm] -\dfrac{x}{9} + y = \dfrac{5}{9} \xrightarrow[\text{Multiply by 9}]{} 9\left(-\dfrac{x}{9} + y\right) = 9\left(\dfrac{5}{9}\right) \xrightarrow[\text{Simplify}]{} \end{cases} \begin{cases} x + 3y = 2 \\ -x + 9y = 5 \end{cases}$$

Step 2:   The coefficients of x are opposites.

Step 3:   The variable x is eliminated when we add the resulting equations.

$$\begin{aligned} x + 3y &= 2 \\ -x + 9y &= 5 \\ \hline 12y &= 7 \end{aligned}$$ In the left column: x + (−x) = 0.

Step 4:   Now solve the resulting equation to find y.

$$12y = 7$$

$$y = \dfrac{7}{12}$$ Divide both sides by 12. This is the y-value of the solution.

Step 5:   We can find x by substituting $\frac{7}{12}$ for y in any equation containing both variables. However, that computation could be complicated, because $\frac{7}{12}$ is a fraction. Instead, we can begin again with the system that is cleared of fractions, but this time, eliminate y. If we multiply both sides of the first equation by -3, this creates the term $-9y$, whose coefficient is opposite that of the $9y$ term in the second equation.

$$\begin{cases} x + 3y = 2 \xrightarrow{\text{Multiply by −3}} -3(x + 3y) = -3(2) \xrightarrow{\text{Simplify}} \\ -x + 9y = 5 \xrightarrow[\text{Unchanged}]{} -x + 9y = 5 \xrightarrow[\text{Unchanged}]{} \end{cases} \begin{cases} -3x - 9y = -6 \\ -x + 9y = 5 \end{cases}$$

When we add the resulting equations, y is eliminated.

$$\begin{aligned} -3x - 9y &= -6 \\ -x + 9y &= 5 \\ \hline -4x \phantom{{}- 9y} &= -1 \end{aligned}$$ In the middle column: 9y + (−9y) = 0.

Now we solve the resulting equation to find x.

$$-4x = -1$$

$$x = \frac{1}{4}$$ Divide both sides by -4. This is the x-value of the solution.

Step 6: The solution is $\left(\frac{1}{4}, \frac{7}{12}\right)$. To verify this, check it in the original equations.

 Self Check 5   Solve the system:
$$\begin{cases} -\dfrac{1}{5}x + y = \dfrac{8}{5} \\ \dfrac{x}{8} + \dfrac{y}{2} = \dfrac{1}{4} \end{cases}$$

Now Try **Problem 61**

4 **Use Elimination to Identify Inconsistent Systems and Dependent Equations.**

We have solved inconsistent systems and systems of dependent equations by substitution and by graphing. We can also solve these systems using the elimination method.

 EXAMPLE 6   Solve the system:
$$\begin{cases} 3x - 2y = 2 \\ -3x + 2y = -12 \end{cases}$$, if possible.

Strategy   We will use the elimination method to solve this system.

Why   The terms $3x$ and $-3x$ are immediately eliminated.

Solution

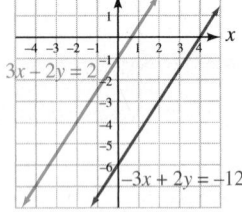

$$\begin{array}{ll} 3x - 2y = 2 & \text{In the left column: } 3x + (-3x) = 0. \\ \underline{-3x + 2y = -12} & \text{In the middle column: } -2y + 2y = 0. \\ \quad\quad\quad 0 = -10 & \end{array}$$

In eliminating x, the variable y is eliminated as well. The resulting false statement, $0 = -10$, indicates that the system has no solution and is inconsistent. The graphs of the equations help to verify this; they are parallel lines.

 Self Check 6   Solve the system:
$$\begin{cases} 2x - 7y = 5 \\ -2x + 7y = 3 \end{cases}$$

Now Try **Problem 65**

EXAMPLE 7   Solve the system:
$$\begin{cases} \dfrac{2x - 5y}{15} = \dfrac{8}{15} \\ -0.2x + 0.5y = -0.8 \end{cases}$$

Strategy   We will begin by clearing the equations of fractions and decimals. Then we will use the elimination method to solve the resulting equivalent system.

Why   In this form, the equations do not contain terms with coefficients that are opposites.

Solution   We can multiply both sides of the first equation by 15 to clear it of fractions and both sides of the second equation by 10 to clear it of decimals.

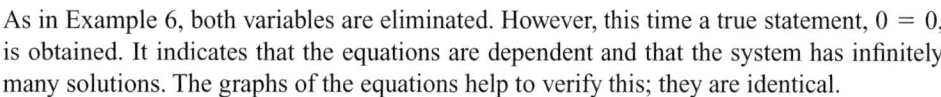

$$\begin{cases} \dfrac{2x - 5y}{15} = \dfrac{8}{15} \\ -0.2x + 0.5y = -0.8 \end{cases} \longrightarrow \begin{matrix} 15\left(\dfrac{2x - 5y}{15}\right) = 15\left(\dfrac{8}{15}\right) \\ 10(-0.2x + 0.5y) = 10(-0.8) \end{matrix} \longrightarrow \begin{cases} 2x - 5y = 8 \\ -2x + 5y = -8 \end{cases}$$

We add the resulting equations to get

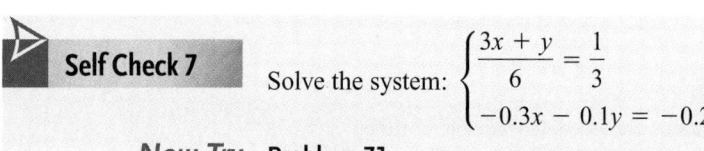

$$\begin{array}{rl} 2x - 5y = & 8 \quad \text{In the left column: } 2x + (-2x) = 0. \\ -2x + 5y = & -8 \quad \text{In the middle column: } -5y + 5y = 0. \\ \hline 0 = & 0 \quad \text{In the right column: } 8 + (-8) = 0. \end{array}$$

As in Example 6, both variables are eliminated. However, this time a true statement, $0 = 0$, is obtained. It indicates that the equations are dependent and that the system has infinitely many solutions. The graphs of the equations help to verify this; they are identical.

To find several of the infinitely many solutions, we can substitute some values of x, say -1, 4, and 9, in either equation and solve for y. The results are: $(-1, -2)$, $(4, 0)$, and $(9, 2)$.

Self Check 7   Solve the system: $\begin{cases} \dfrac{3x + y}{6} = \dfrac{1}{3} \\ -0.3x - 0.1y = -0.2 \end{cases}$

Now Try   **Problem 71**

---

5   **Determine the Most Efficient Method to Use to Solve a Linear System.**

If no method is specified for solving a particular linear system, the following guidelines can be helpful in determining whether to use graphing, substitution, or elimination.

1. If you want to show trends and see the point that the two graphs have in common, then use the **graphing method.** However, this method is not exact and can be lengthy.

2. If one of the equations is solved for one of the variables, or easily solved for one of the variables, use the **substitution method.**

3. If both equations are in standard $Ax + By = C$ form, and no variable has a coefficient of 1 or -1, use the **elimination method.**

4. If the coefficient of one of the variables is 1 or -1, you have a choice. You can write each equation in standard ($Ax + By = C$) form and use elimination, or you can solve for the variable with coefficient 1 or -1 and use substitution.

Here are some examples of suggested approaches:

$$\begin{cases} 2x + 3y = 1 \\ y = 4x - 3 \end{cases} \qquad \begin{cases} 5x + 3y = 9 \\ 8x + 4y = 3 \end{cases} \qquad \begin{cases} 4x - y = -6 \\ 3x + 2y = 1 \end{cases} \qquad \begin{cases} x - 23 = 6y \\ 7x - 9y = -3 \end{cases}$$

   Substitution   Elimination   Elimination   Substitution

Each method that we use to solve systems of equations has advantages and disadvantages.

Method	Advantages	Disadvantages
Graphing	• You see the solutions • The graphs allow you to observe trends	• Inaccurate when the solutions are not integers or are large numbers off the graph
Substitution	• Always gives the exact solutions • Works well if one of the equations is solved for one of the variables, or if it is easy to solve for one of the variables	• You do not see the solution • If no variable has a coefficient of 1 or -1, solving for one of the variables often involves fractions
Elimination	• Always gives the exact solutions • Works well if no variable has a coefficient of 1 or -1	• You do not see the solution • The equations must be written in the form $Ax + By = C$

▷ **ANSWERS TO SELF CHECKS** **1.** $(2, 4)$ **2.** $(-3, -3)$ **3.** $(0, 5)$ **4.** $(-5, 6)$ **5.** $\left(-\frac{22}{9}, \frac{10}{9}\right)$
6. No solution **7.** Infinitely many solutions

STUDY SET
4.3

VOCABULARY

Fill in the blanks.

1. The coefficients of $3x$ and $-3x$ are _____.

2. When the following equations are added, the variable y will be _____.

$$5x - 6y = 10$$
$$-3x + 6y = 24$$

CONCEPTS

3. In the following system, which terms have coefficients that are opposites?

$$\begin{cases} 3x + 7y = -25 \\ 4x - 7y = 12 \end{cases}$$

4. Fill in the blank. The objective of the elimination method is to obtain two equations whose sum will be one equation in one _____.

5. Add each pair of equations.
  a. $\begin{array}{r} 2a + 2b = -6 \\ 3a - 2b = 2 \\ \hline \end{array}$ **b.** $\begin{array}{r} x - 3y = 15 \\ -x - y = -14 \\ \hline \end{array}$

6. a. Multiply both sides of $4x + y = 2$ by 3.

  b. Multiply both sides of $x - 3y = 4$ by -2.

7. If the elimination method is used to solve

$$\begin{cases} 3x + 12y = 4 \\ 6x - 4y = 7 \end{cases}$$

  a. By what would we multiply the first equation to eliminate x?

  b. By what would we multiply the second equation to eliminate y?

8. Suppose the following system is solved using the elimination method and it is found that x is 2. Find the value of y.

$$\begin{cases} 4x + 3y = 11 \\ 3x - 2y = 4 \end{cases}$$

9. What algebraic step should be performed to

  a. Clear $\frac{2}{3}x + 4y = -\frac{4}{5}$ of fractions?

  b. Clear $0.2x - 0.9y = 6.4$ of decimals?

10. a. Suppose $0 = 0$ is obtained when a system is solved by the elimination method. Does the system have a solution? Which of the following is a possible graph of the system?

  b. Suppose $0 = 2$ is obtained when a system is solved by the elimination method. Does the system have a solution? Which of the following is a possible graph of the system?

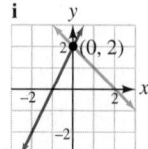
i

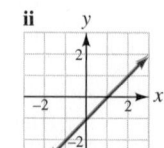
ii

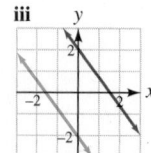
iii

NOTATION

Complete the solution to solve the system.

11. Solve: $\begin{cases} x + y = 5 \\ x - y = -3 \end{cases}$

$$\begin{aligned} x + y &= 5 \\ x - y &= -3 \\ \hline &= 2 \\ x &= \end{aligned}$$

$x + y = 5$ This is the first equation.

$ + y = 5$

$ y = 4$

The solution is (,).

12. Write each equation of the system in standard $Ax + By = C$ form:

$$\begin{cases} 7x + y + 3 = 0 \\ 8x + 4 = -y \end{cases} \rightarrow \begin{cases} \\ \end{cases}$$

GUIDED PRACTICE

Use the elimination method to solve each system. See Example 1.

13. $\begin{cases} x + y = 5 \\ x - y = 1 \end{cases}$

14. $\begin{cases} x - y = 4 \\ x + y = 8 \end{cases}$

15. $\begin{cases} x + y = 1 \\ x - y = 5 \end{cases}$

16. $\begin{cases} x - y = -5 \\ x + y = 1 \end{cases}$

17. $\begin{cases} x + y = -5 \\ -x + y = -1 \end{cases}$

18. $\begin{cases} -x + y = -3 \\ x + y = 1 \end{cases}$

19. $\begin{cases} 4x + 3y = 24 \\ 4x - 3y = -24 \end{cases}$

20. $\begin{cases} -9x + 5y = -9 \\ -9x - 5y = -9 \end{cases}$

21. $\begin{cases} 2s + t = -2 \\ -2s - 3t = -6 \end{cases}$

22. $\begin{cases} -2x + 4y = 12 \\ 2x + 4y = 28 \end{cases}$

23. $\begin{cases} 5x - 4y = 8 \\ -5x - 4y = 8 \end{cases}$

24. $\begin{cases} 2r + s = -8 \\ -2r + 4s = 28 \end{cases}$

Use the elimination method to solve each system. See Example 2.

25. $\begin{cases} x + 3y = -9 \\ x + 8y = -4 \end{cases}$

26. $\begin{cases} x + 7y = -22 \\ x + 9y = -24 \end{cases}$

27. $\begin{cases} 5c + 2d = -5 \\ 6c + 2d = -10 \end{cases}$

28. $\begin{cases} 11c + 3d = -68 \\ 10c + 3d = -64 \end{cases}$

29. $\begin{cases} 7x - y = 10 \\ 8x - y = 13 \end{cases}$

30. $\begin{cases} 6x - y = 4 \\ 9x - y = 10 \end{cases}$

31. $\begin{cases} 3a - b = -9 \\ 4a - b = -17 \end{cases}$

32. $\begin{cases} -7x - y = 22 \\ 4x - y = -44 \end{cases}$

Use the elimination method to solve each system. See Example 3.

33. $\begin{cases} 7x + 4y = 14 \\ 3x - 2y = -20 \end{cases}$

34. $\begin{cases} 5x - 14y = 32 \\ -x - 6y = 20 \end{cases}$

35. $\begin{cases} 7x - 50y = -43 \\ x + 3y = 4 \end{cases}$

36. $\begin{cases} x - 2y = -1 \\ 12x + 11y = 23 \end{cases}$

37. $\begin{cases} 9a + 16b = -36 \\ 7a + 4b = 48 \end{cases}$

38. $\begin{cases} 4a + 7b = -24 \\ 9a + b = 64 \end{cases}$

39. $\begin{cases} 8x + 12y = -22 \\ 3x - 2y = 8 \end{cases}$

40. $\begin{cases} 3x + 2y = 45 \\ 5x - 4y = 20 \end{cases}$

41. $\begin{cases} 6x + 5y + 29 = 0 \\ 2x = 3y - 5 \end{cases}$

42. $\begin{cases} 3x = 20y + 1 \\ 4x + 5y - 33 = 0 \end{cases}$

43. $\begin{cases} c = d - 9 \\ 5c = 3d - 35 \end{cases}$

44. $\begin{cases} a = b + 7 \\ 3a - 15 = 5b \end{cases}$

Use the elimination method to solve each system. See Example 4.

45. $\begin{cases} 4x + 3y = 7 \\ 3x - 2y = -16 \end{cases}$

46. $\begin{cases} 3x - 2y = 20 \\ 2x + 7y = 5 \end{cases}$

47. $\begin{cases} 5a + 8b = 2 \\ 11a - 3b = 25 \end{cases}$

48. $\begin{cases} 7a - 5b = 24 \\ 12a + 8b = 8 \end{cases}$

49. $\begin{cases} 2x + 11y = -10 \\ 5x + 4y = 22 \end{cases}$

50. $\begin{cases} 3x + 4y = 12 \\ 4x + 5y = 17 \end{cases}$

51. $\begin{cases} 7x = 21 - 6y \\ 4x + 5y = 12 \end{cases}$

52. $\begin{cases} -4x = -3y - 13 \\ -6x + 8y = -16 \end{cases}$

53. $\begin{cases} 4x - 7y + 32 = 0 \\ 5x = 4y - 2 \end{cases}$

54. $\begin{cases} 6x = -3y \\ 5x + 15 = 5y \end{cases}$

55. $\begin{cases} 9x + 21 = 3y \\ 4x = 7y + 19 \end{cases}$

56. $\begin{cases} 7x + 11 = 4y \\ 4x = 7y + 22 \end{cases}$

Use the elimination method to solve each system. See Example 5.

57. $\begin{cases} \dfrac{3}{5}s + \dfrac{4}{5}t = 1 \\ -\dfrac{1}{4}s + \dfrac{3}{8}t = 1 \end{cases}$

58. $\begin{cases} \dfrac{1}{2}x + \dfrac{4}{7}y = -1 \\ 5x - \dfrac{4}{5}y = -10 \end{cases}$

59. $\begin{cases} \dfrac{1}{2}s - \dfrac{1}{4}t = 1 \\ \dfrac{1}{3}s + t = 3 \end{cases}$

60. $\begin{cases} \dfrac{3}{5}x + y = 1 \\ \dfrac{4}{5}x - y = -1 \end{cases}$

61. $\begin{cases} x - \dfrac{4}{3}y = \dfrac{1}{3} \\ 2x + \dfrac{3}{2}y = \dfrac{1}{2} \end{cases}$

62. $\begin{cases} x + y = -\dfrac{1}{4} \\ x - \dfrac{y}{2} = -\dfrac{3}{2} \end{cases}$

63. $\begin{cases} 4a + 7b = 2 \\ 9a - 3b = 1 \end{cases}$

64. $\begin{cases} 5a - 7b = 6 \\ 7a - 6b = 8 \end{cases}$

Use the elimination method to solve each system. If there is no solution, or infinitely many solutions, so indicate. See Examples 6 and 7.

65. $\begin{cases} 3x - 5y = -29 \\ 3x - 5y = 15 \end{cases}$

66. $\begin{cases} 2a - 3b = -6 \\ 2a - 3b = 8 \end{cases}$

67. $\begin{cases} 3x - 16 = 5y \\ -3x + 5y - 33 = 0 \end{cases}$

68. $\begin{cases} 2x + 5y - 13 = 0 \\ -2x + 13 = 5y \end{cases}$

69. $\begin{cases} 0.4x - 0.7y = -1.9 \\ -x + \dfrac{7y}{4} = \dfrac{19}{4} \end{cases}$ **70.** $\begin{cases} 0.1x + 2y + 0.2 = 0 \\ -\dfrac{x}{4} - 5y = \dfrac{1}{2} \end{cases}$

71. $\begin{cases} \dfrac{x - 6y}{2} = 7 \\ -x + 6y + 14 = 0 \end{cases}$ **72.** $\begin{cases} \dfrac{-18x + y}{2} = \dfrac{7}{2} \\ 18x = y \end{cases}$

TRY IT YOURSELF

Solve the system by either the substitution or the elimination method, if possible.

73. $\begin{cases} y = -3x + 9 \\ y = x + 1 \end{cases}$ **74.** $\begin{cases} x = 5y - 4 \\ x = 9y - 8 \end{cases}$

75. $\begin{cases} 4x + 6y = 5 \\ 8x - 9y = 3 \end{cases}$ **76.** $\begin{cases} 3a + 4b = 36 \\ 6a - 2b = -21 \end{cases}$

77. $\begin{cases} 6x - 3y = -7 \\ y + 9x = 6 \end{cases}$ **78.** $\begin{cases} 9x + 4y = 31 \\ y - 5 = 6x \end{cases}$

79. $\begin{cases} 4x - 8y = 36 \\ 3x - 6y = 27 \end{cases}$ **80.** $\begin{cases} 2x + 4y = 15 \\ 3x = 8 - 6y \end{cases}$

81. $\begin{cases} x = y \\ 0.1x + 0.2y = 1.0 \end{cases}$ **82.** $\begin{cases} x = y \\ 0.4x - 0.8y = -0.5 \end{cases}$

83. $\begin{cases} 9x - 10y = 0 \\ \dfrac{9x - 3y}{63} = 1 \end{cases}$ **84.** $\begin{cases} 8x - 9y = 0 \\ \dfrac{2x - 3y}{6} = -1 \end{cases}$

85. $\begin{cases} \dfrac{m}{4} + \dfrac{n}{3} = -\dfrac{1}{12} \\ \dfrac{m}{2} - \dfrac{5}{4}n = \dfrac{7}{4} \end{cases}$ **86.** $\begin{cases} \dfrac{x}{2} - \dfrac{y}{3} = -2 \\ \dfrac{x}{3} + \dfrac{2}{3}y = \dfrac{4}{3} \end{cases}$

87. $\begin{cases} 3x + 12y = -12 \\ x = 3y + 10 \end{cases}$ **88.** $\begin{cases} 3x + 2y = 3 \\ y = 2x - 16 \end{cases}$

APPLICATIONS

89. EDUCATION   The graph shows educational trends during the years 1980–2004 for persons 25 years or older in the United States. The equation $9x + 11y = 352$ approximates the percent y that had less than high school completion. The equation $5x - 11y = -198$ approximates the percent y that had a Bachelor's or higher degree. In each case, x is the number of years since 1980. Use the elimination method to determine in what year the percents were equal.

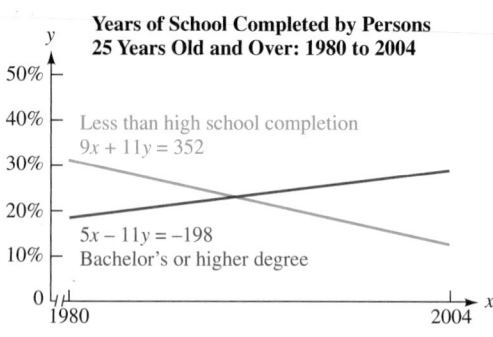

Years of School Completed by Persons 25 Years Old and Over: 1980 to 2004

Less than high school completion $9x + 11y = 352$

$5x - 11y = -198$ Bachelor's or higher degree

Source: U.S. Department of Commerce, Census Bureau

90. NEWSPAPERS   The graph shows the trends in the newspaper publishing industry during the years 1990–2004 in the United States. The equation $37x - 2y = -1,128$ models the number y of morning newspapers published and $31x + y = 1,059$ models the number y of evening newspapers published. In each case, x is the number of years since 1990. Use the elimination method to determine in what year there was an equal number of morning and evening newspapers being published.

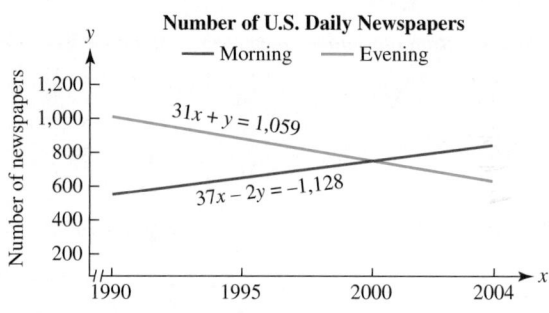

Number of U.S. Daily Newspapers

— Morning   — Evening

$31x + y = 1,059$

$37x - 2y = -1,128$

Source: Editor and Publisher Yearbook data

WRITING

91. Why is the method for solving systems that is discussed in this section called the *elimination method*?

92. If the elimination method is to be used to solve this system, what is wrong with the form in which it is written?

$$\begin{cases} 2x - 5y = -3 \\ -2y + 5x = 10 \end{cases}$$

93. Can the system $\begin{cases} 2x + 5y = 13 \\ -2x - 3y = -5 \end{cases}$ be solved more easily using the elimination method or the substitution method? Explain.

94. Explain the error in the following work.

Solve: $\begin{cases} x + y = 1 \\ x - y = 5 \end{cases}$

$$\begin{array}{r} x + y = 1 \\ +x - y = 5 \\ \hline 2x \quad\quad = 6 \end{array}$$

$$\frac{2x}{2} = \frac{6}{2}$$

$$\boxed{x = 3}$$

The solution is 3.

REVIEW

95. Find an equation of the line with slope $-\frac{11}{6}$ that passes through $(2, -6)$. Write the equation in slope–intercept form.

96. Solve $S = 2\pi rh + 2\pi r^2$ for h.

97. Evaluate: $-10(18 - 4^2)^3$

98. Evaluate: -5^2

CHALLENGE PROBLEMS

Use the elimination method to solve each system.

99. $\begin{cases} \dfrac{x - 3}{2} = \dfrac{11}{6} - \dfrac{y + 5}{3} \\ \dfrac{x + 3}{3} - \dfrac{5}{12} = \dfrac{y + 3}{4} \end{cases}$

100. $\begin{cases} 4(x + 1) = 17 - 3(y - 1) \\ 2(x + 2) + 3(y - 1) = 9 \end{cases}$

SECTION 4.4
Problem Solving Using Systems of Equations

 WORD PROBLEMS

Objectives

1. Assign variables to two unknowns.
2. Use systems to solve geometry problems.
3. Use systems to solve number-value problems.
4. Use systems to solve interest, uniform motion, and mixture problems.

In previous chapters, many applied problems were modeled and solved with an equation in one variable. In this section, the application problems involve two unknowns. It is often easier to solve such problems using a two-variable approach.

 Assign Variables to Two Unknowns.

The following steps are helpful when solving problems involving two unknown quantities.

Problem-Solving Strategy	
	1. **Analyze the problem** by reading it carefully to understand the given facts. Often a diagram or table will help you visualize the facts of the problem.
	2. Pick different variables to represent two unknown quantities. Translate the words of the problem to **form two equations** involving each of the two variables.
	3. **Solve the system** of equations using graphing, substitution, or elimination.
	4. **State the conclusion.**
	5. **Check the results** in the words of the problem.

| EXAMPLE 1 | *Motion Pictures.* Each year, Academy Award winners are presented with Oscars. The 13.5-inch statuette has a base on which a gold-plated figure stands. The figure itself is 7.5 inches taller than its base. Find the height of the figure and the height of the base. |

Analyze the Problem

- The statuette is a total of 13.5 inches tall.
- The figure is 7.5 inches taller than the base.
- Find the height of the figure and the height of the base.

Form Two Equations

Let x = the height of the figure, in inches, and y = the height of the base, in inches. We can translate the words of the problem into two equations, each involving x and y.

The height of the figure	plus	the height of the base	is	13.5 inches.
x	$+$	y	$=$	13.5

> **Caution**
> If two variables are used to represent two unknown quantities, we must form a system of two equations to find the unknowns.

The height of the figure	is	the height of the base	plus	7.5 inches.
x	$=$	y	$+$	7.5

The resulting system is: $\begin{cases} x + y = 13.5 \\ x = y + 7.5 \end{cases}$

Solve the System

Since the second equation is solved for x, we will use substitution to solve the system.

$$\begin{cases} x + y = 13.5 \\ x = y + 7.5 \end{cases}$$

$x + y = 13.5$ This is the first equation of the system.

$y + 7.5 + y = 13.5$ Substitute $y + 7.5$ for x.

$2y + 7.5 = 13.5$ Combine like terms: $y + y = 2y$.

$2y = 6$ Subtract 7.5 from both sides.

$y = 3$ Divide both sides by 2. This is the height of the base.

To find x, substitute 3 for y in the second equation of the system.

$x = y + 7.5$ This is the substitution equation.

$x = 3 + 7.5$ Substitute 3 for y.

$x = 10.5$ This is the height of the figure.

> **Caution**
> In this problem we are to find two unknowns, the height of the figure and height of the base. Remember to give both in the *State the Conclusion* step of the solution.

State the Conclusion

The height of the figure is 10.5 inches and the height of the base is 3 inches.

Check the Results

The sum of 10.5 inches and 3 inches is 13.5 inches, and the 10.5-inch figure is 7.5 inches taller than the 3-inch base. The results check.

Now Try **Problem 17**

② Use Systems to Solve Geometry Problems.

Two angles are said to be **complementary** if the sum of their measures is 90°. Two angles are said to be **supplementary** if the sum of their measures is 180°.

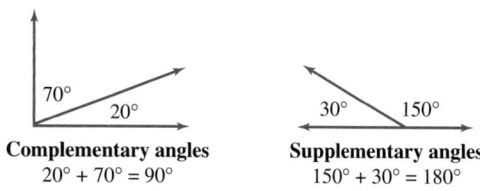

Complementary angles
$20° + 70° = 90°$

Supplementary angles
$150° + 30° = 180°$

EXAMPLE 2   *Angles.*   The difference of the measures of two complementary angles is 6°. Find the measure of each angle.

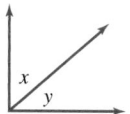

Analyze the Problem

- Since the angles are complementary, the sum of their measures is 90°.
- The word *difference* indicates subtraction. If the measure of the smaller angle is subtracted from the measure of the larger angle, the result will be 6°.
- Find the measure of the larger angle and the measure of the smaller angle.

Form Two Equations
Let x = the measure of the larger angle and y = the measure of the smaller angle. We can translate the words of the problem into two equations, each involving x and y.

The measure of the larger angle	plus	the measure of the smaller angle	is	90°.
x	$+$	y	$=$	90

The measure of the larger angle	minus	the measure of the smaller angle	is	6°.
x	$-$	y	$=$	6

The resulting system is: $\begin{cases} x + y = 90 \\ x - y = 6 \end{cases}$

Solve the System
Since the coefficients of y are opposites, we will use elimination to solve the system.

$$x + y = 90$$
$$\underline{x - y = 6} \quad \text{Add the equations to eliminate } y.$$
$$2x \quad = 96$$
$$x = 48 \quad \text{Divide both sides by 2. This is the measure of the larger angle.}$$

To find y, substitute 48 for x in the first equation of the system.

$$x + y = 90$$

$$48 + y = 90 \quad \text{Substitute 48 for x.}$$

$$y = 42 \quad \text{Subtract 48 from both sides. This is the measure of the smaller angle.}$$

State the Conclusion The measure of the larger angle is 48° and the measure of the smaller angle is 42°.

Check the Results The sum of 48° and 42° is 90°, and the difference is 6°. The results check.

 Now Try Problem 13

EXAMPLE 3 *History.* In 1917, James Montgomery Flagg created the classic *I Want You* poster to help recruiting for World War I. The perimeter of the poster is 114 inches, and its length is 9 inches less than twice its width. Find the length and the width of the poster.

Library of Congress LC-USZC4-3859

Analyze the Problem

- The perimeter of the rectangular poster is 114 inches.
- The length is 9 inches less than twice the width.
- Find the length and the width of the poster.

Form Two Equations Let l = the length of the poster, in inches, and w = the width of the poster, in inches. The perimeter of a rectangle is the sum of two lengths and two widths, as given by the formula $P = 2l + 2w$, so we have

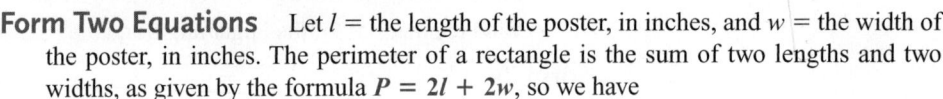

2	times	the length of the poster	plus	2	times	the width of the poster	is	114 inches.
2	·	l	+	2	·	w	=	114

If the length of the poster is 9 inches less than twice the width, we have

The length of the poster	is	2	times	the width of the poster	minus	9 inches.
l	=	2	·	w	−	9

The resulting system is: $\begin{cases} 2l + 2w = 114 \\ l = 2w - 9 \end{cases}$

Solve the System Since the second equation is solved for l, we will use substitution to solve the system.

$$2l + 2w = 114 \quad \text{This is the first equation of the system.}$$

$$2(2w - 9) + 2w = 114 \quad \text{Substitute } 2w - 9 \text{ for } l. \text{ Don't forget the parentheses.}$$

$$4w - 18 + 2w = 114 \quad \text{Distribute the multiplication by 2.}$$

$$6w - 18 = 114 \quad \text{Combine like terms: } 4w + 2w = 6w.$$

$$6w = 132 \quad \text{Add 18 to both sides.}$$

$$w = 22 \quad \text{Divide both sides by 6. This is the width of the poster.}$$

To find l, substitute 22 for w in the second equation of the system.

$$l = 2w - 9$$
$$l = 2(22) - 9$$
$$l = 44 - 9$$
$$l = 35$$ This is the length of the poster.

State the Conclusion   The length of the poster is 35 inches and the width is 22 inches.

Check the Results   The perimeter is $2(35) + 2(22) = 70 + 44 = 114$ inches, and 35 inches is 9 inches less than twice 22 inches. The results check.

 Now Try   **Problem 23**

3   **Use Systems to Solve Number-Value Problems.**

EXAMPLE 4   *Photography.*   At a school, two picture packages are available, as shown in the illustration. Find the cost of a class picture and the cost of an individual wallet-size picture.

Analyze the Problem

- Package 1 contains 1 class picture and 10 wallet-size pictures.

- Package 2 contains 2 class pictures and 15 wallet-size pictures.

- Find the cost of a class picture and the cost of a wallet-size picture.

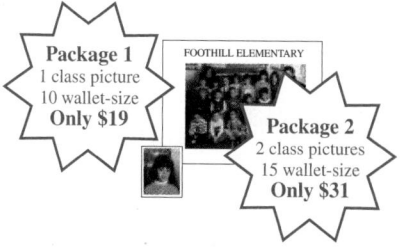

Form Two Equations   Let c = the cost of one class picture and w = the cost of one wallet-size picture. We can use the fact that **Number · value = total value** to write an equation that models the first package. We note that (in dollars) the cost of 1 class picture is $1 \cdot c = c$ and the cost of 10 wallet-size pictures is $10 \cdot w = 10w$.

The cost of 1 class picture	plus	the cost of 10 wallet-size pictures	is	$19.
c	$+$	$10w$	$=$	19

To write an equation that models the second package, we note that (in dollars) the cost of 2 class pictures is $2 \cdot c = 2c$, and the cost of 15 wallet-size pictures is $15 \cdot w = 15w$.

The cost of 2 class pictures	plus	the cost of 15 wallet-size pictures	is	$31.
$2c$	$+$	$15w$	$=$	31

The resulting system is: $\begin{cases} c + 10w = 19 \\ 2c + 15w = 31 \end{cases}$

Solve the System   We can use elimination to solve this system. To eliminate c, we proceed as follows.

$$-2c - 20w = -38 \quad \text{Multiply both sides of } c + 10w = 19 \text{ by } -2.$$
$$\underline{2c + 15w = 31}$$
$$-5w = -7 \quad \text{Add the equations to eliminate } c.$$
$$w = 1.4 \quad \text{Divide both sides by } -5. \text{ This is the cost of a wallet-size picture.}$$

To find c, substitute 1.4 for w in the first equation of the original system.

$$c + 10w = 19$$
$$c + 10(1.4) = 19 \quad \text{Substitute 1.4 for } w.$$
$$c + 14 = 19 \quad \text{Multiply.}$$
$$c = 5 \quad \text{Subtract 14 from both sides. This is the cost of a class picture.}$$

State the Conclusion   A class picture costs $5 and a wallet-size picture costs $1.40.

Check the Results   Package 1 has 1 class picture and 10 wallets: $5 + 10($1.40) = $5 + $14 = $19. Package 2 has 2 class pictures and 15 wallets: 2($5) + 15($1.40) = $10 + $21 = $31. The results check.

 Now Try   **Problem 27**

4 Use Systems to Solve Interest, Uniform Motion, and Mixture Problems.

EXAMPLE 5   *White-Collar Crime.*   Investigators discovered that a small business secretly moved $150,000 out of the country to avoid paying income tax. Some of the money was invested in a Swiss bank account that paid 8% interest annually. The remainder was deposited in a Cayman Islands account, paying 7% annual interest. The investigation also revealed that the combined interest earned the first year was $11,500. How much money was invested in each account?

Analyze the Problem   We are told that an unknown part of the $150,000 was invested at an annual rate of 8% and the rest at 7%. Together, the accounts earned $11,500 in interest.

Form Two Equations   Let x = the amount invested in the Swiss account and y = the amount invested in the Cayman Islands account. Because the total investment was $150,000, we have

The amount invested in the Swiss account	plus	the amount invested in the Cayman Islands account	is	$150,000
x	$+$	y	$=$	150,000

We can use the formula $I = Prt$ to determine that x dollars invested for 1 year at 8% earns $x \cdot 0.08 \cdot 1 = 0.08x$ dollars. Similarly, y dollars invested for 1 year at 7% earns $y \cdot 0.07 \cdot 1 = 0.07y$ dollars. If the total combined interest earned was $11,500, we have

The income on the 8% investment	plus	the income on the 7% investment	is	$11,500.
$0.08x$	$+$	$0.07y$	$=$	11,500

The resulting system is: $\begin{cases} x + y = 150,000 \\ 0.08x + 0.07y = 11,500 \end{cases}$

Solve the System   To solve the system, clear the second equation of decimals. Then eliminate x.

Caution
It is incorrect to let

x = the amount invested in each account

This implies that *equal amounts* were invested in the Swiss and Cayman Island accounts. We do not know that.

$$-8x - 8y = -1,200,000 \qquad \text{Multiply both sides of } x + y = 150,000 \text{ by } -8.$$
$$\underline{8x + 7y = 1,150,000} \qquad \text{Multiply both sides of } 0.08x + 0.07y = 11,500 \text{ by } 100.$$
$$-y = -50,000$$
$$y = 50,000 \qquad \text{Multiply both sides by } -1.$$

To find x, substitute 50,000 for y in the first equation of the original system.

$$x + y = 150,000$$
$$x + \mathbf{50,000} = 150,000 \qquad \text{Substitute 50,000 for } y.$$
$$x = 100,000 \qquad \text{Subtract 50,000 from both sides.}$$

State the Conclusion $100,000 was invested in the Swiss bank account, and $50,000 was invested in the Cayman Islands account.

Check the Results

$$\$100,000 + \$50,000 = \$150,000 \qquad \text{The two investments total \$150,000.}$$
$$0.08(\$100,000) = \$8,000 \qquad \text{The Swiss bank account earned \$8,000.}$$
$$0.07(\$50,000) = \$3,500 \qquad \text{The Cayman Islands account earned \$3,500.}$$

The combined interest is $8,000 + $3,500 = $11,500. The results check.

 Now Try   **Problem 35**

EXAMPLE 6   *Boating.*   A boat traveled 30 miles downstream in 3 hours and made the return trip in 5 hours. Find the speed of the boat in still water and the speed of the current.

Analyze the Problem   Traveling downstream, the speed of the boat will be faster than it would be in still water. Traveling upstream, the speed of the boat will be slower than it would be in still water.

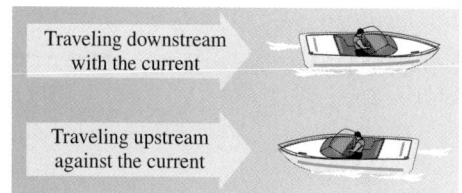

Traveling downstream with the current

Traveling upstream against the current

Form Two Equations   Let s = the speed of the boat in still water and c = the speed of the current. Then the speed of the boat going downstream is $s + c$ and the speed of the boat going upstream is $s - c$. Using the formula $d = rt$, we find that $3(s + c)$ represents the distance traveled downstream and $5(s - c)$ represents the distance traveled upstream. We can organize the facts of the problem in a table.

	Rate ·	Time =	Distance
Downstream	$s + c$	3	$3(s + c)$
Upstream	$s - c$	5	$5(s - c)$

Enter this information first.

↑ Set each of these expressions for distance traveled equal to 30.

Since each trip is 30 miles long, the Distance column of the table helps us to write two equations in two variables. To write each equation in standard form, use the distributive property.

$$\begin{cases} 3(s + c) = 30 \\ 5(s - c) = 30 \end{cases} \xrightarrow[\text{Distribute}]{\text{Distribute}} \begin{cases} 3s + 3c = 30 \\ 5s - 5c = 30 \end{cases}$$

Solve the System To eliminate c, we proceed as follows.

$$\begin{array}{ll} 15s + 15c = 150 & \text{Multiply both sides of } 3s + 3c = 30 \text{ by 5.} \\ \underline{15s - 15c = 90} & \text{Multiply both sides of } 5s - 5c = 30 \text{ by 3.} \\ 30s \qquad\quad = 240 & \\ \qquad\quad s = 8 & \text{Divide both sides by 30. This is the speed of the boat in still water.} \end{array}$$

To find c, it appears that the computations will be easiest if we use $3s + 3c = 30$.

$$\begin{array}{ll} 3s + 3c = 30 & \\ 3(8) + 3c = 30 & \text{Substitute 8 for } s. \\ 24 + 3c = 30 & \text{Multiply.} \\ 3c = 6 & \text{Subtract 24 from both sides.} \\ c = 2 & \text{Divide both sides by 3. This is the speed of the current.} \end{array}$$

State the Conclusion The speed of the boat in still water is 8 mph and the speed of the current is 2 mph.

Check the Results With a 2-mph current, the boat's downstream speed will be $8 + 2 = 10$ mph. In 3 hours, it will travel $10 \cdot 3 = 30$ miles. With a 2-mph current, the boat's upstream speed will be $8 - 2 = 6$ mph. In 5 hours, it will cover $6 \cdot 5 = 30$ miles. The results check.

 Now Try **Problem 41**

EXAMPLE 7 *Medical Technology.* A laboratory technician has one batch of antiseptic that is 40% alcohol and a second batch that is 60% alcohol. She would like to make 8 fluid ounces of solution that is 55% alcohol. How many fluid ounces of each batch should she use?

Analyze the Problem Some 60% solution must be added to some 40% solution to make a 55% solution.

Form Two Equations Let $x =$ the number of ounces to be used from batch 1 and $y =$ the number of ounces to be used from batch 2. The amount of alcohol in each solution is given by

$$\begin{array}{c} \text{Amount of} \\ \text{solution} \end{array} \cdot \begin{array}{c} \text{strength of} \\ \text{solution} \end{array} = \begin{array}{c} \text{amount of} \\ \text{alcohol} \end{array}$$

We can organize the facts of the problem in a table.

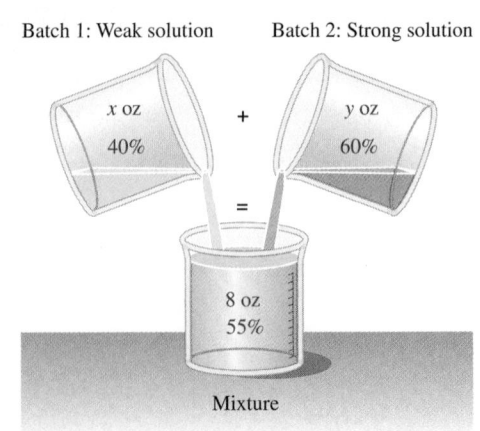

	Amount · Strength = Amount of alcohol		
Batch 1 (too weak)	x	0.40	$0.40x$
Batch 2 (too strong)	y	0.60	$0.60y$
Mixture	8	0.55	$0.55(8)$

One equation comes from information in this column.

40%, 60%, and 55% have been expressed as decimals.

Another equation comes from information in this column.

The information in the table provides two equations.

$$\begin{cases} x + y = 8 \\ 0.40x + 0.60y = 0.55(8) \end{cases}$$

The number of ounces of batch 1 plus the number of ounces of batch 2 equals the total number of ounces in the mixture.

The amount of alcohol in batch 1 plus the amount of alcohol in batch 2 equals the amount of alcohol in the mixture.

Solve the System   We can solve this system by elimination. To eliminate x, we proceed as follows.

$$\begin{array}{rl} -40x - 40y = -320 & \text{Multiply both sides of the first equation by } -40. \\ \underline{40x + 60y = 440} & \text{Multiply both sides of the second equation by 100.} \\ 20y = 120 & \\ y = 6 & \text{Divide both sides by 20. This is the number of fluid ounces of batch 2 needed.} \end{array}$$

To find x, we substitute 6 for y in the first equation of the original system.

$$\begin{array}{rl} x + y = 8 & \\ x + 6 = 8 & \text{Substitute.} \\ x = 2 & \text{Subtract 6 from both sides. This is the number of fluid ounces of batch 1 needed.} \end{array}$$

State the Conclusion   The technician should use 2 fluid ounces of the 40% solution and 6 fluid ounces of the 60% solution.

Check the Results   Note that 2 ounces + 6 ounces = 8 ounces, the required number. Also, the amount of alcohol in the two solutions is equal to the amount of alcohol in the mixture.

Alcohol in batch 1: $0.40x = 0.40(2) = 0.8$ ounces
Alcohol in batch 2: $0.60y = 0.60(6) = 3.6$ ounces $\Big\}$ Total: 4.4 ounces
Alcohol in the mixture: $0.55(8) = 4.4$ ounces

The results check.

 *Now Try*   **Problem 45**

EXAMPLE 8   *Breakfast Cereal.*   One ounce of raisins (by weight) sells for 22¢ and one ounce of bran flakes (by weight) sells for 12¢. How many ounces of each should be used to create a 20-ounce box of raisin bran cereal that can be sold for 15¢ an ounce?

Analyze the Problem   We will use a two-variable approach to solve this dry mixture problem.

Form Two Equations   Let x = the number of ounces of raisins and y = the number of ounces of bran flakes that should be mixed. The value of the mixture and the value of each of its components is given by

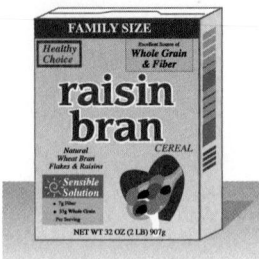

$$\text{Amount} \cdot \text{price} = \text{total value}$$

Thus, the value of x ounces of raisins is $x \cdot 22¢$ or $22x¢$ and the value of y ounces of bran flakes is $y \cdot 12¢$ or $12y¢$. The sum of these values is also equal to the total value of the final mixture, that is $20 \cdot 15¢$ or $300¢$. This information is shown in the table.

	Amount	· Price	=Total value
Raisins	x	22	$22x$
Bran flakes	y	12	$12y$
Mixture	20	15	20(15)

The facts of the problem give the following two equations:

The number of ounces of raisins	plus	the number of ounces of bran flakes	is	20.
x	$+$	y	$=$	20

The value of the raisins	plus	the value of the bran flakes	is	the value of the mixture.
$22x$	$+$	$12y$	$=$	20(15)

Solve the System   To find out how many ounces of raisins and bran flakes are needed we solve the following system:

$$\begin{cases} x + y = 20 \\ 22x + 12y = 300 \quad \text{Multiply: 20(15) = 300.} \end{cases}$$

To solve this system by substitution, we can solve the first equation for x:

$$x + y = 20$$
$$x = 20 - y \quad \text{This is the substitution equation.}$$

Then we substitute $20 - y$ for x in the second equation of the system and solve for y.

$$22x + 12y = 300$$
$$22(20 - y) + 12y = 300 \quad \text{Substitute } 20 - y \text{ for } x.$$
$$440 - 22y + 12y = 300 \quad \text{Distribute the multiplication by 22.}$$
$$440 - 10y = 300 \quad \text{Combine like terms: } -22y + 12y = -10y.$$
$$-10y = -140 \quad \text{Subtract 440 from both sides.}$$
$$y = 14 \quad \text{Divide both sides by } -10. \text{ This is the number of ounces of bran flakes needed.}$$

To find x, we substitute 14 for y in the substitution equation and simplify the right side.

$$x = 20 - y$$
$$= 20 - 14 \quad \text{Substitute 14 for } y.$$
$$= 6 \quad \text{This is the number of ounces of raisins needed.}$$

State the Conclusion To obtain 20 ounces of raisin bran cereal, 6 ounces of raisins and 14 ounces of bran flakes should be combined.

Check the Results When 6 ounces of raisins and 14 ounces of bran flakes are combined, the result is 20 ounces of raisin bran cereal. The 6 ounces of raisins are valued at $6 \cdot 22\cent = 132\cent$ and the 14 ounces of bran flakes are valued at $14 \cdot 12\cent = 168\cent$. The sum of those values, $132\cent + 168\cent = 300\cent$, is the same as the value of the mixture, $20 \cdot 15\cent = 300\cent$. The results check.

Now Try **Problem 49**

STUDY SET
4.4

VOCABULARY

Fill in the blanks.

1. Two angles are said to be _____ if the sum of their measures is 90°. Two angles are said to be _____ if the sum of their measures is 180°.

2. Problems that involve moving vehicles are called uniform _____ problems. Problems that involve combining ingredients are called _____ problems. Problems that involve collections of different items having different values are called number-_____ problems.

CONCEPTS

3. A length of pipe is to be cut into two pieces. The longer piece is to be 1 foot less than twice the shorter piece. Write two equations that model the situation.

20 ft

4. Two angles are complementary. The measure of the larger angle is four times the measure of the smaller angle. Write two equations that model the situation.

5. Two angles are supplementary. The measure of the smaller angle is 25° less than the measure of the larger angle. Write two equations that model the situation.

6. The perimeter of the following Ping-Pong table is 28 feet. The length is 4 feet more than the width. Write two equations that model the situation.

7. Let x = the cost of a chicken taco, in dollars, and y = the cost of a beef taco, in dollars. Write an equation that models the offer shown in the advertisement.

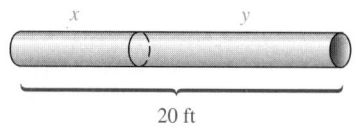

TUESDAY TACO SPECIAL

5 CHICKEN TACOS   *2 BEEF TACOS*

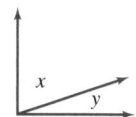

only $10

8. a. Complete the following table.

	Principal ·	Rate ·	Time =	Interest
City Bank	x	5%	1 yr	
USA Savings	y	11%	1 yr	

b. A total of $50,000 was deposited in the two accounts. Use that information to write an equation about the principal.

c. A total of $4,300 was earned by the two accounts. Use that information to write an equation about the interest.

9. For each case below, write an algebraic expression that represents the speed of the canoe in miles per hour if its speed in still water is x mph.

Downstream

Current
c mph

Upstream

Current
c mph

10. Complete the table, which contains information about an airplane flying in windy conditions.

	Rate · Time = Distance		
With wind	$x + y$	3	
Against wind	$x - y$	5	

11. a. If the contents of the two test tubes are poured into a third tube, how much solution will the third tube contain? (mL stands for milliliter. A milliliter is about 15 drops from an eyedropper.)

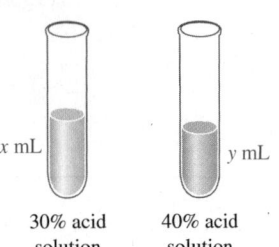

x mL y mL

30% acid
solution   40% acid
solution

b. Which of the following strengths could the mixture possibly be: 27%, 33%, or 44% acid solution?

12. a. Complete the table, which contains information about mixing two salt solutions to get 12 gallons of a 3% salt solution.

	Amount · Strength = Amount of salt		
Weak	x	0.01	
Strong	y	0.06	
Mix			

b. Use the information from the Amount column to write an equation.

c. Use the information from the Amount of salt column to write an equation.

GUIDED PRACTICE

See Example 2.

13. COMPLEMENTARY ANGLES   Two angles are complementary. The measure of one angle is 10° more than three times the measure of the other. Find the measure of each angle.

14. SUPPLEMENTARY ANGLES   Two angles are supplementary. The measure of one angle is 20° less than 19 times the measure of the other. Find the measure of each angle.

15. SUPPLEMENTARY ANGLES   The difference of the measures of two supplementary angles is 80°. Find the measure of each angle.

16. COMPLEMENTARY ANGLES   Two angles are complementary. The measure of one angle is 15° more than one-half of the measure of the other. Find the measure of each angle.

APPLICATIONS

Write a system of two equations in two variables to solve each problem.

17. TREE TRIMMING
When fully extended, the arm on a tree service truck is 51 feet long. If the upper part of the arm is 7 feet shorter than the lower part, how long is each part of the arm?

Upper part

Lower part

TREE SERVICE

18. ALASKA   Most of the 1,422-mile-long Alaskan Highway is actually in Canada. Find the length of the highway that is in Alaska and the length of the highway that is in Canada if it is known that the difference in the lengths is 1,020 miles.

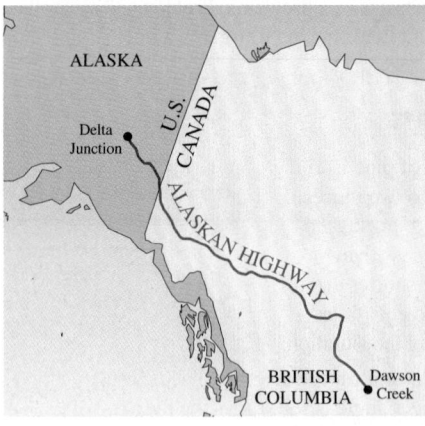

ALASKA

Delta
Junction

U.S.
CANADA

ALASKAN HIGHWAY

BRITISH   Dawson
COLUMBIA   Creek

19. GOVERNMENT   The salaries of the president and vice president of the United States total $608,100 a year. If the president makes $191,900 more than the vice president, find each of their salaries.

20. CAUSES OF DEATH   According to the *National Vital Statistics Reports,* in 2004, the number of Americans who died from heart disease was about 6 times the number who died from accidents. If the total number of deaths from these two causes was approximately 763,000, how many Americans died from each cause in 2004?

21. MONUMENTS The Marine Corps War Memorial in Arlington, Virginia, portrays the raising of the U.S. flag on Iwo Jima during World War II. Find the measures of the two angles shown if the measure of ∠2 is 15° less than twice the measure of ∠1.

Angle 1   Angle 2

22. PHYSICAL THERAPY To rehabilitate her knee, an athlete does leg extensions. Her goal is to regain a full 90° range of motion in this exercise. Use the information in the illustration to determine her current range of motion in degrees and the number of degrees of improvement she still needs to make.

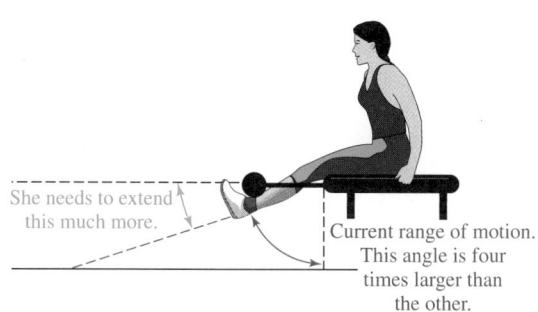

She needs to extend this much more.

Current range of motion. This angle is four times larger than the other.

23. THEATER SCREENS At an IMAX theater, the giant rectangular movie screen has a width 26 feet less than its length. If its perimeter is 332 feet, find the length and the width of the screen.

24. ART In 1770, Thomas Gainsborough painted *The Blue Boy*. The sum of the length and width of the painting is 118 inches. The difference of the length and width is 22 inches. Find the length and width.

© Huntington Library/SuperStock

25. GEOMETRY A 50-meter path surrounds a rectangular garden. The width of the garden is two-thirds its length. Find the length and width.

26. BALLROOM DANCING A rectangular-shaped dance floor has a perimeter of 200 feet. If the floor were 20 feet wider, its width would equal its length. Find the length and width of the dance floor.

27. EMPTY CARTRIDGES A bank recycles its empty printer and copier cartridges. In January, the bank received $40 for recycling 5 printer and 2 copier cartridges. In February, the bank received $57 for recycling 6 printer and 3 copier cartridges. How much is the bank paid for an empty printer cartridge and for an empty copier cartridge?

28. THANKSGIVING DINNER There are a total of 510 calories in 6 ounces of turkey and one slice of pumpkin pie. There are a total of 580 calories in 4 ounces of turkey and two slices of pumpkin pie. How many calories are there in 1 ounce of turkey and in one slice of pumpkin pie?

29. *from Campus to Careers*
            Portrait Photographer

Suppose you are a wedding photographer and you sell:
Package 1: one 10 × 14 and ten 8 × 10 color photos for $239.50
Package 2: one 10 × 14 and five 8 × 10 color photos for $134.50
A newlywed couple buys Package 1 and decides that they want one more 10 × 14 and one more 8 × 10 photograph. At the same prices, what should you charge them for each additional photograph?

© Peter Steiner/Alamy

30. BUYING PAINTING SUPPLIES Two partial receipts for paint supplies are shown. (Assume no sales tax was charged.) Find the cost of one gallon of paint and the cost of one paint brush.

VISTA PAINTS
Dec. 10, 2004

8 gallons latex paint @
3 brushes @
Total $ 270.00

VISTA PAINTS
Dec. 12, 2004

6 gallons latex paint @
2 brushes @
Total $ 200.00

31. COLLECTING STAMPS   Determine the price of an Elvis Presley stamp and a Statue of Liberty stamp given the following information.

 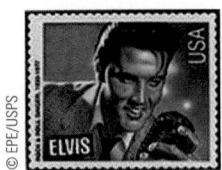

- One Elvis stamp and one Liberty stamp cost a total of 63¢.
- A sheet of 40 Elvis stamps and a sheet of 20 Liberty stamps cost a total of $18.40.   (*Hint:* 63¢ = $0.63)

32. RECYCLING   A boy scout troop earned $24 by recycling a total of 330 beverage containers. The recycling rates are shown below. How many of the small capacity containers and how many of the large capacity containers did they recycle? (*Hint:* 5¢ = $.05 and 10¢ = $0.10)

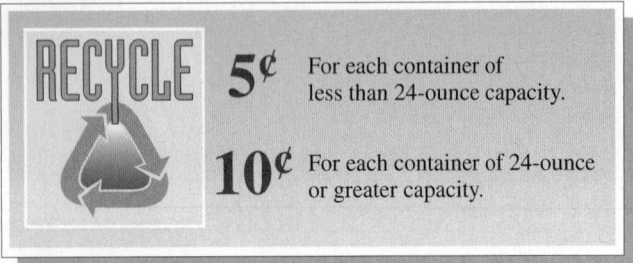

RECYCLE

5¢   For each container of less than 24-ounce capacity.

10¢   For each container of 24-ounce or greater capacity.

33. SELLING ICE CREAM   At a store, ice cream cones cost $1.80 and sundaes cost $3.30. One day the receipts for a total of 148 cones and sundaes were $360.90. How many cones were sold? How many sundaes?

34. BUYING TICKETS   The ticket prices for a movie are shown below. Receipts for one showing were $1,740 for an audience of 190 people. How many general admission tickets and how many senior citizen tickets were sold?

TICKETS
General Admission: $10
Seniors: $6
Showtimes: 5, 8, 11

35. STUDENT LOANS   A college used a $5,000 gift from an alumnus to make two student loans. The first was at 5% annual interest to a nursing student. The second was at 7% to a business major. If the college collected $310 in interest the first year, how much was loaned to each student?

36. FINANCIAL PLANNING   In investing $6,000 of a couple's money, a financial planner put some of it into a savings account paying 6% annual interest. The rest was invested in a riskier mini-mall development plan paying 12% annually. The combined interest earned for the first year was $540. How much money was invested at each rate?

37. INVESTING A BONUS   A businessman invested part of his $40,000 end-of-the-year bonus in an international fund that paid an annual yield of 8%. The rest of the bonus was invested in an offshore bank that paid an annual yield of 9%. Find the amount of each investment if he made a total of $3,415 in interest from them the first year.

38. PENSION FUNDS   A state employees' pension fund invested a total of one million dollars in two accounts that earned 3.5% and 4.5% annual interest. At the end of the year, the total interest earned from the two investments was $39,000. How much was invested at each rate?

39. LOSSES   A CEO deposited part of $22,000 in an account paying 4% interest annually. The rest of the money was invested in a biotech company that, after only one year, caused him to lose 3% of his initial investment in it. Find the amount of each investment if the net interest he earned the first year was only $110.

40. LOTTERY WINNINGS   After winning $60,000 in the lottery, a retired teacher gave $10,000 of it to her grandchildren. She invested part of the remainder in a growth fund that earned 4.4% annually and the rest in certificates of deposit paying a 5.8% annual percentage yield. The interest that she received on these two investments totaled $2,732 at the end of the first year. Find the amount of each investment.

41. THE GULF STREAM   The Gulf stream is a warm ocean current of the North Atlantic Ocean that flows northward, as shown below. Heading north with the Gulf Stream, a cruise ship traveled 300 miles in 10 hours. Against the current, it took 15 hours to make the return trip. Find the speed of the ship in still water and the speed of the current.

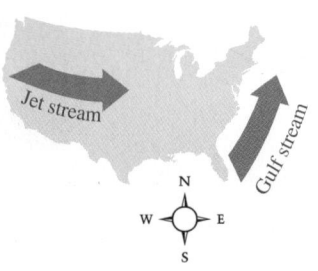

42. THE JET STREAM   The jet stream is a strong wind current that flows across the United States, as shown on the previous page. Flying with the jet stream, a plane flew 3,000 miles in 5 hours. Against the same wind, the trip took 6 hours. Find the speed of the plane in still air and the speed of the wind current.

43. AVIATION An airplane can fly with the wind a distance of 800 miles in 4 hours. However, the return trip against the wind takes 5 hours. Find the speed of the plane in still air and the speed of the wind.

44. BOATING   A boat can travel 24 miles downstream in 2 hours and can make the return trip in 3 hours. Find the speed of the boat in still water and the speed of the current.

45. MARINE BIOLOGY   A marine biologist wants to set up an aquarium containing 3% salt water. He has two tanks on hand that contain 6% and 2% salt water. How much water from each tank must he use to fill a 32-gallon aquarium with a 3% saltwater mixture?

46. COMMEMORATIVE COINS   A foundry has been commissioned to make souvenir coins. The coins are to be made from an alloy that is 40% silver. The foundry has on hand two alloys, one with 50% silver content and one with a 25% silver content. How many kilograms of each alloy should be used to make 20 kilograms of the 40% silver alloy?

47. CLEANING FLOORS   A custodian is going to mix a 4% ammonia solution and a 12% ammonia solution to get 1 gallon (128 fluid ounces) of a 9% ammonia solution. How many fluid ounces of the 4% solution and the 12% solution should be used?

48. MOUTHWASH   A pharmacist has a mouthwash solution that is 6% ethanol alcohol and another that is 18% ethanol alcohol. How many milliliters of each must be mixed to make 750 milliliters of a mouthwash that is 10% ethanol alcohol?

49. COFFEE SALES   A coffee supply store waits until the orders for its special blend reach 100 pounds before making up a batch. Columbian coffee selling for $8.75 a pound is blended with Brazilian coffee selling for $3.75 a pound to make a product that sells for $6.35 a pound. How much of each type of coffee should be used to make the blend that will fill the orders?

50. MIXING NUTS   A merchant wants to mix peanuts with cashews, as shown in the illustration, to get 48 pounds of mixed nuts that will be sold at $8 per pound. How many pounds of each should the merchant use?

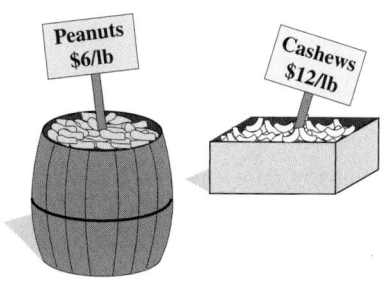

51. GOURMET FOODS   A New York delicatessen sells marinated mushrooms for $12 a pint and stuffed Kalamata olives for $9 a pint. How many pints of each should be used to get 20 pints of a mixture that will sell for $10 a pint?

52. HERBS   Ginger root powder sells for $6.50 a pound and ginkgo leaf powder sells for $9.50 a pound. How many pounds of each should be used to make 15 pounds of a mixture that sells for $7 a pound?

WRITING

53. Explain why a table is helpful in solving uniform motion and mixture problems.

54. A man paid $89 for two shirts and four pairs of socks. If we let x = the cost of a shirt, in dollars, and y = the cost of a pair of socks, in dollars, an equation modeling the purchase is $2x + 4y = 89$. Explain why there is not enough information to determine the cost of a shirt or the cost of a pair of socks.

REVIEW

Graph each inequality. Then describe the graph using interval notation.

55. $x < 4$

56. $x \geq -3$

57. $-1 < x \leq 2$

58. $-2 \leq x \leq 0$

CHALLENGE PROBLEMS

59. On the last scale, how many nails will it take to balance 1 nut?

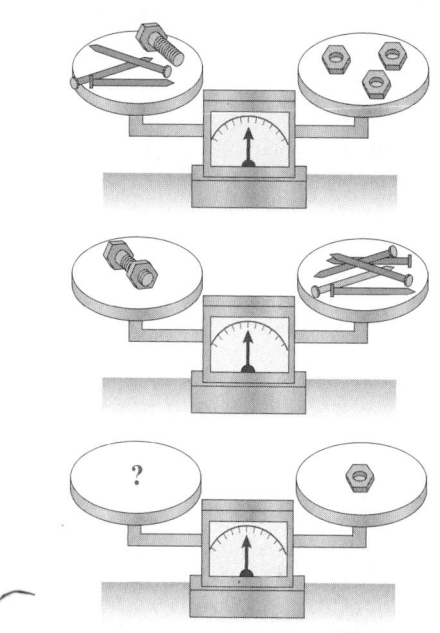

60. FARMING   In a pen of goats and chickens, there are 40 heads and 130 feet. How many goats and chickens are in the pen?

SECTION 4.5
Solving Systems of Linear Inequalities

Objectives

1 Solve a system of linear inequalities by graphing.

2 Solve application problems involving systems of linear inequalities.

In Section 4.1, we solved systems of linear *equations* graphically by finding the point of intersection of two lines. Now we consider systems of linear *inequalities,* such as

$$\begin{cases} x + y \geq -1 \\ x - y \geq 1 \end{cases}$$

To solve systems of linear inequalities, we again find the points of intersection of graphs. In this case, however, we are not looking for an intersection of two lines, but an intersection of two regions.

1 **Solve a System of Linear Inequalities by Graphing.**

A solution of a **system of linear inequalities** is an ordered pair that satisfies each inequality. *To solve a system of linear inequalities* means to find all of its solutions. This can be done by graphing each inequality on the same set of axes and finding the points that are common to every graph in the system.

EXAMPLE 1   Graph the solutions of the system: $\begin{cases} x + y \geq -1 \\ x - y \geq 1 \end{cases}$

Strategy   We will graph the solutions of $x + y \geq -1$ in one color and the solutions of $x - y \geq 1$ in another color on the same coordinate system.

Why   We need to see where the graphs of the two inequalities intersect (overlap).

Solution   To graph $x + y \geq -1$, we begin by graphing the boundary line $x + y = -1$. Since the inequality contains an $\geq$ symbol, the boundary is a solid line. Because the coordinates of the test point $(0, 0)$ satisfy $x + y \geq -1$, we shade (in red) the side of the boundary that contains $(0, 0)$. See part (a) of the figure on the next page.

Graph the boundary: The intercept method

$x + y = -1$

x	y	(x, y)
0	−1	$(0, -1)$
−1	0	$(-1, 0)$

Shading: Check the test point $(0, 0)$

$x + y \geq -1$

$0 + 0 \overset{?}{\geq} -1$ Substitute.

$0 \geq -1$ True

$(0, 0)$ is a solution of $x + y \geq -1$.

In part (b) of the figure, we superimpose the graph of $x - y \geq 1$ on the graph of $x + y \geq -1$ so that we can determine the points that the graphs have in common. To graph $x - y \geq 1$, we graph the boundary $x - y = 1$ as a solid line. Since the test point $(0, 0)$ does not satisfy $x - y \geq 1$, we shade (in blue) the half-plane that does not contain $(0, 0)$.

Graph the boundary: The intercept method

$$x - y = 1$$

x	y	(x, y)
0	−1	$(0, -1)$
1	0	$(1, 0)$

Shading: Check the test point (0, 0)

$$x - y \geq 1$$

$$0 - 0 \overset{?}{\geq} 1 \quad \text{Substitute.}$$

$$0 \geq 1 \quad \text{False}$$

$(0, 0)$ is not a solution of $x - y \geq 1$.

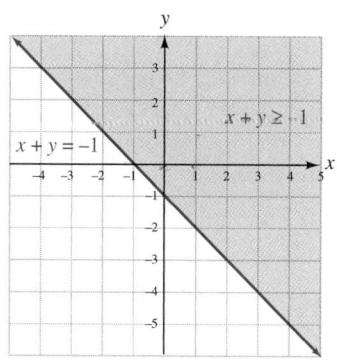

The graph of $x + y \geq -1$ is shaded in red.

(a)

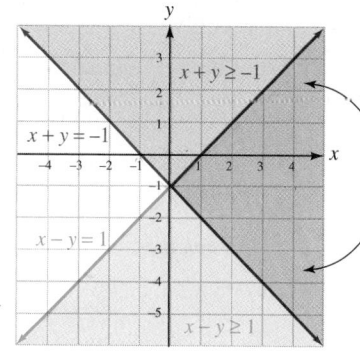

The solutions of the system are shaded in purple. The purple region is the intersection or overlap of the red and blue shaded regions. It includes portions of each boundary.

The graph of $x - y \geq 1$ is shaded in blue. It is drawn over the graph of $x + y \geq -1$.

(b)

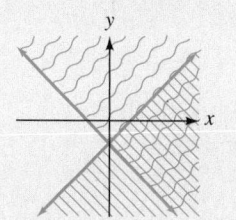

Success Tip

Colored pencils are often used to graph systems of inequalities. A standard pencil can also be used. Just draw different patterns of lines instead of shading.

In part (b) of the figure, the area that is shaded twice represents the solutions of the given system. Any point in the doubly shaded region in purple (including the purple portions of each boundary) has coordinates that satisfy both inequalities.

Since there are infinitely many solutions, we cannot check each of them. However, as an informal check, we can select one ordered pair, say (4, 1), that lies in the doubly shaded region and show that its coordinates satisfy both inequalities of the system.

Check: $x + y \geq -1$ The first inequality. $x - y \geq 1$ The second inequality.

$$4 + 1 \overset{?}{\geq} -1 \qquad\qquad\qquad 4 - 1 \overset{?}{\geq} 1$$

$$5 \geq -1 \quad \text{True} \qquad\qquad\qquad 3 \geq 1 \quad \text{True}$$

The resulting true statements verify that (4, 1) is a solution of the system. If we pick a point that is not in the doubly shaded region, such as (1, 3), (−2, −2), or (0, −4), the coordinates of that point will fail to satisfy one or both of the inequalities.

Self Check 1   Graph the solutions of the system: $\begin{cases} x - y \leq 2 \\ x + y \geq -1 \end{cases}$

Now Try   **Problem 15**

In general, to solve systems of linear inequalities, we will follow these steps.

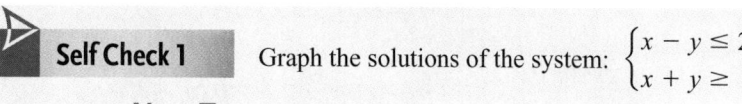

Solving Systems of Linear Inequalities

1. Graph each inequality on the same rectangular coordinate system.

2. Use shading to highlight the intersection of the graphs (the region where the graphs overlap). The points in this region are the solutions of the system.

3. As an informal check, pick a point from the region where the graphs intersect and verify that its coordinates satisfy each inequality of the original system.

EXAMPLE 2   Graph the solutions of the system: $\begin{cases} y > 3x \\ 2x + y < 4 \end{cases}$

Strategy   We will graph the solutions of $y > 3x$ in one color and the solutions of $2x + y < 4$ in another color on the same coordinate system to see where the graphs intersect.

Why   The solution set of the system is the set of all points in the intersection of the two graphs.

Solution   To graph $y > 3x$, we begin by graphing the boundary line $y = 3x$. Since the inequality contains an $>$ symbol, the boundary is a dashed line. Because the boundary passes through $(0, 0)$, we use $(2, 0)$ as the test point instead. Since $(2, 0)$ does not satisfy $y > 3x$, we shade (in red) the half-plane that does not contain $(2, 0)$. See part (a) of the following figure.

Graph the boundary: Slope and y-intercept

$$y = 3x + 0$$

$$m = 3 \qquad b = 0$$

Slope: $\dfrac{3}{1}$ y-intercept: $(0, 0)$

Shading: Check the test point $(2, 0)$

$$y > 3x$$

$$0 \overset{?}{>} 3(2) \qquad \text{Substitute.}$$

$$0 > 6 \qquad \text{False}$$

Since $0 > 6$ is false, $(2, 0)$ is not a solution of $y > 3x$.

In part (b) of the figure, we superimpose the graph of $2x + y < 4$ on the graph of $y > 3x$ to determine the points that the graphs have in common. To graph $2x + y < 4$, we graph the boundary $2x + y = 4$ as a dashed line. Then we shade (in blue) the half-plane that contains $(0, 0)$, because the coordinates of the test point satisfy $2x + y < 4$.

Graph the boundary: The intercept method

$$2x + y = 4$$

x	y	(x, y)
0	4	$(0, 4)$
2	0	$(2, 0)$

Shading: Use the test point $(0, 0)$

$$2x + y < 4$$

$$2(0) + 0 \overset{?}{<} 4 \qquad \text{Substitute.}$$

$$0 < 4 \qquad \text{True}$$

Since $0 < 4$ is true, $(0, 0)$ is a solution of $2x + y < 4$.

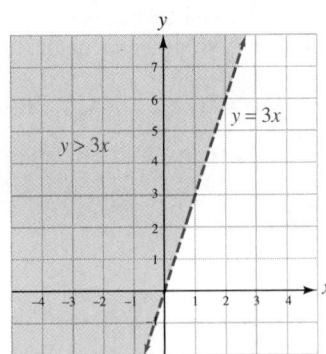

The graph of $y > 3x$ is shaded in red.

(a)

The solutions of the system are shaded in purple. Points on the boundaries are not solutions.

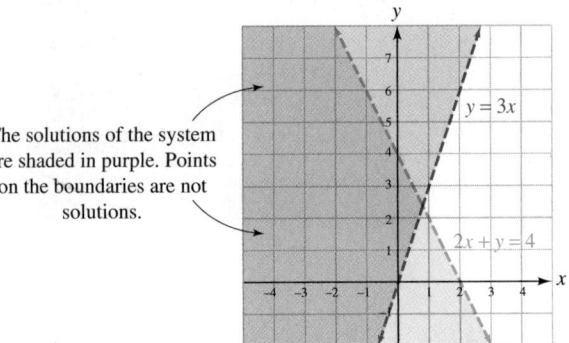

The graph of $2x + y < 4$ is shaded in blue. It is drawn over the graph of $y > 3x$.

(b)

In part (b) of the figure, the area that is shaded twice represents the solutions of the given system. Any point in the doubly shaded region in purple has coordinates that satisfy both inequalities. Pick a point in the region and show that this is true. Note that the region does not include either boundary; points on the boundaries are not solutions of the system.

Self Check 2   Graph the solutions of the system: $\begin{cases} x + 3y < 3 \\ y > \dfrac{1}{3}x \end{cases}$

Now Try   **Problem 23**

EXAMPLE 3   Graph the solutions of the system: $\begin{cases} x \leq 2 \\ y > 3 \end{cases}$

Strategy   We will graph the solutions of $x \leq 2$ in one color and the solutions of $y > 3$ in another color on the same coordinate system to see where the graphs of the two inequalities intersect.

Why   The solution set of the system is the set of all points in the intersection of the two graphs.

Solution   The boundary of the graph of $x \leq 2$ is the line $x = 2$. Since the inequality contains the symbol $\leq$, we draw the boundary as a solid line. The test point $(0, 0)$ makes $x \leq 2$ true, so we shade the side of the boundary that contains $(0, 0)$. See part (a) of the figure on the next page.

Graph the boundary: A table of solutions

$x = 2$

x	y	(x, y)
2	0	(2, 0)
2	2	(2, 2)
2	4	(2, 4)

Shading: Check the test point $(0, 0)$

$x \leq 2$

$0 \leq 2$ True

Since $0 \leq 2$ is true, $(0, 0)$ is a solution of $x \leq 2$.

In part (b) of the figure, the graph of $y > 3$ is superimposed over the graph of $x \leq 2$. The boundary of the graph of $y > 3$ is the line $y = 3$. Since the inequality contains the symbol $>$, we draw the boundary as a dashed line. The test point $(0, 0)$ makes $y > 3$ false, so we shade the side of the boundary that does not contain $(0, 0)$.

Graph the boundary: A table of solutions

$y = 3$

x	y	(x, y)
0	3	(0, 3)
1	3	(1, 3)
4	3	(4, 3)

Shading: Check the test point $(0, 0)$

$y > 3$

$0 > 3$ False

Since $0 > 3$ is false, $(0, 0)$ is not a solution of $y > 3$.

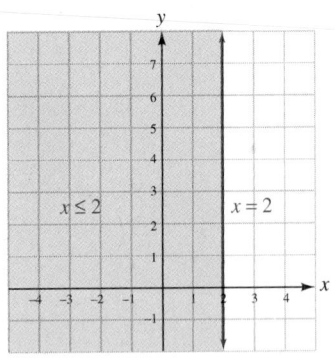

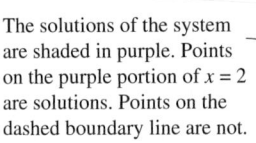

The solutions of the system are shaded in purple. Points on the purple portion of $x = 2$ are solutions. Points on the dashed boundary line are not.

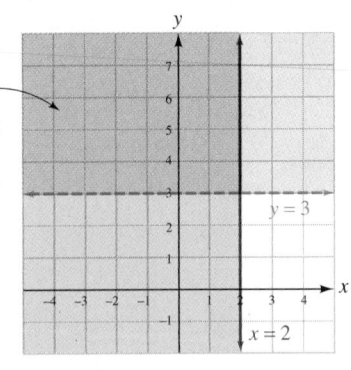

The graph of $x \leq 2$ is shaded in red.

The graph of $y > 3$ is shaded in blue. It is drawn over the graph of $x \leq 2$.

(a)

(b)

The area that is shaded twice represents the solutions of the system of inequalities. Any point in the doubly shaded region in purple has coordinates that satisfy both inequalities, including the purple portion of the $x = 2$ boundary. Pick a point in the region and show that this is true.

Self Check 3 Graph the solutions of the system: $\begin{cases} y \leq 1 \\ x > 2 \end{cases}$

Now Try **Problem 25**

EXAMPLE 4 Graph the solutions of the system: $\begin{cases} x \geq 0 \\ y \geq 0 \\ x + 2y \leq 6 \end{cases}$

Strategy We will graph the solutions of $x \geq 0$, $y \geq 0$, and $x + 2y \leq 6$ on the same coordinate system to see where all three graphs intersect (overlap).

Why The solution set of the system is the set of all points in the intersection of the three graphs.

Solution This is a system of three linear inequalities. If shading is used to graph them on the same set of axes, it can become difficult to interpret the results. Instead, we can draw directional arrows attached to each boundary line in place of the shading.

- The graph of $x \geq 0$ has the boundary $x = 0$ and includes all points on the y-axis and to the right.
- The graph of $y \geq 0$ has the boundary $y = 0$ and includes all points on the x-axis and above.
- The graph of $x + 2y \leq 6$ has the boundary $x + 2y = 6$. Because the coordinates of the origin satisfy $x + 2y \leq 6$, the graph includes all points on and below the boundary.

The solutions of the system are the points that lie on triangle OPQ and the shaded triangular region that it encloses.

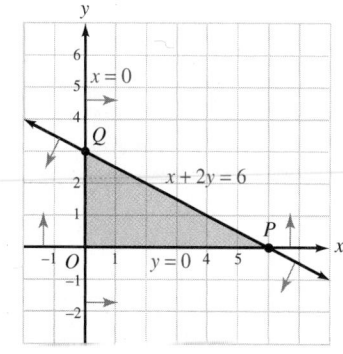

$x + 2y = 6$

x	y	(x, y)
0	3	(0, 3)
6	0	(6, 0)

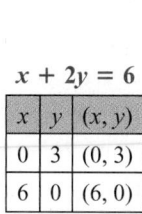 **Self Check 4** Graph the solutions of the system: $\begin{cases} x \leq 1 \\ y \leq 2 \\ 2x - y \leq 4 \end{cases}$

Now Try **Problem 41**

---

2 Solve Application Problems Involving Systems of Linear Inequalities.

EXAMPLE 5 *Landscaping.* A homeowner budgets from $300 to $600 for trees and bushes to landscape his yard. After shopping around, he finds that good trees cost $150 each and mature bushes cost $75 each. What combinations of trees and bushes can he buy?

Analyze the Problem

- At least $300 but not more than $600 is to be spent for trees and bushes.
- Trees cost $150 each and bushes cost $75 each.
- What combination of trees and bushes can he buy?

Form Two Inequalities Let x = the number of trees purchased and y = the number of bushes purchased. We then form the following system of inequalities:

The cost of a tree	times	the number of trees purchased	plus	the cost of a bush	times	the number of bushes purchased	should at least be	$300.
$150	·	x	+	$75	·	y	≥	$300

The cost of a tree	times	the number of trees purchased	plus	the cost of a bush	times	the number of bushes purchased	should not be more than	$600.
$150	·	x	+	$75	·	y	≤	$600

Solve the System To solve the following system of linear inequalities

$$\begin{cases} 150x + 75y \geq 300 \\ 150x + 75y \leq 600 \end{cases}$$

we use the graphing methods discussed in this section. Neither a negative number of trees nor a negative number of bushes can be purchased, so we restrict the graph to Quadrant I.

State the Conclusion   The coordinates of each point highlighted in the graph give a possible combination of the number of trees, x, and the number of bushes, y, that can be purchased. Written as ordered pairs, these possibilities are

(0, 4), (0, 5), (0, 6), (0, 7), (0, 8),
(1, 2), (1, 3), (1, 4), (1, 5), (1, 6),
(2, 0), (2, 1), (2, 2), (2, 3), (2, 4),
(3, 0), (3, 1), (3, 2), (4, 0)

Check the Result   Suppose the homeowner picks the combination of 3 trees and 2 bushes, as represented by (3, 2). Show that this point satisfies both inequalities of the system.

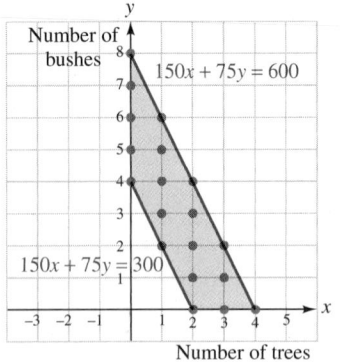

 Now Try   Problem 47

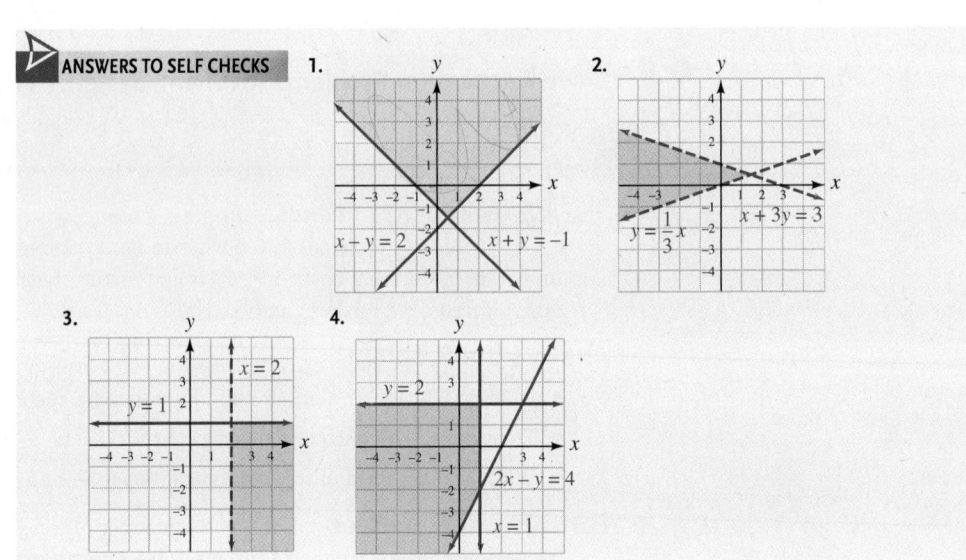

ANSWERS TO SELF CHECKS

STUDY SET
4.5

VOCABULARY

Fill in the blanks.

1. $\begin{cases} x + y > 2 \\ x + y < 4 \end{cases}$ is a system of linear _____.

2. To graph the linear inequality $x + y > 2$, first graph the _____ $x + y = 2$. Then pick the test _____ (0, 0) to determine which half-plane to shade.

3. To find the solutions of a system of two linear inequalities graphically, look for the _____, or overlap, of the two shaded regions.

4. The phrase *should not surpass* can be represented by the inequality symbol ____ and the phrase *must be at least* can be represented by the inequality symbol ____.

CONCEPTS

5. **a.** What is the equation of the boundary line of the graph of $3x - y < 5$?

   b. Is the boundary a solid or dashed line?

6. a. What is the equation of the boundary line of the graph of $y \geq 4x$?

  b. Is the boundary a solid or dashed line?

  c. Why can't $(0, 0)$ be used as a test point to determine what to shade?

7. Find the slope and the y-intercept of the line whose equation is $y = 4x - 3$.

8. Complete the table to find the x- and y-intercepts of the line whose equation is $3x - 2y = 6$.

x	y
0	
	0

9. The boundary of the graph of $2x + y > 4$ is shown.

  a. Does the point $(0, 0)$ make the inequality true?

  b. Should the region above or below the boundary be shaded?

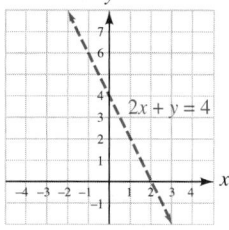

10. Linear inequality 1 is graphed in red and linear inequality 2 is graphed in blue. Determine whether a true or false statement results when

  a. The coordinates of point A are substituted into inequality 1

  b. The coordinates of point A are substituted into inequality 2

  c. The coordinates of point B are substituted into inequality 1

  d. The coordinates of point B are substituted into inequality 2

  e. The coordinates of point C are substituted into inequality 1

  f. The coordinates of point C are substituted into inequality 2

Inequality 1 solutions are in red

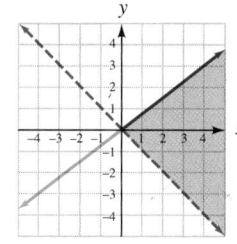

Inequality 2 solutions are in blue

11. The graph of a system of two linear inequalities is shown. Determine whether each point is a solution of the system.

  a. $(4, -2)$

  b. $(1, 3)$

  c. the origin

12. Use a check to determine whether each ordered pair is a solution of the system.

$$\begin{cases} x + 2y \geq -1 \\ x - y < 2 \end{cases}$$

  a. $(1, 4)$ **b.** $(-2, 0)$

13. Match each equation, inequality, or system with the graph of its solution.

  a. $x + y = 2$

  b. $x + y \geq 2$

  c. $\begin{cases} x + y = 2 \\ x - y = 2 \end{cases}$

  d. $\begin{cases} x + y \geq 2 \\ x - y \leq 2 \end{cases}$

  i. 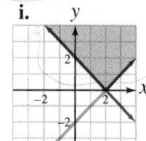   **ii.**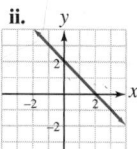

  iii. 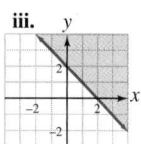   **iv.**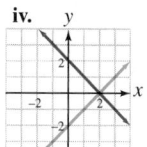

14. Match the system of inequalities with the correct graph.

  a. $\begin{cases} x \geq 2 \\ y < 1 \end{cases}$

  b. $\begin{cases} x > 2 \\ y \leq 1 \end{cases}$

  c. $\begin{cases} x \geq 2 \\ y \geq 1 \end{cases}$

  d. $\begin{cases} x > 2 \\ y > -1 \end{cases}$

  i. 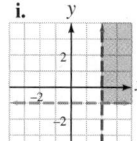   **ii.**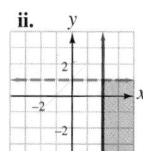

  iii. 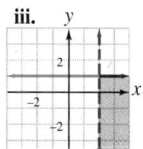   **iv.**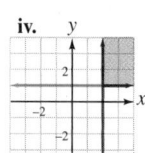

GUIDED PRACTICE

Graph the solutions of each system. See Examples 1–3.

15. $\begin{cases} x + 2y \leq 3 \\ 2x - y \geq 1 \end{cases}$

16. $\begin{cases} 2x + y \geq 3 \\ x - 2y \leq -1 \end{cases}$

17. $\begin{cases} x + y < -1 \\ x - y > -1 \end{cases}$

18. $\begin{cases} x + y > 2 \\ x - y < -2 \end{cases}$

19. $\begin{cases} 2x - 3y \leq 0 \\ y \geq x - 1 \end{cases}$

20. $\begin{cases} y > 2x - 4 \\ y \geq -x - 1 \end{cases}$

21. $\begin{cases} x + y < 2 \\ x + y \leq 1 \end{cases}$

22. $\begin{cases} y > -x + 2 \\ y < -x + 4 \end{cases}$

23. $\begin{cases} y > 2x \\ x + 2y < 6 \end{cases}$

24. $\begin{cases} y \le 2x \\ x + y < 4 \end{cases}$

25. $\begin{cases} x \ge 2 \\ y \le 3 \end{cases}$

26. $\begin{cases} x \ge -1 \\ y > -2 \end{cases}$

27. $\begin{cases} x > 0 \\ y > 0 \end{cases}$

28. $\begin{cases} x \le 0 \\ y < 0 \end{cases}$

29. $\begin{cases} 3x + 4y \ge -7 \\ 2x - 3y \ge 1 \end{cases}$

30. $\begin{cases} 3x + y \le 1 \\ 4x - y \ge -8 \end{cases}$

31. $\begin{cases} 2x + y < 7 \\ y > 2 - 2x \end{cases}$

32. $\begin{cases} 2x + y \ge 6 \\ y \le 4x - 6 \end{cases}$

33. $\begin{cases} 2x - 4y > -6 \\ 3x + y \ge 5 \end{cases}$

34. $\begin{cases} 2x - 3y < 0 \\ 2x + 3y \ge 12 \end{cases}$

35. $\begin{cases} 3x - y + 4 \le 0 \\ 3y > -2x - 10 \end{cases}$

36. $\begin{cases} 3x + 2y - 12 \ge 0 \\ x < -2 + y \end{cases}$

37. $\begin{cases} y \ge x \\ y \le \frac{1}{3}x + 1 \end{cases}$

38. $\begin{cases} y > 3x \\ y \le -x - 1 \end{cases}$

39. $\begin{cases} x + y > 0 \\ y - x < -2 \end{cases}$

40. $\begin{cases} y + 2x \le 0 \\ y \le \frac{1}{2}x + 2 \end{cases}$

Graph the solutions of each system. See Example 4.

41. $\begin{cases} x \ge 0 \\ y \ge 0 \\ x + y \le 3 \end{cases}$

42. $\begin{cases} x - y \le 6 \\ x + 2y \le 6 \\ x \ge 0 \end{cases}$

43. $\begin{cases} x - y < 4 \\ y \le 0 \\ x \ge 0 \end{cases}$

44. $\begin{cases} 2x + y \le 2 \\ y > x \\ x \ge 0 \end{cases}$

APPLICATIONS

45. BIRDS OF PREY Parts (a) and (b) of the illustration show the individual fields of vision for each eye of an owl. In part (c), shade the area where the fields of vision overlap—that is, the area that is seen by both eyes.

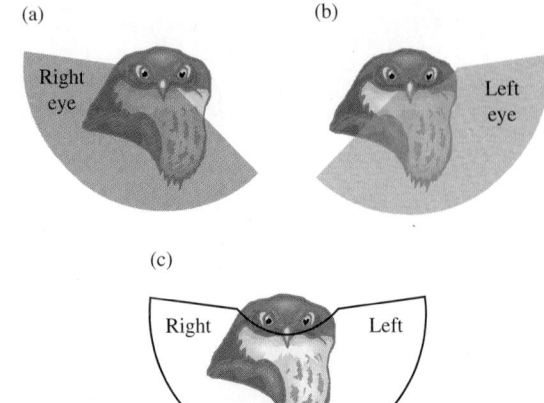

46. EARTH SCIENCE Shade the area of the earth's surface that is north of the Tropic of Capricorn and south of the Tropic of Cancer.

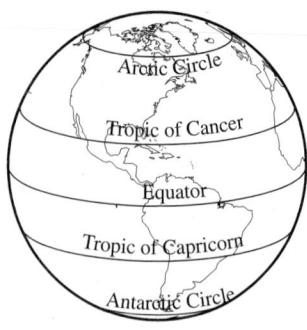

In Problems 47–52, graph each system of inequalities and give two possible solutions. See Example 5.

47. BUYING COMPACT DISCS Melodic Music has compact discs on sale for either $10 or $15. If a customer wants to spend at least $30 but no more than $60 on CDs, graph a system of inequalities showing the possible combinations of $10 CDs ($x$) and $15 CDs ($y$) that the customer can buy.

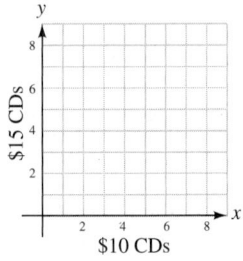

48. BUYING BOATS Boatworks wholesales aluminum boats for $800 and fiberglass boats for $600. Northland Marina wants to make a purchase totaling at least $2,400 but no more than $4,800. Graph a system of inequalities showing the possible combinations of aluminum boats (x) and fiberglass boats (y) that can be ordered.

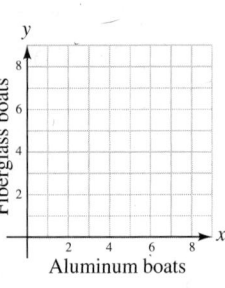

49. BUYING FURNITURE A distributor wholesales desk chairs for $150 and side chairs for $100. Best Furniture wants its order to total no more than $900; Best also wants to order more side chairs than desk chairs. Graph a system of inequalities showing the possible combinations of desk chairs (x) and side chairs (y) that can be ordered.

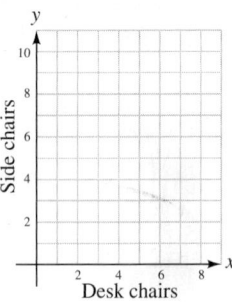

50. ORDERING FURNACE EQUIPMENT   J. Bolden Heating Company wants to order no more than $2,000 worth of electronic air cleaners and humidifiers from a wholesaler that charges $500 for air cleaners and $200 for humidifiers. If Bolden wants more humidifiers than air cleaners, graph a system of inequalities showing the possible combinations of air cleaners (x) and humidifiers (y) that can be ordered.

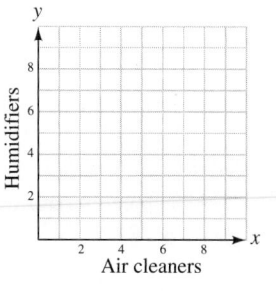

51. PESTICIDES   To eradicate a fruit fly infestation, helicopters sprayed an area of a city that can be described by $y \geq -2x + 1$ (within the city limits). Two weeks later, more spraying was ordered over the area described by $y \geq \frac{1}{4}x - 4$ (within the city limits). Show the part of the city that was sprayed twice.

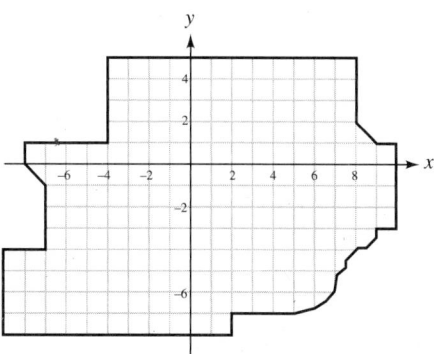

52. REDEVELOPMENT   Refer to the following diagram. A government agency has declared an area of a city east of First Street, north of Second Avenue, south of Sixth Avenue, and west of Fifth Street as eligible for federal redevelopment funds. Describe this area of the city mathematically using a system of four inequalities, if the corner of Central Avenue and Main Street is considered the origin.

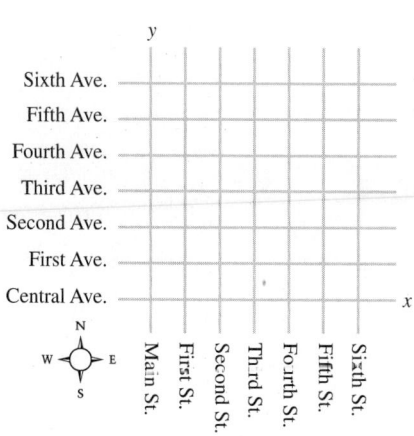

WRITING

53. Explain how to use graphing to solve a system of inequalities.

54. When a solution of a system of linear inequalities is graphed, what does the shading represent?

55. Describe how the graphs of the solutions of these systems are similar and how they differ.

$$\begin{cases} x + y = 4 \\ x - y = 4 \end{cases} \quad \text{and} \quad \begin{cases} x + y \geq 4 \\ x - y \geq 4 \end{cases}$$

56. Explain when a system of inequalities will have no solutions.

REVIEW

Simplify each expression.

57. $8\left(\frac{3}{4}t\right)$

58. $27\left(\frac{2}{3}m\right)$

59. $-\frac{2}{3}(3w - 6)$

60. $\frac{1}{2}(2y - 8)$

61. $-\frac{7}{16}x - \frac{3}{16}x$

62. $-\frac{5}{18}x - \frac{7}{18}x$

63. $60\left(\frac{3}{20}r - \frac{4}{15}\right)$

64. $72\left(\frac{7}{8}f - \frac{8}{9}\right)$

CHALLENGE PROBLEMS

Graph the solutions of each system.

65. $\begin{cases} \dfrac{x}{3} - \dfrac{y}{2} < -3 \\ \dfrac{x}{3} + \dfrac{y}{2} > -1 \end{cases}$

66. $\begin{cases} 3x + y < -2 \\ y > 3(1 - x) \end{cases}$

67. $\begin{cases} 2x + 3y \leq 6 \\ 3x + y \leq 1 \\ x \leq 0 \end{cases}$

68. $\begin{cases} x \geq 0 \\ y \geq 0 \\ 9x + 3y \leq 18 \\ 3x + 6y \leq 18 \end{cases}$

CHAPTER 4
Summary & Review

SECTION 4.1   Solving Systems of Equations by Graphing

DEFINITIONS AND CONCEPTS	EXAMPLES

When two equations are considered at the same time, we say that they form a **system of equations.**

A **solution of a system** of equations in two variables is an ordered pair that satisfies both equations of the system.

Is (4, 3) a solution of the system $\begin{cases} x + y = 7 \\ x - y = 5 \end{cases}$?

To answer this question, we substitute 4 for x and 3 for y in each equation.

$$x + y = 7 \qquad\qquad x - y = 5$$
$$4 + 3 \stackrel{?}{=} 7 \qquad\qquad 4 - 3 \stackrel{?}{=} 5$$
$$7 = 7 \quad \text{True} \qquad\qquad 1 = 5 \quad \text{False}$$

Although (4, 3) satisfies the first equation, it does not satisfy the second. Because it does not satisfy both equations, it is not a solution of the system.

To **solve a system graphically:**

1. Graph each equation on the same coordinate system.

2. Determine the coordinates of the point of intersection of the graphs. That ordered pair is the solution.

3. Check the solution in each equation of the original system.

Use graphing to solve the system: $\begin{cases} y = -2x + 3 \\ x - 2y = 4 \end{cases}$

Step 1: Graph each equation as shown below.

$$y = -2x + 3 \qquad x - 2y = 4$$

$$m = \frac{-2}{1}$$

$$b = 3$$

x	y
0	-2
4	0

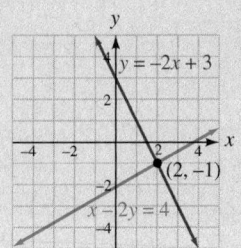

Step 2: It appears that the graphs intersect at the point $(2, -1)$. To verify that this is the solution of the system, substitute 2 for x and -1 for y in each equation.

Step 3: Check

$$x - 2y = 4 \qquad\qquad y = -2x + 3$$
$$2 - 2(-1) \stackrel{?}{=} 4 \qquad\qquad -1 \stackrel{?}{=} -2(2) + 3$$
$$4 = 4 \quad \text{True} \qquad\qquad -1 = -1 \qquad \text{True}$$

Since $(2, -1)$ makes both equations true, it is the solution of the system.

SECTION 4.1   Solving Systems of Equations by Graphing–*continued*

DEFINITIONS AND CONCEPTS

A system of equations that has at least one solution is called a **consistent system.** If the graphs of the equations of the system are parallel lines, the system has no solution and is called an **inconsistent system.**

Equations with different graphs are called **independent equations.** If the graphs of the equations in a system are the same line, the system has infinitely many solutions. The equations are called **dependent equations.**

We can determine the **number of solutions** that a system of two linear equations has by writing each equation in slope-intercept form, $y = mx + b$, and comparing the slopes and y-intercepts.

EXAMPLES

There are three possible outcomes when solving a system of two linear equations by graphing.

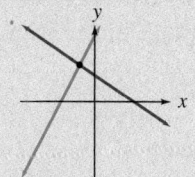

Consistent system
Independent equations
• Exactly one solution
• The lines have different slopes.

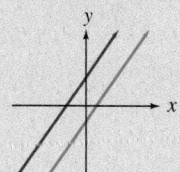

Inconsistent system
Independent equations
• No solution
• The lines have the same slope but different y-intercepts.

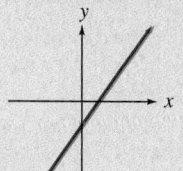

Consistent system
Dependent equations
• Infinitely many solutions
• The lines have the same slope and same y-intercept.

REVIEW EXERCISES

Determine whether the ordered pair is a solution of the system.

1. $(2, -3)$, $\begin{cases} 3x - 2y = 12 \\ 2x + 3y = -5 \end{cases}$

2. $\left(\dfrac{7}{2}, -\dfrac{2}{3}\right)$, $\begin{cases} 3y = 2x - 9 \\ 2x + 3y = 6 \end{cases}$

Use the graphing method to solve each system.

3. $\begin{cases} x + y = 7 \\ 2x - y = 5 \end{cases}$

4. $\begin{cases} 2x + y = 5 \\ y = -\dfrac{x}{3} \end{cases}$

5. $\begin{cases} 3x + 6y = 6 \\ x + 2y - 2 = 0 \end{cases}$

6. $\begin{cases} 6x + 3y = 12 \\ y = -2x + 2 \end{cases}$

7. Find the slope and the y-intercept of the graph of each line in the system $\begin{cases} y = -2x + 1 \\ 8x + 4y = 3 \end{cases}$. Then, use that information to determine the number of solutions of the system.

8. COLLEGE ENROLLMENT   Estimate the point of intersection of the graphs. Explain its significance.

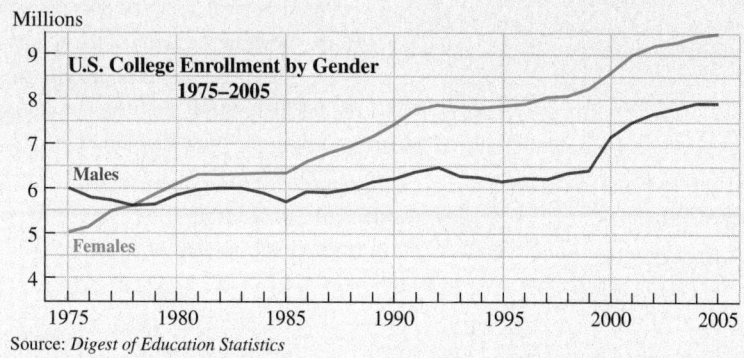

Source: *Digest of Education Statistics*

SECTION 4.2 Solving Systems of Equations by Substitution

DEFINITIONS AND CONCEPTS	EXAMPLES

DEFINITIONS AND CONCEPTS

To solve a system of equations in x and y by **substitution:**

1. Solve one of the equations for either x or y.

2. Substitute the expression for x (or for y) obtained in step 1 into the other equation and solve the equation.

3. Substitute the value of the variable found in step 2 into the substitution equation to find the value of the remaining variable.

4. Check the proposed solution in the equations of the original system.

With the substitution method, the objective is to use an appropriate substitution to obtain *one equation in one variable.*

If in step 2 the variable drops out and a false statement results, the system has **no solution.** If a true statement results, the system has **infinitely many solutions.**

EXAMPLES

Use substitution to solve the system: $\begin{cases} y = x + 7 \\ x + 2y = 5 \end{cases}$

Step 1: The first equation is already solved for y.

Step 2: Substitute $x + 7$ for y in the second equation.

$$x + 2y = 5$$
$$x + 2(x + 7) = 5 \qquad \text{Substitute } x + 7 \text{ for } y.$$
$$x + 2x + 14 = 5 \qquad \text{Distribute.}$$
$$3x + 14 = 5 \qquad \text{Combine like terms.}$$
$$3x = -9 \qquad \text{Subtract 14 from both sides.}$$
$$x = -3 \qquad \text{Divide both sides by 3.}$$

Step 3: $y = x + 7 \qquad$ This is the substitution equation.

$$y = -3 + 7 \qquad \text{Substitute } -3 \text{ for } x.$$
$$y = 4 \qquad \text{Simplify.}$$

The following check verifies that the solution is $(-3, 4)$.

Step 4: Check

$$y = x + 7 \qquad\qquad x + 2y = 5$$
$$4 \overset{?}{=} -3 + 7 \qquad\qquad -3 + 2(4) \overset{?}{=} 5$$
$$4 = 4 \quad \text{True} \qquad\qquad 5 = 5 \quad \text{True}$$

REVIEW EXERCISES

Use the substitution method to solve each system.

9. $\begin{cases} y = 15 - 3x \\ 7y + 3x = 15 \end{cases}$

10. $\begin{cases} x = y \\ 5x - 4y = 3 \end{cases}$

11. $\begin{cases} 6x + 2y = 8 - y + x \\ 3x = 2 - y \end{cases}$

12. $\begin{cases} r = 3s + 7 \\ r = 2s + 5 \end{cases}$

13. $\begin{cases} 9x + 3y - 5 = 0 \\ 3x + y = \dfrac{5}{3} \end{cases}$

14. $\begin{cases} \dfrac{x}{2} + \dfrac{y}{2} = 11 \\ \dfrac{5x}{16} - \dfrac{3y}{16} = \dfrac{15}{8} \end{cases}$

15. When solving a system using the substitution method, suppose you obtain the result $8 = 9$.

 a. How many solutions does the system have?

 b. Describe the graph of the system.

 c. What term is used to describe the system?

16. Fill in the blank. With the substitution method, the objective is to use an appropriate substitution to obtain one equation in _____ variable.

SECTION 4.3 Solving Systems of Equations by Elimination (Addition)

DEFINITIONS AND CONCEPTS

To solve a system of equations in x and y using **elimination (addition):**

1. Write each equation in the standard $Ax + By = C$ form.

2. Multiply one (or both) equations by nonzero quantities to make the coefficients of x (or y) opposites.

3. Add the equations to eliminate the terms involving x (or y).

4. Solve the equation obtained in step 3.

5. Find the value of the other variable by substituting the value of the variable found in step 4 into any equation containing both variables.

6. Check the solution in the equations of the original system.

With the elimination method, the basic objective is to obtain two equations whose sum will be one equation in one variable.

If in step 3 both variables drop out and a false statement results, the system has **no solution.** If a true statement results, the system has **infinitely many solutions.**

EXAMPLES

Use elimination to solve: $\begin{cases} 2x - 3y = 4 \\ 3x + y = -5 \end{cases}$

Step 1: Both equations are written in $Ax + By = C$ form.

Step 2: Multiply the second equation by 3 so that the coefficients of y are opposites.

Step 3:

$$2x - 3y = 4$$
$$\underline{9x + 3y = -15}$$
$$11x \quad\quad = -11 \quad \text{Add the like terms, column by column.}$$

Step 4: Solve for x.

$$11x = -11$$
$$x = -1 \quad \text{Divide both sides by 11.}$$

Step 5: Find y.

$$3x + y = -5 \quad \text{This is the second equation.}$$
$$3(-1) + y = -5 \quad \text{Substitute } -1 \text{ for x.}$$
$$y = -2$$

The solution is $(-1, -2)$.

Step 6: Check

$$2x - 3y = 4 \qquad\qquad 3x + y = -5$$
$$2(-1) - 3(-2) \overset{?}{=} 4 \qquad 3(-1) + (-2) \overset{?}{=} -5$$
$$-2 + 6 \overset{?}{=} 4 \qquad\qquad -3 - 2 \overset{?}{=} -5$$
$$4 = 4 \quad \text{True} \qquad\qquad -5 = -5 \quad \text{True}$$

REVIEW EXERCISES

17. Write each equation of the system $\begin{cases} 4x + 2y - 7 = 0 \\ 3y = 5x + 6 \end{cases}$ in standard $Ax + By = C$ form.

18. Fill in the blank. With the elimination method, the basic objective is to obtain two equations whose sum will be one equation in _____ variable.

Solve each system using the elimination (addition) method.

19. $\begin{cases} 2x + y = 1 \\ 5x - y = 20 \end{cases}$

20. $\begin{cases} x + 8y = 7 \\ x - 4y = 1 \end{cases}$

21. $\begin{cases} 5a + b = 2 \\ 3a + 2b = 11 \end{cases}$

22. $\begin{cases} 11x + 3y = 27 \\ 8x + 4y = 36 \end{cases}$

23. $\begin{cases} 9x + 3y = 15 \\ 3x = 5 - y \end{cases}$

24. $\begin{cases} -\dfrac{a}{4} - \dfrac{b}{3} = \dfrac{1}{12} \\ \dfrac{a}{2} - \dfrac{5b}{4} = \dfrac{7}{4} \end{cases}$

25. $\begin{cases} 0.02x + 0.05y = 0 \\ 0.3x - 0.2y = -1.9 \end{cases}$

26. $\begin{cases} -\dfrac{1}{4}x = 1 - \dfrac{2}{3}y \\ 6x - 18y = 5 - 2y \end{cases}$

For each system, determine which method, substitution or elimination (addition), would be easier to use to solve the system and explain why.

27. $\begin{cases} 6x + 2y = 5 \\ 3x - 3y = -4 \end{cases}$

28. $\begin{cases} x = 5 - 7y \\ 3x - 3y = -4 \end{cases}$

SECTION 4.4 Problem Solving Using Systems of Equations

DEFINITIONS AND CONCEPTS	EXAMPLES

We can solve many types of problems using a system of two linear equations in two variables:

- Geometry problems
- Number-value problems
- Interest problems
- Uniform motion problems
- Liquid and dry mixture problems

To solve problems involving two unknown quantities:

1. **Analyze** the facts of the problem. Make a table or a diagram if it is helpful.

2. Pick different variables to represent the two unknown quantities. **Form two equations** involving those variables.

3. **Solve** the system of equations using graphing, substitution, or elimination.

4. **State** the conclusion.

5. **Check** the result.

Two angles are said to be **complementary** if the sum of their measures is 90°. Two angles are said to be **supplementary** if the sum of their measures is 180°.

The difference of the measures of two supplementary angles is 40°. Find the measure of each angle.

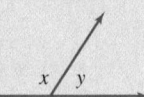

Analyze Since the angles are supplementary, the sum of their measures is 180°. If we subtract the smaller angle from the larger, the result is given to be 40°.

Form Let x = the measure (in degrees) of the larger angle and y = the measure (in degrees) of the smaller angle. Then we have the following two equations.

$$\begin{cases} x + y = 180 & \text{Their sum is 180°.} \\ x - y = 40 & \text{Their difference is 40°.} \end{cases}$$

Solve If we add the equations, we get

$$x + y = 180$$
$$\underline{x - y = 40}$$
$$2x \quad = 220$$
$$x = 110 \quad \text{Divide both sides by 2. This is the measure of the larger angle.}$$

We can use the first equation of the system to find y.

$$x + y = 180$$
$$110 + y = 180 \quad \text{Substitute 110 for x.}$$
$$y = 70 \quad \text{Subtract 110 from both sides. This is the measure of the smaller angle.}$$

State The angles measure 110° and 70°.

Check Angles with measures of 110° and 70° are supplementary (their sum is 180°) and their difference is 40°. The results check.

REVIEW EXERCISES

Write a system of two equations in two variables to solve each problem.

29. ELEVATIONS   The elevation of Las Vegas, Nevada, is 20 times greater than that of Baltimore, Maryland. The sum of their elevations is 2,100 feet. Find the elevation of each city.

30. PAINTING EQUIPMENT   When fully extended, a ladder is 35 feet in length. If the extension is 7 feet shorter than the base, how long is each part of the ladder?

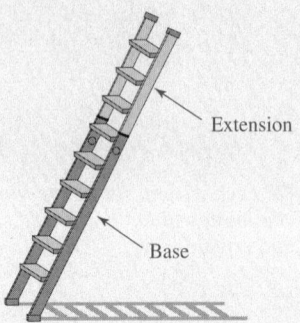

Extension

Base

31. GEOMETRY   Two angles are complementary. The measure of one is 15° more than twice the measure of the other. Find the measure of each angle.

32. CRASH INVESTIGATION   In an effort to protect evidence, investigators used 420 yards of yellow "Police Line—Do Not Cross" tape to seal off a large rectangular-shaped area around an airplane crash site. How much area will the investigators have to search if the width of the rectangle is three-fourths of the length?

33. Complete each table.

a.

	Amount ·	Strength =	Amount of pesticide
Weak	x	0.02	
Strong	y	0.09	
Mixture	100	0.08	

b.

	Rate ·	Time =	Distance
With the wind	$s + w$	5	
Against the wind	$s - w$	7	

c.

	P ·	r ·	$t =$	I
Mack Financial	x	0.11	1	
Union Savings	y	0.06	1	

d.

	Amount	· Price	= Total value
Carmel corn	x	4	
Peanuts	y	8	
Mixture	10	5	

34. CANDY STORE   A merchant wants to mix gummy worms worth $6 per pound and gummy bears worth $3 per pound to make 30 pounds of a mixture worth $4.20 per pound. How many pounds of each type of candy should he use?

35. BOATING   It takes a motorboat 4 hours to travel 56 miles down a river, and 3 hours longer to make the return trip. Find the speed of the current.

36. SHOPPING   Packages containing two bottles of contact lens cleaner and three bottles of soaking solution cost $63.40, and packages containing three bottles of cleaner and two bottles of soaking solution cost $69.60. Find the cost of a bottle of cleaner and a bottle of soaking solution.

37. INVESTING   Carlos invested part of $3,000 in a 10% certificate account and the rest in a 6% passbook account. The total annual interest from both accounts is $270. How much did he invest at 6%?

38. ANTIFREEZE   How much of a 40% antifreeze solution must a mechanic mix with a 70% antifreeze solution if she needs 20 gallons of a 50% antifreeze solution?

SECTION 4.5   Solving Systems of Linear Inequalities

DEFINITIONS AND CONCEPTS

A solution of a **system of linear inequalities** is an ordered pair that satisfies each inequality.

To **solve a system of linear inequalities:**

1. Graph each inequality on the same coordinate system.

2. Use shading to highlight the intersection of the graphs. The points in this region are the solutions of the system.

3. As an informal check, pick a point from the region and verify that its coordinates satisfy each inequality of the original system.

EXAMPLES

Graph the solutions of the system: $\begin{cases} y \le x + 1 \\ y > -1 \end{cases}$

Step 1: Graph each inequality on the same coordinate system as shown.

Step 2: Use shading to highlight where the graphs intersect.

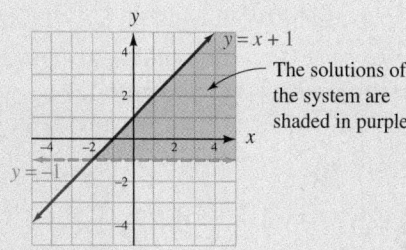

The solutions of the system are shaded in purple.

Step 3: Pick a point from the solution region such as $(1, 0)$ and verify that it satisfies both inequalities.

REVIEW EXERCISES

Solve each system of inequalities.

39. $\begin{cases} 5x + 3y < 15 \\ 3x - y > 3 \end{cases}$

40. $\begin{cases} 3y \le x \\ y > 3x \end{cases}$

41. GIFT SHOPPING   A grandmother wants to spend at least $40 but no more than $60 on school clothes for her grandson. If T-shirts sell for $10 each and pants sell for $20 each, write a system of inequalities that describes the possible numbers of T-shirts x and pairs of pants y that she can buy. Graph the system and give two possible solutions.

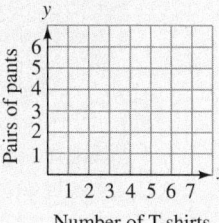

Number of T-shirts

42. Use a check to determine whether each ordered pair is a solution of the system: $\begin{cases} x + 2y \le 3 \\ 2x - y > 1 \end{cases}$

a. $(5, -4)$ **b.** $(-1, -3)$

CHAPTER 4
Test

Determine whether the ordered pair is a solution of the system.

1. $(5, 3)$, $\begin{cases} 3x + 2y = 21 \\ x + y = 8 \end{cases}$

2. $(-2, -1)$, $\begin{cases} 4x + y = -9 \\ 2x - 3y = -7 \end{cases}$

3. Fill in the blanks.

   a. A _____ of a system of linear equations is an ordered pair that satisfies each equation.

   b. A system of equations that has at least one solution is called a _____ system.

   c. A system of equations that has no solution is called an _____ system.

   d. Equations with different graphs are called _____ equations.

   e. A system of _____ equations has an infinite number of solutions.

4. Find the slope and the y-intercept of the graph of each line in the system $\begin{cases} y = 4x - 10 \\ x - 2y = -16 \end{cases}$. Then, use that information to determine the *number of solutions* of the system. **Do not solve the system.**

Solve each system by graphing.

5. $\begin{cases} y = 2x - 1 \\ x - 2y = -4 \end{cases}$

6. $\begin{cases} x + y = 5 \\ y = -x \end{cases}$

The following graph shows two different ways in which a salesperson can be paid according to the number of items she sells each month.

7. What is the point of intersection of the graphs? Explain its significance.

8. Which plan is better for the salesperson if she feels that selling 30 items per month is almost impossible?

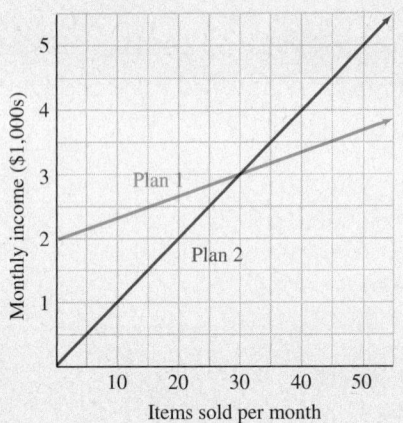

Solve each system by substitution.

9. $\begin{cases} y = x - 1 \\ 2x + y = -7 \end{cases}$

10. $\begin{cases} 3x + 6y = -15 \\ x + 2y = -5 \end{cases}$

Solve each system using elimination.

11. $\begin{cases} 3x - y = 2 \\ 2x + y = 8 \end{cases}$

12. $\begin{cases} 4x + 3y = -3 \\ -3x = -4y + 21 \end{cases}$

Solve each system using substitution or elimination.

13. $\begin{cases} 3x - 5y - 16 = 0 \\ \dfrac{x}{2} - \dfrac{5}{6}y = \dfrac{1}{3} \end{cases}$

14. $\begin{cases} 3a + 4b = -7 \\ 2b - a = -1 \end{cases}$

15. $\begin{cases} y = 3x - 1 \\ y = 2x + 4 \end{cases}$

16. $\begin{cases} 0.6c + 0.5c = 0 \\ 0.02c + 0.09d = 0 \end{cases}$

Write a system of two equations in two variables to solve each problem.

17. CHILD CARE   On a mother's 22-mile commute to work, she drops her daughter off at a child care facility. The first part of the trip is 6 miles less than the second part. How long is each part of her morning commute?

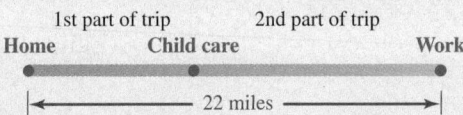

18. VACATIONING   It cost a family of 7 a total of $187 for general admission tickets to the San Diego Zoo. How many adult tickets and how many child tickets were purchased?

19. FINANCIAL PLANNING   A woman invested some money at 8% and some at 9%. The interest for 1 year on the combined investment of $10,000 was $840. How much was invested at each rate?

20. TAILWINDS/HEADWINDS   Flying with a tailwind, a pilot flew an airplane 450 miles in 2.5 hours. Flying into a headwind, the return trip took 3 hours. Find the speed of the plane in calm air and the speed of the wind.

21. TETHER BALL   The angles shown in the illustration are complementary. The measure of the larger angle is 10° more than three times the measure of the smaller angle. Find the measure of each angle.

22. ANTIFREEZE   How many pints of a 5% antifreeze solution and how many pints of a 20% antifreeze solution must be mixed to obtain 12 pints of a 15% solution?

23. SUNSCREEN   A sunscreen selling for $1.50 per ounce is to be combined with another sunscreen selling for $0.80 per ounce. How many ounces of each are needed to make 10 ounces of a sunscreen mix that sells for $1.01 per ounce?

24. Solve the system by graphing.

$$\begin{cases} 2x + 3y \leq 6 \\ x > 2 \end{cases}$$

25. CLOTHES SHOPPING   This system of inequalities describes the number of $20 shirts, x, and $40 pairs of pants, y, a person can buy if he or she plans to spend not less than $80 but not more than $120. Graph the system. Then give three solutions.

$$\begin{cases} 20x + 40y \geq 80 \\ 20x + 40y \leq 120 \end{cases}$$

26. Use a check to determine whether (3, 1) is a solution of the system: $\begin{cases} y \leq 2x - 1 \\ x + 3y > 6 \end{cases}$

GROUP PROJECT

WRITING APPLICATION PROBLEMS

Overview: In Section 4.4, you solved application problems by translating the words of the problem into a system of two equations. In this activity, you will reverse these steps.

Instructions: Form groups of 2 or 3 students. For each type of application, write a problem that could be solved using the given equations. If you need help getting started, refer to the specific problem types in the text. When finished writing the five applications, pick one problem and solve it completely.

A rectangle problem:
$$\begin{cases} 2l + 2w = 320 \\ l = w + 40 \end{cases}$$

An interest problem:
$$\begin{cases} x + y = 75,000 \\ 0.03x + 0.05y = 2,750 \end{cases}$$

A liquid mixture problem:
$$\begin{cases} x + y = 36 \\ 0.50x + 0.20y = 0.30(36) \end{cases}$$

A number-value problem:
$$\begin{cases} 5x + 2y = 23 \\ 3x + 7y = 37 \end{cases}$$

A with–against the wind problem:
$$\begin{cases} 2(x + y) = 600 \\ 3(x - y) = 600 \end{cases}$$

CHAPTER 5

Exponents and Polynomials

© Paul Arthur/Getty Images

from **Campus to Careers**
Sound Engineering Technician

Today's digitally recorded music is crystal clear thanks to the talents of sound engineering technicians. They operate console mixing boards and microphones to record the music, voices, and sound effects that are so much a part of our media-filled lives. The job requires strong mathematical skills and an aptitude for working with electronic equipment. Sound technicians constantly work with numbers as they read meters and graphs, adjust dials and switches, and keep written logs.

Many exciting advancements are taking place in the music industry. In **Problem 85** of **Study Set 5.4,** you will explore the rapid growth of iTunes, a computer program made by Apple Computer that plays, organizes, and enables us to buy music files and much more online.

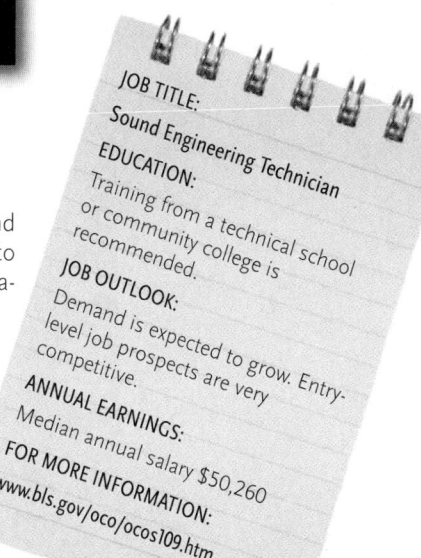

JOB TITLE:
Sound Engineering Technician

EDUCATION:
Training from a technical school or community college is recommended.

JOB OUTLOOK:
Demand is expected to grow. Entry-level job prospects are very competitive.

ANNUAL EARNINGS:
Median annual salary $50,260

FOR MORE INFORMATION:
www.bls.gov/oco/ocos109.htm

375

Study Skills Workshop
Attendance

It's not uncommon for students' enthusiasm to lessen toward the middle of the term. Sometimes their effort and attendance begin to slip. Realize that missing even one class can have a great effect on your grade. Being tardy takes its toll as well. If you are just a few minutes late, or miss an entire class, you risk getting behind. So, keep the following tips in mind.

ARRIVE ON TIME, OR A LITTLE EARLY: When you arrive, get out your note-taking materials and homework. Identify any questions or comments that you plan to ask your instructor once the class starts.

IF YOU MUST MISS A CLASS: Get a set of notes, the homework assignments, and any handouts that the instructor may have provided for the day(s) that you missed.

STUDY THE MATERIAL YOU MISSED: Take advantage of the help that comes with this textbook, such as the CD and the videotapes. Watch the explanations of the material from the section(s) that you missed.

Now Try This

1. Plan ahead! List five possible situations that could cause you to be late to class or miss a class. (Some examples are parking/traffic delays, lack of a babysitter, oversleeping, or job responsibilities.) What can you do ahead of time so that these situations won't cause you to be tardy or absent?

2. Watch one section on the CD or the videotape series that accompanies this book. Take notes as you watch the explanations.

SECTION 5.1
Rules for Exponents

Objectives

 ❶ Identify bases and exponents.

❷ Multiply exponential expressions that have like bases.

❸ Divide exponential expressions that have like bases.

❹ Raise exponential expressions to a power.

❺ Find powers of products and quotients.

In this section, we will use the definition of exponent to develop some rules for simplifying expressions that contain exponents.

❶ **Identify Bases and Exponents.**

Recall that an **exponent** indicates repeated multiplication. It indicates how many times the **base** is used as a factor. For example, 3^5 represents the product of five 3's.

$$\text{Exponent} \rightarrow \overset{\text{5 factors of 3}}{\overbrace{3^5 = 3 \cdot 3 \cdot 3 \cdot 3 \cdot 3}}$$

$$\text{Base} \nearrow$$

In general, we have the following definition.

Natural-Number Exponents

A natural-number exponent tells how many times its base is to be used as a factor. For any number x and any natural number n,

$$x^n = \overset{n \text{ factors of } x}{\overbrace{x \cdot x \cdot x \cdot \,\cdots\, \cdot x}}$$

Expressions of the form x^n are called **exponential expressions.** The base of an exponential expression can be a number, a variable, or a combination of numbers and variables. Some examples are:

Notation

Bases that contain a − sign *must* be written within parentheses.

$$(-2s)^3 \leftarrow \text{Exponent}$$
$$\underset{\text{Base}}{\mid}$$

$$10^5 = 10 \cdot 10 \cdot 10 \cdot 10 \cdot 10 \qquad \text{The base is 10. The exponent is 5. Read as "10 to the fifth power."}$$

$$y^2 = y \cdot y \qquad \text{The base is y. The exponent is 2. Read as "y squared."}$$

$$(-2s)^3 = (-2s)(-2s)(-2s) \qquad \text{The base is } -2s. \text{ The exponent is 3. Read as "negative 2s raised to the third power" or "negative 2s cubed."}$$

$$-8^4 = -(8 \cdot 8 \cdot 8 \cdot 8) \qquad \text{Since the } - \text{ sign is not written within parentheses, the base is 8. The exponent is 4. Read as "the opposite (or the negative) of 8 to the fourth power."}$$

When an exponent is 1, it is usually not written. For example, $4 = 4^1$ and $x = x^1$.

EXAMPLE 1 Identify the base and the exponent in each expression:
a. 8^5 **b.** $7a^3$ **c.** $(7a)^3$

Strategy To identify the base and exponent, we will look for the form ___.

Why The exponent is the small raised number to the right of the base.

Solution
a. In 8^5, the base is 8 and the exponent is 5.
b. $7a^3$ means $7 \cdot a^3$. Thus, the base is a, not $7a$. The exponent is 3.
c. Because of the parentheses in $(7a)^3$, the base is $7a$ and the exponent is 3.

Self Check 1 Identify the base and the exponent:
a. $3y^4$ **b.** $(3y)^4$

Now Try **Problems 13 and 17**

EXAMPLE 2 Write each expression in an equivalent form using an exponent:

a. $\dfrac{p}{3} \cdot \dfrac{p}{3} \cdot \dfrac{p}{3} \cdot \dfrac{p}{3}$ **b.** $5 \cdot t \cdot t \cdot t$

Strategy We will look for repeated factors and count the number of times each appears.

Why We can use an exponent to represent repeated multiplication.

Solution

a. Since there are four repeated factors of $\frac{p}{3}$ in $\frac{p}{3} \cdot \frac{p}{3} \cdot \frac{p}{3} \cdot \frac{p}{3}$, the expression can be written $\left(\frac{p}{3}\right)^4$.

b. Since there are three repeated factors of t in $5 \cdot t \cdot t \cdot t$, the expression can be written $5t^3$.

Self Check 2 Write as an exponential expression:
$(x + y)(x + y)(x + y)(x + y)(x + y)$

Now Try **Problems 25 and 27**

2 **Multiply Exponential Expressions That Have Like Bases.**

To develop a rule for multiplying exponential expressions that have the same base, we consider the product $6^2 \cdot 6^3$. Since 6^2 means that 6 is to be used as a factor two times, and 6^3 means that 6 is to be used as a factor three times, we have

$$6^2 \cdot 6^3 = \overbrace{6 \cdot 6}^{\text{2 factors of 6}} \cdot \overbrace{6 \cdot 6 \cdot 6}^{\text{3 factors of 6}}$$

$$= \overbrace{6 \cdot 6 \cdot 6 \cdot 6 \cdot 6}^{\text{5 factors of 6}}$$

$$= 6^5$$

We can quickly find this result if we keep the common base of 6 and add the exponents on 6^2 and 6^3.

$$6^2 \cdot 6^3 = 6^{2+3} = 6^5$$

This example suggests the following rule for exponents.

Product Rule for Exponents	To multiply exponential expressions that have the same base, keep the common base and add the exponents.

For any number x and any natural numbers m and n,

$$x^m \cdot x^n = x^{m+n}$$ Read as "x to the *m*th power times x to the *n*th power equals x to the *m* plus *n*th power."

EXAMPLE 3 Simplify: **a.** $9^5(9^6)$ **b.** $x^3 \cdot x^4$ **c.** $y^2 y^4 y$
d. $(x + 2)^8(x + 2)^7$ **e.** $(c^2 d^3)(c^4 d^5)$

Strategy In each case, we want to write an equivalent expression using one base and one exponent. We will use the product rule for exponents to do this.

Why The product rule for exponents is used to multiply exponential expressions that have the same base.

Caution

Don't make the mistake of multiplying the bases when using the product rule. Keep the *same* base.

$$9^5(9^6) \neq 81^{11}$$

Caution

Don't make the mistake of "distributing" an exponent over a sum (or difference). *There is no such rule.*

$$(x + 2)^{15} \neq x^{15} + 2^{15}$$

Solution

a. $9^5(9^6) = 9^{5+6} = 9^{11}$ Keep the common base, 9, and add the exponents.

Since 9^{11} is a very large number, we will leave the answer in this form. We won't evaluate it.

b. $x^3 \cdot x^4 = x^{3+4} = x^7$ Keep the common base, x, and add the exponents.

c. $y^2y^4y = y^2y^4y^1$ Write y as y^1.

$= y^{2+4+1}$ Keep the common base, y, and add the exponents.

$= y^7$

d. $(x + 2)^8(x + 2)^7 = (x + 2)^{8+7}$ Keep the common base, x + 2, and add the exponents.

$= (x + 2)^{15}$

e. $(c^2d^3)(c^4d^5) = (c^2c^4)(d^3d^5)$ Use the commutative and associative properties of multiplication to group like bases together.

$= (c^{2+4})(d^{3+5})$ Keep the common base, c, and add the exponents.
Keep the common base, d, and add the exponents.

$= c^6d^8$

 Self Check 3   Simplify: **a.** $7^8(7^7)$ **b.** x^2x^3x
   c. $(y - 1)^5(y - 1)^5$ **d.** $(s^4t^3)(s^4t^4)$

Now Try   Problems 35 and 41

Caution   We cannot use the product rule to simplify expressions like $3^2 \cdot 2^3$, where the bases are not the same. However, we can simplify this expression by doing the arithmetic:

$$3^2 \cdot 2^3 = 9 \cdot 8 = 72$$

EXAMPLE 4   *Geometry.*   Find an expression that represents the area of the rectangle.

x^3 feet

x^5 feet

Strategy   We will multiply the length of the rectangle by its width.

Why   The area of a rectangle is equal to the product of its length and width.

Solution

Area = **length** · **width**   This is the formula for the area of a rectangle.

$= x^5 \cdot x^3$ Substitute x^5 for the length and x^3 for the width.

$= x^{5+3}$ Use the product rule: Keep the common base, x, and add the exponents.

$= x^8$

The area of the rectangle is x^8 square feet, which can be written as x^8 ft^2.

 Now Try   Problem 45

3 **Divide Exponential Expressions That Have Like Bases.**

To develop a rule for dividing exponential expressions that have the same base, we consider the quotient $\frac{4^5}{4^2}$, where the exponent in the numerator is greater than the exponent in the denominator. We can simplify this fraction by removing the common factors of 4 in the numerator and denominator:

$$\frac{4^5}{4^2} = \frac{4 \cdot 4 \cdot 4 \cdot 4 \cdot 4}{4 \cdot 4} = \frac{\overset{1}{\cancel{4}} \cdot \overset{1}{\cancel{4}} \cdot 4 \cdot 4 \cdot 4}{\underset{1}{\cancel{4}} \cdot \underset{1}{\cancel{4}}} = 4^3$$

We can quickly find this result if we keep the common base, 4, and subtract the exponents on 4^5 and 4^2.

$$\frac{4^5}{4^2} = 4^{5-2} = 4^3$$

This example suggests another rule for exponents.

Quotient Rule for Exponents	To divide exponential expressions that have the same base, keep the common base and subtract the exponents.

For any nonzero number x and any natural numbers m and n, where $m > n$,

$$\frac{x^m}{x^n} = x^{m-n}$$ Read as "x to the mth power divided by x to the nth power equals x to the m minus nth power."

EXAMPLE 5 Simplify each expression:

a. $\dfrac{20^{16}}{20^9}$ **b.** $\dfrac{x^9}{x^3}$ **c.** $\dfrac{(7.5n)^{12}}{(7.5n)^{11}}$ **d.** $\dfrac{a^3b^8}{ab^5}$

Strategy In each case, we want to write an equivalent expression using one base and one exponent. We will use the quotient rule for exponents to do this.

Why The quotient rule for exponents is used to divide exponential expressions that have the same base.

Solution

> **Caution**
> Don't make the mistake of dividing the bases when using the quotient rule. Keep the *same* base.
> $$\frac{20^{16}}{20^9} \neq 1^7$$

a. $\dfrac{20^{16}}{20^9} = 20^{16-9}$ Keep the common base, 20, and subtract the exponents.

$\qquad\quad = 20^7$ Since 20^7 is a very large number, we will leave the answer in this form. We won't evaluate it.

b. $\dfrac{x^9}{x^3} = x^{9-3}$ Keep the common base, x, and subtract the exponents.

$\qquad\quad = x^6$

c. $\dfrac{(7.5n)^{12}}{(7.5n)^{11}} = (7.5n)^{12-11}$ Keep the common base, 7.5n, and subtract the exponents.

$\qquad\qquad\quad = (7.5n)^1$

$\qquad\qquad\quad = 7.5n$ Any number raised to the first power is simply that number.

d. $\dfrac{a^3 b^8}{ab^5} = \dfrac{a^3}{a^1} \cdot \dfrac{b^8}{b^5}$ Group the common bases together. Write *a* as a^1.

$= a^{3-1} b^{8-5}$ Keep the common base *a* and subtract the exponents.
                  Keep the common base *b* and subtract the exponents.

$= a^2 b^3$

Self Check 5   Simplify: **a.** $\dfrac{55^{30}}{55^5}$ **b.** $\dfrac{a^5}{a^3}$ **c.** $\dfrac{(8.9t)^8}{(8.9t)^7}$

                            d. $\dfrac{b^{15} c^4}{b^4 c}$

Now Try   **Problems 51 and 57**

EXAMPLE 6   Simplify: $\dfrac{a^3 a^5 a^7}{a^4 a}$

Strategy   We want to write an equivalent expression using one base and one exponent. First, we will use the product rule to simplify the numerator and the denominator. Then, we will use the quotient rule to simplify that result.

Why   The expression involves multiplication and division of exponential expressions that have the same base.

Solution   We simplify the numerator and denominator separately and proceed as follows.

$\dfrac{a^3 a^5 a^7}{a^4 a} = \dfrac{a^{15}}{a^5}$ In the numerator, keep the common base, *a*, and add the exponents. In the denominator, keep the common base, *a*, and add the exponents.

$= a^{15-5}$ Keep the common base, *a*, and subtract the exponents.

$= a^{10}$

Self Check 6   Simplify: $\dfrac{b^2 b^6 b}{b^4 b^4}$

Now Try   **Problem 63**

Recall that like terms are terms with exactly the same variables raised to exactly the same powers. To add or subtract exponential expressions, they must be like terms. To multiply or divide exponential expressions, only the bases need to be the same.

$x^5 + x^2$ These are not like terms; the exponents are different. We cannot add.

$x^2 + x^2 = 2x^2$ These are like terms; we can add. Recall that $x^2 = 1x^2$.

$x^5 \cdot x^2 = x^7$ The bases are the same; we can multiply.

$\dfrac{x^5}{x^2} = x^3$ The bases are the same; we can divide.

4   **Raise Exponential Expressions to a Power.**

To develop another rule for exponents, we consider $(5^3)^4$. Here, an exponential expression, 5^3, is raised to a power. Since 5^3 is the base and 4 is the exponent, $(5^3)^4$ can be written as $5^3 \cdot 5^3 \cdot 5^3 \cdot 5^3$. Because each of the four factors of 5^3 contains three factors of 5, there are $4 \cdot 3$ or 12 factors of 5.

The Language of Algebra
An exponential expression raised to a power, such as $(5^3)^4$, is also called a *power of a power*.

$$\overbrace{}^{\text{12 factors of } x}$$

$$(5^3)^4 = 5^3 \cdot 5^3 \cdot 5^3 \cdot 5^3 = \underbrace{5 \cdot 5 \cdot 5}_{5^3} \cdot \underbrace{5 \cdot 5 \cdot 5}_{5^3} \cdot \underbrace{5 \cdot 5 \cdot 5}_{5^3} \cdot \underbrace{5 \cdot 5 \cdot 5}_{5^3} = 5^{12}$$

We can quickly find this result if we keep the common base of 5 and multiply the exponents.

$$(5^3)^4 = 5^{3 \cdot 4} = 5^{12}$$

This example suggests the following rule for exponents.

Power Rule for Exponents

To raise an exponential expression to a power, keep the base and multiply the exponents. For any number x and any natural numbers m and n,

$$(x^m)^n = x^{m \cdot n} = x^{mn}$$ Read as "the quantity of x to the mth power raised to the nth power equals x to the mnth power."

EXAMPLE 7   Simplify: **a.** $(2^3)^7$ **b.** $[(-6)^2]^5$ **c.** $(z^8)^8$

Strategy   In each case, we want to write an equivalent expression using one base and one exponent. We will use the power rule for exponents to do this.

Why   Each expression is a power of a power.

Solution
a. $(2^3)^7 = 2^{3 \cdot 7} = 2^{21}$ Keep the base, 2, and multiply the exponents.

b. $[(-6)^2]^5 = (-6)^{2 \cdot 5} = (-6)^{10}$ Keep the base, −6, and multiply the exponents. Since $(-6)^{10}$ is a very large number, we will leave the answer in this form.

c. $(z^8)^8 = z^{8 \cdot 8} = z^{64}$ Keep the base, z, and multiply the exponents.

Self Check 7   Simplify: **a.** $(4^6)^5$ **b.** $(y^5)^2$
Now Try   Problems 71 and 73

EXAMPLE 8   Simplify: **a.** $(x^2x^5)^2$ **b.** $(z^2)^4(z^3)^3$

Strategy   In each case, we want to write an equivalent expression using one base and one exponent. We will use the product and power rules for exponents to do this.

Why   The expressions involve multiplication of exponential expressions that have the same base and they involve powers of powers.

Solution
a. $(x^2x^5)^2 = (x^7)^2$ Within the parentheses, keep the common base, x, and add the exponents.
$= x^{14}$ Keep the base, x, and multiply the exponents.

b. $(z^2)^4(z^3)^3 = z^8z^9$ For each power of z raised to a power, keep the base and multiply the exponents.
$= z^{17}$ Keep the common base, z, and add the exponents.

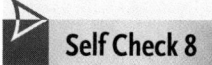

Self Check 8   Simplify: **a.** $(a^4a^3)^3$ **b.** $(a^3)^3(a^4)^2$

Now Try   Problems 77 and 81

5 **Find Powers of Products and Quotients.**

To develop more rules for exponents, we consider the expression $(2x)^3$, which is a *power of the product* of 2 and x, and the expression $\left(\frac{2}{x}\right)^3$, which is a *power of the quotient* of 2 and x.

$$(2x)^3 = 2x \cdot 2x \cdot 2x$$
$$= (2 \cdot 2 \cdot 2)(x \cdot x \cdot x)$$
$$= 2^3x^3$$
$$= 8x^3$$

$$\left(\frac{2}{x}\right)^3 = \frac{2}{x} \cdot \frac{2}{x} \cdot \frac{2}{x}$$ Assume $x \neq 0$.

$$= \frac{2 \cdot 2 \cdot 2}{x \cdot x \cdot x}$$ Multiply the numerators.
                   Multiply the denominators.

$$= \frac{2^3}{x^3}$$

$$= \frac{8}{x^3}$$ Evaluate: $2^3 = 8$.

These examples suggest the following rules for exponents.

Powers of a Product and a Quotient

To raise a product to a power, raise each factor of the product to that power. To raise a quotient to a power, raise the numerator and the denominator to that power.

   For any numbers x and y, and any natural number n,

$$(xy)^n = x^ny^n \qquad \text{and} \qquad \left(\frac{x}{y}\right)^n = \frac{x^n}{y^n}, \quad \text{where } y \neq 0$$

EXAMPLE 9   Simplify: **a.** $(3c)^4$ **b.** $(x^2y^3)^5$ **c.** $\left(-\frac{1}{4}a^3b\right)^2$

Strategy   In each case, we want to write the expression in an equivalent form in which each base is raised to a single power. We will use the power of a product rule for exponents to do this.

Why   Within each set of parentheses is a product, and each of those products is raised to a power.

Solution

a. $(3c)^4 = 3^4c^4$ Raise each factor of the product $3c$ to the 4th power.

   $= 81c^4$ Evaluate: $3^4 = 81$.

b. $(x^2y^3)^5 = (x^2)^5(y^3)^5$ Raise each factor of the product x^2y^3 to the 5th power.

   $= x^{10}y^{15}$ For each power of a power, keep each base, x and y, and multiply the exponents.

c. $\left(-\frac{1}{4}a^3b\right)^2 = \left(-\frac{1}{4}\right)^2(a^3)^2b^2$ Raise each factor of the product $-\frac{1}{4}a^3b$ to the 2nd power.

   $= \frac{1}{16}a^6b^2$ Evaluate: $\left(-\frac{1}{4}\right)^2 = \frac{1}{16}$. Keep the base a and multiply the exponents.

Self Check 9   Simplify: **a.** $(2t)^4$ **b.** $(c^3d^4)^6$
   c. $\left(-\frac{1}{3}ab^5\right)^3$

Now Try   **Problems 85 and 91**

EXAMPLE 10   Simplify: $\dfrac{(a^3b^4)^2}{ab^5}$

Strategy   We want to write the expression in an equivalent form in which each base is raised to a single power. We will use the power of a product rule and the quotient rule for exponents to do this.

Why   The expression involves a power of a product and it is the quotient of exponential expressions that have the same base.

Solution

$$\frac{(a^3b^4)^2}{ab^5} = \frac{(a^3)^2(b^4)^2}{a^1b^5}$$ In the numerator, raise each factor within the parentheses to the second power. In the denominator, write a as a^1.

$$= \frac{a^6b^8}{a^1b^5}$$ In the numerator, for each power of a power, keep each base, a and b, and multiply the exponents.

$$= a^{6-1}b^{8-5}$$ Keep each of the bases, a and b, and subtract the exponents.

$$= a^5b^3$$

Self Check 10   Simplify: $\dfrac{(c^4d^5)^3}{c^2d^3}$

Now Try   **Problem 93**

EXAMPLE 11   Simplify: $\dfrac{(5b)^9}{(5b)^6}$

Strategy   We want to write the expression in an equivalent form in which the base is raised to a single power. We will begin by using the quotient rule for exponents to write an equivalent expression.

Why   The expression involves division of exponential expressions that have the same base, $5b$.

Solution

$$\frac{(5b)^9}{(5b)^6} = (5b)^{9-6}$$ Keep the common base, $5b$, and subtract the exponents.

$$= (5b)^3$$

$$= 5^3b^3$$ Use the power of a product rule: Raise each factor within the parentheses to the 3rd power.

$$= 125b^3$$ Evaluate 5^3.

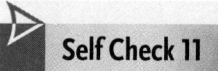

Self Check 11   Simplify: $\dfrac{(-2h)^{20}}{(-2h)^{14}}$

Now Try   Problem 97

EXAMPLE 12   Simplify: **a.** $\left(\dfrac{4}{k}\right)^3$ **b.** $\left(\dfrac{3x^2}{2y^3}\right)^5$

Strategy   We want to write each expression in an equivalent form using each base raised to a single power. We will use the power of a quotient rule for exponents to do this.

Why   Within each set of parentheses is a quotient, and each of those quotients is raised to a power.

Solution

a. Since $\dfrac{4}{k}$ is the quotient of 4 and k, the expression $\left(\dfrac{4}{k}\right)^3$ is a power of a quotient.

$$\left(\frac{4}{k}\right)^3 = \frac{4^3}{k^3} = \frac{64}{k^3} \qquad \text{Raise the numerator and denominator to the 3rd power.}$$
$$\text{Then evaluate } 4^3.$$

b. $\left(\dfrac{3x^2}{2y^3}\right)^5 = \dfrac{(3x^2)^5}{(2y^3)^5}$ Raise the numerator and the denominator to the 5th power.

$$= \frac{3^5(x^2)^5}{2^5(y^3)^5} \qquad \text{In the numerator and denominator, raise each factor within the}$$
$$\text{parentheses to the 5th power.}$$

$$= \frac{243x^{10}}{32y^{15}} \qquad \text{Evaluate } 3^5 \text{ and } 2^5. \text{ For each power of a power, keep the base and multiply}$$
$$\text{the exponents.}$$

Self Check 12   Simplify: **a.** $\left(\dfrac{x}{7}\right)^3$ **b.** $\left(\dfrac{2x^3}{3y^2}\right)^4$

Now Try   Problems 101 and 105

The rules for natural-number exponents are summarized as follows.

Rules for Exponents

If m and n represent natural numbers and there are no divisions by zero, then

Exponent of 1	**Product rule**	**Power rule**
$x^1 = x$	$x^m x^n = x^{m+n}$	$(x^m)^n = x^{mn}$

Quotient rule	**Power of a product**	**Power of a quotient**
$\dfrac{x^m}{x^n} = x^{m-n}$	$(xy)^n = x^n y^n$	$\left(\dfrac{x}{y}\right)^n = \dfrac{x^n}{y^n}$

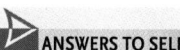

▷ **ANSWERS TO SELF CHECKS**   **1. a.** Base: y, exponent: 4   **b.** Base: $3y$, exponent: 4   **2.** $(x + y)^5$

3. a. 7^{15} **b.** x^6 **c.** $(y - 1)^{10}$ **d.** $s^8 t^7$ **5. a.** 55^{25} **b.** a^2 **c.** $8.9t$ **d.** $b^{11}c^3$ **6.** b

7. a. 4^{30} **b.** y^{10} **8. a.** a^{21} **b.** a^{17} **9. a.** $16t^4$ **b.** $c^{18}d^{24}$ **c.** $-\frac{1}{27}a^3b^{15}$ **10.** $c^{10}d^{12}$ **11.** $64h^6$

12. a. $\dfrac{x^3}{343}$ **b.** $\dfrac{16x^{12}}{81y^8}$

STUDY SET
5.1

VOCABULARY

Fill in the blank.

1. Expressions such as x^4, 10^3, and $(5t)^2$ are called _____ expressions.

2. Match each expression with the proper description.

$$\frac{a^8}{a^2} \qquad (a^4b^2)^5 \qquad \left(\frac{a^6}{a}\right)^3 \qquad (a^8)^4 \qquad a^5 \cdot a^3$$

   a. Product of exponential expressions with the same base
   b. Quotient of exponential expressions with the same base
   c. Power of an exponential expression
   d. Power of a product
   e. Power of a quotient

CONCEPTS

Fill in the blanks.

3. a. $(3x)^4 = $ ▢ $\cdot$ ▢ $\cdot$ ▢ $\cdot$ ▢
   b. $(-5y)(-5y)(-5y) = $ ▢

4. a. $x = x$ ▢   b. $x^m x^n = $ ▢
   c. $(xy)^n = $ ▢   d. $(a^b)^c = $ ▢
   e. $\dfrac{x^m}{x^n} = $ ▢   f. $\left(\dfrac{a}{b}\right)^n = $ ▢

5. To simplify each expression, determine whether you add, subtract, multiply, or divide the exponents.

   a. $\dfrac{x^8}{x^2}$ b. $b^6 \cdot b^9$
   c. $(n^8)^4$ d. $(a^4b^2)^5$

6. a. To simplify $(2y^3z^2)^4$, what factors within the parentheses must be raised to the fourth power?
   b. To simplify $\left(\frac{y^3}{z^2}\right)^4$, what two expressions must be raised to the fourth power?

Simplify each expression, if possible.

7. a. $x^2 + x^2$ b. $x^2 \cdot x^2$
8. a. $x^2 + x$ b. $x^2 \cdot x$
9. a. $x^3 - x^2$ b. $\dfrac{x^3}{x^2}$
10. a. $4^2 \cdot 2^4$ b. $\dfrac{x^3}{y^2}$

NOTATION

Complete each solution to simplify each expression.

11. $(x^4x^2)^3 = ($ ▢ $)^3 = x$ ▢

12. $\dfrac{a^3a^4}{a^2} = \dfrac{a}{a^2}$ ▢ $= a$ ▢ $^{-2} = a$ ▢

GUIDED PRACTICE

Identify the base and the exponent in each expression. See Example 1.

13. 4^3 14. $(-8)^2$

15. x^5 16. $\left(\dfrac{5}{x}\right)^3$

17. $(-3x)^2$ 18. $(2xy)^{10}$

19. $-\dfrac{1}{3}y^6$ 20. $-x^4$

21. $9m^{12}$ 22. $3.14r^4$

23. $(y + 9)^4$ 24. $(z - 2)^3$

Write each expression in an equivalent form using an exponent. See Example 2.

25. $4t \cdot 4t \cdot 4t \cdot 4t$
26. $-5u(-5u)(-5u)(-5u)(-5u)$
27. $-4 \cdot t \cdot t \cdot t \cdot t \cdot t$ 28. $-5 \cdot u \cdot u \cdot u$
29. $\dfrac{t}{2} \cdot \dfrac{t}{2} \cdot \dfrac{t}{2}$ 30. $\dfrac{x}{c} \cdot \dfrac{x}{c} \cdot \dfrac{x}{c} \cdot \dfrac{x}{c}$
31. $(x - y)(x - y)$ 32. $(m + 4)(m + 4)$

Use the product rule for exponents to simplify each expression. Write the results using exponents. See Example 3.

33. $5^3 \cdot 5^4$ 34. $3^4 \cdot 3^6$
35. $a^3 \cdot a^3$ 36. $m^7 \cdot m^7$
37. bb^2b^3 38. aa^3a^5
39. $(y - 2)^5(y - 2)^2$ 40. $(t + 1)^5(t + 1)^3$
41. $(a^2b^3)(a^3b^3)$ 42. $(u^3v^5)(u^4v^5)$
43. $cd^4 \cdot cd$ 44. $ab^3 \cdot ab^4$

Find an expression that represents the area or volume of each figure. Recall that the formula for the volume of a rectangular solid is V = length · width · height. See Example 4.

45.

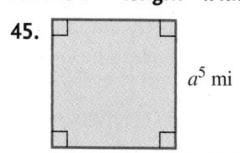

a^5 mi

a^5 mi

46.

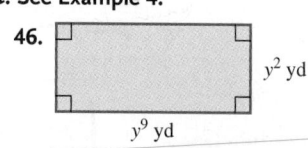

y^2 yd

y^9 yd

47.

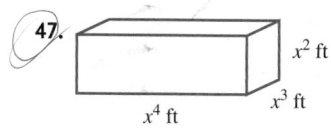

x^2 ft

x^3 ft

x^4 ft

48.

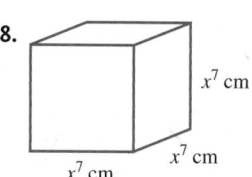

x^7 cm

x^7 cm

x^7 cm

Use the quotient rule for exponents to simplify each expression. Write the results using exponents. See Example 5.

49. $\dfrac{8^{12}}{8^4}$

50. $\dfrac{10^4}{10^2}$

51. $\dfrac{x^{15}}{x^3}$

52. $\dfrac{y^6}{y^3}$

53. $\dfrac{(3.7p)^7}{(3.7p)^2}$

54. $\dfrac{(0.25y)^9}{(0.25y)^3}$

55. $\dfrac{(k-2)^{15}}{(k-2)}$

56. $\dfrac{(m+8)^{20}}{(m+8)}$

57. $\dfrac{c^3d^7}{cd}$

58. $\dfrac{r^8s^9}{rs}$

59. $\dfrac{x^4y^7}{xy^3}$

60. $\dfrac{p^7q^{10}}{p^2q^7}$

Use the product and quotient rules for exponents to simplify each expression. See Example 6.

61. $\dfrac{y^3y^4}{yy^2}$

62. $\dfrac{b^4b^5}{b^2b^3}$

63. $\dfrac{a^2a^3a^4}{a^8}$

64. $\dfrac{h^3h^6h}{h^9}$

65. $\dfrac{t^5t^6t}{t^2t^3}$

66. $\dfrac{m^5m^{12}m}{m^7m^4}$

67. $\dfrac{s^2s^2s^2}{s^3s}$

68. $\dfrac{w^4w^4w^4}{w^2w}$

Use the power rule for exponents to simplify each expression. Write the results using exponents. See Example 7.

69. $(3^2)^4$

70. $(4^3)^3$

71. $[(-4.3)^3]^8$

72. $[(-1.7)^9]^8$

73. $(m^{50})^{10}$

74. $(n^{25})^4$

75. $(y^5)^3$

76. $(b^3)^6$

Use the product and power rules for exponents to simplify each expression. See Example 8.

77. $(x^2x^3)^5$

78. $(y^3y^4)^4$

79. $(p^2p^3)^5$

80. $(r^3r^4)^2$

81. $(t^3)^4(t^2)^3$

82. $(b^2)^5(b^3)^2$

83. $(u^4)^2(u^3)^2$

84. $(v^5)^2(v^3)^4$

Use the power of a product rule for exponents to simplify each expression. See Example 9.

85. $(6a)^2$

86. $(3b)^3$

87. $(5y)^4$

88. $(4t)^4$

89. $(-2r^2s^3)^3$

90. $(-2x^2y^4)^5$

91. $\left(-\dfrac{1}{3}y^2z^4\right)^5$

92. $\left(-\dfrac{1}{4}t^3u^8\right)^2$

Use rules for exponents to simplify each expression. See Example 10.

93. $\dfrac{(ab^2)^3}{a^2b^2}$

94. $\dfrac{(m^3n^4)^3}{m^3n^6}$

95. $\dfrac{(r^4s^3)^4}{r^3s^9}$

96. $\dfrac{(x^2y^5)^5}{x^6y^2}$

Use rules for exponents to simplify each expression. See Example 11.

97. $\dfrac{(6k)^7}{(6k)^4}$

98. $\dfrac{(-3a)^{12}}{(-3a)^{10}}$

99. $\dfrac{(3q)^5}{(3q)^3}$

100. $\dfrac{(ab)^8}{(ab)^4}$

Use the power of a quotient rule for exponents to simplify each expression. See Example 12.

101. $\left(\dfrac{a}{b}\right)^3$

102. $\left(\dfrac{r}{s}\right)^4$

103. $\left(\dfrac{m}{3}\right)^4$

104. $\left(\dfrac{n}{5}\right)^3$

105. $\left(\dfrac{8a^2}{11b^5}\right)^2$

106. $\left(\dfrac{7g^4}{6h^3}\right)^2$

107. $\left(\dfrac{3m^4}{2n^5}\right)^5$

108. $\left(\dfrac{2s^2}{3t^5}\right)^5$

TRY IT YOURSELF

Simplify each expression.

109. $\left(\dfrac{x^2}{y^3}\right)^5$

110. $\left(\dfrac{u^4}{v^2}\right)^6$

111. $y^3y^2y^4$

112. y^4yy^6

113. $\dfrac{15^9}{15^6}$

114. $\dfrac{25^{13}}{25^7}$

115. $\left(\dfrac{y^3y^5}{yy^2}\right)^3$

116. $\left(\dfrac{s^5s^6}{s^2s^2}\right)^4$

117. $(-6a^3b^2)^3$

118. $(-10r^3s^2)^2$

119. $\dfrac{(a^2b^2)^{15}}{(ab)^9}$

120. $\dfrac{(s^3t^3)^4}{(st)^2}$

121. $(n^4n)^3(n^3)^6$

122. $(y^3y)^2(y^2)^2$

123. $\dfrac{(6h)^8}{(6h)^6}$

124. $\dfrac{(-7r)^{10}}{(-7r)^8}$

APPLICATIONS

125. ART HISTORY Leonardo da Vinci's drawing relating a human figure to a square and a circle is shown. Find an expression for the following:

a. The area of the square if the man's height is $5x$ feet.

b. The area of the circle if the waist-to-feet distance is $3a$ feet. Leave π in your answer.

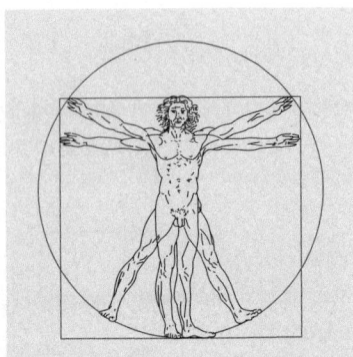

126. PACKAGING A bowling ball fits tightly against all sides of a cardboard box that it is packaged in. Find expressions for the volume of the ball and box. Leave π in your answer.

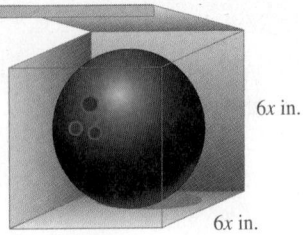

$6x$ in.

$6x$ in.

$6x$ in.

127. CHILDBIRTH Mr. and Mrs. Emory Harrison, of Johnson City, Tennessee, had 13 sons in a row during the 1940s and 1950s. The **probability** of a family of 13 children all being male is $\left(\frac{1}{2}\right)^{13}$. Evaluate this expression.

128. TOYS A Super Ball is dropped from a height of 1 foot and always rebounds to four-fifths of its previous height. The rebound height of the ball after the third bounce is $\left(\frac{4}{5}\right)^3$ feet. Evaluate this expression. Is the third bounce more or less than $\frac{1}{2}$ foot high?

WRITING

129. Explain the mistake in the following work.

$$2^3 \cdot 2^2 = 4^5 = 1{,}024$$

130. Explain why we can simplify $x^4 \cdot x^5$, but cannot simplify $x^4 + x^5$.

REVIEW

Match each equation with its graph below.

131. $y = 2x - 1$ **132.** $y = 3x - 1$
133. $y = 3$ **134.** $x = 3$

a. 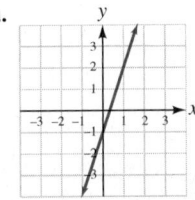 **b.**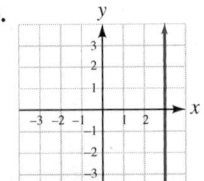

c. 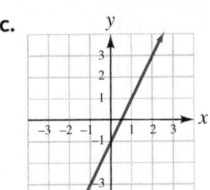 **d.**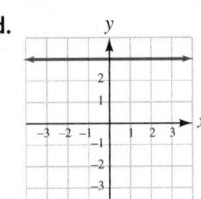

CHALLENGE PROBLEMS

135. Simplify each expression. The variables represent natural numbers.

a. $x^{2m}x^{3m}$ **b.** $(y^{5c})^4$
c. $\dfrac{m^{8x}}{m^{4x}}$ **d.** $(2a^{6y})^4$

136. Is the operation of raising to a power commutative? That is, is $a^b = b^a$? Explain.

SECTION 5.2
Zero and Negative Exponents

Objectives

1. Use the zero exponent rule.
2. Use the negative integer exponent rule.
3. Use exponent rules to change negative exponents in fractions to positive exponents.
4. Use all exponent rules to simplify expressions.

We now extend the discussion of natural-number exponents to include exponents that are zero and exponents that are negative integers.

1. **Use the Zero Exponent Rule.**

To develop the definition of a zero exponent, we will simplify the expression $\frac{5^3}{5^3}$ in two ways and compare the results.

First, we apply the quotient rule for exponents, where we subtract the equal exponents in the numerator and denominator. The result is 5^0. In the second approach, we write 5^3 as $5 \cdot 5 \cdot 5$ and remove the common factors of 5 in the numerator and denominator. The result is 1.

$$\frac{5^3}{5^3} = 5^{3-3} = 5^0 \qquad \frac{5^3}{5^3} = \frac{\overset{1}{\cancel{5}} \cdot \overset{1}{\cancel{5}} \cdot \overset{1}{\cancel{5}}}{\underset{1}{\cancel{5}} \cdot \underset{1}{\cancel{5}} \cdot \underset{1}{\cancel{5}}} = 1$$

These results must be equal.

Since $\frac{5^3}{5^3} = 5^0$ and $\frac{5^3}{5^3} = 1$, we conclude that $5^0 = 1$. This observation suggests the following definition.

Zero Exponents	Any nonzero base raised to the 0 power is 1. For any nonzero real number x, $$x^0 = 1$$

EXAMPLE 1   Simplify. Assume $a \neq 0$:   **a.** $(-8)^0$ **b.** $\left(\dfrac{14}{15}\right)^0$ **c.** $(3a)^0$ **d.** $3a^0$

Strategy   We note that each exponent is 0. To simplify the expressions, we will identify the base and use the zero-exponent rule.

Why   If an expression contains a nonzero base raised to the 0 power, we can replace it with 1.

The Language of Algebra
The zero exponent definition does not define 0^0. This expression is called an *indeterminate form*, which is beyond the scope of this book.

Solution

a. $(-8)^0 = 1$ *Because the base is −8 and the exponent is 0.*

b. $\left(\dfrac{14}{15}\right)^0 = 1$ *Because the base is $\frac{14}{15}$ and the exponent is 0.*

c. $(3a)^0 = 1$ *Because of the parentheses, the base is 3a. The exponent is 0.*

d. $3a^0 = 3 \cdot a^0$ Since there are no parentheses, the base is a, not $3a$. The exponent is 0.

$= 3 \cdot 1$

$= 3$

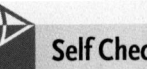 **Self Check 1**   Simplify each expression: **a.** $(0.75)^0$ **b.** $-5c^0d$
c. $(5c)^0$

Now Try   **Problems 13 and 17**

 **2**   **Use the Negative Integer Exponent Rule.**

The Language of Algebra
The *negative integers* are:
$-1, -2, -3, -4, -5 \ldots$

To develop the definition of a negative exponent, we will simplify $\frac{6^2}{6^5}$ in two ways and compare the results.

If we apply the quotient rule for exponents, where we subtract the greater exponent in the denominator from the lesser exponent in the numerator, we get 6^{-3}. In the second approach, we remove the two common factors of 6 to get $\frac{1}{6^3}$.

$$\frac{6^2}{6^5} = 6^{2-5} = 6^{-3} \qquad \frac{6^2}{6^5} = \frac{\overset{1}{\cancel{6}} \cdot \overset{1}{\cancel{6}}}{\cancel{6} \cdot \cancel{6} \cdot 6 \cdot 6 \cdot 6} = \frac{1}{6^3}$$
$$\underset{1 \quad 1}{}$$

⎿——— These must be equal. ———⏌

Since $\frac{6^2}{6^5} = 6^{-3}$ and $\frac{6^2}{6^5} = \frac{1}{6^3}$, we conclude that $6^{-3} = \frac{1}{6^3}$. Note that 6^{-3} is equal to the reciprocal of 6^3. This observation suggests the following definition.

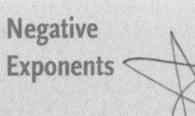 **Negative Exponents**

For any nonzero real number x and any integer n,

$$x^{-n} = \frac{1}{x^n}$$

In words, x^{-n} is the reciprocal of x^n.

From the definition, we see that another way to write x^{-n} is to write its reciprocal and change the sign of the exponent. For example,

$$5^{-4} = \frac{1}{5^4}$$ Think of the reciprocal of 5^{-4}, which is $\frac{1}{5^{-4}}$. Then change the sign of the exponent.

EXAMPLE 2   Express using positive exponents and simplify, if possible:
a. 3^{-2} **b.** y^{-1} **c.** $(-2)^{-3}$ **d.** $5^{-2} - 10^{-2}$

Strategy   Since each exponent is a negative number, we will use the negative exponent rule.

Why   This rule enables us to write an exponential expression that has a negative exponent in an equivalent form using a positive exponent.

Solution

a. $3^{-2} = \dfrac{1}{3^2} = \dfrac{1}{9}$ Write the reciprocal of 3^{-2} and change the sign of the exponent.

b. $y^{-1} = \dfrac{1}{y^1} = \dfrac{1}{y}$ Write the reciprocal of y^{-1} and change the sign of the exponent.

c. $(-2)^{-3} = \dfrac{1}{(-2)^3}$ Because of the parentheses, the base is -2. Write the reciprocal of $(-2)^{-3}$ and change the exponent.

$\qquad\quad = -\dfrac{1}{8}$ Evaluate: $(-2)^3 = -8$.

d. $5^{-2} - 10^{-2} = \dfrac{1}{5^2} - \dfrac{1}{10^2}$ Write the reciprocal of 5^{-2} and 10^{-2} and change the sign of each exponent.

$\qquad\qquad\quad = \dfrac{1}{25} - \dfrac{1}{100}$ Evaluate: $5^2 = 25$ and $10^2 = 100$.

$\qquad\qquad\quad = \dfrac{4}{100} - \dfrac{1}{100}$ Build $\frac{1}{25}$ to have a denominator of 100 so that the fractions can be subtracted: $\frac{1}{25} = \frac{1}{25} \cdot \frac{4}{4} = \frac{4}{100}$.

$\qquad\qquad\quad = \dfrac{3}{100}$

Self Check 2   Express using positive exponents and simplify, if possible:
a. 8^{-2} **b.** x^{-5} **c.** $(-3)^{-3}$ **d.** $2^{-3} - 4^{-2}$

Now Try   **Problems 25, 29, and 37**

EXAMPLE 3   Simplify. Do not use negative exponents in the answer.
a. $9m^{-3}$ **b.** -5^{-2}

Strategy   We note that each exponent is a negative number. We will identify the base for each negative exponent and then use the negative exponent rule.

Why   This rule enables us to write an exponential expression that has a negative exponent in an equivalent form using a positive exponent.

Solution
a. $9m^{-3} = 9 \cdot m^{-3}$ The base is m. The exponent is -3.

$\qquad\quad = 9 \cdot \dfrac{1}{m^3}$ Write the reciprocal of m^{-3} and change the sign of the exponent. Since 9 is not part of the base, it is not part of the reciprocal.

$\qquad\quad = \dfrac{9}{m^3}$ Multiply.

b. $-5^{-2} = -1 \cdot 5^{-2}$ The base is 5. The exponent is -2.

$\qquad\quad = -1 \cdot \dfrac{1}{5^2}$ Write the reciprocal of 5^{-2} and change the sign of the exponent. Since -1 is not part of the base, it is not part of the reciprocal.

$\qquad\quad = -\dfrac{1}{25}$ Evaluate 5^2 and multiply. The result is negative because of the $-$ sign in front of -5^{-2}.

> **Self Check 3**   Simplify. Do not use negative exponents in the answer.
> **a.** $12h^{-9}$ **b.** -2^{-4}
>
> **Now Try**   Problems 41 and 45

3 **Use Exponent Rules to Change Negative Exponents in Fractions to Positive Exponents.**

Negative exponents can appear in the numerator and/or the denominator of a fraction. To develop rules for such situations, we consider the following example.

$$\frac{a^{-4}}{b^{-3}} = \frac{\dfrac{1}{a^4}}{\dfrac{1}{b^3}} = \frac{1}{a^4} \div \frac{1}{b^3} = \frac{1}{a^4} \cdot \frac{b^3}{1} = \frac{b^3}{a^4}$$

We can obtain this result in a simpler way. In $\frac{a^{-4}}{b^{-3}}$, we can move a^{-4} from the numerator to the denominator and change the sign of the exponent, and we can move b^{-3} from the denominator to the numerator and change the sign of the exponent.

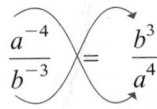

The Language of Algebra
Factors of a numerator or denominator may be moved *across the fraction bar* if we change the sign of their exponent.

This example suggests the following rules.

Changing from Negative to Positive Exponents	A factor can be moved from the denominator to the numerator or from the numerator to the denominator of a fraction if the sign of its exponent is changed. For any nonzero real numbers x and y, and any integers m and n, $$\frac{1}{x^{-n}} = x^n \qquad \text{and} \qquad \frac{x^{-m}}{y^{-n}} = \frac{y^n}{x^m}$$

These rules streamline the process when simplifying fractions involving negative exponents.

EXAMPLE 4   Simplify. Do not use negative exponents in the answer.
a. $\dfrac{1}{d^{-10}}$ **b.** $\dfrac{2^{-3}}{3^{-4}}$ **c.** $\dfrac{-5s^{-2}}{t^{-9}}$

Strategy   We will move any factors in the numerator that have a negative exponent to the denominator. Then we will move any factors in the denominator that have a negative exponent to the numerator.

Why   In this process, the sign of a negative exponent changes to positive.

Solution

a. $\dfrac{1}{d^{-10}} = d^{10}$ Move d^{-10} to the numerator and change the sign of the exponent.

Caution

A common error is to mistake the − sign in −5 for a negative exponent. *It is not an exponent.* The −5 should not be moved to the denominator and have its sign changed.

$$\frac{-5s^{-2}}{t^{-9}} \neq \frac{t^9}{5s^2}$$

b. $\dfrac{2^{-3}}{3^{-4}} = \dfrac{3^4}{2^3}$ Move 2^{-3} to the denominator and change the sign of the exponent. Move 3^{-4} to the numerator and change the sign of the exponent.

$$= \frac{81}{8}$$ Evaluate: $3^4 = 81$ and $2^3 = 8$.

c. $\dfrac{-5s^{-2}}{t^{-9}} = \dfrac{-5t^9}{s^2}$ Since $-5s^{-2}$ has no parentheses, s is the base. Move only s^{-2} to the denominator and change the sign of the exponent. Do not move -5. Move t^{-9} to the numerator and change the sign of the exponent.

Self Check 4 Simplify. Do not use negative exponents in the answer.

a. $\dfrac{1}{w^{-5}}$ **b.** $\dfrac{5^{-2}}{4^{-3}}$ **c.** $\dfrac{-8h^{-6}}{a^{-7}}$

Now Try **Problems 51, 55, and 59**

When a fraction is raised to a negative power, we can use rules for exponents to change the sign of the exponent. For example, we see that

↓ The exponent is the opposite of −3.

$$\left(\frac{a}{2}\right)^{-3} = \frac{a^{-3}}{2^{-3}} = \frac{2^3}{a^3} = \left(\frac{2}{a}\right)^3$$

↑ The base is the reciprocal of $\frac{a}{2}$.

This process can be streamlined using the following rule.

Negative Exponents and Reciprocals

A fraction raised to a power is equal to the reciprocal of the fraction raised to the opposite power.

For any nonzero real numbers x and y, and any integer n,

$$\left(\frac{x}{y}\right)^{-n} = \left(\frac{y}{x}\right)^n$$

EXAMPLE 5 Simplify: $\left(\dfrac{4}{m}\right)^{-2}$

Strategy We want to write this fraction that is raised to a negative power in an equivalent form that involves a positive power. We will use the negative exponent and reciprocal rules to do this.

Why It is usually easier to simplify exponential expressions if the exponents are positive.

Solution

$$\left(\frac{4}{m}\right)^{-2} = \left(\frac{m}{4}\right)^2$$ The base is the fraction $\frac{4}{m}$ and the exponent is −2. Write the reciprocal of the base and change the sign of the exponent.

$$= \frac{m^2}{4^2}$$ Use the power of a quotient rule: Raise the numerator and denominator to the second power.

$$= \frac{m^2}{16}$$ Evaluate: $4^2 = 16$.

Self Check 5 Simplify: $\left(\frac{c}{9}\right)^{-2}$

Now Try Problem 65

4 **Use All Exponent Rules to Simplify Expressions.**

The rules for exponents involving products, powers, and quotients are also true for zero and negative exponents.

Summary of Exponent Rules

If m and n represent integers and there are no divisions by zero, then

Product rule
$$x^m \cdot x^n = x^{m+n}$$

Power rule
$$(x^m)^n = x^{mn}$$

Power of a product
$$(xy)^n = x^n y^n$$

Quotient rule
$$\frac{x^m}{x^n} = x^{m-n}$$

Power of a quotient
$$\left(\frac{x}{y}\right)^n = \frac{x^n}{y^n}$$

Exponents of 0 and 1
$$x^0 = 1 \text{ and } x^1 = x$$

Negative exponent
$$x^{-n} = \frac{1}{x^n}$$

Negative exponents appearing in fractions
$$\frac{1}{x^{-n}} = x^n \qquad \frac{x^{-m}}{y^{-n}} = \frac{y^n}{x^m} \qquad \left(\frac{x}{y}\right)^{-n} = \left(\frac{y}{x}\right)^n$$

The rules for exponents are used to simplify expressions involving products, quotients, and powers. In general, an expression involving exponents is simplified when

- Each base occurs only once
- There are no parentheses
- There are no negative or zero exponents

EXAMPLE 6 Simplify. Do not use negative exponents in the answer.

a. $x^5 \cdot x^{-3}$ **b.** $\dfrac{x^3}{x^7}$ **c.** $(x^3)^{-2}$ **d.** $(2a^3b^{-5})^3$ **e.** $\left(\dfrac{3}{b^5}\right)^{-4}$

Strategy In each case, we want to write an equivalent expression using one base and one positive exponent. We will use rules for exponents to do this.

Why These expressions are not in simplest form. In parts a and b, the base x occurs more than once. In parts c, d, and e, there is a negative exponent.

Solution

a. $x^5 \cdot x^{-3} = x^{5+(-3)} = x^2$ Use the product rule: Keep the base, x, and add exponents.

b. $\dfrac{x^3}{x^7} = x^{3-7}$ Use the quotient rule: Keep the base, x, and subtract the exponents.

$= x^{-4}$ Do the subtraction: 3 − 7 = −4.

$= \dfrac{1}{x^4}$ Write the reciprocal of x^{-4} and change the sign of the exponent.

Success Tip

We can use the negative exponent rule to simplify $(x^3)^{-2}$ in an alternate way:

$$(x^3)^{-2} = \frac{1}{(x^3)^2} = \frac{1}{x^6}$$

c. $(x^3)^{-2} = x^{-6}$ Use the power rule: Keep the base, x, and multiply exponents.

$\qquad\;\; = \dfrac{1}{x^6}$ Write the reciprocal of x^{-6} and change the sign of the exponent.

d. $(2a^3b^{-5})^3 = 2^3(a^3)^3(b^{-5})^3$ Raise each factor of the product $2a^3b^{-5}$ to the 3rd power.

$\qquad\qquad\;\; = 8a^9b^{-15}$ ~~Use the power rule: Multiply exponents.~~

$\qquad\qquad\;\; = \dfrac{8a^9}{b^{15}}$ Move b^{-15} to the denominator and change the sign of the exponent.

e. $\left(\dfrac{3}{b^5}\right)^{-4} = \left(\dfrac{b^5}{3}\right)^4$ The base is the fraction $\frac{3}{b^5}$ and its exponent is -4. Write the reciprocal of the base and change the sign of the exponent.

$\qquad\qquad\; = \dfrac{(b^5)^4}{(3)^4}$ Use the power of a quotient rule: Raise the numerator and denominator to the 4th power.

$\qquad\qquad\; = \dfrac{b^{20}}{81}$ Use the power rule: Keep the base, b, and multiply the exponents 5 and 4. Evaluate: $3^4 = 81$.

  Self Check 6   Simplify. Do not use negative exponents in the answer.

a. $t^8 \cdot t^{-4}$ **b.** $\dfrac{a^3}{a^8}$ **c.** $(n^4)^{-5}$

d. $(4c^2d^{-1})^3$ **e.** $\left(\dfrac{c^4}{2}\right)^{-3}$

Now Try   **Problems 69, 73, 77, 81, and 85**

EXAMPLE 7   Simplify. Do not use negative exponents in the answer.

a. $\dfrac{y^{-4}y^{-3}}{y^{-20}}$ **b.** $\dfrac{7^{-1}a^3b^4}{6^{-2}a^5b^2}$ **c.** $\left(\dfrac{x^{-3}y^2}{xy^{-3}}\right)^2$

Strategy   In each case, we want to write an equivalent expression that uses each base with a positive exponent only once.

Why   These expressions are not in simplest form. The bases occur more than once and the expressions contain negative exponents.

Solution

Caution

We cannot use this approach if the negative exponents occur in a sum or difference of terms. For example:

$$\frac{y^{-4} + y^{-3}}{y^{-20}} \neq \frac{y^{20}}{y^4 + y^3}$$

You will study this situation in more detail in your next algebra course.

a. $\dfrac{y^{-4}y^{-3}}{y^{-20}} = \dfrac{y^{-7}}{y^{-20}}$ In the numerator, use the product rule: Keep the common base, y, and add exponents: $-4 + (-3) = -7$.

$\qquad\qquad = y^{-7-(-20)}$ Use the quotient rule: Keep the common base, y, and subtract exponents.

$\qquad\qquad = y^{13}$ Do the subtraction: $-7 - (-20) = -7 + 20 = 13$.

Alternate solution: To avoid working with negative numbers, we could move each factor across the fraction bar and change the sign of its exponent.

$$\frac{y^{-4}y^{-3}}{y^{-20}} = \frac{y^{20}}{y^4y^3} = \frac{y^{20}}{y^7} = y^{13}$$

Success Tip

Since the rules for exponents can be applied in different orders, there are several equally valid ways to simplify these expressions.

b. $\dfrac{7^{-1}a^3b^4}{6^{-2}a^5b^2} = \dfrac{6^2a^3b^4}{7^1a^5b^2}$

Move 7^{-1} to the denominator. Change the sign of the exponent.
Move 6^{-2} to the numerator. Change the sign of the exponent.

$= \dfrac{36a^{3-5}b^{4-2}}{7}$

Use the quotient rule twice: Keep each base, a and b, and subtract exponents.

$= \dfrac{36a^{-2}b^2}{7}$

$= \dfrac{36b^2}{7a^2}$

Move a^{-2} to the denominator and change the sign of the exponent.

c. $\left(\dfrac{x^{-3}y^2}{xy^{-3}}\right)^2 = [x^{-3-1}y^{2-(-3)}]^2$

Within the parentheses, use the quotient rule twice: Keep each base, x and y, and subtract exponents.

$= (x^{-4}y^5)^2$

$= x^{-8}y^{10}$

Raise each factor within the parentheses to the second power.

$= \dfrac{y^{10}}{x^8}$

Move x^{-8} to the denominator and change the sign of its exponent. y^{10} is not affected.

Alternate solution: To simplify the expression, we can begin on the "outside" by using the power of a quotient rule first.

$$\left(\dfrac{x^{-3}y^2}{xy^{-3}}\right)^2 = \dfrac{(x^{-3})^2(y^2)^2}{x^2(y^{-3})^2} = \dfrac{x^{-6}y^4}{x^2y^{-6}} = \dfrac{y^4y^6}{x^6x^2} = \dfrac{y^{10}}{x^8}$$

 Self Check 7

Simplify. Do not use negative exponents in the answer.

a. $\dfrac{a^{-4}a^{-5}}{a^{-3}}$ **b.** $\dfrac{1^{-4}x^5y^3}{9^{-2}x^3y^6}$ **c.** $\left(\dfrac{c^{-2}d^2}{c^4d^{-3}}\right)^3$

Now Try **Problems 89, 93, and 97**

ANSWERS TO SELF CHECKS 1. **a.** 1 **b.** $-5d$ **c.** 1 2. **a.** $\frac{1}{64}$ **b.** $\frac{1}{x^5}$ **c.** $-\frac{1}{27}$ **d.** $\frac{1}{16}$
3. **a.** $\frac{12}{h^9}$ **b.** $-\frac{1}{16}$ 4. **a.** w^5 **b.** $\frac{64}{25}$ **c.** $-\frac{8a^7}{h^6}$ 5. $\frac{81}{c^2}$ 6. **a.** t^4 **b.** $\frac{1}{a^5}$ **c.** $\frac{1}{n^{20}}$ **d.** $\frac{64c^6}{d^3}$ **e.** $\frac{8}{c^{12}}$
7. **a.** $\frac{1}{a^6}$ **b.** $\frac{81x^2}{y^3}$ **c.** $\frac{d^{15}}{c^{18}}$

STUDY SET
5.2

VOCABULARY

Fill in the blanks.

1. In the expression 5^{-1}, the exponent is a _____ integer.

2. x^{-n} is the _____ of x^n.

CONCEPTS

3. Complete the table.

Expression	Base	Exponent
4^{-2}		
$6x^{-5}$		
$\left(\frac{3}{y}\right)^{-8}$		
-7^{-1}		
$(-2)^{-3}$		
$10a^0$		

4. Complete each rule for exponents.

a. $x^m \cdot x^n = $

b. $x^0 = $

c. $(x^m)^n = $

d. $(xy)^n = $

e. $\left(\dfrac{x}{y}\right)^n = $

f. $x^{-n} = $

g. $\dfrac{1}{x^{-n}} = $

h. $\dfrac{x^{-m}}{y^{-n}} = $

i. $\dfrac{x^m}{x^n} = $

j. $\left(\dfrac{x}{y}\right)^{-n} = $

Complete each table.

5.

x	3^x
2	
1	
0	
-1	
-2	

6.

x	$(-9)^x$
2	
1	
0	
-1	
-2	

7. Fill in the blanks.

a. $2^{-3} = \dfrac{1}{2}$

b. $\dfrac{1}{t^{-6}} = t$

8. A factor can be moved from the denominator to the numerator or from the numerator to the denominator of a fraction if the _____ of its exponent is changed.

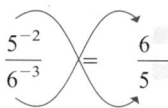

$$\dfrac{5^{-2}}{6^{-3}} \quad = \quad \dfrac{6}{5}$$

9. A fraction raised to a power is equal to the _____ of the fraction raise to the opposite power.

$$\left(\dfrac{3}{d}\right)^{-2} = \left(\dfrac{d}{3}\right)$$

10. Determine whether each statement is true or false.

a. $6^{-2} = -36$

b. $6^{-2} = \dfrac{1}{36}$

c. $\dfrac{x^3}{y^{-2}} = \dfrac{y^2}{x^3}$

d. $\dfrac{-6x^{-5}}{y^{-6}} = \dfrac{y^6}{6x^5}$

NOTATION

Complete each solution to simplify each expression.

11. $(y^5y^3)^{-5} = \left(\quad\right)^{-5} = y \quad = \dfrac{1}{y}$

12. $\left(\dfrac{a^2b}{a^{-3}b^3}\right)^3 = \left(a^{2-}\ b^{1}\ \right)^3$

$$= (a\ \ b\ \ )^3$$

$$= a\ \ b$$

$$= \dfrac{a^{15}}{b}$$

GUIDED PRACTICE

Simplify each expression. See Example 1.

13. 7^0

14. 9^0

15. $\left(\dfrac{1}{4}\right)^0$

16. $\left(\dfrac{3}{8}\right)^0$

17. $2x^0$

18. $8t^0$

19. $15(-6x)^0$

20. $4(-12y)^0$

21. $\left(\dfrac{a^2b^3}{ab^4}\right)^0$

22. $\left(\dfrac{xyz}{x^2y}\right)^0$

23. $\dfrac{5}{2x^0}$

24. $\dfrac{4}{3a^0}$

Express using positive exponents and simplify, if possible. See Example 2.

25. 2^{-2}

26. 7^{-2}

27. 6^{-1}

28. 5^{-1}

29. x^{-9}

30. y^{-3}

31. b^{-5}

32. c^{-4}

33. $(-5)^{-1}$

34. $(-8)^{-1}$

35. $(-10)^{-3}$

36. $(-9)^{-2}$

37. $2^{-2} + 4^{-1}$

38. $-9^{-1} + 9^{-2}$

39. $9^0 - 9^{-1}$

40. $7^{-1} - 7^0$

Simplify. Do not use negative exponents in the answer. See Example 3.

41. $15g^{-6}$

42. $16t^{-3}$

43. $5x^{-3}$

44. $27m^{-3}$

45. -3^{-3}

46. -6^{-3}

47. -8^{-2}

48. -4^{-2}

Simplify. Do not use negative exponents in the answer. See Example 4.

49. $\dfrac{1}{5^{-3}}$

50. $\dfrac{1}{3^{-3}}$

51. $\dfrac{1}{r^{-20}}$

52. $\dfrac{1}{s^{-30}}$

53. $\dfrac{8}{s^{-1}}$

54. $\dfrac{6}{k^{-2}}$

55. $\dfrac{2^{-4}}{3^{-1}}$

56. $\dfrac{7^{-2}}{2^{-3}}$

57. $\dfrac{a^{-5}}{b^{-2}}$

58. $\dfrac{r^{-6}}{s^{-1}}$

59. $\dfrac{-4d^{-1}}{p^{-10}}$

60. $\dfrac{-9m^{-1}}{n^{-30}}$

Simplify. See Example 5.

61. $\left(\dfrac{1}{6}\right)^{-2}$

62. $\left(\dfrac{1}{7}\right)^{-2}$

63. $\left(\dfrac{1}{2}\right)^{-3}$

64. $\left(\dfrac{1}{5}\right)^{-3}$

65. $\left(\dfrac{c}{d}\right)^{-8}$

66. $\left(\dfrac{a}{x}\right)^{-10}$

67. $\left(\dfrac{3}{m}\right)^{-4}$

68. $\left(\dfrac{2}{t}\right)^{-4}$

Simplify. Do not use negative exponents in the answer. See Example 6.

69. $y^8 \cdot y^{-2}$

70. $m^{10} \cdot m^{-6}$

71. $b^{-7} \cdot b^{14}$

72. $c^{-9} \cdot c^{14}$

73. $\dfrac{y^4}{y^5}$

74. $\dfrac{t^7}{t^{10}}$

75. $\dfrac{h^{-5}}{h^2}$

76. $\dfrac{y^{-3}}{y^4}$

77. $(x^4)^{-3}$

78. $(y^{-3})^2$

79. $(b^2)^{-4}$

80. $(n^3)^{-5}$

81. $(6s^4t^{-7})^2$

82. $(11r^{10}s^{-3})^2$

83. $(2u^{-2}v^5)^5$

84. $(3w^{-8}x^3)^4$

85. $\left(\dfrac{4}{x^3}\right)^{-3}$

86. $\left(\dfrac{2}{b^5}\right)^{-2}$

87. $\left(\dfrac{y^4}{3}\right)^{-2}$

88. $\left(\dfrac{p^3}{2}\right)^{-2}$

Simplify. Do not use negative exponents in the answer. See Example 7.

89. $\dfrac{y^{-3}}{y^{-4}y^{-2}}$

90. $\dfrac{x^{-12}}{x^{-3}x^{-4}}$

91. $\dfrac{a^{-5}a^{-9}}{a^{-8}}$

92. $\dfrac{b^{-2}b^{-3}}{b^{-9}}$

93. $\dfrac{2^{-1}a^4b^2}{3^{-2}a^2b^4}$

94. $\dfrac{6^{-2}b^9c^3}{5^{-3}b^4c^8}$

95. $\dfrac{9^{-2}s^6t}{4^{-3}s^4t^5}$

96. $\dfrac{2^{-5}m^{10}n^6}{5^{-2}m^6n^{10}}$

97. $\left(\dfrac{x^2y^{-2}}{x^{-5}y^3}\right)^4$

98. $\left(\dfrac{r^4s^{-3}}{r^{-3}s^7}\right)^3$

99. $\left(\dfrac{y^3z^{-2}}{y^{-4}z^3}\right)^2$

100. $\left(\dfrac{xy^3}{x^{-1}y^{-1}}\right)^3$

TRY IT YOURSELF

Simplify. Do not use negative exponents in the answer. Assume that no variables are 0.

101. $\left(\dfrac{a^4}{2b}\right)^{-3}$

102. $\left(\dfrac{n^8}{9m}\right)^{-2}$

103. $\dfrac{r^{-50}}{r^{-70}}$

104. $\dfrac{m^{-30}}{m^{-40}}$

105. $(5d^{-2})^3$

106. $(9s^{-6})^2$

107. $-15x^0y$

108. $24g^0h^2$

109. $\left(\dfrac{4}{h^{10}}\right)^{-2}$

110. $\left(\dfrac{x^4}{3}\right)^{-4}$

111. $x^{-3} \cdot x^{-3}$

112. $y^{-2} \cdot y^{-2}$

113. $\left(\dfrac{c^3d^{-4}}{c^{-1}d^5}\right)^3$

114. $\left(\dfrac{s^2t^{-8}}{s^{-9}t^2}\right)^4$

115. $\dfrac{2^{-2}g^{-2}h^{-3}}{9^{-1}h^{-3}}$

116. $\dfrac{5^{-1}x^{-2}y^{-3}}{8^{-2}x^{-11}}$

117. $(2x^3y^{-2})^5$

118. $(3u^{-2}v^3)^3$

119. $\dfrac{t(t^{-2})^{-2}}{t^{-5}}$

120. $\dfrac{d(d^{-3})^{-3}}{d^{-7}}$

121. $\dfrac{-4s^{-5}}{t^{-2}}$

122. $\dfrac{-9k^{-8}}{m^{-2}}$

123. $(x^{-4}x^3)^3$

124. $(y^{-2}y)^3$

APPLICATIONS

125. **UNIT COMPARISONS**   Consider the relative sizes of the items below. In the measurement column, write the most appropriate power of 10 from the following list. Each power is used only once. (A meter is slightly longer than one yard.)

10^{-5} 10^{-4} 10^{-3} 10^{-2} 10^{-1} 10^0

smallest   largest

Item	Measurement (meter)
Thickness of a dime	
Height of a bathroom sink	
Length of a pencil eraser	
Thickness of soap bubble film	
Length of a cell phone	
Thickness of a piece of paper	

126. **ELECTRONICS**   The total resistance R of a certain circuit is given by

$$R = \left(\dfrac{1}{R_1} + \dfrac{1}{R_2}\right)^{-1} + R_3$$

Find R if $R_1 = 4$, $R_2 = 2$, and $R_3 = 1$.

WRITING

127. Explain how you would help a friend understand that 2^{-3} is not equal to -8.

128. Explain the error:

$$\dfrac{-5x^{-2}}{y^{-2}} = \dfrac{y^2}{5x^2}$$

Find the slope of the line that passes through the given points.

129. $(1, -4)$ and $(3, -7)$ **130.** $(1, 3)$ and $(3, -1)$

131. Write an equation of the line having slope $\frac{3}{4}$ and y-intercept -5.

132. Find an equation of the line that passes through $(4, 4)$ and $(-6, -6)$. Write the answer in slope–intercept form.

133. Simplify each expression. Do not use negative exponents in the answer. The variable m represents a positive integer.

    a. $r^{5m}r^{-6m}$ **b.** $\dfrac{x^{3m}}{x^{6m}}$

134. Write an expression equivalent to $\left(\dfrac{2x^3y^7}{3z^5}\right)^9$ that involves only *negative* exponents.

SECTION 5.3
Scientific Notation

Objectives

1 Convert from scientific to standard notation.

2 Write numbers in scientific notation.

3 Perform computations with scientific notation.

Scientists often deal with extremely large and small numbers. For example, the distance from the Earth to the sun is approximately 150,000,000 kilometers. The influenza virus, which causes flu symptoms of cough, sore throat, and headache, has a diameter of 0.00000256 inch.

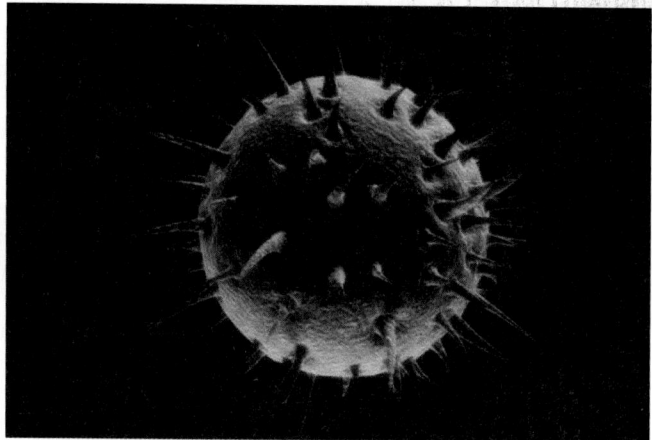

The numbers 150,000,000 and 0.00000256 are written in **standard notation,** which is also called **decimal notation.** Because they contain many zeros, they are difficult to read and cumbersome to work with in calculations. In this section, we will discuss a more convenient form in which we can write such numbers.

 Convert from Scientific to Standard Notation.

Scientific notation provides a compact way of writing very large or very small numbers.

Scientific Notation	A positive number is written in **scientific notation** when it is written in the form $N \times 10^n$, where $1 \le N < 10$ and n is an integer.	

To write numbers in scientific notation, you need to be familiar with **powers of 10,** like those listed in the table below.

Power of 10	10^{-3}	10^{-2}	10^{-1}	10^0	10^1	10^2	10^3
Value	$\dfrac{1}{1,000} = 0.001$	$\dfrac{1}{100} = 0.01$	$\dfrac{1}{10} = 0.1$	1	10	100	1,000

Two examples of numbers written in scientific notation are shown below. Note that each of them is the product of a decimal number (between 1 and 10) and a power of 10.

An integer exponent
$$3.67 \times 10^2 \qquad 2.158 \times 10^{-3}$$
A decimal that is at least 1, but less than 10

A number written in scientific notation can be converted to standard notation by performing the indicated multiplication. For example, to convert 3.67×10^2, we recall that multiplying a decimal by 100 moves the decimal point 2 places to the right.

$$3.67 \times 10^2 = 3.67 \times 100 = 3\,6\,7.$$

To convert 2.158×10^{-3} to standard notation, we recall that dividing a decimal by 1,000 moves the decimal point 3 places to the left.

$$2.158 \times 10^{-3} = 2.158 \times \frac{1}{10^3} = 2.158 \times \frac{1}{1,000} = \frac{2.158}{1,000} = 0.0\,0\,2\,1\,5\,8$$

In 3.67×10^2 and 2.158×10^{-3}, the exponent gives the number of decimal places that the decimal point moves, and the sign of the exponent indicates the direction in which it moves. Applying this observation to several other examples, we have

$5.32 \times 10^6 = 5\,3\,2\,0\,0\,0\,0.$ Move the decimal point 6 places to the right.

$1.95 \times 10^{-5} = 0.0\,0\,0\,0\,1\,9\,5$ Move the decimal point $|-5| = 5$ places to the left.

$9.7 \times 10^0 = 9.7$ There is no movement of the decimal point.

The following procedure summarizes our observations.

Converting from Scientific to Standard Notation	1. If the exponent is positive, move the decimal point the same number of places to the right as the exponent. 2. If the exponent is negative, move the decimal point the same number of places to the left as the absolute value of the exponent.

| **EXAMPLE 1** | Convert to standard notation: **a.** 3.467×10^5 **b.** 8.9×10^{-4} |

Strategy   In each case, we need to identify the exponent on the power of 10 and consider its sign.

Why   The exponent gives the number of decimal places that we should move the decimal point. The sign of the exponent indicates whether it should be moved to the right or the left.

The Language of Algebra
Standard notation is also called *decimal notation*.

Solution

a. Since the exponent in 10^5 is 5, the decimal point moves 5 places to the right.

$3\ 4\ 6\ 7\ 0\ 0.$ *To move 5 places to the right, two placeholder zeros must be written.*

Thus, $3.467 \times 10^5 = 346{,}700$.

b. Since the exponent in 10^{-4} is -4, the decimal point moves 4 places to the left.

$0.0\ 0\ 0\ 8\ 9$ *To move 4 places to the left, three placeholder zeros must be written.*

Thus, $8.9 \times 10^{-4} = 0.00089$

| **Self Check 1** | Convert to standard notation: **a.** 4.88×10^6 **b.** 9.8×10^{-3} |

Now Try   Problems 13 and 17

2   Write Numbers in Scientific Notation.

To write a number in scientific notation ($N \times 10^n$) we first determine N and then n.

| **EXAMPLE 2** | Write each number in scientific notation: **a.** 150,000,000   **b.** 0.00000256   **c.** 432×10^5 |

Strategy   We will write each number as the product of a number between 1 and 10 and a power of 10.

Why   Numbers written in scientific notation have the form $N \times 10^n$.

Solution

a. We must write 150,000,000 (the distance from the Earth to the sun) as the product of a number between 1 and 10 and a power of 10. We note that 1.5 lies between 1 and 10. To obtain 150,000,000, we must move the decimal point in 1.5 exactly 8 places to the right.

$1.5\ 0\ 0\ 0\ 0\ 0\ 0\ 0$

This will happen if we multiply 1.5 by 10^8. Therefore,

$150{,}000{,}000 = 1.5 \times 10^8$ *This is the distance (in kilometers) from the Earth to the sun.*

Notation

Don't apply the negative exponent rule when writing numbers in scientific notation.

2.56×10^{-6} ~~$2.56 \times \dfrac{1}{10^6}$~~

Calculators

Scientific notation

Calculators automatically change into scientific notation when a result is too large or too small to fit the answer display.

$(453.46)^5 = 1.917321395 \times 10^{13}$

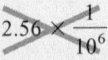

$(0.0005)^{12} = 2.44140625 \times 10^{-40}$

$2.44140625 \quad {}^{-40}$

b. We must write 0.00000256 (the diameter in inches of a flu virus) as the product of a number between 1 and 10 and a power of 10. We note that 2.56 lies between 1 and 10. To obtain 0.00000256, the decimal point in 2.56 must be moved 6 places to the left.

$0\,0\,0\,0\,0\,0\,2.56$

This will happen if we multiply 2.56 by 10^{-6}. Therefore,

$0.00000256 = 2.56 \times 10^{-6}$ *This is the diameter (in inches) of a flu virus.*

c. The number 432×10^5 is not written in scientific notation because 432 is not a number between 1 and 10. To write this number in scientific notation, we proceed as follows:

$432 \times 10^5 = 4.32 \times 10^2 \times 10^5$ *Write 432 in scientific notation.*

$\qquad\qquad = 4.32 \times 10^7$ *Use the product rule to find $10^2 \times 10^5$. Keep the base, 10, and add the exponents.*

Written in scientific notation, 432×10^5 is 4.32×10^7.

▷ **Self Check 2**   Write each number in scientific notation:
a. 93,000,000   **b.** 0.00009055
c. 85×10^{-3}

Now Try   **Problems 31, 35, and 55**

The results from Example 2 illustrate the following forms to use when converting numbers from standard to scientific notation.

For real numbers between 0 and 1: $\Box \times 10^{\text{negative integer}}$
For real numbers at least 1, but less than 10: $\Box \times 10^0$
For real numbers greater than or equal to 10: $\Box \times 10^{\text{positive integer}}$

3   **Perform Computations with Scientific Notation.**

Another advantage of scientific notation becomes apparent when we evaluate products or quotients that involve very large or small numbers. If we express those numbers in scientific notation, we can use rules for exponents to make the calculations easier.

EXAMPLE 3   *Astronomy.*   Except for the sun, the nearest star visible to the naked eye from most parts of the United States is Sirius. Light from Sirius reaches Earth in about 70,000 hours. If light travels at approximately 670,000,000 mph, how far from Earth is Sirius?

Strategy   We can use the formula $d = rt$ to find the distance from Sirius to Earth.

Why   We know the *rate* at which light travels and the *time* it takes to travel from Sirius to the Earth.

Solution   The rate at which light travels is 670,000,000 mph and the time it takes the light to travel from Sirius to Earth is 70,000 hr. To find the distance from Sirius to Earth, we proceed as follows:

$$d = rt$$ This is the formula for distance traveled.

$$d = 670,000,000(70,000)$$ Substitute 670,000,000 for r and 70,000 for t.

$$= (6.7 \times 10^8)(7.0 \times 10^4)$$ Write each number in scientific notation.

$$= (6.7 \cdot 7.0) \times (10^8 \times 10^4)$$ Group the decimals together and the powers of 10 together.

$$= (6.7 \cdot 7.0) \times 10^{8+4}$$ Use the product rule to find $10^8 \times 10^4$. Keep the base, 10, and add exponents.

$$= 46.9 \times 10^{12}$$ Do the multiplication and the addition.

We note that 46.9 is not between 0 and 1, so 46.9×10^{12} is not written in scientific notation. To answer in scientific notation, we proceed as follows.

$$= 4.69 \times 10^1 \times 10^{12}$$ Write 46.9 in scientific notation as 4.69×10^1.

$$= 4.69 \times 10^{13}$$ Use the product rule to find $10^1 \times 10^{12}$. Keep the base, 10, and add the exponents.

Sirius is approximately 4.69×10^{13} or 46,900,000,000,000 miles from Earth.

 Now Try   Problem 61

EXAMPLE 4   ***Atoms.***   As an example of how scientific notation is used in chemistry, we can approximate the weight (in grams) of one atom of the element uranium by evaluating the following expression.

$$\frac{2.4 \times 10^2}{6 \times 10^{23}}$$

Strategy   To simplify, we will divide the numbers and powers of 10 separately.

Why   We can then use the quotient rule for exponents to simplify the calculations.

Solution

$$\frac{2.4 \times 10^2}{6 \times 10^{23}} = \frac{2.4}{6} \times \frac{10^2}{10^{23}}$$ Divide the decimals and the powers of 10 separately.

$$= \frac{2.4}{6} \times 10^{2-23}$$ For the powers of 10, use the quotient rule. Keep the base, 10, and subtract the exponents.

$$= 0.4 \times 10^{-21}$$ Divide the decimals. Subtract the exponents. The result is not in scientific notation form.

$$= 4 \times 10^{-1} \times 10^{-21}$$ Write 0.4 in scientific notation as 4×10^{-1}.

$$= 4 \times 10^{-22}$$ Use the product rule to find $10^{-1} \times 10^{-21}$. Keep the base, 10, and add the exponents.

One atom of uranium weighs 4×10^{-22} gram or 0.0000000000000000000004 g.

 Self Check 4   Find the approximate weight (in grams) of one atom of gold by evaluating: $\frac{1.98 \times 10^2}{6 \times 10^{23}}$

Now Try   Problem 65

Calculators

Entering scientific notation
To evaluate the expression in Example 4 on a scientific calculator, we enter the numbers using the EE key:

2.4 EE 2 ÷ 6 EE 23 =

STUDY SET
5.3

VOCABULARY

Fill in the blanks.

1. 4.84×10^5 is written in _____ notation. 484,000 is written in _____ notation.
2. 10^3, 10^{50}, and 10^{-4} are _____ of 10.

CONCEPTS

Fill in the blanks.

3. When we multiply a decimal by 10^5, the decimal point moves 5 places to the _____. When we multiply a decimal by 10^{-7}, the decimal point moves 7 places to the _____.
4. Describe the procedure for converting a number from scientific notation to standard form.
   a. If the exponent is positive, move the decimal point the same number of places to the _____ as the exponent.
   b. If the exponent is negative, move the decimal point the same number of places to the _____ as the absolute value of the exponent.
5. a. When a real number greater than or equal to 10 is written in scientific notation, the exponent on 10 is a _____ integer.
   b. When a real number between 0 and 1 is written in scientific notation, the exponent on 10 is a _____ integer.
6. The arrows show the movement of a decimal point. By what power of 10 was each decimal multiplied?
   a. 0.000000556 b. $8,041,000,000.$

Fill in the blanks to write each number in scientific notation.

7. a. $7,700 = \times 10^3$ b. $500,000 = \times 10^5$
   c. $114,000,000 = 1.14 \times 10^{}$
8. a. $0.0082 = \times 10^{-3}$
   b. $0.0000001 = \times 10^{-7}$
   c. $0.00003457 = 3.457 \times 10^{}$
9. Write each expression so that the decimal numbers are grouped together and the powers of ten are grouped together.
   a. $(5.1 \times 10^9)(1.5 \times 10^{22})$
   b. $\dfrac{8.8 \times 10^{30}}{2.2 \times 10^{19}}$

10. Simplify each expression.
    a. $10^{24} \times 10^{33}$ b. $\dfrac{10^{50}}{10^{36}}$
    c. $\dfrac{10^{15} \times 10^{27}}{10^{40}}$

NOTATION

11. Fill in the blanks. A positive number is written in scientific notation when it is written in the form $N \times 10^n$, where $ \le N < $ and n is an _____.
12. Express each power of 10 in fraction form and decimal form.
    a. 10^{-3} b. 10^{-6}

GUIDED PRACTICE

Convert each number to standard notation. See Example 1.

13. 2.3×10^2	14. 3.75×10^4
15. 8.12×10^5	16. 1.2×10^3
17. 1.15×10^{-3}	18. 4.9×10^{-2}
19. 9.76×10^{-4}	20. 7.63×10^{-5}
21. 6.001×10^6	22. 9.998×10^5
23. 2.718×10^0	24. 3.14×10^0
25. 6.789×10^{-2}	26. 4.321×10^{-1}
27. 2.0×10^{-5}	28. 7.0×10^{-6}

Write each number in scientific notation. See Example 2.

29. 23,000	30. 4,750
31. 1,700,000	32. 290,000
33. 0.062	34. 0.00073
35. 0.0000051	36. 0.04
37. 5,000,000,000	38. 7,000,000
39. 0.0000003	40. 0.0001
41. 909,000,000	42. 7,007,000,000
43. 0.0345	44. 0.000000567
45. 9	46. 2
47. 11	48. 55
49. 1,718,000,000,000,000,000	
50. 44,180,000,000,000,000,000	
51. 0.0000000000000123	
52. 0.0000000000000000555	
53. 73×10^4	54. 99×10^5
55. 201.8×10^{15}	56. 154.3×10^{17}

57. 0.073×10^{-3}

58. 0.0017×10^{-4}

59. 36.02×10^{-20}

60. 56.29×10^{-30}

Use scientific notation to perform the calculations. Give all answers in scientific notation and standard notation. See Examples 3 and 4.

61. $(3.4 \times 10^2)(2.1 \times 10^3)$

62. $(4.1 \times 10^{-3})(3.4 \times 10^4)$

63. $(8.4 \times 10^{-13})(4.8 \times 10^9)$

64. $(5.5 \times 10^{-15})(2.2 \times 10^{13})$

65. $\dfrac{2.24 \times 10^4}{5.6 \times 10^7}$

66. $\dfrac{2.47 \times 10^5}{3.8 \times 10^{-5}}$

67. $\dfrac{9.3 \times 10^2}{3.1 \times 10^{-2}}$

68. $\dfrac{7.2 \times 10^6}{1.2 \times 10^8}$

69. $\dfrac{0.00000129}{0.0003}$

70. $\dfrac{169,000,000,000}{26,000,000}$

71. $(0.0000000056)(5,500,000)$

72. $(0.000000061)(3,500,000,000)$

73. $\dfrac{96,000}{(12,000)(0.00004)}$

74. $\dfrac{(0.48)(14,400,000)}{96,000,000}$

75. $\dfrac{2,475}{(132,000,000,000,000)(0.25)}$

76. $\dfrac{147,000,000,000,000}{(0.000049)(25)}$

Find each power.

77. $(456.4)^6$

78. $(0.009)^{-6}$

79. 225^{-5}

80. $\left(\dfrac{1}{3}\right)^{-55}$

APPLICATIONS

81. ASTRONOMY   The distance from Earth to Alpha Centauri (the nearest star outside our solar system) is about 25,700,000,000,000 miles. Write this number in scientific notation.

82. WATER   According to the U.S. Geological Survey , the total water supply of the world is 326,000,000,000,000,000,000 gallons. Write this number in scientific notation.

83. EARTH, SUN, MOON   The surface area of Earth is 1.97×10^8 square miles, the surface area of the sun is 1.09×10^{17} square miles, and the surface area of the moon is 1.46×10^7 square miles. Convert each number to standard notation.

84. ATOMS   The number of atoms in 1 gram of iron is approximately 1.08×10^{22}. Convert this number to standard notation.

85. SAND   The mass of one grain of beach sand is approximately 0.00000000045 ounce. Write this number in scientific notation.

86. MOLECULES   The mass of a water molecule is approximately 0.00000000000000000000001056 ounce. Write this number in scientific notation.

87. WAVELENGTHS   Examples of the most common types of electromagnetic waves are given in the table. List the wavelengths in order from shortest to longest.

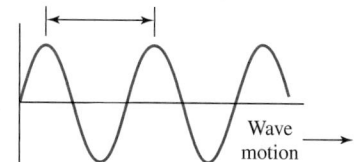

This distance between the two crests of the wave is called the wavelength.

Wave motion

Type	Use	Wavelength (in meters)
visible light	lighting	9.3×10^{-6}
infrared	photography	3.7×10^{-5}
x-ray	medical	2.3×10^{-11}
radio wave	communication	3.0×10^2
gamma ray	treating cancer	8.9×10^{-14}
microwave	cooking	1.1×10^{-2}
ultraviolet	sun lamp	6.1×10^{-8}

88. EXPLORATION   On July 4, 1997, the *Pathfinder*, carrying the rover vehicle called Sojourner, landed on Mars. The distance from Mars to Earth is approximately 3.5×10^7 miles. Use scientific notation to express this distance in feet. (*Hint:* 5,280 feet = 1 mile.)

89. PROTONS   The mass of one proton is approximately 1.7×10^{-24} gram. Use scientific notation to express the mass of 1 million protons.

90. SOUND   The speed of sound in air is approximately 3.3×10^4 centimeters per second. Use scientific notation to express this speed in kilometers per second. (*Hint:* 100 centimeters = 1 meter and 1,000 meters = 1 kilometer.)

91. LIGHT YEARS   One light year is about 5.87×10^{12} miles. Use scientific notation to express this distance in feet. (*Hint:* 5,280 feet = 1 mile.)

92. OIL As of 2006, Saudi Arabia was believed to have crude oil reserves of about 2.643×10^{11} barrels. A barrel contains 42 gallons of oil. Use scientific notation to express Saudi Arabia's oil reserves in gallons. (Source: *infoplease*)

93. INSURED DEPOSITS As of June 2006, the total insured deposits in U.S. banks and savings and loans was approximately 6.4×10^{12} dollars. If this money was invested at a rate of 4% simple annual interest, how much would it earn in 1 year? Use scientific notation to express the answer. (Source: Federal Deposit Insurance Corporation.)

94. CURRENCY As of December 2006, the number of $20 bills in circulation was approximately 5.96×10^9. What was the total value of the currency? Express the answer in scientific notation and standard notation. (Source: The Federal Reserve.)

95. POWERS OF 10 In the United States, we use Latin prefixes in front of "illion" to name extremely large numbers. Write each number in scientific notation.

One million: 1,000,000

One billion: 1,000,000,000

One trillion: 1,000,000,000,000

One quadrillion: 1,000,000,000,000,000

One quintillion: 1,000,000,000,000,000,000

96. SUPERCOMPUTERS As of June 2006, the world's fastest computer was IBM's BlueGene/L System. If it could make 2.81×10^{14} calculations in one second, how many could it make in one minute? Answer in scientific notation.

97. In what situations would scientific notation be more convenient than standard notation?

98. To multiply a number by a power of 10, we move the decimal point. Which way, and how far? Explain.

99. 2.3×10^{-3} contains a negative sign but represents a positive number. Explain.

100. Explain why 237.8×10^8 is not written in scientific notation.

101. If $y = -1$, find the value of $-5y^{55}$.

102. What is the y-intercept of the graph of $y = -3x - 5$?

103. COUNSELING At the end of her first year of practice, a family counselor had 75 clients. At the end of her second year, she had 105 clients. If a linear trend continues, write an equation that gives the number of clients c the counselor will have at the end of t years.

104. Is $(0, -5)$ a solution of $2x + 3y \geq -14$?

105. Consider 2.5×10^{-4}. Answer the following questions in scientific notation form.
   a. What is its opposite?
   b. What is its reciprocal?

106. a. Write the numbers one million and one millionth in scientific notation.
   b. By what number must we multiply one millionth to get one million?

SECTION 5.4
Polynomials

Objectives
1. Know the vocabulary for polynomials.
2. Evaluate polynomials.
3. Graph equations defined by polynomials.

In this section, we will discuss a special type of algebraic expression called a *polynomial*.

1 Know the Vocabulary for Polynomials.

Recall from Chapter 1 that a *term* is a product or quotient of numbers and/or variables. A single number or variable is also a term. Some examples of terms are:

$$14, \quad x, \quad -6y^3, \quad 9cd^2, \quad \text{and} \quad \frac{5}{y}$$

Polynomials

A **polynomial** is a single term or a sum of terms in which all variables have whole-number exponents and no variable appears in a denominator.

Here are some examples of polynomials:

$$3x + 2, \quad 4y^2 - 2y - 3, \quad a^3 + 3a^2b + 3ab^2 + b^3, \quad \text{and} \quad -8xy^2z$$

The polynomial $3x + 2$ is the sum of two terms, $3x$ and 2, and we say it is a **polynomial in one variable, x.** A single number is called a **constant,** and so its last term, 2, is called the **constant term.**

A polynomial is defined as a single term or the sum of several terms. Since $4y^2 - 2y - 3$ can be written as the sum $4y^2 + (-2y) + (-3)$, it has three terms, $4y^2$, $-2y$, and -3. It is written in **descending powers** of y, because the exponents on y decrease from left to right. When a polynomial is written in descending powers, the first term, in this case $4y^2$, is called the **leading term.** The coefficient of the leading term, in this case 4, is called the **leading coefficient.**

A polynomial can have more than one variable. For example, $a^3 + 3a^2b + 3ab^3 + b^3$ is a **polynomial in two variables,** a and b. It has four terms and is written in descending powers of a and **ascending powers** of b. The polynomial $-8xy^2z$ is a polynomial in three variables, $x, y,$ and z, and has only one term.

Polynomials are classified according to the number of terms they have. A polynomial with exactly one term is called a **monomial;** exactly two terms, a **binomial;** and exactly three terms, a **trinomial.** Polynomials with four or more terms have no special names.

Polynomials		
Monomials	Binomials	Trinomials
$-6x$	$9u - 4$	$5t^2 + 4t + 3$
$5.5x^3y^2$	$-29z^4 - z^2$	$27x^3 - 6x^2 - 2x$
11	$18a^2b + 4ab$	$\frac{1}{2}a^2 + 2ab + b^2$

Polynomials and their terms can be described according to the exponents on their variables.

Degree of a Term of a Polynomial

The **degree of a term** of a polynomial in one variable is the value of the exponent on the variable. If a polynomial is in more than one variable, the **degree of a term** is the sum of the exponents on the variables in that term. The **degree of a nonzero constant** is 0.

Here are some examples:

$9x^6$ has degree **6**.

$-2a^4$ has degree **4**.

$47x^2y^{11}$ has degree **13** because $2 + 11 = 13$.

8 has degree **0** since it can be written as $8x^0$.

We determine the *degree of a polynomial* by considering the degrees of each of its terms.

Degree of a Polynomial

The **degree of a polynomial** is the same as the highest degree of any term of the polynomial.

EXAMPLE 1   Use the vocabulary of this section to describe each polynomial:

$$\textbf{a. } d^4 + 9d^2 - 16 \qquad \textbf{b. } \frac{1}{2}x^2 - x$$

$$\textbf{c. } -6y^{14} - 1.5y^9z^9 + 2.5y^8z^{10} + yz^{11}$$

Strategy   First, we will identify the variable(s) in the polynomial and determine whether it is written in ascending or descending powers. Then we will count the number of terms in the polynomial and determine the degree of each term.

Why   The number of terms determines the type of polynomial. The highest degree of any term of the polynomial determines its degree.

Solution

> **The Language of Algebra**
>
> To *descend* means to move from higher to lower. When we write $d^4 + 9d^2 - 16$ as $d^4 + 0d^3 + 9d^2 + 0d^1 - 16d^0$, we more clearly see the *descending* powers of d.

a. $d^4 + 9d^2 - 16$ is a polynomial in one variable that is written in descending powers of d. If we write the subtraction as addition of the opposite, we see that it has 3 terms, $d^4, 9d^2$, and -16, and is therefore a trinomial. The highest degree of any of its terms is 4, so it is of degree 4.

$$d^4 + 9d^2 - 16 = \underset{\underset{\text{1st Term}}{\uparrow}}{d^4} + \underset{\underset{\text{2nd Term}}{\uparrow}}{9d^2} + \underset{\underset{\text{3rd Term}}{\uparrow}}{(-16)}$$

Term	Coefficient	Degree
d^4	1	4
$9d^2$	9	2
-16	-16	0

Degree of the polynomial: **4**

> **Success Tip**
>
> Recall that *terms* are separated by $+$ symbols and that the *coefficient* of a term is the numerical factor of the term.

b. $\frac{1}{2}x^2 - x$ is a polynomial in one variable. It is written in descending powers of x. Since it has two terms, it is a binomial. The highest degree of any of its terms is 2, so it is of degree 2.

Term	Coefficient	Degree
$\frac{1}{2}x^2$	$\frac{1}{2}$	2
$-x$	-1	1

Degree of the polynomial: **2**

c. $-6y^{14} - 1.5y^9z^9 + 2.5y^8z^{10} + yz^{11}$ is a polynomial in two variables, y and z. It is written in descending powers of y and ascending powers of z. It has 4 terms, and therefore has no special name. The highest degree of any term is 18, so it is of degree 18.

Term	Coefficient	Degree
$-6y^{14}$	-6	14
$-1.5y^9z^9$	-1.5	**18**
$2.5y^8z^{10}$	2.5	**18**
yz^{11}	1	12

Degree of the polynomial: **18**

 Self Check 1   Describe each polynomial: **a.** $x^2 + 4x - 16$
   b. $-14s^5t + s^4t^3$

Now Try   **Problems 17 and 35**

2   **Evaluate Polynomials.**

A polynomial can have different values depending on the number that is substituted for its variable (or variables).

EXAMPLE 2   Evaluate $3x^2 + 4x - 5$ for $x = 0$ and $x = -2$.

Strategy   We will substitute the given value for each x in the polynomial and follow the rules for the order of operations.

Why   To *evaluate a polynomial* means to find its numerical value, once we know the value of its variable.

Solution

Caution
Recall that to evaluate $3(-2)^2$, the rules for the order of operations require that we find $(-2)^2$ first, and then multiply that result by 3.

For x = 0:

$$3x^2 + 4x - 5 = 3(0)^2 + 4(0) - 5$$
$$= 3(0) + 4(0) - 5$$
$$= 0 + 0 - 5$$
$$= -5$$

For x = -2:

$$3x^2 + 4x - 5 = 3(-2)^2 + 4(-2) - 5$$
$$= 3(4) + 4(-2) - 5$$
$$= 12 + (-8) - 5$$
$$= -1$$

 Self Check 2   Evaluate $-x^3 + x - 2x + 3$ for $x = -3$.
Now Try   **Problem 54**

EXAMPLE 3   ***Supermarket Displays.*** The polynomial $\frac{1}{3}c^3 + \frac{1}{2}c^2 + \frac{1}{6}c$ gives the number of cans used in a display shaped like a square pyramid, having a square base formed by c cans per side. Find the number of cans used in the display.

Strategy   We will evaluate the polynomial for $c = 4$.

Why   From the illustration, we see that each side of the square base is formed by 4 cans.

Solution

$$\frac{1}{3}c^3 + \frac{1}{2}c^2 + \frac{1}{6}c = \frac{1}{3}(4)^3 + \frac{1}{2}(4)^2 + \frac{1}{6}(4)$$ Substitute 4 for c.

$$= \frac{1}{3}(64) + \frac{1}{2}(16) + \frac{1}{6}(4)$$ Evaluate the exponential expressions first.

$$= \frac{64}{3} + 8 + \frac{2}{3}$$ Do the multiplication, and then simplify: $\frac{4}{6} = \frac{2}{3}$.

$$= 30$$ Add the fractions: $\frac{64}{3} + \frac{2}{3} = \frac{66}{3} = 22$.

There are 30 cans of soup in the display.

 Now Try   **Problem 81**

In the following example, we evaluate a polynomial in two variables.

EXAMPLE 4   Evaluate $3p^2q - 4pq^2$ for $p = 2$ and $q = -3$.

Strategy   We will substitute the given values for each p and q in the polynomial and follow the rules for the order of operations.

Why   To evaluate a polynomial means to find its numerical value, once we know the value of its variables.

Solution

$$
\begin{aligned}
3p^2q - 4pq^2 &= 3(2)^2(-3) - 4(2)(-3)^2 && \text{Substitute 2 for } p \text{ and } -3 \text{ for } q. \\
&= 3(4)(-3) - 4(2)(9) && \text{Find the powers.} \\
&= -36 - 72 && \text{Do the multiplication.} \\
&= -108 && \text{Do the subtraction.}
\end{aligned}
$$

Self Check 4   Evaluate $3a^3b^2 + 2a^2b$ for $a = 2$ and $b = -1$.
Now Try   Problem 61

---

3   **Graph Equations Defined by Polynomials.**

In Chapter 3, we graphed equations such as $y = x$ and $y = 2x - 3$. Recall that these equations are called *linear equations* and that their graphs are straight lines. Note that the right side of the first two equations is a polynomial of degree 1.

$$y = x \qquad y = 2x - 3 \qquad\qquad y = x^2 \qquad\qquad y = x^3 + 1$$

The degree of each polynomial is 1.   The degree of this polynomial is 2.   The degree of this polynomial is 3.

We can also graph equations defined by polynomials with degrees greater than 1.

EXAMPLE 5   Graph: $y = x^2$

Strategy   We will find several solutions of the equation, plot them on a rectangular coordinate system, and then draw a smooth curve passing through the points.

Why   To *graph* an equation in two variables means to make a drawing that represents all of its solutions.

Solution   To find some solutions of this equation, we select several values of x that will make the computations easy. Then we find each corresponding value of y. If $x = -3$, we substitute -3 for x in $y = x^2$ and find y.

$$y = x^2 = (-3)^2 = 9$$

Thus, $(-3, 9)$ is a solution. In a similar manner, we find the corresponding y-values for x-values of $-2, -1, 0, 1, 2,$ and 3. If we plot the ordered pairs listed in the table and join the points with a smooth curve, we get the graph shown on the next page, which is called a **parabola**.

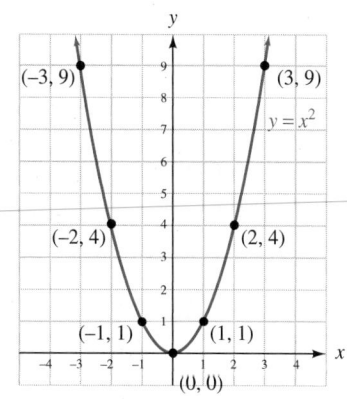

$y = x^2$

x	y	(x, y)
-3	9	$(-3, 9)$
-2	4	$(-2, 4)$
-1	1	$(-1, 1)$
0	0	$(0, 0)$
1	1	$(1, 1)$
2	4	$(2, 4)$
3	9	$(3, 9)$

 Self Check 5   Graph: $y = x^2 - 2$
Now Try   **Problem 69**

EXAMPLE 6   Graph: $y = -x^2 + 2$

Strategy   We will find several solutions of the equation, plot them on a rectangular coordinate system, and then draw a smooth curve passing through the points.

Why   To *graph* an equation in two variables means to make a drawing that represents all of its solutions.

Solution   To make a table of solutions, we select x-values of $-3, -2, -1, 0, 1, 2,$ and 3 and find each corresponding y-value. For example, if $x = -3$, we have

$y = -x^2 + 2$
$y = -(-3)^2 + 2$ Substitute -3 for x.
$y = -(9) + 2$ Evaluate the exponential expression first: $(-3)^2 = 9$.
$y = -7$ Do the addition: $-9 + 2 = -7$.

The ordered pair $(-3, -7)$ is a solution. Six other solutions appear in the table. After plotting each pair, we join the points with a smooth curve to obtain the graph, a parabola opening downward.

$y = -x^2 + 2$

x	y	(x, y)
-3	-7	$(-3, -7)$
-2	-2	$(-2, -2)$
-1	1	$(-1, 1)$
0	2	$(0, 2)$
1	1	$(1, 1)$
2	-2	$(2, -2)$
3	-7	$(3, -7)$

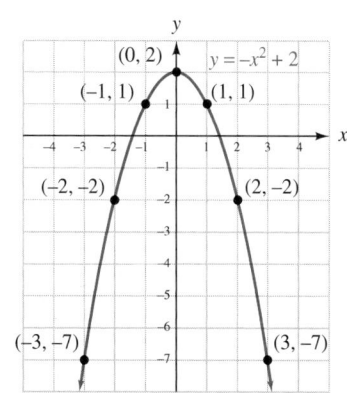

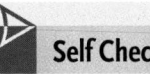

Self Check 6   Graph: $y = -x^2$

Now Try   **Problem 71**

EXAMPLE 7   Graph: $y = x^3 + 1$

Strategy   We will find several solutions of the equation, plot them on a rectangular coordinate system, and then draw a smooth curve passing through the points.

Why   To *graph* an equation in two variables means to make a drawing that represents all of its solutions.

Solution   If we let $x = -2$, we have

$y = x^3 + 1$

$y = (-2)^3 + 1$ Substitute -2 for x.

$y = -8 + 1$ Evaluate the exponential expression first: $(-2)^3 = -8$.

$y = -7$ Do the addition.

The ordered pair $(-2, -7)$ is a solution. This pair and others that satisfy the equation are listed in the table. Plotting the ordered pairs and joining the points with a smooth curve gives us the graph.

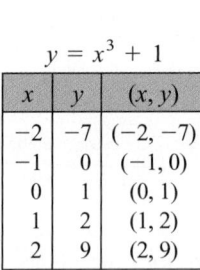

$y = x^3 + 1$

x	y	(x, y)
-2	-7	$(-2, -7)$
-1	0	$(-1, 0)$
0	1	$(0, 1)$
1	2	$(1, 2)$
2	9	$(2, 9)$

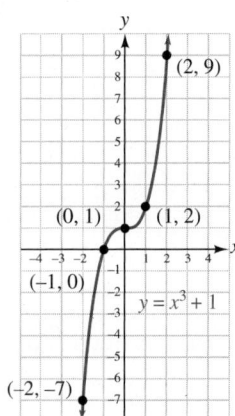

Self Check 7   Graph: $y = x^3 - 1$

Now Try   **Problem 75**

5.
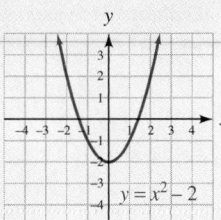

6.
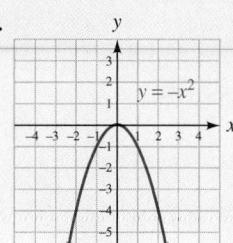

7.
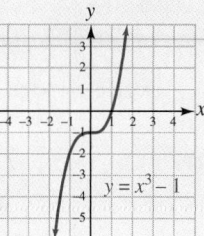

STUDY SET
5.4

VOCABULARY

Fill in the blanks.

1. A _____ is a term or a sum of terms in which all variables have whole-number exponents and no variable appears in a denominator.

2. The _____ of a polynomial are separated by + symbols.

3. $x^3 - 6x^2 + 9x - 2$ is a polynomial in _____ variable, and is written in _____ powers of x and $c^3 + 2c^2d - d^2$ is a polynomial in _____ variables and is written in _____ powers of d.

4. For the polynomial $6x^2 + 3x - 1$, the _____ term is $6x^2$, and the leading _____ is 6. The _____ term is -1.

5. A _____ is a polynomial with exactly one term. A _____ is a polynomial with exactly two terms. A _____ is a polynomial with exactly three terms.

6. The _____ of the term $3x^7$ is 7 because x appears as a factor 7 times: $3 \cdot x \cdot x \cdot x \cdot x \cdot x \cdot x \cdot x$.

7. To _____ the polynomial $x^2 - 2x + 1$ for $x = 6$, we substitute 6 for x and follow the rules for the order of operations.

8. The graph of $y = x^2$ is a cup-shaped curve called a _____.

CONCEPTS

Determine whether each expression is a polynomial.

9. **a.** $x^3 - 5x^2 - 2$ **b.** $x^{-4} - 5x$

     c. $x^2 - \dfrac{1}{2x} + 3$ **d.** $x^3 - 1$

     e. $x^2 - y^2$ **f.** $a^4 + a^3 + a^2 + a$

10. Fill in the blank so that the term has degree 5.

     a. $9x$ **b.** $-\dfrac{2}{3}xy$

Make a term-coefficient-degree table like that shown in Example 1 for each polynomial.

11. $8x^2 + x - 7$

Term	Coefficient	Degree

Degree of the polynomial: _____

12. $y^4 - y^3 + 16y^2 + 3y$

Term	Coefficient	Degree

Degree of the polynomial: _____

13. $8a^6b^3 - 27ab$

Term	Coefficient	Degree

Degree of the polynomial: _____

14. $-1.2c^4 + 2.4c^2d^2 - 3.6d^4$

Term	Coefficient	Degree

Degree of the polynomial:

NOTATION

15. **a.** Write $x - 9 + 3x^2 + 5x^3$ in descending powers of x.

b. Write $-2xy + y^2 + x^2$ in ascending powers of y.

16. Complete the solution. Evaluate $-2x^2 + 3x - 1$ for $x = -2$.

$$-2x^2 + 3x - 1 = -2(\quad)^2 + 3(\quad) - 1$$
$$= -2(\quad) + 3(\quad) - 1$$
$$= \quad + (-6) - 1$$
$$= \quad$$

GUIDED PRACTICE

Classify each polynomial as a monomial, a binomial, a trinomial, or none of these. See Example 1.

17. $3x + 7$ **18.** $3y - 5$

19. $y^2 + 4y + 3$ **20.** $9xy$

21. $\frac{3}{2}z^2$ **22.** $\frac{3}{5}x^4 - \frac{2}{5}x^3 + \frac{3}{5}x - 1$

23. $t - 32$ **24.** $12z^4$

25. $s^2 - 23s + 31$ **26.** $2x^3 - 5x^2 + 6x - 3$

27. $6x^5 - x^4 - 3x^3 + 7$ **28.** x^3

29. $3m^3n - 4m^2n^2 + mn - 1$
30. $4p^3q^2 + 7p^2q^3 + pq^4 - q^5$
31. $2a^2 - 3ab + b^2$ **32.** $a^3b - ab^3$

Find the degree of each polynomial. See Example 1.

33. $3x^4$ **34.** $3x^5$

35. $-2x^2 + 3x + 1$ **36.** $-5x^4 + 3x^2 - 3x$

37. $\frac{1}{3}x - 5$ **38.** $\frac{1}{2}y^3 + 4y^2$

39. $-5r^2s^2 - r^3s + 3$ **40.** $4r^2s^3 - 5r^2s^8$
41. $x^{12} + 3x^2y^3$ **42.** $17ab^5 - 12a^3b$

43. 38 **44.** -24

45. $\frac{3}{2}m^7 - \frac{3}{4}m^{18}$ **46.** $\frac{7}{8}t^{10} - \frac{1}{8}t^{16}$

47. $5.5tw - 6.5t^2w - 7.5t^3$ **48.** $0.4h + 0.6h^4c + 0.6h^5$

Evaluate each expression. See Examples 2 and 3.

49. $x^2 - x + 1$ for   **50.** $x^2 - x + 7$ for
   a. $x = 2$ **a.** $x = 6$
   b. $x = -3$ **b.** $x = -2$

51. $4t^2 + 2t - 8$ for   **52.** $3s^2 - 2s + 8$ for
   a. $t = -1$ **a.** $s = 1$
   b. $t = 0$ **b.** $s = 0$

53. $\frac{1}{2}a^2 - \frac{1}{4}a$ for   **54.** $\frac{1}{3}b^2 - \frac{1}{9}b$ for
   a. $a = 4$ **a.** $b = 9$
   b. $a = -8$ **b.** $b = -9$

55. $-9.2x^2 + x - 1.4$ for   **56.** $-10.3x^2 - x + 6.5$ for
   a. $x = -1$ **a.** $x = -1$
   b. $x = -2$ **b.** $x = -2$

57. $x^3 + 3x^2 + 2x + 4$ for   **58.** $x^3 - 3x^2 - x + 9$ for
   a. $x = 2$ **a.** $x = 3$
   b. $x = -2$ **b.** $x = -3$

59. $y^4 - y^3 + y^2 + 2y - 1$ **60.** $-y^4 + y^3 + y^2 + y + 1$
   for   for
   a. $y = 1$ **a.** $y = 1$
   b. $y = -1$ **b.** $y = -1$

Evaluate each polynomial for $a = -2$ and $b = 3$. See Example 4.

61. $6a^2b$ **62.** $4ab^2$

63. $a^3 + b^3$ **64.** $a^3 - b^3$

65. $a^2 + 5ab - b^2$ **66.** $a^3 - 2ab + b^3$

67. $5ab^3 - ab - b + 10$ **68.** $-a^3b + ab - a - 21$

Construct a table of solutions and then graph the equation. See Examples 5–7.

69. $y = x^2 + 1$ **70.** $y = x^2 - 4$
71. $y = -x^2 - 2$ **72.** $y = -x^2 + 1$
73. $y = 2x^2 - 3$ **74.** $y = -2x^2 + 2$
75. $y = x^3 + 2$ **76.** $y = x^3 + 4$
77. $y = x^3 - 3$ **78.** $y = x^3 - 2$
79. $y = -x^3 - 1$ **80.** $y = -x^3$

APPLICATIONS

81. SUPERMARKETS   A grocer plans to set up a pyramid-shaped display of cantaloupes like that shown in Example 3. If each side of the square base of the display is made of six cantaloupes, how many will be used in the display?

82. PACKAGING   The polynomial $4x^3 - 44x^2 + 120x$ gives the volume (in cubic inches) of the resulting box when a square with sides x inches long is cut from each corner of a box 10 in. $\times$ 12 in. piece of cardboard. Find the volume of a box if 3-inch squares are cut out.

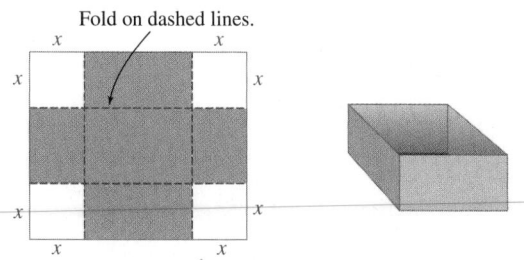

Fold on dashed lines.

83. STOPPING DISTANCE   The number of feet that a car travels before stopping depends on the driver's reaction time and the braking distance, as shown in the illustration. For one driver, the stopping distance is given by the polynomial $0.04v^2 + 0.9v$ where v is the velocity of the car. Find the stopping distance when the driver is traveling at 30 mph.

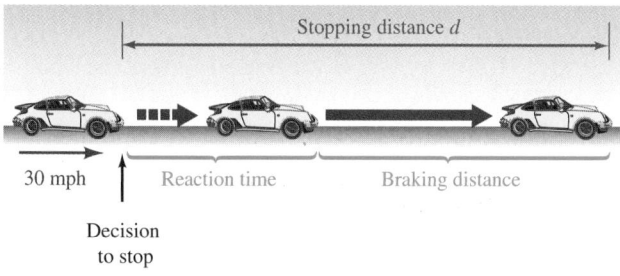

Stopping distance d

30 mph   Reaction time   Braking distance

Decision to stop

84. SUSPENSION BRIDGES   The polynomial $-0.0000001s^4 + 0.0066667s^2 + 400$ approximates the length of the cable between the two vertical towers of a bridge, where s is the sag in the cable (in feet). Estimate the length of the cable if the sag is 24.6 feet.

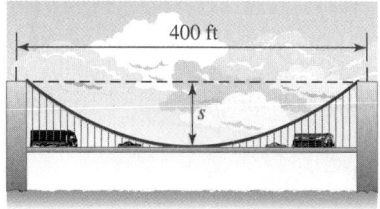

400 ft

85. *from Campus to Careers*
Sound Engineering Technician

© Paul Arthur/Getty Images

Many people involved in the recording industry have been impressed by the success of Apple's iTunes Music Store. The polynomial $1.144x^2 + 5.771x + 0.452$ approximates the number of users (in millions) of iTunes. When $x = 0$, the polynomial estimates the number of users (in millions) as of January 2004. When $x = 1$, it estimates the number of users as of January 2005, and so on. Use the polynomial to estimate the number of iTunes users as of January 2008. (Source: *WebSiteOptimization.com*)

86. ONLINE   When $x = 0$, the polynomial $-0.619x^2 + 11.778x + 12.171$ approximates the percent of the U.S. population that had gone online by 1995. When $x = 1$, it approximates the percent of the U.S. population that had gone online by 1996, and so on. Use the polynomial to approximate the percent of the U.S. population that had gone online by 2006. (Source: *The State of the News Media 2007*)

87. SCIENCE HISTORY   The Italian scientist Galileo Galilei (1564–1642) built an incline plane like that shown to study falling objects. As the ball rolled down, he measured the time it took the ball to travel different distances. Graph the data and then connect the points with a smooth curve.

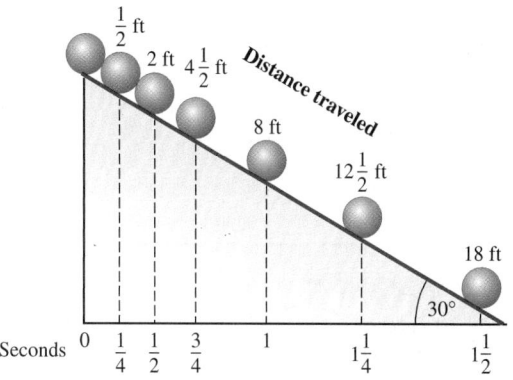

88. DOLPHINS   At a marine park, three trained dolphins jump in unison over an arching stream of water whose path can be described by the equation $y = -0.05x^2 + 2x$. Given the takeoff points for each dolphin, how high must each jump to clear the stream of water?

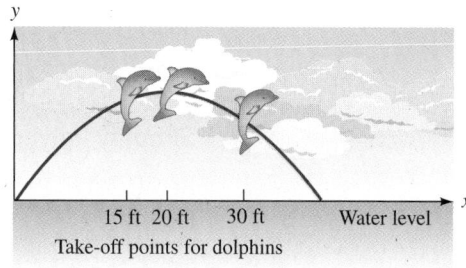

Take-off points for dolphins

WRITING

89. Describe how to determine the degree of a polynomial.
90. List some words that contain the prefixes *mono, bi,* or *tri.*

91. To graph $y = x^2 - 4$, a table of solutions is constructed and a graph is drawn. Explain the error.

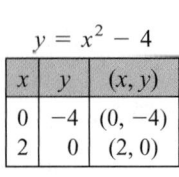

$y = x^2 - 4$

x	y	(x, y)
0	-4	$(0, -4)$
2	0	$(2, 0)$

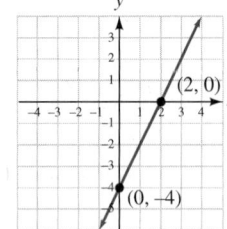

92. The expression $x + y$ is a binomial. Is xy also a binomial? Explain.

Solve each inequality. Write the solution set in interval notation and graph it.

93. $-4(3y + 2) \leq 28$ **94.** $-5 < 3t + 4 \leq 13$

Simplify each expression. Do not use negative exponents in the answer.

95. $(x^2x^4)^3$ **96.** $(a^2)^3(a^3)^2$

97. $\left(\dfrac{y^2y^5}{y^4}\right)^3$ **98.** $\left(\dfrac{2t^3}{t}\right)^{-4}$

CHALLENGE PROBLEMS

99. Find a three-term polynomial of degree 2 whose value will be 1 when it is evaluated for $x = 2$.

100. Graph: $y = 2x^3 - 3x^2 - 11x + 6$

SECTION 5.5
Adding and Subtracting Polynomials

Objectives

① Simplify polynomials by combining like terms.
② Add polynomials.
③ Subtract polynomials.

If we are to add (or subtract) objects, they must be similar. For example, we can add dollars to dollars and inches to inches, but we can't add dollars to inches. If you keep this concept in mind, then adding and subtracting polynomials will be easy. It simply involves combining like terms.

① Simplify Polynomials by Combining Like Terms.

Recall that **like terms** have the same variables with the same exponents. Only the coefficients may differ.

Like terms	*Unlike terms*	
$-7x$ and $15x$	$-7x$ and $15a$	Different variables
$9.4y^3$ and $1.6y^3$	$9.4y^3$ and $1.6y^2$	Different exponents on the same variable
$\dfrac{1}{2}x^5y^2$ and $-\dfrac{1}{3}x^5y^2$	$\dfrac{1}{2}x^5y^2$ and $-\dfrac{1}{3}x^2y^5$	Different exponents on different variables

Also recall that to **combine like terms,** we combine their coefficients and keep the same variables with the same exponents. For example,

$$4y + 5y = (4 + 5)y \qquad 8x^2 - x^2 = (8 - 1)x^2$$
$$= 9y \qquad\qquad\quad = 7x^2$$

The Language of Algebra
Simplifying the sum or difference of like terms is called *combining like terms.*

Polynomials with like terms can be simplified by combining like terms.

| **EXAMPLE 1** | Simplify each polynomial by combining like terms: |

a. $4x^4 + 81x^4$ **b.** $-0.3r - 0.4r + 0.6r$

c. $17x^2y^2 + 2x^2y - 6x^2y^2$ **d.** $\dfrac{3}{4}p^2 + \dfrac{1}{2}q^2 - 7 + \dfrac{1}{3}p^2 - \dfrac{5}{4}q^2 + 4$

Strategy   We will use the distributive property in reverse to add (or subtract) the coeffi-cients of the like terms. We will keep the same variables raised to the same powers.

Why   To *combine like terms* means to add or subtract the like terms in an expression.

Solution

a. $4x^4 + 81x^4 = 85x^4$ Think: $(4 + 81)x^4 = 85x^4$.

b. $-0.3r - 0.4r + 0.6r = -0.1r$ Think: $(-0.3 - 0.4 + 0.6)r = -0.1r$.

c. The first and third terms are like terms.

$$17x^2y^2 + 2x^2y - 6x^2y^2 = 11x^2y^2 + 2x^2y \quad \text{Think: } (17 - 6)x^2y^2 = 11x^2y^2.$$

d. $\dfrac{3}{4}p^2 + \dfrac{1}{2}q^2 - 7 + \dfrac{1}{3}p^2 - \dfrac{5}{4}q^2 + 4$

$$= \left(\dfrac{3}{4} + \dfrac{1}{3}\right)p^2 + \left(\dfrac{1}{2} - \dfrac{5}{4}\right)q^2 - 7 + 4 \qquad \text{Combine like terms.}$$

$$= \left(\dfrac{9}{12} + \dfrac{4}{12}\right)p^2 + \left(\dfrac{2}{4} - \dfrac{5}{4}\right)q^2 - 7 + 4 \qquad \begin{array}{l}\text{Build equivalent fractions:}\\ \frac{3}{4}\cdot\frac{3}{3}=\frac{9}{12},\ \frac{1}{3}\cdot\frac{4}{4}=\frac{4}{12},\ \text{and } \frac{1}{2}\cdot\frac{2}{2}=\frac{2}{4}.\end{array}$$

$$= \dfrac{13}{12}p^2 - \dfrac{3}{4}q^2 - 3 \qquad \text{Do the additions and the subtraction.}$$

Caution   Do not try to clear this expression of fractions by multiplying it by the LCD 12. That strategy works only when we multiply *both sides of an equation* by the LCD.

$$\cancel{12}\left(\dfrac{3}{4}p^2 + \dfrac{1}{2}q^2 - 7 + \dfrac{1}{3}p^2 - \dfrac{5}{4}q^2 + 4\right)$$

| **Self Check 1** | Simplify each polynomial: **a.** $6m^4 + 3m^4$ |

b. $-19x + 21x - x$

c. $1.7s^3t + 0.3s^2t - 0.6s^3t$

d. $\frac{1}{8}c^5 + \frac{1}{3}d^5 - 9 + \frac{5}{4}c^5 - \frac{3}{5}d^5 + 1$

Now Try   **Problems 13, 27, and 33**

  2   **Add Polynomials.**

When adding polynomials horizontally, each polynomial is usually enclosed within parenthe-ses. For example,

$$(3x^2 + 6x + 7) + (2x - 5)$$

is the sum of a trinomial and a binomial. To find the sum, we reorder and regroup the terms using the commutative and associative properties of addition so that like terms are together.

Caution
When combining like terms, the exponents on the variables *stay the same*. Don't incorrectly add the exponents.

$$(3x^2 + 6x + 7) + (2x - 5) = 3x^2 + (6x + 2x) + (7 - 5)$$ The x-terms are together.
The constant terms are
together.

$$= 3x^2 + 8x + 2$$ Combine like terms.

This example suggests the following rule.

Adding Polynomials	To add polynomials, combine their like terms.

EXAMPLE 2   Add the polynomials:

a. $(-6a^3 + 5a^2 - 7a + 9) + (4a^3 - 5a^2 - a - 8)$

b. $\left(\frac{1}{2}m^2 + \frac{2}{3}m + 1\right) + \left(\frac{3}{4}m^2 - \frac{7}{9}m - 4\right)$ **c.** $(16g^2 - h^2) + (4g^2 + 2gh + 10h^2)$

Strategy   We will reorder and regroup to get the like terms together. Then we will combine like terms.

Why   To add polynomials means to combine their like terms.

Solution

Success Tip
Combine the like terms in order: a^3-terms first, a^2-terms second, a-terms third, and constants last. Then the answer will be in descending powers of a.

a. $(-6a^3 + 5a^2 - 7a + 9) + (4a^3 - 5a^2 - a - 8)$

$= (-6a^3 + 4a^3) + (5a^2 - 5a^2) + (-7a - a) + (9 - 8)$ Group like terms together.

$= -2a^3 + 0a^2 + (-8a) + 1$ Combine like terms.

$= -2a^3 - 8a + 1$

b. $\left(\frac{1}{2}m^2 + \frac{2}{3}m + 1\right) + \left(\frac{3}{4}m^2 - \frac{7}{9}m - 4\right)$

$= \left(\frac{1}{2}m^2 + \frac{3}{4}m^2\right) + \left(\frac{2}{3}m - \frac{7}{9}m\right) + (1 - 4)$ Group like terms together.

$= \left(\frac{2}{4}m^2 + \frac{3}{4}m^2\right) + \left(\frac{6}{9}m - \frac{7}{9}m\right) + (1 - 4)$ Build equivalent fractions: $\frac{1}{2} \cdot \frac{2}{2} = \frac{2}{4}$ and $\frac{2}{3} \cdot \frac{3}{3} = \frac{6}{9}$.

$= \frac{5}{4}m^2 - \frac{1}{9}m - 3$ Do the addition and subtraction.

c. $(16g^2 - h^2) + (4g^2 + 2gh + 10h^2)$

$= (16g^2 + 4g^2) + 2gh + (-h^2 + 10h^2)$ Group like terms together.

$= 20g^2 + 2gh + 9h^2$ Combine like terms.

 Self Check 2   Add the polynomials:

a. $(2a^2 - a + 4) + (5a^2 + 6a - 5)$

b. $\left(\frac{3}{2}b^3 + \frac{4}{5}b + 7\right) + \left(\frac{3}{4}b^3 - \frac{11}{10}b - 10\right)$

c. $(7x^2 - 2xy - y^2) + (4x^2 - y^2)$

Now Try   **Problems 41, 43, and 47**

EXAMPLE 3	***Trapezoids.*** Find a polynomial that represents the perimeter of the trapezoid.

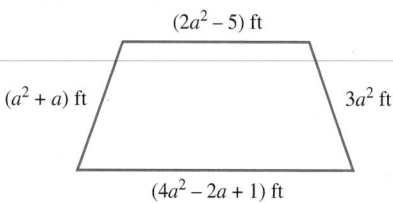

$(2a^2 - 5)$ ft

$(a^2 + a)$ ft $3a^2$ ft

$(4a^2 - 2a + 1)$ ft

Strategy   We will add the polynomials that represent the lengths of the sides of the trapezoid.

Why   To find the perimeter of a figure, we find the distance around the figure by finding the sum of the lengths of its sides.

Solution   To add the four polynomials that represent the lengths of the sides of the trapezoid, we combine their like terms.

$(2a^2 - 5) + (a^2 + a) + (4a^2 - 2a + 1) + 3a^2$

$= (2a^2 + a^2 + 4a^2 + 3a^2) + (a - 2a) + (-5 + 1)$ Reorder and regroup terms.

$= 10a^2 - a - 4$ Combine like terms.

The perimeter of the trapezoid is $(10a^2 - a - 4)$ ft.

> *Now Try*   **Problem 51**

The Language of Algebra
A *trapezoid* is a four-sided figure with exactly two sides parallel.

Polynomials can also be added vertically by aligning like terms in columns.

EXAMPLE 4	Add $4x^2 - 3$ and $3x^2 - 8x + 8$ using vertical form.

Strategy   First, we will write one polynomial underneath the other and draw a horizontal line beneath them. Then we will add the like terms, column by column, and write each result under the line.

Why   *Vertical form* means to use an approach similar to that used in arithmetic to add two numbers.

Solution   When performing vertical addition, any missing term may be written with a coefficient of 0. Since the first polynomial does not have an x-term, we insert a placeholder term $0x$ in the second column so that the constant terms line up in the third column.

x^2-terms ─┐   ┌ x-terms   ┌─ Constants

$$\begin{array}{rrr} 4x^2 & +\,0x & -\,3 \\ 3x^2 & -\,8x & +\,8 \\ \hline 7x^2 & -8x & +5 \end{array}$$

In the x^2-column, find $4x^2 + 3x^2$.

In the x-column, find $0x + (-8x)$.

In the constant column, find $-3 + 8$.

The sum is $7x^2 - 8x + 5$.

Success Tip
In arithmetic, we use vertical form so that we add digits in like place-value columns. In polynomial addition, we combine the like terms in each column.

Hundreds ─┐ Tens ┌─ Ones

403
+388

Self Check 4 Add $4q^2 - 7$ and $2q^2 - 8q + 9$ using vertical form.

Now Try **Problem 53**

3 **Subtract Polynomials.**

Recall from Chapter 1 that we can use the distributive property to find the opposite of several terms enclosed within parentheses. For example, we consider $-(2a^2 - a + 9)$.

$$-(2a^2 - a + 9) = -1(2a^2 - a + 9) \quad \text{Replace the } - \text{ symbol in front of the parentheses with } -1.$$

$$= -2a^2 + a - 9 \quad \text{Use the distributive property to remove parentheses.}$$

This example illustrates the following method of subtracting polynomials.

Subtracting Polynomials	To subtract two polynomials, change the signs of the terms of the polynomial being subtracted, drop the parentheses, and combine like terms.

EXAMPLE 5 Subtract the polynomials: **a.** $(3a^2 - 4a - 6) - (2a^2 - a + 9)$
b. $(-t^3u + 2t^2u - u + 1) - (-3t^2u - u + 8)$

Strategy In each case, we will change the signs of the terms of the polynomial being subtracted, drop the parentheses, and combine like terms.

Why This is the method for subtracting two polynomials.

Solution

a. $(3a^2 - 4a - 6) - (2a^2 - a + 9)$

$$= 3a^2 - 4a - 6 - 2a^2 + a - 9 \quad \text{Change the sign of each term of } 2a^2 - a + 9 \text{ and drop the parentheses.}$$

$$= a^2 - 3a - 15 \quad \text{Combine like terms.}$$

b. $(-t^3u + 2t^2u - u + 1) - (-3t^2u - u + 8)$

$$= -t^3u + 2t^2u - u + 1 + 3t^2u + u - 8 \quad \text{Change the sign of each term of } -3t^2u - u + 8 \text{ and drop the parentheses.}$$

$$= -t^3u + 5t^2u - 7 \quad \text{Combine like terms.}$$

> **Success Tip**
> After some practice, you will be able to reorder and regroup the terms of the polynomials in your head to combine them.

Self Check 5 Subtract the polynomials:
a. $(8a^3 - 5a^2 + 5) - (a^3 - a^2 - 7)$
b. $(x^2y - 2x + y - 2) - (6x + 9y - 2)$

Now Try **Problems 61 and 71**

Polynomials can also be subtracted vertically by aligning like terms in columns.

EXAMPLE 6   Subtract $3x^2 - 2x + 3$ from $2x^2 + 4x - 1$ using vertical form.

Strategy   Since $3x^2 - 2x + 3$ is to be subtracted from $2x^2 + 4x - 1$, we will write $3x^2 - 2x + 3$ underneath $2x^2 + 4x - 1$, change the sign of each term of $3x^2 - 2x + 3$ and add, column-by-column.

Why   *Vertical form* means to arrange the like terms in columns.

Solution

$$
\begin{array}{l}
2x^2 + 4x - 1 \\
\underline{-(3x^2 - 2x + 3)}
\end{array}
\quad
\begin{array}{c}
\xrightarrow[\text{and add}]{\text{Change signs}}
\end{array}
\quad
\begin{array}{l}
2x^2 + 4x - 1 \\
\underline{-3x^2 + 2x - 3} \\
-x^2 + 6x - 4
\end{array}
$$

In the x^2-column, find $2x^2 + (-3x^2)$.

In the x-column, find $4x + 2x$.

In the constant column, find $-1 + (-3)$.

The difference is $-x^2 + 6x - 4$.

 Self Check 6   Subtract $2p^2 + 2p - 8$ from $5p^2 - 6p + 7$ using vertical form.

Now Try   **Problem 73**

EXAMPLE 7   Subtract $1.2a^4 - 0.7a$ from the sum of $0.6a^4 + 1.5a$ and $0.4a^4 - 1.1a$.

Strategy   First, we will translate the words of the problem into mathematical symbols. Then we will perform the indicated operations.

Why   The words of the problem contain the key phrases *subtract from* and *sum*.

Solution   Since $1.2a^4 - 0.7a$, is to be subtracted from the sum, the order must be reversed when we translate to mathematical symbols.

Subtract $1.2a^4 - 0.7a$ from   the sum of $0.6a^4 + 1.5a$ and $0.4a^4 - 1.1a$.

$$[(0.6a^4 + 1.5a) + (0.4a^4 - 1.1a)] - (1.2a^4 - 0.7a)$$ Use brackets [ ] to enclose the sum.

Next, we remove the grouping symbols to obtain

$$= 0.6a^4 + 1.5a + 0.4a^4 - 1.1a - 1.2a^4 + 0.7a$$ Change the sign of each term within $(1.2a^4 - 0.7a)$ and drop the parentheses.

$$= -0.2a^4 + 1.1a$$ Combine like terms.

 Self Check 7   Subtract $-0.2q^2 - 0.2q$ from the sum of $0.1q^2 - 0.6q$ and $0.3q^2 + 0.1q$.

Now Try   **Problem 87**

EXAMPLE 8   *Fireworks.*   Two firework shells are fired upward at the same time from different platforms. The height, after t seconds, of the first shell is $(-16t^2 + 160t + 3)$ feet. The height, after t seconds, of a higher-flying second shell is $(-16t^2 + 200t + 1)$ feet.

a. Find a polynomial that represents the difference in the heights of the shells.

b. In 5 seconds, the first shell reaches its peak and explodes. How much higher is the second shell at that time?

Strategy   To find the difference in their heights, we will subtract the height of the first shell from the height of the higher-flying second shell.

Why   The key word *difference* indicates that we should subtract the polynomials.

Solution

a. Since the height of the higher flying second shell is represented by $-16t^2 + 200t + 1$ and the height of the lower flying shell is represented by $-16t^2 + 160t + 3$, we can find their difference by performing the following subtraction.

$$(-16t^2 + 200t + 1) - (-16t^2 + 160t + 3)$$
$$= -16t^2 + 200t + 1 + 16t^2 - 160t - 3 \qquad \text{Change the sign of each term of } -16t^2 + 160t + 3 \text{ and remove parentheses.}$$
$$= 40t - 2 \qquad \text{Combine like terms.}$$

The difference in the heights of the shells t seconds after being fired is $(40t - 2)$ feet.

b. To find the difference in their heights after 5 seconds, we will evaluate the polynomial found in part (a) at a value of 5 seconds. If we substitute 5 for t, we have

$$40t - 2 = 40(5) - 2 = 200 - 2 = 198$$

When the first shell explodes, the second shell will be 198 feet higher than the first shell.

   Now Try   **Problem 109**

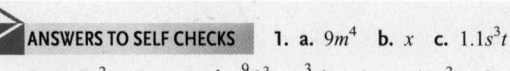
ANSWERS TO SELF CHECKS   **1. a.** $9m^4$ **b.** x **c.** $1.1s^3t + 0.3s^2t$ **d.** $\frac{11}{8}c^5 - \frac{4}{15}d^5 - 8$
   2. a. $7a^2 + 5a - 1$ **b.** $\frac{9}{4}b^3 - \frac{3}{10}b - 3$ **c.** $11x^2 - 2xy - 2y^2$ **4.** $6q^2 - 8q + 2$
   5. a. $7a^3 - 4a^2 + 12$ **b.** $x^2y - 8x - 8y$ **6.** $3p^2 - 8p + 15$ **7.** $0.6q^2 - 0.3q$

STUDY SET
5.5

VOCABULARY

Fill in the blanks.

1. $(b^3 - b^2 - 9b + 1) + (b^3 - b^2 - 9b + 1)$ is the sum of two _____.

2. $(b^2 - 9b + 11) - (4b^2 - 14b)$ is the _____ of a trinomial and a binomial.

3. _____ terms have the same variables with the same exponents

4. The polynomial $2t^4 + 3t^3 - 4t^2 + 5t - 6$ is written in _____ powers of t.

CONCEPTS

Fill in the blanks.

5. To add polynomials, _____ their like terms.

6. To subtract polynomials, _____ the signs of the terms of the polynomial being subtracted, drop parentheses, and combine like terms.

7. Simplify each polynomial, if possible.
   a. $2x^2 + 3x^2$ **b.** $15m^3 - m^3$
   c. $8a^3b - a^3b$ **d.** $6cd + 4c^2d$

8. What is the result when the addition is done in the x-column?

$$4x^2 + x - 12$$
$$\underline{5x^2 - 8x + 23}$$

9. Write without parentheses.
   a. $-(5x^2 - 8x + 23)$ **b.** $-(-5y^4 + 3y^2 - 7)$

10. What is the result when the subtraction is done in the x-column?

$$8x^2 - 7x - 1 8x^2 - 7x - 1$$
$$\underline{-(4x^2 + 6x - 9)} \longrightarrow \underline{-4x^2 - 6x + 9}$$

NOTATION

Fill in the blanks to add (subtract) the polynomials.

11. $(6x^2 + 2x + 3) + (4x^2 - 7x + 1)$
$$= (6x^2 + \boxed{}) + (\boxed{} - 7x) + (3 + \boxed{})$$
$$= \boxed{} - 5x + \boxed{}$$

12. $(6x^2 + 2x + 3) - (4x^2 - 7x + 1)$
$$= 6x^2 + 2x + 3 \boxed{} 4x^2 \boxed{} 7x - 1$$
$$= \boxed{} + 9x + \boxed{}$$

GUIDED PRACTICE

Simplify each polynomial and write it in descending powers of one variable. **See Example 1.**

13. $8t^2 + 4t^2$ 14. $15x^2 + 10x^2$

15. $-32u^3 - 16u^3$ 16. $-25x^3 - 7x^3$

17. $18x^2 - 19x + 2x^2$ 18. $17y^2 - 22y - y^2$

19. $3r^4 - 4r + 7r^4$ 20. $-2b^4 + 7b - 3b^4$

21. $10x^2 - 8x + 9x - 9x^2$ 22. $-3y^2 - y - 6y^2 + 7y$

23. $\frac{1}{5}x^2 - \frac{3}{8}x + \frac{2}{3}x^2 + \frac{1}{4}x$

24. $\frac{6}{7}y^2 + \frac{1}{2}y - \frac{2}{3}y^2 + \frac{1}{5}y$

25. $0.6x^3 + 0.8x^4 + 0.7x^3 + (-0.8x^4)$

26. $1.9m^4 - 2.4m^6 - 3.7m^4 + 2.8m^6$

27. $\frac{1}{2}st + \frac{3}{2}st$ 28. $\frac{2}{5}at + \frac{1}{5}at$

29. $\frac{2}{3}d^2 - \frac{1}{4}c^2 + \frac{5}{6}c^2 - \frac{1}{2}cd + \frac{1}{3}d^2$

30. $\frac{3}{5}s^2 - \frac{2}{5}t^2 - \frac{1}{2}s^2 - \frac{7}{10}st - \frac{3}{10}st$

31. $-4ab + 4ab - ab$ 32. $xy - 4xy - 2xy$

33. $4x^2y + 5 - 6x^3y - 3x^2y + 2x^3y$

34. $5b - 9ab^2 + 10a^3b - 8ab^2 - 9a^3b$

35. $-7cd - 8d^2 - 5cd + 8d^2 - 4c^2$

36. $-3rt - 7t^2 - 6rt + 7t^2 - 6r^2$

Add the polynomials. **See Example 2.**

37. $(3x + 7) + (4x - 3)$

38. $(2y - 3) + (4y + 7)$

39. $(9d^2 + 6d) + (8d - 4d^2)$

40. $(2c^2 - 4c) + (8c - c^2)$

41. $(3q^2 - 5q + 7) + (2q^2 + q - 12)$

42. $(2t^2 + 11t - 15) + (-5t^2 - 13t + 10)$

43. $\left(\frac{2}{3}y^3 + \frac{3}{4}y^2 + \frac{1}{2}\right) + \left(\frac{1}{3}y^3 + \frac{1}{5}y^2 - \frac{1}{6}\right)$

44. $\left(\frac{1}{16}r^6 + \frac{1}{2}r^3 - \frac{11}{12}\right) + \left(\frac{9}{16}r^6 + \frac{9}{4}r^3 + \frac{1}{12}\right)$

45. $(0.3p + 2.1q) + (0.4p - 3q)$

46. $(-0.3r - 5.2s) + (0.8r - 5.2s)$

47. $(2x^2 + xy + 3y^2) + (5x^2 - y^2)$

48. $(-4a^2 - ab + 15b^2) + (5a^2 - b^2)$

Find a polynomial that represents the perimeter of the figure. See Example 3.

49.

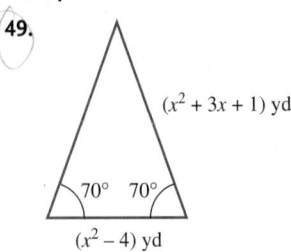

$(x^2 + 3x + 1)$ yd

70° 70°

$(x^2 - 4)$ yd

50.

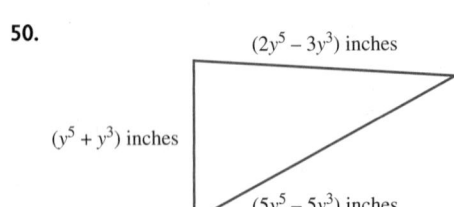

$(2y^5 - 3y^3)$ inches

$(y^5 + y^3)$ inches

$(5y^5 - 5y^3)$ inches

51.

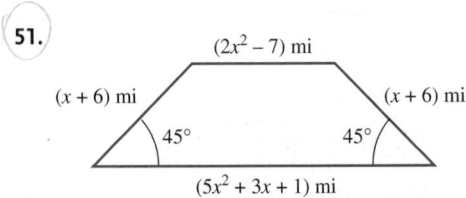

$(2x^2 - 7)$ mi

$(x + 6)$ mi $(x + 6)$ mi

45°   45°

$(5x^2 + 3x + 1)$ mi

52.

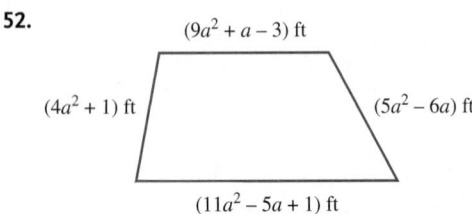

$(9a^2 + a - 3)$ ft

$(4a^2 + 1)$ ft $(5a^2 - 6a)$ ft

$(11a^2 - 5a + 1)$ ft

Use vertical form to add the polynomials. **See Example 4.**

53. $3x^2 + 4x + 5$
$2x^2 - 3x + 6$

54. $6x^3 - 4x^2 + 7$
$7x^3 + 9x^2 + 12$

55. $6a^2 + 7a + 9$
$-9a^2 - 2$

56. $-2c^2 - 3c - 5$
$14c^2 - 1$

57. $z^3 + 6z^2 - 7z + 16$
$9z^3 - 6z^2 + 8z - 18$

58. $3x^3 + 4x^2 - 3x + 5$
$3x^3 - 4x^2 - x - 7$

59. $-3x^3y^2 + 4x^2y - 4x + 9$
$2x^3y^2 + 9x - 3$

60. $3x^2y^2 + 4xy + 25$
$5x^2y^2 - 12$

Subtract the polynomials. **See Example 5.**

61. $(3a^2 - 2a + 4) - (a^2 - 3a + 7)$

62. $(2b^2 + 3b - 5) - (2b^2 - 4b - 9)$

63. $(9a^2 + 3a) - (2a - 4a^2)$

64. $(4b^2 + 3b) - (7b - b^2)$

65. $(-4h^3 + 5h^2 + 15) - (h^3 - 15)$

66. $(-c^5 + 5c^4 - 12) - (2c^5 - c^4)$

67. $\left(\dfrac{3}{8}s^8 - \dfrac{3}{4}s^7\right) - \left(\dfrac{1}{3}s^8 + \dfrac{1}{5}s^7\right)$

68. $\left(\dfrac{5}{6}q^9 - \dfrac{4}{5}q^8\right) - \left(\dfrac{1}{4}q^9 + \dfrac{3}{8}q^8\right)$

69. $(0.03f^2 + 0.25f + 0.91) - (0.17f^2 - 1.18)$

70. $(0.05r^2 - 0.33r) - (0.48\,r^2 + 0.15r + 2.14)$

71. $(5ab + 2b^2) - (2 + ab + b^2)$

72. $(mn + 8n^2) - (6 - 5mn + n^2)$

Use vertical form to subtract the polynomials. **See Example 6.**

73. $3x^2 + 4x + 5$
$-(2x^2 - 2x + 3)$

74. $6y^2 + 4y + 13$
$-(3y^2 - 6y + 7)$

75. $4x^3 + 4x^2 - 3x + 10$
$-(5x^3 - 2x^2 - 4x - 4)$

76. $7m^5 + m^3 + 9m^2 - m$
$-(8m^5 - 2m^3 + m^2 + m)$

77. $0.8x^3 - 2.3x + 0.6$
$-(0.2x^3 - 1.2x^2 - 3.6x + 0.9)$

78. $9.7y^3 + y + 1.1$
$-(6.3y^3 - 4.4y^2 + 2.7y + 8.8)$

79. $3x^3y^2 + 4x^2y + 7x + 12$
$-(-4x^3y^2 + 6x^2y + 9x - 3)$

80. $-2x^2y^2 + 12y^2$
$-(10x^2y^2 + 9xy - 24y^2)$

Perform the operations. **See Example 7.**

81. Subtract $(s^2 + 4s + 2)$ from $(5s^2 - s + 9)$.

82. Subtract $(4p^2 - 4p - 40)$ from $(10p^2 - p - 30)$.

83. Subtract $(-y^5 + 5y^4 - 1.2)$ from $(2y^5 - y^4)$.

84. Subtract $(-4w^3 + 5w^2 + 7.6)$ from $(w^3 - 15w^2)$.

85. Find the sum when $(3x^2 + 4x - 7)$ is added to the sum of $(-2x^2 - 7x + 1)$ and $(-4x^2 + 8x - 1)$.

86. Find the difference when $(32x^2 - 17x + 45)$ is subtracted from the sum of $(23x^2 - 12x - 7)$ and $(-11x^2 + 12x + 7)$.

87. Find the difference when $(t^3 - 2t^2 + 2)$ is subtracted from the sum of $(3t^3 + t^2)$ and $(-t^3 + 6t - 3)$.

88. Find the difference when $(-3z^3 - 4z + 7)$ is subtracted from the sum of $(2z^2 + 3z - 7)$ and $(-4z^3 - 2z - 3)$.

89. $(2x^2 - 3x + 1) - (4x^2 - 3x + 2) + (2x^2 + 3x + 2)$

90. $(-3z^2 - 4z + 7) + (2z^2 + 2z - 1) - (2z^2 - 3z + 7)$

91. $(-2.7t^2 + 2.1t - 1.7) + (3.1t^2 - 2.5t + 2.3) - (1.7t^2 - 1.1t)$

92. $(1.04x^2 - 5.01) + (1.33x - 1.91x^2 + 5.02) - (1.07x^2 - 2.07x)$

TRY IT YOURSELF

Perform the operations.

93. $(-8x^2 - 3x) - (-11x^2 + 6x + 10)$

94. $(5m^2 - 8m + 8) - (-20m^2 + m)$

95. $(3x^2 - 3x - 2) + (3x^2 + 4x - 3)$

96. $(4c^2 + 3c - 2) + (3c^2 + 4c + 2)$

97. $\left(\frac{7}{8}r^4 + \frac{5}{9}r^2 - \frac{9}{4}\right) - \left(-\frac{3}{8}r^4 - \frac{2}{3}r^2 - \frac{1}{4}\right)$

98. $\left(\frac{4}{5}t^4 - \frac{1}{3}t^2 + \frac{1}{2}\right) - \left(-\frac{1}{2}t^4 + \frac{3}{8}t^2 - \frac{1}{16}\right)$

99. $\begin{array}{r} 8c^2 - 4c - 5 \\ -(-c^2 + 2c + 9) \\ \hline \end{array}$

100. $\begin{array}{r} 3t^3 - 4t^2 - 3t + 5 \\ +11t^3 \qquad - 8t - 2 \\ \hline \end{array}$

101. $(10 - 2st - 3s^2t) + (4 - 6st)$

102. $(20 - 4rt - 5r^2t) + (10 - 5rt)$

103. $(12.1h^3 + 9.9h^2) - (7.3h^3 + 1.1h^2)$

104. $(5.7n^3 - 2.1n) - (-6.2n^3 - 3.9n)$

APPLICATIONS

105. GREEK ARCHITECTURE

   a. Find a polynomial that represents the difference in the heights of the columns.

   b. If the columns were stacked one atop the other, to what height would they reach?

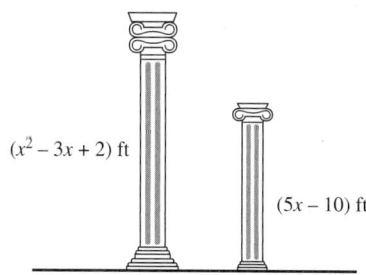

$(x^2 - 3x + 2)$ ft

$(5x - 10)$ ft

106. JETS   Find a polynomial that represents the length of the passenger jet.

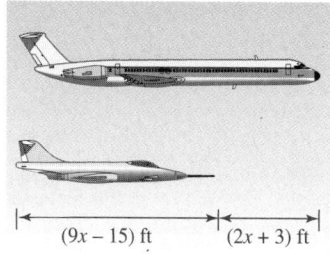

$(9x - 15)$ ft $(2x + 3)$ ft

107. PIÑATAS   Find a polynomial that represents the length of the rope used to hold up the piñata.

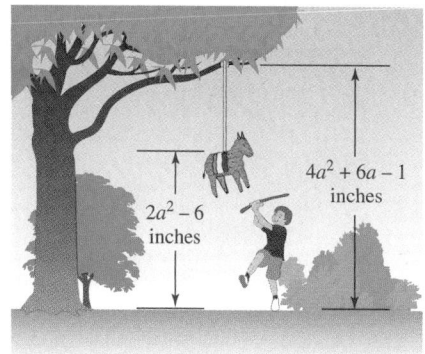

$4a^2 + 6a - 1$ inches

$2a^2 - 6$ inches

108. READING BLUEPRINTS   Find a polynomial that represents

   a. the difference in the length and width of the one-bedroom apartment.

   b. the perimeter of the apartment.

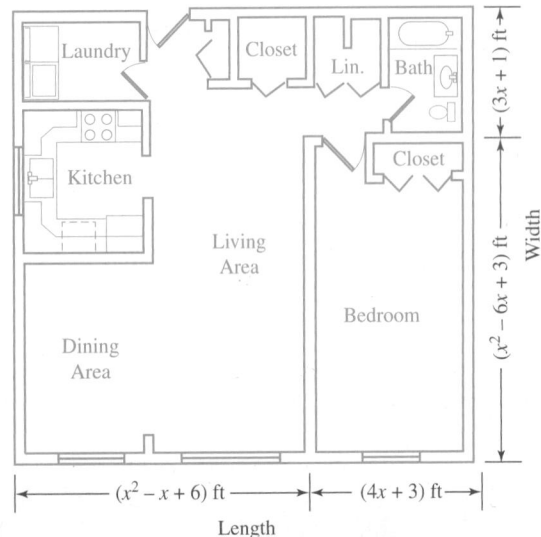

Length

109. NAVAL OPERATIONS   Two warning flares are fired upward at the same time from different parts of a ship. The height of the first flare is $(-16t^2 + 128t + 20)$ feet and the height of the higher-traveling second flare is $(-16t^2 + 150t + 40)$ feet, after t seconds.

   a. Find a polynomial that represents the difference in the heights of the flares.

   b. In 4 seconds, the first flare reaches its peak, explodes, and lights up the sky. How much higher is the second flare at that time?

110. AUTO MECHANICS   Find a polynomial that represents the length of the fan belt shown in the diagram in the next column. The dimensions are in inches. Leave π in your answer.

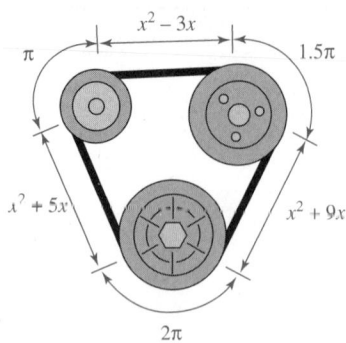

WRITING

111. How do you recognize like terms?

112. Explain why the vertical form used in algebra to add $2x^2 + 4x + 3$ and $5x^2 + 3x + 6$ is similar to the vertical form used in arithmetic to add 243 and 536.

113. Explain the error below.

$$7x^2y + 6x^2y = 13x^4y^2$$

114. Explain the error below.

$$(12x^2 - 4) - (3x^2 - 1) = 12x^2 - 4 - 3x^2 - 1$$
$$= 9x^2 - 5$$

115. A student was asked to simplify $\frac{1}{6}x^2 - 3 + \frac{2}{3}x^2$. Explain the error below:

$$6\left(\frac{1}{6}x^2 - 3 + \frac{2}{3}x^2\right) = x^2 - 18 + 4x^2$$
$$= 5x^2 - 18$$

116. Explain the error below.

Subtract $(2d^2 - d - 3)$ from $(d^2 - 9)$:

$$(2d^2 - d - 3) - (d^2 - 9) = d^2 - d + 6$$

REVIEW

117. What is the sum of the measures of the angles of a triangle?

118. What is the formula for
   a. the area of a circle?
   b. the area of a triangle?

119. Graph: $y = -\frac{1}{2}x + 2$

120. Graph: $2x + 3y = 9$

CHALLENGE PROBLEMS

121. What polynomial must be added to $2x^2 - x + 3$ so that the sum is $6x^2 - 7x - 8$?

122. Is the sum of two trinomials always a trinomial? Explain why or why not.

SECTION 5.6
Multiplying Polynomials

Objectives

1. Multiply monomials.
2. Multiply a polynomial by a monomial.
3. Multiply binomials.
4. Multiply polynomials.

We now discuss multiplying polynomials. We will begin with the simplest case—finding the product of two monomials.

1 Multiply Monomials.

To find the product of two monomials, such as $8x^2$ and $3x^4$, we use the commutative and associative properties of multiplication to reorder and regroup the factors.

$$(8x^2)(3x^4) = (8 \cdot 3)(x^2 \cdot x^4) \quad \text{Group the coefficients together and the variables together.}$$
$$= 24x^6 \quad \text{Simplify: } x^2 \cdot x^4 = x^{2+4} = x^6.$$

This example suggests the following rule.

> **Success Tip**
> In this section, you will see that every polynomial multiplication is a series of monomial multiplications.

> **Multiplying Monomials**
>
> To multiply two monomials, multiply the numerical factors (the coefficients) and then multiply the variable factors.

EXAMPLE 1   Multiply: **a.** $(6r)(r)$ **b.** $3t^4(-2t^5)$

   c. $\left(\dfrac{1}{3}a^2b^3\right)(21ab^2)$ **d.** $-4y^5z^2(2y^3z^3)(3yz)$

Strategy   We will multiply the numerical factors and then multiply the variable factors.

Why   The commutative and associative properties of multiplication enable us to reorder and regroup the factors.

> **Success Tip**
> Notice that we *multiply* the coefficients and we *add* the exponents of the like bases.

Solution

a. $(6r)(r) = 6r^2$ Recall that $r = 1r$. Think: $6 \cdot 1 = 6$ and $r \cdot r = r^2$.

b. $(3t^4)(-2t^5) = -6t^9$ Think: $3(-2) = -6$ and $t^4 \cdot t^5 = t^{4+5} = t^9$.

c. $\left(\dfrac{1}{3}a^2b^3\right)(21ab^2) = 7a^3b^5$ Think: $\frac{1}{3} \cdot 21 = \frac{21}{3} = 7$, $a^2 \cdot a = a^3$, and $b^3 \cdot b^2 = b^5$.

d. $-4y^5z^2(2y^3z^3)(3yz) = -24y^9z^6$ Think: $-4(2)(3) = -24$, $y^5 \cdot y^3 \cdot y = y^9$, and $z^2 \cdot z^3 \cdot z = z^6$.

 Self Check 1   Multiply: **a.** $18t(t)$ **b.** $-10d^8(-6d^3)$

   c. $(16y^{12})\left(\frac{1}{4}y^2\right)$ **d.** $(5a^3b^3)(-6a^3b^4)(ab)$

***Now Try*   Problems 13, 19 and 23**

2 **Multiply a Polynomial by a Monomial.**

We can use the distributive property to find the product of a monomial and a binomial such as $5x$ and $2x + 4$:

$$5x(2x + 4) = 5x(2x) + 5x(4) \qquad \text{Distribute the multiplication by 5x.}$$
$$= 10x^2 + 20x \qquad \text{Multiply the monomials.}$$

This example suggests the following rule.

Multiplying Polynomials by Monomials	To multiply a monomial and a polynomial, multiply each term of the polynomial by the monomial.

EXAMPLE 2   Multiply: **a.** $3a^2(3a^2 - 5a + 2)$
b. $-2xz^3(6x^3z + x^2z^2 - xz^3 + 7z^4)$ **c.** $(-m^4 - 2.5)(4.1m^3)$

Strategy   To find each product, we will multiply each term of the polynomial by the monomial.

Why   We use the distributive property to multiply a monomial and a polynomial.

Solution

a. Multiply each term of $3a^2 - 5a + 2$ by $3a^2$.

$$3a^2(3a^2 - 5a + 2)$$
$$= 3a^2(3a^2) + 3a^2(-5a) + 3a^2(2) \qquad \text{Distribute the multiplication by } 3a^2.$$
$$= 9a^4 - 15a^3 + 6a^2 \qquad \text{Multiply the monomials.}$$

b. Multiply each term of $6x^3z + x^2z^2 - xz^3 + 7z^4$ by $-2xz^3$.

$$-2xz^3(6x^3z + x^2z^2 - xz^3 + 7z^4)$$
$$= -2xz^3(6x^3z) - 2xz^3(x^2z^2) - 2xz^3(-xz^3) - 2xz^3(7z^4)$$
$$= -12x^4z^4 - 2x^3z^5 + 2x^2z^6 - 14xz^7 \qquad \text{Multiply the monomials.}$$

c. Multiply each term of $-m^4 - 2.5$ by $4.1m^3$.

$$(-m^4 - 2.5)(4.1m^3) = -m^4(4.1m^3) - 2.5(4.1m^3) \qquad \text{Distribute the multiplication by } 4.1m^3.$$
$$= -4.1m^7 - 10.25m^3 \qquad \text{Multiply the monomials.}$$

Self Check 2   Multiply: **a.** $5c^2(4c^3 - 9c - 8)$
b. $-s^2t^2(-s^4t^2 + s^3t^3 - s^2t^4 + 7s)$
c. $(w^7 - 2w)6w^5$

Now Try   Problems 29, 37, and 39

EXAMPLE 3   ***Parallelograms.***   Find a polynomial that represents the area of the parallelogram.

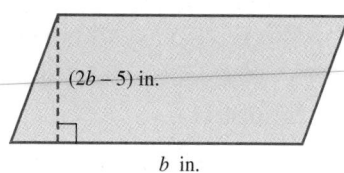

(2b − 5) in.

b in.

Strategy   We will multiply the length of the base of the parallelogram by its height.

Why   The area of a parallelogram is equal to the product of the length of its base and its height.

The Language of Algebra
A *parallelogram* is a four-sided figure whose opposite sides are parallel.

Solution

$Area = \textbf{base} \cdot \text{height}$ This is the formula for the area of a parallelogram.

$\qquad = b(2b - 5)$ b is the length of the base. Substitute 2b − 5 for the height.

$\qquad = 2b^2 - 5b$ Distribute the multiplication by b.

The area of the parallelogram is $(2b^2 - 5b)$ square inches, which can be written as $(2b^2 - 5b)$ in.^{2}.

 Now Try   **Problem 41**

3   **Multiply Binomials.**

The distributive property can also be used to multiply binomials. For example, to multiply $2a + 4$ and $3a + 5$, we think of $2a + 4$ as a single quantity and distribute it over each term of $3a + 5$.

$$(2a + 4)(3a + 5) = (2a + 4)3a + (2a + 4)5$$

$$= (2a + 4)3a + (2a + 4)5$$

$$= (2a)3a + (4)3a + (2a)5 + (4)5 \qquad \text{Distribute the multiplication by 3a and by 5.}$$

$$= 6a^2 + 12a + 10a + 20 \qquad \text{Multiply the monomials.}$$

$$= 6a^2 + 22a + 20 \qquad \text{Combine like terms.}$$

In the third line of the solution, notice that each term of $3a + 5$ has been multiplied by each term of $2a + 4$. This example suggests the following rule.

Multiplying Binomials   To multiply two binomials, multiply each term of one binomial by each term of the other binomial, and then combine like terms.

EXAMPLE 4   Multiply: $(5x - 8)(x + 1)$

Strategy   To find the product, we will multiply $x + 1$ by $5x$ and by -8.

Why   To multiply two binomials, each term of one binomial must be multiplied by each term of the other binomial.

Solution

$$(5x - 8)(x + 1) = 5x(x + 1) - 8(x + 1) \qquad \text{Multiply } x + 1 \text{ by } 5x \text{ and multiply } x + 1 \text{ by } -8.$$
$$= 5x^2 + 5x - 8x - 8 \qquad \text{Distribute the multiplication by 5x.}$$
$$\text{Distribute the multiplication by } -8.$$
$$= 5x^2 - 3x - 8 \qquad \text{Combine like terms.}$$

Self Check 4   Multiply: $(9y + 3)(y - 4)$
Now Try   **Problem 49**

The Language of Algebra
An *acronym* is an abbreviation of several words in such a way that the abbreviation itself forms a word. The *acronym* FOIL helps us remember the order to follow when multiplying two binomials: First, Outer, Inner, Last.

We can use a shortcut method, called the **FOIL method,** to multiply binomials. FOIL is an acronym for **F**irst terms, **O**uter terms, **I**nner terms, **L**ast terms. To use the FOIL method to multiply $2a + 4$ by $3a + 5$, we

1. multiply the **F**irst terms $2a$ and $3a$ to obtain $6a^2$,
2. multiply the **O**uter terms $2a$ and 5 to obtain $10a$,
3. multiply the **I**nner terms 4 and $3a$ to obtain $12a$, and
4. multiply the **L**ast terms 4 and 5 to obtain 20.

Then we simplify the resulting polynomial, if possible.

$$\overset{\text{First}}{\underset{\text{Inner}}{\overset{\text{Outer}}{(2a + 4)(3a + 5)}}} = \overset{F}{2a(3a)} + \overset{O}{2a(5)} + \overset{I}{4(3a)} + \overset{L}{4(5)}$$

$$= 6a^2 + 10a + 12a + 20 \qquad \text{Multiply the monomials.}$$
$$= 6a^2 + 22a + 20 \qquad \text{Combine like terms.}$$

EXAMPLE 5   Multiply: **a.** $(x + 5)(x + 7)$ **b.** $(3x + 4)(2x - 3)$
c. $\left(2r - \dfrac{1}{2}\right)\left(2r + \dfrac{5}{2}\right)$ **d.** $(3a^2 - 7b)(a^2 - b)$

Strategy   We will use the FOIL method.

Why   In each case we are to find the product of two binomials, and the FOIL method is a shortcut for multiplying two binomials.

Solution

a.

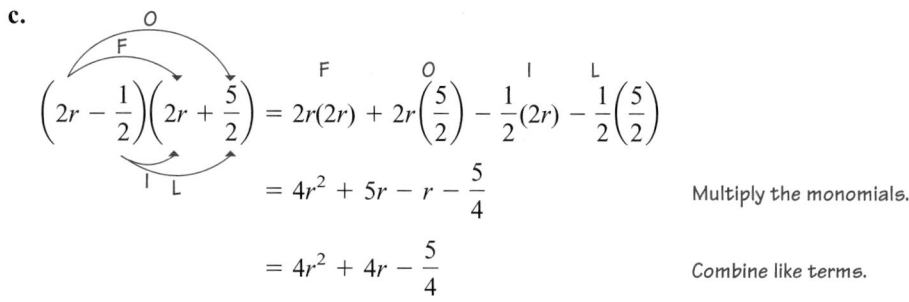

$$(x + 5)(x + 7) = x(x) + x(7) + 5(x) + 5(7)$$
$$= x^2 + 7x + 5x + 35 \qquad \text{Multiply the monomials.}$$
$$= x^2 + 12x + 35 \qquad \text{Combine like terms.}$$

b.

$$(3x + 4)(2x - 3) = 3x(2x) + 3x(-3) + 4(2x) + 4(-3)$$
$$= 6x^2 - 9x + 8x - 12 \qquad \text{Multiply the monomials.}$$
$$= 6x^2 - x - 12 \qquad \text{Combine like terms.}$$

c.

$$\left(2r - \frac{1}{2}\right)\left(2r + \frac{5}{2}\right) = 2r(2r) + 2r\left(\frac{5}{2}\right) - \frac{1}{2}(2r) - \frac{1}{2}\left(\frac{5}{2}\right)$$
$$= 4r^2 + 5r - r - \frac{5}{4} \qquad \text{Multiply the monomials.}$$
$$= 4r^2 + 4r - \frac{5}{4} \qquad \text{Combine like terms.}$$

d.

$$(3a^2 - 7b)(a^2 - b) = 3a^2(a^2) + 3a^2(-b) - 7b(a^2) - 7b(-b)$$
$$= 3a^4 - 3a^2b - 7a^2b + 7b^2 \qquad \text{Multiply the monomials.}$$
$$= 3a^4 - 10a^2b + 7b^2 \qquad \text{Combine like terms.}$$

> **Self Check 5** Multiply: **a.** $(y + 3)(y + 1)$
> **b.** $(2a - 1)(3a + 2)$
> **c.** $\left(4x - \frac{1}{2}\right)\left(4x + \frac{3}{4}\right)$
> **d.** $(5y^3 - 2)(2y^3 - 7)$
> ***Now Try*** **Problems 51, 57, and 59**

④ ## Multiply Polynomials.

To develop a general rule for multiplying any two polynomials, we will find the product of $2x + 3$ and $3x^2 + 3x + 5$. In the solution, the distributive property is used four times.

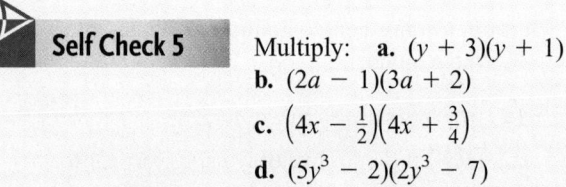

$$(2x + 3)(3x^2 + 3x + 5) = (2x + 3)3x^2 + (2x + 3)3x + (2x + 3)5 \qquad \text{Distribute.}$$
$$= (2x + 3)3x^2 + (2x + 3)3x + (2x + 3)5$$
$$= (2x)3x^2 + (3)3x^2 + (2x)3x + (3)3x + (2x)5 + (3)5 \qquad \text{Distribute.}$$
$$= 6x^3 + 9x^2 + 6x^2 + 9x + 10x + 15 \qquad \text{Multiply the monomials.}$$
$$= 6x^3 + 15x^2 + 19x + 15 \qquad \text{Combine like terms.}$$

In the third line of the solution, note that each term of $3x^2 + 3x + 5$ has been multiplied by each term of $2x + 3$. This example suggests the following rule.

Multiplying Polynomials	To multiply two polynomials, multiply each term of one polynomial by each term of the other polynomial, and then combine like terms.

EXAMPLE 6   Multiply: $(7y + 3)(6y^2 - 8y + 1)$

Strategy   We will multiply each term of the trinomial, $6y^2 - 8y + 1$, by each term of the binomial, $7y + 3$.

Why   To multiply two polynomials, we must multiply each term of one polynomial by each term of the other polynomial.

Solution

$$(7y + 3)(6y^2 - 8y + 1)$$

$$= 7y(6y^2) + 7y(-8y) + 7y(1) + 3(6y^2) + 3(-8y) + 3(1)$$

$$= 42y^3 - 56y^2 + 7y + 18y^2 - 24y + 3 \qquad \text{Multiply the monomials.}$$

$$= 42y^3 - 38y^2 - 17y + 3 \qquad \text{Combine like terms.}$$

  Self Check 6   Multiply: $(3a^2 - 1)(2a^4 - a^2 - a)$

Now Try   **Problem 67**

It is often convenient to multiply polynomials using a vertical form similar to that used to multiply whole numbers.

EXAMPLE 7   Multiply using vertical form: **a.** $(3a^2 - 4a + 7)(2a + 5)$
b. $(6y^3 - 5y + 4)(-4y^2 - 3)$

Strategy   First, we will write one polynomial underneath the other and draw a horizontal line beneath them. Then, we will multiply each term of the upper polynomial by each term of the lower polynomial.

Why   *Vertical form* means to use an approach similar to that used in arithmetic to multiply two numbers.

Solution

Success Tip
Multiplying two polynomials in vertical form is much like multiplying two numbers in arithmetic.

$$
\begin{array}{r}
347 \\
\times\ 25 \\
\hline
1735 \\
+694 \\
\hline
8675
\end{array}
$$

a. Multiply:

$$
\begin{array}{r}
3a^2 - 4a + 7 \\
2a + 5 \\
\hline
15a^2 - 20a + 35 \\
6a^3 - 8a^2 + 14a \\
\hline
6a^3 + 7a^2 - 6a + 35
\end{array}
$$

Multiply $3a^2 - 4a + 7$ by 5.

Multiply $3a^2 - 4a + 7$ by 2a.

In each column, combine like terms.

b. With this method, it is often necessary to leave a space for a missing term to vertically align like terms.

Multiply:

$$
\begin{array}{r}
6y^3 - 5y + 4 \\
-4y^2 \quad\ - 3 \\
\hline
-18y^3 \qquad\quad + 15y - 12 \\
-24y^5 + 20y^3 - 16y^2 \\
\hline
-24y^5 +\ \ 2y^3 - 16y^2 + 15y - 12
\end{array}
$$

Multiply $6y^3 - 5y + 4$ by -3.

Multiply $6y^3 - 5y + 4$ by $-4y^2$.

Leave a space for any missing powers of y. In each column, combine like terms.

Self Check 7   Multiply using vertical form:
a. $(3x + 2)(2x^2 - 4x + 5)$
b. $(-2x^2 + 3)(2x^2 - 4x - 1)$

Now Try   **Problem 77**

When finding the product of three polynomials, we begin by multiplying any two of them, and then we multiply that result by the third polynomial.

EXAMPLE 8   Find the product: $-3a(4a + 1)(a - 7)$

Strategy   We will find the product of $4a + 1$ and $a - 7$ and then multiply that result by $-3a$.

Why   It is wise to perform the most difficult multiplication first. (In this case, that would be the product of the two binomials). Save the simpler multiplication by $-3a$ for last.

Solution

$$
\begin{aligned}
-3a(4a + 1)(a - 7) &= -3a(4a^2 - 28a + a - 7) \\
&= -3a(4a^2 - 27a - 7) \\
&= -12a^3 + 81a^2 + 21a
\end{aligned}
$$

Multiply the two binomials.

Combine like terms within the parentheses.

Distribute the multiplication by $-3a$.

Self Check 8   Find the product: $-2y(y + 3)(3y - 2)$
Now Try   **Problem 81**

ANSWERS TO SELF CHECKS   **1. a.** $18t^2$ **b.** $60d^{11}$ **c.** $4y^{14}$ **d.** $-30a^7b^8$ **2. a.** $20c^5 - 45c^3 - 40c^2$
b. $s^6t^4 - s^5t^5 + s^4t^6 - 7s^3t^2$ **c.** $6w^{12} - 12w^6$ **4.** $9y^2 - 33y - 12$ **5. a.** $y^2 + 4y + 3$
b. $6a^2 + a - 2$ **c.** $16x^2 + x - \frac{3}{8}$ **d.** $10y^6 - 39y^3 + 14$ **6.** $6a^6 - 5a^4 - 3a^3 + a^2 + a$
7. a. $6x^3 - 8x^2 + 7x + 10$ **b.** $-4x^4 + 8x^3 + 8x^2 - 12x - 3$ **8.** $-6y^3 - 14y^2 + 12y$

STUDY SET
5.6

VOCABULARY

Fill in the blanks.

1. $(2x^3)(3x^4)$ is the product of two _____.
2. $(2a - 4)(3a + 5)$ is the product of two _____.
3. In the acronym FOIL, F stands for _____ terms, O for _____ terms, I for _____ terms, and L for _____ terms.
4. $(2a - 4)(3a^2 + 5a - 1)$ is the product of a _____ and a _____.

CONCEPTS

Fill in the blanks.

5. **a.** To multiply two polynomials, multiply _____ term of one polynomial by _____ term of the other polynomial, and then combine like terms.

   b. When multiplying three polynomials, we begin by multiplying _____ two of them, and then we multiply that result by the _____ polynomial.

6. Label each arrow using one of the letters F, O, I, or L. Then fill in the blanks.

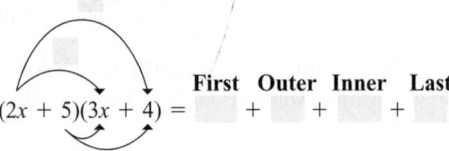

$$\text{First} \quad \text{Outer} \quad \text{Inner} \quad \text{Last}$$
$$(2x + 5)(3x + 4) = \boxed{} + \boxed{} + \boxed{} + \boxed{}$$

7. Simplify each polynomial by combining like terms.
   a. $6x^2 - 8x + 9x - 12$
   b. $5x^4 + 3ax^2 + 5ax^2 + 3a^2$

8. **a.** Add: $(x - 4) + (x + 8)$
   b. Subtract: $(x - 4) - (x + 8)$
   c. Multiply: $(x - 4)(x + 8)$

NOTATION

Complete each solution.

9. $(9n^3)(8n^2) = (9 \cdot \boxed{})(\boxed{} \cdot n^2) = \boxed{}$

10. $7x(3x^2 - 2x + 5) = \boxed{}(3x^2) - \boxed{}(2x) + \boxed{}(5)$
    $$= \boxed{} - 14x^2 + 35x$$

11. $(2x + 5)(3x - 2) = 2x(3x) - \boxed{}(2) + \boxed{}(3x) - \boxed{}(2)$
    $$= 6x^2 - \boxed{} + \boxed{} - 10$$
    $$= 6x^2 + \boxed{} - 10$$

12.
$$
\begin{array}{r}
3x^2 + 4x - 2 \\
2x + 3 \\
\hline
\boxed{} + 12x - 6 \\
6x^3 + 8x^2 - 4x \phantom{{}-6} \\
\hline
\boxed{} + 17x^2 + \boxed{} - 6
\end{array}
$$

GUIDED PRACTICE

Multiply. See Example 1.

13. $5m(m)$ 14. $4s(s)$
15. $(3x^2)(4x^3)$ 16. $(-2a^3)(11a^2)$
17. $(1.2c^3)(5c^3)$ 18. $(2.5h^4)(2h^4)$
19. $(3b^2)(-2b)(4b^3)$ 20. $(3y)(7y^2)(-y^4)$
21. $(2x^2y^3)(4x^3y^2)$ 22. $(-5x^3y^6)(2x^2y^2)$
23. $(8a^5)\left(-\dfrac{1}{4}a^6\right)$ 24. $\left(-\dfrac{2}{3}x^6\right)(9x^3)$

Multiply. See Example 2.

25. $3x(x + 4)$ 26. $3a(a + 2)$
27. $-4t(t^2 - 7)$ 28. $-6s(s^2 - 3)$
29. $9x^2(x^2 - 2x + 6)$ 30. $4y^2(y^2 + 5y - 10)$
31. $-2x^3(3x^2 - x + 1)$ 32. $-4b^3(2b^2 - 2b + 2)$
33. $0.3p^5(0.4p^4 - 6p^2)$ 34. $0.5u^5(0.4u^6 - 0.5u^3)$
35. $\dfrac{5}{8}t^2(t^6 + 8t^2)$ 36. $\dfrac{4}{9}a^2(9a^3 + a^2)$
37. $-4x^2z(3x^2 + z^2 + xz - 1)$
38. $-3x^2y(x^2 + y^2 + xy - 1)$
39. $(x^2 - 12x)(6x^{12})$ 40. $(w^9 - 11w)(2w^7)$

Find a polynomial that represents the area of the parallelogram or rectangle. See Example 3.

41.

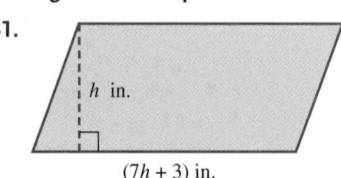

h in.

$(7h + 3)$ in.

42.

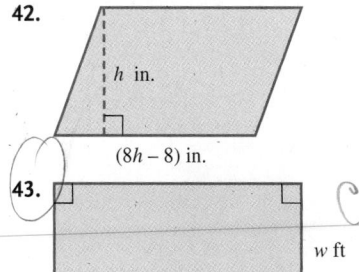

h in.

$(8h - 8)$ in.

43.

w ft

$(4w - 2)$ ft

44. $(8w + 1)$ yd

w yd

Multiply. See Examples 4 and 5.

45. $(y + 3)(y + 5)$ **46.** $(a + 4)(a + 5)$

47. $(t + 4)(t - 3)$ **48.** $(x + 7)(x - 6)$

49. $(m + 6)(m - 9)$ **50.** $(n + 8)(n - 10)$

51. $(4y - 5)(y + 7)$ **52.** $(3x - 4)(x + 5)$

53. $(2x - 3)(6x - 5)$ **54.** $(5x - 3)(2x - 3)$

55. $(3.8y - 1)(2y - 1)$ **56.** $(2.6x - 3)(2x - 1)$

57. $\left(6m - \dfrac{2}{3}\right)\left(3m - \dfrac{4}{3}\right)$ **58.** $\left(8t - \dfrac{1}{2}\right)\left(4t - \dfrac{5}{2}\right)$

59. $(t^2 - 3)(t^2 - 4)$ **60.** $(s^3 - 6)(s^3 - 8)$

61. $(a + b)(a + b)$ **62.** $(m + n)(m + n)$

63. $(3a - 2b)(4a + b)$ **64.** $(2t + 3s)(3t - s)$

Multiply. See Example 6.

65. $(x + 2)(x^2 - 2x + 3)$

66. $(x - 5)(x^2 + 2x - 3)$

67. $(4t + 3)(t^2 + 2t + 3)$

68. $(3x + 1)(2x^2 - 3x + 1)$

69. $(x - 2)(x^2 + 2x + 4)$

70. $(a + 3)(a^2 - 3a + 9)$

71. $(x^2 + 6x + 7)(2x - 5)$

72. $(y^2 - 2y + 1)(4y + 8)$

73. $(-3x + y)(x^2 - 8xy + 16y^2)$

74. $(3x - y)(x^2 + 3xy - y^2)$

75. $(r^2 - r + 3)(r^2 - 4r - 5)$

76. $(w^2 + w - 9)(w^2 - w + 3)$

Multiply using vertical form. See Example 7.

77. $x^2 - 2x + 1$
    $x + 2$

78. $5r^2 + r + 6$
    $2r - 1$

79. $4x^2 + 3x - 4$
    $3x + 2$

80. $x^2 - x + 1$
    $x + 1$

Multiply. See Example 8.

81. $4x(2x + 1)(x - 2)$

82. $5a(3a - 2)(2a + 3)$

83. $-3a(a + b)(a - b)$

84. $-2r(r + s)(r + s)$

85. $(-2a^2)(-3a^3)(3a - 2)$

86. $(3x)(-2x^2)(x + 4)$

87. $(x - 4)(x + 1)(x - 3)$

88. $(x + 6)(x - 2)(x - 4)$

TRY IT YOURSELF

Multiply.

89. $(5x - 2)(6x - 1)$ **90.** $(8x - 1)(3x - 7)$

91. $(3x^2 + 4x - 7)(2x^2)$ **92.** $(2y^2 - 7y - 8)(3y^3)$

93. $(6x^2z^5)(-3xz^3)$ **94.** $(-5r^4t^2)(2r^2t)$

95. $2a^2 + 3a + 1$
    $3a^2 - 2a + 4$

96. $3y^2 + 2y - 4$
    $2y^2 - 4y + 3$

97. $(t + 2s)(9t - 3s)$ **98.** $(4t - u)(3t + u)$

99. $\left(\dfrac{1}{2}a\right)(4a^4)(a^5)$ **100.** $(12b)\left(\dfrac{7}{6}b\right)(b^4)$

101. $4y(y + 3)(y + 7)$ **102.** $2t(t + 8)(t + 10)$

103. $8.2pq(2pq - 3p + 5q)$

104. $5.3ab(2ab + 6a - 3b)$

105. $(x + 6)(x^3 + 5x^2 - 4x - 4)$

106. $(x - 8)(x^3 - 4x^2 - 2x - 2)$

107. $\left(4a - \dfrac{5}{4}r\right)\left(4a + \dfrac{3}{4}r\right)$ **108.** $\left(5c - \dfrac{2}{5}t\right)\left(10c + \dfrac{1}{5}t\right)$

APPLICATIONS

109. STAMPS   Find a polynomial that represents the area of the stamp.

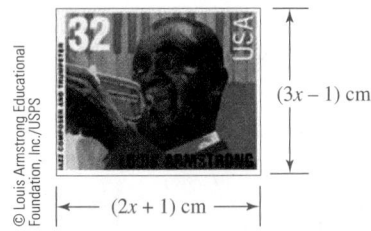

$(3x - 1)$ cm

$(2x + 1)$ cm

110. PARKING   Find a polynomial that represents the total area of the van-accessible parking space and its access aisle.

$(x + 10)$ ft $2x$ ft

111. SUNGLASSES   An ellipse is an oval-shaped curve. The area of an ellipse is approximately $0.785ab$, where a is its length and b is its width. Find a polynomial that represents the approximate area of one of the elliptical-shaped lenses of the sunglasses.

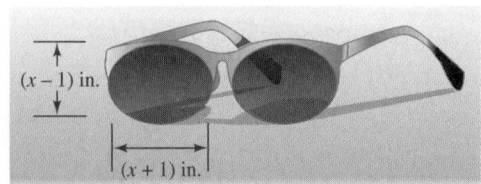

$(x - 1)$ in.

$(x + 1)$ in.

112. GARDENING   Refer to the illustration in the next column.

   a. Find the area of the region planted with corn, tomatoes, beans, and carrots. Add your answers to find the total area of the garden.

   b. Find the length and width of the garden. Multiply your answers to find its area.

   c. How do the answers from parts (a) and (b) for the area of the garden compare?

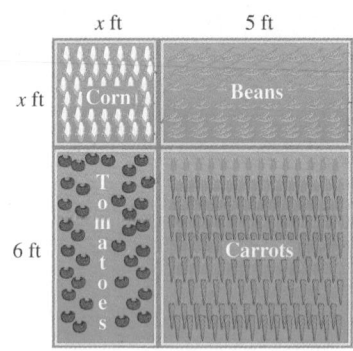

x ft   5 ft

x ft   Corn   Beans

6 ft   Tomatoes   Carrots

113. LUGGAGE   Find a polynomial that represents the volume of the garment bag. (Recall that the formula for the volume of a rectangular solid is $V = lwh$.)

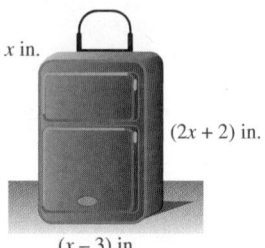

x in.

$(2x + 2)$ in.

$(x - 3)$ in.

114. BASEBALL   Find a polynomial that represents the volume within the batting cage.

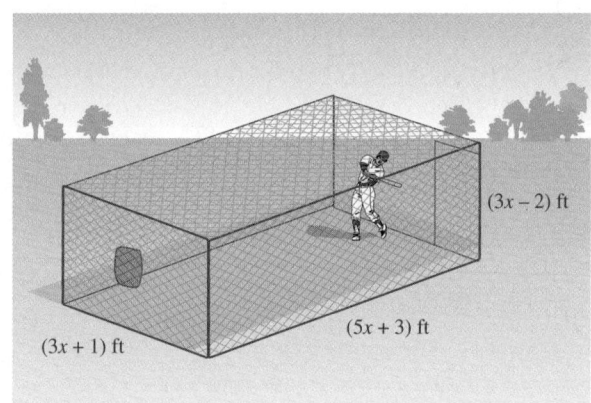

$(3x - 2)$ ft

$(5x + 3)$ ft

$(3x + 1)$ ft

WRITING

115. Is the product of a monomial and a monomial always a monomial? Explain.

116. Explain this diagram.

$$(5x + 6)(7x - 1)$$

117. Explain why the FOIL method cannot be used to find $(3x + 2)(4x^2 - x + 10)$.

118. Explain the error: $(x + 3)(x - 2) = x^2 - 6$

119. Explain why the vertical form used in algebra to multiply $2x^2 + 3x + 1$ and $3x + 2$ is similar to the vertical form used in arithmetic to multiply 231 and 32.

120. Would the OLIF method give the same result as the FOIL method when multiplying two binomials? Explain why or why not.

REVIEW

121. What is the slope of
    a. Line 1?
    b. Line 2?

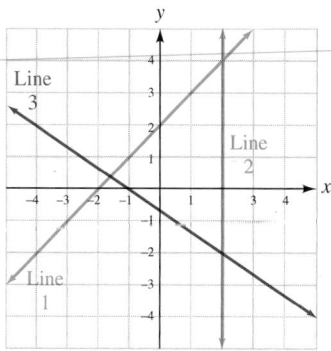

122. What is the slope of
    a. Line 3?
    b. the x-axis?

123. What is the y-intercept of Line 1?

124. What is the x-intercept of Line 1?

CHALLENGE PROBLEMS

125. a. Find each of the following products.
    i. $(x - 1)(x + 1)$
    ii. $(x - 1)(x^2 + x + 1)$
    iii. $(x - 1)(x^3 + x^2 + x + 1)$

    b. Write a product of two polynomials such that the result is $x^5 - 1$.

126. Solve: $(y - 1)(y + 6) = (y - 3)(y - 2) + 8$

SECTION 5.7
Special Products

Objectives

1 Square a binomial.

2 Multiply the sum and difference of the same two terms.

3 Find higher powers of binomials.

4 Simplify expressions containing polynomial multiplication.

Certain products of binomials, called **special products,** occur so often that it is worthwhile to learn their forms.

1   **Square a Binomial.**

To develop a rule to find the *square of a binomial sum,* we consider $(x + y)^2$. We can use the definition of exponent and the procedure for multiplying two binomials to find the product.

$$(x + y)^2 = (x + y)(x + y) \quad \text{In } (x + y)^2, \text{ the base is } (x + y) \text{ and the exponent is 2.}$$
$$= x^2 + xy + xy + y^2 \quad \text{Multiply the binomials.}$$
$$= x^2 + 2xy + y^2 \quad \text{Combine like terms: } xy + xy = 1xy + 1xy = 2xy.$$

Note that the terms of the resulting trinomial are related to the terms of the binomial that was squared.

$$(x + y)^2 = x^2 + 2xy + y^2$$

    The square of the second term, y.
    Twice the product of the first and second terms, x and y.
    The square of the first term, x.

To develop a rule to find the *square of a binomial difference,* we consider $(x - y)^2$.

Success Tip

The illustration can be used to visualize a special product. The area of the large square is $(x + y)(x + y) = (x + y)^2$. The sum of the four smaller areas is $x^2 + xy + xy + y^2$ or $x^2 + 2xy + y^2$. Thus,

$$(x + y)^2 = x^2 + 2xy + y^2$$

$$(x - y)^2 = (x - y)(x - y) \quad \text{In } (x - y)^2, \text{ the base is } (x - y) \text{ and the exponent is 2.}$$
$$= x^2 - 2xy + y^2 \quad \text{Multiply the binomials.}$$
$$\text{Combine like terms: } -xy - xy = -2xy.$$

Again, the terms of the resulting trinomial are related to the terms of the binomial that was squared.

$$(x - y)^2 = x^2 - 2xy + y^2$$

The square of the second term, −y.
Twice the product of the first and second terms, x and −y.
The square of the first term, x.

The observations from these two examples suggest the following **special-product rules.**

Squaring a Binomial

The **square of a binomial** is a trinomial, such that:

- Its first term is the square of the first term of the binomial.
- Its last term is the square of the second term of the binomial.
- Its middle term is twice the product of both terms of the binomial.

$$(A + B)^2 = A^2 + 2AB + B^2 \qquad (A - B)^2 = A^2 - 2AB + B^2$$

EXAMPLE 1 Find each square: **a.** $(t + 9)^2$ **b.** $(8a - 5)^2$
c. $(d + 0.5)^2$ **d.** $\left(c^3 - \dfrac{7}{2}d\right)^2$

The Language of Algebra
When squaring a binomial, the result is called a *perfect-square trinomial.*
$$(t + 9)^2 = \underline{t^2 + 18t + 81}$$
Perfect-square trinomial

Strategy To find each square of a binomial, we will use one of the special-product rules.

Why This approach is faster than using the FOIL method.

Solution

a. $(t + 9)^2$ is the square of a binomial sum. The first term is t and the second term is 9.

$$(t + 9)^2 = \underbrace{t^2}_{\substack{\text{The square of} \\ \text{the first term, } t.}} + \underbrace{2(t)(9)}_{\substack{\text{Twice the product} \\ \text{of both terms.}}} + \underbrace{9^2}_{\substack{\text{The square of the} \\ \text{second term, 9.}}}$$

$$= t^2 + 18t + 81$$

Caution
The square of a binomial is a *trinomial.* A common error when squaring a binomial is to forget the middle term of the product.
$$(8a - 5)^2 \neq 64a^2 + 25$$
Missing −80a

b. $(8a - 5)^2$ is the square of a binomial difference. The first term is $8a$ and the second term is -5.

$$(8a - 5)^2 = \underbrace{(8a)^2}_{\substack{\text{The square of} \\ \text{the first term, 8a.}}} + \underbrace{2(8a)(-5)}_{\substack{\text{Twice the product} \\ \text{of both terms.}}} + \underbrace{(-5)^2}_{\substack{\text{The square of the} \\ \text{second term, −5.}}}$$

$$= 64a^2 - 80a + 25 \quad \text{Use the power of a product rule: } (8a)^2 = 8^2a^2 = 64a^2.$$

c. $(d + 0.5)^2$ is the square of a binomial sum. The first term is d and the second term is 0.5.

$$(d + 0.5)^2 = \underbrace{(d)^2}_{\substack{\text{The square of} \\ \text{the first term, } d.}} + \underbrace{2(d)(0.5)}_{\substack{\text{Twice the product} \\ \text{of both terms.}}} + \underbrace{(0.5)^2}_{\substack{\text{The square of the} \\ \text{second term, 0.5.}}}$$

$$= d^2 + d + 0.25$$

d. $\left(c^3 - \frac{7}{2}d\right)^2$ is the square of a binomial difference. The first term is c^3 and the second term is $-\frac{7}{2}d$.

$$\left(c^3 - \frac{7}{2}d\right)^2 = \underbrace{(c^3)^2}_{\substack{\text{The square of} \\ \text{the first term, } c^3.}} + \underbrace{2(c^3)\left(-\frac{7}{2}d\right)}_{\substack{\text{Twice the product} \\ \text{of both terms.}}} + \underbrace{\left(-\frac{7}{2}d\right)^2}_{\substack{\text{The square of the} \\ \text{second term, } -\frac{7}{2}d.}}$$

$$= c^6 - 7c^3d + \frac{49}{4}d^2 \qquad \text{Use rules for exponents to find } (c^3)^2 \text{ and } \left(-\frac{7}{2}d\right)^2.$$

Self Check 1   Find each square: **a.** $(r + 6)^2$
b. $(7g - 2)^2$ **c.** $(v + 0.8)^2$
d. $\left(w^4 - \frac{3}{2}y\right)^2$

Now Try   **Problems 9, 19, and 29**

2   **Multiply the Sum and Difference of the Same Two Terms.**

A final special product that occurs often has the form $(A + B)(A - B)$. In these products, one binomial is the sum of two terms and the other binomial is the difference of the same two terms. To develop a rule to find such products, consider the following multiplication:

$$(x + y)(x - y) = x^2 - xy + xy - y^2 \qquad \text{Multiply the binomials.}$$
$$= x^2 - y^2 \qquad\qquad\quad \text{Combine like terms: } -xy + xy = 0.$$

Success Tip

We can use the FOIL method to find each of the special products discussed in this section. However, these forms occur so often, it is worthwhile to learn the special-product rules.

Note that when we combined like terms, we added opposites. This will always be the case for products of this type; the sum of the outer and inner products will be 0. The first and last products will be squares.

$$(x + y)(x - y) = x^2 - y^2$$

The square of the second term, y.
The square of the first term, x.

These observations suggest a third **special-product rule.**

Multiplying the Sum and Difference of Two Terms

The product of the sum of two terms and difference of the same two terms is the square of the first term minus the square of the second term.

$$(A + B)(A - B) = A^2 - B^2$$

EXAMPLE 2   Multiply: **a.** $(m + 2)(m - 2)$ **b.** $(3y + 4)(3y - 4)$
c. $\left(b - \frac{2}{3}\right)\left(b + \frac{2}{3}\right)$ **d.** $(t^4 - 6u)(t^4 + 6u)$

Strategy   To find the product of each pair of binomials, we will use the special-product rule for the sum and difference of the same two terms.

Why   This approach is faster than using the FOIL method.

The Language of Algebra
When multiplying the sum and difference of two terms, the result is called a *difference of two squares*.

$$(m + 2)(m - 2) = \underline{m^2 - 4}$$

Difference of two squares

Solution

a. $(m + 2)$ and $(m - 2)$ are the sum and difference of the same two terms, m and 2.

$$(m + 2)(m - 2) = \underbrace{m^2}_{\substack{\text{The square of the} \\ \text{first term, } m.}} - \underbrace{2^2}_{\substack{\text{The square of the} \\ \text{second term, 2.}}}$$

$$= m^2 - 4$$

b. $(3y + 4)$ and $(3y - 4)$ are the sum and difference of the same two terms, $3y$ and 4.

$$(3y + 4)(3y - 4) = \underbrace{(3y)^2}_{\substack{\text{The square of the} \\ \text{first term, 3y.}}} - \underbrace{4^2}_{\substack{\text{The square of the} \\ \text{second term, 4.}}}$$

$$= 9y^2 - 16$$

c. By the commutative property of multiplication, the special-product rule can be written with the factor containing the $-$ symbol first: $(A - B)(A + B) = A^2 - B^2$. Since $\left(b - \frac{2}{3}\right)$ and $\left(b + \frac{2}{3}\right)$ are the difference and sum of the same two terms, b and $\frac{2}{3}$, we have

$$\left(b - \frac{2}{3}\right)\left(b + \frac{2}{3}\right) = \underbrace{b^2}_{\substack{\text{The square of the} \\ \text{first term, } b.}} - \underbrace{\left(\frac{2}{3}\right)^2}_{\substack{\text{The square of the} \\ \text{second term, } \frac{2}{3}.}}$$

$$= b^2 - \frac{4}{9}$$

d. $(t^4 - 6u)$ and $(t^4 + 6u)$ are the difference and sum of the same two terms, t^4 and $6u$.

$$(t^4 - 6u)(t^4 + 6u) = \underbrace{(t^4)^2}_{\substack{\text{The square of the} \\ \text{first term, } t^4.}} - \underbrace{(6u)^2}_{\substack{\text{The square of the} \\ \text{second term, } 6u.}}$$

$$= t^8 - 36u^2$$

 Self Check 2   Multiply: **a.** $(b + 4)(b - 4)$
b. $(5m + 9)(5m - 9)$ **c.** $\left(s - \frac{3}{4}\right)\left(s + \frac{3}{4}\right)$
d. $(c^3 + 2d)(c^3 - 2d)$

Now Try   Problems 35, 37, and 41

3   **Find Higher Powers of Binomials.**

When we find the third, fourth, or even higher powers of a binomial, we say that we are **expanding the binomial.** The special-product rules can be used in such cases. The result is an expression that has more terms than the original binomial.

EXAMPLE 3   Expand: $(x + 1)^3$

Strategy   We will use a special-product rule to find the third power of $x + 1$.

Why   Since $(x + 1)^3$ can be written as $(x + 1)^2(x + 1)$, we can use a special-product rule to quickly find $(x + 1)^2$.

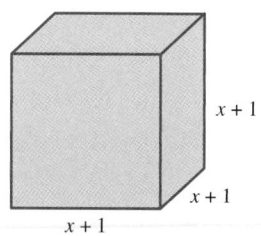

Solution

$$(x + 1)^3 = (x + 1)^2(x + 1)$$

$$= (x^2 + 2x + 1)(x + 1) \qquad \text{Find } (x + 1)^2 \text{ using the rule for the square of a sum.}$$

$$= (x^2 + 2x + 1)(x + 1) \qquad \text{Multiply the binomial and the trinomial.}$$

$$= x^2(x) + x^2(1) + 2x(x) + 2x(1) + 1(x) + 1(1) \qquad \begin{array}{l}\text{Multiply each term of } x + 1 \\ \text{by each term of } x^2 + 2x + 1.\end{array}$$

$$= x^3 + x^2 + 2x^2 + 2x + x + 1 \qquad \text{Multiply the monomials.}$$

$$= x^3 + 3x^2 + 3x + 1 \qquad \text{Combine like terms.}$$

Notation

$(x + 1)^3$ represents the volume of a cube with sides of length $x + 1$ units.

 Self Check 3   Expand: $(n - 3)^3$

         Now Try   **Problem 49**

---

4 **Simplify Expressions Containing Polynomial Multiplication.**

We can use a modified version of the order of operations rule to simplify expressions that involve polynomial addition, subtraction, multiplication, and raising to a power.

Order of Operations with Polynomials

1. If possible, simplify any polynomials within parentheses by combining like terms.
2. Square (or expand) all polynomials raised to powers using the FOIL method or a special-product rule.
3. Perform all polynomial multiplications using the distributive property, the FOIL method, or a special-product rule.
4. Perform all polynomial additions and subtractions by combining like terms.

EXAMPLE 4   Simplify each expression:
  a. $-8(y^2 - 2y + 3) - 4(2y^2 + y - 6)$
  b. $(x + 1)(x - 2) + 3x(x + 3)$
  c. $(3y - 2)^2 - (y - 5)(y + 5)$

Strategy   We will follow the rules for the order of operations to simplify each expression.

Why   If we don't follow the correct order of operations, we can obtain different results that are not equivalent.

Solution

a. The two polynomials within the parentheses do not simplify further and no polynomials are raised to a power. To perform the multiplication, we will use the distributive property twice. Then we will combine like terms.

$$-8(y^2 - 2y + 3) - 4(2y^2 + y - 6) = -8y^2 + 16y - 24 - 8y^2 - 4y + 24 \qquad \text{Distribute.}$$

$$= -16y^2 + 12y \qquad \begin{array}{l}\text{Add and subtract to} \\ \text{combine like terms.}\end{array}$$

b. The three polynomials within parentheses do not simplify further and no polynomials are raised to a power. To perform the multiplications, we use the FOIL method and the distributive property. Then we will combine like terms.

$$(x + 1)(x - 2) + 3x(x + 3)$$

$$= x^2 - x - 2 + 3x^2 + 9x \qquad \text{Use the FOIL method to find } (x + 1)(x - 2). \text{ Distribute the multiplication by } 3x.$$

$$= 4x^2 + 8x - 2 \qquad \text{Combine like terms.}$$

c. The three polynomials within parentheses do not simplify further. To square $3y - 2$, we use a special-product rule. To find the product of $(y - 5)(y + 5)$, we will use the special-product rule for the sum and difference of the same two terms. Then we will combine like terms.

$$(3y - 2)^2 - (y - 5)(y + 5)$$

$$= 9y^2 - 12y + 4 - (y^2 - 25) \qquad \text{Write } y^2 - 25 \text{ within parentheses so that both terms are subtracted.}$$

$$= 9y^2 - 12y + 4 - y^2 + 25 \qquad \text{Change the sign of each term within } (y^2 - 25) \text{ and drop the parentheses.}$$

$$= 8y^2 - 12y + 29 \qquad \text{Combine like terms.}$$

Self Check 4   Simplify each expression:
a. $2(a^2 - 3a) + 5(a^2 + 2a)$
b. $(x - 4)(x + 6) + 5x(2x - 1)$
c. $(a + 9)(a - 9) - (2a - 4)^2$

Now Try   **Problems 61, 63, and 67**

EXAMPLE 5   *Triangles.*   Find a polynomial that represents the area of the triangle.

Strategy   We will multiply one-half, the length of the base, and the height of the triangle.

Why   The area of a triangle is equal to one-half the product of the length of its base and its height.

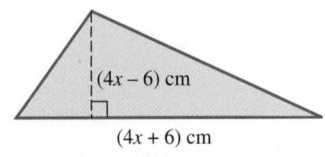
(4x − 6) cm
(4x + 6) cm

Solution   We begin by substituting $4x + 6$ for the length of the base and $4x - 6$ for the height in the formula for the area of a triangle. It is wise to find $(4x + 6)(4x - 6)$ first, using a special-product rule, and then to multiply that result by $\frac{1}{2}$.

$$\text{Area} = \frac{1}{2} \cdot \textbf{base} \cdot \textbf{height}$$

$$= \frac{1}{2}(4x + 6)(4x - 6)$$

$$= \frac{1}{2}(16x^2 - 36) \qquad \text{Use the special product rule for a sum and difference of two terms.}$$

$$= 8x^2 - 18 \qquad \text{Distribute the multiplication by } \frac{1}{2}.$$

Success Tip
Recall that to multiply three polynomials, begin by multiplying any two of them. Then multiply that result by the third polynomial.

The area of the triangle is $(8x^2 - 18)$ square centimeters, which can be written as $(8x^2 - 18)$ cm^2.

 Now Try Problem 73

 ANSWERS TO SELF CHECKS **1. a.** $r^2 + 12r + 36$ **b.** $49g^2 - 28g + 4$ **c.** $v^2 + 1.6v + 0.64$
d. $w^8 - 3w^4y + \frac{9}{4}y^2$ **2. a.** $b^2 - 16$ **b.** $25m^2 - 81$ **c.** $s^2 - \frac{9}{16}$ **d.** $c^6 - 4d^2$
3. $n^3 - 9n^2 + 27n - 27$ **4. a.** $7a^2 + 4a$ **b.** $11x^2 - 3x - 24$ **c.** $-3a^2 + 16a - 97$

STUDY SET
5.7

VOCABULARY

Fill in the blanks.

1. Expressions of the form $(x + y)^2$, $(x - y)^2$, and $(x + y)(x - y)$ occur so frequently in algebra that they are called special _____.

2. $(2x + 3)^2$ is the _____ of a binomial and $(a + 6)(a - 6)$ is the product of the sum and difference of the _____ two terms.

CONCEPTS

3. Complete each special product.
  a. $(x + y)^2 = x^2 + 2xy + y^2$
    └ The _____ of the second term
    └ _____ the product of the first and second terms
    └ The square of the _____ term
  b. $(x + y)(x - y) = x^2 - y^2$
    └ The square of the _____ term
    └ The _____ of the first term

4. Consider the binomial $5x + 4$.
  a. What is the square of its first term?
  b. What is twice the product of its two terms?
  c. What is the square of its second term?

NOTATION

Complete each solution to find the product.

5. $(x + 4)^2 = ^2 + 2(x)() + ^2$
     $= x^2 + + 16$

6. $(6r - 1)^2 = ()^2 2(6r)(1) + (-1)^2$
     $= - + 1$

7. $(s + 5)(s - 5) = ^2 - ^2$
     $= s^2 - $

8. True or false: $(t + 7)(t - 7) = (t - 7)(t + 7)$?

GUIDED PRACTICE

Find each product. **See Example 1.**

9. $(x + 1)^2$ **10.** $(y + 7)^2$

11. $(r + 2)^2$ **12.** $(n + 10)^2$

13. $(m - 6)^2$ **14.** $(b - 1)^2$

15. $(f - 8)^2$ **16.** $(w - 9)^2$

17. $(4x + 5)^2$ **18.** $(6y + 3)^2$

19. $(7m - 2)^2$ **20.** $(9b - 2)^2$

21. $(1 - 3y)^2$ **22.** $(1 - 4a)^2$

23. $(y + 0.9)^2$ **24.** $(d + 0.2)^2$

25. $(a^2 + b^2)^2$ **26.** $(c^2 + d^2)^2$

27. $(r^2 - s^2)^2$ **28.** $(t^2 - u^2)^2$

29. $\left(s + \dfrac{3}{4}\right)^2$ **30.** $\left(y - \dfrac{5}{3}\right)^2$

31. $\left(d^4 + \dfrac{1}{4}\right)^2$ **32.** $\left(q^6 + \dfrac{1}{3}\right)^2$

Find each product. See Example 2.

33. $(x + 3)(x - 3)$

34. $(y + 6)(y - 6)$

35. $(d + 7)(d - 7)$

36. $(t + 2)(t - 2)$

37. $(2p + 7)(2p - 7)$

38. $(5t + 4)(5t - 4)$

39. $(3n + 1)(3n - 1)$

40. $(5a + 4)(5a - 4)$

41. $\left(c + \dfrac{3}{4}\right)\left(c - \dfrac{3}{4}\right)$

42. $\left(m + \dfrac{4}{5}\right)\left(m - \dfrac{4}{5}\right)$

43. $\left(6b + \dfrac{1}{2}\right)\left(6b - \dfrac{1}{2}\right)$

44. $\left(4h + \dfrac{2}{3}\right)\left(4h - \dfrac{2}{3}\right)$

45. $(0.4 - 9m^2)(0.4 + 9m^2)$

46. $(0.3 - 2c^2)(0.3 + 2c^2)$

47. $(5 - 6g)(5 + 6g)$

48. $(6 - c^2)(6 + c^2)$

Expand each binomial. See Example 3.

49. $(x + 4)^3$

50. $(y + 2)^3$

51. $(n - 6)^3$

52. $(m - 5)^3$

53. $(2g - 3)^3$

54. $(3x - 2)^3$

55. $(a + b)^3$

56. $(c - d)^3$

57. $(2m + n)^3$

58. $(p - 2q)^3$

59. $(n - 2)^4$

60. $(c + d)^4$

Perform the operations. See Example 4.

61. $2(x^2 + 7x - 1) - 3(x^2 - 2x + 2)$

62. $5(y^2 - 2y - 6) + 6(2y^2 + 2y - 5)$

63. $2t(t + 2) + (t - 1)(t + 9)$

64. $3y(y + 2) + (y + 1)(y - 1)$

65. $(x + y)(x - y) + x(x + y)$

66. $(3x + 4)(2x - 2) - (2x + 1)(x + 3)$

67. $(5a - 1)^2 - (a - 8)(a + 8)$

68. $(4b + 1)^2 - (b - 7)(b + 7)$

69. $-5d(4d - 1)^2$

70. $-2h(7h - 2)^2$

71. $4d(d^2 + g^3)(d^2 - g^3)$

72. $8y(x^2 + y^2)(x^2 - y^2)$

Find a polynomial that represents the area of the figure. Leave π in your answer. See Example 5.

73.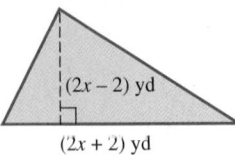
$(2x - 2)$ yd
$(2x + 2)$ yd

74.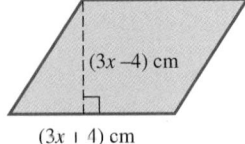
$(3x - 4)$ cm
$(3x + 4)$ cm

75.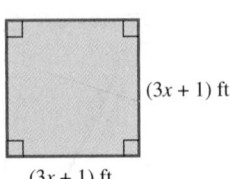
$(3x + 1)$ ft
$(3x + 1)$ ft

76.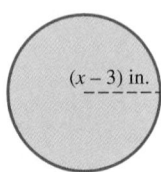
$(x - 3)$ in.

TRY IT YOURSELF

Perform the operations.

77. $(2v^3 - 8)^2$

78. $(8x^4 - 3)^2$

79. $3x(2x + 3)(2x + 3)$

80. $4y(3y + 4)(3y + 4)$

81. $(4f + 0.4)(4f - 0.4)$

82. $(4t + 0.6)(4t - 0.6)$

83. $(r^2 + 10s)^2$

84. $(m^2 + 8n)^2$

85. $2(x + 3) + 4(x - 2)$

86. $3(y - 4) - 5(y + 3)$

87. $(2a - 3b)^2$

88. $(2x + 5y)^2$

89. $(n + 6)(n - 6)$

90. $(a + 12)(a - 12)$

91. $(m + 10)^2 - (m - 8)^2$

92. $(5y - 1)^2 - (y + 7)(y - 7)$

93. $\left(5m - \dfrac{6}{5}\right)^2$

94. $\left(6m - \dfrac{7}{6}\right)^2$

95. $(2e + 1)^3$

96. $(3m - 2n)^3$

97. $(x - 2)^2$

98. $(a + 2)^2$

99. $(3x - 2)^2 + (2x + 1)^2$

100. $(4a - 3)^2 + (a + 6)^2$

101. $(6 - 2d^3)^2$

102. $(6 - 5p^2)^2$

103. $(8x + 3)^2$

104. $(4b - 8)^2$

APPLICATIONS

105. PLAYPENS   Find a polynomial that represents the area of the floor of the playpen.

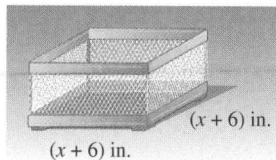

$(x + 6)$ in.
$(x + 6)$ in.

106. STORAGE   Find a polynomial that represents the volume of the cubicle.

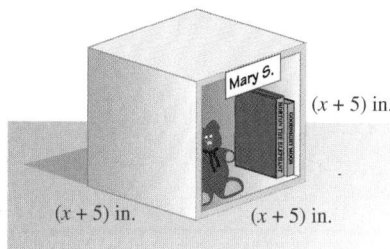

$(x + 5)$ in.
$(x + 5)$ in. $(x + 5)$ in.

107. PAPER TOWELS   The amount of space (volume) occupied by the paper on the roll of paper towels is given by the expression $\pi h(R + r)(R - r)$, where R is the outer radius and r is the inner radius. Perform the indicated multiplication.

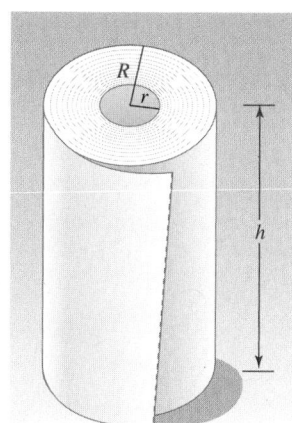

R
r
h

108. SIGNAL FLAGS   Refer to the illustration in the next column. Find a polynomial that represents the area in blue of the maritime signal flag for the letter p. The dimensions are given in centimeters.

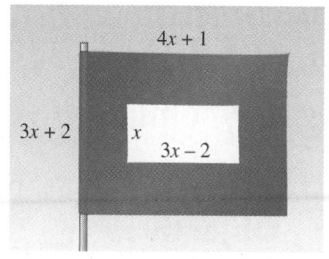

$4x + 1$
$3x + 2$
x
$3x - 2$

WRITING

109. What is a binomial? Explain how to square it.

110. Writing $(x + y)^2$ as $x^2 + y^2$ illustrates a common error. Explain.

111. We can find $(2x + 3)^2$ and $(5y - 6)^2$ using the FOIL method or using special product rules. Explain why the special product rules are faster.

112. **a.** Fill in the blanks: $(xy)^2$ is the _____ of product and $(x + y)^2$ is the _____ of a sum.
  b. Explain why $(xy)^2 \neq (x + y)^2$.

REVIEW

113. Find the prime factorization of 189.

114. Complete each statement. For any nonzero number a,

  a. $\dfrac{a}{a} =$ **b.** $\dfrac{a}{1} =$

  c. $\dfrac{0}{a} =$ **d.** $\dfrac{a}{0}$ is

115. Simplify: $\dfrac{30}{36}$ **116.** Add: $\dfrac{5}{12} + \dfrac{1}{4}$

117. Multiply: $\dfrac{7}{8} \cdot \dfrac{3}{5}$ **118.** Divide: $\dfrac{1}{3} \div \dfrac{4}{5}$

CHALLENGE PROBLEMS

119. **a.** Find two binomials whose product is a binomial.

  b. Find two binomials whose product is a trinomial.

  c. Find two binomials whose product is a four-term polynomial.

120. A special-product rule can be used to find $31 \cdot 29$.

$$31 \cdot 29 = (30 + 1)(30 - 1)$$
$$= 30^2 - 1^2$$
$$= 900 - 1$$
$$= 899$$

Use this method to find $52 \cdot 48$.

SECTION 5.8
Dividing Polynomials

Objectives

1 Divide a monomial by a monomial.

2 Divide a polynomial by a monomial.

3 Divide a polynomial by a polynomial.

In this section, we will conclude our study of operations with polynomials by discussing division of polynomials. To begin, we consider the simplest case, the quotient of two monomials.

1 **Divide a Monomial by a Monomial.**

To divide monomials, we can use the method for simplifying fractions or a method that involves a rule for exponents.

EXAMPLE 1   Divide the monomials: **a.** $\dfrac{21x^5}{7x^2}$ **b.** $\dfrac{10r^6s}{6rs^3}$

Strategy   We can use the rules for simplifying fractions and/or the quotient rule for exponents.

Why   We need to make sure that the numerator and denominator have no common factors other than 1. If that is the case, then the fraction is in *simplest form*.

> **Success Tip**
> In this section, you will see that regardless of the number of terms involved, every polynomial division is a series of monomial divisions.

Solution

By simplifying fractions

a. $\dfrac{21x^5}{7x^2} = \dfrac{3 \cdot \overset{1}{\cancel{7}} \cdot \overset{1}{\cancel{x}} \cdot \overset{1}{\cancel{x}} \cdot x \cdot x \cdot x}{\underset{1}{\cancel{7}} \cdot \underset{1}{\cancel{x}} \cdot \underset{1}{\cancel{x}}}$

$= 3x^3$

b. $\dfrac{10r^6s}{6rs^3} = \dfrac{\overset{1}{\cancel{2}} \cdot 5 \cdot \overset{1}{\cancel{r}} \cdot r \cdot r \cdot r \cdot r \cdot r \cdot \overset{1}{\cancel{s}}}{\underset{1}{\cancel{2}} \cdot 3 \cdot \underset{1}{\cancel{r}} \cdot \underset{1}{\cancel{s}} \cdot s \cdot s}$

$= \dfrac{5r^5}{3s^2}$

Using the rules for exponents

$\dfrac{21x^5}{7x^2} = 3x^{5-2}$ Divide the coefficients.

$= 3x^3$ Subtract the exponents.

$\dfrac{10r^6s}{6rs^3} = \dfrac{5}{3}r^{6-1}s^{1-3}$ Simplify $\frac{10}{6}$.

$= \dfrac{5}{3}r^5s^{-2}$ Subtract exponents.

$= \dfrac{5r^5}{3s^2}$ Move s^{-2} to the denominator and change the sign of the exponent.

 Self Check 1   Divide the monomials: **a.** $\frac{30y^4}{5y^2}$ **b.** $\frac{8c^2d^6}{32c^5d^2}$

Now Try   **Problems 15 and 23**

2 **Divide a Polynomial by a Monomial.**

Recall that to add two fractions with the same denominator, we add their numerators and keep their common denominator.

$$\frac{a}{d} + \frac{b}{d} = \frac{a+b}{d}$$

We can use this rule in reverse to divide polynomials by monomials.

Dividing a Polynomial by a Monomial

To divide a polynomial by a monomial, divide each term of the polynomial by the monomial. If A, B, and D represent monomials, where $D \neq 0$, then

$$\frac{A+B}{D} = \frac{A}{D} + \frac{B}{D}$$

EXAMPLE 2 Divide: **a.** $\dfrac{9x^2 + 6x}{3x}$ **b.** $\dfrac{12a^4b^3 - 18a^3b^2 + 2a^2}{6a^2b^2}$

Strategy We will divide each term of the polynomial in the numerator by the monomial in the denominator.

Why A fraction bar indicates division of the numerator by the denominator.

Solution

a. Here, we have a binomial divided by a monomial.

$$\frac{9x^2 + 6x}{3x} = \frac{9x^2}{3x} + \frac{6x}{3x}$$ Divide each term of the numerator, $9x^2 + 6x$, by the denominator, $3x$.

$$= 3x^{2-1} + 2x^{1-1}$$ Do each monomial division. Divide the coefficients. Keep each base and subtract the exponents.

$$= 3x + 2$$ Recall that $x^0 = 1$.

The Language of Algebra
The names of the parts of a division statement are

Dividend
$$\frac{9x^2 + 6x}{3x} = 3x + 2$$
Divisor   Quotient

Check: We multiply the divisor, $3x$, and the quotient, $3x + 2$. The result should be the dividend, $9x^2 + 6x$.

$$3x(3x + 2) = 9x^2 + 6x$$ The answer checks.

b. Here, we have a trinomial divided by a monomial.

$$\frac{12a^4b^3 - 18a^3b^2 + 2a^2}{6a^2b^2} = \frac{12a^4b^3}{6a^2b^2} - \frac{18a^3b^2}{6a^2b^2} + \frac{2a^2}{6a^2b^2}$$ Divide each term of the numerator by the denominator, $6a^2b^2$.

$$= 2a^{4-2}b^{3-2} - 3a^{3-2}b^{2-2} + \frac{a^{2-2}}{3b^2}$$ Do each monomial division. Simplify: $\frac{2}{6} = \frac{1}{3}$.

$$= 2a^2b - 3a + \frac{1}{3b^2}$$ $b^{2-2} = b^0 = 1$ and $a^{2-2} = a^0 = 1$.

Success Tip
The sum, difference, and product of two polynomials are always polynomials. However, as seen in Example 2b, the quotient of two polynomials is not always a polynomial.

Recall that the variables in a polynomial must have whole-number exponents. Therefore, the result, $2a^2b - 3a + \frac{1}{3b^2}$, is not a polynomial because the last term has a variable in the denominator.

Check:

$$6a^2b^2\left(2a^2b - 3a + \frac{1}{3b^2}\right) = 12a^4b^3 - 18a^3b^2 + 2a^2 \quad \text{The answer checks.}$$

Self Check 2   Divide: **a.** $\dfrac{50h^3 + 15h^2}{5h^2}$

b. $\dfrac{22s^5t^2 - s^4t^3 + 44s^2t^4}{11s^2t^2}$

Now Try   **Problems 29, 37, and 45**

3   **Divide a Polynomial by a Polynomial.**

To divide a polynomial by a polynomial (other than a monomial), we use a method similar to long division in arithmetic.

EXAMPLE 3   Divide $x^2 + 5x + 6$ by $x + 2$.

Strategy   We will use the long division method. The dividend is $x^2 + 5x + 6$ and the divisor is $x + 2$.

Why   Since the divisor has more than one term, we must use the long division method to divide the polynomials.

Solution   We write the division using a long division symbol $\overline{}\big)$ and proceed as follows:

The Language of Algebra
Notice how the instruction *to divide* a polynomial by a binomial translates:

Divide $x^2 + 5x + 6$ by $x + 2$

$x + 2\overline{)x^2 + 5x + 6}$

Step 1: $\overset{\displaystyle x}{\textcircled{x} + 2\overline{)\textcircled{x^2} + 5x + 6}}$
Divide the first term of the dividend by the first term of the divisor: $\frac{x^2}{x} = x$. Write the result, x, above the long division symbol.

Step 2: $\begin{array}{r} x \\ x + 2\overline{)x^2 + 5x + 6} \\ x^2 + 2x \end{array}$
Multiply each term of the divisor by x. Write the result, $x^2 + 2x$, under $x^2 + 5x$, and draw a line. Be sure to align the like terms.

Step 3: $\begin{array}{r} x \\ x + 2\overline{)x^2 + 5x + 6} \\ -(x^2 + 2x) \downarrow \\ \hline 3x + 6 \end{array}$
Subtract $x^2 + 2x$ from $x^2 + 5x$. Work column by column: $x^2 - x^2 = 0$ and $5x - 2x = 3x$.

Bring down the next term, 6.

Success Tip
Notice that this method is much like that used for division of whole numbers.

$\begin{array}{r} 13 \\ 12\overline{)156} \\ -12\downarrow \\ \hline 036 \\ -36 \\ \hline 0 \end{array}$

Hundreds
Tens
Ones

Step 4: $\begin{array}{r} x + 3 \\ \textcircled{x} + 2\overline{)x^2 + 5x + 6} \\ -(x^2 + 2x) \\ \hline \textcircled{3x} + 6 \end{array}$
Divide the first term of $3x + 6$ by the first term of the divisor: $\frac{3x}{x} = 3$. Write $+ 3$ above the long division symbol to form the second term of the quotient.

Step 5: $\begin{array}{r} x + 3 \\ x + 2\overline{)x^2 + 5x + 6} \\ -(x^2 + 2x) \\ \hline 3x + 6 \\ 3x + 6 \end{array}$
Multiply each term of the divisor by 3. Write the result, $3x + 6$, under $3x + 6$ and draw a line. Be sure to align the like terms.

Success Tip

The long division method aligns like terms vertically.

$$
\begin{array}{r}
x + 3 \\
x + 2\overline{)x^2 + 5x + 6} \\
-(x^2 + 2x) \\
\hline
3x + 6 \\
-(3x + 6) \\
\hline
0
\end{array}
$$

x^2-terms ⌐
  x-terms ⌐
    constants ⌐

Step 6:
$$
\begin{array}{r}
x + 3 \\
x + 2\overline{)x^2 + 5x + 6} \\
-(x^2 + 2x) \\
\hline
3x + 6 \\
-(3x + 6) \\
\hline
0
\end{array}
$$
Subtract $3x + 6$ from $3x + 6$. Work vertically: $3x - 3x = 0$ and $6 - 6 = 0$.

This is the remainder.

The quotient is $x + 3$ and the remainder is 0.

Step 7:   Check the result by verifying that $(x + 2)(x + 3)$ is $x^2 + 5x + 6$.

$$
\begin{aligned}
(x + 2)(x + 3) &= x^2 + 3x + 2x + 6 \\
&= x^2 + 5x + 6
\end{aligned}
$$
The answer checks.

▷ **Self Check 3**   Divide $x^2 + 7x + 12$ by $x + 3$.
   Now Try **Problem 49**

The long division method used in algebra can have a remainder just as long division in arithmetic often does.

EXAMPLE 4   Divide: $(6x^2 - 7x - 2) \div (2x - 1)$

Strategy   We will use the long division method. The dividend is $6x^2 - 7x - 2$ and the divisor is $2x - 1$.

Why   Since the divisor has more than one term, we must use the long division method to divide the polynomials.

Solution

Success Tip

The long-division method is a series of four steps that are repeated:
• Divide
• Multiply
• Subtract
• Bring down the next term

Step 1:
$$
\begin{array}{r}
3x \\
2x - 1\overline{)6x^2 - 7x - 2}
\end{array}
$$
Divide the first term of the dividend by the first term of the divisor: $\frac{6x^2}{2x} = 3x$. Write the result, $3x$, above the long division symbol.

Step 2:
$$
\begin{array}{r}
3x \\
2x - 1\overline{)6x^2 - 7x - 2} \\
6x^2 - 3x
\end{array}
$$
Multiply each term of the divisor by $3x$. Write the result, $6x^2 - 3x$, under $6x^2 - 7x$, and draw a line.

Step 3:
$$
\begin{array}{r}
3x \\
2x - 1\overline{)6x^2 - 7x - 2} \\
-(6x^2 - 3x) \downarrow \\
\hline
-4x - 2
\end{array}
$$
Subtract $6x^2 - 3x$ from $6x^2 - 7x$. Work vertically: $6x^2 - 6x^2 = 0$ and $-7x - (-3x) = -7x + 3x = -4x$. Bring down the next term, -2.

Step 4:
$$
\begin{array}{r}
3x - 2 \\
2x - 1\overline{)6x^2 - 7x - 2} \\
-(6x^2 - 3x) \\
\hline
-4x - 2
\end{array}
$$
Divide the first term of $-4x - 2$ by the first term of the divisor: $\frac{-4x}{2x} = -2$. Write -2 above the long division symbol to form the second term of the quotient.

Step 5:
$$2x - 1 \overline{)6x^2 - 7x - 2}$$
$$\underline{-(6x^2 - 3x)}$$
$$-4x - 2$$
$$\underline{-4x + 2}$$

quotient: $3x - 2$

Multiply each term of the divisor by -2. Write the result, $-4x + 2$, under $-4x - 2$, and draw a line.

Step 6:
$$2x - 1 \overline{)6x^2 - 7x - 2}$$
$$\underline{-(6x^2 - 3x)}$$
$$-4x - 2$$
$$\underline{-(-4x + 2)}$$
$$-4$$

quotient: $3x - 2$

Subtract $-4x + 2$ from $-4x - 2$. Work vertically: $-4x - (-4x) = -4x + 4x = 0$ and $-2 - 2 = -4.$

The quotient is $3x - 2$ and the remainder is -4. It is common to write the answer in Quotient $+ \frac{remainder}{divisor}$ form as either

$$3x - 2 + \frac{-4}{2x - 1} \quad \text{or} \quad 3x - 2 - \frac{4}{2x - 1}$$

Step 7: We can check the result using the fact that for any division:

Divisor $\cdot$ quotient $+$ remainder $=$ dividend

$$(2x - 1)(3x - 2) \quad + \quad (-4) \quad = 6x^2 - 4x - 3x + 2 + (-4)$$
$$= 6x^2 - 7x - 2 \qquad \text{The answer checks.}$$

Self Check 4 Divide $(8x^2 + 6x - 3) \div (2x + 3)$. Check the result.

Now Try Problem 63

The division method works best when the terms of the divisor and the dividend are written in descending powers of the variable. If the powers in the dividend or divisor are not in descending order, we use the commutative property of addition to write them that way.

EXAMPLE 5 Divide $4x^2 + 2x^3 + 12 - 2x$ by $x + 3$.

Strategy We will write the dividend in descending powers of x and use the long division method.

Why It is easier to align like terms in columns when the powers of the variable are written in descending order.

Solution

$$
\begin{array}{r}
2x^2 - 2x + 4 \\
x + 3 \overline{\smash{\big)}\ 2x^3 + 4x^2 - 2x + 12} \\
-(2x^3 + 6x^2) \quad\downarrow \\
\hline
-2x^2 - 2x \\
-(-2x^2 - 6x) \quad\downarrow \\
\hline
4x + 12 \\
-(4x + 12) \\
\hline
0
\end{array}
$$

The first division: $\frac{2x^3}{x} = 2x^2$.

The second division: $\frac{-2x^2}{x} = -2x$.

The third division: $\frac{4x}{x} = 4$.

Check: $(x + 3)(2x^2 - 2x + 4) = 2x^3 - 2x^2 + 4x + 6x^2 - 6x + 12$
$$= 2x^3 + 4x^2 - 2x + 12 \qquad \text{The answer checks.}$$

> **Self Check 5**   Divide $x^2 - 10x + 6x^3 + 4$ by $2x - 1$.
>
> *Now Try*   **Problem 65**

When we write the terms of a dividend in descending powers, we must determine whether some powers of the variable are missing. If any are missing, insert such terms with a coefficient of 0 or leave blank spaces for them. This keeps like terms in the same column, which is necessary when performing the subtraction in vertical form.

EXAMPLE 6   Divide: $\dfrac{27x^3 + 1}{3x + 1}$

Strategy   The divisor is $3x + 1$. The dividend, $27x^3 + 1$, does not have an x^2-term or an x-term. We will insert a $0x^2$ term and a $0x$ term as placeholders, and use the long division method.

Why   We insert placeholder terms so that like terms will be aligned in the same column when we subtract.

Solution

$$
\begin{array}{r}
9x^2 - 3x + 1 \\
3x + 1 \overline{\smash{\big)}\ 27x^3 + 0x^2 + 0x + 1} \\
-(27x^3 + 9x^2) \quad\downarrow \\
\hline
-9x^2 + 0x \\
-(-9x^2 - 3x) \quad\downarrow \\
\hline
3x + 1 \\
-(3x + 1) \\
\hline
0
\end{array}
$$

The first division: $\frac{27x^3}{3x} = 9x^2$.

The second division: $\frac{-9x^2}{3x} = -3x$.

The third division: $\frac{3x}{3x} = 1$.

Check: $(3x + 1)(9x^2 - 3x + 1) = 27x^3 - 9x^2 + 3x + 9x^2 - 3x + 1$
$$= 27x^3 + 1 \qquad \text{The answer checks.}$$

> **Self Check 6**   Divide $\dfrac{x^2 - 9}{x - 3}$. Check the result.
>
> *Now Try*   **Problem 73**

| **EXAMPLE 7** | ***Toys.*** The area of an Etch A Sketch screen is represented by the polynomial $(35x^2 + 43x + 12)$ in.^{2}. If the width of the screen is $(5x + 4)$ inches, what expression represents its length? |

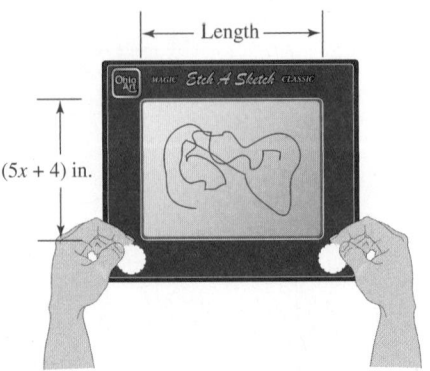

Strategy We will find the length of the screen by dividing its area, $(35x^2 + 43x + 12)$ in.^{2}, by its width, $(5x + 4)$ inches.

Why Recall that the area of a rectangle is given by the formula $A = lw$. If we divide both sides of the formula by w, we see that $l = \frac{A}{w}$.

Solution

$$\text{Length} = \frac{\textbf{area}}{\textbf{width}} \qquad \text{This is the formula for the length of a rectangle.}$$

$$= \frac{35x^2 + 43x + 12}{5x + 4} \qquad \begin{array}{l}\text{Substitute } 35x^2 + 43x + 12 \text{ for the area and } 5x + 4 \text{ for the}\\ \text{width.}\end{array}$$

To divide $35x^2 + 43x + 12$ by $5x + 4$, we use long division.

$$
\begin{array}{r}
7x + 3 \\
5x + 4 \,\overline{)\,35x^2 + 43x + 12} \\
-(35x^2 + 28x) \\
\hline
15x + 12 \\
-(15x + 12) \\
\hline
0
\end{array}
$$

The first division: $\frac{35x^2}{5x} = 7x$.

The second division: $\frac{15x}{5x} = 3$.

The length of the Etch A Sketch screen is $(7x + 3)$ inches.

 Now Try **Problem 101**

STUDY SET
5.8

VOCABULARY

Fill in the blanks.

1. The expression $\dfrac{18x^7}{9x^4}$ is a monomial divided by a _____ .

2. The expression $\dfrac{6x^3y - 4x^2y^2 + 8xy^3 - 2y^4}{2x^4}$ is a _____ divided by a monomial.

3. The expression $\dfrac{x^2 - 8x + 12}{x - 6}$ is a trinomial divided by a _____ .

4.
$$
\begin{array}{r}
x - 2 \\
x - 6 \overline{\smash{)}\, x^2 - 8x - 4} \\
-(x^2 - 6x) \\
\hline
-2x - 4 \\
-(-2x + 12) \\
\hline
-16
\end{array}
$$

CONCEPTS

5. The long division method is a series of four steps that are repeated. Put them in the correct order:

   subtract   multiply   bring down   divide

6. In the following long division, find the answer to the subtraction that must be performed at this stage.

$$
\begin{array}{r}
x \\
x - 7 \overline{\smash{)}\, x^2 - 9x - 6} \\
-(x^2 - 7x)
\end{array}
$$

7. Fill in the blanks: To check an answer of a long division, we use the fact that

   Divisor · _____ + remainder = _____

8. Check to see whether the following result of a long division is correct.

$$
\frac{x^2 + 4x - 20}{x - 3} = x + 7 + \frac{1}{x - 3}
$$

NOTATION

Complete each solution.

9. $\dfrac{28x^5 - x^3 + 5x^2}{7x^2} = \dfrac{28x^5}{7x^2} - \dfrac{x^3}{7x^2} + \dfrac{5x^2}{7x^2}$

$$
= 4x^{} - \frac{x^{3-2}}{} + \frac{5x^{}}{7}
$$

$$
= - \frac{x}{7} + \frac{}{}
$$

10.
$$
\begin{array}{r}
 + 2 \\
x + 2 \overline{\smash{)}\, x^2 + 4x + 5} \\
-(x^2 +) \\
\hline
 + 5 \\
-(2x + 4)
\end{array}
$$

11. Write the polynomial $2x^2 - 1 + 5x^4$ in descending powers of x and insert placeholders for each missing term.

12. True or false: $6x + 4 + \dfrac{-3}{x + 2} = 6x + 4 - \dfrac{3}{x + 2}$

GUIDED PRACTICE

Divide the monomials. **See Example 1.**

13. $\dfrac{x^5}{x^2}$

14. $\dfrac{a^{12}}{a^8}$

15. $\dfrac{45m^{10}}{9m^5}$

16. $\dfrac{24n^{12}}{8n^4}$

17. $\dfrac{12h^8}{9h^6}$

18. $\dfrac{22b^9}{6b^6}$

19. $\dfrac{-3d^4}{15d^8}$

20. $\dfrac{-4x^3}{16x^5}$

21. $\dfrac{10s^2}{s^3}$

22. $\dfrac{16y^3}{y^4}$

23. $\dfrac{8x^3y^2}{40xy^6}$

24. $\dfrac{3y^3z}{18yz^6}$

25. $\dfrac{-16r^3y^2}{-4r^2y^7}$

26. $\dfrac{-35xz^6}{-7x^8z^2}$

27. $\dfrac{-65rs^2}{15r^2s^5}$

28. $\dfrac{112uz^4}{-42u^3z^8}$

Divide the polynomial by the monomial. **See Example 2.**

29. $\dfrac{6x + 9}{3}$

30. $\dfrac{8x + 12}{4}$

31. $\dfrac{9m - 6}{m}$

32. $\dfrac{10n - 6}{n}$

33. $\dfrac{a - a^3 + a^4}{a^4}$

34. $\dfrac{b^2 + b^3 - b^4}{b^4}$

35. $\dfrac{8x^9 - 32x^6}{4x^4}$

36. $\dfrac{30y^8 + 40y^7}{10y^6}$

37. $\dfrac{6h^{12} + 48h^9}{24h^{10}}$

38. $\dfrac{4x^{14} - 36x^8}{36x^{12}}$

39. $\dfrac{-18w^6 - 9}{9w^4}$

40. $\dfrac{-40f^4 + 16}{8f^3}$

41. $\dfrac{9s^8 - 18s^5 + 12s^4}{3s^3}$

42. $\dfrac{16b^{10} + 4b^6 - 20b^4}{4b^2}$

43. $\dfrac{7c^5 + 21c^4 - 14c^3 - 35c}{7c^2}$

44. $\dfrac{12r^{15} - 48r^{12} + r^{10} - 18r^8}{6r^{10}}$

45. $\dfrac{12x^3y^2 - 8x^2y - 4x}{4xy}$

46. $\dfrac{12a^2b^2 - 8a^2b - 4ab}{4ab}$

47. $\dfrac{-25x^2y^3 + 30xy^2 - 5xy}{-5x^2y^2}$

48. $\dfrac{-30a^4b^4 - 15a^3b - 10a^2b^2}{-10a^2b^3}$

Perform each division. See Examples 3 and 4.

49. Divide $x^2 + 8x + 12$ by $x + 2$.

50. Divide $x^2 + 5x + 6$ by $x + 2$.

51. Divide $x^2 - 5x + 6$ by $x - 3$.

52. Divide $x^2 - 12x + 32$ by $x - 4$.

53. $\dfrac{6a^2 + 5a - 6}{2a + 3}$

54. $\dfrac{3b^2 - 5b + 2}{3b - 2}$

55. $\dfrac{3b^2 + 11b + 6}{3b + 2}$

56. $\dfrac{8a^2 + 2a - 3}{2a - 1}$

57. $(x^2 + 6x + 15) \div (x + 5)$

58. $(x^2 + 10x + 30) \div (x + 6)$

59. $a - 5\,\overline{)\,a^2 - 17a + 64}$

60. $b - 2\,\overline{)\,b^2 - 4b + 6}$

61. $\dfrac{2x^2 + 5x + 2}{2x + 3}$

62. $\dfrac{3x^2 - 8x + 3}{3x - 2}$

63. $\dfrac{6x^2 - 11x + 2}{3x - 1}$

64. $\dfrac{4x^2 + 6x - 1}{2x + 1}$

Perform each division. See Example 5.

65. $x + 2\,\overline{)\,3x + 2x^2 - 2}$

66. $x + 3\,\overline{)\,-x + 2x^2 - 21}$

67. $(3 + 11x + 10x^2) \div (5x + 3)$

68. $(6x + 1 + 9x^2) \div (3x + 1)$

69. $2x - 7\,\overline{)\,-x - 21 + 2x^2}$

70. $2x - 1\,\overline{)\,x - 2 + 6x^2}$

71. $3 + 4x\,\overline{)\,3 - 5x^2 - 2x + 4x^3}$

72. $2x + 3\,\overline{)\,7x^2 - 3 + 4x + 2x^3}$

Perform each division. See Example 6.

73. $(a^2 - 25) \div (a + 5)$

74. $(b^2 - 36) \div (b + 6)$

75. $(x^2 - 1) \div (x - 1)$

76. $(x^2 - 9) \div (x + 3)$

77. $\dfrac{4x^2 - 9}{2x + 3}$

78. $\dfrac{25x^2 - 16}{5x - 4}$

79. $\dfrac{81b^2 - 49}{9b - 7}$

80. $\dfrac{16t^2 - 121}{4t + 11}$

81. $\dfrac{x^3 + 1}{x + 1}$

82. $\dfrac{x^3 - 8}{x - 2}$

83. $\dfrac{y^3 + y}{y - 2}$

84. $\dfrac{a^3 + a}{a + 3}$

TRY IT YOURSELF

Perform each division.

85. Divide $y^2 + 13y + 13$ by $y + 1$.

86. Divide $z^2 + 7z + 14$ by $z + 3$.

87. $\dfrac{15a^8b^2 - 10a^2b^5}{5a^3b^2}$

88. $\dfrac{9a^4b^3 - 16a^3b^4}{12a^2b}$

89. $3x + 2\,\overline{)\,2 + 7x + 6x^3 + 10x^2}$

90. $3x - 2\,\overline{)\,4x - 4 + 6x^3 - x^2}$

91. $\dfrac{5x^4 - 10x}{25x^3}$

92. $\dfrac{24x^7 - 32x^2}{16x^3}$

93. $\dfrac{a^3 - 1}{a - 1}$

94. $\dfrac{y^3 + 8}{y + 2}$

95. $\dfrac{6x^3 + x^2 + 2x + 1}{3x - 1}$

96. $\dfrac{3y^3 - 4y^2 + 2y + 3}{y + 3}$

97. $\dfrac{8x^{17}y^{20}}{16x^{15}y^{30}}$

98. $\dfrac{21a^{30}b^{15}}{14a^{40}b^{12}}$

99. $(6m^2 - m - 40) \div (2m + 5)$

100. $(12d^2 - 20d + 3) \div (6d - 1)$

APPLICATIONS

101. FURNACE FILTERS   The area of the furnace filter is $(x^2 - 2x - 24)$ square inches. What expression represents its length?

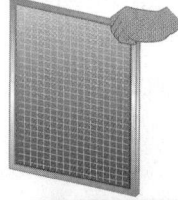

$(x + 4)$ in.

102. MINI-BLINDS   The area covered by the mini-blinds is $(3x^3 - 6x)$ square feet. What expression represents the length of the blinds?

$3x$ ft

103. POOL   The rack shown in the illustration is used to set up the balls for a game of pool. If the perimeter of the rack, in inches, is given by the polynomial $6x^2 - 3x + 9$, what expression represents the approximate length of one side?

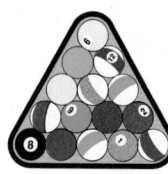

104. COMMUNICATIONS   Telephone poles were installed every $(2x - 3)$ feet along a stretch of railroad track $(8x^3 - 6x^2 + 5x - 21)$ feet long. What expression represents the number of poles that were used?

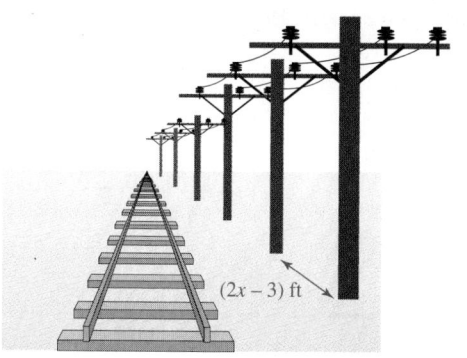

$(2x - 3)$ ft

WRITING

105. Explain how to check the following long division.

$$\begin{array}{r} x + 5 \\ 3x + 5 \overline{)\,3x^2 + 20x - 5} \\ -(3x^2 + 5x) \\ \hline 15x - 5 \\ -(15x + 25) \\ \hline -30 \end{array}$$

106. Explain the difference in the methods used to divide $\frac{x^2 - 3x + 2}{x}$ as compared to $\frac{x^2 - 3x + 2}{x - 2}$.

107. How do you know when to stop the long division method when dividing polynomials?

108. When dividing $x^3 + 1$ by $x + 1$, why is it helpful to write $x^3 + 1$ as $x^3 + 0x^2 + 0x + 1$?

REVIEW

109. Write an equation of the line with slope $-\frac{11}{6}$ that passes through $(2, -6)$. Write the answer in slope–intercept form.

110. Solve $S = 2\pi rh + 2\pi r^2$ for h.

111. Evaluate: $-10(18 - 4^2)^3$

112. Evaluate: -5^2

CHALLENGE PROBLEMS

Perform each division.

113. $\dfrac{6a^3 - 17a^2b + 14ab^2 - 3b^3}{2a - 3b}$

114. $(2x^4 + 3x^3 + 3x^2 - 5x - 3) \div (2x^2 - x - 1)$

115. $(x^6 + 2x^4 - 6x^2 - 9) \div (x^2 + 3)$

116. $\dfrac{6x^{6m}y^{6n} + 15x^{4m}y^{7n} - 24x^{2m}y^{8n}}{3x^{2m}y^n}$

CHAPTER 5
Summary & Review

SECTION 5.1 Rules for Exponents

DEFINITIONS AND CONCEPTS	EXAMPLES
An **exponent** indicates repeated multiplication. It tells how many times the **base** is to be used as a factor. Exponent ⌐ *n* factors of *x* $x^n = \overbrace{x \cdot x \cdot x \cdot \;\cdots\; \cdot x}$ Base ⌐	Identify the base and the exponent in each expression. $2^6 = 2 \cdot 2 \cdot 2 \cdot 2 \cdot 2 \cdot 2$ 2 is the base and 6 is the exponent. $(-xy)^3 = (-xy)(-xy)(-xy)$ $-xy$ is the base and 3 is the exponent. $5t^4 = 5 \cdot t \cdot t \cdot t \cdot t$ t is the base and 4 is the exponent. $8^1 = 8$ 8 is the base and 1 is the exponent.
Rules for Exponents: If *m* and *n* represent integers and there are no divisions by 0, then **Product rule:** $x^m x^n = x^{m+n}$ **Quotient rule:** $\dfrac{x^m}{x^n} = x^{m-n}$ **Power rule:** $(x^m)^n = x^{m \cdot n} = x^{mn}$ **Power of a product rule:** $(xy)^m = x^m y^m$ **Power of a quotient rule:** $\left(\dfrac{x}{y}\right)^n = \dfrac{x^n}{y^n}$	Simplify each expression: $5^2 5^7 = 5^{2+7} = 5^9$ $\dfrac{t^7}{t^3} = t^{7-3} = t^4$ $(6^3)^7 = 6^{3 \cdot 7} = 6^{21}$ $(2p)^5 = 2^5 p^5 = 32p^5$ $\left(\dfrac{s}{4}\right)^4 = \dfrac{s^4}{4^4} = \dfrac{s^4}{256}$

REVIEW EXERCISES

1. Identify the base and the exponent in each expression.

   a. n^{12} **b.** $(2x)^6$

   c. $3r^4$ **d.** $(y-7)^3$

2. Write each expression in an equivalent form using an exponent.

   a. $m \cdot m \cdot m \cdot m \cdot m$ **b.** $-3 \cdot x \cdot x \cdot x \cdot x$

   c. $(x+8)(x+8)$ **d.** $\left(\dfrac{1}{2}pq\right)\left(\dfrac{1}{2}pq\right)\left(\dfrac{1}{2}pq\right)$

Simplify each expression. Assume there are no divisions by 0.

3. $7^4 \cdot 7^8$ **4.** $mmnn$

5. $(y^7)^3$ **6.** $(3x)^4$

7. $\dfrac{b^{12}}{b^3}$ **8.** $-b^3 b^4 b^5$

9. $(-16s^3)^2 s^4$ **10.** $(2.1x^2 y)^2$

11. $[(-9)^3]^5$ **12.** $(a^5)^3 (a^2)^4$

13. $\left(\dfrac{1}{2}x^2 x^3\right)^3$ **14.** $\left(\dfrac{x^7}{3xy}\right)^2$

15. $\dfrac{(m-25)^{16}}{(m-25)^4}$ **16.** $\dfrac{(5y^2 z^3)^3}{(yz)^5}$

17. $\dfrac{a^5 a^4 a^5}{a^2 a}$ **18.** $\dfrac{(cd)^9}{(cd)^4}$

Find an expression that represents the area or the volume of each figure, whichever is appropriate.

19.   **20.**

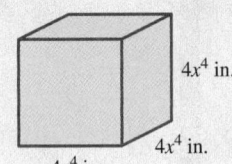

$4x^4$ in.
$4x^4$ in.
$4x^4$ in.

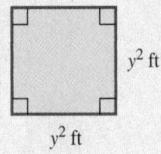

y^2 ft
y^2 ft

SECTION 5.2　Zero and Negative Exponents

DEFINITIONS AND CONCEPTS	EXAMPLES
Rules for exponents: For any nonzero real numbers x and y and any integers m and n,	Simplify each expression. Do not use negative exponents in the answer.

Zero exponent: $x^0 = 1$

$5^0 = 1$

Negative exponents: $x^{-n} = \dfrac{1}{x^n}$

$4^{-2} = \dfrac{1}{4^2} = \dfrac{1}{16}$　　and　　$7c^{-6} = \dfrac{7}{c^6}$

Negative to positive rules:

$$\dfrac{1}{x^{-n}} = x^n \qquad \dfrac{x^{-m}}{y^{-n}} = \dfrac{y^n}{x^m}$$

$\dfrac{1}{t^{-8}} = t^8$　　and　　$\dfrac{2^{-4}}{x^{-6}} = \dfrac{x^6}{2^4} = \dfrac{x^6}{16}$

Negative exponents and reciprocals:

$$\left(\dfrac{x}{y}\right)^{-n} = \left(\dfrac{y}{x}\right)^{n}$$

$\left(\dfrac{x}{10}\right)^{-3} = \left(\dfrac{10}{x}\right)^{3} = \dfrac{10^3}{x^3} = \dfrac{1{,}000}{x^3}$

REVIEW EXERCISES

Simplify each expression. Do not use negative exponents in the answer.

21. x^0

22. $(3x^2y^2)^0$

23. $3x^0$

24. 10^{-3}

25. -5^{-2}

26. $\dfrac{t^4}{t^{10}}$

27. $\dfrac{8}{x^{-5}}$

28. $-6y^4y^{-5}$

29. $\dfrac{7^{-2}}{2^{-3}}$

30. $(x^{-3}x^{-4})^{-2}$

31. $\left(\dfrac{-3r^4r^{-3}}{r^{-3}r^7}\right)^3$

32. $\left(\dfrac{4z^4}{z^3}\right)^{-2}$

33. $\dfrac{3^{-2}c^3d^3}{2^{-3}c^2d^8}$

34. $\dfrac{t^{-30}}{t^{-60}}$

35. $\dfrac{w(w^{-3})^{-4}}{w^{-9}}$

36. $\left(\dfrac{4}{f^4}\right)^{-10}$

SECTION 5.3　Scientific Notation

DEFINITIONS AND CONCEPTS	EXAMPLES
A positive number is written in **scientific notation** when it is written in the form $N \times 10^n$, where $1 \le N < 10$ and n is an integer.	Write each number in scientific notation. $32{,}500 = 3.25 \times 10^4$　　and　　$0.0025 = 2.5 \times 10^{-3}$ 　　　4 decimal places　　　　　　3 decimal places Write each number in standard notation. $1.91 \times 10^5 = 191{,}000$　　and　　$4.7 \times 10^{-6} = 0.0000047$ 　　　5 decimal places　　　　　　6 decimal places
Scientific notation provides an easier way to perform computations involving very large or very small numbers.	Use scientific notation to perform the calculation: $\dfrac{684{,}000{,}000}{456{,}000} = \dfrac{6.84 \times 10^8}{4.56 \times 10^5} = \dfrac{6.84}{4.56} \times \dfrac{10^8}{10^5} = 1.5 \times 10^3 \text{ or } 1{,}500$

REVIEW EXERCISES

Write each number in scientific notation.

37. 720,000,000

38. 9,370,000,000,000,000

39. 0.00000000942

40. 0.00013

41. 0.018×10^{-2}

42. 853×10^{3}

Write each number in standard notation.

43. 1.26×10^{5}

44. 3.919×10^{-8}

45. 2.68×10^{0}

46. 5.76×10^{1}

Evaluate each expression by first writing each number in scientific notation. Express the result in scientific notation and standard notation.

47. $\dfrac{(0.000012)(0.000004)}{0.00000016}$

48. $\dfrac{(4,800,000)(20,000,000)}{600,000}$

49. WORLD POPULATION As of January 2007, the world's population was estimated to be 6.57 billion. Write this number in standard notation and in scientific notation.

50. ATOMS The illustration shows a cross section of an atom. How many nuclei (plural for nucleus), placed end to end, would it take to stretch across the atom?

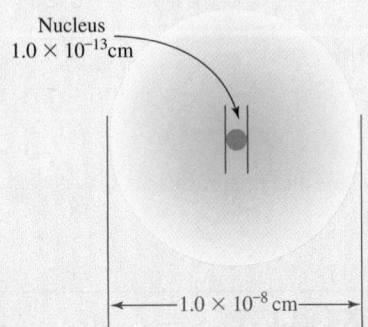

Nucleus
1.0×10^{-13}cm

$\leftarrow\!\!-1.0 \times 10^{-8}$ cm$-\!\!\rightarrow$

SECTION 5.4 Polynomials

DEFINITIONS AND CONCEPTS	EXAMPLES
A **polynomial** is a single term or a sum of terms in which all variables have whole-number exponents and no variable appears in a denominator.	Polynomials: $32,\quad -5x^2y^3,\quad 7p^3 - 14q^3,\quad 4m^2 + 5m - 12$ Not Polynomials: $y^2 - y^{-5},\quad 4x^3 - \dfrac{7}{x} + 3x$
A polynomial with exactly one term is called a **monomial.** A polynomial with exactly two terms is called a **binomial.** A polynomial with exactly three terms is called a **trinomial.**	*Monomials* *Binomials* *Trinomials* $3x^2$ $2y^3 + 3y$ $3p^2 - 7p + 12$ $-12m^3n^2$ $87t - 25$ $4p^2q^3 - 8p^2q^2 + 12p^2q$
The **coefficient of a term** is its numerical factor. The **degree of a term** of a polynomial in one variable is the value of the exponent on the variable. If a polynomial has more than one variable, the **degree of a term** is the sum of the exponents on the variables. The **degree of a nonzero constant** is 0.	*Term* *Coefficient* *Degree of the term* $6a^7$ 6 7 $-7.3x^5y^4$ -7.3 $5 + 4 = 9$ 32 32 0
The **degree of a polynomial** is equal to the highest degree of any term of the polynomial.	*Polynomial* *Degree of the polynomial* $7m^3 - 4m^2 + 5m - 12$ 3 $\dfrac{1}{2}a^4b + \dfrac{3}{4}a^3b^2 - \dfrac{2}{3}a^2b^4$ $2 + 4 = 6$
To **evaluate a polynomial** for a given value, substitute the value for the variable and follow the rules for the order of operations.	Evaluate $3x^2 - 4x + 2$ for $x = 2$. $\begin{aligned} 3x^2 - 4x + 2 &= 3(2)^2 - 4(2) + 2 \quad \text{Substitute 2 for each x.} \\ &= 3(4) - 8 + 2 \quad \text{Evaluate } (2)^2 \text{ first.} \\ &= 6 \end{aligned}$

SECTION 5.4 Polynomials—*continued*

DEFINITIONS AND CONCEPTS	EXAMPLES
We can **graph equations defined by polynomials** such as $y = x^2 - 2$, $y = -x^2$, and $y = x^3 + 1$. The graph of $y = x^2 - 2$ is a cup-shaped curve called a **parabola**.	Graph: $y = x^2 - 2$ Find several solutions of the equation, plot them on a rectangular coordinate system, and then draw a smooth curve passing through the points. 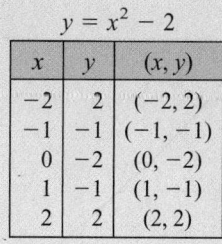 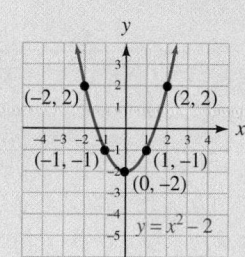

$y = x^2 - 2$

x	y	(x, y)
-2	2	$(-2, 2)$
-1	-1	$(-1, -1)$
0	-2	$(0, -2)$
1	-1	$(1, -1)$
2	2	$(2, 2)$

Select x-values. ⌐
Find y-values. ↑
⌐ Plot points.

REVIEW EXERCISES

51. Consider the polynomial $3x^3 - x^2 + x + 10$.

   a. How many terms does the polynomial have?

   b. What is the lead term?

   c. What is the coefficient of each term?

   d. What is the constant term?

52. Find the degree of each polynomial and classify it as a monomial, binomial, trinomial, or none of these.

   a. $13x^7$ **b.** $-16a^2b$

   c. $5^3x + x^2$ **d.** $-3x^5 + x - 1$

   e. $9xy^2 + 21x^3y^3$ **f.** $4s^4 - 3s^2 + 5s + 4$

53. Evaluate $-x^5 - 3x^4 + 3$ for $x = 0$ and $x = -2$.

54. DIVING The number of inches that the woman deflects the diving board is given by the polynomial $0.1875x^2 - 0.0078125x^3$ where x is the number of feet that she stands from the front anchor point of the board. Find the amount of deflection if she stands on the end of the diving board, 8 feet from the anchor point.

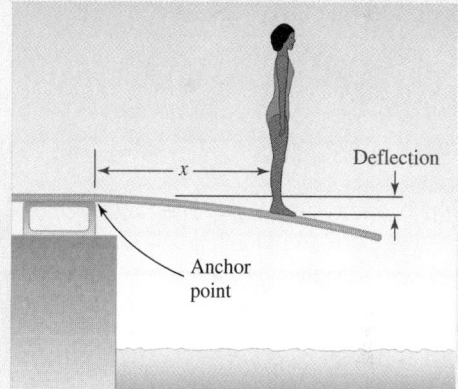

Construct a table of solutions like the one shown here and then graph the equation.

x	-2	-1	0	1	2
y					

55. $y = x^2$ **56.** $y = x^3 + 1$

SECTION 5.5 Adding and Subtracting Polynomials

DEFINITIONS AND CONCEPTS	EXAMPLES
To **simplify a polynomial,** combine like terms.	Simplify: $3r^4 - 4r^3 + 7r^4 + 8r^2$ $\qquad = 10r^4 - 4r^3 + 8r^2$ Combine like terms. Think: $(3 + 7)r^4 = 10r^4$.
To **add polynomials,** combine their like terms.	Add: $(4x^2 + 9x + 4) + (3x^2 - 5x - 1)$ $\qquad = (4x^2 + 3x^2) + (9x - 5x) + (4 - 1)$ Group like terms. $\qquad = 7x^2 + 4x + 3$ Combine like terms.

SECTION 5.5   Adding and Subtracting Polynomials—*continued*

DEFINITIONS AND CONCEPTS	EXAMPLES

To **subtract two polynomials,** change the signs of the terms of the polynomial being subtracted, drop the parentheses, and combine like terms.

Subtract: $(8a^3b - 4ab^2) - (-3a^3b + 9ab^2)$

$= 8a^3b - 4ab^2 + 3a^3b - 9ab^2$ Change the sign of each term of $-3a^3b + 9ab^2$ and drop the parentheses.

$= 11a^3b - 13ab^2$ Combine like terms.

REVIEW EXERCISES

Simplify each polynomial and write the result in descending powers of one variable.

57. $6y^3 + 8y^4 + 7y^3 + (-8y^4)$

58. $4a^2b + 5 - 6a^3b - 3a^2b + 2a^3b + 1$

59. $\frac{5}{6}x^2 + \frac{1}{3}y^2 - \frac{1}{4}x^2 - \frac{3}{4}xy + \frac{2}{3}y^2$

60. $7.6c^5 - 2.1c^3 - 0.9c^5 + 8.1c^4$

Perform the operations.

61. $(2r^6 + 14r^3) + (23r^6 - 5r^3 + 5r)$

62. $(7.1a^2 + 2.2a - 5.8) - (3.4a^2 - 3.9a + 11.8)$

63. $(3r^3s + r^2s^2 - 3rs^3 - 3s^4) + (r^3s - 8r^2s^2 - 4rs^3 + s^4)$

64. $\left(\frac{7}{8}m^4 - \frac{1}{5}m^3\right) - \left(\frac{1}{4}m^4 + \frac{1}{5}m^3\right) - \frac{3}{5}m^3$

65. Find the difference when $(-3z^3 - 4z + 7)$ is subtracted from the sum of $(2z^2 + 3z - 7)$ and $(-4z^3 - 2z - 3)$.

66. GARDENING   Find a polynomial that represents the length of the wooden handle of the shovel.

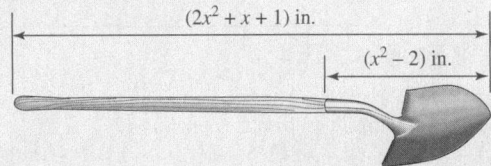

$(2x^2 + x + 1)$ in.

$(x^2 - 2)$ in.

67. Add:

$$3x^2 + 5x + 2$$
$$x^2 - 3x + 6$$

68. Subtract:

$$20x^3 \quad\quad + 12x$$
$$-(12x^3 + 7x^2 - 7x)$$

SECTION 5.6   Multiplying Polynomials

DEFINITIONS AND CONCEPTS	EXAMPLES

To **multiply two monomials,** multiply the numerical factors (the coefficients) and then multiply the variable factors.

Multiply: $(5p^6)(2p^5) = (5 \cdot 2)(p^6 \cdot p^5)$ Group the coefficients together and the variables together.

$= 10p^{11}$ Think: $5 \cdot 2 = 10$ and $p^6 \cdot p^5 = p^{6+5} = p^{11}$.

To **multiply a monomial and a polynomial,** multiply each term of the polynomial by the monomial.

Multiply: $3r^2(2r^4 + 7r^2 - 4)$

$= 3r^2(2r^4) + 3r^2(7r^2) + 3r^2(-4)$ Distribute the multiplication by $3r^2$.

$= 6r^6 + 21r^4 - 12r^2$ Multiply the monomials.

SECTION 5.6 Multiplying Polynomials—*continued*

DEFINITIONS AND CONCEPTS	EXAMPLES
To **multiply two binomials,** use the *FOIL method:* F: First O: Outer I: Inner L: Last	Multiply: $(3m + 4)(2m - 5) = 3m(2m) + 3m(-5) + 4(2m) + 4(-5)$ $\qquad = 6m^2 - 15m + 8m - 20$ Multiply the monomials. $\qquad = 6m^2 - 7m - 20$ Combine like terms.
To **multiply two polynomials,** multiply each term of one polynomial by each term of the other polynomial and then combine like terms.	Multiply: $(a - b)(6a^2 - 4ab + b^2)$ $= a(6a^2) + a(-4ab) + a(b^2) - b(6a^2) - b(-4ab) - b(b^2)$ $= 6a^3 - 4a^2b + ab^2 - 6a^2b + 4ab^2 - b^3$ Multiply the monomials. $= 6a^3 - 10a^2b + 5ab^2 - b^3$ Combine like terms.
When finding the **product of three polynomials,** begin by multiplying any two of them, and then multiply that result by the third polynomial.	Multiply: $-9x^4(x - 1)(x - 7) = -9x^4(x^2 - 7x - x + 7)$ Multiply the two binomials. $= -9x^4(x^2 - 8x + 7)$ Combine like terms within the parentheses. $= -9x^6 + 72x^5 - 63x^4$ Distribute the multiplication by $-9x^4$.

REVIEW EXERCISES

Multiply.

69. $(2x^2)(5x)$

70. $(-6x^4z^3)(x^6z^2)$

71. $5b^3 \cdot 6b^2 \cdot 4b^6$

72. $\dfrac{2}{3}h^5(3h^9 + 12h^6)$

73. $3n^2(3n^2 - 5n + 2)$

74. $x^2y(y^2 - xy)$

75. $2x(3x^4)(x + 2)$

76. $-a^2b^2(-a^4b^2 + a^3b^3 - ab^4 + 7a)$

77. $(x + 3)(x + 2)$

78. $(2x + 1)(x - 1)$

79. $(3t - 3)(2t + 2)$

80. $(3n^4 - 5n^2)(2n^4 - n^2)$

81. $-a^5(a^2 - b)(5a^2 + b)$

82. $6.6(a - 1)(a + 1)$

83. $\left(3t - \dfrac{1}{3}\right)\left(6t + \dfrac{5}{3}\right)$

84. $(5.5 - 6b)(2 - 4b)$

85. $(2a - 3)(4a^2 + 6a + 9)$

86. $(8x^2 + x - 2)(7x^2 + x - 1)$

87. Multiply using vertical form: $\begin{array}{r} 4x^2 - 2x + 1 \\ 2x + 1 \\ \hline \end{array}$

88. Refer to the illustration. Find a polynomial that represents

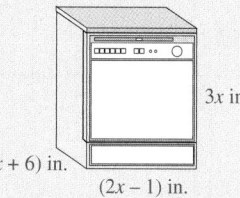

$3x$ in.

$(x + 6)$ in.

$(2x - 1)$ in.

    a. the perimeter of the base of the dishwasher.

    b. the area of the base of the dishwasher.

    c. the volume occupied by the dishwasher.

SECTION 5.7   Special Products

DEFINITIONS AND CONCEPTS	EXAMPLES

The following **special products** occur so often that it is worthwhile to learn their forms.

Square of a binomial:

$$(A + B)^2 = A^2 + 2AB + B^2$$

This is the square of a binomial sum.

$$(A - B)^2 = A^2 - 2AB + B^2$$

This is the square of a binomial difference.

Multiplying the Sum and Difference of the Same Two Terms:

$$(A + B)(A - B) = A^2 - B^2$$

Multiply: $(n + 4)^2 = \underbrace{n^2}_{\substack{\text{The square of}\\ \text{the first term, } n.}} + \underbrace{2(n)(4)}_{\substack{\text{Twice the product}\\ \text{of both terms.}}} + \underbrace{4^2}_{\substack{\text{The square of the}\\ \text{second term, 4.}}}$

$$= n^2 + 8n + 16$$

Multiply: $(5a - 1)^2 = \underbrace{(5a)^2}_{\substack{\text{The square of}\\ \text{the first term, } 5a.}} + \underbrace{2(5a)(-1)}_{\substack{\text{Twice the product}\\ \text{of both terms.}}} + \underbrace{(-1)^2}_{\substack{\text{The square of the}\\ \text{second term, } -1.}}$

$$= 25a^2 - 10a + 1$$

Multiply: $(x + 8)(x - 8) = \underbrace{x^2}_{\substack{\text{The square of}\\ \text{the first term, } x.}} - \underbrace{8^2}_{\substack{\text{The square of the}\\ \text{second term, 8.}}}$

$$= x^2 - 64$$

REVIEW EXERCISES

Find each product.

89. $(a - 3)^2$

90. $(m + 2)^3$

91. $(x + 7)(x - 7)$

92. $(2x - 0.9)(2x + 0.9)$

93. $(2y + 1)^2$

94. $(y^2 + 1)(y^2 - 1)$

95. $(6r^2 + 10s)^2$

96. $-(8a - 3c)^2$

97. $80s(r^2 + s^2)(r^2 - s^2)$

98. $4b(3b - 4)^2$

99. $\left(t - \dfrac{3}{4}\right)^2$

100. $\left(x + \dfrac{4}{3}\right)^2$

Perform the operations.

101. $3(9x^2 + 3x + 7) - 2(11x^2 - 5x + 9)$

102. $(5c - 1)^2 - (c + 6)(c - 6)$

103. GRAPHIC ARTS   A Dr. Martin Luther King poster has his picture with a $\frac{1}{2}$-inch wide blue border around it. The length of the poster is $(x + 3)$ inches and the width is $(x - 1)$ inches. Find a polynomial that represents the area of the *picture* of Dr. King.

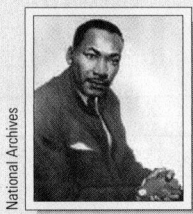

National Archives

104. Find a polynomial that represents the area of the triangle.

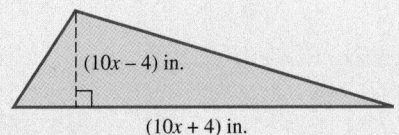

$(10x - 4)$ in.

$(10x + 4)$ in.

SECTION 5.8   Dividing Polynomials

DEFINITIONS AND CONCEPTS	EXAMPLES
To **divide monomials,** use the method for simplifying fractions and/or the quotient rule for exponents.	Divide the monomials:

$$\frac{8p^2q}{4pq^3} = \frac{2 \cdot \overset{1}{\cancel{4}} \cdot \overset{1}{\cancel{p}} \cdot p \cdot \overset{1}{\cancel{q}}}{\overset{}{\cancel{4}} \cdot \underset{1}{\cancel{p}} \cdot \underset{1}{\cancel{q}} \cdot q \cdot q} \quad \text{or} \quad \frac{8p^2q}{4pq^3} = \frac{8}{4}p^{2-1}q^{1-3} \qquad \text{Keep each base and subtract the exponents.}$$

$$= \frac{2p}{q^2} \qquad\qquad\qquad\qquad = 2p^1q^{-2}$$

$$\qquad\qquad\qquad\qquad\qquad\qquad = \frac{2p}{q^2} \qquad \begin{array}{l}\text{Move } q^{-2}\text{ to the}\\\text{denominator and}\\\text{change the sign of}\\\text{the exponent.}\end{array}$$

| To **divide a polynomial by a monomial,** divide each term of the numerator by the denominator | Divide: $\dfrac{9a^2b^4 - 12a^3b^6}{18ab^5} = \dfrac{9a^2b^4}{18ab^5} - \dfrac{12a^3b^6}{18ab^5}$ |

$$= \frac{a}{2b} - \frac{2a^2b}{3} \qquad \text{Perform each monomial division.}$$

| Long division can be used to **divide a polynomial by a polynomial** (other than a monomial). The long-division method is a series of four steps that are repeated: Divide, multiply, subtract, and bring down the next term.

When the division has a remainder, write the answer in the form: Quotient $+\ \dfrac{\text{remainder}}{\text{divisor}}$ | Divide $4x^2 - 4x + 5$ by $2x + 1$. |

$$\begin{array}{r}
2x - 3 \\
2x+1{\overline{\smash{\big)}\,4x^2 - 4x + 5}}\\
\underline{-(4x^2 + 2x)}\\
-6x + 5\\
\underline{-(-6x - 3)}\\
8
\end{array}$$

The first division: $\frac{4x^2}{2x} = 2x$.

The second division: $\frac{-6x}{2x} = -3$.

The remainder is 8.

The result is: $2x - 3 + \dfrac{8}{2x + 1}$

| The long division method works best when the terms of the divisor and the dividend are written in **descending powers of the variable.**

When the dividend has **missing terms,** insert such terms with a coefficient of 0, or leave a blank space. | Set up each long division. |

$$\frac{5x + x^3 + 3 + 3x^2}{x + 1} \qquad\qquad\qquad\qquad \frac{x^2 - 9}{x - 3}$$

The terms of the dividend are not in descending powers of x.

The dividend is missing a term.

$$x+1{\overline{\smash{\big)}\,x^3 + 3x^2 + 5x + 3}} \qquad\qquad x-3{\overline{\smash{\big)}\,x^2 + 0x - 9}}$$

REVIEW EXERCISES

Divide. Do not use negative exponents in the answer.

105. $\dfrac{16n^8}{8n^5}$

106. $\dfrac{-14x^2y}{21xy^3}$

107. $\dfrac{a^{15} - 24a^8}{6a^{12}}$

108. $\dfrac{15a^5b + ab^2 - 25b}{5a^2b}$

109. $x - 1{\overline{\smash{\big)}\,x^2 - 6x + 5}}$

110. $\dfrac{2x^2 + 3 + 7x}{x + 3}$

111. $(15x^2 - 8x - 8) \div (3x + 2)$

112. Divide $25y^2 - 9$ by $5y + 3$.

113. $3x + 1{\overline{\smash{\big)}\,-13x - 4 + 9x^3}}$

114. $2x - 1{\overline{\smash{\big)}\,6x^3 + x^2 + 1}}$

115. Use multiplication to show that $(3y^2 + 11y + 6) \div (y + 3)$ is $3y + 2$.

116. BEDDING   The area of a rectangular-shaped bed sheet is represented by the polynomial $(4x^3 + 12x^2 + x - 12)$ in.^{2}. If the width of the sheet is $(2x + 3)$ inches, find a polynomial that represents its length.

CHAPTER 5
Test

1. Fill in the blanks.

   a. In the expression y^{10}, the _____ is y and the _____ is 10.

   b. We call a polynomial with exactly one term a _____, with exactly two terms a _____, and with exactly three terms a _____.

   c. The _____ of a term of a polynomial in one variable is the value of the exponent on the variable.

   d. $(x + y)^2$, $(x - y)^2$, and $(x + y)(x - y)$ are called _____ products.

2. Use exponents to rewrite $2xxxyyyy$.

Simplify each expression. Do not use negative exponents in the answer.

3. $y^2(yy^3)$

4. $\left(\frac{1}{2}x^3\right)^5 (x^2)^3$

5. $3.5x^0$

6. $2y^{-5}y^2$

7. 5^{-3}

8. $\dfrac{(x + 1)^{15}}{(x + 1)^6}$

9. $\dfrac{(y^{-5})^{-4}}{yy^{-2}}$

10. $\left(\dfrac{a^2b^{-1}}{4a^3b^{-2}}\right)^3$

11. $\left(\dfrac{8}{m^6}\right)^{-2}$

12. $\dfrac{-6a}{b^{-9}}$

13. Find an expression that represents the volume of a cube that has sides of length $10y^4$ inches.

14. ELECTRICITY   One ampere (amp) corresponds to the flow of 6,250,000,000,000,000,000 electrons per second past any point in a direct current (DC) circuit. Write this number in scientific notation.

15. Write 9.3×10^{-5} in standard notation.

16. Evaluate $(2.3 \times 10^{18})(4.0 \times 10^{-15})$. Write the answer in scientific notation and standard notation.

17. Identify $x^4 + 8x^2 - 12$ as a monomial, binomial, or trinomial. Then complete the table.

Term	Coefficient	Degree

Degree of the polynomial

18. Find the degree of the polynomial $3x^3y + 2x^2y^3 - 5xy^2 - 6y$.

19. Complete the table of solutions for $y = x^2 + 2$ and then graph the equation.

x	-2	-1	0	1	2
y					

20. FREE FALL   A visitor standing on the rim of the Grand Canyon drops a rock over the side. The distance (in feet) that the rock is from the canyon floor t seconds after being dropped is given by the polynomial $-16t^2 + 5{,}184$. Find the position of the rock 18 seconds after being dropped. Explain your answer.

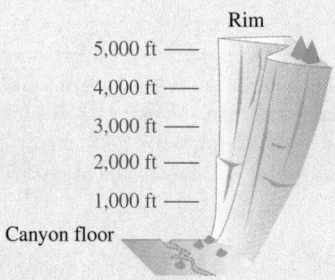

Simplify each polynomial.

21. $\dfrac{3}{5}x^2 + 6 + \dfrac{1}{4}x - 8 - \dfrac{1}{2}x^2 + \dfrac{1}{3}x$

22. $4a^2b + 5 - 6a^3b - 3a^2b + 2a^3b$

Perform the operations.

23. $(12.1h^3 - 9.9h^2 + 9.5) + (7.3h^3 - 1.2h^2 - 10.1)$

24. Subtract $b^3c - 3bc + 12$ from the sum of $6b^3c - 3bc$ and $b^3c - 2bc$.

25. Subtract:

$$-5y^3 + 4y^2 + 3$$
$$\underline{-(-2y^3 - 14y^2 + 17y - 32)}$$

26. Find a polynomial that represents the perimeter of the rectangle.

$(5a^2 + 3a - 1)$ in.

$(a - 9)$ in.

Multiply.

27. $(2x^3y^3)(5x^2y^8)$

28. $9b^3(8b^4)(-b)$

29. $3y^2(y^2 - 2y + 3)$

30. $(x - 5)(3x + 4)$

31. $\left(6t + \dfrac{1}{2}\right)\left(2t - \dfrac{3}{2}\right)$

32. $(2x - 3)(x^2 - 2x + 4)$

33. $(1 + 10c)(1 - 10c)$

34. $(7b^3 - 3t)^2$

35. $(2.2a)(a + 5)(a - 3)$

36. Perform the operations: $(x + y)(x - y) + (x + y)^2$

Divide.

37. $\dfrac{6a^2 - 12b^2}{24ab}$

38. $\dfrac{x^2 + x - 6}{x + 3}$

39. $2x - 1\overline{)\,1 + x^2 + 6x^3}$

40. Find a polynomial that represents the width of a rectangle if its area is represented by the polynomial $(x^2 - 6x + 5)$ ft^2 and the length is $(x - 1)$ ft.

41. Use a check to determine whether $(5m^2 - 29m - 6) \div (5m + 1) = m - 6$.

42. Is $(a + b)^2 = a^2 + b^2$? Show why or why not.

GROUP PROJECT

BINOMIAL MULTIPLICATION AND THE AREA OF RECTANGLES

Overview: In this activity, rectangles are used to visualize binomial multiplication.

Instructions: Form groups of 2 or 3 students. Study the figure. The area of the large rectangle is given by $(x + 2)(x + 3)$. The area of the large rectangle is also the sum of the areas of the four smaller rectangles: $x^2 + 3x + 2x + 6$. Thus,

$$(x + 2)(x + 3) = x^2 + 3x + 2x + 6 = x^2 + 5x + 6$$

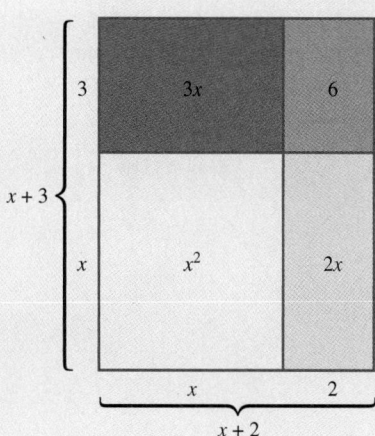

Draw three similar models to represent the following products.

1. $(x + 4)(x + 5)$ 2. $(x + 8)^2$ 3. $x(x + 6)$ 4. $(2x + 1)^2$

Determine the missing number so that the rectangle has the given area.

5. Area: $x^2 + 9x + 14$ 6. Area: $x^2 + 16x + 55$

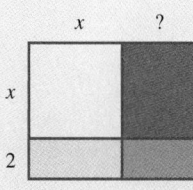

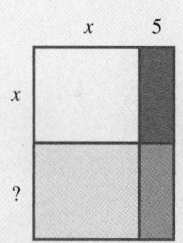

CUMULATIVE REVIEW
Chapters 1–5

1. Use exponents to write the prime factorization of 270. [Section 1.2]

2. a. Use the variables a and b to state the commutative property of addition. [Section 1.4]

   b. Use the variables x, y, and z to state the associative property of multiplication. [Section 1.6]

Evaluate each expression.

3. $3 - 4[-10 - 4(-5)]$ [Section 1.7]

4. $\dfrac{|-45| - 2(-5) + 1^5}{2 \cdot 9 - 2^4}$ [Section 1.7]

Simplify each expression.

5. $27\left(\dfrac{2}{3}x\right)$ [Section 1.9]

6. $3x^2 + 2x^2 - 5x^2$ [Section 1.9]

Solve each equation.

7. $2 - (4x + 7) = 3 + 2(x + 2)$ [Section 2.2]

8. $\dfrac{2}{5}y + 3 = 9$ [Section 2.2]

9. CANDY SALES The circle graph shows how $6.3 billion in seasonal candy sales for 2005 was spent. Find the candy sales for Halloween. [Section 2.3]

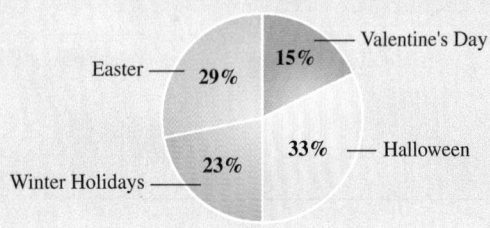

Source: National Confectioners Association

10. AIR CONDITIONING Find the volume of air contained in the duct. Round to the nearest tenth of a cubic foot. [Section 2.4]

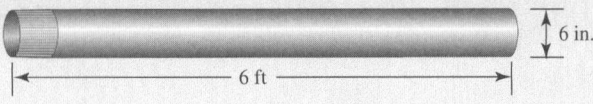

6 in.

6 ft

11. ANGLE OF ELEVATION Find x. [Section 2.5]

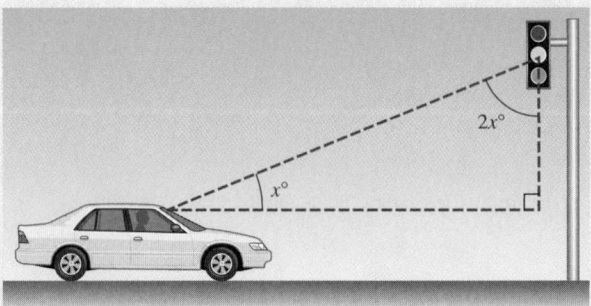

12. LIVESTOCK AUCTION A farmer is going to sell one of her prize hogs at an auction and would like to make $6,000 after paying a 4% commission to the auctioneer. For what selling price will the farmer make this amount of money? [Section 2.5]

13. STOCK MARKET An investment club invested part of $45,000 in a high-yield mutual fund that earned 12% annual simple interest. The remainder of the money was invested in Treasury bonds that earned 6.5% simple annual interest. The two investments earned $4,300 in one year. How much was invested in each account? [Section 2.6]

14. Solve $-4x + 6 > 17$ and graph the solution set. Then describe the graph using interval notation. [Section 2.7]

Graph each equation.

15. $y = 3x$ [Section 3.2]   16. $x = -2$ [Section 3.3]

17. Find the slope of the line passing through $(6, -2)$ and $(-3, 2)$. [Section 3.4]

18. Find the slope and y-intercept of the line. Then write the equation of the line. [Section 3.5]

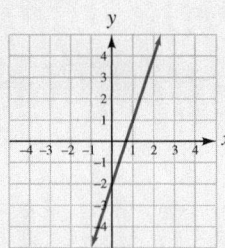

19. Without graphing, determine whether the graphs of $y = \dfrac{3}{2}x - 1$ and $2x + 3y = 10$ are parallel, perpendicular, or neither. [Section 3.5]

20. Write the equation of the line that passes through $(-2, 10)$ with slope -4. Write the result in slope–intercept form. [Section 3.6]

21. Is $(-2, 1)$ a solution of $2x - 3y \geq -6$? [Section 3.7]

22. If $f(x) = 2x^2 + 3x - 9$, find $f(-5)$. [Section 3.8]

23. Is $\left(\dfrac{2}{3}, -1\right)$ a solution of the system $\begin{cases} y = -3x + 1 \\ 3x + 3y = -2 \end{cases}$? [Section 4.1]

24. Solve the system $\begin{cases} 3x + 2y = 14 \\ y = \dfrac{1}{4}x \end{cases}$ by graphing. [Section 4.1]

25. Solve the system $\begin{cases} 2b - 3a = 18 \\ a + 3b = 5 \end{cases}$ by substitution. [Section 4.2]

26. Solve the system $\begin{cases} 8s + 10t = 24 \\ 11s - 3t = -34 \end{cases}$ by elimination (addition). [Section 4.3]

27. VACATIONS   One-day passes to Universal Studios Hollywood cost a family of 5 (2 adults and 3 children) $275. A family of 6 (3 adults and 3 children) paid $336 for their one-day passes. Find the cost of an adult one-day pass and a child's one-day pass to Universal Studios. [Section 4.4]

28. Graph: $\begin{cases} y \leq 2x - 1 \\ x + 3y > 6 \end{cases}$ [Section 4.5]

Simplify. Do not use negative exponents in the answer.

29. $(-3x^2y^4)^2$ [Section 5.1]

30. $(v^5)^2(v^3)^4$ [Section 5.1]

31. $ab^3c^4 \cdot ab^4c^2$ [Section 5.1]

32. $\left(\dfrac{4t^3t^4t^5}{3t^2t^6}\right)^3$ [Section 5.1]

33. $(2y)^{-4}$ [Section 5.2]

34. $\dfrac{a^4b^0}{a^{-3}}$ [Section 5.2]

35. -5^{-2} [Section 5.2]

36. $\left(\dfrac{a}{x}\right)^{-10}$ [Section 5.2]

Write each number in scientific notation.

37. 615,000 [Section 5.3]

38. 0.0000013 [Section 5.3]

39. Graph: $y = x^2$ [Section 5.4]

40. MUSICAL INSTRUMENTS   The amount of deflection of the horizontal beam (in inches) is given by the polynomial $0.01875x^4 - 0.15x^3 + 1.2x$, where x is the distance (in feet) that the gong is hung from one end of the beam. Find the deflection if the gong is hung in the middle of the support. [Section 5.4]

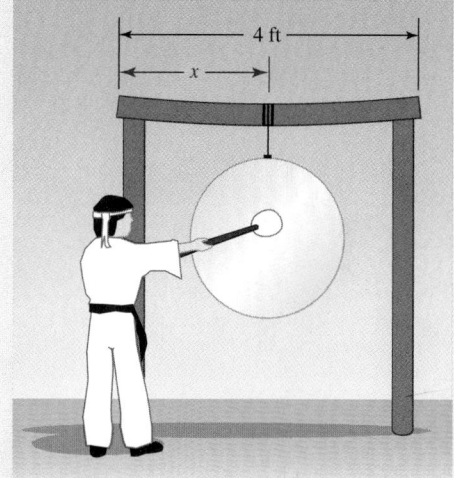

Perform the operations.

41. $(4c^2 + 3c - 2) + (3c^2 + 4c + 2)$ [Section 5.5]

42. Subtract: $17x^4 - 3x^2 - 65x - 12$
$\underline{-(23x^4 + 14x^2 + 3x - 23)}$
[Section 5.5]

43. $(2t + 3s)(3t - s)$ [Section 5.6]

44. $3x(2x + 3)^2$ [Section 5.7]

45. $5x + 3 \overline{)11x + 10x^2 + 3}$ [Section 5.8]

46. $\dfrac{2x - 32}{16x}$ [Section 5.8]

CHAPTER 6

Factoring and Quadratic Equations

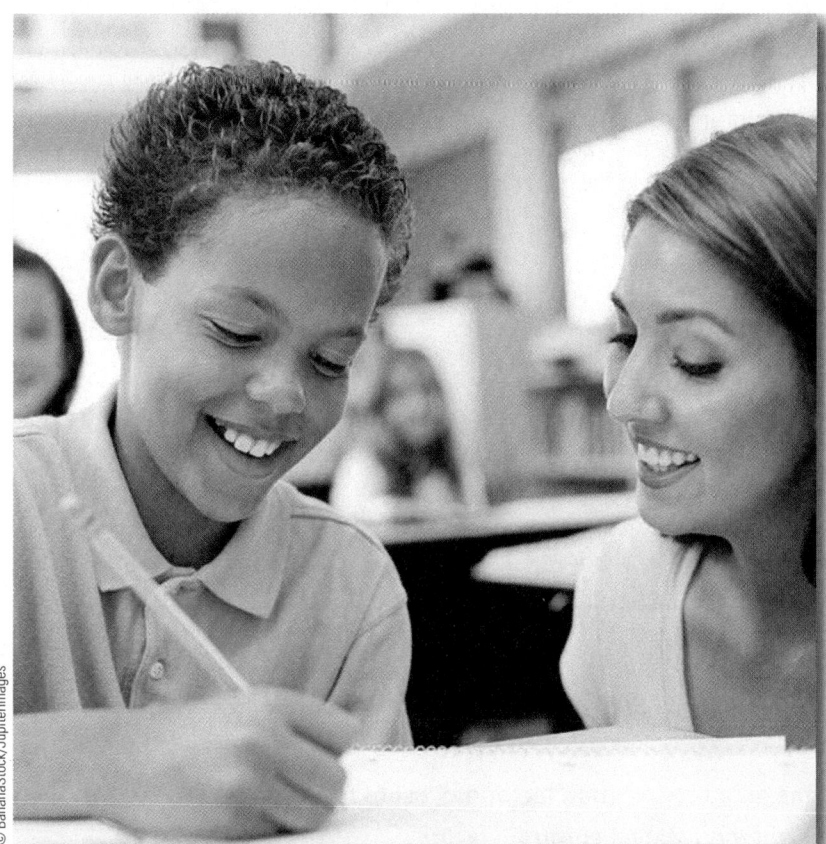

from *Campus to Careers*
Elementary School Teacher

It has been said that a teacher takes a hand, opens a mind, and touches a heart. That is certainly true for the thousands of dedicated elementary school teachers across the country. Elementary school teachers use their training in mathematics in many ways. Besides teaching math on a daily basis, they calculate student grades, analyze test results, and order instructional materials and supplies. They use measurement and geometry for designing bulletin board displays and they construct detailed schedules so that the classroom time is used wisely.

In **Problem 17** of **Study Set 6.8,** you will determine the maximum dimensions of a bulletin board so that it meets the fire code requirements of an elementary school classroom.

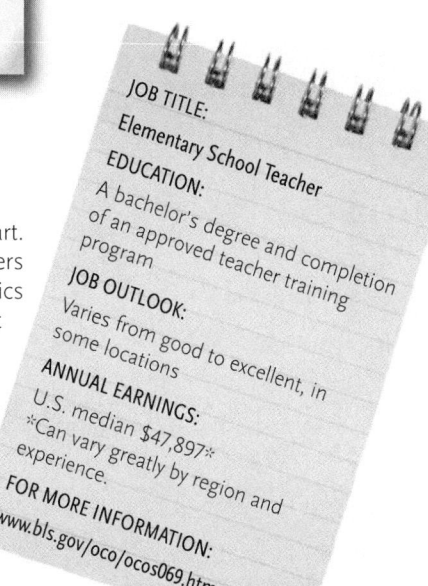

JOB TITLE:
Elementary School Teacher

EDUCATION:
A bachelor's degree and completion of an approved teacher training program

JOB OUTLOOK:
Varies from good to excellent, in some locations

ANNUAL EARNINGS:
U.S. median $47,897*
*Can vary greatly by region and experience.

FOR MORE INFORMATION:
www.bls.gov/oco/ocos069.htm

469

Study Skills Workshop
Reading the Textbook

Reading an algebra textbook is different from reading a newspaper or a novel. Here are two ways that you should be reading this textbook.

SKIMMING FOR AN OVERVIEW: This is a quick way to look at material just *before* it is covered in class. It helps you become familiar with the new vocabulary and notation that will be used by your instructor in the lecture. It lays a foundation.

READING FOR UNDERSTANDING: This in-depth type of reading is done more slowly, with a pencil and paper at hand. Don't skip anything—every word counts! You should do just as much writing as you do reading. Highlight the important points and work each example. If you become confused, stop and reread the material until you understand it.

Now Try This

Choose a section from this chapter and . . .

1. . . . quickly skim it. Write down any terms in bold face type and the titles of any properties, definitions, or strategies that are given in the colored boxes.
2. . . . work each *Self Check* problem. Your solutions should look like those in the *Examples*. Be sure to include your own "author notes" (the sentences in red to the right of each step of a solution).

SECTION 6.1
The Greatest Common Factor; Factoring by Grouping

Objectives

1. Find the greatest common factor of a list of terms.
2. Factor out the greatest common factor.
3. Factor by grouping.

> ***Success Tip***
> On the game show *Jeopardy!*, answers are revealed and contestants respond with the appropriate questions. Factoring is similar. Answers to multiplications are given. You are to respond by telling what factors were multiplied.

In Chapter 5, we learned how to multiply polynomials. For example, to multiply $3x + 5$ by $4x$, we use the distributive property, as shown below.

$$4x(3x + 5) = 4x \cdot 3x + 4x \cdot 5$$
$$= 12x^2 + 20x$$

In this section, we reverse the previous steps and determine what factors were multiplied to obtain $12x^2 + 20x$. We call that process *factoring the polynomial.*

Multiplication: Given the factors, we find a polynomial. ⟶

$$4x(3x + 5) = 12x^2 + 20x$$

⟵ Factoring: Given a polynomial, we find the factors.

To **factor a polynomial** means to express it as a product of two (or more) polynomials. The first step when factoring a polynomial is to determine whether its terms have any common factors.

1 **Find the Greatest Common Factor of a List of Terms.**

To determine whether two or more integers have common factors, it is helpful to write them as products of prime numbers. For example, the prime factorizations of 42 and 90 are given below.

$$42 = \mathbf{2 \cdot 3} \cdot 7 \qquad 90 = \mathbf{2 \cdot 3} \cdot 3 \cdot 5$$

The highlighting shows that 42 and 90 have one factor of 2 and one factor of 3 in common. To find their *greatest common factor* (*GCF*), we multiply the common factors: $2 \cdot 3 = 6$. Thus, the GCF of 42 and 90 is 6.

The Greatest Common Factor (GCF)

The **greatest common factor (GCF)** of a list of integers is the largest common factor of those integers.

Recall from arithmetic that the factors of a number divide the number exactly, leaving no remainder. Therefore, the greatest common factor of two or more integers is the largest natural number that divides each of the integers exactly.

EXAMPLE 1 Find the GCF of each list of numbers: **a.** 21 and 140 **b.** 24, 60, and 96 **c.** 9, 10, and 30

Strategy We will prime factor each number in the list. Then we will identify the common prime factors and find their product.

Why The product of the common prime factors is the GCF of the numbers in the list.

Solution

a. The prime factorization of each number is shown:

$$21 = 3 \cdot 7$$
$$140 = 2 \cdot 2 \cdot 5 \cdot 7 \quad \text{This can be written as } 2^2 \cdot 5 \cdot 7.$$

Since the only prime factor common to 21 and 140 is 7, the GCF of 21 and 140 is 7.

b. To find the GCF of three numbers, we proceed in a similar way by first finding the prime factorization of each number in the list.

$$24 = 2 \cdot 2 \cdot 2 \cdot 3 \quad \text{This can be written as } 2^3 \cdot 3.$$
$$60 = 2 \cdot 2 \cdot 3 \cdot 5 \quad \text{This can be written as } 2^2 \cdot 3 \cdot 5.$$
$$96 = 2 \cdot 2 \cdot 2 \cdot 2 \cdot 2 \cdot 3 \quad \text{This can be written as } 2^5 \cdot 3.$$

The highlighting shows that 24, 60, and 96 have two factors of 2 and one factor of 3 in common. The GCF of 24, 60, and 96 is the product of their common prime factors.

$$\text{GCF} = 2 \cdot 2 \cdot 3 = 2^2 \cdot 3^1 = 12$$

c. Since there are no prime factors common to 9, 10, and 30, their GCF is 1.

$$9 = 3 \cdot 3$$
$$10 = 2 \cdot 5$$
$$30 = 2 \cdot 3 \cdot 5$$

Success Tip Note that the GCF, 7, divides 21 and 140 exactly: $\frac{21}{7} = 3 \quad \frac{140}{7} = 20$

Success Tip The exponent on any factor in a GCF is the *smallest* exponent that appears on that factor in all of the numbers under consideration.

Self Check 1   Find the GCF of each list of numbers: **a.** 24 and 70
b. 22, 25, and 98   **c.** 45, 60, and 75

Now Try   **Problems 17 and 21**

To find the greatest common factor of a list of terms, we can use the following approach.

Strategy for Finding the GCF

1. Write each coefficient as a product of prime factors.
2. Identify the numerical and variable factors common to each term.
3. Multiply the common numerical and variable factors identified in Step 2 to obtain the GCF. If there are no common factors, the GCF is 1.

EXAMPLE 2   Find the GCF of each list of terms: **a.** $12x^2$ and $20x$
b. $9a^5b^2$, $15a^4b^2$,   and $90a^3b^3$

Strategy   We will prime factor each coefficient of each term in the list. Then we will identify the numerical and variable factors common to each term and find their product.

Why   The product of the common factors is the GCF of the terms in the list.

Solution

a. *Step 1:*   We write each coefficient, 12 and 20, as a product of prime factors. Recall that an exponent, as in x^2, indicates repeated multiplication.

$$12x^2 = 2 \cdot 2 \cdot 3 \cdot x \cdot x \quad \text{This can be written as } 2^2 \cdot 3 \cdot x^2.$$
$$20x = 2 \cdot 2 \cdot 5 \cdot x \quad \text{This can be written as } 2^2 \cdot 5 \cdot x.$$

Success Tip
One way to identify common factors is to circle them:

$$12x^2 = 2 \cdot 2 \cdot 3 \cdot x \cdot x$$
$$20x = 2 \cdot 2 \cdot 5 \cdot x$$
$$GCF = 2 \cdot 2 \cdot x = 4x$$

Step 2:   There are two common factors of 2 and one common factor of x.

Step 3:   We multiply the common factors, 2, 2, and x, to obtain the GCF.

$$GCF = 2 \cdot 2 \cdot x = 2^2 \cdot x = 4x$$

b. *Step 1:*   We write the coefficients, 9, 15, and 90, as products of primes. The exponents on the variables represent repeated multiplication.

Success Tip
The exponent on any variable in a GCF is the *smallest* exponent that appears on that variable in all of the terms under consideration.

$$9a^5b^2 = 3 \cdot 3 \cdot a \cdot a \cdot a \cdot a \cdot a \cdot b \cdot b \quad \text{This can be written as } 3^2 \cdot a^5 \cdot b^2.$$
$$15a^4b^2 = 3 \cdot 5 \cdot a \cdot a \cdot a \cdot a \cdot b \cdot b \quad \text{This can be written as } 3 \cdot 5 \cdot a^4 \cdot b^2.$$
$$90a^3b^3 = 2 \cdot 3 \cdot 3 \cdot 5 \cdot a \cdot a \cdot a \cdot b \cdot b \cdot b \quad \text{This can be written as } 2 \cdot 3^2 \cdot 5 \cdot a^3 \cdot b^3.$$

Step 2:   The highlighting shows one common factor of 3, three common factors of a, and two common factors of b.

Step 3: $GCF = 3 \cdot a \cdot a \cdot a \cdot b \cdot b = 3a^3b^2$

Self Check 2   Find the GCF of each list of terms: **a.** $33c$ and $22c^4$
b. $42s^3t^2$, $63s^2t^4$,   and $21s^3t^3$

Now Try   **Problems 29 and 35**

Factor Out the Greatest Common Factor.

The concept of greatest common factor is used to factor polynomials. For example, to factor $12x^2 + 20x$, we note that there are two terms, $12x^2$ and $20x$. We previously determined that the GCF of $12x^2$ and $20x$ is $4x$. With this in mind, we write each term of $12x^2 + 20x$ as a product of the GCF and one other factor. Then we apply the distributive property in reverse: $ab + ac = a(b + c)$.

$$12x^2 + 20x = 4x \cdot 3x + 4x \cdot 5 \qquad \text{Write } 12x^2 \text{ and } 20x \text{ as the product of the GCF, } 4x, \text{ and one other factor.}$$

$$= 4x(3x + 5) \qquad \text{Write an expression so that the multiplication by } 4x \text{ distributes over the terms } 3x \text{ and } 5.$$

We have found that the factored form of $12x^2 + 20x$ is $4x(3x + 5)$. This process is called **factoring out the greatest common factor.**

EXAMPLE 3   Factor: **a.** $8m + 24$ **b.** $35a^3b^2 - 14a^2b^3$
c. $3x^4 - 5x^3 + x^2$

Strategy   First, we will determine the GCF of the terms of the polynomial. Then we will write each term of the polynomial as the product of the GCF and one other factor.

Why   We can then use the distributive property to factor out the GCF.

Solution

a. Since the GCF of $8m$ and 24 is 8, we write $8m$ and 24 as the product of 8 and one other factor.

$$8m + 24 = 8 \cdot m + 8 \cdot 3$$

$$= 8(m + 3) \qquad \text{Factor out the GCF, 8.}$$

To check, we multiply: $8(m + 3) = 8 \cdot m + 8 \cdot 3 = 8m + 24$. Since we obtain the original polynomial, $8m + 24$, the factorization is correct.

Caution   Remember to factor out the greatest common factor, not just a common factor. If we factored out 4 in the previous example, we would get

$$8m + 24 = 4(2m + 6)$$

However, the terms in red within parentheses have a common factor of 2, indicating that the factoring is not complete.

b. First, find the GCF of $35a^3b^2$ and $14a^2b^3$.

$$\left.\begin{array}{l} 35a^3b^2 = 5 \cdot 7 \cdot a \cdot a \cdot a \cdot b \cdot b \\ 14a^2b^3 = 2 \cdot 7 \cdot a \cdot a \cdot b \cdot b \cdot b \end{array}\right\} \quad \text{The GCF is } 7a^2b^2.$$

Now, we write $35a^3b^2$ and $14a^2b^3$ as the product of the GCF, $7a^2b^2$, and one other factor.

$$35a^3b^2 - 14a^2b^3 = 7a^2b^2 \cdot 5a - 7a^2b^2 \cdot 2b$$

$$= 7a^2b^2(5a - 2b) \qquad \text{Factor out the GCF, } 7a^2b^2.$$

We check by multiplying: $7a^2b^2(5a - 2b) = 35a^3b^2 - 14a^2b^3$.

c. We factor out the GCF of the three terms, which is x^2.

$$3x^4 - 5x^3 + x^2 = x^2(3x^2) - x^2(5x) + x^2(1) \qquad \text{Write the last term, } x^2, \text{ as } x^2(1).$$
$$= x^2(3x^2 - 5x + 1) \qquad \text{Factor out the GCF, } x^2.$$

We check by multiplying: $x^2(3x^2 - 5x + 1) = 3x^4 - 5x^3 + x^2$.

Self Check 3   Factor: **a.** $6f + 36$ **b.** $48s^2t^2 - 84s^3t$
     c. $y^6 - 10y^4 - y^3$

Now Try   **Problems 41, 51, and 53**

EXAMPLE 4   Factor -1 from each polynomial: **a.** $-a^3 + 2a^2 - 4$
     b. $6 - x$

Strategy   We will write each term of the polynomial as the product of -1 and one other factor.

Why   We can then use the distributive property to factor out the -1.

Solution

a. $-a^3 + 2a^2 - 4 = (-1)a^3 + (-1)(-2a^2) + (-1)4$

$$= -1(a^3 - 2a^2 + 4) \qquad \text{Factor out } -1.$$
$$= -(a^3 - 2a^2 + 4) \qquad \text{The 1 need not be written.}$$

We check by multiplying: $-(a^3 - 2a^2 + 4) = -a^3 + 2a^2 - 4$.

b. $6 - x = (-1)(-6) + (-1)x$

$$= -1(-6 + x) \qquad \text{Factor out } -1.$$
$$= -(x - 6) \qquad \text{The 1 need not be written. Within the parentheses, write the}$$
$$\qquad\qquad\qquad\qquad \text{binomial with the x-term first.}$$

Self Check 4   Factor -1 from each polynomial:
     a. $-b^4 - 3b^2 + 2$ **b.** $9 - t$

Now Try   **Problems 63 and 69**

EXAMPLE 5   Factor out the opposite of the GCF in $-20m + 30$.

Strategy   First, we will determine the GCF of the terms of the polynomial. Then we will write each term of the polynomial as the product of the opposite of the GCF and one other factor.

Why   We can then use the distributive property to factor out the opposite of the GCF.

Solution   Since the GCF is 10, the opposite of the GCF is -10. We write each term of the polynomial as the product of -10 and another factor. Then we factor out -10.

$$-20m + 30 = (-10)(2m) + (-10)(-3)$$
$$= -10(2m - 3)$$

Note that the leading coefficient of the polynomial within the parentheses is positive.

We check by multiplying: $-10(2m - 3) = -20m + 30$.

 Self Check 5   Factor out the opposite of the GCF in $-44c + 55$.
Now Try   **Problem 75**

EXAMPLE 6   Factor: $x(x + 4) + 3(x + 4)$

Strategy   We will identify the terms of the expression and find their GCF.

Why   We can then use the distributive property to factor out the GCF.

Solution   The expression has two terms: $\underbrace{x(x + 4)}_{\text{The first term}}$ + $\underbrace{3(x + 4)}_{\text{The second term}}$

The GCF of the terms is the binomial $x + 4$, which can be factored out.

$$x(x + 4) + 3(x + 4) = (x + 4)x + (x + 4)3$$

Write each term as the product of $(x + 4)$ and one other factor.

$$= (x + 4)(x + 3)$$

Factor out the common factor, $(x + 4)$.

 **Self Check 6**   Factor: $2y(y - 1) + 7(y - 1)$
Now Try   **Problem 81**

③   **Factor by Grouping.**

Although the terms of many polynomials don't have a common factor, other than 1, it is possible to factor some of them by arranging their terms in convenient groups. This method is called **factoring by grouping.**

EXAMPLE 7   Factor by grouping: **a.** $2x^3 + x^2 + 12x + 6$
b. $5c - 5d + cd - d^2$

Strategy   We will factor out a common factor from the first two terms and a common factor from the last two terms.

Why   This will produce a common binomial factor that can then be factored out.

Solution

a. Except for 1, there is no factor that is common to all four terms. However, the first two terms, $2x^3$ and x^2, have a common factor, x^2, and the last two terms, $12x$ and 6, have a common factor, 6.

$$2x^3 + x^2 \quad + \quad 12x + 6$$

$$x^2(2x + 1) \quad + \quad 6(2x + 1)$$

We now see that $2x^3 + x^2$ and $12x + 6$ have a common binomial factor, $2x + 1$, which can be factored out.

$$2x^3 + x^2 + 12x + 6 = x^2(2x + 1) + 6(2x + 1) \qquad \text{Factor } 2x^3 + x^2 \text{ and } 12x + 6.$$

$$= (2x + 1)(x^2 + 6) \qquad \text{Factor out } 2x + 1.$$

> **Caution**
> Factoring by grouping can be attempted on any polynomial with four or more terms. However, not every such polynomial can be factored in this way.

We can check the factorization by multiplying. The result should be the original polynomial.

$$(2x + 1)(x^2 + 6) = 2x^3 + 12x + x^2 + 6$$

$$= 2x^3 + x^2 + 12x + 6 \qquad \text{Rearrange the terms to get the original polynomial.}$$

> **Caution**
> Don't think that $5(c - d) + d(c - d)$ is in factored form. It is a sum of two terms. To be in factored form, the result must be a product.

b. The first two terms have a common factor, 5, and the last two terms have a common factor, d.

$$5c - 5d + cd - d^2 = 5(c - d) + d(c - d) \qquad \text{Factor out 5 from } 5c - 5d \text{ and } d \text{ from } cd - d^2.$$

$$= (c - d)(5 + d) \qquad \text{Factor out the common binomial factor, } c - d.$$

We can check by multiplying:

$$(c - d)(5 + d) = 5c + cd - 5d - d^2$$

$$= 5c - 5d + cd - d^2$$

▷ **Self Check 7**   Factor by grouping: **a.** $3n^3 + 2n^2 + 9n + 6$
b. $7x - 7y + xy - y^2$

Now Try   **Problems 85 and 87**

Factoring a Four-termed Polynomial by Grouping

1. Group the terms of the polynomial so that the first two terms have a common factor and the last two terms have a common factor.

2. Factor out the common factor from each group.

3. Factor out the resulting common binomial factor. If there is no common binomial factor, regroup the terms of the polynomial and repeat steps 2 and 3.

By the multiplication property of 1, we know that 1 is a factor of every term. We can use this observation to factor certain polynomials by grouping.

EXAMPLE 8   Factor: **a.** $x^3 + 6x^2 + x + 6$ **b.** $x^2 - ax - x + a$

Strategy   We will follow the steps for factoring a four-termed polynomial.

Why   Since the terms of the polynomials do not have a common factor (other than 1), the only option is to attempt to factor these polynomials by grouping.

Solution

a. The first two terms, x^3 and $6x^2$, have a common factor of x^2. The only common factor of the last two terms, x and 6, is 1.

$$x^3 + 6x^2 \ + \ x + 6 \ = x^2(x + 6) + 1(x + 6) \quad \text{Factor out } x^2 \text{ from } x^3 + 6x^2.$$
$$\text{Factor out 1 from } x + 6.$$
$$= (x + 6)(x^2 + 1) \qquad \text{Factor out the common binomial factor, } x + 6.$$

Check the factorization by multiplying.

b. Since x is a common factor of the first two terms, we can factor it out and proceed as follows.

$$x^2 - ax \ - x + a = x(x - a) - x + a \quad \text{Factor out } x \text{ from } x^2 - ax.$$

When factoring four terms by grouping, if the coefficient of the 3rd term is negative, we often factor out a negative coefficient from the last two terms. If we factor -1 from $-x + a$, a common binomial factor $x - a$ appears within the second set of parentheses, which we can factor out.

$$x^2 - ax \ - x + a \ = x(x - a) - 1(x - a) \quad \text{To factor out } -1, \text{ change the sign of } -x$$
$$\text{and } a, \text{ and write } -1 \text{ in front of the parentheses.}$$
$$= (x - a)(x - 1) \qquad \text{Factor out the common factor, } x - a.$$

Check by multiplying.

> **Success Tip**
> When we factor out -1 from the last two terms,
> $$x^2 - ax \ - x + a$$
> $$= x(x - a) - 1(x - a)$$
> the signs of those terms change within the parentheses. The binomials within both sets of parentheses are then identical.

Self Check 8   Factor: **a.** $a^5 + 11a^4 + a + 11$
                          b. $b^2 - bc - b + c$

Now Try   Problems 93 and 95

The next example illustrates that when factoring a polynomial, we should always look for a common factor first.

EXAMPLE 9   Factor: $10k + 10m - 2km - 2m^2$

Strategy   Since all four terms have a common factor of 2, we factor it out first. Then we will factor the resulting polynomial by grouping.

Why   Factoring out the GCF first makes factoring by any method easier.

Solution   After factoring out 2 from all four terms, notice that within the parentheses, the first two terms have a common factor of 5, and the last two terms have a common factor of $-m$.

$$10k + 10m - 2km - 2m^2 = 2(5k + 5m - km - m^2) \qquad \text{Factor out the GCF, 2.}$$

$$= 2[5(k + m) - m(k + m)] \qquad \begin{array}{l}\text{Factor out 5 from } 5k + 5m.\\ \text{Factor out } -m \text{ from}\\ -km - m^2. \text{ This causes the}\\ \text{signs of } -km \text{ and } -m^2 \text{ to}\\ \text{change within the second set}\\ \text{of parentheses.}\end{array}$$

$$= 2[(k + m)(5 - m)] \qquad \begin{array}{l}\text{Factor out the common}\\ \text{binomial factor, } k + m.\end{array}$$

$$= 2(k + m)(5 - m) \qquad \begin{array}{l}\text{Drop the unnecessary}\\ \text{brackets.}\end{array}$$

Check by multiplying.

Self Check 9   Factor: $4t + 4s + 4tz + 4sz$

Now Try   Problem 101

ANSWERS TO SELF CHECKS   **1. a.** 2 **b.** 1 **c.** 15 **2. a.** $11c$ **b.** $21s^2t^2$ **3. a.** $6(f + 6)$
b. $12s^2t(4t - 7s)$ **c.** $y^3(y^3 - 10y - 1)$ **4. a.** $-(b^4 + 3b^2 - 2)$ **b.** $-(t - 9)$ **5.** $-11(4c - 5)$
6. $(y - 1)(2y + 7)$ **7. a.** $(3n + 2)(n^2 + 3)$ **b.** $(x - y)(7 + y)$ **8. a.** $(a + 11)(a^4 + 1)$
b. $(b - c)(b - 1)$ **9.** $4(t + s)(1 + z)$

STUDY SET
6.1

VOCABULARY

Fill in the blanks.

1. To _____ a polynomial means to express it as a product of two (or more) polynomials.

2. GCF stand for _____ _____ _____. When we write $2x + 4$ as $2(x + 2)$, we say that we have _____ out the GCF, 2.

3. To factor $m^3 + 3m^2 + 4m + 12$ by _____, we begin by writing $m^2(m + 3) + 4(m + 3)$.

4. The terms $x(x - 1)$ and $4(x - 1)$ have the common _____ factor $x - 1$.

CONCEPTS

5. Complete each factorization.

  a. $6x = 2 \cdot \ \ \cdot x$ **b.** $35h^2 = 5 \cdot \ \ \cdot h \cdot$

  c. $18y^3z = 2 \cdot$ 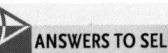 $\cdot 3 \cdot \ \ \cdot y \cdot \ \ \cdot z$

6. a. Find the GCF of $30x^2$ and $105x^3$.

$$30x^2 = 2 \cdot 3 \cdot 5 \cdot x \cdot x$$
$$105x^3 = 3 \cdot 5 \cdot 7 \cdot x \cdot x \cdot x$$

  b. Find the GCF of $12a^2b^2$, $15a^3b$, and $75a^4b^2$.

$$12a^2b^2 = 2 \cdot 2 \cdot 3 \cdot a \cdot a \cdot b \cdot b$$
$$15a^3b = 3 \cdot 5 \cdot a \cdot a \cdot a \cdot b$$
$$75a^4b^2 = 3 \cdot 5 \cdot 5 \cdot a \cdot a \cdot a \cdot a \cdot b \cdot b$$

7. a. Write a binomial such that the GCF of its terms is 2.

  b. Write a trinomial such that the GCF of its terms is x.

8. Check to determine whether each factorization is correct.

  a. $9y^3 + 5y^2 - 15y = 3y(3y^2 + 2y - 5)$

  b. $3s^3 + 2s^2 + 6s + 4 = (3s + 2)(s^2 + 2)$

Fill in the blanks to complete each factorization.

9. $2x^2 + 6x = 2x \cdot x + 2x \cdot 3$

$$= \quad (\quad)$$

10. $3t^3 - t^2 + 15t - 5 = t^2(3t - 1) + 5(3t - 1)$

$$= (\quad)(\quad)$$

11. Consider the polynomial $2k - 8 + hk - 4h$.
   a. How many terms does the polynomial have?
   b. Is there a common factor of all the terms, other than 1?
   c. What is the GCF of the first two terms and what is the GCF of the last two terms?

12. What is the first step in factoring $8y^2 - 16yz - 6y + 12z$?

NOTATION

Complete each factorization.

13. $8m^2 - 32m + 16 = \quad (m^2 - 4m + 2)$

14. $10a^4 - 15a^3 = 5a \quad (2a - 3)$

15. $b^3 - 6b^2 + 2b - 12 = \quad (b - 6) + \quad (b - 6)$

$$= (\quad)(b^2 + 2)$$

16. $12 + 8n - 3m - 2mn = 4(3 + 2n) \quad m(3 + 2n)$

$$= (3 + 2n)(4 \quad m)$$

GUIDED PRACTICE

Find the GCF of each list of numbers. See Example 1.

17. 6, 10
18. 10, 15
19. 18, 24
20. 60, 72
21. 14, 21, 42
22. 16, 24, 48
23. 40, 32, 24
24. 28, 35, 21

Find the GCF of each list of terms. See Example 2.

25. m^4, m^3
26. c^2, c^7
27. $15x, 25$
28. $9a, 21$
29. $20c^2, 12c$
30. $18r, 27r^3$
31. $18a^4, 9a^3, 27a^3$
32. $33m^5, 22m^6, 11m^5$
33. $24a^2, 16a^3b, 40ab$
34. $12r^2, 15rs, 9r^2s^2$
35. $6m^4n, 12m^3n^2, 9m^3n^3$
36. $15c^2d^4, 10c^2d, 40c^3d^3$
37. $4(x + 7), 9(x + 7)$
38. $2(y - 1), 5(y - 1)$
39. $4(p - t), p(p - t)$
40. $a(b + c), 3(b + c)$

Factor out the GCF. See Example 3.

41. $3x + 6$
42. $2y - 10$
43. $18x + 24$
44. $15s - 35$
45. $18m - 9$
46. $24s + 8$
47. $d^2 - 7d$
48. $a^2 + 9a$
49. $15c^3 + 25$
50. $33h^4 - 22$
51. $24a - 16a^2$
52. $18r - 30r^2$

53. $14x^2 - 7x - 7$
54. $27a^2 - 9a + 9$
55. $t^4 + t^3 + 2t^2$
56. $b^4 - b^3 - 3b^2$
57. $ab + ac - ad$
58. $rs - rt + ru$
59. $21x^2y^3 + 3xy^2$
60. $3x^2y^3 - 9x^4y^3$

Factor out −1 from each polynomial. See Example 4.

61. $-a - b$
62. $-x - 2y$
63. $-2x + 5$
64. $-3x + 8$
65. $-3r + 2s - 3$
66. $-6yz + 12xz + 5xy$
67. $-x^2 - x + 16$
68. $-t^2 - 9t + 1$
69. $5 - x$
70. $10 - m$
71. $9 - 4a$
72. $7 - 8b$

Factor each polynomial by factoring out the opposite of the GCF. See Example 5.

73. $-3x^2 - 6x$
74. $-4a^2 - 6a$
75. $-4a^2b + 12a^3$
76. $-25x^4 + 30x^2$
77. $-24x^4 - 48x^3 + 36x^2$
78. $-28a^5 - 42a^4 + 14a^3$
79. $-4a^3b^2 + 14a^2b^2 - 10ab^2$
80. $-30x^4y^3 + 24x^3y^2 - 60x^2y$

Factor each expression. See Example 6.

81. $y(x + 2) + 3(x + 2)$
82. $r(t + v) + 3(t + v)$
83. $m(p - q) - 5(p - q)$
84. $ab(c - 7) - 12(c - 7)$

Factor by grouping. See Example 7.

85. $2x + 2y + ax + ay$
86. $bx + bz + 5x + 5z$
87. $rs - ru + 8sw - 8uw$
88. $12ab - 4ac + 3db - dc$
89. $7m^3 - 2m^2 + 14m - 4$
90. $9s^3 - 2s^2 + 36s - 8$
91. $5x^3 - x^2 + 10x - 2$
92. $6a^3 - a^2 + 18a - 3$

Factor by grouping. See Example 8.

93. $ab + ac + b + c$
94. $xy + 3y^2 + x + 3y$

95. $rs + 4s^2 - r - 4s$ **96.** $tx + tz - x - z$

97. $2ax + 2bx - 3a - 3b$ **98.** $rx + sx - ry - sy$

99. $mp - np - mq + nq$ **100.** $9p - 9q - mp + mq$

Factor by grouping. Remember to factor out the GCF first. See Example 9.

101. $ax^3 - 2ax^2 + 5ax - 10a$
102. $x^3y^2 - 2x^2y^2 + 3xy^2 - 6y^2$
103. $6x^3 - 6x^2 + 12x - 12$
104. $3x^3 - 6x^2 + 15x - 30$

TRY IT YOURSELF

Factor.

105. $h^2(14 + r) + 5(14 + r)$
106. $x(y + 9) - 21(y + 9)$
107. $22a^3 - 33a^2$
108. $39r^3 + 26r^2$
109. $ax + bx - a - b$
110. $2xy + y^2 - 2x - y$
111. $15r^8 - 18r^6 - 30r^5$
112. $24cm - 12cn + 16c$
113. $27mp + 9mq - 9np - 3nq$
114. $4abc + 4ac^2 - 2bc - 2c^2$
115. $-60p^2t^2 - 80pt^3$
116. $-25x^5y^7 + 75x^3y^2$
117. $6x^2 - 2xy - 15x + 5y$
118. $m^3 + 5m^2 + m + 5$
119. $2x^3z - 4x^2z + 32xz - 64z$
120. $4a^2b + 12a^2 - 8ab - 24a$
121. $12uvw^3 - 54uv^2w^2$
122. $14xyz - 16x^2y^2z$
123. $x^3 + x^2 + x + 1$
124. $m^4 + m^3 + 2m + 2$

APPLICATIONS

125. GEOMETRY   The dimensions of the rectangle shown below can be found by factoring the polynomial that represents its area. Find the polynomials that represent the length and the width of the rectangle.

Area = $(x^3 + 4x^2 + 5x + 20)$ ft^2

126. INTERIOR DECORATING   The expression $\pi rs + \pi Rs$ can be used to find the amount of material needed to make the lamp shade shown. Factor the expression.

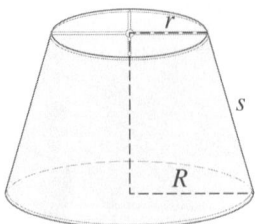

WRITING

127. Explain how to find the GCF of $32a^3$ and $16a^2$.

128. Explain this diagram.

Multiplication $\longrightarrow$

$$3x^2(5x^2 - 6x + 4) = 15x^4 - 18x^3 + 12x^2$$

$\longleftarrow$ Factoring

129. Explain how factorizations of polynomials are checked. Give an example.

130. Explain the error.

Factor out the GCF: $30a^3 - 12a^2 = 6a(5a^2 - 2a)$

REVIEW

131. INSURANCE COSTS   A college student's good grades earned her a student discount on her car insurance premium. What was the percent of decrease, to the nearest percent, if her annual premium was lowered from $1,050 to $925?

132. CALCULATING GRADES   A student has test scores of 68%, 75%, and 79% in a government class. What must she score on the last exam to earn a B (80% or better) in the course?

CHALLENGE PROBLEMS

133. Factor: $6x^{4m}y^n + 21x^{3m}y^{2n} - 15x^{2m}y^{3n}$

134. Factor $ab - b^2 - bc + ac - bc - c^2$ by grouping.

SECTION 6.2
Factoring Trinomials of the Form $x^2 + bx + c$

Objectives

❶ Factor trinomials of the form $x^2 + bx + c$.

❷ Factor trinomials of the form $x^2 + bx + c$ after factoring out the GCF.

❸ Factor trinomials of the form $x^2 + bx + c$ using the grouping method.

In Chapter 5, we learned how to multiply binomials. For example, to multiply $x + 1$ and $x + 2$, we proceed as follows.

$$(x + 1)(x + 2) = x^2 + 2x + x + 2$$
$$= x^2 + 3x + 2$$

To *factor the trinomial* $x^2 + 3x + 2$, we will reverse the multiplication process and determine what factors were multiplied to obtain this result. Since the product of two binomials is often a trinomial, many trinomials factor into two binomials.

Multiplication: Given the binomial factors, we find a trinomial. ⟶

$$(x + 1)(x + 2) = x^2 + 3x + 2$$

⟵ Factoring: Given a trinomial, we find the binomial factors.

The Language of Algebra
Recall that when a polynomial in one variable is written in descending powers of that variable, the coefficient of the first term is called the *leading coefficient*.

To begin the discussion of trinomial factoring, we consider trinomials of the form $x^2 + bx + c$, such as

$$x^2 + 8x + 15, \qquad y^2 - 13y + 12, \qquad a^2 + a - 20, \qquad \text{and} \qquad z^2 - 20z - 21$$

In each case, the **leading coefficient**—the coefficient of the squared variable—is 1.

❶ **Factor Trinomials of the Form $x^2 + bx + c$.**

To develop a method for factoring trinomials, we will find the product of $x + 6$ and $x + 4$ and make some observations about the result.

$$(x + 6)(x + 4) = x \cdot x + x \cdot 4 + 6 \cdot x + 6 \cdot 4 \qquad \text{Use the FOIL method.}$$
$$= x^2 + 4x + 6x + 24$$
$$= x^2 + 10x + 24$$

First term ⎤ Middle term ⎤ Last term

The result is a trinomial, where

The Language of Algebra
If a term of a trinomial is a number only, it is called a *constant term*.

$$x^2 + 10x + 24$$
Constant term

- the first term, x^2, is the product of x and x
- the last term, 24, is the product of 6 and 4
- the coefficient of the middle term, 10, is the sum of 6 and 4

These observations suggest a strategy to use to factor trinomials that have 1 as the leading coefficient.

EXAMPLE 1   Factor: $x^2 + 8x + 15$

Strategy   We will assume that $x^2 + 8x + 15$ is the product of two binomials and we will use a systematic method to find their terms.

Why   Since the terms of $x^2 + 8x + 15$ do not have a common factor (other than 1), the only option available is to try to factor it as the product of two binomials.

Solution   We represent the binomials using two sets of parentheses. Since the first term of the trinomial is x^2, we enter x and x as the first terms of its binomial factors.

$$x^2 + 8x + 15 = \left(x \,\boxed{}\right)\left(x \,\boxed{}\right) \qquad \text{Because } x \cdot x \text{ will give } x^2.$$

The second terms of the binomials must be two integers whose product is 15 and whose sum is 8. Since the integers must have a positive product and a positive sum, we consider only pairs of positive integer factors of 15. The only such pairs, $1 \cdot 15$ and $3 \cdot 5$, are listed in the table. Then we find the sum of each pair and enter each result in the table.

Positive factors of 15	Sum of the positive factors of 15
$1 \cdot 15 = 15$	$1 + 15 = 16$
$3 \cdot 5 = 15$	$3 + 5 = 8$

List all of the pairs of positive integers that multiply to give 15.

Add each pair of factors.

The second row of the table contains the correct pair of integers 3 and 5, whose product is 15 and whose sum is 8. To complete the factorization, we enter 3 and 5 as the second terms of the binomial factors.

$$x^2 + 8x + 15 = (x + 3)(x + 5)$$

We can check the factorization by multiplying:

$$(x + 3)(x + 5) = x^2 + 5x + 3x + 15$$
$$= x^2 + 8x + 15 \qquad \text{This is the original trinomial.}$$

Notation
By the commutative property of multiplication, the order of the binomial factors in a factorization does not matter. Thus, we can also write:

$$x^2 + 8x + 15 = (x + 5)(x + 3)$$

Self Check 1   Factor: $y^2 + 7y + 10$
Now Try   Problem 15

EXAMPLE 2   Factor: $y^2 - 13y + 12$

Strategy   We will assume that $y^2 - 13y + 12$ is the product of two binomials and we will use a systematic method to find their terms.

Why   Since the terms of $y^2 - 13y + 12$ do not have a common factor (other than 1), the only option available is to try to factor it as the product of two binomials.

Solution   We represent the binomials using two sets of parentheses. Since the first term of the trinomial is y^2, the first term of each binomial factor must be y.

$$y^2 - 13y + 12 = \left(y \,\boxed{}\right)\left(y \,\boxed{}\right) \qquad \text{Because } y \cdot y \text{ will give } y^2.$$

$$x^2 + 8x + 15$$
$$5 + 3 = 8$$
$$(x + 5)(x + 3)$$

The second terms of the binomials must be two integers whose product is 12 and whose sum is −13. Since the integers must have a positive product and a negative sum, we only consider pairs of negative integer factors of 12. The possible pairs are listed in the table.

Negative factors of 12	Sum of the negative factors of 12
$-1(-12) = 12$	$-1 + (-12) = -13$
$-2(-6) = 12$	$-2 + (-6) = -8$
$-3(-4) = 12$	$-3 + (-4) = -7$

You can stop listing the factors after finding the correct combination.

The first row of the table contains the correct pair of integers −1 and −12, whose product is 12 and whose sum is −13. To complete the factorization, we enter −1 and −12 as the second terms of the binomial factors.

$$y^2 - 13y + 12 = (y - 1)(y - 12)$$

We check the factorization by multiplying:

$$(y - 1)(y - 12) = y^2 - 12y - y + 12$$
$$= y^2 - 13y + 12 \quad \text{This is the original trinomial.}$$

The Language of Algebra
Make sure you understand this vocabulary: *Many trinomials factor as the product of two binomials.*

Trinomial

Product of two binomials

$$y^2 - 13y + 12 = (y - 1)(y - 12)$$

Self Check 2   Factor: $p^2 - 6p + 8$

Now Try   **Problem 19**

EXAMPLE 3   Factor: $a^2 + a - 20$

Strategy   We will assume that $a^2 + a - 20$ is the product of two binomials and we will use a systematic method to find their terms.

Why   Since the terms of $a^2 + a - 20$ do not have a common factor (other than 1), the only option available is to try to factor it as the product of two binomials.

Solution   We represent the binomials using two sets of parentheses. Since the first term of the trinomial is a^2, the first term of each binomial factor must be a.

$$a^2 + a - 20 = \left(a \,\boxed{}\,\right)\left(a \,\boxed{}\,\right) \quad \text{Because } a \cdot a \text{ will give } a^2.$$

To determine the second terms of the binomials, we must find two integers whose product is −20 and whose sum is 1. Because the integers must have a negative product, their signs must be different. The possible pairs are listed in the table.

It is wise to follow an order when listing the factors in the table so that you don't skip the correct combination. Here, the first factors 1, 2, 4, 5, 10, and 20, are listed from least to greatest.

Factors of −20	Sum of the factors of −20
$1(-20) = -20$	$1 + (-20) = -19$
$2(-10) = -20$	$2 + (-10) = -8$
$4(-5) = -20$	$4 + (-5) = -1$
$5(-4) = -20$	$5 + (-4) = 1$
$10(-2) = -20$	$10 + (-2) = 8$
$20(-1) = -20$	$20 + (-1) = 19$

The fourth row of the table contains the correct pair of integers 5 and -4, whose product is -20 and whose sum is 1. To complete the factorization, we enter 5 and -4 as the second terms of the binomial factors.

$$a^2 + a - 20 = (a + 5)(a - 4)$$

Check the factorization by multiplying.

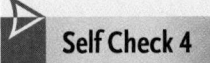

Self Check 3   Factor: $m^2 + m - 42$
Now Try   Problem 27

EXAMPLE 4   Factor: $z^2 - 4z - 21$

Strategy   We will assume that $z^2 - 4z - 21$ is the product of two binomials and we will use a systematic method to find their terms.

Why   Since the terms of $z^2 - 4z - 21$ do not have a common factor (other than 1), the only option available is to try to factor it as the product of two binomials.

Solution   We represent the binomials using two sets of parentheses. Since the first term of the trinomial is z^2, the first term of each binomial factor must be z.

$$z^2 - 4z - 21 = (z \boxed{})(z \boxed{}) \qquad \text{Because } z \cdot z \text{ will give } z^2.$$

To determine the second terms of the binomials, we must find two integers whose product is -21 and whose sum is -4. Because the integers must have a negative product, their signs must be different. The possible pairs are listed in the table.

Factors of -21	Sum of the factors of -21
$1(-21) = -21$	$1 + (-21) = -20$
$3(-7) = -21$	$3 + (-7) = -4$
$7(-3) = -21$	$7 + (-3) = 4$
$21(-1) = -21$	$21 + (-1) = 20$

The second row of the table contains the correct pair of integers 3 and -7, whose product is -21 and whose sum is -4. To complete the factorization, we enter 3 and -7 as the second terms of the binomial factors.

$$z^2 - 4z - 21 = (z + 3)(z - 7)$$

Check by multiplying.

Self Check 4   Factor: $q^2 - 2q - 24$
Now Try   Problem 33

The following guidelines are helpful when factoring trinomials.

Factoring Trinomials Whose Leading Coefficient is 1	To factor a trinomial of the form $x^2 + bx + c$, find two numbers whose product is c and whose sum is b. 1. If c is positive, the numbers have the same sign. 2. If c is negative, the numbers have different signs. Then write the trinomial as a product of two binomials. You can check by multiplying. $$x^2 + bx + c = \left(x \boxed{}\right)\left(x \boxed{}\right)$$ The product of these numbers must be c and their sum must be b.

EXAMPLE 5   Factor: $-h^2 + 2h + 63$

Strategy   We will factor out -1 and then factor the resulting trinomial.

Why   It is easier to factor trinomials that have a positive leading coefficient.

Solution   After factoring out -1, we factor the trinomial within the parentheses.

$$
\begin{aligned}
-h^2 + 2h + 63 &= -1(h^2 - 2h - 63) && \text{Factor out } -1. \\
&= -(h^2 - 2h - 63) && \text{The 1 need not be written.} \\
&= -(h + 7)(h - 9) && \text{Factor } h^2 - 2h - 63.
\end{aligned}
$$

Check:
$$
\begin{aligned}
-(h + 7)(h - 9) &= -(h^2 - 9h + 7h - 63) && \text{Multiply the binomials first.} \\
&= -(h^2 - 2h - 63) && \text{Combine like terms.} \\
&= -h^2 + 2h + 63 && \text{Drop the } - \text{ sign and change the sign} \\
& && \text{of every term within the parentheses.}
\end{aligned}
$$

The result is the original trinomial.

Self Check 5   Factor: $-x^2 + 11x - 28$

Now Try   Problem 37

We can factor trinomials in two variables in a similar way.

EXAMPLE 6   Factor: $x^2 - 4xy - 5y^2$

Strategy   We will assume that $x^2 - 4xy - 5y^2$ is the product of two binomials and we will use a systematic method to find their terms.

Why   Since the terms of $x^2 - 4xy - 5y^2$ do not have a common factor (other than 1), the only option available is to try to factor it as the product of two binomials.

Solution   We represent the binomials using two sets of parentheses. Since the first term of the trinomial is x^2, the first term of each binomial factor must be x. Since the third term contains y^2, the last term of each binomial factor must contain y. To complete the factorization, we need to determine the coefficient of each y-term.

$$x^2 - 4xy - 5y^2 = \left(x \;\boxed{}\; y\right)\left(x \;\boxed{}\; y\right) \quad \text{Because x · x will give } x^2 \text{ and y · y will give } y^2.$$

The coefficients of y must be two integers whose product is -5 and whose sum is -4. Such a pair is 1 and -5. Instead of writing the first factor as $(x + 1y)$, we write it as $(x + y)$, because $1y = y$.

$$x^2 - 4xy - 5y^2 = (x + y)(x - 5y)$$

Check: $(x + y)(x - 5y) = x^2 - 5xy + xy - 5y^2$
$$= x^2 - 4xy - 5y^2 \qquad \text{This is the original trinomial.}$$

 Self Check 6   Factor: $s^2 + 6st - 7t^2$
Now Try   **Problem 45**

2   Factor Trinomials of the Form $x^2 + bx + c$ After Factoring Out the GCF.

If the terms of a trinomial have a common factor, it should be factored out first. A trinomial is **factored completely** when no factor can be factored further.

EXAMPLE 7   Factor completely: $2x^4 + 26x^3 + 80x^2$

Strategy   We will factor out the GCF, $2x^2$, first. Then we will factor the resulting trinomial.

Why   The first step in factoring any polynomial is to factor out the GCF. Factoring out the GCF first makes factoring by any method easier.

Solution   We begin by factoring out the GCF, $2x^2$, from $2x^4 + 26x^3 + 80x^2$.

$$2x^4 + 26x^3 + 80x^2 = 2x^2(x^2 + 13x + 40)$$

Next, we factor $x^2 + 13x + 40$. The integers 8 and 5 have a product of 40 and a sum of 13, so the completely factored form of the given trinomial is

$$2x^4 + 26x^3 + 80x^2 = 2x^2(x + 8)(x + 5) \quad \text{The complete factorization must include } 2x^2.$$

Check: $2x^2(x + 8)(x + 5) = 2x^2(x^2 + 13x + 40)$
$$= 2x^4 + 26x^3 + 80x^2 \qquad \text{This is the original trinomial.}$$

 Self Check 7   Factor completely: $4m^5 + 8m^4 - 32m^3$
Now Try   **Problem 51**

EXAMPLE 8   Factor completely: $-13g^2 + 36g + g^3$

Strategy   We will write the terms of the trinomial in descending powers of g.

Why   It is easier to factor a trinomial if its terms are written in descending powers of one variable.

Solution

$$
\begin{aligned}
-13g^2 + 36g + g^3 &= g^3 - 13g^2 + 36g && \text{Rearrange the terms.} \\
&= g(g^2 - 13g + 36) && \text{Factor out the GCF, } g. \\
&= g(g - 9)(g - 4) && \text{Factor the trinomial.}
\end{aligned}
$$

Check the factorization by multiplying.

> **Caution**
> For multistep factorizations, don't forget to write the GCF in the final factored form.

 Self Check 8   Factor completely: $-12t + t^3 + 4t^2$
         Now Try   **Problem 63**

If a trinomial with integer coefficients cannot be factored using only integers, it is called a **prime trinomial.**

EXAMPLE 9   Factor $x^2 + 2x + 3$, if possible.

Strategy   We will assume that $x^2 + 2x + 3$ is the product of two binomials and we will use a systematic method to find their terms.

Why   Since the terms of $x^2 + 2x + 3$ do not have a common factor (other than 1), the only option available is to try to factor it as the product of two binomials.

Solution   To factor the trinomial, we must find two integers whose product is 3 and whose sum is 2. The possible factorizations are shown in the table.

> **The Language of Algebra**
> When a trinomial is not factorable using only integers, we say it is *prime* and that it does not factor *over* the integers.

Factors of 3	Sum of the factors of 3
$1(3) = 3$	$1 + 3 = 4$
$-1(-3) = 3$	$-1 + (-3) = -4$

Since there are no two integers whose product is 3 and whose sum is 2, the trinomial $x^2 + 2x + 3$ cannot be factored and is a *prime trinomial.*

 Self Check 9   Factor $x^2 - 4x + 6$, if possible.
         Now Try   **Problem 71**

3 **Factor Trinomials of the Form $x^2 + bx + c$ Using the Grouping Method.**

Another way to factor trinomials of the form $x^2 + bx + c$ is to write them as equivalent four-termed polynomials and factor by grouping. To factor $x^2 + 8x + 15$ using this method, we proceed as follows.

1. First, identify b as the coefficient of the x-term, and c as the last term. For trinomials of the form $x^2 + bx + c$, we call c the **key number.**

$$\left. \begin{array}{c} x^2 + bx + c \\ \downarrow \quad\; \downarrow \\ x^2 + 8x + 15 \end{array} \right\} b = 8 \text{ and } c = 15$$

2. Now find two integers whose product is the key number, 15, and whose sum is $b = 8$. Since the integers must have a positive product and a positive sum, we consider only positive factors of 15.

Key number = 15	$b = 8$
Positive factors of 15	Sum of the positive factors of 15
$1 \cdot 15 = 15$	$1 + 15 = 16$
$3 \cdot 5 = 15$	$3 + 5 = 8$

The second row of the table contains the correct pair of integers 3 and 5, whose product is the key number 15 and whose sum is $b = 8$.

3. Express the middle term, $8x$, of the trinomial as the *sum of two terms,* using the integers 3 and 5 found in step 2 as coefficients of the two terms.

$$x^2 + 8x + 15 = x^2 + 3x + 5x + 15 \qquad \text{Express 8x as 3x + 5x.}$$

4. Factor the equivalent four-term polynomial by grouping:

$$x^2 + 3x + 5x + 15 = x(x + 3) + 5(x + 3) \qquad \begin{array}{l}\text{Factor x out of } x^2 + 3x \text{ and 5 out of}\\ 5x + 15.\end{array}$$

$$\qquad\qquad\qquad\quad = (x + 3)(x + 5) \qquad \text{Factor out x + 3.}$$

Check the factorization by multiplying.

The grouping method is an alternative to the method for factoring trinomials discussed earlier in this section. It is especially useful when the constant term, c, has many factors.

Factoring Trinomials of the Form $x^2 + bx + c$ Using Grouping

To factor a trinomial that has a leading coefficient of 1:

1. Identify b and the key number, c.
2. Find two integers whose product is the key number and whose sum is b.
3. Express the middle term, bx, as the sum (or difference) of two terms. Enter the two numbers found in step 2 as coefficients of x in the form shown below. Then factor the equivalent four-term polynomial by grouping.

$$x^2 + \boxed{}\,x + \boxed{}\,x + c$$

The product of these numbers must be c, and their sum must be b.

4. Check the factorization using multiplication.

EXAMPLE 10   Factor by grouping: $a^2 + a - 20$

Strategy   We will express the middle term, a, of the trinomial as the difference of two carefully chosen terms.

Why   We want to produce an equivalent four-term polynomial that can be factored by grouping.

Solution   Since $a^2 + a - 20 = a^2 + 1a - 20$, we identify b as **1** and the key number c as **−20**. We must find two integers whose product is −20 and whose sum is 1. Since the integers must have a negative product, their signs must be different.

<table>
<tr><td colspan="2" align="center">Key number = −20</td><td align="center">$b = 1$</td></tr>
</table>

Factors of −20	Sum of the factors of −20
$1(-20) = -20$	$1 + (-20) = -19$
$2(-10) = -20$	$2 + (-10) = -8$
$4(-5) = -20$	$4 + (-5) = -1$
$5(-4) = -20$	$5 + (-4) = 1$
$10(-2) = -20$	$10 + (-2) = 8$
$20(-1) = -20$	$20 + (-1) = 19$

Success Tip

We could also express the middle term as $-4a + 5a$. We obtain the same binomial factors, but in reverse order.

$a^2 - 4a + 5a - 20$
$= a(a - 4) + 5(a - 4)$
$= (a - 4)(a + 5)$

The fourth row of the table contains the correct pair of integers 5 and −4, whose product is −20 and whose sum is 1. They serve as the coefficients of $5a$ and $-4a$, the two terms that we use to represent the middle term, a, of the trinomial.

$a^2 + a - 20 = a^2 + 5a - 4a - 20$ Express the middle term, a, as 5a − 4a.

$\qquad\qquad = a(a + 5) - 4(a + 5)$ Factor a out of $a^2 + 5a$ and −4 out of −4a − 20.

$\qquad\qquad = (a + 5)(a - 4)$ Factor out a + 5.

Check the factorization by multiplying.

Self Check 10   Factor by grouping: $m^2 + m - 42$

Now Try   Problem 27

EXAMPLE 11   Factor by grouping: $x^2 - 4xy - 5y^2$

Strategy   We will express the middle term, $-4xy$, of the trinomial as the sum of two carefully chosen terms.

Why   We want to produce an equivalent four-term polynomial that can be factored by grouping.

Solution   In $x^2 - 4xy - 5y^2$, we identify b as −4 and the key number c as **−5**. We must find two integers whose product is −5 and whose sum is −4. Such a pair is −5 and 1. They

<table>
<tr><td align="center">Key number = −5</td><td align="center">$b = -4$</td></tr>
<tr><td align="center">Factors</td><td align="center">Sum</td></tr>
<tr><td align="center">$-5(1) = -5$</td><td align="center">$-5 + 1 = -4$</td></tr>
</table>

serve as the coefficients of $-5xy$ and $1xy$, the two terms that we use to represent the middle term, $-4xy$, of the trinomial.

$$x^2 - 4xy - 5y^2 = x^2 - 5xy + 1xy - 5y^2$$

Express the middle term, $-4xy$, as $-5xy + 1xy$. ($1xy - 5xy$ could also be used.)

$$= x(x - 5y) + y(x - 5y)$$

Factor x out of $x^2 - 5xy$ and y out of $1xy - 5y^2$.

$$= (x - 5y)(x + y)$$

Factor out x − 5y.

Check the factorization by multiplying.

 Self Check 11   Factor by grouping: $q^2 - 2qt - 24t^2$

Now Try   **Problem 45**

EXAMPLE 12   Factor completely: $2x^3 - 20x^2 + 18x$

Strategy   We will factor out the GCF, $2x$, first. Then we will factor the resulting trinomial using the grouping method.

Why   The first step in factoring any polynomial is to factor out the GCF.

Solution   We begin by factoring out the GCF, $2x$, from $2x^3 - 20x^2 + 18x$.

$$2x^3 - 20x^2 + 18x = 2x(x^2 - 10x + 9)$$

Key number = 9	$b = -10$
Factors	Sum
$-9(-1) = 9$	$-9 + (-1) = -10$

To factor $x^2 - 10x + 9$ by grouping, we must find two integers whose product is the key number 9 and whose sum is $b = -10$. Such a pair is -9 and -1.

$$x^2 - 10x + 9 = x^2 - 9x - 1x + 9$$

Express −10x as −9x − 1x. (−1x − 9x could also be used.)

$$= x(x - 9) - 1(x - 9)$$

Factor x out of $x^2 − 9x$ and −1 out of −1x + 9.

$$= (x - 9)(x - 1)$$

Factor out x − 9.

The complete factorization of the original trinomial is

$$2x^3 - 20x^2 + 18x = 2x(x - 9)(x - 1)$$

Don't forget to write the GCF, 2x.

Check the factorization by multiplying.

 Self Check 12   Factor completely: $3m^3 - 27m^2 + 24m$

Now Try   **Problem 51**

 ANSWERS TO SELF CHECK   **1.** $(y + 2)(y + 5)$ **2.** $(p - 2)(p - 4)$ **3.** $(m + 7)(m - 6)$
4. $(q + 4)(q - 6)$ **5.** $-(x - 4)(x - 7)$ **6.** $(s + 7t)(s - t)$ **7.** $4m^3(m + 4)(m - 2)$
8. $t(t - 2)(t + 6)$ **9.** Prime trinomial **10.** $(m + 7)(m - 6)$ **11.** $(q + 4t)(q - 6t)$
12. $3m(m - 8)(m - 1)$

STUDY SET
6.2

VOCABULARY

Fill in the blanks.

1. The trinomial $x^2 - x - 12$ _____ as the product of two binomials: $(x - 4)(x + 3)$.
2. A _____ trinomial cannot be factored by using only integers.
3. The _____ coefficient of $x^2 - 3x + 2$ is 1.
4. A trinomial is factored _____ when no factor can be factored further.

CONCEPTS

Fill in the blanks.

5. a. Before attempting to factor a trinomial, be sure that it is written in _____ powers of a variable.
   b. Before attempting to factor a trinomial into two binomials, always factor out any _____ factors first.

6. $x^2 + x - 56 = \left(x \;\boxed{}\right)\left(x \;\boxed{}\right)$

   The product of these numbers must be ____ ,
   and their sum must be ____ .

7. $x^2 + 5x + 3$ cannot be factored because we cannot find two integers whose product is ____ and whose sum is ____ .

8. Complete the following table.

Factors of 8	Sum of the factors of 8
1(8)	
2( )	
−1(−8)	
(−4)	

9. Check to determine whether each factorization is correct.
   a. $x^2 - x - 20 = (x + 5)(x - 4)$
   b. $4a^2 + 12a - 16 = 4(a - 1)(a + 4)$

10. Find two integers whose
    a. product is 10 and whose sum is 7.
    b. product is 8 and whose sum is −6.
    c. product is −6 and whose sum is 1.
    d. product is −9 and whose sum is −8.

11. Consider a trinomial of the form $x^2 + bx + c$.
    a. If c is positive, what can be said about the two integers that should be chosen for the factorization?

    b. If c is negative, what can be said about the two integers that should be chosen for the factorization?

12. Fill in each blank to explain how to factor $x^2 + 7x + 10$ by grouping.

    We express the middle term, $7x$, as the sum of ____ terms:
    $$x^2 + 7x + 10 = x^2 + \boxed{}x + \boxed{}x + 10$$

    The product of these numbers must be ____ ,
    and their sum must be ____ .

NOTATION

13. To factor a trinomial, a student made a table and circled the correct pair of integers, as shown. Complete the factorization of the trinomial.

    $(x \quad)(x \quad)$

Factors	Sum
1(−6)	−5
2(−3)	−1
③(−2)	①
6(−1)	5

14. To factor a trinomial by grouping, a student made a table and circled the correct pair of integers, as shown. Enter the correct coefficients.

    $x^2 + \quad x + \quad x + 16$

    Key number = 16

Factors	Sum
1 · 16	17
②·8	⑩
4 · 4	8

GUIDED PRACTICE

Factor each trinomial. See Examples 1 and 2 or Example 10.

15. $x^2 + 3x + 2$
16. $y^2 + 4y + 3$
17. $z^2 + 7z + 12$
18. $x^2 + 7x + 10$
19. $m^2 - 5m + 6$
20. $n^2 - 7n + 10$
21. $t^2 - 11t + 28$
22. $c^2 - 9c + 8$
23. $r^2 - 9r + 18$
24. $y^2 - 17y + 72$
25. $a^2 - 46a + 45$
26. $r^2 - 37r + 36$

Factor each trinomial. See Examples 3 and 4 or Example 10.

27. $x^2 + 5x - 24$

28. $u^2 + u - 42$

29. $t^2 + 13t - 48$

30. $m^2 + 2m - 48$

31. $a^2 - 6a - 16$

32. $a^2 - 10a - 39$

33. $b^2 - 9b - 36$

34. $x^2 - 3x - 40$

Factor each trinomial. See Example 5.

35. $-x^2 - 7x - 10$

36. $-x^2 + 9x - 20$

37. $-t^2 - t + 30$

38. $-t^2 - 15t + 34$

39. $-r^2 - 3r + 54$

40. $-d^2 - 2d + 63$

41. $-m^2 + 18m - 77$

42. $-n^2 + 14n - 33$

Factor each trinomial. See Example 6 or Example 11.

43. $a^2 + 4ab + 3b^2$

44. $a^2 + 6ab + 5b^2$

45. $x^2 - 6xy - 7y^2$

46. $x^2 + 10xy - 11y^2$

47. $r^2 + sr - 2s^2$

48. $m^2 + mn - 6n^2$

49. $a^2 - 5ab + 6b^2$

50. $p^2 - 7pq + 10q^2$

Factor completely. See Example 7 or Example 12.

51. $2x^2 + 10x + 12$

52. $3y^2 - 21y + 18$

53. $6a^2 - 30a + 24$

54. $4b^2 + 12b - 16$

55. $5a^2 - 25a + 30$

56. $2b^2 - 20b + 18$

57. $-z^3 + 29z^2 - 100z$

58. $-m^3 + m^2 + 56m$

59. $-n^4 + 28n^3 + 60n^2$

60. $-c^5 + 16c^4 + 80c^3$

61. $4x^4 + 16x^3 + 16x^2$

62. $3a^4 + 30a^3 + 75a^2$

Write each trinomial in descending powers of one variable and factor. See Example 8.

63. $80 - 24x + x^2$

64. $y^2 + 100 + 25y$

65. $10y + 9 + y^2$

66. $x^2 - 13 - 12x$

67. $r^2 - 16 + 6r$

68. $u^2 - 12 - u$

69. $4rx + r^2 + 3x^2$

70. $a^2 + 5b^2 + 6ab$

Factor each trinomial, if possible. See Example 9.

71. $u^2 + 10u + 15$

72. $v^2 + 9v + 15$

73. $r^2 + 2r - 4$

74. $r^2 - 9r - 12$

TRY IT YOURSELF

Choose the correct method from Section 6.1 or Section 6.2 to factor completely each of the following.

75. $5x + 15 + xy + 3y$

76. $ab + b + 2a + 2$

77. $26n^2 - 8n$

78. $40c^2 - 12c$

79. $a^2 - 4a - 5$

80. $t^2 - 5t - 50$

81. $-x^2 + 21x + 22$

82. $-r^2 + 14r - 45$

83. $4xy - 4x + 28y - 28$

84. $3xy - 3x + 15y - 15$

85. $24b^4 - 48b^3 + 36b^2$

86. $28n^5 - 42n^4 + 28n^3$

87. $x^2 + 4xy + 4y^2$

88. $m^2 - 8mn + 16n^2$

89. $a^2 - 4ab - 12b^2$

90. $p^2 + pq - 6q^2$

91. $r^2 - 2r + 4$

92. $m^2 + 3m - 20$

93. $t(x + 2) + 7(x + 2)$

94. $r(t - v) + 10(t - v)$

95. $s^4 + 11s^3 - 26s^2$

96. $x^4 + 14x^3 + 45x^2$

97. $15s^3 + 75$

98. $33g^4 - 99$

99. $-13y + y^2 - 14$

100. $-3a + a^2 + 2$

101. $2x^2 - 12x + 16$

102. $6t^2 - 18t - 24$

APPLICATIONS

103. **PETS**   The cage shown on the next page is used for transporting dogs. Its volume is $(x^3 + 12x^2 + 27x)$ in.^{3}. The dimensions of the cage can be found by factoring. If the cage is longer than it is tall and taller than it is wide, find the polynomials that represent its length, width, and height.

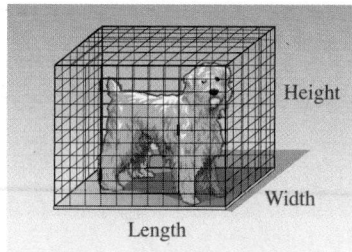

Height

Width

Length

104. PHOTOGRAPHY   A picture cube is a clever way to display 6 photographs in a small amount of space. Suppose the surface area of the entire cube is given by the polynomial $(6s^2 + 12s + 6)$ in.^{2}. Find the polynomial that represents the length of an edge of the cube.

WRITING

105. Explain what it means when we say that a trinomial is the product of two binomials. Give an example.

106. Are $2x^2 - 12x + 16$ and $x^2 - 6x + 8$ factored in the same way? Explain.

107. When factoring $x^2 - 2x - 3$, one student got $(x - 3)(x + 1)$, and another got $(x + 1)(x - 3)$. Are both answers acceptable? Explain.

108. In the partial solution shown below, a student began to factor the trinomial. Write a note to the student explaining his mistake.

Factor: $x^2 - 2x - 63$

$(x -)(x -)$

109. Explain the error in the following factorization.

$$x^3 + 8x^2 + 15x = x(x^2 + 8x + 15)$$
$$= (x + 3)(x + 5)$$

110. Explain why the factorization is not complete.

$$2y^2 - 12y + 16 = 2(y^2 - 6y + 8)$$

REVIEW

Simplify each expression. Write each answer without negative exponents.

111. $\dfrac{x^{12}x^{-7}}{x^3 x^4}$

112. $\dfrac{a^4 a^{-2}}{a^2 a^0}$

113. $(x^{-3}x^{-2})^2$

114. $\left(\dfrac{18a^2 b^3 c^{-4}}{3a^{-1}b^2 c}\right)^{-3}$

CHALLENGE PROBLEMS

Factor completely.

115. $x^2 - \dfrac{6}{5}x + \dfrac{9}{25}$

116. $x^2 - 0.5x + 0.06$

117. $x^{2m} - 12x^m - 45$

118. $x^2(y + 1) - 3x(y + 1) - 70(y + 1)$

119. Find all positive integer values of c that make $n^2 + 6n + c$ factorable.

120. Find all integer values of b that make $x^2 + bx - 44$ factorable.

SECTION 6.3
Factoring Trinomials of the Form $ax^2 + bx + c$

Objectives

1 Factor trinomials using the trial-and-check method.

2 Factor trinomials after factoring out the GCF.

3 Factor trinomials using the grouping method.

In this section we will factor trinomials with leading coefficients other than 1, such as

$$2x^2 + 5x + 3, \qquad 6a^2 - 17a + 5, \qquad \text{and} \qquad 4b^2 + 8bc - 45c^2$$

We can use two methods to factor these trinomials. With the first method, we make educated guesses and then check them using multiplication. The correct factorization is determined through a process of elimination. The second method is an extension of factoring by grouping.

Factor Trinomials Using the Trial-and-Check Method.

> ### EXAMPLE 1   Factor: $2x^2 + 5x + 3$
>
> **Strategy**   We will assume that $2x^2 + 5x + 3$ is the product of two binomials and we will use a systematic method to find their terms.
>
> **Why**   Since the terms of $2x^2 + 5x + 3$ do not have a common factor (other than 1), the only option available is to try to factor it as the product of two binomials.
>
> **Solution**   We represent the binomials using two sets of parentheses. Since the first term of the trinomial is $2x^2$, we enter $2x$ and x as the first terms of the binomial factors.
>
> $$\left(2x\ \boxed{}\right)\left(x\ \boxed{}\right)$$ Because $2x \cdot x$ will give $2x^2$.
>
> The second terms of the binomials must be two integers whose product is 3. Since the coefficients of the terms of $2x^2 + 5x + 3$ are positive, we only consider pairs of positive integer factors of 3. Since there is just one such pair, $1 \cdot 3$, we can enter 1 and 3 as the second terms of the binomials, or we can reverse the order and enter 3 and 1.
>
> $$(2x + 1)(x + 3) \qquad \text{or} \qquad (2x + 3)(x + 1)$$
>
> The first possibility is incorrect, because when we find the outer and inner products and combine like terms, we obtain an incorrect middle term of $7x$.
>
> Outer: 6x
> $$(2x + 1)(x + 3)$$ Multiply and add to find the middle term: $6x + x = 7x$.
> Inner: x
>
> The second possibility is correct, because it gives a middle term of $5x$.
>
> Outer: 2x
> $$(2x + 3)(x + 1)$$ Multiply and add to find the middle term: $2x + 3x = 5x$.
> Inner: 3x
>
> Thus,
>
> $$2x^2 + 5x + 3 = (2x + 3)(x + 1)$$
>
> Check the factorization by multiplying:
>
> $$(2x + 3)(x + 1) = 2x^2 + 2x + 3x + 3$$
> $$= 2x^2 + 5x + 3 \qquad \text{This is the original trinomial.}$$
>
> **Self Check 1**   Factor: $2x^2 + 5x + 2$
> **Now Try**   Problem 19

The Language of Algebra
To *interchange* means to put each in the place of the other. We create all of the possible factorizations by *interchanging* the second terms of the binomials.

$$(2x + 1)(x + 3)$$
$$(2x + 3)(x + 1)$$

> ### EXAMPLE 2   Factor: $6a^2 - 17a + 5$
>
> **Strategy**   We will assume that $6a^2 - 17a + 5$ is the product of two binomials and we will use a systematic method to find their terms.

Why Since the terms of $6a^2 - 17a + 5$ do not have a common factor (other than 1), the only option available is to try to factor it as the product of two binomials.

Solution We represent the binomials using two sets of parentheses. Since the first term is $6a^2$, the first terms of the factors must be $6a$ and a or $3a$ and $2a$.

$$\left(6a\ \boxed{}\right)\left(a\ \boxed{}\right) \quad \text{or} \quad \left(3a\ \boxed{}\right)\left(2a\ \boxed{}\right)$$ Because $6a \cdot a$ or $3a \cdot 2a$ will give $6a^2$.

The second terms of the binomials must be two integers whose product is 5. Since the last term of $6a^2 - 17a + 5$ is positive and the coefficient of the middle term is negative, we only consider negative integer factors of the last term. Since there is just one such pair, $-1(-5)$, we can enter -1 and -5, or we can reverse the order and enter -5 and -1 as second terms of the binomials.

$$(6a - 1)(a - 5) \quad -30a - a = -31a \qquad (6a - 5)(a - 1) \quad -6a - 5a = -11a$$

$$(3a - 1)(2a - 5) \quad -15a - 2a = -17a \qquad (3a - 5)(2a - 1) \quad -3a - 10a = -13a$$

Only the possibility shown in blue gives the correct middle term of $-17a$. Thus,

$$6a^2 - 17a + 5 = (3a - 1)(2a - 5)$$

We check by multiplying: $(3a - 1)(2a - 5) = 6a^2 - 17a + 5$.

Self Check 2   Factor: $6b^2 - 19b + 3$
Now Try   Problem 27

EXAMPLE 3   Factor: $3y^2 - 7y - 6$

Strategy We will assume that $3y^2 - 7y - 6$ is the product of two binomials and we will use a systematic method to find their terms.

Why Since the terms of $3y^2 - 7y - 6$ do not have a common factor (other than 1), the only option available is to try to factor it as the product of two binomials.

Solution Since the first term is $3y^2$, the first terms of the binomial factors must be $3y$ and y.

$$\left(3y\ \boxed{}\right)\left(y\ \boxed{}\right)$$ Because $3y \cdot y$ will give $3y^2$.

The second terms of the binomials must be two integers whose product is -6. There are four such pairs: $1(-6)$, $-1(6)$, $2(-3)$, and $-2(3)$. When these pairs are entered, and then reversed as second terms of the binomials, there are eight possibilities to consider. Four of them can be discarded because they include a binomial whose terms have a common factor. If the terms of $3y^2 - 7y - 6$ do not have a common factor (other than 1), neither can any of its binomial factors.

Success Tip

If the terms of a trinomial do not have a common factor other than 1, the terms of each of its binomial factors will not have a common factor other than 1.

For 1 *and* −6:
$$\overset{-18y}{(3y + 1)(y - 6)} \qquad \text{or} \qquad \cancel{(3y - 6)(y + 1)}$$
$$\underset{y}{}$$
$$-18y + y = -17y \qquad\qquad 3y - 6 \text{ has a common factor of 3.}$$

For −1 *and* 6:
$$\overset{18y}{(3y - 1)(y + 6)} \qquad \text{or} \qquad \cancel{(3y + 6)(y - 1)}$$
$$\underset{-y}{}$$
$$18y - y = 17y \qquad\qquad 3y + 6 \text{ has a common factor of 3.}$$

Success Tip

Reversing the signs within the binomial factors reverses the sign of the middle term. For example, the factors 2 and −3 give the middle term −7y, while −2 and 3 give the middle term 7y.

For 2 *and* −3:
$$\overset{-9y}{(3y + 2)(y - 3)} \qquad \text{or} \qquad \cancel{(3y - 3)(y + 2)}$$
$$\underset{2y}{}$$
$$-9y + 2y = -7y \qquad\qquad 3y - 3 \text{ has a common factor of 3.}$$

For −2 *and* 3:
$$\overset{9y}{(3y - 2)(y + 3)} \qquad \text{or} \qquad \cancel{(3y + 3)(y - 2)}$$
$$\underset{-2y}{}$$
$$9y - 2y = 7y \qquad\qquad 3y + 3 \text{ has a common factor of 3.}$$

Success Tip

All eight possible factorizations are listed. In practice, you will often find the correct factorization without having to examine the entire list of possibilities.

Only the possibility shown in green gives the correct middle term of −7y. Thus,

$$3y^2 - 7y - 6 = (3y + 2)(y - 3)$$

Check the factorization by multiplying.

> **Self Check 3**   Factor: $5t^2 - 23t - 10$
>
> **Now Try**   **Problem 35**

EXAMPLE 4   Factor: $4b^2 + 8bc - 45c^2$

Strategy   We will assume that $4b^2 + 8bc - 45c^2$ is the product of two binomials and we will use a systematic method to find their terms.

Why   Since the terms of $4b^2 + 8bc - 45c^2$ do not have a common factor (other than 1), the only option available is to try to factor it as the product of two binomials.

Notation

The trinomial is in two variables, b and c. It is written in descending powers of b and ascending powers of c.

Solution   Since the first term is $4b^2$, the first terms of the binomial factors must be $4b$ and b or $2b$ and $2b$. Since the last term contains c^2, the second terms of the binomial factors must contain c.

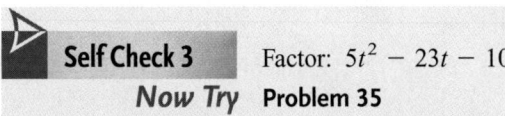

$\left(4b \ \boxed{} \ c\right)\left(b \ \boxed{} \ c\right)$ or $\left(2b \ \boxed{} \ c\right)\left(2b \ \boxed{} \ c\right)$ *Because 4b · b or 2b · 2b gives 4b², and because c · c gives c².*

The coefficients of c must be two integers whose product is −45. Since the coefficient of the last term is negative, the signs of the integers must be different. If we pick factors of $4b$ and b for the first terms, and −1 and 45 for the coefficients of c, the multiplication gives an incorrect middle term of 179bc.

$$180bc$$

$$(4b - c)(b + 45c) \qquad 180bc - bc = 179bc$$

$$-bc$$

If we pick factors of $4b$ and b for the first terms, and 15 and -3 for the coefficients of c, the multiplication gives an incorrect middle term of $3bc$.

$$-12bc$$

$$(4b + 15c)(b - 3c) \qquad -12bc + 15bc = 3bc$$

$$15bc$$

If we pick factors of $2b$ and $2b$ for the first terms, and -5 and 9 for the coefficients of c, we have

$$18bc$$

$$(2b - 5c)(2b + 9c) \qquad 18bc - 10bc = 8bc$$

$$-10bc$$

which gives the correct middle term of $8bc$. Thus,

$$4b^2 + 8bc - 45c^2 = (2b - 5c)(2b + 9c)$$

Check the factorization by multiplying.

Self Check 4   Factor: $4x^2 + 4xy - 3y^2$

Now Try   Problem 43

Because guesswork is often necessary, it is difficult to give specific rules for factoring trinomials with leading coefficients other than 1. However, the following hints are helpful when using the **trial-and-check method.**

Factoring Trinomials with Leading Coefficients Other Than 1

To factor trinomials with leading coefficients other than 1:

1. Factor out any GCF (including -1 if that is necessary to make a positive in a trinomial of the form $ax^2 + bx + c$).

2. Write the trinomial as a product of two binomials. The coefficients of the first terms of each binomial factor must be factors of a, and the last terms must be factors of c.

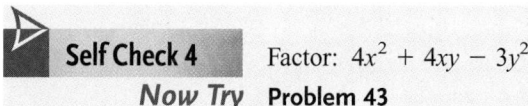

The product of these numbers must be a.

$$ax^2 + bx + c = (\square x \ \square)(\square x \ \square)$$

The product of these numbers must be c.

3. If c is positive, the signs within the binomial factors match the sign of b. If c is negative, the signs within the binomial factors are opposites.

4. Try combinations of first terms and second terms until you find the one that gives the proper middle term. If no combination works, the trinomial is prime.

5. Check by multiplying.

Factor Trinomials After Factoring Out the GCF.

If the terms of a trinomial have a common factor, the GCF (or the opposite of the GCF) should always be factored out first.

EXAMPLE 5   Factor: $2x^2 - 8x^3 + 3x$

Strategy   We will write the trinomial in descending powers of x and factor out the common factor , $-x$.

Why   It is easier to factor trinomials that have a positive leading coefficient.

Solution   Write the trinomial in descending powers of x: $-8x^3 + 2x^2 + 3x$.

$$-8x^3 + 2x^2 + 3x = -x(8x^2 - 2x - 3)$$ Factor out the opposite of the GCF, $-x$.

> **The Language of Algebra**
> When asked to *factor* a polynomial, that means we should *factor completely*.

We now factor $8x^2 - 2x - 3$. Its factorization has the form

$$\left(8x \;\boxed{}\right)\left(x \;\boxed{}\right) \text{ or } \left(2x \;\boxed{}\right)\left(4x \;\boxed{}\right)$$ Because $8x \cdot x$ or $4x \cdot 2x$ gives $8x^2$.

The second terms of the binomials must be two integers whose product is -3. There are two such pairs: $1(-3)$ and $-1(3)$. Since the coefficient of middle term $-2x$ is small, we pick the smaller factors of $8x^2$, which are $2x$ and $4x$, for the first terms and 1 and -3 for the second terms.

$$\overset{\displaystyle -6x}{(2x + 1)(4x - 3)}\underset{\displaystyle 4x}{} \qquad -6x + 4x = -2x$$

This factorization gives the correct middle term of $-2x$. Thus,

$$8x^2 - 2x - 3 = (2x + 1)(4x - 3)$$

We can now give the complete factorization of the original trinomial.

> **Caution**
> For multistep factorization, don't forget to write the GCF (or its opposite) in the final factored form.

$$\begin{aligned}-8x^3 + 2x^2 + 3x &= -x(8x^2 - 2x - 3) \\ &= -x(2x + 1)(4x - 3)\end{aligned}$$

Check the factorization by multiplying.

 Self Check 5   Factor: $12y - 2y^3 - 2y^2$

Now Try   **Problem 53**

---

❸ **Factoring Trinomials Using the Grouping Method.**

Another way to factor a trinomial of the form $ax^2 + bx + c$ is to write it as an equivalent four-termed polynomial and factor it by grouping. For example, to factor $2x^2 + 5x + 3$, we proceed as follows.

1. Identify the values of a, b, and c.

$$\left.\begin{array}{ccc} ax^2 & + bx & + c \\ \downarrow & \downarrow & \downarrow \\ 2x^2 & + 5x & + 3 \end{array}\right\} a = 2, \, b = 5, \text{ and } c = 3$$

Then, find the product ac, called the **key number:** $ac = 2(3) = 6$.

2. Next, find two integers whose product is $ac = 6$ and whose sum is $b = 5$. Since the integers must have a positive product and a positive sum, we consider only positive factors of 6.

Key number = 6 $b = 5$

Positive factors of 6	Sum of the positive factors of 6
$1 \cdot 6 = 6$	$1 + 6 = 7$
$2 \cdot 3 = 6$	$2 + 3 = 5$

The second row of the table contains the correct pair of integers 2 and 3, whose product is 6 and whose sum is 5.

3. Express the middle term, $5x$, of the trinomial as the *sum of two terms,* using the integers 2 and 3 found in step 2 as coefficients of the two terms.

$$2x^2 + 5x + 3 = 2x^2 + 2x + 3x + 3 \quad \text{Express 5x as 2x + 3x.}$$

4. Factor the equivalent four-term polynomial by grouping:

$$2x^2 + 2x + 3x + 3 = 2x(x + 1) + 3(x + 1) \quad \text{Factor 2x out of } 2x^2 + 2x \text{ and 3 out of 3x + 3.}$$

$$= (x + 1)(2x + 3) \quad \text{Factor out x + 1.}$$

Check by multiplying.

Factoring by grouping is especially useful when the leading coefficient, a, and the constant term, c, have many factors.

Factoring Trinomials by Grouping

To factor a trinomial by grouping:

1. Factor out any GCF (including -1 if that is necessary to make a positive in a trinomial of the form $ax^2 + bx + c$).
2. Identify a, b, and c, and find the key number ac.
3. Find two integers whose product is the key number and whose sum is b.
4. Express the middle term, bx, as the sum (or difference) of two terms. Enter the two numbers found in step 3 as coefficients of x in the form shown below. Then factor the equivalent four-term polynomial by grouping.

$$ax^2 + \boxed{}x + \boxed{}x + c$$

The product of these numbers must be ac and their sum must be b.

5. Check the factorization by multiplying.

EXAMPLE 6   Factor by grouping: $10x^2 + 13x - 3$

Strategy   We will express the middle term, $13x$, of the trinomial as the sum of two carefully chosen terms.

Why   We want to produce an equivalent four-term polynomial that can be factored by grouping.

Solution   In $10x^2 + 13x - 3$, we have $a = 10$, $b = 13$, and $c = -3$. The key number is $ac = 10(-3) = -30$. We must find a factorization of -30 such that the sum of the factors is $b = 13$. The possible factor pairs are listed in the table. Since the factors must have a negative product, their signs must be different.

Key number $= -30$ $b = 13$

Factors of -30	Sum of the factors of -30
$1(-30) = -30$	$1 + (-30) = -29$
$2(-15) = -30$	$2 + (-15) = -13$
$3(-10) = -30$	$3 + (-10) = -7$
$5(-6) = -30$	$5 + (-6) = -1$
$6(-5) = -30$	$6 + (-5) = 1$
$10(-3) = -30$	$10 + (-3) = 7$
$\mathbf{15(-2) = -30}$	$15 + (-2) = 13$
$30(-1) = -30$	$30 + (-1) = 29$

It is wise to follow an order when listing the factors in the table so that you don't skip the correct combination. Here, the first factors 1, 2, 3, 5, 6, 10, 15, and 30 are listed from least to greatest.

The seventh row contains the correct pair of numbers 15 and -2, whose product is -30 and whose sum is 13. They serve as the coefficients of $15x$ and $-2x$, the two terms that we use to represent the middle term, $13x$, of the trinomial.

$$10x^2 + 13x - 3 = 10x^2 + \mathbf{15x - 2x} - 3 \qquad \text{Express 13x as 15x − 2x.}$$

Finally, we factor by grouping.

$$10x^2 + 15x - 2x - 3 = 5x(2x + 3) - 1(2x + 3) \qquad \begin{array}{l}\text{Factor out 5x from } 10x^2 + 15x.\\ \text{Factor out } -1 \text{ from } -2x - 3.\end{array}$$

$$= (2x + 3)(5x - 1) \qquad \text{Factor out 2x + 3.}$$

So $10x^2 + 13x - 3 = (2x + 3)(5x - 1)$. Check the factorization by multiplying.

Self Check 6   Factor by grouping: $15a^2 + 17a - 4$

Now Try   Problems 19, 27, and 35

EXAMPLE 7   Factor: $12x^5 - 17x^4 + 6x^3$

Strategy   We will factor out the GCF, x^3, first. Then we will factor the resulting trinomial using the grouping method.

Why   The first step in factoring any polynomial is to factor out the GCF.

Solution   The GCF of the three terms of the trinomial is x^3.

$$12x^5 - 17x^4 + 6x^3 = x^3(12x^2 - 17x + 6)$$

To factor $12x^2 - 17x + 6$, we must find two integers whose product is $12(6) = 72$ and whose sum is -17. Two such numbers are -8 and -9. They serve as the coefficients of $-8x$ and $-9x$, the two terms that we use to represent the middle term, $-17x$, of the trinomial.

Key number $= 72$ $b = -17$

Factors	Sum
$-8(-9) = 72$	$-8 + (-9) = -17$

$$12x^2 - 17x + 6 = 12x^2 - 8x - 9x + 6$$

Express $-17x$ as $-8x - 9x$.
($-9x - 8x$ could also be used.)

$$= 4x(3x - 2) - 3(3x - 2)$$

Factor out $4x$ and factor out -3.

$$= (3x - 2)(4x - 3)$$

Factor out $3x - 2$.

The complete factorization of the original trinomial is

$$12x^5 - 17x^4 + 6x^3 = x^3(3x - 2)(4x - 3)$$ Don't forget to write the GCF, x^3.

Check the factorization by multiplying.

 Self Check 7   Factor: $21a^4 - 13a^3 + 2a^2$
Now Try   **Problem 53**

ANSWERS TO SELF CHECKS   **1.** $(2x + 1)(x + 2)$ **2.** $(6b - 1)(b - 3)$ **3.** $(5t + 2)(t - 5)$
4. $(2x + 3y)(2x - y)$ **5.** $-2y(y + 3)(y - 2)$ **6.** $(3a + 4)(5a - 1)$ **7.** $a^2(7a - 2)(3a - 1)$

STUDY SET
6.3

VOCABULARY

Fill in the blanks.

1. The _____ coefficient of $3x^2 - x - 12$ is 3.

2. Given $5y^2 + 16y + 3 = (5y + 1)(y + 3)$. We say that $5y^2 + 16y + 3$ factors as the product of two _____.

3. The first terms of the binomial factors $(5y + 1)(y + 3)$ are ___ and ___. The second terms of the binomial factors are ___ and ___.

4. To factor $2m^2 + 11m + 12$ by _____, we write it as $2m^2 + 8m + 3m + 12$.

CONCEPTS

5. If $10x^2 - 27x + 5$ is to be factored as the product of two binomials, what are the possible *first terms* of the binomial factors?

6. Complete each sentence.

The product of these
numbers must be ___ .

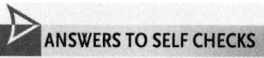

$$5x^2 + 6x - 8 = (\square x \; \square)(\square x \; \square)$$

The product of these
numbers must be ___ .

7. a. Fill in the blanks. When factoring a trinomial, we write it in _____ powers of the variable. Then we factor out any _____ (including -1 if that is necessary to make the lead _____ positive).

b. What is the GCF of the terms of $6s^4 + 33s^3 + 36s^2$?

c. Factor out -1 from $-2d^2 + 19d - 8$.

8. Check to determine whether $(3t - 1)(5t - 6)$ is the correct factorization of $15t^2 - 19t + 6$.

A trinomial has been partially factored. Complete each statement that describes the type of integers we should consider for the blanks.

9. $5y^2 - 13y + 6 = \left(5y \; \boxed{}\right)\left(y \; \boxed{}\right)$
Since the last term of the trinomial is positive and the middle term is negative, the integers must be _____ factors of 6.

10. $5y^2 + 13y + 6 = \left(5y \; \boxed{}\right)\left(y \; \boxed{}\right)$
Since the last term of the trinomial is positive and the middle term is positive, the integers must be _____ factors of 6.

11. $5y^2 - 7y - 6 = \left(5y \; \boxed{}\right)\left(y \; \boxed{}\right)$
Since the last term of the trinomial is negative, the signs of the integers will be _____.

12. $5y^2 + 7y - 6 = \left(5y \; \boxed{}\right)\left(y \; \boxed{}\right)$
Since the last term of the trinomial is negative, the signs of the integers will be _____.

13. Complete the key number table.

Negative factors of 12	Sum of the negative factors of 12
$-1(-12)$	
$-2(\quad)$	
(-4)	

14. Complete the sentence to explain how to factor $3x^2 + 16x + 5$ by grouping.

$$3x^2 + 16x + 5 = 3x^2 + \square\, x + \square\, x + 5$$

The product of these numbers must be ___ and their sum must be ___.

NOTATION

15. a. Suppose we wish to factor $12b^2 + 20b - 9$ by grouping. Identify a, b, and c.
b. What is the key number, ac?

16. To factor $6x^2 + 13x + 6$ by grouping, a student made a table and circled the correct pair of integers, as shown. Enter the correct coefficients.

$$6x^2 + \quad x + \quad x + 6$$

$ac = 36 \quad b = 13$

Factors	Sum
$1 \cdot 36$	37
$2 \cdot 18$	20
$3 \cdot 12$	15
$4 \cdot 9$	13
$6 \cdot 6$	12

Complete each step of the factorization of the trinomial by grouping.

17. $12t^2 + 17t + 6 = 12t^2 + 9t + 8t + 6$
$$= \quad (4t + 3) + \quad (4t + 3)$$
$$= (\quad)(3t + 2)$$

18. $35t^2 - 11t - 6 = 35t^2 + 10t - 21t - 6$
$$= 5t(7t + 2) \quad 3(7t + 2)$$
$$= (\quad)(5t - 3)$$

GUIDED PRACTICE

Factor. See Example 1 or Example 6.

19. $2x^2 + 3x + 1$ **20.** $3x^2 + 4x + 1$

21. $3a^2 + 10a + 3$ **22.** $2b^2 + 7b + 3$

23. $5x^2 + 7x + 2$ **24.** $7t^2 + 12t + 5$

25. $7x^2 + 18x + 11$ **26.** $5n^2 + 12n + 7$

Factor. See Example 2 or Example 6.

27. $4x^2 - 8x + 3$ **28.** $4z^2 - 13z + 3$

29. $8x^2 - 22x + 5$ **30.** $15a^2 - 28a + 5$

31. $15t^2 - 26t + 7$ **32.** $10x^2 - 9x + 2$

33. $6y^2 - 13y + 2$ **34.** $6y^2 - 43y + 7$

Factor. See Example 3 or Example 6.

35. $3x^2 - 2x - 21$ **36.** $3u^2 - 44u - 15$

37. $5m^2 - 7m - 6$ **38.** $5y^2 - 18y - 8$

39. $7y^2 + 55y - 8$ **40.** $7x^2 + 33x - 10$

41. $11y^2 + 7y - 4$ **42.** $13y^2 + 9y - 4$

Factor. See Example 4.

43. $6r^2 + rs - 2s^2$ **44.** $3m^2 + 5mn + 2n^2$

45. $4x^2 + 8xy + 3y^2$ **46.** $4b^2 + 15bc - 4c^2$

47. $8m^2 + 91mn + 33n^2$ **48.** $2m^2 + 17mn - 9n^2$

49. $15x^2 - xy - 6y^2$ **50.** $4a^2 - 15ab + 9b^2$

Factor. See Example 5 or Example 7.

51. $-26x + 6x^2 - 20$ **52.** $-28 + 6a^2 - 2a$

53. $15a + 8a^3 - 26a^2$ **54.** $16r - 40r^2 + 25r^3$

55. $2u^2 - 6v^2 - uv$ **56.** $6a^2 + 6b^2 - 13ab$

57. $36y^2 - 88y + 32$ **58.** $70a^2 - 95a + 30$

59. $130r^2 + 20r - 110$ **60.** $170h^2 - 210h - 260$

61. $-y^3 - 13y^2 - 12y$ **62.** $-2xy^2 - 8xy + 24x$

63. $-6x^4 + 15x^3 + 9x^2$ **64.** $-9y^4 - 3y^3 + 6y^2$

65. $16m^3n + 20m^2n^2 + 6mn^3$ **66.** $-28u^3v^3 + 26u^2v^4 - 6uv^5$

TRY IT YOURSELF

Factor each polynomial completely. If an expression is prime, so indicate.

67. $6t^2 - 7t - 20$ **68.** $6w^2 + 13w + 5$

69. $15p^2 - 2pq - q^2$ **70.** $8c^2 - 10cd + 3d^2$

71. $4t^2 - 16t + 7$

72. $9x^2 - 32x + 15$

73. $8y^2 - 2y - 1$

74. $14y^2 + 11y + 2$

75. $18x^2 + 31x - 10$

76. $20y^2 - 93y - 35$

77. $10u^2 - 13u - 6$

78. $8m^2 + 5m - 10$

79. $3x^2 + x + 6$

80. $2u^2 + 3u + 25$

81. $30r^5 + 63r^4 - 30r^3$

82. $6s^5 - 26s^4 - 20s^3$

83. $6p^2 + pq - q^2$

84. $12m^2 - 11mn + 2n^2$

85. $-12y^2 - 12 + 25y$

86. $-10t^2 + 1 + 3t$

Choose the correct method from Sections 6.1, 6.2, or 6.3 to factor completely each of the following.

87. $m^2 + 3m - 28$

88. $-b^2 - 5b + 24$

89. $6a^3 + 15a^2$

90. $9x^4 + 27x^6$

91. $x^3 - 2x^2 + 5x - 10$

92. $x^3 - x^2 + 2x - 2$

93. $5y^2 + 3 - 8y$

94. $3t^2 + 7 - 10t$

95. $-2x^2 - 10x - 12$

96. $4y^2 + 36y + 72$

97. $12x^3y^3 - 18x^2y^3 + 15x^2y^2$

98. $15c^2d^3 - 25c^3d^2 - 10c^4d^4$

99. $a^2 - 7ab + 10b^2$

100. $x^2 - 13xy + 12y^2$

101. $9u^6 - 71u^5 - 8u^4$

102. $25n^8 - 49n^7 - 2n^6$

APPLICATIONS

103. FURNITURE   The area of a desktop is represented by the trinomial $(4x^2 + 20x - 11)$ in.^{2}. Factor it to find the polynomials that represent its length and width.

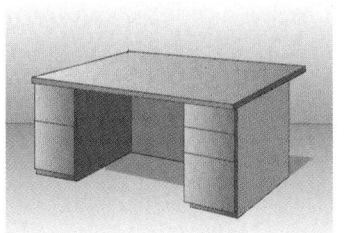

104. STORAGE   The volume of an 8-foot-wide portable storage container is represented by the trinomial $(72x^2 + 120x - 400)$ ft^3. Its dimensions can be determined by factoring the trinomial. Find the polynomials that represent height and the length of the container.

Width
8 ft

SUPER STORAGE

WRITING

105. Two students factor $2x^2 + 20x + 42$ and get two different answers: $(2x + 6)(x + 7)$ and $(x + 3)(2x + 14)$.

Do both answers check? Why don't they agree? Is either answer completely correct? Explain.

106. Why is the process of factoring $6x^2 - 5x - 6$ more complicated than the process of factoring $x^2 - 5x - 6$?

107. Suppose a factorization check of $(3x - 9)(5x + 7)$ gives a middle term $-24x$, but a middle term of $24x$ is actually needed. Explain how to quickly obtain the correct factorization.

108. Suppose we want to factor $2x^2 + 7x - 72$. Explain why $(2x - 1)(x + 72)$ is not a wise choice to try first.

REVIEW

Evaluate each expression.

109. -7^2

110. $(-7)^2$

111. 7^0

112. 7^{-2}

113. $\dfrac{1}{7^{-2}}$

114. $2 \cdot 7^2$

CHALLENGE PROBLEMS

Factor completely.

115. $6a^{10} + 5a^5 - 21$

116. $3x^4y^2 - 29x^2y + 56$

117. $8x^2(c^2 + c - 2) - 2x(c^2 + c - 2) - (c^2 + c - 2)$

118. Find all integer values of b that make $2x^2 + bx - 5$ factorable.

SECTION 6.4
Factoring Perfect-Square Trinomials and the Difference of Two Squares

Objectives

1 Recognize perfect-square trinomials.

2 Factor perfect-square trinomials.

3 Factor the difference of two squares.

In this section, we will discuss a method that can be used to factor two types of trinomials, called *perfect-square trinomials*. We also develop techniques for factoring a type of binomial called the *difference of two squares*.

1 Recognize Perfect-Square Trinomials.

We have seen that the square of a binomial is a trinomial. We have also seen that the special-product rules shown below can be used to quickly find the square of a sum and the square of a difference. The terms of the resulting trinomial are related to the terms of the binomial that was squared.

$$(A + B)^2 = \qquad A^2 \qquad + \qquad 2AB \qquad + \qquad B^2$$

<table>
<tr><td>This is the square
of the first term
of the binomial.</td><td>This is twice the
product of the terms
of the binomial.</td><td>This is the square
of the last term
of the binomial.</td></tr>
</table>

$$(A - B)^2 = \qquad A^2 \qquad - \qquad 2AB \qquad + \qquad B^2$$

> **Success Tip**
> To prepare for this section, it would be helpful to review Section 5.7 Special Products.

Trinomials that are squares of a binomial are called **perfect-square trinomials.** Some examples are

$y^2 + 6y + 9$	Because it is the square of $(y + 3)$: $(y + 3)^2 = y^2 + 6y + 9$
$t^2 - 14t + 49$	Because it is the square of $(t - 7)$: $(t - 7)^2 = t^2 - 14t + 49$
$4m^2 - 20m + 25$	Because it is the square of $(2m - 5)$: $(2m - 5)^2 = 4m^2 - 20m + 25$

EXAMPLE 1 Determine whether the following are perfect-square trinomials:
a. $x^2 + 10x + 25$ **b.** $c^2 - 12c - 36$
c. $25y^2 - 30y + 9$ **d.** $4t^2 + 18t + 81$

Strategy We will compare each trinomial, term-by-term, to one of the special-product forms discussed in Section 5.7.

Why If a trinomial matches one of these forms, it is a perfect-square trinomial.

Solution
a. To determine whether this is a perfect-square trinomial, we note that

$$x^2 + 10x + 25$$

<table>
<tr><td>The first term is
the square of x.</td><td>The middle term is twice
the product of x and 5:
$2 \cdot x \cdot 5 = 10x$.</td><td>The last term is
the square of 5.</td></tr>
</table>

Thus, $x^2 + 10x + 25$ is a perfect-square trinomial.

b. To determine whether this is a perfect-square trinomial, we note that

$$c^2 - 12c - 36$$

The last term, −36, is not
the square of a real number.

Since the last term is negative, $c^2 - 12c - 36$ is not a perfect-square trinomial.

c. To determine whether this is a perfect-square trinomial, we note that

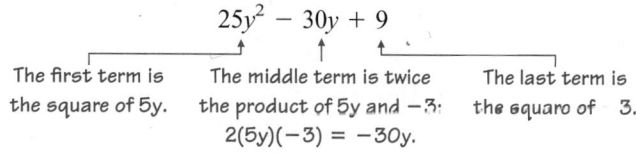

$$25y^2 - 30y + 9$$

The first term is	The middle term is twice	The last term is
the square of 5y.	the product of 5y and −3:	the square of 3.
	$2(5y)(-3) = -30y.$	

Thus, $25y^2 - 30y + 9$ is a perfect-square trinomial.

d. To determine whether this is a perfect-square trinomial, we note that

$$4t^2 + 18t + 81$$

The first term is	The middle term is not	The last term is
the square of 2t.	twice the product of 2t and 9,	the square of 9.
	because $2(2t)(9) = 36t$.	

Thus, $4t^2 + 18t + 81$ is not a perfect-square trinomial.

Self Check 1   Determine whether the following are perfect-square trinomials:
  a. $y^2 + 4y + 4$ **b.** $b^2 - 6b - 9$
  c. $4z^2 + 4z + 4$ **d.** $49x^2 - 28x + 16$

Now Try   Problems 13 and 17

 2   **Factor Perfect-Square Trinomials.**

We can factor perfect-square trinomials using the methods previously discussed in Sections 6.2 and 6.3. However, in many cases, we can factor them more quickly by inspecting their terms and applying the special-product rules in reverse.

| Factoring Perfect-Square Trinomials | $A^2 + 2AB + B^2 = (A + B)^2$ Each of these trinomials factors as the square of a binomial. |
| | $A^2 - 2AB + B^2 = (A - B)^2$ |

When factoring perfect-square trinomials, it is helpful to know the integers that are perfect squares. The number 400, for example, is a perfect-integer square, because $400 = 20^2$.

$1 = 1^2$	$25 = 5^2$	$81 = 9^2$	$169 = 13^2$	$289 = 17^2$
$4 = 2^2$	$36 = 6^2$	$100 = 10^2$	$196 = 14^2$	$324 = 18^2$
$9 = 3^2$	$49 = 7^7$	$121 = 11^2$	$225 = 15^2$	$361 = 19^2$
$16 = 4^2$	$64 = 8^2$	$144 = 12^2$	$256 = 16^2$	$400 = 20^2$

EXAMPLE 2   Factor: **a.** $x^2 + 20x + 100$ **b.** $9x^2 - 30xy + 25y^2$

Strategy   The terms of each trinomial do not have a common factor (other than 1). We will determine whether each is a perfect-square trinomial.

Why   If it is, we can factor it using a special-product rule in reverse.

Solution

a. $x^2 + 20x + 100$ is a perfect-square trinomial, because:

- The first term x^2 is the square of x.
- The last term 100 is the square of 10.
- The middle term is twice the product of x and 10: $2(x)(10) = 20x$.

To find the factorization, we match $x^2 + 20x + 100$ to the proper rule for factoring a perfect-square trinomial.

$$A^2 + 2\ A\quad B + B^2 = (A + B)^2$$
$$x^2 + 20x + 10^2 = x^2 + 2 \cdot x \cdot 10 + 10^2 = (x + 10)^2$$

Therefore, $x^2 + 20x + 10^2 = (x + 10)^2$. Check by finding $(x + 10)^2$.

b. $9x^2 - 30xy + 25y^2$ is a perfect-square trinomial, because:

- The first term $9x^2$ is the square of $3x$: $(3x)^2 = 9x^2$.
- The last term $25y^2$ is the square of $-5y$: $(-5y)^2 = 25y^2$.
- The middle term is twice the product of $3x$ and $-5y$: $2(3x)(-5y) = -30xy$.

We can use these observations to write the trinomial in one of the special-product forms that then leads to its factorization.

$$9x^2 - 30xy + 25y^2 = (3x)^2 - 2(3x)(5y) + (-5y)^2 \quad -2(3x)(5y) = 2(3x)(-5y).$$
$$= (3x - 5y)^2$$

Therefore, $9x^2 - 30xy + 25y^2 = (3x - 5y)^2$. Check by finding $(3x - 5y)^2$.

 Self Check 2   Factor: **a.** $x^2 + 18x + 81$
     b. $16x^2 - 8xy + y^2$

Now Try   **Problems 21 and 31**

> **Success Tip**
> The sign of the middle term of a perfect-square trinomial is the same as the sign of the second term of the squared binomial.
>
> $$A^2 + 2AB + B^2 = (A + B)^2$$
> $$A^2 - 2AB + B^2 = (A - B)^2$$

EXAMPLE 3   Factor completely: $4a^3 - 4a^2 + a$

Strategy   We will factor out the GCF, a, first. Then we will factor the resulting perfect-square trinomial using a special-product rule in reverse.

Why   The first step in factoring any polynomial is to factor out the GCF.

Solution   The terms of $4a^3 - 4a^2 + a$ have the common factor a, which should be factored out first. Within the parentheses, we recognize $4a^2 - 4a + 1$ as a perfect square trinomial of the form $A^2 - 2AB + B^2$, and factor it as such.

$$4a^3 - 4a^2 + a = a(4a^2 - 4a + 1) \quad \text{Factor out } a.$$
$$= a(2a - 1)^2 \quad 4a^2 = (2a)^2, 1 = (-1)^2, \text{ and } -4a = 2(2a)(-1).$$

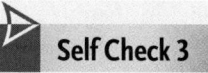

Self Check 3 Factor completely: $49x^3 - 14x^2 + x$

Now Try **Problem 37**

③ Factor the Difference of Two Squares.

The Language of Algebra
The expression $A^2 - B^2$ is a *difference of two squares*, whereas $(A - B)^2$ is the *square of a difference*. They are not equivalent because $(A - B)^2 \neq A^2 - B^2$.

Recall the special-product rule for multiplying the sum and difference of the same two terms:

$$(A + B)(A - B) = A^2 - B^2$$

The binomial $A^2 - B^2$ is called a **difference of two squares,** because A^2 is the square of A and B^2 is the square of B. If we reverse this rule, we obtain a method for factoring a difference of two squares.

$$\overset{\text{Factoring} \longrightarrow}{A^2 - B^2 = (A + B)(A - B)}$$

This pattern is easy to remember if we think of a difference of two squares as the square of a **F**irst quantity minus the square of a **L**ast quantity.

Factoring a Difference of Two Squares

To factor the square of a **F**irst quantity minus the square of a **L**ast quantity, multiply the **F**irst plus the **L**ast by the **F**irst minus the **L**ast.

$$F^2 - L^2 = (F + L)(F - L)$$

EXAMPLE 4 Factor, if possible: **a.** $x^2 - 9$ **b.** $16 - b^2$
c. $n^2 - 45$ **d.** $a^2 + 81$

Strategy The terms of each binomial do not have a common factor (other than 1). The only option available is to attempt to factor each as a difference of two squares.

Why If a binomial is a difference of two squares, we can factor it using a special-product rule in reverse.

Notation
By the commutative property of multiplication, the factors of a difference of two squares can be written in either order. For example, we can write:

$$x^2 - 9 = (x - 3)(x + 3)$$

Solution

a. $x^2 - 9$ is the difference of two squares because it can be written as $x^2 - 3^2$. We can match it to the rule for factoring a difference of two squares to find the factorization.

$$F^2 - L^2 = (F + L)(F - L)$$
$$x^2 - 3^2 = (x + 3)(x - 3) \quad \text{9 is a perfect-integer square: } 9 = 3^2.$$

Therefore, $x^2 - 9 = (x + 3)(x - 3)$.

Check by multiplying: $(x + 3)(x - 3) = x^2 - 9$.

b. $16 - b^2$ is the difference of two squares because $16 - b^2 = 4^2 - b^2$. Therefore,

$$16 - b^2 = (4 + b)(4 - b) \quad \text{16 is a perfect-integer square: } 16 = 4^2.$$

Check by multiplying.

c. Since 45 is not a perfect-integer square, $n^2 - 45$ cannot be factored using integers. It is prime.

d. $a^2 + 81$ can be written $a^2 + 9^2$, and is, therefore, the **sum of two squares.** We might attempt to factor $a^2 + 81$ as $(a + 9)(a + 9)$ or $(a - 9)(a - 9)$. However, the following checks show that neither product is $a^2 + 81$.

$$(a + 9)(a + 9) = a^2 + 18a + 81 \qquad (a - 9)(a - 9) = a^2 - 18a + 81$$

In general, the sum of two squares (with no common factor other than 1) cannot be factored using real numbers. Thus, $a^2 + 81$ is prime.

Self Check 4   Factor, if possible:
a. $c^2 - 4$ **b.** $121 - t^2$
c. $x^2 - 24$ **d.** $s^2 + 36$
Now Try   **Problems 45 and 53**

Terms containing variables such as $25x^2$ and $4y^4$ are perfect squares, because they can be written as the square of a quantity. For example:

$$25x^2 = (5x)^2 \qquad \text{and} \qquad 4y^4 = (2y^2)^2$$

EXAMPLE 5   Factor: **a.** $25x^2 - 49$ **b.** $4y^4 - 121z^2$

Strategy   In each case, the terms of the binomial do not have a common factor (other than 1). To factor them, we will write each binomial in a form that clearly shows it is a difference of two squares.

Why   We can then use a special-product rule in reverse to factor it.

Solution
a. We can write $25x^2 - 49$ in the form $(5x)^2 - 7^2$ and match it to the rule for factoring the difference of two squares:

$$\begin{array}{ccccccc} F^2 & - & L^2 & = & (F & + & L)(F & - & L) \\ \downarrow & & \downarrow & & \downarrow & & \downarrow & \downarrow & & \downarrow \\ (5x)^2 & - & 7^2 & = & (5x & + & 7)(5x & - & 7) \end{array}$$

Therefore, $25x^2 - 49 = (5x + 7)(5x - 7)$. Check by multiplying.

b. We can write $4y^4 - 121z^2$ in the form $(2y^2)^2 - (11z)^2$ and match it to the rule for factoring the difference of two squares:

$$\begin{array}{ccccccc} F^2 & - & L^2 & = & (F & + & L)(F & - & L) \\ \downarrow & & \downarrow & & \downarrow & & \downarrow & \downarrow & & \downarrow \\ (2y^2)^2 & - & (11z)^2 & = & (2y^2 & + & 11z)(2y^2 & - & 11z) \end{array}$$

Therefore, $4y^4 - 121z^2 = (2y^2 + 11z)(2y^2 - 11z)$. Check by multiplying.

Success Tip
Remember that a *difference of two squares* is a binomial. Each term is a square and the terms have different signs. The powers of the variables in the terms must be even.

Self Check 5   Factor: **a.** $16y^2 - 9$
b. $9m^2 - 64n^4$
Now Try   **Problems 57 and 61**

EXAMPLE 6   Factor completely: $8x^2 - 8$

Strategy   We will factor out the GCF, 8, first. Then we will factor the resulting difference of two squares.

Why   The first step in factoring any polynomial is to factor out the GCF.

Solution

$$8x^2 - 8 = 8(x^2 - 1) \qquad \text{The GCF is 8.}$$
$$= 8(x + 1)(x - 1) \qquad \text{Think of } x^2 - 1 \text{ as } x^2 - 1^2 \text{ and factor the difference of two squares.}$$

Check: $\quad 8(x + 1)(x - 1) = 8(x^2 - 1) \qquad$ Multiply the binomials first.
$$= 8x^2 - 8 \qquad \text{Distribute the multiplication by 8.}$$

Self Check 6   Factor completely: $2p^2 - 200$
   Now Try   **Problem 65**

Sometimes we must factor a difference of two squares more than once to completely factor a polynomial.

EXAMPLE 7   Factor completely: $x^4 - 16$

Strategy   The terms of $x^4 - 16$ do not have a common factor (other than 1). To factor this binomial, we will write it in a form that clearly shows it is a difference of two squares.

Why   We can then use a special-product rule in reverse to factor it.

Caution
Factoring a polynomial is complete when no factor can be factored further.

Solution

$$x^4 - 16 = (x^2)^2 - 4^2 \qquad \text{Write } x^4 \text{ as } (x^2)^2 \text{ and 16 as } 4^2.$$
$$= (x^2 + 4)(x^2 - 4) \qquad \text{Factor the difference of two squares.}$$
$$= (x^2 + 4)(x + 2)(x - 2) \qquad \text{Factor } x^2 - 4, \text{ which is itself a difference of two squares. The binomial } x^2 + 4 \text{ is a sum of two squares and does not factor further.}$$

Self Check 7   Factor completely: $a^4 - 81$
   Now Try   **Problem 75**

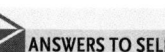
ANSWERS TO SELF CHECKS   **1. a.** Yes **b.** No **c.** No **d.** No **2. a.** $(x + 9)^2$ **b.** $(4x - y)^2$
3. $x(7x - 1)^2$ **4. a.** $(c + 2)(c - 2)$ **b.** $(11 + t)(11 - t)$ **c.** Prime **d.** Prime
5. a. $(4y + 3)(4y - 3)$ **b.** $(3m + 8n^2)(3m - 8n^2)$ **6.** $2(p + 10)(p - 10)$
7. $(a^2 + 9)(a + 3)(a - 3)$

STUDY SET
6.4

VOCABULARY

Fill in the blanks.

1. $x^2 + 6x + 9$ is a _____-square trinomial because it is the square of the binomial $x + 3$.
2. The binomial $x^2 - 25$ is called a _____ of two squares and it factors as $(x + 5)(x - 5)$. The binomial $x^2 + 25$ is a _____ of two squares and since it does not factor using integers, it is _____.

CONCEPTS

Fill in the blanks.

3. Consider $25x^2 + 30x + 9$.
   a. The first term is the square of ___.
   b. The last term is the square of ___.
   c. The middle term is twice the product of ___ and ___.
4. Consider $49x^2 - 28xy + 4y^2$.
   a. The first term is the square of ___.
   b. The last term is the square of ___.
   c. The middle term is twice the product of ___ and ___.
5. a. $x^2 + 2xy + y^2 = ($ ___ $+$ ___ $)^2$
   b. $x^2 - 2xy + y^2 = (x$ ___ $y)^2$
   c. $x^2 - y^2 = (x$ ___ $y)($ ___ $-$ ___ $)$
6. a. $36x^2 = ($ ___ $)^2$ b. $100x^4 = ($ ___ $)^2$
   c. $4x^2 - 9 = ($ ___ $)^2 - ($ ___ $)^2$
7. List the squares of the integers from 1 through 20.

8. Use multiplication to determine if each factorization is correct.
   a. $9y^2 - 12y + 4 = (3y - 2)^2$
   b. $n^2 - 16 = (n + 8)(n - 8)$

NOTATION

Complete each factorization.

9. $x^2 + 10x + 25 = (x + 5)$
10. $9b^2 - 12b + 4 = (3b$ ___ $2)^2$
11. $x^2 - 64 = (x$ ___ $8)(x$ ___ $8)$
12. $16t^2 - 49 = (4t +$ ___ $)(4t -$ ___ $)$

GUIDED PRACTICE

Determine whether each of the following is a perfect-square trinomial. See Example 1.

13. $x^2 + 18x + 81$ 14. $x^2 + 14x + 49$
15. $y^2 + 2y + 4$ 16. $y^2 + 4y + 16$

17. $9n^2 - 30n - 25$ 18. $9a^2 - 48a - 64$
19. $4y^2 - 12y + 9$ 20. $9y^2 - 30y + 25$
$3y$ 5

Factor. See Example 2.

21. $x^2 + 6x + 9$ 22. $x^2 + 10x + 25$

23. $b^2 + 2b + 1$ 24. $m^2 + 12m + 36$

25. $c^2 - 12c + 36$ 26. $d^2 - 10d + 25$

27. $9y^2 - 24y + 16$ 28. $49z^2 - 14z + 1$

29. $9 + 4x^2 + 12x$ 30. $121 + 4x^2 + 44x$

31. $36m^2 + 60mn + 25n^2$ 32. $25a^2 + 30ab + 9b^2$

33. $81x^2 - 72xy + 16y^2$ 34. $9x^2 - 48xy + 64y^2$

35. $49t^2 - 28ts + 4s^2$ 36. $81p^2 - 36pq + 4q^2$

Factor completely. See Example 3.

37. $3u^2 - 18u + 27$ 38. $3v^2 - 42v + 147$

39. $36x^3 + 12x^2 + x$ 40. $4x^4 - 20x^3 + 25x^2$

41. $18a^5 + 84a^4b + 98a^3b^2$ 42. $32b^6 + 80b^5c + 50b^4c^2$

43. $-100t^2 + 20t - 1$ 44. $-81r^2 - 18r - 1$

Factor completely. If a polynomial can't be factored, write "prime." See Example 4.

45. $x^2 - 4$ 46. $x^2 - 9$

47. $x^2 - 16$ 48. $x^2 - 25$

49. $36 - y^2$ 50. $49 - w^2$

51. $-25 + t^2$ 52. $-144 + h^2$

53. $a^2 + b^2$ 54. $121a^2 + b^2$

55. $y^2 - 63$ 56. $x^2 - 27$

Factor. See Example 5.

57. $25t^2 - 64$ 58. $49d^2 - 16$

59. $81y^2 - 1$

60. $400z^2 - 1$

61. $9x^4 - y^2$

62. $4x^2 - z^4$

63. $16c^2 - 49d^4$

64. $36a^2 - 121b^4$

Factor completely. See Example 6.

65. $8x^2 - 32y^2$

66. $2a^2 - 200b^2$

67. $63a^2 - 7$

68. $20x^2 - 5$

69. $x^3 - 144x$

70. $g^3 - 121g$

71. $6x^4 - 6x^2y^2$

72. $4b^2y - 16c^2y$

Factor completely. See Example 7.

73. $81 - s^4$

74. $y^4 - 625$

75. $b^4 - 256$

76. $m^4n^4 - 16$

77. $16t^4 - 16s^4$

78. $2p^4 - 32q^4$

79. $25m^4 - 25$

80. $9 - 9n^4$

TRY IT YOURSELF

Factor completely.

81. $a^4 - 144b^2$

82. $81y^4 - 100z^2$

83. $9x^2y^2 + 30xy + 25$

84. $s^2t^2 - 20st + 100$

85. $t^2 - 20t + 100$

86. $r^2 + 24r + 144$

87. $z^2 - 64$

88. $25 + B^2$

89. $3m^4 - 3n^4$

90. $5a^4 - 80b^4$

91. $25m^2 + 70m + 49$

92. $25x^2 + 20x + 4$

Choose the correct method from Section 6.1, Section 6.2, Section 6.3, or Section 6.4 to factor completely each of the following:

93. $x^2 + x - 42$

94. $rx - sx + r - s$

95. $x^2 - 9$

96. $3a^2 - 4a - 4$

97. $24a^3b - 16a^2b$

98. $20ns^2 - 60nu + 100n$

99. $-2r^2 + 28r - 80$

100. $10s - 39 + s^2$

101. $x^3 + 3x^2 + 4x + 12$

102. $2y^2 - 128z^2$

103. $4b^2 - 20b + 25$

104. $a^2 - 4ab - 12b^2$

APPLICATIONS

105. GENETICS   The Hardy–Weinberg equation, one of the fundamental concepts in population genetics, is $p^2 + 2pq + q^2 = 1$, where p represents the frequency of a certain dominant gene and q represents the frequency of a certain recessive gene. Factor the left side of the equation.

106. SIGNAL FLAGS   The maritime signal flag for the letter X is shown. Find the polynomial that represents the area of the shaded region and express it in factored form.

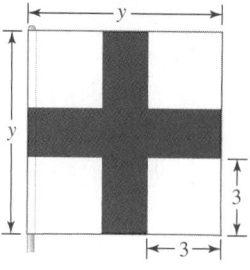

107. PHYSICS   The illustration shows a time-sequence picture of a falling apple. Factor the expression, which gives the difference in the distance fallen by the apple during the time interval from t_1 to t_2 seconds.

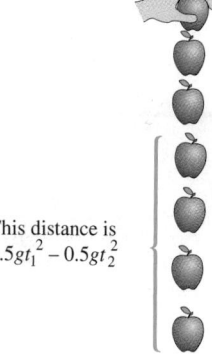

This distance is $0.5gt_1^2 - 0.5gt_2^2$

108. DARTS   A circular dart board has a series of rings around a solid center, called the bullseye. To find the area of the outer grey ring, we can use the formula $A = \pi R^2 - \pi r^2$. Factor the expression on the right side of the equation.

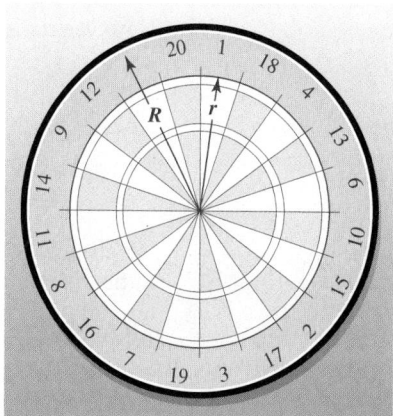

WRITING

109. When asked to factor $x^2 - 25$, one student wrote $(x + 5)(x - 5)$, and another student wrote $(x - 5)(x + 5)$. Are both answers correct? Explain.

110. Explain the error that was made in the following factorization:

$$x^2 - 100 = (x + 50)(x - 50)$$

111. Explain why the following factorization isn't complete.

$$x^4 - 625 = (x^2 + 25)(x^2 - 25)$$

112. Explain why $a^2 + 2a + 1$ is a perfect square trinomial and why $a^2 + 4a + 1$ isn't a perfect square trinomial.

REVIEW

Perform each division.

113. $\dfrac{5x^2 + 10y^2 - 15xy}{5xy}$

114. $\dfrac{-30c^2d^2 - 15c^2d - 10cd^2}{-10cd}$

115. $2a - 1\overline{)a - 2 + 6a^2}$

116. $4b + 3\overline{)4b^3 - 5b^2 - 2b + 3}$

CHALLENGE PROBLEMS

117. For what value of c does $80x^2 - c$ factor as $5(4x + 3)(4x - 3)$?

118. Find all values of b so that $0.16x^2 + bxy + 0.25y^2$ is a perfect square trinomial.

Factor completely.

119. $81x^6 + 36x^3y^2 + 4y^4$

120. $p^2 + p + \dfrac{1}{4}$

121. $c^2 + 1.6c + 0.64$

122. $x^{2n} - y^{4n}$

123. $(x + 5)^2 - y^2$

124. $\dfrac{1}{2} - 2a^2$

125. $c^2 - \dfrac{1}{16}$

126. $t^2 - \dfrac{9}{25}$

SECTION 6.5
Factoring the Sum and Difference of Two Cubes

Objective ❶ Factor the sum and difference of two cubes.

In this section we will discuss how to factor two types of binomials, called the *sum* and the *difference of two cubes.*

❶ Factor the Sum and Difference of Two Cubes.

We have seen that the sum of two squares, such as $x^2 + 4$ or $25a^2 + 9b^2$, cannot be factored. However, the sum of two cubes and the difference of two cubes can be factored.

The sum of two cubes	The difference of two cubes
$x^3 + 8$	$a^3 - 64b^3$

This is x cubed.   This is 2 cubed: $2^3 = 8$.

This is a cubed.   This is 4b cubed: $(4b)^3 = 64b^3$.

To find rules for factoring the sum of two cubes and the difference of two cubes, we need to find the products shown below. Note that each term of the trinomial is multiplied by each term of the binomial.

The Language of Algebra
The expression $x^3 + y^3$ is a *sum of two cubes*, whereas $(x + y)^3$ is the *cube of a sum*. If you expand $(x + y)^3$, you will see that $(x + y)^3 \neq x^3 + y^3$.

$$(x + y)(x^2 - xy + y^2) = x^3 - x^2y + xy^2 + x^2y - xy^2 + y^3$$
$$= x^3 + y^3 \qquad \text{Combine like terms: } -x^2y + x^2y = 0 \text{ and } xy^2 - xy^2 = 0.$$

$$(x - y)(x^2 + xy + y^2) = x^3 + x^2y + xy^2 - x^2y - xy^2 - y^3$$
$$= x^3 - y^3 \qquad \text{Combine like terms.}$$

These results justify the rules for factoring the **sum and difference of two cubes.** They are easier to remember if we think of a sum (or a difference) of two cubes as the cube of a **First** quantity plus (or minus) the cube of the **Last** quantity.

Factoring the Sum and Difference of Two Cubes	To factor the cube of a First quantity plus the cube of a Last quantity, multiply the First plus the Last by the First squared, minus the First times the Last, plus the Last squared. $$F^3 + L^3 = (F + L)(F^2 - FL + L^2)$$ To factor the cube of a First quantity minus the cube of a Last quantity, multiply the First minus the Last by the First squared, plus the First times the Last, plus the Last squared. $$F^3 - L^3 = (F - L)(F^2 + FL + L^2)$$

To factor the sum or difference of two cubes, it's helpful to know the cubes of integers from 1 to 10. The number 216, for example, is a **perfect-integer cube,** because $216 = 6^3$.

$1 = 1^3$	$27 = 3^3$	$125 = 5^3$	$343 = 7^3$	$729 = 9^3$
$8 = 2^3$	$64 = 4^3$	$216 = 6^3$	$512 = 8^3$	$1,000 = 10^3$

EXAMPLE 1   Factor: $x^3 + 8$

Strategy   We will write $x^3 + 8$ in a form that clearly shows it is the sum of two cubes.

Why   We can then use the rule for factoring the sum of two cubes.

Solution $x^3 + 8$ is the sum of two cubes because it can be written as $x^3 + 2^3$. We can match it to the rule for factoring the sum of two cubes to find its factorization.

$$\mathbf{F^3 + L^3 = (F + L)(F^2 - F \; L + L^2)}$$

To write the trinomial factor:
· Square the first term of the binomial factor.
· Multiply the terms of the binomial factor.
· Square the last term of the binomial factor.

$$x^3 + 2^3 = (x + 2)(x^2 - x \cdot 2 + 2^2)$$
$$= (x + 2)(x^2 - 2x + 4) \qquad x^2 - 2x + 4 \text{ does not factor.}$$

Therefore, $x^3 + 8 = (x + 2)(x^2 - 2x + 4)$. We can check by multiplying.

$$(x + 2)(x^2 - 2x + 4) = x^3 + 2x^2 - 2x^2 - 4x + 4x + 8$$
$$= x^3 + 8 \qquad \text{This is the original binomial.}$$

> **Caution**
> A common error is to try to factor $x^2 - 2x + 4$. It is not a perfect square trinomial, because the middle term needs to be $-4x$. Furthermore, it cannot be factored by the methods of Section 6.2. It is prime.

Self Check 1   Factor: $h^3 + 27$
Now Try   **Problem 17**

Terms containing variables such as $64b^3$ and m^6 are also perfect cubes, because they can be written as the cube of a quantity:

$$64b^3 = (4b)^3 \qquad \text{and} \qquad m^6 = (m^2)^3$$

EXAMPLE 2   Factor: $a^3 - 64b^3$

Strategy   We will write $a^3 - 64b^3$ in a form that clearly shows it is the difference of two cubes.

Why   We can then use the rule for factoring the difference of two cubes.

Solution $a^3 - 64b^3$ is the difference of two cubes because it can be written as $a^3 - (4b)^3$. We can match it to the rule for factoring the difference of two cubes to find its factorization.

$$\mathbf{F}^3 - \mathbf{L}^3 = (\mathbf{F} - \mathbf{L})(\mathbf{F}^2 + \mathbf{F}\ \mathbf{L} + \mathbf{L}^2)$$

$$a^3 - (4b)^3 = (a - 4b)[a^2 + a \cdot 4b + (4b)^2]$$
$$= (a - 4b)(a^2 + 4ab + 16b^2) \quad a^2 + 4ab + 16b^2 \text{ does not factor.}$$

Therefore, $a^3 - 64b^3 = (a - 4b)(a^2 + 4ab + 16b^2)$. Check by multiplying.

Self Check 2   Factor: $8c^3 - 1$

Now Try   Problem 37

You should memorize the rules for factoring the sum and the difference of two cubes. Note that the right side of each rule has the form

(a binomial)(a trinomial)

and that there is a relationship between the signs that appear in these forms.

The Sum of Two Cubes

The same sign

$$F^3 + L^3 = (F + L)(F^2 - FL + L^2)$$

Opposite   Always plus
signs

The Difference of Two Cubes

The same sign

$$F^3 - L^3 = (F - L)(F^2 + FL + L^2)$$

Opposite   Always plus
signs

If the terms of a binomial have a common factor, the GCF (or the opposite of the GCF) should always be factored out first.

EXAMPLE 3   Factor: $-2t^5 + 250t^2$

Strategy   We will factor out the common factor, $-2t^2$. We can then factor the resulting binomial as a difference of two cubes.

Why   The first step in factoring any polynomial is to factor out the GCF, or its opposite.

Solution
$$-2t^5 + 250t^2 = -2t^2(t^3 - 125) \qquad \text{Factor out the opposite of the GCF, } -2t^2.$$
$$= -2t^2(t - 5)(t^2 + 5t + 25) \quad \text{Factor } t^3 - 125.$$

Therefore, $-2t^5 + 250t^2 = -2t^2(t - 5)(t^2 + 5t + 25)$. Check by multiplying.

Self Check 3   Factor: $4c^3 + 4d^3$

Now Try   **Problem 43**

ANSWERS TO SELF CHECKS   **1.** $(h + 3)(h^2 - 3h + 9)$ **2.** $(2c - 1)(4c^2 + 2c + 1)$
3. $4(c + d)(c^2 - cd + d^2)$

STUDY SET
6.5

VOCABULARY

Fill in the blanks.

1. $x^3 + 27$ is the _____ of two cubes and $a^3 - 125$ is the difference of two _____.

2. The factorization of $x^3 + 8$ is $(x + 2)(x^2 - 2x + 4)$. The first factor is a binomial and the second is a _____.

CONCEPTS

Fill in the blanks.

3. a. $F^3 + L^3 = (\;\;\; + \;\;\;)(F^2 - FL + L^2)$

  b. $F^3 - L^3 = (F \;\;\; L)(\;\;\; + FL + \;\;\;)$

4. $m^3 + 64$

         ↑   ↑

      This is   This is
      ▨ cubed.   ▨ cubed.

5. $216n^3 - 125$

         ↑   ↑

      This is   This is
      ▨ cubed.   ▨ cubed.

6. a. $x^3 + 64y^3 = (\;\;\;)^3 + (\;\;\;)^3$

  b. $8x^3 - 27 = (\;\;\;)^3 - (\;\;\;)^3$

7. List the first ten positive integer cubes.

8. $(x - 2)(x^2 + 2x + 4)$ is the factorization of what binomial?

9. Use multiplication to determine if the factorization is correct.

  $b^3 + 27 = (b + 3)(b^2 + 3b + 9)$

10. The factorization of $y^3 + 27$ is $(y + 3)(y^2 - 3y + 9)$. Is this factored completely, or does $y^2 - 3y + 9$ factor further?

NOTATION

Complete each factorization.

11. $a^3 + 8 = (a + 2)(a^2 - \;\;\; + 4)$

12. $x^3 - 1 = (x - 1)(x^2 + \;\;\; + 1)$

13. $b^3 + 27 = (\;\;\;)(b^2 - 3b + 9)$

14. $z^3 - 125 = (z - 5)(\;\;\; + 5z + \;\;\;)$

Give an example of each type of expression.

15. a. the sum of two cubes

  b. the cube of a sum

16. a. the difference of two cubes

  b. the cube of a difference

GUIDED PRACTICE

Factor. See Example 1.

17. $y^3 + 125$ **18.** $b^3 + 216$

19. $a^3 + 64$ **20.** $n^3 + 1$

21. $n^3 + 512$ **22.** $t^3 + 729$

23. $8 + t^3$ **24.** $27 + y^3$

25. $a^3 + 1{,}000b^3$
26. $8u^3 + w^3$
27. $125c^3 + 27d^3$
28. $64m^3 + 343n^3$

Factor. See Example 2.

29. $a^3 - 27$ **30.** $r^3 - 8$

31. $m^3 - 343$ **32.** $y^3 - 216$

33. $216 - v^3$ **34.** $125 - t^3$

35. $8s^3 - t^3$

36. $27a^3 - b^3$

37. $1,000a^3 - w^3$

38. $s^3 - 64t^3$

39. $64x^3 - 27y^3$

40. $27x^3 - 1,000y^3$

Factor completely. See Example 3.

41. $2x^3 + 2$

42. $8y^3 + 8$

43. $3d^3 + 81$

44. $2x^3 + 54$

45. $x^4 - 216x$

46. $x^5 - 125x^2$

47. $64m^3x - 8n^3x$

48. $16r^4 - 128rs^3$

TRY IT YOURSELF

Choose the correct method from Section 6.1 through Section 6.5 and factor completely.

49. $x^2 + 8x + 16$

50. $64p^3 - 27$

51. $9r^2 - 16s^2$

52. $-63 - 13x + 6x^2$

53. $xy - ty + sx - st$

54. $12p^2 + 14p - 6$

55. $4p^3 + 32q^3$

56. $56a^4 - 15a^3 + a^2$

57. $16c^3t^2 + 20c^2t^3 + 6ct^4$

58. $-t^2 - 9t + 1$

59. $36e^4 - 36$

60. $3(z + 4) - a(z + 4)$

61. $35a^3b^2 - 14a^2b^3 + 14a^3b^3$

62. $-y^2 - 15y + 34$

63. $36r^2 + 60rs + 25s^2$

64. $16u^2 - 16$

APPLICATIONS

65. MAILING BREAKABLES   Write a polynomial that describes the amount of space in the larger box that must be filled with styrofoam chips if the smaller box containing a glass tea cup is to be placed within the larger box for mailing. Then factor the polynomial.

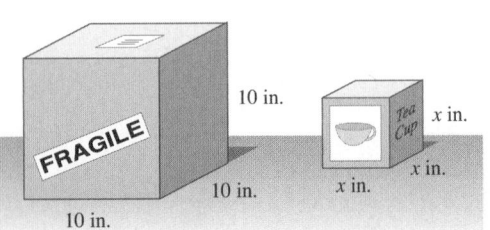

66. MELTING ICE   In one hour, the block of ice shown below had melted to the size shown on the right. Write a polynomial that describes the volume of ice that melted away. Then factor the polynomial.

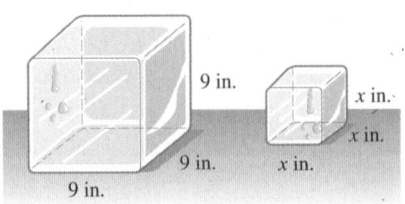

WRITING

67. Explain why $x^3 - 25$ is not a difference of two cubes.

68. Explain this diagram. Then draw a similar diagram for the difference of two cubes.

$$\overbrace{}^{\text{The same}}$$
$$F^3 + L^3 = (F + L)(F^2 - FL + L^2)$$
$$\underbrace{}_{\text{Opposite}} \quad \underbrace{}_{\text{Always plus}}$$

REVIEW

69. When expressed as a decimal, is $\frac{7}{9}$ a terminating or a repeating decimal?

70. Solve: $x + 20 = 4x - 1 + 2x$

71. Write the set of integers.

72. Solve: $2x + 2 = \frac{2}{3}x - 2$

73. Evaluate $2x^2 + 5x - 3$ for $x = -3$.

74. Check to determine whether 4 is a solution of $3(m - 8) + 2m = 4 - (m + 2)$.

CHALLENGE PROBLEMS

75. Consider: $x^6 - 1$

    a. Write the binomial as a difference of two squares. Then factor.

    b. Write the binomial as a difference of two cubes. Then factor.

76. What binomial multiplied by $(a^2b^2 + 7ab + 49)$ gives a difference of two cubes?

Factor completely.

77. $x^6 - y^9$

78. $\dfrac{125}{8}s^3 + \dfrac{1}{27}t^3$

79. $64x^{12} + y^{15}z^{18}$

80. $x^{3m} - y^{3n}$

SECTION 6.6
A Factoring Strategy

The factoring methods discussed so far will be used in the remaining chapters to simplify expressions and solve equations. In such cases, we must determine the factoring method—it will not be specified. This section will give you practice in selecting the appropriate factoring method to use given a randomly chosen polynomial.

The following strategy is helpful when factoring polynomials.

Steps for Factoring a Polynomial

1. Is there a common factor? If so, factor out the GCF, or the opposite of the GCF so that the leading coefficient is positive.

2. How many terms does the polynomial have?

   If it has *two terms,* look for the following problem types:

   a. The difference of two squares

   b. The sum of two cubes

   c. The difference of two cubes

   If it has *three terms,* look for the following problem types:

   a. A perfect-square trinomial

   b. If the trinomial is not a perfect square, use the trial-and-check-method or the grouping method.

   If it has *four or more terms,* try to factor by grouping.

3. Can any factors be factored further? If so, factor them completely.

4. Does the factorization check? Check by multiplying.

EXAMPLE 1   Factor: $2x^4 - 162$

Strategy   We will answer the four questions listed in the *Steps for Factoring a Polynomial.*

Why   The answers to these questions help us determine which factoring techniques to use.

The Language of Algebra
Recall that *to factor a polynomial* means to express it as a product of two (or more) polynomials.

Solution   *Is there a common factor?*   Yes. Factor out the GCF, which is 2.

$$2x^4 - 162 = 2(x^4 - 81)$$

How many terms does it have?   The polynomial within the parentheses, $x^4 - 81$, has two terms. It is a difference of two squares.

$$2x^4 - 162 = 2(x^4 - 81) \qquad \text{Think of } x^4 - 81 \text{ as } (x^2)^2 - 9^2.$$
$$= 2(x^2 + 9)(x^2 - 9)$$

Is it factored completely?   No. $x^2 - 9$ is also the difference of two squares and can be factored.

The Language of Algebra
Remember that the instruction to *factor* means to *factor completely.* A polynomial is *factored completely* when no factor can be factored further.

$$2x^4 - 162 = 2(x^4 - 81)$$
$$= 2(x^2 + 9)(x^2 - 9) \qquad \text{Think of } x^2 - 9 \text{ as } x^2 - 3^2.$$
$$= 2(x^2 + 9)(x + 3)(x - 3) \qquad x^2 + 9 \text{ is a sum of two squares and does not factor.}$$

Therefore, $2x^4 - 162 = 2(x^2 + 9)(x + 3)(x - 3)$.

Does it check?   Yes.

$$2(x^2 + 9)(x + 3)(x - 3) = 2(x^2 + 9)(x^2 - 9) \qquad \text{Multiply } (x + 3)(x - 3) \text{ first.}$$
$$= 2(x^4 - 81) \qquad \text{Multiply } (x^2 + 9)(x^2 - 9).$$
$$= 2x^4 - 162 \qquad \text{This is the original polynomial.}$$

Self Check 1   Factor: $11a^6 - 11a^2$

Now Try   **Problem 21**

EXAMPLE 2   Factor: $-4c^5d^2 - 12c^4d^3 - 9c^3d^4$

Strategy   We will answer the four questions listed in the *Steps for Factoring a Polynomial.*

Why   The answers to these questions help us determine which factoring techniques to use.

Solution   ***Is there a common factor?***   Yes. Factor out the opposite of the GCF, $-c^3d^2$, so that the leading coefficient is positive.

$$-4c^5d^2 - 12c^4d^3 - 9c^3d^4 = -c^3d^2(4c^2 + 12cd + 9d^2)$$

How many terms does it have?   The polynomial within the parentheses has three terms. It is a perfect-square trinomial because $4c^2 = (2c)^2$, $9d^2 = (3d)^2$, and $12cd = 2 \cdot 2c \cdot 3d$.

$$-4c^5d^2 - 12c^4d^3 - 9c^3d^4 = -c^3d^2(4c^2 + 12cd + 9d^2)$$
$$= -c^3d^2(2c + 3d)^2$$

Is it factored completely?   Yes. The binomial $2c + 3d$ does not factor further.

Therefore, $-4c^5d^2 - 12c^4d^3 - 9c^3d^4 = -c^3d^2(2c + 3d)^2$.

Does it check?   Yes.

$$-c^3d^2(2c + 3d)^2 = -c^3d^2(4c^2 + 12cd + 9d^2) \qquad \text{Use a special-product rule.}$$
$$= -4c^5d^2 - 12c^4d^3 - 9c^3d^4 \qquad \text{This is the original polynomial.}$$

Self Check 2   Factor: $-32h^4 - 80h^3 - 50h^2$

Now Try   **Problem 33**

EXAMPLE 3   Factor: $y^4 - 3y^3 + y - 3$

Strategy   We will answer the four questions listed in the *Steps for Factoring a Polynomial.*

Why   The answers to these questions help us determine which factoring techniques to use.

Solution   ***Is there a common factor?***   No. There is no common factor (other than 1).

How many terms does it have?   Since the polynomial has four terms, we will try factoring by grouping.

$$y^4 - 3y^3 + y - 3 = y^3(y - 3) + 1(y - 3)$$ Factor out y^3 from $y^4 - 3y^3$. Factor out 1 from $y - 3$.

$$= (y - 3)(y^3 + 1)$$

Is it factored completely?   No. We can factor $y^3 + 1$ as a sum of two cubes.

$$y^4 - 3y^3 + y - 3 = y^3(y - 3) + 1(y - 3)$$

$$= (y - 3)(y^3 + 1)$$ Think of $y^3 + 1$ as $y^3 + 1^3$.

$$= (y - 3)(y + 1)(y^2 - y + 1)$$ $y^2 - y + 1$ does not factor further.

Therefore, $y^4 - 3y^3 + y - 3 = (y - 3)(y + 1)(y^2 - y + 1)$.

Does it check?   Yes.

$$(y - 3)(y + 1)(y^2 - y + 1) = (y - 3)(y^3 + 1)$$ Multiply the last two factors.

$$= y^4 + y - 3y^3 - 3$$ Use the FOIL method.

$$= y^4 - 3y^3 + y - 3$$ This is the original polynomial.

> **Self Check 3**   Factor: $b^4 + b^3 + 8b + 8$
> **Now Try**   **Problem 37**

EXAMPLE 4   Factor: $32n - 4n^2 + 4n^3$

Strategy   We will answer the four questions listed in the *Steps for Factoring a Polynomial.*

Why   The answers to these questions help us determine which factoring techniques to use.

Solution   ***Is there a common factor?***   Yes. When we write the terms in descending powers of n, we see that the GCF is $4n$.

$$4n^3 - 4n^2 + 32n = 4n(n^2 - n + 8)$$

How many terms does it have?
The polynomial within the parentheses has three terms. It is not a perfect-square trinomial because the last term, 8, is not a perfect-integer square.

Negative factors of 8	Sum of the negative factors of 8
$-1(-8) = 8$	$-1 + (-8) = -9$
$-2(-4) = 8$	$-2 + (-8) = -10$

To factor the trinomial $n^2 - n + 8$, we must find two integers whose product is 8 and whose sum is -1. As we see in the table, there are no such integers. Thus, $n^2 - n + 8$ is prime.

Is it factored completely?   Yes.

Therefore, $4n^3 - 4n^2 + 32n = 4n(n^2 - n + 8)$.   Remember to write the GCF, $4n$, from the first step.

Does it check?   Yes.

$$4n(n^2 - n + 8) = 4n^3 - 4n^2 + 32n$$ This is the original polynomial.

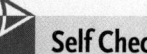

> **Self Check 4**   Factor: $6m^2 - 54m + 6m^3$
> **Now Try**   **Problem 45**

EXAMPLE 5 Factor: $3y^3 - 4y^2 - 4y$

Strategy We will answer the four questions listed in the *Steps for Factoring a Polynomial*.

Why The answers to these questions help us determine which factoring techniques to use.

Solution *Is there a common factor?* Yes. The GCF is y.

$$3y^3 - 4y^2 - 4y = y(3y^2 - 4y - 4)$$

How many terms does it have? The polynomial within the parentheses has three terms. It is not a perfect-square trinomial because the first term, $3y^2$, is not a perfect square.

   If we use grouping to factor $3y^2 - 4y - 4$, the key number is $ac = 3(-4) = -12$. We must find two integers whose product is -12 and whose sum is $b = -4$.

Key number = -12	$b = -4$
Factors of -12	Sum of the factors of -12
$2(-6) = -12$	$2 + (-6) = -4$

From the table, the correct pair is 2 and -6. They serve as the coefficients of $2y$ and $-6y$, the two terms that we use to represent the middle term, $-4y$, of the trinomial.

$$3y^2 - 4y - 4 = 3y^2 + 2y - 6y - 4 \qquad \text{Express } -4y \text{ as } 2y - 6y.$$
$$= y(3y + 2) - 2(3y + 2) \qquad \text{Factor } y \text{ from the first two terms and factor} \\ -2 \text{ the last two terms.}$$
$$= (3y + 2)(y - 2) \qquad \text{Factor out } 3y + 2.$$

The trinomial $3y^2 - 4y - 4$ factors as $(3y + 2)(y - 2)$.

Is it factored completely? Yes. Because $3y + 2$ and $y - 2$ do not factor.

Therefore, $3y^3 - 4y^2 - 4y = y(3y + 2)(y - 2)$. Remember to write the GCF, y, from the first step.

Does it check? Yes.

$$y(3y + 2)(y - 2) = y(3y^2 - 4y - 4) \qquad \text{Multiply the binomials.}$$
$$= 3y^3 - 4y^2 - 4y \qquad \text{This is the original polynomial.}$$

Self Check 5 Factor: $6y^3 + 21y^2 - 12y$

Now Try Problem 67

ANSWERS TO SELF CHECKS **1.** $11a^2(a^2 + 1)(a + 1)(a - 1)$ **2.** $-2h^2(4h + 5)^2$ **3.** $(b + 1)(b + 2)(b^2 - 2b + 4)$ **4.** $6m(m^2 + m - 9)$ **5.** $3y(2y - 1)(y + 4)$

STUDY SET
6.6

VOCABULARY

Fill in the blanks.

1. To factor a polynomial means to express it as a _____ of two (or more) polynomials.
2. A polynomial is factored _____ when each factor is prime.

CONCEPTS

For each of the following polynomials, which factoring method would you use first?

3. $2x^5y - 4x^3y$

4. $9b^2 + 12y - 5$

5. $x^2 + 18x + 81$

6. $ax + ay - x - y$

7. $x^3 + 27$

8. $y^3 - 64$

9. $m^2 + 3mn + 2n^2$

10. $16 - 25z^2$

11. What is the first question that should be asked when using the strategy of this section to factor a polynomial?

12. Use multiplication to determine whether the factorization is correct.

$$5c^3d^2 - 40c^2d^3 + 35cd^4 = 5cd^2(c - 7d)(c - d)$$

NOTATION

Complete each factorization.

13. $6m^3 - 28m^2 + 16m = 2m(3m^2 - \quad + 8)$
    $$= 2m(3m - 2)(\quad - 4)$$

14. $2a^3 + 3a^2 - 2a - 3$
    $$= \quad (2a + 3) - 1(\quad + 3)$$
    $$= (\quad)(a^2 - 1)$$
    $$= (2a + 3)(a + 1)(\quad)$$

TRY IT YOURSELF

The following is a list of random factoring problems. Factor each expression completely. If a expression is not factorable, write "prime." See Examples 1–5.

15. $2b^2 + 8b - 24$

16. $32 - 2t^4$

17. $8p^3q^7 + 4p^2q^3$

18. $8m^2n^3 - 24mn^4$

19. $2 + 24y + 40y^2$

20. $6r^2 + 3rs - 18s^2$

21. $8x^4 - 8$

22. $t - 90 + t^2$

23. $14c - 147 + c^2$

24. $ab^2 - 4a + 3b^2 - 12$

25. $x^2 + 7x + 1$

26. $3a^3 + 24b^3$

27. $-2x^5 + 128x^2$

28. $16 - 40z + 25z^2$

29. $a^2c + a^2d^2 + bc + bd^2$

30. $6t^4 + 14t^3 - 40t^2$

31. $-9x^2y^2 + 6xy - 1$

32. $x^2y^2 - 2x^2 - y^2 + 2$

33. $-20m^3 - 100m^2 - 125m$

34. $5x^3y^3z^4 + 25x^2y^4z^2 - 35x^3y^2z^5$

35. $2c^2 - 5cd - 3d^2$

36. $125p^3 - 64y^3$

37. $p^4 - 2p^3 - 8p + 16$

38. $a^2 + 8a + 3$

39. $a^2(x - a) - b^2(x - a)$

40. $70p^4q^3 - 35p^4q^2 + 49p^5q^2$

41. $a^2b^2 - 144$

42. $-16x^4y^2z + 24x^5y^3z^4 - 15x^2y^3z^7$

43. $2x^3 + 10x^2 + x + 5$

44. $u^2 - 18u + 81$

45. $8v^2 - 14v^3 + v^4$

46. $28 - 3m - m^2$

47. $x^4 - 13x^2 + 36$

48. $81r^4 - 256$

49. $8a^2x^3 - 2b^2x$

50. $12x^2 + 14x - 6$

51. $6x^2 - 14x + 8$

52. $12x^2 - 12$

53. $4x^2y^2 + 4xy^2 + y^2$

54. $81r^4s^2 - 24rs^5$

55. $4m^5 + 500m^2$

56. $ae + bf + af + be$

57. $x^4 - 2x^2 - 8$

58. $6x^2 - x - 16$

59. $4x^2 + 9y^2$

60. $x^4y + 216xy^4$

61. $16a^5 - 54a^2$

62. $25x^2 - 16y^2$

63. $27x - 27y - 27z$

64. $12x^2 + 52x + 35$

65. $xy - ty + xs - ts$

66. $bc + b + cd + d$

67. $35x^8 - 2x^7 - x^6$

68. $x^3 - 25$

69. $5(x - 2) + 10y(x - 2)$

70. $16x^2 - 40x^3 + 25x^4$

71. $49p^2 + 28pq + 4q^2$

72. $x^2y^2 - 6xy - 16$

73. $4t^2 + 36$

74. $r^5 + 3r^3 + 2r^2 + 6$

75. $m^2n^2 - 9m^2 + 3n^2 - 27$

76. $z^2 + 6yz^2 + 9y^2z^2$

WRITING

77. Which factoring method do you find the most difficult? Why?

78. What four questions make up the factoring strategy for polynomials discussed in this section?

79. What does it mean to factor a polynomial?

80. How is a factorization checked?

REVIEW

81. Graph the real numbers -3, 0, 2, and $-\frac{3}{2}$ on a number line.

82. Graph the interval $(-2, 3]$ on a number line.

83. Graph: $y = \frac{1}{2}x + 1$

84. Graph: $y < 2 - 3x$

CHALLENGE PROBLEMS

Factor completely using rational numbers.

85. $x^6 - 4x^3 - 12$

86. $x(x - y) - y(y - x)$

87. $24 - x^3 + 8x^2 - 3x$

88. $25b^2 + 14b + \frac{49}{25}$

89. $x^9 + y^6$

90. $\frac{1}{4} - \frac{u^2}{81}$

SECTION 6.7
Solving Quadratic Equations by Factoring

Objectives

1 Define quadratic equations.

2 Solve quadratic equations using the zero-factor property.

3 Solve third-degree equations by factoring.

The factoring methods that we have discussed have many applications in algebra. In this section, we will use factoring to solve *quadratic equations*. These equations are different from those that we solved in Chapter 2. They contain a term where the variable is raised to the second power, such as x^2 or t^2.

1 Define Quadratic Equations.

In a linear, or first degree equation, such as $2x + 3 = 8$, the exponent on the variable is an unwritten 1. A quadratic, or second degree equation, has a term in which the exponent on the variable is 2, and has no other terms of higher degree.

Quadratic Equations	A **quadratic equation** is an equation that can be written in the **standard form**
	$$ax^2 + bx + c = 0$$
	where a, b, and c represent real numbers, and $a \neq 0$.

Some examples of quadratic equations are

$$x^2 - 2x - 63 = 0, \qquad x^2 - 25 = 0, \qquad \text{and} \qquad 2x^2 + 3x = 2$$

The first two equations are in standard form. To write the third equation in standard form, we subtract 2 from both sides to get $2x^2 + 3x - 2 = 0$.

② Solve Quadratic Equations Using the Zero-Factor Property.

To **solve a quadratic equation,** we find all values of the variable that make the equation true. The methods that we used to solve linear equations in Chapter 2 cannot be used to solve a quadratic equation, because we cannot isolate the variable on one side of the equation. However, we can often solve quadratic equations using factoring and the following property of real numbers.

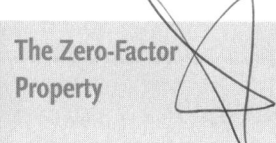

The Zero-Factor Property

When the product of two real numbers is 0, at least one of them is 0.
> If a and b represent real numbers, and
> $$\text{if } ab = 0, \text{ then } \quad a = 0 \quad \text{or} \quad b = 0$$

EXAMPLE 1   Solve: $(4x - 1)(x + 6) = 0$

Strategy   We will set $4x - 1$ equal to 0 and $x + 6$ equal to 0 and solve each equation.

Why   If the product of $4x - 1$ and $x + 6$ is 0, then, by the zero-factor property, $4x - 1$ must equal 0, or $x + 6$ must equal 0.

Solution   If $(4x - 1)(x + 6) = 0$ is to be a true statement, then either

$$4x - 1 = 0 \qquad \text{or} \qquad x + 6 = 0$$

Now we solve each of these linear equations.

$$
\begin{array}{lll}
4x - 1 = 0 & & \text{or} \qquad x + 6 = 0 \\
4x = 1 & \text{Add 1 to both sides.} & \qquad x = -6 \quad \text{Subtract 6 from both sides.} \\
x = \dfrac{1}{4} & \text{Divide both sides by 4.} &
\end{array}
$$

The results must be checked separately to see whether each of them produces a true statement. We substitute $\frac{1}{4}$ and then -6 for x in the original equation and evaluate the left side.

Check $\dfrac{1}{4}$:

$$(4x - 1)(x + 6) = 0$$
$$\left[4\left(\frac{1}{4}\right) - 1\right]\left(\frac{1}{4} + 6\right) \stackrel{?}{=} 0$$
$$(1 - 1)\left(\frac{25}{4}\right) \stackrel{?}{=} 0$$
$$0\left(\frac{25}{4}\right) \stackrel{?}{=} 0$$
$$0 = 0 \quad \text{True}$$

Check -6:

$$(4x - 1)(x + 6) = 0$$
$$[4(-6) - 1](-6 + 6) \stackrel{?}{=} 0$$
$$(-24 - 1)(0) \stackrel{?}{=} 0$$
$$-25(0) \stackrel{?}{=} 0$$
$$0 = 0 \quad \text{True}$$

The resulting true statements indicate that $(4x - 1)(x + 6) = 0$ has two solutions: $\frac{1}{4}$ and -6. Recall from Chapter 2 that the *solution set* of an equation is the set of all numbers that make the equation true. Thus, the solution set is $\left\{-6, \frac{1}{4}\right\}$.

> **Self Check 1**   Solve: $(x - 12)(5x + 6) = 0$
> **Now Try**   **Problem 15**

In Example 1, the left side of $(4x - 1)(x + 6) = 0$ is in factored form and the right side is 0, so we can immediately use the zero-factor property. However, to solve many quadratic equations, we must factor before using the zero-factor property.

EXAMPLE 2   Solve: $x^2 - 2x - 63 = 0$

Strategy   We will factor the trinomial on the left side of the equation and use the zero-factor property.

Why   To use the zero-factor property, we need one side of the equation to be factored completely and the other side to be 0.

Solution

$x^2 - 2x - 63 = 0$	This is the equation to solve.
$(x + 7)(x - 9) = 0$	Factor the trinomial, $x^2 - 2x - 63$.
$x + 7 = 0$ or $x - 9 = 0$	Set each factor equal to 0.
$x = -7$ $\quad$ $x = 9$	Solve each linear equation.

> **Success Tip**
> When you see the word *solve* in this example, you probably think of steps from Chapter 2 such as combining like terms, distributing, or doing something to both sides. However, to solve this quadratic equation, we begin by factoring $x^2 - 2x - 63$.

To check the results, we substitute -7 and then 9 for x in the original equation and evaluate the left side.

Check -7:
$$x^2 - 2x - 63 = 0$$
$$(-7)^2 - 2(-7) - 63 \stackrel{?}{=} 0$$
$$49 - (-14) - 63 \stackrel{?}{=} 0$$
$$63 - 63 \stackrel{?}{=} 0$$
$$0 = 0 \quad \text{True}$$

Check 9:
$$x^2 - 2x - 63 = 0$$
$$(9)^2 - 2(9) - 63 \stackrel{?}{=} 0$$
$$81 - 18 - 63 \stackrel{?}{=} 0$$
$$63 - 63 \stackrel{?}{=} 0$$
$$0 = 0 \quad \text{True}$$

The solutions of $x^2 - 2x - 63 = 0$ are -7 and 9, and the solution set is $\{-7, 9\}$.

> **Self Check 2**   Solve: $x^2 + 5x + 6 = 0$
> **Now Try**   **Problem 27**

The previous examples suggest the following strategy to solve quadratic equations by factoring.

The Factoring Method for Solving a Quadratic Equation	1. Write the equation in standard form: $ax^2 + bx + c = 0$ or $0 = ax^2 + bx + c$.
	2. Factor completely.
	3. Use the zero-factor property to set each factor equal to 0.
	4. Solve each resulting linear equation.
	5. Check the results in the original equation.

With this method, we factor *expressions* to solve *equations*.

EXAMPLE 3   Solve: $x^2 - 25 = 0$

Strategy   We will factor the binomial on the left side of the equation and use the zero-factor property.

Why   To use the zero-factor property, we need one side of the equation to be factored completely and the other side to be 0.

Solution   We factor the difference of two squares on the left side of the equation and proceed as follows.

$x^2 - 25 = 0$	This is the equation to solve.
$(x + 5)(x - 5) = 0$	Factor the difference of two squares, $x^2 - 25$.
$x + 5 = 0$ or $x - 5 = 0$	Set each factor equal to 0.
$x = -5$ $x = 5$	Solve each linear equation.

Notation
Although $x^2 - 25 = 0$ is missing a term involving x, it is a quadratic equation in standard form $ax^2 + bx + c = 0$, where $a = 1$, $b = 0$, and $c = -25$.

Check each result by substituting it into the original equation.

Check -5:	*Check* 5:
$x^2 - 25 = 0$	$x^2 - 25 = 0$
$(-5)^2 - 25 \stackrel{?}{=} 0$	$5^2 - 25 \stackrel{?}{=} 0$
$25 - 25 \stackrel{?}{=} 0$	$25 - 25 \stackrel{?}{=} 0$
$0 = 0$ True	$0 = 0$ True

The solutions of $x^2 - 25 = 0$ are -5 and 5, and the solution set is $\{-5, 5\}$.

Self Check 3   Solve: $x^2 - 49 = 0$

Now Try   **Problem 35**

EXAMPLE 4   Solve: $6x^2 = 12x$

Strategy   We will subtract $12x$ from both sides of the equation to get 0 on the right side. Then we will factor the resulting binomial and use the zero-factor property.

Why   To use the zero-factor property, we need one side of the equation to be factored completely and the other side to be 0.

Solution   The equation is not in standard form, $ax^2 + bx + c = 0$. To get 0 on the right side, we proceed as follows.

$$6x^2 = 12x$$ This is the equation to solve.

$$6x^2 - 12x = 12x - 12x$$ Use the subtraction property of equality to get 0 on the right side: Subtract 12x from both sides.

$$6x^2 - 12x = 0$$ Combine like terms: 12x − 12x = 0. This equation is in standard form.

To solve this equation, we factor the left side and proceed as follows.

$$6x(x - 2) = 0$$ Factor out the GCF, 6x.

$$6x = 0 \quad \text{or} \quad x - 2 = 0$$ Set each factor equal to 0.

$$x = \frac{0}{6} \qquad\qquad x = 2$$ Solve each equation.

$$x = 0$$

The solutions are 0 and 2 and the solution set is $\{0, 2\}$. Check each solution in the original equation, $6x^2 = 12x$.

> **Self Check 4**   Solve: $5x^2 = 25x$
> **Now Try**   Problem 47

EXAMPLE 5   Solve: $2x^2 - 2 = -3x$

Strategy   We will add $3x$ to both sides of the equation to get 0 on the right side. Then we will factor the resulting trinomial and use the zero-factor property.

Why   To use the zero-factor property, we need one side of the equation to be factored completely and the other side to be 0.

Solution   The equation is not in standard form, $ax^2 + bx + c = 0$. To get 0 on the right side, we proceed as follows.

$$2x^2 - 2 = -3x$$ This is the equation to solve.

$$2x^2 + 3x - 2 = -3x + 3x$$ Use the addition property of equality to get 0 on the right side: Add 3x to both sides.

$$2x^2 + 3x - 2 = 0$$ Combine like terms: −3x + 3x = 0. This equation is in standard form.

$$(2x - 1)(x + 2) = 0$$ Factor the trinomial.

$$2x - 1 = 0 \quad \text{or} \quad x + 2 = 0$$ Set each factor equal to 0.

$$2x = 1 \qquad\qquad x = -2$$ Solve each equation.

$$x = \frac{1}{2}$$

The solutions are $\frac{1}{2}$ and -2 and the solution set is $\left\{-2, \frac{1}{2}\right\}$. Check each solution in the original equation, $2x^2 - 2 = -3x$.

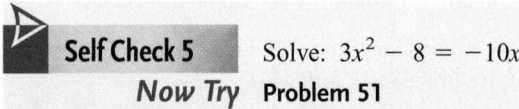

> **Self Check 5**   Solve: $3x^2 - 8 = -10x$
> **Now Try**   Problem 51

Unlike linear equations, quadratic equations have two solutions. In some cases, however, the two solutions are the same number.

EXAMPLE 6 Solve: $x(9x - 12) = -4$

Strategy To write the equation in standard form, we will distribute the multiplication by x and add 4 to both sides. Then we will factor the resulting trinomial and use the zero-factor property.

Why To use the zero-factor property, we need one side of the equation to be factored completely and the other side to be 0.

Solution

$x(9x - 12) = -4$ This is the equation to solve.

$9x^2 - 12x = -4$ Distribute the multiplication by x.

$9x^2 - 12x + 4 = -4 + 4$ To get 0 on the right side, add 4 to both sides.

$9x^2 - 12x + 4 = 0$ Combine like terms: $-4 + 4 = 0$. This equation is in standard form.

$(3x - 2)(3x - 2) = 0$ Factor the trinomial, $9x^2 - 12x + 4$.

$3x - 2 = 0$ or $3x - 2 = 0$ Set each factor equal to 0.

$3x = 2$ | $3x = 2$ Solve each equation.

$x = \frac{2}{3}$ | $x = \frac{2}{3}$

After solving both equations, we see that $\frac{2}{3}$ is a *repeated solution*. Thus, the solution set is $\left\{\frac{2}{3}\right\}$. Check this result by substituting it into the original equation.

Caution

To use the zero-factor property, one side of the equation must be 0. In this example, it would be incorrect to set each factor equal to -4.

~~$x = -4$ or $9x - 12 = -4$~~

If the product of two numbers is -4, one of them does not have to be -4. For example, $2(-2) = -4$.

Self Check 6 Solve: $x(4x + 12) = -9$

Now Try **Problem 59**

3 Solve Third-Degree Equations by Factoring.

Some equations involving polynomials with degrees higher than 2 can also be solved by using the factoring method. In such cases, we use an extension of the zero-factor property: When the product of two *or more* real numbers is 0, at least one of them is 0.

EXAMPLE 7 Solve: $6x^3 + 12x = 17x^2$

Strategy This equation is not quadratic, because it contains a term involving x^3. However, we can solve it by using factoring. First we get 0 on the right side by subtracting $17x^2$ from both sides. Then we factor the polynomial on the left side and use an extension of the zero-factor property.

Why   To use the zero-factor property, we need one side of the equation to be factored completely and the other side to be 0.

The Language of Algebra
Since the highest degree of any term in $6x^3 + 12x = 17x^2$ is 3, it is called a *third-degree* equation. Note that it has three solutions.

Solution

$$6x^3 + 12x = 17x^2 \qquad \text{This is the equation to solve.}$$
$$6x^3 - 17x^2 + 12x = 17x^2 - 17x^2 \qquad \text{To get 0 on the right side, subtract } 17x^2 \text{ from both sides.}$$
$$6x^3 - 17x^2 + 12x = 0 \qquad \text{Combine like terms: } 17x^2 - 17x^2 = 0.$$
$$x(6x^2 - 17x + 12) = 0 \qquad \text{Factor out the GCF, } x.$$
$$x(2x - 3)(3x - 4) = 0 \qquad \text{Factor the trinomial, } 6x^2 - 17x + 12.$$

If $x(2x - 3)(3x - 4) = 0$, then at least one of the factors is equal to 0.

$$x = 0 \qquad \text{or} \qquad 2x - 3 = 0 \qquad \text{or} \qquad 3x - 4 = 0 \quad \text{Set each factor equal to 0.}$$
$$2x = 3 \qquad\qquad 3x = 4 \quad \text{Solve each equation.}$$
$$x = \frac{3}{2} \qquad\qquad x = \frac{4}{3}$$

The solutions are 0, $\frac{3}{2}$, and $\frac{4}{3}$ and the solution set is $\left\{0, \frac{4}{3}, \frac{3}{2}\right\}$. Check each solution in the original equation, $6x^3 + 12x = 17x^2$.

 Self Check 7   Solve: $10x^3 + x^2 = 2x$
Now Try   Problem 67

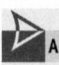 **ANSWERS TO SELF CHECKS**   **1.** $12, -\frac{6}{5}$ **2.** $-2, -3$ **3.** $-7, 7$ **4.** $0, 5$ **5.** $\frac{2}{3}, -4$ **6.** $-\frac{3}{2}$
7. $0, \frac{2}{5}, -\frac{1}{2}$

STUDY SET
6.7

VOCABULARY

Fill in the blanks.

1. $2x^2 + 3x - 1 = 0$ and $x^2 - 36 = 0$ are examples of _____ equations.

2. $ax^2 + bx + c = 0$ is called the _____ form of a quadratic equation.

3. The _____ property states that if the product of two numbers is 0, at least one of them is 0: If $ab = 0$, then $a =$ ▢ or $b =$ ▢.

4. Since the highest degree of any term in $x^3 - 5x^2 - 6x = 0$ is 3, it is called a _____-degree equation.

CONCEPTS

5. Which of the following are quadratic equations?
  a. $x^2 + 2x - 10 = 0$ **b.** $2x - 10 = 0$
  c. $x^2 = 15x$ **d.** $x^3 + x^2 + 2x = 0$

6. Write each equation in the standard form $ax^2 + bx + c = 0$.
  a. $x^2 + 2x = 6$ **b.** $x^2 = 5x$
  c. $3x(x - 8) = -9$ **d.** $4x^2 = 25$

7. Set $5x + 4$ equal to 0 and solve for x.

8. What step should be performed first to solve $x^2 - 6x - 16 = 0$?

9. What step (or steps) should be performed first before factoring is used to solve each equation?

    a. $x^2 + 7x = -6$

    b. $x(x + 7) = 3$

10. Check to determine whether the given number is a solution of the given quadratic equation.

    a. $x^2 - 4x = 0$; 4

    b. $x^2 - 2x - 7 = 0$; -2

NOTATION

Complete each solution to solve the equation.

11. $(x - 1)(x + 7) = 0$

$$x - 1 = \boxed{} \quad \text{or} \quad \boxed{} = 0$$

$$x = 1 \qquad\qquad x = \boxed{}$$

12. $7y^2 + 14y = 0$

$$\boxed{}(y + 2) = 0$$

$$7y = 0 \qquad y + 2 = 0$$

$$y = \boxed{} \qquad y = -2$$

13. $p^2 - p - 6 = 0$

$$(\boxed{} - 3)(p + 2) = 0$$

$$\boxed{} = 0 \quad \text{or} \quad p + 2 = \boxed{}$$

$$p = \boxed{} \qquad p = \boxed{}$$

14. $4y^2 - 25 = 0$

$$(2y + \boxed{})(2y - \boxed{}) = 0$$

$$2y + 5 = \boxed{} \quad \text{or} \quad \boxed{} = 0$$

$$2y = \boxed{} \qquad 2y = 5$$

$$y = -\frac{5}{2} \qquad y = \boxed{}$$

GUIDED PRACTICE

Solve each equation. See Example 1.

15. $(x - 3)(x - 2) = 0$ **16.** $(x + 2)(x + 3) = 0$

17. $(x + 7)(x - 7) = 0$ **18.** $(x - 8)(x + 8) = 0$

19. $6x(2x - 5) = 0$ **20.** $5x(5x + 7) = 0$

21. $-7a(3a + 10) = 0$ **22.** $-6t(2t - 9) = 0$

23. $t(t - 6)(t + 8) = 0$ **24.** $n(n + 1)(n - 6) = 0$

25. $(x - 1)(x + 2)(x - 3) = 0$ **26.** $(x + 2)(x + 3)(x - 4) = 0$

Solve each equation. See Example 2.

27. $x^2 - 13x + 12 = 0$ **28.** $x^2 + 7x + 6 = 0$

29. $x^2 - 4x - 21 = 0$ **30.** $x^2 + 2x - 15 = 0$

31. $x^2 - 9x + 8 = 0$ **32.** $x^2 - 14x + 45 = 0$

33. $a^2 + 8a + 15 = 0$ **34.** $a^2 - 17a + 60 = 0$

Solve each equation. See Example 3.

35. $x^2 - 81 = 0$ **36.** $x^2 - 36 = 0$

37. $t^2 - 25 = 0$ **38.** $m^2 - 49 = 0$

39. $4x^2 - 1 = 0$ **40.** $9y^2 - 1 = 0$

41. $9y^2 - 49 = 0$ **42.** $16z^2 - 25 = 0$

Solve each equation. See Example 4.

43. $w^2 = 7w$ **44.** $x^2 = 5x$

45. $s^2 = 16s$ **46.** $p^2 = 20p$

47. $4y^2 = 12y$ **48.** $5m^2 = 15m$

49. $3x^2 = -8x$ **50.** $3s^2 = -4s$

Solve each equation. See Example 5.

51. $3x^2 + 5x = 2$ **52.** $3x^2 + 14x = -8$

53. $2x^2 + x = 3$ **54.** $2x^2 - 5x = -2$

55. $5x^2 + 1 = 6x$ **56.** $6x^2 + 1 = 5x$

57. $2x^2 - 3x = 20$ **58.** $2x^2 - 3x = 14$

Solve each equation. See Example 6.

59. $4r(r + 7) = -49$ **60.** $5m(5m + 8) = -16$

61. $9a(a - 3) = 3a - 25$ **62.** $3x(3x + 10) = 6x - 16$

63. $z(z - 7) = -12$ **64.** $p(p + 1) = 6$

65. $(n + 8)(n - 3) = -30$ **66.** $(2s + 5)(s + 1) = -1$

Solve each equation. See Example 7.

67. $x^3 + 3x^2 + 2x = 0$ **68.** $x^3 - 7x^2 + 10x = 0$

69. $k^3 - 27k - 6k^2 = 0$ **70.** $j^3 - 22j - 9j^2 = 0$

71. $x^3 - 6x^2 = -9x$ **72.** $m^3 - 8m^2 + 16m = 0$

73. $2x^3 = 2x(x + 2)$ **74.** $x^3 + 7x^2 = x^2 - 9x$

TRY IT YOURSELF

Solve each equation.

75. $4x^2 = 81$

76. $9y^2 = 64$

77. $x^2 - 16x + 64 = 0$

78. $h^2 + 2h + 1 = 0$

79. $(2s - 5)(s + 6) = 0$

80. $h(3h - 4)(h + 1) = 0$

81. $3b^2 - 30b = 6b - 60$

82. $2m^2 - 8m = 2m - 12$

83. $k^3 + k^2 - 20k = 0$

84. $n^3 - 6n^2 + 8n = 0$

85. $x^2 - 100 = 0$

86. $z^2 - 25 = 0$

87. $3y^2 - 14y - 5 = 0$

88. $4y^2 - 11y - 3 = 0$

89. $(x - 2)(x^2 - 8x + 7) = 0$

90. $(x - 1)(x^2 + 5x + 6) = 0$

91. $4a^2 + 1 = 8a + 1$

92. $3b^2 - 6 = 12b - 6$

93. $2b(6b + 13) = -12$

94. $5f(5f - 16) = -15$

95. $3a^3 + 4a^2 + a = 0$

96. $10b^3 - 15b^2 - 25b = 0$

97. $-15x^2 + 2 + 7x = 0$

98. $-8x^2 + 3 - 10x = 0$

99. $4p^2 - 121 = 0$

100. $q^2 - \dfrac{1}{4} = 0$

101. $d(8d - 9) = -1$

102. $6n^3 - 6n = 0$

WRITING

103. Explain the zero-factor property.

104. Find the error in the following solution.

$$x(x + 1) = 6$$
$$x = 6 \quad \text{or} \quad x + 1 = 6$$
$$x = 5$$

The solutions are 6 and 5.

105. A student solved $x^2 - 5x + 6 = 0$ and obtained two solutions: 2 and 3. Explain the error in his check.

Check:
$$x^2 - 5x + 6 = 0$$
$$2^2 - 5(3) + 6 \stackrel{?}{=} 0$$
$$4 - 15 + 6 \stackrel{?}{=} 0$$
$$-5 = 0 \quad \text{False}$$

2 is not a solution.   3 is not a solution.

106. In this section, we solved quadratic equations by factoring. Did we always obtain two different solutions? Explain.

107. What is wrong with the step used to solve $x^2 = 2x$ shown below?

$$x^2 = 2x$$
$$\frac{x^2}{x} = \frac{2x}{x}$$
$$x = 2$$

The solution is 2.

108. Explain the error in the following solution.

Factor: $x^2 - 5x + 6$

$$(x - 2)(x - 3) = 0$$
$$x - 2 = 0 \quad \text{or} \quad x - 3 = 0$$
$$x = 2 \quad \quad x = 3$$

The solutions are 2 and 3.

REVIEW

109. EXERCISE   A doctor advises a patient to exercise at least 15 minutes but less than 30 minutes per day. Use a compound inequality to express the range of these times t in minutes.

110. SNACKS   A bag of peanuts is worth $0.30 less than the same size bag of cashews. Equal amounts of peanuts and cashews are used to make 40 bags of a mixture that is worth $1.05 per bag. How much is a bag of cashews worth?

CHALLENGE PROBLEMS

Solve each equation.

111. $x^4 - 625 = 0$

112. $2a^3 + a^2 - 32a - 16 = 0$

113. $(x - 3)^2 = 2x + 9$

114. $(x + 3)^2 = (2x - 1)^2$

SECTION 6.8
Applications of Quadratic Equations

Objectives

1 Solve problems involving geometric figures.

2 Solve problems involving consecutive integers.

3 Solve problems using the Pythagorean theorem.

4 Solve problems given the quadratic equation model.

In Chapter 2, we solved mixture, investment, and uniform motion problems. To model those situations, we used *linear equations* in one variable. We will now consider situations that are modeled by *quadratic equations*.

1 **Solve Problems Involving Geometric Figures.**

We can use the five-step problem solving strategy and the factoring method for solving quadratic equations to find the dimensions of certain figures, given their area.

© Morgan Art Foundation Limited/Art Resource, NY/ARS

EXAMPLE 1 *Painting.* In 2002, the pop art painting *American Sweetheart*, by artist Robert Indiana, sold for $614,500. The area of the rectangular painting is 32 square feet. Find the dimensions of the painting if it is twice as long as it is wide.

Analyze the Problem

• The area of the painting is 32 ft².

• The length is twice as long as the width.

• Find the length and width.

Form an Equation Since the length is related to the width, let w = the width of the painting in feet. Then $2w$ = the length of the painting. To form an equation, we use the formula for the area of a rectangle, $A = lw$, where $A = 32$.

The area of the rectangle	equals	the length	times	the width.
32	=	$2w$	·	w

Solve the Equation

$$32 = 2w \cdot w$$

$$32 = 2w^2$$ Multiply 2w and w. Note that this is a quadratic equation.

$$0 = 2w^2 - 32$$ To get 0 on the left side, subtract 32 from both sides.

$$0 = 2(w^2 - 16)$$ Factor out the GCF, 2.

$$0 = 2(w + 4)(w - 4)$$ Factor the difference of two squares, $w^2 - 16$.

$$w + 4 = 0 \quad \text{or} \quad w - 4 = 0$$ Since 2 cannot equal 0, discard that possibility. Set each factor that contains a variable equal to 0.

$$\cancel{w = -4} \quad \bigg| \quad w = 4$$ Solve each equation.

The Language of Algebra
When solving real-world application problems, we *discard* or *reject* any solutions of equations that do not make sense, such as a negative width of a painting.

2w

w

State the Conclusion The solutions of the equation are −4 and 4. Since *w* represents the width of the picture, and the width cannot be negative, we discard −4. Thus, the width of the picture is 4 feet and the length is 2 · 4 = 8 feet.

Check the Result A rectangle with dimensions 4 feet by 8 feet has an area of 32 ft^2, and the length is twice the width. The answers check.

 Now Try Problem 13

© Nicholas Pitt/Alamy

EXAMPLE 2 **Windmills.** The height of a triangular canvas sail of a windmill is 1 foot less than twice the length of its base. If the sail has an area of 22.5 ft^2, find the length of the base and the height.

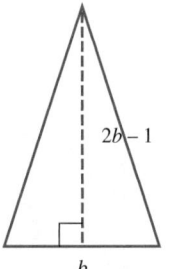

Analyze the Problem

- The height is 1 ft less than twice the length of the base.
- The area is 22.5 ft^2.
- Find the length of the base and the height.

Form an Equation Since the height is related to the length of the base, we let b = the length of the base of the sail in feet. Then $2b - 1$ = the height of the sail. To form an equation, we use the formula for the area of a triangle: $A = \frac{1}{2}bh$, where $A = 22.5$.

The area of the triangle	equals	one-half	times	the length of the base	times	the height.
22.5	=	$\frac{1}{2}$	·	b	·	$(2b - 1)$

Solve the Equation

$$22.5 = \frac{1}{2}b(2b - 1)$$

$$2 \cdot 22.5 = 2 \cdot \frac{1}{2}b(2b - 1)$$ To clear the equation of the fraction, multiply both sides by 2.

$$45 = b(2b - 1)$$ Multiply: $2 \cdot 22.5 = 45$ and $2 \cdot \frac{1}{2} = 1$.

$$45 = 2b^2 - b$$ Distribute the multiplication by b. Note that this is a quadratic equation.

$$0 = 2b^2 - b - 45$$ To get 0 on the left side, subtract 45 from both sides.

$$0 = (2b + 9)(b - 5)$$ Factor the trinomial.

$2b + 9 = 0$ or $b - 5 = 0$ Set each factor equal to 0.

$2b = -9$ $b = 5$ Solve each equation.

$b = -\frac{9}{2}$ (crossed out)

> **Caution**
> A common error is to incorrectly "distribute" the 2 on the right side of the equation.
>
>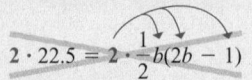
> $$2 \cdot 22.5 = 2 \cdot \frac{1}{2}b(2b - 1)$$ (crossed out)

State the Conclusion The solutions of the equation are $-\frac{9}{2}$ and 5. Since b represents the length of the base of the sail, and it cannot be negative, we discard $-\frac{9}{2}$. The length of the base is then 5 feet, and the height is $2(5) - 1 = 9$ feet.

Check the Result   A triangle with height 9 feet and base 5 feet has area $\frac{1}{2}(9)(5) = 22.5$ ft^2, and the height is 1 foot less than twice the base. The answers check.

 Now Try   **Problem 19**

②   Solve Problems Involving Consecutive Integers.

Consecutive integers are integers that follow one another, such as 15 and 16. When solving consecutive integer problems, if we let $x =$ the first integer, then:

- two consecutive integers are x and $x + 1$
- two consecutive even integers are x and $x + 2$
- two consecutive odd integers are x and $x + 2$

> **EXAMPLE 3**   *Women's Tennis.*   In the 1998 Australian Open, sisters Venus and Serena Williams played against each other for the first time as professionals. Venus was victorious over her younger sister. At that time, their ages were consecutive integers whose product was 272. How old were Venus and Serena when they met in this match?

Analyze the Problem

- Venus is older than Serena.
- Their ages were consecutive integers.
- The product of their ages was 272.
- Find Venus' and Serena's age when they played this match.

Form an Equation   Let $x =$ Serena's age when she played in the 1998 Australian Open. Since their ages were consecutive integers, and since Venus is older, we let $x + 1 =$ Venus' age. The word *product* indicates multiplication.

Serena's age	times	Venus' age	was	272.
x	$\cdot$	$(x + 1)$	$=$	272

Solve the Equation

$$x(x + 1) = 272$$
$$x^2 + x = 272 \qquad \text{Distribute the multiplication by x. Note that this is a quadratic equation.}$$
$$x^2 + x - 272 = 0 \qquad \text{Subtract 272 from both sides to make the right side 0.}$$
$$(x + 17)(x - 16) = 0 \qquad \text{Factor } x^2 + x - 272. \text{ Two numbers whose product is } -272 \text{ and whose sum is 1 are 17 and } -16.$$
$$x + 17 = 0 \quad \text{or} \quad x - 16 = 0 \qquad \text{Set each factor equal to 0.}$$
$$x = -17 \qquad\qquad x = 16 \qquad \text{Solve each equation.}$$

State the Conclusion   The solutions of the equation are -17 and 16. Since x represents Serena's age, and it cannot be negative, we discard -17. Thus, Serena Williams was 16 years old and Venus Williams was $16 + 1 = 17$ years old when they played against each other for the first time as professionals.

Check the Result Since 16 and 17 are consecutive integers, and since $16 \cdot 17 = 272$, the answers check.

 Now Try **Problem 23**

③ **Solve Problems Using the Pythagorean Theorem.**

Caution
It doesn't matter which leg of the right triangle is labeled *a* and which is labeled *b*. However, the hypotenuse must be labeled *c*.

A **right triangle** is a triangle that has a 90° (right) angle. The longest side of a right triangle is the **hypotenuse,** which is the side opposite the right angle. The remaining two sides are the **legs** of the triangle. The **Pythagorean theorem** provides a formula relating the lengths of the three sides of a right triangle.

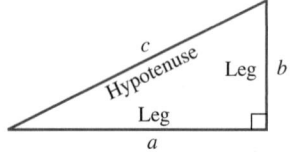

The Pythagorean Theorem

If *a* and *b* are the lengths of the legs of a right triangle and *c* is the length of the hypotenuse, then

$$a^2 + b^2 = c^2$$

In a right triangle, the sum of the squares of the lengths of the two legs is equal to the square of the length of the hypotenuse.

Pythagoras

EXAMPLE 4 *Right Triangles.* The longer leg of a right triangle is 3 units longer than the shorter leg. If the hypotenuse is 6 units longer than the shorter leg, find the lengths of the sides of the triangle.

Analyze the Problem We begin by drawing a right triangle and labeling the legs and the hypotenuse.

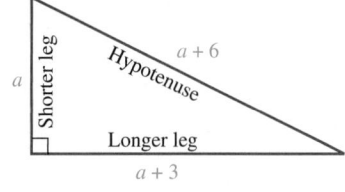

Form an Equation Let a = length of the shorter leg. Then the length of the hypotenuse is $a + 6$ and the length of the longer leg is $a + 3$. By the Pythagorean theorem, we have

$\left(\begin{array}{c}\text{The length of}\\ \text{the shorter leg}\end{array}\right)^2$	plus	$\left(\begin{array}{c}\text{the length of}\\ \text{the longer leg}\end{array}\right)^2$	equals	$\left(\begin{array}{c}\text{the length of the}\\ \text{hypotenuse}\end{array}\right)^2$
a^2	$+$	$(a+3)^2$	$=$	$(a+6)^2$

The Language of Algebra
A *theorem* is a mathematical statement that can be proved. The *Pythagorean theorem* is named after *Pythagoras,* a Greek mathematician who lived about 2,500 years ago. He is thought to have been the first to prove the theorem.

Solve the Equation

$$a^2 + (a + 3)^2 = (a + 6)^2$$

$a^2 + a^2 + 6a + 9 = a^2 + 12a + 36$ Find $(a + 3)^2$ and $(a + 6)^2$.

$2a^2 + 6a + 9 = a^2 + 12a + 36$ On the left side: $a^2 + a^2 = 2a^2$.

$a^2 - 6a - 27 = 0$ To get 0 on the right side, subtract a^2, $12a$, and 36 from both sides. This is a quadratic equation.

$(a - 9)(a + 3) = 0$ Factor the trinomial.

$a - 9 = 0$ or $a + 3 = 0$ Set each factor equal to 0.

$a = 9$ $\cancel{a = -3}$ Solve each equation.

State the Conclusion   Since a side cannot have a negative length, we discard the solution −3. Thus, the shorter leg is 9 units long, the hypotenuse is $9 + 6 = 15$ units long, and the longer leg is $9 + 3 = 12$ units long.

Check the Result   The longer leg, 12, is 3 units longer than the shorter leg, 9. The hypotenuse, 15, is 6 units longer than the shorter leg, 9, and the side lengths satisfy the Pythagorean theorem. So the results check.

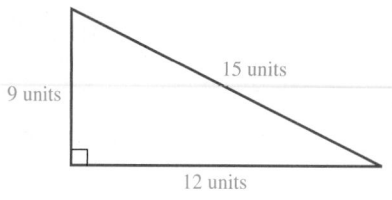

15 units

9 units

12 units

$$9^2 + 12^2 \stackrel{?}{=} 15^2$$
$$81 + 144 \stackrel{?}{=} 225$$
$$225 = 225$$

   Now Try   **Problem 31**

---

4   **Solve Problems Given the Quadratic Equation Model.**

A quadratic equation can be used to describe the height of an object that is projected upward, such as a ball thrown into the air or an arrow shot into the sky.

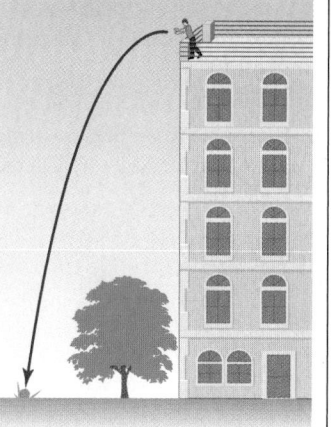

EXAMPLE 5   ***College Pranks.***   A student uses rubber tubing to launch a water balloon from the roof of his dormitory. The height h (in feet) of the balloon, t seconds after being launched, is given by the formula $h = -16t^2 + 48t + 64$. After how many seconds will the balloon hit the ground?

Analyze the Problem   When the water balloon hits the ground, its height will be 0 feet. To find the time that it takes for the balloon to hit the ground, we set h equal to 0, and solve the quadratic equation for t.

Form an Equation $h = -16t^2 + 48t + 64$

$0 = -16t^2 + 48t + 64$ Substitute 0 for the height, h. This is a quadratic equation.

Success Tip

Note that the common factor, −16, divides −16, 48, and 64 exactly:

$$\frac{-16}{-16} = 1 \qquad \frac{48}{-16} = -3$$
$$\frac{64}{-16} = -4$$

Solve the Equation

$0 = -16t^2 + 48t + 64$

$0 = -16(t^2 - 3t - 4)$ Factor out the opposite of the GCF, −16.

$0 = -16(t + 1)(t - 4)$ Factor the trinomial.

$t + 1 = 0$ or $t - 4 = 0$ Since −16 cannot equal 0, discard that possibility. Set each factor that contains a variable equal to 0.

$t = -1$ | $t = 4$ Solve each equation.

State the Conclusion   The equation has two solutions, -1 and 4. Since t represents time, and, in this case, time cannot be negative, we discard -1. The second solution, 4, indicates that the balloon hits the ground 4 seconds after being launched.

Check the Result   Check this result by substituting 4 for t in $h = -16t^2 + 48t + 64$. You should get $h = 0$.

 Now Try   Problem 37

STUDY SET
6.8

VOCABULARY

Fill in the blanks.

1. Integers that follow one another, such as 6 and 7, are called _____ integers.

2. A _____ triangle is a triangle that has a 90° angle.

3. The longest side of a right triangle is the _____. The remaining two sides are the _____ of the triangle.

4. The _____ theorem is a formula that relates the lengths of the three sides of a right triangle.

CONCEPTS

5. A rectangle has an area of 40 in.^{2}. The length is 3 inches longer than the width. Which rectangle below meets these conditions?

   i.

   4 in.

   10 in.

   ii.

   5 in.

   8 in.

6. A triangle has an area of 15 ft^2. The height is 7 feet less than twice the length of the base. Which triangle below meets these conditions?

   i.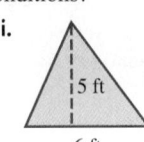

   5 ft

   6 ft

   ii.

   3 ft

   10 ft

7. Multiply both sides of the equation by 2. *Do not solve.*

   $10 = \frac{1}{2}b(b + 5)$

8. Fill in the blanks.
   a. If the length of the hypotenuse of a right triangle is c and the lengths of the other two legs are a and b, then _____ $= c^2$.
   b. In a right triangle, the sum of the _____ of the lengths of the two legs is equal to the square of the length of the _____.

9. a. What kind of triangle is shown?
   b. What are the lengths of the legs of the triangle?
   c. How much longer is the hypotenuse than the shorter leg?

   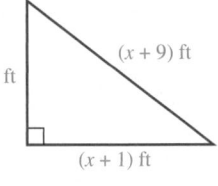

   $(x + 9)$ ft

   x ft

   $(x + 1)$ ft

10. A ball is thrown into the air. Its height h in feet, t seconds after being released, is given by the formula $h = -16t^2 + 24t + 6$. When the ball hits the ground, what is the value of h?

NOTATION

Complete the solution to solve the equation.

11. $0 = -16t^2 + 32t + 48$

   $0 = \boxed{}(t^2 - 2t - 3)$

   $0 = -16(t - 3)(t + \boxed{})$

   $t - 3 = \boxed{}$ or $t + 1 = \boxed{}$

   $t = \boxed{}$ | $t = \boxed{}$

12. Fill in the blanks.
    a. Consecutive integers can be represented by x and _____.
    b. Consecutive odd integers can be represented by x and _____.
    c. Consecutive even integers can be represented by x and _____.

APPLICATIONS

Geometry Problems

13. FLAGS   The length of the flag of Australia is twice as long as the width. Find the dimensions of an Australian flag if its area is 18 ft².

14. BILLIARDS   Pool tables are rectangular, and their length is twice the width. Find the dimensions of a pool table if it occupies 50 ft² of floor space.

15. X-RAYS.   A rectangular-shaped x-ray film has an area of 80 square inches. The length is 2 inches longer than the width. Find its width and length.

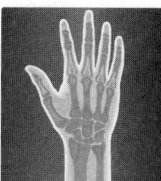

16. INSULATION   The area of the rectangular slab of foam insulation in the illustration is 36 square meters. Find the dimensions of the slab.

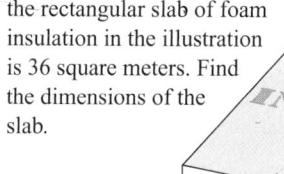

w meters

(2*w* + 1) meters

17. *from* **Campus to Careers**
    Bulletin Boards

Suppose you are an elementary school teacher. You want to order a rectangular bulletin board to mount on a classroom wall that has an area of 90 square feet. Fire code requirements allow for no more than 30% of a classroom wall to be covered by a bulletin board. If the length of the board to be three times as long as the width, what are the dimensions of the largest bulletin board that meets fire code?

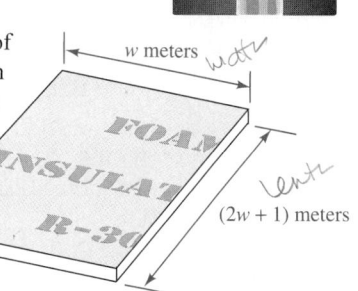
© BananaStock/Jupiterimages

18. TUBING   Refer to the diagram in the next column. A piece of cardboard in the shape of a parallelogram is twisted to form the tube. The parallelogram has an area of 60 square inches. If its height *h* is 7 inches more than the length of the base *b*, what is the length of the base? (*Hint:* The formula for the area of a parallelogram is $A = bh$.)

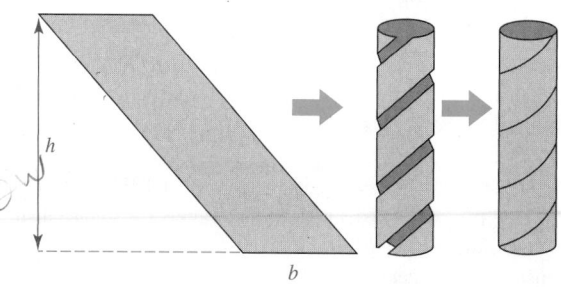

h

b

19. JEANS   The height of the triangular-shaped logo on a pair of jeans is 1 centimeter less than the length of its base. If the area of the logo is 15 square centimeters, find the length of the base and the height.

20. SHUFFLEBOARD   The area of the numbered triangle on a shuffle board court is 27 ft². Its height is 3 feet more than the length of the base. Find the length of the base and the height.

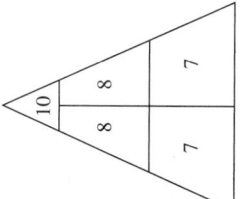

21. SAILBOATS   Refer to the diagram of a sail shown here. The length of the *luff* is 3 times longer than the length of the *foot* of the sail. Find the length of the foot and the length of the luff.

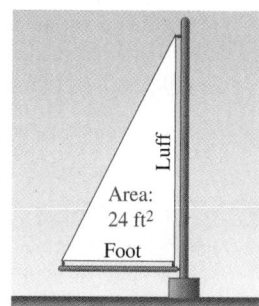

Luff

Area: 24 ft²

Foot

22. DESIGNING TENTS   The length of the base of the triangular sheet of canvas above the door of a tent is 2 feet more than twice its height. The area is 30 square feet. Find the height and the length of the base of the triangle.

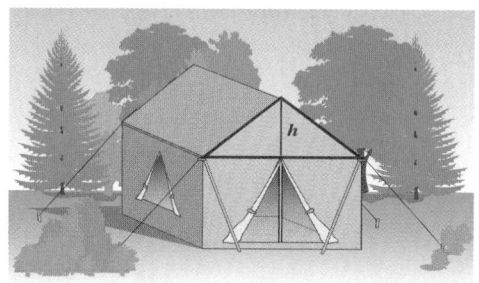

Consecutive Integer Problems

23. NASCAR   The car numbers of drivers Kasey Kahne and Scott Riggs are consecutive positive integers whose product is 90. If Kahne's car number is the smaller, what is the number of each car.

24. BASEBALL   Catcher Thurman Munson and pitcher Whitey Ford are two of the sixteen New York Yankees who have had their uniform numbers retired. Their uniform numbers are consecutive integers whose product is 240. If Munson's was the smaller number, determine the uniform number of each player.

25. CUSTOMER SERVICE   At a pharmacy, customers take a ticket to reserve their turn for service. If the product of the ticket number now being served and the next ticket number to be served is 156, what number is now being served?

26. HISTORY   Delaware was the first state to enter the Union and Hawaii was the 50th. If we order the positions of entry for the rest of the states, we find that Kentucky entered the Union right after Vermont, and the product of their order-of-entry numbers is 210. Use the given information to complete these statements:

Kentucky was the ____ th state to enter the Union.

Vermont was the ____ th state to enter the Union.

27. PLOTTING POINTS   The x-coordinate and y-coordinate of a point in quadrant I are consecutive odd integers whose product is 143. Find the coordinates of the point.

28. PRESIDENTS   George Washington was born on 2-22-1732 (February 22, 1732). He died in 1799 at the age of 67. The month in which he died and the day of the month on which he died are consecutive even integers whose product is 168. When did Washington die?

Pythagorean Theorem Problems

29. HIGH-ROPES ADVEN-TURES COURSES   A builder of a high-ropes adventure course wants to secure a pole by attaching a support cable from the anchor stake 8 yards from its base to a point 6 yards up the pole. How long should the cable be?

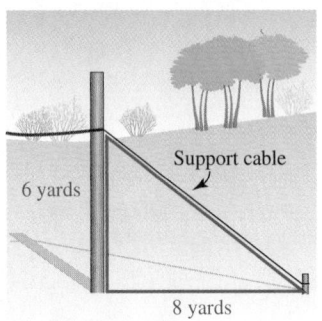

30. WIND DAMAGE   A tree was blown over in a wind storm. Find x. Then find the height of the tree when it was standing upright.

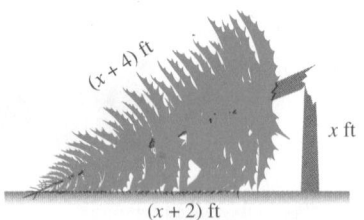

31. MOTO X   Find x, the height of the landing ramp.

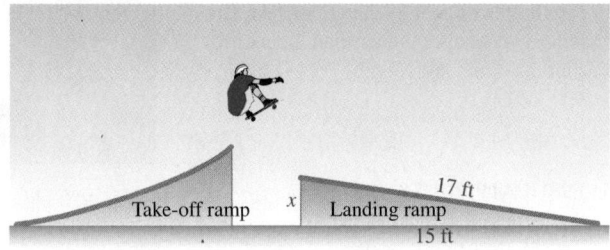

32. GARDENING TOOLS   The dimensions (in millimeters) of the teeth of a pruning saw blade are given in the illustration. Find each length.

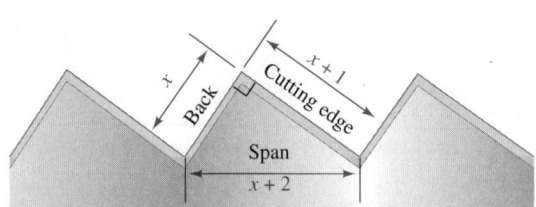

33. BOATING   The inclined ramp of the boat launch is 8 meters longer than the rise of the ramp. The run is 7 meters longer than the rise. How long are the three sides of the ramp?

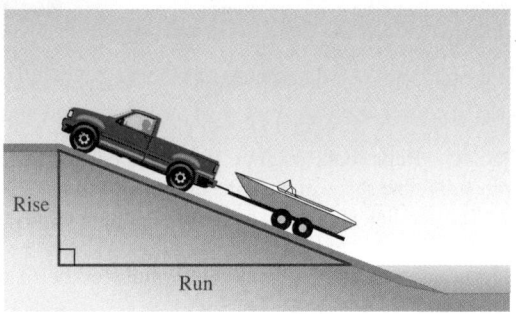

34. CAR REPAIRS   To create some space to work under the front end of a car, a mechanic drives it up steel ramps. A ramp is 1 foot longer than the back, and the base is 2 feet longer than the back of the ramp. Find the length of each side of the ramp.

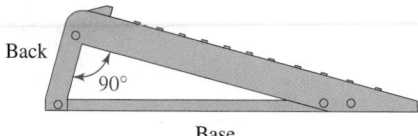

Back
90°
Base

Quadratic Equation Model Problems

35. THRILL RIDES   At the peak of a roller coaster ride, a rider's sunglasses fly off his head. The height h (in feet) of the glasses, t seconds after he loses them, is given by $h = -16t^2 + 64t + 80$. After how many seconds will the glasses hit the ground?   (*Hint:* Factor out -16.)

36. PARADES   A celebrity on the top of a parade float is tossing pieces of candy to the people on the street below. The height h (in feet) of a piece of candy, t seconds after being thrown, is given by $h = -16t^2 + 16t + 32$. After how many seconds will the candy hit the ground?   (*Hint:* Factor out -16.)

37. SOFTBALL   A pitcher can throw a fastball underhand at 63 feet per second (about 45 mph). If she throws a ball into the air with that velocity, its height h in feet, t seconds after being released, is given by $h = -16t^2 + 63t + 4$. After the ball is thrown, in how many seconds will it hit the ground?   (*Hint:* Factor out -16.)

38. OFFICIATING   Before a football game, a coin toss is used to determine which team will kick off. The height h (in feet) of a coin above the ground t seconds after being flipped up into the air is given by $h = -16t^2 + 22t + 3$. How long does a team captain have to call heads or tails if it must be done while the coin is in the air?   (*Hint:* Factor out -1.)

39. DOLPHINS   Refer to the illustration. The height h in feet reached by a dolphin t seconds after breaking the surface of the water is given by $h = -16t^2 + 32t$. How long will it take the dolphin to jump out of the water and touch the trainer's hand?

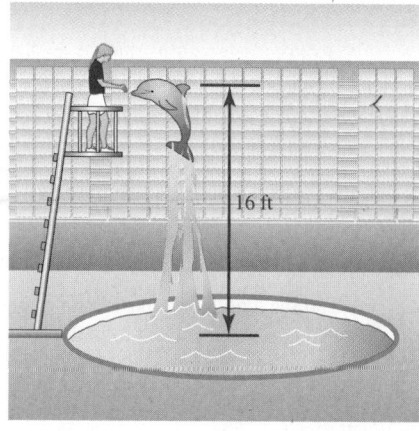

16 ft

40. EXHIBITION DIVING   In Acapulco, Mexico, men diving from a cliff to the water 64 feet below are quite a tourist attraction. A diver's height h above the water (in feet), t seconds after diving, is given by $h = -16t^2 + 64$. How long does a dive last?

41. CHOREOGRAPHY   For the finale of a musical, 36 dancers are to assemble in a triangular-shaped series of rows, where each row has one more dancer than the previous row. The illustration shows the beginning of such a formation. The relationship between the number of rows r and the number of dancers d is given by $d = \frac{1}{2}r(r + 1)$. Determine the number of rows in the formation.

42. CRAFTS   The illustration shows how a wall hanging can be created by stretching yarn from peg to peg across a wooden ring. The relationship between the number of pegs p placed evenly around the ring and the number of yarn segments s that criss-cross the ring is given by the formula $s = \frac{p(p - 3)}{2}$. How many pegs are needed if the designer wants 27 segments to criss-cross the ring?   (*Hint:* Multiply both sides of the equation by 2.)

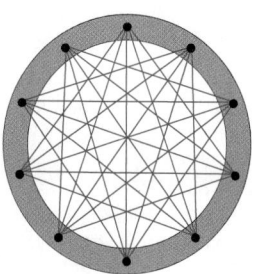

WRITING

43. A student was asked to solve the following problem: *The length of a rectangular room is 2 feet more than twice the width. If the area of the room is 60 square feet, find its dimensions.* Here is the student's solution:

Since $10 \cdot 6 = 60$, the length of the room is 10 feet and the width is 6 feet.

Explain why his solution is incorrect.

44. Suppose that to find the length of the base of a triangle, you write a quadratic equation and solve it to find $b = 6$ or $b = -8$. Explain why one solution should be discarded.

45. What error is apparent in the following illustration?

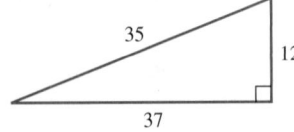

46. When naming the legs of a right triangle, explain why it doesn't matter which leg you label a and which leg you label b.

REVIEW

Find each special product.

47. $(5b - 2)^2$

48. $(2a + 3)^2$

49. $(s^2 + 4)^2$

50. $(m^2 - 1)^2$

51. $(9x + 6)(9x - 6)$

52. $(5b + 2)(5b - 2)$

CHALLENGE PROBLEMS

53. POOL BORDERS The owners of a 10-meter wide by 25-meter long rectangular swimming pool want to surround the pool with a crushed-stone border of uniform width. They have enough stone to cover 74 square meters. How wide should they make the border?

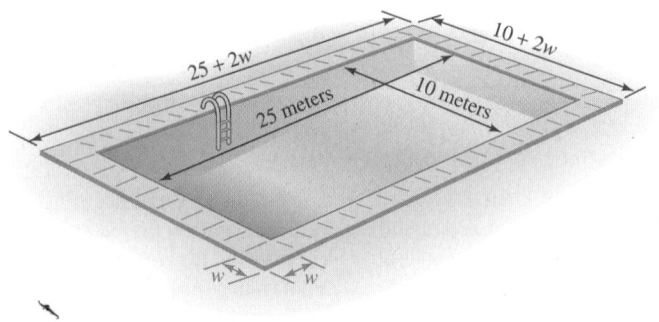

54. Find h.

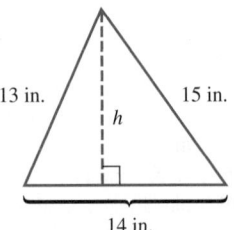

CHAPTER 6
Summary & Review

SECTION 6.1 The Greatest Common Factor; Factoring by Grouping

DEFINITIONS AND CONCEPTS	EXAMPLES
Factoring is multiplication reversed. To **factor a polynomial** means to express it as a product of two (or more) polynomials.	Multiplication: Given the factors, we find a polynomial. ⟶ $2x(5x + 3) = 10x^2 + 6x$ ⟵ Factoring: Given a polynomial, we find the factors.
A natural number is in **prime-factored form** when it is written as the product of prime numbers.	The prime-factored form of 28 is $2 \cdot 2 \cdot 7 = 2^2 \cdot 7$.

SECTION 6.1   The Greatest Common Factor; Factoring by Grouping—*continued*

DEFINITIONS AND CONCEPTS	EXAMPLES
To find the **greatest common factor, GCF,** of a list of terms 1. Write each coefficient as a product of prime factors. 2. Identify the numerical and variable factors common to each term. 3. Multiply the common numerical and variable factors identified in Step 2 to obtain the GCF. If there are no common factors, the GCF is 1.	Find the GCF of $35x^4$, $63x^3$, and $42x^2$. $35x^4 = 5 \cdot 7 \cdot x \cdot x \cdot x \cdot x$ $63x^3 = 3 \cdot 3 \cdot 7 \cdot x \cdot x \cdot x$ $\Big\}$ GCF $= 7 \cdot x \cdot x = 7x^2$ $42x^2 = 2 \cdot 3 \cdot 7 \cdot x \cdot x$
The first step of factoring a polynomial is to see whether the terms of the polynomial have a common factor. If they do, **factor out the GCF.**	Factor: $35x^4 + 63x^3 - 42x^2$ $= 7x^2(5x^2 + 9x - 6)$ Factor out the GCF, $7x^2$. Use multiplication to check the factorization: $7x^2(5x^2 + 9x - 6) = 35x^4 + 63x^3 - 42x^2$ This is the original polynomial.
If a polynomial has four terms, try **factoring by grouping.** 1. Group the terms of the polynomial so that the first two terms have a common factor and the last two terms have a common factor. 2. Factor out the common factor from each group. 3. Factor out the resulting common binomial factor. If there is no common binomial factor, regroup the terms of the polynomial and repeat steps 2 and 3.	Factor: $ax - bx + ay - by$ $= x(a - b) + y(a - b)$ Factor out x from $ax - bx$ and y from $ay - by$. $= (a - b)(x + y)$ Factor out the common binomial factor, $(a - b)$.

REVIEW EXERCISES

Find the prime-factorization of each number.

1. 35

2. 96

Find the GCF of each list.

3. 28 and 35

4. $36a^4$, $54a^3$, and $126a^6$

Factor.

5. $3x + 9y$

6. $5ax^2 + 15a$

7. $7s^5 + 14s^3$

8. $\pi ab - \pi ac$

9. $24x^3 + 60x^2 - 48x$

10. $x^5y^3z^2 + xy^5z^3 - xy^3z^2$

11. $-5ab^2 + 10a^2b - 15ab$

12. $4(x - 2) - x(x - 2)$

Factor out −1.

13. $-a - 7$

14. $-4t^2 + 3t - 1$

Factor.

15. $2c + 2d + ac + ad$

16. $3xy + 18x - 5y - 30$

17. $2a^3 + 2a^2 - a - 1$

18. $4m^2n + 12m^2 - 8mn - 24m$

SECTION 6.2   Factoring Trinomials of the Form $x^2 + bx + c$

DEFINITIONS AND CONCEPTS	EXAMPLES

Many trinomials factor as the product of two binomials. To **factor a trinomial** of the form $x^2 + bx + c$, whose **leading coefficient is 1,** find two integers whose product is c and whose sum is b.

$$x^2 + bx + c = \left(x \boxed{}\right)\left(x \boxed{}\right)$$

The product of these numbers must be c and their sum must be b.

Use the FOIL method to check the factorization.

Factor: $p^2 + 7p + 12$

$$= \left(p \boxed{}\right)\left(p \boxed{}\right)$$
$$= (p + 3)(p + 4)$$

Positive factors of 12	Sum of positive factors of 12
$1 \cdot 12 = 12$	$1 + 12 = 13$
$2 \cdot 6 = 12$	$2 + 6 = 8$
$3 \cdot 4 = 12$	$3 + 4 = 7$

Check: $(p + 3)(p + 4) = p^2 + 4p + 3p + 12$
$$= p^2 + 7p + 12$$

Before factoring a trinomial, write it in **descending powers** of one variable. Also, factor out -1 if that is necessary to make the **leading coefficient positive.**

Factor: $7q - q^2 - 6$

$$= -q^2 + 7q - 6 \qquad \text{Write the terms in descending powers of } q.$$
$$= -(q^2 - 7q + 6) \qquad \text{Factor out } -1.$$
$$= -(q - 1)(q - 6) \qquad \text{Factor the trinomial.}$$

If a trinomial cannot be factored using only integers, it is called a **prime trinomial.**

$t^2 + 2t - 5$ is a prime trinomial because there are no two integers whose product is -5 and whose sum is 2.

The GCF should always be factored out first. A trinomial is **factored completely** when no factor can be factored further.

Use multiplication to check the factorization.

Factor completely: $3m^3 - 6m^2 - 24m$

$$= 3m(m^2 - 2m - 8) \qquad \text{Factor out the GCF, } 3m, \text{ first.}$$
$$= 3m(m - 4)(m + 2) \qquad \text{Factor the trinomial.}$$

To factor a trinomial of the form $x^2 + bx + c$ by **grouping,** write it as a equivalent four-term polynomial:

$$x^2 + \boxed{}\, x + \boxed{}\, x + c$$

The product of these numbers must be c, and their sum must be b.

Then factor the four-term polynomial by grouping.

Use the FOIL method to check the factorization.

Factor by grouping: $p^2 + 7p + 12$

We must find two numbers whose product is $c = 12$ and whose sum is $b = 7$. Two such numbers are 4 and 3. They serve as the coefficients of $4p$ and $3p$, the two terms that we use to represent the middle term, $7p$, of the trinomial.

$$p^2 + 7p + 12 = p^2 + 4p + 3p + 12 \qquad \text{Express } 7p \text{ as } 4p + 3p.$$
$$= p(p + 4) + 3(p + 4) \qquad \text{Factor } p \text{ out of } p^2 + 4p \text{ and 3 out of } 3p + 12.$$
$$= (p + 4)(p + 3) \qquad \text{Factor out } (p + 4).$$

REVIEW EXERCISES

19. What is the leading coefficient of $x^2 + 8x - 9$?

20. Complete the table.

Factors of 6	Sum of the factors of 6
1(6)	
2(3)	
−1(−6)	
−2(−3)	

Factor each trinomial, if possible.

21. $x^2 + 2x - 24$

22. $x^2 - 18x - 40$

23. $x^2 - 14x + 45$

24. $t^2 + 10t + 15$

25. $-y^2 + 15y - 56$

26. $10y + 9 + y^2$

27. $c^2 + 3cd - 10d^2$

28. $-3mn + m^2 + 2n^2$

29. Explain how we can check to determine whether $(x - 4)(x + 5)$ is the factorization of $x^2 + x - 20$.

30. Explain why $x^2 + 7x + 11$ is prime.

Completely factor each trinomial.

31. $5a^5 + 45a^4 - 50a^3$

32. $-4x^2y - 4x^3 + 24xy^2$

SECTION 6.3 Factoring Trinomials of the Form $ax^2 + bx + c$

DEFINITIONS AND CONCEPTS	EXAMPLES

We can use the **trial-and-check method** to factor trinomials with **leading coefficients other than 1.** Write the trinomial as the product of two binomials and determine four integers.

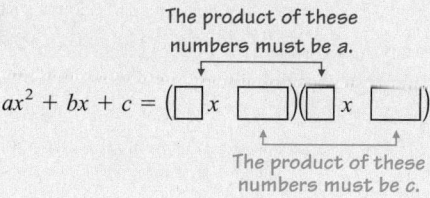

The product of these
numbers must be a.

$$ax^2 + bx + c = (\boxed{}\,x\; \boxed{})(\boxed{}\,x\; \boxed{})$$

The product of these
numbers must be c.

Use the FOIL method to check the factorization.

Factor: $2x^2 - 5x - 12$

Since the first term is $2x^2$, the first terms of the binomial factors must be $2x$ and x.

$$(2x\; \boxed{})(x\; \boxed{}) \quad \text{Because } 2x \cdot x \text{ will give } 2x^2$$

The second terms of the binomials must be two integers whose product is -12. There are six such pairs:

$$12(-1), \quad 6(-2), \quad 3(-4), \quad 4(-3), \quad 2(\;6), \quad \text{and} \quad 1(-12)$$

The pair in blue gives the correct middle term when we use the FOIL method to check:

Outer: $-8x$

$$(2x + 3)(x - 4) \quad \text{Combine like terms: } -8x + 3x = -5x.$$

Inner: $3x$

Thus, $2x^2 - 5x - 12 = (2x + 3)(x - 4)$.

To factor $ax^2 + bx + c$ by **grouping,** write it as an equivalent four-term polynomial:

$$ax^2 + \boxed{}\,x + \boxed{}\,x + c$$

The product of these numbers must
be ac, and their sum must be b.

Then factor the four-term polynomial by grouping.

Use the FOIL method to check your work.

Factor by grouping: $2x^2 - 5x - 12$

We must find two numbers whose product is $ac = 2(-12) = -24$ and whose sum is $b = -5$. Two such numbers are -8 and 3. They serve as the coefficients of $-8x$ and $3x$, the two terms that we use to represent the middle term, $-5x$, of the trinomial.

$$\begin{aligned} 2x^2 - 5x - 12 &= 2x^2 - 8x + 3x - 12 \quad \text{Express } -5x \text{ as } -8x + 3x. \\ &= 2x(x - 4) + 3(x - 4) \\ &= (x - 4)(2x + 3) \quad \text{Factor out } (x - 4). \end{aligned}$$

REVIEW EXERCISES

Factor each trinomial completely, if possible.

33. $2x^2 - 5x - 3$

34. $35y^2 + 11y - 10$

35. $-3x^2 + 13x + 30$

36. $-33p^2 - 6p + 18p^3$

37. $4b^2 - 17bc + 4c^2$

38. $7y^2 + 7y - 18$

39. ENTERTAINING The rectangular-shaped area occupied by a table setting is $(12x^2 - x - 1)$ square inches. Factor the polynomial to find the binomials that represent the length and width of the table setting.

40. In the following work, a student began to factor $5x^2 - 8x + 3$. Explain his mistake.

$$(5x - \quad)(x + \quad)$$

SECTION 6.4 Factoring Perfect-Square Trinomials and the Difference of Two Squares

DEFINITIONS AND CONCEPTS	EXAMPLES
Trinomials that are squares of a binomial are called **perfect-square trinomials.** We can factor perfect-square trinomials by applying the special-product rules in reverse. $A^2 + 2AB + B^2 = (A + B)^2$ $A^2 - 2AB + B^2 = (A - B)^2$	Factor: $g^2 + 8g + 16$ and $m^2 - 18mn + 81n^2$ We match each trinomial to a special-product form shown in the left column. $g^2 + 8g + 16 = g^2 + 2 \cdot 4 \cdot g + 4^2 = (g + 4)^2$ $m^2 - 18mn + 81n^2 = m^2 - 2 \cdot m \cdot 9n + (-9n)^2 = (m - 9n)^2$
To factor the **difference of two squares,** use the rule $F^2 - L^2 = (F + L)(F - L)$ It will be helpful to review the table of **squares of integers** shown on page 507.	Factor: $25b^2 - 36$ $\qquad = (5b)^2 - 6^2$ This is a difference of two squares. $\qquad = (5b + 6)(5b - 6)$
In general, the **sum of two squares** (with no common factor other than 1) cannot be factored using real numbers.	$x^2 + 100$ and $36y^2 + 49$ are prime polynomials.

REVIEW EXERCISES

Factor completely, if possible.

41. $x^2 + 10x + 25$ **42.** $9y^2 + 16 - 24y$ **47.** $x^2y^2 - 400$ **48.** $8at^2 - 32a$

43. $-z^2 + 2z - 1$ **44.** $25a^2 + 20ab + 4b^2$ **49.** $c^4 - 256$ **50.** $h^2 + 36$

45. $x^2 - 9$ **46.** $49t^2 - 121y^2$

SECTION 6.5 Factoring the Sum and Difference of Two Cubes

DEFINITIONS AND CONCEPTS	EXAMPLES
To factor the **sum** and **difference of two cubes,** use the following rules. $F^3 + L^3 = (F + L)(F^2 - FL + L^2)$ $F^3 - L^3 = (F - L)(F^2 + FL + L^2)$ It will be helpful to review the table of **cubes of integers** shown on page 515.	Factor: $p^3 + 64$ and $125a^3 - 27b^3$ We match each binomial to a factoring rule shown in the left column. $p^3 + 64 = p^3 + 4^3$ This is a sum of two cubes. $\qquad = (p + 4)(p^2 - p \cdot 4 + 4^2)$ $\qquad = (p + 4)(p^2 - 4p + 16)$ $125a^3 - 27b^3 = (5a)^3 - (3b)^3$ This is a difference of two cubes. $\qquad = (5a - 3b)[(5a)^2 + 5a \cdot 3b + (3b)^2]$ $\qquad = (5a - 3b)(25a^2 + 15ab + 27b^2)$

REVIEW EXERCISES

Factor each polynomial completely.

51. $b^3 + 1$ **53.** $p^3 + 125q^3$

52. $x^3 - 216$ **54.** $16x^5 - 54x^2y^3$

SECTION 6.6   A Factoring Strategy

DEFINITIONS AND CONCEPTS

To factor a random polynomial, use the **factoring strategy** discussed in Section 6.6 on page 519.

Remember that the instruction to factor means to **factor completely**. A polynomial is factored completely when no factor can be factored further.

EXAMPLES

Factor: $a^5 + 8a^2 + 4a^3 + 32$

Is there a common factor?   No. There is no common factor (other than 1).

How many terms does it have?   Since the polynomial has four terms, try factoring by grouping.

$$a^5 + 8a^2 + 4a^3 + 32 = a^2(a^3 + 8) + 4(a^3 + 8) \quad \text{Factor } a^2 \text{ from } a^5 + 8a^2 \text{ and 4 from } 4a^3 + 32.$$

$$= (a^3 + 8)(a^2 + 4) \quad \text{Factor out } a^3 + 8.$$

Is it factored completely?   No. We can factor $a^3 + 8$ as a sum of two cubes.

$$a^5 + 8a^2 + 4a^3 + 32$$

$$= (a^3 + 8)(a^2 + 4) \quad a^2 + 4 \text{ is prime.}$$

$$= (a + 2)(a^2 - 2a + 4)(a^2 + 4) \quad a^2 - 2a + 4 \text{ is prime.}$$

Does it check?   Use multiplication to check.

REVIEW EXERCISES

Factor each polynomial completely, if possible.

55. $14y^3 + 6y^4 - 40y^2$ **56.** $5s^2t + 5s^2u^2 + 5tv + 5u^2v$ **61.** $2t^3 + 10$ **62.** $121p^2 + 36q^2$

57. $j^4 - 16$ **58.** $-3j^3 - 24$ **63.** $x^2z + 64y^2z + 16xyz$ **64.** $18c^3d^2 - 12c^3d - 24c^2d$

59. $400x + 400 - m^2x - m^2$ **60.** $12w^4 - 36w^3 + 27w^2$

SECTION 6.7   Solving Quadratic Equations by Factoring

DEFINITIONS AND CONCEPTS

A **quadratic equation** is an equation that can be written in the **standard form** $ax^2 + bx + c = 0$, where a, b, and c are real numbers and $a \neq 0$.

The Zero-Factor Property

If the product of two or more numbers is 0, then at least one of the numbers is 0.

To use the **factoring method to solve a quadratic equation:**

1. Write the equation in standard form: $ax^2 + bx + c = 0$ or $0 = ax^2 + bx + c$

2. Factor completely.

3. Use the *zero-factor property* to set each factor equal to 0.

4. Solve each resulting linear equation.

5. Check each result in the original equation.

EXAMPLES

Examples of quadratic equations are:

$$5x^2 + 25x = 0, \quad 4a^2 - 9 = 0, \quad \text{and} \quad y^2 - 13y = 6$$

If $(x + 2)(x - 3) = 0$ then,

$$x + 2 = 0 \quad \text{or} \quad x - 3 = 0.$$

Solve: $\quad 5x^2 + 25x = 0 \qquad$ Solve: $\qquad 4a^2 - 9 = 0$

$$5x(x + 5) = 0 \qquad\qquad (2a + 3)(2a - 3) = 0$$

$5x = 0 \quad$ or $\quad x + 5 = 0 \qquad 2a + 3 = 0 \quad$ or $\quad 2a - 3 = 0$

$x = 0 \quad \mid \quad x = -5 \qquad\qquad 2a = -3 \quad \mid \quad 2a = 3$

The solutions are 0 and -5.

$$a = -\frac{3}{2} \quad \Big| \quad a = \frac{3}{2}$$

The solutions are $-\frac{3}{2}$ and $\frac{3}{2}$.

Check each result in the original equation.

SECTION 6.7   Solving Quadratic Equations by Factoring—*continued*

DEFINITIONS AND CONCEPTS	EXAMPLES
To use the zero-factor property to solve a quadratic equation, we need one side of the equation to be factored completely and the other side to be 0.	Solve: $5y^2 - 13y = 6$ $5y^2 - 13y - 6 = 6 - 6$ To get 0 on the right side, subtract 6 from both sides. $5y^2 - 13y - 6 = 0$ $(5y + 2)(y - 3) = 0$ $5y + 2 = 0$ or $y - 3 = 0$ $5y = -2$ $y = 3$ $y = -\dfrac{2}{5}$ The solutions are $-\frac{2}{5}$ and 3. Check each result in the original equation.

REVIEW EXERCISES

Solve each equation by factoring.

65. $8x(x - 6) = 0$
66. $(4x - 7)(x + 1) = 0$
67. $x^2 + 2x = 0$
68. $x^2 - 9 = 0$
69. $144x^2 - 25 = 0$
70. $a^2 - 7a + 12 = 0$
71. $2t^2 + 28t + 98 = 0$
72. $2x - x^2 = -24$
73. $5a^2 - 6a + 1 = 0$
74. $2p^3 = 2p(p + 2)$

SECTION 6.8   Applications of Quadratic Equations

DEFINITIONS AND CONCEPTS	EXAMPLES
To solve application problems, use the **five-step problem solving strategy:** 1. Analyze the problem 2. Form an equation 3. Solve the equation 4. State the conclusion 5. Check the result	Find two consecutive positive integers whose product is 72. **Analyze the Problem** *Consecutive integers* are integers that follow each other. The word *product* indicates multiplication. **Form an Equation** Let $x =$ the smaller positive integer. Then $x + 1 =$ the larger integer.

The smaller integer	times	the larger integer	equals	72.
x	$\cdot$	$(x + 1)$	$=$	72

Solve the Equation

$x(x + 1) = 72$

$x^2 + x - 72 = 0$ Remove parentheses. To get 0 on the right side, subtract 72 from both sides.

$(x + 9)(x - 8) = 0$ Factor the trinomial.

$x + 9 = 0$ or $x - 8 = 0$ Set each factor equal to 0.
$x = -9$ $x = 8$ Solve each linear equation.

State the Conclusion   Since we are looking for positive integers, the solution -9 must be discarded. Thus, the smaller integer is 8 and the larger integer is $x + 1 = 9$.

Check the Result   The integers 8 and 9 are consecutive positive integers and their product is 72.

SECTION 6.8 Applications of Quadratic Equations—*continued*

DEFINITIONS AND CONCEPTS	EXAMPLES
The Pythagorean Theorem: If a and b are the lengths of the legs of a right triangle and c is the length of the hypotenuse, then $$a^2 + b^2 = c^2$$	To show that a triangle with sides of 5, 12, and 13 units is a right triangle, we verify that $5^2 + 12^2 = 13^2$. $5^2 + 12^2 \overset{?}{=} 13^2$ $25 + 144 \overset{?}{=} 169$ $169 = 169$ True 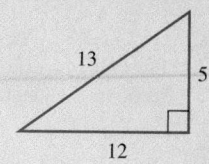

REVIEW EXERCISES

75. CONSTRUCTION   The face of the triangular concrete panel has an area of 45 square meters, and its base is 3 meters longer than twice its height. Find the length of its base.

76. ACADEMY AWARDS   Meryl Streep has now surpassed Katherine Hepburn as the most-nominated actress for Oscars. The number of times Streep has been nominated and the number of times Hepburn has been nominated are consecutive even integers whose product is 168. How many times was each actress nominated?

77. TIGHTROPE WALKERS   A circus performer intends to walk up a taut cable shown in the illustration to a platform at the top of a pole. How high above the ground is the platform?

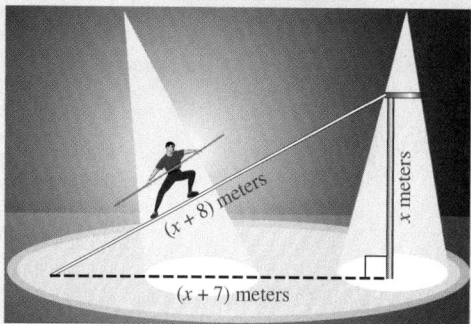

78. BALLOONING   A hot-air balloonist accidentally dropped his camera overboard while traveling at a height of 1,600 ft. The height h in feet of the camera t seconds after being dropped is given by $h = -16t^2 + 1,600$. In how many seconds will the camera hit the ground?

CHAPTER 6
Test

1. Fill in the blanks.
   a. The letters GCF stand for _____ _____ _____.
   b. To factor a polynomial means to express it as a _____ of two (or more) polynomials.
   c. The _____ theorem provides a formula relating the lengths of the three sides of a right triangle.
   d. $y^2 - 25$ is a _____ of two squares.
   e. The trinomial $x^2 + x - 6$ factors as the product of two _____: $(x + 3)(x - 2)$.

2. a. Find the prime factorizations of 45 and 30.

   b. Find the greatest common factor of $45x^4$ and $30x^3$.

Factor completely. If an expression cannot be factored, write "prime."

3. $4x + 16$

4. $q^2 - 81$

5. $30a^2b^3 - 20a^3b^2 + 5ab$

6. $x^2 + 9$

7. $2x(x + 1) + 3(x + 1)$

8. $x^2 + 4x + 3$

9. $-x^2 + 9x + 22$

10. $60x^2 - 32x^3 + x^4$

11. $9a - 9b + ax - bx$

12. $2a^2 + 5a - 12$

13. $18x^2 + 60xy + 50y^2$

14. $x^3 + 8$

15. $20m^8 - 15m^6$ **16.** $3a^3 - 81$

17. $16x^4 - 81$ **18.** $a^3 + 5a^2 + a + 5$

19. CHECKERS   The area of a square checkerboard is represented by $(25x^2 - 40x + 16)$ in.^{2}. Find the polynomial that represents the length of a side of the checkerboard.

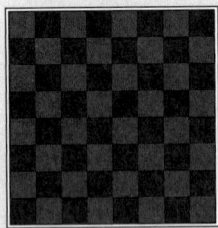

20. Factor $x^2 - 3x - 54$. Show a check of your answer.

Solve each equation.

21. $(x + 3)(x - 2) = 0$ **22.** $x^2 - 25 = 0$

23. $36x^2 - 6x = 0$ **24.** $x^2 + 6x = -9$

25. $6x^2 + x - 1 = 0$ **26.** $a(a - 7) = 18$

27. $x^3 + 7x^2 = -6x$

28. DRIVING SAFETY   All cars have a blind spot where it is difficult for the driver to see a car behind and to the right. The area of the rectangular blind spot shown is 54 ft^2. Its length is 3 feet longer than its width. Find its dimensions.

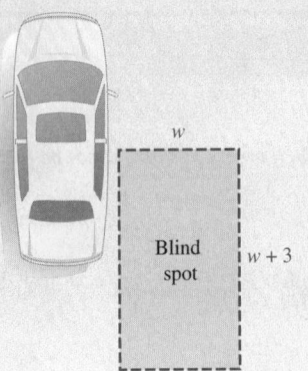

29. ROCKETRY   The height h, in feet, of a toy rocket t seconds after being launched is given by $h = -16t^2 + 80t$. After how many seconds will the rocket hit the ground?

30. ATV'S   The area of a triangular-shaped safety flag on an all-terrain vehicle is 33 in.^{2}. Its height is 1 inch less than twice the length of the base. Find the length of the base and the height of the flag.

31. Find two consecutive positive integers whose product is 156.

32. Find the length of the hypotenuse of the right triangle shown.

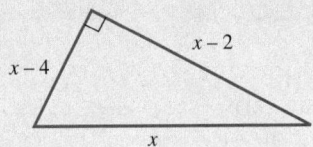

33. What is a quadratic equation? Give an example.

34. If the product of two numbers is 0, what conclusion can be drawn about the numbers?

GROUP PROJECT

FACTORING MODELS

Overview: In this activity, you will construct geometric models to find factorizations of several trinomials.

Instructions: Form groups of 2 or 3 students.

1. Copy and cut out each of the following figures. On each figure, write its area.

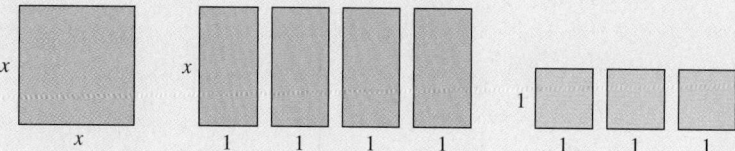

Write a trinomial that represents the *sum* of the areas of the eight figures by combining any like terms: _____ + _____ + _____

2. Now assemble the eight figures to form the large rectangle shown below.

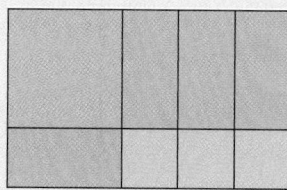

Write an expression that represents the *length* of the rectangle: _____ + _____

Write an expression that represents the *width* of the rectangle: _____ + _____

Express the area of the rectangle as the product of its length and width:

(_____)(_____)

3. The set of figures used in step 1 and the set of figures used in step 2 are the same. Therefore, the expressions for the areas must be equal. Set your answers from steps 1 and 2 equal to find the factorization of the trinomial $x^2 + 4x + 3$.

$$\frac{\rule{3cm}{0.4pt}}{\text{Answer from step 1}} = \frac{\rule{3cm}{0.4pt}}{\text{Answer from step 2}}$$

4. Make a new model to find the factorization of $x^2 + 5x + 4$.   (*Hint:* You will need to make one more 1-by-x figure and one more 1-by-1 figure.)

_____ = _____

5. Make a new model to find the factorization of $2x^2 + 5x + 2$.   (*Hint:* You will need to make one more x-by-x figure.)

_____ = _____

CHAPTER 7

Rational Expressions and Equations

© BananaStock/SuperStock

from Campus to Careers
Recreation Director

People of all ages enjoy participating in activities, such as arts and crafts, camping, sports, and the performing arts. Recreation directors plan, organize, and oversee these activities in local playgrounds, camps, community centers, religious organizations, theme parks, and tourist attractions. The job of recreation director requires mathematical skills such as budgeting, scheduling, and forecasting trends.

Problem 33 of **Study Set 7.7** involves an area of responsibility for many recreation directors—swimming pools.

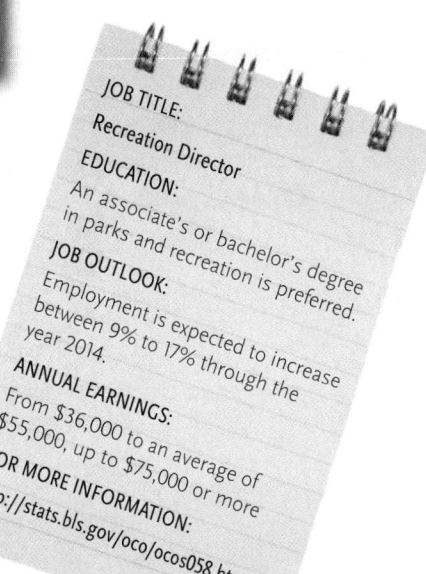

JOB TITLE:
Recreation Director

EDUCATION:
An associate's or bachelor's degree in parks and recreation is preferred.

JOB OUTLOOK:
Employment is expected to increase between 9% to 17% through the year 2014.

ANNUAL EARNINGS:
From $36,000 to an average of $55,000, up to $75,000 or more

FOR MORE INFORMATION:
http://stats.bls.gov/oco/ocos058.htm

Study Skills Workshop
Study Groups

Study groups give students an opportunity to ask their classmates questions, share ideas, compare lecture notes, and review for tests. If something like this interests you, here are some suggestions.

GROUP SIZE: A study group should be small—from 3 to 6 people is best.

TIME AND PLACE: You should meet regularly in a place where you can spread out and talk without disturbing others.

GROUND RULES: The study group will be more effective if, early on, you agree on some rules to follow.

Now Try This

Would you like to begin a study group? If so, you need to answer the following questions.

Who will be in your group? Where will your group meet? How often will it meet? For how long will each session last? Will you have a group leader? What will be the leader's responsibilities? What will you try to accomplish each session? How will the members prepare for each meeting? Will you follow a set agenda each session? How will the members share contact information? When will you discuss ways to improve the study sessions?

SECTION 7.1
Simplifying Rational Expressions

Objectives

1. Evaluate rational expressions.
2. Find numbers that cause a rational expression to be undefined.
3. Simplify rational expressions.
4. Simplify rational expressions that have factors that are opposites.

Fractions that are the quotient of two integers are *rational numbers*. Examples are $\frac{1}{2}$ and $\frac{9}{5}$. Fractions such as

$$\frac{3}{2y}, \qquad \frac{x}{x+2}, \qquad \text{and} \qquad \frac{2a^2 - 8a}{a^2 - 6a + 8}$$

that are the quotient of two polynomials are called **rational expressions.**

Rational Expressions

A **rational expression** is an expression of the form $\frac{A}{B}$, where A and B are polynomials and B does not equal 0.

1 Evaluate Rational Expressions.

Rational expressions can have different values depending on the number that is substituted for the variable.

EXAMPLE 1 Evaluate $\dfrac{2x - 1}{x^2 + 1}$ for $x = -3$ and for $x = 0$.

Strategy We will replace each x in the rational expression with the given value of the variable. Then we will evaluate the numerator and denominator separately, and simplify, if possible.

Why Recall from Chapter 1 that to *evaluate an expression* means to find its numerical value, once we know the value of its variable.

Solution

For $x = -3$:

$$\frac{2x - 1}{x^2 + 1} = \frac{2(-3) - 1}{(-3)^2 + 1}$$

$$= \frac{-6 - 1}{9 + 1}$$

$$= -\frac{7}{10}$$

For $x = 0$:

$$\frac{2x - 1}{x^2 + 1} = \frac{2(0) - 1}{(0)^2 + 1}$$

$$= \frac{0 - 1}{0 + 1}$$

$$= -1$$

Self Check 1 Evaluate $\frac{2x - 1}{x^2 + 1}$ for $x = 7$.

Now Try **Problem 13**

---

2 Find Numbers that Cause a Rational Expression to be Undefined.

The fraction bar in a rational expression indicates division. Since division by 0 is undefined, we must make sure that the denominator of a rational expression is not equal to 0.

EXAMPLE 2 Find all real numbers for which the rational expression is undefined: **a.** $\dfrac{7x}{x - 5}$ **b.** $\dfrac{3x - 2}{x^2 - x - 6}$ **c.** $\dfrac{8}{x^2 + 1}$

Strategy To find the real numbers for which each rational expression is undefined, we will find the values of the variable that make the *denominator* 0.

Why We don't need to examine the numerator of the rational expression; it can be any value, including 0. It's a denominator of 0 that makes a rational expression undefined, because a denominator of 0 indicates division by 0.

Solution

a. The denominator of $\frac{7x}{x - 5}$ will be 0 if we replace x with 5.

$$\frac{7x}{x - 5} = \frac{7(5)}{5 - 5} = \frac{35}{0}$$

Since $\frac{35}{0}$ is undefined, the rational expression $\frac{7x}{x - 5}$ is undefined for $x = 5$.

b. $\frac{3x - 2}{x^2 - x - 6}$ will be undefined for values of x that make the denominator 0. To find these values, we set $x^2 - x - 6$ equal to 0, and solve for x.

$$x^2 - x - 6 = 0 \qquad \text{Set the denominator of } \tfrac{3x-2}{x^2-x-6} \text{ equal to 0.}$$

$$(x - 3)(x + 2) = 0 \qquad \text{To solve the quadratic equation, factor the trinomial.}$$

$$x - 3 = 0 \quad \text{or} \quad x + 2 = 0 \qquad \text{Set each factor equal to 0.}$$

$$x = 3 \qquad \qquad \quad x = -2 \qquad \text{Solve each equation.}$$

Since 3 and -2 make the denominator 0, the rational expression $\frac{3x - 2}{x^2 - x - 6}$ is undefined for $x = 3$ and $x = -2$.

The Language of Algebra
Another way that Example 2 could be phrased is: State the *restrictions* on the variable. For $\frac{3x-2}{x^2-x-6}$, we can state the *restrictions* by writing $x \neq 3$ and $x \neq -2$.

<table>
<tr><td align="center">For x = 3:</td><td align="center">For x = −2:</td></tr>
</table>

$$\frac{3x - 2}{x^2 - x - 6} = \frac{3(3) - 2}{3^2 - 3 - 6} \qquad\qquad \frac{3x - 2}{x^2 - x - 6} = \frac{3(-2) - 2}{(-2)^2 - (-2) - 6}$$

$$= \frac{9 - 2}{9 - 3 - 6} \qquad\qquad\qquad = \frac{-6 - 2}{4 + 2 - 6}$$

$$= \frac{7}{0} \quad \begin{array}{l}\text{This expression}\\\text{is undefined.}\end{array} \qquad\qquad = \frac{-8}{0} \quad \begin{array}{l}\text{This expression}\\\text{is undefined.}\end{array}$$

c. No matter what real number is substituted for x, the denominator, $x^2 + 1$, will not be equal to 0. (A number squared plus 1 cannot equal 0.) Thus, no real numbers make $\frac{8}{x^2 + 1}$ undefined.

Self Check 2   Find all real numbers for which the rational expression is undefined: **a.** $\frac{x}{x + 9}$ **b.** $\frac{9x + 7}{x^2 - 25}$ **c.** $\frac{4 - x}{x^2 + 64}$

Now Try **Problems 25 and 33**

---

③ Simplify Rational Expressions.

Success Tip
Unless otherwise stated, we will now assume that the variables in rational expressions represent real numbers for which the denominator is not zero.

In Section 1.2, we simplified fractions by removing a factor equal to 1. For example, to simplify $\frac{6}{15}$, we factor 6 and 15, and then remove the factor $\frac{3}{3}$.

$$\frac{6}{15} = \frac{2 \cdot 3}{5 \cdot 3} = \frac{2}{5} \cdot \frac{3}{3} = \frac{2}{5} \cdot 1 = \frac{2}{5}$$

To streamline this process, we can replace $\frac{3}{3}$ in $\frac{2 \cdot 3}{5 \cdot 3}$ with the equivalent fraction $\frac{1}{1}$.

$$\frac{6}{15} = \frac{2 \cdot 3}{5 \cdot 3} = \frac{2 \cdot \overset{1}{\cancel{3}}}{5 \cdot \underset{1}{\cancel{3}}} = \frac{2}{5} \qquad \text{We are removing } \frac{3}{3} = 1.$$

We can simplify rational expressions in a similar manner using a procedure that is based on the following property.

The Fundamental Property of Rational Expressions

If A, B, and C are polynomials, and B and C are not 0,

$$\frac{AC}{BC} = \frac{A}{B}$$

A rational expression is **simplified** if its numerator and denominator have no common factors other than 1. To simplify a rational expression, follow these steps.

Simplifying Rational Expressions

1. Factor the numerator and denominator completely to determine their common factors.
2. Remove factors equal to 1 by replacing each pair of factors common to the numerator and denominator with the equivalent fraction $\frac{1}{1}$.
3. Multiply the remaining factors in the numerator and in the denominator.

EXAMPLE 3 Simplify: $\dfrac{21x^3y}{14x^2y^2}$

Strategy We will write the numerator and denominator in factored form and then remove pairs of factors that are equal to 1.

Why The rational expression is simplified when the numerator and denominator have no common factor other than 1.

Solution

$$\frac{21x^3y}{14x^2y^2} = \frac{3 \cdot 7 \cdot x \cdot x \cdot x \cdot y}{2 \cdot 7 \cdot x \cdot x \cdot y \cdot y}$$

Factor the numerator and the denominator.

$$= \frac{3 \cdot \overset{1}{\cancel{7}} \cdot \overset{1}{\cancel{x}} \cdot \overset{1}{\cancel{x}} \cdot x \cdot \overset{1}{\cancel{y}}}{2 \cdot \underset{1}{\cancel{7}} \cdot \underset{1}{\cancel{x}} \cdot \underset{1}{\cancel{x}} \cdot \underset{1}{\cancel{y}} \cdot y}$$

Simplify by replacing $\frac{7}{7}$, $\frac{x}{x}$, and $\frac{y}{y}$ with the equivalent fraction $\frac{1}{1}$. This removes the factor $\frac{7 \cdot x \cdot x \cdot y}{7 \cdot x \cdot x \cdot y}$, which is equal to 1.

$$= \frac{3x}{2y}$$

Multiply the remaining factors in the numerator: $3 \cdot 1 \cdot 1 \cdot 1 \cdot x \cdot 1 = 3x.$
Multiply the remaining factors in the denominator: $2 \cdot 1 \cdot 1 \cdot 1 \cdot 1 \cdot y = 2y.$

We say that $\frac{21x^3y}{14x^2y^2}$ simplifies to $\frac{3x}{2y}$.

An alternate approach is to use rules for exponents to simplify the rational expression.

$$\frac{21x^3y}{14x^2y^2} = \frac{3 \cdot \overset{1}{\cancel{7}} \cdot x^{3-2}y^{1-2}}{2 \cdot \underset{1}{\cancel{7}}} = \frac{3x^1y^{-1}}{2} = \frac{3x}{2y}$$

To divide exponential expressions with the same base, keep the base and subtract the exponents.

Self Check 3 Simplify: $\frac{32a^3b^2}{24ab^4}$

Now Try Problem 39

To simplify rational expressions, we often make use of the factoring methods discussed in Chapter 6.

EXAMPLE 4 Simplify: **a.** $\dfrac{30t - 6}{36}$ **b.** $\dfrac{x^2 + 13x + 12}{x^2 + 12x}$

Strategy We will begin by factoring the numerator and denominator. Then we will remove any factors common to the numerator and denominator.

Why   We need to make sure that the numerator and denominator have no common factor other than 1. When this is the case, the rational expression is simplified.

Solution

a. $\dfrac{30t - 6}{36} = \dfrac{6(5t - 1)}{6 \cdot 6}$ Factor the numerator: The GCF is 6.
Factor the denominator.

$= \dfrac{\overset{1}{\cancel{6}}(5t - 1)}{\underset{1}{\cancel{6}} \cdot 6}$ Simplify by removing a factor equal to 1. Replace $\frac{6}{6}$ with $\frac{1}{1}$.

$= \dfrac{5t - 1}{6}$ Multiply the remaining factors in the numerator: $1 \cdot (5t - 1) = 5t - 1$.
Multiply the remaining factors in the denominator: $1 \cdot 6 = 6$.

The Language of Algebra
Sometimes, a common factor of the numerator and denominator has two or more terms. Here, we remove the common *binomial* factor, $x + 12$.

b. $\dfrac{x^2 + 13x + 12}{x^2 + 12x} = \dfrac{(x + 1)(x + 12)}{x(x + 12)}$ Factor the numerator.
Factor the denominator: The GCF is x.

$= \dfrac{(x + 1)\overset{1}{\cancel{(x + 12)}}}{x\underset{1}{\cancel{(x + 12)}}}$ Simplify by replacing $\frac{x + 12}{x + 12}$ with the equivalent fraction $\frac{1}{1}$. This removes the factor $\frac{x + 12}{x + 12} = 1$.

$= \dfrac{x + 1}{x}$ This rational expression cannot be simplified further.

▷ **Self Check 4**   Simplify: a. $\dfrac{4t - 20}{12}$ b. $\dfrac{x^2 - x - 6}{x^2 - 3x}$

Now Try   **Problems 43 and 47**

Caution   When simplifying rational expressions, we can only remove factors common to the entire numerator and denominator. *It is incorrect to remove any terms common to the numerator and denominator.*

$\dfrac{\overset{1}{\cancel{x}} + 1}{\underset{1}{\cancel{x}}}$ $\dfrac{a^2 - 3a + \overset{1}{\cancel{2}}}{a + \underset{1}{\cancel{2}}}$ $\dfrac{\overset{1}{\cancel{y^2}} - 36}{\underset{1}{\cancel{y^2}} - y - 7}$

x is a term of $x + 1$. 2 is a term of $a^2 - 3a + 2$ and a term of $a + 2$. y^2 is a term of $y^2 - 36$ and a term of $y^2 - y - 7$.

EXAMPLE 5   Simplify: a. $\dfrac{3x^2 - 8x - 3}{2x^5 - 18x^3}$ b. $\dfrac{(x - y)^4}{x^2 - 2xy + y^2}$

Strategy   We will begin by factoring the numerator and denominator using the methods discussed in Chapter 5. Then we will remove any factors common to the numerator and denominator.

The Language of Algebra
When a rational expression is simplified, the result is an *equivalent expression*. This means that $\frac{3x^2 - 8x - 3}{2x^5 - 18x^3}$ has the same value as $\frac{3x + 1}{2x^3(x + 3)}$ for all values of x, except those that make either denominator 0.

Why   We need to make sure that the numerator and denominator have no common factor other than 1. When this is the case, then the rational expression is simplified.

Solution

a. $\dfrac{3x^2 - 8x - 3}{2x^5 - 18x^3} = \dfrac{(3x + 1)(x - 3)}{2x^3(x^2 - 9)}$ Factor the trinomial in the numerator.
Factor the denominator: The GCF is $2x^3$.

$= \dfrac{(3x + 1)(x - 3)}{2x^3(x + 3)(x - 3)}$ In the denominator, factor the difference of two squares, $x^2 - 9$.

$$= \frac{(3x + 1)(x - 3)}{2x^3(x + 3)(x - 3)}$$

Simplify by replacing $\frac{x-3}{x-3}$ with the equivalent fraction $\frac{1}{1}$. This removes the factor $\frac{x-3}{x-3} = 1$.

$$= \frac{3x + 1}{2x^3(x + 3)}$$

It is not necessary to perform the multiplication $2x^3(x + 3)$ in the result. It is usually more convenient to leave the denominator in factored form.

b. $\dfrac{(x - y)^4}{x^2 - 2xy + y^2} = \dfrac{(x - y)^4}{(x - y)^2}$

In the denominator, factor the perfect square trinomial $x^2 - 2xy + y^2$.

$$= \frac{(x - y)(x - y)(x - y)(x - y)}{(x - y)(x - y)}$$

Write the repeated multiplication indicated by each exponent.

$$= \frac{(x - y)(x - y)(x - y)(x - y)}{(x - y)(x - y)}$$

Simplify by replacing each $\frac{x-y}{x-y}$ with $\frac{1}{1}$.

$$= (x - y)^2$$

Use an exponent to write the repeated multiplication in the numerator.

Self Check 5 Simplify: **a.** $\dfrac{4x^2 - 4x - 15}{8x^3 - 50x}$

b. $\dfrac{(a + 3b)^5}{a^2 + 6ab + 9b^2}$

Now Try **Problems 51 and 53**

EXAMPLE 6 Simplify: $\dfrac{5(x + 3) - 5}{7(x + 3) - 7}$

Strategy We will begin by simplifying the numerator, $5(x + 3) - 5$, and the denominator, $7(x + 3) - 7$, separately. Then we will factor each result and remove any common factors.

Why We cannot immediately remove $x + 3$ because it is not a factor of the *entire* numerator and the *entire* denominator.

Solution

$$\frac{5(x + 3) - 5}{7(x + 3) - 7} = \frac{5x + 15 - 5}{7x + 21 - 7}$$

Use the distributive property in the numerator and in the denominator.

$$= \frac{5x + 10}{7x + 14}$$

Combine like terms: $15 - 5 = 10$ and $21 - 7 = 14$.

$$= \frac{5(x + 2)}{7(x + 2)}$$

Factor the numerator: The GCF is 5.
Factor the denominator: The GCF is 7.

$$= \frac{5(x + 2)}{7(x + 2)}$$

Simplify by replacing $\frac{x+2}{x+2}$ with the equivalent fraction $\frac{1}{1}$. This removes the factor $\frac{x+2}{x+2} = 1$.

$$= \frac{5}{7}$$

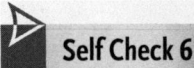

Self Check 6   Simplify: $\dfrac{4(x-2)+4}{3(x-2)+3}$

Now Try   **Problem 55**

④ **Simplify Rational Expressions That Have Factors That Are Opposites.**

If the terms of two polynomials are the same, except that they are opposite in sign, the polynomials are **opposites.** For example, the following pairs of polynomials are opposites.

$$2a - 1 \quad \text{and} \quad 1 - 2a \qquad\qquad -3x^2 - x + 5 \quad \text{and} \quad 3x^2 + x - 5$$

Compare terms: $2a$ and $-2a$; -1 and 1.   Compare terms: $-3x^2$ and $3x^2$; $-x$, and x; 5 and -5.

We have seen that the quotient of two real numbers that are opposites is always -1:

$$\dfrac{2}{-2} = -1 \qquad \dfrac{-78}{78} = -1 \qquad \dfrac{3.5}{-3.5} = -1$$

Likewise, the quotient of two binomials that are opposites is always -1.

EXAMPLE 7   Simplify: $\dfrac{2a-1}{1-2a}$

Strategy   We will rearrange the terms of the numerator, $2a - 1$, and factor out -1.

Why   This step is useful when the numerator and denominator contain factors that are opposites, such as $2a - 1$ and $1 - 2a$. It produces a common factor that can be removed.

Solution

$$\dfrac{2a-1}{1-2a} = \dfrac{-1+2a}{1-2a} \qquad \text{Think of the numerator, } 2a - 1 \text{, as } 2a + (-1) \text{. Then change the order of the terms: } 2a + (-1) = -1 + 2a.$$

$$= \dfrac{-1(1-2a)}{1-2a} \qquad \text{Factor out } -1 \text{ from the two terms of the numerator.}$$

$$= \dfrac{-1(\overset{1}{\cancel{1-2a}})}{\underset{1}{\cancel{1-2a}}} \qquad \text{Simplify by replacing } \tfrac{1-2a}{1-2a} \text{ with the equivalent fraction } \tfrac{1}{1}. \text{ This removes the factor } \tfrac{1-2a}{1-2a} = 1.$$

$$= \dfrac{-1}{1} \qquad \text{Multiply the remaining factors in the numerator.}$$

$$= -1 \qquad \text{Any number divided by 1 is itself.}$$

Self Check 7   Simplify: $\dfrac{3p-2}{2-3p}$

Now Try   **Problem 59**

In general, we have this fact.

The Quotient of Opposites   The quotient of any nonzero polynomial and its opposite is -1.

For each of the following rational expressions, the numerator and denominator are opposites. Thus, each expression is equal to −1.

$$\frac{x-6}{6-x} = -1 \qquad \frac{2a-9b}{9b-2a} = -1 \qquad \frac{-3x^2-x+5}{3x^2+x-5} = -1$$

This fact can be used to simplify certain rational expressions by removing a factor equal to −1. If a factor of the numerator is the opposite of a factor of the denominator, we can replace them with the equivalent fraction $\frac{-1}{1}$, as shown in the following example.

EXAMPLE 8   Simplify, if possible: **a.** $\dfrac{y^2-1}{3-3y}$ **b.** $\dfrac{t+8}{t-8}$

Strategy   We will begin by factoring the numerator and denominator. Then we look for common factors, or factors that are opposites, and remove them.

Why   We need to make sure that the numerator and denominator have no common factor (or opposite factors) other than 1. When this is the case, then the rational expression is simplified.

Solution

a. $\dfrac{y^2-1}{3-3y} = \dfrac{(y+1)(y-1)}{3(1-y)}$ Factor the numerator.
Factor the denominator.

$$= \dfrac{(y+1)\overset{-1}{\cancel{(y-1)}}}{3\underset{1}{\cancel{(1-y)}}}$$ Since $y-1$ and $1-y$ are opposites, simplify by replacing $\frac{y-1}{1-y}$ with the equivalent fraction $\frac{-1}{1}$. This removes the factor $\frac{y-1}{1-y} = -1$.

$$= \dfrac{-(y+1)}{3}$$

This result may be written in several other equivalent forms.

$\dfrac{-(y+1)}{3} = -\dfrac{y+1}{3}$ The − symbol in −(y + 1) can be written in the front of the fraction, and the parentheses can be dropped.

$\dfrac{-(y+1)}{3} = \dfrac{-y-1}{3}$ The − symbol in −(y + 1) represents a factor of −1. Distribute the multiplication by −1 in the numerator.

$\dfrac{-(y+1)}{3} = \dfrac{y+1}{-3}$ The − symbol in −(y + 1) can be applied to the denominator. However, we don't usually use this form.

b. The binomials $t+8$ and $t-8$ are not opposites because their first terms do not have opposite signs. Thus, $\frac{t+8}{t-8}$ does not simplify.

Self Check 8   Simplify, if possible: **a.** $\dfrac{m^2-100}{10m-m^2}$
b. $\dfrac{2x-3}{2x+3}$

Now Try   Problem 63

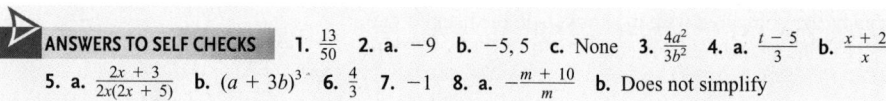

ANSWERS TO SELF CHECKS   **1.** $\frac{13}{50}$ **2. a.** -9 **b.** $-5, 5$ **c.** None   **3.** $\frac{4a^2}{3b^2}$ **4. a.** $\frac{t-5}{3}$ **b.** $\frac{x+2}{x}$
5. a. $\frac{2x+3}{2x(2x+5)}$ **b.** $(a+3b)^3$ **6.** $\frac{4}{3}$ **7.** -1 **8. a.** $\frac{-m+10}{m}$ **b.** Does not simplify

STUDY SET
7.1

VOCABULARY

Fill in the blanks.

1. A quotient of two polynomials, such as $\frac{x^2 + x}{x^2 - 3x}$, is called a _____ expression.

2. To simplify a rational expression, we remove common _____ of the numerator and denominator.

3. Because of the division by 0, the expression $\frac{8}{0}$ is _____.

4. The binomials $x - 15$ and $15 - x$ are called _____, because their terms are the same, except that they are opposite in sign.

CONCEPTS

5. When we simplify $\frac{x^2 + 5x}{4x + 20}$, the result is $\frac{x}{4}$. These equivalent expressions have the same value for all real numbers, except $x = -5$. Show that they have the same value for $x = 1$.

6. Determine whether each pair of polynomials are opposites. Write *yes* or *no*.
   a. $y + 7$ and $y - 7$
   b. $b - 20$ and $20 - b$
   c. $x^2 + 2x - 1$ and $-x^2 - 2x - 1$

7. Simplify each expression, if possible.
   a. $\frac{x - 8}{x - 8}$ b. $\frac{x - 8}{8 - x}$
   c. $\frac{x + 8}{8 + x}$ d. $\frac{x + 8}{x}$

8. Simplify each expression.
   a. $\frac{(x + 2)(x - 2)}{(x + 1)(x + 2)}$ b. $\frac{y(y - 2)}{9(2 - y)}$
   c. $\frac{(2m + 7)(m - 5)}{(2m + 7)}$ d. $\frac{x \cdot x}{x \cdot x(x - 30)}$

NOTATION

Complete the solution to simplify the rational expression.

9. $\dfrac{x^2 + 2x + 1}{x^2 + 4x + 3} = \dfrac{(x + 1)(\quad + 1)}{(x + 3)(x + \quad)}$

   $= \dfrac{(x + 1)\overset{1}{\cancel{(x + 1)}}}{(x + 3)\underset{1}{}}$

   $= \dfrac{x + 1}{}$

10. In the following table, a student's answers to three homework problems are compared with the answers in the back of the book. Are the answers equivalent?

Answer	Book's answer	Equivalent?
$\frac{-3}{x + 3}$	$-\frac{3}{x + 3}$	
$\frac{-x + 4}{6x + 1}$	$\frac{-(x - 4)}{6x + 1}$	
$\frac{x + 7}{(x - 4)(x + 2)}$	$\frac{x + 7}{(x + 2)(x - 4)}$	

GUIDED PRACTICE

Evaluate each expression for x = 6. See Example 1.

11. $\dfrac{x - 2}{x - 5}$ 12. $\dfrac{3x - 2}{x - 2}$

13. $\dfrac{x^2 - 4x - 12}{x^2 + x - 2}$ 14. $\dfrac{x^2 - 1}{x^3 - 1}$

15. $\dfrac{-x + 1}{x^2 - 5x - 6}$ 16. $\dfrac{-2x^2 - 3}{x - 6}$

Evaluate each expression for y = −3. See Example 1.

17. $\dfrac{y + 5}{3y - 2}$ 18. $\dfrac{2y + 9}{y^2 + 25}$

19. $\dfrac{y^2 + 9}{9 - y^2}$ 20. $\dfrac{-y - 11}{y^2 + 2y - 3}$

21. $\dfrac{-y}{y^2 - y + 6}$ 22. $\dfrac{y^3}{3y^2 + 1}$

Find all real numbers for which the rational expression is undefined. See Example 2.

23. $\dfrac{x + 5}{8x}$ 24. $\dfrac{4x - 1}{6x}$

25. $\dfrac{15}{x - 2}$ 26. $\dfrac{5x}{x + 5}$

27. $\dfrac{15x + 2}{x^2 + 6}$ 28. $\dfrac{x^2 - 4x}{x^2 + 4}$

29. $\dfrac{x + 1}{2x - 1}$ 30. $\dfrac{-6x}{3x - 1}$

31. $\dfrac{30x}{x^2 - 36}$ 32. $\dfrac{2x - 15}{x^2 - 49}$

33. $\dfrac{15}{x^2 + x - 2}$ 34. $\dfrac{x - 20}{x^2 + 2x - 8}$

Simplify each expression. See Example 3.

35. $\dfrac{45}{9a}$ 36. $\dfrac{48}{16y}$

37. $\dfrac{6x^2}{4x^2}$ 38. $\dfrac{9x}{6x}$

39. $\dfrac{42c^3 d}{18cd^3}$ **40.** $\dfrac{49m^4 n^5}{35mn^6}$

41. $\dfrac{36a^3 b^8}{44ab^9}$ **42.** $\dfrac{45x^2 y^3}{20xy^4}$

Simplify each expression, if possible. See Examples 4–5.

43. $\dfrac{6x + 3}{3y}$ **44.** $\dfrac{4x + 12}{2y}$

45. $\dfrac{x + 3}{3x + 9}$ **46.** $\dfrac{2x - 14}{x - 7}$

47. $\dfrac{x^2 - 4}{x^2 - 6x + 8}$ **48.** $\dfrac{y^2 - 25}{y^2 - 3y - 10}$

49. $\dfrac{2x^2}{x + 2}$ **50.** $\dfrac{5y^2}{y + 5}$

51. $\dfrac{4b^2 + 4b + 1}{(2b + 1)^3}$ **52.** $\dfrac{9y^2 - 12y + 4}{(3y - 2)^3}$

53. $\dfrac{m^2 - 2mn + n^2}{7m^2 - 7n^2}$ **54.** $\dfrac{11c^2 - 11d^2}{c^2 - 2cd + d^2}$

Simplify each expression. See Example 6.

55. $\dfrac{10(c - 3) + 10}{3(c - 3) + 3}$ **56.** $\dfrac{6(d + 3) - 6}{7(d + 3) - 7}$

57. $\dfrac{6(x + 3) - 18}{3x - 18}$ **58.** $\dfrac{4(t - 1) + 4}{4t + 4}$

Simplify each expression. See Examples 7–8.

59. $\dfrac{2x - 7}{7 - 2x}$ **60.** $\dfrac{18 - d}{d - 18}$

61. $\dfrac{3 - 4t}{8t - 6}$ **62.** $\dfrac{5t - 1}{3 - 15t}$

63. $\dfrac{2 - a}{a^2 - a - 2}$ **64.** $\dfrac{4 - b}{b^2 - 5b + 4}$

65. $\dfrac{25 - 5m}{m^2 - 25}$ **66.** $\dfrac{36 - 6h}{h^2 - 36}$

TRY IT YOURSELF

Simplify each expression, if possible.

67. $\dfrac{a^3 - a^2}{a^4 - a^3}$ **68.** $\dfrac{2c^4 + 2c^3}{4c^5 + 4c^4}$

69. $\dfrac{4 - x^2}{x^2 - x - 2}$ **70.** $\dfrac{81 - y^2}{y^2 + 10y + 9}$

71. $\dfrac{6x - 30}{5 - x}$ **72.** $\dfrac{6t - 42}{7 - t}$

73. $\dfrac{x^2 + 3x + 2}{x^2 + x - 2}$ **74.** $\dfrac{x^2 + x - 6}{x^2 - x - 2}$

75. $\dfrac{15x^2 y}{5xy^2}$ **76.** $\dfrac{12xz}{4xz^2}$

77. $\dfrac{x(x - 8) + 16}{16 - x^2}$ **78.** $\dfrac{x^2 - 3(2x - 3)}{9 - x^2}$

79. $\dfrac{4c + 4d}{d + c}$ **80.** $\dfrac{a + b}{5b + 5a}$

81. $\dfrac{3x^2 - 27}{2x^2 - 5x - 3}$ **82.** $\dfrac{2x^2 - 8}{3x^2 - 5x - 2}$

83. $\dfrac{-3x^2 + 10x + 77}{x^2 - 4x - 21}$ **84.** $\dfrac{-2x^2 + 5x + 3}{x^2 + 2x - 15}$

85. $\dfrac{16a^2 - 1}{4a + 4}$ **86.** $\dfrac{25m^2 - 1}{5m + 5}$

87. $\dfrac{8u^2 - 2u - 15}{4u^4 + 5u^3}$ **88.** $\dfrac{6n^2 - 7n + 2}{3n^3 - 2n^2}$

89. $\dfrac{(2x + 3)^4}{4x^2 + 12x + 9}$ **90.** $\dfrac{(3y - 2)^5}{9y^2 - 12y + 4}$

91. $\dfrac{6a + 3(a + 2) + 12}{a + 2}$ **92.** $\dfrac{2y + 4(y - 1) - 2}{y - 1}$

93. $\dfrac{15x - 3x^2}{25y - 5xy}$ **94.** $\dfrac{18c - 2c^2}{81d - 9cd}$

APPLICATIONS

95. ORGAN PIPES   The number of vibrations n per second of an organ pipe is given by the formula $n = \dfrac{512}{L}$ where L is the length of the pipe in feet. How many times per second will a 6-foot pipe vibrate?

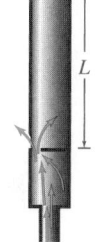

96. RAISING TURKEYS   The formula $T = \dfrac{2,000m}{m + 1}$ gives the number T of turkeys on a poultry farm m months after the beginning of the year. How many turkeys will there be on the farm by the end of July?

97. MEDICAL DOSAGES   The formula $c = \dfrac{4t}{t^2 + 1}$ gives the concentration c (in milligrams per liter) of a certain dosage of medication in a patient's blood stream t hours after the medication is administered. Suppose the patient received the medication at noon. Find the concentration of medication in his blood at the following times later that afternoon.

98. MANUFACTURING   If a company produces x child car seats, the average cost c (in dollars) to produce one car seat is given by the formula $c = \dfrac{50x + 50{,}000}{x}$. Find the company's average production cost if 1,000 are produced.

WRITING

99. Explain why $\dfrac{x - 7}{7 - x} = -1$.

100. Explain why $\dfrac{x - 3}{x + 4}$ is undefined for $x = -4$ but defined for $x = 3$.

101. Explain the error in the following work:

$$\frac{x}{x + 2} = \frac{\overset{1}{\cancel{x}}}{\underset{1}{\cancel{x}} + 2} = \frac{1}{3}$$

102. Explain why there are no values for x for which $\dfrac{x - 7}{x^2 + 49}$ is undefined.

REVIEW

State each property using the variables a, b, and when necessary, c.

103. a. The associative property of addition

   b. The commutative property of multiplication

104. a. The distributive property

   b. The zero-factor property

CHALLENGE PROBLEMS

Simplify each expression.

105. $\dfrac{(x^2 + 2x + 1)(x^2 - 2x + 1)}{(x^2 - 1)^2}$

106. $\dfrac{2x^2 + 2x - 12}{x^3 + 3x^2 - 4x - 12}$

107. $\dfrac{x^3 - 27}{x^3 - 9x}$

108. $\dfrac{b^3 + a^3}{a^2 - ab + b^2}$

SECTION 7.2
Multiplying and Dividing Rational Expressions

Objectives
1. Multiply rational expressions.
2. Divide rational expressions.
3. Convert units of measurement.

 *will Be easier than 7.1*

In this section, we will extend the rules for multiplying and dividing fractions to problems involving multiplication and division of rational expressions.

1   **Multiply Rational Expressions.**

Recall that to multiply fractions, we multiply their numerators and multiply their denominators. For example,

$$\frac{4}{7} \cdot \frac{3}{5} = \frac{4 \cdot 3}{7 \cdot 5} \qquad \text{Multiply the numerators and multiply the denominators.}$$

$$= \frac{12}{35}$$

We use the same procedure to multiply rational expressions.

Multiplying Rational Expressions	To multiply rational expressions, multiply their numerators and their denominators. Then, if possible, factor and simplify.
	For any two rational expressions, $\dfrac{A}{B}$ and $\dfrac{C}{D}$,
	$$\frac{A}{B} \cdot \frac{C}{D} = \frac{AC}{BD}$$

EXAMPLE 1   Multiply: **a.** $\dfrac{x+1}{x} \cdot \dfrac{9}{4x^2}$ **b.** $\dfrac{35x^3}{17y} \cdot \dfrac{y}{5x}$

Strategy   To find the product, we will use the rule for multiplying rational expressions. In the process, we must be prepared to factor the numerators and denominators so that any common factors can be removed.

Why   We want to give the result in simplified form, which requires that the numerator and denominator have no common factors other than 1.

Solution

a. $\dfrac{x+1}{x} \cdot \dfrac{9}{4x^2} = \dfrac{9(x+1)}{4x^3}$ *Multiply the numerators.*
Multiply the denominators.

Since the numerator and denominator do not share any common factors, $\dfrac{9(x+1)}{4x^3}$ cannot be simplified. We can leave the numerator in factored form, or we can distribute the multiplication by 9 and write the result as $\dfrac{9x+9}{4x^3}$.

b. $\dfrac{35x^3}{17y} \cdot \dfrac{y}{5x} = \dfrac{35x^3 \cdot y}{17y \cdot 5x}$ *Multiply the numerators.*
Multiply the denominators.

It is obvious that the numerator and denominator of $\dfrac{35x^3 \cdot y}{17y \cdot 5x}$ have several common factors, such as 5, x and y. These common factors become more apparent when we factor the numerator and denominator completely.

$$\dfrac{35x^3 \cdot y}{17y \cdot 5x} = \dfrac{5 \cdot 7 \cdot x \cdot x \cdot x \cdot y}{17 \cdot y \cdot 5 \cdot x}$$ *Factor $35x^3$.*

$$= \dfrac{\overset{1}{\cancel{5}} \cdot 7 \cdot \overset{1}{\cancel{x}} \cdot x \cdot x \cdot \overset{1}{\cancel{y}}}{17 \cdot \underset{1}{\cancel{y}} \cdot \underset{1}{\cancel{5}} \cdot \underset{1}{\cancel{x}}}$$ *Simplify by replacing $\frac{5}{5}$, $\frac{x}{x}$, and $\frac{y}{y}$ with the equivalent fraction $\frac{1}{1}$. This removes the factor $\frac{5 \cdot x \cdot y}{5 \cdot x \cdot y} = 1$.*

$$= \dfrac{7x^2}{17}$$ *Multiply the remaining factors in the numerator.*
Multiply the remaining factors in the denominator.

> **Caution**
> When multiplying rational expressions, always write the result in simplest form by removing any factors common to the numerator and denominator.

Self Check 1   Multiply and simplify the result, if possible:

a. $\dfrac{y}{y+6} \cdot \dfrac{12}{y-4}$ **b.** $\dfrac{a^4}{8b} \cdot \dfrac{24b}{11a^3}$

Now Try   **Problems 11 and 17**

EXAMPLE 2   Multiply: **a.** $\dfrac{x+3}{2x+4} \cdot \dfrac{6}{x^2-9}$

b. $\dfrac{8x^2-8x}{x^2+x-56} \cdot \dfrac{3x^2-22x+7}{x-x^2}$

Strategy   To find the product, we will use the rule for multiplying rational expressions. In the process, we need to factor the monomials, binomials, or trinomials that are not prime, so that any common factors can be removed.

Why   We want to give the result in simplified form, which requires that the numerator and denominator have no common factor other than 1.

Solution

a. $\dfrac{x+3}{2x+4} \cdot \dfrac{6}{x^2-9} = \dfrac{(x+3)6}{(2x+4)(x^2-9)}$ Multiply the numerators and multiply the denominators.

$= \dfrac{(x+3) \cdot 3 \cdot 2}{2(x+2)(x+3)(x-3)}$ Factor 6. Factor out the GCF, 2, from $2x + 4$. Factor the difference of two squares, $x^2 - 9$.

$= \dfrac{\overset{1}{(x+3)} \cdot 3 \cdot \overset{1}{2}}{\underset{1}{2}(x+2)\underset{1}{(x+3)}(x-3)}$ Simplify by replacing $\frac{x+3}{x+3}$ and $\frac{2}{2}$ with $\frac{1}{1}$. This removes the factor $\frac{2 \cdot (x+3)}{2 \cdot (x+3)} = 1$.

$= \dfrac{3}{(x+2)(x-3)}$ Multiply the remaining factors in the numerator.
Multiply the remaining factors in the denominator.

b. $\dfrac{8x^2-8x}{x^2+x-56} \cdot \dfrac{3x^2-22x+7}{x-x^2}$

$= \dfrac{(8x^2-8x)(3x^2-22x+7)}{(x^2+x-56)(x-x^2)}$ Multiply the numerators and multiply the denominators.

$= \dfrac{8x(x-1)(3x-1)(x-7)}{(x+8)(x-7)x(1-x)}$ Factor all four polynomials.

$= \dfrac{8x\overset{1}{(x-1)}(3x-1)\overset{-1}{(x-7)}}{(x+8)\underset{1}{(x-7)}\underset{1}{x}\underset{1}{(1-x)}}$ Simplify. Since $x - 1$ and $1 - x$ are opposites, replace $\frac{x-1}{1-x}$ with $\frac{-1}{1}$. This removes the factor $\frac{x-1}{1-x} = -1$.

$= \dfrac{-8(3x-1)}{x+8}$ Multiply the remaining factors in the numerator.
Multiply the remaining factors in the denominator.

The result can also be written as $-\dfrac{8(3x-1)}{x+8}$.

 Self Check 2   Multiply and simplify the result, if possible:

a. $\dfrac{3n-9}{3n+2} \cdot \dfrac{9n^2-4}{6}$

b. $\dfrac{m^2-4m-5}{2m-m^2} \cdot \dfrac{2m^2-4m}{3m^2-14m-5}$

Now Try   Problems 25 and 33

EXAMPLE 3   Multiply: **a.** $63x\left(\dfrac{1}{7x}\right)$ **b.** $5a\left(\dfrac{3a-1}{a}\right)$

Strategy   We will write each of the monomials, $63x$ and $5a$, as rational expressions with denominator 1. (Remember, any number divided by 1 remains unchanged.) Then we will use the rule for multiplying rational expressions.

Why   Writing $63x$ and $5a$ over 1 is helpful during the multiplication process when we multiply numerators and multiply denominators.

Solution

a. $63x\left(\dfrac{1}{7x}\right) = \dfrac{63x}{1}\left(\dfrac{1}{7x}\right)$ Write $63x$ as a fraction: $63x = \frac{63x}{1}$.

$= \dfrac{63x \cdot 1}{1 \cdot 7 \cdot x}$ Multiply the numerators and multiply the denominators.

$$= \frac{\overset{1}{9} \cdot \overset{1}{\cancel{7}} \cdot \cancel{x} \cdot 1}{1 \cdot \cancel{7} \cdot \cancel{x}} \quad \text{Write 63x in factored form as } 9 \cdot 7 \cdot x. \text{ Then simplify by removing a}$$
$$\underset{1 \quad 1}{} \qquad \text{factor equal to 1: } \tfrac{7x}{7x}.$$

$$= 9 \qquad\qquad \text{Because } \tfrac{9}{1} = 9.$$

b. $5a\left(\dfrac{3a - 1}{a}\right) = \dfrac{5a}{1}\left(\dfrac{3a - 1}{a}\right) \quad$ Write 5a as a fraction: $5a = \tfrac{5a}{1}$.

$$= \frac{\overset{1}{5\cancel{a}}(3a - 1)}{1 \cdot \underset{1}{\cancel{a}}} \qquad \text{Multiply the numerators and multiply the denominators. Then} \\ \text{simplify by removing a factor equal to 1: } \tfrac{a}{a}.$$

$$= 5(3a - 1)$$

Note that $5(3a - 1)$ can be written as $15a - 5$.

Self Check 3   Multiply and simplify the result, if possible: **a.** $36b\left(\frac{1}{6b}\right)$
b. $4x\left(\frac{x + 3}{x}\right)$

Now Try   **Problems 37 and 41**

2   ## Divide Rational Expressions.

Recall that one number is the **reciprocal** of another if their product is 1. To find the reciprocal of a fraction, we invert its numerator and denominator. We have seen that to divide fractions, we multiply the first fraction by the reciprocal of the second fraction.

$$\frac{4}{7} \div \frac{3}{5} = \frac{4}{7} \cdot \frac{5}{3} \qquad \text{Invert } \tfrac{3}{5} \text{ and change the division to a multiplication.}$$

$$= \frac{20}{21} \qquad \text{Multiply the numerators and multiply the denominators.}$$

We use the same procedure to divide rational expressions.

Dividing Rational Expressions   To divide two rational expressions, multiply the first by the reciprocal of the second. Then, if possible, we factor and simplify.

For any two rational expressions, $\frac{A}{B}$ and $\frac{C}{D}$, where $\frac{C}{D} \neq 0$,

$$\frac{A}{B} \div \frac{C}{D} = \frac{A}{B} \cdot \frac{D}{C} = \frac{AD}{BC}$$

EXAMPLE 4   Divide: **a.** $\dfrac{a}{13} \div \dfrac{17}{26}$ **b.** $\dfrac{9x}{35y} \div \dfrac{15x^2}{14}$

Strategy   We will use the rule for dividing rational expressions. After multiplying by the reciprocal, we will factor the monomials that are not prime, and remove any common factors of the numerator and denominator.

Why   We want to give the result in simplified form, which requires that the numerator and denominator have no common factor other than 1.

Solution

a. $\dfrac{a}{13} \div \dfrac{17}{26} = \dfrac{a}{13} \cdot \dfrac{26}{17}$ Multiply by the reciprocal of $\frac{17}{26}$.

$= \dfrac{a \cdot 2 \cdot 13}{13 \cdot 17}$ Multiply the numerators and denominators. Then factor 26 as $2 \cdot 13$.

$= \dfrac{a \cdot 2 \cdot \cancel{13}}{\cancel{13} \cdot 17}$ Simplify by removing common factors of the numerator and denominator.

$= \dfrac{2a}{17}$ Multiply the remaining factors in the numerator.
Multiply the remaining factors in the denominator.

b. $\dfrac{9x}{35y} \div \dfrac{15x^2}{14} = \dfrac{9x}{35y} \cdot \dfrac{14}{15x^2}$ Multiply by the reciprocal of $\frac{15x^2}{14}$.

$= \dfrac{3 \cdot 3 \cdot x \cdot 2 \cdot 7}{5 \cdot 7 \cdot y \cdot 3 \cdot 5 \cdot x \cdot x}$ Multiply the numerators and denominators. Then factor 9, 35, 14, and $15x^2$.

$= \dfrac{3 \cdot \cancel{3} \cdot \cancel{x} \cdot 2 \cdot \cancel{7}}{5 \cdot \cancel{7} \cdot y \cdot \cancel{3} \cdot 5 \cdot \cancel{x} \cdot x}$ Simplify by removing factors equal to 1.

$= \dfrac{6}{25xy}$ Multiply the remaining factors in the numerator.
Multiply the remaining factors in the denominator.

Self Check 4   Divide and simplify the result, if possible: $\dfrac{8a}{3b} \div \dfrac{16a^2}{9b^2}$

Now Try   **Problems 45 and 49**

EXAMPLE 5   Divide: $\dfrac{x^2 + x}{3x - 15} \div \dfrac{(x + 1)^2}{6x - 30}$

Strategy   To find the quotient, we will use the rule for dividing rational expressions. After multiplying by the reciprocal, we will factor the binomials that are not prime, and remove any common factors of the numerator and denominator.

Why   We want to give the result in simplified form, which requires that the numerator and denominator have no common factor other than 1.

Solution

$\dfrac{x^2 + x}{3x - 15} \div \dfrac{(x + 1)^2}{6x - 30}$

$= \dfrac{x^2 + x}{3x - 15} \cdot \dfrac{6x - 30}{(x + 1)^2}$ Multiply by the reciprocal of $\frac{(x + 1)^2}{6x - 30}$.

$= \dfrac{x(x + 1) \cdot 2 \cdot 3(x - 5)}{3(x - 5)(x + 1)(x + 1)}$ Multiply the numerators and multiply the denominators. Then factor the binomials. Write $(x + 1)^2$ as repeated multiplication.

$= \dfrac{x(\cancel{x + 1}) \cdot 2 \cdot \cancel{3}(\cancel{x - 5})}{\cancel{3}(\cancel{x - 5})(\cancel{x + 1})(x + 1)}$ Simplify by removing common factors of the numerator and denominator.

$$= \frac{2x}{x + 1}$$ Multiply the remaining factors in the numerator.
Multiply the remaining factors in the denominator.

Self Check 5 Divide and simplify the result, if possible:
$$\frac{z^2 - 9}{z^2 + 4z + 3} \div \frac{z^2 - 3z}{(z + 1)^2}$$

Now Try **Problem 59**

EXAMPLE 6 Divide: $\dfrac{2x^2 - 3xy - 2y^2}{2x + y} \div (4y^2 - x^2)$

Strategy We begin by writing $4y^2 - x^2$ as a rational expression by inserting a denominator 1. Then we will use the rule for dividing rational expressions.

Why Writing $4y^2 - x^2$ over 1 is helpful when we invert its numerator and denominator to find its reciprocal.

Solution

$$\frac{2x^2 - 3xy - 2y^2}{2x + y} \div (4y^2 - x^2)$$

$$= \frac{2x^2 - 3xy - 2y^2}{2x + y} \div \frac{4y^2 - x^2}{1}$$ Write $4y^2 - x^2$ as a fraction with a denominator of 1.

$$= \frac{2x^2 - 3xy - 2y^2}{2x + y} \cdot \frac{1}{4y^2 - x^2}$$ Multiply by the reciprocal of $\frac{4y^2 - x^2}{1}$.

$$= \frac{(2x + y)(x - 2y) \cdot 1}{(2x + y)(2y + x)(2y - x)}$$ Multiply the numerators and denominators. Then factor $2x^2 - 3xy - 2y^2$ and $4y^2 - x^2$.

$$= \frac{\overset{1}{\cancel{(2x + y)}}(x \overset{-1}{\cancel{-2y)}} \cdot 1}{\underset{1}{\cancel{(2x + y)}}(2y + x)\underset{1}{\cancel{(2y - x)}}}$$ Since $x - 2y$ and $2y - x$ are opposites, simplify by replacing $\frac{x - 2y}{2y - x}$ with $\frac{-1}{1}$.

$$= \frac{-1}{2y + x}$$

Note that $\frac{-1}{2y + x}$ can be written as $-\frac{1}{2y + x}$.

Self Check 6 Divide and simplify the result, if possible: $(b - a) \div \dfrac{a^2 - b^2}{a^2 + ab}$

Now Try **Problem 67**

Convert Units of Measurement.

We can use the concepts discussed in this section to make conversions from one unit of measure to another. *Unit conversion factors* play an important role in this process. A **unit conversion factor** is a fraction that has a value of 1. For example, we can use the fact that 1 square yard = 9 square feet to form two unit conversion factors:

$$\frac{1 \text{ yd}^2}{9 \text{ ft}^2} = 1 \quad \substack{\text{Read as "1 square yard} \\ \text{per 9 square feet."}} \qquad \frac{9 \text{ ft}^2}{1 \text{ yd}^2} = 1 \quad \substack{\text{Read as "9 square} \\ \text{feet per 1 square yard."}}$$

Since a unit conversion factor is equal to 1, multiplying a measurement by a unit conversion factor does not change the measurement, it only changes the units of measure.

EXAMPLE 7 *Carpeting.* A roll of carpeting is 12 feet wide and 150 feet long. Find the number of square yards of carpeting on the roll.

Strategy We will begin by determining the number of square feet of carpeting on the roll. Then we will multiply that result by a unit conversion factor.

Why A properly chosen unit conversion factor can convert the number of square feet of carpeting on the roll to the number of square yards on the roll.

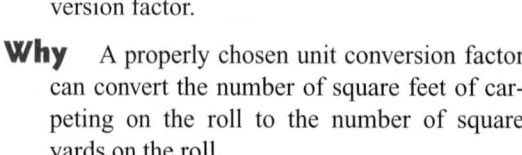

Solution When unrolled, the carpeting forms a rectangular shape with an area of $12 \cdot 150 = 1{,}800$ square feet. We will multiply $1{,}800 \text{ ft}^2$ by a unit conversion factor such that the units of ft^2 are removed and the units of yd^2 are introduced. Since $1 \text{ yd}^2 = 9 \text{ ft}^2$, we will use $\frac{1 \text{yd}^2}{9 \text{ft}^2}$.

$$\frac{1{,}800 \text{ ft}^2}{1 \text{ roll}} = \frac{1{,}800 \text{ ft}^2}{1 \text{ roll}} \cdot \frac{1 \text{ yd}^2}{9 \text{ ft}^2}$$ Multiply by a unit conversion factor that relates yd^2 to ft^2.

$$= \frac{1{,}800 \text{ ft}^2}{1 \text{ roll}} \cdot \frac{1 \text{ yd}^2}{9 \text{ ft}^2}$$ Remove the units of ft^2 that are common to the numerator and denominator.

$$= \frac{200 \text{ yd}^2}{1 \text{ roll}}$$ Divide 1,800 by 9 to get 200.

There are 200 yd^2 of carpeting on the roll.

 Self Check 7 Convert $5{,}400 \text{ ft}^2$ to square yards.

Now Try **Problem 71**

EXAMPLE 8 *The Speed of Light.* The speed with which light moves through space is about 186,000 miles per second. Express this speed in miles per minute.

Strategy The speed of light can be expressed as $\frac{186{,}000 \text{ mi}}{1 \text{ sec}}$. We will multiply that fraction by a unit conversion factor.

Why A properly chosen unit conversion factor can convert the number of miles traveled per second to the number of miles traveled per minute.

Solution We will multiply $\frac{186{,}000 \text{ mi}}{1 \text{ sec}}$ by a unit conversion factor such that the units of seconds are removed and the units of minutes are introduced. Since $60 \text{ seconds} = 1$ minute, we will use $\frac{60 \text{ sec}}{1 \text{ min}}$.

$$\frac{186{,}000 \text{ mi}}{1 \text{ sec}} = \frac{186{,}000 \text{ mi}}{1 \text{ sec}} \cdot \frac{60 \text{ sec}}{1 \text{ min}}$$ Multiply by a unit conversion factor that relates seconds to minutes.

$$= \frac{186,000 \text{ mi}}{1 \text{ sec}} \cdot \frac{60 \text{ sec}}{1 \text{ min}}$$ Remove the units of seconds that are common to the numerator and denominator.

$$= \frac{11,160,000 \text{ mi}}{1 \text{ min}}$$ Multiply 186,000 and 60 to get 11,160,000.

The speed of light is about 11,160,000 miles per minute.

 Self Check 8   A mosquito flaps it wings about 600 times per second. How many times is that per minute?

Now Try   **Problem 75**

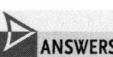 **ANSWERS TO SELF CHECKS**   1. a. $\frac{12y}{(y+6)(y-4)}$ b. $\frac{3a}{11}$ 2. a. $\frac{(n-3)(3n-2)}{2}$ b. $-\frac{2(m+1)}{3m+1}$
3. a. 6 b. $4x + 12$ 4. $\frac{3b}{2a}$ 5. $\frac{z+1}{z}$ 6. $-a$ 7. 600 yd² 8. 36,000 flaps per minute

STUDY SET
7.2

VOCABULARY

Fill in the blanks.

1. The _____ of $\frac{x^2 + 6x + 1}{10x}$ is $\frac{10x}{x^2 + 6x + 1}$.

2. A ____ conversion factor is a fraction that is equal to 1, such as $\frac{3 \text{ ft}}{1 \text{ yd}}$.

CONCEPTS

Fill in the blanks.

3. a. To multiply rational expressions, multiply their _____ and multiply their _____. To divide two rational expressions, multiply the first by the _____ of the second. In symbols,
   b. $\frac{A}{B} \cdot \frac{C}{D} = $ —— and $\frac{A}{B} \div \frac{C}{D} = \frac{A}{B} \cdot $ ——

Simplify each expression.

4. $\frac{(x+7) \cdot 2 \cdot 5}{5(x+1)(x+7)(x-9)}$

5. $\frac{y \cdot y \cdot y(15-y)}{y(y-15)(y+1)}$

6. a. Write $3x + 5$ in fractional form.
   b. What is the reciprocal of $18x$?

7. Find the product of the rational expression and its reciprocal.
   $$\frac{3}{x+2} \cdot \frac{x+2}{3}$$

8. Use the fact that 1 tablespoon = 3 teaspoons to write two unit conversion factors.

NOTATION

9. What units are common to the numerator and denominator?
   $$\frac{45 \text{ ft}}{1} \cdot \frac{1 \text{ yd}}{3 \text{ ft}}$$

10. a. What fact is indicated by the unit conversion factor $\frac{1 \text{ day}}{24 \text{ hours}}$?
    b. Fill in the blank: $\frac{1 \text{ day}}{24 \text{ hours}} = $ ▢.

GUIDED PRACTICE

Multiply, and then simplify, if possible. See Example 1.

11. $\frac{3}{7} \cdot \frac{y}{2}$ 12. $\frac{2}{7} \cdot \frac{z}{3}$

13. $\frac{y+2}{y} \cdot \frac{3}{y^2}$ 14. $\frac{4}{a+1} \cdot \frac{a}{7}$

15. $\frac{35n}{12} \cdot \frac{16}{7n^2}$ 16. $\frac{11m}{21} \cdot \frac{14}{55m^3}$

17. $\frac{2x^2y}{3xy} \cdot \frac{3xy^2}{2}$ 18. $\frac{2x^2z}{z} \cdot \frac{5x}{z}$

Multiply, and then simplify, if possible. See Example 2.

19. $\frac{x+5}{5} \cdot \frac{x}{x+5}$ 20. $\frac{a-9}{9} \cdot \frac{8a}{a-9}$

21. $\dfrac{x-2}{x} \cdot \dfrac{2x}{2-x}$

22. $\dfrac{y-3}{y} \cdot \dfrac{3y}{3-y}$

23. $\dfrac{2x+6}{x+3} \cdot \dfrac{3}{4x}$

24. $\dfrac{3y-9}{y-3} \cdot \dfrac{y}{3y^2}$

25. $\dfrac{(x+1)^2}{x+2} \cdot \dfrac{x+2}{x+1}$

26. $\dfrac{(y-3)^2}{y-5} \cdot \dfrac{y-5}{y-3}$

27. $\dfrac{x^2-x}{x} \cdot \dfrac{3x-6}{3-3x}$

28. $\dfrac{5z-10}{z+2} \cdot \dfrac{3}{6-3z}$

29. $\dfrac{x^2+x-6}{5x} \cdot \dfrac{5x-10}{x+3}$

30. $\dfrac{z^2+4z-5}{5z-5} \cdot \dfrac{5z}{z+5}$

31. $\dfrac{m^2-2m-3}{2m+4} \cdot \dfrac{m^2-4}{m^2+3m+2}$

32. $\dfrac{p^2-p-6}{3p-9} \cdot \dfrac{2p^2-5p-3}{p^2-3p}$

33. $\dfrac{6a^2}{a^2+6a+9} \cdot \dfrac{(a+3)^4}{4a^5}$

34. $\dfrac{9b^3}{b^2-8b+16} \cdot \dfrac{(b-4)^4}{15b^8}$

Multiply, and then simplify, if possible. See Example 3.

35. $7m\left(\dfrac{5}{m}\right)$

36. $9p\left(\dfrac{10}{p}\right)$

37. $15x\left(\dfrac{x+1}{5x}\right)$

38. $30t\left(\dfrac{t-7}{10t}\right)$

39. $12y\left(\dfrac{5y-8}{6y}\right)$

40. $16x\left(\dfrac{3x+8}{4x}\right)$

41. $24\left(\dfrac{3a-5}{2a}\right)$

42. $28\left(\dfrac{8-3t}{4t}\right)$

Divide, and then simplify, if possible. See Example 4.

43. $\dfrac{2}{y} \div \dfrac{4}{3}$

44. $\dfrac{3}{a} \div \dfrac{9}{5}$

45. $\dfrac{3a}{25} \div \dfrac{1}{5}$

46. $\dfrac{3y}{8} \div \dfrac{3}{2}$

47. $\dfrac{x^3}{18y} \div \dfrac{x}{6y}$

48. $\dfrac{21x}{z^2} \div \dfrac{7x^3}{z^5}$

49. $\dfrac{27p^4}{35q} \div \dfrac{9p}{21q}$

50. $\dfrac{12}{25s^5} \div \dfrac{10}{15s^2}$

Divide, and then simplify, if possible. See Example 5.

51. $\dfrac{9a-18}{28} \div \dfrac{9a^3}{35}$

52. $\dfrac{3x+6}{40} \div \dfrac{3x^2}{24}$

53. $\dfrac{x^2-4}{3x+6} \div \dfrac{2-x}{x+2}$

54. $\dfrac{x^2-9}{5x+15} \div \dfrac{3-x}{x+3}$

55. $\dfrac{m^2+m-20}{m} \div \dfrac{4-m}{m}$

56. $\dfrac{n^2+4n-21}{n} \div \dfrac{3-n}{n}$

57. $\dfrac{t^2+5t-14}{t} \div \dfrac{t-2}{t}$

58. $\dfrac{r^2+12r+11}{r} \div \dfrac{r+11}{r}$

59. $\dfrac{x^2-2x-35}{3x^2+27x} \div \dfrac{3x^2+17x+10}{18x^2+12x}$

60. $\dfrac{x^2-x-6}{2x^2+9x+10} \div \dfrac{x^2-25}{2x^2+15x+25}$

61. $\dfrac{36c^2-49d^2}{3d^3} \div \dfrac{12c+14d}{d^4}$

62. $\dfrac{25y^2-16z^2}{2yz} \div \dfrac{10y-8z}{y^2}$

Divide, and then simplify, if possible. See Example 6.

63. $\dfrac{x^2-1}{3x-3} \div (x+1)$

64. $\dfrac{x^2-16}{x-4} \div (3x+12)$

65. $\dfrac{n^2-10n+9}{n-9} \div (n-1)$

66. $\dfrac{r^2-11r+18}{r-9} \div (r-2)$

67. $\dfrac{2r-3s}{12} \div (4r^2-12rs+9s^2)$

68. $\dfrac{3m+n}{18} \div (9m^2+6mn+n^2)$

69. $24n^2 \div \dfrac{18n^3}{n-1}$

70. $12m \div \dfrac{16m^2}{m+4}$

Complete each unit conversion. See Examples 7 and 8.

71. $\dfrac{150 \text{ yards}}{1} \cdot \dfrac{3 \text{ feet}}{1 \text{ yard}} = ?$

72. $\dfrac{60 \text{ inches}}{1} \cdot \dfrac{1 \text{ feet}}{12 \text{ inches}} = ?$

73. $\dfrac{6 \text{ pints}}{1} \cdot \dfrac{1 \text{ gallon}}{8 \text{ pints}} = ?$

74. $\dfrac{4 \text{ cups}}{1} \cdot \dfrac{1 \text{ gallon}}{16 \text{ cups}} = ?$

75. $\dfrac{30 \text{ miles}}{1 \text{ hour}} \cdot \dfrac{1 \text{ hour}}{60 \text{ minute}} = ?$

76. $\dfrac{300 \text{ meters}}{3 \text{ months}} \cdot \dfrac{12 \text{ months}}{1 \text{ year}} = ?$

77. $\dfrac{30 \text{ meters}}{1 \text{ seconds}} \cdot \dfrac{60 \text{ seconds}}{1 \text{ minutes}} = ?$

78. $\dfrac{288 \text{ inches}^2}{1 \text{ year}} \cdot \dfrac{1 \text{ feet}^2}{144 \text{ inches}^2} = ?$

TRY IT YOURSELF

Perform the operations and simplify, if possible.

79. $\dfrac{b^2-5b+6}{b^2-10b+16} \div \dfrac{b^2+2b}{b^2-6b-16}$

80. $\dfrac{m^2+m-6}{m^2-6m+9} \div \dfrac{m^2-4}{m^2-9}$

81. $\dfrac{5x+5}{25} \cdot \dfrac{5}{(x+1)^3}$

82. $\dfrac{7t-7}{28} \cdot \dfrac{4}{(t-1)^4}$

83. $10h\left(\dfrac{5h-3}{2h}\right)$

84. $33r\left(\dfrac{5r+4}{11r}\right)$

85. $\dfrac{n^2 - 9}{n^2 - 3n} \div \dfrac{n + 3}{n^2 - n}$

86. $\dfrac{b^2 - b}{b + 2} \div \dfrac{b^2 - 2b}{b^2 - 4}$

87. $\dfrac{10r^2s}{6rs^2} \cdot \dfrac{3r^3}{2rs}$

88. $\dfrac{3a^3b}{25cd^3} \cdot \dfrac{5cd^2}{6ab}$

89. $\dfrac{7}{3p^3} \cdot \dfrac{p + 2}{p}$

90. $\dfrac{5t^2}{11} \cdot \dfrac{2t}{t - 5}$

91. $\dfrac{5x^2 + 13x - 6}{x + 3} \div \dfrac{5x^2 - 17x + 6}{x - 2}$

92. $\dfrac{3p^2 + 5p - 2}{p^3 + 2p^2} \div \dfrac{6p^2 + 13p - 5}{2p^3 + 5p^2}$

93. $\dfrac{4x^2 - 12xy + 9y^2}{x^3y^2} \cdot \dfrac{x^3y}{4x^2 - 9y^2}$

94. $\dfrac{ab^4}{25a^2 - 16b^2} \cdot \dfrac{25a^2 - 40ab + 16b^2}{a^2b^4}$

APPLICATIONS

95. GEOMETRY   Find the area of the rectangle.

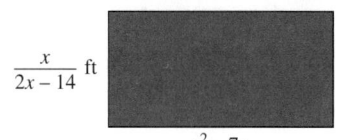
$\dfrac{x}{2x - 14}$ ft

$\dfrac{x^2 - 7x}{5}$ ft

96. MOTION   The table contains algebraic expressions for the rate an object travels and the time traveled at that rate. Complete the table.

Rate (mph)	Time (hr)	Distance (mi)
$\dfrac{k^2 + k - 6}{k - 3}$	$\dfrac{k^2 - 9}{k^2 - 4}$	

97. TALKING   According to the *Sacramento Bee* newspaper, the number of words an average man speaks a day is about 12,000. How many words does an average man speak in 1 year?   (*Hint:* 365 days = 1 year.)

98. CLASSROOM SPACE   The recommended size of an elementary school classroom in the United States is approximately 900 square feet. Convert this to square yards.

99. NATURAL LIGHT   According to the University of Georgia School Design and Planning Laboratory, the basic classroom should have at least 72 square feet of windows for natural light. Convert this to square yards.

100. TRUCKING   A cement truck holds 9 cubic yards of concrete. How many cubic feet of concrete does it hold?   (*Hint:* 27 cubic feet = 1 cubic yard.)

101. BEARS   The maximum speed a grizzly bear can run is about 30 miles per hour. What is its maximum speed in miles per minute?

102. FUEL ECONOMY   Use the information that follows to determine the miles per fluid ounce of gasoline for city and for highway driving for the Dodge Dakota Pickup.   (*Hint:* 1 gallon = 128 fluid ounces.)

2007 Dodge Dakota Pickup
Fuel Economy

Fuel Type	Regular
MPG (city)	16
MPG (highway)	20

103. TV TRIVIA   On the comedy television series *Green Acres* (1965–1971), New York socialites Oliver Wendell Douglas (played by Eddie Albert) and his wife, Lisa Douglas (played by Eva Gabor), move from New York to purchase a 160-acre farm in Hooterville. Convert this to square miles.   (*Hint:* 1 square mile = 640 acres.)

104. CAMPING   The capacity of backpacks is usually given in cubic inches. Convert a backpack capacity of 5,400 cubic inches to cubic feet.   (*Hint:* 1 cubic foot = 1,728 cubic inches.)

WRITING

105. Explain how to multiply rational expressions.

106. To divide rational expressions, you must first know how to multiply rational expressions. Explain why.

107. Explain why 60 miles per hour and 1 mile per minute are the same speed.

108. Explain why the unit conversion factor $\dfrac{1 \text{ ft}}{12 \text{ in}}$ is equal to 1.

REVIEW

109. HARDWARE   A brace has a length that is 2 inches less than twice the width of the shelf that it supports. The brace is anchored to the wall 8 inches below the shelf. Find the width of the shelf and the length of the brace.

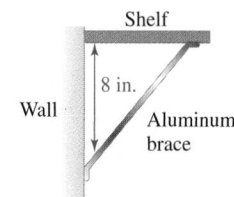
Shelf
8 in.
Wall
Aluminum brace

110. Solve $A = \frac{1}{2}h(b + d)$ for b.

CHALLENGE PROBLEMS

Perform the operations. Simplify, if possible.

111. $\dfrac{c^3 - 2c^2 + 5c - 10}{c^2 - c - 2} \cdot \dfrac{c^3 + c^2 - 5c - 5}{c^4 - 25}$

112. $\dfrac{x^3 - y^3}{x^3 + y^3} \div \dfrac{x^3 + x^2y + xy^2}{x^2y - xy^2 + y^3}$

113. $\dfrac{-x^3 + x^2 + 6x}{3x^3 + 21x^2} \div \left(\dfrac{2x + 4}{3x^2} \div \dfrac{2x + 14}{x^2 - 3x} \right)$

114. $\dfrac{x^2 - y^2}{2x^4 - 2x^3} \div \left(\dfrac{x - y}{2x^2} \div \dfrac{x + y}{x^2 + 2xy + y^2} \right)$

SECTION 7.3
Adding and Subtracting with Like Denominators; Least Common Denominators

Objectives

 Add and subtract rational expressions that have the same denominator.

2 Find the least common denominator.

3 Build rational expressions into equivalent expressions.

In this section, we extend the rules for adding and subtracting fractions to problems involving addition and subtraction of rational expressions.

1 **Add and Subtract Rational Expressions That Have the Same Denominator.**

Recall from Chapter 1 that to add (or subtract) fractions that have the same denominator, we add (or subtract) their numerators and write the sum (or difference) over the common denominator. For example,

The Language of Algebra
We can describe $\frac{3}{7}$ and $\frac{2}{7}$ as having the *same* denominator, *common* denominators, or *like* denominators.

$$\frac{3}{7} + \frac{2}{7} = \frac{3 + 2}{7} \qquad \text{and} \qquad \frac{18}{25} - \frac{9}{25} = \frac{18 - 9}{25}$$

$$= \frac{5}{7} \qquad\qquad\qquad = \frac{9}{25}$$

We use the same procedure to add and subtract rational expressions with like denominators.

Adding and Subtracting Rational Expressions That Have the Same Denominator

To add (or subtract) rational expressions that have same denominator, add (or subtract) their numerators and write the sum (or difference) over the common denominator. Then, if possible, factor and simplify.

If $\frac{A}{D}$ and $\frac{B}{D}$ are rational expressions,

$$\frac{A}{D} + \frac{B}{D} = \frac{A + B}{D} \qquad \text{and} \qquad \frac{A}{D} - \frac{B}{D} = \frac{A - B}{D}$$

EXAMPLE 1   Add: **a.** $\dfrac{x}{8} + \dfrac{3x}{8}$ **b.** $\dfrac{4s - 9}{9t} + \dfrac{7}{9t}$

Strategy   We will add the numerators and write the sum over the common denominator. Then, if possible, we will factor and simplify.

Why   This is the rule for adding rational expressions, such as these, that have the same denominator.

Caution
We *do not* add rational expressions by adding numerators and adding denominators!

$$\frac{x}{8} + \frac{3x}{8} \neq \frac{4x}{16}$$

The same caution applies when subtracting rational expressions.

Solution
a. The given rational expressions have the same denominator, 8.

$$\frac{x}{8} + \frac{3x}{8} = \frac{x + 3x}{8}$$

$$= \frac{4x}{8} \qquad \text{Combine like terms in the numerator: } x + 3x = 4x.$$
$$\text{This result can be simplified.}$$

$$= \frac{\overset{1}{\cancel{4}} \cdot x}{2 \cdot \underset{1}{\cancel{4}}} \qquad \text{Factor 8 as } 4 \cdot 2. \text{ Then simplify by removing a factor equal to 1.}$$

$$= \frac{x}{2}$$

b. The given rational expressions have the same denominator, $9t$.

$$\frac{4s - 9}{9t} + \frac{7}{9t} = \frac{4s - 9 + 7}{9t} \qquad \text{Add the numerators. Write the sum over the common denominator, } 9t.$$

$$= \frac{4s - 2}{9t} \qquad \text{Combine like terms in the numerator: } -9 + 7 = -2.$$

To attempt to simplify the result, we factor the numerator to get $\frac{2(2s - 1)}{9t}$. Since the numerator and denominator do not have any common factors, $\frac{4s - 2}{9t}$ cannot be simplified. Thus,

$$\frac{4s - 9}{9t} + \frac{7}{9t} = \frac{4s - 2}{9t}$$

Self Check 1   Add and simplify the result, if possible: **a.** $\frac{2x}{15} + \frac{4x}{15}$

b. $\frac{3m - 8}{23n} + \frac{2}{23n}$

Now Try   **Problems 17 and 21**

EXAMPLE 2   Add: **a.** $\dfrac{3x + 21}{5x + 10} + \dfrac{8x + 1}{5x + 10}$ **b.** $\dfrac{x^2 + 9x - 7}{2x(x - 6)} + \dfrac{x^2 - 9x}{(x - 6)2x}$

Strategy   We will add the numerators and write the sum over the common denominator. Then, if possible, we will factor and simplify.

Why   This is the rule for adding rational expressions that have the same denominator.

Solution

a. $\dfrac{3x + 21}{5x + 10} + \dfrac{8x + 1}{5x + 10} = \dfrac{3x + 21 + 8x + 1}{5x + 10}$ Add the numerators. Write the sum over the common denominator, $5x + 10$.

$$= \frac{11x + 22}{5x + 10} \qquad \begin{array}{l}\text{Combine like terms in the numerator:} \\ 3x + 8x = 11x \text{ and } 21 + 1 = 22.\end{array}$$

$$= \frac{11(\overset{1}{\cancel{x + 2}})}{5(\underset{1}{\cancel{x + 2}})} \qquad \begin{array}{l}\text{Factor the numerator: The GCF is 11. Factor} \\ \text{the denominator: The GCF is 5. Then simplify} \\ \text{by removing a factor equal to 1.}\end{array}$$

$$= \frac{11}{5}$$

b. By the commutative property of multiplication, $2x(x - 6) = (x - 6)2x$. Therefore, the denominators are the same. We add the numerators and write the sum over the common denominator.

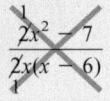

$$\frac{x^2 + 9x - 7}{2x(x - 6)} + \frac{x^2 - 9x}{(x - 6)2x} = \frac{x^2 + 9x - 7 + x^2 - 9x}{2x(x - 6)}$$

$$= \frac{2x^2 - 7}{2x(x - 6)}$$

Combine like terms in the numerator: $x^2 + x^2 = 2x^2$ and $9x - 9x = 0$.

Since the numerator, $2x^2 - 7$, does not factor, $\frac{2x^2 - 7}{2x(x - 6)}$ is in simplest form.

Self Check 2   Add and simplify the result, if possible: **a.** $\frac{m + 3}{3m - 9} + \frac{m - 9}{3m - 9}$

b. $\frac{c^2 - c}{(c - 1)(c + 2)} + \frac{c^2 - 10c}{(c + 2)(c - 1)}$

Now Try   **Problems 25 and 27**

EXAMPLE 3   Subtract: $\dfrac{x + 6}{x^2 + 4x - 5} - \dfrac{1}{x^2 + 4x - 5}$

Strategy   We will subtract the numerators and write the sum over the common denominator. Then, if possible, we will factor and simplify.

Why   This is the rule for subtracting rational expressions that have the same denominator.

Solution

$$\frac{x + 6}{x^2 + 4x - 5} - \frac{1}{x^2 + 4x - 5} = \frac{x + 6 - 1}{x^2 + 4x - 5}$$

Subtract the numerators. Write the difference over the common denominator, $x^2 + 4x - 5$.

$$= \frac{x + 5}{x^2 + 4x - 5}$$

Combine like terms in the numerator: $6 - 1 = 5$.

$$= \frac{\overset{1}{\cancel{x + 5}}}{\underset{1}{\cancel{(x + 5)}}(x - 1)}$$

Factor the denominator. Then simplify by removing a factor equal to 1.

$$= \frac{1}{x - 1}$$

Self Check 3   Subtract and simplify the result, if possible:

$$\frac{n - 3}{n^2 - 16} - \frac{1}{n^2 - 16}$$

Now Try   **Problem 37**

EXAMPLE 4   Subtract: **a.** $\dfrac{x^2 + 10x}{x + 3} - \dfrac{4x - 9}{x + 3}$

b. $\dfrac{x^2}{(x + 7)(x - 8)} - \dfrac{-x^2 + 14x}{(x + 7)(x - 8)}$

Strategy   We will use the rule for subtracting rational expressions that have the same denominators. In both cases, it is important to note that the numerator of the second fraction has *two* terms.

Why   We must make sure that entire numerator (not just the first term) of the second fraction is subtracted.

Solution

a. To subtract the numerators, each term of $4x - 9$ must be subtracted from $x^2 + 10x$.

This $-$ symbol applies to the entire numerator $4x - 9$.

This numerator is written within parentheses to make sure that we subtract both of its terms.

$$\frac{x^2 + 10x}{x + 3} - \frac{4x - 9}{x + 3} = \frac{x^2 + 10x - (4x - 9)}{x + 3}$$

Subtract the numerators. Write the difference over the common denominator.

$$= \frac{x^2 + 10x - 4x + 9}{x + 3}$$

In the numerator, use the distributive property: $-(4x - 9) = -1(4x - 9) = -4x + 9$.

$$= \frac{x^2 + 6x + 9}{x + 3}$$

Combine like terms in the numerator: $10x - 4x = 6x$.

$$= \frac{(x + 3)(x + 3)}{x + 3}$$

To see if the result simplifies, factor the numerator.

$$= \frac{\overset{1}{(\cancel{x + 3})}(x + 3)}{\underset{1}{\cancel{x + 3}}}$$

Simplify by removing a factor equal to 1.

$$= x + 3$$

b. We subtract the numerators and write the difference over the common denominator.

$$\frac{x^2}{(x + 7)(x - 8)} - \frac{-x^2 + 14x}{(x + 7)(x - 8)} = \frac{x^2 - (-x^2 + 14x)}{(x + 7)(x - 8)}$$

Write the second numerator within parentheses.

$$= \frac{x^2 + x^2 - 14x}{(x + 7)(x - 8)}$$

Use the distributive property: $-(-x^2 + 14x) = x^2 - 14x$.

$$= \frac{2x^2 - 14x}{(x + 7)(x - 8)}$$

In the numerator, combine like terms: $x^2 + x^2 = 2x^2$.

In an attempt to simplify, we can factor $2x^2 - 14x$ as $2x(x - 7)$. However, the numerator and denominator have no common factors. The result is in simplest form.

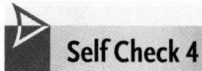 **Self Check 4**   Subtract and simplify the result, if possible:

a. $\dfrac{x^2 + 3x}{x - 1} - \dfrac{5x - 1}{x - 1}$

b. $\dfrac{3y^2}{(y + 3)(y - 3)} - \dfrac{-3y^2 + y}{(y + 3)(y - 3)}$

Now Try Problems 41 and 47

❷ Find the Least Common Denominator.

We will now discuss two skills that are needed for adding and subtracting rational expressions that have unlike denominators. To begin, let's consider

$$\frac{11}{8x} + \frac{7}{18x^2}$$

To add these expressions, we must express them as equivalent expressions with a common denominator. The **least common denominator (LCD)** is usually the easiest one to use. The least common denominator of several rational expressions can be found as follows.

Finding the LCD	1. Factor each denominator completely.
	2. The LCD is a product that uses each different factor obtained in step 1 the greatest number of times it appears in any one factorization.

EXAMPLE 5 Find the LCD of each pair of rational expressions:

a. $\dfrac{11}{8x}$ and $\dfrac{7}{18x^2}$ **b.** $\dfrac{20}{x}$ and $\dfrac{4x}{x-9}$

Strategy We will begin by factoring completely the denominator of each rational expression. Then we will form a product using each factor the greatest number of times it appears in any one factorization.

Why Since the LCD must contain the factors of each denominator, we need to write each denominator in factored form.

Solution

a. $8x = 2 \cdot 2 \cdot 2 \cdot x$ Prime factor 8.

$18x^2 = 2 \cdot 3 \cdot 3 \cdot x \cdot x$ Prime factor 18. Factor x^2.

The factorizations of $8x$ and $18x^2$ contain the factors 2, 3, and x. The LCD of $\dfrac{11}{8x}$ and $\dfrac{7}{18x^2}$ should contain each factor of $8x$ and $18x^2$ the greatest number of times it appears in any one factorization.

> The greatest number of times the factor 2 appears is three times.
> The greatest number of times the factor 3 appears is twice.
> The greatest number of times the factor x appears is twice.

$$\text{LCD} = 2 \cdot 2 \cdot 2 \cdot 3 \cdot 3 \cdot x \cdot x$$
$$= 72x^2$$

The LCD for $\dfrac{11}{8x}$ and $\dfrac{7}{18x^2}$ is $72x^2$.

b. Since the denominators of $\dfrac{20}{x}$ and $\dfrac{4x}{x-9}$ are completely factored, the factor x appears once and the factor $x - 9$ appears once. Thus, the LCD is $x(x - 9)$.

Self Check 5 Find the LCD of each pair of rational expressions:

a. $\dfrac{y+7}{6y^3}$ and $\dfrac{7}{75y}$ **b.** $\dfrac{a-3}{a+3}$ and $\dfrac{21}{a}$

Now Try Problems 53 and 59

EXAMPLE 6 Find the LCD of each pair of rational expressions:

a. $\dfrac{x}{7x+7}$ and $\dfrac{x-2}{5x+5}$ **b.** $\dfrac{6-x}{x^2+8x+16}$ and $\dfrac{15x}{x^2-16}$

Strategy We will begin by factoring completely each binomial and trinomial in the denominators of the rational expressions. Then we will form a product using each factor the greatest number of times it appears in any one factorization.

Why Since the LCD must contain the factors of each denominator, we need to write each denominator in factored form.

Solution

a. Factor each denominator completely.

$$7x + 7 = 7(x + 1) \quad \text{The GCF is 7.}$$

$$5x + 5 = 5(x + 1) \quad \text{The GCF is 5.}$$

The factorizations of $7x + 7$ and $5x + 5$ contain the factors 7, 5, and $x + 1$. The LCD of $\frac{x}{7x + 7}$ and $\frac{x - 2}{5x + 5}$ should contain each factor of $7x + 7$ and $5x + 5$ the greatest number of times it appears in any one factorization.

> The greatest number of times the factor 7 appears is once.
> The greatest number of times the factor 5 appears is once.
> The greatest number of times the factor $x + 1$ appears is once.

$$\text{LCD} = 7 \cdot 5 \cdot (x + 1) = 35(x + 1)$$

b. Factor each denominator completely.

$$x^2 + 8x + 16 = (x + 4)(x + 4) \quad \text{Factor the trinomial.}$$

$$x^2 - 16 = (x + 4)(x - 4) \quad \text{Factor the difference of two squares.}$$

The factorizations of $x^2 + 8x + 16$ and $x^2 - 16$ contain the factors $x + 4$ and $x - 4$.

> The greatest number of times the factor $x + 4$ appears is twice.
> The greatest number of times the factor $x - 4$ appears is once.

$$\text{LCD} = (x + 4)(x + 4)(x - 4) = (x + 4)^2(x - 4)$$

Self Check 6   Find the LCD of each pair of rational expressions:

a. $\dfrac{x^3}{x^2 - 6x}$ and $\dfrac{25x}{2x - 12}$

b. $\dfrac{m + 1}{m^2 - 9}$ and $\dfrac{6m^2}{m^2 - 6m + 9}$

Now Try **Problems 63 and 69**

❸ Build Rational Expressions into Equivalent Expressions.

Recall from Chapter 1 that writing a fraction as an equivalent fraction with a larger denominator is called **building the fraction.** For example, to write $\frac{3}{5}$ as an equivalent fraction with a denominator of 35, we multiply it by 1 in the form of $\frac{7}{7}$:

$$\frac{3}{5} = \frac{3}{5} \cdot \frac{7}{7} = \frac{21}{35} \quad \begin{array}{l} \text{Multiply the numerators.} \\ \text{Multiply the denominators.} \end{array}$$

To add and subtract rational expressions with different denominators, we must write them as equivalent expressions having a common denominator. To do so, we build rational expressions.

Building Rational Expressions   To build a rational expression, multiply it by 1 in the form of $\frac{c}{c}$, where c is any nonzero number or expression.

EXAMPLE 7 Write each rational expression as an equivalent expression with the indicated denominator: **a.** $\dfrac{7}{15n}$, denominator $30n^3$

b. $\dfrac{6x}{x+4}$, denominator $(x+4)(x-4)$

Strategy We will begin by asking, "By what must we multiply the given denominator to get the required denominator?"

Why The answer to that question helps us determine the form of 1 to be used to build an equivalent rational expression.

Solution

a. We need to multiply the denominator of $\dfrac{7}{15n}$ by $2n^2$ to obtain a denominator of $30n^3$. It follows that $\dfrac{2n^2}{2n^2}$ is the form of 1 that should be used to build an equivalent expression.

$$\frac{7}{15n} = \frac{7}{15n} \cdot \frac{2n^2}{2n^2} \quad \text{Multiply the given rational expression by 1, in the form of } \tfrac{2n^2}{2n^2}.$$

$$= \frac{14n^2}{30n^3} \quad \begin{array}{l}\text{Multiply the numerators.}\\ \text{Multiply the denominators.}\end{array}$$

> **The Language of Algebra**
> We say that $\frac{7}{15n}$ and $\frac{14n^2}{30n^3}$ are *equivalent expressions* because they have the same value for all values of *n*, except those that make either denominator 0.

b. We need to multiply the denominator of $\dfrac{6x}{x+4}$ by $x-4$ to obtain a denominator of $(x+4)(x-4)$. It follows that $\dfrac{x-4}{x-4}$ is the form of 1 that should be used to build an equivalent expression.

$$\frac{6x}{x+4} = \frac{6x}{x+4} \cdot \frac{x-4}{x-4} \quad \text{Multiply the given rational expression by 1, in the form of } \tfrac{x-4}{x-4}.$$

$$= \frac{6x(x-4)}{(x+4)(x-4)} \quad \begin{array}{l}\text{Multiply the numerators.}\\ \text{Multiply the denominators.}\end{array}$$

$$= \frac{6x^2-24x}{(x+4)(x-4)} \quad \begin{array}{l}\text{In the numerator, distribute the multiplication by 6x.}\\ \text{Leave the denominator in factored form.}\end{array}$$

Notation To get this answer, we multiplied the factors in the numerator to obtain a polynomial in unfactored form: $6x^2 - 24x$. However, we left the denominator in factored form. This approach is beneficial in the next section when we add and subtract rational expressions with unlike denominators.

Self Check 7 Write each rational expression as an equivalent expression with the indicated denominator: **a.** $\dfrac{7}{20m^2}$, denominator $60m^3$
b. $\dfrac{2c}{c+1}$, denominator $(c+1)(c+3)$

Now Try **Problem 77**

EXAMPLE 8 Write $\dfrac{x+1}{x^2+6x}$ as an equivalent expression with a denominator of $x(x+6)(x+2)$.

Strategy We will begin by factoring the denominator of $\dfrac{x+1}{x^2+6x}$. Then we will compare the factors of x^2+6x to those of $x(x+6)(x+2)$.

Why   This comparison will enable us to answer the question, "By what must we multiply $x^2 + 6x$ to obtain $x(x + 6)(x + 2)$?"

Solution   We factor the denominator to determine what factors are missing.

$$\frac{x + 1}{x^2 + 6x} = \frac{x + 1}{x(x + 6)} \qquad \text{Factor out the GCF, x, from } x^2 + 6x.$$

It is now apparent that we need to multiply the denominator by $x + 2$ to obtain a denominator of $x(x + 6)(x + 2)$. It follows that $\frac{x + 2}{x + 2}$ is the form of 1 that should be used to build an equivalent expression.

$$\frac{x + 1}{x^2 + 6x} = \frac{x + 1}{x(x + 6)} \cdot \frac{x + 2}{x + 2} \qquad \begin{array}{l}\text{Multiply the given rational expression by 1, in the form of}\\ \frac{x + 2}{x + 2}.\end{array}$$

$$= \frac{(x + 1)(x + 2)}{x(x + 6)(x + 2)} \qquad \begin{array}{l}\text{Multiply the numerators.}\\ \text{Multiply the denominators.}\end{array}$$

$$= \frac{x^2 + 3x + 2}{x(x + 6)(x + 2)} \qquad \begin{array}{l}\text{In the numerator, use the FOIL method to multiply}\\ (x + 1)(x + 2).\\ \text{Leave the denominator in factored form.}\end{array}$$

> **Notation**
> When building rational expressions, write the numerator of the result as a polynomial in unfactored form. Write the denominator in factored form.

  Self Check 8   Write $\frac{x - 3}{x^2 - 4x}$ as an equivalent expression with a denominator of $x(x - 4)(x + 8)$.

Now Try   **Problem 83**

ANSWERS TO SELF CHECKS   **1. a.** $\frac{2x}{5}$ **b.** $\frac{3m - 6}{23n}$ or $\frac{3(m - 2)}{23n}$ **2. a.** $\frac{2}{3}$ **b.** $\frac{2c^2 - 11c}{(c - 1)(c + 2)}$ or $\frac{c(2c - 11)}{(c - 1)(c + 2)}$
3. $\frac{1}{n + 4}$ **4. a.** $x - 1$ **b.** $\frac{6y^2 - y}{(y + 3)(y - 3)}$ **5. a.** $150y^3$ **b.** $a(a + 3)$ **6. a.** $2x(x - 6)$
b. $(m + 3)(m - 3)^2$ **7. a.** $\frac{21m}{60m^3}$ **b.** $\frac{2c^2 + 6c}{(c + 1)(c + 3)}$ **8.** $\frac{(x - 3)(x + 8)}{x(x - 4)(x + 8)}$

STUDY SET
7.3

VOCABULARY

Fill in the blanks.

1. The rational expressions $\frac{7}{6n}$ and $\frac{n + 1}{6n}$ have the common _____ 6n.

2. The _____ _____ denominator of $\frac{x - 8}{x + 6}$ and $\frac{6 - 5x}{x}$ is $x(x + 6)$.

3. To _____ a rational expression, we multiply it by a form of 1. For example: $\frac{2}{n^2} \cdot \frac{8}{8} = \frac{16}{8n^2}$

4. $\frac{2}{n^2}$ and $\frac{16}{8n^2}$ are _____ expressions. They have the same value for all values of n, except for $n = 0$.

CONCEPTS

Fill in the blanks.

5. To add or subtract rational expressions that have the same denominator, add or subtract the _____, and write the sum or difference over the common _____. In symbols, $\frac{A}{D} + \frac{B}{D} = \underline{\quad}$ and $\frac{A}{D} - \frac{B}{D} = \underline{\quad}$

6. To find the least common denominator of several rational expressions, _____ each denominator completely. The LCD is a product that uses each different factor the _____ number of times it appears in any one factorization.

7. The sum of two rational expressions is $\frac{4x + 4}{5(x + 1)}$. Factor the numerator and then simplify the result.

8. Factor each denominator completely.

a. $\dfrac{17}{40x^2}$ **b.** $\dfrac{x+25}{2x^2-6x}$

9. Consider the following factorizations.

$$18x - 36 = 2 \cdot 3 \cdot 3 \cdot (x-2)$$

$$3x - 6 = 3(x-2)$$

a. What is the greatest number of times the factor 3 appears in any one factorization?

b. What is the greatest number of times the factor $x-2$ appears in any one factorization?

10. Fill in the blanks. To write $\dfrac{x}{x-9}$ as an equivalent rational expression with a denominator of $3x(x-9)$, we need to multiply the denominator by ___ . It follows that ___ is the form of 1 that should be used to build $\dfrac{x}{x-9}$.

NOTATION

Complete the solution.

11. $\dfrac{5}{x} - \dfrac{x-1}{x} = \dfrac{5-(\boxed{})}{x}$

$= \dfrac{5 \boxed{} x \boxed{} 1}{x}$

$= \dfrac{\boxed{} - x}{x}$

12. The type of multiplication that is used to build rational expressions is shown below. Fill in the blanks.

a. $\dfrac{4x}{5} \cdot \dfrac{2}{2} = \dfrac{\boxed{}}{10}$ **b.** $\dfrac{3}{t} \cdot \dfrac{t-2}{t-2} = \dfrac{\boxed{}}{t(t-2)}$

c. $\dfrac{m+1}{m-3} \cdot \dfrac{m-5}{m-5} = \dfrac{\boxed{}}{(m-3)(m-5)}$

GUIDED PRACTICE

Add and simplify the result, if possible. See Examples 1 and 2.

13. $\dfrac{9}{x} + \dfrac{2}{x}$ **14.** $\dfrac{4}{s} + \dfrac{4}{s}$

15. $\dfrac{x}{18} + \dfrac{5}{18}$ **16.** $\dfrac{7}{10} + \dfrac{3y}{10}$

17. $\dfrac{x}{9} + \dfrac{2x}{9}$ **18.** $\dfrac{5x}{7} + \dfrac{9x}{7}$

19. $\dfrac{a-5}{3a^3} + \dfrac{5}{3a^3}$ **20.** $\dfrac{b^3-8}{10b^4} + \dfrac{8}{10b^4}$

21. $\dfrac{x+3}{2y} + \dfrac{x+5}{2y}$ **22.** $\dfrac{y+2}{10z} + \dfrac{y+4}{10z}$

23. $\dfrac{2}{r^2-3r-10} + \dfrac{r}{r^2-3r-10}$

24. $\dfrac{1}{h^2-4h-5} + \dfrac{h}{h^2-4h-5}$

25. $\dfrac{3x-5}{x-2} + \dfrac{6x-13}{x-2}$ **26.** $\dfrac{8x-7}{x+3} + \dfrac{2x+37}{x+3}$

27. $\dfrac{a^2+a}{4a^2-8a} + \dfrac{2a^2-7a}{4a^2-8a}$ **28.** $\dfrac{3b^2+16b}{6b^2+9b} + \dfrac{7b^2-b}{6b^2+9b}$

Subtract and simplify the result, if possible. See Example 3.

29. $\dfrac{2x}{25} - \dfrac{x}{25}$ **30.** $\dfrac{16c}{11} - \dfrac{4c}{11}$

31. $\dfrac{35t}{99} - \dfrac{13t}{99}$ **32.** $\dfrac{44y}{72} - \dfrac{35y}{72}$

33. $\dfrac{m-1}{6m^2} - \dfrac{5}{6m^2}$ **34.** $\dfrac{c+7}{4c^4} - \dfrac{3}{4c^4}$

35. $\dfrac{17a}{2a+4} - \dfrac{7a}{2a+4}$ **36.** $\dfrac{10b}{3b-18} - \dfrac{4b}{3b-18}$

37. $\dfrac{t}{t^2+t-2} - \dfrac{1}{t^2+t-2}$

38. $\dfrac{r}{r^2-2r-3} - \dfrac{3}{r^2-2r-3}$

39. $\dfrac{11w+6}{3w(w-9)} - \dfrac{11w}{3w(w-9)}$

40. $\dfrac{y+8}{2y(y-14)} - \dfrac{y}{2y(y-14)}$

Subtract and simplify the result, if possible. See Example 4.

41. $\dfrac{3y-2}{2y+6} - \dfrac{2y-5}{2y+6}$ **42.** $\dfrac{5x+8}{3x+15} - \dfrac{3x-2}{3x+15}$

43. $\dfrac{6x^2}{3x+2} - \dfrac{11x+10}{3x+2}$ **44.** $\dfrac{8a^2}{2a+5} - \dfrac{4a^2+25}{2a+5}$

45. $\dfrac{6x-5}{3xy} - \dfrac{3x-5}{3xy}$ **46.** $\dfrac{7x+7}{5y} - \dfrac{2x+7}{5y}$

47. $\dfrac{2-p}{p^2-p} - \dfrac{-p+2}{p^2-p}$ **48.** $\dfrac{2-7n}{n^2+5} - \dfrac{-7n+2}{n^2+5}$

49. $\dfrac{8}{9-3x^2} - \dfrac{-6x+8}{9-3x^2}$ **50.** $\dfrac{5}{10-5t^2} - \dfrac{-15t+5}{10-5t^2}$

51. $\dfrac{-4x}{3x^2-7x+2} - \dfrac{-3x-2}{3x^2-7x+2}$

52. $\dfrac{-3c}{5c^2-16c+3} - \dfrac{-2c-3}{5c^2-16c+3}$

Find the LCD of each pair of rational expressions. See Examples 5 and 6.

53. $\dfrac{1}{2x}, \dfrac{9}{6x}$ **54.** $\dfrac{4}{9y}, \dfrac{11}{3y}$

55. $\dfrac{33}{15a^3}, \dfrac{9}{10a}$ **56.** $\dfrac{m-21}{12m^4}, \dfrac{m+1}{18m}$

57. $\dfrac{35}{3a^2b}, \dfrac{23}{a^2b^3}$ **58.** $\dfrac{27}{c^3d}, \dfrac{17}{2c^2d^3}$

59. $\dfrac{8}{c}, \dfrac{8-c}{c+2}$ **60.** $\dfrac{d^2-5}{d+9}, \dfrac{d-3}{d}$

61. $\dfrac{3x+1}{3x-1}, \dfrac{3x}{3x+1}$

62. $\dfrac{b+1}{b-1}, \dfrac{b}{b+1}$

63. $\dfrac{b-9}{4b+8}, \dfrac{b}{6}$

64. $\dfrac{b^2-b}{10b-15}, \dfrac{11b}{10}$

65. $\dfrac{6-k}{2k+4}, \dfrac{11}{8k}$

66. $\dfrac{5m+6}{4m+12}, \dfrac{7}{6m}$

67. $\dfrac{-2x}{x^2-1}, \dfrac{5x}{x+1}$

68. $\dfrac{7-y^2}{y^2-4}, \dfrac{y-49}{y+2}$

69. $\dfrac{4x-5}{x^2-4x-5}, \dfrac{3x+1}{x^2-25}$

70. $\dfrac{44}{s^2-9}, \dfrac{s+9}{s^2-s-6}$

71. $\dfrac{5n^2-16}{2n^2+13n+20}, \dfrac{3n^2}{n^2+8n+16}$

72. $\dfrac{4y+25}{y^2+10y+25}, \dfrac{y^2-7}{2y^2+17y+35}$

Build each rational expression into an equivalent expression with the given denominator. See Examples 7 and 8.

73. $\dfrac{5}{r}$; $10r$

74. $\dfrac{4}{y}$; $7y$

75. $\dfrac{8}{x}$; $x^2 y$

76. $\dfrac{7}{y}$; xy^2

77. $\dfrac{9}{4b}$; $12b^2$

78. $\dfrac{7}{6c}$; $30c^2$

79. $\dfrac{3x}{x+1}$; $(x+1)^2$

80. $\dfrac{5y}{y-2}$; $(y-2)^2$

81. $\dfrac{x-2}{x}$; $x(x+3)$

82. $\dfrac{y-4}{y}$; $y(y-9)$

83. $\dfrac{t+5}{4t+8}$; $20(t+2)$

84. $\dfrac{x+7}{3x-15}$; $6(x-5)$

85. $\dfrac{y+3}{y^2-5y+6}$; $4y(y-2)(y-3)$

86. $\dfrac{3x-4}{x^2+3x+2}$; $8x(x+1)(x+2)$

87. $\dfrac{12-h}{h^2-81}$; $3(h+9)(h-9)$

88. $\dfrac{m^2}{m^2-100}$; $9(m+10)(m-10)$

TRY IT YOURSELF

Perform the operations. Then simplify, if possible.

89. $\dfrac{3t}{t^2-8t+7} - \dfrac{3}{t^2-8t+7}$

90. $\dfrac{10x}{x^2-2x+1} - \dfrac{10}{x^2-2x+1}$

91. $\dfrac{c}{c^2-d^2} - \dfrac{d}{c^2-d^2}$

92. $\dfrac{b}{b^2-4} - \dfrac{2}{b^2-4}$

93. $\dfrac{11n}{(n+4)(n-2)} - \dfrac{4n-1}{(n-2)(n+4)}$

94. $\dfrac{1}{(t-1)(t+1)} - \dfrac{6-t}{(t+1)(t-1)}$

95. $\dfrac{11}{36y} + \dfrac{9}{36y}$

96. $\dfrac{13}{24w} + \dfrac{17}{24w}$

97. $\dfrac{3x^2}{x+1} - \dfrac{-x+2}{x+1}$

98. $\dfrac{8b^2}{3b-2} - \dfrac{-b^2+4}{3b-2}$

99. $\dfrac{5r-27}{3r^2-9r} + \dfrac{4r}{3r^2-9r}$

100. $\dfrac{9a}{5a^2+25a} + \dfrac{a+50}{5a^2+25a}$

APPLICATIONS

101. GEOMETRY   What is the difference of the length and width of the rectangle?

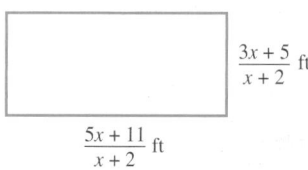

$\dfrac{3x+5}{x+2}$ ft

$\dfrac{5x+11}{x+2}$ ft

102. GEOMETRY   What is the perimeter of the rectangle in Problem 101?

WRITING

103. Explain how to add fractions with the same denominator.

104. Explain how to find a least common denominator.

105. Explain the error in the following solution:

$$\dfrac{2x+3}{x+5} - \dfrac{x+2}{x+5} = \dfrac{2x+3-x+2}{x+5}$$

$$= \dfrac{x+5}{x+5}$$

$$= 1$$

106. Explain the error in the following solution:

$$\dfrac{y+4}{y} - \dfrac{1}{y} = \dfrac{y+4-1}{y+y}$$

$$= \dfrac{y+3}{2y}$$

107. Explain why the LCD of $\dfrac{5}{h^2}$ and $\dfrac{3}{h}$ is h^2 and not h^3.

108. Explain how multiplication by 1 is used to build a rational expression.

Give the formula for . . .

109. a. simple interest

 b. the area of a triangle

 c. the perimeter of a rectangle

110. a. the slope of a line

 b. distance traveled

 c. the area of a circle

CHALLENGE PROBLEMS

Perform the operations. Simplify the result, if possible.

111. $\dfrac{3xy}{x-y} - \dfrac{x(3y-x)}{x-y} - \dfrac{x(x-y)}{x-y}$

112. $\dfrac{9t^3 - 12t^2}{27t^3 - 64} - \dfrac{-3t+4}{27t^3 - 64}$

113. $\dfrac{2a^2 + 2}{a^3 + 8} + \dfrac{a^3 + a}{a^3 + 8}$

114. Find the LCD of these rational expressions.

$$\frac{2}{a^3 + 8}, \frac{a}{a^2 - 4}, \frac{2a+5}{a^3 - 8}$$

SECTION 7.4
Adding and Subtracting with Unlike Denominators

Objectives

1 Add and subtract rational expressions that have unlike denominators.

2 Add and subtract rational expressions that have denominators that are opposites.

We have discussed a method for finding the least common denominator (LCD) of two rational expressions. We have also built rational expressions into equivalent expressions having a given denominator. We will now use these skills to add and subtract rational expressions with unlike denominators.

 Add and Subtract Rational Expressions That Have Unlike Denominators.

The following steps summarize how to add or subtract rational expressions that have different denominators.

Adding and Subtracting Rational Expressions That Have Unlike Denominators	1. Find the LCD.
	2. Rewrite each rational expression as an equivalent expression with the LCD as the denominator. To do so, build each fraction using a form of 1 that involves any factor(s) needed to obtain the LCD.
	3. Add or subtract the numerators and write the sum or difference over the LCD.
	4. Simplify the result, if possible.

EXAMPLE 1 Add: $\dfrac{9x}{7} + \dfrac{3x}{5}$

Strategy We will use the procedure for adding rational expressions that have unlike denominators. The first step is to determine the LCD.

Why If we are to add (or subtract) fractions, their denominators must be the same. Since the denominators of these rational expressions are different, we cannot add them in their present form.

sevenths $\dfrac{9x}{7} + \dfrac{3x}{5}$ fifths

Not the same number

Solution

Step 1: The denominators are 7 and 5. The LCD is $7 \cdot 5 = 35$.

Step 2: We need to multiply the denominator of $\frac{9x}{7}$ by 5 and we need to multiply the denominator of $\frac{3x}{5}$ by 7 to obtain the LCD, 35. It follows that $\frac{5}{5}$ and $\frac{7}{7}$ are the forms of 1 that should be used to write the equivalent rational expressions.

> **Caution**
>
> In Step 2, *don't simplify* $\frac{45x}{35}$ and $\frac{21x}{35}$, because that will take you back to the original problem.

$$\dfrac{9x}{7} + \dfrac{3x}{5} = \dfrac{9x}{7} \cdot \dfrac{5}{5} + \dfrac{3x}{5} \cdot \dfrac{7}{7}$$ Build the rational expressions so that each has a denominator of 35.

$$= \dfrac{45x}{35} + \dfrac{21x}{35}$$ Multiply the numerators.
Multiply the denominators.

Step 3: $= \dfrac{45x + 21x}{35}$ Add the numerators. Write the sum over the common denominator.

$$= \dfrac{66x}{35}$$ Combine like terms in the numerator: $45x + 21x = 66x$.

Step 4: Since 66 and 35 have no common factor other than 1, the result cannot be simplified.

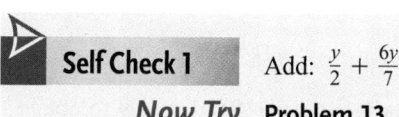

Self Check 1 Add: $\frac{y}{2} + \frac{6y}{7}$

Now Try **Problem 13**

EXAMPLE 2 Subtract: $\dfrac{13}{18b^2} - \dfrac{1}{24b}$

Strategy We will use the procedure for subtracting rational expressions that have unlike denominators. The first step is to determine the LCD.

Why If we are to subtract fractions, their denominators must be the same. Since the denominators of these rational expressions are different, we cannot subtract them in their present form.

Solution

Step 1: To find the LCD, we form a product that uses each different factor of $18b^2$ and $24b$ the greatest number of times it appears in any one factorization.

$$\left.\begin{array}{l} 18b^2 = 2 \cdot 3 \cdot 3 \cdot b \cdot b \\ 24b = 2 \cdot 2 \cdot 2 \cdot 3 \cdot b \end{array}\right\} \text{LCD} = 2 \cdot 2 \cdot 2 \cdot 3 \cdot 3 \cdot b \cdot b = 72b^2$$

Step 2: We need to multiply $18b^2$ by 4 to obtain $72b^2$, and $24b$ by $3b$ to obtain $72b^2$. It follows that we should use $\frac{4}{4}$ and $\frac{3b}{3b}$ to build the equivalent rational expressions.

Notation

When checking your answers with those in the back of the book, remember that the results can often be presented in several equivalent forms. For example, this result can also be expressed as

$$\frac{-3b + 52}{72b^2}$$

$$\frac{13}{18b^2} - \frac{1}{24b} = \frac{13}{18b^2} \cdot \frac{4}{4} - \frac{1}{24b} \cdot \frac{3b}{3b}$$

Build the rational expressions so that each has a denominator of $72b^2$.

$$= \frac{52}{72b^2} - \frac{3b}{72b^2}$$

Multiply the numerators.
Multiply the denominators.

Step 3: $\qquad = \frac{52 - 3b}{72b^2}$

Subtract the numerators. Write the difference over the common denominator.

Step 4: Since $52 - 3b$ does not factor, the result cannot be simplified.

Self Check 2 Subtract: $\frac{5}{21z^2} - \frac{3}{28z}$

Now Try Problem 29

EXAMPLE 3 Add: $\dfrac{3}{2x + 18} + \dfrac{27}{x^2 - 81}$

Strategy We use the procedure for adding rational expressions when the denominators are binomials. The first step is to find the LCD.

Why Since the denominators are different, we cannot add these rational expressions in their present form.

Solution After factoring the denominators, we see that the greatest number of times each of the factors 2, $x + 9$, and $x - 9$ appear in any one of the factorizations is once.

$$\left.\begin{array}{l} 2x + 18 = 2(x + 9) \\ x^2 - 81 = (x + 9)(x - 9) \end{array}\right\} \text{LCD} = 2(x + 9)(x - 9)$$

Since we need to multiply $2(x + 9)$ by $x - 9$ to obtain the LCD and $(x + 9)(x - 9)$ by 2 to obtain the LCD, $\frac{x - 9}{x - 9}$ and $\frac{2}{2}$ are the forms of 1 to use to build the equivalent rational expressions.

$$\frac{3}{2x + 18} + \frac{27}{x^2 - 81} = \frac{3}{2(x + 9)} + \frac{27}{(x + 9)(x - 9)}$$

Write each denominator in factored form.

$$= \frac{3}{2(x + 9)} \cdot \frac{x - 9}{x - 9} + \frac{27}{(x + 9)(x - 9)} \cdot \frac{2}{2}$$

Build the expressions so that each has a denominator of $2(x + 9)(x - 9)$.

Success Tip

To build the first rational expression, we use the distributive property to multiply the numerators. Note that we don't multiply out the denominators.

$$\frac{3}{2(x + 9)} \cdot \frac{x - 9}{x - 9}$$

The result is $\frac{3x - 27}{2(x + 9)(x - 9)}$.

$$= \frac{3x - 27}{2(x + 9)(x - 9)} + \frac{54}{2(x + 9)(x - 9)}$$

Multiply numerators.
Multiply denominators.

Although it is not required, the factors of each denominator are written in the same order.

$$= \frac{3x - 27 + 54}{2(x + 9)(x - 9)}$$

Add the numerators. Write the sum over the common denominator.

$$= \frac{3x + 27}{2(x + 9)(x - 9)}$$

Combine like terms in the numerator:
$-27 + 54 = 27$.

Caution
Always write the result in simplest form by removing any factors common to the numerator and denominator.

$$= \frac{\overset{1}{3(\cancel{x+9})}}{\underset{1}{2(\cancel{x+9})(x-9)}}$$

$$= \frac{3}{2(x-9)}$$

Factor the numerator. Then simplify the expression by removing a factor equal to 1.

Self Check 3   Add: $\frac{2}{5x+25} + \frac{4}{x^2-25}$

Now Try   **Problem 37**

EXAMPLE 4   Subtract: $\dfrac{x}{x-1} - \dfrac{x-6}{x-4}$

Strategy   We use the same procedure for subtracting rational expressions when the denominators are binomials. The first step is to find the LCD.

Why   Since the denominators are different, we cannot subtract these rational expressions in their present form.

Solution   The denominators of $\frac{x}{x-1}$ and $\frac{x-6}{x-4}$ are completely factored. The factor $x-1$ appears once and the factor $x-4$ appears once. Thus, the LCD $= (x-1)(x-4)$.

We need to multiply the first denominator by $x-4$ to obtain the LCD and the second denominator by $x-1$ to obtain the LCD. It follows that $\frac{x-4}{x-4}$ and $\frac{x-1}{x-1}$ are the forms of 1 to use to build the equivalent rational expressions.

Success Tip
To build the second rational expression, we use the FOIL method to multiply the numerators: Note that we don't multiply out the denominators.

$$\frac{x-6}{x-4} \cdot \frac{x-1}{x-1}$$

The result is $\frac{x^2-7x+6}{(x-4)(x-1)}$.

$$\frac{x}{x-1} - \frac{x-6}{x-4} = \frac{x}{x-1} \cdot \frac{x-4}{x-4} - \frac{x-6}{x-4} \cdot \frac{x-1}{x-1}$$

Build the rational expressions so that each has a denominator of $(x-1)(x-4)$.

$$= \frac{x^2-4x}{(x-1)(x-4)} - \frac{x^2-7x+6}{(x-4)(x-1)}$$

Multiply the numerators. Multiply the denominators.

By the commutative property of multiplication, these are like denominators.

$$= \frac{x^2-4x - (x^2-7x+6)}{(x-1)(x-4)}$$

Subtract the numerators. Remember the parentheses. Write the difference over the common denominator.

$$= \frac{x^2-4x - x^2+7x-6}{(x-1)(x-4)}$$

In the numerator, use the distributive property: $-(x^2-7x+6) = -1(x^2-7x+6) = -x^2+7x-6$.

$$= \frac{3x-6}{(x-1)(x-4)}$$

Combine like terms in the numerator: $x^2-x^2=0$ and $-4x+7x=3x$.

The numerator factors as $3(x-2)$. Since the numerator and denominator have no common factor, the result is in simplest form.

Self Check 4   Subtract: $\dfrac{x}{x+9} - \dfrac{x-7}{x+8}$

Now Try   **Problem 45**

| **EXAMPLE 5** | Subtract: $\dfrac{m}{m^2 + 5m + 6} - \dfrac{2}{m^2 + 3m + 2}$ |

Strategy   We use the same procedure for subtracting rational expressions when the denominators are trinomials. The first step is to find the LCD.

Why   Since the denominators are different, we cannot subtract these rational expressions in their present form.

Solution   Factor each denominator and form the LCD.

$$\left.\begin{array}{l} m^2 + 5m + 6 = (m + 2)(m + 3) \\ m^2 + 3m + 2 = (m + 2)(m + 1) \end{array}\right\} \text{LCD} = (m + 2)(m + 3)(m + 1)$$

Examining the factored forms, we see that the first denominator must be multiplied by $m + 1$, and the second must be multiplied by $m + 3$ to obtain the LCD. To build the expressions, we will use $\dfrac{m + 1}{m + 1}$ and $\dfrac{m + 3}{m + 3}$.

$$\dfrac{m}{m^2 + 5m + 6} - \dfrac{2}{m^2 + 3m + 2}$$

$$= \dfrac{m}{(m + 2)(m + 3)} - \dfrac{2}{(m + 2)(m + 1)}$$
Write each denominator in factored form.

$$= \dfrac{m}{(m + 2)(m + 3)} \cdot \dfrac{m + 1}{m + 1} - \dfrac{2}{(m + 2)(m + 1)} \cdot \dfrac{m + 3}{m + 3}$$
Build each expression, so that it has a denominator of $(m + 2)(m + 3)(m + 1)$.

$$= \dfrac{m^2 + m}{(m + 2)(m + 3)(m + 1)} - \dfrac{2m + 6}{(m + 2)(m + 1)(m + 3)}$$
Multiply numerators. Multiply denominators.

By the commutative property of multiplication, these are like denominators.

$$= \dfrac{m^2 + m - (2m + 6)}{(m + 2)(m + 3)(m + 1)}$$
Subtract the numerators. Remember the parentheses. Write the difference over the common denominator.

$$= \dfrac{m^2 + m - 2m - 6}{(m + 2)(m + 3)(m + 1)}$$
Use the distributive property: $-(2m + 6) = -1(2m + 6) = -2m - 6$.

$$= \dfrac{m^2 - m - 6}{(m + 2)(m + 3)(m + 1)}$$
Combine like terms in the numerator.

$$= \dfrac{\overset{1}{(m - 3)(\cancel{m + 2})}}{(\cancel{m + 2})(m + 3)(m + 1)}$$
Factor the numerator and simplify the expression by removing a factor equal to 1.

$$= \dfrac{m - 3}{(m + 3)(m + 1)}$$

 Self Check 5   Subtract: $\dfrac{b}{b^2 - 2b - 8} - \dfrac{6}{b^2 + b - 20}$

Now Try   **Problem 57**

| **EXAMPLE 6** | Add: $\dfrac{4b}{a - 5} + b$ |

Strategy   We will begin by writing the second addend, b, as $\dfrac{b}{1}$ and then find the LCD.

Why   To add b to the rational expression, $\frac{4b}{a-5}$, we must rewrite b as a rational expression.

Solution   The LCD of $\frac{4b}{a-5}$ and $\frac{b}{1}$ is $1(a-5)$, or simply $a-5$. Since we must multiply the denominator of $\frac{b}{1}$ by $a-5$ to obtain the LCD, we will use $\frac{a-5}{a-5}$ to write an equivalent rational expression.

$$\frac{4b}{a-5} + b = \frac{4b}{a-5} + \frac{b}{1} \cdot \frac{a-5}{a-5} \qquad \text{Build } \tfrac{b}{1} \text{ so that it has a denominator of } a-5.$$

$$= \frac{4b}{a-5} + \frac{ab-5b}{a-5} \qquad \begin{array}{l}\text{Multiply numerators: } b(a-5) = ab - 5b.\\ \text{Multiply denominators: } 1(a-5) = a-5.\end{array}$$

$$= \frac{4b+ab-5b}{a-5} \qquad \begin{array}{l}\text{Add the numerators. Write the sum over the common}\\ \text{denominator.}\end{array}$$

$$= \frac{ab-b}{a-5} \qquad \text{Combine like terms in the numerator: } 4b - 5b = -b.$$

Although the numerator factors as $b(a-1)$, the numerator and denominator do not have a common factor. Therefore, the result is in simplest form.

> ▷ **Self Check 6**   Add: $\dfrac{10y}{n+4} + y$
>
> *Now Try*   **Problem 65**

Success Tip

Since the denominator of $\frac{4b}{a-5}$ is the LCD, we *do not* need to build it by multiplying by it a form of 1. The step below is unnecessary.

---

2   **Add and Subtract Rational Expressions That Have Denominators That Are Opposites.**

Recall that two polynomials are **opposites** if their terms are the same but they are opposite in sign. For example, $x-4$ and $4-x$ are opposites. If we multiply one of these binomials by -1, the subtraction is reversed, and the result is the other binomial.

$$-1(x-4) = -x+4 \qquad\qquad -1(4-x) = -4+x$$

$$= 4-x \quad \begin{array}{l}\text{Write the}\\ \text{expression}\\ \text{with 4 first.}\end{array} \qquad\qquad\qquad = x-4 \quad \begin{array}{l}\text{Write the}\\ \text{expression}\\ \text{with x first.}\end{array}$$

These results suggest a general fact.

Multiplying by -1	When a polynomial is multiplied by -1, the result is its opposite.

This fact can be used when adding or subtracting rational expressions whose denominators are opposites.

> **EXAMPLE 7**   Subtract: $\dfrac{x}{x-7} - \dfrac{1}{7-x}$

Strategy   We note that the denominators are opposites. Either can serve as the LCD; we will choose $x-7$. To obtain a common denominator, we will multiply $\frac{1}{7-x}$ by $\frac{-1}{-1}$.

Why   When $7-x$ is multiplied by -1, the subtraction is reversed, and the result is $x-7$.

Solution We must multiply the denominator of $\frac{1}{7-x}$ by -1 to obtain the LCD. It follows that $\frac{-1}{-1}$ should be the form of 1 that is used to write an equivalent rational expression.

$$\frac{x}{x-7} - \frac{1}{7-x} = \frac{x}{x-7} - \frac{1}{7-x} \cdot \frac{-1}{-1} \quad \text{Build } \frac{1}{7-x} \text{ so that it has a denominator of } x - 7.$$

$$= \frac{x}{x-7} - \frac{-1}{-7+x} \quad \begin{array}{l}\text{Multiply the numerators.} \\ \text{Multiply the denominators.}\end{array}$$

$$= \frac{x}{x-7} - \frac{-1}{x-7} \quad \begin{array}{l}\text{Rewrite the second denominator:} \\ -7 + x = x - 7.\end{array}$$

$$= \frac{x - (-1)}{x-7} \quad \begin{array}{l}\text{Subtract the numerators. Remember the} \\ \text{parentheses. Write the difference over the} \\ \text{common denominator.}\end{array}$$

$$= \frac{x+1}{x-7} \quad \text{Simplify the numerator.}$$

The result does not simplify.

 Self Check 7 Add: $\frac{n}{n-8} + \frac{12}{8-n}$

Now Try Problem 71

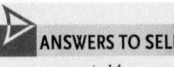 **ANSWERS TO SELF CHECK** 1. $\frac{19y}{14}$ 2. $\frac{20-9z}{84z^2}$ 3. $\frac{2}{5(x-5)}$ 4. $\frac{6x+63}{(x+9)(x+8)}$ 5. $\frac{b+3}{(b+2)(b+5)}$
6. $\frac{ny+14y}{n+4}$ 7. $\frac{n-12}{n-8}$

STUDY SET
7.4

VOCABULARY

Fill in the blanks.

1. $\frac{x}{x-7}$ and $\frac{1}{x-7}$ have like denominators. $\frac{x+5}{x-7}$ and $\frac{4x}{x+7}$ have _____ denominators.

2. Two polynomials are _____ if their terms are the same, but are opposite in sign.

CONCEPTS

3. Write each denominator in factored form.

  a. $\frac{x+1}{20x^2}$

  b. $\frac{3x^2-4}{x^2+4x-12}$

4. The factorizations of the denominators of two rational expressions are given. Find the LCD.

  a. $12a = 2 \cdot 2 \cdot 3 \cdot a$
     $18a^2 = 2 \cdot 3 \cdot 3 \cdot a \cdot a$

  b. $x^2 - 36 = (x+6)(x-6)$
     $3x - 18 = 3(x-6)$

5. What is the LCD for $\frac{x-1}{x+6}$ and $\frac{1}{x+3}$?

6. The LCD for $\frac{1}{9n^2}$ and $\frac{37}{15n^3}$ is $3 \cdot 3 \cdot 5 \cdot n \cdot n \cdot n = 45n^3$. If we want to add these rational expressions, what form of 1 should be used

  a. to build $\frac{1}{9n^2}$?

  b. to build $\frac{37}{15n^3}$?

Fill in the blanks.

7. To build $\frac{x}{x+2}$ so that it has a denominator of $5(x+2)$, we multiply it by 1 in the form of _____.

8. To build $\frac{8x}{2-x}$ so that it has a denominator of $x-2$, we multiply it by 1 in the form of _____.

NOTATION

Complete the solution.

9. $\dfrac{2}{5} + \dfrac{7}{3x} = \dfrac{2}{5} \cdot \dfrac{}{3x} + \dfrac{7}{3x} \cdot \dfrac{}{5}$

$= \dfrac{6x}{} + \dfrac{35}{}$

$= \dfrac{6x + }{15x}$

10. Are the student's answers and the book's answers equivalent?

Student's answer	Book's answer	Equivalent?
$\dfrac{m^2 + 2m}{(m-1)(m-4)}$	$\dfrac{m^2 + 2m}{(m-4)(m-1)}$	
$\dfrac{-5x^2 - 7}{4x(x+3)}$	$\dfrac{-5x^2 - 7}{4x(x+3)}$	
$\dfrac{-2x}{x-y}$	$\dfrac{2x}{x-y}$	

GUIDED PRACTICE

Perform the operations. Simplify, if possible. See Example 1.

11. $\dfrac{x}{3} + \dfrac{2x}{7}$

12. $\dfrac{y}{4} + \dfrac{3y}{5}$

13. $\dfrac{5y}{6} + \dfrac{5y}{3}$

14. $\dfrac{4x}{3} + \dfrac{x}{6}$

15. $\dfrac{7}{8} - \dfrac{4}{t}$

16. $\dfrac{5}{3} - \dfrac{2}{m}$

17. $\dfrac{4b}{3} - \dfrac{5b}{12}$

18. $\dfrac{21y}{12} - \dfrac{7y}{6}$

Perform the operations. Simplify, if possible. See Example 2.

19. $\dfrac{7}{m^2} - \dfrac{2}{m}$

20. $\dfrac{6}{n^2} - \dfrac{2}{n}$

21. $\dfrac{3}{x^2} + \dfrac{17}{x}$

22. $\dfrac{7}{c} + \dfrac{14}{c^2}$

23. $\dfrac{3}{5p} - \dfrac{5}{10p}$

24. $\dfrac{15}{16a} - \dfrac{3}{4a}$

25. $\dfrac{1}{6t} - \dfrac{11}{8t}$

26. $\dfrac{3}{10a} - \dfrac{13}{15a}$

27. $\dfrac{11}{5x} - \dfrac{5}{6x}$

28. $\dfrac{5}{9y} - \dfrac{1}{4y}$

29. $\dfrac{1}{6c^4} + \dfrac{8}{9c^2}$

30. $\dfrac{7}{8b^2} + \dfrac{5}{6b^3}$

Perform the operations. Simplify, if possible. See Example 3.

31. $\dfrac{1}{2a + 4} + \dfrac{5}{a^2 - 4}$

32. $\dfrac{5}{p^2 - 9} + \dfrac{2}{3p + 9}$

33. $\dfrac{2}{3a - 2} + \dfrac{5}{9a^2 - 4}$

34. $\dfrac{2}{5b - 3} + \dfrac{5}{25b^2 - 9}$

35. $\dfrac{4}{a + 2} - \dfrac{7}{a^2 + 4a + 4}$

36. $\dfrac{9}{b^2 - 2b + 1} - \dfrac{2}{b - 1}$

37. $\dfrac{6}{5m^2 - 5m} - \dfrac{3}{5m - 5}$

38. $\dfrac{9}{2c^2 - 2c} - \dfrac{5}{2c - 2}$

Perform the operations. Simplify, if possible. See Example 4.

39. $\dfrac{9}{t + 3} + \dfrac{8}{t + 2}$

40. $\dfrac{2}{m - 3} + \dfrac{7}{m - 4}$

41. $\dfrac{3x}{2x - 1} - \dfrac{2x}{2x + 3}$

42. $\dfrac{2y}{5y - 1} - \dfrac{2y}{3y + 2}$

43. $\dfrac{1}{5x} + \dfrac{7x}{x + 5}$

44. $\dfrac{10h}{h - 3} + \dfrac{7}{9h}$

45. $\dfrac{x}{x + 1} + \dfrac{x - 1}{x}$

46. $\dfrac{t - 2}{t} + \dfrac{t}{t + 3}$

47. $\dfrac{s + 7}{s + 3} - \dfrac{s - 3}{s + 7}$

48. $\dfrac{t + 5}{t - 5} - \dfrac{t - 5}{t + 5}$

49. $\dfrac{3m}{m - 2} - \dfrac{m - 3}{m + 5}$

50. $\dfrac{2x}{x + 2} - \dfrac{x + 1}{x - 3}$

Perform the operations. Simplify, if possible. See Example 5.

51. $\dfrac{4}{s^2 + 5s + 4} + \dfrac{s}{s^2 + 2s + 1}$

52. $\dfrac{d}{d^2 + 6d + 5} - \dfrac{3}{d^2 + 5d + 4}$

53. $\dfrac{5}{x^2 - 9x + 8} - \dfrac{3}{x^2 - 6x - 16}$

54. $\dfrac{3}{t^2 + t - 6} + \dfrac{1}{t^2 + 3t - 10}$

55. $\dfrac{2}{a^2 + 4a + 3} + \dfrac{1}{a + 3}$

56. $\dfrac{1}{c + 6} + \dfrac{4}{c^2 + 8c + 12}$

57. $\dfrac{8}{y^2 - 16} - \dfrac{7}{y^2 - y - 12}$

58. $\dfrac{6}{s^2 - 9} - \dfrac{5}{s^2 - s - 6}$

Perform the operations. Simplify, if possible. See Example 6.

59. $\dfrac{8}{x} + 6$

60. $\dfrac{2}{y} + 7$

61. $\dfrac{9}{x - 4} + x$

62. $\dfrac{9}{m + 4} + 9$

63. $b - \dfrac{3}{a^2}$

64. $c - \dfrac{5}{3b}$

65. $\dfrac{x + 2}{x + 1} - 5$

66. $\dfrac{y + 8}{y - 8} - 4$

Perform the operations. Simplify, if possible. See Example 7.

67. $\dfrac{7}{a - 4} + \dfrac{5}{4 - a}$

68. $\dfrac{4}{b - 6} + \dfrac{b}{6 - b}$

69. $\dfrac{c}{7c - d} - \dfrac{d}{d - 7c}$

70. $\dfrac{a}{5a - 3b} - \dfrac{b}{3b - 5a}$

71. $\dfrac{3d - 3}{d - 9} - \dfrac{3d}{9 - d}$

72. $\dfrac{2x + 2}{x - 2} - \dfrac{2x}{2 - x}$

73. $\dfrac{g}{g^2 - 4} + \dfrac{2}{4 - g^2}$

74. $\dfrac{h}{h^2 - 49} + \dfrac{7}{49 - h^2}$

TRY IT YOURSELF

Perform the operations and simplify, if possible.

75. $\dfrac{j}{j^2 + 9j + 20} - \dfrac{4}{j^2 + 7j + 12}$

76. $\dfrac{r}{r^2 + 5r + 6} - \dfrac{2}{r^2 + 3r + 2}$

77. $\dfrac{10}{x - 1} + y$

78. $\dfrac{3}{s - 8} + t$

79. $\dfrac{b}{b + 1} - \dfrac{b - 1}{b + 2}$

80. $\dfrac{x}{x - 2} - \dfrac{x + 2}{x + 3}$

81. $\dfrac{y}{y - 1} - \dfrac{4}{1 - y}$

82. $\dfrac{1}{t - 7} - \dfrac{t}{7 - t}$

83. $\dfrac{n}{5} - \dfrac{n - 2}{15}$

84. $\dfrac{m}{9} - \dfrac{m + 1}{27}$

85. $\dfrac{y + 2}{5y^2} + \dfrac{y + 4}{15y}$

86. $\dfrac{x + 3}{x^2} + \dfrac{x + 5}{2x}$

87. $\dfrac{x}{x - 2} + \dfrac{4 + 2x}{x^2 - 4}$

88. $\dfrac{y}{y + 3} - \dfrac{2y - 6}{y^2 - 9}$

89. $\dfrac{7}{3a} + \dfrac{1}{a - 2}$

90. $\dfrac{5}{9x} + \dfrac{4}{x + 6}$

Perform the operations and simplify, if possible. Be careful to apply the correct method because these problems involve addition, subtraction, multiplication, and division.

91. a. $\dfrac{5}{2x} + \dfrac{4x}{15}$ b. $\dfrac{5}{2x} \cdot \dfrac{4x}{15}$

92. a. $\dfrac{2a + 4}{3} - \dfrac{9}{a + 2}$ b. $\dfrac{2a + 4}{3} \cdot \dfrac{9}{a + 2}$

93. a. $\dfrac{t}{t - 5} - \dfrac{t}{t^2 - 25}$ b. $\dfrac{t}{t - 5} \div \dfrac{t}{t^2 - 25}$

94. a. $\dfrac{1}{m + 2} - \dfrac{2}{m^2 + 4m + 4}$ b. $\dfrac{1}{m + 2} \div \dfrac{2}{m^2 + 4m + 4}$

APPLICATIONS

95. Find the total height of the funnel.

96. What is the difference between the diameter of the opening at the top of the funnel and the diameter of its spout?

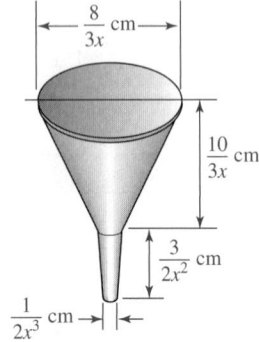

WRITING

97. Explain the error:

$$\dfrac{3}{x} + \dfrac{8}{y} = \dfrac{3 + 8}{x + y} = \dfrac{11}{x + y}$$

98. Explain how to add two rational expressions with unlike denominators.

99. When will the LCD of two rational expressions be the product of the denominators of those rational expressions? Give an example.

100. Explain how multiplication by $\dfrac{-1}{-1}$ is used in this section.

REVIEW

101. Find the slope and y-intercept of the graph of $y = 8x + 2$.

102. Find the slope and y-intercept of the graph of $3x + 4y = -36$.

103. What is the slope of the graph of $y = 2$?

104. Is the graph of the equation $x = 0$ the x-axis or the y-axis?

CHALLENGE PROBLEMS

Perform the operations and simplify the result, if possible.

105. $\dfrac{a}{a - 1} - \dfrac{2}{a + 2} + \dfrac{3(a - 2)}{a^2 + a - 2}$

106. $\dfrac{2x}{x^2 - 3x + 2} + \dfrac{2x}{x - 1} - \dfrac{x}{x - 2}$

107. $\dfrac{1}{a + 1} + \dfrac{a^2 - 7a + 10}{2a^2 - 2a - 4} \cdot \dfrac{2a^2 - 50}{a^2 + 10a + 25}$

108. $1 - \dfrac{(x - 2)^2}{(x + 2)^2}$

SECTION 7.5
Simplifying Complex Fractions

Objectives

1 Simplify complex fractions using division.

2 Simplifying complex fractions using the LCD.

A rational expression whose numerator and/or denominator contain fractions is called a **complex rational expression** or a **complex fraction.** The expression above the main fraction bar of a complex fraction is the numerator, and the expression below the main fraction bar is the denominator. Two examples of complex fractions are:

$$\dfrac{\dfrac{5x}{3}}{\dfrac{2x}{9}}$$ ← Numerator of complex fraction → $$\dfrac{\dfrac{1}{2} - \dfrac{1}{x}}{\dfrac{x}{3} + \dfrac{1}{5}}$$

← Main fraction bar →

← Denominator of complex fraction →

In this section, we will discuss two methods for simplifying complex fractions. To **simplify a complex fraction** means to write it in the form $\frac{A}{B}$, where A and B are polynomials that have no common factors.

1 Simplify Complex Fractions Using Division.

One method for simplifying complex fractions uses the fact that the main fraction bar indicates division.

**Simplifying Complex Fractions
Method 1:
Using Division**

1. Add or subtract in the numerator and/or denominator so that the numerator is a single fraction and the denominator is a single fraction.
2. Perform the indicated division by multiplying the numerator of the complex fraction by the reciprocal of the denominator.
3. Simplify the result, if possible.

EXAMPLE 1   Simplify: $\dfrac{\dfrac{5x^2}{3}}{\dfrac{2x^3}{9}}$

Strategy   We will perform the division indicated by the main fraction bar using the procedure for dividing rational expressions from Section 7.2.

Why   We can skip the first step of method 1 and immediately divide because the numerator and the denominator of the complex fraction are already single fractions.

Solution

$$\dfrac{\dfrac{5x^2}{3}}{\dfrac{2x^3}{9}} = \dfrac{5x^2}{3} \div \dfrac{2x^3}{9}$$ Write the division indicated by the main fraction bar using a ÷ symbol.

The Language of Algebra
The second step of this method could also be phrased: Perform the division by *inverting the denominator of the complex fraction and multiplying.*

$$= \frac{5x^2}{3} \cdot \frac{9}{2x^3}$$ To divide rational expressions, multiply the first by the reciprocal of the second.

$$= \frac{5x^2 \cdot 9}{3 \cdot 2x^3}$$ Multiply the numerators.
Multiply the denominators.

$$= \frac{5 \cdot \overset{1}{\cancel{x}} \cdot \overset{1}{\cancel{x}} \cdot \overset{1}{\cancel{3}} \cdot 3}{\underset{1}{\cancel{3}} \cdot 2 \cdot \underset{1}{\cancel{x}} \cdot \underset{1}{\cancel{x}} \cdot x}$$ Factor 9 as 3 · 3. Then simplify by removing factors equal to 1.

$$= \frac{15}{2x}$$ Multiply the remaining factors in the numerator.
Multiply the remaining factors in the denominator.

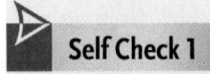 **Self Check 1**   Simplify: $\dfrac{\frac{7y^3}{8}}{\frac{21y^2}{20}}$

Now Try   **Problem 19**

EXAMPLE 2   Simplify: $\dfrac{\frac{1}{2} - \frac{1}{x}}{\frac{x}{3} + \frac{1}{5}}$

Strategy   We will simplify the expressions above and below the main fraction bar separately to write $\frac{1}{2} - \frac{1}{x}$ and $\frac{x}{3} + \frac{1}{5}$ as single fractions. Then we will perform the indicated division.

Why   The numerator and the denominator of the complex fraction must be written as single fractions before dividing.

Solution   To write the numerator as a single fraction, we build $\frac{1}{2}$ and $\frac{1}{x}$ to have an LCD of $2x$, and then subtract. To write the denominator as a single fraction, we build $\frac{x}{3}$ and $\frac{1}{5}$ to have an LCD of 15, and then add.

$$\frac{\frac{1}{2} - \frac{1}{x}}{\frac{x}{3} + \frac{1}{5}} = \frac{\frac{1}{2} \cdot \frac{x}{x} - \frac{1}{x} \cdot \frac{2}{2}}{\frac{x}{3} \cdot \frac{5}{5} + \frac{1}{5} \cdot \frac{3}{3}}$$
← The LCD for the numerator is $2x$. Build each fraction so that each has a denominator of $2x$.
← The LCD for the denominator is 15. Build each fraction so that each has a denominator of 15.

$$= \frac{\frac{x}{2x} - \frac{2}{2x}}{\frac{5x}{15} + \frac{3}{15}}$$ Multiply the numerators and multiply the denominators.

$$= \frac{\frac{x-2}{2x}}{\frac{5x+3}{15}}$$ Subtract in the numerator and add in the denominator of the complex fraction.

Now that the numerator and the denominator of the complex fraction are single fractions, we perform the indicated division.

$$\frac{\dfrac{x-2}{2x}}{\dfrac{5x+3}{15}} = \frac{x-2}{2x} \div \frac{5x+3}{15}$$ Write the division indicated by the main fraction bar using a ÷ symbol.

$$= \frac{x-2}{2x} \cdot \frac{15}{5x+3}$$ Multiply by the reciprocal of $\frac{5x+3}{15}$.

$$= \frac{15(x-2)}{2x(5x+3)}$$ Multiply the numerators.
Multiply the denominators.

Since the numerator and denominator have no common factor, the result does not simplify.

Notation
The result after simplifying a complex fraction can often have several equivalent forms. The result for Example 2 could be written:
$$\frac{15x-30}{2x(5x+3)}$$

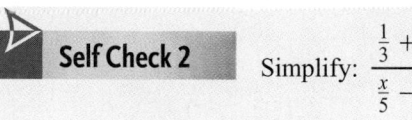

Self Check 2   Simplify: $\dfrac{\frac{1}{3}+\frac{1}{x}}{\frac{x}{5}-\frac{1}{2}}$

Now Try   **Problem 27**

EXAMPLE 3   Simplify: $\dfrac{\frac{6}{x}+y}{\frac{6}{y}+x}$

Strategy   We will simplify the expressions above and below the main fraction bar separately to write $\frac{6}{x}+y$ and $\frac{6}{y}+x$ as single fractions. Then we will perform the indicated division.

Why   The numerator and the denominator of the complex fraction must be written as single fractions before dividing.

Solution   To write $\frac{6}{x}+y$ as a single fraction, we build y into a fraction with a denominator of x and add. To write $\frac{6}{y}+x$ as a single fraction, we build x into a fraction with a denominator of y and add.

$$\frac{\frac{6}{x}+y}{\frac{6}{y}+x} = \frac{\frac{6}{x}+\frac{y}{1}\cdot\frac{x}{x}}{\frac{6}{y}+\frac{x}{1}\cdot\frac{y}{y}}$$ ← Write y as $\frac{y}{1}$. The LCD for the numerator is x. Build $\frac{y}{1}$ so that it has a denominator of x.
← Write x as $\frac{x}{1}$. The LCD for the denominator is y. Build $\frac{x}{1}$ so that it has a denominator of y.

$$= \frac{\frac{6}{x}+\frac{xy}{x}}{\frac{6}{y}+\frac{xy}{y}}$$ Multiply in the numerator and multiply in the denominators.

$$= \frac{\frac{6+xy}{x}}{\frac{6+xy}{y}}$$ Add in the numerator and in the denominator of the complex fraction.

Now that the numerator and the denominator of the complex fraction are single fractions, we can perform the division.

Success Tip

Simplifying using division (method 1) works well when a complex fraction is written, or can be easily written, as a quotient of two single rational expressions.

$$\frac{\dfrac{6+xy}{x}}{\dfrac{6+xy}{y}} = \frac{6+xy}{x} \div \frac{6+xy}{y}$$ Write the division indicated by the main fraction bar using a ÷ symbol.

$$= \frac{6+xy}{x} \cdot \frac{y}{6+xy}$$ Multiply by the reciprocal of $\frac{6+xy}{y}$.

$$= \frac{y(6+xy)}{x(6+xy)}$$ Multiply the numerators. Multiply the denominators.

$$= \frac{y(6+\cancel{xy})}{x(6+\cancel{xy})}$$ Simplify the result by removing a factor equal to 1.

$$= \frac{y}{x}$$

 Self Check 3   Simplify: $\dfrac{\frac{2}{a}-b}{\frac{2}{b}-a}$

Now Try   **Problem 39**

❷ Simplify Complex Fractions Using the LCD.

A second method for simplifying complex fractions uses the concepts of LCD and multiplication by a form of 1. The multiplication by 1 produces a simpler, equivalent expression, which will not contain fractions in its numerator or denominator.

Simplifying Complex Fractions
Method 2: Multiplying by the LCD

1. Find the LCD of all fractions within the complex fraction.
2. Multiply the complex fraction by 1 in the form $\frac{\text{LCD}}{\text{LCD}}$.
3. Perform the operations in the numerator and denominator. No fractional expressions should remain within the complex fraction.
4. Simplify the result, if possible.

We will use method 2 to rework Example 2.

EXAMPLE 4   Simplify: $\dfrac{\frac{1}{2}-\frac{1}{x}}{\frac{x}{3}+\frac{1}{5}}$

Strategy   Using method 1 to simplify this complex fraction, we worked with $\frac{1}{2}-\frac{1}{x}$ and $\frac{x}{3}+\frac{1}{5}$ separately. With method 2, we will use the LCD of *all four* fractions within the complex fraction.

Why   Multiplying a complex fraction by 1 in the form of $\frac{\text{LCD}}{\text{LCD}}$ clears its numerator and denominator of fractions.

Solution The denominators of all the fractions within the complex fraction are 2, x, 3, and 5. Thus, their LCD is $2 \cdot x \cdot 3 \cdot 5 = 30x$.

We now multiply the complex fraction by a factor equal to 1, using the LCD: $\frac{30x}{30x} = 1$.

Success Tip
With method 2, each term of the numerator and each term of the denominator of the complex fraction is multiplied by the LCD. Arrows can be helpful in showing this.

$$\frac{\dfrac{1}{2} - \dfrac{1}{x}}{\dfrac{x}{3} + \dfrac{1}{5}} = \frac{\dfrac{1}{2} - \dfrac{1}{x}}{\dfrac{x}{3} + \dfrac{1}{5}} \cdot \frac{30x}{30x}$$

$$= \frac{\left(\dfrac{1}{2} - \dfrac{1}{x}\right)30x}{\left(\dfrac{x}{3} + \dfrac{1}{5}\right)30x}$$
← Multiply the numerators.
← Multiply the denominators.

$$= \frac{\dfrac{1}{2}(30x) - \dfrac{1}{x}(30x)}{\dfrac{x}{3}(30x) + \dfrac{1}{5}(30x)}$$
← In the numerator, distribute the multiplication by 30x.
← In the denominator, distribute the multiplication by 30x.

$$= \frac{15x - 30}{10x^2 + 6x}$$
Perform each of the four multiplications by 30x. Notice that no fractional expressions remain within the complex fraction.

To attempt to simplify the result, factor the numerator and denominator. Since they do not have a common factor, the result is in simplest form.

$$\frac{15x - 30}{10x^2 + 6x} = \frac{15(x - 2)}{2x(5x + 3)}$$

Self Check 4 Use method 2 to simplify: $\dfrac{\dfrac{1}{4} - \dfrac{1}{x}}{\dfrac{x}{5} + \dfrac{1}{3}}$

Now Try **Problem 49**

EXAMPLE 5 Simplify: $\dfrac{\dfrac{1}{8} - \dfrac{1}{y}}{\dfrac{8 - y}{4y^2}}$

Strategy Using method 1, we would work with $\frac{1}{8} - \frac{1}{y}$ and $\frac{8 - y}{4y^2}$ separately. With method 2, we use the LCD of all three fractions within the complex fraction.

Why Multiplying a complex fraction by 1 in the form of $\frac{\text{LCD}}{\text{LCD}}$ clears its numerator and denominator of fractions.

Solution The denominators of all fractions within the complex fraction are 8, y, and $4y^2$. Therefore, the LCD is $8y^2$ and we multiply the complex fraction by a factor equal to 1, using the LCD: $\frac{8y^2}{8y^2} = 1$.

$$\frac{\dfrac{1}{8} - \dfrac{1}{y}}{\dfrac{8 - y}{4y^2}} = \frac{\dfrac{1}{8} - \dfrac{1}{y}}{\dfrac{8 - y}{4y^2}} \cdot \frac{8y^2}{8y^2}$$

Success Tip
When simplifying a complex fraction, the same result will be obtained regardless of the method used. See Example 2.

$$= \frac{\left(\dfrac{1}{8} - \dfrac{1}{y}\right)8y^2}{\left(\dfrac{8-y}{4y^2}\right)8y^2}$$ ← Multiply the numerators.

← Multiply the denominators.

$$= \frac{\dfrac{1}{8}(8y^2) - \dfrac{1}{y}(8y^2)}{\left(\dfrac{8-y}{4y^2}\right)(8y^2)}$$ Distribute the multiplication by $8y^2$.

$$= \frac{y^2 - 8y}{(8-y)2}$$ Perform each of the three multiplications by $8y^2$.

$$= \frac{\overset{-1}{y(\cancel{y-8})}}{(\cancel{8-y})2}$$ In the numerator, factor out the GCF, y. Since y − 8 and 8 − y are opposites, simplify by replacing $\frac{y-8}{8-y}$ with $\frac{-1}{1}$.

$$= -\frac{y}{2}$$

 Self Check 5 Simplify: $\dfrac{\dfrac{10-n}{5n^2}}{\dfrac{1}{10} - \dfrac{1}{n}}$

Now Try **Problem 55**

EXAMPLE 6 Simplify: $\dfrac{1}{1 + \dfrac{1}{x+1}}$

Strategy Although either method can be used, we will use method 2 to simplify this complex fraction.

Why Method 2 is often easier when the complex fraction contains a sum or difference.

Solution The only fraction within the complex fraction has the denominator $x + 1$. Therefore, the LCD is $x + 1$. We multiply the complex fraction by a factor equal to 1, using the LCD: $\frac{x+1}{x+1} = 1$.

$$\frac{1}{1 + \dfrac{1}{x+1}} = \frac{1}{1 + \dfrac{1}{x+1}} \cdot \frac{x+1}{x+1}$$

$$= \frac{1(x+1)}{\left(1 + \dfrac{1}{x+1}\right)(x+1)}$$ Multiply the numerators.
Multiply the denominators.

$$= \frac{1(x+1)}{1(x+1) + \dfrac{1}{x+1}(x+1)}$$ In the denominator, distribute the multiplication by x + 1.

$$= \frac{x+1}{x+1+1}$$ Perform each of the three multiplications by x + 1.

$$= \frac{x + 1}{x + 2}$$ Combine like terms in the denominator.

The result does not simplify.

 Self Check 6   Simplify: $\dfrac{2}{\dfrac{1}{x + 2} + 2}$

Now Try   **Problem 63**

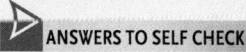 ANSWERS TO SELF CHECK   **1.** $\frac{5y}{6}$ **2.** $\frac{10(x + 3)}{3x(2x - 5)}$ **3.** $\frac{b}{a}$ **4.** $\frac{15(x - 4)}{4x(3x + 5)}$ **5.** $-\frac{2}{n}$ **6.** $\frac{2(x + 2)}{2x + 5}$

STUDY SET
7.5

VOCABULARY

Fill in the blanks.

1. The expression $\dfrac{\frac{2}{3} - \frac{1}{x}}{\frac{x - 3}{4}}$ is called a _____ rational expression
or a _____ fraction.

2. In a complex fraction, the numerator is above the _____ fraction bar and the _____ is below it.

CONCEPTS

Fill in the blanks.

3. Method 1: To _____ a complex fraction, write its numerator and denominator as single fractions. Then perform the indicated _____ by multiplying the numerator of the complex fraction by the _____ of the denominator.

4. Method 2: To _____ a complex fraction, find the LCD of ____ fractions within the complex fraction. Multiply the complex fraction by 1 in the form _____. Then perform the operations.

5. Consider: $\dfrac{\frac{x - 3}{4}}{\frac{1}{12} - \frac{x}{6}}$

   a. What is the numerator of the complex fraction? Is it a single fraction?

   b. What is the denominator of the complex fraction? Is it a single fraction?

6. Consider the complex fraction: $\dfrac{\frac{1}{y} - \frac{1}{3}}{\frac{5}{6} + \frac{1}{y}}$

   a. What is the LCD of all fractions in the complex fraction?

   b. To simplify the complex fraction using method 2, it should be multiplied by what form of 1?

NOTATION

Fill in the blanks to simplify each complex fraction.

7. $\dfrac{\frac{12}{y^2}}{\frac{4}{y^3}} = \dfrac{12}{y^2} \quad \dfrac{4}{y^3}$

8. $\dfrac{\left(\frac{1}{5} - \frac{1}{a}\right)}{\left(\frac{a}{4} + \frac{2}{a}\right)} \cdot \dfrac{20a}{20a} = \dfrac{\frac{1}{5}(\quad) - \frac{1}{a}(\quad)}{\frac{a}{4}(\quad) + \frac{2}{a}(\quad)}$

$$= \dfrac{4a - \quad}{\quad + \quad}$$

GUIDED PRACTICE

Simplify each complex fraction. See Example 1.

9. $\dfrac{\frac{2}{3}}{\frac{3}{4}}$

10. $\dfrac{\frac{3}{5}}{\frac{2}{7}}$

11. $\dfrac{\frac{x}{2}}{\frac{6}{5}}$

12. $\dfrac{\frac{9}{4}}{\frac{7}{x}}$

13. $\dfrac{\frac{x}{y}}{\frac{1}{x}}$

14. $\dfrac{\frac{y}{x}}{\frac{x}{xy}}$

15. $\dfrac{\dfrac{n}{8}}{\dfrac{1}{n^2}}$

16. $\dfrac{\dfrac{1}{m}}{\dfrac{m^3}{15}}$

17. $\dfrac{\dfrac{4a}{11}}{\dfrac{6a}{55}}$

18. $\dfrac{\dfrac{14}{15m}}{\dfrac{21}{25m}}$

19. $\dfrac{-\dfrac{x^4}{30}}{\dfrac{7x}{15}}$

20. $\dfrac{-\dfrac{5x^2}{24}}{\dfrac{x^5}{56}}$

21. $\dfrac{\dfrac{10x}{x-3}}{\dfrac{6}{x-3}}$

22. $\dfrac{\dfrac{18a}{a-4}}{\dfrac{12}{a-4}}$

23. $\dfrac{\dfrac{4t-8}{t^2}}{\dfrac{8t-16}{t^5}}$

24. $\dfrac{\dfrac{9m-27}{m^6}}{\dfrac{2m-6}{m^8}}$

Simplify each complex fraction. See Examples 2 or 4.

25. $\dfrac{\dfrac{1}{2}+\dfrac{3}{4}}{\dfrac{3}{2}+\dfrac{1}{4}}$

26. $\dfrac{\dfrac{2}{3}-\dfrac{5}{2}}{\dfrac{2}{3}-\dfrac{3}{2}}$

27. $\dfrac{\dfrac{1}{4}+\dfrac{1}{y}}{\dfrac{y}{3}-\dfrac{1}{2}}$

28. $\dfrac{\dfrac{2}{x}-\dfrac{1}{3}}{\dfrac{2}{3}+\dfrac{x}{5}}$

29. $\dfrac{\dfrac{1}{y}-\dfrac{5}{2}}{\dfrac{3}{y}}$

30. $\dfrac{\dfrac{1}{6}-\dfrac{5}{s}}{\dfrac{2}{s}}$

31. $\dfrac{\dfrac{4}{c}-\dfrac{c}{6}}{\dfrac{2}{c}}$

32. $\dfrac{\dfrac{10}{n}-\dfrac{n}{4}}{\dfrac{8}{n}}$

33. $\dfrac{\dfrac{2}{s}-\dfrac{2}{s^2}}{\dfrac{4}{s^3}+\dfrac{4}{s^2}}$

34. $\dfrac{\dfrac{2}{x^3}-\dfrac{2}{x}}{\dfrac{4}{x}+\dfrac{8}{x^2}}$

35. $\dfrac{\dfrac{1}{a^2b}-\dfrac{5}{ab}}{\dfrac{3}{ab}-\dfrac{7}{ab^2}}$

36. $\dfrac{\dfrac{3}{ab^2}+\dfrac{6}{a^2b}}{\dfrac{6}{a}-\dfrac{9}{b^2}}$

Simplify each complex fraction. See Examples 3 or 5.

37. $\dfrac{\dfrac{2}{3}+1}{\dfrac{1}{3}+1}$

38. $\dfrac{\dfrac{3}{5}-2}{\dfrac{2}{5}-2}$

39. $\dfrac{\dfrac{1}{x}-3}{\dfrac{5}{x}+2}$

40. $\dfrac{\dfrac{1}{y}+3}{\dfrac{3}{y}-2}$

41. $\dfrac{\dfrac{2}{x}+2}{\dfrac{4}{x}+2}$

42. $\dfrac{\dfrac{3}{x}-3}{\dfrac{9}{x}-3}$

43. $\dfrac{\dfrac{3y}{x}-y}{y-\dfrac{y}{x}}$

44. $\dfrac{\dfrac{y}{x}+3y}{y+\dfrac{2y}{x}}$

45. $\dfrac{4-\dfrac{1}{8h}}{12+\dfrac{3}{4h}}$

46. $\dfrac{12+\dfrac{1}{3b}}{12-\dfrac{1}{b^2}}$

47. $\dfrac{1-\dfrac{9}{d^2}}{2+\dfrac{6}{d}}$

48. $\dfrac{1-\dfrac{16}{a^2}}{\dfrac{12}{a}+3}$

Simplify each complex fraction. See Examples 4 and 5.

49. $\dfrac{\dfrac{1}{6}-\dfrac{2}{x}}{\dfrac{1}{6}+\dfrac{1}{x}}$

50. $\dfrac{\dfrac{3}{4}+\dfrac{1}{y}}{\dfrac{5}{6}-\dfrac{1}{y}}$

51. $\dfrac{\dfrac{a}{7}-\dfrac{7}{a}}{\dfrac{1}{a}+\dfrac{1}{7}}$

52. $\dfrac{\dfrac{t}{9}-\dfrac{9}{t}}{\dfrac{1}{t}+\dfrac{1}{9}}$

53. $\dfrac{\dfrac{m}{n}+\dfrac{n}{m}}{\dfrac{m}{n}-\dfrac{n}{m}}$

54. $\dfrac{\dfrac{2a}{b}-\dfrac{b}{a}}{\dfrac{2a}{b}+\dfrac{b}{a}}$

55. $\dfrac{\dfrac{d+2}{2}}{\dfrac{d}{3}-\dfrac{d}{4}}$

56. $\dfrac{\dfrac{d^2}{4}+\dfrac{4d}{5}}{\dfrac{d+1}{2}}$

57. $\dfrac{\dfrac{2}{c^2}}{\dfrac{1}{c}+\dfrac{5}{4}}$

58. $\dfrac{\dfrac{7}{s^2}}{\dfrac{1}{s}+\dfrac{10}{3}}$

59. $\dfrac{\dfrac{2}{x}}{\dfrac{2}{y}-\dfrac{4}{x}}$

60. $\dfrac{\dfrac{2y}{3}}{\dfrac{2y}{3}-\dfrac{8}{y}}$

Simplify each complex fraction. See Example 6.

61. $\dfrac{\dfrac{1}{x+1}}{1+\dfrac{1}{x+1}}$

62. $\dfrac{\dfrac{1}{x-1}}{1-\dfrac{1}{x-1}}$

63. $\dfrac{\dfrac{x}{x+2}}{\dfrac{x}{x+2}+x}$

64. $\dfrac{\dfrac{2}{x-2}}{\dfrac{2}{x-2}-1}$

65. $\dfrac{3+\dfrac{3}{x-1}}{3-\dfrac{3}{x-1}}$

66. $\dfrac{2-\dfrac{2}{x+1}}{2+\dfrac{2}{x+1}}$

67. $\dfrac{m-\dfrac{1}{2m+1}}{1-\dfrac{m}{2m+1}}$

68. $\dfrac{1-\dfrac{r}{2r+1}}{r-\dfrac{1}{2r+1}}$

TRY IT YOURSELF

Simplify each complex fraction.

69. $\dfrac{\dfrac{1}{p}+\dfrac{1}{q}}{\dfrac{1}{p}}$

70. $\dfrac{\dfrac{m}{n}+1}{1-\dfrac{m}{n}}$

71. $\dfrac{\dfrac{40x^2}{1}}{\dfrac{20x}{9}}$

72. $\dfrac{\dfrac{18n^2}{1}}{\dfrac{6n}{13}}$

73. $\dfrac{\dfrac{1}{c}+\dfrac{1}{2}}{\dfrac{1}{c^2}-\dfrac{1}{4}}$

74. $\dfrac{\dfrac{1}{m}-\dfrac{1}{n}}{\dfrac{m}{n}-\dfrac{n}{m}}$

75. $\dfrac{\dfrac{1}{r+1}+1}{\dfrac{3}{r-1}+1}$

76. $\dfrac{5+\dfrac{1}{n+7}}{4-\dfrac{2}{n+7}}$

77. $\dfrac{\dfrac{b^2-81}{18a^2}}{\dfrac{4b-36}{9a}}$

78. $\dfrac{\dfrac{8x-64}{y}}{\dfrac{x^2-64}{y^2}}$

79. $\dfrac{1+\dfrac{6}{t}+\dfrac{8}{t^2}}{1+\dfrac{1}{t}-\dfrac{12}{t^2}}$

80. $\dfrac{1-p+\dfrac{2}{p}}{\dfrac{6}{p^2}+\dfrac{1}{p}-1}$

81. $\dfrac{1}{\dfrac{1}{x}+\dfrac{1}{y}}$

82. $\dfrac{1}{\dfrac{b}{a}-\dfrac{a}{b}}$

83. $\dfrac{-\dfrac{25}{16x^2}}{\dfrac{15}{32x^5}}$

84. $\dfrac{\dfrac{21}{8g^3}}{-\dfrac{35}{16g^8}}$

APPLICATIONS

85. SLOPE We can use the slope formula to find the slope of a line that passes through $\left(\frac{1}{2},\frac{1}{3}\right)$ and $\left(\frac{3}{4},\frac{5}{8}\right)$:

$$m = \dfrac{\dfrac{5}{8}-\dfrac{1}{3}}{\dfrac{3}{4}-\dfrac{1}{2}}$$

Simplify the complex fraction to find *m*.

86. PITCHING The earned run average (ERA) is a statistic that gives the average number of earned runs a pitcher allows. For a softball pitcher, this is based on a six-inning game. The formula for ERA is

$$\text{ERA} = \dfrac{\dfrac{\text{earned runs}}{\text{innings pitched}}}{6}$$

Simplify the complex fraction on the right side of the formula.

87. ELECTRONICS In electronic circuits, resistors are tiny components that limit the flow of an electric current. An important formula about two resistors in a circuit is

$$\text{Total resistance} = \dfrac{1}{\dfrac{1}{R_1}+\dfrac{1}{R_2}}$$

(Recall that R_1 is read as R sub one.)

Simplify the complex fraction on the right side of the formula.

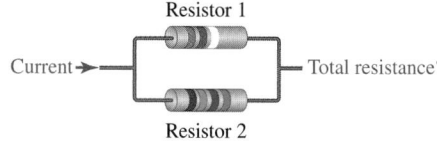

88. DATA ANALYSIS Use the data in the table to find the average measurement for the three-trial experiment.

	Trial 1	Trial 2	Trial 3
Measurement	$\frac{k}{2}$	$\frac{k}{3}$	$\frac{k}{2}$

WRITING

89. What is a complex fraction? Give several examples.

90. Explain how to use method 1 to simplify: $\dfrac{1+\frac{1}{x}}{3-\frac{1}{x}}$

91. Explain how to use method 2 to simplify the expression in Problem 90.

92. a. List an advantage and a disadvantage of using method 1 to simplify a complex fraction.

b. List an advantage and a disadvantage of using method 2 to simplify a complex fraction.

REVIEW

Simplify each expression. Write each answer without using parentheses or negative exponents.

93. $(8x)^0$

94. $\left(-\dfrac{3r}{4r^3}\right)^4$

95. $\left(\dfrac{4x^3}{5x^{-3}}\right)^{-2}$

96. $\left(\dfrac{12xy^{-3}}{3x^{-2}y^2}\right)^{-2}$

CHALLENGE PROBLEMS

Simplify.

97. $\dfrac{\dfrac{h}{h^2 + 3h + 2}}{\dfrac{4}{h + 2} - \dfrac{4}{h + 1}}$

98. $\dfrac{\dfrac{2}{b^2 - 1} - \dfrac{3}{ab - a}}{\dfrac{3}{ab - a} - \dfrac{2}{b^2 - 1}}$

99. $a + \dfrac{a}{1 + \dfrac{a}{a + 1}}$

100. $\dfrac{y^{-2} + 1}{y^{-2} - 1}$

SECTION 7.6
Solving Rational Equations

Objectives

1 Solve rational equations.

2 Solve for a specified variable in a formula.

In Chapter 2, we solved equations such as $\frac{1}{6}x + \frac{5}{2} = \frac{1}{3}$ by multiplying both sides by the LCD. With this approach, the equation that results is equivalent to the original equation, but easier to solve because it is cleared of fractions.

In this section, we will extend the fraction-clearing strategy to solve another type of equation, called a *rational equation*.

Rational Equations

A **rational equation** is an equation that contains one or more rational expressions.

Rational equations often have a variable in a denominator. Some examples are:

$$\frac{2x}{3} = \frac{x}{6} + \frac{3}{2} \qquad \frac{2}{x} + \frac{1}{4} = \frac{5}{2x} \qquad \frac{11x}{x - 5} = 6 + \frac{55}{x - 5}$$

1 **Solve Rational Equations.**

To solve a rational equation, we find all the values of the variable that make the equation true. Any value of the variable that makes a denominator in a rational equation equal to 0 cannot be a solution of the equation. Such a number must be rejected, because division by 0 is undefined.

The following steps can be used to solve rational equations.

Strategy for Solving Rational Equations

1. Determine which numbers cannot be solutions of the equation.
2. Multiply both sides of the equation by the LCD of all rational expressions in the equation. This clears the equation of fractions.
3. Solve the resulting equation.
4. Check all possible solutions in the original equation.

EXAMPLE 1   Solve: $\dfrac{2x}{3} = \dfrac{x}{6} + \dfrac{3}{2}$

Strategy   We will use the multiplication property of equality to clear this rational equation of fractions by multiplying both sides by the LCD.

Why   Equations that contain only integers are usually easier to solve than equations that contain fractions.

Solution   There are no restrictions on x, because no value of x ever makes a denominator 0. Since the denominators are 3, 6, and 2, we multiply both sides of the equation by the LCD, 6.

<table>
<tr><td>$\dfrac{2x}{3} = \dfrac{x}{6} + \dfrac{3}{2}$</td><td></td></tr>
<tr><td>$6\left(\dfrac{2x}{3}\right) = 6\left(\dfrac{x}{6} + \dfrac{3}{2}\right)$</td><td>Multiply both sides of the equation by the LCD of $\frac{2x}{3}, \frac{x}{6},$ and $\frac{3}{2}$, which is 6.</td></tr>
<tr><td>$6\left(\dfrac{2x}{3}\right) = 6\left(\dfrac{x}{6}\right) + 6\left(\dfrac{3}{2}\right)$</td><td>Distribute the multiplication by 6.</td></tr>
<tr><td>$2 \cdot \overset{1}{\cancel{3}}\left(\dfrac{2x}{\underset{1}{\cancel{3}}}\right) = \overset{1}{\cancel{6}}\left(\dfrac{x}{\underset{1}{\cancel{6}}}\right) + \overset{1}{\cancel{2}} \cdot 3\left(\dfrac{3}{\underset{1}{\cancel{2}}}\right)$</td><td>Perform the three multiplications by 6 by first removing common factors of the numerator and denominator. Try to do this step in your head.</td></tr>
<tr><td>$4x = x + 9$</td><td>Simplify. Note that the fractions have been cleared.</td></tr>
<tr><td>$3x = 9$</td><td>To eliminate x on the right side, subtract x from both sides.</td></tr>
<tr><td>$x = 3$</td><td>To undo the multiplication by 3, divide both sides by 3.</td></tr>
</table>

To check, we replace each x with 3 in the original equation.

$\dfrac{2x}{3} = \dfrac{x}{6} + \dfrac{3}{2}$ This is the original equation.

$\dfrac{2(3)}{3} \overset{?}{=} \dfrac{3}{6} + \dfrac{3}{2}$ Substitute 3 for x.

$2 \overset{?}{=} \dfrac{1}{2} + \dfrac{3}{2}$ Simplify: $\dfrac{\overset{1}{2(3)}}{\underset{}{3}} = 2$ and $\dfrac{3}{6} = \dfrac{1}{2}$.

$2 = 2$ $\dfrac{1}{2} + \dfrac{3}{2} = \dfrac{4}{2} = 2.$

Since we obtain a true statement, 3 is the solution of $\dfrac{2x}{3} = \dfrac{x}{6} + \dfrac{3}{2}$. The solution set is $\{3\}$.

   Self Check 1   Solve: $\dfrac{3x}{5} = \dfrac{x}{2} + \dfrac{1}{10}$

Now Try   **Problem 15**

Caution

Always enclose the left and right sides of an equation within parentheses when multiplying both sides by the LCD.

$\left(\dfrac{2x}{3}\right) = \left(\dfrac{x}{6} + \dfrac{3}{2}\right)$

Notation

Here is an alternate way to remove common factors when multiplying by the LCD:

$\overset{2}{\cancel{6}}\left(\dfrac{2x}{\underset{1}{\cancel{3}}}\right)$ and $\overset{3}{\cancel{6}}\left(\dfrac{3}{\underset{1}{\cancel{2}}}\right)$

EXAMPLE 2   Solve: $\dfrac{2}{x} + \dfrac{1}{4} = \dfrac{5}{2x}$

Strategy   This equation contains two rational expressions that have a variable in their denominator. We begin by asking, "What value(s) of x make either denominator 0?" Then we will clear the equation of fractions by multiplying both sides by the LCD.

Why If a number makes the denominator of a rational expression 0, that number cannot be a solution of the equation because division by 0 is undefined.

Solution If x is 0, the denominators of $\frac{2}{x}$ and $\frac{5}{2x}$ are 0 and the expressions would be undefined. Therefore, 0 cannot be a solution.

Since the denominators are x, 4, and $2x$, we multiply both sides of the equation by the LCD, $4x$, to clear the equation of fractions.

$$\frac{2}{x} + \frac{1}{4} = \frac{5}{2x}$$

$$4x\left(\frac{2}{x} + \frac{1}{4}\right) = 4x\left(\frac{5}{2x}\right)$$

Write each side of the equation within parentheses, and then multiply both sides by 4x.

$$4x\left(\frac{2}{x}\right) + 4x\left(\frac{1}{4}\right) = 4x\left(\frac{5}{2x}\right)$$

Distribute the multiplication by 4x.

$$\overset{1}{4x}\left(\frac{2}{\overset{}{x}}\right) + \overset{1}{4}x\left(\frac{1}{\overset{}{4}}\right) = 2\cdot\overset{1}{2}\cdot\overset{1}{x}\left(\frac{5}{\overset{}{2\cdot x}}\right)$$

On the right side, factor 4x as 2 · 2 · x. Perform the three multiplications by 4x by first removing common factors of each numerator and denominator. Try to do this step in your head.

$$8 + x = 10$$

Simplify. Note that the fractions have been cleared.

$$x = 2$$

To solve the resulting equation, subtract 8 from both sides.

The solution of $\frac{2}{x} + \frac{1}{4} = \frac{5}{2x}$ is 2. The solution set is {2}. Check by substituting 2 for each x in the original equation.

Caution
After multiplying both sides by the LCD and simplifying, the equation should not contain any fractions. If it does, check for an algebraic error, or perhaps your LCD is incorrect.

Success Tip
Don't confuse procedures. To *simplify the expression* $\frac{2}{x} + \frac{1}{4}$, we build each fraction to have the LCD $4x$, add the numerators, and write the sum over the LCD. To *solve the equation* $\frac{2}{x} + \frac{1}{4} = \frac{5}{2x}$, we multiply both sides by the LCD $4x$ to eliminate the denominators.

 Self Check 2 Solve: $\frac{1}{6} + \frac{4}{3x} = \frac{5}{x}$

Now Try **Problem 23**

EXAMPLE 3 Solve: $y - \frac{12}{y} = 4$

Strategy Since the only denominator is y, we will multiply both sides of the equation by y.

Why Multiplying both sides by y will clear the equation of the fraction, $\frac{12}{y}$.

Solution If y is 0, the denominator of $\frac{12}{y}$ is 0 and the fraction would be undefined. Therefore, 0 cannot be a solution.

$$y - \frac{12}{y} = 4$$

$$y\left(y - \frac{12}{y}\right) = y(4)$$

Write each side of the equation within parentheses and then multiply both sides by the LCD, y.

$$y(y) - y\left(\frac{12}{y}\right) = y(4)$$

Distribute the multiplication by y.

$$y^2 - 12 = 4y$$

Simplify: $y\left(\frac{12}{y}\right) = 12$. Note that the fraction has been cleared.

Caution
By the multiplication property of equality, each term on both sides of the equation must be multiplied by the LCD. Here, it would be incorrect to only multiply the second term, $\frac{12}{y}$, by the LCD.

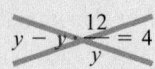

We can solve the resulting quadratic equation using the factoring method.

$$y^2 - 4y - 12 = 0 \qquad \text{Subtract 4y from both sides to get 0 on the right side.}$$

$$(y - 6)(y + 2) = 0 \qquad \text{Factor the trinomial.}$$

$$y - 6 = 0 \quad \text{or} \quad y + 2 = 0 \qquad \text{Set each factor equal to 0.}$$

$$y = 6 \qquad\qquad y = -2 \qquad \text{Solve each equation.}$$

There are two possible solutions, 6 and -2, to check.

Check y = 6:

$$y - \frac{12}{y} = 4$$

$$6 - \frac{12}{6} \stackrel{?}{=} 4$$

$$6 - 2 \stackrel{?}{=} 4$$

$$4 = 4 \quad \text{True}$$

Check y = -2:

$$y - \frac{12}{y} = 4 \qquad \text{This is the original equation.}$$

$$-2 - \frac{12}{-2} \stackrel{?}{=} 4$$

$$-2 - (-6) \stackrel{?}{=} 4$$

$$4 = 4 \quad \text{True}$$

The solutions of $y - \frac{12}{y} = 4$ are 6 and -2.

 Self Check 3 · Solve: $x - \frac{24}{x} = -5$

Now Try **Problem 29**

EXAMPLE 4 Solve: $\dfrac{11x}{x - 5} = 6 + \dfrac{55}{x - 5}$

Strategy Since both denominators are $x - 5$, we multiply both sides by the LCD, $x - 5$.

Why This will clear the equation of fractions.

Solution If x is 5, the denominators of $\frac{11x}{x - 5}$ and $\frac{55}{x - 5}$ are 0, and the expressions are undefined. Therefore, 5 cannot be a solution of the equation.

> **Caution**
> Even if you do not make an arithmetic or algebraic error when solving a rational equation, a possible solution may not check.

$$\frac{11x}{x - 5} = 6 + \frac{55}{x - 5}$$

$$(x - 5)\left(\frac{11x}{x - 5}\right) = (x - 5)\left(6 + \frac{55}{x - 5}\right) \qquad \text{Write each side of the equation within parentheses and then multiply both sides by } x - 5.$$

$$(\overset{1}{\cancel{x - 5}})\left(\frac{11x}{\cancel{x - 5}}\right)\underset{1}{} = (x - 5)6 + (\overset{1}{\cancel{x - 5}})\left(\frac{55}{\cancel{x - 5}}\right)\underset{1}{} \qquad \text{Distribute the multiplication by } x - 5. \text{ Remove the common binomial factor } (x - 5) \text{ of the numerator and denominator.}$$

$$11x = (x - 5)6 + 55 \qquad \text{Simplify. Note that the fractions have been cleared.}$$

$$11x = 6x - 30 + 55 \qquad \text{To solve the resulting equation, distribute the multiplication by 6.}$$

$$11x = 6x + 25 \qquad \text{Combine like terms: } -30 + 55 = 25.$$

$$5x = 25 \qquad \text{To eliminate 6x on the right side, subtract 6x from both sides.}$$

$$x = 5 \qquad \text{To undo the multiplication by 5, divide both sides by 5.}$$

> **The Language of Algebra**
> *Extraneous* means not a vital part. Mathematicians speak of *extraneous* solutions. Rock groups don't want *extraneous* sounds (like feedback) coming from their amplifiers. Artists erase *extraneous* marks on their sketches.

We have determined that 5 makes both denominators in the original equation 0. Therefore, 5 cannot be a solution. Since 5 is the only possible solution, and it must be rejected, it follows that $\frac{11x}{x-5} = 6 + \frac{55}{x-5}$ has no solution. The solution set is written as { } or $\varnothing$.

When solving an equation, a possible solution that does not satisfy the original equation is called an **extraneous solution.** In this example, 5 is an extraneous solution.

Self Check 4   Solve $\frac{9x}{x-6} = 3 + \frac{54}{x-6}$, if possible.

Now Try   **Problem 39**

EXAMPLE 5   Solve: $\dfrac{x+5}{x+3} + \dfrac{1}{x^2+2x-3} = 1$

Strategy   We will multiply both sides by the LCD of the two rational expressions in the equation. But first, we must factor the second denominator.

Why   To determine the restrictions on the variable and to find the LCD, we need to write $x^2 + 2x - 3$ in factored form.

Solution   Since the trinomial $x^2 + 2x - 3$ factors as $(x+3)(x-1)$, we can write the given equation as:

$$\frac{x+5}{x+3} + \frac{1}{(x+3)(x-1)} = 1 \qquad \begin{array}{l}\text{If } x \text{ is } -3, \text{ the first denominator is 0. If } x \text{ is } -3 \text{ or 1, the}\\ \text{second denominator is 0.}\end{array}$$

We see that -3 and 1 cannot be solutions of the equation, because they make rational expressions in the equation undefined.

Since the denominators are $x + 3$ and $(x+3)(x-1)$, we multiply both sides of the equation by the LCD, $(x+3)(x-1)$, to clear the fractions.

$$(x+3)(x-1)\left[\frac{x+5}{x+3} + \frac{1}{(x+3)(x-1)}\right] = (x+3)(x-1)[1] \qquad \text{Write each side within brackets [\ ].}$$

$$\overset{1}{(x+3)}(x-1)\frac{x+5}{\underset{1}{x+3}} + \overset{1}{(x+3)}\overset{1}{(x-1)}\frac{1}{\underset{1}{(x+3)}\underset{1}{(x-1)}} = (x+3)(x-1)1 \qquad \begin{array}{l}\text{Distribute the multiplication by}\\ (x+3)(x-1) \text{ and remove common}\\ \text{factors.}\end{array}$$

$$(x-1)(x+5) + 1 = (x+3)(x-1) \qquad \text{Simplify.}$$

The Language of Algebra
We say that the LCD is a *multiplier* that clears a rational equation of fractions.

To solve the resulting equation, we multiply the binomials on the left side and the right side, and proceed as follows.

$$x^2 + 4x - 5 + 1 = x^2 + 2x - 3 \qquad \text{Find } (x-1)(x+5) \text{ and } (x+3)(x-1).$$

$$x^2 + 4x - 4 = x^2 + 2x - 3 \qquad \text{Combine like terms: } -5 + 1 = -4.$$

$$4x - 4 = 2x - 3 \qquad \text{Subtract } x^2 \text{ from both sides.}$$

$$2x - 4 = -3 \qquad \begin{array}{l}\text{To eliminate } 2x \text{ on the right side, subtract } 2x \text{ from both}\\ \text{sides.}\end{array}$$

$$2x = 1 \qquad \text{To undo the subtraction of 4, add 4 to both sides.}$$

$$x = \frac{1}{2} \qquad \text{To undo the multiplication by 2, divide both sides by 2.}$$

A check will show that $\frac{1}{2}$ is the solution of the original equation.

Self Check 5 Solve: $\dfrac{1}{x+3} + \dfrac{1}{x-3} = \dfrac{5}{x^2-9}$

Now Try **Problem 47**

② Solve for a Specified Variable in a Formula.

Many formulas are expressed as rational equations. To solve such formulas for a specified variable, we use the same steps, in the same order, as we do when solving rational equations having only one variable.

EXAMPLE 6 ***Determining a Child's Dosage.*** The formula $C = \dfrac{AD}{A+12}$ is called **Young's rule.** It is a way to find the approximate child's dose C of a medication, where A is the age of the child in years and D is the recommended dosage for an adult. Solve the formula for D.

Strategy As we have done in the previous examples, we will begin by multiplying both sides of the equation by the LCD to clear it of the fraction.

Why To isolate D on the right side of the equation, we must first isolate the term AD on that side. That calls for clearing the right side of the denominator $A + 12$.

Solution

$$C = \frac{AD}{A+12}$$

$$(A+12)(C) = (A+12)\left(\frac{AD}{A+12}\right) \qquad \text{Write each side of the formula within parentheses, and then multiply both sides by the LCD, } A+12.$$

$$(A+12)C = AD \qquad \text{Simplify the right side: } (\overset{1}{\cancel{A+12}})\left(\frac{AD}{\underset{1}{\cancel{A+12}}}\right).$$

$$AC + 12C = AD \qquad \text{Distribute the multiplication by } C.$$

$$\frac{AC+12C}{A} = D \qquad \text{To undo the multiplication by } A \text{ on the right side and isolate } D, \text{ divide both sides by } A.$$

Solving Young's rule for D, we have $D = \dfrac{AC+12C}{A}$.

Self Check 6 Solve $R = \dfrac{eS}{T-10}$ for S.

Now Try **Problem 61**

© Natalee Hazelwood/Alamy

EXAMPLE 7 ***Photography.*** The design of a camera lens uses the formula $\dfrac{1}{f} = \dfrac{1}{p} + \dfrac{1}{q}$, where f is the focal length of the lens, p is the distance from the lens to the object, and q is the distance from the lens to the image. Solve the formula for q.

Strategy   We will begin by multiplying both sides of the equation by the LCD.

Why   It will be easier to isolate q if there are no fractions.

Solution

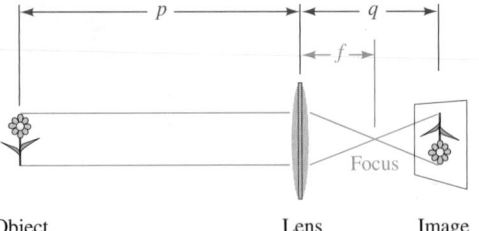

Object   Lens   Image

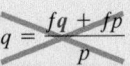

Caution

A common error, is to divide both sides of $pq = fq + fp$ by p to solve for q:

$$q = \frac{fq + fp}{p}$$

The formula is not solved for q because q appears on *both sides* of the equation.

$$\frac{1}{f} = \frac{1}{p} + \frac{1}{q}$$

$$fpq\left(\frac{1}{f}\right) = fpq\left(\frac{1}{p} + \frac{1}{q}\right)$$
Write each side of the formula within parentheses and then multiply both sides by the LCD, fpq.

Distribute the multiplication by fpq and then remove the common factors of each numerator and denominator.

$$pq = fq + fp$$
Simplify.

If we subtract fq from both sides, all terms that contain q will be on the left side.

$$pq - fq = fp$$
Subtract fq from both sides.

$$q(p - f) = fp$$
Factor out the GCF, q, from the two terms on the left side.

$$\frac{q(p - f)}{p - f} = \frac{fp}{p - f}$$
To undo the multiplication by $(p - f)$ and isolate q, divide both sides by $p - f$.

$$q = \frac{fp}{p - f}$$
Simplify the left side: $= q$.

Solving the formula for q, we have $q = \frac{fp}{p - f}$.

Self Check 7   Solve the formula in Example 7 for p.

Now Try   **Problem 63**

ANSWERS TO SELF CHECK   **1.** 1   **2.** 22   **3.** 3, -8 **4.** No solution   **5.** $\frac{5}{2}$ **6.** $S = \frac{RT - 10R}{e}$

7. $p = \frac{fq}{q - f}$

STUDY SET
7.6

VOCABULARY

Fill in the blanks.

1. Equations that contain one or more rational expressions, such as $\frac{x}{x + 2} = 4 + \frac{10}{x + 2}$, are called _____ equations.

2. To *solve* a rational equation we find all the values of the variable that make the equation _____.

3. To *clear* a rational equation of fractions, _____ both sides by the LCD of all rational expressions in the equation.

4. When solving a rational equation, if we obtain a number that does not satisfy the original equation, the number is called an _____ solution.

CONCEPTS

5. Is 5 a solution of the given rational equation?

a. $\dfrac{1}{x-1} = 1 - \dfrac{3}{x-1}$

b. $\dfrac{x}{x-5} = 3 + \dfrac{5}{x-5}$

6. A student was asked to solve a rational equation. The first step of his solution is as follows:

$$12x\left(\dfrac{5}{x} + \dfrac{2}{3}\right) = 12x\left(\dfrac{7}{4x}\right)$$

a. What equation was he asked to solve?

b. What LCD is used to clear the equation of fractions?

7. Consider the rational equation $\dfrac{x}{x-3} = \dfrac{1}{x} + \dfrac{2}{x-3}$.

a. What values of x make a denominator 0?

b. What values of x make a rational expression undefined?

c. What numbers can't be solutions of the equation?

8. A student solved a rational equation and found 8 to be a possible solution. When she checked 8, she obtained $\dfrac{3}{0} = \dfrac{1}{0} + \dfrac{2}{3}$. What conclusion can be drawn?

By what should both sides of the equation be multiplied to clear it of fractions?

9. a. $\dfrac{1}{y} = 20 - \dfrac{5}{y}$ **b.** $\dfrac{x}{x^2-4} = \dfrac{4}{x-2}$

10. a. $\dfrac{x}{5} = \dfrac{3x}{10} + \dfrac{7}{2x}$ **b.** $\dfrac{2x}{x-6} = 4 + \dfrac{1}{x-6}$

11. Perform each multiplication.

a. $4x\left(\dfrac{3}{4x}\right)$ **b.** $(x+6)(x-2)\left(\dfrac{3}{x-2}\right)$

12. Fill in the blanks.

$$\underbrace{8x\left(\dfrac{3}{4x}\right)}_{} = \underbrace{8x\left(\dfrac{1}{8x}\right)}_{} + \underbrace{8x\left(\dfrac{5}{4}\right)}_{}$$

NOTATION

Complete the solution to solve the equation.

13.
$$\dfrac{2}{a} + \dfrac{1}{2} = \dfrac{7}{2a}$$

$$\left(\dfrac{2}{a} + \dfrac{1}{2}\right) = \left(\dfrac{7}{2a}\right)$$

$$\left(\dfrac{2}{a}\right) + \left(\dfrac{1}{2}\right) = \left(\dfrac{7}{2a}\right)$$

$$\boxed{} + a =$$

$$4 + a - 4 = 7 - \boxed{}$$

$$a =$$

14. Can $5x\left(\dfrac{2}{x} + \dfrac{4}{5}\right)$ be written as $5x \cdot \dfrac{2}{x} + \dfrac{4}{5}$? Explain.

GUIDED PRACTICE

Solve each equation and check the result. If an equation has no solution, so indicate. See Examples 1 and 2.

15. $\dfrac{2}{3} = \dfrac{1}{2} + \dfrac{x}{6}$ **16.** $\dfrac{7}{4} = \dfrac{x}{8} + \dfrac{5}{2}$

17. $\dfrac{s}{12} - \dfrac{s}{2} = \dfrac{5s}{4}$ **18.** $\dfrac{n}{18} - \dfrac{n}{6} = \dfrac{4n}{3}$

19. $\dfrac{x}{18} = \dfrac{1}{3} - \dfrac{x}{2}$ **20.** $\dfrac{x}{4} = \dfrac{1}{2} - \dfrac{3x}{20}$

21. $\dfrac{5}{3k} + \dfrac{1}{k} = -2$ **22.** $\dfrac{3}{4h} + \dfrac{2}{h} = 1$

23. $\dfrac{1}{4} - \dfrac{5}{6} = \dfrac{1}{a}$ **24.** $\dfrac{5}{9} - \dfrac{1}{3} = \dfrac{1}{b}$

25. $\dfrac{1}{8} + \dfrac{2}{b} - \dfrac{1}{12} = 0$ **26.** $\dfrac{1}{14} + \dfrac{2}{n} - \dfrac{2}{21} = 0$

Solve each equation and check the result. If an equation has no solution, so indicate. See Example 3.

27. $x + \dfrac{8}{x} = 6$ **28.** $z - \dfrac{16}{z} = 6$

29. $\dfrac{10}{t} - t = 3$ **30.** $\dfrac{7}{p} - p = -6$

31. $\dfrac{20}{c} + c = -9$ **32.** $d = 4 + \dfrac{21}{d}$

33. $4 + \dfrac{15}{p} = 3p$ **34.** $2x = 6 + \dfrac{8}{x}$

Solve each equation and check the result. If an equation has no solution, so indicate. See Example 4.

35. $\dfrac{2}{y+1} + 5 = \dfrac{12}{y+1}$ **36.** $\dfrac{3}{p+6} - 2 = \dfrac{7}{p+6}$

37. $\dfrac{x}{x-5} - \dfrac{5}{x-5} = 3$

38. $\dfrac{3}{y-2} + 1 = \dfrac{3}{y-2}$

39. $\dfrac{a^2}{a+2} - a = \dfrac{4}{a+2}$

40. $\dfrac{z^2}{z+1} + 2 = \dfrac{1}{z+1}$

41. $\dfrac{5a}{a+1} - 4 = \dfrac{3}{a+1}$ **42.** $\dfrac{4}{b-3} = \dfrac{b+5}{b-3} - 5$

43. $\dfrac{z-4}{z-3} = \dfrac{z+2}{z+1}$ **44.** $\dfrac{a+2}{a+8} = \dfrac{a-3}{a-2}$

45. $\dfrac{2}{3-t} = \dfrac{-t}{t+3}$ **46.** $\dfrac{n}{n+1} = \dfrac{6}{n+7}$

Solve each equation and check the result. If an equation has no solution, so indicate. See Example 5.

47. $\dfrac{2x}{x^2 + x - 2} + \dfrac{2}{x + 2} = 1$

48. $\dfrac{4x}{x^2 + 2x - 3} + \dfrac{3}{x + 3} = 1$

49. $\dfrac{4}{y^2 - 4} = \dfrac{1}{y - 2} + \dfrac{1}{y + 2}$

50. $\dfrac{2w}{w^2 - 9} = \dfrac{1}{w + 3} - \dfrac{4}{w - 3}$

51. $\dfrac{3}{x - 2} + \dfrac{1}{x} = \dfrac{6x + 4}{x^2 - 2x}$

52. $\dfrac{x}{x - 1} - \dfrac{12}{x^2 - x} = \dfrac{-1}{x - 1}$

53. $\dfrac{2}{4m + 12} - \dfrac{m + 1}{3m + 9} = \dfrac{m}{2m + 6}$

54. $\dfrac{5}{4y - 4} + \dfrac{y - 2}{2y - 2} = \dfrac{y}{5y - 5}$

Solve each formula for the indicated variable. See Examples 6 and 7.

55. $\dfrac{P}{n} = rt$ for P

56. $\dfrac{F}{m} = a$ for F

57. $\dfrac{a}{b} = \dfrac{c}{d}$ for d

58. $\dfrac{pc}{s} = \dfrac{t}{r}$ for c

59. $h = \dfrac{2A}{b + d}$ for A

60. $T = \dfrac{3R}{M - n}$ for R

61. $I = \dfrac{E}{R + r}$ for r

62. $\dfrac{S}{k + h} = E$ for k

63. $\dfrac{1}{a} + \dfrac{1}{b} = 1$ for a

64. $\dfrac{1}{a} - \dfrac{1}{b} = 1$ for b

65. $\dfrac{5}{x} - \dfrac{4}{y} = \dfrac{5}{z}$ for x

66. $\dfrac{2}{c} + \dfrac{2}{d} = \dfrac{1}{h}$ for c

67. $\dfrac{1}{r} + \dfrac{1}{s} = \dfrac{1}{t}$ for r

68. $\dfrac{1}{x} - \dfrac{1}{y} = \dfrac{1}{z}$ for x

69. $F = \dfrac{L^2}{6d} + \dfrac{d}{2}$ for L^2

70. $H = \dfrac{J^3}{cd} - \dfrac{K^3}{d}$ for J^3

TRY IT YOURSELF

Solve each equation and check the result. If an equation has no solution, so indicate.

71. $\dfrac{1}{3} + \dfrac{2}{x - 3} = 1$

72. $\dfrac{3}{5} + \dfrac{7}{x + 2} = 2$

73. $\dfrac{7}{q^2 - q - 2} + \dfrac{1}{q + 1} = \dfrac{3}{q - 2}$

74. $\dfrac{3}{x - 1} - \dfrac{1}{x + 9} = \dfrac{18}{x^2 + 8x - 9}$

75. $\dfrac{1}{8} + \dfrac{2}{y} = \dfrac{1}{y} + \dfrac{1}{10}$

76. $\dfrac{7}{10} + \dfrac{4}{c} = \dfrac{1}{c} + \dfrac{11}{15}$

77. $4 - \dfrac{8}{x + 1} = \dfrac{8x}{x + 1}$

78. $\dfrac{x}{x - 2} = \dfrac{2}{x - 2} + 2$

79. $\dfrac{3}{x + 1} = \dfrac{x - 2}{x + 1} + \dfrac{x - 2}{2}$

80. $\dfrac{2}{x - 1} + \dfrac{x - 2}{3} = \dfrac{4}{x - 1}$

81. $\dfrac{3}{x} + 2 = 3$

82. $\dfrac{2}{x} + 9 = 11$

83. $\dfrac{3}{5d} + \dfrac{4}{3} = \dfrac{9}{10d}$

84. $\dfrac{2}{3d} + \dfrac{1}{4} = \dfrac{11}{6d}$

85. $\dfrac{n}{n^2 - 9} + \dfrac{n + 8}{n + 3} = \dfrac{n - 8}{n - 3}$

86. $\dfrac{7}{x - 5} = \dfrac{40}{x^2 - 25} + \dfrac{3}{x + 5}$

87. $y + \dfrac{2}{3} = \dfrac{2y - 12}{3y - 9}$

88. $1 - \dfrac{3}{b} = \dfrac{-8b}{b^2 + 3b}$

89. $\dfrac{a - 1}{7} - \dfrac{a - 2}{14} = \dfrac{1}{2}$

90. $\dfrac{3x - 1}{6} - \dfrac{x + 3}{2} = \dfrac{3x + 4}{3}$

For each expression, perform the indicated operations and then simplify, if possible. Solve each equation and check the result.

91. a. $\dfrac{a}{3} + \dfrac{3}{5} + \dfrac{a}{15}$

    b. $\dfrac{a}{3} + \dfrac{3}{5} = \dfrac{a}{15}$

92. a. $\dfrac{1}{6x} - \dfrac{2}{x - 6}$

    b. $\dfrac{1}{6x} = \dfrac{2}{x - 6}$

93. a. $\dfrac{x}{x - 2} - \dfrac{1}{x - 3}$

    b. $\dfrac{x}{x - 2} - \dfrac{1}{x - 3} = 1$

94. a. $\dfrac{u^2 + 1}{u^2 - u} - \dfrac{u}{u - 1}$

    b. $\dfrac{u^2 + 1}{u^2 - u} - \dfrac{u}{u - 1} = \dfrac{1}{u}$

APPLICATIONS

95. MEDICINE   Radioactive tracers are used for diagnostic work in nuclear medicine. The *effective half-life H* of a radioactive material in an organism is given by the formula $H = \dfrac{RB}{R + B}$ where R is the radioactive half-life and B is the biological half-life of the tracer. Solve the formula for R.

96. CHEMISTRY Charles's law describes the relationship between the volume and temperature of a gas that is kept at a constant pressure. It can be expressed as $\frac{V_1}{V_2} = \frac{T_1}{T_2}$ where V_1 and V_2 are variables representing two different volumes, and T_1 and T_2 are variables representing two different temperatures. (Recall that the notation V_1 is read as *V sub one.*) Solve for V_2.

97. ELECTRONICS Most electronic circuits require resistors to make them work properly. Resistors are components that limit current. An important formula about resistors in a circuit is $\frac{1}{r} = \frac{1}{r_1} + \frac{1}{r_2}$. Solve for r.

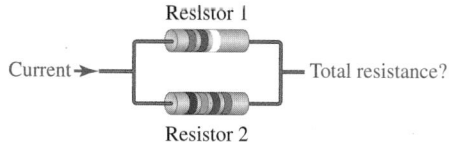

Resistor 1

Current → Total resistance?

Resistor 2

98. MATHEMATICAL FORMULAS To quickly find the sum $\frac{1}{2} + \frac{1}{4} + \frac{1}{8} + \frac{1}{16} + \frac{1}{32} + \frac{1}{64} + \frac{1}{128}$, mathematicians use the formula $S = \frac{a(1 - r^n)}{1 - r}$. Solve the formula for a.

WRITING

99. Explain how the multiplication property of equality is used to solve rational equations. Give an example.

100. When solving rational equations, how do you know whether a solution is extraneous?

101. What is meant by clearing a rational equation of fractions? Give an example.

102. Explain the difference between the procedure used to simplify $\frac{1}{x} + \frac{1}{3}$ and the procedure used to solve $\frac{1}{x} + \frac{1}{3} = \frac{1}{2}$.

REVIEW

103. UNIFORMS A cheerleading squad had their school mascot embroidered on the front of their uniform sweaters. They were charged $18.50 per sweater plus a one time set up fee of $75. If the project cost $445, how many sweaters were embroidered?

104. GEOMETRY The vertex angle of an isosceles triangle is 46°. Find the measure of each base angle.

CHALLENGE PROBLEMS

105. Solve: $x^{-2} + 2x^{-1} + 1 = 0$

106. ENGINES A formula that is used in the design and testing of diesel engines is $E = 1 - \frac{T_4 - T_1}{a(T_3 - T_2)}$. Solve the formula for T_1.

SECTION 7.7
Problem Solving Using Rational Equations

Objectives

 1 Solve number problems.

2 Solve uniform motion problems.

3 Solve shared-work problems.

4 Solve investment problems.

We will now use the five-step problem-solving strategy to solve application problems from a variety of areas, including banking, petroleum engineering, sports, and travel. In each case, we will use a rational equation to model the situation. We begin with an example in which we find an unknown number.

1 Solve Number Problems.

EXAMPLE 1 *Number Problem.* If the same number is added to both the numerator and the denominator of the fraction $\frac{3}{5}$, the result is $\frac{4}{5}$. Find the number.

Analyze the Problem

- Begin with the fraction $\frac{3}{5}$.

- Add the same number to the numerator and to the denominator.

- The result is $\frac{4}{5}$.

- Find the number.

Form an Equation

Let $n =$ the unknown number. To form an equation, add the unknown number to the numerator and to the denominator of $\frac{3}{5}$. Then set the result equal to $\frac{4}{5}$.

$$\frac{3 + n}{5 + n} = \frac{4}{5}$$

Solve the Equation

To solve this rational equation, we begin by clearing it of fractions.

$$\frac{3 + n}{5 + n} = \frac{4}{5}$$

$$\overset{1}{5(5 + n)}\left(\frac{3 + n}{\underset{1}{5 + n}}\right) = \overset{1}{5}(5 + n)\left(\frac{4}{\underset{1}{5}}\right)$$ Multiply both sides by the LCD, $5(5 + n)$. Then remove common factors of the numerator and denominator.

$$5(3 + n) = (5 + n)4$$ Simplify.

$$15 + 5n = 20 + 4n$$ Distribute the multiplication by 5 and by 4.

$$15 + n = 20$$ To isolate the variable term of the left side, subtract $4n$ from both sides.

$$n = 5$$ To undo the addition of 15, subtract 15 from both sides.

State the Conclusion

The number is 5.

Check the Result

When we add 5 to both the numerator and denominator of $\frac{3}{5}$, we get

$$\frac{3 + 5}{5 + 5} = \frac{8}{10} = \frac{4}{5}$$

The result checks.

 Now Try **Problem 13**

---

2   Solve Uniform Motion Problems.

Recall that we use the distance formula $d = rt$ to solve motion problems. The relationship between distance, rate, and time can be expressed in another way by solving for t.

$$d = rt$$ Distance = rate · time.

$$\frac{d}{r} = \frac{rt}{r}$$ To undo the multiplication by r and isolate t, divide both sides by r.

$$\frac{d}{r} = t$$ Simplify the right side: $\dfrac{\overset{1}{\cancel{r}} \cdot t}{\underset{1}{\cancel{r}}} = t.$

$$t = \frac{d}{r}$$

This result suggests an alternate form of the distance formula, Time $= \frac{\text{distance}}{\text{rate}}$, that is used to solve the next example.

EXAMPLE 2   ***Runners.***   A coach can run 10 miles in the same amount of time as his best student-athlete can run 12 miles. If the student runs 1 mile per hour (mph) faster than the coach, find the running speeds of the coach and the student.

Analyze the Problem

- The coach runs 10 miles in the same time that the student runs 12 miles.
- The student runs 1 mph faster than the coach.
- Find the speed that each runs.

Form an Equation   Since the student's speed is 1 mph faster than the coach's, let $r =$ the speed that the coach can run. Then, $r + 1 =$ the speed that the student can run. The expressions for the rates are entered in the Rate column of the table. The distances run by the coach and by the student are entered in the Distance column of the table.

Using $t = \frac{d}{r}$, we find that the time it takes the coach to run 10 miles, at a rate of r mph, is $\frac{10}{r}$ hours. Similarly, we find that the time it takes the student to run 12 miles, at a rate of $(r + 1)$ mph, is $\frac{12}{r+1}$ hours. These expressions are entered in the Time column of the table.

	Rate	· Time	= Distance
Coach	r	$\frac{10}{r}$	10
Student	$r + 1$	$\frac{12}{r+1}$	12

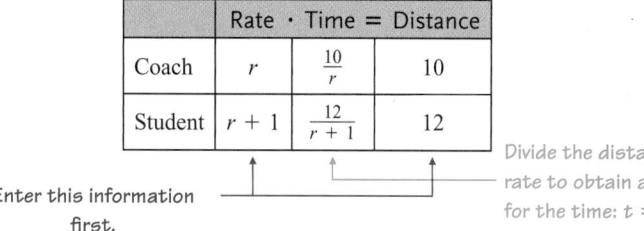

Enter this information first.

Divide the distance by the rate to obtain an expression for the time: $t = \frac{d}{r}$.

The time it takes the coach to run 10 miles	is the same as	the time it takes the student to run 12 miles.
$\frac{10}{r}$	$=$	$\frac{12}{r+1}$

Solve the Equation   To solve this rational equation, we begin by clearing it of fractions.

$$\frac{10}{r} = \frac{12}{r+1}$$

$$\overset{1}{\cancel{r}}(r+1)\left(\frac{10}{\underset{1}{\cancel{r}}}\right) = r(r\cancel{+1})\overset{1}{\left(\frac{12}{\underset{1}{\cancel{r+1}}}\right)} \qquad \text{Multiply both sides by the LCD, } r(r+1). \text{ Then remove common factors of the numerator and denominator.}$$

$$(r+1)10 = 12r \qquad \text{Simplify.}$$

$$10r + 10 = 12r \qquad \text{Distribute the multiplication by 10.}$$

$$10 = 2r \qquad \text{To isolate the variable term on the right, subtract 10}r \text{ from both sides.}$$

$$5 = r \qquad \text{To undo the multiplication by 2, divide both sides by 2.}$$

If $r = 5$, then $r + 1 = 6$.

State the Conclusion The coach's running speed is 5 mph and the student's running speed is 6 mph.

Check the Result The coach will run 10 miles in $\frac{10 \text{ miles}}{5 \text{ mph}} = 2$ hours. The student will run 12 miles in $\frac{12 \text{ miles}}{6 \text{ mph}} = 2$ hours. The times are the same; the results check.

 Now Try **Problem 23**

3 **Solve Shared-Work Problems.**

Problems in which two or more people (or machines) work together to complete a job are called *shared-work problems*. To solve such problems, we must determine the **rate of work** for each person (or machine) involved. For example, suppose it takes you 4 hours to clean your house. Your rate of work can be expressed as $\frac{1}{4}$ of the job is completed per hour. If someone else takes 5 hours to clean the same house, they complete $\frac{1}{5}$ of the job per hour. In general, a rate of work can be determined in the following way.

Rate of Work	If a job can be completed in t units of time, the rate of work can be expressed as: $\frac{1}{t}$ of the job is completed per unit of time.

To solve shared-work problems, we must also determine what fractional part of a job is completed. To do this, we use the formula

Work completed = rate of work · time worked   or $W = rt$

EXAMPLE 3 *Payroll.* At the end of a pay period, it takes the president of a company 15 minutes to sign all of her employees' payroll checks. What fractional part of the job is completed if the president signs checks for 10 minutes?

Strategy We will begin by finding the president's check-signing rate. Then we can use the formula $W = rt$ to find the part of the job that is completed.

Why We know the time worked is 10 minutes. To use the work formula to find what part of the job is completed, we also need to know the president's work rate.

Solution If all of the checks can be signed in 15 minutes, the president's work rate is $\frac{1}{15}$ job per minute. Substituting into the work formula, we have

$W = rt$

$= \dfrac{1}{15} \cdot 10$ Substitute $\frac{1}{15}$ for r, the work rate, and 10 for t, the time worked.

$= \dfrac{10}{15}$ Multiply.

$= \dfrac{2}{3}$ Simplify by removing a common factor of 5.

In 10 minutes, the president will complete $\frac{2}{3}$ of the job of signing the payroll checks.

5. It takes a night security officer 45 minutes to check each of the doors in an office building to make sure they are locked. What is the officer's rate of work?

6. It takes an elementary school teacher 4 hours to make out the semester report cards. What part of the job does she complete in x hours?

7. a. Solve $d = rt$ for t.

   b. Solve $I = Prt$ for P.

8. Complete the table.

	r	$\cdot$	t	$=$	d
Snowmobile	r				4
4 × 4 truck	$r - 5$				3

9. Complete the table.

	Rate	$\cdot$	Time	$=$	Work completed
1st printer	$\frac{1}{15}$		x		
2nd printer	$\frac{1}{8}$		x		

10. Complete the table.

	P	$\cdot$	r	$\cdot$	t	$=$	I
City savings bank			r		1		50
Credit union			$r - 0.02$		1		75

NOTATION

11. Write $\frac{55}{9}$ days using a mixed number.

12. a. Write 9% as a decimal.

   b. Write 0.035 as a percent.

GUIDED PRACTICE

Solve each of these number problems. **See Example 1.**

13. If the same number is added to both the numerator and the denominator of $\frac{2}{5}$, the result is $\frac{2}{3}$. Find the number.

14. If the same number is subtracted from both the numerator and the denominator of $\frac{11}{13}$, the result is $\frac{3}{4}$. Find the number.

15. If the denominator of $\frac{3}{4}$ is increased by a number, and the numerator is doubled, the result is 1. Find the number.

16. If a number is added to the numerator of $\frac{7}{8}$, and the same number is subtracted from the denominator, the result is 2. Find the number.

17. If a number is added to the numerator of $\frac{3}{4}$, and twice as much is added to the denominator, the result is $\frac{4}{7}$. Find the number.

18. If a number is added to the numerator of $\frac{5}{7}$, and twice as much is subtracted from the denominator, the result is 8. Find the number.

19. The sum of a number and its reciprocal is $\frac{13}{6}$. Find each number.

20. The sum of the reciprocals of two consecutive even integers is $\frac{7}{24}$. Find each integer.

APPLICATIONS

21. COOKING   If the same number is added to both the numerator and the denominator of the amount of butter used in the following recipe for toffee, the result is the amount of brown sugar to be used. Find the number.

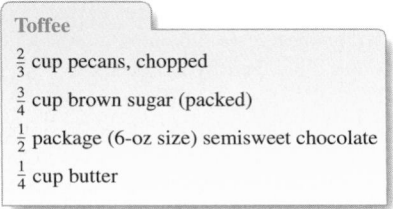

Toffee
$\frac{2}{3}$ cup pecans, chopped
$\frac{3}{4}$ cup brown sugar (packed)
$\frac{1}{2}$ package (6-oz size) semisweet chocolate
$\frac{1}{4}$ cup butter

22. TAPE MEASURES   If the same number is added to both the numerator and the denominator of the first measurement, the result is the second measurement. Find the number.

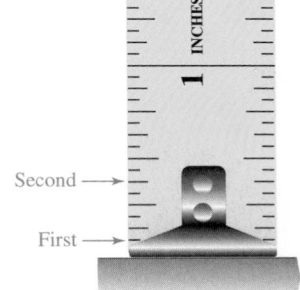

Second →
First →

23. TOUR DE FRANCE   Maurice Garin of France won the first Tour de France bicycle road race in 1903. In 2005, American Lance Armstrong won his seventh consecutive Tour de France. Armstrong's average speed in 2005 was 10 mph faster than Garin's in 1903. In the time it took Garin to ride 80 miles, Armstrong could have ridden 130 miles. Find each cyclist average speed.

24. PHYSICAL FITNESS   A woman can bicycle 28 miles in the same time as it takes her to walk 8 miles. She can ride 10 mph faster than she can walk. How fast can she walk?

25. PACKAGING FRUIT   The diagram on the next page shows how apples are processed for market. Although the second conveyor belt is shorter, an apple spends the same amount of time on each belt because the second conveyor moves 1 foot per second slower than the first. Determine the speed of each conveyor belt.

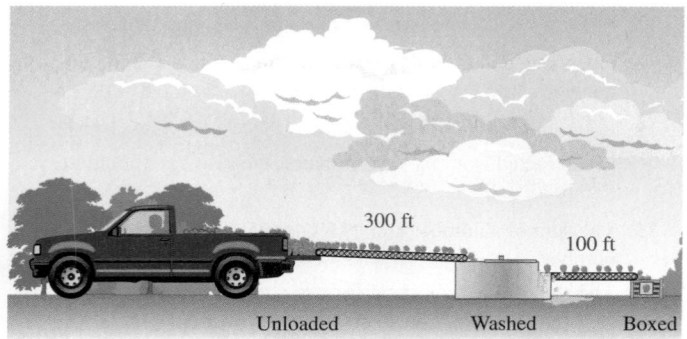

300 ft

100 ft

Unloaded   Washed   Boxed

	Rate · Time = Distance		
Downwind	$255 + x$		300
Upwind	$255 - x$		210

26. COMPARING TRAVEL A plane can fly 300 miles in the same time as it takes a car to go 120 miles. If the car travels 90 mph slower than the plane, find the speed of the plane.

27. BIRDS IN FLIGHT Although flight speed is dependent upon the weather and the wind, in general, a Canada goose can fly about 10 mph faster than a great blue heron. In the same time that a Canada goose travels 120 miles, a great blue heron travels 80 miles. Find their flying speeds.

28. FAST CARS The top speed of a Dodge Charger SRT8 is 33 mph less than the top speed of a Chevrolet Corvette Z06. At their top speeds, a Corvette can travel 6 miles in the same time that a Charger can travel 5 miles. Find the top speed of each car.

29. WIND SPEED When a plane flies downwind, the wind pushes the plane so that its speed is the *sum* of the speed of the plane in still air and the speed of the wind. Traveling upwind, the wind pushes against the plane so that its speed is the *difference* of the speed of the plane in still air and the speed of the wind. Suppose a plane that travels 255 mph in still air can travel 300 miles downwind in the same time as it takes to travel 210 miles upwind. Complete the following table and find the speed of the wind, represented by x.

30. BOATING A boat that travels 18 mph in still water can travel 22 miles downstream in the same time as it takes to travel 14 miles upstream. Find the speed of the current in the river. (See problem 29.)

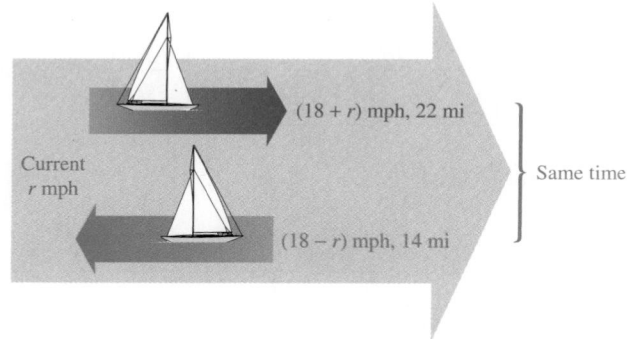

Current
r mph

$(18 + r)$ mph, 22 mi

$(18 - r)$ mph, 14 mi

Same time

31. ROOFING HOUSES A homeowner estimates that it will take her 7 days to roof her house. A professional roofer estimates that he could roof the house in 4 days. How long will it take if the homeowner helps the roofer?

32. HOLIDAY DECORATING One crew can put up holiday decorations in the mall in 8 hours. A second crew can put up the decorations in 10 hours. How long will it take if both crews work together to decorate the mall?

33. *from Campus to Careers*
Recreation Director

Suppose you are a recreation director at a summer camp. The water in the camp swimming pool was drained out for the winter and it is now time to refill the pool. One pipe can fill the empty pool in 12 hours and another can fill the empty pool in 18 hours. Suppose both pipes are opened at 8:00 A.M. and you have scheduled a swimming activity for 2:00 P.M. that day. Will the pool be filled by then?

34. GROUNDSKEEPING It takes a groundskeeper 45 minutes to prepare a softball field for a game. It takes his assistant 55 minutes to prepare the same field. How long will it take if they work together to prepare the field?

35. FILLING A POOL One inlet pipe can fill an empty pool in 4 hours, and a drain can empty the pool in 8 hours. How long will it take the pipe to fill the pool if the drain is left open?

36. SEWAGE TREATMENT A sludge pool is filled by two inlet pipes. One pipe can fill the pool in 15 days, and the other can fill it in 21 days. However, if no sewage is added, continuous waste removal will empty the pool in 36 days. How long will it take the two inlet pipes to fill an empty sludge pool?

37. GRADING PAPERS On average, it takes a teacher 30 minutes to grade a set of quizzes. It takes her teacher's aide twice as long to do the same grading. How long will it take if they work together to grade a set of quizzes?

38. DOG KENNELS It takes the owner/operator of a dog kennel 6 hours to clean all of the cages. It takes his assistant 2 hours more than that to clean the same cages. How long will it take if they work together?

39. PRINTERS It takes a printer 6 hours to print the class schedules for all of the students enrolled in a community college. A faster printer can print the schedules in 4 hours. How long will it take the two printers working together to print $\frac{3}{4}$ of the class schedules?

40. OFFICE WORK In 5 hours, a secretary can address 100 envelopes. Another secretary can address 100 envelopes in 6 hours. How long would it take the secretaries, working together, to address 300 envelopes. (*Hint:* Think of addressing 300 envelopes as three 100-envelope jobs.)

41. COMPARING INVESTMENTS An amount of money invested for 1 year in tax-free bonds will earn $300. In a certain credit union account, that same amount of money will only earn $200 interest in a year, because the interest paid is 2% less than that paid by the bonds. Find the rate of interest paid by each investment.

42. COMPARING INVESTMENTS An amount of money invested for 1 year in a savings account will earn $1,500. That same amount of money, invested in a mini-mall development will earn $6,500 interest in a year, because the interest paid is 10% more than that paid by the savings account. Find the rate of interest paid by each investment.

43. COMPARING INVESTMENTS Two certificates of deposit (CDs) pay interest at rates that differ by 1%. Money invested for 1 year in the first CD earns $175 interest. The same principal invested in the second CD earns $200. Find the two rates of interest.

44. COMPARING INTEREST RATES Two bond funds pay interest at rates that differ by 2%. Money invested for 1 year in the first fund earns $315 interest. The same amount invested in the second fund earns $385. Find the lower rate of interest.

45. In Example 4, one inlet pipe could fill an oil tank in 7 days, and another could fill the same tank in 9 days. We were asked to find how long it would take if both pipes were used. Explain why each of the following approaches is incorrect.

The time it would take to fill the tank

- is the *sum* of the lengths of time it takes each pipe to fill the tank: 7 days + 9 days = 16 days.
- is the *difference* in the lengths of time it takes each pipe to fill the tank: 9 days − 7 days = 2 days.
- is the *average* of the lengths of time it takes each pipe to fill the tank:

$$\frac{7 \text{ days} + 9 \text{ days}}{2} = \frac{16 \text{ days}}{2} = 8 \text{ days.}$$

46. Write a shared-work problem that can be modeled by the equation:

$$\frac{x}{3} + \frac{x}{4} = 1$$

REVIEW

47. Solve using substitution: $\begin{cases} x + y = 4 \\ y = 3x \end{cases}$

48. Solve using elimination (addition): $\begin{cases} 5x - 4y = 19 \\ 3x + 2y = 7 \end{cases}$

49. Use a check to determine whether $\frac{21}{5}$ is a solution of: $x + 20 = 4x - 1 + 2x$

50. Solve: $4x^2 + 8x = 0$

51. Evaluate $2x^2 + 5x - 3$ for $x = -3$.

52. Solve $T - R = ma$ for R.

CHALLENGE PROBLEMS

53. RIVER TOURS A river boat tour begins by going 60 miles upstream against a 5-mph current. There, the boat turns around and returns with the current. What still-water speed should the captain use to complete the tour in 5 hours?

54. TRAVEL TIME A company president flew 680 miles one way in the corporate jet, but returned in a smaller plane that could fly only half as fast. If the total travel time was 6 hours, find the speeds of the planes.

55. SALES A dealer bought some radios for a total of $1,200. She gave away 6 radios as gifts, sold the rest for $10 more than she paid for each radio, and broke even. How many radios did she buy?

56. FURNACE REPAIRS A repairman purchased several furnace-blower motors for a total cost of $210. If his cost per motor had been $5 less, he could have purchased one additional motor. How many motors did he buy at the regular rate?

SECTION 7.8
Proportions and Similar Triangles

Objectives

1 Write ratios and rates in simplest form.

2 Solve proportions.

3 Use proportions to solve problems.

4 Use proportions to solve problems involving similar triangles.

In this section, we will discuss a problem-solving tool called a *proportion*. A proportion is a type of rational equation that involves two *ratios* or two *rates*.

1 **Write Ratios and Rates in Simplest Form.**

Ratios enable us to compare numerical quantities. Here are some examples.

• To prepare fuel for a lawnmower, gasoline is mixed with oil in a 50-to-1 ratio.
• In the stock market, winning stocks might outnumber losers by a ratio of 7 to 4.
• Gold is combined with other metals in the ratio of 14 to 10 to make 14-karat jewelry.

Ratios	A **ratio** is the quotient of two numbers or the quotient of two quantities that have the same units.

There are three ways to write a ratio: as a fraction, using the word *to,* or with a colon. For example, the comparison of the number of winning stocks to the number of losing stocks mentioned earlier can be written as

$$\frac{7}{4}, \qquad 7 \text{ to } 4, \qquad \text{or} \qquad 7{:}4$$

Each of these forms can be read as "the ratio of 7 to 4."

EXAMPLE 1   Translate each phrase into a ratio written in fractional form:
a. The ratio of 5 to 9   **b.** 12 ounces to 2 pounds

Strategy   To translate. we need to identify the number (or quantity) before the word *to* and the number (or quantity) after it.

Why   The number before the word *to* is the numerator of the ratio and the number after it is the denominator.

Solution

a. The ratio of 5 to 9 is written $\frac{5}{9}$.

b. To write a ratio of two quantities with the same units, we must express 2 pounds in terms of ounces. Since 1 pound = 16 ounces, 2 pounds = 32 ounces. The ratio of 12 ounces to 32 ounces can be simplified so that no units appear in the final form.

$$\frac{12 \text{ ounces}}{32 \text{ ounces}} = \frac{\overset{1}{3} \cdot \overset{1}{\cancel{4}} \text{ } \cancel{\text{ounces}}}{\cancel{4} \cdot 8 \text{ } \cancel{\text{ounces}}}_{1 \quad 1} = \frac{3}{8}$$

Self Check 1   Translate each phrase into a ratio written in fractional form:
a. The ratio of 15 to 2   **b.** 12 hours to 2 days

Now Try   **Problem 25**

A quotient that compares quantities with different units is called a **rate.** For example, if the 495-mile drive from New Orleans to Dallas takes 9 hours, the average rate of speed is the quotient of the miles driven and the length of time the trip takes.

$$\text{Average rate of speed} = \frac{495 \text{ miles}}{9 \text{ hours}} = \frac{\overset{1}{\cancel{9}} \cdot 55 \text{ miles}}{\underset{1}{\cancel{9}} \cdot 1 \text{ hours}} = \frac{55 \text{ miles}}{1 \text{ hour}}$$

Rates   A **rate** is a quotient of two quantities that have different units.

   Solve Proportions.

If two ratios or two rates are equal, we say that they are *in proportion.*

Proportion   A **proportion** is a mathematical statement that two ratios or two rates are equal.

Some examples of proportions are

$$\frac{1}{2} = \frac{3}{6} \qquad \frac{3 \text{ waiters}}{7 \text{ tables}} = \frac{9 \text{ waiters}}{21 \text{ tables}} \qquad \frac{a}{b} = \frac{c}{d}$$

- The proportion $\frac{1}{2} = \frac{3}{6}$ can be read as "1 is to 2 as 3 is to 6."

- The proportion $\frac{3 \text{ waiters}}{7 \text{ tables}} = \frac{9 \text{ waiters}}{21 \text{ tables}}$ can be read as "3 waiters is to 7 tables as 9 waiters is to 21 tables."

- The proportion $\frac{a}{b} = \frac{c}{d}$ can be read as "a is to b as c is to d."

Each of the four numbers in a proportion is called a **term.** The first and fourth terms are called the **extremes,** and the second and third terms are called the **means.**

First term ⟶ $\dfrac{a}{b} = \dfrac{c}{d}$ ⟵ Third term   *a and d are the extremes. b and c are the means.*
Second term ⟶ $\phantom{\dfrac{a}{b}}$ ⟵ Fourth term

For the proportion $\frac{a}{b} = \frac{c}{d}$, we can show that the product of the extremes, *ad*, is equal to the product of the means, *bc*, by multiplying both sides of the proportion by *bd*, and observing that $ad = bc$.

$$\frac{a}{b} = \frac{c}{d}$$

$$\frac{1}{\cancel{bd}} \cdot \frac{a}{\cancel{b}} = \cancel{bd} \cdot \frac{c}{\cancel{d}} \atop {1}$$ To clear the fractions, multiply both sides by the LCD, *bd*. Remove common factors of the numerator and denominator.

$$ad = bc$$ Simplify: $\frac{b}{b} = 1$ and $\frac{d}{d} = 1$.

Since $ad = bc$, the product of the extremes equals the product of the means.

The same products ad and bc can be found by multiplying diagonally in the proportion $\frac{a}{b} = \frac{c}{d}$. We call ad and bc **cross products**.

$$ad \qquad\qquad bc$$
$$\frac{a}{b} \diagtimes \frac{c}{d}$$

The Fundamental Property of Proportions	In a proportion, the product of the extremes is equal to the product of the means.
	If $\frac{a}{b} = \frac{c}{d}$, then $ad = bc$ and if $ad = bc$, then $\frac{a}{b} = \frac{c}{d}$.

EXAMPLE 2   Determine whether each equation is a proportion:

a. $\dfrac{3}{7} = \dfrac{9}{21}$ **b.** $\dfrac{8}{3} = \dfrac{13}{5}$

Strategy   We will check to see whether the product of the extremes is equal to the product of the means.

Why   If the product of the extremes equals the product of the means, the equation is a proportion. If the cross products are not equal, the equation is not a proportion.

Solution

a. The product of the extremes is $3 \cdot 21 = 63$. The product of the means is $7 \cdot 9 = 63$. Since the cross products are equal, $\frac{3}{7} = \frac{9}{21}$ is a proportion.

$$3 \cdot 21 = 63 \qquad 7 \cdot 9 = 63$$
$$\frac{3}{7} \diagtimes \frac{9}{21}$$ Each cross product is 63.

b. The product of the extremes is $8 \cdot 5 = 40$. The product of the means is $3 \cdot 13 = 39$. Since the cross products are not equal, the equation is not a proportion: $\frac{8}{3} \neq \frac{13}{5}$.

$$8 \cdot 5 = 40 \qquad 3 \cdot 13 = 39$$
$$\frac{8}{3} \diagtimes \frac{13}{5}$$ One cross product is 40 and the other is 39.

Caution

We cannot remove common factors "across" an = symbol.

When this is done, the original proportion, $\frac{3}{7} = \frac{9}{21}$, which we found to be true, is made false: $\frac{1}{7} = \frac{9}{7}$.

Self Check 2   Determine whether the equation $\frac{6}{13} = \frac{24}{53}$ is a proportion.

Now Try   **Problems 29**

The fundamental property of proportions provides us with a way to solve proportions.

EXAMPLE 3   Solve: $\dfrac{3}{2} = \dfrac{9}{x}$

Strategy   To solve for x, we will set the cross products equal.

Why   This equation is a proportion, and in a proportion the product of the means equals the product of the extremes.

Solution

$$\frac{3}{2} = \frac{9}{x} \qquad \text{This is the given proportion.}$$

$$3 \cdot x = 2 \cdot 9 \qquad \text{Find each cross product and set them equal.}$$

$$3x = 18 \qquad \text{Do the multiplications.}$$

$$\frac{3x}{3} = \frac{18}{3} \qquad \text{To isolate } x, \text{ divide both sides by 3.}$$

$$x = 6$$

> **Caution**
> Remember that a cross product is the product of the means or extremes of a proportion. It would be incorrect to try to compute cross products to solve $\dfrac{12}{18} = \dfrac{4}{x} + \dfrac{1}{2}$ because there is more than one term on the right side of the equation. It is *not* a proportion.

Check: To check the result, we substitute 6 for x in $\dfrac{3}{2} = \dfrac{9}{x}$ and find the cross products.

$$3 \cdot 6 = 18 \qquad 2 \cdot 9 = 18$$

$$\frac{3}{2} \overset{?}{=} \frac{9}{6} \qquad \text{Each cross product is 18.}$$

Since the cross products are equal, the solution of $\dfrac{3}{2} = \dfrac{9}{x}$ is 6. The solution set is $\{6\}$.

 Self Check 3   Solve: $\dfrac{15}{x} = \dfrac{25}{40}$

Now Try   **Problem 35**

EXAMPLE 4   Solve: $\dfrac{a}{2} = \dfrac{4}{a-2}$

Strategy   To solve for a, we will set the cross products equal.

Why   Since this equation is a proportion, the product of the means equals the product of the extremes.

Solution

$$\frac{a}{2} = \frac{4}{a-2} \qquad \text{This is the given proportion.}$$

$$a(a-2) = 2 \cdot 4 \qquad \text{Find each cross product and set them equal. Don't forget to write the parentheses.}$$

$$a^2 - 2a = 8 \qquad \text{On the left hand side, distribute the multiplication by } a. \text{ This is a quadratic equation.}$$

$$a^2 - 2a - 8 = 0 \qquad \text{To get 0 on the right side of the equation, subtract 8 from both sides.}$$

$$(a+2)(a-4) = 0 \qquad \text{Factor } a^2 - 2a - 8.$$

> **Success Tip**
> Since proportions are rational equations, they can also be solved by multiplying both sides by the LCD. Here, an alternate approach is to multiply both sides by $2(a-2)$:
>
> $$2(a-2)\left(\frac{a}{2}\right) = 2(a-2)\left(\frac{4}{a-2}\right)$$

$$a + 2 = 0 \quad \text{or} \quad a - 4 = 0 \quad \text{Set each factor equal to 0.}$$
$$a = -2 \quad | \quad a = 4 \quad \text{Solve each equation.}$$

The solutions are -2 and 4. Verify this using a check.

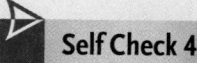

Self Check 4   Solve: $\dfrac{6}{c} = \dfrac{c - 1}{5}$

Now Try   **Problem 47**

③ Use Proportions to Solve Problems.

We can use proportions to solve many problems. If we are given a ratio (or rate) comparing two quantities, the words of the problem can be translated into a proportion, and we can solve it to find the unknown.

EXAMPLE 5   *Grocery Shopping.*   If 6 apples cost $1.38, how much will 16 apples cost?

Analyze the Problem   We know the cost of 6 apples; we are to find the cost of 16 apples.

Form a Proportion   Let $c =$ the cost of 16 apples. If we compare the number of apples to their cost, the two ratios must be equal.

6 apples is to $1.38 as 16 apples is to $c.

$$\text{Number of apples} \longrightarrow \dfrac{6}{1.38} = \dfrac{16}{c} \longleftarrow \text{Number of apples}$$
$$\text{Cost} \longrightarrow \qquad\qquad \longleftarrow \text{Cost}$$

The Language of Algebra
Remember that the word *to* separates the numerator and denominator of a ratio. If the units are written outside the ratio, we can write 6 apples is *to* $1.38 as

$$\dfrac{6}{1.38}$$

Solve the Proportion   We drop the units, find each cross product, set them equal, and then solve the resulting equation for c.

$$6 \cdot c = 1.38(16) \quad \text{In a proportion, the product of the extremes equals the product of the means.}$$

$$6c = 22.08 \quad \text{Multiply: } 1.38(16) = 22.08.$$

$$\dfrac{6c}{6} = \dfrac{22.08}{6} \quad \text{To undo the multiplication by 6 and isolate } c, \text{ divide both sides by 6.}$$

$$c = 3.68 \quad \text{Recall that } c \text{ represents the cost of 16 apples.}$$

State the Conclusion   Sixteen apples will cost $3.68.

Check the Result   We can use estimation to check the result. 16 apples are about 3 times as many as 6 apples, which cost $1.38. If we multiply $1.38 by 3, we get an estimate of the cost of 16 apples: $1.38 \cdot 3 = \$4.14$. The result, $3.68, seems reasonable.

Self Check 5   If 9 tickets to a concert cost $112.50, how much will 15 tickets cost?

Now Try   **Problem 71**

CAUTION   When solving problems using proportions, we must make sure that the units of both numerators are the same and the units of both denominators are the same. In Example 5, it would be incorrect to write

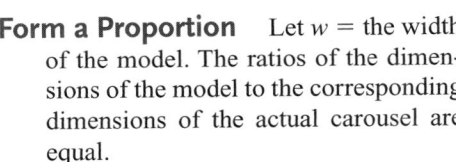

Cost of 6 apples ⟶ $\dfrac{1.38}{6} = \dfrac{16}{c}$ ⟵ 16 apples
6 apples ⟶ ⟵ Cost of 16 apples

EXAMPLE 6   *Miniatures.*   A **scale** is a ratio (or rate) that compares the size of a model, drawing, or map with the size of an actual object. The scale indicates that 1 inch on the model carousel is equivalent to 160 inches on the actual carousel. How wide should the model be if the actual carousel is 35 feet wide?

Analyze the Problem   We are asked to determine the width of the miniature carousel if a ratio of 1 inch to 160 inches is used. We would like the width of the model to be given in inches, not feet, so we will express the 35-foot width of the actual carousel as $35 \cdot 12 = 420$ inches.

Carousel ratio
1 inch:160 inches

?

Form a Proportion   Let w = the width of the model. The ratios of the dimensions of the model to the corresponding dimensions of the actual carousel are equal.

1 inch is to 160 inches as w inches is to 420 inches.

Model ⟶ $\dfrac{1}{160} = \dfrac{w}{420}$ ⟵ Model
Actual size ⟶ ⟵ Actual size

Solve the Proportion   We drop the units, find each cross product, set them equal, and then solve the resulting equation for w.

$420 = 160w$ In a proportion, the product of the extremes is equal to the product of the means.

$\dfrac{420}{160} = \dfrac{160w}{160}$ To undo the multiplication by 160 and isolate w, divide both sides by 160.

$2.625 = w$ Recall that w represents the width of the model.

State the Conclusion   The width of the miniature carousel should be 2.625 in., or $2\frac{5}{8}$ in.

Check the Result   A width of $2\frac{5}{8}$ in. is approximately 3 in. When we write the ratio of the model's approximate width to the width of the actual carousel, we get $\dfrac{3}{420} = \dfrac{1}{140}$, which is about $\dfrac{1}{160}$. The answer seems reasonable.

Now Try   **Problem 85**

When shopping, *unit prices* can be used to compare costs of different sizes of the same brand to determine the best buy. The **unit price** gives the cost per unit, such as cost per ounce, cost per pound, or cost per sheet. We can find the unit price of an item using a proportion.

EXAMPLE 7   ***Comparison Shopping.***   Which size of toothpaste is the better buy?

 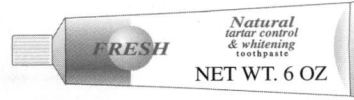

NET WT. 4 OZ   NET WT. 6 OZ

$2.19 $2.79

Solution   To find the unit price for each tube, we let x = the price of 1 ounce of toothpaste. Then we set up and solve the following proportions.

For the 4-ounce tube:

Ounces $\longrightarrow$ $\dfrac{4}{2.19} = \dfrac{1}{x}$ $\longleftarrow$ Ounce
Price $\longrightarrow$ $\phantom{\dfrac{4}{2.19}}$ $\longleftarrow$ Price

$$4x = 2.19$$

$$x = \frac{2.19}{4}$$

$$x \approx 0.55$$ The unit price is approximately $0.55.

For the 6-ounce tube:

Ounces $\longrightarrow$ $\dfrac{6}{2.79} = \dfrac{1}{x}$ $\longleftarrow$ Ounce
Price $\longrightarrow$ $\phantom{\dfrac{6}{2.79}}$ $\longleftarrow$ Price

$$6x = 2.79$$

$$x = \frac{2.79}{6}$$

$$x \approx 0.47$$ The unit price is approximately $0.47.

The price of 1 ounce of toothpaste from the 4-ounce tube is about 55¢. The price for 1 ounce of toothpaste from the 6-ounce tube is about 47¢. Since the 6-ounce tube has the lower unit price, it is the better buy.

Self Check 7   Which is the better buy: 3 pounds of hamburger for $6.89 or 5 pounds for $12.49?

Now Try   **Problem 89**

4   **Use Proportions to Solve Problems Involving Similar Triangles.**

If two angles of one triangle have the same measures as two angles of a second triangle, the triangles have the same shape. Triangles with the same shape, but not necessarily the same size, are called **similar triangles.** In the following figure, $\triangle ABC \sim \triangle DEF$. (Read the symbol $\sim$ as "is similar to.")

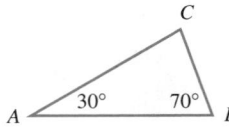 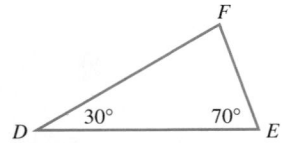

Property of Similar Triangles

If two triangles are **similar,** all pairs of corresponding sides are in proportion.

For the similar triangles previously shown, the following proportions are true.

$$\frac{AB}{DE} = \frac{BC}{EF}, \qquad \frac{BC}{EF} = \frac{CA}{FD}, \qquad \text{and} \qquad \frac{CA}{FD} = \frac{AB}{DE}$$ Read AB as "the length of segment AB."

EXAMPLE 8 **Finding the Height of a Tree.** A tree casts a shadow 18 feet long at the same time as a woman 5 feet tall casts a shadow 1.5 feet long. Find the height of the tree.

Analyze the Problem The figure shows the similar triangles determined by the tree and its shadow and the woman and her shadow. Since the triangles are similar, the lengths of their corresponding sides are in proportion. We can use this fact to find the height of the tree.

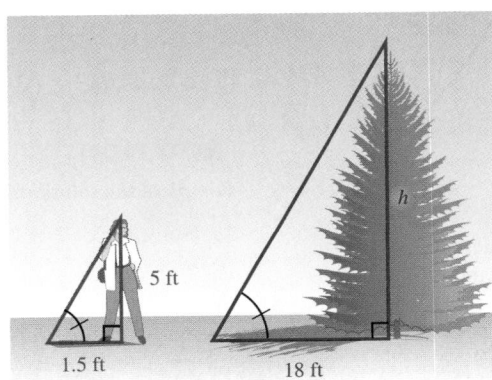

Each triangle has a right angle. Since the sun's rays strike the ground at the same angle, the angles highlighted with a tick mark have the same measure. Therefore, two angles of the smaller triangle have the same measures as two angles of the larger triangle; the triangles are similar.

Success Tip

Similar triangles do not have to be positioned the same. When they are placed differently, be careful to match their corresponding letters correctly. Here, $\triangle RST \sim \triangle MNO$.

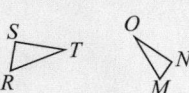

Form a Proportion If we let h = the height of the tree, we can find h by solving the following proportion.

$$\frac{h}{5} = \frac{18}{1.5} \qquad \frac{\text{Height of the tree}}{\text{Height of the woman}} = \frac{\text{Length of shadow of the tree}}{\text{Length of shadow of the woman}}$$

Solve the Proportion

$$1.5h = 5(18) \qquad \text{In a proportion, the product of the extremes equals the product of the means.}$$

$$1.5h = 90 \qquad \text{Multiply.}$$

$$\frac{1.5h}{1.5} = \frac{90}{1.5} \qquad \text{To undo the multiplication by 1.5 and isolate } h, \text{ divide both sides by 1.5.}$$

$$h = 60 \qquad \text{Do the decimal division, } 1.5\overline{)90}, \text{ to get 60.}$$

State the Conclusion The tree is 60 feet tall.

Check the Result $\frac{18}{1.5} = 12$ and $\frac{60}{5} = 12$. Since the ratios are the same, the result checks.

  Now Try **Problems 55 and 97**

ANSWERS TO SELF CHECKS **1. a.** $\frac{15}{2}$ **b.** $\frac{1}{4}$ **2.** No **3.** 24 **4.** $-5, 6$ **5.** $187.50 **7.** 3 lb for $6.89

STUDY SET
7.8

VOCABULARY

Fill in the blanks.

1. A _____ is the quotient of two numbers or the quotient of two quantities with the same units. A _____ is a quotient of two quantities that have different units.

2. A _____ is a mathematical statement that two ratios or two rates are equal.

3. In $\frac{50}{3} = \frac{x}{9}$, the terms 50 and 9 are called the _____ and the terms 3 and x are called the _____ of the proportion.

4. The _____ products for the proportion $\frac{5}{2} = \frac{6}{x}$ are $5x$ and 12.

5. Examples of _____ prices are $1.65 per gallon, 17¢ per day, and $50 per foot.

6. Two triangles with the same shape, but not necessarily the same size, are called _____ triangles.

CONCEPTS

7. Fill in the blanks: In a proportion, the product of the extremes is _____ to the product of the means. In symbols,

   If $\dfrac{a}{b} = \dfrac{c}{d}$, then $\boxed{} = \boxed{}$.

8. Is 45 a solution of $\frac{5}{3} = \frac{75}{x}$?

9. SNACKS   In a sample of 25 bags of potato chips, 2 were found to be underweight. Complete the following proportion that could be used to find the number of underweight bags that would be expected in a shipment of 1,000 bags of potato chips.

   Number of bags $\longrightarrow$ $\dfrac{\boxed{}}{\boxed{}} = \dfrac{\boxed{}}{\boxed{}}$ $\longleftarrow$ Number of bags
   Number underweight $\longrightarrow$ $\quad\quad\quad\quad$ $\longleftarrow$ Number underweight

10. MINIATURES   A model of the Seattle Space Needle is to be made using a scale of 2 inches to 35 feet. Complete the following proportion to determine the height h of the model.

    $\dfrac{2}{35} = \dfrac{\boxed{}}{\boxed{}}$

605 ft

11. KLEENEX   Complete the following proportion that can be used to find the unit price of facial tissue if a box of 85 tissues sells for $2.19.

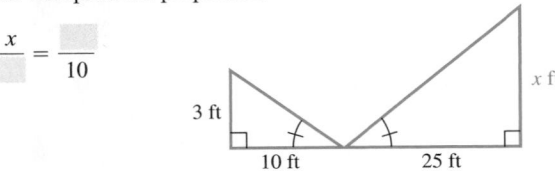

Price $\longrightarrow$ $\dfrac{\boxed{}}{85} = \dfrac{x}{\boxed{}}$ $\longleftarrow$ Price
Number of sheets $\longrightarrow$ $\quad\quad\quad$ $\longleftarrow$ Number of sheets

12. The two triangles shown in the following illustration are similar. Complete the proportion.

    $\dfrac{x}{\boxed{}} = \dfrac{\boxed{}}{10}$

    3 ft   10 ft   25 ft x ft

NOTATION

Complete the solution.

13. Solve for x: $\dfrac{12}{18} = \dfrac{x}{24}$

    $12 \cdot 24 = 18 \cdot \boxed{}$

    $\boxed{} = 18x$

    $\dfrac{288}{\boxed{}} = \dfrac{18x}{\boxed{}}$

    $16 = x$

14. Write the ratio of 25 to 4 in two other forms.

15. Fill in the blanks: The proportion $\dfrac{20}{1.6} = \dfrac{100}{8}$ can be read: 20 is to 1.6 ____ 100 is ____ 8.

16. Fill in the blank: We read $\triangle XYZ \sim \triangle MNO$ as: triangle XYZ is _____ to triangle MNO.

GUIDED PRACTICE

Translate each ratio into a fraction in simplest form. **See Example 1.**

17. 4 boxes to 15 boxes
18. 2 miles to 9 miles
19. 18 watts to 24 watts
20. 11 cans to 121 cans
21. 30 days to 24 days
22. 45 people to 30 people
23. 90 minutes to 3 hours
24. 20 inches to 2 feet
25. 8 quarts to 4 gallons
26. 6 feet to 12 yards
27. 6,000 feet to 1 mile (*Hint:* 1 mi = 5,280 ft)
28. 5 tons to 4,000 pounds (*Hint:* 1 ton = 2,000 lb)

Determine whether each equation is a true proportion. **See Example 2.**

29. $\dfrac{7}{3} = \dfrac{14}{6}$
30. $\dfrac{7}{16} = \dfrac{3}{7}$
31. $\dfrac{5}{8} = \dfrac{12}{19.4}$
32. $\dfrac{9}{32} = \dfrac{4.5}{16}$

Solve each proportion. See Example 3.

33. $\dfrac{2}{3} = \dfrac{x}{6}$

34. $\dfrac{3}{6} = \dfrac{x}{8}$

35. $\dfrac{63}{g} = \dfrac{9}{2}$

36. $\dfrac{27}{x} = \dfrac{9}{4}$

37. $\dfrac{x+1}{5} = \dfrac{3}{15}$

38. $\dfrac{x-1}{7} = \dfrac{2}{21}$

39. $\dfrac{5-x}{17} = \dfrac{13}{34}$

40. $\dfrac{4-x}{13} = \dfrac{11}{26}$

41. $\dfrac{15}{7b+5} = \dfrac{5}{2b+1}$

42. $\dfrac{8}{3n+6} = \dfrac{16}{3n-3}$

43. $\dfrac{8x}{3} = \dfrac{11x+9}{4}$

44. $\dfrac{3x}{16} = \dfrac{x+2}{5}$

Solve each proportion. See Example 4.

45. $\dfrac{2}{3x} = \dfrac{x}{6}$

46. $\dfrac{y}{4} = \dfrac{4}{y}$

47. $\dfrac{b-5}{3} = \dfrac{2}{b}$

48. $\dfrac{2}{q} = \dfrac{q-3}{2}$

49. $\dfrac{a-4}{a} = \dfrac{15}{a+4}$

50. $\dfrac{s}{s-5} = \dfrac{s+5}{24}$

51. $\dfrac{t+3}{t+5} = \dfrac{-1}{2t}$

52. $\dfrac{5h}{14h+3} = \dfrac{1}{h}$

Each pair of triangles is similar. Find the missing side length. See Example 8.

53.

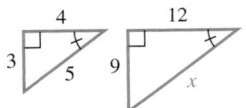

54.

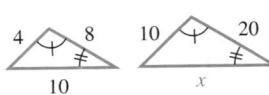

55.

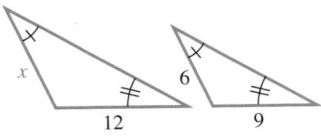

56.

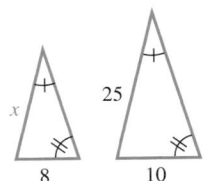

TRY IT YOURSELF

Solve each proportion.

57. $\dfrac{x-1}{x+1} = \dfrac{2}{3x}$

58. $\dfrac{2}{x+6} = \dfrac{-2x}{5}$

59. $\dfrac{x+1}{4} = \dfrac{3x}{8}$

60. $\dfrac{x-1}{9} = \dfrac{2x}{3}$

61. $\dfrac{y-4}{y+1} = \dfrac{y+3}{y+6}$

62. $\dfrac{r-6}{r-8} = \dfrac{r+1}{r-4}$

63. $\dfrac{c}{10} = \dfrac{10}{c}$

64. $\dfrac{-6}{r} = \dfrac{r}{-6}$

65. $\dfrac{m}{3} = \dfrac{4}{m+1}$

66. $\dfrac{n}{2} = \dfrac{5}{n+3}$

67. $\dfrac{3}{3b+4} = \dfrac{2}{5b-6}$

68. $\dfrac{2}{4d-1} = \dfrac{3}{2d+1}$

APPLICATIONS

69. GEAR RATIOS   Write each ratio in two ways: as a fraction in simplest form and using a colon.

a. The number of teeth of the larger gear to the number of teeth of the smaller gear

b. The number of teeth of the smaller gear to the number of teeth of the larger gear

70. FACULTY–STUDENT RATIOS   At a college, there are 300 faculty members and 2,850 students. Find the rate of faculty to students. (This is often referred to as the faculty-to-student ratio, even though the units are different.)

71. SHOPPING FOR CLOTHES   If shirts are on sale at two for $25, how much do five shirts cost?

72. COMPUTING A PAYCHECK   Billie earns $412 for a 40-hour week. If she missed 10 hours of work last week, how much did she get paid?

73. COOKING   A recipe for spaghetti sauce requires four 16-ounce bottles of ketchup to make 2 gallons of sauce. How many bottles of ketchup are needed to make 10 gallons of sauce?

74. MIXING PERFUME   A perfume is to be mixed in the ratio of 3 drops of pure essence to 7 drops of alcohol. How many drops of pure essence should be mixed with 56 drops of alcohol?

75. CPR   A first aid handbook states that when performing cardiopulmonary resuscitation on an adult, the ratio of chest compressions to breaths should be 5:2. If 210 compressions were administered to an adult patient, how many breaths should have been given?

76. COOKING   A recipe for wild rice soup follows. Find the amounts of chicken broth, rice, and flour needed to make 15 servings.

Wild Rice Soup	
A sumptuous side dish with a nutty flavor	
3 cups chicken broth	1 cup light cream
$\frac{2}{3}$ cup uncooked rice	2 tablespoons flour
$\frac{1}{4}$ cup sliced onions	$\frac{1}{8}$ teaspoon pepper
$\frac{1}{2}$ cup shredded carrots	Serves: 6

77. **NUTRITION**   The table shows the nutritional facts about a 10-oz chocolate milkshake sold by a fast-food restaurant. Use the information to complete the table for the 16-oz shake. Round to the nearest unit when an answer is not exact.

	Calories	Fat (gm)	Protein (gm)
10-oz chocolate milkshake	355	8	9
16-oz chocolate milkshake			

78. **ENGINEERING**   A portion of a bridge is shown. Use the fact that $\frac{AB}{BC}$ is in proportion to $\frac{FE}{ED}$ to find *FE*.

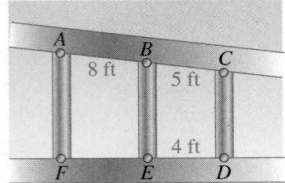

79. **QUALITY CONTROL**   Out of a sample of 500 men's shirts, 17 were rejected because of crooked collars. How many crooked collars would you expect to find in a run of 15,000 shirts?

80. **PHOTO ENLARGEMENTS**   The 3-by-5 photo is to be blown up to the larger size. Find *x*.

81. **MIXING FUEL**   The instructions on a can of oil intended to be added to lawnmower gasoline are shown below. Are these instructions correct?   (*Hint:* There are 128 ounces in 1 gallon.)

Recommended	Gasoline	Oil
50 to 1	6 gal	16 oz

82. **DRIVER'S LICENSES**   Of the 50 states, Alabama has one of the highest ratios of licensed drivers to residents. If the ratio is 399:500 and the population of Alabama is about 4,500,000, how many residents of that state have a driver's license?

83. **CAPTURE–RELEASE METHOD**   To estimate the ground squirrel population on his acreage, a farmer trapped, tagged, and then released a dozen squirrels. Two weeks later, the farmer trapped 35 squirrels and noted that 3 were tagged. Use this information to estimate the number of ground squirrels on his acreage.

84. **CONCRETE**   A 2:3 concrete mix means that for every two parts of sand, three parts of gravel are used. How much sand should be used in a mix composed of 25 cubic feet of gravel?

85. **MODEL RAILROADS**   An HO scale model railroad engine is 6 inches long. If the HO scale is 1 to 87, how long is a real engine, in inches? In feet?

86. **MODEL RAILROADS**   An N scale model railroad caboose is 4.5 inches long. If the N scale is 1 to 160, how long is a real caboose, in inches? In feet?

87. **BLUEPRINTS**   The scale for the drawing shown means that a $\frac{1}{4}$-inch length $\left(\frac{1}{4}''\right)$ on the drawing corresponds to an actual size of 1 foot ($1'$-$0''$). Suppose the length of the kitchen is $2\frac{1}{2}$ inches on the drawing. How long is the actual kitchen?

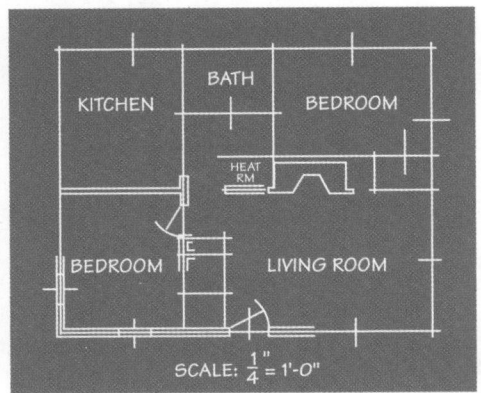

88. **THE TITANIC**   A 1:144 scale model of the *Titanic* is to be built. If the ship was 882 feet long, find the length of the model.

For each of the following purchases, determine the better buy. **See Example 7.**

89. Trumpet lessons: 45 minutes for $25 or 60 minutes for $35

90. Memory for a computer: 128 megabytes for $26 or 512 megabytes for $110

91. Business cards: 100 for $9.99 or 150 for $12.99

92. Dog food: 20 pounds for $7.49 or 44 pounds for $14.99

93. Soft drinks: 6-pack for $1.50 or a case (24 cans) for $6.25

94. Donuts: A dozen for $6.24 or a baker's dozen (13) for $6.65

95.

96.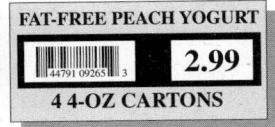

FAT-FREE PEACH YOGURT **4.79** 6 4-OZ CARTONS
FAT-FREE PEACH YOGURT **2.99** 4 4-OZ CARTONS

97. HEIGHT OF A TREE A tree casts a shadow of 26 feet at the same time as a 6-foot man casts a shadow of 4 feet. Find the height of the tree.

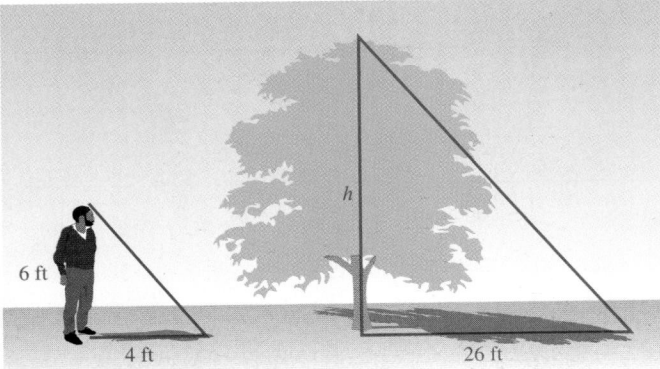

98. HEIGHT OF A BUILDING A man places a mirror on the ground and sees the reflection of the top of a building, as shown. The two triangles in the illustration are similar. Find the height, h, of the building.

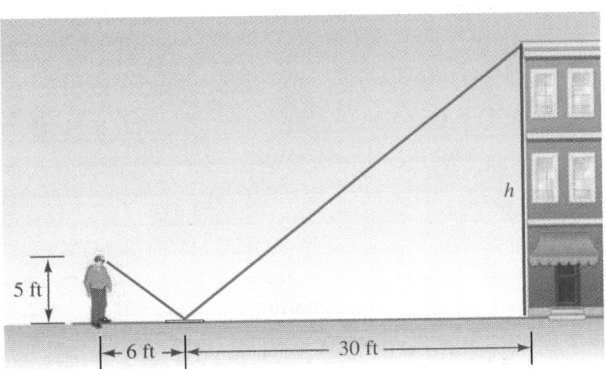

99. SURVEYING To find the width of a river, a surveyor laid out the following similar triangles. Find w.

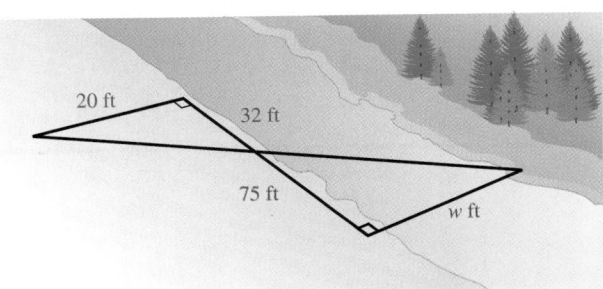

100. FLIGHT PATHS An airplane ascends 100 feet as it flies a horizontal distance of 1,000 feet. How much altitude will it gain as it flies a horizontal distance of 1 mile? (*Hint:* 5,280 feet=1 mile.)

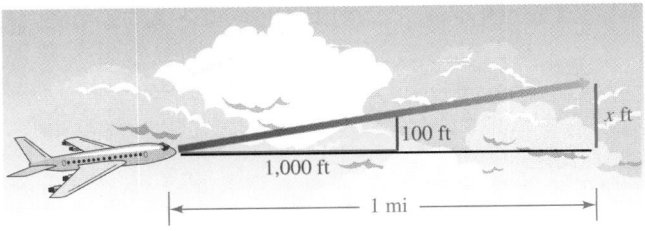

WRITING

101. Explain how to solve the equation $\frac{7}{6} = \frac{2}{x}$ and how to simplify the expression $\frac{7}{6} \cdot \frac{2}{x}$.

102. Explain why the concept of cross products cannot immediately be used to solve the equation:

$$\frac{x}{3} - \frac{3x}{4} = \frac{1}{12}$$

103. What are similar triangles?

104. What is a unit price? Give an example.

REVIEW

105. Change $\frac{9}{10}$ to a percent.

106. Change $33\frac{1}{3}\%$ to a fraction.

107. Find 30% of 1,600.

108. SHOPPING Maria bought a dress for 25% off the original price of $98. How much did the dress cost?

CHALLENGE PROBLEMS

109. Suppose $\frac{a}{b} = \frac{c}{d}$. Write three other proportions using $a, b, c,$ and d.

110. Verify that $\frac{3}{5} = \frac{12}{20} = \frac{3 + 12}{5 + 20}$. Is the following rule always true? Explain.

$$\frac{a}{b} = \frac{c}{d} = \frac{a + c}{b + d}$$

CHAPTER 7
Summary & Review

SECTION 7.1 Simplifying Rational Expressions

DEFINITIONS AND CONCEPTS	EXAMPLES
A **rational expression** is an expression of the form $\frac{A}{B}$, where A and B are polynomials and B does not equal 0.	Rational expressions: $\frac{8}{7t}$, $\frac{a}{a-3}$, and $\frac{4x^2-16x}{x^2-6x+8}$
To **evaluate a rational expression,** we substitute the values of its variables and simplify.	Evaluate $\frac{3x+1}{x-2}$ for $x = 5$. $$\frac{3x+1}{x-2} = \frac{3(5)+1}{5-2} = \frac{16}{3} \quad \text{Substitute 5 for } x.$$
To find the real numbers for which a **rational expression is undefined,** find the values of the variable that make the denominator 0.	For which real numbers is $\frac{11}{2x-3}$ undefined? $$2x - 3 = 0 \quad \text{Set the denominator equal to 0 and solve for } x.$$ $$2x = 3$$ $$x = \frac{3}{2} \quad \text{The expression is undefined for } x = \frac{3}{2}.$$
To **simplify a rational expression:** 1. Factor the numerator and the denominator completely. 2. Remove factors equal to 1. 3. Multiply the remaining factors in the numerator and denominator and simplify, if possible.	Simplify: $\dfrac{x^2-4}{x^2-7x+10} = \dfrac{(x+2)\overset{1}{\cancel{(x-2)}}}{(x-5)\underset{1}{\cancel{(x-2)}}}$ Factor and simplify. $$= \frac{x+2}{x-5}$$
The quotient of any nonzero expression and its **opposite** is -1.	$\dfrac{2t-3}{3-2t} = -1$ Because $2t-3$ and $3-2t$ are opposites.

REVIEW EXERCISES

1. Find the values of x for which the rational expression $\frac{x-1}{x^2-16}$ is undefined.

2. Evaluate $\frac{x^2-1}{x-5}$ for $x = -2$.

Simplify each rational expression, if possible. Assume that no denominators are zero.

3. $\dfrac{3x^2}{6x^3}$

4. $\dfrac{5xy^2}{2x^2y^2}$

5. $\dfrac{x^2}{x^2+x}$

6. $\dfrac{a^2-4}{a+2}$

7. $\dfrac{3p-2}{2-3p}$

8. $\dfrac{8-x}{x^2-5x-24}$

9. $\dfrac{2x^2-16x}{2x^2-18x+16}$

10. $\dfrac{x^2+x-2}{x^2-x-2}$

11. $\dfrac{x^2-2xy+y^2}{(x-y)^3}$

12. $\dfrac{4(t+3)+8}{3(t+3)+6}$

13. Explain the error in the following work: $\dfrac{x+1}{x} = \dfrac{\cancel{x}+1}{\cancel{x}} = \dfrac{2}{1} = 2.$

14. DOSAGES   Cowling's rule is a formula that can be used to determine the dosage of a prescription medication for children. If C is the proper child's dosage, D is an adult dosage, and A is the child's age in years, then $C = \frac{D(A+1)}{24}$. Find the daily dosage of an antibiotic for an 11-year-old child if the adult daily dosage is 300 milligrams.

SECTION 7.2   Multiplying and Dividing Rational Expressions

DEFINITIONS AND CONCEPTS	EXAMPLES
To **multiply rational expressions,** multiply their numerators and multiply their denominators. $$\frac{A}{B} \cdot \frac{C}{D} = \frac{AC}{BD}$$ Then simplify, if possible.	Multiply: $\dfrac{4b}{b+2} \cdot \dfrac{7}{b} = \dfrac{4b \cdot 7}{(b+2)b}$ $= \dfrac{4\overset{1}{\cancel{b}} \cdot 7}{(b+2)\underset{1}{\cancel{b}}}$ Simplify. $= \dfrac{28}{b+2}$
To find the **reciprocal** of a rational expression, invert its numerator and denominator.	The reciprocal of $\dfrac{c}{c+7}$ is $\dfrac{c+7}{c}$.
To **divide rational expressions,** multiply the first expression by the reciprocal of the second. $$\frac{A}{B} \div \frac{C}{D} = \frac{A}{B} \cdot \frac{D}{C} = \frac{AD}{BC}$$ Then simplify, if possible.	Divide: $\dfrac{t}{t+1} \div \dfrac{8}{t^2+t} = \dfrac{t}{t+1} \cdot \dfrac{t^2+t}{8}$ $= \dfrac{t \cdot t\overset{1}{(\cancel{t+1})}}{\underset{1}{(\cancel{t+1})}8}$ Factor and simplify. $= \dfrac{t^2}{8}$
A **unit conversion factor** is a fraction that has a value of 1.	$\dfrac{1 \text{ yd}^2}{9 \text{ ft}^2} = 1$ and $\dfrac{1 \text{ mi}}{5{,}280 \text{ ft}} = 1$

REVIEW EXERCISES

Multiply and simplify, if possible.

15. $\dfrac{3xy}{2x} \cdot \dfrac{4x}{2y^2}$

16. $56x\left(\dfrac{12}{7x}\right)$

17. $\dfrac{x^2-1}{x^2+2x} \cdot \dfrac{x}{x+1}$

18. $\dfrac{x^2+x}{3x-15} \cdot \dfrac{6x-30}{x^2+2x+1}$

Divide and simplify, if possible.

19. $\dfrac{3x^2}{5x^2y} \div \dfrac{6x}{15xy^2}$

20. $\dfrac{x^2-x-6}{1-2x} \div \dfrac{x^2-2x-3}{2x^2+x-1}$

21. Determine whether the given fraction is a unit conversion factor.

   a. $\dfrac{1 \text{ ft}}{12 \text{ in.}}$

   b. $\dfrac{60 \text{ min}}{1 \text{ day}}$

   c. $\dfrac{2{,}000 \text{ lb}}{1 \text{ ton}}$

   d. $\dfrac{1 \text{ gal}}{4 \text{ qt}}$

22. TRAFFIC SIGNS Convert the speed limit on the sign from miles per hour to miles per minute.

SPEED LIMIT 20 mph

SECTION 7.3   Adding and Subtracting with Like Denominators; Least Common Denominators

DEFINITIONS AND CONCEPTS	EXAMPLES
To **add (or subtract) rational expressions** that have the same denominator, add (or subtract) their numerators and write the sum (or difference) over their common denominator. $$\frac{A}{D} + \frac{B}{D} = \frac{A+B}{D} \qquad \frac{A}{D} - \frac{B}{D} = \frac{A-B}{D}$$	Add: $\dfrac{2b}{3b-9} + \dfrac{b}{3b-9} = \dfrac{2b+b}{3b-9}$ $= \dfrac{\overset{1}{\cancel{3}}b}{\underset{1}{\cancel{3}}(b-3)}$ Factor and simplify. $= \dfrac{b}{b-3}$

SECTION 7.3 *—continued*

DEFINITIONS AND CONCEPTS	EXAMPLES
Then simplify, if possible.	Subtract: $\dfrac{x+1}{x} - \dfrac{x-1}{x} = \dfrac{x+1-(x-1)}{x}$ *Don't forget the parentheses.* $= \dfrac{x+1-x+1}{x}$ $= \dfrac{2}{x}$ *Combine like terms.*
To find the **LCD** of several fractions, factor each denominator completely. Form a product using each different factor the greatest number of times it appears in any one factorization.	Find the LCD of $\dfrac{3}{x^3-x^2}$ and $\dfrac{x}{x^2-1}$. $\left.\begin{array}{l} x^3 - x^2 = x \cdot x \cdot (x-1) \\ x^2 - 1 = (x+1)(x-1) \end{array}\right\}$ LCD $= x \cdot x \cdot (x-1)(x+1)$
To **build an equivalent rational expression,** multiply the given expression by 1 in the form of $\dfrac{c}{c}$ where $c \neq 0$.	$\dfrac{7}{4t} = \dfrac{7}{4t} \cdot \dfrac{3t}{3t}$ and $\dfrac{x+1}{x-7} = \dfrac{x+1}{x-7} \cdot \dfrac{x-1}{x-1}$ $= \dfrac{21t}{12t^2}$ $\qquad\qquad = \dfrac{(x+1)(x-1)}{(x-7)(x-1)}$ $\qquad\qquad\qquad\qquad = \dfrac{x^2-1}{x^2-8x+7}$

REVIEW EXERCISES

Add or subtract and simplify, if possible.

23. $\dfrac{13}{15d} - \dfrac{8}{15d}$

24. $\dfrac{x}{x+y} + \dfrac{y}{x+y}$

25. $\dfrac{3x}{x-7} - \dfrac{x-2}{x-7}$

26. $\dfrac{a}{a^2-2a-8} + \dfrac{2}{a^2-2a-8}$

Find the LCD of each pair of rational expressions.

27. $\dfrac{12}{x}, \dfrac{1}{9}$

28. $\dfrac{1}{2x^3}, \dfrac{5}{8x}$

29. $\dfrac{7}{m}, \dfrac{m+2}{m-8}$

30. $\dfrac{x}{5x+1}, \dfrac{5x}{5x-1}$

31. $\dfrac{6-a}{a^2-25}, \dfrac{a^2}{a-5}$

32. $\dfrac{4t+25}{t^2+10t+25}, \dfrac{t^2-7}{2t^2+17t+35}$

Build each rational expression into an equivalent fraction having the denominator shown in red.

33. $\dfrac{9}{a}, 7a$

34. $\dfrac{2y+1}{x-9}, x(x-9)$

35. $\dfrac{b+7}{3b-15}, 6(b-5)$

36. $\dfrac{9r}{r^2+6r+5}, (r+1)(r-4)(r+5)$

SECTION 7.4   Adding and Subtracting with Unlike Denominators

DEFINITIONS AND CONCEPTS	EXAMPLES

DEFINITIONS AND CONCEPTS

To **add (or subtract) rational expressions** with unlike denominators:

1. Find the LCD.

2. Write each rational expression as an equivalent expression whose denominator is the LCD.

3. Add (or subtract) the numerators and write the sum (or difference) over the LCD.

4. Simplify the resulting rational expression if possible.

EXAMPLES

Add: $\dfrac{4x}{x} + \dfrac{2}{x-1} = \dfrac{4x}{x} \cdot \dfrac{x-1}{x-1} + \dfrac{2}{x-1} \cdot \dfrac{x}{x}$ The LCD is $x(x-1)$.

$= \dfrac{4x(x-1)}{x(x-1)} + \dfrac{2x}{x(x-1)}$

$= \dfrac{4x^2 - 4x + 2x}{x(x-1)}$ Distribute the multiplication by $4x$.

$= \dfrac{4x^2 - 2x}{x(x-1)}$ Combine like terms.

$= \dfrac{2\overset{1}{\cancel{x}}(2x-1)}{\underset{1}{\cancel{x}}(x-1)}$ Factor and simplify.

$= \dfrac{2(2x-1)}{x-1}$

When a polynomial is multiplied by -1, the result is its opposite. This fact is used when adding or subtracting rational expressions whose **denominators are opposites.**

Add: $\dfrac{c}{c-4} + \dfrac{1}{4-c} = \dfrac{c}{c-4} + \dfrac{1}{4-c} \cdot \dfrac{-1}{-1}$

$= \dfrac{c}{c-4} + \dfrac{-1}{c-4}$ $-1(4-c) = c-4$

$= \dfrac{c-1}{c-4}$

REVIEW EXERCISES

Add or subtract and simplify, if possible.

37. $\dfrac{1}{7} - \dfrac{1}{a}$

38. $\dfrac{x}{x-1} + \dfrac{1}{x}$

39. $\dfrac{2t+2}{t^2 + 2t + 1} - \dfrac{1}{t+1}$

40. $\dfrac{x+2}{2x} - \dfrac{2-x}{x^2}$

41. $\dfrac{6}{b-1} - \dfrac{b}{1-b}$

42. $\dfrac{8}{c} + 6$

43. $\dfrac{n+7}{n+3} - \dfrac{n-3}{n+7}$

44. $\dfrac{4}{t+2} - \dfrac{7}{(t+2)^2}$

45. $\dfrac{6}{a^2-9} - \dfrac{5}{a^2-a-6}$

46. $\dfrac{2}{3y-6} + \dfrac{3}{4y+8}$

47. Working on a homework assignment, a student added two rational expressions and obtained $\dfrac{-5n^3 - 7}{3n(n+6)}$. The answer given in the back of the book was $-\dfrac{5n^3 + 7}{3n(n+6)}$. Are the answers equivalent?

48. DIGITAL VIDEO CAMERAS   Find the perimeter and the area of the LED screen of the camera.

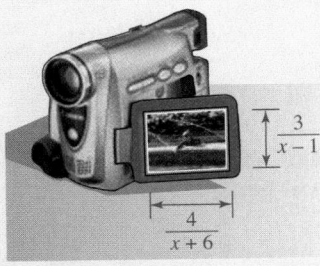

$\dfrac{3}{x-1}$

$\dfrac{4}{x+6}$

SECTION 7.5   Simplifying Complex Fractions

DEFINITIONS AND CONCEPTS	EXAMPLES
Complex fractions contain fractions in their numerators and/or their denominators.	Complex fractions: $\dfrac{\frac{2}{t}}{\frac{5}{4t}}$ and $\dfrac{\frac{3}{m}+\frac{m}{4}}{\frac{m}{2}}$
To **simplify a complex fraction:** **Method 1** Write the numerator and the denominator as single rational expressions and perform the indicated division.	Simplify: $\dfrac{\frac{3}{m}+\frac{m}{2}}{\frac{m}{4}}=\dfrac{\frac{3}{m}\cdot\frac{2}{2}+\frac{m}{2}\cdot\frac{m}{m}}{\frac{m}{4}}$ In the numerator, build to have an LCD of 2m. $=\dfrac{\frac{6}{2m}+\frac{m^2}{2m}}{\frac{m}{4}}$ The main fraction bar indicates division. $=\dfrac{\frac{6+m^2}{2m}}{\frac{m}{4}}$ Add the fractions in the numerator. $=\dfrac{(6+m^2)\cdot\overset{1}{\cancel{2}}\cdot 2}{\underset{1}{\cancel{2}m\cdot m}}$ Multiply by the reciprocal of $\frac{m}{4}$. Factor and simplify. $=\dfrac{12+2m^2}{m^2}$ Distribute the multiplication by 2.
Method 2 Determine the LCD of all the rational expressions in the complex fraction and multiply the complex fraction by 1, written in the form $\frac{\text{LCD}}{\text{LCD}}$.	Simplify: $\dfrac{\frac{3}{m}+\frac{m}{2}}{\frac{m}{4}}=\dfrac{\frac{3}{m}+\frac{m}{2}}{\frac{m}{4}}\cdot\dfrac{4m}{4m}$ The LCD for all the rational expressions is 4m. $=\dfrac{\frac{3}{m}\cdot 4m+\frac{m}{2}\cdot 4m}{\frac{m}{4}\cdot 4m}$ In the numerator, distribute the multiplication by 4m. $=\dfrac{12+2m^2}{m^2}$ Perform each multiplication by 4m.

REVIEW EXERCISES
Simplify each complex fraction.

49. $\dfrac{\frac{n^4}{30}}{\frac{7n}{15}}$

50. $\dfrac{\frac{r^2-81}{18s^2}}{\frac{4r-36}{9s}}$

51. $\dfrac{\frac{1}{y}+1}{\frac{1}{y}-1}$

52. $\dfrac{\frac{7}{a^2}}{\frac{1}{a}+\frac{10}{3}}$

53. $\dfrac{\frac{2}{x-1}+\frac{x-1}{x+1}}{\frac{1}{x^2-1}}$

54. $\dfrac{\frac{1}{x^2y}-\frac{5}{xy}}{\frac{3}{xy}-\frac{7}{xy^2}}$

SECTION 7.6   Solving Rational Equations

DEFINITIONS AND CONCEPTS	EXAMPLES

To **solve a rational equation** we use the multiplication property of equality to clear the equation of fractions. Use these steps:

1. Determine which numbers cannot be solutions.

2. Multiply both sides of the equation by the LCD of the rational expressions contained in the equation.

3. Solve the resulting equation.

4. Check all possible solutions in the *original* equation. A possible solution that does not satisfy the original equation is called an **extraneous solution.**

Solve:

$$\frac{y}{y-2} - 1 = \frac{1}{y}$$ Since no denominators can be 0, $y \neq 2$ and $y \neq 0$.

$$y(y-2)\left(\frac{y}{y-2} - 1\right) = y(y-2)\left(\frac{1}{y}\right)$$ The LCD is $y(y-2)$.

$$y(y-2)\left(\frac{y}{y-2}\right) - y(y-2)1 = y(y-2)\left(\frac{1}{y}\right)$$ Simplify.

$$y \cdot y - y(y-2) = (y-2) \cdot 1$$

$$y^2 - y^2 + 2y = y - 2$$

$$2y = y - 2$$ Combine like terms.

$$y = -2$$

REVIEW EXERCISES

Solve each equation and check the result. If an equation has no solution, so indicate.

55. $\dfrac{3}{x} = \dfrac{2}{x-1}$

56. $\dfrac{a}{a-5} = 3 + \dfrac{5}{a-5}$

57. $\dfrac{2}{3t} + \dfrac{1}{t} = \dfrac{5}{9}$

58. $a = \dfrac{3a-50}{4a-24} - \dfrac{3}{4}$

59. $\dfrac{4}{x+2} - \dfrac{3}{x+3} = \dfrac{6}{x^2+5x+6}$

60. $\dfrac{3}{x+1} - \dfrac{x-2}{2} = \dfrac{x-2}{x+1}$

61. ENGINEERING   The efficiency E of a Carnot engine is given by the following formula. Solve it for T_1.

$$E = 1 - \frac{T_2}{T_1}$$

62. Solve for y: $\dfrac{1}{x} = \dfrac{1}{y} + \dfrac{1}{z}$

SECTION 7.7   Problem Solving Using Rational Equations

DEFINITIONS AND CONCEPTS	EXAMPLES

To solve application problems, follow these steps:

1. Analyze the problem.

2. Form an equation.

3. Solve the equation.

4. State the conclusion.

5. Check the result.

Rate of Work: If a job can be completed in t units of time, the rate of work can be expressed as $\frac{1}{t}$ of the job is completed per unit of time.

Shared work problems:
Work completed = rate of work · time worked

WASHING CARS   Working alone, Carlos can wash the family SUV in 30 minutes. Victor, his brother, can wash the same SUV in 20 minutes working alone. How long will it take them if they wash the SUV together?

Analyze the Problem   Let $x =$ the number of minutes it will take Carlos and Victor, working together, to wash the SUV. Enter the data in a table.

	Rate	· Time	= Work Completed
Carlos	$\frac{1}{30}$	x	$\frac{x}{30}$
Victor	$\frac{1}{20}$	x	$\frac{x}{20}$

Form an equation   The part of the job done by Carlos plus the part of the job done by Victor equals 1 job completed.

$$\frac{x}{30} + \frac{x}{20} = 1$$

SECTION 7.7   Problem Solving Using Rational Equations—continued

DEFINITIONS AND CONCEPTS	EXAMPLES
	Solve the equation

$$60\left(\frac{x}{30}+\frac{x}{20}\right)=60(1) \quad \text{Multiply both sides by the LCD, 60.}$$

$$60\left(\frac{x}{30}\right)+60\left(\frac{x}{20}\right)=60(1) \quad \text{On the left side, distribute the multiplication by 60.}$$

$$2x+3x=60 \quad \text{Perform each multiplication by 60.}$$

$$5x=60 \quad \text{Combine like terms.}$$

$$x=\frac{60}{5} \quad \text{Divide both sides by 5.}$$

$$x=12$$

State the Conclusion   Working together, it will take Carlos and Victor 12 minutes to wash the family SUV.

Check the Result   In 12 minutes, Carlos will do $\frac{12}{30}=\frac{24}{60}$ of the job and Victor will do $\frac{12}{20}=\frac{36}{60}$ of the job. Together they will do $\frac{24}{60}+\frac{36}{60}=\frac{60}{60}$ or 1 whole job. The result checks.

Uniform motion problems: Time $=\dfrac{\text{distance}}{\text{rate}}$

See Example 2 in Section 7.7.

Investment problems: Principal $=\dfrac{\text{Interest}}{\text{rate}\cdot\text{time}}$

See Example 5 in Section 7.7.

REVIEW EXERCISES

63. NUMBER PROBLEM   If a number is subtracted from the denominator of $\frac{4}{5}$ and twice as much is added to the numerator, the result is 5. Find the number.

64. EXERCISE   A woman can bicycle 30 miles in the same time that it takes her to jog 10 miles. If she can ride 10 mph faster than she can jog, how fast can she jog?

65. HOUSE CLEANING   A maid can clean a house in 4 hours. What is her rate of work?

66. HOUSE PAINTING   If a homeowner can paint a house in 14 days and a professional painter can paint it in 10 days, how long will it take if they work together?

67. INVESTMENTS   In 1 year, a student earned $100 interest on money she deposited at a savings and loan. She later learned that the money would have earned $120 if she had deposited it at a credit union, because the credit union paid 1% more interest at the time. Find the rate she received from the savings and loan.

68. WIND SPEED   A plane flies 400 miles downwind in the same amount of time as it takes to travel 320 miles upwind. If the plane can fly at 360 mph in still air, find the velocity of the wind.

SECTION 7.8   Proportions and Similar Triangles

DEFINITIONS AND CONCEPTS	EXAMPLES
A **ratio** is the quotient of two numbers with the same units.	Ratios: $\frac{2}{3}$, $\frac{1}{50}$, and 2:3
A **rate** is the quotient of two quantities with different units.	Rates: $\frac{4\text{ oz}}{6\text{ lb}}$, $\frac{525\text{ mi}}{15\text{ hr}}$, and $\frac{\$1.95}{2\text{ lb}}$

SECTION 7.8   Proportions and Similar Triangles—*continued*

DEFINITIONS AND CONCEPTS	EXAMPLES
A **proportion** is a statement that two ratios or two rates are equal. In the proportion $\frac{a}{b} = \frac{c}{d}$, a and d are the **extremes** and b and c are the **means.** In any proportion, the product of the extremes is equal to the product of the means. (The **cross products** are equal.) To **solve a proportion,** set the product of the extremes equal to the product of the means and solve the resulting equation.	A proportion: $\frac{4}{9} = \frac{28}{63}$ Extremes: 4 and 63 Means: 9 and 28 A proportion: $\frac{4}{9} = \frac{28}{63}$ Cross product: $4 \cdot 63 = 252$ Cross product: $9 \cdot 28 = 252$ Solve the proportion: $\frac{3}{2} = \frac{x}{10}$ $3 \cdot 10 = 2 \cdot x$ *Set the cross products equal.* $30 = 2x$ $15 = x$ *Solve for x.*
Triangles with the same shape but not necessarily the same size are called **similar triangles.** The lengths of the corresponding sides of two similar triangles are in proportion.	In these similar triangles: $\frac{a}{d} = \frac{b}{e} = \frac{c}{f}$ 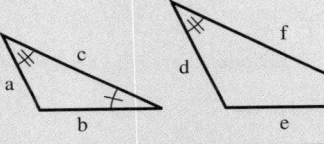
A **scale** is a ratio (or rate) that compares the size of a model to the size of an actual object. **Unit prices** can be used to compare costs of different sizes of the same brand to determine the best buy.	See Example 6 in Section 7.8. For the same item, a cost of $\frac{\$1.95}{1\ \text{lb}}$ is a better buy than a cost of $\frac{\$1.99}{1\ \text{lb}}$. See Example 7 in Section 7.8.

REVIEW EXERCISES

Determine whether each equation is a true proportion.

69. $\dfrac{4}{7} = \dfrac{20}{34}$

70. $\dfrac{5}{7} = \dfrac{30}{42}$

Solve each proportion.

71. $\dfrac{3}{x} = \dfrac{6}{9}$

72. $\dfrac{x}{3} = \dfrac{x}{5}$

73. $\dfrac{x-2}{5} = \dfrac{x}{7}$

74. $\dfrac{2x}{x+4} = \dfrac{3}{x-1}$

75. DENTISTRY   The diagram below was displayed in a dentist's office. According to the diagram, if the dentist has 340 adult patients, how many will develop gum disease?

3 out of 4 adults will develop gum disease.

76. UTILITY POLES   A telephone pole casts a shadow 12 feet long at the same time that a man 6 feet tall casts a shadow of 3.6 feet. How tall is the pole?

77. PORCELAIN FIGURINES   A model of a flutist, standing and playing at a music stand, was made using a 1/12th scale. If the scale model is 5.5 inches tall, how tall is the flutist?

78. COMPARISON SHOPPING   Which is the better buy for recordable compact discs: 150 for $60 or 250 for $98?

CHAPTER 7
Test

1. Fill in the blanks.

  a. A quotient of two polynomials, such as $\frac{x+7}{x^2+2x}$, is called a _____ expression.

  b. Two triangles with the same shape, but not necessarily the same size, are called _____ triangles.

  c. A _____ is a mathematical statement that two ratios or two rates are equal.

  d. To _____ a rational expression, we multiply it by a form of 1. For example, $\frac{2}{5x} \cdot \frac{8}{8} = \frac{16}{40x}$.

  e. To simplify $\frac{x-3}{(x+3)(x-3)}$, we remove common _____ of the numerator and denominator.

2. MEMORY   The formula $n = \frac{35 + 5d}{d}$ approximates the number of words n that a certain person can recall d days after memorizing a list of 50 words. How many words will the person remember in 1 week?

For what real numbers are each rational expression undefined?

3. $\dfrac{6x-9}{5x}$ **4.** $\dfrac{x}{x^2+x-6}$

5. THE INTERNET   A dial-up modem transmits up to 56K bits per second (K is an abbreviation for one thousand). Convert this to bits per minute.

6. Explain the error: $\dfrac{x+5}{5} = \dfrac{x + \cancel{5}^{1}}{\cancel{5}_{1}}$.

$$= x + 1$$

Simplify each rational expression.

7. $\dfrac{48x^2y}{54xy^2}$ **8.** $\dfrac{7m-49}{7-m}$

9. $\dfrac{2x^2-x-3}{4x^2-9}$ **10.** $\dfrac{3(x+2)-3}{6x+5-(3x+2)}$

Find the LCD of each pair of rational expressions.

11. $\dfrac{19}{3c^2d}, \dfrac{6}{c^2d^3}$ **12.** $\dfrac{4n+25}{n^2-4n-5}, \dfrac{6n}{n^2-25}$

Perform the operations. Simplify, if possible.

13. $\dfrac{12x^2y}{15xy} \cdot \dfrac{25y^2}{16x}$ **14.** $\dfrac{x^2+3x+2}{3x+9} \cdot \dfrac{x+3}{x^2-4}$

15. $\dfrac{x-x^2}{3x^2+6x} \div \dfrac{3x-3}{3x^3+6x^2}$ **16.** $\dfrac{a^2-16}{a-4} \div (6a+24)$

17. $\dfrac{3y+7}{2y+3} - \dfrac{-3y-2}{2y+3}$ **18.** $\dfrac{2n}{5m} - \dfrac{n}{2}$

19. $\dfrac{x+1}{x} + \dfrac{x-1}{x+1}$ **20.** $\dfrac{a+3}{a-1} - \dfrac{a+4}{1-a}$

21. $\dfrac{9}{c-4} + c$ **22.** $\dfrac{6}{t^2-9} - \dfrac{5}{t^2-t-6}$

Simplify each complex fraction.

23. $\dfrac{\dfrac{3m-9}{8m}}{\dfrac{5m-15}{32}}$ **24.** $\dfrac{\dfrac{3}{as^2} + \dfrac{6}{a^2s}}{\dfrac{6}{a} - \dfrac{9}{s^2}}$

Solve each equation. If an equation has no solution, so indicate.

25. $\dfrac{1}{3} + \dfrac{4}{3y} = \dfrac{5}{y}$ **26.** $\dfrac{9n}{n-6} = 3 + \dfrac{54}{n-6}$

27. $\dfrac{7}{q^2-q-2} + \dfrac{1}{q+1} = \dfrac{3}{q-2}$

28. $\dfrac{2}{3} = \dfrac{2c-12}{3c-9} - c$

29. $\dfrac{y}{y-1} = \dfrac{y-2}{y}$

30. Solve for B: $H = \dfrac{RB}{R+B}$

31. HEALTH RISKS   A medical newsletter states that a "healthy" waist-to-hip ratio for men is 19:20 or less. Does the patient shown in the illustration fall within the "healthy" range?

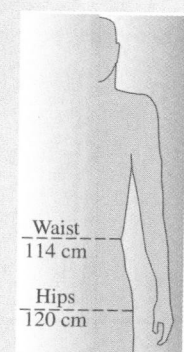

Waist
$\overline{114 \text{ cm}}$

Hips
$\overline{120 \text{ cm}}$

32. CURRENCY EXCHANGE RATES Preparing for a visit to London, a New York resident exchanged 3,500 U.S. dollars for British pounds. (A pound is the basic monetary unit of Great Britain.) If the exchange rate was 100 U.S. dollars for 51 British pounds, how many British pounds did the traveler receive?

33. TV TOWERS A television tower casts a shadow 114 feet long at the same time that a 6-foot-tall television reporter casts a shadow of 4 feet. Find the height of the tower.

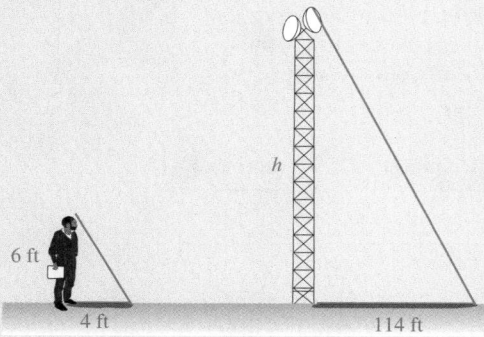

6 ft

h

4 ft   114 ft

34. COMPARISON SHOPPING Which is the better buy for fabric softener: 80 sheets for $3.89 or 120 sheets for $6.19?

35. CLEANING HIGHWAYS One highway worker can pick up all the trash on a strip of highway in 7 hours, and his helper can pick up the trash in 9 hours. How long will it take them if they work together?

36. PHYSICAL FITNESS A man roller-blades at a rate 6 miles per hour faster than he jogs. In the same time it takes him to roller-blade 5 miles he can jog 2 miles. How fast does he jog?

37. NUMBER PROBLEM If a number is subtracted from the numerator of $\frac{5}{8}$ and twice as much is added to the denominator, the result is $\frac{1}{4}$. Find the number.

38. Explain what it means to clear the following equation of fractions.

$$\frac{u}{u-1} + \frac{1}{u} = \frac{u^2+1}{u^2-u}$$

Why is this a helpful first step in solving the equation?

GROUP PROJECT

WHAT IS π?

Overview: In this activity, you will discover an important fact about the ratio of the circumference to the diameter of a circle.

Instructions: Form groups of 2 or 3 students. With a piece of string or a cloth tape measure, find the circumference and the diameter of objects that are circular in shape. You can measure anything that is round: for example, a coin, the top of a can, a tire, or a wastepaper basket. Enter your results in a table, as shown below. Convert each measurement to a decimal, and then use a calculator to determine a decimal approximation of the ratio of the circumference C to diameter d.

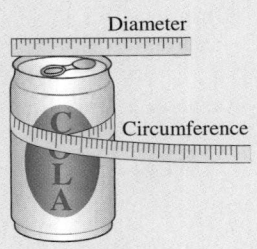

Object	Circumference	Diameter	$\frac{C}{d}$ (approx.)
A quarter	$2\frac{15}{16}$ in. = 2.9375 in.	$\frac{15}{16}$ in. = 0.9375 in.	3.13333

Since early history, mathematicians have known that the ratio of the circumference to the diameter of a circle is the same for any size circle, approximately 3. Today, following centuries of study, we know that this ratio is exactly 3.141592653589. . . .

$$\frac{C}{d} = 3.141592653589\ldots$$

The Greek letter π (pi) is used to represent the ratio of circumference to diameter:

$$\pi = \frac{C}{d}, \qquad \text{where } \pi = 3.141592653589\ldots$$

Are the ratios in your table numerically close to π? Give some reasons why they aren't exactly 3.141592653589 . . . in each case.

CUMULATIVE REVIEW
Chapters 1–7

1. Determine whether each statement is true or false. [Section 1.3]

   a. Every integer is a whole number.

   b. 0 is not a rational number.

   c. π is an irrational number.

   d. The set of integers is the set of whole numbers and their opposites.

2. Insert the proper symbol, $<$ or $>$, in the blank to make a true statement.

$$|2 - 4| \quad \underline{\quad} \quad -(-6) \text{ [Section 1.5]}$$

3. Evaluate: $9^2 - 3[45 - 3(6 + 4)]$ [Section 1.7]

4. Find the average (mean) test score of a student in a history class with scores of 80, 73, 61, 73, and 98. [Section 1.7]

5. Simplify: $8(c + 7) - 2(c - 3)$ [Section 1.9]

6. Solve: $\frac{4}{5}d = -4$ [Section 2.1]

7. Solve: $2 - 3(x - 5) = 4(x - 1)$ [Section 2.2]

8. GRAND KING SIZE BEDS   Because Americans are taller compared to 100 years ago, bed manufacturers are making larger models. Find the percent of increase in sleeping area of the new grand king size bed compared to the standard king size. [Section 2.3]

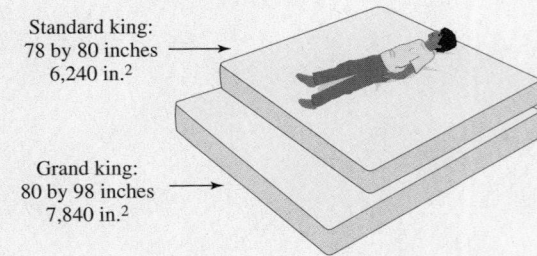

Standard king:
78 by 80 inches
6,240 in.²

Grand king:
80 by 98 inches
7,840 in.²

9. Solve $A - c = 2B + r$ for B. [Section 2.4]

10. Change 40°C to degrees Fahrenheit. [Section 2.4]

11. Find the volume of a pyramid that has a square base, measuring 6 feet on a side, and whose height is 20 feet. [Section 2.4]

12. BLENDING TEA   One grade of tea (worth \$6.40 per pound) is to be mixed with another grade (worth \$4 per pound) to make 20 pounds of a mixture that will be worth \$5.44 per pound. How much of each grade of tea must be used? [Section 2.6]

13. SPEED OF A PLANE   Two planes are 6,000 miles apart and their speeds differ by 200 mph. If they travel toward each other and meet in 5 hours, find the speed of the slower plane. [Section 2.6]

14. Solve $7x + 2 \geq 4x - 1$. Write the solution set in interval notation and graph it. [Section 2.7]

15. Graph: $y = 2x - 3$ [Section 3.2]

16. Find the slope of the line passing through $(-1, 3)$ and $(3, -1)$. [Section 3.4]

17. CUTTING STEEL   The graph shows the amount of wear (in millimeters) on a cutting blade for a given length of a cut (in meters). Find the rate of change in the length of the cutting blade. [Section 3.4]

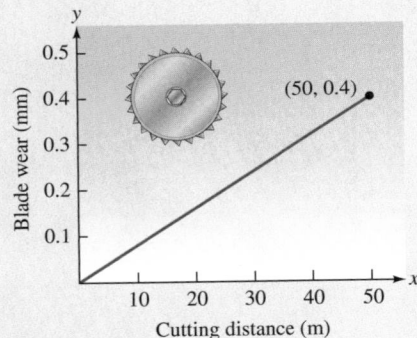

18. What is the slope of a line perpendicular to the line $y = -\frac{7}{8}x - 6$? [Section 3.5]

19. Write an equation of the line that has slope 3 and passes through the point $(1, 5)$. Write the answer in slope-intercept form. [Section 3.6]

20. Graph: $3x - 2y \leq 6$ [Section 3.7]

21. If $f(x) = -3x^2 - 6x$, find $f(-2)$. [Section 3.8]

22. Fill in the blanks. The set of all possible input values for a function is called the _____ and the set of all output values is called the _____. [Section 3.8]

23. Solve the system $\begin{cases} x + y = 1 \\ y = x + 5 \end{cases}$ by graphing. [Section 4.1]

24. Solve the system: $\begin{cases} x = 3y - 1 \\ 2x - 3y = 4 \end{cases}$ [Section 4.2]

25. Solve the system: $\begin{cases} 2x + 3y = -1 \\ 3x + 5y = -2 \end{cases}$ [Section 4.3]

26. POKER   After a night of cards, a poker player finished with some red chips (worth \$5 ech) and some blue chips (worth \$10 each). He received \$190 when he cashed in the 23 chips. How many of each colored chip did he finish with? [Section 4.4]

Simplify each expression. Write each answer without using negative exponents.

27. $x^4 x^3$ [Section 5.1]

28. $(x^2 x^3)^5$ [Section 5.1]

29. $\left(\dfrac{y^3 y}{2yy^2}\right)^3$ [Section 5.1]

30. $\left(\dfrac{-2a}{b}\right)^5$ [Section 5.1]

31. $(a^{-2}b^3)^{-4}$ [Section 5.2]

32. $\dfrac{9b^0 b^3}{3b^{-3} b^4}$ [Section 5.2]

33. Write 290,000 in scientific notation. [Section 5.3]

34. What is the degree of the polynomial $5x^3 - 4x + 16$ [Section 5.4]

35. Graph: $y = -x^3$ [Section 5.4]

36. CONCENTRIC CIRCLES   The area of the ring between the two concentric circles of radius r and R is given by the formula

$$A = \pi(R + r)(R - r)$$

Do the multiplication on the right-hand side of the equation. [Section 5.7]

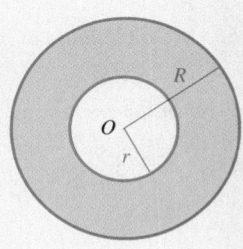

Perform the operations.

37. $(3x^2 - 3x - 2) + (3x^2 + 4x - 3)$ [Section 5.5]

38. $\left(\dfrac{1}{16}t^3 + \dfrac{1}{2}t^2 - \dfrac{1}{6}t\right) - \left(\dfrac{9}{16}t^3 + \dfrac{9}{4}t^2 - \dfrac{1}{12}t\right)$ [Section 5.5]

39. $(2x^2 y^3)(3x^2 y^2)$ [Section 5.6]

40. $(2y - 5)(3y + 7)$ [Section 5.6]

41. $-4x^2 z(3x^2 - z)$ [Section 5.6]

42. $(3a - 4)^2$ [Section 5.7]

43. $\dfrac{6x + 9}{3}$ [Section 5.8]

44. $2x + 3\overline{)2x^3 + 7x^2 + 4x - 3}$ [Section 5.8]

Factor each polynomial completely, if possible.

45. $k^3 t - 3k^2 t$ [Section 6.1]

46. $2ab + 2ac + 3b + 3c$ [Section 6.1]

47. $u^2 - 18u + 81$ [Section 6.2]

48. $-r^2 + 2 + r$ [Section 6.2]

49. $u^2 + 10u + 15$
[Section 6.2]

50. $6x^2 - 63 - 13x$
[Section 6.3]

51. $2a^2 - 200b^2$
[Section 6.4]

52. $b^3 + 125$
[Section 6.5]

Solve each equation by factoring.

53. $5x^2 + x = 0$
[Section 6.7]

54. $6x^2 - 5x = -1$
[Section 6.7]

55. COOKING   The electric griddle shown has a cooking surface of 160 square inches. Find the length and the width of the griddle. [Section 6.7]

$w + 6$

w

56. For what values of x is the rational expression $\frac{3x^2}{x^2 - 25}$ undefined? [Section 7.1]

Perform the operations. Simplify, if possible.

57. $\dfrac{2x^2 - 8x}{x^2 - 6x + 8}$
[Section 7.1]

58. $\dfrac{x^2 - 16}{4 - x} \div \dfrac{3x + 12}{x^3}$
[Section 7.2]

59. $\dfrac{8m^2}{2m + 5} - \dfrac{4m^2 + 25}{2m + 5}$
[Section 7.3]

60. $\dfrac{4}{x - 3} + \dfrac{5}{3 - x}$
[Section 7.4]

61. $\dfrac{m}{m^2 + 5m + 6} - \dfrac{2}{m^2 + 3m + 2}$ [Section 7.4]

62. Simplify: $\dfrac{2 - \dfrac{2}{x + 1}}{2 + \dfrac{2}{x}}$ [Section 7.5]

Solve each equation.

63. $\dfrac{7}{5x} - \dfrac{1}{2} = \dfrac{5}{6x} + \dfrac{1}{3}$ [Section 7.6]

64. $\dfrac{u}{u - 1} + \dfrac{1}{u} = \dfrac{u^2 + 1}{u^2 - u}$ [Section 7.6]

65. DRAINING A TANK   If one outlet pipe can drain a tank in 24 hours, and another pipe can drain the tank in 36 hours, how long will it take for both pipes to drain the tank? [Section 7.7]

66. HEIGHT OF A TREE   A tree casts a shadow of 29 feet at the same time as a vertical yardstick casts a shadow of 2.5 feet. Find the height of the tree. [Section 7.8]

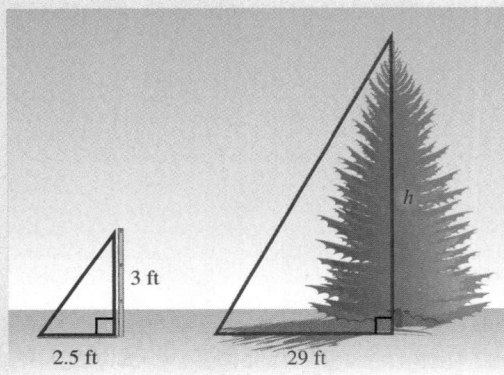

3 ft

h

2.5 ft

29 ft

CHAPTER 8

Transition to Intermediate Algebra

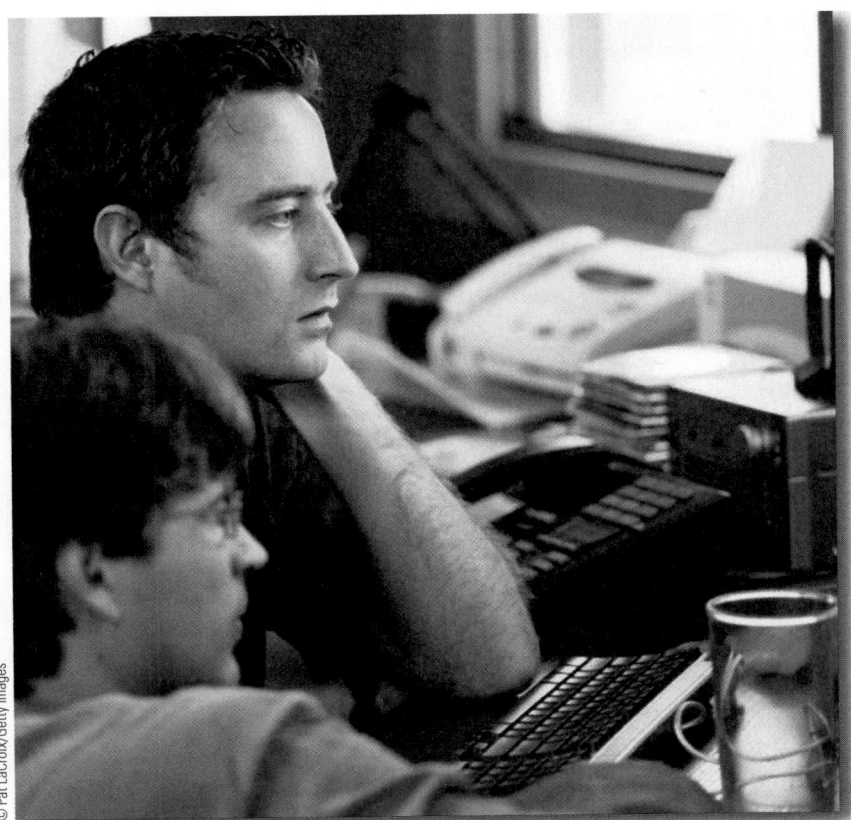

© Pat LaCroix/Getty Images

from Campus to Careers
Webmaster

If you use the Internet, then you have seen firsthand what webmasters do. They design and maintain websites for individuals and companies on the World Wide Web. The job of webmaster requires excellent computer and technical skills. A background in business, art, and design is also helpful. Since webmasters are often called on to be troubleshooters when technical difficulties arise, those considering entering the field are encouraged to study math to strengthen their problem-solving abilities.

In **Problem 28** of **Study Set 8.9**, you will use the language of variation to describe various aspects of Internet usage and website design.

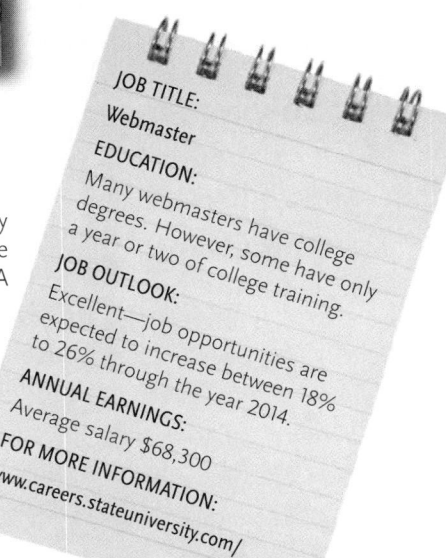

JOB TITLE:
Webmaster

EDUCATION:
Many webmasters have college degrees. However, some have only a year or two of college training.

JOB OUTLOOK:
Excellent—job opportunities are expected to increase between 18% to 26% through the year 2014.

ANNUAL EARNINGS:
Average salary $68,300

FOR MORE INFORMATION:
www.careers.stateuniversity.com/

Since a student's level of effort is significantly influenced by his or her attitude, you should strive to maintain a positive mental outlook for the entire term. Here are three suggestions to help you do that:

PERSONAL REMINDERS: From time-to-time, remind yourself of the ways in which you will benefit by passing the course.

DON'T DWELL ON THE NEGATIVE: Counterproductive feelings of stress and math anxiety can often be overcome with extra preparation, support services, and even relaxation techniques.

ACCOMPLISH GOALS AND EARN REWARDS: Reward yourself after studying, learning a difficult concept, or completing a homework assignment. The reward can be small, like listening to music, reading a novel, playing a sport, or spending some time with friends.

Now Try This

1. List six ways in which you will benefit by passing this course. For example, it will get you one step closer to a college degree or it will improve your problem-solving abilities.

2. List three ways in which you can to respond to feelings of stress or math anxiety, should they arise during the term.

3. List some simple ways that you can reward yourself when you complete one of the class goals that you set for yourself.

SECTION 8.1
Review of Solving Linear Equations, Formulas, and Linear Inequalities

Objectives

1. Use properties of equality to solve linear equations.
2. Identify identities and contradictions.
3. Solve formulas for a specified variable.
4. Solve linear inequalities.
5. Use linear equations and inequalities to solve application problems.

In this section, we will review how to solve equations and formulas. We will then extend our review to include inequalities.

1 Use Properties of Equality to Solve Linear Equations.

Recall that an **equation** is a statement indicating that two expressions are equal. The set of numbers that satisfy an equation is called its **solution set,** and the elements in the solution set are called **solutions** of the equation. To **solve an equation,** we must find all of its solutions.

In this section we will solve *linear equations in one variable.*

Linear Equations

A **linear equation in one variable** can be written in the form $ax + b = c$, where a, b, and c are real numbers, and $a \neq 0$.

Some examples of linear equations in one variable are

$$5a + 2 = 8, \qquad \frac{7}{3}y = -28, \qquad \text{and} \qquad 3(2x - 1) = 2x + 9$$

We can solve a linear equation in one variable by using the following **properties of equality** to replace it with simpler **equivalent equations** that have the same solution set. We continue this process until the variable is isolated on one side of the $=$ symbol.

1. Adding the same number to, or subtracting the same number from, both sides of an equation does not change its solution.

2. Multiplying or dividing both sides of an equation by the same nonzero number does not change its solution.

Success Tip
You may want to review the properties of equality on pages 110, 111, 113, and 115.

EXAMPLE 1   Solve: $3(2x - 1) = 2x + 9$

Strategy   First, we will use the distributive property to remove parentheses on the left side of the equation. Then, to eliminate $2x$ from the right side, we will subtract $2x$ from both sides.

Why   To solve for x, all the terms containing x must be on the same side of the equation.

Solution

$3(2x - 1) = 2x + 9$	This is the equation to solve.
$6x - 3 = 2x + 9$	Distribute the multiplication by 3.
$6x - 3 - 2x = 2x + 9 - 2x$	To eliminate 2x from the right side, subtract 2x from both sides.
$4x - 3 = 9$	Combine like terms.
$4x - 3 + 3 = 9 + 3$	To isolate the variable term 4x, undo the subtraction of 3 on the left side by adding 3 to both sides.
$4x = 12$	Combine like terms.
$x = 3$	To isolate x, undo the multiplication by 4 by dividing both sides by 4.

Check: We substitute 3 for x in the original equation to see whether it satisfies the equation.

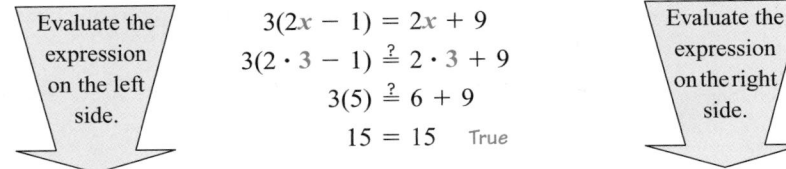

$$3(2x - 1) = 2x + 9$$
$$3(2 \cdot 3 - 1) \overset{?}{=} 2 \cdot 3 + 9$$
$$3(5) \overset{?}{=} 6 + 9$$
$$15 = 15 \quad \text{True}$$

Since 3 satisfies the original equation, it is the solution. The solution set is $\{3\}$.

> ▷
> | **Self Check 1** | Solve: $2(3x - 2) = 3x - 13$ |
> | *Now Try* | **Problem 21** |

In general, we will follow these steps to solve linear equations in one variable. Not every step is needed to solve every equation.

Strategy for Solving Linear Equations in One Variable

1. **Clear the equation of fractions or decimals:** Multiply both sides by the LCD to clear fractions or multiply both sides by a power of 10 to clear decimals.
2. **Simplify each side of the equation:** Use the distributive property to remove parentheses and combine like terms on each side.
3. **Isolate the variable term on one side:** Add (or subtract) to get the variable term on one side of the equation and a number on the other using the addition (or subtraction) property of equality.
4. **Isolate the variable:** Multiply (or divide) to isolate the variable using the multiplication (or division) property of equality.
5. **Check the result:** Substitute the possible solution for the variable in the original equation to see if a true statement results.

EXAMPLE 2   Solve: $\dfrac{1}{3}(6x + 15) = \dfrac{3}{2}(x + 2) - 2$

Strategy   We will follow the strategy for solving equations.

Why   This is the most efficient way to solve a linear equation in one variable.

Solution

Step 1: We can clear the equation of fractions by multiplying both sides by the least common denominator (LCD) of $\frac{1}{3}$ and $\frac{3}{2}$. The LCD of these fractions is the smallest number that can be divided by both 2 and 3 exactly. That number is 6.

> ### Success Tip
> Before multiplying both sides of an equation by the LCD, frame the left side and frame the right side with parentheses or brackets.

$$\frac{1}{3}(6x + 15) = \frac{3}{2}(x + 2) - 2$$ This is the equation to solve.

$$6\left[\frac{1}{3}(6x + 15)\right] = 6\left[\frac{3}{2}(x + 2) - 2\right]$$ To eliminate the fractions, multiply both sides by the LCD, 6.

$$2(6x + 15) = 6 \cdot \frac{3}{2}(x + 2) - 6 \cdot 2$$ On the left side, perform the multiplication: $6 \cdot \frac{1}{3} = 2$. On the right side, distribute the multiplication by 6.

$$2(6x + 15) = 9(x + 2) - 12$$ Multiply on the right side.

Step 2: Use the distributive property to remove parentheses and combine like terms on the right side.

$$12x + 30 = 9x + 18 - 12$$ Distribute the multiplication by 2 and the multiplication by 9.

$$12x + 30 = 9x + 6$$

Step 3: To get the variable term on the left side and the constant on the right side, subtract $9x$ and 30 from both sides.

$$12x + 30 - 9x - 30 = 9x + 6 - 9x - 30$$
$$3x = -24 \qquad \text{On each side, combine like terms.}$$

Step 4: The coefficient of the variable is 3. To isolate x, we undo the multiplication by 3 by dividing both sides by 3.

$$\frac{3x}{3} = \frac{-24}{3}$$
$$x = -8$$

Step 5: Check by substituting -8 for x in the original equation and evaluating each side.

$$\frac{1}{3}(6x + 15) = \frac{3}{2}(x + 2) - 2$$

$$\frac{1}{3}[6(-8) + 15] \stackrel{?}{=} \frac{3}{2}(-8 + 2) - 2$$

$$\frac{1}{3}(-48 + 15) \stackrel{?}{=} \frac{3}{2}(-6) - 2$$

$$\frac{1}{3}(-33) \stackrel{?}{=} -9 - 2$$

$$-11 = -11 \qquad \text{True}$$

The solution is -8 and the solution set is $\{-8\}$.

Self Check 2	Solve: $\frac{1}{3}(2x - 2) = \frac{1}{4}(5x + 1) + 2$
Now Try	**Problem 29**

2 **Identify Identities and Contradictions.**

The equations discussed so far are called **conditional equations.** For these equations, some numbers satisfy the equation and others do not. An **identity** is an equation that is satisfied by every number for which both sides of the equation are defined. A **contradiction** is an equation that is never true.

EXAMPLE 3	Solve: **a.** $-2(x - 1) - 4 = -4(1 + x) + 2x + 2$
	b. $-6.2(-x - 1) - 4 = 4.2x - (-2x)$

Strategy   In each case, we will follow the strategy for solving equations.

Why   This is the most efficient way to solve a linear equation in one variable.

Solution

a. Since there are no fractions to clear, we will begin by using the distributive property to remove the parentheses on the left and right sides of the equation.

$$-2(x - 1) - 4 = -4(1 + x) + 2x + 2 \qquad \text{This is the equation to solve.}$$
$$-2x + 2 - 4 = -4 - 4x + 2x + 2 \qquad \text{Use the distributive property.}$$
$$-2x - 2 = -2x - 2 \qquad \text{On each side, combine like terms.}$$
$$-2x - 2 + 2x = -2x - 2 + 2x \qquad \text{To attempt to isolate the variable on one side of the equation, add 2x to both sides.}$$

$$-2 = -2 \qquad \text{True}$$

The terms involving x drop out. The resulting true statement indicates that the original equation is true for every value of x. The solution set is the set of real numbers denoted $\mathbb{R}$. The equation is an identity.

b.

$-6.2(-x - 1) - 4 = 4.2x - (-2x)$	This is the equation to solve.
$-6.2(-x - 1) - 4 = 4.2x + 2x$	Simplify: $-(-2x) = 2x$.
$10[-6.2(-x - 1) - 4] = 10(4.2x + 2x)$	To remove the decimals, multiply both sides by 10.
$-62(-x - 1) - 40 = 42x + 20x$	Distribute the multiplication by 10.
$62x + 62 - 40 = 42x + 20x$	On the left side, remove parentheses.
$62x + 22 = 62x$	On each side, combine like terms.
$62x + 22 - 62x = 62x - 62x$	To attempt to isolate the variable on one side of the equation, subtract 62x from both sides.
$22 = 0$	False

The terms involving x drop out. The resulting false statement indicates that no value for x makes the original equation true. The solution set contains no elements and can be denoted as the **empty set** { } or the **null set** $\varnothing$. The equation is a contradiction.

Self Check 3   Solve: **a.** $3(a + 1) - (20 + a) = 5(a - 1) - 3(a + 4)$

b. Solve: $0.3(a + 4) + 0.2 = 0.2(a - 1) + 0.1a + 1.9$

Now Try   **Problems 31 and 35**

③   Solve Formulas for a Specified Variable.

Real-world applications sometimes call for a formula solved for one variable to be solved for a different variable. *To solve a formula for a specified variable* means to isolate that variable on one side of the formula and write all other quantities on the other side.

EXAMPLE 4   For simple interest, the formula $A = P + Prt$ gives the amount of money in an account at the end of a specific time t. A represents the amount, P the principal, and r the rate of interest. Solve the formula for t.

Strategy   To solve for t, we treat it as if it were the only variable in the equation. To isolate t, we will use the same strategy that we used to solve linear equations in one variable.

Why   We can solve this formula as if it were an equation in one variable because all the other variables, A, P, and r, are treated as if they were numbers (constants).

Solution

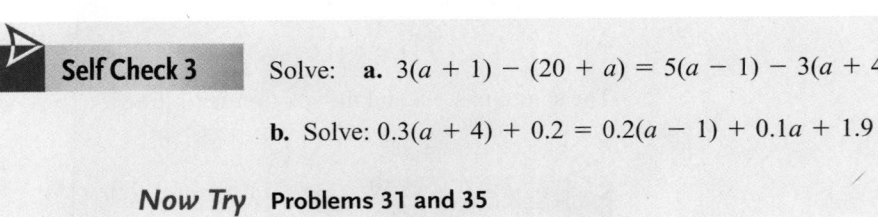

To solve for t, we will isolate t on this side of the equation.

$A = P + Prt$	
$A - P = Prt$	To isolate the term involving t, subtract P from both sides.
$\dfrac{A - P}{Pr} = \dfrac{Prt}{Pr}$	To isolate t, divide both sides by Pr (or multiply both sides by $\frac{1}{Pr}$).

$$\frac{A - P}{Pr} = t \qquad \text{On the right side, remove the common factors: } \frac{\cancel{P}r\cancel{t}}{\cancel{P}r\cancel{t}}.$$

$$t = \frac{A - P}{Pr} \qquad \text{Write the equation with } t \text{ on the left side.}$$

Self Check 4   Solve $A = P + Prt$ for r.
Now Try   **Problem 39**

EXAMPLE 5   The formula for the area of a trapezoid is $A = \frac{1}{2}h(b_1 + b_2)$. Solve the formula for b_1.

Strategy   To solve for b_1, we will treat it as if it were the only variable in the equation. To isolate b_1, we will use the same strategy that we used to solve linear equations in one variable.

Why   We can solve the formula as if it were an equation in one variable because all the other variables, A, h, and b_2, are treated as if it they were numbers (constants).

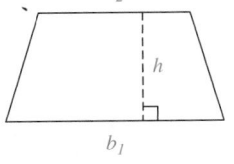

Trapezoid

Solution

To solve for b_1, we will isolate b_1 on this side of the equation.

$$A = \frac{1}{2}h(b_1 + b_2)$$

$$2 \cdot A = 2 \cdot \frac{1}{2}h(b_1 + b_2) \qquad \text{Multiply both sides by 2 to clear the equation of the fraction.}$$

$$2A = h(b_1 + b_2) \qquad \text{Simplify each side of the equation.}$$

$$2A = hb_1 + hb_2 \qquad \text{Distribute the multiplication by } h.$$

$$2A - hb_2 = hb_1 \qquad \begin{array}{l}\text{Subtract } hb_2 \text{ from both sides to isolate the variable term } hb_1 \\ \text{on the right side. This step is done mentally.}\end{array}$$

$$\frac{2A - hb_2}{h} = \frac{hb_1}{h} \qquad \begin{array}{l}\text{To isolate } b_1, \text{ undo the multiplication by } h \text{ by dividing both} \\ \text{sides by } h.\end{array}$$

$$\frac{2A - b_2h}{h} = b_1 \qquad \text{On the right side, remove the common factor of } h: \frac{\cancel{h}b_1}{\cancel{h}}.$$

$$b_1 = \frac{2A - hb_2}{h} \qquad \text{Reverse the sides of the equation so that } b_1 \text{ is on the left.}$$

When solving formulas for a specified variable, there is often more than one way to express the result. In this case, we could perform the division by h on the right side term-by-term: $b_1 = \frac{2A}{h} - \frac{hb_2}{h}$. After removing the common factor of h in the numerator and denominator of the second fraction, we obtain the following equivalent form of the result: $b_1 = \frac{2A}{h} - b_2$.

Caution   Do not try to simplify the result in the following way. It is incorrect because h is not a factor of the entire numerator.

$$b_1 = \frac{2A - \cancel{h}b_2}{\cancel{h}}$$

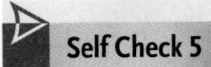

Self Check 5   Solve $S = \frac{180(t - 2)}{7}$ for t.

Now Try   Problem 47

4   **Solve Linear Inequalities.**

Inequalities are statements indicating that two quantities are unequal. They usually contain one or more of the following symbols.

Inequality Symbols		
$<$ is less than	$>$ is greater than	$\neq$ is not equal to
$\leq$ is less than or equal to	$\geq$ is greater than or equal to	

In this section, we will work with **linear inequalities** in one variable.

Linear Inequalities

A **linear inequality** in one variable (say, x) is any inequality that can be expressed in one of the following forms, where a, b, and c represent real numbers and $a \neq 0$.

$$ax + b < c \qquad ax + b \leq c \qquad ax + b > c \qquad ax + b \geq c$$

Some examples of linear inequalities are

$$3x \leq 0, \qquad 3(2x - 9) < 9, \qquad \text{and} \qquad -12x - 8 \geq 16$$

To **solve a linear inequality** means to find all values that, when substituted for the variable, make the inequality true. The set of all solutions of an inequality is called its **solution set.** The solution set of an inequality can be graphed as an **interval** on the number line. We can also write a solution set using **interval notation** and **set-builder notation.** We use the following properties to solve inequalities.

Success Tip
You may want to review the properties of inequality on pages 176 and 177.

1. Adding the same number to, or subtracting the same number from, both sides of an inequality does not change the solutions.
2. Multiplying or dividing both sides of an inequality by the same positive number does not change the solutions.
3. If we multiply or divide both sides of an inequality by a negative number, the direction of the inequality symbol must be *reversed* for the inequalities to have the same solutions.

After applying one of these properties, the resulting inequality is equivalent to the original one. Like equivalent equations, **equivalent inequalities** have the same solution set.

EXAMPLE 6   Solve $3(2x - 9) < 9$. Write the solution set in interval notation and graph it.

Strategy   We will use the properties of inequalities and the same strategy for solving equations to isolate x on one side of the inequality.

Why   Once we have obtained an equivalent inequality, with the variable isolated on one side, the solution set will be obvious.

Solution

$3(2x - 9) < 9$ This is the inequality to solve.

$6x - 27 < 9$ Distribute the multiplication by 3.

$6x < 36$ To isolate the variable term 6x, undo the subtraction of 27 on the left side by adding 27 to both sides.

$x < 6$ To isolate x, undo the multiplication by 6 by dividing both sides by 6.

The solution set is the interval $(-\infty, 6)$, whose graph is shown. We can also write the solution set using set-builder notation: $\{x \mid x < 6\}$. We read this notation as "the set of all real numbers x such that x is less than 6."

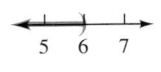

Since the solution set contains infinitely many real numbers, we cannot check all of them to see whether they satisfy the original inequality. However, as an informal check, we can pick one number in the graph, near the endpoint, such as 5, and see whether it satisfies the inequality. We can also pick one number not in the graph, but near the endpoint, such as 7, and see whether it fails to satisfy the inequality.

Check a value in the graph: x = 5	*Check a value not in the graph: x = 7*
$3(2x - 9) < 9$	$3(2x - 9) < 9$
$3[2(5) - 9] \overset{?}{<} 9$	$3[2(7) - 9] \overset{?}{<} 9$
$3(10 - 9) \overset{?}{<} 9$	$3(14 - 9) \overset{?}{<} 9$
$3(1) \overset{?}{<} 9$	$3(5) \overset{?}{<} 9$
$3 < 9$ True	$15 < 9$ False

Since 5 satisfies $3(2x - 9) < 9$ and 7 does not, the solution set appears to be correct.

Self Check 6   Solve $2(3x + 2) > -44$. Write the solution set in interval notation and graph it.

Now Try   **Problem 53**

EXAMPLE 7   Solve $-12x - 8 \le 16$. Write the solution set in interval notation and graph it.

Strategy   We will use the properties of inequalities and the same strategy for solving equations to isolate x on one side of the inequality.

Why   This will give the solution set of the inequality.

Solution   Once we have obtained an equivalent inequality, with the variable isolated on one side, the solution set will be obvious.

$-12x - 8 \le 16$

$-12x \le 24$ To undo the subtraction of 8, add 8 to both sides.

$x \ge -2$ To undo the multiplication by -12, divide both sides by -12. Because we are dividing by a negative number, we reverse the $\le$ symbol.

Success Tip
We must remember to reverse the inequality symbol every time we multiply or divide both sides by a negative number.

The solution set is $\{x \mid x \ge -2\}$ or the interval $[-2, \infty)$, whose graph is shown.

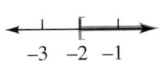

Self Check 7   Solve: $-6x + 6 \leq 0$
Now Try   **Problem 59**

EXAMPLE 8   Solve $3a - 4 < 3(a + 5)$. Write the solution set in interval notation and graph it.

Strategy   We will use the properties of inequalities and the same strategy for solving equations to isolate a on one side of the inequality.

Why   To solve for a, all the terms involving a need to be on one side of the inequality.

Solution

$$3a - 4 < 3(a + 5)$$
$$3a - 4 < 3a + 15 \qquad \text{Distribute the multiplication by 3.}$$
$$3a - 4 - 3a < 3a + 15 - 3a \quad \text{Subtract 3a from both sides.}$$
$$-4 < 15 \qquad \text{True}$$

The terms involving a drop out. The resulting true statement indicates that the original inequality is true for all values of a. Therefore, the solution set is the set of real numbers, denoted $(-\infty, \infty)$ or $\mathbb{R}$, and its graph is as shown.

Success Tip
If the variables drop out when solving an inequality and the result is false, the solution set contains no elements and is denoted by $\varnothing$.

Self Check 8   Solve: $-8n + 10 \geq 1 - 2(4n - 2)$
Now Try   **Problem 67**

---

5   **Use Linear Equations and Inequalities to Solve Application Problems.**

We can use equations and inequalities to solve real-world problems.

EXAMPLE 9   *Travel Promotions.*   The price of a 7-day Alaskan cruise, normally $2,752 per person, is reduced by $1.75 per person for large groups traveling together. How large a group is needed for the price to be $2,500 per person?

Analyze the Problem   For a group of 20 people, the cost is reduced by $1.75 for each person and the $2,752 price is reduced by 20($1.75) = $35.

The per-person price of the cruise = $2,752 − 20($1.75)

For a group of 30 people, the $2,752 cost is reduced by 30($1.75) = $52.50.

The per-person price of the cruise = $2,752 − 30($1.75)

Form an Equation   If we let x = the group size necessary for the price of the cruise to be $2,500 per person, we can form the following equation:

The per-person price of the cruise	is	$2,752	minus	the number of people in the group	times	$1.75.
2,500	=	2,752	−	x	·	1.75

Solve the Equation

$$2{,}500 = 2{,}752 - 1.75x$$

$2{,}500 - \textbf{2{,}752} = 2{,}752 - 1.75x - \textbf{2{,}752}$ Subtract 2,752 from both sides.

$-252 = -1.75x$ Simplify each side.

$144 = x$ To isolate x, divide both sides by −1.75.

State the Conclusion   If 144 people travel together, the price will be $2,500 per person.

Check the Result   For 144 people, the cruise cost of $2,752 will be reduced by $144(\$1.75) = \$252$. If we subtract, $\$2{,}752 - \$252 = \$2{,}500$. The answer checks.

   Now Try   **Problem 103**

To determine whether to use an equation or an inequality to solve a problem, look for key words and phrases. For example, phrases like *does not exceed, is no more than,* and *is at least* translate into inequalities.

EXAMPLE 10   ***Communication.***   A satellite phone is a mobile phone that sends and receives calls using satellites instead of landlines or cellular broadcasting towers. The advantage of a satellite phone is that it can complete calls from anywhere, such as the Sahara desert, the top of Mount Everst, or an African jungle. If a satellite telephone company charges callers $5.50 for the first three minutes and 88¢ for each additional minute, for how many minutes can a call last if the cost is not to exceed $20?

Analyze the Problem   We are given the rate at which a call is billed. Since the cost of a call is not to exceed $20, the cost must be *less than or equal to* $20. This phrase indicates that we should write an inequality to find how long a call can last.

Form an Inequality   We will let $x =$ the total number of minutes that a call can last. Then the cost of a call will be $5.50 for the first three minutes plus 88¢ times the number of additional minutes, where the number of *additional* minutes is $x - 3$ (the total number of minutes minus the first 3 minutes). With this information, we can form an inequality.

The cost of the first three minutes	plus	the cost of the additional minutes	is not to exceed	$20.
5.50	+	$0.88(x - 3)$	$\leq$	20

Solve the Inequality   To simplify the computations, we first clear the inequality of decimals.

$5.50 + 0.88(x - 3) \leq 20$ Write 88¢ as $0.88.

$550 + 88(x - 3) \leq 2{,}000$ To eliminate the decimals, multiply both sides by 100.

$550 + 88x - 264 \leq 2{,}000$ Distribute the multiplication by 88.

$88x + 286 \leq 2{,}000$ Combine like terms.

$88x \leq 1{,}714$ Subtract 286 from both sides.

$x \leq 19.47727273 \ldots$ Divide both sides by 88.

State the Conclusion　Since the phone company doesn't bill for part of a minute, the longest time a call can last is 19 minutes. If a call lasts for $x = 19.47727273\ldots$ minutes, it will be charged as a 20-minute call, and the cost will be $5.50 + $0.88(17) = $20.46.

Check the Result　If the call lasts 19 minutes, the cost will be $5.50 + $0.88(16) = $19.58. This is less than $20. The result checks.

 Now Try　Problem 109

ANSWERS TO SELF CHECKS　**1.** -3　**2.** -5　**3. a.** All real numbers, $\mathbb{R}$; identity　**b.** No solution, $\varnothing$; contradiction　**4.** $r = \frac{A - P}{Pt}$　**5.** $t = \frac{7S + 360}{180}$　**6.** $(-8, \infty)$ ⟵—┤—┬—┬—⟶ -9 -8 -7
7. $[1, \infty)$ ⟵—┬—├—┬—⟶ 0 1 2　**8.** $(-\infty, \infty)$ ⟵—┬—┬—┬—⟶ -1 0 1

STUDY SET
8.1

VOCABULARY

Fill in the blanks.

1. An _____ is a statement indicating that two expressions are equal.

2. Any number that makes an equation true when substituted for its variable is said to _____ the equation. Such numbers are called _____ of the equation.

3. $2x + 1 = 4$ and $5(y - 3) = 8$ are examples of _____ equations in one variable.

4. If two equations have the same solution set, they are called _____ equations.

5. An _____ is an equation that is satisfied by every number for which both sides of the equation are defined.

6. An equation that is never true is called a _____.

7. $<$, $>$, $\le$, and $\ge$ are _____ symbols.

8. To _____ an inequality means to find all values of the variable that make the inequality true.

CONCEPTS

Fill in the blanks.

9. a. Adding the _____ number to, or subtracting the same number from, both sides of an equation does not change its solution.

b. Multiplying or dividing both sides of an equation by the _____ nonzero number does not change its solution.

10. If we multiply both sides of an inequality by a negative number, the direction of the inequality must be _____ for the inequalities to have the same solutions.

11. Determine whether -5 is a solution of the following equation and inequality.
　a. $5(2x + 7) = 2x - 4$　　**b.** $3x + 6 \le -9$

12. The solution set of a linear inequality in x is graphed on the right. Determine whether a true or false statement results when 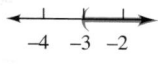 -4 -3 -2
　a. -4 is substituted for x.
　b. -3 is substituted for x.
　c. 0 is substituted for x.

NOTATION

13. Match each interval with its graph.
　a. $(-\infty, -1]$　　**i.** ⟵—┬—┤—┬—⟶ 0 1 2
　b. $(-\infty, 1)$　　**ii.** ⟵—┬—├—┬—⟶ -2 -1 0
　c. $[-1, \infty)$　　**iii.** ⟵—┬—┤—┬—⟶ -2 -1 0

14. a. Suppose that when solving a linear inequality, the variable drops out, and the result is $6 \le 10$. Write the solution set in interval notation and graph it.

b. Suppose that when solving a linear inequality, the variable drops out, and the result is $7 < -1$. What symbol is used to represent the solution set?

GUIDED PRACTICE

Solve each equation. Check the result. See Example 1.

15. $4x + 1 = 13$

16. $4x - 8 = 16$

17. $3(x + 1) = 15$

18. $-2(x + 5) = 30$

19. $2x + 6(2x + 3) = -10$

20. $3(2y - 4) - 6 = 3y$

21. $7(a + 2) = 4a + 17$

22. $5(5 - a) = 37 - 2a$

Solve each equation. See Example 2.

23. $\frac{1}{2}x - 4 = -1 + 2x$

24. $2x + 3 = \frac{2}{3}x - 1$

25. $\frac{x}{2} - \frac{x}{3} = 4$

26. $\frac{x}{2} + \frac{x}{3} = 10$

27. $\frac{1}{6}(x + 12) + 1 = \frac{x}{3}$

28. $\frac{3}{2}(y + 4) = \frac{20 - y}{2}$

29. $\frac{4}{5}(x + 5) = \frac{7}{8}(3x + 23) - 7$

30. $\frac{2}{3}(2x + 2) + 4 = \frac{1}{6}(5x + 29)$

Solve each equation. If an equation is an identity or a contradiction, so indicate. See Example 3.

31. $2x - 6 = -2x + 4(x - 2)$

32. $-3x = -2x + 1 - (5 + x)$

33. $2y + 1 = 5(0.2y + 1) - (4 - y)$

34. $4(2 - 3t) + 6t = -6t + 8$

35. $\frac{7}{2}(y - 1) + \frac{1}{2} = \frac{1}{2}(7y - 6)$

36. $2(x - 3) = \frac{3}{2}(x - 4) + \frac{x}{2}$

37. $0.3(x - 4) + 0.6 = -0.2(x + 4) + 0.5x$

38. $0.5(y + 2) + 0.7 - 0.3y = 0.2(y + 9)$

Solve each formula for the specified variable. See Examples 4 and 5.

39. $P = 2l + 2w$ for w

40. $P = 2l + 2w$ for l

41. $V = \frac{1}{3}Bh$ for B

42. $A = \frac{1}{2}bh$ for b

43. $T - W = ma$ for W

44. $G = U - TS + PV$ for S

45. $z = \frac{x - \mu}{\sigma}$ for x

46. $P = L + \frac{s}{f}i$ for s

47. $S = \frac{n(a + l)}{2}$ for l

48. $h = 48t + \frac{1}{2}at^2$ for a

49. $l = a + (n - 1)d$ for d

50. $P = 2(w + h + l)$ for h

Solve each inequality. Write the solution set in interval notation and then graph it. See Example 6.

51. $5x - 3 > 7$

52. $7x - 9 < 5$

53. $9a + 11 \leq 29$

54. $3b - 26 \geq 4$

55. $3(z - 2) \leq 2(z + 7)$

56. $5(3 + z) > -3(z + 3)$

57. $2x + 4 + 6x > 2 - 3x + 2$

58. $5x + 6 + 2x \geq 2 - x + 4$

Solve each inequality. Write the solution set in interval notation and then graph it. See Example 7.

59. $-3x - 1 \leq 5$

60. $-2x + 6 \geq 16$

61. $-5t + 3 \leq 5$

62. $-9t + 6 \geq 16$

63. $-7y + 5 > -5y - 1$

64. $8 - 9y \geq -y$

65. $t + 1 - 3t \geq t - 20$

66. $a + 4 - 10a > a - 16$

Solve each inequality. Write the solution set in interval notation and then graph it. See Example 8 and the Success Tip in the margin.

67. $2(5x - 6) > 4x - 15 + 6x$

68. $3(4x - 2) > 14x - 7 - 2x$

69. $\frac{3b + 7}{3} \leq \frac{2b - 9}{2}$

70. $-\frac{5x}{4} > \frac{3 - 5x}{4}$

TRY IT YOURSELF

Solve each equation. If an equation is an identity or a contradiction, so indicate.

71. $2r - 5 = 1 - r$

72. $5s - 13 = s - 1$

73. $0.2(a - 5) - 0.1(3a + 1) = 0$

74. $0.8(3a - 5) - 0.4(2a + 3) = 1.2$

75. $-\frac{4}{5}s = 2$

76. $-\frac{9}{8}s = 3$

77. $\frac{1}{2}(3y + 2) - \frac{5}{8} = \frac{3}{4}y$

78. $-\frac{3}{4}(4c - 3) + \frac{7}{8}c = \frac{19}{16}$

79. $8x + 3(2 - x) = 5x + 6$

80. $2(2a + 1) - 1 = 4a + 1$

81. $12 + 3(x - 4) - 21 = 5[5 - 4(4 - x)]$

82. $1 + 3[-2 + 6(4 - 2x)] = -(x + 3)$

83. $\dfrac{3 + p}{3} - 4p = 1 - \dfrac{p + 7}{2}$

84. $\dfrac{4 - t}{2} - \dfrac{3t}{5} = 2 + \dfrac{t + 1}{3}$

85. $5x + 10 = 5x$

86. $4(t - 2) - t = -(9 - 3t)$

87. $9.8 - 16r = -15.7 - r$

88. $15s + 8.1 - 2s = 8.1 - s$

89. $-4[p - (3 - p)] = 3(6p - 2)$

90. $2[5(4 - a) + 2(a - 1)] = 3 - a$

Solve each inequality. Write the solution set in interval notation and then graph it.

91. $-3(a + 2) > 2(a + 1)$

92. $-4(y - 1) < y + 8$

93. $\dfrac{x - 7}{2} - \dfrac{x - 1}{5} \le -\dfrac{x}{4}$

94. $\dfrac{3a + 1}{3} - \dfrac{4 - 3a}{5} \le -\dfrac{1}{15}$

95. $5(2n + 2) - n > 3n - 3(1 - 2n)$

96. $-1 + 4(y - 1) + 2y \le \dfrac{1}{2}(12y - 30) + 15$

97. $0.4x + 0.4 \le 0.1x + 0.85$

98. $0.05 + 0.8x \le 0.5x - 0.7$

99. $\dfrac{1}{2}y + 2 \ge \dfrac{1}{3}y - 4$

100. $\dfrac{1}{4}x - \dfrac{1}{3} \le x + 2$

101. $7 < \dfrac{5}{3}a - 3$

102. $5 > \dfrac{7}{2}a - 9$

APPLICATIONS

103. SPRING TOURS   A group of junior high students will be touring Washington, D.C. Their chaperons will have the $1,810 cost of the tour reduced by $15.50 for each student they supervise. How many students will a chaperon have to supervise so that his or her cost to take the tour will be $1,500?

104. MACHINING   Each pass through a lumber plane shaves off 0.015 inch of thickness from a board. How many times must a board, originally 0.875 inch thick, be run through the planer if a board of thickness 0.74 inch is desired?

105. MOVING EXPENSES   To help move his furniture, a man rents a truck for $41.50 per day plus 35¢ per mile. If he has budgeted $150 for transportation expenses, how many miles will he be able to drive the truck if the move takes 1 day?

106. COMPUTING SALARIES   A student working for a delivery company earns $57.50 per day plus $4.75 for each package she delivers. How many deliveries must she make each day to earn $200 a day?

107. FENCING PENS   A man has 150 feet of fencing to build the two-part pen shown in the illustration. If one end is a square and the other a rectangle, find the outside dimensions of the pen.

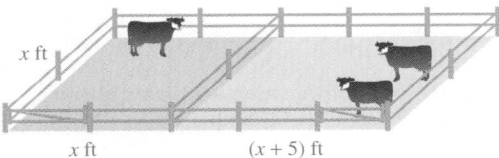

108. FENCING PASTURES   A farmer has 624 feet of fencing to enclose a pasture. Because a river runs along one side, fencing will be needed on only three sides. Find the dimensions of the pasture if its length is double its width.

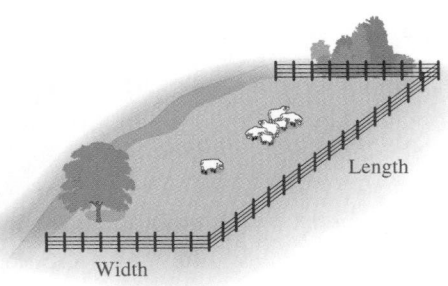

109. FUND-RAISING   A school PTA wants to rent a dunking tank for its annual school fund-raising carnival. The cost is $85.00 for the first 3 hours and then $19.50 for each additional hour or part thereof. How long can the tank be rented if up to $185 is budgeted for this expense?

110. AVERAGING GRADES   A student has scores of 70, 77, and 85 on three government exams. What score does she need on a fourth exam to give her an average of 80 or better?

111. WORK SCHEDULES   A student works two part-time jobs. He earns $8 an hour for working at the college library and $15 an hour for construction work. To save time for study, he limits his work to 25 hours a week. If he enjoys the work at the library more, how many hours can he work at the library and still earn at least $300 a week?

112. SCHEDULING EQUIPMENT   An excavating company charges $300 an hour for the use of a backhoe and $500 an hour for the use of a bulldozer. (Part of an hour counts as a full hour.) The company employs one operator for 40 hours per week to operate the machinery. If the company wants to bring in at least $18,500 each week from equipment rental, how many hours per week can it schedule the operator to use a backhoe?

113. VIDEO GAME SYSTEMS   A student who can afford to spend up to $1,000 sees the ad shown in the illustration. If she decides to buy the video game system, find the greatest number of video games that she can also purchase. (Disregard sales tax.)

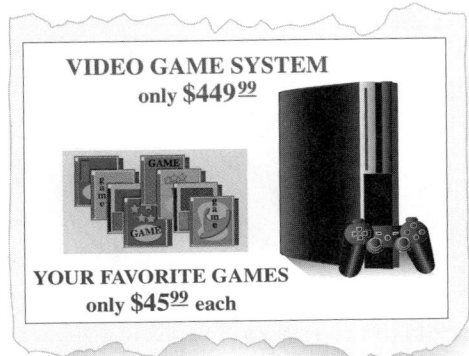

114. INVESTMENTS   If a woman has invested $10,000 at 8% annual interest, how much more must she invest at 9% so that her annual income will exceed $1,250?

WRITING

115. Explain why the equation $x = x + 1$ doesn't have a real-number solution.

116. Explain what is wrong with the following statement:

> *When solving inequalities involving negative numbers, the direction of the inequality symbol must be reversed.*

REVIEW

Simplify each expression. Write answers using only positive exponents.

117. $\left(\dfrac{t^3 t^5 t^{-6}}{t^2 t^{-4}}\right)^{-3}$

118. $\left(\dfrac{a^{-2} b^3 a^5 b^{-2}}{a^6 b^{-5}}\right)^{-4}$

CHALLENGE PROBLEMS

119. Find the value of k that makes 4 a solution of the following linear equation in x.

$$k + 3x - 6 = 3kx - k + 16$$

120. Solve: $0.75(x - 5) - \dfrac{4}{5} = \dfrac{1}{6}(3x + 1) + 3.2$

121. Consider the following "solution" of the inequality $\frac{1}{3} > \frac{1}{x}$ where it appears that the solution set is the interval $(3, \infty)$.

$$\frac{1}{3} > \frac{1}{x}$$

$$3x\left(\frac{1}{3}\right) > 3x\left(\frac{1}{x}\right)$$

$$x > 3$$

a. Show that $x = -1$ makes the original inequality true.

b. If $x = -1$ makes the original inequality true, there must be an error in the "solution." Where is it?

122. MEDICAL PLANS   A college provides its employees with a choice of the two medical plans shown in the following table. For what size hospital bills is Plan 2 better for the employee than Plan 1?   (*Hint:* The cost to the employee includes both the deductible payment and the employee's coinsurance payment.)

Plan 1	Plan 2
Employee pays $100	Employee pays $200
Plan pays 70% of the rest	Plan pays 80% of the rest

SECTION 8.2
Solving Compound Inequalities

Objectives

1 Find the intersection and the union of two sets.

2 Solve compound inequalities containing the word *and.*

3 Solve double linear inequalities.

4 Solve compound inequalities containing the word *or.*

A label on a first-aid cream warns the user about the temperature at which the medication should be stored. A careful reading reveals that the storage instructions consist of two parts:

> The storage temperature should be at least 59°F

and

> the storage temperature should be at most 77°F

DIRECTIONS: Clean the affected area thoroughly. Apply a small amount of this product (an amount equal to the surface area of the tip of a finger) on the area 1 to 3 times daily. Do not use in eyes. **Store at 59° to 77°F.** Do not use longer than 1 week. Keep this and all drugs out of the reach of children.

When the word *and* or the word *or* is used to connect pairs of inequalities, we call the statement a *compound inequality.* To solve compound inequalities, we need to know how to find the *intersection* and *union* of two sets.

 Find the Intersection and the Union of Two Sets.

Just as operations such as addition and multiplication are performed on real numbers, operations can also be performed on sets. The operation of intersection of two sets produces a new third set that consists of all of the elements that the two given sets have in common.

The Intersection of Two Sets	The **intersection of set A and set B,** written $A \cap B$, is the set of all elements that are common to set A and set B.

The operation of union of two sets produces a third set that is a combination of all of the elements of the two given sets.

The Union of Two Sets	The **union of set A and set B,** written $A \cup B$, is the set of elements that belong to set A or set B or both.

Venn diagrams can be used to illustrate the intersection and union of sets. The area shown in purple in figure (a) represents $A \cap B$ and the area shown in both shades of red in figure (b) represents $A \cup B$.

(a)   (b)

EXAMPLE 1   Let $A = \{0, 1, 2, 3, 4, 5, 6\}$ and $B = \{-4, -2, 0, 2, 4\}$.
 a. Find $A \cap B$.   b. Find $A \cup B$.

Strategy   In part (a), we will find the elements that sets A and B have in common, and in part (b), we will find the elements that are in one set or the other.

Why   The symbol $\cap$ means intersection, and the symbol $\cup$ means union.

Solution
 a. Since the numbers 0, 2, and 4 are common to both sets A and B, we have

$$A \cap B = \{0, 2, 4\}$$

 b. Since the numbers in either or both sets are $-4, -2, 0, 1, 2, 3, 4, 5,$ and 6, we have

$$A \cup B = \{-4, -2, 0, 1, 2, 3, 4, 5, 6\}$$

 Self Check 1   Let $C = \{8, 9, 10, 11\}$ and $D = \{3, 6, 9, 12, 15\}$.
 a. Find $C \cap D$.   b. Find $C \cup D$.

Now Try **Problems 17 and 21**

---

② **Solve Compound Inequalities Containing the Word *And*.**

When two inequalities are joined with the word *and*, we call the statement a **compound inequality.** Some examples are

$$x \geq -3 \quad \text{and} \quad x \leq 6$$

$$\frac{x}{2} + 1 > 0 \quad \text{and} \quad 2x - 3 < 5$$

$$x + 3 \leq 2x - 1 \quad \text{and} \quad 3x - 2 < 5x - 4$$

The solution set of a compound inequality containing the word *and* includes all numbers that make both of the inequalities true. That is, it is the intersection of their solution sets. We can find the solution set of the compound inequality $x \geq -3$ and $x \leq 6$, for example, by graphing the solution sets of each inequality on the same number line and looking for the numbers common to both graphs.

In the following figure, the graph of the solution set of $x \geq -3$ is shown in red, and the graph of the solution set of $x \leq 6$ is shown in blue.

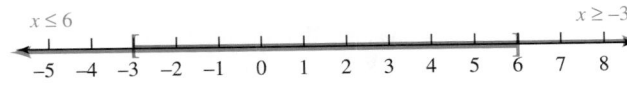

The figure below shows the graph of the solution of the compound inequality $x \geq -3$ and $x \leq 6$. The purple shaded interval, where the red and blue graphs intersect, represents the numbers that are common to the graphs of $x \geq -3$ and $x \leq 6$.

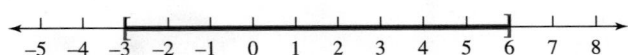

The solution set of $x \geq -3$ and $x \leq 6$ is the **bounded interval** $[-3, 6]$, where the brackets indicate that the endpoints, -3 and 6, are included. It represents all real numbers between -3 and 6, including -3 and 6. Intervals such as this, which contain both endpoints, are called **closed intervals.**

Since the solution set of $x \geq -3$ and $x \leq 6$ is the intersection of the solution sets of the two inequalities, we can write

$$[-3, \infty) \cap (-\infty, 6] = [-3, 6]$$

The solution set of the compound inequality $x \geq -3$ and $x \leq 6$ can be expressed in several ways:

1. As a graph:

2. In words: all real numbers between -3 and 6, including -3 and 6

3. In interval notation: $[-3, 6]$

4. Using set-builder notation: $\{x \mid x \geq -3 \text{ and } x \leq 6\}$

EXAMPLE 2   Solve $\dfrac{x}{2} + 1 > 0$ and $2x - 3 < 5$. Graph the solution set and write it using interval notation and set-builder notation.

Strategy   We will solve each inequality separately. Then we will graph the two solution sets on the same number line and determine their intersection.

Why   The solution set of a compound inequality containing the word *and* is the intersection of the solution sets of the two inequalities.

Solution   In each case, we can use properties of inequality to isolate the variable on one side of the inequality.

$$\frac{x}{2} + 1 > 0 \qquad \text{and} \qquad 2x - 3 < 5 \qquad \text{This is the compound inequality to solve.}$$

$$\frac{x}{2} > -1 \qquad\qquad\qquad 2x < 8$$

$$x > -2 \qquad\qquad\qquad\quad x < 4$$

Next, we graph the solutions of each inequality on the same number line and determine their intersection.

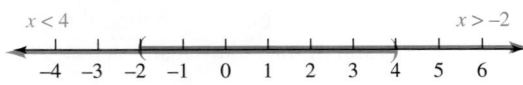

We see that the intersection of the graphs is the set of all real numbers between -2 and 4. The solution set of the compound inequality is the interval $(-2, 4)$, whose graph is shown below. This bounded interval, which does not include either endpoint, is called an **open interval.** Written using set-builder notation, the solution set is $\{x \mid x > -2 \text{ and } x < 4\}$.

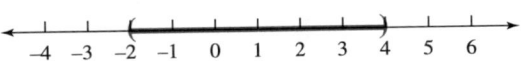

Self Check 2   Solve $3x > -18$ and $\frac{x}{5} - 1 \le 1$. Graph the solution set and write it using interval notation.

Now Try   **Problem 27**

The solution of the compound inequality in the Self Check of Example 2 is the interval $(-6, 10]$. A bounded interval such as this, which includes only one endpoint, is called a **half-open interval.** The following chart shows the various types of bounded intervals, along with the inequalities and interval notation that describe them.

Intervals			
	Open intervals	The interval (a, b) includes all real numbers x such that $a < x < b$.	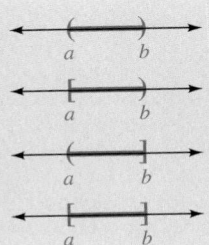
	Half-open intervals	The interval $[a, b)$ includes all real numbers x such that $a \le x < b$.	
		The interval $(a, b]$ includes all real numbers x such that $a < x \le b$.	
	Closed intervals	The interval $[a, b]$ includes all real numbers x such that $a \le x \le b$.	

EXAMPLE 3   Solve $x + 3 \le 2x - 1$ and $3x - 2 < 5x - 4$. Graph the solution set and write it using interval notation and set-builder notation.

Strategy   We will solve each inequality separately. Then we will graph the two solution sets on the same number line and determine their intersection.

Why   The solution set of a compound inequality containing the word *and* is the intersection of the solution sets of the two inequalities.

Solution   In each case, we can use properties of inequality to isolate the variable on one side.

$$x + 3 \le 2x - 1 \qquad \text{and} \qquad 3x - 2 < 5x - 4 \qquad \text{This is the compound inequality to solve.}$$
$$4 \le x \qquad\qquad\qquad\qquad 2 < 2x$$
$$x \ge 4 \qquad\qquad\qquad\qquad 1 < x$$
$$\qquad\qquad\qquad\qquad\qquad x > 1$$

The graph of $x \ge 4$ is shown below in red and the graph of $x > 1$ is shown below in blue.

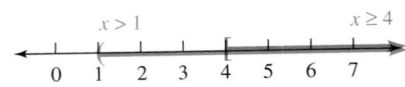

Only those values of x where $x \ge 4$ and $x > 1$ are in the solution set of the compound inequality. Since all numbers greater than or equal to 4 are also greater than 1, the solutions are the numbers x where $x \geq 4$. The solution set is the interval $[4, \infty)$, whose graph is shown below. Written using set-builder notation, the solution set is $\{x \mid x \geq 4\}$.

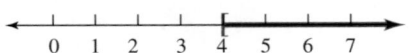

Self Check 3   Solve $2x + 3 < 4x + 2$ and $3x + 1 < 5x + 3$. Graph the solution set and write it using interval notation.

Now Try   **Problem 29**

EXAMPLE 4   Solve $x - 1 > -3$ and $2x < -8$, if possible.

Strategy   We will solve each inequality separately. Then we will graph the two solution sets on the same number line and determine their intersection, if any.

Why   The solution set of a compound inequality containing the word *and* is the intersection of the solution sets of the two inequalities.

Solution   In each case, we can use properties of inequality to isolate the variable on one side.

$$x - 1 > -3 \quad \text{and} \quad 2x < -8 \quad \textit{This is the compound inequality to solve.}$$
$$x > -2 \qquad\qquad\quad x < -4$$

The graphs of the solution sets shown below do not intersect. Since there are no numbers that make both parts of the original compound inequality true, $x - 1 > -3$ and $2x < -8$ has no solution.

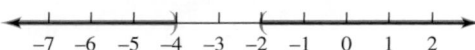

The solution set of the compound inequality is the empty set, which can be written as $\varnothing$.

Self Check 4   Solve $2x - 3 < x - 2$ and $0 < x - 3.5$, if possible.

Now Try   **Problem 31**

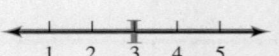
---

3   **Solve Double Linear Inequalities.**

Inequalities that contain exactly two inequality symbols are called **double inequalities.** An example is

$$-3 \leq 2x + 5 < 7 \quad \textit{Read as "−3 is less than or equal to 2x + 5 and 2x + 5 is less than 7."}$$

Any double linear inequality can be written as a compound inequality containing the word *and*. In general, the following is true.

| **Double Linear Inequalities** | The compound inequality $c < x < d$ is equivalent to $c < x$ and $x < d$. |

EXAMPLE 5   Solve $-3 \le 2x + 5 < 7$. Graph the solution set and write it using interval notation and set-builder notation.

Strategy   We will solve the double inequality by applying properties of inequality to *all three of its parts* to isolate x in the middle.

Why   This double inequality $-3 \le 2x + 5 < 7$ means that $-3 \le 2x + 5$ and $2x + 5 < 7$. We can solve it more easily by leaving it in its original form.

Solution

$-3 \le 2x + 5 < 7$	This is the double inequality to solve.
$-3 - 5 \le 2x + 5 - 5 < 7 - 5$	To undo the addition of 5, subtract 5 from all three parts.
$-8 \le 2x < 2$	Perform the subtractions.
$\dfrac{-8}{2} \le \dfrac{2x}{2} < \dfrac{2}{2}$	To isolate x, undo the multiplication by 2 by dividing all three parts by 2.
$-4 \le x < 1$	Perform the divisions.

The solution set of the double linear inequality is the half-open interval $[-4, 1)$, whose graph is shown below. Written using set-builder notation, the solution set is $\{x \mid -4 \le x < 1\}$.

 Self Check 5   Solve $-5 \le 3x - 8 \le 7$. Graph the solution set and write it using interval notation.

Now Try   **Problem 35**

> *Notation*
> Note that the two inequality symbols in $-3 \le 2x + 5 < 7$ point in the same direction and point to the smaller number.

Caution   When multiplying or dividing all three parts of a double inequality by a negative number, don't forget to reverse the direction of both inequalities. As an example, we solve $-15 < -5x \le 25$.

$-15 < -5x \le 25$	
$\dfrac{-15}{-5} > \dfrac{-5x}{-5} \ge \dfrac{25}{-5}$	Divide all three parts by -5 to isolate x in the middle. Reverse both inequality signs.
$3 > x \ge -5$	Perform the divisions.
$-5 \le x < 3$	Write an equivalent double inequality with the smaller number, -5, on the left.

4   Solve Compound Inequalities Containing the Word *Or.*

A warning on the water temperature gauge of a commercial dishwasher cautions the operator to shut down the unit if

The water temperature goes below 140°

or

The water temperature goes above 160°

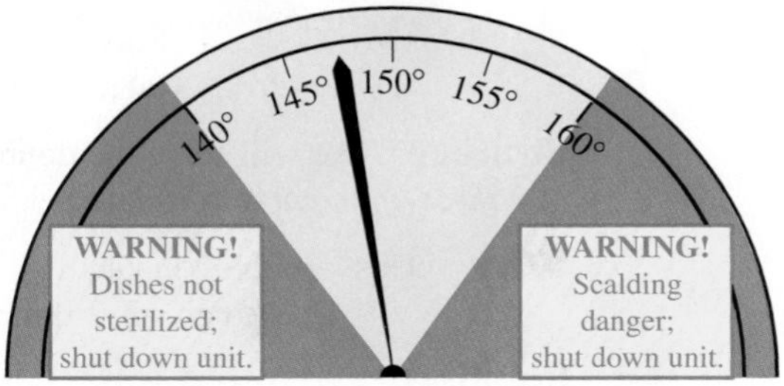

When two inequalities are joined with the word *or,* we also call the statement a compound inequality. Some examples are

$x < 140$ or $x > 160$

$x \le -3$ or $x \ge 2$

$\frac{x}{3} > \frac{2}{3}$ or $-(x - 2) > 3$

Caution
It is incorrect to write the statement $x \le -3$ or $x \ge 2$ as the double inequality $2 \le x \le -3$, because that would imply that $2 \le -3$, which is false.

The solution set of a compound inequality containing the word *or* includes all numbers that make one or the other or both inequalities true. That is, it is the union of their solution sets. We can find the solution set of $x \le -3$ or $x \ge 2$, for example, by drawing the graphs of each inequality on the same number line.

In the following figure, the graph of the solution set of $x \le -3$ is shown in red, and the graph of the solution set of $x \ge 2$ is shown in blue.

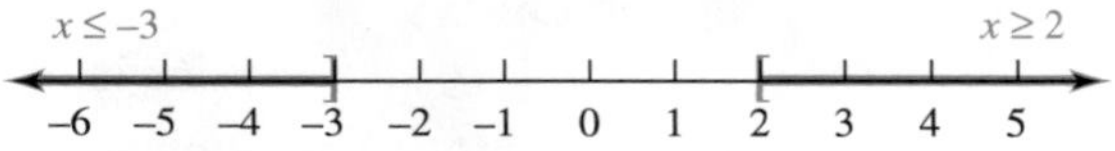

The figure below shows the graph of the solution set of $x \le -3$ or $x \ge 2$. This graph is a union of the graph of $x \le -3$ with the graph of $x \ge 2$.

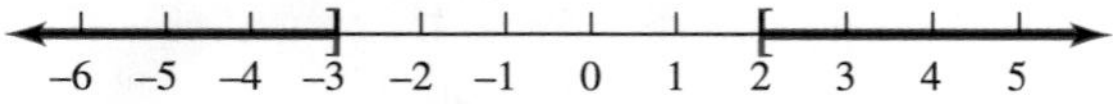

For the compound inequality $x \le -3$ or $x \ge 2$, we can write the solution set as the union of two intervals:

$$(-\infty, -3] \cup [2, \infty)$$

We can express the solution set of the compound inequality $x \le -3$ or $x \ge 2$ in several ways:

1. As a graph: 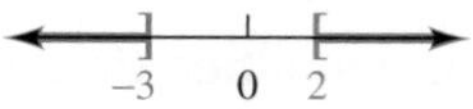

2. In words: all real numbers less than or equal to -3 *or* greater than or equal to 2
3. As the union of two intervals: $(-\infty, -3] \cup [2, \infty)$
4. Using set-builder notation: $\{x \mid x \le -3 \text{ or } x \ge 2\}$

EXAMPLE 6   Solve $\frac{x}{3} > \frac{2}{3}$ or $-(x - 2) > 3$. Graph the solution set and write it using interval notation and set-builder notation.

Strategy   We will solve each inequality separately. Then we will graph the two solution sets on the same number line to show their union.

Why   The solution set of a compound inequality containing the word *or* is the union of the solution sets of the two inequalities.

Solution   To solve each inequality, we proceed as follows:

$$\frac{x}{3} > \frac{2}{3} \quad \text{or} \quad -(x - 2) > 3 \qquad \text{This is the compound inequality to solve.}$$

$$x > 2 \qquad\qquad\qquad -x + 2 > 3$$
$$-x > 1$$
$$x < -1$$

Next, we graph the solutions of each inequality on the same number line and determine their union.

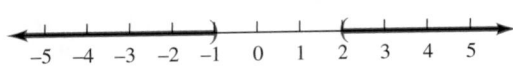

The union of the two solution sets consists of all real numbers less than -1 or greater than 2. The solution set of the compound inequality is the union of two intervals: $(-\infty, -1) \cup (2, \infty)$. Its graph appears below. Written using set-builder notation, the solution set is $\{x \mid x > 2 \text{ or } x < -1\}$.

The Language of Algebra
The meaning of the word *or* in a compound inequality differs from our everyday use of the word. For example, when we say, "I will go shopping today *or* tomorrow," we mean that we will go one day or the other, but *not* both. With compound inequalities, *or* includes one possibility, or the other, or both.

(number line graph from −5 to 5 showing solution set)

Self Check 6   Solve $\frac{x}{2} > 2$ or $-3(x - 2) > 0$. Graph the solution set and write it using interval notation.

Now Try   Problem 41

EXAMPLE 7   Solve $x + 3 \geq -3$ or $-x > 0$. Graph the solution set and write it using interval notation and set-builder notation.

Strategy   We will solve each inequality separately. Then we will graph the two solution sets on the same number line to show their union.

Why   The solution set of a compound inequality containing the word *or* is the union of the solution sets of the two inequalities.

Solution   To solve each inequality, we proceed as follows:

$$x + 3 \geq -3 \quad \text{or} \quad -x > 0 \qquad \text{This is the compound inequality to solve.}$$
$$x \geq -6 \qquad\qquad x < 0$$

We graph the solution set of each inequality on the same number line and determine their union.

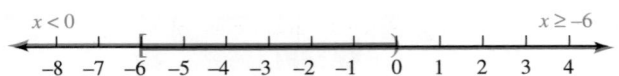

Since the entire number line is shaded, all real numbers satisfy the original compound inequality and the solution set is denoted as $(-\infty, \infty)$ or $\mathbb{R}$. Its graph is shown below. Written using set-builder notation, the solution set is $\{x \mid x \text{ is a real number}\}$.

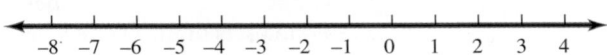

Self Check 7 Solve $x - 1 < 5$ or $-2x \leq 10$. Graph the solution set and write it using interval notation

Now Try Problem 43

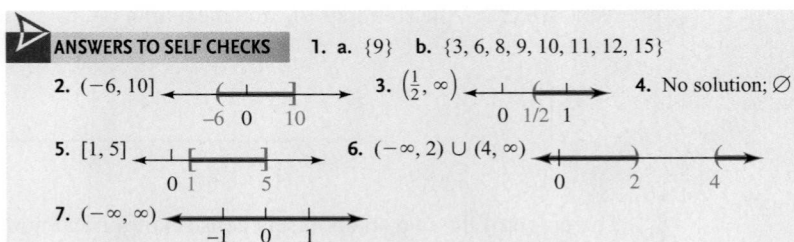

ANSWERS TO SELF CHECKS **1. a.** $\{9\}$ **b.** $\{3, 6, 8, 9, 10, 11, 12, 15\}$
2. $(-6, 10]$ **3.** $\left(\frac{1}{2}, \infty\right)$ **4.** No solution; $\varnothing$
5. $[1, 5]$ **6.** $(-\infty, 2) \cup (4, \infty)$
7. $(-\infty, \infty)$

STUDY SET
8.2

VOCABULARY

Fill in the blanks.

1. The _____ of two sets is the set of elements that are common to both sets and the _____ of two sets is the set of elements that are in one set, or the other, or both.

2. $x \geq 3$ and $x < 4$ is a _____ inequality.

3. $-6 < x + 1 \leq 1$ is a _____ linear inequality.

4. (2, 8) is an example of an open _____, $[-4, 0]$ is an example of a _____ interval, and (0, 9] is an example of a half-_____ interval.

CONCEPTS

Fill in the blanks.

5. a. The solution set of a compound inequality containing the word *and* includes all numbers that make _____ inequalities true.

b. The solution set of a compound inequality containing the word *or* includes all numbers that make _____, or the other, or _____ inequalities true.

6. The double inequality $4 < 3x + 5 \leq 15$ is equivalent to $4 < 3x + 5$ _____ $3x + 5 \leq 15$.

7. a. When solving a compound inequality containing the word *and,* the solution set is the _____ of the solution sets of the inequalities.

b. When solving a compound inequality containing the word *or,* the solution set is the _____ of the solution sets of the inequalities.

8. When multiplying or dividing all three parts of a double inequality by a negative number, the direction of both inequality symbols must be _____.

9. In each case, determine whether -3 is a solution of the compound inequality.

a. $\frac{x}{3} + 1 \geq 0$ and $2x - 3 < -10$

b. $2x \leq 0$ or $-3x < -5$

10. In each case, determine whether -3 is a solution of the double linear inequality.

a. $-1 < -3x + 4 < 12$

b. $-1 < -3x + 4 < 14$

11. Use interval notation, if possible, to describe the intersection of each pair of graphs.

a.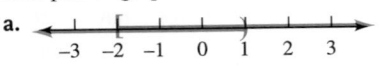

b.

c.

12. Use interval notation to describe the union of each pair of graphs.

a.

b.

c.

NOTATION

13. Fill in the blanks: We read $\cup$ as _____ and $\cap$ as _____.

14. Match each interval with its corresponding graph.

a. $[2, 3)$ i.

b. $(2, 3)$ ii.

c. $[2, 3]$ iii.

15. What set is represented by the interval notation $(-\infty, \infty)$? Graph it.

16. a. Graph: $(-\infty, 2) \cup [3, \infty)$

b. Graph: $(-\infty, 3) \cap [-2, \infty)$

GUIDED PRACTICE

Let $A = \{0, 1, 2, 3, 4, 5, 6\}$, $B = \{4, 6, 8, 10\}$, $C = \{-3, -1, 0, 1, 2\}$, and $D = \{-3, 1, 2, 5, 8\}$.
Find each set. See Example 1.

17. $A \cap B$ 18. $A \cap D$

19. $C \cap D$ 20. $B \cap C$

21. $B \cup C$ 22. $A \cup C$

23. $A \cup D$ 24. $C \cup D$

Solve each compound inequality, if possible. Graph the solution set (if one exists) and write it using interval notation. See Examples 2–4.

25. $x > -2$ and $x \leq 5$

26. $x \leq -4$ and $x \geq -7$

27. $2x - 1 > 3$ and $x + 8 \leq 11$

28. $5x - 3 \geq 2$ and $6 \geq 4x - 3$

29. $6x + 1 < 5x - 3$ and $\dfrac{x}{2} + 9 \leq 6$

30. $\dfrac{2}{3}x + 1 > -9$ and $\dfrac{3}{4}x - 1 > -10$

31. $x + 2 < -\dfrac{1}{3}x$ and $-6x < 9x$

32. $\dfrac{3}{2}x + \dfrac{1}{5} < 5$ and $2x + 1 > 9$

Solve each double inequality. Graph the solution set and write it using interval notation. See Example 5.

33. $4 \leq x + 3 \leq 7$ 34. $-5.3 \leq x - 2.3 \leq -1.3$

35. $0.9 < 2x - 0.7 < 1.5$ 36. $7 < 3x - 2 < 25$

Solve each compound inequality. Graph the solution set and write it using interval notation. See Examples 6 and 7.

37. $x \leq -2$ or $x > 6$

38. $x \geq -1$ or $x \leq -3$

39. $x - 3 < -4$ or $-x + 2 < 0$

40. $4x < -12$ or $\dfrac{x}{2} > 4$

41. $3x + 2 < 8$ or $2x - 3 > 11$

42. $3x + 4 < -2$ or $3x + 4 > 10$

43. $2x > x + 3$ or $\dfrac{x}{8} + 1 < \dfrac{13}{8}$

44. $2(x + 2) < x - 11$ or $-\dfrac{x}{5} < 20$

TRY IT YOURSELF

Solve each compound inequality, if possible. Graph the solution set (if one exists) and write it using interval notation.

45. $-4(x + 2) \geq 12$ or $3x + 8 < 11$

46. $4.5x - 1 < -10$ or $6 - 2x \geq 12$

47. $2.2x < -19.8$ and $-4x < 40$

48. $\dfrac{1}{2}x \leq 2$ and $0.75x \geq -6$

49. $-2 < -b + 3 < 5$

50. $2 < -t - 2 < 9$

51. $4.5x - 2 > 2.5$ or $\dfrac{1}{2}x \leq 1$

52. $0 < x$ or $3x - 5 > 4x - 7$

53. $5(x - 2) \geq 0$ and $-3x < 9$

54. $x - 1 \leq 2(x + 2)$ and $x \leq 2x - 5$

55. $-x < -2x$ and $3x > 2x$

56. $-\dfrac{x}{4} > -2.5$ and $9x > 2(4x + 5)$

57. $-6 < -3(x - 4) \leq 24$

58. $-4 \leq -2(x + 8) < 8$

59. $2x + 1 \geq 5$ and $-3(x + 1) \geq -9$

60. $2(-2) \le 3x - 1$ and $3x - 1 \le -1 - 3$

61. $\dfrac{4.5x - 12}{2} < x$ or $-15.3 > -3(x - 1.4)$

62. $y + 0.52 < 1.05y$ or $9.8 - 15y > -15.7$

63. $\dfrac{x}{0.7} + 5 > 4$ and $-4.8 \le \dfrac{3x}{-0.125}$

64. $5(x + 1) \le 4(x + 3)$ and $x + 12 < -3$

65. $-24 < \dfrac{3}{2}x - 6 \le -15$

66. $-4 > \dfrac{2}{3}x - 2 > -6$

67. $\dfrac{x}{3} - \dfrac{x}{4} > \dfrac{1}{6}$ or $\dfrac{x}{2} + \dfrac{2}{3} \le \dfrac{3}{4}$

68. $\dfrac{a}{2} + \dfrac{7}{4} > 5$ or $\dfrac{3}{8} + \dfrac{a}{3} \le \dfrac{5}{12}$

69. $0 \le \dfrac{4 - x}{3} \le 2$

70. $-2 \le \dfrac{5 - 3x}{2} \le 2$

71. $x \le 6 - \dfrac{1}{2}x$ and $\dfrac{1}{2}x + 1 \ge 3$

72. $3\left(x + \dfrac{2}{3}\right) \le -7$ and $2(x + 2) \ge -2$

APPLICATIONS

73. BABY FURNITURE   Refer to the illustration. A company manufactures various sizes of play yard cribs having perimeters between 128 and 192 inches, inclusive.

a. Complete the double inequality that describes the range of the perimeters of the play yard shown.

$$\boxed{} \le 4s \le \boxed{}$$

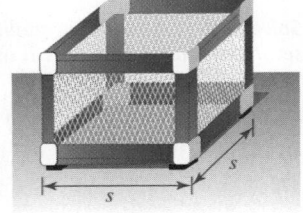

b. Solve the double inequality to find the range of the side lengths of the play yard.

74. TRUCKING   The distance that a truck can travel in 8 hours, at a constant rate of r mph, is given by $8r$. A trucker wants to travel at least 350 miles, and company regulations don't allow him to exceed 450 miles in one 8-hour shift.

a. Complete the double inequality that describes the mileage range of the truck.

$$\boxed{} \le 8r \le \boxed{}$$

b. Solve the double inequality to find the range of the average rate (speed) of the truck for the 8-hour trip.

75. Suppose that you are an HVAC technician and that you are repairing a thermostat in a commercial building. During business hours as shown in figure (a), the *Temp range* control is set at 5. This means that the heater comes on when the room temperature gets 5 degrees below the *Temp setting* and the air conditioner comes on when the room temperature gets 5 degrees above the *Temp setting.*

a. Use interval notation to describe the temperature range when neither the heater nor the air conditioner will come on during business hours.

b. After business hours, the Temp range setting is changed to save energy. See figure (b). Use interval notation to describe the after business hours temperature range when neither the heater nor the air conditioner will be come on.

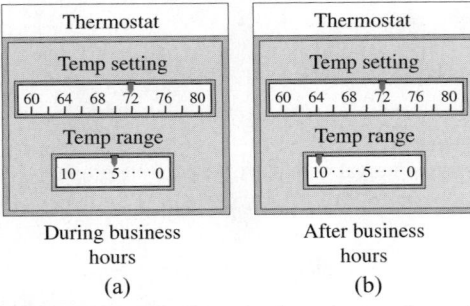

During business   After business
    hours   hours
     (a)   (b)

76. TREATING FEVERS   Use the flow chart to determine what action should be taken for a 13-month-old child who has had a 99.8° temperature for 3 days and is not suffering any other symptoms. T represents the child's temperature, A the child's age in months, and S the number of hours the child has experienced the symptoms.

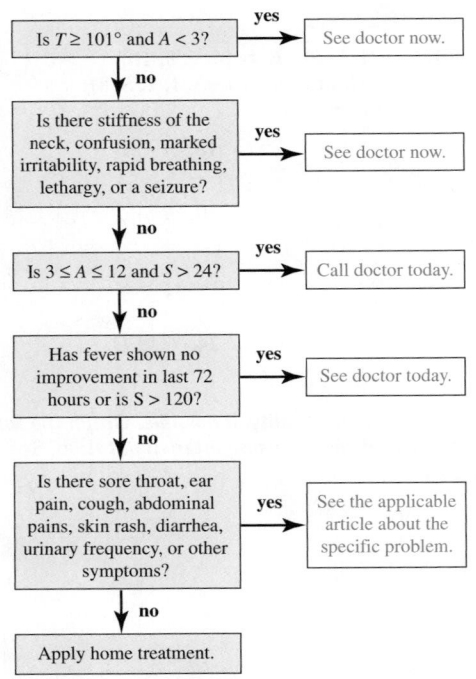

Based on information from *Take Care of Yourself* (Addison-Wesley, 1993)

77. U.S. HEALTH CARE   Refer to the following graph. Let *P* represent the percent of children covered by private insurance, *M* the percent covered by Medicare/Medicaid, and *N* the percent not covered. For what years are the following true?

a. $P \geq 65$ and $M \geq 20$

b. $P > 63$ or $M \geq 25$

c. $M \geq 25$ and $N \leq 10$

d. $M \geq 26$ or $N < 11$

U.S. Health Care Coverage for Persons Under 18 Years of Age (in percent)

Private insurance ☐   Medicaid ☐
Not covered ☐

Year	Private insurance	Medicaid	Not covered
2001	66.7	21.2	11.0
2002	63.5	24.8	10.9
2003	63.0	26.0	9.8
2004	63.2	26.4	9.2

Source: U.S. Department of Health and Human Services

78. POLLS   For each response to the poll question shown below, the *margin of error* is +/− (read as "plus or minus") 3.4%. This means that for the statistical methods used to do the polling, the actual response could be as much as 3.4 points more or 3.4 points less than shown. Use interval notation to describe the possible interval (in percent) for each response.

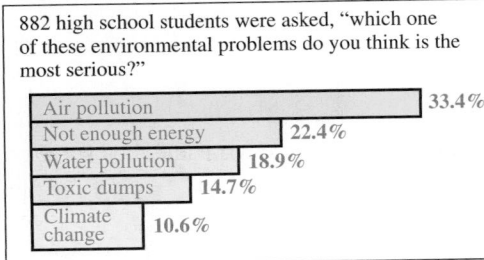

882 high school students were asked, "which one of these environmental problems do you think is the most serious?"

Air pollution 33.4%
Not enough energy 22.4%
Water pollution 18.9%
Toxic dumps 14.7%
Climate change 10.6%

Source: Zogby International

79. STREET INTERSECTIONS   Refer to the illustration in the next column.

a. Shade the area that represents the intersection of the two streets shown in the illustration.

b. Shade the area that represents the union of the two streets.

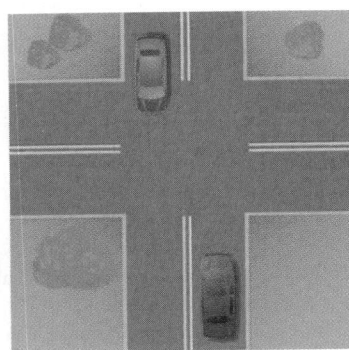

80. TRAFFIC SIGNS   The pair of signs shown below are a real-life example of which concept discussed in this section?

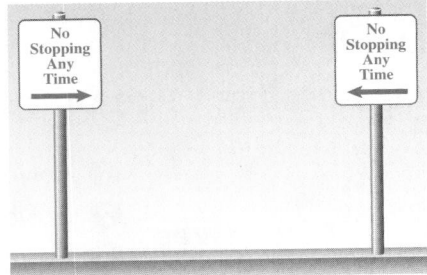

WRITING

81. Explain how to find the union and how to find the intersection of $(-\infty, 5)$ and $(-2, \infty)$ graphically.

82. Explain why the double inequality

$$2 < x < 8$$

can be written in the equivalent form

$$2 < x \text{ and } x < 8$$

83. Explain the meaning of the notation $(2, 3)$ for each type of graph.

a.

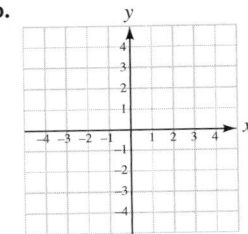

b.

84. The meaning of the word *or* in a compound inequality differs from our everyday use of the word. Explain the difference.

85. Describe each set in words.

   a. $(-3, 3)$

   b. $[7, 12]$

   c. $(-\infty, 5] \cup (6, \infty)$

86. What is incorrect about the double inequality
$3 < -3x + 4 < -3$?

REVIEW

87. INVESTMENTS   Equal amounts are invested at 6%, 7%, and 8% annual interest. The three investments yield a total of $2,037 annual interest. Find the total amount of money invested.

88. MEDICATIONS   A doctor prescribes an ointment that is 2% hydrocortisone. A pharmacist has 1% and 5% concentrations in stock. How many ounces of each should the pharmacist use to make a 1-ounce tube?

CHALLENGE PROBLEMS

Solve each compound inequality. Graph the solution set and write it in interval notation.

89. $-5 < \dfrac{x+2}{-2} < 0$ or $2x + 10 \geq 30$

90. $-2 \leq \dfrac{x-4}{3} \leq 0$ and $\dfrac{x-5}{2} \geq -3$

SECTION 8.3
Solving Absolute Value Equations and Inequalities

Objectives

  1   Solve equations of the form $|X| = k$.

  2   Solve equations with two absolute values.

  3   Solve inequalities of the form $|X| < k$.

  4   Solve inequalities of the form $|X| > k$.

Many quantities studied in mathematics, science, and engineering are expressed as positive numbers. To guarantee that a quantity is positive, we often use absolute value. In this section, we will consider equations and inequalities involving the absolute value of an algebraic expression. Some examples are

$$|3x - 2| = 5, \qquad |2x - 3| < 9, \qquad \text{and} \qquad \left| \frac{3-x}{5} \right| \geq 6$$

To solve these *absolute value equations* and *inequalities,* we write and then solve equivalent compound equations and inequalities.

1   **Solve Equations of the Form $|X| = k$.**

Recall that the absolute value of a real number is its distance from 0 on a number line. To solve the **absolute value equation** $|x| = 5$, we must find all real numbers x whose distance from 0 on the number line is 5. There are two such numbers: 5 and -5. It follows that the solutions of $|x| = 5$ are 5 and -5 and the solution set is $\{5, -5\}$.

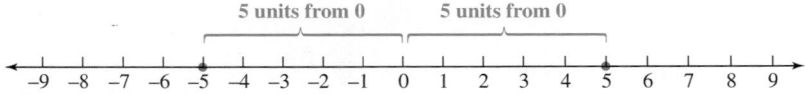

The results from this example suggest the following approach for solving absolute value equations.

Solving Absolute Value Equations	For any positive number k and any algebraic expression X: To solve $|X| = k$, solve the equivalent compound equation $$X = k \quad \text{or} \quad X = -k$$

The statement $X = k$ or $X = -k$ is called a **compound equation** because it consists of two equations joined with the word *or*.

EXAMPLE 1	Solve: **a.** $|x| = 8$ **b.** $|s| = 0.003$

Strategy   To solve each of these absolute value equations, we will write and solve an equivalent compound equation.

Why   We can use this approach because an equation of the form $|x| = k$, where k is positive, is equivalent to $x = k$ or $x = -k$.

The Language of Algebra
When we say that the absolute value equation and a compound equation are *equivalent,* we mean that they have the same solution(s).

Solution

a. The absolute value equation $|x| = 8$ is equivalent to the compound equation

$$x = 8 \quad \text{or} \quad x = -8$$

Therefore, the solutions of $|x| = 8$ are 8 and -8 and the solution set is $\{8, -8\}$.

b. The absolute value equation $|s| = 0.003$ is equivalent to the compound equation

$$s = 0.003 \quad \text{or} \quad s = -0.003$$

Therefore, the solutions of $|s| = 0.003$ are 0.003 and -0.003 and the solution set is $\{0.003, -0.003\}$.

 Self Check 1   Solve: **a.** $|y| = 24$ **b.** $|x| = \frac{1}{2}$

Now Try   **Problems 17 and 19**

The equation-solving procedure discussed in Example 1 can often be used when the expression within absolute value bars is more complicated than a single variable.

EXAMPLE 2	Solve: **a.** $|3x - 2| = 5$ **b.** $|10 - x| = -40$

Strategy   To solve the first equation, we will write and then solve an equivalent compound equation. We will solve the second equation by inspection.

Why   Both equations are of the form $|X| = k$. However, the standard method for solving absolute value equations cannot be applied to $|10 - x| = -40$ because k is negative.

Solution

a. The absolute value equation $|3x - 2| = 5$ is equivalent to the compound equation

$$3x - 2 = 5 \quad \text{or} \quad 3x - 2 = -5$$

Now we solve each equation for x:

$$3x - 2 = 5 \quad \text{or} \quad 3x - 2 = -5$$
$$3x = 7 \qquad\qquad\quad 3x = -3$$
$$x = \frac{7}{3} \qquad\qquad\quad x = -1$$

The results must be checked separately to see whether each of them produces a true statement. We substitute $\frac{7}{3}$ for x and then -1 for x in the original equation.

Check: **For** $x = \dfrac{7}{3}$ **For** $x = -1$

$$|3x - 2| = 5 \qquad\qquad\qquad |3x - 2| = 5$$
$$\left|3\left(\frac{7}{3}\right) - 2\right| \overset{?}{=} 5 \qquad\qquad |3(-1) - 2| \overset{?}{=} 5$$
$$|7 - 2| \overset{?}{=} 5 \qquad\qquad\qquad |-3 - 2| \overset{?}{=} 5$$
$$|5| \overset{?}{=} 5 \qquad\qquad\qquad\quad |-5| \overset{?}{=} 5$$
$$5 = 5 \quad \text{True} \qquad\qquad\qquad 5 = 5 \quad \text{True}$$

The resulting true statements indicate that the equation has two solutions: $\frac{7}{3}$ and -1.

b. Since an absolute value can never be negative, there are no real numbers x that make $|10 - x| = -40$ true. The equation has no solution and the solution set is $\varnothing$.

Self Check 2   Solve: **a.** $|2x - 3| = 7$ **b.** $\left|\dfrac{x}{4} - 1\right| = -3$

Now Try   **Problems 23 and 29**

Caution   When solving absolute value equations (or inequalities), isolate the absolute value expression on one side *before* writing the equivalent compound statement.

EXAMPLE 3   Solve: $\left|\dfrac{2}{3}x + 3\right| + 4 = 10$

Strategy   We will first isolate $\left|\frac{2}{3}x + 3\right|$ on the left side of the equation and then write and solve an equivalent compound equation.

Why   After isolating the absolute value expression on the left, the resulting equation will have the desired form $|X| = k$.

Solution

$$\left|\frac{2}{3}x + 3\right| + 4 = 10 \quad \text{\small This is the equation to solve.}$$

$$\left|\frac{2}{3}x + 3\right| = 6 \quad \begin{array}{l}\text{\small To isolate the absolute value expression, subtract 4} \\ \text{\small from both sides. The equation is in the form } |X| = k.\end{array}$$

With the absolute value now isolated, we can solve $\left|\frac{2}{3}x + 3\right| = 6$ by writing and solving an equivalent compound equation:

$$\frac{2}{3}x + 3 = 6 \quad \text{or} \quad \frac{2}{3}x + 3 = -6$$

Now we solve each equation for x:

$$\frac{2}{3}x + 3 = 6 \quad \text{or} \quad \frac{2}{3}x + 3 = -6$$

$$\frac{2}{3}x = 3 \qquad\qquad \frac{2}{3}x = -9$$

$$2x = 9 \qquad\qquad 2x = -27$$

$$x = \frac{9}{2} \qquad\qquad x = -\frac{27}{2}$$

Verify that both solutions, $\frac{9}{2}$ and $-\frac{27}{2}$, check by substituting them into the original equation.

 Self Check 3   Solve: $|0.4x - 2| - 0.6 = 0.4$

Now Try **Problem 35**

EXAMPLE 4   Solve: $3\left|\dfrac{1}{2}x - 5\right| - 4 = -4$

Strategy   We will first isolate $\left|\dfrac{1}{2}x - 5\right|$ on the left side of the equation and then write and solve an equivalent compound equation.

Why   After isolating the absolute value expression, the resulting equation will have the desired form $|X| = k$.

Solution

$$3\left|\frac{1}{2}x - 5\right| - 4 = -4 \qquad \text{This is the equation to solve.}$$

$$3\left|\frac{1}{2}x - 5\right| = 0 \qquad \text{Add 4 to both sides.}$$

$$\left|\frac{1}{2}x - 5\right| = 0 \qquad \begin{array}{l}\text{To isolate the absolute value expression, divide both sides by 3. The}\\ \text{equation is in the form } |X| = k.\end{array}$$

Since 0 is the only number whose absolute value is 0, the expression $\frac{1}{2}x - 5$ must be 0, and we have

$$\frac{1}{2}x - 5 = 0$$

$$\frac{1}{2}x = 5 \qquad \text{Add 5 to both sides.}$$

$$x = 10 \qquad \text{To solve for x, multiply both sides by 2.}$$

The solution is 10. Verify that it satisfies the original equation.

> **Success Tip**
> To solve most absolute value equations, we must consider two cases. However, if an absolute value is equal to **0**, we need only consider one: the case when the expression within the absolute value bars is equal to **0**.

 Self Check 4   Solve: $-5\left|\dfrac{2}{3}x + 4\right| + 1 = 1$

Now Try **Problem 41**

In Section 2.6 we discussed absolute value functions and their graphs. If we are given an output of an absolute value function, we can work in reverse to find the corresponding input(s).

EXAMPLE 5 Let $f(x) = |x + 4|$. For what value(s) of x is $f(x) = 20$?

Strategy We will substitute 20 for $f(x)$ and solve for x.

Why In the equation, there are two unknowns, x and $f(x)$. If we replace $f(x)$ with 20, we can solve the resulting absolute value equation for x.

Solution

$$f(x) = |x + 4|$$
$$20 = |x + 4| \quad \text{Substitute 20 for } f(x).$$

To solve $20 = |x + 4|$, we write and then solve an equivalent compound equation:

$$20 = x + 4 \quad \text{or} \quad -20 = x + 4$$

Now we solve each equation for x:

$$20 = x + 4 \quad \text{or} \quad -20 = x + 4$$
$$16 = x \quad \quad \quad -24 = x$$

To check, substitute 16 and then -24 for x in $f(x) = |x + 4|$, and verify that $f(x) = 20$ in each case.

Self Check 5 For what value(s) of x is $f(x) = 11$?

Now Try **Problem 77**

---

2 Solve Equations with Two Absolute Values.

Equations can contain two absolute value expressions. To develop a strategy to solve them, consider the following examples.

$$|3| = |3| \quad \text{or} \quad |-3| = |-3| \quad \text{or} \quad |3| = |-3| \quad \text{or} \quad |-3| = |3|$$

↑　↑　　　　　　↑　　↑　　　　　　↑　　↑　　　　　　↑　　↑

The same number.　　　The same number.　　These numbers are opposites.　　These numbers are opposites.

These four possible cases are really just two cases: *For two expressions to have the same absolute value, they must either be equal or be opposites of each other.* This observation suggests the following approach for solving equations having two absolute value expressions.

Solving Equations with Two Absolute Values

For any algebraic expressions X and Y:

To solve $|X| = |Y|$, solve the compound equation $X = Y$ or $X = -Y$.

EXAMPLE 6 Solve: $|5x + 3| = |3x + 25|$

Strategy To solve this equation, we will write and then solve an equivalent compound equation.

Why   We can use this approach because the equation is of the form $|X| = |Y|$.

Solution   The equation $|5x + 3| = |3x + 25|$, with the two absolute value expressions, is equivalent to the following compound equation:

The expressions within the
absolute value symbols are equal

The expressions within the
absolute value symbols are opposites

$$5x + 3 = 3x + 25 \qquad \text{or} \qquad 5x + 3 = -(3x + 25)$$
$$2x = 22 \qquad\qquad\qquad 5x + 3 = -3x - 25$$
$$x = 11 \qquad\qquad\qquad 8x = -28$$
$$x = -\frac{28}{8}$$
$$x = -\frac{7}{2}$$

Verify that both solutions, 11 and $-\frac{7}{2}$, check by substituting them into the original equation.

Self Check 6   Solve: $|2x - 3| = |4x + 9|$

Now Try   **Problem 49**

③   **Solve Inequalities of the Form $|X| < k$.**

To solve the **absolute value inequality** $|x| < 5$, we must find all real numbers x whose distance from 0 on the number line is less than 5. From the graph, we see that there are many such numbers. For example, -4.999, -3, -2.4, $-1\frac{7}{8}$, $-\frac{3}{4}$, 0, 1, 2.8, 3.001, and 4.999 all meet this requirement. We conclude that the solution set is all numbers between -5 and 5, which can be written $(-5, 5)$.

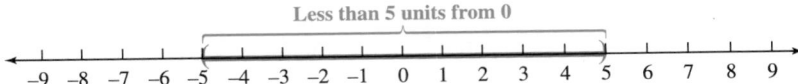

Less than 5 units from 0

Since x is between -5 and 5, it follows that $|x| < 5$ is equivalent to $-5 < x < 5$. This observation suggests the following approach for solving absolute value inequalities of the form $|X| < k$ and $|X| \le k$.

Solving $|X| < k$ and $|X| \le k$

For any positive number k and any algebraic expression X:

To solve $|X| < k$, solve the equivalent double inequality $-k < X < k$.

To solve $|X| \le k$, solve the equivalent double inequality $-k \le X \le k$.

EXAMPLE 7   Solve $|2x - 3| < 9$ and graph the solution set.

Strategy   To solve this absolute value inequality, we will write and solve an equivalent double inequality.

Why   We can use this approach because the inequality is of the form $|X| < k$, and k is positive.

Solution The absolute value inequality $|2x - 3| < 9$ is equivalent to the double inequality

$$-9 < 2x - 3 < 9$$

which we can solve for x:

$$-9 < 2x - 3 < 9$$
$$-6 < 2x < 12 \qquad \text{To isolate the variable term } 2x, \text{ add 3 to all three parts.}$$
$$-3 < x < 6 \qquad \text{To isolate } x, \text{ divide all parts by 2.}$$

Any number between -3 and 6 is in the solution set. This is the interval $(-3, 6)$; its graph is shown on the right.

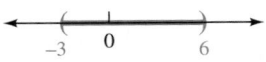

 Self Check 7 Solve $|3x + 2| < 4$ and graph the solution set.
Now Try **Problems 57 and 61**

Because it is related to distance, absolute value can be used to describe the amount of error involved when measurements are taken.

EXAMPLE 8 ***Tolerances.*** When manufactured parts are inspected by a quality control engineer, they are classified as acceptable if each dimension falls within a given *tolerance range* of the dimensions listed on the blueprint. For the bracket shown in the figure, the distance between the two drilled holes is given as 2.900 inches. Because the tolerance is ± 0.015 inch, this distance can be as much as 0.015 inch longer or 0.015 inch shorter, and the part will be considered acceptable. The acceptable distance d between holes can be represented by the absolute value inequality $|d - 2.900| \le 0.015$. Solve the inequality and explain the result.

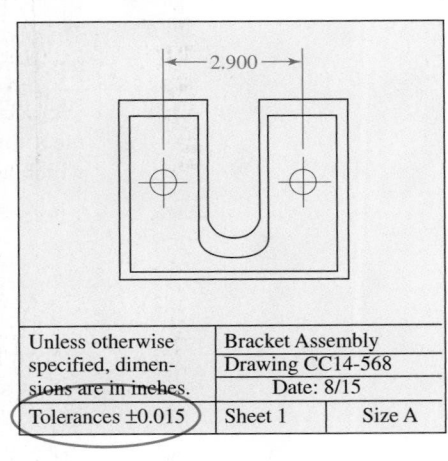

Unless otherwise specified, dimensions are in inches.	Bracket Assembly	
	Drawing CC14-568	
	Date: 8/15	
Tolerances ±0.015	Sheet 1	Size A

Strategy To solve $|d - 2.900| \le 0.015$, we will write and solve an equivalent double inequality.

Why We can use this approach because the inequality is of the form $|X| \le k$, and k is positive.

Solution The absolute value inequality $|d - 2.900| \le 0.015$ is equivalent to the double inequality

$$-0.015 \le d - 2.900 \le 0.015$$

which we can solve for d:

$$-0.015 \le d - 2.900 \le 0.015$$
$$2.885 \le d \le 2.915 \qquad \text{To isolate } d, \text{ add 2.900 to all three parts.}$$

The solution set is the interval [2.885, 2.915]. This means that the distance between the two holes should be between 2.885 and 2.915 inches, inclusive. If the distance is less than 2.885 inches or more than 2.915 inches, the part should be rejected.

 Now Try **Problem 103**

EXAMPLE 9   Solve: $|4x - 5| < -2$

Strategy   We will solve this inequality by inspection.

Why   The inequality $|4x - 5| < -2$ is of the form $|X| < k$. However, the standard method for solving absolute value inequalities cannot be used because k is negative.

Solution   Since $|4x - 5|$ is always greater than or equal to 0 for any real number x, this absolute value inequality has no solution. The solution set is $\varnothing$.

 Self Check 9   Solve: $|6x + 24| < -51$
Now Try   **Problem 65**

---

4   **Solve Inequalities of the Form $|X| > k$.**

To solve the absolute value inequality $|x| > 5$, we must find all real numbers x whose distance from 0 on the number line is greater than 5. From the following graph, we see that there are many such numbers. For example, -5.001, -6, -7.5, and $-8\frac{3}{8}$, as well as 5.001, 6.2, 7, 8, and $9\frac{1}{2}$ all meet this requirement. We conclude that the solution set is all numbers less than -5 or greater than 5, which can be written as the union of two intervals: $(-\infty, -5) \cup (5, \infty)$.

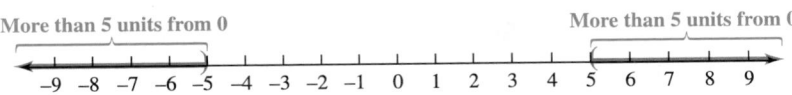

Since x is less than -5 or greater than 5, it follows that $|x| > 5$ is equivalent to $x < -5$ or $x > 5$. This observation suggests the following approach for solving absolute value inequalities of the form $|X| > k$ and $|X| \geq k$.

Solving $|X| > k$ and $|X| \geq k$

For any positive number k and any algebraic expression X:

To solve $|X| > k$, solve the equivalent compound inequality $X < -k$ or $X > k$.
To solve $|X| \geq k$, solve the equivalent compound inequality $X \leq -k$ or $X \geq k$.

EXAMPLE 10   Solve $\left| \dfrac{3 - x}{5} \right| \geq 6$ and graph the solution set.

Strategy   To solve this absolute value inequality, we will write and solve an equivalent compound inequality

Why   We can use this approach because the equation is of the form $|X| \geq k$, and k is positive.

Solution The absolute value inequality $\left|\frac{3-x}{5}\right| \geq 6$ is equivalent to the compound inequality

$$\frac{3-x}{5} \leq -6 \quad \text{or} \quad \frac{3-x}{5} \geq 6$$

Now we solve each inequality for x:

$$\frac{3-x}{5} \leq -6 \quad \text{or} \quad \frac{3-x}{5} \geq 6$$

$3 - x \leq -30$	$3 - x \geq 30$ To clear the fraction, multiply both sides by 5.
$-x \leq -33$	$-x \geq 27$ Subtract 3 from both sides.
$x \geq 33$	$x \leq -27$ To isolate x, divide both sides by −1 and reverse the direction of the inequality symbol.

The solution set is the union of two intervals: $(-\infty, -27] \cup [33, \infty)$. Its graph appears on the right.

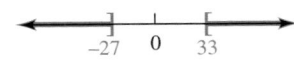

 Self Check 10 Solve $\left|\frac{2-x}{4}\right| \geq 1$ and graph the solution set.

Now Try **Problems 69 and 71**

EXAMPLE 11 Solve $\left|\frac{2}{3}x - 2\right| - 3 > 6$ and graph the solution set.

Strategy We will first isolate $\left|\frac{2}{3}x - 2\right|$ on the left side of the inequality and write and solve an equivalent compound inequality.

Why After isolating the absolute value expression, the resulting inequality will have the desired form $|X| > k$.

Solution We add 3 to both sides to isolate the absolute value on the left side.

$$\left|\frac{2}{3}x - 2\right| - 3 > 6 \quad \text{This is the inequality to solve.}$$

$$\left|\frac{2}{3}x - 2\right| > 9 \quad \text{Add 3 to both sides to isolate the absolute value.}$$

To solve this absolute value inequality, we write and solve an equivalent compound inequality:

$$\frac{2}{3}x - 2 < -9 \quad \text{or} \quad \frac{2}{3}x - 2 > 9$$

$\frac{2}{3}x < -7$	$\frac{2}{3}x > 11$ Add 2 to both sides.
$2x < -21$	$2x > 33$ Multiply both sides by 3.
$x < -\frac{21}{2}$	$x > \frac{33}{2}$ To isolate x, divide both sides by 2.

The solution set is the union of two intervals: $\left(-\infty, -\frac{21}{2}\right) \cup \left(\frac{33}{2}, \infty\right)$. Its graph appears on the right.

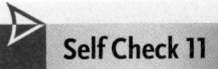 **Self Check 11** Solve $\left|\frac{3}{4}x + 2\right| - 1 > 3$ and graph the solution set.

Now Try **Problem 73**

EXAMPLE 12 Solve $\left|\frac{x}{8} - 1\right| \geq -4$ and graph the solution set.

Strategy We will solve this inequality by inspection.

Why The inequality $\left|\frac{x}{8} - 1\right| \geq -4$ is of the form $|x| \geq k$. However, the standard method for solving absolute value inequalities cannot be used because k is negative.

Solution Since $\left|\frac{x}{8} - 1\right|$ is always greater than or equal to 0 for any real number x, this absolute value inequality is true for all real numbers. The solution set is the interval $(-\infty, \infty)$ or $\mathbb{R}$. Its graph appears on the right.

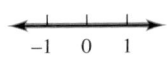

 Self Check 12 Solve $|-x - 9| > -0.5$ and graph the solution set.

Now Try **Problem 75**

The following summary shows how we can interpret absolute value in three ways. Assume $k > 0$.

Geometric description	*Graphic description*	*Algebraic description*
1. $|x| = k$ means that x is k units from 0 on the number line.	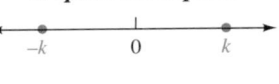	$|x| = k$ is equivalent to $x = k$ or $x = -k$.
2. $|x| < k$ means that x is less than k units from 0 on the number line.	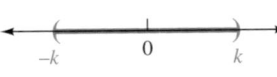	$|x| < k$ is equivalent to $-k < x < k$.
3. $|x| > k$ means that x is more than k units from 0 on the number line.	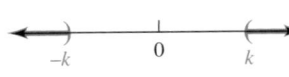	$|x| > k$ is equivalent to $x > k$ or $x < -k$.

Using Your Calculator ***Solving Absolute Value Equations and Inequalities***

We can solve absolute value equations and inequalities with a graphing calculator. For example, to solve $|2x - 3| = 9$, we graph the equations $y = |2x - 3|$ and $y = 9$ on the same coordinate system, as shown in the figure. The equation $|2x - 3| = 9$ will be true for all x-coordinates of points that lie on *both* graphs. Using the TRACE or the INTERSECT feature, we can see that the graphs intersect at the points $(-3, 9)$ and $(6, 9)$. Thus, the solutions of the absolute value equation are -3 and 6.

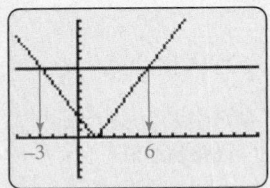

The inequality $|2x - 3| < 9$ will be true for all x-coordinates of points that lie on the graph of $y = |2x - 3|$ and *below* the graph of $y = 9$. We see that these values of x are between -3 and 6. Thus, the solution set is the interval $(-3, 6)$.

The inequality $|2x - 3| > 9$ will be true for all x-coordinates of points that lie on the graph of $y = |2x - 3|$ and *above* the graph of $y = 9$. We see that these values of x are less than -3 or greater than 6. Thus, the solution set is the union of two intervals: $(-\infty, -3) \cup (6, \infty)$.

ANSWERS TO SELF CHECKS   **1. a.** $24, -24$ **b.** $\frac{1}{2}, -\frac{1}{2}$ **2. a.** $5, -2$ **b.** No solution, $\varnothing$

3. $7.5, 2.5$ **4.** -6 **5.** $7, -15$ **6.** $-1, -6$ **7.** $\left(-2, \frac{2}{3}\right)$ **9.** No solution

10. $(-\infty, -2] \cup [6, \infty)$

11. $(-\infty, -8) \cup \left(\frac{8}{3}, \infty\right)$ **12.** $(-\infty, \infty)$

STUDY SET
8.3

VOCABULARY

Fill in the blanks.

1. The _____ _____ of a number is its distance from 0 on a number line.

2. $|2x - 1| = 10$ is an absolute value _____ and $|2x - 1| > 10$ is an absolute value _____.

3. To _____ the absolute value in $|3 - x| - 4 = 5$, we add 4 to both sides.

4. When we say that the absolute value equation and a compound equation are equivalent, we mean that they have the same _____.

5. When two equations are joined by the word *or*, such as $x + 1 = 5$ or $x + 1 = -5$, we call the statement a _____ equation.

6. $f(x) = |6x - 2|$ is called an absolute value _____.

CONCEPTS

Fill in the blanks.

7. To solve these absolute value equations and inequalities, we write and solve equivalent _____ equations and inequalities.

8. For two expressions to have the same absolute value, they must either be equal or _____ of each other.

9. Consider the following real numbers:
$-3, -2.01, -2, -1.99, -1, 0, 1, 1.99, 2, 2.01, 3$
  a. Which of them make $|x| = 2$ true?
  b. Which of them make $|x| < 2$ true?
  c. Which of them make $|x| > 2$ true?

10. Determine whether -3 is a solution of the given equation or inequality.
  a. $|x - 1| = 4$ **b.** $|x - 1| > 4$
  c. $|x - 1| \leq 4$ **d.** $|5 - x| = |x + 12|$

11. For each absolute value equation, write an equivalent compound equation.
  a. $|x - 7| = 8$ is equivalent to

$x - 7 = \boxed{}$ or $x - 7 = \boxed{}$

  b. $|x + 10| = |x - 3|$ is equivalent to

$x + 10 = \boxed{}$ or $x + 10 = \boxed{}$

12. For each absolute value inequality, write an equivalent compound inequality.
  a. $|x + 5| < 1$ is equivalent to

$\boxed{} < x + 5 < \boxed{}$

  b. $|x - 6| \geq 3$ is equivalent to

$x - 6 \leq \boxed{}$ or $x - 6 \geq \boxed{}$

13. For each absolute value equation or inequality, write an equivalent compound equation or inequality. *Do not solve.*

    a. $|x| = 8$ **b.** $|x| \geq 8$

    c. $|x| \leq 8$ **d.** $|5x - 1| = |x + 3|$

14. Perform the necessary steps to isolate the absolute value expression on one side of the equation. *Do not solve.*

    a. $|3x + 2| - 7 = -5$

    b. $6 + |5x - 19| \leq 40$

NOTATION

15. Match each equation or inequality with its graph.

    a. $|x| = 1$ **i.**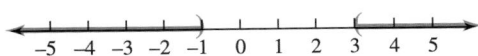

    b. $|x| > 1$ **ii.**

    c. $|x| < 1$ **iii.**

16. Describe the set graphed below using interval notation.

GUIDED PRACTICE

Solve each equation. See Example 1.

17. $|x| = 23$ **18.** $|x| = 90$

19. $|x| = \dfrac{3}{4}$ **20.** $|x| = 6.95$

Solve each equation. See Example 2.

21. $|x - 5| = 8$ **22.** $|x - 7| = 4$

23. $|3x + 2| = 16$ **24.** $|5x - 3| = 22$

25. $\left|\dfrac{x}{5}\right| = 10$ **26.** $\left|\dfrac{x}{7}\right| = 2$

27. $|2x + 3.6| = 9.8$ **28.** $|4x - 24.8| = 32.4$

29. $|50.4 - 3x| = -1$ **30.** $\left|75 - \dfrac{1}{3}x\right| = -1$

31. $\left|\dfrac{7}{2}x + 3\right| = -5$ **32.** $|x - 2.1| = -16.3$

Solve each equation. See Example 3.

33. $|x - 3| - 19 = 3$ **34.** $|x - 10| + 30 = 50$

35. $|3x - 7| + 8 = 22$ **36.** $|6x - 3| + 7 = 28$

37. $|3 - 4x| + 1 = 6$ **38.** $|8 - 5x| - 8 = 10$

39. $\left|\dfrac{7}{8}x + 5\right| - 2 = 7$ **40.** $\left|\dfrac{3}{4}x + 4\right| - 5 = 11$

Solve each equation. See Example 4.

41. $\left|\dfrac{1}{5}x + 2\right| - 8 = -8$ **42.** $\left|\dfrac{1}{9}x + 4\right| + 25 = 25$

43. $2|3x + 24| = 0$ **44.** $8\left|\dfrac{2x}{3} + 10\right| = 0$

45. $-5|2x - 9| + 14 = 14$ **46.** $-10|16x + 4| - 3 = -3$

47. $6 - 3|10x + 5| = 6$ **48.** $15 - |12x + 12| = 15$

Solve each equation. See Example 6.

49. $|5x - 12| = |4x - 16|$ **50.** $|4x - 7| = |3x - 21|$

51. $|10x| = |x - 18|$ **52.** $|6x| = |x + 45|$

53. $|2 - x| = |3x + 2|$ **54.** $|4x + 3| = |9 - 2x|$

55. $|5x - 7| = |4(x + 1)|$ **56.** $|2x + 1| = |3(x + 1)|$

Solve each inequality. Graph the solution set and write it using interval notation. See Examples 7 and 9.

57. $|x| < 4$ **58.** $|x| < 9$

59. $|x + 9| \leq 12$ **60.** $|x - 8| \leq 12$

61. $\left|\dfrac{x}{4} + 3\right| \leq 3$ **62.** $\left|\dfrac{x}{10} - 1\right| \leq 1$

63. $|3x - 2| < 10$ **64.** $|4 - 3x| \leq 13$

65. $|5x - 12| < -5$ **66.** $|3x + 2| \leq -3$

67. $|3.4x| + 19.7 \leq 19.6$ **68.** $|1.9x| - 3.1 < -3.2$

Solve each inequality. Graph the solution set and write it using interval notation. See Examples 10–12.

69. $|x| > 3$

70. $|x| > 7$

71. $|x - 12| > 24$

72. $|x + 5| \geq 7$

73. $|5x - 1| - 2 \geq 0$

74. $|6x - 3| - 5 \geq 0$

75. $|4x + 3| \geq -5$

76. $|7x + 2| \geq -8$

See Examples 5, 7, and 10.

77. Let $f(x) = |x + 3|$. For what value(s) of x is $f(x) = 3$?

78. Let $g(x) = |2 - x|$. For what value(s) of x is $g(x) = 2$?

79. Let $f(x) = |2(x - 1) + 4|$. For what value(s) of x is $f(x) < 4$?

80. Let $h(x) = \left|\frac{x}{5} - \frac{1}{2}\right|$. For what value(s) of x is $h(x) > \frac{9}{10}$?

TRY IT YOURSELF

Solve each equation and inequality. For the inequalities, graph the solution set and write it using interval notation.

81. $|3x + 2| + 1 > 15$

82. $|2x - 5| - 5 > 20$

83. $6\left|\dfrac{x - 2}{3}\right| \leq 24$

84. $8\left|\dfrac{x - 2}{3}\right| > 32$

85. $-7 = 2 - |0.3x - 3|$

86. $-1 = 1 - |0.1x + 8|$

87. $|2 - 3x| \geq -8$

88. $|-1 - 2x| > 5$

89. $|7x + 12| = |x - 6|$

90. $|8 - x| = |x + 2|$

91. $3|2 - 3x| + 2 \leq 2$

92. $|15x - 45| + 7 \leq 7$

93. $-14 = |x - 3|$

94. $-75 = |x + 4|$

95. $\dfrac{6}{5} = \left|\dfrac{3x}{5} + \dfrac{x}{2}\right|$

96. $\dfrac{11}{12} = \left|\dfrac{x}{3} - \dfrac{3x}{4}\right|$

97. $-|2x - 3| < -7$

98. $-|3x + 1| < -8$

99. $|0.5x + 1| < -23$

100. $15 \geq 7 - |1.4x + 9|$

APPLICATIONS

101. TEMPERATURE RANGES   The temperatures on a sunny summer day satisfied the inequality $|t - 78°| \leq 8°$, where t is a temperature in degrees Fahrenheit. Solve this inequality and express the range of temperatures as a double inequality.

102. OPERATING TEMPERATURES   A car CD player has an operating temperature of $|t - 40°| < 80°$, where t is a temperature in degrees Fahrenheit. Solve the inequality and express this range of temperatures as an interval.

103. AUTO MECHANICS   On most cars, the bottoms of the front wheels are closer together than the tops, creating a *camber angle*. This lessens road shock to the steering system. (See the illustration.) The specifications for a certain car state that the camber angle c of its wheels should be $0.6° \pm 0.5°$.

    a. Express the range with an inequality containing absolute value symbols.

    b. Solve the inequality and express this range of camber angles as an interval.

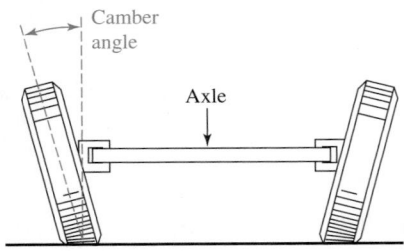

104. STEEL PRODUCTION   A sheet of steel is to be 0.250 inch thick with a tolerance of 0.025 inch.

    a. Express this specification with an inequality containing absolute value symbols, using x to represent the thickness of a sheet of steel.

    b. Solve the inequality and express the range of thickness as an interval.

105. ERROR ANALYSIS   In a lab, students measured the percent of copper p in a sample of copper sulfate. The students know that copper sulfate is actually 25.46% copper by mass. They are to compare their results to the actual value and find the amount of *experimental error*. Which measurements shown in the illustration satisfy the absolute value inequality $|p - 25.46| \leq 1.00$?

Lab 4	Section A
Title:	
"Percent copper (Cu) in copper sulfate $(CuSO_4 \cdot 5H_2O)$"	

Results

	% Copper
Trial #1:	22.91%
Trial #2:	26.45%
Trial #3:	26.49%
Trial #4:	24.76%

106. ERROR ANALYSIS See Exercise 105. Which measurements satisfy the absolute value inequality $|p - 25.46| > 1.00$?

WRITING

107. Explain the error.

Solve: $|x| + 2 = 6$

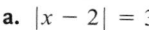

$x + 2 = 6$ or $x + 2 = -6$

$x = 4$ | $x = -8$

108. Explain why the equation $|x - 4| = -5$ has no solution.

109. Explain the differences between the solution sets of $|x| < 8$ and $|x| > 8$.

110. Explain how to use the graph in the illustration to solve the following.

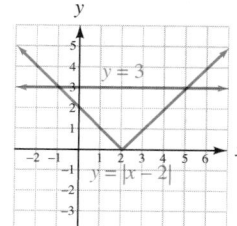

   a. $|x - 2| = 3$

   b. $|x - 2| \le 3$

   c. $|x - 2| \ge 3$

REVIEW

111. RAILROAD CROSSINGS
The warning sign in the illustration is to be painted on the street in front of a railroad crossing. If y is 30° more than twice x, find x and y.

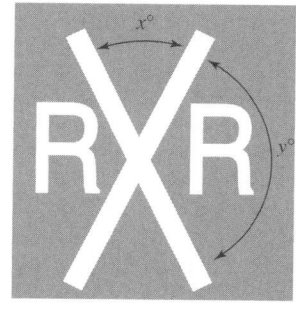

112. STATUE OF LIBERTY   From the foundation of the large pedestal on which it sits to the top of the torch, the Statue of Liberty National Monument measures 305 feet. The pedestal is 3 feet taller than the statue. Find the height of the pedestal and the height of the statue.

CHALLENGE PROBLEMS

113. a. For what values of k does $|x| + k = 0$ have exactly two solutions?

   b. For what values of k does $|x| + k = 0$ have exactly one solution?

114. Solve: $2^{|2x - 3|} = 64$

SECTION 8.4
Review of Factoring Methods: GCF, Grouping, Trinomials

Objectives

  1 Factor out the greatest common factor.

  2 Factor by grouping.

  3 Use factoring to solve formulas for a specified variable.

  4 Factor trinomials.

  5 Use substitution to factor trinomials.

  6 Use the grouping method to factor trinomials.

In Chapter 5, we discussed how to factor polynomials. In this section, we will review that material.

 Factor Out the Greatest Common Factor.

Recall that *when we factor a polynomial, we write a sum of terms as a product of factors.* To perform the most basic type of factoring, we determine whether the terms of the given polynomial have any common factors. This process, called **factoring out the greatest common factor,** is based on the distributive property.

| **EXAMPLE 1** | Factor: $3xy^2z^3 + 6xz^2 - 9xyz^4$ |

Strategy We will determine the GCF of the terms of the polynomial. Then we will write each term of the polynomial as the product of the GCF and one other factor.

Why We can then use the distributive property to factor out the GCF.

Solution We begin by factoring each term:

$$\left. \begin{array}{l} 3xy^2z^3 = 3 \cdot x \cdot y \cdot y \cdot z \cdot z \cdot z \\ 6xz^2 = 2 \cdot 3 \cdot x \cdot z \cdot z \\ 9xyz^4 = 3 \cdot 3 \cdot x \cdot y \cdot z \cdot z \cdot z \cdot z \end{array} \right\} \text{GCF} = 3 \cdot x \cdot z \cdot z = 3xz^2$$

Since each term has one factor of 3, one factor of x, and two factors of z, and there are no other common factors, $3xz^2$ is the greatest common factor of the three terms. We write each term as the product of the GCF, $3xz^2$, and one other factor and proceed as follows:

$$3xy^2z^3 + 6xz^2 - 9xyz^4 = 3xz^2 \cdot y^2z + 3xz^2 \cdot 2 - 3xz^2 \cdot 3yz^2$$

$$= 3xz^2(y^2z + 2 - 3yz^2) \qquad \text{Factor out the GCF, } 3xz^2.$$

We can check the factorization using multiplication.

$$3xz^2(y^2z + 2 - 3yz^2) = 3xy^2z^3 + 6xz^2 - 9xyz^4 \quad \text{This is the original polynomial.}$$

| **Self Check 1** | Factor: $6a^2b^2 - 4ab^3 + 2ab^2$ |
| *Now Try* | **Problem 25** |

| **EXAMPLE 2** | Factor the opposite of the greatest common factor from $-6u^2v^3 + 8u^3v^2$. |

Strategy We will determine the GCF of the terms of the polynomial. Then we will write each term as the product of the opposite of the GCF and one other factor.

Why We can then use the distributive property to factor out the opposite of the GCF.

Solution Because the greatest common factor of the two terms is $2u^2v^2$, the opposite of the greatest common factor is $-2u^2v^2$. To factor out $-2u^2v^2$, we proceed as follows:

$$-6u^2v^3 + 8u^3v^2 = -2u^2v^2 \cdot 3v - (-2u^2v^2)4u$$

$$= -2u^2v^2(3v - 4u) \qquad \text{Note that the leading coefficient of the polynomial within the parentheses is positive.}$$

| **Self Check 2** | Factor out the opposite of the GCF: $-3p^3q + 6p^2q^2$ |
| *Now Try* | **Problem 31** |

A polynomial that cannot be factored is called a **prime polynomial** or an **irreducible polynomial.**

EXAMPLE 3   Factor $9x + 16$, if possible.

Strategy   First, we will determine the GCF of the terms of the polynomial.

Why   If the terms have no common factors (other than 1), this polynomial does not factor.

Solution   We factor each term of $9x + 16$:

$$9x = 3 \cdot 3 \cdot x \qquad 16 = 2 \cdot 2 \cdot 2 \cdot 2$$

Since there are no common factors other than 1, this polynomial cannot be factored. It is a prime polynomial.

    Self Check 3   Factor: $6a^3 + 7b^2$
Now Try   **Problem 35**

A common factor can have more than one term.

EXAMPLE 4   Factor: $a(x - y + z) - b(x - y + z) + 3(x - y + z)$

Strategy   We will factor out the trinomial $x - y + z$ from each term.

Why $x - y + z$ is the GCF of each term of the given expression.

Solution

$$a(x - y + z) - b(x - y + z) + 3(x - y + z) = (x - y + z)(a - b + 3)$$

    Self Check 4   Factor: $c^2(y^2 + 1) + d^2(y^2 + 1)$
Now Try   **Problem 41**

---

2   **Factor by Grouping.**

Sometimes polynomials having four or more terms can be factored by removing common factors from groups of terms. This process is called **factoring by grouping.**

EXAMPLE 5   Factor: $2c - 2d + cd - d^2$

Strategy   Since the four terms of the polynomial do not have a common factor (other than 1), we will attempt to factor the polynomial by grouping. We will factor out a common factor from the first two terms and from the last two terms.

Why   This will produce a common binomial factor that can be factored out.

Success Tip
You may want to review the steps in the process of factoring by grouping on page 476.

Solution   The first two terms have a common factor, 2, and the last two terms have a common factor, d.

$$2c - 2d + cd - d^2 = 2(c - d) + d(c - d) \quad \text{Factor out 2 from } 2c - 2d \text{ and } d \text{ from } cd - d^2.$$

$$= (c - d)(2 + d) \quad \text{Factor out the common binomial factor, } c - d.$$

We check by multiplying:

$$(c - d)(2 + d) = 2c + cd - 2d - d^2$$
$$= 2c - 2d + cd - d^2 \quad \text{Rearrange the terms to get the original polynomial.}$$

Self Check 5   Factor: $7m - 7n + mn - n^2$
Now Try   Problems 43 and 47

To factor a polynomial, it is often necessary to factor more than once. When factoring a polynomial, *always look for a common factor first.*

EXAMPLE 6   Factor: $3x^3y - 4x^2y^2 - 6x^2y + 8xy^2$

Strategy   Since all four terms have a common factor of xy, we factor it out first. Then we will attempt to factor the resulting polynomial by grouping.

Why   Factoring out the GCF first makes factoring by any method easier.

Solution   We begin by factoring out the common factor of xy.

$$3x^3y - 4x^2y^2 - 6x^2y + 8xy^2 = xy(3x^2 - 4xy - 6x + 8y)$$

We can now factor $3x^2 - 4xy - 6x + 8y$ by grouping:

$$3x^3y - 4x^2y^2 - 6x^2y + 8xy^2$$
$$= xy(3x^2 - 4xy - 6x + 8y)$$
$$= xy[x(3x - 4y) - 2(3x - 4y)] \quad \text{Factor x from } 3x^2 - 4xy \text{ and } -2 \text{ from } -6x + 8y.$$
$$= xy(3x - 4y)(x - 2) \quad \text{Factor out } 3x - 4y.$$

Because xy, $3x - 4y$, and $x - 2$ are prime, no further factoring can be done; the factorization is complete.

Caution
The instruction "Factor" means to factor the given expression completely. Each factor of a completely factored expression will be prime.

Self Check 6   Factor: $3a^3b + 3a^2b - 2a^2b^2 - 2ab^2$
Now Try   Problem 51

3   **Use Factoring to Solve Formulas for a Specified Variable.**

Factoring is often required to solve a formula for one of its variables.

EXAMPLE 7   *Electronics.*   The formula $r_1r_2 = rr_2 + rr_1$ is used in electronics to relate the combined resistance, r, of two resistors wired in parallel. The variable r_1 represents the resistance of the first resistor, and the variable r_2 represents the resistance of the second. Solve for r_2.

Strategy   To isolate r_2 on one side of the equation, we will get all the terms involving r_2 on the left side and all the terms not involving r_2 on the right side.

Why   To *solve a formula for a specified variable* means to isolate that variable on one side of the equation, with all other variables and constants on the opposite side.

Solution

   We want to isolate this variable on one side of the equation.

$$r_1r_2 = rr_2 + rr_1$$

$$r_1r_2 - rr_2 = rr_1$$ To eliminate rr_2 on the right side, subtract rr_2 from both sides.

$$r_2(r_1 - r) = rr_1$$ On the left side, factor out the GCF r_2 from $r_1r_2 - rr_2$.

$$\frac{\overset{1}{r_2(\cancel{r_1 - r})}}{\underset{1}{\cancel{r_1 - r}}} = \frac{rr_1}{r_1 - r}$$ To isolate r_2 on the left side, divide both sides by $r_1 - r$.

$$r_2 = \frac{rr_1}{r_1 - r}$$

  Self Check 7   Solve $f_1f_2 = ff_1 + ff_2$ for f_1.

Now Try   **Problem 57**

4 **Factor Trinomials.**

Recall that many trinomials factor as the product of two binomials.

EXAMPLE 8   Factor: $x^2 - 6x + 8$

Strategy   We will assume that this trinomial is the product of two binomials. We must find the terms of the binomials.

Why   Since the terms of $x^2 - 6x + 8$ do not have a common factor (other than 1), the only option is to try to factor it as the product of two binomials.

Success Tip
You may want to review the procedure for factoring trinomials with a leading coefficient of 1 on page 485.

Solution   We represent the binomials using two sets of parentheses. Since the first term of the trinomial is x^2, we enter x and x as the first terms of the binomial factors.

$$x^2 - 6x + 8 = \left(x \;\boxed{}\;\right)\left(x \;\boxed{}\;\right)$$ Because $x \cdot x$ will give x^2.

The second terms of the binomials must be two integers whose product is 8 and whose sum is -6. We list all possible integer-pair factors of 8 in the table.

Factors of 8	Sum of the factors of 8
$1(8) = 8$	$1 + 8 = 9$
$2(4) = 8$	$2 + 4 = 6$
$-1(-8) = 8$	$-1 + (-8) = -9$
$-2(-4) = 8$	$-2 + (-4) = -6$

⟵ This is the pair to choose.

The fourth row of the table contains the correct pair of integers -2 and -4, whose product is 8 and whose sum is -6. To complete the factorization, we enter -2 and -4 as the second terms of the binomial factors.

$$x^2 - 6x + 8 = (x - 2)(x - 4)$$

Check: We can verify the factorization by multiplication:

$$(x - 2)(x - 4) = x^2 - 4x - 2x + 8 \quad \text{Use the FOIL method.}$$
$$= x^2 - 6x + 8 \quad \text{This is the original trinomial.}$$

Self Check 8   Factor: $a^2 - 7a + 12$

Now Try   **Problem 63**

EXAMPLE 9   Factor: $2a^2 + 4ab - 30b^2$

Strategy   We will factor out the GCF, 2, first. Then we will factor the resulting trinomial.

Why   The first step in factoring any polynomial is to factor out the GCF. Factoring out the GCF first makes factoring by any method easier.

Solution   Each term in this trinomial has a common factor of 2, which can be factored out.

$$2a^2 + 4ab - 30b^2 = 2(a^2 + 2ab - 15b^2)$$

Next, we factor $a^2 + 2ab - 15b^2$. Since the first term of the trinomial is a^2, the first term of each binomial factor must be a. Since the third term contains b^2, the last term of each binomial factor must contain b. To complete the factorization, we need to determine the coefficient of each b-term.

$$a^2 + 2ab - 15b^2 = \left(a \boxed{} b\right)\left(a \boxed{} b\right) \quad \text{Because } a \cdot a \text{ will give } a^2 \text{ and } b \cdot b \text{ will give } b^2.$$

The coefficients of b must be two integers whose product is -15 and whose sum is 2. We list the factors of -15 and find the pair whose sum is 2.

This is the pair to choose.
↓
$$1(-15) \qquad 3(-5) \qquad 5(-3) \qquad 15(-1)$$

The only factorization where the sum of the factors is 2 (which is the coefficient of the middle term of $a^2 + 2ab - 15b^2$) is $5(-3)$. Thus,

$$2a^2 + 4ab - 30b^2 = 2(a^2 + 2ab - 15b^2)$$
$$= 2(a + 5b)(a - 3b)$$

Verify this result by multiplication.

Self Check 9   Factor: $3p^2 + 6pq - 24q^2$

Now Try   **Problem 67**

There are more combinations of coefficients to consider when factoring trinomials with leading coefficients other than 1. Because it is not easy to give specific rules for factoring such trinomials, we will use a method called the **trial-and-check method.**

EXAMPLE 10   Factor: $3p^2 - 4p - 4$

Strategy   We will assume that this trinomial is the product of two binomials. To find their terms, we will make educated guesses and then check them using multiplication.

Why   Since the terms of the trinomial do not have a common factor (other than 1), the only option is to try to factor it as the product of two binomials.

Solution   To factor the trinomial, we note that the first terms of the binomial factors must be $3p$ and p to give the first term of $3p^2$.

$$3p^2 - 4p - 4 = \left(3p \; \boxed{}\right)\left(p \; \boxed{}\right) \qquad \text{Because } 3p \cdot p \text{ will give } 3p^2.$$

The second terms of the binomials must be two integers whose product is -4. There are three such pairs: $1(-4)$, $-1(4)$, and $-2(2)$. When these pairs are entered, and then reversed, as second terms of the binomials, there are six possibilities to consider.

For 1 and -4:

$(3p + 1)(p - 4)$
$-12p + p = -11p$

$(3p - 4)(p + 1)$
$3p + (-4p) = -p$

For -1 and 4:

$(3p - 1)(p + 4)$
$12p + (-p) = 11p$

$(3p + 4)(p - 1)$
$-3p + 4p = p$

For -2 and 2:

$(3p - 2)(p + 2)$
$6p + (-2p) = 4p$

$(3p + 2)(p - 2)$
$-6p + 2p = -4p$

Of these possibilities, only the one in blue gives the required middle term of $-4p$. Thus,

$$3p^2 - 4p - 4 = (3p + 2)(p - 2)$$

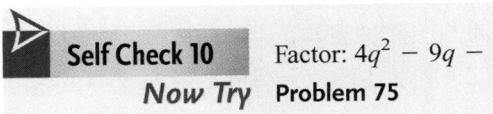

Self Check 10   Factor: $4q^2 - 9q - 9$

Now Try   Problem 75

Success Tip
You may want to review the procedure for factoring trinomials using the trial-and-check method on page 497.

Notation
By the commutative property of multiplication, the factors of a trinomial can be written in either order. Thus, we could also write:
$$3p^2 - 4p - 4 = (p - 2)(3p + 2)$$

EXAMPLE 11   Factor: $6y^3 + 13x^2y^3 + 6x^4y^3$

Strategy   We write the expression in descending powers of x.

Why   It is easier to factor a trinomial if its terms are written in descending powers of one variable.

Solution   We write the expression in descending powers of x and factor out the greatest common factor, y^3.

$$6y^3 + 13x^2y^3 + 6x^4y^3 = 6x^4y^3 + 13x^2y^3 + 6y^3$$
$$= y^3(6x^4 + 13x^2 + 6)$$

To factor $6x^4 + 13x^2 + 6$, we examine its terms.

- Since the first term is $6x^4$, the first terms of the binomial factors must be either $2x^2$ and $3x^2$ or x^2 and $6x^2$.

$$6x^4 + 13x^2 + 6 = \left(2x^2 \;\boxed{}\right)\left(3x^2 \;\boxed{}\right) \text{ or } \left(x^2 \;\boxed{}\right)\left(6x^2 \;\boxed{}\right)$$

- Since the signs of the middle term and the last term of the trinomial are positive, the signs within each binomial factor will be positive.

- Since the product of the last terms of the binomial factors must be 6, we must find two numbers whose product is 6 that will lead to a middle term of $13x^2$.

After trying some combinations, we find the one that works.

$$6x^4y^3 + 13x^2y^3 + 6y^3 = y^3(6x^4 + 13x^2 + 6)$$
$$= y^3(2x^2 + 3)(3x^2 + 2)$$

  Self Check 11   Factor: $4b + 11a^2b + 6a^4b$

Now Try   **Problem 87**

5   **Use Substitution to Factor Trinomials.**

For more complicated expressions, especially those involving a quantity within parentheses, a substitution sometimes helps to simplify the factoring process.

EXAMPLE 12   Factor: $(x + y)^2 + 7(x + y) + 12$

Strategy   We will use a substitution where we will replace each expression $x + y$ with the variable z and factor the resulting trinomial.

Why   The resulting trinomial will be easier to factor because it will be in only one variable, z.

Solution   If we use the substitution $z = x + y$, we obtain

$$(x + y)^2 + 7(x + y) + 12 = z^2 + 7z + 12 \qquad \text{Replace } x + y \text{ with } z.$$
$$= (z + 4)(z + 3) \qquad \text{Factor the trinomial.}$$

To find the factorization of $(x + y)^2 + 7(x + y) + 12$, we substitute $x + y$ for each z in the expression $(z + 4)(z + 3)$.

$$(z + 4)(z + 3) = (x + y + 4)(x + y + 3)$$

Thus, $(x + y)^2 + 7(x + y) + 12 = (x + y + 4)(x + y + 3)$

> **Self Check 12** Factor: $(a + b)^2 - 3(a + b) - 10$
> **Now Try** **Problem 95**

6 **Use the Grouping Method to Factor Trinomials.**

Another way to factor trinomials is to write them as equivalent four-termed polynomials and factor by grouping.

EXAMPLE 13 Factor by grouping: **a.** $x^2 + 8x + 15$ **b.** $10x^2 + 13xy - 3y^2$

Strategy In each case, we will express the middle term of the trinomial as the sum of two terms.

Why We want to produce an equivalent four-termed polynomial that can be factored by grouping.

Solution

a. Since $x^2 + 8x + 15 = 1x^2 + 8x + 15$, we identify a as 1, b as 8, and c as 15. The key number is $ac = 1(15) = 15$. We must find two integers whose product is the key number 15 and whose sum is $b = 8$. Since the integers must have a positive product and a positive sum, we consider only positive factors of 15.

Key number = 15	$b = 8$
Positive factors of 15	Sum of the factors of 15
$1 \cdot 15 = 15$	$1 + 15 = 16$
$3 \cdot 5 = 15$	$3 + 5 = 8$

The second row of the table contains the correct pair of integers 3 and 5, whose product is 15 and whose sum is 8.

We can express the middle term, $8x$, of the trinomial as the *sum of two terms,* using the integers 3 and 5 as coefficients of the two terms and factor the equivalent four-termed polynomial by grouping:

$$x^2 + 8x + 15 = x^2 + 3x + 5x + 15 \quad \text{Express 8x as 3x + 5x.}$$
$$x^2 + 3x + 5x + 15 = x(x + 3) + 5(x + 3) \quad \text{Factor x out of } x^2 + 3x \text{ and 5 out of 5x + 15.}$$
$$= (x + 3)(x + 5) \quad \text{Factor out the GCF, x + 3.}$$

Check the factorization by multiplying.

b. In $10x^2 + 13xy - 3y^2$, we have $a = 10$, $b = 13$, and $c = -3$. The key number is $ac = 10(-3) = -30$. We must find a factorization of -30 such that the sum of the factors is $b = 13$. Since the factors must have a negative product, their signs must be different. The possible factor pairs are listed in the table.

Key number = −30	$b = 13$
Factors of −30	Sum of the factors of −30
$1(-30) = -30$	$1 + (-30) = -29$
$2(-15) = -30$	$2 + (-15) = -13$
$3(-10) = -30$	$3 + (-10) = -7$
$5(-6) = -30$	$5 + (-6) = -1$
$6(-5) = -30$	$6 + (-5) = 1$
$10(-3) = -30$	$10 + (-3) = 7$
$15(-2) = -30$	$15 + (-2) = 13$
$30(-1) = -30$	$30 + (-1) = 29$

The seventh row contains the correct pair of numbers 15 and -2, whose product is -30 and whose sum is 13. They serve as the coefficients of two terms, $15xy$ and $-2xy$, that we place between $10x^2$ and $-3y^2$.

$$10x^2 + 13xy - 3y^2 = 10x^2 + 15xy - 2xy - 3y^2 \quad \text{Express } 13xy \text{ as } 15xy - 2xy.$$

We factor the resulting four-term polynomial by grouping.

$$10x^2 + 15xy - 2xy - 3y^2 = 5x(2x + 3y) - y(2x + 3y) \quad \begin{array}{l}\text{Factor out } 5x \text{ from}\\ 10x^2 + 15xy. \text{ Factor out}\\ -y \text{ from } -2xy - 3y^2.\end{array}$$

$$= (2x + 3y)(5x - y) \quad \begin{array}{l}\text{Factor out the GCF,}\\ 2x + 3y.\end{array}$$

Thus, $10x^2 + 13xy - 3y^2 = (2x + 3y)(5x - y)$. Check by multiplying.

Self Check 13   Factor by grouping: **a.** $m^2 + 13m + 42$
                     b. $15a^2 + 17ab - 4b^2$

Now Try   **Problems 63 and 75**

ANSWERS TO SELF CHECKS   **1.** $2ab^2(3a - 2b + 1)$ **2.** $-3p^2q(p - 2q)$ **3.** A prime polynomial
4. $(y^2 + 1)(c^2 + d^2)$ **5.** $(m - n)(7 + n)$ **6.** $ab(3a - 2b)(a + 1)$ **7.** $f_1 = \frac{ff_2}{f_2 - f}$
8. $(a - 4)(a - 3)$ **9.** $3(p + 4q)(p - 2q)$ **10.** $(4q + 3)(q - 3)$ **11.** $b(2a^2 + 1)(3a^2 + 4)$
12. $(a + b + 2)(a + b - 5)$ **13. a.** $(m + 7)(m + 6)$ **b.** $(3a + 4b)(5a - b)$

STUDY SET
8.4

VOCABULARY

Fill in the blanks.

1. When we write $2x + 4$ as $2(x + 2)$, we say that we have
_____ $2x + 4$.

2. When we factor a polynomial, we write a sum of terms as a
_____ of factors.

3. The abbreviation GCF stands for _____ _____ _____.

4. If a polynomial cannot be factored, it is called a _____
polynomial or an irreducible polynomial.

5. To factor $ab + 6a + 2b + 12$ by _____, we begin by
factoring out a from the first two terms and 2 from the last two
terms.

6. The trinomial $4a^2 - 5a - 6$ is written in _____ powers
of a.

7. The _____ coefficient of $x^2 - 3x + 2$ is 1, the _____
of the middle term is -3, and the last term is ☐.

8. The statement $x^2 - x - 12 = (x - 4)(x + 3)$ shows that
$x^2 - x - 12$ factors into the _____ of two binomials.

CONCEPTS

9. The prime factorizations of three terms are shown here. Find
their GCF.

$2 \cdot 2 \cdot 3 \cdot x \cdot x \cdot y \cdot y \cdot y$

$2 \cdot 3 \cdot 3 \cdot x \cdot y \cdot y \cdot y \cdot y$

$2 \cdot 3 \cdot 3 \cdot 7 \cdot x \cdot x \cdot x \cdot y \cdot y$

10. Use multiplication to determine whether $(3t - 1)(5t - 6)$ is the
correct factorization of $15t^2 - 19t + 6$.

11. Complete the table.

Factors of 8	Sum of the factors of 8
$1(8) = 8$	
$2(4) = 8$	
$-1(-8) = 8$	
$-2(-4) = 8$	

12. Find two integers whose
   a. product is 10 and whose sum is 7.
   b. product is 8 and whose sum is -6.
   c. product is -6 and whose sum is 1.
   d. product is -9 and whose sum is -8.

13. Complete the key number table.

Key number $= 12$ $b = -7$

Negative factors of 12	Sum of the factors of 12
$-1(-12) = 12$	
$-3(-4) = 12$	

14. Use the substitution $x = a + b$ to rewrite the trinomial $6(a + b)^2 - 17(a + b) - 3$.

NOTATION

Complete each factorization.

15. $15c^3d^4 - 25c^2d^4 + 5c^3d^6 = \quad (3c - 5 + cd^2)$

16. $x^3 - x^2 + 2x - 2 = \quad (x - 1) + \quad (x - 1)$
$= (\quad)(x^2 + 2)$

17. $6m^2 + 7m - 3 = (\quad - 1)(2m + \quad)$

18. $2y^2 + 10y + 12 = \quad (y^2 + 5y + 6)$
$= 2(y + \quad)(\quad + 2)$

GUIDED PRACTICE

Factor each polynomial. See Example 1.

19. $2x^2 - 6x$

20. $3y^3 + 3y^2$

21. $15x^2y - 10x^2y^2$

22. $63x^3y^2 + 81x^2y^4$

23. $27z^3 + 12z^2 + 3z$

24. $25t^6 - 10t^3 + 5t^2$

25. $24s^3 - 12s^2t + 6st^2$

26. $18y^2z^2 + 12y^2z^3 - 24y^4z^3$

Factor each polynomial by factoring out the opposite of the GCF. See Example 2.

27. $-8a - 16$

28. $-6b - 30$

29. $-6x^2 - 3xy$

30. $-15y^3 - 25y^2$

31. $-18a^2b + 12ab^2$

32. $-21t^5 + 28t^3$

33. $-8a^4c^8 + 28a^3c^8 - 20a^2c^9$ **34.** $-30x^{10}y^3 + 24x^9y^2 - 60x^8y^2$

Factor each polynomial, if possible. See Example 3.

35. $11x^3 - 12y$

36. $14s^3 + 15t^6$

37. $23a^2b^3 + 4x^3y^2$

38. $18p^3q^2 - 5t^5$

Factor. See Example 4.

39. $(x + y)u + (x + y)v$

40. $4(x + y) + t(x + y)$

41. $5(a - b + c) - t(a - b + c)$

42. $(a - b - c)r - (a - b - c)s$

Factor by grouping. See Example 5.

43. $ax + bx + ay + by$

44. $ar - br + as - bs$

45. $x^2 + yx - x - y$

46. $d^2 + cd + c + d$

47. $t^3 - 3t^2 - 7t + 21$

48. $b^3 - 4b^2 - 3b + 12$

49. $a^2 - 4b + ab - 4a$

50. $3c - cd + 3d - c^2$

Factor. See Example 6.

51. $6x^3 - 6x^2 + 12x - 12$

52. $3x^3 - 6x^2 + 15x - 30$

53. $28a^3b^3c + 14a^3c - 4b^3c - 2c$

54. $12x^3z + 12xy^2z - 8x^2yz - 8y^3z$

Solve for the specified variable or expression. See Example 7.

55. $2g = ch + dh$ for h

56. $d_1d_2 = fd_2 + fd_1$ for f

57. $r_1r_2 = rr_2 + rr_1$ for r_1

58. $rx - ty = by$ for y

59. $b^2x^2 + a^2y^2 = a^2b^2$ for a^2

60. $b^2x^2 + a^2y^2 = a^2b^2$ for b^2

61. $Sn = (n - 2)180$ for n

62. $S(1 - r) = a - lr$ for r

Factor. See Example 8 or 13.

63. $x^2 - 5x + 6$

64. $y^2 + 7y + 6$

65. $x^2 + x - 30$

66. $c^2 + 3c - 28$

Factor. See Example 9.

67. $3x^2 + 12xy - 63y^2$

68. $2y^2 + 4yz - 48z^2$

69. $6a^2 - 30ab + 24b^2$

70. $4b^2 + 12bc - 16c^2$

71. $n^4 - 28n^3t - 60n^2t^2$

72. $c^4 - 16c^3d - 80c^2d^2$

73. $-3x^2 + 15xy - 18y^2$

74. $-2y^2 - 16yt + 40t^2$

Factor. See Example 10 or 13.

75. $5x^2 + 13x + 6$

76. $5x^2 + 18x + 9$

77. $7a^2 + 12a + 5$

78. $7a^2 + 36a + 5$

79. $11y^2 + 32y - 3$

80. $2y^2 - 9y - 18$

81. $8x^2 - 22x + 5$

82. $4z^2 - 13z + 3$

83. $6y^2 - 13y + 6$

84. $6x^2 - 11x + 3$

85. $15b^2 + 4b - 4$

86. $8a^2 + 6a - 9$

Factor each expression. See Example 11.

87. $30x^4 - 25x^2 - 20$

88. $14x^4 + 77x^2 + 84$

89. $32x^4 - 96x^2 + 72$

90. $20a^4 + 60a^2 + 45$

91. $64h^5 - 4h + 24h^3$

92. $9x^5 - 24x + 30x^3$

93. $-3a^4 - 5a^2b^2 - 2b^4$

94. $-2x^4 + 3x^2y^2 + 5y^4$

Factor by using a substitution. See Example 12.

95. $(a + b)^2 - 2(a + b) - 24$

96. $(x - y)^2 + 3(x - y) - 10$

97. $(x + a)^2 + 2(x + a) + 1$

98. $(a + b)^2 - 2(a + b) + 1$

TRY IT YOURSELF

Factor completely. Factor out all common factors first including −1 if the first term is negative. If an expression is prime, so indicate.

99. $3(m + n + p) + x(m + n + p)$

100. $x(x - y - z) + y(x - y - z)$

101. $-63u^3v^6 + 28u^2v^7 - 21u^3v^3$

102. $-56x^4y^3 - 72x^3y^4 + 80xy^2$

103. $b^4x^2 - 12b^2x^2 + 35x^2$

104. $c^3x^4 + 11c^3x^2 - 42c^3$

105. $1 - m + mn - n$

106. $a^2x^2 - 10 - 2x^2 + 5a^2$

107. $-x^2 + 4xy + 21y^2$

108. $-a^2 - 4ab + 5b^2$

109. $a^2x + bx - a^2 - b$

110. $x^2y - ax - xy + a$

111. $4y^2 + 4y + 1$

112. $9x^2 + 6x + 1$

113. $b^2 + 8b + 18$

114. $x^2 + 4x - 28$

115. $13r + 3r^2 - 10$

116. $-r + 3r^2 - 10$

117. $y^3 - 12 + 3y - 4y^2$

118. $h^3 - 8 + h - 8h^2$

119. $2y^5 - 26y^3 + 60y$

120. $2y^5 - 26y^3 + 84y$

121. $14(q - r)^2 - 17(q - r) - 6$

122. $8(h + s)^2 + 34(h + s) + 35$

APPLICATIONS

123. CRAYONS   The amount of colored wax used to make the crayon shown in the illustration can be found by computing its volume using the formula

$$V = \pi r^2 h_1 + \frac{1}{3}\pi r^2 h_2$$

Factor the expression on the right side of this equation.

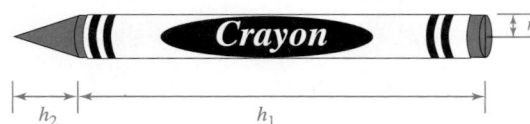

124. PACKAGING   The amount of cardboard needed to make the following cereal box can be found by finding the area A, which is given by the formula

$$A = 2wh + 4wl + 2lh$$

where w is the width, h the height, and l the length. Solve the equation for the width.

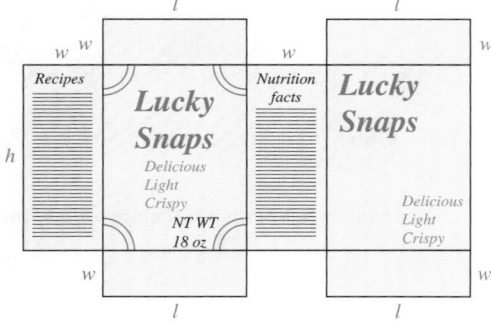

125. ICE   The surface area of the ice cube is represented by the expression $6x^2 + 36x + 54$. Use factoring to find the length of an edge of the cube.

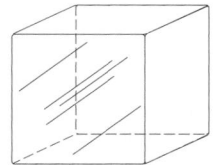

126. CHECKERS   The area of the checkerboard is represented by the expression is $25x^2 - 40x + 16$. Use factoring to find the length of each side.

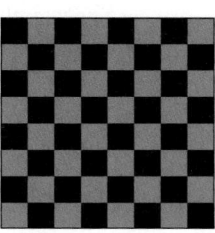

127. Explain the error in the following solution.

Solve for r_1: $r_1 r_2 = rr_2 + rr_1$

$$\frac{r_1 r_2}{r_2} = \frac{rr_2 + rr_1}{r_2}$$

$$\cancel{r_1} = \frac{rr_2 + rr_1}{r_2}$$

128. Explain the error.

Factor: $2x^2 - 4x - 6 = \cancel{(2x + 2)(x - 3)}$

129. INVESTMENTS   Equal amounts are invested in each of three accounts paying 7%, 8%, and 10.5% annually. If one year's combined interest income is $1,249.50, how much is invested in each account?

130. SEARCH AND RESCUE   Two search-and-rescue teams leave base at the same time looking for a lost boy. The first team, on foot, heads north at 2 mph and the other, on horseback, south at 4 mph. How long will it take them to search a distance of 21 miles between them?

Factor out the specified factor.

131. x^2 from $x^{n+2} + x^{n+3}$

132. y^n from $2y^{n+2} - 3y^{n+3}$

Factor. Assume that n is a natural number.

133. $x^{2n} + 2x^n + 1$ **134.** $2a^{6n} - 3a^{3n} - 2$

135. $x^{4n} + 2x^{2n}y^{2n} + y^{4n}$ **136.** $6x^{2n} + 7x^n - 3$

SECTION 8.5

Review of Factoring Methods: The Difference of Two Squares; the Sum and Difference of Two Cubes

Objectives

1 Factor the difference of two squares.

2 Factor the sum and difference of two cubes.

We will now review how to factor the difference of two squares and the sum and difference of two cubes.

1 **Factor the Difference of Two Squares.**

Recall that the difference of the squares of two quantities factors into the product of two binomials.

Difference of Two Squares	$x^2 - y^2 = (x + y)(x - y)$

Success Tip

To factor differences of squares, it is helpful to know these **perfect-integer squares**: 1, 4, 9, 16, 25, 36, 49, 64, 81, 100, 121, 144.

If we think of the difference of two squares as the square of a **F**irst quantity minus the square of a **L**ast quantity, we have the formula

$$F^2 - L^2 = (F + L)(F - L)$$

and we say: *To factor the square of a **F**irst quantity minus the square of a **L**ast quantity, we multiply the **F**irst plus the **L**ast by the **F**irst minus the **L**ast.*

EXAMPLE 1   Factor: $49x^2 - 16$

Strategy   The terms of this binomial do not have a common factor (other than 1). The only option is to attempt to factor it as a difference of two squares.

Why   If a binomial is a difference of two squares, we can factor it using a special product.

Solution $49x^2 - 16$ is the difference of two squares because it can be written as $(7x)^2 - (4)^2$. We can match it to the rule for factoring a difference of two squares to find the factorization.

$$F^2 - L^2 = (F + L)(F - L)$$
$$(7x)^2 - (4)^2 = (7x + 4)(7x - 4)$$

We can verify this result using the FOIL method to do the multiplication.

$$
\begin{aligned}
(7x + 4)(7x - 4) &= 49x^2 - 28x + 28x - 16 \\
&= 49x^2 - 16 \qquad \text{This is the original binomial.}
\end{aligned}
$$

> **Success Tip**
> Always verify a factorization by doing the indicated multiplication. The result should be the original polynomial.

 Self Check 1   Factor: $81p^2 - 25$
Now Try   **Problems 11, 17, and 23**

EXAMPLE 2   Factor: $x^4 - 1$

Strategy   The terms of $x^4 - 1$ do not have a common factor (other than 1). To factor this binomial, we will write it in a form that shows it is a difference of two squares.

Why   We can then use a special-product rule to factor it.

Solution   Because the binomial is the difference of the squares of x^2 and 1, it factors into the sum of x^2 and 1 and the difference of x^2 and 1.

$$
\begin{aligned}
x^4 - 1 &= (x^2)^2 - (1)^2 \\
&= (x^2 + 1)(x^2 - 1)
\end{aligned}
$$

> **Caution**
> The binomial $x^2 + 1$ is the **sum of two squares**. In general, after removing any common factor, a sum of two squares cannot be factored using real numbers.

The factor $x^2 + 1$ is the sum of two quantities and is prime. However, the factor $x^2 - 1$ is the difference of two squares and can be factored as $(x + 1)(x - 1)$. Thus,

$$
\begin{aligned}
x^4 - 1 &= (x^2 + 1)(x^2 - 1) \\
&= (x^2 + 1)(x + 1)(x - 1)
\end{aligned}
$$

 Self Check 2   Factor: $a^4 - 81$
Now Try   **Problem 29**

EXAMPLE 3   Factor: $(x + y)^4 - z^4$

Strategy   We will use a substitution to factor this difference of two squares.

Why   For more complicated expressions, especially those involving a quantity within parentheses, a substitution often helps simplify the factoring process.

Solution   If we use the substitution $a = x + y$, we obtain

$$(x + y)^4 - z^4 = a^4 - z^4 \qquad \text{Replace } x + y \text{ with } a.$$
$$= (a^2 + z^2)(a^2 - z^2) \qquad \text{Factor the difference of two squares.}$$
$$= (a^2 + z^2)(a + z)(a - z) \qquad \text{Factor } a^2 - z^2.$$

To find the factorization of $(x + y)^4 - z^4$, we substitute $x + y$ for each a in the expression $(a^2 + z^2)(a + z)(a - z)$.

$$(a^2 + z^2)(a + z)(a - z) = [(x + y)^2 + z^2](x + y + z)(x + y - z)$$

Thus, $(x + y)^4 - z^4 = [(x + y)^2 + z^2](x + y + z)(x + y - z)$.
If we square the binomial within the brackets, we have

$$(x + y)^4 - z^4 = [x^2 + 2xy + y^2 + z^2](x + y + z)(x + y - z)$$

> ### Caution
> When factoring a polynomial, be sure to factor it completely. Always check to see whether any of the factors of your result can be factored further.

 Self Check 3   Factor: $(a - b)^4 - c^4$

Now Try   **Problem 35**

EXAMPLE 4   Factor: $2x^4y - 32y$

Strategy   We will factor out the GCF of $2y$ and factor the resulting difference of two squares.

Why   The first step in factoring any polynomial is to factor out the GCF.

Solution
$$2x^4y - 32y = 2y(x^4 - 16) \qquad \text{Factor out the GCF, which is } 2y.$$
$$= 2y(x^2 + 4)(x^2 - 4) \qquad \text{Factor } x^4 - 16.$$
$$= 2y(x^2 + 4)(x + 2)(x - 2) \qquad \text{Factor } x^2 - 4.$$

 Self Check 4   Factor: $3a^4 - 3$

Now Try   **Problems 37 and 41**

EXAMPLE 5   Factor: **a.** $x^2 - y^2 + x - y$ **b.** $x^2 + 6x + 9 - z^2$

Strategy   The terms of each expression do not have a common factor (other than 1) and traditional factoring by grouping will not work. Instead, in part (a), we will group only the first two terms of the polynomial and in part (b) we will group the first three terms.

Why   Hopefully, those steps will produce equivalent expressions that can be factored.

Solution   We group the first two terms, factor them as a difference of two squares, and look for a common factor.

a. $x^2 - y^2 + x - y = (x + y)(x - y) + (x - y)$ Factor $x^2 - y^2$. The terms of the resulting expression have a common binomial
factor, $x - y$, that can be factored out.

$= (x - y)(x + y + 1)$ Factor out the GCF, $x - y$.

b. We group the first three terms and factor that trinomial to get:

$x^2 + 6x + 9 - z^2 = (x + 3)(x + 3) - z^2$ $x^2 + 6x + 9$ is a perfect-square
trinomial.

$= (x + 3)^2 - z^2$ The expression that results is a
difference of two squares.

$= (x + 3 + z)(x + 3 - z)$ Factor the difference of two squares.

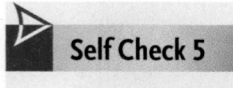

Self Check 5   Factor: **a.** $a^2 - b^2 + a + b$
b. $a^2 + 4a + 4 - b^2$

Now Try   Problems 45 and 49

---

2   **Factor the Sum and Difference of Two Cubes.**

Recall that the sum and difference of two cubes factor as the product of a binomial and a
trinomial.

Sum and Difference of Two Cubes	
$x^3 + y^3 = (x + y)(x^2 - xy + y^2)$	
$x^3 - y^3 = (x - y)(x^2 + xy + y^2)$	

Success Tip
The formulas for factoring the
sum and difference of two cubes
were developed on page 513.

If we think of the sum of two cubes as the sum of the cube of a **First** quantity plus the
cube of a **Last** quantity, we have the formula

$$F^3 + L^3 = (F + L)(F^2 - FL + L^2)$$

*To factor the cube of a **First** quantity plus the cube of a **Last** quantity, we multiply the sum of
the **First** and **Last** by*

- *the **First** squared*
- *minus the **First** times the **Last***
- *plus the **Last** squared*

The formula for the difference of two cubes is

$$F^3 - L^3 = (F - L)(F^2 + FL + L^2)$$

*To factor the cube of a **First** quantity minus the cube of a **Last** quantity, we multiply the
difference of the **First** and **Last** by*

- *the **First** squared*
- *plus the **First** times the **Last***
- *plus the **Last** squared*

EXAMPLE 6   Factor: $a^3 + 8$

Strategy   We will write the binomial in a form that shows it is the sum of two cubes.

Why   We can then use the rule for factoring the sum of two cubes.

Solution   Since $a^3 + 8$ can be written as $a^3 + 2^3$, it is a sum of two cubes, which factors as follows:

$$F^3 + L^3 = (F + L)(F^2 - FL + L^2)$$

$$a^3 + 2^3 = (a + 2)(a^2 - a\,2 + 2^2)$$

$$= (a + 2)(a^2 - 2a + 4) \quad a^2 - 2a + 4 \text{ does not factor.}$$

Therefore, $a^3 + 8 = (a + 2)(a^2 - 2a + 4)$. We can check by multiplying.

$$(a + 2)(a^2 - 2a + 4) = a^3 - 2a^2 + 4a + 2a^2 - 4a + 8$$

$$= a^3 + 8 \qquad \text{This is the original binomial.}$$

Self Check 6   Factor: $p^3 + 27$

Now Try   **Problem 53**

> **Caution**
> In Example 6, a common error is to try to factor $a^2 - 2a + 4$. It is not a perfect-square trinomial, because the middle term needs to be $-4a$. Furthermore, it cannot be factored by the methods of Section 8.4. It is prime.

EXAMPLE 7   Factor: $27a^3 - 64b^6$

Strategy   We will write the binomial in a form that shows it is the difference of two cubes.

Why   We can then use the rule for factoring the difference of two cubes.

Solution   Since $27a^3 - 64b^6$ can be written as $(3a)^3 - (4b^2)^3$, it is a difference of two cubes, which factors as follows:

$$F^3 - L^3 = (F - L)(F^2 + F\,L + L^2)$$

$$(3a)^3 - (4b^2)^3 = (3a - 4b^2)[(3a)^2 + (3a)(4b^2) + (4b^2)^2]$$

$$= (3a - 4b^2)(9a^2 + 12ab^2 + 16b^4)$$

Thus, $27a^3 - 64b^6 = (3a - 4b^2)(9a^2 + 12ab^2 + 16b^4)$.

Self Check 7   Factor: $8c^6 - 125d^3$

Now Try   **Problem 57**

EXAMPLE 8   Factor: $a^3 - (c + d)^3$

Strategy   To factor this expression, we will use the rule for factoring the difference of two cubes.

Why   The terms a^3 and $(c + d)^3$ are perfect cubes.

Success Tip

We could also use the substitution $x = c + d$, factor $a^3 - x^3$ as $(a - x)(a^2 + ax + x^2)$, and then replace each x with $c + d$.

Solution

$$F^3 - L^3 = (F - L)(F^2 + F L + L^2)$$

$$a^3 - (c + d)^3 = [a - (c + d)][a^2 + a(c + d) + (c + d)^2]$$

Now we simplify the expressions inside both sets of brackets.

$$a^3 - (c + d)^3 = (a - c - d)(a^2 + ac + ad + c^2 + 2cd + d^2)$$

Self Check 8   Factor: $(p + q)^3 - r^3$

Now Try   **Problem 61**

EXAMPLE 9   Factor: $x^6 - 64$

Strategy   This binomial is both the difference of two squares and the difference of two cubes. We will write it in a form that shows it is a difference of two squares to begin the factoring process.

Why   It is easier to factor it as the difference of two squares first.

Solution
$$x^6 - 64 = (x^3)^2 - 8^2$$
$$= (x^3 + 8)(x^3 - 8)$$

Each of these factors can be factored further. One is the sum of two cubes and the other is the difference of two cubes.

$$x^6 - 64 = (x + 2)(x^2 - 2x + 4)(x - 2)(x^2 + 2x + 4)$$

Self Check 9   Factor: $1 - x^6$

Now Try   **Problem 65**

EXAMPLE 10   Factor: $2a^5 + 250a^2$

Strategy   We will factor out the GCF of $2a^2$ and factor the resulting sum of two cubes.

Why   The first step in factoring any polynomial is to factor out the GCF.

Solution   We first factor out the common factor of $2a^2$ to obtain

$$2a^5 + 250a^2 = 2a^2(a^3 + 125)$$

Then we factor $a^3 + 125$ as the sum of two cubes to obtain

$$2a^5 + 250a^2 = 2a^2(a + 5)(a^2 - 5a + 25)$$

Self Check 10   Factor: $3x^5 + 24x^2$

Now Try   **Problem 71**

STUDY SET
8.5

VOCABULARY

Fill in the blanks.

1. When the polynomial $4x^2 - 25$ is written as $(2x)^2 - (5)^2$, we see that it is the difference of two _____.

2. When the polynomial $8x^3 + 125$ is written as $(2x)^3 + (5)^3$, we see that it is the sum of two _____.

CONCEPTS

3. a. Write the first ten perfect-integer squares.

   b. Write the first ten perfect-integer cubes.

4. a. Use multiplication to verify that the sum of two squares $x^2 + 25$ does not factor as $(x + 5)(x + 5)$.

   b. Use multiplication to verify that the difference of two squares $x^2 - 25$ factors as $(x + 5)(x - 5)$.

5. Complete each factorization.
   a. $F^2 - L^2 = (F + L)(\qquad)$
   b. $F^3 + L^3 = (F + L)(\qquad)$
   c. $F^3 - L^3 = (F - L)(\qquad)$

6. Factor each binomial.
   a. $5p^2 + 20$
   b. $5p^2 - 20$
   c. $5p^3 + 20$
   d. $5p^3 + 40$

NOTATION

7. Give an example of each.
   a. a difference of two squares
   b. a square of a difference
   c. a sum of two squares
   d. a sum of two cubes
   e. a cube of a sum

8. Fill in the blanks.
   a. $36y^2 - 49m^4 = (\quad)^2 - (\quad)^2$
   b. $125h^3 - 27k^6 = (\quad)^3 - (\quad)^3$

GUIDED PRACTICE

Factor, if possible. See Example 1.

9. $x^2 - 16$

10. $y^2 - 49$

11. $9y^2 - 64$

12. $16x^2 - 81$

13. $144 - c^2$

14. $25 - t^2$

15. $100m^2 - 1$

16. $144x^2 - 1$

17. $81a^2 - 49b^2$

18. $64r^2 - 121s^2$

19. $x^2 + 25$

20. $a^2 + 36$

21. $9r^4 - 121s^2$

22. $81a^4 - 16b^2$

23. $16t^2 - 25w^4$

24. $9r^2 - 25s^4$

25. $100r^2s^4 - t^4$

26. $400x^2z^4 - a^4$

27. $36x^4y^2 - 49z^6$

28. $4a^2b^4 - 9d^6$

Factor completely. See Example 2.

29. $x^4 - y^4$
30. $16n^4 - 1$
31. $16a^4 - 81b^4$
32. $81m^4 - 256n^4$

Factor. See Example 3.

33. $(x + y)^2 - z^2$

34. $a^2 - (b - c)^2$

35. $(r - s)^2 - t^4$

36. $(m + n)^2 - p^4$

Factor each expression. Factor out any GCF first. See Example 4.

37. $2x^2 - 288$

38. $8x^2 - 72$

39. $3x^3 - 243x$

40. $2x^3 - 32x$

41. $5ab^4 - 5a$

42. $3ac^4 - 243a$

43. $64b - 4b^5$

44. $1,250n - 2n^5$

Factor by first grouping the appropriate terms. See Example 5.

45. $c^2 - d^2 + c + d$

46. $s^2 - t^2 + s - t$

47. $a^2 - b^2 + 2a - 2b$

48. $m^2 - n^2 + 3m + 3n$

49. $x^2 + 12x + 36 - y^2$

50. $x^2 - 6x + 9 - 4y^2$

51. $x^2 - 2x + 1 - 9z^2$

52. $x^2 + 10x + 25 - 16z^2$

Factor each sum of cubes. See Example 6.

53. $a^3 + 125$

54. $b^3 + 64$

55. $8r^3 + s^3$

56. $27t^3 + u^3$

Factor each difference of cubes. See Example 7.

57. $64t^6 - 27v^3$

58. $125m^3 - x^6$

59. $x^3 - 216y^6$

60. $8c^6 - 343w^3$

Factor. See Example 8.

61. $(a - b)^3 + 27$

62. $(b - c)^3 - 1,000$

63. $64 - (a + b)^3$

64. $1 - (x + y)^3$

Factor each expression completely. Factor a difference of two squares first. See Example 9.

65. $x^6 - 1$

66. $x^6 - y^6$

67. $x^{12} - y^6$

68. $a^{12} - 64$

Factor each sum or difference of cubes. Factor out the GCF first. See Example 10.

69. $5x^3 + 625$

70. $2x^3 - 128$

71. $4x^5 - 256x^2$

72. $2x^6 + 54x^3$

TRY IT YOURSELF

Factor each expression, if possible.

73. $64a^3 - 125b^6$

74. $8x^6 - 27y^3$

75. $288b^2 - 2b^6$

76. $98x - 2x^5$

77. $x^2 - y^2 + 8x + 8y$

78. $5m - 5n + m^2 - n^2$

79. $x^9 + y^9$

80. $x^6 + y^6$

81. $144a^2t^2 - 169b^6$

82. $25x^6 - 81y^2z^2$

83. $100a^2 + 9b^2$

84. $25s^4 + 16t^2$

85. $81c^4d^4 - 16t^4$

86. $256x^4 - 81y^4$

87. $128u^2v^3 - 2t^3u^2$

88. $56rs^2t^3 + 7rs^2v^6$

89. $y^2 - (2x - t)^2$

90. $(15 - r)^2 - s^2$

91. $x^2 + 20x + 100 - 9z^2$

92. $49a^2 - b^2 - 14b - 49$

93. $(c - d)^3 + 216$

94. $1 - (x + y)^3$

95. $\dfrac{1}{36} - y^4$

96. $\dfrac{4}{81} - m^4$

97. $m^6 - 64$

98. $y^6 - 1$

99. $(a + b)x^3 + 27(a + b)$

100. $(c - d)r^3 - (c - d)s^3$

101. $x^9 - y^{12}z^{15}$

102. $r^{12} + s^{18}t^{24}$

APPLICATIONS

103. CANDY   To find the amount of chocolate used in the outer coating of the malted-milk ball shown, we can find the volume V of the chocolate shell using the formula

$$V = \frac{4}{3}\pi r_1{}^3 - \frac{4}{3}\pi r_2{}^3$$

Factor the expression on the right side of the formula.

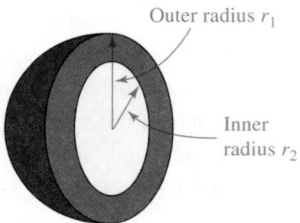

Outer radius r_1

Inner radius r_2

104. MOVIE STUNTS   The function that gives the distance a stuntwoman is above the ground t seconds after she falls over the side of a 144-foot tall building is:

$$h(t) = 144 - 16t^2$$

Factor the right side.

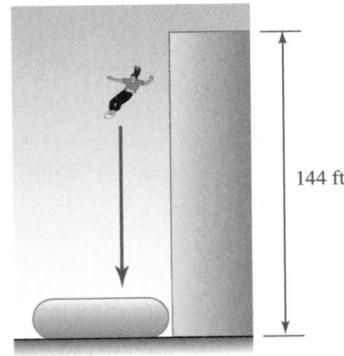

144 ft

WRITING

105. Explain how the patterns used to factor the sum and difference of two cubes are similar and how they differ.

106. Explain why the factorization is not complete.

Factor: $1 - t^8 = (1 + t^4)(1 - t^4)$

107. Explain the error.

Factor: $4g^2 - 16 = (2g + 4)(2g - 4)$

108. When asked to factor $81t^2 - 16$, one student answered $(9t - 4)(9t + 4)$, and another answered $(9t + 4)(9t - 4)$. Explain why both students are correct.

REVIEW

For each of the following purchases, determine the better buy.

109. Flute lessons: 45 minutes for $25 or 1 hour for $35.

110. Tissue paper: 15 sheets for $1.39 or a dozen sheets for $1.10.

CHALLENGE PROBLEMS

Factor. Assume all variables represent natural numbers.

111. $4x^{2n} - 9y^{2n}$

112. $25 - x^{6n}$

113. $a^{3b} - c^{3b}$

114. $27x^{3n} + y^{3n}$

115. $x^{32} - y^{32}$

116. Find the error in this proof that $2 = 1$.

$$x = y$$
$$x^2 = xy$$
$$x^2 - y^2 = xy - y^2$$
$$(x + y)(x - y) = y(x - y)$$
$$\frac{(x + y)(x - y)}{(x - y)} = \frac{y(x - y)}{(x - y)}$$
$$x + y = y$$
$$y + y = y$$
$$2y = y$$
$$\frac{2y}{y} = \frac{y}{y}$$
$$2 = 1$$

SECTION 8.6
Review of Rational Expressions and Rational Equations

Objectives

1. Simplify rational expressions.
2. Multiply and divide rational expressions.
3. Add and subtract rational expressions.
4. Simplify complex fractions.
5. Solve rational equations.

Recall that rational expressions are algebraic fractions with polynomial numerators and denominators.

Rational Expressions

A **rational expression** is an expression of the form $\frac{A}{B}$, where A and B are polynomials and $B \neq 0$.

Caution
Since division by 0 is undefined, the value of a polynomial in the denominator of a rational expression cannot be 0. For example, m cannot be -2 in the expression $\frac{5m + n}{8m + 16}$ because the denominator would be 0.

Some examples of rational expressions are

$$\frac{8y^3z^5}{6y^4z^3}, \quad \frac{3x}{x - 7}, \quad \frac{5m + n}{8m + 16}, \quad \text{and} \quad \frac{6a^2 - 13a + 6}{3a^2 + a - 2}$$

To add, subtract, multiply, and divide rational expressions we use the same rules that we use for performing those operations with arithmetic fractions.

1 Simplify Rational Expressions.

To **simplify a rational expression** we use the following strategy.

Simplifying Rational Expressions

1. Factor the numerator and denominator completely to determine their common factors.
2. Remove factors equal to 1 by replacing each pair of factors common to the numerator and denominator with the equivalent fraction $\frac{1}{1}$.
3. Multiply the remaining factors in the numerator and in the denominator.

EXAMPLE 1   Simplify: $\dfrac{8y^3z^5}{6y^4z^3}$

Strategy   We will begin by writing the numerator and denominator in factored form. Then we will remove any factors common to the numerator and denominator.

Why   The rational expression is simplified when the numerator and denominator have no common factors other than 1.

Solution

$$\frac{8y^3z^5}{6y^4z^3} = \frac{\overset{1}{2} \cdot 4 \cdot \overset{1}{y} \cdot \overset{1}{y} \cdot \overset{1}{y} \cdot \overset{1}{z} \cdot \overset{1}{z} \cdot \overset{1}{z} \cdot z \cdot z}{\underset{1}{2} \cdot 3 \cdot \underset{1}{y} \cdot \underset{1}{y} \cdot \underset{1}{y} \cdot y \cdot \underset{1}{z} \cdot \underset{1}{z} \cdot \underset{1}{z}}$$

Replace $\frac{2}{2}$, $\frac{y}{y}$, and $\frac{z}{z}$ with $\frac{1}{1}$.

This removes the factor $\frac{2 \cdot y \cdot y \cdot y \cdot z \cdot z \cdot z}{2 \cdot y \cdot y \cdot y \cdot z \cdot z \cdot z} = 1$.

$$= \frac{4z^2}{3y}$$

   Self Check 1   Simplify: $\dfrac{10k}{25k^2}$

Now Try   Problem 15

The fractions in Example 1 and the Self Check can also be simplified by using the rules of exponents:

$$\frac{8y^3z^5}{6y^4z^3} = \frac{\overset{1}{2} \cdot 4 \cdot y^{3-4} \cdot z^{5-3}}{\underset{1}{2} \cdot 3} \qquad \frac{10k}{25k^2} = \frac{\overset{1}{5} \cdot 2 \cdot k^{1-2}}{\underset{1}{5} \cdot 5}$$

To divide exponential expressions with the same base, keep the base and subtract the exponents.

$$= \frac{4 \cdot y^{-1} \cdot z^2}{3} \qquad\qquad = \frac{2 \cdot k^{-1}}{5}$$

$$= \frac{4z^2}{3y} \qquad\qquad\qquad = \frac{2}{5k}$$

EXAMPLE 2   Simplify: $\dfrac{2x^2 + 11x + 12}{3x^2 + 11x - 4}$

Strategy   We will begin by factoring the numerator and denominator completely. Then we will remove any factors common to the numerator and denominator.

Why   We need to make sure that the numerator and denominator have no common factors other than 1. If that is the case, then the rational expression is simplified.

Solution

$$\dfrac{2x^2 + 11x + 12}{3x^2 + 11x - 4} = \dfrac{(2x + 3)\overset{1}{\cancel{(x + 4)}}}{(3x - 1)\underset{1}{\cancel{(x + 4)}}} \qquad \text{Remove a factor equal to 1: } \tfrac{x + 4}{x + 4} = 1.$$

$$= \dfrac{2x + 3}{3x - 1} \qquad \text{This expression does not simplify further.}$$

> **Caution**
> In Example 2, do not remove the x's in the result $\tfrac{2x + 3}{3x - 1}$. The x in the numerator is not a factor of the entire numerator. Likewise, the x in the denominator is not a factor of the entire denominator.
>
>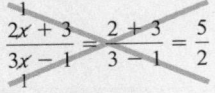

> **Self Check 2**   Simplify: $\dfrac{2x^2 + 5x + 2}{3x^2 + 5x - 2}$
>
> **Now Try**   Problem 21

Recall that if the terms of two polynomials are the same except that they are opposite in sign, the polynomials are opposites.

The Quotient of Opposites

The quotient of any nonzero polynomial and its opposite is -1.

EXAMPLE 3   Simplify: $\dfrac{3x^2 - 10xy - 8y^2}{4y^2 - xy}$

Strategy   We will begin by factoring the numerator and denominator. Then we look for common factors, or factors that are opposites, and remove them.

Why   We need to make sure that the numerator and denominator have no common factors other than 1. If that is the case, then the rational expression is simplified.

Solution   We factor the numerator and denominator. Because $x - 4y$ and $4y - x$ are opposites, their quotient is -1.

> **Caution**
> A $-$ symbol in front of a fraction may be placed in the numerator or the denominator, but not to both. For example,
>
> $$-\dfrac{3x + 2y}{y} \neq \dfrac{-3x - 2y}{-y}$$

$$\dfrac{3x^2 - 10xy - 8y^2}{4y^2 - xy} = \dfrac{(3x + 2y)\overset{-1}{\cancel{(x - 4y)}}}{y\underset{1}{\cancel{(4y - x)}}} \qquad \begin{array}{l}\text{Since } x - 4y \text{ and } 4y - x \text{ are opposites,}\\[2pt]\text{simplify by replacing } \tfrac{x - 4y}{4y - x} \text{ with the equivalent}\\[2pt]\text{fraction } \tfrac{-1}{1} = -1.\end{array}$$

$$= \dfrac{-(3x + 2y)}{y}$$

This result can also be written as $-\dfrac{3x + 2y}{y}$ or $\dfrac{-3x - 2y}{y}$.

> **Self Check 3**   Simplify: $\dfrac{2a^2 - 3ab - 9b^2}{3b^2 - ab}$
>
> **Now Try**   Problem 25

Multiply and Divide Rational Expressions.

Recall that to multiply two fractions, we multiply the numerators and multiply the denominators. We use the same strategy to multiply rational expressions.

Multiplying Rational Expressions	To multiply rational expressions, multiply their numerators and their denominators. Then, if possible, factor and simplify.

For any two rational expressions, $\frac{A}{B}$ and $\frac{C}{D}$,

$$\frac{A}{B} \cdot \frac{C}{D} = \frac{AC}{BD}$$

EXAMPLE 4   Multiply: $\dfrac{x^2 - 6x + 9}{20x} \cdot \dfrac{5x^2}{x^2 - 9}$

Strategy   To find the product, we will use the rule for multiplying rational expressions. In the process, we must be prepared to factor the numerators and denominators so that any common factors can be removed.

Why   We want to give the result in simplified form.

Solution

$$\dfrac{x^2 - 6x + 9}{20x} \cdot \dfrac{5x^2}{x^2 - 9} = \dfrac{(x^2 - 6x + 9)5x^2}{20x(x^2 - 9)} \quad \begin{array}{l}\text{Multiply the numerators.}\\ \text{Multiply the denominators.}\end{array}$$

$$= \dfrac{(x - 3)(x - 3)5 \cdot x \cdot x}{4 \cdot 5 \cdot x(x + 3)(x - 3)} \quad \begin{array}{l}\text{Factor the numerator.}\\ \text{Factor the denominator.}\end{array}$$

$$= \dfrac{\overset{1}{\cancel{(x - 3)}}(x - 3)\overset{1}{\cancel{5}} \cdot \overset{1}{\cancel{x}} \cdot x}{4 \cdot \underset{1}{\cancel{5}} \cdot \underset{1}{\cancel{x}}(x + 3)\underset{1}{\cancel{(x - 3)}}} \quad \begin{array}{l}\text{Simplify by removing common factors}\\ \text{of the numerator and denominator.}\end{array}$$

$$= \dfrac{x(x - 3)}{4(x + 3)}$$

> *Caution*
> When multiplying rational expressions, always write the result in simplest form by removing any factors common to the numerator and denominator.

We could distribute in the numerator and/or denominator and write the result as $\frac{x^2 - 3x}{4(x + 3)}$ or $\frac{x^2 - 3x}{4x + 12}$. Check with your instructor to see which form of the result he or she prefers.

 Self Check 4   Multiply: $\dfrac{a^2 + 6a + 9}{18a} \cdot \dfrac{3a^3}{a + 3}$

     Now Try   **Problem 29**

EXAMPLE 5   Multiply: $(2x - x^2) \cdot \dfrac{x}{x^2 - xb - 2x + 2b}$

Strategy   We will write $2x - x^2$ as a rational expression with denominator 1. (Remember, any number divided by 1 remains unchanged.) Then we will use the rule for multiplying rational expressions.

Why   Writing $2x - x^2$ as $\frac{2x - x^2}{1}$ is helpful during the multiplication process when we multiply numerators and multiply denominators.

Solution

$$(2x - x^2) \cdot \frac{x}{x^2 - xb - 2x + 2b}$$

$$= \frac{2x - x^2}{1} \cdot \frac{x}{x^2 - xb - 2x + 2b} \qquad \text{Write } 2x - x^2 \text{ as } \tfrac{2x - x^2}{1}.$$

$$= \frac{(2x - x^2)x}{1(x^2 - xb - 2x + 2b)} \qquad \begin{array}{l}\text{Multiply the numerators.}\\ \text{Multiply the denominators.}\end{array}$$

$$= \frac{x(2 - x)x}{1[x(x - b) - 2(x - b)]} \qquad \begin{array}{l}\text{Factor out } x \text{ in the numerator. In the denominator,}\\ \text{begin factoring by grouping.}\end{array}$$

$$= \frac{x(2 - x)x}{1(x - b)(x - 2)} \qquad \begin{array}{l}\text{In the denominator, complete the factoring by}\\ \text{grouping. The brackets [ ] are no longer needed.}\end{array}$$

$$= \frac{x \overset{-1}{\cancel{(2 - x)}} x}{1(x - b)\underset{1}{\cancel{(x - 2)}}} \qquad \begin{array}{l}\text{The quotient of any nonzero quantity and its}\\ \text{opposite is } -1\text{: } \tfrac{2 - x}{x - 2} = -1.\end{array}$$

$$= \frac{-x^2}{x - b}$$

Since the $-$ symbol can be written in front of the fraction, this result can also be written as

$$-\frac{x^2}{x - b}$$

Self Check 5   Multiply: $\dfrac{x^2 + 5x + 6}{4x + 8 - x^2 - 2x}(x^2 - 4x)$

Now Try   **Problem 35**

> **Success Tip**
> We would obtain the same answer if we had factored the numerators and denominators first and simplified before we multiplied.

Recall that to divide fractions, we multiply the first fraction by the reciprocal of the second fraction. We use the same strategy to divide rational expressions.

Dividing Rational Expressions

To divide two rational expressions, multiply the first by the reciprocal of the second. Then, if possible, factor and simplify.

For any two rational expressions, $\frac{A}{B}$ and $\frac{C}{D}$, where $\frac{C}{D} \neq 0$,

$$\frac{A}{B} \div \frac{C}{D} = \frac{A}{B} \cdot \frac{D}{C} = \frac{AD}{BC}$$

EXAMPLE 6   Divide: $\dfrac{x^3 + 8}{4x + 4} \div \dfrac{x^2 - 2x + 4}{2x^2 - 2}$

Strategy   We will write the division $\frac{x^3 + 8}{4x + 4} \div \frac{x^2 - 2x + 4}{2x^2 - 2}$ as the equivalent multiplication $\frac{x^3 + 8}{4x + 4} \cdot \frac{2x^2 - 2}{x^2 - 2x + 4}$.

Why   We can then perform the multiplication as in Examples 4 and 5.

Solution

$$\frac{x^3 + 8}{4x + 4} \div \frac{x^2 - 2x + 4}{2x^2 - 2}$$

$$= \frac{x^3 + 8}{4x + 4} \cdot \frac{2x^2 - 2}{x^2 - 2x + 4}$$

Multiply the first rational expression by the reciprocal of the second.

$$= \frac{(x^3 + 8)(2x^2 - 2)}{(4x + 4)(x^2 - 2x + 4)}$$

Multiply the numerators.
Multiply the denominators.

$$= \frac{(x + 2)(\overset{1}{\cancel{x^2 - 2x + 4}})2(\overset{1}{\cancel{x + 1}})(x - 1)}{2 \cdot \cancel{2}(\underset{1}{\cancel{x + 1}})(\underset{1}{\cancel{x^2 - 2x + 4}})}$$

Factor $x^3 + 8$, $2x^2 - 2$, and $4x + 4$. The polynomial $x^2 - 2x + 4$ does not factor. Write 4 as $2 \cdot 2$. Then simplify.

$$= \frac{(x + 2)(x - 1)}{2}$$

 Self Check 6   Divide: $\dfrac{x^3 - 8}{9x - 9} \div \dfrac{x^2 + 2x + 4}{3x^2 - 3x}$

Now Try   **Problem 43**

EXAMPLE 7   Simplify: $\dfrac{x^2 + 2x - 3}{6x^2 + 5x + 1} \div \dfrac{2x^2 - 2}{2x^2 - 5x - 3} \cdot \dfrac{6x^2 + 4x - 2}{x^2 - 2x - 3}$

Strategy   We will consider the division first by multiplying the first rational expression by the reciprocal of the second. Then we will find the product of the three rational expressions.

Why   By the rules for the order of operations, we must perform division and multiplication in order from left to right.

Solution   Since multiplications and divisions are done in order from left to right, we begin by focusing on the division. We introduce grouping symbols to emphasize this. To divide the expressions within the parentheses, we invert $\dfrac{2x^2 - 2}{2x^2 - 5x - 3}$ and multiply.

$$\left(\frac{x^2 + 2x - 3}{6x^2 + 5x + 1} \div \frac{2x^2 - 2}{2x^2 - 5x - 3} \right) \frac{6x^2 + 4x - 2}{x^2 - 2x - 3} = \left(\frac{x^2 + 2x - 3}{6x^2 + 5x + 1} \cdot \frac{2x^2 - 5x - 3}{2x^2 - 2} \right) \frac{6x^2 + 4x - 2}{x^2 - 2x - 3}$$

Next, we multiply the three rational expressions and simplify the result.

$$= \frac{(x^2 + 2x - 3)(2x^2 - 5x - 3)(6x^2 + 4x - 2)}{(6x^2 + 5x + 1)(2x^2 - 2)(x^2 - 2x - 3)}$$

$$= \frac{(x + 3)(\overset{1}{\cancel{x - 1}})(\overset{1}{\cancel{2x + 1}})(\overset{1}{\cancel{x - 3}})2(3x - 1)(\overset{1}{\cancel{x + 1}})}{(3x + 1)(\underset{1}{\cancel{2x + 1}})\cancel{2}(x + 1)(\underset{1}{\cancel{x - 1}})(\underset{1}{\cancel{x - 3}})(\underset{1}{\cancel{x + 1}})}$$

$$= \frac{(x + 3)(3x - 1)}{(3x + 1)(x + 1)}$$

 Self Check 7   Simplify: $\dfrac{x^2 - 25}{4x^2 + 12x + 9} \div \dfrac{x^2 - 5x}{3x - 1} \cdot \dfrac{2x + 3}{3x^2 + 14x - 5}$

Now Try   **Problem 47**

3   **Add and Subtract Rational Expressions.**

To add (or subtract) fractions with like denominators, we add (or subtract) the numerators and keep the same denominator. We use the same strategy to add (or subtract) rational expressions with like denominators.

Adding and Subtracting Rational Expressions That Have the Same Denominator	To add (or subtract) rational expressions that have same denominator, add (or subtract) their numerators and write the sum (or difference) over the common denominator. Then, if possible, factor and simplify. If $\frac{A}{D}$ and $\frac{B}{D}$ are rational expressions, $$\frac{A}{D} + \frac{B}{D} = \frac{A+B}{D} \quad \text{and} \quad \frac{A}{D} - \frac{B}{D} = \frac{A-B}{D}$$

EXAMPLE 8   Add $\dfrac{a^2}{a^2-36} + \dfrac{6a}{a^2-36}$ and simplify the result.

Strategy   We will add the numerators and write the sum over the common denominator.

Why   This is the rule for adding rational expressions that have the *same* denominator.

Solution

$$\frac{a^2}{a^2-36} + \frac{6a}{a^2-36} = \frac{a^2+6a}{a^2-36}$$ Add the numerators. Write the sum over the common denominator, $a^2 - 36$.

We can factor the binomials in the numerator and the denominator.

$$\frac{a^2}{a^2-36} + \frac{6a}{a^2-36} = \frac{a(a+6)}{(a+6)(a-6)}$$

$$= \frac{\overset{1}{a(\cancel{a+6})}}{(\cancel{a+6})(a-6)}$$ Simplify.

$$= \frac{a}{a-6}$$

> **Caution**
> When adding or subtracting rational expressions, always write the result in simplest form by removing any factors common to the numerator and denominator.

Self Check 8   Add: $\dfrac{2b}{b^2-4} + \dfrac{b^2}{b^2-4}$

***Now Try*   Problem 55**

To add or subtract rational expressions with unlike denominators, we build them to rational expressions with the same denominator.

Building Rational Expressions	To build a rational expression, multiply it by 1 in the form of $\frac{c}{c}$, where c is any nonzero number or expression.

When adding (or subtracting) rational expressions with unlike denominators, we will write the rational expressions with the smallest common denominator possible, called the **least** (or lowest) **common denominator (LCD)**. To find the least common denominator of several rational expressions, we follow these steps.

Finding the LCD	1. Factor each denominator completely.
	2. The LCD is a product that uses each different factor obtained in step 1 the greatest number of times it appears in any one factorization.

EXAMPLE 9   Add: $\dfrac{5a}{24b} + \dfrac{11a}{18b^2}$

Strategy   We will find the LCD of these rational expressions. Then we will build the rational expressions so each one has the LCD as its denominator.

Why   Since the denominators are different, we cannot add these rational expressions in their present form.

Solution   To find the LCD, we write each denominator as the product of prime numbers and variables.

$$24b = 2 \cdot 2 \cdot 2 \cdot 3 \cdot b = 2^3 \cdot 3 \cdot b$$
$$18b^2 = 2 \cdot 3 \cdot 3 \cdot b \cdot b = 2 \cdot 3^2 \cdot b^2$$

Then we form a product using each of these factors the greatest number of times it appears in any one factorization.

> The greatest number of times the factor 2 appears is three times.
> The greatest number of times the factor 3 appears is twice.
> The greatest number of times the factor b appears is twice.

$$LCD = 2 \cdot 2 \cdot 2 \cdot 3 \cdot 3 \cdot b \cdot b$$
$$= 72b^2$$

We now multiply the numerator and denominator of each rational expression by whatever it takes to build their denominators to $72b^2$.

$$\frac{5a}{24b} + \frac{11a}{18b^2} = \frac{5a}{24b} \cdot \frac{3b}{3b} + \frac{11a}{18b^2} \cdot \frac{4}{4} \qquad \text{Build each rational expression.}$$

$$= \frac{15ab}{72b^2} + \frac{44a}{72b^2} \qquad \begin{array}{l}\text{Multiply the numerators.}\\ \text{Multiply the denominators.}\end{array}$$

$$= \frac{15ab + 44a}{72b^2} \qquad \begin{array}{l}\text{Add the numerators. Write the sum over the}\\ \text{common denominator. The result does not simplify.}\end{array}$$

    Self Check 9   Add: $\dfrac{3}{28z^3} + \dfrac{5x}{21z}$

Now Try   **Problem 61**

EXAMPLE 10   Subtract: $\dfrac{x+1}{x^2 - 2x + 1} - \dfrac{x-4}{x^2 - 1}$

Strategy   We will factor each denominator, find the LCD, and build the rational expressions so each one has the LCD as its denominator.

Why   Since the denominators are different, we cannot subtract these rational expressions in their present form.

Solution We factor each denominator to find the LCD:

$$x^2 - 2x + 1 = (x-1)(x-1) = (x-1)^2$$
$$x^2 - 1 = (x+1)(x-1)$$

The greatest number of times the factor $x-1$ appears is twice.
The greatest number of times the factor $x+1$ appears is once.

The LCD is $(x-1)^2(x+1)$ or $(x-1)(x-1)(x+1)$.

We now write each rational expression with its denominator in factored form. Then we multiply each numerator and denominator by the missing factor, so that each rational expression has a denominator of $(x-1)(x-1)(x+1)$.

$$\frac{x+1}{x^2-2x+1} - \frac{x-4}{x^2-1}$$

$$= \frac{x+1}{(x-1)(x-1)} - \frac{x-4}{(x+1)(x-1)}$$

Write each denominator in factored form.

$$= \frac{x+1}{(x-1)(x-1)} \cdot \frac{x+1}{x+1} - \frac{x-4}{(x+1)(x-1)} \cdot \frac{x-1}{x-1}$$

Build each rational expression.

$$= \frac{x^2+2x+1}{(x-1)(x-1)(x+1)} - \frac{x^2-5x+4}{(x+1)(x-1)(x-1)}$$

Multiply the numerators using the FOIL method. Multiply the denominators.

This numerator is written within parentheses to make sure we subtract all three of its terms.

$$= \frac{x^2+2x+1-(x^2-5x+4)}{(x-1)(x-1)(x+1)}$$

Subtract the numerators. Write the difference over the denominator.

$$= \frac{x^2+2x+1-x^2+5x-4}{(x-1)(x-1)(x+1)}$$

In the numerator, subtract the trinomials.

$$= \frac{7x-3}{(x-1)(x-1)(x+1)}$$

Combine like terms. The result does not simplify.

$$= \frac{7x-3}{(x-1)^2(x+1)}$$

Write $(x-1)(x-1)$ as $(x-1)^2$.

Success Tip
To build each rational expression, we use the FOIL method to multiply the numerators. Note that we don't multiply out the denominators. For example, to build the second rational expression, we have:

$$\frac{x-4}{(x+1)(x-1)} \cdot \frac{x-1}{x-1}$$

The result is:

$$\frac{x^2-5x+4}{(x+1)(x-1)(x-1)}$$

Self Check 10 Subtract: $\dfrac{a+2}{a^2-4a+4} - \dfrac{a-3}{a^2-4}$

Now Try Problem 77

4 Simplify Complex Fractions.

A rational expression whose numerator and/or denominator contain rational expressions is called a **complex rational expression** (or a **complex fraction**). The expression above the main fraction bar of a complex fraction is the numerator, and the expression below the main fraction bar is the denominator. Two examples are:

Success Tip
Review Method 1 and Method 2 for simplifying complex fractions on pages 591 and 594.

$$\frac{\dfrac{3a}{b}}{\dfrac{6ac}{b^2}} \qquad \frac{\dfrac{1}{a^2-3a+2}}{\dfrac{3}{a-2}-\dfrac{2}{a-1}}$$

← Numerator
← Main fraction bar
← Denominator

The first method to **simplify complex fractions** uses the fact that the main fraction bar indicates division. With the second method, we multiply the numerator and denominator of the complex fraction by the LCD of all fractions in the complex fraction.

EXAMPLE 11   Simplify: $\dfrac{\dfrac{3a}{b}}{\dfrac{6ac}{b^2}}$

Strategy   We will perform the division indicated by the main fraction bar.

Why   We know how to divide rational expressions.

Solution

> **Success Tip**
> Simplifying using division works well when a complex fraction is written as a quotient of two single rational expressions.

$$\dfrac{\dfrac{3a}{b}}{\dfrac{6ac}{b^2}} = \dfrac{3a}{b} \div \dfrac{6ac}{b^2}$$ The main fraction bar of the complex fraction indicates division.

$$= \dfrac{3a}{b} \cdot \dfrac{b^2}{6ac}$$ To divide rational expressions, multiply the first by the reciprocal of the second.

$$= \dfrac{3a \cdot b^2}{b \cdot 6ac}$$ Multiply the numerators.
Multiply the denominators.

$$= \dfrac{\overset{1}{\cancel{3}} \cdot \overset{1}{\cancel{a}} \cdot \overset{1}{\cancel{b}} \cdot b}{\underset{1}{\cancel{b}} \cdot 2 \cdot \underset{1}{\cancel{3}} \cdot \underset{1}{\cancel{a}} \cdot c}$$ Factor the numerator and denominator. Then simplify by removing common factors of the numerator and denominator.

$$= \dfrac{b}{2c}$$ Multiply the remaining factors in the numerator.
Multiply the remaining factors in the denominator.

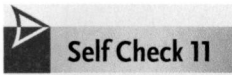 **Self Check 11**   Simplify: $\dfrac{\dfrac{2x}{y^2}}{\dfrac{6xz}{y}}$

Now Try   **Problem 79**

EXAMPLE 12   Simplify: $\dfrac{\dfrac{1}{a^2 - 3a + 2}}{\dfrac{3}{a - 2} - \dfrac{2}{a - 1}}$

Strategy   We will factor $a^2 - 3a + 2$ and determine the LCD for all the fractions appearing in the complex fraction. Then we will use Method 2 to simplify.

Why   Method 2 works well when the complex fraction has sums and/or differences in the numerator or denominator.

Solution   After factoring $a^2 - 3a + 2$, we will see that the LCD of the fractions in the complex fraction is $(a - 2)(a - 1)$. We multiply the numerator and the denominator by the LCD.

Success Tip
Simplifying using the LCD works well when the complex fraction has sums and/or differences in the numerator or denominator.

$$\frac{\dfrac{1}{a^2 - 3a + 2}}{\dfrac{3}{a - 2} - \dfrac{2}{a - 1}} = \frac{\dfrac{1}{(a - 2)(a - 1)}}{\dfrac{3}{a - 2} - \dfrac{2}{a - 1}}$$

Factor $a^2 - 3a + 2$.

$$= \frac{\dfrac{1}{(a - 2)(a - 1)}}{\dfrac{3}{a - 2} - \dfrac{2}{a - 1}} \cdot \frac{(a - 2)(a - 1)}{(a - 2)(a - 1)}$$

Multiply the complex fraction by a factor equal to 1 in the form of $\frac{LCD}{LCD}$.

$$= \frac{\left[\dfrac{1}{(a - 2)(a - 1)}\right](a - 2)(a - 1)}{\left(\dfrac{3}{a - 2} - \dfrac{2}{a - 1}\right)(a - 2)(a - 1)}$$

Multiply the numerators. Multiply the denominators.

$$= \frac{\dfrac{(a - 2)(a - 1)}{(a - 2)(a - 1)}}{\dfrac{3(a - 2)(a - 1)}{a - 2} - \dfrac{2(a - 2)(a - 1)}{a - 1}}$$

Perform the multiplication in the numerator. In the denominator, distribute the LCD, $(a - 2)(a - 1)$.

$$= \frac{1}{3(a - 1) - 2(a - 2)}$$

Simplify each of the three rational expressions highlighted in blue.

$$= \frac{1}{3a - 3 - 2a + 4}$$

In the denominator, remove parentheses.

$$= \frac{1}{a + 1}$$

Combine like terms.

 **Self Check 12**   Simplify: $\dfrac{\dfrac{b}{b + 4} + \dfrac{3}{b + 3}}{\dfrac{b}{b^2 + 7b + 12}}$

Now Try   **Problem 87**

The Language of Algebra
After multiplying a complex fraction by $\frac{LCD}{LCD}$ and performing the multiplications, the numerator and denominator of the complex fraction will be cleared of fractions.

5   **Solve Rational Equations.**

If an equation contains one or more rational expressions, it is called a **rational equation.** Recall that we use a fraction-clearing strategy to solve rational equations.

EXAMPLE 13   Solve: $1 + \dfrac{8a}{a^2 + 3a} = \dfrac{3}{a}$

Strategy   We will find the LCD of the rational expressions in the equation and multiply both sides by the LCD.

Why   This will clear the equation of fractions.

Success Tip
You may want to review the procedure for solving rational equations on page 600.

Solution Since the binomial $a^2 + 3a$ factors as $a(a + 3)$, we can write the given equation as:

$$1 + \frac{8a}{a(a + 3)} = \frac{3}{a} \qquad \text{Factor } a^2 + 3a.$$

We see that 0 and −3 cannot be solutions of the equation, because they make rational expressions in the equation undefined.

We can clear the equation of fractions by multiplying both sides by $a(a + 3)$, which is the LCD of the two rational expressions.

$$a(a + 3)\left[1 + \frac{8a}{a(a + 3)}\right] = a(a + 3)\left(\frac{3}{a}\right) \qquad \text{Multiply both sides by the LCD.}$$

$$a(a + 3)1 + a(a + 3)\frac{8a}{a(a + 3)} = a(a + 3)\left(\frac{3}{a}\right) \qquad \begin{array}{l}\text{On the left side, distribute the}\\\text{multiplication by } a(a + 3).\end{array}$$

$$a(a + 3)1 + \overset{1}{\cancel{a}}\overset{1}{\cancel{(a + 3)}}\frac{8a}{\underset{1}{\cancel{a}}\underset{1}{\cancel{(a + 3)}}} = \overset{1}{\cancel{a}}(a + 3)\left(\frac{3}{\underset{1}{\cancel{a}}}\right) \qquad \begin{array}{l}\text{Remove common factors of the}\\\text{numerator and denominator.}\end{array}$$

$$a^2 + 3a + 8a = 3a + 9 \qquad \begin{array}{l}\text{Simplify each side. The resulting}\\\text{quadratic equation does not contain}\\\text{any fractions.}\end{array}$$

$$a^2 + 8a - 9 = 0 \qquad \begin{array}{l}\text{To get 0 on the right side, subtract } 3a\\\text{and 9 from both sides.}\end{array}$$

$$(a + 9)(a - 1) = 0 \qquad \text{Factor the left side.}$$

$$a + 9 = 0 \quad \text{or} \quad a - 1 = 0 \qquad \text{Set each factor equal to 0.}$$

$$a = -9 \qquad\qquad a = 1 \qquad \text{Solve each equation.}$$

The solutions are −9 and 1. Verify that both satisfy the original equation.

Self Check 13 Solve: $1 + \frac{2}{2b + 1} = \frac{5}{2b^2 + b}$

Now Try Problem 95

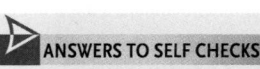

ANSWERS TO SELF CHECKS 1. $\frac{2}{5k}$ 2. $\frac{2x + 1}{3x - 1}$ 3. $-\frac{2a + 3b}{b}$ or $\frac{-2a - 3b}{b}$ 4. $\frac{a^2(a + 3)}{6}$ 5. $-x(x + 3)$
6. $\frac{x(x - 2)}{3}$ 7. $\frac{1}{x(2x + 3)}$ 8. $\frac{b}{b - 2}$ 9. $\frac{9 + 20xz^2}{84z^3}$ 10. $\frac{9a - 2}{(a - 2)^2(a + 2)}$ 11. $\frac{1}{3yz}$ 12. $\frac{b^2 + 6b + 12}{b}$
13. $-\frac{5}{2}, 1$

STUDY SET
8.6

VOCABULARY

Fill in the blanks.

1. A quotient of two polynomials, such as $\frac{x^2 + x}{x^2 - 3x}$, is called a _____ expression.

2. To simplify a rational expression, we remove factors _____ to the numerator and denominator.

3. The quotient of _____ is −1. For example, $\frac{x - 8}{8 - x} = -1$.

4. In the rational expression $\frac{(x + 2)(3x - 1)}{(x + 2)(4x + 2)}$, $x + 2$ is a common _____ of the numerator and the denominator.

5. The least _____ _____ of $\frac{x-8}{x+6}$ and $\frac{6-5x}{x}$ is $x(x+6)$.

6. To _____ a rational expression, we multiply it by a form of 1. For example, $\frac{2}{n^2} \cdot \frac{8}{8} = \frac{16}{8n^2}$.

CONCEPTS

Fill in the blanks.

7. To multiply rational expressions, multiply their _____ and multiply their _____. To divide two rational expressions, multiply the first by the _____ of the second.

8. To add or subtract rational expressions that have the same denominator, add or subtract the _____ and write the sum or difference over the common _____.

9. To find the least common denominator of several rational expressions, _____ each denominator completely. The LCD is a product that uses each different factor the _____ number of times it appears in any one factorization.

10. The expression $4 - y$ must be multiplied by _____ to obtain $y - 4$.

11. Consider the following factorizations.

$$18x - 36 = 2 \cdot 3 \cdot 3 \cdot (x - 2)$$
$$3x^2 - 3x - 6 = 3(x - 2)(x + 1)$$

a. What is the greatest number of times the factor 3 appears in any one factorization?

b. What is the greatest number of times the factor $x - 2$ appears in any one factorization?

12. The LCD for $\frac{2x+1}{x^2+5x+6}$ and $\frac{3x}{x^2-4}$ is

$$LCD = (x + 2)(x + 3)(x - 2)$$

If we want to subtract these rational expressions, what form of 1 should be used

a. to build $\frac{2x+1}{x^2+5x+6}$?

b. to build $\frac{3x}{x^2-4}$?

NOTATION

13. a. Write $5x^2 + 35x$ as a fraction.

b. What is the reciprocal of $5x^2 + 35x$?

14. Fill in the blank: The fraction $\dfrac{\frac{a}{b}}{\frac{c}{d}}$ is equivalent to $\frac{a}{b} \quad \frac{c}{d}$.

GUIDED PRACTICE

Simplify each rational expression. See Example 1.

15. $\dfrac{15a^2}{25a^8}$

16. $\dfrac{12x}{16x^7}$

17. $\dfrac{24x^3y^4}{54x^4y^3}$

18. $\dfrac{15a^5b^4}{21a^2b^5}$

Simplify each rational expression. See Example 2.

19. $\dfrac{5x^2 - 10x}{x^2 - 4x + 4}$

20. $\dfrac{x^2 + 6x + 9}{2x^2 + 6x}$

21. $\dfrac{6x^2 - 7x - 5}{2x^2 + 5x + 2}$

22. $\dfrac{6x^2 + x - 2}{8x^2 + 2x - 3}$

Simplify each rational expression. See Example 3.

23. $\dfrac{4 - x^2}{x^2 - x - 2}$

24. $\dfrac{x^2 - 2x - 15}{25 - x^2}$

25. $\dfrac{p^3 + p^2q - 2pq^2}{pq^2 + p^2q - 2p^3}$

26. $\dfrac{m^3 - mn^2}{mn^2 + m^2n - 2m^3}$

Perform the operations and simplify, if possible. See Example 4.

27. $\dfrac{10a^2}{3b^4} \cdot \dfrac{12b^3}{5a^2}$

28. $\dfrac{16c^3}{5d^2} \cdot \dfrac{25d}{12c}$

29. $\dfrac{3p^2}{6p + 24} \cdot \dfrac{p^2 - 16}{6p}$

30. $\dfrac{y^2 + 6y + 9}{15y} \cdot \dfrac{3y^2}{2y + 6}$

31. $\dfrac{x^2 + 2x + 1}{9x} \cdot \dfrac{2x^2 - 2x}{2x^2 - 2}$

32. $\dfrac{a + 6}{a^2 - 16} \cdot \dfrac{3a - 12}{3a + 18}$

33. $\dfrac{2x^2 - x - 3}{x^2 - 1} \cdot \dfrac{x^2 + x - 2}{2x^2 + x - 6}$

34. $\dfrac{2p^2 - 5p - 3}{p^2 - 9} \cdot \dfrac{2p^2 + 5p - 3}{2p^2 + 5p + 2}$

Perform the operations and simplify, if possible. See Example 5.

35. $(6a - a^2) \cdot \dfrac{a^3}{a^3 - 6a^2 + 3a - 18}$

36. $(10n - n^2) \cdot \dfrac{n^6}{n^3 - 10n^2 - 2n + 20}$

37. $(x^2 + x - 2cx - 2c) \cdot \dfrac{x^2 + 3x + 2}{4c^2 - x^2}$

38. $(2ax - 10x - a + 5) \cdot \dfrac{x}{x - 2x^2}$

Perform the operations and simplify, if possible. See Example 6.

39. $\dfrac{m^2n}{4} \div \dfrac{mn^3}{6}$

40. $\dfrac{a^4b}{14} \div \dfrac{a^3b^2}{21}$

41. $\dfrac{x^2 - 16}{x^2 - 25} \div \dfrac{x + 4}{x - 5}$

42. $\dfrac{a^2 - 9}{a^2 - 49} \div \dfrac{a + 3}{a + 7}$

43. $\dfrac{5c + 1}{6} \div \dfrac{125c^3 + 1}{6c + 6}$

44. $\dfrac{6m - 8}{9m^3} \div \dfrac{27m^3 - 64}{9m + 9}$

45. $\dfrac{3n^2 + 5n - 2}{12n^2 - 13n + 3} \div \dfrac{n^2 + 3n + 2}{4n^2 + 5n - 6}$

46. $\dfrac{8y^2 - 14y - 15}{6y^2 - 11y - 10} \div \dfrac{4y^2 - 9y - 9}{3y^2 - 7y - 6}$

75. $\dfrac{6}{5d^2 - 5d} - \dfrac{3}{5d - 5}$

76. $\dfrac{9}{2r^2 - 2r} - \dfrac{5}{2r - 2}$

Perform the operations and simplify, if possible. See Example 7.

47. $\dfrac{6a^2 - 7a - 3}{2a^2 - 2} \div \dfrac{4a^2 - 12a + 9}{a^2 - 1} \cdot \dfrac{2a^2 - a - 3}{3a^2 - 2a - 1}$

48. $\dfrac{x^2 - x - 12}{x^2 + x - 2} \div \dfrac{x^2 - 6x + 8}{x^2 - 3x - 10} \cdot \dfrac{x^2 - 3x + 2}{x^2 - 2x - 15}$

49. $\dfrac{4x^2 - 10x + 6}{x^4 - 3x^3} \cdot \dfrac{x - 3}{2 - 2x} \div \dfrac{2x - 3}{2x^3}$

50. $\dfrac{2x^2 - 2x - 4}{x^2 + 2x - 8} \cdot \dfrac{3x^2 + 15x}{x + 1} \div \dfrac{100 - 4x^2}{x^2 - x - 20}$

Perform the operations and simplify, if possible. See Example 8.

51. $\dfrac{3}{a + 7} - \dfrac{a}{a + 7}$

52. $\dfrac{x}{x + 4} + \dfrac{5}{x + 4}$

53. $\dfrac{5x}{x + 1} + \dfrac{3}{x + 1} - \dfrac{2x}{x + 1}$

54. $\dfrac{4}{a + 4} - \dfrac{2a}{a + 4} + \dfrac{3a}{a + 4}$

55. $\dfrac{3x}{x^2 - 9} - \dfrac{9}{x^2 - 9}$

56. $\dfrac{9x}{x^2 - 1} - \dfrac{9}{x^2 - 1}$

57. $\dfrac{3y - 2}{2y + 6} - \dfrac{2y - 5}{2y + 6}$

58. $\dfrac{5x + 8}{3x + 15} - \dfrac{3x - 2}{3x + 15}$

Perform the operations and simplify, if possible. See Example 9.

59. $\dfrac{3}{4x} + \dfrac{2}{3x}$

60. $\dfrac{2}{5a} + \dfrac{3}{2a}$

61. $\dfrac{8}{9y^2} + \dfrac{1}{6y^4}$

62. $\dfrac{5}{6a^3} + \dfrac{7}{8a^2}$

63. $\dfrac{3}{4ab^2} - \dfrac{5}{2a^2b}$

64. $\dfrac{1}{5xy^3} - \dfrac{2}{15x^2y}$

65. $\dfrac{y - 7}{y^2} - \dfrac{y + 7}{2y}$

66. $\dfrac{x + 5}{xy} - \dfrac{x - 1}{x^2y}$

Perform the operations and simplify, if possible. See Example 10.

67. $\dfrac{3}{x + 2} + \dfrac{5}{x - 4}$

68. $\dfrac{2}{a + 4} - \dfrac{6}{a + 3}$

69. $\dfrac{x + 2}{x + 5} - \dfrac{x - 3}{x + 7}$

70. $\dfrac{7}{x + 3} + \dfrac{4x}{x + 6}$

71. $\dfrac{x}{x^2 + 5x + 6} + \dfrac{x}{x^2 - 4}$

72. $\dfrac{x}{x^2 + 2x + 1} + \dfrac{x}{x^2 - 1}$

73. $\dfrac{4}{x^2 - 2x - 3} - \dfrac{x}{3x^2 - 7x - 6}$

74. $\dfrac{5}{x^2 + 5x - 6} - \dfrac{x}{2x^2 + 7x - 30}$

77. $\dfrac{m}{m^2 + 9m + 20} - \dfrac{4}{m^2 + 7m + 12}$

78. $\dfrac{x + 3}{2x^2 - 5x + 2} - \dfrac{3x - 1}{x^2 - x - 2}$

Simplify each complex fraction. See Example 11.

79. $\dfrac{\dfrac{4x}{y}}{\dfrac{6xz}{y^2}}$

80. $\dfrac{\dfrac{5t^4}{9x}}{\dfrac{2t}{18x}}$

81. $\dfrac{\dfrac{t}{x^2 - y^2}}{\dfrac{t}{x + y}}$

82. $\dfrac{\dfrac{x^2 + 5x + 6}{3xy}}{\dfrac{x^2 - 9}{6xy}}$

Simplify each complex fraction. See Example 12.

83. $\dfrac{\dfrac{1}{a} + \dfrac{1}{b}}{\dfrac{1}{a}}$

84. $\dfrac{\dfrac{1}{b}}{\dfrac{1}{a} - \dfrac{1}{b}}$

85. $\dfrac{\dfrac{y}{x} - \dfrac{x}{y}}{\dfrac{1}{x} + \dfrac{1}{y}}$

86. $\dfrac{\dfrac{y}{x} - \dfrac{x}{y}}{\dfrac{1}{y} - \dfrac{1}{x}}$

87. $\dfrac{\dfrac{h}{h^2 + 3h + 2}}{\dfrac{4}{h + 2} - \dfrac{4}{h + 1}}$

88. $\dfrac{\dfrac{3}{z - 3} + \dfrac{2}{z - 2}}{\dfrac{5z}{z^2 - 5z + 6}}$

89. $\dfrac{\dfrac{2}{y - 1} - \dfrac{2}{y}}{\dfrac{3}{y - 1} - \dfrac{1}{1 - y}}$

90. $\dfrac{\dfrac{1}{x} - \dfrac{4}{x - 1}}{\dfrac{3}{x - 1} + \dfrac{2}{x}}$

Solve each equation. See Example 13.

91. $\dfrac{3}{y} + \dfrac{7}{2y} = 13$

92. $\dfrac{2}{x} + \dfrac{1}{2} = \dfrac{7}{2x}$

93. $\dfrac{2}{x} + \dfrac{1}{2} = \dfrac{9}{4x} - \dfrac{1}{2x}$

94. $\dfrac{7}{5x} - \dfrac{1}{2} = \dfrac{5}{6x} + \dfrac{1}{3}$

95. $\dfrac{a}{2} = \dfrac{a - 6}{3a - 9} - \dfrac{1}{3}$

96. $\dfrac{b}{5} = \dfrac{b - 14}{2b - 16} - \dfrac{1}{2}$

97. $\dfrac{2}{5x - 5} + \dfrac{x - 2}{15} = \dfrac{4}{5x - 5}$

98. $\dfrac{3}{2x + 4} = \dfrac{x - 2}{2} + \dfrac{x - 5}{2x + 4}$

TRY IT YOURSELF

In the following problems, simplify each expression by performing the indicated operations and solve each equation.

99. $\dfrac{p^3 - q^3}{q^2 - p^2} \cdot \dfrac{q^2 + pq}{p^3 + p^2 q + pq^2}$

100. $\dfrac{x^3 + y^3}{x^3 - y^3} \div \dfrac{x^2 - xy + y^2}{x^2 + xy + y^2}$

101. $\dfrac{t}{t^2 + 5t + 6} - \dfrac{2}{t^2 + 3t + 2}$

102. $\dfrac{2a}{a^2 - 2a - 8} + \dfrac{3}{a^2 - 5a + 4}$

103. $\dfrac{\dfrac{2}{x+3} - \dfrac{1}{x-3}}{\dfrac{3}{x^2 - 9}}$

104. $\dfrac{2 + \dfrac{1}{x^2 - 1}}{1 + \dfrac{1}{x-1}}$

105. $\dfrac{4}{m^2 - 9} + \dfrac{5}{m^2 - m - 12} = \dfrac{7}{m^2 - 7m + 12}$

106. $\dfrac{34}{x^2} + \dfrac{13}{20x} = \dfrac{3}{2x}$

107. $\dfrac{2}{x-1} - \dfrac{2x}{x^2 - 1} - \dfrac{x}{x^2 + 2x + 1}$

108. $\dfrac{x-2}{x^2 - 3x} + \dfrac{2x-1}{x^2 + 3x} - \dfrac{2}{x^2 - 9}$

109. $(2x^2 - 15x + 25) \div \dfrac{2x^2 - 3x - 5}{x + 1}$

110. $(x^2 - 6x + 9) \div \dfrac{x^2 - 9}{x + 3}$

111. $\dfrac{y^3 - x^3}{2x^2 + 2xy + x + y} \cdot \dfrac{2x^2 - 5x - 3}{yx - 3y - x^2 + 3x}$

112. $\dfrac{ax + ay + bx + by}{x^3 - 27} \cdot \dfrac{x^2 + 3x + 9}{xc + xd + yc + yd}$

113. $\dfrac{24n^4}{16n^4 + 24n^3}$

114. $\dfrac{18m^4}{36m^4 - 9m^3}$

115. $1 + x - \dfrac{x}{x - 5}$

116. $2 - x + \dfrac{3}{x - 9}$

117. $\dfrac{ax + by + ay + bx}{a^2 - b^2}$

118. $\dfrac{3x^2 - 3y^2}{x^2 + 2y + 2x + yx}$

119. $\dfrac{3}{s - 2} + \dfrac{s - 14}{2s^2 - 3s - 2} - \dfrac{4}{2s + 1} = 0$

120. $\dfrac{1}{y^2 - 2y - 3} + \dfrac{1}{y^2 - 4y + 3} - \dfrac{1}{y^2 - 1} = 0$

121. $\dfrac{5}{x + 4} + \dfrac{1}{x + 4} = x - 1$

122. $\dfrac{5}{3x + 12} - \dfrac{1}{9} = \dfrac{x - 1}{3x}$

123. $\dfrac{\dfrac{ac - ad - c + d}{a^3 - 1}}{\dfrac{c^2 - 2cd + d^2}{a^2 + a + 1}}$

124. $\dfrac{\dfrac{tx + ty - 2x - 2y}{x^2 + 2xy + y^2}}{\dfrac{t^3 - 8}{15x + 15y}}$

125. $\dfrac{5x}{x - 3} + \dfrac{4x}{3 - x}$

126. $\dfrac{8x}{x - 4} - \dfrac{10x}{4 - x}$

APPLICATIONS

127. PHYSICS The following table contains data from a physics experiment. k_1 and k_2 are constants. Complete the table.

Trial	Rate (m/sec)	Time (sec)	Distance (m)
1	$\dfrac{k_1^2 + 3k_1 + 2}{k_1 - 3}$	$\dfrac{k_1^2 - 3k_1}{k_1 + 1}$	
2	$\dfrac{k_2^2 + 6k_2 + 5}{k_2 + 1}$		$k_2^2 + 11k_2 + 30$

128. DRAFTING Among the tools used in drafting are $45° - 45° - 90°$ and $30° - 60° - 90°$ triangles. Find the perimeter of each triangle and express each result as a rational expression.

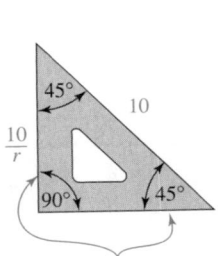

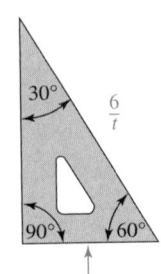

For a $45°$–$45°$–$90°$ triangle, these two sides are the same length.

For a $30°$–$60°$–$90°$ triangle, this side is half as long as the hypotenuse.

129. THE AMAZON The Amazon River flows in a general eastern direction to the Atlantic Ocean. In Brazil, when the river is at low stage, the rate of flow is about 5 mph. Suppose that a river guide can canoe in still water at a rate of r mph.

a. Complete the table to find rational expressions that represent the time it would take the guide to canoe 3 miles downriver and to canoe 3 miles upriver.

b. Find the difference in the times for the trips upriver and downriver. Express the result as a rational expression.

	Rate (mph)	Time (hr)	Distance (mi)
Downriver	$r + 5$		3
Upriver	$r - 5$		3

130. ENGINEERING   The stiffness of the shaft shown below is given by the formula

$$k = \cfrac{1}{\cfrac{1}{k_1} + \cfrac{1}{k_2}}$$

where k_1 and k_2 are the individual stiffnesses of each section. Simplify the complex fraction.

WRITING

131. A student compared his answer, $\frac{a - 3b}{2b - a}$, with the answer, $\frac{3b - a}{a - 2b}$, in the back of the text. Is the student's work correct? Explain.

132. Explain the error that is made in the following work:

$$\frac{3x^2 + 1}{3y} = \frac{\cancel{3}x^2 + 1}{\cancel{3}y} = \frac{x^2 + 1}{y}$$

133. Write some comments to the student who wrote the following solution, explaining his misunderstanding.

$$\text{Multiply: } \frac{1}{x} \cdot \frac{3}{2} = \frac{1 \cdot 2}{x \cdot 2} \cdot \frac{3 \cdot x}{2 \cdot x}$$

$$= \frac{2}{2x} \cdot \frac{3x}{2x}$$

$$= \frac{6x}{2x}$$

134. Explain how to find the least common denominator of a set of rational expressions.

REVIEW

Write a system of two equations in two variables to solve each problem.

135. INTEGER PROBLEMS   The sum of two integers is 38, and their difference is 12. Find the integers.

136. INTEGER PROBLEMS   Twice one integer plus another integer is 21. If the first integer plus 3 times the second is 33, find the integers.

CHALLENGE PROBLEMS

137. Simplify: $\dfrac{a^6 - 64}{(a^2 + 2a + 4)(a^2 - 2a + 4)}$

138. Add: $x^{-1} + x^{-2} + x^{-3} + x^{-4} + x^{-5}$

139. Simplify: $[(x^{-1} + 1)^{-1} + 1]^{-1}$

140. Find two rational expressions, each with denominator $x^2 + 5x + 6$, such that their sum is $\frac{1}{x + 2}$.

SECTION 8.7
Review of Linear Equations in Two Variables

Objectives

1 Graph linear equations using point plotting and the intercept method.

2 Find the slope of a line.

3 Write equations of lines.

4 Recognize parallel and perpendicular lines.

5 Write a linear equation model.

6 Find the midpoint of a line segment.

Recall the definition of a *linear equation in two variables*.

Standard (General) Form of a Linear Equation	A **linear equation in two variables** is an equation that can be written in the form $$Ax + By = C$$ where A, B, and C are real numbers and A and B are not both 0.

Some other examples of linear equations are

$$y = 4x - 7, \qquad 2x - 5y = 10, \qquad y = 3, \qquad \text{and} \qquad x = -2$$

Linear equations can be graphed in several ways. Generally, the form in which an equation is written determines the method that we use to graph it.

1 **Graph Linear Equations Using Point Plotting and the Intercept Method.**

The *graph* of a linear equation is a mathematical picture of the infinitely many solutions of the equation. To graph linear equations solved for *y*, we can select values of *x* and calculate the corresponding values of *y*.

EXAMPLE 1   Graph: $y = -\dfrac{1}{2}x + 3$

Strategy   We will find three solutions of the equation, plot them on a rectangular coordinate system, and then draw a straight line passing through the points.

Why   To *graph* an equation in two variables means to make a drawing that represents all of its solutions.

Solution   To find three solutions of this equation, we select three values for *x* that will make the computations easy. Then we find each corresponding value of *y*. For example, if *x* is -2, we have

$$y = -\frac{1}{2}x + 3$$

$$y = -\frac{1}{2}(-2) + 3 \qquad \text{Substitute } -2 \text{ for } x.$$

$$y = 1 + 3 \qquad\qquad \text{Evaluate the right side.}$$

$$y = 4$$

Thus, $(-2, 4)$ is a solution. In a similar manner, we find corresponding *y*-values for *x*-values of 0 and 2 and enter the solutions in the table below.

When we plot the ordered-pair solutions on a rectangular coordinate system, we see that they lie in a straight line. Using a straight edge or ruler, we then draw a straight line through the points because the graph of any solution of $y = -\frac{1}{2}x + 3$ will lie on this line. Furthermore, every point of this line represents a solution. We call the line the *graph of the equation*. It represents all of the solutions of $y = -\frac{1}{2}x + 3$.

> **Success Tip**
> When selecting *x*-values for a table of solutions, a rule of thumb is to choose a negative number, a positive number, and 0. Here, we choose *x*-values that are multiples of 2 to make the computations easier when multiplying *x* by $-\frac{1}{2}$.
> As a result, the corresponding *y*-values are integers, and not difficult-to-plot fractions.

$y = -\dfrac{1}{2}x + 3$

x	y	(x, y)
-2	4	$(-2, 4)$
0	3	$(0, 3)$
2	2	$(2, 2)$

↑ ↑ ↑
Select x. Find y. Plot (x, y).

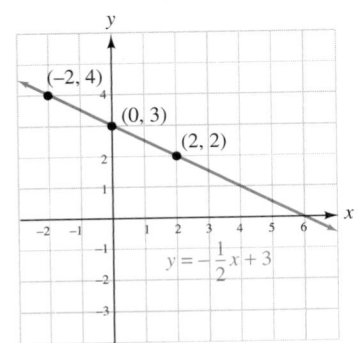

Plot the ordered pairs.

Draw a straight line through the points.
This is the *graph of the equation*.

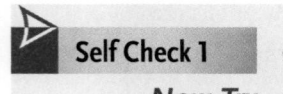

Self Check 1   Graph: $y = \frac{1}{3}x + 1$

Now Try   **Problem 23**

In Example 1, the graph intersected the y-axis at the point $(0, 3)$, called the y-intercept, and it intersected the x-axis at the point $(6, 0)$, called the x-intercept. In general, we have the following definitions.

Intercepts of a Line	The **y-intercept** of a line is the point $(0, b)$, where the line intersects the y-axis. To find b, substitute 0 for x in the equation of the line and solve for y.
	The **x-intercept** of a line is the point $(a, 0)$, where the line intersects the x-axis. To find a, substitute 0 for y in the equation of the line and solve for x.

Plotting the x- and y-intercepts of a graph and drawing a line through them is called the **intercept method of graphing a line.** This method is useful when graphing linear equations written in the standard (general) form $Ax + By = C$.

EXAMPLE 2   Graph $2x - 5y = 10$ by finding the intercepts.

Strategy   We will let $x = 0$ to find the y-intercept of the graph and then let $y = 0$ to find the x-intercept.

Why   Since two points determine a line, the y-intercept and x-intercept are enough information to graph this linear equation.

Solution   To find the y-intercept of the graph, we substitute 0 for x and solve for y. To find the x-intercept of the graph, we substitute 0 for y and solve for x.

Success Tip
The exponent on each variable of a linear equation is an understood 1. For example, $2x - 5y = 10$ can be thought of as $2x^1 - 5y^1 = 10$.

y-intercept:		*x-intercept:*	
$2x - 5y = 10$		$2x - 5y = 10$	
$2(0) - 5y = 10$	Substitute 0 for x.	$2x - 5(0) = 10$	Substitute 0 for y.
$-5y = 10$		$2x = 10$	
$y = -2$		$x = 5$	

The y-intercept is $(0, -2)$. The x-intercept is $(5, 0)$.

Although two points provide enough information to draw the graph of the equation, it is a good idea to find and plot a third point as a check. If the three points do not lie on a line, then at least one of them is in error.

To find the coordinates of a third point, we can substitute any convenient number (such as -5) for x and solve for y:

The Language of Algebra
For any two points, exactly one line passes through them. We say two points *determine* a line.

$2x - 5y = 10$	This is the equation to graph.
$2(-5) - 5y = 10$	Substitute −5 for x.
$-10 - 5y = 10$	
$-5y = 20$	Add 10 to both sides.
$y = -4$	To isolate y, divide both sides by −5.

The line will also pass through the point $(-5, -4)$. We plot the intercepts and the check point, draw a straight line through them, and label the line as $2x - 5y = 10$.

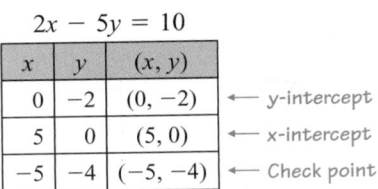

$$2x - 5y = 10$$

x	y	(x, y)	
0	−2	(0, −2)	← y-intercept
5	0	(5, 0)	← x-intercept
−5	−4	(−5, −4)	← Check point

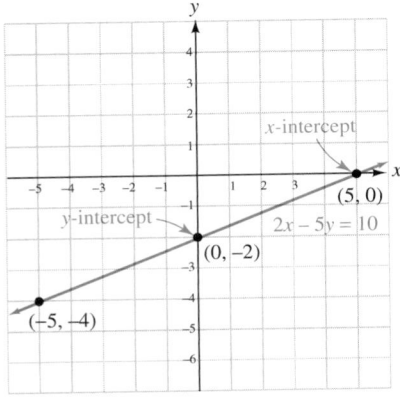

The Language of Algebra
Be careful with pronunciation: The point where a line *intersects* the x- or y-axis is called an *intercept*.

Self Check 2   Graph $5x + 15y = -15$ by finding the y- and x-intercepts.

Now Try   **Problem 27**

Equations such as $y = 3$ and $x = -2$ are linear equations, because they can be written in the general form $Ax + By = C$.

$y = 3$ is equivalent to $0x + 1y = 3$

$x = -2$ is equivalent to $1x + 0y = -2$

EXAMPLE 3   Graph: **a.** $y = 4$ **b.** $x = -2$

Strategy   To find three ordered-pair solutions of $y = 4$, we will select three values for x and use 4 for y each time. To find three ordered-pair solutions of $x = -2$, we will select three values for y and use -2 for x each time.

Why   The first equation requires that $y = 4$ and the second equation requires that $x = -2$.

Solution
a. Since the equation $y = 4$ does not contain x, the numbers chosen for x have no effect on y. The value of y is always 4.
    After plotting the ordered pairs shown in the table on the next page, and drawing a straight line through them, we see that the graph is a horizontal line, parallel to the x-axis, with a y-intercept of $(0, 4)$. The line has no x-intercept.

b. Since the equation $x = -2$ does not contain y, the value of y can be any number.
    After plotting the ordered pairs shown in the table on the next page, and drawing a straight line through them, we see that the graph is a vertical line, parallel to the y-axis, with an x-intercept of $(-2, 0)$. The line has no y-intercept.

Success Tip

The graph of $y = 0$ is the x-axis and the graph of $x = 0$ is the y-axis.

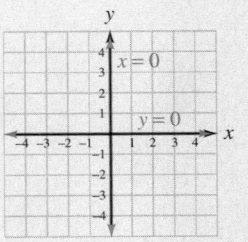

$y = 4$

x	y	(x, y)
-3	4	$(-3, 4)$
0	4	$(0, 4)$
2	4	$(2, 4)$

↑ ↑
Choose   Each
any   value of
number y must
for x.   be 4.

$x = -2$

x	y	(x, y)
-2	-3	$(-2, -3)$
-2	0	$(-2, 0)$
-2	2	$(-2, 2)$

↑ ↑
Each   Choose
value of   any
x must   number
be -2.   for y.

Self Check 3   Graph: **a.** $x = 4$ **b.** $y = -3$

Now Try Problems 35 and 37

The results of Example 3 suggest the following facts.

Equations of Horizontal and Vertical Lines

The equation $y = b$ represents the horizontal line that intersects the y-axis at $(0, b)$.

The equation $x = a$ represents the vertical line that intersects the x-axis at $(a, 0)$.

The graph of the equation $y = 0$ has special significance; it is the x-axis. Similarly, the graph of the equation $x = 0$ is the y-axis.

Using Your Calculator

Generating Tables of Solutions and Graphing Lines

If an equation in x and y is solved for y, we can use a graphing calculator to generate a table of solutions. The instructions in this discussion are for a TI-84 or TI-84 plus graphing calculator. For specific details about other brands, please consult the owner's manual.

To construct a table of solutions for $2x - 5y = 10$, we first solve for y.

$2x - 5y = 10$

$\qquad -5y = -2x + 10$ Subtract 2x from both sides.

$\qquad\qquad y = \dfrac{2}{5}x - 2$ Divide both sides by −5 and simplify.

To enter $y = \frac{2}{5}x - 2$, we press $\boxed{Y =}$ and enter $(2/5)x - 2$, as shown in figure (a) on the next page. (Ignore the subscript 1 on y; it is not relevant at this time.)

To enter the x-values that are to appear in the table, we press $\boxed{\text{2nd}}$ $\boxed{\text{TBLSET}}$ and enter the first value for x on the line labeled TblStart =. In figure (b), -5 has been entered on this line. Other values for x that are to appear in the table are determined by setting an increment value on the line labeled ΔTbl =. Figure (b) shows that an increment of 1 was entered. This means that each x-value in the table will be 1 unit larger than the previous x-value.

The final step is to press the keys [2nd] [TABLE]. This displays a table of solutions, as shown in figure (c).

To see the graph of $y = \frac{2}{5}x - 2$, we press [GRAPH], as shown in figure (d).

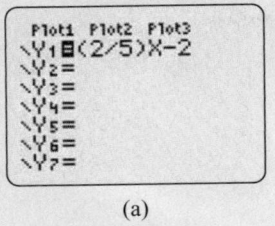
(a)

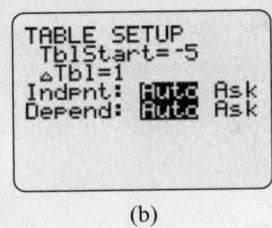
(b)

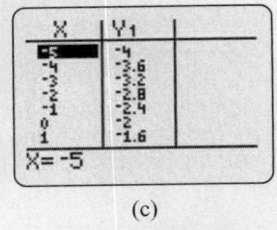
(c)

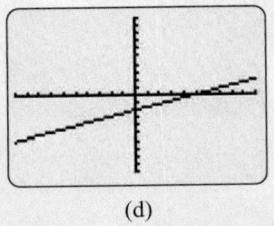
(d)

2 **Find the Slope of a Line.**

Recall that the *slope* of a line is a measure of the steepness of the line.

Slope of a Line

The **slope** of a line passing through points (x_1, y_1) and (x_2, y_2) is

$$m = \frac{\text{vertical change}}{\text{horizontal change}} = \frac{\text{change in } y}{\text{change in } x} = \frac{y_2 - y_1}{x_2 - x_1} \quad \text{where } x_2 \neq x_1$$

The Language of Algebra
The symbol Δ is the letter delta from the Greek alphabet.

The change in y (denoted as Δy and read as "delta y") is the **rise** of the line between two points on the line. The change in x (denoted as Δx and read as "delta x") is the **run.** Using this terminology, we can define slope as the ratio of the rise to the run:

$$m = \frac{\Delta y}{\Delta x} = \frac{\text{rise}}{\text{run}} \quad \text{where } \Delta x \neq 0$$

EXAMPLE 4 Find the slope of the line graphed in figure (a) below.

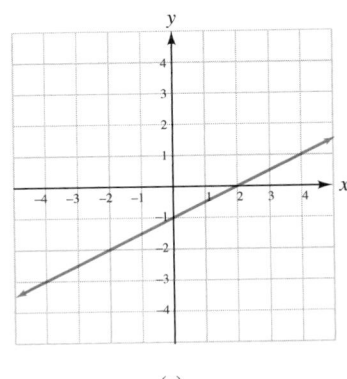
(a)

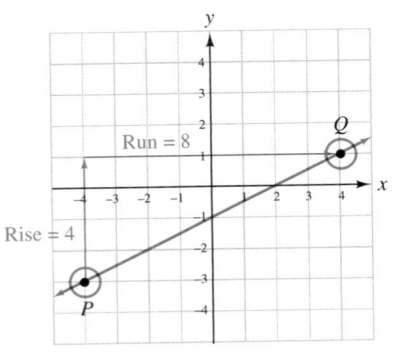
(b)

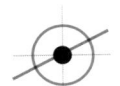
Pick two points on the line that also lie on the intersection of two grid lines.

Strategy We will pick two points on the line, construct a slope triangle, and find the rise and run. Then we will write the ratio of rise to run and simplify the result, if possible.

Why The slope of a line is the ratio of the rise to the run.

Solution We begin by choosing two points on the line, P and Q, as shown in figure (b). One way to move from P to Q is to start at P, move upward, a rise of 4 grid squares, and then to the right, a run of 8 grid squares, to reach Q. These steps create a right triangle called a **slope triangle.**

$$m = \frac{\text{rise}}{\text{run}} = \frac{4}{8} = \frac{1}{2} \quad \text{\textit{Simplify the fraction. The result is positive.}}$$

The slope of the line is $\frac{1}{2}$.

Self Check 4 Find the slope of the line shown in the previous graph using two points different from those used in the solution of Example 4.

Now Try **Problems 39 and 41**

EXAMPLE 5 Find the slope of the line passing through $(-2, 4)$ and $(3, -4)$.

Strategy We will use the slope formula to find the slope.

Why We know the coordinates of two points on the line.

Solution We can let $(x_1, y_1) = (-2, 4)$ and $(x_2, y_2) = (3, -4)$. Then we have

$$m = \frac{y_2 - y_1}{x_2 - x_1} \quad \text{\textit{This is the slope formula.}}$$

$$= \frac{-4 - 4}{3 - (-2)} \quad \text{\textit{Substitute −4 for } } y_2 \text{\textit{, 4 for}}$$
$$\text{\textit{y_1, 3 for x_2, and −2 for x_1.}}$$

$$= \frac{-8}{5}$$

$$= -\frac{8}{5} \quad \text{\textit{Write $\frac{-8}{5}$ with the − sign in front of the fraction. The result is negative.}}$$

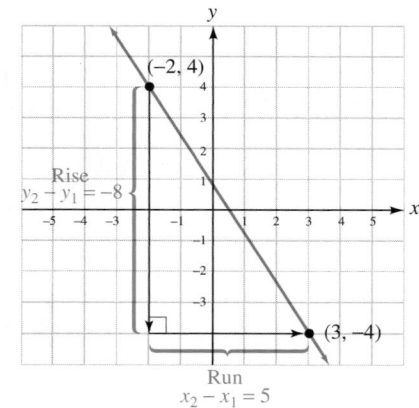

The slope of the line is $-\frac{8}{5}$.

The graph of the line passing through $(-2, 4)$ and $(3, -4)$ is shown above. Notice that we obtain the same result when the slope of the line is found graphically.

$$m = \frac{\text{rise}}{\text{run}} = \frac{-8}{5} = -\frac{8}{5}$$

Self Check 5 Find the slope of the line passing through $(-3, 6)$ and $(4, -8)$.

Now Try **Problem 43**

To classify the slope of a line as positive or negative, follow the line from left to right. If a line rises, its slope is positive. If a line drops, its slope is negative. If a line is horizontal, its slope is 0. If a line is vertical, it has undefined slope.

Success Tip When drawing a slope triangle, movement upward or to the right is positive. Movement downward or to the left is negative.

Success Tip When calculating slope, it doesn't matter which point we call (x_1, y_1) and which point we call (x_2, y_2). We will obtain the same result in Example 5 if we let $(x_1, y_1) = (3, -4)$ and $(x_2, y_2) = (-2, 4)$.

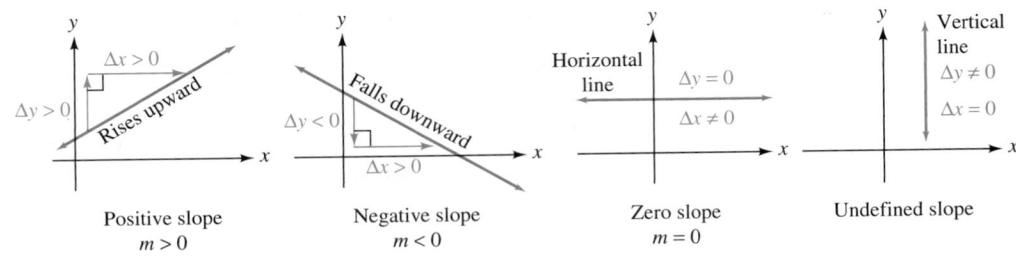

Positive slope
$m > 0$

Negative slope
$m < 0$

Zero slope
$m = 0$

Undefined slope

Caution
Undefined and **0** do not mean the same thing. A horizontal line has a defined slope; it is **0**. A vertical line does not have a defined slope; we say its slope is undefined.

3   Write Equations of Lines.

Recall the following two forms of the equation of a line.

Equations of Lines

Point–slope form: The equation of the line passing through (x_1, y_1) and with slope m is

$$y - y_1 = m(x - x_1)$$

Slope–intercept form: The equation of the line with slope m and y-intercept $(0, b)$ is

$$y = mx + b$$

EXAMPLE 6   Find an equation of the line passing through $(-5, 4)$ and $(8, -6)$. Write the equation in slope–intercept form.

Strategy   We will use the point–slope form, $y - y_1 = m(x - x_1)$, to write an equation of the line.

Why   We know the coordinates of a point on the line and we can calculate the unknown slope of the line using the slope formula.

Solution   First we find the slope of the line.

$$m = \frac{y_2 - y_1}{x_2 - x_1}$$ This is the slope formula.

$$= \frac{-6 - 4}{8 - (-5)}$$ Substitute -6 for y_2, 4 for y_1, 8 for x_2, and -5 for x_1.

$$= -\frac{10}{13}$$ This is the slope of the line.

Since the line passes through $(-5, 4)$ and $(8, -6)$, we can choose either point and substitute its coordinates into the point–slope form. If we select $(-5, 4)$, we substitute -5 for x_1, 4 for y_1, and $-\frac{10}{13}$ for m and proceed as follows.

Success Tip
In Example 6, either of the given points can be used as (x_1, y_1) when writing the point–slope equation. Looking ahead, we usually choose the point whose coordinates will make the computations the easiest.

$$y - y_1 = m(x - x_1)$$ This is the point–slope form.

$$y - 4 = -\frac{10}{13}[x - (-5)]$$ Substitute $-\frac{10}{13}$ for m, -5 for x_1, and 4 for y_1.

$$y - 4 = -\frac{10}{13}(x + 5)$$ Simplify the expression within the brackets. This equation is in point–slope form.

To write this equation in slope–intercept form, we solve for y.

$$y - 4 = -\frac{10}{13}x - \frac{50}{13}$$ To remove the parentheses, distribute $-\frac{10}{13}$.

$$y = -\frac{10}{13}x + \frac{2}{13}$$ To isolate y, add 4 in the form of $\frac{52}{13}$ to both sides and simplify: $-\frac{50}{13} + \frac{52}{13} = \frac{2}{13}$.

The equation of the line in slope–intercept form is $y = -\frac{10}{13}x + \frac{2}{13}$.

Self Check 6   Find an equation of the line passing through $(-2, 5)$ and $(4, -3)$. Write the equation in slope–intercept form.

Now Try   **Problem 51**

4   **Recognize Parallel and Perpendicular Lines.**

Recall these two facts from Chapter 3:

- Different lines having the same slope are parallel.
- If the slopes of two lines are negative reciprocals, the lines are perpendicular.

EXAMPLE 7   **a.** Show that the lines represented by $2y = -3x + 12$ and $6x + 4y = 7$ are parallel.

b. Show that the lines represented by $3(x - 3y) = 10$ and $-3x = y + 5$ are perpendicular.

Strategy   We will write each equation in slope–intercept form and compare their slopes.

Why   If the slopes are equal, the lines are parallel. If the slopes are negative reciprocals, the lines are perpendicular.

Solution

a. We solve each equation for y to see whether the lines are distinct (different) and whether their slopes are equal.

$$2y = -3x + 12 \qquad\qquad\qquad 6x + 4y = 7$$

$$y = -\frac{3}{2}x + 6 \quad m = -\frac{3}{2} \text{ and } b = 6 \qquad\qquad 4y = -6x + 7$$

$$y = -\frac{3}{2}x + \frac{7}{4} \quad m = -\frac{3}{2} \text{ and } b = \frac{7}{4}$$

Since the values of b in these equations are different $\left(6 \text{ and } \frac{7}{4}\right)$, the lines have different y-intercepts and are distinct. Since the slope of each line is $-\frac{3}{2}$, they are parallel.

b. We solve each equation for y to see whether the slopes of their straight-line graphs are negative reciprocals.

$$3(x - 3y) = 10 \qquad\qquad\qquad -3x = y + 5$$

$$3x - 9y = 10 \qquad\qquad\qquad -3x - 5 = y$$

$$-9y = -3x + 10 \qquad\qquad\qquad y = -3x - 5 \quad m = -3 \text{ and } b = -5$$

$$y = \frac{1}{3}x - \frac{10}{9} \quad m = \frac{1}{3} \text{ and } b = -\frac{10}{9}$$

Since the slopes are negative reciprocals $\left(\frac{1}{3} \text{ and } -3\right)$, the lines are perpendicular.

Self Check 7
a. Are the lines represented by $3x - 2y = 4$ and $2x = 5(y + 1)$ parallel?
b. Are the lines represented by $3x + 2y = 6$ and $2x - 3y = 6$ perpendicular?

Now Try **Problems 59 and 61**

EXAMPLE 8 Find an equation of the line that passes through $(-2, 5)$ and is parallel to the line $y = 8x - 3$. Write the equation in slope–intercept form.

Strategy We will use the point–slope form to write an equation of the line.

Why We know that the line passes through $(-2, 5)$. We can use the fact that the lines are parallel to determine the unknown slope of the desired line.

Solution Since the slope of the line represented by $y = 8x - 3$ is the coefficient of x, the slope is 8. Since the desired equation is to have a graph that is parallel to the graph of $y = 8x - 3$, its slope must also be 8.

We substitute -2 for x_1, 5 for y_1, and 8 for m in the point–slope form and simplify.

$y - y_1 = m(x - x_1)$

$y - 5 = 8[x - (-2)]$ Substitute 5 for y_1, 8 for m, and -2 for x_1.

$y - 5 = 8(x + 2)$ Simplify within the brackets. This equation is in point–slope form. To write the equation in slope–intercept form, we must isolate y.

$y - 5 = 8x + 16$ To remove parentheses, distribute 8.

$y = 8x + 21$ To isolate y, add 5 to both sides.

The equation in slope–intercept form is $y = 8x + 21$.

> **Success Tip**
> If this problem had called for the equation of a line that is perpendicular to $y = 8x - 3$, the slope of the desired line would be the negative reciprocal of 8, which is $-\frac{1}{8}$.

Self Check 8 Write an equation of the line that is parallel to the line $y = 8x - 3$ and passes through $(0, 0)$.

Now Try **Problem 69**

5 **Write a Linear Equation Model.**

Linear models can be used to describe certain types of financial gain or loss. For example, **straight-line depreciation** is used when aging equipment declines in value and **straight-line appreciation** is used when property or collectibles increase in value.

EXAMPLE 9   *Accounting.*   After purchasing a new drill press, a machine shop owner had his accountant prepare a depreciation worksheet for tax purposes. See the illustration.

a. Assuming straight-line depreciation, write an equation that gives the value v of the drill press after x years of use.

b. Find the value of the drill press after $2\frac{1}{2}$ years of use.

c. What is the economic meaning of the v-intercept of the line?

d. What is the economic meaning of the slope of the line?

> **Depreciation Worksheet**
>
> Drill press $1,970
> (new)
>
> Salvage value $270
> (in 10 years)

Strategy   We will read the problem with the hope that we can find information about the slope of the line and the coordinates of a point (or points) that lie on the line.

Why   If we know the slope of the line and the coordinates of a point (or points) that lie on the line, we can write its equation using slope–intercept form or point–slope form.

Solution

a. The facts presented in the worksheet can be expressed as ordered pairs of the form

$$(x, v)$$

Number of years of use ⌐   ⌐ Value of the drill press

• When purchased, the new $1,970 drill press had been used 0 years: (0, 1,970).

• After 10 years of use, the value of the drill press will be $270: (10, 270).

A sketch showing these ordered pairs and the line of depreciation is helpful in visualizing the situation.

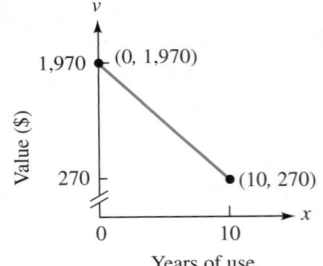

Since we know two points that lie on the line, we can write its equation. As we saw in Example 6, the first step is to find the slope of the line.

$$m = \frac{v_2 - v_1}{x_2 - x_1}$$ This is the slope formula using the variables x and v.

$$= \frac{270 - 1,970}{10 - 0}$$ Let $(x_1, v_1) = (0, 1,970)$ and $(x_2, v_2) = (10, 270)$ and substitute.

$$= \frac{-1,700}{10}$$ Simplify.

$$= -170$$ This is the slope of the line.

To find the equation of the line, we substitute -170 for m, 0 for x_1, and 1,970 for v_1 in the point–slope form and simplify.

$$v - v_1 = m(x - x_1)$$ This is the point–slope form using the variables x and v.

$$v - 1,970 = -170(x - 0)$$ Substitute.

$$v = -170x + 1,970$$ To isolate v, add 1,970 to both sides.

The value v of the drill press after x years of use is given by the straight-line depreciation model $v = -170x + 1,970$.

b. To find the value of the drill press after $2\frac{1}{2}$ years of use, we substitute 2.5 for x in the depreciation equation and find v.

$$v = -170x + 1,970$$ This is the straight-line depreciation equation.

$$= -170(2.5) + 1,970$$ Substitute.

The Language of Algebra
In this problem, the value of the drill press *depreciates*. The value of an item can also *appreciate*, which means to increase in value over time. Certain types of art, antiques, and jewelry *appreciate* quickly.

$$= -425 + 1,970$$

$$= 1,545$$

In $2\frac{1}{2}$ years, the drill press will be worth $1,545.

c. From the sketch, we see that the v-intercept of the graph of the depreciation line is $(0, 1,970)$. This gives the original cost of the drill press, $1,970.

d. Each year, the value of the drill press decreases by $170, because the slope of the line is -170. The slope of the line is the *annual depreciation rate*.

 Now Try **Problem 87**

6 **Find the Midpoint of a Line Segment.**

If point M in the figure on the left lies midway between points $P(x_1, y_1)$ and $Q(x_2, y_2)$, point M is called the **midpoint** of segment PQ. To find the coordinates of M, we average the x-coordinates and average the y-coordinates of P and Q.

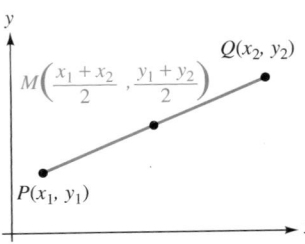

The Midpoint Formula	The **midpoint** of the line segment with endpoints at (x_1, y_1) and (x_2, y_2) is the point with coordinates of $$\left(\frac{x_1 + x_2}{2}, \frac{y_1 + y_2}{2}\right)$$

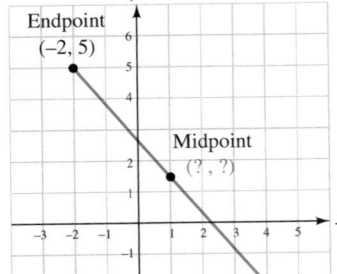

EXAMPLE 10   Find the midpoint of the line segment with endpoints $(-2, 5)$ and $(4, -2)$.

Strategy   To find the coordinates of the midpoint, we find the average of the x-coordinates and the average of the y-coordinates of the endpoints.

Why   This is what is called for by the expressions $\frac{x_1 + x_2}{2}$ and $\frac{y_1 + y_2}{2}$ of the midpoint formula.

Solution   We can let $(x_1, y_1) = (-2, 5)$ and $(x_2, y_2) = (4, -2)$. After substituting these values into the expressions for the x- and y-coordinates in the midpoint formula, we will evaluate each expression to find the coordinates of the midpoint.

$$\frac{x_1 + x_2}{2} = \frac{-2 + 4}{2}$$

$$= \frac{2}{2}$$

$$= 1$$

$$\frac{y_1 + y_2}{2} = \frac{5 + (-2)}{2}$$

$$= \frac{3}{2}$$

Thus, the midpoint is $\left(1, \frac{3}{2}\right)$.

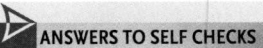

Self Check 10 Find the midpoint of the line segment with endpoints $(-1, 8)$ and $(5, 2)$.

Now Try **Problem 77**

ANSWERS TO SELF CHECKS

1.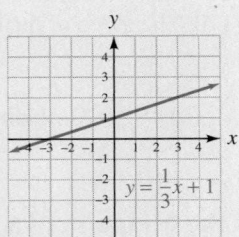

2.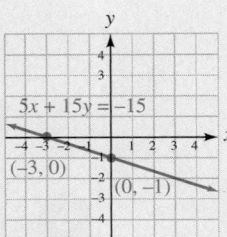

3.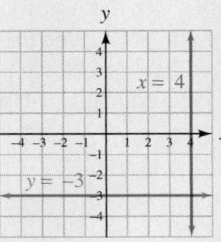

4. $\frac{1}{2}$ 5. -2 6. $y = -\frac{4}{3}x + \frac{7}{3}$ 7. **a.** No **b.** Yes 8. $y = 8x$

10. $(2, 5)$

STUDY SET
8.7

VOCABULARY

Fill in the blanks.

1. $y = 3x - 1$ is a linear equation in _____ variables, x and y.

2. A solution of an equation in two variables is an ordered _____ of numbers that makes the equation a true statement.

3. Solutions of equations in two variables can be listed in a _____ of solutions.

4. The line that represents all of the solutions of a linear equation is called the _____ of the equation.

5. The point where a graph intersects the y-axis is called the _____ and the point where it intersects the x-axis is called the _____.

6. _____ is defined as the change in y divided by the change in x.

CONCEPTS

Fill in the blanks.

7. The graph of any equation of the form $x = a$ is a _____ line. The graph of any equation of the form $y = b$ is a _____ line.

8. The formula to compute slope is $m =$ _____.

9. The slope of a _____ line is 0. A _____ line has no defined slope.

10. The graph of the equation $y = 0$ is the _____. The graph of the equation $x = 0$ is the _____.

11. _____ lines have the same slope. The slopes of _____ lines are negative reciprocals.

12. The point–slope form of the equation of a line is _____.

13. The slope–intercept form of the equation of a line is _____.

14. The midpoint of a line segment joining (x_1, y_1) and (x_2, y_2) is given by the formula _____.

15. Refer to the graph.

   a. What is the x-intercept and what is the y-intercept of the line?

   b. If the coordinates of point P are substituted into the equation of the line that is graphed here, will a true or a false statement result?

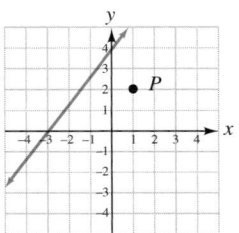

16. Use the graph to determine three solutions of $2x + 3y = 9$.

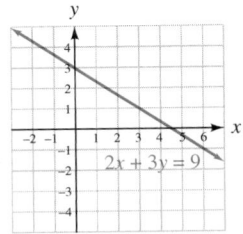

NOTATION

Fill in the blanks.

17. The symbol x_1 is read as "x _____."

18. The symbol Δ is the letter _____ from the Greek alphabet. The change in x is written as [ ].

GUIDED PRACTICE

Graph each equation. See Example 1.

19. $y = x - 2$

20. $y = -x + 4$

21. $y = x$

22. $y = -2x$

23. $y = 2x - 3$

24. $y = -3x + 2$

25. $y = -\dfrac{1}{3}x - 1$

26. $y = -\dfrac{1}{2}x + \dfrac{5}{2}$

Graph each equation using the intercept method. Label the intercepts on each graph. See Example 2.

27. $3x + 4y = 12$

28. $4x - 3y = 12$

29. $3y = 6x - 9$

30. $2x = 4y - 10$

31. $3x + 4y - 8 = 0$

32. $-2y - 3x + 9 = 0$

33. $3x = 4y - 11$

34. $-5x + 3y = 11$

Graph each equation. See Example 3.

35. $x = 3$

36. $y = -4$

37. $y - 2 = 0$

38. $x + 1 = 0$

Find the slope of each line. See Example 4.

39.

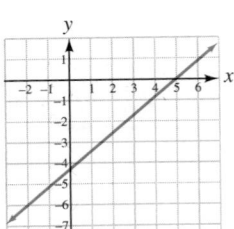

40.

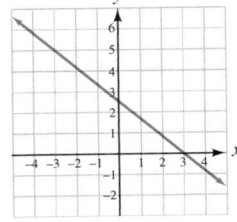

41.

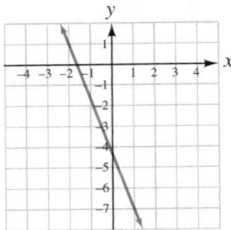

42.

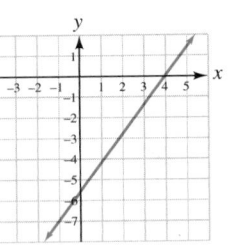

Find the slope of the line that passes through the given points, if possible. See Example 5.

43. $(1, 4), (3, 9)$

44. $(2, 2), (9, 6)$

45. $(-1, 8), (6, 1)$

46. $(-5, -8), (3, 8)$

47. $(3, -1), (-6, 2)$

48. $(0, -8), (-5, 0)$

49. $(-7, 5), (-7, 2)$

50. $(3, -5), (3, 8)$

Find the equation of the line with the given properties. Write the equation in slope–intercept form. See Example 6.

51. Passes through $(4, 0)$ and $(6, -8)$

52. Passes through $(-2, -5)$ and $(1, 4)$

53. Passes through $(-4, 5)$ and $(2, -6)$

54. Passes through $(3, 4)$ and $(0, -3)$

55. Slope 5; passes through $(4, -5)$

56. Slope -7; passes through $(-3, -7)$

57. Slope 3; y-intercept $(0, 17)$

58. Slope -2; y-intercept $(0, 11)$

Determine whether the graphs of each pair of equations are parallel, perpendicular, or neither. See Example 7.

59. $y = 3x + 4, y = 3x - 7$ par

60. $y = 4x - 13, y = \dfrac{1}{4}x + 13$

61. $x + y = 2, y = x + 5$

62. $x = y + 2, y = x + 3$

63. $3x + 6y = 1, y = \dfrac{1}{2}x$

64. $2x + 3y = 9, 3x - 2y = 5$

65. $y = 3, x = 4$

66. $y = -3, y = -7$

Find an equation of the line that passes through the given point and is parallel to the given line. Write the equation in slope–intercept form. See Example 8.

67. $(0, 0), y = 4x - 7$

68. $(0, 0), x = -3y - 12$

69. $(2, 5), 4x - y = 7$

70. $(-6, 3), y + 3x = -12$

Find an equation of the line that passes through the given point and is perpendicular to the given line. Write the equation in slope–intercept form. **See Example 8 and the Success Tip in the margin.**

71. $(0, 0)$, $x = -3y - 12$

72. $(0, 0)$, $y = 4x - 7$

73. $(2, 5)$, $4x - y = 7$

74. $(-6, 3)$, $y + 3x = -12$

Find the midpoint of a line segment with the given endpoints. **See Example 10.**

75. $(0, 0)$, $(6, 8)$ **76.** $(10, 12)$, $(0, 0)$

77. $(6, 8)$, $(12, 16)$ **78.** $(10, 4)$, $(2, -2)$

79. $(-2, -8)$, $(3, -8)$ **80.** $(-5, -2)$, $(7, 3)$

81. $(7, 1)$, $(-10, 4)$ **82.** $(-4, -3)$, $(4, -8)$

APPLICATIONS

83. HIGHWAY GRADES   Find the *grade* of the road shown below by expressing the slope as a percent. (*Hint:* 1 mi = 5,280 ft.)

84. DECK DESIGNS   See the illustration. Find the slopes of the cross-brace and the supports. Is the cross-brace perpendicular to either support?

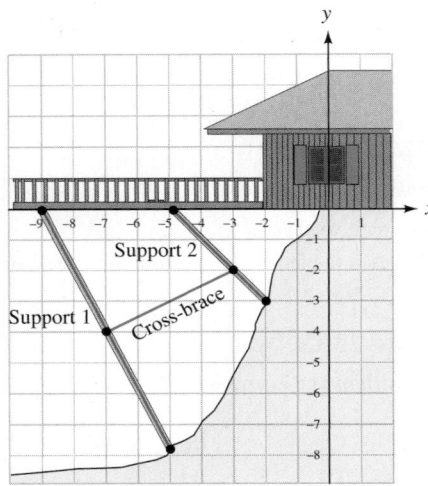

85. U.S. MUSIC SALES   The following line graph models the approximate number of CDs that were shipped for sale in the United States from 1980 through 2005.

   a. Find the rate of increase (the slope) in the number of CDs shipped from 1990–2000.

   b. Find the rate of decrease (the slope) in the number of CDs shipped from 2000–2005.

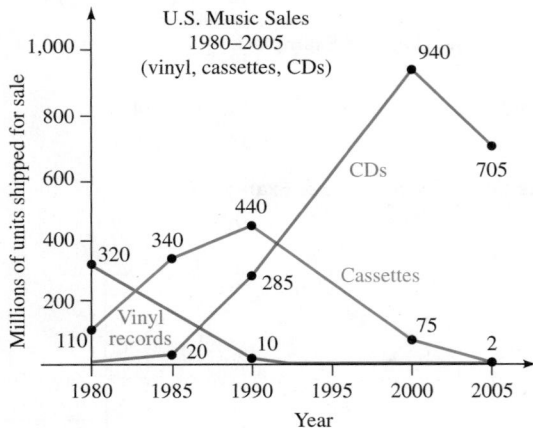

86. CRAIGSLIST   Find the straight-line depreciation equation for the TV in the following ad found on Craigslist (an online website featuring free classified advertisements).

---

46 inch SHARP high definition LCDTV - $1,000

Reply to: <u>sale-001234@saleslist.org</u>
Date: 2007-10-30, 7:20PM EST

3-year-old TV with surround sound & vision, remote.
New $3,295. Asking $1,000.

---

87. SALVAGE VALUES   A truck was purchased for $19,984. Its salvage value at the end of 8 years is expected to be $1,600. Find the straight-line depreciation equation.

88. REAL ESTATE   Refer to the illustration below.
   a. Use the information given in the description of the property to write a straight-line appreciation equation for the house.

   b. What will be the predicted value of the home when it is 25 years old?

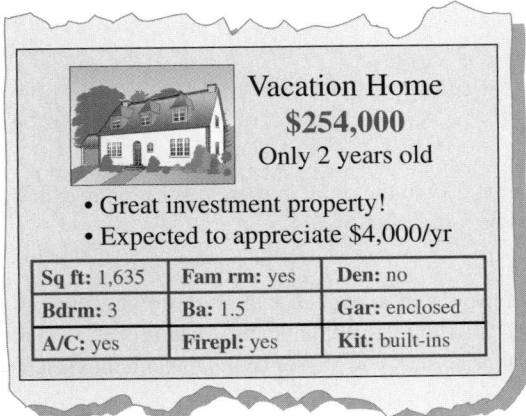

Vacation Home
$254,000
Only 2 years old

• Great investment property!
• Expected to appreciate $4,000/yr

Sq ft: 1,635	Fam rm: yes	Den: no
Bdrm: 3	Ba: 1.5	Gar: enclosed
A/C: yes	Firepl: yes	Kit: built-ins

89. CRIMINOLOGY   City growth and the number of burglaries for a certain city are related by a linear equation. Records show that 575 burglaries were reported in a year when the local population was 77,000 and that the rate of increase in the number of burglaries was 1 for every 100 new residents.
   a. Using the variables p for population and B for burglaries, write an equation (in slope–intercept form) that police can use to predict future burglary statistics.
   b. How many burglaries can be expected when the population reaches 110,000?

90. COLLEGE COSTS   According to the *College Board*, in 1980, the average tuition and fees at a private college were $8,850 a year. Since then, the annual cost has increased by about $514 per year.
   a. Write a linear model in slope–intercept form that gives the cost c to attend a private college t years after 1980.

   b. Use the model to predict what the average tuition and fees at a private college will be in the year 2050.

WRITING

91. Explain how to graph a line using the intercept method.
92. When graphing a line by plotting points, why is it a good practice to find three solutions instead of two?
93. A student was asked to determine the slope of the graph of the line $y = 6x - 4$. His answer was $m = 6x$. Explain his error.
94. Explain why a vertical line has no defined slope.

REVIEW

Evaluate each expression.

95. $-5^2 - 5 - 5(-5)$

96. $\dfrac{1}{3}\left(\dfrac{1}{6}\right) - \left(-\dfrac{1}{3}\right)^2$

97. $\dfrac{|-25| - 2(-5)}{2^4 - 9}$

98. $\dfrac{3[-9 + 2(7 - 3)]}{(8 - 5)(9 - 7)}$

99. Solve $P = 2l + 2w$ for w.

100. Solve $T_f = T_a(1 - F)$ for T_a.

CHALLENGE PROBLEMS

101. Graph: $\dfrac{1}{5}x = 6 - \dfrac{3}{10}y$

102. The line passing through $(1, 3)$ and $(-2, 7)$ is perpendicular to the line passing through points $(4, b)$ and $(8, -1)$. Without graphing, find b.

103. a. Solve $Ax + By = C$ for y, and thereby show that the slope of its graph is $-\dfrac{A}{B}$ and its y-intercept is $\left(0, \dfrac{C}{B}\right)$.

   b. Show that the x-intercept of the graph of $Ax + By = C$ is $\left(\dfrac{C}{A}, 0\right)$.

104. If $(6, -5)$ is the midpoint of segment PQ and the coordinates of point Q are $(-5, -8)$, find the coordinates of point P.

SECTION 8.8
Functions

Objectives

1. Identify functions.
2. Use function notation.
3. Find the domain and range of a function.
4. Graph linear functions.
5. Use the vertical line test.
6. Find function values, domains, and ranges graphically.
7. Graph nonlinear functions.
8. Graph functions using translations and reflections.

In Section 3.8, we introduced the concept of function. In this section, we will review that material and extend it to functions that are nonlinear.

1. **Identify Functions.**

Recall that a **relation** is a set of ordered pairs and that a *function* is a special type of relation.

Functions

A **function** is a set of ordered pairs (a relation) in which to each first component, there corresponds exactly one second component.

The set of first components is called the **domain of the function** and the set of second components is called the **range of the function.**

Since we will often work with sets of ordered pairs of the form (x, y), it is helpful to define a function using the variables x and y.

y Is a Function of x

Given a relation in x and y, if to each value of x in the domain there corresponds exactly one value of y in the range, then y is said to be a function of x.

In the previous definition, since y depends on x, we call x the **independent variable** and y the **dependent variable**. The set of all possible values that can be used for the independent variable is the **domain** of the function, and the set of all values of the dependent variable is the **range** of the function.

EXAMPLE 1 In each case, determine whether the relation defines y to be a function of x.

a.

x   y

5 ──→ 4
7 ──→ 6
11 ──→ 10

b.

x	y
8	2
1	4
8	3
9	9

c. $\{(-2, 3), (-1, 3), (0, 3), (1, 3)\}$

Strategy In each case, we will determine whether there is more than one value of y that corresponds to a single value of x.

Why   If to any *x*-value there corresponds more than one *y*-value, then *y* is not a function of *x*.

Solution

a. The arrow diagram defines a function because to each *value of x* there corresponds exactly one value of *y*: 5→4, 7→6, and 11→10.

b. The table does not define a function, because to the *x*-value 8 there corresponds more than one *y*-value. In the first row, to the *x*-value 8, there corresponds the *y*-value 2. In the third row, to the same *x*-value 8, there corresponds a different *y*-value, 3.

c. Since to each value of *x*, there corresponds exactly one value of *y*, the set of ordered pairs defines *y* to be a function of *x*. In this case, the same *y*-value, 3, corresponds to each *x*-value. This example illustrates an important fact: *Two different ordered pairs of a function can have the same y-value, but they cannot have the same x-value.*

Success Tip

Every function is, by definition, a relation. However, not every relation is a function, as we see in part (b).

Self Check 1   In each case, determine whether the relation defines *y* to be a function of *x*.

a.

x y

b.

x	y
-1	-60
0	55
3	0

c. $\{(4, -1), (9, 2), (16, 15), (4, 4)\}$

Now Try   **Problems 15, 17, and 19**

The Language of Algebra

We can also think of a function as a rule or correspondence that *assigns* exactly one range value to each domain value.

A function can also be defined by an equation. For example, $y = \frac{1}{2}x + 3$ sets up a rule in which to each value of *x* there corresponds exactly one value of *y*. To find the *y*-value (called an **output**) that corresponds to the *x*-value 4 (called an **input**), we substitute 4 for *x* and evaluate the right side of the equation.

$$y = \frac{1}{2}x + 3$$

$$= \frac{1}{2}(4) + 3 \quad \text{Substitute 4 for x. The input is 4.}$$

$$= 2 + 3$$

$$= 5 \quad \text{This is the output.}$$

For the function $y = \frac{1}{2}x + 3$, a *y*-value of 5 corresponds to an *x*-value of 4.

Not all equations define functions, as we will see in the next example.

EXAMPLE 2   Determine whether each equation defines *y* to be a function of *x*.
a. $y = 2x - 5$ **b.** $y^2 = x$

Strategy   In each case, we will determine whether there is more than one value of *y* that corresponds to a single value of *x*.

Why   If to any *x*-value there corresponds more than one *y*-value, then *y* is not a function of *x*.

Solution

a. To find the output value y that corresponds to an input value x, we multiply x by 2 and subtract 5. Since this arithmetic gives one result, to each value of x there corresponds exactly one value of y. Thus, $y = 2x - 5$ defines y to be a function of x.

b. The equation $y^2 = x$ does not define y to be a function of x, because more than one value of y corresponds to a single value of x. For example, if x is 16, then the equation becomes $y^2 = 16$ and y can be either 4 or -4. This is because $4^2 = 16$ and $(-4)^2 = 16$.

x	y
16	4
16	-4

 Self Check 2   Determine whether each equation defines y to be a function of x.

   a. $y = -x + 1$ **b.** $\left|\dfrac{1}{2}y\right| = x$

Now Try   Problems 23 and 25

---

2   **Use Function Notation.**

A special notation is used to name functions that are defined by equations.

Function Notation	The notation $y = f(x)$ indicates that the variable y is a function of x.

Since $y = f(x)$, the equations $y = \frac{1}{2}x + 3$ and $f(x) = \frac{1}{2}x + 3$ are equivalent. We read $f(x) = \frac{1}{2}x + 3$ as "f of x is equal to one-half of x plus 3."

EXAMPLE 3   Let $f(x) = 4x + 3$. Find: **a.** $f(-1)$ **b.** $f(r + 1)$

Strategy   We will substitute -1 and $r + 1$ for each x in $f(x) = 4x + 3$ and evaluate (simplify) the right side.

Why   Whatever expression appears within the parentheses in $f(\ )$ is to be substituted for each x in $f(x) = 4x + 3$.

Solution

a. To find $f(-1)$, we replace x with -1:

$$f(x) = 4x + 3$$
$$f(-1) = 4(-1) + 3$$
$$= -1$$

b. To find $f(r + 1)$, we replace x with $r + 1$:

$$f(x) = 4x + 3$$
$$f(r + 1) = 4(r + 1) + 3$$
$$= 4r + 4 + 3$$
$$= 4r + 7 \quad \text{ordered pair}$$

$$(r+1, 4r+7)$$

 Self Check 3   Let $f(x) = -2x - 1$. Find: **a.** $f(2)$ **b.** $f(-t)$

Now Try   Problems 33, 45, and 51

(handwritten:)
$f(-t) = -2(-t) - 1$
$f(t) = 2t - 1$
$(-t, 2t-1)$

$f(x) = -2x - 1$ $f(2) = -2(2) - 1$
$f(2) = -5$ $(2, -5)$

If we are given an output of a function, we can find the corresponding input(s).

EXAMPLE 4   Let $f(x) = 8x + 1$. For what value(s) of x is $f(x) = -23$?

Strategy   We will substitute -23 for $f(x)$ in the equation $f(x) = 8x + 1$ and solve for x.

Why   In the equation, there are two unknowns, x and $f(x)$. If we replace $f(x)$ with -23, we can use equation-solving techniques to find x.

Solution
$$f(x) = 8x + 1$$
$$-23 = 8x + 1 \qquad \text{Substitute } -23 \text{ for } f(x).$$
$$-24 = 8x \qquad \text{Subtract 1 from both sides.}$$
$$-3 = x \qquad \text{Divide both sides by 8.}$$

To check, we can substitute -3 for x and verify that $f(-3) = -23$.

$$f(x) = 8x + 1$$
$$f(-3) = 8(-3) + 1$$
$$= -24 + 1$$
$$= -23$$

Self Check 4   For what value(s) of x is $f(x) = -15$?

Now Try   Problem 53

---

③   **Find the Domain and Range of a Function.**

We can think of a function as a machine that takes some input x and turns it into some output $f(x)$, as shown in figure (a). The machine shown in figure (b) turns the input -6 into the output -11. The set of numbers that we put into the machine is the *domain* of the function, and the set of numbers that come out is the *range*.

The Language of Algebra
The last two letters in the word *domain* help us remember that it is the set of all *inputs* of a function.

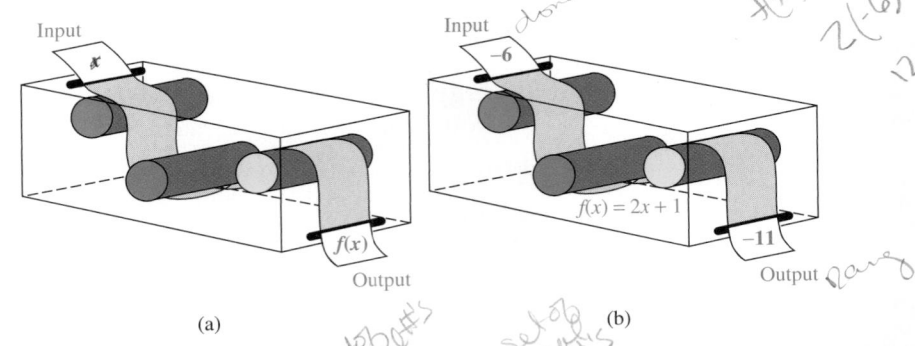

(a)   (b)

EXAMPLE 5   Find the domain and range of each function:
   a. $\{(-2, 4), (0, 6), (2, 8)\}$ **b.** $f(x) = 3x + 1$ **c.** $f(x) = \dfrac{1}{x - 2}$

Strategy   To find the domain, we must find the set of all possible numbers that are permissible inputs for x. To find the range, we must find the set of all possible outputs y.

Why The possible input set for x is the domain. The set of possible outputs y is the range.

Solution

a. The three ordered pairs set up a correspondence between x (the input) and y (the output), where a single value of y is assigned to each x.

- The domain is the set of first components: $\{-2, 0, 2\}$.
- The range is the set of second components: $\{4, 6, 8\}$.

b. We can evaluate $3x + 1$ for any real-number input x. So the domain of the function is *the set of real numbers*. Since the output y can be any real number, the range is the set of real numbers, which can be represented by the symbol $\mathbb{R}$.

c. To find the domain of $f(x) = \dfrac{1}{x - 2}$, we must exclude any real-number inputs for which the fraction $\dfrac{1}{x - 2}$ is undefined. Since 2 would make the denominator of the fraction 0, it must be excluded. Since any real number except 2 can be substituted for x in the equation $f(x) = \dfrac{1}{x - 2}$, the domain is *the set of all real numbers except 2*.

Since a fraction with a numerator of 1 cannot be 0, the range is the set of all real numbers except 0.

Self Check 5   Find the domain and range of each function:
a. $\{(-3, 5), (-2, 7), (1, 11)\}$
b. $f(x) = \dfrac{2}{x + 3}$

Now Try   **Problems 57, 59, and 61**

---

4   **Graph Linear Functions.**

We have seen that a function assigns to each value of x a single value $f(x)$. The input-output pairs that a function generates can be plotted on a rectangular coordinate system to get the graph of the function.

EXAMPLE 6   Graph the function: $f(x) = \dfrac{1}{2}x + 3$

Strategy   To graph the function, we can think of $f(x)$ as y and use the same methods that we used to graph linear equations in Section 2.2.

Why   The notation $f(x) = \frac{1}{2}x + 3$ is another way to write $y = \frac{1}{2}x + 3$.

Solution   We begin by constructing a table of function values. To make a table, we select several values for x and find the corresponding values of $f(x)$. If $x = -2$, we have

$$f(x) = \frac{1}{2}x + 3 \qquad \text{This is the function to graph.}$$

$$f(-2) = \frac{1}{2}(-2) + 3 \qquad \text{Substitute } -2 \text{ for each } x.$$

$$= -1 + 3 \qquad \text{Evaluate the right side.}$$

$$= 2$$

Thus, $f(-2) = 2$ and the ordered pair $(-2, 2)$ lies on the graph of f.

In a similar way, we find the corresponding values of $f(x)$ for x-values of 0 and 2 and record them in the table. Then we plot the ordered pairs and draw a straight line through the points to get the graph of $f(x) = \frac{1}{2}x + 3$.

This axis can be labeled y or $f(x)$.

$f(x) = \frac{1}{2}x + 3$

x	$f(x)$
-2	2	→ $(-2, 2)$
0	3	→ $(0, 3)$
2	4	→ $(2, 4)$

↑ Select x. ↑ Find f(x). ↑ Plot the point.

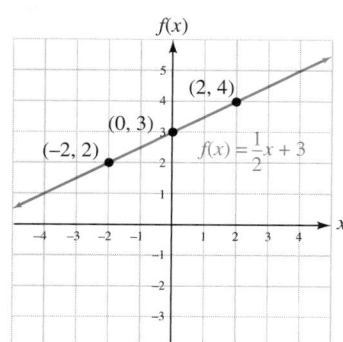

Self Check 6 Graph: $f(x) = -3x - 2$

Now Try Problem 65

The function $f(x) = \frac{1}{2}x + 3$ that was graphed in Example 6 is called a *linear function*. In general, a **linear function** is a function that can be written in the form $f(x) = mx + b$. Its graph is a straight line with slope m and y-intercept $(0, b)$.

5 Use the Vertical Line Test.

Some graphs define functions and some do not. If every vertical line that intersects a graph does so exactly once, the graph represents a function, because to each value of x there corresponds exactly one value of y. If any vertical line that intersects a graph does so more than once, the graph cannot represent a function, because to one value of x there would correspond more than one value of y.

The Vertical Line Test

If a vertical line intersects a graph in more than one point, the graph is not the graph of a function.

The graph shown in red in figure (a) is not the graph of a function because the vertical line intersects the graph at more than one point. The points of intersection indicate that two values of y (2.5 and -2.5) correspond to the x-value 3.

The graph shown in red in figure (b) represents a function, because no vertical line intersects the graph at more than one point. Several vertical lines are drawn to illustrate this.

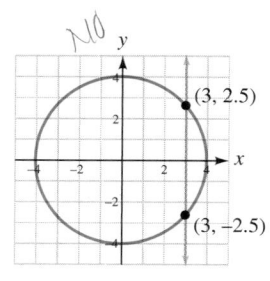

(a)

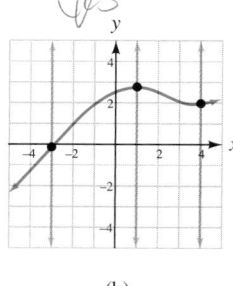

(b)

Find Function Values, Domains, and Ranges Graphically.

Since a graph is often the best way to describe a function, we need to know how to interpret graphs of functions. From the graph of a function, we can determine function values.

EXAMPLE 7   Refer to the graph of function f in figure (a).   **a.** Find $f(-3)$.
b. Find the value of x for which $f(x) = -2$.

Strategy   In each case, we will use the information provided by the function notation to locate a specific point on the graph and determine its x- and y-coordinates.

Why   Once we locate the specific point, one of its coordinates will equal the value that we are asked to find.

Solution

a. To find $f(-3)$, we need to find the y-coordinate of the point on the graph of f whose x-coordinate is -3. If we draw a vertical line through -3 on the x-axis, as shown in figure (b), the line intersects the graph of f at $(-3, 5)$. Therefore, 5 corresponds to -3, and it follows that $f(-3) = 5$.

b. To find the input value x that has an output value $f(x) = -2$, we draw a horizontal line through -2 on the y-axis, as shown in figure (c) and note that it intersects the graph of f at $(4, -2)$. Since -2 corresponds to 4, it follows that $f(x) = -2$ if $x = 4$.

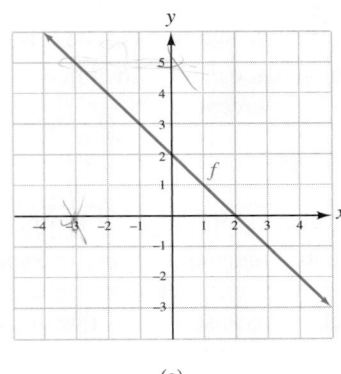   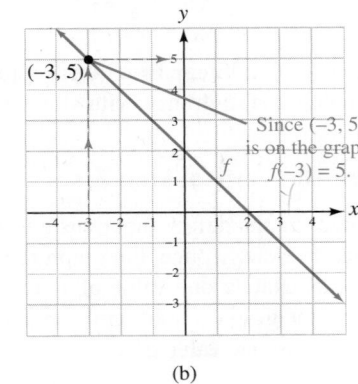   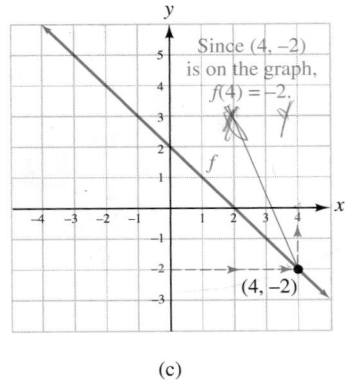

  (a)   (b)   (c)

▷ **Self Check 7**   Refer to the graph of function g.
a. Find $g(-3)$.
b. Find the x-value for which $g(x) = 4$.

Now Try   Problem 77

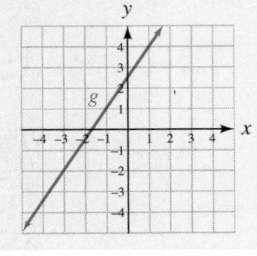

We can determine the domain and range of a function from its graph. For example, to find the domain of the linear function graphed in figure (a) on the next page, we *project* the graph onto the x-axis. Because the graph of the function extends indefinitely to the left and to the right, the projection includes all the real numbers. Therefore, the domain of the function is the set of real numbers.

To determine the range of the same linear function, we project the graph onto the y-axis, as shown in figure (b). Because the graph of the function extends indefinitely upward and downward, the projection includes all the real numbers. Therefore, the range of the function is the set of real numbers.

Project the graph onto the x-axis.

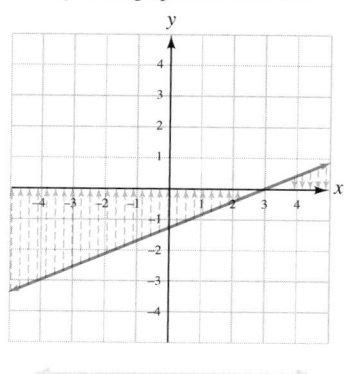

Domain: all real numbers

(a)

Project the graph onto the y-axis.

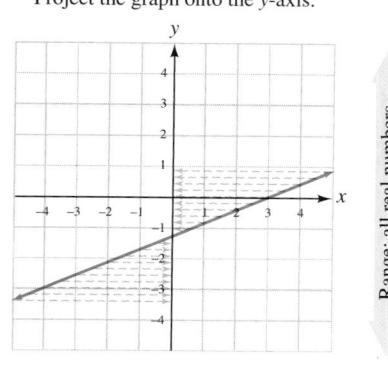

Range: all real numbers

(b)

The Language of Algebra
Think of the *projection* of a graph on an axis as the *shadow* that the graph makes on the axis.

7   **Graph Nonlinear Functions.**

We have seen that the graph of a linear function is a line. We will now consider several examples of **nonlinear functions** whose graphs are not lines. We will begin with $f(x) = x^2$, called the **squaring function.**

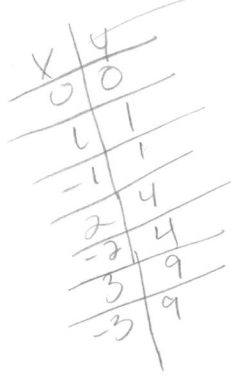

EXAMPLE 8   Graph $f(x) = x^2$ and find its domain and range.

Strategy   We will graph the function by creating a table of function values and plotting the corresponding ordered pairs.

Why   After drawing a smooth curve though the plotted points, we will have the graph.

Solution   To graph the function, we select several x-values and find the corresponding values of $f(x)$. For example, if we select -3 for x, we have

$f(x) = x^2$ This is the function to graph.

$f(-3) = (-3)^2$ Substitute -3 for each x.

$\quad\quad = 9$

All equation will be Square parabola 𝖸

Since $f(-3) = 9$, the ordered pair $(-3, 9)$ lies on the graph of f. In a similar manner, we find the corresponding values of $f(x)$ for six other x-values and list the ordered pairs in the table of values. Then we plot the points and draw a smooth curve through them to get the graph, called a **parabola.**

$f(x) = x^2$

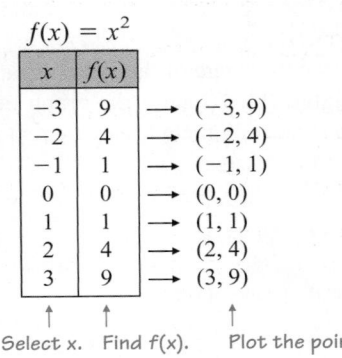

x	$f(x)$		
-3	9	→	$(-3, 9)$
-2	4	→	$(-2, 4)$
-1	1	→	$(-1, 1)$
0	0	→	$(0, 0)$
1	1	→	$(1, 1)$
2	4	→	$(2, 4)$
3	9	→	$(3, 9)$

↑   ↑   ↑

Select x.   Find f(x).   Plot the point.

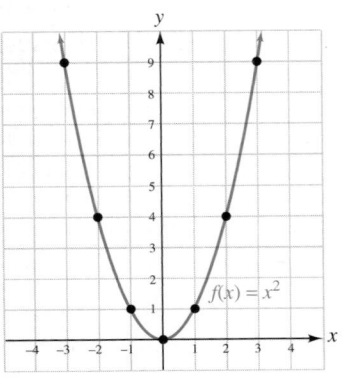

Because the graph extends indefinitely to the left and to the right, the projection of the graph onto the *x*-axis includes all the real numbers. See figure (a). This means that the domain of the squaring function is the set of real numbers.

Project the graph onto the *x*-axis.

Project the graph onto the *y*-axis.

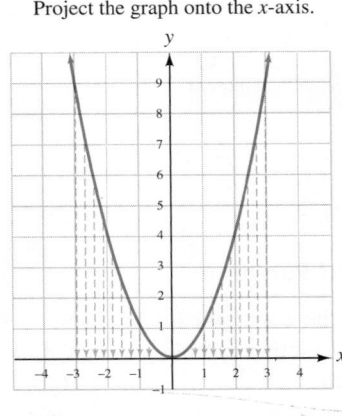

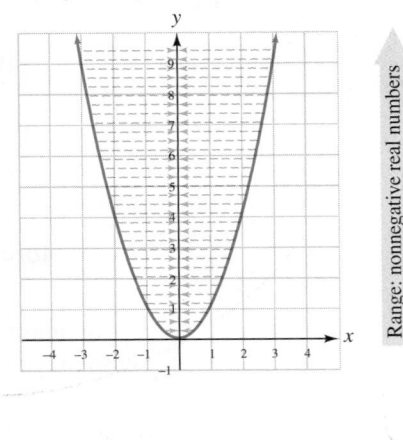

Range: nonnegative real numbers

Domain: all real numbers

(a)   (b)

Because the graph extends upward indefinitely from the point $(0, 0)$, the projection of the graph on the *y*-axis includes only positive real numbers and 0. See figure (b) above. This means that the range of the squaring function is the set of nonnegative real numbers.

Self Check 8   Graph $g(x) = x^2 - 2$ and find its domain and range. Compare the graph to the graph of $f(x) = x^2$.

Now Try Problem 81

Another nonlinear function is $f(x) = x^3$, called the **cubing function.**

EXAMPLE 9   Graph $f(x) = x^3$ and find its domain and range.

Strategy   We will graph the function by creating a table of function values and plotting the corresponding ordered pairs.

Why   After drawing a smooth curve though the plotted points, we will have the graph.

Solution   To graph the function, we select several values for x and find the corresponding values of $f(x)$. For example, if we select -2 for x, we have

$$f(x) = x^3$$
$$f(-2) = (-2)^3 \quad \text{Substitute } -2 \text{ for each } x.$$
$$= -8$$

Since $f(-2) = -8$, the ordered pair $(-2, -8)$ lies on the graph of f. In a similar manner, we find the corresponding values of $f(x)$ for four other x-values and list the ordered pairs in the table. Then we plot the points and draw a smooth curve through them to get the graph.

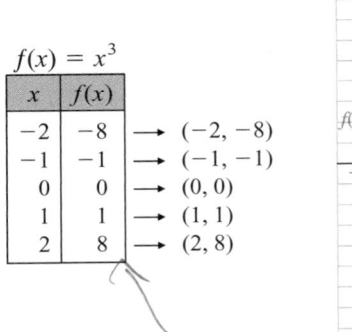

$$f(x) = x^3$$

x	$f(x)$
-2	-8
-1	-1
0	0
1	1
2	8

$\rightarrow (-2, -8)$
$\rightarrow (-1, -1)$
$\rightarrow (0, 0)$
$\rightarrow (1, 1)$
$\rightarrow (2, 8)$

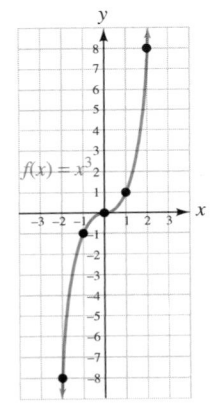

Because the graph of the function extends indefinitely to the left and to the right, the projection includes all the real numbers. Therefore, the domain of the cubing function is the set of real numbers.

Because the graph of the function extends indefinitely upward and downward, the projection includes all the real numbers. Therefore, the range of the cubing function is the set of real numbers.

Self Check 9   Graph $g(x) = x^3 + 1$ and find its domain and range. Compare the graph to the graph of $f(x) = x^3$.

add 1 to this will be up 1 on graph if it would be down 1 or graph

Now Try   Problem 83

A third nonlinear function is $f(x) = |x|$, called the **absolute value function**.

EXAMPLE 10   Graph $f(x) = |x|$ and find its domain and range.

Strategy   We will graph the function by creating a table of function values and plotting the corresponding ordered pairs.

Why   After drawing lines though the plotted points, we will have the graph.

Solution   To graph the function, we select several x-values and find the corresponding values for $f(x)$. For example, if we choose -3 for x, we have

$$f(x) = |x|$$
$$f(-3) = |-3| \quad \text{Substitute } -3 \text{ for each } x.$$
$$= 3$$

Since $f(-3) = 3$, the ordered pair $(-3, 3)$ lies on the graph of f. In a similar manner, we find the corresponding values of $f(x)$ for six other x-values and list the ordered pairs in the table. Then we plot the points and connect them to get the following V-shaped graph.

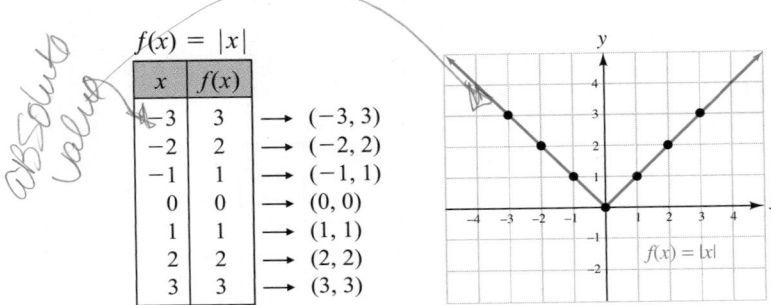

$f(x) = |x|$

x	$f(x)$	
-3	3	$\rightarrow (-3, 3)$
-2	2	$\rightarrow (-2, 2)$
-1	1	$\rightarrow (-1, 1)$
0	0	$\rightarrow (0, 0)$
1	1	$\rightarrow (1, 1)$
2	2	$\rightarrow (2, 2)$
3	3	$\rightarrow (3, 3)$

Because the graph extends indefinitely to the left and to the right, the projection of the graph onto the x-axis includes all the real numbers. Thus, the domain of the absolute value function is the set of real numbers.

Because the graph extends upward indefinitely from the point $(0, 0)$, the projection of the graph on the y-axis includes only positive real numbers and 0. Thus, the range of the absolute value function is the set of nonnegative real numbers.

▷ **Self Check 10**   Graph $g(x) = |x - 2|$ and find its domain and range. Compare the graph to the graph of $f(x) = |x|$.

Now Try   **Problem 85**

Using Your Calculator   *Graphing Functions*

We can graph nonlinear functions with a graphing calculator. For example, to graph $f(x) = x^2$ in a standard window of $[-10, 10]$ for x and $[-10, 10]$ for y, we first press $\boxed{Y =}$. Then we enter the function by typing $x \mathbin{^\wedge} 2$ (or x followed by $\boxed{x^2}$) and press the $\boxed{\text{GRAPH}}$ key. We will obtain the graph shown in figure (a) on the next page.

To graph $f(x) = x^3$, we enter the function by typing $x \wedge 3$ and then press the $\boxed{\text{GRAPH}}$ key to obtain the graph in figure (b). To graph $f(x) = |x|$, we enter the function by selecting abs from the NUM option within the MATH menu, typing x, and pressing the $\boxed{\text{GRAPH}}$ key to obtain the graph in figure (c).

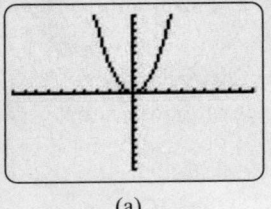   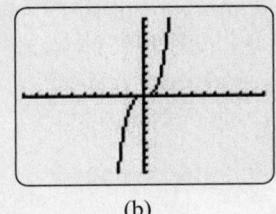   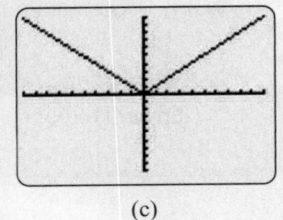   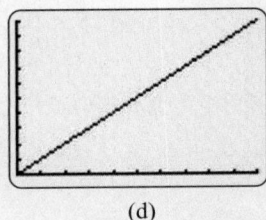

(a)   (b)   (c)   (d)

When using a graphing calculator, we must be sure that the viewing window does not show a misleading graph. For example, if we graph $f(x) = |x|$ in the window [0, 10] for x and [0, 10] for y, we will obtain a misleading graph that looks like a line. See figure (d). This is not correct. The proper graph is the V-shaped graph shown in figure (c). One of the challenges of using graphing calculators is finding an appropriate viewing window.

8 Graph Functions Using Translations and Reflections.

Examples 8, 9, and 10 and their Self Checks suggest that the graphs of different functions may be identical except for their positions in the coordinate plane. For example, the figure on the right shows the graph of $f(x) = x^2 + k$ for three different values of k. If $k = 0$, we get the graph of $f(x) = x^2$. If $k = 3$, we get the graph of $f(x) = x^2 + 3$, which is identical to the graph of $f(x) = x^2$ except that it is shifted 3 units upward. If $k = -4$, we get the graph of $f(x) = x^2 - 4$, which is identical to the graph of $f(x) = x^2$ except that it is shifted 4 units downward. These shifts are called **vertical translations.**

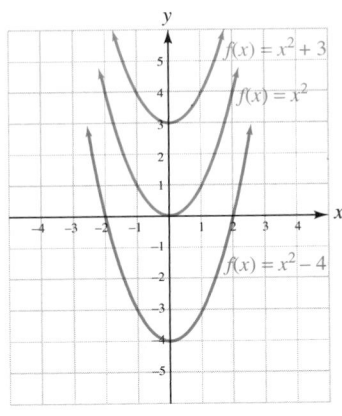

In general, we can make these observations.

Vertical Translations

If f is a function and k represents a positive number, then

- The graph of $y = f(x) + k$ is identical to the graph of $y = f(x)$ except that it is translated k units upward.
- The graph of $y = f(x) - k$ is identical to the graph of $y = f(x)$ except that it is translated k units downward.

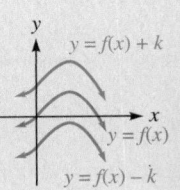

EXAMPLE 11   Graph: $g(x) = |x| + 2$

Strategy   We will graph $g(x) = |x| + 2$ by translating (shifting) the graph of $f(x) = |x|$ upward 2 units.

Why   The addition of 2 in $g(x) = |x| + 2$ causes a vertical shift of the graph of the absolute value function 2 units upward.

Solution   Each point used to graph $f(x) = |x|$, which is shown in gray, is shifted 2 units upward to obtain the graph of $g(x) = |x| + 2$, which is shown in red.

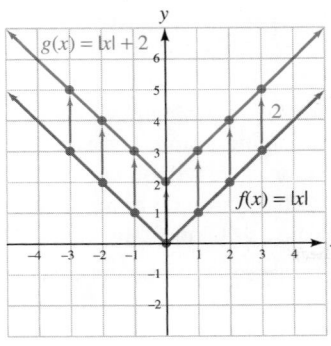

To graph $g(x) = |x| + 2$, translate each point on the graph of $f(x) = |x|$ up 2 units.

Self Check 11   Graph: $g(x) = |x| - 3$

Now Try   **Problem 89**

The figure on the right shows the graph of $f(x) = (x + h)^2$ for three different values of h. If $h = 0$, we get the graph of $f(x) = x^2$. The graph of $f(x) = (x - 3)^2$ is identical to the graph of $f(x) = x^2$ except that it is shifted 3 units to the right. The graph of $f(x) = (x + 2)^2$ is identical to the graph of $f(x) = x^2$ except that it is shifted 2 units to the left. These shifts are called **horizontal translations**.

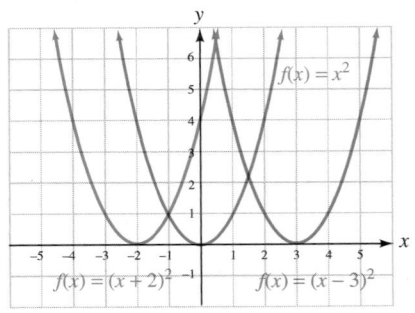

In general, we can make these observations.

Horizontal Translations

If f is a function and h is a positive number, then

- The graph of $y = f(x - h)$ is identical to the graph of $y = f(x)$ except that it is translated h units to the right.
- The graph of $y = f(x + h)$ is identical to the graph of $y = f(x)$ except that it is translated h units to the left.

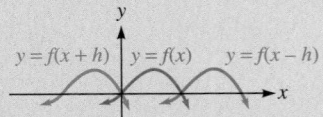

EXAMPLE 12   Graph: $g(x) = (x + 3)^3$

Strategy   We will graph $g(x) = (x + 3)^3$ by translating (shifting) the graph of $f(x) = x^3$ to the left 3 units.

Why   The addition of 3 to x in $g(x) = (x + 3)^3$ causes a horizontal shift of the graph of the cubing function 3 units to the left.

To graph $g(x) = (x + 3)^3$, translate each point on the graph of $f(x) = x^3$ to the left 3 units.

Solution   Each point used to graph $f(x) = x^3$, which is shown in gray, is shifted 3 units to the left to obtain the graph of $g(x) = (x + 3)^3$, which is shown in red.

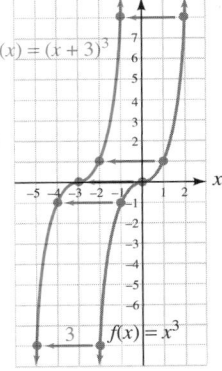

 Self Check 12   Graph: $g(x) = (x - 2)^2$

Now Try   **Problem 91**

The graphs of some functions involve horizontal and vertical translations.

EXAMPLE 13   Graph: $g(x) = (x - 5)^2 - 2$

Strategy   To graph $g(x) = (x - 5)^2 - 2$, we will perform two translations by shifting the graph of $f(x) = x^2$ to the right 5 units and then 2 units downward.

Why   The subtraction of 5 from x in $g(x) = (x - 5)^2 - 2$ causes a horizontal shift of the graph of the squaring function 5 units to the right and the subtraction of 2 causes a vertical shift of the graph 2 units downward.

Solution   Each point used to graph $f(x) = x^2$, which is shown in gray, is shifted 5 units to the right and 2 units downward to obtain the graph of $g(x) = (x - 5)^2 - 2$, which is shown in red.

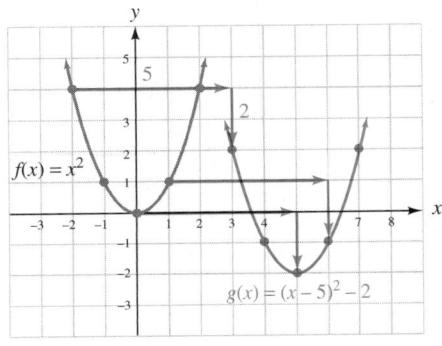

To graph $g(x) = (x - 5)^2 - 2$, translate each point on the graph of $f(x) = x^2$ to the right 5 units and then 2 units downward.

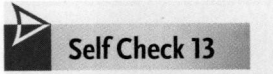 **Self Check 13** Graph: $g(x) = |x + 2| - 3$

Now Try Problem 97

The following figure shows a table of values for $f(x) = x^2$ and for $g(x) = -x^2$. We note that for a given value of x, the corresponding y-value in the tables are opposites. When graphed, we see that the $-$ sign in $g(x) = -x^2$ has the effect of flipping the graph of $f(x) = x^2$ over the x-axis so that the parabola opens downward. We say that the graph of $g(x) = -x^2$ is a **reflection** of the graph of $f(x) = x^2$ about the x-axis.

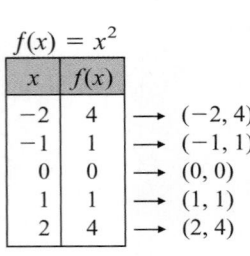

$f(x) = x^2$

x	$f(x)$	
-2	4	→ $(-2, 4)$
-1	1	→ $(-1, 1)$
0	0	→ $(0, 0)$
1	1	→ $(1, 1)$
2	4	→ $(2, 4)$

$g(x) = -x^2$

x	$g(x)$	
-2	-4	→ $(-2, -4)$
-1	-1	→ $(-1, -1)$
0	0	→ $(0, 0)$
1	-1	→ $(1, -1)$
2	-4	→ $(2, -4)$

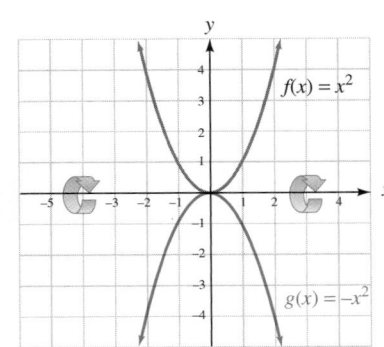

Reflection of a Graph

The graph of $y = -f(x)$ is the graph of $y = f(x)$ reflected about the x-axis.

EXAMPLE 14 Graph: $g(x) = -x^3$

Strategy We will graph $g(x) = -x^3$ by reflecting the graph of $f(x) = x^3$ about the x-axis.

Why Because of the $-$ sign in $g(x) = -x^3$, the y-coordinate of each point on the graph of function g is the opposite of the y-coordinate of the corresponding point on the graph $f(x) = x^3$.

Solution To graph $g(x) = -x^3$, we use the graph of $f(x) = x^3$ from Example 9. First, we reflect the portion of the graph of $f(x) = x^3$ in quadrant I to quadrant IV, as shown. Then we reflect the portion of the graph of $f(x) = x^3$ in quadrant III to quadrant II.

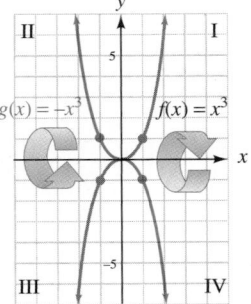

 Self Check 14 Graph: $g(x) = -|x|$

Now Try Problem 107

ANSWERS TO SELF CHECKS   **1. a.** No; $(0, 2), (0, 3)$ **b.** Yes   **c.** No; $(4, -1), (4, 4)$ **2. a.** Yes
b. No; $(1, 2), (1, -2)$ **3. a.** -5 **b.** $2t - 1$ **4.** -2 **5. a.** $\{-3, -2, 1\}, \{5, 7, 11\}$
b. D: the set of all real numbers except -3; R: the set of all real numbers except 0

6.

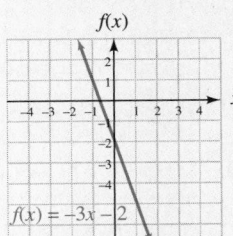

$f(x)$

$f(x) = -3x - 2$

7. a. -2 **b.** 1

8. D: the set of real numbers;
R: the set of real numbers
greater than or equal to -2;
the graph has the same shape,
but is 2 units lower.

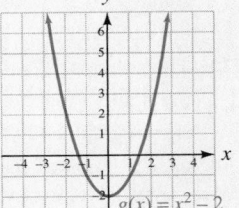

$g(x) = x^2 - 2$

9. D: the set of real numbers;
R: the set of real numbers;
the graph has the same shape,
but is 1 unit higher.

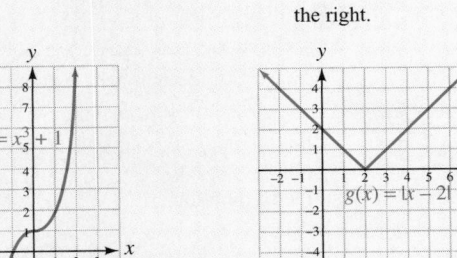

$g(x) = x^3 + 1$

10. D: the set of real numbers;
R: the set of nonnegative real
numbers; the graph has the
same shape, but is 2 units to
the right.

$g(x) = |x - 2|$

11.

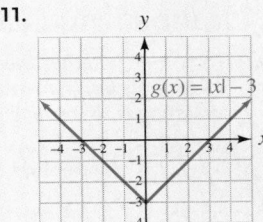

$g(x) = |x| - 3$

12.

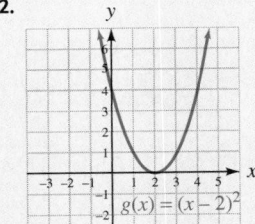

$g(x) = (x - 2)^2$

13.

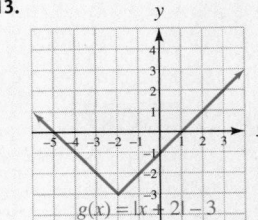

$g(x) = |x + 2| - 3$

14.

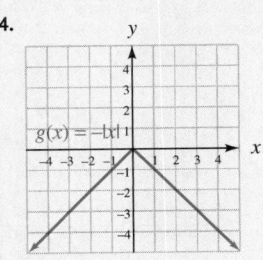

$g(x) = -|x|$

STUDY SET
8.8

VOCABULARY

Fill in the blanks.

1. A _____ is a set of ordered pairs (a relation) in which to each first component, there corresponds exactly one second component.

2. We can think of a function as a machine that takes input x and turns it into some _____ $f(x)$. The set of numbers that we put into the machine is the _____ of the function, and the set of numbers that come out is the is the _____.

3. The function $f(x) = x^2$ is called the _____ function. Its graph is a cup-like shape called a _____.

4. The function $f(x) = x^3$ is called the _____ function.

5. The function $f(x) = |x|$ is called the _____ _____ function.

6. We can use the _____ line test to determine if a graph represents a function.

CONCEPTS

7. The arrow diagram describes a function.
   a. What is the domain of the function?
   b. What is the range of the function?

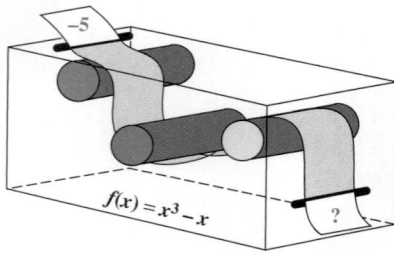

8. For the given input, what value will the function machine output?

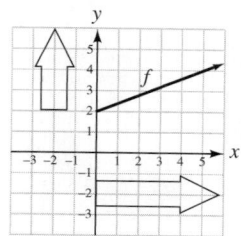

9. Consider the graph of the function f.
   a. Label each arrow in the illustration with the appropriate term: *domain* or *range*.
   b. Give the domain and range of f.

10. Translate each point plotted on the graph 5 units to the left and then up 1 unit.

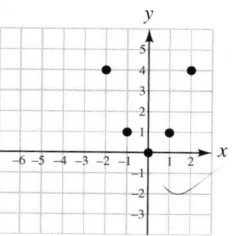

NOTATION

Fill in the blanks.

11. We read $f(x) = 5x - 6$ as "f ▢ x is $5x$ minus 6."

12. Since $y = $ ▢, the equations $y = 3x + 2$ and $f(x) = 3x + 2$ are equivalent.

13. The notation $f(2) = 7$ indicates that when the x-value ▢ is input into a function rule, the output is ▢. This fact can be shown graphically by plotting the ordered pair (▢ , ▢).

14. a. The graph of $f(x) = (x + 4)^3$ is the same as the graph of $f(x) = x^3$ except that it is shifted ___ units to the ____.
    b. The graph of $f(x) = x^3 - 2$ is the same as the graph of $f(x) = x^3$ except that it is shifted ___ units _____.
    c. The graph of $f(x) = x^2 + 5$ is the same as the graph of $f(x) = x^2$ except that it is shifted ___ units ____.

GUIDED PRACTICE

Determine whether the relation defines y to be a function of x. If it does not, find two ordered pairs where more than one value of y corresponds to a single value of x. See Example 1.

15.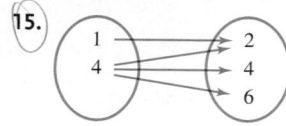

16.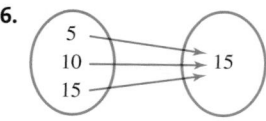

17. $\{(-2, 7), (-1, 10), (0, 13), (1, 16)\}$

18. $\{(-2, 4), (-3, 8), (-3, 12), (-4, 16)\}$

19.

x	y
-4	6
-1	0
0	-3
2	4
-1	2

20.

x	y
1	1
2	2
3	3
4	4

Determine whether each equation defines y to be a function of x. If it does not, find two ordered pairs where more than one value of y corresponds to a single value of x. See Example 2.

21. $y = 2x + 3$　　　　22. $y = 4x - 1$

23. $y = 4x^2$　　　　24. $y^2 = x$

25. $y^4 = x$

26. $y = \dfrac{1}{x}$

27. $xy = 9$

28. $y = |x|$

29. $y = \dfrac{1}{x^2}$

30. $x + 1 = |y|$

31. $x = |y|$

32. $xy = -4$

For each function, find $f(3)$ and $f(-1)$. See Example 3.

33. $f(x) = 5x + 7$

34. $f(x) = 3x + 3$

35. $f(x) = 9 - 2x$

36. $f(x) = 12 + 3x$

For each function, find $g(2)$ and $g(t)$. See Example 3.

37. $g(x) = 2x^2 - x$

38. $g(x) = 5x^2 + 2x$

39. $g(x) = x^3 - 1$

40. $g(x) = x^3$

Find $h(2)$ and $h(-2)$. See Example 3.

41. $h(x) = |x| + 2$

42. $h(x) = |x| - 5$

43. $h(x) = x^2 - 2$

44. $h(x) = x^2 + 3$

45. $h(x) = \dfrac{1}{x + 3}$

46. $h(x) = \dfrac{3}{x - 4}$

47. $h(x) = \dfrac{x}{x - 3}$

48. $h(x) = \dfrac{x}{x^2 + 2}$

Find $g(w)$ and $g(w + 1)$. See Example 3.

49. $g(x) = 3x - 5$

50. $g(x) = 2x - 7$

51. $g(x) = x^2 + 9$

52. $g(x) = 4 - x^2$

See Example 4.

53. Let $f(x) = -2x + 5$. For what value(s) of x is $f(x) = 5$?

54. Let $f(x) = -2x + 5$. For what value(s) of x is $f(x) = -7$?

55. Let $f(x) = \dfrac{3}{2}x - 2$. For what value(s) of x is $f(x) = -\dfrac{1}{2}$?

56. Let $f(x) = \dfrac{3}{2}x - 2$. For what value(s) of x is $f(x) = \dfrac{2}{3}$?

Find the domain and range of each function. See Example 5.

57. $\{(-2, 3), (4, 5), (6, 7)\}$

58. $\{(0, 2), (1, 2), (3, 4)\}$

59. $s(x) = 3x + 6$

60. $h(x) = \dfrac{4}{5}x - 8$

61. $f(x) = \dfrac{1}{x - 4}$

62. $f(x) = \dfrac{5}{x + 1}$

63. $s(x) = |x - 7|$

64. $f(x) = x^2$

Graph each linear function. See Example 6.

65. $f(x) = 2x - 1$

66. $f(x) = -x + 2$

67. $f(x) = \dfrac{2}{3}x - 2$

68. $f(x) = -\dfrac{3}{2}x - 3$

Use the vertical line test to determine whether the given graph represents a function. If it does not, find two ordered pairs where more than one value of y corresponds to a single value of x. See Objective 5.

69.
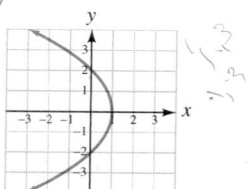

70.
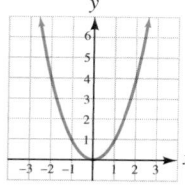

71.
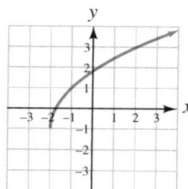

72.
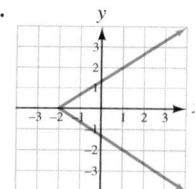

73.
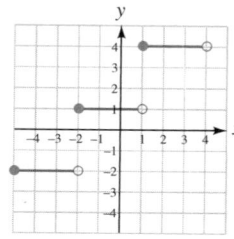

74.
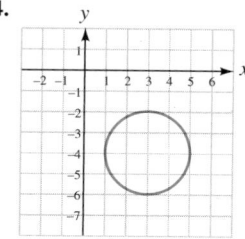

75.
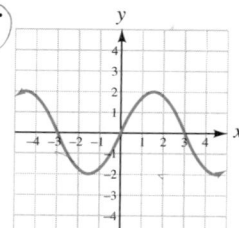

76.
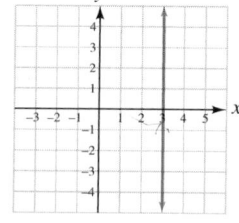

Refer to the given graph to find each value. See Example 7.

77. a. $f(-2)$

b. $f(0)$

c. The value of x for which $f(x) = 4$.

d. The value of x for which $f(x) = -2$.

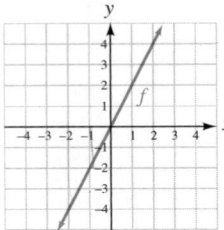

78. a. $g(-2)$

b. $g(0)$

c. The value of x for which $g(x) = 3$.

d. The values of x for which $g(x) = -1$.

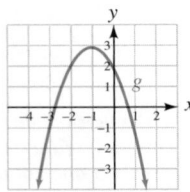

79. a. $s(-3)$

b. $s(3)$

c. The values of x for which $s(x) = 0$.

d. The values of x for which $s(x) = 3$.

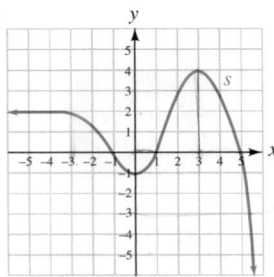

80. a. $h(-3)$

b. $h(4)$

c. The value(s) of x for which $h(x) = 1$.

d. The value of x for which $h(x) = 0$.

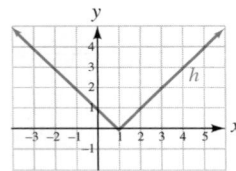

Graph each function by creating a table of function values and plotting points. Give the domain and range of the function. See Examples 8, 9, and 10.

81. $f(x) = x^2 + 2$

82. $f(x) = x^2 - 4$

83. $f(x) = x^3 - 3$

84. $f(x) = x^3 + 2$

85. $f(x) = |x - 1|$

86. $f(x) = |x + 4|$

87. $f(x) = (x + 4)^2$

88. $f(x) = (x - 1)^3$

For each of the following functions, first sketch the graph of its associated function, $f(x) = x^2$, $f(x) = x^3$, or $f(x) = |x|$. Then draw the graph of function g using a translation and give its domain and range. See Examples 11 and 12.

89. $g(x) = |x| - 2$

90. $g(x) = |x + 2|$

91. $g(x) = (x + 1)^3$

92. $g(x) = x^3 + 5$

93. $g(x) = x^2 - 3$

94. $g(x) = (x - 6)^2$

95. $g(x) = |x| + 1$

96. $g(x) = x^3 + 4$

For each of the following functions, first sketch the graph of its associated function, $f(x) = x^2$, $f(x) = x^3$, or $f(x) = |x|$. Then draw the graph of function g using a translation. See Example 13.

97. $g(x) = |x - 2| - 1$

98. $g(x) = (x + 2)^2 - 1$

99. $g(x) = (x + 1)^3 - 2$

100. $g(x) = |x + 4| + 3$

101. $g(x) = (x - 2)^2 + 4$

102. $g(x) = (x - 4)^2 + 3$

103. $g(x) = |x + 3| + 5$

104. $g(x) = (x - 3)^2 - 2$

For each of the following functions, first sketch the graph of its associated function, $f(x) = x^2$, $f(x) = x^3$, or $f(x) = |x|$. Then draw the graph of function g using a translation and/or a reflection. See Example 14.

105. $g(x) = -x^3$

106. $g(x) = -|x|$

107. $g(x) = -x^2$

108. $g(x) = -(x + 1)^2$

109. $g(x) = -|x + 5|$

110. $g(x) = -(x + 4)^3$

111. $g(x) = -x^2 + 3$

112. $g(x) = -|x| - 4$

Graph each function using window settings of $[-4, 4]$ for x and $[-4, 4]$ for y. The graph is not what it appears to be. Pick a better viewing window and find a better representation of the true graph. See Using Your Calculator: Graphing Functions.

113. $f(x) = x^2 + 8$

114. $f(x) = x^3 - 8$

115. $f(x) = |x + 5|$

116. $f(x) = |x - 5|$

117. $f(x) = (x - 6)^2$

118. $f(x) = (x + 9)^2$

119. $f(x) = x^3 + 8$

120. $f(x) = x^3 - 12$

APPLICATIONS

121. DECONGESTANTS The temperature in degrees Celsius that is equivalent to a temperature in degrees Fahrenheit is given by the linear function $C(F) = \frac{5}{9}(F - 32)$. Use this function to find the low and high temperature extremes, in degrees Celsius, in which a bottle of decongestant should be stored.

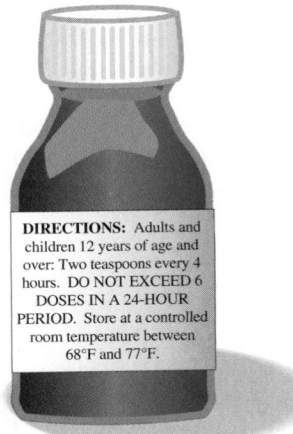

DIRECTIONS: Adults and children 12 years of age and over: Two teaspoons every 4 hours. DO NOT EXCEED 6 DOSES IN A 24-HOUR PERIOD. Store at a controlled room temperature between 68°F and 77°F.

122. CONCESSIONAIRES A baseball club pays a vendor $125 per game for selling bags of peanuts for $4.75 each.

  a. Write a linear function that describes the income the vendor makes for the baseball club during a game if she sells b bags of peanuts.

  b. Find the income the baseball club will make if the vendor sells 110 bags of peanuts during a game.

123. HOME CONSTRUCTION In a proposal to some clients, a housing contractor listed the following costs:

Fees, permits, miscellaneous	$12,000
Construction, per square foot	$95

  a. Write a linear function that the clients could use to determine the cost of building a home having f square feet.

  b. Find the cost to build a home having 1,950 square feet.

124. CHEMICAL REACTIONS When students mixed solutions of acetone and chloroform, they found that heat was generated. However, as time passed, the mixture cooled down. The graph shows data points of the form (time, temperature) taken by the students.

  a. The linear function $T(t) = -\frac{t}{240} + 30$ models the relationship between the elapsed time t since the solutions were combined and the temperature $T(t)$ of the mixture. Graph the function.

  b. Predict the temperature of the mixture immediately after the two solutions are combined.

  c. Is $T(180)$ more or less than the temperature recorded by the students for $t = 300$?

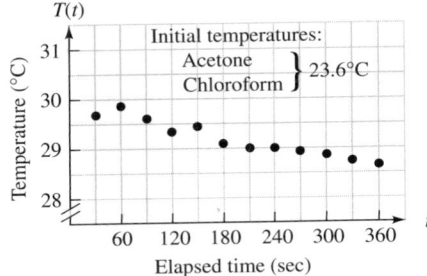

WRITING

125. What is a function?

126. Explain why -4 isn't in the domain of $f(x) = \frac{1}{x + 4}$.

127. What does it mean to vertically translate a graph?

128. Explain why the correct choice of window settings is important when using a graphing calculator.

REVIEW

Solve each system of equations.

129. $\begin{cases} 3x + 4y = -24 \\ 5x + 12y = -72 \end{cases}$

130. $\begin{cases} 5x + 2y = 11 \\ 7x + 6y = 9 \end{cases}$

CHALLENGE PROBLEMS

Graph each function.

131. $f(x) = \begin{cases} |x| & \text{for } x \geq 0 \\ x^3 & \text{for } x < 0 \end{cases}$

132. $f(x) = \begin{cases} x^2 & \text{for } x \geq 0 \\ |x| & \text{for } x < 0 \end{cases}$

SECTION 8.9
Variation

Objectives

1. Solve problems involving direct variation.
2. Solve problems involving inverse variation.
3. Solve problems involving joint variation.
4. Solve problems involving combined variation.

In this section, we introduce four types of *variation models,* each of which expresses a special relationship between two or more quantities. We will use these models to solve problems involving travel, lighting, geometry, and highway construction.

1 Solve Problems Involving Direct Variation.

To introduce direct variation, we consider the formula for the circumference of a circle

$$C = \pi D$$

where C is the circumference, D is the diameter, and $\pi \approx 3.14159$. If we double the diameter of a circle, we determine another circle with a larger circumference C_1 such that

$$C_1 = \pi(2D) = 2\pi D = 2C$$

Thus, doubling the diameter results in doubling the circumference. Likewise, if we triple the diameter, we will triple the circumference.

In the formula $C = \pi D$, we say that the variables C and D *vary directly,* or that they are *directly proportional.* This is because C is always found by multiplying D by a constant. In this example, the constant π is called the *constant of variation* or the *constant of proportionality.*

Direct Variation

The words "y varies directly as x" or "y is directly proportional to x" means that $y = kx$ for some nonzero constant k. The constant k is called the **constant of variation** or the **constant of proportionality.**

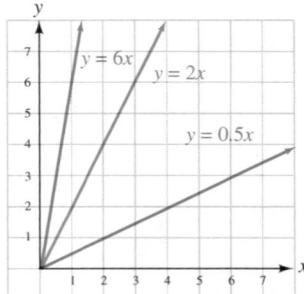

Since the formula for direct variation ($y = kx$) defines a linear function, its graph is always a line with a y-intercept at the origin. The graph of $y = kx$ where $x \geq 0$ appears in the margin for three positive values of k.

One example of direct variation is Hooke's law from physics. Hooke's law states that the distance a spring will stretch varies directly as the force that is applied to it.

If d represents a distance and f represents a force, this verbal model of Hooke's law can be expressed as

$$d = kf \quad \text{The direct variation model can be read as "d is directly proportional to f."}$$

where k is the constant of variation. Suppose we know that a certain spring stretches 10 inches when a weight of 6 pounds is attached (see the figure). We can find k as follows:

$$d = kf$$
$$10 = k(6) \quad \text{Substitute 10 for d and 6 for f.}$$
$$\frac{5}{3} = k$$

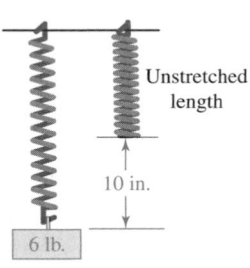

Unstretched length

10 in.

6 lb.

To find the force required to stretch the spring a distance of 35 inches, we can solve the equation $d = kf$ for f, with $d = 35$ and $k = \frac{5}{3}$.

$$d = kf$$

$$35 = \frac{5}{3}f \qquad \text{Substitute 35 for } d \text{ and } \tfrac{5}{3} \text{ for } k.$$

$$105 = 5f \qquad \text{Multiply both sides by 3.}$$

$$21 = f \qquad \text{Divide both sides by 5.}$$

Thus, the force required to stretch the spring a distance of 35 inches is 21 pounds.

EXAMPLE 1   **Currency Exchange.**   The currency calculator shown here converts from U.S. dollars to Japanese yen. When exchanging these currencies, the number of yen received is directly proportional to the number of dollars to be exchanged. How many yen will an exchange of $1,200 bring?

convert

US Dollar USD

amount

500

into

Japanese Yen JPY

amount

57,250

Strategy   We will use a direct variation model to solve this problem.

Why   The words *the number of yen received is directly proportional to the number of dollars to be exchanged* indicate that this type of model should be used.

Solution   The verbal model can be represented by the following equation

$$y = kd \qquad \text{This is a direct variation model.}$$

where y is the number of yen, k is the constant of variation, *never changes*, and d is the number of dollars. From the illustration, we see that an exchange of $500 brings 57,250 yen. To find k, we substitute 500 for d and 57,250 for y, and then we solve for k.

$$y = kd$$

$$57{,}250 = k(500)$$

$$114.5 = k \qquad \text{To isolate } k, \text{ divide both sides by 500.}$$

To find how many yen an exchange of $1,200 will bring, we substitute 114.5 for k and 1,200 for d in the direct variation model, and then we evaluate the right side.

$$y = kd$$

$$y = 114.5(1{,}200) \qquad \frac{57{,}250}{500} \qquad \frac{y2}{1200}$$

$$y = 137{,}400$$

An exchange of $1,200 will bring 137,400 yen.

Self Check 1   When exchanging currencies, the number of British pounds received is directly proportional to the number of U.S. dollars to be exchanged. If $800 converts to 392 pounds, how many pounds will be received if $1,500 is exchanged?

Now Try   **Problem 31**

We can use the following steps to solve variation problems.

Solving Variation Problems	To solve a variation problem:
	1. Translate the verbal model into an equation.
	2. Substitute the first set of values into the equation from step 1 to determine the value of k.
	3. Substitute the value of k into the equation from step 1.
	4. Substitute the remaining set of values into the equation from step 3 and solve for the unknown.

2 **Solve Problems Involving Inverse Variation.**

In the formula $w = \frac{12}{l}$, w gets smaller as l gets larger, and w gets larger as l gets smaller. Since these variables vary in opposite directions in a predictable way, we say that the variables **vary inversely,** or that they are **inversely proportional.** The constant 12 is the constant of variation.

Inverse Variation	The words "y varies inversely as x" or "y is inversely proportional to x" mean that $y = \frac{k}{x}$ for some nonzero constant k. The constant k is called the **constant of variation.**

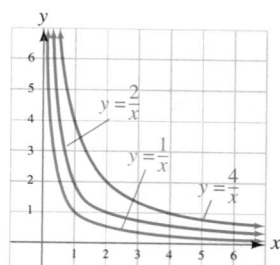

The formula for inverse variation, $y = \frac{k}{x}$, defines a rational function whose graph will have the x- and y-axes as asymptotes. The graph of $y = \frac{k}{x}$ where $x > 0$ appears in the margin for three positive values of k.

In an elevator, the amount of floor space per person varies inversely as the number of people in the elevator. If f represents the amount of floor space per person and n the number of people in the elevator, the relationship between f and n can be expressed by the following equation.

$$f = \frac{k}{n}$$ This inverse variation model can also be read as "f is inversely proportional to n."

Success Tip

For any inverse variation equation of the form $y = \frac{k}{x}$, where $k > 0$: as x increases, y decreases.

The figure shows 6 people in an elevator; each has 8.25 square feet of floor space. To determine how much floor space each person would have if 15 people were in the elevator, we begin by determining k.

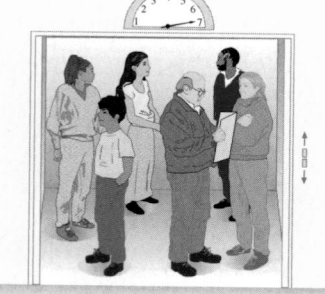

$$f = \frac{k}{n}$$

$$8.25 = \frac{k}{6}$$ Substitute 8.25 for f and 6 for n.

$$k = 49.5$$ Multiply both sides by 6 to solve for k.

To find the amount of floor space per person if 15 people are in the elevator, we proceed as follows:

Success Tip

If we multiply both sides of $y = \frac{k}{x}$ by x, we get $xy = k$. Thus, for the inverse variation model, k is simply the product of one pair of values of x and y. (Assume $x \neq 0$.)

$$f = \frac{k}{n}$$

$$f = \frac{49.5}{15}$$ Substitute 49.5 for k and 15 for n.

$$f = 3.3$$ Do the division.

If 15 people were in the elevator, each would have 3.3 square feet of floor space.

EXAMPLE 2 *Photography.* The intensity I of light received from a light source varies inversely as the square of the distance from the light source. If a photographer, 16 feet away from his subject, has a light meter reading of 4 foot-candles of luminance, what will the meter read if the photographer moves in for a close-up 4 feet away from the subject?

Strategy We will use the inverse variation model of the form $I = \frac{k}{d^2}$, where I represents the intensity and d^2 represents the square of the distance from the light source.

Why The words *intensity varies inversely as the square of the distance* indicate that this type of model should be used.

Solution

$$I = \frac{k}{d^2}$$
This inverse variation model can also be read as "I is inversely proportional to d^2."

To find k, we substitute 4 for I and 16 for d and solve for k.

> ### Success Tip
> The constant of variation is usually positive, because most real-life applications involve only positive quantities. However, the definitions of *direct, inverse, joint,* and *combined variation* allow for a negative constant of variation.

$$I = \frac{k}{d^2}$$

$$4 = \frac{k}{16^2}$$

$$4 = \frac{k}{256}$$

$$1{,}024 = k$$
To isolate k, multiply both sides by 256.

To find the intensity when the photographer is 4 feet away from the subject, we substitute 4 for d and 1,024 for k and simplify.

$$I = \frac{k}{d^2}$$

$$I = \frac{1{,}024}{4^2}$$

$$= 64$$

The intensity at 4 feet is 64 foot-candles.

 Self Check 2 Find the intensity when the photographer is 8 feet away from the subject.

Now Try Problem 33

---

3 **Solve Problems Involving Joint Variation.**

There are times when one variable varies as the product of several variables. For example, the area of a triangle varies directly with the product of its base and height:

$$A = \frac{1}{2}bh$$

Such variation is called *joint variation.*

Joint Variation

If one variable varies directly as the product of two or more variables, the relationship is called **joint variation**. If y varies jointly with x and z, then $y = kxz$. The nonzero constant k is called the **constant of variation**.

EXAMPLE 3   *Force of the Wind.*   The force of the wind on a billboard varies jointly as the area of the billboard and the square of the wind velocity. When the wind is blowing at 20 mph, the force on a billboard 30 feet wide and 18 feet high is 972 pounds. Find the force on a billboard having an area of 300 square feet caused by a 40-mph wind.

Strategy   We will use the joint variation model $f = kAv^2$, where f represents the force of the wind, A represents the area of the billboard, and v^2 represents the square of the velocity of the wind.

Why   The words *the force of the wind on a billboard varies jointly as the area of the billboard and the square of the wind velocity* indicate that this type of model should be used.

Solution

$$f = kAv^2$$ The joint variation model can also be read as "f is directly proportional to the product of A and v^2."

Since the billboard is 30 feet wide and 18 feet high, it has an area of $30 \cdot 18 = 540$ square feet. We can find k by substituting 972 for f, 540 for A, and 20 for v.

$$f = kAv^2$$
$$972 = k(540)(20)^2$$
$$972 = k(216{,}000) \qquad \text{Evaluate: } (20)^2 = 400. \text{ Then do the multiplication.}$$
$$0.0045 = k \qquad \text{Divide both sides by 216,000 to solve for } k.$$

To find the force exerted on a 300-square-foot billboard by a 40-mph wind, we use the formula $f = 0.0045Av^2$ and substitute 300 for A and 40 for v.

$$f = 0.0045Av^2$$
$$= 0.0045(300)(40)^2$$
$$= 2{,}160$$

The 40-mph wind exerts a force of 2,160 pounds on the billboard.

 Now Try   **Problem 37**

④ Solve Problems Invoving Combined Variation.

Many applied problems involve a combination of direct and inverse variation. Such variation is called **combined variation.**

EXAMPLE 4   *Highway Construction.*   The time it takes to build a highway varies directly as the length of the road, and inversely as the number of workers. If it takes 100 workers 4 weeks to build 2 miles of highway, how long will it take 80 workers to build 10 miles of highway?

Strategy   We will use the combined variation model $t = \frac{kl}{w}$, where t represents the time in days, l represents the length of road built in miles, and w represents the number of workers.

Why   The words *the time it takes to build a highway varies directly as the length of the road, and inversely with the number of workers* indicate that this type of model should be used.

Solution   The relationship between these variables can be expressed by the equation

$$t = \frac{kl}{w} \qquad \text{This is a combined variation model.}$$

We substitute 4 for t, 100 for w, and 2 for l to find k:

$$4 = \frac{k(2)}{100}$$

$$400 = 2k \qquad \text{Multiply both sides by 100.}$$

$$200 = k \qquad \text{Divide both sides by 2 to solve for } k.$$

We now substitute 80 for w, 10 for l, and 200 for k in the equation $t = \frac{kl}{w}$ and simplify:

$$t = \frac{kl}{w}$$

$$t = \frac{200(10)}{80}$$

$$= 25$$

It will take 25 weeks for 80 workers to build 10 miles of highway.

Self Check 4   How long will it take 60 workers to build 6 miles of highway?

Now Try   **Problem 43**

ANSWERS TO SELF CHECKS   **1.** 735 British pounds   **2.** 16 foot-candles   **4.** 20 weeks

STUDY SET
8.9

VOCABULARY

Fill in the blanks.

1. The equation $y = kx$ defines _____ variation: As x increases, y _____.

2. The equation $y = \frac{k}{x}$ defines _____ variation: As x increases, y _____.

3. A constant is a _____.

4. The equation $y = kxz$ represents _____ variation.

5. The equation $y = \frac{kx}{z}$ means that y varies _____ with x and _____ with z.

CONCEPTS

Determine whether direct or inverse variation applies and sketch a possible graph for the situation.

6.

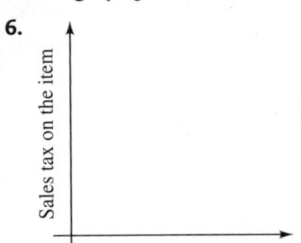

Sales tax on the item (vertical axis) / Price of an item (horizontal axis)

7.

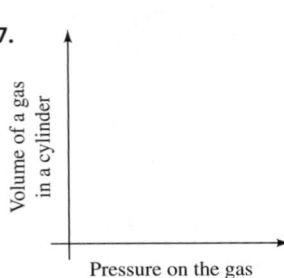

Volume of a gas in a cylinder (vertical axis) / Pressure on the gas (horizontal axis)

8.

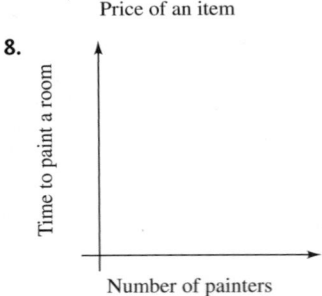

Time to paint a room (vertical axis) / Number of painters (horizontal axis)

9.

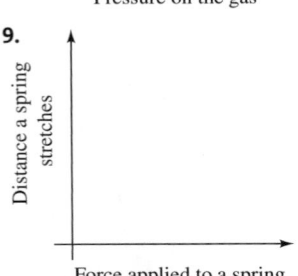

Distance a spring stretches (vertical axis) / Force applied to a spring (horizontal axis)

NOTATION

10. Determine whether the equation defines direct variation.

a. $y = kx$ *Direct*

b. $y = k + x$

c. $y = \dfrac{k}{x}$ *inversely*

d. $m = kc$ *direct*

11. Determine whether each equation defines inverse variation.

a. $y = kx$

b. $y = \dfrac{k}{x}$

c. $y = \dfrac{x}{k}$

d. $d = \dfrac{k}{g}$

GUIDED PRACTICE

Express each verbal model in symbols. See Objectives 1 and 2.

12. A varies directly as the square of p.

13. t varies directly as s.

14. z varies inversely as the cube of t.

15. v varies inversely as the square of r.

Express each verbal model in symbols. See Objectives 3 and 4.

16. C varies jointly as x, y, and z.

17. d varies jointly as r and t.

18. P varies directly as the square of a and inversely as the cube of j.

19. M varies inversely as the cube of n and jointly as x and the square of z.

Express each variation model in words. In each equation, k is the constant of variation. See Objectives 1 and 2.

20. $r = kt$

21. $A = kr^3$

A varies directly as the cube of r

22. $b = \dfrac{k}{h}$

23. $d = \dfrac{k}{W^4}$

Express each variation model in words. In each equation, k is the constant of variation. See Objectives 3 and 4.

24. $U = krs^2t$

25. $L = kmn$

26. $P = \dfrac{km}{n}$

27. $R = \dfrac{kL}{d^2}$

APPLICATIONS

28. Campus to Careers

Webmaster

The language of variation is often used to describe various aspects of the Internet and websites. Determine whether each statement, generally speaking, is true or false.

© Pat LaCroix/Getty Images

a. The dollar amount of sales that an Internet website receives is inversely proportional to amount of Internet traffic that visits the website.

b. The download time of an Internet website varies directly with the bandwidth being used.

c. Search engines like Google place a value on a website that is directly proportional to the number of sites that link to it.

Solve each problem by writing a variation model.

29. GRAVITY The force of gravity acting on an object varies directly as the mass of the object. The force on a mass of 5 kilograms is 49 newtons. What is the force acting on a mass of 12 kilograms?

30. FREE FALL An object in free fall travels a distance s that is directly proportional to the square of the time t. If an object falls 1,024 feet in 8 seconds, how far will it fall in 10 seconds?

31. FINDING DISTANCE The distance that a car can go varies directly as the number of gallons of gasoline it consumes. If a car can go 288 miles on 12 gallons of gasoline, how far can it go on a full tank of 18 gallons?

32. FARMING The number of days that a given number of bushels of corn will last when feeding cattle varies inversely as the number of animals. If x bushels will feed 25 cows for 10 days, how long will the feed last for 10 cows?

33. ORGAN PIPES Refer to the illustration on the next page. The frequency of vibration of air in an organ pipe is inversely proportional to the length of the pipe. If a pipe 2 feet long vibrates 256 times per second, how many times per second will a 6-foot pipe vibrate?

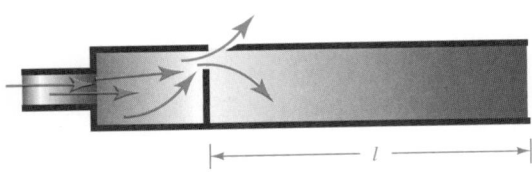

34. GAS PRESSURE Under constant temperature, the volume occupied by a gas varies inversely to the pressure applied. If the gas occupies a volume of 20 cubic inches under a pressure of 6 pounds per square inch, find the volume when the gas is subjected to a pressure of 10 pounds per square inch.

35. REAL ESTATE The following table shows the listing price for three homes in the same general locality. Write the variation model (direct or inverse) that describes the relationship between the listing price and the number of square feet of a house in this area.

Number of square feet	Listing price
1,720	$180,600
1,205	$126,525
1,080	$113,400

36. TRUCKING COSTS The costs of a trucking company vary jointly as the number of trucks in service and the number of hours they are used. When 4 trucks are used for 6 hours each, the costs are $1,800. Find the costs of using 10 trucks, each for 12 hours.

37. OIL STORAGE The number of gallons of oil that can be stored in a cylindrical tank varies jointly as the height of the tank and the square of the radius of its base. The constant of proportionality is 23.5. Find the number of gallons that can be stored in the cylindrical tank shown.

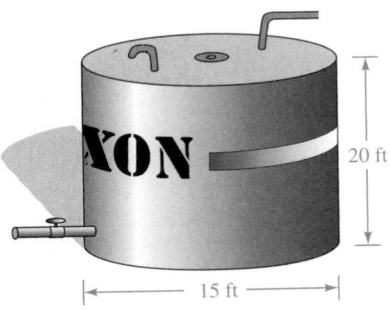

15 ft
20 ft

38. ELECTRONICS The voltage (in volts) measured across a resistor is directly proportional to the current (in amperes) flowing through the resistor. The constant of variation is the **resistance** (in ohms). If 6 volts is measured across a resistor carrying a current of 2 amperes, find the resistance.

39. ELECTRONICS The power (in watts) lost in a resistor (in the form of heat) varies directly as the square of the current (in amperes) passing through it. The constant of proportionality is the resistance (in ohms). What power is lost in a 5-ohm resistor carrying a 3-ampere current?

40. STRUCTURAL ENGINEERING The deflection of a beam is inversely proportional to its width and the cube of its depth. If the deflection of a 4-inch wide by 4-inch deep beam is 1.1 inches, find the deflection of a 2-inch wide by 8-inch deep beam positioned as in the illustration.

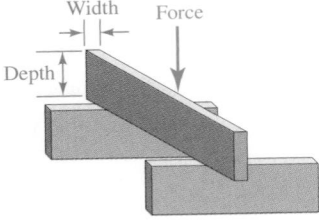

41. STRUCTURAL ENGINEERING Find the deflection of the beam in Exercise 40 when the beam is positioned as in the illustration.

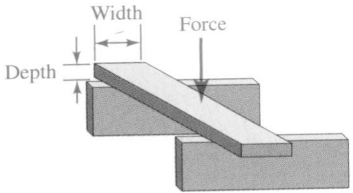

42. TENSION IN A STRING When playing with a Skip-It toy, a child swings a weighted ball on the end of a string in a circular motion around one leg while jumping over the revolving string with the other leg. See the illustration. The tension T in the string is directly proportional to the square of the speed s of the ball and inversely proportional to the radius r of the circle. If the tension in the string is 6 pounds when the speed of the ball is 6 feet per second and the radius is 3 feet, find the tension when the speed is 8 feet per second and the radius is 2.5 feet.

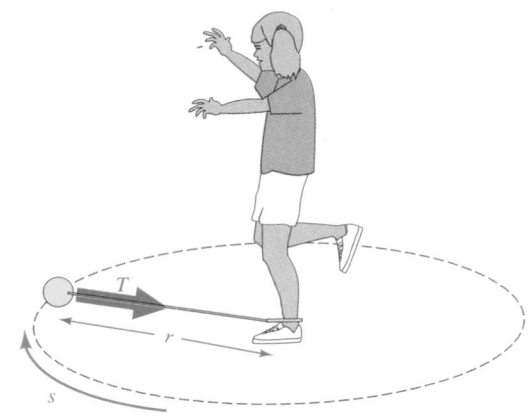

43. GAS PRESSURE   The pressure of a certain amount of gas is directly proportional to the temperature (measured on the Kelvin scale) and inversely proportional to the volume. A sample of gas at a pressure of 1 atmosphere occupies a volume of 1 cubic meter at a temperature of 273 Kelvin. When heated, the gas expands to twice its volume, but the pressure remains constant. To what temperature is it heated?

WRITING

44. Distinguish between a *ratio* and a *proportion*.

45. Give examples of two quantities from everyday life that vary directly and two quantities that vary inversely.

REVIEW

46. Write 35,000 in scientific notation.

47. Write 0.00035 in scientific notation.

48. Write 2.5×10^{-3} in standard notation.

49. Write 2.5×10^4 in standard notation.

CHALLENGE PROBLEMS

50. As the cost of a purchase that is less than $5 increases, the amount of change received from a five-dollar bill decreases. Is this inverse variation? Explain.

51. You've probably heard of Murphy's first law:

> If anything can go wrong, it will.

Another of Murphy's laws is:

> The chances of a piece of bread falling with the grape-jelly side down varies directly with the cost of the carpet.

Write one of your own witty sayings using the phrase *varies directly*.

CHAPTER 8
Summary & Review

SECTION 8.1   Review of Solving Linear Equations, Formulas, and Linear Inequalities

DEFINITIONS AND CONCEPTS	EXAMPLES

Strategy for Solving Linear Equations in One Variable

1. Clear the equation of fractions or decimals.
2. Simplify each side of the equation by removing all sets of parentheses and combining like terms.
3. Isolate the variable term on one side of the equation.
4. Isolate the variable.
5. Check the result in the original equation.

Solve: $\dfrac{x-1}{6} + x = \dfrac{2}{3} - \dfrac{x+2}{6}$

$6\left(\dfrac{x-1}{6} + x\right) = 6\left(\dfrac{2}{3} - \dfrac{x+2}{6}\right)$ Multiply both sides by 6 to clear the fractions.

$x - 1 + 6x = 4 - (x + 2)$ Simplify. Don't forget the parentheses.

$x - 1 + 6x = 4 - x - 2$ Remove parentheses.

$7x - 1 = 2 - x$ Combine like terms on each side.

$7x - 1 + x = 2 - x + x$ To eliminate $-x$ on the right side, add x to both sides.

$8x - 1 = 2$ Combine like terms on each side.

$8x - 1 + 1 = 2 + 1$ To isolate the variable term 8x, add 1 to both sides.

$8x = 3$ Simplify each side.

$\dfrac{8x}{8} = \dfrac{3}{8}$ Isolate the variable x by dividing both sides by 8.

$x = \dfrac{3}{8}$

The solution is $\frac{3}{8}$ and the solution set is $\left\{\frac{3}{8}\right\}$. Check this result to verify that it satisfies the *original* equation.

An **identity** is an equation that is satisfied by every number for which both sides are defined.

When we solve $x + 5 + x = 2x + 5$, the variables drop out and we obtain a true statement $5 = 5$. All real numbers are solutions. The solution set is the set of real numbers denoted $\mathbb{R}$.

A **contradiction** is an equation that is never true.

When we solve $y + 2 = y$, the variables drop out and we obtain a false statement $2 = 0$. The equation has no solutions. The solution set contains no elements and can be denoted as the **empty set** { } or the **null set** $\varnothing$.

To **solve a formula for a specified variable** means to isolate that variable on one side of the equation, with all other variables and constants on the opposite side.

Solve $F = \dfrac{mMg}{r^2}$ for M.

$Fr^2 = mMg$ To clear the fraction, multiply both sides by r^2.

$\dfrac{Fr^2}{mg} = M$ To isolate M, divide both sides by mg.

$M = \dfrac{Fr^2}{mg}$ Write M on the left side.

SECTION 8.1   Review of Solving Linear Equations, Formulas, and Linear Inequalities—*continued*

DEFINITIONS AND CONCEPTS	EXAMPLES

To **solve a linear inequality** in one variable we use **properties of inequality** to find the values of its variable that make the inequality true.

If we multiply or divide both sides of an inequality by a negative number, the direction of the **inequality symbol must be reversed** for the inequalities to have the same solutions.

The set of all solutions of an inequality is called its **solution set**.

Solve: $-5x + 7 > 22$

$$-5x + 7 - 7 > 22 - 7 \quad \text{Subtract 7 from both sides.}$$
$$-5x > 15$$
$$\frac{-5x}{-5} < \frac{15}{-5} \quad \text{Divide both sides by } -5 \text{ and reverse the direction of the inequality symbol.}$$
$$x < -3$$

The solution set is:

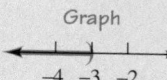

Graph	Interval notation	Set-builder notation
	$(-\infty, -3)$	$\{x \mid x < -3\}$

$-4 \quad -3 \quad -2$

REVIEW EXERCISES

Solve each equation. If an equation is an identity or a contradiction, so indicate.

1. $5x + 12 = 0$

2. $-3x - 7 + x = 6x + 20 - 5x$

3. $4(y - 1) = 28$

4. $2 - 13(x - 1) = 4 - 6x$

5. $\frac{8}{3}(x - 5) = \frac{2}{5}(x - 4)$

6. $\frac{3y}{4} - 14 = -\frac{y}{3} - 1$

7. $2x + 4 = 2(x + 3) - 2$

8. $3x - 2 - x = 2(x - 4)$

9. $-\frac{5}{4}p = 10$

10. $\frac{4t + 1}{3} - \frac{t + 5}{6} = \frac{t - 3}{6}$

Solve each formula for the indicated variable.

11. $V = \pi r^2 h$ for h

12. $v = \frac{1}{6}ab(x + y)$ for x

Solve each inequality. Give each solution set in interval notation and graph it.

13. $0.3x - 0.4 \geq 1.2 - 0.1x$

14. $\frac{7}{4}(x + 3) < \frac{3}{8}(x - 3)$

15. $-16 < -\frac{4}{5}x$

16. $5(2n + 2) - n > 3n - 3(1 - 2n)$

17. CARPENTRY   A carpenter wants to cut a 20-foot rafter so that one piece is 3 times as long as the other. Where should he cut the board?

18. GEOMETRY   A rectangle is 4 meters longer than it is wide. If the perimeter of the rectangle is 28 meters, find its length and width.

SECTION 8.2   Solving Compound Inequalities

DEFINITIONS AND CONCEPTS	EXAMPLES

The **intersection** of two sets A and B, written $A \cap B$, is the set of all elements that are common to set A and set B.

The **union** of two sets A and B, written $A \cup B$, is the set of all elements that are in set A, set B, or both.

Let $A = \{-2, 0, 3, 5\}$ and $B = \{-3, 0, 5, 7\}$.

$A \cap B = \{0, 5\}$ The intersection contains the elements that the sets have in common.

$A \cup B = \{-3, -2, 0, 3, 5, 7\}$ The union contains the elements that are in one or the other set, or both.

SECTION 8.2   Solving Compound Inequalities—*continued*

DEFINITIONS AND CONCEPTS	EXAMPLES
When the word *and* or the word *or* is used to connect pairs of inequalities, we call the statement a **compound inequality**. The solution set of a **compound inequality containing the word *and*** includes all numbers that make both of the inequalities true. That is, it is the intersection of their solution sets.	Solve: $2x - 1 \le 5$ and $5x + 1 > 4$ We solve each inequality separately. Then we graph the two solution sets on the same number line and determine their intersection. $\begin{array}{cc} 2x - 1 \le 5 & \text{and} \quad 5x + 1 > 4 \\ 2x \le 6 & 5x > 3 \\ x \le 3 & x > \dfrac{3}{5} \end{array}$ 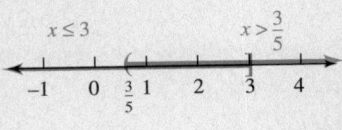 The purple-shaded interval is where the red and blue graphs overlap. Thus, the solution set is: Interval notation: $\left(\dfrac{3}{5}, 3\right]$
Inequalities that contain exactly two inequality symbols are called **double inequalities**. Any double linear inequality can be written as a compound inequality containing the word *and*. For example: $c < x < d$ is equivalent to $c < x$ and $x < d$	Solve: $-7 \le 3x - 1 < 5$ We apply properties of inequality to *all three of its parts* to isolate x in the middle. $\begin{aligned} -7 &\le 3x - 1 < 5 \\ -7 + 1 &\le 3x - 1 + 1 < 5 + 1 \qquad \text{Add 1 to all three parts.} \\ -6 &\le 3x < 6 \\ \dfrac{-6}{3} &\le \dfrac{3x}{3} < \dfrac{6}{3} \qquad\qquad \text{Divide each part by 3.} \\ -2 &\le x < 2 \end{aligned}$ The solution set is: Interval notation: $[-2, 2)$
The solution set of a **compound inequality containing the word *or*** includes all numbers that make one or the other, or both, inequalities true. That is, it is the union of their solution sets.	Solve: $2x - 1 > 5$ or $-(5x - 7) \ge 2$ We solve each inequality separately. Then we graph the two solution sets on the same number line to show their union. $\begin{array}{cc} 2x - 1 > 5 & \text{or} \quad -(5x - 7) \ge 2 \\ 2x > 6 & -5x + 7 \ge 2 \\ x > 3 & -5x \ge -5 \\ & x \le 1 \end{array}$ The solution set is: Interval notation: $(-\infty, 1] \cup (3, \infty)$ This is the union of two intervals.

REVIEW EXERCISES

Let $A = \{-6, -3, 0, 3, 6\}$ and $B = \{-5, -3, 3, 8\}$.

19. Find $A \cap B$.

20. Find $A \cup B$.

Determine whether -4 is a solution of the compound inequality.

21. $x < 0$ and $x > -5$

22. $x + 3 < -3x - 1$ and $4x - 3 > 3x$

Graph each set.

23. $(-3, 3) \cup [1, 6]$

24. $(-\infty, 2] \cap [1, 4)$

Solve each compound inequality. Graph the solution set and write it using interval notation.

25. $-2x > 8$ and $x + 4 \ge -6$

26. $5(x + 2) \le 4(x + 1)$ and $11 + x < 0$

27. $\frac{2}{5}x - 2 < -\frac{4}{5}$ and $\frac{x}{-3} < -1$

28. $4\left(x - \frac{1}{4}\right) \le 3x - 1$ and $x \ge 0$

Solve each double inequality. Graph the solution set and write it using interval notation.

29. $3 < 3x + 4 < 10$ **30.** $-2 \le \frac{5 - x}{2} \le 2$

Determine whether −4 is a solution of the compound inequality.

31. $x < 1.6$ or $x > -3.9$

32. $x + 1 < 2x - 1$ or $4x - 3 > 3x$

Solve each compound inequality. Graph the solution set and write it using interval notation.

33. $x + 1 < -4$ or $x - 4 > 0$ **34.** $\frac{x}{2} + 3 > -2$ or $4 - x > 4$

35. RUGS   A manufacturer makes a line of decorator rugs that are 4 feet wide and of varying lengths x (in feet). The floor area covered by the rugs ranges from 17 ft² to 25 ft². Write and then solve a double linear inequality to find the range of the lengths of the rugs.

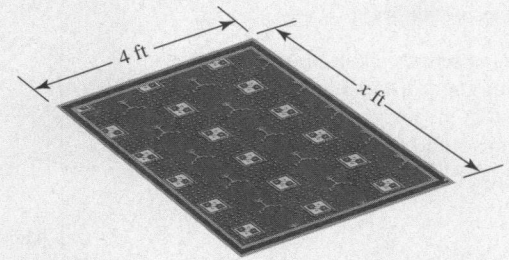

36. Match each word in Column I with *two* associated items in Column II.

Column I	Column II
a. or	**i.** ∩
	ii. ∪
b. and	**iii.** intersection
	iv. union

SECTION 8.3   Solving Absolute Value Equations and Inequalities

DEFINITIONS AND CONCEPTS	EXAMPLES
To **solve absolute value equations** of the form $\lvert X \rvert = k$, where $k > 0$, solve the equivalent **compound equation** $\quad X = k \quad$ or $\quad X = -k$ If k is negative, then $\lvert X \rvert = k$ has no solution.	Solve: $\lvert 2x + 1 \rvert = 7$ This absolute value equation is equivalent to the following compound equation, which we can solve: $\quad 2x + 1 = 7 \quad$ or $\quad 2x + 1 = -7$ $\qquad 2x = 6 \qquad\qquad\quad 2x = -8$ $\qquad\quad x = 3 \qquad\qquad\qquad x = -4$ This equation has two solutions: 3 and −4. The solution set is $\{-4, 3\}$. Solve: $\lvert 4x - 5 \rvert = -3$ Since an absolute value can never be negative, there are no real numbers x that make $\lvert 4x - 5 \rvert = -3$ true. The equation has no solution and the solution set is $\varnothing$.
To **solve absolute value equations** of the form $\lvert X \rvert = \lvert Y \rvert$, solve the compound equation $\quad X = Y \quad$ or $\quad X = -Y$	Solve: $\lvert 3x - 2 \rvert = \lvert 2x + 4 \rvert$ This equation is equivalent to the following compound equation, which we can solve: $\quad 3x - 2 = 2x + 4 \quad$ or $\quad 3x - 2 = -(2x + 4)$ $\qquad x - 2 = 4 \qquad\qquad\quad 3x - 2 = -2x - 4$ $\qquad\quad x = 6 \qquad\qquad\qquad 5x - 2 = -4$ $\qquad\qquad\qquad\qquad\qquad\qquad 5x = -2$ $\qquad\qquad\qquad\qquad\qquad\qquad\;\; x = -\frac{2}{5}$ This equation has two solutions: 6 and $-\frac{2}{5}$. The solution set is $\left\{-\frac{2}{5}, 6\right\}$.

SECTION 8.3   Solving Absolute Value Equations and Inequalities—*continued*

DEFINITIONS AND CONCEPTS	EXAMPLES
To **solve absolute value inequalities** of the form $\lvert X \rvert < k$, where $k > 0$, solve the equivalent double inequality $-k < X < k$. Use a similar approach to solve $\lvert X \rvert \le k$.	Solve: $\lvert 4x - 3 \rvert < 9$ This inequality is equivalent to the following double inequality, which we can solve: $$-9 < 4x - 3 < 9$$ $$-6 < 4x < 12 \qquad \text{Add 3 to all three parts.}$$ $$-\frac{3}{2} < x < 3 \qquad \text{Divide each part by 4 and simplify.}$$ The solution set is: Interval notation: $\left(-\frac{3}{2}, 3\right)$
To **solve absolute value inequalities** of the form $\lvert X \rvert \ge k$, where $k > 0$, solve the equivalent compound inequality $X \le -k$ or $X \ge k$. Use a similar approach to solve $X > k$.	Solve: $\lvert 3x + 1 \rvert \ge 7$ This inequality is equivalent to the following compound inequality, which we can solve: $$3x + 1 \le -7 \quad \text{or} \quad 3x + 1 \ge 7$$ $$3x \le -8 \qquad\qquad 3x \ge 6$$ $$x \le -\frac{8}{3} \qquad\qquad x \ge 2$$ Interval notation: $\left(-\infty, -\frac{8}{3}\right] \cup [2, \infty)$ This is the union of two intervals.

REVIEW EXERCISES

Solve each absolute value equation.

37. $\lvert 4x \rvert = 8$

38. $2\lvert 3x + 1 \rvert - 1 = 19$

39. $\left\lvert \dfrac{3}{2}x - 4 \right\rvert - 10 = -1$

40. $\left\lvert \dfrac{2 - x}{3} \right\rvert = -4$

41. $\lvert -4(2x - 6) \rvert = 0$

42. $\left\lvert \dfrac{3}{8} + \dfrac{x}{3} \right\rvert = \dfrac{5}{12}$

43. $\lvert 3x + 2 \rvert = \lvert 2x - 3 \rvert$

44. $\left\lvert \dfrac{2(1 - x) + 1}{2} \right\rvert = \left\lvert \dfrac{3x - 2}{3} \right\rvert$

Solve each absolute value inequality. Graph the solution set and write it using interval notation.

45. $\lvert x \rvert \le 3$

46. $\lvert 2x + 7 \rvert < 3$

47. $2\lvert 5 - 3x \rvert \le 28$

48. $\left\lvert \dfrac{2}{3}x + 14 \right\rvert + 6 < 6$

49. $\lvert x \rvert > 1$

50. $\left\lvert \dfrac{1 - 5x}{3} \right\rvert \ge 7$

51. $\lvert 3x - 8 \rvert - 4 > 0$

52. $\left\lvert \dfrac{3}{2}x - 14 \right\rvert \ge 0$

53. Explain why $\lvert 0.04x - 8.8 \rvert < -2$ has no solution.

54. Explain why the solution set of $\left\lvert \dfrac{3x}{50} + \dfrac{1}{45} \right\rvert \ge -\dfrac{4}{5}$ is the set of all real numbers.

55. PRODUCE Before packing, freshly picked tomatoes are weighed on the scale shown. Tomatoes having a weight w (in ounces) that falls within the highlighted range are sold to grocery stores.

a. Complete the following absolute value inequality that expresses the acceptable weight range:

$|w - \quad | \leq \quad$

b. Solve the inequality from part (a) and express the acceptable weight range using interval notation.

56. Let $f(x) = \frac{1}{3}|6x| - 1$. For what value(s) of x is $f(x) = 5$?

SECTION 8.4   Review of Factoring Methods: GCF, Grouping, Trinomials

DEFINITIONS AND CONCEPTS	EXAMPLES
The first step of factoring a polynomial is to see whether the terms of the polynomial have a common factor. If they do, **factor out the GCF.**	Factor: $14a^4 + 35a^3 - 56a^2 = 7a^2(2a^2 + 5a - 8)$ *Factor out the GCF, $7a^2$.* Use multiplication to check the factorization: $7a^2(2a^2 + 5a - 8) = 14a^4 + 35a^3 - 56a^2$ *This is the original polynomial.*
If an expression has four or more terms, try to factor the expression by **grouping.**	Factor: $ax - 2x + 3a - 6$ $ax - 2x + 3a - 6 = \boxed{ax - 2x} + \boxed{3a - 6}$ *Group the terms.* $= x(a - 2) + 3(a - 2)$ *Factor x from ax − 2x.* *Factor 3 from 3a − 6.* $= (a - 2)(x + 3)$ *Factor out the GCF, a − 2.*
Many trinomials factor as the product of two binomials. To **factor a trinomial** of the form $x^2 + bx + c$, whose **leading coefficient is 1,** find two integers whose product is c and whose sum is b.	Factor: $p^2 + 14p + 45$ We must find two integers whose product is 45 and whose sum is 14. Since $5 \cdot 9 = 45$ and $5 + 9 = 14$, two such numbers are 5 and 9, and we have $p^2 + 14p + 45 = (p + 5)(p + 9)$ **Check:** $(p + 5)(p + 9) = p^2 + 9p + 5p + 45 = p^2 + 14p + 45$
We can use the **trial-and-check method** to factor trinomials with **leading coefficients other than 1.** Write the trinomial as the product of two binomials and determine four integers.	Factor: $2x^2 - 5x - 12$ Since the first term is $2x^2$, the first terms of the binomial factors must be $2x$ and x. $(2x \boxed{})(x \boxed{})$ *Because 2x · x will give $2x^2$.* The second terms of the binomials must be two integers whose product is -12. There are six such pairs: $1(-12), 2(-6), 3(-4), 4(-3), 6(-2),$ and $12(-1)$ The pair in blue gives the correct middle term, $-5x$, when we use the FOIL method to check: *Outer: −8x* $(2x + 3)(x - 4)$ $-8x + 3x = -5x$ *Inner: 3x* Thus, $2x^2 - 5x - 12 = (2x + 3)(x - 4)$.

SECTION 8.4   Review of Factoring Methods: GCF, Grouping, Trinomials—*continued*

DEFINITIONS AND CONCEPTS	EXAMPLES
To factor $ax^2 + bx + c$ by **grouping,** write it as an equivalent four-term polynomial: $$ax^2 + \boxed{}x + \boxed{}x + c$$ The product of these numbers must be ac, and their sum must be b. Then factor the four-term polynomial by grouping. Use the FOIL method to check.	Factor by grouping: $2x^2 - 5x - 12$ We must find two integers whose product is $ac = 2(-12) = -24$ and whose sum is $b = -5$. Two such numbers are -8 and 3. They serve as the coefficients of $-8x$ and $3x$, the two terms that we use to represent the middle term, $-5x$, of the trinomial. $$\begin{aligned} 2x^2 - 5x - 12 &= 2x^2 - 8x + 3x - 12 \quad \text{Express } -5x \text{ as } -8x + 3x. \\ &= 2x(x - 4) + 3(x - 4) \\ &= (x - 4)(2x + 3) \quad \text{Factor out } (x - 4). \end{aligned}$$

REVIEW EXERCISES

Factor, if possible.

57. $z^2 - 11z + 30$

58. $x^4 + 4x^2 + x^2y + 4y$

59. $4a^2 - 5a + 1$

60. $27x^3y^3z^3 + 81x^4y^5z^2 - 90x^2y^3z^7$

61. $15b^2 + 4b - 4$

62. $-x^2 - 3x + 28$

63. $15x^2 - 57xy - 12y^2$

64. $w^8 - w^4 - 90$

65. $r^2y - ar - ry + a + r - 1$

66. $49a^6 + 84a^3b^2 + 36b^4$

67. $3b^2 + 2b + 1$

68. $2a^4 + 4a^3 - 6a^2$

69. Use a substitution to factor: $(s + t)^2 - 2(s + t) + 1$

70. Solve $m_1m_2 = mm_2 + mm_1$ for m_1.

SECTION 8.5   The Difference of Two Squares; the Sum and Difference of Two Cubes

DEFINITIONS AND CONCEPTS	EXAMPLES
The **difference of two squares:** To factor the square of a First quantity minus the square of a Last quantity, multiply the First plus the Last by the First minus the Last. $$F^2 - L^2 = (F + L)(F - L)$$	Factor: $x^2y^2 - 100$ $$\begin{aligned} x^2y^2 - 100 &= (xy)^2 - 10^2 \quad \text{This is a difference of two squares.} \\ &= (xy + 10)(xy - 10) \end{aligned}$$
In general, the **sum of two squares** (with no common factor other than 1) cannot be factored using real numbers.	$x^2 + 100$ and $36y^2 + 49z^4$ are prime polynomials.
The **sum of two cubes:** To factor the cube of a First quantity plus the cube of a Last quantity, multiply the First plus the Last by the First squared, minus the First times the Last, plus the Last squared. $$F^3 + L^3 = (F + L)(F^2 - FL + L^2)$$	Factor: $y^3 + 27z^6$ $$\begin{aligned} y^3 + 27z^6 &= y^3 + (3z^2)^3 \quad \text{This is a sum of two cubes.} \\ &= (y + 3z^2)[y^2 - y \cdot 3z^2 + (3z^2)^2] \\ &= (y + 3z^2)(y^2 - 3yz^2 + 9z^4) \end{aligned}$$
The **difference of two cubes:** To factor the cube of a First quantity minus the cube of a Last quantity, multiply the First minus the Last by the First squared, plus the First times the Last, plus the Last squared. $$F^3 - L^3 = (F - L)(F^2 + FL + L^2)$$	Factor: $125s^3 - 64$ $$\begin{aligned} 125s^3 - 64 &= (5s)^3 - 4^3 \quad \text{This is a difference of two cubes.} \\ &= (5s - 4)[(5s)^2 + 5s \cdot 4 + 4^2] \\ &= (5s - 4)(25s^2 + 20s + 16) \end{aligned}$$

REVIEW EXERCISES

Factor, if possible.

71. $z^2 - 16$

72. $x^2y^4 - 64z^6$

73. $a^2b^2 + c^2$

74. $c^2 - (a + b)^2$

75. $32a^4c - 162b^4c$

76. $k^2 + 2k + 1 - 9m^2$

77. $m^2 - n^2 - m - n$

78. $t^3 + 64$

79. $8a^3 - 125b^9$

80. SPANISH ROOF TILE   The amount of clay used to make a roof tile is given by

$$V = \frac{\pi}{2}r_1{}^2h - \frac{\pi}{2}r_2{}^2h$$

Factor the right side of the formula completely.

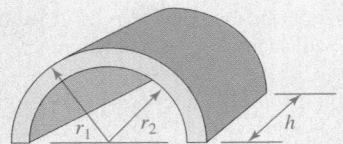

SECTION 8.6   Review of Rational Expressions and Rational Equations

DEFINITIONS AND CONCEPTS	EXAMPLES
To simplify a rational expression: 1. Factor the numerator and denominator completely. 2. Remove factors equal to 1 by replacing each pair of factors common to the numerator and denominator with the equivalent fraction $\frac{1}{1}$. 3. Multiply the remaining factors in the numerator and in the denominator.	Simplify: $\dfrac{x^2 - 4}{2x + 4} = \dfrac{\overset{1}{\cancel{(x + 2)}}(x - 2)}{2\underset{1}{\cancel{(x + 2)}}} = \dfrac{x - 2}{2}$ Simplify: $\dfrac{2a^3 - 5a^2 - 12a}{2a^3 - 11a^2 + 12a} = \dfrac{\overset{1}{\cancel{a}}(2a + 3)\overset{1}{\cancel{(a - 4)}}}{\underset{1}{\cancel{a}}(2a - 3)\underset{1}{\cancel{(a - 4)}}} = \dfrac{2a + 3}{2a - 3}$
To multiply rational expressions, multiply the numerators and multiply the denominators. $$\frac{A}{B} \cdot \frac{C}{D} = \frac{AC}{BD}$$ Then simplify, if possible.	Multiply, and then simplify, if possible. $\dfrac{x^2 - 4}{x + 3} \cdot \dfrac{3x + 9}{x + 2} = \dfrac{(x^2 - 4)(3x + 9)}{(x + 3)(x + 2)}$ Multiply the numerators. Multiply the denominators. $= \dfrac{\overset{1}{\cancel{(x + 2)}}(x - 2) \cdot 3 \cdot \overset{1}{\cancel{(x + 3)}}}{\underset{1}{\cancel{(x + 3)}}\underset{1}{\cancel{(x + 2)}}}$ Factor completely and then simplify. $= 3(x - 2)$
To divide rational expressions, multiply the first by the reciprocal of the second. $$\frac{A}{B} \div \frac{C}{D} = \frac{A}{B} \cdot \frac{D}{C} = \frac{AD}{BC}$$ Then simplify, if possible.	Divide, and then simplify, if possible. $\dfrac{x^2 + 4x + 3}{x^2 + 3x} \div \dfrac{3}{x} = \dfrac{x^2 + 4x + 3}{x^2 + 3x} \cdot \dfrac{x}{3}$ Multiply the first rational expression by the reciprocal of the second. $= \dfrac{(x^2 + 4x + 3) \cdot x}{(x^2 + 3x) \cdot 3}$ Multiply the numerators. Multiply the denominators. $= \dfrac{(x + 1)\overset{1}{\cancel{(x + 3)}} \cdot \overset{1}{\cancel{x}}}{\underset{1}{\cancel{x}}\underset{1}{\cancel{(x + 3)}} \cdot 3}$ Factor completely and then simplify. $= \dfrac{x + 1}{3}$ Multiply the remaining factors in the numerator. Multiply the remaining factors in the denominator.

SECTION 8.6   Review of Rational Expressions and Rational Equations—*continued*

DEFINITIONS AND CONCEPTS	EXAMPLES
To **add (or subtract) two rational expressions with like denominators,** add (or subtract) the numerators and keep the common denominator. Then, if possibe, factor and simplify.	Add: $$\frac{x^2-26}{x-5}+\frac{1}{x-5}=\frac{x^2-26+1}{x-5}$$ Add the numerators. Write the sum over the common denominator, $x-5$. $$=\frac{x^2-25}{x-5}$$ Combine like terms. $$=\frac{(x+5)\overset{1}{\cancel{(x-5)}}}{\underset{1}{\cancel{x-5}}}$$ To simplify the result, factor the numerator and remove the factor common to the numerator and denominator. $$=x+5$$
To **add or subtract rational expressions with unlike denominators,** find the LCD and express each rational expression with a denominator that is the LCD. Add (or subtract) the resulting fractions and simplify the result, if possible.	Subtract: $$\frac{2x}{x+5}-\frac{1}{x}=\frac{2x}{x+5}\cdot\frac{x}{x}-\frac{1}{x}\cdot\frac{x+5}{x+5}$$ Build each rational expression to have the LCD of $x(x+5)$. $$=\frac{2x^2}{x(x+5)}-\frac{x+5}{x(x+5)}$$ Multiply the numerators. Multiply the denominators. $$=\frac{2x^2-(x+5)}{x(x+5)}$$ Subtract the numerators. Write the difference over the common denominator. $$=\frac{2x^2-x-5}{x(x+5)}$$ The result does not simplify.
Two methods are used to simplify **complex fractions.** **Method 1:** Write the numerator and denominator as single fractions. Then divide the fractions and simplify. This method works well when a complex fraction is written, or can be easily written, as a quotient of two single rational expressions.	Simplify: $$\frac{\frac{4x^2}{y^3}}{\frac{14x}{y}}=\frac{4x^2}{y^3}\div\frac{14x}{y}$$ The main fraction bar of the complex fraction indicates division. $$=\frac{4x^2}{y^3}\cdot\frac{y}{14x}$$ To divide rational expressions, multiply the first by the reciprocal of the second. $$=\frac{4x^2\cdot y}{y^3\cdot 14x}$$ Multiply the numerators. Multiply the denominators. $$=\frac{\overset{1}{\cancel{2}}\cdot 2\cdot\overset{1}{\cancel{x}}\cdot x\cdot\overset{1}{\cancel{y}}}{\underset{1}{\cancel{y}}\cdot y\cdot y\cdot\underset{1}{\cancel{2}}\cdot 7\cdot\underset{1}{\cancel{x}}}$$ Factor the numerator and denominator. Then simplify by removing common factors of the numerator and denominator. $$=\frac{2x}{7y^2}$$ Multiply the remaining factors in the numerator. Multiply the remaining factors in the denominator.

SECTION 8.6 Review of Rational Expressions and Rational Equations—*continued*

DEFINITIONS AND CONCEPTS	EXAMPLES

Method 2: Determine the LCD of all the rational expressions in the complex fraction and multiply the complex fraction by 1, written in the form $\frac{LCD}{LCD}$.

This method works well when the complex fraction has sums and/or differences in the numerator or denominator.

Simplify:

$$\frac{\dfrac{1}{x} - y}{\dfrac{5}{2x}} = \frac{\dfrac{1}{x} - y}{\dfrac{5}{2x}} \cdot \frac{2x}{2x}$$

The LCD of all the rational expressions in the complex fraction is 2x. Multiply the complex fraction by 1 in the form $\frac{2x}{2x}$.

$$= \frac{\left(\dfrac{1}{x} - y\right)2x}{\left(\dfrac{5}{2x}\right)2x}$$

Multiply the numerators.
Multiply the denominators.

$$= \frac{\dfrac{1}{x} \cdot 2x - y \cdot 2x}{5}$$

In the numerator, distribute the multiplication by 2x. In the denominator, perform the multiplication by 2x.

$$= \frac{2 - 2xy}{5}$$

In the numerator, perform each multiplication by 2x.

To solve a rational equation:

1. Factor all denominators.

2. Determine which numbers cannot be solutions of the equation.

3. Multiply both sides of the equation by the LCD of all rational expressions in the equation.

4. Use the distributive property to remove parentheses, remove any factors equal to 1, and write the result in simplified form.

5. Solve the resulting equation.

6. Check all possible solutions in the original equation.

All possible solutions of a rational equation must be checked. Multiplying both sides of an equation by a quantity that contains a variable can lead to **extraneous solutions.**

Solve: $\dfrac{3}{2} + \dfrac{1}{a - 4} = \dfrac{5}{2a - 8}$

If we factor the last denominator, the equation can be written as:

$$\frac{3}{2} + \frac{1}{a - 4} = \frac{5}{2(a - 4)}$$

We see that 4 cannot be a solution of the equation, because it makes at least one of the rational expressions in the equation undefined.

We can clear the equation of fractions by multiplying both sides by $2(a - 4)$, which is the LCD of the three rational expressions.

$$2(a - 4)\left(\frac{3}{2} + \frac{1}{a - 4}\right) = 2(a - 4)\left[\frac{5}{2(a - 4)}\right]$$ Multiply both sides by the LCD.

$$2(a - 4)\left(\frac{3}{2}\right) + 2(a - 4)\left(\frac{1}{a - 4}\right) = 2(a - 4)\left[\frac{5}{2(a - 4)}\right]$$ Distribute.

$$\overset{1}{2}(a - 4)\left(\frac{3}{\underset{1}{2}}\right) + 2(a \overset{1}{-} 4)\left(\frac{1}{a \underset{1}{-} 4}\right) = \overset{1}{2}(a \overset{1}{-} 4)\left[\frac{5}{\underset{1}{2}(a \underset{1}{-} 4)}\right]$$ Remove common factors.

$$(a - 4)3 + 2 = 5$$ Simplify.

$$3a - 12 + 2 = 5$$ Distribute.

$$3a - 10 = 5$$ Combine like terms.

$$3a = 15$$

$$a = 5$$

The solution is 5. Verify that it satisfies the original equation.

REVIEW EXERCISES

Simplify each rational expression.

81. $\dfrac{62x^2y}{144xy^2}$

82. $\dfrac{2m - 2n}{n - m}$

Perform the operations and simplify, if possible.

83. $\dfrac{3x^3y^4}{c^2d} \cdot \dfrac{c^3d^2}{21x^5y^4}$

84. $\dfrac{2a^2 - 5a - 3}{a^2 - 9} \div \dfrac{2a^2 + 5a + 2}{2a^2 + 5a - 3}$

85. $\dfrac{m^2 + 3m + 9}{m^2 + mp + mr + pr} \div \dfrac{m^3 - 27}{am + ar + bm + br}$

86. $\dfrac{x^3 + 3x^2 + 2x}{2x^2 - 2x - 12} \cdot \dfrac{3x^2 - 3x}{x^3 - 3x^2 - 4x} \div \dfrac{x^2 + 3x + 2}{2x^2 - 4x - 16}$

87. $\dfrac{d^2}{c^3 - d^3} + \dfrac{c^2 + cd}{c^3 - d^3}$

88. $\dfrac{4}{t - 3} + \dfrac{6}{3 - t}$

89. $\dfrac{5x}{14z^2} + \dfrac{y^2}{16z}$

90. $\dfrac{4}{3xy - 6y} - \dfrac{4}{10 - 5x}$

91. $\dfrac{y + 7}{y + 3} - \dfrac{y - 3}{y + 7}$

92. $\dfrac{2x}{x + 1} + \dfrac{3x}{x + 2} + \dfrac{4x}{x^2 + 3x + 2}$

Simplify each complex fraction.

93. $\dfrac{\dfrac{4a^3b^2}{9c}}{\dfrac{14a^3b}{9c^4}}$

94. $\dfrac{\dfrac{p^2 - 9}{6pt}}{\dfrac{p^2 + 5p + 6}{3pt}}$

95. $\dfrac{1 - \dfrac{1}{x} - \dfrac{2}{x^2}}{1 + \dfrac{4}{x} + \dfrac{3}{x^2}}$

96. $\dfrac{\dfrac{2}{b^2 - 1} - \dfrac{3}{ab - a}}{\dfrac{3}{ab - a} - \dfrac{2}{b^2 - 1}}$

Solve each equation. If a solution is extraneous, so indicate.

97. $\dfrac{4}{x} - \dfrac{1}{10} = \dfrac{7}{2x}$

98. $\dfrac{3}{y} - \dfrac{2}{y + 1} = \dfrac{1}{2}$

99. $\dfrac{2}{3x + 15} - \dfrac{1}{18} = \dfrac{1}{3x + 12}$

100. $\dfrac{3}{x + 2} = \dfrac{1}{2 - x} + \dfrac{2}{x^2 - 4}$

101. $\dfrac{x + 3}{x - 5} + \dfrac{2x^2 + 6}{x^2 - 7x + 10} = \dfrac{3x}{x - 2}$

102. $\dfrac{5a}{a - 3} - 7 = \dfrac{15}{a - 3}$

SECTION 8.7   Review of Linear Equations in Two Variables

DEFINITIONS AND CONCEPTS	EXAMPLES

DEFINITIONS AND CONCEPTS

The **standard** or **general form** of a **linear equation** in two variables is $Ax + By = C$, where A, B, and C are real numbers and A and B are not both zero.

The **graph of an equation** is the graph of all points on the rectangular coordinate system whose coordinates satisfy the equation.

The graph of a linear equation in two variables is a line.

To find the **y-intercept** of a line, substitute 0 for x in the equation and solve for y. To find the **x-intercept** of a line, substitute 0 for y in the equation and solve for x.

Plotting the x- and y-intercepts of a graph and drawing a line through them is called the **intercept method for graphing a line.**

EXAMPLES

Graph $2x - 5y = 10$ using the intercept method.

To find the x-intercept, substitute 0 for y and solve for x.

$$2x - 5y = 10$$
$$2x - 5(0) = 10$$
$$2x = 10$$
$$x = 5$$

To find the y-intercept, substitute 0 for x and solve for y.

$$2x - 5y = 10$$
$$2(0) - 5y = 10$$
$$-5y = 10$$
$$y = -2$$

The x-intercept is $(5, 0)$. The y-intercept is $(0, -2)$.

We can find a third point on the line as a check point and draw a line through the three points to get the graph.

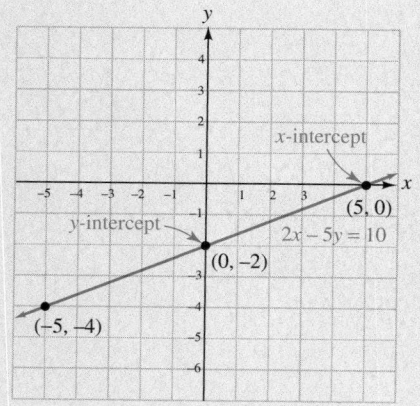

SECTION 8.7   Review of Linear Equations in Two Variables—*continued*

DEFINITIONS AND CONCEPTS	EXAMPLES
The **slope** m of a line is a ratio that compares the vertical and horizontal change as we move along the line from one point to another. $$m = \frac{\text{vertical change}}{\text{horizontal change}} = \frac{\text{rise}}{\text{run}} = \frac{\Delta y}{\Delta x}$$ The **slope formula:** $$m = \frac{y_2 - y_1}{x_2 - x_1} \quad \text{if } x_1 \neq x_2$$ Horizontal lines have a slope of 0. Vertical lines have no defined slope.	Find the slope of the line passing through $(-5, -2)$ and $(7, -14)$. $$m = \frac{y_2 - y_1}{x_2 - x_1} = \frac{-14 - (-2)}{7 - (-5)} = \frac{-12}{12} = -1$$
Equations of a line: **Point–slope form:** $$y - y_1 = m(x - x_1)$$ **Slope–intercept form:** $$y = mx + b$$ **Standard (general) form:** $$Ax + By = C$$ **Horizontal line:** **Vertical line:** $$x = a \qquad\qquad y = b$$	Find an equation of a line with slope 2 that passes through $(4, 5)$. Write the equation in slope–intercept form. $$y - y_1 = m(x - x_1)$$ $$y - 5 = 2(x - 4) \quad \text{Substitute 5 for } y_1, \text{4 for } x_1, \text{ and 2 for } m.$$ To write the equation in slope–intercept form, we solve for y. $$y - 5 = 2x - 8$$ $$y = 2x - 8 + 5$$ $$y = 2x - 3$$ To write the equation in standard form, we proceed as follows: $$y = 2x - 3 \quad \text{This is slope–intercept form.}$$ $$3 = 2x - y \quad \text{Add 3 and subtract } y \text{ from both sides.}$$ $$2x - y = 3 \quad \text{This is standard form.}$$
Slopes can be used to identify **parallel** and **perpendicular lines.**	The graphs of the lines $y = 7x + 5$ and $y = 7x - 3$ are parallel because each line has slope 7. The graphs of the lines $y = 6x + 1$ and $y = -\frac{1}{6}x$ are perpendicular because their slopes are 6 and $-\frac{1}{6}$, which are negative reciprocals.
The **midpoint** of a line segment with endpoints (x_1, y_1) and (x_2, y_2) is the point with coordinates $$\left(\frac{x_1 + x_2}{2}, \frac{y_1 + y_2}{2} \right)$$	Find the midpoint of the segment joining $(-3, 7)$ and $(5, -8)$. We let $(x_1, y_1) = (-3, 7)$ and $(x_2, y_2) = (5, -8)$ and substitute the coordinates into the midpoint formula. $$\left(\frac{x_1 + x_2}{2}, \frac{y_1 + y_2}{2} \right) = \left(\frac{-3 + 5}{2}, \frac{7 + (-8)}{2} \right)$$ $$= \left(\frac{2}{2}, -\frac{1}{2} \right)$$ $$= \left(1, -\frac{1}{2} \right) \quad \text{This is the midpoint.}$$

REVIEW EXERCISES

Graph each equation.

103. $y = -\dfrac{1}{3}x - 1$ **104.** $x = -2$

Graph each equation using the intercept method.

105. $2x + y = 4$ **106.** $3x - 4y - 8 = 0$

107. Find the slope of the graph of $2x - 3y = 18$.

108. Find the slope of lines l_1 and l_2 in the illustration.

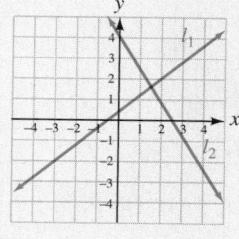

Find the slope of the line passing through the given points.

109. $(2, 5)$ and $(5, 8)$ **110.** $(3, -2)$ and $(-6, 12)$

111. $(-2, 4)$ and $(8, 4)$ **112.** $(-5, -4)$ and $(-5, 8)$

Determine whether the lines with the given slopes are parallel, perpendicular, or neither.

113. $m_1 = 4$, $m_2 = -\dfrac{1}{4}$ **114.** $m_1 = 0.5$, $m_2 = \dfrac{1}{2}$

Find an equation of the line with the given properties. Write the equation in slope–intercept form.

115. Slope of 3; passing through $(-8, 5)$

116. Passing through $(-2, 4)$ and $(6, -9)$

Find an equation of the line with the given properties. Write the equation in slope–intercept form.

117. Passes through $(-3, -5)$; parallel to the graph of $3x - 2y = 7$

118. Passes through $(-3, -5)$; perpendicular to the graph of $3x - 2y = 7$

119. ACCOUNTING   A business purchases a copy machine for $8,700 and will depreciate it on a straight-line basis over the next 5 years. At the end of its useful life, it will be sold as scrap for $100. Find its depreciation equation.

120. Find the midpoint of the line segment joining $(-3, 5)$ and $(6, 11)$.

SECTION 8.8   Functions

DEFINITIONS AND CONCEPTS	EXAMPLES
A **function** is a set of ordered pairs (a relation) in which to each first component there corresponds exactly one second component. Since we often work with sets of ordered pairs of the form (x, y), it is helpful to define a function using the variables x and y: **y is a function of x:** Given a relation in x and y, if to each value of x in the domain there corresponds exactly one value of y in the range, then y is said to be a function of x. Since y depends on x, we call x the **independent variable** and y the **dependent variable**.	The arrow diagram does not define y as a function of x because to the x-value 4 there corresponds more than one y-value: 2 and 6. 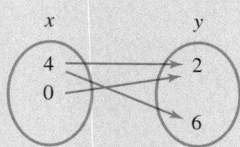 The equation $x = \lvert y \rvert$ does not define y as a function of x because more than one value of y corresponds to a single value of x. If x is 2, for example, the equation becomes $2 = \lvert y \rvert$ and y can be either 2 or -2.
The **function notation** $y = f(x)$ indicates that the variable y is a function of x. It is read as "f of x." Think of a function as a machine that takes some **input** x and turns it into some **output** $f(x)$, called a **function value**.	If $f(x) = 2x + 1$, find $f(-2)$ and $f(n + 1)$. $f(x) = 2x + 1$ $f(x) = 2x + 1$ $f(-2) = 2(-2) + 1$ $f(n + 1) = 2(n + 1) + 1$ $= -4 + 1$ $= 2n + 2 + 1$ $= -3$ $= 2n + 3$ Thus, $f(-2) = -3$. Thus, $f(n + 1) = 2n + 3$.

SECTION 8.8 Functions—*continued*

DEFINITIONS AND CONCEPTS	EXAMPLES
The input-output pairs that a function generates can be written as ordered pairs and plotted on a rectangular coordinate system to give the **graph of the function.** A **linear function** is a function that can be written in the form $f(x) = mx + b$. The graph of a linear function is a straight line.	Graph the linear function $f(x) = -4x - 2$. We make a table of values, plot the points, and draw the graph. 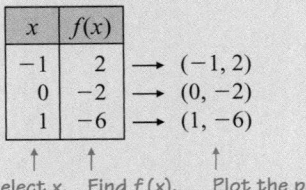 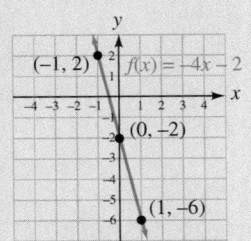 Select x. Find f(x). Plot the point.
The **domain** of a function is the set of input values. The **range** is the set of output values.	Find the domain of: $f(x) = \dfrac{10}{x + 3}$ The number -3 cannot be substituted for x, because that would make the denominator equal to 0. Since any real number except -3 can be substituted for x, the domain is *the set of all real numbers except* -3.
The **vertical line test:** If a vertical line intersects a graph in more than one point, the graph is not the graph of a function.	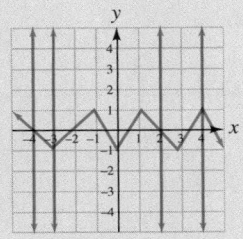 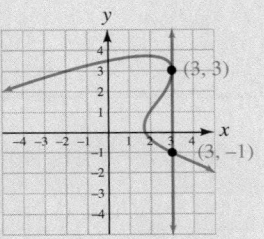 The graph of a function Not the graph of a function
These three basic functions are used so often in algebra that you should memorize their names and their graphs. They are called **nonlinear functions** because their graphs are not lines.	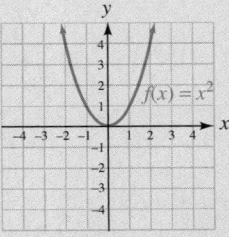 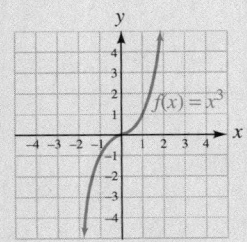 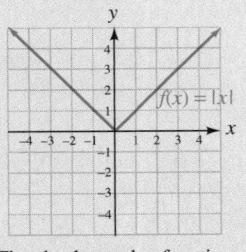 The squaring function The cubing function The absolute value function
We can find the domain and range of a function from its graph. The **domain of a function** is the *projection* of its graph onto the *x*-axis. **The range of a function** is the *projection* of its graph onto the *y*-axis.	Find the domain and range of function f. 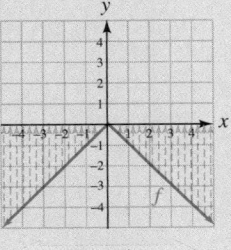 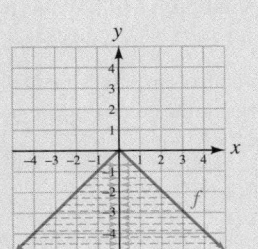 Domain R: The set of nonpositive real numbers D: The set of real numbers

SECTION 8.8 Functions–*continued*

DEFINITIONS AND CONCEPTS	EXAMPLES
A **vertical translation** shifts a graph upward or downward. A **horizontal translation** shifts a graph left or right. A **reflection** "flips" a graph about the x-axis.	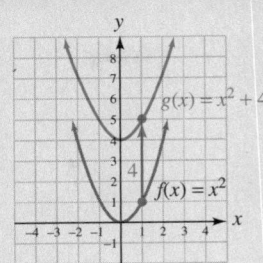 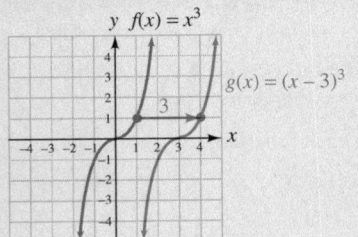 To graph $g(x) = x^2 + 4$, translate each point on the graph of $f(x) = x^2$ up 4 units. To graph $g(x) = (x - 3)^3$, translate each point on the graph of $f(x) = x^3$ to the right 3 units.

REVIEW EXERCISES

121. Fill in the blanks.

   a. A _____ is a set of ordered pairs (a relation) in which to each first component there corresponds exactly one second component.

   b. Given a relation in x and y, if to each value of x in the domain there corresponds exactly one value of y in the range, y is said to be a _____ of x. We call x the independent _____ and y the _____ variable.

122. Determine whether the relation defines y as a function of x. If it does not, find two ordered pairs where more than one value of y corresponds to a single value of x.

   a. x y **b.**

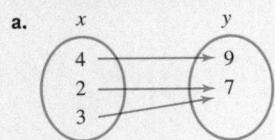

x	y
-1	8
0	5
4	1
-1	9

   c. $\{(14, 6), (-1, 14), (6, 0), (-3, 8)\}$

Determine whether each equation determines y to be a function of x. If it does not, find two ordered pairs where more than one value of y corresponds to a single value of x.

123. $y = 6x - 4$ **124.** $|y| = x$

Let $f(x) = 3x + 2$ and $g(a) = \dfrac{a^2 - 4a + 4}{2}$. Find each function value.

125. $f(-3)$ **126.** $g(8)$

127. $g(-2)$ **128.** $f(t)$

129. Let $f(x) = -5x + 7$. For what value x of is $f(x) = -8$?

130. Let $g(x) = \frac{3}{4}x - 1$. For what value of x is $g(x) = 0$?

Find the domain and range of each function.

131. $f(x) = 4x - 1$

132. $f(x) = \dfrac{4}{2 - x}$

Determine whether each graph represents a function. If it does not, find two ordered pairs where more than one value of y corresponds to a single value of x.

133.   **134.**

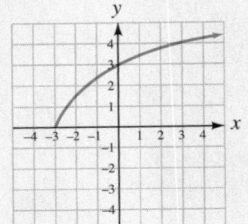   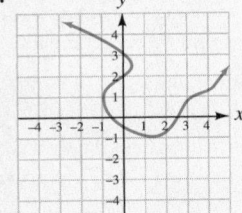

135. Graph: $f(x) = \dfrac{2}{3}x - 2$

136. Use the graph in the illustration to find each value.

   a. $f(-2)$

   b. $f(3)$

   c. The value(s) for which $f(x) = -1$

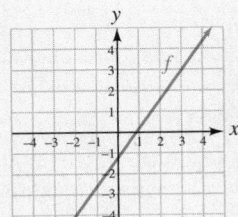

137. Fill in the blanks.

    a. The graph of $f(x) = x^2 + 6$ is the same as the graph of $f(x) = x^2$ except that it is shifted ___ units ___.

    b. The graph of $f(x) = (x + 6)^2$ is the same as the graph of $f(x) = x^2$ except that it is shifted ___ units to the ___.

138. Graph $f(x) = |x + 2|$, $g(x) = |x| - 3$, and $h(x) = -|x|$ on the same coordinate system.

Graph each function.

139. $f(x) = x^2 - 3$ **140.** $f(x) = (x - 2)^3 + 1$

Give the domain and range of each function.

141.

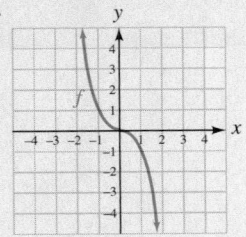

142.

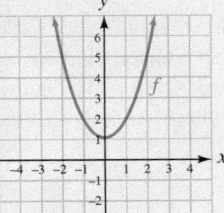

SECTION 8.9   Variation

DEFINITIONS AND CONCEPTS	EXAMPLES
The words *y varies directly as x* or *y is directly proportional to x* mean that $y = kx$ for some nonzero constant k, called the **constant of variation**. The words *y varies inversely as x* or *y is inversely proportional to x* mean that $y = \frac{k}{x}$ for some nonzero constant k.	The distance d that a spring stretches *varies directly* as the force f attached to the spring: $d = kf$. If the voltage in an electric circuit is kept constant, the current I *varies inversely* as the resistance R: $$I = \frac{k}{R}$$
Strategy for Solving Variation Problems 1. Translate the verbal model into an equation. 2. Substitute the first set of values into the equation from step 1 to determine the value of k. 3. Substitute the value of k into the equation from step 1. 4. Substitute the remaining set of values into the equation from step 3 and solve for the unknown.	Suppose d varies inversely as h. If $d = 5$ when $h = 4$, find d when $h = 10$. 1. The words d *varies inversely as* h translate to $d = \frac{k}{h}$. 2. If we substitute 5 for d and 4 for h, we have $$5 = \frac{k}{4}$$ $$20 = k \quad \text{To find } k, \text{ multiply both sides by 4.}$$ 3. Since $k = 20$, the inverse variation equation is $d = \frac{20}{h}$. 4. To answer the final question, we substitute 10 for h in the inverse variation model: $$d = \frac{20}{10} = 2$$
Joint variation: One variable varies as the product of several variables. For example, $y = kxz$ (k is a constant).	The number of gallons g of oil that can be stored in a cylindrical tank *varies jointly* as the height h of the tank and the square of the radius r of its base: $g = khr^2$.
Combined variation: A combination of direct and inverse variation. For example, $$y = \frac{kx}{z} \quad (k \text{ is a constant})$$	The gravitational force F between two objects with masses m_1 and m_2 *varies directly* as the product of their masses and *inversely* as the square of the distance d between them: $$F = \frac{km_1m_2}{d^2}$$

REVIEW EXERCISES

143. PROPERTY TAX The property tax in a certain county varies directly as assessed valuation. If a tax of $1,575 is charged on a single-family home assessed at $90,000, determine the property tax on an apartment complex assessed at $312,000.

144. ELECTRICITY For a fixed voltage, the current in an electrical circuit varies inversely as the resistance in the circuit. If a certain circuit has a current of $2\frac{1}{2}$ amps when the resistance is 150 ohms, find the current in the circuit when the resistance is doubled.

145. Assume that y varies jointly with x and z. Find the constant of variation if $x = 24$ when $y = 3$ and $z = 4$.

146. HURRICANE WINDS The wind force on a vertical surface varies jointly as the area of the surface and the square of the wind's velocity. If a 10-mph wind exerts a force of 1.98 pounds on the sign shown here, find the force on the sign when the wind is blowing at 80 mph.

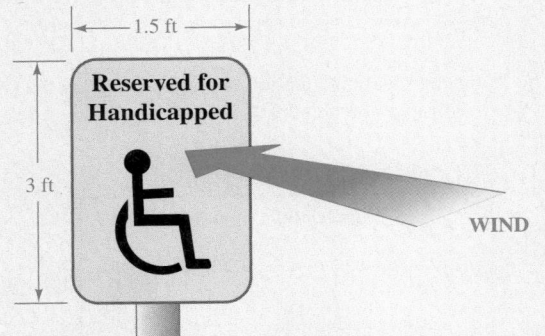

147. Does the graph below show direct or inverse variation?

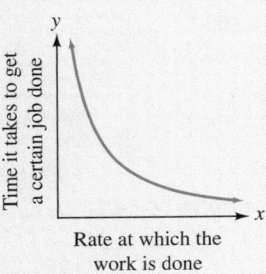

148. Assume that x_1 varies directly with the third power of t and inversely with x_2. Find the constant of variation if $x_1 = 1.6$ when $t = 8$ and $x_2 = 64$.

CHAPTER 8
TEST

1. Fill in the blanks.

  a. To _____ an equation means to find all of the values of the variable that make the equation true.

  b. $<, >, \leq,$ and $\geq$ are _____ symbols.

  c. The statement $x^2 - x - 12 = (x - 4)(x + 3)$ shows that the trinomial $x^2 - x - 12$ factors as the product of two _____.

  d. The _____ of $\frac{x + 1}{x - 7}$ is $\frac{x - 7}{x + 1}$.

  e. Given a relation in x and y, if to each value of x in the domain there corresponds exactly one value of y in the range, y is said to be a _____ of x. We call x the independent _____ and y the _____ variable.

2. Use a check to determine whether 6.7 is a solution of $1.6y + (-3) = y + 1.02$.

Solve each equation.

3. $t + 18 = 5t - 3 + t$

4. $\frac{2}{3}(2s + 2) = \frac{1}{6}(5s + 29) - 4$

5. $6 - (x - 3) - 5x = 3[1 - 2(x + 2)]$

6. Solve $n = \frac{360}{180 - a}$ for a.

7. CALCULATORS The viewing window of a calculator has a perimeter of 26 centimeters and is 5 centimeters longer than it is wide. Find the dimensions of the window.

8. AVERAGING GRADES Use the information from the gradebook to determine what score Karen Nelson-Sims needs on the fifth exam so that her exam average exceeds 80.

Sociology 101 8:00-10:00 pm MW	Exam 1	Exam 2	Exam 3	Exam 4	Exam 5
Nelson-Sims, Karen	70	79	85	88	

Solve each inequality. Write the solution set in interval notation and graph it.

9. $-2(2x + 3) \geq 14$

10. $-2 < \dfrac{x - 4}{3} < 4$

11. $3x \geq -2x + 5$ and $7 \geq 4x - 2$

12. $3x < -9$ or $-\dfrac{x}{4} < -2$

13. $|2x - 4| > 22$

14. $2|3(x - 2)| \leq 4$

Solve each equation.

15. $|2x + 3| - 19 = 0$

16. $|3x + 4| = |x + 12|$

Factor, if possible.

17. $12a^3b^2c - 3a^2b^2c^2 + 6abc^3$

18. $4y^4 - 64$

19. $b^3 + 125$

20. $6u^2 + 9u - 6$

21. $ax - xy + ay - y^2$

22. $25m^8 - 60m^4n + 36n^2$

23. $144b^2 + 25$

24. $x^2 + 6x + 9 - y^2$

25. $64a^3 - 125b^6$

26. $(x - y)^2 + 3(x - y) - 10$

Simplify each rational expression.

27. $\dfrac{3y - 6z}{2z - y}$

28. $\dfrac{2x^2 + 7xy + 3y^2}{4xy + 12y^2}$

Perform the operations and simplify, if possible.

29. $\dfrac{x^3 + y^3}{4} \div \dfrac{x^2 - xy + y^2}{2x + 2y}$

30. $\dfrac{xu + 2u + 3x + 6}{u^2 - 9} \cdot \dfrac{13u - 39}{x^2 + 3x + 2}$

31. $\dfrac{-3t + 4}{t^2 + t - 20} + \dfrac{5t + 6}{t^2 + t - 20}$

32. $\dfrac{a + 3}{a^2 - a - 2} - \dfrac{a - 4}{a^2 - 2a - 3}$

Simplify each complex fraction.

33. $\dfrac{\dfrac{2u^2w^2}{v^2}}{\dfrac{4uw^4}{uv}}$

34. $\dfrac{\dfrac{4}{3k} + \dfrac{k}{k + 1}}{\dfrac{k}{k + 1} - \dfrac{3}{k}}$

35. Find an equation of the line that passes through $(-2, 1)$ and is parallel to the graph of $y = -\dfrac{3}{2}x - 7$. Write the equation in slope–intercept form.

36. Find the slope and the y-intercept of the graph of $2x + 9 = -6y$.

37. Find the slope of the line passing through $(-3, 5)$ and $(4, -6)$.

38. ACCOUNTING After purchasing a new color copier, a business owner had his accountant prepare a depreciation worksheet for tax purposes. (See the illustration.)

   a. Assuming straight-line depreciation, write an equation that gives the value v of the copier after x years of use.

   b. If the depreciation equation is graphed, explain the significance of its v-intercept.

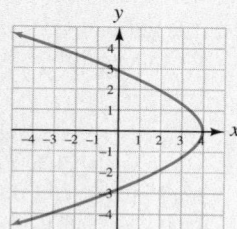

Depreciation Worksheet

Color copier . . . . . . $4,000 (new)

Salvage value . . . . . . $400 (in 6 years)

39. Find the x- and y-intercepts of the graph of $2x - 5y = 10$. Then graph the equation.

40. Graph: $y = -2$

41. Does the table define y as a function of x?

x	y
−3	4
4	−3
1	4
2	5

42. Determine whether the graph represents a function.

43. Let $g(x) = x^2 - 2x - 1$. Find: $g(0)$

44. Let $f(x) = -\frac{4}{5}x - 12$. For what value of x is $f(x) = 4$?

45. Graph: $g(x) = -|x + 2|$

46. Give the domain and range of function f graphed on the right.

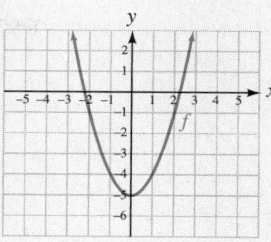

Refer to the graph of function f.

47. Find: $f(-2)$

48. Find the value of x for which $f(x) = 3$.

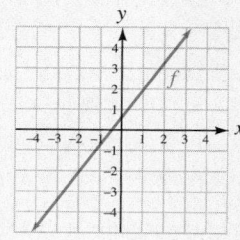

49. Assume that y varies directly with x. If $x = 30$ when $y = 4$, find y when $x = 9$.

50. SOUND   Sound intensity (loudness) varies inversely as the square of the distance from the source. If a rock band has a sound intensity of 100 decibels 30 feet away from the amplifier, find the sound intensity 60 feet away from the amplifier.

GROUP PROJECT

VENN DIAGRAMS

Overview: In this activity, we will discuss several of the fundamental concepts of what is known as *set theory*.

Instructions: *Venn diagrams* are a convenient way to visualize relationships between sets and operations on sets. They were invented by the English mathematician John Venn (1834–1923). To draw a Venn diagram, we begin with a large rectangle, called the *universal set*. Ovals or circles are then drawn in the interior of the rectangle to represent subsets of the universal set.

Form groups of 2 or 3 students. Study the following figures, which illustrate three set operations: union, intersection, and complement.

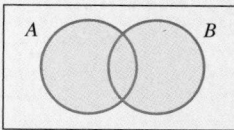
$A \cup B$
The shaded region is the *union* of set A and set B.

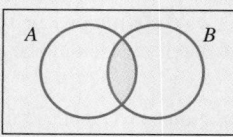
$A \cap B$
The shaded region is the *intersection* of set A and set B.

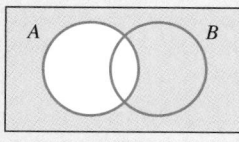
$\overline{A}$
The shaded region is the complement of set A.

For each of the following exercises, sketch the following blank Venn diagram and then shade the indicated region.

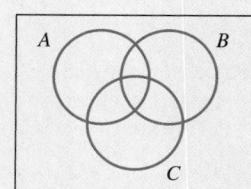

1. $A \cup B$

2. $A \cap B$

3. $A \cap C$

4. $A \cup C$

5. $A \cup B \cup C$

6. $A \cap B \cap C$

7. $(B \cup C) \cap A$

8. $C \cup (A \cap B)$

9. $\overline{A}$

10. $\overline{B} \cup \overline{C}$

11. $\overline{B} \cap \overline{C}$

12. $\overline{A \cup B}$

CUMULATIVE REVIEW
Chapters 1–8

1. Determine whether each statement is true or false. [Section 1.3]
   a. All whole numbers are integers.
   b. π is a rational number.
   c. A real number is either rational or irrational.

2. Evaluate: $\dfrac{-3(3+2)^2 - (-5)}{17 - |-22|}$ [Section 1.7]

3. Simplify: $3p - 6(p+z) + p$ [Section 1.9]

4. Solve: $2 - (4x + 7) = 3 + 2(x + 2)$ [Section 2.2]

5. BACKPACKS Pediatricians advise that children should not carry more than 20% of their own body weight in a backpack. According to this warning, how much weight can a fifth-grade girl who weighs 85 pounds safely carry in her backpack? [Section 2.3]

6. SURFACE AREA The total surface area A of a box with dimensions l, w, and h is given by the formula $A = 2lw + 2wh + 2lh$. If $A = 202$ square inches, $l = 9$ inches, and $w = 5$ inches, find h. [Section 2.4]

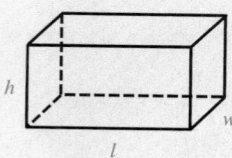

7. SEARCH AND RESCUE Two search and rescue teams leave base at the same time, looking for a lost boy. The first team, on foot, heads north at 2 mph and the other, on horseback, south at 4 mph. How long will it take them to search a distance of 21 miles between them? [Section 2.6]

8. BLENDING COFFEE A store sells regular coffee for $8 a pound and gourmet coffee for $14 a pound. Using 40 pounds of the gourmet coffee, the owner makes a blend to put on sale for $10 a pound. How many pounds of regular coffee should he use? [Section 2.6]

9. Solve: $3 - 3x \geq 6 + x$. Graph the solution set. Then describe the graph using interval notation. [Section 2.7]

10. Is $(-6, -7)$ a solution of $4x - 3y = -4$? [Section 3.1]

Graph each equation.

11. $y = \dfrac{1}{2}x$ [Section 3.2]

12. $3x - 4y = 12$ [Section 3.3]

13. $x = 5$ [Section 3.3]

14. $y = 2x^2 - 3$ [Section 5.4]

15. SHOPPING On the graph in the next column, the line approximates the growth in retail sales for U.S. shopping centers during the years 1994–2005. Find the rate of increase in sales by finding the slope of the line. [Section 3.4]

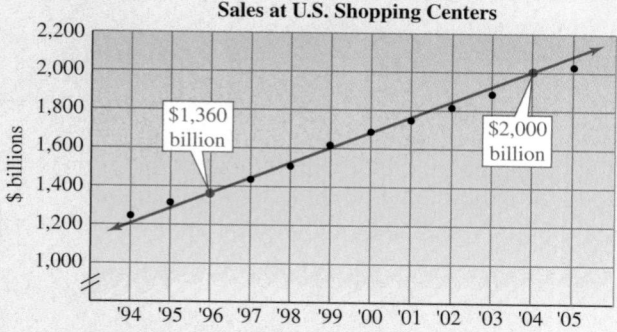

Sales at U.S. Shopping Centers

Based on data from International Council of Shopping Centers

16. What is the slope of the line defined by each equation? [Section 3.5]
   a. $y = 3x - 7$ b. $2x + 3y = -10$

17. Find the equation of the line passing through $(-2, 5)$ and $(4, 8)$. Write the answer in slope–intercept form. [Section 3.6]

18. If $f(x) = 2x^2 - 3x + 1$, find $f(-3)$. [Section 3.8]

19. BOATING The graph shows the vertical distance from a point on the tip of a propeller to the centerline as the propeller spins. Is this the graph of a function? [Section 3.8]

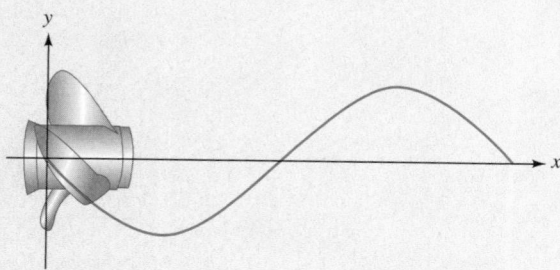

20. Solve the system $\begin{cases} x + y = 4 \\ y = x + 6 \end{cases}$ by graphing. [Section 4.1]

Solve each system of equations.

21. $\begin{cases} x = y + 4 \\ 2x + y = 5 \end{cases}$ [Section 4.2]

22. $\begin{cases} 3s + 4t = 5 \\ 2s - 3t = -8 \end{cases}$ [Section 4.3]

23. FINANCIAL PLANNING In investing $6,000 of a couple's money, a financial planner put some of it into a savings account paying 6% annual interest. The rest was invested in a riskier mini-mall development plan paying 12% annually. The combined interest earned for the first year was $540. How much money was invested at each rate? Use two variables to solve this problem. [Section 4.4]

24. Graph: $\begin{cases} 3x + 2y \geq 6 \\ x + 3y \leq 6 \end{cases}$ [Section 4.5]

Simplify. Use only positive exponents in your answers.

25. $(x^5)^2(x^7)^3$ [Section 5.1]   **26.** $\left(\dfrac{a^3b}{c^4}\right)^5$ [Section 5.1]

27. $4^{-3} \cdot 4^{-2} \cdot 4^5$ [Section 5.2]   **28.** $(2a^{-2}b^3)^{-4}$ [Section 5.2]

29. ASTRONOMY   The **parsec**, a unit of distance used in astronomy, is 3×10^{16} meters. The distance to Betelgeuse, a star in the constellation Orion, is 1.6×10^2 parsecs. Use scientific notation to express this distance in meters. [Section 5.3]

30. Write 0.0000000043 in scientific notation. [Section 5.3]

Perform the operations.

31. $(3a^2 - 2a + 4) - (a^2 - 3a + 7)$ [Section 5.5]

32. $0.3p^5(0.4p^4 - 6p^2)$ [Section 5.6]

33. $(-3t + 2s)(2t - 3s)$ [Section 5.6]

34. $(4b - 8)^2$ [Section 5.7]

35. $\left(6b + \dfrac{1}{2}\right)\left(6b - \dfrac{1}{2}\right)$ [Section 5.7]

36. $x + 2\overline{)2x^2 + 3x - 2}$ [Section 5.8]

Factor completely, if possible.

37. $12x^2y - 6xy^2 + 9xy^3$ [Section 6.1]

38. $2x^2 + 2xy - 3x - 3y$ [Section 6.1]

39. $x^2 + 7x + 10$ [Section 6.2]

40. $6a^2 - 7a - 20$ [Section 6.3]

41. $6 + 3x^2 + x$ [Section 6.3]

42. $25a^2 - 70ab + 49b^2$ [Section 6.4]

43. $a^3 + 8b^3$ [Section 6.5]

44. $2x^5 - 32x$ [Section 6.6]

Solve each equation.

45. $x^2 + 3x + 2 = 0$ [Section 6.7]

46. $5x^2 = 10x$ [Section 6.7]

47. $6x^2 - x = 2$ [Section 6.7]

48. $a^2 - 25 = 0$ [Section 6.7]

49. CHILDREN'S STICKERS   A rectangular-shaped sticker has an area of 20 cm². The width is 1 cm shorter than the length. Find the length of the sticker. [Section 6.8]

A   **is for alligator**

50. Simplify: $\dfrac{x^2 + 2x + 1}{x^2 - 1}$ [Section 7.1]

Perform the operations. Simplify, if possible.

51. $\dfrac{p^2 - p - 6}{3p - 9} \div \dfrac{p^2 + 6p + 9}{p^2 - 9}$ [Section 7.2]

52. $\dfrac{12x^2}{7 - x} \cdot \dfrac{x - 7}{20x^3}$ [Section 7.2]

53. $\dfrac{13}{15a} - \dfrac{8}{15a}$ [Section 7.3]

54. $\dfrac{x + 2}{x + 5} - \dfrac{x - 3}{x + 7}$ [Section 7.4]

55. $\dfrac{1}{6b^4} - \dfrac{8}{9b^2}$ [Section 7.4]

56. $\dfrac{\dfrac{1}{x} + \dfrac{1}{y}}{\dfrac{1}{x} - \dfrac{1}{y}}$ [Section 7.5]

57. Solve: $\dfrac{7}{a^2 - a - 2} + \dfrac{1}{a + 1} = \dfrac{3}{a - 2}$ [Section 7.6]

58. FILLING A POOL   An inlet pipe can fill an empty swimming pool in 5 hours, and another inlet pipe can fill the pool in 4 hours. How long will it take both pipes to fill the pool? [Section 7.7]

59. ONLINE SALES   A company found that, on average, it made 9 online sales transactions for every 500 hits on its Internet Web site. If the company's Web site had 360,000 hits in one year, how many sales transactions did it have that year? [Section 7.8]

60. The triangles shown below are similar. Find x.   [Section 7.8]

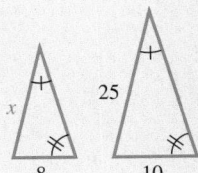

Solve each compound inequality. Graph the solution set and write it using interval notation.

61. $3x + 2 < 8$ or $2x - 3 > 11$ [Section 8.2]

62. $3x + 4 < -2$ or $3x + 4 > 10$ [Section 8.2]

Solve each equation.

63. $2|4x - 3| + 1 = 19$ [Section 8.3]

64. $|2x - 1| = |3x + 4|$ [Section 8.3]

Solve each inequality. Graph the solution set and write it using interval notation.

65. $|3x - 2| \leq 4$ [Section 8.3]

66. $|2x + 3| - 1 > 4$ [Section 8.3]

67. Use the substitution to factor $(x + y)^2 + 7(x + y) + 12$. [Section 8.4]

68. Factor: $(a - b)^4 - c^4$ [Section 8.5]

69. Factor: $a^2 - b^2 + a + b$ [Section 8.5]

70. Factor: $a^2 + 4a + 4 - b^2$ [Section 8.5]

71. Multiply, and then simplify if possible:

$$(2ax - 10x - a + 5) \cdot \frac{x}{x - 2x^2} \quad [\text{Section 8.6}]$$

72. Add: $\dfrac{x}{x - 4} + \dfrac{5}{4 - x}$ [Section 8.6]

73. Perform the operations and simplify, if possible:

$$\frac{2x^2 - 2x - 4}{x^2 + 2x - 8} \cdot \frac{3x^2 + 15x}{x + 1} \div \frac{100 - 4x^2}{x^2 - x - 20} \quad [\text{Section 8.6}]$$

74. Determine whether the graphs of $2x + 3y = 9$ and $3x - 2y = 5$ are parallel, perpendicular, or neither. [Section 8.7]

75. Find the domain and range of the relation
$\{(1, -12), (-6, 8), (5, 8), (0, 0), (1, 4)\}$. [Section 8.8]

76. Let $f(x) = \frac{1}{3}x + 4$. For what value of x is $f(x) = 2$? [Section 8.8]

77. Find the domain of the function $f(x) = \frac{1}{x - 2}$. [Section 8.8]

78. Graph $f(x) = |x| - 2$ and give its domain and range. [Section 8.8]

79. Suppose w varies directly as x. If $w = 1.2$ when $x = 4$, find w when $x = 30$. [Section 8.9]

80. FARMING   The number of days a given number of bushels of corn will last when feeding chickens varies inversely with the number of animals. If a certain number of bushels will feed 300 chickens for 4 days, how long will the feed last for 1,200 chickens? [Section 8.9]

CHAPTER 9

Radical Expressions and Equations

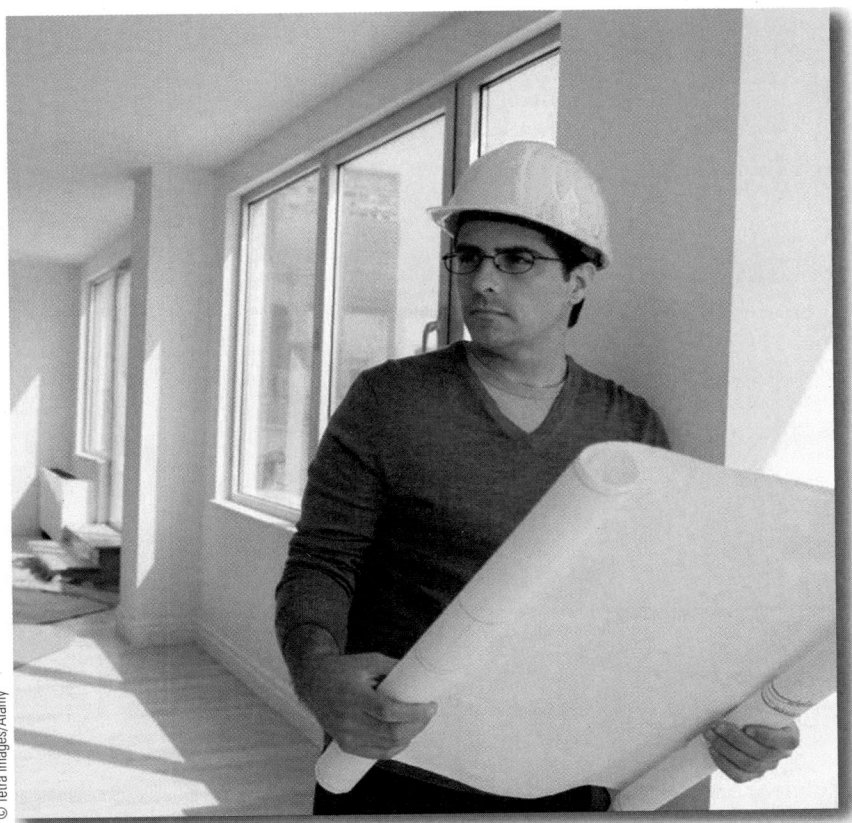

from *Campus to Careers*
General Contractor

The growing popularity of remodeling has created a boom for general contractors. If it's an additional bedroom you need or a makeover of a dated kitchen or bathroom, they can provide design and construction expertise, as well as knowledge of local building code requirements. From the planning stages of a project through its completion, general contractors use mathematics every step of the way.

In **Problem 135** of **Study Set 9.2,** you will use concepts from this chapter to examine the movement of construction materials through a tight hallway.

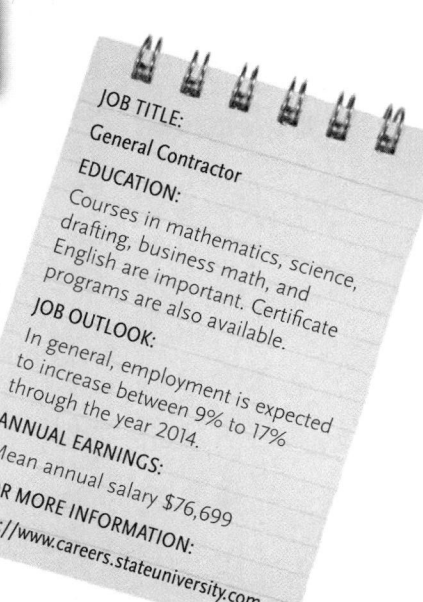

JOB TITLE:
General Contractor

EDUCATION:
Courses in mathematics, science, drafting, business math, and English are important. Certificate programs are also available.

JOB OUTLOOK:
In general, employment is expected to increase between 9% to 17% through the year 2014.

ANNUAL EARNINGS:
Mean annual salary $76,699

FOR MORE INFORMATION:
http://www.careers.stateuniversity.com

Study Skills Workshop
Don't Just Memorize

Many students attempt to learn algebra by rote memorization. Unfortunately, as the term progresses, they find that does not work. When they encounter problem types slightly different from those that they have memorized, they experience great difficulty. Remember, memorization only provides a superficial grasp of the concepts. When learning a new algebraic procedure, it is most important that you:

UNDERSTAND "WHY": Be able to explain the purpose for each step in the procedure and why they are applied in that order.

UNDERSTAND "WHEN": Be able to explain what types of problems are solved using the procedure and what types are not.

Now Try This

1. Choose five problems in the Guided Practice section of a Study Set to solve. Write a *Strategy* and *Why* statement for each solution in your own words.

2. Select a procedure that is introduced in this chapter and explain when it should be used. Write a *Caution* statement that warns of a possible pitfall when using it. Give an example of an application problem that can be solved using the procedure.

SECTION 9.1
Radical Expressions and Radical Functions

Objectives

1. Find square roots.
2. Find square roots of expressions containing variables.
3. Graph the square root function.
4. Find cube roots.
5. Graph the cube root function.
6. Find nth roots.

In this section, we will reverse the squaring process and learn how to find *square roots* of numbers. Then we will generalize the concept of root and consider cube roots, fourth roots, and so on. We will also discuss a new family of functions, called *radical functions*.

1. Find Square Roots.

When we raise a number to the second power, we are squaring it, or finding its **square.**

- The square of 5 is 25 because $5^2 = 25$.
- The square of -5 is 25, because $(-5)^2 = 25$.

We can reverse the squaring process to find **square roots** of numbers. For example, to find the square roots of 25, we ask ourselves "What number, when squared, is equal to 25?" There are two possible answers.

- 5 is a square root of 25, because $5^2 = 25$.
- -5 is a square root of 25, because $(-5)^2 = 25$.

In general, we have the following definition.

Square Root of a	The number b is a **square root** of the number a if $b^2 = a$.

Every positive number has two square roots, one positive and one negative. For example, the two square roots of 9 are 3 and -3, and the two square roots of 144 are 12 and -12. The number 0 is the only real number with exactly one square root. In fact, it is its own square root, because $0^2 = 0$.

The Language of Algebra
We can read $\sqrt{9}$ as "the square root of 9" or as "radical 9."

A **radical symbol** $\sqrt{}$ represents the **positive** or **principal square root** of a number. Since 3 is the positive square root of 9, we can write

$$\sqrt{9} = 3$$

The symbol $-\sqrt{}$ represents the **negative square root** of a number. It is the opposite of the principal square root. Since -12 is the negative square root of 144, we can write

$$-\sqrt{144} = -12 \quad \text{Read as "the negative square root of 144 is } -12\text{" or "the opposite of the square root of 144 is } -12\text{."}$$

Square Root Notation	If a is a positive real number,

1. $\sqrt{a}$ represents the **positive** or **principal square root** of a. It is the positive number we square to get a.
2. $-\sqrt{a}$ represents the **negative square root** of a. It is the opposite of the principal square root of a: $-\sqrt{a} = -1 \cdot \sqrt{a}$.
3. The principal square root of 0 is 0: $\sqrt{0} = 0$.

The number or variable expression under a radical symbol is called the **radicand.** Together, the radical symbol and radicand are called a **radical.** An algebraic expression containing a radical is called a **radical expression.**

Radical symbol
$$\underbrace{\sqrt{81}}_{\text{Radical}} \leftarrow \text{Radicand}$$

To evaluate square root radical expressions, it is helpful to memorize the whole numbers that are perfect squares.

$1 = 1^2$	$25 = 5^2$	$81 = 9^2$	$169 = 13^2$	$289 = 17^2$
$4 = 2^2$	$36 = 6^2$	$100 = 10^2$	$196 = 14^2$	$324 = 18^2$
$9 = 3^2$	$49 = 7^2$	$121 = 11^2$	$225 = 15^2$	$361 = 19^2$
$16 = 4^2$	$64 = 8^2$	$144 = 12^2$	$256 = 16^2$	$400 = 20^2$

EXAMPLE 1 Evaluate each square root:

a. $\sqrt{81}$ b. $-\sqrt{225}$ c. $\sqrt{\dfrac{49}{4}}$ d. $\sqrt{0.36}$

Strategy In each case, we will determine what positive number, when squared, produces the radicand.

Why The symbol $\sqrt{}$ indicates that the positive square root of the number written under it should be found.

Solution

a. $\sqrt{81} = 9$ *Because $9^2 = 81$.* b. $-\sqrt{225} = -15$ *Because $-\sqrt{225} = -1 \cdot \sqrt{225}$.*

c. $\sqrt{\dfrac{49}{4}} = \dfrac{7}{2}$ *Because $\left(\dfrac{7}{2}\right)^2 = \dfrac{49}{4}$.* d. $\sqrt{0.36} = 0.6$ *Because $(0.6)^2 = 0.36$.*

Self Check 1 Evaluate each square root: a. $\sqrt{64}$ b. $-\sqrt{1}$

c. $\sqrt{\dfrac{1}{16}}$ d. $\sqrt{0.09}$

Now Try **Problems 21, 25, and 27**

A table of square roots

n	$\sqrt{n}$
1	1.000
2	1.414
3	1.732
4	2.000
5	**2.236**
6	2.449
7	2.646
8	2.828
9	3.000
10	3.162

A number such as 81, 225, $\frac{1}{4}$, and 0.36, that is the square of some rational number, is called a **perfect square.** In Example 1, we saw that the square root of a perfect square is a rational number.

If a positive number is not a perfect square, its square root is irrational. For example, $\sqrt{5}$ is an irrational number because 5 is not a perfect square. Since $\sqrt{5}$ is irrational, its decimal representation is nonterminating and nonrepeating. We can find an approximate value of $\sqrt{5}$ using the square root key $\boxed{\sqrt{}}$ on a calculator or from the table of square roots found in Appendix I.

$$\sqrt{5} \approx 2.236067978$$

Caution Square roots of negative numbers are not real numbers. For example, $\sqrt{-9}$ is not a real number, because no real number squared equals -9. Square roots of negative numbers come from a set called the **imaginary numbers,** which we will discuss later in this chapter. If we attempt to evaluate $\sqrt{-9}$ using a calculator, we will get an error message.

> **Caution**
> Although they look similar, these radical expressions have very different meanings.
>
> $-\sqrt{9} = -3$
>
> $\sqrt{-9}$ is not a real number.

Scientific calculator

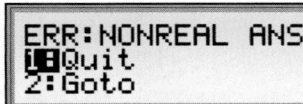

Graphing calculator

We summarize three important facts about square roots as follows.

Square Roots	1. If a is a perfect square, then $\sqrt{a}$ is rational.
	2. If a is a positive number that is not a perfect square, then $\sqrt{a}$ is irrational.
	3. If a is a negative number, then $\sqrt{a}$ is not a real number.

② Find Square Roots of Expressions Containing Variables.

If $x \neq 0$, the positive number x^2 has x and $-x$ for its two square roots. To denote the positive square root of $\sqrt{x^2}$, we must know whether x is positive or negative.

If x is positive, we can write

$$\sqrt{x^2} = x \qquad \sqrt{x^2} \text{ represents the positive square root of } x^2, \text{ which is } x.$$

If x is negative, then $-x$ is positive and we can write

$$\sqrt{x^2} = -x \qquad \sqrt{x^2} \text{ represents the positive square root of } x^2, \text{ which is } -x.$$

If we don't know whether x is positive or negative, we can use absolute value symbols to guarantee that $\sqrt{x^2}$ is positive.

| Definition of x^2 | For any real number x, |
| | $$\sqrt{x^2} = |x|$$ |

We use this definition to *simplify* square root radical expressions.

EXAMPLE 2   Simplify: **a.** $\sqrt{16x^2}$ **b.** $\sqrt{x^2 + 2x + 1}$
c. $\sqrt{m^4}$ **d.** $\sqrt{49r^8}$

Strategy   In each case, we will determine what positive expression, when squared, produces the radicand.

Why   The symbol $\sqrt{}$ indicates that the positive square root of the expression written under it should be found.

Solution   If x, m, and r can be any real number, we have

a. $\sqrt{16x^2} = \sqrt{(4x)^2}$ Write the radicand $16x^2$ as $(4x)^2$.

$\qquad\quad = |4x|$ Because $(|4x|)^2 = 16x^2$. Since x could be negative, absolute value symbols are needed.

$\qquad\quad = 4|x|$ Since 4 is a positive constant in the product 4x, we can write it outside the absolute value symbols.

b. $\sqrt{x^2 + 2x + 1} = \sqrt{(x + 1)^2}$ Factor the radicand: $x^2 + 2x + 1 = (x + 1)^2$.

$\qquad\qquad\qquad = |x + 1|$ Since x + 1 can be negative (for example, when x = −5, x + 1 is −4), absolute value symbols are needed.

c. $\sqrt{m^4} = m^2$ Because $(m^2)^2 = m^4$. Since $m^2 \geq 0$, no absolute value symbols are needed.

d. $\sqrt{49r^8} = 7r^4$ Because $(7r^4)^2 = 49r^8$. Since $r^4 \geq 0$, no absolute value symbols are needed.

> **Self Check 2**   Simplify: **a.** $\sqrt{25a^2}$ **b.** $\sqrt{16a^4}$
> **c.** $\sqrt{x^2 - 18x + 81}$ **d.** $\sqrt{100d^8}$
>
> *Now Try*   **Problems 37, 39, 43, and 45**

If we know that x is positive in parts (a) and (b) of Example 2, we don't need to use absolute value symbols. For example, if $x > 0$, then

$$\sqrt{16x^2} = 4x \qquad \text{If } x \text{ is positive, } 4x \text{ is positive.}$$

$$\sqrt{x^2 + 2x + 1} = x + 1 \qquad \text{If } x \text{ is positive, } x + 1 \text{ is positive.}$$

3   **Graph the Square Root Function.**

Since there is one principal square root for every nonnegative real number x, the equation $f(x) = \sqrt{x}$ determines a function, called a **square root function.** Square root functions belong to a larger family of functions known as **radical functions.**

EXAMPLE 3   Graph $f(x) = \sqrt{x}$ and find the domain and range of the function.

Strategy   We will graph the function by creating a table of function values and plotting the corresponding ordered pairs.

Why   After drawing a smooth curve though the plotted points, we will have the graph.

Solution   To graph the function, we select several values for x and find the corresponding values of $f(x)$. We begin with $x = 0$, since 0 is the smallest input for which $\sqrt{x}$ is defined.

$$f(x) = \sqrt{x}$$
$$f(0) = \sqrt{0} \qquad \text{Substitute 0 for } x.$$
$$f(0) = 0$$

We enter 0 for x and 0 for $f(x)$ in the table. Then we let $x = 1, 4, 9,$ and 16, and list each corresponding function value in the table. After plotting the ordered pairs, we draw a smooth curve through the points to get the graph shown in figure (a). Since the equation defines a function, its graph passes the vertical line test.

We can use a graphing calculator to get the graph shown in figure (b). From either graph, we can see that the domain and the range are the set of nonnegative real numbers. Expressed in interval notation, the domain is $[0, \infty)$, and the range is $[0, \infty)$.

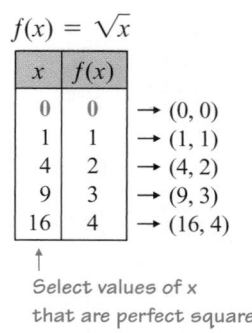

$f(x) = \sqrt{x}$

x	$f(x)$	
0	0	→ (0, 0)
1	1	→ (1, 1)
4	2	→ (4, 2)
9	3	→ (9, 3)
16	4	→ (16, 4)

↑
Select values of x
that are perfect squares.

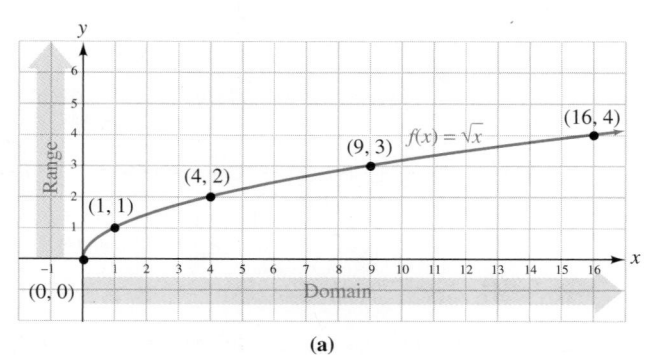

(a)

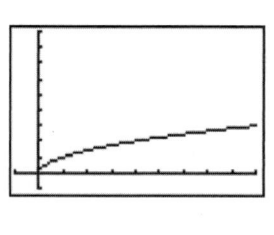

(b)

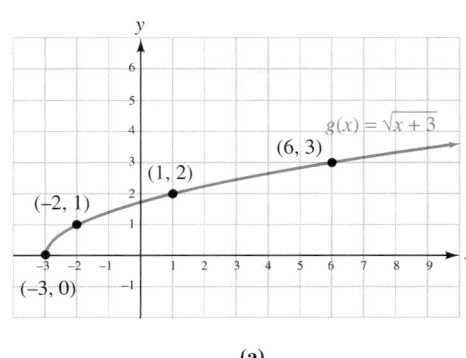

> **Self Check 3**   Graph: $g(x) = \sqrt{x} + 2$. Then give its domain and range and compare it with the graph of $f(x) = \sqrt{x}$.
>
> **Now Try**   **Problem 53**

EXAMPLE 4   Consider the function $g(x) = \sqrt{x + 3}$.
a. Find its domain.   **b.** Graph the function.   **c.** Find its range.

Strategy   We will determine the domain algebraically by finding all the values of x for which $x + 3 \geq 0$.

Why   Since the expression $\sqrt{x + 3}$ is not a real number when $x + 3$ is negative, we must require that $x + 3 \geq 0$

Solution
a. To determine the domain of the function, we solve the following inequality:

$$x + 3 \geq 0 \qquad \text{Because we cannot find the square root of a negative number.}$$

$$x \geq -3 \qquad \text{To solve for x, subtract 3 from both sides.}$$

The x-inputs must be real numbers greater than or equal to -3. Thus, the domain of $g(x) = \sqrt{x + 3}$ is the interval $[-3, \infty)$.

b. To graph the function, we construct a table of function values. We begin by selecting $x = -3$, since -3 is the smallest input for which $\sqrt{x + 3}$ is defined.

$$g(x) = \sqrt{x + 3}$$
$$g(-3) = \sqrt{-3 + 3}$$
$$= \sqrt{0}$$
$$= 0$$

We enter -3 for x and 0 for $g(x)$ in the table. Then we let $x = -2, 1,$ and 6 and list each corresponding function value in the table. After plotting the ordered pairs, we draw a smooth curve through the points to get the graph shown in figure (a). In figure (b), we see that the graph of $g(x) = \sqrt{x + 3}$ is the graph of $f(x) = \sqrt{x}$, translated 3 units to the left.

$g(x) = \sqrt{x + 3}$

x	$g(x)$	
-3	0	$\rightarrow (-3, 0)$
-2	1	$\rightarrow (-2, 1)$
1	2	$\rightarrow (1, 2)$
6	3	$\rightarrow (6, 3)$

↑
Select values of x that make x + 3 a perfect square.

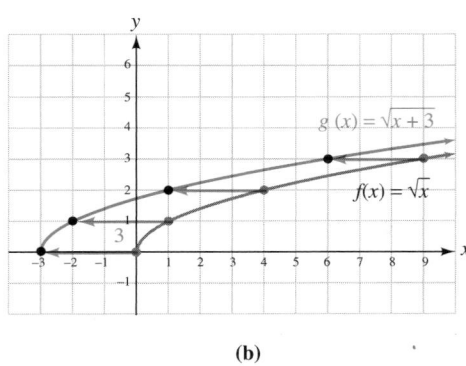

(a)   (b)

c. From the graph, we see that the range of $g(x) = \sqrt{x + 3}$ is $[0, \infty)$.

Self Check 4 Consider $h(x) = \sqrt{x - 2}$. **a.** Find its domain.
b. Graph the function. **c.** Find its range.

Now Try **Problem 55**

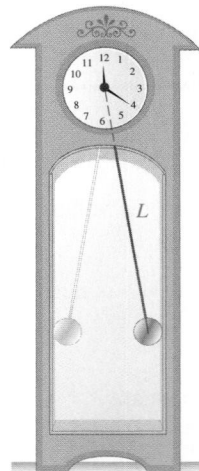

EXAMPLE 5 *Pendulums.* The **period of a pendulum** is the time required for the pendulum to swing back and forth to complete one cycle. The period (in seconds) is a function of the pendulum's length L (in feet) and is given by

$$f(L) = 2\pi\sqrt{\frac{L}{32}}$$

Find the period of the 5-foot-long pendulum of a clock. Round the result to the nearest tenth.

Strategy To find the period of the pendulum we will find $f(5)$.

Why The notation $f(5)$ represents the period (in seconds) of a pendulum whose length L is 5 feet.

Solution

$$f(L) = 2\pi\sqrt{\frac{L}{32}}$$

$$f(5) = 2\pi\sqrt{\frac{5}{32}} \qquad \text{Substitute 5 for L.}$$

$$\approx 2.483647066 \qquad \text{Use a calculator to find an approximation.}$$

The period is approximately 2.5 seconds.

Notation

$2\pi\sqrt{\dfrac{5}{32}}$ means $2 \cdot \pi \cdot \sqrt{\dfrac{5}{32}}$

Self Check 5 Find the period of a pendulum that is 3 feet long. Round the result to the nearest hundredth

Now Try **Problem 105**

Using Your Calculator *Evaluating A Square Root Function*

To solve Example 5 with a graphing calculator, we graph the function $f(x) = 2\pi\sqrt{\frac{x}{32}}$, as in figure (a). We then trace and move the cursor toward an x-value of 5 until we see the coordinates shown in figure (b). The pendulum's period is given by the y-value shown on the screen. By zooming in, we can get better results.

After entering $Y_1 = 2\pi\sqrt{\frac{x}{32}}$, we can also use the TABLE mode to find $f(5)$. See figure (c).

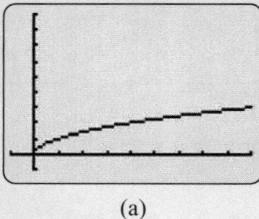

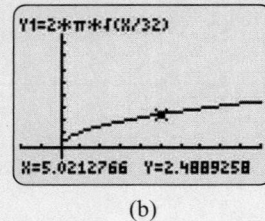

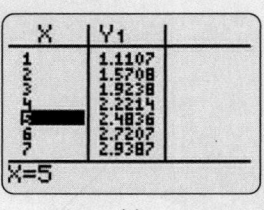

(a)   (b)   (c)

Find Cube Roots.

When we raise a number to the third power, we are cubing it, or finding its **cube.** We can reverse the cubing process to find **cube roots** of numbers. To find the cube root of 8, we ask "What number, when cubed, is equal to 8?" It follows that 2 is a cube root of 8, because $2^3 = 8$.

In general, we have this definition.

Cube Root of *a*	The number *b* is a **cube root** of the real number *a* if $b^3 = a$.

All real numbers have one real cube root. A positive number has a positive cube root, a negative number has a negative cube root, and the cube root of 0 is 0.

Cube Root Notation	The **cube root of *a*** is denoted by $\sqrt[3]{a}$. By definition,
	$\sqrt[3]{a} = b$ if $b^3 = a$

Earlier, we determined that the cube root of 8 is 2. In symbols, we can write: $\sqrt[3]{8} = 2$. The number 3 is called the **index.**

> **Notation**
> For the square root symbol $\sqrt{}$, the unwritten index is understood to be 2.
> $$\sqrt{a} = \sqrt[2]{a}$$

Index
$\downarrow$
$\sqrt[3]{8}$

A number such as 125, $\frac{1}{64}$, -27, and -8, that is the cube of some rational number, is called a **perfect cube.** To simplify cube root radical expressions, we look for perfect cubes and apply the following definition.

Definition of $\sqrt[3]{x^3}$	For any real number *x*,
	$$\sqrt[3]{x^3} = x$$

To simplify cube root radical expressions, it is helpful to memorize the whole numbers that are perfect cubes.

$1 = 1^3$	$27 = 3^3$	$125 = 5^3$	$343 = 7^3$	$729 = 9^3$
$8 = 2^3$	$64 = 4^3$	$216 = 6^3$	$512 = 8^3$	$1{,}000 = 10^3$

EXAMPLE 6 Simplify: **a.** $\sqrt[3]{125}$ **b.** $\sqrt[3]{\dfrac{1}{64}}$ **c.** $\sqrt[3]{-27x^3}$ **d.** $\sqrt[3]{-8a^3b^6}$

Strategy In each case, we will determine what number or expression, when cubed, produces the radicand.

Why The symbol $\sqrt[3]{}$ indicates that the cube root of the number written under it should be found.

> **Success Tip**
> Since every real number has exactly one real cube root, absolute value symbols are not to be used when simplifying cube roots.

Solution

a. $\sqrt[3]{125} = 5$ Because $5^3 = 5 \cdot 5 \cdot 5 = 125$.

b. $\sqrt[3]{\dfrac{1}{64}} = \dfrac{1}{4}$ Because $\left(\frac{1}{4}\right)^3 = \frac{1}{4} \cdot \frac{1}{4} \cdot \frac{1}{4} = \frac{1}{64}$.

c. $\sqrt[3]{-27x^3} = -3x$ Because $(-3x)^3 = (-3x)(-3x)(-3x) = -27x^3$.

d. $\sqrt[3]{-8a^3b^6} = -2ab^2$ Because $(-2ab^2)^3 = (-2ab^2)(-2ab^2)(-2ab^2) = -8a^3b^6$.

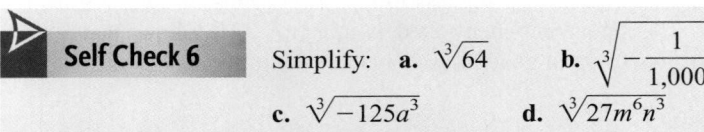

Self Check 6   Simplify: **a.** $\sqrt[3]{64}$ **b.** $\sqrt[3]{-\dfrac{1}{1,000}}$

c. $\sqrt[3]{-125a^3}$ **d.** $\sqrt[3]{27m^6n^3}$

Now Try   **Problems 59, 61, and 67**

5 **Graph the Cube Root Function.**

Since there is one cube root for every real number x, the equation $f(x) = \sqrt[3]{x}$ defines a function, called the **cube root function.** Like square root functions, cube root functions belong to the family of radical functions.

EXAMPLE 7   Consider $f(x) = \sqrt[3]{x}$.   **a.** Graph the function.
b. Find its domain and range.   **c.** Graph: $g(x) = \sqrt[3]{x} - 2$

Strategy   We will graph the function by creating a table of function values and plotting the corresponding ordered pairs.

Why   After drawing a smooth curve though the plotted points, we will have the graph. The answers to parts (b) and (c) can then be determined from the graph.

Solution

a. To graph the function, we select several values for x and find the corresponding values of $f(x)$. We begin with $x = -8$.

$$f(x) = \sqrt[3]{x}$$
$$f(-8) = \sqrt[3]{-8}   \text{Substitute } -8 \text{ for } x.$$
$$f(-8) = -2$$

We enter -8 for x and -2 for $f(x)$ in the table. Then we let $x = -1, 0, 1,$ and 8, and list each corresponding function value in the table. After plotting the ordered pairs, we draw a smooth curve through the points to get the graph shown in figure (a).

$f(x) = \sqrt[3]{x}$

x	$f(x)$	
-8	-2	$\rightarrow (-8, -2)$
-1	-1	$\rightarrow (-1, -1)$
0	0	$\rightarrow (0, 0)$
1	1	$\rightarrow (1, 1)$
8	2	$\rightarrow (8, 2)$

↑
Select values of x that are perfect cubes.

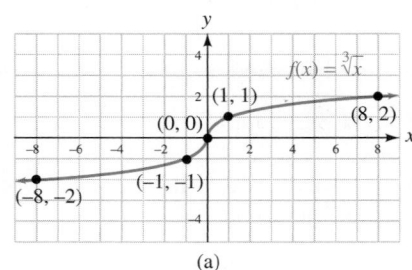

(a)

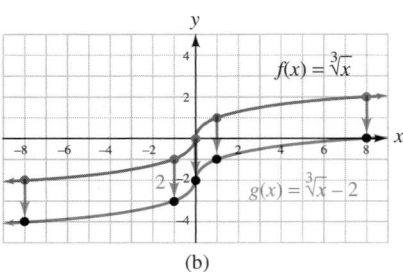

(b)

b. From the graph in figure (a), we see that the domain and the range of function f are the set of real numbers. Thus, the domain is $(-\infty, \infty)$ and the range is $(-\infty, \infty)$.

c. Refer to figure (b). The graph of $g(x) = \sqrt[3]{x} - 2$ is the graph of $f(x) = \sqrt[3]{x}$, translated 2 units downward.

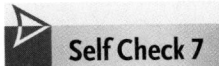

Self Check 7	Consider $f(x) = \sqrt[3]{x} + 1$. **a.** Graph the function.
	b. Find its domain and range.
Now Try	**Problem 69**

6   **Find *n*th Roots.**

Just as there are square roots and cube roots, there are fourth roots, fifth roots, sixth roots, and so on. In general, we have the following definition.

nth Roots of a

The ***nth root of a*** is denoted by $\sqrt[n]{a}$, and

$$\sqrt[n]{a} = b \qquad \text{if} \qquad b^n = a$$

The number n is called the **index** (or **order**) of the radical. If n is an even natural number, a must be positive or zero, and b must be positive.

When n is an odd natural number, the expression $\sqrt[n]{x}$, where $n > 1$, represents an **odd root**. Since every real number has just one real nth root when n is odd, we don't need absolute value symbols when finding odd roots. For example,

$$\sqrt[5]{243} = \sqrt[5]{3^5} = 3 \qquad \text{Because } 3^5 = 243.$$

$$\sqrt[7]{-128x^7} = \sqrt[7]{(-2x)^7} = -2x \quad \text{Because } (-2x)^7 = -128x^7.$$

When n is an even natural number, the expression $\sqrt[n]{x}$, where $x > 0$, represents an **even root**. In this case, there will be one positive and one negative real nth root. For example, the real sixth roots of 729 are 3 and -3, because $3^6 = 729$ and $(-3)^6 = 729$. When finding even roots, we can use absolute value symbols to guarantee that the nth root is positive.

$$\sqrt[4]{(-3)^4} = |-3| = 3 \qquad\qquad \text{We could also simplify this as follows:}$$
$$\sqrt[4]{(-3)^4} = \sqrt[4]{81} = 3 .$$

$$\sqrt[6]{729x^6} = \sqrt[6]{(3x)^6} = |3x| = 3|x| \quad \text{The absolute value symbols guarantee}$$
$$\text{that the sixth root is positive.}$$

In general, we have the following rules.

Rules for $\sqrt[n]{x^n}$

If x is a real number and $n > 1$, then

If n is an odd natural number, $\sqrt[n]{x^n} = x$.

If n is an even natural number, $\sqrt[n]{x^n} = |x|$.

EXAMPLE 8	Evaluate each radical expression, if possible: **a.** $\sqrt[4]{625}$
	b. $\sqrt[4]{-1}$ **c.** $\sqrt[5]{-32}$ **d.** $\sqrt[6]{\dfrac{1}{64}}$ **e.** $\sqrt[7]{10^7}$

Strategy   In each case, we will determine what number, when raised to the fourth, fifth, sixth, or seventh power, produces the radicand.

Why   The symbols $\sqrt[4]{}$, $\sqrt[5]{}$, $\sqrt[6]{}$, and $\sqrt[7]{}$ indicate that the fourth, fifth, sixth, or seventh root of the number written under it should be found.

Solution

a. $\sqrt[4]{625} = 5$, because $5^4 = 625$.   Read $\sqrt[4]{625}$ as "the fourth root of 625."

b. $\sqrt[4]{-1}$ is not a real number.   An even root of a negative number is not a real number.

c. $\sqrt[5]{-32} = -2$, because $(-2)^5 = -32$.   Read $\sqrt[5]{-32}$ as "the fifth root of −32."

d. $\sqrt[6]{\dfrac{1}{64}} = \dfrac{1}{2}$, because $\left(\dfrac{1}{2}\right)^6 = \dfrac{1}{64}$.   Read $\sqrt[6]{\frac{1}{64}}$ as "the sixth root of $\frac{1}{64}$."

e. $\sqrt[7]{10^7} = 10$, because $10^7 = 10^7$.   Read $\sqrt[7]{10^7}$ as "the seventh root of 10^7."

Caution
When n is even $(n > 1)$ and $x < 0$, $\sqrt[n]{x}$ is not a real number. For example, $\sqrt[4]{-81}$ is not a real number, because no real number raised to the fourth power is −81.

Self Check 8   Evaluate, if possible: a. $\sqrt[4]{\dfrac{1}{81}}$ b. $\sqrt[5]{10^5}$

c. $\sqrt[6]{-64}$

Now Try   **Problems 73, 77, and 79**

Using Your Calculator

Finding Roots

The square root key $\boxed{\sqrt{}}$ on a scientific calculator can be used to evaluate square roots. To evaluate roots with an index greater than 2, we can use the root key $\boxed{\sqrt[x]{y}}$. For example, the function $r(V) = \sqrt[3]{\dfrac{3V}{4\pi}}$ gives the radius of a sphere with volume V. To find the radius of the spherical propane tank shown on the left, we substitute 113 for V to get

PROPANE
Capacity 113 ft³

$$r(V) = \sqrt[3]{\dfrac{3V}{4\pi}}$$

$$r(113) = \sqrt[3]{\dfrac{3(113)}{4\pi}}$$

To evaluate a root, we enter the radicand and press the root key $\boxed{\sqrt[x]{y}}$ followed by the index of the radical, which in this case is 3.

$\boxed{3}\boxed{\times}\,113\,\boxed{\div}\,\boxed{(}\,\boxed{4}\boxed{\times}\boxed{\pi}\,\boxed{)}\,\boxed{=}\,\boxed{\text{2nd}}\,\boxed{\sqrt[x]{y}}\,\boxed{3}\,\boxed{=}$ ⟨ 2.999139118 ⟩

To evaluate the cube root of $\dfrac{3(113)}{4\pi}$ with a graphing calculator, we enter

$\boxed{\text{MATH}}\,\boxed{4}\,\boxed{(}\,\boxed{3}\boxed{\times}\,113\,\boxed{)}\,\boxed{\div}\,\boxed{(}\,\boxed{4}\boxed{\times}\,\boxed{\text{2nd}}\,\boxed{\pi}\,\boxed{)}\,\boxed{)}\,\boxed{\text{ENTER}}$

```
³√((3*113)/(4*π))
>
              2.999139118
```

The radius of the propane tank is about 3 feet.

EXAMPLE 9   Simplify each radical expression. Assume that x can be any real number.   **a.** $\sqrt[5]{x^5}$ **b.** $\sqrt[4]{16x^4}$ **c.** $\sqrt[6]{(x+4)^6}$ **d.** $\sqrt[4]{81x^8}$

Strategy   When the index n is odd, we will determine what expression, when raised to the nth power produces the radicand. When the index n is even, we will determine what positive expression, when raised to the nth power produces the radicand.

Why   This is the definition of nth root.

Solution

a. $\sqrt[5]{x^5} = x$ Since n is odd, absolute value symbols aren't needed.

b. $\sqrt[4]{16x^4} = |2x| = 2|x|$ Since n is even and x can be negative, absolute value symbols are needed to guarantee that the result is positive.

c. $\sqrt[6]{(x+4)^6} = |x+4|$ Absolute value symbols are needed to guarantee that the result is positive.

d. $\sqrt[4]{81x^8} = 3x^2$ Because $(3x^2)^4 = 81x^8$. Since $3x^2 \geq 0$ for any value of x, no absolute value symbols are needed.

Self Check 9   Simplify. Assume all variables are unrestricted.   **a.** $\sqrt[6]{x^6}$
b. $\sqrt[5]{(a+5)^5}$ **c.** $\sqrt[4]{16a^8}$

Now Try   Problems 81, 83, 85, and 87

If we know that x is positive in parts (b) and (c) of Example 9, we don't need to use absolute value symbols. For example, if $x > 0$, then

$$\sqrt[4]{16x^4} = 2x$$ If x is positive, $2x$ is positive.

$$\sqrt[6]{(x+4)^6} = x+4$$ If x is positive, $x+4$ is positive.

We summarize the definitions concerning $\sqrt[n]{x}$ as follows.

Summary of the Definitions of $\sqrt[n]{x}$

If n is a natural number greater than 1 and x is a real number,

If $x > 0$, then $\sqrt[n]{x}$ is the positive number such that $\left(\sqrt[n]{x}\right)^n = x$.

If $x = 0$, then $\sqrt[n]{x} = 0$.

If $x < 0$ $\begin{cases} \text{and } n \text{ is odd, then } \sqrt[n]{x} \text{ is the negative number such that } \left(\sqrt[n]{x}\right)^n = x. \\ \text{and } n \text{ is even, then } \sqrt[n]{x} \text{ is not a real number.} \end{cases}$

ANSWERS TO SELF CHECKS 1. a. 8 b. -1 c. $\frac{1}{4}$ d. 0.3 2. a. $5|a|$ b. $4a^2$ c. $|x-9|$ d. $10d^4$

3.
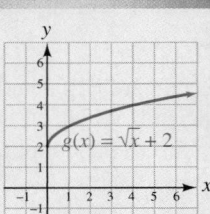

D: $[0, \infty)$;
R: $[2, \infty)$;
the graph is
2 units higher

4. a. $[2, \infty)$ b.
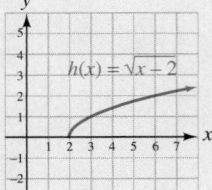
c. $[0, \infty)$

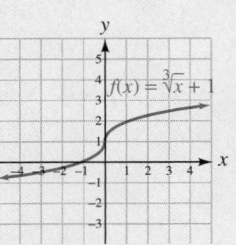

5. 1.92 sec 6. a. 4 b. $-\frac{1}{10}$ c. $-5a$ d. $3m^2n$ 7. a.

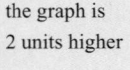

b. D: $(-\infty, \infty)$; R: $(-\infty, \infty)$ 8. a. $\frac{1}{3}$ b. 10 c. Not a real number 9. a. $|x|$ b. $a+5$ c. $2a^2$

STUDY SET
9.1

VOCABULARY

Fill in the blanks.

1. $5x^2$ is the _____ root of $25x^4$ because $(5x^2)^2 = 25x^4$. The _____ root of 216 is 6 because $6^3 = 216$.

2. The symbol $\sqrt{}$ is called a _____ symbol or a _____ root symbol.

3. In the expression $\sqrt[3]{27x^6}$, the _____ is 3 and $27x^6$ is the _____.

4. When we write $\sqrt{b^4} = b^2$, we say that we have _____ the radical expression.

5. When n is an odd number, $\sqrt[n]{x}$ represents an _____ root. When n is an _____ number, $\sqrt[n]{x}$ represents an even root.

6. $f(x) = \sqrt{x}$ and $g(x) = \sqrt[3]{x}$ are _____ functions.

CONCEPTS

Fill in the blanks.

7. b is a square root of a if $b^2 = $.

8. $\sqrt{0} = $ and $\sqrt[3]{0} = $.

9. The number 25 has _____ square roots. The principal square root of 25 is .

10. $\sqrt{-4}$ is not a real number, because no real number _____ equals -4.

11. $\sqrt[3]{x} = y$ if $y^3 = $.

12. $\sqrt{x^2} = $ and $\sqrt[3]{x^3} = $

13. The graph of $g(x) = \sqrt{x} + 3$ is the graph of $f(x) = \sqrt{x}$ translated _____ units _____.

14. The graph of $g(x) = \sqrt{x+5}$ is the graph of $f(x) = \sqrt{x}$ translated _____ units to the _____.

15. The graph of a square root function f is shown. Find each of the following.
    a. $f(11)$ b. $f(2)$
    c. The value(s) of x for which $f(x) = 2$
    d. The domain and range of f

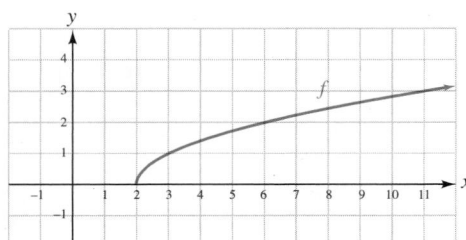

16. The graph of a cube root function f is shown on the next page. Find each of the following.
    a. $f(-8)$ b. $f(0)$
    c. The value(s) of x for which $f(x) = -2$
    d. The domain and range of f

NOTATION

Translate each sentence into mathematical symbols.

17. The square root of x squared is the absolute value of x.

18. The cube root of x cubed is x.

19. f of x equals the square root of the quantity x minus five.

20. The fifth root of negative thirty-two is negative two.

GUIDED PRACTICE

Evaluate each square root, if possible, without using a calculator. See Objective 1 and Example 1.

21. $\sqrt{100}$ **22.** $\sqrt{49}$

23. $-\sqrt{64}$ **24.** $-\sqrt{1}$

25. $\sqrt{\dfrac{1}{9}}$ **26.** $\sqrt{\dfrac{4}{25}}$

27. $\sqrt{0.25}$ **28.** $\sqrt{0.16}$

29. $\sqrt{-81}$ **30.** $-\sqrt{-49}$

31. $\sqrt{121}$ **32.** $\sqrt{144}$

Use a calculator to find each square root. Give each answer to four decimal places. See Objective 1.

33. $\sqrt{12}$ **34.** $\sqrt{340}$

35. $\sqrt{679.25}$ **36.** $\sqrt{0.0063}$

Simplify each expression. Assume that all variables are unrestricted and use absolute value symbols when necessary. See Example 2.

37. $\sqrt{4x^2}$ **38.** $\sqrt{64t^2}$

39. $\sqrt{81h^4}$ **40.** $\sqrt{36y^4}$

41. $\sqrt{36s^6}$ **42.** $\sqrt{9y^6}$

43. $\sqrt{144m^8}$ **44.** $\sqrt{4n^8}$

45. $\sqrt{y^2 - 2y + 1}$ **46.** $\sqrt{b^2 - 14b + 49}$

47. $\sqrt{a^4 + 6a^2 + 9}$ **48.** $\sqrt{x^4 + 10x^2 + 25}$

49. Let $f(x) = \sqrt{x - 4}$. Find each function value. See Example 3.

  a. $f(8)$ **b.** $f(29)$

50. Let $f(x) = \sqrt{x^2 + 1}$. Find each function value. Use a calculator to approximate each answer to four decimal places. See Example 3.

  a. $f(4)$ **b.** $f(2.35)$

51. Let $g(x) = \sqrt[3]{x - 4}$. Find each function value. See Example 7.

  a. $g(12)$ **b.** $g(-23)$

52. Let $g(x) = \sqrt[3]{x^2 + 1}$. Find each function value. Use a calculator to approximate each answer to four decimal places. See Using Your Calculator: Finding Roots.

  a. $g(6)$ **b.** $g(21.57)$

Complete each table and graph the function. Find the domain and range. See Example 3.

53. $f(x) = -\sqrt{x}$

x	y
0	
1	
4	
9	
16	

54. $f(x) = \sqrt{x} + 2$

x	y
0	
1	
4	
9	
16	

Find the domain of each function, graph it, and find its range. See Example 4.

55. $f(x) = \sqrt{x + 4}$ **56.** $f(x) = \sqrt{x - 1}$

Simplify each cube root. See Example 6.

57. $\sqrt[3]{1}$ **58.** $\sqrt[3]{8}$

59. $\sqrt[3]{-125}$ **60.** $\sqrt[3]{-27}$

61. $\sqrt[3]{\dfrac{8}{27}}$ **62.** $\sqrt[3]{\dfrac{125}{64}}$

63. $\sqrt[3]{64}$ **64.** $\sqrt[3]{1,000}$

65. $\sqrt[3]{-216a^3}$ **66.** $\sqrt[3]{-512x^3}$

67. $\sqrt[3]{-1,000p^6q^3}$ **68.** $\sqrt[3]{-343a^6b^3}$

Complete each table and graph the function. Give the domain and range. See Example 7.

69. $f(x) = \sqrt[3]{x} - 3$

x	y
-8	
-1	
0	
1	
8	

70. $f(x) = -\sqrt[3]{x}$

x	y
-8	
-1	
0	
1	
8	

Graph each function. Give the domain and range. See Example 7.

71. $f(x) = \sqrt[3]{x} - 3$

72. $f(x) = \sqrt[3]{x} + 3$

Evaluate each radical expression, if possible, without using a calculator. See Example 8.

73. $\sqrt[4]{81}$

74. $\sqrt[6]{64}$

75. $-\sqrt[5]{243}$

76. $-\sqrt[4]{625}$

77. $\sqrt[6]{-256}$

78. $\sqrt[6]{-729}$

79. $\sqrt[7]{-\dfrac{1}{128}}$

80. $\sqrt[5]{-\dfrac{243}{32}}$

Simplify each radical expression. Assume all variables are unrestricted. See Example 9.

81. $\sqrt[5]{32a^5}$

82. $\sqrt[5]{-32x^5}$

83. $\sqrt[4]{81a^4}$

84. $\sqrt[8]{t^8}$

85. $\sqrt[6]{k^{12}}$

86. $\sqrt[6]{64b^6}$

87. $\sqrt[4]{(m+4)^8}$

88. $\sqrt[4]{(x-7)^8}$

TRY IT YOURSELF

Simplify each radical expression, if possible. Assume all variables are unrestricted.

89. $\sqrt[3]{64s^9t^6}$

90. $\sqrt[3]{1,000a^6b^6}$

91. $-\sqrt{49b^8}$

92. $-\sqrt{144t^4}$

93. $-\sqrt[5]{-\dfrac{1}{32}}$

94. $-\sqrt[5]{-243}$

95. $\sqrt[3]{-125m^6}$

96. $\sqrt[3]{-216z^9} \quad -6z^3$

97. $\sqrt[6]{64a^6b^6}$

98. $\sqrt{169p^4q^2}$

99. $\sqrt[7]{(x+2)^7}$

100. $\sqrt[6]{(x+4)^6}$

101. $\sqrt[4]{-81}$

102. $\sqrt[6]{-1}$

103. $\sqrt{n^2 + 12n + 36}$

104. $\sqrt{s^2 - 20s + 100}$

APPLICATIONS

Use a calculator to solve each problem. Round answers to the nearest tenth.

105. EMBROIDERY   The radius *r* of a circle is given by the formula

$$r = \sqrt{\dfrac{A}{\pi}}$$

where *A* is its area. Find the diameter of the embroidery hoop if there are 38.5 in.^{2} of stretched fabric on which to embroider.

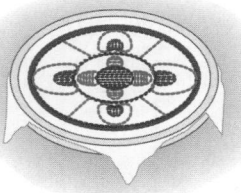

106. PENDULUMS   Find the period of a pendulum with length 1 foot. *See Example 5.*

107. SHOELACES   The formula $S = 2\left[H + L + (p-1)\sqrt{H^2 + V^2}\right]$ can be used to calculate the correct shoelace length for the criss-cross lacing pattern shown in the illustration, where *p* represents the number of *pairs* of eyelets. Find the correct shoelace length if *H* (horizontal distance) = 50 millimeters, *L* (length of end) = 250 millimeters, and *V* (vertical distance) = 20 millimeters. (*Source:* Ian's Shoelace Site at www.fieggen.com)

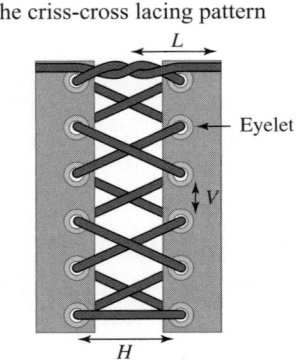

108. BASEBALL   The length of a diagonal of a square is given by the function $d(s) = \sqrt{2s^2}$, where *s* is the length of a side of the square. Find the distance from home plate to second base on a softball diamond and on a baseball diamond. The illustration gives the dimensions of each type of infield.

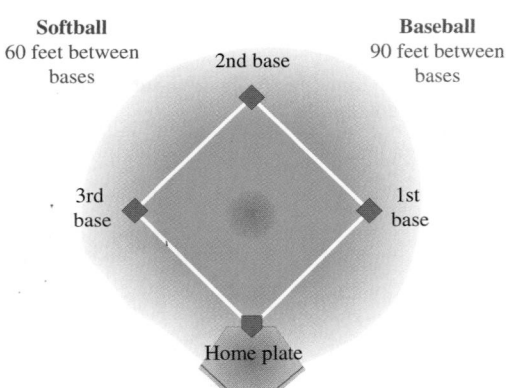

109. PULSE RATES   The approximate pulse rate (in beats per minute) of an adult who is *t* inches tall is given by the function

$$p(t) = \dfrac{590}{\sqrt{t}}$$

The *Guinness Book of World Records 2008* lists Leonid Stadnyk of Ukraine as the tallest living man, at 8 feet, 5.5 inches. Find his approximate pulse rate as predicted by the function.

110. THE GRAND CANYON   The time t (in seconds) that it takes for an object to fall a distance of s feet is given by the formula

$$t = \frac{\sqrt{s}}{4}$$

In some places, the Grand Canyon is one mile (5,280 feet) deep. How long would it take a stone dropped over the edge of the canyon to hit bottom?

111. BIOLOGY   Scientists will place five rats inside a clear plastic hemisphere and control the environment to study the rats' behavior. The function

$$d(V) = \sqrt[3]{12\left(\frac{V}{\pi}\right)}$$

gives the diameter of a hemisphere with volume V. Use the function to determine the diameter of the base of the hemisphere, if each rat requires 125 cubic feet of living space.

112. AQUARIUMS   The function

$$s(g) = \sqrt[3]{\frac{g}{7.5}}$$

determines how long (in feet) an edge of a cube-shaped tank must be if it is to hold g gallons of water. What dimensions should a cube-shaped aquarium have if it is to hold 1,250 gallons of water?

113. COLLECTIBLES   The *effective rate of interest r* earned by an investment is given by the formula

$$r = \sqrt[n]{\frac{A}{P}} - 1$$

where P is the initial investment that grows to value A after n years. Determine the effective rate of interest earned by a collector on a Lladró porcelain figurine purchased for $800 and sold for $950 five years later.

114. LAW ENFORCEMENT   The graphs of the two radical functions shown in the illustration can be used to estimate the speed (in mph) of a car involved in an accident. Suppose a police accident report listed skid marks to be 220 feet long but failed to give the road conditions. Estimate the possible speeds the car was traveling prior to the brakes being applied.

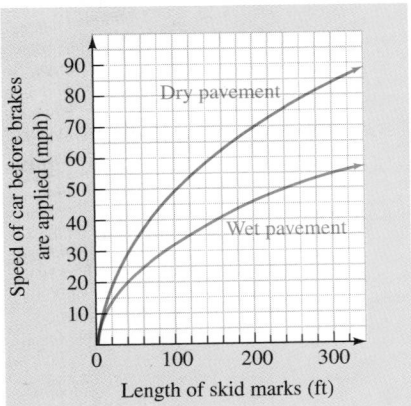

WRITING

115. If x is any real number, that is, if x is unrestricted, then $\sqrt{x^2} = x$ is not correct. Explain.

116. Explain why $\sqrt{36}$ is just 6, and not -6.

117. Explain what is wrong with the graph in the illustration if it is supposed to be the graph of $f(x) = \sqrt{x}$.

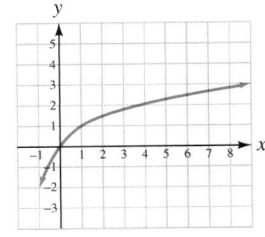

118. Explain how to estimate the domain and range of the radical function shown below.

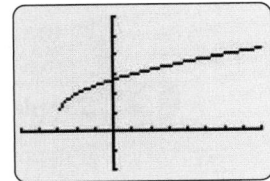

REVIEW

Perform the operations and simplify when possible.

119. $\dfrac{x^2 - x - 6}{x^2 - 2x - 3} \cdot \dfrac{x^2 - 1}{x^2 + x - 2}$

120. $\dfrac{x^2 - 3xy - 4y^2}{x^2 + cx - 2yx - 2cy} \div \dfrac{x^2 - 2xy - 3y^2}{x^2 + cx - 4yx - 4cy}$

121. $\dfrac{3}{m + 1} + \dfrac{3m}{m - 1}$

122. $\dfrac{2x + 3}{3x - 1} - \dfrac{x - 4}{2x + 1}$

CHALLENGE PROBLEMS

123. Graph $f(x) = -\sqrt{x - 2} + 3$ and find the domain and range.

124. Simplify $\sqrt{9a^{16} + 12a^8 b^{25} + 4b^{50}}$ and assume that $a > 0$ and $b > 0$.

SECTION 9.2
Rational Exponents

Objectives

1. Simplify expressions of the form $a^{1/n}$.
2. Simplify expressions of the form $a^{m/n}$.
3. Convert between radicals and rational exponents.
4. Simplify expressions with negative rational exponents.
5. Use rules for exponents to simplify expressions.
6. Simplify radical expressions.

In this section, we will extend the definition of exponent to include rational (fractional) exponents. We will see how expressions such as $9^{1/2}$, $\left(\frac{1}{16}\right)^{3/4}$, and $(-32x^5)^{-2/5}$ can be simplified by writing them in an equivalent radical form using two new rules for exponents.

Simplify Expressions of the Form $a^{1/n}$.

It is possible to raise numbers to fractional powers. To give meaning to rational exponents, we first consider $\sqrt{7}$. Because $\sqrt{7}$ is the positive number whose square is 7, we have

The Language of Algebra
Rational exponents are also called *fractional exponents*.

$$\left(\sqrt{7}\right)^2 = 7$$

We now consider the notation $7^{1/2}$. If rational exponents are to follow the same rules as integer exponents, the square of $7^{1/2}$ must be 7, because

$$(7^{1/2})^2 = 7^{1/2 \cdot 2} \quad \text{Keep the base and multiply the exponents.}$$
$$= 7^1 \quad \text{Do the multiplication: } \tfrac{1}{2} \cdot 2 = 1.$$
$$= 7$$

Since the square of $7^{1/2}$ and the square of $\sqrt{7}$ are both equal to 7, we define $7^{1/2}$ to be $\sqrt{7}$. Similarly,

$$7^{1/3} = \sqrt[3]{7}, \qquad 7^{1/4} = \sqrt[4]{7}, \qquad \text{and} \qquad 7^{1/5} = \sqrt[5]{7}$$

In general, we have the following definition.

The Definition of $x^{1/n}$

A **rational exponent** of $\frac{1}{n}$ indicates the *n*th root of its base.

If *n* represents a positive integer greater than 1 and $\sqrt[n]{x}$ represents a real number,

$$x^{1/n} = \sqrt[n]{x}$$

We can use this definition to simplify exponential expressions that have rational exponents with a numerator of 1. For example, to simplify $8^{1/3}$, we write it as an equivalent expression in radical form and proceed as follows:

$$\underbrace{8}_{\text{Radicand}}{}^{\overbrace{1/3}^{\text{Index}}} = \sqrt[3]{8} = 2$$

The base of the exponential expression, 8, is the radicand of the radical expression. The denominator of the fractional exponent, 3, is the index of the radical.

Thus, $8^{1/3} = 2$.

EXAMPLE 1 Evaluate: **a.** $9^{1/2}$ **b.** $(-64)^{1/3}$ **c.** $16^{1/4}$ **d.** $-\left(\dfrac{1}{32}\right)^{1/5}$

Strategy First, we will identify the base and the exponent of the exponential expression. Then we will write the expression in an equivalent radical form using the rule for rational exponents $x^{1/n} = \sqrt[n]{x}$.

Why We can then use the methods from Section 9.1 to evaluate the resulting square root, cube root, fourth root, and fifth root.

Solution

a. $9^{1/2} = \sqrt{9} = 3$ *Because the denominator of the exponent is 2, find the square root of the base, 9.*

b. $(-64)^{1/3} = \sqrt[3]{-64} = -4$ *Because the denominator of the exponent is 3, find the cube root of the base, −64.*

c. $16^{1/4} = \sqrt[4]{16} = 2$ *Because the denominator of the exponent is 4, find the fourth root of the base, 16.*

d. $-\left(\dfrac{1}{32}\right)^{1/5} = -\sqrt[5]{\dfrac{1}{32}} = -\dfrac{1}{2}$ *Because the denominator of the exponent is 5, find the fifth root of the base, $\frac{1}{32}$.*

Self Check 1 Evaluate: **a.** $16^{1/2}$ **b.** $\left(-\dfrac{27}{8}\right)^{1/3}$

  c. $-(81)^{1/4}$ **d.** $1^{1/5}$

Now Try Problems 17, 21, 23, and 25

As with radicals, when n is an *odd natural number* in the expression $x^{1/n}$, where $n > 1$, there is exactly one real nth root, and we don't need to use absolute value symbols.

When n is an *even natural number,* there are two nth roots. Since we want the expression $x^{1/n}$ to represent the positive nth root, we must often use absolute value symbols to guarantee that the simplified result is positive. Thus, if n is even,

$$(x^n)^{1/n} = |x|$$

When n is even and x is negative, the expression $x^{1/n}$ is not a real number.

EXAMPLE 2 Simplify each expression, if possible. Assume that the variables can be any real number. **a.** $(-27x^3)^{1/3}$ **b.** $(256a^8)^{1/8}$

  c. $[(y + 4)^2]^{1/2}$ **d.** $(25b^4)^{1/2}$ **e.** $(-256x^4)^{1/4}$

Strategy We will write each exponential expression in an equivalent radical form using the rule for rational exponents $x^{1/n} = \sqrt[n]{x}$.

Why We can then use the methods of Section 9.1 to simplify the resulting radical expression.

Solution

a. $(-27x^3)^{1/3} = \sqrt[3]{-27x^3} = -3x$ *Because $(-3x)^3 = -27x^3$. Since n is odd, no absolute value symbols are needed.*

b. $(256a^8)^{1/8} = \sqrt[8]{256a^8} = 2|a|$ *Because $(2|a|)^8 = 256a^8$. Since n is even and a can be any real number, 2a can be negative. Thus, absolute value symbols are needed.*

c. $[(y + 4)^2]^{1/2} = \sqrt{(y + 4)^2} = |y + 4|$ Because $|y + 4|^2 = (y + 4)^2$. Since n is even and y can be any real number, $y + 4$ can be negative. Thus, absolute value symbols are needed.

d. $(25b^4)^{1/2} = \sqrt{25b^4} = 5b^2$ Because $(5b^2)^2 = 25b^4$. Since $b^2 \geq 0$, no absolute value symbols are needed.

e. $(-256x^4)^{1/4} = \sqrt[4]{-256x^4}$, which is not a real number   Because no real number raised to the 4th power is -256.

Self Check 2   Simplify each expression, if possible. Assume that the variables can be any real number.
    a. $(-8n^3)^{1/3}$ **b.** $(625a^4)^{1/4}$
    c. $(b^4)^{1/2}$ **d.** $(-64b^{12})^{1/6}$

Now Try   Problems 27, 35, 37, and 39

If we were told that the variables represent positive real numbers in parts (b) and (c) of Example 2, the absolute value symbols in the answers would not be needed.

$(256a^8)^{1/8} = 2a$ If a represents a positive real number, then $2a$ is positive.

$[(y + 4)^2]^{1/2} = y + 4$ If y represents a positive real number, then $y + 4$ is positive.

We summarize the cases as follows.

Summary of the Definitions of $x^{1/n}$

If n is a natural number greater than 1 and x is a real number,

If $x > 0$, then $x^{1/n}$ is the real number such that $(x^{1/n})^n = x$.

If $x = 0$, then $x^{1/n} = 0$.

If $x < 0$ $\begin{cases} \text{and } n \text{ is odd, then } x^{1/n} \text{ is the negative number such that } (x^{1/n})^n = x. \\ \text{and } n \text{ is even, then } x^{1/n} \text{ is not a real number.} \end{cases}$

2 Simplify Expressions of the Form $a^{m/n}$.

We can extend the definition of $x^{1/n}$ to include fractional exponents with numerators other than 1. For example, since $8^{2/3}$ can be written as $(8^{1/3})^2$, we have

$8^{2/3} = (8^{1/3})^2$

$\qquad = (\sqrt[3]{8})^2$ Write $8^{1/3}$ in radical form.

$\qquad = 2^2$ Find the cube root first: $\sqrt[3]{8} = 2$.

$\qquad = 4$ Then find the power.

Thus, we can simplify $8^{2/3}$ by finding the second power of the cube root of 8.

The numerator of the rational exponent is the power.

$8^{2/3} = (\sqrt[3]{8})^2$ The base of the exponential expression is the radicand.

The denominator of the rational exponent is the index of the radical.

We can also simplify $8^{2/3}$ by taking the cube root of 8 squared.

$$8^{2/3} = (8^2)^{1/3}$$

$$= 64^{1/3} \qquad \text{Find the power first: } 8^2 = 64.$$

$$= \sqrt[3]{64} \qquad \text{Write } 64^{1/3} \text{ in radical form.}$$

$$= 4 \qquad \text{Now find the cube root.}$$

In general, we have the following definition.

The Definition of $x^{m/n}$	If m and n represent positive integers ($n \neq 1$) and $\sqrt[n]{x}$ represents a real number, $$x^{m/n} = \left(\sqrt[n]{x}\right)^m \qquad \text{and} \qquad x^{m/n} = \sqrt[n]{x^m}$$

Because of the previous definition, we can interpret $x^{m/n}$ in two ways:

1. $x^{m/n}$ means the nth root of the mth power of x.
2. $x^{m/n}$ means the mth power of the nth root of x.

We can use this definition to evaluate exponential expressions that have rational exponents with a numerator that is not 1. To avoid large numbers, we usually find the root of the base first and then calculate the power using the rule $x^{m/n} = \left(\sqrt[n]{x}\right)^m$.

EXAMPLE 3   Evaluate: **a.** $32^{2/5}$ **b.** $81^{3/4}$ **c.** $(-64)^{2/3}$ **d.** $-\left(\dfrac{1}{25}\right)^{3/2}$

Strategy   First, we will identify the base and the exponent of the exponential expression. Then we will write the expression in an equivalent radical form using the rule for rational exponents $x^{m/n} = \left(\sqrt[n]{x}\right)^m$.

Why   We know how to evaluate square roots, cube roots, fourth roots, and fifth roots.

Solution

a. To evaluate $32^{2/5}$, we write it in an equivalent radical form. The denominator of the rational exponent is the same as the index of the corresponding radical. The numerator of the rational exponent indicates the power to which the radical base is raised.

$$32^{2/5} = \left(\sqrt[5]{32}\right)^2 = (2)^2 = 4 \qquad \text{Because the exponent is 2/5, find the fifth root of the base, 32, to get 2. Then find the second power of 2.}$$

$$\text{b. } 81^{3/4} = \left(\sqrt[4]{81}\right)^3 = (3)^3 = 27 \qquad \text{Because the exponent is 3/4, find the fourth root of the base, 81, to get 3. Then find the third power of 3.}$$

c. For $(-64)^{2/3}$, the base is -64.

$$(-64)^{2/3} = \left(\sqrt[3]{-64}\right)^2 = (-4)^2 = 16 \qquad \text{Because the exponent is 2/3, find the cube root of the base, } -64, \text{ to get } -4. \text{ Then find the second power of } -4.$$

d. For $-\left(\dfrac{1}{25}\right)^{3/2}$, the base is $\dfrac{1}{25}$, not $-\dfrac{1}{25}$.

Power
Root

$$-\left(\frac{1}{25}\right)^{3/2} = -\left(\sqrt[2]{\frac{1}{25}}\right)^{3} = -\left(\frac{1}{5}\right)^{3} = -\frac{1}{125}$$

Because the exponent is 3/2, find the square root of the base, $\frac{1}{25}$, to get $\frac{1}{5}$. Then find the third power of $\frac{1}{5}$.

 Self Check 3   Evaluate: **a.** $16^{3/2}$ **b.** $125^{4/3}$

c. $(-216)^{2/3}$ **d.** $-\left(\dfrac{1}{32}\right)^{4/5}$

Now Try **Problems 43, 47, and 49**

EXAMPLE 4   Simplify each expression. All variables represent positive real numbers.   **a.** $(36m^4)^{3/2}$ **b.** $(-8x^3)^{4/3}$ **c.** $-(x^5y^5)^{2/5}$

Strategy   We will write each exponential expression in an equivalent radical form using the rule for rational exponents $x^{m/n} = \left(\sqrt[n]{x}\right)^m$.

Why   We can then use the methods of Section 9.1 to simplify the resulting radical expression.

Solution

Power
Root

a. $(36m^4)^{3/2} = \left(\sqrt[2]{36m^4}\right)^3 = (6m^2)^3 = 216m^6$

Because the exponent is 3/2, find the square root of the base, $36m^4$, to get $6m^2$. Then find the third power of $6m^2$.

Power
Root

b. $(-8x^3)^{4/3} = \left(\sqrt[3]{-8x^3}\right)^4 = (-2x)^4 = 16x^4$

Because the exponent is 4/3, find the cube root of the base, $-8x^3$, to get $-2x$. Then find the fourth power of $-2x$.

Power
Root

c. $-(x^5y^5)^{2/5} = -\left(\sqrt[5]{x^5y^5}\right)^2 = -(xy)^2 = -x^2y^2$

Because the exponent is 2/5, find the fifth root of the base, x^5y^5, to get xy. Then find the second power of xy.

 Self Check 4   Simplify each expression. All variables represent positive real numbers.   **a.** $(4c^4)^{3/2}$ **b.** $(-27m^3n^3)^{2/3}$

c. $-(32a^{10})^{3/5}$

Now Try **Problems 51 and 53**

Using Your Calculator   *Rational Exponents*

We can evaluate expressions containing rational exponents using the exponential key $\boxed{y^x}$ or $\boxed{x^y}$ on a scientific calculator. For example, to evaluate $10^{2/3}$, we enter

10 $\boxed{y^x}$ $\boxed{(}$ 2 $\boxed{\div}$ 3 $\boxed{)}$ $\boxed{=}$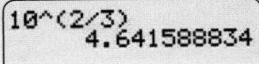

Note that parentheses were used when entering the power. Without them, the calculator would interpret the entry as $10^2 \div 3$.

To evaluate the exponential expression using a graphing calculator, we use the $\boxed{\wedge}$ key, which raises a base to a power. Again, we use parentheses when entering the power.

10 $\boxed{\wedge}$ $\boxed{(}$ 2 $\boxed{\div}$ 3 $\boxed{)}$ $\boxed{\text{ENTER}}$

```
10^(2/3)
        4.641588834
```

To the nearest hundredth, $10^{2/3} \approx 4.64$.

3 **Convert Between Radicals and Rational Exponents.**

We can use the rules for rational exponents to convert expressions from radical form to exponential form, and vice versa.

EXAMPLE 5   Write $\sqrt{5xyz}$ as an exponential expression with a rational exponent.

Strategy   We will use the first rule for rational exponents in reverse: $\sqrt[n]{x} = x^{1/n}$.

Why   We are given a radical expression and we want to write an equivalent exponential expression.

Solution   The radicand is $5xyz$, so the base of the exponential expression is $5xyz$. The index of the radical is an understood 2, so the denominator of the fractional exponent is 2.

$$\sqrt{5xyz} = (5xyz)^{1/2}$$ Recall: $\sqrt[2]{5xyz} = \sqrt{5xyz}$.

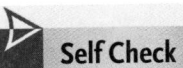   **Self Check 5**   Write $\sqrt[6]{7ab}$ as an exponential expression with a rational exponent.

Now Try   **Problem 59**

Rational exponents appear in formulas used in many disciplines, such as science and engineering.

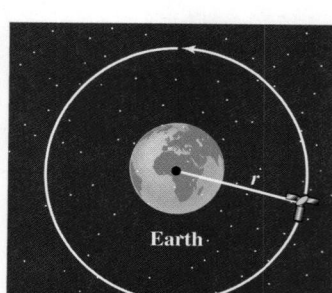

Earth

EXAMPLE 6   *Satellites.*   The formula

$$r = \left(\frac{GMP^2}{4\pi^2}\right)^{1/3}$$

gives the orbital radius (in meters) of a satellite circling Earth, where G and M are constants and P is the time in seconds for the satellite to make one complete revolution. Write the formula using a radical.

Strategy   We will use the first rule for rational exponents: $x^{1/n} = \sqrt[n]{x}$.

Why   We are given an exponential expression involving a rational exponent with a numerator of 1 and we want to write an equivalent radical expression.

Solution   The fractional exponent $\frac{1}{3}$, with a numerator of 1 and a denominator of 3, indicates that we are to find the cube root of the base of the exponential expression. So we have

$$r = \sqrt[3]{\frac{GMP^2}{4\pi^2}}$$

 Now Try   **Problem 63**

4   **Simplify Expressions with Negative Rational Exponents.**

To be consistent with the definition of negative integer exponents, we define $x^{-m/n}$ as follows.

Definition of $x^{-m/n}$	If m and n are positive integers, $\frac{m}{n}$ is in simplified form, and $x^{1/n}$ is a real number, then

$$x^{-m/n} = \frac{1}{x^{m/n}} \qquad \text{and} \qquad \frac{1}{x^{-m/n}} = x^{m/n} \quad (x \neq 0)$$

From the definition, we see that another way to write $x^{-m/n}$ is to write its reciprocal and change the sign of the exponent.

EXAMPLE 7   Simplify each expression. Assume that x can represent any nonzero real number.   **a.** $64^{-1/2}$ **b.** $(-16)^{-5/4}$

c. $-625^{-3/4}$ **d.** $(-32x^5)^{-2/5}$ **e.** $\dfrac{1}{25^{-3/2}}$

Strategy   We will use one of the rules $x^{-m/n} = \dfrac{1}{x^{m/n}}$ or $\dfrac{1}{x^{-m/n}} = x^{m/n}$ to write the reciprocal of each exponential expression and change the exponent's sign to positive.

Why   If we can produce an equivalent expression having a positive rational exponent, we can use the methods of this section to simplify it.

Solution

a.

Reciprocal

$$64^{-1/2} = \frac{1}{64^{1/2}} = \frac{1}{\sqrt{64}} = \frac{1}{8}$$

Change sign

Because the exponent is negative, write the reciprocal of $64^{-1/2}$, and change the sign of the exponent.

Caution

A negative exponent does not indicate a negative number. For example,

$$64^{-1/2} = \frac{1}{8}$$

b. $(-16)^{-5/4}$ is not a real number because $(-16)^{1/4}$ is not a real number.

c. In $-625^{-3/4}$, the base is 625.

$$-625^{-3/4} = -\frac{1}{625^{3/4}} = -\frac{1}{\left(\sqrt[4]{625}\right)^3} = -\frac{1}{(5)^3} = -\frac{1}{125}$$

d. $(-32x^5)^{-2/5} = \frac{1}{(-32x^5)^{2/5}} = \frac{1}{\left(\sqrt[5]{-32x^5}\right)^2} = \frac{1}{(-2x)^2} = \frac{1}{4x^2}$

Caution

A base of 0 raised to a negative power is undefined. For example, $0^{-2} = \frac{1}{0^2}$ is undefined because we cannot divide by 0.

e. $\dfrac{1}{25^{-3/2}} = 25^{3/2} = (\sqrt{25})^3 = (5)^3 = 125$

Because the exponent is negative, write the reciprocal of $\frac{1}{25^{-3/2}}$, and change the sign of the exponent.

Self Check 7   Simplify. Assume that a can represent any nonzero real number.

a. $9^{-1/2}$ **b.** $(36)^{-3/2}$

c. $(-27a^3)^{-2/3}$ **d.** $-\dfrac{1}{81^{-3/4}}$

Now Try   Problems **67, 73,** and **79**

5   **Use the Rules for Exponents to Simplify Expressions.**

We can use the rules for exponents to simplify many expressions with fractional exponents. If all variables represent positive real numbers, absolute value symbols are not needed.

EXAMPLE 8   Simplify each expression. All variables represent positive real numbers. Write all answers using positive exponents only.

a. $5^{2/7}5^{3/7}$ **b.** $(11^{2/7})^3$ **c.** $(a^{2/3}b^{1/2})^6$ **d.** $\dfrac{a^{8/3}a^{1/3}}{a^2}$

Strategy   We will use the product, power, and quotient rules for exponents to simplify each expression.

Why   The familiar rules for exponents discussed in Chapter 5 are valid for rational exponents.

Solution

a. $5^{2/7}5^{3/7} = 5^{2/7 + 3/7}$ Use the rule $x^m x^n = x^{m+n}$.

$\qquad\quad = 5^{5/7}$ Add: $\frac{2}{7} + \frac{3}{7} = \frac{5}{7}$.

b. $(11^{2/7})^3 = 11^{(2/7)(3)}$ Use the rule $(x^m)^n = x^{mn}$.

$\qquad\quad = 11^{6/7}$ Multiply: $\frac{2}{7}(3) = \frac{6}{7}$.

c. $(a^{2/3}b^{1/2})^6 = (a^{2/3})^6(b^{1/2})^6$ Use the rule $(xy)^n = x^n y^n$.

$\qquad\qquad = a^{12/3}b^{6/2}$ Use the rule $(x^m)^n = x^{mn}$ twice.

$\qquad\qquad = a^4 b^3$ Simplify the exponents.

d. $\dfrac{a^{8/3}a^{1/3}}{a^2} = a^{8/3 + 1/3 - 2}$ Use the rules $x^m x^n = x^{m+n}$ and $\dfrac{x^m}{x^n} = x^{m-n}$.

$\qquad\qquad = a^{8/3 + 1/3 - 6/3}$ To establish an LCD, write -2 as $-\dfrac{6}{3}$.

$\qquad\qquad = a^{3/3}$ Simplify: $\dfrac{8}{3} + \dfrac{1}{3} - \dfrac{6}{3} = \dfrac{3}{3}$.

$\qquad\qquad = a$ Simplify: $\dfrac{3}{3} = 1$.

Self Check 8   Simplify. All variables represent positive real numbers.
a. $2^{1/5}2^{2/5}$ **b.** $(12^{1/3})^4$
c. $(x^{1/3}y^{3/2})^6$ **d.** $\dfrac{x^{5/3}x^{2/3}}{x^{1/3}}$

Now Try   **Problems 83, 87, and 91**

EXAMPLE 9   Perform each multiplication and simplify when possible. Assume all variables represent positive real numbers. Write all answers using positive exponents only.
a. $a^{4/5}(a^{1/5} + a^{3/5})$ **b.** $x^{1/2}(x^{-1/2} - x^{1/2})$

Strategy   We will use the distributive property and multiply each term within the parentheses by the term outside the parentheses.

Why   The first expression has the form $a(b + c)$ and the second has the form $a(b - c)$.

Solution

a. $a^{4/5}(a^{1/5} + a^{3/5}) = a^{4/5}a^{1/5} + a^{4/5}a^{3/5}$ Use the distributive property.

$\qquad\qquad = a^{4/5 + 1/5} + a^{4/5 + 3/5}$ Use the rule $x^m x^n = x^{m+n}$.

$\qquad\qquad = a^{5/5} + a^{7/5}$ Add the exponents.

$\qquad\qquad = a + a^{7/5}$ We cannot add these terms because they are not like terms.

b. $x^{1/2}(x^{-1/2} - x^{1/2}) = x^{1/2}x^{-1/2} - x^{1/2}x^{1/2}$ Use the distributive property.

$\qquad\qquad = x^{1/2 + (-1/2)} - x^{1/2 + 1/2}$ Use the rule $x^m x^n = x^{m+n}$.

$\qquad\qquad = x^0 - x^1$ Add the exponents.

$\qquad\qquad = 1 - x$ Simplify: $x^0 = 1$.

Self Check 9   Simplify: $t^{5/8}(t^{3/8} - t^{-5/8})$. Assume t represents a positive real number.

Now Try   **Problem 97**

---

6   Simplify Radical Expressions.

We can simplify many radical expressions by using the following steps.

Using Rational Exponents to Simplify Radicals	1. Change the radical expression into an exponential expression.
	2. Simplify the rational exponents.
	3. Change the exponential expression back into a radical.

EXAMPLE 10 Simplify: **a.** $\sqrt[4]{3^2}$ **b.** $\sqrt[8]{x^6}$ **c.** $\sqrt[9]{27x^6y^3}$ **d.** $\sqrt[5]{\sqrt[3]{t}}$

Strategy We will write each radical expression as an equivalent exponential expression and use rules for exponents to simplify it. Then we will change that result back into a radical.

Why When the given expression is written in an equivalent exponential form, we can use rules for exponents and our arithmetic skills with fractions to simplify the exponents.

Solution

a. $\sqrt[4]{3^2} = (3^2)^{1/4}$ *Change the radical to an exponential expression.*

$= 3^{2/4}$ *Use the rule $(x^m)^n = x^{mn}$.*

$= 3^{1/2}$ *Simplify the fractional exponent: $\frac{2}{4} = \frac{1}{2}$.*

$= \sqrt{3}$ *Change back to radical form.*

b. $\sqrt[8]{x^6} = (x^6)^{1/8}$ *Change the radical to an exponential expression.*

$= x^{6/8}$ *Use the rule $(x^m)^n = x^{mn}$.*

$= x^{3/4}$ *Simplify the fractional exponent: $\frac{6}{8} = \frac{3}{4}$.*

$= (x^3)^{1/4}$ *Write $\frac{3}{4}$ as $3\left(\frac{1}{4}\right)$.*

$= \sqrt[4]{x^3}$ *Change back to radical form.*

c. $\sqrt[9]{27x^6y^3} = (3^3x^6y^3)^{1/9}$ *Write 27 as 3^3 and change the radical to an exponential expression.*

$= 3^{3/9}x^{6/9}y^{3/9}$ *Raise each factor to the $\frac{1}{9}$ power by multiplying the fractional exponents.*

$= 3^{1/3}x^{2/3}y^{1/3}$ *Simplify each fractional exponent.*

$= (3x^2y)^{1/3}$ *Use the rule $(xy)^n = x^ny^n$.*

$= \sqrt[3]{3x^2y}$ *Change back to radical form.*

d. $\sqrt[5]{\sqrt[3]{t}} = \sqrt[5]{t^{1/3}}$ *Change the radical $\sqrt[3]{t}$ to exponential notation.*

$= (t^{1/3})^{1/5}$ *Change the radical $\sqrt[5]{t^{1/3}}$ to exponential notation.*

$= t^{1/15}$ *Use the rule $(x^m)^n = x^{mn}$. Multiply: $\frac{1}{3} \cdot \frac{1}{5} = \frac{1}{15}$.*

$= \sqrt[15]{t}$ *Change back to radical form.*

Self Check 10 Simplify: **a.** $\sqrt[6]{3^3}$ **b.** $\sqrt[4]{49x^2y^2}$

c. $\sqrt[3]{\sqrt[4]{m}}$

Now Try **Problems 99, 105, and 107**

STUDY SET
9.2

VOCABULARY

Fill in the blanks.

1. The expressions $4^{1/2}$ and $(-8)^{-2/3}$ have _____ exponents.

2. In the exponential expression $27^{4/3}$, 27 is the _____ and 4/3 is the _____.

3. In the radical expression $\sqrt[4]{16x^8}$, 4 is the _____, and $16x^8$ is the _____.

4. $32^{4/5}$ means the fourth _____ of the fifth _____ of 32.

CONCEPTS

5. Complete the table by writing the given expression in the alternate form.

Radical form	Exponential form
$\sqrt[5]{25}$	
	$(-27)^{2/3}$
$\left(\sqrt[4]{16}\right)^{-3}$	
	$81^{3/2}$
$-\sqrt{\dfrac{9}{64}}$	

6. In your own words, explain the three rules for rational exponents illustrated in the diagrams below.

a. $(-32)^{1/5} = \sqrt[5]{-32}$

b. $125^{4/3} = (\sqrt[3]{125})^4$

c. $8^{-1/3} = \dfrac{1}{8^{1/3}}$

7. Graph each number on the number line.

$$\left\{ 8^{2/3}, (-125)^{1/3}, -16^{-1/4}, 4^{3/2}, -\left(\frac{9}{100}\right)^{-1/2} \right\}$$

8. a. Evaluate $25^{3/2}$ by writing in the form $(25^{1/2})^3$.
   b. Evaluate $25^{3/2}$ by writing in the form $(25^3)^{1/2}$.
   c. Which way was easier?

Complete each rule for exponents.

9. $x^{1/n} =$ [ ]

10. $x^{m/n} =$ [ ] $= \sqrt[n]{x^m}$

11. $x^{-m/n} =$ [ ]

12. $\dfrac{1}{x^{-m/n}} =$ [ ]

NOTATION

Complete each solution.

13. Simplify:
$$(100a^4)^{3/2} = \left(\sqrt{}\right)^3$$
$$= \left(\right)^3$$
$$= 1,000a^6$$

14. Simplify:
$$(m^{1/3}n^{1/2})^6 = \left(\right)^6 (n^{1/2})^6$$
$$= m^{} n^{6/2}$$
$$= m^2 n^3$$

GUIDED PRACTICE

Evaluate each expression. See Example 1.

15. $125^{1/3}$
16. $8^{1/3}$
17. $81^{1/4}$
18. $625^{1/4}$
19. $32^{1/5}$
20. $0^{1/5}$
21. $(-216)^{1/3}$
22. $(-1,000)^{1/3}$
23. $-16^{1/4}$
24. $-125^{1/3}$
25. $\left(\dfrac{1}{4}\right)^{1/2}$
26. $\left(\dfrac{1}{16}\right)^{1/2}$

Simplify each expression, if possible. Assume that the variables can be any real number, and use absolute value symbols when necessary. See Example 2.

27. $(4x^4)^{1/2}$
28. $(25a^8)^{1/2}$
29. $(x^2)^{1/2}$
30. $(x^3)^{1/3}$
31. $(m^4)^{1/2}$
32. $(a^4)^{1/4}$
33. $(-64p^8)^{1/2}$
34. $(-16q^4)^{1/2}$
35. $(-27n^9)^{1/3}$
36. $(-64t^9)^{1/3}$
37. $(16x^4)^{1/4}$
38. $(-x^4)^{1/4}$
39. $(-64x^8)^{1/8}$
40. $(243x^{10})^{1/5}$
41. $[(x+1)^6]^{1/6}$
42. $[(x+5)^8]^{1/8}$

Evaluate each expression. See Example 3.

43. $36^{3/2}$

44. $27^{2/3}$

45. $16^{3/4}$

46. $-100^{3/2}$

47. $\left(-\dfrac{1}{216}\right)^{2/3}$

48. $\left(\dfrac{4}{9}\right)^{3/2}$

49. $-4^{5/2}$

50. $(-125)^{4/3}$

Simplify each expression. All variables represent positive real numbers. See Example 4.

51. $(25x^4)^{3/2}$

52. $(27a^3b^3)^{2/3}$

53. $(-8x^6y^3)^{2/3}$

54. $(-32x^{10}y^5)^{4/5}$

55. $(81x^4y^8)^{3/4}$

56. $\left(\dfrac{1}{16}x^8y^4\right)^{3/4}$

57. $-\left(\dfrac{x^5}{32}\right)^{4/5}$

58. $-\left(\dfrac{27}{64y^6}\right)^{2/3}$

Change each radical to an exponential expression. See Example 5.

59. $\sqrt[5]{8abc}$

60. $\sqrt[7]{7p^2q}$

61. $\sqrt[3]{a^2 - b^2}$

62. $\sqrt{x^2 + y^2}$

Change each exponential expression to a radical. See Example 6.

63. $(6x^3y)^{1/4}$

64. $(7a^2b^2)^{1/5}$

65. $(2s^2 - t^2)^{1/2}$

66. $(x^3 + y^3)^{1/3}$

Simplify each expression. All variables represent positive real numbers. See Example 7.

67. $4^{-1/2}$

68. $49^{-1/2}$

69. $125^{-1/3}$

70. $8^{-1/3}$

71. $16^{-3/2}$

72. $(16)^{-5/4}$

73. $-(1,000y^3)^{-2/3}$

74. $-(81c^4)^{-3/2}$

75. $\left(-\dfrac{27}{8}\right)^{-4/3}$

76. $\left(\dfrac{25}{49}\right)^{-3/2}$

77. $\left(\dfrac{16}{81y^4}\right)^{-3/4}$

78. $\left(-\dfrac{8x^3}{27}\right)^{-1/3}$

79. $\dfrac{1}{32^{-1/5}}$

80. $\dfrac{1}{64^{-1/6}}$

81. $\dfrac{1}{9^{-5/2}}$

82. $\dfrac{1}{16^{-5/2}}$

Simplify each expression. Write the answers without negative exponents. All variables represent positive real numbers. See Example 8.

83. $9^{3/7}9^{2/7}$

84. $4^{2/5}4^{2/5}$

85. $6^{-2/3}6^{-4/3}$

86. $5^{1/3}5^{-5/3}$

87. $(m^{2/3}m^{1/3})^6$

88. $(b^{3/5}b^{2/5})^8$

89. $(a^{1/2}b^{1/3})^{3/2}$

90. $(mn^{-2/3})^{-3/5}$

91. $\dfrac{3^{4/3}3^{1/3}}{3^{2/3}}$

92. $\dfrac{2^{5/6}2^{1/3}}{2^{1/2}}$

93. $\dfrac{a^{3/4}a^{3/4}}{a^{1/2}}$

94. $\dfrac{b^{4/5}b^{4/5}}{b^{3/5}}$

Perform the multiplications. All variables represent positive real numbers. See Example 9.

95. $y^{1/3}(y^{2/3} + y^{5/3})$

96. $y^{2/5}(y^{-2/5} + y^{3/5})$

97. $x^{3/5}(x^{7/5} - x^{-3/5} + 1)$

98. $x^{4/3}(x^{2/3} + 3x^{5/3} - 4)$

Use rational exponents to simplify each radical. All variables represent positive real numbers. See Example 10.

99. $\sqrt[4]{5^2}$

100. $\sqrt[6]{7^3}$

101. $\sqrt[9]{11^3}$

102. $\sqrt[12]{13^4}$

103. $\sqrt[6]{p^3}$

104. $\sqrt[8]{q^2}$

105. $\sqrt[10]{x^2y^2}$

106. $\sqrt[6]{x^2y^2}$

107. $\sqrt[9]{\sqrt{c}}$

108. $\sqrt[4]{\sqrt{x}}$

109. $\sqrt[5]{\sqrt[3]{7m}}$

110. $\sqrt[3]{\sqrt[4]{21x}}$

Use a calculator to evaluate each expression. Round to the nearest hundredth. See Using Your Calculator: Rational Exponents.

111. $15^{1/3}$

112. $(50.5)^{1/4}$

113. $(1.045)^{2/5}$

114. $(-1,000)^{3/5}$

TRY IT YOURSELF

Simplify each expression. All variables represent positive real numbers.

115. $(25y^2)^{1/2}$

116. $(-27x^3)^{1/3}$

117. $-\left(\dfrac{a^4}{81}\right)^{3/4}$

118. $-\left(\dfrac{b^8}{625}\right)^{3/4}$

119. $\dfrac{p^{8/5}p^{7/5}}{p^2}$

120. $\dfrac{c^{2/3}c^{2/3}}{c^{1/3}}$

121. $(-27x^6)^{-1/3}$

122. $(16a^4)^{-1/2}$

123. $n^{1/5}(n^{2/5} - n^{-1/5})$

124. $t^{4/3}(t^{5/3} + t^{-4/3})$

125. $\dfrac{1}{9^{-5/2}}$

126. $\dfrac{1}{16^{-5/2}}$

127. $\sqrt[4]{25b^2}$

128. $\sqrt[9]{8x^6}$

129. $-(8a^3b^6)^{-2/3}$

130. $-(25s^4t^6)^{-3/2}$

APPLICATIONS

131. BALLISTIC PENDULUMS   The formula

$$v = \frac{m + M}{m}(2gh)^{1/2}$$

gives the velocity (in ft/sec) of a bullet with weight m fired into a block with weight M, that raises the height of the block h feet after the collision. The letter g represents a constant, 32. Find the velocity of the bullet to the nearest ft/sec.

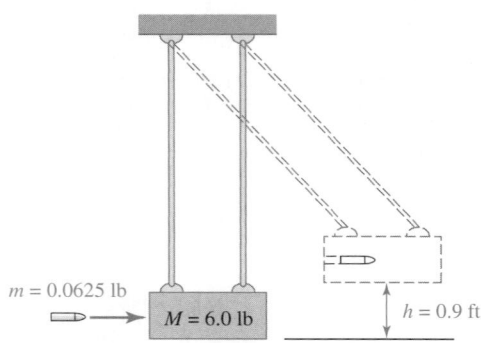

$m = 0.0625$ lb

$M = 6.0$ lb

$h = 0.9$ ft

132. GEOGRAPHY   The formula

$$A = [s(s - a)(s - b)(s - c)]^{1/2}$$

gives the area of a triangle with sides of length a, b, and c, where s is one-half of the perimeter. Estimate the area of Virginia (to the nearest square mile) using the data given in the illustration.

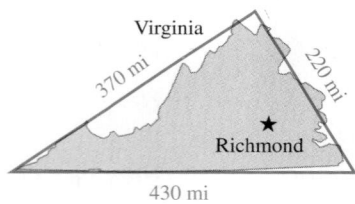

Virginia

370 mi

220 mi

★ Richmond

430 mi

133. RELATIVITY   One concept of relativity theory is that an object moving past an observer at a speed near the speed of light appears to have a larger mass because of its motion. If the mass of the object is m_0 when the object is at rest relative to the observer, its mass m will be given by the formula

$$m = m_0\left(1 - \frac{v^2}{c^2}\right)^{-1/2}$$

when it is moving with speed v (in miles per second) past the observer. The variable c is the speed of light, 186,000 mi/sec. If a proton with a rest mass of 1 unit is accelerated by a nuclear accelerator to a speed of 160,000 mi/sec, what mass will the technicians observe it to have? Round to the nearest hundredth.

134. LOGGING   The width w and height h of the strongest rectangular beam that can be cut from a cylindrical log of radius a are given by

$$w = \frac{2a}{3}(3^{1/2}) \quad \text{and} \quad h = a\left(\frac{8}{3}\right)^{1/2}$$

Find the width, height, and cross-sectional area of the strongest beam that can be cut from a log with *diameter* 4 feet. Round to the nearest hundredth.

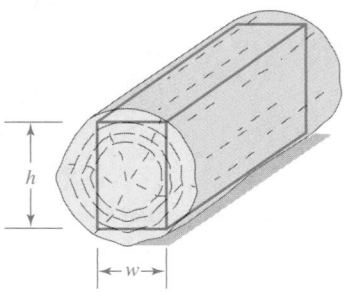

135. *from Campus to Careers*
              General Contractor

The length L of the longest board that can be carried horizontally around the right-angle corner of two intersecting hallways is given by the formula

$$L = (a^{2/3} + b^{2/3})^{3/2}$$

where a and b represent the widths of the hallways. Find the longest shelf that a carpenter can carry around the corner if $a = 40$ in. and $b = 64$ in. Give your result in inches and in feet. In each case, round to the nearest tenth.

© Tetra Images/Alamy

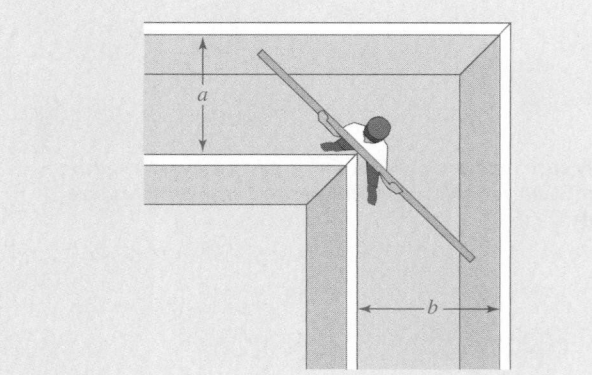

136. CUBICLES   The area of the base of a cube is given by the function $A(V) = V^{2/3}$, where V is the volume of the cube. In a preschool room, 18 children's cubicles like the one shown are placed on the floor around the room. Estimate how much floor space is lost to the cubicles. Give your answer in square inches and in square feet.

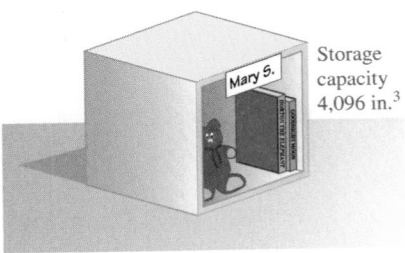

WRITING

137. What is a rational exponent? Give some examples.

138. Explain how the root key $\boxed{\sqrt[x]{y}}$ on a scientific calculator can be used in combination with other keys to evaluate the expression $16^{3/4}$.

REVIEW

139. COMMUTING TIME   The time it takes a car to travel a certain distance varies inversely with its rate of speed. If a certain trip takes 3 hours at 50 miles per hour, how long will the trip take at 60 miles per hour?

140. BANKRUPTCY   After filing for bankruptcy, a company was able to pay its creditors only 15 cents on the dollar. If the company owed a lumberyard \$9,712, how much could the lumberyard expect to be paid?

CHALLENGE PROBLEMS

141. The fraction $\frac{2}{4}$ is equal to $\frac{1}{2}$. Is $16^{2/4}$ equal to $16^{1/2}$? Explain.

142. Explain how would you evaluate an expression with a mixed-number exponent? For example, what is $8^{1\frac{1}{3}}$? What is $25^{2\frac{1}{2}}$?

SECTION 9.3
Simplifying and Combining Radical Expressions

Objectives

1 Use the product rule to simplify radical expressions.
2 Use prime factorization to simplify radical expressions.
3 Use the quotient rule to simplify radical expressions.
4 Add and subtract radical expressions.

In algebra, it is often helpful to replace an expression with a simpler equivalent expression. This is certainly true when working with radicals. In most cases, radical expressions should be written in simplified form. We use two rules for radicals to do this.

1 Use the Product Rule to Simplify Radical Expressions.

To introduce the product rule for radicals, we will find $\sqrt{4 \cdot 25}$ and $\sqrt{4}\sqrt{25}$, and compare the results.

Square root of a product
$$\sqrt{4 \cdot 25} = \sqrt{100}$$
$$= 10$$

Product of square roots
$$\sqrt{4}\sqrt{25} = 2 \cdot 5$$
$$= 10$$

In each case, the answer is 10. Thus, $\sqrt{4 \cdot 25} = \sqrt{4}\sqrt{25}$.

Notation
The products $\sqrt{4}\sqrt{25}$ and $\sqrt[3]{8}\sqrt[3]{27}$ can also be written using a raised dot:

$\sqrt{4} \cdot \sqrt{25}$ $\sqrt[3]{8} \cdot \sqrt[3]{27}$

Similarly, we will find $\sqrt[3]{8 \cdot 27}$ and $\sqrt[3]{8}\sqrt[3]{27}$ and compare the results.

Cube root of a product	**Product of cube roots**
$\sqrt[3]{8 \cdot 27} = \sqrt[3]{216}$	$\sqrt[3]{8}\sqrt[3]{27} = 2 \cdot 3$
$= 6$	$= 6$

In each case, the answer is 6. Thus, $\sqrt[3]{8 \cdot 27} = \sqrt[3]{8}\sqrt[3]{27}$. These results illustrate the *product rule for radicals*.

The Product Rule for Radicals

The *n*th root of the product of two numbers is equal to the product of their *n*th roots. If $\sqrt[n]{a}$ and $\sqrt[n]{b}$ are real numbers,

$$\sqrt[n]{ab} = \sqrt[n]{a}\sqrt[n]{b}$$

Caution   The product rule for radicals applies to the *n*th root of a product. There is no such property for sums or differences. For example,

$$\sqrt{9 + 4} \neq \sqrt{9} + \sqrt{4} \qquad\qquad \sqrt{9 - 4} \neq \sqrt{9} - \sqrt{4}$$
$$\sqrt{13} \neq 3 + 2 \qquad\qquad\qquad \sqrt{5} \neq 3 - 2$$
$$\sqrt{13} \neq 5 \qquad\qquad\qquad\qquad \sqrt{5} \neq 1$$

Thus, $\sqrt{a + b} \neq \sqrt{a} + \sqrt{b}$ and $\sqrt{a - b} \neq \sqrt{a} - \sqrt{b}$.

The product rule for radicals can be used to simplify radical expressions. When a radical expression is written in **simplified form,** each of the following is true.

Simplified Form of a Radical Expression

1. Each factor in the radicand is to a power that is less than the index of the radical.
2. The radicand contains no fractions or negative numbers.
3. No radicals appear in the denominator of a fraction.

To simplify radical expressions, we must often factor the radicand using two natural-number factors. To simplify square-root, cube-root, and fourth-root radicals, it is helpful to memorize the following lists.

Perfect squares: **1, 4, 9, 16, 25, 36, 49, 64, 81, 100, 121, 144, 169, 196, 225, . . .**

Perfect cubes: **1, 8, 27, 64, 125, 216, 343, 512, 729, 1,000, . . .**

Perfect-fourth powers: **1, 16, 81, 256, 625, . . .**

EXAMPLE 1   Simplify: **a.** $\sqrt{12}$ **b.** $\sqrt{98}$ **c.** $\sqrt[3]{54}$ **d.** $-\sqrt[4]{48}$

Strategy   We will factor each radicand into two factors, one of which is a perfect square, perfect cube, or perfect-fourth power, depending on the index of the radical. Then we can use the product rule for radicals to simplify the expression.

Why   Factoring the radicand in this way leads to a square root, cube root, or fourth root of a perfect square, perfect cube, or perfect-fourth power that we can easily simplify.

Solution

a. To simplify $\sqrt{12}$, we first factor 12 so that one factor is the largest perfect square that divides 12. Since 4 is the largest perfect-square factor of 12, we write 12 as $4 \cdot 3$, use the product rule for radicals, and simplify.

$$\sqrt{12} = \sqrt{4 \cdot 3} \qquad \text{Write 12 as } 12 = 4 \cdot 3.$$

Write the perfect-square factor first.

$$= \sqrt{4}\sqrt{3} \qquad \text{The square root of a product is equal to the product of the square roots.}$$

$$= 2\sqrt{3} \qquad \text{Evaluate } \sqrt{4}. \text{ Read as "2 times the square root of 3" or as "2 radical 3."}$$

We say that $2\sqrt{3}$ is the simplified form of $\sqrt{12}$.

b. The largest perfect-square factor of 98 is 49. Thus,

$$\sqrt{98} = \sqrt{49 \cdot 2} \qquad \text{Write 98 in factored form: } 98 = 49 \cdot 2.$$

$$= \sqrt{49}\sqrt{2} \qquad \text{The square root of a product is equal to the product of the square roots: } \sqrt{49 \cdot 2} = \sqrt{49}\sqrt{2}.$$

$$= 7\sqrt{2} \qquad \text{Evaluate } \sqrt{49}.$$

c. Since the largest perfect-cube factor of 54 is 27, we have

$$\sqrt[3]{54} = \sqrt[3]{27 \cdot 2} \qquad \text{Write 54 as } 27 \cdot 2.$$

$$= \sqrt[3]{27}\sqrt[3]{2} \qquad \text{The cube root of a product is equal to the product of the cube roots: } \sqrt[3]{27 \cdot 2} = \sqrt[3]{27}\sqrt[3]{2}.$$

$$= 3\sqrt[3]{2} \qquad \text{Evaluate } \sqrt[3]{27}.$$

d. The largest perfect-fourth power factor of 48 is 16. Thus,

$$-\sqrt[4]{48} = -\sqrt[4]{16 \cdot 3} \qquad \text{Write 48 as } 16 \cdot 3.$$

$$= -\sqrt[4]{16}\sqrt[4]{3} \qquad \text{The fourth root of a product is equal to the product of the fourth roots: } \sqrt[4]{16 \cdot 3} = \sqrt[4]{16} \cdot \sqrt[4]{3}.$$

$$= -2\sqrt[4]{3} \qquad \text{Evaluate } \sqrt[4]{16}.$$

Self Check 1 Simplify: **a.** $\sqrt{20}$ **b.** $\sqrt[3]{24}$ **c.** $\sqrt[5]{128}$

Now Try Problems 13, 17, and 19

Variable expressions can also be perfect squares, perfect cubes, perfect-fourth powers, and so on. For example,

$$\text{Perfect squares: } x^2, x^4, x^6, x^8, x^{10}, \ldots$$
$$\text{Perfect cubes: } x^3, x^6, x^9, x^{12}, x^{15}, \ldots$$
$$\text{Perfect-fourth powers: } x^4, x^8, x^{12}, x^{16}, x^{20}, \ldots$$

EXAMPLE 2 Simplify: **a.** $\sqrt{m^9}$ **b.** $\sqrt{128a^5}$ **c.** $\sqrt[3]{-24x^5}$ **d.** $\sqrt[5]{a^9 b^5}$
All variables represent positive real numbers.

Strategy We will factor each radicand into two factors, one of which is a perfect *n*th power.

Success Tip
In Example 1, a radical of a product is written as a product of radicals:

$$\sqrt[n]{ab} = \sqrt[n]{a}\sqrt[n]{b}$$

Why   We can then apply the rule *the nth root of a product is the product of the nth roots* to simplify the radical expression.

Solution

a. The largest perfect-square factor of m^9 is m^8.

$$\sqrt{m^9} = \sqrt{m^8 \cdot m} \qquad \text{Write } m^9 \text{ in factored form as } m^8 \cdot m.$$
$$= \sqrt{m^8}\sqrt{m} \qquad \text{Use the product rule for radicals.}$$
$$= m^4\sqrt{m} \qquad \text{Simplify } \sqrt{m^8}.$$

b. Since the largest perfect-square factor of 128 is 64 and the largest perfect-square factor of a^5 is a^4, the largest perfect-square factor of $128a^5$ is $64a^4$. We write $128a^5$ as $64a^4 \cdot 2a$ and proceed as follows:

$$\sqrt{128a^5} = \sqrt{64a^4 \cdot 2a} \qquad \text{Write } 128a^5 \text{ in factored form as } 64a^4 \cdot 2a.$$
$$= \sqrt{64a^4}\sqrt{2a} \qquad \text{Use the product rule for radicals.}$$
$$= 8a^2\sqrt{2a} \qquad \text{Simplify } \sqrt{64a^4}.$$

c. We write $-24x^5$ as $-8x^3 \cdot 3x^2$ and proceed as follows:

$$\sqrt[3]{-24x^5} = \sqrt[3]{-8x^3 \cdot 3x^2} \qquad 8x^3 \text{ is the largest perfect-cube factor of } 24x^5. \text{ Since the radicand is negative, we factor it using } -8x^3.$$
$$= \sqrt[3]{-8x^3}\sqrt[3]{3x^2} \qquad \text{Use the product rule for radicals.}$$
$$= -2x\sqrt[3]{3x^2} \qquad \text{Simplify } \sqrt[3]{-8x^3}.$$

d. The largest perfect-fifth power factor of a^9 is a^5, and b^5 is a perfect-fifth power.

$$\sqrt[5]{a^9b^5} = \sqrt[5]{a^5b^5 \cdot a^4} \qquad a^5b^5 \text{ is the largest perfect-fifth power factor of } a^9b^5.$$
$$= \sqrt[5]{a^5b^5}\sqrt[5]{a^4} \qquad \text{Use the product rule for radicals.}$$
$$= ab\sqrt[5]{a^4} \qquad \text{Simplify } \sqrt[5]{a^5b^5}.$$

The Language of Algebra
Perfect-fifth powers of a are
$$a^5, a^{10}, a^{15}, a^{20}, a^{25}, \ldots$$

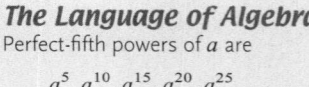

Self Check 2   Simplify. All variables represent positive real numbers.
a. $\sqrt{98b^3}$ **b.** $\sqrt[3]{-54y^5}$
c. $\sqrt[4]{t^8u^{15}}$

Now Try   Problems 21, 29, and 31

2   **Use Prime Factorization to Simplify Radical Expressions.**

When simplifying radical expressions, prime factorization can be helpful in determining how to factor the radicand.

EXAMPLE 3 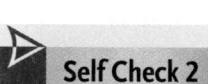   Simplify. All variables represent positive real numbers.
a. $\sqrt{150}$ **b.** $\sqrt[3]{297b^4}$ **c.** $\sqrt[4]{224s^8t^7}$

Strategy   In each case, the way to factor the radicand is not obvious. Another approach is to prime-factor the coefficient of the radicand and look for groups of like factors.

Why   Identifying groups of like factors of the radicand leads to a factorization of the radicand that can be easily simplified.

Solution

a. $\sqrt{150} = \sqrt{2 \cdot 3 \cdot 5 \cdot 5}$ Write 150 in prime-factored form.

$\quad\quad\quad\;\; = \sqrt{2 \cdot 3}\sqrt{5 \cdot 5}$ Group the pair of like factors together and use the product rule for radicals.

$\quad\quad\quad\;\; = \sqrt{2 \cdot 3}\sqrt{5^2}$ Write $5 \cdot 5$ as 5^2.

$\quad\quad\quad\;\; = \sqrt{6} \cdot 5$ Evaluate $\sqrt{5^2}$.

$\quad\quad\quad\;\; = 5\sqrt{6}$ Write the factor 5 first.

b. $\sqrt[3]{297b^4} = \sqrt[3]{3 \cdot 3 \cdot 3 \cdot 11 \cdot b^3 \cdot b}$ Write 297 in prime-factored form. The largest perfect-cube factor of b^4 is b^3.

$\quad\quad\quad\quad\;\; = \sqrt[3]{3 \cdot 3 \cdot 3 \cdot b^3}\sqrt[3]{11b}$ Group the three like factors of 3 together and use the product rule for radicals.

$\quad\quad\quad\quad\;\; = \sqrt[3]{3^3 b^3}\sqrt[3]{11b}$ Write $3 \cdot 3 \cdot 3$ as 3^3.

$\quad\quad\quad\quad\;\; = 3b\sqrt[3]{11b}$ Simplify $\sqrt[3]{3^3 b^3}$.

c. $\sqrt[4]{224 s^8 t^7} = \sqrt[4]{2 \cdot 2 \cdot 2 \cdot 2 \cdot 2 \cdot 7 \cdot s^8 \cdot t^4 \cdot t^3}$ Write 224 in prime-factored form. The largest perfect-fourth power factor of t^7 is t^4.

$\quad\quad\quad\quad\;\; = \sqrt[4]{2 \cdot 2 \cdot 2 \cdot 2 \cdot s^8 \cdot t^4}\sqrt[4]{2 \cdot 7 \cdot t^3}$ Group the four like factors of 2 together and use the product rule for radicals.

$\quad\quad\quad\quad\;\; = \sqrt[4]{2^4 s^8 t^4}\sqrt[4]{2 \cdot 7 \cdot t^3}$ Write $2 \cdot 2 \cdot 2 \cdot 2$ as 2^4.

$\quad\quad\quad\quad\;\; = 2s^2 t\sqrt[4]{14t^3}$ Simplify $\sqrt[4]{2^4 s^8 t^4}$.

 Self Check 3   Simplify: **a.** $\sqrt{275}$ **b.** $\sqrt[3]{189 c^4 d^3}$

Now Try   Problems 33 and 39

---

3   **Use the Quotient Rule to Simplify Radical Expressions.**

To introduce the quotient rule for radicals, we will find $\sqrt{\dfrac{100}{4}}$ and $\dfrac{\sqrt{100}}{\sqrt{4}}$ and compare the results.

Square root of a quotient

$$\sqrt{\frac{100}{4}} = \sqrt{25}$$
$$= 5$$

Quotient of square roots

$$\frac{\sqrt{100}}{\sqrt{4}} = \frac{10}{2}$$
$$= 5$$

Since the answer is 5 in each case, $\sqrt{\dfrac{100}{4}} = \dfrac{\sqrt{100}}{\sqrt{4}}$.

Similarly, we will find $\sqrt[3]{\dfrac{64}{8}}$ and $\dfrac{\sqrt[3]{64}}{\sqrt[3]{8}}$, and compare the results.

Cube root of a quotient

$$\sqrt[3]{\frac{64}{8}} = \sqrt[3]{8}$$
$$= 2$$

Quotient of cube roots

$$\frac{\sqrt[3]{64}}{\sqrt[3]{8}} = \frac{4}{2}$$
$$= 2$$

Since the answer is 2 in each case, $\sqrt[3]{\dfrac{64}{8}} = \dfrac{\sqrt[3]{64}}{\sqrt[3]{8}}$. These results illustrate the *quotient rule for radicals*.

The Quotient Rule for Radicals	The *n*th root of the quotient of two numbers is equal to the quotient of their *n*th roots. If $\sqrt[n]{a}$ and $\sqrt[n]{b}$ are real numbers, then $$\sqrt[n]{\dfrac{a}{b}} = \dfrac{\sqrt[n]{a}}{\sqrt[n]{b}} \qquad (b \neq 0)$$

EXAMPLE 4   Simplify each expression: **a.** $\sqrt{\dfrac{7}{64}}$ **b.** $\sqrt{\dfrac{15}{49x^2}}$ **c.** $\sqrt[3]{\dfrac{10x^2}{27y^6}}$

All variables represent positive real numbers.

Strategy   In each case, the radical is not in simplified form because the radicand contains a fraction. To write each of these expressions in simplified form, we will use the quotient rule for radicals.

Why   Writing these expressions in $\dfrac{\sqrt[n]{a}}{\sqrt[n]{b}}$ form leads to square roots of perfect squares and cube roots of perfect cubes that we can easily simplify.

Solution
a. We can use the quotient rule for radicals to simplify each expression.

$$\sqrt{\dfrac{7}{64}} = \dfrac{\sqrt{7}}{\sqrt{64}} \qquad \text{The square root of a quotient is equal to the quotient of the square roots.}$$

$$= \dfrac{\sqrt{7}}{8} \qquad \text{Evaluate } \sqrt{64}.$$

> **Success Tip**
> In Example 4, a radical of a quotient is written as a quotient of radicals:
> $$\sqrt[n]{\dfrac{a}{b}} = \dfrac{\sqrt[n]{a}}{\sqrt[n]{b}}$$

b. $\sqrt{\dfrac{15}{49x^2}} = \dfrac{\sqrt{15}}{\sqrt{49x^2}}$ The square root of a quotient is equal to the quotient of the square roots.

$$= \dfrac{\sqrt{15}}{7x} \qquad \text{Simplify the denominator: } \sqrt{49x^2} = 7x.$$

c. $\sqrt[3]{\dfrac{10x^2}{27y^6}} = \dfrac{\sqrt[3]{10x^2}}{\sqrt[3]{27y^6}}$ The cube root of a quotient is equal to the quotient of the cube roots.

$$= \dfrac{\sqrt[3]{10x^2}}{3y^2} \qquad \text{Simplify the denominator.}$$

 Self Check 4   Simplify. All variables represent positive real numbers.

a. $\sqrt[3]{\dfrac{25}{27}}$ **b.** $\sqrt{\dfrac{11}{36a^2}}$ **c.** $\sqrt[4]{\dfrac{a^3}{625y^{12}}}$

Now Try   Problems 41, 43, and 51

EXAMPLE 5 Simplify each expression. All variables represent positive real numbers. **a.** $\dfrac{\sqrt{45xy^2}}{\sqrt{5x}}$ **b.** $\dfrac{\sqrt[3]{-432x^5}}{\sqrt[3]{8x}}$

Strategy We will use the quotient rule for radicals in reverse: $\dfrac{\sqrt[n]{a}}{\sqrt[n]{b}} = \sqrt[n]{\dfrac{a}{b}}$.

Why When the radicands are written under a single radical symbol, the result is a rational expression. Our hope is that the rational expression can be simplified.

Solution

a. We can write the quotient of the square roots as the square root of a quotient.

$$\frac{\sqrt{45xy^2}}{\sqrt{5x}} = \sqrt{\frac{45xy^2}{5x}} \qquad \text{Use the quotient rule for radicals. Note that the resulting radicand is a rational expression.}$$

$$= \sqrt{9y^2} \qquad \text{Simplify the radicand: } \frac{45xy^2}{5x} = \frac{\cancel{5}\cdot 9\cdot \cancel{x}\cdot y^2}{\cancel{5}\cdot \cancel{x}} = 9y^2.$$

$$= 3y \qquad \text{Simplify the radical.}$$

b. We can write the quotient of the cube roots as the cube root of a quotient.

$$\frac{\sqrt[3]{-432x^5}}{\sqrt[3]{8x}} = \sqrt[3]{-\frac{432x^5}{8x}} \qquad \text{Use the quotient rule for radicals. Note that the resulting radicand is a rational expression.}$$

$$= \sqrt[3]{-54x^4} \qquad \text{Simplify the radicand: } -\frac{432x^5}{8x} = -54x^4.$$

$$= \sqrt[3]{-27x^3 \cdot 2x} \qquad 27x^3 \text{ is the largest perfect cube that divides } 54x^4.$$

$$= \sqrt[3]{-27x^3}\sqrt[3]{2x} \qquad \text{Use the product rule for radicals.}$$

$$= -3x\sqrt[3]{2x} \qquad \text{Simplify: } \sqrt[3]{-27x^3} = -3x.$$

Self Check 5 Simplify each expression. All variables represent positive real numbers. **a.** $\dfrac{\sqrt{50ab^2}}{\sqrt{2a}}$ **b.** $\dfrac{\sqrt[3]{-2{,}000x^5v^3}}{\sqrt[3]{2x}}$

Now Try Problems 55 and 59

---

4 Add and Subtract Radical Expressions.

Radical expressions with the same index and the same radicand are called **like** or **similar** radicals. For example, $3\sqrt{2}$ and $2\sqrt{2}$ are like radicals. However,

- $3\sqrt{5}$ and $4\sqrt{2}$ are not like radicals, because the radicands are different.
- $3\sqrt[4]{5}$ and $2\sqrt[3]{5}$ are not like radicals, because the indices are different.

For an expression with two or more radical terms, we should attempt to combine like radicals, if possible. For example, to simplify the expression $3\sqrt{2} + 2\sqrt{2}$, we use the distributive property to factor out $\sqrt{2}$ and simplify.

Success Tip

In Example 5, a quotient of radicals is written as a radical of a quotient.

$$\frac{\sqrt[n]{a}}{\sqrt[n]{b}} = \sqrt[n]{\frac{a}{b}}$$

Success Tip

Combining like radicals is similar to combining like terms.

$$3\sqrt{2} + 2\sqrt{2} = 5\sqrt{2}$$

$$3x + 2x = 5x$$

$$3\sqrt{2} + 2\sqrt{2} = (3 + 2)\sqrt{2}$$
$$= 5\sqrt{2}$$

Radicals with the same index but different radicands can often be written as like radicals. For example, to simplify the expression $\sqrt{75} - \sqrt{27}$, we simplify both radicals first and then combine the like radicals.

$$\sqrt{75} - \sqrt{27} = \sqrt{25 \cdot 3} - \sqrt{9 \cdot 3} \qquad \text{Write 75 and 27 in factored form.}$$
$$= \sqrt{25}\sqrt{3} - \sqrt{9}\sqrt{3} \qquad \text{Use the product rule for radicals.}$$
$$= 5\sqrt{3} - 3\sqrt{3} \qquad \text{Evaluate } \sqrt{25} \text{ and } \sqrt{9}.$$
$$= (5 - 2)\sqrt{3} \qquad \text{Factor out } \sqrt{3}.$$
$$= 3\sqrt{3}$$

As the previous examples suggest, we can add or subtract radicals as follows.

Adding and Subtracting Radicals

To add or subtract radicals, simplify each radical, if possible, and combine like radicals.

EXAMPLE 6   Simplify: **a.** $2\sqrt{12} - 3\sqrt{48}$ **b.** $\sqrt[3]{16} + \sqrt[3]{54} - \sqrt[3]{24}$

Strategy   Since the radicals in each part are unlike radicals, we cannot add or subtract them in their current form. However, we will simplify the radicals and hope that like radicals result.

Why   Like radicals can be combined.

Solution

a. We begin by simplifying each radical expression:

$$2\sqrt{12} - 3\sqrt{48} = 2\sqrt{4 \cdot 3} - 3\sqrt{16 \cdot 3}$$
$$= 2\sqrt{4}\sqrt{3} - 3\sqrt{16}\sqrt{3}$$
$$= 2(2)\sqrt{3} - 3(4)\sqrt{3}$$
$$= 4\sqrt{3} - 12\sqrt{3} \qquad \text{Both expressions have the same index and radicand.}$$
$$= (4 - 12)\sqrt{3} \qquad \text{Combine like radicals.}$$
$$= -8\sqrt{3}$$

b. We begin by simplifying each radical expression:

$$\sqrt[3]{16} + \sqrt[3]{54} - \sqrt[3]{24} = \sqrt[3]{8 \cdot 2} + \sqrt[3]{27 \cdot 2} - \sqrt[3]{8 \cdot 3}$$
$$= \sqrt[3]{8}\sqrt[3]{2} + \sqrt[3]{27}\sqrt[3]{2} - \sqrt[3]{8}\sqrt[3]{3}$$
$$= 2\sqrt[3]{2} + 3\sqrt[3]{2} - 2\sqrt[3]{3}$$

Now we combine the two radical expressions that have the same index and radicand.

$$\sqrt[3]{16} + \sqrt[3]{54} - \sqrt[3]{24} = (2 + 3)\sqrt[3]{2} - 2\sqrt[3]{3} \qquad \text{Combine like radicals.}$$
$$= 5\sqrt[3]{2} - 2\sqrt[3]{3}$$

Caution Even though the expressions $5\sqrt[3]{2}$ and $2\sqrt[3]{3}$ have the same index, we cannot combine them, because their radicands are different. Neither can we combine radical expressions having the same radicand but a different index. For example, the expression $\sqrt[3]{2} + \sqrt[4]{2}$ cannot be simplified.

Self Check 6 Simplify: **a.** $3\sqrt{75} - 2\sqrt{12} + 2\sqrt{48}$
b. $\sqrt[3]{24} - \sqrt[3]{16} + \sqrt[3]{54}$

Now Try **Problems 67 and 71**

EXAMPLE 7 Simplify: $\sqrt[3]{16x^4} + \sqrt[3]{54x^4} - \sqrt[3]{-128x^4}$

Strategy Since the radicals are unlike radicals, we cannot add or subtract them in their current form. However, we will simplify the radicals and hope that like radicals result.

Why Like radicals can be combined.

Solution We begin by simplifying each radical expression.

$$\sqrt[3]{16x^4} + \sqrt[3]{54x^4} - \sqrt[3]{-128x^4}$$
$$= \sqrt[3]{8x^3 \cdot 2x} + \sqrt[3]{27x^3 \cdot 2x} - \sqrt[3]{-64x^3 \cdot 2x}$$
$$= \sqrt[3]{8x^3}\sqrt[3]{2x} + \sqrt[3]{27x^3}\sqrt[3]{2x} - \sqrt[3]{-64x^3}\sqrt[3]{2x}$$
$$= 2x\sqrt[3]{2x} + 3x\sqrt[3]{2x} + 4x\sqrt[3]{2x} \quad \text{All three radicals have the same index and radicand.}$$
$$= (2x + 3x + 4x)\sqrt[3]{2x} \quad \text{Combine like radicals.}$$
$$= 9x\sqrt[3]{2x} \quad \text{Within the parentheses, combine like terms.}$$

Self Check 7 Simplify: $\sqrt{32x^3} + \sqrt{50x^3} - \sqrt{18x^3}$
Now Try **Problems 81 and 83**

ANSWERS TO SELF CHECKS **1. a.** $2\sqrt{5}$ **b.** $2\sqrt[3]{3}$ **c.** $2\sqrt[5]{4}$ **2. a.** $7b\sqrt{2b}$ **b.** $-3y\sqrt[3]{2y^2}$
c. $t^2u^3\sqrt[4]{u^3}$ **3. a.** $5\sqrt{11}$ **b.** $3cd\sqrt[3]{7c}$ **4. a.** $\dfrac{\sqrt[3]{25}}{3}$ **b.** $\dfrac{\sqrt{11}}{6a}$ **c.** $\dfrac{\sqrt[4]{a^3}}{5y^3}$ **5. a.** $5b$ **b.** $-10xv\sqrt[3]{x}$
6. a. $19\sqrt{3}$ **b.** $2\sqrt[3]{3} + \sqrt[3]{2}$ **7.** $6x\sqrt{2x}$

STUDY SET
9.3

VOCABULARY

Fill in the blanks.

1. Radical expressions such as $\sqrt[3]{4}$ and $6\sqrt[3]{4}$ with the same index and the same radicand are called _____ radicals.

2. Numbers such as 1, 4, 9, 16, 25, and 36 are called perfect _____. Numbers such as 1, 8, 27, 64, and 125 are called perfect _____. Numbers such as 1, 16, 81, 256, and 625 are called perfect-fourth _____.

3. The largest perfect-square _____ of 27 is 9. The largest _____-cube factor of 16 is 8.

4. To _____ $\sqrt{24}$ means to write it as $2\sqrt{6}$.

CONCEPTS

Fill in the blanks.

5. The product rule for radicals: $\sqrt[n]{ab} =$ _____. In words, the nth root of the _____ of two numbers is equal to the product of their nth _____.

6. The quotient rule for radicals: $\sqrt[n]{\dfrac{a}{b}} =$ _____. In words, the nth root of the _____ of two numbers is equal to the quotient of their nth _____.

7. Consider the expressions

$$\sqrt{4 \cdot 5} \quad \text{and} \quad \sqrt{4}\sqrt{5}$$

Which expression is

a. the square root of a product?

b. the product of square roots?

c. How are these two expressions related?

8. Consider the expressions

$$\frac{\sqrt[3]{a}}{\sqrt[3]{x^2}} \quad \text{and} \quad \sqrt[3]{\frac{a}{x^2}}$$

Which expression is

a. the cube root of a quotient?

b. the quotient of cube roots?

c. How are these two expressions related?

9. **a.** Write two radical expressions that have the same radicand but a different index. Can the expressions be added?

b. Write two radical expressions that have the same index but a different radicand. Can the expressions be added?

10. Fill in the blanks.

a. $5\sqrt{6} + 3\sqrt{6} = (\quad + \quad)\sqrt{6} = \quad \sqrt{6}$

b. $9\sqrt[3]{n} - 2\sqrt[3]{n} = (\quad - \quad)\sqrt[3]{n} = 7$

NOTATION

Complete each solution.

11. Simplify:

$$\sqrt[3]{32k^4} = \sqrt[3]{\qquad \cdot 4k}$$
$$= \sqrt[3]{\qquad}\ \sqrt[3]{4k}$$
$$= 2k\sqrt[3]{\qquad}$$

12. Simplify:

$$\frac{\sqrt{80s^2t^4}}{\sqrt{5s^2}} = \sqrt{\frac{80s^2t^4}{\qquad}}$$
$$= \sqrt{\qquad}$$
$$= 4t^2$$

GUIDED PRACTICE

Simplify each expression. See Example 1.

13. $\sqrt{50}$ 14. $\sqrt{28}$

15. $\sqrt{45}$ 16. $\sqrt{54}$

17. $\sqrt[3]{32}$ 18. $\sqrt[3]{40}$

19. $\sqrt[4]{48}$ 20. $\sqrt[4]{32}$

Simplify each radical expression. All variables represent positive real numbers. See Example 2.

21. $\sqrt{75a^2}$ 22. $\sqrt{50x^2}$

23. $\sqrt{32b}$ 24. $\sqrt{80c}$

25. $\sqrt{128a^3b^5}$ 26. $\sqrt{75b^8c}$

27. $\sqrt{300xy}$ 28. $\sqrt{200x^2y}$

29. $\sqrt[3]{-54x^6}$ 30. $\sqrt[3]{-81a^3}$

31. $\sqrt[4]{32x^{12}y^4}$ 32. $\sqrt[5]{64x^{10}y^5}$

Simplify each radical expression. All variables represent positive real numbers. See Example 3.

33. $\sqrt{242}$ 34. $\sqrt{363}$

35. $\sqrt{112a^3}$ 36. $\sqrt{147a^5}$

37. $-\sqrt[5]{96a^4}$ 38. $-\sqrt[7]{256t^6}$

39. $\sqrt[3]{405x^{12}y^4}$ 40. $\sqrt[3]{280a^5b^6}$

Simplify each radical expression. All variables represent positive real numbers. See Example 4.

41. $\sqrt{\dfrac{11}{9}}$

42. $\sqrt{\dfrac{3}{4}}$

43. $\sqrt[3]{\dfrac{7}{64}}$

44. $\sqrt[3]{\dfrac{4}{125}}$

45. $\sqrt[4]{\dfrac{3}{625}}$

46. $\sqrt[5]{\dfrac{2}{243}}$

47. $\sqrt[5]{\dfrac{3x^{10}}{32}}$

48. $\sqrt[6]{\dfrac{5y^{12}}{64}}$

49. $\sqrt{\dfrac{z^2}{16x^2}}$

50. $\sqrt{\dfrac{b^4}{64a^8}}$

51. $\sqrt[4]{\dfrac{5x}{16z^4}}$

52. $\sqrt[3]{\dfrac{11a^2}{125b^6}}$

Simplify each expression. All variables represent positive real numbers. See Example 5.

53. $\dfrac{\sqrt{500}}{\sqrt{5}}$

54. $\dfrac{\sqrt{128}}{\sqrt{2}}$

55. $\dfrac{\sqrt{98x^3}}{\sqrt{2x}}$

56. $\dfrac{\sqrt{75y^5}}{\sqrt{3y}}$

57. $\dfrac{\sqrt[3]{48x^7}}{\sqrt[3]{6x}}$

58. $\dfrac{\sqrt[3]{64y^8}}{\sqrt[3]{8y^2}}$

59. $\dfrac{\sqrt[3]{189a^5}}{\sqrt[3]{7a}}$

60. $\dfrac{\sqrt[3]{243x^8}}{\sqrt[3]{9x}}$

Simplify by combining like radicals. All variables represent positive real numbers. See Objective 4 and Example 6.

61. $5\sqrt{7} + 3\sqrt{7}$

62. $11\sqrt{3} + 2\sqrt{3}$

63. $20\sqrt[3]{4} - 15\sqrt[3]{4}$

64. $30\sqrt[3]{6} - 10\sqrt[3]{6}$

65. $\sqrt{8} + \sqrt{2}$

66. $\sqrt{45} + \sqrt{20}$

67. $\sqrt{98} - \sqrt{50} - \sqrt{72}$

68. $\sqrt{20} + \sqrt{125} - \sqrt{80}$

69. $\sqrt[3]{32} - \sqrt[3]{108}$

70. $\sqrt[3]{80} - \sqrt[3]{10{,}000}$

71. $2\sqrt[3]{125} - 5\sqrt[3]{64}$

72. $3\sqrt[3]{27} + 12\sqrt[3]{216}$

73. $14\sqrt[4]{32} - 15\sqrt[4]{2}$

74. $23\sqrt[4]{3} + \sqrt[4]{48}$

75. $\sqrt{80} + \sqrt{45} - \sqrt{27}$

76. $\sqrt{63} + \sqrt{72} - \sqrt{28}$

Simplify by combining like radicals. All variables represent positive real numbers. See Example 7.

77. $4\sqrt{2x} + 6\sqrt{2x}$

78. $6\sqrt[3]{5y} + 3\sqrt[3]{5y}$

79. $8\sqrt[5]{7a^2} - 7\sqrt[5]{7a^2}$

80. $10\sqrt[6]{12xy} - \sqrt[6]{12xy}$

81. $\sqrt{18t} + \sqrt{300t} - \sqrt{243t}$

82. $\sqrt{80m} - \sqrt{128m} + \sqrt{288m}$

83. $2\sqrt[3]{16} - \sqrt[3]{54} - 3\sqrt[3]{128}$

84. $\sqrt[3]{250} - 4\sqrt[3]{5} + \sqrt[3]{16}$

85. $2\sqrt[3]{64a} + 2\sqrt[3]{8a}$

86. $3\sqrt[4]{x^4y} - 2\sqrt[4]{x^4y}$

87. $\sqrt[4]{64} + 5\sqrt[4]{4} - \sqrt[4]{324}$

88. $\sqrt[4]{48} - \sqrt[4]{243} - \sqrt[4]{768}$

TRY IT YOURSELF

Simplify each expression. All variables represent positive real numbers.

89. $\sqrt[6]{m^{11}}$

90. $\sqrt[6]{n^{13}}$

91. $\sqrt{8y^7} + \sqrt{32y^7} - \sqrt{2y^7}$

92. $\sqrt{y^5} - \sqrt{9y^5} - \sqrt{25y^5}$

93. $\sqrt{\dfrac{125n^5}{64n}}$

94. $\sqrt{\dfrac{72q^7}{25q^3}}$

95. $\sqrt[5]{x^6y^2} + \sqrt[5]{32x^6y^2} + \sqrt[5]{x^6y^2}$

96. $\sqrt[3]{xy^4} + \sqrt[3]{8xy^4} - \sqrt[3]{27xy^4}$

97. $\sqrt[4]{208m^4n}$

98. $\sqrt[4]{128p^8q^3}$

99. $\sqrt[3]{\dfrac{a^7}{64a}}$

100. $\sqrt[3]{\dfrac{b^3c^8}{125c^5}}$

101. $\sqrt[5]{64t^{11}}$

102. $\sqrt[5]{243r^{22}}$

103. $\sqrt[3]{24x} + \sqrt[3]{3x}$

104. $\sqrt[3]{16y} + \sqrt[3]{128y}$

APPLICATIONS

First give the exact answer, expressed as a simplified radical expression. Then give an approximation, rounded to the nearest tenth.

105. UMBRELLAS The surface area of a cone is given by the formula $S = \pi r \sqrt{r^2 + h^2}$, where r is the radius of the base and h is its height. Use this formula to find the number of square feet of waterproof cloth used to make the umbrella shown.

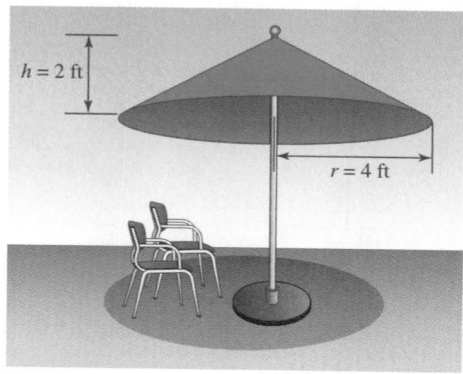

106. STRUCTURAL ENGINEERING Engineers have determined that two additional supports need to be added to strengthen the truss shown. Find the length L of each new support using the formula

$$L = \sqrt{\frac{b^2}{2} + \frac{c^2}{2} - \frac{a^2}{4}}$$

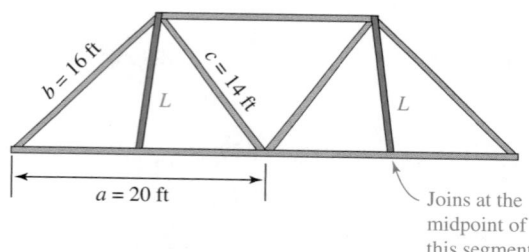

Joins at the midpoint of this segment

107. BLOW DRYERS The current I (in amps), the power P (in watts), and the resistance R (in ohms) are related by the formula $I = \sqrt{\frac{P}{R}}$. What current is needed for a 1,200-watt hair dryer if the resistance is 16 ohms?

108. COMMUNICATIONS SATELLITES Engineers have determined that a spherical communications satellite needs to have a capacity of 565.2 cubic feet to house all of its operating systems. The volume V of a sphere is related to its radius r by the formula $r = \sqrt[3]{\frac{3V}{4\pi}}$. What radius must the satellite have to meet the engineer's specification? Use 3.14 for π.

109. DUCTWORK The following pattern is laid out on a sheet of galvanized tin. Then it is cut out with snips and bent to make an air conditioning duct connection. Find the total length of the cut that must be made with the tin snips. (All measurements are in inches.)

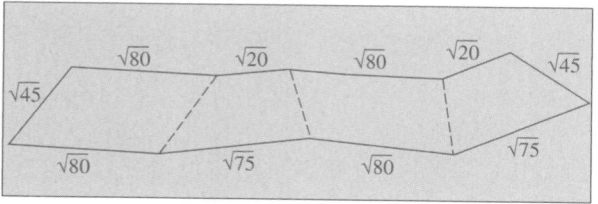

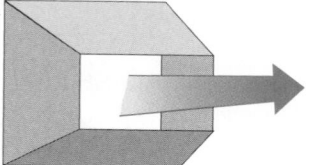

110. OUTDOOR COOKING The diameter of a circle is given by the function $d(A) = 2\sqrt{\frac{A}{\pi}}$, where A is the area of the circle. Find the difference between the diameters of the barbecue grills.

Cooking area 147π in.²

Cooking area 48π in.²

WRITING

111. Explain why each expression is not in simplified form.

   a. $\sqrt[3]{9x^4}$ **b.** $\sqrt{\dfrac{24m}{25}}$ **c.** $\dfrac{\sqrt[4]{c^3}}{\sqrt[4]{16}}$

112. How are the procedures used to simplify $3x + 4x$ and $3\sqrt{x} + 4\sqrt{x}$ similar?

113. Explain the mistake in the student's solution shown below.
Simplify: $\sqrt[3]{54}$

$$\sqrt[3]{54} = \sqrt[3]{27 + 27}$$
$$= \sqrt[3]{27} + \sqrt[3]{27}$$
$$= 3 + 3$$
$$= 6$$

114. Explain how the graphs of $Y_1 = 3\sqrt{24x} + \sqrt{54x}$ (on the left) and $Y_1 = 9\sqrt{6x}$ (on the right) can be used to verify the simplification $3\sqrt{24x} + \sqrt{54x} = 9\sqrt{6x}$. In each graph, settings of $[-5, 20]$ for x and $[-5, 100]$ for y were used.

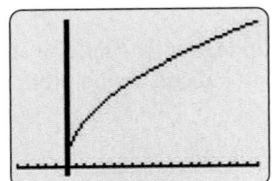 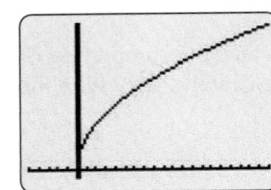

REVIEW

Perform each operation.

115. $3x^2y^3(-5x^3y^{-4})$

116. $(2x^2 - 9x - 5) \cdot \dfrac{x}{2x^2 + x}$

117. $2p - 5\overline{)6p^2 - 7p - 25}$

118. $\dfrac{xy}{\dfrac{1}{x} - \dfrac{1}{y}}$

CHALLENGE PROBLEMS

119. Can you find any numbers a and b such that
$$\sqrt{a + b} = \sqrt{a} + \sqrt{b}?$$

120. Find the sum:
$$\sqrt{3} + \sqrt{3^2} + \sqrt{3^3} + \sqrt{3^4} + \sqrt{3^5}$$

SECTION 9.4
Multiplying and Dividing Radical Expressions

Objectives

 1 Multiply radical expressions.

2 Find powers of radical expressions.

3 Rationalize denominators.

4 Rationalize denominators that have two terms.

5 Rational numerators.

In this section, we will discuss the methods we can use to multiply and divide radical expressions.

1 **Multiply Radical Expressions.**

We have used the *product rule for radicals* to write radical expressions in simplified form. We can also use this rule to multiply radical expressions that have the same index.

The Product Rule for Radicals	The product of the *n*th roots of two nonnegative numbers is equal to the *n*th root of the product of those numbers.

If $\sqrt[n]{a}$ and $\sqrt[n]{b}$ are real numbers,

$$\sqrt[n]{a} \cdot \sqrt[n]{b} = \sqrt[n]{a \cdot b}$$

EXAMPLE 1   Multiply and then simplify, if possible:

a. $\sqrt{5}\sqrt{10}$ **b.** $3\sqrt{6}\left(2\sqrt{3}\right)$ **c.** $-2\sqrt[3]{7x}\cdot 6\sqrt[3]{49x^2}$

Strategy   In each expression, we will use the product rule for radicals to multiply factors of the form $\sqrt[n]{a}$ and $\sqrt[n]{b}$.

Why   The product rule for radicals is used to multiply radicals that have the same index.

Solution

a. $\sqrt{5}\sqrt{10} = \sqrt{5\cdot 10}$ Use the product rule for radicals.

$\qquad\qquad = \sqrt{50}$ Multiply under the radical. Note that $\sqrt{50}$ can be simplified.

$\qquad\qquad = \sqrt{25\cdot 2}$ Prepare to simplify: factor 50.

$\qquad\qquad = 5\sqrt{2}$ Simplify: $\sqrt{25\cdot 2} = \sqrt{25}\sqrt{2} = 5\sqrt{2}$.

b. We use the commutative and associative properties of multiplication to multiply the integer factors and the radicals separately. Then we simplify any radicals in the product, if possible.

$3\sqrt{6}\left(2\sqrt{3}\right) = 3(2)\sqrt{6}\sqrt{3}$ Multiply the integer factors, 3 and 2, and multiply the radicals.

$\qquad\qquad = 6\sqrt{18}$ Use the product rule for radicals.

$\qquad\qquad = 6\sqrt{9}\sqrt{2}$ Simplify: $\sqrt{18} = \sqrt{9\cdot 2} = \sqrt{9}\sqrt{2}$.

$\qquad\qquad = 6(3)\sqrt{2}$ Evaluate: $\sqrt{9} = 3$.

$\qquad\qquad = 18\sqrt{2}$ Multiply.

c. $-2\sqrt[3]{7x}\cdot 6\sqrt[3]{49x^2} = -2(6)\sqrt[3]{7x}\sqrt[3]{49x^2}$ Write the integer factors together and the radicals together.

$\qquad\qquad = -12\sqrt[3]{7x\cdot 49x^2}$ Multiply the integer factors, -2 and 6, and multiply the radicals.

$\qquad\qquad = -12\sqrt[3]{7x\cdot 7^2 x^2}$ Write 49 as 7^2.

$\qquad\qquad = -12\sqrt[3]{7^3 x^3}$ Prepare to simplify: write $7x\cdot 7^2 x^2$ as $7^3 x^3$.

$\qquad\qquad = -12(7x)$ Simplify: $\sqrt[3]{7^3 x^3} = 7x$.

$\qquad\qquad = -84x$ Multiply.

Self Check 1   Multiply. All variables represent positive real numbers.

a. $\sqrt{7}\sqrt{14}$

b. $-2\sqrt[3]{2}\left(5\sqrt[3]{12}\right)$ **c.** $\sqrt[4]{4x^3}\cdot 9\sqrt[4]{8x^2}$

Now Try   Problems 11, 19, and 21

Recall that to multiply a polynomial by a monomial, we use the distributive property. We use the same technique to multiply a radical expression that has two or more terms by a radical expression that has only one term.

EXAMPLE 2   Multiply and then simplify, if possible: $3\sqrt{3}\left(4\sqrt{8} - 5\sqrt{10}\right)$

Strategy   We will use the distributive property and multiply each term within the parentheses by the term outside the parentheses.

Why   The given expression has the form $a(b - c)$.

Solution

$$3\sqrt{3}\left(4\sqrt{8} - 5\sqrt{10}\right)$$

$$= 3\sqrt{3} \cdot 4\sqrt{8} - 3\sqrt{3} \cdot 5\sqrt{10} \qquad \text{Distribute the multiplication by } 3\sqrt{3}.$$

$$= 12\sqrt{24} - 15\sqrt{30} \qquad \text{Multiply the integer factors and use the product rule to multiply the radicals.}$$

$$= 12\sqrt{4}\sqrt{6} - 15\sqrt{30} \qquad \text{Simplify: } \sqrt{24} = \sqrt{4 \cdot 6} = \sqrt{4}\sqrt{6}.$$

$$= 12(2)\sqrt{6} - 15\sqrt{30} \qquad \text{Evaluate: } \sqrt{4} = 2.$$

$$= 24\sqrt{6} - 15\sqrt{30}$$

Self Check 2   Multiply and then simplify, if possible: $4\sqrt{2}\left(3\sqrt{5} - 2\sqrt{8}\right)$

Now Try   **Problems 31 and 33**

Recall that to multiply two binomials, we multiply each term of one binomial by each term of the other binomial and simplify. We multiply two radical expressions, each having two terms, in the same way.

EXAMPLE 3   Multiply and then simplify, if possible:
a. $\left(\sqrt{7} + \sqrt{2}\right)\left(\sqrt{7} - 3\sqrt{2}\right)$
b. $\left(\sqrt[3]{x^2} - 4\sqrt[3]{5}\right)\left(\sqrt[3]{x} + \sqrt[3]{2}\right)$

Strategy   As with binomials, we will multiply each term within the first set of parentheses by each term within the second set of parentheses.

Why   This is an application of the FOIL method for multiplying binomials.

Solution

a. $\left(\sqrt{7} + \sqrt{2}\right)\left(\sqrt{7} - 3\sqrt{2}\right)$

$$\overset{\text{F}}{= \sqrt{7}\sqrt{7}} \;\; \overset{\text{O}}{- 3\sqrt{7}\sqrt{2}} \;\; \overset{\text{I}}{+ \sqrt{2}\sqrt{7}} \;\; \overset{\text{L}}{- 3\sqrt{2}\sqrt{2}} \qquad \text{Use the FOIL method.}$$

$$= 7 - 3\sqrt{14} + \sqrt{14} - 3(2) \qquad \text{Perform each multiplication.}$$

$$= 7 - 2\sqrt{14} - 6 \qquad \text{Combine like radicals.}$$

$$= 1 - 2\sqrt{14} \qquad \text{Combine like terms.}$$

b. $\left(\sqrt[3]{x^2} - 4\sqrt[3]{5}\right)\left(\sqrt[3]{x} + \sqrt[3]{2}\right)$

$$= \sqrt[3]{x^2}\sqrt[3]{x} + \sqrt[3]{x^2}\sqrt[3]{2} - 4\sqrt[3]{5}\sqrt[3]{x} - 4\sqrt[3]{5}\sqrt[3]{2} \qquad \text{Use the FOIL method.}$$

$$= \sqrt[3]{x^3} + \sqrt[3]{2x^2} - 4\sqrt[3]{5x} - 4\sqrt[3]{10} \qquad \text{Perform each multiplication.}$$

$$= x + \sqrt[3]{2x^2} - 4\sqrt[3]{5x} - 4\sqrt[3]{10} \qquad \text{Simplify the first term.}$$

Self Check 3   Multiply and then simplify, if possible:

   a. $\left(\sqrt{5} + 2\sqrt{3}\right)\left(\sqrt{5} - \sqrt{3}\right)$

   b. $\left(\sqrt[3]{a} + 9\sqrt[3]{2}\right)\left(\sqrt[3]{a^2} - \sqrt[3]{3}\right)$

Now Try   **Problems 37 and 41**

②   **Find Powers of Radical Expressions.**

To find the power of a radical expression, such as $\left(\sqrt{5}\right)^2$ or $\left(\sqrt[3]{2}\right)^3$, we can use the definition of exponent and the product rule for radicals.

$$\left(\sqrt{5}\right)^2 = \sqrt{5}\sqrt{5} \qquad\qquad\qquad \left(\sqrt[3]{2}\right)^3 = \sqrt[3]{2} \cdot \sqrt[3]{2} \cdot \sqrt[3]{2}$$

$$= \sqrt{25} \qquad\qquad\qquad\qquad\qquad = \sqrt[3]{8}$$

$$= 5 \qquad\qquad\qquad\qquad\qquad\qquad = 2$$

These results illustrate the following property of radicals.

The *n*th Power of the *n*th Root	If $\sqrt[n]{a}$ is a real number, $$\left(\sqrt[n]{a}\right)^n = a$$

EXAMPLE 4   Find: **a.** $\left(\sqrt{5}\right)^2$ **b.** $\left(2\sqrt[3]{7x^2}\right)^3$ **c.** $\left(\sqrt{m+1} + 2\right)^2$

Strategy   In part (a), we will use the definition of square root. In part (b), we will use a power rule for exponents. In part (c), we will use the FOIL method.

Why   Part (a) is the square of a square root, part (b) has the form $(xy)^n$, and part (c) has the form $(x + y)^2$.

Solution

a. $\left(\sqrt{5}\right)^2 = 5$ Because the square of the square root of 5 is 5.

b. We can use the power of a product rule for exponents to find $\left(2\sqrt[3]{7x^2}\right)^3$.

$$\left(2\sqrt[3]{7x^2}\right)^3 = 2^3\left(\sqrt[3]{7x^2}\right)^3 \qquad \text{Raise each factor of } 2\sqrt[3]{7x^2} \text{ to the 3rd power.}$$

$$= 8(7x^2) \qquad\qquad \text{Evaluate: } 2^3 = 8. \text{ Use } \left(\sqrt[n]{a}\right)^n = a.$$

$$= 56x^2$$

c. We can use the FOIL method to find the product.

$$\left(\sqrt{m+1} + 2\right)^2 = \left(\sqrt{m+1} + 2\right)\left(\sqrt{m+1} + 2\right)$$

$$= \left(\sqrt{m+1}\right)^2 + 2\sqrt{m+1} + 2\sqrt{m+1} + 2 \cdot 2$$

$$= m + 1 + 2\sqrt{m+1} + 2\sqrt{m+1} + 4 \qquad \text{Use } \left(\sqrt[n]{a}\right)^n = a.$$

$$= m + 4\sqrt{m+1} + 5 \qquad\qquad\qquad\qquad \text{Combine like terms.}$$

Success Tip

Since $\left(\sqrt{m+1} + 2\right)^2$ has the form $(x + y)^2$, we could also use a special-product rule to find this square of a sum quickly.

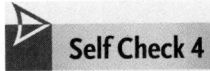

Self Check 4   Find: **a.** $\left(\sqrt{11}\right)^2$ **b.** $\left(3\sqrt[3]{4y}\right)^3$
c. $\left(\sqrt{x-8}-5\right)^2$

Now Try   Problems 43, 49, and 51

3 **Rationalize Denominators.**

We have seen that when a radical expression is written in simplified form, each of the following statements is true.

1. Each factor in the radicand is to a power that is less than the index of the radical.
2. The radicand contains no fractions or negative numbers.
3. No radicals appear in the denominator of a fraction.

We now consider radical expressions that do not satisfy requirements 2 or 3. We will introduce an algebraic technique, called *rationalizing the denominator,* that is used to write such expressions in an equivalent simplified form.

To divide radical expressions, we **rationalize the denominator** of a fraction to replace the denominator with a rational number. For example, to divide $\sqrt{5}$ by $\sqrt{3}$, we write the division as the fraction

$$\frac{\sqrt{5}}{\sqrt{3}}$$ This radical expression is not in simplified form, because a radical appears in the denominator.

We want to find a fraction equivalent to $\frac{\sqrt{5}}{\sqrt{3}}$ that does not have a radical in its denominator. If we multiply $\frac{\sqrt{5}}{\sqrt{3}}$ by $\frac{\sqrt{3}}{\sqrt{3}}$, the denominator becomes $\sqrt{3} \cdot \sqrt{3} = 3$, a rational number.

$$\frac{\sqrt{5}}{\sqrt{3}} = \frac{\sqrt{5}}{\sqrt{3}} \cdot \frac{\sqrt{3}}{\sqrt{3}}$$ To build an equivalent fraction, multiply by $\frac{\sqrt{3}}{\sqrt{3}} = 1$.

$$= \frac{\sqrt{15}}{3}$$ Multiply the numerators: $\sqrt{5} \cdot \sqrt{3} = \sqrt{15}$.
Multiply the denominators: $\sqrt{3} \cdot \sqrt{3} = \left(\sqrt{3}\right)^2 = 3$.

Thus, $\frac{\sqrt{5}}{\sqrt{3}} = \frac{\sqrt{15}}{3}$. These equivalent fractions represent the same number, but have different forms. Since there is no radical in the denominator, and $\sqrt{15}$ is in simplest form, the expression $\frac{\sqrt{15}}{3}$ is in simplified form.

EXAMPLE 5   Rationalize the denominator: **a.** $\sqrt{\dfrac{20}{7}}$ **b.** $\dfrac{4}{\sqrt[3]{2}}$

Strategy   We look at each denominator and ask, "By what must we multiply it to obtain a rational number?" Then we will multiply each expression by a carefully chosen form of 1.

Why   We want to produce an equivalent expression that does not have a radical in its denominator.

Solution

a. This radical expression is not in simplified form, because the radicand contains a fraction. We begin by writing the square root of the quotient as the quotient of two square roots:

$$\sqrt{\frac{20}{7}} = \frac{\sqrt{20}}{\sqrt{7}} \qquad \text{Use the division property of radicals: } \sqrt[n]{\frac{a}{b}} = \frac{\sqrt[n]{a}}{\sqrt[n]{b}}.$$

To rationalize the denominator, we proceed as follows:

$$\frac{\sqrt{20}}{\sqrt{7}} = \frac{\sqrt{20}}{\sqrt{7}} \cdot \frac{\sqrt{7}}{\sqrt{7}} \qquad \text{To build an equivalent fraction, multiply by } \frac{\sqrt{7}}{\sqrt{7}} = 1.$$

$$= \frac{\sqrt{140}}{7} \qquad \begin{array}{l} \text{Multiply the numerators.} \\ \text{Multiply the denominators: } \sqrt{7} \cdot \sqrt{7} = \left(\sqrt{7}\right)^2 = 7. \end{array}$$

$$= \frac{2\sqrt{35}}{7} \qquad \text{Simplify: } \sqrt{140} = \sqrt{4 \cdot 35} = \sqrt{4}\sqrt{35} = 2\sqrt{35}.$$

> **Caution**
>
> Do not attempt to remove a common factor of 7 from the numerator and denominator of $\frac{2\sqrt{35}}{7}$. The numerator, $2\sqrt{35}$, does not have a factor of 7.
>
> $$\frac{2\sqrt{35}}{7} = \frac{2 \cdot \sqrt{5 \cdot 7}}{7}$$

b. This expression is not in simplified form because a radical appears in the denominator of a fraction. Here, we must rationalize a denominator that is a cube root. We multiply the numerator and the denominator by a number that will give a perfect cube under the radical. Since $2 \cdot 4 = 8$ is a perfect cube, $\sqrt[3]{4}$ is such a number.

$$\frac{4}{\sqrt[3]{2}} = \frac{4}{\sqrt[3]{2}} \cdot \frac{\sqrt[3]{4}}{\sqrt[3]{4}} \qquad \text{To build an equivalent fraction, multiply by } \frac{\sqrt[3]{4}}{\sqrt[3]{4}} = 1.$$

$$= \frac{4\sqrt[3]{4}}{\sqrt[3]{8}} \qquad \begin{array}{l} \text{Multiply the numerators. Multiply the denominators.} \\ \text{This radicand is now a perfect cube.} \end{array}$$

$$= \frac{4\sqrt[3]{4}}{2} \qquad \text{Evaluate the denominator: } \sqrt[3]{8} = 2.$$

$$= 2\sqrt[3]{4} \qquad \text{Simplify the fraction: } \frac{4\sqrt[3]{4}}{2} = \frac{\overset{1}{\cancel{2}} \cdot 2\sqrt[3]{4}}{\underset{1}{\cancel{2}}} = 2\sqrt[3]{4}.$$

> **Caution**
>
> Multiplying $\frac{4}{\sqrt[3]{2}}$ by $\frac{\sqrt[3]{2}}{\sqrt[3]{2}}$ does not rationalize the denominator.
>
> $$\frac{4}{\sqrt[3]{2}} \cdot \frac{\sqrt[3]{2}}{\sqrt[3]{2}} = \frac{4\sqrt[3]{2}}{\sqrt[3]{4}}$$
>
> Since 4 is not a perfect cube, this radical does not simplify.

> **Self Check 5**   Rationalize the denominator: **a.** $\sqrt{\dfrac{8}{5}}$ **b.** $\dfrac{5}{\sqrt[3]{3}}$
>
> **Now Try**   Problems 57, 59, and 63

EXAMPLE 6   Rationalize the denominator: $\dfrac{\sqrt{5xy^2}}{\sqrt{xy^3}}$

Strategy   We will begin by using the quotient rule for radicals in reverse $\dfrac{\sqrt[n]{a}}{\sqrt[n]{b}} = \sqrt[n]{\dfrac{a}{b}}$.

Why   When the radicands are written under a single radical symbol, the result is a rational expression. Our hope is that the rational expression can be simplified, which could possibly make rationalizing the denominator easier.

Solution There are two methods we can use to rationalize the denominator. In each method, we simplify the expression first.

Method 1

$$\frac{\sqrt{5xy^2}}{\sqrt{xy^3}} = \sqrt{\frac{5xy^2}{xy^3}}$$

$$= \sqrt{\frac{5}{y}}$$

$$= \frac{\sqrt{5}}{\sqrt{y}}$$

$$= \frac{\sqrt{5}}{\sqrt{y}} \cdot \frac{\sqrt{y}}{\sqrt{y}} \quad \text{Multiply outside the radical.}$$

$$= \frac{\sqrt{5y}}{y}$$

Method 2

$$\frac{\sqrt{5xy^2}}{\sqrt{xy^3}} = \sqrt{\frac{5xy^2}{xy^3}}$$

$$= \sqrt{\frac{5}{y}}$$

$$= \sqrt{\frac{5}{y} \cdot \frac{y}{y}} \quad \text{Multiply within the radical.}$$

$$= \frac{\sqrt{5y}}{\sqrt{y^2}}$$

$$= \frac{\sqrt{5y}}{y}$$

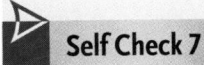

Self Check 6 Rationalize the denominator: $\dfrac{\sqrt{4ab^3}}{\sqrt{2a^2b^2}}$

Now Try **Problems 67 and 73**

EXAMPLE 7 Rationalize the denominator: $\dfrac{11}{\sqrt{20q^5}}$

Strategy We will simplify the radical expression in the denominator before rationalizing the denominator.

Why We could begin by multiplying $\dfrac{11}{\sqrt{20q^5}}$ by $\dfrac{\sqrt{20q^5}}{\sqrt{20q^5}}$. However, to work with smaller numbers and simpler radical expressions, it is easier if we simplify $\sqrt{20q^5}$ first, and then rationalize the denominator.

Solution

$$\frac{11}{\sqrt{20q^5}} = \frac{11}{\sqrt{4q^4 \cdot 5q}} \qquad \text{To prepare to simplify } \sqrt{20q^5}, \text{ factor } 20q^5 \text{ as } 4q^4 \cdot 5q.$$

$$= \frac{11}{2q^2\sqrt{5q}} \qquad \text{Simplify: } \sqrt{4q^4 \cdot 5q} = \sqrt{4q^4}\sqrt{5q} = 2q^2\sqrt{5q}.$$

$$= \frac{11}{2q^2\sqrt{5q}} \cdot \frac{\sqrt{5q}}{\sqrt{5q}} \qquad \text{To rationalize the denominator, multiply by } \frac{\sqrt{5q}}{\sqrt{5q}} = 1.$$

$$= \frac{11\sqrt{5q}}{2q^2(5q)} \qquad \begin{array}{l}\text{Multiply the numerators.}\\ \text{Multiply the denominators: } \sqrt{5q} \cdot \sqrt{5q} = \left(\sqrt{5q}\right)^2 = 5q.\end{array}$$

$$= \frac{11\sqrt{5q}}{10q^3} \qquad \text{Multiply in the denominator.}$$

Self Check 7 Rationalize the denominator: $\dfrac{7}{\sqrt{18c^3}}$

Now Try **Problems 75 and 81**

EXAMPLE 8 Rationalize each denominator: **a.** $\dfrac{5}{\sqrt[3]{6n^2}}$ **b.** $\dfrac{\sqrt[4]{2}}{\sqrt[4]{9a}}$

Strategy In part (a), we will examine the radicand in the denominator and ask, "By what must we multiply it to obtain a perfect cube?" In part (b), we will examine the radicand in the denominator and ask, "By what must we multiply it to obtain a perfect-fourth power?"

Why The answers to those questions will determine what form of 1 we use to rationalize each denominator.

Solution

a. To rationalize the denominator $\sqrt[3]{6n^2}$, we need the radicand to be a perfect cube. Since $6n^2 = 6 \cdot n \cdot n$, the radicand needs two more factors of 6 and one more factor of n. It follows that we should multiply the given expression by $\dfrac{\sqrt[3]{36n}}{\sqrt[3]{36n}}$.

$$\dfrac{5}{\sqrt[3]{6n^2}} = \dfrac{5}{\sqrt[3]{6n^2}} \cdot \dfrac{\sqrt[3]{36n}}{\sqrt[3]{36n}} \qquad \text{Multiply by a form of 1 to rationalize the denominator.}$$

$$= \dfrac{5\sqrt[3]{36n}}{\sqrt[3]{216n^3}} \quad \longleftarrow \quad \begin{array}{l}\text{Multiply the numerators. Multiply the denominators.}\\ \text{This radicand is now a perfect cube.}\end{array}$$

$$= \dfrac{5\sqrt[3]{36n}}{6n} \qquad \text{Simplify the denominator: } \sqrt[3]{216n^3} = 6n.$$

b. To rationalize the denominator $\sqrt[4]{9a}$, we need the radicand to be a perfect-fourth power. Since $9a = 3 \cdot 3 \cdot a$, the radicand needs two more factors of 3 and three more factors of a. It follows that we should multiply the given expression by $\dfrac{\sqrt[4]{9a^3}}{\sqrt[4]{9a^3}}$.

$$\dfrac{\sqrt[4]{2}}{\sqrt[4]{9a}} = \dfrac{\sqrt[4]{2}}{\sqrt[4]{9a}} \cdot \dfrac{\sqrt[4]{9a^3}}{\sqrt[4]{9a^3}} \qquad \text{Multiply by a form of 1 to rationalize the denominator.}$$

$$= \dfrac{\sqrt[4]{18a^3}}{\sqrt[4]{81a^4}} \quad \longleftarrow \quad \begin{array}{l}\text{Multiply the numerators. Multiply the denominators.}\\ \text{This radicand is now a perfect-fourth power.}\end{array}$$

$$= \dfrac{\sqrt[4]{18a^3}}{3a} \qquad \text{Simplify the denominator: } \sqrt[4]{81a^4} = 3a.$$

Self Check 8 Rationalize each denominator:

a. $\dfrac{27}{\sqrt[3]{100a}}$ **b.** $\dfrac{\sqrt[4]{3}}{\sqrt[4]{4y^2}}$

Now Try **Problems 83 and 87**

4 **Rationalize Denominators That Have Two Terms.**

So far, we have rationalized denominators that have only one term. We will now discuss a method to rationalize denominators that have two terms.

One-termed denominators

$$\dfrac{\sqrt{5}}{\sqrt{3}}, \quad \dfrac{11}{\sqrt{20q^5}}, \quad \dfrac{4}{\sqrt[3]{2}}$$

Two-termed denominators

$$\dfrac{1}{\sqrt{2} + 1}, \quad \dfrac{\sqrt{x} + \sqrt{2}}{\sqrt{x} - \sqrt{2}}$$

To rationalize the denominator of $\dfrac{1}{\sqrt{2} + 1}$, for example, we multiply the numerator and denominator by $\sqrt{2} - 1$, because the product $\left(\sqrt{2} + 1\right)\left(\sqrt{2} - 1\right)$ contains no radicals.

$$\left(\sqrt{2} + 1\right)\left(\sqrt{2} - 1\right) = \left(\sqrt{2}\right)^2 - (1)^2 \quad \text{Use a special-product rule.}$$
$$= 2 - 1$$
$$= 1$$

Radical expressions that involve the sum and difference of the same two terms, such as $\sqrt{2} + 1$ and $\sqrt{2} - 1$, are called **conjugates**.

EXAMPLE 9   Rationalize the denominator: **a.** $\dfrac{1}{\sqrt{2} + 1}$ **b.** $\dfrac{\sqrt{x} + \sqrt{2}}{\sqrt{x} - \sqrt{2}}$

Strategy   In each part, we will rationalize the denominator by multiplying the numerator and the denominator by the conjugate of the denominator.

Why   Multiplying each denominator by its conjugate will produce a new denominator that does not contain radicals.

Solution

a. To find a fraction equivalent to $\dfrac{1}{\sqrt{2} + 1}$ that does not have a radical in its denominator, we multiply $\dfrac{1}{\sqrt{2} + 1}$ by a form of 1 that uses the conjugate of $\sqrt{2} + 1$.

$$\frac{1}{\sqrt{2} + 1} = \frac{1}{\sqrt{2} + 1} \cdot \frac{\sqrt{2} - 1}{\sqrt{2} - 1}$$

$$= \frac{\sqrt{2} - 1}{\left(\sqrt{2}\right)^2 - (1)^2} \qquad \text{Multiply the numerators.}$$
$$\qquad\qquad\qquad\quad \text{Multiply the denominators using a}$$
$$\qquad\qquad\qquad\quad \text{special-product rule.}$$

$$= \frac{\sqrt{2} - 1}{2 - 1}$$

$$= \frac{\sqrt{2} - 1}{1}$$

$$= \sqrt{2} - 1$$

b. We multiply the numerator and denominator by $\sqrt{x} + \sqrt{2}$, which is the conjugate of $\sqrt{x} - \sqrt{2}$, and simplify.

$$\frac{\sqrt{x} + \sqrt{2}}{\sqrt{x} - \sqrt{2}} = \frac{\sqrt{x} + \sqrt{2}}{\sqrt{x} - \sqrt{2}} \cdot \frac{\sqrt{x} + \sqrt{2}}{\sqrt{x} + \sqrt{2}}$$

$$= \frac{x + \sqrt{2x} + \sqrt{2x} + 2}{\left(\sqrt{x}\right)^2 - \left(\sqrt{2}\right)^2} \qquad \text{Multiply the numerators.}$$
$$\qquad\qquad\qquad\qquad\quad \text{Multiply the denominators.}$$

$$= \frac{x + \sqrt{2x} + \sqrt{2x} + 2}{x - 2}$$

$$= \frac{x + 2\sqrt{2x} + 2}{x - 2} \qquad \text{In the numerator, combine like radicals.}$$

 Self Check 9   Rationalize the denominator: $\dfrac{\sqrt{x} - \sqrt{2}}{\sqrt{x} + \sqrt{2}}$

Now Try   Problems 91 and 99

Rationalize Numerators.

In calculus, we sometimes have to rationalize a numerator by multiplying the numerator and denominator of the fraction by the conjugate of the numerator.

EXAMPLE 10   Rationalize the numerator: $\dfrac{\sqrt{x}-3}{\sqrt{x}}$

Strategy   To rationalize the numerator, we will multiply the numerator and the denominator by the conjugate of the numerator.

Why   After rationalizing the numerator, we can simplify the expression. Although the result will not be in simplified form, this nonsimplified form is often desirable in calculus.

Solution   We multiply the numerator and denominator by $\sqrt{x}+3$, which is the conjugate of the numerator.

$$\frac{\sqrt{x}-3}{\sqrt{x}}=\frac{\sqrt{x}-3}{\sqrt{x}}\cdot\frac{\sqrt{x}+3}{\sqrt{x}+3}$$ Multiply by a form of 1 to rationalize the numerator.

$$=\frac{\left(\sqrt{x}\right)^2-(3)^2}{x+3\sqrt{x}}$$ Multiply the numerators using a special-product rule. Multiply the denominators.

$$=\frac{x-9}{x+3\sqrt{x}}$$

 Self Check 10   Rationalize the numerator: $\dfrac{\sqrt{x}+3}{\sqrt{x}}$

Now Try   **Problem 105**

ANSWERS TO SELF CHECKS 1. a. $7\sqrt{2}$ b. $-20\sqrt[3]{3}$ c. $18x\sqrt[4]{2x}$ 2. $12\sqrt{10}-32$
3. a. $-1+\sqrt{15}$ b. $a-\sqrt[3]{3a}+9\sqrt[3]{2a^2}-9\sqrt[3]{6}$ 4. a. 11 b. $108y$ c. $x-10\sqrt{x-8}+17$
5. a. $\frac{2\sqrt{10}}{5}$ b. $\frac{5\sqrt[3]{9}}{3}$ 6. $\frac{\sqrt{2ab}}{a}$ 7. $\frac{7\sqrt{2c}}{6c^2}$ 8. a. $\frac{27\sqrt[3]{10a^2}}{10a}$ b. $\frac{\sqrt[4]{12y^2}}{2y}$ 9. $\frac{x-2\sqrt{2x}+2}{x-2}$ 10. $\frac{x-9}{x-3\sqrt{x}}$

STUDY SET
9.4

VOCABULARY

Fill in the blanks.

1. To multiply $\left(\sqrt{3}+\sqrt{2}\right)\left(\sqrt{3}-2\sqrt{2}\right)$, we can use the _____ method.

2. To multiply $2\sqrt{5}\left(3\sqrt{8}+\sqrt{3}\right)$, use the _____ property.

3. The denominator of the fraction $\frac{4}{\sqrt{5}}$ is an _____ number.

4. The _____ of $\sqrt{x}+1$ is $\sqrt{x}-1$.

5. To obtain a _____-cube radicand in the denominator of $\frac{\sqrt[3]{7}}{\sqrt[3]{5n}}$, we multiply the fraction by $\frac{\sqrt[3]{25n^2}}{\sqrt[3]{25n^2}}$.

6. To _____ the denominator of $\frac{4}{\sqrt{5}}$, we multiply the fraction by $\frac{\sqrt{5}}{\sqrt{5}}$.

CONCEPTS

7. Perform each operation, if possible.

   a. $4\sqrt{6} + 2\sqrt{6}$ **b.** $4\sqrt{6}\left(2\sqrt{6}\right)$

   c. $3\sqrt{2} - 2\sqrt{3}$ **d.** $3\sqrt{2}\left(-2\sqrt{3}\right)$

8. Perform each operation, if possible.

   a. $5 + 6\sqrt[3]{6}$ **b.** $5\left(6\sqrt[3]{6}\right)$

   c. $\dfrac{30\sqrt[3]{15}}{5}$ **d.** $\dfrac{\sqrt[3]{15}}{5}$

NOTATION

Fill in the blanks.

9. Multiply:

$$5\sqrt{8} \cdot 7\sqrt{6} = 5(7)\sqrt{8}$$
$$= 35\sqrt{}$$
$$= 35\sqrt{ \cdot 3}$$
$$= 35()\sqrt{3}$$
$$= 140\sqrt{3}$$

10. Rationalize the denominator:

$$\frac{9}{\sqrt[3]{4a^2}} = \frac{9}{\sqrt[3]{4a^2}} \cdot \frac{\sqrt[3]{2a}}{}$$
$$= \frac{9\sqrt[3]{2a}}{\sqrt[3]{}}$$
$$= \frac{9\sqrt[3]{2a}}{}$$

GUIDED PRACTICE

Multiply and simplify, if possible. All variables represent positive real numbers. See Example 1.

11. $\sqrt{3}\sqrt{15}$ **12.** $\sqrt{5}\sqrt{15}$

13. $2\sqrt{3}\sqrt{6}$ **14.** $-3\sqrt{11}\sqrt{33}$

15. $\left(3\sqrt[3]{9}\right)\left(2\sqrt[3]{3}\right)$ **16.** $\left(2\sqrt[3]{16}\right)\left(-\sqrt[3]{4}\right)$

17. $\sqrt[3]{2} \cdot \sqrt[3]{12}$ **18.** $\sqrt[3]{3} \cdot \sqrt[3]{18}$

19. $6\sqrt{ab^3}\left(8\sqrt{ab}\right)$ **20.** $3\sqrt{8x}\left(2\sqrt{2x^3y}\right)$

21. $\sqrt[4]{5a^3}\sqrt[4]{125a^2}$ **22.** $\sqrt[4]{2r^3}\sqrt[4]{8r^2}$

23. $-4\sqrt[3]{5r^2s}\left(5\sqrt[3]{2r}\right)$ **24.** $-\sqrt[3]{3xy^2}\left(-\sqrt[3]{9x^3}\right)$

25. $\sqrt{x(x+3)}\sqrt{x^3(x+3)}$ **26.** $\sqrt{y^2(x+y)}\sqrt{(x+y)^3}$

Multiply and simplify, if possible. All variables represent positive real numbers. See Example 2.

27. $3\sqrt{5}\left(4 - \sqrt{5}\right)$ **28.** $2\sqrt{7}\left(3 - \sqrt{7}\right)$

29. $\sqrt{2}\left(4\sqrt{6} + 2\sqrt{7}\right)$ **30.** $-\sqrt{3}\left(\sqrt{7} - \sqrt{15}\right)$

31. $-2\sqrt{5x}\left(4\sqrt{2x} - 3\sqrt{3}\right)$ **32.** $3\sqrt{7t}\left(2\sqrt{7t} + 3\sqrt{3t^2}\right)$

33. $\sqrt[3]{2}\left(4\sqrt[3]{4} + \sqrt[3]{12}\right)$ **34.** $\sqrt[3]{3}\left(2\sqrt[3]{9} + \sqrt[3]{18}\right)$

Multiply and simplify, if possible. All variables represent positive real numbers. See Example 3.

35. $\left(\sqrt{2} + 1\right)\left(\sqrt{2} - 3\right)$

36. $\left(2\sqrt{3} + 1\right)\left(\sqrt{3} - 1\right)$

37. $\left(\sqrt{3x} - \sqrt{2y}\right)\left(\sqrt{3x} + \sqrt{2y}\right)$

38. $\left(\sqrt{3m} + \sqrt{2n}\right)\left(\sqrt{3m} - \sqrt{2n}\right)$

39. $\left(2\sqrt[3]{4} - 3\sqrt[3]{2}\right)\left(3\sqrt[3]{4} + 2\sqrt[3]{10}\right)$

40. $\left(4\sqrt[3]{9} - 3\sqrt[3]{3}\right)\left(4\sqrt[3]{3} + 2\sqrt[3]{6}\right)$

41. $\left(\sqrt[3]{5z} + \sqrt[3]{3}\right)\left(\sqrt[3]{5z} + 2\sqrt[3]{3}\right)$

42. $\left(\sqrt[3]{3p} - 2\sqrt[3]{2}\right)\left(\sqrt[3]{3p} + \sqrt[3]{2}\right)$

Square or cube each quantity and simplify the result, if possible. See Example 4.

43. $\left(\sqrt{7}\right)^2$ **44.** $\left(\sqrt{11}\right)^2$

45. $\left(\sqrt[3]{12}\right)^3$ **46.** $\left(\sqrt[3]{9}\right)^3$

47. $\left(3\sqrt{2}\right)^2$ **48.** $\left(2\sqrt{5}\right)^2$

49. $\left(-2\sqrt[3]{2x^2}\right)^3$ **50.** $\left(-3\sqrt[3]{10y^3}\right)^3$

51. $\left(3\sqrt{2r} - 2\right)^2$

52. $\left(2\sqrt{3t} + 5\right)^2$

53. $\left(\sqrt{3x} + \sqrt{3}\right)^2$

54. $\left(\sqrt{5x} - \sqrt{3}\right)^2$

Rationalize each denominator. See Example 5.

55. $\sqrt{\dfrac{2}{7}}$ **56.** $\sqrt{\dfrac{5}{3}}$

57. $\sqrt{\dfrac{8}{3}}$ **58.** $\sqrt{\dfrac{8}{7}}$

59. $\dfrac{4}{\sqrt{6}}$ **60.** $\dfrac{8}{\sqrt{10}}$

61. $\dfrac{1}{\sqrt[3]{2}}$ **62.** $\dfrac{2}{\sqrt[3]{6}}$

63. $\dfrac{3}{\sqrt[3]{9}}$ **64.** $\dfrac{2}{\sqrt[3]{a}}$

65. $\dfrac{1}{\sqrt[4]{4}}$ **66.** $\dfrac{1}{\sqrt[5]{2}}$

Rationalize each denominator. All variables represent positive real numbers. See Example 6.

67. $\dfrac{\sqrt{10y^2}}{\sqrt{2y^3}}$ **68.** $\dfrac{\sqrt{15b^2}}{\sqrt{5b^3}}$

69. $\dfrac{\sqrt{48x^2}}{\sqrt{8x^2y}}$ **70.** $\dfrac{\sqrt{9xy}}{\sqrt{3x^2y}}$

71. $\dfrac{\sqrt[3]{12t^3}}{\sqrt[3]{54t^2}}$ **72.** $\dfrac{\sqrt[3]{15m^4}}{\sqrt[3]{12m^3}}$

73. $\dfrac{\sqrt[3]{4a^6}}{\sqrt[3]{2a^5b}}$ **74.** $\dfrac{\sqrt[3]{9x^5y^4}}{\sqrt[3]{3x^5y^5}}$

Rationalize each denominator. All variables represent positive real numbers. See Example 7.

75. $\dfrac{23}{\sqrt{50p^5}}$ **76.** $\dfrac{11}{\sqrt{75s^5}}$

77. $\dfrac{7}{\sqrt{24b^3}}$ **78.** $\dfrac{13}{\sqrt{32n^3}}$

79. $\sqrt[3]{\dfrac{5}{16}}$ **80.** $\sqrt[3]{\dfrac{2}{81}}$

81. $\sqrt[3]{\dfrac{4}{81}}$ **82.** $\sqrt[3]{\dfrac{7}{16}}$

Rationalize each denominator. All variables represent positive real numbers. See Example 8.

83. $\dfrac{19}{\sqrt[3]{5c^2}}$ **84.** $\dfrac{1}{\sqrt[3]{4m^2}}$

85. $\dfrac{\sqrt[3]{3}}{\sqrt[3]{2r}}$ **86.** $\dfrac{\sqrt[3]{7}}{\sqrt[3]{100s}}$

87. $\dfrac{\sqrt[4]{2}}{\sqrt[4]{3t^2}}$ **88.** $\dfrac{\sqrt[4]{3}}{\sqrt[4]{5b^3}}$

89. $\dfrac{25}{\sqrt[4]{8a}}$ **90.** $\dfrac{4}{\sqrt[4]{9t}}$

Rationalize each denominator. All variables represent positive real numbers. See Example 9.

91. $\dfrac{\sqrt{2}}{\sqrt{5}+3}$ **92.** $\dfrac{\sqrt{3}}{\sqrt{3}-2}$

93. $\dfrac{2}{\sqrt{x}+1}$ **94.** $\dfrac{3}{\sqrt{x}-2}$

95. $\dfrac{\sqrt{7}-\sqrt{2}}{\sqrt{2}+\sqrt{7}}$ **96.** $\dfrac{\sqrt{3}+\sqrt{2}}{\sqrt{3}-\sqrt{2}}$

97. $\dfrac{3\sqrt{2}-5\sqrt{3}}{2\sqrt{3}-3\sqrt{2}}$ **98.** $\dfrac{3\sqrt{6}+5\sqrt{5}}{2\sqrt{5}-3\sqrt{6}}$

99. $\dfrac{\sqrt{x}-\sqrt{y}}{\sqrt{x}+\sqrt{y}}$ **100.** $\dfrac{\sqrt{x}+\sqrt{y}}{\sqrt{x}-\sqrt{y}}$

101. $\dfrac{2z-1}{\sqrt{2z}-1}$
(*Hint:* Do not perform the multiplication of the numerators.)

102. $\dfrac{3t-1}{\sqrt{3t}+1}$
(*Hint:* Do not perform the multiplication of the numerators.)

Rationalize each numerator. All variables represent positive real numbers. See Example 10.

103. $\dfrac{\sqrt{x}+3}{x}$ **104.** $\dfrac{2+\sqrt{x}}{5x}$

105. $\dfrac{\sqrt{x}+\sqrt{y}}{\sqrt{x}}$ **106.** $\dfrac{\sqrt{x}-\sqrt{y}}{\sqrt{x}+\sqrt{y}}$

TRY IT YOURSELF

The following problems involve addition, subtraction, and multiplication of radical expressions, as well as rationalizing the denominator. Perform the operations and simplify, if possible. All variables represent positive real numbers.

107. $\sqrt{x}\left(\sqrt{14x}+\sqrt{2}\right)$ **108.** $2\sqrt[3]{16}-3\sqrt[3]{128}-\sqrt[3]{54}$

109. $\left(10\sqrt[3]{2x}\right)^3$ **110.** $\dfrac{\sqrt{3}}{\sqrt{98x^2}}$

111. $\left(3p+\sqrt{5}\right)^2$ **112.** $\sqrt{288t}+\sqrt{80t}-\sqrt{128t}$

113. $\sqrt{\dfrac{72m^8}{25m^3}}$

114. $\left(\sqrt{14x}+\sqrt{3}\right)\left(\sqrt{14x}-\sqrt{3}\right)$

115. $\sqrt[4]{3n^2}\sqrt[4]{27n^3}$ **116.** $\dfrac{\sqrt{y}-2}{\sqrt{y}+3}$

117. $\dfrac{\sqrt[3]{x}}{\sqrt[3]{9}}$

118. $\sqrt[5]{\dfrac{2}{243}}$

APPLICATIONS

119. **STATISTICS** An example of a normal distribution curve, or *bell-shaped* curve, is shown. A fraction that is part of the equation that models this curve is $\dfrac{1}{\sigma\sqrt{2\pi}}$, where σ is a letter from the Greek alphabet. Rationalize the denominator of the fraction.

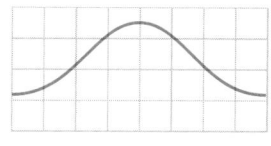

120. **ANALYTIC GEOMETRY** The length of the perpendicular segment drawn from $(-2, 2)$ to the line with equation $2x - 4y = 4$ is given by

$$L = \dfrac{|2(-2) + (-4)(2) + (-4)|}{\sqrt{(2)^2 + (-4)^2}}$$

Find L. Express the result in simplified radical form. Then give an approximation to the nearest tenth.

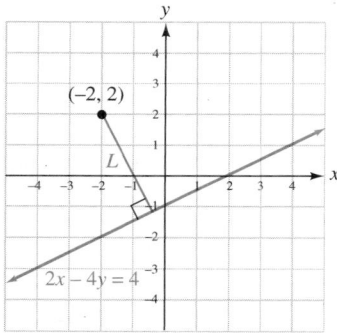

121. **TRIGONOMETRY** In trigonometry, we must often find the ratio of the lengths of two sides of right triangles. Use the information in the illustration to find the ratio

$$\dfrac{\text{length of side } AC}{\text{length of side } AB}$$

Write the result in simplified radical form.

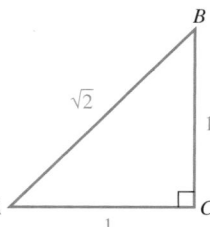

122. **ENGINEERING** Refer to the illustration in the next column that shows a block connected to two walls by springs. A measure of how fast the block will oscillate when the spring system is set in motion is given by the formula

$$\omega = \sqrt{\dfrac{k_1 + k_2}{m}}$$

where k_1 and k_2 indicate the stiffness of the springs and m is the mass of the block. Rationalize the right side and restate the formula.

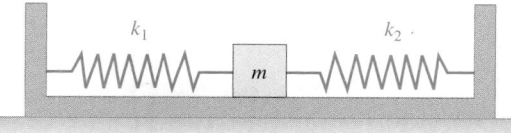

WRITING

123. Consider $\dfrac{\sqrt{3}}{\sqrt{7}} = \dfrac{\sqrt{3}}{\sqrt{7}} \cdot \dfrac{\sqrt{7}}{\sqrt{7}}$. Explain why the expressions on the left side and the right side of the equation are equal.

124. To rationalize the denominator of $\dfrac{\sqrt[4]{12}}{\sqrt[4]{3}}$, why wouldn't we multiply the numerator and denominator by $\dfrac{\sqrt[4]{3}}{\sqrt[4]{3}}$?

125. Explain why $\dfrac{\sqrt[3]{12}}{\sqrt[3]{5}}$ is not in simplified form.

126. Explain why $\sqrt{\dfrac{3a}{11k}}$ is not in simplified form.

127. Explain why $\sqrt{m} \cdot \sqrt{m} = m$ but $\sqrt[3]{m} \cdot \sqrt[3]{m} \neq m$. Assume that m represents a positive number.

128. Explain why the product of $\sqrt{m} + 3$ and $\sqrt{m} - 3$ does not contain a radical.

REVIEW

Solve each equation.

129. $\dfrac{8}{b - 2} + \dfrac{3}{2 - b} = -\dfrac{1}{b}$

130. $\dfrac{2}{x - 2} + \dfrac{1}{x + 1} = \dfrac{1}{(x + 1)(x - 2)}$

CHALLENGE PROBLEMS

131. Multiply: $\sqrt{2} \cdot \sqrt[3]{2}$. (*Hint:* Keep in mind two things. The indices (plural for *index*) must be the same to use the product rule for radicals, and radical expressions can be written using rational exponents.)

132. Show that $\dfrac{\sqrt[3]{a^2} + \sqrt[3]{a}\sqrt[3]{b} + \sqrt[3]{b^2}}{\sqrt[3]{a^2} + \sqrt[3]{a}\sqrt[3]{b} + \sqrt[3]{b^2}}$ can be used to rationalize the denominator of $\dfrac{1}{\sqrt[3]{a} - \sqrt[3]{b}}$.

SECTION 9.5
Solving Radical Equations

Objectives

1 Solve equations containing one radical.

2 Solve equations containing two radicals.

3 Solve formulas containing radicals.

When we solve equations containing fractions, we clear them of the fractions by multiplying both sides by the LCD. To solve equations containing radical expressions, we take a similar approach. The first step is to clear them of the radicals. To do this, we raise both sides to a power.

 Solve Equations Containing One Radical.

Radical equations contain a radical expression with a variable in the radicand. Some examples are

$$\sqrt{x + 3} = 4 \qquad \sqrt[3]{x^3 + 7} = x + 1 \qquad \sqrt{x} + \sqrt{x + 2} = 2$$

To solve equations containing radicals, we will use the *power rule*.

The Power Rule	If we raise two equal quantities to the same power, the results are equal quantities. If x, y, and n are real numbers and $x = y$, then $$x^n = y^n$$

If both sides of an equation are raised to the same power, all solutions of the original equation are also solutions of the new equation. However, the resulting equation might not be equivalent to the original equation. For example, if we square both sides of the equation

(1) $x = 3$

with a solution set of $\{3\}$, we obtain the equation

(2) $x^2 = 9$

with a solution set of $\{3, -3\}$.

Equations 1 and 2 are not equivalent, because they have different solution sets. The solution -3 of Equation 2 does not satisfy Equation 1. Since raising both sides of an equation to the same power can produce an equation with proposed solutions that don't satisfy the original equation, we must always check each proposed solution in the original equation and discard any **extraneous solutions.**

When we use the power rule to solve square root radical equations, it produces expressions of the form $\left(\sqrt{a}\right)^2$. We have seen that when this expression is simplified, the radical symbol is removed.

> **Caution**
>
> Recall that a similar caution about checking proposed solutions was given in Chapter 6 when we solved rational equations, such as:
>
> $$\frac{11}{x - 5} = 6 + \frac{55}{x - 5}$$

The Square of a Square Root	For any nonnegative real number a,	$(\sqrt{a})^2 = a$

EXAMPLE 1   Solve: $\sqrt{x + 3} = 4$

Strategy   We will use the power rule and square both sides of the equation.

Why   Squaring both sides will produce, on the left side, the expression $\left(\sqrt{x + 3}\right)^2$ that simplifies to $x + 3$. This step clears the equation of the radical.

Solution

$$\sqrt{x + 3} = 4 \qquad \text{This is the equation to solve.}$$

$$\left(\sqrt{x + 3}\right)^2 = (4)^2 \qquad \text{To clear the equation of the square root, square both sides.}$$

$$x + 3 = 16 \qquad \text{Perform the operations on each side.}$$

$$x = 13 \qquad \text{Solve the resulting equation by subtracting 3 from both sides.}$$

We must check the proposed solution 13 to see whether it satisfies the original equation.

The Language of Algebra
When we square both sides of an equation, we are *raising both sides to the second power.*

The Language of Algebra
Proposed solutions are also called *potential* or *possible* solutions.

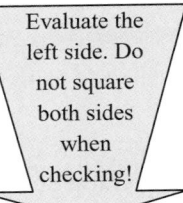
Evaluate the left side. Do not square both sides when checking!

Check: $\sqrt{x + 3} = 4$ This is the original equation.

$\sqrt{13 + 3} \overset{?}{=} 4$ Substitute 13 for x.

$\sqrt{16} \overset{?}{=} 4$

$4 = 4$ True

Since 13 satisfies the original equation, it is the solution. The solution set is $\{13\}$.

Self Check 1   Solve: $\sqrt{a - 2} = 3$

***Now Try*   Problems 13 and 17**

The method used in Example 1 to solve a radical equation containing a square root can be generalized, as follows.

Solving an Equation Containing Radicals	1. Isolate one radical expression on one side of the equation.
	2. Raise both sides of the equation to the power that is the same as the index of the radical.
	3. Solve the resulting equation. If it still contains a radical, go back to step 1.
	4. Check the results to eliminate extraneous solutions.

EXAMPLE 2   **Amusement Park Rides.** The distance d in feet that an object will fall in t seconds is given by the formula

$$t = \sqrt{\frac{d}{16}}$$

If the designers of the amusement park attraction want the riders to experience 3 seconds of vertical free fall, what length of vertical drop is needed?

Strategy   We will begin by substituting 3 for the time t in the formula.

Why   We can then solve the resulting radical equation in one variable to find the unknown distance d.

Solution

$$t = \sqrt{\frac{d}{16}}$$

$$3 = \sqrt{\frac{d}{16}} \qquad \text{Substitute 3 for } t. \text{ Here the radical is isolated on the right side.}$$

$$(3)^2 = \left(\sqrt{\frac{d}{16}}\right)^2 \qquad \text{To clear the equation of the square root, square both sides.}$$

$$9 = \frac{d}{16} \qquad \text{Perform the operations on each side.}$$

$$144 = d \qquad \text{Solve the resulting equation by multiplying both sides by 16.}$$

The amount of vertical drop needs to be 144 feet.

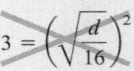

Caution

When using the power rule, don't forget to raise both sides to the same power. For this example, a common error would be to write

$$3 = \left(\sqrt{\frac{d}{16}}\right)^2$$

▷ **Self Check 2**   How long a vertical drop is needed if the riders are to free fall for 3.5 seconds?

Now Try   **Problem 97**

EXAMPLE 3   Solve: $\sqrt{3x + 1} + 1 = x$

Strategy   Since 1 is outside the square root symbol, there are two terms on the left side of the equation. To isolate the radical, we will subtract 1 from both sides.

Why   This will put the equation in a form where we can square both sides to clear the radical.

Solution

Notation

In Section 9.2, we saw that radical expressions can be written with rational exponents. The equation in this example could also be written as

$$(3x + 1)^{1/2} + 1 = x$$

$$\sqrt{3x + 1} + 1 = x \qquad \text{This is the equation to solve.}$$

$$\sqrt{3x + 1} = x - 1 \qquad \text{To isolate the radical on the left side, subtract 1 from both sides.}$$

$$\left(\sqrt{3x + 1}\right)^2 = (x - 1)^2 \qquad \text{Square both sides to eliminate the square root.}$$

$$3x + 1 = x^2 - 2x + 1 \qquad \text{On the right side, use the FOIL method:}$$
$$(x - 1)^2 = (x - 1)(x - 1) = x^2 - x - x + 1 = x^2 - 2x + 1.$$

$$0 = x^2 - 5x$$

To get 0 on the left side, subtract $3x$ and 1 from both sides. This is a quadratic equation.

$$0 = x(x - 5)$$

Factor out the GCF, x.

$$x = 0 \quad \text{or} \quad x - 5 = 0$$

Set each factor equal to 0.

$$x = 0 \quad | \quad x = 5$$

We must check each proposed solution to see whether it satisfies the original equation.

This is the check for 0:

$$\sqrt{3x + 1} + 1 = x$$

$$\sqrt{3(0) + 1} + 1 \overset{?}{=} 0$$

$$\sqrt{1} + 1 \overset{?}{=} 0$$

$$2 = 0 \quad \text{False}$$

This is the check for 5:

$$\sqrt{3x + 1} + 1 = x \qquad \text{This is the original equation.}$$

$$\sqrt{3(5) + 1} + 1 \overset{?}{=} 5$$

$$\sqrt{16} + 1 \overset{?}{=} 5$$

$$5 = 5 \quad \text{True}$$

The proposed solution 0 does not check; it must be discarded. Since the only solution is 5, the solution set is $\{5\}$.

Success Tip

Even if you are certain that no algebraic mistakes were made when solving a radical equation, you must still check your solutions. Raising both sides to a power can introduce extraneous solutions that must be discarded.

Self Check 3   Solve: $\sqrt{4x + 1} + 1 = x$

Now Try   **Problems 21 and 25**

Using Your Calculator

Solving Radical Equations

To find solutions for $\sqrt{3x + 1} + 1 = x$ with a graphing calculator, we graph the functions $f(x) = \sqrt{3x + 1} + 1$ and $g(x) = x$, as in figure (a). We then trace to find the approximate x-coordinate of their intersection point, as in figure (b). After repeated zooms, we will see that $x = 5$.

We can also use the INTERSECT feature to approximate the point of intersection of the graphs. See figure (c). The intersection point of (5, 5), with x-coordinate 5, implies that 5 is a solution of the radical equation.

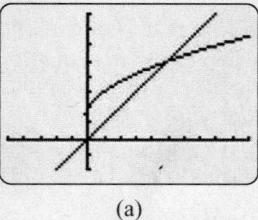 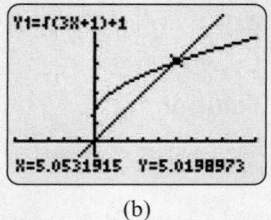 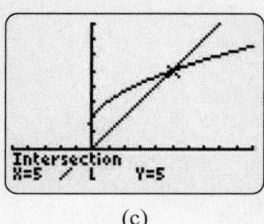

(a)   (b)   (c)

EXAMPLE 4   Solve: $\sqrt{3x} + 8 = 2$

Strategy   Since 8 is outside the square root symbol, there are two terms on the left side of the equation. To isolate the radical, we will subtract 8 from both sides.

Why   This will put the equation in a form where we can square both sides to clear the radical.

Success Tip

After isolating the radical, we obtained the equation $\sqrt{3x} = -6$. Since $\sqrt{3x}$ cannot be negative, we immediately know that $\sqrt{3x} + 8 = 2$ has no solution.

Solution

$$\sqrt{3x} + 8 = 2 \qquad \text{This is the equation to solve.}$$

$$\sqrt{3x} = -6 \qquad \text{To isolate the radical on the left side, subtract 8 from both sides.}$$

$$\left(\sqrt{3x}\right)^2 = (-6)^2 \qquad \text{Square both sides to eliminate the square root.}$$

$$3x = 36$$

$$x = 12 \qquad \text{To solve the resulting equation, divide both sides by 3.}$$

We check the proposed solution 12 in the original equation.

$$\sqrt{3x} + 8 = 2$$

$$\sqrt{3(12)} + 8 \stackrel{?}{=} 2 \qquad \text{Substitute 12 for x.}$$

$$\sqrt{36} + 8 \stackrel{?}{=} 2$$

$$6 + 8 \stackrel{?}{=} 2$$

$$14 = 2 \qquad \text{False}$$

Since 12 does not satisfy the original equation, it is extraneous. The equation has no solution. The solution set is $\varnothing$.

Self Check 4   Solve: $\sqrt{a - 9} + 3 = 0$

Now Try   **Problem 29**

The power rule can be used to solve radical equations that involve cube roots, fourth roots, fifth roots, and so on.

EXAMPLE 5   Solve: $\sqrt[3]{x^3 + 7} = x + 1$

Strategy   Note that the index of the radical is 3. We will use the power rule and cube both sides of the equation.

Why   Cubing both sides will produce, on the left side, the expression $\left(\sqrt[3]{x^3 + 7}\right)^3$ that simplifies to $x^3 + 7$. This step clears the equation of the radical.

Solution

$$\sqrt[3]{x^3 + 7} = x + 1 \qquad \text{This is the equation to solve.}$$

$$\left(\sqrt[3]{x^3 + 7}\right)^3 = (x + 1)^3 \qquad \text{Cube both sides to eliminate the cube root.}$$

$$x^3 + 7 = x^3 + 3x^2 + 3x + 1 \qquad \begin{array}{l}\text{Perform the operations on each side. On the right:}\\ (x + 1)^3 = (x + 1)(x + 1)(x + 1).\end{array}$$

$$0 = 3x^2 + 3x - 6 \qquad \begin{array}{l}\text{To get 0 on the left side, subtract } x^3 \text{ and 7 from}\\ \text{both sides. This is a quadratic equation.}\end{array}$$

$$0 = x^2 + x - 2 \qquad \text{Divide both sides by 3.}$$

$$0 = (x + 2)(x - 1) \qquad \text{Factor the trinomial.}$$

$$x + 2 = 0 \quad \text{or} \quad x - 1 = 0 \qquad \text{Set each factor equal to 0.}$$

$$x = -2 \quad \vert \quad x = 1$$

Success Tip

After raising both sides of a radical equation to a power, we use $\left(\sqrt[n]{a}\right)^n = a$ to simplify one side. For example:

$$\left(\sqrt[3]{x^3 + 7}\right)^3 = x^3 + 7$$

Notation

Since $\sqrt[3]{x^3 + 7}$ can be written as $(x^3 + 7)^{1/3}$, this equation could also be written as $(x^3 + 7)^{1/3} = x + 1$.

We check each proposed solution, -2 and 1, to see whether they satisfy the original equation.

Check:

$$\sqrt[3]{x^3 + 7} = x + 1$$

$$\sqrt[3]{(-2)^3 + 7} \overset{?}{=} -2 + 1$$

$$\sqrt[3]{-8 + 7} \overset{?}{=} -1$$

$$\sqrt[3]{-1} \overset{?}{=} -1$$

$$-1 = -1 \quad \text{True}$$

$$\sqrt[3]{x^3 + 7} = x + 1$$

$$\sqrt[3]{1^3 + 7} \overset{?}{=} 1 + 1$$

$$\sqrt[3]{1 + 7} \overset{?}{=} 2$$

$$\sqrt[3]{8} \overset{?}{=} 2$$

$$2 = 2 \quad \text{True}$$

Both -2 and 1 satisfy the original equation. Thus, the solution set is $\{-2, 1\}$.

 Self Check 5   Solve: $\sqrt[3]{x^3 + 8} = x + 2$

Now Try   **Problems 33 and 37**

EXAMPLE 6   Let $f(x) = \sqrt[4]{2x + 1}$. For what value(s) of x is $f(x) = 5$?

Strategy   We will substitute 5 for $f(x)$ and solve the equation $5 = \sqrt[4]{2x + 1}$. To do so, we will raise both sides of the equation to the fourth power.

Why   Raising both sides to the fourth power will produce, on the right side, the expression $\left(\sqrt[4]{2x + 1}\right)^4$ that simplifies to $2x + 1$. This step clears the equation of the radical.

Solution   To find the value(s) where $f(x) = 5$, we substitute 5 for $f(x)$ and solve for x.

$$f(x) = \sqrt[4]{2x + 1}$$

$$5 = \sqrt[4]{2x + 1} \qquad \text{This is the equation to solve.}$$

Since the equation contains a fourth root, we raise both sides to the fourth power to solve for x.

$$(5)^4 = \left(\sqrt[4]{2x + 1}\right)^4 \qquad \text{Use the power rule to eliminate the radical.}$$

$$625 = 2x + 1 \qquad \text{Perform the operations on each side.}$$

$$624 = 2x \qquad \text{To solve the resulting equation, subtract 1 from both sides.}$$

$$312 = x \qquad \text{Divide both sides by 2.}$$

If $x = 312$, then $f(x) = 5$. Verify this by evaluating $f(312)$.

 Self Check 6   Let $g(x) = \sqrt[5]{10x + 1}$. For what value(s) of x is $g(x) = 1$?

Now Try   **Problem 41**

---

2   **Solve Equations Containing Two Radicals.**

To solve an equation containing two radicals, we want to have one radical on the left side and one radical on the right side.

EXAMPLE 7   Solve: $\sqrt{5x + 9} = 2\sqrt{3x + 4}$

Strategy   We will square both sides to clear the equation of both radicals.

Why   We can immediately square both sides since each radical is isolated on one side of the equation.

Solution

$$\sqrt{5x + 9} = 2\sqrt{3x + 4}$$ This is the equation to solve.

$$\left(\sqrt{5x + 9}\right)^2 = \left(2\sqrt{3x + 4}\right)^2$$ Square both sides to eliminate the radicals.

$$5x + 9 = 2^2\left(\sqrt{3x + 4}\right)^2$$ Simplify on the left. On the right, raise each factor of the product $2\sqrt{3x + 4}$ to the second power.

$$5x + 9 = 4(3x + 4)$$ Perform the operations on the right.

$$5x + 9 = 12x + 16$$ To solve the resulting equation, distribute the multiplication by 4.

$$-7 = 7x$$ Subtract 5x and 16 from both sides.

$$-1 = x$$ Divide both sides by 7.

We check the solution by substituting -1 for x in the original equation.

$$\sqrt{5x + 9} = 2\sqrt{3x + 4}$$

$$\sqrt{5(-1) + 9} \stackrel{?}{=} 2\sqrt{3(-1) + 4}$$ Substitute -1 for x.

$$\sqrt{4} \stackrel{?}{=} 2\sqrt{1}$$

$$2 = 2$$ True

The solution is -1 and the solution set is $\{-1\}$.

> **Self Check 7**   Solve: $\sqrt{x - 4} = 2\sqrt{x - 16}$
>
> **Now Try**   **Problem 47**

> **Caution**
>
> When finding $\left(2\sqrt{3x + 4}\right)^2$, remember to square both 2 and $\sqrt{3x + 4}$ to get:
>
> $$2^2\left(\sqrt{3x + 4}\right)^2$$
>
> This is an application of the power of a product rule for exponents:
>
> $$(xy)^n = x^n y^n$$

When more than one radical appears in an equation, we must often use the power rule more than once.

EXAMPLE 8   Solve: $\sqrt{x} + \sqrt{x + 2} = 2$

Strategy   We will isolate $\sqrt{x + 2}$ on the left side of the equation and square both sides to eliminate it. After simplifying the resulting equation, we will isolate the remaining radical term and square both sides a second time to eliminate it.

Why   Each time that we square both sides, we are able to clear the equation of one radical.

Solution

$$\sqrt{x} + \sqrt{x + 2} = 2$$ This is the equation to solve.

$$\sqrt{x + 2} = 2 - \sqrt{x}$$ To isolate $\sqrt{x + 2}$, subtract $\sqrt{x}$ from both sides.

$$\left(\sqrt{x + 2}\right)^2 = \left(2 - \sqrt{x}\right)^2$$ Square both sides to eliminate the square root on the left side.

(1) $$x + 2 = \left(2 - \sqrt{x}\right)^2$$ Perform the operation on the left side.

To square the expression on the right side, we can use FOIL or a special-product rule:

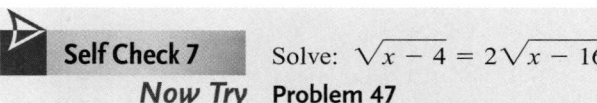

$$\left(2 - \sqrt{x}\right)^2 = \left(2 - \sqrt{x}\right)\left(2 - \sqrt{x}\right) = 4 - 2\sqrt{x} - 2\sqrt{x} + x = 4 - 4\sqrt{x} + x$$

> **Caution**
>
> When finding $\left(2 - \sqrt{x}\right)^2$, remember to use FOIL or a special-product rule. Do not just square the first term and the second term.
>
> $$\left(2 - \sqrt{x}\right)^2 \neq 2^2 - \left(\sqrt{x}\right)^2$$

Equation 1 can now be written as

$$x + 2 = 4 - 4\sqrt{x} + x$$

Since the equation still contains a radical, we need to square both sides again. Before doing that, we must isolate the radical on one side.

$2 = 4 - 4\sqrt{x}$	Subtract x from both sides.
$-2 = -4\sqrt{x}$	To isolate the radical term, subtract 4 from both sides.
$\dfrac{1}{2} = \sqrt{x}$	To isolate the radical, divide both sides by -4.
$\left(\dfrac{1}{2}\right)^2 = \left(\sqrt{x}\right)^2$	To eliminate the radical, square both sides again.
$\dfrac{1}{4} = x$	Perform the operations on each side.

Check:

$$\sqrt{x} + \sqrt{x + 2} = 2 \qquad \text{This is the original equation.}$$

$$\sqrt{\frac{1}{4}} + \sqrt{\frac{1}{4} + 2} \overset{?}{=} 2 \qquad \text{Substitute } \tfrac{1}{4} \text{ for x.}$$

$$\frac{1}{2} + \sqrt{\frac{9}{4}} \overset{?}{=} 2 \qquad \text{Think of 2 as } \tfrac{8}{4} \text{ and add: } \tfrac{1}{4} + \tfrac{8}{4} = \tfrac{9}{4}.$$

$$\frac{1}{2} + \frac{3}{2} \overset{?}{=} 2 \qquad \text{Evaluate } \sqrt{\tfrac{9}{4}}.$$

$$2 = 2 \qquad \text{True}$$

The result $\frac{1}{4}$ checks. The solution set is $\left\{\frac{1}{4}\right\}$.

Self Check 8   Solve: $\sqrt{a} + \sqrt{a + 3} = 3$

Now Try **Problems 53 and 61**

Using Your Calculator   **Solving Radical Equations**

To find solutions for $\sqrt{x} + \sqrt{x + 2} = 4$ (an equation similar to Example 8) with a graphing calculator, we graph the functions $f(x) = \sqrt{x} + \sqrt{x + 2}$ and $g(x) = 4$. We then trace to find an approximation of the x-coordinate of their intersection point, as in figure (a). From the figure, we can see that $x \approx 2.98$. We can zoom to get better results.

Figure (b) shows that the INTERSECT feature gives the approximate coordinates of the point of intersection of the two graphs as (3.06, 4). Therefore, an approximate solution of the radical equation is 3.06. Check its reasonableness.

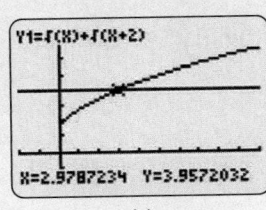
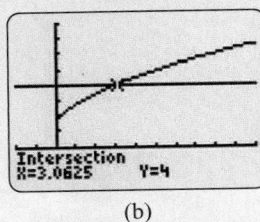

(a)   (b)

③ **Solve Formulas Containing Radicals.**

To *solve a formula for a variable* means to isolate that variable on one side of the equation, with all other quantities on the other side.

EXAMPLE 9 **Depreciation Rates.** Some office equipment that is now worth V dollars originally cost C dollars 3 years ago. The rate r at which it has depreciated is given by

$$r = 1 - \sqrt[3]{\dfrac{V}{C}}$$

Solve the formula for C.

Strategy To isolate the radical, we will subtract 1 from both sides. We can then eliminate the radical by cubing both sides.

Why Cubing both sides will produce, on the right, the expression $\left(\sqrt[3]{\dfrac{V}{C}}\right)^3$ that simplifies to $\dfrac{V}{C}$. This step clears the equation of the radical.

Solution We begin by isolating the cube root on the right side of the equation.

$$r = 1 - \sqrt[3]{\dfrac{V}{C}} \qquad \text{This is the depreciation model.}$$

$$r - 1 = -\sqrt[3]{\dfrac{V}{C}} \qquad \text{Subtract 1 from both sides to isolate the radical.}$$

$$(r - 1)^3 = \left(-\sqrt[3]{\dfrac{V}{C}}\right)^3 \qquad \text{To eliminate the radical, cube both sides.}$$

$$(r - 1)^3 = -\dfrac{V}{C} \qquad \text{Simplify the right side.}$$

$$C(r - 1)^3 = -V \qquad \text{To clear the equation of the fraction, multiply both sides by } C.$$

$$C = -\dfrac{V}{(r - 1)^3} \qquad \text{To isolate } C, \text{ divide both sides by } (r - 1)^3.$$

 Self Check 9 A formula used in statistics to determine the size of a sample to obtain a desired degree of accuracy is

$$E = z_0 \sqrt{\dfrac{pq}{n}}$$

Solve the formula for n.

Now Try **Problem 69**

ANSWERS TO SELF CHECKS **1.** 11 **2.** 196 ft **3.** 6, 0 is extraneous **4.** 18 is extraneous, no solution, ∅
5. 0, −2 **6.** 0 **7.** 20 **8.** 1 **9.** $n = \dfrac{z_0^{\,2}pq}{E^2}$

STUDY SET
9.5

VOCABULARY

Fill in the blanks.

1. Equations such as $\sqrt{x + 4} - 4 = 5$ and $\sqrt[3]{x + 1} = 12$ are called _____ equations.

2. When solving equations containing radicals, try to _____ one radical expression on one side of the equation.

3. Squaring both sides of an equation can introduce _____ solutions.

4. To _____ a proposed solution means to substitute it into the original equation and see whether a true statement results.

CONCEPTS

5. Fill in the blanks: The power rule states that if x, y, and n are real numbers and $x = y$, then

   $$x^{\square} = y$$

6. Determine whether 6 is a solution of each radical equation.
   a. $\sqrt{x + 3} = x - 3$
   b. $\sqrt[3]{5x - 3} + 9 = x$

7. What is the first step in solving each equation?
   a. $\sqrt{x + 11} = 5$
   b. $\sqrt[3]{5x + 4} + 3 = 30$
   c. $\sqrt{x + 8} - \sqrt{2x + 9} = 1$

8. Simplify each expression.
   a. $\left(\sqrt{x}\right)^2$
   b. $\left(\sqrt{x - 5}\right)^2$
   c. $\left(4\sqrt{2x}\right)^2$
   d. $\left(\sqrt[3]{4x - 8}\right)^3$
   e. $\left(\sqrt[4]{8x}\right)^4$

9. Find: $\left(\sqrt{x} - 3\right)^2$

10. Solve $\sqrt{x - 2} + 2 = 4$ using the graphs below.

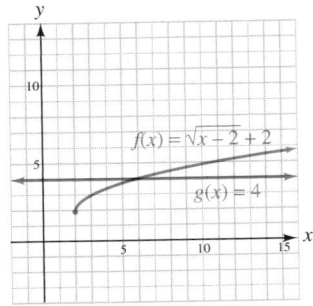

NOTATION

Complete each solution.

11. Solve: $\sqrt{3x + 3} - 1 = 5$

    $$\sqrt{3x + 3} = \square$$
    $$\left(\sqrt{3x + 3}\right)^{\square} = (6)^{\square}$$
    $$\square = 36$$
    $$3x = \square$$
    $$x = \square$$

    Does the proposed solution check?

12. Fill in the blanks. Write each radical equation using a rational exponent.
    a. $\sqrt{x + 10} + 5 = 15$ can be written $(x + 10)^{\square} + 5 = 15$
    b. $\sqrt[3]{2t + 4} = x - 1$ can be written $(2t + 4)^{\square} = x - 1$

GUIDED PRACTICE

Solve each equation. Write all proposed solutions. Cross out those that are extraneous. See Example 1.

13. $\sqrt{a - 3} = 1$
14. $\sqrt{x - 10} = 1$
15. $\sqrt{4x + 5} = 5$
16. $\sqrt{5x - 6} = 2$
17. $\sqrt{6x + 13} = 3$
18. $\sqrt{6x + 1} = 5$
19. $\sqrt{\frac{1}{3}x - 2} = 8$
20. $\sqrt{\frac{1}{2}x + 3} = 6$

Solve each equation. Write all proposed solutions. Cross out those that are extraneous. See Example 3 and the Notation feature in the margin.

21. $\sqrt{2x + 11} + 2 = x$
22. $\sqrt{2a - 3} + 3 = a$
23. $\sqrt{2r - 3} + 9 = r$
24. $\sqrt{-x + 2} + 2 = x$
25. $(3t + 7)^{1/2} - t = 1$
26. $(t + 3)^{1/2} - t = 1$
27. $(9 - a)^{1/2} - a = 3$
28. $(4 - a)^{1/2} - a = 2$

Solve each equation. Write all proposed solutions. Cross out those that are extraneous. See Example 4.

29. $\sqrt{5x} + 10 = 8$
30. $\sqrt{3x} + 5 = 2$
31. $\sqrt{5 - x} + 10 = 9$
32. $1 = 2 + \sqrt{4x + 75}$

Solve each equation. See Example 5 and the Notation feature in the margin.

33. $\sqrt[3]{7n - 1} = 3$
34. $\sqrt[3]{12m + 4} = 4$
35. $\sqrt[3]{x^3 - 7} = x - 1$
36. $\sqrt[3]{b^3 - 63} = b - 3$

37. $(m^3 + 26)^{1/3} = m + 2$ **38.** $(x^3 + 56)^{1/3} = x + 2$

39. $(5r + 14)^{1/3} = 4$ **40.** $(2b + 29)^{1/3} = 3$

See Example 6.

41. Let $f(x) = \sqrt[4]{3x + 1}$. For what value(s) of x is $f(x) = 4$?

42. Let $f(x) = \sqrt{2x^2 - 7x}$. For what value(s) of x is $f(x) = 2$?

43. Let $f(x) = \sqrt[3]{3x - 6}$. For what value(s) of x is $f(x) = -3$?

44. Let $f(x) = \sqrt[5]{4x - 4}$. For what value(s) of x is $f(x) = -2$?

Solve each equation. See Example 7.

45. $\sqrt{3x + 12} = \sqrt{5x - 12}$ **46.** $\sqrt{m + 4} = \sqrt{2m - 5}$

47. $2\sqrt{4x + 1} = \sqrt{x + 4}$ **48.** $\sqrt{6 - 2x} = 4\sqrt{x - 3}$

49. $\sqrt{6t + 9} = 3\sqrt{t}$ **50.** $\sqrt{12x + 24} = 6\sqrt{x}$

51. $(34x + 26)^{1/3} = 4(x - 1)^{1/3}$ **52.** $(a^2 + 2a)^{1/3} = 2(a - 1)^{1/3}$

Solve each equation. Write all proposed solutions. Cross out those that are extraneous. See Example 8.

53. $\sqrt{x - 5} + \sqrt{x} = 5$ **54.** $\sqrt{x - 7} + \sqrt{x} = 7$

55. $\sqrt{z + 3} - \sqrt{z} = 1$ **56.** $\sqrt{x + 12} + \sqrt{x} = 6$

57. $\sqrt{x + 5} + \sqrt{x - 3} = 4$ **58.** $\sqrt{b + 7} - \sqrt{b - 5} = 2$

59. $\sqrt{r + 16} + \sqrt{r + 9} = 7$ **60.** $\sqrt{x + 8} - \sqrt{x - 4} = 2$

61. $3 = \sqrt{y + 4} - \sqrt{y + 7}$ **62.** $3 = \sqrt{u - 3} - \sqrt{u}$

63. $2 = \sqrt{2u + 7} - \sqrt{u}$ **64.** $1 = \sqrt{4s + 5} - \sqrt{2s + 2}$

Solve each equation for the specified variable or expression. See Example 9.

65. $v = \sqrt{2gh}$ for h **66.** $d = 1.4\sqrt{h}$ for h

67. $T = 2\pi\sqrt{\dfrac{l}{32}}$ for l **68.** $d = \sqrt[3]{\dfrac{12V}{\pi}}$ for V

69. $r = \sqrt[3]{\dfrac{A}{P}} - 1$ for A **70.** $r = \sqrt[3]{\dfrac{A}{P}} - 1$ for P

71. $L_A = L_B\sqrt{1 - \dfrac{v^2}{c^2}}$ for v^2 **72.** $R_1 = \sqrt{\dfrac{A}{\pi} - R_2{}^2}$ for A

TRY IT YOURSELF

Solve each equation. Write all proposed solutions. Cross out those that are extraneous.

73. $2\sqrt{x} = \sqrt{5x - 16}$ **74.** $3\sqrt{x} = \sqrt{3x + 54}$

75. $n = (n^3 + n^2 - 1)^{1/3}$ **76.** $(m^4 + m^2 - 25)^{1/4} = m$

77. $\sqrt{y + 2} + y = 4$ **78.** $\sqrt{22y + 86} - y = 9$

79. $\sqrt[3]{x + 8} = -2$ **80.** $\sqrt[3]{x + 4} = -1$

81. $1 = \sqrt{x + 5} - \sqrt{x}$ **82.** $2 = \sqrt{x + 8} - \sqrt{x}$

83. $x = \dfrac{\sqrt{12x - 5}}{2}$ **84.** $x = \dfrac{\sqrt{16x - 12}}{2}$

85. $(n^2 + 6n + 3)^{1/2} = (n^2 - 6n - 3)^{1/2}$

86. $(m^2 - 12m - 3)^{1/2} = (m^2 + 12m + 3)^{1/2}$

87. $\sqrt{x - 5} - \sqrt{x + 3} = 4$ **88.** $\sqrt{x + 8} - \sqrt{x - 4} = -2$

89. $\sqrt[4]{10y + 6} = 2\sqrt[4]{y}$ **90.** $\sqrt[4]{21a + 39} = 3\sqrt[4]{a - 1}$

91. $\sqrt{-5x + 24} = 6 - x$ **92.** $-s - 3 = 2\sqrt{5 - s}$

93. $\sqrt{2x + 5} = 1$ **94.** $\sqrt{3x + 10} = 1$

95. $\sqrt{6x + 2} - \sqrt{5x + 3} = 0$ **96.** $\sqrt{5x + 2} - \sqrt{x + 10} = 0$

APPLICATIONS

97. HIGHWAY DESIGN   A curved road will accommodate traffic traveling s mph if the radius of the curve is r feet, according to the formula $s = 3\sqrt{r}$. If engineers expect 40-mph traffic, what radius should they specify? Give the result to the nearest foot.

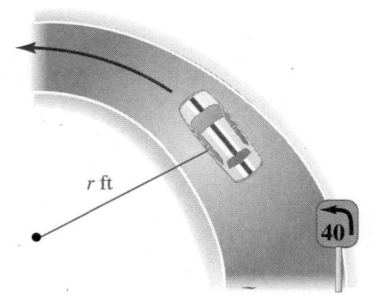

98. FORESTRY The taller a lookout tower, the farther an observer can see. That distance d (called the *horizon distance,* measured in miles) is related to the height h of the observer (measured in feet) by the formula $d = 1.22\sqrt{h}$. How tall must a lookout tower be to see the edge of the forest, 25 miles away? (Round to the nearest foot.)

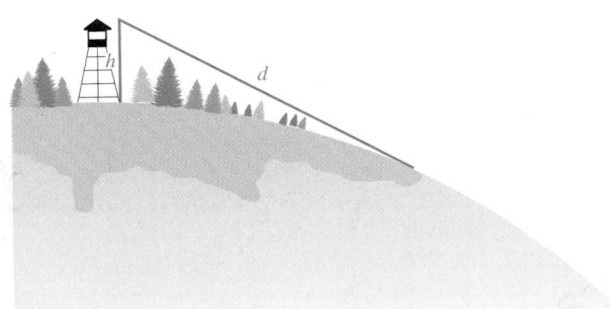

99. WIND POWER The power generated by a windmill is related to the velocity of the wind by the formula

$$v = \sqrt[3]{\frac{P}{0.02}}$$

where P is the power (in watts) and v is the velocity of the wind (in mph). Find how much power the windmill is generating when the wind is 29 mph.

100. DIAMONDS The *effective rate of interest r* earned by an investment is given by the formula

$$r = \sqrt[n]{\frac{A}{P}} - 1$$

where P is the initial investment that grows to value A after n years. If a diamond buyer got \$4,000 for a 1.73-carat diamond that he had purchased 4 years earlier, and earned an annual rate of return of 6.5% on the investment, what did he originally pay for the diamond?

101. THEATER PRODUCTIONS The ropes, pulleys, and sandbags shown in the illustration are part of a mechanical system used to raise and lower scenery for a stage play. For the scenery to be in the proper position, the following formula must apply:

$$w_2 = \sqrt{w_1{}^2 + w_3{}^2}$$

If $w_2 = 12.5$ lb and $w_3 = 7.5$ lb, find w_1.

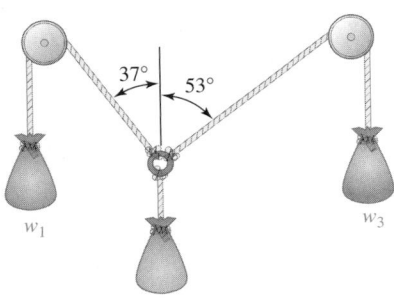

102. CARPENTRY During construction, carpenters often brace walls as shown in the illustration, where the length L of the brace is given by the formula

$$L = \sqrt{f^2 + h^2}$$

If a carpenter nails a 10-ft brace to the wall 6 feet above the floor, how far from the base of the wall should he nail the brace to the floor?

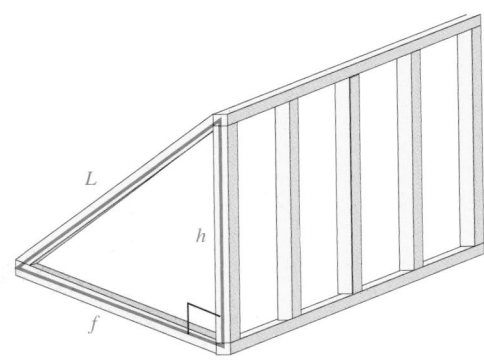

103. SUPPLY AND DEMAND The number of wrenches that will be produced at a given price can be predicted by the formula $s = \sqrt{5x}$, where s is the supply (in thousands) and x is the price (in dollars). The demand d for wrenches can be predicted by the formula $d = \sqrt{100 - 3x^2}$. Find the equilibrium price—that is, find the price at which supply will equal demand.

104. SUPPLY AND DEMAND The number of mirrors that will be produced at a given price can be predicted by the formula $s = \sqrt{23x}$, where s is the supply (in thousands) and x is the price (in dollars). The demand d for mirrors can be predicted by the formula $d = \sqrt{312 - 2x^2}$. Find the equilibrium price—that is, find the price at which supply will equal demand.

WRITING

105. What is wrong with the work shown below?

Solve: $\sqrt{x+1} - 3 = 8$
$$\sqrt{x+1} = 11$$
$$\left(\sqrt{x+1}\right)^2 = 11$$
$$x + 1 = 11$$
$$x = 10$$

106. The first step of a student's solution is shown below. What is a better way to begin the solution?

Solve: $\sqrt{x} + \sqrt{x+22} = 12$
$$\left(\sqrt{x} + \sqrt{x+22}\right)^2 = 12^2$$

107. Explain why it is immediately apparent that $\sqrt{8x - 7} = -2$ has no solution.

108. Explain the error in the following work.

Solve: ~~$\sqrt{2y + 1} = \sqrt{y + 7} + 3$~~

~~$\left(\sqrt{2y + 1}\right)^2 = \left(\sqrt{y + 7} + 3\right)^2$~~

~~$2y + 1 = y + 7 + 9$~~

109. To solve $\sqrt{2x + 7} = \sqrt{x}$ we need only square both sides once. To solve $\sqrt{2x + 7} = \sqrt{x} + 2$ we have to square both sides twice. Why does the second equation require more work?

110. Explain how $\sqrt{2x - 1} = x$ can be solved graphically.

111. Explain how the table can be used to solve $\sqrt{4x - 3} - 2 = \sqrt{2x - 5}$ if $Y_1 = \sqrt{4x - 3} - 2$ and $Y_2 = \sqrt{2x - 5}$.

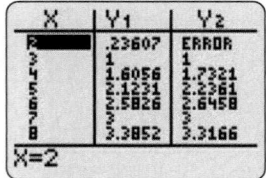

112. Explain how to use the graph of $f(x) = \sqrt[3]{x - 0.5} - 1$, shown in the illustration, to approximate the solution of $\sqrt[3]{x - 0.5} = 1$.

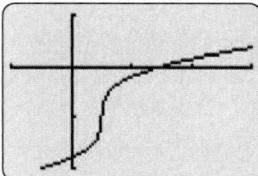

REVIEW

113. LIGHTING   The intensity of light from a lightbulb varies inversely as the square of the distance from the bulb. If you are 5 feet away from a bulb and the intensity is 40 foot-candles, what will the intensity be if you move 20 feet away from the bulb?

114. COMMITTEES   What type of variation is shown in the illustration? As the number of people on this committee increased, what happened to its effectiveness?

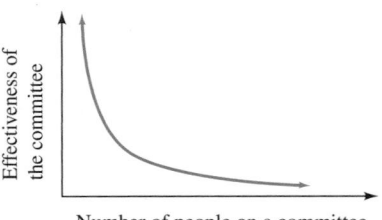

115. TYPESETTING   If 12-point type is 0.166044 inch tall, how tall is 30-point type?

116. GUITAR STRINGS   The frequency of vibration of a string varies directly as the square root of the tension and inversely as the length of the string. Suppose a string 2.5 feet long, under a tension of 16 pounds, vibrates 25 times per second. Find k, the constant of proportionality.

CHALLENGE PROBLEMS

Solve each equation. Write all proposed solutions. Cross out those that are extraneous.

117. $\sqrt[3]{2x} = \sqrt{x}$

118. $\sqrt[4]{x} = \sqrt{\dfrac{x}{4}}$

119. $\sqrt{x + 2} + \sqrt{2x} = \sqrt{18 - x}$

120. $\sqrt{8 - x} - \sqrt{3x - 8} = \sqrt{x - 4}$

SECTION 9.6
Geometric Applications of Radicals

Objectives

1 Use the Pythagorean theorem to solve problems.

2 Solve problems involving 45°–45°–90° triangles.

3 Solve problems involving 30°–60°–90° triangles.

4 Use the distance formula to solve problems.

We will now consider applications of square roots in geometry. Then we will find the distance between two points on a rectangular coordinate system, using a formula that contains a square root. We begin by considering an important theorem about right triangles.

Use the Pythagorean Theorem to Solve Problems.

If we know the lengths of two legs of a right triangle, we can find the length of the **hypotenuse** (the side opposite the 90° angle) by using the **Pythagorean theorem.**

The Pythagorean Theorem	If a and b are the lengths of the legs of a right triangle and c is the length of the hypotenuse, $$a^2 + b^2 = c^2$$	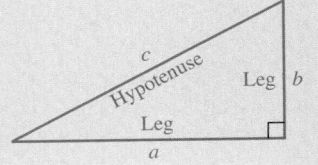

In words, the Pythagorean theorem is expressed as follows:

In any right triangle, the square of the hypotenuse is equal to the sum of the squares of the two legs.

Suppose the right triangle shown in the margin has legs of length 3 and 4 units. To find the length of the hypotenuse, we use the Pythagorean theorem.

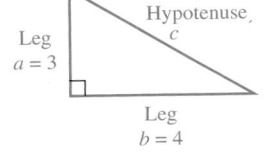

$$a^2 + b^2 = c^2$$
$$3^2 + 4^2 = c^2 \qquad \text{Substitute 3 for } a \text{ and 4 for } b.$$
$$9 + 16 = c^2$$
$$25 = c^2$$

To find c, we ask "What number, when squared, is equal to 25?" There are two such numbers: the positive square root of 25 and the negative square root of 25. Since c represents the length of the hypotenuse, and it cannot be negative, it follows that c is the positive square root of 25.

$$\sqrt{25} = c \qquad \text{Recall that a radical symbol } \sqrt{} \text{ is used to represent the positive, or principal square root of a number.}$$
$$5 = c$$

The length of the hypotenuse is 5 units.

The Language of Algebra
A *theorem* is a mathematical statement that can be proved. The Pythagorean theorem is named after Pythagoras, a Greek mathematician who lived about 2,500 years ago. He is thought to have been the first to prove the theorem.

EXAMPLE 1   *Firefighting.* To fight a fire, the forestry department plans to clear a rectangular firebreak around the fire, as shown in the illustration on the next page. Crews are equipped with mobile communications that have a 3,000-yard range. Can crews at points A and B remain in radio contact?

Strategy   We will use the Pythagorean theorem to find the distance between points A and B.

Why   If this distance is less than 3,000 yards, they can communicate. If it is greater than 3,000 yards, they cannot communicate.

Solution   The line segments connecting points A, B, and C form a right triangle. To find the distance c from point A to point B, we can use the Pythagorean theorem, substituting 2,400 for a and 1,000 for b and solving for c.

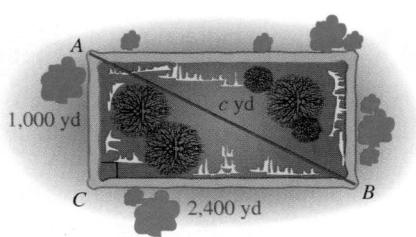

$$a^2 + b^2 = c^2$$

$$2,400^2 + 1,000^2 = c^2$$

$$5,760,000 + 1,000,000 = c^2$$

$$6,760,000 = c^2$$

$$\sqrt{6,760,000} = c$$ If $c^2 = 6{,}760{,}000$, then c must be a square root of 6,760,000. Because c represents a length, it must be the positive square root of 6,760,000.

$$2,600 = c$$ Use a calculator to find the square root.

The two crews are 2,600 yards apart. Because this distance is less than the 3,000-yard range of the radios, they can communicate by radio.

> ▷ **Self Check 1**   Can the crews communicate if $b = 1{,}500$ yards?
>
> **Now Try**   Problems 15 and 55

Caution
When using the Pythagorean theorem $a^2 + b^2 = c^2$, we can let a represent the length of either leg of the right triangle. We then let b represent the length of the other leg. The variable c must always represent the length of the hypotenuse.

2   **Solve Problems Involving 45°– 45°– 90° Triangles.**

An **isosceles right triangle** is a right triangle with two legs of equal length. Isosceles right triangles have angle measures of 45°, 45°, and 90°. If we know the length of one leg of an isosceles right triangle, we can use the Pythagorean theorem to find the length of the hypotenuse. Since the triangle shown in the margin is a right triangle, we have

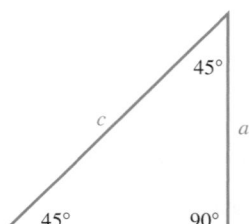

$$c^2 = a^2 + b^2$$

$$c^2 = a^2 + a^2$$ Both legs are a units long, so replace b with a.

$$c^2 = 2a^2$$ Combine like terms.

$$c = \sqrt{2a^2}$$ If $c^2 = 2a^2$, then c must be a square root of $2a^2$. Because c represents a length, it must be the positive square root of $2a^2$.

$$c = a\sqrt{2}$$ Simplify the radical: $\sqrt{2a^2} = \sqrt{2}\sqrt{a^2} = \sqrt{2}a = a\sqrt{2}$.

Thus, *in an isosceles right triangle, the length of the hypotenuse is $\sqrt{2}$ times the length of one leg.*

EXAMPLE 2   If one leg of an isosceles right triangle is 10 feet long, find the exact length of the hypotenuse. Then approximate the length to two decimal places.

Strategy   We will multiply the length of the known leg by $\sqrt{2}$.

Why   The length of the hypotenuse of an isosceles right triangle is $\sqrt{2}$ times the length of one leg.

Solution   Since the length of the hypotenuse is the length of a leg times $\sqrt{2}$, we have

$$c = 10\sqrt{2}$$

The exact length of the hypotenuse is $10\sqrt{2}$ feet. If we approximate to two decimal places, the length is 14.14 feet.

Self Check 2   Find the exact length of the hypotenuse of an isosceles right triangle if one leg is 12 meters long. Then approximate the length to two decimal places.

Now Try   **Problem 23**

If the length of the hypotenuse of an isosceles right triangle is known, we can use the Pythagorean theorem to find the length of each leg.

EXAMPLE 3   Find the exact length of each leg of the following isosceles right triangle shown in the margin. Then approximate the lengths to two decimal places.

Strategy   We will use the Pythagorean theorem to form an equation that we can solve to find the unknown length of a leg of the triangle.

Why   We use the Pythagorean theorem because the triangle is a right triangle and we are given the length of its hypotenuse.

Solution   We use the Pythagorean theorem.

$$c^2 = a^2 + b^2$$

$$25^2 = a^2 + a^2 \qquad \text{Since both legs are } a \text{ units long, substitute } a \text{ for } b.$$
$$\text{The hypotenuse is 25 units long. Substitute 25 for } c.$$

$$25^2 = 2a^2 \qquad \text{Combine like terms.}$$

$$\frac{625}{2} = a^2 \qquad \text{To isolate } a^2, \text{ square 25 and then divide both sides by 2.}$$

$$\sqrt{\frac{625}{2}} = a \qquad \text{If } a^2 = \frac{625}{2}, \text{ then } a \text{ must be the positive square root of } \frac{625}{2}.$$

$$\frac{\sqrt{625}}{\sqrt{2}} \cdot \frac{\sqrt{2}}{\sqrt{2}} = a \qquad \text{Write } \sqrt{\frac{625}{2}} \text{ as } \frac{\sqrt{625}}{\sqrt{2}}. \text{ Then rationalize the denominator.}$$

$$\frac{25\sqrt{2}}{2} = a \qquad \text{In the numerator, evaluate the radical: } \sqrt{625} = 25.$$
$$\text{Multiply the denominators: } \sqrt{2} \cdot \sqrt{2} = 2.$$

The exact length of each leg is $\frac{25\sqrt{2}}{2}$ units. If we approximate to two decimal places, the length is 17.68 units.

Self Check 3   Find the exact length of each leg of an isosceles right triangle if the length of the hypotenuse is 9 inches. Then approximate the lengths to two decimal places.

Now Try   **Problem 27**

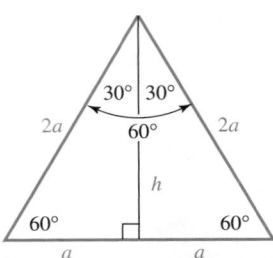

 Solve Problems Involving 30°–60°–90° Triangles.

From geometry, we know that an **equilateral triangle** is a triangle with three sides of equal length and three 60° angles. Each side of the equilateral triangle shown in the margin is $2a$ units long. If an **altitude** (height) is drawn to its base, the altitude divides the base into two segments of equal length and divides the equilateral triangle into two 30°–60°–90° triangles. From the figure, we can see that the shorter leg of each 30°–60°–90° triangle (the side *opposite* the 30° angle) is a units long. Thus,

The length of the hypotenuse of a 30°–60°–90° triangle is twice as long as the shorter leg.

We can discover another relationship between the legs of a 30°–60°–90° triangle if we find the length of the altitude h in the figure. We begin by applying the Pythagorean theorem to one of the 30°–60°–90° triangles.

$$a^2 + b^2 = c^2$$
$$a^2 + h^2 = (2a)^2 \quad \text{The altitude is } h \text{ units long, so replace } b \text{ with } h.$$
$$\qquad\qquad\qquad \text{The hypotenuse is } 2a \text{ units long, so replace } c \text{ with } 2a.$$
$$a^2 + h^2 = 4a^2 \quad (2a)^2 = (2a)(2a) = 4a^2.$$
$$h^2 = 3a^2 \quad \text{Subtract } a^2 \text{ from both sides.}$$
$$h = \sqrt{3a^2} \quad \text{If } h^2 = 3a^2, \text{ then } h \text{ must be the positive square root of } 3a^2.$$
$$h = a\sqrt{3} \quad \text{Simplify the radical: } \sqrt{3a^2} = \sqrt{3}\sqrt{a^2} = a\sqrt{3}.$$

We see that the altitude—the longer leg of the 30°–60°–90° triangle—is $\sqrt{3}$ times as long as the shorter leg. Thus,

The length of the longer leg of a 30°–60°–90° triangle is $\sqrt{3}$ times the length of the shorter leg.

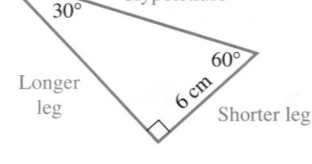

| **EXAMPLE 4** | Find the length of the hypotenuse and the length of the longer leg of the 30°–60°–90° triangle shown in the margin. |

Strategy To find the length of the hypotenuse, we will multiply the length of the shorter leg by 2. To find the length of the longer leg, we will multiply the length of the shorter leg by $\sqrt{3}$.

Why These side-length relationships are true for any 30°–60°–90° triangle.

Solution Since the length of the hypotenuse of a 30°–60°–90° triangle is twice as long as the shorter leg, and the length of the shorter leg is 6 cm, the hypotenuse is $2 \cdot 6 = 12$ cm.

Since the length of the longer leg is $\sqrt{3}$ times the length of the shorter leg, and the length of the shorter leg is 6 cm, the longer leg is $6\sqrt{3}$ cm (about 10.39 cm).

Self Check 4 Find the length of the hypotenuse and the longer leg of a 30°–60°–90° triangle if the shorter leg is 8 centimeters long.

***Now Try* Problem 31**

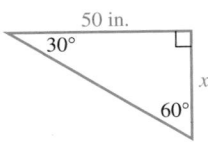

50 in.

30°

60°

x

EXAMPLE 5   Find the length of the hypotenuse and the length of the shorter leg of the 30°–60°–90° triangle shown in the margin. Then approximate the lengths to two decimal places.

Strategy   We will find the length of the shorter leg first.

Why   Once we know the length of the shorter leg, we can multiply it by 2 to find the length of the hypotenuse.

Solution   If we let x = the length in inches of the shorter leg of the triangle, we can form an equation by translating the following statement:

The length of the longer leg of a 30°–60°–90° triangle	is	$\sqrt{3}$	times	the length of the shorter leg.
50	=	$\sqrt{3}$	$\cdot$	x

To find the length of the shorter leg, we solve the equation for x.

$$50 = \sqrt{3}x$$

$$\frac{50}{\sqrt{3}} = \frac{\sqrt{3}x}{\sqrt{3}} \qquad \text{To isolate } x, \text{ divide both sides by } \sqrt{3}.$$

$$\frac{50}{\sqrt{3}} = x$$

The length of the shorter leg is exactly $\frac{50}{\sqrt{3}}$ inches. To write this number in simplified radical form, we rationalize the denominator.

$$\frac{50}{\sqrt{3}} \cdot \frac{\sqrt{3}}{\sqrt{3}} = \frac{50\sqrt{3}}{3}$$

Thus, the length of the shorter leg is exactly $\frac{50\sqrt{3}}{3}$ inches (about 28.87 inches).

Since the length of the hypotenuse of a 30°–60°–90° triangle is twice as long as the shorter leg, the hypotenuse is $2 \cdot \frac{50\sqrt{3}}{3} = \frac{100\sqrt{3}}{3}$ inches (about 57.74 inches).

 Self Check 5   Find the length of the hypotenuse and the shorter leg of a 30°–60°–90° triangle if the longer leg is 15 feet long. Then approximate the lengths to two decimal places.

Now Try   Problem 35

EXAMPLE 6   *Stretching Exercises.*   A doctor prescribed the exercise shown in figure (a) on the next page. The patient was instructed to raise his leg to an angle of 60° and hold the position for 10 seconds. If the patient's leg is 36 inches long, how high off the floor will his foot be when his leg is held at the proper angle?

Strategy   This situation is modeled by a 30°–60°–90° triangle. We will begin by finding the length of the shorter leg.

Why Once we know the length of the shorter leg, we can easily find the length of the longer leg, which represents the distance the patient's foot is off the ground.

Solution In figure (b), we see right triangle *ABC*, which models the situation. Since the length of the hypotenuse is twice as long as the side opposite the 30° angle, side *AC* is half as long as the hypotenuse. Since the hypotenuse is given to be 36 inches long, side *AC* must be 18 inches long.

Since the length of the longer leg (the leg opposite the 60° angle) is $\sqrt{3}$ times the length of the shorter leg (side *AC*), side *BC* is $18\sqrt{3}$, or about 31 inches long. So the patient's foot will be about 31 inches from the floor when his leg is in the proper position.

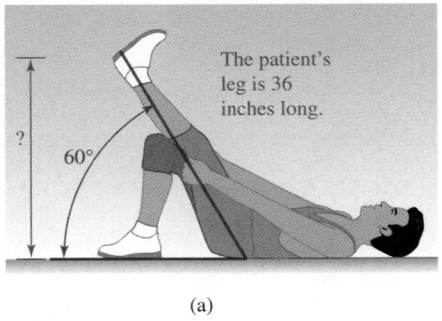

The patient's leg is 36 inches long.

(a)

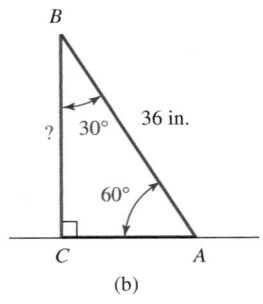

(b)

Now Try **Problems 39 and 61**

④ **Use the Distance Formula to Solve Problems.**

With the *distance formula*, we can find the distance between any two points graphed on a rectangular coordinate system.

To find the distance *d* between points $P(x_1, y_1)$ and $Q(x_2, y_2)$ shown in the figure below, we construct the right triangle *PRQ*. The distance between *P* and *R* is $|x_2 - x_1|$, and the distance between *R* and *Q* is $|y_2 - y_1|$. We apply the Pythagorean theorem to the right triangle *PRQ* to get

$$d^2 = |x_2 - x_1|^2 + |y_2 - y_1|^2$$
$$= (x_2 - x_1)^2 + (y_2 - y_1)^2 \quad \text{Because } |x_2 - x_1|^2 = (x_2 - x_1)^2 \text{ and } |y_2 - y_1|^2 = (y_2 - y_1)^2.$$

Because *d* represents the distance between two points, it must be equal to the positive square root of $(x_2 - x_1)^2 + (y_2 - y_1)^2$.

$$d = \sqrt{(x_2 - x_1)^2 + (y_2 - y_1)^2}$$

We call this result the *distance formula*.

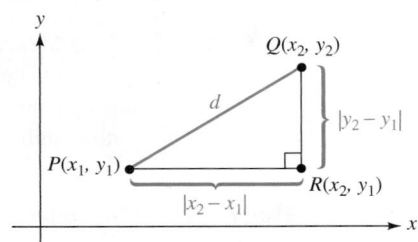

Distance Formula   The distance d between two points with coordinates (x_1, y_1) and (x_2, y_2) is given by

$$d = \sqrt{(x_2 - x_1)^2 + (y_2 - y_1)^2}$$

EXAMPLE 7   Find the distance between the points $(-2, 3)$ and $(4, -5)$.

Strategy   We will use the distance formula.

Why   We know the x- and y-coordinates of both points.

Solution   To find the distance, we can use the distance formula by substituting 4 for x_2, -2 for x_1, -5 for y_2, and 3 for y_1.

$$
\begin{aligned}
d &= \sqrt{(x_2 - x_1)^2 + (y_2 - y_1)^2} \\
&= \sqrt{[4 - (-2)]^2 + (-5 - 3)^2} \\
&= \sqrt{(4 + 2)^2 + (-5 - 3)^2} \\
&= \sqrt{6^2 + (-8)^2} \\
&= \sqrt{36 + 64} \\
&= \sqrt{100} \\
&= 10
\end{aligned}
$$

The distance between the points is 10 units.

Self Check 7   Find the distance between $(-2, -2)$ and $(3, 10)$.

Now Try   Problems 47 and 49

EXAMPLE 8   ***Robotics.***   Robots are used to weld parts of an automobile chassis on an automated production line. To do so, an imaginary coordinate system is superimposed on the side of the vehicle, and the robot is programmed to move to specific positions to make each weld. (See the figure below, which is scaled in inches.) How far does the tip of the welder unit move from position 1 to position 2?

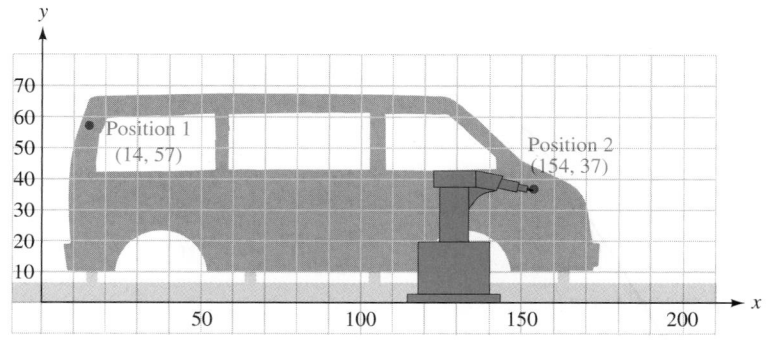

Strategy   We will use the distance formula to find how far the tip of the welder unit moves.

Why We know the *x*- and *y*-coordinates of position 1 and position 2.

Solution To find the distance *d* that the tip of the welder unit moves from position 1 at (14, 57) to position 2 at (154, 37), we proceed as follows:

$$d = \sqrt{(x_2 - x_1)^2 + (y_2 - y_1)^2} \qquad \text{This is the distance formula.}$$

$$d = \sqrt{(154 - 14)^2 + (37 - 57)^2} \qquad \text{Substitute 154 for } x_2, \text{ 14 for } x_1, \text{ 37 for } y_2, \text{ and 57 for } y_1.$$

$$= \sqrt{140^2 + (-20)^2}$$

$$= \sqrt{20{,}000} \qquad \text{Evaluate: } 140^2 + (-20)^2 = 19{,}600 + 400 = 20{,}000.$$

$$= 100\sqrt{2} \qquad \text{Simplify: } \sqrt{20{,}000} = \sqrt{100 \cdot 100 \cdot 2} = 100\sqrt{2}.$$

The tip of the welder unit moves $100\sqrt{2}$ inches (about 141.42 inches) from position 1 to position 2.

Now Try Problem 67

ANSWERS TO SELF CHECKS   **1.** Yes   **2.** $12\sqrt{2}$ m ≈ 16.97 m   **3.** $\frac{9\sqrt{2}}{2}$ in. ≈ 6.36 in.
4. 16 cm, $8\sqrt{3}$ cm   **5.** $10\sqrt{3}$ in. ≈ 17.32 in., $5\sqrt{3}$ in. ≈ 8.66 in.   **7.** 13

STUDY SET
9.6

VOCABULARY

Fill in the blanks.

1. In a right triangle, the side opposite the 90° angle is called the
_____.

2. An _____ right triangle is a right triangle with two legs of equal length.

3. The _____ theorem states that in any right triangle, the square of the hypotenuse is equal to the sum of the squares of the lengths of the two legs.

4. An _____ triangle has three sides of equal length and three 60° angles.

CONCEPTS

Fill in the blanks.

5. If *a* and *b* are the lengths of the legs of a right triangle and *c* is the length of the hypotenuse, then ⬚ + ⬚ = ⬚.

6. In any right triangle, the square of the hypotenuse is equal to the _____ of the squares of the two _____.

7. In an isosceles right triangle, the length of the hypotenuse is times the length of one leg.

8. The shorter leg of a 30°–60°–90° triangle is _____ as long as the hypotenuse.

9. The length of the longer leg of a 30°–60°–90° triangle is ⬚ times the length of the shorter leg.

10. In a 30°–60°–90° triangle, the shorter leg is opposite the _____ angle, and the longer leg is opposite the _____ angle.

11. The formula to find the distance between points (x_1, y_1) and (x_2, y_2) is $d = \sqrt{\text{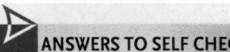} \quad + \quad}$.

12. Solve for *c*, where *c* represents the length of the hypotenuse of a right triangle. Simplify the result, if possible.
   a. $c^2 = 64$
   b. $c^2 = 15$
   c. $c^2 = 24$

NOTATION

Complete each solution.

13. Evaluate. Approximate to two decimal places.

$$\sqrt{(-1-3)^2 + [2-(-4)]^2} = \sqrt{(-4)^2 + [\ \ ]^2}$$
$$= \sqrt{}$$
$$= \sqrt{} \cdot 13$$
$$= \sqrt{13}$$
$$\approx $$

14. Solve $8^2 + 4^2 = c^2$ and assume $c > 0$.

$$ + 16 = c^2$$
$$ = c^2$$
$$\sqrt{} = $$
$$\sqrt{} \cdot 5 = c$$
$$\sqrt{5} = c$$
$$c 8.94$$

GUIDED PRACTICE

The lengths of two sides of the right triangle ABC are given. Find the length of the missing side. See Example 1.

15. $a = 6$ ft and $b = 8$ ft
16. $a = 5$ in. and $b = 12$ in.
17. $a = 8$ ft and $b = 15$ ft
18. $a = 24$ yd and $b = 7$ yd
19. $b = 9$ ft and $c = 41$ ft
20. $b = 18$ m and $c = 82$ m
21. $a = 10$ cm and $c = 26$ cm
22. $a = 14$ in. and $c = 50$ in.

Find the missing side lengths in each triangle. Give the exact answer and then an approximation to two decimal places when appropriate. See Example 2.

23.

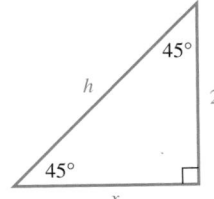

24.

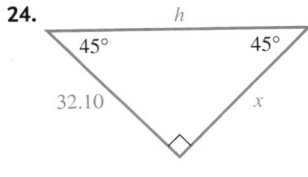

25. One leg of an isosceles right triangle is 3.2 feet long. Find the length of its hypotenuse. Give the exact answer and then an approximation to two decimal places.

26. One side of a square is $5\frac{1}{2}$ in. long. Find the length of its diagonal. Give the exact answer and then an approximation to two decimal places.

Find the missing side lengths in each triangle. Give the exact answer and then an approximation to two decimal places. See Example 3.

27.

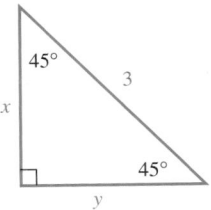

28.

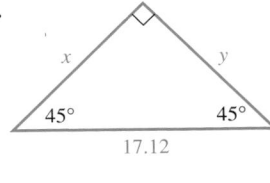

29. PHOTOGRAPHS The diagonal of a square photograph measures 10 inches. Find the length of one of its sides. Give the exact answer and then an approximation to two decimal places.

30. PARKING LOTS The diagonal of a square parking lot is approximately 1,414 feet long.
  a. Find the length of one side of the parking lot. Round to the nearest foot.
  b. Find the approximate area of the parking lot.

Find the missing side lengths in each triangle. Give the exact answer and then an approximation to two decimal places, when appropriate. See Example 4.

31.

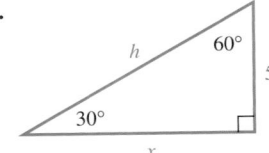

32.

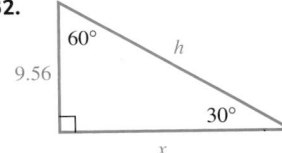

33. In a 30°–60°–90° triangle, the length of the leg opposite the 30° angle is 75 cm. Find the length of the leg opposite the 60° angle and the length of the hypotenuse. Give the exact answer and then an approximation to two decimal places, when appropriate.

34. In a 30°–60°–90° triangle, the length of the shorter leg is $5\sqrt{2}$ inches. Find the length of the hypotenuse and the length of the longer leg. Give the exact answer and then an approximation to two decimal places.

Find the missing lengths in each triangle. Give the exact answer and then an approximation to two decimal places. See Example 5.

35.

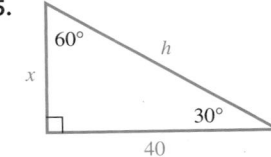

36.

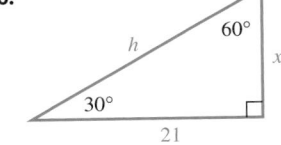

37. In a 30°–60°–90° right triangle, the length of the leg opposite the 60° angle is 55 millimeters. Find the length of the leg opposite the 30° and the length of the hypotenuse. Give the exact answer and then an approximation to two decimal places.

38. In a 30°–60°–90° right triangle, the length of the longer leg is 24 yards. Find the length of the hypotenuse and the length of the shorter leg. Give the exact answer and then an approximation to two decimal places.

Find the missing lengths in each triangle. Give the exact answer and then an approximation to two decimal places, when appropriate. See Example 6.

39.

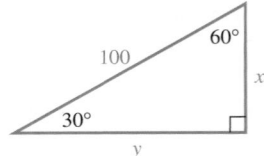

40.

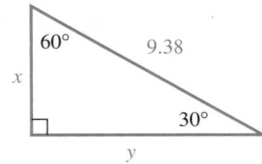

41. In a 30°–60°–90° right triangle, the length of the hypotenuse is 1.5 feet. To the nearest hundredth, find the length of the shorter leg and the length of the longer leg. Give the exact answer and then an approximation to two decimal places, when appropriate.

42. In a 30°–60°–90° right triangle, the length of the hypotenuse is $12\sqrt{3}$ inches. Find the length of the leg opposite the 30° angle and the length of the leg opposite the 60° angle. Give the exact answer and then an approximation to two decimal places, when appropriate.

Find the distance between each pair of points. See Example 7.

43. (0, 0), (3, −4)

44. (0, 0), (−12, 16)

45. (−2, −8), (3, 4)

46. (−5, −2), (7, 3)

47. (6, 8), (12, 16)

48. (10, 4), (2, −2)

49. (−2, 1), (3, 4)

50. (2, −3), (4, −8)

51. (−1, −6), (3, −4)

52. (−3, 5), (−5, −5)

53. (−2, −1), (−5, 8)

54. (4, 7), (−4, −5)

APPLICATIONS

55. SOCCER The allowable length of a rectangular soccer field used for international adult matches can be from 100 to 110 meters and the width can be from 64 to 75 meters.

   a. Find the length of the diagonal of the field that has the minimum allowable length and minimum allowable width. Give an approximation to two decimal places.

   b. Find the length of the diagonal of the field that has the maximum allowable length and maximum allowable width. Give the exact answer and an approximation to two decimal places.

56. Find the exact length of the diagonal (in blue) of one of the *faces* of the cube shown below.

57. Find the exact length of the diagonal (in green) of the cube shown below.

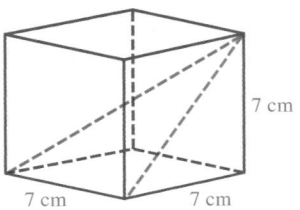

58. Use the distance formula to show that a triangle with vertices (−2, 4), (2, 8), and (6, 4) is isosceles.

59. WASHINGTON, D.C. The square in the map shows the 100-square-mile site selected by George Washington in 1790 to serve as a permanent capital for the United States. In 1847, the part of the district lying on the west bank of the Potomac was returned to Virginia. Find the exact coordinates of each corner of the original square that outlined the District of Columbia and approximations to two decimal places.

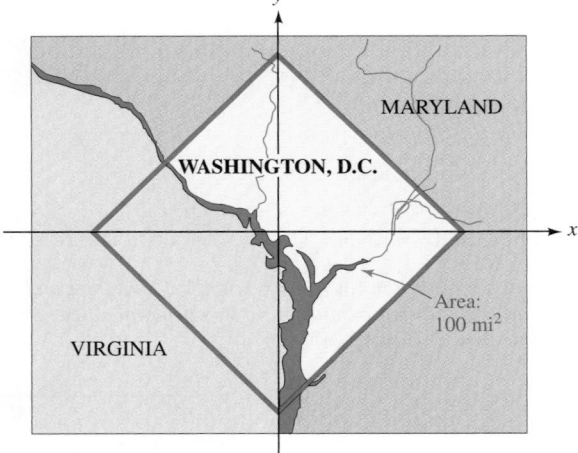

60. PAPER AIRPLANES   The illustration gives the directions for making a paper airplane from a square piece of paper with sides 8 inches long. Find the length *l* of the plane when it is completed. Give the exact answer and an approximation to two decimal places.

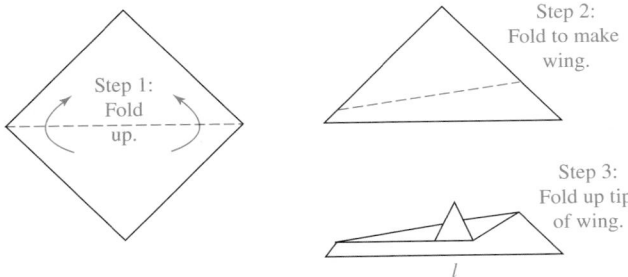

Step 1: Fold up.

Step 2: Fold to make wing.

Step 3: Fold up tip of wing.

l

61. HARDWARE   The sides of a regular hexagonal nut are 10 millimeters long. Find the height *h* of the nut. Give the exact answer and an approximation to two decimal places.

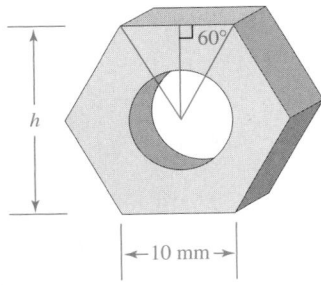

60°

h

←10 mm→

62. IRONING BOARDS   Find the height *h* of the ironing board shown in the illustration. Give the exact answer and an approximation to two decimal places.

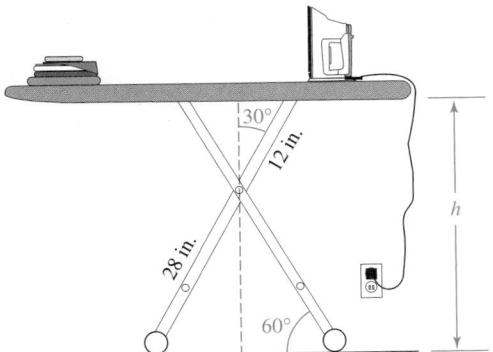

30°

12 in.

28 in.

60°

h

63. BASEBALL   A baseball diamond is a square, 90 feet on a side. If the third baseman fields a ground ball 10 feet directly behind third base, how far must he throw the ball to throw a runner out at first base? Give the exact answer and an approximation to two decimal places.

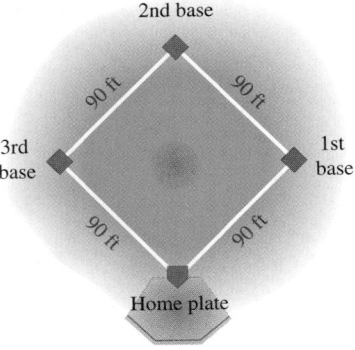

2nd base

90 ft

90 ft

3rd base

1st base

90 ft

90 ft

Home plate

64. BASEBALL   A shortstop fields a grounder at a point one-third of the way from second base to third base. How far will he have to throw the ball to make an out at first base? Give the exact answer and an approximation to two decimal places.

65. CLOTHESLINES   A pair of damp jeans are hung on a clothesline to dry. They pull the center down 1 foot. By how much is the line stretched? Give an approximation to two decimal places.

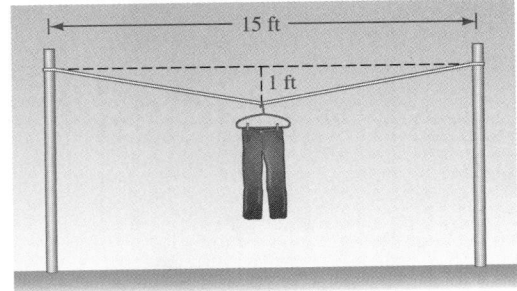

15 ft

1 ft

66. FIREFIGHTING The base of the 37-foot ladder is 9 feet from the wall. Will the top reach a window ledge that is 35 feet above the ground? Verify your result.

37 ft

h ft

9 ft

67. ART HISTORY A figure displaying some of the characteristics of Egyptian art is shown in the illustration. Use the distance formula to find the following dimensions of the drawing. Round your answers to two decimal places.

a. From the foot to the eye

b. From the belt to the hand holding the staff

c. From the shoulder to the symbol held in the hand

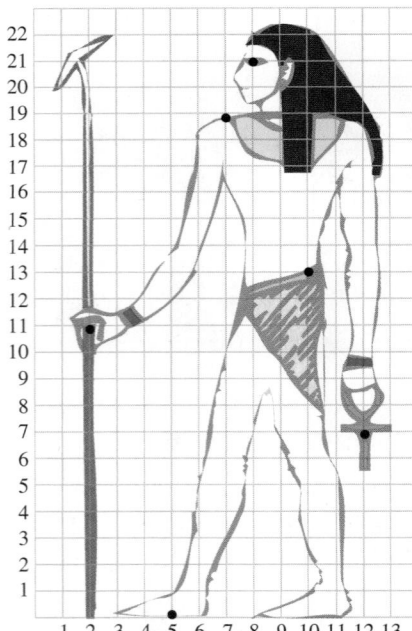

68. PACKAGING The diagonal *d* of a rectangular box with dimensions $a \times b \times c$ is given by

$$d = \sqrt{a^2 + b^2 + c^2}$$

Will the umbrella fit in the shipping carton in the illustration? Verify your result.

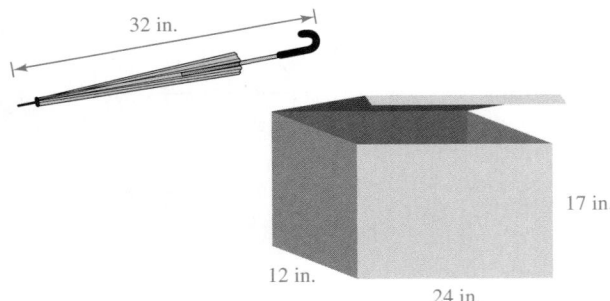

32 in.

17 in.

12 in.

24 in.

69. PACKAGING An archaeologist wants to ship a 34-inch femur bone. Will it fit in a 4-inch-tall box that has a 24-inch-square base? (See Exercise 68.) Verify your result.

70. TELEPHONE SERVICE The telephone cable in the illustration runs from *A* to *B* to *C* to *D*. How much cable is required to run from *A* to *D* directly?

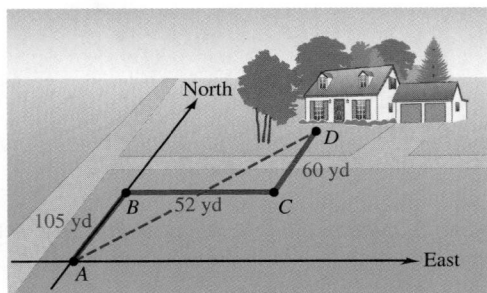

North

D

60 yd

105 yd   *B*   52 yd   *C*

A   East

WRITING

71. State the Pythagorean theorem in words.

72. List the facts that you learned about special right triangles in this section.

73. When the lengths of the sides of a certain triangle are substituted into the equation of the Pythagorean theorem, the result is a false statement. Explain why.

$$a^2 + b^2 = c^2$$
$$2^2 + 4^2 = 5^2$$
$$4 + 16 = 25$$
$$20 = 25 \quad \text{False}$$

74. Explain how the distance formula and the Pythagorean theorem can be used to show that a triangle with vertices $(2, 3)$, $(-3, 4)$, and $(1, -2)$ is a right triangle.

REVIEW

75. DISCOUNT BUYING A repairman purchased some washing-machine motors for a total of $224. When the unit cost decreased by $4, he was able to buy one extra motor for the same total price. How many motors did he buy originally?

76. AVIATION An airplane can fly 650 miles with the wind in the same amount of time as it can fly 475 miles against the wind. If the wind speed is 40 mph, find the speed of the plane in still air.

77. Find the mean of 16, 6, 10, 4, 5, 13

78. Find the median of 16, 6, 10, 4, 5

CHALLENGE PROBLEMS

79. Find the length of the diagonal of the cube.

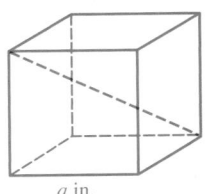

a in.

80. Show that the length of the diagonal of the rectangular solid shown is $\sqrt{a^2 + b^2 + c^2}$ cm.

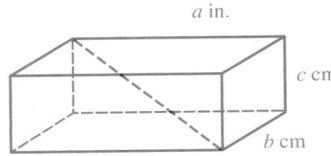

c cm
b cm
a cm

Find the distance between each pair of points.

81. $\left(\sqrt{48}, \sqrt{150}\right)$ and $\left(\sqrt{12}, \sqrt{24}\right)$

82. $\left(\sqrt{8}, -\sqrt{20}\right)$ and $\left(\sqrt{50}, -\sqrt{45}\right)$

SECTION 9.7
Complex Numbers

Objectives

1 Express square roots of negative numbers in terms of i.
2 Write complex numbers in the form $a + bi$.
3 Add and subtract complex numbers.
4 Multiply complex numbers.
5 Divide complex numbers.
6 Perform operations involving powers of i.

Recall that the square root of a negative number is not a real number. However, an expanded number system, called the *complex number system,* gives meaning to square roots of negative numbers, such as $\sqrt{-9}$ and $\sqrt{-25}$. To define complex numbers, we use a number that is denoted by the letter i.

1 **Express Square Roots of Negative Numbers in Terms of i.**

Some equations do not have real-number solutions. For example, $x^2 = -1$ has no real-number solutions because the square of a real number is never negative. To provide a solution to this equation, mathematicians have defined the number i so that $i^2 = -1$.

The Number i

The **imaginary number i** is defined as

$$i = \sqrt{-1}$$

From the definition, it follows that $i^2 = -1$.

This definition enables us to write the square root of any negative number in terms of i.

We can use extensions of the product and quotient rules for radicals to write the square root of a negative number as the product of a real number and i.

EXAMPLE 1 Write each expression in terms of i:

a. $\sqrt{-9}$ b. $\sqrt{-7}$ c. $-\sqrt{-18}$ d. $\sqrt{-\dfrac{24}{49}}$

Strategy We will write each radicand as the product of -1 and a positive number. Then we will apply the appropriate rules for radicals.

Why We want our work to produce a factor of $\sqrt{-1}$ so that we can replace it with i.

Solution After factoring the radicand, we use an extension of the product rule for radicals.

a. $\sqrt{-9} = \sqrt{-1 \cdot 9} = \sqrt{-1}\sqrt{9} = i \cdot 3 = 3i$ Replace $\sqrt{-1}$ with i.

b. $\sqrt{-7} = \sqrt{-1 \cdot 7} = \sqrt{-1}\sqrt{7} = i\sqrt{7}$ or $\sqrt{7}i$ Replace $\sqrt{-1}$ with i.

c. $-\sqrt{-18} = -\sqrt{-1 \cdot 9 \cdot 2} = -\sqrt{-1}\sqrt{9}\sqrt{2} = -i \cdot 3 \cdot \sqrt{2} = -3i\sqrt{2}$ or $-3\sqrt{2}i$

d. After factoring the radicand, use an extension of the product and quotient rules for radicals.

$$\sqrt{-\frac{24}{49}} = \sqrt{-1 \cdot \frac{24}{49}} = \frac{\sqrt{-1 \cdot 24}}{\sqrt{49}} = \frac{\sqrt{-1}\sqrt{4}\sqrt{6}}{\sqrt{49}} = \frac{2i\sqrt{6}}{7} \text{ or } \frac{2\sqrt{6}}{7}i$$

Self Check 1 Write each expression in terms of i:

a. $\sqrt{-25}$ b. $-\sqrt{-19}$ c. $\sqrt{-45}$

d. $\sqrt{-\dfrac{50}{81}}$

Now Try Problems 19, 21, and 27

The results from Example 1 illustrate a rule for simplifying square roots of negative numbers.

Square Root of a Negative Number

For any positive real number b,
$$\sqrt{-b} = i\sqrt{b}$$

To justify this rule, we use the fact that $\sqrt{-1} = i$.

$$\sqrt{-b} = \sqrt{-1 \cdot b}$$
$$= \sqrt{-1}\sqrt{b}$$
$$= i\sqrt{b}$$

Write Complex Numbers in the Form $a + bi$.

The imaginary number i is used to define *complex numbers*.

Complex Numbers	A **complex number** is any number that can be written in the form $a + bi$, where a and b are real numbers and $i = \sqrt{-1}$.
	Complex numbers of the form $a + bi$, where $b \neq 0$, are also called **imaginary numbers.***

Notation

It is acceptable to use $a - bi$ as a substitute for the form $a + bi$. For example:

$$6 - 9i = 6 + (-9)i.$$

For a complex number written in the **standard form** $a + bi$, we call a the **real part** and b the **imaginary part**. Some examples of complex numbers written in standard form are

$$2 + 11i \qquad 6 - 9i \qquad -\frac{1}{2} + 0i \qquad 0 + i\sqrt{3}$$

Two complex numbers $a + bi$ and $c + di$ are equal if and only if $a = c$ and $b = d$. Thus, $0.5 + 0.9i = \frac{1}{2} + \frac{9}{10}i$ because $0.5 = \frac{1}{2}$ and $0.9 = \frac{9}{10}$.

EXAMPLE 2   Write each number in the form $a + bi$:
  a. 6 **b.** $\sqrt{-64}$ **c.** $-2 + \sqrt{-63}$

Strategy   We will determine a, the real part, and we will simplify the radical (if necessary) to determine the bi part.

Why   We can put the two parts together to produce the desired $a + bi$ form.

Solution

a. $6 = 6 + 0i$ *The real part is 6. The imaginary part is 0.*

b. $\sqrt{-64} = 0 + 8i$ *The real part is 0. Simplify:* $\sqrt{-64} = \sqrt{-1}\sqrt{64} = 8i.$

c. $-2 + \sqrt{-63} = -2 + 3i\sqrt{7}$ *The real part is −2.*
  Simplify: $\sqrt{-63} = \sqrt{-1}\sqrt{63} = \sqrt{-1}\sqrt{9}\sqrt{7} = 3i\sqrt{7}.$

 Self Check 2   Write each number in the form $a + bi$:
  a. -18 **b.** $\sqrt{-36}$
  c. $1 + \sqrt{-24}$

Now Try   Problems 29 and 33

Success Tip

Just as real numbers are either rational or irrational, but not both, complex numbers are either real or imaginary, but not both.

The following illustration shows the relationship between the real numbers, the imaginary numbers, and the complex numbers.

Complex numbers							
Real numbers				Imaginary numbers			
-6	$\frac{5}{16}$	-1.75	π	$9 + 7i$	$-2i$	$\frac{1}{4} - \frac{3}{4}i$	
$48 + 0i$	0	$-\sqrt{10}$	$-\frac{7}{2}$	$0.56i$	$\sqrt{-10}$	$6 + i\sqrt{3}$	

*Some textbooks define imaginary numbers as complex numbers with $a = 0$ and $b \neq 0$.

Add and Subtract Complex Numbers.

Adding and subtracting complex numbers is similar to adding and subtracting polynomials.

Addition and Subtraction of Complex Numbers	1. To add complex numbers, add their real parts and add their imaginary parts. 2. To subtract complex numbers, add the opposite of the complex number being subtracted.

EXAMPLE 3   Perform each operation. Write the answers in the form $a + bi$.
a. $(8 + 4i) + (12 + 8i)$ **b.** $(-6 + 4i) - (3 + 2i)$
c. $\left(7 - \sqrt{-16}\right) + \left(9 + \sqrt{-4}\right)$

Strategy   To add the complex numbers, we will add their real parts and add their imaginary parts. To subtract the complex numbers, we will add the opposite of the complex number to be subtracted.

Why   We perform the indicated operations as if the complex numbers were polynomials with i as a variable.

Solution

a. $(8 + 4i) + (12 + 8i) = (8 + 12) + (4 + 8)i$

The sum of the imaginary parts
The sum of the real parts

$$= 20 + 12i$$

Add

b. $(-6 + 4i) - (3 + 2i) = (-6 + 4i) + (-3 - 2i)$ To find the opposite, change the sign of each term of $3 + 2i$.

the opposite

$$= [-6 + (-3)] + [4 + (-2)]i \quad \text{Add the real parts. Add the imaginary parts.}$$

$$= -9 + 2i$$

Success Tip
Always change complex numbers to $a + bi$ form before performing any arithmetic.

c. $\left(7 - \sqrt{-16}\right) + \left(9 + \sqrt{-4}\right)$
$\quad = (7 - 4i) + (9 + 2i)$ Write $\sqrt{-16}$ and $\sqrt{-4}$ in terms of i.
$\quad = (7 + 9) + (-4 + 2)i$ Add the real parts. Add the imaginary parts.
$\quad = 16 - 2i$ Write $16 + (-2i)$ in the form $16 - 2i$.

 Self Check 3   Perform the operations. Write the answers in the form $a + bi$.
a. $(3 - 5i) + (-2 + 7i)$
b. $\left(3 - \sqrt{-25}\right) - \left(-2 + \sqrt{-49}\right)$
Now Try   Problems 37 and 43

Multiply Complex Numbers.

Since imaginary numbers are not real numbers, some properties of real numbers do not apply to imaginary numbers. For example, we cannot use the product rule for radicals to multiply two imaginary numbers.

Caution If a and b are both negative, then $\sqrt{a}\sqrt{b} \neq \sqrt{ab}$. For example, if $a = -4$ and $b = -9$,

$$\sqrt{-4}\sqrt{-9} = \sqrt{-4(-9)} = \sqrt{36} = 6 \qquad \sqrt{-4}\sqrt{-9} = 2i(3i) = 6i^2 = 6(-1) = -6$$

EXAMPLE 4 Multiply: $\sqrt{-2}\sqrt{-20}$

Strategy To multiply the imaginary numbers, we will first write $\sqrt{-2}$ and $\sqrt{-20}$ in $i\sqrt{b}$ form. Then we will use the product rule for radicals.

Why We cannot immediately use the product rule for radicals because it does not apply when both radicands are negative.

Solution

$$
\begin{aligned}
\sqrt{-2}\sqrt{-20} &= \left(i\sqrt{2}\right)\left(2i\sqrt{5}\right) && \text{Simplify: } \sqrt{-20} = i\sqrt{20} = 2i\sqrt{5}. \\
&= 2i^2\sqrt{2\cdot 5} && \text{Multiply: } i\cdot 2i = 2i^2. \text{ Use the product rule for radicals.} \\
&= 2i^2\sqrt{10} \\
&= 2(-1)\sqrt{10}\cdot && \text{Replace } i^2 \text{ with } -1. \\
&= -2\sqrt{10} && \text{Multiply.}
\end{aligned}
$$

 Self Check 4 Multiply: $\sqrt{-3}\sqrt{-32}$

Now Try **Problem 47**

Multiplying complex numbers is similar to multiplying polynomials.

EXAMPLE 5 Multiply. Write the answers in the form $a + bi$.
a. $6(2 + 9i)$ **b.** $-5i(4 - 8i)$

Strategy We will use the distributive property to find the products.

Why We perform the indicated operations as if the complex numbers were polynomials with i as a variable.

Solution

a. $6(2 + 9i) = 6(2) + 6(9i)$ Use the distributive property.

$\qquad\qquad\quad = 12 + 54i$ Perform each multiplication.

b. $-5i(4 - 8i) = -5i(4) - (-5i)8i$ Use the distributive property.

$\qquad\qquad\quad\; = -20i + 40i^2$ Perform each multiplication.

$\qquad\qquad\quad\; = -20i + 40(-1)$ Replace i^2 with -1.

Caution
A common mistake is to replace i with -1. Remember, $i \neq -1$. By definition, $i = \sqrt{-1}$ and $i^2 = -1$.

$$= -20i - 40 \qquad \text{Multiply.}$$

$$= -40 - 20i \qquad \text{Write the real part, } -40, \text{ as the first term.}$$

Self Check 5   Multiply. Write the answers in the form $a + bi$.
a. $-2(-9 - i)$ **b.** $10i(7 + 4i)$

Now Try   **Problems 49 and 55**

EXAMPLE 6   Multiply. Write the answers in the form $a + bi$.
a. $(2 + 3i)(3 - 2i)$ **b.** $(-4 + 2i)(2 + i)$

Strategy   We will use the FOIL method to multiply the two complex numbers.

Why   We perform the indicated operations as if the complex numbers were binomials with i as a variable.

Solution

> ***Success Tip***
>
> i is not a variable, but you can think of it as one when adding, subtracting, and multiplying. For example:
>
> $$-4i + 9i = 5i$$
> $$6i - 2i = 4i$$
> $$i \cdot i = i^2$$
>
> Remember that the expression i^2 simplifies to -1.

a. $(2 + 3i)(3 - 2i) = \overset{\text{F}}{6} - \overset{\text{O}}{4i} + \overset{\text{I}}{9i} - \overset{\text{L}}{6i^2}$ Use the FOIL method.

$$= 6 + 5i - 6(-1) \qquad \begin{array}{l}\text{Combine the imaginary terms: } -4i + 9i = 5i. \\ \text{Replace } i^2 \text{ with } -1.\end{array}$$

$$= 6 + 5i + 6 \qquad \text{Simplify the last term.}$$

$$= 12 + 5i \qquad \text{Combine like terms.}$$

b. $(-4 + 2i)(2 + i) = -8 - 4i + 4i + 2i^2$ Use the FOIL method.

$$= -8 + 0i + 2(-1) \qquad \begin{array}{l}\text{Combine like terms: } -4i + 4i = 0i. \text{ Replace } i^2 \\ \text{with } -1.\end{array}$$

$$= -8 + 0i - 2 \qquad \text{Multiply.}$$

$$= -10 + 0i \qquad \text{Combine like terms.}$$

Self Check 6   Multiply. Write the answers in the form $a + bi$.
$(-2 + 3i)(3 - 2i)$

Now Try   **Problem 59**

---

⑤   Divide Complex Numbers.

Before we can discuss division of complex numbers, we must introduce an important fact about *complex conjugates*.

Complex Conjugates	The complex numbers $a + bi$ and $a - bi$ are called **complex conjugates**.

For example,

- $7 + 4i$ and $7 - 4i$ are complex conjugates.
- $5 - i$ and $5 + i$ are complex conjugates.
- $-6i$ and $6i$ are complex conjugates, because $-6i = 0 - 6i$ and $6i = 0 + 6i$.

In general, the product of the complex number $a + bi$ and its complex conjugate $a - bi$ is the real number $a^2 + b^2$, as the following work shows:

$$(a + bi)(a - bi) = a^2 - abi + abi - b^2i^2 \quad \text{Use the FOIL method.}$$
$$= a^2 - b^2(-1) \quad -abi + abi = 0. \text{ Replace } i^2 \text{ with } -1.$$
$$= a^2 + b^2$$

EXAMPLE 7 Find the product of $3 + 5i$ and its complex conjugate.

Strategy The complex conjugate of $3 + 5i$ is $3 - 5i$. We will find their product by using the FOIL method.

Why We perform the indicated operations as if the complex numbers were binomials with i as a variable.

Solution We can find the product as follows:

$$(3 + 5i)(3 - 5i) = 9 - 15i + 15i - 25i^2 \quad \text{Use the FOIL method.}$$
$$= 9 - 25i^2 \quad \text{Combine like terms: } -15i + 15i = 0.$$
$$= 9 - 25(-1) \quad \text{Replace } i^2 \text{ with } -1.$$
$$= 9 + 25$$
$$= 34$$

The product of $3 + 5i$ and its conjugate $3 - 5i$ is the real number 34.

Self Check 7 Multiply: $(2 + 3i)(2 - 3i)$

Now Try Problem 65

Recall that to divide *radical expressions,* we rationalized the denominator. We will use a similar approach to divide complex numbers. To divide two complex numbers when the divisor has two terms, we use the following strategy.

Division of Complex Numbers

To divide complex numbers, multiply the numerator and denominator by the complex conjugate of the denominator.

EXAMPLE 8 Divide. Write the answers in the form $a + bi$.

a. $\dfrac{3}{6 + i}$ b. $\dfrac{1 + 2i}{3 - 4i}$

Strategy We will build each fraction by multiplying it by a form of 1 that uses the conjugate of the denominator.

Why This step produces a *real number* in the denominator so that the result can then be written in the form $a + bi$.

Solution

a. We want to build a fraction equivalent to $\frac{3}{6+i}$ that does not have i in the denominator. To make the denominator, $6 + i$, a real number, we need to multiply it by its complex conjugate, $6 - i$. It follows that $\frac{6-i}{6-i}$ should be the form of 1 that is used to build $\frac{3}{6+i}$.

$$\frac{3}{6+i} = \frac{3}{6+i} \cdot \frac{6-i}{6-i}$$

To build an equivalent fraction, multiply by $\frac{6-i}{6-i} = 1$.

$$= \frac{18 - 3i}{36 - 6i + 6i - i^2}$$

To multiply the numerators, distribute the multiplication by 3. Use the FOIL method to multiply the denominators.

$$= \frac{18 - 3i}{36 - (-1)}$$

Combine like terms: $-6i + 6i = 0$. Replace i^2 with -1. Note that the denominator no longer contains i.

$$= \frac{18 - 3i}{37}$$

Simplify the denominator. This notation represents the difference of two fractions that have the common denominator 37: $\frac{18}{37}$ and $\frac{3i}{37}$.

$$= \frac{18}{37} - \frac{3}{37}i$$

Write the complex number in the form $a + bi$.

b. We can make the denominator of $\frac{1+2i}{3-4i}$ a real number by multiplying it by the complex conjugate of $3 - 4i$, which is $3 + 4i$. It follows that $\frac{3+4i}{3+4i}$ should be the form of 1 that is used to build $\frac{1+2i}{3-4i}$.

$$\frac{1+2i}{3-4i} = \frac{1+2i}{3-4i} \cdot \frac{3+4i}{3+4i}$$

To build an equivalent fraction, multiply by $\frac{3+4i}{3+4i} = 1$.

$$= \frac{3 + 4i + 6i + 8i^2}{9 + 12i - 12i - 16i^2}$$

Use the FOIL method to multiply the numerators and the denominators.

$$= \frac{3 + 10i + 8(-1)}{9 - 16(-1)}$$

Combine like terms in the numerator and denominator. Replace i^2 with -1. The denominator is now a real number.

$$= \frac{3 + 10i - 8}{9 + 16}$$

Simplify the numerator and denominator.

$$= \frac{-5 + 10i}{25}$$

Combine like terms in the numerator and denominator.

$$= \frac{\overset{1}{\cancel{5}}(-1 + 2i)}{\underset{1}{\cancel{5} \cdot 5}}$$

Factor out 5 in the numerator and remove the common factor of 5.

$$= \frac{-1 + 2i}{5}$$

Simplify. This notation represents the sum of two fractions that have the common denominator 5.

$$= -\frac{1}{5} + \frac{2}{5}i$$

Write the complex number in the form $a + bi$.

Self Check 8 Divide. Write the answers in the form $a + bi$.

a. $\frac{6}{5+2i}$ **b.** $\frac{2-4i}{5-3i}$

Now Try **Problems 71 and 77**

EXAMPLE 9   Divide and write the answer in the form $a + bi$: $\dfrac{4 + \sqrt{-16}}{2 + \sqrt{-4}}$

Strategy   We will begin by writing $\sqrt{-16}$ and $\sqrt{-4}$ in $i\sqrt{b}$ form.

Why   To perform any computations, the numerator and denominator should be written in the form $a + bi$.

Solution

$$\frac{4 + \sqrt{-16}}{2 + \sqrt{-4}} = \frac{4 + 4i}{2 + 2i}$$ Simplify: $\sqrt{-16} = \sqrt{-1}\sqrt{16} = 4i$ and $\sqrt{-4} = \sqrt{-1}\sqrt{4} = 2i$.

$$= \frac{2(\overset{1}{\cancel{2 + 2i}})}{\underset{1}{\cancel{2 + 2i}}}$$ Factor out 2 in the numerator and remove the common factor of $2 + 2i$.

$$= 2$$

$$= 2 + 0i$$ Write 2 in the form $a + bi$.

>
> **Self Check 9**   Divide and write the answer in the form $a + bi$: $\dfrac{21 - \sqrt{-49}}{3 - \sqrt{-1}}$
> **Now Try**   **Problem 81**

EXAMPLE 10   Divide and write the result in the form $a + bi$: $\dfrac{7}{2i}$

Strategy   We will use $\dfrac{-2i}{-2i}$ as the form of 1 to build $\dfrac{7}{2i}$.

Why   Since the denominator $2i$ can be expressed as $0 + 2i$, its conjugate is $0 - 2i$. However, instead of building with $\dfrac{0 - 2i}{0 - 2i}$ we will drop the zeros and just use $\dfrac{-2i}{-2i}$.

Solution

$$\frac{7}{2i} = \frac{7}{2i} \cdot \frac{-2i}{-2i}$$ To build an equivalent fraction, multiply by $\dfrac{-2i}{-2i} = 1$.

$$= \frac{-14i}{-4i^2}$$ Multiply the numerators and multiply the denominators.

$$= \frac{-14i}{-4(-1)}$$ Replace i^2 with -1. The denominator is now a real number.

$$= \frac{-14i}{4}$$ Simplify the denominator.

$$= -\frac{7i}{2}$$ Simplify the fraction: $-\dfrac{\overset{1}{\cancel{2}} \cdot 7i}{\underset{1}{\cancel{2}} \cdot 2}$.

$$= 0 - \frac{7}{2}i$$ Write in the form $a + bi$.

Success Tip

In this example, the denominator of $\frac{7}{2i}$ is of the form bi. In such cases, we can eliminate i in the denominator by simply multiplying by $\frac{i}{i}$.

$$\frac{7}{2i} = \frac{7}{2i} \cdot \frac{i}{i} = \frac{7i}{2i^2} = -\frac{7}{2}i$$

> **Self Check 10**   Divide and write the answer in the form $a + bi$: $\dfrac{3}{4i}$
> **Now Try**   **Problem 85**

6 Perform Operations Involving Powers of i.

The powers of i produce an interesting pattern:

$$i = \sqrt{-1} = i \qquad\qquad i^5 = i^4 i = 1i = i$$
$$i^2 = \left(\sqrt{-1}\right)^2 = -1 \qquad i^6 = i^4 i^2 = 1(-1) = -1$$
$$i^3 = i^2 i = -1i = -i \qquad i^7 = i^4 i^3 = 1(-i) = -i$$
$$i^4 = i^2 i^2 = (-1)(-1) = 1 \qquad i^8 = i^4 i^4 = (1)(1) = 1$$

Success Tip
Note that the powers of i cycle through four possible outcomes:

The pattern continues: $i, -1, -i, 1, \ldots$.

Larger powers of i can be simplified by using the fact that $i^4 = 1$. For example, to simplify i^{29}, we note that 29 divided by 4 gives a quotient of 7 and a remainder of 1. Thus, $29 = 4 \cdot 7 + 1$ and

$$i^{29} = i^{4 \cdot 7 + 1} \qquad 4 \cdot 7 = 28.$$
$$= (i^4)^7 \cdot i^1 \qquad \text{Use the rules for exponents } x^{m \cdot n} = (x^m)^n \text{ and } x^{m+n} = x^m \cdot x^n.$$
$$= 1^7 \cdot i \qquad \text{Simplify: } i^4 = 1.$$
$$= i \qquad \text{Simplify: } 1 \cdot i = i.$$

The result of this example illustrates the following fact.

Powers of i	If n is a natural number that has a remainder of r when divided by 4, then $$i^n = i^r$$

EXAMPLE 11   Simplify: i^{55}

Strategy   We will examine the remainder when we divide the exponent 55 by 4.

Why   The remainder determines the power to which i is raised in the simplified form.

Solution   We divide 55 by 4 and get a remainder of 3. Therefore,

$$i^{55} = i^3 = -i$$

$$\begin{array}{r} 13\text{R }3 \\ 4\overline{)55} \\ -4 \\ \hline 15 \\ -12 \\ \hline 3 \end{array}$$

Success Tip
If we divide the natural number exponent n of a power of i by 4, the remainder indicates the simplified form of i^n:

$R = 1$: i
$R = 2$: -1
$R = 3$: $-i$
$R = 0$: 1

Self Check 11   Simplify: i^{62}

Now Try   Problem 91

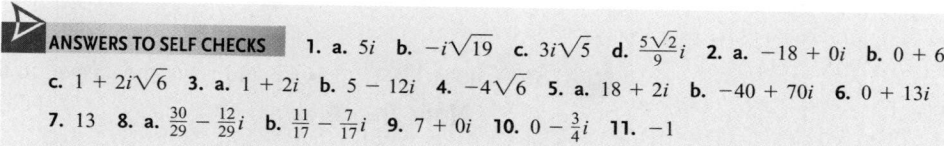

ANSWERS TO SELF CHECKS   **1. a.** $5i$ **b.** $-i\sqrt{19}$ **c.** $3i\sqrt{5}$ **d.** $\frac{5\sqrt{2}}{9}i$ **2. a.** $-18 + 0i$ **b.** $0 + 6i$ **c.** $1 + 2i\sqrt{6}$ **3. a.** $1 + 2i$ **b.** $5 - 12i$ **4.** $-4\sqrt{6}$ **5. a.** $18 + 2i$ **b.** $-40 + 70i$ **6.** $0 + 13i$ **7.** 13 **8. a.** $\frac{30}{29} - \frac{12}{29}i$ **b.** $\frac{11}{17} - \frac{7}{17}i$ **9.** $7 + 0i$ **10.** $0 - \frac{3}{4}i$ **11.** -1

STUDY SET
9.7

VOCABULARY

Fill in the blanks.

1. The _____ number i is defined as $i = \sqrt{-1}$. We call i^{25} a _____ of i.

2. A _____ number is any number that can be written in the form $a + bi$, where a and b are real numbers and $i = \sqrt{-1}$.

3. For the complex number $2 + 5i$, we call 2 the _____ part and 5 the _____ part.

4. $6 + 3i$ and $6 - 3i$ are called complex _____.

CONCEPTS

Fill in the blanks.

5. **a.** $i = $
   b. $i^2 = $
   c. $i^3 = $
   d. $i^4 = $
   e. In general, the powers of i cycle through _____ possible outcomes.

6. Simplify:
$$\sqrt{-36} = \sqrt{ \cdot 36} = \sqrt{}\sqrt{36} = 6$$

7. To add (or subtract) complex numbers, add (or subtract) their _____ parts and add (or subtract) their _____ parts.

8. To multiply two complex numbers, such as $(2 + 3i)(3 + 5i)$, we can use the _____ method.

9. To divide $6 + 7i$ by $1 - 8i$, we multiply $\frac{6 + 7i}{1 - 8i}$ by a 1 in the form of ____.

10. Give the complex conjugate of each number.
    a. $2 - 3i$
    b. 2
    c. $-3i$

11. Complete the illustration. Label the real numbers, the imaginary numbers, the complex numbers, the rational numbers, and the irrational numbers.

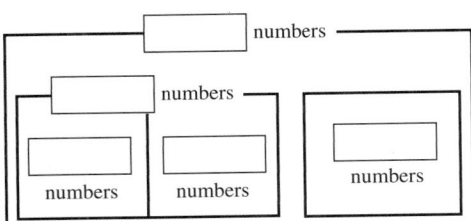

12. Determine whether each statement is true or false.
    a. Every complex number is a real number.
    b. Every real number is a complex number.
    c. i is a real number.
    d. The square root of a negative number is an imaginary number.

NOTATION

Complete each solution.

13. $(3 + 2i)(3 - i) = \boxed{} - 3i + \boxed{} - 2i^2$
$$= 9 + 3i + \boxed{}$$
$$= \boxed{} + 3i$$

14. $\dfrac{3}{2 - i} = \dfrac{3}{2 - i} \cdot \dfrac{\boxed{}}{\boxed{}}$
$$= \dfrac{6 + \boxed{}}{4 - \boxed{}}$$
$$= \dfrac{6 + 3i}{\boxed{}}$$
$$= \dfrac{}{\boxed{}} + \dfrac{3}{5}i$$

15. Determine whether each statement is true or false.
    a. $\sqrt{6}i = i\sqrt{6}$
    b. $\sqrt{8}i = \sqrt{8i}$
    c. $\sqrt{-25} = -\sqrt{25}$
    d. $-i = i$

16. Write each number in the form $a + bi$.
    a. $\dfrac{9 + 11i}{4}$
    b. $\dfrac{1 - i}{18}$

GUIDED PRACTICE

Express each number in terms of i. See Example 1.

17. $\sqrt{-9}$
18. $\sqrt{-4}$
19. $\sqrt{-7}$
20. $\sqrt{-11}$
21. $\sqrt{-24}$
22. $\sqrt{-28}$
23. $-\sqrt{-72}$
24. $-\sqrt{-24}$
25. $5\sqrt{-81}$
26. $6\sqrt{-49}$
27. $\sqrt{-\dfrac{25}{9}}$
28. $-\sqrt{-\dfrac{121}{144}}$

Write each number in the form $a + bi$. See Example 2.

29. **a.** 5
    b. $\sqrt{-49}$
30. **a.** -43
    b. $\sqrt{-169}$

31. a. $1 + \sqrt{-25}$

32. a. $21 + \sqrt{-16}$

b. $-3 + \sqrt{-8}$

b. $-9 + \sqrt{-12}$

33. a. $76 - \sqrt{-54}$

34. a. $88 - \sqrt{-98}$

b. $-7 + \sqrt{-19}$

b. $-2 + \sqrt{-35}$

35. a. $-6 - \sqrt{-9}$

36. a. $-45 - \sqrt{-81}$

b. $3 + \sqrt{-6}$

b. $8 + \sqrt{-7}$

Perform the operations. Write all answers in the form a + bi. See Example 3.

37. $(3 + 4i) + (5 - 6i)$

38. $(8 + 3i) + (-7 - 2i)$

39. $(6 - i) + (9 + 3i)$

40. $(5 + 3i) - (6 - 9i)$

41. $(7 - 3i) - (4 + 2i)$

42. $(5 - 4i) - (3 + 2i)$

43. $\left(8 + \sqrt{-25}\right) - \left(7 + \sqrt{-4}\right)$

44. $\left(-7 + \sqrt{-81}\right) - \left(-2 - \sqrt{-64}\right)$

Multiply. See Example 4.

45. $\sqrt{-1}\sqrt{-36}$

46. $\sqrt{-9}\sqrt{-100}$

47. $\sqrt{-2}\sqrt{-12}$

48. $\sqrt{-3}\sqrt{-45}$

Multiply. Write all answers in the form a + bi. See Example 5.

49. $3(2 - 9i)$

50. $-4(3 + 4i)$

51. $7(5 - 4i)$

52. $-5(3 + 2i)$

53. $2i(7 - 3i)$

54. $i(8 + 2i)$

55. $-5i(5 - 5i)$

56. $2i(7 + 2i)$

Multiply. Write all answers in the form a + bi. See Example 6.

57. $(2 + i)(3 - i)$

58. $(4 - i)(2 + i)$

59. $(3 - 2i)(2 + 3i)$

60. $(3 - i)(2 + 3i)$

61. $(4 + i)(3 - i)$

62. $(1 - 5i)(1 - 4i)$

63. $(2 + i)^2$

64. $(3 - 2i)^2$

Find the product of the given complex number and its conjugate. See Example 7.

65. $2 + 6i$

66. $5 + 2i$

67. $-4 - 7i$

68. $-10 - 9i$

Divide. Write all answers in the form a + bi. See Example 8.

69. $\dfrac{9}{5 + i}$

70. $\dfrac{4}{2 - i}$

71. $\dfrac{11i}{4 - 7i}$

72. $\dfrac{2i}{3 + 8i}$

73. $\dfrac{3 - 2i}{4 - i}$

74. $\dfrac{6 - i}{2 + i}$

75. $\dfrac{7 + 4i}{2 - 5i}$

76. $\dfrac{2 + 3i}{2 - 3i}$

77. $\dfrac{7 + 3i}{4 - 2i}$

78. $\dfrac{5 - 3i}{4 + 2i}$

79. $\dfrac{1 - 3i}{3 + i}$

80. $\dfrac{3 + 5i}{1 - i}$

Divide. Write all answers in the form a + bi. See Example 9.

81. $\dfrac{8 + \sqrt{-144}}{2 + \sqrt{-9}}$

82. $\dfrac{3 + \sqrt{-36}}{1 + \sqrt{-4}}$

83. $\dfrac{-4 - \sqrt{-4}}{2 + \sqrt{-1}}$

84. $\dfrac{-5 - \sqrt{-25}}{1 + \sqrt{-1}}$

Divide. Write all answers in the form a + bi. See Example 10.

85. $\dfrac{5}{3i}$

86. $\dfrac{3}{8i}$

87. $-\dfrac{2}{7i}$

88. $-\dfrac{8}{5i}$

Simplify each expression. See Example 11.

89. i^{21}

90. i^{19}

91. i^{27}

92. i^{22}

93. i^{100}

94. i^{97}

95. i^{42}

96. i^{200}

TRY IT YOURSELF

Perform the operations. Write all answers in the form a + bi.

97. $(3 - i) - (-1 + 10i)$

98. $(14 + 4i) - (-9 - i)$

99. $\left(2 - \sqrt{-16}\right)\left(3 + \sqrt{-4}\right)$

100. $\left(3 - \sqrt{-4}\right)\left(4 - \sqrt{-9}\right)$

101. $(-6 - 9i) + (4 + 3i)$

102. $(-3 + 11i) + (-1 - 6i)$

103. $\dfrac{-2i}{3 + 2i}$

104. $\dfrac{-4i}{2 - 6i}$

105. $6i(2 - 3i)$

106. $-9i(4 - 6i)$

107. $\dfrac{4}{5i^{35}}$

108. $\dfrac{3}{2i^{17}}$

109. $\left(2 + i\sqrt{2}\right)\left(3 - i\sqrt{2}\right)$

110. $\left(5 + i\sqrt{3}\right)\left(2 - i\sqrt{3}\right)$

111. $\dfrac{5 + 9i}{1 - i}$

112. $\dfrac{5 - i}{3 + 2i}$

113. $(4 - 8i)^2$

114. $(7 - 3i)^2$

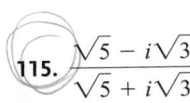

115. $\dfrac{\sqrt{5} - i\sqrt{3}}{\sqrt{5} + i\sqrt{3}}$

116. $\dfrac{\sqrt{3} + i\sqrt{2}}{\sqrt{3} - i\sqrt{2}}$

APPLICATIONS

117. FRACTALS Complex numbers are fundamental in the creation of the intricate geometric shape shown below, called a *fractal*. The process of creating this image is based on the following sequence of steps, which begins by picking any complex number, which we will call z.

1. Square z, and then add that result to z.
2. Square the result from step 1, and then add it to z.
3. Square the result from step 2, and then add it to z.

If we begin with the complex number i, what is the result after performing steps 1, 2, and 3?

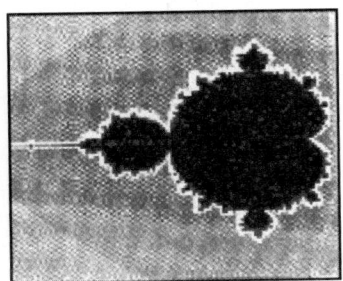

118. ELECTRONICS The impedance Z in an AC (alternating current) circuit is a measure of how much the circuit impedes (hinders) the flow of current through it. The impedance is related to the voltage V and the current I by the formula

$$V = IZ$$

If a circuit has a current of $(0.5 + 2.0i)$ amps and an impedance of $(0.4 - 3.0i)$ ohms, find the voltage.

WRITING

119. What is an imaginary number? What is a complex number?

120. The method used to divide complex numbers is similar to the method used to divide radical expressions. Explain why. Give an example.

121. Explain the error. Then find the correct result.

   a. Add: $\sqrt{-16} + \sqrt{-9} = \sqrt{-25}$

   b. Multiply: $\sqrt{-2}\sqrt{-3} = \sqrt{-2(-3)} = \sqrt{6}$

122. Determine whether the pair of complex numbers are equal. Explain your reasoning.

   a. $4 - \dfrac{2}{5}i,\ \dfrac{8}{2} - 0.4i$

   b. $0.25 + 0.7i,\ \dfrac{1}{4} + \dfrac{7}{10}i$

REVIEW

123. WIND SPEEDS A plane that can fly 200 mph in still air makes a 330-mile flight with a tail wind and returns, flying into the same wind. Find the speed of the wind if the total flying time is $3\dfrac{1}{3}$ hours.

124. FINDING RATES A student drove a distance of 135 miles at an average speed of 50 mph. How much faster would she have to drive on the return trip to save 30 minutes of driving time?

CHALLENGE PROBLEMS

125. Simplify: $\left(i^{349}\right)^{-i^{456}}$

126. Simplify $(2 + 3i)^{-2}$ and write the result in the form $a + bi$.

CHAPTER 9
Summary & Review

SECTION 9.1 Radical Expressions and Radical Functions

DEFINITIONS AND CONCEPTS	EXAMPLES
The number b is a **square root of a** if $b^2 = a$.	7 is a square root of 49 because $7^2 = 49$. -7 is a square root of 49 because $(-7)^2 = 49$.
A **radical symbol** $\sqrt{\ }$ represents the **positive** or **principal square root** of a number. For any real number x, $\sqrt{x^2} = \|x\|$ The symbol $-\sqrt{\ }$ represents the **negative square root** of a number.	Simplify: $\sqrt{25} = 5$ because $5^2 = 25$. $\sqrt{36x^2} = \|6x\| = 6\|x\|$ because $(\|6x\|)^2 = 36x^2$. $\sqrt{\dfrac{r^8}{100}} = \dfrac{r^4}{10}$ because $\left(\dfrac{r^4}{10}\right)^2 = \dfrac{r^8}{100}$. Since $\frac{r^4}{10} \geq 0$, no absolute value symbols are needed. $-\sqrt{81} = -9$ because $(-9)^2 = 81$.
A function of the form $f(x) = \sqrt{x}$ is called a **square root function**.	Find the domain of $f(x) = \sqrt{x-2}$. Since the expression $\sqrt{x-2}$ is not a real number when $x-2$ is negative, we must require that $x - 2 \geq 0$. It follows that x must be greater than or equal to 2. Thus, the domain of $f(x)$ is $[2, \infty)$.
The **cube root** of x is denoted as $\sqrt[3]{x}$ and is defined as $\sqrt[3]{x} = y$ if $y^3 = x$ A function of the form $f(x) = \sqrt[3]{x}$ is called a **cube root function**.	Simplify: $\sqrt[3]{8} = 2$ because $2^3 = 8$. $\sqrt[3]{-64} = -4$ because $(-4)^3 = -64$.
The **nth root of x** is denoted as $\sqrt[n]{x}$. If x is a real number and $n > 1$, then: $\begin{cases} \text{If } n \text{ is an odd natural number, } \sqrt[n]{x^n} = x. \\ \text{If } n \text{ is an even natural number, } \sqrt[n]{x^n} = \|x\|. \end{cases}$	Simplify: $\sqrt[4]{81x^4} = \|3x\| = 3\|x\|$ because $(\|3x\|)^4 = 81x^4$. $\sqrt[5]{32a^{10}} = 2a^2$ because $(2a^2)^5 = 32a^{10}$. $\sqrt[6]{(m-4)^6} = \|m-4\|$ because $(\|m-4\|)^6 = (m-4)^6$.

REVIEW EXERCISES

Simplify each expression, if possible. Assume that x and y can be any real number.

1. $\sqrt{49}$

2. $-\sqrt{121}$

3. $\sqrt{\dfrac{225}{49}}$

4. $\sqrt{-4}$

5. $\sqrt{100a^{12}}$

6. $\sqrt{25x^2}$

7. $\sqrt{x^8}$

8. $\sqrt{x^2 + 4x + 4}$

9. $\sqrt[3]{-27}$

10. $-\sqrt[3]{216}$

11. $\sqrt[3]{64x^6y^3}$

12. $\sqrt[3]{\dfrac{x^9}{125}}$

13. $\sqrt[6]{64}$

14. $\sqrt[5]{-32}$

15. $\sqrt[4]{256x^8y^4}$

16. $\sqrt[15]{(x+1)^{15}}$

17. $-\sqrt[4]{\dfrac{1}{16}}$

18. $\sqrt[4]{-16}$

19. $\sqrt[6]{-1}$

20. $\sqrt[3]{0}$

21. GEOMETRY   The side of a square with area A square feet is given by the function $s(A) = \sqrt{A}$. Find the length of one side of a square that has an area of 169 ft^2.

22. SURFACE AREA OF A CUBE   The total surface area of a cube is related to its volume V by the function $A(V) = 6\sqrt[3]{V^2}$. Find the surface area of a cube with a volume of 8 cm^3.

Graph each function. Find the domain and range.

23. $f(x) = \sqrt{x}$

24. $f(x) = \sqrt[3]{x}$

25. $f(x) = \sqrt{x + 2}$

26. $f(x) = -\sqrt[3]{x} + 3$

SECTION 9.2   Rational Exponents

DEFINITIONS AND CONCEPTS	EXAMPLES
To simplify exponential expressions involving **rational (fractional) exponents,** use the following rules to write the expressions in an equivalent radical form. $x^{1/n} = \sqrt[n]{x}$ $x^{m/n} = \left(\sqrt[n]{x}\right)^m = \sqrt[n]{x^m}$	Simplify. All variables represent positive real numbers. $25^{1/2} = \sqrt{25} = 5 \qquad \left(\dfrac{256}{d^4}\right)^{1/4} = \sqrt[4]{\dfrac{256}{d^4}} = \dfrac{4}{d}$ $8^{2/3} = \left(\sqrt[3]{8}\right)^2 = (2)^2 = 4 \qquad (t^{10})^{6/5} = \left(\sqrt[5]{t^{10}}\right)^6 = (t^2)^6 = t^{12}$
To be consistent with the definition of negative integer exponents, we define $x^{-m/n}$ as follows. $x^{-m/n} = \dfrac{1}{x^{m/n}}$ $\dfrac{1}{x^{-m/n}} = x^{m/n}$	Simplify. All variables represent positive real numbers. $(125)^{-2/3} = \dfrac{1}{(125)^{2/3}} = \dfrac{1}{\left(\sqrt[3]{125}\right)^2} = \dfrac{1}{25}$ $\dfrac{1}{(-32x^5)^{-3/5}} = (-32x^5)^{3/5} = \left(\sqrt[5]{-32x^5}\right)^3 = -8x^3$
The **rules for exponents** can be used to simplify expressions with rational (fractional) exponents.	Simplify: $\dfrac{p^{5/3}p^{8/3}}{p^4} = p^{5/3 + 8/3 - 4} = p^{5/3 + 8/3 - 12/3} = p^{1/3}$
We can write certain radical expressions as an equivalent exponential expression and use rules for exponents to simplify it. Then we can change that result back into a radical.	Simplify: $\sqrt[4]{9} = \sqrt[4]{3^2} = 3^{2/4} = 3^{1/2} = \sqrt{3}$

REVIEW EXERCISES

Write each expression in radical form.

27. $t^{1/2}$

28. $(5xy^3)^{1/4}$

Simplify each expression, if possible. Assume that all variables represent positive real numbers.

29. $25^{1/2}$

30. $-36^{1/2}$

31. $(-36)^{1/2}$

32. $1^{1/5}$

33. $\left(\dfrac{9}{x^2}\right)^{1/2}$

34. $(-8)^{1/3}$

35. $625^{1/4}$

36. $(81c^4d^4)^{1/4}$

37. $9^{3/2}$

38. $8^{-2/3}$

39. $-49^{5/2}$

40. $\dfrac{1}{100^{-1/2}}$

41. $\left(\dfrac{4}{9}\right)^{-3/2}$

42. $\dfrac{1}{25^{5/2}}$

43. $(25x^2y^4)^{3/2}$

44. $(8u^6v^3)^{-2/3}$

Perform the operations. Write answers without negative exponents. Assume that all variables represent positive real numbers.

45. $5^{1/4}5^{1/2}$

46. $a^{3/7}a^{-2/7}$

47. $(k^{4/5})^{10}$

48. $\dfrac{3^{5/6}3^{1/3}}{3^{1/2}}$

Perform the multiplications. Assume all variables represent positive real numbers.

49. $u^{1/2}(u^{1/2} - u^{-1/2})$

50. $v^{2/3}(v^{1/3} + v^{4/3})$

Use rational exponents to simplify each radical. All variables represent positive real numbers.

51. $\sqrt[4]{a^2}$

52. $\sqrt[3]{\sqrt{c}}$

53. VISIBILITY The distance d in miles a person in an airplane can see to the horizon on a clear day is given by the formula $d = 1.22a^{1/2}$, where a is the altitude of the plane in feet. Find d.

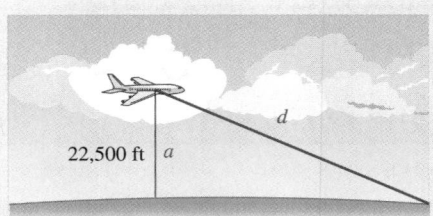

22,500 ft a

54. Substitute the x- and y-coordinates of each point labeled in the graph into the equation

$$x^{2/3} + y^{2/3} = 32$$

Show that each one satisfies the equation.

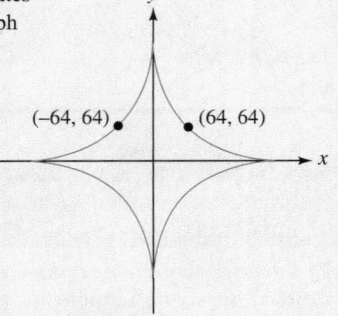

$(-64, 64)$ $(64, 64)$

SECTION 9.3   Simplifying and Combining Radical Expressions

DEFINITIONS AND CONCEPTS	EXAMPLES
Product rule for radicals: $$\sqrt[n]{ab} = \sqrt[n]{a}\sqrt[n]{b}$$ The product rule for radicals can be used to simplify radical expressions. **Simplified form** of a radical: 1. Except for 1, the radicand has no perfect-square factors. 2. No fraction appears in the radicand. 3. No radical appears in the denominator.	Simplify: $\sqrt{98} = \sqrt{49 \cdot 2}$ Write 98 as the product of it greatest perfect-square factor and one other factor. $\quad = \sqrt{49}\sqrt{2}$ The square root of a product is equal to the product of the square roots. $\quad = 7\sqrt{2}$ Evaluate $\sqrt{49}$. Simplify: $\sqrt[3]{16x^4} = \sqrt[3]{8x^3 \cdot 2x}$ Write $16x^4$ as the product of it greatest perfect-cube factor and one other factor. $\quad = \sqrt[3]{8x^3}\sqrt[3]{2x}$ The cube root of a product is equal to the product of the cube roots. $\quad = 2x\sqrt[3]{2x}$ Simplify $\sqrt[3]{8x^3}$.
Quotient rule for radicals: $$\sqrt[n]{\dfrac{a}{b}} = \dfrac{\sqrt[n]{a}}{\sqrt[n]{b}}$$	Simplify: $\sqrt{\dfrac{10}{25x^4}} = \dfrac{\sqrt{10}}{\sqrt{25x^4}} = \dfrac{\sqrt{10}}{5x^2}$ Simplify: $\sqrt[3]{\dfrac{16y^3}{125a^3}} = \dfrac{\sqrt[3]{16y^3}}{\sqrt[3]{125a^3}} = \dfrac{\sqrt[3]{8y^3}\sqrt[3]{2}}{5a} = \dfrac{2y\sqrt[3]{2}}{5a}$
Radical expressions with the same index and radicand are called **like radicals.** Like radicals can be combined by addition and subtraction. To **combine like radicals** we use the distributive property in reverse.	Add: $3\sqrt{6} + 5\sqrt{6} = (3 + 5)\sqrt{6} = 8\sqrt{6}$ Subtract: $8\sqrt[4]{2y} - 9\sqrt[4]{2y} = (8 - 9)\sqrt[4]{2y} = -\sqrt[4]{2y}$
If a sum or difference involves unlike radicals, make sure that each one is written in simplified form. After doing so, like radicals may result that can be combined.	Simplify: $\sqrt[3]{54a^4} - \sqrt[3]{16a^4} = \sqrt[3]{27a^3 \cdot 2a} - \sqrt[3]{8a^3 \cdot 2a}$ $\quad = \sqrt[3]{27a^3}\sqrt[3]{2a} - \sqrt[3]{8a^3}\sqrt[3]{2a}$ $\quad = 3a\sqrt[3]{2a} - 2a\sqrt[3]{2a}$ $\quad = (3a - 2a)\sqrt[3]{2a}$ $\quad = a\sqrt[3]{2a}$

REVIEW EXERCISES

Simplify each expression. All variables represent positive real numbers.

55. $\sqrt{80}$

56. $\sqrt[3]{54}$

57. $\sqrt[4]{160}$

58. $\sqrt[5]{-96}$

59. $\sqrt{8x^5}$

60. $\sqrt[4]{r^{17}}$

61. $\sqrt[3]{-27j^7k}$

62. $\sqrt[3]{-16x^5y^4}$

63. $\sqrt{\dfrac{m}{144n^{12}}}$

64. $\sqrt{\dfrac{17xy}{64a^4}}$

65. $\dfrac{\sqrt[5]{64x^8}}{\sqrt[5]{2x^3}}$

66. $\dfrac{\sqrt[5]{243x^{16}}}{\sqrt[5]{x}}$

Simplify and combine like radicals. All variables represent positive real numbers.

67. $\sqrt{2} + 2\sqrt{2}$

68. $6\sqrt{20} - \sqrt{5}$

69. $2\sqrt[3]{3} - \sqrt[3]{24}$

70. $-\sqrt[4]{32a^5} - 2\sqrt[4]{162a^5}$

71. $2x\sqrt{8} + 2\sqrt{200x^2} + \sqrt{50x^2}$

72. $\sqrt[3]{54x^3} - 3\sqrt[3]{16x^3} + 4\sqrt[3]{128x^3}$

73. $2\sqrt[4]{32t^3} - 8\sqrt[4]{6t^3} + 5\sqrt[4]{2t^3}$

74. $10\sqrt[4]{16x^9} - 8x^2\sqrt[4]{x} + 5\sqrt[4]{x^5}$

75. Explain the error in each simplification.

  a. $2\sqrt{5x} + 3\sqrt{5x} = 5\sqrt{10x}$

  b. $30 + 30\sqrt[4]{2} = 60\sqrt[4]{2}$

  c. $7\sqrt[3]{y^2} - 5\sqrt[3]{y^2} = 2$

  d. $6\sqrt{11ab} - 3\sqrt{5ab} = 3\sqrt{6ab}$

76. SEWING   A corner of fabric is folded over to form a collar and stitched down as shown. From the dimensions given in the figure, determine the exact number of inches of stitching that must be made. Then give an approximation to one decimal place. (All measurements are in inches.)

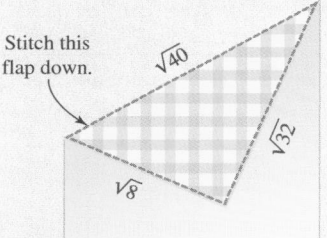

Stitch this flap down. $\sqrt{40}$ $\sqrt{32}$ $\sqrt{8}$

SECTION 9.4 Multiplying and Dividing Radical Expressions

DEFINITIONS AND CONCEPTS	EXAMPLES
We can use the product rule for radicals to **multiply radical expressions** that have the same index: $$\sqrt[n]{a}\,\sqrt[n]{b} = \sqrt[n]{ab}$$	Multiply and then simplify, if possible: $$\sqrt{6}\sqrt{8} = \sqrt{6 \cdot 8} = \sqrt{48} = \sqrt{16 \cdot 3} = \sqrt{16}\sqrt{3} = 4\sqrt{3}$$ $$\sqrt[3]{9x^4}\,\sqrt[3]{3x^2} = \sqrt[3]{9x^4 \cdot 3x^2} = \sqrt[3]{27x^6} = 3x^2$$
We can use the **distributive property** to multiply a radical expression with two or more terms by a radical expression with one term.	Multiply and then simplify, if possible: $$2\sqrt{3}\left(4\sqrt{5} - 5\sqrt{2}\right) = 2\sqrt{3} \cdot 4\sqrt{5} - 2\sqrt{3} \cdot 5\sqrt{2}$$ $$= 2 \cdot 4\sqrt{3 \cdot 5} - 2 \cdot 5\sqrt{3 \cdot 2}$$ $$= 8\sqrt{15} - 10\sqrt{6}$$
We can use the **FOIL method** to multiply a radical expression with two terms by another radical expression with two terms.	Multiply and then simplify, if possible: $$\begin{array}{cccc} & \text{F} & \text{O} & \text{I} & \text{L} \end{array}$$ $$\left(\sqrt[3]{x} - \sqrt[3]{3}\right)\left(\sqrt[3]{x} + \sqrt[3]{9}\right) = \sqrt[3]{x}\sqrt[3]{x} + \sqrt[3]{x}\sqrt[3]{9} - \sqrt[3]{3}\sqrt[3]{x} - \sqrt[3]{3}\sqrt[3]{9}$$ $$= \sqrt[3]{x^2} + \sqrt[3]{9x} - \sqrt[3]{3x} - \sqrt[3]{27}$$ $$= \sqrt[3]{x^2} + \sqrt[3]{9x} - \sqrt[3]{3x} - 3$$

SECTION 9.4   Multiplying and Dividing Radical Expressions—*continued*

DEFINITIONS AND CONCEPTS	EXAMPLES
If a radical appears in a denominator of a fraction, or if a radicand contains a fraction, we can write the radical in simplest form by **rationalizing the denominator.** To rationalize a denominator, we multiply the given expression by a carefully chosen form of 1.	Rationalize the denominator: $\sqrt{\dfrac{10}{3}} = \dfrac{\sqrt{10}}{\sqrt{3}} \cdot \dfrac{\sqrt{3}}{\sqrt{3}}$ $\qquad$ $\sqrt[3]{\dfrac{5}{2p^2}} = \dfrac{\sqrt[3]{5}}{\sqrt[3]{2p^2}} \cdot \dfrac{\sqrt[3]{4p}}{\sqrt[3]{4p}}$ $\qquad\quad = \dfrac{\sqrt{30}}{3}$ $\qquad\qquad\qquad = \dfrac{\sqrt[3]{20p}}{\sqrt[3]{8p^3}}$ $\qquad\qquad\qquad\qquad\qquad\qquad\quad = \dfrac{\sqrt[3]{20p}}{2p}$
Radical expressions that involve the sum and difference of the same two terms are called **conjugates.**	Conjugates: $\sqrt{2x} + 3$ and $\sqrt{2x} - 3$
To **rationalize a two-termed denominator** of a fraction, multiply the numerator and the denominator by the **conjugate** of the denominator.	Rationalize the denominator: $\dfrac{\sqrt{x} - 2}{\sqrt{x} + 2} = \dfrac{\sqrt{x} - 2}{\sqrt{x} + 2} \cdot \dfrac{\sqrt{x} - 2}{\sqrt{x} - 2}$ $\qquad\quad = \dfrac{\sqrt{x}\sqrt{x} - 2\sqrt{x} - 2\sqrt{x} + 4}{\sqrt{x}\sqrt{x} - 2\sqrt{x} + 2\sqrt{x} - 4}$ $\qquad\quad = \dfrac{x - 4\sqrt{x} + 4}{x - 4}$

REVIEW EXERCISES

Simplify each expression. All variables represent positive real numbers.

77. $\sqrt{7}\sqrt{7}$

78. $\left(2\sqrt{5}\right)\left(3\sqrt{2}\right)$

79. $\left(-2\sqrt{8}\right)^2$

80. $2\sqrt{6}\sqrt{15}$

81. $\sqrt{9x}\sqrt{x}$

82. $\left(\sqrt[3]{x+1}\right)^3$

83. $-\sqrt[3]{2x^2}\sqrt[3]{4x^8}$

84. $\sqrt[5]{9} \cdot \sqrt[5]{27}$

85. $3\sqrt{7t}\left(2\sqrt{7t} + 3\sqrt{3t^2}\right)$

86. $-\sqrt[4]{4x^5y^{11}}\sqrt[4]{8x^9y^3}$

87. $\left(\sqrt{3b} + \sqrt{3}\right)^2$

88. $\left(\sqrt[3]{3p} - 2\sqrt[3]{2}\right)\left(\sqrt[3]{3p} + \sqrt[3]{2}\right)$

Rationalize each denominator. All variables represent positive real numbers.

89. $\dfrac{10}{\sqrt{3}}$

90. $\sqrt{\dfrac{3}{5xy}}$

91. $\dfrac{\sqrt[3]{6u}}{\sqrt[3]{u^5}}$

92. $\dfrac{\sqrt[4]{a}}{\sqrt[4]{3b^2}}$

93. $\dfrac{2}{\sqrt{2} - 1}$

94. $\dfrac{4\sqrt{x} - 2\sqrt{z}}{\sqrt{z} + 4\sqrt{x}}$

95. Rationalize the numerator: $\dfrac{\sqrt{a} - \sqrt{b}}{\sqrt{a}}$

96. VOLUME   The formula relating the radius r of a sphere and its volume V is $r = \sqrt[3]{\dfrac{3V}{4\pi}}$. Write the radical in simplest form.

SECTION 9.5   Solving Radical Equations

DEFINITIONS AND CONCEPTS	EXAMPLES

We can use the **power rule** to solve equations containing radicals.

> If $x = y$, then $x^n = y^n$.

To **solve equations containing radicals:**

1. Isolate one radical expression on one side of the equation.

2. Raise both sides of the equation to the power that is the same as the index.

3. Solve the resulting equation. If it still contains a radical, go back to step 1.

4. Check the solutions to eliminate **extraneous** solutions.

Solve each equation.

$$\sqrt{2x - 2} + 1 = x$$
$$\sqrt{2x - 2} = x - 1$$
$$\left(\sqrt{2x - 2}\right)^2 = (x - 1)^2$$
$$2x - 2 = x^2 - 2x + 1$$
$$0 = x^2 - 4x + 3$$
$$0 = (x - 3)(x - 1)$$
$$x - 3 = 0 \quad \text{or} \quad x - 1 = 0$$
$$x = 3 \quad \mid \quad x = 1$$

The solutions are 3 and 1. Verify that each satisfies the original equation.

$$\sqrt[3]{x + 2} = 3$$
$$\left(\sqrt[3]{x + 2}\right)^3 = 3^3$$
$$x + 2 = 27$$
$$x = 25$$

The solution is 25. Verify that it satisfies the original equation.

When **more than one radical** appears in an equation, we must often use the power rule more than once.

Solve:

$$\sqrt{x} + \sqrt{x + 5} = 5$$

$$\sqrt{x + 5} = 5 - \sqrt{x} \qquad \text{To isolate each radical, subtract } \sqrt{x} \text{ from both sides.}$$

$$\left(\sqrt{x + 5}\right)^2 = \left(5 - \sqrt{x}\right)^2 \qquad \text{To eliminate one radical, square both sides.}$$

$$x + 5 = 25 - 10\sqrt{x} + x \qquad \text{Perform the operations on each side.}$$

$$-20 = -10\sqrt{x} \qquad \text{To isolate the radical term, subtract 25 and } x \text{ from both sides.}$$

$$2 = \sqrt{x} \qquad \text{To isolate the radical, divide both sides by } -10.$$

$$(2)^2 = \left(\sqrt{x}\right)^2 \qquad \text{To eliminate the radical, square both sides again.}$$

$$4 = x$$

The solution is 4. Verify that it satisfies the original equation.

REVIEW EXERCISES

Solve each equation. Write all proposed solutions. Cross out those that are extraneous.

97. $\sqrt{7x - 10} - 1 = 11$

98. $u = \sqrt{25u - 144}$

99. $2\sqrt{y - 3} = \sqrt{2y + 1}$

100. $\sqrt{z + 1} + \sqrt{z} = 2$

101. $\sqrt[3]{x^3 + 56} - 2 = x$

102. $a = \sqrt{a^2 + 5a - 35}$

103. $(x + 2)^{1/2} - (4 - x)^{1/2} = 0$ **104.** $\sqrt{b^2 + b} = \sqrt{3 - b^2}$

105. $\sqrt[4]{8x - 8} + 2 = 0$

106. $\sqrt{2m + 4} - \sqrt{m + 3} = 1$

107. Let $f(x) = \sqrt{2x^2 - 7x}$. For what value(s) of x is $f(x) = 2$?

108. Using the graphs of $f(x) = \sqrt{2x - 3}$ and $g(x) = -2x + 5$, estimate the solution of

$$\sqrt{2x - 3} = -2x + 5$$

Check the result.

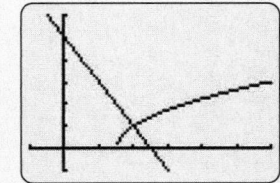

Solve each equation for the specified variable.

109. $r = \sqrt{\dfrac{A}{P} - 1}$ for P

110. $h = \sqrt[3]{\dfrac{12I}{b}}$ for I

SECTION 9.6 Geometric Applications of Radicals

DEFINITIONS AND CONCEPTS	EXAMPLES
The Pythagorean theorem: If a and b are the lengths of the **legs** of a right triangle and c is the length of the **hypotenuse**, then $a^2 + b^2 = c^2$. 	Find the length of the third side of the right triangle. $a^2 + b^2 = c^2$ This is the Pythagorean theorem. $6^2 + b^2 = 10^2$ Substitute 6 for a and 10 for c. $36 + b^2 = 100$ $b^2 = 64$ To isolate b^2, subtract 36 from both sides. $b = \sqrt{64}$ Since b must be positive, find the positive square root of 64. $b = 8$ The length of the third side of the triangle is 8 ft.
In an **isosceles right triangle,** the length of the hypotenuse is $\sqrt{2}$ times the length of one leg.	If the length of one leg of an isosceles right triangle is 7 feet, the length of the hypotenuse is $7\sqrt{2}$ feet.
The hypotenuse of a **30°–60°–90° triangle** is twice as long as the shorter leg (the leg opposite the 30° angle.) The length of the longer leg (the leg opposite the 60° angle) is $\sqrt{3}$ times the length of the shorter leg.	If the shorter leg of a 30°–60°–90° triangle is 9 inches long: • The length of the hypotenuse is $2 \cdot 9 = 18$ inches. • The length of the longer leg is $\sqrt{3} \cdot 9 = 9\sqrt{3}$ inches. 30°–60°–90° Triangle
The distance d between two points with coordinates (x_1, y_1) and (x_2, y_2) is given by **the distance formula:** $d = \sqrt{(x_2 - x_1)^2 + (y_2 - y_1)^2}$	The distance between points $(-2, 3)$ and $(1, 7)$ is $\begin{aligned} d &= \sqrt{(x_2 - x_1)^2 + (y_2 - y_1)^2} \\ &= \sqrt{[1 - (-2)]^2 + (7 - 3)^2} \\ &= \sqrt{3^2 + 4^2} \\ &= \sqrt{9 + 16} \\ &= \sqrt{25} \\ &= 5 \end{aligned}$

REVIEW EXERCISES

111. CARPENTRY The gable end of the roof shown below is divided in half by a vertical brace, 8 feet in height. Find the length of the roof line.

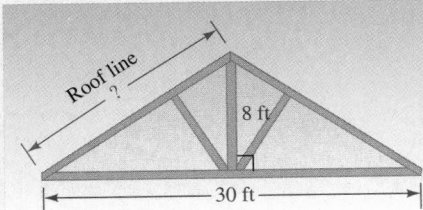

112. SAILING A technique called *tacking* allows a sailboat to make progress into the wind. A sailboat follows the course shown below. Find *d*, the distance the boat advances into the wind after tacking.

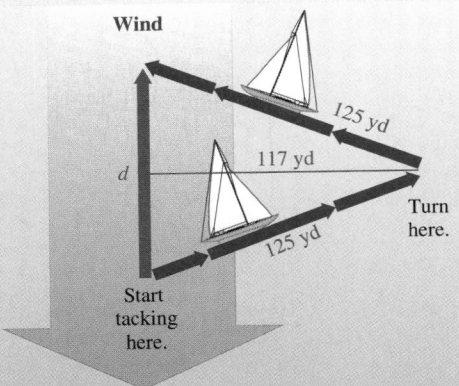

For problems 113–118, give the exact answer and then an approximation to two decimal places, when appropriate.

113. Find the length of the hypotenuse of an isosceles right triangle if the length of one leg is 7 meters.

114. The length of the hypotenuse of an isosceles right triangle is 15 yards. Find the length of one leg of the triangle.

115. The length of the hypotenuse of a 30°–60°–90° triangle is 12 centimeters. Find the length of each leg.

116. In a 30°–60°–90° triangle, the length of the longer leg is 60 feet. Find the length of the hypotenuse and the length of the shorter leg.

117. Find *x* and *y*.

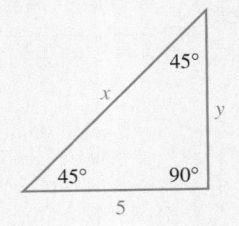

118. Find *x* and *y*.

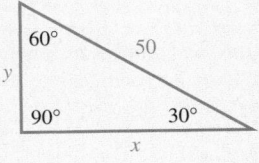

Find the distance between the points.

119. $(1, 3)$ and $(6, -9)$

120. $(-4, 6)$ and $(-2, 8)$

SECTION 9.7   Complex Numbers

DEFINITIONS AND CONCEPTS	EXAMPLES
The **imaginary number** *i* is defined as $$i = \sqrt{-1}$$ From the definition, it follows that $i^2 = -1$.	Write each expression in terms of i : $$\sqrt{-81} = \sqrt{-1 \cdot 81} \qquad \sqrt{-24} = \sqrt{-1 \cdot 24}$$ $$= \sqrt{-1}\sqrt{81} \qquad\quad = \sqrt{-1}\sqrt{24}$$ $$= i \cdot 9 \qquad\qquad\quad = i\sqrt{4}\sqrt{6}$$ $$= 9i \qquad\qquad\quad = 2i\sqrt{6} \ \text{ or } \ 2\sqrt{6}i$$
A **complex number** is any number that can be written in the form $a + bi$, where *a* and *b* are real numbers and $i = \sqrt{-1}$. We call *a* the **real part** and *b* the **imaginary part**.	Complex numbers: $5 + 3i$ 5 is the real part and 3 is the imaginary part. $16 = 16 + 0i$ 16 is the real part and 0 is the imaginary part. $9i = 0 + 9i$ 0 is the real part and 9 is the imaginary part.

SECTION 9.7 Complex Numbers—*continued*

DEFINITIONS AND CONCEPTS	EXAMPLES
Adding and subtracting complex numbers is similar to adding and subtracting polynomials. To **add two complex numbers,** add their real parts and add their imaginary parts.	Add. Write the answer in the form $a + bi$. $(7 - 5i) + (3 + 9i) = (7 + 3) + (-5 + 9)i$ Add the real parts. Add the imaginary parts. $= 10 + 4i$
To **subtract two complex numbers,** add the opposite of the complex number being subtracted.	Subtract. Write the answer in the form $a + bi$. $(8 - i) - (-1 + 6i) = (8 - i) + (1 - 6i)$ Add the opposite of $-1 + 6i$. $= (8 + 1) + [-1 + (-6)]i$ Add the real parts. Add the imaginary parts. $= 9 - 7i$
Multiplying complex numbers is similar to multiplying polynomials.	Multiply. Write the answers in the form $a + bi$. $3i(6 - 4i) = 18i - 12i^2$ $(4 + 7i)(2 - i) = 8 - 4i + 14i - 7i^2$ $= 18i - 12(-1)$ $= 8 + 10i - 7(-1)$ $= 18i + 12$ $= 8 + 10i + 7$ $= 12 + 18i$ $= 15 + 10i$
The complex numbers $a + bi$ and $a - bi$ are called **complex conjugates.**	The complex numbers $7 - 2i$ and $7 + 2i$ are complex conjugates.
To **divide complex numbers,** multiply the numerator and denominator by the complex conjugate of the denominator. The process is similar to rationalizing denominators.	Divide. Write the answers in the form $a + bi$. $\dfrac{3}{1 + i} \cdot \dfrac{1 - i}{1 - i} = \dfrac{3(1 - i)}{1 - i + i - i^2}$ $\dfrac{6 + i}{2 - i} \cdot \dfrac{2 + i}{2 + i} = \dfrac{12 + 6i + 2i + i^2}{4 + 2i - 2i - i^2}$ $= \dfrac{3(1 - i)}{1 - (-1)}$ $= \dfrac{12 + 8i + (-1)}{4 - (-1)}$ $= \dfrac{3 - 3i}{2}$ $= \dfrac{11 + 8i}{5}$ $= \dfrac{3}{2} - \dfrac{3}{2}i$ $= \dfrac{11}{5} + \dfrac{8}{5}i$
The **powers of i** cycle through four possible outcomes: i, -1, $-i$, and 1. $i^1 = \boxed{i} = i^5 = i^9 = \dots$ $i^2 = \boxed{-1} = i^6 = i^{10} = \dots$ $i^3 = \boxed{-i} = i^7 = i^{11} = \dots$ $i^4 = \boxed{1} = i^8 = i^{12} = \dots$	Simplify: i^{66} We divide 66 by 4 to get a remainder of 2. Thus, $i^{66} = i^2 = -1$. $\begin{array}{r} 16 \text{ R2} \\ 4\overline{)66} \\ \underline{-4} \\ 26 \\ \underline{-24} \\ 2 \end{array}$

REVIEW EXERCISES

Write each expression in terms of i.

121. $\sqrt{-25}$ **122.** $\sqrt{-18}$

123. $-\sqrt{-6}$ **124.** $\sqrt{-\dfrac{9}{64}}$

125. Complete the diagram.

Complex numbers

_____ numbers	_____ numbers

126. Determine whether each statement is true or false.
  a. Every real number is a complex number.
  b. $3 - 4i$ is a complex number.
  c. $\sqrt{-4}$ is a real number.
  d. i is a real number.

Give the complex conjugate of each number.

127. a. $3 + 6i$ **128. a.** $-1 - 7i$
   b. $19i$ **b.** $-i$

Perform the operations. Write all answers in the form a + bi.

129. $(3 + 4i) + (5 - 6i)$ **130.** $\left(7 - \sqrt{-9}\right) - \left(4 + \sqrt{-4}\right)$

131. $3i(2 - i)$ **132.** $(2 - 7i)(-3 + 4i)$

133. $\sqrt{-3} \cdot \sqrt{-9}$ **134.** $(9i)^2$

135. $\dfrac{5 + 14i}{2 + 3i}$ **136.** $\dfrac{3}{11i}$

Simplify each expression.

137. i^{42} **138.** i^{97}

CHAPTER 9
TEST

1. Fill in the blanks.
  a. The symbol $\sqrt{\ }$ is called a _____ symbol.
  b. The _____ number i is defined as $i = \sqrt{-1}$.
  c. Squaring both sides of an equation can introduce _____ solutions.
  d. An _____ right triangle is a right triangle with two legs of equal length.
  e. To _____ the denominator of $\dfrac{4}{\sqrt{5}}$, we multiply the fraction by $\dfrac{\sqrt{5}}{\sqrt{5}}$.
  f. A _____ number is any number that can be written in the form $a + bi$, where a and b are real numbers and $i = \sqrt{-1}$.

2. a. State the product rule for radicals.

  b. State the quotient rule for radicals.

3. Graph $f(x) = \sqrt{x} - 1$. Find the domain and range of the function.

4. DIVING   Refer to the illustration in the next column. The velocity v of an object in feet per second after it has fallen a distance of d feet is approximated by the function $v(d) = \sqrt{64.4d}$. Olympic diving platforms are 10 meters tall (approximately 32.8 feet). Estimate the velocity at which a diver hits the water from this height. Round to the nearest foot per second.

32.8 feet

5. Use the graph to find each of the following.
  a. $f(-1)$
  b. $f(8)$
  c. The value(s) of x for which $f(x) = 1$
  d. The domain and range of f

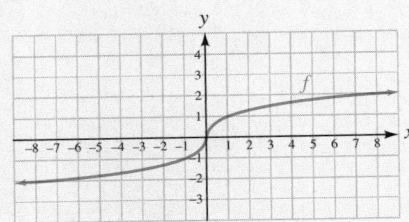

6. Explain why $\sqrt[4]{-16}$ is not a real number.

Simplify each expression. All variables represent positive real numbers. Write answers without using negative exponents.

7. $(49x^4)^{1/2}$

8. $-27^{2/3}$

9. $36^{-3/2}$

10. $\left(-\dfrac{8}{125n^6}\right)^{-2/3}$

11. $\dfrac{2^{5/3}2^{1/6}}{2^{1/2}}$

12. $(a^{2/3})^{1/6}$

Simplify each expression. The variables are unrestricted.

13. $\sqrt{x^2}$

14. $\sqrt{y^2 - 10y + 25}$

Simplify each expression. All variables represent positive real numbers.

15. $\sqrt[3]{-64x^3y^6}$

16. $\sqrt{\dfrac{4a^2}{9}}$

17. $\sqrt[5]{(t + 8)^5}$

18. $\sqrt{540x^3y^5}$

19. $\dfrac{\sqrt[3]{24x^{15}y^4}}{\sqrt[3]{y}}$

20. $\sqrt[4]{32}$

Perform the operations and simplify. All variables represent positive real numbers.

21. $2\sqrt{48y^5} - 3y\sqrt{12y^3}$

22. $2\sqrt[3]{40} - \sqrt[3]{5,000} + 4\sqrt[3]{625}$

23. $\sqrt[4]{243z^{13}} + z\sqrt[4]{48z^9}$

24. $-2\sqrt{xy}\left(3\sqrt{x} + \sqrt{xy^3}\right)$

25. $\left(3\sqrt{2} + \sqrt{3}\right)\left(2\sqrt{2} - 3\sqrt{3}\right)$

26. $\left(\sqrt[3]{2a} + 9\right)^2$

27. $\dfrac{8}{\sqrt{10}}$

28. $\dfrac{3t - 1}{\sqrt{3t} - 1}$

29. $\sqrt[3]{\dfrac{9}{4a}}$

30. Rationalize the numerator: $\dfrac{\sqrt{5} + 3}{-4\sqrt{2}}$

Solve each equation. Write all proposed solutions. Cross out those that are extraneous.

31. $4\sqrt{x} = \sqrt{x + 1}$

32. $\sqrt[3]{6n + 4} - 4 = 0$

33. $1 = \sqrt{u - 3} + \sqrt{u}$

34. $(2m^2 - 9)^{1/2} = m$

35. $\sqrt{t - 2} - t + 2 = 0$

36. $\sqrt{x - 8} + 10 = 0$

37. $\sqrt[4]{15 - a} = \sqrt[4]{13 - 2a}$

38. Solve $r = \sqrt[3]{\dfrac{GMt^2}{4\pi^2}}$ for G.

Find the missing side lengths in each triangle. Give the exact answer and then an approximation to two decimal places, when appropriate.

39.

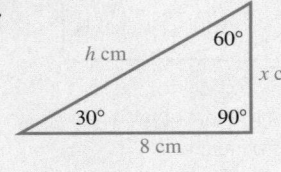

40.

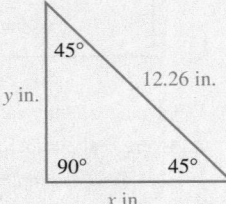

41. Find the distance between $(-2, 5)$ and $(22, 12)$.

42. SHIPPING CRATES   The diagonal brace on the shipping crate in the illustration is 53 inches. Find the height h of the crate.

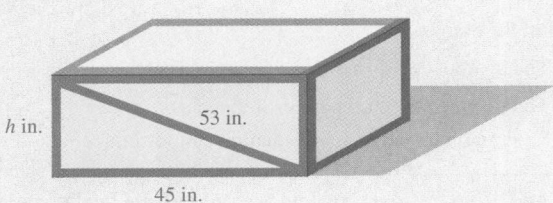

43. Express $\sqrt{-45}$ in terms of i.

44. Simplify: i^{106}

Perform the operations. Write all answers in the form $a + bi$.

45. $(9 + 4i) + (-13 + 7i)$

46. $\left(3 - \sqrt{-9}\right) - \left(-1 + \sqrt{-16}\right)$

47. $15i(3 - 5i)$

48. $(8 + 10i)(-7 - i)$

49. $\dfrac{1}{i\sqrt{2}}$

50. $\dfrac{2 + i}{3 - i}$

GROUP PROJECT

A SPIRAL OF ROOTS

Overview: In this activity, you will create a visual representation of a collection of square roots.

Instructions: Form groups of 2 or 3 students. You will need a piece of unlined paper, a protractor, a ruler, and a pencil. Begin by drawing an isosceles right triangle with legs of length 1 inch in the middle of the paper. (See the illustration.) Use the Pythagorean theorem to determine the length of the hypotenuse. Draw a second right triangle using the hypotenuse of the first right triangle as one leg. Draw its second leg with length 1 inch. Find the length of the hypotenuse of the second triangle.

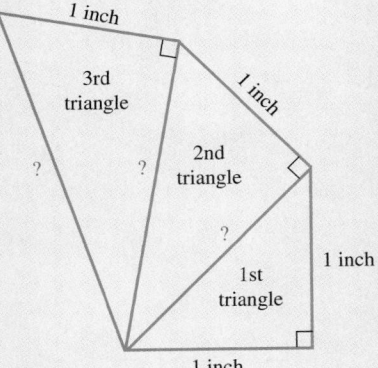

Continue creating right triangles, using the previous hypotenuse as one leg and drawing a new second leg of length 1 inch each time. Calculate the length of each resulting hypotenuse. When the figure begins to spiral onto itself, you may stop the process. Make a list of the lengths of each hypotenuse. What pattern do you see?

GRAPHING IN THREE DIMENSIONS

Overview: In this activity, you will find the distance between two points that lie in three-dimensional space.

Instructions: Form groups of 2 or 3 students. In a three-dimensional Cartesian coordinate system, the positive x-axis is horizontal and pointing toward the viewer (out of the page), the positive y-axis is also horizontal and pointing to the right, and the positive z-axis is vertical, pointing up. A point is located by plotting an ordered triple of numbers (x, y, z). In the illustration, the point $(3, 2, 4)$ is plotted.

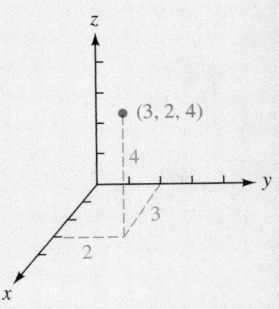

In three dimensions, the distance formula is

$$d = \sqrt{(x_2 - x_1)^2 + (y_2 - y_1)^2 + (z_2 - z_1)^2}$$

1. Copy the illustration shown. Then plot the point $(1, 4, 3)$. Use the distance formula to find the distance between these two points.
2. Draw another three-dimensional coordinate system and plot the points $(-3, 3, -4)$ and $(2, -3, 2)$. Use the distance formula to find the distance between these two points.

CHAPTER 10

Quadratic Equations, Functions, and Inequalities

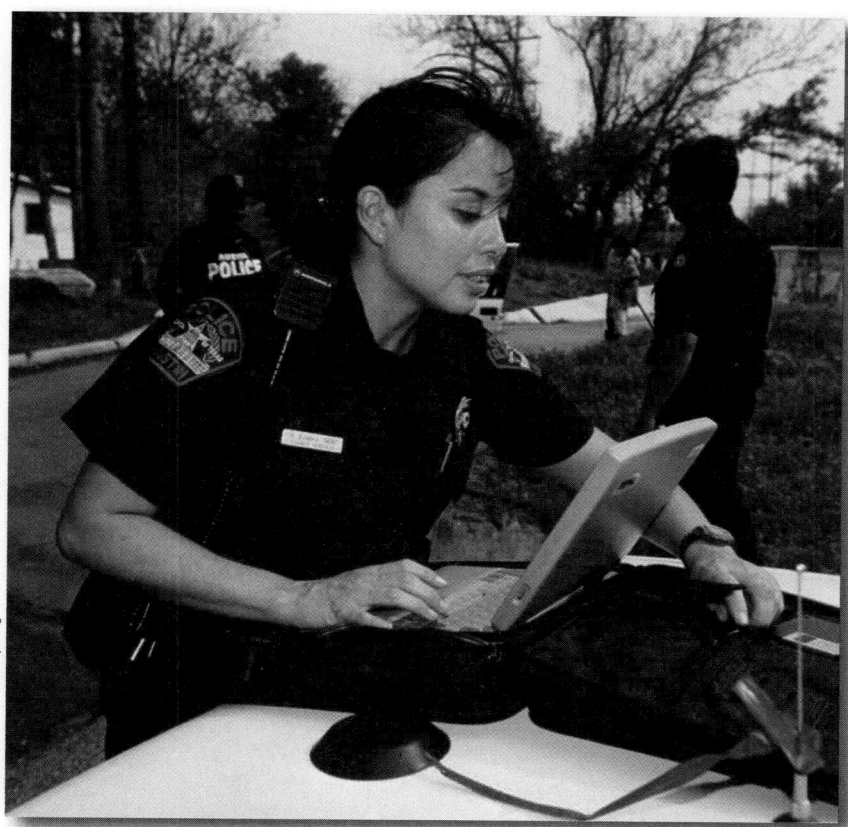

from **Campus to Careers**
Police Patrol Officer

The responsibilities of a police patrol officer are extremely broad. Quite often, he or she must make a split-second decision while under enormous pressure. One internet vocational website cautions anyone considering such a career to "pay attention in your mathematics and science classes. Those classes help sharpen your ability to think things through and solve problems—an important part of police work."

Police patrol officers often use yellow "DO NOT CROSS" barricade tape to keep the public from entering a crime scene. In **problem 89** of **Study Set 10.4,** you will determine the maximum rectangular area that can be sealed off using a 300-foot roll of barricade tape.

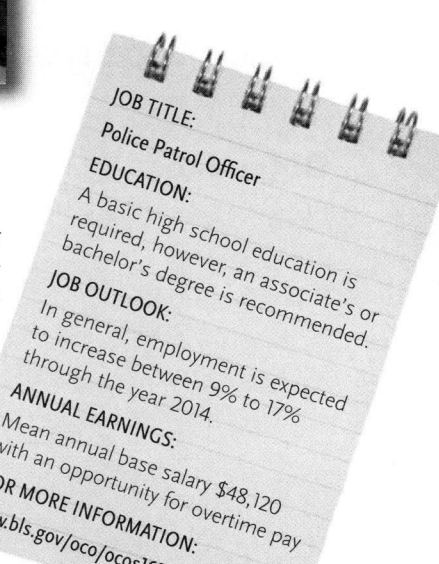

JOB TITLE:
Police Patrol Officer

EDUCATION:
A basic high school education is required, however, an associate's or bachelor's degree is recommended.

JOB OUTLOOK:
In general, employment is expected to increase between 9% to 17% through the year 2014.

ANNUAL EARNINGS:
Mean annual base salary $48,120 with an opportunity for overtime pay

FOR MORE INFORMATION:
www.bls.gov/oco/ocos160.htm

Study Skills Workshop
Organizing Your Notebook

If you're like most students, your algebra notebook could probably use some attention at this stage of the course. You will definitely appreciate a well-organized notebook when it comes time to study for the final exam. Here are some suggestions to put it in tip-top shape.

ORGANIZE YOUR NOTEBOOK INTO SECTIONS: Create a separate section in the notebook for each chapter (or unit of study) that your class has covered this term.

ORGANIZE THE PAPERS WITHIN EACH SECTION: One recommended order is to begin each section with your class notes, followed by your completed homework assignments, then any study sheets or handouts, and, finally, all graded quizzes and tests.

Now Try This

1. Organize your algebra notebook using the guidelines given above.
2. Write a Table of Contents to place at the beginning of your notebook. List each chapter (or unit of study) and include the dates over which the material was covered.
3. Compare your completed notebook with those of other students in your class. Have you overlooked any important items that would be useful when studying for the final exam?

SECTION 10.1
The Square Root Property and Completing the Square

Objectives

❶ Use the square root property to solve quadratic equations.
❷ Solve quadratic equations by completing the square.
❸ Use quadratic equations to solve application problems.

Recall that a *quadratic equation* is an equation of the form $ax^2 + bx + c = 0$, where a, b, and c are real numbers and $a \neq 0$. We have solved quadratic equations using factoring and the zero-factor property.

EXAMPLE 1   Solve: $6x^2 - 7x - 3 = 0$

Strategy   We will factor the trinomial on the left side of the equation and use the zero-factor property to solve for x.

Why   To use the zero-factor property, we need one side of the equation to be factored completely and the other side to be 0.

Solution

$$6x^2 - 7x - 3 = 0$$

$$(2x - 3)(3x + 1) = 0 \qquad \text{Factor the trinomial.}$$

$$2x - 3 = 0 \quad \text{or} \quad 3x + 1 = 0 \qquad \text{Set each factor equal to 0.}$$

$$x = \frac{3}{2} \qquad \qquad x = -\frac{1}{3} \qquad \text{Solve each linear equation.}$$

Self Check 1 Solve: $4x^2 - 4x - 3 = 0$

Now Try Problem 21

Many expressions do not factor as easily as $6x^2 - 7x - 3$. For example, it would be difficult to solve $2x^2 + 4x + 1 = 0$ by factoring, because $2x^2 + 4x + 1$ cannot be factored by using only integers. We will now develop a more general method that enables us to solve any quadratic equation. It is based on the *square root property*.

 Use the Square Root Property to Solve Quadratic Equations.

To develop general methods for solving quadratic equations, we first consider the equation $x^2 = c$. If $c \geq 0$, we can find the real solutions of $x^2 = c$ as follows:

$$x^2 = c$$

$$x^2 - c = 0 \qquad \text{Subtract } c \text{ from both sides.}$$

$$x^2 - \left(\sqrt{c}\right)^2 = 0 \qquad \text{Replace } c \text{ with } \left(\sqrt{c}\right)^2, \text{ since } c = \left(\sqrt{c}\right)^2.$$

$$\left(x + \sqrt{c}\right)\left(x - \sqrt{c}\right) = 0 \qquad \text{Factor the difference of two squares.}$$

$$x + \sqrt{c} = 0 \quad \text{or} \quad x - \sqrt{c} = 0 \qquad \text{Set each factor equal to 0.}$$

$$x = -\sqrt{c} \qquad \qquad x = \sqrt{c} \qquad \text{Solve each linear equation.}$$

The solutions of $x^2 = c$ are $\sqrt{c}$ and $-\sqrt{c}$.

The Square Root Property

For any nonnegative real number c, if $x^2 = c$, then

$$x = \sqrt{c} \qquad \text{or} \qquad x = -\sqrt{c}$$

EXAMPLE 2 Solve: $x^2 - 12 = 0$

Strategy We will add 12 to both sides of the equation and use the square root property to solve for x.

Why After adding 12 to both sides, the resulting equivalent equation will have the desired form $x^2 = c$.

Notation
We can use **double-sign notation** $\pm$ to write the solutions in compact form as $\pm 2\sqrt{3}$. Read $\pm$ as "positive or negative."

Solution

$$x^2 - 12 = 0 \qquad \text{This is the equation to solve.}$$

$$x^2 = 12 \qquad \text{To isolate } x^2 \text{ on the left side, add 12 to both sides.}$$

$$x = \sqrt{12} \quad \text{or} \quad x = -\sqrt{12} \qquad \text{Use the square root property.}$$

$$x = 2\sqrt{3} \qquad \qquad x = -2\sqrt{3} \qquad \text{Simplify: } \sqrt{12} = \sqrt{4}\sqrt{3} = 2\sqrt{3}.$$

Check: $x^2 - 12 = 0$ $x^2 - 12 = 0$
$$\left(2\sqrt{3}\right)^2 - 12 \stackrel{?}{=} 0 \qquad\qquad \left(-2\sqrt{3}\right)^2 - 12 \stackrel{?}{=} 0$$
$$12 - 12 \stackrel{?}{=} 0 \qquad\qquad\qquad 12 - 12 \stackrel{?}{=} 0$$
$$0 = 0 \quad \text{True} \qquad\qquad\qquad 0 = 0 \quad \text{True}$$

The exact solutions are $2\sqrt{3}$ and $-2\sqrt{3}$ and the solution set is $\{2\sqrt{3}, -2\sqrt{3}\}$. We can use a calculator to approximate the solutions. To the nearest hundredth, they are ± 3.46.

 Self Check 2   Solve: $x^2 - 18 = 0$
Now Try   **Problems 23 and 27**

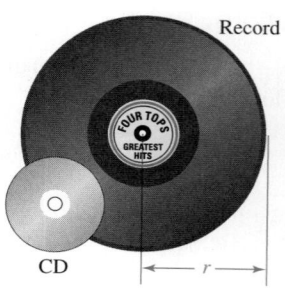

Record

CD $\leftarrow r \rightarrow$

EXAMPLE 3   *Phonograph Records.* Before compact discs, music was recorded on thin vinyl discs. The discs used for long-playing records had a surface area of about 111 square inches per side and were played at $33\frac{1}{3}$ revolutions per minute on a turntable. Find the radius of a long-playing record.

Strategy   The area A of a circle with radius r is given by the formula $A = \pi r^2$. We will find the radius of a record by substituting 111 for A and dividing both sides by π. Then we will use the square root property to solve for r.

Why   After substituting 111 for A and dividing both sides by π, the resulting equivalent equation will have the desired form $r^2 = c$.

Solution
$$A = \pi r^2 \qquad\qquad \text{This is the formula for the area of a circle.}$$
$$111 = \pi r^2 \qquad\qquad \text{Substitute 111 for A.}$$
$$\frac{111}{\pi} = r^2 \qquad\qquad \text{To undo the multiplication by } \pi, \text{ divide both sides by } \pi.$$
$$r = \sqrt{\frac{111}{\pi}} \quad \text{or} \quad r = -\sqrt{\frac{111}{\pi}} \qquad \text{Use the square root property. Since the radius of the record cannot be negative, discard the second solution.}$$

The radius of a long-playing record is $\sqrt{\frac{111}{\pi}}$ inches—to the nearest tenth, 5.9 inches.

 Now Try   **Problem 99**

EXAMPLE 4   Use the square root property to solve $(x - 1)^2 = 16$.

Strategy   Instead of a variable squared on the left side, we have a quantity squared. We still use the square root property to solve the equation.

Why   We want to eliminate the square on the binomial, so that we can eventually isolate the variable on one side of the equation.

Solution

$$(x - 1)^2 = 16$$ This is the equation to solve.

$$x - 1 = \pm\sqrt{16}$$ Use the square root property.

$$x - 1 = \pm 4$$ Simplify: $\sqrt{16} = 4$.

$$x = 1 \pm 4$$ To isolate x, add 1 to both sides.

$$x = 1 + 4 \quad \text{or} \quad x = 1 - 4$$ To find one solution, use $+$. To find the other, use $-$.

$$x = 5 \quad | \quad x = -3$$ Add (subtract).

Verify that 5 and -3 satisfy the original equation.

Self Check 4   Use the square root property to solve $(x + 2)^2 = 9$.

Now Try   Problems 31 and 35

> **Notation**
> It is standard practice to write the addition of the 1 in front of the $\pm$ symbol.
>
> $$x = 1 \pm 4$$
>
> not
>
> $$x = \pm 4 + 1$$
>
> We read 1 ± 4 as "one plus or minus four."

Some quadratic equations have solutions that are not real numbers.

EXAMPLE 5   Solve: $4x^2 + 25 = 0$

Strategy   We will subtract 25 from both sides of the equation and divide both sides by 4. Then we will use the square root property to solve for x.

Why   After subtracting 25 from both sides and dividing both sides by 4, the resulting equivalent equation will have the desired form $x^2 = c$.

Solution

$$4x^2 + 25 = 0$$ This is the equation to solve.

$$x^2 = -\frac{25}{4}$$ To isolate x^2, subtract 25 from both sides and divide both sides by 4.

$$x = \pm\sqrt{-\frac{25}{4}}$$ Use the square root property.

Since

$$\sqrt{-\frac{25}{4}} = \sqrt{-1 \cdot \frac{25}{4}} = \sqrt{-1}\frac{\sqrt{25}}{\sqrt{4}} = \frac{5}{2}i$$

we have

$$x = \pm\frac{5}{2}i$$

Since the solutions are $\frac{5}{2}i$ and $-\frac{5}{2}i$, the solution set is $\left\{\frac{5}{2}i, -\frac{5}{2}i\right\}$.

Check:

$$4x^2 + 25 = 0$$
$$4\left(\frac{5}{2}i\right)^2 + 25 \overset{?}{=} 0$$
$$4\left(\frac{25}{4}\right)i^2 + 25 \overset{?}{=} 0$$
$$25(-1) + 25 \overset{?}{=} 0$$
$$0 = 0 \quad \text{True}$$

$$4x^2 + 25 = 0$$
$$4\left(-\frac{5}{2}i\right)^2 + 25 \overset{?}{=} 0$$
$$4\left(\frac{25}{4}\right)i^2 + 25 \overset{?}{=} 0$$
$$25(-1) + 25 \overset{?}{=} 0$$
$$0 = 0 \quad \text{True}$$

> **The Language of Algebra**
> The $\pm$ symbol is often seen in political polls. A candidate with 48% ($\pm 4\%$) support could be between $48 + 4 = 52\%$ and $48 - 4 = 44\%$.

Self Check 5   Solve: $16x^2 + 49 = 0$

Now Try   **Problem 41**

---

2   **Solve Quadratic Equations by Completing the Square.**

When the polynomial in a quadratic equation doesn't factor easily, we can solve the equation by *completing the square*. This method is based on the following special products:

$$x^2 + 2bx + b^2 = (x + b)^2 \qquad \text{and} \qquad x^2 - 2bx + b^2 = (x - b)^2$$

In each of these perfect-square trinomials, the third term is the square of one-half of the coefficient of x.

> **The Language of Algebra**
> Recall that trinomials that are the square of a binomial are called *perfect-square trinomials*.

- In $x^2 + 2bx + b^2$, the coefficient of x is $2b$. If we find $\frac{1}{2} \cdot 2b$, which is b, and square it, we get the third term, b^2.

- In $x^2 - 2bx + b^2$, the coefficient of x is $-2b$. If we find $\frac{1}{2}(-2b)$, which is $-b$, and square it, we get the third term: $(-b)^2 = b^2$.

We can use these observations to change certain binomials into perfect-square trinomials. For example, to change $x^2 + 12x$ into a perfect-square trinomial, we find one-half of the coefficient of x, square the result, and add the square to $x^2 + 12x$.

$$x^2 + 12x + \boxed{}$$

Find one-half of the coefficient of x.   Add the square to the binomial.

$$\frac{1}{2} \cdot 12 = 6 \qquad 6^2 = 36$$

Square the result.

We obtain the perfect-square trinomial $x^2 + 12x + 36$, which factors as $(x + 6)^2$. By adding 36 to $x^2 + 12x$, we say that we have *completed the square on* $x^2 + 12x$.

> **Completing the Square**
>
> To complete the square on $x^2 + bx$, add the square of one-half of the coefficient of x:
>
> $$x^2 + bx + \left(\frac{1}{2}b\right)^2$$

EXAMPLE 6   Complete the square and factor the resulting perfect-square trinomial: **a.** $x^2 + 10x$ **b.** $x^2 - 11x$

Strategy   We will add the square of one-half of the coefficient of x to the given binomial.

Why   Adding such a term will change the binomial into a perfect-square trinomial that will factor.

> **Caution**
> Realize that when we complete the square on a binomial, we are not writing an equivalent trinomial expression. Thus, it would be incorrect to use an $=$ symbol between the two.
>
> $\cancel{x^2 + 10x} = \cancel{x^2 + 10x + 25}$

Solution

a. To make $x^2 + 10x$ a perfect-square trinomial, we find one-half of 10, square it, and add the result to $x^2 + 10x$.

$$x^2 + 10x + 25 \qquad \tfrac{1}{2} \cdot 10 = 5 \text{ and } 5^2 = 25. \text{ Add 25 to the binomial.}$$

This trinomial factors as $(x + 5)^2$.

b. To make $x^2 - 11x$ a perfect-square trinomial, we find one-half of -11, square it, and add the result to $x^2 - 11x$.

$$x^2 - 11x + \frac{121}{4} \qquad \tfrac{1}{2}(-11) = -\tfrac{11}{2} \text{ and } \left(-\tfrac{11}{2}\right)^2 = \tfrac{121}{4}. \text{ Add } \tfrac{121}{4} \text{ to the binomial.}$$

This trinomial factors as $\left(x - \frac{11}{2}\right)^2$.

Self Check 6   Complete the square on $a^2 - 5a$ and factor the resulting trinomial.

Now Try   **Problems 47 and 49**

To solve an equation of the form $ax^2 + bx + c = 0$ by completing the square, we use the following steps.

Completing the Square to Solve a Quadratic Equation in x

1. If the coefficient of x^2 is 1, go to step 2. If it is not, make it 1 by dividing both sides of the equation by the coefficient of x^2.
2. Get all variable terms on one side of the equation and constants on the other side.
3. Complete the square by finding one-half of the coefficient of x, squaring the result, and adding the square to both sides of the equation.
4. Factor the perfect-square trinomial as the square of a binomial.
5. Solve the resulting equation using the square root property.
6. Check your answers in the original equation.

EXAMPLE 7   Solve $x^2 + 8x + 7 = 0$ by completing the square.

Strategy   We will begin by subtracting 7 from both sides of the equation. Then we will proceed to complete the square to solve for x.

Why   We subtract 7 from both sides to isolate the variable terms, x^2 and $8x$, on the left side of the equation and the constant term on the right side.

Solution
Step 1: In this example, the coefficient of x^2 is 1.

Step 2: To prepare to complete the square, we subtract 7 from both sides.

$$x^2 + 8x + 7 = 0 \qquad \text{This is the equation to solve.}$$
$$x^2 + 8x = -7$$

Step 3: The coefficient of x is 8, one-half of 8 is 4, and $4^2 = 16$. To complete the square, we add 16 to both sides.

$$x^2 + 8x + 16 = 16 - 7$$
(1) $x^2 + 8x + 16 = 9 \qquad \text{Simplify: } 16 - 7 = 9.$

Step 4: Since the left side of Equation 1 is a perfect-square trinomial, we can factor it to get $(x + 4)^2$.

$$x^2 + 8x + 16 = 9$$
(2) $(x + 4)^2 = 9$

Step 5: We can solve Equation 2 by using the square root property.

$$x + 4 = \pm\sqrt{9}$$
$$x + 4 = \pm 3 \qquad \text{Simplify: } \sqrt{9} = 3.$$
$$x = -4 \pm 3 \qquad \text{To isolate } x, \text{ subtract 4 from both sides.}$$

This result represents two solutions. To find the first solution we add 3, and to find the second solution, we subtract 3.

$$x = -4 + 3 \quad \text{or} \quad x = -4 - 3 \qquad \text{± represents + or −.}$$
$$x = -1 \qquad\qquad\quad x = -7 \qquad \text{Add (subtract).}$$

Step 6: The solutions are −1 and −7. Check each one in the original equation.

Self Check 7   Solve $x^2 + 12x + 11 = 0$ by completing the square.
Now Try   **Problem 51**

If the coefficient of the squared variable (called the **leading coefficient**) of a quadratic equation is not 1, we must make it 1 before we can complete the square.

EXAMPLE 8   Solve $6x^2 + 5x - 6 = 0$ by completing the square.

Strategy   We will begin by dividing both sides of the equation by 6.

Why   This will create a leading coefficient that is 1 so that we can proceed to complete the square to solve the equation.

Solution

Step 1: To make the coefficient of x^2 equal to 1, we divide both sides of the equation by 6.

$$6x^2 + 5x - 6 = 0 \qquad \text{This is the equation to solve.}$$
$$\frac{6x^2}{6} + \frac{5}{6}x - \frac{6}{6} = \frac{0}{6} \qquad \text{Divide both sides by 6, term-by-term.}$$
$$x^2 + \frac{5}{6}x - 1 = 0 \qquad \text{Simplify.}$$

Step 2: To have the constant term on one side of the equation and the variable terms on the other, add 1 to both sides.

$$x^2 + \frac{5}{6}x = 1$$

Step 3: The coefficient of x is $\frac{5}{6}$, one-half of $\frac{5}{6}$ is $\frac{5}{12}$, and $\left(\frac{5}{12}\right)^2 = \frac{25}{144}$. To complete the square, we add $\frac{25}{144}$ to both sides.

$$x^2 + \frac{5}{6}x + \frac{25}{144} = 1 + \frac{25}{144}$$
$$(3) \quad x^2 + \frac{5}{6}x + \frac{25}{144} = \frac{169}{144} \qquad \text{Simplify the right side: } 1 + \frac{25}{144} = \frac{144}{144} + \frac{25}{144} = \frac{169}{144}.$$

Step 4: Factor the left side of Equation 3.

(4) $\left(x + \dfrac{5}{12}\right)^2 = \dfrac{169}{144}$ $\qquad$ $x^2 + \dfrac{5}{6}x + \dfrac{25}{144}$ is a perfect-square trinomial.

Step 5: We can solve Equation 4 by using the square root property.

$$x + \frac{5}{12} = \pm\sqrt{\frac{169}{144}}$$

$$x + \frac{5}{12} = \pm\frac{13}{12} \qquad \text{Simplify: } \pm\sqrt{\tfrac{169}{144}} = \pm\tfrac{13}{12}.$$

$$x = -\frac{5}{12} \pm \frac{13}{12} \qquad \text{To isolate x, subtract } \tfrac{5}{12} \text{ from both sides.}$$

$$x = -\frac{5}{12} + \frac{13}{12} \quad \text{or} \quad x = -\frac{5}{12} - \frac{13}{12}$$

$$x = \frac{8}{12} \qquad\qquad x = -\frac{18}{12} \qquad \text{Add (subtract) the fractions.}$$

$$x = \frac{2}{3} \qquad\qquad x = -\frac{3}{2} \qquad \text{Simplify each fraction.}$$

Step 6: Check each solution in the original equation.

Self Check 8 $\qquad$ Solve: $3x^2 + 2x - 8 = 0$

Now Try **Problem 59**

EXAMPLE 9 $\qquad$ Solve: $2x^2 + 4x + 1 = 0$

Strategy $\quad$ We will follow the steps for solving a quadratic equation by completing the square.

Why $\quad$ Since the trinomial $2x^2 + 4x + 1$ cannot be factored using only integers, solving the equation by completing the square is our only option at this time.

Solution

$$2x^2 + 4x + 1 = 0 \qquad \text{This is the equation to solve.}$$

$$x^2 + 2x + \frac{1}{2} = 0 \qquad \text{Divide both sides by 2 to make the coefficient of } x^2 \text{ equal to 1.}$$

$$x^2 + 2x \quad\;\; = -\frac{1}{2} \qquad \text{Subtract } \tfrac{1}{2} \text{ from both sides.}$$

$$x^2 + 2x + 1 = -\frac{1}{2} + 1 \qquad \text{Square one-half of the coefficient of x and add it to both sides.}$$

$$(x + 1)^2 = \frac{1}{2} \qquad \text{Factor and combine like terms.}$$

$$x + 1 = \pm\sqrt{\frac{1}{2}} \qquad \text{Use the square root property.}$$

$$x = -1 \pm \sqrt{\frac{1}{2}} \qquad \text{To isolate x, subtract 1 from both sides.}$$

To write $\sqrt{\dfrac{1}{2}}$ in simplified radical form, we write it as a quotient of square roots and then rationalize the denominator.

$$x = -1 + \frac{\sqrt{2}}{2} \quad \text{or} \quad x = -1 - \frac{\sqrt{2}}{2} \qquad \sqrt{\tfrac{1}{2}} = \frac{\sqrt{1}}{\sqrt{2}} = \frac{1 \cdot \sqrt{2}}{\sqrt{2}\sqrt{2}} = \frac{\sqrt{2}}{2}.$$

We can express each solution in an alternate form if we write -1 as a fraction with a denominator of 2.

$$x = -\frac{2}{2} + \frac{\sqrt{2}}{2} \quad \text{or} \quad x = -\frac{2}{2} - \frac{\sqrt{2}}{2} \qquad \text{Write } -1 \text{ as } -\tfrac{2}{2}.$$

$$x = \frac{-2 + \sqrt{2}}{2} \qquad\qquad x = \frac{-2 - \sqrt{2}}{2} \qquad \text{Add (subtract) the numerators and keep the common denominator 2.}$$

> **Caution**
>
> Recall that to simplify a fraction, we remove common factors of the numerator and denominator. Since -2 is a term of the numerator of $\dfrac{-2 + \sqrt{2}}{2}$, no further simplification of this expression can be made.

The exact solutions are $\dfrac{-2 + \sqrt{2}}{2}$ and $\dfrac{-2 - \sqrt{2}}{2}$, or simply, $\dfrac{-2 \pm \sqrt{2}}{2}$. We can use a calculator to approximate them. To the nearest hundredth, they are -0.29 and -1.71.

Self Check 9   Solve: $3x^2 + 6x + 1 = 0$

Now Try   **Problem 63**

Using Your Calculator

Checking Solutions of Quadratic Equations

We can use a graphing calculator to check the solutions of the quadratic equation $2x^2 + 4x + 1 = 0$ found in Example 9. After entering $Y_1 = 2x^2 + 4x + 1$, we call up the home screen by pressing $\boxed{\text{2nd}}$ $\boxed{\text{QUIT}}$. Then we press $\boxed{\text{VARS}}$, arrow to Y-VARS, press $\boxed{\text{ENTER}}$, and enter 1 to get the display shown in figure (a). We evaluate $2x^2 + 4x + 1$ for $x = \dfrac{-2 + \sqrt{2}}{2}$ by entering the solution using function notation, as shown in figure (b). When $\boxed{\text{Enter}}$ is pressed, the result of 0 is confirmation that $x = \dfrac{-2 + \sqrt{2}}{2}$ is a solution of the equation.

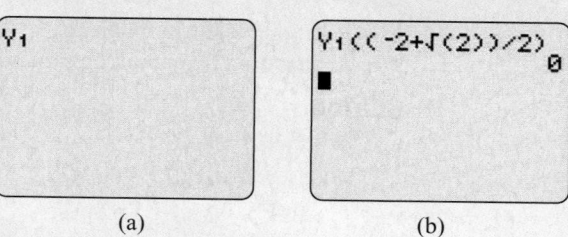

(a)   (b)

In the next example, the solutions of the equation are two complex numbers that contain i.

EXAMPLE 10   Solve: $3x^2 + 2x + 2 = 0$

Strategy   We will follow the steps for solving a quadratic equation by completing the square.

Why   Since the trinomial $3x^2 + 2x + 2$ cannot be factored using only integers, solving the equation by completing the square is our only option at this time.

Success Tip

Examples 7 and 8 can be solved by factoring. However, Examples 9 and 10 cannot. This illustrates that completing the square can be used to solve any quadratic equation.

Solution

$$3x^2 + 2x + 2 = 0$$ This is the equation to solve.

$$x^2 + \frac{2}{3}x + \frac{2}{3} = \frac{0}{3}$$ Divide both sides by 3 to make the coefficient of x^2 equal to 1.

$$x^2 + \frac{2}{3}x \qquad = -\frac{2}{3}$$ Subtract $\frac{2}{3}$ from both sides.

$$x^2 + \frac{2}{3}x + \frac{1}{9} = \frac{1}{9} - \frac{2}{3}$$ $\frac{1}{2} \cdot \frac{2}{3} = \frac{1}{3}$ and $\left(\frac{1}{3}\right)^2 = \frac{1}{9}$. Add $\frac{1}{9}$ to both sides.

$$\left(x + \frac{1}{3}\right)^2 = -\frac{5}{9}$$ Factor the left side and combine terms: $\frac{1}{9} - \frac{2}{3} = \frac{1}{9} - \frac{6}{9} = -\frac{5}{9}$.

$$x + \frac{1}{3} = \pm\sqrt{-\frac{5}{9}}$$ Use the square root property.

$$x = -\frac{1}{3} \pm \sqrt{-\frac{5}{9}}$$ To isolate x, subtract $\frac{1}{3}$ from both sides.

Since

$$\sqrt{-\frac{5}{9}} = \sqrt{-1 \cdot \frac{5}{9}} = \sqrt{-1}\frac{\sqrt{5}}{\sqrt{9}} = \frac{\sqrt{5}}{3}i$$

we have

$$x = -\frac{1}{3} \pm \frac{\sqrt{5}}{3}i$$

The solutions are $-\frac{1}{3} + \frac{\sqrt{5}}{3}i$ and $-\frac{1}{3} - \frac{\sqrt{5}}{3}i$.

Notation

The solutions are written in complex number form $a + bi$. They could also be written as

$$\frac{-1 \pm i\sqrt{5}}{3}$$

Self Check 10   Solve: $x^2 + 4x + 6 = 0$

Now Try   **Problem 67**

❸   **Use Quadratic Equations to Solve Application Problems.**

EXAMPLE 11   *Graduation Announcements.*   To create the announcement shown, a graphic artist must follow two design requirements:

- A border of uniform width should surround the text.
- Equal areas should be devoted to the text and to the border.

To meet these requirements, how wide should the border be?

Analyze the Problem   The text occupies $4 \cdot 3 = 12$ in.^{2} of space. The border must also have an area of 12 in.^{2}.

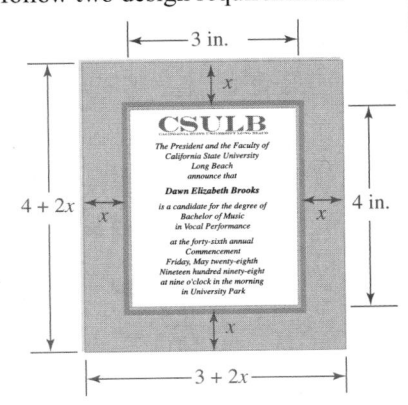

Form an Equation If we let $x =$ the width of the border in inches, the length of the announcement is $(4 + 2x)$ inches and the width is $(3 + 2x)$ inches. We can now form an equation.

The area of the announcement	minus	the area of the text	equals	the area of the border.
$(4 + 2x)(3 + 2x)$	$-$	12	$=$	12

Solve the Equation

$$(4 + 2x)(3 + 2x) - 12 = 12$$

$$12 + 8x + 6x + 4x^2 - 12 = 12 \quad \text{On the left side, use the FOIL method.}$$

$$4x^2 + 14x = 12 \quad \text{Combine like terms.}$$

$$2x^2 + 7x - 6 = 0 \quad \text{Subtract 12 from both sides. Then divide both sides of } 4x^2 + 14x - 12 = 0 \text{ by 2.}$$

Since the trinomial on the left side does not factor, we will solve the equation by completing the square.

$$x^2 + \frac{7}{2}x - 3 = 0 \qquad \text{Divide both sides by 2 so that the coefficient of } x^2 \text{ is 1.}$$

$$x^2 + \frac{7}{2}x = 3 \qquad \text{Add 3 to both sides.}$$

$$x^2 + \frac{7}{2}x + \frac{49}{16} = 3 + \frac{49}{16} \qquad \text{One-half of } \frac{7}{2} \text{ is } \frac{7}{4}. \text{ Square } \frac{7}{4}, \text{ which is } \frac{49}{16}, \text{ and add it to both sides.}$$

$$\left(x + \frac{7}{4}\right)^2 = \frac{97}{16} \qquad \text{On the left side, factor the trinomial. On the right side, } 3 = \frac{3 \cdot 16}{1 \cdot 16} = \frac{48}{16} \text{ and } \frac{48}{16} + \frac{49}{16} = \frac{97}{16}.$$

$$x + \frac{7}{4} = \pm\sqrt{\frac{97}{16}} \qquad \text{Use the square root property.}$$

$$x = -\frac{7}{4} \pm \frac{\sqrt{97}}{4} \qquad \text{Subtract } \frac{7}{4} \text{ from both sides and simplify: } \sqrt{\frac{97}{16}} = \frac{\sqrt{97}}{\sqrt{16}} = \frac{\sqrt{97}}{4}.$$

$$x = \frac{-7 + \sqrt{97}}{4} \quad \text{or} \quad x = \frac{-7 - \sqrt{97}}{4} \qquad \text{Write each expression as a single fraction.}$$

State the Conclusion The width of the border should be $\dfrac{-7 + \sqrt{97}}{4} \approx 0.71$ inch. $\left(\text{We discard the solution } \dfrac{-7 - \sqrt{97}}{4}, \text{ since it is negative.}\right)$

Check the Result If the border is 0.71 inch wide, the announcement has an area of about $5.42 \cdot 4.42 \approx 23.96$ in.^{2}. If we subtract the area of the text from the area of the announcement, we get $23.96 - 12 = 11.96$ in.^{2}. This represents the area of the border, which was to be 12 in.^{2}. The answer seems reasonable.

 Now Try **Problem 101**

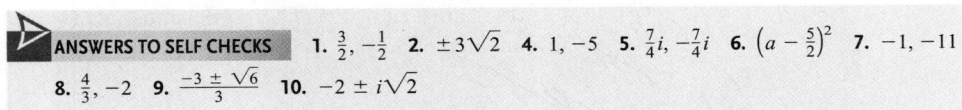

STUDY SET
10.1

VOCABULARY

Fill in the blanks.

1. An equation of the form $ax^2 + bx + c = 0$, where $a \neq 0$, is called a _____ equation.
2. $x^2 + 6x + 9$ is called a _____-square trinomial because it factors as $(x + 3)^2$.
3. When we add 16 to $x^2 + 8x$, we say that we have completed the _____ on $x^2 + 8x$.
4. The _____ coefficient of $5x^2 - 2x + 7$ is 5 and the _____ term is 7.

CONCEPTS

Fill in the blanks.

5. For any nonnegative number c, if $x^2 = c$, then $x =$ _____ or $x =$ _____.
6. To complete the square on $x^2 + 10x$, add the square of _____ of the coefficient of x.
7. Find one-half of the given number and square the result.
   a. 12   **b.** -5
8. Fill in the blanks to complete the square. Then factor the resulting perfect-square trinomial.
   a. $x^2 + 8x +$ ▢ $= (x +$ ▢ $)^2$
   b. $x^2 - 9x +$ ▢ $= \left(x -$ ▢ $\right)^2$
9. What is the first step to solve each equation by completing the square? **Do not solve.**
   a. $x^2 + 9x + 7 = 0$
   b. $4x^2 + 5x - 16 = 0$
10. Determine whether $-2 + \sqrt{2}$ is a solution of $x^2 + 4x + 2 = 0$.
11. Determine whether each statement is true or false.
    a. Any quadratic equation can be solved by the factoring method.
    b. Any quadratic equation can be solved by completing the square.
12. **a.** Write an expression that represents the width of the larger rectangle shown in red.
    b. Write an expression that represents the length of the larger rectangle shown in red.

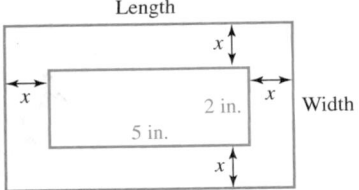

NOTATION

13. We read $8 \pm \sqrt{3}$ as "eight _____ ___ _____ the square root of 3."
14. When solving a quadratic equation, a student obtains $x = \dfrac{-5 \pm \sqrt{7}}{3}$.
    a. How many solutions are represented by this notation? List them.
    b. Approximate the solutions to the nearest hundredth.

GUIDED PRACTICE

Use factoring to solve each equation. **See Example 1.**

15. $6x^2 + 12x = 0$	**16.** $5x^2 + 11x = 0$
17. $y^2 - 25 = 0$	**18.** $y^2 - 16 = 0$
19. $r^2 + 6r + 8 = 0$	**20.** $x^2 + 9x + 20 = 0$
21. $2z^2 = -2 + 5z$	**22.** $3x^2 = 8 - 10x$

Use the square root property to solve each equation. **See Example 2.**

23. $t^2 - 81 = 0$	**24.** $w^2 - 49 = 0$
25. $x^2 = 36$	**26.** $x^2 = 144$
27. $z^2 - 50 = 0$	**28.** $u^2 - 24 = 0$
29. $3x^2 - 16 = 0$	**30.** $5x^2 - 49 = 0$

Use the square root property to solve each equation. **See Example 4.**

31. $(x + 5)^2 = 9$	**32.** $(x - 1)^2 = 4$
33. $(t + 4)^2 = 16$	**34.** $(s - 7)^2 = 9$
35. $(x + 5)^2 - 3 = 0$	**36.** $(x + 3)^2 - 7 = 0$
37. $(a - 2)^2 = 8$	**38.** $(c - 2)^2 = 12$

Use the square root property to solve each equation. **See Example 5.**

39. $p^2 = -16$	**40.** $q^2 = -25$
41. $4m^2 + 81 = 0$	**42.** $9n^2 + 121 = 0$
43. $(x - 3)^2 = -5$	**44.** $(x + 2)^2 = -3$
45. $(y - 1)^2 + 4 = 0$	**46.** $(t + 4)^2 + 49 = 0$

Complete the square and factor the resulting perfect-square trinomial. **See Example 6.**

47. $x^2 + 24x$
48. $y^2 - 18y$
49. $a^2 - 7a$
50. $b^2 + 11b$

Use completing the square to solve each equation. **See Example 7.**

51. $x^2 + 2x - 8 = 0$	**52.** $x^2 + 6x + 5 = 0$
53. $k^2 - 8k + 12 = 0$	**54.** $p^2 - 4p + 3 = 0$
55. $g^2 + 5g - 6 = 0$	**56.** $s^2 + 5s - 14 = 0$
57. $x^2 - 3x - 4 = 0$	**58.** $x^2 - 7x + 12 = 0$

Use completing the square to solve each equation. See Example 8.

59. $2x^2 - x - 1 = 0$

60. $2x^2 - 5x + 2 = 0$

61. $12t^2 - 5t - 3 = 0$

62. $5m^2 + 13m - 6 = 0$

Use completing the square to solve each equation. See Example 9.

63. $3x^2 - 12x + 1 = 0$

64. $6x^2 - 12x + 1 = 0$

65. $2x^2 + 5x - 2 = 0$

66. $2x^2 - 8x + 5 = 0$

Use completing the square to solve each equation. See Example 10.

67. $p^2 + 2p + 2 = 0$

68. $x^2 - 6x + 10 = 0$

69. $y^2 + 8y + 18 = 0$

70. $n^2 + 10n + 28 = 0$

TRY IT YOURSELF

Solve each equation.

71. $(3x - 1)^2 = 25$

72. $(5x - 2)^2 = 64$

73. $3x^2 - 6x = 1$

74. $2x^2 - 6x = -3$

75. $x^2 + 8x + 6 = 0$

76. $x^2 + 6x + 4 = 0$

77. $6x^2 + 72 = 0$

78. $5x^2 + 40 = 0$

79. $x^2 - 2x = 17$

80. $x^2 + 10x = 7$

81. $m^2 - 7m + 3 = 0$

82. $m^2 - 5m + 3 = 0$

83. $7h^2 = 35$

84. $9n^2 = 99$

85. $\dfrac{7x + 1}{5} = -x^2$

86. $\dfrac{3}{8}x^2 = \dfrac{1}{8} - x$

87. $t^2 + t + 3 = 0$

88. $b^2 - b + 5 = 0$

89. $(8x + 5)^2 = 24$

90. $(3y - 2)^2 = 18$

91. $r^2 - 6r - 27 = 0$

92. $s^2 - 6s - 40 = 0$

93. $4p^2 + 2p + 3 = 0$

94. $3m^2 - 2m + 3 = 0$

APPLICATIONS

95. MOVIE STUNTS　According to the *Guinness Book of World Records*, stuntman Dan Koko fell a distance of 312 feet into an airbag after jumping from the Vegas World Hotel and Casino. The distance d in feet traveled by a free-falling object in t seconds is given by the formula $d = 16t^2$. To the nearest tenth of a second, how long did the fall last?

96. ACCIDENTS　The height h (in feet) of an object that is dropped from a height of s feet is given by the formula $h = s - 16t^2$, where t is the time the object has been falling. A 5-foot-tall woman on a sidewalk looks directly overhead and sees a window washer drop a bottle from four stories up. How long does she have to get out of the way? Round to the nearest tenth. (A story is 12 feet.)

97. GEOGRAPHY　The surface area S of a sphere is given by the formula $S = 4\pi r^2$, where r is the radius of the sphere. An almanac lists the surface area of the Earth as 196,938,800 square miles. Assuming the Earth to be spherical, what is its radius to the nearest mile?

98. FLAGS　In 1912, an order by President Taft fixed the width and length of the U.S. flag in the ratio 1 to 1.9. If 100 square feet of cloth are to be used to make a U.S. flag, estimate its dimensions to the nearest $\frac{1}{4}$ foot.

99. AUTOMOBILE ENGINES　As the piston shown moves upward, it pushes a cylinder of a gasoline/air mixture that is ignited by the spark plug. The formula that gives the volume of a cylinder is $V = \pi r^2 h$, where r is the radius and h the height. Find the radius of the piston (to the nearest hundredth of an inch) if it displaces 47.75 cubic inches of gasoline/air mixture as it moves from its lowest to its highest point.

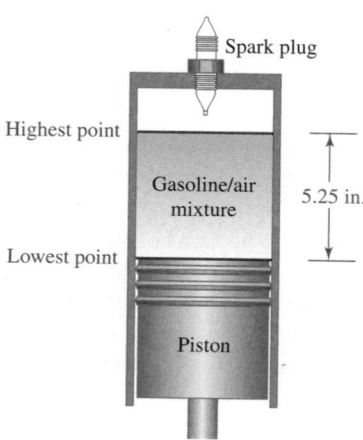

100. INVESTMENTS　If P dollars are deposited in an account that pays an annual rate of interest r, then in n years, the amount of money A in the account is given by the formula $A = P(1 + r)^n$. A savings account was opened on January 3, 2006, with a deposit of $10,000 and closed on January 2, 2008, with an ending balance of $11,772.25. Find the rate of interest.

101. PICTURE FRAMING The matting around the picture has a uniform width. How wide is the matting if its area equals the area of the picture? Round to the nearest hundredth of an inch.

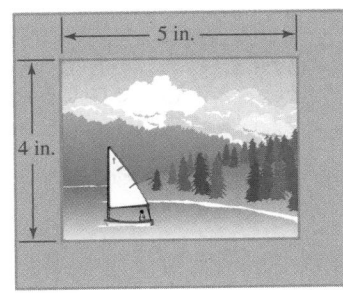

102. SWIMMING POOLS In the advertisement shown, how wide will the free concrete decking be if a uniform width is constructed around the perimeter of the pool? Round to the nearest hundredth of a yard. (*Hint:* Note the difference in units. Convert the dimensions of the pool to yards.)

SAHARA POOL & SPA
SUMMER SPECIAL
This 18 ft x 30 ft pool: only $28,500
Buy now and receive 28 square yards of concrete decking *FREE!*

103. DIMENSIONS OF A RECTANGLE A rectangle is 4 feet longer than it is wide, and its area is 20 square feet. Find its dimensions to the nearest tenth of a foot.

104. DIMENSIONS OF A TRIANGLE The height of a triangle is 4 meters longer than twice its base. Find the base and height if the area of the triangle is 10 square meters. Round to the nearest hundredth of a meter.

WRITING

105. Give an example of a perfect-square trinomial. Why do you think the word "perfect" is used to describe it?

106. Explain why completing the square on $x^2 + 5x$ is more difficult than completing the square on $x^2 + 4x$.

107. Explain the error in the work shown below.

a. $\dfrac{4 \pm \sqrt{3}}{8} = \dfrac{\overset{1}{\cancel{4}} \pm \sqrt{3}}{\underset{1}{\cancel{4}} \cdot 2} = \dfrac{1 \pm \sqrt{3}}{2}$

b. $\dfrac{1 \pm \sqrt{5}}{5} = \dfrac{1 \pm \overset{1}{\cancel{5}}}{\underset{1}{\cancel{5}}} = \dfrac{1 \pm 1}{1}$

108. Explain the steps involved in expressing $8 \pm \dfrac{\sqrt{15}}{2}$ as a single fraction with denominator 2.

REVIEW

Simplify each expression. All variables represent positive real numbers.

109. $\sqrt[3]{40a^3b^6}$ **110.** $\sqrt[8]{x^{24}}$

111. $\sqrt[4]{\dfrac{16}{625}}$ **112.** $\sqrt{175a^2b^3}$

CHALLENGE PROBLEMS

113. What number must be added to $x^2 + \sqrt{3}x$ to make a perfect-square trinomial?

114. Solve $x^2 + \sqrt{3}x - \dfrac{1}{4} = 0$ by completing the square.

Solve for the specified variable. Assume that all variables represent positive numbers. Express all radicals in simplified form.

115. $E = mc^2$ for c **116.** $A = \pi r^2$ for r

SECTION 10.2
The Quadratic Formula

Objectives

1. Derive the quadratic formula.
2. Solve quadratic equations using the quadratic formula.
3. Write equivalent equations to make quadratic formula computations easier.
4. Use the quadratic formula to solve application problems.

We can solve quadratic equations by completing the square, but the work is often tedious. In this section, we will develop a formula, called the *quadratic formula,* that enables us solve quadratic equations with less effort.

Derive the Quadratic Formula.

To develop a formula that will produce the solutions of a quadratic equation, we start with the **general quadratic equation** $ax^2 + bx + c = 0$ with $a > 0$, and solve it for x by completing the square.

$$ax^2 + bx + c = 0$$

$$\frac{ax^2}{a} + \frac{bx}{a} + \frac{c}{a} = \frac{0}{a} \qquad \text{\textit{Divide both sides by } a \textit{ so that the coefficient of } x^2 \textit{ is 1.}}$$

$$x^2 + \frac{b}{a}x + \frac{c}{a} = 0 \qquad \text{\textit{Simplify: } } \frac{ax^2}{a} = x^2. \text{ \textit{Write } } \frac{bx}{a} \text{ \textit{ as } } \frac{b}{a}x.$$

$$x^2 + \frac{b}{a}x \qquad\quad = -\frac{c}{a} \qquad \text{\textit{Subtract } } \frac{c}{a} \text{ \textit{ from both sides so that only the terms involving } x \textit{ are on}}$$
$$\text{\textit{the left side of the equation.}}$$

We can complete the square on $x^2 + \frac{b}{a}x$ by adding the square of one-half of the coefficient of x. Since the coefficient of x is $\frac{b}{a}$, we have $\frac{1}{2} \cdot \frac{b}{a} = \frac{b}{2a}$ and $\left(\frac{b}{2a}\right)^2 = \frac{b^2}{4a^2}$.

$$x^2 + \frac{b}{a}x + \frac{b^2}{4a^2} = -\frac{c}{a} + \frac{b^2}{4a^2} \qquad \text{\textit{To complete the square, add } } \frac{b^2}{4a^2} \text{ \textit{ to both sides.}}$$

$$x^2 + \frac{b}{a}x + \frac{b^2}{4a^2} = -\frac{4ac}{4aa} + \frac{b^2}{4a^2} \qquad \text{\textit{Multiply } } -\frac{c}{a} \text{ \textit{ by } } \frac{4a}{4a}. \text{ \textit{Now the fractions on the}}$$
$$\text{\textit{right side have the common denominator } } 4a^2.$$

$$\left(x + \frac{b}{2a}\right)^2 = \frac{b^2 - 4ac}{4a^2} \qquad \text{\textit{On the left side, factor. On the right side, add the}}$$
$$\text{\textit{fractions.}}$$

$$x + \frac{b}{2a} = \pm\sqrt{\frac{b^2 - 4ac}{4a^2}} \qquad \text{\textit{Use the square root property.}}$$

$$x + \frac{b}{2a} = \pm\frac{\sqrt{b^2 - 4ac}}{\sqrt{4a^2}} \qquad \text{\textit{The square root of a quotient is the quotient of}}$$
$$\text{\textit{square roots.}}$$

$$x + \frac{b}{2a} = \pm\frac{\sqrt{b^2 - 4ac}}{2a} \qquad \text{\textit{Since } } a > 0, \sqrt{4a^2} = 2a.$$

$$x = -\frac{b}{2a} \pm \frac{\sqrt{b^2 - 4ac}}{2a} \qquad \text{\textit{To isolate } x, \textit{ subtract } } \frac{b}{2a} \text{ \textit{ from both sides.}}$$

$$x = \frac{-b \pm \sqrt{b^2 - 4ac}}{2a} \qquad \text{\textit{Combine the fractions.}}$$

This result is called the *quadratic formula*. To develop this formula, we assumed that a was positive. If a is negative, similar steps are used, and we obtain the same result.

The Language of Algebra
To *derive* means to obtain by reasoning. To *derive* the quadratic formula means to solve $ax^2 + bx + c = 0$ for x, using the series of steps shown here, to obtain

$$x = \frac{-b \pm \sqrt{b^2 - 4ac}}{2a}$$

Quadratic Formula

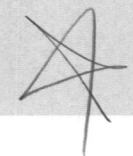

The solutions of $ax^2 + bx + c = 0$, with $a \neq 0$, are

$$x = \frac{-b \pm \sqrt{b^2 - 4ac}}{2a}$$

2   **Solve Quadratic Equations Using the Quadratic Formula.**

In the next example, we will use the quadratic formula to solve a quadratic equation.

EXAMPLE 1   Solve $2x^2 - 5x - 3 = 0$ by using the quadratic formula.

Strategy   We will begin by comparing $2x^2 - 5x - 3 = 0$ to the standard form $ax^2 + bx + c = 0$.

Why   To use the quadratic formula, we need to identify the values of a, b, and c.

Solution

$$2x^2 - 5x - 3 = 0 \quad \text{This is the equation to solve.}$$
$$\uparrow \qquad \uparrow \qquad \uparrow$$
$$ax^2 + bx + c = 0$$

We see that $a = 2$, $b = -5$, and $c = -3$. To find the solutions of the equation, we substitute these values into the quadratic formula and evaluate the right side.

$$x = \frac{-b \pm \sqrt{b^2 - 4ac}}{2a} \qquad \text{This is the quadratic formula.}$$

$$x = \frac{-(-5) \pm \sqrt{(-5)^2 - 4(2)(-3)}}{2(2)} \qquad \text{Substitute 2 for } a, \ -5 \text{ for } b, \text{ and } -3 \text{ for } c.$$

$$x = \frac{5 \pm \sqrt{25 - (-24)}}{4} \qquad \text{Simplify: } -(-5) = 5. \text{ Evaluate the power and multiply within the radical. Multiply in the denominator.}$$

$$x = \frac{5 \pm \sqrt{49}}{4} \qquad \text{Simplify within the radical.}$$

$$x = \frac{5 \pm 7}{4} \qquad \text{Simplify: } \sqrt{49} = 7.$$

To find the first solution, we evaluate the expression using the $+$ symbol. To find the second solution, we evaluate the expression using the $-$ symbol.

$$x = \frac{5 + 7}{4} \qquad \text{or} \qquad x = \frac{5 - 7}{4}$$

$$x = \frac{12}{4} \qquad\qquad\qquad x = \frac{-2}{4}$$

$$x = 3 \qquad\qquad\qquad\quad x = -\frac{1}{2}$$

The solutions are 3 and $-\frac{1}{2}$ and the solution set is $\left\{3, -\frac{1}{2}\right\}$. Check each solution in the original equation.

Self Check 1   Solve $4x^2 - 7x - 2 = 0$ by using the quadratic formula.

Now Try   **Problem 13**

To solve a quadratic equation in x using the quadratic formula, we follow these steps.

Solving a Quadratic Equation in x Using the Quadratic Formula	1. Write the equation in standard form: $ax^2 + bx + c = 0$.
	2. Identify a, b, and c.
	3. Substitute the values for a, b, and c in the quadratic formula
	$$x = \frac{-b \pm \sqrt{b^2 - 4ac}}{2a}$$
	and evaluate the right side to obtain the solutions.

EXAMPLE 2 Solve: $2x^2 = -4x - 1$

Strategy We will write the equation in standard form $ax^2 + bx + c = 0$. Then we will identify the values of a, b, and c, and substitute these values into the quadratic formula.

Why The quadratic equation must be in standard form to identify the values of a, b, and c.

Solution To write the equation in standard form, we need to have all nonzero terms on the left side and 0 on the right side.

$$2x^2 = -4x - 1 \qquad \text{This is the equation to solve.}$$
$$2x^2 + 4x + 1 = 0 \qquad \text{To get 0 on the right side, add 4x and 1 to both sides.}$$

In this equation, $a = 2$, $b = 4$, and $c = 1$.

$$x = \frac{-b \pm \sqrt{b^2 - 4ac}}{2a} \qquad \text{This is the quadratic formula.}$$

$$x = \frac{-4 \pm \sqrt{4^2 - 4(2)(1)}}{2(2)} \qquad \text{Substitute 2 for } a,\ 4 \text{ for } b, \text{ and 1 for } c.$$

$$x = \frac{-4 \pm \sqrt{16 - 8}}{4} \qquad \text{Evaluate the expression within the radical. Multiply in the denominator.}$$

$$x = \frac{-4 \pm \sqrt{8}}{4}$$

$$x = \frac{-4 \pm 2\sqrt{2}}{4} \qquad \text{Simplify: } \sqrt{8} = \sqrt{4 \cdot 2} = 2\sqrt{2}.$$

Success Tip
Perhaps you noticed that Example 1 could be solved by factoring. However, that is not the case for Example 2. These observations illustrate that the quadratic formula can be used to solve any quadratic equation.

We can write the solutions in simpler form by factoring out 2 from the terms in the numerator and removing the common factor of 2 in the numerator and denominator.

Notation
The solutions can also be written as
$$-\frac{2}{2} \pm \frac{\sqrt{2}}{2} = -1 \pm \frac{\sqrt{2}}{2}$$

$$x = \frac{-4 \pm 2\sqrt{2}}{4} = \frac{2(-2 \pm \sqrt{2})}{4} = \frac{\overset{1}{\cancel{2}}(-2 \pm \sqrt{2})}{\underset{1}{\cancel{2} \cdot 2}} = \frac{-2 \pm \sqrt{2}}{2}$$

The solutions are $\frac{-2 + \sqrt{2}}{2}$ and $\frac{-2 - \sqrt{2}}{2}$ and the solution set is $\left\{ \frac{-2 + \sqrt{2}}{2}, \frac{-2 - \sqrt{2}}{2} \right\}$. We can approximate the solutions using a calculator. To two decimal places, they are -0.29 and -1.71.

▷ **Self Check 2** Solve $3x^2 = 2x + 3$. Approximate the solutions to two decimal places.

Now Try Problem 25

The solutions to the next example are imaginary numbers.

EXAMPLE 3   Solve: $x^2 + x = -1$

Strategy   We will write the equation in standard form $ax^2 + bx + c = 0$. Then we will identify the values of a, b, and c, and substitute these values into the quadratic formula.

Why   The quadratic equation must be in standard form to identify the values of a, b, and c.

Solution   To write $x^2 + x = -1$ in standard form, we add 1 to both sides, to get

$$x^2 + x + 1 = 0 \quad \text{This is the equation to solve.}$$

In this equation, $a = 1$, $b = 1$, and $c = 1$:

$$x = \frac{-b \pm \sqrt{b^2 - 4ac}}{2a}$$

$$x = \frac{-1 \pm \sqrt{1^2 - 4(1)(1)}}{2(1)} \quad \text{Substitute 1 for } a, \text{ 1 for } b, \text{ and 1 for } c.$$

$$x = \frac{-1 \pm \sqrt{1 - 4}}{2} \quad \text{Evaluate the expression within the radical.}$$

$$x = \frac{-1 \pm \sqrt{-3}}{2}$$

$$x = \frac{-1 \pm i\sqrt{3}}{2} \qquad \sqrt{-3} = \sqrt{-1 \cdot 3} = \sqrt{-1}\sqrt{3} = i\sqrt{3}.$$

> **Notation**
> The solutions are written in complex number form $a + bi$. They could also be written as
> $$\frac{-1 \pm i\sqrt{3}}{2}$$

The   solutions   are $-\frac{1}{2} + \frac{\sqrt{3}}{2}i$ and $-\frac{1}{2} - \frac{\sqrt{3}}{2}i$ and   the   solution   set   is $\left\{ -\frac{1}{2} + \frac{\sqrt{3}}{2}i, -\frac{1}{2} - \frac{\sqrt{3}}{2}i \right\}$

 Self Check 3   Solve: $a^2 + 3a + 5 = 0$
Now Try   **Problem 29**

---

3   **Write Equivalent Equations to Make Quadratic Formula Computations Easier.**

When solving a quadratic equation by the quadratic formula, we can often simplify the computations by solving a simpler, but equivalent equation.

EXAMPLE 4   For each equation below, write an equivalent equation so that the quadratic formula computations will be simpler.

a. $-2x^2 + 4x - 1 = 0$ **b.** $x^2 + \frac{4}{5}x - \frac{1}{3} = 0$

c. $20x^2 - 60x - 40 = 0$ **d.** $0.03x^2 - 0.04x - 0.01 = 0$

Strategy   We will multiply both sides of each equation by a carefully chosen number.

Why   In each case, the objective is to find an equivalent equation whose values of a, b, and c are easier to work with than those of the given equation.

Solution

a. It is often easier to solve a quadratic equation using the quadratic formula if a is positive. If we multiply (or divide) both sides of $-2x^2 + 4x - 1 = 0$ by -1, we obtain an equivalent equation with $a > 0$.

$$-2x^2 + 4x - 1 = 0 \qquad \text{Here, } a = -2.$$
$$-1(-2x^2 + 4x - 1) = -1(0)$$
$$2x^2 - 4x + 1 = 0 \qquad \text{Now } a = 2.$$

> **Success Tip**
>
> Unlike completing the square, the quadratic formula does not require the leading coefficient to be 1.

b. For $x^2 + \frac{4}{5}x - \frac{1}{3} = 0$, two coefficients are fractions: $b = \frac{4}{5}$ and $c = -\frac{1}{3}$. We can multiply both sides of the equation by their least common denominator, 15, to obtain an equivalent equation having coefficients that are integers.

$$x^2 + \frac{4}{5}x - \frac{1}{3} = 0 \qquad \text{Here, } a = 1, b = \frac{4}{5}, \text{ and } c = -\frac{1}{3}.$$
$$15\left(x^2 + \frac{4}{5}x - \frac{1}{3}\right) = 15(0)$$
$$15x^2 + 12x - 5 = 0 \qquad \text{Now } a = 15, b = 12, \text{ and } c = -5.$$

c. For $20x^2 - 60x - 40 = 0$, the coefficients 20, -60, and -40 have a common factor of 20. If we divide both sides of the equation by their GCF, we obtain an equivalent equation having smaller coefficients.

$$20x^2 - 60x - 40 = 0 \qquad \text{Here, } a = 20, b = -60, \text{ and } c = -40.$$
$$\frac{20x^2}{20} - \frac{60x}{20} - \frac{40}{20} = \frac{0}{20}$$
$$x^2 - 3x - 2 = 0 \qquad \text{Now } a = 1, b = -3, \text{ and } c = -2.$$

d. For $0.03x^2 - 0.04x - 0.01 = 0$, all three coefficients are decimals. We can multiply both sides of the equation by 100 to obtain an equivalent equation having coefficients that are integers.

$$0.03x^2 - 0.04x - 0.01 = 0 \qquad \text{Here, } a = 0.03, b = -0.04, \text{ and } c = -0.01.$$
$$100(0.03x^2 - 0.04x - 0.01) = 100(0)$$
$$3x^2 - 4x - 1 = 0 \qquad \text{Now } a = 3, b = -4, \text{ and } c = -1.$$

 Self Check 4   For each equation, write an equivalent equation so that the quadratic formula computations will be simpler.

a. $-6x^2 + 7x - 9 = 0$

b. $\frac{1}{3}x^2 - \frac{2}{3}x - \frac{5}{6} = 0$

c. $44x^2 + 66x - 99 = 0$

d. $0.08x^2 - 0.07x - 0.02 = 0$

Now Try **Problems 37 and 39**

---

4   ### Use the Quadratic Formula to Solve Application Problems.

A variety of real-world applications can be modeled by quadratic equations. However, such equations are often difficult or even impossible to solve using the factoring method. In those cases, we can use the quadratic formula to solve the equation.

EXAMPLE 5   *Shortcuts.*   Instead of using the hallways, students are wearing a path through a planted quad area to walk 195 feet directly from the classrooms to the cafeteria. If the length of the hallway from the office to the cafeteria is 105 feet longer than the hallway from the office to the classrooms, how much walking are the students saving by taking the shortcut?

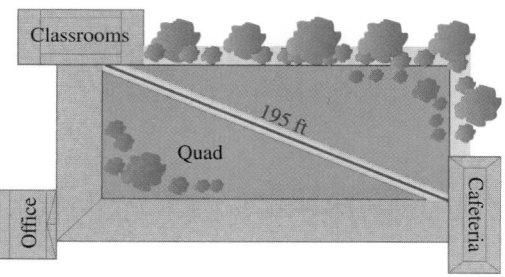

Analyze the Problem   The two hallways and the shortcut form a right triangle with a hypotenuse 195 feet long. We will use the Pythagorean theorem to solve this problem.

Form an Equation   If we let x = the length (in feet) of the hallway from the classrooms to the office, then the length of the hallway from the office to the cafeteria is $(x + 105)$ feet. Substituting these lengths into the Pythagorean theorem, we have

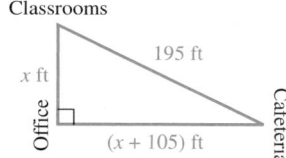

$$a^2 + b^2 = c^2$$ This is the Pythagorean theorem.

$$x^2 + (x + 105)^2 = 195^2$$ Substitute x for a, (x + 105) for b, and 195 for c.

$$x^2 + x^2 + 105x + 105x + 11{,}025 = 38{,}025$$ Find $(x + 105)^2$.

$$2x^2 + 210x + 11{,}025 = 38{,}025$$ Combine like terms.

$$2x^2 + 210x - 27{,}000 = 0$$ To get 0 on the right side, subtract 38,025 from both sides.

$$x^2 + 105x - 13{,}500 = 0$$ The coefficients have a common factor of 2: divide both sides by 2.

Solve the Equation   To solve $x^2 + 105x - 13{,}500 = 0$, we will use the quadratic formula with $a = 1$, $b = 105$, and $c = -13{,}500$.

$$x = \frac{-b \pm \sqrt{b^2 - 4ac}}{2a}$$

$$x = \frac{-105 \pm \sqrt{105^2 - 4(1)(-13{,}500)}}{2(1)}$$

$$x = \frac{-105 \pm \sqrt{65{,}025}}{2}$$ Simplify: $105^2 - 4(1)(-13{,}500) = 11{,}025 + 54{,}000 = 65{,}025$.

$$x = \frac{-105 \pm 255}{2}$$ Use a calculator: $\sqrt{65{,}025} = 255$.

$$x = \frac{150}{2} \quad \text{or} \quad x = \frac{-360}{2}$$

$$x = 75 \qquad\qquad x = -180$$ Since the length of the hallway can't be negative, discard the solution -180.

State the Conclusion   The length of the hallway from the classrooms to the office is 75 feet. The length of the hallway from the office to the cafeteria is $75 + 105 = 180$ feet. Instead of using the hallways, a distance of $75 + 180 = 255$ feet, the students are taking the 195-foot shortcut to the cafeteria, a savings of $(255 - 195)$, or 60 feet.

Check the Result   The length of the 180-foot hallway is 105 feet longer than the length of the 75-foot hallway. The sum of the squares of the lengths of the hallways is $75^2 + 180^2 = 38{,}025$. This equals the square of the length of the 195-foot shortcut: $195^2 = 38{,}025$. The result checks.

 Now Try **Problem 73**

EXAMPLE 6   *Mass Transit.*   A bus company has 4,000 passengers daily, each currently paying a 75¢ fare. For each 15¢ fare increase, the company estimates that it will lose 50 passengers. If the company needs to bring in $6,570 per day to stay in business, what fare must be charged to produce this amount of revenue?

Analyze the Problem   To understand how a fare increase affects the number of passengers, let's consider what happens if there are two fare increases. We organize the data in a table. The fares are expressed in terms of dollars.

Number of increases	New fare	Number of passengers
One $0.15 increase	$0.75 + $0.15(1) = $0.90	$4{,}000 - 50(1) = 3{,}950$
Two $0.15 increases	$0.75 + $0.15(2) = $1.05	$4{,}000 - 50(2) = 3{,}900$

In general, the new fare will be the old fare ($0.75) plus the number of fare increases times $0.15. The number of passengers who will pay the new fare is 4,000 minus 50 times the number of $0.15 fare increases.

Form an Equation   If we let $x =$ the number of $0.15 fare increases necessary to bring in $6,570 daily, then $(0.75 + 0.15x)$ is the fare that must be charged. The number of passengers who will pay this fare is $4{,}000 - 50x$. We can now form an equation.

The bus fare	times	the number of passengers who will pay that fare	equals	$6,570.
$(0.75 + 0.15x)$	$\cdot$	$(4{,}000 - 50x)$	$=$	$6{,}570$

Solve the Equation

$$(0.75 + 0.15x)(4{,}000 - 50x) = 6{,}570$$

$$3{,}000 - 37.5x + 600x - 7.5x^2 = 6{,}570 \qquad \text{Multiply the binomials.}$$

$$-7.5x^2 + 562.5x + 3{,}000 = 6{,}570 \qquad \text{Combine like terms: } -37.5x + 600x = 562.5x.$$

$$-7.5x^2 + 562.5x - 3{,}570 = 0 \qquad \text{To get 0 on the right side, subtract 6,570 from both sides.}$$

$$7.5x^2 - 562.5x + 3{,}570 = 0 \qquad \text{Multiply both sides by } -1 \text{ so that the value of } a, 7.5, \text{ is positive.}$$

To solve this equation, we will use the quadratic formula.

$$x = \frac{-b \pm \sqrt{b^2 - 4ac}}{2a}$$

$$x = \frac{-(-562.5) \pm \sqrt{(-562.5)^2 - 4(7.5)(3,570)}}{2(7.5)}$$ Substitute 7.5 for a, -562.5 for b, and 3,570 for c.

$$x = \frac{562.5 \pm \sqrt{209,306.25}}{15}$$ Simplify: $(-562.5)^2 - 4(7.5)(3,570) =$ 316,406.25 $-$ 107.100 $=$ 209,306.25.

$$x = \frac{562.5 \pm 457.5}{15}$$ Use a calculator: $\sqrt{209,306.25} = 457.5$.

$$x = \frac{1,020}{15} \quad \text{or} \quad x = \frac{105}{15}$$

$$x = 68 \qquad \qquad x = 7$$

State the Conclusion   If there are 7 fifteen-cent increases in the fare, the new fare will be $0.75 + $0.15(7) = $1.80. If there are 68 fifteen-cent increases in the fare, the new fare will be $0.75 + $0.15(68) = $10.95. Although this fare would bring in the necessary revenue, a $10.95 bus fare is unreasonable, so we discard it.

Check the Result   A fare of $1.80 will be paid by [4,000 − 50(7)] = 3,650 bus riders. The amount of revenue brought in would be $1.80(3,650) = $6,570. The result checks.

 Now Try   **Problem 83**

EXAMPLE 7   *Lawyers.*   The number of lawyers in the United States is approximated by the function $N(x) = 45x^2 + 21,000x + 560,000$, where $N(x)$ is the number of lawyers and x is the number of years after 1980. In what year does this model indicate that the United States had one million lawyers? (Based on data from the American Bar Association)

Strategy   We will substitute 1,000,000 for $N(x)$ in the equation and solve for x.

Why   The value of x will give the number of years after 1980 that the United States had 1,000,000 lawyers.

Solution
$$N(x) = 45x^2 + 21,000x + 560,000$$
$$1,000,000 = 45x^2 + 21,000x + 560,000$$ Replace $N(x)$ with 1,000,000.
$$0 = 45x^2 + 21,000x - 440,000$$ To get 0 on the left side, subtract 1,000,000 from both sides.

We can simplify the computations by dividing both sides of the equation by 5, which is the greatest common factor of 45, 21,000, and 440,000.

$$9x^2 + 4,200x - 88,000 = 0$$ Divide both sides by 5.

We solve this equation using the quadratic formula.

$$x = \frac{-b \pm \sqrt{b^2 - 4ac}}{2a}$$

$$x = \frac{-4,200 \pm \sqrt{(4,200)^2 - 4(9)(-88,000)}}{2(9)}$$ Substitute 9 for a, 4,200 for b, and $-88,000$ for c.

$$x = \frac{-4,200 \pm \sqrt{20,808,000}}{18}$$ Evaluate the expression within the radical.

$$x \approx \frac{362}{18} \quad \text{or} \quad x \approx \frac{-8{,}762}{18} \quad \text{Use a calculator.}$$

$$x \approx 20.1 \quad \bcancel{x \approx -486.8} \quad \text{Since the model is defined only for positive values of } x, \text{ we discard the second solution.}$$

In 20.1 years after 1980, or in early 2000, the model predicts that United States had approximately 1,000,000 lawyers.

 Now Try Problem 87

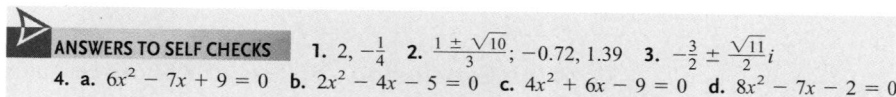 **ANSWERS TO SELF CHECKS** **1.** $2, -\frac{1}{4}$ **2.** $\frac{1 \pm \sqrt{10}}{3}$; $-0.72, 1.39$ **3.** $-\frac{3}{2} \pm \frac{\sqrt{11}}{2}i$
4. a. $6x^2 - 7x + 9 = 0$ **b.** $2x^2 - 4x - 5 = 0$ **c.** $4x^2 + 6x - 9 = 0$ **d.** $8x^2 - 7x - 2 = 0$

STUDY SET
10.2

VOCABULARY

Fill in the blanks.

1. A _____ equation in one variable is any equation that can be written in the standard form $ax^2 + bx + c = 0$.

2. $x = \frac{-b \pm \sqrt{b^2 - 4ac}}{2a}$ is called the _____ formula.

CONCEPTS

3. Write each equation in standard form.
 a. $x^2 + 2x = -5$ **b.** $3x^2 = -2x + 1$

4. For each quadratic equation, find the values of a, b, and c.
 a. $x^2 + 5x + 6 = 0$ **b.** $8x^2 - x = 10$

5. Determine whether each statement is true or false.
 a. Any quadratic equation can be solved by using the quadratic formula.
 b. Any quadratic equation can be solved by completing the square.

6. What is wrong with the beginning of the solution shown below?

 Solve: $x^2 - 3x = 2$

 $a = 1 \quad b = -3 \quad c = 2$

Evaluate each expression.

7. $\frac{-2 \pm \sqrt{2^2 - 4(1)(-8)}}{2(1)}$

8. $\frac{-(-1) \pm \sqrt{(-1)^2 - 4(2)(-4)}}{2(2)}$

9. A student used the quadratic formula to solve a quadratic equation and obtained $x = \frac{-2 \pm \sqrt{3}}{2}$.
 a. How many solutions does the equation have? What are they exactly?
 b. Graph the solutions on a number line.

10. Simplify each of the following.
 a. $\frac{3 \pm 6\sqrt{2}}{3}$ **b.** $\frac{-12 \pm 4\sqrt{7}}{8}$

NOTATION

11. On a quiz, students were asked to write the quadratic formula. What is wrong with each answer shown below?
 a. $x = -b \pm \frac{\sqrt{b^2 - 4ac}}{2a}$
 b. $x = \frac{-b\sqrt{b^2 - 4ac}}{2a}$

12. When reading $\frac{-b \pm \sqrt{b^2 - 4ac}}{2a}$, we say, "The _____ of b, plus or _____ the square _____ of b _____ minus times a times c, all _____ $2a$."

GUIDED PRACTICE

Use the quadratic formula to solve each equation. **See Example 1.**

13. $x^2 - 3x + 2 = 0$ **14.** $x^2 + 3x + 2 = 0$

15. $x^2 + 12x = -36$

16. $x^2 - 18x + 81 = 0$

17. $2x^2 + x - 3 = 0$

18. $6x^2 - x - 1 = 0$

19. $12t^2 - 5t - 2 = 0$

20. $12z^2 + 5z - 3 = 0$

Solve each equation. Approximate the solutions to two decimal places. See Example 2.

21. $x^2 = x + 7$

22. $t^2 = t + 4$

23. $5x^2 + 5x = -1$

24. $2x^2 + 7x = -1$

25. $3y^2 + 1 = -6y$

26. $4w^2 + 1 = -6w$

27. $4m^2 = 4m + 19$

28. $3y^2 = 12y - 4$

Solve each equation. See Example 3.

29. $2x^2 + x + 1 = 0$

30. $2x^2 + 3x + 5 = 0$

31. $3x^2 - 2x + 1 = 0$

32. $3x^2 - 2x + 5 = 0$

33. $x^2 - 2x + 2 = 0$

34. $x^2 - 4x + 8 = 0$

35. $4a^2 + 4a + 5 = 0$

36. $4b^2 + 4b + 17 = 0$

For each equation, write an equivalent quadratic equation that will be easier to solve. Do not solve the equation. See Example 4.

37. a. $-5x^2 + 9x - 2 = 0$

   b. $1.6t^2 + 2.4t - 0.9 = 0$

38. a. $\frac{1}{8}x^2 + \frac{1}{2}x - \frac{3}{4} = 0$

   b. $33y^2 + 99y - 66 = 0$

39. a. $45x^2 + 30x - 15 = 0$

   b. $\frac{1}{3}m^2 - \frac{1}{2}m - \frac{1}{3} = 0$

40. a. $0.6t^2 - 0.1t - 0.2 = 0$

   b. $-a^2 - 15a + 12 = 0$

TRY IT YOURSELF

Solve each equation.

41. $x^2 - \frac{14}{15}x = \frac{8}{15}$

42. $x^2 = -\frac{5}{4}x + \frac{3}{2}$

43. $3x^2 - 4x = -2$

44. $2x^2 + 3x = -3$

45. $-16y^2 - 8y + 3 = 0$

46. $-16x^2 - 16x - 3 = 0$

47. $2x^2 - 3x - 1 = 0$

48. $3x^2 - 9x - 2 = 0$

49. $-x^2 + 10x = 18$

50. $-x^2 - 6x - 2 = 0$

51. $x^2 - 6x = 391$

52. $x^2 - 27x - 280 = 0$

53. $x^2 + 5x - 5 = 0$

54. $x^2 - 3x - 27 = 0$

55. $9h^2 - 6h + 7 = 0$

56. $5x^2 = 2x - 1$

57. $50x^2 + 30x - 10 = 0$

58. $120b^2 + 120b - 40 = 0$

59. $0.6x^2 + 0.03 - 0.4x = 0$

60. $2x^2 + 0.1x = 0.04$

61. $\frac{1}{8}x^2 - \frac{1}{2}x + 1 = 0$

62. $\frac{1}{2}x^2 + 3x + \frac{13}{2} = 0$

63. $\frac{a^2}{10} - \frac{3a}{5} + \frac{7}{5} = 0$

64. $\frac{c^2}{4} + c + \frac{11}{4} = 0$

65. $\frac{x^2}{2} + \frac{5}{2}x = -1$

66. $\frac{x^2}{8} - \frac{x}{4} = \frac{1}{2}$

67. $900x^2 - 8{,}100x = 1{,}800$

68. $14x^2 - 21x = 49$

Solve each equation. Approximate the solutions to two decimal places.

69. $\frac{1}{4}x^2 - \frac{1}{6}x - \frac{1}{6} = 0$

70. $81x^2 + 12x - 80 = 0$

71. $0.7x^2 - 3.5x - 25 = 0$

72. $-4.5x^2 + 0.2x + 3.75 = 0$

APPLICATIONS

73. CROSSWALKS   Refer to the illustration on the next page. Instead of using the Main Street and First Avenue crosswalks to get from Nordstroms to Best Buy, a shopper uses the diagonal crosswalk to walk 97 feet directly from one corner to the other. If the length of the Main Street crosswalk is 7 feet longer than the First Avenue crosswalk, how much walking does the shopper save by using the diagonal crosswalk?

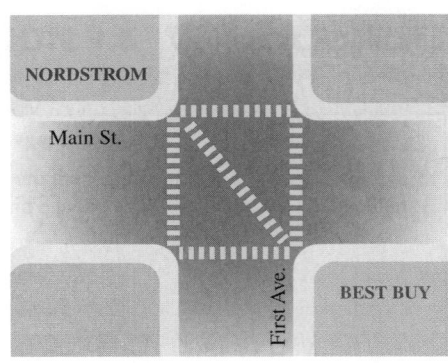

74. **BADMINTON** The person who wrote the instructions for setting up the badminton net shown below forgot to give the specific dimensions for securing the pole. How long is the support string?

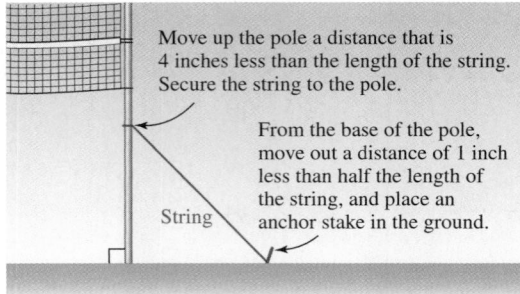

Move up the pole a distance that is 4 inches less than the length of the string. Secure the string to the pole.

From the base of the pole, move out a distance of 1 inch less than half the length of the string, and place an anchor stake in the ground.

String

75. **RIGHT TRIANGLES** The hypotenuse of a right triangle is 2.5 units long. The longer leg is 1.7 units longer than the shorter leg. Find the lengths of the sides of the triangle.

76. **TELEVISIONS** The screen size of a television is measured diagonally from one corner to the opposite corner. In 2007, Sharp developed the world's largest TV screen to date—a 108-inch flat-panel liquid crystal display (LCD). Find the width and the height of the rectangular screen if the width were 41 inches greater than the height. Round to the nearest inch.

77. **IMAX SCREENS** The largest permanent movie screen is in the Panasonic Imax theater at Darling Harbor, Sydney, Australia. The rectangular screen has an area of 11,349 square feet. Find the dimensions of the screen if it is 20 feet longer than it is wide.

78. **WORLD'S LARGEST LED SCREEN** A huge suspended LED screen is the centerpiece of The Place, a popular mall in Beijing, China. Find the length and width of the rectangular screen if the length is 10 meters more than 8 times its width, and the viewable area is 7,500 square meters.

© Directphoto.org/Alamy

79. **PARKS** Central Park is one of New York's best-known landmarks. Rectangular in shape, its length is 5 times its width. When measured in miles, its perimeter numerically exceeds its area by 4.75. Find the dimensions of Central Park if we know that its width is less than 1 mile.

80. **HISTORY** One of the important cities of the ancient world was Babylon. Greek historians wrote that the city was square-shaped. Measured in miles, its area numerically exceeded its perimeter by about 124. Find its dimensions. (Round to the nearest tenth.)

81. **POLYGONS** A five-sided polygon, called a *pentagon,* has 5 diagonals. The number of diagonals d of a polygon of n sides is given by the formula

$$d = \frac{n(n-3)}{2}$$

Find the number of sides of a polygon if it has 275 diagonals.

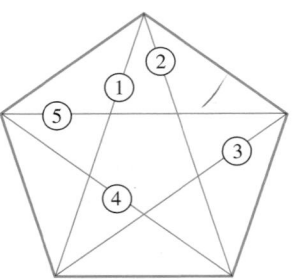

82. **METAL FABRICATION** A box with no top is to be made by cutting a 2-inch square from each corner of a square sheet of metal. After bending up the sides, the volume of the box is to be 220 cubic inches. Find the length of a side of the square sheet of metal that should be used in the construction of the box. Round to the nearest hundredth.

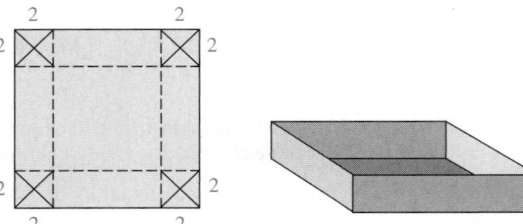

83. **DANCES** Tickets to a school dance cost $4 and the projected attendance is 300 people. It is further projected that for every 10¢ increase in ticket price, the average attendance will decrease by 5. At what ticket price will the receipts from the dance be $1,248?

84. **TICKET SALES** A carnival usually sells three thousand 75¢ ride tickets on a Saturday. For each 15¢ increase in price, management estimates that 80 fewer tickets will be sold. What increase in ticket price will produce $2,982 of revenue on Saturday?

85. **MAGAZINE SALES** The *Gazette's* profit is $20 per year for each of its 3,000 subscribers. Management estimates that the profit per subscriber will increase by 1¢ for each additional subscriber over the current 3,000. How many subscribers will bring a total profit of $120,000?

86. INVESTMENT RATES A woman invests $1,000 in a fund for which interest is compounded annually at a rate r. After one year, she deposits an additional $2,000. After two years, the balance in the account is

$$\$1,000(1 + r)^2 + \$2,000(1 + r)$$

If this amount is $3,368.10, find r.

87. U.S. EMPLOYMENT TRENDS The percent of men ages 55–64 in the labor force is approximated by the function $P(x) = 0.03x^2 - 1.45x + 83.5$, where $P(x)$ is the percent and x is the number of years after 1970. The model is valid for the years 1970–2007; thus $0 \le x \le 37$. When does the model indicate that 75% of the men ages 55–64 were part of the labor force? (Based on data from the *Statistical Abstract of the United States,* 2007)

88. WATER The water usage by residents of Santa Barbara, California, is approximated by the function $g(x) = -0.36x^2 + 7.86x + 80.8$, where $g(x)$ is the average number of gallons used each day per person and x is the number of years after 1990. The model is valid for the years 1990–2005; thus, $0 \le x \le 15$. In what year does the model indicate that the daily usage was 115 gallons per person? (Based on data from the City of Santa Barbara Public Works Department)

WRITING

89. Explain why the quadratic formula, in most cases, is easier to use to solve a quadratic equation than is the method of completing the square.

90. On an exam, a student was asked to solve the equation $-4w^2 - 6w - 1 = 0$. Her first step was to multiply both sides of the equation by -1. She then used the quadratic formula to solve $4w^2 + 6w + 1 = 0$ instead. Is this a valid approach? Explain.

REVIEW

Change each radical to an exponential expression.

91. $\sqrt{n}$

92. $\sqrt[7]{8r^2s}$

93. $\sqrt[4]{3b}$

94. $3\sqrt[3]{c^2 - d^2}$

Write each expression in radical form.

95. $t^{1/3}$

96. $(3m^2n^2)^{1/5}$

97. $(3t)^{1/4}$

98. $(c^2 + d^2)^{1/2}$

CHALLENGE PROBLEMS

All of the equations we have solved so far have had rational-number coefficients. However, the quadratic formula can be used to solve quadratic equations with irrational or even imaginary coefficients. Solve each equation.

99. $x^2 + 2\sqrt{2}x - 6 = 0$

100. $\sqrt{2}x^2 + x - \sqrt{2} = 0$

101. $x^2 - 3ix - 2 = 0$

102. $100ix^2 + 300x - 200i = 0$

SECTION 10.3
The Discriminant and Equations That Can Be Written in Quadratic Form

Objectives

1 Use the discriminant to determine number and type of solutions.

2 Solve equations that are quadratic in form.

3 Solve problems involving quadratic equations.

We have seen that solutions of the quadratic equation $ax^2 + bx + c = 0$ with $a \ne 0$ are given by the formula

$$x = \frac{-b \pm \sqrt{b^2 - 4ac}}{2a}$$

discriminant of - no real solution will have imaginary # i   complex #'s

In this section, we will examine the radicand within the quadratic formula to distinguish or "*discriminate*" among the three types of solutions—rational, irrational, or imaginary.

 Use the Discriminant to Determine Number and Type of Solutions.

The expression $b^2 - 4ac$ that appears under the radical symbol in the quadratic formula is called the **discriminant.** The discriminant can be used to predict what kind of solutions a quadratic equation has without solving it.

The Discriminant

under square root sign (handwritten)

For a quadratic equation of the form $ax^2 + bx + c = 0$ with real-number coefficients and $a \neq 0$, the expression $b^2 - 4ac$ is called the **discriminant** and can be used to determine the number and type of the solutions of the equation.

Discriminant: $b^2 - 4ac$	Number and type of solutions
Positive	Two different real numbers
0 . . . . . *will have 1 real #*	One repeated solution, a rational number
Negative	Two different imaginary numbers that are complex conjugates

Discriminant: $b^2 - 4ac$	Number and type of solutions
A perfect square	Two different rational numbers
Positive and not a perfect square	Two different irrational numbers

---

EXAMPLE 1   Determine the number and type of solutions for each equation:
a. $x^2 + x + 1 = 0$ **b.** $3x^2 + 5x + 2 = 0$

Strategy   We will identify the values of a, b, and c in each equation. Then we will use those values to compute $b^2 - 4ac$, the discriminant.

Why   Once we know whether the discriminant is positive, 0, or negative, and whether it is a perfect square, we can determine the number and type of the solutions of the equation.

Solution

a. For $x^2 + x + 1 = 0$, the discriminant is:

$$b^2 - 4ac = 1^2 - 4(1)(1) \quad \text{Substitute: } a = 1, b = 1, \text{ and } c = 1.$$
$$= -3 \quad \text{The result is a negative number.}$$

Since $b^2 - 4ac < 0$, the solutions of $x^2 + x + 1 = 0$ are two different imaginary numbers that are complex conjugates.

b. For $3x^2 + 5x + 2 = 0$, the discriminant is:

$$b^2 - 4ac = 5^2 - 4(3)(2) \quad \text{Substitute: } a = 3, b = 5, \text{ and } c = 2.$$
$$= 25 - 24$$
$$= 1 \quad \text{The result is a positive number.}$$

Since $b^2 - 4ac > 0$ and $b^2 - 4ac$ is a perfect square, the solutions of $3x^2 + 5x + 2 = 0$ are two different rational numbers.

> **Success Tip**
> The discriminant can also be used to determine factorability. The trinomial $ax^2 + bx + c$ with integer coefficients and $a \neq 0$ will factor as the product of two binomials with integer coefficients if the value of $b^2 - 4ac$ is a perfect square. If $b^2 - 4ac = 0$, the factors will be the same.

Self Check 1   Determine the number and type of solutions for
a. $x^2 + x - 1 = 0$
b. $3x^2 + 4x + 2 = 0$

Now Try   Problems 11, 13, and 15

2   **Solve Equations That Are Quadratic in Form.**

We have discussed four methods that are used to solve quadratic equations. The table on the next page shows some advantages and disadvantages of each method.

Method	Advantages	Disadvantages	Examples
Factoring and the zero-factor property	It can be very fast. When each factor is set equal to 0, the resulting equations are usually easy to solve.	Some polynomials may be difficult to factor and others impossible.	$x^2 - 2x - 24 = 0$ $4a^2 - a = 0$
Square root property	It is the fastest way to solve equations of the form $ax^2 = n$ ($n =$ a number) or $(ax + b)^2 = n$.	It only applies to equations that are in these forms.	$x^2 = 27$ $(2y + 3)^2 = 25$
Completing the square*	It can be used to solve any quadratic equation. It works well with equations of the form $x^2 + bx = n$, where b is even.	It involves more steps than the other methods. The algebra can be cumbersome if the leading coefficient is not 1.	$x^2 + 4x + 1 = 0$ $t^2 - 14t - 9 = 0$
Quadratic formula	It can be used to solve any quadratic equation.	It involves several computations where sign errors can be made. Often the result must be simplified.	$x^2 + 3x - 33 = 0$ $4s^2 - 10s + 5 = 0$

*The quadratic formula is just a condensed version of completing the square and is usually easier to use. However, you need to know how to complete the square because it is used in more advanced mathematics courses.

To determine the most efficient method for a given equation, we can use the following strategy.

$$ax^2 + bx + c = 0$$

Strategy for Solving Quadratic Equations

1. See whether the equation is in a form such that the **square root method** is easily applied.
2. See whether the equation is in a form such that the **completing the square method** is easily applied.
3. If neither Step 1 nor Step 2 is reasonable, write the equation in $ax^2 + bx + c = 0$ form.
4. See whether the equation can be solved using the **factoring method.**
5. If you can't factor, solve the equation by the **quadratic formula.**

Many nonquadratic equations can be written in quadratic form ($ax^2 + bx + c = 0$) and solved using the techniques discussed in previous sections. For example, a careful inspection of the equation $x^4 - 5x^2 + 4 = 0$ leads to the following observations:

The leading term, x^4, is the square of the expression x^2 in the middle term: $x^4 = (x^2)^2$.

$$x^4 - 5x^2 + 4 = 0$$

The last term is a constant.

Equations that contain an expression, the same expression squared, and a constant term are said to be *quadratic in form*. One method used to solve such equations is to make a substitution.

EXAMPLE 2   Solve: $x^4 - 3x^2 - 4 = 0$

Strategy   Since the leading term, x^4, is the square of the expression x^2 in the middle term, we will substitute y for x^2.

Why   Our hope is that such a substitution will produce an equation that we can solve using one of the methods previously discussed.

Solution   If we write x^4 as $(x^2)^2$, the equation takes the form

$$(x^2)^2 - 3x^2 - 4 = 0$$

and it is said to be *quadratic in x^2*. We can solve this equation by letting $y = x^2$.

$$y^2 - 3y - 4 = 0 \quad \text{Replace each } x^2 \text{ with } y.$$

We can solve this quadratic equation by factoring.

$$(y - 4)(y + 1) = 0 \qquad \text{Factor } y^2 - 3y - 4.$$
$$y - 4 = 0 \quad \text{or} \quad y + 1 = 0 \qquad \text{Set each factor equal to 0.}$$
$$y = 4 \quad \Big| \quad\quad y = -1$$

These are not the solutions for x. To find x, we reverse the substitution by replacing each y with x^2 and proceed as follows:

$$x^2 = 4 \qquad \text{or} \quad x^2 = -1 \qquad \text{Substitute } x^2 \text{ for } y.$$
$$x = \pm\sqrt{4} \quad\Big| \quad x = \pm\sqrt{-1} \quad \text{Use the square root property.}$$
$$x = \pm 2 \quad\Big| \quad\quad x = \pm i$$

This equation has four solutions: $2, -2, i,$ and $-i$. Check each one in the original equation.

Self Check 2	Solve: $x^4 - 5x^2 - 36 = 0$
Now Try	**Problem 23**

EXAMPLE 3   Solve: $x - 7\sqrt{x} + 12 = 0$

Strategy   Since the leading term, x, is the square of the expression $\sqrt{x}$ in the middle term, we will substitute y for $\sqrt{x}$.

Why   Our hope is that such a substitution will produce an equation that we can solve using one of the methods previously discussed.

Solution   We examine the leading term and the middle term.

The leading term, x, is the square of the expression $\sqrt{x}$ in middle term: $x = \left(\sqrt{x}\right)^2.$ $x - 7\sqrt{x} + 12 = 0$ ↖The last term is a constant.

If we write x as $\left(\sqrt{x}\right)^2$, the equation takes the form

$$\left(\sqrt{x}\right)^2 - 7\sqrt{x} + 12 = 0$$

and it is said to be *quadratic in $\sqrt{x}$*. We can solve this equation by letting $y = \sqrt{x}$ and factoring.

$$y^2 - 7y + 12 = 0 \quad \text{Replace each } \sqrt{x} \text{ with } y.$$
$$(y - 3)(y - 4) = 0 \quad \text{Factor } y^2 - 7y + 12.$$
$$y - 3 = 0 \quad \text{or} \quad y - 4 = 0 \quad \text{Set each factor equal to 0.}$$
$$y = 3 \quad \Big| \quad\quad y = 4$$

To find x, we reverse the substitution and replace each y with $\sqrt{x}$. Then we solve the resulting radical equations by squaring both sides.

$$\sqrt{x} = 3 \quad \text{or} \quad \sqrt{x} = 4$$
$$x = 9 \quad | \quad x = 16$$

The solutions are 9 and 16. Check each solution in the original equation.

Self Check 3   Solve: $x + \sqrt{x} - 6 = 0$
Now Try **Problem 27**

EXAMPLE 4   Solve: $2m^{2/3} - 2 = 3m^{1/3}$

Strategy   We will write the equation in descending powers of m and look for a possible substitution to make.

Why   Our hope is that a substitution will produce an equation that we can solve using one of the methods previously discussed.

Solution   After writing the equation in descending powers of m, we see that

$$2m^{2/3} - 3m^{1/3} - 2 = 0$$

is *quadratic in* $m^{1/3}$, because $m^{2/3} = (m^{1/3})^2$. We will use the substitution $y = m^{1/3}$ to write this equation in quadratic form.

$$2m^{2/3} - 3m^{1/3} - 2 = 0$$
$$2(m^{1/3})^2 - 3m^{1/3} - 2 = 0 \qquad \text{Write } m^{2/3} \text{ as } (m^{1/3})^2.$$
$$2y^2 - 3y - 2 = 0 \qquad \text{Replace each } m^{1/3} \text{ with } y.$$
$$(2y + 1)(y - 2) = 0 \qquad \text{Factor } 2y^2 - 3y - 2.$$
$$2y + 1 = 0 \quad \text{or} \quad y - 2 = 0 \quad \text{Set each factor equal to 0.}$$
$$y = -\frac{1}{2} \quad \bigg| \quad y = 2$$

To find m, we reverse the substitution and replace each y with $m^{1/3}$. Then we solve the resulting equations by cubing both sides.

$$m^{1/3} = -\frac{1}{2} \quad \text{or} \quad m^{1/3} = 2$$
$$(m^{1/3})^3 = \left(-\frac{1}{2}\right)^3 \quad \bigg| \quad (m^{1/3})^3 = (2)^3 \quad \begin{array}{l}\text{Recall that } m^{1/3} = \sqrt[3]{m}. \text{ To solve for } m, \text{ cube both}\\ \text{sides.}\end{array}$$
$$m = -\frac{1}{8} \quad \bigg| \quad m = 8$$

The solutions are $-\frac{1}{8}$ and 8. Check each solution in the original equation.

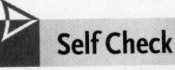

Self Check 4   Solve: $a^{2/3} = -3a^{1/3} + 10$
Now Try **Problem 31**

EXAMPLE 5   Solve: $(4t + 2)^2 - 30(4t + 2) + 224 = 0$

Strategy   Since the leading term, $(4t + 2)^2$, is the square of the expression $4t + 2$ in the middle term, we will substitute y for $4t + 2$.

Why   Our hope is that such a substitution will produce an equation that we can solve using one of the methods previously discussed.

Solution   This equation is *quadratic in* $4t + 2$. If we make the substitution $y = 4t + 2$, we have

$$y^2 - 30y + 224 = 0$$

which can be solved by using the quadratic formula.

$$y = \frac{-b \pm \sqrt{b^2 - 4ac}}{2a}$$

$$y = \frac{-(-30) \pm \sqrt{(-30)^2 - 4(1)(224)}}{2(1)} \qquad \text{Substitute 1 for } a, \ -30 \text{ for } b, \text{ and } 224 \text{ for } c.$$

$$y = \frac{30 \pm \sqrt{900 - 896}}{2} \qquad \text{Simplify within the radical.}$$

$$y = \frac{30 \pm 2}{2} \qquad\qquad \sqrt{900 - 896} = \sqrt{4} = 2.$$

$$y = 16 \quad \text{or} \quad y = 14$$

To find t, we reverse the substitution and replace y with $4t + 2$. Then we solve for t.

$$4t + 2 = 16 \quad \text{or} \quad 4t + 2 = 14$$
$$4t = 14 \qquad\qquad 4t = 12$$
$$t = 3.5 \qquad\qquad t = 3$$

Verify that 3.5 and 3 satisfy the original equation.

Self Check 5   Solve: $(n + 3)^2 - 6(n + 3) = -8$

Now Try   **Problem 35**

EXAMPLE 6   Solve: $15a^{-2} - 8a^{-1} + 1 = 0$

Strategy   We will write the equation with positive exponents and look for a possible substitution to make.

Why   Our hope is that a substitution will produce an equation that we can solve using one of the methods previously discussed.

Solution   When we write the terms $15a^{-2}$ and $-8a^{-1}$ using positive exponents, we see that this equation is *quadratic in* $\frac{1}{a}$.

> **Success Tip**
> We could also solve this equation by multiplying both sides by the LCD, a^2.

$$\frac{15}{a^2} - \frac{8}{a} + 1 = 0 \qquad \text{Think of this equation as } 15 \cdot \left(\frac{1}{a}\right)^2 - 8 \cdot \frac{1}{a} + 1 = 0.$$

If we let $y = \frac{1}{a}$, the resulting quadratic equation can be solved by factoring.

$$15y^2 - 8y + 1 = 0 \quad \text{Substitute } y^2 \text{ for } \frac{1}{a^2} \text{ and } y \text{ for } \frac{1}{a}.$$

$$(5y - 1)(3y - 1) = 0 \quad \text{Factor } 15y^2 - 8y + 1 = 0.$$

$$5y - 1 = 0 \quad \text{or} \quad 3y - 1 = 0$$

$$y = \frac{1}{5} \qquad\qquad y = \frac{1}{3}$$

To find a, we reverse the substitution and replace each y with $\frac{1}{a}$. Then we proceed as follows:

$$\frac{1}{a} = \frac{1}{5} \quad \text{or} \quad \frac{1}{a} = \frac{1}{3}$$

$$5 = a \qquad\qquad 3 = a \quad \text{Solve the proportions.}$$

The solutions are 5 and 3. Check each solution in the original equation.

Self Check 6   Solve: $28c^{-2} - 3c^{-1} - 1 = 0$

Now Try   **Problem 41**

3 **Solve Problems Involving Quadratic Equations.**

EXAMPLE 7   ***Household Appliances.***   The illustration shows a temperature control on a washing machine. When the *warm* setting is selected, both the hot and cold water pipes open to fill the tub in 2 minutes 15 seconds. When the *cold* setting is chosen, the tub fills 45 seconds faster than when the *hot* setting is used. How long does it take to fill the washing machine with hot water?

Electronic Temperature Control

Water Temp

Analyze the Problem   It is helpful to organize the facts of this shared-work problem in a table.

Form an Equation   Let $x =$ the number of seconds it takes to fill the tub with hot water. Since the cold water inlet fills the tub in 45 seconds less time, $x - 45 =$ the number of seconds it takes to fill the tub with cold water. The hot and cold water inlets will be open for the same time: 2 minutes 15 seconds, or 135 seconds.

To determine the work completed by each inlet, multiply the rate by the time.

	Rate	· Time =	Work completed
Hot water	$\dfrac{1}{x}$	135	$\dfrac{135}{x}$
Cold water	$\dfrac{1}{x - 45}$	135	$\dfrac{135}{x - 45}$

Enter this information first.   Multiply to get each of these entries: $W = rt$.

Success Tip

An alternate way to form an equation is to note that what the hot water inlet can do in 1 second plus what the cold water inlet can do in 1 second equals what they can do together in 1 second:

$$\frac{1}{x} + \frac{1}{x - 45} = \frac{1}{135}$$

In shared-work problems, 1 represents one whole job completed. So we have,

The fraction of tub filled with hot water	plus	the fraction of the tub filled with cold water	equals	1 tub filled.
$\dfrac{135}{x}$	$+$	$\dfrac{135}{x - 45}$	$=$	1

Solve the Equation

$$\frac{135}{x} + \frac{135}{x - 45} = 1$$

$$x(x - 45)\left(\frac{135}{x} + \frac{135}{x - 45}\right) = x(x - 45)(1)$$

Multiply both sides by the LCD $x(x - 45)$ to clear the rational equation of fractions.

$$x(x - 45)\frac{135}{x} + x(x - 45)\frac{135}{x - 45} = x(x - 45)(1)$$

Distribute the multiplication by $x(x - 45)$.

$$135(x - 45) + 135x = x(x - 45)$$

Simplify each side.

$$135x - 6{,}075 + 135x = x^2 - 45x$$

Distribute the multiplication by 135 and by x.

$$270x - 6{,}075 = x^2 - 45x$$

Combine like terms.

$$0 = x^2 - 315x + 6{,}075$$

To get 0 on the left side, subtract 270x from both sides, and add 6,075 to both sides.

To solve this equation, we will use the quadratic formula, with $a = 1$, $b = -315$, and $c = 6{,}075$.

$$x = \frac{-b \pm \sqrt{b^2 - 4ac}}{2a}$$

$$x = \frac{-(-315) \pm \sqrt{(-315)^2 - 4(1)(6{,}075)}}{2(1)}$$

Substitute 1 for *a*, −315 for *b*, and 6,075 for *c*.

$$x = \frac{315 \pm \sqrt{99{,}225 - 24{,}300}}{2}$$

Simplify within the radical.

$$x = \frac{315 \pm \sqrt{74{,}925}}{2}$$

$$x \approx \frac{589}{2} \quad \text{or} \quad x \approx \frac{41}{2}$$

$$x \approx 294 \quad \quad \cancel{x \approx 21}$$

State the Conclusion We can discard the solution of 21 seconds, because this would imply that the cold water inlet fills the tub in a negative number of seconds $(21 - 45 = -24)$. Therefore, the hot water inlet fills the washing machine tub in about 294 seconds, which is 4 minutes 54 seconds.

Check the Result Use estimation to check the result.

 Now Try Problem 83

STUDY SET
10.3

VOCABULARY

Fill in the blanks.

1. For the quadratic equation $ax^2 + bx + c = 0$, the _____ is $b^2 - 4ac$.

2. We can solve $x - 2\sqrt{x} - 8 = 0$ by making the _____ $y = \sqrt{x}$.

CONCEPTS

Consider the quadratic equation $ax^2 - bx + c = 0$, where a, b, and c represent rational numbers, and fill in the blanks.

3. If $b^2 - 4ac < 0$, the solutions of the equation are two different imaginary numbers that are complex _____.

4. If $b^2 - 4ac = $, the equation has one repeated rational-number solution.

5. If $b^2 - 4ac$ is a perfect square, the solutions of the equation are two different _____ numbers.

6. If $b^2 - 4ac$ is positive and not a perfect square, the solutions of the equation are two different _____ numbers.

7. For each equation, determine the substitution that should be made to write the equation in quadratic form.

   a. $x^4 - 12x^2 + 27 = 0$ Let $y = $

   b. $x - 13\sqrt{x} + 40 = 0$ Let $y = $

   c. $x^{2/3} + 2x^{1/3} - 3 = 0$ Let $y = $

   d. $x^{-2} - x^{-1} - 30 = 0$ Let $y = $

   e. $(x + 1)^2 - (x + 1) - 6 = 0$ Let $y = $

8. Fill in the blanks.

   a. $x^4 = \left(\quad\right)^2$ b. $x = \left(\quad\right)^2$

   c. $x^{2/3} = \left(\quad\right)^2$ d. $\dfrac{1}{x^2} = \left(\quad\right)^2$

NOTATION

Complete the solution.

9. To find the type of solutions for the equation $x^2 + 5x + 6 = 0$, we compute the discriminant.

   $$b^2 - \quad = \quad^2 - 4(1)\left(\quad\right)$$
   $$= 25 - \quad$$
   $$= 1$$

   Since a, b, and c are rational numbers and the value of the discriminant is a perfect square, the solutions are two different _____ numbers.

10. Fill in the blanks to write each equation in quadratic form.

    a. $x^4 - 2x^2 - 15 = 0 \rightarrow \left(\quad\right)^2 - 2\quad - 15 = 0$

    b. $x - 2\sqrt{x} + 3 = 0 \rightarrow \left(\quad\right)^2 - 2\quad + 3 = 0$

    c. $8m^{2/3} - 10m^{1/3} - 3 = 0 \rightarrow 8\left(\quad\right)^2 - 10\quad - 3 = 0$

GUIDED PRACTICE

Use the discriminant to determine the number and type of solutions for each equation. Do not solve. See Example 1.

11. $4x^2 - 4x + 1 = 0$ 12. $6x^2 - 5x - 6 = 0$

13. $5x^2 + x + 2 = 0$ 14. $3x^2 + 10x - 2 = 0$

15. $2x^2 = 4x - 1$ 16. $9x^2 = 12x - 4$

17. $x(2x - 3) = 20$ 18. $x(x - 3) = -10$

19. $3x^2 - 10 = 0$ 20. $5x^2 - 24 = 0$

21. $x^2 - \dfrac{14}{15}x = \dfrac{8}{15}$ 22. $x^2 = -\dfrac{5}{4}x + \dfrac{3}{2}$

Solve each equation. See Example 2.

23. $x^4 - 17x^2 + 16 = 0$ 24. $x^4 - 10x^2 + 9 = 0$

25. $x^4 + 5x^2 - 36 = 0$ 26. $x^4 - 15x^2 - 16 = 0$

Solve each equation. See Example 3.

27. $x - 13\sqrt{x} + 40 = 0$ 28. $x - 9\sqrt{x} + 18 = 0$

29. $2x + \sqrt{x} - 3 = 0$ 30. $2x - \sqrt{x} - 1 = 0$

Solve each equation. See Example 4.

31. $a^{2/3} - 2a^{1/3} = 3$ 32. $r^{2/3} + 4r^{1/3} = 5$

33. $x^{2/3} + 2x^{1/3} - 8 = 0$ 34. $x^{2/3} - 7x^{1/3} + 12 = 0$

Solve each equation. See Example 5.

35. $(c + 1)^2 - 4(c + 1) + 3 = 0$

36. $(a - 5)^2 - 4(a - 5) - 21 = 0$

37. $2(2x + 1)^2 - 7(2x + 1) + 6 = 0$

38. $3(2 - x)^2 + 10(2 - x) - 8 = 0$

Solve each equation. See Example 6.

39. $m^{-2} + m^{-1} - 6 = 0$

40. $t^{-2} + t^{-1} - 42 = 0$

41. $8x^{-2} - 10x^{-1} - 3 = 0$

42. $2x^{-2} - 5x^{-1} - 3 = 0$

Solve each equation. See Example 7.

43. $1 - \dfrac{5}{x} = \dfrac{10}{x^2}$

44. $1 - \dfrac{3}{x} = \dfrac{5}{x^2}$

45. $\dfrac{1}{2} + \dfrac{1}{b} = \dfrac{1}{b - 7}$

46. $\dfrac{1}{4} - \dfrac{1}{n} = \dfrac{1}{n + 3}$

TRY IT YOURSELF

Solve each equation.

47. $2x - \sqrt{x} = 3$

48. $3x + 4\sqrt{x} = 4$

49. $x^{-2} + 2x^{-1} - 3 = 0$

50. $x^{-2} + 2x^{-1} - 8 = 0$

51. $x^4 + 19x^2 + 18 = 0$

52. $t^4 + 4t^2 - 5 = 0$

53. $(k - 7)^2 + 6(k - 7) + 10 = 0$

54. $(d + 9)^2 - 4(d + 9) + 8 = 0$

55. $\dfrac{2}{x - 1} + \dfrac{1}{x + 1} = 3$

56. $\dfrac{3}{x - 2} - \dfrac{1}{x + 2} = 5$

57. $x - 6x^{1/2} = -8$

58. $x - 5x^{1/2} + 4 = 0$

59. $(y^2 - 9)^2 + 2(y^2 - 9) - 99 = 0$

60. $(a^2 - 4)^2 - 4(a^2 - 4) - 32 = 0$

61. $x^{-4} - 2x^{-2} + 1 = 0$

62. $4x^{-4} + 1 = 5x^{-2}$

63. $t^4 + 3t^2 = 28$

64. $3h^4 + h^2 - 2 = 0$

65. $2x^{2/5} - 5x^{1/5} = -3$

66. $2x^{2/5} + 3x^{1/5} = -1$

67. $9\left(\dfrac{3m + 2}{m}\right)^2 - 30\left(\dfrac{3m + 2}{m}\right) + 25 = 0$

68. $4\left(\dfrac{c - 7}{c}\right)^2 - 12\left(\dfrac{c - 7}{c}\right) + 9 = 0$

69. $\dfrac{3}{a - 1} = 1 - \dfrac{2}{a}$

70. $1 + \dfrac{4}{x} = \dfrac{3}{x^2}$

71. $\left(8 - \sqrt{a}\right)^2 + 6\left(8 - \sqrt{a}\right) - 7 = 0$

72. $\left(10 - \sqrt{t}\right)^2 - 4\left(10 - \sqrt{t}\right) - 45 = 0$

73. $x + \dfrac{2}{x - 2} = 0$

74. $x + \dfrac{x + 5}{x - 3} = 0$

75. $3x + 5\sqrt{x} + 2 = 0$

76. $3x - 4\sqrt{x} + 1 = 0$

77. $x^4 - 6x^2 + 5 = 0$

78. $2x^4 - 26x^2 + 24 = 0$

79. $8(t + 1)^{-2} - 30(t + 1)^{-1} + 7 = 0$

80. $2(s - 2)^{-2} + 3(s - 2)^{-1} - 5 = 0$

81. $\dfrac{1}{x + 2} + \dfrac{24}{x + 3} = 13$

82. $\dfrac{3}{x} + \dfrac{4}{x + 1} = 2$

APPLICATIONS

83. CROWD CONTROL   After a performance at a county fair, security guards have found that the grandstand area can be emptied in 6 minutes if both the east and west exits are opened. If just the east exit is used, it takes 4 minutes longer to clear the grandstand than it does if just the west exit is opened. How long does it take to clear the grandstand if everyone must file through the west exit? Round to the nearest tenth of a minute.

84. PAPER ROUTES   When a father, in a car, and his son, on a bicycle, work together to distribute the morning newspaper, it takes them 35 minutes to complete the route. Working alone, it takes the son 25 minutes longer than the father. To the nearest minute, how long does it take the son to cover the route on his bicycle?

85. ASSEMBLY LINES   A newly manufactured product traveled 300 feet on a high-speed conveyor belt at a rate of r feet per second. It could have traveled the 300 feet in 3 seconds less time if the speed of the conveyor belt was increased by 5 feet per second. Find r.

86. BICYCLING   Tina bicycles 160 miles at the rate of r mph. The same trip would have taken 2 hours longer if she had decreased her speed by 4 mph. Find r.

87. ARCHITECTURE   A **golden rectangle** is one of the most visually appealing of all geometric forms. The Parthenon, built by the Greeks in the 5th century B.C., fits into a golden rectangle if its ruined triangular pediment is included. See the illustration. In a golden rectangle, the length l and width w must satisfy the equation

$$\frac{l}{w} = \frac{w}{l - w}$$

If a rectangular billboard is to have a width of 20 feet, what should its length be so that it is a golden rectangle? Round to the nearest tenth.

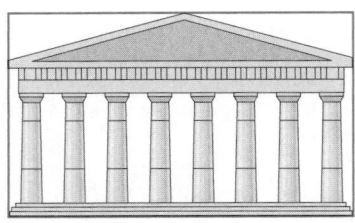

88. ENTRY DOOR DESIGNS   An architect needs to determine the height h of the window shown in the illustration. The radius r, the width w, and the height h of the circular-shaped window are related by the formula

$$r = \frac{4h^2 + w^2}{8h}$$

If w is to be 34 inches and r is to be 18 inches, find h to the nearest tenth of an inch.

WRITING

89. Describe how to predict what type of solutions the equation $3x^2 - 4x + 5 = 0$ will have.

90. What error is made in the following solution?

Solve: $x^4 - 12x^2 + 27 = 0$

Let $y = x^2$

$$y^2 - 12y + 27 = 0$$
$$(y - 9)(y - 3) = 0$$

$$y - 9 = 0 \quad \text{or} \quad y - 3 = 0$$
$$y = 9 \qquad \qquad y = 3$$

The solutions of $x^4 - 12x^2 + 27 = 0$ are 9 and 3.

REVIEW

91. Write an equation of the vertical line that passes through $(3, 4)$.

92. Find an equation of the line that passes through $(-1, -6)$ and $(-2, -1)$. Write the equation in slope–intercept form.

93. Write an equation of the line with slope $\frac{2}{3}$ that passes through the origin.

94. Find an equation of the line that passes through $(2, -3)$ and is perpendicular to the line whose equation is $y = \frac{x}{5} + 6$. Write the equation in slope–intercept form.

CHALLENGE PROBLEMS

95. Solve: $x^6 + 17x^3 + 16 = 0$

96. Find the real-number solutions of $x^4 - 3x^2 - 2 = 0$. Rationalize the denominators of the solutions.

SECTION 10.4
Quadratic Functions and Their Graphs

Objectives

1 Graph functions of the form $f(x) = ax^2$.

2 Graph functions of the form $f(x) = ax^2 + k$.

3 Graph functions of the form $f(x) = a(x - h)^2$.

4 Graph functions of the form $f(x) = a(x - h)^2 + k$.

5 Graph functions of the form $f(x) = ax^2 + bx + c$ by completing the square.

6 Find the vertex using $-\frac{b}{2a}$.

7 Determine minimum and maximum values.

8 Solve quadratic equations graphically.

In this section, we will discuss methods for graphing *quadratic functions*.

Quadratic Functions	A **quadratic function** is a second-degree polynomial function that can be written in the form $$f(x) = ax^2 + bx + c$$ where a, b, and c are real numbers and $a \neq 0$.

Quadratic functions are often written in another form, called **standard form,**

$$f(x) = a(x - h)^2 + k$$

Notation

Since $y = f(x)$, quadratic functions can also be written as $y = ax^2 + bx + c$ and $y = a(x - h)^2 + k$.

where a, h, and k are real numbers and $a \neq 0$. This form is useful because a, h, and k give us important information about the graph of the function. To develop a strategy for graphing quadratic functions written in standard form, we will begin by considering the simplest case, $f(x) = ax^2$.

1 **Graph Functions of the Form $f(x) = ax^2$.**

One way to graph quadratic functions is to plot points.

EXAMPLE 1   Graph: **a.** $f(x) = x^2$ **b.** $g(x) = 3x^2$ **c.** $s(x) = \dfrac{1}{3}x^2$

Strategy   We can make a table of values for each function, plot each point, and connect them with a smooth curve.

Why   At this time, this method is our only option.

Solution   After graphing each curve, we see that the graph of $g(x) = 3x^2$ is narrower than the graph of $f(x) = x^2$, and the graph of $s(x) = \frac{1}{3}x^2$ is wider than the graph of $f(x) = x^2$. For $f(x) = ax^2$, the smaller the value of $|a|$, the wider the graph.

$f(x) = x^2$

x	$f(x)$
-2	4
-1	1
0	0
1	1
2	4

$g(x) = 3x^2$

x	$g(x)$
-2	12
-1	3
0	0
1	3
2	12

——The values of $g(x)$ increase faster than the values of $f(x)$, making its graph steeper. ——

$s(x) = \dfrac{1}{3}x^2$

x	$s(x)$
-2	$\frac{4}{3}$
-1	$\frac{1}{3}$
0	0
1	$\frac{1}{3}$
2	$\frac{4}{3}$

——The values of $s(x)$ increase more slowly than the values of $f(x)$, making its graph flatter. ——

Now Try   Problem 15

EXAMPLE 2 Graph: $f(x) = -3x^2$

Strategy We make a table of values for the function, plot each point, and connect them with a smooth curve.

Why At this time, this method is our only option.

Solution After graphing the curve, we see that it opens downward and has the same shape as the graph of $g(x) = 3x^2$ that was graphed in Example 1.

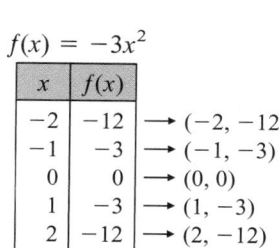

$f(x) = -3x^2$

x	$f(x)$	
-2	-12	$\rightarrow (-2, -12)$
-1	-3	$\rightarrow (-1, -3)$
0	0	$\rightarrow (0, 0)$
1	-3	$\rightarrow (1, -3)$
2	-12	$\rightarrow (2, -12)$

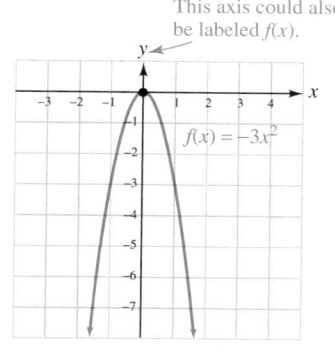

This axis could also be labeled $f(x)$.

$f(x) = -3x^2$

Self Check 2 Graph: $f(x) = -\frac{1}{3}x^2$

Now Try Problem 17

The graphs of functions of the form $f(x) = ax^2$ are **parabolas.** The lowest point on a parabola that opens upward, or the highest point on a parabola that opens downward, is called the **vertex** of the parabola. The vertical line, called an **axis of symmetry,** that passes through the vertex divides the parabola into two congruent halves. If we fold the paper along the axis of symmetry, the two sides of the parabola will match.

The Language of Algebra

An axis of symmetry divides a parabola into two matching sides. The sides are said to be *mirror images* of each other.

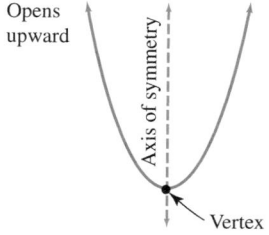

Opens upward

Axis of symmetry

Vertex

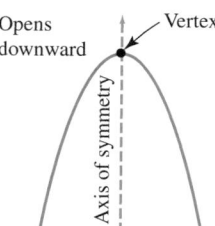

Opens downward

Vertex

Axis of symmetry

The results from Examples 1 and 2 confirm the following facts.

The Graph of
$f(x) = ax^2$

The graph of $f(x) = ax^2$ is a parabola opening upward when $a > 0$ and downward when $a < 0$, with vertex at the point $(0, 0)$ and axis of symmetry the line $x = 0$.

2 **Graph Functions of the Form $f(x) = ax^2 + k$.**

> **EXAMPLE 3**   Graph: **a.** $f(x) = 2x^2$ **b.** $g(x) = 2x^2 + 3$
>   **c.** $s(x) = 2x^2 - 3$
>
> **Strategy**   We make a table of values for each function, plot each point, and connect them with a smooth curve.
>
> **Why**   At this time, this method is our only option.
>
> **Solution**   After graphing the curves, we see that the graph of $g(x) = 2x^2 + 3$ is identical to the graph of $f(x) = 2x^2$, except that it has been translated 3 units upward. The graph of $s(x) = 2x^2 - 3$ is identical to the graph of $f(x) = 2x^2$, except that it has been translated 3 units downward. In each case, the axis of symmetry is the line $x = 0$.

$f(x) = 2x^2$

x	$f(x)$
-2	8
-1	2
0	0
1	2
2	8

$g(x) = 2x^2 + 3$

x	$g(x)$
-2	11
-1	5
0	3
1	5
2	11

$s(x) = 2x^2 - 3$

x	$s(x)$
-2	5
-1	-1
0	-3
1	-1
2	5

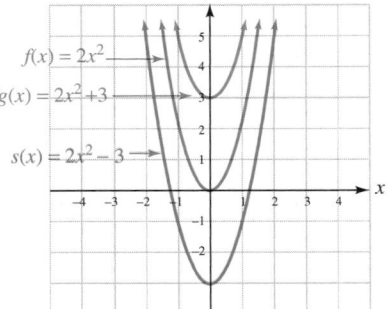

—For each x-value, g(x) is 3 more than f(x).—

—For each x-value, s(x) is 3 less than f(x).—

  Now Try   **Problem 19**

The results of Example 3 confirm the following facts.

The Graph of $f(x) = ax^2 + k$	The graph of $f(x) = ax^2 + k$ is a parabola having the same shape as $f(x) = ax^2$ but translated k units upward if k is positive and $|k|$ units downward if k is negative. The vertex is at the point $(0, k)$, and the axis of symmetry is the line $x = 0$.

3 **Graph Functions of the Form $f(x) = a(x - h)^2$.**

> **EXAMPLE 4**   Graph: **a.** $f(x) = 2x^2$ **b.** $g(x) = 2(x - 3)^2$
>   **c.** $s(x) = 2(x + 3)^2$
>
> **Strategy**   We make a table of values for each function, plot each point, and connect them with a smooth curve.
>
> **Why**   At this time, this method is our only option.
>
> **Solution**   We note that the graph of $g(x) = 2(x - 3)^2$ on the next page is identical to the graph of $f(x) = 2x^2$, except that it has been translated 3 units to the right. The graph of $s(x) = 2(x + 3)^2$ is identical to the graph of $f(x) = 2x^2$, except that it has been translated 3 units to the left.

$f(x) = 2x^2$

x	$f(x)$
-2	8
-1	2
0	0
1	2
2	8

$g(x) = 2(x - 3)^2$

x	$g(x)$
1	8
2	2
3	0
4	2
5	8

$s(x) = 2(x + 3)^2$

x	$s(x)$
-5	8
-4	2
-3	0
-2	2
-1	8

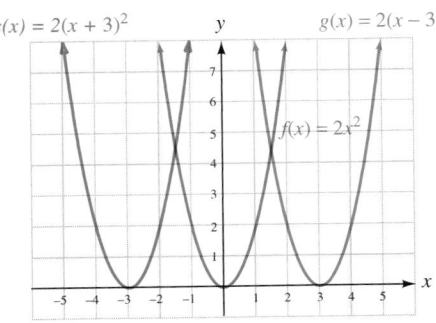

 When an x-value is increased by 3, the function's outputs are the same.

 When an x-value is decreased by 3, the function's outputs are the same.

> **Now Try** **Problem 21**

The results of Example 4 confirm the following facts.

The Graph of $f(x) = a(x - h)^2$	The graph of $f(x) = a(x - h)^2$ is a parabola having the same shape as $f(x) = ax^2$ but translated h units to the right if h is positive and $|h|$ units to the left if h is negative. The vertex is at the point $(h, 0)$, and the axis of symmetry is the line $x = h$.

4 **Graph Functions of the Form** $f(x) = a(x - h)^2 + k$.

The results of Examples 1–4 suggest a general strategy for graphing quadratic functions that are written in the form $f(x) = a(x - h)^2 + k$.

Graphing a Quadratic Function in Standard Form	The graph of the quadratic function $$f(x) = a(x - h)^2 + k \quad \text{where } a \neq 0$$ is a parabola with vertex at (h, k). The axis of symmetry is the line $x = h$. The parabola opens upward when $a > 0$ and downward when $a < 0$.

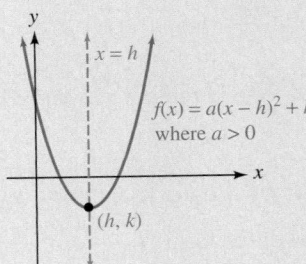

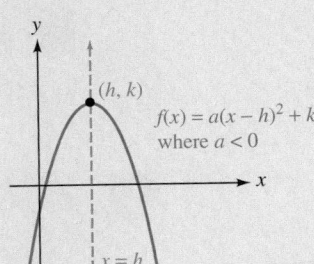

EXAMPLE 5 Graph: $f(x) = 2(x - 3)^2 - 4$. Label the vertex and draw the axis of symmetry.

Strategy We will determine whether the graph opens upward or downward and find its vertex and axis of symmetry. Then we will plot some points and complete the graph.

Why This method will be more efficient than plotting many points.

Solution The graph of $f(x) = 2(x - 3)^2 - 4$ is identical to the graph of $g(x) = 2(x - 3)^2$, except that it has been translated 4 units downward. The graph of $g(x) = 2(x - 3)^2$ is identical to the graph of $s(x) = 2x^2$, except that it has been translated 3 units to the right.

We can learn more about the graph of $f(x) = 2(x - 3)^2 - 4$ by determining a, h, and k.

$$f(x) = 2(x - 3)^2 - 4 \atop f(x) = a(x - h)^2 + k \Bigg\} \ a = 2, h = 3, \text{ and } k = -4$$

Upward/downward: Since $a = 2$ and $2 > 0$, the parabola opens upward.

Vertex: The vertex of the parabola is $(h, k) = (3, -4)$, as shown below.

Axis of symmetry: Since $h = 3$, the axis of symmetry is the line $x = 3$, as shown below.

Plotting points: We can construct a table of values to determine several points on the parabola. Since the x-coordinate of the vertex is 3, we choose the x-values of 4 and 5, find $f(4)$ and $f(5)$, and record the results in a table. Then we plot $(4, -2)$ and $(5, 4)$, and use symmetry to locate two other points on the parabola: $(2, -2)$ and $(1, 4)$. Finally, we draw a smooth curve through the points to get the graph.

$$f(x) = 2(x - 3)^2 - 4$$

x	$f(x)$	
4	-2	→ $(4, -2)$
5	4	→ $(5, 4)$

The x-coordinate of the vertex is 3. Choose values for x close to 3 and on the same side of the axis of symmetry.

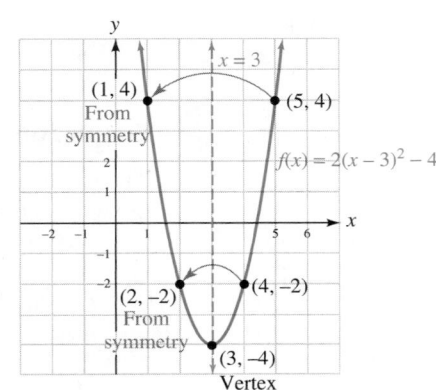

Self Check 5 Graph: $f(x) = 2(x - 1)^2 - 2$. Label the vertex and draw the axis of symmetry.

Now Try Problems 25 and 31

 **Graph Functions of the Form $f(x) = ax^2 + bx + c$ by Completing the Square.**

To graph functions of the form $f(x) = ax^2 + bx + c$, we can complete the square to write the function in standard form $f(x) = a(x - h)^2 + k$.

EXAMPLE 6 Determine the vertex and the axis of symmetry of the graph of $f(x) = x^2 + 8x + 21$. Will the graph open upward or downward?

Strategy To find the vertex and the axis of symmetry, we will complete the square on x and write the equation of the function in standard form.

Why Once the equation is written in standard form, we can determine the values of a, h, and k. The coordinates of the vertex will be (h, k) and the equation of the axis of symmetry will by $x = h$. The graph will open upward if $a > 0$ or downward if $a < 0$.

Solution To determine the vertex and the axis of symmetry of the graph, we complete the square on the right side so we can write the function in $f(x) = a(x - h)^2 + k$ form.

$$f(x) = x^2 + 8x + 21$$

$$f(x) = (x^2 + 8x \quad) + 21 \qquad \text{Prepare to complete the square on } x \text{ by writing parentheses around } x^2 + 8x.$$

To complete the square on $x^2 + 8x$, we note that one-half of the coefficient of x is $\frac{1}{2} \cdot 8 = 4$, and $4^2 = 16$. If we add 16 to $x^2 + 8x$, we obtain a perfect-square trinomial within the parentheses. Since this step adds 16 to the right side, we must also subtract 16 from the right side so that it remains in an equivalent form.

Add 16 to the right side.   Subtract 16 from the right side to counteract the addition of 16.

$$f(x) = (x^2 + 8x + 16) + 21 - 16$$

$$f(x) = (x + 4)^2 + 5 \qquad \text{Factor } x^2 + 8x + 16 \text{ and combine like terms.}$$

The function is now written in standard form, and we can determine a, h, and k.

The standard form requires a minus symbol here.

$$f(x) = \left[(x - (-4))\right]^2 + 5 \qquad \text{Write } x + 4 \text{ as } x - (-4) \text{ to determine } h.$$

$$a = 1 \quad h = -4 \quad k = 5$$

The vertex is $(h, k) = (-4, 5)$ and the axis of symmetry is the line $x = -4$. Since $a = 1$ and $1 > 0$, the parabola opens upward.

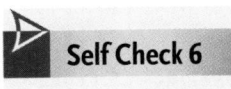

Self Check 6   Determine the vertex and the axis of symmetry of the graph of $f(x) = x^2 + 4x + 10$. Will the graph open upward or downward?

Now Try   **Problem 39**

Success Tip
When a number is added to and that same number is subtracted from one side of an equation, the value of that side of the equation remains the same.

EXAMPLE 7   Graph: $f(x) = 2x^2 - 4x - 1$

Strategy We will complete the square on x and write the equation of the function in standard form, $f(x) = a(x - h)^2 + k$.

Why When the equation is in standard form, we can identify the values of a, h, and k from the equation. This information will help us sketch the graph.

Solution Recall that to complete the square on $2x^2 - 4x$, the coefficient of x^2 must be equal to 1. Therefore, we factor 2 from $2x^2 - 4x$.

$$f(x) = 2x^2 - 4x - 1$$

$$f(x) = 2(x^2 - 2x \quad) - 1$$

To complete the square on $x^2 - 2x$, we note that one-half of the coefficient of x is $\frac{1}{2}(-2) = -1$, and $(-1)^2 = 1$. If we add 1 to $x^2 - 2x$, we obtain a perfect-square trinomial within the parentheses. Since this step adds 2 to the right side, we must also subtract 2 from the right side so that it remains in an equivalent form.

By the distributive property, when 1 is added to the expression within the parentheses, $2 \cdot 1 = 2$ is added to the right side.

Subtract 2 to counteract the addition of 2.

$$f(x) = 2(x^2 - 2x + 1) - 1 - 2$$

$$f(x) = 2(x - 1)^2 - 3 \qquad \text{Factor } x^2 - 2x + 1 \text{ and combine like terms.}$$

We see that $a = 2$, $h = 1$, and $k = -3$. Thus, the vertex is at the point $(1, -3)$, and the axis of symmetry is $x = 1$. Since $a = 2$ and $2 > 0$, the parabola opens upward. We plot the vertex and axis of symmetry as shown below.

Finally, we construct a table of values, plot the points, use symmetry to plot the corresponding points, and then draw the graph.

Success Tip

To find additional points on the graph, select values of x that are close to the x-coordinate of the vertex.

$$f(x) = 2x^2 - 4x - 1$$
$$\text{or}$$
$$f(x) = 2(x - 1)^2 - 3$$

x	$f(x)$	
2	-1	$\rightarrow (2, -1)$
3	5	$\rightarrow (3, 5)$

↑
The x-coordinate of the vertex is 1. Choose values for x close to 1 and on the same side of the axis of symmetry.

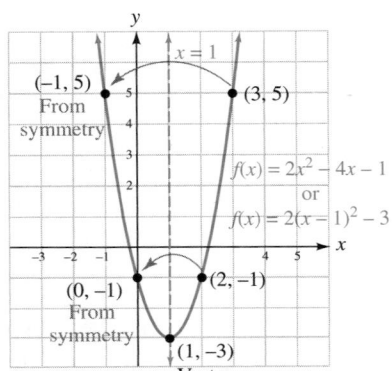

Self Check 7   Graph: $f(x) = 3x^2 - 12x + 8$.

Now Try   Problem 45

6   Find the Vertex Using $-\frac{b}{2a}$.

Because of symmetry, if a parabola has two x-intercepts, the x-coordinate of the vertex is exactly midway between them. We can use this fact to derive a formula to find the vertex of a parabola.

In general, if a parabola has two x-intercepts, they can be found by solving $0 = ax^2 + bx + c$ for x. We can use the quadratic formula to find the solutions. They are

$$x = \frac{-b - \sqrt{b^2 - 4ac}}{2a} \quad \text{and} \quad x = \frac{-b + \sqrt{b^2 - 4ac}}{2a}$$

Thus, the parabola's x-intercepts are $\left(\dfrac{-b - \sqrt{b^2 - 4ac}}{2a}, 0\right)$ and $\left(\dfrac{-b + \sqrt{b^2 - 4ac}}{2a}, 0\right)$.

Since the x-value of the vertex of a parabola is halfway between the two x-intercepts, we can find this value by finding the average, or $\frac{1}{2}$ of the sum of the x-coordinates of the x-intercepts.

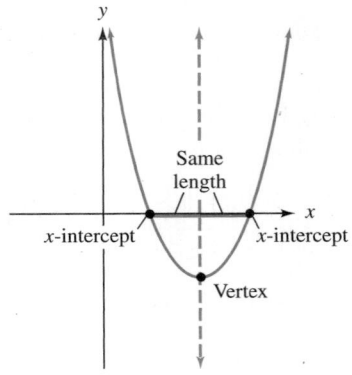

$$x = \frac{1}{2}\left(\frac{-b - \sqrt{b^2 - 4ac}}{2a} + \frac{-b + \sqrt{b^2 - 4ac}}{2a}\right)$$

$$x = \frac{1}{2}\left(\frac{-b - \sqrt{b^2 - 4ac} + (-b) + \sqrt{b^2 - 4ac}}{2a}\right) \quad \begin{array}{l}\text{Add the numerators and keep the}\\\text{common denominator.}\end{array}$$

$$x = \frac{1}{2}\left(\frac{-2b}{2a}\right) \quad \begin{array}{l}\text{Combine like terms: } -b + (-b) = -2b\\\text{and } -\sqrt{b^2 - 4ac} + \sqrt{b^2 - 4ac} = 0.\end{array}$$

$$x = -\frac{b}{2a} \quad \begin{array}{l}\text{Remove the common factor of 2 in the}\\\text{numerator and denominator and simplify.}\end{array}$$

This result is true even if the graph has no x-intercepts.

Formula for the Vertex of a Parabola

The vertex of the graph of the quadratic function $f(x) = ax^2 + bx + c$ is

$$\left(-\frac{b}{2a}, f\left(-\frac{b}{2a}\right)\right)$$

and the axis of symmetry of the parabola is the line $x = -\frac{b}{2a}$.

EXAMPLE 8 Find the vertex of the graph of $f(x) = 2x^2 - 4x - 1$.

Strategy We will determine the values of a and b and substitute into the formula for the vertex of a parabola.

Why It is easier to find the coordinates of the vertex using the formula than it is to complete the square on $2x^2 - 4x - 1$.

Solution The function is written in $f(x) = ax^2 + bx + c$ form, where $a = 2$ and $b = -4$. To find the vertex of its graph, we compute

$$-\frac{b}{2a} = -\frac{-4}{2(2)}$$

$$= -\frac{-4}{4}$$

$$= 1 \quad \begin{array}{l}\text{This is the}\\\text{x-coordinate}\\\text{of the vertex.}\end{array}$$

$$f\left(-\frac{b}{2a}\right) = f(1)$$

$$= 2(1)^2 - 4(1) - 1$$

$$= -3 \quad \begin{array}{l}\text{This is the}\\\text{y-coordinate}\\\text{of the vertex.}\end{array}$$

The vertex is the point $(1, -3)$. This agrees with the result we obtained in Example 7 by completing the square.

 Self Check 8 Find the vertex of the graph of $f(x) = 3x^2 - 12x + 8$.

Now Try **Problem 55**

Using Your Calculator *Finding the Vertex*

We can use a graphing calculator to graph the function $f(x) = 2x^2 + 6x - 3$ and find the coordinates of the vertex and the axis of symmetry of the parabola. If we enter the function, we will obtain the graph shown in figure (a).

We then trace to move the cursor to the lowest point on the graph, as shown in figure (b). By zooming in, we can see that the vertex is the point $(-1.5, -7.5)$, or $\left(-\frac{3}{2}, -\frac{15}{2}\right)$, and that the line $x = -\frac{3}{2}$ is the axis of symmetry.

Some calculators have an fmin or fmax feature that can also be used to find the vertex.

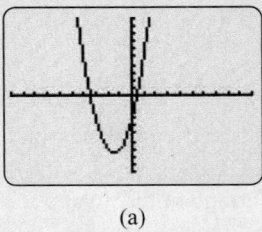

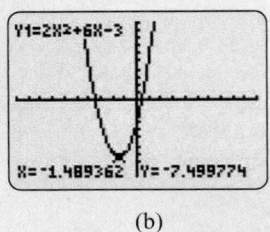

(a)

(b)

We can determine much about the graph of $f(x) = ax^2 + bx + c$ from the coefficients a, b, and c. This information is summarized as follows:

Graphing a Quadratic Function
$f(x) = ax^2 + bx + c$

- Determine whether the parabola opens upward or downward by finding the value of a.
- The x-coordinate of the vertex of the parabola is $x = -\frac{b}{2a}$.
- To find the y-coordinate of the vertex, substitute $-\frac{b}{2a}$ for x and find $f\left(-\frac{b}{2a}\right)$.
- The axis of symmetry is the vertical line passing through the vertex.
- The y-intercept is determined by the value of $f(x)$ when $x = 0$: the y-intercept is $(0, c)$.
- The x-intercepts (if any) are determined by the values of x that make $f(x) = 0$. To find them, solve the quadratic equation $ax^2 + bx + c = 0$.

EXAMPLE 9   Graph: $f(x) = -2x^2 - 8x - 8$

Strategy   We will follow the steps for graphing a quadratic function.

Why   This is the most efficient way to graph a general quadratic function.

Solution

Step 1:   Determine whether the parabola opens upward or downward. The function is in the form $f(x) = ax^2 + bx + c$, with $a = -2$, $b = -8$, and $c = -8$. Since $a < 0$, the parabola opens downward.

Step 2:   Find the vertex and draw the axis of symmetry. To find the coordinates of the vertex, we compute

$$x = -\frac{b}{2a}$$

$$x = -\frac{-8}{2(-2)} \qquad \text{Substitute } -2 \text{ for } a \text{ and } -8 \text{ for } b.$$

$$= -2 \qquad \text{This is the } x\text{-coordinate of the vertex.}$$

$$f\left(-\frac{b}{2a}\right) = f(-2)$$

$$= -2(-2)^2 - 8(-2) - 8$$

$$= -8 + 16 - 8$$

$$= 0 \qquad \text{This is the } y\text{-coordinate of the vertex.}$$

Success Tip

An easy way to remember the vertex formula is to note that $x = \frac{-b}{2a}$ is part of the quadratic formula:

$$x = \frac{-b \pm \sqrt{b^2 - 4ac}}{2a}$$

The vertex of the parabola is the point $(-2, 0)$. This point is in blue on the graph. The axis of symmetry is the line $x = -2$.

Step 3: *Find the x- and y-intercepts.* Since $c = -8$, the *y*-intercept of the parabola is $(0, -8)$. The point $(-4, -8)$, two units to the left of the axis of symmetry, must also be on the graph. We plot both points in black on the graph.

To find the *x*-intercepts, we set $f(x)$ equal to 0 and solve the resulting quadratic equation.

$$f(x) = -2x^2 - 8x - 8$$

$0 = -2x^2 - 8x - 8$ Set f(x) = 0.

$0 = x^2 + 4x + 4$ Divide both sides by -2.

$0 = (x + 2)(x + 2)$ Factor the trinomial.

$x + 2 = 0$ or $x + 2 = 0$ Set each factor equal to 0.

$x = -2$ $x = -2$

Since the solutions are the same, the graph has only one *x*-intercept: $(-2, 0)$. This point is the vertex of the parabola and has already been plotted.

Step 4: *Plot another point.* Finally, we find another point on the parabola. If $x = -3$, then $f(-3) = -2$. We plot $(-3, -2)$ and use symmetry to determine that $(-1, -2)$ is also on the graph. Both points are in green.

Step 5: Draw a smooth curve through the points, as shown.

$$f(x) = -2x^2 - 8x - 8$$

x	$f(x)$
-3	-2	$\longrightarrow (-3, -2)$

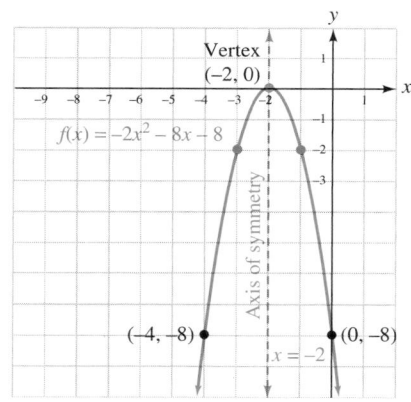

 Now Try   **Problems 59 and 69**

7 **Determine Minimum and Maximum Values.**

It is often useful to know the smallest or largest possible value a quantity can assume. For example, companies try to minimize their costs and maximize their profits. If the quantity is expressed by a quadratic function, the *y*-coordinate of the vertex of the graph of the function gives its minimum or maximum value.

EXAMPLE 10   ***Minimizing Costs.*** A glassworks that makes lead crystal vases has daily production costs given by the function $C(x) = 0.2x^2 - 10x + 650$, where *x* is the number of vases made each day. How many vases should be produced to minimize the per-day costs? What will the costs be?

Strategy   We will find the vertex of the graph of the quadratic function.

Why   The *x*-coordinate of the vertex indicates the number of vases to make to keep costs at a minimum, and the *y*-coordinate indicates the minimum cost.

Solution   The graph of $C(x) = 0.2x^2 - 10x + 650$ is a parabola opening upward. The vertex is the lowest point on the graph. To find the vertex, we compute

$$-\frac{b}{2a} = -\frac{-10}{2(0.2)} \qquad b = -10 \text{ and } a = 0.2. \qquad f\left(-\frac{b}{2a}\right) = f(25)$$

$$= -\frac{-10}{0.4} \qquad\qquad\qquad\qquad = 0.2(25)^2 - 10(25) + 650$$

$$= 25 \qquad\qquad\qquad\qquad\qquad = 525$$

The vertex is (25, 525), and it indicates that the costs are a minimum of $525 when 25 vases are made daily.

To solve this problem with a graphing calculator, we graph the function $C(x) = 0.2x^2 - 10x + 650$. By using TRACE and ZOOM, we can locate the vertex of the graph. The coordinates of the vertex indicate that the minimum cost is $525 when the number of vases produced is 25.

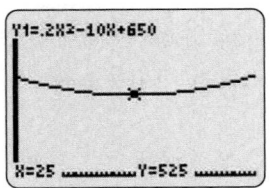

 Now Try   **Problem 85**

8   **Solve Quadratic Equations Graphically.**

When solving quadratic equations graphically, we must consider three possibilities. If the graph of the associated quadratic function has two *x*-intercepts, the quadratic equation has two real-number solutions. Figure (a) shows an example of this. If the graph has one *x*-intercept, as shown in figure (b), the equation has one repeated real-number solution. Finally, if the graph does not have an *x*-intercept, as shown in figure (c), the equation does not have any real-number solutions.

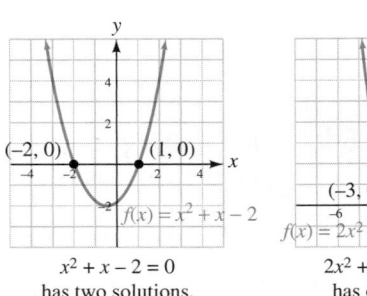

$x^2 + x - 2 = 0$
has two solutions,
−2 and 1.

(a)

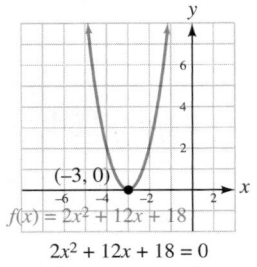

$2x^2 + 12x + 18 = 0$
has one repeated
solution, −3.

(b)

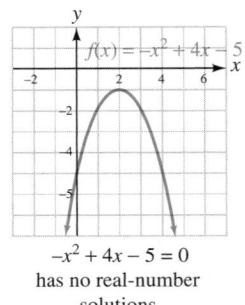

$-x^2 + 4x - 5 = 0$
has no real-number
solutions.

(c)

Using Your Calculator	***Solving Quadratic Equations Graphically***

We can use a graphing calculator to find approximate solutions of quadratic equations. For example, the solutions of $0.7x^2 + 2x - 3.5 = 0$ are the numbers x that will make $y = 0$ in the quadratic function $f(x) = 0.7x^2 + 2x - 3.5$. To approximate these numbers, we graph the quadratic function and read the x-intercepts from the graph using the ZERO feature. (The ZERO feature can be found by pressing $\boxed{2^{nd}}$, CALC, and then 2.) In the figure, we see that the x-coordinate of the left-most x-intercept of the graph is given as -4.082025. This means that an approximate solution of the equation is -4.08. To find the positive x-intercept, we use similar steps.

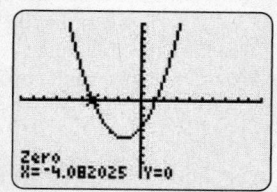

▷ **ANSWERS TO SELF CHECKS**

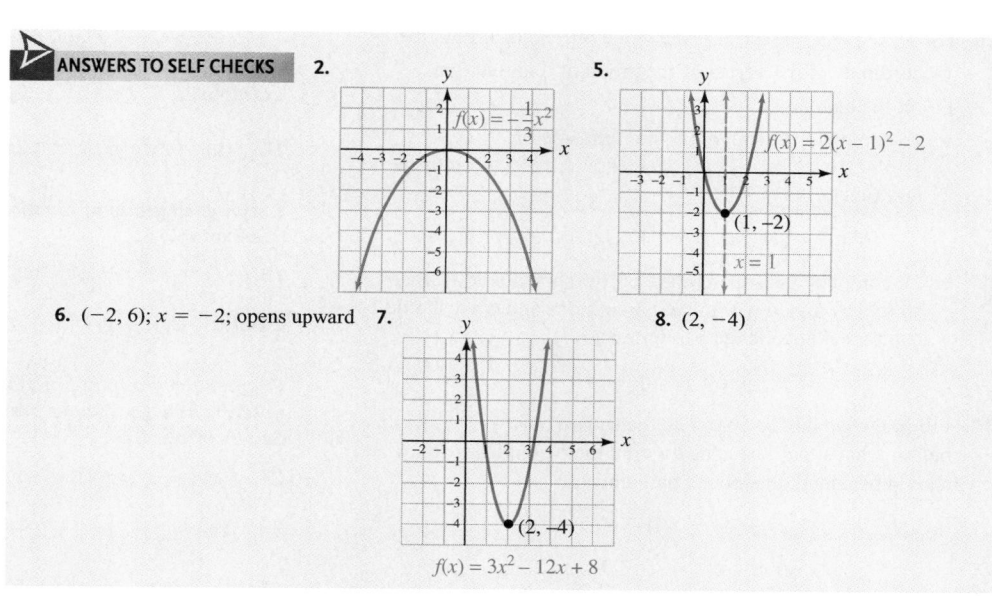

2. $f(x) = -\frac{1}{3}x^2$

5. $f(x) = 2(x - 1)^2 - 2$; $(1, -2)$; $x = 1$

6. $(-2, 6)$; $x = -2$; opens upward   **7.**

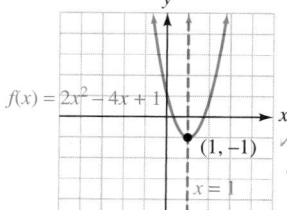

$f(x) = 3x^2 - 12x + 8$; $(2, -4)$

8. $(2, -4)$

STUDY SET
10.4

VOCABULARY

Refer to the graph. Fill in the blanks.

1. $f(x) = 2x^2 - 4x + 1$ is called a _____ function. Its graph is a cup-shaped figure called a _____.

2. The lowest point on the graph is $(1, -1)$. This is called the _____ of the parabola.

3. The vertical line $x = 1$ divides the parabola into two halves. This line is called the _____ ___ _____.

$f(x) = 2x^2 - 4x + 1$; $(1, -1)$; $x = 1$

4. $f(x) = a(x - h)^2 + k$ is called the _____ form of the equation of a quadratic function.

CONCEPTS

5. Refer to the graph.
   a. What are the x-intercepts of the graph?
   b. What is the y-intercept of the graph?
   c. What is the vertex?
   d. What is the axis of symmetry?
   e. What is the domain and the range of the function?

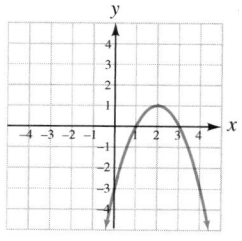

6. The vertex of a parabola is at $(1, -3)$, its y-intercept is $(0, -2)$, and it passes through the point $(3, 1)$, as shown in the illustration. Use the axis of symmetry shown in blue to help determine two other points on the parabola.

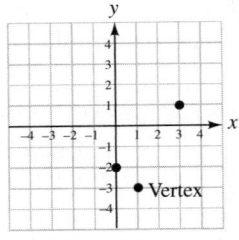

7. Draw the graph of a quadratic function using the given facts about its graph.

- Opens upward
- y-intercept: $(0, -3)$
- Vertex: $(-1, -4)$
- x-intercepts: $(-3, 0)$, $(1, 0)$

x	$f(x)$
2	5

8. For $f(x) = -x^2 + 6x - 7$, the value of $-\frac{b}{2a}$ is 3. Find the y-coordinate of the vertex of the graph of this function.

9. Fill in the blanks.

a. To complete the square on the right side of $f(x) = 2x^2 + 12x + 11$, what should be factored from the first two terms?

$$f(x) = \boxed{}(x^2 + 6x) + 11$$

b. To complete the square on $x^2 + 6x$ shown below, what should be added within the parentheses and what should be subtracted outside the parentheses?

$$f(x) = 2(x^2 + 6x + \boxed{}) + 11 - \boxed{}$$

10. Fill in the blanks. To complete the square on $x^2 + 4x$ shown below, what should be added within the parentheses and what should be added outside the parentheses?

$$f(x) = -5(x^2 + 4x + \boxed{}) + 7 + \boxed{}$$

11. Use the graph of $f(x) = \frac{1}{10}x^2 - \frac{1}{5}x - \frac{3}{2}$, shown below, to estimate the solutions of the equation $\frac{1}{10}x^2 - \frac{1}{5}x - \frac{3}{2} = 0$.

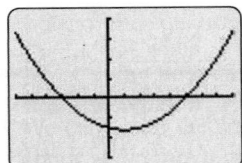

12. Three quadratic equations are to be solved graphically. The graphs of their associated quadratic functions are shown here. Determine which graph indicates that the equation has

a. two real solutions.

b. one repeated real solution.

c. no real solutions.

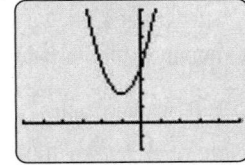

(i)

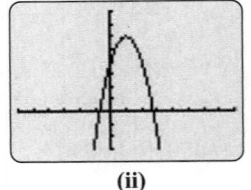

(ii)

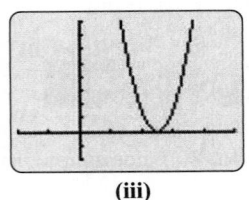

(iii)

NOTATION

13. The function $f(x) = 2(x + 1)^2 + 6$ is written in the form $f(x) = a(x - h)^2 + k$. Is $h = -1$ or is $h = 1$? Explain.

14. Consider the function $f(x) = 2x^2 + 4x - 8$.

a. What are a, b, and c?

b. Find $-\frac{b}{2a}$.

GUIDED PRACTICE

Graph each group of functions on the same coordinate system. See Example 1.

15. $f(x) = x^2$, $g(x) = 2x^2$, $s(x) = \frac{1}{2}x^2$

16. $f(x) = x^2$, $g(x) = 4x^2$, $s(x) = \frac{1}{4}x^2$

Graph each pair of functions on the same coordinate system. See Example 2.

17. $f(x) = 2x^2$, $g(x) = -2x^2$ **18.** $f(x) = \frac{1}{2}x^2$, $g(x) = -\frac{1}{2}x^2$

Graph each group of functions on the same coordinate system. See Example 3.

19. $f(x) = 4x^2$, $g(x) = 4x^2 + 3$, $s(x) = 4x^2 - 2$

20. $f(x) = \frac{1}{3}x^2$, $g(x) = \frac{1}{3}x^2 + 4$, $s(x) = \frac{1}{3}x^2 - 3$

Graph each group of functions on the same coordinate system and describe how the graphs are similar and how they are different. See Example 4.

21. $f(x) = 3x^2$, $g(x) = 3(x + 2)^2$, $s(x) = 3(x - 3)^2$

22. $f(x) = \frac{1}{2}x^2$, $g(x) = \frac{1}{2}(x + 3)^2$, $s(x) = \frac{1}{2}(x - 2)^2$

Find the vertex and the axis of symmetry of the graph of each function. Do not graph the equation, but determine whether the graph will open upward or downward. See Example 5.

23. $f(x) = (x - 1)^2 + 2$

24. $f(x) = 2(x - 2)^2 - 1$

25. $f(x) = -2(x + 3)^2 - 4$

26. $f(x) = -3(x + 1)^2 + 3$

27. $f(x) = -0.5(x - 7.5)^2 + 8.5$

28. $f(x) = -\frac{3}{2}\left(x + \frac{1}{4}\right)^2 + \frac{7}{8}$

29. $f(x) = 2x^2 - 4$

30. $f(x) = 3x^2 - 3$

Determine the vertex and the axis of symmetry of the graph of each function. Then plot several points and complete the graph. See Example 5.

31. $f(x) = (x - 3)^2 + 2$ **32.** $f(x) = (x + 1)^2 - 2$

33. $f(x) = -(x - 2)^2$ **34.** $f(x) = -(x + 2)^2$

35. $f(x) = -2(x + 3)^2 + 4$

36. $f(x) = -2(x - 2)^2 - 4$

37. $f(x) = \dfrac{1}{2}(x + 1)^2 - 3$

38. $f(x) = \dfrac{1}{3}(x - 1)^2 + 2$

Determine the vertex and the axis of symmetry of the graph of each function. Will the graph open upward or downward? See Example 6.

39. $f(x) = x^2 + 4x + 5$

40. $f(x) = x^2 - 4x - 1$

41. $f(x) = -x^2 + 6x - 15$

42. $f(x) = -x^2 - 6x + 3$

Complete the square to write each function in $f(x) = a(x - h)^2 + k$ form. Determine the vertex and the axis of symmetry of the graph of the function. Then plot several points and complete the graph. See Examples 6 and 7.

43. $f(x) = x^2 + 2x - 3$

44. $f(x) = x^2 + 6x + 5$

45. $f(x) = 4x^2 + 24x + 37$

46. $f(x) = 3x^2 - 12x + 10$

47. $f(x) = x^2 + x - 6$

48. $f(x) = x^2 - x - 6$

49. $f(x) = -4x^2 + 16x - 10$

50. $f(x) = -2x^2 + 4x + 3$

51. $f(x) = 2x^2 + 8x + 6$

52. $f(x) = 3x^2 - 12x + 9$

53. $f(x) = -x^2 - 8x - 17$

54. $f(x) = -x^2 + 6x - 8$

Use the vertex formula to find the vertex of the graph of each function. See Example 8.

55. $f(x) = x^2 + 2x - 5$

56. $f(x) = -x^2 + 4x - 5$

57. $f(x) = 2x^2 - 3x + 4$

58. $f(x) = 2x^2 - 7x - 4$

Find the x- and y-intercepts of the graph of the quadratic function. See Example 9.

59. $f(x) = x^2 - 2x - 35$

60. $f(x) = -x^2 - 10x - 21$

61. $f(x) = -2x^2 + 4x$

62. $f(x) = 3x^2 + 6x - 9$

Determine the coordinates of the vertex of the graph of each function using the vertex formula. Then determine the x- and y-intercepts of the graph. Finally, plot several points and complete the graph. See Example 9.

63. $f(x) = x^2 + 4x + 4$

$x = \dfrac{-b}{2a}$

64. $f(x) = x^2 - 6x + 9$

65. $f(x) = -x^2 + 2x - 1$

66. $f(x) = -x^2 - 2x - 1$

67. $f(x) = x^2 - 2x$

68. $f(x) = x^2 + x$

69. $f(x) = 2x^2 - 8x + 6$

70. $f(x) = 3x^2 - 12x + 12$

71. $f(x) = -6x^2 - 12x - 8$

72. $f(x) = -2x^2 + 8x - 10$

73. $f(x) = 4x^2 - 12x + 9$

74. $f(x) = 4x^2 + 4x - 3$

Use a graphing calculator to find the coordinates of the vertex of the graph of each quadratic function. Round to the nearest hundredth. See Using Your Calculator: Finding the Vertex.

75. $f(x) = 2x^2 - x + 1$

76. $f(x) = x^2 + 5x - 6$

77. $f(x) = -x^2 + x + 7$

78. $f(x) = 2x^2 - 3x + 2$

Use a graphing calculator to solve each equation. If an answer is not exact, round to the nearest hundredth. See Using Your Calculator: Solving Quadratic Equations Graphically.

79. $x^2 + x - 6 = 0$

80. $2x^2 - 5x - 3 = 0$

81. $0.5x^2 - 0.7x - 3 = 0$

82. $2x^2 - 0.5x - 2 = 0$

APPLICATIONS

83. CROSSWORD PUZZLES
Darken the appropriate squares to the right of the dashed red line so that the puzzle has symmetry with respect to that line.

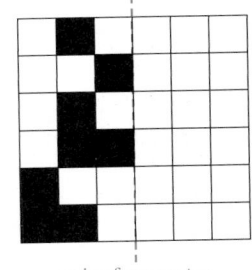

axis of symmetry

84. GRAPHIC ARTS   Draw an axis of symmetry over the letter shown here.

85. OPERATING COSTS The cost C in dollars of operating a certain concrete-cutting machine is related to the number of minutes n the machine is run by the function

$$C(n) = 2.2n^2 - 66n + 655$$

For what number of minutes is the cost of running the machine a minimum? What is the minimum cost?

86. WATER USAGE The height (in feet) of the water level in a reservoir over a 1-year period is modeled by the function

$$H(t) = 3.3(t - 9)^2 + 14$$

where $t = 1$ represents January, $t = 2$ represents February, and so on. How low did the water level get that year, and when did it reach the low mark?

87. FIREWORKS A fireworks shell is shot straight up with an initial velocity of 120 feet per second. Its height s after t seconds is given by the equation $s = 120t - 16t^2$. If the shell is designed to explode when it reaches its maximum height, how long after being fired, and at what height, will the fireworks appear in the sky?

88. BALLISTICS From the top of the building in the illustration, a ball is thrown straight up with an initial velocity of 32 feet per second. The equation

$$s = -16t^2 + 32t + 48$$

gives the height s of the ball t seconds after it is thrown. Find the maximum height reached by the ball and the time it takes for the ball to hit the ground.

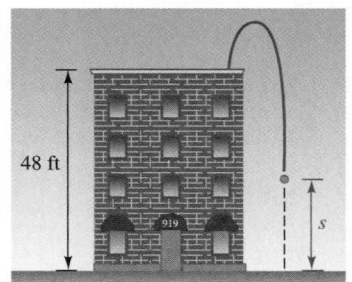

48 ft

89. from Campus to Careers
   Police Patrol Officer

Suppose you are a police patrol officer and you have a 300-foot-long roll of yellow "DO NOT CROSS" barricade tape to seal off an automobile accident, as shown in the illustration. What dimensions should you use to seal off the maximum rectangular area around the collision? What is the maximum area?

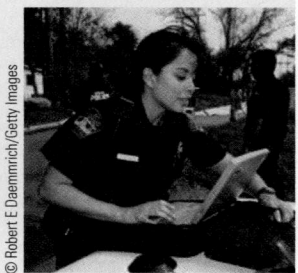

© Robert E Daemmrich/Getty Images

POLICE LINE   DO NOT CROSS

90. RANCHING See the illustration. A farmer wants to fence in three sides of a rectangular field with 1,000 feet of fencing. The other side of the rectangle will be a river. If the enclosed area is to be maximum, find the dimensions of the field.

1,000 ft

91. MILITARY HISTORY The function

$$N(x) = -0.0534x^2 + 0.337x + 0.97$$

gives the number of active-duty military personnel in the United States Army (in millions) for the years 1965–1972, where $x = 0$ corresponds to 1965, $x = 1$ corresponds to 1966, and so on. For this period, when was the army's personnel strength level at its highest, and what was it? Historically, can you explain why?

92. SCHOOL ENROLLMENT After peaking in 1970, school enrollment in the United States fell during the 1970s and 1980s. The total annual enrollment (in millions) in U.S. elementary and secondary schools for the years 1975–1996 is given by the model

$$E(x) = 0.058x^2 - 1.162x + 50.604$$

where $x = 0$ corresponds to 1975, $x = 1$ corresponds to 1976, and so on. For this period, when was enrollment the lowest? What was the enrollment?

93. MAXIMIZING REVENUE The revenue R received for selling x stereos is given by the formula

$$R = -\frac{x^2}{5} + 80x - 1,000$$

How many stereos must be sold to obtain the maximum revenue? Find the maximum revenue.

94. MAXIMIZING REVENUE When priced at $30 each, a toy has annual sales of 4,000 units. The manufacturer estimates that each $1 increase in price will decrease sales by 100 units. Find the unit price that will maximize total revenue. (*Hint:* Total revenue = price · the number of units sold.)

WRITING

95. Use the example of a stream of water from a drinking fountain to explain the concepts of the vertex and the axis of symmetry of a parabola. Draw a picture.

96. What are some quantities that are good to maximize? What are some quantities that are good to minimize?

97. A mirror is held against the y-axis of the graph of a quadratic function. What fact about parabolas does this illustrate?

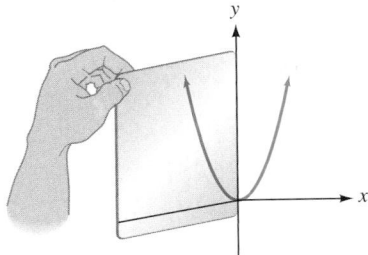

98. The vertex of a quadratic function $f(x) = ax^2 + bx + c$ is given by the formula $\left(-\frac{b}{2a}, f\left(-\frac{b}{2a}\right)\right)$. Explain what is meant by the notation $f\left(-\frac{b}{2a}\right)$.

99. A table of values for $f(x) = 2x^2 - 4x + 3$ is shown. Explain why it appears that the vertex of the graph of f is the point $(1, 1)$.

X	Y₁
.25	2.125
.5	1.5
.75	1.125
1	1
1.25	1.125
1.5	1.5
1.75	2.125

X=.25

100. The illustration shows the graph of the quadratic function $f(x) = -4x^2 + 12x$ with domain $[0, 3]$. Explain how the value of $f(x)$ changes as the value of x increases from 0 to 3.

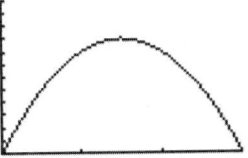

REVIEW

Simplify each expression. Assume all variables represent positive numbers.

101. $\dfrac{\sqrt{3}}{\sqrt{50}}$

102. $\dfrac{3}{\sqrt[3]{9}}$

103. $3\left(\sqrt{5b} - \sqrt{3}\right)^2$

104. $-2\sqrt{5b}\left(4\sqrt{2b} - 3\sqrt{3}\right)$

CHALLENGE PROBLEMS

105. Find a number between 0 and 1 such that the difference of the number and its square is a maximum.

106. Determine a quadratic function whose graph has x-intercepts of $(2, 0)$ and $(-4, 0)$.

SECTION 10.5
Quadratic and Other Nonlinear Inequalities

Objectives
1. Solve quadratic inequalities.
2. Solve rational inequalities.
3. Graph nonlinear inequalities in two variables.

We have previously solved *linear* inequalities in one variable such as $2x + 3 > 8$ and $6x - 7 < 4x - 9$. To find their solution sets, we used properties of inequalities to isolate the variable on one side of the inequality.

In this section, we will solve *quadratic* inequalities in one variable such as $x^2 + x - 6 < 0$ and $x^2 + 4x \geq 5$. We will use an interval testing method on the number line to determine their solution sets.

1 Solve Quadratic Inequalities.

Recall that a quadratic equation can be written in the form $ax^2 + bx + c = 0$. If we replace the = symbol with an inequality symbol, we have a quadratic inequality.

Quadratic Inequalities

A **quadratic inequality** can be written in one of the standard forms

$$ax^2 + bx + c < 0 \qquad ax^2 + bx + c > 0 \qquad ax^2 + bx + c \leq 0 \qquad ax^2 + bx + c \geq 0$$

where a, b, and c are real numbers and $a \neq 0$.

To solve a quadratic inequality in one variable, we will use the following steps to find the values of the variable that make the inequality true.

Solving Quadratic Inequalities

1. Write the inequality in standard form and solve its related quadratic equation.
2. Locate the solutions (called **critical numbers**) of the related quadratic equation on a number line.
3. Test each interval on the number line created in step 2 by choosing a test value from the interval and determining whether it satisfies the inequality. The solution set includes the interval(s) whose test value makes the inequality true.
4. Determine whether the endpoints of the intervals are included in the solution set.

EXAMPLE 1 Solve: $x^2 + x - 6 < 0$

Strategy We will solve the related quadratic equation $x^2 + x - 6 = 0$ by factoring to determine the critical numbers. These critical numbers will separate the number line into intervals.

Why We can test each interval to see whether numbers in the interval are in the solution set of the inequality.

Solution The expression $x^2 + x - 6$ can be positive, negative, or 0, depending on what value is substituted for x. Solutions of the inequality are x-values that make $x^2 + x - 6$ less than 0. To find them, we will follow the steps for solving quadratic inequalities.

Step 1: *Solve the related quadratic equation.* For the quadratic inequality $x^2 + x - 6 < 0$, the related quadratic equation is $x^2 + x - 6 = 0$.

$$x^2 + x - 6 = 0$$
$$(x + 3)(x - 2) = 0 \qquad \text{Factor the trinomial.}$$
$$x + 3 = 0 \quad \text{or} \quad x - 2 = 0 \qquad \text{Set each factor equal to 0.}$$
$$x = -3 \qquad \qquad x = 2 \qquad \text{Solve each equation.}$$

The solutions of $x^2 + x - 6 = 0$ are -3 and 2. These solutions are the critical numbers.

Step 2: *Locate the critical numbers on a number line.* When we highlight -3 and 2 on a number line, they separate the number line into three intervals:

Step 3: *Test each interval.* To determine whether the numbers in $(-\infty, -3)$ are solutions of the inequality, we choose a number from that interval, substitute it for x, and see whether it satisfies $x^2 + x - 6 < 0$. *If one number in that interval satisfies the inequality, all numbers in that interval will satisfy the inequality.*

If we choose -4 from $(-\infty, -3)$, we have:

$$x^2 + x - 6 < 0 \qquad \text{This is the original inequality.}$$
$$(-4)^2 + (-4) - 6 \overset{?}{<} 0 \qquad \text{Substitute } -4 \text{ for x.}$$
$$16 + (-4) - 6 \overset{?}{<} 0$$
$$6 < 0 \qquad \text{False}$$

Since -4 does not satisfy the inequality, none of the numbers in $(-\infty, -3)$ are solutions.
    To test the second interval, $(-3, 2)$, we choose $x = 0$.

$$x^2 + x - 6 < 0 \qquad \textit{This is the original inequality.}$$
$$0^2 + 0 - 6 \overset{?}{<} 0 \qquad \textit{Substitute 0 for x.}$$
$$-6 < 0 \qquad \textit{True}$$

Since 0 satisfies the inequality, all of the numbers in $(-3, 2)$ are solutions.
    To test the third interval, $(2, \infty)$, we choose $x = 3$.

$$x^2 + x - 6 < 0 \qquad \textit{This is the original inequality.}$$
$$3^2 + 3 - 6 \overset{?}{<} 0 \qquad \textit{Substitute 3 for x.}$$
$$9 + 3 - 6 \overset{?}{<} 0$$
$$6 < 0 \qquad \textit{False}$$

Since 3 does not satisfy the inequality, none of the numbers in $(2, \infty)$ are solutions.

> **Success Tip**
>
> If a quadratic inequality contains $\le$ or $\ge$, the endpoints of the intervals are included in the solution set. If the inequality contains $<$ or $>$, they are not.

Step 4: *Are the endpoints included?* From the interval testing, we see that only numbers from $(-3, 2)$ satisfy $x^2 + x - 6 < 0$. The endpoints -3 and 2 are not included in the solution set because they do not satisfy the inequality. (Recall that -3 and 2 make $x^2 + x - 6$ equal to 0.) The solution set is the interval $(-3, 2)$ as graphed on the right.

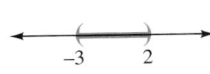

Self Check 1   Solve: $x^2 + x - 12 < 0$
Now Try   **Problem 15**

EXAMPLE 2   Solve: $x^2 + 4x \ge 5$

Strategy   This inequality is not in standard form because it does not have 0 on the right side. We will write it in standard form and solve its related quadratic equation to find any critical numbers. These critical numbers will separate the number line into intervals.

Why   We can then test each interval to see whether numbers in the interval are in the solution set of the inequality.

Solution   To get 0 on the right side, we subtract 5 from both sides.

$$x^2 + 4x \ge 5 \qquad \textit{This is the inequality to solve.}$$
$$x^2 + 4x - 5 \ge 0 \qquad \textit{Write the inequality in the equivalent form } ax^2 + bx + c \ge 0.$$

We can solve the related quadratic equation $x^2 + 4x - 5 = 0$ by factoring.

$$x^2 + 4x - 5 = 0$$
$$(x + 5)(x - 1) = 0 \qquad\qquad \textit{Factor the trinomial.}$$
$$x + 5 = 0 \quad \text{or} \quad x - 1 = 0 \qquad \textit{Set each factor equal to 0.}$$
$$x = -5 \qquad\qquad x = 1$$

The critical numbers -5 and 1 separate the number line into three intervals. We pick a test value from each interval to see whether it satisfies $x^2 + 4x - 5 \ge 0$.

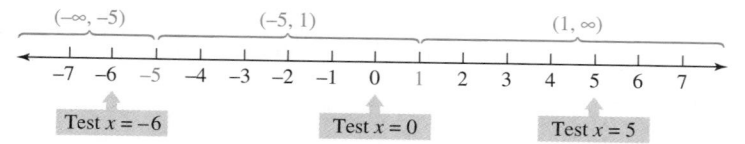

$(-\infty, -5)$ $(-5, 1)$ $(1, \infty)$

Test $x = -6$ Test $x = 0$ Test $x = 5$

$x^2 + 4x - 5 \geq 0$	$x^2 + 4x - 5 \geq 0$	$x^2 + 4x - 5 \geq 0$
$(-6)^2 + 4(-6) - 5 \overset{?}{\geq} 0$	$0^2 + 4(0) - 5 \overset{?}{\geq} 0$	$5^2 + 4(5) - 5 \overset{?}{\geq} 0$
$7 \geq 0$ True	$-5 \geq 0$ False	$40 \geq 0$ True

The numbers in the intervals $(-\infty, -5)$ and $(1, \infty)$ satisfy the inequality. Since the endpoints -5 and 1 also satisfy $x^2 + 4x - 5 \geq 0$, they are included in the solution set. (Recall that -5 and 1 make $x^2 + 4x - 5$ equal to 0.) Thus, the solution set is the union of two intervals: $(-\infty, -5] \cup [1, \infty)$. The graph of the solution set is shown on the right.

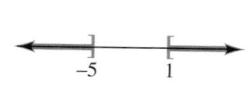

 Self Check 2   Solve: $x^2 + 3x \geq 40$

Now Try   **Problem 19**

② Solve Rational Inequalities.

Rational inequalities in one variable such as $\frac{9}{x} < 8$ and $\frac{x^2 + x - 2}{x - 4} \geq 0$ can also be solved using the interval testing method.

Solving Rational Inequalities

1. Write the inequality in standard form with a single quotient on the left side and 0 on the right side. Then solve its related rational equation.
2. Set the denominator equal to zero and solve that equation.
3. Locate the solutions (called *critical numbers*) found in steps 1 and 2 on a number line.
4. Test each interval on the number line created in step 3 by choosing a test value from the interval and determining whether it satisfies the inequality. The solution set includes the interval(s) whose test value makes the inequality true.
5. Determine whether the endpoints of the intervals are included in the solution set. Exclude any values that make the denominator 0.

EXAMPLE 3   Solve: $\frac{9}{x} < 8$

Strategy   This rational inequality is not in standard form because it does not have 0 on the right side. We will write it in standard form and solve its related rational equation to find any critical numbers. These critical numbers will separate the number line into intervals.

Why   We can test each interval to see whether numbers in the interval are in the solution set of the inequality.

Solution   To get 0 on the right side, we subtract 8 from both sides. We then find a common denominator to write the left side as a single quotient.

$$\frac{9}{x} < 8 \qquad \text{This is the inequality to solve.}$$

$$\frac{9}{x} - 8 < 0 \qquad \text{Subtract 8 from both sides.}$$

$$\frac{9}{x} - 8 \cdot \frac{x}{x} < 0 \qquad \text{To write the left side as a single quotient, build 8 to a fraction with denominator } x.$$

$$\frac{9}{x} - \frac{8x}{x} < 0$$

$$\frac{9 - 8x}{x} < 0 \qquad \text{Subtract the numerators and keep the common denominator, } x.$$

Now we solve the related rational equation.

$$\frac{9 - 8x}{x} = 0$$

$$9 - 8x = 0 \qquad \text{If } x \neq 0, \text{ we can clear the equation of the fraction by multiplying both sides by } x.$$

$$-8x = -9 \qquad \text{Subtract 9 from both sides.}$$

$$x = \frac{9}{8} \qquad \text{This is a critical number.}$$

If we set the denominator of $\frac{9 - 8x}{x}$ equal to 0, we obtain a second critical number, $x = 0$. When graphed, the critical numbers 0 and $\frac{9}{8}$ separate the number line into three intervals. We pick a test value from each interval to see whether it satisfies $\frac{9 - 8x}{x} < 0$.

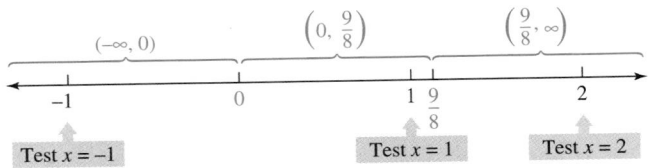

$$\frac{9 - 8x}{x} < 0 \qquad\qquad \frac{9 - 8x}{x} < 0 \qquad\qquad \frac{9 - 8x}{x} < 0$$

$$\frac{9 - 8(-1)}{-1} \overset{?}{<} 0 \qquad \frac{9 - 8(1)}{1} \overset{?}{<} 0 \qquad \frac{9 - 8(2)}{2} \overset{?}{<} 0$$

$$-17 < 0 \quad \text{True} \qquad 1 < 0 \quad \text{False} \qquad -\frac{7}{2} < 0 \quad \text{True}$$

The numbers in the intervals $(-\infty, 0)$ and $\left(\frac{9}{8}, \infty\right)$ satisfy the inequality. We do not include the endpoint 0 in the solution set, because it makes the denominator of the original inequality 0. Neither do we include $\frac{9}{8}$, because it does not satisfy $\frac{9 - 8x}{x} < 0$. (Recall that $\frac{9}{8}$ makes $\frac{9 - 8x}{x}$ equal to 0.) Thus, the solution set is the union of two intervals: $(-\infty, 0) \cup \left(\frac{9}{8}, \infty\right)$. Its graph is shown on the right.

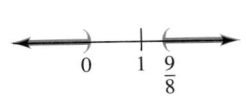

Self Check 3 Solve: $\frac{3}{x} < 5$

Now Try **Problem 23**

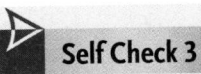

EXAMPLE 4   Solve: $\dfrac{x^2 + x - 2}{x - 4} \geq 0$

Strategy   This inequality is in standard form. We will solve its related rational equation to find any critical numbers. These critical numbers will separate the number line into intervals.

Why   We can test each interval to see whether numbers in the interval are in the solution set of the inequality.

Solution   To solve the related rational equation, we proceed as follows:

$$\frac{x^2 + x - 2}{x - 4} = 0$$

$$x^2 + x - 2 = 0 \qquad \text{If x} \neq 4, \text{ we can clear the equation of the fraction by multiplying both sides by x } - 4.$$

$$(x + 2)(x - 1) = 0 \qquad \text{Factor the trinomial.}$$

$$x + 2 = 0 \quad \text{or} \quad x - 1 = 0 \qquad \text{Set each factor equal to 0.}$$

$$x = -2 \quad \mid \quad x = 1 \qquad \text{These are critical numbers.}$$

If we set the denominator of $\dfrac{x^2 + x - 2}{x - 4}$ equal to 0, we see that $x = 4$ is also a critical number. When graphed, the critical numbers, -2, 1, and 4, separate the number line into four intervals. We pick a test value from each interval to see whether it satisfies $\dfrac{x^2 + x - 2}{x - 4} \geq 0$.

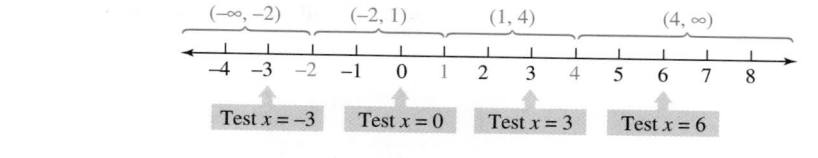

$$\frac{(-3)^2 + (-3) - 2}{-3 - 4} \overset{?}{\geq} 0 \qquad \frac{0^2 + 0 - 2}{0 - 4} \overset{?}{\geq} 0 \qquad \frac{3^2 + 3 - 2}{3 - 4} \overset{?}{\geq} 0 \qquad \frac{6^2 + 6 - 2}{6 - 4} \overset{?}{\geq} 0$$

$$-\frac{4}{7} \geq 0 \qquad\qquad \frac{1}{2} \geq 0 \qquad\qquad -10 \geq 0 \qquad\qquad 20 \geq 0$$

      False   True   False   True

The numbers in the intervals $(-2, 1)$ and $(4, \infty)$ satisfy the inequality. We include the endpoints -2 and 1 in the solution set because they satisfy the inequality. We do not include 4 because it makes the denominator of the inequality 0. Thus, the solution set is the union of two intervals $[-2, 1] \cup (4, \infty)$, as graphed on the right.

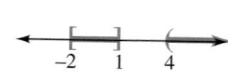

Self Check 4   Solve: $\dfrac{x + 2}{x^2 - 2x - 3} \geq 0$

Now Try   Problem 27

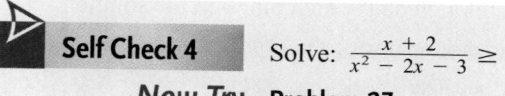

EXAMPLE 5   Solve: $\dfrac{3}{x - 1} < \dfrac{2}{x}$

Strategy   We will subtract $\frac{2}{x}$ from both sides to get 0 on the right side and solve the resulting related rational equation to find any critical numbers. These critical numbers will separate the number line into intervals.

Why We can test each interval to see whether numbers in the interval are in the solution set of the inequality.

Solution

$$\frac{3}{x-1} < \frac{2}{x} \quad \text{This is the inequality to solve.}$$

$$\frac{3}{x-1} - \frac{2}{x} < 0 \quad \text{Subtract } \tfrac{2}{x} \text{ from both sides.}$$

$$\frac{3}{x-1} \cdot \frac{x}{x} - \frac{2}{x} \cdot \frac{x-1}{x-1} < 0 \quad \text{To get a single quotient on the left side, build each rational expression to have the common denominator } x(x-1).$$

$$\frac{3x - 2x + 2}{x(x-1)} < 0 \quad \text{Subtract the numerators and keep the common denominator.}$$

$$\frac{x + 2}{x(x-1)} < 0 \quad \text{Combine like terms.}$$

The only solution of the related rational equation $\frac{x+2}{x(x-1)} = 0$ is -2. Thus, -2 is a critical number. When we set the denominator equal to 0 and solve $x(x-1) = 0$, we find two more critical numbers, 0 and 1. These three critical numbers create four intervals to test.

Success Tip

When the endpoints of an interval are consecutive integers, such as with the third interval $(0, 1)$, we cannot choose an integer as a test value. For these cases, choose a fraction or decimal that lies within the interval.

$(-\infty, -2)$ $(-2, 0)$ $(0, 1)$ $(1, \infty)$

Test $x = -3$ Test $x = -1$ Test $x = 0.5$ Test $x = 3$

$$\frac{-3+2}{-3(-3-1)} \overset{?}{<} 0 \qquad \frac{-1+2}{-1(-1-1)} \overset{?}{<} 0 \qquad \frac{0.5+2}{0.5(0.5-1)} \overset{?}{<} 0 \qquad \frac{3+2}{3(3-1)} \overset{?}{<} 0$$

$$\frac{-1}{-3(-4)} \overset{?}{<} 0 \qquad \frac{1}{-1(-2)} \overset{?}{<} 0 \qquad \frac{2.5}{0.5(-0.5)} \overset{?}{<} 0 \qquad \frac{5}{3(2)} \overset{?}{<} 0$$

$$-\frac{1}{12} < 0 \qquad \frac{1}{2} < 0 \qquad -10 < 0 \qquad \frac{5}{6} < 0$$

True   False   True   False

The numbers 0 and 1 are not included in the solution set because they make the denominator 0, and the number -2 is not included because it does not satisfy the inequality. The solution set is the union of two intervals $(-\infty, -2) \cup (0, 1)$, as graphed on the right.

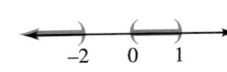

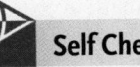 **Self Check 5**   Solve: $\frac{2}{x+1} > \frac{1}{x}$

Now Try Problem 31

Using Your Calculator   *Solving Inequalities Graphically*

We can solve $x^2 + 4x \geq 5$ (Example 2) graphically by writing the inequality as $x^2 + 4x - 5 \geq 0$ and graphing the quadratic function $f(x) = x^2 + 4x - 5$, as shown in figure (a). The solution set of the inequality will be those values of x for which the graph lies on or above the x-axis. We can trace to determine that this is the union of two intervals: $(-\infty, -5] \cup [1, \infty)$.

To solve $\frac{3}{x-1} < \frac{2}{x}$ (Example 5) graphically, we first write the inequality in the form $\frac{x+2}{x(x-1)} < 0$ and then graph the rational function $f(x) = \frac{x+2}{x(x-1)}$, as shown in figure (b). The solution of the inequality will be those values of x for which the graph lies below the axis.

We can trace to see that the graph is below the x-axis when x is less than -2. Since we cannot see the graph in the interval $0 < x < 1$, we redraw the graph using window settings of $[-1, 2]$ for x and $[-25, 10]$ for y, as shown in figure (c).

Now we see that the graph is below the x-axis in the interval $(0, 1)$. Thus, the solution set of the inequality is the union of the two intervals: $(-\infty, -2) \cup (0, 1)$.

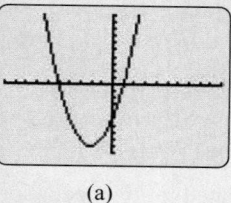   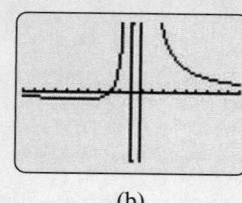   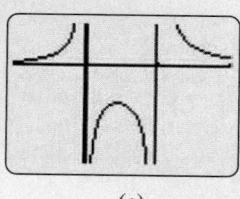

     (a)   (b)   (c)

3   **Graph Nonlinear Inequalities in Two Variables.**

We have previously graphed linear inequalities in two variables such as $y > 3x + 2$ and $2x - 3y \le 6$ using the following steps.

Graphing Inequalities in Two Variables

1. Graph the related equation to find the boundary line of the region. If the inequality allows equality (the symbol is either $\le$ or $\ge$), draw the boundary as a solid line. If equality is not allowed ($<$ or $>$), draw the boundary as a dashed line.

2. Pick a test point that is on one side of the boundary line. (Use the origin if possible.) Replace x and y in the original inequality with the coordinates of that point. If the inequality is satisfied, shade the side that contains that point. If the inequality is not satisfied, shade the other side of the boundary.

We use the same procedure to graph *nonlinear* inequalities in two variables.

EXAMPLE 6   Graph: $y < -x^2 + 4$

Strategy   We will graph the related equation $y = -x^2 + 4$ to establish a boundary parabola. Then we will determine which side of the boundary parabola represents the solution set of the inequality.

Why   To *graph a nonlinear inequality* in two variables means to draw a "picture" of the ordered pairs (x, y) that make the inequality true.

Solution   The graph of the boundary $y = -x^2 + 4$ is a parabola opening downward, with vertex at $(0, 4)$ and axis of symmetry $x = 0$ (the y-axis). Since the inequality contains an $<$ symbol and equality is not allowed, we draw the parabola using a dashed curve.

To determine which region to shade, we pick the test point $(0, 0)$ and substitute its coordinates into the inequality. We shade the region containing $(0, 0)$ because its coordinates satisfy $y < -x^2 + 4$.

Graph the boundary

$$y = -x^2 + 4$$

Compare to $y = a(x - h)^2 + k$

$a = -1$: Opens downward

$h = 0$ and $k = 4$: Vertex $(0, 4)$

Axis of symmetry $x = 0$

x	y
1	3
2	0

Shading: Use the test point (0, 0)

$$y < -x^2 + 4$$
$$0 \overset{?}{<} -0^2 + 4$$
$$0 < 4 \qquad \text{True}$$

Since $0 < 4$ is true, $(0, 0)$ is a solution of $y < -x^2 + 4$.

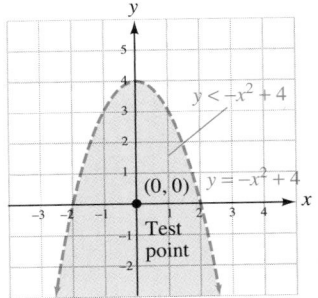

 Self Check 6   Graph: $y \geq -x^2 + 4$

Now Try   **Problem 35**

EXAMPLE 7   Graph: $x \leq |y|$

Strategy   We will graph the related equation $x = |y|$ to establish a boundary. Then we will determine which side of the boundary represents the solution set of the inequality.

Why   To *graph a nonlinear inequality* in two variables means to draw a "picture" of the ordered pairs (x, y) that make the inequality true.

Solution   To graph the boundary, $x = |y|$, we construct a table of solutions, as shown in figure (a). In figure (b), the boundary is graphed using a solid line because the inequality contains a $\leq$ symbol and equality is permitted. Since the origin is on the graph, we cannot use it as a test point. However, any other point, such as $(1, 0)$, will do. We substitute 1 for x and 0 for y into the inequality to get

$$x \leq |y|$$
$$1 \overset{?}{\leq} |0|$$
$$1 \leq 0 \qquad \text{False}$$

Since $1 \leq 0$ is a false statement, the point $(1, 0)$ does not satisfy the inequality and is not part of the graph. Thus, the graph of $x \leq |y|$ is to the left of the boundary. The complete graph is shown in figure (c).

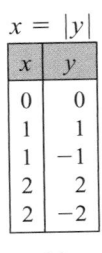

$x = |y|$

x	y
0	0
1	1
1	−1
2	2
2	−2

(a)

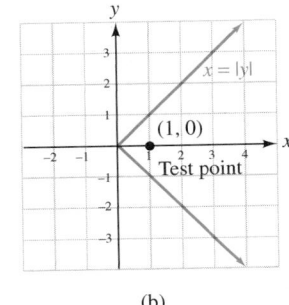

(b)

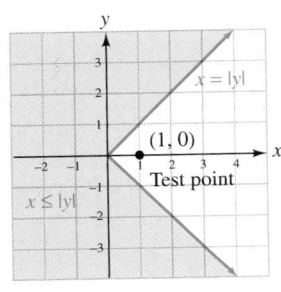

(c)

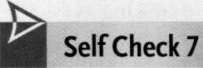 **Self Check 7**   Graph: $x \geq -|y|$
Now Try   **Problem 39**

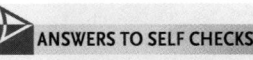 **ANSWERS TO SELF CHECKS**   **1.** $(-4, 3)$

2. $(-\infty, -8] \cup [5, \infty)$ **3.** 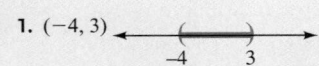 $(-\infty, 0) \cup \left(\frac{3}{5}, \infty\right)$

4. $[-2, -1) \cup (3, \infty)$ **5.** $(-1, 0) \cup (1, \infty)$

6.   **7.**

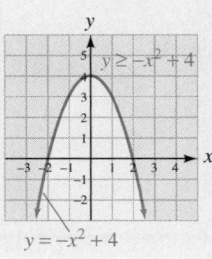

$y \geq -x^2 + 4$

$y = -x^2 + 4$

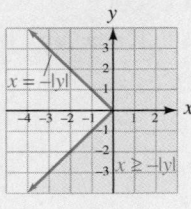

$x = -|y|$

$x \geq -|y|$

STUDY SET
10.5

VOCABULARY

Fill in the blanks.

1. $x^2 + 3x - 18 < 0$ is an example of a _____ inequality in one variable.

2. $\frac{x-1}{x^2 - x - 20} \leq 0$ is an example of a _____ inequality in one variable.

3. $y \leq x^2 - 4x + 3$ is an example of a nonlinear inequality in _____ variables.

4. The set of real numbers greater than 3 can be represented using the _____ notation $(3, \infty)$.

CONCEPTS

5. The critical numbers of a quadratic inequality are highlighted in red on the number line shown below. Use interval notation to represent each interval that must be tested to solve the inequality.

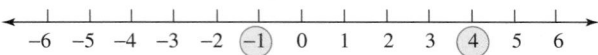

6. Graph each of the following solution sets.

  a. $(-2, 4)$ **b.** $(-\infty, -2) \cup (3, 5]$

7. The graph of the solution set of a rational inequality in one variable is shown. Determine whether each of the following numbers is a solution of the inequality.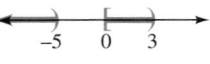

  a. -10 **b.** -5
  c. 0 **d.** 4

8. What are the critical numbers for each inequality?

  a. $x^2 - 2x - 48 \geq 0$ **b.** $\dfrac{x-3}{x(x+4)} > 0$

9. **a.** The results after interval testing for a quadratic inequality containing a $>$ symbol are shown below. (The critical numbers are highlighted in red.) What is the solution set?

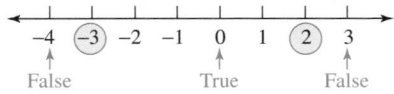

  b. The results after interval testing for a quadratic inequality containing a $\leq$ symbol are shown below. (The critical numbers are highlighted in red.) What is the solution set?

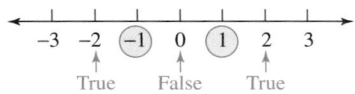

10. Fill in the blank to complete this important fact about the interval testing method discussed in this section: *If one number in an interval satisfies the inequality, ____ numbers in that interval will satisfy the inequality.*

11. a. When graphing the solution of $y \le x^2 + 2x + 1$, should the boundary be solid or dashed?

　　b. Does the test point $(0, 0)$ satisfy the inequality?

12. a. Estimate the solution of $x^2 - x - 6 > 0$ using the graph of $y = x^2 - x - 6$ shown in figure (a) below.

　　b. Estimate the solution of $\frac{x-3}{x} \le 0$ using the graph of $y = \frac{x-3}{x}$ shown in figure (b) below.

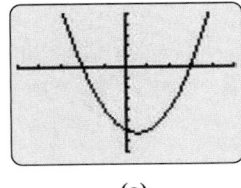 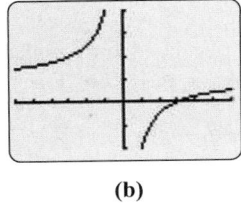

(a)　　　　　　　　　　**(b)**

NOTATION

13. Write the quadratic inequality $x^2 - 6x \ge 7$ in standard form.

14. The solution set of a rational inequality consists of the intervals $(-1, 4]$ and $(7, \infty)$. When writing the solution set, what symbol is used between the two intervals?

GUIDED PRACTICE

Solve each inequality. Write the solution set in interval notation and graph it. See Example 1.

15. $x^2 - 5x + 4 < 0$　　　　**16.** $x^2 + 2x - 8 < 0$

17. $x^2 - 8x + 15 > 0$　　　　**18.** $x^2 - 3x - 4 > 0$

Solve each inequality. Write the solution set in interval notation and graph it. See Example 2.

19. $x^2 - x \ge 42$　　　　　　**20.** $x^2 - x \ge 72$

21. $x^2 + x \le 12$　　　　　　**22.** $x^2 - 8x \le -15$

Solve each inequality. Write the solution set in interval notation and graph it. See Example 3.

23. $\dfrac{1}{x} < 2$　　　　　　　**24.** $\dfrac{1}{x} < 3$

25. $\dfrac{5}{x} \ge -3$　　　　　　**26.** $\dfrac{4}{x} \ge 8$

Solve each inequality. Write the solution set in interval notation and graph it. See Example 4.

27. $\dfrac{x^2 - x - 12}{x - 1} < 0$　　　**28.** $\dfrac{x^2 + x - 6}{x - 4} \ge 0$

29. $\dfrac{6x^2 - 5x + 1}{2x + 1} \ge 0$　　　**30.** $\dfrac{6x^2 + 11x + 3}{3x - 1} < 0$

Solve each inequality. Write the solution set in interval notation and graph it. See Example 5.

31. $\dfrac{3}{x - 2} < \dfrac{4}{x}$　　　　**32.** $\dfrac{-6}{x + 1} \ge \dfrac{1}{x}$

33. $\dfrac{7}{x - 3} \ge \dfrac{2}{x + 4}$　　　**34.** $\dfrac{-5}{x - 4} < \dfrac{3}{x + 1}$

Graph each inequality. See Example 6.

35. $y < x^2 + 1$　　　　　　**36.** $y > x^2 - 3$
37. $y \le x^2 + 5x + 6$　　　　**38.** $y \ge x^2 + 5x + 4$

Graph each inequality. See Example 7.

39. $y < |x + 4|$　　　　　　**40.** $y \le |x - 3|$
41. $y \ge -|x| + 2$　　　　　**42.** $y > |x| - 2$

Use a graphing calculator to solve each inequality. Write the solution set in interval notation. See Using Your Calculator: Solving Inequalities Graphically.

43. $x^2 - 2x - 3 < 0$　　　　**44.** $x^2 + x - 6 > 0$

45. $\dfrac{x + 3}{x - 2} > 0$　　　　　**46.** $\dfrac{3}{x} < 2$

TRY IT YOURSELF

Solve each inequality. Write the solution set in interval notation and graph it.

47. $\dfrac{x}{x + 4} \le \dfrac{1}{x + 1}$

48. $\dfrac{x}{x + 9} \ge \dfrac{1}{x + 1}$

49. $x^2 \ge 9$　　　　　　　　**50.** $x^2 \ge 16$

51. $x^2 + 6x \ge -9$　　　　　**52.** $x^2 + 8x < -16$

53. $\dfrac{x^2 + x - 2}{x - 3} > 0$　　　**54.** $\dfrac{x - 2}{x^2 - 1} > 0$

55. $2x^2 - 50 < 0$

56. $3x^2 - 243 < 0$

57. $\dfrac{2x - 3}{3x + 1} < 0$

58. $\dfrac{x - 5}{x + 1} < 0$

59. $x^2 - 6x + 9 < 0$

60. $x^2 + 4x + 4 > 0$

61. $\dfrac{5}{x + 1} > \dfrac{3}{x - 4}$

62. $\dfrac{3}{x - 2} \le -\dfrac{2}{x + 3}$

APPLICATIONS

63. BRIDGES   If an x-axis is superimposed over the roadway of the Golden Gate Bridge, with the origin at the center of the bridge, the length L in feet of a vertical support cable can be approximated by the formula

$$L = \frac{1}{9{,}000}x^2 + 5$$

For the Golden Gate Bridge, $-2{,}100 < x < 2{,}100$. For what intervals along the x-axis are the vertical cables more than 95 feet long?

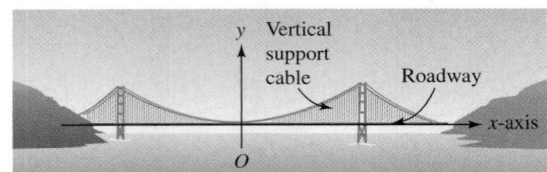

64. MALLS   The number of people n in a mall is modeled by the formula

$$n = -100x^2 + 1{,}200x$$

where x is the number of hours since the mall opened. If the mall opened at 9 A.M., when were there 2,000 or more people in it?

WRITING

65. How are critical numbers used when solving a quadratic inequality in one variable?

66. Explain how to graph $y \ge x^2$.

67. The graph of $f(x) = x^2 - 3x + 4$ is shown below. Explain why the quadratic inequality $x^2 - 3x + 4 < 0$ has no solution.

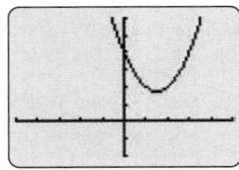

68. Describe the following solution set of a rational inequality in words: $(-\infty, 4] \cup (6, 7)$.

REVIEW

Translate each statement into an equation.

69. x varies directly with y.

70. y varies inversely with t.

71. t varies jointly with x and y.

72. d varies directly with t and inversely with u^2.

CHALLENGE PROBLEMS

73. a. Solve: $x^2 - x - 12 > 0$

  b. Find a rational inequality in one variable that has the same solution set as the quadratic inequality in part (a).

74. a. Solve: $\dfrac{1}{x} < 1$

  b. Now incorrectly "solve" $\dfrac{1}{x} < 1$ by multiplying both sides by x to clear it of the fraction. What part of the solution set is not obtained with this incorrect approach?

CHAPTER 10
Summary & Review

SECTION 10.1   The Square Root Property and Completing the Square

DEFINITIONS AND CONCEPTS	EXAMPLES
We can use the **square root property** to solve equations of the form $x^2 = c$, where $c > 0$. The two solutions are $$x = \sqrt{c} \quad \text{or} \quad x = -\sqrt{c}$$ We can write $x = \sqrt{c}$ or $x = -\sqrt{c}$ in more compact form using **double-sign notation**: $$x = \pm\sqrt{c}$$	Solve: $x^2 = 24$ $\quad x = \sqrt{24} \quad$ or $\quad x = -\sqrt{24} \quad$ Use the square root property. $\quad x = \pm\sqrt{24} \qquad\qquad$ Use double-sign notation. $\quad x = \pm 2\sqrt{6} \qquad\qquad$ Simplify $\sqrt{24}$. The solutions are $2\sqrt{6}$ and $-2\sqrt{6}$. Solve: $(x - 3)^2 = -81$ $\quad x - 3 = \pm\sqrt{-81} \qquad$ Use the square root property and double-sign notation. $\quad x = 3 \pm \sqrt{-81} \qquad$ To isolate x, add 3 to both sides. $\quad x = 3 \pm 9i \qquad\qquad$ Simplify the radical expression. The solutions are $3 + 9i$ and $3 - 9i$.
To **complete the square** on $x^2 + bx$, add the square of one-half of the coefficient of x. $$x^2 + bx + \left(\frac{1}{2}b\right)^2$$	Complete the square on $x^2 + 8x$ and factor the resulting perfect-square trinomial. $\quad x^2 + 8x + 16 \qquad$ The coefficient of x is 8. To complete the square: $\frac{1}{2} \cdot 8 = 4$ and $4^2 = 16$. Add 16 to the binomial. Now we factor: $x^2 + 8x + 16 = (x + 4)^2$
To **solve a quadratic equation in x by completing the square:** 1. If necessary, divide both sides of the equation by the coefficient of x^2 to make its coefficient 1. 2. Get all variable terms on one side of the equation and all constants on the other side. 3. Complete the square. 4. Factor the perfect-square trinomial. 5. Solve the resulting equation by using the square root property. 6. Check your answers in the original equation.	Solve: $3x^2 - 12x + 6 = 0$ $\quad \dfrac{3x^2}{3} - \dfrac{12x}{3} + \dfrac{6}{3} = \dfrac{0}{3} \quad$ To make the leading coefficient 1, divide both sides by 3. $\quad x^2 - 4x + 2 = 0 \qquad$ Do the divisions. $\quad x^2 - 4x \quad\;\; = -2 \qquad$ Subtract 2 from both sides so that the constant term, -2, is on the right side. $\quad x^2 - 4x + 4 = -2 + 4 \quad$ The coefficient of x is -4. To complete the square: $\frac{1}{2}(-4) = -2$ and $(-2)^2 = 4$. Add 4 to both sides. $\quad (x - 2)^2 = 2 \qquad$ Factor the perfect-square trinomial on the left side. Add on the right side. $\quad x - 2 = \pm\sqrt{2} \qquad$ Use the square root property. $\quad x = 2 \pm \sqrt{2} \qquad$ To isolate x, add 2 to both sides. The solutions are $2 + \sqrt{2}$ and $2 - \sqrt{2}$.

REVIEW EXERCISES

Solve each equation by factoring.

1. $x^2 + 9x + 20 = 0$ **2.** $6x^2 + 17x + 5 = 0$

Solve each equation using the square root property.

3. $x^2 = 28$ **4.** $(t + 2)^2 = 36$

5. $a^2 + 25 = 0$ **6.** $5x^2 - 49 = 0$

7. Solve $A = \pi r^2$ for r. Assume all variables represent positive numbers. Express the result in simplified radical form.

8. Complete the square on $x^2 - x$ and then factor the resulting perfect-square trinomial.

Solve each equation by completing the square.

9. $x^2 + 6x + 8 = 0$ **10.** $2x^2 - 6x + 3 = 0$

11. $6a^2 - 12a = -1$ **12.** $x^2 - 2x = -13$

13. Explain why completing the square on $x^2 + 7x$ is more difficult than completing the square on $x^2 + 6x$.

14. Explain the error: $\dfrac{2 \pm \sqrt{7}}{2} = \dfrac{\overset{1}{\cancel{2}} \pm \sqrt{7}}{\underset{1}{\cancel{2}}}$

15. a. Write an expression that represents the width of the larger rectangle shown in red.

b. Write an expression that represents the length of the larger rectangle shown in red.

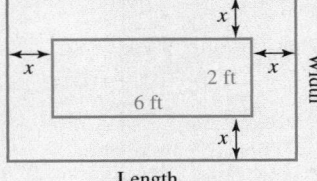

16. HAPPY NEW YEAR As part of a New Year's Eve celebration, a huge ball is to be dropped from the top of a 605-foot-tall building at the proper moment so that it strikes the ground at exactly 12:00 midnight. The distance d in feet traveled by a free-falling object in t seconds is given by the formula $d = 16t^2$. To the nearest second, when should the ball be dropped from the building?

SECTION 10.2   The Quadratic Formula

DEFINITIONS AND CONCEPTS	EXAMPLES

DEFINITIONS AND CONCEPTS

To **solve a quadratic equation in x using the quadratic formula:**

1. Write the equation in standard form: $ax^2 + bx + c = 0$.

2. Identify a, b, and c.

3. Substitute the values for a, b, and c in the quadratic formula

$$x = \frac{-b \pm \sqrt{b^2 - 4ac}}{2a}$$

and evaluate the right side to obtain the solutions.

EXAMPLES

Solve: $3x^2 - 2x - 2 = 0$

Here, $a = 3$, $b = -2$, and $c = -2$.

$$x = \frac{-b \pm \sqrt{b^2 - 4ac}}{2a}$$ This is the quadratic formula.

$$x = \frac{-(-2) \pm \sqrt{(-2)^2 - 4(3)(-2)}}{2(3)}$$ Substitute 3 for a, -2 for b, and -2 for c.

$$x = \frac{2 \pm \sqrt{4 - (-24)}}{6}$$ Evaluate the power and multiply within the radical. Multiply in the denominator.

$$x = \frac{2 \pm \sqrt{28}}{6}$$ Add the opposite: $4 - (-24) = 4 + 24 = 28$.

$$x = \frac{2 \pm 2\sqrt{7}}{6}$$ Simplify the radical: $\sqrt{28} = \sqrt{4}\sqrt{7} = 2\sqrt{7}$.

$$x = \frac{\overset{1}{\cancel{2}}(1 \pm \sqrt{7})}{\underset{1}{\cancel{2} \cdot 3}}$$ Factor out the GCF, 2, from the two terms in the numerator. In the denominator, factor 6 as $2 \cdot 3$. Remove the common factor, 2.

$$x = \frac{1 \pm \sqrt{7}}{3}$$

The exact solutions are $\dfrac{1 + \sqrt{7}}{3}$ and $\dfrac{1 - \sqrt{7}}{3}$. We can use a calculator to approximate them. To two decimal places, they are 1.22 and -0.55.

When solving a quadratic equation using the quadratic formula, we can often **simplify the computations** by solving an equivalent equation that does not involve fractions or decimals, and whose leading coefficient is positive.

Before solving . . .	**do this . . .**	**to get this**
$-3x^2 + 5x - 1 = 0$	Multiply both sides by -1	$3x^2 - 5x + 1 = 0$
$x^2 + \dfrac{7}{8}x - \dfrac{1}{2} = 0$	Multiply both sides by 8	$8x^2 + 7x - 4 = 0$
$60x^2 - 40x + 90 = 0$	Divide both sides by 10	$6x^2 - 4x + 9 = 0$
$0.05x^2 + 0.16x + 0.71 = 0$	Multiply both sides by 100	$5x^2 + 16x + 71 = 0$

REVIEW EXERCISES

Solve each equation using the quadratic formula.

17. $2x^2 + 13x = 7$

18. $-x^2 + 10x - 18 = 0$

19. $x^2 - 10x = 0$

20. $3y^2 = 26y - 2$

21. $\frac{1}{3}p^2 + \frac{1}{2}p + \frac{1}{2} = 0$

22. $3{,}000t^2 - 4{,}000t = -2{,}000$

23. $0.5x^2 + 0.3x - 0.1 = 0$

24. $x^2 - 3x - 27 = 0$

25. TUTORING   A private tutoring company charges $20 for a 1-hour session. Currently, 300 students are tutored each week. Since the company is losing money, the owner has decided to increase the price. For each 50¢ increase, she estimates that 5 fewer students will participate. If the company needs to bring in $6,240 per week to stay in business, what price must be charged for a 1-hour tutoring session to produce this amount of revenue?

26. POSTERS   The specifications for a poster of Cesar Chavez call for a 615-square-inch photograph to be surrounded by a green border. The borders on the top and bottom of the poster are to be twice as wide as those on the sides. Find the width of each border.

35 in.

← 23 in. →

© Hulton-Deutsch Collection/CORBIS

27. ACROBATS   To begin his routine on a trapeze, an acrobat is catapulted upward as shown in the illustration. His distance *d* (in feet) from the arena floor during this maneuver is given by the formula $d = -16t^2 + 40t + 5$, where *t* is the time in seconds since being launched. If the trapeze bar is 25 feet in the air, at what two times will he be able to grab it? Round to the nearest tenth.

28. TRIANGLES   The length of the longer leg of a right triangle exceeds the length of the shorter leg by 23 inches and the length of the hypotenuse is 65 inches. Find the length of each leg of the triangle.

SECTION 10.3   The Discriminant and Equations That Can Be Written in Quadratic Form

DEFINITIONS AND CONCEPTS	EXAMPLES
The **discriminant** predicts the type of solutions of $ax^2 + bx + c = 0$, where *a*, *b*, and *c* are real numbers and $a \neq 0$: 1. If $b^2 - 4ac > 0$, there are two different real-number solutions. If $b^2 - 4ac$ is a perfect square, there are two different rational-number solutions. If $b^2 - 4ac$ is not a perfect square, there are two different irrational-number solutions. 2. If $b^2 - 4ac = 0$, there is one repeated solution, a rational number. 3. If $b^2 - 4ac < 0$, there are two different imaginary-number solutions that are complex conjugates.	In the quadratic equation $2x^2 - 5x - 3 = 0$, we have $a = 2$, $b = -5$, and $c = -3$. So the value of the discriminant is $$b^2 - 4ac = (-5)^2 - 4(2)(-3) = 25 + 24 = 49$$ Since the value of the discriminant is positive and a perfect square, the equation $2x^2 - 5x - 3 = 0$ has two different rational-number solutions.

SECTION 10.3   The Discriminant and Equations That Can Be Written in Quadratic Form–*continued*

DEFINITIONS AND CONCEPTS	EXAMPLES
Equations that contain an expression, the same expression squared, and a constant term are said to be **quadratic in form.** One method used to solve such equations is to make a **substitution.**	Solve: $x^{2/3} - 6x^{1/3} + 5 = 0$ The equation can be written in quadratic form: $(x^{1/3})^2 - 6x^{1/3} + 5 = 0$ We substitute y for $x^{1/3}$ and use factoring to solve the resulting quadratic equation $y^2 - 6y + 5 = 0$. $(y - 1)(y - 5) = 0$ Let $y = x^{1/3}$. $y = 1$ or $y = 5$ Now we reverse the substitution $y = x^{1/3}$ and solve for x. $x^{1/3} = 1$ or $x^{1/3} = 5$ $(x^{1/3})^3 = (1)^3$ $(x^{1/3})^3 = (5)^3$ $x = 1$ $x = 125$ The solutions are 1 and 125. Check both in the original equation.

REVIEW EXERCISES

Use the discriminant to determine the number and type of solutions for each equation.

29. $3x^2 + 4x - 3 = 0$

30. $4x^2 - 5x + 7 = 0$

31. $3x^2 - 4x + \dfrac{4}{3} = 0$

32. $m(2m - 3) = 20$

Solve each equation.

33. $x - 13\sqrt{x} + 12 = 0$

34. $a^{2/3} + a^{1/3} - 6 = 0$

35. $3x^4 + x^2 - 2 = 0$

36. $\dfrac{6}{x + 2} + \dfrac{6}{x + 1} = 5$

37. $(x - 7)^2 + 6(x - 7) + 10 = 0$

38. $m^{-4} - 2m^{-2} + 1 = 0$

39. $4\left(\dfrac{x + 1}{x}\right)^2 + 12\left(\dfrac{x + 1}{x}\right) + 9 = 0$

40. $2m^{2/5} - 5m^{1/5} + 2 = 0$

41. WEEKLY CHORES   Working together, two sisters can do the yard work at their house in 45 minutes. When the older girl does it all herself, she can complete the job in 20 minutes less time than it takes the younger girl working alone. How long does it take the older girl to do the yard work?

42. ROAD TRIPS   A woman drives her automobile 150 miles at a rate of r mph. She could have gone the same distance in 2 hours less time if she had increased her speed by 20 mph. Find r.

SECTION 10.4   Quadratic Functions and Their Graphs

DEFINITIONS AND CONCEPTS	EXAMPLES
A **quadratic function** is a second-degree polynomial function of the form $f(x) = ax^2 + bx + c$	Quadratic functions: $f(x) = 2x^2 - 3x + 5$, $g(x) = -x^2 + 4x$, and $s(x) = \dfrac{1}{4}x^2 - 10$

SECTION 10.4   Quadratic Functions and Their Graphs–*continued*

DEFINITIONS AND CONCEPTS	EXAMPLES

The graph of the quadratic function $f(x) = a(x - h)^2 + k$ where $a \neq 0$ is a **parabola** with **vertex** at (h, k). The **axis of symmetry** is the line $x = h$. The parabola opens upward when $a > 0$ and downward when $a < 0$.

Graph: $f(x) = 2(x + 1)^2 - 8$

$$f(x) = 2[x - (-1)]^2 - 8$$
$$\qquad\quad\uparrow \qquad\quad \uparrow \qquad \uparrow$$
$$f(x) = a(x - \quad h)^2 + k$$

We see that $a = 2$, $h = -1$, and $k = -8$. The graph is a parabola with vertex $(h, k) = (-1, -8)$ and axis of symmetry $x = -1$. Since a is positive, the parabola opens upward.

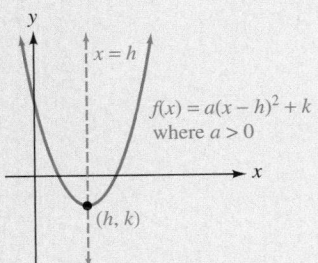

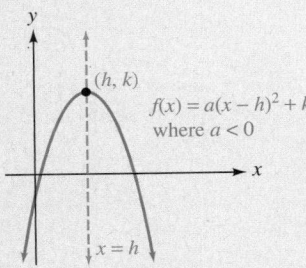

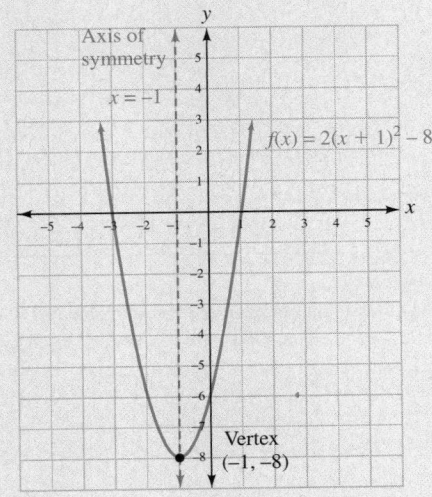

The **vertex** of the graph of $f(x) = ax^2 + bx + c$ is

$$\left(-\frac{b}{2a}, f\left(-\frac{b}{2a} \right) \right)$$

and the axis of symmetry is the line

$$x = -\frac{b}{2a}$$

The y-coordinate of the vertex of the graph of a quadratic function gives the **minimum or maximum value** of the function.

Graph: $f(x) = -x^2 + 3x + 4$

Here, $a = -1$, $b = 3$, and $c = 4$.

- Since $a < 0$, the graph opens downward.

- The x-coordinate of the vertex of the graph is

$$-\frac{b}{2a} = -\frac{3}{2(-1)} = \frac{3}{2}$$

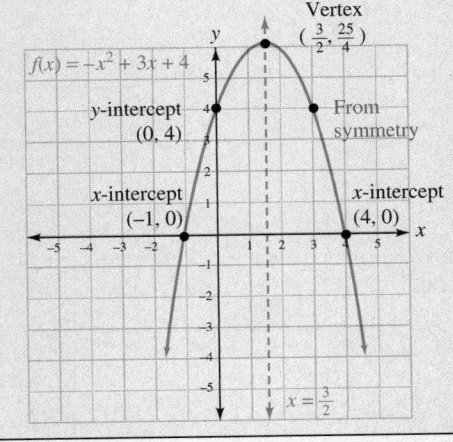

SECTION 10.4   Quadratic Functions and Their Graphs—*continued*

DEFINITIONS AND CONCEPTS	EXAMPLES
The **y-intercept** is determined by the value of $f(x)$ when $x = 0$: the y-intercept is $(0, c)$. To find the **x-intercepts**, let $f(x) = 0$ and solve $ax^2 + bx + c = 0$.	To find the y-coordinate of the vertex, we substitute $\frac{3}{2}$ for x in the function. $$f(x) = -x^2 + 3x + 4 = -\left(\frac{3}{2}\right)^2 + 3\left(\frac{3}{2}\right) + 4 = \frac{25}{4}$$ The vertex of the parabola is the point $\left(\frac{3}{2}, \frac{25}{4}\right)$. • The y-intercept is the value of the function when $x = 0$. Thus, the y-intercept is $(0, 4)$. • To find the x-intercepts, we solve: $\quad -x^2 + 3x + 4 = 0$ $\quad x^2 - 3x - 4 = 0$ Multiply both sides by -1. $\quad (x + 1)(x - 4) = 0$ Factor. $\quad x = -1$ or $x = 4$ The x-intercepts are $(-1, 0)$ and $(4, 0)$.

REVIEW EXERCISES

43. HOSPITALS The annual number of in-patient admissions to U.S. community hospitals for the years 1980–2004 can be modeled by the quadratic function $A(x) = 0.03x^2 - 0.82x + 36.31$, where $A(x)$ is the number of admissions in millions and x is the number of years after 1980. Use the function to estimate the number of in-patient admissions for the year 1992. Round to the nearest tenth of one million.

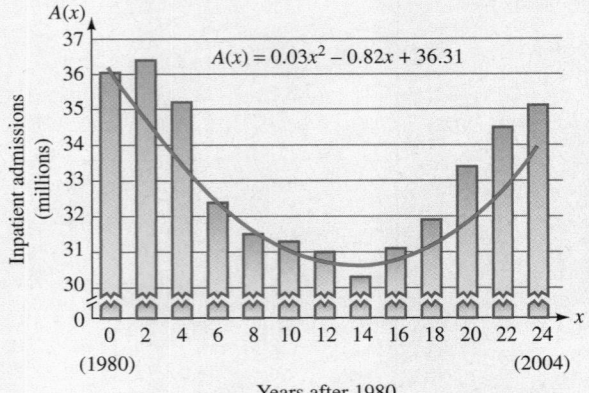

(Source: American Hospital Association)

44. Fill in the blanks. The graph of the quadratic function $f(x) = a(x - h)^2 + k$ is a parabola with vertex at (,). The axis of symmetry is the line $x = h$. The parabola opens upward when $a > 0$ and downward when $a < 0$.

Graph each pair of functions on the same coordinate system.

45. $f(x) = 2x^2$, $g(x) = 2x^2 - 3$

46. $f(x) = -\frac{1}{4}x^2$, $g(x) = -\frac{1}{4}(x + 2)^2$

47. Find the vertex and the axis of symmetry of the graph of $f(x) = -2(x - 1)^2 + 4$. Then plot several points and complete the graph.

48. Complete the square to write $f(x) = 4x^2 + 16x + 9$ in the form $f(x) = a(x - h)^2 + k$. Determine the vertex and the axis of symmetry of the graph. Then plot several points and complete the graph.

49. Find the vertex of the graph of $f(x) = -2x^2 + 4x - 8$ using the vertex formula.

50. First determine the coordinates of the vertex and the axis of symmetry of the graph of $f(x) = x^2 + x - 2$ using the vertex formula. Then determine the x- and y-intercepts of the graph. Finally, plot several points and complete the graph.

51. FARMING The number of farms in the United States for the years 1870–1970 is approximated by

$$N(x) = -1{,}526x^2 + 155{,}652x + 2{,}500{,}200$$

where $x = 0$ represents 1870, $x = 1$ represents 1871, and so on. For this period, when was the number of U.S. farms a maximum? How many farms were there?

52. Estimate the solutions of $-3x^2 - 5x + 2 = 0$ from the graph of $f(x) = -3x^2 - 5x + 2$, shown here.

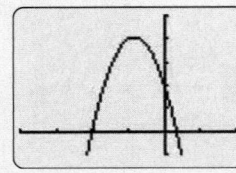

SECTION 10.5   Quadratic and Other Nonlinear Inequalities

DEFINITIONS AND CONCEPTS	EXAMPLES
To solve a quadratic inequality, get 0 on the right side and solve the related quadratic equation. Then locate the **critical numbers** on a number line, test each interval, and check the endpoints.	To solve $x^2 - x - 6 \geq 0$, we solve the related quadratic equation $x^2 - x - 6 = 0$. $x^2 - x - 6 = 0$ $(x - 3)(x + 2) = 0$ Factor. $x = 3$ or $x = -2$ These are the critical numbers that divide the number line into three intervals. 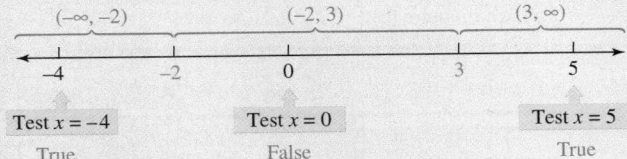 After testing each interval and noting that 3 and -2 satisfy the inequality, we see that the solution set is $(-\infty, -2] \cup [3, \infty)$.
To solve a rational inequality, get 0 on the right side and solve the related rational equation. Then locate the **critical numbers** (including any values that make the denominator 0) on a number line, test each interval, and check the endpoints.	To solve $\frac{x+1}{x-4} < 0$, we solve the related rational equation $\frac{x+1}{x-4} = 0$ to obtain the solution $x = -1$, which is a critical number. Another critical number is $x = 4$, the value that makes the denominator 0. These critical numbers divide the number line into three intervals. 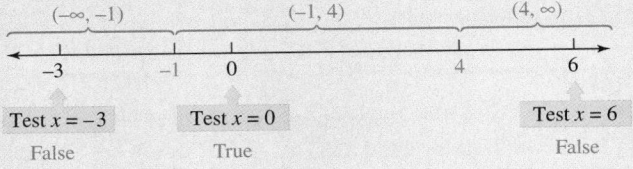 After testing each interval and noting that -1 and 4 do not satisfy the inequality, we see that the solution set is the interval $(-1, 4)$.
To graph a nonlinear inequality in two variables, first graph the boundary. Then use a test point to determine which side of the boundary to shade.	This is the graph of $y \leq x^2 + 5x + 4$. Since the inequality contains the symbol $\leq$, and equality is allowed, we draw the parabola determined by $y = x^2 + 5x + 4$ using a solid line. We shade the region containing the test point $(0, 0)$ because its coordinates satisfy $y \leq x^2 + 5x + 4$. 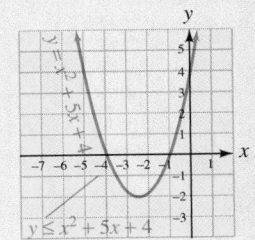

REVIEW EXERCISES

Solve each inequality. Write the solution set in interval notation and graph it.

53. $x^2 + 2x - 35 > 0$ **54.** $x^2 \leq 81$

55. $\dfrac{3}{x} \leq 5$ **56.** $\dfrac{2x^2 - x - 28}{x - 1} > 0$

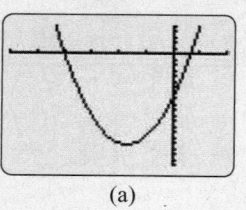 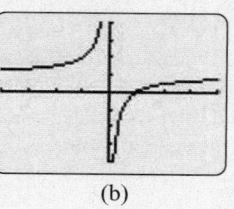

(a)   (b)

57. Estimate the solution set of $3x^2 + 10x - 8 \leq 0$ from the graph of $f(x) = 3x^2 + 10x - 8$ shown in figure (a).

58. Estimate the solution set of $\frac{x-1}{x} > 0$ from the graph of $f(x) = \frac{x-1}{x}$ shown in figure (b).

Graph each inequality.

59. $y < \dfrac{1}{2}x^2 - 1$ **60.** $y \geq -|x|$

CHAPTER 10
Test

1. Fill in the blanks.

  a. An equation of the form $ax^2 + bx + c = 0$, where $a \neq 0$, is called a _____ equation.

  b. When we add 81 to $x^2 + 18x$, we say that we have _____ the _____ on $x^2 + 18x$.

  c. The lowest point on a parabola that opens upward, or the highest point on a parabola that opens downward, is called the _____ of the parabola.

  d. $\dfrac{x-5}{x^2 - x - 56} > 0$ is an example of a _____ inequality in one variable.

  e. $y \leq x^2 - 4x + 3$ is an example of a _____ inequality in two variables.

2. Solve $x^2 - 63 = 0$ using the square root property. Approximate the solutions to the nearest hundredth.

Solve each equation using the square root property.

3. $(a + 7)^2 = 50$ **4.** $m^2 + 4 = 0$

5. Add a number to make $x^2 + 11x$ a perfect-square trinomial. Then factor the result

6. Solve $4x^2 - 16x + 15 = 0$ by completing the square.

Use the quadratic formula to solve each equation.

7. $4x^2 + 4x - 1 = 0$

8. $\dfrac{1}{8}t^2 - \dfrac{1}{4}t = \dfrac{1}{2}$

9. $-t^2 + 4t - 13 = 0$ **10.** $0.01x^2 = -0.08x - 0.15$

Solve each equation by any method.

11. $2y - 3\sqrt{y} + 1 = 0$ **12.** $3 = m^{-2} - 2m^{-1}$

13. $x^4 - x^2 - 12 = 0$

14. $4\left(\dfrac{x+2}{3x}\right)^2 - 4\left(\dfrac{x+2}{3x}\right) - 3 = 0$

15. $\dfrac{1}{n+2} = \dfrac{1}{3} - \dfrac{1}{n}$

16. $5a^{2/3} + 11a^{1/3} = -2$

17. Solve $E = mc^2$ for c. Assume that all variables represent positive numbers. Express any radical in simplified form.

18. Use the discriminant to determine the number and type of solutions for each equation.

  a. $3x^2 + 5x + 17 = 0$

  b. $9m^2 - 12m = -4$

19. TABLECLOTHS   In 1990, Sportex of Highland, Illinois, made what was at the time the world's longest tablecloth. Find the dimensions of the rectangular tablecloth if it covered an area of 6,759 square feet and its length was 8 feet more than 332 times its width.

20. COOKING   Working together, a chef and his assistant can make a pastry dessert in 25 minutes. When the chef makes it himself, it takes him 8 minutes less time than it takes his assistant working alone. How long does it take the chef to make the dessert?

21. DRAWING   An artist uses four equal-sized right triangles to block out a perspective drawing of an old hotel. See the illustration on the next page. For each triangle, the leg on the horizontal line is 14 inches longer than the leg on the center line. The length of each hypotenuse is 26 inches. On the centerline of the drawing, what is the length of the segment extending from the ground to the top of the building?

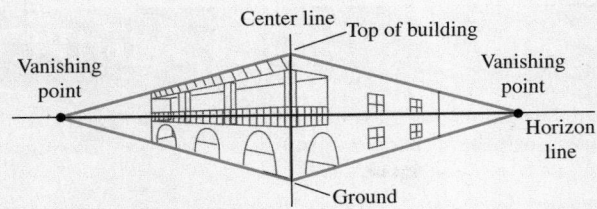

Center line — Top of building
Vanishing point
Vanishing point
Horizon line
Ground

22. **ANTHROPOLOGY**   Anthropologists refer to the shape of the human jaw as a *parabolic dental arcade*. Which function is the best mathematical model of the parabola shown in the illustration?

i. $f(x) = -\dfrac{3}{8}(x-4)^2 + 6$

ii. $f(x) = -\dfrac{3}{8}(x-6)^2 + 4$

iii. $f(x) = -\dfrac{3}{8}x^2 + 6$

iv. $f(x) = \dfrac{3}{8}x^2 + 6$

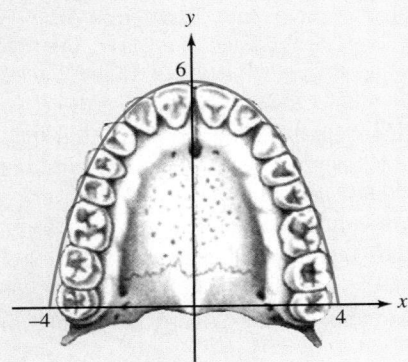

23. Find the vertex and the axis of symmetry of the graph of $f(x) = -3(x-1)^2 + 2$. Then plot several points and complete the graph.

24. Complete the square to write the function $f(x) = 5x^2 + 10x - 1$ in the form $f(x) = a(x-h)^2 + k$. Determine the vertex and the axis of symmetry of the graph. Then plot several points and complete the graph.

25. First determine the coordinates of the vertex and the axis of symmetry of the graph of $f(x) = 2x^2 + x - 1$ using the vertex formula. Then determine the x- and y-intercepts of the graph. Finally, plot several points and complete the graph.

26. **DISTRESS SIGNALS**   A flare is fired directly upward into the air from a boat that is experiencing engine problems. The height of the flare (in feet) above the water, t seconds after being fired, is given by the formula $h = -16t^2 + 112t + 15$. If the flare is designed to explode when it reaches its highest point, at what height will this occur?

Solve each inequality. Write the solution set in interval notation and then graph it.

27. $x^2 - 2x > 8$

28. $\dfrac{x-2}{x+3} \le 0$

29. **WATER USAGE**   The average amount of water used per month by a single-family residential customer in Tucson, Arizona, for the year 2004 is modeled by the function $W(m) = -235m^2 + 2{,}095m + 6{,}540$, where $W(m)$ is the number of gallons and m is the number of months *after March*. Use the function to approximate the average number of gallons of water used in July, which is typically Tucson's warmest month. (Based on data from the City of Tucson Water Department)

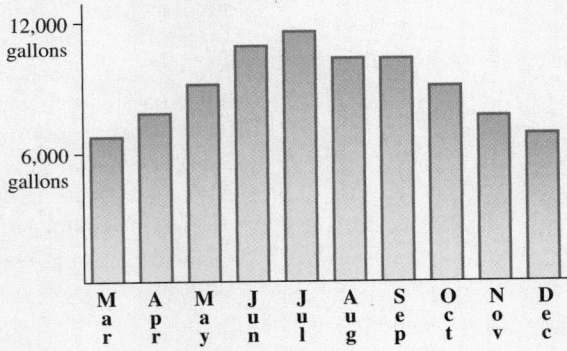

**Tucson: Average Water Usage by Month
Single-Family Residental Customer, 2004**

30. Graph: $y \le -x^2 + 3$

31. The graph of a quadratic function of the form $f(x) = ax^2 + bx + c$ is shown. Estimate the solutions of the corresponding quadratic equation $ax^2 + bx + c = 0$.

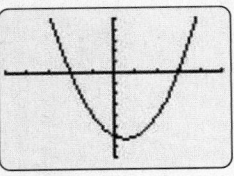

32. See Exercise 31. Estimate the solution of the quadratic inequality $ax^2 + bx + c \le 0$.

GROUP PROJECT

PICTURE FRAMING

Overview: When framing pictures, mats are often used to enhance the images and give them a sense of depth. In this activity, you will use the quadratic formula to design the matting for several pictures.

Instructions: Form groups of 3 students. Each person in your group is to bring a picture to class. You can use a picture from a magazine or newspaper, a picture postcard, or a photograph that is no larger than 5 in. × 7 in. You will also need a pair of scissors, a ruler, glue, and three pieces of construction paper (12 in. × 18 in.).

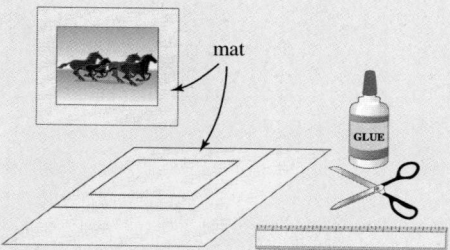

Select one of the pictures and find its area. A mat of *uniform* width is to be placed around the picture. The area of the mat should equal the area of the picture. To determine the proper width of the matting, follow the steps of Example 11 in Section 10.1. However, use the quadratic formula, instead of completing the square, to solve the equation. Once you have determined the proper width, cut out the mat from the construction paper and glue it to the picture.

Then, choose another picture and find its area. Determine the uniform width that a matting should have so that its area is double that of the picture. Cut out the proper-size matting from the construction paper and glue it to the second picture.

Finally, find the area of the third picture and determine the uniform width that a matting should have so that its area is one-half that of the picture. Cut out the proper-size matting from the construction paper and glue it to the third picture.

Is one size matting more visually appealing than another? Discuss this among the members of your group.

CUMULATIVE REVIEW
Chapters 1–10

1. Solve: $3(x + 2) - 2 = -(5 + x) + x$ [Section 2.2]

2. PHARMACISTS How many liters of a 1% glucose solution should a pharmacist mix with 2 liters of a 5% glucose solution to obtain a 2% glucose solution? [Section 2.6]

3. Find the x-intercept and the y-intercept of the graph of the linear equation $5x - 3y = 6$. [Section 3.3]

4. SHORTAGE OF NURSES Use the data in the graph to find the projected rates of change in the supply and demand for registered nurses (RNs) in the United States for the years 2010–2020. [Section 3.4]

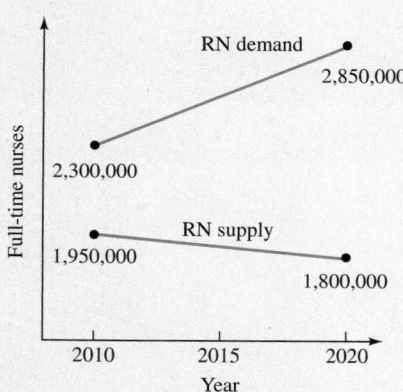

(Source: American Hospital Association)

Find an equation of the line with the given properties. Write the equation in slope–intercept form.

5. Slope 3, passes through $(-2, -4)$ [Section 3.5]

6. Parallel to the graph of $2x + 3y = 6$ and passes through $(0, -2)$ [Section 3.6]

7. Solve the system by graphing:

$$\begin{cases} y = -\dfrac{5}{2}x + \dfrac{1}{2} \\ 2x - \dfrac{3}{2}y = 5 \end{cases}$$ [Section 4.1]

8. Use substitution to solve the system:

$$\begin{cases} x - y = -5 \\ 3x - 2y = -7 \end{cases}$$ [Section 4.2]

9. Graph the solution set of the system:

$$\begin{cases} 3x + 2y > 6 \\ x + 3y \leq 2 \end{cases}$$ [Section 4.5]

10. Simplify: $\left(\dfrac{2a^2 b^3 c^{-4}}{5a^{-2} b^{-1} c^3} \right)^{-3}$ [Section 5.2]

11. Write each number in scientific notation and perform the operations. Give the answer in scientific notation and in standard notation:

$$\dfrac{(1{,}280{,}000{,}000)(2{,}700{,}000)}{240{,}000}$$ [Section 5.3]

12. Simplify: $\dfrac{3}{5}s^2 - \dfrac{2}{5}t^2 - \dfrac{1}{2}s^2 - \dfrac{7}{10}st - \dfrac{3}{10}st$ [Section 5.5]

Perform the indicated operations.

13. $(-8.9t^3 - 2.4t) - (2.1t^3 + 0.8t^2 - t)$ [Section 5.5]

14. $(2a - b)(4a^2 + 2ab + b^2)$ [Section 5.6]

Factor each expression.

15. $12uvw^3 - 18uv^2w^2$ [Section 6.1]

16. $x^2 + 4y - xy - 4x$ [Section 6.1]

17. $30a^4 - 4a^3 - 16a^2$ [Section 6.3]

18. $49s^6 - 84s^3n^2 + 36n^4$ [Section 6.4]

19. $x^4 - 16y^4$ [Section 6.4]

20. $8x^6 + 125y^3$ [Section 6.5]

Solve each equation.

21. $(m + 4)(2m + 3) - 22 = 10m$ [Section 6.7]

22. $6a^3 - 2a = a^2$ [Section 6.7]

23. Simplify: $\dfrac{6x^2 - 7x - 5}{2x^2 + 5x + 2}$ [Section 7.1]

24. Divide: $\dfrac{x^3 + y^3}{x^3 - y^3} \div \dfrac{x^2 - xy + y^2}{x^2 + xy + y^2}$ [Section 7.2]

25. Perform the operations: $\dfrac{1}{x + y} - \dfrac{1}{x - y} + \dfrac{2y}{x^2 - y^2}$ [Section 7.3]

26. Simplify: $\dfrac{\dfrac{1}{r^2 + 4r + 4}}{\dfrac{r}{r + 2} + \dfrac{r}{r + 2}}$ [Section 7.4]

27. Divide: $\dfrac{24x^6y^7 - 12x^5y^{12} + 36xy}{48x^2y^3}$

[Section 7.5]

28. Divide: $3a - 4 \overline{)15a^3 - 29a^2 + 16}$

[Section 7.5]

29. Solve: $\dfrac{x-4}{x-3} + \dfrac{x-2}{x-3} = x - 3$
[Section 7.7]

30. Solve for R: $\dfrac{1}{R} = \dfrac{1}{R_1} + \dfrac{1}{R_2} + \dfrac{1}{R_3}$
[Section 7.7]

31. SINKS   A sink has two faucets, one for cold water and one for hot water. It can be filled by the cold-water faucet in 30 seconds and by the hot-water faucet in 45 seconds. How long will it take to fill the sink if both faucets are opened? [Section 7.7]

32. SNOW REMOVAL   A state highway department uses a 7-to-2 sand-to-salt mix in the winter months for spreading across roadways covered with snow and ice. If they have 6 tons of salt in storage, how many tons of sand should be added to obtain the proper mix? [Section 7.8]

Solve each inequality. Write the solution set in interval notation and then graph it.

33. $5(-2x + 2) > 20 - x$ [Section 8.1]

34. $5x - 3 \geq 2$ and $6 \geq 4x - 3$ [Section 8.2]

35. $|2x - 5| \geq 25$ [Section 8.3]

36. Factor: $x^2 + 10x + 25 - y^8$ [Section 8.5]

37. TIDES   The illustration shows the graph of a function f, which gives the height of the tide for a 24-hour period in Seattle, Washington. (Note that military time is used on the x-axis: 3 A.M. = 3, noon = 12, 9 P.M. = 21, and so on.)
[Section 8.8]

   a. Find the domain of the function.

   b. Find $f(6)$.

   c. What information does $f(12)$ give?

   d. Estimate the values of x for which $f(x) = 0$.

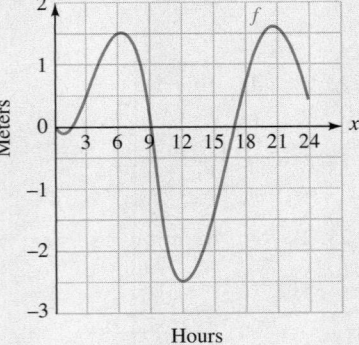

38. Find the domain of the function $f(x) = \dfrac{x}{x + 20}$. [Section 8.8]

39. DELIVERIES   The costs of a delivery company vary jointly with the number of trucks in service and the number of hours they are used. When 8 trucks are used for 12 hours each, the costs are $3,600. Find the costs of using 20 trucks, each for 12 hours. [Section 8.9]

40. Graph the function $f(x) = \sqrt{x - 2}$ and give its domain and range. [Section 9.1]

Simplify each expression.

41. $\sqrt[3]{-27x^3}$ [Section 9.1]

42. $\sqrt{48t^3}$ [Section 9.1]

43. $64^{-2/3}$ [Section 9.2]

44. $\dfrac{x^{5/3}x^{1/2}}{x^{3/4}}$ [Section 9.2]

45. $-3\sqrt[4]{32} - 2\sqrt[4]{162} + 5\sqrt[4]{48}$ [Section 9.3]

46. $3\sqrt{2}\left(2\sqrt{3} - 4\sqrt{12}\right)$ [Section 9.4]

47. $\dfrac{\sqrt{x} + 2}{\sqrt{x} - 1}$ [Section 9.4]

48. $\dfrac{5}{\sqrt[3]{x}}$ [Section 9.4]

Solve each equation.

49. $5\sqrt{x + 2} = x + 8$ [Section 9.5]

50. $\sqrt{x} + \sqrt{x + 2} = 2$ [Section 9.5]

51. Find the length of the hypotenuse of the right triangle in figure (a). [Section 9.6]

52. Find the length of the hypotenuse of the right triangle in figure (b). [Section 9.6]

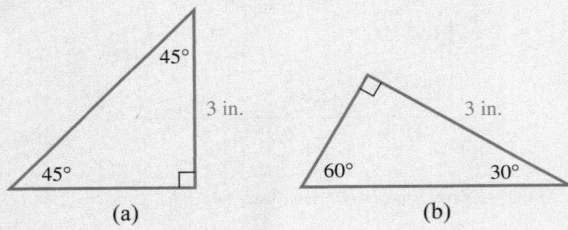

(a)   (b)

53. Find the distance between $(-2, 6)$ and $(4, 14)$. [Section 9.6]

54. Simplify: i^{43} [Section 9.7]

Perform the indicated operations. Write each result in a + bi form.

55. $\left(-7 + \sqrt{-81}\right) - \left(-2 - \sqrt{-64}\right)$ [Section 9.7]

56. $\dfrac{5}{3 - i}$ [Section 9.7]

57. $(2 + i)^2$ [Section 9.7]

58. $\dfrac{-4}{6i^7}$ [Section 9.7]

Solve each equation.

59. $x^2 = 28$ [Section 10.1]

60. $(x - 19)^2 = -5$ [Section 10.1]

61. Use the method of completing the square to solve $2x^2 - 6x + 3 = 0$. [Section 10.1]

62. Use the quadratic formula to solve $a^2 - \dfrac{2}{5}a = -\dfrac{1}{5}$. [Section 10.2]

63. COMMUNITY GARDENS Residents of a community can work their own 16-ft × 24-ft plot of city-owned land if they agree to the following conditions:

- The area of the garden cannot exceed 180 square feet.
- A path of uniform width must be maintained around the garden.

Find the dimensions of the largest possible garden. [Section 10.2]

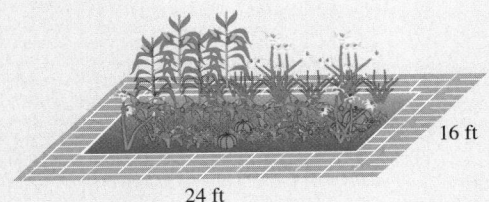

16 ft

24 ft

64. SIDEWALKS A 170-meter-long sidewalk from the mathematics building M to the student center C is shown in red in the illustration. However, students prefer to walk directly from M to C, across a lawn. How long are the two segments of the existing sidewalk? [Section 10.2]

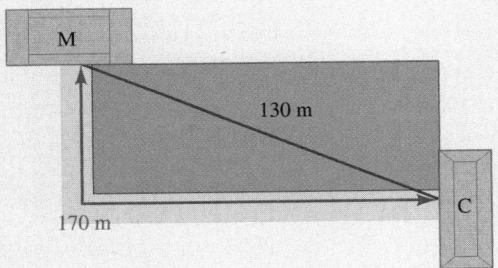

Solve each equation.

65. $t^{2/3} - t^{1/3} = 6$ [Section 10.3]

66. $x^{-4} - 2x^{-2} + 1 = 0$ [Section 10.3]

67. First determine the vertex and the axis of symmetry of the graph of $f(x) = -x^2 - 4x$ using the vertex formula. Then determine the x- and y-intercepts of the graph. Finally, plot several points and complete the graph. [Section 10.4]

Solve each inequality. Write the solution set in interval notation and then graph it.

68. $x^2 - 81 < 0$ [Section 10.5]

69. $\dfrac{1}{x+1} \geq \dfrac{x}{x+4}$ [Section 10.5]

70. a. The graph of $f(x) = 16x^2 + 24x + 9$ is shown below. Estimate the solution(s) of $16x^2 + 24x + 9 = 0$. [Section 10.5]

   b. Use the graph to determine the solution of $16x^2 + 24x + 9 < 0$. [Section 10.5]

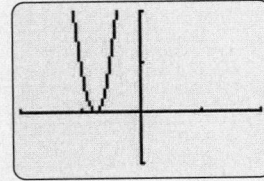

CHAPTER 11

Exponential and Logarithmic Functions

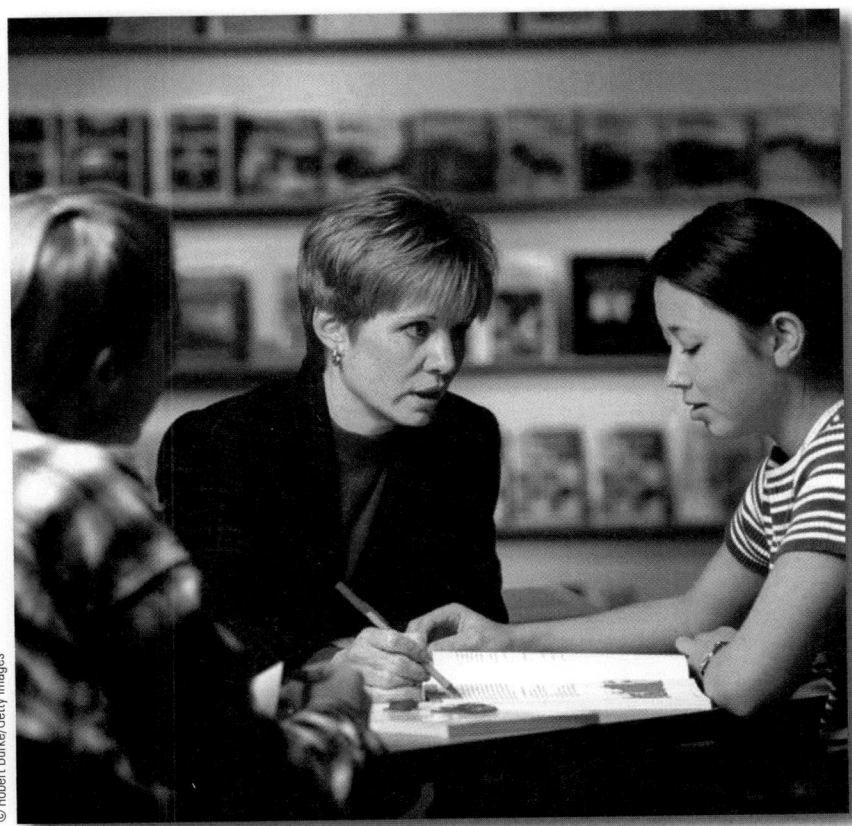

from *Campus to Careers*
Social Worker

For those with a desire to help improve other people's lives, social work is one career option to consider. Social workers offer guidance and counseling to people in crisis. They must be critical thinkers—able to use their logic and reasoning to brainstorm alternative solutions to problems faced by their clients. Social workers use their mathematical skills to construct family budgets, plan personnel schedules, gather and interpret data, and comprehend the statistical methods used in research studies.

Social workers often use occupational test results when counseling their clients about employment options. In **Problem 49** of **Study Set 11.4,** you will use concepts from this chapter to interpret the "learning curve" of a factory trainee.

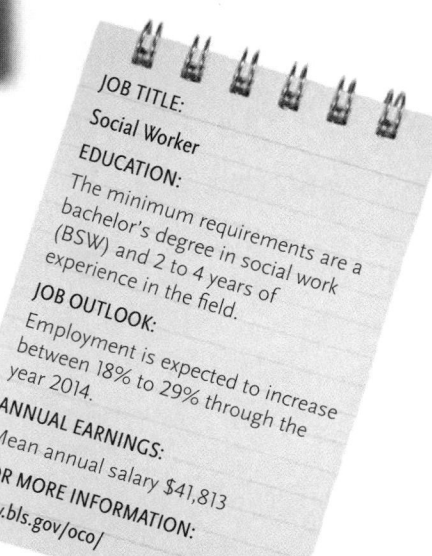

JOB TITLE:
Social Worker

EDUCATION:
The minimum requirements are a bachelor's degree in social work (BSW) and 2 to 4 years of experience in the field.

JOB OUTLOOK:
Employment is expected to increase between 18% to 29% through the year 2014.

ANNUAL EARNINGS:
Mean annual salary $41,813

FOR MORE INFORMATION:
www.bls.gov/oco/

Study Skills Workshop
Participating in Class

One of the keys to success in algebra is to learn as much as you can in class. To get the most out of class meetings, and to make them more enjoyable, you should participate in the following ways:

ASK QUESTIONS: During class, clear up any questions that may arise from your homework assignments or from your instructor's lectures. Also, pay close attention when other students ask questions. You never know when you might face the same difficulty.

ANSWER QUESTIONS: Many instructors direct questions to the class while lecturing. Take advantage of this opportunity to increase your knowledge by attempting to answer all such questions from your instructor.

INTERACT WITH CLASSMATES: Before class begins and after class ends, regularly discuss the material that you are studying with fellow classmates.

Now Try This

1. List the reasons why you do not feel comfortable asking questions in class.
2. While working on your next homework assignment, write down any questions that occur to you so that you will not forget to ask them in class.
3. Exchange a written question about a homework problem with a classmate. See if you can answer each other's question.

SECTION 11.1
Algebra and Composition of Functions

Objectives

1. Add, subtract, multiply, and divide functions.
2. Find the composition of functions.
3. Use graphs to evaluate functions.
4. Define the identity function.
5. Use composite functions to solve problems.

Just as it is possible to perform arithmetic operations on real numbers, it is possible to perform those operations on functions. We call the process of adding, subtracting, multiplying, and dividing functions the *algebra of functions*.

 Add, Subtract, Multiply, and Divide Functions.

The sum, difference, product, and quotient of two functions are themselves functions.

Operations on Functions

If the domains and ranges of functions f and g are subsets of the real numbers, then
The **sum** of f and g, denoted as $f + g$, is defined by

$$(f + g)(x) = f(x) + g(x)$$

The **difference** of f and g, denoted as $f - g$, is defined by

$$(f - g)(x) = f(x) - g(x)$$

The **product** of f and g, denoted as $f \cdot g$, is defined by

$$(f \cdot g)(x) = f(x)g(x)$$

The **quotient** of f and g, denoted as f / g, is defined by

$$(f/g)(x) = \frac{f(x)}{g(x)} \quad \text{where } g(x) \neq 0$$

The domain of each of these functions is the set of real numbers x that are in the domain of both f and g. In the case of the quotient, there is the further restriction that $g(x) \neq 0$.

EXAMPLE 1   Let $f(x) = 2x^2 + 1$ and $g(x) = 5x - 3$. Find each function and its domain: **a.** $f + g$ **b.** $f - g$

Strategy   We will add and subtract the functions as if they were binomials.

Why   We add because of the plus symbol in $f + g$, and we subtract because of the minus symbol in $f - g$.

Solution

a. $(f + g)(x) = f(x) + g(x)$

$\qquad\qquad = (2x^2 + 1) + (5x - 3)$ Replace f(x) with $2x^2$ + 1 and g(x) with 5x − 3.

$\qquad\qquad = 2x^2 + 1 + 5x - 3$ Drop the parentheses.

$\qquad\qquad = 2x^2 + 5x - 2$ Combine like terms.

The domain of $f + g$ is the set of real numbers that are in the domain of both f and g. Since the domain of both f and g is the interval $(-\infty, \infty)$, the domain of $f + g$ is the interval $(-\infty, \infty)$.

b. $(f - g)(x) = f(x) - g(x)$

$\qquad\qquad = (2x^2 + 1) - (5x - 3)$

$\qquad\qquad = 2x^2 + 1 - 5x + 3$ Change the sign of each term of 5x − 3 and drop the parentheses.

$\qquad\qquad = 2x^2 - 5x + 4$ Combine like terms.

Since the domain of both f and g is $(-\infty, \infty)$, the domain of $f - g$ is the interval $(-\infty, \infty)$.

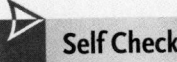

Self Check 1   Let $f(x) = 3x - 2$ and $g(x) = 2x^2 + 3x$. Find:
a. $f + g$
b. $f - g$

Now Try   **Problems 21 and 23**

EXAMPLE 2   Let $f(x) = 2x^2 + 1$ and $g(x) = 5x - 3$. Find each function and its domain: **a.** $f \cdot g$ **b.** f/g

Strategy   We will multiply and divide the functions as if they were binomials.

Why   We multiply because of the raised dot in $f \cdot g$ and we divide because of the fraction bar in f / g.

Solution

a. $(f \cdot g)(x) = f(x) \cdot g(x)$

$\qquad\qquad = (2x^2 + 1)(5x - 3)$ *Replace f(x) with 2x² + 1 and g(x) with 5x − 3.*

$\qquad\qquad = 10x^3 - 6x^2 + 5x - 3$ *Multiply the binomials.*

The domain of $f \cdot g$ is the set of real numbers that are in the domain of both f and g. Since the domain of both f and g is the interval $(-\infty, \infty)$, the domain of $f \cdot g$ is the interval $(-\infty, \infty)$.

b. $(f/g)(x) = \dfrac{f(x)}{g(x)}$

$\qquad\qquad = \dfrac{2x^2 + 1}{5x - 3}$

Since the denominator of the fraction cannot be 0, $x \neq \frac{3}{5}$. Thus, the domain of f/g is the union of two intervals: $\left(-\infty, \frac{3}{5}\right) \cup \left(\frac{3}{5}, \infty\right)$.

 Self Check 2   Let $f(x) = 2x^2 - 3$ and $g(x) = x^2 + 1$. Find:

a. $f \cdot g$

b. f/g

Now Try   **Problems 25 and 27**

There is a relationship that can be seen between the graphs of two functions and the graph of their sum (or difference) function. For example, in the following illustration, the graph of $f + g$, which gives the total number of elementary and secondary students in the United States, can be found by adding the graph of f, which gives the number of elementary students, to the graph of g, which gives the number of secondary students. For any given x-value, we simply add the two corresponding y-values to get the graph of the sum function $f + g$.

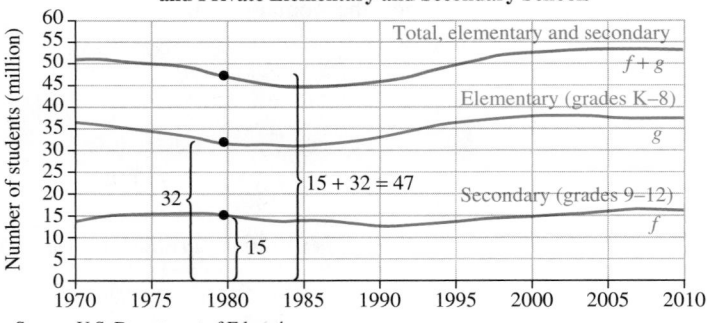

Actual and Projected Enrollment in U.S. Public and Private Elementary and Secondary Schools

Source: U.S. Department of Education

2   Find the Composition of Functions.

We have seen that a function can be represented by a machine: We put in a number from the domain, and a number from the range comes out. For example, if we put the number 2 into the machine shown in figure (a), the number $f(2) = 8$ comes out. In general, if we put x into the machine shown in figure (b), the value $f(x)$ comes out.

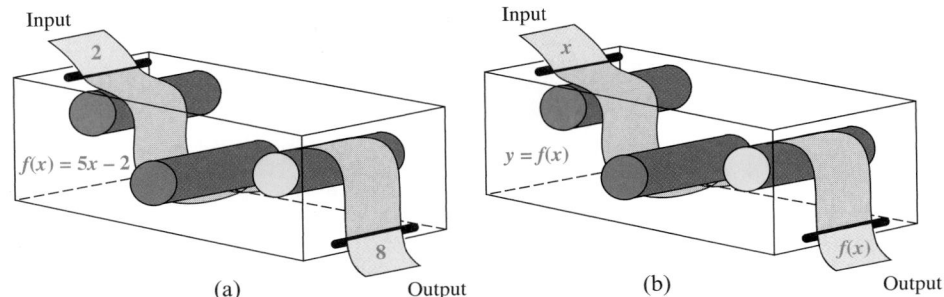

(a)   Output   (b)   Output

Often one quantity is a function of a second quantity that depends, in turn, on a third quantity. For example, the cost of a car trip is a function of the gasoline consumed. The amount of gasoline consumed, in turn, is a function of the number of miles driven. Such chains of dependence can be analyzed mathematically as **compositions of functions.**

Suppose that $y = f(x)$ and $y = g(x)$ define two functions. Any number x in the domain of g will produce the corresponding value $g(x)$ in the range of g. If $g(x)$ is in the domain of function f, then $g(x)$ can be substituted into f, and a corresponding value $f(g(x))$ will be determined. This two-step process defines a new function, called a **composite function,** denoted by $f \circ g$. (This is read as "f composed with g" or "the composition of f and g" or "f circle g.")

The function machines shown below illustrate the composition $f \circ g$. When we put a number into the function g, a value $g(x)$ comes out. The value $g(x)$ then goes into function f, which transforms $g(x)$ into $f(g(x))$. (This is read as "f of g of x.") If the function machines for g and f were connected to make a single machine, that machine would be named $f \circ g$.

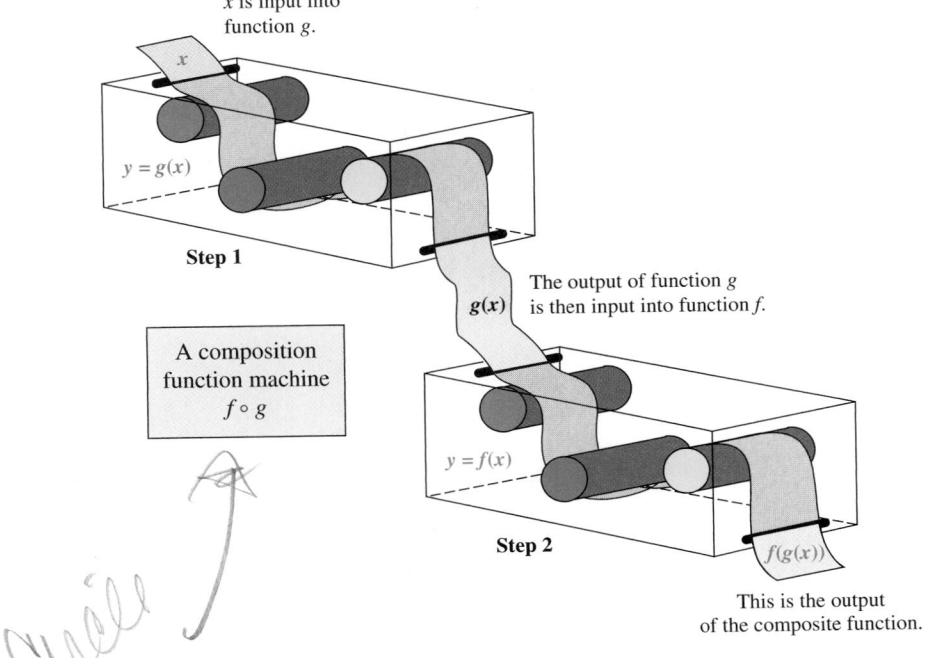

x is input into function g.

Step 1

A composition function machine $f \circ g$

The output of function g is then input into function f.

Step 2

This is the output of the composite function.

To be in the domain of the composite function $f \circ g$, a number x has to be in the domain of g and the output of g must be in the domain of f. Thus, the domain of $f \circ g$ consists of those numbers x that are in the domain of g, and for which $g(x)$ is in the domain of f.

Composite Functions

The **composite function** $f \circ g$ is defined by

$$(f \circ g)(x) = f(g(x))$$

Caution

$(f \overset{\downarrow}{\circ} g)(x)$
　　　Composition

does not mean

$(f \overset{\downarrow}{\cdot} g)(x)$
　　　Multiplication

If $f(x) = 4x$ and $g(x) = 3x + 2$, to find $f \circ g$ and $g \circ f$, we proceed as follows.

$$(f \circ g)(x) = f(g(x)) \qquad (g \circ f)(x) = g(f(x))$$
$$= f(3x + 2) \qquad\qquad = g(4x)$$
$$= 4(3x + 2) \qquad\qquad = 3(4x) + 2$$
$$= 12x + 8 \qquad\qquad = 12x + 2$$

Different results

The different results illustrate that the composition of functions is not commutative. Ususally, we will find that $(f \circ g)(x) \neq (g \circ f)(x)$.

EXAMPLE 3 Let $f(x) = 2x + 1$ and $g(x) = x - 4$. Find:
a. $(f \circ g)(9)$ **b.** $(f \circ g)(x)$ **c.** $(g \circ f)(-2)$

Strategy In part (a), we will find $f(g(9))$. In part (b), we will find $f(g(x))$. In part (c), we will find $g(f(-2))$.

Why To evaluate a composition function written with the circle $\circ$ notation, we rewrite it using nested parentheses: $(f \circ g)(x) = f(g(x))$.

The Language of Algebra

$f(\underbrace{g(x)})$

We call these nested parentheses.

Solution

a. $(f \circ g)(9)$ means $f(g(9))$. In figure (a) on the next page, function g receives the number 9, subtracts 4, and releases the number $g(9) = 5$. Then 5 goes into the f function, which doubles 5 and adds 1. The final result, 11, is the output of the composite function $f \circ g$:

Read as "f of g of 9."

$$(f \circ g)(9) = f(g(9)) \quad \text{Change from } \circ \text{ notation to nested parentheses.}$$
$$= f(5) \quad \text{Evaluate: } g(9) = 9 - 4 = 5.$$
$$= 2(5) + 1 \quad \text{Evaluate f(5) using f(x) = 2x + 1.}$$
$$= 11$$

$2(x-4)+1 = 2(9-4)+1$
$2(5)+1 = 11$

Thus, $(f \circ g)(9) = 11$.

Notation

The notation $f \circ g$ can also be read as "f circle g." Remember, it means that the function g is applied first and function f is applied second.

b. $(f \circ g)(x)$ means $f(g(x))$. In figure (a) on the next page, function g receives the number x, subtracts 4, and releases the number $x - 4$. Then $x - 4$ goes into the f function, which doubles $x - 4$ and adds 1. The final result, $2x - 7$, is the output of the composite function $f \circ g$.

Read as "f of g of x."

$$(f \circ g)(x) = f(g(x)) \quad \text{Change from } \circ \text{ notation to nested parentheses.}$$
$$= f(x - 4) \quad \text{We are given g(x) = x - 4.}$$
$$= 2(x - 4) + 1 \quad \text{Find f(x - 4) using f(x) = 2x + 1.}$$
$$= 2x - 8 + 1$$
$$= 2x - 7$$

Thus, $(f \circ g)(x) = 2x - 7$.

c. $(g \circ f)(-2)$ means $g(f(-2))$. In figure (b) below, function f receives the number -2, doubles it and adds 1, and releases -3 into the g function. Function g subtracts 4 from -3 and outputs a final result of -7. Thus,

Read as "*g* of *f* of −2."
↓

$$(g \circ f)(-2) = g(f(-2)) \qquad \text{Change from } \circ \text{ notation to nested parentheses.}$$
$$= g(-3) \qquad \text{Evaluate } f(-2) \text{ using } f(x) = 2x + 1.$$
$$= -3 - 4 \qquad \text{Evaluate } g(-3) \text{ using } g(x) = x - 4.$$
$$= -7$$

Thus, $(g \circ f)(-2) = -7$.

9   *x*

$g(x) = x - 4$

Step 1:
Function g is
applied first.

$f \circ g$

5    $x - 4$

$f(x) = 2x + 1$

Step 2:
Apply function f.

11    $2x - 7$

(a)

−2

$f(x) = 2x + 1$

Step 1:
Function f is
applied first.

$g \circ f$

−3

$g(x) = x - 4$

Step 2:
Apply function g.

−7

(b)

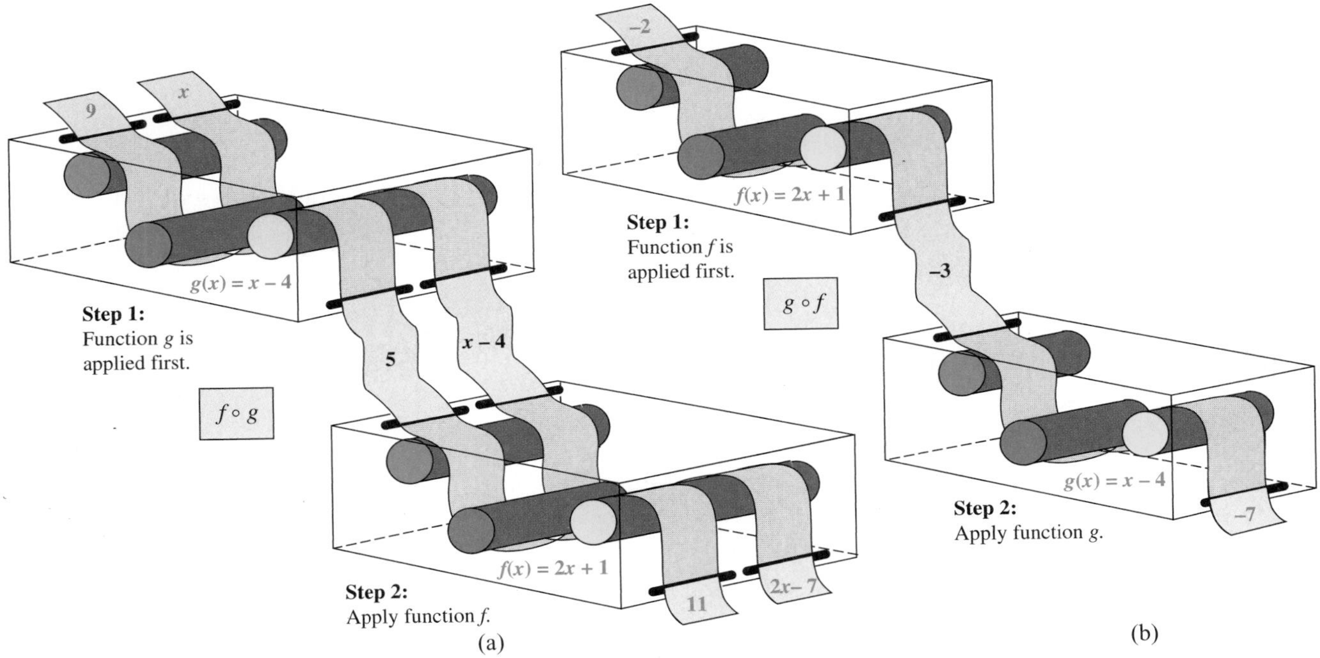

▷ **Self Check 3** Let $f(x) = x^3$ and $g(x) = 6 - x$. Find:
a. $(f \circ g)(8)$ **b.** $(g \circ f)(1)$ **c.** $(g \circ f)(x)$

Now Try **Problems 37, 39, and 45**

3 **Use Graphs to Evaluate Functions.**

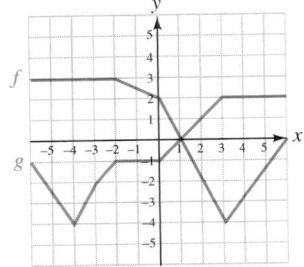

EXAMPLE 4 Refer to the graphs of functions f and g on the left to find each of the following.
a. $(f + g)(-4)$
b. $(f \cdot g)(2)$
c. $(f \circ g)(-3)$

Strategy We will express the sum, product, and composition functions in terms of the functions from which they are formed.

Why   We can evaluate sum, product, and composition functions at a given x-value by evaluating each function from which they are formed at that x-value.

Solution

a. $(f + g)(-4) = f(-4) + g(-4)$

$\qquad\qquad\quad = 3 + (-4)$

$\qquad\qquad\quad = -1$

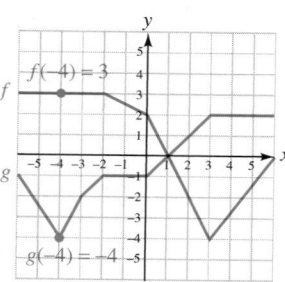

b. $(f \cdot g)(2) = f(2) \cdot g(2)$

$\qquad\qquad\quad = -2 \cdot 1$

$\qquad\qquad\quad = -2$

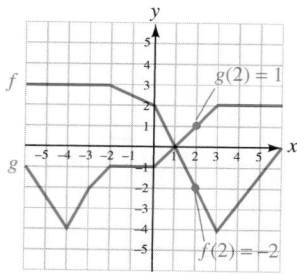

c. $(f \circ g)(-3) = f(g(-3))$

$\qquad\qquad\quad\; = f(-2)$

$\qquad\qquad\quad\; = 3$

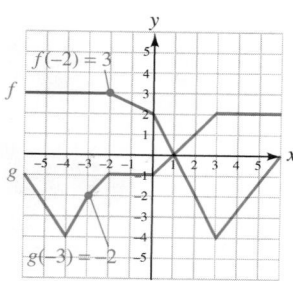

 Self Check 4   Refer to the graph above to find each of the following.

$\qquad$ **a.** $(f - g)(3)$ $\qquad$ **b.** $\left(\dfrac{f}{g}\right)(-2)$ $\qquad$ **c.** $(g \circ f)(3)$

Now Try   **Problem 65**

---

④   Define the Identity Function.

The **identity function** is defined by the equation $I(x) = x$. Under this function, the value that is assigned to any real number x is x itself. For example $I(2) = 2$, $I(-3) = -3$, and $I(7.5) = 7.5$. If f is any function, the composition of f with the identity function is just the function f:

$$(f \circ I)(x) = (I \circ f)(x) = f(x)$$

EXAMPLE 5 Let f be any function and let I be the identity function, $I(x) = x$.
Show that   **a.** $(f \circ I)(x) = f(x)$ **b.** $(I \circ f)(x) = f(x)$

Strategy In part (a), we will find $f(I(x))$ and in part (b), we will find $I(f(x))$.

Why To find a composition function written with the circle $\circ$ notation, we rewrite it using the nested parentheses notation.

Solution

a. $(f \circ I)(x)$ means $f(I(x))$. Because $I(x) = x$, we have

$$(f \circ I)(x) = f(I(x)) = f(x)$$

b. $(I \circ f)(x)$ means $I(f(x))$. Because I passes any number through unchanged, we have

$$(I \circ f)(x) = I(f(x)) = f(x)$$

The Language of Algebra
The *identity* function pairs each real number with itself such that each output is *identical* to its corresponding input.

 Now Try Problems 69 and 71

5 Use Composite Functions to Solve Problems.

EXAMPLE 6 **Biological Research.** A specimen is stored in refrigeration at a temperature of 15° Fahrenheit. Biologists remove the specimen and warm it at a controlled rate of 3°F per hour. Express its Celsius temperature as a function of the time t since it was removed from refrigeration.

Strategy We will express the Fahrenheit temperature of the specimen as a function of the time t since it was removed from refrigeration. Then we will express the Celsius temperature of the specimen as a function of its Fahrenheit temperature and find the composition of the two functions.

Why The Celsius temperature of the specimen is a function of its Fahrenheit temperature. Its Fahrenheit temperature is a function of the time since it was removed from refrigeration. This chain of dependence suggests that we write a composition of functions.

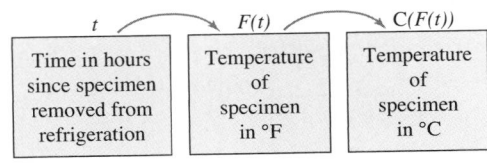

Solution The temperature of the specimen is 15°F when the time $t = 0$. Because it warms at a rate of 3°F per hour, its initial temperature of 15°F increases by $3t$°F in t hours. The Fahrenheit temperature at time t of the specimen is given by the function

$$F(t) = 3t + 15$$

The Celsius temperature C is a function of this Fahrenheit temperature F, given by the function

$$C(F) = \frac{5}{9}(F - 32)$$

To express the specimen's Celsius temperature as a function of *time,* we find the composite function $(C \circ F)(t)$.

$$(C \circ F)(t) = C(F(t))$$

$$= C(3t + 15) \qquad \text{Substitute } 3t + 15 \text{ for } F(t).$$

$$= \frac{5}{9}[(3t + 15) - 32] \qquad \text{Substitute } 3t + 15 \text{ for } F \text{ in } \frac{5}{9}(F - 32).$$

$$= \frac{5}{9}(3t - 17) \qquad \text{Simplify within the brackets.}$$

$$= \frac{15}{9}t - \frac{85}{9} \qquad \text{Distribute the multiplication by } \frac{5}{9}.$$

$$= \frac{5}{3}t - \frac{85}{9}$$

The composite function, $C(t) = \frac{5}{3}t - \frac{85}{9}$, gives the temperature of the specimen in degrees Celsius *t* hours after it is removed from refrigeration.

 Now Try   **Problem 75**

ANSWERS TO SELF CHECKS   1. a. $(f + g)(x) = 2x^2 + 6x - 2$ b. $(f - g)(x) = -2x^2 - 2$
2. a. $(f \cdot g)(x) = 2x^4 - x^2 - 3$ b. $(f/g)(x) = \frac{2x^2 - 3}{x^2 + 1}$ 3. a. -8 b. 5 c. $(g \circ f)(x) = 6 - x^3$
4. a. -6 b. -3 c. -4

STUDY SET
11.1

VOCABULARY

Fill in the blanks.

1. The _____ of *f* and *g*, denoted as $f + g$, is defined by $(f + g)(x) = $ [_____] and the _____ of *f* and *g*, denoted as $f - g$, is defined by $(f - g)(x) = $ [_____].

2. The _____ of *f* and *g*, denoted as $f \cdot g$, is defined by $(f \cdot g)(x) = $ [_____] and the _____ of *f* and *g*, denoted as f/g, is defined by $(f/g)(x) = $ [____].

3. The _____ of the function $f + g$ is the set of real numbers *x* that are in the domain of both *f* and *g*.

4. The _____ function $f \circ g$ is defined by $(f \circ g)(x) = $ [_____].

5. Under the _____ function, the value that is assigned to any real number *x* is *x* itself: $I(x) = x$.

6. When reading the notation $f(g(x))$, we say "*f* ___ *g* ___ *x*."

CONCEPTS

7. Fill in the blanks.
   a. $(f \circ g)(3) = f(\quad)$
   b. To find $f(g(3))$, we first find _____ and then substitute that value for *x* in $f(x)$.

8. a. If $f(x) = 3x + 1$ and $g(x) = 1 - 2x$, find $f(g(3))$ and $g(f(3))$.
   b. Is the composition of functions commutative?

9. Fill in the three blanks in the drawing of the function machines that show how to compute $g(f(-2))$.

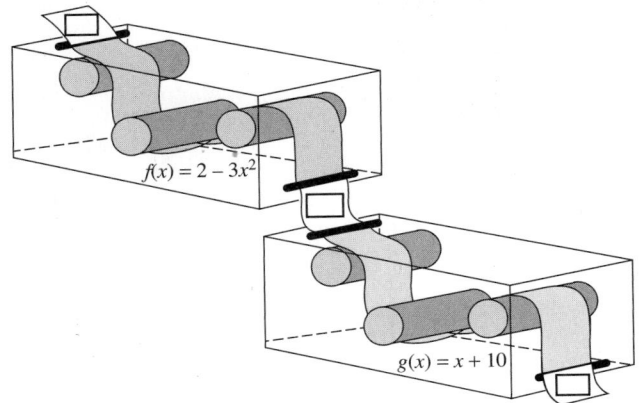

$f(x) = 2 - 3x^2$

$g(x) = x + 10$

10. Complete the table of values for the identity function, $I(x) = x$. Then graph it.

x	$I(x)$
-3	
-2	
-1	
0	
1	
2	
3	

NOTATION

Complete each solution.

11. Let $f(x) = 3x - 1$ and $g(x) = 2x + 3$. Find $f \cdot g$.

$(f \cdot g)(x) = f(x) \cdot \ \ \ \ $

$\qquad = \ \ \ \ \ \ (2x + 3)$

$\qquad = 6x^2 + \ \ \ - \ \ \ - 3$

$(f \cdot g)(x) = 6x^2 + 7x - 3$

12. Let $f(x) = 3x - 1$ and $g(x) = 2x + 3$. Find $f \circ g$.

$(f \circ g)(x) = f(\ \ \ \ )$

$\qquad = f(\ \ \ \ \ \ )$

$\qquad = 3(\ \ \ \ \ \ ) - 1$

$\qquad = \ \ \ + \ \ \ - 1$

$(f \circ g)(x) = 6x + 8$

GUIDED PRACTICE

Let $f(x) = 3x$ and $g(x) = 4x$. Find each function and its domain.
See Examples 1 and 2.

13. $f + g$

14. $f - g$

15. $g - f$

16. $g + f$

17. $f \cdot g$

18. f/g

19. g/f

20. $g \cdot f$

Let $f(x) = 2x + 1$ and $g(x) = x - 3$. Find each function and its domain. See Examples 1 and 2.

21. $f + g$

22. $f - g$

23. $g - f$

24. $g + f$

25. $f \cdot g$

26. f/g

27. g/f

28. $g \cdot f$

Let $f(x) = 3x - 2$ and $g(x) = 2x^2 + 1$. Find each function and its domain. See Examples 1 and 2.

29. $f - g$

30. $f + g$

31. f/g

32. $f \cdot g$

Let $f(x) = x^2 - 1$ and $g(x) = x^2 - 4$. Find each function and its domain.

33. $f - g$

34. $f + g$

35. g/f

36. $g \cdot f$

Let $f(x) = 2x + 1$ and $g(x) = x^2 - 1$. Find each of the following. See Example 3.

37. $(f \circ g)(2)$

38. $(g \circ f)(2)$

39. $(g \circ f)(-3)$

40. $(f \circ g)(-3)$

41. $(f \circ g)(0)$

42. $(g \circ f)(0)$

43. $(f \circ g)\left(\dfrac{1}{2}\right)$

44. $(g \circ f)\left(\dfrac{1}{3}\right)$

45. $(f \circ g)(x)$

46. $(g \circ f)(x)$

47. $(g \circ f)(2x)$

48. $(f \circ g)(2x)$

Let $f(x) = 3x - 2$ and $g(x) = x^2 + x$. Find each of the following. See Example 3.

49. $(f \circ g)(4)$

50. $(g \circ f)(4)$

51. $(g \circ f)(-3)$

52. $(f \circ g)(-3)$

53. $(g \circ f)(0)$

54. $(f \circ g)(0)$

55. $(g \circ f)(x)$

56. $(f \circ g)(x)$

Let $f(x) = \frac{1}{x}$ **and** $g(x) = \frac{1}{x^2}$. **Find each of the following.**

57. $(f \circ g)(4)$

58. $(f \circ g)(6)$

59. $(g \circ f)\left(\frac{1}{3}\right)$

60. $(g \circ f)\left(\frac{1}{10}\right)$

61. $(g \circ f)(8x)$

62. $(f \circ g)(5x)$

63. If $f(x) = x + 1$ and $g(x) = 2x - 5$, show that $(f \circ g)(x) \neq (g \circ f)(x)$.

64. If $f(x) = x^2 + 1$ and $g(x) = 3x^2 - 2$, show that $(f \circ g)(x) \neq (g \circ f)(x)$.

See Example 4.

65. Refer to graphs. Find:

  a. $(f + g)(-5)$

  b. $(f - g)(3)$

  c. $(f \cdot g)(-3)$

  d. $(f/g)(0)$

  e. $(f \circ g)(3)$

  f. $(g \circ f)(2)$

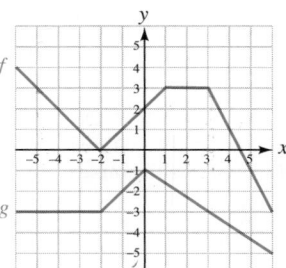

66. Refer to graphs. Find:

  a. $(g + f)(2)$

  b. $(g - f)(-5)$

  c. $(g \cdot f)(1)$

  d. $(g/f)(-6)$

  e. $(g \circ f)(4)$

  f. $(f \circ g)(6)$

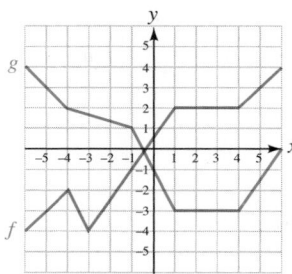

67. Use the tables of values for functions f and g to find each of the following.

x	$f(x)$
1	3
5	8

x	$g(x)$
1	4
5	0

  a. $(f + g)(1)$ **b.** $(f - g)(5)$

  c. $(f \cdot g)(1)$ **d.** $(g/f)(5)$

68. Use the table of values for functions f and g to find each of the following.

x	$f(x)$
2	5
4	7

x	$g(x)$
1	2
5	-3

  a. $(f \circ g)(1)$ **b.** $(g \circ f)(2)$

Let $f(x) = 3x + 1$ **and find each composition. See Example 5.**

69. $(f \circ I)(x)$

70. $(I \circ f)(x)$

Let $f(x) = -2x - 5$ **and find each composition. See Example 5.**

71. $(I \circ f)(x)$

72. $(f \circ I)(x)$

APPLICATIONS

73. SAT SCORES   Refer to the following illustration. The graph of function m gives the average score on the mathematics portion of the SAT college entrance exam, the graph of function v gives the average score on the verbal portion, and x represents the number of years since 1990. (In 2006, the SAT Test was renamed the SAT Reasoning Test and it now includes three parts: mathematics, critical reading, and writing.)

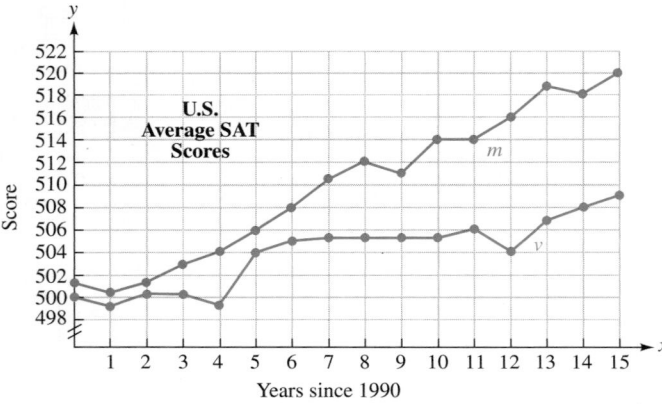

Source: *The World Almanac*, 1999 and Infoplease.com

  a. Find $(m + v)(3)$ and explain what information about SAT scores it gives.

  b. Find $(m - v)(6)$ and explain what information about SAT scores it gives.

  c. Find: $(m + v)(15)$

  d. Find: $(m - v)(15)$

74. BACHELOR'S DEGREES   Refer to the following illustration. The graph of function *m* gives the actual and projected number of bachelor's degrees awarded to men, and the graph of function *w* gives the actual and projected number of bachelor's degrees awarded to women.

**Actual and Projected Numbers
for Bachelor's Degrees Awarded in the U.S.**

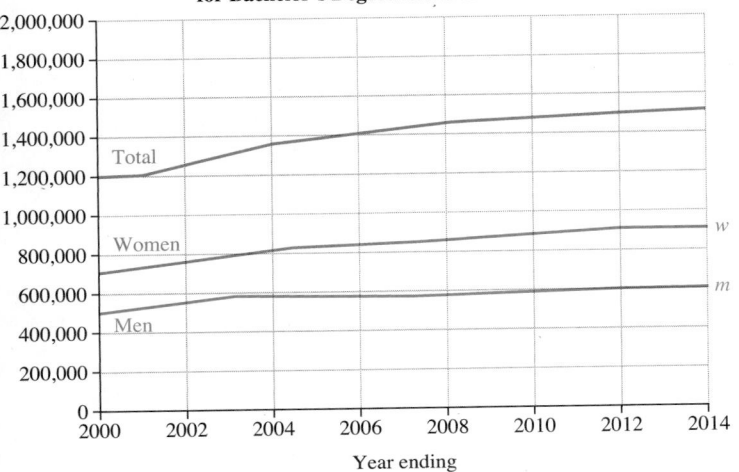

Source: U.S. Department of Education

a. Estimate $(w + m)(2000)$ and explain what information about bachelor's degrees it gives.

b. Estimate $(w - m)(2000)$ and explain what information about bachelor's degrees it gives.

c. Estimate: $(w + m)(2012)$

d. Estimate: $(w - m)(2012)$

75. METALLURGY   A molten alloy must be cooled slowly to control crystallization. When removed from the furnace, its temperature is 2,700°F, and it will be cooled at 200° per hour. Write a composition function that expresses the Celsius temperature as a function of the number of hours *t* since cooling began. (*Hint*: $C = \frac{5}{9}(F - 32)$.)

76. WEATHER FORECASTING   A high-pressure area promises increasingly warmer weather for the next 48 hours. The temperature is now 34° Celsius and is expected to rise 1° every 6 hours. Write a composition function that expresses the Fahrenheit temperature as a function of the number of hours from now. (*Hint*: $F = \frac{9}{5}C + 32$.)

77. VACATION MILEAGE COSTS

a. Use the following graphs to determine the cost of the gasoline consumed if a family drove 500 miles on a summer vacation.

b. Write a composition function that expresses the cost of the gasoline consumed on the vacation as a function of the miles driven.

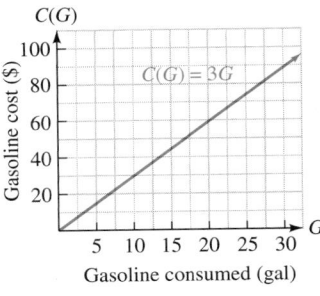

78. HALLOWEEN COSTUMES   The tables on the back of a pattern package can be used to determine the number of yards of material needed to make a rabbit costume for a child.

a. How many yards of material are needed if the child's chest measures 29 inches?

b. In this exercise, one quantity is a function of a second quantity that depends, in turn, on a third quantity. Explain this dependence.

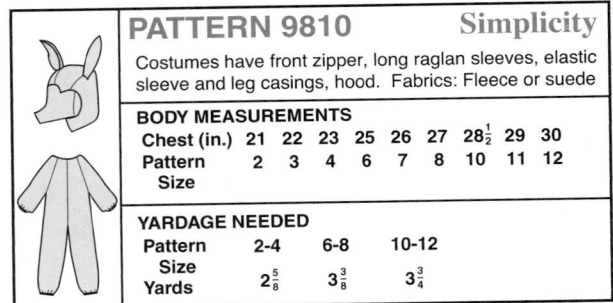

| **PATTERN 9810** | | | | | | | | | | **Simplicity** |

Costumes have front zipper, long raglan sleeves, elastic sleeve and leg casings, hood. Fabrics: Fleece or suede

BODY MEASUREMENTS

Chest (in.)	21	22	23	25	26	27	$28\frac{1}{2}$	29	30
Pattern Size	2	3	4	6	7	8	10	11	12

YARDAGE NEEDED

Pattern Size	2-4	6-8	10-12
Yards	$2\frac{5}{8}$	$3\frac{3}{8}$	$3\frac{3}{4}$

WRITING

79. Exercise 77 illustrates a chain of dependence between the cost of the gasoline, the gasoline consumed, and the miles driven. Describe another chain of dependence that could be represented by a composition function.

80. In this section, what operations are performed on functions? Give an example of each.

81. Write out in words how to say each of the following:

$$(f \circ g)(2) \qquad g(f(-8))$$

82. If $Y_1 = f(x)$ and $Y_2 = g(x)$, explain how to use the following tables to find $g(f(2))$.

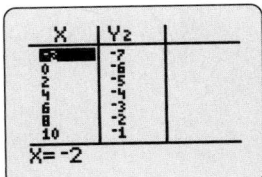

REVIEW

Simplify each complex fraction.

83. $\dfrac{\dfrac{ac - ad - c + d}{a^3 - 1}}{\dfrac{c^2 - 2cd + d^2}{a^2 + a + 1}}$

84. $\dfrac{2 + \dfrac{1}{x^2 - 1}}{1 + \dfrac{1}{x - 1}}$

CHALLENGE PROBLEMS

Fill in the blanks.

85. If $f(x) = x^2$ and $g(x) = \underline{}$, then $(f \circ g)(x) = 4x^2 + 20x + 25$.

86. If $f(x) = \sqrt{3x}$ and $g(x) = \underline{}$, then $(g \circ f)(x) = 9x^2 + 7$.

Refer to the following graphs of functions f and g.

87. Graph the sum function $f + g$ on the given coordinate system.

88. Graph the difference function $f - g$ on the given coordinate system.

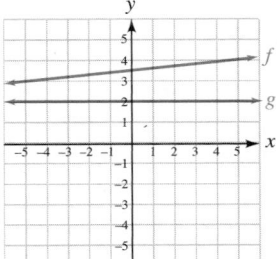

SECTION 11.2
Inverse Functions

Objectives

① Determine whether a function is a one-to-one function.

② Use the horizontal line test to determine whether a function is one-to-one.

③ Find the equation of the inverse of a function.

④ Find the composition of a function and its inverse.

⑤ Graph a function and its inverse.

In the previous section, we created new functions from given functions by using the operations of arithmetic and composition. Another way to create new functions is to find the *inverse of a function*.

① **Determine Whether a Function Is a One-to-One Function.**

In figure (a) below, the arrow diagram defines a function f. If we reverse the arrows as shown in figure (b), we obtain a new correspondence where the range of f becomes the domain of the new correspondence, and the domain of f becomes the range. The new correspondence is a function because to each member of the domain, there corresponds exactly one member of the range. We call this new correspondence the **inverse** of f, or f inverse.

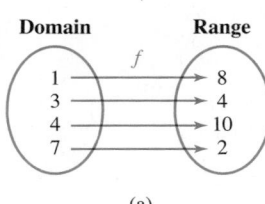

(a)

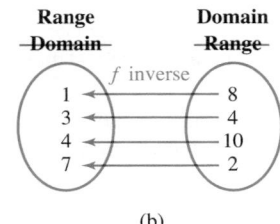

(b)

This reversing process does not always produce a function. For example, if we reverse the arrows in function *g* defined by the diagram in figure (a) below, the resulting correspondence is not a function. This is because to the number 2 in the domain, there corresponds two members of the range: 8 and 4.

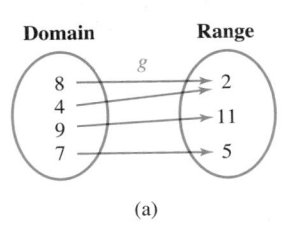
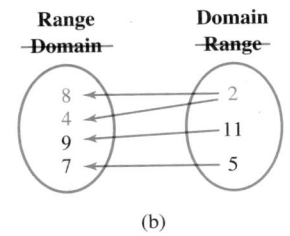

(a)   (b)

The question that arises is, "What must be true of an original function to guarantee that the reversing process produces a function?" The answer is: *the original function must be one-to-one.*

We have seen that in a function, each input determines exactly one output. For some functions, different inputs determine different outputs, as in figure (a). For other functions, different inputs might determine the *same* output, as in figure (b). When a function has the property that different inputs determine different outputs, as in figure (a), we say the function is *one-to-one.*

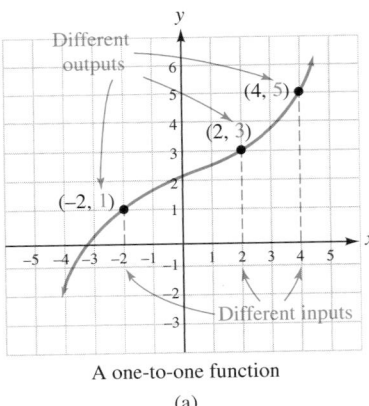

A one-to-one function
(a)

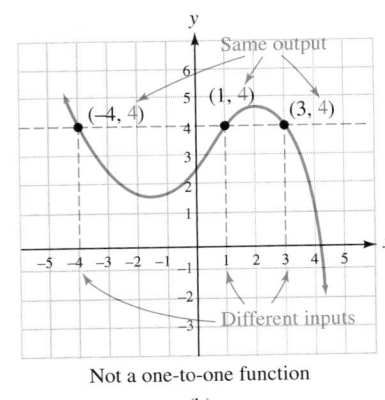

Not a one-to-one function
(b)

One-to-One Functions	A function is called a **one-to-one function** if different inputs determine different outputs.

EXAMPLE 1   Determine whether each function is one-to-one.   **a.** $f(x) = x^2$
b. $f(x) = x^3$

Strategy   We will determine whether different inputs have different outputs.

Why   If different inputs have different outputs, the function is one-to-one. If different inputs have the same output, the function is not one-to-one.

Solution

a. Since two different inputs, -3 and 3, have the same output 9, $f(x) = x^2$ is not a one-to-one function.

$$f(-3) = (-3)^2 = 9 \text{ and } f(3) = 3^2 = 9$$

x	$f(x)$
-3	9
3	9

The output 9 does not correspond to exactly one input.

b. Since different numbers have different cubes, each input of $f(x) = x^3$ determines a different output. This function is one-to-one.

Self Check 1   Determine whether each function is one-to-one. If not, find an output that corresponds to more than one input.
a. $f(x) = 2x + 3$ **b.** $f(x) = x^2$

Now Try   Problems 19 and 21

2   **Use the Horizontal Line Test to Determine Whether a Function Is One-to-One.**

To determine whether a function is one-to-one, it is often easier to view its graph rather than its defining equation. If two (or more) points on the graph of a function have the same y-coordinate, the function is not one-to-one. This observation suggests the following *horizontal line test*.

The Horizontal Line Test   A function is one-to-one if each horizontal line that intersects its graph does so exactly once.

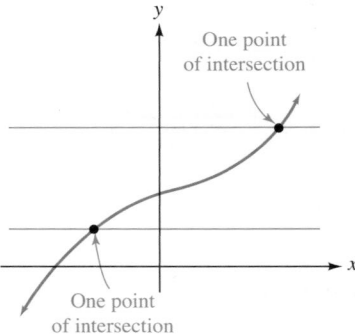

One point of intersection

One point of intersection

A one-to-one function

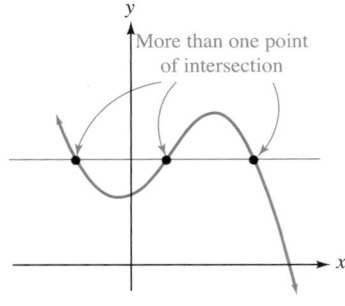

More than one point of intersection

Not a one-to-one function

EXAMPLE 2   Use the horizontal line test to determine whether the following graphs of functions represent one-to-one functions.

Strategy   We will draw horizontal lines through the graph of the function and see how many times each line intersects the graph.

Why   If each horizontal line intersects the graph of the function exactly once, the graph represents a one-to-one function. If any horizontal line intersects the graph of the function more than once, the graph does not represent a one-to-one function.

Solution

a. Because we can draw a horizontal line that intersects the graph of the function shown in figure (a) twice, the graph does not represent a one-to-one function.

b. Because every horizontal line that intersects the graph of the function in figure (b) does so exactly once, the graph represents a one-to-one function.

Success Tip

Recall that we use the vertical line test to determine whether a graph represents a function. We use the horizontal line test to determine whether the function that is graphed is one-to-one.

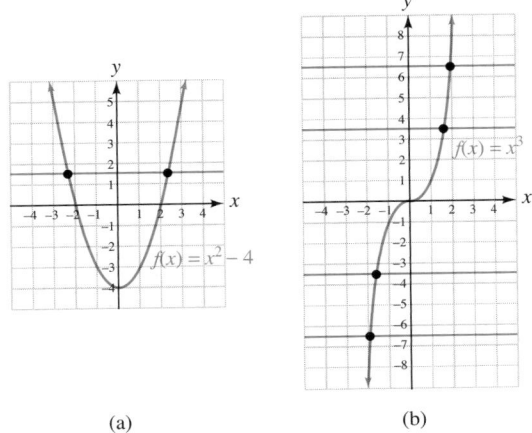

(a)   (b)

Self Check 2   Determine whether the following graphs represent one-to-one functions.

a.   **b.**

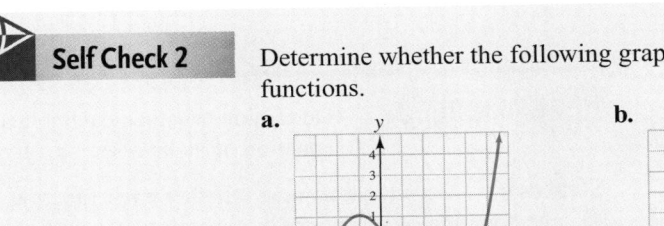

Now Try   Problems 27 and 29

③   Find the Equation of the Inverse of a Function.

If f is the one-to-one function defined by the arrow diagram in figure (a), it turns the number 1 into 10, 2 into 20, and 3 into 30. The ordered pairs that define f can be listed in a table. Since the inverse of f must turn 10 back into 1, 20 back into 2, and 30 back into 3, it consists of the ordered pairs shown in the table in figure (b).

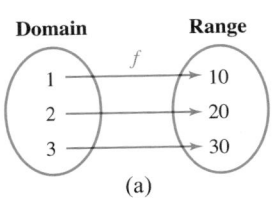

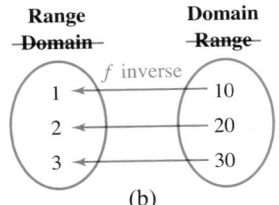

(a)   (b)

The x- and y-coordinates are interchanged.

We note that the domain of f and the range of its inverse is $\{1, 2, 3\}$. The range of f and the domain of its inverse is $\{10, 20, 30\}$.

This example suggests that to form the inverse of a function f, we simply interchange the coordinates of each ordered pair that determines f. When the inverse of a function is also a function, we call it **f inverse** and denote it with the symbol f^{-1}. The symbol $f^{-1}(x)$ is read as "the inverse of $f(x)$" or "f inverse of x."

The Inverse of a Function	If f is a one-to-one function consisting of ordered pairs of the form (x, y), the **inverse of f**, denoted f^{-1}, is the one-to-one function consisting of all ordered pairs of the form (y, x).

When a one-to-one function is defined by an equation, we use the following method to find the equation of its inverse.

Finding the Equation of the Inverse of a Function	If a function is one-to-one, we find its inverse as follows:

1. If the function is written using function notation, replace $f(x)$ with y.
2. Interchange the variables x and y.
3. Solve the resulting equation for y.
4. Substitute $f^{-1}(x)$ for y.

EXAMPLE 3   Determine whether each function is one-to-one. If so, find the equation of its inverse.   **a.** $f(x) = 4x + 2$ **b.** $f(x) = x^3$

Strategy   We will determine whether each function is one-to-one. If it is, we can find the equation of its inverse by replacing $f(x)$ with y, interchanging x and y, and solving for y.

Why   The reason for interchanging the variables is this: If a one-to-one function takes an input x into an output y, by definition, its inverse function has the reverse effect.

Solution

a. We recognize $f(x) = 4x + 2$ as a linear function whose graph is a straight line with slope 4 and y-intercept $(0, 2)$. Since such a graph would pass the horizontal line test, we conclude that f is one-to-one.

To find the inverse, we proceed as follows:

$$f(x) = 4x + 2$$
$$y = 4x + 2 \qquad \text{Replace } f(x) \text{ with } y.$$
$$x = 4y + 2 \qquad \text{Interchange the variables } x \text{ and } y.$$
$$x - 2 = 4y \qquad \text{To isolate the term } 4y, \text{ subtract 2 from both sides.}$$
$$\frac{x - 2}{4} = y \qquad \text{To solve for } y, \text{ divide both sides by 4.}$$
$$y = \frac{x - 2}{4} \qquad \text{Write the equation with } y \text{ on the left side.}$$

To denote that this equation is the inverse of function f, we replace y with $f^{-1}(x)$.

$$f^{-1}(x) = \frac{x - 2}{4}$$

b. In Example 2, we used the horizontal line test to determine that $f(x) = x^3$ is a one-to-one function. See figure (b) in Example 2.
To find its inverse, we proceed as follows:

$$f(x) = x^3$$
$$y = x^3 \qquad \text{Replace } f(x) \text{ with } y.$$
$$x = y^3 \qquad \text{Interchange the variables } x \text{ and } y.$$
$$\sqrt[3]{x} = y \qquad \text{To solve for } y, \text{ take the cube root of both sides.}$$
$$y = \sqrt[3]{x} \qquad \text{Write the equation with } y \text{ on the left side}$$

Replacing y with $f^{-1}(x)$, we have

$$f^{-1}(x) = \sqrt[3]{x}$$

> **Caution**
> Only one-to-one functions have inverse functions.

Self Check 3 Determine whether each function is one-to-one. If it is, find the equation of its inverse. **a.** $f(x) = -5x - 3$ **b.** $f(x) = x^5$

Now Try Problems 35 and 45

 4 **Find the Composition of a Function and Its Inverse.**

To emphasize a relationship between a function and its inverse, we substitute some number x, such as $x = 3$, into the function $f(x) = 4x + 2$ of Example 3(a). The corresponding value of y that is produced is

$$f(3) = 4(3) + 2 = 14 \qquad \text{f determines the ordered pair (3, 14).}$$

If we substitute 14 into the inverse function, $f^{-1}(x) = \frac{x-2}{4}$, the corresponding value of y that is produced is

$$f^{-1}(14) = \frac{14-2}{4} = 3 \qquad f^{-1} \text{ determines the ordered pair (14, 3).}$$

Thus, the function f turns 3 into 14, and the inverse function f^{-1} turns 14 back into 3.
In general, the composition of a function and its inverse function is the identity function, $I(x) = x$, such that any input x has the output x. This fact can be stated symbolically as follows.

> **The Composition of Inverse Functions**
> For any one-to-one function f and its inverse, f^{-1},
> $$(f \circ f^{-1})(x) = x \quad \text{and} \quad (f^{-1} \circ f)(x) = x$$

We can use this property to determine whether two functions are inverses.

EXAMPLE 4 Show that $f(x) = 4x + 2$ and $f^{-1}(x) = \frac{x-2}{4}$ are inverses.

Strategy We will find the composition of $f(x)$ and $f^{-1}(x)$ in both directions and show that the result is x.

Why   Only when the result of the composition is x in both directions are the functions inverses.

Solution   To show that $f(x) = 4x + 2$ and $f^{-1}(x) = \frac{x - 2}{4}$ are inverses, we must show that for each composition, an input of x gives an output of x.

$$(f \circ f^{-1})(x) = f(f^{-1}(x))$$

$$= f\left(\frac{x - 2}{4}\right)$$

$$= 4\left(\frac{x - 2}{4}\right) + 2$$

$$= x - 2 + 2$$

$$= x$$

$$(f^{-1} \circ f)(x) = f^{-1}(f(x))$$

$$= f^{-1}(4x + 2)$$

$$= \frac{4x + 2 - 2}{4}$$

$$= \frac{4x}{4}$$

$$= x$$

Because $(f \circ f^{-1})(x) = x$ and $(f^{-1} \circ f)(x) = x$, the functions are inverses.

Self Check 4   Show that $f(x) = x - 4$ and $g(x) = x + 4$ are inverses.

Now Try   Problem 55

---

5   **Graph a Function and Its Inverse.**

If a point (a, b) is on the graph of function f, it follows that the point (b, a) is on the graph of f^{-1}, and vice versa. There is a geometric relationship between a pair of points whose coordinates are interchanged. For example, in the graph, we see that the line segment between $(1, 3)$ and $(3, 1)$ is perpendicular to and cut in half by the line $y = x$. We say that $(1, 3)$ and $(3, 1)$ are mirror images of each other with respect to $y = x$.

Since each point on the graph of f^{-1} is a mirror image of a point on the graph of f, and vice versa, the graphs of f and f^{-1} must be mirror images of each other with respect to $y = x$.

> **Success Tip**
> Recall that the line $y = x$ passes through points whose x- and y-coordinates are equal: $(-1, -1)$, $(0, 0)$, $(1, 1)$, $(2, 2)$, and so on.

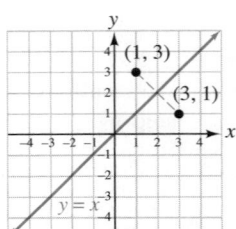

EXAMPLE 5   Find the equation of the inverse of $f(x) = -\frac{3}{2}x + 3$. Then graph f and its inverse on one coordinate system.

Strategy   We will determine whether the function has an inverse. If so, we will replace $f(x)$ with y, interchange x and y, and solve for y to obtain the equation of the inverse.

Why   The reason for interchanging the variables is this: If a one-to-one function takes an input x into an output y, by definition, its inverse function has the reverse effect.

Solution   Since $f(x) = -\frac{3}{2}x + 3$ is a linear function, it is one-to-one and has an inverse. To find the inverse function, we replace $f(x)$ with y, and interchange x and y to obtain

$$x = -\frac{3}{2}y + 3$$

Then we solve for y to get

$$x - 3 = -\frac{3}{2}y \qquad \text{Subtract 3 from both sides.}$$

$$-\frac{2}{3}x + 2 = y \qquad \text{To isolate y, multiply both sides by } -\frac{2}{3}.$$

When we replace y with $f^{-1}(x)$, we have $f^{-1}(x) = -\frac{2}{3}x + 2$.

To graph f and f^{-1}, we construct tables of values and plot points. Because the functions are inverses of each other, their graphs are mirror images about the line $y = x$

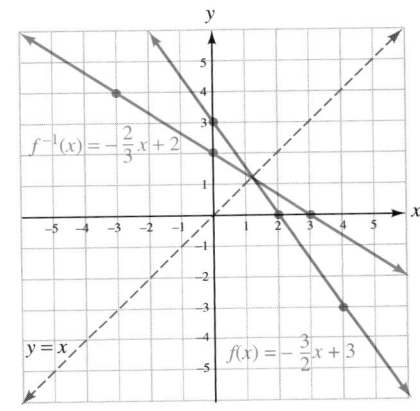

$$f(x) = -\frac{3}{2}x + 3 \qquad\qquad f^{-1}(x) = -\frac{2}{3}x + 2$$

x	$f(x)$	
0	3	$\to (0, 3)$
2	0	$\to (2, 0)$
4	-3	$\to (4, -3)$

x	$f^{-1}(x)$	
3	0	$\to (3, 0)$
0	2	$\to (0, 2)$
-3	4	$\to (-3, 4)$

Success Tip

To graph f^{-1}, we don't need to construct a table of values. We can simply interchange the coordinates of the ordered pairs in the table for f and use them to graph f^{-1}.

Self Check 5   Find the inverse of $f(x) = \frac{2}{3}x - 2$. Then graph the function and its inverse on one coordinate system.

Now Try   **Problem 59**

Using Your Calculator

Graphing the Inverse of a Function

We can use a graphing calculator to check the result found in Example 5. First, we enter $f(x) = -\frac{3}{2}x + 3$ and then enter what we believe to be the inverse function, $f^{-1}(x) = -\frac{2}{3}x + 2$, as well as the equation $y = x$. See figure (a). Before graphing, we adjust the display so that the graphing grid will be composed of squares. The axis of symmetry is then at a 45° angle to the positive x-axis.

In figure (b), it appears that the two graphs are symmetric about the line $y = x$. Although it is not definitive, this visual check does help to validate the result of Example 5.

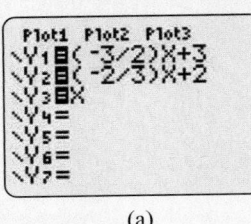

(a)

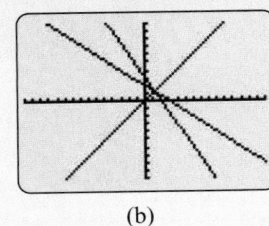

(b)

EXAMPLE 6 Graph the inverse of function f shown in figure (a).

Strategy We will find the coordinates of several points on curve f in figure (a). After interchanging the coordinates of these points, we will plot them as shown in figure (b).

Why The reason for interchanging the coordinates is this: If (a, b) is a point on the graph of a one-to-one function, then the point (b, a) is on the graph of its inverse.

Solution In figure (a), we see that the points $(-5, -3)$, $(-2, -1)$, $(0, 2)$, $(3, 3)$, $(5, 4)$, and $(7, 5)$ lie on the graph of function f. To graph the inverse, we interchange their coordinates, and plot them, as shown in figure (b). Then we graph the line $y = x$ and use symmetry to draw a smooth curve through those points to get the graph of f^{-1}.

> **The Language of Algebra**
> We can also say that the graphs of f and f^{-1} are *reflections* of each other about the line $y = x$, or they are *symmetric about* $y = x$.

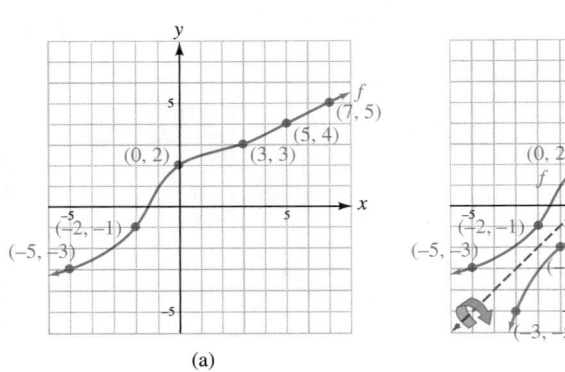

(a)   (b)

Now Try Problem 59

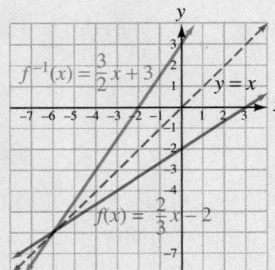

STUDY SET
11.2

VOCABULARY

Fill in the blanks.

1. A function is called a _____ function if different inputs determine different outputs.
2. The _____ line test can be used to determine whether the graph of a function represents a one-to-one function.
3. The functions f and f^{-1} are _____.
4. The graphs of a function and its inverse are mirror _____ of each other with respect to $y = x$. We also say that their graphs are _____ with respect to the line $y = x$.

CONCEPTS

Fill in the blanks.

5. If any horizontal line that intersects the graph of a function does so more than once, the function is not _____.
6. To find the inverse of the function $f(x) = 2x - 3$, we begin by replacing $f(x)$ with y, and then we _____ x and y.
7. If f is a one-to-one function, the domain of f is the _____ of f^{-1}, and the range of f is the _____ of f^{-1}.
8. If a function turns an input of 2 into an output of 5, the inverse function will turn an input of 5 into the output ___.
9. If f is a one-to-one function, and if $f(1) = 6$, then $f^{-1}(6) = $ ___ ?
10. If the point $(9, -4)$ is on the graph of the one-to-one function f, then the point (___ , ___) is on the graph of f^{-1}.
11. **a.** Is the correspondence defined by the arrow diagram a one-to-one function?

    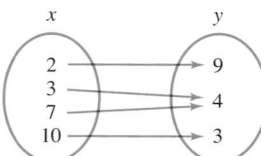

    b. Is the correspondence defined by the table a one-to-one function?

x	$f(x)$
-2	4
-1	1
0	0
2	4
3	9

12. Is the inverse of a one-to-one function always a function?

13. Use the table of values of the one-to-one function f to complete a table of values for f^{-1}.

x	$f(x)$
-4	-2
0	0
8	4

x	$f^{-1}(x)$
-2	
0	
4	

14. Redraw the graph of function f. Then graph f^{-1} and the axis of symmetry on the same coordinate system.

    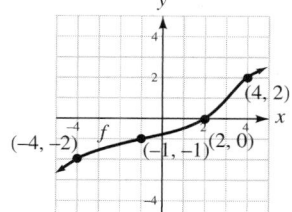

NOTATION

Complete each solution.

15. Find the inverse of $f(x) = 2x - 3$.

$$\boxed{} = 2x - 3$$
$$x = \boxed{} - 3$$
$$x + \boxed{} = 2y$$
$$\frac{x + 3}{2} = \boxed{}$$

The inverse of $f(x) = 2x - 3$ is $\boxed{\phantom{f^{-1}}}(x) = \dfrac{x + 3}{2}$.

16. Find the inverse of $f(x) = \sqrt[3]{x} + 2$.

$$\boxed{} = \sqrt[3]{x} + 2$$
$$x = \sqrt[3]{\boxed{}} + 2$$
$$x - \boxed{} = \sqrt[3]{y}$$
$$(x - 2)^3 = \boxed{}$$

The inverse of $f(x) = \sqrt[3]{x} + 2$ is $\boxed{\phantom{f^{-1}}}(x) = (x - 2)^3$.

17. The symbol f^{-1} is read as "the ~~inverse~~ of f" or "f ~~inverse~~"

18. Explain the difference in the meaning of the -1 in the notation $f^{-1}(x)$ as compared with x^{-1}.

GUIDED PRACTICE

Determine whether each function is one-to-one. See Example 1.

19. $f(x) = 2x$

20. $f(x) = |x|$

21. $f(x) = x^4$

22. $f(x) = x^3 + 1$

23. $f(x) = -x^2 + 3x$

24. $f(x) = \frac{2}{3}x + 8$

25. $\{(1, 1), (2, 1), (3, 1), (4, 1)\}$

26. $\{(3, 2), (2, 1), (1, 0)\}$

Each graph represents a function. Use the horizontal line test to determine whether the function is one-to-one. See Example 2.

27.
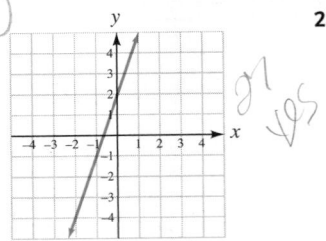

28.
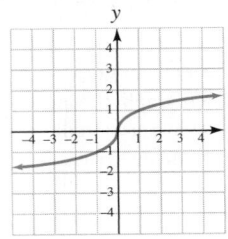

29.
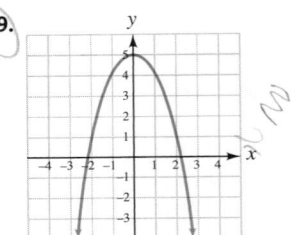

30.
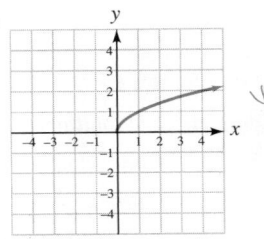

31.
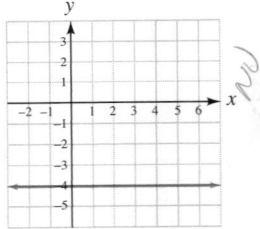

32.
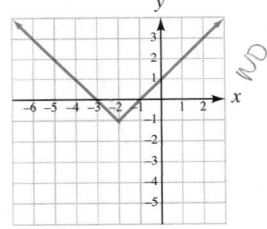

33.
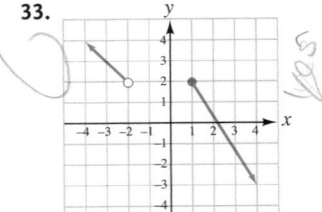

34.
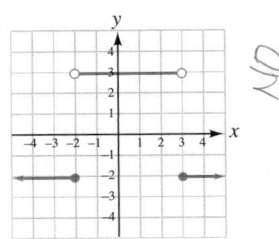

Each of the following functions is one-to-one. Find the inverse of each function and express it using $f^{-1}(x)$ notation. See Example 3.

35. $f(x) = 2x + 4$

36. $f(x) = 5x - 1$

37. $f(x) = \frac{x}{5} + \frac{4}{5}$

38. $f(x) = \frac{x}{3} - \frac{1}{3}$

39. $f(x) = \frac{x - 4}{5}$

40. $f(x) = \frac{2x + 6}{3}$

41. $f(x) = \frac{2}{x - 3}$

42. $f(x) = \frac{3}{x + 1}$

43. $f(x) = \frac{4}{x}$

44. $f(x) = \frac{1}{x}$

45. $f(x) = x^3 + 8$

46. $f(x) = x^3 - 4$

47. $f(x) = \sqrt[3]{x}$

48. $f(x) = \sqrt[3]{x - 5}$

49. $f(x) = (x + 10)^3$

50. $f(x) = (x - 9)^3$

51. $f(x) = 2x^3 - 3$

52. $f(x) = \frac{3}{x^3} - 1$

53. $f(x) = \frac{x^7}{2}$

54. $f(x) = \frac{x^9}{4}$

Show that each pair of functions are inverses. See Example 4.

55. $f(x) = 2x + 9, \ f^{-1}(x) = \frac{x - 9}{2}$

56. $f(x) = 5x - 1, \ f^{-1}(x) = \frac{x + 1}{5}$

57. $f(x) = \frac{2}{x - 3}, \ f^{-1}(x) = \frac{2}{x} + 3$

58. $f(x) = \sqrt[3]{x - 6}, \ f^{-1}(x) = x^3 + 6$

Find the inverse of each function. Then graph the function and its inverse on one coordinate system. Show the line of symmetry on the graph. See Examples 5 and 6.

59. $f(x) = 2x$

60. $f(x) = -3x$

61. $f(x) = 4x + 3$

62. $f(x) = \frac{x}{3} + \frac{1}{3}$

63. $f(x) = -\frac{2}{3}x + 3$

64. $f(x) = -\frac{1}{3}x + \frac{4}{3}$

65. $f(x) = x^3$

66. $f(x) = x^3 + 1$

67. $f(x) = x^2 - 1 \ (x \geq 0)$

68. $f(x) = x^2 + 1 \ (x \geq 0)$

APPLICATIONS

69. INTERPERSONAL RELATIONSHIPS Feelings of anxiety in a relationship can increase or decrease, depending on what is going on in the relationship. The graph shows how a person's anxiety might vary as a relationship develops over time.

  a. Is this the graph of a function? Is its inverse a function?

  b. Does each anxiety level correspond to exactly one point in time? Use the dashed lined labeled *Maximum threshold* to explain.

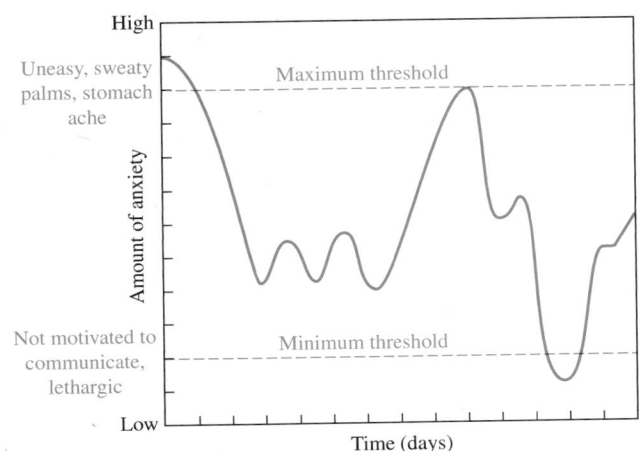

Source: Gudykunst, *Building Bridges: Interpersonal Skills for a Changing World* (Houghton Mifflin, 1994)

70. LIGHTING LEVELS The ability of the eye to see detail increases as the level of illumination increases. This relationship can be modeled by a function *E*, whose graph is shown here.

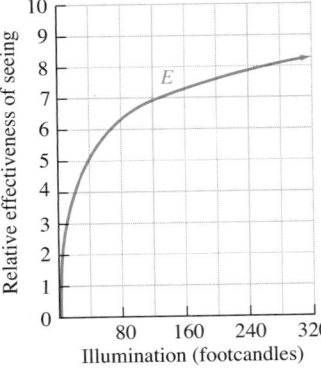

  a. From the graph, determine $E(240)$.

  b. Is function *E* one-to-one? Does *E* have an inverse?

  c. If the effectiveness of seeing in an office is 7, what is the illumination in the office? How can this question be asked using inverse function notation?

WRITING

71. In your own words, what is a one-to-one function?

72. Two functions are graphed on the square grid on the right along with the line $y = x$. Explain why the functions cannot be inverses of each other.

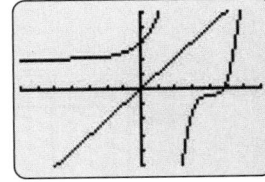

73. Explain how the graph of a one-to-one function can be used to draw the graph of its inverse function.

74. **a.** Explain the purpose of the vertical line test.

  b. Explain the purpose of the horizontal line test.

75. In the illustration, a function *f* and its inverse f^{-1} have been graphed on the same coordinate system. Explain what concept can be demonstrated by folding the graph paper on the dashed line.

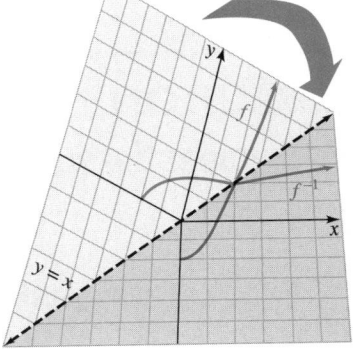

76. Write in words how to read the notation.

  a. $f^{-1}(x) = \dfrac{1}{2}x - 3$

  b. $(f \circ f^{-1})(x) = x$

REVIEW

Simplify. Write the result in a + bi form.

77. $3 - \sqrt{-64}$

78. $(2 - 3i) + (4 + 5i)$

79. $(3 + 4i)(2 - 3i)$

80. $\dfrac{6 + 7i}{3 - 4i}$

81. $(6 - 8i)^2$

82. i^{100}

CHALLENGE PROBLEMS

83. Find the inverse of $f(x) = \dfrac{x + 1}{x - 1}$.

84. Using the functions of Exercise 83, show that $(f \circ f^{-1})(x) = x$ and $(f^{-1} \circ f)(x) = x$.

85. A table of values for a function f is shown in figure (a). A table of values for f^{-1} is shown in figure (b). Use the tables to find $f^{-1}(f(4))$ and $f(f^{-1}(2))$.

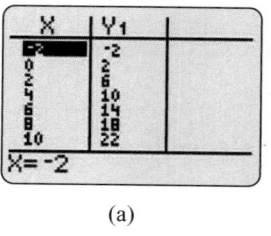

(a)

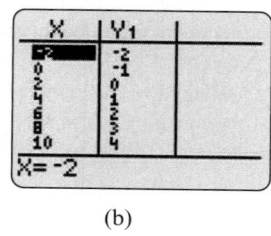

(b)

86. a. The graph of a one-to-one function lies entirely in quadrant I. In what quadrant does the graph of its inverse lie?

b. The graph of a one-to-one function lies entirely in quadrant II. In what quadrant does the graph of its inverse lie?

c. The graph of a one-to-one function lies entirely in quadrant III. In what quadrant does the graph of its inverse lie?

d. The graph of a one-to-one function lies entirely in quadrant IV. In what quadrant does the graph of its inverse lie?

SECTION 11.3
Exponential Functions

Objectives

1 Simplify expressions containing irrational exponents.

2 Define exponential functions.

3 Graph exponential functions.

4 Use exponential functions in applications involving growth or decay.

The graph in figure (a) below shows the balance in a bank account in which $10,000 was invested in 2006 at 9%, compounded monthly. The graph shows that in the year 2016, the value of the account will be approximately $25,000, and in the year 2036, the value will be approximately $147,000. The rapidly rising red curve is the graph of a function called an *exponential function*.

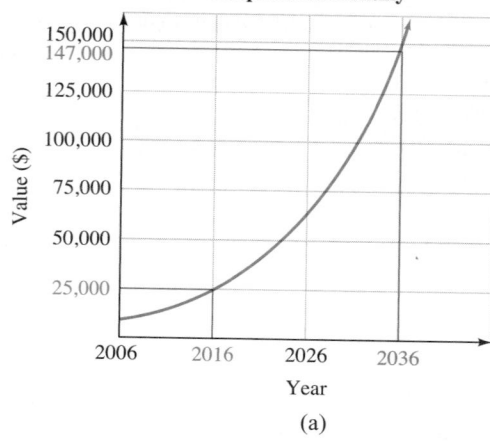

(a)

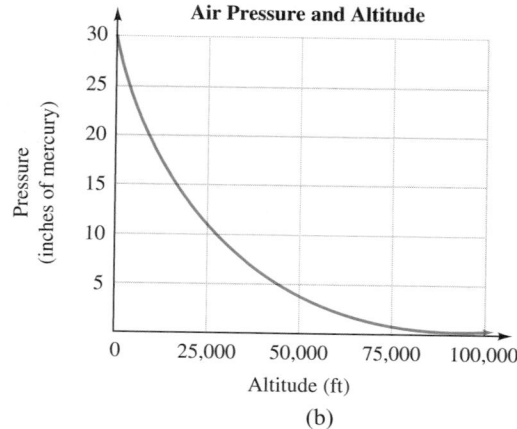

(b)

If you have ever climbed a high mountain or gone up in an airplane that does not have a pressurized cabin, you have probably felt the effects of low air pressure. The graph in figure

(b) on the previous page shows the how atmospheric pressure decreases with increasing altitude. The rapidly falling red curve is also the graph of an *exponential function.*

Exponential functions are used to model many other situations, such as population growth, the spread of an epidemic, the temperature of a heated object as it cools, and radioactive decay. Before we can discuss exponential functions in more detail, we must define irrational exponents.

1 Simplify Expressions Containing Irrational Exponents.

We have discussed expressions of the form b^x, where x is a rational number.

$8^{1/2}$ means "the square root of 8."

$5^{1/3}$ means "the cube root of 5."

$3^{-2/5} = \dfrac{1}{3^{2/5}}$ means "the reciprocal of the fifth root of 3^2."

To give meaning to b^x when x is an irrational number, we consider the expression

$5^{\sqrt{2}}$ where $\sqrt{2}$ is the irrational number $1.414213562\ldots$

We can successively approximate $5^{\sqrt{2}}$ by the following rational powers:

$$5^{1.4}, \qquad 5^{1.41}, \qquad 5^{1.414}, \qquad 5^{1.4142}, \qquad 5^{1.41421}, \ldots$$

Using concepts from advanced mathematics, it can be shown that there is exactly one number that these powers approach. We define $5^{\sqrt{2}}$ to be that number. This process can be used to approximate $5^{\sqrt{2}}$ to as many decimal places as desired.

Any other positive irrational exponent can be defined in the same manner, and negative irrational exponents can be defined using reciprocals. Thus, if b is positive, b^x has meaning for any real number x.

We can use a calculator to obtain a very good approximation of an exponential expression with an irrational exponent.

Using Your Calculator

Evaluating Exponential Expressions

To find the value of $5^{\sqrt{2}}$ with a scientific calculator, we enter these numbers and press these keys:

5 $\boxed{y^x}$ 2 $\boxed{\sqrt{}}$ = $\boxed{9.738517742}$

With a graphing calculator, we enter these numbers and press these keys:

5 $\boxed{\wedge}$ $\boxed{\text{2nd}}$ $\boxed{\sqrt{}}$ 2 $\boxed{)}$ $\boxed{\text{ENTER}}$ 5^√(2)
                                                          9.738517742

It can be shown that all of the familiar rules of exponents are also true for irrational exponents.

EXAMPLE 1   Simplify: **a.** $\left(5^{\sqrt{2}}\right)^{\sqrt{2}}$ **b.** $b^{\sqrt{3}} \cdot b^{\sqrt{12}}$

Strategy   We will use the power rule $(x^m)^n = x^{mn}$ to simplify part (a) and the product rule $x^m x^n = x^{m+n}$ to simplify part (b).

Why   These rules for exponents hold true for irrational exponents.

Solution

a. $\left(5^{\sqrt{2}}\right)^{\sqrt{2}} = 5^{\sqrt{2}\sqrt{2}}$ Keep the base and multiply the exponents.

$= 5^2$ Multiply: $\sqrt{2}\sqrt{2} = \sqrt{4} = 2$.

$= 25$

b. $b^{\sqrt{3}} \cdot b^{\sqrt{12}} = b^{\sqrt{3}+\sqrt{12}}$ Keep the base and add the exponents.

$= b^{\sqrt{3}+2\sqrt{3}}$ Simplify: $\sqrt{12} = \sqrt{4}\sqrt{3} = 2\sqrt{3}$.

$= b^{3\sqrt{3}}$ Combine like radicals. $\sqrt{3} + 2\sqrt{3} = 3\sqrt{3}$.

Self Check 1   Simplify: **a.** $\left(3^{\sqrt{2}}\right)^{\sqrt{8}}$ **b.** $b^{\sqrt{2}} \cdot b^{\sqrt{18}}$

Now Try   Problems 15 and 17

② Define Exponential Functions.

If $b > 0$ and $b \neq 1$, the function $f(x) = b^x$ is called an **exponential function.** Since x can be any real number, its domain is the set of real numbers. This is the interval $(-\infty, \infty)$.

Because b is positive, the value of $f(x)$ is positive, and the range is the set of positive numbers. This is the interval $(0, \infty)$.

Since $b \neq 1$, an exponential function cannot be the constant function $f(x) = 1^x$, in which $f(x) = 1$ for every real number x.

Exponential Functions

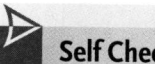

An **exponential function with base b** is defined by the equations

$$f(x) = b^x \qquad \text{or} \qquad y = b^x$$

where $b > 0$, $b \neq 1$, and x is a real number. The domain of $f(x) = b^x$ is the interval $(-\infty, \infty)$, and the range is the interval $(0, \infty)$.

③ Graph Exponential Functions.

Since the domain and range of $f(x) = b^x$ are sets of real numbers, we can graph exponential functions on a rectangular coordinate system.

EXAMPLE 2   Graph: $f(x) = 2^x$

Strategy   We will graph the function by creating a table of function values and plotting the corresponding ordered pairs.

Why   After drawing a smooth curve though the plotted points, we will have the graph.

Notation

We have previously graphed the linear function $f(x) = 2x$ and the squaring function $f(x) = x^2$. For the exponential function $f(x) = 2^x$, note that the variable is in the exponent.

Solution   To graph $f(x) = 2^x$, we choose several values for x and find the corresponding values of $f(x)$. If x is -1, we have

$f(x) = 2^x$

$f(-1) = 2^{-1}$ Substitute -1 for x.

$= \dfrac{1}{2}$

The point $\left(-1, \frac{1}{2}\right)$ is on the graph of $f(x) = 2^x$. In a similar way, we find the corresponding values of $f(x)$ for x values of 0, 1, 2, 3, and 4 and list them in a table. Then we plot the ordered pairs and draw a smooth curve through them.

$f(x) = 2^x$

x	$f(x)$	
-1	$\frac{1}{2}$	$\rightarrow \left(-1, \frac{1}{2}\right)$
0	1	$\rightarrow (0, 1)$
1	2	$\rightarrow (1, 2)$
2	4	$\rightarrow (2, 4)$
3	8	$\rightarrow (3, 8)$
4	16	$\rightarrow (4, 16)$

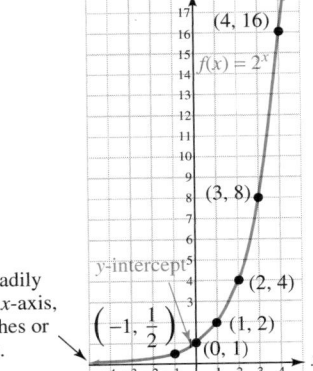

The graph steadily approaches the x-axis, but never touches or crosses it.

The Language of Algebra

We have encountered the word *asymptote* earlier, when we graphed rational functions. Recall that an asymptote is not part of the graph. It is a line that the graph approaches and, in this case, never touches.

From the graph, we can see that the domain of $f(x) = 2^x$ is the interval $(-\infty, \infty)$ and the range is the interval $(0, \infty)$. Since the graph passes the horizontal line test, the function is one-to-one.

Note that as x decreases, the values of $f(x)$ decrease and approach 0. Thus, the x-axis is a horizontal asymptote of the graph. The graph does not have an x-intercept, the y-intercept is $(0, 1)$, and the graph passes through the point $(1, 2)$.

> **Self Check 2** Graph: $g(x) = 4^x$
>
> **Now Try** **Problem 23**

 EXAMPLE 3   Graph: $f(x) = \left(\frac{1}{3}\right)^x$

Strategy   We will graph the function by creating a table of function values and plotting the corresponding ordered pairs.

Why   After drawing a smooth curve through the plotted points, we will have the graph.

Solution   If $x = -2$, we have

$$f(x) = \left(\frac{1}{3}\right)^x$$

$$f(-2) = \left(\frac{1}{3}\right)^{-2}$$

$$= \left(\frac{3}{1}\right)^2 \qquad \text{Recall: } \left(\frac{x}{y}\right)^{-n} = \left(\frac{y}{x}\right)^n.$$

$$= 9$$

The point $(-2, 9)$ is on the graph of $f(x) = \left(\frac{1}{3}\right)^x$. In a similar way, we find the corresponding values of $f(x)$ for other x-values and list them in a table.

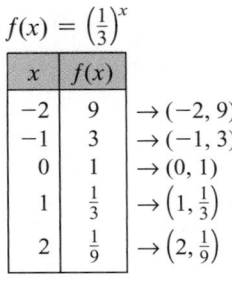

$$f(x) = \left(\frac{1}{3}\right)^x$$

x	$f(x)$	
-2	9	$\rightarrow (-2, 9)$
-1	3	$\rightarrow (-1, 3)$
0	1	$\rightarrow (0, 1)$
1	$\frac{1}{3}$	$\rightarrow \left(1, \frac{1}{3}\right)$
2	$\frac{1}{9}$	$\rightarrow \left(2, \frac{1}{9}\right)$

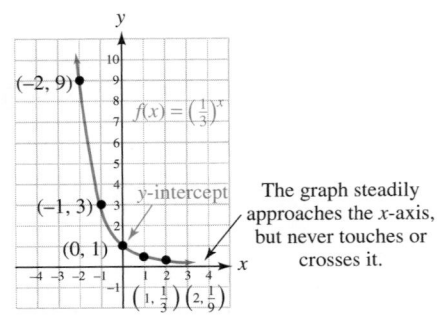

The graph steadily approaches the x-axis, but never touches or crosses it.

From the graph, we can verify that the domain of $f(x) = \left(\frac{1}{3}\right)^x$ is the interval $(-\infty, \infty)$ and the range is the interval $(0, \infty)$. Since the graph passes the horizontal line test, the function is one-to-one.

Note that as x increases, the values of $f(x)$ decrease and approach 0. Thus, the x-axis is a horizontal asymptote of the graph. The graph does not have an x-intercept, the y-intercept is $(0, 1)$, and the graph passes through the point $\left(1, \frac{1}{3}\right)$.

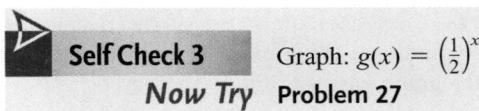

Self Check 3   Graph: $g(x) = \left(\frac{1}{2}\right)^x$

Now Try   **Problem 27**

Examples 2 and 3 illustrate the following properties of exponential functions.

Properties of Exponential Functions	The domain of the exponential function $f(x) = b^x$ is the interval $(-\infty, \infty)$.
	The range is the interval $(0, \infty)$.
	The graph has a y-intercept of $(0, 1)$.
	The x-axis is an asymptote of the graph.
	The graph of $f(x) = b^x$ passes through the point $(1, b)$.

In Example 2 (where $b = 2$), the values of y increase as the values of x increase. Since the graph rises as we move to the right, we call the function an *increasing function*. When $b > 1$, the larger the value of b, the steeper the curve.

In Example 3 $\left(\text{where } b = \frac{1}{3}\right)$, the values of y decrease as the values of x increase. Since the graph drops as we move to the right, we call the function a *decreasing function*. When $0 < b < 1$, the smaller the value of b, the steeper the curve.

In general, the following is true.

Increasing and Decreasing Functions	If $b > 1$, then $f(x) = b^x$ is an **increasing function.** If $0 < b < 1$, then $f(x) = b^x$ is a **decreasing function.**	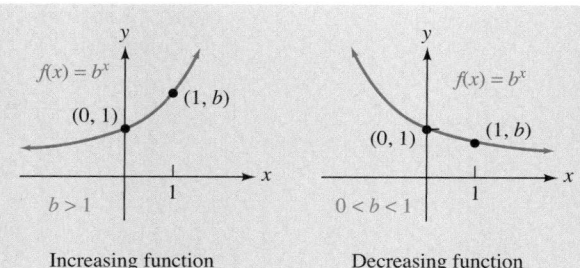

An exponential function with base b is either increasing (for $b > 1$) or decreasing ($0 < b < 1$). Since different real numbers x determine different values of b^x, exponential functions are one-to-one.

Using Your Calculator

Graphing Exponential Functions

To use a graphing calculator to graph $f(x) = \left(\frac{2}{3}\right)^x$ and $g(x) = \left(\frac{3}{2}\right)^x$, we enter the right sides of the equations after the symbols $Y_1 =$ and $Y_2 =$. The screen will show the following equations.

$$Y_1 = (2/3)\text{^}X$$
$$Y_2 = (3/2)\text{^}X$$

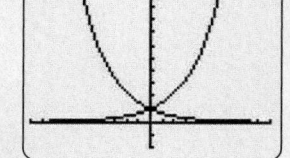

If we press ⏹GRAPH⏹, we will obtain the display shown.

We note that the graph of $f(x) = \left(\frac{2}{3}\right)^x$ passes through $(0, 1)$. Since $\frac{2}{3} < 1$, the function is decreasing. The graph of $g(x) = \left(\frac{3}{2}\right)^x$ also passes through $(0, 1)$. Since $\frac{3}{2} > 1$, the function is increasing. Since both graphs pass the horizontal line test, each function is one-to-one.

The graphs of many exponential functions are translations of basic graphs.

EXAMPLE 4   Graph each function by using a translation:

    a. $g(x) = 2^x - 4$ **b.** $g(x) = \left(\frac{1}{3}\right)^{x+3}$

Strategy   We will graph $g(x) = 2^x - 4$ by translating the graph of $f(x) = 2^x$ downward 4 units. We will graph $g(x) = \left(\frac{1}{3}\right)^{x+3}$ by translating the graph of $f(x) = \left(\frac{1}{3}\right)^x$ to the left 3 units.

Why   The subtraction of 4 in $g(x) = 2^x - 4$ causes a vertical shift of the graph of the base-2 exponential function 4 units downward. The addition of 3 to x in $g(x) = \left(\frac{1}{3}\right)^{x+3}$ causes a horizontal shift of the graph of the base-$\frac{1}{3}$ exponential function 3 units to the left.

Solution

a. The graph of $g(x) = 2^x - 4$ will be the same shape as the graph of $f(x) = 2^x$, except it is shifted 4 units downward.

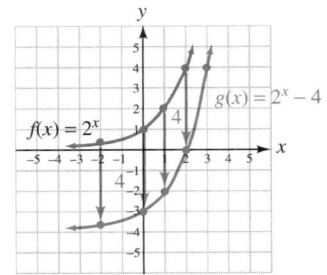

To graph $g(x) = 2^x - 4$, translate each point on the graph of $f(x) = 2^x$ down 4 units.

b. The graph of $g(x) = \left(\frac{1}{3}\right)^{x+3}$ will be the same shape as the graph of $f(x) = \left(\frac{1}{3}\right)^x$, except it is shifted 3 units to the left.

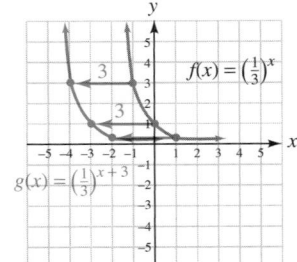

To graph $g(x) = \left(\frac{1}{3}\right)^{x+3}$, translate each point on the graph of $f(x) = \left(\frac{1}{3}\right)^x$ to the left 3 units.

 Self Check 4   Graph each function by using a translation:

a. $g(x) = \left(\frac{1}{4}\right)^x + 2$ **b.** $g(x) = 4^{x-3}$

Now Try   **Problems 31 and 33**

---

4   **Use Exponential Functions in Applications Involving Growth or Decay.**

EXAMPLE 5   ***Professional Baseball Salaries.*** The exponential function $s(t) = 170{,}000(1.12)^t$ approximates the average salary of a major league baseball player, where t is the number of years after 1980 and $0 \le t \le 25$. Sketch the graph of the function. (Source: Baseball Almanac)

Strategy   We will graph the function by creating a table of function values and plotting the corresponding ordered pairs.

Why   After drawing a smooth curve though the plotted points, we will have the graph.

Solution   The function values for $t = 0$ and $t = 5$ are computed as follows:

$t = 0$ *(the year 1980)*

$s(0) = 170{,}000(1.12)^0$

$\quad\quad = 170{,}000(1)$ Any nonzero number to the 0 power is 1.

$\quad\quad = 170{,}000$

$t = 5$ *(the year 1985)*

$s(5) = 170{,}000(1.12)^5$

$\quad\quad \approx 299{,}598$ Use a calculator.

The Language of Algebra

The word *exponential* is used in many settings to describe rapid growth. For example, we hear that the processing power of computers is growing *exponentially*.

To approximate $s(5)$, use the keystrokes 170000 $\boxed{\times}$ 1.12 $\boxed{y^x}$ 5 $\boxed{=}$ on a scientific calculator and 170000 $\boxed{\times}$ 1.12 $\boxed{\wedge}$ 5 $\boxed{\text{ENTER}}$ on a graphing calculator.

In a similar way, we find the corresponding values of $s(t)$ for t-values of 10, 15, 20, and 25 and list them in a table. Then we plot the ordered pairs and draw a smooth curve through them to get the graph.

t	$s(t)$
0	170,000
5	299,598
10	527,994
15	930,506
20	1,639,870
25	2,890,011

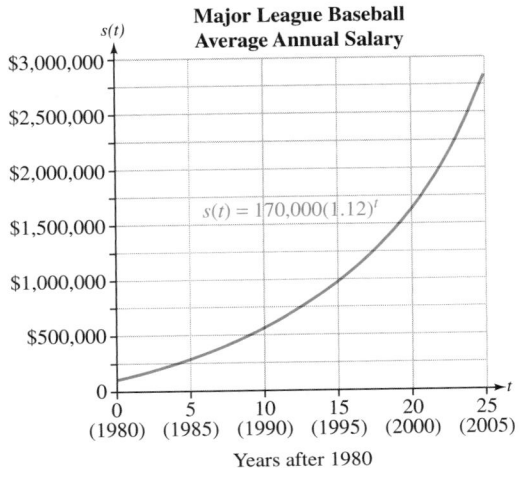

Major League Baseball
Average Annual Salary

$s(t) = 170,000(1.12)^t$

Years after 1980

 Now Try   **Problem 45**

Using Your Calculator

Graphing Exponential Functions

To use a graphing calculator to graph the exponential function $s(t) = 170,000(1.12)^t$, we enter the right side of the equation after the symbol $Y_1 =$ and replace the variable t with x. The display will show the equation

$Y_1 = 170000(1.12\wedge X)$

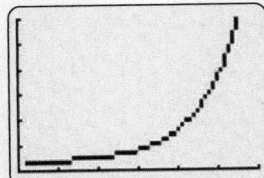

With window settings $[0, 30]$ for x and Xscale $= 5$ and $[0, 3000000]$ for y and Yscale $= 500000$, we obtain the display shown when we press $\boxed{\text{GRAPH}}$.

Exponential functions can be used to calculate compound interest. The following discussion gives some insight into why this is so.

If an amount of money P, called the **principal**, is deposited in an account paying an annual interest rate r, we can find the amount A in the account at the end of t years by using the formula

$$A = P + Prt \quad \text{or} \quad A = P(1 + rt)$$

Suppose that we deposit $500 in such an account that pays interest every 6 months. Then $P = 500$, and after 6 months $\left(\frac{1}{2}\text{ year}\right)$, the amount in the account will be

$$A = 500(1 + rt)$$
$$= 500\left(1 + r \cdot \frac{1}{2}\right) \quad \text{Substitute } \tfrac{1}{2} \text{ for } t.$$
$$= 500\left(1 + \frac{r}{2}\right)$$

The account will begin the second 6-month period with a value of $500\left(1 + \frac{r}{2}\right)$. After the second 6-month period, the amount in the account will be

$$A = P(1 + rt)$$
$$A = \left[500\left(1 + \frac{r}{2}\right)\right]\left(1 + r \cdot \frac{1}{2}\right) \quad \text{Substitute } 500\left(1 + \tfrac{r}{2}\right) \text{ for } P \text{ and } \tfrac{1}{2} \text{ for } t.$$
$$= 500\left(1 + \frac{r}{2}\right)\left(1 + \frac{r}{2}\right)$$
$$= 500\left(1 + \frac{r}{2}\right)^2$$

At the end of a third 6-month period, the amount in the account will be

$$A = 500\left(1 + \frac{r}{2}\right)^3$$

In this discussion, the earned interest is deposited back in the account and also earns interest, and we say that the account is earning **compound interest.** The preceding example suggests the following formula for compound interest.

Formula for Compound Interest

If P is deposited in an account and interest is paid k times a year at an annual rate r, the amount A in the account after t years is given by

$$A = P\left(1 + \frac{r}{k}\right)^{kt}$$

EXAMPLE 6   *Educational Savings Plan.*   To save for college, parents of a newborn child invest $12,000 in a mutual fund at 10% interest, compounded quarterly.

a. Find a function for the amount in the account after t years.

b. If the quarterly interest paid is continually reinvested, how much money will be in the account when the child is 18 years old?

Strategy   To write a function for the amount in the account after t years, we will substitute the given values for P, r, and k into the compound interest formula.

Why   The resulting equation will involve only two variables, A and t. Then we can write that equation using function notation.

Solution

a. When we substitute 12,000 for P, 0.10 for r, and 4 for k in the formula for compound interest, the resulting formula involves only two variables, A and t.

$$A = P\left(1 + \frac{r}{k}\right)^{kt}$$

$$A = 12{,}000\left(1 + \frac{0.10}{4}\right)^{4t}$$ Since the interest is compounded quarterly, $k = 4$.
Express $r = 10\%$ as a decimal.

Since the value of A depends on the value of t, we can express this relationship using function notation.

$$A(t) = 12{,}000\left(1 + \frac{0.10}{4}\right)^{4t}$$

$$A(t) = 12{,}000(1 + 0.025)^{4t}$$ Evaluate: $\frac{0.10}{4} = 0.025$.

$$A(t) = 12{,}000(1.025)^{4t}$$ This exponential function has a base of 1.025.

b. To find how much money will be in the account when the child is 18 years old, we need to find $A(18)$.

$$A(18) = 12{,}000(1.025)^{4(18)}$$ Substitute 18 for t.

$$= 12{,}000(1.025)^{72}$$

$$= 12{,}000(1.025)^{72}$$

$$\approx 71{,}006.74$$ Use a scientific calculator and press these keys:
12000 $\times$ 1.025 y^x 72 $=$.

When the child is 18 years old, the account will contain $71,006.74.

Self Check 6   How much money would be in the account after 18 years if the parents initially invested $20,000?

Now Try   **Problem 49**

Using Your Calculator

Solving Investment Problems

Suppose $1 is deposited in an account earning 6% annual interest, compounded monthly. To use a graphing calculator to estimate how much will be in the account in 100 years, we can substitute 1 for P, 0.06 for r, and 12 for k in the formula and simplify.

$$A = P\left(1 + \frac{r}{k}\right)^{kt} = 1\left(1 + \frac{0.06}{12}\right)^{12t} = (1.005)^{12t}$$

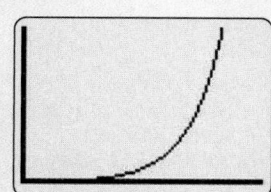

We now graph the function $A(t) = (1.005)^{12t}$ using window settings of [0, 120] and [0, 400] with Xscale = 1 and Yscale = 1 to obtain the graph shown. We can then trace and zoom to estimate that $1 grows to be approximately $397 in 100 years. From the graph, we can see that the money grows slowly in the early years and rapidly in the later years.

Examples 5 and 6 are applications illustrating exponential growth. In the next example, we see an application of exponential decay.

EXAMPLE 7   ***Internet Access.***   The exponential function $c(t) = 6.2(0.9)^t$ approximates the ratio of students to instructional computers with Internet access in U.S. public schools, where t is the number of years after 2000 and $0 \le t \le 5$. Use the function to answer the following questions.

a. The National Center for Educational Statistics did not survey the public schools in 2004. What ratio does the exponential function give for 2004?

b. If the current trend continues, what ratio of students to instructional computers with Internet access does the function predict for the year 2014?

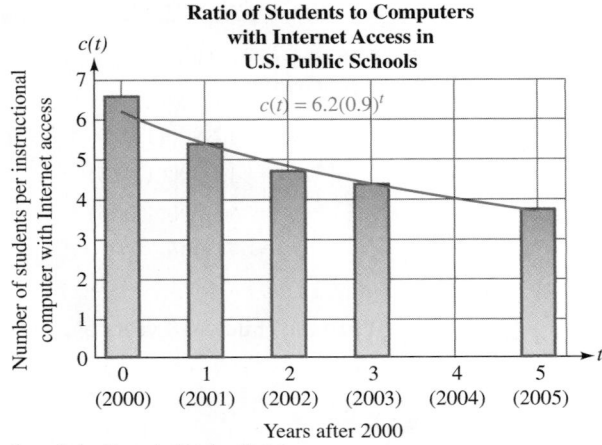

Ratio of Students to Computers with Internet Access in U.S. Public Schools

$c(t) = 6.2(0.9)^t$

Years after 2000

Source: National Center for Educational Statistics

Strategy   We will substitute 4 and 14 into the function $c(t) = 6.2(0.9)^t$ for t.

Why   The year 2004 is 4 years after 2000 and the year 2014 is 14 years after 2000.

Solution

a. To determine the ratio for the year 2004 as given by the exponential function, we evaluate it for $t = 4$.

$$c(4) = 6.2(0.9)^4$$
$$\approx 4.1 \qquad \text{Use a calculator.}$$

In the year 2004, the exponential function predicts the ratio was approximately 4.1 students to each computer with Internet access.

b. To predict the ratio for the year 2014, we evaluate the function for $t = 14$.

$$c(14) = 6.2(0.9)^{14}$$
$$\approx 1.4 \qquad \text{Use a calculator.}$$

If the trend continues, in 2014, the ratio will be approximately 1.4 students to each computer with Internet access.

> **Success Tip**
> $c(t) = 6.2(0.9)^t$ is a decreasing function because the base, 0.9, is such that $0 < 0.9 < 1$.

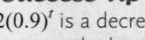 **Now Try** Problem 55

▶ **ANSWER TO SELF CHECKS**   **1. a.** 81,   **b.** $b^{4\sqrt{2}}$

2.

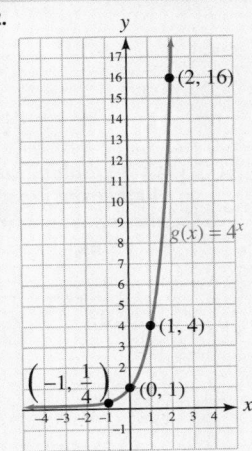

3.

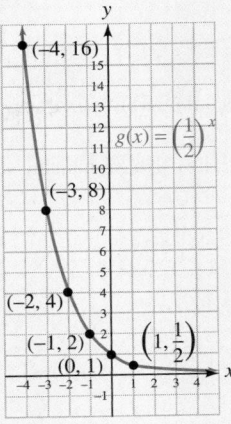

4. a.

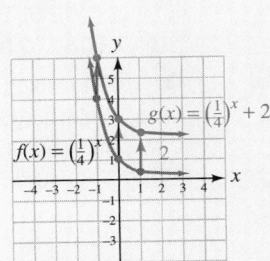

b.

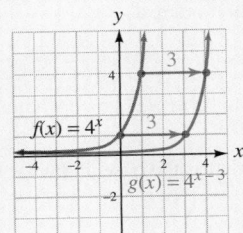

6. $118,344.56

STUDY SET
11.3

VOCABULARY

Refer to the graph of $f(x) = 3^x$.

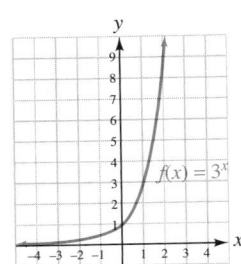

1. What type of function is $f(x) = 3^x$?

2. What is the domain of the function?

3. What is the range of the function?

4. a. What is the y-intercept of the graph?

   b. What is the x-intercept of the graph?

5. Is the function one-to-one?

6. What is an asymptote of the graph?

7. Is f an increasing or a decreasing function?

8. The graph passes through the point $(1, y)$. What is y?

CONCEPTS

9. Evaluate each expression.

   a. 3^{-2}

   b. $\left(\frac{1}{2}\right)^4$

   c. $\left(\frac{1}{5}\right)^{-2}$

10. Evaluate each expression using a calculator. Round to the nearest tenth.

   a. $20{,}000(1.036)^{52}$

   b. $92(0.88)^6$

11. Match each situation to the exponential graph that best models it.

   a. The number of cell phone subscribers in the world over the past 5 years

   b. The level of caffeine in the bloodstream after drinking a cup of coffee

   c. The amount of money in a bank account earning interest compounded quarterly

   d. The number of rabbits in a population with a high birth rate

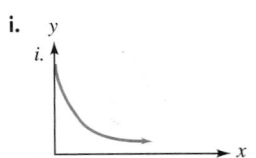   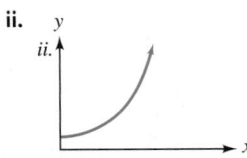

12. What formula is used to determine the amount of money in a savings account earning compound interest?

NOTATION

13. For an exponential function of the form $f(x) = b^x$, what are the restrictions on b?

14. In $A(t) = 16{,}000\left(1 + \dfrac{0.05}{365}\right)^{365t}$, what is the base and what is the exponent?

GUIDED PRACTICE

Simplify each expression. Write answers using positive exponents.
See Example 1.

15. $\left(2^{\sqrt{3}}\right)^{\sqrt{3}}$

16. $\left(3^{\sqrt{5}}\right)^{\sqrt{5}}$

17. $7^{\sqrt{3}}\,7^{\sqrt{12}}$

18. $3^{\sqrt{2}}\,3^{\sqrt{18}}$

19. $\dfrac{3^{2\sqrt{7}}}{3^{\sqrt{7}}}$

20. $\dfrac{5^{6\sqrt{2}}}{5^{4\sqrt{2}}}$

21. $5^{-\sqrt{5}}$

22. $4^{-\sqrt{5}}$

Graph each function. See Examples 2 and 3.

23. $f(x) = 3^x$

24. $f(x) = 6^x$

25. $f(x) = 5^x$

26. $f(x) = 7^x$

27. $f(x) = \left(\dfrac{1}{4}\right)^x$

28. $f(x) = \left(\dfrac{1}{5}\right)^x$

29. $f(x) = \left(\dfrac{1}{6}\right)^x$

30. $f(x) = \left(\dfrac{1}{8}\right)^x$

Graph each function by plotting points or using a translation.
See Example 4.

31. $g(x) = 3^x - 2$

32. $g(x) = 2^x + 1$

33. $g(x) = 2^{x+1}$

34. $g(x) = 3^{x-1}$

35. $g(x) = 4^{x-1} + 2$

36. $g(x) = 4^{x+1} - 2$

37. $g(x) = -2^x$

38. $g(x) = -3^x$

Use a graphing calculator to graph each function. Determine whether the function is an increasing or a decreasing function. See Using Your Calculator: Graphing Exponential Functions.

39. $f(x) = \dfrac{1}{2}(3^{x/2})$

40. $f(x) = -3(2^{x/3})$

41. $f(x) = 2(3^{-x/2})$

42. $f(x) = -\dfrac{1}{4}(2^{-x/2})$

APPLICATIONS

43. WORLD POPULATION   See the following graph.

   a. Estimate when the world's population reached $\frac{1}{2}$ billion and when it reached 1 billion.

   b. Estimate the world's population in the year 2000.

   c. What type of function does it appear could be used to model the world's population growth?

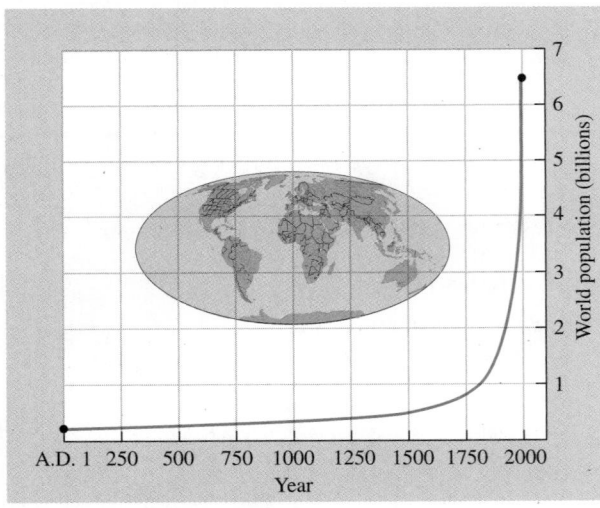

Source: United Nations Population Division

44. GLOBAL WARMING The following graph from the United States Environmental Protection Agency shows the projected sea level changes due to anticipated global warming.

a. What type of function does it appear could be used to model the sea level change?

b. When were the earliest instrumental records of sea level change made?

c. For the year 2100, what is the upper-end projection for sea level change? What is the lower-level projection?

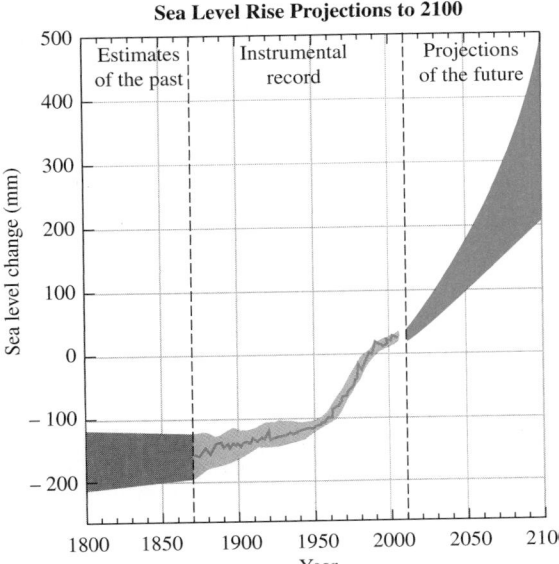

Sea Level Rise Projections to 2100

Source: United States Environmental Protection Agency

45. COMPUTER VIRUSES Suppose the number of computers infected by the spread of a virus through an e-mail is described by the exponential function $c(t) = 5(1.03)^t$, where t is the number of minutes since the first infected e-mail was opened. Graph the function.

46. SALVAGE VALUE A small business purchased a computer for $5,000. It is expected that its value each year will be 75% of its value the preceding year. The value (in dollars) of the computer, t years after its purchase, is given by the exponential function $v(t) = 5,000(0.75)^t$. Graph the function.

47. DIVING *Bottom time* is the time a scuba diver spends descending plus the actual time spent at a certain depth. Graph the bottom time limits given in the table.

Bottom time limits			
Depth (ft)	Bottom time (min)	Depth (ft)	Bottom time (min)
30	no limit	80	40
35	310	90	30
40	200	100	25
50	100	110	20
60	60	120	15
70	50	130	10

48. VALUE OF A CAR The graph shows how the value of the average car depreciates as a percent of its original value over a 10-year period. It also shows the yearly maintenance costs as a percent of the car's value.

a. When is the car worth half of its purchase price?

b. When is the car worth a quarter of its purchase price?

c. When do the average yearly maintenance costs surpass the value of the car?

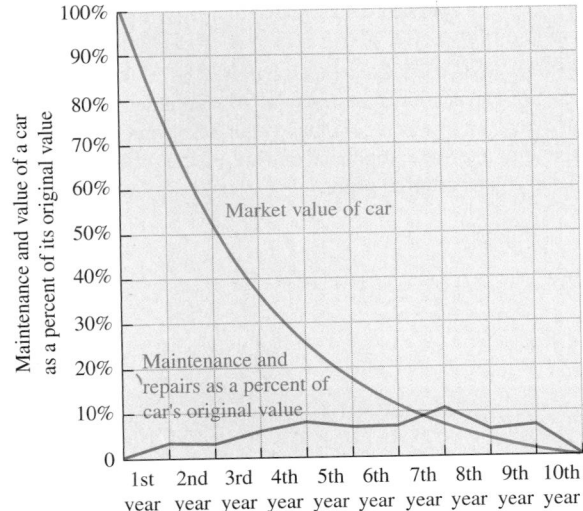

Source: U.S. Department of Transportation

In Exercises 49–54, assume that there are no deposits or withdrawals.

49. COMPOUND INTEREST   An initial deposit of $10,000 earns 8% interest, compounded quarterly. How much will be in the account after 10 years?

50. COMPOUND INTEREST   An initial deposit of $10,000 earns 8% interest, compounded monthly. How much will be in the account after 10 years?

51. COMPARING INTEREST RATES   How much more interest could $1,000 earn in 5 years, compounded quarterly, if the annual interest rate were $5\frac{1}{2}$% instead of 5%?

52. COMPARING SAVINGS PLANS   Which institution in the ads provides the better investment?

Fidelity Savings & Loan
Earn 5.25%
compounded monthly

Union Trust
Money Market Account
paying 5.35%
compounded annually

53. COMPOUND INTEREST   If $1 had been invested on July 4, 1776, at 5% interest, compounded annually, what would it be worth on July 4, 2076?

54. FREQUENCY OF COMPOUNDING $10,000 is invested in each of two accounts, both paying 6% annual interest. In the first account, interest compounds quarterly, and in the second account, interest compounds daily. Find the difference between the accounts after 20 years.

55. GUITARS   The frets on the neck of a guitar are placed so that pressing a string against them determines the strings' vibrating length. The exponential function $f(n) = 650(0.94)^n$ gives the vibrating length (in millimeters) of a string on a certain guitar for the fret number n. Find the length of the vibrating string when a guitarist holds down a string at the 7th fret.

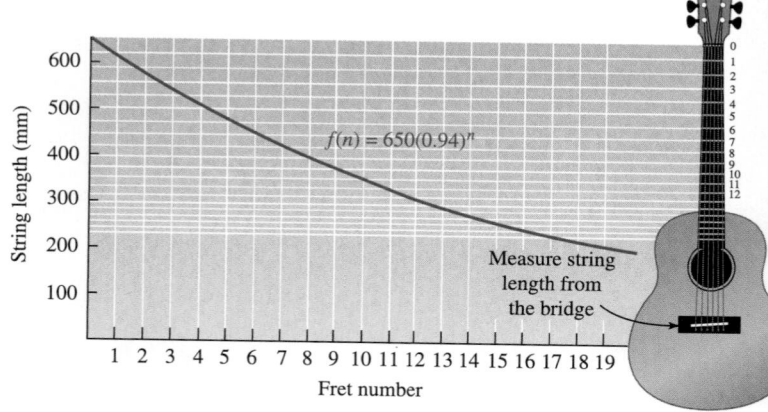

$f(n) = 650(0.94)^n$

Measure string length from the bridge

String length (mm)

Fret number

56. BACTERIAL CULTURES   A colony of 6 million bacteria was determined to be growing in the culture medium shown in illustration (a). If the population P of bacteria after t hours is given by the function $P(t) = 6,000,000(2.3)^t$, find the population in the culture later in the day using the information given in illustration (b).

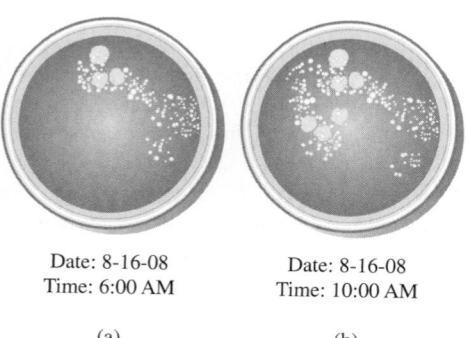

Date: 8-16-08
Time: 6:00 AM

Date: 8-16-08
Time: 10:00 AM

(a)   (b)

57. RADIOACTIVE DECAY   Five hundred grams of a radio-active material decays according to the formula $A = 500\left(\frac{2}{3}\right)^t$, where t is measured in years. Find the amount present in 10 years. Round to the nearest one-tenth of a gram.

58. DISCHARGING A BATTERY   The charge remaining in a battery decreases as the battery discharges. The charge C (in coulombs) after t days is given by the function $C(t) = 0.0003(0.7)^t$. Find the charge after 5 days.

59. POPULATION GROWTH   The population of North Rivers is decreasing exponentially according to the formula $P = 3,745(0.93)^t$, where t is measured in years from the present date. Find the population in 6 years, 9 months.

60. THE LOUISIANA PURCHASE   In 1803, the United States negotiated the Louisiana Purchase with France. The country doubled its territory by adding 827,000 square miles of land for $15 million. If the land appreciated at the rate of 6% each year, what would one square mile of land be worth in 2005?

WRITING

61. If world population is increasing exponentially, why is there cause for concern?

62. How do the graphs of $f(x) = 3^x$ and $g(x) = \left(\frac{1}{3}\right)^x$ differ? How are they similar?

63. A snowball rolling downhill grows *exponentially* with time. Explain what this means. Sketch a simple graph that models the situation.

64. Explain why the graph of $f(x) = 3^x$ gets closer and closer to the *x*-axis as the values of x decrease. Does the graph ever cross the *x*-axis? Explain why or why not.

65. Describe the graphs of $f(x) = x^2$ and $g(x) = 2^x$ in words.

66. Write a paragraph explaining the concept that is illustrated in the following graph.

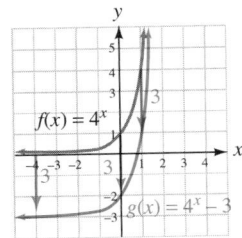

REVIEW

In Exercises 67–70, refer to the illustration below in which lines r and s are parallel.

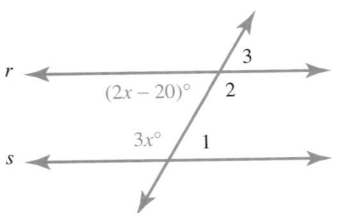

67. Find x.

68. Find the measure of $\angle 1$.

69. Find the measure of $\angle 2$.

70. Find the measure of $\angle 3$.

CHALLENGE PROBLEMS

71. In the definition of the exponential function, b could not be negative. Why?

72. Graph $f(x) = 3^x$. Then use the graph to estimate the value of $3^{1.5}$.

73. Graph $y = x^{1/2}$ and $y = \left(\frac{1}{2}\right)^x$ on the same set of coordinate axes. Estimate the coordinates of any point(s) that the graphs have in common.

74. Find the value of b that would cause the graph of $f(x) = b^x$ to look like the graph below.

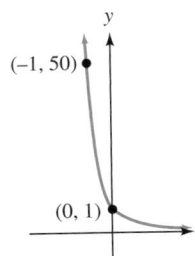

SECTION 11.4
Base-e Exponential Functions

Objectives

1. Define e and identify the formula for exponential growth/decay.
2. Define the natural exponential function.
3. Graph the natural exponential function.
4. Use base-e exponential functions in applications involving growth or decay.

Any positive number can be used as a base of an exponential function. However, some bases are used more often than others. An exponential function that has many applications is one whose base is an irrational number represented by the letter e. We will now discuss one way of arriving at this important number e.

 Define e and Identify the Formula for Exponential Growth/Decay.

If a bank pays interest twice a year, we say that interest is compounded semiannually. If it pays interest four times a year, we say that interest is compounded quarterly. If it pays interest continuously (infinitely many times in a year), we say that interest is compounded continuously.

To develop the formula for continuous compound interest, we start with the formula

$$A = P\left(1 + \frac{r}{k}\right)^{kt}$$ This is the formula for compound interest: r is the annual rate and k is the number of times per year interest is paid.

Notation
Swiss born Leonhard Euler (1707–1783) is said to have published more than any mathematician in history. He had a great influence on the notation that we use today. Through his work, the symbol e came into common use.

and let $rn = k$. Since r and k are positive numbers, so is n.

$$A = P\left(1 + \frac{r}{rn}\right)^{rnt}$$

We can then simplify the fraction $\frac{r}{rn}$ and use the commutative property of multiplication to change the order of the exponents.

$$A = P\left(1 + \frac{1}{n}\right)^{nrt}$$

Finally, we can use a property of exponents to write the formula as

(1) $$A = P\left[\left(1 + \frac{1}{n}\right)^{n}\right]^{rt}$$ Use the property $a^{mn} = (a^m)^n$.

n	$\left(1 + \frac{1}{n}\right)^{n}$
1	2
2	2.25
4	2.44140625 ...
12	2.61303529 ...
365	2.71456748 ...
1,000	2.71692393 ...
100,000	2.71826830 ...
1,000,000	2.71828137 ...

To find the value of $\left(1 + \frac{1}{n}\right)^{n}$, we evaluate it for several values of n, as shown in the table. The results suggest that as n gets larger, the value of $\left(1 + \frac{1}{n}\right)^{n}$ approaches the number 2.71828 .... This number is called e, which has the following value.

$$e = 2.718281828459 \ldots$$

Like π, the number e is irrational. Its decimal representation is nonterminating and non-repeating. Rounded to four decimal places, $e \approx 2.7183$.

If we replace $\left(1 + \frac{1}{n}\right)^n$ in Equation 1 with e, we will get the formula for exponential growth.

$$A = P\left[\left(1 + \frac{1}{n}\right)^n\right]^{rt}$$

$$A = Pe^{rt} \qquad \text{Substitute } e \text{ for } \left(1 + \frac{1}{n}\right)^n.$$

Formula for Exponential Growth/Decay

If a quantity P increases or decreases at an annual rate r, compounded continuously, the amount A after t years is given by

$$A = Pe^{rt}$$

If time is measured in years, r is called the **annual growth rate**. If r is negative, the growth represents a decrease.

For a given quantity P, say 10,000, and a given rate r, say 5%, we can write the formula for exponential growth using function notation:

$$A(t) = 10{,}000e^{0.05t}$$

 Define the Natural Exponential Function.

Of all possible bases for an exponential function, e is the most convenient for problems involving growth or decay. Since these situations occur often in natural settings, we call $f(x) = e^x$ the *natural exponential function*.

The Natural Exponential Function

The function defined by $f(x) = e^x$ is the **natural exponential function** (or the **base-e exponential function**) where $e = 2.71828\ldots$. The domain of $f(x) = e^x$ is the interval $(-\infty, \infty)$. The range is the interval $(0, \infty)$.

The $\boxed{e^x}$ key on a calculator is used to evaluate the natural exponential function.

Using Your Calculator

The Natural Exponential Function Key

To compute the amount to which $12,000 will grow if invested for 18 years at 10% annual interest, compounded continuously, we substitute 12,000 for P, 0.10 for r, and 18 for t in the formula for continuous compound interest and simplify.

$$A = Pe^{rt} = 12{,}000e^{0.10(18)} = 12{,}000e^{1.8} \qquad \text{Write 10\% as 0.10.}$$

To evaluate this expression using a scientific calculator, we enter

$$1.8 \;\boxed{e^x}\;\boxed{\times}\; 12000 \;\boxed{=}$$

$$\boxed{72595.76957}$$

Using a graphing calculator, we enter

$$12000 \;\boxed{\times}\;\boxed{2nd}\;\boxed{e^x}\;1.8\;\boxed{)}\;\boxed{\text{ENTER}}$$

$$\boxed{\begin{array}{l}12000*e\char`^(1.8)\\ \qquad\qquad 72595.76957\end{array}}$$

After 18 years, the account will contain $72,595.77. This is $1,589.03 more than the result in Example 6 in the Section 11.3, where interest was compounded quarterly.

EXAMPLE 1   *Investing.* If $25,000 accumulates interest at an annual rate of 8%, compounded continuously, find the balance in the account in 50 years.

Strategy   We will substitute 25,000 for P, 0.08 for r, and 50 for t in the formula $A = Pe^{rt}$ and calculate the value of A.

Why   The words *compounded continuously* indicate that we should use the exponential growth/decay formula.

Solution

$$A = Pe^{rt} \qquad \text{This is the formula for continuous compound interest.}$$

$$A = 25{,}000e^{0.08(50)} \qquad \text{Write 8\% as 0.08.}$$

$$= 25{,}000e^{4}$$

$$\approx 1{,}364{,}953.75 \qquad \text{Use a calculator.}$$

In 50 years, the balance will be $1,364,953.75—more than a million dollars.

    Self Check 1   Find the balance in 60 years.

Now Try   Problems 19 and 37

---

3   **Graph the Natural Exponential Function.**

To graph $f(x) = e^x$, we construct a table of function values by choosing several values for x and finding the corresponding values of $f(x)$. For example, $x = -2$, we have

$$f(x) = e^x$$

$$f(-2) = e^{-2}$$

$$= 0.135335283 \ldots \qquad \text{Use a calculator.}$$

$$\approx 0.1 \qquad \text{Round to the nearest tenth.}$$

We enter $(-2, 0.1)$ in the table. Similarly, we find $f(-1)$, $f(0)$, $f(1)$, and $f(2)$, enter each result in the table, and plot the ordered pairs. We draw a smooth curve through the points to get the graph.

From the graph, we can verify that the domain of $f(x) = e^x$ is the interval $(-\infty, \infty)$ and the range is the interval $(0, \infty)$. Since the graph passes the horizontal line test, the function is one-to-one.

Note that as x decreases, the values of $f(x)$ decrease and approach 0. Thus, the x-axis is a horizontal asymptote of the graph. The graph does not have an x-intercept, the y-intercept is $(0, 1)$, and the graph passes through the point $(1, e)$.

$$f(x) = e^x$$

x	$f(x)$	
-2	$\frac{1}{e^2} \approx 0.1$	$\rightarrow (-2, 0.1)$
-1	$\frac{1}{e^1} \approx 0.4$	$\rightarrow (-1, 0.4)$
0	$e^0 = 1$	$\rightarrow (0, 1)$
1	$e^1 = e$	$\rightarrow (1, e)$
2	$e^2 \approx 7.4$	$\rightarrow (2, 7.4)$

The outputs can be found using the $\boxed{e^x}$ key on a calculator. Some are rounded to the nearest tenth to make point-plotting easier.

The graph steadily approaches the x-axis, but never touches or crosses it.

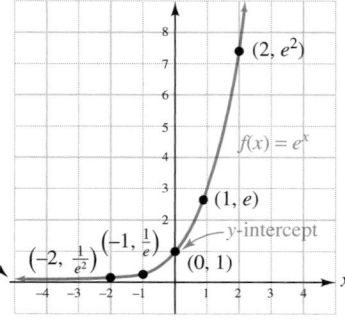

To graph more complicated natural exponential functions, point-plotting can be tedious. In such cases, we will use a graphing calculator.

Using Your Calculator

Graphing Exponential Functions

The figure shows the calculator graph of $f(x) = 3e^{-x/2}$. To graph this function, we enter the right side of the equation after the symbol $Y_1 =$. The display will show the equation

$$Y_1 = 3(e^{\wedge}(-X/2))$$

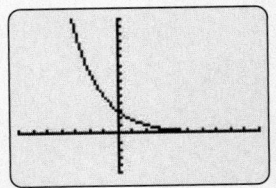

The graphs of many functions are translations of the natural exponential function.

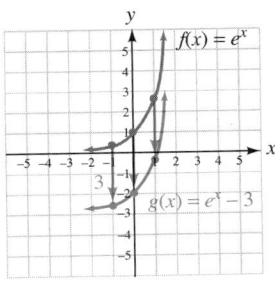

To graph $g(x) = e^x - 3$, translate each point on the graph of $f(x) = e^x$ down 3 units.

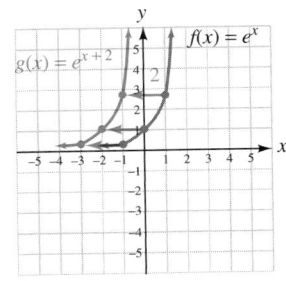

To graph $g(x) = e^{x+2}$, translate each point on the graph of $f(x) = e^x$ to the left 2 units.

We can illustrate the effects of vertical and horizontal translations of the natural exponential function by using a graphing calculator.

Using Your Calculator

Translations of the Natural Exponential Function

Figure (a) shows the calculator graphs of $f(x) = e^x$, $g(x) = e^x + 5$, and $h(x) = e^x - 3$. To graph them, we enter the right sides of the equations after $Y_1 =$, $Y_2 =$, and $Y_3 =$. The display will show:

$$Y_1 = e^{\wedge}(X) \qquad Y_2 = e^{\wedge}(X) + 5 \qquad Y_3 = e^{\wedge}(X) - 3$$

The graph of $g(x) = e^x + 5$ is 5 units above the graph of $f(x) = e^x$ and the graph of $h(x) = e^x - 3$ is 3 units below the graph of $f(x) = e^x$.

Figure (b) shows the calculator graphs of $f(x) = e^x$, $g(x) = e^{x+5}$, and $h(x) = e^{x-3}$. The graph of $g(x) = e^{x+5}$ is 5 units to the left of the graph of $f(x) = e^x$ and the graph of $h(x) = e^{x-3}$ is 3 units to the right of the graph of $f(x) = e^x$.

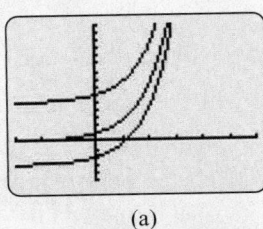

(a)

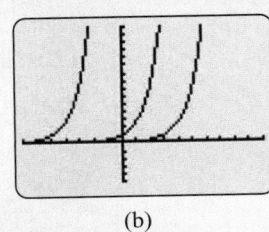

(b)

Use Base-*e* Exponential Functions in Applications Involving Growth or Decay.

An equation based on the natural exponential function provides a model for **population growth**. In the **Malthusian model for population growth,** the future population of a colony is related to the present population by the formula $A = Pe^{rt}$. (Note that this is the same formula as for continuous compound interest.)

EXAMPLE 2   *City Planning.*   The population of a city is currently 15,000, but economic conditions are causing the population to decrease 3% each year. If this trend continues, find the population in 30 years.

Strategy   We will substitute 15,000 for P, -0.03 for r, and 30 for t in the formula $A = Pe^{rt}$ and calculate the value of A.

Why   Since the population is decreasing 3% each year, the annual growth rate is -3%, or -0.03.

Solution

> **Success Tip**
> For quantities that are decreasing, remember to enter a negative value for r, the annual rate, in the formula $A = Pe^{rt}$.

$$A = Pe^{rt} \qquad \text{This is the model for population growth/decay.}$$
$$A = 15,000e^{-0.03(30)} \qquad \text{Substitute.}$$
$$= 15,000e^{-0.9}$$
$$\approx 6,099 \qquad \text{Use a calculator and round to the nearest whole number.}$$

In 30 years, the expected population will be 6,099.

 Self Check 2   Find the population in 50 years.

Now Try   **Problems 23 and 47**

The English economist Thomas Robert Malthus (1766–1834) was a pioneer in studying population. He believed that poverty and starvation were unavoidable because the human population tends to grow exponentially but the food supply tends to grow linearly.

EXAMPLE 3   *Food Shortages.*   Suppose that a country with a population of 1,000 people is growing exponentially according to the function

$$P(t) = 1,000e^{0.02t} \qquad \text{The annual growth rate is } 2\% = 0.02.$$

where t is in years. Furthermore, assume that the food supply F, measured in adequate food per day per person, is growing linearly according to the function

$$F(t) = 30.625t + 2,000 \quad (t \text{ is time in years})$$

In how many years will the population outstrip the food supply?

Strategy   We will use a graphing calculator to graph these two functions and find the point where the graphs intersect.

Why   This will be the point where the food supply is exactly adequate to feed the population. Beyond this point, there will be a shortage of food.

Solution We can use a graphing calculator with window settings of [0, 100] for x and [0, 10,000] for y. After graphing the functions, we obtain figure (a). The food supply is modeled by the straight line, and the population is modeled by the curved line. If we trace, as in figure (b), we can find the point where the two graphs intersect. From the graph, we can see that the food supply will be adequate for about 71 years. At that time, the population of approximately 4,200 people will begin to have problems.

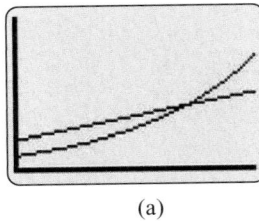
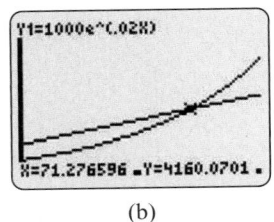

(a)   (b)

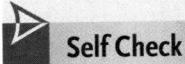 **Self Check 3** Suppose that the population grows at a 2.5% rate. Use a graphing calculator to determine for how many years the food supply will be adequate.

Now Try **Problem 35**

EXAMPLE 4 *Baking.* A mother takes a cake out of the oven and sets it on a rack to cool. The function $T(t) = 68 + 220e^{-0.18t}$ gives the cake's temperature in degrees Fahrenheit after it has cooled for t minutes. If her children will be home from school in 20 minutes, will the cake have cooled enough for the children to eat it? (Assume that 80°F, or cooler, would be a comfortable eating temperature.)

Strategy We will substitute 20 for t in the function $T(t) = 68 + 220e^{-0.18t}$.

Why The variable t represents the number of minutes the cake has cooled.

Solution When the children arrive home, the cake will have cooled for 20 minutes. To find the temperature of the cake at that time, we need to find $T(20)$.

$$T(t) = 68 + 220e^{-0.18t}$$
$$T(20) = 68 + 220e^{-0.18(20)} \quad \text{Substitute 20 for } t.$$
$$= 68 + 220e^{-3.6}$$
$$\approx 74.0 \quad \text{Use a calculator.}$$

When the children return home, the temperature of the cake will be about 74°, and it can be eaten.

 Now Try **Problem 55**

 ANSWERS TO SELF CHECKS **1.** $3,037,760.44 **2.** 3,347 **3.** About 51 years

STUDY SET
11.4

VOCABULARY

Refer to the graph of $f(x) = e^x$.

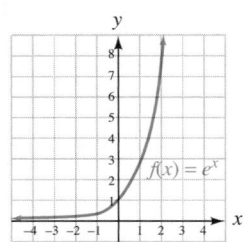

1. What is the name of the function $f(x) = e^x$?

2. What is the domain of the function?
3. What is the range of the function?
4. **a.** What is the y-intercept of the graph?
   b. What is the x-intercept of the graph?
5. Is the function one-to-one?
6. What is an asymptote of the graph?
7. Is f an increasing or a decreasing function?
8. The graph passes through the point $(1, y)$. What is y?

CONCEPTS

Fill in the blanks.

9. In _____ compound interest, the number of compoundings is infinitely large.
10. The formula for exponential growth/decay is $A =$ ▢e▢ .
11. To two decimal places, the value of e is ▢ .
12. If n gets larger and larger, the value of $\left(1 + \frac{1}{n}\right)^n$ approaches the value of ▢ .
13. Graph each irrational number on the number line: $\left\{\pi, e, \sqrt{2}\right\}$.

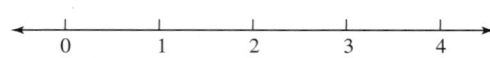

14. Complete the table of values. Round to the nearest hundredth.

x	-2	-1	0	1	2
e^x					

15. The function $f(x) = e^x$ is graphed and the TRACE feature is used, as shown here. What is the y-coordinate of the point on the graph having an x-coordinate of 1? What is the symbol that represents this number?

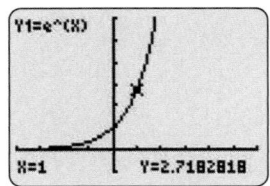

16. The illustration shows a table of values for $f(x) = e^x$. As x decreases, what happens to the values of $f(x)$ listed in the Y_1 column? Will the value of $f(x)$ ever be 0 or negative?

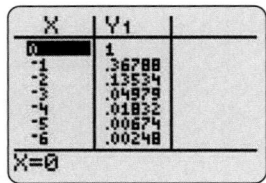

NOTATION

Find A using the formula $A = Pe^{rt}$ given the following values of P, r, and t. Round to the nearest tenth.

17. $P = 1,000$, $r = 0.09$, and $t = 10$

$A = $ ▢$e^{(0.09)(\ \ )}$

$= 1,000e$ ▢

$\approx$ ▢   *Use a calculator.*

18. $P = 50,000$, $r = -0.12$, and $t = 50$

$A = 50,000e^{(\ \ \ \ )(50)}$

$= 50,000e$ ▢

$\approx$ ▢   *Use a calculator.*

GUIDED PRACTICE

Find A using the formula $A = Pe^{rt}$ given the following values of P, r, and t. Round to the nearest hundredth. See Example 1.

19. $P = 5,000$, $r = 8\%$, $t = 20$ years
20. $P = 15,000$, $r = 6\%$, $t = 40$ years
21. $P = 20,000$, $r = 10.5\%$, $t = 50$ years
22. $P = 25,000$, $r = 6.5\%$, $t = 100$ years

Find A using the formula A = Pe^rt given the following values of P, r, and t. Round to the nearest hundredth. See Example 2.

23. $P = 15{,}895$, $r = -2\%$, $t = 16$ years

24. $P = 33{,}999$, $r = -4\%$, $t = 21$ years

25. $P = 565$, $r = -0.5\%$, $t = 8$ years

26. $P = 110$, $r = -0.25\%$, $t = 9$ years

Graph each function. See Objective 3.

27. $f(x) = e^x$

28. $f(x) = -e^x$

29. $f(x) = e^x + 1$

30. $f(x) = e^x - 2$

31. $y = e^{x+3}$

32. $y = e^{x-5}$

33. $f(x) = 2e^x$

34. $f(x) = \frac{1}{2}e^x$

APPLICATIONS

 In Exercises 35 and 36, use a graphing calculator to solve each problem.

35. THE MALTHUSIAN MODEL   In Example 3, suppose that better farming methods changed the function for food growth to $F(t) = 31t + 2{,}000$. How long would the food supply be adequate?

36. THE MALTHUSIAN MODEL   In Example 3, suppose that a birth-control program changed the function for population growth to $P(t) = 1{,}000e^{0.01t}$. How long would the food supply be adequate?

In Exercises 37–42, assume that there are no deposits or withdrawals.

37. CONTINUOUS COMPOUND INTEREST   An initial investment of $5,000 earns 8.2% interest, compounded continuously. What will the investment be worth in 12 years?

38. CONTINUOUS COMPOUND INTEREST   An initial investment of $2,000 earns 8% interest, compounded continuously. What will the investment be worth in 15 years?

39. COMPARISON OF COMPOUNDING METHODS   An initial deposit of $5,000 grows at an annual rate of 8.5% for 5 years. Compare the final balances resulting from annual compounding and continuous compounding.

40. COMPARISON OF COMPOUNDING METHODS   An initial deposit of $30,000 grows at an annual rate of 8% for 20 years. Compare the final balances resulting from annual compounding and continuous compounding.

41. DETERMINING THE INITIAL DEPOSIT   An account now contains $11,180 and has been accumulating interest at 7% annual interest, compounded continuously, for 7 years. Find the initial deposit.

42. DETERMINING THE PREVIOUS BALANCE   An account now contains $3,610 and has been accumulating interest at 8% annual interest, compounded continuously. How much was in the account 4 years ago?

43. POPULATION OF THE UNITED STATES   Graph the U.S. census population figures shown in the table (in millions). What type of function does it appear could be used to model the population?

Year	Population	Year	Population
1790	3.9	1900	76.0
1800	5.3	1910	92.2
1810	7.2	1920	106.0
1820	9.6	1930	123.2
1830	12.9	1940	132.1
1840	17.0	1950	151.3
1850	23.1	1960	179.3
1860	31.4	1970	203.3
1870	38.5	1980	226.5
1880	50.1	1990	248.7
1890	62.9	2000	281.4

44. OZONE CONCENTRATIONS   A *Dobson* unit is the most basic measure used in ozone research. Roughly 300 Dobson units are equivalent to the height of 2 pennies stacked on top of each other. Suppose the ozone layer thickness (in Dobsons) over a certain city is modeled by the function $A(t) = 300e^{-0.0011t}$, where t is the number of years after 1990. Estimate how thick the ozone layer will be in 2015.

45. POPULATION OF THE UNITED STATES   The exponential function $A(t) = 123e^{0.0117t}$ approximates the population of the United States (in millions), where t is the number of years after 1930. Use the function to estimate the U.S. population for these important dates:
- 1937 The Golden Gate Bridge is completed
- 1941 The United States enters World War II
- 1955 Rosa Parks refuses to give up her seat on a Montgomery, Alabama, bus
- 1969 Astronaut Neil Armstrong walks on the moon
- 1974 President Nixon resigns
- 1986 The *Challenger* space shuttle explodes
- 1997 *The Simpsons* becomes the longest running cartoon television series in history

46. WORLD POPULATION GROWTH   The population of Earth is approximately 6.6 billion people and is growing at an annual rate of 1.167%. Use the exponential growth model to find the world population in 30 years.

47. HIGHS AND LOWS Liberia, located on the west coast of Africa, has one of the greatest population growth rates in the world. Bulgaria, in southeastern Europe, has one of the smallest. Use an exponential growth/decay model to complete the table.

Country	Population 2007	Annual growth rate	Estimated population 2020
Liberia	3,195,931	4.84%	
Bulgaria	7,322,858	−0.84%	

Source: CIA World Factbook

48. DISINFECTANTS The exponential function $A(t) = 2,000,000e^{-0.588t}$ approximates the number of germs on a table top, t minutes after disinfectant was sprayed on it. Estimate the germ count on the table 5 minutes after it is sprayed.

49. from Campus to Careers
Social Worker

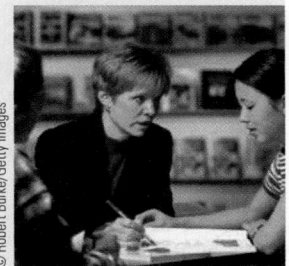

Social workers often use occupational test results when counseling their clients about employment options. The "learning curve" below shows that as a factory trainee assembled more chairs, the assembly time per chair generally decreased. If company standards required an average assembly time of 10 minutes or less, how many chairs did the trainee have to assemble before meeting company standards?

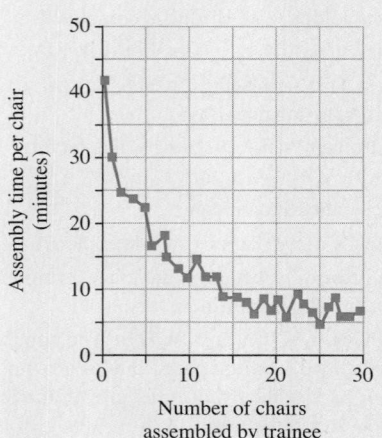

Number of chairs
assembled by trainee

50. ANTS Shortly after an explorer ant discovers a food source, a recruitment process begins in which numerous additional ants travel to the source. The number of ants at the source grows exponentially according to the function $a(t) = 1.36\left(\frac{e}{2.5}\right)^t$, where t is the number of minutes since the explorer discovered the food. How many ants will be at the source in 40 minutes?

51. EPIDEMICS The spread of hoof-and-mouth disease through a herd of cattle can be modeled by the function

$$P(t) = 2e^{0.27t} \quad (t \text{ is in days})$$

If a rancher does not quickly treat the two cows that now have the disease, how many cattle will have the disease in 12 days?

52. OCEANOGRAPHY The width w (in millimeters) of successive growth spirals of the sea shell *Catapulus voluto*, shown below, is given by the exponential function

$$w(n) = 1.54e^{0.503n}$$

where n is the spiral number. Find the width, to the nearest tenth of a millimeter, of the sixth spiral.

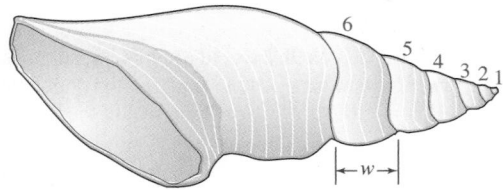

53. HALF-LIFE OF A DRUG The quantity of a prescription drug in the bloodstream of a patient t hours after it is administered can be modeled by an exponential function. (See the graph.) Determine the time it takes to eliminate half of the initial dose from the body from the graph.

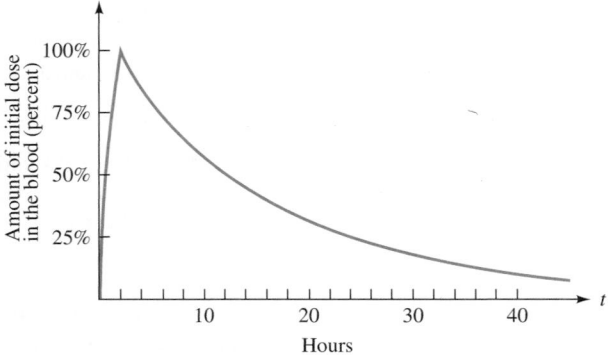

Hours

54. MEDICINE The concentration of a certain prescription drug in an organ after t minutes is modeled by the function

$$f(t) = 0.08(1 - e^{-0.1t})$$

where $f(t)$ is the concentration at time t. Find the concentration of the drug at 30 minutes.

55. SKYDIVING Before the parachute opens, a skydiver's velocity in meters per second is modeled by the function

$$f(t) = 50(1 - e^{-0.2t})$$

where $f(t)$ is the velocity at time t. Find the velocity after 20 seconds of free fall.

56. FREE FALL After t seconds a certain falling object has a velocity $f(t)$ modeled by the function $f(t) = 50(1 - e^{-0.3t})$. Which is falling faster after 2 seconds—the object or the skydiver in Exercise 55?

WRITING

57. Explain why the graph of $y = e^x - 5$ is five units below the graph of $y = e^x$.

58. A feature article in a newspaper stated that the sport of snowboarding was growing *exponentially*. Explain what the author of the article meant by that.

59. As of 2007, the population growth rate for Russia was -0.37% annually. What are some of the consequences for a country that has a negative population growth?

60. What is e?

REVIEW

Simplify each expression. Assume that all variables represent positive numbers.

61. $\sqrt{240x^5}$

62. $\sqrt[3]{-125x^5y^4}$

63. $4\sqrt{48y^3} - 3y\sqrt{12y}$

64. $\sqrt[4]{48z^5} + \sqrt[4]{768z^5}$

CHALLENGE PROBLEMS

65. Without using a calculator, determine whether the statement $e^e > e^3$ is true or false. Explain your reasoning.

66. Graph the function defined by the equation

$$f(x) = \frac{e^x + e^{-x}}{2}$$

from $x = -2$ to $x = 2$. The graph will look like a parabola, but it is not. The graph, called a **catenary**, is important in the design of power distribution networks, because it represents the shape of a uniform flexible cable whose ends are suspended from the same height. The function is called the **hyperbolic cosine function.**

67. If $e^{t+5} = ke^t$, find k.

68. If $e^{5t} = k^t$, find k.

SECTION 11.5
Logarithmic Functions

Objectives

  1 Define logarithm.

2 Write logarithmic equations as exponential equations.

3 Write exponential equations as logarithmic equations.

4 Evaluate logarithmic expressions.

5 Graph logarithmic functions.

6 Use logarithmic functions in applications.

In this section, we will discuss inverses of exponential functions. These functions are called *logarithmic functions,* and they can be used to solve problems from fields such as electronics, seismology (the study of earthquakes), and business.

1 **Define Logarithm.**

The graph of the exponential function $f(x) = 2^x$ is shown in red on the next page. Since it passes the horizontal line test, it is a one-to-one function and has an inverse. To graph f^{-1}, we interchange the coordinates of the ordered pairs in the table, plot those points, and draw a smooth curve through them, as shown in blue. As expected, the graphs of f and f^{-1} are symmetric with respect to the line $y = x$.

$f(x) = 2^x$

To graph f^{-1}, interchange each pair of coordinates.

x	$f(x)$		
-3	$\frac{1}{8}$	$\to \left(-3, \frac{1}{8}\right)$	$\left(\frac{1}{8}, -3\right)$
-2	$\frac{1}{4}$	$\to \left(-2, \frac{1}{4}\right)$	$\left(\frac{1}{4}, -2\right)$
-1	$\frac{1}{2}$	$\to \left(-1, \frac{1}{2}\right)$	$\left(\frac{1}{2}, -1\right)$
0	1	$\to (0, 1)$	$(1, 0)$
1	2	$\to (1, 2)$	$(2, 1)$
2	4	$\to (2, 4)$	$(4, 2)$
3	8	$\to (3, 8)$	$(8, 3)$

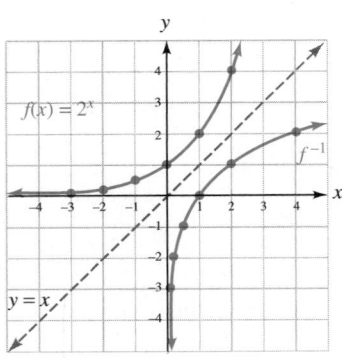

To write an equation for the inverse of $f(x) = 2^x$, we proceed as follows:

$$f(x) = 2^x$$
$$y = 2^x \qquad \text{Replace } f(x) \text{ with } y.$$
$$x = 2^y \qquad \text{Interchange the variables } x \text{ and } y.$$

We cannot solve the equation for y because we have not discussed methods for solving equations with a variable in an exponent. However, we can translate the relationship $x = 2^y$ into words:

$$y = \text{ the power to which we raise 2 to get } x$$

If we substitute $f^{-1}(x)$ for y, we see that

$$f^{-1}(x) = \text{ the power to which we raise 2 to get } x$$

If we define the symbol $\log_2 x$ to mean *the power to which we raise 2 to get x*, we can write the equation for the inverse as

$$f^{-1}(x) = \log_2 x \qquad \text{Read } \log_2 x \text{ as "the logarithm, base 2, of } x \text{" or "log, base 2, of } x.\text{"}$$

We have found that the inverse of the exponential function $f(x) = 2^x$ is $f^{-1}(x) = \log_2 x$. To find the inverse of exponential functions with other bases, such as $f(x) = 3^x$ and $f(x) = 10^x$, we define logarithm in the following way.

Definition of Logarithm

For all positive numbers b, where $b \neq 1$, and all positive numbers x,

$$y = \log_b x \quad \text{is equivalent to} \quad x = b^y$$

This definition guarantees that any pair (x, y) that satisfies the logarithmic equation $y = \log_b x$ also satisfies the exponential equation $x = b^y$. Because of this relationship, a statement written in logarithmic form can be written in an equivalent exponential form, and vice versa. The following diagram will help you remember the respective positions of the exponent and base in each form.

Exponent

$$y = \log_b x \qquad x = b^y$$

Base

② Write Logarithmic Equations as Exponential Equations.

The following table shows several pairs of equivalent equations.

Logarithmic equation	Exponential equation
$\log_2 8 = 3$	$2^3 = 8$
$\log_3 81 = 4$	$3^4 = 81$
$\log_4 4 = 1$	$4^1 = 4$
$\log_5 \dfrac{1}{125} = -3$	$5^{-3} = \dfrac{1}{125}$

EXAMPLE 1   Write each logarithmic equation as an exponential equation:

a. $\log_4 64 = 3$ **b.** $\log_7 \sqrt{7} = \dfrac{1}{2}$ **c.** $\log_6 \dfrac{1}{36} = -2$

Strategy   To write an equivalent exponential equation, we will determine which number will serve as the base and which will serve as the exponent.

Why   We can then use the definition of logarithm to move from one form to the other: $\log_b x = y$ is equivalent to $x = b^y$.

Solution

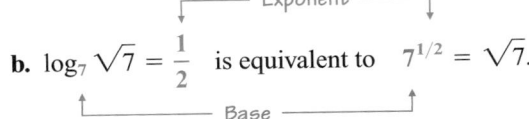

a. $\log_4 64 = 3$ is equivalent to $4^3 = 64$.

b. $\log_7 \sqrt{7} = \dfrac{1}{2}$ is equivalent to $7^{1/2} = \sqrt{7}$.

c. $\log_6 \dfrac{1}{36} = -2$ is equivalent to $6^{-2} = \dfrac{1}{36}$.

 Self Check 1   Write $\log_2 128 = 7$ as an exponential equation.
Now Try   Problem 21

③ Write Exponential Equations as Logarithmic Equations.

EXAMPLE 2   Write each exponential equation as a logarithmic equation:

a. $8^0 = 1$ **b.** $6^{1/3} = \sqrt[3]{6}$ **c.** $\left(\dfrac{1}{4}\right)^2 = \dfrac{1}{16}$

Strategy   To write an equivalent logarithmic equation, we will determine which number will serve as the base and where we will place the exponent.

Why   We can then use the definition of logarithm to move from one form to the other: $x = b^y$ is equivalent to $\log_b x = y$.

Solution

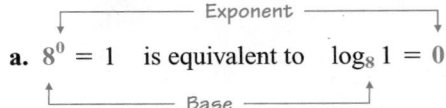

a. $8^0 = 1$ is equivalent to $\log_8 1 = 0$

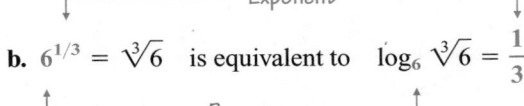

b. $6^{1/3} = \sqrt[3]{6}$ is equivalent to $\log_6 \sqrt[3]{6} = \dfrac{1}{3}$

c. $\left(\dfrac{1}{4}\right)^2 = \dfrac{1}{16}$ is equivalent to $\log_{1/4} \dfrac{1}{16} = 2$

 Self Check 2   Write $9^{-1} = \frac{1}{9}$ as a logarithmic equation.

Now Try   Problem 29

Certain logarithmic equations can be solved by writing them as exponential equations.

EXAMPLE 3   Solve each equation for x:

a. $\log_x 25 = 2$ **b.** $\log_3 x = -3$ **c.** $\log_{1/2} \dfrac{1}{16} = x$

Strategy   To solve each logarithmic equation, we will instead write and solve an equivalent exponential equation.

Why   The resulting exponential equation is easier to solve because the variable term is often isolated on one side.

Solution

a. Since $\log_x 25 = 2$ is equivalent to $x^2 = 25$, we can solve $x^2 = 25$ to find x.

$$x^2 = 25$$
$$x = \pm\sqrt{25} \quad \text{Use the square root property.}$$
$$x = \pm 5$$

Success Tip
Recall that equivalent equations have the same solutions.

In the expression $\log_x 25$, the base of the logarithm is x. Because the base must be positive, we discard -5 and we have

$$x = 5$$

To check the solution of 5, we verify that $\log_5 25 = 2$.

b. Since $\log_3 x = -3$ is equivalent to $3^{-3} = x$, we can solve $3^{-3} = x$ to find x.

$$3^{-3} = x$$

$$\frac{1}{3^3} = x$$

$$x = \frac{1}{27}$$

To check the solution of $\frac{1}{27}$, we verify that $\log_3 \frac{1}{27} = -3$.

c. Since $\log_{1/2} \frac{1}{16} = x$ is equivalent to $\left(\frac{1}{2}\right)^x = \frac{1}{16}$, we can solve $\left(\frac{1}{2}\right)^x = \frac{1}{16}$ to find x.

$$\left(\frac{1}{2}\right)^x = \frac{1}{16}$$

$$\left(\frac{1}{2}\right)^x = \left(\frac{1}{2}\right)^4 \quad \text{Write } \tfrac{1}{16} \text{ as a power of } \tfrac{1}{2} \text{ to match the bases: } \tfrac{1}{2} \cdot \tfrac{1}{2} \cdot \tfrac{1}{2} \cdot \tfrac{1}{2} = \tfrac{1}{16}.$$

$$x = 4 \qquad \text{Since the bases are the same, and since exponential functions are one-to-one, the exponents must be equal.}$$

To check the solution of 4, we verify that $\log_{1/2} \frac{1}{16} = 4$.

> **Success Tip**
> To solve this equation, we note that if the bases are equal, the exponents must be equal.
>
> $$\left(\frac{1}{2}\right)^x = \left(\frac{1}{2}\right)^4$$

Self Check 3   Solve each equation for x:
a. $\log_x 49 = 2$ **b.** $\log_{1/3} x = 2$ **c.** $\log_6 216 = x$

Now Try   Problems 37, 39, and 41

---

4   **Evaluate Logarithmic Expressions.**

In the previous examples, we have seen that the logarithm of a number is an exponent. In fact,

   $\log_b x$ *is the exponent to which b is raised to get x.*

Translating this statement into symbols, we have

   $$b^{\log_b x} = x$$

EXAMPLE 4   Evaluate each logarithmic expression:

a. $\log_8 64$ **b.** $\log_3 \frac{1}{3}$ **c.** $\log_4 2$

Strategy   After identifying the base, we will ask "To what power must the base be raised to get the other number?"

Why   That power is the value of the logarithmic expression.

Solution
a. $\log_8 64 = 2$ Ask: "To what power must we raise 8 to get 64?"
Since $8^2 = 64$, the answer is the 2nd power.

b. $\log_3 \frac{1}{3} = -1$ Ask: "To what power must we raise 3 to get $\frac{1}{3}$?"
Since $3^{-1} = \frac{1}{3}$, the answer is the -1 power.

c. $\log_4 2 = \dfrac{1}{2}$ Ask: "To what power must we raise 4 to get 2?"

Since $\sqrt{4} = 4^{1/2} = 2$, the answer is the $\frac{1}{2}$ power.

Self Check 4   Evaluate each expression: **a.** $\log_9 81$ **b.** $\log_4 \dfrac{1}{16}$

c. $\log_9 3$

Now Try   **Problems 61 and 63**

The Language of Algebra
London professor Henry Briggs (1561–1630) and Scottish lord John Napier (1550–1617) are credited with developing the concept of common logarithms. Their tables of logarithms were useful tools at that time for those performing large calculations.

For computational purposes and in many applications, we will use base-10 logarithms (also called **common logarithms**). When the base b is not indicated in the notation $\log x$, we assume that $b = 10$:

$\log x$ means $\log_{10} x$

The table below shows several pairs of equivalent statements involving base-10 logarithms.

Logarithmic form	*Exponential form*
$\log 100 = 2$	$10^2 = 100$ Read $\log 100$ as "log of 100."
$\log \dfrac{1}{10} = -1$	$10^{-1} = \dfrac{1}{10}$
$\log 1 = 0$	$10^0 = 1$

In general, we have

$\log_{10} 10^x = x$

EXAMPLE 5   Evaluate each logarithmic expression, if possible:

a. $\log 1{,}000$ **b.** $\log \dfrac{1}{100}$ **c.** $\log 10$ **d.** $\log(-10)$

Strategy   After identifying the base, we will ask "To what power must 10 be raised to get the other number?"

Why   That power is the value of the logarithmic expression.

Solution

a. $\log 1{,}000 = 3$ Ask: "To what power must we raise 10 to get 1,000?"

Since $10^3 = 1{,}000$, the answer is: the 3rd power.

b. $\log \dfrac{1}{100} = -2$ Ask: "To what power must we raise 10 to get $\frac{1}{100}$?"

Since $10^{-2} = \frac{1}{100}$, the answer is: the -2 power.

c. $\log 10 = 1$ Ask: "To what power must we raise 10 to get 10?"

Since $10^1 = 10$, the answer is: the 1st power.

d. To find $\log(-10)$, we must find a power of 10 such that $10^? = -10$. There is no such number. Thus, $\log(-10)$ is undefined.

Self Check 5   Evaluate each expression: **a.** $\log 10{,}000$

b. $\log \dfrac{1}{1{,}000}$ **c.** $\log 0$

Now Try   **Problems 65 and 67**

Many logarithmic expressions cannot be evaluated by inspection. For example, to find log 2.34, we ask, "To what power must we raise 10 to get 2.34?" This answer isn't obvious. In such cases, we use a calculator.

Using Your Calculator   *Evaluating Logarithms*

To find log 2.34 with a scientific calculator we enter

2.34 $\boxed{\text{LOG}}$ $\boxed{.369215857}$

On some calculators, the $\boxed{10^x}$ key also serves as the $\boxed{\text{LOG}}$ key when $\boxed{\text{2nd}}$ or $\boxed{\text{SHIFT}}$ is pressed. This is because $f(x) = 10^x$ and $f(x) = \log x$ are inverses.

To use a graphing calculator, we enter

$\boxed{\text{LOG}}$ 2.34 $\boxed{)}$ $\boxed{\text{ENTER}}$ $\boxed{\begin{array}{l} \text{log(2.34)} \\ \qquad\quad .369215857 \end{array}}$

To four decimal places, log 2.34 = 0.3692. This means, $10^{0.3692} \approx 2.34$.

If we attempt to evaluate logarithmic expressions such as log 0, or the logarithm of a negative number, such as log(-5), an error message like the following will be displayed.

$\boxed{\text{Error}}$ $\boxed{\begin{array}{l} \texttt{ERR:DOMAIN} \\ \texttt{1:QUIT} \\ \texttt{2:Go to} \end{array}}$ $\boxed{\begin{array}{l} \texttt{ERR:NONREAL ANS} \\ \texttt{1:QUIT} \\ \texttt{2:Go to} \end{array}}$

EXAMPLE 6   Solve log x = 0.3568 and round to four decimal places.

Strategy   To solve this logarithmic equation, we will instead write and solve an equivalent exponential equation.

Why   The resulting exponential equation is easier to solve because the variable term is isolated on one side.

Solution   The equation log x = 0.3568 is equivalent to $10^{0.3568} = x$. Since we cannot determine $10^{0.3568}$ by inspection, we will use a calculator to find an approximate solution. We enter

10 $\boxed{y^x}$ .3568 $\boxed{=}$

The display reads $\boxed{2.274049951}$. To four decimal places,

$x = 2.2740$

If your calculator has a $\boxed{10^x}$ key, enter .3568 and press it to get the same result. The solution is 2.2740. To check, use your calculator to verify that log 2.2740 ≈ 0.3568.

 Self Check 6   Solve log x = 1.87737 and round to four decimal places.

Now Try   **Problem 77**

Graph Logarithmic Functions.

Because an exponential function defined by $f(x) = b^x$ is one-to-one, it has an inverse function that is defined by $x = b^y$. When we write $x = b^y$ in the equivalent form $y = \log_b x$, the result is called a *logarithmic function*.

Logarithmic Functions	If $b > 0$ and $b \neq 1$, the **logarithmic function with base b** is defined by the equations

$$f(x) = \log_b x \quad \text{or} \quad y = \log_b x$$

The domain of $f(x) = \log_b x$ is the interval $(0, \infty)$ and the range is the interval $(-\infty, \infty)$.

Since every logarithmic function is the inverse of a one-to-one exponential function, logarithmic functions are one-to-one.

We can plot points to graph logarithmic functions. For example, to graph $f(x) = \log_2 x$, we construct a table of function values, plot the resulting ordered pairs, and draw a smooth curve through the points to get the graph, as shown in figure (a). To graph $f(x) = \log_{1/2} x$, we use the same method, as shown in figure (b).

$f(x) = \log_2 x$

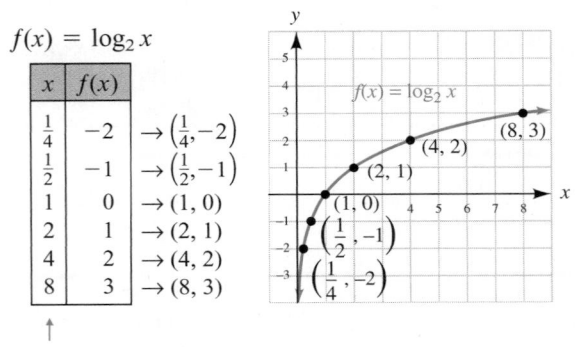

x	$f(x)$	
$\frac{1}{4}$	-2	$\rightarrow \left(\frac{1}{4}, -2\right)$
$\frac{1}{2}$	-1	$\rightarrow \left(\frac{1}{2}, -1\right)$
1	0	$\rightarrow (1, 0)$
2	1	$\rightarrow (2, 1)$
4	2	$\rightarrow (4, 2)$
8	3	$\rightarrow (8, 3)$

Because the base of the function is 2, choose values for x that are integer powers of 2.

(a)

$f(x) = \log_{1/2} x$

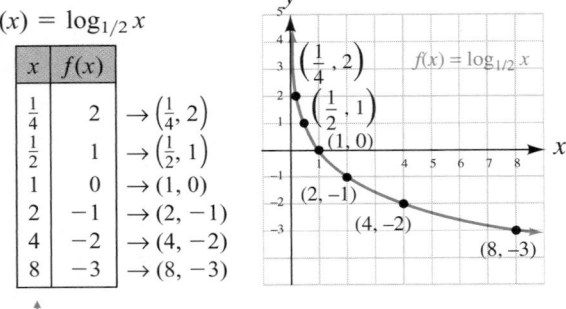

x	$f(x)$	
$\frac{1}{4}$	2	$\rightarrow \left(\frac{1}{4}, 2\right)$
$\frac{1}{2}$	1	$\rightarrow \left(\frac{1}{2}, 1\right)$
1	0	$\rightarrow (1, 0)$
2	-1	$\rightarrow (2, -1)$
4	-2	$\rightarrow (4, -2)$
8	-3	$\rightarrow (8, -3)$

Because the base of the function is $\frac{1}{2}$, choose values for x that are integers powers of $\frac{1}{2}$.

(b)

The graphs of all logarithmic functions are similar to those shown below. If $b > 1$, the logarithmic function is increasing, as in figure (a). If $0 < b < 1$, the logarithmic function is decreasing, as in figure (b).

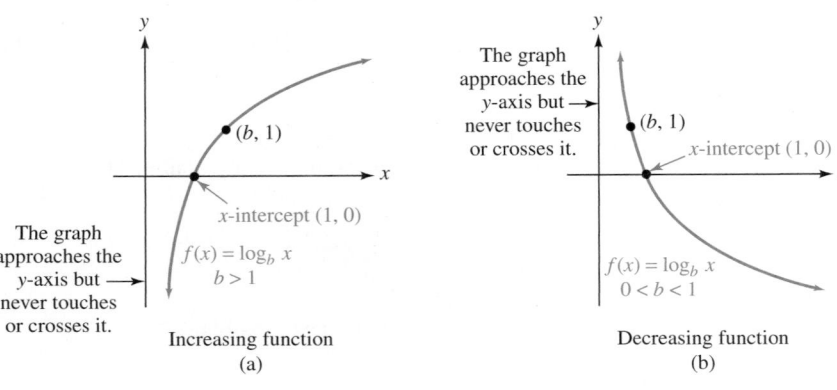

The graph approaches the y-axis but never touches or crosses it.

$(b, 1)$

x-intercept $(1, 0)$

$f(x) = \log_b x$
$b > 1$

Increasing function
(a)

The graph approaches the y-axis but never touches or crosses it.

$(b, 1)$

x-intercept $(1, 0)$

$f(x) = \log_b x$
$0 < b < 1$

Decreasing function
(b)

Properties of Logarithmic Functions	The graph of $f(x) = \log_b x$ (or $y = \log_b x$) has the following properties. 1. It passes through the point $(1, 0)$. 2. It passes through the point $(b, 1)$. 3. The y-axis (the line $x = 0$) is an asymptote. 4. The domain is the interval $(0, \infty)$ and the range is the interval $(-\infty, \infty)$.

Caution

Since the domain of the logarithmic function is the set of positive real numbers, it is impossible to find the logarithm of 0 or the logarithm of a negative number. For example,

$$\log_2(-4) \quad \text{and} \quad \log_2 0$$

are undefined.

The exponential and logarithmic functions are inverses of each other, so their graphs have symmetry about the line $y = x$. The graphs of $f(x) = \log_b x$ and $g(x) = b^x$ are shown in figure (a) when $b > 1$ and in figure (b) when $0 < b < 1$.

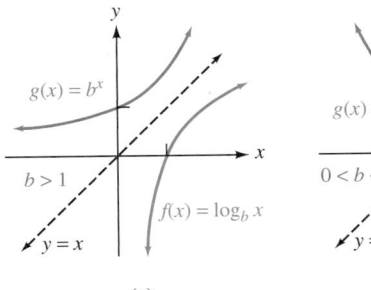
(a)

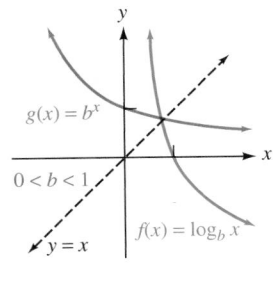
(b)

The graphs of many functions involving logarithms are translations of the basic logarithmic graphs.

EXAMPLE 7   Graph each function by using a translation:
    a. $g(x) = 3 + \log_2 x$ **b.** $g(x) = \log_{1/2}(x - 1)$

Strategy   We will graph $g(x) = 3 + \log_2 x$ by translating the graph of $f(x) = \log_2 x$ upward 3 units. We will graph $g(x) = \log_{1/2}(x - 1)$ by translating the graph of $f(x) = \log_{1/2} x$ to the right 1 unit.

Why   The addition of 3 in $g(x) = 3 + \log_2 x$ causes a vertical shift of the graph of the base-2 logarithmic function 3 units upward. The subtraction of 1 from x in $g(x) = \log_{1/2}(x - 1)$ causes a horizontal shift of the graph of the base-$\frac{1}{2}$ logarithmic function 1 unit to the right.

Solution

a. The graph of $g(x) = 3 + \log_2 x$ will be the same shape as the graph of $f(x) = \log_2 x$, except that it is shifted 3 units upward.

Notation

Since $y = f(x)$, we can write

$$f(x) = \log_2 x$$

as

$$y = \log_2 x$$

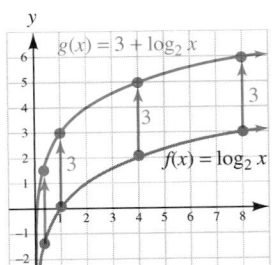

To graph $g(x) = 3 + \log_2 x$, translate each point on the graph of $f(x) = \log_2 x$ up 3 units.

b. The graph of $g(x) = \log_{1/2}(x - 1)$ will be the same shape as the graph of $f(x) = \log_{1/2} x$, except it is shifted 1 unit to the right.

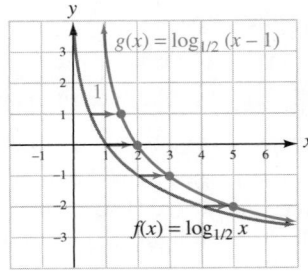

To graph $g(x) = \log_{1/2}(x - 1)$, translate each point on the graph of $f(x) = \log_{1/2} x$ to the right 1 unit.

 Self Check 7   Graph each function by using a translation.

a. $g(x) = (\log_3 x) - 2$ **b.** $g(x) = \log_{1/3}(x + 2)$

Now Try   **Problems 89 and 91**

To graph more complicated logarithmic functions, a graphing calculator is a useful tool.

Using Your Calculator   ***Graphing Logarithmic Functions***

To use a calculator to graph the logarithmic function $f(x) = -2 + \log_{10}\frac{x}{2}$, we enter the right side of the equation after the symbol $Y_1 =$. The display will show the equation

$$Y_1 = -2 + \log(X/2)$$

If we use window settings of $[-1, 5]$ for x and $[-4, 1]$ for y and press the $\boxed{\text{GRAPH}}$ key, we will obtain the graph shown.

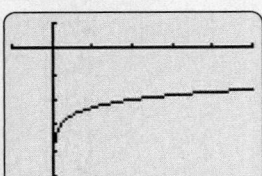

 Use Logarithmic Functions in Applications.

Logarithmic functions, like exponential functions, can be used to model certain types of growth and decay. Common logarithms are used in electrical engineering to express the voltage gain (or loss) of an electronic device such as an amplifier. The unit of gain (or loss), called the **decibel,** is defined by a logarithmic relation.

Decibel Voltage Gain

If E_O is the output voltage of a device and E_I is the input voltage, the decibel voltage gain of the device (db gain) is given by

$$\text{db gain} = 20 \log \frac{E_O}{E_I}$$

© Image Source Pink/Getty Images

EXAMPLE 8 **db Gain.** If the input to an amplifier is 0.5 volt and the output is 40 volts, find the decibel voltage gain of the amplifier.

Strategy We will substitute into the formula for db gain and evaluate the right side using a calculator.

Why We can use this formula to find the db gain because we are given the input voltage E_I and the output voltage E_O.

Solution We can find the decibel voltage gain by substituting 0.5 for E_I and 40 for E_O into the formula for db gain:

$$\text{db gain} = 20 \log \frac{E_O}{E_I}$$

$$\text{db gain} = 20 \log \frac{40}{0.5}$$

$$= 20 \log 80 \qquad \text{Divide: } \frac{40}{0.5} = 80.$$

$$\approx 38 \qquad \text{Use a calculator: } 20 \log 80 \text{ means } 20 \cdot \log 80.$$

The amplifier provides a 38-decibel voltage gain.

Self Check 8 If the input to an amplifier is 0.6 volt and the output is 40 volts, find the decibel voltage gain of the amplifier.

Now Try **Problem 97**

In seismology, common logarithms are used to measure the intensity of earthquakes on the **Richter scale.** The intensity of an earthquake is given by the following logarithmic function.

Richter Scale

If R is the intensity of an earthquake, A is the amplitude (measured in micrometers) of the ground motion, and P is the period (the time of one oscillation of the Earth's surface measured in seconds), then

$$R = \log \frac{A}{P}$$

EXAMPLE 9 **Earthquakes.** Find the measure on the Richter scale of an earthquake with an amplitude of 5,000 micrometers (0.5 centimeter) and a period of 0.1 second.

Strategy We will substitute into the formula for intensity of an earthquake and evaluate the right side using a calculator.

Why We can use this formula to find the intensity of the earthquake because we are given the amplitude A and the period P.

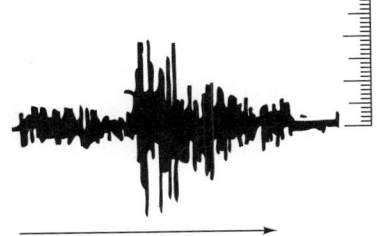

Time

Solution   We substitute 5,000 for A and 0.1 for P in the Richter scale formula and proceed as follows:

$$R = \log \frac{A}{P}$$

$$R = \log \frac{5,000}{0.1}$$

$$= \log 50,000 \qquad \text{Divide: } \frac{5,000}{0.1} = 50,000.$$

$$\approx 4.698970004 \qquad \text{Use a calculator.}$$

The earthquake measures about 4.7 on the Richter scale.

The Language of Algebra
The Richter scale was developed in 1935 by Charles F. Richter of the California Institute of Technology.

Self Check 9   Find the measure on the Richter scale of an earthquake with an amplitude of 4,000 micrometers (0.4 centimeter) and a period of 0.2 second.

Now Try   **Problem 101**

ANSWERS TO SELF CHECKS   1. $2^7 = 128$ 2. $\log_9 \frac{1}{9} = -1$ 3. a. 7   b. $\frac{1}{9}$ c. 3
4. a. 2   b. -2 c. $\frac{1}{2}$ 5. a. 4   b. -3,   c. Undefined   6. 75.3998
7. a.   b.   8. About 36 db   9. 4.3

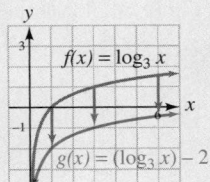

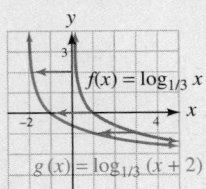

STUDY SET
11.5

VOCABULARY

Refer to the graph of $f(x) = \log_4 x$.

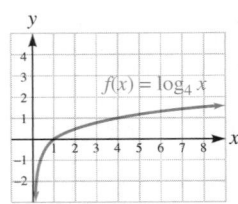

1. What type of function is $f(x) = \log_4 x$?
2. What is the domain of the function?
3. What is the range of the function?

4. a. What is the y-intercept of the graph?
   b. What is the x-intercept of the graph?
5. Is f a one-to-one function?
6. What is an asymptote of the graph?
7. Is f an increasing or a decreasing function?
8. The graph passes through the point $(4, y)$. What is y?

CONCEPTS

Fill in the blanks.

9. The equation $y = \log_b x$ is equivalent to the exponential equation $\boxed{} = \boxed{}$.
10. $\log_b x$ is the _____ to which b is raised to get x.
11. The functions $f(x) = \log_{10} x$ and $f(x) = 10^x$ are _____ functions.
12. The inverse of an exponential function is called a _____ function.

Complete the table of values, where possible.

13. $f(x) = \log x$

x	$f(x)$
100	
$\frac{1}{100}$	

14. $f(x) = \log_5 x$

x	$f(x)$
25	
$\frac{1}{25}$	

15. $f(x) = \log_6 x$

Input	Output
6	
-6	
0	

16. $f(x) = \log_8 x$

Input	Output
8	
-8	
0	

17. a. Use a calculator to complete the table of values for $f(x) = \log x$. Round to the nearest hundredth.

x	$f(x)$
0.5	
1	
2	
4	
6	
8	
10	

b. Graph $f(x) = \log x$. Note that the units on the x- and y-axes are different.

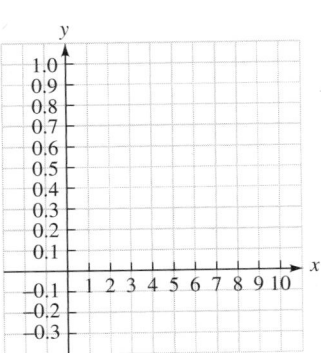

18. For each function, determine its inverse, $f^{-1}(x)$.

   a. $f(x) = 10^x$ **b.** $f(x) = 3^x$

   c. $f(x) = \log x$ **d.** $f(x) = \log_2 x$

NOTATION

Fill in the blanks.

19. a. $\log x = \log \quad x$ **b.** $\log_{10} 10^x =$

20. a. We read $\log_5 25$ as "log,   5,   25."

   b. We read $\log x$ as "   of x."

GUIDED PRACTICE

Write each logarithmic equation as an exponential equation. See Example 1.

21. $\log_3 81 = 4$ **22.** $\log_7 7 = 1$

23. $\log_{10} 10 = 1$ **24.** $\log_{10} 100 = 2$

25. $\log_4 \dfrac{1}{64} = -3$ **26.** $\log_6 \dfrac{1}{36} = -2$

27. $\log_5 \sqrt{5} = \dfrac{1}{2}$ **28.** $\log_8 \sqrt[3]{8} = \dfrac{1}{3}$

Write each exponential equation as a logarithmic equation. See Example 2.

29. $8^2 = 64$ **30.** $10^3 = 1{,}000$

31. $4^{-2} = \dfrac{1}{16}$ **32.** $3^{-4} = \dfrac{1}{81}$

33. $\left(\dfrac{1}{2}\right)^{-5} = 32$ **34.** $\left(\dfrac{1}{3}\right)^{-3} = 27$

35. $x^y = z$ **36.** $m^n = p$

Solve for x. See Example 3.

37. $\log_x 81 = 2$ **38.** $\log_x 9 = 2$
39. $\log_8 x = 2$ **40.** $\log_7 x = 0$
41. $\log_5 125 = x$ **42.** $\log_4 16 = x$
43. $\log_5 x = -2$ **44.** $\log_3 x = -4$

45. $\log_{36} x = -\dfrac{1}{2}$ **46.** $\log_{27} x = -\dfrac{1}{3}$

47. $\log_x 0.01 = -2$ **48.** $\log_x 0.001 = -3$
49. $\log_{27} 9 = x$ **50.** $\log_{12} x = 0$
51. $\log_x 5^3 = 3$ **52.** $\log_x 5 = 1$

53. $\log_{100} x = \dfrac{3}{2}$ **54.** $\log_x \dfrac{1}{1{,}000} = -\dfrac{3}{2}$

55. $\log_x \dfrac{1}{64} = -3$ **56.** $\log_x \dfrac{1}{100} = -2$

57. $\log_8 x = 0$ **58.** $\log_4 8 = x$

59. $\log_x \dfrac{\sqrt{3}}{3} = \dfrac{1}{2}$ **60.** $\log_x \dfrac{9}{4} = 2$

Evaluate each logarithmic expression. See Examples 4 and 5.

61. $\log_2 8$

62. $\log_3 9$

63. $\log_4 16$

64. $\log_6 216$

65. $\log 1{,}000{,}000$

66. $\log 100{,}000$

67. $\log \dfrac{1}{10}$

68. $\log \dfrac{1}{10{,}000}$

69. $\log_{1/2} \dfrac{1}{32}$

70. $\log_{1/3} \dfrac{1}{81}$

71. $\log_9 3$

72. $\log_{125} 5$

Use a calculator to find each value. Give answers to four decimal places. See Using Your Calculator: Evaluating Logarithms.

73. $\log 3.25$

74. $\log 0.57$

75. $\log 0.00467$

76. $\log 375.876$

Use a calculator to solve each equation. Round answers to four decimal places. See Example 6.

77. $\log x = 3.7813$

78. $\log x = 2.8945$

79. $\log x = -0.7630$

80. $\log x = -1.3587$

81. $\log x = -0.5$

82. $\log x = -0.926$

83. $\log x = -1.71$

84. $\log x = 1.4023$

Graph each function. Determine whether each function is an increasing or a decreasing function. See Objective 5.

85. $f(x) = \log_3 x$

86. $f(x) = \log_{1/3} x$

87. $y = \log_{1/2} x$

88. $y = \log_4 x$

Graph each function by plotting points or by using a translation. (The basic logarithmic functions graphed in Exercises 85–88 will be helpful.) See Example 7.

89. $f(x) = 3 + \log_3 x$

90. $f(x) = (\log_{1/3} x) - 1$

91. $y = \log_{1/2}(x - 2)$

92. $y = \log_4(x + 2)$

Graph each pair of inverse functions on the same coordinate system. Draw the axis of symmetry. See Objective 1.

93. $f(x) = 6^x$
   $f^{-1}(x) = \log_6 x$

94. $f(x) = 3^x$
   $f^{-1}(x) = \log_3 x$

95. $f(x) = 5^x$
   $f^{-1}(x) = \log_5 x$

96. $f(x) = 8^x$
   $f^{-1}(x) = \log_8 x$

APPLICATIONS

97. INPUT VOLTAGE   Find the db gain of an amplifier if the input voltage is 0.71 volt when the output voltage is 20 volts.

98. OUTPUT VOLTAGE   Find the db gain of an amplifier if the output voltage is 2.8 volts when the input voltage is 0.05 volt.

99. db GAIN   Find the db gain of the amplifier shown below.

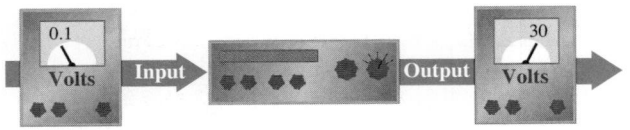

100. db GAIN   An amplifier produces an output of 80 volts when driven by an input of 0.12 volt. Find the amplifier's db gain.

101. THE RICHTER SCALE   An earthquake has amplitude of 5,000 micrometers and a period of 0.2 second. Find its measure on the Richter scale.

102. EARTHQUAKES   Find the period of an earthquake with amplitude of 80,000 micrometers that measures 6 on the Richter scale.

103. EARTHQUAKES   An earthquake with a period of $\frac{1}{4}$ second measures 4 on the Richter scale. Find its amplitude.

104. EARTHQUAKES   In 1985, Mexico City experienced an earthquake of magnitude 8.1 on the Richter scale. In 1989, the San Francisco Bay area was rocked by an earthquake measuring 7.1. By what factor must the amplitude of an earthquake change to increase its severity by 1 point on the Richter scale? (Assume that the period remains constant.)

105. CHILDREN'S HEIGHT   The function $h(A) = 29 + 48.8 \log(A + 1)$ gives the percent of the adult height a male child A years old has attained. If a boy is 9 years old, what percent of his adult height will he have reached?

106. DEPRECIATION   In business, equipment is often depreciated using the double declining-balance method. In this method, a piece of equipment with a life expectancy of N years, costing $\$C$, will depreciate to a value of $\$V$ in n years, where n is given by the formula

$$n = \frac{\log V - \log C}{\log\left(1 - \frac{2}{N}\right)}$$

A computer that cost $\$37{,}000$ has a life expectancy of 5 years. If it has depreciated to a value of $\$8{,}000$, how old is it?

107. INVESTING   If $\$P$ is invested at the end of each year in an annuity earning annual interest at a rate r, the amount in the account will be $\$A$ after n years, where

$$n = \frac{\log\left[\dfrac{Ar}{P} + 1\right]}{\log(1 + r)}$$

If $\$1{,}000$ is invested each year in an annuity earning 12% annual interest, how long will it take for the account to be worth $\$20{,}000$?

108. GROWTH OF MONEY   If $\$5{,}000$ is invested each year in an annuity earning 8% annual interest, how long will it take for the account to be worth $\$50{,}000$? (See Exercise 107.)

WRITING

109. Explain the mathematical relationship between $f(x) = \log x$ and $g(x) = 10^x$.

110. Explain why it is impossible to find the logarithm of a negative number.

111. A table of solutions for $f(x) = \log x$ is shown below. As x decreases and gets close to 0, what happens to the values of $f(x)$?

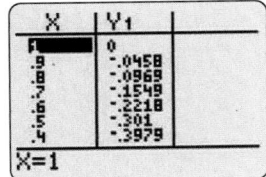

112. What question should be asked when evaluating the expression $\log_4 16$?

REVIEW

Solve each equation.

113. $\sqrt[3]{6x + 4} = 4$

114. $\sqrt{3x + 4} = \sqrt{7x + 2}$

115. $\sqrt{a + 1} - 1 = 3a$

116. $3 - \sqrt{t - 3} = \sqrt{t}$

CHALLENGE PROBLEMS

117. Without graphing, determine the domain of the function $f(x) = \log_5 (x^2 - 1)$. Express the result in interval notation.

118. Evaluate: $\log_6(\log_5(\log_4 1{,}024))$

SECTION 11.6
Base-e Logarithmic Functions

Objectives

1 Define base-*e* logarithms.
2 Evaluate natural logarithmic expressions.
3 Graph the natural logarithmic function.
4 Use natural logarithmic functions in applications.

We have seen the importance of e in modeling the growth and decay of natural events. Just as $f(x) = e^x$ is called the natural exponential function, its inverse, the base-e logarithmic function, is called the *natural logarithmic function*. Natural logarithmic functions have many applications. They play a very important role in advanced mathematics courses, such as calculus.

1 Define Base-*e* Logarithms.

Of all possible bases for a logarithmic function, e is the most convenient for problems involving growth or decay. Since these situations occur often in natural settings, base-e logarithms are called **natural logarithms** or **Napierian logarithms** after John Napier (1550–1617). They are usually written as $\ln x$ rather than $\log_e x$:

$\ln x$ means $\log_e x$ Read ln x letter-by-letter as "ℓ ... n ... of x."

In general, the logarithm of a number is an exponent. For natural logarithms,

ln x is the exponent to which e is raised to get x.

Translating this statement into symbols, we have

$e^{\ln x} = x$

Caution
Because of the font used to print the natural log of x, some students initially misread the notation as ln x. In handwriting, ln x should look like

Evaluate Natural Logarithmic Expressions.

EXAMPLE 1   Evaluate each natural logarithmic expression:

   a. $\ln e$ **b.** $\ln \dfrac{1}{e^2}$ **c.** $\ln 1$ **d.** $\ln \sqrt{e}$

Strategy   Since the base is e in each case, we will ask "To what power must e be raised to get the given number?"

Why   That power is the value of the logarithmic expression.

Solution

a. $\ln e = 1$ Ask: "To what power must we raise e to get e?"

Since $e^1 = e$, the answer is: the 1st power.

b. $\ln \dfrac{1}{e^2} = -2$ Ask: "To what power must we raise e to get $\frac{1}{e^2}$?"

Since $e^{-2} = \frac{1}{e^2}$, the answer is: the -2 power.

c. $\ln 1 = 0$ Ask: "To what power must we raise e to get 1?"

Since $e^0 = 1$, the answer is: the 0 power.

d. $\ln \sqrt{e} = \dfrac{1}{2}$ Ask: "To what power must we raise e to get $\sqrt{e}$?"

Since $e^{1/2} = \sqrt{e}$, the answer is: the $\frac{1}{2}$ power.

▷ **Self Check 1**   Evaluate each expression:

   a. $\ln e^3$ **b.** $\ln \dfrac{1}{e}$ **c.** $\ln \sqrt[3]{e}$

Now Try   Problems 15, 19, and 21

Many natural logarithmic expressions are not as easy to evaluate as those in the previous example. For example, to find $\ln 2.34$, we ask, "To what power must we raise e to get 2.34?" The answer isn't obvious. In such cases, we use a calculator.

Using Your Calculator   *Evaluating Base-e (Natural) Logarithms*

To find $\ln 2.34$ with a scientific calculator, we enter

   2.34 $\boxed{\text{LN}}$ $\boxed{.850150929}$

On some calculators, the $\boxed{e^x}$ key also serves as the $\boxed{\text{LN}}$ key when $\boxed{\text{2nd}}$ or $\boxed{\text{SHIFT}}$ is pressed. This is because $f(x) = e^x$ and $g(x) = \ln x$ are inverses.

To use a graphing calculator, we enter

   $\boxed{\text{LN}}$ 2.34 $\boxed{)}$ $\boxed{\text{ENTER}}$ $\boxed{\begin{array}{l}\texttt{ln(2.34)}\\ \texttt{.8501509294}\end{array}}$

To four decimal places, $\boxed{\ln}\ 2.34 = 0.8502$. This means that $e^{0.8502} \approx 2.34$.

If we attempt to evaluate logarithmic expressions such as $\ln 0$, or the logarithm of a negative number, such as $\ln(-5)$, then one of the following error statements will be are displayed.

$\boxed{\text{Error}}$ $\boxed{\begin{array}{l}\texttt{ERR:DOMAIN}\\ \texttt{1:QUIT}\\ \texttt{2:Go to}\end{array}}$ $\boxed{\begin{array}{l}\texttt{ERR:NONREAL ANS}\\ \texttt{1:QUIT}\\ \texttt{2:Go to}\end{array}}$

Certain natural logarithmic equations can be solved by writing them as natural exponential equations.

EXAMPLE 2   Solve each equation: **a.** $\ln x = 1.335$ and **b.** $\ln x = -5.5$. Give each result to four decimal places.

Strategy   To solve this logarithmic equation, we will instead write and solve an equivalent exponential equation.

Why   The resulting exponential equation is easier to solve because the variable term is isolated on one side.

Solution

a. Since the base of the natural logarithmic function is e, the logarithmic equation $\ln x = 1.335$ is equivalent to exponential equation $e^{1.335} = x$. To use a scientific calculator to find x, enter:

$$1.335 \; e^x$$

The display will read 3.799995946. To four decimal places,

$$x = 3.8000$$

The solution is 3.8000. To check, use your calculator to verify that $\ln 3.8000 \approx 1.335$.

b. The equation $\ln x = -5.5$ is equivalent to $e^{-5.5} = x$. To use a scientific calculator to find x, enter:

$$5.5 \; \boxed{+/-} \; e^x$$

The display will read 0.004086771. To four decimal places,

$$x = 0.0041$$

The solution is 0.0041. To check, use your calculator to verify that $\ln 0.0041 \approx -5.5$.

 Self Check 2   Solve each equation. Give each result to four decimal places.
   a. $\ln x = 1.9344$ **b.** $-3 = \ln x$

Now Try   **Problems 35 and 39**

---

③ Graph the Natural Logarithmic Function.

Because the natural exponential function defined by $f(x) = e^x$ is one-to-one, it has an inverse function that is defined by $x = e^y$. When we write $x = e^y$ in the equivalent form $y = \ln x$, the result is called the *natural logarithmic function*.

The Natural Logarithmic Function

The **natural logarithmic function** with base e is defined by the equations

$$f(x) = \ln x \quad \text{or} \quad y = \ln x, \text{ where } \ln x = \log_e x.$$

The domain of $f(x) = \ln x$ is the interval $(0, \infty)$, and the range is the interval $(-\infty, \infty)$.

Since the natural logarithmic function is the inverse of the one-to-one natural exponential function, the natural logarithmic function is one-to-one.

To graph $f(x) = \ln x$, we can construct a table of function values, plot the resulting ordered pairs, and draw a smooth curve through the points to get the graph shown in figure (a). Figure (b) shows the calculator graph of $f(x) = \ln x$.

$f(x) = \ln x$ To plot these ordered pairs, use a calculator to approximate x.

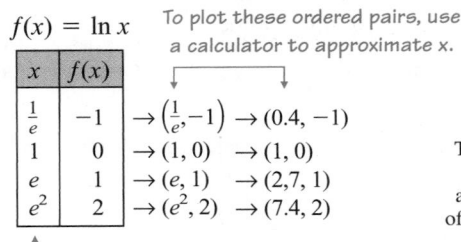

x	$f(x)$	
$\frac{1}{e}$	-1	$\rightarrow \left(\frac{1}{e}, -1\right) \rightarrow (0.4, -1)$
1	0	$\rightarrow (1, 0) \quad \rightarrow (1, 0)$
e	1	$\rightarrow (e, 1) \quad \rightarrow (2.7, 1)$
e^2	2	$\rightarrow (e^2, 2) \rightarrow (7.4, 2)$

Since the base of the natural logarithmic function is e, choose x-values that are integer powers of e.

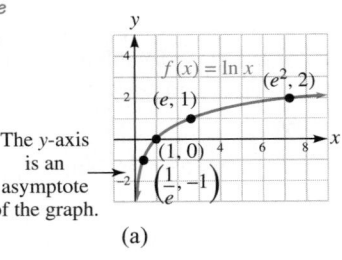

The y-axis is an asymptote of the graph.

(a)

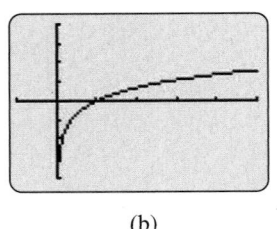

(b)

The natural exponential function and the natural logarithm function are inverse functions. The figure shows that their graphs are symmetric to the line $y = x$.

$f(x) = e^x$

$f^{-1}(x) = \ln x$

$y = x$

Using Your Calculator

Graphing Base-e Logarithmic Functions

Many graphs of logarithmic functions involve translations of the graph of $f(x) = \ln x$. For example, the figure below shows calculator graphs of the functions $f(x) = \ln x$, $g(x) = (\ln x) + 2$, and $h(x) = (\ln x) - 3$.

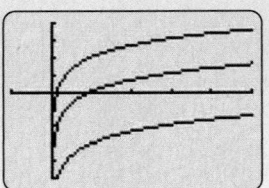

The graph of $g(x) = (\ln x) + 2$ is 2 units above the graph of $f(x) = \ln x$.

The graph of $h(x) = (\ln x) - 3$ is 3 units below the graph of $f(x) = \ln x$.

The next figure shows the calculator graph of the functions $f(x) = \ln x$, $g(x) = \ln (x - 2)$, and $h(x) = \ln (x + 3)$.

The graph of $h(x) = \ln (x + 3)$ is 3 units to the left of the graph of $f(x) = \ln x$.

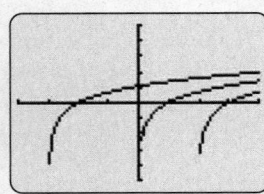

The graph of $g(x) = \ln (x - 2)$ is 2 units to the right of the graph of $f(x) = \ln x$.

4 Use Natural Logarithmic Functions in Applications.

If a population grows exponentially at a certain annual rate, the time required for the population to double is called the **doubling time.** It is given by the following formula.

Formula for Doubling Time	If r is the annual rate, compounded continuously, and t is the time required for a population to double, then $$t = \frac{\ln 2}{r}$$

© Grant Faint/Getty Images

EXAMPLE 3   *Doubling Time.*   The population of the Earth is growing at the approximate rate of 1.17% per year. If this rate continues, how long will it take for the population to double?

Strategy   We will substitute 1.17% for r in the formula for doubling time and evaluate the right side using a calculator.

Why   We can use this formula because we are given the annual rate of continuous compounding.

Solution   Since the population is growing at the rate of 1.17% per year, we substitute 0.0117 for r in the formula for doubling time and simplify.

$$t = \frac{\ln 2}{r}$$

$$t = \frac{\ln 2}{0.0117}$$

$$\approx 59.24334877 \quad \text{Use a calculator. Find ln 2 first, then divide the result by 0.0117.}$$

The population of the Earth will double in about 59 years.

Self Check 3   If the population's annual growth rate could be reduced to 1% per year, what would be the doubling time?

Now Try   **Problem 47**

EXAMPLE 4   *Doubling Time.*   How long will it take $1,000 to double at an annual rate of 8%, compounded continuously?

Strategy   We will substitute 8% for r in the formula for doubling time and evaluate the right side using a calculator. In this case, the information that the original amount is $1,000 is unnecessary information.

Why   We can use this formula because we are given the annual rate of continuous compounding.

Solution   We substitute 0.08 for r and proceed as follows:

$$t = \frac{\ln 2}{r} \qquad \text{This is the formula for doubling time.}$$

$$t = \frac{\ln 2}{0.08}$$

$$\approx 8.664339757 \qquad \text{Use a calculator. Find In 2 first, then divide the result by 0.08.}$$

It will take about $8\frac{2}{3}$ years for the money to double.

 Self Check 4   How long will it take at 9%, compounded continuously?

Now Try   **Problem 50**

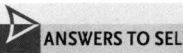 **ANSWERS TO SELF CHECKS**   **1. a.** 3   **b.** -1 **c.** $\frac{1}{3}$ **2. a.** 6.9199   **b.** 0.0498   **3.** About 46 years
4. About 7.7 years

STUDY SET
11.6

VOCABULARY

Fill in the blanks.

1. $f(x) = \ln x$ is called the _____ logarithmic function. The base
is ▢.

2. If a population grows exponentially at a certain annual rate,
the time required for the population to double is called the
_____ _____.

CONCEPTS

3. a. Use a calculator to complete the table of values for
$f(x) = \ln x$. Round to the nearest hundredth.

x	$f(x)$
0.5	
1	
2	
4	
6	
8	
10	

b. Graph $f(x) = \ln x$. Note that the units on the x- and y-axes are
different.

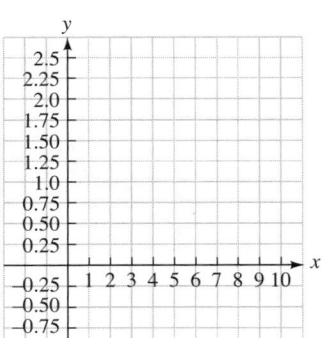

4. What is the inverse of the natural logarithmic function
$f(x) = \ln x$?

Fill in the blanks.

5. The domain of the function $f(x) = \ln x$ is the interval ▢
and the range of the function is the interval ▢.

6. The graph of $f(x) = \ln x$ has the x-intercept ($\,$, 0). The y-axis
is an _____ of the graph.

7. To find $\ln e^2$, we ask, "To what power must we raise ▢ to get
e^2?" Since the answer is the 2nd power, $\ln e^2 = $ ▢.

8. The logarithmic equation $\ln x = 1.5318$ is equivalent to the exponential equation $\boxed{} = \boxed{}$.

9. The illustration shows the graph of $f(x) = \ln x$, as well as a vertical translation of that graph. Using the notation $g(x)$ for the translation, write the defining equation for the function.

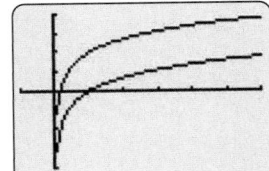

10. In the illustration, $f(x) = \ln x$ was graphed, and the TRACE feature was used. What is the x-coordinate of the point on the graph having a y-coordinate of 1? What is the name given this number?

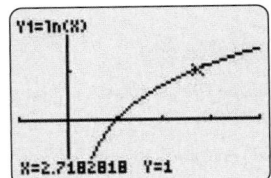

NOTATION

Fill in the blanks.

11. We read $\ln x$ letter-by-letter as "$\boxed{}$ ... $\boxed{}$ ... of x."

12. **a.** $\ln 2$ means $\log_{\boxed{}} 2$.

   b. $\log 2$ means $\log_{\boxed{}} 2$.

13. To evaluate a base-10 logarithm with a calculator, use the $\boxed{}$ key. To evaluate the base-e logarithm, use the $\boxed{}$ key.

14. If a population grows exponentially at a rate r, the time it will take the population to double is given by the formula $t = \boxed{}$.

GUIDED PRACTICE

Evaluate each expression without using a calculator.
See Example 1.

15. $\ln e^5$

16. $\ln e^2$

17. $\ln e^6$

18. $\ln e^4$

19. $\ln \dfrac{1}{e}$

20. $\ln \dfrac{1}{e^3}$

21. $\ln \sqrt[4]{e}$

22. $\ln \sqrt[5]{e}$

23. $\ln \sqrt[3]{e^2}$

24. $\ln \sqrt[4]{e^3}$

25. $\ln e^{-7}$

26. $\ln e^{-10}$

Use a calculator to evaluate each expression, if possible. Express all answers to four decimal places. See Using Your Calculator: Evaluating Base-e Logarithms.

27. $\ln 35.15$

28. $\ln 0.675$

29. $\ln 0.00465$

30. $\ln 378.96$

31. $\ln 1.72$

32. $\ln 2.7$

33. $\ln (-0.1)$

34. $\ln (-10)$

Solve each equation. Express all answers to four decimal places. See Example 2.

35. $\ln x = 1.4023$

36. $\ln x = 2.6490$

37. $\ln x = 4.24$

38. $\ln x = 0.926$

39. $\ln x = -3.71$

40. $\ln x = -0.28$

41. $1.001 = \ln x$

42. $\ln x = -0.001$

Use a graphing calculator to graph each function. See Objective 2. See Using Your Calculator: Base-e Logarithmic Functions.

43. $f(x) = \ln\left(\dfrac{1}{2}x\right)$

44. $f(x) = \ln x^2$

45. $f(x) = \ln (-x)$

46. $f(x) = \ln (3x)$

APPLICATIONS

Use a calculator to solve each problem.

47. THE PEACH STATE   Chattahoochee County, Georgia, grew by 13.2% between 2005 and 2006, making it the fastest-growing county in the United States at that time. If the growth rate remains constant, how long will it take for the population of the county to double?

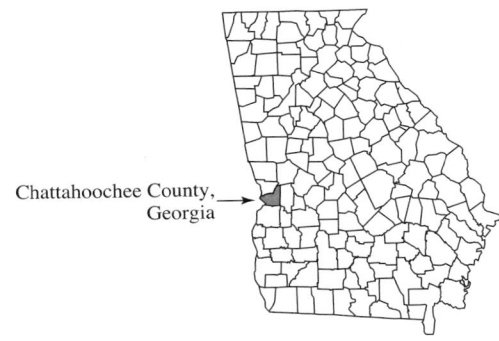

Chattahoochee County, Georgia

48. THE SILVER STATE   Nevada is the one of the fastest growing states in the United States. Determine the number of years it would take for the population of each of Nevada's four largest cities to double in size using the 2005 data below.

Rank	City	Population 2005	Annual Rate of Increase
1.	Las Vegas	569,838	3.7%
2.	Henderson	241,134	4.8%
3.	Reno	206,735	3.8%
4.	North Las Vegas	180,219	9.2%

49. THE NORTH STAR STATE   Minnesota's population increased by 40,362 persons, or 0.8%, between July 2005 and July 2006. If the growth rate remains constant, how long will it take for the population of the state to double?

50. DOUBLING MONEY   How long will it take $1,000 to double if it is invested at an annual rate of 5% compounded continuously?

51. POPULATION GROWTH   A population growing continuously at an annual rate r will triple in a time t given by the formula

$$t = \frac{\ln 3}{r}$$

How long will it take the population of a town to triple if it is growing at the rate of 12% per year?

52. TRIPLING MONEY   Find the length of time for $25,000 to triple when it is invested at 6% annual interest, compounded continuously. See Exercise 51.

53. FORENSIC MEDICINE   To estimate the number of hours t that a murder victim had been dead, a coroner used the formula

$$t = \frac{1}{0.25} \ln \frac{98.6 - T_s}{82 - T_s}$$

where T_s is the temperature of the surroundings where the body was found. If the crime took place in an apartment where the thermostat was set at 70°F, approximately how long ago did the murder occur?

54. MAKING JELL-O   After the contents of a package of JELL-O are combined with boiling water, the mixture is placed in a refrigerator whose temperature remains a constant 42°F. Estimate the number of hours t that it will take for the JELL-O to cool to 50°F using the formula

$$t = -\frac{1}{0.9} \ln \frac{50 - T_r}{200 - T_r}$$

where T_r is the temperature of the refrigerator.

WRITING

55. Explain the difference between the functions $f(x) = \log x$ and $g(x) = \ln x$.

56. How are the functions $f(x) = \ln x$ and $g(x) = e^x$ related?

57. Explain why $\ln e = 1$.

58. Why is $f(x) = \ln x$ called the natural logarithmic function?

59. A table of values for $f(x) = \ln x$ is shown below. Explain why ERROR appears in the Y_1 column for the first three entries.

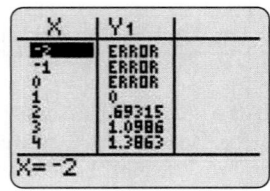

60. The graphs of $f(x) = \ln x$, $g(x) = e^x$, and $y = x$ are shown below. Describe the relationship between the graphs in words.

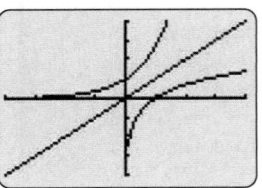

REVIEW

Write the equation of the required line in slope–intercept form, if possible.

61. Parallel to $y = 5x - 8$ and passing through the origin

62. Having a slope of 7 and a y-intercept of 3

63. Passing through the point $(3, 2)$ and perpendicular to the line $y = \frac{2}{3}x - 12$

64. Parallel to the line $3x + 2y = 9$ and passing through the point $(-3, 5)$

65. Vertical line through the point $(2, 3)$

66. Horizontal line through the point $(2, 3)$

CHALLENGE PROBLEMS

67. Use the formula $P = P_0 e^{rt}$ to verify that P will be twice P_0 when $t = \frac{\ln 2}{r}$.

68. Use the formula $P = P_0 e^{rt}$ to verify that P will be three times as large as P_0 when $t = \frac{\ln 3}{r}$.

69. Find a formula to find how long it will take money to quadruple.

70. Use a graphing calculator to graph the function $f(x) = \frac{1}{1 + e^{-2x}}$. Describe its graph in words.

SECTION 11.7
Properties of Logarithms

Objectives

1. Use the four basic properties of logarithms.
2. Use the product rule for logarithms.
3. Use the quotient rule for logarithms.
4. Use the power rule for logarithms.
5. Write logarithmic expressions as a single logarithm.
6. Use the change-of-base formula.
7. Use properties of logarithms to solve application problems.

Since a logarithm is an exponent, we would expect there to be properties of logarithms just as there are properties of exponents. In this section, we will introduce seven properties of logarithms and use them to simplify and expand logarithmic expressions.

1 **Use the Four Basic Properties of Logarithms.**

The first four properties of logarithms follow directly from the definition of logarithm.

Properties of Logarithms

For all positive numbers b, where $b \neq 1$,

1. $\log_b 1 = 0$ **2.** $\log_b b = 1$ **3.** $\log_b b^x = x$ **4.** $b^{\log_b x} = x$ $(x > 0)$

We can use the definition of logarithm to prove that these properties are true.

1. $\log_b 1 = 0$, because $b^0 = 1$.
2. $\log_b b = 1$, because $b^1 = b$.
3. $\log_b b^x = x$, because $b^x = b^x$.
4. $b^{\log_b x} = x$, because $\log_b x$ is the exponent to which b is raised to get x.

Properties 3 and 4 also indicate that the composition of the exponential and logarithmic functions (in both directions) is the identity function. This is expected, because the exponential and logarithmic functions are inverse functions.

EXAMPLE 1   Simplify each expression:

a. $\log_5 1$ **b.** $\log_3 3$ **c.** $\ln e^3$ **d.** $6^{\log_6 7}$

Strategy   We will compare each logarithmic expression to the left side of the previous four properties of logarithms.

Why   When we get a match, the property will provide the answer.

Solution

a. By property 1, $\log_5 1 = 0$, because $5^0 = 1$.

b. By property 2, $\log_3 3 = 1$, because $3^1 = 3$.

c. By property 3, $\ln e^3 = 3$, because $e^3 = e^3$.

d. By property 4, $6^{\log_6 7} = 7$, because $\log_6 7$ is the power to which 6 is raised to get 7.

 Self Check 1   Simplify: **a.** $\log_4 1$ **b.** $\log_4 4$ **c.** $\log_2 2^4$
    d. $5^{\log_5 2}$

Now Try   **Problems 19, 21, 23, and 27**

2   **Use the Product Rule for Logarithms.**

The next property of logarithms is related to the product rule for exponents: $x^m \cdot x^n = x^{m+n}$.

> **The Product Rule for Logarithms**
>
> The logarithm of a product is equal to the sum of the logarithms.
>   For all positive real numbers M, N, and b, where $b \neq 1$,
>
> $$\log_b MN = \log_b M + \log_b N$$

EXAMPLE 2   Write each expression as a sum of logarithms. Then simplify, if possible.   **a.** $\log_2 (2 \cdot 7)$ **b.** $\log 100x$ **c.** $\log_5 125yz$

Strategy   In each case, we will use the product rule for logarithms.

Why   We use the product rule because each of the logarithmic expressions has the form $\log_b MN$.

Solution

a. $\log_2 (2 \cdot 7) = \log_2 2 + \log_2 7$ The log of a product is the sum of the logs.

    $\qquad\qquad\quad = 1 + \log_2 7$ Simplify: By property 2, $\log_2 2 = 1$.

b. Recall that $100x$ means $100 \cdot x$.

    $\log 100x = \log 100 + \log x$ The log of a product is the sum of the logs.

    $\qquad\quad = 2 + \log x$ Simplify: By property 3, $\log 100 = \log 10^2 = 2$.

c. We can write $125yz$ as $(125y)z$.

    $\log_5 125yz = \log_5 (125y)z$ Group the first two factors together.

    $\qquad\quad = \log_5 (125y) + \log_5 z$ The log of a product is the sum of the logs.

    $\qquad\quad = \log_5 125 + \log_5 y + \log_5 z$ The log of a product is the sum of the logs.

    $\qquad\quad = 3 + \log_5 y + \log_5 z$ Simplify: By property 3, $\log_5 125 = \log_5 5^3 = 3$.

> **Notation**
>
> To avoid any confusion, we can use parentheses when writing the logarithm of a product:
>
> $\log 100x = \log (100x)$

 Self Check 2   Write each expression as the sum of logarithms. Then simplify, if possible.   **a.** $\log_3 (3 \cdot 4)$ **b.** $\log 1{,}000y$
    c. $\log_5 25cd$

Now Try   **Problems 31 and 35**

PROOF To prove the product rule for logarithms, we let $x = \log_b M$, $y = \log_b N$, and use the definition of logarithm to write each equation in exponential form.

$$M = b^x \quad \text{and} \quad N = b^y$$

Then $MN = b^x b^y$, and a property of exponents gives

$$MN = b^{x+y} \quad \text{Keep the base and add the exponents: } b^x b^y = b^{x+y}.$$

We write this exponential equation in logarithmic form as

$$\log_b MN = x + y$$

Substituting the values of x and y completes the proof.

$$\log_b MN = \log_b M + \log_b N$$

Caution By the product rule, the logarithm of a *product* is equal to the *sum* of the logarithms. The logarithm of a sum or a difference usually does not simplify. In general,

$$\log_b (M + N) \neq \log_b M + \log_b N \quad \text{and} \quad \log_b (M - N) \neq \log_b M - \log_b N$$

For example,

$$\log_2 (2 + 7) \neq \log_2 2 + \log_2 7 \quad \text{and} \quad \log (100 - y) \neq \log 100 - \log y$$

Using Your Calculator **Verifying Properties of Logarithms**

We can use a calculator to illustrate the product rule for logarithms by showing that

$$\ln (3.7 \cdot 15.9) = \ln 3.7 + \ln 15.9$$

We calculate the left and right sides of the equation separately and compare the results. To use a scientific calculator to find $\ln (3.7 \cdot 15.9)$, we enter

3.7 $\boxed{\times}$ 15.9 $\boxed{=}$ $\boxed{\text{LN}}$ $\boxed{4.074651929}$

To find $\ln 3.7 + \ln 15.9$, we enter

3.7 $\boxed{\text{LN}}$ $\boxed{+}$ 15.9 $\boxed{\text{LN}}$ $\boxed{=}$ $\boxed{4.074651929}$

Since the left and right sides are equal, the equation $\ln (3.7 \cdot 15.9) = \ln 3.7 + \ln 15.9$ is true.

③ Use the Quotient Rule for Logarithms.

The next property of logarithms is related to the quotient rule for exponents: $\frac{x^m}{x^n} = x^{m-n}$.

The Quotient Rule for Logarithms

The logarithm of a quotient is equal to the difference of the logarithms. For all positive real numbers M, N, and b, where $b \neq 1$,

$$\log_b \frac{M}{N} = \log_b M - \log_b N$$

The proof of the quotient rule for logarithms is similar to the proof for the product rule for logarithms.

EXAMPLE 3   Write each expression as a difference of logarithms. Then simplify, if possible.   **a.** $\ln \dfrac{10}{7}$ **b.** $\log_4 \dfrac{x}{64}$

Strategy   In both cases, we will apply the quotient rule for logarithms.

Why   We use the quotient rule because each of the logarithmic expressions has the form $\log_b \dfrac{M}{N}$.

Solution

a. $\ln \dfrac{10}{7} = \ln 10 - \ln 7$ The log of a quotient is the difference of the logs.

b. $\log_4 \dfrac{x}{64} = \log_4 x - \log_4 64$ The log of a quotient is the difference of the logs.

$\qquad\qquad = \log_4 x - 3$ Simplify: $\log_4 64 = \log_4 4^3 = 3$.

 Self Check 3   Write each expression as a difference of logarithms. Then simplify, if possible.

a. $\log_6 \dfrac{6}{5}$ **b.** $\ln \dfrac{y}{100}$

Now Try   **Problem 39**

Caution   By the quotient rule, the logarithm of a *quotient* is equal to the *difference* of the logarithms. The logarithm of a quotient is not the quotient of the logarithms:

$$\log_b \frac{M}{N} \neq \frac{\log_b M}{\log_b N}$$

For example,

$$\ln \frac{10}{7} \neq \frac{\ln 10}{\ln 7} \quad \text{and} \quad \log_4 \frac{x}{64} \neq \frac{\log_4 x}{\log_4 64}$$

In the next example, the product and quotient rules for logarithms are used in combination to rewrite an expression.

EXAMPLE 4   Write $\log \dfrac{xy}{10z}$ as the sum and/or difference of logarithms of a single quantity. Then simplify, if possible.

Strategy   We will use the quotient rule for logarithms and then the product rule.

Why   We use the quotient rule because $\log \dfrac{xy}{10z}$ has the form $\log_b \dfrac{M}{N}$. We later use the product rule because the numerator and denominator of $\dfrac{xy}{10z}$ contain products.

Solution   We begin by applying the quotient rule for logarithms.

$$\log \frac{xy}{10z} = \log xy - \log 10z \qquad \text{The log of a quotient is the difference of the logs.}$$

$$= \log x + \log y - (\log 10 + \log z) \qquad \text{The log of a product is the sum of the logs.}$$

Write parentheses here so that the sum, $\log 10 + \log z$, is subtracted.

$$= \log x + \log y - \log 10 - \log z \qquad \text{Change the sign of each term of } \log 10 + \log z \text{ and drop the parentheses.}$$

$$= \log x + \log y - 1 - \log z \qquad \text{Simplify: } \log 10 = 1.$$

 Self Check 4   Write $\log_b \dfrac{x}{yz}$ as the sum and/or difference of logarithms of a single quantity. Then simplify, if possible.

Now Try   **Problem 45**

4   **Use the Power Rule for Logarithms.**

The next property of logarithms is related to the power rule for exponents: $(x^m)^n = x^{mn}$.

The Power Rule for Logarithms	The logarithm of a power is equal to the power times the logarithm. For all real positive numbers M and b, where $b \neq 1$, and any real number p, $$\log_b M^p = p \log_b M$$

EXAMPLE 5   Write each logarithm without an exponent or a square root:

a. $\log_5 6^2$ **b.** $\log \sqrt{10}$

Strategy   In each case, we will use the power rule for logarithms.

Why   We use the power rule because $\log_5 6^2$ has the form $\log_b M^p$, as will $\log \sqrt{10}$ if we write $\sqrt{10}$ as $10^{1/2}$.

Solution

a. $\log_5 6^2 = 2 \log_5 6$ The log of a power is equal to the power times the log.

b. $\log \sqrt{10} = \log(10)^{1/2}$ Write $\sqrt{10}$ using a fractional exponent: $\sqrt{10} = (10)^{1/2}$.

$$= \frac{1}{2} \log 10 \qquad \text{The log of a power is equal to the power times the log.}$$

$$= \frac{1}{2} \qquad \text{Simplify: } \log 10 = 1.$$

 Self Check 5   Write each logarithm without an exponent or a cube root:

a. $\ln x^4$ **b.** $\log_2 \sqrt[3]{3}$

Now Try   **Problems 51 and 53**

PROOF   To prove the power rule, we let $x = \log_b M$, write the expression in exponential form, and raise both sides to the pth power:

$$M = b^x$$
$$(M)^p = (b^x)^p \quad \text{Raise both sides to the } p\text{th power.}$$
$$M^p = b^{px} \quad \text{Keep the base and multiply the exponents.}$$

Using the definition of logarithms gives

$$\log_b M^p = px$$

Substituting the value for x completes the proof.

$$\log_b M^p = p \log_b M$$

EXAMPLE 6   Write each logarithm as the sum and/or difference of logarithms of a single quantity: **a.** $\log_b x^2 y^3 z$ **b.** $\ln \dfrac{y^3 \sqrt{x}}{z}$

Strategy   In part (a), we will use the product rule and the power rules for logarithms. In part (b), we will use the quotient rule, the product rule, and the power rule for logarithms.

Why   In part (a), we first use the product rule because the expression has the form $\log_b MN$. In part (b), we first use the quotient rule because the expression has the form $\log_b \dfrac{M}{N}$.

Solution
a.   We recognize that $\log_b x^2 y^3 z$ is the logarithm of a product.

$$\log_b x^2 y^3 z = \log_b x^2 + \log_b y^3 + \log_b z \quad \text{The log of a product is the sum of the logs.}$$
$$= 2 \log_b x + 3\log_b y + \log_b z \quad \text{The log of a power is the power times the log.}$$

The Language of Algebra
In Examples 2, 3, 4, and 6, we use properties of logarithms to *expand* logarithmic expressions.

b.   The expression $\ln \dfrac{y^3 \sqrt{x}}{z}$ is the logarithm of a quotient.

$$\ln \dfrac{y^3 \sqrt{x}}{z} = \ln y^3 \sqrt{x} - \ln z \quad \text{The log of a quotient is the difference of the logs.}$$
$$= \ln y^3 + \ln \sqrt{x} - \ln z \quad \text{The log of a product is the sum of the logs.}$$
$$= \ln y^3 + \ln x^{1/2} - \ln z \quad \text{Write } \sqrt{x} \text{ as } x^{1/2}.$$
$$= 3 \ln y + \frac{1}{2} \ln x - \ln z \quad \text{The log of a power is the power times the log.}$$

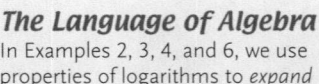 **Self Check 6**   Expand: $\log \sqrt[4]{\dfrac{x^3 y}{z}}$

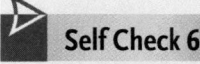 **Now Try**   **Problems 59 and 61**

5 Write Logarithmic Expressions as a Single Logarithm.

EXAMPLE 7	Write each logarithmic expression as one logarithm:

$$\textbf{a. } 3 \log_5 x + \frac{1}{2} \log_5 y \qquad \textbf{b. } \frac{1}{2} \log_b (x - 2) - \log_b y + 3 \log_b z$$

Strategy   In part (a), we will use the power rule and product rule for logarithms in reverse. In part (b), we will use the power rule, the quotient rule, and the product rule for logarithms in reverse.

Why   We use the power rule because we see expressions of the form $p \log_b M$. The $+$ symbol between logarithmic terms suggests that we use the product rule and the $-$ symbol between such terms suggests that we use the quotient rule.

Solution

a. We begin by using the power rule on both terms of the expression.

$$3 \log_5 x + \frac{1}{2} \log_5 y = \log_5 x^3 + \log_5 y^{1/2} \qquad \text{A power times a log is the log of the power.}$$

$$= \log_5 (x^3 \cdot y^{1/2}) \qquad \text{The sum of two logs is the log of the product.}$$

$$= \log_5 x^3 y^{1/2}$$

$$= \log_5 x^3 \sqrt{y} \qquad \text{Write } y^{1/2} \text{ as } \sqrt{y}.$$

b. The first and third terms of this expression can be rewritten using the power rule of logarithms.

$$\frac{1}{2} \log_b (x - 2) - \log_b y + 3 \log_b z$$

$$= \log_b (x - 2)^{1/2} - \log_b y + \log_b z^3 \qquad \text{A power times a log is the log of the power.}$$

$$= \log_b \frac{(x - 2)^{1/2}}{y} + \log_b z^3 \qquad \text{The difference of two logs is the log of the quotient.}$$

$$= \log_b \frac{\sqrt{x - 2}}{y} + \log_b z^3 \qquad \text{Write } (x - 2)^{1/2} \text{ as } \sqrt{x - 2}.$$

$$= \log_b \left(\frac{\sqrt{x - 2}}{y} \cdot z^3 \right) \qquad \text{The sum of two logs is the log of the product.}$$

$$= \log_b \frac{z^3 \sqrt{x - 2}}{y}$$

> **The Language of Algebra**
> In these examples, we use properties of logarithms to *condense* the given expression into a single logarithmic expression. To *condense* means to make more compact. Summer school is a *condensed* version of the regular semester.

 Self Check 7   Write the expression as one logarithm:

$$2 \log_a x + \frac{1}{2} \log_a y - 2 \log_a (x - y)$$

Now Try   **Problems 71 and 75**

The properties of logarithms can be used when working with numerical values.

EXAMPLE 8   Find approximations for each logarithm given that log 2 ≈ 0.3010 and log 3 ≈ 0.4771:   **a.** log 6   **b.** log 18

Strategy   We will express 6 and 18 using factors of 2 and 3 and then use properties of logarithms to simplify each resulting expression.

Why   We express 6 and 18 using factors of 2 and 3 because we are given values of log 2 and log 3.

Solution

a. log 6 = log (2 · 3)   Write 6 using the factors 2 and 3.

   = log 2 + log 3   The log of a product is the sum of the logs.

   ≈ 0.3010 + 0.4771   Substitute the value of each logarithm.

   ≈ 0.7781

b. log 18 = log (2 · 3²)   Write 18 using the factors 2 and 3.

   = log 2 + log 3²   The log of a product is the sum of the logs.

   = log 2 + 2 log 3   The log of a power is the power times the log.

   ≈ 0.3010 + 2(0.4771)   Substitute the value of each logarithm.

   ≈ 1.2552

Self Check 8   Find approximations for each logarithm.   **a.** log 1.5   **b.** log 0.75

Now Try   **Problems 79 and 81**

We summarize the properties of logarithms as follows.

Properties of Logarithms

If b, M, and N are positive real numbers, $b \neq 1$, and p is any real number,

1. $\log_b 1 = 0$
2. $\log_b b = 1$
3. $\log_b b^x = x$
4. $b^{\log_b x} = x$
5. $\log_b MN = \log_b M + \log_b N$
6. $\log_b \dfrac{M}{N} = \log_b M - \log_b N$
7. $\log_b M^p = p \log_b M$

6   **Use the Change-of-Base Formula.**

Most calculators can find common logarithms and natural logarithms. If we need to find a logarithm with some other base, we use a conversion formula.

   If we know the base-a logarithm of a number, we can find its logarithm to some other base b by using a formula called the **change-of-base formula.**

Change-of-Base Formula

For any logarithmic bases a and b, and any positive real number x,

$$\log_b x = \frac{\log_a x}{\log_a b}$$

We can use any positive number other than 1 for base b in the change-of-base formula. However, we usually use 10 or e because of the capabilities of a standard calculator.

EXAMPLE 9   Find: $\log_3 5$

Strategy   To evaluate this base-3 logarithm, we will substitute into the change-of-base formula.

Why   We assume that the reader does not have a calculator that evaluates base-3 logarithms (at least not directly). Thus, the only alternative is to change the base.

Solution   To find $\log_3 5$, we substitute 3 for b, 10 for a, and 5 for x in the change-of-base formula and simplify:

$$\log_b x = \frac{\log_a x}{\log_a b}$$

$$\log_3 5 = \frac{\log_{10} 5}{\log_{10} 3} \qquad \text{Substitute: } b = 3, x = 5, \text{ and } a = 10.$$

$$\approx 1.464973521 \qquad \text{Use a scientific calculator and enter } 5 \boxed{\log} \div 3 \boxed{\log} = .$$

To four decimal places, $\log_3 5 = 1.4650$.

We can also use the natural logarithm function (base e) in the change-of-base formula to find a base-3 logarithm.

$$\log_b x = \frac{\log_a x}{\log_a b}$$

$$\log_3 5 = \frac{\log_e 5}{\log_e 3} \qquad \text{Substitute: } b = 3, x = 5, \text{ and } a = e.$$

$$\log_3 5 = \frac{\ln 5}{\ln 3} \qquad \text{Write } \log_e 5 \text{ as ln 5 and } \log_e 3 \text{ as ln 3.}$$

$$\approx 1.464973521 \qquad \text{Use a calculator.}$$

We obtain the same result.

Self Check 9   Find $\log_5 3$ to four decimal places.

Now Try   Problem 87

PROOF   To prove the change-of-base formula, we begin with the equation $\log_b x = y$.

$$y = \log_b x$$

$$x = b^y \qquad \text{Change the equation from logarithmic to exponential form.}$$

$$\log_a x = \log_a b^y \qquad \text{Take the base-}a \text{ logarithm of both sides.}$$

$$\log_a x = y \log_a b \qquad \text{The log of a power is the power times the log.}$$

$$y = \frac{\log_a x}{\log_a b} \qquad \text{Divide both sides by } \log_a b.$$

$$\log_b x = \frac{\log_a x}{\log_a b} \qquad \text{Refer to the first equation and substitute } \log_b x \text{ for } y.$$

7   Use Properties of Logarithms to Solve Application Problems.

In chemistry, common logarithms are used to express the acidity of solutions. The more acidic a solution, the greater the concentration of hydrogen ions. This concentration is indicated by the *pH scale*, or *hydrogen ion index*. The pH of a solution is defined as follows.

pH of a Solution	If [H$^+$] is the hydrogen ion concentration in gram-ions per liter, then
	$$pH = -\log[H^+]$$

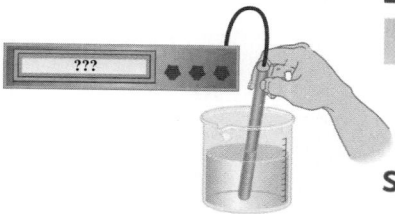

EXAMPLE 10   *pH Meters.*   One of the most accurate ways to measure pH is with a probe and meter. What reading should the meter give for pure water if water has a hydrogen ion concentration [H$^+$] of approximately 10^{-7} gram-ions per liter?

Strategy   We will substitute into the formula for pH and use the power rule for logarithms to simplify the right side.

Why   After substituting 10^{-7} for H$^+$ in $-\log[H^+]$, the resulting expression will have the form $\log_b M^p$.

Solution   Since pure water has approximately 10^{-7} gram-ions per liter, its pH is

$pH = -\log[H^+]$ This is the formula for pH.

$pH = -\log 10^{-7}$

$\quad = -(-7)\log 10$ The log of a power is the power times the log.

$\quad = -(-7) \cdot 1$ Simplify: $\log 10 = 1$.

$\quad = 7$

The meter should give a reading of 7.

 Now Try   **Problem 99**

EXAMPLE 11   *Hydrogen Ion Concentration.*   Find the hydrogen ion concentration of seawater if its pH is 8.5.

Strategy   To find the hydrogen ion concentration, we will substitute 8.5 for pH in the formula $pH = -\log[H^+]$ and solve the resulting equation for [H$^+$].

Why   After substituting for pH, the resulting logarithmic equation can be solved by solving an equivalent exponential equation.

Solution

$pH = -\log[H^+]$ This is the formula for pH.

$8.5 = -\log[H^+]$ Substitute 8.5 for pH.

$-8.5 = \log[H^+]$ Multiply both sides by -1.

$[H^+] = 10^{-8.5}$ Write the equation in the equivalent exponential form.

© Garry Gay/Alamy

We can use a calculator to find that

$$[H^+] \approx 3.2 \times 10^{-9} \text{ gram-ions per liter}$$

 Now Try **Problem 101**

STUDY SET
11.7

VOCABULARY

Fill in the blanks.

1. The expression $\log_3 4x$ is the logarithm of a _____.

2. The expression $\log_2 \frac{5}{x}$ is the logarithm of a _____.

3. The expression $\log 4^x$ is the logarithm of a _____.

4. In the expression $\log_5 4$, the number 5 is the _____ of the logarithm.

CONCEPTS

Fill in the blanks.

5. $\log_b 1 =$

6. $\log_b b =$

7. $\log_b MN = \log_b \quad + \log_b$

8. $b^{\log_b x} =$

9. $\log_b \frac{M}{N} = \log_b M \quad \log_b N$

10. $\log_b M^p = p \log_b$

11. $\log_b b^x =$

12. $\log_b (A + B) \quad \log_b A + \log_b B$

13. $\log_b \frac{M}{N} \quad \frac{\log_b M}{\log_b N}$

14. $\log_b AB \quad \log_b A + \log_b B$

15. $\log_b x = \frac{\log_a x}{}$

16. pH =

NOTATION

Complete each solution.

17. $\log_b rst = \log_b (\quad)t$
$= \log_b (rs) + \log_b$
$= \log_b \quad + \log_b \quad + \log_b t$

18. $\log \frac{r}{st} = \log r - \log (\quad)$
$= \log r - (\log \quad + \log t)$
$= \log r - \log s$

GUIDED PRACTICE

In this Study Set, assume that all variables represent positive numbers and $b \neq 1$.

Evaluate each expression. See Example 1.

19. $\log_6 1$ **20.** $\log_9 9$

21. $\log_4 4^7$ **22.** $\ln e^8$

23. $5^{\log_5 10}$ **24.** $8^{\log_8 10}$

25. $\log_5 5^2$ **26.** $\log_4 4^2$

27. $\ln e$ **28.** $\log_7 1$

29. $\log_3 3^7$ **30.** $5^{\log_5 8}$

Write each logarithm as a sum. Then simplify, if possible. See Example 2.

31. $\log_2 (4 \cdot 5)$ **32.** $\log_3 (27 \cdot 5)$

33. $\log 25y$ **34.** $\log xy$

35. $\log 100pq$

36. $\log 1,000rs$

37. $\log 5xyz$

38. $\log 10abc$

Write each logarithm as a difference. Then simplify, if possible.
See Example 3.

39. $\log \dfrac{100}{9}$

40. $\ln \dfrac{27}{e}$

41. $\log_6 \dfrac{x}{36}$

42. $\log_8 \dfrac{y}{8}$

Write each logarithm as the sum and/or difference of logarithms of a single quantity. Then simplify, if possible. See Example 4.

43. $\log \dfrac{7c}{2}$

44. $\log \dfrac{9t}{4}$

45. $\log \dfrac{10x}{y}$

46. $\log_2 \dfrac{ab}{4}$

47. $\ln \dfrac{exy}{z}$

48. $\ln \dfrac{5p}{e}$

49. $\log_8 \dfrac{1}{8m}$

50. $\log_6 \dfrac{1}{36r}$

Write each logarithm without an exponent or a radical symbol. Then simplify, if possible. See Example 5.

51. $\ln y^7$

52. $\ln z^9$

53. $\log \sqrt{5}$

54. $\log \sqrt[3]{7}$

55. $\log e^{-3}$

56. $\log e^{-1}$

57. $\log_7 \left(\sqrt[5]{100}\right)^3$

58. $\log_2 \left(\sqrt{10}\right)^5$

Write each logarithm as the sum and/or difference of logarithms of a single quantity. Then simplify, if possible. See Example 6.

59. $\log xyz^2$

60. $\log 4xz^2$

61. $\log_2 \dfrac{2\sqrt[3]{x}}{y}$

62. $\log_3 \dfrac{\sqrt[4]{x}}{yz}$

63. $\log x^3 y^2$

64. $\log xy^2 z^3$

65. $\log_b \sqrt{xy}$

66. $\log_b x^3 \sqrt{y}$

67. $\log_a \dfrac{\sqrt[3]{x}}{\sqrt[4]{yz}}$

68. $\log_b \sqrt[4]{\dfrac{x^3 y^2}{z^4}}$

69. $\ln x\sqrt{z}$

70. $\ln \sqrt{xy}$

Write each logarithmic expression as one logarithm. See Example 7.

71. $\log_2 (x+1) + 9\log_2 x$

72. $2\log x + \dfrac{1}{2}\log y$

73. $\log_3 x + \log_3 (x+2) - \log_3 8$

74. $-2\log x - 3\log y + \log z$

75. $-3\log_b x - 2\log_b y + \dfrac{1}{2}\log_b z$

76. $3\log_b (x+1) - 2\log_b (x+2) + \log_b x$

77. $\ln \left(\dfrac{x}{z}+x\right) - \ln \left(\dfrac{y}{z}+y\right)$

78. $\ln (xy+y^2) - \ln (xz+yz) + \ln z$

Assume that $\log_b 4 = 0.6021$, $\log_b 7 = 0.8451$, and $\log_b 9 = 0.9542$. Use these values to evaluate each logarithm. See Example 8.

79. $\log_b 28$

80. $\log_b \dfrac{7}{4}$

81. $\log_b \dfrac{4}{63}$

82. $\log_b 36$

83. $\log_b \dfrac{63}{4}$

84. $\log_b 2.25$

85. $\log_b 64$

86. $\log_b 49$

Use the change-of-base formula to find each logarithm to four decimal places. See Example 9.

87. $\log_3 7$

88. $\log_7 3$

89. $\log_{1/3} 3$

90. $\log_{1/2} 6$

91. $\log_3 8$

92. $\log_5 10$

93. $\log_{\sqrt{2}} \sqrt{5}$

94. $\log_\pi e$

Use a calculator to verify that each equation is true. See Using Your Calculator: Verifying Properties of Logarithms.

95. $\log (2.5 \cdot 3.7) = \log 2.5 + \log 3.7$

96. $\ln (2.25)^4 = 4\ln 2.25$

97. $\ln \dfrac{11.3}{6.1} = \ln 11.3 - \ln 6.1$

98. $\log \sqrt{24.3} = \dfrac{1}{2}\log 24.3$

APPLICATIONS

99. pH OF A SOLUTION Find the pH of a solution with a hydrogen ion concentration of 1.7×10^{-5} gram-ions per liter.

100. pH OF PICKLES The hydrogen ion concentration of sour pickles is 6.31×10^{-4}. Find the pH.

101. HYDROGEN ION CONCENTRATION Find the hydrogen ion concentration of a saturated solution of calcium hydroxide whose pH is 13.2.

102. AQUARIUMS To test for safe pH levels in a freshwater aquarium, a test strip is compared with the scale shown below. Find the corresponding range in the hydrogen ion concentration.

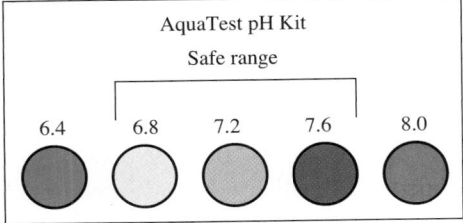

AquaTest pH Kit

Safe range

6.4   6.8   7.2   7.6   8.0

WRITING

103. Explain the difference between a logarithm of a product and the product of logarithms.

104. How can the LOG key on a calculator be used to find $\log_2 7$?

Explain why each statement is false.

105. $\log xy = (\log x)(\log y)$

106. $\log ab = \log a + 1$

107. $\log_b (A - B) = \dfrac{\log_b A}{\log_b B}$

108. $\dfrac{\log_b A}{\log_b B} = \log_b A - \log_b B$

REVIEW

Consider the line that passes through $P(-2, 3)$ and $Q(4, -4)$.

109. Find the slope of line PQ.

110. Find the distance between P and Q.

111. Find the midpoint of line segment PQ.

112. Write the equation in slope–intercept form of line PQ.

CHALLENGE PROBLEMS

113. Explain why $e^{\ln x} = x$.

114. If $\log_b 3x = 1 + \log_b x$, find b.

115. Show that $\log_{b^2} x = \dfrac{1}{2}\log_b x$.

116. Show that $e^{x \ln a} = a^x$.

SECTION 11.8
Exponential and Logarithmic Equations

Objectives

1. Solve exponential equations.
2. Solve logarithmic equations.
3. Solve radioactive decay problems.
4. Solve population growth problems.

An **exponential equation** contains a variable in one of its exponents. Some examples of exponential equations are

$$3^{x+1} = 81, \qquad 6^{x-3} = 2^x, \qquad \text{and} \qquad e^{0.9t} = 8$$

A **logarithmic equation** is an equation with a logarithmic expression that contains a variable. Some examples of logarithmic equations are

$$\log 5x = 3, \qquad \log (3x + 2) = \log (2x - 3), \qquad \text{and} \qquad \log_2 7 - \log_2 x = 5$$

In this section, we will learn how to solve exponential and logarithmic equations.

1. Solve Exponential Equations.

If both sides of an exponential equation can be expressed as a power of the same base, we can use the following property to solve it.

Exponent Property of Equality	If two exponential expressions with the same base are equal, their exponents are equal. For any real number b, where $b \neq -1, 0,$ or 1, $$b^x = b^y \quad \text{is equivalent to} \quad x = y$$

EXAMPLE 1   Solve: $3^{x+1} = 81$

Strategy   We will express the right side of the equation as a power of 3.

Why   We can then use the exponent property of equality to set the exponents equal and solve for x.

Solution

$3^{x+1} = 81$ This is the equation to solve.

$3^{x+1} = 3^4$ Write 81 as a power of 3: $81 = 3^4$.

$x + 1 = 4$ If two exponential expressions with the same base are equal, their exponents are equal.

$x = 3$

The solution is 3 and the solution set is $\{3\}$. To check this result, we substitute 3 for x in the original equation.

Check: $3^{x+1} = 81$

$3^{3+1} \stackrel{?}{=} 81$

$3^4 \stackrel{?}{=} 81$

$81 = 81$ True

    Self Check 1   Solve: $5^{3x-4} = 25$

Now Try   **Problem 21**

EXAMPLE 2   Solve: $2^{x^2+2x} = \frac{1}{2}$

Strategy   We will express the right side of the equation as a power of 2.

Why   We can then use the exponent property of equality to set the exponents equal and solve for x.

Solution

$2^{x^2+2x} = \frac{1}{2}$ This is the equation to solve.

$2^{x^2+2x} = 2^{-1}$ Write $\frac{1}{2}$ as a power of 2: $\frac{1}{2} = 2^{-1}$.

$x^2 + 2x = -1$ If two exponential expressions with the same base are equal, their exponents are equal.

$x^2 + 2x + 1 = 0$ Add 1 to both sides.

$(x + 1)(x + 1) = 0$ Factor the trinomial.

$x + 1 = 0$ or $x + 1 = 0$ Set each factor equal to 0.

$x = -1$ | $x = -1$

We see that the two solutions are the same. Thus, -1 is a repeated solution and the solution set is $\{-1\}$. Verify that -1 satisfies the original equation.

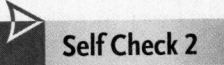

Self Check 2   Solve: $3^{x^2-2x} = \frac{1}{3}$

Now Try   **Problem 25**

Using Your Calculator

Solving Exponential Equations Graphically

To use a graphing calculator to approximate the solutions of $2^{x^2+2x} = \frac{1}{2}$ (see Example 2), we can subtract $\frac{1}{2}$ from both sides of the equation to get

$$2^{x^2+2x} - \frac{1}{2} = 0$$

and graph the corresponding function

$$f(x) = 2^{x^2+2x} - \frac{1}{2}$$

as shown in figure (a).

   The solutions of $2^{x^2+2x} - \frac{1}{2} = 0$ are the x-coordinates of the x-intercepts of the graph of $f(x) = 2^{x^2+2x} - \frac{1}{2}$. Using the ZERO feature, we see in figure (a) that the graph has only one x-intercept, $(-1, 0)$. Therefore, -1 is the only solution of $2^{x^2+2x} - \frac{1}{2} = 0$.

   We can also solve $2^{x^2+2x} = \frac{1}{2}$ using the INTERSECT feature found on most graphing calculators. After graphing $Y_1 = 2^{x^2+2x}$ and $Y_2 = \frac{1}{2}$, we select INTERSECT, which approximates the coordinates of the point of intersection of the two graphs. From the display shown in figure (b), we can conclude that the solution is -1. Verify this by checking.

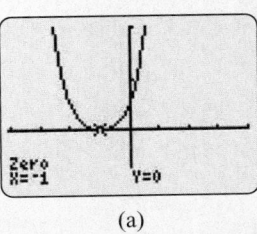   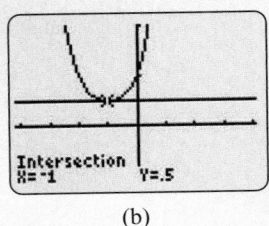

(a)   (b)

When it is difficult or impossible to write each side of an exponential equation as a power of the same base, we can often use the following property of logarithms to solve the equation.

Logarithm Property of Equality

If two positive numbers are equal, the logarithms base-b of the numbers are equal. For any positive number b, where $b \neq 1$, and positive numbers x and y,

$$\log_b x = \log_b y \qquad \text{is equivalent to} \qquad x = y$$

EXAMPLE 3   Solve: $3^x = 5$

Strategy   We will take the base-10 logarithm of both sides of the equation.

Why   We can then use the logarithm property of equality to move the variable x from its current position as an exponent to a position as a factor.

Solution   Unlike Example 1, where we solved $3^{x+1} = 81$, it is not possible to write each side of $3^x = 5$ as a power of the same base 3. Instead, we use the logarithm property of equality and *take the logarithm of each side* to solve the equation. Although any base logarithm can be chosen, the computations with a calculator are usually simplest if we use a common or natural logarithm.

$$3^x = 5 \qquad \text{This is the equation to solve.}$$

$$\log 3^x = \log 5 \qquad \text{Take the common logarithm of each side.}$$

$$x \log 3 = \log 5 \qquad \text{The log of a power is the power times the log: } \log 3^x = x \log 3.$$
$$\text{Note that the variable } x \text{ is now a factor of } x \log 3.$$

$$x = \frac{\log 5}{\log 3} \qquad \text{To isolate } x, \text{ divide both sides by } \log 3. \text{ This is the exact solution.}$$

$$x \approx 1.464973521 \qquad \text{Use a calculator.}$$

The exact solution is $\dfrac{\log 5}{\log 3}$. To four decimal places, the solution is 1.4650.

We can also take the natural logarithm of each side of the equation to solve for x.

$$3^x = 5$$

$$\ln 3^x = \ln 5 \qquad \text{Take the natural logarithm of each side.}$$

$$x \ln 3 = \ln 5 \qquad \text{Use the power rule of logarithms: } \ln 3^x = x \ln 3.$$

$$x = \frac{\ln 5}{\ln 3} \qquad \text{To isolate } x, \text{ divide both sides by } \ln 3.$$

$$x \approx 1.464973521 \qquad \text{Use a calculator.}$$

The result is the same using the natural logarithm. To check the approximate solution, we substitute 1.4650 for x in 3^x and see if $3^{1.4650}$ is close to 5.

> ***Check:*** $3^x = 5$
> $$3^{1.4650} \stackrel{?}{=} 5$$
> $$5.000145454 \approx 5 \qquad \text{Use a calculator: Enter 3 } \boxed{y^x} \text{ 1.4650 } \boxed{=} .$$

Self Check 3   Solve $5^x = 4$ and give the answer to four decimal places.

Now Try   **Problem 29**

EXAMPLE 4   Solve: $6^{x-3} = 2^x$

Strategy   We will take the common logarithm of both sides of the equation.

Why   We can then move the expression $x - 3$ from its current position as an exponent to a position as a factor.

Solution

$$6^{x-3} = 2^x \qquad \text{This is the equation to solve.}$$

$$\log 6^{x-3} = \log 2^x \qquad \text{Take the common logarithm of each side.}$$

$$(x - 3)\log 6 = x \log 2 \qquad \text{The log of a power is the power times the log. Note that the expression } x - 3 \text{ is now a factor of } (x - 3)\log 6.$$

$$x \log 6 - 3 \log 6 = x \log 2 \qquad \text{Distribute the multiplication by } \log 6.$$

$$x \log 6 - x \log 2 = 3 \log 6 \qquad \text{On both sides, add } 3 \log 6 \text{ and subtract } x \log 2.$$

$$x(\log 6 - \log 2) = 3 \log 6 \qquad \text{Factor out } x \text{ on the left side.}$$

$$x = \frac{3 \log 6}{\log 6 - \log 2} \qquad \text{To isolate } x, \text{ divide both sides by } \log 6 - \log 2.$$

$$x \approx 4.892789261 \qquad \text{Use a calculator.}$$

> **The Language of Algebra**
> $\frac{3 \log 6}{\log 6 - \log 2}$ is the *exact* solution
> of $6^{x-3} = 2^x$. An *approximate*
> solution is 4.8928.

To four decimal places, the solution is 4.8928. To check the approximate solution, we substitute 4.8928 for each x in $6^{x-3} = 2^x$. The resulting values on the left and right sides of the equation should be approximately equal.

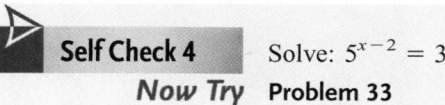

Self Check 4   Solve: $5^{x-2} = 3^x$

Now Try **Problem 33**

EXAMPLE 5   Solve: $e^{0.9t} = 10$

Strategy   We will take the natural logarithm of both sides of the equation.

Why   We can then move the expression $0.9t$ from its current position as an exponent to a position as a factor.

Solution   The exponential expression on the left side has base e. In such cases, the computations are easier when we take the natural logarithm of each side.

$$e^{0.9t} = 10 \qquad \text{This is the equation to solve.}$$

$$\ln e^{0.9t} = \ln 10 \qquad \text{Take the natural logarithm of each side.}$$

$$0.9t \ln e = \ln 10 \qquad \text{Use the power rule of logarithms: } \ln e^{0.9t} = 0.9t \ln e. \text{ Note that the expression } 9t \text{ is now a factor of } 0.9t \ln e.$$

$$0.9t \cdot 1 = \ln 10 \qquad \text{Simplify: } \ln e = 1.$$

$$0.9t = \ln 10$$

$$t = \frac{\ln 10}{0.9} \qquad \text{To isolate } t, \text{ divide both sides by } 0.9.$$

$$t \approx 2.558427881 \qquad \text{Use a calculator.}$$

To four decimal places, the solution is 2.5584. Verify this using a check.

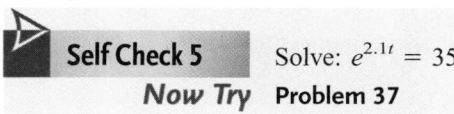

Self Check 5   Solve: $e^{2.1t} = 35$

Now Try **Problem 37**

Solve Logarithmic Equations.

A **logarithmic equation** is an equation containing a variable in a logarithmic expression. We can solve many logarithmic equations using properties of logarithms.

EXAMPLE 6 Solve: $\log 5x = 3$

Strategy Recall that $\log 5x = \log_{10} 5x$. To solve $\log 5x = 3$, we will instead write and solve an equivalent base-10 exponential equation.

Why The resulting exponential equation is easier to solve because the variable term is isolated on one side.

Solution

$\log 5x = 3$ This is the equation to solve.

$10^3 = 5x$ Write the equivalent base-10 exponential equation.

$1{,}000 = 5x$ Simplify: $10^3 = 1{,}000$.

$200 = x$ To isolate x, divide both sides by 5.

The solution is 200 and the solution set is $\{200\}$.

> **Caution**
> Always check your solutions to a logarithmic equation.

Check:
$\log 5x = 3$
$\log 5(200) \overset{?}{=} 3$ Substitute 200 for x.
$\log 1{,}000 \overset{?}{=} 3$ Multiply 5(200) = 1,000.
$3 = 3$ Evaluate: $\log 1{,}000 = \log 10^3 = 3$.

 Self Check 6 Solve: $\log_2 (x - 3) = -1$

Now Try Problem 41

EXAMPLE 7 Solve: $\log(3x + 2) = \log(2x - 3)$

Strategy We will use the logarithmic property of equality to see that $3x + 2 = 2x - 3$.

Why We can use the logarithm property of equality because the given equation, $\log(3x + 2) = \log(2x - 3)$, has the form $\log_b x = \log_b y$.

Solution

$\log(3x + 2) = \log(2x - 3)$ This is the equation to solve.

$3x + 2 = 2x - 3$ If the logarithms of two numbers are equal, the numbers are equal.

$x + 2 = -3$ Subtract 2x from both sides.

$x = -5$ Subtract 2 from both sides.

> **Caution**
> Don't make this error of trying to "distribute" log:
> $\log (3x + 2)$
> The notation **log** is not a number, it is the name of a function and cannot be distributed.

Check:
$\log(3x + 2) = \log(2x - 3)$
$\log[3(-5) + 2] \overset{?}{=} \log[2(-5) - 3]$ Substitute −5 for x.
$\log(-13) \overset{?}{=} \log(-13)$ Evaluate within brackets. Recall that $\log(-13)$ is undefined.

Since the logarithm of a negative number does not exist, the proposed solution of −5 must be discarded. This equation has no solutions. Its solution set is $\varnothing$.

Self Check 7   Solve: $\log (5x + 14) = \log (7x - 2)$

Now Try   **Problem 49**

EXAMPLE 8   Solve: $\log x + \log (x - 3) = 1$

Strategy   We will use the product rule for logarithms in reverse: The sum of two logarithms is equal to the logarithm of a product. Then we will write and solve an equivalent exponential equation.

Why   We use the product rule for logarithms because the left side of the equation, $\log x + \log (x - 3)$, has the form $\log_b M + \log_b N$.

Solution

$\log x + \log (x - 3) = 1$	This is the equation to solve.
$\log x(x - 3) = 1$	On the left side, use the product rule for logarithms.
$\log_{10} x(x - 3) = 1$	The base of the logarithm is 10.
$x(x - 3) = 10^1$	Write an equivalent base-10 exponential equation.
$x^2 - 3x - 10 = 0$	Distribute the multiplication by x, and then subtract 10 from both sides.
$(x + 2)(x - 5) = 0$	Factor the trinomial.
$x + 2 = 0$ or $x - 5 = 0$	Set each factor equal to 0.
$x = -2$ | $x = 5$	

Check: The number -2 is not a solution because it does not satisfy the equation (a negative number does not have a logarithm). We will check the other result, 5.

$\log x + \log (x - 3) = 1$	
$\log 5 + \log (5 - 3) \overset{?}{=} 1$	Substitute 5 for x.
$\log 5 + \log 2 \overset{?}{=} 1$	
$\log 10 \overset{?}{=} 1$	Use the product rule of logarithms: $\log 5 + \log 2 = \log (5 \cdot 2) = \log 10$.
$1 = 1$	Evaluate: $\log 10 = 1$.

Since 5 satisfies the equation, it is the solution.

> **Caution**
> The proposed solutions of a logarithmic equation must be checked to see whether they produce undefined logarithms in the original equation.

Self Check 8   Solve: $\log x + \log (x + 3) = 1$

Now Try   **Problem 53**

Using Your Calculator   *Solving Logarithmic Equations Graphically*

To use a graphing calculator to approximate the solutions of the logarithmic equation $\log x + \log (x - 3) = 1$ (see Example 8), we can subtract 1 from both sides of the equation to get

$$\log x + \log (x - 3) - 1 = 0$$

and graph the corresponding function

$$f(x) = \log x + \log (x - 3) - 1$$

as shown in figure (a). Since the solution of the equation is the x-value that makes $f(x) = 0$, the solution is the x-coordinate of the x-intercept of the graph. We can use the ZERO feature to find that this x-value is 5.

We can also solve $\log x + \log (x - 3) = 1$ using the INTERSECT feature. After graphing $Y_1 = \log x + \log (x - 3)$ and $Y_2 = 1$, we select INTERSECT, which approximates the coordinates of the point of intersection of the two graphs. From the display shown in figure (b), we can conclude that the solution is 5.

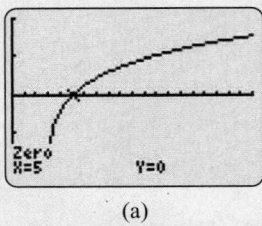 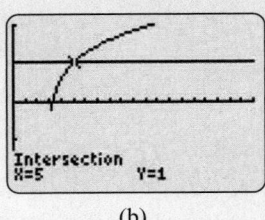

(a)   (b)

EXAMPLE 9   Solve: $\log_2 7 - \log_2 x = 5$

Strategy   We will use the quotient rule for logarithms in reverse: The difference of two logarithms is equal to the logarithm of a quotient. Then we will write and solve an equivalent exponential equation.

Why   We use the quotient rule for logarithms because the left side of the equation, $\log_2 7 - \log_2 x$, has the form $\log_b M - \log_b N$.

Solution

$$\log_2 7 - \log_2 x = 5 \qquad \text{This is the equation to solve.}$$

$$\log_2 \frac{7}{x} = 5 \qquad \text{On the left side, use the quotient rule for logarithms.}$$

$$\frac{7}{x} = 2^5 \qquad \text{Write an equivalent base-2 exponential equation.}$$

$$\frac{7}{x} = 32 \qquad \text{Evaluate: } 2^5 = 32.$$

$$7 = 32x \qquad \text{To clear the equation of the fraction, multiply both sides by x.}$$

$$\frac{7}{32} = x \qquad \text{To isolate x, divide both sides by 32.}$$

The solution is $\frac{7}{32}$. Verify that it satisfies the original equation.

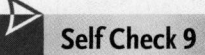 **Self Check 9**   Solve: $\log_2 9 - \log_2 x = 4$

Now Try   **Problem 57**

3   Solve Radioactive Decay Problems.

Experiments have determined the time it takes for half of a sample of a radioactive material to decompose. This time is a constant, called the material's **half-life.**

When living organisms die, the oxygen–carbon dioxide cycle common to all living things ceases, and carbon-14, a radioactive isotope with a half-life of 5,700 years, is no longer absorbed. By measuring the amount of carbon-14 present in an ancient object, archaeologists can estimate the object's age by using the radioactive decay formula.

Radioactive Decay Formula	If A is the amount of radioactive material present at time t, A_0 was the amount present at $t = 0$, and h is the material's half-life, then $$A = A_0 2^{-t/h}$$

© Images&Stories/Alamy

EXAMPLE 10   *Carbon-14 Dating.*   How old is a piece of wood that retains only one-third of its original carbon-14 content?

Strategy   If A_0 is the original carbon-14 content, then today's content $A = \frac{1}{3}A_0$. We will substitute $\frac{A_0}{3}$ for A and 5,700 for h in the radioactive decay formula and solve for t.

Why   The value of t is the estimated age of the piece of wood.

Solution   To find the time t when $A = \frac{1}{3}A_0$, we substitute $\frac{A_0}{3}$ for A and 5,700 for h in the radioactive decay formula and solve for t:

$A = A_0 2^{-t/h}$	This is the radioactive decay model.
$\dfrac{A_0}{3} = A_0 2^{-t/5,700}$	The half-life of carbon-14 is 5,700 years.
$1 = 3(2^{-t/5,700})$	Divide both sides by A_0 and multiply both sides by 3.
$\log 1 = \log 3(2^{-t/5,700})$	Take the common logarithm of both sides.
$0 = \log 3 + \log 2^{-t/5,700}$	$\log 1 = 0$, and use the product rule for logarithms.
$-\log 3 = -\dfrac{t}{5,700} \log 2$	Subtract $\log 3$ from both sides and use the power rule of logarithms.
$5,700\left(\dfrac{\log 3}{\log 2}\right) = t$	Multiply both sides by $-\dfrac{5,700}{\log 2}$.
$t \approx 9,034.286254$	Use a calculator.

> **Notation**
> The initial amount of radioactive material is represented by A_0, and it is read as "A sub 0."

The piece of wood is approximately 9,000 years old.

   Self Check 10   How old is a piece of wood that retains 25% of its original carbon-14 content?

Now Try   **Problem 97**

   ## 4   Solve Population Growth Problems.

When there is sufficient food and space available, populations of living organisms tend to increase exponentially according to the following growth model.

| **Exponetial Growth Model** | If P is the population at some time t, P_0 is the initial population at $t = 0$, and k depends on the rate of growth, then
$$P = P_0 e^{kt}$$ |

EXAMPLE 11   *Population Growth.*   The bacteria in a laboratory culture increased from an initial population of 500 to 1,500 in 3 hours. How long will it take for the population to reach 10,000?

Strategy   We will substitute 500 for P_0, 1,500 for P, and 3 for t into the exponential growth model and solve for k:

Why   Once we know the value of k, we can substitute 10,000 for P, 500 for P_0, and the value of k into the exponential growth model and solve for the time t.

Solution

$P = P_0 e^{kt}$	This is the population growth formula.
$1,500 = 500(e^{k3})$	Substitute 1,500 for P, 500 for P_0, and 3 for t.
$3 = e^{3k}$	Divide both sides by 500.
$3k = \ln 3$	Write the equivalent base-e logarithmic equation.
$k = \dfrac{\ln 3}{3}$	Divide both sides by 3.

> **Notation**
> The initial population of bacteria is represented by P_0, and it is read as "P sub 0."

To find when the population will reach 10,000, we substitute 10,000 for P, 500 for P_0, and $\frac{\ln 3}{3}$ for k in the equation $P = P_0 e^{kt}$ and solve for t:

$P = P_0 e^{kt}$	
$10,000 = 500 e^{[(\ln 3)/3]t}$	
$20 = e^{[(\ln 3)/3]t}$	Divide both sides by 500.
$\left(\dfrac{\ln 3}{3}\right)t = \ln 20$	Write the equivalent base-e logarithmic equation.
$t = \dfrac{3 \ln 20}{\ln 3}$	Multiply both sides by $\frac{3}{\ln 3}$.
≈ 8.180499084	Use a calculator.

The culture will reach 10,000 bacteria in about 8 hours.

 Self Check 11   How long will it take the population to reach 20,000?

Now Try   Problem 109

VOCABULARY

Fill in the blanks.

1. An equation with a variable in its exponent, such as $3^{2x} = 8$, is called a(n) _____ equation.
2. An equation with a logarithmic expression that contains a variable, such as $\log_5 (2x - 3) = \log_5 (x + 4)$, is a(n) _____ equation.

CONCEPTS

Fill in the blanks.

3. **a.** The exponent property of equality: If two exponential expressions with the same base are equal, their exponents are _____.

    $b^x = b^y$ is equivalent to _____.

   b. The logarithm property of equality: If the logarithms base-b of two numbers are equal, the numbers are _____.

    $\log_b x = \log_b y$ is equivalent to _____.

4. The right side of the exponential equation $5^{x-3} = 125$ can be written as a power of ___.
5. If $6^{4x} = 6^{-2}$, then $4x =$ ___.
6. **a.** Write the equivalent base-10 exponential equation for $\log (x + 1) = 2$.

   b. Write the equivalent base-e exponential equation for $\ln (x + 1) = 2$.

Fill in the blanks.

7. To solve $5^x = 2$, we can take the _____ of both sides of the equation to get $\log 5^x = \log 2$.
8. The power rule for logarithms provides a way of moving the variable x from its position as an _____ to a position as a factor of $x \log 5$.
9. If the power rule for logarithms is used on the left side of the equation $\log 7^x = 12$, the resulting equation is ___ $\log 7 = 12$.
10. If $e^{x+2} = 4$, then $\ln e^{x+2} =$ ___.
11. Perform a check to determine whether -2 is a solution of $5^{2x+3} = \frac{1}{5}$.
12. Perform a check to determine whether 4 is a solution of $\log_5 (x + 1) = 2$.
13. Use a calculator to determine whether 2.5646 is an approximate solution of $2^{2x+1} = 70$.
14. How do we solve $x \ln 3 = \ln 5$ for x?

15. **a.** Find $\frac{\log 8}{\log 5}$. Round to four decimal places.

    b. Find $\frac{2 \ln 12}{\ln 9}$. Round to four decimal places.

16. Does $\frac{\log 7}{\log 3} = \log 7 - \log 3$?
17. Complete each formula.

    a. Radioactive decay: $A =$ ___ .

    b. Population growth: $P =$ ___ .

18. Use the graphs to estimate the solution of each equation.

    a. $2^x = 3^{-x+3}$

    b. $3 \log (x - 1) = 2 \log x$

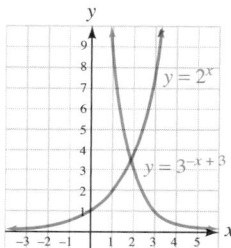
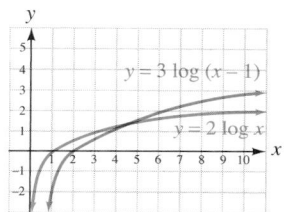

NOTATION

Complete each solution.

19. Solve: $2^x = 7$.

$$2^x = \log 7$$
$$x \quad = \log 7$$
$$x = \frac{\log 7}{\log 2}$$
$$x \approx$$

20. Solve: $\log_2 (2x - 3) = \log_2 (x + 4)$.

$$\quad = x + 4$$
$$x = 7$$

GUIDED PRACTICE

Solve each equation. **See Example 1.**

21. $6^{x-2} = 36$

22. $3^{x+1} = 27$

23. $5^{4x} = \frac{1}{125}$

24. $8^{-2x+1} = \frac{1}{64}$

Solve each equation. See Example 2.

25. $2^{x^2 - 2x} = 8$

26. $3^{x^2 - 3x} = 81$

27. $3^{x^2 + 4x} = \dfrac{1}{81}$

28. $7^{x^2 + 3x} = \dfrac{1}{49}$

Solve each equation. Give answers to four decimal places. See Example 3.

29. $4^x = 5$

30. $7^x = 12$

31. $13^{x-1} = 2$

32. $5^{x+1} = 3$

Solve each equation. Give answers to four decimal places. See Example 4.

33. $2^{x+1} = 3^x$

34. $6^x = 7^{x-4}$

35. $5^{x-3} = 3^{2x}$

36. $8^{3x} = 9^{x+1}$

Solve each equation. Give answers to four decimal places. See Example 5.

37. $e^{2.9x} = 4.5$

38. $e^{3.3t} = 9.1$

39. $e^{-0.2t} = 14.2$

40. $e^{-0.7x} = 6.2$

Solve each equation. See Example 6.

41. $\log 2x = 4$

42. $\log 5x = 4$

43. $\log_3 (x - 3) = 2$

44. $\log_4 (2x - 1) = 3$

45. $\log (7 - x) = 2$

46. $\log (2 - x) = 3$

47. $\log \dfrac{1}{8} x = -2$

48. $\log \dfrac{1}{5} x = -3$

Solve each equation. See Example 7.

49. $\log (3 - 2x) = \log (x + 24)$

50. $\log (3x + 5) = \log (2x + 6)$

51. $\ln (3x + 1) = \ln (x + 7)$

52. $\ln (x^2 + 4x) = \ln (x^2 + 16)$

Solve each equation. See Example 8.

53. $\log x + \log (x - 48) = 2$

54. $\log x + \log (x + 9) = 1$

55. $\log_5 (4x - 1) + \log_5 x = 1$

56. $\log_2 (x - 7) + \log_2 x = 3$

Solve each equation. See Example 9.

57. $\log 5 - \log x = 1$

58. $\log 11 - \log x = 2$

59. $\log_3 4x - \log_3 7 = 2$

60. $\log_2 5x - \log_2 3 = 4$

TRY IT YOURSELF

Solve each equation. Give approximate solutions to four decimal places.

61. $\log 2x = \log 4$

62. $\log 3x = \log 9$

63. $\ln x = 1$

64. $\ln x = 5$

65. $7^{x^2} = 10$

66. $8^{x^2} = 11$

67. $\log (x + 90) + \log x = 3$

68. $\log (x - 90) + \log x = 3$

69. $3^{x-6} = 81$

70. $5^{x+4} = 125$

71. $\log \dfrac{4x + 1}{2x + 9} = 0$

72. $\log \dfrac{2 - 5x}{2(x + 8)} = 0$

73. $15 = 9^{x+1}$

74. $29 = 5^{x-6}$

75. $\log x^2 = 2$

76. $\log x^3 = 3$

77. $\log (x - 6) - \log (x - 2) = \log \dfrac{5}{x}$

78. $\log (3 - 2x) - \log (x + 9) = 0$

79. $\log_3 x = \log_3 \left(\dfrac{1}{x} \right) + 4$

80. $\log_5 (7 + x) + \log_5 (8 - x) - \log_5 2 = 2$

81. $2 \log_2 x = 3 + \log_2 (x - 2)$

82. $2 \log_3 x - \log_3 (x - 4) = 2 + \log_3 2$

83. $\log (7y + 1) = 2 \log (y + 3) - \log 2$

84. $2 \log (y + 2) = \log (y + 2) - \log 12$

85. $e^{3x} = 9$

86. $e^{4x} = 60$

87. $\dfrac{\log (5x + 6)}{2} = \log x$

88. $\dfrac{1}{2} \log (4x + 5) = \log x$

Use a graphing calculator to solve each equation. If an answer is not exact, round to the nearest tenth. See Using Your Calculator: Solving Exponential Equations Graphically or Solving Logarithmic Equations Graphically.

89. $2^{x+1} = 7$

90. $3^{x-1} = 2^x$

91. $3^x - 10 = 3^{-x}$

92. $2^x - 8 = 5 + 2^{-x}$

93. $\log x + \log (x - 15) = 2$

94. $\log x + \log (x + 3) = 1$

95. $\ln (2x + 5) - \ln 3 = \ln (x - 1)$

96. $2 \log (x^2 + 4x) = 1$

APPLICATIONS

97. TRITIUM DECAY   The half-life of tritium is 12.4 years. How long will it take for 25% of a sample of tritium to decompose?

98. RADIOACTIVE DECAY   In 2 years, 20% of a radioactive element decays. Find its half-life.

99. THORIUM DECAY   An isotope of thorium, written as ^{227}Th, has a half-life of 18.4 days. How long will it take for 80% of the sample to decompose?

100. LEAD DECAY   An isotope of lead, written as ^{201}Pb, has a half-life of 8.4 hours. How many hours ago was there 30% more of the substance?

101. CARBON-14 DATING   A bone fragment analyzed by archaeologists contains 60% of the carbon-14 that it is assumed to have had initially. How old is it?

102. CARBON-14 DATING   Only 10% of the carbon-14 in a small wooden bowl remains. How old is the bowl?

103. COMPOUND INTEREST   If $500 is deposited in an account paying 8.5% annual interest, compounded semiannually, how long will it take for the account to increase to $800?

104. CONTINUOUS COMPOUND INTEREST   In Exercise 103, how long will it take if the interest is compounded continuously?

105. COMPOUND INTEREST   If $1,300 is deposited in a savings account paying 9% interest, compounded quarterly, how long will it take the account to increase to $2,100?

106. COMPOUND INTEREST   A sum of $5,000 deposited in an account grows to $7,000 in 5 years. Assuming annual compounding, what interest rate is being paid?

107. RULE OF SEVENTY   A rule of thumb for finding how long it takes an investment to double is called the **rule of seventy.** To apply the rule, divide 70 by the interest rate written as a percent. At 5%, an investment takes $\frac{70}{5} = 14$ years to double. At 7%, it takes $\frac{70}{7} = 10$ years. Explain why this formula works.

108. BACTERIAL GROWTH   A bacterial culture grows according to the function

$$P(t) = P_0 a^t$$

If it takes 5 days for the culture to triple in size, how long will it take to double in size?

109. RODENT CONTROL   The rodent population in a city is currently estimated at 30,000. If it is expected to double every 5 years, when will the population reach 1 million?

110. POPULATION GROWTH   The population of a city is expected to triple every 15 years. When can the city planners expect the present population of 140 persons to double?

111. BACTERIA CULTURE   A bacteria culture doubles in size every 24 hours. By how much will it have increased in 36 hours?

112. OCEANOGRAPHY   The intensity I of a light a distance x meters beneath the surface of a lake decreases exponentially. Use the illustration to find the depth at which the intensity will be 20%.

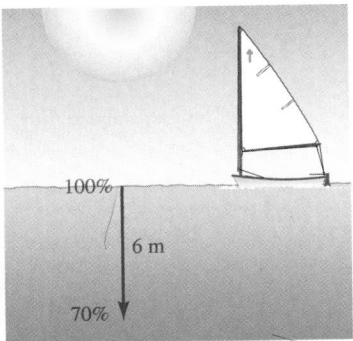

113. NEWTON'S LAW OF COOLING   Water initially at 100°C is left to cool in a room at temperature 60°C. After 3 minutes, the water temperature is 90°. The water temperature T is a function of time t given by the following formula. Find k.

$$T = 60 + 40e^{kt}$$

114. NEWTON'S LAW OF COOLING   Refer to Exercise 113 and find the time for the water temperature to reach 70°C.

WRITING

115. Explain how to solve the equation $2^{x+1} = 31$.

116. Explain how to solve the equation $2^{x+1} = 32$.

117. Write a justification for each step of the solution.

$$15^x = 9 \qquad \text{This is the equation to solve.}$$
$$\log 15^x = \log 9 \qquad \underline{\hspace{3cm}}.$$
$$x \log 15 = \log 9 \qquad \underline{\hspace{3cm}}.$$
$$x = \frac{\log 9}{\log 15} \qquad \underline{\hspace{3cm}}.$$

118. What is meant by the term *half-life*?

REVIEW

119. Find the length of leg AC.

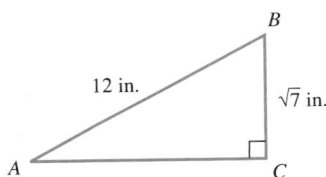

120. The amount of medicine a patient should take is often proportional to his or her weight. If a patient weighing 83 kilograms needs 150 milligrams of medicine, how much will be needed by a person weighing 99.6 kilograms?

CHALLENGE PROBLEMS

121. Without solving the following equation, find the values of x that cannot be a solution:

$$\log (x - 3) - \log (x^2 + 2) = 0$$

122. Solve: $x^{\log x} = 10,000$

123. Solve: $\dfrac{\log_2 (6x - 8)}{\log_2 x} = 2$

124. Solve: $\dfrac{\log (3x - 4)}{\log x} = 2$

CHAPTER 11
Summary & Review

SECTION 11.1 Algebra and Composition of Functions

DEFINITIONS AND CONCEPTS	EXAMPLES

DEFINITIONS AND CONCEPTS

Just as it is possible to perform arithmetic operations on real numbers, it is possible to perform those operations on functions.

The **sum, difference, product,** and **quotient functions** are defined as:

$$(f + g)(x) = f(x) + g(x)$$
$$(f - g)(x) = f(x) - g(x)$$
$$(f \cdot g)(x) = f(x)g(x)$$
$$(f/g)(x) = \frac{f(x)}{g(x)}, \quad \text{with } g(x) \neq 0$$

EXAMPLES

Let $f(x) = 2x + 1$ and $g(x) = x^2$.

$(f + g)(x) = f(x) + g(x)$ $\quad$ $(f - g)(x) = f(x) - g(x)$
$\qquad = 2x + 1 + x^2$ $\qquad\qquad = 2x + 1 - x^2$
$\qquad = x^2 + 2x + 1$ $\qquad\qquad = -x^2 + 2x + 1$

$(f \cdot g)(x) = f(x) \cdot g(x)$ $\quad$ $(f/g)(x) = \dfrac{f(x)}{g(x)}$

$\qquad = (2x + 1)x^2$ $\qquad\qquad = \dfrac{2x + 1}{x^2}$

$\qquad = 2x^3 + x^2$

DEFINITIONS AND CONCEPTS

Often one quantity is a function of a second quantity that depends, in turn, on a third quantity. Such chains of dependence can be modeled by a **composition of functions.**

Composition of functions:

$$(f \circ g)(x) = f(g(x))$$

EXAMPLES

Let $f(x) = 4x - 9$ and $g(x) = x^3$. Find $(f \circ g)(2)$ and $(f \circ g)(x)$.

$(f \circ g)(2) = f(g(2))$ $\quad$ *Change to nested parentheses notation.*
$\qquad = f(8)$ $\qquad$ *Evaluate: $g(2) = 2^3 = 8$.*
$\qquad = 4(8) - 9$ $\qquad$ *Evaluate $f(8)$ using $f(x) = 4x - 9$.*
$\qquad = 23$

$(f \circ g)(x) = f(g(x)) = f(x^3) = 4x^3 - 9$

REVIEW EXERCISES

Let $f(x) = 2x$ and $g(x) = x + 1$. Find each function and its domain.

1. $f + g$

2. $f - g$

3. $f \cdot g$

4. f/g

Let $f(x) = x^2 + 2$ and $g(x) = 2x + 1$. Find each of the following.

5. $(f \circ g)(-1)$

6. $(g \circ f)(0)$

7. $(f \circ g)(x)$

8. $(g \circ f)(x)$

9. Use the graphs of functions f and g to find each of the following.
   a. $(f + g)(2)$
   b. $(f \cdot g)(-4)$
   c. $(f \circ g)(4)$
   d. $(g \circ f)(6)$

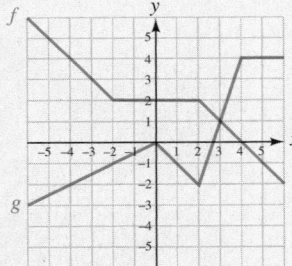

10. MILEAGE COSTS The function $f(m) = \frac{m}{8}$ gives the number of gallons of fuel consumed if a bus travels m miles. The function $C(f) = 3.25f$ gives the cost (in dollars) of f gallons of fuel. Write a composition function that expresses the cost of the fuel consumed as a function of the number of miles driven.

SECTION 11.2 Inverse Functions

DEFINITIONS AND CONCEPTS	EXAMPLES
A function is called a **one-to-one function** if different inputs determine different outputs.	The function $f(x) = 3x - 5$ is a one-to-one function because different inputs have different outputs. Since two different inputs, -2 and 2, have the same output 16, the function $f(x) = x^4$ is not one-to-one.
Horizontal line test: A function is one-to-one if every horizontal line intersects the graph of the function at most once.	The function $f(x) = \lvert x + 1\rvert$ is a not one-to-one function because we can draw a horizontal line that intersects its graph twice.
To find the inverse of a function, replace $f(x)$ with y, interchange the variables x and y, solve for y, and replace y with $f^{-1}(x)$.	To find the inverse of the one-to-one function $f(x) = 2x + 1$, we proceed as follows: $f(x) = 2x + 1$ $y = 2x + 1$ Replace f(x) with y. $x = 2y + 1$ Interchange the variables x and y. $\dfrac{x-1}{2} = y$ Solve for y. $f^{-1}(x) = \dfrac{x-1}{2}$ Replace y with f⁻¹(x).
If a point (a, b) is on the graph of function f, it follows that the point (b, a) is on the graph of f^{-1}, and vice versa. The graph of a function and its inverse are **symmetric about the line $y = x$.**	The graphs of $f(x) = 2x + 1$ and $f^{-1}(x) = \frac{x-1}{2}$ are symmetric about the line $y = x$ as shown in the illustration.
For any one-to-one function f and its inverse, f^{-1}, $(f \circ f^{-1})(x) = x$ and $(f^{-1} \circ f)(x) = x$	The composition of $f(x) = 2x + 1$ and its inverse $f^{-1}(x) = \frac{x-1}{2}$ is the identity function $f(x) = x$. $(f \circ f^{-1})(x) = f(f^{-1}(x)) = f\left(\dfrac{x-1}{2}\right) = 2\left(\dfrac{x-1}{2}\right) + 1 = x - 1 + 1 = x$ $(f^{-1} \circ f)(x) = f^{-1}(f(x)) = f^{-1}(2x + 1) = \dfrac{2x + 1 - 1}{2} = \dfrac{2x}{2} = x$

REVIEW EXERCISES

In Exercises 11–16, determine whether the function is one-to-one.

11. $f(x) = x^2 + 3$

12. $f(x) = \dfrac{1}{3}x - 8$

13. $\{(3, 4), (5, 10), (10, -1), (6, 6)\}$

14.

x	$f(x)$
0	-5
2	10
4	-5
6	15

15.

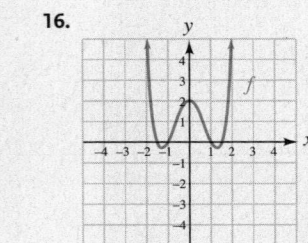

16.

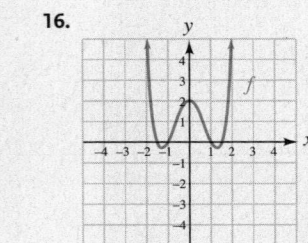

17. Use the table of values of the one-to-one function f to complete a table of values for f^{-1}.

x	$f(x)$
-6	-6
-1	-3
7	12
20	3

x	$f^{-1}(x)$
-6	
-3	
12	
3	

18. Given the graph of function f, graph f^{-1} on the same coordinate axes. Label the axis of symmetry.

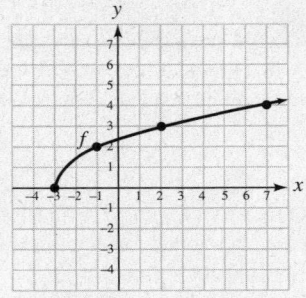

Find the inverse of each function.

19. $f(x) = 6x - 3$

20. $f(x) = \dfrac{4}{x - 1}$

21. $f(x) = (x + 2)^3$

22. $f(x) = \dfrac{x}{6} - \dfrac{1}{6}$

23. Find the inverse of $f(x) = \sqrt[3]{x - 1}$. Then graph the function and its inverse on one coordinate system. Show the axis of symmetry on the graph.

24. Use composition to show that $f(x) = 5 - 4x$ and $f^{-1}(x) = -\dfrac{x - 5}{4}$ are inverse functions.

SECTION 11.3   Exponential Functions

DEFINITIONS AND CONCEPTS	EXAMPLES
An **exponential function** with base b is defined by the equation $\quad f(x) = b^x$, with $b > 0$, $b \neq 1$ Properties of an exponential function $f(x) = b^x$: The **domain** is the interval $(-\infty, \infty)$. The **range** is the interval $(0, \infty)$. Its graph has a **y-intercept** of $(0, 1)$. The x-axis is an **asymptote** of its graph. The graph **passes through** the point $(1, b)$. If $b > 1$, then $f(x) = b^x$ is an **increasing function.** If $0 < b < 1$, then $f(x) = b^x$ is a **decreasing function.**	The graphs of $f(x) = 2^x$ and $g(x) = \left(\dfrac{1}{2}\right)^x$ are shown below. 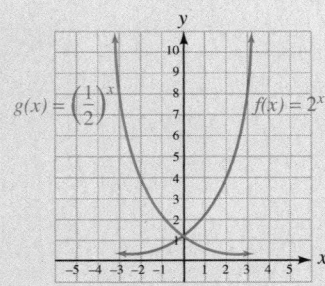 Since the base 2 is greater than 1, the function $f(x) = 2^x$ is an increasing function. Since the base $\dfrac{1}{2}$ is such that $0 < \dfrac{1}{2} < 1$, the function $g(x) = \left(\dfrac{1}{2}\right)^x$ is a decreasing function.

SECTION 11.3 Exponential Functions—*continued*

DEFINITIONS AND CONCEPTS	EXAMPLES

Exponential functions are used to model many situations, such as population **growth,** the spread of an epidemic, the temperature of a heated object as it cools, and radioactive **decay.**

Exponential functions are suitable models for describing **compound interest:**

If P is the deposit, and interest is paid k times a year at an annual rate r, the amount A in the account after t years is given by

$$A = P\left(1 + \frac{r}{k}\right)^{kt}$$

If \$15,000 is deposited in an account paying an annual interest rate of 7.5%, compounded monthly, how much will be in the account in 60 years?

$$A(t) = 15,000\left(1 + \frac{0.075}{12}\right)^{12t}$$

To write the formula in function notation, substitute for P, r, and k.

$$A(60) = 15,000\left(1 + \frac{0.075}{12}\right)^{12(60)}$$

Substitute 60 for t.

$$= 15,000\left(1 + \frac{0.075}{12}\right)^{720}$$

$$\approx 1,331,479.52$$

Use a calculator.

In 60 years, the account will contain \$1,331,479.52.

REVIEW EXERCISES

Use properties of exponents to simplify each expression.

25. $5^{\sqrt{6}} \cdot 5^{3\sqrt{6}}$ **26.** $\left(2^{\sqrt{14}}\right)^{\sqrt{2}}$

Graph each function and give the domain and the range. Label the y-intercept.

27. $f(x) = 3^x$

28. $f(x) = \left(\frac{1}{3}\right)^x$

29. $f(x) = \left(\frac{1}{2}\right)^x - 2$

30. $f(x) = 3^{x-1}$

31. In Exercise 30, what is the asymptote of the graph of $f(x) = 3^{x-1}$?

32. COAL PRODUCTION   The table gives the number of tons of coal produced in the United States for the years 1800–1920. Graph the data. What type of function does it appear could be used to model coal production over this period?

Year	Tons	Year	Tons
1800	108,000	1870	40,429,000
1810	178,000	1880	79,407,000
1820	881,000	1890	157,771,000
1830	1,334,000	1900	269,684,000
1840	2,474,000	1910	501,596,000
1850	8,356,000	1920	658,265,000
1860	20,041,000		

Source: *World Book Encyclopedia*

33. COMPOUND INTEREST   How much will \$10,500 become if it earns 9% annual interest, compounded quarterly, for 60 years?

34. DEPRECIATION   The value (in dollars) of a certain model car is given by the function $V(t) = 12,000\left(10^{-0.155t}\right)$, where t is the number of years from the present. Find the value of the car in 5 years.

SECTION 11.4 Base-e Exponential Functions

DEFINITIONS AND CONCEPTS	EXAMPLES
Of all possible bases for an exponential function, e is the most convenient for problems involving growth or decay. $e = 2.718281828459 \ldots$ The function defined by $f(x) = e^x$ is the **natural exponential function.**	From the graph, we see that the domain of the natural exponential function is $(-\infty, \infty)$ and the range is $(0, \infty)$. 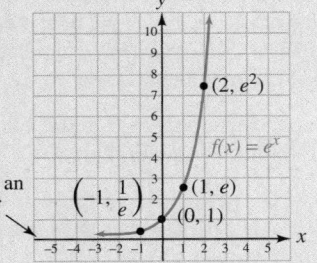 The x-axis is an asymptote of the graph.
Exponential growth/decay: If a quantity increases or decreases at an annual rate r, **compounded continuously,** the amount A after t years is given by $A = Pe^{rt}$ If r is negative, the amount decreases.	If \$30,000 accumulates interest at an annual rate of 9%, compounded continuously, find the amount in the account after 25 years. $A = Pe^{rt}$ This is the formula for continuous compound interest. $= 30{,}000e^{0.09 \cdot 25}$ Substitute 30,000 for P, 0.09 for r, and 25 for t. $= 30{,}000e^{2.25}$ $\approx 284{,}632.08$ Use a calculator. In 30 years, the account will contain \$284,632.08. Suppose the population of a city of 50,000 people is decreasing exponentially according to the function $P(t) = 50{,}000e^{-0.003t}$, where t is measured in years from the present date. Find the expected population of the city in 20 years. $P(t) = 50{,}000e^{-0.003t}$ Since r is negative, this is the exponential decay model. $P(20) = 50{,}000e^{-0.003(20)}$ Substituite 20 for t. $= 50{,}000e^{-0.06}$ $\approx 47{,}088$ Use a calculator. After 20 years, the expected population will be about 47,088 people.

REVIEW EXERCISES

Graph each function, and give the domain and the range.

35. $f(x) = e^x + 1$

36. $f(x) = e^{x-3}$

37. INTEREST COMPOUNDED CONTINUOUSLY   If \$10,500 accumulates interest at an annual rate of 9%, compounded continuously, how much will be in the account in 60 years?

38. THE GRAND CANYON STATE   In 2006, Arizona ended Nevada's 19-year reign as the nation's fastest growing state. The population of Arizona at the time was 6,166,318 with an annual growth rate of 3.6%. Predict the population of Arizona in 2016, assuming the growth rate remains the same.

39. MORTGAGE RATES   There was the housing boom in the 1980s as the baby boomers (those born from 1946–1964) bought their homes. The average annual interest rate in percent on a 30-year fixed-rate home mortgage for the years 1980–1996 can be approximated by the function $r(t) = 13.9e^{-0.035t}$, where t is the number of years since 1980. To the nearest hundredth of a percent, what does this model predict was the 30-year fixed rate in 1980? In 1985? In 1990?

40. MEDICAL TESTS   A radioactive dye is injected into a patient as part of a test to detect heart disease. The amount of dye remaining in his bloodstream t hours after the injection is given by the function $f(t) = 10e^{-0.27t}$. How can you determine from the function that the amount of dye in the bloodstream is decreasing?

SECTION 11.5   Logarithmic Functions

DEFINITIONS AND CONCEPTS	EXAMPLES
Definition of logarithm: If $b > 0$, $b \neq 1$, and x is positive, then $\qquad y = \log_b x \quad$ means $\quad x = b^y$	*Logarithmic form* $\qquad$ *Exponential form* $\log_5 125 = 3 \quad$ is equivalent to $\quad 5^3 = 125$ $\log_2 \dfrac{1}{8} = -3 \quad$ is equivalent to $\quad 2^{-3} = \dfrac{1}{8}$
$\log_b x$ is the exponent to which b is raised to get x.	To evaluate $\log_4 16$ we ask: "To what power must we raise 4 to get 16?" Since $4^2 = 16$, the answer is: the 2nd power. Thus, $\qquad \log_4 16 = 2$
For computational purposes and in many applications, we use base-10 logarithms, called **common logarithms.** $\qquad \log x \quad$ means $\quad \log_{10} x$	$\log \dfrac{1}{1,000} = -3 \quad$ because $\quad 10^{-3} = \dfrac{1}{1,000}$
If $b > 0$ and $b \neq 1$, the **logarithmic function with base b** is defined by $f(x) = \log_b x$. The domain is $(0, \infty)$ and the range is $(-\infty, \infty)$. If $b > 1$, then $f(x) = \log_b x$ is an increasing function. If $0 < b < 1$, then $f(x) = \log_b x$ is a decreasing function.	The graph of the logarithmic function $f(x) = \log_2 x$. From the graph, we see that $f(x) = \log_2 x$ is an increasing function.
The exponential function $f(x) = b^x$ and the logarithmic function $f(x) = \log_b x$ are inverses of each other.	$f(x) = 3^x$ and $f^{-1}(x) = \log_3 x$ are inverses of each other. Their graphs are symmetric about the line $y = x$.
Logarithmic functions, like exponential functions, can be used to **model** certain types of growth and decay. *Decibel voltage gain:* $\qquad \text{db gain} = 20 \log \dfrac{E_O}{E_I}$ *The Richter scale:* $\qquad R = \log \dfrac{A}{P}$	If the input to an amplifier is 0.4 volt and the output is 30 volts, find the decibel voltage gain. $\text{db gain} = 20 \log \dfrac{E_O}{E_I}$ $\qquad = 20 \log \dfrac{30}{0.4} \qquad$ Substitute 30 for E_O and 0.4 for E_I. $\qquad \approx 37.50122527 \qquad$ Use a calculator. The db gain is about 38 decibels.

REVIEW EXERCISES

41. Give the domain and range of $f(x) = \log x$.

42. Explain why a student got the following message when she used a calculator to evaluate log 0.

Error

43. Write the statement $\log_4 64 = 3$ in exponential form.

44. Write the statement $7^{-1} = \frac{1}{7}$ in logarithmic form.

Evaluate, if possible.

45. $\log_3 9$

46. $\log_9 \frac{1}{81}$

47. $\log_{1/2} 1$

48. $\log_5 (-25)$

49. $\log_6 \sqrt{6}$

50. $\log 1,000$

Solve for x.

51. $\log_2 x = 5$

52. $\log_3 x = -4$

53. $\log_x 16 = 2$

54. $\log_x \frac{1}{100} = -2$

55. $\log_9 3 = x$

56. $\log_{27} 3 = x$

Use a calculator to find the value of x to four decimal places.

57. $\log 4.51 = x$

58. $\log x = 1.43$

Graph each function and its inverse on the same coordinate system. Draw the axis of symmetry.

59. $f(x) = \log_4 x$ and $g(x) = 4^x$

60. $f(x) = \log_{1/3} x$ and $g(x) = \left(\frac{1}{3}\right)^x$

Graph each function. Label the x-intercept.

61. $f(x) = \log (x - 2)$

62. $f(x) = 3 + \log x$

63. ELECTRICAL ENGINEERING Find the db gain of an amplifier with an output of 18 volts and an input of 0.04 volt.

64. EARTHQUAKES An earthquake had a period of 0.3 second and an amplitude of 7,500 micrometers. Find its measure on the Richter scale.

SECTION 11.6 Base-*e* Logarithmic Functions

DEFINITIONS AND CONCEPTS	EXAMPLES
Of all possible bases for a logarithmic function, e is the most convenient for problems involving growth or decay. Since these situations occur often in natural settings, base-e logarithms are called **natural logarithms**: $\ln x$ means $\log_e x$	$\ln 5.7$ means $\log_e 5.7$
$\ln x$ is the exponent to which e is raised to get x.	To evaluate $\ln \frac{1}{e^4}$ we ask: "To what power must we raise e to get $\frac{1}{e^4}$?" Since $e^{-4} = \frac{1}{e^4}$, the answer is: the -4th power. Thus, $$\ln \frac{1}{e^4} = -4$$
The **natural logarithmic function** with base e is defined by $$f(x) = \ln x$$ The domain is the interval $(0, \infty)$ and the range is the interval $(-\infty, \infty)$.	The graph of the natural logarithmic function $$f(x) = \ln x$$ From the graph, we see that $f(x) = \ln x$ is an increasing function. The y-axis is an asymptote of the graph. 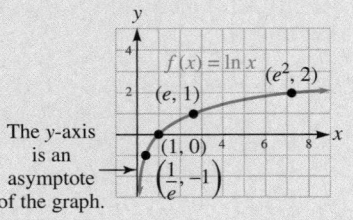

SECTION 11.6   Base-*e* Logarithmic Functions—*continued*

DEFINITIONS AND CONCEPTS	EXAMPLES
The natural exponential function $f(x) = e^x$ and the natural logarithmic function $f^{-1}(x) = \ln x$ are **inverses** of each other.	The graphs are symmetric about the line $y = x$. 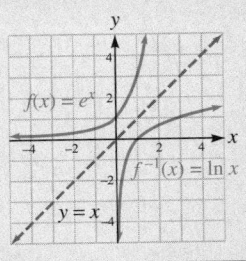
If a population grows exponentially at a certain annual rate r, the time required for the population to double is called the **doubling time**. It is given by the formula: $$t = \frac{\ln 2}{r}$$	The population of a town is growing at a rate of 3% per year. If this rate continues, how long will it take the population to double? We substitute 0.03 for r and use a calculator to perform the computation. $$t = \frac{\ln 2}{r} = \frac{\ln 2}{0.03} \approx 23.10490602$$ The population will double in about 23.1 years.

REVIEW EXERCISES

Evaluate each expression, if possible. Do not use a calculator.

65. $\ln e$

66. $\ln e^2$

67. $\ln \frac{1}{e^5}$

68. $\ln \sqrt{e}$

69. $\ln(-e)$

70. $\ln 0$

71. $\ln 1$

72. $\ln e^{-7}$

Use a calculator to evaluate each expression. Express all answers to four decimal places.

73. $\ln 452$

74. $\ln 0.85$

Solve each equation. Express all answers to four decimal places.

75. $\ln x = 2.336$

76. $\ln x = -8.8$

77. Explain the difference between the functions $f(x) = \log x$ and $g(x) = \ln x$.

78. What function is the inverse of $f(x) = \ln x$?

Graph each function.

79. $f(x) = 1 + \ln x$

80. $f(x) = \ln(x + 1)$

81. POPULATION GROWTH   How long will it take the population of Mexico to double if the growth rate is currently about 1.153%?

82. BOTANY   The height (in inches) of a certain plant is approximated by the function $H(a) = 13 + 20.03 \ln a$, where a is its age in years. How tall will it be when it is 19 years old?

SECTION 11.7 Properties of Logarithms

DEFINITIONS AND CONCEPTS	EXAMPLES

Properties of logarithms: If M, N, and b are positive real numbers, $b \neq 1$
 	
 1. $\log_b 1 = 0$ 2. $\log_b b = 1$
 3. $\log_b b^x = x$ 4. $b^{\log_b x} = x$
 	
 5. *Product rule for logarithms:*
 	
 $\log_b MN = \log_b M + \log_b N$
 	
 6. *Quotient rule for logarithms:*
 	
 $\log_b \dfrac{M}{N} = \log_b M - \log_b N$
 	
 7. *Power rule for logarithms:*
 	
 $\log_b M^P = p \log_b M$
	Apply a property of logarithms and then simplify, if possible.
	
	 1. $\log_3 1 = 0$ 2. $\log_7 7 = 1$
	 3. $\log_5 5^3 = 3$ 4. $9^{\log_9 10} = 10$
	 5. $\log_2(6 \cdot 8) = \log_2 6 + \log_2 8$
	              $= \log_2 6 + 3$
	
	 6. $\log_2 \dfrac{8}{6} = \log_2 8 - \log_2 6$
	          $= 3 - \log_2 6$
	
	 7. $\log_2 7^3 = 3 \log_2 7$

Properties of logarithms can be used to **expand** logarithmic expressions.	Write $\log_3 (x^2 y^3)$ as the sum and/or difference of logarithms of a single quantity. $\log_3 (x^2 y^3) = \log_3 x^2 + \log_3 y^3$ The log of a product is the sum of the logs. $= 2 \log_3 x + 3 \log_3 y$ The log of a power is the power times the log.

Properties of logarithms can be used to **condense** certain logarithmic expressions.	Write $3 \ln x - \frac{1}{2} \ln y$ as a single logarithm.
	
	 $3 \ln x - \dfrac{1}{2} \ln y = \ln x^3 - \ln y^{1/2}$ A power times a log is the log of the power.
	
	             $= \ln \dfrac{x^3}{y^{1/2}}$ The difference of two logs is the log of the quotient.
	
	             $= \ln \dfrac{x^3}{\sqrt{y}}$ Write $y^{1/2}$ as $\sqrt{y}$.

If we need to find a logarithm with some base other than 10 or e, we can use a conversion formula. **Change-of-base formula:** $\log_b x = \dfrac{\log_a x}{\log_a b}$	Find $\log_7 6$ to four decimal places. $\log_7 6 = \dfrac{\log 6}{\log 7} \approx 0.920782221$ To four decimal places, $\log_7 6 = 0.9208$. To check, verify that $7^{0.9208}$ is approximately 6.

In chemistry, common logarithms are used to express the acidity of solutions.
 	
 pH scale:
 	
 $\text{pH} = -\log[H^+]$
	Find the pH of a liquid with a hydrogen ion concentration of 10^{-8} gram-ions per liter.
	
	 $\text{pH} = -\log[H^+]$
	      $= -\log 10^{-8}$ Substitute 10^{-8} for $[H^+]$.
	      $= -(-8) \log 10$ The log of a power is the power times the log.
	      $= 8$ Simplify: $\log 10 = 1$.

REVIEW EXERCISES

Simplify each expression.

83. $\log_2 1$ **84.** $\log_9 9$

85. $\log 10^3$ **86.** $7^{\log_7 4}$

Write each logarithm as the sum and/or difference of logarithms of a single quantity. Then simplify, if possible.

87. $\log_3 27x$

88. $\log \frac{100}{x}$

89. $\log_5 \sqrt{27}$

90. $\log_b 10ab$

Write each logarithm as the sum and/or difference of logarithms of a single quantity.

91. $\log_b \frac{x^2 y^3}{z}$

92. $\ln \sqrt{\frac{x}{yz^2}}$

Write each logarithmic expression as one logarithm.

93. $3 \log_2 x - 5 \log_2 y + 7 \log_2 z$

94. $-3 \log_b y - 7 \log_b z + \frac{1}{2} \log_b (x + 2)$

Assume that $\log_b 5 = 1.1609$ and $\log_b 8 = 1.5000$ and find each value to four decimal places.

95. $\log_b 40$ **96.** $\log_b 64$

97. Find $\log_5 17$ to four decimal places.

98. pH OF GRAPEFRUIT The pH of grapefruit juice is about 3.1. Find its hydrogen ion concentration.

SECTION 11.8 Exponential and Logarithmic Equations

DEFINITIONS AND CONCEPTS	EXAMPLES
An **exponential equation** contains a variable in one of its exponents. If both sides of an exponential equation can be expressed as a power of the same base, we can use the following property to solve it: $\quad b^x = b^y$ is equivalent to $x = y$	Solve: $3^{x+2} = 27$ We express the right side of the equation as a power of 3. $\quad 3^{x+2} = 3^3$ Write 27 as 3^3. $\quad x + 2 = 3$ If two exponential expressions with the same base are equal, their exponents are equal. $\quad\quad x = 1$ The solution is 1. Check it in the original equation.
When it is difficult to write each side of an exponential equation as a power of the same base, **take the logarithm of each side.**	Solve $4^x = 7$ and give the answer to four decimal places. We take the base-10 logarithm of both sides of the equation. $\quad \log 4^x = \log 7$ $\quad x \log 4 = \log 7$ The log of a power is the power times the log. $\quad x = \dfrac{\log 7}{\log 4}$ To isolate x, divide both sides by log 4. $\quad x \approx 1.4037$ Use a calculator. To four decimal places, the solution is 1.4037. To check the approximate solution, we substitute 1.4037 for x in $4^x = 7$ and use a calculator to evaluate the left side.

SECTION 11.8   Exponential and Logarithmic Equations—*continued*

DEFINITIONS AND CONCEPTS	EXAMPLES
A **logarithmic equation** is an equation containing a variable in a logarithmic expression. Certain logarithmic equations can be solved using the following property: $\log_b x = \log_b y$ is equivalent to $x = y$	Solve: $\log(4x - 3) = \log(2x + 7)$ $\quad \log(4x - 3) = \log(2x + 7)$ $\qquad\quad 4x - 3 = 2x + 7$ *If the logarithms of two numbers are equal,* $\qquad\qquad\qquad\qquad\qquad$ *the numbers are equal.* $\qquad\qquad\quad 2x = 10$ $\qquad\qquad\qquad x = 5$ The solution is 5. Check it in the original equation.
To solve some logarithmic equations, we instead write and solve an equivalent exponential equation.	Solve: $\log_4(x + 1) = 2$ We will write the equivalent base-4 exponential equation. $\qquad \log_4(x + 1) = 2$ $\qquad\qquad x + 1 = 4^2$ $\qquad\qquad x + 1 = 16$ $\qquad\qquad\quad x = 15$ The solution is 15. Check it in the original equation.
When there is sufficient food and space available, populations of living organisms tend to increase exponentially according to the following **growth model**. ***Population growth*** $\quad P = P_0 e^{kt}$	Find the number of bacteria in a culture of 1,000 bacteria if are they are allowed to reproduce for 5 hours. Assume $k = \frac{\ln 3}{3}$. $\quad P = P_0 e^{k\,t}$ *This is the population growth model.* $\qquad = 1{,}000 e^{\frac{\ln 3}{3} \cdot 5}$ *Substitute.* $\qquad \approx 6{,}240$ *Use a calculator.* In 5 hours, there will be approximately 6,240 bacteria.

REVIEW EXERCISES

Solve each equation. Give approximate answers to four decimal places.

99. $5^{x+6} = 25$ **100.** $2^{x^2+4x} = \frac{1}{8}$

101. $3^x = 7$ **102.** $2^x = 3^{x-1}$

103. $e^x = 7$ **104.** $e^{-0.4t} = 25$

Solve each equation.

105. $\log(x - 4) = 2$ **106.** $\ln(2x - 3) = \ln 15$

107. $\log x + \log(29 - x) = 2$

108. $\log_2 x + \log_2(x - 2) = 3$

109. $\frac{\log(7x - 12)}{\log x} = 2$

110. $\log_2(x + 2) + \log_2(x - 1) = 2$

111. $\log x + \log(x - 5) = \log 6$

112. $\log 3 - \log(x - 1) = -1$

113. Evaluate both sides of the statement $\frac{\log 8}{\log 15} \neq \log 8 - \log 15$ to show that the sides are indeed not equal.

114. CARBON-14 DATING   A wooden statue found in Egypt has a carbon-14 content that is two-thirds of that found in living wood. If the half-life of carbon-14 is 5,700 years, how old is the statue?

115. ANTS   The number of ants in a colony is estimated to be 800. If the ant population is expected to triple every 14 days, how long will it take for the population to reach one million?

116. The approximate coordinates of the points of intersection of the graphs of $f(x) = \log x$ and $g(x) = 1 - \log(7 - x)$ are shown in parts (a) and (b) of the illustration. Use the graphs to estimate the solutions of the logarithmic equation $\log x = 1 - \log(7 - x)$. Then check your answers.

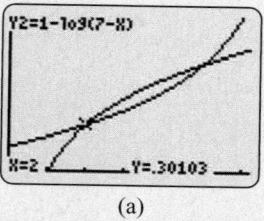

(a)

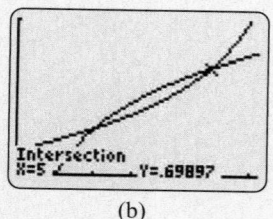

(b)

CHAPTER 11
Test

1. Fill in the blanks.

   a. A _____ function is denoted by $f \circ g$.

   b. $f(x) = e^x$ is the _____ exponential function.

   c. In _____ compound interest, the number of compoundings is infinitely large.

   d. The functions $f(x) = \log_{10} x$ and $f(x) = 10^x$ are _____ functions.

   e. $f(x) = \log_4 x$ is a _____ function.

2. Write out in words how to say each of the following:

   a. $f \circ g$

   b. $g(f(8))$

Let $f(x) = x + 9$ and $g(x) = 4x^2 - 3x + 2$. Find each function and give its domain.

3. $f + g$

4. g/f

Let $f(x) = 2x^2 + 3$ and $g(x) = 4x - 8$. Find each composition.

5. $(g \circ f)(-3)$

6. $(f \circ g)(x)$

Use the tables of values for functions f and g to find each of the following.

7. a. $(f \cdot g)(9)$

    b. $(f \circ g)(-3)$

x	$f(x)$
9	−1
10	17

x	$g(x)$
−3	10
9	16

8. Refer to the graphs of functions f and g to find each of the following.

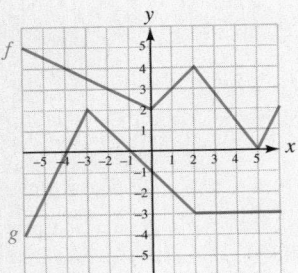

   a. $(g/f)(-4)$

   b. $(f \circ g)(1)$

   c. $(f + g)(2)$

   d. $(f \cdot g)(0)$

   e. $(g - f)(1)$

Determine whether each function is one-to-one.

9. $f(x) = x$

10.

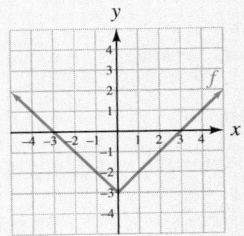

11. Find the inverse of $f(x) = -\frac{1}{3}x$ and then graph f and its inverse on the same coordinate axes.

12. Find the inverse of $f(x) = (x - 15)^3$.

13. Use composition to show that $f(x) = 4x + 4$ and $f^{-1}(x) = \frac{x - 4}{4}$ are inverse functions.

14. Consider the following graph of the function f.

   a. Is f a one-to-one function?

   b. Is its inverse a function?

   c. What is $f^{-1}(260)$? What information does it give?

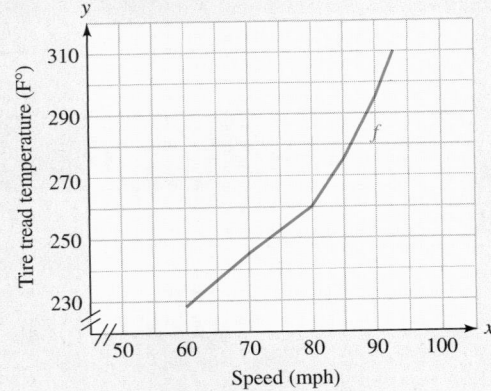

Relationship Between Car Speed and Tire Temperature

Graph each function and give the domain and the range.

15. $f(x) = 2^x + 1$

16. $f(x) = 3^{-x}$

17. RADIOACTIVE DECAY A radioactive material decays according to the formula $A = A_0(2)^{-t}$. How much of a 3-gram sample will be left in 6 years?

18. COMPOUND INTEREST An initial deposit of $1,000 earns 6% interest, compounded twice a year. How much will be in the account in one year?

19. Graph $f(x) = e^x$. Label the y-intercept and the asymptote of the graph.

20. POPULATION GROWTH   As of July 2007, the population of India was estimated to be 1,129,866,154, with an annual growth rate of 1.606%. If the growth rate remains the same, how large will the population be in 30 years?

21. Write the statement $\log_6 \frac{1}{36} = -2$ in exponential form.

22. a. What are the domain and range of the function $f(x) = \log x$?
    b. What is the inverse of $f(x) = \log x$?

Evaluate each logarithmic expression, if possible.

23. $\log_5 25$

24. $\log_9 \frac{1}{81}$

25. $\log (-100)$

26. $\ln \frac{1}{e^6}$

27. $\log_4 2$

28. $\log_{1/3} 1$

Solve for x.

29. $\log_x 32 = 5$

30. $\log_8 x = \frac{4}{3}$

31. $\log_3 x = -3$

32. $\ln x = 1$

Graph each function.

33. $f(x) = -\log_3 x$

34. $f(x) = \ln x$

35. CHEMISTRY   pH Find the pH of a solution with a hydrogen ion concentration of 3.7×10^{-7}. (*Hint:* pH $= -\log [H^+]$.)

36. ELECTRONICS   Find the db gain of an amplifier when $E_O = 60$ volts and $E_I = 0.3$ volt. *Hint:* db gain $= 20 \log \frac{E_O}{E_I}$.

37. Use a calculator to find x to four decimal places: $\log x = -1.06$

38. Use the change-of-base formula to find $\log_7 3$ to four decimal places.

39. Write the expression $\log_b a^2bc^3$ as the sum and/or difference of logarithms of a single quantity. Then simplify, if possible.

40. Write the expression $\frac{1}{2} \ln (a + 2) + \ln b - 3 \ln c$ as a logarithm of a single quantity.

Solve each equation. Give approximate answers to four decimal places.

41. $5^x = 3$

42. $3^{x-1} = 27$

43. $\ln (5x + 2) = \ln (2x + 5)$

44. $\log x + \log (x - 9) = 1$

45. The illustration shows the graphs of $y = \frac{1}{2} \ln (x - 1)$ and $y = \ln 2$ and the approximate coordinates of their point of intersection. Estimate the solution of the logarithmic equation $\frac{1}{2} \ln (x - 1) = \ln 2$. Then check the result.

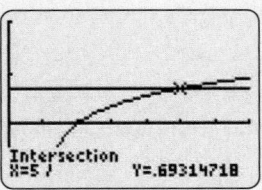

46. INSECTS   The number of insects attracted to a bright light is currently 5. If the number is expected to quadruple every 6 minutes, how long will it take for the number to reach 500?

GROUP PROJECT

THE NUMBER *e*

Overview: In this activity, you will use a calculator to find progressively more accurate approximations of *e*.

Instructions: Form groups of two students. Each student will need a scientific calculator.

Begin by finding an approximation of *e* using the $\boxed{e^x}$ key on your calculator. Copy the table shown on the next page, and write the number displayed on the calculator screen at the top of the table.

The value of *e* can be calculated to any degree of accuracy by adding the terms of the following pattern:

$$e = 1 + 1 + \frac{1}{2} + \frac{1}{2 \cdot 3} + \frac{1}{2 \cdot 3 \cdot 4} + \frac{1}{2 \cdot 3 \cdot 4 \cdot 5} + \cdots$$

The more terms that are added, the closer the sum will be to *e*.

You are to add as many terms as necessary until you obtain a sum that matches the value of e ,given by the $\boxed{e^x}$ key on your calculator. Work together as a team. One member of the group should compute the fractional form of the term to be added. (See the middle column of the table.) The other member should take that information and calculate the cumulative sum. (See the right column of the table.)

How many terms must be added so that the cumulative sum approximation and the $\boxed{e^x}$ key approximation match in each decimal place?

Approximation of e found using the $\boxed{e^x}$ key: $e \approx$ _____

Number of terms in the sum	Term (Expressed as a fraction)	Cumulative sum (An approximation of e)
1	1	1
2	1	2
3	$\frac{1}{2}$	2.5
4	$\frac{1}{2 \cdot 3} = \frac{1}{6}$	2.666666667
$\vdots$	$\vdots$	$\vdots$

CHAPTER 12

More on Systems of Equations

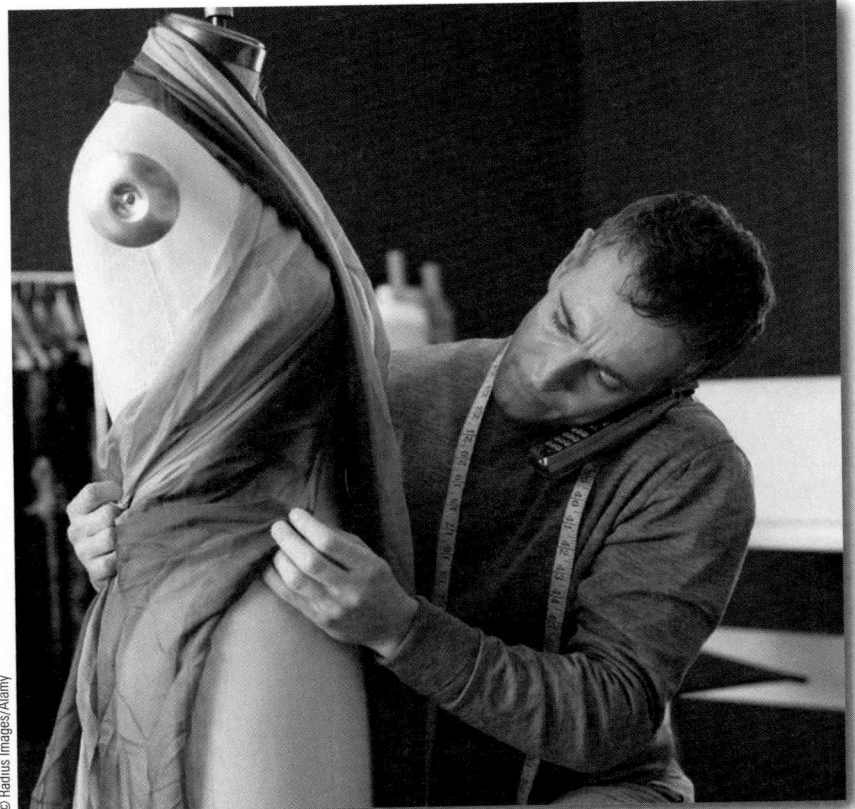

© Radius Images/Alamy

from *Campus to Careers*
Fashion Designer

Fashion designers help create the billions of clothing articles, shoes, and accessories purchased every year by consumers. Fashion design relies heavily on mathematical skills, including knowledge of lines, angles, curves, and measurement. Designers also use mathematics in the manufacturing and marketing parts of the industry as they calculate labor costs and determine the markups and markdowns involved in retail pricing.

In **Problem 11** of **Study Set 12.3** we will examine the production side of fashion design as we determine the number of coats, shirts, and slacks that can be made with the available labor.

JOB TITLE:
Fashion Designer

EDUCATION:
Many community colleges and vocational schools provide training for the fashion industry.

JOB OUTLOOK:
The best opportunities will be in designing clothing sold in department stores and retail chains.

ANNUAL EARNINGS:
$69,270

FOR MORE INFORMATION:
www.collegeboard.com/csearch/majors_careers/profiles/

Preparing for a Final Exam

Final exams can be stressful for many students because the number of topics to study can seem overwhelming. Here are some suggestions to help reduce the stress and prepare you for the test.

GET ORGANIZED: Gather all of your notes, study sheets, homework assignments, and especially all of your returned tests to review.

TALK WITH YOUR INSTRUCTOR: Ask your instructor to list the topics that may appear on the final and those that won't be covered.

MANAGE YOUR TIME: Adjust your daily schedule 1 week before the final so that it includes extended periods of study time.

Now Try This

1. Review your old tests. Make a list of the test problems that you are still unsure about and see a tutor or your instructor to get help.

2. Make a practice final exam that includes one or more of each type of problem that may appear on the test.

3. Make a detailed study plan. Determine when, where, and what you will study each day for 1 week before the final.

SECTION 12.1
Solving Systems of Equations in Two Variables

Objectives

1. Determine whether an ordered pair is a solution of a system.
2. Solve systems of linear equations by graphing.
3. Use graphing to identify inconsistent systems and dependent equations.
4. Solve systems of linear equations by substitution.
5. Solve systems of linear equation by the elimination (addition) method.
6. Use substitution and elimination to identify inconsistent systems and dependent equations.
7. Solve problems using systems of equations.

In this section, we will review graphical and algebraic methods for solving systems of two linear equations in two variables.

Determine Whether an Ordered Pair Is a Solution of a System.

When two equations with the same variables are considered simultaneously (at the same time), we say that they form a **system of equations.** We will use a left brace { when writing a system of equations. An example is

$$\begin{cases} 2x + 5y = -1 \\ x - y = -4 \end{cases}$$ Read as "the system of equations $2x + 5y = -1$ and $x - y = -4$."

A **solution of a system** of equations in two variables is an ordered pair that satisfies both equations of the system.

EXAMPLE 1   Determine whether $(-3, 1)$ is a solution of each system of equations.   **a.** $\begin{cases} 2x + 5y = -1 \\ x - y = -4 \end{cases}$ **b.** $\begin{cases} 5y = 2 - x \\ y = 3x \end{cases}$

Strategy   We will substitute the x- and y-coordinates of $(-3, 1)$ for the corresponding variables in both equations of the system.

Why   If both equations are satisfied (made true) by the x- and y-coordinates, the ordered pair is a solution of the system.

Solution

a. To determine whether $(-3, 1)$ is a solution, we substitute -3 for x and 1 for y in each equation.

Check: $2x + 5y = -1$ First equation. $x - y = -4$ Second equation.

$2(-3) + 5(1) \overset{?}{=} -1$ $-3 - 1 \overset{?}{=} -4$

$-6 + 5 \overset{?}{=} -1$ $-4 = -4$ True

$-1 = -1$ True

Since $(-3, 1)$ satisfies both equations, it is a solution of the system.

b. Again, we substitute -3 for x and 1 for y in each equation.

Check: $5y = 2 - x$ First equation. $y = 3x$ Second equation.

$5(1) \overset{?}{=} 2 - (-3)$ $1 \overset{?}{=} 3(-3)$

$5 \overset{?}{=} 2 + 3$ $1 = -9$ False

$5 = 5$ True

Although $(-3, 1)$ satisfies the first equation, it does not satisfy the second. Because it does not satisfy both equations, $(-3, 1)$ is not a solution of the system.

Self Check 1   Determine whether $(6, -2)$ is a solution of $\begin{cases} x - 2y = 10 \\ y = 3x - 20 \end{cases}$.

Now Try   Problem 13

Solve Systems of Linear Equations by Graphing.

To solve a system of two linear equations in two variables by graphing, we use the following steps.

The Graphing Method	1. Graph each equation on the same rectangular coordinate system.
	2. Determine the coordinates of the point of intersection of the graphs. That ordered pair is the solution of the system.
	3. If the graphs have no point in common, the system has no solution.
	4. Check the proposed solution in each equation of the original system.

When a system of equations (as in Example 2) has at least one solution, the system is called a **consistent system.**

EXAMPLE 2   Solve the system by graphing: $\begin{cases} x + 2y = 4 \\ 2x - y = 3 \end{cases}$

Strategy   We will graph both equations on the same coordinate system.

Why   The graph of a linear equation is a picture of its solutions. If both equations are graphed on the same coordinate system, we can see whether they have any common solutions.

Solution   We graph both equations as shown in the figure.

$x + 2y = 4$

x	y	(x, y)
4	0	$(4, 0)$
0	2	$(0, 2)$
-2	3	$(-2, 3)$

$2x - y = 3$

x	y	(x, y)
$\frac{3}{2}$	0	$\left(\frac{3}{2}, 0\right)$
0	-3	$(0, -3)$
-1	-5	$(-1, -5)$

Use the intercept method to graph each line. Let $y = 0$ and find x. Let $x = 0$ and find y. Find a third point on the line by selecting a convenient value of x and finding y.

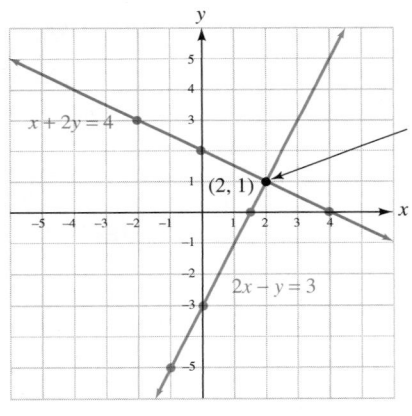

The point of intersection gives the solution of the system.

Although infinitely many ordered pairs (x, y) satisfy $x + 2y = 4$, and infinitely many ordered pairs (x, y) satisfy $2x - y = 3$, only the coordinates of the point where the graphs intersect satisfy both equations. From the graph, it appears that the intersection point has coordinates $(2, 1)$. To verify that it is the solution, we substitute 2 for x and 1 for y in both equations and verify that $(2, 1)$ satisfies each one.

> **Success Tip**
> When determining the coordinates of a point of intersection from a graph, realize that they are simply estimates. Only after algebraically checking a proposed solution can we be assured that it is an actual solution.

Check:

First equation.	Second equation.
$x + 2y = 4$	$2x - y = 3$
$2 + 2(1) \stackrel{?}{=} 4$	$2(2) - 1 \stackrel{?}{=} 3$
$2 + 2 \stackrel{?}{=} 4$	$4 - 1 \stackrel{?}{=} 3$
$4 = 4$ True	$3 = 3$ True

Since $(2, 1)$ makes both equations true, it is the solution of the system. The solution set is $\{(2, 1)\}$.

▷ **Self Check 2**   Solve the system by graphing: $\begin{cases} x - 3y = -5 \\ 2x + y = 4 \end{cases}$

Now Try   **Problem 19**

3   **Use Graphing to Identify Inconsistent Systems and Dependent Equations.**

When a system has no solution (as in Example 3), it is called an **inconsistent system**.

EXAMPLE 3   Solve the system by graphing, if possible: $\begin{cases} 2x + 3y = 6 \\ 4x + 6y = 24 \end{cases}$

Strategy   We will graph each equation on one set of coordinate axes and hope to determine the intersection point.

Why   If there is one, the coordinates of the intersection point will be the solution of the system.

Solution   Using the intercept method, we graph both equations, as shown below.

$2x + 3y = 6$

x	y	(x, y)
3	0	$(3, 0)$
0	2	$(0, 2)$
-3	4	$(-3, 4)$

$4x + 6y = 24$

x	y	(x, y)
6	0	$(6, 0)$
0	4	$(0, 4)$
-3	6	$(-3, 6)$

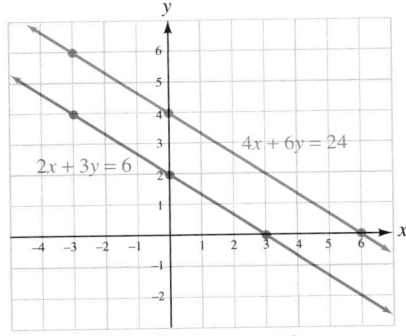

Parallel lines—no solution

In this example, the graphs are parallel because the slopes of the lines are equal and they have different y-intercepts. We can see that the slope of each line is $-\frac{2}{3}$ by writing each equation in slope-intercept form.

$$2x + 3y = 6 \qquad\qquad 4x + 6y = 24$$
$$3y = -2x + 6 \qquad\qquad 6y = -4x + 24$$
$$y = -\frac{2}{3}x + 2 \qquad\qquad y = -\frac{2}{3}x + 4$$

Because the lines are parallel, there is no point of intersection. Such a system has *no solution* and it is called an **inconsistent system**. The solution set is the empty set, which is written ∅.

▷ **Self Check 3**   Solve the system by graphing, if possible: $\begin{cases} 3y - 2x = 6 \\ 2x - 3y = 6 \end{cases}$

Now Try   **Problem 23**

When the equations of a system have different graphs (as in Examples 2 and 3), the equations are called **independent equations.** Two equations with the same graph are called **dependent equations.**

EXAMPLE 4   Solve the system by graphing: $\begin{cases} y = \dfrac{1}{2}x + 2 \\ 2x + 8 = 4y \end{cases}$

Strategy   We will graph both equations on the same coordinate system.

Why   If both equations are graphed on the same coordinate system, we can see whether they have any common solutions.

Solution   We graph each equation on one set of coordinate axes, as shown below.

Graph by using the slope and y-intercept.

$$y = \frac{1}{2}x + 2$$

$$m = \frac{1}{2} \qquad b = 2$$

$$\text{Slope} = \frac{1}{2} \qquad y\text{-intercept: } (0, 2)$$

Graph by using the intercept method.

$$2x + 8 = 4y$$

x	y	(x, y)
-4	0	$(-4, 0)$
0	2	$(0, 2)$
2	3	$(2, 3)$

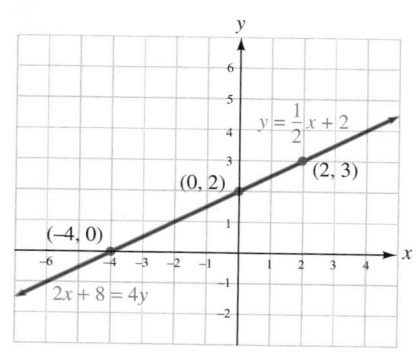

The same line — infinitely many solutions

The graphs appear to be identical. We can verify this by writing the second equation in slope–intercept form and observing that it is the same as the first equation.

$$y = \frac{1}{2}x + 2 \quad \text{First equation.}$$

$$2x + 8 = 4y \quad \text{Second equation.}$$

$$\frac{2x}{4} + \frac{8}{4} = \frac{4y}{4} \quad \text{Divide both sides by 4.}$$

$$\frac{1}{2}x + 2 = y$$

We see that the equations of the system are equivalent. Because they are different forms of the same equation, they are called **dependent equations.**

Since the graphs are the same line, they have infinitely many points in common. The coordinates of each of those points satisfy both equations of the system. In cases like this, we say that there are *infinitely many solutions.* The solution set can be written using set-builder notation as

$$\left\{ (x, y) \mid y = \frac{1}{2}x + 2 \right\} \quad \text{Read as, "the set of all ordered pairs (x, y), such that } y = \tfrac{1}{2}x + 2.\text{"}$$

We can also express the solution set using the second equation of the system in the set-builder notation: $\{(x, y) \mid 2x + 8 = 4y\}$.

Some instructors prefer that the set-builder notation use an equation in standard form with coefficients that are integers having no common factor other than 1. Such an equation that is equivalent to $y = \frac{1}{2}x + 2$ and $2x + 8 = 4y$ is $x - 2y = -4$. The set-builder notation solution for this example could, therefore, be written as $\{(x, y) \mid x - 2y = -4\}$.

From the graph, it appears that three of the infinitely many solutions are $(-4, 0)$, $(0, 2)$, and $(2, 3)$. Check each of them to verify that both equations of the system are satisfied.

 **Self Check 4** Solve the system by graphing: $\begin{cases} 2x - y = 4 \\ y = 2x - 4 \end{cases}$

Now Try **Problem 25**

We now summarize the possibilities that can occur when two linear equations, each in two variables, are graphed.

Solving a System of Equations by the Graphing Method

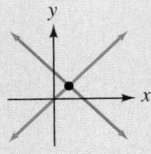 If the lines are different and intersect, the equations are independent, and the system is consistent. **One solution exists.** It is the point of intersection.

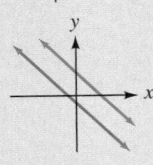 If the lines are different and parallel, the equations are independent, and the system is inconsistent. **No solution exists.**

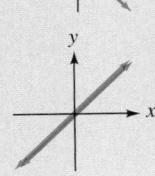 If the lines coincide, the equations are dependent, and the system is consistent. **Infinitely many solutions exist.** Any point on the line is a solution.

If each equation in one system is equivalent to a corresponding equation in another system, the systems are called **equivalent.**

EXAMPLE 5 Solve the system by graphing: $\begin{cases} \dfrac{3}{2}x - y = \dfrac{5}{2} \\ x + \dfrac{1}{2}y = 4 \end{cases}$

Strategy We will use the multiplication property of equality to clear both equations of fractions and solve the resulting equivalent system by graphing.

Why It is usually easier to solve systems that do not contain fractions.

Solution We multiply both sides of $\frac{3}{2}x - y = \frac{5}{2}$ by 2 to eliminate the fractions and obtain the equation $3x - 2y = 5$. We multiply both sides of $x + \frac{1}{2}y = 4$ by 2 to eliminate the fraction and obtain the equation $2x + y = 8$.

The original system   *An equivalent system*

$$\begin{cases} \dfrac{3}{2}x - y = \dfrac{5}{2} \xrightarrow{\text{Multiply by 2}} 2\left(\dfrac{3}{2}x - y\right) = 2\left(\dfrac{5}{2}\right) \xrightarrow{\text{Simplify}} \begin{cases} 3x - 2y = 5 \\ \\ x + \dfrac{1}{2}y = 4 \xrightarrow{\text{Multiply by 2}} 2\left(x + \dfrac{1}{2}y\right) = 2(4) \xrightarrow{\text{Simplify}} \end{cases} 2x + y = 8 \end{cases}$$

Since the new system is equivalent to the original system, they have the same solution. If we graph the equations of the new system, it appears that the point where the lines intersect is (3, 2).

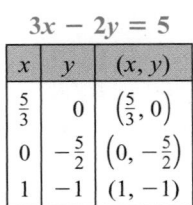

$3x - 2y = 5$

x	y	(x, y)
$\frac{5}{3}$	0	$\left(\frac{5}{3}, 0\right)$
0	$-\frac{5}{2}$	$\left(0, -\frac{5}{2}\right)$
1	-1	$(1, -1)$

$2x + y = 8$

x	y	(x, y)
4	0	$(4, 0)$
0	8	$(0, 8)$
1	6	$(1, 6)$

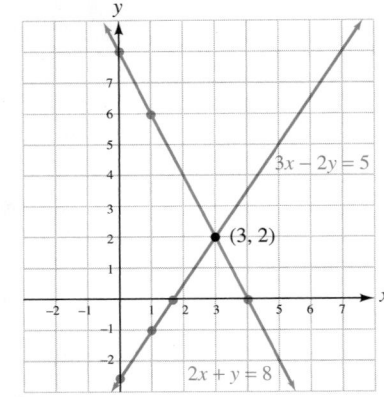

> ### Caution
> When checking the solution of a system of equations, always substitute the values of the variables into the original equations.

To verify that (3, 2) is the solution, we substitute 3 for x and 2 for y in each equation of the original system.

Check: $\frac{3}{2}x - y = \frac{5}{2}$ *First equation.* $x + \frac{1}{2}y = 4$ *Second equation.*

$\frac{3}{2}(3) - 2 \stackrel{?}{=} \frac{5}{2}$ $3 + \frac{1}{2}(2) \stackrel{?}{=} 4$

$\frac{9}{2} - \frac{4}{2} \stackrel{?}{=} \frac{5}{2}$ $3 + 1 \stackrel{?}{=} 4$

$\frac{5}{2} = \frac{5}{2}$ *True* $4 = 4$ *True*

 Self Check 5   Solve the system by graphing: $\begin{cases} \dfrac{1}{2}x + \dfrac{1}{2}y = -1 \\ \dfrac{1}{3}x - \dfrac{1}{2}y = -4 \end{cases}$

Now Try **Problem 29**

Using Your Calculator   *Solving Systems by Graphing*

The graphing method is limited to equations with two variables. Systems with three or more variables cannot be solved graphically. Also, it is often difficult to find exact solutions graphically. However, the TRACE and ZOOM capabilities of graphing calculators enable us to get very good approximations of such solutions.

To solve the system $\begin{cases} 3x + 2y = 12 \\ 2x - 3y = 12 \end{cases}$

with a graphing calculator, we must first solve each equation for y so that we can enter the equations into the calculator. After solving for y, we obtain the following equivalent system:

$$\begin{cases} y = -\dfrac{3}{2}x + 6 \\ y = \dfrac{2}{3}x - 4 \end{cases}$$

If we use window settings of $[-10, 10]$ for x and for y, the graphs of the equations will look like those in figure (a). If we zoom in on the intersection point of the two lines and trace, we will get an approximate solution like the one shown in figure (b). To get better results, we can do more zooms. We would then find that, to the nearest hundredth, the solution is $(4.63, -0.94)$. Verify that this is reasonable.

We can also find the intersection of two lines by using the INTERSECT feature found on most graphing calculators. To locate INTERSECT, press 2nd , CALC , 5, followed by ENTER . After graphing the lines and using INTERSECT, we obtain a graph similar to figure (c). The display shows the approximate coordinates of the point of intersection.

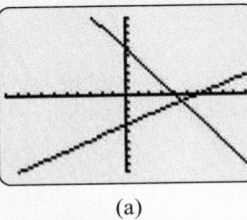
(a)

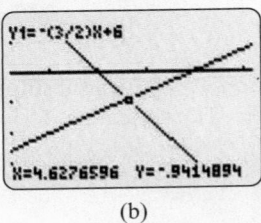
(b)

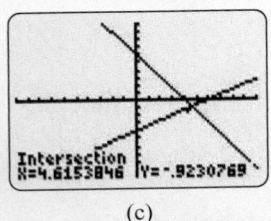
(c)

4 **Solve Systems of Linear Equations by Substitution.**

The graphing method enables us to visualize the process of solving systems of equations. However, it can be difficult to determine the exact coordinates of the point of intersection. We now review an algebraic method that we can use to find the exact solutions of systems of equations.

The *substitution method* works well for solving systems where one equation is solved, or can be easily solved, for one of the variables. To solve a system of two linear equations in x and y by the substitution method, we can follow these steps.

The Substitution Method

1. Solve one of the equations for either x or y—preferably a variable with a coefficient of 1 or -1. If this is already done, go to step 2. (We call this equation the **substitution equation.**)
2. Substitute the expression for x or for y obtained in step 1 into the other equation and solve that equation.
3. Substitute the value of the variable found in step 2 into the substitution equation to find the value of the remaining variable.
4. Check the proposed solution in each equation of the original system. Write the solution as an ordered pair.

> **EXAMPLE 6**   Solve the system by substitution: $\begin{cases} 4x + y = 13 \\ -2x + 3y = -17 \end{cases}$

Strategy   We will use the substitution method. Since the system does not contain an equation solved for x or y, we must choose an equation and solve it for x or y. It is easiest to solve for y in the first equation, because y has a coefficient of 1.

Why   Solving $4x + y = 13$ for x or solving $-2x + 3y = -17$ for x or y would involve working with cumbersome fractions.

Solution

Step 1: We solve the first equation for y, because y has a coefficient of 1.

$$4x + y = 13$$
$$y = -4x + 13 \quad \text{To isolate y, subtract 4x from both sides.}$$
$$\text{This is the substitution equation.}$$

> **The Language of Algebra**
> Since substitution involves algebra and not graphing, it is called an *algebraic* method for solving a system.

Because y and $-4x + 13$ are equal, we can substitute $-4x + 13$ for y in the second equation of the system.

$$y = \boxed{-4x + 13} \qquad -2x + 3y = -17$$

> **Success Tip**
> With this method, the objective is to use an appropriate substitution to obtain *one* equation in *one* variable.

Step 2: We then substitute $-4x + 13$ for y in the second equation to eliminate the variable y from that equation. The result will be an equation containing only one variable, x.

$$-2x + 3y = -17 \quad \text{This is the second equation of the system.}$$
$$-2x + 3(-4x + 13) = -17 \quad \text{Substitute } -4x + 13 \text{ for y.}$$
$$\text{Write the parentheses so that the multiplication by 3 is distributed over both terms of } -4x + 13.$$
$$-2x - 12x + 39 = -17 \quad \text{Distribute the multiplication by 3.}$$
$$-14x + 39 = -17 \quad \text{Combine like terms.}$$
$$-14x = -56 \quad \text{Subtract 39 from both sides.}$$
$$x = 4 \quad \text{To solve for x, divide both sides by } -14.$$
$$\text{This is the x-value of the solution.}$$

> **The Language of Algebra**
> The phrase *back-substitute* can also be used to describe step 3 of the substitution method. To find y, we *back-substitute* 4 for x in the equation $y = -4x + 13$.

Step 3: To find y, we substitute 4 for x in the substitution equation and evaluate the right side.

$$y = -4x + 13 \quad \text{This is the substitution equation.}$$
$$= -4(4) + 13 \quad \text{Substitute 4 for x.}$$
$$= -16 + 13$$
$$= -3 \quad \text{This is the y-value of the solution.}$$

Step 4: To verify that $(4, -3)$ satisfies both equations, we substitute 4 for x and -3 for y into each equation of the original system and simplify.

Check:

$$4x + y = 13 \quad \text{First equation.} \qquad\qquad -2x + 3y = -17 \quad \text{Second equation.}$$
$$4(4) + (-3) \stackrel{?}{=} 13 \qquad\qquad -2(4) + 3(-3) \stackrel{?}{=} -17$$
$$16 - 3 \stackrel{?}{=} 13 \qquad\qquad -8 - 9 \stackrel{?}{=} -17$$
$$13 = 13 \quad \text{True} \qquad\qquad -17 = -17 \quad \text{True}$$

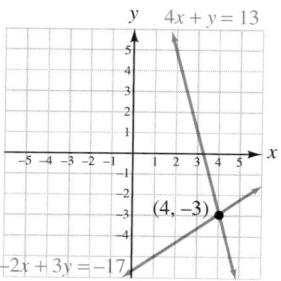
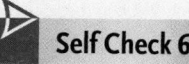

Since $(4, -3)$ satisfies both equations of the system, it is the solution of the system. The solution set is $\{(4, -3)\}$. The graphs of the equations of the system help to verify this. They appear to intersect at $(4, -3)$, as shown on the left.

Self Check 6   Solve the system by substitution: $\begin{cases} x + 3y = 9 \\ 2x - y = -10 \end{cases}$

Now Try   **Problem 41**

⑤   Solve Systems of Linear Equations by the Elimination (Addition) Method.

Recall that with the elimination (addition) method, we combine the equations of the system in a way that will eliminate terms involving one of the variables.

The Elimination (Addition) Method

1. Write both equations of the system in standard (general) form: $Ax + By = C$.
2. If necessary, multiply one or both of the equations by a nonzero number chosen to make the coefficients of x (or the coefficients of y) opposites.
3. Add the equations to eliminate the terms involving x (or y).
4. Solve the equation resulting from step 3.
5. Find the value of the remaining variable by substituting the solution found in step 4 into any equation containing both variables. Or, repeat steps 2–4 to eliminate the other variable.
6. Check the proposed solution in each equation of the original system. Write the solution as an ordered pair.

EXAMPLE 7   Solve: $\begin{cases} \dfrac{4}{3}x + \dfrac{1}{2}y = -\dfrac{2}{3} \\ 0.3x + 0.4y = 1 \end{cases}$

Strategy   We will find an equivalent system without fractions or decimals and use the elimination method to solve it.

Why   It's usually easier to solve a system of equations that involves only integers.

Solution

Step 1: To clear the first equation of the fractions, we multiply both sides by 6. To clear the second equation of decimals, we multiply both sides by 10.

$$\begin{cases} \dfrac{4}{3}x + \dfrac{1}{2}y = -\dfrac{2}{3} \\ 0.3x + 0.4y = 1 \end{cases} \xrightarrow[\text{Multiply by 10}]{\text{Multiply by 6}} \begin{array}{l} 6\left(\dfrac{4}{3}x + \dfrac{1}{2}y\right) = 6\left(-\dfrac{2}{3}\right) \\ 10(0.3x + 0.4y) = 10(1) \end{array} \xrightarrow[\text{Simplify}]{\text{Simplify}} \begin{cases} 8x + 3y = -4 \\ 3x + 4y = 10 \end{cases}$$

Step 2: To make the y-terms drop out when we add the equations, we multiply both sides of $8x + 3y = -4$ by 4 and both sides of $3x + 4y = 10$ by -3 to get

$$\begin{cases} 8x + 3y = -4 \\ 3x + 4y = 10 \end{cases} \xrightarrow[\text{Multiply by } -3]{\text{Multiply by 4}} \begin{cases} 32x + 12y = -16 \\ -9x - 12y = -30 \end{cases}$$

Step 3: When these equations are added, the y-terms drop out.

$$32x + 12y = -16$$
$$\underline{-9x - 12y = -30}$$ Add the like terms, column by column: $32x + (-9x) = 23x$,
$$23x \qquad\;\; = -46$$ $12y + (-12y) = 0$, and $-16 + (-30) = -46$.

Step 4: We solve the resulting equation to find x.

$$23x = -46$$
$$x = -2 \quad \text{To solve for x, divide both sides by 23. This is the x-value of the solution.}$$

Step 5: To find y, we can substitute -2 for x in either of the equations of the original system or either of the equations of the equivalent system. It appears the computations will be the simplest if we use $3x + 4y = 10$.

$$3x + 4y = 10 \quad \text{This is the second equation of the equivalent system.}$$
$$3(-2) + 4y = 10 \quad \text{Substitute } -2 \text{ for x.}$$
$$-6 + 4y = 10 \quad \text{Simplify.}$$
$$4y = 16 \quad \text{To isolate the variable term, add 6 to both sides.}$$
$$y = 4 \quad \text{To solve for y, divide both sides by 4. This is the y-value of the solution.}$$

Step 6: The solution is $(-2, 4)$ and the solution set is $\{(-2, 4)\}$. Verify that the solution checks using the original equations.

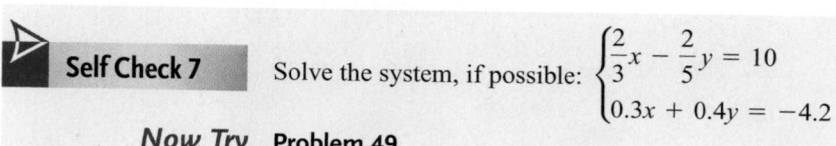

Self Check 7   Solve the system, if possible: $\begin{cases} \dfrac{2}{3}x - \dfrac{2}{5}y = 10 \\ 0.3x + 0.4y = -4.2 \end{cases}$

Now Try   **Problem 49**

6   **Use Substitution and Elimination to Identify Inconsistent Systems and Dependent Equations.**

EXAMPLE 8   Solve $\begin{cases} y = 2x + 4 \\ 8x - 4y = 7 \end{cases}$, if possible.

Strategy   We will use the substitution method to solve this system.

Why   The substitution method works well when one of the equations of the system (in this case, $y = 2x + 4$) is solved for a variable.

Solution   Since the first equation is solved for y, we will use the substitution method.

$$y = 2x + 4 \quad \text{This is the substitution equation.}$$
$$8x - 4y = 7 \quad \text{This is the second equation of the system.}$$
$$8x - 4(2x + 4) = 7 \quad \text{Substitute } 2x + 4 \text{ for y.}$$

We can now try to solve this equation for x:

$$8x - 8x - 16 = 7 \quad \text{Distribute the multiplication by } -4.$$
$$-16 = 7 \quad \text{Simplify the left side: } 8x - 8x = 0.$$

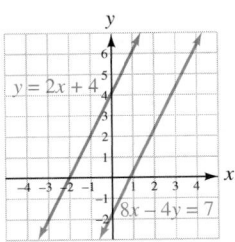

Here, the terms involving x drop out, and we get $-16 = 7$. This false statement indicates that the system has *no solution* and is, therefore, inconsistent. The solution set is $\varnothing$. The graphs of the equations of the system help to verify this. They appear to be parallel lines, as shown on the left.

> **Self Check 8**   Solve the system, if possible: $\begin{cases} 5x - 15y = 2 \\ x = 3y - 1 \end{cases}$
>
> *Now Try*   **Problem 51**

EXAMPLE 9   Solve: $\begin{cases} 2(2x + 3y) = 12 \\ -2x = 3y - 6 \end{cases}$

Strategy   We will write each equation in standard (general) form $Ax + By = C$ and use the elimination (addition) method to solve the resulting equivalent system.

Why   In their current form, the equations do not contain terms with coefficients that are opposites.

Solution   To write the first equation in standard (general) form, we use the distributive property. To write the second equation in standard form, we subtract $3y$ from both sides.

The first equation

$$2(2x + 3y) = 12$$
$$4x + 6y = 12$$

The second equation

$$-2x = 3y - 6$$
$$-2x - 3y = -6$$

We now copy $4x + 6y = 12$ and multiply both sides of $-2x - 3y = -6$ by 2 to get

$$\begin{aligned} 4x + 6y &= 12 \\ -4x - 6y &= -12 \\ \hline 0 &= 0 \end{aligned}$$

When we add like terms, column by column, the result is $0x + 0y = 0$, which simplifies to $0 = 0$.

Here, both the x- and y-terms drop out. The resulting true statement $0 = 0$ indicates that the equations are dependent and that the system has an *infinitely many solutions*. The solution set is written using set-builder notation as $\{(x, y) \mid 4x + 6y = 12\}$ and is read as "the set of all ordered pairs (x, y) such that $4x + 6y = 12$."

Note that the equations of the system are dependent equations, because when the second equation is multiplied by -2, it becomes the first equation. The graphs of these equations are, therefore, the same line. To find some of the infinitely many solutions of the system, we can substitute 0, 3, and -3 for x in either equation to obtain $(0, 2)$, $(3, 0)$, and $(-3, 4)$.

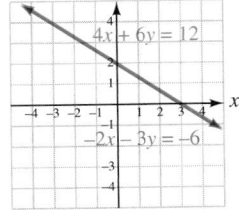

> **Self Check 9**   Solve the system: $\begin{cases} 2x = 5(y + 2) \\ -4x + 10y = -20 \end{cases}$
>
> *Now Try*   **Problem 53**

Solve Problems Using Systems of Equations.

> **EXAMPLE 10**   *Retail Sales.*   An electronics store carries two brands of car audio CD receivers. The one made by Pioneer sells for $129 and the other, made by Panasonic, sells for $189. If the store receipts from the sale of 36 car CD receivers totaled $5,604, how many of each brand were sold?
>
> **Analyze the Problem**   We will let $x =$ the number of Pioneer CD receivers that sold for $129 and $y =$ the number of Panasonic CD receivers that sold for $189. Thus, the receipts from the sale of the Pioneer receivers were $129x$, and the receipts from the sale of the Panasonic receivers were $189y$.
>
> **Form Two Equations**   We can translate the words of the problem into two equations, each involving x and y.
>
> | The number of Pioneer CD receivers sold | plus | the number of Panasonic CD receivers sold | is | the total number of CD receivers sold. |
> |---|---|---|---|---|
> | x | $+$ | y | $=$ | 36 |
>
> | The receipts from the sale of the Pioneer CD receivers | plus | the receipts from the sale of the Panasonic CD receivers | is | the total receipts. |
> |---|---|---|---|---|
> | $129x$ | $+$ | $189y$ | $=$ | 5,604 |
>
> **Solve the System**   We can solve the following system for x and y to find out how many of each type of CD receiver were sold.
>
> $$\begin{cases} x + y = 36 \\ 129x + 189y = 5,604 \end{cases}$$
>
> We multiply both sides of $x + y = 36$ by -189, add the resulting equation to $129x + 189y = 5,604$, and solve for x:
>
> $$\begin{array}{rl} -189x - 189y = & -6,804 \\ \underline{129x + 189y = 5,604} & \text{Add the like terms, column by column.} \\ -60x = -1,200 & \\ x = 20 & \text{Divide both sides by } -60. \end{array}$$
>
> To find y, we substitute 20 for x in the first equation and solve for y:
>
> $$\begin{array}{rl} x + y = 36 & \text{This is the first equation of the original system.} \\ 20 + y = 36 & \text{Substitute 20 for } x. \\ y = 16 & \text{To solve for } y, \text{ subtract 20 from both sides.} \end{array}$$
>
> **State the Conclusion**   The store sold 20 of the Pioneer car audio CD receivers and 16 of the Panasonic car audio CD receivers.
>
> **Check the Result**   If 20 of Pioneer receivers were sold and 16 of the Panasonic receivers were sold, a total of 36 receivers were sold. Since the value of Pioneer receivers sold is $20(\$129) = \$2,580$, and the value of the Panasonic receivers sold is $16(\$189) = \$3,024$, the total receipts were $\$2,580 + \$3,024 = \$5,604$. The results check.

Success Tip
We could also solve this system using the substitution method. If we subtract y from both sides of the first equation, we obtain the substitution equation $x = 36 - y$.

 Now Try   **Problem 97**

▷ **ANSWERS TO SELF CHECKS** **1.** Yes

2. (1, 2)   **3.** No solution, ∅; inconsistent system   **4.** Infinitely many solutions,

$\{(x, y) \mid y = 2x - 4\}$
or $\{(x, y) \mid 2x - y = 4\}$;
dependent equations

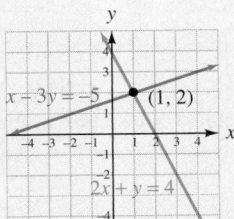
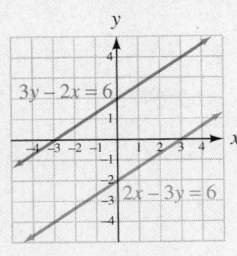
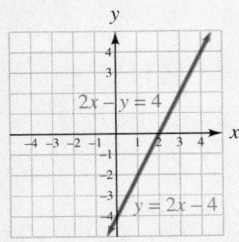

5. (−6, 4)   **6.** (−3, 4) **7.** (6, −15) **8.** No solution, ∅; inconsistent system

9. Infinitely many solutions, $\{(x, y) \mid 2x - 5y = 10\}$; dependent equations

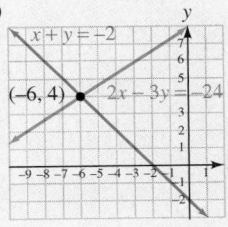

STUDY SET
12.1

VOCABULARY

Fill in the blanks.

1. The pair of equations $\begin{cases} x - y = -1 \\ 2x - y = 1 \end{cases}$ is called a _____ of equations.

2. Because the ordered pair (2, 3) satisfies both equations in problem 1, it is a _____ of the system of equations.

3. When the graphs of the equations of a system are identical lines, the equations are called dependent and the system has _____ many solutions.

4. A system that has no solution is called an _____ system.

CONCEPTS

5. Refer to the illustration. Determine whether a true or a false statement is obtained when the coordinates of

a. point *A* are substituted into the equation for line l_1.

b. point *B* are substituted into the equation for line l_1.

c. point *C* are substituted into the equation for line l_1.

d. point *C* are substituted into the equation for line l_2.

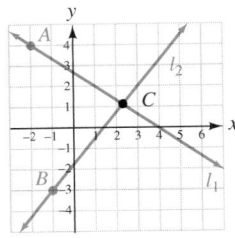

6. Refer to the illustration.

a. How many ordered pairs satisfy the equation $3x + y = 3$? Name three.

b. How many ordered pairs satisfy the equation $\frac{2}{3}x - y = -3$? Name three.

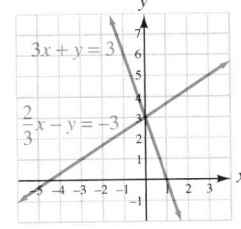

c. How many ordered pairs satisfy both equations? Name it or them.

7. If the system $\begin{cases} 4x - 3y = 7 \\ 3x - 2y = 6 \end{cases}$ is to be solved using the elimination method, by what constant should each equation be multiplied if

a. the *x*-terms are to drop out?

b. the *y*-terms are to drop out?

8. Consider the system: $\begin{cases} \frac{2}{3}x - \frac{y}{6} = \frac{16}{9} \\ 0.03x + 0.02y = 0.03 \end{cases}$

a. What step should be performed to clear the first equation of fractions?

b. What step should be performed to clear the second equation of decimals?

NOTATION

Complete each solution.

9. Solve: $\begin{cases} y = 3x - 7 \\ x + y = 5 \end{cases}$

10. Solve: $\begin{cases} 6x + 2y = 0 \\ x - 2y = 7 \end{cases}$

$x + () = 5$

$x + 3x - 7 = $

$x - 7 = 5$

$4x = $

$x = 3$

$y = 3x - 7$

$y = 3() - 7$

$y = $

The solution is (, 2).

$6x + 2y = 0$

$x - 2y = 7$

$7x = $

$x = $

$x - 2y = 7$

$ - 2y = 7$

$-2y = $

$y = $

The solution is (, -3).

GUIDED PRACTICE

Determine whether the ordered pair is a solution of the system of equations. See Example 1.

11. $(-4, 3)$; $\begin{cases} 4x - y = -19 \\ 3x + 2y = -6 \end{cases}$

12. $(-1, 2)$; $\begin{cases} 3x - y = -5 \\ x - y = -4 \end{cases}$

13. $(2, -3)$; $\begin{cases} y + 2 = \dfrac{1}{2}x \\ 3x + 2y = 0 \end{cases}$

14. $(1, 2)$; $\begin{cases} 2x - y = 0 \\ y = \dfrac{1}{2}x + \dfrac{3}{2} \end{cases}$

15. $\left(\dfrac{1}{2}, \dfrac{1}{3}\right)$; $\begin{cases} 2x + 3y = 2 \\ 4x - 9y = 1 \end{cases}$

16. $\left(-\dfrac{3}{4}, \dfrac{2}{3}\right)$; $\begin{cases} 4x + 3y = -1 \\ 4x - 3y = -5 \end{cases}$

17. $(-0.2, 0.5)$; $\begin{cases} 2x + 5y = 2.1 \\ 5x + y = -0.5 \end{cases}$

18. $(2.1, -3.2)$; $\begin{cases} x + y = -1.1 \\ 2x - 3y = 13.8 \end{cases}$

Solve each system by graphing. See Example 2.

19. $\begin{cases} x + y = 6 \\ x - y = 2 \end{cases}$

20. $\begin{cases} x - y = 4 \\ 2x + y = 5 \end{cases}$

21. $\begin{cases} y = -2x + 1 \\ x - 2y = -7 \end{cases}$

22. $\begin{cases} 3x - y = -3 \\ y = -2x - 7 \end{cases}$

Solve each system by graphing, if possible. If a system is inconsistent or if the equations are dependent, state this. See Examples 3 and 4.

23. $\begin{cases} 3x - 3y = 4 \\ x - y = 4 \end{cases}$

24. $\begin{cases} 5x + 2y = 6 \\ -10x - 4y = -12 \end{cases}$

25. $\begin{cases} x = 3 - 2y \\ 2x + 4y = 6 \end{cases}$

26. $\begin{cases} 3x = 5 - 2y \\ 3x + 2y = 7 \end{cases}$

Solve each system by graphing. See Example 5.

27. $\begin{cases} \dfrac{1}{6}x = \dfrac{1}{3}y + \dfrac{1}{2} \\ y = x \end{cases}$

28. $\begin{cases} x = y + 3 \\ \dfrac{1}{4}x - \dfrac{1}{6}y = \dfrac{1}{3} \end{cases}$

29. $\begin{cases} \dfrac{1}{3}x - \dfrac{7}{6}y = \dfrac{1}{2} \\ \dfrac{1}{5}y = \dfrac{1}{3}x + \dfrac{7}{15} \end{cases}$

30. $\begin{cases} \dfrac{3}{5}x + \dfrac{1}{4}y = -\dfrac{11}{10} \\ \dfrac{1}{8}x = \dfrac{13}{24} + \dfrac{1}{3}y \end{cases}$

Use a graphing calculator to solve each system. Give all answers to the nearest hundredth. See Using Your Calculator: Solving Systems by Graphing.

31. $\begin{cases} y = 3.2x - 1.5 \\ y = -2.7x - 3.7 \end{cases}$

32. $\begin{cases} y = -0.45x + 5 \\ y = 5.55x - 13.7 \end{cases}$

33. $\begin{cases} 1.7x + 2.3y = 3.2 \\ y = 0.25x + 8.95 \end{cases}$

34. $\begin{cases} 2.75x = 12.9y - 3.79 \\ 7.1x - y = 35.76 \end{cases}$

Solve each system by substitution. See Example 6.

35. $\begin{cases} y = 3x \\ x + y = 8 \end{cases}$

36. $\begin{cases} y = x + 2 \\ x + 2y = 16 \end{cases}$

37. $\begin{cases} x = 2 + y \\ 2x + y = 13 \end{cases}$

38. $\begin{cases} x = -5 + y \\ 3x - 2y = -7 \end{cases}$

39. $\begin{cases} x + 2y = 6 \\ 3x - y = -10 \end{cases}$

40. $\begin{cases} 2x - y = -21 \\ 4x + 5y = 7 \end{cases}$

41. $\begin{cases} 5x + 3y = -26 \\ 3x + y = -14 \end{cases}$

42. $\begin{cases} 3x + 5y = 4 \\ 5x + y = 14 \end{cases}$

Solve each system by elimination (addition). See Example 7.

43. $\begin{cases} x - y = 7 \\ x + y = 11 \end{cases}$

44. $\begin{cases} a + b = 5 \\ a - b = 11 \end{cases}$

45. $\begin{cases} 2s + 3t = -8 \\ 2s - 3t = -8 \end{cases}$

46. $\begin{cases} x + 2y = -21 \\ x - 2y = 11 \end{cases}$

47. $\begin{cases} 3x + 4y = -24 \\ 5x + 12y = -72 \end{cases}$

48. $\begin{cases} 5x + 2y = 11 \\ 7x + 6y = 9 \end{cases}$

49. $\begin{cases} \dfrac{5}{6}x + \dfrac{1}{2}y = 12 \\ 0.3x + 0.5y = 5.6 \end{cases}$

50. $\begin{cases} \dfrac{1}{3}x + \dfrac{1}{2}y = \dfrac{31}{6} \\ 0.3x + 0.2y = 3.9 \end{cases}$

Solve each system, if possible. If a system is inconsistent or if the equations are dependent, state this. See Examples 8 and 9.

51. $\begin{cases} 6x + 3y = 18 \\ y = -2x + 5 \end{cases}$

52. $\begin{cases} 8x - 4y = 16 \\ 2x - 4 = y \end{cases}$

53. $\begin{cases} 3x - y = 5 \\ 21x = 7(y + 5) \end{cases}$

54. $\begin{cases} 4x + 8y = 15 \\ x = 2(2 - y) \end{cases}$

TRY IT YOURSELF

Solve each system by graphing, if possible. If a system is inconsistent or if the equations are dependent, state this. (Hint: Several coordinates of points of intersection are fractions.)

55. $\begin{cases} 4x - 3y = 5 \\ y = -2x \end{cases}$

56. $\begin{cases} 2x + 2y = -1 \\ 3x + 4y = 0 \end{cases}$

57. $\begin{cases} y = -\dfrac{5}{2}x + \dfrac{1}{2} \\ 2x - \dfrac{3}{2}y = 5 \end{cases}$

58. $\begin{cases} \dfrac{5}{2}x + 3y = 6 \\ y = -\dfrac{5}{6}x + 2 \end{cases}$

59. $\begin{cases} x = \dfrac{11 - 2y}{3} \\ y = \dfrac{11 - 6x}{4} \end{cases}$

60. $\begin{cases} x = \dfrac{1 - 3y}{4} \\ y = \dfrac{12 + 3x}{2} \end{cases}$

61. $\begin{cases} x = 13 - 4y \\ 3x = 4 + 2y \end{cases}$

62. $\begin{cases} 3x = 7 - 2y \\ 2x = 2 + 4y \end{cases}$

63. $\begin{cases} x = 2 \\ y = -\dfrac{1}{2}x + 2 \end{cases}$

64. $\begin{cases} y = -2 \\ y = \dfrac{2}{3}x - \dfrac{4}{3} \end{cases}$

65. $\begin{cases} x + 3y = 6 \\ y = -\dfrac{1}{3}x + 2 \end{cases}$

66. $\begin{cases} 2x - y = -4 \\ 2y = 4x - 6 \end{cases}$

Solve each system by any method, if possible. If a system is inconsistent or if the equations are dependent, so indicate.

67. $\begin{cases} 2x + 3y = 8 \\ 3x - 2y = -1 \end{cases}$

68. $\begin{cases} x = \dfrac{3}{2}y + 5 \\ 2x - 3y = 8 \end{cases}$

69. $\begin{cases} 4(x - 2) = -9y \\ 2(x - 3y) = -3 \end{cases}$

70. $\begin{cases} 2(2x + 3y) = 5 \\ 8x = 3(1 + 3y) \end{cases}$

71. $\begin{cases} 0.3a + 0.1b = 0.5 \\ \dfrac{4}{3}a + \dfrac{1}{3}b = 3 \end{cases}$

72. $\begin{cases} 0.9p + 0.2q = 1.2 \\ \dfrac{2}{3}p + \dfrac{1}{9}q = 1 \end{cases}$

73. $\begin{cases} \dfrac{x}{2} + \dfrac{y}{2} = 6 \\ \dfrac{x}{3} + \dfrac{y}{3} = 4 \end{cases}$

74. $\begin{cases} \dfrac{x}{2} - \dfrac{y}{3} = -4 \\ \dfrac{x}{2} + \dfrac{y}{9} = 0 \end{cases}$

75. $\begin{cases} x = \dfrac{2}{3}y \\ y = 4x + 5 \end{cases}$

76. $\begin{cases} 5x - 2y = 19 \\ y = \dfrac{1 - 3x}{4} \end{cases}$

77. $\begin{cases} 3x - 4y = 9 \\ x + 2y = 8 \end{cases}$

78. $\begin{cases} 3x - 2y = -10 \\ 6x + 5y = 25 \end{cases}$

79. $\begin{cases} x - \dfrac{4y}{5} = 4 \\ \dfrac{y}{3} = \dfrac{x}{2} - \dfrac{5}{2} \end{cases}$

80. $\begin{cases} 3x - 2y = \dfrac{9}{2} \\ \dfrac{x}{2} - \dfrac{3}{4} = 2y \end{cases}$

81. $\begin{cases} \dfrac{2}{3}x - \dfrac{1}{4}y = -8 \\ 0.5x - 0.375y = -9 \end{cases}$

82. $\begin{cases} 0.5x + 0.5y = 6 \\ \dfrac{x}{2} - \dfrac{y}{2} = -2 \end{cases}$

83. $\begin{cases} \dfrac{3}{2}p + \dfrac{1}{3}q = 2 \\ \dfrac{2}{3}p + \dfrac{1}{9}q = 1 \end{cases}$

84. $\begin{cases} a + \dfrac{b}{3} = \dfrac{5}{3} \\ \dfrac{a + b}{3} = 3 - a \end{cases}$

85. $\begin{cases} \dfrac{m - n}{5} + \dfrac{m + n}{2} = 6 \\ \dfrac{m - n}{2} - \dfrac{m + n}{4} = 3 \end{cases}$

86. $\begin{cases} \dfrac{r - 2}{5} + \dfrac{s + 3}{2} = 5 \\ \dfrac{r + 3}{2} + \dfrac{s - 2}{3} = 6 \end{cases}$

Solve each system. To do so, substitute a for $\frac{1}{x}$ and b for $\frac{1}{y}$ and solve for a and b. Then find x and y using the fact that $a = \frac{1}{x}$ and $b = \frac{1}{y}$.

87. $\begin{cases} \dfrac{1}{x} + \dfrac{1}{y} = \dfrac{5}{6} \\ \dfrac{1}{x} - \dfrac{1}{y} = \dfrac{1}{6} \end{cases}$

88. $\begin{cases} \dfrac{1}{x} + \dfrac{1}{y} = \dfrac{9}{20} \\ \dfrac{1}{x} - \dfrac{1}{y} = \dfrac{1}{20} \end{cases}$

89. $\begin{cases} \dfrac{1}{x} + \dfrac{2}{y} = -1 \\ \dfrac{2}{x} - \dfrac{1}{y} = -7 \end{cases}$

90. $\begin{cases} \dfrac{3}{x} - \dfrac{2}{y} = -30 \\ \dfrac{2}{x} - \dfrac{3}{y} = -30 \end{cases}$

APPLICATIONS

91. HEARING TESTS   See the illustration. At what frequency and decibel level were the hearing test results the same for the left and right ear? Write your answer as an ordered pair.

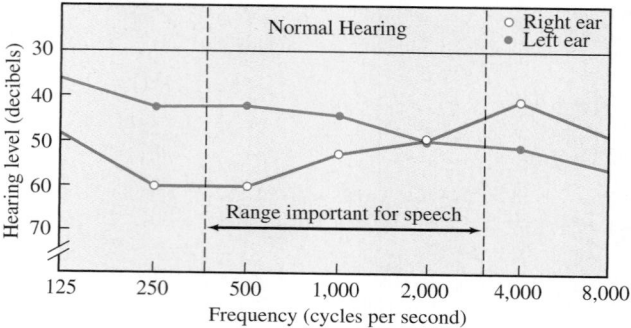

92. THE INTERNET

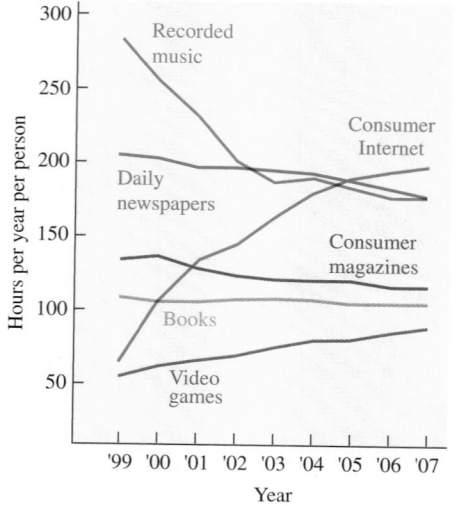

Source: Veronis Suhler Stevenson

The graph in the previous column shows the growing importance of the Internet in the daily lives of Americans. Determine when the time spent on the following activities was the same. Approximately how many hours per year were spent on each?

a. Internet and reading magazines

b. Internet and reading newspapers

c. Internet and reading books

d. Reading newspapers and listening to recorded music

93. SUPPLY AND DEMAND   The demand function, graphed below, describes the relationship between the price x of a certain camera and the demand for the camera.

a. The supply function, $S(x) = \frac{25}{4}x - 525$, describes the relationship between the price x of the camera and the number of cameras the manufacturer is willing to supply. Graph this function on the illustration.

b. For what price will the supply of cameras equal the demand?

c. As the price of the camera is increased, what happens to supply and what happens to demand?

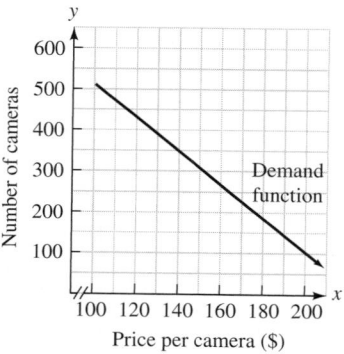

94. COST AND REVENUE   The function $C(x) = 200x + 400$ gives the cost for a college to offer x sections of an introductory class in CPR (cardiopulmonary resuscitation). The function $R(x) = 280x$ gives the amount of revenue the college brings in when offering x sections of CPR.

a. Find the *break-even point* (where cost = revenue) by graphing each function on the same coordinate system.

b. How many sections does the college need to offer to make a profit on the CPR training course?

95. BUSINESS   Estimate the break-even point (where cost = revenue) on the graph on the next page. Explain why it is called the *break-even point.*

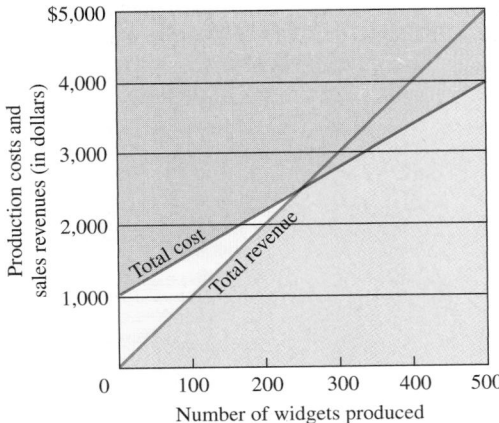

96. NAVIGATION   The paths of two ships are tracked on the same coordinate system. One ship is following a path described by the equation $2x + 3y = 6$, and the other is following a path described by the equation $y = \frac{2}{3}x - 3$.

   a. Is there a possibility of a collision?

   b. What are the coordinates of the danger point?

   c. Is a collision a certainty?

In Exercises 97–112, write a system of two equations in two variables to solve each problem.

97. TICKET SALES   The ticket prices for a Halloween haunted house were $5 for adults and $3 for children. On a day when a total of 390 tickets were purchased, the receipts were $1,470. How many of each type of ticket were sold?

98. ICE CREAM SALES   At a snack shop, ice cream cones cost $2.10 and sundaes cost $3.20. One day the receipts for a total of 140 cones and sundaes were $360. How many cones were sold? How many sundaes?

99. ADVERTISING   Use the information in the ad to find the cost of a 15-second and the cost of a 30-second radio commercial on radio station KLIZ.

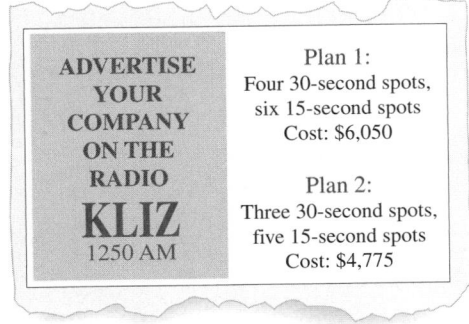

ADVERTISE YOUR COMPANY ON THE RADIO

KLIZ
1250 AM

Plan 1:
Four 30-second spots, six 15-second spots
Cost: $6,050

Plan 2:
Three 30-second spots, five 15-second spots
Cost: $4,775

100. CONCERTS   According to StubHub.com, in 2006, two tickets to a Rolling Stones concert and two tickets to a Jimmy Buffet concert cost, on average, a total of $792. At those prices, four tickets to see the Stones and two tickets to see Jimmy Buffet cost $1,320. What was the average cost of a Rolling Stones ticket and a Jimmy Buffet ticket in 2006?

101. GEOMETRY   An acute angle is an angle with measure less than 90°. In a right triangle, the measure of one acute angle is 15° greater than two times the measure of the other acute angle. Find the measure of each acute angle.

102. NEW YORK CITY   The triangular-shaped Flatiron Building in Manhattan has a perimeter of 499 feet at its base. It is bordered on each side by a street. The 5th Avenue front of the building is 198 feet long. The Broadway front is 43 feet more than twice as long as the East 22nd Street front. Find the length of the Broadway front and East 22nd Street front. (Source: New York Public Library)

© Bill Ross/Corbis

103. INVESTMENT CLUBS   Part of $8,000 was invested by an investment club at 10% interest and the rest at 12%. If the annual income from these investments is $900, how much was invested at each rate?

104. RETIREMENT INCOME   A retired couple invested part of $12,000 at 6% interest and the rest at 7.5%. If their annual income from these investments is $810, how much was invested at each rate?

105. SNOWMOBILING   A man rode a snowmobile at the rate of 20 mph and then skied cross country at the rate of 4 mph. During the 6-hour trip, he traveled 48 miles. How long did he snowmobile, and how long did he ski?

106. SALMON   It takes a salmon 40 minutes to swim 10,000 feet upstream and 8 minutes to swim that same portion of a river downstream. Find the speed of the salmon in still water and the speed of the current.

107. PRODUCTION PLANNING   A manufacturer builds racing bikes and mountain bikes, with the per unit manufacturing costs shown in the table. The company has budgeted $26,150 for materials and $31,800 for labor. How many bicycles of each type can be built?

Model	Cost of materials	Cost of labor
Racing	$110	$120
Mountain	$140	$180

108. FARMING A farmer keeps some animals on a strict diet. Each animal is to receive 15 grams of protein and 7.5 grams of carbohydrates. The farmer uses two food mixes, with nutrients as shown in the table. How many grams of each mix should be used to provide the correct nutrients for each animal?

Mix	Protein	Carbohydrates
Mix A	12%	9%
Mix B	15%	5%

109. COSMETOLOGY A beauty shop specializing in permanents has fixed costs of $2,101.20 per month. The owner estimates that the cost for each permanent is $23.60, which covers labor, chemicals, and electricity. If her shop can give as many permanents as she wants at a price of $44 each, how many must be given each month for her to break even?

110. MIXING CANDY How many pounds of each candy shown in the illustration must be mixed to obtain 60 pounds of candy that would be worth $4 per pound?

Gummy Bears
$3.50/lb

Jelly Beans
$5.50/lb

111. DERMATOLOGY Tests of an antibacterial face-wash cream showed that a mixture containing 0.3% Triclosan (active ingredient) gave the best results. How many grams of cream from each tube should be used to make an equal-size tube of the 0.3% cream?

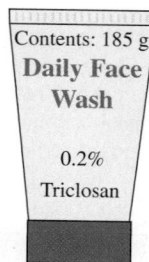

Contents: 185 g
Daily Face Wash

0.2%
Triclosan

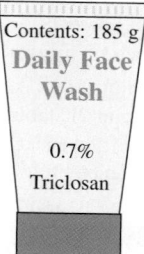

Contents: 185 g
Daily Face Wash

0.7%
Triclosan

112. MIXING SOLUTIONS How many ounces of the two alcohol solutions in the illustration must be mixed to obtain 100 ounces of a 12.2% solution?

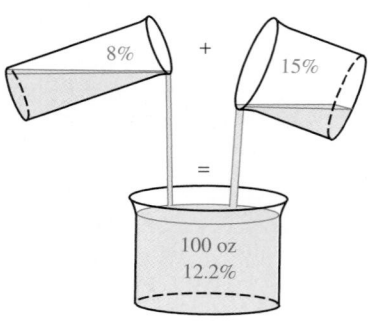

8%   +   15%

=

100 oz
12.2%

WRITING

113. Which method would you use to solve the system? Explain.

$$\begin{cases} y - 1 = 3x \\ 3x + 2y = 12 \end{cases}$$

114. Which method would you use to solve the system? Explain.

$$\begin{cases} 2x + 4y = 9 \\ 3x - 5y = 20 \end{cases}$$

115. When solving a system, what advantages are there with the substitution and elimination methods compared with the graphing method?

116. When using the elimination (addition) method, how can you tell whether

   a. a system of linear equations has no solution?

   b. a system of linear equations has infinitely many solutions?

REVIEW

Solve each formula for the specified variable.

117. $\dfrac{V_2}{V_1} = \dfrac{P_1}{P_2}$ for P_1

118. $\dfrac{1}{r} = \dfrac{1}{r_1} + \dfrac{1}{r_2}$ for r

119. $S = \dfrac{a - lr}{1 - r}$ for r

120. $P = \dfrac{Q_1}{Q_2 - Q_1}$ for Q_1

CHALLENGE PROBLEMS

121. If the solution of the system $\begin{cases} Ax + By = -2 \\ Bx - Ay = -26 \end{cases}$ is $(-3, 5)$, find the values of A and B.

122. Solve $\begin{cases} 2ab - 3cd = 1 \\ 3ab - 2cd = 1 \end{cases}$ and assume that b and d are constants.

SECTION 12.2
Solving Systems of Equations in Three Variables

Objectives

1. Determine whether an ordered triple is a solution of a system.
2. Solve systems of three linear equations in three variables.
3. Solve systems of equations with missing variable terms.
4. Identify inconsistent systems and dependent equations.

In previous sections, we solved systems of linear equations in two variables. We will now extend this discussion to consider systems of linear equations in *three* variables.

1 Determine Whether an Ordered Triple is a Solution of a System.

The equation $x - 5y + 7z = 10$, where each variable is raised to the first power, is an example of a linear equation in three variables. In general, we have the following definition.

Standard (General) Form

A **linear equation in three variables** is an equation that can be written in the form

$$Ax + By + Cz = D$$

where A, B, C, and D are real numbers and A, B, and C are not all 0.

A solution of a linear equation in three variables is an **ordered triple** of numbers of the form (x, y, z) whose coordinates satisfy the equation. For example, $(2, 0, 1)$ is a solution of $x + y + z = 3$ because a true statement results when we substitute 2 for x, 0 for y, and 1 for z: $2 + 0 + 1 = 3$.

A **solution of a system of three linear equations** in three variables is an ordered triple that satisfies each equation of the system.

EXAMPLE 1   Determine whether $(-4, 2, 5)$ is a solution of the system:

$$\begin{cases} 2x + 3y + 4z = 18 \\ 3x + 4y + z = 1 \\ x + y + 3z = 13 \end{cases}$$

Strategy   We will substitute the x-, y-, and z-coordinates of $(-4, 2, 5)$ for the corresponding variables in each equation of the system.

Why   If each equation is satisfied by the x-, y-, and z-coordinates, the ordered triple is a solution of the system.

Solution   We substitute -4 for x, 2 for y, and 5 for z in each equation.

The first equation

$2x + 3y + 4z = 18$
$2(-4) + 3(2) + 4(5) \stackrel{?}{=} 18$
$-8 + 6 + 20 \stackrel{?}{=} 18$
$18 = 18$ True

The second equation

$3x + 4y + z = 1$
$3(-4) + 4(2) + 5 \stackrel{?}{=} 1$
$-12 + 8 + 5 \stackrel{?}{=} 1$
$1 = 1$ True

The third equation

$x + y + 3z = 13$
$-4 + 2 + 3(5) \stackrel{?}{=} 13$
$-4 + 2 + 15 \stackrel{?}{=} 13$
$13 = 13$ True

Since $(-4, 2, 5)$ satisfies each equation, it is a solution of the system.

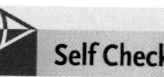

Self Check 1   Determine whether $(6, -3, 1)$ is a solution of the system:

$$\begin{cases} x - y + z = 10 \\ x + 4y - z = -7 \\ 3x - y + 4z = 24 \end{cases}$$

Now Try   **Problem 11**

The Language of Algebra
Recall that when a system of equations has at least one solution, the system is called a *consistent* system, and if a system has no solution, the system is called *inconsistent*.

The graph of an equation of the form $Ax + By + Cz = D$ is a flat surface called a **plane.** A system of three linear equations with three variables is consistent or inconsistent, depending on how the three planes corresponding to the three equations intersect. The following illustration shows some of the possibilities. A system of three linear equations in three variables can have exactly one solution, no solution, or infinitely many solutions.

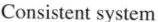

Consistent system   Consistent system   Inconsistent systems

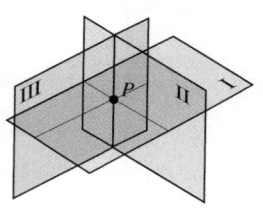   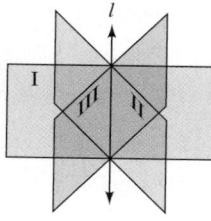   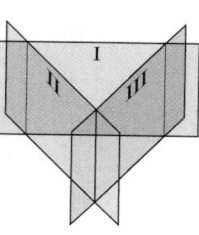

The three planes intersect at a single point P: one solution

The three planes have a line l in common: infinitely many solutions

The three planes have no point in common: no solutions

(a)   (b)   (c)

 Solve Systems of Three Linear Equations in Three Variables.

To **solve a system of three linear equations** in three variables means to find all of the solutions of the system. Solving such a system by graphing is not practical because we would need a coordinate system with three axes.

The substitution method is useful to solve systems of three equations where one or more equations have only two variables. However, the best way to solve systems of three linear equations in three variables is usually the elimination method.

Solving a System of Three Linear Equations by Elimination

1. Write each equation in standard form $Ax + By + Cz = D$ and clear any decimals or fractions.
2. Pick any two equations and eliminate a variable.
3. Pick a different pair of equations and eliminate the same variable as in step 1.
4. Solve the resulting pair of two equations in two variables.
5. To find the value of the third variable, substitute the values of the two variables found in step 4 into any equation containing all three variables and solve the equation.
6. Check the proposed solution in all three of the original equations. Write the solution as an ordered triple.

EXAMPLE 2 Solve the system: $\begin{cases} 2x + y + 4z = 12 \\ x + 2y + 2z = 9 \\ 3x - 3y - 2z = 1 \end{cases}$

Strategy Since the coefficients of the z-terms are opposites in the second and third equations, we will add the left and right sides of those equations to eliminate z. Then we will chose another pair of equations and eliminate z again.

Why The result will be a system of two equations in x and y that we can solve by elimination.

Solution

Step 1: We can skip step 1 because each equation is written in standard form and there are no fractions or decimals to clear. We will number each equation and move to step 2.

(1) $\quad \begin{cases} 2x + y + 4z = 12 \\ \textbf{(2)} \quad x + 2y + 2z = 9 \\ \textbf{(3)} \quad 3x - 3y - 2z = 1 \end{cases}$

Step 2: If we pick equations 2 and 3 and add them, the variable z is eliminated.

(2) $\quad x + 2y + 2z = 9$
(3) $\quad \underline{3x - 3y - 2z = 1}$
(4) $\quad 4x - y = 10 \quad$ *This equation does not contain z.*

Step 3: We now pick a different pair of equations (equations 1 and 3) and eliminate z again. If each side of equation 3 is multiplied by 2, and the resulting equation is added to equation 1, z is eliminated.

(1) $\quad 2x + y + 4z = 12$
$\quad\quad \underline{6x - 6y - 4z = 2} \quad$ *This is 2(3x − 3y − 2z) = 2(1).*
(5) $\quad 8x - 5y = 14 \quad$ *This equation does not contain z.*

Step 4: Equations 4 and 5 form a system of two equations in x and y.

(4) $\quad \begin{cases} 4x - y = 10 \\ \textbf{(5)} \quad 8x - 5y = 14 \end{cases}$

To solve this system, we multiply equation 4 by -5 and add the resulting equation to equation 5 to eliminate y.

$\quad\quad -20x + 5y = -50 \quad$ *This is −5(4x − y) = −5(10).*
(5) $\quad \underline{8x - 5y = 14}$
$\quad\quad -12x = -36$
$\quad\quad\quad\quad x = 3 \quad\quad$ *Divide both sides by −12. This is the x-value of the solution.*

To find y, we substitute 3 for x in any equation containing x and y (such as equation 5) and solve for y:

(5) $\quad 8x - 5y = 14$
$\quad\quad 8(3) - 5y = 14 \quad\quad$ *Substitute 3 for x.*
$\quad\quad 24 - 5y = 14 \quad\quad$ *Simplify.*
$\quad\quad\quad -5y = -10 \quad\quad$ *Subtract 24 from both sides.*
$\quad\quad\quad\quad y = 2 \quad\quad$ *Divide both sides by −5. This is the y-value of the solution.*

Step 5: To find z, we substitute 3 for x and 2 for y in any equation containing x, y, and z (such as equation 1) and solve for z:

(1) $2x + y + 4z = 12$
$\quad\ 2(3) + 2 + 4z = 12$ <small>Substitute 3 for x and 2 for y.</small>
$\quad\qquad 8 + 4z = 12$ <small>Simplify.</small>
$\quad\qquad\qquad 4z = 4$ <small>Subtract 8 from both sides.</small>
$\quad\qquad\qquad\ \ z = 1$ <small>Divide both sides by 4. This is the z-value of the solution.</small>

Step 6: To verify that the solution is (3, 2, 1), we substitute 3 for x, 2 for y, and 1 for z in the three equations of the original system. The solution set is written as $\{(3, 2, 1)\}$. Since this system has a solution, it is a consistent system.

> ▷ **Self Check 2**   Solve the system: $\begin{cases} 2x - 3y + 2z = -7 \\ x + 4y - z = 10 \\ 3x + 2y + z = 4 \end{cases}$
>
> ***Now Try***   **Problem 15**

❸   **Solve Systems of Equations with Missing Variable Terms.**

When one or more of the equations of a system is missing a variable term, the elimination of a variable that is normally performed in step 2 of the solution process can be skipped.

EXAMPLE 3   Solve the system: $\begin{cases} 3x = 6 - 2y + z \\ -y - 2z = -8 - x \\ x = 1 - 2z \end{cases}$

Strategy   Since the third equation does not contain the variable y, we will work with the first and second equations to obtain another equation that does not contain y.

Why   Then we can use the elimination method to solve the resulting system of two equations in x and z.

Solution
Step 1: We use the addition property of equality to write each equation in the standard form $Ax + By + Cz = D$ and number each equation.

(1) $\begin{cases} 3x + 2y - z = 6 \end{cases}$ <small>Add 2y and subtract z from both sides of 3x = 6 − 2y + z.</small>
(2) $\begin{cases} x - y - 2z = -8 \end{cases}$ <small>Add x to both sides of −y − 2z = −8 − x.</small>
(3) $\begin{cases} x + 2z = 1 \end{cases}$ <small>Add 2z to both sides of x = 1 − 2z.</small>

Step 2: Since equation 3 does not have a y-term, we can skip to step 3, where we will find another equation that does not contain a y-term.

Step 3: If each side of equation 2 is multiplied by 2 and the resulting equation is added to equation 1, y is eliminated.

(1) $3x + 2y - \ z = \quad 6$
$\qquad\ \underline{2x - 2y - 4z = -16}$ <small>This is 2(x − y − 2z) = 2(−8).</small>
(4) $5x \qquad\quad -5z = -10$

Step 4: Equations 3 and 4 form a system of two equations in x and z:

(3) $\begin{cases} x + 2z = 1 \\ 5x - 5z = -10 \end{cases}$
(4)

To solve this system, we multiply equation 3 by -5 and add the resulting equation to equation 4 to eliminate x:

$$-5x - 10z = -5 \quad \text{This is } -5(x + 2z) = -5(1).$$

(4) $\dfrac{5x - 5z = -10}{}$

$$-15z = -15$$

$$z = 1 \qquad \text{Divide both sides by } -15. \text{ This is the } z\text{-value of the solution.}$$

To find x, we substitute 1 for z in equation 3.

(3) $x + 2z = 1$

$$x + 2(1) = 1 \qquad \text{Substitute 1 for } z.$$

$$x + 2 = 1 \qquad \text{Multiply.}$$

$$x = -1 \qquad \text{Subtract 2 from both sides.}$$

Step 5: To find y, we substitute -1 for x and 1 for z in equation 1:

(1) $3x + 2y - z = 6$

$$3(-1) + 2y - 1 = 6 \qquad \text{Substitute } -1 \text{ for } x \text{ and 1 for } z.$$

$$-3 + 2y - 1 = 6 \qquad \text{Multiply.}$$

$$2y = 10 \qquad \text{Simplify and add 4 to both sides.}$$

$$y = 5 \qquad \text{Divide both sides by 2.}$$

The solution of the system is $(-1, 5, 1)$ and the solution set is $\{(-1, 5, 1)\}$.

Step 6: Check the proposed solution in all three of the original equations.

> **Success Tip**
> We don't have to find the values of the variables in alphabetical order. In Step 2, choose the variable that is the easiest to eliminate. In this example, the value of z is found first.

Self Check 3   Solve the system: $\begin{cases} x + 2y = 1 + z \\ 2x = 3 + y - z \\ x + z = 3 \end{cases}$

Now Try   **Problem 23**

EXAMPLE 4   Solve the system: $\begin{cases} x - y + 4z = -30 & \textbf{(1)} \\ x + 2y = 200 & \textbf{(2)} \\ y + z = 30 & \textbf{(3)} \end{cases}$

Strategy   Since the second equation does not contain the variable z, we will work with the first and third equations to obtain another equation that does not contain z.

Why   Then we can use the elimination method to solve the resulting system of two equations in x and y.

Solution

Step 1: Each of the equations is written in standard form, however, two of the equations are missing variable terms.

Step 2: Since equation 2 does not have a z-term, we can skip to step 3, where we will find another equation that does not contain a z-term.

Step 3: If each side of the equation 3 is multiplied by -4 and the resulting equation is added to the equation 1, z is eliminated.

(1) $x - y + 4z = -30$

$\underline{ -4y - 4z = -120}$ This is $-4(y + z) = -4(30).$

(4) $x - 5y = -150$

Step 4: Equations 2 and 4 form a system of two equations in x and y.

(2) $\begin{cases} x + 2y = 200 \\ x - 5y = -150 \end{cases}$
(4)

To solve this system, we multiply equation 4 by -1 and add the resulting equation to equation 2 to eliminate x.

(2) $x + 2y = 200$

$\underline{ -x + 5y = 150}$ This is $-1(x - 5y) = -1(-150).$

$7y = 350$

$y = 50$ To find y, divide both sides by 7.

We then find the value of x to complete step 4 and the value of z (step 5) in the same way as in Examples 2 and 3.

Step 6: Check that the solution is $(100, 50, -20)$.

> ▷ **Self Check 4**   Solve the system: $\begin{cases} x + y + 3z = 35 \\ x + 3y = -20 \\ 2y + z = -35 \end{cases}$
>
> **Now Try**   **Problem 29**

In the following example, we will solve the same system from Example 4 using a different approach.

EXAMPLE 5   Use substitution to solve the system: $\begin{cases} x - y + 4z = -30 \\ x + 2y = 200 \\ y + z = 30 \end{cases}$

Strategy   We will solve the second equation for x and the third equation for z. This creates two substitution equations. Then we will substitute the results for x and for z in the first equation.

Why   These substitutions will produce one equation in the variable y.

Solution   The method we will use is much like solving systems of two equations by substitution. However, we must find two substitution equations, instead of one.

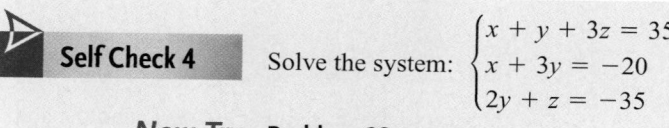

$\begin{cases} x - y + 4z = -30 \\ x + 2y = 200 \\ y + z = 30 \end{cases}$ Solve for x Solve for z $\begin{cases} x - y + 4z = -30 & \text{(1)} \\ x = 200 - 2y & \text{(2)} \\ z = 30 - y & \text{(3)} \end{cases}$

Since the variable x is isolated in equation 2, we will substitute $200 - 2y$ for x in equation 1, and since the variable z is isolated in equation 3, we will substitute $30 - y$ for z in equation 1. These substitutions will eliminate x and z from equation 1, leaving an equation in one variable, y.

$$x - y + 4z = -30 \qquad \text{This is equation 1.}$$

$$200 - 2y - y + 4(30 - y) = -30 \qquad \text{Substitute } 200 - 2y \text{ for x and } 30 - y \text{ for z.}$$

$$200 - 2y - y + 120 - 4y = -30 \qquad \text{Distribute the multiplication by 4.}$$

$$320 - 7y = -30 \qquad \text{On the left side, combine like terms.}$$

$$-7y = -350 \qquad \text{Subtract 320 from both sides.}$$

$$y = 50 \qquad \text{To solve for y, divide both sides by } -7.$$

As expected, this is the same value for y that we obtained using elimination in Example 4. We can substitute 50 for y in equation 2 to find that $x = 100$ and 50 for y in equation 3 to find that $z = -20$. Using elimination or substitution, we find that the solution is $(100, 50, -20)$.

 Self Check 5   Use substitution to solve the system: $\begin{cases} x + y - 4z = -54 \\ x - y = -6 \\ 3y + z = 12 \end{cases}$

Now Try   Problem 31

 Identify Inconsistent Systems and Dependent Equations.

EXAMPLE 6   If possible, solve the system: $\begin{cases} 2a + b - 3c = -3 & \textbf{(1)} \\ 3a - 2b + 4c = 2 & \textbf{(2)} \\ 4a + 2b - 6c = -7 & \textbf{(3)} \end{cases}$

Strategy   Since the coefficients of the b-terms are opposites in the second and third equations, we will add the left and right sides of those equations to eliminate b. Then we will choose another pair of equations and eliminate b again.

Why   The result will be a system of two equations in a and c that we can attempt to solve by elimination.

Solution   We add equations 2 and 3 of the system to eliminate b.

$$\begin{array}{ll} \textbf{(2)} & 3a - 2b + 4c = 2 \\ \textbf{(3)} & \underline{4a + 2b - 6c = -7} \\ \textbf{(4)} & 7a - 2c = -5 \end{array}$$

We can multiply both sides of equation 1 by 2 and add the resulting equation to equation 2 to eliminate b again:

$$\begin{array}{ll} & 4a + 2b - 6c = -6 \qquad \text{This is } 2(2a + b - 3c) = 2(-3). \\ \textbf{(2)} & \underline{3a - 2b + 4c = 2} \\ \textbf{(5)} & 7a - 2c = -4 \end{array}$$

Equations 4 and 5 form a system in a and c.

$$\begin{array}{ll} \textbf{(4)} & \begin{cases} 7a - 2c = -5 \\ \textbf{(5)} & 7a - 2c = -4 \end{cases} \end{array}$$

If we multiply both sides of equation 5 by -1 and add the result to equation 4, the terms involving a and c are both eliminated.

Success Tip

If you obtain a false statement at any time in the solution process, you need not proceed. Such an outcome indicates that the system has no solution.

(4) $7a - 2c = -5$

$\underline{-7a + 2c = 4}$ This is $-1(7a - 2c) = -1(-4)$.

$0 = -1$

The false statement $0 = -1$ indicates that the system has *no solution* and is, therefore, inconsistent. The solution set is $\varnothing$.

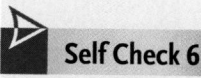

Self Check 6   Solve the system, if possible: $\begin{cases} 2a + b - 3c = 8 \\ 3a - 2b + 4c = 10 \\ 4a + 2b - 6c = -5 \end{cases}$

Now Try   **Problem 35**

EXAMPLE 7   Solve the system: $\begin{cases} \dfrac{4}{5}x - y + z = \dfrac{53}{5} \\ x - 2y - z = 8 \\ 0.2x - 0.3y + 0.1z = 2.3 \end{cases}$

Strategy   We will find an equivalent system without fractions or decimals and use elimination to solve it.

Why   It's easier to solve a system of equations that involves only integers.

Solution   To clear equation 1 of fractions, we multiply both sides by the LCD of the fractions, which is 5. To clear equation 3 of decimals, we multiply both sides by 10.

(1) $\begin{cases} \dfrac{4}{5}x - y + z = \dfrac{53}{5} \\ \textbf{(2)} \quad x - 2y - z = 8 \\ \textbf{(3)} \quad 0.2x - 0.3y + 0.1z = 2.3 \end{cases}$
$\xrightarrow[\text{Unchanged}]{\text{Multiply by 5}}$
$\xrightarrow{\text{Multiply by 10}}$
$\begin{cases} 4x - 5y + 5z = 53 \quad \textbf{(4)} \\ x - 2y - z = 8 \quad\quad\; \textbf{(2)} \\ 2x - 3y + z = 23 \quad\; \textbf{(5)} \end{cases}$

If each side of equation 2 is multiplied by 5 and the resulting equation is added to equation 4, the variable z is eliminated.

(4) $4x - 5y + 5z = 53$

$\underline{5x - 10y - 5z = 40}$ This is $5(x - 2y - z) = 5(8)$.

(6) $9x - 15y = 93$

If we add equations 2 and 5, the variable z is eliminated again.

(2) $x - 2y - z = 8$

(5) $\underline{2x - 3y + z = 23}$

(7) $3x - 5y = 31$

Equations 6 and 7 form a system in x and y. If each side of equation 7 is multiplied by -3 and the resulting equation is added to equation 6, both x and y are eliminated.

(6) $9x - 15y = 93$

$\underline{-9x + 15y = -93}$ This is $-3(3x - 5y) = -3(31)$.

$0 = 0$

The resulting true statement, $0 = 0$, indicates that the equations of the system are dependent, and that the system has *infinitely many solutions*. This is the case when all three equations of a system represent the same plane or when the intersection of the planes is a line.

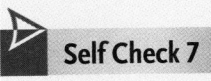 **Self Check 7**   Solve the system:
$$\begin{cases} x - 2y - z = 1 \\ x + \dfrac{4}{3}y + z = \dfrac{5}{3} \\ 0.02x + 0.01y + 0.01z = 0.03 \end{cases}$$

Now Try   **Problem 37**

When the equations in a system of two equations with two variables are dependent, the system has infinitely many solutions. This is not always true for systems of three equations with three variables. In fact, a system can have dependent equations and still be inconsistent. The following illustration shows the different possibilities.

Consistent system

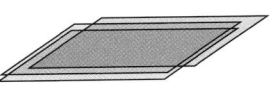

When three planes coincide, the equations are dependent, and there are infinitely many solutions.

(a)

Consistent system

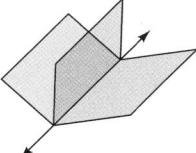

When three planes intersect in a common line, the equations are dependent, and there are infinitely many solutions.

(b)

Inconsistent system

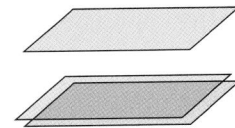

When two planes coincide and are parallel to a third plane, the system is inconsistent, and there are no solutions.

(c)

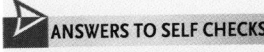 **ANSWERS TO SELF CHECKS**   **1.** No   **2.** $(2, 1, -4)$ **3.** $(1, 1, 2)$ **4.** $(40, -20, 5)$ **5.** $(-6, 0, 12)$
6. No solution, $\varnothing$; inconsistent system   **7.** Infinitely many solutions; dependent equations

---

STUDY SET
12.2

VOCABULARY

Fill in the blanks.

1. $\begin{cases} 2x + y - 3z = 0 \\ 3x - y + 4z = 5 \\ 4x + 2y - 6z = 0 \end{cases}$ is called a _____ of three linear equations in three variables. Each equation is written in _____ $Ax + By + Cz = D$ form.

2. If the first two equations of the system in Exercise 1 are added, the variable y is _____.

3. Solutions of a system of three equations in three variables, x, y, and z, are written in the form (x, y, z) and are called ordered _____.

4. The graph of the equation $2x + 3y + 4z = 5$ is a flat surface called a _____.

5. When three planes coincide, the equations of the system are _____, and there are infinitely many solutions.

6. When three planes intersect in a line, the system will have _____ many solutions.

CONCEPTS

7. For each graph of a system of three equations, determine whether the solution set contains one solution, infinitely many solutions, or no solution.

a.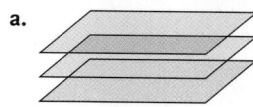

b.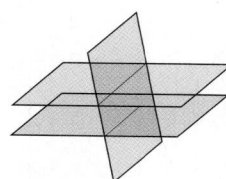

8. Consider the system:
$$\begin{cases} \text{(1)} & -2x + y + 4z = 3 \\ \text{(2)} & x - y + 2z = 1 \\ \text{(3)} & x + y - 3z = 2 \end{cases}$$

   a. What is the result if equation 1 and equation 2 are added?

   b. What is the result if equation 2 and equation 3 are added?

   c. What variable was eliminated in the steps performed in parts (a) and (b)?

NOTATION

9. For the following system, clear the equations of any fractions or decimals and write each equation in $Ax + By + Cz = D$ form.

$$\begin{cases} x + y = 3 - 4z \\ 0.7x - 0.2y + 0.8z = 1.5 \\ \dfrac{x}{2} + \dfrac{y}{3} - \dfrac{z}{6} = \dfrac{2}{3} \end{cases} \longrightarrow \Bigg\{$$

10. What is the purpose of the numbers shown in red in front of the equations below?

$$\begin{cases} \text{(1)} & x + y - z = 6 \\ \text{(2)} & 2x - y + z = 3 \\ \text{(3)} & 5x + 3y - z = -2 \end{cases}$$

GUIDED PRACTICE

Determine whether the ordered triple is a solution of the system. See Example 1.

11. $(2, 1, 1),$ $\begin{cases} x - y + z = 2 \\ 2x + y - z = 4 \\ 2x - 3y + z = 2 \end{cases}$

12. $(-3, 2, -1),$ $\begin{cases} 3x + y - z = -6 \\ 2x + 2y + 3z = -1 \\ x + y + 2z = 1 \end{cases}$

13. $(6, -7, -5),$ $\begin{cases} 3x - 2y - z = 37 \\ x - 3y = 27 \\ 2x + 7y + 2z = -48 \end{cases}$

14. $(-4, 0, 9),$ $\begin{cases} x + 2y - 3z = -31 \\ 2x + 6z = 46 \\ 3x - y = -12 \end{cases}$

Solve each system. See Example 2.

15. $\begin{cases} x + y + z = 4 \\ 2x + y - z = 1 \\ 2x - 3y + z = 1 \end{cases}$

16. $\begin{cases} x + y + z = 4 \\ x - y + z = 2 \\ x - y - 2z = -1 \end{cases}$

17. $\begin{cases} 3x + 2y - 5z = 3 \\ 4x - 2y - 3z = -10 \\ 5x - 2y - 2z = -11 \end{cases}$

18. $\begin{cases} 5x + 4y + 2z = -2 \\ 3x + 4y - 3z = -27 \\ 2x - 4y - 7z = -23 \end{cases}$

19. $\begin{cases} 2x + 6y + 3z = 9 \\ 5x - 3y - 5z = 3 \\ 4x + 3y + 2z = 15 \end{cases}$

20. $\begin{cases} 4x - 3y + 5z = 23 \\ 2x - 5y - 3z = 13 \\ -4x - 6y + 7z = 7 \end{cases}$

21. $\begin{cases} 4x - 5y - 8z = -52 \\ 2x - 3y - 4z = -26 \\ 3x + 7y + 8z = 31 \end{cases}$

22. $\begin{cases} 2x + 6y + 3z = -20 \\ 5x - 3y - 5z = 47 \\ 4x + 3y + 2z = 4 \end{cases}$

Solve each system. See Example 3.

23. $\begin{cases} 3x + 3z = 6 - 4y \\ 7x - 5z = 46 + 2y \\ 4x = 31 - z \end{cases}$

24. $\begin{cases} 5x + 6z = 4y - 21 \\ 9x + 2y = 3z - 47 \\ 3x + y = -19 \end{cases}$

25. $\begin{cases} 2x + z = -2 + y \\ 8x - 3y = -2 \\ 6x - 2y + 3z = -4 \end{cases}$

26. $\begin{cases} 3y + z = -1 \\ -x + 2z = -9 + 6y \\ 9y + 3z = -9 + 2x \end{cases}$

Solve each system using elimination. See Example 4.

27. $\begin{cases} x + y + 3z = 35 \\ -x - 3y = 20 \\ 2y + z = -35 \end{cases}$

28. $\begin{cases} x + 2y + 3z = 11 \\ 5x - y = 13 \\ 2x - 3z = -11 \end{cases}$

29. $\begin{cases} 3x + 2y - z = 7 \\ 6x - 3y = -2 \\ 3y - 2z = 8 \end{cases}$

30. $\begin{cases} 2x + y = 4 \\ -x - 2y + 8z = 7 \\ -y + 4z = 5 \end{cases}$

Solve each system using substitution. See Example 5.

31. $\begin{cases} r + s - 3t = 21 \\ r + 4s = 9 \\ 5s + t = -4 \end{cases}$

32. $\begin{cases} r - s + 6t = 12 \\ r + 6s = -28 \\ 7s + t = -26 \end{cases}$

33. $\begin{cases} x - 8z = -30 \\ 3x + y - 4z = 5 \\ y + 7z = 30 \end{cases}$

34. $\begin{cases} x + 6z = -36 \\ 5x + 3y - 2z = -20 \\ y + 4z = -20 \end{cases}$

Solve each system. If a system is inconsistent or if the equations are dependent, state this. See Examples 6 and 7.

35. $\begin{cases} 7a + 9b - 2c = -5 \\ 5a + 14b - c = -11 \\ 2a - 5b - c = 3 \end{cases}$

36. $\begin{cases} 3x + 4y + z = 10 \\ x - 2y + z = -3 \\ 2x + y + z = 5 \end{cases}$

37. $\begin{cases} 7x - y - z = 10 \\ x - 3y + z = 2 \\ x + 2y - z = 1 \end{cases}$

38. $\begin{cases} 2a - b + c = 6 \\ -5a - 2b - 4c = -30 \\ a + b + c = 8 \end{cases}$

TRY IT YOURSELF

Solve each system, if possible. If a system is inconsistent or if the equations are dependent, state this.

39. $\begin{cases} 2a + 3b - 2c = 18 \\ 5a - 6b + c = 21 \\ 4b - 2c - 6 = 0 \end{cases}$

40. $\begin{cases} r - s + t = 4 \\ r + 2s - t = -1 \\ r + s - 3t = -2 \end{cases}$

41. $\begin{cases} 2x + 2y - z = 2 \\ x + 3z - 24 = 0 \\ y = 7 - 4z \end{cases}$

42. $\begin{cases} r - 3t = -11 \\ r + s + t = 13 \\ s - 4t = -12 \end{cases}$

43. $\begin{cases} b + 2c = 7 - a \\ a + c = 2(4 - b) \\ 2a + b + c = 9 \end{cases}$

44. $\begin{cases} 0.02a = 0.02 - 0.03b - 0.01c \\ 4a + 6b + 2c - 5 = 0 \\ a + c = 3 + 2b \end{cases}$

45. $\begin{cases} 2x + y - z = 1 \\ x + 2y + 2z = 2 \\ 4x + 5y + 3z = 3 \end{cases}$

46. $\begin{cases} 2x + 2y + 3z = 10 \\ 3x + y - z = 0 \\ x + y + 2z = 6 \end{cases}$

47. $\begin{cases} 0.4x + 0.3z = 0.4 \\ 2y - 6z = -1 \\ 4(2x + y) = 9 - 3z \end{cases}$

48. $\begin{cases} a + b + c = 180 \\ \dfrac{a}{4} + \dfrac{b}{2} + \dfrac{c}{3} = 60 \\ 2b + 3c - 330 = 0 \end{cases}$

49. $\begin{cases} r + s + 4t = 3 \\ 3r + 7t = 0 \\ 3s + 5t = 0 \end{cases}$

50. $\begin{cases} x - y = 3 \\ 2x - y + z = 1 \\ x + z = -2 \end{cases}$

51. $\begin{cases} 0.5a + 0.3b = 2.2 \\ 1.2c - 8.5b = -24.4 \\ 3.3c + 1.3a = 29 \end{cases}$

52. $\begin{cases} 4a - 3b = 1 \\ 6a - 8c = 1 \\ 2b - 4c = 0 \end{cases}$

53. $\begin{cases} 2x + 3y = 6 - 4z \\ 2x = 3y + 4z - 4 \\ 4x + 6y + 8z = 12 \end{cases}$

54. $\begin{cases} -x + 5y - 7z = 0 \\ 4x + y - z = 0 \\ x + y - 4z = 0 \end{cases}$

55. $\begin{cases} a + b = 2 + c \\ a = 3 + b - c \\ -a + b + c - 4 = 0 \end{cases}$

56. $\begin{cases} 0.1x - 0.3y + 0.4z = 0.2 \\ 2x + y + 2z = 3 \\ 4x - 5y + 10z = 7 \end{cases}$

57. $\begin{cases} x + \dfrac{1}{3}y + z = 13 \\ \dfrac{1}{2}x - y + \dfrac{1}{3}z = -2 \\ x + \dfrac{1}{2}y - \dfrac{1}{3}z = 2 \end{cases}$

58. $\begin{cases} x - \dfrac{1}{5}y - z = 9 \\ \dfrac{1}{4}x + \dfrac{1}{5}y - \dfrac{1}{2}z = 5 \\ 2x + y + \dfrac{1}{6}z = 12 \end{cases}$

APPLICATIONS

59. GRAPHS OF SYSTEMS   Explain how each of the following pictures could be thought of as an example of the graph of a system of three equations. Then describe the solution, if there is any.

a.

b.

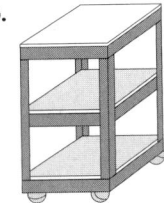

c.

d.

60. ZOOLOGY Refer to the illustration below. An X-ray of a mouse revealed a cancerous tumor located at the intersection of the coronal, sagittal, and transverse planes. From this description, would you expect the tumor to be at the base of the tail, on the back, in the stomach, on the tip of the right ear, or in the mouth of the mouse?

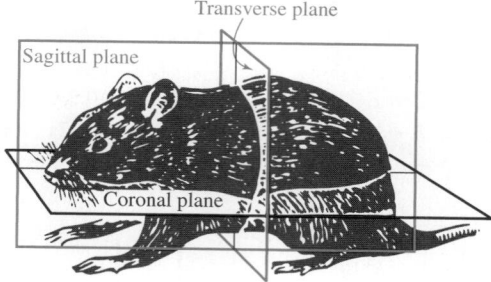

Transverse plane
Sagittal plane
Coronal plane

61. NBA RECORDS The three highest one-game point totals by one player in a National Basketball Association game are shown below.

Pts	Player, team	Date
x	Wilt Chamberlain, Philadelphia	3/2/1962
y	Kobe Bryant, Los Angeles	1/22/2006
z	Wilt Chamberlain, Philadelphia	12/8/1961

Solve the following system to find x, y, and z.

$$\begin{cases} x + y + z = 259 \\ x - y = 19 \\ x - z = 22 \end{cases}$$

62. BICYCLE FRAMES The angle measures of the triangular part of the bicycle frame shown can be found by solving the following system. Find x, y, and z.

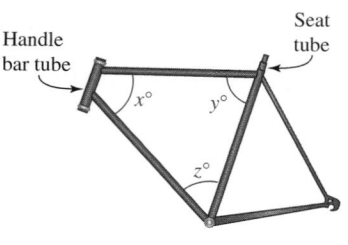

Handle bar tube

Seat tube

$$\begin{cases} x + y + z = 180 \\ x + y = 120 \\ y + z = 135 \end{cases}$$

WRITING

63. Explain how a system of three equations in three variables can be reduced to a system of two equations in two variables.

64. What makes a system of three equations with three variables inconsistent?

65. What does the graph of a linear equation in three variables such as $2x - 3y + 9z = 10$ look like?

66. What situation discussed in this section looks like two walls of a room and the floor meeting in a corner?

REVIEW

Graph each of the basic functions.

67. $f(x) = |x|$

68. $g(x) = x^2$

69. $h(x) = x^3$

70. $S(x) = x$

CHALLENGE PROBLEMS

Solve each system.

71. $$\begin{cases} w + x + y + z = 3 \\ w - x + y + z = 1 \\ w + x - y + z = 1 \\ w + x + y - z = 3 \end{cases}$$

72. $$\begin{cases} \dfrac{1}{x} + \dfrac{1}{y} + \dfrac{1}{z} = 3 \\ \dfrac{2}{x} + \dfrac{1}{y} - \dfrac{1}{z} = 0 \\ \dfrac{1}{x} - \dfrac{2}{y} + \dfrac{4}{z} = 21 \end{cases}$$

SECTION 12.3
Problem Solving Using Systems of Three Equations

Objectives

① Assign variables to three unknowns.

② Use systems to solve curve-fitting problems.

Problems that involve three unknown quantities can be solved using a strategy similar to that for solving problems involving two unknowns. To solve such problems, we will write three equations in three variables to model the situation and then we will use the methods of Section 12.2 to solve the system formed by the three equations.

1   **Assign Variables to Three Unknowns.**

EXAMPLE 1   ***Tool Manufacturing.*** A company makes three types of hammers, which are marketed as "good," "better," and "best." The cost of manufacturing each type of hammer is $4, $6, and $7, respectively, and the hammers sell for $6, $9, and $12. Each day, the cost of manufacturing 100 hammers is $520, and the daily revenue from their sale is $810. How many hammers of each type are manufactured?

Analyze the Problem   We need to find how many of each type of hammer are manufactured daily. Since there are three unknowns, we must write three equations to find them.

Form Three Equations   Let x = the number of good hammers, y = the number of better hammers, and z = the number of best hammers. We know that

The cost of manufacturing
- the good hammers is $4x$ ($4 times x hammers).
- the better hammers is $6y$ ($6 times y hammers).
- the best hammers is $7z$ ($7 times z hammers).

The revenue received by selling
- the good hammers is $6x$ ($6 times x hammers).
- the better hammers is $9y$ ($9 times y hammers).
- the best hammers is $12z$ ($12 times z hammers).

We can use the facts of the problem to write three equations.

The number of good hammers	plus	the number of better hammers	plus	the number of best hammers	is	the total number of hammers.
x	$+$	y	$+$	z	$=$	100

The cost of good hammers	plus	the cost of better hammers	plus	the cost of best hammers	is	the total cost.
$4x$	$+$	$6y$	$+$	$7z$	$=$	520

The revenue from good hammers	plus	the revenue from better hammers	plus	the revenue from best hammers	is	the total revenue.
$6x$	$+$	$9y$	$+$	$12z$	$=$	810

Solve the System   To find how many hammers of each type are manufactured, we must solve the following system of three equations in three variables:

$$\begin{cases} (1) & x + y + z = 100 \\ (2) & 4x + 6y + 7z = 520 \\ (3) & 6x + 9y + 12z = 810 \end{cases}$$

If we multiply equation 1 by -7 and add the result to equation 2, we get

$-7x - 7y - 7z = -700$ This is $-7(x + y + z) = -7(100)$.

(2) $\underline{4x + 6y + 7z = 520}$

(4) $-3x - y = -180$

If we multiply equation 1 by -12 and add the result to equation 3, we get

$$-12x - 12y - 12z = -1{,}200 \qquad \text{This is } -12(x + y + z) = -12(100).$$

$$
\begin{array}{rl}
\textbf{(3)} & 6x + 9y + 12z = 810 \\
\hline
\textbf{(5)} & -6x - 3y = -390
\end{array}
$$

We can multiply equation 4 by -3 and add it to equation 5 to eliminate y.

$$
\begin{array}{rl}
 & 9x + 3y = 540 \qquad \text{This is } -3(-3x - y) = -3(-180). \\
\textbf{(5)} & \underline{-6x - 3y = -390} \\
 & 3x = 150
\end{array}
$$

$$x = 50 \qquad \text{To solve for } x, \text{ divide both sides by 3.}$$
$$ \text{This is the number of good hammers manufactured.}$$

To find y, we substitute 50 for x in equation 4:

$$-3x - y = -180$$
$$-3(50) - y = -180 \qquad \text{Substitute 50 for } x.$$
$$-150 - y = -180$$
$$-y = -30 \qquad \text{Add 150 to both sides.}$$
$$y = 30 \qquad \text{To solve for } y, \text{ divide both sides by } -1.$$
$$ \text{This is the number of better hammers manufactured.}$$

To find z, we substitute 50 for x and 30 for y in equation 1:

$$x + y + z = 100$$
$$50 + 30 + z = 100$$
$$z = 20 \qquad \text{To solve for } z, \text{ subtract 80 from both sides.}$$
$$ \text{This is the number of best hammers manufactured.}$$

State the Conclusion   Each day, the company manufactures 50 good hammers, 30 better hammers, and 20 best hammers.

Check the Result   If the company manufactures **50** good hammers, **30** better hammers, and **20** best hammers each day, that is a total of **50** + **30** + **20** = 100 hammers. The cost of manufacturing the three types of hammers is $4(**50**) + $6(**30**) + $7(**20**) = $200 + $180 + $140 or $520. The revenue from the sale of the hammers is $6(**50**) + $9(**30**) + $12(**20**) = $300 + $270 + $240 or $810. The results check.

 Now Try   **Problem 7**

EXAMPLE 2   *The Olympics.*   The three countries that won the most medals in the 2004 summer Olympic games were the United States, Russia, and China. Together they won a total of 258 medals, with the U.S. medal count 11 more than Russia's and Russia's medal count 29 more than China's. Find the number of medals won by each country.

Analyze the Problem   We need to find how many medals the United States, Russia, and China won. Since there are three unknowns, we must write three equations to find them.

Form Three Equations   Let x = the number of medals won by the United States, y = the number of medals won by Russia, and z = the number of medals won by China.

We can use the facts of the problem to write three equations.

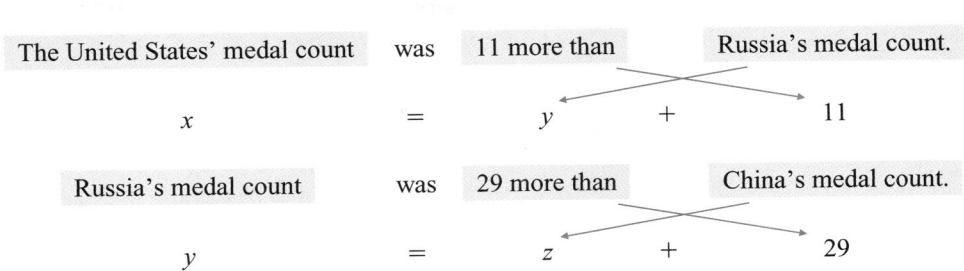

The number of medals won by the United States	plus	the number of medals won by Russia	plus	the number of medals won by China	was	258.
x	$+$	y	$+$	z	$=$	258

The United States' medal count	was	11 more than	Russia's medal count.
x	$=$	y	$+$	11

Russia's medal count	was	29 more than	China's medal count.
y	$=$	z	$+$	29

Solve the System   We can use substitution to solve the resulting system of three equations. If we solve equation 3 for z, then the resulting equation 4 and equation 2 can serve as substitution equations.

(1) $\begin{cases} x + y + z = 258 \\ (2) \quad x = y + 11 \\ (3) \quad y = z + 29 \end{cases}$
$\xrightarrow[\text{Unchanged}]{\text{Unchanged}}$
$\xrightarrow{\text{Solve for } z}$
$\begin{cases} x + y + z = 258 \quad \text{(1)} \\ x = \boxed{y + 11} \quad \text{(2)} \\ z = \boxed{y - 29} \quad \text{(4)} \end{cases}$

When we substitute for x and z in equation 1, we obtain an equation in one variable, y.

$$x + y + z = 258 \qquad \text{This is equation 1.}$$
$$y + 11 + y + y - 29 = 258 \qquad \text{Substitute } y + 11 \text{ for } x \text{ and } y - 29 \text{ for } z.$$
$$3y - 18 = 258 \qquad \text{On the left side, combine like terms.}$$
$$3y = 276 \qquad \text{Add 18 to both sides.}$$
$$y = 92 \qquad \text{To solve for } y, \text{ divide both sides by 3.}$$
$$\text{This is the number of medals won by Russia.}$$

To find x, we substitute 92 for y in equation 2. To find z, we substitute 92 for y in equation 4.

$$x = y + 11 \qquad \text{This is equation 2.} \qquad\qquad z = y - 29 \qquad \text{This is equation 4.}$$
$$= 92 + 11 \qquad\qquad\qquad\qquad\qquad\qquad = 92 - 29$$
$$= 103 \qquad \text{This is the U.S. medal count.} \qquad = 63 \qquad \text{This is China's medal count.}$$

State the Conclusion   In the 2004 summer Olympics, the United States won 103 medals, Russia won 92, and China won 63.

Check the Result   The sum of $103 + 92 + 63$ is 258. Furthermore, 103 is 11 more than 92, and 92 is 29 more than 63. The results check.

 Now Try   **Problem 13**

2   **Use Systems to Solve Curve-Fitting Problems.**

The process of determining an equation whose graph contains given points is called **curve fitting**.

EXAMPLE 3   The equation of a parabola opening upward or downward is of the form $y = ax^2 + bx + c$. Find the equation of the parabola graphed on the left by determining the values of a, b, and c.

Strategy   We will substitute the x- and y-coordinates of three points that lie on the graph into the equation $y = ax^2 + bx + c$. This will produce a system of three equations in three variables that we can solve to find a, b, and c.

Why   Once we know a, b, and c, we can write the equation.

Solution   Since the parabola passes through the points $(-1, 5)$, $(1, 1)$, and $(2, 2)$, each pair of coordinates must satisfy the equation $y = ax^2 + bx + c$. If we substitute each pair into $y = ax^2 + bx + c$, we will get a system of three equations in three variables.

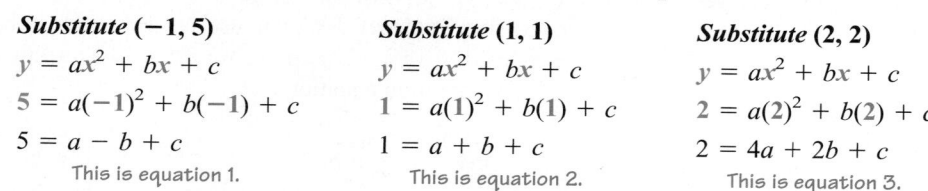

Substitute $(-1, 5)$
$y = ax^2 + bx + c$
$5 = a(-1)^2 + b(-1) + c$
$5 = a - b + c$
 This is equation 1.

Substitute $(1, 1)$
$y = ax^2 + bx + c$
$1 = a(1)^2 + b(1) + c$
$1 = a + b + c$
 This is equation 2.

Substitute $(2, 2)$
$y = ax^2 + bx + c$
$2 = a(2)^2 + b(2) + c$
$2 = 4a + 2b + c$
 This is equation 3.

The three equations above give the system, which we can solve to find a, b, and c.

$$\begin{cases} \textbf{(1)} & a - b + c = 5 \\ \textbf{(2)} & a + b + c = 1 \\ \textbf{(3)} & 4a + 2b + c = 2 \end{cases}$$

If we add equations 1 and 2, we obtain

$$\begin{array}{r} a - b + c = 5 \\ \underline{a + b + c = 1} \\ \textbf{(4)} \quad 2a \qquad + 2c = 6 \end{array}$$

If we multiply equation 1 by 2 and add the result to equation 3, we get

$$\begin{array}{r} 2a - 2b + 2c = 10 \\ \textbf{(3)} \quad \underline{4a + 2b + \ c = \ 2} \\ \textbf{(5)} \quad 6a \qquad + 3c = 12 \end{array}$$

We can then divide both sides of equation 4 by 2 to get equation 6 and divide both sides of equation 5 by 3 to get equation 7. We now have the system

$$\begin{cases} \textbf{(6)} & a + c = 3 \\ \textbf{(7)} & 2a + c = 4 \end{cases}$$

To eliminate c, we multiply equation 6 by -1 and add the result to equation 7. We get

$$\begin{array}{r} -a - c = -3 \quad \text{This is } -1(a + c) = -1(3). \\ \underline{2a + c = \ \ 4} \\ a \qquad = \ 1 \end{array}$$

To find c, we can substitute 1 for a in equation 6 and find that $c = 2$. To find b, we can substitute 1 for a and 2 for c in equation 2 and find that $b = -2$.

After we substitute these values of a, b, and c into the equation $y = ax^2 + bx + c$, we have the equation of the parabola.

Success Tip
If a point lies on the graph of an equation, it is a solution of the equation, and the coordinates of the point satisfy the equation.

$$y = ax^2 + bx + c$$
$$y = 1x^2 - 2x + 2$$
$$y = x^2 - 2x + 2$$

 Now Try **Problem 25**

STUDY SET
12.3

VOCABULARY

Fill in the blanks.

1. If a point lies on the graph of an equation, it is a solution of the equation, and the coordinates of the point _____ the equation.

2. The process of determining an equation whose graph contains given points is called curve _____.

CONCEPTS

Write a system of three equations in three variables that models the situation. Do not solve the system.

3. DESSERTS   A bakery makes three kinds of pies: chocolate cream, which sells for $5; apple, which sells for $6; and cherry, which sells for $7. The cost to make the pies is $2, $3, and $4, respectively. Let x = the number of chocolate cream pies made daily, y = the number of apple pies made daily, and z = the number of cherry pies made daily.

- Each day, the bakery makes 50 pies.
- Each day, the revenue from the sale of the pies is $295.
- Each day, the cost to make the pies is $145.

4. FAST FOODS   Let x = the number of calories in a Big Mac hamburger, y = the number of calories in a small order of French fries, and z = the number of calories in a medium Coca-Cola.

- The total number of calories in a Big Mac hamburger, a small order of French fries, and a medium Coke is 1,000.
- The number of calories in a Big Mac is 260 more than in a small order of French fries.
- The number of calories in a small order of French fries is 40 more than in a medium Coke. (Source: McDonald's USA)

5. What equation results when the coordinates of the point $(2, -3)$ are substituted into $y = ax^2 + bx + c$?

6. The equation $y = 5x^2 - 6x + 1$ is written in the form $y = ax^2 + bx + c$. What are a, b, and c?

APPLICATIONS

7. MAKING STATUES   An artist makes three types of ceramic statues (large, medium, and small) at a monthly cost of $650 for 180 statues. The manufacturing costs for the three types are $5, $4, and $3. If the statues sell for $20, $12, and $9, respectively, how many of each type should be made to produce $2,100 in monthly revenue?

8. PUPPETS   A toy company makes a total of 500 puppets in three sizes during a production run. The small puppets cost $5 to make and sell for $8 each, the standard size puppets cost $10 to make and sell for $16 each, and the super-size puppets cost $15 to make and sell for $25. The total cost to make the puppets is $4,750 and the revenue from their sale is $7,700. How many small, standard, and super-size puppets are made during a production run?

9. NUTRITION   A dietitian is to design a meal that will provide a patient with exactly 14 grams (g) of fat, 9 g of carbohydrates, and 9 g of protein. She is to use a combination of the three foods listed in the table. If one ounce of each of the foods has the nutrient content shown in the table, how many ounces of each food should be used?

Food	Fat	Carbohydrates	Protein
A	2 g	1 g	2 g
B	3 g	2 g	1 g
C	1 g	1 g	2 g

(g stands for gram)

10. NUTRITIONAL PLANNING One ounce of each of three foods has the vitamin and mineral content shown in the table. How many ounces of each must be used to provide exactly 22 milligrams (mg) of niacin, 12 mg of zinc, and 20 mg of vitamin C?

Food	Niacin	Zinc	Vitamin C
A	1 mg	1 mg	2 mg
B	2 mg	1 mg	1 mg
C	2 mg	1 mg	2 mg

(mg stands for milligram)

11. *from Campus to Careers*
Fashion Designer

A clothing manufacturer makes coats, shirts, and slacks. The time required for cutting, sewing, and packaging each item is shown in the table. How many of each should be made to use all available labor hours?

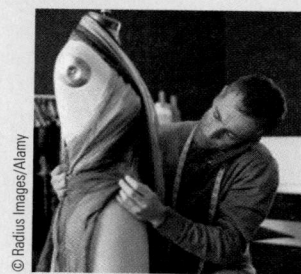

© Radius Images/Alamy

	Coats	Shirts	Slacks	Time available
Cutting	20 min	15 min	10 min	115 hr
Sewing	60 min	30 min	24 min	280 hr
Packaging	5 min	12 min	6 min	65 hr

12. SCULPTING A wood sculptor carves three types of statues with a chainsaw. The number of hours required for carving, sanding, and painting a totem pole, a bear, and a deer are shown in the table. How many of each should be produced to use all available labor hours?

	Totem pole	Bear	Deer	Time available
Carving	2 hr	2 hr	1 hr	14 hr
Sanding	1 hr	2 hr	2 hr	15 hr
Painting	3 hr	2 hr	2 hr	21 hr

13. NFL RECORDS Jerry Rice, who played the majority of his career with the San Francisco 49ers and the Oakland Raiders, holds the all-time record for touchdown (TD) passes caught. Here are some interesting facts about this feat.

- He caught 30 more TD passes from Steve Young than he did from Joe Montana.
- He caught 39 more TD passes from Joe Montana than he did from Rich Gannon.
- He caught a total of 156 TD passes from Young, Montana, and Gannon.

Determine the number of touchdown passes Rice has caught from Young, from Montana, and from Gannon.

14. HOT DOGS In 12 minutes, the top three finishers in the 2007 Nathan's Hot Dog Eating Contest consumed a total of 178 hot dogs. The winner, Joey Chestnut, ate 3 more hot dogs than the runner-up, Takeru Kobayashi. Pat Bertoletti finished a distant third, 14 hot dogs behind Kobayashi. How many hot dogs did each person eat?

15. EARTH'S ATMOSPHERE Use the information in the circle graph to determine what percent of Earth's atmosphere is nitrogen, is oxygen, and is other gases.

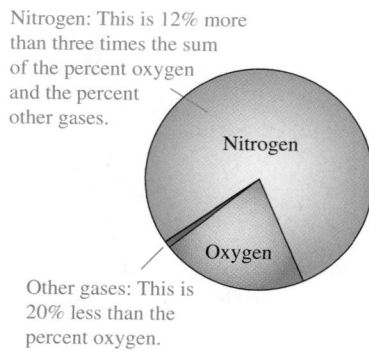

Nitrogen: This is 12% more than three times the sum of the percent oxygen and the percent other gases.

Other gases: This is 20% less than the percent oxygen.

16. DECEASED CELEBRITIES Between October 2005 and October 2006, the estates of Kurt Cobain, Elvis Presley, and Charles M. Schultz (Snoopy cartoonist) earned a total of $127 million. Together, the Presley and Schultz estates earned $27 million more than the Cobain estate. If the Schultz estate had earned $15 million more, it would equal the value of the Cobain estate. Use this information to label the vertical axis of the graph below. (Source: Forbes.com)

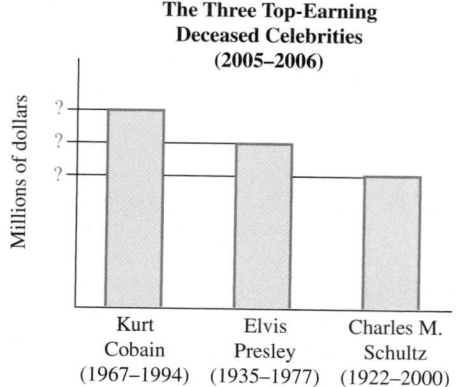

The Three Top-Earning Deceased Celebrities (2005–2006)

17. TRIANGLES The sum of the measures of the angles of any triangle is 180°. In $\triangle ABC$, $\angle A$ measures 100° less than the sum of the measures of $\angle B$ and $\angle C$, and the measure of $\angle C$ is 40° less than twice the measure of $\angle B$. Find the measure of each angle of the triangle.

18. QUADRILATERALS A quadrilateral is a four-sided polygon. The sum of the measures of the angles of any quadrilateral is 360°. In the illustration below, the measures of $\angle A$ and $\angle B$ are the same. The measure of $\angle C$ is 20° greater than the measure of $\angle A$, and the measure of $\angle D$ is 60° less than $\angle B$. Find the measure of $\angle A$, $\angle B$, and $\angle C$.

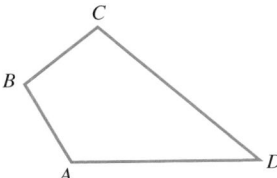

19. TV HISTORY *X-Files, Will & Grace,* and *Seinfeld* are three of the most popular television shows of all time. The total number of episodes of these three shows is 575. There are 21 more episodes of *X-Files* than *Seinfeld,* and the difference between the number of episodes of *Will & Grace* and *Seinfeld* is 14. Find the number of episodes of each show.

20. TRAFFIC LIGHTS At a traffic light, one cycle through green-yellow-red lasts for 80 seconds. The green light is on eight times longer than the yellow light, and the red light is on eleven times longer than the yellow light. For how long is each colored light on during one cycle?

21. POTPOURRI The owner of a home decorating shop wants to mix dried rose petals selling for $6 per pound, dried lavender selling for $5 per pound, and buckwheat hulls selling for $4 per pound to get 10 pounds of a mixture that would sell for $5.50 per pound. She wants to use twice as many pounds of rose petals as lavender. How many pounds of each should she use?

22. MIXING NUTS The owner of a candy store wants to mix some peanuts worth $3 per pound, some cashews worth $9 per pound, and some Brazil nuts worth $9 per pound to get 50 pounds of a mixture that will sell for $6 per pound. She uses 15 fewer pounds of cashews than peanuts. How many pounds of each did she use?

23. PIGGY BANKS When a child breaks open her piggy bank, she finds a total of 64 coins, consisting of nickels, dimes, and quarters. The total value of the coins is $6. If the nickels were dimes, and the dimes were nickels, the value of the coins would be $5. How many nickels, dimes, and quarters were in the piggy bank?

24. THEATER SEATING The illustration in the next column shows the cash receipts and the ticket prices from two sold-out Sunday performances of a play. Find the number of seats in each of the three sections of the 800-seat theater.

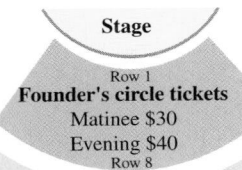

Sunday Ticket Receipts	
Matinee	$13,000
Evening	$23,000

25. ASTRONOMY Comets have elliptical orbits, but the orbits of some comets are so large that they are indistinguishable from parabolas. Find an equation of the form $y = ax^2 + bx + c$ for the parabola that closely describes the orbit of the comet shown in the illustration.

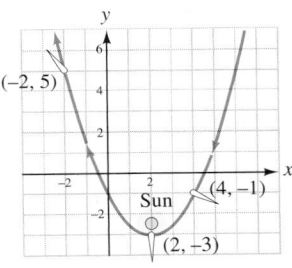

26. CURVE FITTING Find an equation of the form $y = ax^2 + bx + c$ for the parabola shown in the illustration.

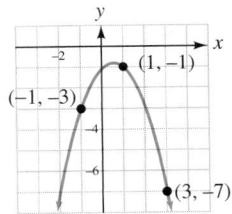

27. WALKWAYS A circular sidewalk is to be constructed in a city park. The walk is to pass by three particular areas of the park, as shown in the illustration. If an equation of a circle is of the form $x^2 + y^2 + Cx + Dy + E = 0$, find an equation that describes the path of the sidewalk by determining C, D, and E.

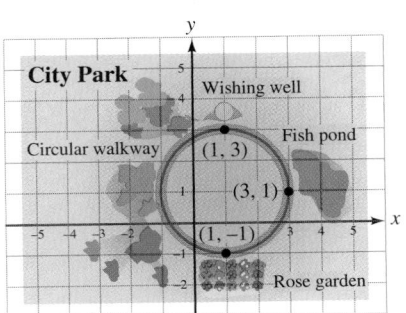

28. CURVE FITTING   The equation of a circle is of the form
$$x^2 + y^2 + Cx + Dy + E = 0.$$
Find an equation of the circle shown in the illustration by determining C, D, and E.

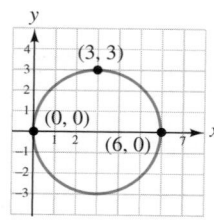

33. $xy = 9$

34. $y = |x|$

35. $x + 1 = |y|$

36. $y = \dfrac{1}{x^2}$

37. $y^2 = x$

38. $x = |y|$

WRITING

29. Explain why the following problem does not give enough information to answer the question: The sum of three integers is 48. If the first integer is doubled, the sum is 60. Find the integers.

30. Write an application problem that can be solved using a system of three equations in three variables.

REVIEW

Determine whether each equation defines y to be a function of x. If it does not, find two ordered pairs where more than one value of y corresponds to a single value of x.

31. $y = \dfrac{1}{x}$

32. $y^4 = x$

CHALLENGE PROBLEMS

39. DIGITS PROBLEM   The sum of the digits of a three-digit number is 8. Twice the hundreds digit plus the tens digit is equal to the ones digit. If the digits of the number are reversed, the new number is 82 more than twice the original number. What is the three-digit number?

40. PURCHASING PETS   A pet store owner spent $100 to buy 100 animals. He bought at least one iguana, one guinea pig, and one mouse, but no other kinds of animals. If an iguana cost $10.00, a guinea pig cost $3.00, and a mouse cost $0.50, how many of each did he buy?

SECTION 12.4
Solving Systems of Equations Using Matrices

Objectives

1. Define a matrix and determine its order.
2. Write the augmented matrix for a system.
3. Perform elementary row operations on matrices.
4. Use matrices to solve a system of two equations.
5. Use matrices to solve a system of three equations.
6. Use matrices to identify inconsistent systems and dependent equations.

In this section, we will discuss another way to solve systems of linear equations. This technique uses a mathematical tool called a *matrix* in a series of steps that are based on the elimination (addition) method.

1   **Define a Matrix and Determine Its Order.**

Another way to solve systems of equations involves rectangular arrays of numbers called *matrices* (plural of matrix).

Matrices

A **matrix** is any rectangular array of numbers arranged in rows and columns, written within brackets.

Some examples of matrices are

$$A = \begin{bmatrix} 1 & -3 & 8 \\ 2 & 5 & -1 \end{bmatrix} \begin{matrix} \leftarrow \text{Row 1} \\ \leftarrow \text{Row 2} \end{matrix} \qquad B = \begin{bmatrix} 1 & 4 & -2 & -4 \\ 6 & -2 & 6 & 1 \\ 3 & 8 & -3 & 12 \end{bmatrix} \begin{matrix} \leftarrow \text{Row 1} \\ \leftarrow \text{Row 2} \\ \leftarrow \text{Row 3} \end{matrix}$$

$$\begin{matrix} \uparrow & \uparrow & \uparrow \\ \text{Column} & \text{Column} & \text{Column} \\ 1 & 2 & 3 \end{matrix} \qquad\qquad \begin{matrix} \uparrow & \uparrow & \uparrow & \uparrow \\ \text{Column} & \text{Column} & \text{Column} & \text{Column} \\ 1 & 2, & 3 & 4 \end{matrix}$$

Each number in a matrix is called an **element** or an **entry** of the matrix. A matrix with m rows and n columns has **order** $m \times n$, which is read as "m by n." Because matrix A has two rows and three columns, its order is 2×3. The order of matrix B is 3×4 because it has three rows and four columns.

②   Write the Augmented Matrix for a System.

To show how to use matrices to solve systems of linear equations, we consider the system

$$\begin{cases} x - y = 4 \\ 2x + y = 5 \end{cases}$$

which can be represented by the following matrix, called an **augmented matrix:**

$$\begin{bmatrix} 1 & -1 & \vdots & 4 \\ 2 & 1 & \vdots & 5 \end{bmatrix} \qquad \text{The constants appear to the right of the dashed line.}$$

Each row of the augmented matrix represents one equation of the system. The first two columns of the augmented matrix are determined by the coefficients of x and y in the equations of the system. The last column is determined by the constants in the equations.

$$\begin{bmatrix} 1 & -1 & \vdots & 4 \\ 2 & 1 & \vdots & 5 \end{bmatrix} \qquad \begin{matrix} \text{This row represents the equation } x - y = 4. \\ \text{This row represents the equation } 2x + y = 5. \end{matrix}$$

$$\begin{matrix} \text{Coefficients} & \text{Coefficients} & \text{Constants} \\ \text{of } x & \text{of } y \end{matrix}$$

EXAMPLE 1   Represent each system using an augmented matrix:

a. $\begin{cases} 3x + y = 11 \\ x - 8y = 0 \end{cases}$ **b.** $\begin{cases} 2a + b - 3c = -3 \\ 9a + 4c = 2 \\ a - b - 6c = -7 \end{cases}$

Strategy   We will write the coefficients of the variables and the constants from each equation in rows to form a matrix. The coefficients are written to the left of a vertical dashed line and constants to the right.

Why   In an augmented matrix, each row represents one equation of the system.

Solution   Since the equations of each system are written in standard form, we can easily write the corresponding augmented matrices.

a. $\begin{cases} 3x + y = 11 & \leftrightarrow \\ x - 8y = 0 & \leftrightarrow \end{cases} \begin{bmatrix} 3 & 1 & \vdots & 11 \\ 1 & -8 & \vdots & 0 \end{bmatrix}$

b. $\begin{cases} 2a + b - 3c = -3 & \leftrightarrow \\ 9a + 4c = 2 & \leftrightarrow \\ a - b - 6c = -7 & \leftrightarrow \end{cases} \begin{bmatrix} 2 & 1 & -3 & \vdots & -3 \\ 9 & 0 & 4 & \vdots & 2 \\ 1 & -1 & -6 & \vdots & -7 \end{bmatrix}$ In the second row, 0 is entered as the coefficient of the missing *b*-term.

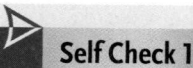

Self Check 1   Represent each system using an augmented matrix:

a. $\begin{cases} 2x - 4y = 9 \\ 5x - y = -2 \end{cases}$

b. $\begin{cases} a + b - c = -4 \\ -2b + 7c = 0 \\ 10a + 8b - 4c = 5 \end{cases}$

Now Try   Problem 13

③   Perform Elementary Row Operations on Matrices.

The Language of Algebra
Two matrices are *equivalent* if they represent systems that have the same solution set.

To solve a 2×2 system of equations using matrices, we transform the augmented matrix into an equivalent matrix that has 1's down its main diagonal and a 0 below the 1 in the first column. A matrix written in this form is said to be in **row echelon form.** We can easily determine the solution of the associated system of equations when an augmented matrix is written in this form.

$$\begin{bmatrix} 1 & a & b \\ 0 & 1 & c \end{bmatrix}$$ *a, b,* and *c* represent real numbers.

Main diagonal

To write an augmented matrix in row echelon form, we use three operations called *elementary row operations.*

Elementary Row Operations

Type 1: Any two rows of a matrix can be interchanged.
Type 2: Any row of a matrix can be multiplied by a nonzero constant.
Type 3: Any row of a matrix can be changed by adding a nonzero constant multiple of another row to it.

None of these row operations affect the solution of a given system of equations. The changes to the augmented matrix produce equivalent matrices that correspond to systems with the same solution.

- A type 1 row operation corresponds to interchanging two equations of the system.
- A type 2 row operation corresponds to multiplying both sides of an equation by a nonzero constant.
- A type 3 row operation corresponds to adding a nonzero multiple of one equation to another.

EXAMPLE 2   Perform the following elementary row operations:

$$A = \begin{bmatrix} 2 & 4 & -3 \\ 1 & -8 & 0 \end{bmatrix} \qquad B = \begin{bmatrix} 1 & -1 & 2 \\ 4 & -8 & 0 \end{bmatrix} \qquad C = \begin{bmatrix} 2 & 1 & -8 & 4 \\ 0 & 1 & 4 & -2 \\ 0 & 0 & -6 & 24 \end{bmatrix}$$

a. Type 1: Interchange rows 1 and 2 of matrix A.

b. Type 2: Multiply row 3 of matrix C by $-\frac{1}{6}$.

c. Type 3: To the numbers in row 2 of matrix B, add the results of multiplying each number in row 1 by -4.

Strategy   We will perform elementary row operations on each matrix as if we were performing those operations on the equations of a system.

Why The rows of an augmented matrix correspond to the equations of a system.

Solution

a. Interchanging rows 1 and 2 of matrix $A = \begin{bmatrix} 2 & 4 & \vdots & -3 \\ 1 & -8 & \vdots & 0 \end{bmatrix}$ gives $\begin{bmatrix} 1 & -8 & \vdots & 0 \\ 2 & 4 & \vdots & -3 \end{bmatrix}$.

We can represent the instruction to interchange rows 1 and 2 with the symbol $R_1 \leftrightarrow R_2$.

b. We multiply each number in row 3 of matrix $C = \begin{bmatrix} 2 & 1 & -8 & \vdots & 4 \\ 0 & 1 & 4 & \vdots & -2 \\ 0 & 0 & -6 & \vdots & 24 \end{bmatrix}$ by $-\frac{1}{6}$. Note

that rows 1 and 2 remain unchanged.

$$\begin{bmatrix} 2 & 1 & -8 & \vdots & 4 \\ 0 & 1 & 4 & \vdots & -2 \\ 0 & 0 & 1 & \vdots & -4 \end{bmatrix}$$ We can represent the instruction to multiply the third row by $-\frac{1}{6}$ with the symbol $-\frac{1}{6}R_3$.

c. We multiply each number from the first row of matrix $B = \begin{bmatrix} 1 & -1 & \vdots & 2 \\ 4 & -8 & \vdots & 0 \end{bmatrix}$ by -4 to get

$$-4 \quad 4 \quad -8 \quad \text{This is } -4R_1.$$

We then add these numbers to the entries in row 2 of matrix B. (Note that row 1 remains unchanged.)

$$\begin{bmatrix} 1 & -1 & \vdots & 2 \\ 4 + (-4) & -8 + 4 & \vdots & 0 + (-8) \end{bmatrix}$$ This procedure is represented by $-4R_1 + R_2$, which means "Multiply row 1 by -4 and add the result to row 2."

After simplifying the bottom row, we have the matrix $\begin{bmatrix} 1 & -1 & \vdots & 2 \\ 0 & -4 & \vdots & -8 \end{bmatrix}$.

Success Tip

This elementary row operation corresponds to the following elimination method step for solving a system:

$$\begin{array}{r} 4x - 8y = 0 \\ -4x + 4y = -8 \\ \hline -4y = -8 \end{array}$$

 Self Check 2 Use the augmented matrices of Example 2 and perform the following:

a. Interchange the rows of matrix B.

b. To the numbers in row 1 of matrix A, add the results of multiplying each number in row 2 by -2.

c. Multiply row 1 of matrix C by $\frac{1}{2}$.

Now Try Problems 17, 19, and 23

4 Use Matrices to Solve a System of Two Equations.

We can solve a system of two linear equations using a series of elementary row operations.

EXAMPLE 3 Use matrices to solve the system: $\begin{cases} 2x + y = 5 \\ x - y = 4 \end{cases}$

Strategy We will represent the system with an augmented matrix and use a series of elementary row operations to produce an equivalent matrix in row echelon form.

Why When the resulting row echelon form matrix is written as a system of two equations, we will know the value of one variable, and the value of the other can be found using substitution.

Solution   We can represent the system with the following augmented matrix:

$$\left[\begin{array}{cc:c} 2 & 1 & 5 \\ 1 & -1 & 4 \end{array}\right]$$

First, we want to get a 1 in the top row of the first column where the shaded 2 is. This can be done by applying a type 1 row operation and interchanging rows 1 and 2.

$$\left[\begin{array}{cc:c} 1 & -1 & 4 \\ 2 & 1 & 5 \end{array}\right] \quad R_1 \leftrightarrow R_2$$

To get a 0 in the first column where the shaded 2 is, we use a type 3 row operation and multiply each entry in row 1 by -2 to get

$$-2 \quad 2 \quad -8$$

and add these numbers to the entries in row 2.

$$\left[\begin{array}{cc:c} 1 & -1 & 4 \\ 2+(-2) & 1+2 & 5+(-8) \end{array}\right] \quad \text{This is } -2R_1 + R_2.$$

After simplifying the bottom row, we have

$$\left[\begin{array}{cc:c} 1 & -1 & 4 \\ 0 & 3 & -3 \end{array}\right]$$

To get a 1 in the bottom row of the second column where the shaded 3 is, we use a type 2 row operation and multiply row 2 by $\frac{1}{3}$.

$$\left[\begin{array}{cc:c} 1 & -1 & 4 \\ 0 & 1 & -1 \end{array}\right] \quad \frac{1}{3}R_2$$

This augmented matrix represents the system of equations

$$\begin{cases} 1x - 1y = 4 \\ 0x + 1y = -1 \end{cases}$$

Writing the equations without the coefficients of 1 and -1 and dropping the $0x$ term, we have

(1) $\begin{cases} x - y = 4 \\ **(2)** \quad y = -1 \end{cases}$

From equation 2, we see that $y = -1$. We can back substitute -1 for y in equation 1 to find x.

$$\begin{aligned} x - y &= 4 \\ x - (-1) &= 4 \quad &\text{Substitute } -1 \text{ for } y. \\ x + 1 &= 4 \quad &\text{Simplify: } -(-1) = 1. \\ x &= 3 \quad &\text{To solve for } x, \text{ subtract 1 from both sides.} \end{aligned}$$

The solution of the system is $(3, -1)$ and the solution set is $\{(3, -1)\}$. Verify that this ordered pair satisfies both equations in the original system.

Self Check 3   Use matrices to solve the system: $\begin{cases} 2x + y = -9 \\ x - 3y = 13 \end{cases}$

Now Try   Problem 27

When a system of linear equations has one solution, we can use the following steps to solve it.

Solving Systems of Linear Equations Using Matrices	1. Write an augmented matrix for the system. 2. Use elementary row operations to transform the augmented matrix into a matrix in row echelon form with 1's down its main diagonal and 0's under the 1's. 3. When step 2 is complete, write the resulting system and use *back-substitution* to find the solution. 4. Check the proposed solution in the equations of the original system.

5   **Use Matrices to Solve a System of Three Equations.**

To show how to use matrices to solve systems of three linear equations containing three variables, we consider the following system that can be represented by the augmented matrix to its right.

$$\begin{cases} 3x + y + 5z = 8 \\ 2x + 3y - z = 6 \\ x + 2y + 2z = 10 \end{cases} \qquad \begin{bmatrix} 3 & 1 & 5 & | & 8 \\ 2 & 3 & -1 & | & 6 \\ 1 & 2 & 2 & | & 10 \end{bmatrix}$$

To solve the 3 × 3 system of equations, we transform the augmented matrix into a matrix with 1's down its main diagonal and 0's below its main diagonal.

$$\begin{bmatrix} 1 & a & b & | & c \\ 0 & 1 & d & | & e \\ 0 & 0 & 1 & | & f \end{bmatrix} \qquad a, b, c, \ldots, f \text{ represent real numbers.}$$

Main diagonal

EXAMPLE 4   Use matrices to solve the system:

$$\begin{cases} 3x + y + 5z = 8 \\ 2x + 3y - z = 6 \\ x + 2y + 2z = 10 \end{cases}$$

Strategy   We will represent the system with an augmented matrix and use a series of elementary row operations to produce an equivalent matrix in row echelon form.

Why   When the resulting matrix in row echelon form is written as a system of three equations, the value of one variable is known, and the values of the other two can be found using substitution.

Solution   This system can be represented by the augmented matrix

$$\begin{bmatrix} 3 & 1 & 5 & | & 8 \\ 2 & 3 & -1 & | & 6 \\ 1 & 2 & 2 & | & 10 \end{bmatrix}$$

To get a 1 in the first column where the shaded 3 is, we perform a type 1 row operation by interchanging rows 1 and 3.

$$\begin{bmatrix} 1 & 2 & 2 & | & 10 \\ 2 & 3 & -1 & | & 6 \\ 3 & 1 & 5 & | & 8 \end{bmatrix} \qquad R_1 \leftrightarrow R_3$$

Success Tip

Follow this order in getting 1's and 0's in the proper positions of the augmented matrix.

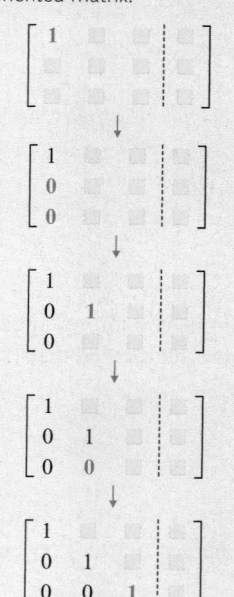

To get a 0 where the shaded 2 is, we perform a type 3 row operation by multiplying each entry in row 1 by -2 to get

$$-2 \qquad -4 \qquad -4 \qquad -20$$

and add these numbers to the entries in row 2.

$$\begin{bmatrix} 1 & 2 & 2 & \vdots & 10 \\ 2 + (-2) & 3 + (-4) & -1 + (-4) & \vdots & 6 + (-20) \\ 3 & 1 & 5 & \vdots & 8 \end{bmatrix}$$ This is $-2R_1 + R_2$.

After simplifying the second row, we have

$$\begin{bmatrix} 1 & 2 & 2 & \vdots & 10 \\ 0 & -1 & -5 & \vdots & -14 \\ 3 & 1 & 5 & \vdots & 8 \end{bmatrix}$$

To get a 0 where the shaded 3 is, we perform another type 3 row operation by multiplying the entries in row 1 by -3 and adding the results to row 3.

$$\begin{bmatrix} 1 & 2 & 2 & \vdots & 10 \\ 0 & -1 & -5 & \vdots & -14 \\ 0 & -5 & -1 & \vdots & -22 \end{bmatrix}$$ $-3R_1 + R_3$

To get a 1 where the shaded -1 is, we perform a type 2 row operation by multiplying row 2 by -1.

$$\begin{bmatrix} 1 & 2 & 2 & \vdots & 10 \\ 0 & 1 & 5 & \vdots & 14 \\ 0 & -5 & -1 & \vdots & -22 \end{bmatrix}$$ $-1R_2$

To get a 0 where the shaded -5 is, we perform a type 3 row operation by multiplying the entries in row 2 by 5 and adding the results to row 3.

$$\begin{bmatrix} 1 & 2 & 2 & \vdots & 10 \\ 0 & 1 & 5 & \vdots & 14 \\ 0 & 0 & 24 & \vdots & 48 \end{bmatrix}$$ $5R_2 + R_3$

To get a 1 where the shaded 24 is, we perform a type 2 row operation by multiplying row 3 by $\frac{1}{24}$.

$$\begin{bmatrix} 1 & 2 & 2 & \vdots & 10 \\ 0 & 1 & 5 & \vdots & 14 \\ 0 & 0 & 1 & \vdots & 2 \end{bmatrix}$$ Note the 1s along the main diagonal
$\frac{1}{24}R_3$

The final augmented matrix represents the system

$$\begin{cases} 1x + 2y + 2z = 10 \\ 0x + 1y + 5z = 14 \\ 0x + 0y + 1z = 2 \end{cases} \text{ which can be written as } \begin{cases} x + 2y + 2z = 10 & \textbf{(1)} \\ \phantom{x + 2y + {}} y + 5z = 14 & \textbf{(2)} \\ \phantom{x + 2y + 2y + {}} z = 2 & \textbf{(3)} \end{cases}$$

From equation 3, we can see that $z = 2$. To find y, we back substitute 2 for z in equation 2 and solve for y:

$$y + 5z = 14 \qquad \text{This is equation 2.}$$
$$y + 5(2) = 14 \qquad \text{Substitute 2 for } z.$$
$$y + 10 = 14$$
$$y = 4 \qquad \text{To solve for } y, \text{ subtract 10 from both sides.}$$

Thus, $y = 4$. To find x, we back substitute 2 for z and 4 for y in equation 1 and solve for x:

$$x + 2y + 2z = 10 \quad \text{This is equation 1.}$$
$$x + 2(4) + 2(2) = 10 \quad \text{Substitute 2 for z and 4 for y.}$$
$$x + 8 + 4 = 10$$
$$x + 12 = 10$$
$$x = -2 \quad \text{To solve for x, subtract 12 from both sides.}$$

Thus, $x = -2$. The solution of the given system is $(-2, 4, 2)$ and the solution set is $\{(-2, 4, 2)\}$. Verify that this ordered triple satisfies each equation of the original system.

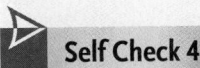 **Self Check 4**   Use matrices to solve the system:
$$\begin{cases} 2x - y + z = 5 \\ x + y - z = -2 \\ -x + 2y + 2z = 1 \end{cases}$$

Now Try   **Problem 29**

6 **Use Matrices to Identify Inconsistent Systems and Dependent Equations.**

In the next example, we will see how to recognize inconsistent systems and systems of dependent equations when matrices are used to solve them.

EXAMPLE 5   If possible, use matrices to solve the system.

a. $\begin{cases} x + y = -1 \\ -3x - 3y = -5 \end{cases}$ b. $\begin{cases} 2x - y = 4 \\ -6x + 3y = -12 \end{cases}$

Strategy   We will represent the system with an augmented matrix and use a series of elementary row operations to produce an equivalent matrix in row echelon form.

Why   When the resulting matrix in row echelon form is written as a system of two equations, we can determine whether the system is consistent or inconsistent and whether the equations are dependent or independent.

Solution

a. The system $\begin{cases} x + y = -1 \\ -3x - 3y = -5 \end{cases}$ can be represented by the augmented matrix

$$\left[\begin{array}{cc|c} 1 & 1 & -1 \\ -3 & -3 & -5 \end{array}\right]$$

Since the matrix has a 1 in the top row of the first column, we proceed to get a 0 where the shaded -3 is by multiplying row 1 by 3 and adding the results to row 2.

$$\left[\begin{array}{cc|c} 1 & 1 & -1 \\ 0 & 0 & -8 \end{array}\right] \quad 3R_1 + R_2$$

This augmented matrix represents the system

$$\begin{cases} x + y = -1 \\ 0 + 0 = -8 \end{cases}$$

The equation in the second row of this system simplifies to $0 = -8$. This false statement indicates that the system is inconsistent and has no solution. The solution set is $\varnothing$.

b. The system $\begin{cases} 2x - y = 4 \\ -6x + 3y = -12 \end{cases}$ can be represented by the augmented matrix

$$\begin{bmatrix} 2 & -1 & \vdots & 4 \\ -6 & 3 & \vdots & -12 \end{bmatrix}$$

To get a 1 where the shaded 2 is, we perform a type 2 row operation by multiplying row 1 by $\frac{1}{2}$.

$$\begin{bmatrix} 1 & -\dfrac{1}{2} & \vdots & 2 \\ -6 & 3 & \vdots & -12 \end{bmatrix} \quad \frac{1}{2}R_1$$

To get a 0 where the shaded -6 is, we perform a type 3 row operation by multiplying the entries in row 1 by 6 and adding the results to the entries in row 2.

$$\begin{bmatrix} 1 & -\dfrac{1}{2} & \vdots & 2 \\ 0 & 0 & \vdots & 0 \end{bmatrix} \quad 6R_1 + R_2$$

This augmented matrix represents the system

$$\begin{cases} x - \dfrac{1}{2}y = 2 \\ 0 + 0 = 0 \end{cases}$$

The equation in the second row of the system simplifies to $0 = 0$. This true statement indicates that the equations are dependent and that the system has infinitely many solutions. The solution set is $\{(x, y) \mid 2x - y = 4\}$.

Self Check 5   If possible, use matrices to solve the system.

a. $\begin{cases} 4x - 8y = 9 \\ x - 2y = -5 \end{cases}$

b. $\begin{cases} x - 3y = 6 \\ -4x + 12y = -24 \end{cases}$

Now Try **Problems 33 and 35**

ANSWERS TO SELF CHECKS

1. a. $\begin{bmatrix} 2 & -4 & \vdots & 9 \\ 5 & -1 & \vdots & -2 \end{bmatrix}$ **b.** $\begin{bmatrix} 1 & 1 & -1 & \vdots & -4 \\ 0 & -2 & 7 & \vdots & 0 \\ 10 & 8 & -4 & \vdots & 5 \end{bmatrix}$ **2. a.** $\begin{bmatrix} 4 & -8 & \vdots & 0 \\ 1 & -1 & \vdots & 2 \end{bmatrix}$

b. $\begin{bmatrix} 0 & 20 & \vdots & -3 \\ 1 & -8 & \vdots & 0 \end{bmatrix}$ **c.** $\begin{bmatrix} 1 & \dfrac{1}{2} & -4 & \vdots & 2 \\ 0 & 1 & 4 & \vdots & -2 \\ 0 & 0 & -6 & \vdots & 24 \end{bmatrix}$ **3.** $(-2, -5)$ **4.** $(1, -1, 2)$ **5. a.** No solution, $\varnothing$;

inconsistent system **b.** $\{(x, y) \mid x - 3y = 6\}$, infinitely many solutions; dependent equations

STUDY SET
12.4

VOCABULARY

Fill in the blanks.

1. A _____ is a rectangular array of numbers written within brackets.

2. Each number in a matrix is called an _____ or entry of the matrix.

3. If the order of a matrix is 3×4, it has 3 _____ and 4 _____ . We read 3×4 as "3 ____ 4."

4. Elementary _____ operations are used to produce equivalent matrices that lead to the solution of a system.

5. A matrix that represents the equations of a system is called an _____ matrix.

6. The matrix $\begin{bmatrix} 1 & 3 & \vdots & -2 \\ 0 & 1 & \vdots & 4 \end{bmatrix}$, with 1's down its main _____ and a 0 below the 1 in the first column, is in row echelon form.

CONCEPTS

7. For each matrix, determine the number of rows and the number of columns.

   a. $\begin{bmatrix} 4 & 6 & \vdots & -1 \\ 1 & 9 & \vdots & -3 \end{bmatrix}$

   b. $\begin{bmatrix} 1 & -2 & 3 & \vdots & 1 \\ 0 & 1 & 6 & \vdots & 4 \\ 0 & 0 & 1 & \vdots & \frac{1}{3} \end{bmatrix}$

8. Fill in the blanks to complete each elementary row operation:

   a. Type 1: Any two rows of a matrix can be _____.

   b. Type 2: Any row of a matrix can be _____ by a nonzero constant.

   c. Type 3: Any row of a matrix can be changed by _____ a nonzero constant multiple of another row to it.

9. Matrices were used to solve a system. The final augmented matrix is shown. Fill in the blanks.

   a. $\begin{bmatrix} 1 & 1 & \vdots & 10 \\ 0 & 1 & \vdots & 6 \end{bmatrix}$ represents $\begin{cases} \square + \square = 10 \\ \square = 6 \end{cases}$

   Therefore, $y = \square$. Using back substitution, we find that $x = \square$.

   b. $\begin{bmatrix} 1 & -2 & 1 & \vdots & -16 \\ 0 & 1 & 2 & \vdots & 8 \\ 0 & 0 & 1 & \vdots & 4 \end{bmatrix}$ represents $\begin{cases} \square - 2y + \square = -16 \\ y + \square = 8 \\ \square = 4 \end{cases}$

   Therefore, $z = \square$. Using back substitution, we find that $y = \square$ and $x = \square$.

10. a. Which augmented matrix shown below indicates that its associated system of equations has no solution?

    b. Which augmented matrix indicates that the equations of its associated system are dependent?

    i. $\begin{bmatrix} 1 & 2 & \vdots & -4 \\ 0 & 0 & \vdots & 0 \end{bmatrix}$ ii. $\begin{bmatrix} 1 & 3 & \vdots & 6 \\ 0 & 1 & \vdots & 0 \end{bmatrix}$ iii. $\begin{bmatrix} 1 & 2 & \vdots & -4 \\ 0 & 0 & \vdots & 2 \end{bmatrix}$

NOTATION

11. Explain what each symbolism means.

    a. $R_1 \leftrightarrow R_2$

    b. $\frac{1}{2}R_1$

    c. $6R_2 + R_3$

12. Complete the solution.

   Solve using matrices: $\begin{cases} 4x - y = 14 \\ x + y = 6 \end{cases}$

   $\begin{bmatrix} 4 & \square & \vdots & 14 \\ 1 & 1 & \vdots & 6 \end{bmatrix}$

   $\begin{bmatrix} \square & 1 & \vdots & 6 \\ 4 & -1 & \vdots & 14 \end{bmatrix} \quad R_1 \leftrightarrow R_2$

   $\begin{bmatrix} 1 & 1 & \vdots & 6 \\ 0 & \square & \vdots & -10 \end{bmatrix} \quad -4R_1 + R_2$

   $\begin{bmatrix} 1 & 1 & \vdots & 6 \\ 0 & 1 & \vdots & 2 \end{bmatrix} \quad -\frac{1}{5}R_2$

   This matrix represents the system:

   $\begin{cases} x + y = 6 \\ \quad\ \square = 2 \end{cases}$

   The solution is ($\square$, 2).

GUIDED PRACTICE

Represent each system using an augmented matrix. See Example 1.

13. $\begin{cases} x + 2y = 6 \\ 3x - y = -10 \end{cases}$

14. $\begin{cases} x + y + z = 4 \\ 2x + y - z = 1 \\ 2x - 3y = 1 \end{cases}$

For each augmented matrix, give the system of equations that it represents. See Example 3.

15. $\begin{bmatrix} 1 & 6 & \vdots & 7 \\ 0 & 1 & \vdots & 4 \end{bmatrix}$

16. $\begin{bmatrix} 1 & -2 & 9 & \vdots & 1 \\ 0 & 1 & 4 & \vdots & 0 \\ 0 & 0 & 1 & \vdots & -7 \end{bmatrix}$

Perform each of the following elementary row operations on the augmented matrix $\begin{bmatrix} -3 & 1 & \vdots & -6 \\ 1 & -4 & \vdots & 4 \end{bmatrix}$. See Example 2.

17. $R_1 \leftrightarrow R_2$

18. $5R_2$

19. $-\dfrac{1}{3}R_1$

20. $3R_2 + R_1$

Perform each of the following elementary row operations on the augmented matrix $\begin{bmatrix} 3 & 6 & -9 & | & 0 \\ 1 & 5 & -2 & | & 1 \\ -2 & 2 & -2 & | & 5 \end{bmatrix}$. *See Example 2.*

21. $R_2 \leftrightarrow R_3$

22. $-\dfrac{1}{2}R_3$

23. $-R_1 + R_2$

24. $2R_2 + R_3$

Use matrices to solve each system of equations. See Example 3.

25. $\begin{cases} x + y = 2 \\ x - y = 0 \end{cases}$ **26.** $\begin{cases} x + y = 3 \\ x - y = -1 \end{cases}$

27. $\begin{cases} 2x + y = 1 \\ x + 2y = -4 \end{cases}$ **28.** $\begin{cases} 5x - 4y = 10 \\ x - 7y = 2 \end{cases}$

Use matrices to solve each system of equations. See Example 4.

29. $\begin{cases} x + y + z = 6 \\ x + 2y + z = 8 \\ x + y + 2z = 7 \end{cases}$ **30.** $\begin{cases} x + y + z = 6 \\ x + 2y + z = 8 \\ x + y + 2z = 9 \end{cases}$

31. $\begin{cases} 3x + y - 3z = 5 \\ x - 2y + 4z = 10 \\ x + y + z = 13 \end{cases}$ **32.** $\begin{cases} 2x + y - 3z = -1 \\ 3x - 2y - z = -5 \\ x - 3y - 2z = -12 \end{cases}$

Use matrices to solve each system of equations. If the equations of a system are dependent or if a system is inconsistent, state this. See Example 5.

33. $\begin{cases} x - 3y = 9 \\ -2x + 6y = 18 \end{cases}$ **34.** $\begin{cases} -6x + 12y = 10 \\ 2x - 4y = 8 \end{cases}$

35. $\begin{cases} -4x - 4y = -12 \\ x + y = 3 \end{cases}$ **36.** $\begin{cases} 5x - 15y = 10 \\ x - 3y = 2 \end{cases}$

TRY IT YOURSELF

Use matrices to solve each system of equations. If the equations of a system are dependent or if a system is inconsistent, state this.

37. $\begin{cases} 2x + 3y - z = -8 \\ x - y - z = -2 \\ -4x + 3y + z = 6 \end{cases}$ **38.** $\begin{cases} 2a + b + 3c = 3 \\ -2a - b + c = 5 \\ 4a - 2b + 2c = 2 \end{cases}$

39. $\begin{cases} 2x - y = -1 \\ x - 2y = 1 \end{cases}$ **40.** $\begin{cases} 2x - y = 0 \\ x + y = 3 \end{cases}$

41. $\begin{cases} 3x + 4y = -12 \\ 9x - 2y = 6 \end{cases}$ **42.** $\begin{cases} 2x - 3y = 16 \\ -4x + y = -22 \end{cases}$

43. $\begin{cases} 2x + y - z = 1 \\ x + 2y + 2z = 2 \\ 4x + 5y + 3z = 3 \end{cases}$ **44.** $\begin{cases} x - y = 1 \\ 2x - z = 0 \\ 2y - z = -2 \end{cases}$

45. $\begin{cases} 8x - 2y = 4 \\ 4x - y = 2 \end{cases}$ **46.** $\begin{cases} 9x - 3y = 6 \\ 3x - y = 8 \end{cases}$

47. $\begin{cases} 2x + y - 2z = 6 \\ 4x - y + z = -1 \\ 6x - 2y + 3z = -5 \end{cases}$ **48.** $\begin{cases} 2x - 3y + 3z = 14 \\ 3x + 3y - z = 2 \\ -2x + 6y + 5z = 9 \end{cases}$

49. $\begin{cases} 6x + y - z = -2 \\ x + 2y + z = 5 \\ 5y - z = 2 \end{cases}$ **50.** $\begin{cases} 2x + 3y - 2z = 18 \\ 5x - 6y + z = 21 \\ 4y - 2z = 6 \end{cases}$

51. $\begin{cases} 5x + 3y = 4 \\ 3y - 4z = 4 \\ x + z = 1 \end{cases}$ **52.** $\begin{cases} y + 2z = -2 \\ x + y = 1 \\ 2x - z = 0 \end{cases}$

APPLICATIONS

53. DIGITAL PHOTOGRAPHY
A digital camera stores the black and white photograph shown on the right as a 512×512 matrix. Each element of the matrix corresponds to a small dot of grey scale shading, called a *pixel,* in the picture. How many elements does a 512×512 matrix have?

54. DIGITAL IMAGING   A scanner stores a black and white photograph as a matrix that has a total of 307,200 elements. If the matrix has 480 rows, how many columns does it have?

Write a system of equations to solve each problem. Use matrices to solve the system.

55. COMPLENTARY ANGLES   One angle measures 46° more than the measure of its complement. Find the measure of each angle.   (*Hint:* The sum of the measures of complementary angles is 90°.)

56. SUPPLEMENTARY ANGLES   One angle measures 14° more than the measure of its supplement. Find the measure of each angle.   (*Hint:* The sum of the measures of supplementary angles is 180°.)

57. TRIANGLES   In the illustration, $\angle B$ measures 25° more than the measure of $\angle A$, and the measure of $\angle C$ is 5° less than twice the measure of $\angle A$. Find the measure of each angle of the triangle.

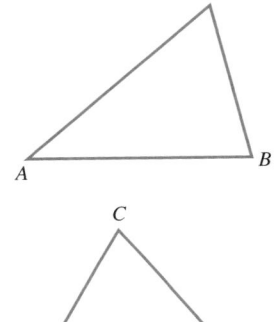

58. TRIANGLES   In the illustration, $\angle A$ measures 10° less than the measure of $\angle B$, and the measure of $\angle B$ is 10° less than the measure of $\angle C$. Find the measure of each angle of the triangle.

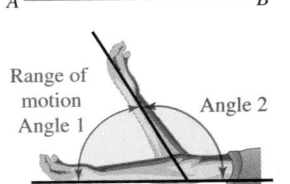

59. PHYSICAL THERAPY   After an elbow injury, a volleyball player has restricted movement of her arm. Her range of motion (the measure of $\angle 1$) is 28° less than the measure of $\angle 2$. Find the measure of each angle.

Range of motion
Angle 1   Angle 2

60. ICE SKATING   Three circles are traced out by a figure skater during her performance. If the centers of the circles are the given distances apart, find the radius of each circle.

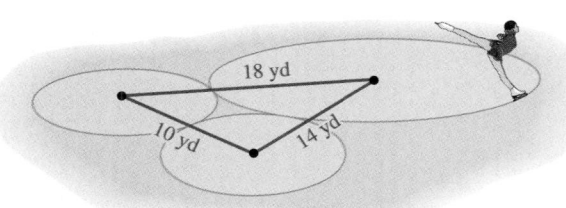

18 yd
10 yd   14 yd

WRITING

61. For the system $\begin{cases} 2x - 3y = 5 \\ 4x + 8 = y \end{cases}$, explain what is wrong with writing its corresponding augmented matrix as $\begin{bmatrix} 2 & -3 & | & 5 \\ 4 & 8 & | & 1 \end{bmatrix}$. How should it be written?

62. Explain what is meant by the phrase *back substitution*. Give an example of how it was used in this section.

63. Explain how a type 3 row operation is similar to the elimination (addition) method of solving a system of equations.

64. If the system represented by the following augmented matrix has no solution, what do you know about k? Explain your answer.

$$\begin{bmatrix} 1 & 1 & 0 & | & 1 \\ 0 & 0 & 1 & | & 2 \\ 0 & 0 & 0 & | & k \end{bmatrix}$$

REVIEW

65. What is the formula used to find the slope of a line, given two points on the line?

66. What is the form of the equation of a horizontal line? Of a vertical line?

67. What is the point–slope form of the equation of a line?

68. What is the slope–intercept form of the equation of a line?

CHALLENGE PROBLEMS

Use matrices to solve the system.

69. $\begin{cases} x^2 + y^2 + z^2 = 14 \\ 2x^2 + 3y^2 - 2z^2 = -7 \\ x^2 - 5y^2 + z^2 = 8 \end{cases}$

70. $\begin{cases} w + x + y + z = 0 \\ w - 2x + y - 3z = -3 \\ 2w + 3x + y - 2z = -1 \\ 2w - 2x - 2y + z = -12 \end{cases}$

SECTION 12.5
Solving Systems of Equations Using Determinants

Objectives

1. Evaluate 2×2 and 3×3 determinants.
2. Use Cramer's rule to solve systems of two equations.
3. Use Cramer's rule to solve systems of three equations.

In this section, we will discuss another method for solving systems of linear equations. With this method, called *Cramer's rule,* we work with combinations of the coefficients and the constants of the equations written as *determinants.*

1 **Evaluate 2×2 and 3×3 Determinants.**

An idea related to the concept of matrix is the **determinant.** A determinant is a number that is associated with a **square matrix,** a matrix that has the same number of rows and columns. For any square matrix A, the symbol $|A|$ represents the determinant of A. To write a determinant, we put the elements of a square matrix between two vertical lines.

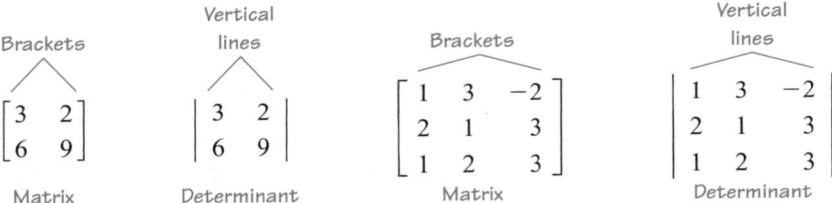

Brackets

$$\begin{bmatrix} 3 & 2 \\ 6 & 9 \end{bmatrix}$$

Matrix

Vertical lines

$$\begin{vmatrix} 3 & 2 \\ 6 & 9 \end{vmatrix}$$

Determinant

Brackets

$$\begin{bmatrix} 1 & 3 & -2 \\ 2 & 1 & 3 \\ 1 & 2 & 3 \end{bmatrix}$$

Matrix

Vertical lines

$$\begin{vmatrix} 1 & 3 & -2 \\ 2 & 1 & 3 \\ 1 & 2 & 3 \end{vmatrix}$$

Determinant

Like matrices, determinants are classified according to the number of rows and columns they contain. The determinant above, on the left, is a 2×2 determinant. The other is a 3×3 determinant.

The determinant of a 2×2 matrix is the number that is equal to the product of the numbers on the main diagonal minus the product of the numbers on the other diagonal.

$$\begin{vmatrix} a & b \\ c & d \end{vmatrix}$$
Main diagonal

$$\begin{vmatrix} a & b \\ c & d \end{vmatrix}$$
Other diagonal

Value of a 2×2 Determinant

If a, b, c, and d are numbers, the **determinant** of the matrix $\begin{bmatrix} a & b \\ c & d \end{bmatrix}$ is:

$$\begin{vmatrix} a & b \\ c & d \end{vmatrix} = ad - bc$$

EXAMPLE 1   Evaluate each determinant: **a.** $\begin{vmatrix} 3 & 2 \\ 6 & 9 \end{vmatrix}$ **b.** $\begin{vmatrix} -20 & 1 \\ -8 & 4 \end{vmatrix}$

Strategy   We will find the product of the numbers on the main diagonal and the product of the numbers along the other diagonal and subtract the results.

Why   The value of a determinant of the form $\begin{vmatrix} a & b \\ c & d \end{vmatrix}$ is $ad - bc$.

Caution
Although the symbols used to write a determinant are a pair of vertical lines just like the vertical lines used for absolute value, the two concepts are unrelated. In part (b), notice that the value of the determinant is negative.

Solution   To evaluate the determinant, we proceed as follows:

a.   This is always minus.

$$\begin{vmatrix} 3 & 2 \\ 6 & 9 \end{vmatrix} = 3(9) - 2(6) = 27 - 12 = 15$$

b. $\begin{vmatrix} -20 & 1 \\ -8 & 4 \end{vmatrix} = -20(4) - 1(-8) = -80 - (-8) = -80 + 8 = -72$

Self Check 1   Evaluate: $\begin{vmatrix} 4 & -3 \\ 2 & 1 \end{vmatrix}$

Now Try   **Problems 15 and 17**

A 3×3 determinant is evaluated by **expanding by minors.** The following definition shows how we can evaluate a 3×3 determinant by expanding by minors along the first row.

Value of a 3×3 Determinant

Minor of a_1 Minor of b_1 Minor of c_1

$$\begin{vmatrix} a_1 & b_1 & c_1 \\ a_2 & b_2 & c_2 \\ a_3 & b_3 & c_3 \end{vmatrix} = a_1 \begin{vmatrix} b_2 & c_2 \\ b_3 & c_3 \end{vmatrix} - b_1 \begin{vmatrix} a_2 & c_2 \\ a_3 & c_3 \end{vmatrix} + c_1 \begin{vmatrix} a_2 & b_2 \\ a_3 & b_3 \end{vmatrix}$$

To find the minor of a_1, we cross out the elements of the determinant that are in the same row and column as a_1:

$$\begin{vmatrix} a_1 & b_1 & c_1 \\ a_2 & b_2 & c_2 \\ a_3 & b_3 & c_3 \end{vmatrix}$$ The minor of a_1 is $\begin{vmatrix} b_2 & c_2 \\ b_3 & c_3 \end{vmatrix}$.

To find the minor of b_1, we cross out the elements of the determinant that are in the same row and column as b_1:

$$\begin{vmatrix} a_1 & b_1 & c_1 \\ a_2 & b_2 & c_2 \\ a_3 & b_3 & c_3 \end{vmatrix}$$ The minor of b_1 is $\begin{vmatrix} a_2 & c_2 \\ a_3 & c_3 \end{vmatrix}$.

To find the minor of c_1, we cross out the elements of the determinant that are in the same row and column as c_1:

$$\begin{vmatrix} a_1 & b_1 & c_1 \\ a_2 & b_2 & c_2 \\ a_3 & b_3 & c_3 \end{vmatrix}$$ The minor of c_1 is $\begin{vmatrix} a_2 & b_2 \\ a_3 & b_3 \end{vmatrix}$.

EXAMPLE 2   Evaluate the determinant: $\begin{vmatrix} 1 & 3 & -2 \\ 2 & 0 & 3 \\ 1 & 2 & 3 \end{vmatrix}$

Strategy   We will expand the determinant along the first row using the numbers in the first row and their corresponding minors.

Why   We can then evaluate the resulting 2×2 determinants and simplify.

Solution   To evaluate the determinant, we can use the first row and expand the determinant by minors:

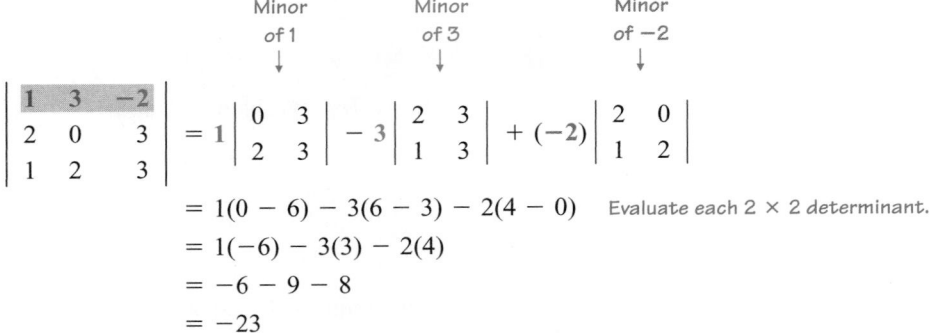

$$= 1(0 - 6) - 3(6 - 3) - 2(4 - 0) \quad \text{Evaluate each } 2 \times 2 \text{ determinant.}$$
$$= 1(-6) - 3(3) - 2(4)$$
$$= -6 - 9 - 8$$
$$= -23$$

▷ **Self Check 2**   Evaluate: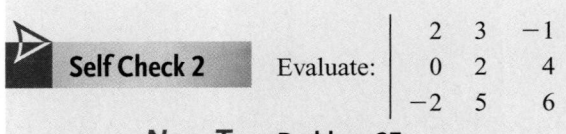

Now Try   **Problem 27**

We can evaluate a 3×3 determinant by expanding it by the minors of any row or column. We will get the same value. To determine the signs between the terms of the expansion of a 3×3 determinant, we use the following array of signs.

Array of Signs for a 3 × 3 Determinant			
+	−	+	
−	+	−	This array of signs is often called the checkerboard pattern.
+	−	+	

To remember the sign pattern, note that there is a + sign in the upper left position and that the signs alternate for all of the positions that follow.

EXAMPLE 3   Evaluate the determinant $\begin{vmatrix} 1 & 3 & -2 \\ 2 & 0 & 3 \\ 1 & 2 & 3 \end{vmatrix}$ by expanding by the minors of the middle column. (This is the determinant of Example 2.)

Strategy   We will expand the determinant using the numbers in the middle column and their corresponding minors. We will use the sign pattern − + − between the terms of the expansion.

Why   We can then evaluate the resulting 2×2 determinants and simplify.

Solution   To evaluate the determinant, we proceed as follows:

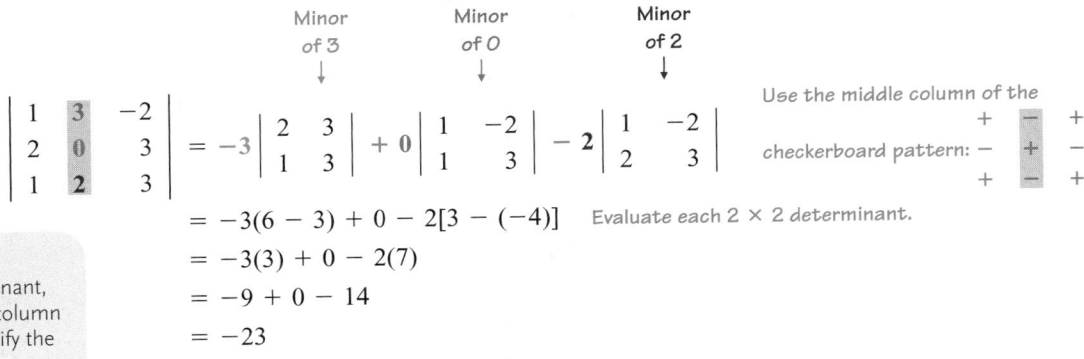

$$\begin{vmatrix} 1 & 3 & -2 \\ 2 & 0 & 3 \\ 1 & 2 & 3 \end{vmatrix} = -3\begin{vmatrix} 2 & 3 \\ 1 & 3 \end{vmatrix} + 0\begin{vmatrix} 1 & -2 \\ 1 & 3 \end{vmatrix} - 2\begin{vmatrix} 1 & -2 \\ 2 & 3 \end{vmatrix}$$

Use the middle column of the checkerboard pattern:

$$= -3(6-3) + 0 - 2[3 - (-4)] \quad \text{Evaluate each 2 × 2 determinant.}$$
$$= -3(3) + 0 - 2(7)$$
$$= -9 + 0 - 14$$
$$= -23$$

Success Tip

When evaluating a determinant, expanding along a row or column that contains 0's can simplify the computations.

As expected, we get the same value as in Example 2.

Self Check 3   Evaluate $\begin{vmatrix} 2 & 3 & -1 \\ 0 & 2 & 4 \\ -2 & 5 & 6 \end{vmatrix}$ by expanding by the minors of the first column.

Now Try   **Problem 31**

Using Your Calculator   *Evaluating Determinants*

It is possible to use a graphing calculator to evaluate determinants. For example, to evaluate the determinant in Example 3, we first enter the matrix by pressing the MATRIX key, selecting EDIT, and pressing the ENTER key. Next, we enter the dimensions and the elements of the matrix to get figure (a). We then press 2nd QUIT to clear the screen, press MATRIX , select MATH, and press 1 to get figure (b). We then press MATRIX , select NAMES, press 1, and press) and ENTER to get the value of the determinant. Figure (c) shows that the value of the determinant is -23.

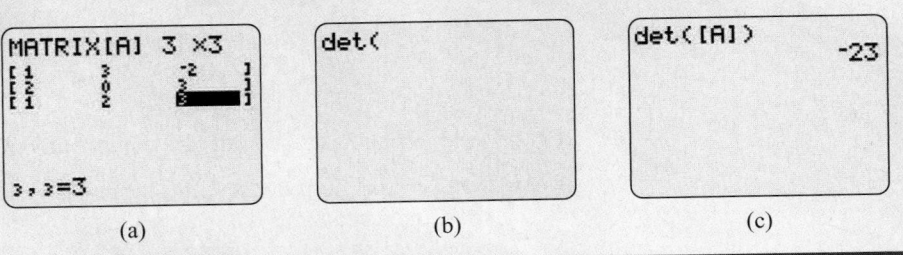

(a)   (b)   (c)

---

2   **Use Cramer's Rule to Solve Systems of Two Equations.**

The method of using determinants to solve systems of linear equations is called **Cramer's rule,** named after the 18th-century Swiss mathematician Gabriel Cramer. To develop Cramer's rule, we consider the system

$$\begin{cases} ax + by = e \\ cx + dy = f \end{cases}$$

where x and y are variables and a, b, c, d, e, and f are constants.

If we multiply both sides of the first equation by d and multiply both sides of the second equation by $-b$, we can add the equations and eliminate y:

$$\begin{array}{ll} adx + bdy = ed & \text{This is } d(ax + by) = d(e). \\ \underline{-bcx - bdy = -bf} & \text{This is } -b(cx + dy) = -b(f). \\ adx - bcx \quad\quad = ed - bf \end{array}$$

To solve for x, we use the distributive property to write $adx - bcx$ as $(ad - bc)x$ on the left side and divide each side by $ad - bc$:

$$(ad - bc)x = ed - bf$$

$$x = \frac{ed - bf}{ad - bc} \quad\quad \text{where } ad - bc \neq 0$$

We can find y in a similar way. After eliminating the variable x, we get

$$y = \frac{af - ec}{ad - bc} \quad\quad \text{where } ad - bc \neq 0$$

Note that the denominator for both x and y is

$$\begin{vmatrix} a & b \\ c & d \end{vmatrix} = ad - bc$$

The numerators can be expressed as determinants also:

$$x = \frac{ed - bf}{ad - bc} = \frac{\begin{vmatrix} e & b \\ f & d \end{vmatrix}}{\begin{vmatrix} a & b \\ c & d \end{vmatrix}} \quad\quad \text{and} \quad\quad y = \frac{af - ec}{ad - bc} = \frac{\begin{vmatrix} a & e \\ c & f \end{vmatrix}}{\begin{vmatrix} a & b \\ c & d \end{vmatrix}}$$

Cramer's Rule for Two Equations in Two Variables

The solution of the system $\begin{cases} ax + by = e \\ cx + dy = f \end{cases}$ is given by

$$x = \frac{D_x}{D} = \frac{\begin{vmatrix} e & b \\ f & d \end{vmatrix}}{\begin{vmatrix} a & b \\ c & d \end{vmatrix}} \quad\quad \text{and} \quad\quad y = \frac{D_y}{D} = \frac{\begin{vmatrix} a & e \\ c & f \end{vmatrix}}{\begin{vmatrix} a & b \\ c & d \end{vmatrix}}$$

If every determinant is 0, the system is consistent, but the equations are dependent.

If $D = 0$ and D_x or D_y is nonzero, the system is inconsistent. If $D \neq 0$, the system is consistent, and the equations are independent.

The following observations are helpful when memorizing the three determinants of Cramer's rule.

- The denominator determinant, D, is formed by using the coefficients a, b, c, and d of the variables in the equations.

$$D = \begin{vmatrix} a & b \\ c & d \end{vmatrix}$$

x-term $\longrightarrow$ $\llcorner$ y-term
coefficients   coefficients

- The numerator determinants, D_x and D_y, are the same as the denominator determinant, D, except that the column of coefficients of the variable for which we are solving is replaced with the column of constants e and f.

$$D_x = \begin{vmatrix} e & b \\ f & d \end{vmatrix} \qquad\qquad D_y = \begin{vmatrix} a & e \\ c & f \end{vmatrix}$$

Replace the x-term coefficients with the constants.   Replace the y-term coefficients with the constants.

EXAMPLE 4   Use Cramer's rule to solve the system: $\begin{cases} 4x - 3y = 6 \\ -2x + 5y = 4 \end{cases}$

Strategy   We will evaluate three determinants, D, D_x, and D_y.

Why   The x-value of the solution of the system is the quotient of D_x and D and the y-value of the solution is the quotient of two determinants, D_y and D.

Solution   The denominator determinant D is made up of the coefficients of x and y:

$$D = \begin{vmatrix} 4 & -3 \\ -2 & 5 \end{vmatrix}$$

To solve for x, we form the numerator determinant D_x from D by replacing its first column (the coefficients of x) with the column of constants (6 and 4).

To solve for y, we form the numerator determinant D_y from D by replacing the second column (the coefficients of y) with the column of constants (6 and 4).

To find the values of x and y, we evaluate each determinant:

$$x = \frac{D_x}{D} = \frac{\begin{vmatrix} 6 & -3 \\ 4 & 5 \end{vmatrix}}{\begin{vmatrix} 4 & -3 \\ -2 & 5 \end{vmatrix}} = \frac{6(5) - (-3)(4)}{4(5) - (-3)(-2)} = \frac{30 + 12}{20 - 6} = \frac{42}{14} = 3$$

$$y = \frac{D_y}{D} = \frac{\begin{vmatrix} 4 & 6 \\ -2 & 4 \end{vmatrix}}{\begin{vmatrix} 4 & -3 \\ -2 & 5 \end{vmatrix}} = \frac{4(4) - 6(-2)}{14} = \frac{16 + 12}{14} = \frac{28}{14} = 2$$

The solution of the system is (3, 2). Verify that it satisfies both equations.

Self Check 4   Use Cramer's rule to solve the system: $\begin{cases} 2x - 3y = -16 \\ 3x + 5y = 14 \end{cases}$

Now Try   Problem 39

Success Tip

We can now solve systems of linear equations in five ways:
- graphing
- substitution
- elimination (addition)
- matrices
- Cramer's rule

In the next example, we will see how to recognize inconsistent systems when Cramer's rule is used to solve them.

EXAMPLE 5   Use Cramer's rule to solve $\begin{cases} 7x = 8 - 4y \\ 2y = 3 - \frac{7}{2}x \end{cases}$, if possible.

Strategy   We will try to evaluate three determinants, D, D_x, and D_y.

Why   The x-value of the solution of the system is the quotient of D_x and D, and the y-value of the solution is the quotient of two determinants, D_y and D.

Solution   Before we can form the required determinants, the equations of the system must be written in standard form: $Ax + By = C$.

$$\begin{cases} 7x = 8 - 4y \\ 2y = 3 - \dfrac{7}{2}x \end{cases} \qquad \begin{cases} 7x + 4y = 8 \\ 7x + 4y = 6 \end{cases}$$

Add 4y to both sides.

Multiply both sides by 2 and add 7x to both sides.

When we attempt to use Cramer's rule to solve this system for x, we obtain

$$x = \frac{D_x}{D} = \frac{\begin{vmatrix} 8 & 4 \\ 6 & 4 \end{vmatrix}}{\begin{vmatrix} 7 & 4 \\ 7 & 4 \end{vmatrix}} = \frac{32 - 24}{28 - 28} = \frac{8}{0}, \text{ which is undefined.}$$

Since the denominator determinant D is 0 and the numerator determinant D_x is not 0, the system is inconsistent. It has no solution and the solution set is $\varnothing$.

We can see directly from the system that it is inconsistent. For any values of x and y, it is impossible that 7 times x plus 4 times y could be both 8 and 6.

Self Check 5   Use Cramer's rule to solve the system $\begin{cases} 3x = 8 - 4y \\ y = \dfrac{5}{2} - \dfrac{3}{4}x \end{cases}$, if possible.

Now Try   Problem 47

---

3   **Use Cramer's Rule to Solve Systems of Three Equations.**

Cramer's rule can be extended to solve systems of three linear equations with three variables.

Cramer's Rule for Three Equations in Three Variables

The solution of the system $\begin{cases} ax + by + cz = j \\ dx + ey + fz = k \\ gx + hy + iz = l \end{cases}$ is given by

$$x = \frac{D_x}{D}, \qquad y = \frac{D_y}{D}, \qquad \text{and} \qquad z = \frac{D_z}{D}, \qquad \text{where}$$

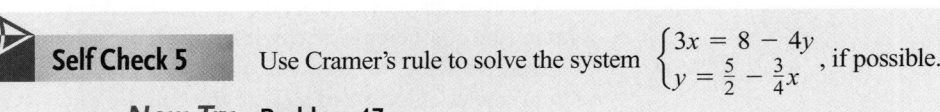

$$D = \begin{vmatrix} a & b & c \\ d & e & f \\ g & h & i \end{vmatrix} \text{ Only the coefficients of the variables.} \qquad D_x = \begin{vmatrix} j & b & c \\ k & e & f \\ l & h & i \end{vmatrix} \text{ Replace the x-term coefficients with the constants.}$$

$$D_y = \begin{vmatrix} a & j & c \\ d & k & f \\ g & l & i \end{vmatrix} \text{ Replace the y-term coefficients with the constants.} \qquad D_z = \begin{vmatrix} a & b & j \\ d & e & k \\ g & h & l \end{vmatrix} \text{ Replace the z-term coefficients with the constants.}$$

If every determinant is 0, the system is consistent, but the equations are dependent.

If $D = 0$ and D_x or D_y or D_z is nonzero, the system is inconsistent. If $D \neq 0$, the system is consistent, and the equations are independent.

EXAMPLE 6   Use Cramer's rule to solve the system: $\begin{cases} 2x + y + 4z = 12 \\ x + 2y + 2z = 9 \\ 3x - 3y - 2z = 1 \end{cases}$

Strategy   We will evaluate four determinants, D, D_x, D_y, and D_z.

Why   The x-value of the solution of the system is the quotient of D_x and D, the y-value of the solution is the quotient of two determinants, D_y and D, and the z-value of the solution is the quotient of two determinants, D_z and D.

Solution   The denominator determinant D is the determinant formed by the coefficients of the variables. The numerator determinants, D_x, D_y, and D_z, are formed by replacing the coefficients of the variable being solved for by the column of constants. We form the quotients for x, y, and z and evaluate each determinant by expanding by minors about the first row:

$$x = \frac{D_x}{D} = \frac{\begin{vmatrix} 12 & 1 & 4 \\ 9 & 2 & 2 \\ 1 & -3 & -2 \end{vmatrix}}{\begin{vmatrix} 2 & 1 & 4 \\ 1 & 2 & 2 \\ 3 & -3 & -2 \end{vmatrix}} = \frac{12\begin{vmatrix} 2 & 2 \\ -3 & -2 \end{vmatrix} - 1\begin{vmatrix} 9 & 2 \\ 1 & -2 \end{vmatrix} + 4\begin{vmatrix} 9 & 2 \\ 1 & -3 \end{vmatrix}}{2\begin{vmatrix} 2 & 2 \\ -3 & -2 \end{vmatrix} - 1\begin{vmatrix} 1 & 2 \\ 3 & -2 \end{vmatrix} + 4\begin{vmatrix} 1 & 2 \\ 3 & -3 \end{vmatrix}} = \frac{12(2) - 1(-20) + 4(-29)}{2(2) - 1(-8) + 4(-9)} = \frac{-72}{-24} = 3$$

$$y = \frac{D_y}{D} = \frac{\begin{vmatrix} 2 & 12 & 4 \\ 1 & 9 & 2 \\ 3 & 1 & -2 \end{vmatrix}}{\begin{vmatrix} 2 & 1 & 4 \\ 1 & 2 & 2 \\ 3 & -3 & -2 \end{vmatrix}} = \frac{2\begin{vmatrix} 9 & 2 \\ 1 & -2 \end{vmatrix} - 12\begin{vmatrix} 1 & 2 \\ 3 & -2 \end{vmatrix} + 4\begin{vmatrix} 1 & 9 \\ 3 & 1 \end{vmatrix}}{-24} = \frac{2(-20) - 12(-8) + 4(-26)}{-24} = \frac{-48}{-24} = 2$$

$$z = \frac{D_z}{D} = \frac{\begin{vmatrix} 2 & 1 & 12 \\ 1 & 2 & 9 \\ 3 & -3 & 1 \end{vmatrix}}{\begin{vmatrix} 2 & 1 & 4 \\ 1 & 2 & 2 \\ 3 & -3 & -2 \end{vmatrix}} = \frac{2\begin{vmatrix} 2 & 9 \\ -3 & 1 \end{vmatrix} - 1\begin{vmatrix} 1 & 9 \\ 3 & 1 \end{vmatrix} + 12\begin{vmatrix} 1 & 2 \\ 3 & -3 \end{vmatrix}}{-24} = \frac{2(29) - 1(-26) + 12(-9)}{-24} = \frac{-24}{-24} = 1$$

The solution of this system is $(3, 2, 1)$. Verify that it satisfies the three original equations.

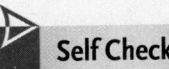 **Self Check 6**   Use Cramer's rule to solve the system: $\begin{cases} x + y + 2z = 6 \\ 2x - y + z = 9 \\ x + y - 2z = -6 \end{cases}$

Now Try   Problem 51

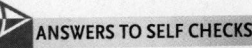 **ANSWERS TO SELF CHECKS**   **1.** 10   **2.** −44   **3.** −44   **4.** $(-2, 4)$
5. No solution, $\varnothing$; inconsistent system   **6.** $(2, -2, 3)$

STUDY SET
12.5

VOCABULARY

Fill in the blanks.

1. $\begin{vmatrix} 4 & 9 \\ -6 & 1 \end{vmatrix}$ is a _____. The numbers 4 and 1 lie along its main _____.

2. A determinant is a number that is associated with a _____ matrix.

3. The _____ of b_1 in $\begin{vmatrix} a_1 & b_1 & c_1 \\ a_2 & b_2 & c_2 \\ a_3 & b_3 & c_3 \end{vmatrix}$ is $\begin{vmatrix} a_2 & c_2 \\ a_3 & c_3 \end{vmatrix}$.

4. _____ rule uses determinants to solve systems of linear equations.

CONCEPTS

Fill in the blanks.

5. $\begin{vmatrix} a & b \\ c & d \end{vmatrix} = \blacksquare - \blacksquare$

6. To find the minor of 5, we cross out the elements of the determinant that are in the same row and column as __.

$\begin{vmatrix} 3 & 5 & 1 \\ 6 & -2 & 2 \\ 8 & -1 & 4 \end{vmatrix}$

7. In evaluating the determinant below, about what row or column was it expanded?

$\begin{vmatrix} 5 & 1 & -1 \\ 8 & 7 & 4 \\ 9 & 7 & 6 \end{vmatrix} = -1\begin{vmatrix} 8 & 7 \\ 9 & 7 \end{vmatrix} - 4\begin{vmatrix} 5 & 1 \\ 9 & 7 \end{vmatrix} + 6\begin{vmatrix} 5 & 1 \\ 8 & 7 \end{vmatrix}$

8. Set up the denominator determinant D for the system
$\begin{cases} 3x + 4y = 7 \\ 2x - 3y = 5 \end{cases}$

9. Set up the denominator determinant D for the system
$\begin{cases} x + 2y = -8 \\ 3x + y - z = -2. \\ 8x + 4y - z = 6 \end{cases}$

10. For the system $\begin{cases} 3x + 2y = 1 \\ 4x - y = 3 \end{cases}$, $D_x = -7$, $D_y = 5$, and $D = -11$. Find the solution of the system.

11. For the system $\begin{cases} 2x + 3y - z = -8 \\ x - y - z = -2 \\ -4x + 3y + z = 6 \end{cases}$, $D_x = -28$, $D_y = -14$, $D_z = 14$, and $D = 14$. Find the solution.

12. Fill in the blank. If the denominator determinant D for a system of equations is 0, the equations of the system are dependent or the system is _____.

NOTATION

Complete the evaluation of each determinant.

13. $\begin{vmatrix} 5 & -2 \\ -2 & 6 \end{vmatrix} = 5(\ ) - (-2)(-2)$
$= \blacksquare - 4$
$= 26$

14. $\begin{vmatrix} 2 & 1 & 3 \\ 3 & 4 & 2 \\ 1 & 5 & 3 \end{vmatrix}$
$= 2\begin{vmatrix} 4 & \blacksquare \\ 5 & 3 \end{vmatrix} - 1\begin{vmatrix} 3 & 2 \\ \blacksquare & 3 \end{vmatrix} + 3\begin{vmatrix} 3 & 4 \\ 1 & \blacksquare \end{vmatrix}$
$= 2(\blacksquare - 10) - 1(9 - \blacksquare) + 3(15 - \blacksquare)$
$= 2(2) - 1(\ ) + \blacksquare(11)$
$= 4 - 7 + \blacksquare$
$= 30$

GUIDED PRACTICE

Evaluate each determinant. See Example 1.

15. $\begin{vmatrix} 2 & 3 \\ 2 & 5 \end{vmatrix}$

16. $\begin{vmatrix} 3 & 2 \\ 2 & 4 \end{vmatrix}$

17. $\begin{vmatrix} -9 & 7 \\ 4 & -2 \end{vmatrix}$

18. $\begin{vmatrix} -1 & 2 \\ 3 & -4 \end{vmatrix}$

19. $\begin{vmatrix} 5 & 20 \\ 10 & 6 \end{vmatrix}$

20. $\begin{vmatrix} 10 & 15 \\ 15 & 5 \end{vmatrix}$

21. $\begin{vmatrix} -6 & -2 \\ 15 & 4 \end{vmatrix}$

22. $\begin{vmatrix} 3 & -2 \\ 12 & -8 \end{vmatrix}$

23. $\begin{vmatrix} -9 & -1 \\ -10 & -5 \end{vmatrix}$

24. $\begin{vmatrix} -7 & -7 \\ -6 & -4 \end{vmatrix}$

25. $\begin{vmatrix} 8 & 8 \\ -9 & -9 \end{vmatrix}$

26. $\begin{vmatrix} 20 & -3 \\ 20 & -3 \end{vmatrix}$

Evaluate each determinant. See Examples 2 and 3.

27. $\begin{vmatrix} 3 & 2 & 1 \\ 4 & 1 & 2 \\ 5 & 3 & 1 \end{vmatrix}$

28. $\begin{vmatrix} 6 & 2 & 3 \\ 1 & 5 & 4 \\ 2 & 3 & 5 \end{vmatrix}$

29. $\begin{vmatrix} 1 & -2 & 3 \\ -2 & 1 & 1 \\ -3 & -2 & 1 \end{vmatrix}$

30. $\begin{vmatrix} 1 & 1 & 2 \\ 2 & 1 & -2 \\ 3 & 1 & 3 \end{vmatrix}$

31. $\begin{vmatrix} -2 & 5 & 1 \\ 0 & 3 & 4 \\ -1 & 2 & 6 \end{vmatrix}$ **32.** $\begin{vmatrix} 4 & -1 & 2 \\ 6 & -1 & 0 \\ 1 & -3 & 4 \end{vmatrix}$

33. $\begin{vmatrix} 1 & -4 & 1 \\ 3 & 0 & -2 \\ 3 & 1 & -2 \end{vmatrix}$ **34.** $\begin{vmatrix} 8 & -3 & 1 \\ 1 & 0 & 2 \\ 3 & -9 & 4 \end{vmatrix}$

35. $\begin{vmatrix} 1 & 2 & 1 \\ -3 & 7 & 3 \\ -4 & 3 & -5 \end{vmatrix}$ **36.** $\begin{vmatrix} 1 & 4 & 7 \\ 2 & 5 & 8 \\ 3 & 6 & 9 \end{vmatrix}$

37. $\begin{vmatrix} 1 & 2 & 0 \\ 0 & 1 & 2 \\ 0 & 0 & 1 \end{vmatrix}$ **38.** $\begin{vmatrix} 1 & 0 & 1 \\ 0 & 1 & 0 \\ 1 & 1 & 1 \end{vmatrix}$

Use Cramer's rule to solve each system of equations. See Example 4.

39. $\begin{cases} x + y = 6 \\ x - y = 2 \end{cases}$ **40.** $\begin{cases} x - y = 4 \\ 2x + y = 5 \end{cases}$

41. $\begin{cases} x + 2y = -21 \\ x - 2y = 11 \end{cases}$ **42.** $\begin{cases} 5x + 2y = 11 \\ 7x + 6y = 9 \end{cases}$

43. $\begin{cases} 3x - 4y = 9 \\ x + 2y = 8 \end{cases}$ **44.** $\begin{cases} 2x + 2y = -1 \\ 3x + 4y = 0 \end{cases}$

45. $\begin{cases} 2x + 3y = 31 \\ 3x + 2y = 39 \end{cases}$ **46.** $\begin{cases} 5x + 3y = 72 \\ 3x + 5y = 56 \end{cases}$

Use Cramer's rule to solve each system of equations, if possible. If a system is inconsistent or if the equations are dependent, so indicate. See Example 5.

47. $\begin{cases} 3x + 2y = 11 \\ 6x + 4y = 11 \end{cases}$ **48.** $\begin{cases} 5x - 4y = 20 \\ 10x - 8y = 30 \end{cases}$

49. $\begin{cases} \dfrac{5}{6}x = 2 - y \\ 10x + 12y = 24 \end{cases}$ **50.** $\begin{cases} 16x - 8y = 32 \\ x - 2 = \dfrac{y}{2} \end{cases}$

Use Cramer's rule to solve each system of equations. See Example 6.

51. $\begin{cases} x + y + z = 4 \\ x + y - z = 0 \\ x - y + z = 2 \end{cases}$ **52.** $\begin{cases} x + y + z = 4 \\ x - y + z = 2 \\ x - y - z = 0 \end{cases}$

53. $\begin{cases} 3x + 2y - z = -8 \\ 2x - y + 7z = 10 \\ 2x + 2y - 3z = -10 \end{cases}$ **54.** $\begin{cases} x + 2y + 2z = 10 \\ 2x + y + 2z = 9 \\ 2x + 2y + z = 1 \end{cases}$

Use Cramer's rule to solve each system of equations, if possible. If a system is inconsistent or if the equations are dependent, so indicate.

55. $\begin{cases} 2x + y + z = 5 \\ x - 2y + 3z = 10 \\ x + y - 4z = -3 \end{cases}$ **56.** $\begin{cases} x + y + 2z = 7 \\ x + 2y + z = 8 \\ 2x + y + z = 9 \end{cases}$

57. $\begin{cases} y = \dfrac{-2x + 1}{3} \\ 3x - 2y = 8 \end{cases}$ **58.** $\begin{cases} 2x + 3y = -1 \\ x = \dfrac{y - 9}{4} \end{cases}$

59. $\begin{cases} 4x - 3y = 1 \\ 6x - 8z = 1 \\ 2y - 4z = 0 \end{cases}$ **60.** $\begin{cases} 4x + 3z = 4 \\ 2y - 6z = -1 \\ 8x + 4y + 3z = 9 \end{cases}$

61. $\begin{cases} 2x + y - z - 1 = 0 \\ x + 2y + 2z - 2 = 0 \\ 4x + 5y + 3z - 3 = 0 \end{cases}$ **62.** $\begin{cases} 2x - y + 4z + 2 = 0 \\ 5x + 8y + 7z = -8 \\ x + 3y + z + 3 = 0 \end{cases}$

63. $\begin{cases} 3x - 16 = 5y \\ -3x + 5y - 33 = 0 \end{cases}$ **64.** $\begin{cases} 2x + 5y - 13 = 0 \\ -2x + 13 = 5y \end{cases}$

65. $\begin{cases} x + y = 1 \\ \dfrac{1}{2}y + z = \dfrac{5}{2} \\ x - z = -3 \end{cases}$ **66.** $\begin{cases} \dfrac{1}{2}x + y + z + \dfrac{3}{2} = 0 \\ x + \dfrac{1}{2}y + z - \dfrac{1}{2} = 0 \\ x + y + \dfrac{1}{2}z + \dfrac{1}{2} = 0 \end{cases}$

67. $\begin{cases} 2x + 3y = 0 \\ 4x - 6y = -4 \end{cases}$ **68.** $\begin{cases} 4x - 3y = -1 \\ 8x + 3y = 4 \end{cases}$

69. $\begin{cases} 2x + 3y + 4z = 6 \\ 2x - 3y - 4z = -4 \\ 4x + 6y + 8z = 12 \end{cases}$ **70.** $\begin{cases} x - 3y + 4z - 2 = 0 \\ 2x + y + 2z - 3 = 0 \\ 4x - 5y + 10z - 7 = 0 \end{cases}$

Evaluate each determinant. See Using Your Calculator: Evaluating Determinants.

71. $\begin{vmatrix} 25 & -36 & 44 \\ -11 & 21 & 54 \\ 37 & -31 & 19 \end{vmatrix}$ **72.** $\begin{vmatrix} 13 & -27 & 62 \\ -38 & 27 & -52 \\ 10 & -300 & 42 \end{vmatrix}$

73. $\begin{vmatrix} -280 & 191 & -356 \\ -211 & -102 & -422 \\ 400 & -213 & -333 \end{vmatrix}$ **74.** $\begin{vmatrix} 4.1 & 2.2 & -3.3 \\ 2.7 & -5.9 & 6.8 \\ 2.3 & 5.3 & 0.6 \end{vmatrix}$

APPLICATIONS

Write a system of equations to solve each problem. Then use Cramer's rule to solve the system.

75. SIGNALING   Refer to the illustration below. A system of sending signals uses two flags held in various positions to represent letters of the alphabet. The illustration shows how the letter U is signaled. Find x and y, if y is to be 30° more than x.

76. INVESTING A student wants to earn \$1,320 interest in the first year by investing \$20,000 in the three stocks listed in the table. Because HiTech is a high-risk investment, he wants to invest three times as much in SaveTel and OilCo combined as he invests in HiTech. How much should he invest in each stock?

Stock	Rate of return
HiTech	10%
SaveTel	5%
OilCo	6%

WRITING

77. Explain the difference between a matrix and a determinant. Give an example of each.

78. When evaluating $\begin{vmatrix} 4 & -1 & 2 \\ 6 & -1 & 0 \\ 1 & -3 & 4 \end{vmatrix}$, why is it helpful to expand by the minors of the numbers in the third column?

79. Explain how to find the minor of an element of a determinant.

80. Explain how to find x when solving a system of three linear equations in x, y, and z by Cramer's rule. Use the words *coefficients* and *constants* in your explanation.

81. Explain how the following checkerboard pattern is used when evaluating a 3 × 3 determinant.

$$\begin{matrix} + & - & + \\ - & + & - \\ + & - & + \end{matrix}$$

82. Briefly describe each of the five methods of this chapter that can be used to solve a system of two linear equations in two variables.

REVIEW

83. Are the lines $y = 2x - 7$ and $x - 2y = 7$ perpendicular?

84. Are the lines $y = 2x - 7$ and $2x - y = 10$ parallel?

85. How are the graphs of $f(x) = x^2$ and $g(x) = x^2 - 2$ related?

86. Is the graph of a circle the graph of a function?

87. The graph of a line passes through $(0, -3)$. Is this the x-intercept or the y-intercept of the line?

88. What is the name of the function $f(x) = |x|$?

89. For the function $y = 2x^2 + 6x + 1$, what is the independent variable and what is the dependent variable?

90. If $f(x) = x^3 - x$, what is $f(-1)$?

CHALLENGE PROBLEMS

91. Show that $\begin{vmatrix} x & y & 1 \\ -2 & 3 & 1 \\ 3 & 5 & 1 \end{vmatrix} = 0$ represents the equation of the line passing through $(-2, 3)$ and $(3, 5)$.

92. Show that $\dfrac{1}{2}\begin{vmatrix} 0 & 0 & 1 \\ 3 & 0 & 1 \\ 0 & 4 & 1 \end{vmatrix}$ represents the area of the triangle with vertices at $(0, 0)$, $(3, 0)$, and $(0, 4)$.

CHAPTER 12
Summary & Review

SECTION 12.1   Solving Systems of Equations in Two Variables

DEFINITIONS AND CONCEPTS	EXAMPLES
When two equations are considered at the same time, we say that they form a **system of equations.** A **solution of a system** of equations in two variables is an ordered pair that satisfies both equations of the system.	The ordered pair $(2, -3)$ is a solution of the system $\begin{cases} x + y = -1 \\ x - 2y = 8 \end{cases}$ because its coordinates, $x = 2$ and $y = -3$, satisfy both equations. $x + y = -1$ First equation. $x - 2y = 8$ Second equation. $2 + (-3) \stackrel{?}{=} -1$ $2 - 2(-3) \stackrel{?}{=} 8$ $-1 = -1$ True $8 = 8$ True
To **solve a system graphically:** 1. Graph each equation on the same rectangular coordinate system. 2. Determine the coordinates of the point where the graphs intersect. That ordered pair is the solution. 3. If the graphs have no point in common, the system has no solution. 4. Check the proposed solution in each equation of the original system.	To solve the system $\begin{cases} x + y = -1 \\ x - 2y = 8 \end{cases}$ by graphing, we graph each equation as shown in the illustration below. 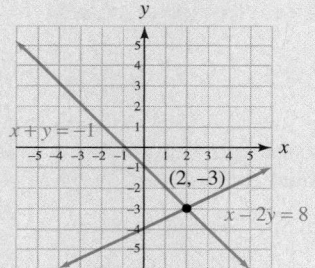 The graphs appear to intersect at the point $(2, -3)$. The check shown above verifies that $(2, -3)$ is the solution of the system.
A system of equations that has at least one solution is called a **consistent system.** If the graphs are parallel lines, the system has no solution, and it is called an **inconsistent system.** Equations with different graphs are called **independent equations.** If the graphs are the same line, the system has infinitely many solutions. The equations are called **dependent equations.**	Since the system shown above has a solution, it is a *consistent system.* Since the graphs are different, the equations are *independent.* ***Consistent system***　　***Inconsistent system***　　***Dependent equations*** 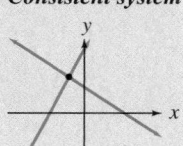　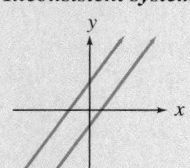　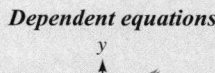

SECTION 12.1   Solving Systems of Equations in Two Variables—*continued*

DEFINITIONS AND CONCEPTS	EXAMPLES

DEFINITIONS AND CONCEPTS

To solve a system of two linear equations in x and y by the **substitution method:**

1. Solve one equation for either x or y. This is called the *substitution equation.*

2. Substitute the resulting expression for that variable into the other equation and solve it.

3. Substitute the value of the variable found in step 2 into the substitution equation and solve that equation.

4. Check the proposed solution in each of the original equations. Write the solution as an ordered pair.

If in step 2 the variable drops out and a false statement results, the system has **no solution.** If a true statement results, the system has **infinitely many solutions** and we can use **set-builder notation** to write the solution set.

EXAMPLES

Use substitution to solve the system: $\begin{cases} x + y = -1 \\ x - 2y = 8 \end{cases}$

Step 1: We will solve $x + y = -1$ for y.

$y = -x - 1$ This is the substitution equation.

Step 2: We substitute $-x - 1$ for y in the second equation and solve for x.

$x - 2(-x - 1) = 8$

$x + 2x + 2 = 8$ Distribute.

$3x = 6$ Combine like terms and subtract 2 from both sides.

$x = 2$ Divide both sides by 3.

Step 3: We substitute 2 for x in the substitution equation and solve for y.

$y = -x - 1$

$y = -2 - 1$

$y = -3$

Step 4: The solution is $(2, -3)$. Verify this by checking it in each of the original equations.

DEFINITIONS AND CONCEPTS

To solve a system of two linear equations in x and y by the **elimination (addition) method:**

1. Write both equations of the system in standard (general) form: $Ax + By = C$.

2. If necessary, multiply one or both of the equations by a nonzero number chosen to make the coefficients of x (or the coefficients of y) opposites.

3. Add the equations to eliminate the terms involving x (or y).

4. Solve the equation resulting from step 3.

5. Find the value of the remaining variable by substituting the solution found in step 4 into any equation containing both variables. Or, repeat steps 2–4 to eliminate the other variable.

6. Check the proposed solution in each equation of the original system. Write the solution as an ordered pair.

If in step 3 both variables drop out and a false statement results, the system has **no solution.** If a true statement results, the system has **infinitely many solutions** and we can use **set-builder notation** to write the solution set.

EXAMPLES

Use elimination to solve the system: $\begin{cases} x + y = -1 \\ x - 2y = 8 \end{cases}$

Step 1: Since both equations are in standard form, we move to step 2.

Step 2: We can multiply both sides of the first equation by 2 to get the coefficients of y to be opposites.

$$\begin{matrix} \textbf{(1)} & \begin{cases} x + y = -1 & \xrightarrow{\text{Multiply by 2}} & \begin{cases} 2x + 2y = -2 & \textbf{(3)} \\ \textbf{(2)} & x - 2y = 8 & \xrightarrow{\text{Unchanged}} & x - 2y = 8 & \textbf{(2)} \end{cases} \end{cases} \end{matrix}$$

Step 3: We add equations 3 and 2 to eliminate y.

$2x + 2y = -2$

$\underline{x - 2y = 8}$ Add like terms, column-by-column.

$3x = 6$

Step 4: Since the resulting equation has only one variable, we can solve it for x.

$3x = 6$

$x = 2$ Divide both sides by 3.

Step 5: To find y, we can substitute 2 for x in equation 1.

$x + y = -1$

$2 + y = -1$

$y = -3$ Subtract 2 from both sides.

Step 6: The solution is $(2, -3)$. Verify this by checking it in each of the original equations.

REVIEW EXERCISES

1. See the illustration.

   a. Give three points that satisfy the equation $2x + y = 5$.

   b. Give three points that satisfy the equation $x - y = 4$.

   c. Find the solution of $\begin{cases} 2x + y = 5 \\ x - y = 4 \end{cases}$.

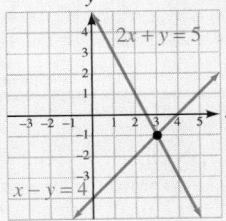

2. **TELEVISION**   The graphs below show the percent of time that Americans spent viewing broadcast television networks (such as CBS, NBC, ABC) and cable television networks (such as TNT, NICK, ESPN, USA). Explain the importance of the point of intersection of the graphs.

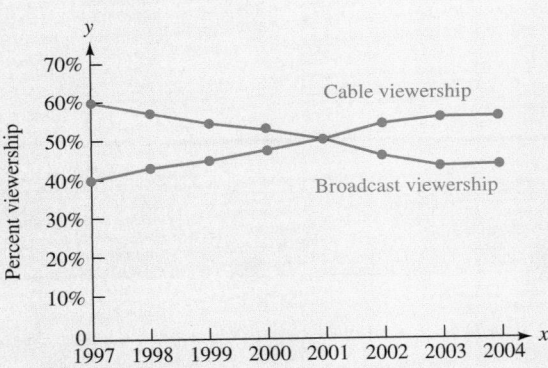

Source: CSFB Media & Entertainment Stock Source

Determine whether the ordered pair is a solution of the system of equations.

3. $\left(-1, \frac{1}{2}\right)$, $\begin{cases} x + 2y = 0 \\ x + 4y = 1 \end{cases}$

4. $(13, 23)$, $\begin{cases} 3a - 2b + 7 = 0 \\ -2a + b = -4 \end{cases}$

Solve each system by the graphing method, if possible. If a system is inconsistent or if the equations are dependent, state this.

5. $\begin{cases} 2x + y = 11 \\ -x + 2y = 7 \end{cases}$

6. $\begin{cases} y = -\frac{3}{2}x \\ 2x - 3y + 13 = 0 \end{cases}$

7. $\begin{cases} \frac{1}{2}x + \frac{1}{3}y = 2 \\ y = 6 - \frac{3}{2}x \end{cases}$

8. $\begin{cases} \frac{x}{3} - \frac{y}{2} = 1 \\ 6x - 9y = 3 \end{cases}$

Use the graphs in the illustration to solve each equation. Check each answer.

9. $2(2 - x) + x = x$

10. $2(2 - x) + x = 5$

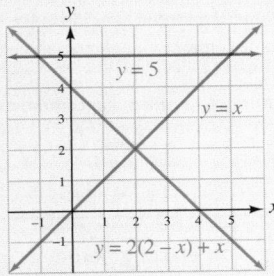

Solve each system using the substitution method.

11. $\begin{cases} x = y - 4 \\ 2x + 3y = 7 \end{cases}$

12. $\begin{cases} y = 2x + 5 \\ 3x - 5y = -4 \end{cases}$

13. $\begin{cases} 0.1x + 0.2y = 1.1 \\ 2x - y = 2 \end{cases}$

14. $\begin{cases} x = -2 - 3y \\ -2x - 6y = 4 \end{cases}$

Solve each system by elimination (addition), if possible. If a system is inconsistent or if the equations are dependent, so indicate.

15. $\begin{cases} x + y = -2 \\ 2x + 3y = -3 \end{cases}$

16. $\begin{cases} 2x - 3y = 5 \\ 2x - 3y = 8 \end{cases}$

17. $\begin{cases} x + \frac{1}{2}y = 7 \\ -2x = 3y - 6 \end{cases}$

18. $\begin{cases} y = \dfrac{x - 3}{2} \\ x = \dfrac{2y + 7}{2} \end{cases}$

19. To solve $\begin{cases} 5x - 2y = 19 \\ 3x + 4y = 1 \end{cases}$, which method, elimination or substitution, would you use? Explain.

20. Estimate the solution of the system $\begin{cases} y = -\frac{2}{3}x \\ 2x - 3y = -4 \end{cases}$ from the graphs in the illustration. Then solve the system algebraically.

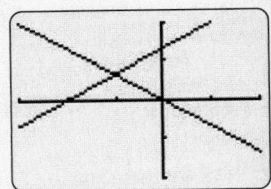

Write a system of two equations in two variables to solve each problem.

21. LIBRARY CARDS   Residents of a city are charged $3 for a library card; nonresidents are charged $7. On a day when a total of 11 library cards were purchased, the receipts were $41. How many resident and nonresident library cards were sold?

22. MAPS   Refer to the illustration. The distance between Austin and Houston is 4 miles less than twice the distance between Austin and San Antonio. The round trip from Houston to Austin to San Antonio and back to Houston is 442 miles. Determine the mileages between Austin and Houston and between Austin and San Antonio.

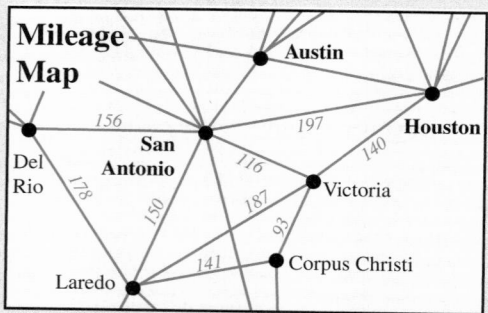

23. RIVERBOATS   A Mississippi riverboat travels 30 miles downstream in three hours and then makes the return trip upstream in five hours. Find the speed of the riverboat in still water and the speed of the current.

24. MIXING SOLUTIONS   How many fluid ounces of 6% sucrose solution and how many ounces of an 18% sucrose solution must be mixed to make 750 ounces of a 10% sucrose solution?

25. INVESTING   One year, a couple invested a total of $10,000 in two projects. The first investment, a mini-mall, made a 6% profit. The other investment, a skateboard park, made a 12% profit. If their investments made $960, how much was invested at each rate?

26. COOKING   Two teaspoons and five tablespoons is a total of 85 milliliters of liquid. Five teaspoons and two tablespoons is a total of 55 milliliters of liquid. Find the number of milliliters in one teaspoon and the number of milliliters in one tablespoon.

SECTION 12.2   Solving Systems of Equations in Three Variables

DEFINITIONS AND CONCEPTS	EXAMPLES
The graph of an equation of the form $Ax + By + Cz = D$ is a flat surface called a **plane**. A **solution of a system of three linear equations** in three variables is an **ordered triple** that satisfies each equation of the system.	The ordered triple $(4, 0, -3)$ is a solution of $\begin{cases} x + y - z = 7 \\ x - y + z = 1 \\ 2x + y + z = 5 \end{cases}$ because its coordinates, $x = 4$, $y = 0$, and $z = -3$, satisfy each equation:

$$
\begin{array}{ccc}
x + y - z = 7 & x - y + z = 1 & 2x + y + z = 5 \\
4 + 0 - (-3) \overset{?}{=} 7 & 4 - 0 + (-3) \overset{?}{=} 1 & 2(4) + 0 + (-3) \overset{?}{=} 5 \\
7 = 7 & 1 = 1 & 5 = 5 \\
\text{True} & \text{True} & \text{True}
\end{array}
$$

To **solve a system of three linear equations** by elimination: **1.** Write each equation in standard form $Ax + By + Cz = D$ and clear any decimals or fractions. **2.** Pick any two equations and eliminate a variable.	Solve the system: $\begin{cases} x + 2y - z = 1 & \textbf{(1)} \\ 2x - y + z = 6 & \textbf{(2)} \\ x + 3y - z = 2 & \textbf{(3)} \end{cases}$ **Step 1:** Each equation is written in standard form. **Step 2:** To eliminate z, we add equations 1 and 2. $\begin{array}{ll} \textbf{(1)} & x + 2y - z = 1 \\ \textbf{(2)} & 2x - y + z = 6 \\ \hline \textbf{(4)} & 3x + y \quad\;\; = 7 \end{array}$

SECTION 12.2   Solving Systems of Equations in Three Variables—*continued*

DEFINITIONS AND CONCEPTS	EXAMPLES

3. Pick a different pair of equations and eliminate the same variable as in step 2.

4. Solve the resulting pair of two equations in two variables.

5. To find the value of the third variable, substitute the values of the two variables found in step 4 into any equation containing all three variables and solve the equation.

6. Check the proposed solution in all three of the original equations. Write the solution as an ordered triple.

If at any time in the elimination process the variables drop out and a false statement results, the system has **no solution.** If a true statement results, the system has **infinitely many solutions.**

Step 3: To eliminate z again, we add equations 2 and 3.

$$(2) \quad 2x - y + z = 6$$
$$(3) \quad \underline{x + 3y - z = 2}$$
$$(5) \quad 3x + 2y \quad\;\; = 8$$

Step 4: Equations 4 and 5 form a system of two equations in x and y. To solve this system, we multiply equation 4 by -1 and add the resulting equation to equation 5 to eliminate x.

$$-3x - \;\; y = -7 \quad \text{This is } -1(3x + y) = -1(7).$$
$$(5) \quad \underline{3x + 2y = 8}$$
$$y = 1$$

To find x, we substitute 1 for y in any equation containing x and y (such as equation 4) and solve for x:

$$3x + y = 7 \quad \text{This is equation 4.}$$
$$3x + 1 = 7 \quad \text{Substitute 1 for } y.$$
$$x = 2 \quad \text{Solve for x.}$$

Step 5: To find z, we substitute 2 for x and 1 for y in any equation containing x, y, and z (such as equation 2) and solve for z:

$$2x - y + z = 6 \quad \text{This is equation 2.}$$
$$2(2) - 1 + z = 6 \quad \text{Substitute for x and y.}$$
$$4 - 1 + z = 6$$
$$z = 3 \quad \text{Solve for z.}$$

Step 6: The solution is $(2, 1, 3)$. Verify this by checking it in each of the original equations.

REVIEW EXERCISES

27. Determine whether $(2, -1, 1)$ is a solution of the system:

$$\begin{cases} x - y + z = 4 \\ x + 2y - z = -1 \\ x + y - 3z = -1 \end{cases}$$

28. A system of three linear equations in three variables is graphed on the right. Does the system have a solution? If so, how many solutions does it have?

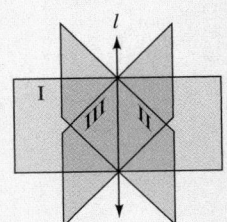

Solve each system, if possible. If a system is inconsistent or if the equations are dependent, state this.

29. $\begin{cases} x - 2y + 3z = -7 \\ -x + 3y + 2z = -8 \\ 2x - y - z = 7 \end{cases}$

30. $\begin{cases} x + y + z = 4 \\ x - 2y - z = 1 \\ 2x - y - 2z = -1 \end{cases}$

31. $\begin{cases} x + y - z = -3 \\ x + z = 2 \\ 2x - y = 3 - 2z \end{cases}$

32. $\begin{cases} b - 4c = 2 \\ a - b + 2c = 1 \\ 2a - 2b = -2 - 5c \end{cases}$

33. $\begin{cases} x + 2z = 10 \\ 3x + 2y - 3z = 8 \\ y + 4z = 6 \end{cases}$

34. $\begin{cases} x + 3y + z = 14 \\ x - 5y = -19 \\ 3y + z = 13 \end{cases}$

35. $\begin{cases} 2x + 3y + z = -5 \\ -x + 2y - z = -6 \\ 3x + y + 2z = 4 \end{cases}$

36. $\begin{cases} 3x + 3y + 6z = -6 \\ -x - y - 2z = 2 \\ 2x + 2y + 4z = -4 \end{cases}$

SECTION 12.3   Problem Solving Using Systems of Three Equations

DEFINITIONS AND CONCEPTS	EXAMPLES

Problems that involve **three unknown quantities** can be solved using a strategy similar to that for solving problems involving two unknowns.

BATTERIES   A hardware store sells three types of batteries: AA size for \$1 each, C size for \$1.50 each, and D size for \$2.00 each. One Saturday, the store sold 25 batteries for a total of \$34. If the number of C batteries that were sold was four less than the number of AA batteries that were sold, how many of each size battery were sold?

Analyze   To find the three unknowns we will write a system of three equations in three variables.

Form   Let A = the number of AA batteries sold, C = the number of C batteries sold, and D = the number of D batteries sold. The given information leads to three equations:

$\begin{cases} A + C + D = 25 \\ 1A + 1.50C + 2D = 34 \\ C = A - 4 \end{cases}$ The total number of batteries sold was 25.
The total value of the batteries sold was \$34.
The number of C batteries sold was 4 less than AA batteries sold.

If we multiply the second equation by 10 to clear the decimal and write the third equation in standard form, we have the system:

(1)
(2) $\begin{cases} A + C + D = 25 \\ 10A + 15C + 20D = 340 \\ -A + C = -4 \end{cases}$
(3)

Solve   Since equation 3 does not contain a D-term, we will find another equation that does not contain a D-term. If each side of equation 1 is multiplied by -20 and the resulting equation is added to equation 2, D is eliminated, and we obtain

$\quad\quad -20A - 20C - 20D = -500$ This is $-20(A + C + D) = -20(25)$.
(2) $\quad\underline{\;\;10A + 15C + 20D = \quad 340\;}$
(4) $\quad -10A - \;\;\;5C \quad\quad\quad = -160$

Equations 3 and 4 form a system of two equations in A and C that can be solved in the usual manner. (The remaining work is left to the reader.)

$$\begin{cases} -10A - 5C = -160 \\ -A + C = -4 \end{cases}$$

State   There were 12 AA batteries, 8 C batteries, and 5 D batteries sold.

Check   Verify that these results are correct by checking them in the words of the problem.

REVIEW EXERCISES

37. TEDDY BEARS   A toy company produces three sizes of teddy bears. Each day, the total cost to produce the bears is $850, the total time needed to stuff them is 480 minutes, and the total time needed to sew them is 1,260 minutes. Use the information in the table to determine how many of each type of teddy bear are produced daily.

Size of teddy bear	Production cost	Stuffing time	Sewing time
Small	$3	2 min	6 min
Medium	$5	3 min	8 min
Large	$10	5 min	12 min

38. VETERINARY MEDICINE   The daily requirements of a balanced diet for an animal are shown in the nutritional pyramid. The number of grams per cup of nutrients in three food mixes are shown in the table. How many cups of each mix should be used to meet the daily requirements for protein, carbohydrates, and essential fatty acids in the animal's diet?

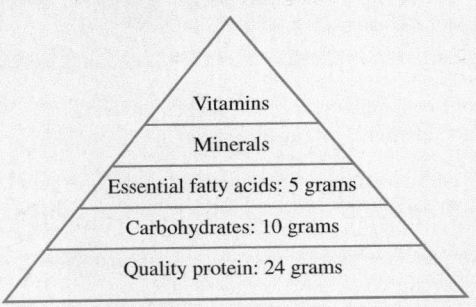

	Grams per cup		
	Protein	Carbohydrates	Fatty acids
Mix A	5	2	1
Mix B	6	3	2
Mix C	8	3	1

39. FINANCIAL PLANNING   A financial planner invested $22,000 in three accounts, paying 5%, 6%, and 7% annual interest. She invested $2,000 more at 6% than at 5%. If the total interest earned in one year was $1,370, how much was invested at each rate?

40. BALLISTICS   The path of a thrown object is a parabola with an equation of $y = ax^2 + bx + c$. The parabola passes through the points $(0, 0)$, $(8, 12)$, and $(12, 15)$. Find a, b, and c.

SECTION 12.4   Solving Systems of Equations Using Matrices

DEFINITIONS AND CONCEPTS	EXAMPLES
A **matrix** is a rectangular array of numbers. Each number in a matrix is called an **element** or an **entry** of the matrix. A matrix with m rows and n columns has **order** $m \times n$.	A 2×3 matrix: $\begin{bmatrix} 2 & -7 & 5 \\ -3 & 4 & 1 \end{bmatrix}$ A 3×3 matrix: $\begin{bmatrix} 5 & -3 & 12 \\ 4 & 7 & -5 \\ 1 & -4 & 2 \end{bmatrix}$
A system of linear equations can be represented by an **augmented matrix.** Each row of the augmented matrix represents one equation of the system.	The system of equations $\begin{cases} 3x + 5y = 12 \\ 2x - 7y = -5 \end{cases}$ can be represented by the augmented matrix $\begin{bmatrix} 3 & 5 & \vdots & 12 \\ 2 & -7 & \vdots & -5 \end{bmatrix}$.

SECTION 12.4 Solving Systems of Equations Using Matrices—*continued*

DEFINITIONS AND CONCEPTS	EXAMPLES

Systems of linear equations can be solved by using matrices and **elementary row operations:**

1. Any two rows can be interchanged.

2. Any row can be multiplied by a nonzero constant.

3. Any row can be changed by adding a nonzero constant multiple of another row to it.

To solve a system of two linear equations in two unknowns using matrices, we transform the augmented matrix into an equivalent matrix that has 1's down its main diagonal and a 0 below the 1 in the first column. A matrix written in this form is said to be in **row echelon form.**

$$\begin{bmatrix} 1 & a & | & b \\ 0 & 1 & | & c \end{bmatrix}$$ *a, b,* and *c* represent real numbers.

Main diagonal

Matrices can also be used to solve systems of three linear equations containing three variables.

Solve the system using matrices: $\begin{cases} 2x - 3y = 0 \\ x + 2y = 7 \end{cases}$

This system is represented by the augmented matrix

$$\begin{bmatrix} 2 & -3 & | & 0 \\ 1 & 2 & | & 7 \end{bmatrix}$$

We can write this matrix in row echelon form by performing the following elementary row operations

$$\begin{bmatrix} 1 & 2 & | & 7 \\ 2 & -3 & | & 0 \end{bmatrix}$$ Interchange row 1 and row 2. In symbols: $R_1 \leftrightarrow R_2$.

$$\begin{bmatrix} 1 & 2 & | & 7 \\ 0 & -7 & | & -14 \end{bmatrix}$$ Multiply row 1 by -2 and add to row 2: In symbols: $-2R_1 + R_2$.

$$\begin{bmatrix} 1 & 2 & | & 7 \\ 0 & 1 & | & 2 \end{bmatrix}$$ Multply row 2 by $-\frac{1}{7}$. In symbols: $-\frac{1}{7}R_2$.

This augmented matrix represents the system: $\begin{cases} x + 2y = 7 \\ y = 2 \end{cases}$

Thus $y = 2$ and by back substitution, $x = 3$. The solution of the system is (3, 2).

REVIEW EXERCISES

Represent each system of equations using an augmented matrix.

41. $\begin{cases} 5x + 4y = 3 \\ x - y = -3 \end{cases}$

42. $\begin{cases} x + 2y + 3z = 6 \\ x - 3y - z = 4 \\ 6x + y - 2z = -1 \end{cases}$

43. Perform each of the following elementary row operations on the augmented matrix: $\begin{bmatrix} 6 & 12 & | & -6 \\ 1 & 3 & | & -2 \end{bmatrix}$.

a. $R_1 \leftrightarrow R_2$

b. $\frac{1}{6}R_1$

c. $-6R_2 + R_1$

44. Perform each of the following elementary row operations on the augmented matrix: $\begin{bmatrix} 2 & -1 & 1 & | & 3 \\ 1 & 1 & 0 & | & -1 \\ 3 & -1 & -2 & | & 7 \end{bmatrix}$.

a. $R_1 \leftrightarrow R_2$

b. $3R_2$

c. $-2R_2 + R_1$

Solve each system using matrices, if possible. If a system is inconsistent or if the equations are dependent, state this.

45. $\begin{cases} x - y = 4 \\ 3x + 7y = -18 \end{cases}$

46. $\begin{cases} x + 2y - 3z = 5 \\ x + y + z = 0 \\ 3x + 4y + 2z = -1 \end{cases}$

47. $\begin{cases} 16x - 8y = 32 \\ -2x + y = -4 \end{cases}$

48. $\begin{cases} x + 2y - z = 4 \\ x + 3y + 4z = 1 \\ 2x + 4y - 2z = 3 \end{cases}$

SECTION 12.5 Solving Systems of Equations Using Determinants

DEFINITIONS AND CONCEPTS	EXAMPLES
A determinant is a number that is associated with a **square matrix,** a matrix that has the same number of rows and columns.	A 2 × 2 determinant: $\begin{vmatrix} 3 & -3 \\ -4 & 5 \end{vmatrix}$ A 3 × 3 determinant: $\begin{vmatrix} 3 & 8 & 3 \\ 7 & 2 & 2 \\ 1 & 5 & 1 \end{vmatrix}$

To evaluate a 2 × 2 determinant:

$$\begin{vmatrix} a & b \\ c & d \end{vmatrix} = ad - bc$$

Evaluate:

$$\begin{vmatrix} 3 & -3 \\ -4 & 5 \end{vmatrix} = 3(5) - (-3)(-4) = 15 - 12 = 3$$

To evaluate a 3 × 3 determinant, we **expand it by minors** along any row or column.

Evaluate:

Minor of 3   Minor of 8   Minor of 3

$$\begin{vmatrix} 3 & 8 & 3 \\ 7 & 2 & 2 \\ 1 & 5 & 1 \end{vmatrix} = 3\begin{vmatrix} 2 & 2 \\ 5 & 1 \end{vmatrix} - 8\begin{vmatrix} 7 & 2 \\ 1 & 1 \end{vmatrix} + 3\begin{vmatrix} 7 & 2 \\ 1 & 5 \end{vmatrix}$$

$$= 3(-8) - 8(5) + 3(33) = -24 - 40 + 99 = 35$$

Cramer's rule can be used to solve systems of two linear equations in two variables.

Cramer's rule can be extended to solve systems of **three linear equations** with three variables.

Use Cramer's rule to solve: $\begin{cases} 2x - 3y = 0 \\ x + 2y = 7 \end{cases}$

The denominator determinant is D: $\begin{vmatrix} 2 & -3 \\ 1 & 2 \end{vmatrix} = 4 - (-3) = 7$

The numerator determinant for x is D_x: $\begin{vmatrix} 0 & -3 \\ 7 & 2 \end{vmatrix} = 0 - (-3)(7) = 21$

The numerator determinant for y is D_y: $\begin{vmatrix} 2 & 0 \\ 1 & 7 \end{vmatrix} = 14 - 0 = 14$

Thus, we have:

$$x = \frac{D_x}{D} = \frac{\begin{vmatrix} 0 & -3 \\ 7 & 2 \end{vmatrix}}{\begin{vmatrix} 2 & -3 \\ 1 & 2 \end{vmatrix}} = \frac{21}{7} = 3 \qquad y = \frac{D_y}{D} = \frac{\begin{vmatrix} 2 & 0 \\ 1 & 7 \end{vmatrix}}{\begin{vmatrix} 2 & -3 \\ 1 & 2 \end{vmatrix}} = \frac{14}{7} = 2$$

The solution of the system is (3, 2).

REVIEW EXERCISES

Evaluate each determinant.

49. $\begin{vmatrix} 2 & 3 \\ -4 & 3 \end{vmatrix}$

50. $\begin{vmatrix} -3 & -4 \\ 5 & -6 \end{vmatrix}$

51. $\begin{vmatrix} -1 & 2 & -1 \\ 2 & -1 & 3 \\ 1 & -2 & 2 \end{vmatrix}$

52. $\begin{vmatrix} 3 & -2 & 2 \\ 1 & -2 & -2 \\ 2 & 1 & -1 \end{vmatrix}$

Use Cramer's rule to solve each system, if possible. If a system is inconsistent or if the equations are dependent, state this.

53. $\begin{cases} 3x + 4y = 10 \\ 2x - 3y = 1 \end{cases}$

54. $\begin{cases} -6x - 4y = -6 \\ 3x + 2y = 5 \end{cases}$

55. $\begin{cases} x + 2y + z = 0 \\ 2x + y + z = 3 \\ x + y + 2z = 5 \end{cases}$

56. $\begin{cases} 2x + 3y + z = 2 \\ x + 3y + 2z = 7 \\ x - y - z = -7 \end{cases}$

CHAPTER 12
TEST

1. Fill in the blanks.

a. $\begin{cases} 2x - 7y = 1 \\ 4x - y = -8 \end{cases}$ is called a _____ of linear equations.

b. The matrix $\begin{bmatrix} -10 & 3 \\ 4 & 9 \end{bmatrix}$ has 2 _____ and 2 _____.

c. Solutions of a system of three equations in three variables, x, y, and z, are written in the form (x, y, z) and are called ordered _____.

d. The graph of the equation $2x + 3y + 4z = 5$ is a flat surface called a _____.

e. A _____ is a rectangular array of numbers written within brackets.

2. Solve the system by graphing: $\begin{cases} 2x + y = 5 \\ y = 2x - 3 \end{cases}$

3. Determine whether $\left(-\frac{1}{2}, -\frac{2}{3}\right)$ is a solution of the system:
$\begin{cases} 10x - 12y = 3 \\ 18x - 15y = 1 \end{cases}$.

4. POLITICS Explain the importance of the point of intersection of the graphs shown below.

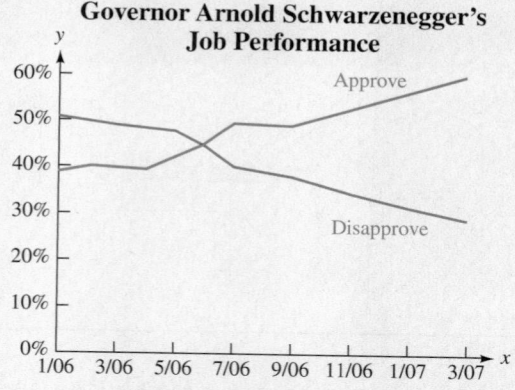

Governor Arnold Schwarzenegger's Job Performance

Source: Field Research Corporation

5. Use the vocabulary of this chapter to describe each system of two linear equations in two variables graphed below. Does the system have a solution (or solutions)?

a.

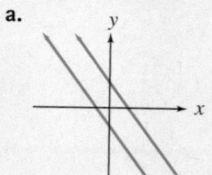

b.

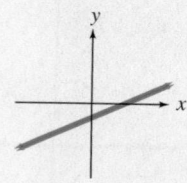

6. Use the graphs in the illustration to solve
$3(x - 2) - 2(-2 + x) = 1$.

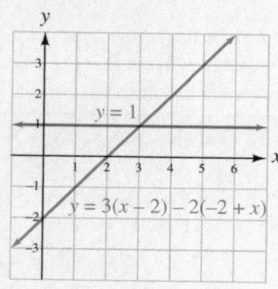

7. Use substitution to solve the system: $\begin{cases} 2x - 4y = 14 \\ x + 2y = 7 \end{cases}$

8. Use elimination (addition) to solve the system:
$\begin{cases} 2c + 3d = -5 \\ 3c - 2d = 12 \end{cases}$

Solve each system by any method, if possible. If a system is inconsistent or if the equations are dependent, state this.

9. $\begin{cases} 3(x + y) = x - 3 \\ -y = \dfrac{2x + 3}{3} \end{cases}$

10. $\begin{cases} 0.6x + 0.5y = 1.2 \\ x - \dfrac{4}{9}y + \dfrac{5}{9} = 0 \end{cases}$

Write a system of two equations in two variables to solve each problem.

11. TRAFFIC SIGNS In the sign, find x and y, if y is 15 more than x.

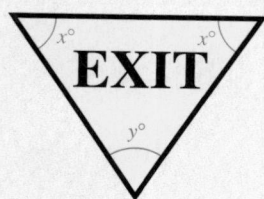

12. ANTIFREEZE How much of a 40% antifreeze solution must a mechanic mix with an 80% antifreeze solution if 20 gallons of a 50% antifreeze solution are needed?

13. WEDDING PICTURES A professional photographer offers two different packages for wedding pictures. Use the information in the advertisement to determine the cost of one 8 × 10-inch photograph and the cost of one 5 × 7-inch photograph.

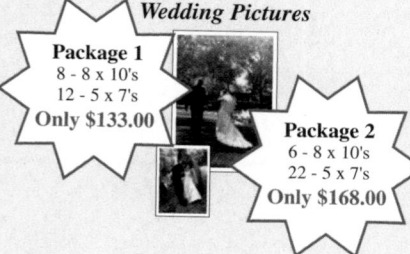

Wedding Pictures

Package 1
8 - 8 x 10's
12 - 5 x 7's
Only $133.00

Package 2
6 - 8 x 10's
22 - 5 x 7's
Only $168.00

14. Determine whether $\left(-1, -\frac{1}{2}, 5\right)$ is a solution of:

$$\begin{cases} x - 2y + z = 5 \\ 2x + 4y = -4 \\ -6y + 4z = 22 \end{cases}$$

15. Solve the system: $\begin{cases} x + y + z = 4 \\ x + y - z = 6 \\ 2x - 3y + z = -1 \end{cases}$

16. Solve the system: $\begin{cases} z - 2y = 1 \\ x + y + z = 1 \\ x + 5y = 4 \end{cases}$

17. MOVIE TICKETS   The receipts for one showing of a movie were $410 for an audience of 100 people. The ticket prices are given in the table. If twice as many children's tickets as general admission tickets were purchased, how many of each type of ticket were sold?

Ticket prices	
Children	$3
General admission	$6
Seniors	$5

18. Let $A = \begin{bmatrix} 1 & 7 & | & -3 \\ 3 & -1 & | & 13 \end{bmatrix}$. Write the matrix obtained when the elementary row operations $-3R_1 + R_2$ are performed on matrix A.

Use matrices to solve each system, if possible. If a system is inconsistent or if the equations are dependent, state this.

19. $\begin{cases} x + y = 4 \\ 2x - y = 2 \end{cases}$

20. $\begin{cases} x - 3y + 2z = 1 \\ x - 2y + 3z = 5 \\ 2x - 6y + 4z = 3 \end{cases}$

Evaluate each determinant.

21. $\begin{vmatrix} 2 & -3 \\ -4 & 5 \end{vmatrix}$

22. $\begin{vmatrix} 1 & 2 & 0 \\ 2 & 0 & 3 \\ 1 & -2 & 2 \end{vmatrix}$

23. Use Cramer's rule to solve the system:

$$\begin{cases} x - y = -6 \\ 3x + y = -6 \end{cases}$$

24. Solve the following system for z only, using Cramer's rule.

$$\begin{cases} x + y + z = 4 \\ x + y - z = 6 \\ 2x - 3y + z = -1 \end{cases}$$

GROUP PROJECT

Overview: In this activity, you are to interpret a graph that contains a break point and submit your observations in writing in the form of a financial report.

Instructions: Form groups of 2 or 3 students. Suppose you are a financial analyst for a coat hanger company The setup cost of a machine that makes wooden coat hangers is $400. After setup, it costs $1.50 to make each hanger (the unit cost). Management is considering the purchase of a new machine that can manufacture the same type of coat hanger at a cost of $1.25 per hanger. If the setup cost of the new machine is $500, find the number of coat hangers that the company would need to manufacture to make the cost the same using either machine. This is called the **break point**.

Then write a brief report that could be given to company managers, explaining their options concerning the purchase of the new machine. Under what conditions should they keep the machine currently in use? Under what conditions should they buy the new machine?

Overview: In this activity, you will explore the advantages and disadvantages of several methods for solving a system of linear equations.

Instructions: Form groups of 5 students. Have each member of your group solve the system

$$\begin{cases} x - y = 4 \\ 2x + y = 5 \end{cases}$$

in a different way. The methods to use are graphing, substitution, elimination, matrices, and Cramer's rule. Have each person briefly explain to the group his or her method of solution. After everyone has presented a solution, discuss the advantages and drawbacks of each method. Then rank the five methods, from most desirable to least desirable.

CHAPTERS 1–12
Cumulative Review

1. True or false: [Section 1.3]

   a. Every rational number can be written as a ratio of two integers.

   b. The set of real numbers corresponds to all points on the number line.

   c. The whole numbers and their opposites form the set of integers.

2. Evaluate: $-4 + 2[-7 - 3(-9)]$ [Section 1.7]

3. Solve $\frac{5}{6}k = 10$ and check the result. [Section 2.2]

4. Solve $-(3a + 1) + a = 2$ and check the result. [Section 2.2]

5. Solve $T = 2r + 2t$ for r. [Section 2.4]

6. LOOSE CHANGE   The Coinstar machines that are in many grocery stores count unsorted coins and print out a voucher that can be exchanged for cash at the checkout stand. However, to use this service, a processing fee is charged. If a boy turned in a jar of coins worth $50 and received a voucher for $45.55, what was the processing fee (expressed as a percent) charged by Coinstar? [Section 2.3]

7. Solve $5x + 7 < 2x + 1$ and graph the solution set. Then use interval notation to describe the solution. [Section 2.7]

8. Graph the line: $y = -3$ [Section 3.3]

9. Graph the line passing through $(-2, -1)$ and having slope $\frac{4}{3}$. [Section 3.4]

10. TV NEWS   The graph in red approximates the evening news viewership on all networks for the years 1995–2005. Find the rate of decrease over this period of time. [Section 3.4]

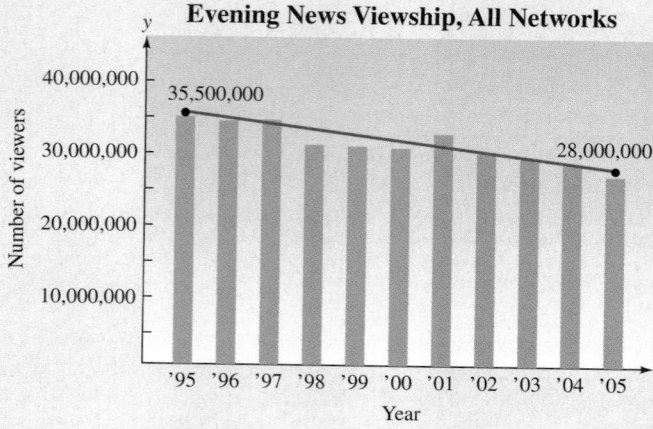

Evening News Viewship, All Networks

Source: The State of the News Media, 2006

11. Find an equation of the line whose graph has slope $\frac{1}{4}$ and passes through $(8, 1)$. Write the equation in slope–intercept form. [Section 3.5]

12. Are the graphs of $y = 4x + 9$ and $x + 4y = -10$ parallel, perpendicular, or neither? [Section 3.5]

Solve each system using the method indicated.

13. Graphing [Section 4.1]

$$\begin{cases} x + y = 1 \\ y = x + 5 \end{cases}$$

14. Substitution [Section 4.2]

$$\begin{cases} y = 2x + 5 \\ x + 2y = -5 \end{cases}$$

15. Elimination (addition) [Section 4.3]

$$\begin{cases} \dfrac{3}{5}s + \dfrac{4}{5}t = 1 \\ -\dfrac{1}{4}s + \dfrac{3}{8}t = 1 \end{cases}$$

16. MIXING CANDY   How many pounds of each candy must be mixed to obtain 48 pounds of candy that would be worth $4 per pound? Use two variables to solve this problem. [Section 4.4]

Sweet-n-sour
$3/lb

Bit-O-Honey
$6/lb

17. AVIATION   With the wind, a plane can fly 3,000 miles in 5 hours. Against the wind, the trip takes 6 hours. Find the airspeed of the plane (the speed in still air) and the speed of the wind. Use two variables to solve this problem. [Section 4.4]

18. Solve the system of linear inequalities. [Section 4.5]

$$\begin{cases} 3x + 4y \geq -7 \\ 2x - 3y \geq 1 \end{cases}$$

Simplify each expression. Write each answer without using parentheses or negative exponents.

19. $y^3(y^2y^4)^3$ [Section 5.1]

20. $\left(\dfrac{21x^{-2}y^2z^{-2}}{7x^3y^{-1}}\right)^{-2}$ [Section 5.2]

21. FIVE-CARD POKER   The odds against being dealt the hand shown are about 2.6×10^6 to 1. Express the odds using standard notation. [Section 5.3]

22. Write 0.00073 in scientific notation. [Section 5.3]

Perform the operations.

23. $4(4x^3 + 2x^2 - 3x - 8) - 5(2x^3 - 3x + 8)$
   [Section 5.5]

24. $(-2a^3)(3a^2)$ [Section 5.6]

25. $(2b - 1)(3b + 4)$ [Section 5.6]

26. $(3x + y)(2x^2 - 3xy + y^2)$ [Section 5.6]

27. $(2x + 5y)^2$ [Section 5.7]

28. $(9m^2 - 1)(9m^2 + 1)$ [Section 5.7]

29. $\dfrac{12a^3b - 9a^2b^2 + 3ab}{6a^2b}$ [Section 5.8]

30. $x - 3\overline{)2x^2 - 3 - 5x}$ [Section 5.8]

Factor each expression completely.

31. $6a^2 - 12a^3b + 36ab$
   [Section 6.1]

32. $2x + 2y + ax + ay$
   [Section 6.1]

33. $25t^2 - 16$
   [Section 6.4]

34. $b^3 + 125$
   [Section 6.5]

Solve each equation.

35. $3x^2 + 8x = 0$ [Section 6.7]   **36.** $15x^2 - 2 = 7x$ [Section 6.7]

37. HEIGHT OF A TRIANGLE   The triangle shown has an area of 22.5 square inches. Find its height. [Section 6.8]

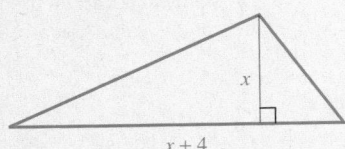

38. For what value(s) of x is $\dfrac{5}{x^2 - 2x}$ undefined? [Section 7.1]

Simplify each expression.

39. $\dfrac{3x^2 - 27}{x^2 + 3x - 18}$ [Section 7.1]   **40.** $\dfrac{a - 15}{15 - a}$ [Section 7.1]

Perform the operations.

41. $\dfrac{x^2 - x - 6}{2x^2 + 9x + 10} \div \dfrac{x^2 - 25}{2x^2 + 15x + 25}$ [Section 7.2]

42. $\dfrac{1}{s^2 - 4s - 5} + \dfrac{s}{s^2 - 4s - 5}$ [Section 7.3]

43. $\dfrac{x + 5}{xy} - \dfrac{x - 1}{x^2 y}$
   [Section 7.4]

44. $\dfrac{x}{x - 2} + \dfrac{3x}{x^2 - 4}$
   [Section 7.4]

Simplify each complex fraction.

45. $\dfrac{\dfrac{9m - 27}{m^6}}{\dfrac{2m - 6}{m^8}}$
   [Section 7.5]

46. $\dfrac{\dfrac{5}{y} + \dfrac{4}{y + 1}}{\dfrac{4}{y} - \dfrac{5}{y + 1}}$
   [Section 7.5]

47. Solve: $\dfrac{7}{q^2 - q - 2} + \dfrac{1}{q + 1} = \dfrac{3}{q - 2}$ [Section 7.6]

48. ROOFING   A homeowner estimates that it will take him 7 days to roof his house. A professional roofer estimates that he can roof the house in 4 days. How long will it take if the homeowner helps the roofer? [Section 7.7]

49. LOSING WEIGHT   If a person cuts his or her daily calorie intake by 100, it will take 350 days for that person to lose 10 pounds. How long will it take for the person to lose 25 pounds? [Section 7.8]

50. $\triangle ABC$ and $\triangle DEC$ are similar triangles. Find x. [Section 7.8]

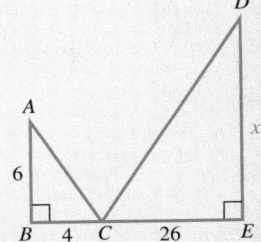

Solve each inequality. Write the solution set in interval notation and graph it.

51. $\left|\dfrac{x - 2}{3}\right| - 4 \le 0$ [Section 8.2]

52. $3x + 2 < 8$ or $2x - 3 > 11$ [Section 8.3]

Factor completely.

53. $x^2 + 4x + 4 - y^2$ [Section 8.4]

54. $b^4 - 17b^2 + 16$ [Section 8.5]

55. If $f(x) = 3x^2 + 3x - 8$, find $f(-1)$. [Section 8.8]

56. Graph $f(x) = (x - 1)^3$ and determine the domain and range of f. [Section 8.8]

57. BOATING   The graph on the next page shows the vertical distance from a point on the tip of a propeller to the centerline as the propeller spins. Is this the graph of a function? [Section 8.8]

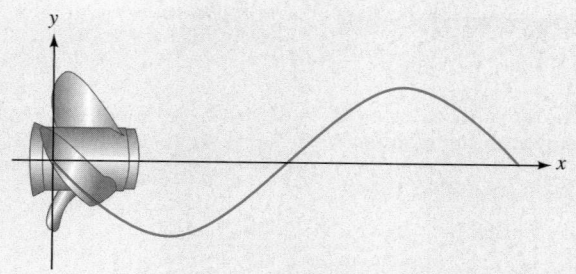

58. GEARS The speed of a gear varies inversely with the number of teeth. If a gear with 10 teeth makes 3 revolutions per second, how many revolutions per second will a gear with 25 teeth make? [Section 8.9]

59. Simplify $\sqrt{4x^2}$ and assume that the variable is unrestricted. [Section 9.1]

60. Simplify: $(-8)^{-4/3}$ [Section 9.2]

Simplify each expression. All variables represent positive numbers.

61. $\sqrt{100a^6b^4}$ [Section 9.3]   **62.** $\sqrt[4]{16x^7y^4}$ [Section 9.3]

63. $3\sqrt{24} + \sqrt{54}$ [Section 9.3]

64. $\sqrt[5]{x^6y^2} + \sqrt[5]{32x^6y^2} + \sqrt[5]{x^6y^2}$ [Section 9.3]

65. $\sqrt{\dfrac{72x^3}{y^2}}$ [Section 9.4]

66. $\sqrt[3]{\dfrac{27m^3}{8n^6}}$ [Section 9.4]

Rationalize the denominator.

67. $\dfrac{2}{\sqrt[3]{a}}$ [Section 9.4]   **68.** $\dfrac{\sqrt{x} - \sqrt{y}}{\sqrt{x} + \sqrt{y}}$ [Section 9.4]

69. Solve: $2 + \sqrt{u} = \sqrt{2u + 7}$ [Section 9.5]

70. STORAGE CUBES The diagonal distance across the face of each of the stacking cubes is 15 inches. What is the height of the entire storage arrangement? Round to the nearest tenth of an inch. [Section 9.6]

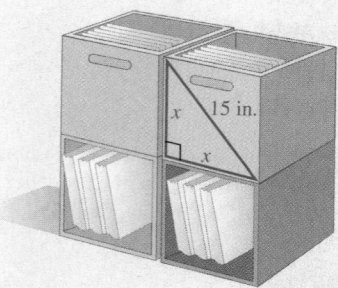

Write each expression in terms of i.

71. $\sqrt{-49}$ [Section 9.7]   **72.** $\sqrt{-54}$ [Section 9.7]

Perform the operations. Express each answer in the form $a + bi$.

73. $(2 + 3i) - (1 - 2i)$ [Section 9.7]

74. $(7 - 4i) + (9 + 2i)$ [Section 9.7]

75. $(3 - 2i)(4 - 3i)$ [Section 9.7]

76. $\dfrac{3 - i}{2 + i}$ [Section 9.7]

77. Solve $x^2 + 8x + 12 = 0$ by completing the square. [Section 10.1]

78. Solve $4x^2 - x - 2 = 0$ using the quadratic formula. Give the exact solutions, and then approximate each to the nearest hundredth. [Section 10.2]

Solve each equation. Express the solutions in the form $a + bi$.

79. $x^2 + 16 = 0$ [Section 10.2]   **80.** $x^2 - 4x = -5$ [Section 10.2]

81. Solve: $a^{2/3} + a^{1/3} - 6 = 0$ [Section 10.3]

82. Graph the quadratic equation $y = 2x^2 + 8x + 6$. Find the vertex, the x- and y-intercepts, and the axis of symmetry of the graph. [Section 10.4]

83. Let $f(x) = 3x - 2$ and $g(x) = x^2 + x$. Find $(f \circ g)(-3)$. [Section 11.1]

84. Find the inverse of $f(x) = -\dfrac{3}{2}x + 3$. [Section 11.2]

Graph each function. Determine the domain and range.

85. $f(x) = 5^x$ [Section 11.3]   **86.** $f(x) = \ln x$ [Section 11.6]

87. Give an approximate value of e to the nearest tenth. [Section 11.4]

88. WORLD POPULATION GROWTH The population of the Earth is approximately 6.6 billion people and is growing at an annual rate of 1.167%. Use the exponential growth model to find the world population in 20 years. [Section 11.4]

Find each value of x.

89. $\log_x 5 = 1$ [Section 11.5]   **90.** $\log_8 x = 2$ [Section 11.5]

Evaluate each expression.

91. $\log_9 \dfrac{1}{81}$ [Section 11.5]   **92.** $\ln e$ [Section 11.6]

93. Write $\ln \dfrac{y^3\sqrt{x}}{z}$ as the sum and/or difference of logarithms of a single quantity. Then simplify, if possible. [Section 11.7]

94. Write $2 \log x - 3 \log y + \log z$ as one logarithm. [Section 11.7]

Solve each equation. Give answers to four decimal places when necessary.

95. $5^{x-3} = 3^{2x}$ [Section 11.8]

96. $\log(x + 90) = 3 - \log x$ [Section 11.8]

97. Solve: $\begin{cases} x - y + z = 4 \\ x + 2y - z = -1 \\ x + y - 3z = -2 \end{cases}$ [Section 12.2]

98. TRIANGLES   The sum of the measures of the angles of any triangle is 180°. In $\triangle DEF$, $\angle D$ measures 100° less than the sum of the measures of $\angle E$ and $\angle F$, and the measure of $\angle F$ is 40° less than twice the measure of $\angle E$. Find the measure of each angle of the triangle. [Section 12.3]

99. Solve by using matrices: $\begin{cases} 2x + y = 1 \\ x + 2y = -4 \end{cases}$ [Section 12.4]

100. Solve by using Cramer's rule: $\begin{cases} 3x - 4y = 9 \\ x + 2y = 8 \end{cases}$ [Section 12.5]

CHAPTER 13

Conic Sections; More Graphing

© Rich LaSalle/Getty Images

from *Campus to Careers*
Traffic Engineer

Traffic engineers design roads, streets, and highways for the safe and efficient movement of people and goods. They use traffic flow formulas to determine what kinds of roads are needed and then find economical ways to construct and operate them. During the planning stages, traffic engineers make detailed drawings and graphs of the project. Because highway and street construction is often publicly funded, they make budgets, submit bid proposals, and perform cost analysis studies to make sure that highway tax money is spent wisely.

In **Problem 85** of **Study Set 13.1,** you will design two sections of a freeway that are to be joined with a curve that is one-quarter of a circle.

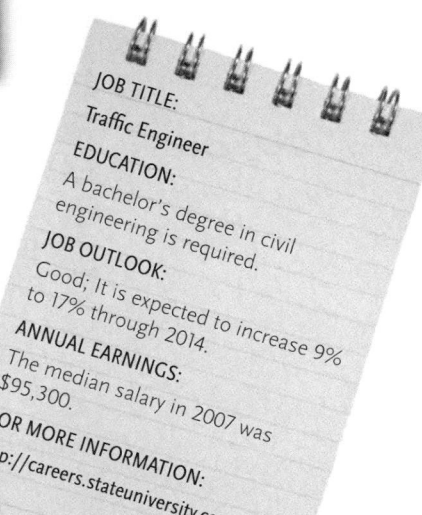

JOB TITLE:
Traffic Engineer

EDUCATION:
A bachelor's degree in civil engineering is required.

JOB OUTLOOK:
Good; It is expected to increase 9% to 17% through 2014.

ANNUAL EARNINGS:
The median salary in 2007 was $95,300.

FOR MORE INFORMATION:
http://careers.stateuniversity.com

Study Skills Workshop
Preparing for Your Next Math Course

Before moving on to a new mathematics course, it's worthwhile to take some time to reflect on your effort and performance in this course.

Now Try This

As this course draws to a close, here are some questions to ask yourself.

1. How was my attendance?
2. Was I organized? Did I have the right materials?
3. Did I follow a regular schedule?
4. Did I pay attention in class and take good notes?
5. Did I spend the appropriate amount of time on homework?
6. How did I prepare for tests? Did I have a test-taking strategy?
7. Was I part of a study group? If not, why not? If so, was it worthwhile?
8. Did I ever seek extra help from a tutor or from my instructor?
9. In what topics was I the strongest? In what topics was I the weakest?
10. If I had it to do over, would I do anything differently?

SECTION 13.1
The Circle and the Parabola

Objectives

 1 Identify conic sections and some of their applications.

2 Graph equations of circles written in standard form.

3 Write the equation of a circle, given its center and radius.

4 Convert the general form of the equation of a circle to standard form.

5 Solve problems involving circles.

6 Convert the general form of the equation of a parabola to standard form to graph it.

We have previously graphed first-degree equations in two variables such as $y = 3x + 8$ and $4x - 3y = 12$. Their graphs are lines. In this section, we will graph second-degree equations in two variables such as $x^2 + y^2 = 25$ and $x = -3y^2 - 12y - 13$. The graphs of these equations are *conic sections*.

1 Identify Conic Sections and Some of Their Applications.

The curves formed by the intersection of a plane with an infinite right-circular cone are called **conic sections.** Those curves have four basic shapes, called **circles, parabolas, ellipses,** and **hyperbolas,** as shown on the next page.

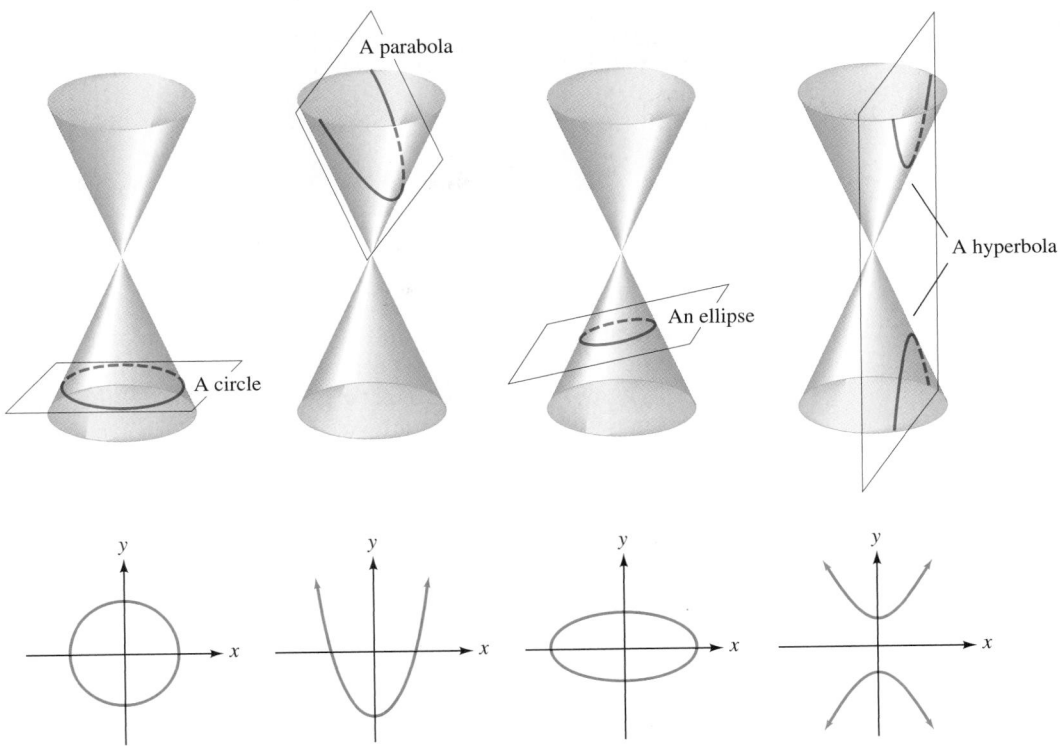

Conic sections have many applications. For example, everyone is familiar with circular wheels and gears, pizza cutters, and hula hoops.

Parabolas can be rotated to generate dish-shaped surfaces called **paraboloids.** Any light or sound placed at the **focus** of a paraboloid is reflected outward in parallel paths. This property makes parabolic surfaces ideal for flashlight and headlight reflectors. It also makes parabolic surfaces good antennas, because signals captured by such antennas are concentrated at the focus. Parabolic mirrors are capable of concentrating the rays of the sun at a single point, thereby generating tremendous heat. This property is used in the design of solar furnaces.

Any object thrown upward and outward travels in a parabolic path. An example of this is a stream of water flowing from a drinking fountain. In architecture, many arches are parabolic in shape, because this gives them strength. Cables that support suspension bridges hang in the shape of a parabola.

The Language of Algebra
Conic sections are often simply called *conics.*

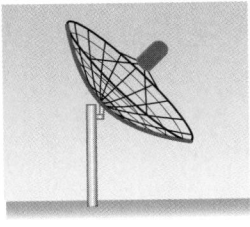

Radar dish

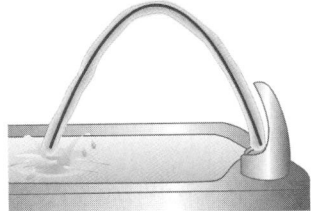

Stream of water

Support cables

Ellipses have optical and acoustical properties that are useful in architecture and engineering. Many arches are portions of an ellipse, because the shape is pleasing to the eye. The planets and many comets have elliptical orbits. Certain gears have elliptical shapes to provide nonuniform motion.

Arches

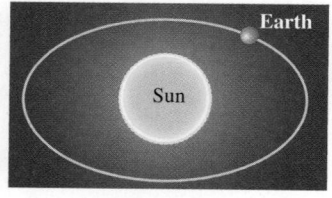

Earth's orbit

Gears

Hyperbolas serve as the basis of a navigational system known as LORAN (LOng RAnge Navigation). They are also used to find the source of a distress signal, are the basis for the design of hypoid gears, and describe the orbits of some comets.

A sonic shock wave created by a jet aircraft has the shape of a cone. In level flight, the sound wave intersects the ground as one branch of a hyperbola, as shown below. People in different places along the curve on the ground hear and feel the sonic boom at the same time.

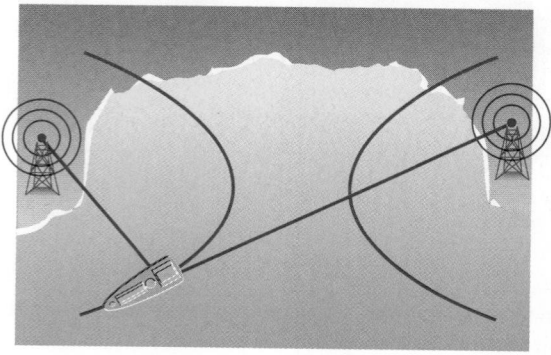

Navigation

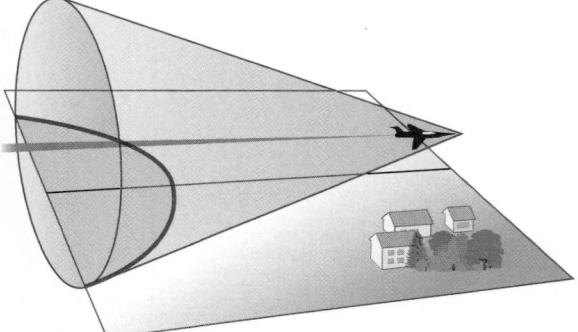

Sonic boom

② Graph Equations of Circles Written in Standard Form.

Every conic section can be represented by a second-degree equation in x and y. To find the equation of a circle, we use the following definition.

Definition of a Circle	A **circle** is the set of all points in a plane that are a fixed distance from a fixed point called its **center.** The fixed distance is called the **radius** of the circle.

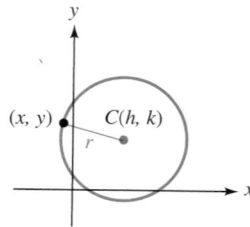

If we let (h, k) be the center of a circle and (x, y) be some point on a circle that is graphed on a rectangular coordinate system, the distance from (h, k) to (x, y) is the radius r of the circle. We can use the distance formula to find r.

$$r = \sqrt{(x - h)^2 + (y - k)^2}$$

We can square both sides to eliminate the radical and obtain

$$r^2 = (x - h)^2 + (y - k)^2$$

This result is called the *standard form of the equation of a circle* with radius r and center at (h, k).

Equation of a Circle	The **standard form of the equation of a circle** with radius r and center at (h, k) is $$(x - h)^2 + (y - k)^2 = r^2$$

EXAMPLE 1   Find the center and the radius of each circle and then graph it:
a. $(x - 4)^2 + (y - 1)^2 = 9$ **b.** $x^2 + y^2 = 25$
c. $(x + 3)^2 + y^2 = 12$

Strategy   We will compare each equation to the standard form of the equation of a circle, $(x - h)^2 + (y - k)^2 = r^2$, and identify h, k, and r.

Why   The center of the circle is the point with coordinates (h, k) and the radius of the circle is r.

Solution

a. The color highlighting shows how to compare the given equation to the standard form to find h, k, and r.

$$(x - 4)^2 + (y - 1)^2 = 9$$
$$\uparrow \qquad \uparrow \qquad \uparrow$$
$$(x - h)^2 + (y - k)^2 = r^2$$

$h = 4$, $k = 1$, and $r^2 = 9$. Since the radius of a circle must be positive, $r = 3$.

The center of the circle is $(h, k) = (4, 1)$ and the radius is 3.
To plot four points on the circle, we move up, down, left, and right 3 units from the center, as shown in figure (a). Then we draw a circle through the points to get the graph of $(x - 4)^2 + (y - 1)^2 = 9$, as shown in figure (b).

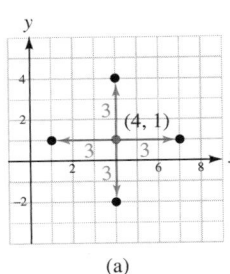
(a)

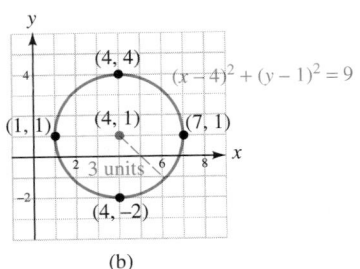
(b)

b. To find h and k, we will write $x^2 + y^2 = 25$ in the following way:

$$(x - 0)^2 + (y - 0)^2 = 25$$
$$\uparrow \qquad \uparrow \qquad \uparrow$$
$$(x - h)^2 + (y - k)^2 = r^2$$

$h = 0$, $k = 0$, and $r^2 = 25$. Since the radius must be positive, $r = 5$.

The center of the circle is at $(0, 0)$ and the radius is 5.
To plot four points on the circle, we move up, down, left, and right 5 units from the center. Then we draw a circle through the points to get the graph of $x^2 + y^2 = 25$, as shown.

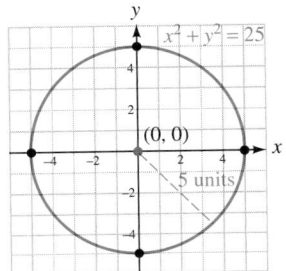

c. To find h, we will write $x + 3$ as $x - (-3)$.

Standard form requires a minus symbol here.
↓

$$[x - (-3)]^2 + (y - 0)^2 = 12$$
$$\uparrow \qquad \uparrow \qquad \uparrow$$
$$(x - h)^2 + (y - k)^2 = r^2$$

$h = -3$, $k = 0$, and $r^2 = 12$.

Since $r^2 = 12$, we have

$$r = \pm\sqrt{12} = \pm 2\sqrt{3} \qquad \text{Use the square root property.}$$

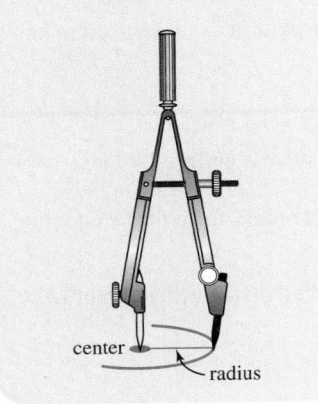

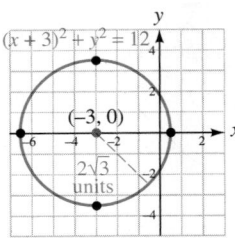

$(x + 3)^2 + y^2 = 12$

$(-3, 0)$

$2\sqrt{3}$ units

Since the radius can't be negative, $r = 2\sqrt{3}$. The center of the circle is at $(-3, 0)$ and the radius is $2\sqrt{3}$.

To plot four points on the circle, we move up, down, left, and right $2\sqrt{3} \approx 3.5$ units from the center. We then draw a circle through the points to get the graph of $(x + 3)^2 + y^2 = 12$, as shown on the left.

 Self Check 1   Find the center and the radius of each circle and then graph it:

     a. $(x - 3)^2 + (y + 4)^2 = 4$ **b.** $x^2 + y^2 = 8$

Now Try   **Problems 15, 19, and 21**

③ Write the Equation of a Circle, Given its Center and Radius.

Because a circle is determined by its center and radius, that information is all we need to know to write its equation.

EXAMPLE 2   Write the equation of the circle with radius 9 and center at $(6, -5)$.

Strategy   We substitute 9 for r, 6 for h, and -5 for k in the standard form of the equation of a circle, $(x - h)^2 + (y - k)^2 = r^2$.

Why   When writing the standard form, the center is represented by the ordered pair (h, k) and the radius as r.

Notation
Standard form can be written

$$(x - 6)^2 + (y + 5)^2 = 9^2$$
or
$$(x - 6)^2 + (y + 5)^2 = 81$$

Solution

$$(x - h)^2 + (y - k)^2 = r^2$$
$$(x - 6)^2 + [y - (-5)]^2 = 9^2 \quad \text{Substitute 6 for } h, -5 \text{ for } k, \text{ and 9 for } r.$$
$$(x - 6)^2 + (y + 5)^2 = 9^2 \quad \text{Write } y - (-5) \text{ as } y + 5.$$

If we express 9^2 as 81, we have

$$(x - 6)^2 + (y + 5)^2 = 81$$

 Self Check 2   Write the equation of the circle with radius 10 and center at $(-7, 1)$.

Now Try   **Problems 23, 27, and 31**

---

④ Convert the General Form of the Equation of a Circle to Standard Form.

In Example 2, the result was written in standard form: $(x - 6)^2 + (y + 5)^2 = 81$. If we square $x - 6$ and $y + 5$, we obtain a different form for the equation of the circle.

$$(x - 6)^2 + (y + 5)^2 = 9^2$$
$$x^2 - 12x + 36 + y^2 + 10y + 25 = 81 \quad \text{Square each binomial.}$$
$$x^2 - 12x + y^2 + 10y - 20 = 0 \quad \text{Subtract 81 from both sides. Combine like terms.}$$
$$x^2 + y^2 - 12x + 10y - 20 = 0 \quad \text{Rearrange the terms, writing the squared terms first.}$$

Success Tip
This example illustrates an important fact: The equation of a circle contains both x^2 and y^2 terms on the same side of the equation with equal coefficients.

This result is written in the *general form of the equation of a circle*.

Equation of a Circle

The **general form of the equation of a circle** is

$$x^2 + y^2 + Dx + Ey + F = 0$$

We can convert from the general form to the standard form of the equation of a circle by completing the square.

EXAMPLE 3   Write the equation $x^2 + y^2 - 4x + 2y - 11 = 0$ in standard form and graph it.

Strategy   We will rearrange the terms to write the equation in the form $x^2 - 4x + y^2 + 2y = 11$ and complete the square on x and y.

Why   Standard form contains the expressions $(x - h)^2$ and $(y - k)^2$. We can obtain a perfect-square trinomial that factors as $(x - 2)^2$ by completing the square on $x^2 - 4x$. We can complete the square on $y^2 + 2y$ to obtain an expression of the form $(y + 1)^2$.

Solution   To write the equation in standard form, we complete the square twice.

$$x^2 + y^2 - 4x + 2y - 11 = 0$$

$$x^2 - 4x + y^2 + 2y = 11 \qquad \text{Write the x-terms together, the y-terms together, and add 11 to both sides.}$$

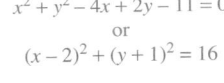
$$x^2 + y^2 - 4x + 2y - 11 = 0$$
or
$$(x - 2)^2 + (y + 1)^2 = 16$$

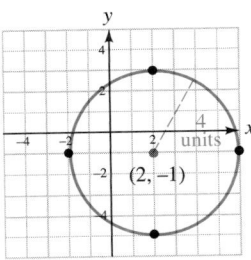

To complete the square on $x^2 - 4x$, we note that $\frac{1}{2}(-4) = -2$ and $(-2)^2 = 4$. To complete the square on $y^2 + 2y$, we note that $\frac{1}{2}(2) = 1$ and $1^2 = 1$. We add **4** and **1** to both sides of the equation.

$$x^2 - 4x + 4 + y^2 + 2y + 1 = 11 + 4 + 1$$

$$(x - 2)^2 + (y + 1)^2 = 16 \qquad \text{Factor } x^2 - 4x + 4 \text{ and } y^2 + 2y + 1.$$

The equation can also be written as $(x - 2)^2 + (y + 1)^2 = 4^2$.

We can determine the circle's center and radius by comparing this equation to the standard form of the equation of a circle, $(x - h)^2 + (y - k)^2 = r^2$. We see that $h = 2$, $k = -1$, and $r = 4$. We can use the center, $(h, k) = (2, -1)$ and the radius $r = 4$, to graph the circle as shown on the left.

Self Check 3   Write the equation $x^2 + y^2 + 12x - 6y - 4 = 0$ in standard form and graph it.

Now Try   **Problem 35**

Using Your Calculator   *Graphing Circles*

Since the graphs of circles fail the vertical line test, their equations do not represent functions. It is more difficult to use a graphing calculator to graph equations that are not functions. For example, to graph the circle described by $(x - 1)^2 + (y - 2)^2 = 4$, we must split the equation into two functions and graph each one separately. We begin by solving the equation for y.

$$(x - 1)^2 + (y - 2)^2 = 4$$

$$(y - 2)^2 = 4 - (x - 1)^2 \qquad \text{Subtract } (x - 1)^2 \text{ from both sides.}$$

$$y - 2 = \pm \sqrt{4 - (x - 1)^2} \qquad \text{Use the square root property.}$$

$$y = 2 \pm \sqrt{4 - (x - 1)^2} \qquad \text{Add 2 to both sides.}$$

This equation defines two functions. If we graph

$$y = 2 + \sqrt{4 - (x - 1)^2} \quad \text{and} \quad y = 2 - \sqrt{4 - (x - 1)^2}$$

we get the distorted circle shown in figure (a). To get a better circle, we can use the graphing calculator's square window feature, which gives an equal unit distance on both the *x*- and *y*-axes. (Press ZOOM, 5, ENTER.) Using this feature, we get the circle shown in figure (b). Sometimes the two arcs will not connect because of approximations made by the calculator at each endpoint.

The graph of
$y = 2 + \sqrt{4 - (x - 1)^2}$
is the top half of the circle.

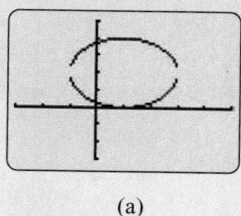

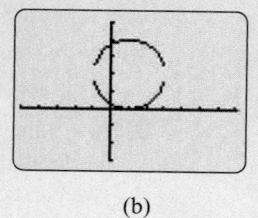

The graph of
$y = 2 - \sqrt{4 - (x - 1)^2}$
is the bottom half of the circle.

(a)

(b)

5 Solve Problems Involving Circles.

EXAMPLE 4 *Radio Translators.* The broadcast area of a television station is bounded by the circle $x^2 + y^2 = 3{,}600$, where *x* and *y* are measured in miles. A translator station picks up the signal and retransmits it from the center of a circular area bounded by

$$(x + 30)^2 + (y - 40)^2 = 1{,}600$$

Find the location of the translator and the greatest distance from the main transmitter that the signal can be received.

Strategy Refer to the figure below. We will find two distances: the distance from the TV station transmitter to the translator and the distance from the translator to the outer edge of its coverage.

Why The greatest distance of reception from the main transmitter is the sum of those two distances.

Solution The coverage of the TV station is bounded by $x^2 + y^2 = 60^2$, a circle centered at the origin with a radius of 60 miles, as shown in yellow in the figure. Because the translator is at the center of the circle $(x + 30)^2 + (y - 40)^2 = 1{,}600$, it is located at $(-30, 40)$, a point 30 miles west and 40 miles north of the TV station. The radius of the translator's coverage is $\sqrt{1{,}600}$, or 40 miles.

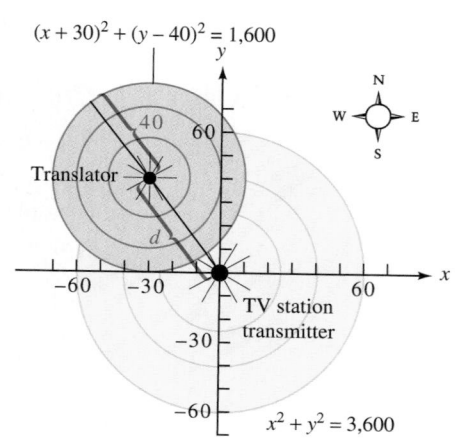

As shown in the figure, the greatest distance of reception is the sum of *d*, the distance from the translator to the television station, and 40 miles, the radius of the translator's coverage.

To find d, we use the distance formula to find the distance between the origin, $(x_1, y_1) = (0, 0)$, and $(x_2, y_2) = (-30, 40)$.

$$d = \sqrt{(x_2 - x_1)^2 + (y_2 - y_1)^2}$$ *The distance formula was introduced in Section 7.6.*

$$d = \sqrt{(-30 - 0)^2 + (40 - 0)^2}$$

$$d = \sqrt{(-30)^2 + 40^2}$$

$$= \sqrt{900 + 1{,}600}$$

$$= \sqrt{2{,}500}$$

$$= 50$$

The translator is located 50 miles from the television station, and it broadcasts the signal 40 miles. The greatest reception distance from the main transmitter signal is, therefore, $50 + 40$, or 90 miles.

 Now Try   **Problem 83**

 6   **Convert the General Form of the Equation of a Parabola to Standard Form to Graph It.**

Another type of conic section is the parabola.

Definition of a Parabola

A **parabola** is the set of all points in a plane that are equidistant from a fixed point, called the **focus,** and a fixed line, called the **directrix.**

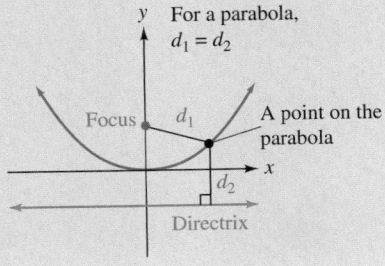

We have previously discussed parabolas whose graphs open upward or downward. Parabolas can also open to the right and to the left, but they do not define functions because their graphs fail the vertical line test.

The two general forms of the equation of a parabola are similar.

Equation of a Parabola

The **general forms of the equation of a parabola** are:

1. $y = ax^2 + bx + c$ The graph opens upward if $a > 0$ and downward if $a < 0$.
2. $x = ay^2 + by + c$ The graph opens to the right if $a > 0$ and to the left if $a < 0$.

Recall from Chapter 8 that equations written in the standard form $y = a(x - h)^2 + k$ represent parabolas with vertex at (h, k) and axis of symmetry $x = h$. They open upward when $a > 0$ and downward when $a < 0$.

EXAMPLE 5   Write $y = -2x^2 + 12x - 15$ in standard form and graph it.

Strategy   We will complete the square on x to write the equation in standard form, $y = a(x - h)^2 + k$.

Why   Standard form contains the expression $(x - h)^2$. We can obtain a perfect-square trinomial that factors into that form by completing the square on x.

Solution   Because the equation is not in standard form, the coordinates of the vertex are not obvious. To write the equation in standard form, we complete the square on x.

$$y = -2x^2 + 12x - 15$$

$$y = -2(x^2 - 6x \qquad) - 15 \qquad \text{Factor out } -2 \text{ from } -2x^2 + 12x.$$

This step adds $-2 \cdot 9$ Add 18 to counteract
or -18 to this side.   the addition of -18.

$$y = -2(x^2 - 6x + 9) - 15 + 18 \qquad \text{Complete the square on } x^2 - 6x.$$

$$y = -2(x - 3)^2 + 3 \qquad \text{Factor } x^2 - 6x + 9 \text{ and combine like terms.}$$

This equation is written in the form $y = a(x - h)^2 + k$, where $a = -2$, $h = 3$, and $k = 3$. Thus, the graph of the equation is a parabola that opens downward with vertex at $(3, 3)$ and an axis of symmetry $x = 3$. We can construct a table of solutions and use symmetry to plot several points on the parabola. Then we draw a smooth curve through the points to get the graph of $y = -2x^2 + 12x - 15$, as shown below.

Success Tip

Recall that we can find the x-coordinate of the vertex using

$$x = -\frac{b}{2a} = -\frac{12}{2(-2)} = 3$$

To find the y-coordinate, substitute:

$$y = -2(3)^2 + 12(3) - 15$$
$$= 3$$

The vertex is at $(3, 3)$.

$$y = -2x^2 + 12x - 15$$

x	y
1	−5	→ (1, −5)
2	1	→ (2, 1)

↑

Because the x-coordinate of the vertex is 3, choose values for x that are close to 3 on the same side of the axis of symmetry.

(3, 3)

(2, 1)   (4, 1)

$y = -2x^2 + 12x - 15$
or
$y = -2(x - 3)^2 + 3$

(1, −5)   (5, −5)

$x = 3$

Self Check 5   Write $y = 2x^2 + 4x + 5$ in standard form and graph it.

Now Try   Problem 39

The *standard form* for the equation of a parabola that opens to the right or left is similar to $y = a(x - h)^2 + k$, except that the variables, x and y, exchange positions as do the constants, h and k.

Standard Form of the Equation of a Parabola

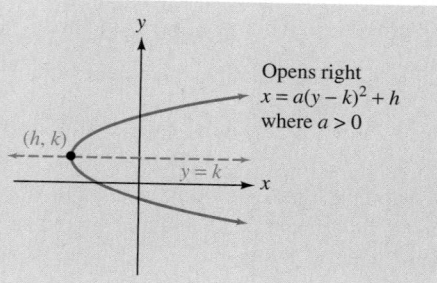

Opens right
$x = a(y - k)^2 + h$
where $a > 0$

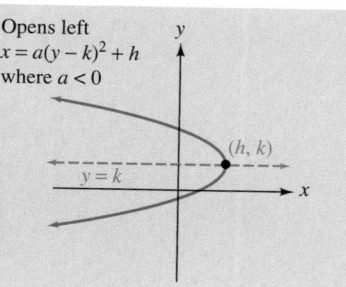

Opens left
$x = a(y - k)^2 + h$
where $a < 0$

EXAMPLE 6   Graph: $x = \dfrac{1}{2}y^2$

Strategy   We will compare the equation to the standard form of the equation of a parabola to find a, h, and k.

Why   Once we know these values, we can locate the vertex of the graph. We also know whether the parabola will open to the left or to the right.

Solution   This equation is written in the form $x = a(y - k)^2 + h$, where $a = \frac{1}{2}$, $k = 0$, and $h = 0$. The graph of the equation is a parabola that opens to the right with vertex at $(0, 0)$ and an axis of symmetry $y = 0$.

To construct a table of solutions, we choose values of y and find their corresponding values of x. For example, if $y = 1$, we have

$$x = \frac{1}{2}y^2$$

$$x = \frac{1}{2}(1)^2 \quad \text{Substitute 1 for } y.$$

$$x = \frac{1}{2}$$

The point $\left(\frac{1}{2}, 1\right)$ is on the parabola.

We plot the ordered pairs from the table and use symmetry to plot three more points on the parabola. Then we draw a smooth curve through the points to get the graph of $x = \frac{1}{2}y^2$, as shown below.

$x = \frac{1}{2}y^2$

x	y
$\frac{1}{2}$	1
2	2
8	4

$\rightarrow \left(\frac{1}{2}, 1\right)$
$\rightarrow (2, 2)$
$\rightarrow (8, 4)$

↑
Because the y-coordinate of the vertex is
0, choose values for y that are close to 0
on the same side of the axis of symmetry.

Self Check 6   Graph: $x = -\dfrac{2}{3}y^2$

Now Try   **Problem 43**

EXAMPLE 7   Write $x = -3y^2 - 12y - 13$ in standard form and graph it.

Strategy   We will complete the square on y to write the equation in standard form, $x = a(y - k)^2 + h$.

Why   Standard form contains the expression $(y - k)^2$. We can obtain a perfect-square trinomial that factors into that form by completing the square on y.

Solution   To write the equation in standard form, we complete the square.

$$x = -3y^2 - 12y - 13$$
$$x = -3(y^2 + 4y \qquad) - 13 \qquad \text{Factor out } -3 \text{ from } -3y^2 - 12y.$$
$$x = -3(y^2 + 4y + 4) - 13 + 12 \qquad \text{Complete the square on } y^2 + 4y. \text{ Then add 12 to the}$$
$$\text{right side to counteract } -3 \cdot 4 = -12.$$
$$x = -3(y + 2)^2 - 1 \qquad \text{Factor } y^2 + 4y + 4 \text{ and combine like terms.}$$

This equation is in the standard form $x = a(y - k)^2 + h$, where $a = -3$, $k = -2$, and $h = -1$. The graph of the equation is a parabola that opens to the left with vertex at $(-1, -2)$ and an axis of symmetry $y = -2$.

We can construct a table of solutions and use symmetry to plot several points on the parabola. Then we draw a smooth curve through the points to get the graph of $x = -3y^2 - 12y - 13$, as shown below.

$$x = -3y^2 - 12y - 13$$
$$\text{or}$$
$$x = -3(y + 2)^2 - 1$$

x	y
-4	-1	$\rightarrow (-4, -1)$
-13	0	$\rightarrow (-13, 0)$

↑
Choose values for y, and find the corresponding x-values.

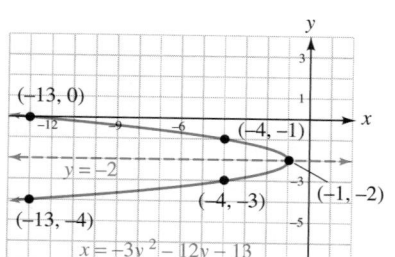

 Self Check 7   Write $x = 3y^2 - 6y - 1$ in standard form and graph it.

Now Try   **Problem 49**

 ANSWERS TO SELF CHECKS   **1. a.**

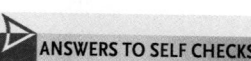

b.

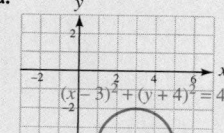

2. $(x + 7)^2 + (y - 1)^2 = 100$ **3.**

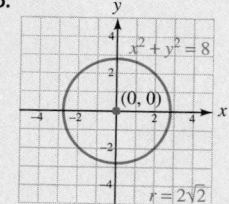

5.

6.

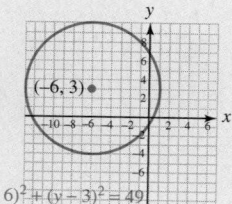

7.

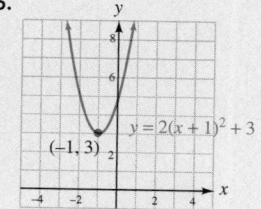

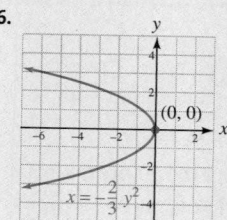

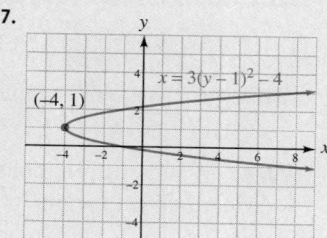

STUDY SET
13.1

VOCABULARY

Fill in the blanks.

1. The curves formed by the intersection of a plane with an infinite right-circular cone are called _____ _____.

2. Give the name of each curve shown below.

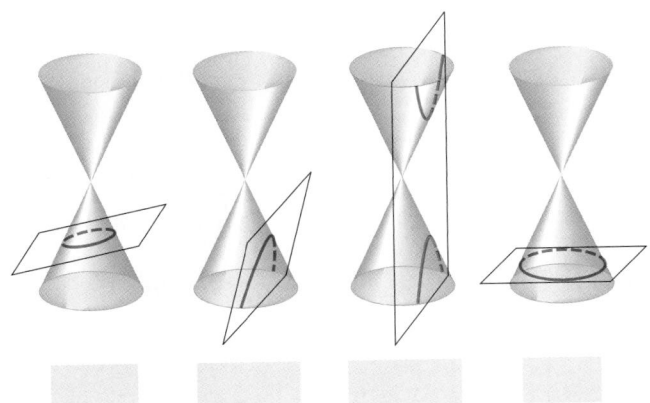

3. A _____ is the set of all points in a plane that are a fixed distance from a fixed point called its center. The fixed distance is called the _____.

4. A parabola is the set of all points in a plane that are equidistant from a fixed point and a fixed _____.

CONCEPTS

5. **a.** Write the standard form of the equation of a circle.

   b. Write the standard form of the equation of a circle with the center at the origin.

6. **a.** Find the center and the radius of the circle graphed on the right.

   b. Write the equation of the circle.

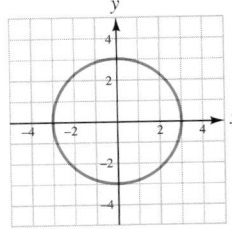

7. **a.** Find the center and the radius of the circle graphed on the right.

   b. Write the equation of the circle.

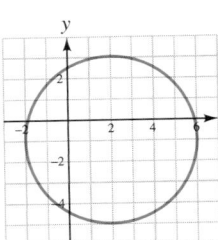

8. Fill in the blanks. To complete the square on $x^2 + 2x$ and on $y^2 - 6y$, what numbers must be added to each side of the equation?

$$x^2 + 2x + y^2 - 6y = 2$$

$$x^2 + 2x + \boxed{} + y^2 - 6y + \boxed{} = 2 + \boxed{} + \boxed{}$$

9. **a.** What is the standard form of the equation of a parabola opening upward or downward?

   b. What is the standard form of the equation of a parabola opening to the right or left?

10. Fill in the blanks.

   a. To complete the square on the right side, what should be factored from the first two terms?

   $$x = 4y^2 + 16y + 9$$

   $$x = \boxed{}(y^2 + 4y) + 9$$

   b. To complete the square on $y^2 + 4y$, what should be added within the parentheses, and what should be subtracted outside the parentheses?

   $$x = 4(y^2 + 4y + \boxed{}) + 9 - \boxed{}$$

11. Determine whether the graph of each equation is a circle or a parabola.

   a. $x^2 + y^2 - 6x + 8y - 10 = 0$

   b. $y^2 - 2x + 3y - 9 = 0$

   c. $x^2 + 5x - y = 0$

   d. $x^2 + 12x + y^2 = 0$

12. Draw a parabola using the given facts.
   - Opens right
   - Passes through $(-2, 1)$
   - Vertex $(-3, 2)$
   - x-intercept $(1, 0)$

NOTATION

13. Find h, k, and r: $(x - 6)^2 + (y + 2)^2 = 9$

14. **a.** Find a, h, and k: $y = 6(x - 5)^2 - 9$

   b. Find a, h, and k: $x = -3(y + 2)^2 + 1$

GUIDED PRACTICE

Find the center and radius of each circle and graph it. See Example 1.

15. $x^2 + y^2 = 9$

16. $x^2 + y^2 = 16$

17. $x^2 + (y + 3)^2 = 1$

18. $(x + 4)^2 + y^2 = 1$

19. $(x + 3)^2 + (y - 1)^2 = 16$

20. $(x - 1)^2 + (y + 4)^2 = 9$

21. $x^2 + y^2 = 6$

22. $x^2 + y^2 = 10$

Write the equation of a circle in standard form with the following properties. See Example 2.

23. Center at the origin; radius 1

24. Center at the origin; radius 4

25. Center at $(6, 8)$; radius 5

26. Center at $(5, 3)$; radius 2

27. Center at $(-2, 6)$; radius 12

28. Center at $(5, -4)$; radius 6

29. Center at $(0, 0)$; radius $\dfrac{1}{4}$

30. Center at $(0, 0)$; radius $\dfrac{1}{3}$

31. Center at $\left(\dfrac{2}{3}, -\dfrac{7}{8}\right)$; radius $\sqrt{2}$

32. Center at $(-0.7, -0.2)$; radius $\sqrt{11}$

33. Center at the origin; diameter $4\sqrt{2}$

34. Center at the origin; diameter $8\sqrt{3}$

Write each equation of a circle in standard form and graph it. Give the coordinates of its center and give the radius. See Example 3.

35. $x^2 + y^2 - 2x + 4y = -1$

36. $x^2 + y^2 + 6x - 4y = -12$

37. $x^2 + y^2 + 4x + 2y = 4$

38. $x^2 + y^2 + 8x + 2y = -13$

Write each equation of a parabola in standard form and graph it. Give the coordinates of the vertex. See Example 5.

39. $y = 2x^2 - 4x + 5$ 40. $y = x^2 + 4x + 5$

41. $y = -x^2 - 2x + 3$ 42. $y = -2x^2 - 4x$

Graph each equation of a parabola. Give the coordinates of the vertex. See Example 6.

43. $x = y^2$ 44. $x = 2y^2$

45. $x = 2(y + 1)^2 + 3$ 46. $x = 3(y - 2)^2 - 1$

Write each equation of a parabola in standard form and graph it. Give the coordinates of the vertex. See Example 7.

47. $x = y^2 - 2y + 5$ 48. $x = y^2 + 6y + 8$

49. $x = -3y^2 + 18y - 25$ 50. $x = -2y^2 + 4y + 1$

Use a graphing calculator to graph each equation. (Hint: Solve for y and graph two functions.) See Using Your Calculator: Graphing Circles.

51. $x^2 + y^2 = 7$ 52. $x^2 + y^2 = 5$

53. $(x + 1)^2 + y^2 = 16$ 54. $x^2 + (y - 2)^2 = 4$

Use a graphing calculator to graph each equation. (Hint: Solve for y and graph two functions when necessary.)

55. $x = 2y^2$ 56. $x = y^2 - 4$

57. $x^2 - 2x + y = 6$ 58. $x = -2(y - 1)^2 + 2$

TRY IT YOURSELF

Write each equation in standard form, if it is not already so, and graph it. If the graph is a circle, give the coordinates of its center and its radius. If the graph is a parabola, give the coordinates of its vertex.

59. $x = y^2 - 6y + 4$ 60. $x = y^2 - 8y + 13$

61. $(x - 2)^2 + y^2 = 25$ 62. $x^2 + (y - 3)^2 = 25$

63. $x^2 + y^2 - 6x + 8y + 18 = 0$

64. $x^2 + y^2 - 4x + 4y - 3 = 0$

65. $y = 4x^2 - 16x + 17$ 66. $y = 4x^2 - 32x + 63$

67. $(x - 1)^2 + (y - 3)^2 = 15$ 68. $(x + 1)^2 + (y + 1)^2 = 8$

69. $x = -y^2 + 1$ 70. $x = -y^2 - 5$

71. $(x - 2)^2 + (y - 4)^2 = 36$ 72. $(x - 3)^2 + (y - 2)^2 = 36$

73. $x = -\dfrac{1}{4}y^2$ 74. $x = 4y^2$

75. $x = -6(y - 1)^2 + 3$ 76. $x = -6(y + 1)^2 - 4$

77. $x^2 + y^2 + 2x - 8 = 0$ 78. $x^2 + y^2 - 4y = 12$

79. $x = \dfrac{1}{2}y^2 + 2y$ 80. $x = -\dfrac{1}{3}y^2 - 2y$

81. $y = -4(x + 5)^2 + 5$ 82. $y = -4(x - 4)^2 - 4$

APPLICATIONS

83. BROADCAST RANGES Radio stations applying for licensing may not use the same frequency if their broadcast areas overlap. One station's coverage is bounded by $x^2 + y^2 - 8x - 20y + 16 = 0$, and the other's by $x^2 + y^2 + 2x + 4y - 11 = 0$. May they be licensed for the same frequency?

84. MESHING GEARS For design purposes, the large gear is described by the circle $x^2 + y^2 = 16$. The smaller gear is a circle centered at $(7, 0)$ and tangent to the larger circle. Find the equation of the smaller gear.

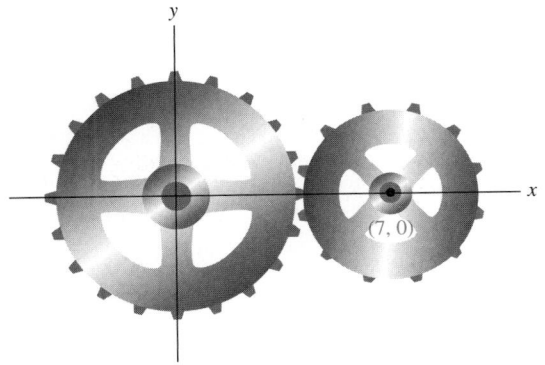

85. from Campus to Careers
   Traffic Engineer

Suppose you are a traffic engineer and you are designing two sections of a new freeway so that they join with a curve that is one-quarter of a circle, as shown. The equation of the circle is $x^2 + y^2 - 10x - 12y + 52 = 0$, where distances are measured in miles.

a. How far from City Hall will the new freeway intersect State Street?

b. How far from City Hall will the new freeway intersect Highway 60?

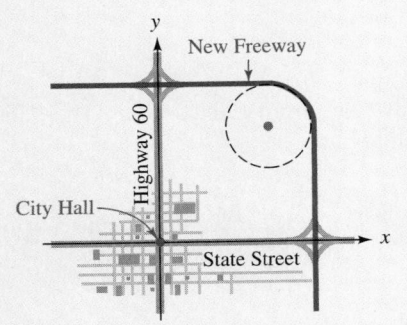

86. WALKWAYS The walkway shown is bounded by the two circles $x^2 + y^2 = 2,500$ and $(x - 10)^2 + y^2 = 900$, measured in feet. Find the largest and the smallest width of the walkway.

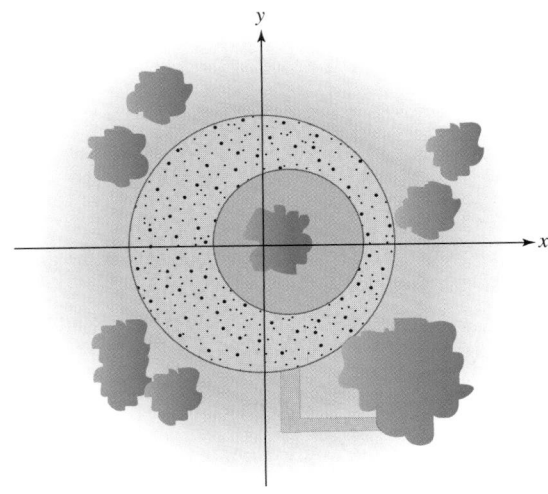

87. PROJECTILES The cannonball in the illustration follows the parabolic path $y = 30x - x^2$. How far short of the castle does it land?

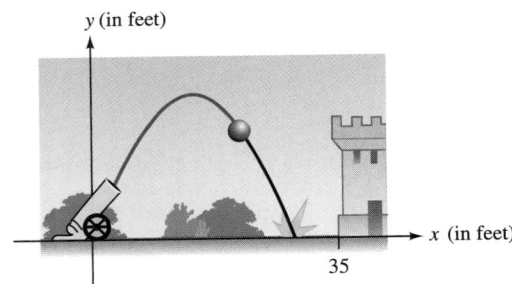

88. PROJECTILES In Exercise 87, how high does the cannonball get?

89. COMETS If the orbit of the comet is approximated by the equation $2y^2 - 9x = 18$, how far is it from the sun at the vertex V of the orbit? Distances are measured in astronomical units (AU).

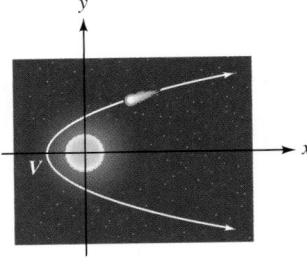

90. SATELLITE ANTENNAS   The cross section of the satellite antenna in the illustration is a parabola given by the equation $y = \frac{1}{16}x^2$, with distances measured in feet. If the dish is 8 feet wide, how deep is it?

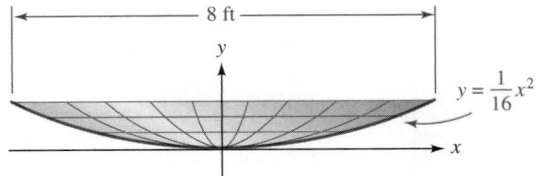

WRITING

91. Explain how to decide from its equation whether the graph of a parabola opens up, down, right, or left.

92. From the equation of a circle, explain how to determine the radius and the coordinates of the center.

93. On the day of an election, the following warning was posted in front of a school. Explain what it means.

> *No electioneering within a 1,000-foot radius of this polling place.*

94. What is meant by the *turning radius* of a truck?

REVIEW

Solve each equation.

95. $|3x - 4| = 11$

96. $\left|\dfrac{4 - 3x}{5}\right| = 12$

97. $|3x + 4| = |5x - 2|$

98. $|6 - 4x| = |x + 2|$

CHALLENGE PROBLEMS

99. Could the intersection of a plane with an infinite right-circular cone as shown on page 939 be a single point? If so, draw a picture that illustrates this.

100. Under what conditions will the graph of $x = a(y - k)^2 + h$ have no y-intercepts?

101. Write the equation of a circle with a diameter whose endpoints are at $(-2, -6)$ and $(8, 10)$.

102. Write the equation of a circle with a diameter whose endpoints are at $(-5, 4)$ and $(7, -3)$.

SECTION 13.2
The Ellipse

Objectives

  1 Define an ellipse.
  2 Graph ellipses centered at the origin.
  3 Graph ellipses centered at (h, k).
  4 Solve problems involving ellipses.

A third conic section is an oval-shaped curve called an *ellipse*. Ellipses can be nearly round and look almost like a circle, or they can be long and narrow. In this section, we will learn how to construct ellipses and how to graph equations that represent ellipses.

1 **Define an Ellipse.**

To define a circle, we considered a fixed distance from a fixed point. The definition of an ellipse involves *two* distances from *two* fixed points.

Definition of an Ellipse	An **ellipse** is the set of all points in a plane for which the sum of the distances from two fixed points is a constant.

The figure on the next page illustrates that any point on an ellipse is a constant distance $d_1 + d_2$ from two fixed points, each of which is called a **focus**. Midway between the **foci** is the **center** of the ellipse.

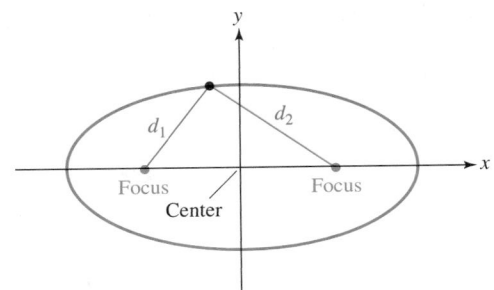

We can construct an ellipse by placing two thumbtacks fairly close together to serve as foci. We then tie each end of a piece of string to a thumbtack, catch the loop with the point of a pencil, and (keeping the string taut) draw the ellipse.

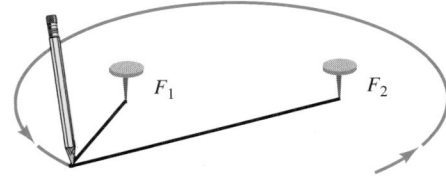

2 **Graph Ellipses Centered at the Origin.**

The definition of an ellipse can be used to develop the standard equation of an ellipse. To learn more about the derivation, see Problem 70 in the Challenge Problem section of the Study Set.

Equation of an Ellipse Centered at the Origin	The **standard form of the equation of an ellipse** that is symmetric with respect to both axes and centered at $(0, 0)$ is $$\frac{x^2}{a^2} + \frac{y^2}{b^2} = 1 \quad \text{where } a > 0 \text{ and } b > 0$$

To graph an ellipse centered at the origin, it is helpful to know the intercepts of the graph. To find the x-intercepts of the graph of

$$\frac{x^2}{a^2} + \frac{y^2}{b^2} = 1$$

we let $y = 0$ and solve for x.

$$\frac{x^2}{a^2} + \frac{0^2}{b^2} = 1 \qquad \text{Substitute 0 for } y.$$

$$\frac{x^2}{a^2} + 0 = 1 \qquad \text{Simplify: } \frac{0^2}{b^2} = 0.$$

$$x^2 = a^2 \qquad \text{Simplify and multiply both sides by } a^2.$$

$$x = \pm a \qquad \text{Use the square root property.}$$

The x-intercepts are $(a, 0)$ and $(-a, 0)$.

To find the y-intercepts of the graph, we can let $x = 0$ and solve for y.

$$\frac{0^2}{a^2} + \frac{y^2}{b^2} = 1 \qquad \text{Substitute 0 for } x.$$

$$0 + \frac{y^2}{b^2} = 1 \qquad \text{Simplify: } \frac{0^2}{a^2} = 0.$$

$$y^2 = b^2 \qquad \text{Simplify and multiply both sides by } b^2.$$

$$y = \pm b \qquad \text{Use the square root property.}$$

The y-intercepts are $(0, b)$ and $(0, -b)$.

In general, we have the following results.

For $\frac{x^2}{a^2} + \frac{y^2}{b^2} = 1$, if $a > b$, the ellipse is horizontal, as shown in figure (a). If $b > a$, the ellipse is vertical, as shown in figure (b). The points V_1 and V_2 are called the **vertices** of the ellipse. The line segment joining the vertices is called the **major axis,** and its midpoint is called the **center** of the ellipse. The line segment whose endpoints are on the ellipse and that is perpendicular to the major axis at the center is called the **minor axis** of the ellipse.

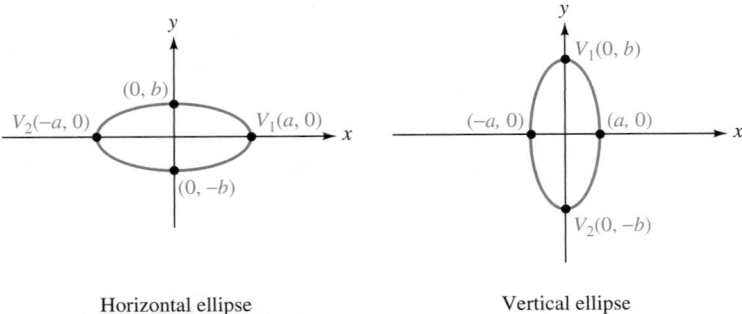

Horizontal ellipse
(a)

Vertical ellipse
(b)

EXAMPLE 1   Graph: $\dfrac{x^2}{36} + \dfrac{y^2}{9} = 1$

Strategy   This equation is in standard $\frac{x^2}{a^2} + \frac{y^2}{b^2} = 1$ form. We will identify a and b.

Why   Once we know a and b, we can determine the intercepts of the graph of the ellipse.

Solution   The color highlighting shows how to compare the given equation to the standard form to find a and b.

$$\frac{x^2}{36} + \frac{y^2}{9} = 1 \qquad \frac{x^2}{a^2} + \frac{y^2}{b^2} = 1$$

Since $a^2 = 36$, it follows that $a = 6$.   Since $b^2 = 9$, it follows that $b = 3$.

The x-intercepts are $(a, 0)$ and $(-a, 0)$, or $(6, 0)$ and $(-6, 0)$. The y-intercepts are $(0, b)$ and $(0, -b)$, or $(0, 3)$ and $(0, -3)$. Using these four points as a guide, we draw an oval-shaped curve through them, as shown in figure (a). The result is a horizontal ellipse.

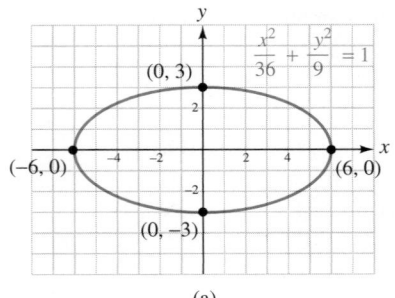

(a)

$$\frac{x^2}{36} + \frac{y^2}{9} = 1$$

x	y	
2	$\pm 2\sqrt{2}$	$\rightarrow (2, \pm 2\sqrt{2})$
4	$\pm \sqrt{5}$	$\rightarrow (4, \pm \sqrt{5})$

↑
Approximate the radicals to graph.

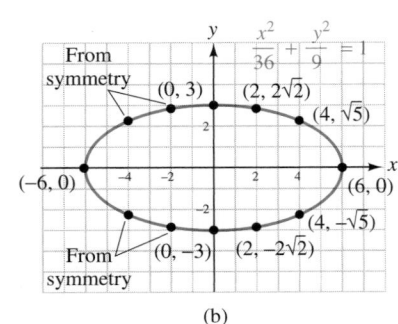

(b)

To increase the accuracy of the graph, we can find additional ordered pairs that satisfy the equation and plot them. For example, if $x = 2$, we have

$$\frac{2^2}{36} + \frac{y^2}{9} = 1 \qquad \text{Substitute 2 for x in the equation of the ellipse.}$$

$$36\left(\frac{4}{36} + \frac{y^2}{9}\right) = 36(1) \qquad \text{To clear the fractions, multiply both sides by the LCD, 36.}$$

$$4 + 4y^2 = 36 \qquad \text{Distribute the multiplication by 36 and simplify.}$$

$$y^2 = 8 \qquad \text{Subtract 4 from both sides and divide both sides by 4.}$$

$$y = \pm\sqrt{8} \qquad \text{Use the square root property.}$$

$$y = \pm 2\sqrt{2} \qquad \text{Simplify the radical.}$$

Since two values of y, $2\sqrt{2}$ and $-2\sqrt{2}$, correspond to the x-value 2, we have found two points on the ellipse: $\left(2, 2\sqrt{2}\right)$ and $\left(2, -2\sqrt{2}\right)$.

In a similar way, we can find the corresponding values of y for the x-value 4. In figure (b) we record these ordered pairs in a table, plot them, use symmetry with respect to the y-axis to plot four other points, and draw the graph of the ellipse.

 Self Check 1   Graph: $\frac{x^2}{49} + \frac{y^2}{25} = 1$

Now Try   Problem 17

EXAMPLE 2   Graph: $16x^2 + y^2 = 16$

Strategy   We will write the equation in standard $\frac{x^2}{a^2} + \frac{y^2}{b^2} = 1$ form.

Why   When the equation is in standard form, we will be able to identify the center and the intercepts of the graph of the ellipse.

Solution   The given equation is not in standard form. To write it in standard form with 1 on the right side, we divide both sides by 16.

$$16x^2 + y^2 = 16$$

$$\frac{16x^2}{16} + \frac{y^2}{16} = \frac{16}{16} \qquad \text{Divide both sides by 16.}$$

$$\frac{x^2}{1} + \frac{y^2}{16} = 1 \qquad \text{Simplify: } \tfrac{16x^2}{16} = x^2 = \tfrac{x^2}{1} \text{ and } \tfrac{16}{16} = 1.$$

> **Success Tip**
> Although the term $\frac{16x^2}{16}$ simplifies to x^2, we write it as the fraction $\frac{x^2}{1}$ so that it has the form $\frac{x^2}{a^2}$.

To determine a and b, we can write the equation in the form

$$\frac{x^2}{1^2} + \frac{y^2}{4^2} = 1 \qquad \text{To find a, write 1 as } 1^2. \text{ To find b, write 16 as } 4^2.$$

Since a^2 (the denominator of x^2) is 1^2, it follows that $a = 1$, and since b^2 (the denominator of y^2) is 4^2, it follows that $b = 4$. Thus, the x-intercepts of the graph are $(1, 0)$ and $(-1, 0)$ and the y-intercepts are $(0, 4)$ and $(0, -4)$. We use these four points as guides to sketch the graph of the ellipse, as shown. The result is a vertical ellipse.

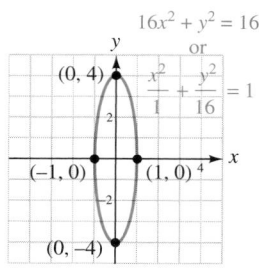

> **Self Check 2**   Graph: $9x^2 + y^2 = 9$
>
> **Now Try**   **Problem 21**

❸ Graph Ellipses Centered at (*h*, *k*).

Not all ellipses are centered at the origin. As with the graphs of circles and parabolas, the graph of an ellipse can be translated horizontally and vertically.

The Equation of an Ellipse Centered at (*h*, *k*)	The **standard form of the equation of a horizontal or vertical ellipse** centered at (h, k) is $$\frac{(x-h)^2}{a^2} + \frac{(y-k)^2}{b^2} = 1 \quad \text{where } a > 0 \text{ and } b > 0$$ For a horizontal ellipse, a is the distance from the center to a vertex. For a vertical ellipse, b is the distance from the center to a vertex.

EXAMPLE 3   Graph: $\dfrac{(x-2)^2}{16} + \dfrac{(y+3)^2}{25} = 1$

Strategy   The equation is in standard $\dfrac{(x-h)^2}{a^2} + \dfrac{(y-k)^2}{b^2} = 1$ form. We will identify h, k, a, and b.

Why   If we know h, k, a, and b, we can graph the ellipse.

Solution   To determine h, k, a, and b, we write the equation in the form

$$\frac{(x-2)^2}{4^2} + \frac{[y-(-3)]^2}{5^2} = 1 \qquad \begin{array}{l}\text{To find } k, \text{ write } y + 3 \text{ as } y - (-3).\\ \text{To find } a, \text{ write } 16 \text{ as } 4^2. \text{ To find } b, \text{ write } 25 \text{ as } 5^2.\end{array}$$

We find the center of the ellipse in the same way we would find the center of a circle, by examining $(x-2)^2$ and $(y+3)^2$. Since $h = 2$ and $k = -3$, this is the equation of an ellipse centered at $(h, k) = (2, -3)$. From the denominators, 4^2 and 5^2, we find that $a = 4$ and $b = 5$. Because $b > a$, it is a vertical ellipse.

We first plot the center, as shown below. Since b is the distance from the center to a vertex for a vertical ellipse, we can locate the vertices by counting 5 units above and 5 units below the center. The vertices are the points $(2, 2)$ and $(2, -8)$.

To locate two more points on the ellipse, we use the fact that a is 4 and count 4 units to the left and to the right of the center. We see that the points $(-2, -3)$ and $(6, -3)$ are also on the graph.

Using these four points as guides, we draw the graph shown below.

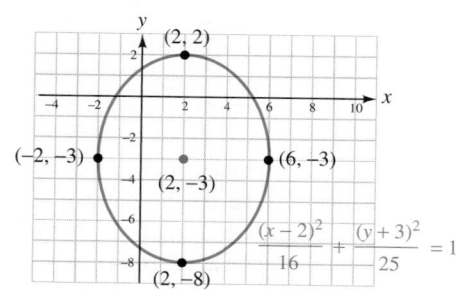

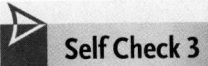

Self Check 3   Graph: $\dfrac{(x-1)^2}{9} + \dfrac{(y+2)^2}{16} = 1$

Now Try   **Problem 25**

Using Your Calculator   *Graphing Ellipses*

To use a graphing calculator to graph the equation from Example 3,

$$\frac{(x-2)^2}{16} + \frac{(y+3)^2}{25} = 1$$

we clear the equation of fractions and solve for y.

$$25(x-2)^2 + 16(y+3)^2 = 400 \qquad \text{Multiply both sides by 400.}$$

$$16(y+3)^2 = 400 - 25(x-2)^2 \qquad \begin{array}{l}\text{Subtract } 25(x-2)^2 \text{ from}\\ \text{both sides.}\end{array}$$

$$(y+3)^2 = \frac{400 - 25(x-2)^2}{16} \qquad \text{Divide both sides by 16.}$$

$$y + 3 = \pm\frac{\sqrt{400 - 25(x-2)^2}}{4} \qquad \text{Use the square root property.}$$

$$y = -3 \pm \frac{\sqrt{400 - 25(x-2)^2}}{4} \qquad \text{Subtract 3 from both sides.}$$

The previous equation represents two functions. On a calculator, we can graph them in a square window to get the ellipse shown below.

$$y = -3 + \frac{\sqrt{400 - 25(x-2)^2}}{4} \qquad \text{and} \qquad y = -3 - \frac{\sqrt{400 - 25(x-2)^2}}{4}$$

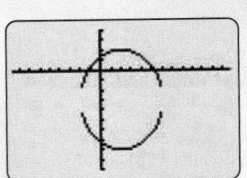

As we saw with circles, the two portions of the ellipse do not quite connect. This is because the graphs are nearly vertical there.

EXAMPLE 4   Graph: $4(x-2)^2 + 9(y-1)^2 = 36$

Strategy   We will write the equation in standard $\dfrac{(x-h)^2}{a^2} + \dfrac{(y-k)^2}{b^2} = 1$ form. Then we will identify h, k, a, and b.

Why   If we know h, k, a, and b, we can graph the ellipse.

Solution   This equation is not in standard form. To write it in standard form with 1 on the right side, we divide both sides by 36.

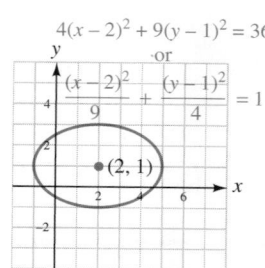

$$4(x-2)^2 + 9(y-1)^2 = 36$$
·or·
$$\frac{(x-2)^2}{9} + \frac{(y-1)^2}{4} = 1$$

$$4(x - 2)^2 + 9(y - 1)^2 = 36$$

$$\frac{4(x-2)^2}{36} + \frac{9(y-1)^2}{36} = \frac{36}{36} \qquad \text{Divide both sides by 36.}$$

$$\frac{(x-2)^2}{9} + \frac{(y-1)^2}{4} = 1 \qquad \text{Simplify: } \frac{4}{36} = \frac{1}{9}, \frac{9}{36} = \frac{1}{4}, \text{ and } \frac{36}{36} = 1.$$

This is the standard form of the equation of a horizontal ellipse, centered at (2, 1), with $a = 3$ and $b = 2$. The graph of the ellipse is shown in the margin.

 Self Check 4   Graph: $12(x - 1)^2 + 3(y + 1)^2 = 48$

Now Try   **Problem 29**

 4   **Solve Problems Involving Ellipses.**

EXAMPLE 5   *Landscape Design.* A landscape architect is designing an elliptical pool that will fit in the center of a 20-by-30-foot rectangular garden, leaving 5 feet of clearance on all sides, as shown in the illustration below. Find the equation of the ellipse.

Strategy   We will establish a coordinate system with its origin at the center of the garden. Then we will determine the *x*- and *y*-intercepts of the edge of the pool.

Why   If we know the *x*- and *y*-intercepts of the graph of the edge of the elliptical pool, we can use that information to write its equation.

Solution   We place the rectangular garden in the coordinate system shown below. To maintain 5 feet of clearance at the ends of the ellipse, the *x*-intercepts must be the points (10, 0) and (−10, 0). Similarly, the *y*-intercepts are the points (0, 5) and (0, −5).

Since the ellipse is centered at the origin, its equation has the form

$$\frac{x^2}{a^2} + \frac{y^2}{b^2} = 1$$

with $a = 10$ and $b = 5$. Thus, the equation of the boundary of the pool is

$$\frac{x^2}{100} + \frac{y^2}{25} = 1$$

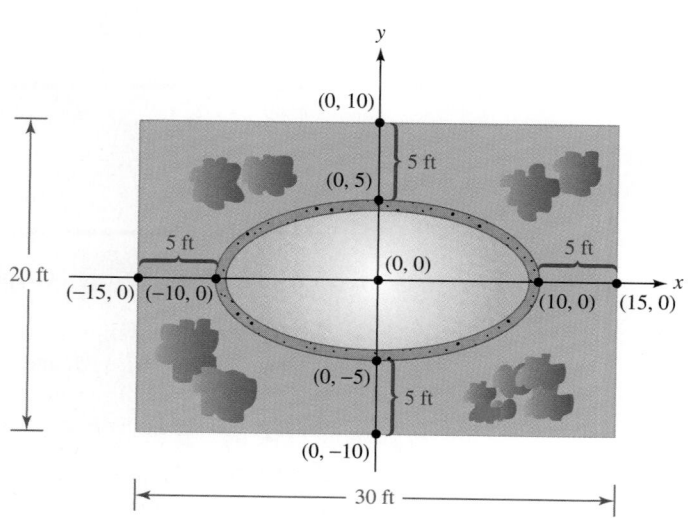

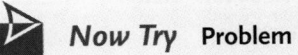

Now Try Problem 53

Ellipses, like parabolas, have reflective properties that are used in many practical applications. For example, any light or sound originating at one focus of an ellipse is reflected by the interior of the figure to the other focus.

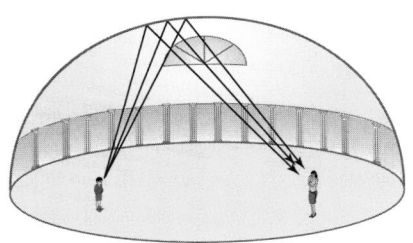

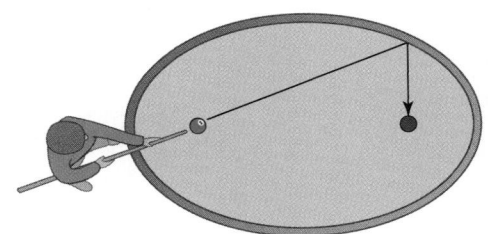

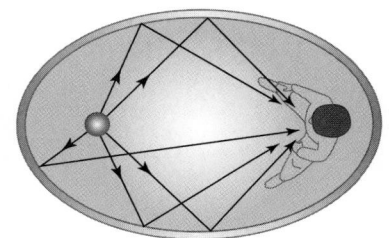

Whispering galleries

In an elliptical dome, even the slightest whisper made by a person standing at one focus can be heard by a person standing at the other focus.

Elliptical billiards tables

When a ball is shot from one focus, it will rebound off the side of the table into a pocket located at the other focus.

Treatment for kidney stones

The patient is positioned in an elliptical tank of water so that the kidney stone is at one focus. High-intensity sound waves generated at another focus are reflected to the stone to shatter it.

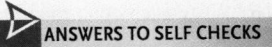

ANSWERS TO SELF CHECKS

1.

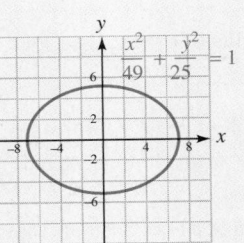

$$\frac{x^2}{49} + \frac{y^2}{25} = 1$$

2.

$9x^2 + y^2 = 9$ or

$$\frac{x^2}{1} + \frac{y^2}{9} = 1$$

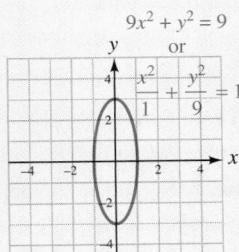

3.

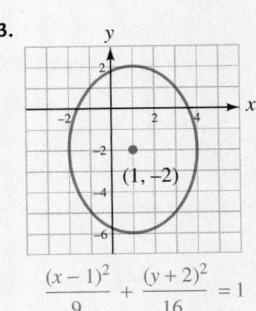

$(1, -2)$

$$\frac{(x-1)^2}{9} + \frac{(y+2)^2}{16} = 1$$

4.

$12(x-1)^2 + 3(y+1)^2 = 48$ or

$$\frac{(x-1)^2}{4} + \frac{(y+1)^2}{16} = 1$$

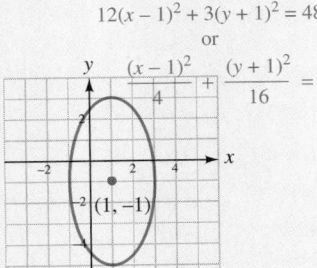

$(1, -1)$

STUDY SET
13.2

VOCABULARY

Fill in the blanks.

1. The curve graphed below is an _____.

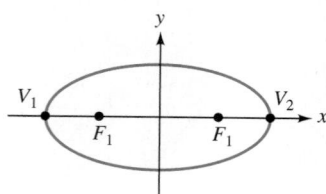

2. An _____ is the set of all points in a plane for which the sum of the distances from two fixed points is a constant.

3. In the graph above, F_1 and F_2 are the _____ of the ellipse. Each one is called a _____ of the ellipse.

4. In the graph above, V_1 and V_2 are the _____ of the ellipse. Each one is called a _____ of the ellipse.

5. The line segment joining the vertices of an ellipse is called the _____ axis of the ellipse.

6. The midpoint of the major axis of an ellipse is the _____ of the ellipse.

CONCEPTS

7. Write the standard form of the equation of an ellipse centered at the origin and symmetric to both axes.

8. Write the standard form of the equation of a horizontal or vertical ellipse centered at (h, k).

9. Find the x- and the y-intercepts of the graph of $\frac{x^2}{a^2} + \frac{y^2}{b^2} = 1$.

10. **a.** Find the center of the ellipse graphed on the right. What are a and b?

    b. Is the ellipse horizontal or vertical?

    c. Find the equation of the ellipse.

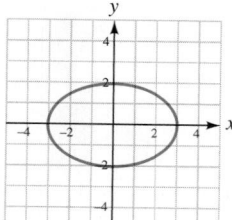

11. **a.** Find the center of the ellipse graphed on the right. What are a and b?

    b. Is the ellipse horizontal or vertical?

    c. Find the equation of the ellipse.

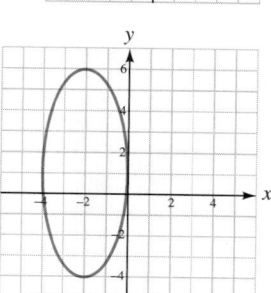

12. Find two points on the graph of $\frac{x^2}{16} + \frac{y^2}{4} = 1$ by letting $x = 2$ and finding the corresponding values of y.

13. Divide both sides of the equation by 64 and write the equation in standard form:

$$4(x - 1)^2 + 64(y + 5)^2 = 64$$

14. Determine whether the graph of each equation is a circle, a parabola, or an ellipse.

    a. $x = y^2 - 2y + 10$

    b. $\dfrac{x^2}{49} + \dfrac{y^2}{64} = 1$

    c. $(x - 3)^2 + (y + 4)^2 = 25$

    d. $2(x - 1)^2 + 8(y + 5)^2 = 32$

NOTATION

15. Find h, k, a, and b: $\dfrac{(x + 8)^2}{100} + \dfrac{(y - 6)^2}{144} = 1$

16. Write each denominator in the equation $\dfrac{x^2}{81} + \dfrac{y^2}{49} = 1$ as the square of a number.

GUIDED PRACTICE

Graph each equation. See Example 1.

17. $\dfrac{x^2}{25} + \dfrac{y^2}{4} = 1$ 18. $\dfrac{x^2}{16} + \dfrac{y^2}{9} = 1$

19. $\dfrac{x^2}{4} + \dfrac{y^2}{9} = 1$ 20. $\dfrac{x^2}{16} + \dfrac{y^2}{25} = 1$

Graph each equation. See Example 2.

21. $x^2 + 9y^2 = 9$ 22. $25x^2 + 9y^2 = 225$

23. $16x^2 + 4y^2 = 64$ 24. $4x^2 + 9y^2 = 36$

Graph each equation. See Example 3.

25. $\dfrac{(x - 2)^2}{9} + \dfrac{(y - 1)^2}{4} = 1$ 26. $\dfrac{(x - 1)^2}{9} + \dfrac{(y - 3)^2}{4} = 1$

27. $\dfrac{(x + 2)^2}{64} + \dfrac{(y - 2)^2}{100} = 1$ 28. $\dfrac{(x - 6)^2}{36} + \dfrac{(y + 6)^2}{144} = 1$

Graph each equation. See Example 4.

29. $(x + 1)^2 + 4(y + 2)^2 = 4$

30. $25(x + 1)^2 + 9y^2 = 225$

31. $16(x - 2)^2 + 4(y + 4)^2 = 256$

32. $4(x - 2)^2 + 9(y - 4)^2 = 144$

Use a graphing calculator to graph each equation. See Using Your Calculator: Graphing Ellipses.

33. $\dfrac{x^2}{9} + \dfrac{y^2}{4} = 1$

34. $x^2 + 16y^2 = 16$

35. $\dfrac{x^2}{4} + \dfrac{(y-1)^2}{9} = 1$

36. $\dfrac{(x+1)^2}{9} + \dfrac{(y-2)^2}{4} = 1$

TRY IT YOURSELF

Write each equation in standard form, if it is not already so, and graph it. The problems include equations that describe circles, parabolas, and ellipses.

37. $(x+1)^2 + (y-2)^2 = 16$

38. $(x-3)^2 + (y+1)^2 = 25$

39. $\dfrac{x^2}{16} + \dfrac{y^2}{1} = 1$

40. $\dfrac{x^2}{1} + \dfrac{y^2}{9} = 1$

41. $x = \dfrac{1}{2}(y-1)^2 - 2$

42. $x = -\dfrac{1}{2}(y+4)^2 + 5$

43. $x^2 + y^2 - 25 = 0$

44. $x^2 = 36 - y^2$

45. $x^2 = 100 - 4y^2$

46. $x^2 = 36 - 4y^2$

47. $y = -3x^2 - 24x - 43$

48. $y = 5x^2 - 60x + 173$

49. $x^2 + y^2 - 2x + 4y - 4 = 0$

50. $x^2 + y^2 + 4x + 6y + 9 = 0$

51. $9(x-1)^2 + 4(y+2)^2 = 36$

52. $16(x-5)^2 + 25(y-4)^2 = 400$

APPLICATIONS

53. FITNESS EQUIPMENT   With elliptical cross-training equipment, the feet move through the natural elliptical pattern that one experiences when walking, jogging, or running. Write the equation of the elliptical pattern shown below.

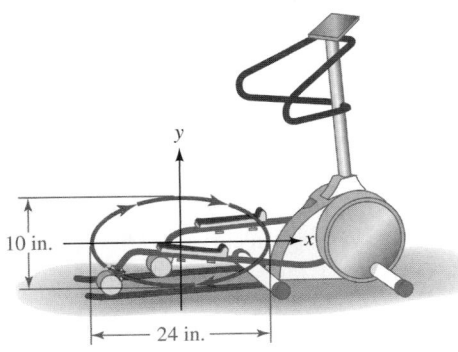

54. DESIGNING AN UNDERPASS   The arch of an underpass is a part of an ellipse. Find the equation of the ellipse.

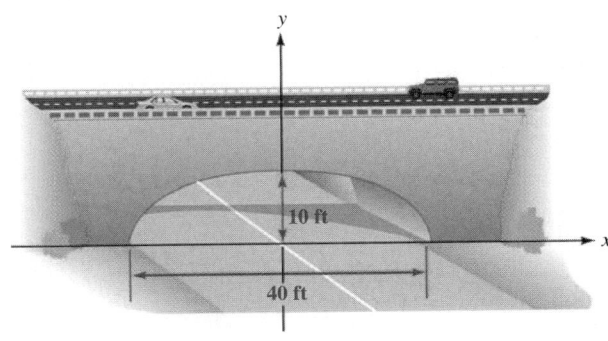

55. CALCULATING CLEARANCE   Find the height of the elliptical arch in Exercise 54 at a point 10 feet from the center of the roadway that passes under the arch.

56. POOL TABLES   Find the equation of the outer edge of the elliptical pool table shown below.

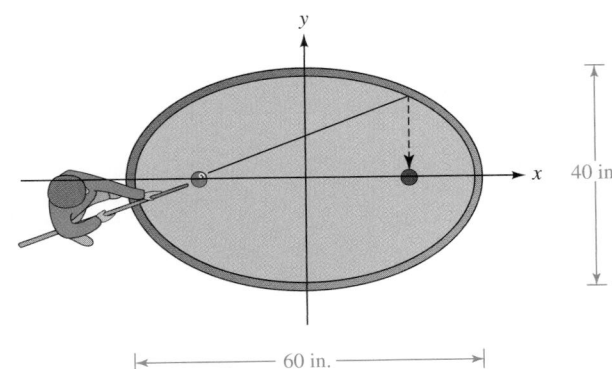

57. AREA OF AN ELLIPSE   The area A bounded by the following ellipse is given by $A = \pi ab$.

$$\dfrac{x^2}{a^2} + \dfrac{y^2}{b^2} = 1$$

Find the area bounded by the ellipse described by $9x^2 + 16y^2 = 144$.

58. AREA OF A TRACK   The elliptical track shown in the figure is bounded by the ellipses $4x^2 + 9y^2 = 576$ and $9x^2 + 25y^2 = 900$. Find the area of the track. (See Exercise 57.)

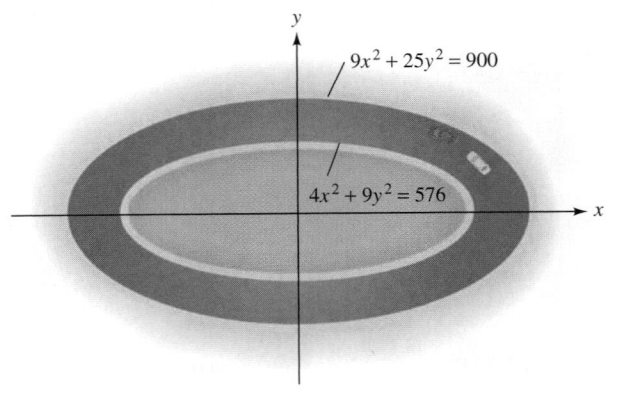

WRITING

59. What is an ellipse?

60. Explain the difference between the focus of an ellipse and the vertex of an ellipse.

61. Compare the graphs of $\frac{x^2}{81} + \frac{y^2}{64} = 1$ and $\frac{x^2}{64} + \frac{y^2}{81} = 1$. Do they have any similarities?

62. What are the reflective properties of an ellipse?

REVIEW

Find each product.

63. $3x^{-2}y^2(4x^2 + 3y^{-2})$

64. $(2a^{-2} - b^{-2})(2a^{-2} + b^{-2})$

Simplify each expression.

65. $\dfrac{x^{-2} + y^{-2}}{x^{-2} - y^{-2}}$

66. $\dfrac{2x^{-3} - 2y^{-3}}{4x^{-3} + 4y^{-3}}$

CHALLENGE PROBLEMS

67. What happens to the graph of the equation $\frac{x^2}{a^2} + \frac{y^2}{b^2} = 1$ when $a = b$?

68. Graph: $9x^2 + 4y^2 = 1$

69. Write the equation $9x^2 + 4y^2 - 18x + 16y = 11$ in the standard form of the equation of an ellipse.

70. Let the foci of an ellipse be $(c, 0)$ and $(-c, 0)$. Suppose that the sum of the distances from any point (x, y) on the ellipse to the two foci is the constant $2a$. Show that the equation for the ellipse is

$$\frac{x^2}{a^2} + \frac{y^2}{a^2 - c^2} = 1$$

Then let $b^2 = a^2 - c^2$ to obtain the standard form of the equation of an ellipse.

SECTION 13.3
The Hyperbola

Objectives

1. Define a hyperbola.
2. Graph hyperbolas centered at the origin.
3. Graph hyperbolas centered at (h, k).
4. Graph equations of the form $xy = k$.
5. Solve problems involving hyperbolas.

The final conic section that we will discuss, the *hyperbola,* is a curve that has two branches. In this section, we will learn how to graph equations that represent hyperbolas.

1   **Define a Hyperbola.**

Ellipses and hyperbolas have completely different shapes, but their definitions are similar. Instead of the *sum* of distances, the definition of a hyperbola involves a *difference* of distances.

Definition of a Hyperbola	A **hyperbola** is the set of all points in a plane for which the difference of the distances from two fixed points is a constant.

The figure below illustrates that any point P on the hyperbola is a constant distance $d_1 - d_2$ from two fixed points, each of which is called a **focus**. Midway between the **foci** is the **center** of the hyperbola.

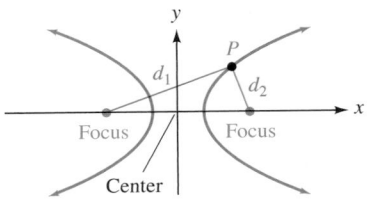
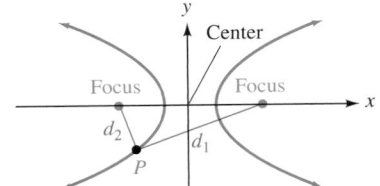

② Graph Hyperbolas Centered at the Origin.

The graph of the equation

$$\frac{x^2}{25} - \frac{y^2}{9} = 1$$

is a hyperbola. To graph the equation, we make a table of solutions that satisfy the equation, plot each point, and join them with a smooth curve.

$$\frac{x^2}{25} - \frac{y^2}{9} = 1$$

x	y	
-7	± 2.9	$\longrightarrow (-7, \pm 2.9)$
-6	± 2.0	$\longrightarrow (-6, \pm 2.0)$
-5	0	$\longrightarrow (-5, 0)$
5	0	$\longrightarrow (5, 0)$
6	± 2.0	$\longrightarrow (6, \pm 2.0)$
7	± 2.9	$\longrightarrow (7, \pm 2.9)$

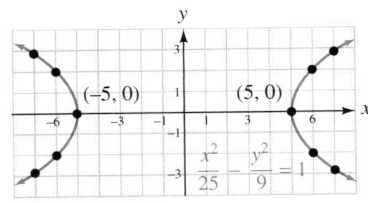

This graph is centered at the origin and intersects the x-axis at $(5, 0)$ and $(-5, 0)$. We also note that the graph does not intersect the y-axis.

It is possible to draw a hyperbola without plotting points. For example, if we want to graph the hyperbola with an equation of

$$\frac{x^2}{a^2} - \frac{y^2}{b^2} = 1$$

we first find the x- and y-intercepts. To find the x-intercepts, we let $y = 0$ and solve for x:

$$\frac{x^2}{a^2} - \frac{0^2}{b^2} = 1$$

$$x^2 = a^2$$

$$x = \pm a \quad \text{Use the square root property.}$$

The hyperbola crosses the x-axis at the points $V_1(a, 0)$ and $V_2(-a, 0)$, called the **vertices** of the hyperbola.

To attempt to find the y-intercepts, we let $x = 0$ and solve for y:

$$\frac{0^2}{a^2} - \frac{y^2}{b^2} = 1$$

$$y^2 = -b^2$$

$$y = \pm\sqrt{-b^2}$$

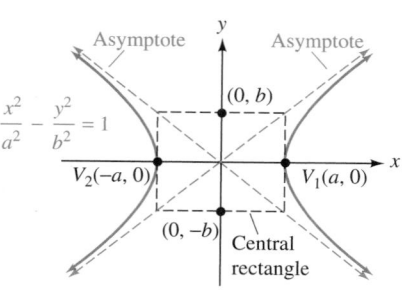

Since b^2 is always positive, $\sqrt{-b^2}$ is an imaginary number. This means that the hyperbola does not intersect the y-axis.

If we construct a rectangle, called the **central rectangle,** whose sides pass horizontally through $\pm b$ on the y-axis and vertically through $\pm a$ on the x-axis, the extended diagonals of the rectangle will be **asymptotes** of the hyperbola. As the hyperbola gets farther away from the origin, its branches get closer and closer to the asymptotes. The asymptotes are not part of the hyperbola, but they serve as a guide when drawing its graph. Since the slopes of the diagonals are $\frac{b}{a}$ and $-\frac{b}{a}$, the equations of the asymptotes are

> **The Language of Algebra**
> The central rectangle is also called the *fundamental rectangle.*

$$y = \frac{b}{a}x \quad \text{and} \quad y = -\frac{b}{a}x$$

Standard Form of the Equation of a Hyperbola Centered at the Origin and Intersecting the x-Axis	Any equation that can be written in the form $$\frac{x^2}{a^2} - \frac{y^2}{b^2} = 1$$ has a graph that is a hyperbola centered at the origin. The x-intercepts are the vertices $V_1(a, 0)$ and $V_2(-a, 0)$. There are no y-intercepts. The asymptotes of the hyperbola are the extended diagonals of the central rectangle, and their equations are $y = \frac{b}{a}x$ and $y = -\frac{b}{a}x$.	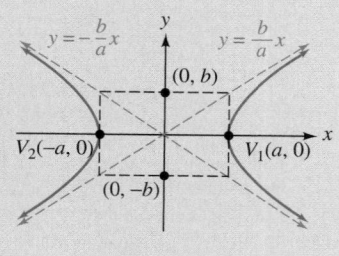

The branches of the hyperbola in previous discussions open to the left and to the right. It is possible for hyperbolas to have different orientations with respect to the x- and y-axes. For example, the branches of a hyperbola can open upward and downward. In that case, the following equation applies.

Standard Form of the Equation of a Hyperbola Centered at the Origin and Intersecting the y-Axis	Any equation that can be written in the form $$\frac{y^2}{a^2} - \frac{x^2}{b^2} = 1$$ has a graph that is a hyperbola centered at the origin. The y-intercepts are the vertices $V_1(0, a)$ and $V_2(0, -a)$. There are no x-intercepts. The asymptotes of the hyperbola are the extended diagonals of the central rectangle, and their equations are $y = \frac{a}{b}x$ and $y = -\frac{a}{b}x$.	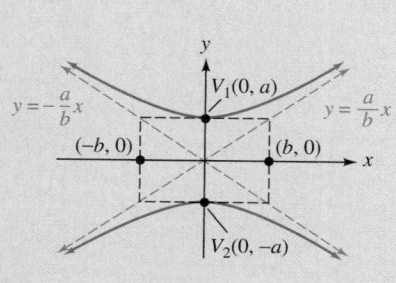

EXAMPLE 1   Graph: $\dfrac{x^2}{9} - \dfrac{y^2}{16} = 1$

Strategy   This equation is in standard $\dfrac{x^2}{a^2} - \dfrac{y^2}{b^2} = 1$ form. We will identify a and b.

Why   We can use a and b to find the vertices of the graph of the hyperbola and the location of the central rectangle.

Solution   The color highlighting shows how to compare the given equation with the standard form to find a and b.

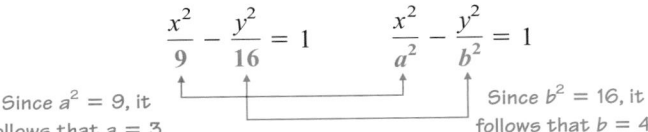

$$\dfrac{x^2}{9} - \dfrac{y^2}{16} = 1 \qquad \dfrac{x^2}{a^2} - \dfrac{y^2}{b^2} = 1$$

Since $a^2 = 9$, it follows that $a = 3$.   Since $b^2 = 16$, it follows that $b = 4$.

This is the standard form of the equation of a hyperbola, centered at the origin, that opens left and right. The x-intercepts are $(a, 0)$ and $(-a, 0)$, or $(3, 0)$ and $(-3, 0)$. They are also the vertices of the hyperbola.

To construct the central rectangle, we use the values of $a = 3$ and $b = 4$. The rectangle passes through $(3, 0)$ and $(-3, 0)$ on the x-axis, and $(0, 4)$ and $(0, -4)$ on the y-axis. We draw extended diagonal dashed lines through the rectangle to obtain the asymptotes and write their equations: $y = \frac{4}{3}x$ and $y = -\frac{4}{3}x$. Then we draw a smooth curve through each vertex that gets close to the asymptotes.

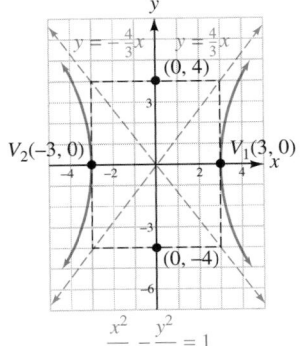

Self Check 1   Graph: $\dfrac{x^2}{25} - \dfrac{y^2}{4} = 1$

Now Try   **Problem 17**

EXAMPLE 2   Graph: $9y^2 - 4x^2 = 36$

Strategy   We will write the equation in standard $\dfrac{y^2}{a^2} - \dfrac{x^2}{b^2} = 1$ form.

Why   When the equation is in standard form, we will be able to identify the center and the vertices of the graph of the hyperbola and the location of the central rectangle.

Solution   To write the equation in standard form, we divide both sides by 36.

$$9y^2 - 4x^2 = 36$$

$$\dfrac{9y^2}{36} - \dfrac{4x^2}{36} = \dfrac{36}{36} \qquad \text{To get a 1 on the right side, divide both sides by 36.}$$

$$\dfrac{y^2}{4} - \dfrac{x^2}{9} = 1 \qquad \text{Simplify each fraction.}$$

This is the standard form of the equation of a hyperbola, centered at the origin, that opens up and down. The color highlighting shows how we compare the resulting equation to the standard form to find a and b.

$$\dfrac{y^2}{4} - \dfrac{x^2}{9} = 1 \qquad \dfrac{y^2}{a^2} - \dfrac{x^2}{b^2} = 1$$

Since $a^2 = 4$, it follows that $a = 2$.   Since $b^2 = 9$, it follows that $b = 3$.

Success Tip
The positive variable term in the standard form equation determines whether a hyperbola is vertical or horizontal. In this example, the positive variable term involves y, so the hyperbola is vertical.

$$\dfrac{y^2}{4} - \dfrac{x^2}{9} = 1$$

The *y*-intercepts are $(0, a)$ and $(0, -a)$, or $(0, 2)$ and $(0, -2)$. They are also the vertices of the hyperbola.

Since $a = 2$ and $b = 3$, the central rectangle passes through $(0, 2)$ and $(0, -2)$, and $(3, 0)$ and $(-3, 0)$. We draw its extended diagonals and sketch the hyperbola.

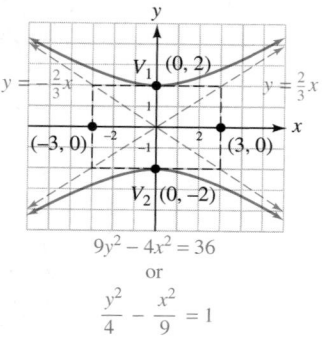

$$9y^2 - 4x^2 = 36$$
or
$$\frac{y^2}{4} - \frac{x^2}{9} = 1$$

 Self Check 2   Graph: $16y^2 - x^2 = 16$
Now Try   **Problem 21**

We can determine whether an equation, when graphed, will be a circle, a parabola, an ellipse, or a hyperbola by examining its variable terms.

$x^2 + y^2 = 16$ — With the variable terms on the same side of the equation, we see that the coefficients of the squared terms are the same. The graph is a circle.

$4x^2 + 9y^2 = 144$ — With the variable terms on the same side of the equation, we see that the coefficients of the squared terms are different, but have the same sign. The graph is an ellipse.

$4x^2 - 9y^2 = 144$ — With the variable terms on the same side of the equation, we see that the coefficients of the squared terms have different signs. The graph is a hyperbola.

$x = y^2 + y - 16$ — Since one variable is squared and the other is not, the graph is a parabola.

Using Your Calculator

Graphing Hyperbolas

To graph $\frac{x^2}{9} - \frac{y^2}{16} = 1$ from Example 1 using a graphing calculator, we follow the same procedure that we used for circles and ellipses. To write the equation as two functions, we solve for *y* to get $y = \pm\frac{\sqrt{16x^2 - 144}}{3}$. Then we graph the following two functions in a square window setting to get the graph of the hyperbola shown below.

$$y = \frac{\sqrt{16x^2 - 144}}{3} \quad \text{and} \quad y = -\frac{\sqrt{16x^2 - 144}}{3}$$

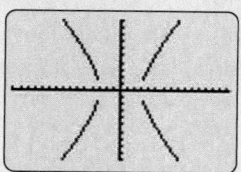

Graph Hyperbolas Centered at (h, k).

If a hyperbola is centered at a point with coordinates (h, k), the following equations apply.

Standard Form of the Equation of a Hyperbola Centered at (h, k)

Any equation that can be written in the form

$$\frac{(x - h)^2}{a^2} - \frac{(y - k)^2}{b^2} = 1$$

is a hyperbola that has its center at (h, k) and opens left and right.
Any equation of the form

$$\frac{(y - k)^2}{a^2} - \frac{(x - h)^2}{b^2} = 1$$

is a hyperbola that has its center at (h, k) and opens up and down.

EXAMPLE 3   Graph:

a. $\dfrac{(x - 3)^2}{16} - \dfrac{(y + 1)^2}{4} = 1$ **b.** $\dfrac{(y - 2)^2}{9} - \dfrac{(x - 1)^2}{9} = 1$

Strategy   We will write each equation in a form that makes it easy to identify h, k, a, and b.

Why   If we know h, k, a, and b, we can graph the hyperbola and the central rectangle.

Solution

a. We can write the given equation as

$$\frac{(x - 3)^2}{4^2} - \frac{[y - (-1)]^2}{2^2} = 1$$

To find k, write $y + 1$ as $y - (-1)$.
To find a, write 16 as 4^2. To find b, write 4 as 2^2.

> **Caution**
> When sketching the graph of a hyperbola, the branches should not touch the asymptotes.

Because the term involving x is positive, the hyperbola opens left and right. We find the center by examining $(x - 3)^2$ and $[y - (-1)]^2$. Since $h = 3$ and $k = -1$, the hyperbola is centered at $(h, k) = (3, -1)$. From the denominators, 4^2 and 2^2, we find that $a = 4$ and $b = 2$. Thus, its vertices are located 4 units to the right and left of the center, at $(7, -1)$ and $(-1, -1)$. Since $b = 2$, we can count 2 units above and

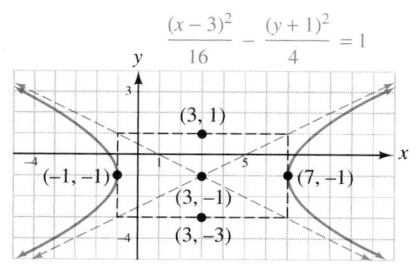

below the center to locate points $(3, 1)$ and $(3, -3)$. With these four points, we can draw the central rectangle along with its extended diagonals (the asymptotes). We can then sketch the hyperbola, as shown.

b. We can write the given equation as

$$\frac{(y - 2)^2}{3^2} - \frac{(x - 1)^2}{3^2} = 1$$

Because the term involving y is positive, the hyperbola opens up and down. We find its center by examining $(y - 2)^2$ and $(x - 1)^2$. Since $k = 2$ and $h = 1$, the hyperbola is centered at $(h, k) = (1, 2)$. From the denominators, 3^2 and 3^2, we find that $a = 3$ and $b = 3$, and we use that information to draw the central rectangle and its extended diagonals (the asymptotes), as shown.

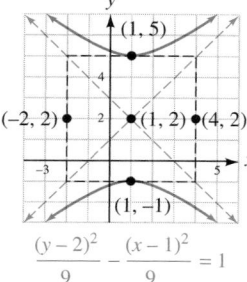

Self Check 3   Graph: **a.** $\dfrac{(x+2)^2}{9} - \dfrac{(y-1)^2}{4} = 1$ **b.** $\dfrac{(y+1)^2}{1} - \dfrac{(x+1)^2}{4} = 1$

Now Try   **Problems 25 and 27**

4 **Graph Equations of the Form** $xy = k$.

There is a special type of hyperbola (also centered at the origin) that does not intersect either the x- or the y-axis. These hyperbolas have equations of the form $xy = k$, where $k \neq 0$.

EXAMPLE 4   Graph: $xy = -8$

Strategy   We will make a table of solutions, plot the points, and connect the points with a smooth curve.

Why   Since this equation cannot be written in standard form, we cannot use the methods used in the previous examples.

Solution   To make a table of solutions, we can solve the equation $xy = -8$ for y:

$$y = \frac{-8}{x}$$

Then we choose several values for x, find the corresponding values of y, and record the results in the table below. We plot the ordered pairs and join them with a smooth curve to obtain the graph of the hyperbola.

The Language of Algebra
The asymptotes of this hyperbola are the x- and y-axes. A hyperbola for which the asymptotes are perpendicular is called a **rectangular hyperbola**.

$xy = -8$ or $y = \dfrac{-8}{x}$

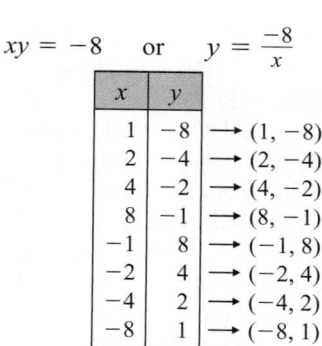

x	y	
1	-8	$\rightarrow (1, -8)$
2	-4	$\rightarrow (2, -4)$
4	-2	$\rightarrow (4, -2)$
8	-1	$\rightarrow (8, -1)$
-1	8	$\rightarrow (-1, 8)$
-2	4	$\rightarrow (-2, 4)$
-4	2	$\rightarrow (-4, 2)$
-8	1	$\rightarrow (-8, 1)$

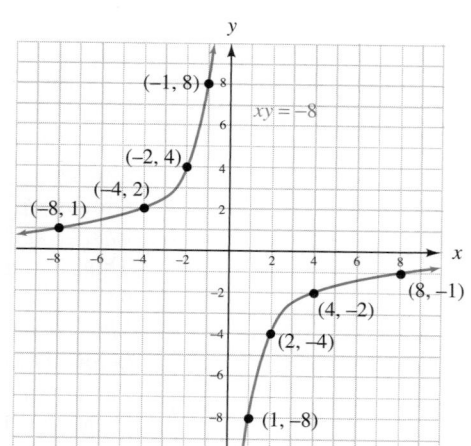

 **Self Check 4**   Graph: $xy = 6$

Now Try   **Problem 33**

The result in Example 4 illustrates the following general equation.

| Equations of Hyperbolas of the Form $xy = k$ | Any equation of the form $xy = k$, where $k \neq 0$, has a graph that is a **hyperbola**, which does not intersect either the x- or y-axis. |

 5   Solve Problems Involving Hyperbolas.

EXAMPLE 5   *Atomic Structure.*   In an experiment that led to the discovery of the atomic structure of matter, Lord Rutherford (1871–1937) shot high-energy alpha particles toward a thin sheet of gold. Many of them were reflected, and Rutherford showed the existence of the nucleus of a gold atom. An alpha particle is repelled by the nucleus at the origin; it travels along the hyperbolic path given by $4x^2 - y^2 = 16$. How close does the particle come to the nucleus?

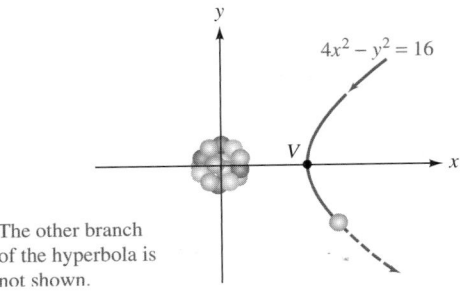

$4x^2 - y^2 = 16$

The other branch of the hyperbola is not shown.

Strategy   We will write the equation in standard form and find the coordinates of point V.

Why   The distance from the origin to point V is the closest the particle comes to the nucleus.

Solution   To find the distance from the nucleus at the origin, we must find the coordinates of the vertex V. To do so, we write the equation of the particle's path in standard form:

$$4x^2 - y^2 = 16$$

$$\frac{4x^2}{16} - \frac{y^2}{16} = \frac{16}{16} \qquad \text{Divide both sides by 16.}$$

$$\frac{x^2}{4} - \frac{y^2}{16} = 1 \qquad \text{Simplify.}$$

$$\frac{x^2}{2^2} - \frac{y^2}{4^2} = 1 \qquad \text{To determine } a \text{ and } b, \text{ write 4 as } 2^2 \text{ and 16 as } 4^2.$$

This equation is in the form

$$\frac{x^2}{a^2} - \frac{y^2}{b^2} = 1$$

with $a = 2$. Thus, the vertex of the path is $(2, 0)$. The particle is never closer than 2 units from the nucleus.

 Now Try   **Problem 61**

▷ **ANSWERS TO SELF CHECKS**

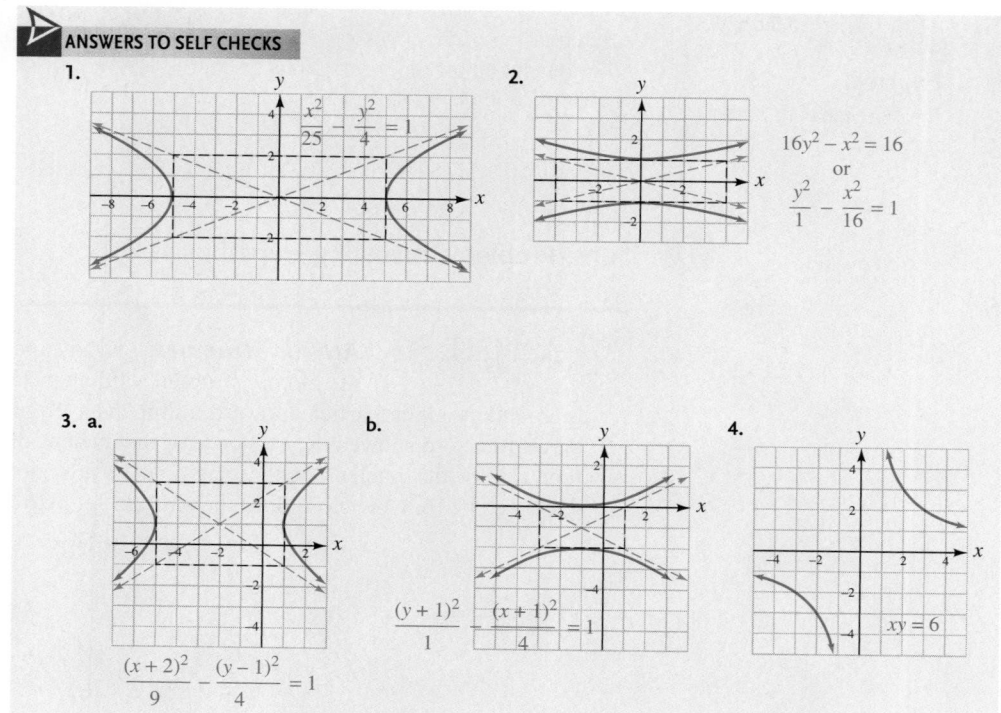

1.

$$\frac{x^2}{25} - \frac{y^2}{4} = 1$$

2.

$$16y^2 - x^2 = 16$$
or
$$\frac{y^2}{1} - \frac{x^2}{16} = 1$$

3. a.

$$\frac{(x + 2)^2}{9} - \frac{(y - 1)^2}{4} = 1$$

b.

$$\frac{(y + 1)^2}{1} - \frac{(x + 1)^2}{4} = 1$$

4.

$$xy = 6$$

STUDY SET
13.3

VOCABULARY

Fill in the blanks.

1. The two-branch curve graphed on the right is a _____.

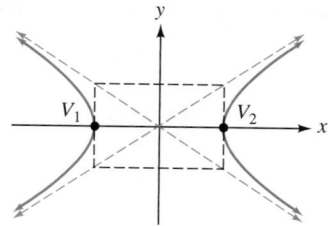

2. A _____ is the set of all points in a plane for which the difference of the distances from two fixed points is a constant.

3. In the graph above, V_1 and V_2 are the _____ of the hyperbola.

4. In the graph above, the figure drawn using dashed black lines is called the _____ _____.

5. The extended _____ of the central rectangle are asymptotes of the hyperbola.

6. To write $9x^2 - 4y^2 = 36$ in _____ form, we divide both sides by 36.

CONCEPTS

7. Write the standard form of the equation of a hyperbola centered at the origin that opens left and right.

8. Write the standard form of the equation of a hyperbola centered at (h, k) that opens up and down.

9. Write the standard form of the equation of a hyperbola centered at (h, k) that opens left and right.

10. **a.** Find the center of the hyperbola graphed below. What are a and b?

   b. Find the x-intercepts of the graph. What are the y-intercepts of the graph?

   c. Find the equation of the hyperbola.

   d. Find the equations of the asymptotes.

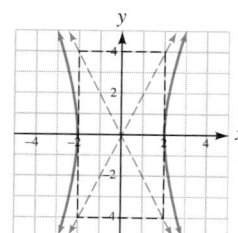

11. a. Find the center of the hyperbola graphed on the right. What are a and b?

    b. Find the equation of the hyperbola.

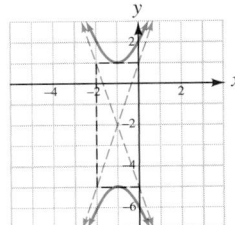

12. a. Fill in the blank: An equation of the form $xy = k$, where $k \neq 0$, has a graph that is a _____ that does not intersect either the x-axis or the y-axis.

    b. Complete the table of solutions for $xy = 10$.

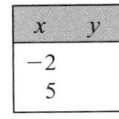

x	y
-2	
5	

13. Divide both sides of the equation by 100 and write the equation in standard form:

$$100(x + 1)^2 - 25(y - 5)^2 = 100$$

14. Determine whether the graph of the equation will be a circle, a parabola, an ellipse, or a hyperbola.

    a. $x^2 + y^2 = 10$

    b. $9y^2 - 16x^2 = 144$

    c. $x = y^2 - 3y + 6$

    d. $4x^2 + 25y^2 = 100$

NOTATION

15. Find h, k, a, and b: $\dfrac{(x - 5)^2}{25} - \dfrac{(y + 11)^2}{36} = 1$

16. Write each denominator in the equation $\dfrac{x^2}{36} - \dfrac{y^2}{81} = 1$ as the square of a number.

GUIDED PRACTICE

Graph each hyperbola. See Example 1.

17. $\dfrac{x^2}{9} - \dfrac{y^2}{4} = 1$

18. $\dfrac{x^2}{4} - \dfrac{y^2}{4} = 1$

19. $\dfrac{y^2}{4} - \dfrac{x^2}{9} = 1$

20. $\dfrac{y^2}{4} - \dfrac{x^2}{64} = 1$

Graph each hyperbola. See Example 2.

21. $y^2 - 4x^2 = 16$

22. $9y^2 - 25x^2 = 225$

23. $25x^2 - y^2 = 25$

24. $9x^2 - 4y^2 = 36$

Graph each hyperbola. See Example 3.

25. $\dfrac{(x - 2)^2}{9} - \dfrac{y^2}{16} = 1$

26. $\dfrac{(x + 2)^2}{16} - \dfrac{(y - 3)^2}{25} = 1$

27. $\dfrac{(y + 1)^2}{1} - \dfrac{(x - 2)^2}{4} = 1$

28. $\dfrac{(y - 2)^2}{4} - \dfrac{(x + 1)^2}{1} = 1$

29. $\dfrac{(x + 1)^2}{9} - \dfrac{(y + 1)^2}{9} = 1$

30. $\dfrac{(x - 2)^2}{16} - \dfrac{(y - 1)^2}{16} = 1$

31. $\dfrac{(y - 3)^2}{25} - \dfrac{x^2}{25} = 1$

32. $\dfrac{(y - 1)^2}{9} - \dfrac{x^2}{9} = 1$

Graph each equation. See Example 4.

33. $xy = 8$

34. $xy = 4$

35. $xy = -10$

36. $xy = -12$

Use a graphing calculator to graph each equation. See Using Your Calculator: Graphing Hyperbolas.

37. $\dfrac{x^2}{9} - \dfrac{y^2}{4} = 1$

38. $y^2 - 16x^2 = 16$

39. $\dfrac{x^2}{4} - \dfrac{(y - 1)^2}{9} = 1$

40. $\dfrac{(y + 1)^2}{9} - \dfrac{(x - 2)^2}{4} = 1$

TRY IT YOURSELF

Write each equation in standard form, if it is not already so, and graph it. The problems include equations that describe circles, parabolas, ellipses, and hyperbolas.

41. $(x + 1)^2 + (y - 2)^2 = 16$

42. $(x - 3)^2 + (y + 4)^2 = 1$

43. $9x^2 - 49y^2 = 441$

44. $25y^2 - 16x^2 = 400$

45. $4(x + 1)^2 + 9(y + 1)^2 = 36$

46. $16x^2 + 25(y - 3)^2 = 400$

47. $4(x + 3)^2 - (y - 1)^2 = 4$

48. $(x + 5)^2 - 16y^2 = 16$

49. $xy = -6$

50. $xy = 10$

51. $x = \dfrac{1}{2}(y - 1)^2 - 2$

52. $x = -\dfrac{1}{4}(y - 3)^2 + 2$

53. $\dfrac{y^2}{25} - \dfrac{(x - 2)^2}{4} = 1$

54. $\dfrac{y^2}{36} - \dfrac{(x + 2)^2}{4} = 1$

55. $y = -x^2 + 6x - 4$

56. $y = x^2 - 2x + 5$

57. $\dfrac{x^2}{1} + \dfrac{y^2}{36} = 1$

58. $\dfrac{x^2}{4} + \dfrac{y^2}{16} = 1$

59. $x^2 + y^2 + 4x - 6y - 23 = 0$

60. $x^2 + y^2 + 8x - 2y - 8 = 0$

APPLICATIONS

61. ALPHA PARTICLES   The particle in the illustration on the next page approaches the nucleus at the origin along the path $9y^2 - x^2 = 81$ in the coordinate system shown. How close does the particle come to the nucleus?

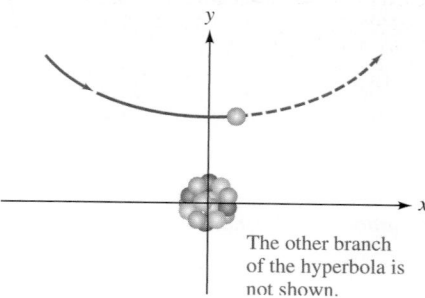

The other branch
of the hyperbola is
not shown.

62. LORAN By determining the difference of the distances between the ship in the illustration and two radio transmitters, the LORAN navigation system places the ship on the hyperbola $x^2 - 4y^2 = 576$ in the coordinate system shown. If the ship is 5 miles out to sea, find its coordinates.

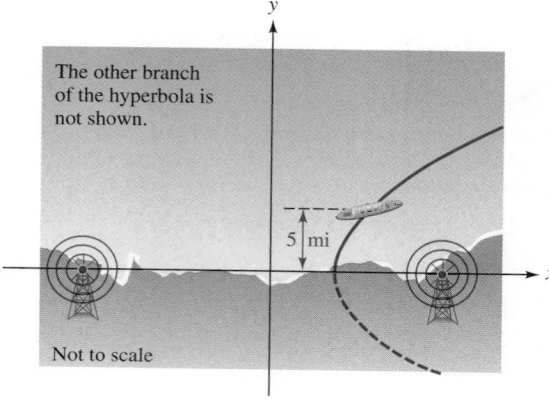

The other branch
of the hyperbola is
not shown.

5 mi

Not to scale

63. SONIC BOOM The position of a sonic boom caused by the faster-than-sound aircraft is one branch of the hyperbola $y^2 - x^2 = 25$ in the coordinate system shown. How wide is the hyperbola 5 miles from its vertex?

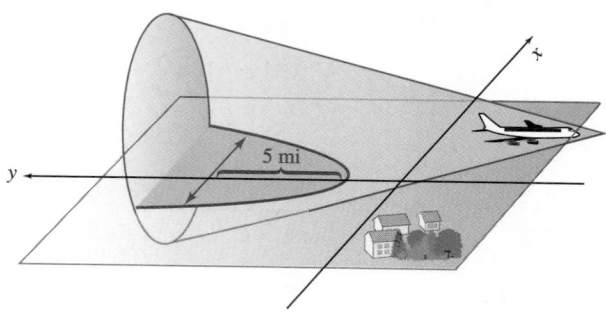

5 mi

64. FLUIDS See the illustration in the next column. Two glass plates in contact at the left, and separated by about 5 millimeters on the right, are dipped in beet juice, which rises by capillary action to form a hyperbola. The hyperbola is modeled by an equation of the form $xy = k$. If the curve passes through the point (12, 2), what is k?

65. What is a hyperbola?

66. Compare the graphs of $\frac{x^2}{81} - \frac{y^2}{64} = 1$ and $\frac{y^2}{81} - \frac{x^2}{64} = 1$. Do they have any similarities?

67. Explain how to determine the dimensions of the central rectangle that is associated with the graph of

$$\frac{x^2}{36} - \frac{y^2}{25} = 1$$

68. Explain why the graph of the following hyperbola has no y-intercept.

$$\frac{x^2}{a^2} - \frac{y^2}{b^2} = 1$$

REVIEW

Find each value of x.

69. $\log_8 x = 2$

70. $\log_{25} x = \frac{1}{2}$

71. $\log_{1/2} \frac{1}{8} = x$

72. $\log_{12} x = 0$

73. $\log_x \frac{9}{4} = 2$

74. $\log_6 216 = x$

75. $\log_x 1,000 = 3$

76. $\log_2 \sqrt{2} = x$

CHALLENGE PROBLEMS

77. Write the equation $x^2 - y^2 - 2x + 4y = 12$ in standard form to show that it describes a hyperbola.

78. Write the equation $x^2 - 4y^2 + 2x - 8y = 7$ in standard form to show that it describes a hyperbola.

79. Write the equation $36x^2 - 25y^2 - 72x - 100y = 964$ in standard form to show that it describes a hyperbola.

80. Write an equation of a hyperbola whose graph has the following characteristics:
   • vertices ($\pm 1, 0$)
   • equations of asymptotes: $y = \pm 5x$

81. Graph: $16x^2 - 25y^2 = 1$

82. Show that the equations of the extended diagonals of the fundamental rectangle of the hyperbola

$$\frac{x^2}{a^2} - \frac{y^2}{b^2} = 1 \text{ are } y = \frac{b}{a}x \text{ and } y = -\frac{b}{a}x$$

SECTION 13.4
Solving Nonlinear Systems of Equations

Objectives

1 Solve systems by graphing.
2 Solve systems by substitution.
3 Solve systems by elimination (addition).

In Chapter 3, we discussed how to solve systems of linear equations by the graphing, substitution, and elimination methods. In this section, we will use these methods to solve systems where at least one of the equations is nonlinear.

1 **Solve Systems by Graphing.**

A solution of a **nonlinear system of equations** is an ordered pair of real numbers that satisfies all of the equations in the system. The **solution set of a nonlinear system** is the set of all such ordered pairs. One way to solve a system of two equations in two variables is to graph the equations on the same rectangular coordinate system.

EXAMPLE 1 Solve $\begin{cases} x^2 + y^2 = 25 \\ 2x + y = 10 \end{cases}$ by graphing.

Strategy We will graph both equations on the same coordinate system.

Why If the equations are graphed on the same coordinate system, we can see whether they have any common solutions.

Solution The graph of $x^2 + y^2 = 25$ is a circle with center at the origin and radius of 5. The graph of $2x + y = 10$ is a line. Depending on whether the line is a **secant** (intersecting the circle at two points) or a **tangent** (intersecting the circle at one point) or does not intersect the circle at all, there are two, one, or no solutions to the system, respectively.

After graphing the circle and the line, it appears that the points of intersection are (5, 0) and (3, 4). To verify that they are solutions of the system, we need to check each one.

Check:

For (5, 0)

$2x + y = 10$	$x^2 + y^2 = 25$
$2(5) + 0 \overset{?}{=} 10$	$5^2 + 0^2 \overset{?}{=} 25$
$10 = 10$	$25 = 25$
True	True

For (3, 4)

$2x + y = 10$	$x^2 + y^2 = 25$
$2(3) + 4 \overset{?}{=} 10$	$3^2 + 4^2 \overset{?}{=} 25$
$10 = 10$	$25 = 25$
True	True

The ordered pair (5, 0) satisfies both equations of the system, and so does (3, 4). Thus, there are two solutions, (5, 0) and (3, 4), and the solution set is $\{(5, 0), (3, 4)\}$.

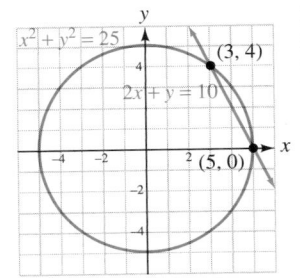

Success Tip
It is helpful to sketch the possibilities before solving the system:

Secant line

2 points of intersection: (2 real solutions)

Tangent line

1 point of intersection: (1 real solution)

No points of intersection: (0 real solutions)

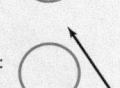

Self Check 1 Solve $\begin{cases} x^2 + y^2 = 25 \\ y = -2x - 5 \end{cases}$ by graphing.

Now Try **Problem 15**

Using Your Calculator *Solving Systems of Equations*

To solve Example 1 with a graphing calculator, we graph the circle and the line on one set of coordinate axes. See figure (a). We then trace to find the coordinates of the intersection points of the graphs. See figures (b) and (c).

We can zoom for better results.

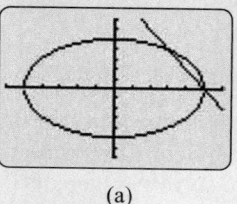   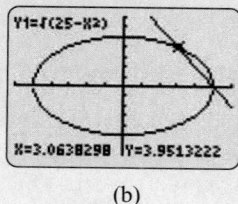   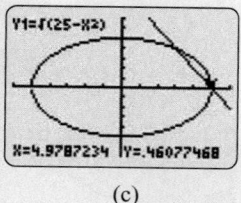

(a)   (b)   (c)

---

2 **Solve Systems by Substitution.**

When solving a system by graphing, it is often difficult to determine the coordinates of the intersection points. A more precise algebraic method called the **substitution method** can be used to solve certain systems involving nonlinear equations.

EXAMPLE 2 Solve $\begin{cases} x^2 + y^2 = 2 \\ 2x - y = 1 \end{cases}$ by substitution.

Strategy We will solve the second equation for y and substitute the result for y in the first equation.

Why We can solve the resulting equation for x and then back substitute to find y.

Solution This system has one second-degree equation and one first-degree equation. We can solve this type of system by substitution. Solving the linear equation for y gives

$$2x - y = 1$$
$$-y = -2x + 1 \quad \text{Subtract 2x from both sides.}$$
$$y = 2x - 1 \quad \text{Multiply both sides by } -1. \text{ We call this the substitution equation.}$$

Because y and $2x - 1$ are equal, we can substitute $2x - 1$ for y in the first equation of the system.

$$y = \boxed{(2x - 1)} \qquad\qquad x^2 + y^2 = 2$$

Then we solve the resulting quadratic equation for x.

$$x^2 + y^2 = 2$$
$$x^2 + (2x - 1)^2 = 2 \quad \text{Substitute 2x - 1 for y.}$$
$$x^2 + 4x^2 - 4x + 1 = 2 \quad \text{Use a special-product rule to find } (2x - 1)^2.$$

Success Tip
With this method, the objective is to use an appropriate substitution to obtain *one* equation in *one* variable.

$$5x^2 - 4x - 1 = 0$$ To get 0 on the right side, subtract 2
from both sides and then combine like terms.

$$(5x + 1)(x - 1) = 0$$ Factor.

$$5x + 1 = 0 \quad \text{or} \quad x - 1 = 0$$ Set each factor equal to 0.

$$x = -\frac{1}{5} \qquad\qquad x = 1$$

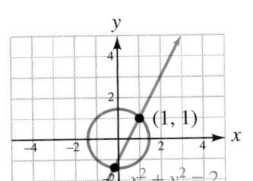

If we substitute $-\frac{1}{5}$ for x in the equation $y = 2x - 1$, we get $y = -\frac{7}{5}$. If we substitute 1 for x in $y = 2x - 1$, we get $y = 1$. Thus, the system has two solutions, $\left(-\frac{1}{5}, -\frac{7}{5}\right)$ and $(1, 1)$. Verify that each ordered pair satisfies both equations of the original system.

The graph in the margin confirms that the system has two solutions, and that one of them is $(1, 1)$. However, it would be virtually impossible to determine from the graph that the coordinates of the second point of intersection are $\left(-\frac{1}{5}, -\frac{7}{5}\right)$.

Self Check 2   Solve $\begin{cases} x^2 + y^2 = 10 \\ y = x + 2 \end{cases}$ by substitution.

Now Try   Problem 23

EXAMPLE 3   Solve: $\begin{cases} 4x^2 + 9y^2 = 5 \\ y = x^2 \end{cases}$

Strategy   Since $y = x^2$, we will substitute y for x^2 in the first equation.

Why   This will give an equation in one variable that we can solve for y. We can then find x by back substitution.

Solution   We can solve this system by substitution.

$$4x^2 + 9y^2 = 5 \qquad \overset{\frown}{} \quad y = x^2$$

When we substitute y for x^2 in the first equation, the result is a quadratic equation in y.

$$4x^2 + 9y^2 = 5$$
$$4y + 9y^2 = 5 \qquad \text{Substitute } y \text{ for } x^2.$$
$$9y^2 + 4y - 5 = 0 \qquad \text{To get 0 on the right side, subtract 5 from both sides.}$$
$$(9y - 5)(y + 1) = 0 \qquad \text{Factor } 9y^2 + 4y - 5.$$
$$9y - 5 = 0 \quad \text{or} \quad y + 1 = 0 \qquad \text{Set each factor equal to 0.}$$
$$y = \frac{5}{9} \qquad\qquad y = -1$$

Since $y = x^2$, the values of x are found by solving the equations

$$x^2 = \frac{5}{9} \quad \text{or} \quad \cancel{x^2 = -1}$$

Because $x^2 = -1$ has no real solutions, this possibility is discarded. The solutions of $x^2 = \frac{5}{9}$ are

$$x = \sqrt{\frac{5}{9}} = \frac{\sqrt{5}}{\sqrt{9}} = \frac{\sqrt{5}}{3} \quad \text{or} \quad x = -\sqrt{\frac{5}{9}} = -\frac{\sqrt{5}}{\sqrt{9}} = -\frac{\sqrt{5}}{3}$$

Success Tip

$4x^2 + 9y^2 = 5$ is the equation of an ellipse centered at $(0, 0)$, and $y = x^2$ is the equation of a parabola with vertex at $(0, 0)$, opening upward. We would expect two solutions.

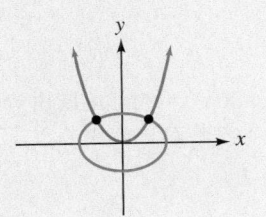

Caution

In this section, we are solving for only the real values of x and y.

Thus, the solutions of the system are

$$\left(\frac{\sqrt{5}}{3}, \frac{5}{9}\right) \quad \text{and} \quad \left(-\frac{\sqrt{5}}{3}, \frac{5}{9}\right)$$

 Self Check 3   Solve: $\begin{cases} x^2 + y^2 = 20 \\ y = x^2 \end{cases}$

Now Try   Problem 27

3 **Solve Systems by Elimination (Addition).**

Another method for solving nonlinear system of equations is the **elimination** or **addition method.** With this method, we combine the equations in a way that will eliminate the terms of one of the variables.

EXAMPLE 4   Solve: $\begin{cases} 3x^2 + 2y^2 = 36 \\ 4x^2 - y^2 = 4 \end{cases}$

Strategy   We will multiply both sides of the second equation by 2 and add the result to the first equation.

Why   This will eliminate the y^2-terms and produce an equation that we can solve for x.

Solution   To solve this system of two second-degree equations, we can use either the substitution or the elimination method. We will use the elimination method because the y^2-terms can be eliminated by multiplying the second equation by 2 and adding it to the first equation.

$$\begin{cases} 3x^2 + 2y^2 = 36 \\ 4x^2 - y^2 = 4 \end{cases} \xrightarrow[\text{Multiply by 2}]{\text{Unchanged}} \begin{cases} 3x^2 + 2y^2 = 36 \\ 8x^2 - 2y^2 = 8 \end{cases}$$

We add the two equations on the right to eliminate y^2 and solve the resulting equation for x:

$$11x^2 = 44$$
$$x^2 = 4$$
$$x = 2 \quad \text{or} \quad x = -2$$

To find y, we can substitute 2 for x and then -2 for x into any equation containing both variables. It appears that the computations will be simplest if we use $3x^2 + 2y^2 = 36$.

For x = 2

$$3x^2 + 2y^2 = 36$$
$$3(2)^2 + 2y^2 = 36$$
$$12 + 2y^2 = 36$$
$$2y^2 = 24$$
$$y^2 = 12$$
$$y = \sqrt{12} \quad \text{or} \quad y = -\sqrt{12}$$
$$y = 2\sqrt{3} \quad | \quad y = -2\sqrt{3}$$

For x = -2

$$3x^2 + 2y^2 = 36$$
$$3(-2)^2 + 2y^2 = 36$$
$$12 + 2y^2 = 36$$
$$2y^2 = 24$$
$$y^2 = 12$$
$$y = \sqrt{12} \quad \text{or} \quad y = -\sqrt{12}$$
$$y = 2\sqrt{3} \quad | \quad y = -2\sqrt{3}$$

The four solutions of this system are

$$\left(2, 2\sqrt{3}\right), \qquad \left(2, -2\sqrt{3}\right), \qquad \left(-2, 2\sqrt{3}\right), \qquad \text{and} \qquad \left(-2, -2\sqrt{3}\right)$$

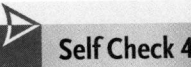

Self Check 4   Solve: $\begin{cases} x^2 + 4y^2 = 16 \\ x^2 - y^2 = 1 \end{cases}$

Now Try   **Problem 31**

ANSWERS TO SELF CHECKS   **1.** $(-4, 3), (0, -5)$

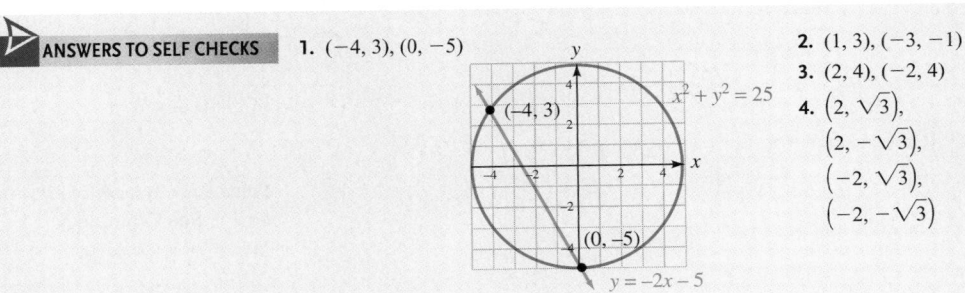

2. $(1, 3), (-3, -1)$
3. $(2, 4), (-2, 4)$
4. $\left(2, \sqrt{3}\right),$ $\left(2, -\sqrt{3}\right),$ $\left(-2, \sqrt{3}\right),$ $\left(-2, -\sqrt{3}\right)$

STUDY SET
13.4

VOCABULARY

Fill in the blanks.

1. $\begin{cases} 4x^2 + 6y^2 = 24 \\ 9x^2 - y^2 = 9 \end{cases}$ is a _____ of two nonlinear equations.

2. The graph of $2x + y = 10$ is a _____ and the graph of $x^2 + y^2 = 25$ is a _____.

3. When solving a system by graphing, it is often difficult to determine the coordinates of the points of _____ of the graphs.

4. Two algebraic methods for solving systems of nonlinear equations are the _____ method and the _____ method.

5. A _____ is a line that intersects a circle at two points.

6. A _____ is a line that intersects a circle at one point.

CONCEPTS

7. **a.** A line can intersect an ellipse in at most _____ points.
   b. An ellipse can intersect a parabola in at most _____ points.
   c. An ellipse can intersect a circle in at most _____ points.
   d. A hyperbola can intersect a circle in at most _____ points.

8. Determine whether $(1, -1)$ is a solution of the system
$$\begin{cases} 2x + y - 1 = 0 \\ x^2 - y^2 = 3 \end{cases}.$$

9. Find the solutions of the system
$$\begin{cases} x^2 + 4y^2 = 25 \\ x^2 - 2y^2 = 1 \end{cases}$$
that is graphed on the right.

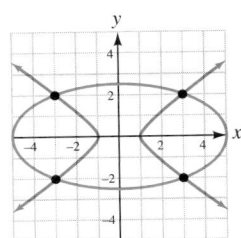

10. Find a substitution equation that can be used to solve the system $\begin{cases} x^2 + y^2 = 9 \\ 2x - y = 3 \end{cases}$.

11. Consider the system $\begin{cases} 6x^2 + y^2 = 9 \\ 3x^2 + 4y^2 = 36 \end{cases}$.

    a. If the y^2-terms are to be eliminated, by what should the first equation be multiplied?

    b. If the x^2-terms are to be eliminated, by what should the second equation be multiplied?

12. Suppose you begin to solve the system $\begin{cases} x^2 + y^2 = 10 \\ 4x^2 + y^2 = 13 \end{cases}$ and find that x is ± 1. Use the first equation to find the corresponding y-values for $x = 1$ and $x = -1$. State the solutions as ordered pairs.

NOTATION

Complete each solution to solve the system.

13. Solve: $\begin{cases} x^2 + y^2 = 5 \\ y = 2x \end{cases}$

$x^2 + y^2 = 5$ This is the first equation.

$x^2 + (\quad)^2 = 5$

$x^2 + 4x^2 = \blacksquare$

$\blacksquare x^2 = 5$

$x^2 = \blacksquare$

$x = 1$ or $x = -1$

If $x = 1$, then $y = 2(\blacksquare) = 2$.   Use the second equation.

If $x = -1$, then $y = 2(\blacksquare) = -2$.

The solutions are $(1, 2)$ and $\left(-1, \blacksquare\right)$.

14. Solve: $\begin{cases} y = x^2 + 2 \\ y = -x^2 + 4 \end{cases}$

$2y = \blacksquare$ Add the equations.

$y = \blacksquare$

If $y = 3$, then

$\blacksquare = x^2 + 2$ This is the first equation.

$1 = x^2$

$\blacksquare 1 = x$

The solutions are

$\left(1, \blacksquare\right)$ and $\left(\blacksquare, 3\right)$

GUIDED PRACTICE

Solve each system of equations by graphing. See Example 1.

15. $\begin{cases} x^2 + y^2 = 9 \\ y - x = 3 \end{cases}$

16. $\begin{cases} x^2 + y^2 = 16 \\ y - x = -4 \end{cases}$

17. $\begin{cases} 9x^2 + 16y^2 = 144 \\ 9x^2 - 16y^2 = 144 \end{cases}$

18. $\begin{cases} x^2 + 9y^2 = 9 \\ 9y^2 - x^2 = 9 \end{cases}$

19. $\begin{cases} y = x^2 - 4x \\ x^2 + y = 0 \end{cases}$

20. $\begin{cases} x^2 - y = 0 \\ y = -x^2 + 4x \end{cases}$

21. $\begin{cases} x^2 + 4y^2 = 4 \\ x = 2y^2 - 2 \end{cases}$

22. $\begin{cases} 4x^2 + y^2 = 4 \\ y = 2x^2 - 2 \end{cases}$

Solve each system of equations by substitution for real values of x and y. See Examples 2 and 3.

23. $\begin{cases} x^2 + y^2 = 5 \\ x + y = 3 \end{cases}$

24. $\begin{cases} x^2 - x - y = 2 \\ 4x - 3y = 0 \end{cases}$

25. $\begin{cases} y = x^2 + 6x + 7 \\ 2x + y = -5 \end{cases}$

26. $\begin{cases} 2x + y = 1 \\ x^2 + y = 4 \end{cases}$

27. $\begin{cases} x^2 + y^2 = 13 \\ y = x^2 - 1 \end{cases}$

28. $\begin{cases} x^2 + y^2 = 10 \\ y = 3x^2 \end{cases}$

29. $\begin{cases} x^2 + y^2 = 30 \\ y = x^2 \end{cases}$

30. $\begin{cases} x^2 + y^2 = 20 \\ y = x^2 \end{cases}$

Solve each system of equations by elimination for real values of x and y. See Example 4.

31. $\begin{cases} x^2 + y^2 = 20 \\ x^2 - y^2 = -12 \end{cases}$

32. $\begin{cases} x^2 + y^2 = 13 \\ x^2 - y^2 = 5 \end{cases}$

33. $\begin{cases} 9x^2 - 7y^2 = 81 \\ x^2 + y^2 = 9 \end{cases}$

34. $\begin{cases} x^2 + y^2 = 25 \\ 2x^2 - 3y^2 = 5 \end{cases}$

35. $\begin{cases} 2x^2 + y^2 = 6 \\ x^2 - y^2 = 3 \end{cases}$

36. $\begin{cases} x^2 + y^2 = 36 \\ 49x^2 + 36y^2 = 1{,}764 \end{cases}$

37. $\begin{cases} x^2 - y^2 = -5 \\ 3x^2 + 2y^2 = 30 \end{cases}$

38. $\begin{cases} 6x^2 + 8y^2 = 182 \\ 8x^2 - 3y^2 = 24 \end{cases}$

Solve each system. See Using Your Calculator: Solving Systems of Equations.

39. $\begin{cases} x^2 - 6x - y = -5 \\ x^2 - 6x + y = -5 \end{cases}$

40. $\begin{cases} x^2 - y^2 = -5 \\ 3x^2 + 2y^2 = 30 \end{cases}$

TRY IT YOURSELF

Solve each system of equations for real values of x and y.

41. $\begin{cases} 2x^2 - 3y^2 = 5 \\ 3x^2 + 4y^2 = 16 \end{cases}$

42. $\begin{cases} 2x^2 - y^2 + 2 = 0 \\ 3x^2 - 2y^2 + 5 = 0 \end{cases}$

43. $\begin{cases} y = x^2 - 4 \\ x^2 - y^2 = -16 \end{cases}$

44. $\begin{cases} y - x = 0 \\ 4x^2 + y^2 = 10 \end{cases}$

45. $\begin{cases} 3y^2 = xy \\ 2x^2 + xy - 84 = 0 \end{cases}$

46. $\begin{cases} x^2 + y^2 = 10 \\ 2x^2 - 3y^2 = 5 \end{cases}$

47. $\begin{cases} y^2 = 40 - x^2 \\ y = x^2 - 10 \end{cases}$

48. $\begin{cases} 25x^2 + 9y^2 = 225 \\ 5x + 3y = 15 \end{cases}$

49. $\begin{cases} 3x - y = -3 \\ 25y^2 - 9x^2 = 225 \end{cases}$

50. $\begin{cases} x - 2y = 2 \\ 9x^2 - 4y^2 = 36 \end{cases}$

51. $\begin{cases} x^2 - y = 0 \\ x^2 - 4x + y = 0 \end{cases}$

52. $\begin{cases} xy = -\dfrac{9}{2} \\ 3x + 2y = 6 \end{cases}$

53. $\begin{cases} x^2 - 2y^2 = 6 \\ x^2 + 2y^2 = 2 \end{cases}$

54. $\begin{cases} x^2 + 9y^2 = 1 \\ x^2 - 9y^2 = 3 \end{cases}$

55. $\begin{cases} y = x^2 - 4 \\ 6x - y = 13 \end{cases}$

56. $\begin{cases} y = x + 1 \\ x^2 - y^2 = 1 \end{cases}$

57. $\begin{cases} x^2 + y^2 = 4 \\ 9x^2 + y^2 = 9 \end{cases}$

58. $\begin{cases} 2x^2 - 6y^2 + 3 = 0 \\ 4x^2 + 3y^2 = 4 \end{cases}$

59. $\begin{cases} xy = \dfrac{1}{6} \\ y + x = 5xy \end{cases}$

60. $\begin{cases} xy = \dfrac{1}{12} \\ y + x = 7xy \end{cases}$

61. $\begin{cases} x^2 = 4 - y \\ y = x^2 + 2 \end{cases}$

62. $\begin{cases} 3x + 2y = 10 \\ y = x^2 - 5 \end{cases}$

63. $\begin{cases} x^2 - y^2 = 4 \\ x + y = 4 \end{cases}$

64. $\begin{cases} x - y = -1 \\ y^2 - 4x = 0 \end{cases}$

APPLICATIONS

Use a nonlinear system of equations to solve each problem.

65. INTEGER PROBLEM   The product of two integers is 32, and their sum is 12. Find the integers.

66. NUMBER PROBLEM   The sum of the squares of two numbers is 221, and the sum of the numbers is 9. Find the numbers.

67. ARCHERY   See the illustration. An arrow shot from the base of a hill follows the parabolic path $y = -\frac{1}{6}x^2 + 2x$, with distances measured in meters. The inclined hill has a slope of $\frac{1}{3}$ and can therefore be modeled by the equation $y = \frac{1}{3}x$. Find the coordinates of the point of impact of the arrow and then its distance from the archer.

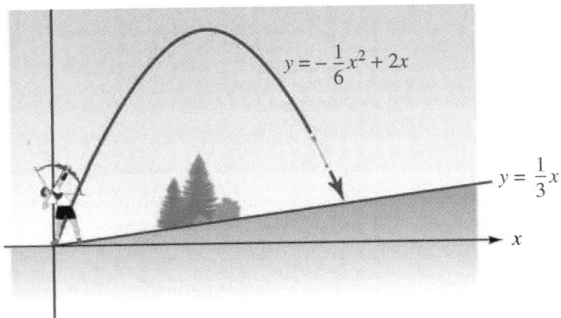

68. GEOMETRY   The area of a rectangle is 63 square centimeters, and its perimeter is 32 centimeters. Find the dimensions of the rectangle.

69. FENCING PASTURES   The rectangular pasture shown here is to be fenced in along a riverbank. If 260 feet of fencing is to enclose an area of 8,000 square feet, find the dimensions of the pasture.

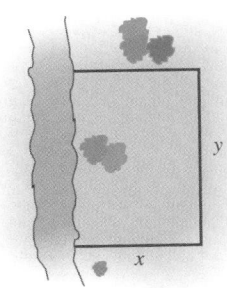

70. DRIVING RATES   Jim drove 306 miles. Jim's brother made the same trip at a speed 17 mph slower than Jim did and required an extra $1\frac{1}{2}$ hours. What was Jim's rate and time?

71. INVESTING   Grant receives $225 annual income from one investment. Jeff invested $500 more than Grant, but at an annual rate of 1% less. Jeff's annual income is $240. What are the amount and rate of Grant's investment?

72. INVESTING   Carol receives $67.50 annual income from one investment. John invested $150 more than Carol at an annual rate of $1\frac{1}{2}$% more. John's annual income is $94.50. What are the amount and rate of Carol's investment?   (*Hint:* There are two answers.)

WRITING

73. a. Describe the benefits of the graphical method for solving a system of equations.

b. Describe the drawbacks of the graphical method.

74. Explain why the elimination method, not the substitution method, is the better method to solve the system

$$\begin{cases} 4x^2 + 9y^2 = 52 \\ 9x^2 + 4y^2 = 52 \end{cases}$$

REVIEW

Solve each equation.

75. $\log 5x = 4$

76. $\log 3x = \log 9$

77. $\dfrac{\log(8x - 7)}{\log x} = 2$

78. $\log x + \log(x + 9) = 1$

CHALLENGE PROBLEMS

79. a. The graphs of the two independent equations of a system are parabolas. How many solutions might the system have?

   b. The graphs of the two independent equations of a system are hyperbolas. How many solutions might the system have?

80. Solve the system for real solutions: $\begin{cases} \dfrac{1}{x} + \dfrac{2}{y} = 1 \\ \dfrac{2}{x} - \dfrac{1}{y} = \dfrac{1}{3} \end{cases}$

81. Solve the system for real solutions: $\begin{cases} \dfrac{1}{x} + \dfrac{3}{y} = 4 \\ \dfrac{2}{x} - \dfrac{1}{y} = 7 \end{cases}$

82. Solve the system $\begin{cases} x^2 - y^2 = 16 \\ x^2 + y^2 = 9 \end{cases}$ over the complex numbers.

CHAPTER 13
Summary & Review

SECTION 13.1   The Circle and the Parabola

DEFINITIONS AND CONCEPTS	EXAMPLES

A **circle** is the set of all points in a plane that are a fixed distance from a fixed point called its **center.** The fixed distance is called the **radius** of the circle.

Standard forms of the equation of a circle:

$x^2 + y^2 = r^2$ Center $(0, 0)$, radius r

$(x - h)^2 + (y - k)^2 = r^2$ Center (h, k), radius r

The graph of the equation $x^2 + y^2 = 16$, which can be written $x^2 + y^2 = 4^2$, is a circle with center at $(0, 0)$ and a radius of 4.

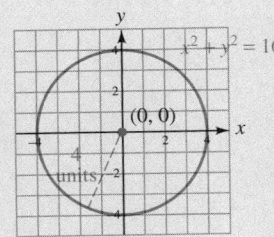 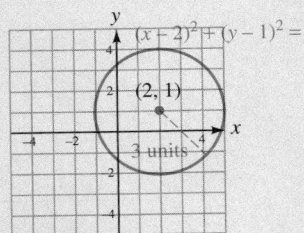

The graph of the equation $(x - 2)^2 + (y - 1)^2 = 9$, which can be written $(x - 2)^2 + (y - 1)^2 = 3^2$, is a circle with center at $(2, 1)$ and a radius of 3.

Because a circle is determined by its center and radius, that information is all we need to know to write its equation.

Write the equation of a circle centered at $(4, -3)$ and with a radius of 5.

In this problem, $h = 4$, $k = -3$, and $r = 5$. We substitute these values into the standard form of the equation of a circle and simplify.

$$(x - h)^2 + (y - k)^2 = r^2$$
$$(x - 4)^2 + [y - (-3)]^2 = 5^2$$
$$(x - 4)^2 + (y + 3)^2 = 25$$

A **parabola** is the set of all points in a plane that are equidistant from a fixed point, called the **focus,** and a fixed line, called the **directrix.**

General forms of the equation of a parabola:

$y = ax^2 + bx + c$ $a > 0$: up; $a < 0$: down

$x = ay^2 + by + c$ $a > 0$: right; $a < 0$: left

Standard forms of the equation of a parabola:

$y = a(x - h)^2 + k$ $a > 0$: up; $a < 0$: down
Vertex at (h, k) Axis of symmetry is $x = h$

$x = a(y - k)^2 + h$ $a > 0$: right; $a < 0$: left
Vertex at (h, k) Axis of symmetry is $y = k$

The equation $x = 2y^2 - 4y + 5$ is the equation of a parabola that opens to the right. To find its vertex and axis of symmetry, we complete the square on x and write the equation in standard form.

$x = 2y^2 - 4y + 5$

$x = 2(y^2 - 2y\quad) + 5$ Factor out 2.

$x = 2(y^2 - 2y + 1) + 5 - 2$ Complete the square.

$x = 2(y - 1)^2 + 3$ Factor and simplify.

From the standard form, we see that $h = 3$ and $k = 1$. Thus, the vertex is at $(3, 1)$ and the axis of symmetry is $y = 1$. To construct a table of solutions, we choose values of y and find their corresponding values of x.

$x = 2(y - 1)^2 + 3$

x	y	
5	2	→ $(5, 2)$
11	3	→ $(11, 3)$

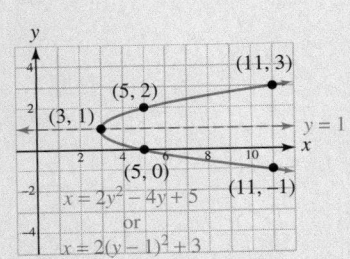

REVIEW EXERCISES

Graph each equation.

1. $x^2 + y^2 = 16$

2. $(x - 4)^2 + (y + 3)^2 = 4$

3. Write the equation in standard form and graph it.

$$x^2 + y^2 + 4x - 2y = 4$$

4. ART HISTORY   Leonardo da Vinci's *Vitruvian Man* (1492) is one of the most famous pen-and-ink drawings of all time. Use the coordinate system that is superimposed on the drawing to write the equation of the circle in standard form.

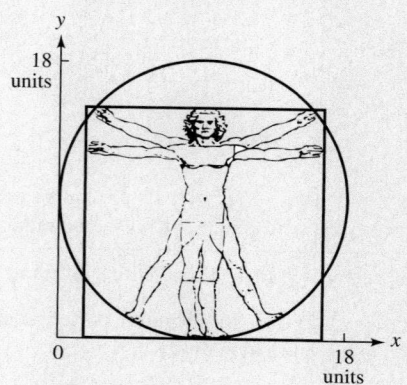

5. Find the center and the radius of the circle whose equation is $(x + 6)^2 + y^2 = 24$.

6. Fill in the blanks: A circle is the set of all points in a plane that are a fixed distance from a point called its _____. The fixed distance is called the _____ of the circle.

Graph each parabola and give the coordinates of the vertex.

7. $x = y^2$

8. $x = 2(y + 1)^2 - 2$

Write each equation in standard form and graph it.

9. $x = -3y^2 + 12y - 7$

10. $y = x^2 + 8x + 11$

11. The axis of symmetry, the vertex, and two additional points on the graph of a parabola are shown. Find the coordinates of two other points on the parabola.

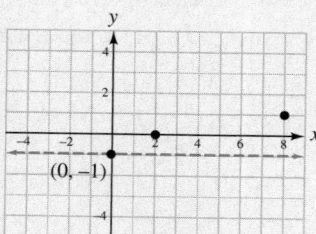

12. LONG JUMP   The equation describing the flight path of the long jumper is $y = -\frac{5}{121}(x - 11)^2 + 5$. Show that she will land at a point 22 feet away from the take-off board.

Take-off board ← ——————— 22 ft ——————— → Landing

SECTION 13.2   The Ellipse

DEFINITIONS AND CONCEPTS	EXAMPLES

An **ellipse** is the set of all points in a plane for which the sum of the distances from two fixed points is a constant.

Standard forms of the equation of an ellipse:

$$\frac{x^2}{a^2} + \frac{y^2}{b^2} = 1 \quad \text{Center } (0, 0)$$

$$\frac{(x - h)^2}{a^2} + \frac{(y - k)^2}{b^2} = 1 \quad \text{Center } (h, k)$$

The equation $\frac{x^2}{9} + \frac{y^2}{4} = 1$, which can be written $\frac{x^2}{3^2} + \frac{y^2}{2^2} = 1$, represents an ellipse that is centered at the origin. Here, $a = 3$ and $b = 2$.

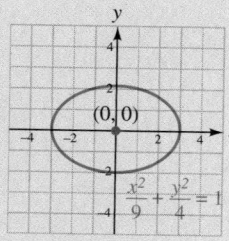

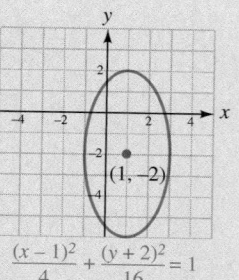

The equation $\frac{(x - 1)^2}{4} + \frac{(y + 2)^2}{16} = 1$, which can be written $\frac{(x - 1)^2}{2^2} + \frac{(y + 2)^2}{4^2} = 1$, represents an ellipse that is centered at $(1, -2)$. Here, $a = 2$ and $b = 4$.

SECTION 13.2 The Ellipse–*continued*

DEFINITIONS AND CONCEPTS	EXAMPLES

To write the equation $25x^2 + 16y^2 = 400$ in standard form, divide both sides by 400 and simplify.

$$25x^2 + 16y^2 = 400$$

$$\frac{25x^2}{400} + \frac{16y^2}{400} = \frac{400}{400} \qquad \text{To get 1 on the right side, divide both sides by 400.}$$

$$\frac{x^2}{16} + \frac{y^2}{25} = 1 \qquad \text{Simplify each fraction.}$$

This result represents an ellipse that is centered at $(0, 0)$, with $a = 4$ and $b = 5$.

REVIEW EXERCISES

Graph each ellipse.

13. $\dfrac{x^2}{16} + \dfrac{y^2}{9} = 1$

14. $\dfrac{(x-2)^2}{4} + \dfrac{(y-1)^2}{25} = 1$

15. $4(x + 1)^2 + 9(y - 1)^2 = 36$

16. Consider the equation $\dfrac{x^2}{144} + \dfrac{y^2}{1} = 1$. Write each term on the left side with a denominator that is the square of a number.

17. Determine whether the graph of each equation is a circle, a parabola, or an ellipse.

   a. $(x - 1)^2 + (y + 9)^2 = 100$

   b. $\dfrac{x^2}{49} + \dfrac{y^2}{121} = 1$

   c. $x = y^2 - 2y + 6$

   d. $16(x - 4)^2 + 4(y + 8)^2 = 16$

18. SALAMI   When a delicatessen slices a cylindrical salami at an angle, the results are elliptical pieces that are larger than circular pieces. See the illustration in the next column. Write the equation of the shape of the slice of salami shown if it was centered at the origin of a coordinate system.

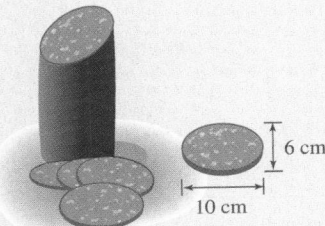

6 cm

10 cm

19. Fill in the blanks: An _____ is the set of all points in a plane for which the sum of the distances from two fixed points is a constant. Each of the fixed points is called a _____.

20. CONSTRUCTION   Sketch the path of the sound when a person, standing at one focus, whispers something in the whispering gallery dome shown below.

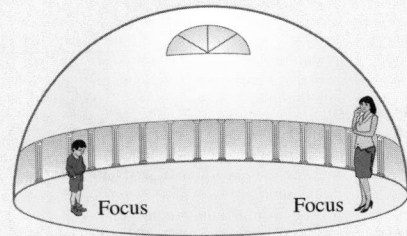

Focus   Focus

SECTION 13.3   The Hyperbola

DEFINITIONS AND CONCEPTS	EXAMPLES

A **hyperbola** is the set of all points in a plane for which the difference of the distances from two fixed points is a constant.

Standard forms of the equation of a hyperbola:

$\dfrac{x^2}{a^2} - \dfrac{y^2}{b^2} = 1$ Center (0, 0), opens left and right

$\dfrac{y^2}{a^2} - \dfrac{x^2}{b^2} = 1$ Center (0, 0), opens up and down

$\dfrac{(x-h)^2}{a^2} - \dfrac{(y-k)^2}{b^2} = 1$ Center (h, k), opens left and right

$\dfrac{(y-k)^2}{a^2} - \dfrac{(x-h)^2}{b^2} = 1$ Center (h, k), opens up and down

The equation $\dfrac{x^2}{4} - \dfrac{y^2}{9} = 1$, which can be written $\dfrac{x^2}{2^2} - \dfrac{y^2}{3^2} = 1$, represents a hyperbola, centered at (0, 0), that opens left and right. Here, $a = 2$ and $b = 3$.

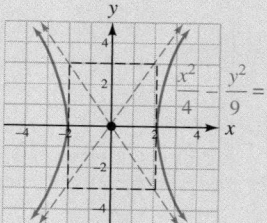
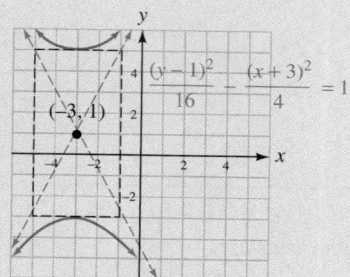

The equation $\dfrac{(y-1)^2}{16} - \dfrac{(x+3)^2}{4} = 1$, which can be written $\dfrac{(y-1)^2}{4^2} - \dfrac{(x+3)^2}{2^2} = 1$, represents a hyperbola, centered at $(-3, 1)$, that opens up and down. Here, $a = 4$ and $b = 2$.

To write the equation $25y^2 - 9x^2 = 225$ in standard form, divide both sides by 225 and simplify.

$$25y^2 - 9x^2 = 225$$

$$\dfrac{25y^2}{225} - \dfrac{9x^2}{225} = \dfrac{225}{225}$$ To get 1 on the right side, divide both sides by 225.

$$\dfrac{y^2}{9} - \dfrac{x^2}{25} = 1$$ Simplify each fraction.

This result represents a hyperbola centered at the origin that opens up and down. Here, $a = 3$ and $b = 5$.

REVIEW EXERCISES

Graph each hyperbola.

21. $\dfrac{y^2}{9} - \dfrac{x^2}{1} = 1$

22. $9(x-1)^2 - 4(y+1)^2 = 36$

23. $\dfrac{(y-2)^2}{25} - \dfrac{(x+1)^2}{25} = 1$

24. $xy = 9$

25. ELECTROSTATIC REPULSION Two similarly charged particles are shot together for an almost head-on collision, as in the illustration. They repel each other and travel the two branches of the hyperbola given by $x^2 - 4y^2 = 4$ on the given coordinate system. How close do they get?

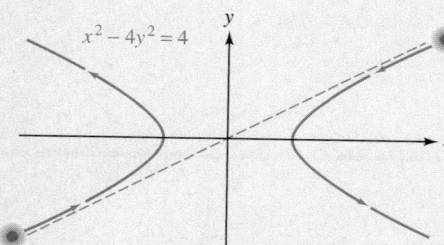

26. Determine whether the graph of each equation will be a circle, parabola, ellipse, or hyperbola.

a. $\dfrac{(x-4)^2}{16} + \dfrac{y^2}{49} = 1$

b. $16(x+3)^2 - 4(y-1)^2 = 64$

c. $x = -4y^2 - y + 1$

d. $x^2 + 2x + y^2 - 4y = 40$

SECTION 13.4   Solving Nonlinear Systems of Equations

DEFINITIONS AND CONCEPTS	EXAMPLES
A **nonlinear system of equations** is a system that contains at least one nonlinear equation. Systems of nonlinear equations are solved by **graphing**, by **substitution**, or by **elimination (addition)**.	To solve the nonlinear system $\begin{cases} x^2 + y^2 = 20 \\ y = x^2 \end{cases}$ by graphing, we graph the equations on the same rectangular coordinate system, and determine the coordinates of the points of intersection of the graphs. 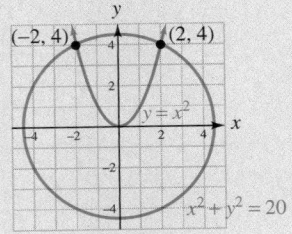 Since the points of intersection of the graphs are $(-2, 4)$ and $(2, 4)$, the solutions of the system are $(-2, 4)$ and $(2, 4)$.

With the **substitution method**, the objective is to use an appropriate substitution to obtain *one* equation in *one* variable.

To use substitution to solve the nonlinear system $\begin{cases} y = 3x - 5 \\ x^2 + y^2 = 5 \end{cases}$, we substitute $3x - 5$ for y in the second equation and solve for x.

$$x^2 + y^2 = 5$$
$$x^2 + (3x - 5)^2 = 5 \qquad \text{This is a quadratic equation in } x.$$
$$x^2 + 9x^2 - 30x + 25 = 5$$
$$10x^2 - 30x + 20 = 0 \qquad \text{Combine terms and subtract 5 from both sides.}$$
$$x^2 - 3x + 2 = 0 \qquad \text{Divide both sides by 10.}$$
$$(x - 2)(x - 1) = 0 \qquad \text{Factor.}$$
$$x - 2 = 0 \quad \text{or} \quad x - 1 = 0 \qquad \text{Set each factor equal to 0.}$$
$$x = 2 \qquad \qquad x = 1$$

If $x = 2$, then $y = 3x - 5 = 3(2) - 5 = 1$.
If $x = 1$, then $y = 3x - 5 = 3(1) - 5 = -2$.

The two solutions of the system are $(2, 1)$ and $(1, -2)$.

With the **elimination (addition) method**, we combine the equations in a way that will eliminate the terms of one of the variables.

To use elimination to solve the nonlinear system $\begin{cases} x^2 - y = 0 \\ x + y = 0 \end{cases}$, we add the equations to get $x^2 + x = 0$. Then we factor this result to get $x = 0$ or $x = -1$. We can substitute these values into the second equation to find y.

If $x = 0$: $\quad x + y = 0 \qquad$ This is the second equation.
$\qquad \qquad \quad 0 + y = 0 \qquad$ Substitute 0 for x.
$\qquad \qquad \qquad \quad y = 0$

If $x = -1$: $\quad x + y = 0 \qquad$ This is the second equation.
$\qquad \qquad \quad -1 + y = 0 \qquad$ Substitute −1 for x.
$\qquad \qquad \qquad \quad y = 1$

The two solutions of the system are $(0, 0)$ and $(-1, 1)$.

REVIEW EXERCISES

27. Determine whether $\left(-\sqrt{11}, -3\right)$ is a solution of the system:
$$\begin{cases} x^2 + y^2 = 20 \\ x^2 - y^2 = 2 \end{cases}$$

28. The graphs of $y^2 - x^2 = 9$ and $x^2 + y^2 = 9$ are shown. Estimate the solutions of the system
$$\begin{cases} y^2 - x^2 = 9 \\ x^2 + y^2 = 9 \end{cases}$$

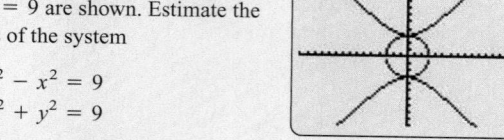

29. Solve the system $\begin{cases} xy = 4 \\ y = 2x - 2 \end{cases}$ by graphing.

30. Determine the maximum number of solutions there could be for a system of equations consisting of the given curves.

   a. A line and an ellipse   **b.** Two hyperbolas

   c. An ellipse and a circle   **d.** A parabola and a circle

31. Suppose the x-coordinate of both points of intersection of the circle, represented by $x^2 + y^2 = 1$, and the hyperbola, defined by $4y^2 - x^2 = 4$, is 0. Without graphing, determine the y-coordinates of both points of intersection. Express the answers as ordered-pair solutions.

32. Find a substitution equation that can be used to solve
$$\begin{cases} x^2 + y^2 = 16 \\ 3x - y = 1 \end{cases}$$ Do not solve the system.

Solve each system for real values of x and y.

33. $\begin{cases} y^2 - x^2 = 16 \\ y + 4 = x^2 \end{cases}$

34. $\begin{cases} y = -x^2 + 2 \\ x^2 - y - 2 = 0 \end{cases}$

35. $\begin{cases} x^2 + 2y^2 = 12 \\ 2x - y = 2 \end{cases}$

36. $\begin{cases} 3x^2 + y^2 = 52 \\ x^2 - y^2 = 12 \end{cases}$

37. $\begin{cases} \dfrac{x^2}{16} + \dfrac{y^2}{12} = 1 \\ \dfrac{x^2}{1} - \dfrac{y^2}{3} = 1 \end{cases}$

38. $\begin{cases} xy = 4 \\ \dfrac{x^2}{1} + \dfrac{y^2}{2} = 9 \end{cases}$

39. $\begin{cases} y = -x^2 + 1 \\ x + y = 5 \end{cases}$

40. $\begin{cases} x = y^2 - 3 \\ x = y^2 - 3y \end{cases}$

CHAPTER 13
Test

1. Fill in the blanks.

   a. The curves formed by the intersection of a plane with an infinite right-circular cone are called _____ sections.

   b. A circle is the set of all points in a plane that are a fixed distance from a point called its _____. The fixed distance is called the _____ of the circle.

   c. The standard form for the equation of a(n) _____ centered at the origin that opens left and right is $\dfrac{x^2}{a^2} - \dfrac{y^2}{b^2} = 1$.

   d. $\begin{cases} y = x^2 + x - 4 \\ x^2 + y^2 = 36 \end{cases}$ is a(n) _____ system of equations.

   e. The standard form for the equation of a(n) _____ centered at the origin is $\dfrac{x^2}{a^2} + \dfrac{y^2}{b^2} = 1$.

2. Find the center and the radius of the circle represented by the equation $x^2 + y^2 = 100$ and graph it.

3. Find the center and the radius of the circle represented by the equation $x^2 + y^2 + 4x - 6y = 5$.

4. TV HISTORY In the early days of television, stations broadcast a black-and-white test pattern like that shown here during the early morning hours. Use the given coordinate system to write an equation of the large, bold circle in the center of the pattern.

Graph each equation.

5. $(x + 2)^2 + (y - 1)^2 = 9$

6. $x = y^2 - 4y + 3$

7. $y = -2x^2 - 4x + 5$

8. $xy = -4$

9. $9x^2 + 4y^2 = 36$

10. $\dfrac{(x - 2)^2}{9} - \dfrac{y^2}{1} = 1$

11. Write the equation in standard form of the ellipse graphed here.

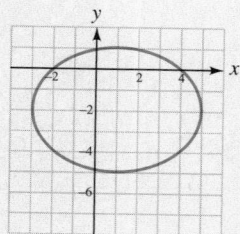

12. LIGHT The cross section of a parabolic mirror is given by the equation $x = \frac{1}{10}y^2$, with distances measured in inches. If the dish is 10 inches wide, how deep is it?

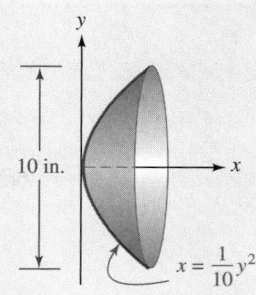

10 in.

$x = \frac{1}{10}y^2$

13. Give an example of the reflective properties of an ellipse. Include a drawing and label it completely.

14. Find the center and the length and width of the central rectangle of the graph of $(x + 1)^2 - (y - 1)^2 = 4$.

15. Find the equation in standard form of the hyperbola graphed here.

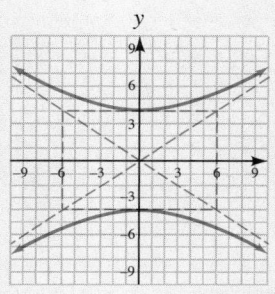

16. Determine whether the graph of each equation will be a circle, a parabola, an ellipse, or a hyperbola.

  a. $25x^2 + 100y^2 = 400$

  b. $9x^2 - y^2 = 9$

  c. $x^2 + 8x + y^2 - 16y - 1 = 0$

  d. $x = 8y^2 - 9y + 4$

Solve each system graphically.

17. $\begin{cases} x^2 + y^2 = 25 \\ y - x = 1 \end{cases}$

Solve each system for real values of x and y.

18. $\begin{cases} 2x - y = -2 \\ x^2 + y^2 = 16 + 4y \end{cases}$

19. $\begin{cases} 5x^2 - y^2 - 3 = 0 \\ x^2 + 2y^2 = 5 \end{cases}$

20. $\begin{cases} xy = -\dfrac{9}{2} \\ 3x + 2y = 6 \end{cases}$

21. $\begin{cases} y = x + 1 \\ x^2 - y^2 = 1 \end{cases}$

22. $\begin{cases} x^2 + 3y^2 = 6 \\ x^2 + y = 8 \end{cases}$

GROUP PROJECT

PARABOLAS

Overview: In this activity, you will construct several models of parabolas.

Instructions: Form groups of 2 or 3 students. You will need a T-square, string, paper, pencil, and a thumbtack. To construct a parabola, secure one end of a piece of string that is as long as the T-square to a large piece of paper using a brad or thumbtack, as shown below. Attach the other end of the string to the upper end of the T-square. Hold the string taut against the T-square with a pencil and slide the T-square along the edge of the table. As the T-square moves, the pencil will trace a parabola.

Each point on the parabola is the same distance away from a given point as it is from a given line. With this model, what is the given point, and what is the given line?

Make other models by moving the fixed point closer and farther away from the edge of the table. How is the shape of the parabola affected?

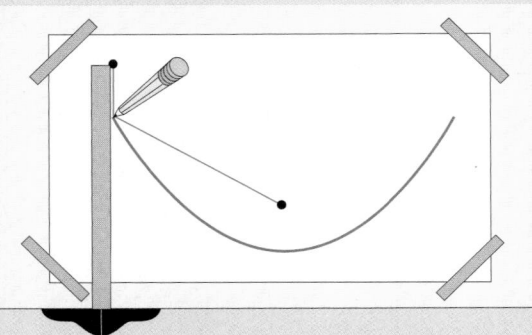

ELLIPSES

Overview: In this activity, you will construct several ellipses.

Instructions: Form groups of 2 or 3 students. You will need two thumbtacks, a pencil, and a length of string with a loop tied at one end. To construct an ellipse, place two thumbtacks (or brads) fairly close together, as shown in the illustration. Catch the loop of the string with the point of the pencil and, keeping the string taut, draw the ellipse.

Make several models by moving one of the thumbtacks farther away and then closer to the other thumbtack. How does the shape of the ellipse change?

For each point on the ellipse, the sum of the distances of the point from two given points is a constant. With this method of construction, what are the two points? What is the constant distance?

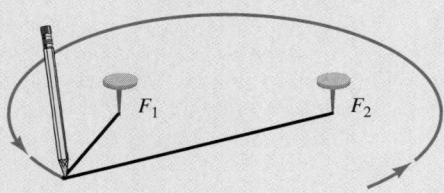

CHAPTER 14

Miscellaneous Topics

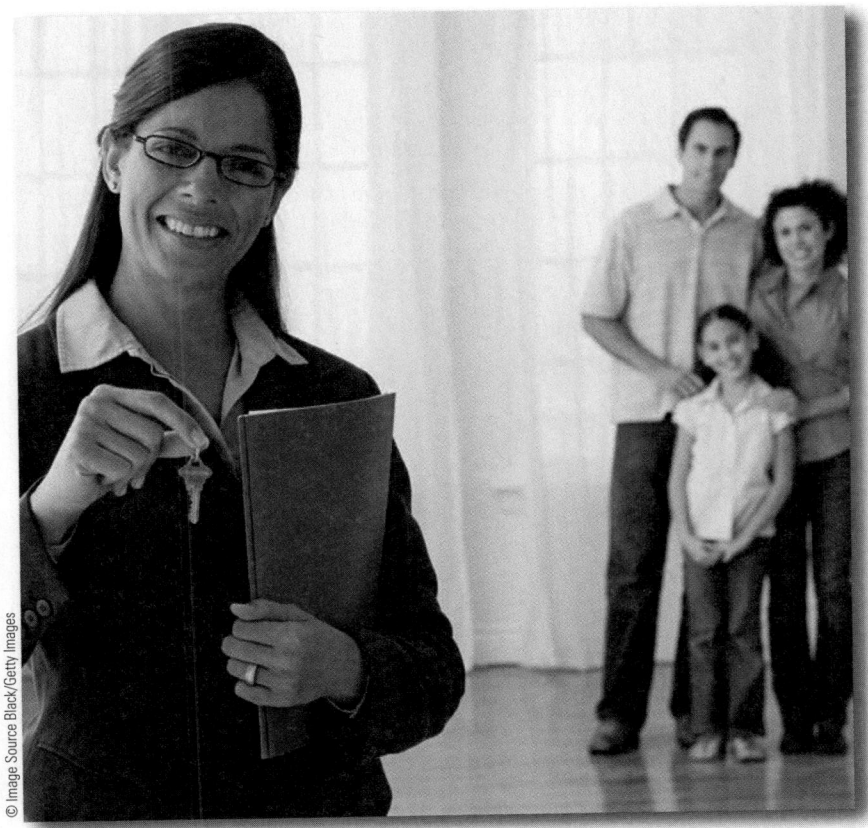

© Image Source Black/Getty Images

from Campus to Careers
Real Estate Sales Agent

Buying a house is probably the biggest purchase that most people will make in their lives. The complex process of purchasing a home is much easier with the help of a real estate agent. Real estate agents use their mathematical skills in many ways. They compute square footage, appraise property, calculate commissions, and write offer sheets. Technology is widely used in the real estate industry. Most sales agents use computers to locate and list available properties and identify sources of financing.

In **Problem 81** of **Study Set 14.3,** you will find what the value of a $250,000 home in Seattle will be in 10 years if property values continue to increase at the current rate.

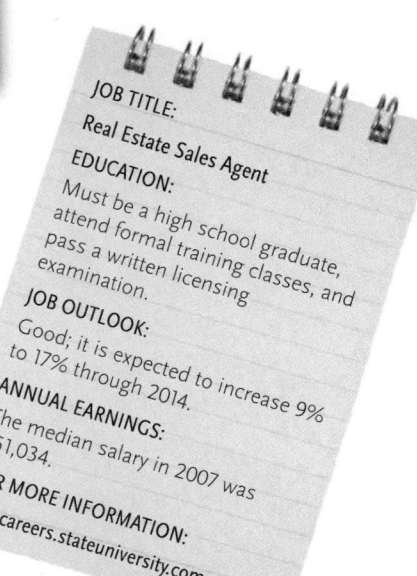

JOB TITLE:
Real Estate Sales Agent

EDUCATION:
Must be a high school graduate, attend formal training classes, and pass a written licensing examination.

JOB OUTLOOK:
Good; it is expected to increase 9% to 17% through 2014.

ANNUAL EARNINGS:
The median salary in 2007 was $51,034.

FOR MORE INFORMATION:
www.careers.stateuniversity.com

Exploring Careers

Ultimately, your choice of career will determine the math course(s) that you need to take after Intermediate Algebra. Before the end of this term, it would be wise to have at least a general idea of your career goals.

HOW DO YOU DECIDE?: Seek the advice of a counselor, visit your school's career center, search the Internet, or read books that will help you discover your interests and possible related careers.

ONCE YOU'VE DECIDED: Talk to your counselor and consult the appropriate college catalogs to develop a long-term plan that will put you on the correct educational path.

Now Try This

1. Do you have a career goal in mind? If so, what is it?
2. Take at least two personality tests and two career-choice tests. A list of tests offered online can be found at http://academic.cengage.com/math/tussy/.
3. Visit a counselor to discuss which classes you should take during your next term and beyond. Make a list of classes that your counselor suggests that you take.

SECTION 14.1
The Binomial Theorem

Objectives

1. Raise binomials to powers.
2. Use Pascal's triangle to expand binomials.
3. Use factorial notation.
4. Use the binomial theorem to expand binomials.
5. Find a specific term of a binomial expansion.

We have discussed how to raise binomials to positive-integer powers. For example, we have seen that

The Language of Algebra
Recall that two-term polynomial expressions such as $a + b$ and $3u - 2v$ are called *binomials*.

$$(a + b)^2 = a^2 + 2ab + b^2$$

and that $(a + b)^3 = (a + b)(a + b)^2$

$$= (a + b)(a^2 + 2ab + b^2)$$
$$= a^3 + 2a^2b + ab^2 + a^2b + 2ab^2 + b^3$$
$$= a^3 + 3a^2b + 3ab^2 + b^3$$

In this section, we will learn how to raise binomials to positive-integer powers without performing the multiplications.

① Raise Binomials to Powers.

To see how to raise binomials to nonnegative-integer powers, we consider the following binomial expansions of $a + b$.

$(a + b)^0 =$	1	1 term
$(a + b)^1 =$	$a + b$	2 terms
$(a + b)^2 =$	$a^2 + 2ab + b^2$	3 terms
$(a + b)^3 =$	$a^3 + 3a^2b + 3ab^2 + b^3$	4 terms
$(a + b)^4 =$	$a^4 + 4a^3b + 6a^2b^2 + 4ab^3 + b^4$	5 terms
$(a + b)^5 =$	$a^5 + 5a^4b + 10a^3b^2 + 10a^2b^3 + 5ab^4 + b^5$	6 terms
$(a + b)^6 =$	$a^6 + 6a^5b + 15a^4b^2 + 20a^3b^3 + 15a^2b^4 + 6ab^5 + b^6$	7 terms

Several patterns appear in these expansions:

1. Each expansion has one more term than the power of the binomial.

2. For each term of an expansion, the sum of the exponents on a and b is equal to the exponent of the binomial being expanded. For example, in the expansion of $(a + b)^5$, the sum of the exponents in each term is 5:

$$4 + 1 = 5 \quad\quad 3 + 2 = 5 \quad\quad 2 + 3 = 5 \quad\quad 1 + 4 = 5$$
$$(a + b)^5 = a^5 \;+\; 5\overbrace{a^4b} \;+\; 10\overbrace{a^3b^2} \;+\; 10\overbrace{a^2b^3} \;+\; 5\overbrace{ab^4} \;+\; b^5$$

3. The first term in each expansion is a, raised to the power of the binomial, and the last term in each expansion is b, raised to the power of the binomial.

4. The exponents on a decrease by one in each successive term, ending with $a^0 = 1$ in the last term. The exponents on b, beginning with $b^0 = 1$ in the first term, increase by one in each successive term. For example, the expansion of $(a + b)^4$ could be written as

$$a^4b^0 + 4a^3b^1 + 6a^2b^2 + 4a^1b^3 + a^0b^4$$

Thus, the variables have the pattern

$$a^n, \quad a^{n-1}b, \quad a^{n-2}b^2, \quad \ldots, \quad ab^{n-1}, \quad b^n$$

5. The coefficients of each expansion begin with 1, increase through some values, and then decrease through those same values, back to 1.

② Use Pascal's Triangle to Expand Binomials.

To see another pattern, we write the coefficients of each expansion of $a + b$ in a triangular array:

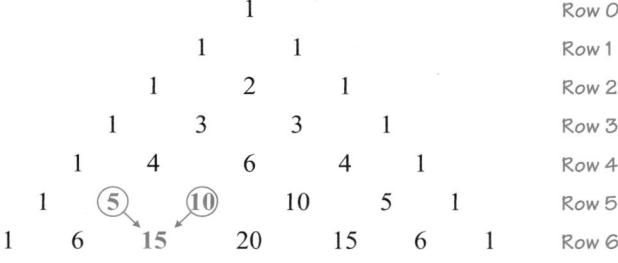

						1							Row 0
					1		1						Row 1
				1		2		1					Row 2
			1		3		3		1				Row 3
		1		4		6		4		1			Row 4
	1		⑤		⑩		10		5		1		Row 5
1		6		15		20		15		6		1	Row 6

In this array, called **Pascal's triangle**, each entry between the 1's is the sum of the closest pair of numbers in the line immediately above it. For example, the first 15 in the bottom row is the sum of the 5 and 10 immediately above it. Pascal's triangle continues with the same pattern forever. The next two lines are

$$\begin{array}{ccccccccccccccc} 1 && 7 && 21 && 35 && 35 && 21 && 7 && 1 \end{array}$$ Row 7

$$\begin{array}{ccccccccccccccccc} 1 && 8 && 28 && 56 && 70 && 56 && 28 && 8 && 1 \end{array}$$ Row 8

EXAMPLE 1   Expand: $(x + y)^5$

Strategy   We will use the pattern shown on the previous page for raising binomials to powers and Pascal's triangle.

Why   The pattern provides the variable expressions in the expansion and Pascal's triangle provides their coefficients.

Solution   The first term in the expansion is x^5, and the exponents on x decrease by one in each successive term. A y first appears in the second term, and the exponents on y increase by one in each successive term, concluding when the term y^5 is reached. Thus, the variable expressions in the expansion are

$$x^5, \quad x^4y, \quad x^3y^2, \quad x^2y^3, \quad xy^4, \quad y^5$$

Since the exponent of the binomial that is being expanded is 5, the coefficients of these variables are found in row 5 of Pascal's triangle.

$$\begin{array}{cccccc} 1 & 5 & 10 & 10 & 5 & 1 \end{array}$$ Remember, the 1 at the top of Pascal's triangle is labeled row 0.

Combining this information gives the following expansion:

$$(x + y)^5 = x^5 + 5x^4y + 10x^3y^2 + 10x^2y^3 + 5xy^4 + y^5$$

> **Self Check 1**   Expand: $(x + y)^4$
>
> ***Now Try***   **Problem 25**

EXAMPLE 2   Expand: $(u - v)^4$

Strategy   We will use the pattern shown on the previous page for raising binomials to powers and Pascal's triangle.

Why   The pattern provides the variable expressions in the expansion and Pascal's triangle provides their coefficients.

Solution   We note that $(u - v)^4$ can be written as $[u + (-v)]^4$. The variable expressions in this expansion are

$$u^4, \quad u^3(-v), \quad u^2(-v)^2, \quad u(-v)^3, \quad (-v)^4$$

and the coefficients are given in row 4 of Pascal's triangle:

$$\begin{array}{ccccc} 1 & 4 & 6 & 4 & 1 \end{array}$$ Remember, the 1 at the top of Pascal's triangle is labeled row 0.

Thus, the required expansion is

$$(u - v)^4 = u^4 + 4u^3(-v) + 6u^2(-v)^2 + 4u(-v)^3 + (-v)^4$$

The Language of Algebra
To *alternate* means to change back and forth. In this expansion, the signs + and − alternate.

Now we simplify each term. When $-v$ is raised to an even power, the sign is positive, and when $-v$ is raised to an odd power, the sign is negative. This causes the signs of the terms in the expansion to alternate between + and −.

$$(u - v)^4 = u^4 - 4u^3v + 6u^2v^2 - 4uv^3 + v^4$$

Self Check 2   Expand: $(x - y)^5$

Now Try   **Problem 27**

3   **Use Factorial Notation.**

Although Pascal's triangle gives the coefficients of the terms in a binomial expansion, it is not the easiest way to expand a binomial. To develop a better way, we introduce **factorial notation.** The symbol $n!$ (read as "*n* **factorial**") is defined as follows.

Factorial Notation

$n!$ is the product of consecutively decreasing natural numbers from n to 1.
  For any natural number n,

$$n! = n(n - 1)(n - 2)(n - 3) \cdot \cdots \cdot 3 \cdot 2 \cdot 1$$

Zero factorial is defined as $0! = 1$.

EXAMPLE 3   Evaluate each expression: **a.** $4!$ **b.** $6!$ **c.** $3! \cdot 2!$ **d.** $5! \cdot 0!$

Strategy   We will use the definition of $n!$.

Why   The definition explains how to evaluate factorials.

Solution
a. $4! = 4 \cdot 3 \cdot 2 \cdot 1 = 24$ Read as "4 factorial."
b. $6! = 6 \cdot 5 \cdot 4 \cdot 3 \cdot 2 \cdot 1 = 720$ Read as "6 factorial."
c. $3! \cdot 2! = (3 \cdot 2 \cdot 1) \cdot (2 \cdot 1) = 6 \cdot 2 = 12$ Find each factorial and multiply the results.
d. $5! \cdot 0! = (5 \cdot 4 \cdot 3 \cdot 2 \cdot 1) \cdot 1 = 120$ Simplify: $0! = 1$.

Self Check 3   Evaluate each expression:
                   a. $7!$ **b.** $4! \cdot 3!$ **c.** $1! \cdot 0!$

Now Try   **Problems 31 and 37**

Using Your Calculator   *Factorials*

We can find factorials using a calculator. For example, to find $12!$ with a scientific calculator, we enter

12 $\boxed{x!}$ (You may have to use a $\boxed{\text{2nd}}$ or $\boxed{\text{SHIFT}}$ key first.) $\boxed{479001600}$

To find 12! on a graphing calculator, we enter

12 [MATH] [▶] to PRB 4 [ENTER]

$$12! \\ \quad 479001600$$

The following property follows from the definition of factorial.

Factorial Property

For any natural number n,

$$n(n - 1)! = n!$$

We can use this property to evaluate many expressions involving factorials.

EXAMPLE 4 Evaluate each expression: **a.** $\dfrac{6!}{5!}$ **b.** $\dfrac{10!}{8!(10 - 8)!}$

Strategy   We will use the factorial property to partially expand the factorial and then we will simplify the fraction.

Why   By using this approach, we can avoid difficult multiplications and divisions.

Solution

a. If we write 6! as $6 \cdot 5!$, we can simplify the fraction by removing the common factor 5! in the numerator and denominator.

$$\frac{6!}{5!} = \frac{6 \cdot 5!}{5!} = \frac{6 \cdot \overset{1}{\cancel{5!}}}{\underset{1}{\cancel{5!}}} = 6 \qquad \text{Simplify: } \tfrac{5!}{5!} = 1.$$

b. We subtract within the parentheses, write 10! as $10 \cdot 9 \cdot 8!$, and simplify.

$$\frac{10!}{8!(10 - 8)!} = \frac{10!}{8! \cdot 2!} = \frac{10 \cdot 9 \cdot \overset{1}{\cancel{8!}}}{\underset{1}{\cancel{8!}} \cdot 2!} = \frac{5 \cdot \overset{1}{\cancel{2}} \cdot 9}{\underset{1}{\cancel{2}} \cdot 1} = 45 \qquad \begin{array}{l}\text{Simplify: } \tfrac{8!}{8!} = 1. \text{ Factor 10 as}\\ 5 \cdot 2 \text{ and simplify: } \tfrac{2}{2} = 1.\end{array}$$

 Self Check 4   Evaluate each expression: **a.** $\dfrac{4!}{3!}$ **b.** $\dfrac{7!}{5!(7 - 5)!}$

Now Try   **Problems 39 and 47**

 Use the Binomial Theorem to Expand Binomials.

The following theorem summarizes our observations about binomial expansions and our work with factorials. Known as the *binomial theorem,* it is usually the best way to expand a binomial.

The Binomial Theorem

For any positive integer n,

$$(a + b)^n = a^n + \frac{n!}{1!(n-1)!}a^{n-1}b + \frac{n!}{2!(n-2)!}a^{n-2}b^2 + \frac{n!}{3!(n-3)!}a^{n-3}b^3$$

$$+ \cdots + \frac{n!}{r!(n-r)!}a^{n-r}b^r + \cdots + b^n$$

In the binomial theorem, the exponents on the variables follow the familiar pattern:

- The sum of the exponents on a and b in each term is n.
- The exponents on a decrease by 1 in each successive term.
- The exponents on b increase by 1 in each successive term.

The method of finding the coefficients involves factorials. Except for the first and last terms, the numerator of each coefficient is $n!$. If the exponent on b in a particular term is r, the denominator of the coefficient of that term is $r!(n-r)!$.

EXAMPLE 5   Use the binomial theorem to expand $(a + b)^3$.

Strategy   We will substitute 3 for n in the binomial theorem and simplify.

Why   The binomial theorem is the fastest way to expand expressions of the form $(a + b)^n$.

Solution

$$(a + b)^3 = a^3 + \frac{3!}{1!(3-1)!}a^2b + \frac{3!}{2!(3-2)!}ab^2 + b^3$$

$$= a^3 + \frac{3!}{1!\cdot 2!}a^2b + \frac{3!}{2!\cdot 1!}ab^2 + b^3$$

$$= a^3 + \frac{3\cdot \overset{1}{\cancel{2!}}}{1!\cdot \underset{1}{\cancel{2!}}}a^2b + \frac{3\cdot \overset{1}{\cancel{2!}}}{\underset{1}{\cancel{2!}}\cdot 1!}ab^2 + b^3 \quad \text{Write 3! as } 3\cdot 2! \text{ to simplify the fractions.}$$

$$= a^3 + 3a^2b + 3ab^2 + b^3$$

Self Check 5   Use the binomial theorem to expand $(a + b)^4$.

Now Try   **Problem 57**

We can find expansions of binomials in variables other than a and b by making substitutions into the binomial theorem.

EXAMPLE 6   Use the binomial theorem to expand $(x - y)^4$.

Strategy   First, we will write $(x - y)^4$ as $[x + (-y)]^4$. Then we will use the binomial theorem with $a = x$, $b = -y$, and $n = 4$.

Why   To directly substitute into the binomial theorem, the difference within the parentheses, $x - y$, must be expressed as a sum.

Solution

$$(x - y)^4 = [x + (-y)]^4$$

$$= x^4 + \frac{4!}{1!(4-1)!}x^3(-y) + \frac{4!}{2!(4-2)!}x^2(-y)^2 + \frac{4!}{3!(4-3)!}x(-y)^3 + (-y)^4$$

$$= x^4 - \frac{4!}{1! \cdot 3!}x^3y + \frac{4!}{2! \cdot 2!}x^2y^2 - \frac{4!}{3! \cdot 1!}xy^3 + y^4$$

$$= x^4 - \frac{4 \cdot \overset{1}{3!}}{1! \cdot 3!}x^3y + \frac{4 \cdot 3 \cdot \overset{1}{2!}}{2! \cdot 2 \cdot 1}x^2y^2 - \frac{4 \cdot \overset{1}{3!}}{3! \cdot 1!}xy^3 + y^4 \qquad \text{Write 4! as } 4 \cdot 3! \text{ and as}$$
$$\qquad\qquad\qquad\qquad\qquad\qquad\qquad\qquad\qquad\qquad\qquad\qquad\qquad\qquad 4 \cdot 3 \cdot 2! \text{ to simplify the}$$
$$\qquad\qquad\qquad\qquad\qquad\qquad\qquad\qquad\qquad\qquad\qquad\qquad\qquad\qquad \text{fractions.}$$

$$= x^4 - 4x^3y + 6x^2y^2 - 4xy^3 + y^4 \qquad\qquad\qquad \text{Note the alternating signs.}$$

 Self Check 6   Use the binomial theorem to expand $(x - y)^3$.

Now Try   **Problem 59**

EXAMPLE 7   Use the binomial theorem to expand $(3u - 2v)^4$.

Strategy   We will write the expansion of $(a + b)^4$. Then we will substitute for a and b to find the expansion of $(3u - 2v)^4$.

Why   For binomials with more complicated terms, the computations are often easier if the general expansion is written first, followed by the appropriate substitutions.

Solution   We can use the binomial theorem to expand $(a + b)^4$.

$$(a + b)^4 = a^4 + \frac{4!}{1!(4-1)!}a^3b + \frac{4!}{2!(4-2)!}a^2b^2 + \frac{4!}{3!(4-3)!}ab^3 + b^4$$

$$= a^4 + 4a^3b + 6a^2b^2 + 4ab^3 + b^4$$

If we write $(3u - 2v)^4$ as $[3u + (-2v)]^4$, we see that the expressions $3u$ and $-2v$ can be substituted for a and b respectively in the expansion of $(a + b)^4$.

$$(3u - 2v) = (3u)^4 + 4(3u)^3(-2v) + 6(3u)^2(-2v)^2 + 4(3u)(-2v)^3 + (-2v)^4$$

$$= 81u^4 - 216u^3v + 216u^2v^2 - 96uv^3 + 16v^4$$

 Self Check 7   Use the binomial theorem to expand $(4a - 5b)^3$.

Now Try   **Problem 69**

⑤ **Find a Specific Term of a Binomial Expansion.**

To find a specific term of a binomial expansion, we don't need to write out the entire expansion. The binomial theorem and the pattern of the terms suggest the following method for finding a single term of an expansion.

Finding a Specific Term of a Binomial Expansion	The $(r + 1)$st term of the expansion of $(a + b)^n$ is $$\frac{n!}{r!(n-r)!}a^{n-r}b^r$$

EXAMPLE 8 Find the 4th term of the expansion of $(a + b)^9$.

Strategy We will determine n and r and substitute into the formula for finding a specific term of a binomial expansion.

Why We will use the formula because it enables us to find the fourth term of the expansion without us having to write out all the terms of the expansion.

Solution To use the formula for finding a specific term of $(a + b)^9$, we must determine n and r. Since $r + 1 = 4$ in the fourth term, $r = 3$ and since this binomial is raised to the 9th power, $n = 9$.

We substitute 3 for r and 9 for n into the formula to find the fourth term.

$$\frac{n!}{r!(n - r)!}a^{n-r}b^r = \frac{9!}{3!(9 - 3)!}a^{9-3}b^3$$

$$= \frac{9!}{3!6!}a^6b^3 \qquad \text{Evaluate: } \frac{9!}{3!6!} = \frac{9 \cdot 8 \cdot 7 \cdot \overset{1}{\cancel{6!}}}{3 \cdot 2 \cdot 1 \cdot \underset{1}{\cancel{6!}}} = 84.$$

$$= 84a^6b^3$$

 Self Check 8 Find the 3rd term of the expansion of $(a + b)^9$.

Now Try Problem 77

EXAMPLE 9 Find the 6th term of the expansion of $\left(x^2 - \frac{y}{2}\right)^7$.

Strategy We will determine n, r, a, and b and substitute these values into the formula for finding a specific term of a binomial expansion.

Why We will use the formula because it enables us to find the sixth term of the expansion without us having to write out all the terms of the expansion.

Solution To use the formula for finding a specific term of $\left(x^2 - \frac{y}{2}\right)^7$, we must determine n, r, a, and b. In the sixth term, $r + 1 = 6$. So $r = 5$. By comparing $\left(x^2 - \frac{y}{2}\right)^7$ to $(a + b)^n$, we see that $a = x^2$, $b = -\frac{y}{2}$, and $n = 7$. We substitute these values into the formula as follows.

$$\frac{n!}{r!(n - r)!}a^{n-r}b^r = \frac{7!}{5!(7 - 5)!}(x^2)^{7-5}\left(-\frac{y}{2}\right)^5$$

$$= \frac{7!}{5!2!}(x^2)^2\left(-\frac{y^5}{32}\right) \qquad \text{Evaluate: } \frac{7!}{5!2!} = \frac{7 \cdot 6 \cdot \overset{1}{\cancel{5!}}}{\underset{1}{\cancel{5!}} \cdot 2 \cdot 1} = 21.$$

$$= -\frac{21}{32}x^4y^5$$

 Self Check 9 Find the 5th term of the expansion of $\left(c^2 - \frac{d}{3}\right)^7$.

Now Try Problem 87

> ◢ **ANSWERS TO SELF CHECKS**
> 1. $x^4 + 4x^3y + 6x^2y^2 + 4xy^3 + y^4$
> 2. $x^5 - 5x^4y + 10x^3y^2 - 10x^2y^3 + 5xy^4 - y^5$ 3. a. 5,040 b. 144 c. 1 4. a. 4 b. 21
> 5. $a^4 + 4a^3b + 6a^2b^2 + 4ab^3 + b^4$ 6. $x^3 - 3x^2y + 3xy^2 - y^3$ 7. $64a^3 - 240a^2b + 300ab^2 - 125b^3$
> 8. $36a^7b^2$ 9. $\frac{35}{81}c^6d^4$

STUDY SET
14.1

VOCABULARY

Fill in the blanks.

1. The two-term polynomial expression $a + b$ is called a _____.

2. $a^4 + 4a^3b + 6a^2b^2 + 4ab^3 + b^4$ is the binomial _____ of $(a + b)^4$.

3. We can use the _____ theorem to raise binomials to positive-integer powers without doing the actual multiplication.

4. The array of numbers that gives the coefficients of the terms of a binomial expansion is called _____ triangle.

5. $n!$ (read as "n _____") is the product of consecutive _____ natural numbers from n to 1.

6. In the expansion $a^3 - 3a^2b + 3ab^2 - b^3$, the signs _____ between + and −.

CONCEPTS

Fill in the blanks.

7. The binomial expansion of $(m + n)^6$ has ____ more term than the power of the binomial.

8. For each term of the expansion of $(a + b)^8$, the sum of the exponents of a and b is ▮.

9. The first term of the expansion of $(r + s)^{20}$ is r▮ and the last term is s▮.

10. In the expansion of $(m - n)^{15}$, the exponents on m _____ and the exponents on n _____.

11. The coefficients of the terms of the expansion of $(c + d)^{20}$ begin with ▮, increase through some values, and then decrease through those same values, back to ▮.

12. Complete Pascal's Triangle:

```
                        1
                    1       1
                1       2      ▮
            1       ▮      3       1
        1       ▮      6       4       1
    1       5      10      10       5       1
  1       ▮     15       ▮      15      6       1
1      7      21     35        ▮     21      7      1
1    8     28     56     70       56      ▮     8     ▮
```

13. $n \cdot (\ ▮\ -\ ▮\ )! = n!$

14. $8! = 8 \cdot\ ▮\ !$

15. $0! = ▮$

16. According to the binomial theorem, the third term of the expansion of $(a + b)^n$ is $\dfrac{▮!}{▮!(n - 2)!}a^{n-2}b▮$.

17. The coefficient of the fourth term of the expansion of $(a + b)^9$ is 9! divided by 3!$(\ ▮\ -\ ▮\ )!$.

18. The exponent on a in the fourth term of the expansion of $(a + b)^6$ is ▮ and the exponent on b is ▮.

19. The exponent on a in the fifth term of the expansion of $(a + b)^6$ is ▮ and the exponent on b is ▮.

20. The expansion of $(a - b)^4$ is

$$a^4\ ▮\ 4a^3b\ ▮\ 6a^2b^2\ ▮\ 4ab^3\ ▮\ b^4$$

21. $(x + y)^3$

$$= x▮\ +\ \frac{▮!}{1!(3 - 1)!}x^2▮\ +\ \frac{3!}{▮!(3 - 2)!}xy▮\ +\ y▮$$

22. Fill in the blanks.

   a. The $(r + 1)$st term of the expansion of $(a + b)^n$ is $\dfrac{n!}{r!(n - ▮)!}a^{▮-r}b▮$.

   b. To use this formula to find the 6th term of the expansion of $\left(m + \frac{n}{2}\right)^8$, we note that $r = ▮$, $n = ▮$, $a = ▮$, and $b = ▮$.

NOTATION

Fill in the blanks.

23. $n! = n(\ ▮\ -\ ▮\ )(n - 2) \cdot\ \cdots\ \cdot 3 \cdot 2 \cdot 1$

24. The symbol 5! is read as "_____ _____" and it means $5 \cdot\ ▮\ \cdot\ ▮\ \cdot\ ▮\ \cdot\ ▮$.

GUIDED PRACTICE

Use Pascal's triangle to expand each binomial. See Examples 1 and 2.

25. $(a + b)^3$
26. $(m + p)^4$
27. $(m - p)^5$
28. $(a - b)^3$

Evaluate each expression. See Examples 3 and 4.

29. 3!

30. 7!

31. $5!$

32. $6!$

33. $3! + 4!$

34. $4! + 4!$

35. $3!(4!)$

36. $2!(3!)$

37. $8(7!)$

38. $4!(5)$

39. $\dfrac{49!}{47!}$

40. $\dfrac{101!}{100!}$

41. $\dfrac{9!}{11!}$

42. $\dfrac{13!}{10!}$

43. $\dfrac{9!}{7!0!}$

44. $\dfrac{7!}{5!0!}$

45. $\dfrac{5!}{1!(5-1)!}$

46. $\dfrac{15!}{14!(15-14)!}$

47. $\dfrac{5!}{3!(5-3)!}$

48. $\dfrac{6!}{4!(6-4)!}$

49. $\dfrac{5!(8-5)!}{4! \cdot 7!}$

50. $\dfrac{6! \cdot 7!}{(8-3)!(7-4)!}$

51. $\dfrac{7!}{5!(7-5)!}$

52. $\dfrac{8!}{6!(8-6)!}$

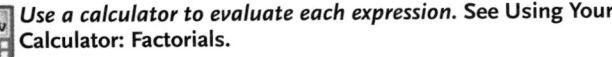 **Use a calculator to evaluate each expression. See Using Your Calculator: Factorials.**

53. $11!$

54. $13!$

55. $20!$

56. $55!$

Use the binomial theorem to expand each expression. See Examples 5 and 6.

57. $(m + n)^4$

58. $(a - b)^4$

59. $(c - d)^5$

60. $(c + d)^5$

61. $(a - b)^9$

62. $(a + b)^7$

63. $(s + t)^6$

64. $(s - t)^6$

Use the binomial theorem to expand each expression. See Example 7.

65. $(2x + y)^3$

66. $(x + 2y)^3$

67. $(2t - 3)^5$

68. $(2b + 1)^4$

69. $(5m - 2n)^4$

70. $(2m + 3n)^5$

71. $\left(\dfrac{x}{3} + \dfrac{y}{2}\right)^3$

72. $\left(\dfrac{x}{2} - \dfrac{y}{3}\right)^3$

73. $\left(\dfrac{x}{3} - \dfrac{y}{2}\right)^4$

74. $\left(\dfrac{x}{2} + \dfrac{y}{3}\right)^4$

75. $(c^2 - d^2)^5$

76. $(u^2 - v^3)^5$

Find the indicated term of each binomial expansion. See Examples 8 and 9.

77. $(x + y)^8$; 3rd term

78. $(x + y)^9$; 7th term

79. $(r + s)^6$; 5th term

80. $(r + s)^7$; 5th term

81. $(x - 1)^{13}$; 3rd term

82. $(x - 1)^{10}$; 5th term

83. $(x - 3y)^4$; 2nd term

84. $(3x - y)^5$; 3rd term

85. $(2x - 3y)^5$; 5th term

86. $(3x - 2y)^4$; 2nd term

87. $\left(\dfrac{c}{2} - \dfrac{d}{3}\right)^4$; 2nd term

88. $\left(\dfrac{c}{3} + \dfrac{d}{2}\right)^5$; 4th term

89. $(2t - 5)^7$; 4th term

90. $(2t - 3)^6$; 6th term

91. $(a^2 + b^2)^6$; 2nd term

92. $(a^2 + b^2)^7$; 6th term

WRITING

93. Describe how to construct Pascal's triangle.

94. Explain why the signs alternate in the expansion of $(x - y)^9$.

95. Explain why the third term of the expansion of $(m + 3n)^9$ could not be $324m^7n^3$.

96. Using your own words, write a definition of $n!$.

REVIEW

Assume that x, y, z, and b represent positive numbers. Use the properties of logarithms to write each expression as the logarithm of a single quantity.

97. $2 \log x + \dfrac{1}{2} \log y$

98. $-2 \log x - 3 \log y + \log z$

99. $\ln(xy + y^2) - \ln(xz + yz) + \ln z$

100. $\log_2 (x + 1) - \log_2 x$

CHALLENGE PROBLEMS

101. Find the constant term in the expansion of $\left(x + \dfrac{1}{x}\right)^{10}$.

102. Find the coefficient of a^5 in the expansion of $\left(a - \dfrac{1}{a}\right)^9$.

103. **a.** If we applied the pattern of the coefficients to the coefficient of the first term in a binomial expansion, the coefficient would be $\dfrac{n!}{0!(n - 0)!}$. Show that this expression is 1.

b. If we applied the pattern of the coefficients to the coefficient of the last term in a binomial expansion, the coefficient would be $\dfrac{n!}{n!(n - n)!}$. Show that this expression is 1.

104. Expand $(i - 1)^7$, where $i = \sqrt{-1}$.

SECTION 14.2
Arithmetic Sequences and Series

Objectives

❶ Find terms of a sequence given the general term.
❷ Find terms of an arithmetic sequence by identifying the first term and the common difference.
❸ Find arithmetic means.
❹ Find the sum of the first n terms of an arithmetic sequence.
❺ Solve application problems involving arithmetic sequences.
❻ Use summation notation.

The word *sequence* is used in everyday conversation when referring to an ordered list. For example, a history instructor might discuss the sequence of events that led up to the sinking of the *Titanic*. In mathematics, a **sequence** is a list of numbers written in a specific order.

❶ **Find Terms of a Sequence Given the General Term.**

Each number in a sequence is called a **term** of the sequence. **Finite sequences** contain a finite number of terms and **infinite sequences** contain infinitely many terms. Two examples of sequences are:

Finite sequence: 1, 5, 9, 13, 17, 21, 25

Infinite sequence: 3, 6, 9, 12, 15, . . .   The . . . indicates that the sequence goes on forever.

Sequences are defined formally using the terminology of functions.

Finite and Infinite Sequences

A **finite sequence** is a function whose domain is the set of natural numbers $\{1, 2, 3, 4, . . . , n\}$ for some natural number n.

An **infinite sequence** is a function whose domain is the set of natural numbers: $\{1, 2, 3, 4, . . .\}$.

Instead of using $f(x)$ notation, we use a_n (read as "a sub n") notation to write the value of a sequence at a number n. For the infinite sequence introduced earlier, we have:

1st term	2nd term	3rd term	4th term	5th term
3,	6,	9,	12,	15, . . .
a_1	a_2	a_3	a_4	a_5

To specifically describe all the terms of a sequence, we can write a formula for a_n, called the **general term** of the sequence. For the sequence 3, 6, 9, 12, 15, . . . , we note that $a_1 = 3 \cdot 1$, $a_2 = 3 \cdot 2$, $a_3 = 3 \cdot 3$, and so on. In general, the nth term of the sequence is found by multiplying n by 3.

$a_n = 3n$ Read a_n as "a sub n."

We can use this formula to find any term of the sequence. For example, to find the 12th term, we substitute 12 for n.

$$a_{12} = 3(12) = 36$$

EXAMPLE 1   Given an infinite sequence with $a_n = 2n - 3$, find each of the following: **a.** the first four terms   **b.** a_{50}

Strategy   We will substitute 1, 2, 3, 4, and 50 for n in the formula that defines the sequence.

Why   To find the first term of the sequence, we let $n = 1$. To find the second term, let $n = 2$, and so on.

Solution

a. $a_1 = 2(1) - 3 = -1$ Substitute 1 for n. $a_2 = 2(2) - 3 = 1$ Substitute 2 for n.

$a_3 = 2(3) - 3 = 3$ Substitute 3 for n. $a_4 = 2(4) - 3 = 5$ Substitute 4 for n.

The first four terms of the sequence are -1, 1, 3, and 5.

b. To find a_{50}, the 50th term of the sequence, we let $n = 50$ in the formula for the nth term:

$$a_{50} = 2(50) - 3 = 97$$

 Self Check 1   Given an infinite sequence with $a_n = 3n + 5$, find each of the following: **a.** the first three terms   **b.** a_{100}

Now Try **Problem 17**

EXAMPLE 2   Find the first four terms of the sequence whose general term is $a_n = \dfrac{(-1)^n}{2^n}$.

Strategy   We will substitute 1, 2, 3, and 4 for n in the formula that defines the sequence.

Why   To find the first term of the sequence, we let $n = 1$. To find the second term, let $n = 2$, and so on.

Solution

$$a_1 = \frac{(-1)^1}{2^1} = -\frac{1}{2} \qquad (-1)^1 = -1 \qquad a_2 = \frac{(-1)^2}{2^2} = \frac{1}{4} \qquad (-1)^2 = 1$$

$$a_3 = \frac{(-1)^3}{2^3} = \frac{-1}{8} = -\frac{1}{8} \quad (-1)^3 = -1 \qquad a_4 = \frac{(-1)^4}{2^4} = \frac{1}{16} \quad (-1)^4 = 1$$

The first four terms of the sequence are $-\frac{1}{2}$, $\frac{1}{4}$, $-\frac{1}{8}$, and $\frac{1}{16}$.

> **Success Tip.**
>
> The factor $(-1)^n$ in $a_n = \frac{(-1)^n}{2^n}$ causes the signs of the terms to alternate between positive (when n is even) and negative (when n is odd).

 Self Check 2   Find the first four terms of the sequence whose general term is $a_n = \dfrac{(-1)^n}{n}$.

Now Try **Problem 25**

Find Terms of an Arithmetic Sequence by Identifying the First Term and the Common Difference.

A sequence where each term is found by adding the same number to the previous term is called an *arithmetic sequence*. Two examples are

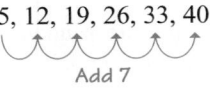

5, 12, 19, 26, 33, 40   This is a finite arithmetic sequence where each term is found by adding 7 to the previous term.

Add 7

3, 1, −1, −3, −5, −7, . . .   This is an infinite arithmetic sequence where each term is found by adding −2 to the previous term.

Add −2

Arithmetic Sequence

An **arithmetic sequence** is a sequence of the form

$$a_1, \quad a_1 + d, \quad a_1 + 2d, \quad a_1 + 3d, \quad \ldots, \quad a_1 + (n - 1)d, \ldots$$

where a_1 is the **first term** and d is the **common difference**. The nth term is given by

$$a_n = a_1 + (n - 1)d$$

We note that the second term of an arithmetic sequence has an addend of $1d$, the third term has an addend of $2d$, the fourth term has an addend of $3d$, and the nth term has an addend of $(n - 1)d$. We also note that the *difference between any two consecutive terms in an arithmetic sequence is d*.

EXAMPLE 3   An arithmetic sequence has a first term 5 and a common difference 4. Write the first five terms of the sequence and find the 25th term.

Strategy   To find the first five terms, we will write the first term and add 4 to each successive term until we produce five terms. To find the 25th term, we will substitute 5 for a_1, 4 for d, and 25 for n in the formula $a_n = a_1 + (n - 1)d$.

Why   The same number is added to each term of an arithmetic sequence to get the next term. However, successively adding 4 to find the 25th term would be time consuming. Using the formula is faster.

Solution   Since the first term is 5 and the common difference is 4, the first five terms are

5, 9, 13, 17, 21

Add 4

Since the first term is $a_1 = 5$ and the common difference is $d = 4$, the arithmetic sequence is defined by the formula

$$a_n = 5 + (n - 1)4 \qquad \text{In } a_n = a_1 + (n - 1)d, \text{ substitute 5 for } a_1 \text{ and 4 for } d.$$

To find the 25th term, we substitute 25 for n and simplify.

$$a_{25} = 5 + (25 - 1)4$$
$$= 5 + (24)4$$
$$= 101$$

The 25th term is 101.

Self Check 3   Write the first five terms of an arithmetic sequence with a first term of 10 and a common difference of 8. Then find the 30th term.

Now Try   **Problem 29**

EXAMPLE 4   The first three terms of an arithmetic sequence are 3, 8, and 13. Find the 100th term.

Strategy   We can use the first three terms to find the common difference. Then we will know the first term of the sequence and the common difference.

Why   Once we know the first term and the common difference, we can use the formula $a_n = a_1 + (n - 1)d$ to find the 100th term by letting $n = 100$.

Solution   The common difference d is the difference between any two successive terms. Since $a_1 = 3$ and $a_2 = 8$, we can find d using subtraction.

$$d = a_2 - a_1 = 8 - 3 = 5 \quad \text{Also note that } a_3 - a_2 = 13 - 8 = 5.$$

To find 100th term, we substitute 3 for a_1, 5 for d, and 100 for n in the formula for the nth term.

$$a_n = a_1 + (n - 1)d$$
$$a_{100} = 3 + (100 - 1)5$$
$$= 3 + (99)5$$
$$= 498$$

Success Tip

The common difference d of an arithmetic sequence is defined to be

$$d = a_{n+1} - a_n$$

Self Check 4   The first three terms of an arithmetic sequence are -3, 6, and 15. Find the 99th term.

Now Try   **Problem 37**

EXAMPLE 5   The first term of an arithmetic sequence is 12 and the 50th term is 3,099. Write the first six terms of the sequence.

Strategy   We will find the common difference by substituting 3,099 for a_n, 12 for a_1, and 50 for n in the formula $a_n = a_1 + (n - 1)d$.

Why   Once we know the first term and the common difference, we can successively add the common difference to each term to produce the first six terms.

Solution   Since the 50th term of the sequence is 3,099, we substitute 3,099 for a_{50}, 12 for a_1, and 50 for n in the formula $a_n = a_1 + (n - 1)d$ and solve for d.

$$a_{50} = a_1 + (n - 1)d$$
$$3{,}099 = 12 + (50 - 1)d \quad \text{Substitute 3,099 for } a_{50}, \text{ 12 for } a_1, \text{ and 50 for } n.$$
$$3{,}099 = 12 + 49d \quad \text{Simplify.}$$
$$3{,}087 = 49d \quad \text{Subtract 12 from both sides.}$$
$$63 = d \quad \text{Divide both sides by 49.}$$

Since the first term is 12 and the common difference is 63, the first six terms are

$$12, 75, 138, 201, 264, 327 \quad \text{Add 63 to a term to get the next term.}$$

Self Check 5   The first term of an arithmetic sequence is 15 and the 12th term is 92. Write the first four terms of the sequence.

Now Try   **Problem 41**

3   **Find Arithmetic Means.**

If numbers are inserted between two numbers a and b to form an arithmetic sequence, the inserted numbers are called **arithmetic means** between a and b. If a single number is inserted, it is called **the arithmetic mean** between a and b.

EXAMPLE 6   Insert two arithmetic means between 6 and 27.

Strategy   Because two arithmetic means are to be inserted between 6 and 27, we will consider a sequence of four terms, with a first term of 6 and a fourth term of 27. We will then use $a_n = a_1 + (n-1)d$ to find the common difference d.

Why   Once we know the first term and the common difference, we can add the common difference to find the two unknown terms.

Solution   The first term is $a_1 = 6$ and the fourth term is $a_4 = 27$. We must find the common difference so that the terms

$$
\begin{array}{cccc}
6, & 6+d, & 6+2d, & 27 \\
\uparrow & \uparrow & \uparrow & \uparrow \\
a_1 & a_2 & a_3 & a_4
\end{array}
$$

form an arithmetic sequence. To find the common difference d, we substitute 6 for a_1, 4 for n, and 27 for a_4 in the formula for the 4th term:

$$a_4 = a_1 + (n-1)d \quad \text{This gives the 4th term of any arithmetic sequence.}$$
$$27 = 6 + (4-1)d \quad \text{Substitute.}$$
$$27 = 6 + 3d \quad \text{Simplify.}$$
$$21 = 3d \quad \text{Subtract 6 from both sides.}$$
$$7 = d \quad \text{Divide both sides by 3.}$$

To find the two arithmetic means between 6 and 27, we add the common difference 7, as shown:

$$
\begin{array}{ll}
6 + d = 6 + 7 & \quad \text{or} \quad 6 + 2d = 6 + 2(7) \\
\qquad = 13 \quad \text{This is } a_2. & \qquad\qquad = 6 + 14 \\
& \qquad\qquad = 20 \quad \text{This is } a_3.
\end{array}
$$

Two arithmetic means between 6 and 27 are 13 and 20.

Self Check 6   Insert two arithmetic means between 8 and 44.

Now Try   **Problem 45**

Find the Sum of the First *n* Terms of an Arithmetic Sequence.

To develop a formula for finding the sum of the first *n* terms of an arithmetic sequence, we let S_n represent the sum of the first *n* terms of an arithmetic sequence:

$$S_n = \quad a_1 \quad + \quad [a_1 + d] \quad + \quad [a_1 + 2d] \quad + \cdots + \quad [a_1 + (n - 1)d]$$

We write the same sum again, but in reverse order:

$$S_n = [a_1 + (n - 1)d] \quad + \quad [a_1 + (n - 2)d] \quad + \quad [a_1 + (n - 3)d] \quad + \cdots + \quad a_1$$

Adding these equations together, term by term, we get

$$2S_n = [2a_1 + (n - 1)d] + [2a_1 + (n - 1)d] + [2a_1 + (n - 1)d] + \cdots + [2a_1 + (n - 1)d]$$

Because there are *n* equal terms on the right side of the preceding equation, we can write

(1) $2S_n = n[2a_1 + (n - 1)d]$

(2) $2S_n = n[a_1 + a_1 + (n - 1)d]$ Write $2a_1$ as $a_1 + a_1$.

$\qquad 2S_n = n(a_1 + a_n)$ Substitute a_n for $a_1 + (n - 1)d$.

$\qquad S_n = \dfrac{n(a_1 + a_n)}{2}$ Divide both sides by 2.

This reasoning establishes the following formula.

Sum of the First *n* Terms of an Arithmetic Sequence	The sum of the first *n* terms of an arithmetic sequence is given by the formula $$S_n = \frac{n(a_1 + a_n)}{2}$$ where a_1 is the first term, a_n is the *n*th (or last) term, and *n* is the number of terms in the sequence.

EXAMPLE 7   Find the sum of the first 40 terms of the arithmetic sequence: 4, 10, 16, . . .

Strategy   We know the first term is 4 and we can find the common difference *d*. We will substitute these values into the formula $a_n = a_1 + (n - 1)d$ to find the last term to be added, a_{40}.

Why   To use the formula $S_n = \dfrac{n(a_1 + a_n)}{2}$ to find the sum of the first 40 terms, we need to know the first term, a_1, and the last term, a_{40}.

Solution   We can substitute 4 for a_1, 40 for *n*, and $10 - 4 = 6$ for *d* into $a_n = a_1 + (n - 1)d$ to get $a_{40} = 4 + (40 - 1)6 = 238$. We then substitute these values into the formula for S_{40}:

$$S_n = \frac{n(a_1 + a_{40})}{2}$$

$$S_{40} = \frac{40(4 + 238)}{2}$$ Substitute: $a_1 = 4$, $n = 40$, and $a_{40} = 238$.

$$= 20(242)$$

$$= 4{,}840$$

The sum of the first 40 terms is 4,840.

Self Check 7   Find the sum of the first 50 terms of the arithmetic sequence: 3, 8, 13, . . .

Now Try   **Problem 53**

5 **Solve Application Problems Involving Arithmetic Sequences.**

EXAMPLE 8   *Halftime Performances.* Each row of a formation formed by the members of a college marching band has one more person in it than the previous row. If 4 people are in the front row and 21 are in the 18th (and last) row, how many band members are there?

Strategy   To find the number of band members, we will write an arithmetic sequence to model the situation and find the sum of its terms.

Why   We can use an arithmetic sequence to model this situation because each row has one more person in it than the previous one. Thus, the common difference is 1.

Solution   When we list the number of band members in each row of the formation, we get the arithmetic sequence 4, 5, 6, . . . , 21, where $a_1 = 4$, $d = 1$, $n = 18$, and $a_{18} = 21$. We can use the formula $S_n = \dfrac{n(a_1 + a_n)}{2}$ to find the sum of the terms of the sequence.

$$S_{18} = \frac{18(4 + 21)}{2} = \frac{18(25)}{2} = 225$$

There are 225 members of the marching band.

Self Check 8   How many band members would it take to form a 10-row formation, if the first row has 5 people in it, the second row has 7 people, the third row has 9 people, and so on?

Now Try   **Problem 95**

6 **Use Summation Notation.**

When the commas between the terms of a sequence are replaced with + signs, we call the sum a **series.** The sum of the terms of an arithmetic sequence is called an **arithmetic series.** Some examples are

$4 + 8 + 12 + 16 + 20 + 24$ Since this series has a limited number of terms, it is a finite arithmetic series.

$5 + 8 + 11 + 14 + 17 + \cdots$ Since this series has infinitely many terms, it is an infinite arithmetic series.

When the general term of a sequence is known, we can use a special notation to write a series. This notation, called **summation notation,** involves the Greek letter Σ (sigma). The expression

$$\sum_{k=1}^{4} 3k$$ Read as "the summation of 3k as k runs from 1 to 4."

designates the sum of all terms obtained if we successively substitute the numbers 1, 2, 3, and 4 for k, called the **index of the summation.** Thus, we have

$$\overset{k=1 \quad k=2 \quad k=3 \quad k=4}{\downarrow \qquad \downarrow \qquad \downarrow \qquad \downarrow}$$

$$\sum_{k=1}^{4} 3k = 3(1) + 3(2) + 3(3) + 3(4)$$

$$= 3 + 6 + 9 + 12$$

$$= 30$$

EXAMPLE 9   Write the series associated with each summation and find the sum:

a. $\displaystyle\sum_{k=1}^{3} (2k + 1)$ **b.** $\displaystyle\sum_{k=2}^{8} k^2$

Strategy   In part (a), we will substitute 1, 2, and 3 for k and add the resulting numbers. In part (b), we will substitute 2, 3, 4, 5, 6, 7, and 8 for k and add the resulting numbers.

Why   Think of k as a counter that begins with the number written at the bottom of the notation and successively increases by 1 until it reaches the number written at the top.

Solution
a. We substitute the integers 1, 2, and 3 for k and find the sum.

$$\sum_{k=1}^{3} (2k + 1) = [2(1) + 1] + [2(2) + 1] + [2(3) + 1]$$

$$= 3 + 5 + 7$$

$$= 15$$

b. We substitute the integers from 2 to 8 for k and find the sum.

$$\sum_{k=2}^{8} k^2 = 2^2 + 3^2 + 4^2 + 5^2 + 6^2 + 7^2 + 8^2$$

$$= 4 + 9 + 16 + 25 + 36 + 49 + 64$$

$$= 203$$

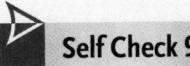 **Self Check 9**   Find the sum: $\displaystyle\sum_{k=1}^{4} (2k^2 - 2)$

Now Try   **Problems 57 and 61**

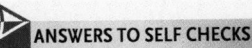**ANSWERS TO SELF CHECKS**   **1. a.** 8, 11, 14   **b.** 305   **2.** $-1, \frac{1}{2}, -\frac{1}{3}, \frac{1}{4}$ **3.** 10, 18, 26, 34, 42; 242

4. 879   **5.** 15, 22, 29, 36   **6.** 20, 32   **7.** 6,275   **8.** 140   **9.** 52

STUDY SET
14.2

VOCABULARY

Fill in the blanks.

1. A _____ is a function whose domain is the set of natural numbers.

2. A sequence with an unlimited number of terms is called a(n) _____ sequence. A sequence with a specific number of terms is called a(n) _____ sequence.

3. Each term of a(n) _____ sequence is found by adding the same number to the previous term.

4. 5, 15, 25, 35, 45, 55, . . . is an example of a(n) _____ sequence. The first _____ is 5 and the common _____ is 10.

5. If a single number is inserted between a and b to form an arithmetic sequence, the number is called the arithmetic _____ between a and b.

6. The sum of the terms of an arithmetic sequence is called an arithmetic _____.

CONCEPTS

7. Write the first three terms of an arithmetic sequence if $a_1 = 1$ and $d = 6$.

8. Given the arithmetic sequence 4, 7, 10, 13, 16, 19, . . . , find a_5 and d.

9. **a.** Write the formula for a_n, the general term of an arithmetic sequence.

   b. Write the formula for S_n, the sum of the first n terms of an arithmetic sequence.

10. An infinite arithmetic sequence is of the form

    $$a_1, a_1 + d, \qquad , a_1 + 3d, \qquad , \ldots$$

NOTATION

Fill in the blanks.

11. The notation a_n represents the _____ term of a sequence.

12. To find the common difference of an arithmetic sequence, we use the formula $d = a_{} - a_{}$.

13. The symbol Σ is the Greek letter _____.

14. In the symbol $\displaystyle\sum_{k=1}^{5} (2k - 5)$, k is called the _____ of summation.

15. We read $\displaystyle\sum_{k=1}^{10} 3k$ as "the _____ of $3k$ as k _____ from 1 to 10."

16. $\displaystyle\sum_{k=1}^{5} k = \boxed{} + \boxed{} + \boxed{} + \boxed{} + \boxed{}$

GUIDED PRACTICE

Write the first five terms of each sequence and then find the specified term. See Example 1.

17. $a_n = 4n - 1$, a_{40}

18. $a_n = 5n - 3$, a_{25}

19. $a_n = -3n + 1$, a_{30}

20. $a_n = -6n + 2$, a_{15}

21. $a_n = -n^2$, a_{20}

22. $a_n = -n^3$, a_{10}

23. $a_n = \dfrac{n - 1}{n}$, a_{12}

24. $a_n = \dfrac{n + 1}{2n}$, a_{100}

Write the first four terms of each sequence. See Example 2.

25. $a_n = \dfrac{(-1)^n}{3^n}$

26. $a_n = \dfrac{(-1)^n}{4^n}$

27. $a_n = (-1)^n(n + 6)$

28. $a_n = (-1)^n(7n)$

Write the first five terms of each arithmetic sequence with the given properties and find the specified term. See Example 3.

29. First term: 3, common difference: 2; find the 10th term.

30. First term: -2, common difference 3; find the 20th term.

31. First term -5, common difference: -3; find the 15th term.

32. First term: 8, common difference: -5; find the 25th term.

33. First term: 7, common difference: 12; find the 30th term.

34. First term: -1, common difference: 4; find the 55th term.

35. First term: -7; common difference: -2; find the 15th term.

36. First term: 8, common difference -3; find the 25th term.

The first three terms of an arithmetic sequence are shown below. Find the specified term. See Example 4.

37. 1, 4, 7, . . . ; 30th term

38. 2, 6, 10, . . . ; 28th term

39. $-5, -1, 3, \ldots$; 17th term

40. $-7, -1, 5, \ldots$; 15th term

Write the first five terms of the arithmetic sequence with the following properties. See Example 5.

41. The first term is 5 and the fifth term is 29.

42. The first term is 4 and the sixth term is 39.

43. The first term is -4 and the sixth term is -39.

44. The first term is -5 and the fifth term is -37.

Insert the given number of arithmetic means between the numbers. See Example 6.

45. Two arithmetic means between 2 and 11

46. Four arithmetic means between 5 and 25

47. Four arithmetic means between 10 and 20

48. Three arithmetic means between 20 and 80

49. Three arithmetic means between 20 and 30

50. Two arithmetic means between 10 and 19

51. One arithmetic mean between -4.5 and 7

52. One arithmetic mean between -6.5 and 8.5

For each arithmetic sequence, find the sum of the specified number of terms. See Example 7.

53. The first 35 terms of 5, 9, 13, $\ldots$

54. The first 50 terms of 7, 12, 17, $\ldots$

55. The first 40 terms of $-5, -1, 3, \ldots$

56. The first 25 terms of 2, $-3, -8, \ldots$

Write the series associated with each summation. See Example 9.

57. $\displaystyle\sum_{k=1}^{4} (3k)$

58. $\displaystyle\sum_{k=1}^{4} (k-9)$

59. $\displaystyle\sum_{k=2}^{4} k^2$

60. $\displaystyle\sum_{k=3}^{5} (-2k)$

Find each sum. See Example 9.

61. $\displaystyle\sum_{k=1}^{4} (6k)$

62. $\displaystyle\sum_{k=2}^{5} (3k)$

63. $\displaystyle\sum_{k=3}^{4} k^3$

64. $\displaystyle\sum_{k=2}^{4} (-k^2)$

65. $\displaystyle\sum_{k=3}^{4} (k^2 + 3)$

66. $\displaystyle\sum_{k=2}^{6} (k^2 + 1)$

67. $\displaystyle\sum_{k=4}^{4} (2k + 4)$

68. $\displaystyle\sum_{k=3}^{3} (3k^2 - 7)$

69. $\displaystyle\sum_{k=2}^{5} (5k)$

70. $\displaystyle\sum_{k=2}^{5} (3k - 5)$

71. $\displaystyle\sum_{k=4}^{6} (4k - 1)$

72. $\displaystyle\sum_{k=3}^{5} (k^3)$

73. Find the common difference of the arithmetic sequence with a first term of 40 if its 44th term is 556.

74. Find the first term of the arithmetic sequence with a common difference of -5 if its 23rd term is -625.

75. Find the sum of the first 12 terms of the arithmetic sequence if its second term is 7 and its third term is 12.

76. Find the sum of the first 16 terms of the arithmetic sequence if its second term is 5 and its fourth term is 9.

77. Find the first five terms of the arithmetic sequence if the common difference is 7 and the sixth term is -83.

78. Find the first five terms of the arithmetic sequence if the common difference is 3 and the seventh term is 12.

79. Find the first six terms of the arithmetic sequence if the common difference is -3 and the ninth term is 10.

80. Find the first six terms of the arithmetic sequence if the common difference is -5 and the tenth term is -27.

81. The first three terms of an arithmetic sequence are 5, 12, and 19. Find the 200th term.

82. The first three terms of an arithmetic sequence are 10, 14, and 18. Find the 500th term.

83. Find the sum of the first 50 natural numbers.

84. Find the sum of the first 100 natural numbers.

85. Find the 37th term of the arithmetic sequence with a second term of -4 and a third term of -9.

86. Find the 40th term of the arithmetic sequence with a second term of 6 and a fourth term of 16.

87. Find the first term of the arithmetic sequence with a common difference of 11 if its 27th term is 263.

88. Find the common difference of the arithmetic sequence with a first term of -164 if its 36th term is -24.

89. Find the 15th term of the arithmetic sequence $\frac{1}{2}, \frac{1}{4}, 0$

90. Find the 14th term of the arithmetic sequence $\frac{2}{3}, \frac{1}{2}, \frac{1}{3}, \ldots$

91. Find the sum of the first 50 odd natural numbers.

92. Find the sum of the first 50 even natural numbers.

93. SAVING MONEY   Yasmeen puts \$60 into a safety deposit box. After each succeeding month, she puts \$50 more in the box. Write the first six terms of an arithmetic sequence that gives the monthly amounts in her savings, and find her savings after 10 years.

94. INSTALLMENT LOANS   Maria borrowed \$10,000, interest-free, from her mother. She agreed to pay back the loan in monthly installments of \$275. Write the first six terms of an arithmetic sequence that shows the balance due after each month, and find the balance due after 17 months.

95. DESIGNING PATIOS Refer to the illustration. Each row of bricks in a triangular patio floor is to have one more brick than the previous row, ending with the longest row of 150 bricks. How many bricks will be needed?

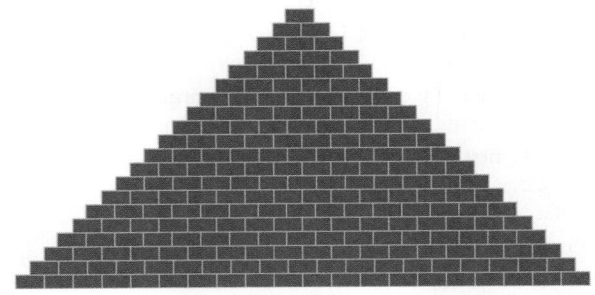

96. LOGGING Logs are stacked so that the bottom row has 30 logs, the next row has 29 logs, the next row has 28 logs, and so on.

a. If there are 20 rows in the stack, how many logs are in the top row?

b. How many logs are in the stack?

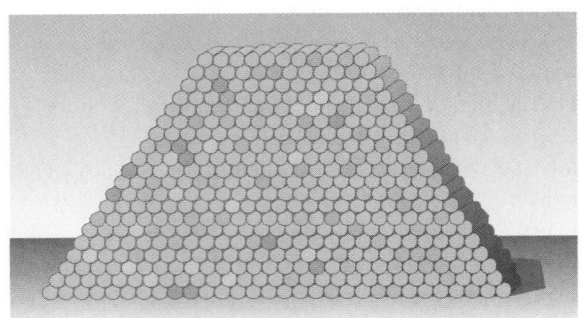

97. HOLIDAY SONGS A popular song of European origin lists the gifts received from someone's "true love" over a 12-day span: a partridge in a pear tree, two turtle doves, three French hens, four calling birds, five gold rings, six geese a-laying, seven swans a-swimming, eight maids a-milking, nine ladies dancing, ten lords a-leaping, eleven pipers piping, and twelve drummers drumming. Use a formula from this section to determine the total number of gifts in this list.

98. INTERIOR ANGLES The sums of the angles of several polygons are given in the table in the next column. Assuming that the pattern continues, complete the table.

Figure	Number of sides	Sum of angles
Triangle	3	180°
Quadrilateral	4	360°
Pentagon	5	540°
Hexagon	6	720°
Octagon	8	
Dodecagon	12	

WRITING

99. Explain why 1, 4, 8, 13, 19, 26, . . . is not an arithmetic sequence.

100. What is the difference between a sequence and a series?

101. What is the difference between a_n and S_n?

102. How is the symbol Σ used in this section?

REVIEW

Assume that x, y, z, and b represent positive numbers. Use the properties of logarithms to write each expression in terms of the logarithms of x, y, and z.

103. $\log_2 \dfrac{2x}{y}$

104. $\ln x\sqrt{z}$

105. $\log x^3 y^2$

106. $\log x^3 y^{1/2}$

CHALLENGE PROBLEMS

Write the summation notation for each sum.

107. Write the summation notation for
$$1 + 4 + 9 + 16 + 25$$

108. Write the summation notation for $3 + 4 + 5 + 6$ without using $k = 1$ in your answer.

109. For what value of x will $x - 2$, $2x + 4$, and $5x - 8$, in that order, form an arithmetic sequence?

110. For what value of x will the arithmetic mean of $x + 4$ and $x + 8$ be 5?

SECTION 14.3
Geometric Sequences and Series

Objectives

1. Find terms of a geometric sequence by identifying the first term and the common ratio.
2. Find geometric means.
3. Find the sum of the first n terms of a geometric sequence.
4. Define and use infinite geometric series.
5. Solve application problems involving geometric sequences.

We have seen that the same number is added to each term of an arithmetic sequence to get the next term. In this section, we will consider another type of sequence where we multiply each term by the same number to get the next term. This type of sequence is called a *geometric sequence*. Two examples are

2, 8, 32, 128, . . .   *This is an infinite geometric sequence where each term is found by multiplying the previous term by 4.*

Multiply by 4

$27, 9, 3, 1, \dfrac{1}{3}, \dfrac{1}{9}$ *This is a finite geometric sequence where each term is found by multiplying the previous term by $\frac{1}{3}$.*

Multiply by $\frac{1}{3}$

1   **Find Terms of a Geometric Sequence by Identifying the First Term and the Common Ratio.**

Each term of a geometric sequence is found by multiplying the previous term by the same number.

Geometric Sequence

A **geometric sequence** is a sequence of the form

$$a_1, \ a_1r, \ a_1r^2, \ a_1r^3, \ldots, \ a_1r^{n-1}, \ldots$$

where a_1 is the **first term** and r is the **common ratio.** The nth term is given by

$$a_n = a_1r^{n-1}$$

We note that the second term of a geometric sequence has a factor of r^1, the third term has a factor of r^2, the fourth term has a factor of r^3, and the nth term has a factor of r^{n-1}. We also note that *r is the quotient obtained when any term is divided by the previous term.*

EXAMPLE 1   A geometric sequence has a first term 5 and a common ratio 3.
a. Write the first five terms of the sequence.   **b.** Find the 9th term.

Strategy   In part (a), we will write the first term and successively multiply each term by 3 until we produce five terms. In part (b), we will substitute 5 for a_1, 3 for r, and 9 for n in the formula for the nth term of a geometric sequence $a_n = a_1r^{n-1}$.

Why To find the terms of a geometric sequence, we multiply each term by the same number to get the next term. To answer part (b), successively multiplying by 3 to find the 9th term would be time consuming. Using the formula is faster.

Solution

a. Because the first term is $a_1 = 5$ and the common ratio is $r = 3$, the first five terms are

$$5, \quad 5(3), \quad 5(3^2), \quad 5(3^3), \quad 5(3^4) \qquad \text{Each term is found by multiplying}$$
$$\uparrow \qquad \uparrow \qquad \uparrow \qquad \uparrow \qquad \uparrow \qquad \text{the previous term by 3.}$$
$$a_1 \qquad a_2 \qquad a_3 \qquad a_4 \qquad a_5$$

or

$$5, 15, 45, 135, 405$$

b. The nth term is $a_n = a_1 r^{n-1}$ with $a_1 = 5$ and $r = 3$. Because we want the ninth term, we let $n = 9$:

$$a_n = a_1 r^{n-1}$$
$$a_9 = 5(3)^{9-1}$$
$$= 5(3)^8$$
$$= 5(6,561)$$
$$= 32,805$$

Self Check 1   A geometric sequence has a first term 3 and a common ratio 4.
   a. Write the first four terms.
   b. Find the 8th term.

Now Try   **Problem 15**

EXAMPLE 2   The first three terms of a geometric sequence are 16, 4, and 1. Find the 7th term.

Strategy   We can use the first three terms to find the common ratio. Then we will know the first term and the common ratio.

Why   Once we know the first term, a_1, and the common ratio, r, we can use the formula $a_n = a_1 r^{n-1}$ to find the seventh term by letting $n = 7$.

Solution   The common ratio r is the ratio between any two successive terms. Since $a_1 = 16$ and $a_2 = 4$, we can find r as follows:

$$r = \frac{a_2}{a_1} = \frac{4}{16} = \frac{1}{4} \qquad \text{Also note that } \frac{a_3}{a_2} = \frac{1}{4}.$$

To find the seventh term, we substitute 16 for a_1, $\frac{1}{4}$ for r, and 7 for n in the formula for the nth term and simplify:

$$a_n = a_1 r^{n-1} \qquad \text{This is the formula for the nth term of a geometric sequence.}$$
$$a_7 = 16\left(\frac{1}{4}\right)^{7-1} \qquad \text{Substitute 16 for a_1, 7 for n, and $\frac{1}{4}$ for r.}$$
$$= 16\left(\frac{1}{4}\right)^6$$

$$= 16\left(\frac{1}{4{,}096}\right)$$

$$= \frac{1}{256}$$

Self Check 2   The first three terms of a geometric sequence are 25, 5, and 1. Find the 7th term.

Now Try   **Problem 19**

EXAMPLE 3   Find the first five terms of the geometric sequence with a first term of 2, a third term of 32, and a common ratio that is positive.

Strategy   We will substitute $a_1 = 2$, $a_3 = 32$, and $n = 3$ into the formula for the nth term of a geometric sequence $a_n = a_1 r^{n-1}$ and solve for r.

Why   Once we know the common ratio, we can successively multiply each term by the common ratio to produce the first five terms.

Solution   We will substitute 3 for n, 2 for a_1, 32 for a_3 and solve for r.

$a_n = a_1 r^{n-1}$ This is the formula for the nth term of a geometric sequence.

$a_3 = 2r^{3-1}$ Substitute 3 for n and 2 for a_1.

$32 = 2r^2$ Substitute 32 for a_3 and simplify.

$16 = r^2$ Divide both sides by 2.

$\pm 4 = r$ Use the square root property.

Since r is given to be positive, $r = 4$. The first five terms are produced by multiplying by the common ratio:

$$2,\ 2 \cdot 4,\ 2 \cdot 4^2,\ 2 \cdot 4^3,\ 2 \cdot 4^4$$

or

$$2,\ 8,\ 32,\ 128,\ 512$$

Self Check 3   Find the first five terms of the geometric sequence with a first term of -2, a fourth term of -54.

Now Try   **Problem 23**

---

2   **Find Geometric Means.**

If numbers are inserted between two numbers a and b to form a geometric sequence, the inserted numbers are called **geometric means** between a and b. If a single number is inserted, that number is called **the geometric mean** between a and b.

EXAMPLE 4   Insert two geometric means between 7 and 1,512.

Strategy   Because two geometric means are to be inserted between 7 and 1,512, we will consider a sequence of four terms, with a first term of 7 and a fourth term of 1,512. We will then use $a_n = a_1 r^{n-1}$ to find the common ratio r.

Why   Once we know the first term and the common ratio, we can multiply by the common ratio to find the two unknown terms.

Solution   In this example, the first term is $a_1 = 7$, and the fourth term is $a_4 = 1,512$. To find the common ratio r so that the terms

$$
\begin{array}{cccc}
7, & 7r, & 7r^2, & 1{,}512 \\
\uparrow & \uparrow & \uparrow & \uparrow \\
a_1 & a_2 & a_3 & a_4
\end{array}
$$

form a geometric sequence, we substitute 4 for n and 7 for a_1 in the formula for the nth term of a geometric sequence and solve for r.

$$a_n = a_1 r^{n-1}$$ This is the formula for the nth term of a geometric sequence.

$$a_4 = 7r^{4-1}$$ Substitute 4 for n and 7 for a_1.

$$1{,}512 = 7r^3$$ Substitute 1,512 for a_4.

$$216 = r^3$$ Divide both sides by 7.

$$6 = r$$ Take the cube root of both sides.

To find the two geometric means between 7 and 1,512, we multiply by the common ratio 6, as shown:

$$7r = 7(6) = 42 \quad \text{and} \quad 7r^2 = 7(6)^2 = 7(36) = 252$$

The numbers 7, 42, 252, and 1,512 are the first four terms of a geometric sequence.

Self Check 4   Insert three positive geometric means between 1 and 16.

Now Try   **Problem 27**

EXAMPLE 5   Find the geometric mean between 2 and 20.

Strategy   Because one geometric mean is to be inserted between 2 and 20, we will consider a sequence of three terms, with a first term of 2 and a third term of 20. We will then use $a_n = a_1 r^{n-1}$ to find the common ratio r.

Why   Once we know the first term and the common ratio, we can multiply by the common ratio to find the unknown term.

Solution   We want to find the middle term of the three-termed geometric sequence

$$
\begin{array}{ccc}
2, & 2r, & 20 \\
\uparrow & \uparrow & \uparrow \\
a_1 & a_2 & a_3
\end{array}
$$

with $a_1 = 2$, $a_3 = 20$, and $n = 3$. To find r, we substitute these values into the formula for the nth term of a geometric sequence:

$$a_n = a_1 r^{n-1}$$ This is the formula for the nth term of a geometric sequence.

$$a_3 = 2r^{3-1}$$ Substitute 3 for n and 2 for a_1.

$$20 = 2r^2 \qquad \text{Substitute 20 for } a_3.$$
$$10 = r^2 \qquad \text{Divide both sides by 2.}$$
$$\pm\sqrt{10} = r \qquad \text{Use the square root property.}$$

Because r can be either $\sqrt{10}$ or $-\sqrt{10}$, there are two values for the geometric mean. They are

$$2r = 2\sqrt{10} \qquad \text{and} \qquad 2r = -2\sqrt{10}$$

The sets of numbers 2, $2\sqrt{10}$, 20 and 2, $-2\sqrt{10}$, 20 both form geometric sequences. The common ratio of the first sequence is $\sqrt{10}$, and the common ratio of the second sequence is $-\sqrt{10}$.

Self Check 5   Find the positive geometric mean between 2 and 200.

Now Try   **Problem 31**

3   **Find the Sum of the First *n* Terms of a Geometric Sequence.**

There is a formula that gives the sum of the first n terms of a geometric sequence. To develop this formula, we let S_n represent the sum of the first n terms of a geometric sequence.

(1) $S_n = a_1 + a_1 r + a_1 r^2 + a_1 r^3 + \cdots + a_1 r^{n-1}$

> *Success Tip*
> If the common factor of a_1 in the numerator of $\frac{a_1 - a_1 r^n}{1 - r}$ is factored out, the formula can be written in a different form:
> $$S_n = \frac{a_1(1 - r^n)}{1 - r}$$

We multiply both sides of Equation 1 by r to get

(2) $S_n r = \qquad a_1 r + a_1 r^2 + a_1 r^3 + \cdots + a_1 r^{n-1} + a_1 r^n$

We now subtract Equation 2 from Equation 1 and solve for S_n:

$$S_n - S_n r = a_1 - a_1 r^n$$
$$S_n(1 - r) = a_1 - a_1 r^n \qquad \text{Factor out } S_n \text{ from the left side.}$$
$$S_n = \frac{a_1 - a_1 r^n}{1 - r} \qquad \text{Divide both sides by } 1 - r.$$

This reasoning establishes the following formula.

Sum of the First *n* Terms of a Geometric Sequence

The sum of the first n terms of a geometric sequence is given by the formula

$$S_n = \frac{a_1 - a_1 r^n}{1 - r} \qquad \text{or} \qquad S_n = \frac{a_1(1 - r^n)}{1 - r} \qquad \text{where } r \neq 1$$

where S_n is the sum, a_1 is the first term, r is the common ratio, and n is the number of terms.

EXAMPLE 6   Find the sum of the first six terms of the geometric sequence: 250, 50, 10, . . .

Strategy   We will find the common ratio r.

Why   If we know the first term and the common ratio, we can use the formula $S_n = \frac{a_1 - a_1 r^n}{1 - r}$ to find the sum of the first six terms by letting $n = 6$.

Solution   The common ratio r is the ratio between any two successive terms. Since $a_1 = 250$ and $a_2 = 50$, we can find r as follows:

$$r = \frac{a_2}{a_1} = \frac{50}{250} = \frac{1}{5} \qquad \text{Also note that } \frac{a_3}{a_2} = \frac{10}{50} = \frac{1}{5}.$$

In this sequence, $a_1 = 250$, $r = \frac{1}{5}$, and $n = 6$. We substitute these values into the formula for the sum of the first n terms of a geometric sequence and simplify:

$$S_n = \frac{a_1 - a_1 r^n}{1 - r}$$

$$S_6 = \frac{250 - 250\left(\frac{1}{5}\right)^6}{1 - \frac{1}{5}}$$

$$= \frac{250 - 250\left(\frac{1}{15{,}625}\right)}{\frac{4}{5}}$$

$$= \left(250 - \frac{250}{15{,}625}\right) \cdot \frac{5}{4} \qquad \text{Multiply the numerator by the reciprocal of the denominator.}$$

$$= 312.48 \qquad \text{Use a calculator.}$$

The sum of the first six terms is 312.48.

 Self Check 6   Find the sum of the first five terms of the geometric sequence 100, 20, 4, . . .

Now Try   **Problem 35**

④   Define and Use Infinite Geometric Series.

When we add the terms of a geometric sequence, we form a **geometric series.** If we form the sum of the terms of an infinite geometric sequence, we get a series called an **infinite geometric series.** For example, if the common ratio r is 3, we have

Infinite geometric sequence	*Infinite geometric series*
2, 6, 18, 54, 162, . . .	$2 + 6 + 18 + 54 + 162 + \cdots$

As the number of terms of this series gets larger, the value of the series gets larger. We can see that this is true by forming some **partial sums.**

The first partial sum of the series is $S_1 = 2$.

The second partial sum of the series is $S_2 = 2 + 6 = 8$.

The third partial sum of the series is $S_3 = 2 + 6 + 18 = 26$.

The fourth partial sum of the series is $S_4 = 2 + 6 + 18 + 54 = 80$.

The Language of Algebra
The word *partial* means only a part, not total. Have you ever seen a *partial* eclipse of the moon?

We can see that as the number of terms gets infinitely large, the value of this series gets infinitely large.

The values of some infinite geometric series get closer and closer to a specific number as the number of terms approaches infinity. One such series is

$$\frac{3}{2} + \frac{3}{4} + \frac{3}{8} + \frac{3}{16} + \frac{3}{32} + \frac{3}{64} + \cdots \quad \text{Here, } r = \tfrac{1}{2}.$$

To see that this is true, we form some partial sums.

The first partial sum is $S_1 = \dfrac{3}{2} = 1.5$

The second partial sum is $S_2 = \dfrac{3}{2} + \dfrac{3}{4} = \dfrac{9}{4} = 2.25$

The third partial sum is $S_3 = \dfrac{3}{2} + \dfrac{3}{4} + \dfrac{3}{8} = \dfrac{21}{8} = 2.625$

The fourth partial sum is $S_4 = \dfrac{3}{2} + \dfrac{3}{4} + \dfrac{3}{8} + \dfrac{3}{16} = \dfrac{45}{16} = 2.8125$

The fifth partial sum is $S_5 = \dfrac{3}{2} + \dfrac{3}{4} + \dfrac{3}{8} + \dfrac{3}{16} + \dfrac{3}{32} = \dfrac{93}{32} = 2.90625$

The sixth partial sum is $S_6 = \dfrac{3}{2} + \dfrac{3}{4} + \dfrac{3}{8} + \dfrac{3}{16} + \dfrac{3}{32} + \dfrac{3}{64} = \dfrac{189}{64} = 2.953125$

As the number of terms in this series gets larger, the values of the partial sums approach the number 3. We say that 3 is the **limit** of S_n as n approaches infinity, and we say that 3 is the **sum of the infinite geometric series.**

To develop a formula for finding the sum of an infinite geometric series, we consider the formula that gives the sum of the first n terms.

$$S_n = \frac{a_1 - a_1 r^n}{1 - r} \quad \text{where } r \neq 1$$

If $|r| < 1$ and a_1 is constant, the term $a_1 r^n$ in the above formula approaches 0 as n becomes very large. For example,

$$a_1\left(\frac{1}{2}\right)^1 = \frac{1}{2}a_1, \quad a_1\left(\frac{1}{2}\right)^2 = \frac{1}{4}a_1, \quad a_1\left(\frac{1}{2}\right)^3 = \frac{1}{8}a_1$$

and so on. Thus, when n is very large, the value of $a_1 r^n$ is negligible, and the term $a_1 r^n$ in the above formula can be ignored. This reasoning justifies the following formula.

Sum of the Terms of an Infinite Geometric Series

If a_1 is the first term and r is the common ratio of an infinite geometric sequence, and $|r| < 1$, the sum of the terms of the corresponding series is given by

$$S = \frac{a_1}{1 - r}$$

EXAMPLE 7 Find the sum of the terms of the infinite geometric series $125 + 25 + 5 + \cdots$

Strategy We will identify a_1 and find r.

Why To use the formula $S = \frac{a_1}{1 - r}$ to find the sum, we need to know the first term, a_1, and the common ratio, r.

Solution In this geometric series, $a_1 = 125$ and $r = \frac{25}{125} = \frac{1}{5}$. Since $|r| = \left|\frac{1}{5}\right| = \frac{1}{5} < 1$, we can find the sum of the series. We do this by substituting 125 for a_1 and $\frac{1}{5}$ for r in the formula $S = \frac{a_1}{1-r}$ and simplifying:

$$S = \frac{a_1}{1-r} = \frac{125}{1-\frac{1}{5}} = \frac{125}{\frac{4}{5}} = 125 \cdot \frac{5}{4} = \frac{625}{4}$$

The sum of the series $125 + 25 + 5 + \cdots$ is $\frac{625}{4} = 156.25$.

Self Check 7   Find the sum of the infinite geometric series:
$$100 + 20 + 4 + \cdots$$

Now Try   **Problem 43**

EXAMPLE 8   Find the sum of the infinite geometric series:
$$64 + (-4) + \frac{1}{4} + \cdots$$

Strategy   We will identify a_1 and find r.

Why   To use the formula $S = \frac{a_1}{1-r}$ to find the sum, we need to know the first term, a_1, and the common ratio, r.

Solution   In this geometric series, $a_1 = 64$ and $r = \frac{-4}{64} = -\frac{1}{16}$. Since $|r| = \left|-\frac{1}{16}\right| = \frac{1}{16} < 1$, we can find the sum of the series. We substitute 64 for a_1 and $-\frac{1}{16}$ for r in the formula $S = \frac{a_1}{1-r}$ and simplify:

$$S = \frac{a_1}{1-r} = \frac{64}{1-\left(-\frac{1}{16}\right)} = \frac{64}{\frac{17}{16}} = 64 \cdot \frac{16}{17} = \frac{1{,}024}{17}$$

The sum of the geometric series $64 + (-4) + \frac{1}{4} + \cdots$ is $\frac{1{,}024}{17}$.

Self Check 8   Find the sum of the infinite geometric series:
$$81 + (-27) + 9 + \ldots$$

Now Try   **Problem 51**

EXAMPLE 9   Change $0.\overline{8}$ to a common fraction.

Strategy   First, we will show that the decimal $0.888\ldots$ can be represented by an infinite geometric series whose terms are fractions. Then we will identify a_1, find the common ratio r, and find the sum.

Why   When we use the formula $S = \frac{a_1}{1-r}$ to find the sum, the result will be the required common fraction.

Solution   The decimal $0.\overline{8}$ can be written as the infinite geometric series

$$0.\overline{8} = 0.888\ldots$$
$$= 0.8 + 0.08 + 0.008 + \cdots$$
$$= \frac{8}{10} + \frac{8}{100} + \frac{8}{1,000} + \cdots$$

Here, $a_1 = \frac{8}{10}$ and $r = \frac{1}{10}$. Because $|r| = \left|\frac{1}{10}\right| = \frac{1}{10} < 1$, we can find the sum as follows:

$$S = \frac{a_1}{1-r} = \frac{\frac{8}{10}}{1 - \frac{1}{10}} = \frac{\frac{8}{10}}{\frac{9}{10}} = \frac{8}{10} \cdot \frac{10}{9} = \frac{8}{9}$$

Thus, $0.\overline{8} = \frac{8}{9}$. Long division will verify that $\frac{8}{9} = 0.888\ldots$.

> **Self Check 9**   Change $0.\overline{6}$ to a common fraction.
>
> ***Now Try***   **Problem 55**

---

5   **Solve Application Problems Involving Geometric Sequences.**

EXAMPLE 10   *Inheritances.*   A father decides to give his son part of his inheritance early. Each year, on the son's birthday, the father will pay the son 15% of what remains in the inheritance fund. If the fund initially begins with $100,000, how much money will be left in the fund after 20 years of payments?

Strategy   We will model the facts of the problem using a geometric sequence with a first term of 100,000 and a common ratio of 0.85.

Why   One of the terms of the sequence will represent the amount of money left in the inheritance fund after 20 years of payments.

Solution   If 15% of the money in the inheritance fund is given to the son each year, 85% of that amount remains after each payment. To find the amount of money that remains in the fund after a payment is made, we multiply the amount that was in the fund by 0.85. Over the years, the amounts of money that are left in the fund after a payment form a geometric sequence.

Amount of money remaining in the inheritance fund

The fund begins with:	$100,000$	$\leftarrow a_1$
After first payment:	$100,000(0.85) = 100,000(0.85)^1$	$\leftarrow a_2$
After second payment:	$[100,000(0.85)^1](0.85) = 100,000(0.85)^2$	$\leftarrow a_3$
After third payment:	$[100,000(0.85)^2](0.85) = 100,000(0.85)^3$	$\leftarrow a_4$
After fourth payment:	$[100,000(0.85)^3](0.85) = 100,000(0.85)^4$	$\leftarrow a_5$
After 20th payment:	???	$\leftarrow a_{21}$

The amount of money remaining in the inheritance fund after 20 years is represented by the 21st term of a geometric sequence, where $a_1 = 100{,}000$, $r = 0.85$, and $n = 21$.

$$a_n = a_1 r^{n-1} \qquad \text{This is the formula for the nth term.}$$
$$a_{21} = a_1 r^{21-1} \qquad \text{Substitute 21 for n.}$$
$$= 100{,}000(0.85)^{21-1} \qquad \text{Substitute 100,000 for } a_1 \text{ and 0.85 for } r.$$
$$= 100{,}000(0.85)^{20}$$
$$\approx 3{,}876 \qquad \text{Use a calculator. Round to the nearest dollar.}$$

In 20 years, approximately \$3,876 of the inheritance fund will be left.

Self Check 10   How much money will be left in the inheritance fund after 30 years of payments?

Now Try   **Problem 79**

EXAMPLE 11   *Testing Steel.*   One way to measure the hardness of a steel anvil is to drop a ball bearing onto the face of the anvil. The bearing should rebound $\frac{4}{5}$ of the distance from which it was dropped. If a bearing is dropped from a height of 10 inches onto a hard forged steel anvil, and if it could bounce forever, what total distance would the bearing travel?

Strategy   We will show that the facts in the problem can be modeled by an infinite geometric sequence.

Why   The sum of the terms of the infinite geometric sequence will give the total distance the bearing will travel.

Solution   The total distance the ball bearing travels is the sum of two motions, falling and rebounding. The bearing falls 10 inches, then rebounds $\frac{4}{5} \cdot 10 = 8$ inches, and falls 8 inches, and rebounds $\frac{4}{5} \cdot 8 = \frac{32}{5}$ inches, and falls $\frac{32}{5}$ inches, and rebounds $\frac{4}{5} \cdot \frac{32}{5} = \frac{128}{25}$ inches, and so on.

The distance the ball falls is given by the sum

$$10 + 8 + \frac{32}{5} + \frac{128}{25} + \cdots \qquad \text{This is an infinite geometric series with } a_1 = 10 \text{ and } r = \frac{4}{5}.$$

The distance the ball rebounds is given by the sum

$$8 + \frac{32}{5} + \frac{128}{25} + \cdots \qquad \text{This is an infinite geometric series with } a_1 = 8 \text{ and } r = \frac{4}{5}.$$

Since each of these is an infinite geometric series with $|r| < 1$, we can use the formula $S = \frac{a_1}{1 - r}$ to find each sum.

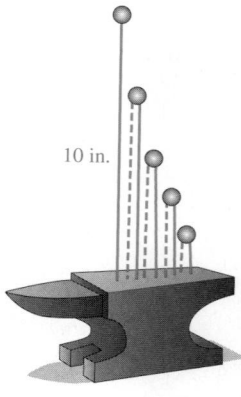

10 in.

Falling: $\dfrac{10}{1 - \dfrac{4}{5}} = \dfrac{10}{\dfrac{1}{5}} = 50$ in.   Rebounding: $\dfrac{8}{1 - \dfrac{4}{5}} = \dfrac{8}{\dfrac{1}{5}} = 40$ in.

The total distance the bearing travels is 50 inches + 40 inches = 90 inches.

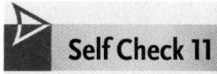

Self Check 11 If a bearing was dropped from a height of 15 inches onto a hard forged steel anvil, and if it could bounce forever, what total distance would it travel?

Now Try **Problem 85**

ANSWERS TO SELF CHECKS **1. a.** 3, 12, 48, 192 **b.** 49,152 **2.** $\frac{1}{625}$ **3.** $-2, -6, -18, -54, -162$
4. 2, 4, 8 **5.** 20 **6.** 124.96 **7.** 125 **8.** $\frac{243}{4}$ **9.** $\frac{2}{3}$ **10.** About $763 **11.** 135 in.

STUDY SET
14.3

VOCABULARY

Fill in the blanks.

1. Each term of a _____ sequence is found by multiplying the previous term by the same number.
2. 8, 16, 32, 64, 128, . . . is an example of a _____ sequence. The first _____ is 8 and the common _____ is 2.
3. If a single number is inserted between a and b to form a geometric sequence, the number is called the geometric _____ between a and b.
4. The sum of the terms of a geometric sequence is called a geometric _____. The sum of the terms of an infinite geometric sequence is called an _____ geometric series.

CONCEPTS

5. Write the first three terms of a geometric sequence if $a_1 = 16$ and $r = \frac{1}{4}$.
6. Given the geometric sequence 1, 2, 4, 8, 16, 32, . . . , find a_5 and r.
7. Write the formula for a_n, the general term of a geometric sequence.
8. **a.** Write the formula for S_n, the sum of the first n terms of a geometric sequence.
   b. Write the formula for S, the sum of the terms of an infinite geometric sequence, where $|r| < 1$.
9. Which of the following values of r satisfy $|r| < 1$?
   a. $r = \frac{2}{3}$ **b.** $r = -3$
   c. $r = 6$ **d.** $r = -\frac{1}{5}$
10. Write $0.\overline{7}$ as an infinite geometric series:

$$0.\overline{7} = 0.7 + 0.07 + 0.007 + \cdots$$
$$= \frac{7}{} + \frac{7}{} + \frac{7}{} + \cdots$$

NOTATION

Fill in the blanks.

11. An infinite geometric sequence is of the form
    $$a_1, \ a_1 r, \ a_1 \boxed{}, \ a_1 r^3, \ \boxed{}, \ldots$$
12. The first four terms of the sequence defined by $a_n = 4(3)^{n-1}$ are $\boxed{}$, $\boxed{}$, $\boxed{}$, $\boxed{}$.
13. To find the common ratio of a geometric sequence, we use the formula $r = \dfrac{a_{n+1}}{a_{\boxed{}}}$.
14. S_8 represents the sum of the first _____ terms of a geometric sequence.

GUIDED PRACTICE

Write the first five terms of each geometric sequence with the given properties and then find the specified term. **See Example 1.**

15. First term: 3, common ratio: 2; find the 9th term
16. First term: -2, common ratio: 2; find the 8th term
17. First term: -5, common ratio: $\frac{1}{5}$; find the 8th term
18. First term: 8, common ratio: $\frac{1}{2}$; find the 10th term

Find the specified term of the geometric sequence with the following properties. **See Example 2.**

19. The first three terms are 2, 6, 18; find the 7th term
20. The first three terms are 50, 100, 200; find the 10th term
21. The first three terms are $\frac{1}{2}, -\frac{5}{2}, \frac{25}{2}$; find the 6th term
22. The first three terms are $\frac{1}{4}, -\frac{3}{4}, \frac{9}{4}$; find the 9th term

Write the first five terms of the geometric sequence with the following properties. See Example 3.

23. First term: 2, $r > 0$, third term: 18

24. First term: 2, $r < 0$, third term: 50

25. First term: 3, fourth term: 24

26. First term: -3, fourth term: -192

Find the geometric means to be inserted in each geometric sequence. See Example 4.

27. Insert three positive geometric means between 2 and 162.

28. Insert four geometric means between 3 and 96.

29. Insert four geometric means between -4 and $-12,500$.

30. Insert three geometric means (two positive and one negative) between -64 and $-1,024$.

Find a geometric mean to be inserted in each geometric sequence. See Example 5.

31. Find the geometric mean between 2 and 128.

32. Find the geometric mean between 3 and 243.

33. Find the geometric mean between 10 and 20.

34. Find the geometric mean between 5 and 15.

For each geometric sequence, find the sum of the specified number of terms. See Example 6.

35. The first 6 terms of 2, 6, 18, . . .

36. The first 6 terms of 2, -6, 18, . . .

37. The first 5 terms of 2, -6, 18, . . .

38. The first 5 terms of 2, 6, 18, . . .

39. The first 8 terms of 3, -6, 12, . . .

40. The first 8 terms of 3, 6, 12, . . .

41. The first 7 terms of 3, 6, 12, . . .

42. The first 7 terms of 3, -6, 12, . . .

Find the sum of each infinite geometric series, if possible. See Examples 7 and 8.

43. $8 + 4 + 2 + \cdots$

44. $12 + 6 + 3 + \cdots$

45. $54 + 18 + 6 + \cdots$

46. $45 + 15 + 5 + \cdots$

47. $-\frac{27}{2} + (-9) + (-6) + \cdots$

48. $-112 + (-28) + (-7) + \cdots$

49. $\frac{9}{2} + 6 + 8 + \cdots$

50. $\frac{18}{25} + \frac{6}{5} + 2 + \cdots$

51. $12 + (-6) + 3 + \cdots$

52. $8 + (-4) + 2 + \cdots$

53. $-45 + 15 + (-5) + \cdots$

54. $-54 + 18 + (-6) + \cdots$

Write each decimal in fraction form. Then check the answer by performing long division. See Example 9.

55. $0.\overline{1}$

56. $0.\overline{2}$

57. $0.\overline{3}$

58. $0.\overline{4}$

59. $0.\overline{12}$

60. $0.\overline{21}$

61. $0.\overline{75}$

62. $0.\overline{57}$

TRY IT YOURSELF

63. Find the common ratio of the geometric sequence with a first term -8 and a sixth term $-1,944$.

64. Find the common ratio of the geometric sequence with a first term 12 and a sixth term $\frac{3}{8}$.

65. Write the first five terms of the geometric sequence if its first term is -64, $r < 0$, and its fifth term is -4.

66. Write the first five terms of the geometric sequence if its first term is -64, $r > 0$, and its fifth term is -4.

67. Find a geometric mean, if possible, between -50 and 10.

68. Find a negative geometric mean, if possible, between -25 and -5.

69. Find the 10th term of the geometric sequence with $a_1 = 7$ and $r = 2$.

70. Find the 12th term of the geometric sequence with $a_1 = 64$ and $r = \frac{1}{2}$.

71. Write the first four terms of the geometric sequence if its first term is -64 and its sixth term is -2.

72. Write the first four terms of the geometric sequence if its first term is -81 and its sixth term is $\frac{1}{3}$.

73. Find the sum of the terms of the geometric sequence $3, \frac{3}{4}, \frac{3}{16}, \frac{3}{64}, \ldots$

74. Find the sum of the terms of the geometric sequence $1, -\frac{1}{2}, \frac{1}{4}, -\frac{1}{8}, \ldots$

75. Find the first term of the geometric sequence with a common ratio -3 and an eighth term -81.

76. Find the first term of the geometric sequence with a common ratio 2 and a tenth term 384.

77. Find the sum of the first five terms of the geometric sequence if its first term is 3 and the common ratio is 2.

78. Find the sum of the first five terms of the geometric sequence if its first term is 5 and the common ratio is -6.

APPLICATIONS

Use a calculator to help solve each problem.

79. DECLINING SAVINGS   John has $10,000 in a safety deposit box. Each year, he spends 12% of what is left in the box. How much will be in the box after 15 years?

80. SAVINGS GROWTH   Sally has $5,000 in a savings account earning 12% annual interest. How much will be in her account 10 years from now? (Assume that Sally makes no deposits or withdrawals.)

81. *from Campus to Careers*
    Real Estate Sales Agent

Suppose you are a real estate sales agent and you are working with a client who is considering buying a $250,000 house in Seattle as an investment. If the property values in that area have a track record of increasing at a rate of 8% per year, what will the house be worth 10 years from now?

© Image Source Black/Getty Images

82. BOAT DEPRECIATION A boat that cost $5,000 when new depreciates at a rate of 9% per year. How much will the boat be worth in 5 years?

83. INSCRIBED SQUARES Each inscribed square in the illustration joins the midpoints of the next larger square. The area of the first square, the largest, is 1 square unit. Find the area of the 12th square.

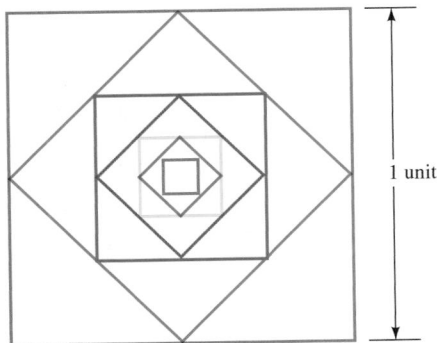

1 unit

84. GENEALOGY The following family tree spans 3 generations and lists 7 people. How many names would be listed in a family tree that spans 10 generations?

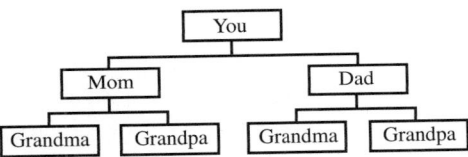

85. BOUNCING BALLS On each bounce, the rubber ball in the illustration rebounds to a height one-half of that from which it fell. Find the total vertical distance the ball travels.

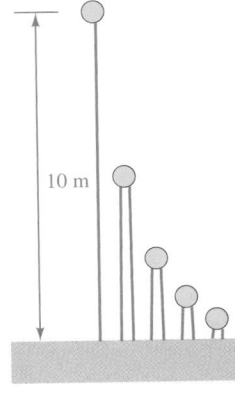

10 m

86. BOUNCING BALLS A golf ball is dropped from a height of 12 feet. On each bounce, it returns to a height that is two-thirds of the distance it fell. Find the total vertical distance the ball travels.

87. PEST CONTROL To reduce the population of a destructive moth, biologists release 1,000 sterilized male moths each day into the environment. If 80% of these moths alive one day survive until the next, then after a long time the population of sterile males is the sum of the infinite geometric series

$$1{,}000 + 1{,}000(0.8) + 1{,}000(0.8)^2 + 1{,}000(0.8)^3 + \cdots$$

Find the long-term population.

88. PENDULUMS On its first swing to the right, a pendulum swings through an arc of 96 inches. Each successive swing, the pendulum travels $\frac{99}{100}$ as far as on the previous swing. Determine the total distance that the pendulum will travel by the time it comes to rest.

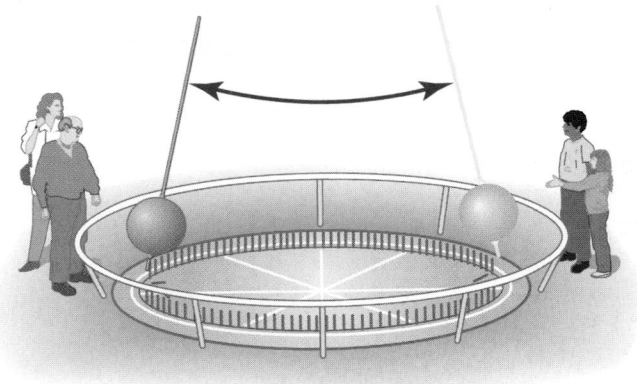

WRITING

89. Describe the real numbers that satisfy $|r| < 1$.

90. Why must the absolute value of the common ratio be less than 1 before an infinite geometric sequence can have a sum?

91. Explain the difference between an arithmetic sequence and a geometric sequence.

92. Why is $1 - \frac{1}{2} + \frac{1}{4} - \frac{1}{8} + \frac{1}{16} - \frac{1}{32} + \cdots$ called an alternating infinite geometric series?

REVIEW

Solve each inequality. Write the solution set using interval notation.

93. $x^2 - 5x - 6 \le 0$

94. $a^2 - 7a + 12 \ge 0$

95. $\dfrac{x-4}{x+3} > 0$

96. $\dfrac{t^2 + t - 20}{t+2} < 0$

CHALLENGE PROBLEMS

97. If $f(x) = 1 + x + x^2 + x^3 + x^4 + \cdots$, find $f\left(\frac{1}{2}\right)$ and $f\left(-\frac{1}{2}\right)$.

98. Find the sum:

$$\frac{1}{\sqrt{3}} + \frac{1}{3} + \frac{1}{3\sqrt{3}} + \frac{1}{9} + \cdots$$

99. If $a > b > 0$, which is larger: the arithmetic mean between a and b or the geometric mean between a and b?

100. Is there a geometric mean between -5 and 5?

CHAPTER 14
Summary & Review

SECTION 14.1   The Binomial Theorem

DEFINITIONS AND CONCEPTS	EXAMPLES
Pascal's triangle gives the coefficients of the terms of the expansion of $(a + b)^n$.	Pascal's triangle:

Pascal's triangle:

$$
\begin{array}{ccccccccccccc}
 & & & & & 1 & & & & & & \text{Row 0} \\
 & & & & 1 & & 1 & & & & & \text{Row 1} \\
 & & & 1 & & 2 & & 1 & & & & \text{Row 2} \\
 & & 1 & & 3 & & 3 & & 1 & & & \text{Row 3} \\
 & 1 & & 4 & & 6 & & 4 & & 1 & & \text{Row 4} \\
1 & & 5 & & 10 & & 10 & & 5 & & 1 & \text{Row 5}
\end{array}
$$

$(x + y)^5 = 1x^5 + 5x^4y + 10x^3y^2 + 10x^2y^3 + 5xy^4 + 1y^5$

DEFINITIONS AND CONCEPTS	EXAMPLES
The symbol $n!$ (*n factorial*) is the product of consecutively decreasing natural numbers from n to 1. $$n! = n(n - 1)(n - 2) \cdot \; \dots \; \cdot 2 \cdot 1$$	Evaluate each expression: $$4! = 4 \cdot 3 \cdot 2 \cdot 1 = 24 \qquad 3! \cdot 2! = 3 \cdot 2 \cdot 1 \cdot 2 \cdot 1 = 12$$ $$\frac{6!}{5!} = \frac{6 \cdot \overset{1}{\cancel{5!}}}{\underset{1}{\cancel{5!}}} = 6 \qquad 1! = 1 \quad \text{and} \quad 0! = 1$$
The **binomial theorem** is usually the best way to expand a binomial. $$(a + b)^n = a^n + \frac{n!}{1!(n - 1)!}a^{n-1}b +$$ $$\frac{n!}{2!(n - 2)!}a^{n-2}b^2 + \; \dots \; + b^n$$	Use the binomial theorem to expand $(p + q)^4$, where $n = 4$. $$(p + q)^4 = p^4 + \frac{4!}{1!(4 - 1)!}p^3q + \frac{4!}{2!(4 - 2)!}p^2q^2 + \frac{4!}{3!(4 - 3)!}pq^3 + q^4$$ $$= p^4 + 4p^3q + 6p^2q^2 + 4pq^3 + q^4$$
To find a specific term of an expansion: The $(r + 1)$st term of the expansion of $(a + b)^n$ is $$\frac{n!}{r!(n - r)!}a^{n-r}b^r$$ Remember that r is always 1 less than the number of the term that you are finding.	Find the third term of the expansion of $(a + b)^5$. In the third term of $(a + b)^5$, $n = 5$ and $r = 2$. $$\frac{5!}{2!(5 - 2)!}a^{5-2}b^2 = \frac{5 \cdot 4 \cdot \overset{1}{\cancel{3!}}}{2 \cdot 1 \cdot \underset{1}{\cancel{3!}}}a^3b^2 = 10a^3b^2$$

REVIEW EXERCISES

1. Complete Pascal's triangle. List the row that gives the coefficients for the expansion of $(a + b)^5$.

$$
\begin{array}{ccccccccccccc}
 & & & & & 1 & & & & & \\
 & & & & 1 & & & & & & \\
 & & & 1 & & 2 & & 1 & & & \\
 & & 1 & & & & 3 & & & & \\
 & & 4 & & 6 & & 4 & & 1 & & \\
 1 & & 5 & & & & 10 & & 5 & & 1 \\
1 & & 6 & & 15 & & 20 & & & & 6 \quad 1
\end{array}
$$

2. Consider the expansion of $(a + b)^{12}$.
  a. How many terms does the expansion have?
  b. For each term, what is the sum of the exponents on a and b?
  c. What is the first term? What is the last term?
  d. How do the exponents on a and b change from term to term?

Evaluate each expression.

3. $4! \cdot 3!$

4. $\dfrac{5!}{3!}$

5. $\dfrac{6!}{2!(6 - 2)!}$

6. $\dfrac{12!}{3!(12 - 3)!}$

7. $(n - n)!$

8. $\dfrac{8!}{7!}$

Use the binomial theorem to find each expansion.

9. $(x + y)^5$

10. $(x - y)^9$

11. $(4x - y)^3$

12. $\left(\dfrac{c}{2} + \dfrac{d}{3}\right)^4$

Find the specified term in each expansion.

13. $(x + y)^4$; third term

14. $(x - y)^6$; fourth term

15. $(3x - 4y)^3$; second term

16. $(u^2 - v^3)^5$; fifth term

SECTION 14.2   Arithmetic Sequences and Series

DEFINITIONS AND CONCEPTS	EXAMPLES
A **sequence** is a list of numbers written in a specific order.	**Finite sequence:** 1, 4, 7, 10, 13 **Infinite sequence:** 2, 6, 10, 14, 18, …
To specifically describe all the terms of a sequence, we can write a formula for a_n, called the **general term** of the sequence.	Find the first four terms of the sequence described by $a_n = 5n + 1$. We substitute 1, 2, 3, and 4, for n in the formula: $a_1 = 5(1) + 1 = 6 \qquad a_2 = 5(2) + 1 = 11$ $a_3 = 5(3) + 1 = 16 \qquad a_4 = 5(4) + 1 = 21$ The first four terms are 6, 11, 16, and 21.
A sequence where each term is found by adding the same number to the previous term is called an **arithmetic sequence.** An arithmetic sequence has the form $a_1, a_1 + d, a_1 + 2d, \dots, a_1 + (n-1)d, \dots$ where a_1 is the first term and d is the common difference. The ***n*th term of an arithmetic sequence** is given by $a_n = a_1 + (n-1)d$. The **common difference** d of an arithmetic sequence is the difference between any two consecutive terms: $d = a_{n+1} - a_n$.	Find the 12th term of the arithmetic sequence $-4, -1, 2, 5, 8, \dots$. First, we find the common difference d. It is the difference between any two successive terms: $d = a_2 - a_1 = -1 - (-4) = 3$ Then we substitute into the formula for the nth term: $a_n = a_1 + (n-1)d$ $a_{12} = -4 + (12-1)3$ In this sequence, $n = 12$, $a_1 = -4$, and $d = 3$. $= -4 + (11)3$ $= -4 + 33$ $= 29$ The 12th term of the sequence is 29.
If numbers are inserted between two given numbers a and b to form an arithmetic sequence, the inserted numbers are **arithmetic means** between a and b.	Find three arithmetic means between -6 and 14. Since there will be five terms, $n = 5$. We also know that $a_1 = -6$ and $a_5 = 14$. We will substitute these values in the formula for the nth term and solve for d. $a_5 = a_1 + (n-1)d$ $14 = -6 + (5-1)d$ $14 = -6 + 4d$ $20 = 4d$ $5 = d$ Since the common difference is 5, we successively add 5 to the terms to get the sequence $-6, -1, 4, 9, 14$. Thus, the three arithmetic means are $-1, 4,$ and 9.

SECTION 14.2   Arithmetic Sequences and Series—*continued*

DEFINITIONS AND CONCEPTS	EXAMPLES

DEFINITIONS AND CONCEPTS

When the commas between the terms of a sequence are replaced with + signs, we call the sum a **series**.

The **sum of the first n terms of an arithmetic sequence** is given by

$$S_n = \frac{n(a_1 + a_n)}{2}$$

EXAMPLES

Find the sum of the first 12 terms of the arithmetic sequence $-4, -1, 2, 5, 8, \ldots$.

Here $a_1 = -4$ and $n = 12$. On the previous page, we found that for this sequence $a_{12} = 29$. We substitute these values into the formula for S_n and simplify.

$$S_n = \frac{n(a_1 + a_n)}{2}$$

$$S_{12} = \frac{12(-4 + 29)}{2}$$

$$= \frac{300}{2}$$

$$= 150$$

The sum of the first 12 terms is 150.

Summation notation involves the Greek letter sigma Σ. It designates the sum of terms called a **series**.

$$\sum_{n=1}^{4} (3k - 2) = [3(1) - 2] + [3(2) - 2] + [3(3) - 2] + [3(4) - 2]$$

$$= 1 + 4 + 7 + 10$$

$$= 22$$

(with labels $k = 1$, $k = 2$, $k = 3$, $k = 4$ above the respective terms)

REVIEW EXERCISES

17. Find the first four terms of the sequence defined by $a_n = 2n - 4$.

18. Find the first five terms of the sequence defined by $a_n = \frac{(-1)^n}{n + 1}$.

19. Find the 50th term of the sequence defined by $a_n = 100 - \frac{n}{2}$.

20. Find the eighth term of an arithmetic sequence whose first term is 7 and whose common difference is 5.

21. Write the first five terms of the arithmetic sequence whose ninth term is 242 and whose seventh term is 212.

22. The first three terms of an arithmetic sequence are 6, −6, and −18. Find the 101st term.

23. Find the common difference of an arithmetic sequence if its 1st term is −515 and the 23rd term is −625.

24. Find two arithmetic means between 8 and 25.

25. Find the sum of the first ten terms of the sequence 9, 6.5, 4, ...

26. Find the sum of the first 28 terms of an arithmetic sequence if the second term is 6 and the sixth term is 22.

Find each sum.

27. $\sum_{k=4}^{6} \frac{1}{2}k$

28. $\sum_{k=2}^{5} 7k^2$

29. $\sum_{k=1}^{4} (3k - 4)$

30. $\sum_{k=10}^{10} 36k$

31. What is the sum of the first 200 natural numbers?

32. SEATING   The illustration shows the first 2 of a total of 30 rows of seats in an amphitheater. The number of seats in each row forms an arithmetic sequence. Find the total number of seats.

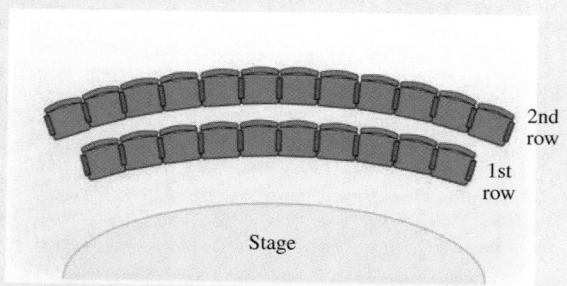

2nd row
1st row
Stage

SECTION 14.3 Geometric Sequences and Series

DEFINITIONS AND CONCEPTS	EXAMPLES

DEFINITIONS AND CONCEPTS

Each term of a **geometric sequence** is found by multiplying the previous term by the same number.

A geometric sequence has the form:

$$a_1, a_1r, a_1r^2, a_1r^3, \ldots, a_1r^{n-1}, \ldots$$

where a_1 is the first term and r is the common ratio.

The **nth term of a geometric sequence** is given by $a_n = a_1r^{n-1}$.

The **common ratio r** of a geometric sequence is the quotient obtained when any term is divided by the previous term:

$$r = \frac{a_{n+1}}{a_n}$$

If numbers are inserted between a and b to form a geometric sequence, the inserted numbers are **geometric means** between a and b.

The **sum of the first n terms of a geometric sequence** is given by:

$$S_n = \frac{a_1 - a_1r^n}{1 - r} \quad r \neq 1$$

EXAMPLES

Find the 8th term of the geometric sequence 8, 4, 2, ...

First, we find the common ratio r. It is the quotient obtained when any term is divided by the previous term:

$$r = \frac{a_2}{a_1} = \frac{4}{8} = \frac{1}{2}$$

Then we substitute into the formula for the nth term:

$$a_n = a_1r^{n-1}$$
$$a_8 = 8\left(\frac{1}{2}\right)^{8-1} \quad \text{Substitute: } n = 8, \ a_1 = 8, \text{ and } r = \frac{1}{2}.$$
$$= 8\left(\frac{1}{2}\right)^7$$
$$= \frac{8}{128} \quad \text{Evaluate: } \left(\frac{1}{2}\right)^7 = \frac{1}{128}.$$
$$= \frac{1}{16} \quad \text{Simplify the fraction: Remove a common factor of 8 in the numerator and denominator.}$$

Find two geometric means between 6 and 162.

Since there will be four terms, $n = 4$. We also know that $a_1 = 6$ and $a_4 = 162$. We will substitute these values in the formula for the nth term and solve for r.

$$a_n = a_1r^{n-1}$$
$$162 = 6r^{4-1}$$
$$27 = r^3 \quad \text{Divide both sides by 6.}$$
$$3 = r \quad \text{Take the cube root of both sides.}$$

Since the common ratio is 3, we successively multiply the terms by 3 to get the sequence 6, 18, 54, 162. Thus, the two geometric means are 18 and 54.

Find the sum of the first five terms of the geometric sequence 2, 6, 18, ...

Here $a_1 = 2$, $r = 3$, and $n = 5$. We substitute these values into the formula for S_n and simplify.

$$S_n = \frac{a_1 - a_1(r)^n}{1 - r}$$
$$S_5 = \frac{2 - (2)(3)^5}{1 - 3}$$
$$= \frac{2 - 2(243)}{-2} \quad \text{Evaluate: } (3)^5 = 243.$$
$$= \frac{2 - 486}{-2}$$
$$= \frac{-484}{-2}$$
$$= 242$$

The sum of the first five terms is 242.

SECTION 14.3   Geometric Sequences and Series—*continued*

DEFINITIONS AND CONCEPTS	EXAMPLES
If we form the sum of the terms of an infinite geometric sequence, we get a series called an **infinite geometric series.** The sum of an infinite geometric series is given by: $$S = \frac{a_1}{1 - r} \qquad \text{where } |r| < 1$$	Find the sum of the infinite geometric series: $12 + 8 + \frac{16}{3} + \cdots$ Here $a_1 = 12$ and $r = \frac{8}{12} = \frac{2}{3}$. (Note that $r < 1$) We substitute these values into the formula for the sum of an infinite series. $$S = \frac{a_1}{1 - r} = \frac{12}{1 - \frac{2}{3}} = \frac{12}{\frac{1}{3}} = 12 \cdot \frac{3}{1} = 36$$ The sum is 36.

REVIEW EXERCISES

33. Find the sixth term of a geometric sequence with a first term of $\frac{1}{8}$ and a common ratio of 2.

34. Write the first five terms of the geometric sequence whose fourth term is 3 and whose fifth term is $\frac{3}{2}$.

35. Find the first term of a geometric sequence if it has a common ratio of -3 and the ninth term is 243.

36. Find two geometric means between -6 and 384.

37. Find the sum of the first seven terms of the sequence 162, 54, 18, ...

38. Find the sum of the first eight terms of the sequence: $\frac{1}{8}, -\frac{1}{4}, \frac{1}{2}, \ldots$

39. FEEDING BIRDS   Tom has 50 pounds of birdseed stored in his garage. Each month, he uses 25% of what is left in the bag to feed the birds in his yard. How much birdseed will be left in 12 months?

40. Find the sum of the infinite geometric series: $25 + 20 + 16 + \cdots$

41. Change the decimal $0.\overline{05}$ to a common fraction.

42. WHAM-O TOYS   Tests have found that 1998 Superballs rebound $\frac{9}{10}$ of the distance from which they are dropped. If a Superball is dropped from a height of 10 feet, and if it could bounce forever, what total distance would it travel?

CHAPTER 14
Test

1. Fill in the blanks.

   a. The array of numbers that gives the coefficients of the terms of a binomial expansion is called _____ triangle.

   b. In the expansion $a^3 - 3a^2b + 3ab^2 - b^3$, the signs _____ between $+$ and $-$.

   c. Each term of an _____ sequence is found by adding the same number to the previous term.

   d. The sum of the terms of an arithmetic sequence is called an arithmetic _____.

   e. Each term of a _____ sequence is found by multiplying the previous term by the same number.

2. Find the first 4 terms of the sequence defined by: $a_n = -6n + 8$

3. Find the first 5 terms of the sequence defined by: $a_n = \frac{(-1)^n}{n^3}$.

4. Evaluate: $\frac{10!}{6!(10 - 6)!}$

5. Expand: $(a - b)^6$.

6. Find the third term in the expansion of $(x^2 + 2y)^4$.

7. Find the tenth term of an arithmetic sequence whose first 3 terms are 3, 10, and 17.

8. Find the sum of the first 12 terms of the sequence: $-2, 3, 8, \ldots$

9. Find two arithmetic means between 2 and 98.

10. Find the common difference of an arithmetic sequence if the second term is $\frac{5}{4}$ and the 17th term is 5.

11. Find the sum of the first 27 terms of an arithmetic sequence if the 4th term is -11 and the 20th term is -75.

12. PLUMBING   Plastic pipe is stacked so that the bottom row has 25 pipes, the next row has 24 pipes, the next row has 23 pipes, and so on until there is 1 pipe at the top of the stack. If a worker removes the top 15 rows of pipe, how many pieces of pipe will be left in the stack?

13. FALLING OBJECTS   If an object is in free fall, the sequence 16, 48, 80, . . . represents the distance in feet that object falls during the 1st second, during the 2nd second, during the 3rd second, and so on. How far will the object fall during the first 10 seconds?

14. Evaluate: $\displaystyle\sum_{k=1}^{3} (2k - 3)$

15. Find the seventh term of the geometric sequence whose first 3 terms are $-\frac{1}{9}, -\frac{1}{3}$, and -1.

16. Find the sum of the first 6 terms of the sequence:
$\frac{1}{27}, \frac{1}{9}, \frac{1}{3}, \cdots$

17. Find the first term of a geometric sequence if the common ratio is $-\frac{2}{3}$ and the fourth term is $-\frac{16}{9}$.

18. Find two geometric means between 3 and 648.

19. Find the sum of infinite geometric series: $9 + 3 + 1 + \cdots$

20. DEPRECIATION   A yacht that cost \$1,500,000 when new depreciates at a rate of 8% per year. How much will the yacht be worth in 10 years?

21. PENDULUMS   On its first swing to the right, a pendulum swings through an arc of 60 inches. Each successive swing, the pendulum travels $\frac{97}{100}$ as far as on the previous swing. Determine the total distance the pendulum will travel by the time it comes to rest.

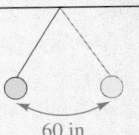

60 in

22. Change the decimal $0.\overline{7}$ to a common fraction.

GROUP PROJECT

THE LANGUAGE OF ALGEBRA

Overview: This activity will help you review for the final exam.

Instructions: Form groups of two students. Match each instruction in column I with the most appropriate item in column II. Each letter in column II is used only once.

Column I

_____ 1. Use the FOIL method.

_____ 2. Apply a rule for exponents to simplify.

_____ 3. Add the rational expressions.

_____ 4. Rationalize the denominator.

_____ 5. Factor completely.

_____ 6. Evaluate the expression for $a = -1$ and $b = -6$.

_____ 7. Express in lowest terms.

_____ 8. Solve for t.

_____ 9. Combine like terms.

_____ 10. Remove parentheses.

_____ 11. Solve the system by graphing.

_____ 12. Find $f(g(x))$.

_____ 13. Solve using the quadratic formula.

_____ 14. Identify the base and the exponent.

_____ 15. Write without a radical symbol.

_____ 16. Write the equation of the line having the given slope and y-intercept.

_____ 17. Solve the inequality.

_____ 18. Complete the square to make a perfect-square trinomial.

_____ 19. Find the slope of the line passing through the given points.

_____ 20. Use a property of logarithms to simplify.

_____ 21. Set each factor equal to zero and solve for x.

_____ 22. State the solution of the compound inequality using interval notation.

_____ 23. Find the inverse function, $h^{-1}(x)$.

_____ 24. Write using scientific notation.

_____ 25. Write the logarithmic statement in exponential form.

_____ 26. Find the sum of the first 6 terms of the sequence.

Column II

a. 2,300,000,000

b. e^3

c. $f(x) = x^2 + 1$ and $g(x) = 5 - 3x$

d. $-2x(3x^2 - 4x + 8)$

e. $4x - 7 > -3x - 7$

f. $\begin{cases} 2x = y - 5 \\ x + y = -1 \end{cases}$

g. $(x^2 - 5)(x^2 + 3)$

h. $(x + 2)(x - 10) = 0$

i. $\dfrac{x - 1}{2x^2} + \dfrac{x + 1}{8x}$

j. $\sqrt{4x^2}$

k. $\ln 6 + \ln x$

l. $\dfrac{10}{\sqrt{6} - \sqrt{2}}$

m. $2x - 8 + 6y - 14$

n. $(3, -2)$ and $(0, -5)$

o. $x^4 \cdot x^3$

p. $\dfrac{4x^2y}{16xy}$

q. $h(x) = 10^x$

r. $\log_2 8 = 3$

s. $m = \dfrac{2}{3}$ and passes through $(0, 2)$

t. $2, 6, 18, \ldots$

u. $3y^3 - 243b^6$

v. $x + 7 \geq 0$ and $-x < -1$

w. $x^2 - 3x - 4 = 0$

x. $Rt = cd + 2t$

y. $-2\pi a^2 - 3b^3$

z. $x^2 + 4x$

CUMULATIVE REVIEW
Chapters 1–14

1. Give the elements of the set $\left\{-\frac{4}{3}, \pi, 5.6, \sqrt{2}, 0, -23, e, 7i\right\}$ that belong to each of the following sets. [Section 1.3]

   a. Whole numbers

   b. Rational numbers

   c. Irrational numbers

   d. Real numbers

2. Solve: $6[x - (2 - x)] = -4(8x + 3)$ [Section 2.2]

3. Solve $A = \frac{1}{2}h(b_1 + b_2)$ for b_2. [Section 2.4]

4. MARTIAL ARTS   Find the measure of each angle of the triangle shown in the illustration. [Section 2.5]

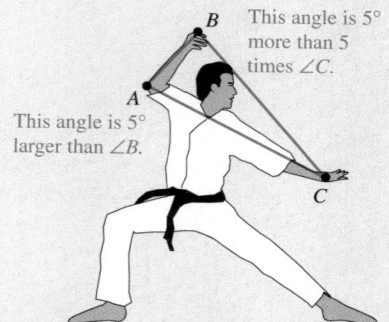

B This angle is 5° more than 5 times $\angle C$.

A This angle is 5° larger than $\angle B$.

C

5. FINANCIAL PLANNING   Anna has some money to invest. Her financial planner tells her that if she can come up with $3,000 more, she will qualify for an 11% annual interest rate. Otherwise, she will have to invest the money at 7.5% annual interest. The financial planner urges her to invest the larger amount, because the 11% investment would yield twice as much annual income as the 7.5% investment. How much does she originally have on hand to invest? [Section 2.6]

6. BOATING   Use the following graph to determine the average rate of change in the sound level of the engine of a boat in relation to the number of revolutions per minute (rpm) of the engine. [Section 3.4]

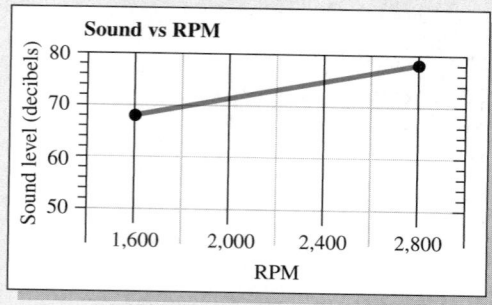

Sound vs RPM

Sound level (decibels) — 50, 60, 70, 80

RPM — 1,600, 2,000, 2,400, 2,800

7. Decide whether the graphs of the equations are parallel or perpendicular. [Section 3.5]

   a. $3x - 4y = 12,\ y = \frac{3}{4}x - 5$

   b. $y = 3x + 4,\ x = -3y + 4$

8. SALVAGE VALUES   A truck was purchased for $28,000. Its salvage value at the end of 6 years is expected to be $7,600. Find the straight-line depreciation equation. [Section 3.6]

Find an equation of the line with the given properties. Write the equation in slope–intercept form.

9. $m = -2$, passing through $(0, 5)$ [Section 2.4]

10. Passing through $(8, -5)$ and $(-5, 4)$ [Section 2.4]

11. Explain why the graph does not represent a function. [Section 3.8]

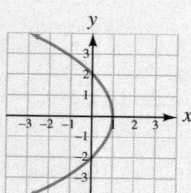

12. If $f(x) = 3x^5 - 2x^2 + 1$, find $f(-1)$ and $f(a)$. [Section 3.8]

13. Use substitution to solve: $\begin{cases} 2x - y = -21 \\ 4x + 5y = 7 \end{cases}$ [Section 4.3]

14. Use addition to solve: $\begin{cases} 4y + 5x - 7 = 0 \\ \frac{10}{7}x - \frac{4}{9}y = \frac{17}{21} \end{cases}$ [Section 4.3]

15. MIXING COFFEE   How many pounds of regular coffee (selling for $4 per pound) and how many pounds of Brazilian coffee (selling for $11.50 per pound) must be combined to get 40 pounds of a mixture worth $6 per pound? [Section 4.4]

16. Graph the solution set: $\begin{cases} 3x - 2y \le 6 \\ y < -x + 2 \end{cases}$ [Section 4.5]

Simplify each expression. Write answers using positive exponents.

17. $(x^2)^5 y^7 y^3 x^{-2} y^0$

[Section 5.2]

18. $\left(\dfrac{3x^5 y^2}{6x^5 y^{-2}}\right)^{-4}$

[Section 5.2]

19. Write 173,000,000,000,000 and 0.000000046 in scientific notation. [Section 5.3]

20. Write each number in scientific notation and perform the indicated operations. Give the answer in scientific notation and in standard notation. [Section 5.3]

$$\frac{(0.00024)(96,000,000)}{(640,000,000)(0.025)}$$

21. Simplify the polynomial: $\frac{9}{4}rt^2 - \frac{5}{3}rt - \frac{1}{2}rt^2 + \frac{5}{6}rt$

[Section 5.5]

22. Find the sum when $(3x^2 + 4x - 7)$ is added to the sum of $(-2x^2 - 7x + 1)$ and $(-4x^2 + 8x - 1)$. [Section 5.5]

Perform the indicated operations and simplify, if possible.

23. $(-2x^2y^3 + 6xy + 5y^2) - (-4x^2y^3 - 7xy + 2y^2)$
    [Section 5.5]
24. $(x - 3y)(x^2 + 3xy + 9y^2)$ [Section 5.6]
25. $(2m^5 - 7)(3m^5 - 1)$ [Section 5.6]
26. $(9ab^2 - 4)^2$ [Section 5.7]

Factor the expression completely.

27. $3x^3y - 4x^2y^2 - 6x^2y + 8xy^2$ [Section 6.1]
28. $b^3 - 4b^2 - 3b + 12$ [Section 6.1]
29. $12y^2 + 23y + 10$ [Section 6.3]
30. $256x^4y^4 - z^8$ [Section 6.4]
31. $27t^3 + u^3$ [Section 6.4]
32. $a^4b^2 - 20a^2b^2 + 64b^2$ [Section 6.6]

33. Solve: $3x^2 + 5x - 2 = 0$ [Section 6.7]
34. Solve: $(x + 7)^2 = -2(x + 7) - 1$ [Section 6.7]

35. Solve: $x^3 + 8x^2 = 9x$ [Section 6.7]
36. PAINTING When it is spread out, a rectangular-shaped painting tarp covers an area of 84 square feet. Its length is 1 foot longer than five times its width. Find its width and length. [Section 6.8]

37. GEOMETRY The longer leg of a right triangle is 2 units longer than the shorter leg. If the hypotenuse is 4 units longer than the shorter leg, find the lengths of the sides of the triangle. [Section 6.8]

38. Simplify: $\dfrac{6x^2 + 13x + 6}{6x^2 + 5x - 6}$ [Section 7.1]

Perform the indicated operations and simplify, if possible.

39. $\dfrac{p^3 - q^3}{q^2 - p^2} \cdot \dfrac{q^2 + pq}{p^3 + p^2q + pq^2}$ [Section 7.2]

40. $\dfrac{2x + 1}{x^4 - 81} + \dfrac{2 - x}{x^4 - 81}$ [Section 7.3]

41. $\dfrac{2}{a - 2} + \dfrac{3}{a + 2} - \dfrac{a - 1}{a^2 - 4}$ [Section 7.4]

42. $\dfrac{\dfrac{y}{x} - \dfrac{x}{y}}{\dfrac{1}{x} + \dfrac{1}{y}}$ [Section 7.5]

43. Solve: $\dfrac{1}{a + 5} = \dfrac{1}{3a + 6} - \dfrac{a + 2}{a^2 + 7a + 10}$ [Section 7.6]

44. Solve $\dfrac{1}{R} = \dfrac{1}{R_1} + \dfrac{1}{R_2} + \dfrac{1}{R_3}$ for R. [Section 7.6]

45. PRINTING PAYCHECKS It takes a printer 6 hours to print the payroll checks for all of the employees of a large company. A faster printer can print the paychecks in 4 hours. How long will it take the two printers working together to print all of the paychecks? [Section 7.7]

46. CAPTURE-RELEASE METHOD To estimate the ground squirrel population on his acreage, a farmer trapped, tagged, and then released two dozen squirrels. Two weeks later, the farmer trapped 31 squirrels and noted that 8 were tagged. Use this information to estimate the number of ground squirrels on his acreage. [Section 7.8]

47. Solve: $3(x - 4) + 6 = -2(x + 4) + 5x$ [Section 8.1]

48. WORK SCHEDULES A student works two part-time jobs. She earns $12 an hour for working at the college book store and $22.50 an hour working graveyard shift at an airport. To save time for study, she limits her work to 25 hours a week. If she enjoys the work at the bookstore more, how many hours can she work at the bookstore and still earn at least $450 a week? [Section 8.1]

Give the solution in interval notation and graph the solution set.

49. Solve: $4.5x - 1 < -10$ or $6 - 2x \geq 12$
    [Section 8.2]
50. Solve: $|5 - 3x| - 14 \leq 0$ [Section 8.3]
51. Use a substitution to factor: $14(x - t)^2 - 17(x - t) - 6$
    [Section 8.4]

52. Solve for λ: $\dfrac{A\lambda}{2} + 1 = 2d + 3\lambda$ [Section 8.5]

53. Factor: $x^2 + 10x + 25 - 16z^2$ [Section 8.5]

54. Does $x = |y|$ define a function? [Section 8.8]

55. Use the graph of function h to find each of the following.
    [Section 8.8]
    a. $h(-3)$
    b. $h(4)$
    c. The value(s) of x for which $h(x) = 1$
    d. The value(s) of x for which $h(x) = 0$

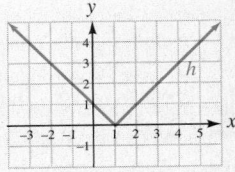

56. Find the domain of the function $f(x) = \frac{8}{x^2 - 25}$. [Section 8.8]

57. LIGHT   The intensity of a light source is inversely proportional to the square of the distance from the source. If the intensity is 18 lumens at a distance of 4 feet, what is the intensity when the distance is 12 feet? [Section 8.9]

58. Graph: $f(x) = \sqrt{x} + 2$. Give the domain and range of the function. [Section 9.1]

59. Evaluate: $\left(\frac{25}{49}\right)^{-3/2}$ [Section 9.2]

60. Simplify: $\sqrt{112a^3 b^5}$ [Section 9.3]

Simplify each expression. All variables represent positive numbers.

61. $\sqrt{98} + \sqrt{8} - \sqrt{32}$ [Section 9.3]

62. $12\sqrt[3]{648x^4} + 3\sqrt[3]{81x^4}$ [Section 9.3]

63. $(2\sqrt{7} + 1)(\sqrt{7} - 1)$ [Section 9.4]

64. $3(\sqrt{5x} - \sqrt{3})^2$ [Section 9.4]

Rationalize each denominator.

65. $\dfrac{\sqrt[3]{4}}{\sqrt[3]{b}}$ [Section 9.4]

66. $\dfrac{3t - 1}{\sqrt{3t} + 1}$ [Section 9.4]

Solve each equation.

67. $2x = \sqrt{16x - 12}$ [Section 9.5]

68. $\sqrt[3]{12m + 4} = 4$ [Section 9.5]

69. $\sqrt{x + 3} - \sqrt{3} = \sqrt{x}$ [Section 9.5]

70. CHANGING DIAPERS   The following illustration shows how to put a diaper on a baby. If the diaper is a square with sides 16 inches long, what is the largest waist size that this diaper can wrap around, assuming an overlap of 1 inch to pin the diaper? [Section 9.6]

71. Express $\sqrt{-25}$ in terms of i. [Section 9.7]

72. Simplify: i^{42} [Section 9.7]

Write each expression in a + bi form.

73. $(-7 + 9i) - (-2 - 8i)$ [Section 9.7]

74. $\dfrac{2 - 5i}{2 + 5i}$ [Section 9.7]

75. Solve: $t^2 = 24$ [Section 10.1]

76. Solve $m^2 + 10m - 7 = 0$ by completing the square. [Section 10.1]

Solve each equation.

77. $4w^2 + 6w + 1 = 0$ [Section 10.2]

78. $3x^2 - 4x = -2$ [Section 10.2]

79. $2(2x + 1)^2 - 7(2x + 1) + 6 = 0$ [Section 10.3]

80. $x^4 + 19x^2 + 18 = 0$ [Section 10.3]

81. TIRE WEAR   Refer to the graph. [Section 10.4]

  a. What type of function does it appear would model the relationship between the inflation of a tire and the percent of service it gives?

  b. At what percent(s) of inflation will a tire offer only 90% of its possible service?

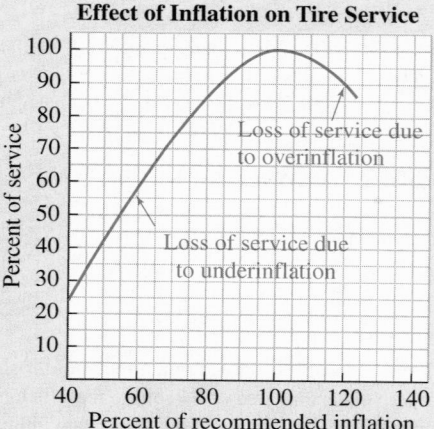

Effect of Inflation on Tire Service

Loss of service due to overinflation

Loss of service due to underinflation

Percent of service

Percent of recommended inflation

82. Graph $f(x) = -6x^2 - 12x - 8$ using the vertex formula. Then determine the x- and y-intercepts of the graph. Finally, plot several points and complete the graph. [Section 10.4]

83. Solve $x^2 - 8x \le -15$. Write the solution set in interval notation and graph it. [Section 10.5]

84. If $f(x) = x^2 - 2$ and $g(x) = 2x + 1$, find $(f \circ g)(x)$. [Section 11.1]

85. Find the inverse function of $f(x) = 2x^3 - 1$. [Section 11.2]

86. Graph $f(x) = \left(\frac{1}{2}\right)^x$ and give the domain and range of the function. [Section 11.3]

87. Graph $f(x) = e^x$ and its inverse on the same coordinate system. [Section 11.4]

88. POPULATION GROWTH In 2007, the population of Mexico was about 109 million and the annual growth rate was 1.153%. If the growth rate remained the same, what would the population of Mexico be 25 years later? [Section 11.4]

Find x.

89. $\log 1{,}000 = x$ [Section 11.5]

90. $\log_8 64 = x$ [Section 11.5]

91. $\log_3 x = -3$ [Section 11.5]

92. $\log_x 25 = 2$ [Section 11.5]

93. $\ln e = x$ [Section 11.6]

94. $\ln \dfrac{1}{e} = x$ [Section 11.6]

95. Find $\ln 0$, if possible. [Section 11.6]

96. Use the properties of logarithms to simplify $\log_6 \dfrac{36}{x^3}$. [Section 11.7]

97. Write the expression $\frac{1}{2}\ln x + \ln y - \ln z$ as a single logarithm. [Section 11.7]

98. BACTERIA GROWTH The bacteria in a laboratory culture increased from an initial population of 200 to 600 in 4 hours. How long will it take the population to reach 8,000? [Section 11.8]

Solve each equation. Round to four decimal places when necessary.

99. $5^{4x} = \dfrac{1}{125}$ [Section 11.8]

100. $2^{x+2} = 3^x$ [Section 11.8]

101. $\log x + \log(x + 9) = 1$ [Section 11.8]

102. $\log_3 x = \log_3 \left(\dfrac{1}{x}\right) + 4$ [Section 11.8]

103. Solve: $\begin{cases} b + 2c = 7 - a \\ a + c = 8 - 2b \\ 2a + b + c = 9 \end{cases}$ [Section 12.2]

104. PUPPETS A toy company makes a total of 500 puppets in three sizes during a production run. The small puppets cost $5 to make and sell for $8 each, the standard-size puppets cost $10 to make and sell for $16 each, and the super-size puppets cost $15 to make and sell for $25 each. The total cost to make the puppets is $4,750 and the revenue from their sale is $7,700. How many small, standard-size, and super-size puppets are made during a production run? [Section 12.3]

105. Use matrices to solve the system: $\begin{cases} 2x + y = 1 \\ x + 2y = -4 \end{cases}$ [Section 12.4]

106. Use Cramer's rule to solve: $\begin{cases} 2(x + y) + 1 = 0 \\ 3x + 4y = 0 \end{cases}$ [Section 12.5]

107. Write the equation of the circle that has its center at $(1, -3)$ and a radius of 2. Graph the equation. [Section 13.1]

108. Complete the square to write the equation $y^2 + 4x - 6y = -1$ in $x = a(y - k)^2 + h$ form. Determine the vertex and the axis of symmetry of the graph. Then plot several points and complete the graph. [Section 13.1]

109. Graph: $\dfrac{(x + 1)^2}{4} + \dfrac{(y - 3)^2}{16} = 1$ [Section 13.2]

110. Write the equation in standard form and graph it: $(x - 2)^2 - 9y^2 = 9$ [Section 13.3]

111. Use the binomial theorem to expand $(3a - b)^4$. [Section 14.1]

112. Find the seventh term of the expansion of $(2x - y)^8$. [Section 14.1]

113. Evaluate: $\dfrac{12!}{10!(12 - 10)!}$ [Section 14.1]

114. Find the 20th term of an arithmetic sequence with a first term -11 and a common difference 6. [Section 14.2]

115. Find the sum of the first 20 terms of an arithmetic sequence with a first term 6 and a common difference 3. [Section 14.2]

116. Evaluate: $\displaystyle\sum_{k=3}^{5} (2k + 1)$ [Section 14.2]

117. Find the seventh term of a geometric sequence with a first term $\frac{1}{27}$ and a common ratio 3. [Section 14.3]

118. BOAT DEPRECIATION How much will a $9,000 boat be worth after 9 years if it depreciates 12% per year? [Section 14.3]

119. Find the sum of the first ten terms of the sequence: $\dfrac{1}{64}, \dfrac{1}{32}, \dfrac{1}{16}, \dots$ [Section 14.3]

120. Find the sum of the infinite series: $9 + 3 + 1 + \cdots$ [Section 14.3]

APPENDIX 1
Roots and Powers

n	n^2	$\sqrt{n}$	n^3	$\sqrt[3]{n}$	n	n^2	$\sqrt{n}$	n^3	$\sqrt[3]{n}$
1	1	1.000	1	1.000	51	2,601	7.141	132,651	3.708
2	4	1.414	8	1.260	52	2,704	7.211	140,608	3.733
3	9	1.732	27	1.442	53	2,809	7.280	148,877	3.756
4	16	2.000	64	1.587	54	2,916	7.348	157,464	3.780
5	25	2.236	125	1.710	55	3,025	7.416	166,375	3.803
6	36	2.449	216	1.817	56	3,136	7.483	175,616	3.826
7	49	2.646	343	1.913	57	3,249	7.550	185,193	3.849
8	64	2.828	512	2.000	58	3,364	7.616	195,112	3.871
9	81	3.000	729	2.080	59	3,481	7.681	205,379	3.893
10	100	3.162	1,000	2.154	60	3,600	7.746	216,000	3.915
11	121	3.317	1,331	2.224	61	3,721	7.810	226,981	3.936
12	144	3.464	1,728	2.289	62	3,844	7.874	238,328	3.958
13	169	3.606	2,197	2.351	63	3,969	7.937	250,047	3.979
14	196	3.742	2,744	2.410	64	4,096	8.000	262,144	4.000
15	225	3.873	3,375	2.466	65	4,225	8.062	274,625	4.021
16	256	4.000	4,096	2.520	66	4,356	8.124	287,496	4.041
17	289	4.123	4,913	2.571	67	4,489	8.185	300,763	4.062
18	324	4.243	5,832	2.621	68	4,624	8.246	314,432	4.082
19	361	4.359	6,859	2.668	69	4,761	8.307	328,509	4.102
20	400	4.472	8,000	2.714	70	4,900	8.367	343,000	4.121
21	441	4.583	9,261	2.759	71	5,041	8.426	357,911	4.141
22	484	4.690	10,648	2.802	72	5,184	8.485	373,248	4.160
23	529	4.796	12,167	2.844	73	5,329	8.544	389,017	4.179
24	576	4.899	13,824	2.884	74	5,476	8.602	405,224	4.198
25	625	5.000	15,625	2.924	75	5,625	8.660	421,875	4.217
26	676	5.099	17,576	2.962	76	5,776	8.718	438,976	4.236
27	729	5.196	19,683	3.000	77	5,929	8.775	456,533	4.254
28	784	5.292	21,952	3.037	78	6,084	8.832	474,552	4.273
29	841	5.385	24,389	3.072	79	6,241	8.888	493,039	4.291
30	900	5.477	27,000	3.107	80	6,400	8.944	512,000	4.309
31	961	5.568	29,791	3.141	81	6,561	9.000	531,441	4.327
32	1,024	5.657	32,768	3.175	82	6,724	9.055	551,368	4.344
33	1,089	5.745	35,937	3.208	83	6,889	9.110	571,787	4.362
34	1,156	5.831	39,304	3.240	84	7,056	9.165	592,704	4.380
35	1,225	5.916	42,875	3.271	85	7,225	9.220	614,125	4.397
36	1,296	6.000	46,656	3.302	86	7,396	9.274	636,056	4.414
37	1,369	6.083	50,653	3.332	87	7,569	9.327	658,503	4.431
38	1,444	6.164	54,872	3.362	88	7,744	9.381	681,472	4.448
39	1,521	6.245	59,319	3.391	89	7,921	9.434	704,969	4.465
40	1,600	6.325	64,000	3.420	90	8,100	9.487	729,000	4.481
41	1,681	6.403	68,921	3.448	91	8,281	9.539	753,571	4.498
42	1,764	6.481	74,088	3.476	92	8,464	9.592	778,688	4.514
43	1,849	6.557	79,507	3.503	93	8,649	9.644	804,357	4.531
44	1,936	6.633	85,184	3.530	94	8,836	9.695	830,584	4.547
45	2,025	6.708	91,125	3.557	95	9,025	9.747	857,375	4.563
46	2,116	6.782	97,336	3.583	96	9,216	9.798	884,736	4.579
47	2,209	6.856	103,823	3.609	97	9,409	9.849	912,673	4.595
48	2,304	6.928	110,592	3.634	98	9,604	9.899	941,192	4.610
49	2,401	7.000	117,649	3.659	99	9,801	9.950	970,299	4.626
50	2,500	7.071	125,000	3.684	100	10,000	10.000	1,000,000	4.642

APPENDIX 2

Synthetic Division

1 Synthetic division

2 The remainder theorem

3 The factor theorem

We have discussed how to divide polynomials by polynomials using a long division process. We will now discuss a shortcut method, called **synthetic division,** that we can use to divide a polynomial by a binomial of the form $x - k$.

1 **Synthetic Division.**

To see how synthetic division works, we consider $(4x^3 - 5x^2 - 11x + 20) \div (x - 2)$. On the left below is the long division, and on the right is the same division with the variables and their exponents removed. The various powers of x can be remembered without actually writing them, because the exponents of the terms in the divisor, dividend, and quotient were written in descending order.

$$
\begin{array}{r}
4x^2 + 3x\ - 5 \\
x - 2)\overline{4x^3 - 5x^2 - 11x + 20} \\
\underline{4x^3 - 8x^2} \\
3x^2 - 11x \\
\underline{3x^2 - \ 6x} \\
-5x + 20 \\
\underline{-5x + 10} \\
10 \quad \text{(remainder)}
\end{array}
$$

$$
\begin{array}{r}
4\quad 3 - \ 5 \\
1 - 2)\overline{4 - 5 - 11 \quad 20} \\
\underline{4 - 8} \\
3 - 11 \\
\underline{3 - \ 6} \\
-5 \quad 20 \\
\underline{-5 \quad 10} \\
10 \quad \text{(remainder)}
\end{array}
$$

The numbers printed in color need not be written, because they are duplicates of the numbers above them. Thus, we can write the division in the form shown below on the left. We can shorten the process further by compressing the work vertically and eliminating the 1 (the coefficient of x in the divisor) as shown below on the right.

The Language of Algebra

Synthetic means devised to imitate something natural. You've probably heard of *synthetic* fuels or *synthetic* fibers. *Synthetic* division imitates the long division process.

$$
\begin{array}{r}
4\quad 3 - 5 \\
1 - 2)\overline{4 - 5 - 11 \quad 20} \\
\underline{- 8} \\
3 \\
\underline{- 6} \\
- 5 \\
\underline{10} \\
10
\end{array}
$$

$$
\begin{array}{r}
4 \quad 3 \quad -5 \\
-2)\overline{4 \quad -5 \quad -11 \quad 20} \\
\underline{-8 \quad -6 \quad 10} \\
3 \quad -5 \quad 10
\end{array}
$$

If we write the 4 in the quotient on the bottom line, that line gives the coefficients of the quotient and the remainder. If we eliminate the top line, the division appears as follows:

$$\begin{array}{r|rrrr} -2 & 4 & -5 & -11 & 20 \\ & & -8 & -6 & 10 \\ \hline & 4 & 3 & -5 & 10 \end{array}$$

The bottom line was obtained by subtracting the middle line from the top line. If we replace the -2 in the divisor by 2, the division process will reverse the signs of every entry in the middle line, and then the bottom line can be obtained by addition. This gives the final form of the synthetic division.

$$\begin{array}{r|rrrr} 2 & 4 & -5 & -11 & 20 \\ & & 8 & 6 & -10 \\ \hline & 4 & 3 & -5 & 10 \end{array}$$

These are the coefficients of the dividend.

These are the coefficients of the quotient and the remainder.

$$4x^2 + 3x - 5 + \frac{10}{x-2}$$

Read the result from the bottom row.

Thus,

$$\frac{4x^3 - 5x^2 - 11x + 20}{x-2} = 4x^2 + 3x - 5 + \frac{10}{x-2}$$

The Language of Algebra
Synthetic division is used to divide a polynomial by a binomial of the form $x - k$. We call k the **synthetic divisor.** In this example, we are dividing by $x - 2$, so k is 2.

EXAMPLE 1   Use synthetic division to find $(6x^2 + 5x - 2) \div (x - 5)$.

Solution   We write the coefficients in the dividend and the 5 in the divisor in the following form:

Since we are dividing the polynomial by $x - 5$, the synthetic divisor is 5.

$$\longrightarrow \quad \begin{array}{r|rrr} 5 & 6 & 5 & -2 \end{array}$$ ← This represents the dividend $6x^2 + 5x - 2$.

Then we follow these steps:

$$\begin{array}{r|rrr} 5 & 6 & 5 & -2 \\ & \downarrow \\ \hline & 6 \end{array}$$ Begin by bringing down the 6.

$$\begin{array}{r|rrr} 5 & 6 & 5 & -2 \\ & & 30 \\ \hline & 6 \end{array}$$ Multiply 5 by 6 to get 30.

$$\begin{array}{r|rrr} 5 & 6 & 5 & -2 \\ & & 30 \\ \hline & 6 & 35 \end{array}$$ Add 5 and 30 to get 35.

$$\begin{array}{r|rrr} 5 & 6 & 5 & -2 \\ & & 30 & 175 \\ \hline & 6 & 35 \end{array}$$ Multiply 35 by 5 to get 175.

$$\begin{array}{r|rrr} 5 & 6 & 5 & -2 \\ & & 30 & 175 \\ \hline & 6 & 35 & 173 \end{array}$$ Add -2 and 175 to get 173.

Success Tip
In his process, numbers below the line are multiplied by the synthetic divisor and that product is carried above the line to the next column. Numbers above the horizontal line are added.

The numbers 6 and 35 represent the quotient $6x + 35$, and 173 is the remainder. Thus,

$$\frac{6x^2 + 5x - 2}{x - 5} = 6x + 35 + \frac{173}{x - 5}$$

 Self Check 1   Divide $5x^2 - 4x + 2$ by $x - 3$.

EXAMPLE 2   Use synthetic division to find $\dfrac{x^3 + x^2 - 1}{x - 3}$.

Solution   We begin by writing

$$\underline{3|\ 1\quad 1\quad 0\quad -1}$$ *Write 0 for the coefficient of x, the missing term.*

and complete the division as follows.

$3|\ 1\quad 1\quad 0\quad -1$	$3|\ 1\quad 1\quad\ 0\quad -1$	$3|\ 1\quad 1\quad\ 0\quad\ -1$
$\qquad\quad 3$	$\qquad\quad 3\quad 12$	$\qquad\quad 3\quad 12\quad 36$
$\overline{1\quad 4}$	$\overline{1\quad 4\quad 12}$	$\overline{1\quad 4\quad 12\quad 35}$
Multiply, then add.	*Multiply, then add.*	*Multiply, then add.*

Thus,

$$\frac{x^3 + x^2 - 1}{x - 3} = x^2 + 4x + 12 + \frac{35}{x - 3}$$

 Self Check 2   Use synthetic division to find $\dfrac{x^3 + 3x - 90}{x - 4}$.

EXAMPLE 3   Use synthetic division to divide $5x^2 + 6x^3 + 2 - 4x$ by $x + 2$.

Solution   First, we write the dividend with the exponents in descending order.

$$6x^3 + 5x^2 - 4x + 2$$

Then we write the divisor in $x - k$ form: $x - (-2)$. Thus, $k = -2$. Using synthetic division, we begin by writing

This represents division by x + 2.

$$\underline{-2|\ 6\qquad 5\quad -4\qquad 2}$$

and complete the division.

Notation

Because the remainder is negative, we can also write the result as

$$6x^2 - 7x + 10 + \frac{-18}{x + 2}$$

$$
\begin{array}{r|rrrr}
-2 & 6 & 5 & -4 & 2 \\
   &   & -12 & 14 & -20 \\
\hline
   & 6 & -7 & 10 & -18
\end{array}
$$ *The remainder is negative.*

Thus,

$$\frac{5x^2 + 6x^3 + 2 - 4x}{x + 2} = 6x^2 - 7x + 10 - \frac{18}{x + 2}$$

> **Self Check 3**   Divide $2x - 4x^2 + 3x^3 - 3$ by $x + 1$.

2 The Remainder Theorem.

Synthetic division is important because of the **remainder theorem.**

Remainder Theorem	If a polynomial $P(x)$ is divided by $x - k$, the remainder is $P(k)$.

It follows from the remainder theorem that we can evaluate polynomials using synthetic division. We illustrate this in the following example.

EXAMPLE 4   Let $P(x) = 2x^3 - 3x^2 - 2x + 1$. Find **a.** $P(3)$ and **b.** the remainder when $P(x)$ is divided by $x - 3$.

Notation
Naming the function with the letter P, instead of f, stresses that we are working with a polynomial function.

Solution

a. To find $P(3)$ we evaluate the function for $x = 3$.

$$P(3) = 2(3)^3 - 3(3)^2 - 2(3) + 1 \quad \text{Substitute 3 for x.}$$
$$= 2(27) - 3(9) - 6 + 1$$
$$= 54 - 27 - 6 + 1$$
$$= 22$$

Thus, $P(3) = 22$.

Success Tip
It is often easier to find $P(k)$ by using synthetic division than by substituting k for x in $P(x)$. This is especially true if k is a decimal.

b. We can use synthetic division to find the remainder when $P(x)$ is divided by $x - 3$.

$$
\begin{array}{r|rrrr}
3 & 2 & -3 & -2 & 1 \\
  &   & 6 & 9 & 21 \\
\hline
  & 2 & 3 & 7 & 22
\end{array}
\quad P(x) = 2x^3 - 3x^2 - 2x + 1
$$

Thus, the remainder is 22.

The same results in parts a and b show that rather than substituting 3 for x in $P(x) = 2x^3 - 3x^2 - 2x + 1$, we can divide $2x^3 - 3x^2 - 2x + 1$ by $x - 3$ to find $P(3)$.

> **Self Check 4**   Let $P(x) = 5x^3 - 3x^2 + x + 6$. Find **a.** $P(1)$ and **b.** use synthetic division to find the remainder when $P(x)$ is divided by $x - 1$.

3 The Factor Theorem.

If two quantities are multiplied, each is called a **factor** of the product. Thus, $x - 2$ is a factor of $6x - 12$, because $6(x - 2) = 6x - 12$. A theorem, called the **factor theorem,** tells us how to find one factor of a polynomial if the remainder of a certain division is 0.

Factor Theorem	If $P(x)$ is a polynomial in x, then
	$P(k) = 0$ if and only if $x - k$ is a factor of $P(x)$

If $P(x)$ is a polynomial in x and if $P(k) = 0$, k is called a **zero of the polynomial function.**

EXAMPLE 5   Let $P(x) = 3x^3 - 5x^2 + 3x - 10$. Show that   **a.** $P(2) = 0$ and   **b.** $x - 2$ is a factor of $P(x)$.

Solution

a. We can use the remainder theorem to evaluate $P(2)$ by dividing $P(x)$ by $x - 2$.

$$
\begin{array}{r|rrrr}
2 & 3 & -5 & 3 & -10 \\
  &   & 6 & 2 & 10 \\
\hline
  & 3 & 1 & 5 & 0
\end{array}
\qquad P(x) = 3x^2 - 5x^2 + 3x - 10
$$

The Language of Algebra
The phrase *if and only if* in the factor theorem means:

If $P(2) = 0$, then $x - 2$ is a factor of $P(x)$

and

If $x - 2$ is a factor of $P(x)$, then $P(2) = 0$.

The remainder in this division is 0. By the remainder theorem, the remainder is $P(2)$. Thus, $P(2) = 0$, and 2 is a zero of the polynomial.

b. Because the remainder is 0, the numbers 3, 1, and 5 in the synthetic division in part a represent the quotient $3x^2 + x + 5$. Thus,

$$
\underbrace{(x - 2)}_{\text{Divisor}} \cdot \underbrace{(3x^2 + x + 5)}_{\text{quotient}} + \underbrace{0}_{\text{remainder}} = \underbrace{3x^3 - 5x^2 + 3x - 10}_{\text{the dividend, } P(x)}
$$

or

$$
(x - 2)(3x^2 + x + 5) = 3x^3 - 5x^2 + 3x - 10
$$

Thus, $x - 2$ is a factor of $3x^3 - 5x^2 + 3x - 10$.

 Self Check 5   Let $P(x) = x^3 - 4x^2 + x + 6$. Show that $x + 1$ is a factor of $P(x)$ using synthetic division.

The result in Example 5 is true, because the remainder, $P(2)$, is 0. If the remainder had not been 0, then $x - 2$ would not have been a factor of $P(x)$.

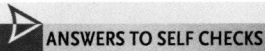 **ANSWERS TO SELF CHECKS**   **1.** $5x + 11 + \dfrac{35}{x-3}$ **2.** $x^2 + 4x + 19 - \dfrac{14}{x-4}$
3. $3x^2 - 7x + 9 - \dfrac{12}{x+1}$ **4.** 9   **5.** Since $P(-1) = 0$, $x + 1$ is a factor of $P(x)$.

STUDY SET
Appendix 2

VOCABULARY

Fill in the blanks.

1. The method of dividing $x^2 + 2x - 9$ by $x - 4$ shown below is called _____ division.

$$
\begin{array}{r|rrr}
4 & 1 & 2 & -9 \\
  &   & 4 & 24 \\
\hline
  & 1 & 6 & 15
\end{array}
$$

2. Synthetic division is used to divide a polynomial by a _____ of the form $x - k$.

3. In Exercise 1, the synthetic _____ is 4.

4. By the _____ theorem, if a polynomial $P(x)$ is divided by $x - k$, the remainder is $P(k)$.

5. The factor _____ tells us how to find one factor of a polynomial if the remainder of a certain division is 0.

6. If $P(x)$ is a polynomial and if $P(k) = 0$, then k is called a _____ of the polynomial.

CONCEPTS

7. a. What division is represented below?

  b. What is the answer?

$$
\begin{array}{r|rrrr}
-2 & 5 & 0 & 1 & -3 \\
   &   & -10 & 20 & -42 \\
\hline
   & 5 & -10 & 21 & -45
\end{array}
$$

Fill in the blanks.

8. In the synthetic division process, numbers below the line are _____ by the synthetic divisor and that product is carried above the line to the next column. Numbers above the horizontal line are _____.

9. Rather than substituting 8 for x in $P(x) = 6x^3 - x^2 - 17x + 9$, we can divide the polynomial _____ by _____ to find $P(8)$.

10. For $P(x) = x^3 - 4x^2 + x + 6$, suppose we know that $P(3) = 0$. Then _____ is a factor of $x^3 - 4x^2 + x + 6$.

NOTATION

Complete each synthetic division.

11. Divide $6x^3 + x^2 - 23x + 2$ by $x - 2$.

$$
\begin{array}{r|rrrr}
 & 6 & & -23 & 2 \\
            &   &   &   & 6 \\
\hline
            &   & 13 & 3 &
\end{array}
$$

12. Divide $2x^3 - 4x^2 - 25x + 15$ by $x + 3$.

$$
\begin{array}{r|rrrr}
 & 2 & -4 & & 15 \\
            &   &   & 30 &   \\
\hline
            &   &   &   & 0
\end{array}
$$

GUIDED PRACTICE

Use synthetic division to perform each division.

13. $\dfrac{x^2 + x - 2}{x - 1}$

14. $\dfrac{x^2 + x - 6}{x - 2}$

15. $\dfrac{x^2 - 7x + 12}{x - 4}$

16. $\dfrac{x^2 - 6x + 5}{x - 5}$

17. $\dfrac{x^2 + 8 + 6x}{x + 4}$

18. $\dfrac{x^2 - 15 - 2x}{x + 3}$

19. $\dfrac{x^2 - 5x + 14}{x + 2}$

20. $\dfrac{x^2 + 13x + 42}{x + 6}$

21. $\dfrac{3x^3 - 10x^2 + 5x - 6}{x - 3}$

22. $\dfrac{2x^3 - 9x^2 + 10x - 3}{x - 3}$

23. $\dfrac{2x^3 - 6 - 5x}{x - 2}$

24. $\dfrac{4x^3 - 1 + 5x^2}{x + 2}$

25. $\dfrac{5x^2 + 4 + 6x^3}{x + 1}$

26. $\dfrac{4 - 3x^2 + x}{x - 4}$

27. $\dfrac{t^3 + t^2 + t + 2}{t + 1}$

28. $\dfrac{m^3 - m^2 - m - 1}{m - 1}$

29. $\dfrac{a^5 - 1}{a - 1}$

30. $\dfrac{b^4 - 81}{b - 3}$

31. $\dfrac{-5x^5 + 4x^4 + 30x^3 + 2x^2 + 20x + 3}{x - 3}$

32. $\dfrac{-6c^5 + 14c^4 + 38c^3 + 4c^2 + 25c - 36}{c - 4}$

33. $\dfrac{8t^3 - 4t^2 + 2t - 1}{t - \dfrac{1}{2}}$

34. $\dfrac{9a^3 + 3a^2 - 21a - 7}{a + \dfrac{1}{3}}$

35. $\dfrac{x^4 - x^3 - 56x^2 - 2x + 16}{x - 8}$

36. $\dfrac{x^4 - 9x^3 + x^2 - 7x - 20}{x - 9}$

Use a calculator and synthetic division to perform each division.

37. $\dfrac{7.2x^2 - 2.1x + 0.5}{x - 0.2}$

38. $\dfrac{2.7x^2 + x - 5.2}{x + 1.7}$

39. $\dfrac{9x^3 - 25}{x + 57}$

40. $\dfrac{0.5x^3 + x}{x - 2.3}$

Let $P(x) = 2x^3 - 4x^2 + 2x - 1$. Evaluate $P(x)$ by substituting the given value of x into the polynomial and simplifying. Then evaluate the polynomial by using the remainder theorem and synthetic division.

41. $P(1)$ **42.** $P(2)$

43. $P(-2)$ **44.** $P(-1)$

45. $P(3)$ **46.** $P(-4)$

47. $P(0)$ **48.** $P(4)$

Let $Q(x) = x^4 - 3x^3 + 2x^2 + x - 3$. Evaluate $Q(x)$ by substituting the given value of x into the polynomial and simplifying. Then evaluate the polynomial by using the remainder theorem and synthetic division.

49. $Q(-1)$ **50.** $Q(1)$

51. $Q(2)$ **52.** $Q(-2)$

53. $Q(3)$ **54.** $Q(0)$

55. $Q(-3)$ **56.** $Q(-4)$

Use the remainder theorem and synthetic division to find $P(k)$.

57. $P(x) = x^3 - 4x^2 + x - 2; k = 2$

58. $P(x) = x^3 - 3x^2 + x + 1; k = 1$

59. $P(x) = 2x^3 + x + 2; k = 3$

60. $P(x) = x^3 + x^2 + 1; k = -2$

61. $P(x) = x^4 - 2x^3 + x^2 - 3x + 2; k = -2$

62. $P(x) = x^5 + 3x^4 - x^2 + 1; k = -1$

63. $P(x) = 3x^5 + 1; k = -\dfrac{1}{2}$

64. $P(x) = 5x^7 - 7x^4 + x^2 + 1; k = 2$

Use the factor theorem and determine whether the first expression is a factor of $P(x)$.

65. $x - 3; P(x) = x^3 - 3x^2 + 5x - 15$

66. $x + 1; P(x) = x^3 + 2x^2 - 2x - 3$
   (*Hint:* Write $x + 1$ as $x - (-1)$.)

67. $x + 2; P(x) = 3x^2 - 7x + 4$
   (*Hint:* Write $x + 2$ as $x - (-2)$.)

68. $x; P(x) = 7x^3 - 5x^2 - 8x$
   (*Hint:* $x = x - 0$.)

WRITING

69. When dividing a polynomial by a binomial of the form $x - k$, synthetic division is considered to be faster than long division. Explain why.

70. Let $P(x) = x^3 - 6x^2 - 9x + 4$. You now know two ways to find $P(6)$. What are they? Which method do you prefer?

71. Explain the factor theorem.

72. What is a *zero* of a polynomial function?

REVIEW

Evaluate each expression for $x = -3, y = -5$, and $z = 0$.

73. $x^2 z(y^3 - z)$ **74.** $|y^3 - z|$

75. $\dfrac{x - y^2}{2y - 1 + x}$ **76.** $\dfrac{2y + 1}{x} - x$

CHALLENGE PROBLEMS

Suppose that
$$P(x) = x^{100} - x^{99} + x^{98} - x^{97} + \cdots + x^2 - x + 1.$$

77. Find the remainder when $P(x)$ is divided by $x - 1$.

78. Find the remainder when $P(x)$ is divided by $x + 1$.

APPENDIX 3
Answers to Selected Exercises

Study Set Section 1.1 (page 8)

1. sum, difference, product, quotient   **3.** constant   **5.** equation
7. horizontal   **9. a.** Equation   **b.** Algebraic Expression
11. a. Algebraic Expression   **b.** Equation
13. Addition, multiplication, division; t
15.

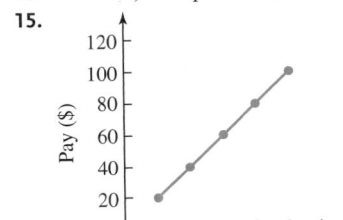

17. is, not, equal, to
19. $5 \cdot 6, 5(6)$ **21.** $4x$
23. $2w$ **25.** $\frac{32}{x}$ **27.** $\frac{55}{5}$
29. 300   **31.** 15-year-old machinery is worth \$35,000.
33. The product of 8 and 2 equals 16.
35. The difference of 11 and 9 equals 2.
37. The sum of x and 2 equals 10.
39. The quotient of 66 and 11 equals 6.   **41.** $p = 100 - d$
43. $7d = h$ **45.** $s = 3c$ **47.** $w = e + 1,200$ **49.** $p = r - 600$
51. $\frac{l}{4} = m$ **53.** 390, 400, 405   **55.** 1,300; 1,200; 1,100
57. $\frac{e}{12}$ **59.** $2c$
61. 90, 75, 60, 45, 30, 15, 0

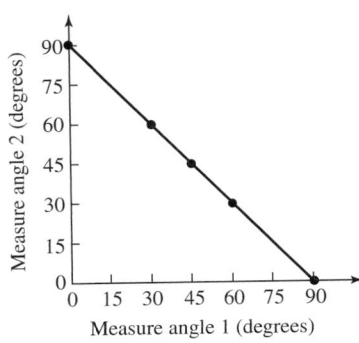

Study Set Section 1.2 (page 22)

1. multiplied   **3.** prime-factored   **5.** equivalent
7. least or lowest   **9. a.** 1   **b.** a **c.** $\frac{a \cdot c}{b \cdot d}$ **d.** $\frac{a \cdot d}{b \cdot c}$ **e.** $\frac{a + b}{d}$
f. $\frac{a - b}{d}$ **11. a.** 1   **b.** 1   **13. a.** $\frac{5}{5}, \frac{25}{30}$ **b.** $2, 7, \frac{2}{7}$
15. $3 \cdot 5 \cdot 5$ **17.** $2 \cdot 2 \cdot 7$ **19.** $3 \cdot 3 \cdot 3 \cdot 3$ **21.** $3 \cdot 3 \cdot 13$
23. $2 \cdot 2 \cdot 5 \cdot 11$ **25.** $2 \cdot 3 \cdot 11 \cdot 19$ **27.** $\frac{5}{48}$ **29.** $\frac{21}{55}$ **31.** $\frac{15}{8}$
33. $\frac{42}{25}$ **35.** $\frac{3}{9}$ **37.** $\frac{24}{54}$ **39.** $\frac{35}{5}$ **41.** $\frac{35}{7}$ **43.** $\frac{1}{3}$ **45.** $\frac{6}{7}$
47. $\frac{3}{8}$ **49.** Lowest terms   **51.** $\frac{2}{3}$ **53.** $\frac{4}{25}$ **55.** $\frac{6}{5}$ **57.** $\frac{4}{7}$
59. $\frac{5}{24}$ **61.** $\frac{22}{35}$ **63.** $\frac{41}{45}$ **65.** $\frac{3}{20}$ **67.** 24   **69.** 4   **71.** $\frac{7}{9}$

73. $\frac{7}{20}$ **75.** $32\frac{2}{3}$ **77.** $2\frac{1}{2}$ **79.** $\frac{5}{9}$ **81.** $5\frac{19}{48}$ **83.** $\frac{19}{15}$
85. 70   **87.** $13\frac{3}{4}$ **89.** $\frac{1}{2}$ **91.** $\frac{14}{5}$ **93.** $\frac{8}{5}$ **95.** $\frac{3}{35}$ **97.** $\frac{9}{4}$
99. $\frac{3}{10}$ **101.** $1\frac{9}{11}$ **103.** $\frac{1}{7}$ **105. a.** $\frac{7}{32}$ in.   **b.** $\frac{3}{32}$ in.
107. $5\frac{3}{4}$ lb   **109.** $40\frac{1}{2}$ in.   **115.** 150, 180

Study Set Section 1.3 (page 32)

1. whole   **3.** integers   **5.** Negatives; positives   **7.** rational
9. irrational   **11.** real   **13. a.** $-\$15$ million
b. $\frac{5}{16}$ in. or $+\frac{5}{16}$ in.   **15. a.** -20 **b.** $\frac{2}{3}$ **17.** -14 and -4
19. square, root   **21.** is, approximately, equal, to   **23.** Greek
25. $-4, -5$
27.

	5	0	-3	$\frac{7}{8}$	0.17	$-9\frac{1}{4}$	$\sqrt{2}$	π
Real	✓	✓	✓	✓	✓	✓	✓	✓
Irrational							✓	✓
Rational	✓	✓	✓	✓	✓	✓		
Integer	✓	✓	✓					
Whole	✓	✓						
Natural	✓							

29. True   **31.** False   **33.** True   **35.** True   **37.** $>$ **39.** $>$
41. $<$ **43.** $>$ **45.** $<$ **47.** $>$ **49.** 0.625   **51.** $0.0\overline{3}$
53. $0.01\overline{6}$ **55.** 0.42
57.

$$-\frac{35}{8} \quad -\pi \quad -1\frac{1}{2} \; -0.333... \; \sqrt{2} \quad\quad 3 \quad 4.25$$

number line from -5 to 5

59.

number line from -5 to 5

61. 83   **63.** $\frac{4}{3}$ **65.** 11   **67.** 6.1   **69.** $>$
71. $<$ **73.** $=$ **75.** $=$ **77.** $<$ **79.** $>$
81. Natural, whole, integers: 9; rational: $9, \frac{15}{16}, 3\frac{1}{8}, 1.765$; irrational: $2\pi, 3\pi, \sqrt{89}$; real: all   **83.** Arrow 1; $|-6| > |5|$
85. a. 2000; $-\$81$ billion   **b.** 1990; $-\$40$ billion   **93.** $\frac{4}{9}$
95. $2\frac{5}{23}$

Study Set Section 1.4 (page 41)

1. sum   **3.** commutative, associative   **5. a.** 6   **b.** -9.2
7. a. Negative   **b.** Positive   **9. a.** $1 + (-5)$ **b.** $-80.5 + 15$
c. $20 + 4$ **d.** $3 + 2.1$ **11.** Step 1: Commutative Property of Addition; Step 2: Associative Property of Addition

A-9

13. a. $x + y = y + x$ **b.** $(x + y) + z = x + (y + z)$ **15.** -9
17. -17 **19.** -74 **21.** -10.3 **23.** $-\frac{17}{12}$ **25.** $-\frac{7}{20}$ **27.** -3
29. 39   **31.** 0   **33.** 2.25   **35.** $-\frac{4}{15}$ **37.** $\frac{3}{8}$ **39.** 16
41. -15 **43.** -21 **45.** -26 **47.** 0.67   **49.** 195   **51.** 215
53. -112 **55.** $1\frac{2}{3}$ **57.** 15.4   **59.** 9   **61.** -5 **63.** 1
65. -1.7 **67.** 70   **69.** -6.6 **71.** $-\frac{1}{8}$ **73.** -14 **75.** 0
77. -22.1 **79.** 2,167   **81.** 68   **83.** $-\frac{15}{28}$ **85.** -0.9
87. 2,150 m   **89.** Woods: -18, Kite: -6, Tolles: -5, Watson: -4
91. 1,242.86   **93.** $-\$99,650,000$ **95.** $-1, 3$
97. 79 feet above sea level   **99.** $-\$23.2$ million
103. True   **105.** -9 and 3

Study Set Section 1.5 (page 49)
1. Subtraction   **3.** range   **5. a.** -12 **b.** $\frac{1}{5}$ **c.** -2.71
d. 0   **7. a.** 8   **b.** -8 **9.** No; $7 + (-8) \neq 15$
11. a. $1 - (-7) = 8$ **b.** $-(-2) = 2$ **c.** $-|-3| = -3$
d. $2 - 6 = -4$ **13.** 55   **15.** x **17.** -25 **19.** $-\frac{3}{16}$
21. -3 **23.** -13 **25.** -10 **27.** 11   **29.** -6 **31.** 1
33. 2   **35.** 40   **37.** 5   **39.** -69 **41.** -88 **43.** 12
45. -1.1 **47.** -3.5 **49.** -2.31 **51.** $-\frac{1}{2}$ **53.** $-\frac{5}{16}$
55. $-\frac{5}{12}$ **57.** 22   **59.** -25 **61.** -11 **63.** -50 **65.** -7
67. -1 **69.** 256   **71.** 0   **73.** -2.1 **75.** $\frac{47}{56}$ **77.** 3
79. -47.5 **81.** 149   **83.** -171 **85.** 4.63   **87.** $-\frac{19}{12}$
89. 160°F   **91.** 21   **93.** Orlando: $-40,000$; LaGuardia: 37,000
95. 1,030 ft   **97. a.** iii   **b.** $-\$116.1$ billion
99. $-68, -78, -147$ **105.** $2 \cdot 3 \cdot 5$ **107.** True

Study Set Section 1.6 (page 59)
1. product, quotient   **3.** associative   **5. a.** positive   **b.** negative
7. a. $-3, 3, -9$ **b.** $0, 8, 0$ **9. a.** a **b.** 1   **c.** 0
d. Undefined   **11. a.** $8 \cdot 5$ **b.** $(-2 \cdot 6)9$ **c.** $\frac{1}{5}$ **d.** 1
13. a. NEG   **b.** Not possible to tell   **c.** POS   **d.** NEG
15. $-4(-5) = 20$ **17.** -4 **19.** -16 **21.** -60 **23.** -66
25. -0.48 **27.** $-\frac{1}{4}$ **29.** 7   **31.** 54   **33.** 9   **35.** -441
37. 2.4   **39.** $\frac{1}{12}$ **41.** 0   **43.** 66   **45.** -720 **47.** -861
49. -216 **51.** 16   **53.** $\frac{9}{7}$; 1   **55.** $-\frac{1}{13}$; 1   **57.** 10   **59.** 3
61. -4 **63.** -17 **65.** -1 **67.** 1   **69.** -9 **71.** -0.005
73. 0   **75.** Undefined   **77.** $-\frac{5}{12}$ **79.** $\frac{15}{4}$ **81.** -4.7
83. -520 **85.** $\frac{1}{24}$ **87.** -11 **89.** $1\frac{1}{2}$ **91.** 30.24   **93.** $-\frac{3}{8}$
95. $-\frac{3}{20}$ **97.** 30.3   **99.** $\frac{15}{16}$ **101.** -67 **103.** 6   **105.** $\$8,000$
107. $-72°$ **109. a.** ii   **b.** 36 lb   **111.** $-51°F$
113. $-\$1,100, -\$400, -\$1,100$ **115. a.** $5, -10$ **b.** $2.5, -5$
c. $7.5, -15$ **d.** $10, -20$ **121.** -5 **123.** $1.08\overline{3}$

Study Set Section 1.7 (page 71)
1. base, exponent, power   **3.** exponent   **5.** order
7. a. Subtraction   **b.** Division   **c.** Addition   **d.** Power
9. a. Parentheses, brackets, braces, absolute value symbols, fraction bar
b. Innermost: parentheses; outermost: brackets   **11. a.** -5 **b.** 5
13. $3, 9, 27, 54, -73$ **15.** 8^3 **17.** $7^3 12^2$ **19.** x^3 **21.** $r^4 s^2$
23. 49   **25.** 216   **27.** 625   **29.** 0.01   **31.** $-\frac{1}{64}$ **33.** $\frac{8}{27}$
35. $36, -36$ **37.** $64, -64$ **39.** -17 **41.** 30   **43.** 43

45. 8   **47.** -34 **49.** -118 **51.** 86   **53.** -8 **55.** -44
57. 0   **59.** -148 **61.** 100   **63.** -32 **65.** 53   **67.** -86
69. -392 **71.** 3   **73.** -19 **75.** $\frac{1}{2}$ **77.** 0   **79.** $-\frac{8}{9}$
81. 13   **83.** -8 **85.** -31 **87.** 11   **89.** 1   **91.** -500
93. -376 **95.** 12   **97.** 39   **99.** Undefined   **101.** -110
103. -54 **105.** $\frac{1}{8}$ **107.** 10   **109.** -1
111. 2^2 square units, 3^2 square units, 4^2 square units
113. About 6   **115. a.** $\$11,875$ **b.** $\$95$ **117.** 81 in.
123. $-17, -5$

Study Set Section 1.8 (page 82)
1. expressions   **3.** terms   **5.** coefficient   **7.** $7, 14, 21, 7w$
9. $(12 - h)$ in.   **11.** $(x + 20)$ ounces   **13. a.** $b - 15$
b. $p + 15$ **15.** $5, 25, 45$ **17. a.** $8y$ **b.** $2cd$ **c.** Commutative
19. a. 4   **b.** $3, 11, -1, 9$ **21.** Term   **23.** Factor   **25.** $l + 15$
27. $50x$ **29.** $\frac{w}{l}$ **31.** $P + \frac{2}{3}p$ **33.** $k^2 - 2,005$ **35.** $2a - 1$
37. $\frac{1,000}{n}$ **39.** $2p + 90$ **41.** $3(35 + h + 300)$ **43.** $p - 680$
45. $4d - 15$ **47.** $2(200 + t)$ **49.** $|a - 2|$ **51.** $0.1d$ or $\frac{1}{10}d$
53. Three-fourths of r **55.** 50 less than t
57. The product of x, y, and z **59.** Twice m, increased by 5
61. $(x + 2)$ in.   **63.** $(36 - x)$ in.   **65.** $60h$ **67.** $\frac{i}{12}$ **69.** $\$8x$
71. $49x¢$ **73.** $\$2t$ **75.** $\$25(x + 2)$ **77.** 2   **79.** 13   **81.** 20
83. -12 **85.** -5 **87.** $-\frac{1}{5}$ **89.** 17   **91.** 36   **93.** 255
95. 8   **97.** $-1, -2, -28$ **99.** $41, 11, 2$ **101.** $150, -450$
103. $0, 0, 5$ **105. a.** Let x = weight of the Element, $2x - 340 =$
weight of the Hummer   **b.** 6,400 lb   **107. a.** Let x = age of
Apple; $x + 80$ = age of IBM; $x - 9$ = age of Dell
b. IBM: 112; Dell: 23   **113.** 60   **115.** $\frac{8}{27}$

Study Set Section 1.9 (page 93)
1. simplify   **3.** distributive   **5.** like   **7. a.** $4, 9, 36$
b. Associative Property of Multiplication   **9. a.** $+$ **b.** $-$ **c.** $-$
d. $+$ **11. a.** $10x$ **b.** Can't be simplified   **c.** $-42x$
d. Can't be simplified   **e.** $18x$ **f.** $3x + 5$ **13. a.** $6(h - 4)$
b. $-(z + 16)$ **15.** $12t$ **17.** $63m$ **19.** $-35q$ **21.** $300t$
23. $11.2x$ **25.** $60c$ **27.** $-96m$ **29.** g **31.** $5x$ **33.** $6y$
35. $5x + 15$ **37.** $-12x - 27$ **39.** $9x + 10$ **41.** $0.4x - 1.6$
43. $36c - 42$ **45.** $-78c + 18$ **47.** $30t + 90$ **49.** $4a - 1$
51. $24t + 16$ **53.** $2w - 4$ **55.** $56y + 32$
57. $50a - 75b + 25$ **59.** $-x + 7$ **61.** $5.6y - 7$ **63.** $3x, -2x$
65. $-3m^3, -m^3$ **67.** $10x$ **69.** 0   **71.** $20b^2$ **73.** r **75.** $28y$
77. $-s^3$ **79.** $-3.6c$ **81.** $0.4r$ **83.** $\frac{4}{5}t$ **85.** $-\frac{5}{8}x$
87. $-6y - 10$ **89.** $-2x + 5$ **91.** $9m^2 + 6m - 4$
93. $4x^2 - 3x + 9$ **95.** $7z - 15$ **97.** $s^2 - 12$ **99.** $-41r + 130$
101. $8x - 9$ **103.** $12c + 34$ **105.** $-10r$ **107.** $-20r$
109. $3a$ **111.** $9r - 16$ **113.** $-6x$ **115.** $c - 13$
117. $a^3 - 8$ **119.** $12x$ **121.** $(4x + 8)$ ft   **125.** 2

Chapter 1 Review (page 96)
1. 1 hr; 100 cars   **2.** 100   **3.** 7 P.M.   **4.** 12 A.M. (midnight)
5. The difference of 15 and 3 equals 12.
6. The sum of 15 and 3 equals 18.
7. The quotient of 15 and 3 equals 5.
8. The product of 15 and 3 equals 45.   **9. a.** $4 \cdot 9$; $4(9)$
b. $\frac{9}{3}$ **10. a.** $8b$ **b.** Prt **11. a.** Equation

b. Expression   **12.** 10, 15, 25   **13. a.** $2 \cdot 12, 3 \cdot 8$ (Answers may vary)   **b.** $2 \cdot 2 \cdot 6$ (Answers may vary)   **c.** 1, 2, 3, 4, 6, 8, 12, 24
14. Equivalent   **15.** $2 \cdot 3^3$ **16.** $3 \cdot 7^2$ **17.** $5 \cdot 7 \cdot 11$
18. Prime   **19.** $\frac{4}{7}$ **20.** $\frac{4}{3}$ **21.** $\frac{40}{64}$ **22.** $\frac{36}{3}$ **23.** 90   **24.** 210
25. $\frac{7}{64}$ **26.** $\frac{5}{21}$ **27.** $\frac{16}{45}$ **28.** $3\frac{1}{4}$ **29.** $\frac{2}{5}$ **30.** $\frac{5}{22}$ **31.** $\frac{59}{60}$
32. $\frac{5}{18}$ **33.** $52\frac{1}{2}$ million   **34.** $\frac{17}{96}$ in.   **35. a.** 0
b. $\{ \ldots, -2, -1, 0, 1, 2, \ldots \}$ **36.** -206 ft   **37. a.** $<$
b. $>$ **38. a.** $\frac{7}{10}$ **b.** $\frac{14}{3}$ **39.** 0.004   **40.** $0.7\overline{72}$
41.

$$-\frac{17}{4} \quad -2 \quad 0.333\ldots \quad \frac{7}{8} \quad \sqrt{2} \quad \pi \quad 3.75$$

(number line from -5 to 5)

42. Natural: 8; whole: 0, 8; integers: $0, -12, 8$; rational: $-\frac{4}{5}$, 99.99, 0, $-12, 4\frac{1}{2}, 0.666\ldots, 8$; irrational: $\sqrt{2}$; real: all
43. False   **44.** False   **45.** True   **46.** True   **47.** $>$ **48.** $<$
49. -82 **50.** 12   **51.** -7 **52.** 0   **53.** -11 **54.** -12.3
55. $-\frac{3}{16}$ **56.** 11   **57. a.** Commutative Property of Addition
b. Associative Property of Addition   **c.** Addition Property of Opposites (Inverse Property of Addition)   **d.** Addition Property of 0 (Identity Property of Addition)   **58.** 118°F   **59. a.** -10 **b.** 3
60. a. $\frac{9}{16}$ **b.** -4 **61.** -19 **62.** $-\frac{14}{15}$ **63.** 5   **64.** 5.7
65. -10 **66.** -29 **67.** 65,233 ft; $65,233 + (-36,205) = 29,028$
68. 428 B.C.; (-428); $-428 + 81 = -347$ **69.** -56 **70.** 1
71. 12   **72.** -12 **73.** 6.36   **74.** -2 **75.** $-\frac{2}{15}$ **76.** 0
77. High: 3, low: -4.5 **78. a.** Associative Property of Multiplication   **b.** Commutative Property of Multiplication
c. Multiplication Property of 1 (Identity Property of Multiplication)
d. Inverse Property of Multiplication   **79.** -1 **80.** -17 **81.** 3
82. $-\frac{6}{5}$ **83.** Undefined   **84.** -4.5 **85.** 0, 18, 0   **86.** $-\$360$
87. a. 8^5 **b.** $9\pi r^2$ **88. a.** 81   **b.** $-\frac{8}{27}$ **c.** 32   **d.** 50
89. 17   **90.** -36 **91.** -169 **92.** 23   **93.** -420 **94.** $-\frac{7}{19}$
95. 113   **96.** Undefined   **97. a.** $(-9)^2 = 81$ **b.** $-9^2 = -81$
98. \$20   **99. a.** 3   **b.** 1   **100. a.** $16, -5, 25$ **b.** $\frac{1}{2}$, 1
101. $h + 25$ **102.** $3s - 15$ **103.** $\frac{1}{2}t - 6$
104. $|2 - a^2|$ **105.** $(n + 4)$ in.   **106.** $(b - 4)$ in.   **107.** $10d$
108. $(x - 5)$ years   **109.** $30, 10d$ **110.** $0, 19, -16$ **111.** 40
112. -36 **113.** $-28w$ **114.** $24x$ **115.** $2.08f$ **116.** r
117. $5x + 15$ **118.** $-2x - 3 + y$ **119.** $3c - 6$
120. $12.6c + 29.4$ **121.** $9p$ **122.** $-7m$ **123.** $4n$
124. $-p - 18$ **125.** $0.1k^2$ **126.** $8a^3 - 1$ **127.** w
128. $4h - 15$ **129.** $(4x + 4)$ ft   **130. a.** x **b.** $-x$
c. $4x + 1$ **d.** $4x - 1$

Chapter 1 Test (page 105)

1. a. equivalent   **b.** product   **c.** reciprocal   **d.** like, terms
e. undefined   **2. a.** \$24   **b.** 5 hr   **3.** 3, 20, 70
4. $2 \cdot 2 \cdot 3 \cdot 3 \cdot 5 = 2^2 \cdot 3^2 \cdot 5$ **5.** $\frac{2}{5}$ **6.** $\frac{3}{2} = 1\frac{1}{2}$ **7.** $\frac{27}{35}$ **8.** $6\frac{11}{15}$
9. \$3.57   **10.** $0.8\overline{3}$

11.

$$-3.75 \quad -3 \quad -1\frac{1}{4} \quad 0.5 \quad \sqrt{2} \quad \frac{7}{2}$$

(number line from -5 to 5)

12. a. True   **b.** False   **c.** True   **d.** True   **13.** A real number is any number that is either a rational or an irrational number.

---

14. a. $>$ **b.** $<$ **c.** $<$ **d.** $>$
15. A gain of 0.6 of a rating point   **16.** -2 **17.** $\frac{3}{8}$ **18. a.** -6
b. $-6 + (-4) = -10$ **19. a.** 14   **b.** $14(-9) = -126$
20. -30 **21.** -2.44 **22.** 0   **23.** $-\frac{27}{125}$ **24.** 0   **25.** -3
26. 50   **27.** 14   **28. a.** Associative Property of Addition
b. The Distributive Property   **c.** Commutative Property of Multiplication   **d.** Inverse Property of Multiplication
e. Identity Property of Addition   **29. a.** 9^5 **b.** $3x^2z^3$ **30.** 170
31. -12 **32.** -100 **33.** -351 **34.** 36   **35.** $4, 17, -59$
36. $2w - 7$ **37. a.** $x - 2$ **b.** $25q$¢   **38.** 3; 5   **39.** $-20x$
40. $224t$ **41.** $-4a + 4$ **42.** $-5.9d^3$ **43.** $14x + 3$
44. $(18x + 6)$ ft

Study Set Section 2.1 (page 117)

1. equation   **3.** solve   **5.** equivalent   **7. a.** $x + 6$ **b.** Neither
c. No   **d.** Yes   **9. a.** c, c **b.** c, c **11. a.** x **b.** y **c.** t
d. h **13.** 5, 5, 50, 50, $\frac{2}{=}$, 45, 50   **15. a.** Is possibly equal to
b. Yes   **17.** No   **19.** No   **21.** No   **23.** No   **25.** Yes
27. No   **29.** No   **31.** Yes   **33.** Yes   **35.** Yes   **37.** 71
39. 18   **41.** -0.9 **43.** 3   **45.** $\frac{8}{9}$ **47.** 3   **49.** $-\frac{1}{25}$
51. -2.3 **53.** 45   **55.** 0   **57.** 21   **59.** -2.64 **61.** 20
63. 15   **65.** -6 **67.** 4   **69.** 4   **71.** 7   **73.** 1   **75.** -6
77. 20   **79.** 0.5   **81.** -18 **83.** $-\frac{4}{21}$ **85.** 13   **87.** 2.5
89. $-\frac{8}{3}$ **91.** $\frac{13}{20}$ **93.** 4   **95.** -5 **97.** -200 **99.** 95
101. 65°   **103.** \$6,000,000   **109.** 0   **111.** $45 - x$

Study Set Section 2.2 (page 127)

1. equation   **3.** identity   **5.** subtraction, multiplication
7. a. $-2x - 8 = -24$ **b.** $-20 = 3x - 16$ **9. a.** $12x$ **b.** $2x$
11. 10   **13.** $+7, +7, 2, 2, 14, \frac{2}{=}, 28, 21, 14$ **15.** 6   **17.** 5
19. -7 **21.** 18   **23.** 16   **25.** 12   **27.** $\frac{10}{3}$ **29.** $-\frac{5}{2}$ **31.** 5
33. -0.25 **35.** 2.9   **37.** -4 **39.** $\frac{11}{5}$ **41.** -1 **43.** -6
45. 0.04   **47.** -6 **49.** -11 **51.** 7   **53.** -11 **55.** 1
57. $\frac{9}{2}$ **59.** 3   **61.** -20 **63.** 6   **65.** $\frac{2}{15}$ **67.** $-\frac{12}{5}$
69. $\frac{27}{5}$ **71.** 5   **73.** 200   **75.** 1,000   **77.** 200   **79.** $\frac{5}{4}$
81. -1 **83.** 1   **85.** 80   **87.** All real numbers
89. No solution   **91.** No solution   **93.** All real numbers
95. $\frac{1}{4}$ **97.** 30   **99.** -11 **101.** No solution   **103.** 1
105. $\frac{52}{9}$ **107.** -6 **109.** -5 **115.** Commutative Property of Multiplication   **117.** Associative Property of Addition

Study Set Section 2.3 (page 135)

1. Percent   **3.** multiplication, is   **5.** $\frac{51}{100}$, 0.51, 51%   **7. a.** 2,449
b. 2,449, what, 14,792   **9. a.** $12 = 0.40 \cdot x$ **b.** $99 = x \cdot 200$
c. $x = 0.66 \cdot 3$ **11. a.** 0.35   **b.** 0.085   **c.** 1.5   **d.** 0.0275
13. 312   **15.** 26%   **17.** 300   **19.** 46.2   **21.** 2.5%   **23.** 1,464
25. 0.48 oz   **27. a.** \$925 billion   **b.** \$600 billion   **29.** \$10.45
31. \$24.20   **33.** 60%, 40%   **35.** 19%   **37.** No (66%)
39. 120   **41. a.** 5 g; 25%   **b.** 20 g   **43.** 2000–2001, about 9%
45. 12%   **47.** 3%   **49.** \$75   **51.** \$300   **53.** \$95,000
55. \$25,600   **61.** $\frac{12}{5} = 2\frac{2}{5}$ **63.** No

Study Set Section 2.4 (page 147)

1. formula   **3.** volume   **5. a.** $d = rt$ **b.** $r = c + m$
c. $p = r - c$ **d.** $I = Prt$ **7.** 11,176,920 mi, 65,280 ft
9. $Ax, -Ax, B, B, B$ **11.** \$240 million   **13.** \$931   **15.** 3.5%
17. \$6,000   **19.** 2.5 mph   **21.** 4.5 hours   **23.** 185°C
25. −454°F   **27.** 20 in.   **29.** 1,885 mm³   **31.** $c = r - m$
33. $b = P - a - c$ **35.** $R = \frac{E}{I}$ **37.** $l = \frac{V}{wh}$ **39.** $r = \frac{C}{2\pi}$
41. $h = \frac{3V}{B}$ **43.** $f = \frac{s}{w}$ **45.** $r = \frac{T - 2t}{2}$ **47.** $x = \frac{C - By}{A}$
49. $m = \frac{2K}{v^2}$ **51.** $c = 3A - a - b$ **53.** $t = T - 18E$
55. $r^2 = \frac{s}{4\pi}$ **57.** $v^2 = \frac{2Kg}{w}$ **59.** $r^3 = \frac{3V}{4\pi}$
61. $M = 4.2B + 19.8$ **63.** $h = \frac{S - 2\pi r^2}{2\pi r}$
65. $y = -3x + 9$ **67.** $y = \frac{1}{3}x + 3$ **69.** $y = -\frac{3}{4}x - 4$
71. $b = \frac{2A}{h} - d$ or $b = \frac{2A - hd}{h}$ **73.** $c = \frac{72 - 8w}{7}$
75. 87, 89, 91
77.

Income Statement (dollars amounts in millions)	Quarter ending Sep 04	Quarter ending Sep 05
Revenue	1,806.2	1,886.0
Cost of goods sold	1,543.4	1,638.9
Operating profit	262.8	247.1

79. 14 in.   **81.** 50 in.   **83.** 25 in., 2.5 in.   **85.** 18.1 in.²
87. 2,463 ft²   **89.** 3,150 cm²   **91.** 6 in.   **93.** 8 ft   **95.** 348 ft³
97. 254 in.²   **99.** $D = \frac{L - 3.25r - 3.25R}{2}$ **105.** 137.76
107. 15%

Study Set Section 2.5 (page 158)

1. consecutive   **3.** vertex, base
5.

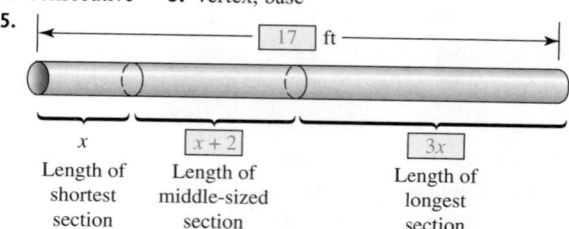

Length of shortest section — x
Length of middle-sized section — $x + 2$
Length of longest section — $3x$

7. \$0.03x   **9.** 180°   **11. a.** $x + 1$ **b.** $x + 2$ **13.** 4 ft, 8 ft
15. 102 mi, 108 mi, 114 mi, 120 mi   **17.** 7.3 ft, 10.7 ft
19. 250 calories in ice cream, 600 calories in pie   **21.** 7   **23.** 580
25. 20   **27.** \$50,000   **29.** \$5,250   **31.** Ronaldo: 15, Mueller: 14
33. *Friends:* 236; *Leave It to Beaver:* 234   **35.** Jan. 8, 10, 12
37. Width: 27 ft, length: 78 ft   **39.** 21 in. by 30.25 in.
41. 7 ft, 7 ft, 11 ft   **43.** 20°   **45.** 42.5°, 70°, 67.5°
47. 22°, 68°   **53.** −24   **55.** $-\frac{40}{37}$ **57.** 1

Study Set Section 2.6 (page 169)

1. investment, motion   **3.** $30,000 - x$ **5.** $r - 150$
7. $35t + 45t = 80, 35t, t, 45t, 80$
9. a. $0.50(6) + 0.25x = 0.30(6 + x), 0.50(6), 0.25x, 6 + x,$
$0.30(6 + x)$ **b.** $0.06x + 0.03(10 - x) = 0.05(10), 0.06x, 10 - x,$
$0.03(10 - x), 0.05(10)$ **11.** 0.06, 0.152   **13.** 4   **15.** 6,000
17. \$15,000 at 4%; \$10,000 at 7%   **19.** Silver: \$1,500; gold: \$2,000
21. \$26,000   **23.** 822: \$9,000; 721: \$6,000   **25.** \$4,900

27. 2 hr   **29.** $\frac{1}{4}$ hr = 15 min   **31.** 1 hr   **33.** 4 hr   **35.** 55 mph
37. 50 gal   **39.** 4%: 5 gal; 1%: 10 gal
41. 32 ounces of 8%; 32 ounces of 22%   **43.** 6 gal   **45.** 50 lb
47. 20 scoops   **49.** 15   **51.** \$4.25   **53.** 17   **55.** 90
57. 40 pennies, 20 dimes, 60 nickels
59. 2-pointers: 50; 3-pointers: 4   **63.** $-50x + 125$
65. $-3x + 3$ **67.** $16y - 16$

Study Set Section 2.7 (page 183)

1. inequality   **3.** interval   **5. a.** same   **b.** positive
c. negative   **7.** $x > 32$ **9. a.** $\le$ **b.** ∞ **c.** [or]
d. $>$ **11.** 5, 5, 12, 4, 4, 3   **13. a.** Yes   **b.** No
15. a. No   **b.** Yes   **17.** $(-\infty, 5)$

19. $(-3, 1]$ **21.** $x < -1, (-\infty, -1)$

23. $-7 < x \le 2, (-7, 2]$
25. $(3, \infty)$
27. $[10, \infty)$
29. $(-\infty, 6)$
31. $(-\infty, 48]$
33. $[2, \infty)$
35. $[3, \infty)$
37. $(7, \infty)$
39. $(-\infty, 0.4]$
41. $[16, \infty)$
43. $(-\infty, 0)$
45. $[-10, \infty)$
47. $(-\infty, -2)$
49. $(-5, \infty)$
51. $(-\infty, 1.5]$
53. $(-\infty, 20]$
55. $(0, \infty)$
57. $\left(\frac{5}{4}, \infty\right)$

59. $\left(-\infty, \frac{3}{2}\right]$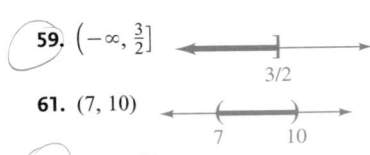
3/2

61. $(7, 10)$
7   10

63. $[-10, 0]$
−10   0

65. $[-6, 10]$
−6   10

67. $[2, 3)$
2   3

69. $(-3, 6]$
−3   6

71. $(-5, -2)$
−5   −2

73. $\left[\frac{9}{4}, \infty\right)$
9/4

75. $(-\infty, -40]$
−40

77. $(-2, 1]$
−2   1

79. $(-\infty, 2]$
2

81. $(-\infty, -27)$
−27

83. $\left(-\infty, \frac{1}{8}\right]$
1/8

85. $[-13, \infty)$
−13

87. $(6, \infty)$
6

89. $[-32, 48]$
−32   48

91. $\left(-\infty, -\frac{11}{4}\right)$
−11/4

93. $\left[-\frac{3}{8}, \infty\right)$
−3/8

95. $(-\infty, -1]$
−1

97. $\left(\frac{6}{7}, \infty\right)$
6/7

99. 98% or better   **101.** More than 27 mpg   **103.** 19 ft or less
105. More than 5 ft   **107.** 40 or less   **109.** 12.5 in. or less
111. a. $26 \text{ lb} \le w \le 31 \text{ lb}$ **b.** $12 \text{ lb} \le w \le 14 \text{ lb}$
c. $18.5 \text{ lb} \le w \le 20.5 \text{ lb}$ **d.** $11 \text{ lb} \le w \le 13 \text{ lb}$ **115.** 1, −3, 6

Chapter 2 Review (page 187)

1. Yes   **2.** No   **3.** No   **4.** No   **5.** Yes   **6.** Yes
7. equation   **8.** True   **9.** 21   **10.** 32   **11.** −20.6
12. 107   **13.** 24   **14.** 2   **15.** −9   **16.** −7.8   **17.** 0
18. $-\frac{16}{5}$ **19.** 2   **20.** −30.6   **21.** 30   **22.** −19   **23.** 4
24. 1   **25.** $\frac{5}{4}$ **26.** $\frac{47}{13}$ **27.** 6   **28.** $-\frac{22}{75}$ **29.** 5   **30.** 1

31. Identity; all real numbers   **32.** Contradiction; no solution
33. a. Percent   **b.** discount   **c.** commission   **34.** 192.4
35. 142.5   **36.** 12%
37. Broadband: 139.3 million; dial-up: 45.3 million
38. $26.74   **39.** No   **40.** $450   **41.** $150   **42.** 1,567%
43. $176   **44.** $11,800   **45.** 8 min   **46.** 4.5%
47. 1,949°F   **48. a.** 168 in.   **b.** 1,440 in.^{2} **c.** 4,320 in.^{3}
49. 76.5 m^2 **50.** 144 in.^{2} **51. a.** 50.27 cm
b. 201 cm^2 **52.** 9.4 ft^3 **53.** 120 ft^3 **54.** 381.70 in.^{3}
55. $h = \frac{A}{2\pi r}$ **56.** $G = 3A - 3BC + K$ **57.** $t = \frac{4C}{s} + d$
58. $y = \frac{3}{4}x + 4$ **59.** 8 ft   **60.** 200   **61.** $2,500,000
62. Labonte: 43; Petty: 45
63. 24.875 in. × 29.875 in. $\left(24\frac{7}{8}\text{ in.} \times 29\frac{7}{8}\text{ in.}\right)$
64. 76.5°, 76.5°   **65.** $16,000 at 7%, $11,000 at 9%   **66.** 20
67. $1\frac{2}{3}$ hr = 1 hr 40 min   **68.** 12, 4   **69.** 10 lb of each   **70.** 2 gal

71. $(-\infty, 1)$
1

72. $(-\infty, 12]$
12

73. $\left(\frac{5}{4}, \infty\right)$
5/4

74. $[3, \infty)$
3

75. $(-\infty, 40]$
40

76. $(9, \infty)$
9

77. $(6, 11)$
6   11

78. $\left(-\frac{7}{2}, \frac{3}{2}\right]$
−7/2   3/2

79. $2.40 \text{ g} \le w \le 2.53 \text{ g}$ **80.** 48 inches or less

Chapter 2 Test (page 194)

1. a. solve   **b.** Percent   **c.** circumference   **d.** inequality
e. multiplication, equality   **2.** No   **3.** 2   **4.** −5   **5.** 22
6. $-\frac{1}{4}$ **7.** 1,336   **8.** All real numbers (an identity)   **9.** $\frac{7}{4}$
10. −4   **11.** No solution (a contradiction)   **12.** 0   **13.** 12.16
14. $76,000   **15.** 6%   **16.** $30   **17.** $295   **18.** −10°C
19. 393 in.^{3} **20.** $r = \frac{A - P}{Pt}$ **21.** 20 in.^{2} **22.** 22 min, 8 min
23. $120,000   **24.** 380 mi, 280 mi   **25.** Green: 16 lb; herbal: 4 lb
26. 412, 413   **27.** $\frac{3}{5}$ hr   **28.** 10 liters   **29.** 68°   **30.** $5,250

31. $[-3, \infty)$
−3

32. $(-\infty, 6.4)$
6.4

33. $[-7, 4)$
−7   4

34. 180 words

Study Set Section 3.1 (page 204)

1. ordered   **3.** axis, axis, origin   **5.** rectangular
7. a. origin, left, up   **b.** origin, right, down   **9. a.** I and II

b. II and III   **c.** IV   **11.** $(3, 5)$ is an ordered pair, $3(5) = 3 \cdot 5$
13. Yes   **15.** Horizontal
17.

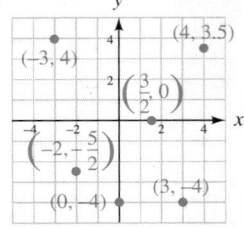

19. $(4, 3), (0, 4), (-5, 0), (-4, -5), (3, -3)$
21. a. 60 beats/min   **b.** 10 min
23. a. 5 min and 50 min after starting   **b.** 20 min
25. a. 2 hr   **b.** $-1,000$ ft   **27. a.** It ascends (rises) 500 ft
b. -500 ft   **29.** Rivets: $(-6, 0), (-2, 0), (2, 0), (6, 0)$; welds:
$(-4, 3), (0, 3), (4, 3)$; anchors: $(-6, -3), (6, -3)$
31. $(G, 2), (G, 3), (G, 4)$ **33. a.** 8
b. It represents the patient's left side.   **35.** $(2, 4)$; 12 sq. units
37.

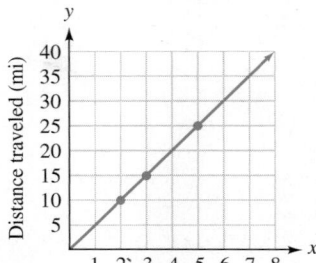

a. 20 mi   **b.** 6 gal
c. 35 mi

39.

a. A 3-yr-old copier is worth
$7,000.   **b.** $1,000

c. 6 yr   **45.** $h = \frac{3(AC + T)}{2}$
47. -1

Study Set Section 3.2 (page 217)

1. two   **3.** table   **5.** linear   **7. a.** 2   **b.** Yes   **c.** No
d. Infinitely many   **9.** solution, point
11. a. $-5, 0, 5$ (Answers may vary)
b. $-10, 0, 10$ (Answers may vary)   **13.** $6, -2, 2, 6$
15. a. $y^1 = \frac{1}{2}x^1 + 7$ **b.** The exponent on x is not 1.   **17.** Yes
19. No   **21.** Yes   **23.** Yes   **25.** No   **27.** No   **29.** 11
31. 4   **33.** 13   **35.** $-\frac{8}{7}$

37.

x	y	(x, y)
8	12	(8, 12)
6	8	(6, 8)

39.

x	y	(x, y)
-5	-13	$(-5, -13)$
-1	-1	$(-1, -1)$

41.

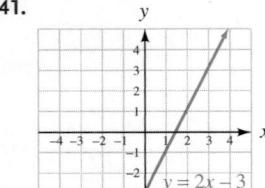

43.

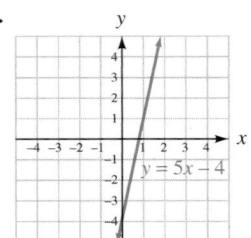

45.

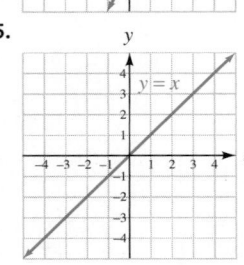

47.

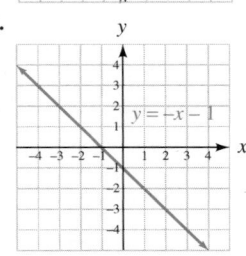

49.

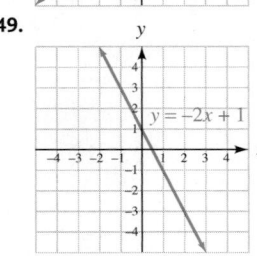

51.

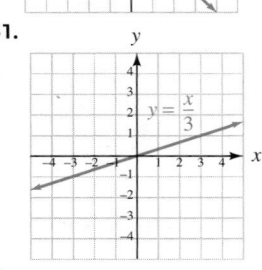

53.

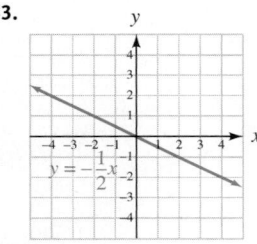

55.

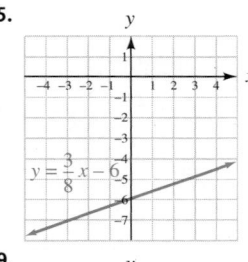

57.

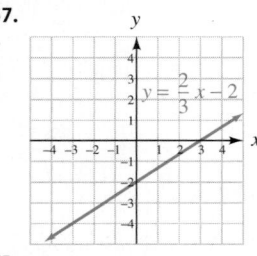

59.

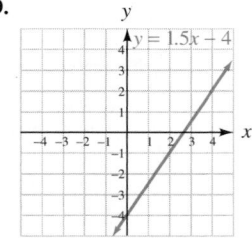

61.

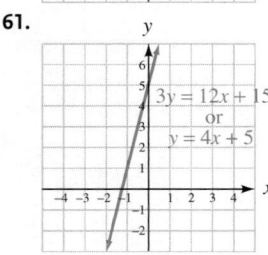

63.

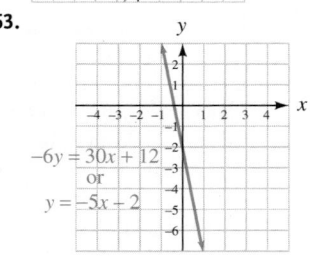

65.

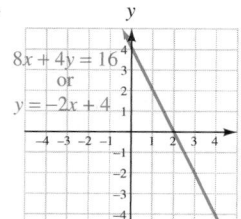

67.

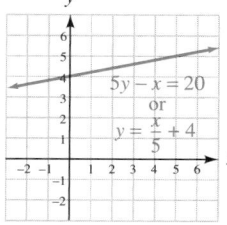

69.

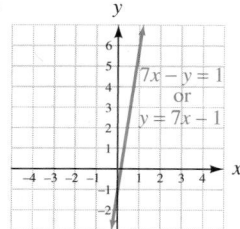

71.

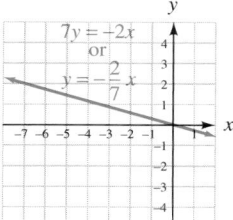

73.

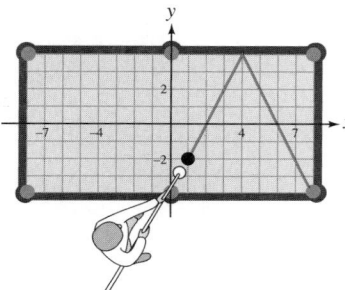

75. 125 hr

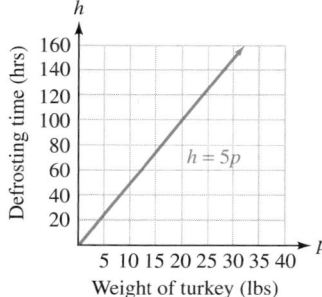

77. 3 oz

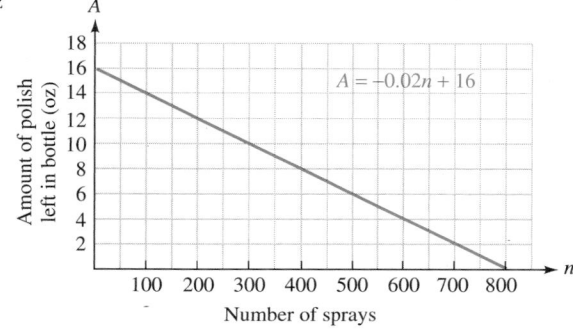

79. About $95

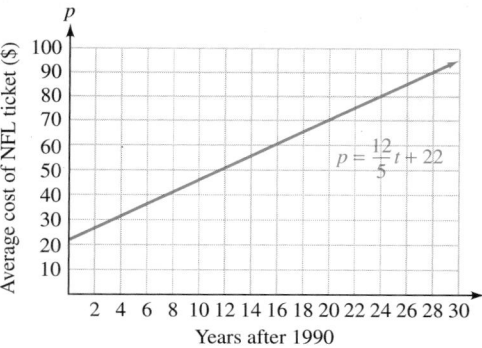

81. About 180

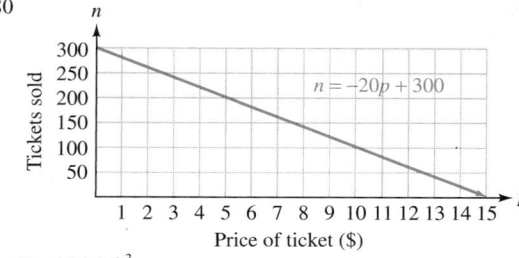

91. $5 + 4c$ **93.** 904.8 ft^3

Study Set Section 3.3 (page 229)

1. x-intercept **3.** horizontal, vertical **5. a.** $0, y$ **b.** $0, x$
7. a. y-intercept: $(0, 40)$; $40,000
b. x-intercept: $(30, 0)$; 30 years after purchase **9.** $y = 0; x = 0$
11. x-intercept: $(4, 0)$, y-intercept: $(0, 3)$
13. x-intercept: $(-5, 0)$, y-intercept: $(0, -4)$
15. y-intercept: $(0, 2)$
17. x-intercept: $\left(-2\frac{1}{2}, 0\right)$; y-intercept: $\left(0, \frac{2}{3}\right)$ (Answers may vary)
19. $(3, 0); (0, 8)$ **21.** $(4, 0); (0, -14)$ **23.** $(-2, 0); \left(0, -\frac{10}{3}\right)$
25. $\left(\frac{3}{2}, 0\right); (0, 9)$

27.

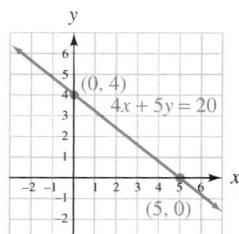

29.

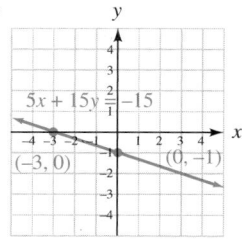

31.

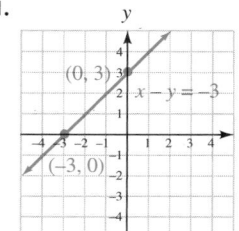

33.

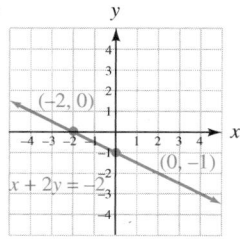

35.
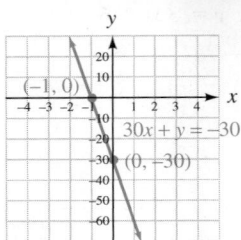

(−1, 0)

$30x + y = -30$

(0, −30)

37.
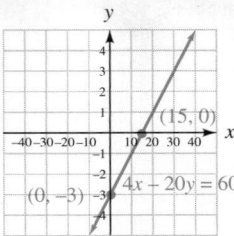

(15, 0)

(0, −3) $4x - 20y = 60$

59. $y = 2$
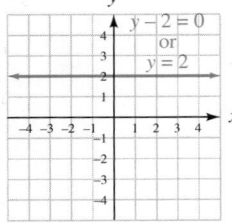

$y - 2 = 0$
or
$y = 2$

39.
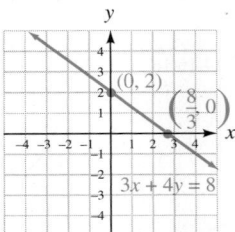

(0, 2)

$\left(\frac{8}{3}, 0\right)$

$3x + 4y = 8$

41.
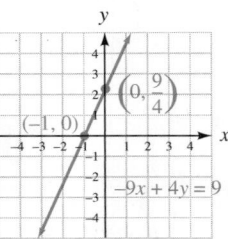

$\left(0, \frac{9}{4}\right)$

(−1, 0)

$-9x + 4y = 9$

61. $x = 1.5$
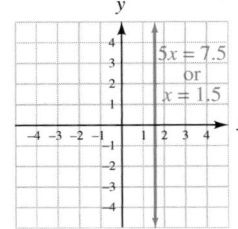

$5x = 7.5$
or
$x = 1.5$

43.
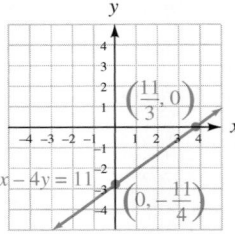

$\left(\frac{11}{3}, 0\right)$

$3x - 4y = 11$

$\left(0, -\frac{11}{4}\right)$

45.
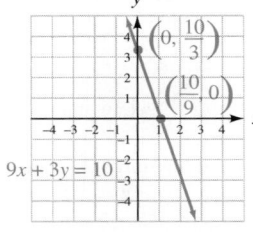

$\left(0, \frac{10}{3}\right)$

$\left(\frac{10}{9}, 0\right)$

$9x + 3y = 10$

63.
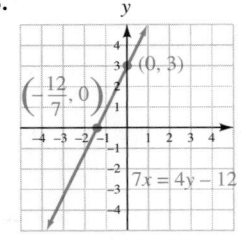

(0, 3)

$\left(-\frac{12}{7}, 0\right)$

$7x = 4y - 12$

65.
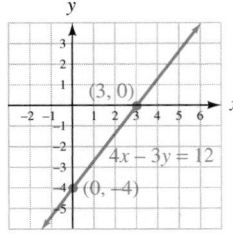

(3, 0)

$4x - 3y = 12$

(0, −4)

47.
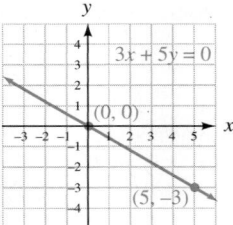

$3x + 5y = 0$

(0, 0)

(5, −3)

49.
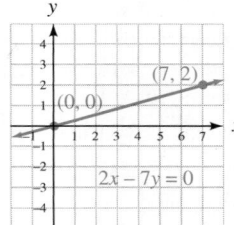

(7, 2)

(0, 0)

$2x - 7y = 0$

67.
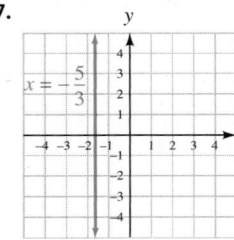

$x = -\frac{5}{3}$

69.
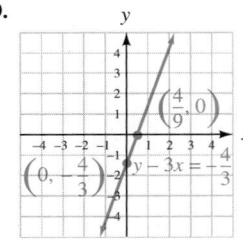

$\left(\frac{4}{9}, 0\right)$

$\left(0, -\frac{4}{3}\right)$ $y - 3x = -\frac{4}{3}$

51.
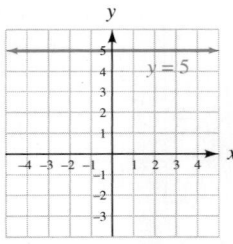

$y = 5$

53.
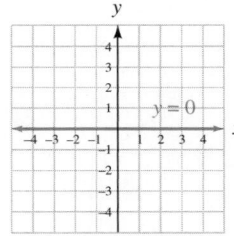

$y = 0$

71.
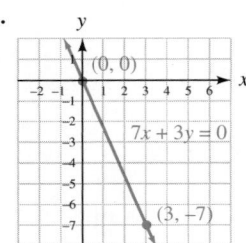

(0, 0)

$7x + 3y = 0$

(3, −7)

73.
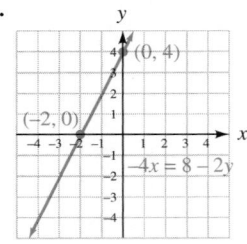

(0, 4)

(−2, 0)

$-4x = 8 - 2y$

55.
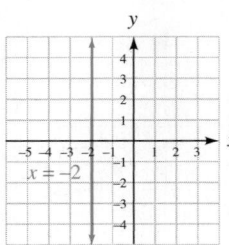

$x = -2$

57.
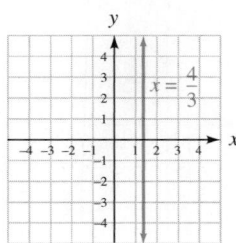

$x = \frac{4}{3}$

75.
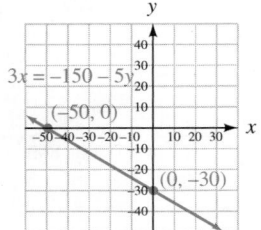

$3x = -150 - 5y$

(−50, 0)

(0, −30)

77. $y = -1$

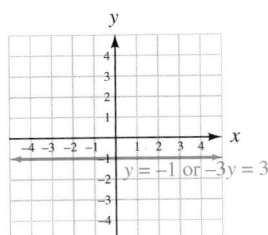

79. a. About $-270°C$ **b.** 0 milliliters

81.

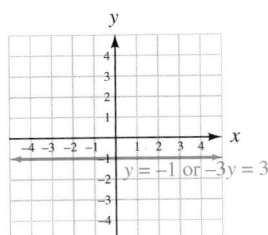

a. If only shrubs are purchased, he can buy 200.
b. If only trees are purchased, he can buy 100.   **87.** $\frac{1}{5}$ **89.** $2x - 6$

Study Set Section 3.4 (page 242)

1. slope, ratio   **3.** change   **5. a.** Line 2   **b.** Line 1   **c.** Line 4
d. Line 3   **7. a.** Line 1   **b.** Line 1   **9.** same   **11. a.** 0
b. Undefined   **c.** $\frac{1}{4}$ **d.** 2   **13.** 40%   **15. a.** $-\frac{1}{6}$ **b.** $\frac{8}{7}$
c. 1   **17. a.** $m = \frac{y_2 - y_1}{x_2 - x_1}$ **b.** sub, sub, over (divided by), two, one
19. 1   **21.** $\frac{2}{3}$ **23.** $\frac{4}{3}$ **25.** -2 **27.** 0   **29.** $-\frac{1}{5}$ **31.** $\frac{1}{2}$
33. 1   **35.** -3 **37.** $\frac{5}{4}$ **39.** $-\frac{1}{2}$ **41.** $\frac{3}{5}$ **43.** 0
45. Undefined   **47.** $-\frac{2}{3}$ **49.** -4.75 **51.** 0   **53.** $\frac{7}{5}$ **55.** $-\frac{2}{5}$
57. $m = \frac{3}{4}$ **59.** $m = 0$ **61.** 0   **63.** 0   **65.** Undefined
67. Undefined   **69.** 0   **71.** Undefined   **73.** Parallel
75. Perpendicular   **77.** Neither   **79.** Perpendicular   **81.** Parallel
83. Neither   **85.** $\frac{5}{9}$ **87.** $-\frac{2}{3}$ **89.** -1 **91.** $\frac{1}{2}$ **93.** $-\frac{2}{5}$
95. $\frac{1}{20}$; 5%   **97.** 4%, 8%, 12%   **99.** Front: $\frac{3}{2}$; side: $\frac{3}{5}$
101. -875 gal per hr   **103.** 380 lb per yr   **109.** 40 lb licorice;
20 lb gumdrops

Study Set Section 3.5 (page 254)

1. slope–intercept   **3. a.** No   **b.** No   **c.** Yes   **d.** No
5. a. $y = 2x + 8$ **b.** $y = -5x - 3$ **c.** $y = \frac{x}{3} - 1$
d. $y = \frac{9}{5}x + 4$ **7.** $-2x, 5y, 5, 5, 5, -\frac{2}{5}, 3, -\frac{2}{5}, (0, 3)$ **9.** $-2, -3$
11. $4, (0, 2)$ **13.** $-5, (0, -8)$ **15.** $4, (0, -9)$ **17.** $-1, (0, 11)$
19. $-20, (0, 1)$ **21.** $\frac{1}{2}, (0, 6)$ **23.** $\frac{1}{4}, \left(0, -\frac{1}{2}\right)$ **25.** $-5, (0, 0)$
27. $\frac{2}{3}, (0, 0)$ **29.** $1, (0, 0)$ **31.** $0, (0, -2)$ **33.** $0, \left(0, -\frac{2}{5}\right)$
35. $-1, (0, 8)$ **37.** $\frac{1}{6}, (0, -1)$ **39.** $-2, (0, 7)$ **41.** $-\frac{3}{2}, (0, 1)$
43. $-\frac{2}{3}, (0, 2)$ **45.** $\frac{3}{5}, (0, -3)$ **47.** $\frac{4}{3}, (0, -4)$ **49.** $1, \left(0, -\frac{11}{6}\right)$

51. $y = 5x - 3$

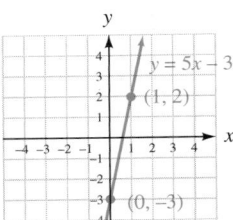

53. $y = -3x + 6$

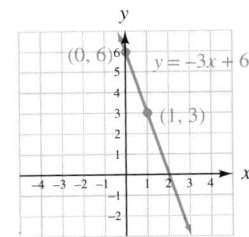

55. $y = \frac{1}{4}x - 2$

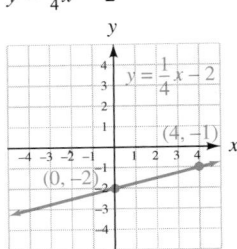

57. $y = -\frac{8}{3}x + 5$

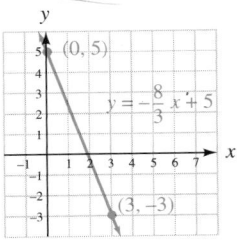

59. $y = 5x - 1$ **61.** $y = -2x + 3$ **63.** $y = \frac{4}{5}x - 2$
65. $y = -\frac{5}{3}x + 2$
67. $3, (0, 3)$ **69.** $-\frac{1}{2}, (0, 2)$

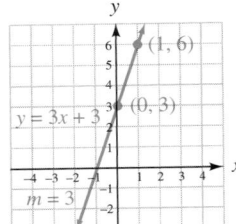

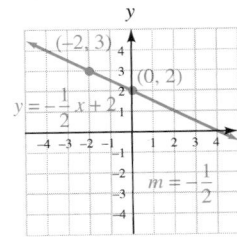

71. $-3, (0, 0)$ **73.** $-4, (0, -4)$

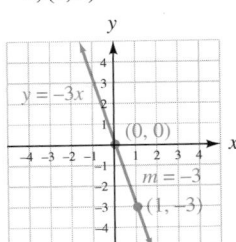

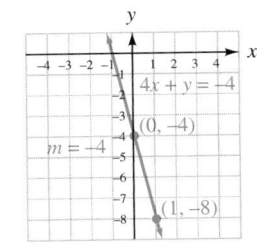

75. $-\frac{3}{4}, (0, 4)$ **77.** $2, (0, -1)$

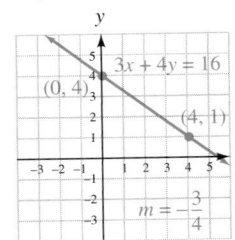

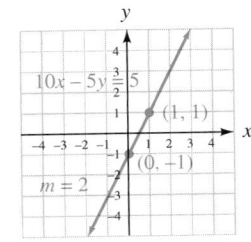

79. Parallel   **81.** Perpendicular   **83.** Neither   **85.** Perpendicular
87. Parallel   **89.** Perpendicular   **91. a.** $c = 2,000h + 5,000$
b. $21,000   **93.** $F = 5t - 10$ **95.** $c = -20m + 500$

97. a. $c = 5x + 20$

b and c.

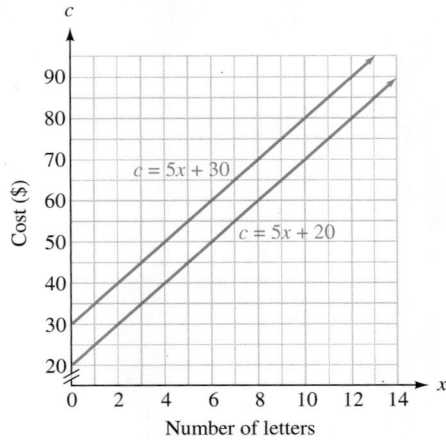

99. $c = 0.89t + 16.63$ **103.** 42 ft, 45 ft, 48 ft, 51 ft

Study Set Section 3.6 (page 264)

1. point–slope, sub, times, minus, one   **3. a.** point–slope
b. slope–intercept   **5. a.** $(-2, -3)$ **b.** $\frac{5}{6}$ **c.** $y + 3 = \frac{5}{6}(x + 2)$
7. $(67, 170), (79, 220)$ **9.** $5, -1, +, 2, 3$
11. point–slope, slope–intercept   **13.** $y - 1 = 3(x - 2)$
15. $y + 1 = \frac{4}{5}(x + 5)$ **17.** $y = 2x - 1$ **19.** $y = -5x - 37$
21. $y = -3x$ **23.** $y = \frac{1}{5}x - 1$ **25.** $y = -\frac{4}{3}x + 4$
27. $y = -\frac{11}{6}x - \frac{7}{3}$ **29.** $y = 2x + 5$ **31.** $y = -\frac{1}{2}x + 1$
33. $y = 5$ **35.** $y = \frac{1}{10}x + \frac{1}{2}$ **37.** $x = -8$ **39.** $y = \frac{1}{2}x$
41. $x = 4$ **43.** $y = 5$

45.

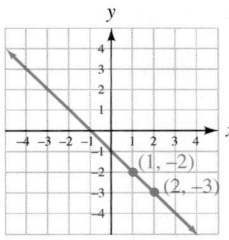

47.

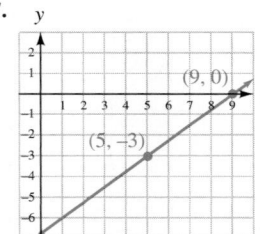

49.

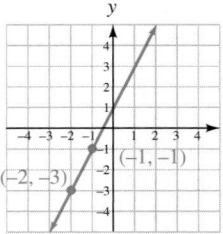

51.

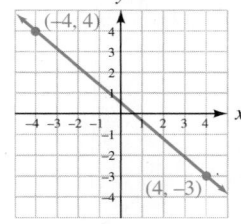

53. $y = \frac{1}{4}x - \frac{5}{4}$ **55.** $y = 12$ **57.** $y = -\frac{2}{3}x + 2$
59. $y = 8x + 4$ **61.** $x = -3$ **63.** $y = 7x$ **65.** $y = -4x - 9$
67. $y = \frac{2}{7}x - 2$ **69.** $y = \frac{1}{10}x$ **71.** $x = -\frac{1}{8}$
73. $h = 3.9r + 28.9$ **75.** $y = -\frac{2}{5}x + 4, y = -7x + 70, x = 10$
77. a. $y = -40m + 920$ **b.** 440 yd^3 **79.** $l = \frac{25}{4}r + \frac{1}{4}$
81. a. $y = -\frac{3}{10}x + \frac{283}{10}$ or $y = -0.3x + 28.3$ **b.** 16.3 gal
87. 17 in. by 39 in.

Study Set Section 3.7 (page 275)

1. inequality   **3.** satisfies   **5.** half-planes   **7.** Yes
9. dashed, solid
11. The half-plane opposite that in which the test point lies
13. a. Yes   **b.** No   **c.** No   **d.** Yes   **15. a.** Is less than
b. Is greater than or equal to   **c.** Is less than or equal to
d. Is possibly greater than   **17.** $=, <$ **19.** Yes   **21.** No
23. No   **25.** Yes

27.

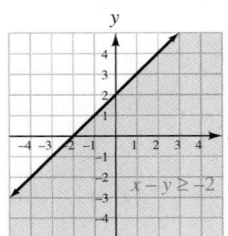

29.

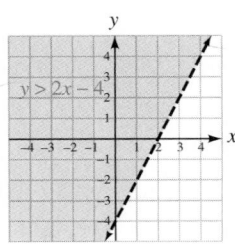

31.

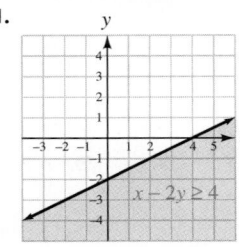

33.

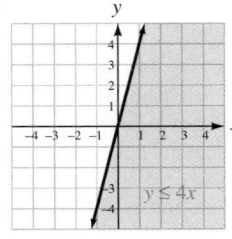

35.

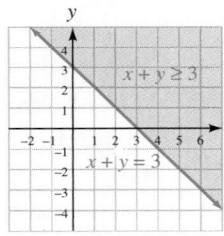

37.

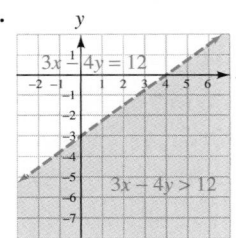

39.

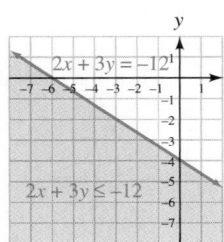

41.

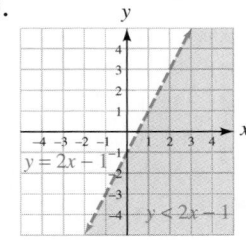

43.

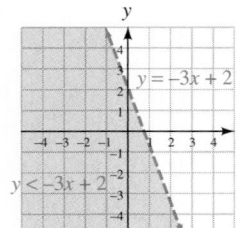

45.

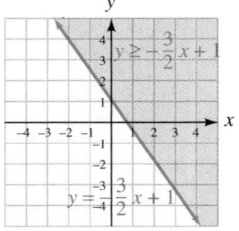

47.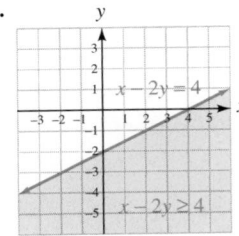

49.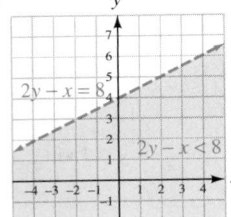

51.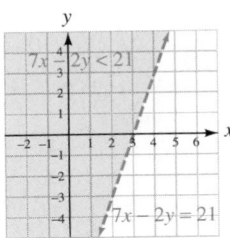

53.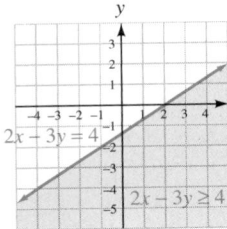

55.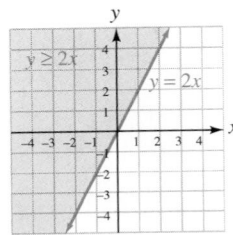

57.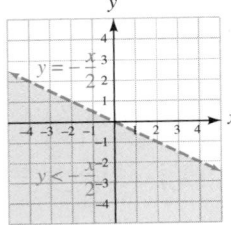

59.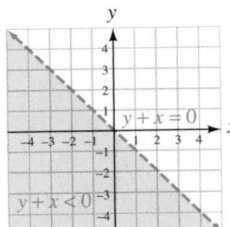

61.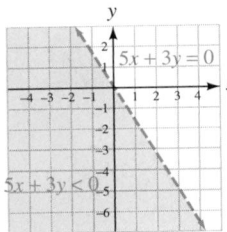

63.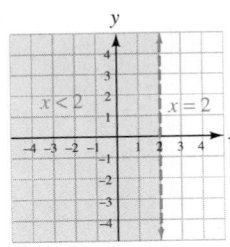

65.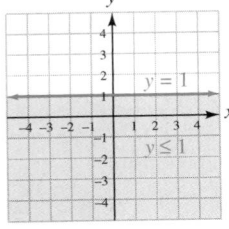

67.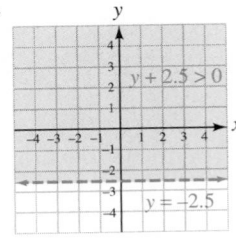

69.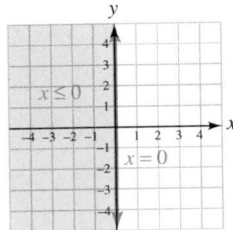

71. No   **73.** ii

75. $(10, 10), (20, 10), (10, 20)$; Answers may vary

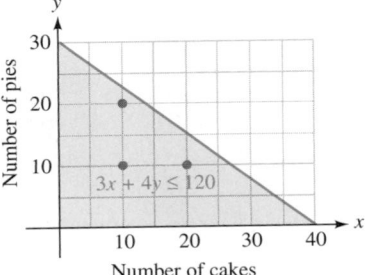

77. $(40, 30), (30, 40), (40, 20)$; Answers may vary

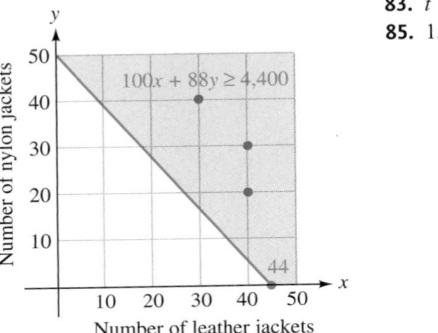

83. $t = \dfrac{A - P}{Pr}$

85. $15x + 22$

Study Set Section 3.8 (page 288)

1. relation   **3.** domain, range   **5.** value

7.

Domain	Range
1990	3.80
1992	4.25
1994	
1996	4.75
1998	5.15
2000	
2002	
2004	
2006	
2008	6.55

9. 33   **11.** of   **13.** 4, 5, (4, 5)
15. Domain: $\{-6, -1, 6, 8\}$;
range: $\{-10, -5, -1, 2\}$
17. Domain: $\{-8, 0, 6\}$; range: $\{9, 50\}$
19. Yes; domain: $\{10, 20, 30\}$;
range: $\{20, 40, 60\}$
21. No; (4, 2), (4, 4), (4, 6)
23. Yes; domain: $\{1, 2, 3, 4, 5\}$;
range: $\{7, 8, 15, 16, 23\}$
25. No; $(-1, 0), (-1, 2)$
27. No; $(3, 4), (3, -4)$ or
$(4, 3), (4, -3)$

29. Yes; domain: $\{-3, 1, 5, 6\}$; range: $\{-8, 0, 4, 9\}$ **31. a.** 3
b. -9 **c.** 0   **d.** 199   **33. a.** 0.32   **b.** 18   **c.** 2,000,000
d. $\frac{1}{32}$ **35. a.** 7   **b.** 14   **c.** 0   **d.** 1   **37. a.** 0   **b.** 990
c. -24 **d.** 210   **39. a.** 36   **b.** 0   **c.** 9   **d.** 4
41. 1.166
43.

x	f(x)
−2	4
−1	1
0	−2
1	−5

45.

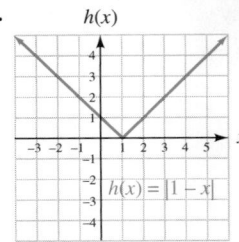

x	h(x)
-2	3
-1	2
0	1
1	0
2	1
3	2
4	3

47.

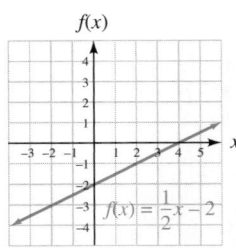

49.

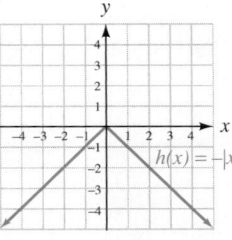

51. Yes   **53.** No; (3, 4), (3, −1) (Answers may vary)
55. No; (0, 2), (0, −4) (Answers may vary)
57. No; (3, 0), (3, 1) (Answers may vary)
59. $f(x) = |x|$ **61.** $900   **63.** 78.5 ft², 1,256.6 ft²   **69.** 80 lb

Chapter 3 Review (page 293)

1.

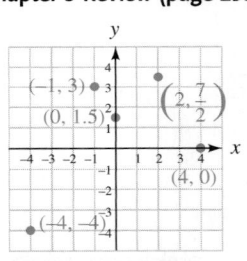

2. (158, 21.5)   **3.** Quadrant III
4. (0, 0)   **5.** (1, 4); 36 square units
6. a. 2,500; week 2   **b.** 1,000
c. 1st week and 5th week
7. Yes

8.

x	y	(x, y)
-2	-6	(-2, -6)
-8	3	(-8, 3)

9. $y = x^2 + 1$ and $y - x^3 = 0$ **10. a.** True   **b.** False

11.

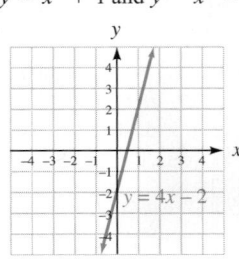

12.

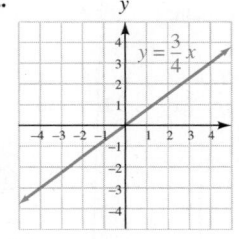

13.

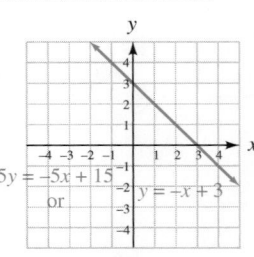

14.

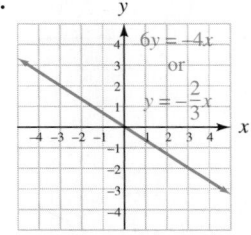

15. About $190 new

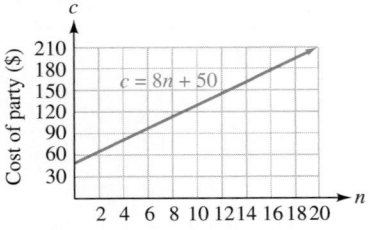

16. a. False   **b.** True   **17.** (−3, 0), (0, 2.5)
18. (0, 25,000); the equipment was originally valued at $25,000.
(10, 0); in 10 years, the sound equipment had no value.
19. x-intercept: (−2, 0); y-intercept: (0, 4)

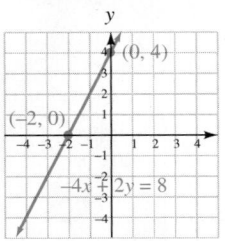

20. x-intercept: $\left(\frac{13}{5}, 0\right)$; y-intercept: $\left(0, -\frac{13}{4}\right)$

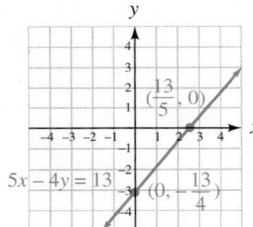

21.

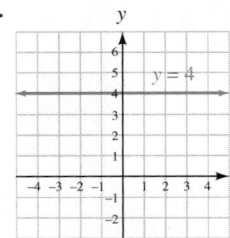

22.

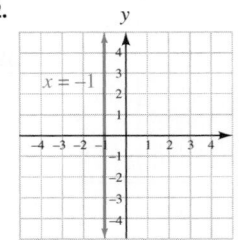

23. $\frac{1}{4}$ **24.** $-\frac{7}{8}$ **25.** −7   **26.** $-\frac{3}{2}$
27. b. Negative slope   **d.** Undefined slope   **c.** 0 slope   **a.** Positive slope

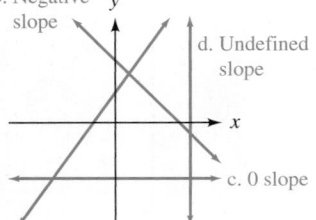

28. $\frac{3}{4}$ **29.** 8.3%   **30. a.** −4.5 million people per yr
b. 4.05 million people per yr   **31.** They are neither.   **32.** $-\frac{7}{5}$
33. $m = \frac{3}{4}$; y-intercept: (0, −2)   **34.** $m = -4$; y-intercept: (0, 0)

35. $m = \frac{1}{8}$; *y*-intercept: $(0, 10)$ **36.** $m = -\frac{7}{5}$; *y*-intercept: $\left(0, -\frac{21}{5}\right)$

37. $y = 4x - 1$ **38.** $y = \frac{3}{2}x - 3$

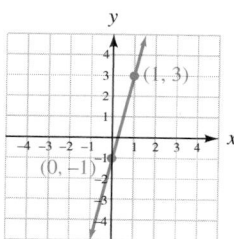

39. $m = 3$; *y*-intercept: $(0, -5)$ **40. a.** $c = 300w + 75{,}000$
b. $90{,}600$

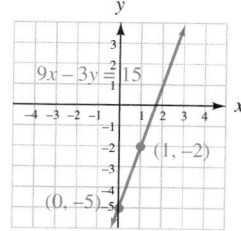

41. Parallel   **42.** Perpendicular
43. $y = 3x + 2$ **44.** $y = -\frac{1}{2}x - 3$

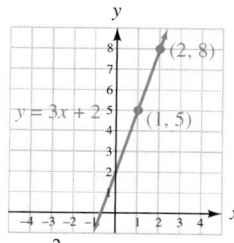   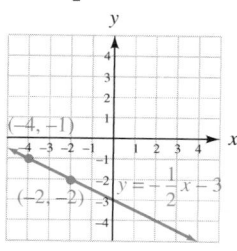

45. $y = \frac{2}{3}x + 5$ **46.** $y = -8$ **47.** $f = -35x + 450$
48. a. $y = 1{,}200x + 19{,}800$ **b.** $\$43{,}800$ **49. a.** Yes
b. Yes   **c.** Yes   **d.** No   **50.** $=, >$
51.   **52.**

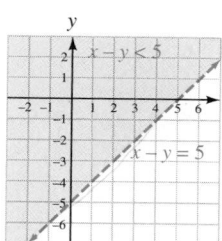   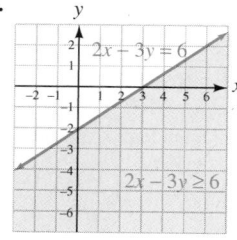

53.   **54.**

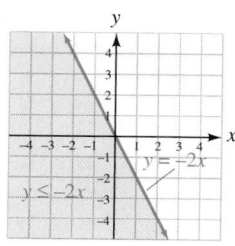   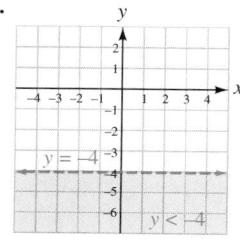

55. a. True   **b.** False   **c.** False

56. $(2, 4), (5, 3), (6, 2)$; Answers may vary

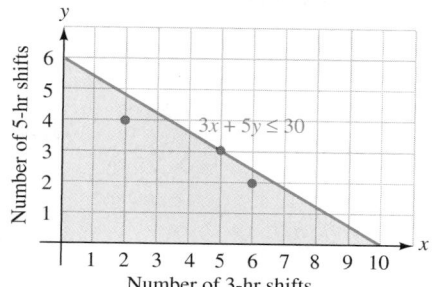

57. Domain: $\{-5, 0, 4, 7\}$; range: $\{-11, -3, 4, 9\}$
58. Domain: $\{-6, 1, 2, 15\}$; range: $\{-8, -2, 9\}$
59. Yes; domain: $\{1, 4, 8\}$; range: $\{0, 6, 9\}$
60. Yes; domain: $\{2, 3, 5, 6\}$; range: $\{1, 4\}$
61. Yes; domain: $\{3, 5, 7, 9\}$; range: $\{9, 25, 49, 81\}$
62. No; $(-1, 2), (-1, 4)$ **63.** domain, range   **64.** $f(x)$ **65.** -3
66. 0   **67.** 21   **68.** $-\frac{7}{4}$ **69.** -5 **70.** 37   **71.** -2
72. -8 **73.** No; $(1, 0.5), (1, 4)$, (Answers may vary)   **74.** Yes
75.

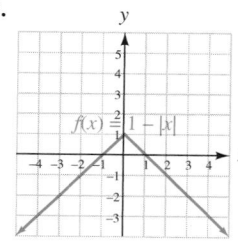

x	f(x)
0	1
1	0
2	-1
-1	0
-2	-1
-3	-2

76. $1{,}004.8 \text{ in.}^3$

Chapter 3 Test (page 302)

1. a. axis, axis   **b.** solution   **c.** linear   **d.** slope   **e.** function
2. 10   **3.** 60   **4.** 1 day before and the 3rd day of the holiday
5. 50 dogs were in the kennel when the holiday began.
6.

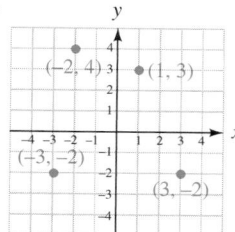

7. $A(2, 4), B(-3, 3), C(-2, -3), D(4, -3)$ **8. a.** III   **b.** IV
9. Yes   **10.**

x	y	(x, y)
2	1	(2, 1)
-6	3	(-6, 3)

11. a. False   **b.** True   **12.**

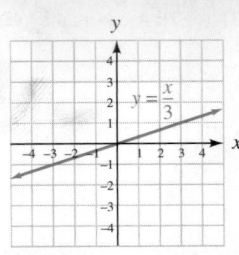

13. x-intercept: $(3, 0)$; y-intercept: $(0, -2)$

14.

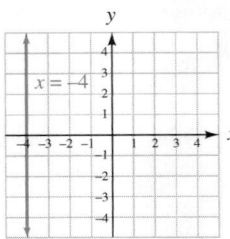

15. $\frac{8}{7}$ **16.** -1 **17.** 0

18. 10%   **19.** Perpendicular

20. Parallel   **21.** -15 ft per mi

22. 25 ft per mi

23.

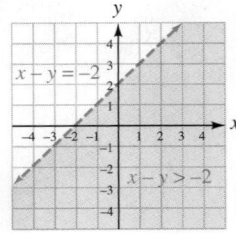

24.

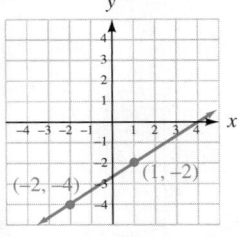

25. $m = -\frac{1}{2}$; $(0, 4)$ **26.** $y = 7x + 19$ **27.** $y = -2x - 5$
28. a. $v = -1,500x + 15,000$ **b.** \$3,000   **29.** Yes

30. $y = -\frac{1}{5}T + 41$

31.

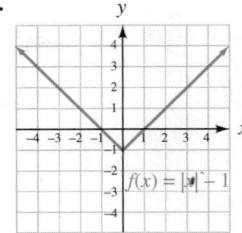

32. Domain: $\{-4, 0, 1, 5\}$;
range: $\{-8, 3, 12\}$
33. Yes; domain: $\{1, 2, 3, 4\}$;
range: $\{1, 2, 3, 4\}$
34. No; $(-3, 9)$, $(-3, -7)$;
35. No; $(2, 3.5)$, $(2, -3.5)$; (Answers
may vary)   **36.** Yes; domain:
$\{5, 10, 15, 20, 25\}$; range: $\{12\}$
37. -13 **38.** 756
39. $C(45) = 28.50$; it costs \$28.50 to
make 45 calls.

40.

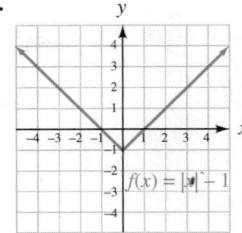

20. $7x - 12$ **21.** 6   **22.** 2.9   **23.** 9   **24.** -19 **25.** $\frac{1}{7}$
26. 1   **27.** $-\frac{55}{6}$ **28.** No solution, contradiction   **29.** -99
30. $-\frac{1}{4}$ **31.** 1,100   **32.** $h = \frac{S - 2\pi r^2}{2\pi r}$ **33.** $3\frac{1}{8}$ in., $\frac{39}{64}$ in.^{2}
34. $45°$

35.

	% acid	Liters	Amount of acid
50% solution	0.50	x	$0.50x$
25% solution	0.25	$13 - x$	$0.25(13 - x)$
30% mixture	0.30	13	$0.30(13)$

36. 7.5 hr   **37.** 80 lb candy corn, 120 lb gumdrops
38. $(-\infty, 48]$ **39.** $(0, \infty)$

40. I and II   **41.** No

42.

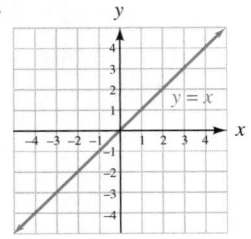

43.

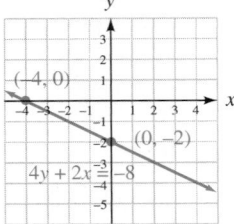

44. 0   **45.** $-\frac{10}{7}$ **46.** $\frac{7}{12}$ **47.** $\frac{2}{3}$, $(0, 2)$ **48.** $y = -2x + 1$
49. $y + 9 = -\frac{7}{8}(x - 2)$; $y = -\frac{7}{8}x - \frac{29}{4}$ **50.** Yes

51.

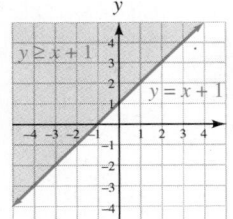

52.

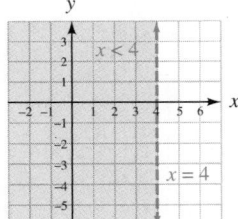

53. 78   **54.** No

Study Set Section 4.1 (page 316)

1. system   **3.** intersection   **5.** consistent, inconsistent
7. a. True   **b.** True   **9. a.** $-5, 2$ **b.** $3, 3, (0, -2)$ **11.** No
solution; independent   **13.** Yes   **15.** Yes   **17.** No   **19.** No
21. No   **23.** Yes   **25.** $(3, 2)$

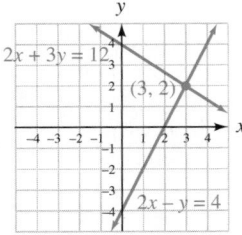

Cumulative Review Chapters 1–3 (page 305)

1. $2^2 \cdot 3^3$ **2.** 0.004   **3. a.** True   **b.** True   **c.** True   **4.** -15
5. -0.77 **6.** -945 **7.** 30   **8.** 2   **9.** 32   **10.** $500 - x$
11. $3, -2$ **12. a.** $2x + 8$ **b.** $-2x + 8$ **13.** $4a + 10$
14. $-63t$ **15.** $4b^2$ **16.** 0   **17.** 4   **18.** $-160a$ **19.** $-3y$

27. $(-1, 5)$

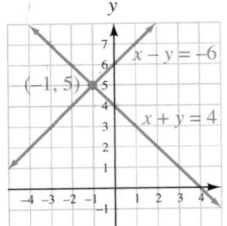

29. $(-2, 0)$

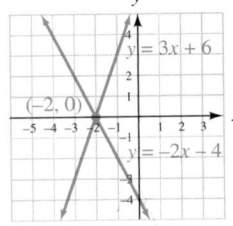

47. $(3, -1)$

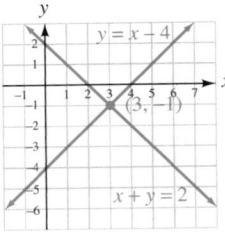

49. $(-6, 1)$

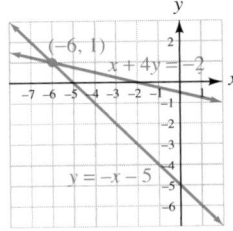

31. Infinitely many solutions

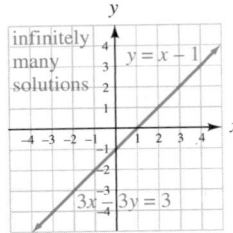

33. No solution

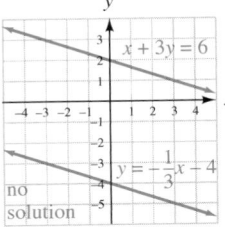

51. $(3, 0)$

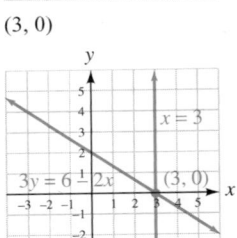

53. $(-2, -3)$

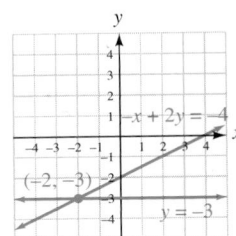

35. $(4, -6)$

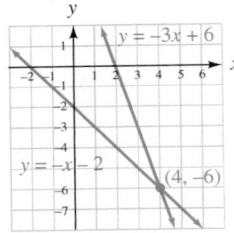

37. $(5, -2)$

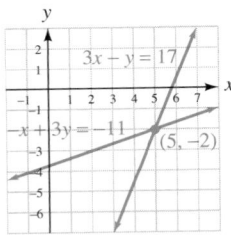

55. $(4, -4)$

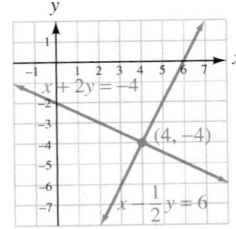

39. $(1, 1)$

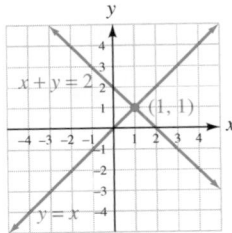

41. $(-4, 0)$

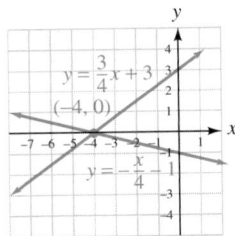

57. 1 solution   **59.** Same line; infinitely many solutions   **61.** No solution   **63.** 1 solution   **65.** $(1, 3)$ **67.** No solution
69. 1994; about 4,100   **71. a.** Houston, New Orleans, St. Augustine
b. St. Louis, Memphis, New Orleans   **c.** New Orleans   **73. a.** The incumbent; 7%   **b.** November 2   **c.** The challenger; 3
75. 10 mi   **83.** $[-3, \infty)$

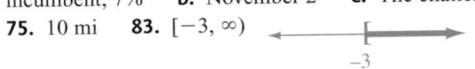

85. $[-8, \infty)$

87. $(-\infty, 16)$

43. No solution

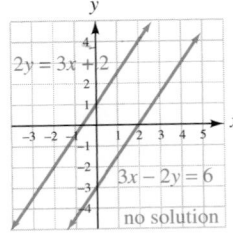

45. Infinitely many solutions

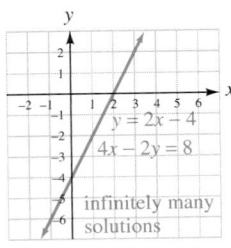

Study Set Section 4.2 (page 327)
1. substituting   **3.** $y = -3x$ **5.** $x + 3(x - 4) = 8$
7. Substitute 3 for a in the second equation.   **9. a.** No   **b.** ii
11. $3x, 4, -2, -2, -6, (-2, -6)$ **13.** $(2, 4)$ **15.** $(3, 0)$
17. $(-1, -1)$ **19.** $(-10, 2)$ **21.** $(-3, -1)$ **23.** $\left(\frac{1}{2}, \frac{1}{3}\right)$
25. $(4, -2)$ **27.** $(-4, -9)$ **29.** $(3, 2)$ **31.** $(-5, 5)$
33. $\left(\frac{2}{3}, -\frac{1}{3}\right)$ **35.** $\left(-4, \frac{5}{4}\right)$ **37.** $(-2, 3)$ **39.** $(-4, -6)$
41. $(3, -2)$ **43.** $(-5, -1)$ **45.** $(-6, 4)$ **47.** $\left(\frac{1}{5}, 4\right)$
49. $\left(10, \frac{15}{2}\right)$ **51.** $(9, 11)$ **53.** $(-4, -1)$ **55.** $(4, 2)$
57. No solution   **59.** Infinitely many solutions   **61.** No solution
63. Infinitely many solutions   **65.** $(3, -2)$ **67.** $(1, 1)$
69. No solution   **71.** $\left(\frac{1}{3}, \frac{2}{3}\right)$ **73.** Infinitely many solutions

75. $(-10, -24)$ **77.** $\left(\frac{1}{2}, 2\right)$ **79.** $\left(-1, \frac{2}{3}\right)$ **81.** Angle of approach: 40°; angle of departure: 37°   **87.** $3^3 \cdot 7$ **89.** 5/6   **91.** 21/40

Study Set Section 4.3 (page 338)

1. opposites   **3.** $7y$ and $-7y$ **5. a.** $5a = -4$ **b.** $-4y = 1$
7. a. -2 **b.** 3   **9. a.** Multiply both sides by 15.   **b.** Multiply both sides by 10.   **11.** $2x, 1, 1, (1, 4)$ **13.** $(3, 2)$ **15.** $(3, -2)$
17. $(-2, -3)$ **19.** $(0, 8)$ **21.** $(-3, 4)$ **23.** $(0, -2)$
25. $(-12, 1)$ **27.** $(-5, 10)$ **29.** $(3, 11)$ **31.** $(-8, -15)$
33. $(-2, 7)$ **35.** $(1, 1)$ **37.** $(12, -9)$ **39.** $\left(1, -\frac{5}{2}\right)$
41. $(-4, -1)$ **43.** $(-4, 5)$ **45.** $(-2, 5)$ **47.** $(2, -1)$
49. $(6, -2)$ **51.** $(3, 0)$ **53.** $(6, 8)$ **55.** $(-4, -5)$ **57.** $(-1, 2)$
59. $(3, 2)$ **61.** $\left(\frac{7}{25}, -\frac{1}{25}\right)$ **63.** $\left(\frac{13}{75}, \frac{14}{75}\right)$ **65.** No solution
67. No solution   **69.** Infinitely many solutions   **71.** Infinitely many solutions   **73.** $(2, 3)$ **75.** $\left(\frac{3}{4}, \frac{1}{3}\right)$ **77.** $\left(\frac{1}{3}, 3\right)$
79. Infinitely many solutions   **81.** $\left(\frac{10}{3}, \frac{10}{3}\right)$ **83.** $(10, 9)$
85. $(1, -1)$ **87.** $(4, -2)$ **89.** 1991   **95.** $y = -\frac{11}{6}x - \frac{7}{3}$
97. -80

Study Set Section 4.4 (page 351)

1. complementary, supplementary   **3.** $x + y = 20, y = 2x - 1$
5. $x + y = 180, y = x - 25$ **7.** $5x + 2y = 10$ **9.** $x + c, x - c$
11. a. $(x + y)$ mL   **b.** 33%   **13.** 20°, 70°   **15.** 50°, 130°
17. 22 ft, 29 ft   **19.** President: $400,000; vice president: $208,100
21. 65°, 115°   **23.** 96 ft, 70 ft   **25.** 15 m, 10 m   **27.** Printer: $2; copier: $15   **29.** $29.50 for a 10 × 14; $21.00 for an 8 × 10
31. Elvis: 29¢; Lucy: 34¢   **33.** 85, 63   **35.** Nursing: $2,000; business: $3,000   **37.** International fund: $18,500; offshore bank: $21,500   **39.** 4% account: $11,000; biotech: $11,000   **41.** 25 mph, 5 mph   **43.** 180 mph, 20 mph   **45.** 8 gal 6% salt water, 24 gal 2% salt water   **47.** 4% solution: 48 oz; 12% solution: 80 oz   **49.** 52 lb $8.75, 48 lb $3.75   **51.** $\frac{20}{3} = 6\frac{2}{3}$ pints of mushrooms; $\frac{40}{3} = 13\frac{1}{3}$ pints of olives   **55.** $(-\infty, 4)$

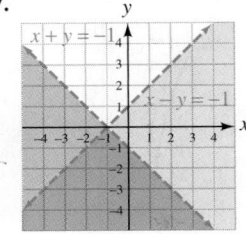

57. $(-1, 2]$

Study Set Section 4.5 (page 362)

1. inequalities   **3.** intersection   **5. a.** $3x - y = 5$ **b.** Dashed
7. Slope: $4 = \frac{4}{1}$, y-intercept: $(0, -3)$ **9. a.** No   **b.** Above
11. a. Yes   **b.** No   **c.** No   **13. a.** ii   **b.** iii   **c.** iv   **d.** i
15.

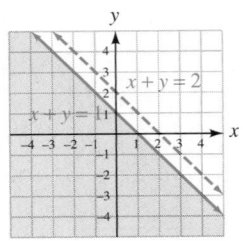

17.

19.

21.

23.

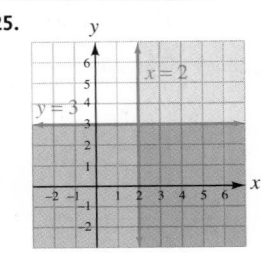

25.

27.

29.

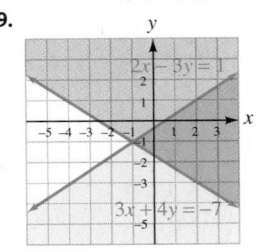

31.

33.

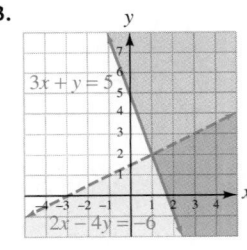

35.

37.

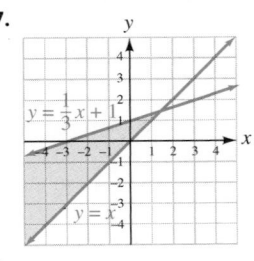

39.

41.

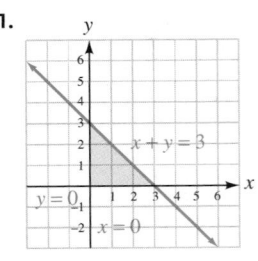

43.

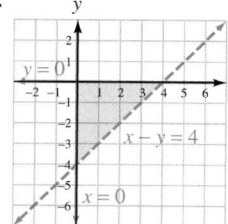

45. (c)

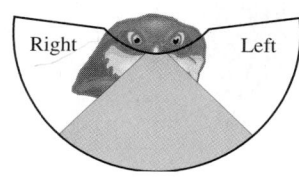

47. 1 $10 CD and 2 $15 CDs; 4 $10 CDs and 1 $15 CD

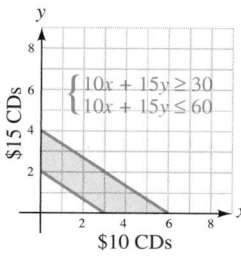

49. 2 desk chairs and 4 side chairs; 1 desk chair and 5 side chairs

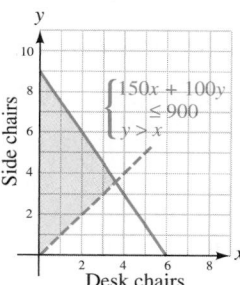

51.

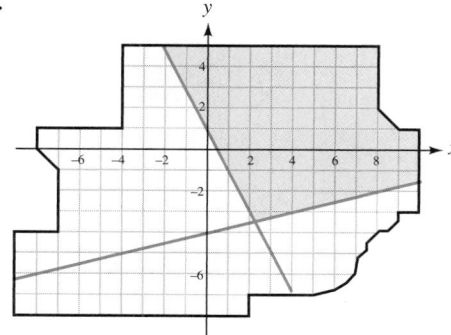

57. $6t$

59. $-2w + 4$

61. $-\frac{5}{8}x$

63. $9r - 16$

Chapter 4 Review (page 366)

1. Yes   **2.** No

3. $(4, 3)$ **4.** $(3, -1)$

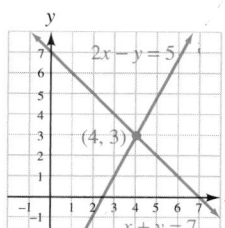
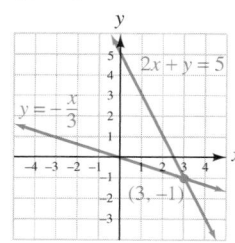

5. Infinitely many solutions   **6.** No solution

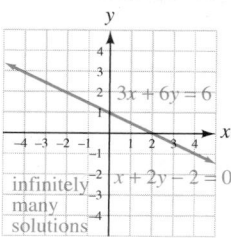
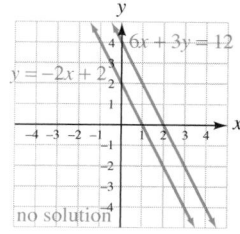

7. No solution   **8.** $(1978, 5{,}600{,}000)$; In 1978, the same number of males as females were enrolled in college, about 5.6 million of each.

9. $(5, 0)$ **10.** $(3, 3)$ **11.** $\left(-\frac{1}{2}, \frac{7}{2}\right)$ **12.** $(1, -2)$ **13.** Infinitely many solutions   **14.** $(12, 10)$ **15. a.** No solution   **b.** Two parallel lines   **c.** Inconsistent system   **16.** one

17. $\begin{cases} 4x + 2y = 7 \\ 5x - 3y = -6 \end{cases}$ **18.** one   **19.** $(3, -5)$ **20.** $\left(3, \frac{1}{2}\right)$

21. $(-1, 7)$ **22.** $(0, 9)$ **23.** Infinitely many solutions

24. $(1, -1)$ **25.** $(-5, 2)$ **26.** No solution   **27.** Elimination; no variables have a coefficient of 1 or -1.   **28.** Substitution; equation 1 is solved for x.   **29.** Las Vegas; 2,000 ft; Baltimore: 100 ft

30. Base: 21 ft; extension: 14 ft   **31.** $65°, 25°$ **32.** $10{,}800$ yd^2

33. a. $0.02x$, $0.09y$, $0.08(100)$ **b.** $5(s + w)$, $7(s - w)$

c. $0.11x$, $0.06y$ **d.** $4x$, $8y$, $10(5)$ **34.** 12 lb worms, 18 lb bears

35. 3 mph   **36.** $16.40, $10.20   **37.** $750

38. $13\frac{1}{3}$ gal 40%, $6\frac{2}{3}$ gal 70%

39.   **40.**

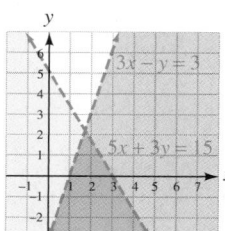
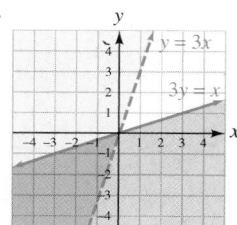

41. $10x + 20y \geq 40$, $10x + 20y \leq 60$; $(3, 1)$: 3 shirts and 1 pair of pants; $(1, 2)$: 1 shirt and 2 pairs of pants (Answers may vary)

42. a. Yes   **b.** No

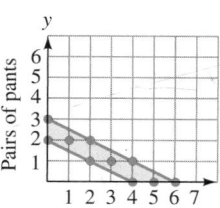

Chapter 4 Test (page 372)

1. Yes   **2.** No   **3. a.** solution   **b.** consistent   **c.** inconsistent

d. independent   **e.** dependent   **4.** Since the lines have different slopes, they will intersect at one point. The system has 1 solution.

5. (2, 3)

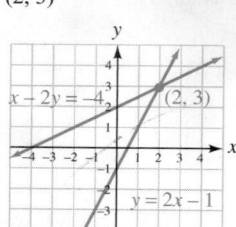

6. No solution

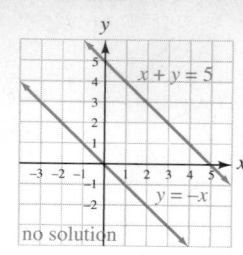

7. (30, 3,000); if 30 items are sold, the salesperson gets paid the same by both plans, $3,000.   **8.** Plan 1   **9.** $(-2, -3)$ **10.** Infinitely many solutions   **11.** (2, 4)   **12.** $(-3, 3)$ **13.** No solution   **14.** $(-1, -1)$ **15.** (5, 14)   **16.** (0, 0)   **17.** 8 mi, 14 mi   **18.** 3 adult tickets; 4 child tickets   **19.** $6,000, $4,000   **20.** 165 mph, 15 mph   **21.** Larger: 70°, smaller: 20°   **22.** 5%: 4 pints; 20%: 8 pints   **23.** $1.50 sunscreen: 3 oz; $0.80 sunscreen: 7 oz

24.

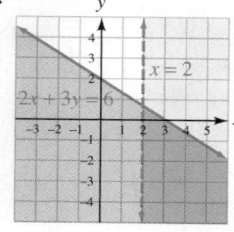

25. (1, 2), (2, 2), (3, 1) (Answers may vary)   **26.** No

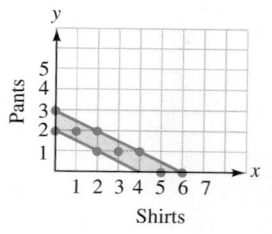

Study Set Section 5.1 (page 386)

1. exponential   **3. a.** $3x \cdot 3x \cdot 3x \cdot 3x$ **b.** $(-5y)^3$
5. a. Subtract   **b.** Add   **c.** Multiply   **d.** Multiply   **7. a.** $2x^2$
b. x^4 **9. a.** Doesn't simplify   **b.** x **11.** x^6, ^{18} **13.** Base 4;
exponent 3   **15.** Base x, exponent 5   **17.** Base $-3x$, exponent 2
19. Base y, exponent 6   **21.** Base m, exponent 12
23. Base $y + 9$, exponent 4   **25.** $(4t)^4$ **27.** $-4t^5$ **29.** $\left(\frac{t}{2}\right)^3$
31. $(x - y)^2$ **33.** 5^7 **35.** a^6 **37.** b^6 **39.** $(y - 2)^7$
41. a^5b^6 **43.** c^2d^5 **45.** a^{10} mi^2 **47.** x^9 ft^3 **49.** 8^8 **51.** x^{12}
53. $(3.7p)^5$ **55.** $(k - 2)^{14}$ **57.** c^2d^6 **59.** x^3y^4 **61.** y^4
63. a **65.** t^7 **67.** s^2 **69.** 3^8 **71.** $(-4.3)^{24}$ **73.** m^{500}
75. y^{15} **77.** x^{25} **79.** p^{25} **81.** t^{18} **83.** u^{14} **85.** $36a^2$
87. $625y^4$ **89.** $-8r^6s^9$ **91.** $-\frac{1}{243}y^{10}z^{20}$ **93.** ab^4 **95.** $r^{13}s^3$
97. $216k^3$ **99.** $9q^2$ **101.** $\frac{a^3}{b^3}$ **103.** $\frac{m^4}{81}$ **105.** $\frac{64a^4}{121b^{10}}$
107. $\frac{243m^{20}}{32n^{25}}$ **109.** $\frac{x^{10}}{y^{15}}$ **111.** y^9 **113.** 15^3 **115.** y^{15}
117. $-216a^9b^6$ **119.** $a^{21}b^{21}$ **121.** n^{33} **123.** $36h^2$
125. a. $25x^2$ ft^2 **b.** $9a^2\pi$ ft^2 **127.** $\frac{1}{8,192}$ **131.** c   **133.** d

Study Set Section 5.2 (page 396)

1. negative   **3.**

Expression	Base	Exponent
4^{-2}	4	-2
$6x^{-5}$	x	-5
$\left(\frac{3}{y}\right)^{-8}$	$\frac{3}{y}$	-8
-7^{-1}	7	-1
$(-2)^{-3}$	-2	-3
$10a^0$	a	0

5.

x	3^x
2	9
1	3
0	1
-1	$\frac{1}{3}$
-2	$\frac{1}{9}$

7. a. 3   **b.** 6   **9.** reciprocal, 2   **11.** y^8, -40, 40   **13.** 1
15. 1   **17.** 2   **19.** 15   **21.** 1   **23.** $\frac{5}{2}$ **25.** $\frac{1}{4}$ **27.** $\frac{1}{6}$ **29.** $\frac{1}{x^9}$
31. $\frac{1}{b^5}$ **33.** $-\frac{1}{5}$ **35.** $-\frac{1}{1,000}$ **37.** $\frac{1}{2}$ **39.** $\frac{8}{9}$ **41.** $\frac{15}{g^6}$ **43.** $\frac{5}{x^3}$
45. $-\frac{1}{27}$ **47.** $\frac{1}{64}$ **49.** 125   **51.** r^{20} **53.** $8s$ **55.** $\frac{3}{16}$ **57.** $\frac{b^2}{a^5}$
59. $-\frac{4p^{10}}{d}$ **61.** 36   **63.** 8   **65.** $\frac{d^8}{c^8}$ **67.** $\frac{m^4}{81}$ **69.** y^6 **71.** b^7
73. $\frac{1}{y}$ **75.** $\frac{1}{h^7}$ **77.** $\frac{1}{x^{12}}$ **79.** $\frac{1}{b^8}$ **81.** $\frac{36s^8}{t^{14}}$ **83.** $\frac{32v^{25}}{u^{10}}$ **85.** $\frac{x^9}{64}$
87. $\frac{9}{y^8}$ **89.** y^3 **91.** $\frac{1}{a^6}$ **93.** $\frac{9a^2}{2b^2}$ **95.** $\frac{64s^2}{81t^4}$ **97.** $\frac{x^{28}}{y^{20}}$ **99.** $\frac{y^{14}}{z^{10}}$
101. $\frac{8b^3}{a^{12}}$ **103.** r^{20} **105.** $\frac{125}{d^6}$ **107.** $-15y$ **109.** $\frac{h^{20}}{16}$ **111.** $\frac{1}{x^6}$
113. $\frac{c^{12}}{d^{27}}$ **115.** $\frac{9}{4g^2}$ **117.** $\frac{32x^{15}}{y^{10}}$ **119.** t^{10} **121.** $-\frac{4t^2}{s^5}$ **123.** $\frac{1}{x^3}$

125.

Item	Measurement (meter)
Thickness of a dime	10^{-3}
Height of a bathroom sink	10^0
Length of a pencil eraser	10^{-2}
Thickness of soap bubble film	10^{-5}
Length of a cell phone	10^{-1}
Thickness of a piece of paper	10^{-4}

129. $-\frac{3}{2}$ **131.** $y = \frac{3}{4}x - 5$

Study Set Section 5.3 (page 404)

1. scientific, standard   **3.** right, left   **5. a.** positive   **b.** negative
7. a. 7.7   **b.** 5.0   **c.** 8   **9. a.** $(5.1 \times 1.5)(10^9 \times 10^{22})$
b. $\frac{8.8}{2.2} \times \frac{10^{30}}{10^{19}}$ **11.** 1, 10, integer   **13.** 230   **15.** 812,000
17. 0.00115   **19.** 0.000976   **21.** 6,001,000   **23.** 2.718
25. 0.06789   **27.** 0.00002   **29.** 2.3×10^4 **31.** 1.7×10^6
33. 6.2×10^{-2} **35.** 5.1×10^{-6} **37.** 5.0×10^9
39. 3.0×10^{-7} **41.** 9.09×10^8 **43.** 3.45×10^{-2}
45. 9.0×10^0 **47.** 1.1×10^1 **49.** 1.718×10^{18}
51. 1.23×10^{-14} **53.** 7.3×10^5 **55.** 2.018×10^{17}
57. 7.3×10^{-5} **59.** 3.602×10^{-19} **61.** 7.14×10^5; 714,000
63. 4.032×10^{-3}; 0.004032   **65.** 4.0×10^{-4}; 0.0004
67. 3.0×10^4; 30,000   **69.** 4.3×10^{-3}; 0.0043
71. 3.08×10^{-2}; 0.0308   **73.** 2.0×10^5; 200,000
75. 7.5×10^{-11}; 0.000000000075   **77.** $9.038030748 \times 10^{15}$
79. $1.734152992 \times 10^{-12}$ **81.** 2.57×10^{13} mi   **83.** 197,000,000
mi^2; 109,000,000,000,000,000 mi^2; 14,600,000 mi^2
85. 4.5×10^{-10} oz   **87.** g, x, u, v, i, m, r   **89.** 1.7×10^{-18} g
91. 3.09936×10^{16} ft   **93.** 2.56×10^{11} dollars
95. 1.0×10^6, 1.0×10^9, 1.0×10^{12}, 1.0×10^{15}, 1.0×10^{18}
101. 5   **103.** $c = 30t + 45$

Study Set Section 5.4 (page 413)

1. polynomial   **3.** one, descending, two, ascending   **5.** monomial, binomial, trinomial   **7.** evaluate   **9. a.** Yes   **b.** No   **c.** No   **d.** Yes   **e.** Yes   **f.** Yes   **11.**

Term	Coefficient	Degree
$8x^2$	8	2
x	1	1
-7	-7	0
Degree of the polynomial 2		

13.

Term	Coefficient	Degree
$8a^6b^3$	8	9
$-27ab$	-27	2
Degree of the polynomial 9		

15. a. $5x^3 + 3x^2 + x - 9$ **b.** $x^2 - 2xy + y^2$ **17.** Binomial   **19.** Trinomial   **21.** Monomial   **23.** Binomial   **25.** Trinomial

27. None of these   **29.** None of these   **31.** Trinomial   **33.** 4th   **35.** 2nd   **37.** 1st   **39.** 4th   **41.** 12th   **43.** 0th   **45.** 18th   **47.** 3rd   **49. a.** 3   **b.** 13   **51. a.** -6 **b.** -8 **53. a.** 7   **b.** 34   **55. a.** -11.6 **b.** -40.2 **57. a.** 28   **b.** 4   **59. a.** 2   **b.** 0   **61.** 72   **63.** 19   **65.** -35 **67.** -257

69. 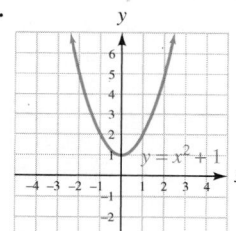   **71.**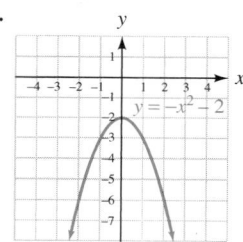

73. 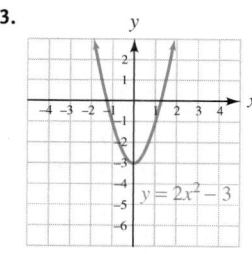   **75.**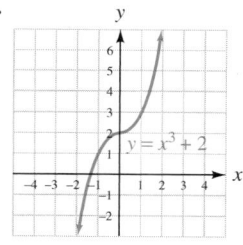

77. 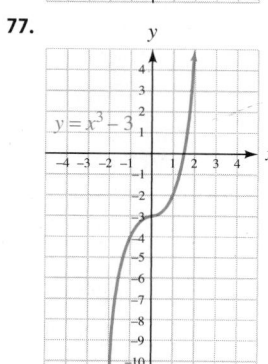   **79.**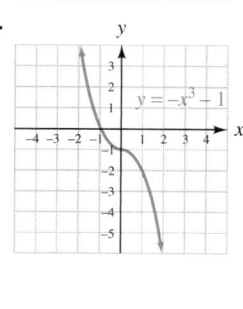

81. 91   **83.** 63 ft   **85.** About 42 million

87.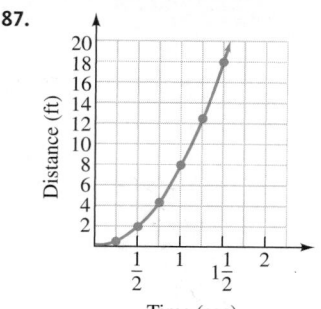

93. $[-3, \infty)$ 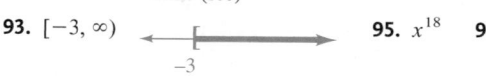   **95.** x^{18} **97.** y^9

Study Set Section 5.5 (page 423)

1. polynomials   **3.** Like   **5.** combine   **7. a.** $5x^2$ **b.** $14m^3$ **c.** $7a^3b$ **d.** $6cd + 4c^2d$ **9. a.** $-5x^2 + 8x - 23$ **b.** $5y^4 - 3y^2 + 7$ **11.** $4x^2, 2x, 1, 10x^2, 4$ **13.** $12t^2$ **15.** $-48u^3$ **17.** $20x^2 - 19x$ **19.** $10r^4 - 4r$ **21.** $x^2 + x$ **23.** $\frac{13}{15}x^2 - \frac{1}{8}x$ **25.** $1.3x^3$ **27.** $2st$ **29.** $\frac{7}{12}c^2 - \frac{1}{2}cd + d^2$ **31.** $-ab$ **33.** $-4x^3y + x^2y + 5$ **35.** $-4c^2 - 12cd$ **37.** $7x + 4$ **39.** $5d^2 + 14d$ **41.** $5q^2 - 4q - 5$ **43.** $y^3 + \frac{19}{20}y^2 + \frac{1}{3}$ **45.** $0.7p - 0.9q$ **47.** $7x^2 + xy + 2y^2$ **49.** $(3x^2 + 6x - 2)$ yd   **51.** $(7x^2 + 5x + 6)$ mi   **53.** $5x^2 + x + 11$ **55.** $-3a^2 + 7a + 7$ **57.** $10z^3 + z - 2$ **59.** $-x^3y^2 + 4x^2y + 5x + 6$ **61.** $2a^2 + a - 3$ **63.** $13a^2 + a$ **65.** $-5h^3 + 5h^2 + 30$ **67.** $\frac{1}{24}s^8 - \frac{19}{20}s^7$ **69.** $-0.14f^2 + 0.25f + 2.09$ **71.** $b^2 + 4ab - 2$ **73.** $x^2 + 6x + 2$ **75.** $-x^3 + 6x^2 + x + 14$ **77.** $0.6x^3 + 1.2x^2 + 1.3x - 0.3$ **79.** $7x^3y^2 - 2x^2y - 2x + 15$ **81.** $4s^2 - 5s + 7$ **83.** $3y^5 - 6y^4 + 1.2$ **85.** $-3x^2 + 5x - 7$ **87.** $t^3 + 3t^2 + 6t - 5$ **89.** $3x + 1$ **91.** $-1.3t^2 + 0.7t + 0.6$ **93.** $3x^2 - 9x - 10$ **95.** $6x^2 + x - 5$ **97.** $\frac{5}{4}r^4 + \frac{11}{9}r^2 - 2$ **99.** $9c^2 - 6c - 14$ **101.** $-3s^2t - 8st + 14$ **103.** $4.8h^3 + 8.8h^2$ **105. a.** $(x^2 - 8x + 12)$ ft   **b.** $(x^2 + 2x - 8)$ ft   **107.** $(2a^2 + 6a + 5)$ in.   **109. a.** $(22t + 20)$ ft   **b.** 108 ft   **117.** $180°$ **119.**

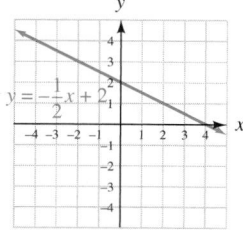

Study Set Section 5.6 (page 434)

1. monomials   **3.** first, outer, inner, last   **5. a.** each, each   **b.** any, third   **7. a.** $6x^2 + x - 12$ **b.** $5x^4 + 8ax^2 + 3a^2$ **9.** $8, n^3, 72n^5$ **11.** $2x, 5, 5, 4x, 15x, 11x$ **13.** $5m^2$ **15.** $12x^5$ **17.** $6c^6$ **19.** $-24b^6$ **21.** $8x^5y^5$ **23.** $-2a^{11}$ **25.** $3x^2 + 12x$ **27.** $-4t^3 + 28t$ **29.** $9x^4 - 18x^3 + 54x^2$ **31.** $-6x^5 + 2x^4 - 2x^3$ **33.** $0.12p^9 - 1.8p^7$ **35.** $\frac{5}{8}t^8 + 5t^4$ **37.** $-12x^4z - 4x^2z^3 - 4x^3z^2 + 4x^2z$ **39.** $6x^{14} - 72x^{13}$ **41.** $(7h^2 + 3h)$ in.^{2} **43.** $(4w^2 - 2w)$ ft^2 **45.** $y^2 + 8y + 15$

47. $t^2 + t - 12$ **49.** $m^2 - 3m - 54$ **51.** $4y^2 + 23y - 35$
53. $12x^2 - 28x + 15$ **55.** $7.6y^2 - 5.8y + 1$
57. $18m^2 - 10m + \frac{8}{9}$ **59.** $t^4 - 7t^2 + 12$ **61.** $a^2 + 2ab + b^2$
63. $12a^2 - 5ab - 2b^2$ **65.** $x^3 - x + 6$
67. $4t^3 + 11t^2 + 18t + 9$ **69.** $x^3 - 8$
71. $2x^3 + 7x^2 - 16x - 35$ **73.** $-3x^3 + 25x^2y - 56xy^2 + 16y^3$
75. $r^4 - 5r^3 + 2r^2 - 7r - 15$ **77.** $x^3 - 3x + 2$
79. $12x^3 + 17x^2 - 6x - 8$ **81.** $8x^3 - 12x^2 - 8x$
83. $-3a^3 + 3ab^2$ **85.** $18a^6 - 12a^5$ **87.** $x^3 - 6x^2 + 5x + 12$
89. $30x^2 - 17x + 2$ **91.** $6x^4 + 8x^3 - 14x^2$ **93.** $-18x^3z^8$
95. $6a^4 + 5a^3 + 5a^2 + 10a + 4$ **97.** $9t^2 + 15st - 6s^2$ **99.** $2a^{10}$
101. $4y^3 + 40y^2 + 84y$ **103.** $16.4p^2q^2 - 24.6p^2q + 41pq^2$
105. $x^4 + 11x^3 + 26x^2 - 28x - 24$ **107.** $16a^2 - 2ar - \frac{15}{16}r^2$
109. $(6x^2 + x - 1)\ \text{cm}^2$ **111.** $(0.785x^2 - 0.785)\ \text{in.}^2$
113. $(2x^3 - 4x^2 - 6x)\ \text{in.}^3$ **121. a.** 1   **b.** Undefined
123. $(0, 2)$

Study Set Section 5.7 (page 443)

1. products   **3. a.** square, Twice, first   **b.** second, square
5. $x, 4, 4, 8x$ **7.** $s, 5, 25$ **9.** $x^2 + 2x + 1$ **11.** $r^2 + 4r + 4$
13. $m^2 - 12m + 36$ **15.** $f^2 - 16f + 64$
17. $16x^2 + 40x + 25$ **19.** $49m^2 - 28m + 4$ **21.** $1 - 6y + 9y^2$
23. $y^2 + 1.8y + 0.81$ **25.** $a^4 + 2a^2b^2 + b^4$ **27.** $r^4 - 2r^2s^2 + s^4$
29. $s^2 + \frac{3}{2}s + \frac{9}{16}$ **31.** $d^8 + \frac{1}{2}d^4 + \frac{1}{16}$ **33.** $x^2 - 9$
35. $d^2 - 49$ **37.** $4p^2 - 49$ **39.** $9n^2 - 1$ **41.** $c^2 - \frac{9}{16}$
43. $36b^2 - \frac{1}{4}$ **45.** $0.16 - 81m^4$ **47.** $25 - 36g^2$
49. $x^3 + 12x^2 + 48x + 64$ **51.** $n^3 - 18n^2 + 108n - 216$
53. $8g^3 - 36g^2 + 54g - 27$ **55.** $a^3 + 3a^2b + 3ab^2 + b^3$
57. $8m^3 + 12m^2n + 6mn^2 + n^3$ **59.** $n^4 - 8n^3 + 24n^2 - 32n + 16$
61. $-x^2 + 20x - 8$ **63.** $3t^2 + 12t - 9$ **65.** $2x^2 + xy - y^2$
67. $24a^2 - 10a + 65$ **69.** $-80d^3 + 40d^2 - 5d$
71. $4d^5 - 4dg^6$ **73.** $(2x^2 - 2)\ \text{yd}^2$ **75.** $(9x^2 + 6x + 1)\ \text{ft}^2$
77. $4v^6 - 32v^3 + 64$ **79.** $12x^3 + 36x^2 + 27x$ **81.** $16f^2 - 0.16$
83. $r^4 + 20r^2s + 100s^2$ **85.** $6x - 2$ **87.** $4a^2 - 12ab + 9b^2$
89. $n^2 - 36$ **91.** $36m + 36$ **93.** $25m^2 - 12m + \frac{36}{25}$
95. $8e^3 + 12e^2 + 6e + 1$ **97.** $x^2 - 4x + 4$ **99.** $13x^2 - 8x + 5$
101. $36 - 24d^3 + 4d^6$ **103.** $64x^2 + 48x + 9$
105. $(x^2 + 12x + 36)\ \text{in.}^2$ **107.** $\pi hR^2 - \pi hr^2$ **113.** $3^3 \cdot 7$
115. $\frac{5}{6}$ **117.** $\frac{21}{40}$

Study Set Section 5.8 (page 453)

1. monomial   **3.** binomial   **5.** Divide, multiply, subtract, bring
down   **7.** quotient, dividend   **9.** $7x^2, x^3, 7x^2, 5, 2, 7, 2, 2, 4x^3, \frac{5}{7}$
11. $5x^4 + 0x^3 + 2x^2 + 0x - 1$ **13.** x^3 **15.** $5m^5$ **17.** $\frac{4h^2}{3}$
19. $-\frac{1}{5d^4}$ **21.** $\frac{10}{s}$ **23.** $\frac{x^2}{5y^4}$ **25.** $\frac{4r}{y^5}$ **27.** $-\frac{13}{3rs^3}$ **29.** $2x + 3$
31. $9 - \frac{6}{m}$ **33.** $\frac{1}{a^3} - \frac{1}{a} + 1$ **35.** $2x^5 - 8x^2$ **37.** $\frac{h^2}{4} + \frac{2}{h}$
39. $-2w^2 - \frac{1}{w^4}$ **41.** $3s^5 - 6s^2 + 4s$ **43.** $c^3 + 3e^2 - 2c - \frac{5}{c}$
45. $3x^2y - 2x - \frac{1}{y}$ **47.** $5y - \frac{6}{x} + \frac{1}{xy}$ **49.** $x + 6$ **51.** $x - 2$
53. $3a - 2$ **55.** $b + 3$ **57.** $x + 1 + \frac{10}{x+5}$
59. $a - 12 + \frac{4}{a-5}$ **61.** $x + 1 + \frac{-1}{2x+3}$ **63.** $2x - 3 + \frac{-1}{3x-1}$
65. $2x - 1$ **67.** $2x + 1$ **69.** $x + 3$ **71.** $x^2 - 2x + 1$
73. $a - 5$ **75.** $x + 1$ **77.** $2x - 3$ **79.** $9b + 7$

81. $x^2 - x + 1$ **83.** $y^2 + 2y + 5 + \frac{10}{y-2}$ **85.** $y + 12 + \frac{1}{y+1}$
87. $3a^5 - \frac{2b^3}{a}$ **89.** $2x^2 + 2x + 1$ **91.** $\frac{x}{5} - \frac{2}{5x^2}$ **93.** $a^2 + a + 1$
95. $2x^2 + x + 1 + \frac{2}{3x-1}$ **97.** $\frac{x^2}{2y^{10}}$ **99.** $3m - 8$
101. $(x - 6)\ \text{in.}$ **103.** $(2x^2 - x + 3)\ \text{in.}$ **109.** $y = -\frac{11}{6}x - \frac{7}{3}$
111. -80

Chapter 5 Review (page 456)

1. a. Base n, exponent 12   **b.** Base $2x$, exponent 6   **c.** Base r,
exponent 4   **d.** Base $y - 7$, exponent 3   **2. a.** m^5 **b.** $-3x^4$
c. $(x + 8)^2$ **d.** $\left(\frac{1}{2}pq\right)^3$ **3.** 7^{12} **4.** m^2n^2 **5.** y^{21} **6.** $81x^4$
7. b^9 **8.** $-b^{12}$ **9.** $256s^{10}$ **10.** $4.41x^4y^2$ **11.** $(-9)^{15}$
12. a^{23} **13.** $\frac{1}{8}x^{15}$ **14.** $\frac{x^{12}}{9y^2}$ **15.** $(m - 25)^{12}$ **16.** $125yz^4$
17. a^{11} **18.** c^5d^5 **19.** $64x^{12}\ \text{in.}^3$ **20.** $y^4\ \text{ft}^2$ **21.** 1   **22.** 1
23. 3   **24.** $\frac{1}{1,000}$ **25.** $-\frac{1}{25}$ **26.** $\frac{1}{t^6}$ **27.** $8x^5$ **28.** $-\frac{6}{y}$
29. $\frac{8}{49}$ **30.** x^{14} **31.** $-\frac{27}{r^9}$ **32.** $\frac{1}{16z^2}$ **33.** $\frac{8c}{9d^5}$ **34.** t^{30}
35. w^{22} **36.** $\frac{f^{40}}{4^{10}}$ **37.** 7.2×10^8 **38.** 9.37×10^{15}
39. 9.42×10^{-9} **40.** 1.3×10^{-4} **41.** 1.8×10^{-4}
42. 8.53×10^5 **43.** 126,000   **44.** 0.00000003919   **45.** 2.68
46. 57.6   **47.** 3.0×10^{-4}; 0.0003   **48.** 1.6×10^8; 160,000,000
49. 6,570,000,000; 6.57×10^9 **50.** $1.0 \times 10^5 = 100,000$
51. a. 4   **b.** $3x^3$ **c.** 3, -1, 1, 10   **d.** 10   **52. a.** 7th,
monomial   **b.** 3rd, monomial   **c.** 2nd, binomial   **d.** 5th,
trinomial   **e.** 6th, binomial   **f.** 4th, none of these   **53.** 3, -13
54. 8 in.

55.   **56.**

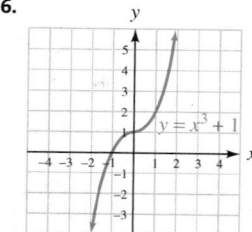

57. $13y^3$ **58.** $-4a^3b + a^2b + 6$ **59.** $\frac{7}{12}x^2 - \frac{3}{4}xy + y^2$
60. $6.7c^5 + 8.1c^4 - 2.1c^3$ **61.** $25r^6 + 9r^3 + 5r$
62. $3.7a^2 + 6.1a - 17.6$ **63.** $4r^3s - 7r^2s^2 - 7rs^3 - 2s^4$
64. $\frac{5}{8}m^4 - m^3$ **65.** $-z^3 + 2z^2 + 5z - 17$ **66.** $(x^2 + x + 3)\ \text{in.}$
67. $4x^2 + 2x + 8$ **68.** $8x^3 - 7x^2 + 19x$ **69.** $10x^3$
70. $-6x^{10}z^5$ **71.** $120b^{11}$ **72.** $2h^{14} + 8h^{11}$
73. $9n^4 - 15n^3 + 6n^2$ **74.** $x^2y^3 - x^3y^2$ **75.** $6x^6 + 12x^5$
76. $a^6b^4 - a^5b^5 + a^3b^6 - 7a^3b^2$ **77.** $x^2 + 5x + 6$
78. $2x^2 - x - 1$ **79.** $6t^2 - 6$ **80.** $6n^8 - 13n^6 + 5n^4$
81. $-5a^9 + 4a^7b + a^5b^2$ **82.** $6.6a^2 - 6.6$ **83.** $18t^2 + 3t - \frac{5}{9}$
84. $24b^2 - 34b + 11$ **85.** $8a^3 - 27$
86. $56x^4 + 15x^3 - 21x^2 - 3x + 2$ **87.** $8x^3 + 1$
88. a. $(6x + 10)\ \text{in.}$ **b.** $(2x^2 + 11x - 6)\ \text{in.}^2$
c. $(6x^3 + 33x^2 - 18x)\ \text{in.}^3$ **89.** $a^2 - 6a + 9$
90. $m^3 + 6m^2 + 12m + 8$ **91.** $x^2 - 49$ **92.** $4x^2 - 0.81$
93. $4y^2 + 4y + 1$ **94.** $y^4 - 1$ **95.** $36r^4 + 120r^2s + 100s^2$
96. $-64a^2 + 48ac - 9c^2$ **97.** $80r^4s - 80s^5$
98. $36b^3 - 96b^2 + 64b$ **99.** $t^2 - \frac{3}{2}t + \frac{9}{16}$ **100.** $x^2 + \frac{8}{3}x + \frac{16}{9}$
101. $5x^2 + 19x + 3$ **102.** $24c^2 - 10c + 37$ **103.** $(x^2 - 4)\ \text{in.}^2$

104. $(50x^2 - 8)$ in.^{2} **105.** $2n^3$ **106.** $-\frac{2x}{3y^2}$ **107.** $\frac{a^3}{6} - \frac{4}{a^4}$

108. $3a^3 + \frac{b}{5a} - \frac{5}{a^2}$ **109.** $x - 5$ **110.** $2x + 1$

111. $5x - 6 + \frac{4}{3x + 2}$ **112.** $5y - 3$ **113.** $3x^2 - x - 4$

114. $3x^2 + 2x + 1 + \frac{2}{2x - 1}$

115. $(y + 3)(3y + 2) = 3y^2 + 11y + 6$ **116.** $(2x^2 + 3x - 4)$ in.

Chapter 5 Test (page 464)

1. a. base, exponent **b.** monomial, binomial, trinomial **c.** degree
d. special **2.** $2x^3y^4$ **3.** y^6 **4.** $\frac{1}{32}x^{21}$ **5.** 3.5 **6.** $\frac{2}{y^3}$

7. $\frac{1}{125}$ **8.** $(x + 1)^9$ **9.** y^{21} **10.** $\frac{b^3}{64a^3}$ **11.** $\frac{m^{12}}{64}$ **12.** $-6ab^9$

13. $1{,}000y^{12}$ in.^{3} **14.** 6.25×10^{18} **15.** 0.000093

16. 9.2×10^3; 9,200 **17.** Trinomial

Term	Coefficient	Degree
x^4	1	4
$8x^2$	8	2
-12	-12	0
Degree of the polynomial 4		

18. 5th degree **19.**

x	-2	-1	0	1	2
y	6	3	2	3	6

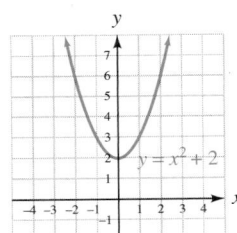

$y = x^2 + 2$

20. 0 ft; the rock hits the canyon floor 18 seconds after being dropped.

21. $\frac{1}{10}x^2 + \frac{7}{12}x - 2$ **22.** $-4a^3b + a^2b + 5$

23. $19.4h^3 - 11.1h^2 - 0.6$ **24.** $6b^3c - 2bc - 12$

25. $-3y^3 + 18y^2 - 17y + 35$ **26.** $(10a^2 + 8a - 20)$ in.

27. $10x^5y^{11}$ **28.** $-72b^8$ **29.** $3y^4 - 6y^3 + 9y^2$

30. $3x^2 - 11x - 20$ **31.** $12t^2 - 8t - \frac{3}{4}$

32. $2x^3 - 7x^2 + 14x - 12$ **33.** $1 - 100c^2$

34. $49b^6 - 42b^3t + 9t^2$ **35.** $2.2a^3 + 4.4a^2 - 33a$

36. $2x^2 + 2xy$ **37.** $\frac{a}{4b} - \frac{b}{2a}$ **38.** $x - 2$

39. $3x^2 + 2x + 1 + \frac{2}{2x - 1}$ **40.** $(x - 5)$ ft

41. Yes; $(5m + 1)(m - 6) = 5m^2 - 29m - 6$

42. No; $(a + b)^2 = a^2 + 2ab + b^2$

Cumulative Review Chapters 1–5 (page 466)

1. $2 \cdot 3^3 \cdot 5$ **2. a.** $a + b = b + a$ **b.** $(xy)z = x(yz)$ **3.** -37
4. 28 **5.** $18x$ **6.** 0 **7.** -2 **8.** 15 **9.** $\$2.079$ billion
10. 1.2 ft^3 **11.** 30 **12.** $\$6{,}250$ **13.** Mutual fund: $\$25{,}000$;
bonds: $\$20{,}000$ **14.**

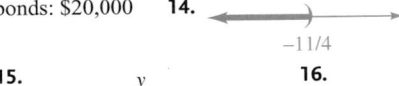

$-11/4$

15.

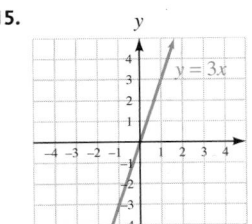

$y = 3x$

16.

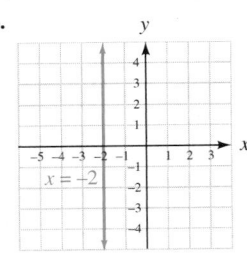

$x = -2$

17. $-\frac{4}{9}$ **18.** $m = 3, (0, -2); y = 3x - 2$ **19.** Perpendicular
20. $y = -4x + 2$ **21.** No **22.** 26 **23.** No
24. $(4, 1)$

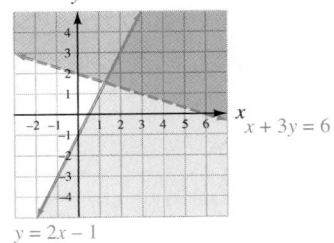

$3x + 2y = 14$

$y = \frac{1}{4}x$ $(4, 1)$

25. $(-4, 3)$ **26.** $(-2, 4)$
27. Adult: $\$61$; child: $\$51$

28.

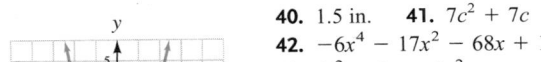

$x + 3y = 6$

$y = 2x - 1$

29. $9x^4y^8$ **30.** v^{22}

31. $a^2b^7c^6$ **32.** $\frac{64t^{12}}{27}$ **33.** $\frac{1}{16y^4}$ **34.** a^7 **35.** $-\frac{1}{25}$ **36.** $\frac{x^{10}}{a^{10}}$
37. 6.15×10^5 **38.** 1.3×10^{-6}
39.

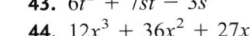

$y = x^2$

40. 1.5 in. **41.** $7c^2 + 7c$
42. $-6x^4 - 17x^2 - 68x + 11$
43. $6t^2 + 7st - 3s^2$
44. $12x^3 + 36x^2 + 27x$
45. $2x + 1$ **46.** $\frac{1}{8} - \frac{2}{x}$

Study Set Section 6.1 (page 478)

1. factor **3.** grouping **5. a.** 3 **b.** $7, h$ **c.** $3, y, y$
7. a. $2x + 4$ (Answers may vary) **b.** $x^3 + x^2 + x$ (Answers may
vary) **9.** $2x, x + 3$ **11. a.** 4 **b.** No **c.** $2; h$ **13.** 8
15. $b^2, 2, b - 6$ **17.** 2 **19.** 6 **21.** 7 **23.** 8 **25.** m^3
27. 5 **29.** $4c$ **31.** $9a^3$ **33.** $8a$ **35.** $3m^3n$ **37.** $x + 7$
39. $p - t$ **41.** $3(x + 2)$ **43.** $6(3x + 4)$ **45.** $9(2m - 1)$
47. $d(d - 7)$ **49.** $5(3c^3 + 5)$ **51.** $8a(3 - 2a)$
53. $7(2x^2 - x - 1)$ **55.** $t^2(t^2 + t + 2)$ **57.** $a(b + c - d)$
59. $3xy^2(7xy + 1)$ **61.** $-(a + b)$ **63.** $-(2x - 5)$
65. $-(3r - 2s + 3)$ **67.** $-(x^2 + x - 16)$ **69.** $-(-5 + x)$ or
$-(x - 5)$ **71.** $-(-9 + 4a)$ or $-(4a - 9)$ **73.** $-3x(x + 2)$
75. $-4a^2(b - 3a)$ **77.** $-12x^2(2x^2 + 4x - 3)$
79. $-2ab^2(2a^2 - 7a + 5)$ **81.** $(x + 2)(y + 3)$
83. $(p - q)(m - 5)$ **85.** $(x + y)(2 + a)$ **87.** $(s - u)(r + 8w)$
89. $(7m - 2)(m^2 + 2)$ **91.** $(5x - 1)(x^2 + 2)$ **93.** $(b + c)(a + 1)$
95. $(r + 4s)(s - 1)$ **97.** $(2x - 3)(a + b)$ **99.** $(m - n)(p - q)$
101. $a(x - 2)(x^2 + 5)$ **103.** $6(x^2 + 2)(x - 1)$
105. $(14 + r)(h^2 + 5)$ **107.** $11a^2(2a - 3)$ **109.** $(a + b)(x - 1)$
111. $3r^5(5r^3 - 6r - 10)$ **113.** $3(3p + q)(3m - n)$
115. $-20pt^2(3p + 4t)$ **117.** $(3x - y)(2x - 5)$
119. $2z(x - 2)(x^2 + 16)$ **121.** $6uvw^2(2w - 9v)$
123. $(x + 1)(x^2 + 1)$ **125.** $(x^2 + 5)$ ft; $(x + 4)$ ft **131.** 12%

Study Set Section 6.2 (page 491)

1. factors **3.** leading **5. a.** descending **b.** common **7.** 3, 5
9. a. No **b.** Yes **11. a.** They are both positive or both negative.
b. One will be positive, the other negative. **13.** $+3, -2$
15. $(x + 2)(x + 1)$ **17.** $(z + 4)(z + 3)$ **19.** $(m - 3)(m - 2)$
21. $(t - 7)(t - 4)$ **23.** $(r - 3)(r - 6)$ **25.** $(a - 45)(a - 1)$
27. $(x + 8)(x - 3)$ **29.** $(t - 3)(t + 16)$ **31.** $(a - 8)(a + 2)$
33. $(b - 12)(b + 3)$ **35.** $-(x + 5)(x + 2)$ **37.** $-(t + 6)(t - 5)$
39. $-(r + 9)(r - 6)$ **41.** $-(m - 7)(m - 11)$
43. $(a + 3b)(a + b)$ **45.** $(x - 7y)(x + y)$ **47.** $(r + 2s)(r - s)$
49. $(a - 3b)(a - 2b)$ **51.** $2(x + 3)(x + 2)$ **53.** $6(a - 4)(a - 1)$
55. $5(a - 3)(a - 2)$ **57.** $-z(z - 4)(z - 25)$
59. $-n^2(n - 30)(n + 2)$ **61.** $4x^2(x + 2)(x + 2) = 4x^2(x + 2)^2$
63. $(x - 4)(x - 20)$ **65.** $(y + 9)(y + 1)$ **67.** $(r - 2)(r + 8)$
69. $(r + 3x)(r + x)$ **71.** Prime **73.** Prime **75.** $(x + 3)(5 + y)$
77. $2n(13n - 4)$ **79.** $(a - 5)(a + 1)$ **81.** $-(x - 22)(x + 1)$
83. $4(y - 1)(x + 7)$ **85.** $12b^2(2b^2 - 4b + 3)$
87. $(x + 2y)(x + 2y) = (x + 2y)^2$ **89.** $(a - 6b)(a + 2b)$
91. Prime **93.** $(x + 2)(t + 7)$ **95.** $s^2(s + 13)(s - 2)$
97. $15(s^3 + 5)$ **99.** $(y - 14)(y + 1)$ **101.** $2(x - 2)(x - 4)$
103. $(x + 9)$ in., x in., $(x + 3)$ in. **111.** $\frac{1}{x^2}$ **113.** $\frac{1}{x^{10}}$

Study Set Section 6.3 (page 502)

1. leading **3.** $5y, y, 1, 3$ **5.** $10x$ and x, $5x$ and $2x$
7. a. descending, GCF, coefficient **b.** $3s^2$ **c.** $-(2d^2 - 19d + 8)$
9. negative **11.** different **13.** $-13; -6, -8; -3, -7$
15. a. 12, 20, -9 **b.** -108 **17.** $3t, 2, 4t + 3$
19. $(2x + 1)(x + 1)$ **21.** $(3a + 1)(a + 3)$ **23.** $(5x + 2)(x + 1)$
25. $(7x + 11)(x + 1)$ **27.** $(2x - 3)(2x - 1)$
29. $(4x - 1)(2x - 5)$ **31.** $(5t - 7)(3t - 1)$ **33.** $(6y - 1)(y - 2)$
35. $(3x + 7)(x - 3)$ **37.** $(5m + 3)(m - 2)$ **39.** $(7y - 1)(y + 8)$
41. $(11y - 4)(y + 1)$ **43.** $(3r + 2s)(2r - s)$
45. $(2x + 3y)(2x + y)$ **47.** $(8m + 3n)(m + 11n)$
49. $(5x + 3y)(3x - 2y)$ **51.** $2(3x + 2)(x - 5)$
53. $a(2a - 5)(4a - 3)$ **55.** $(2u + 3v)(u - 2v)$
57. $4(9y - 4)(y - 2)$ **59.** $10(13r - 11)(r + 1)$
61. $-y(y + 12)(y + 1)$ **63.** $-3x^2(2x + 1)(x - 3)$
65. $2mn(4m + 3n)(2m + n)$ **67.** $(2t - 5)(3t + 4)$
69. $(3p - q)(5p + q)$ **71.** $(2t - 1)(2t - 7)$
73. $(4y + 1)(2y - 1)$ **75.** $(18x - 5)(x + 2)$ **77.** Prime
79. Prime **81.** $3r^3(5r - 2)(2r + 5)$ **83.** $(3p - q)(2p + q)$
85. $-(4y - 3)(3y - 4)$ **87.** $(m + 7)(m - 4)$ **89.** $3a^2(2a + 5)$
91. $(x - 2)(x^2 + 5)$ **93.** $(5y - 3)(y - 1)$ **95.** $-2(x + 2)(x + 3)$
97. $3x^2y^2(4xy - 6y + 5)$ **99.** $(a - 5b)(a - 2b)$
101. $u^4(9u + 1)(u - 8)$ **103.** $(2x + 11)$ in., $(2x - 1)$ in.
109. -49 **111.** 1 **113.** 49

Study Set Section 6.4 (page 510)

1. perfect **3. a.** $5x$ **b.** 3 **c.** $5x, 3$ **5. a.** x, y **b.** $-$
c. $+, x, y$ **7.** 1, 4, 9, 16, 25, 36, 49, 64, 81, 100, 121, 144, 169, 196,
225, 256, 289, 324, 361, 400 **9.** 2 **11.** $+, -$ **13.** Yes
15. No **17.** No **19.** Yes **21.** $(x + 3)^2$ **23.** $(b + 1)^2$
25. $(c - 6)^2$ **27.** $(3y - 4)^2$ **29.** $(2x + 3)^2$ **31.** $(6m + 5n)^2$
33. $(9x - 4y)^2$ **35.** $(7t - 2s)^2$ **37.** $3(u - 3)^2$ **39.** $x(6x + 1)^2$
41. $2a^3(3a + 7b)^2$ **43.** $-(10t - 1)^2$ **45.** $(x + 2)(x - 2)$
47. $(x + 4)(x - 4)$ **49.** $(6 + y)(6 - y)$ **51.** $(t + 5)(t - 5)$
53. Prime **55.** Prime **57.** $(5t + 8)(5t - 8)$

59. $(9y + 1)(9y - 1)$ **61.** $(3x^2 + y)(3x^2 - y)$
63. $(4c + 7d^2)(4c - 7d^2)$ **65.** $8(x + 2y)(x - 2y)$
67. $7(3a + 1)(3a - 1)$ **69.** $x(x + 12)(x - 12)$
71. $6x^2(x + y)(x - y)$ **73.** $(9 + s^2)(3 + s)(3 - s)$
75. $(b^2 + 16)(b + 4)(b - 4)$ **77.** $16(t^2 + s^2)(t + s)(t - s)$
79. $25(m^2 + 1)(m + 1)(m - 1)$ **81.** $(a^2 + 12b)(a^2 - 12b)$
83. $(3xy + 5)^2$ **85.** $(t - 10)^2$ **87.** $(z + 8)(z - 8)$
89. $3(m^2 + n^2)(m + n)(m - n)$ **91.** $(5m + 7)^2$
93. $(x + 7)(x - 6)$ **95.** $(x + 3)(x - 3)$ **97.** $8a^2b(3a - 2)$
99. $-2(r - 10)(r - 4)$ **101.** $(x + 3)(x^2 + 4)$ **103.** $(2b - 5)^2$
105. $(p + q)^2$ **107.** $0.5g(t_1 + t_2)(t_1 - t_2)$ **113.** $\frac{x}{y} + \frac{2y}{x} - 3$
115. $3a + 2$

Study Set Section 6.5 (page 515)

1. sum., cubes **3. a.** F, L **b.** $-, F^2, L^2$ **5.** $6n, 5$ **7.** 1, 8, 27,
64, 125, 216, 343, 512, 729, 1,000 **9.** No **11.** $2a$ **13.** $b + 3$
15. a. $x^3 + 8$ (Answers may vary.) **b.** $(x + 8)^3$
17. $(y + 5)(y^2 - 5y + 25)$ **19.** $(a + 4)(a^2 - 4a + 16)$
21. $(n + 8)(n^2 - 8n + 64)$ **23.** $(2 + t)(4 - 2t + t^2)$
25. $(a + 10b)(a^2 - 10ab + 100b^2)$
27. $(5c + 3d)(25c^2 - 15cd + 9d^2)$ **29.** $(a - 3)(a^2 + 3a + 9)$
31. $(m - 7)(m^2 + 7m + 49)$ **33.** $(6 - v)(36 + 6v + v^2)$
35. $(2s - t)(4s^2 + 2st + t^2)$ **37.** $(10a - w)(100a^2 + 10aw + w^2)$
39. $(4x - 3y)(16x^2 + 12xy + 9y^2)$ **41.** $2(x + 1)(x^2 - x + 1)$
43. $3(d + 3)(d^2 - 3d + 9)$ **45.** $x(x - 6)(x^2 + 6x + 36)$
47. $8x(2m - n)(4m^2 + 2mn + n^2)$ **49.** $(x + 4)^2$
51. $(3r + 4s)(3r - 4s)$ **53.** $(x - t)(y + s)$
55. $4(p + 2q)(p^2 - 2pq + 4q^2)$ **57.** $2ct^2(4c + 3t)(2c + t)$
59. $36(e^2 + 1)(e + 1)(e - 1)$ **61.** $7a^2b^2(5a - 2b + 2ab)$
63. $(6r + 5s)^2$ **65.** $(1,000 - x^3)$ in.^{3}; $(10 - x)(100 + 10x + x^2)$
69. Repeating **71.** $\{\ldots, -3, -2, -1, 0, 1, 2, 3, \ldots\}$ **73.** 0

Study Set Section 6.6 (page 521)

1. product **3.** Factor out the GCF **5.** Perfect-square trinomial
7. Sum of two cubes **9.** Trinomial factoring **11.** Is there a
common factor? **13.** $14m, m$ **15.** $2(b + 6)(b - 2)$
17. $4p^2q^3(2pq^4 + 1)$ **19.** $2(2y + 1)(10y + 1)$
21. $8(x^2 + 1)(x + 1)(x - 1)$ **23.** $(c + 21)(c - 7)$ **25.** Prime
27. $-2x^2(x - 4)(x^2 + 4x + 16)$ **29.** $(c + d^2)(a^2 + b)$
31. $-(3xy - 1)^2$ **33.** $-5m(2m + 5)^2$ **35.** $(2c + d)(c - 3d)$
37. $(p - 2)^2(p^2 + 2p + 4)$ **39.** $(x - a)(a + b)(a - b)$
41. $(ab + 12)(ab - 12)$ **43.** $(x + 5)(2x^2 + 1)$
45. $v^2(v^2 - 14v + 8)$ **47.** $(x + 2)(x - 2)(x + 3)(x - 3)$
49. $2x(2ax + b)(2ax - b)$ **51.** $2(3x - 4)(x - 1)$
53. $y^2(2x + 1)^2$ **55.** $4m^2(m + 5)(m^2 - 5m + 25)$
57. $(x + 2)(x - 2)(x^2 + 2)$ **59.** Prime
61. $2a^2(2a - 3)(4a^2 + 6a + 9)$ **63.** $27(x - y - z)$
65. $(x - t)(y + s)$ **67.** $x^6(7x + 1)(5x - 1)$
69. $5(x - 2)(1 + 2y)$ **71.** $(7p + 2q)^2$ **73.** $4(t^2 + 9)$
75. $(n + 3)(n - 3)(m^2 + 3)$
81.

$$-\frac{3}{2}$$

83.

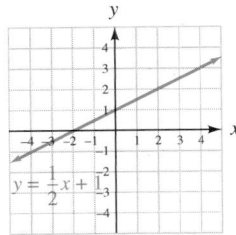

$$y = \frac{1}{2}x + 1$$

Study Set Section 6.7 (page 528)

1. quadratic   **3.** zero-factor, 0, 0   **5. a.** Yes   **b.** No   **c.** Yes
d. No   **7.** $-\frac{4}{5}$ **9. a.** Add 6 to both sides.   **b.** Distribute the
multiplication by x and subtract 3 from both sides.   **11.** $0, x + 7, -7$
13. $p, p - 3, 0, 3, -2$ **15.** $3, 2$ **17.** $-7, 7$ **19.** $0, \frac{5}{2}$
21. $0, -\frac{10}{3}$ **23.** $0, 6, -8$ **25.** $1, -2, 3$
27. $12, 1$ **29.** $-3, 7$ **31.** $8, 1$ **33.** $-3, -5$ **35.** $-9, 9$
37. $-5, 5$ **39.** $-\frac{1}{2}, \frac{1}{2}$ **41.** $-\frac{7}{3}, \frac{7}{3}$ **43.** $0, 7$ **45.** $0, 16$
47. $0, 3$ **49.** $0, -\frac{8}{3}$ **51.** $-2, \frac{1}{3}$ **53.** $-\frac{3}{2}, 1$ **55.** $\frac{1}{5}, 1$
57. $-\frac{5}{2}, 4$ **59.** $-\frac{7}{2}$ **61.** $\frac{5}{3}$ **63.** $3, 4$ **65.** $-3, -2$
67. $0, -1, -2$ **69.** $0, 9, -3$ **71.** $0, 3$ **73.** $0, -1, 2$
75. $-\frac{9}{2}, \frac{9}{2}$ **77.** 8 **79.** $\frac{5}{2}, -6$ **81.** $2, 10$ **83.** $0, -5, 4$
85. $-10, 10$ **87.** $-\frac{1}{3}, 5$ **89.** $2, 7, 1$ **91.** $0, 2$ **93.** $-\frac{2}{3}, -\frac{3}{2}$
95. $0, -1, -\frac{1}{3}$ **97.** $\frac{2}{3}, -\frac{1}{5}$ **99.** $-\frac{11}{2}, \frac{11}{2}$ **101.** $\frac{1}{8}, 1$
109. $15 \text{ min} \leq t < 30 \text{ min}$

Study Set Section 6.8 (page 536)

1. consecutive   **3.** hypotenuse, legs   **5.** ii.   **7.** $20 = b(b + 5)$
9. a. A right triangle   **b.** x ft; $(x + 1)$ ft   **c.** 9 ft
11. $-16, 1, 0, 0, 3, -1$ **13.** Width: 3 ft; length: 6 ft   **15.** 8 in.,
10 in.   **17.** 3 ft by 9 ft   **19.** Base: 6 cm; height: 5 cm
21. Foot: 4 ft; luff: 12 feet   **23.** Kahne: 9; Riggs: 10   **25.** 12
27. $(11, 13)$ **29.** 10 yd   **31.** 8 ft   **33.** 5 m, 12 m, 13 m
35. 5 sec   **37.** 4 sec   **39.** 1 sec   **41.** 8   **47.** $25b^2 - 20b + 4$
49. $s^4 + 8s^2 + 16$ **51.** $81x^2 - 36$

Chapter 6 Review (page 540)

1. $5 \cdot 7$ **2.** $2^5 \cdot 3$ **3.** 7   **4.** $18a^3$ **5.** $3(x + 3y)$
6. $5a(x^2 + 3)$ **7.** $7s^3(s^2 + 2)$ **8.** $\pi a(b - c)$
9. $12x(2x^2 + 5x - 4)$ **10.** $xy^3z^2(x^4 + y^2z - 1)$
11. $-5ab(b - 2a + 3)$ **12.** $(x - 2)(4 - x)$ **13.** $-(a + 7)$
14. $-(4t^2 - 3t + 1)$ **15.** $(c + d)(2 + a)$ **16.** $(y + 6)(3x - 5)$
17. $(a + 1)(2a^2 - 1)$ **18.** $4m(n + 3)(m - 2)$ **19.** 1
20.

Factors of 6	Sum of the factors of 6
1(6)	7
2(3)	5
-1(-6)	-7
-2(-3)	-5

21. $(x + 6)(x - 4)$
22. $(x - 20)(x + 2)$
23. $(x - 5)(x - 9)$
24. Prime
25. $-(y - 8)(y - 7)$
26. $(y + 9)(y + 1)$
27. $(c + 5d)(c - 2d)$
28. $(m - 2n)(m - n)$ **29.** Multiply   **30.** There are no two
integers whose product is 11 and whose sum is 7.
31. $5a^3(a + 10)(a - 1)$ **32.** $-4x(x + 3y)(x - 2y)$
33. $(2x + 1)(x - 3)$ **34.** $(7y + 5)(5y - 2)$
35. $-(3x + 5)(x - 6)$ **36.** $3p(6p + 1)(p - 2)$

37. $(4b - c)(b - 4c)$ **38.** Prime   **39.** $(4x + 1)$ in., $(3x - 1)$ in.
40. The signs of the second terms must be negative.   **41.** $(x + 5)^2$
42. $(3y - 4)^2$ **43.** $-(z - 1)^2$ **44.** $(5a + 2b)^2$
45. $(x + 3)(x - 3)$ **46.** $(7t + 11y)(7t - 11y)$
47. $(xy + 20)(xy - 20)$ **48.** $8a(t + 2)(t - 2)$
49. $(c^2 + 16)(c + 4)(c - 4)$ **50.** Prime
51. $(b + 1)(b^2 - b + 1)$ **52.** $(x - 6)(x^2 + 6x + 36)$
53. $(p + 5q)(p^2 - 5pq + 25q^2)$
54. $2x^2(2x - 3y)(4x^2 + 6xy + 9y^2)$ **55.** $2y^2(3y - 5)(y + 4)$
56. $5(t + u^2)(s^2 + v)$ **57.** $(j^2 + 4)(j + 2)(j - 2)$
58. $-3(j + 2)(j^2 - 2j + 4)$ **59.** $(x + 1)(20 + m)(20 - m)$
60. $3w^2(2w - 3)^2$ **61.** $2(t^3 + 5)$ **62.** Prime   **63.** $z(x + 8y)^2$
64. $6c^2d(3cd - 2c - 4)$ **65.** $0, 6$ **66.** $\frac{7}{4}, -1$ **67.** $0, -2$
68. $-3, 3$ **69.** $-\frac{5}{12}, \frac{5}{12}$ **70.** $3, 4$ **71.** -7 **72.** $6, -4$
73. $1, \frac{1}{5}$ **74.** $0, -1, 2$ **75.** 15 m   **76.** Streep: 14, Hepburn: 12
77. 5m   **78.** 10 sec

Chapter 6 Test (page 547)

1. a. greatest, common, factor   **b.** product   **c.** Pythagorean
d. difference   **e.** binomials   **2. a.** $45 = 3^2 \cdot 5$; $30 = 2 \cdot 3 \cdot 5$
b. $15x^3$ **3.** $4(x + 4)$ **4.** $(q + 9)(q - 9)$
5. $5ab(6ab^2 - 4a^2b + 1)$ **6.** Prime   **7.** $(x + 1)(2x + 3)$
8. $(x + 3)(x + 1)$ **9.** $-(x - 11)(x + 2)$ **10.** $x^2(x - 30)(x - 2)$
11. $(a - b)(9 + x)$ **12.** $(2a - 3)(a + 4)$ **13.** $2(3x + 5y)^2$
14. $(x + 2)(x^2 - 2x + 4)$ **15.** $5m^6(4m^2 - 3)$
16. $3(a - 3)(a^2 + 3a + 9)$ **17.** $(4x^2 + 9)(2x + 3)(2x - 3)$
18. $(a + 5)(a^2 + 1)$ **19.** $(5x - 4)$ in.
20. $(x - 9)(x + 6) = x^2 + 6x - 9x - 54 = x^2 - 3x - 54$
21. $-3, 2$ **22.** $-5, 5$ **23.** $0, \frac{1}{6}$ **24.** -3 **25.** $\frac{1}{3}, -\frac{1}{2}$
26. $9, -2$ **27.** $0, -1, -6$ **28.** 6 ft by 9 ft   **29.** 5 sec
30. Base: 6 in.; height: 11 in.   **31.** 12, 13   **32.** 10
33. A quadratic equation is an equation that can be written in the form
$ax^2 + bx + c = 0$; $x^2 - 2x + 1 = 0$. (Answers may vary.)
34. At least one of them is 0.

Study Set Section 7.1 (page 560)

1. rational   **3.** undefined   **5.** $\frac{6}{24} = \frac{1}{4}$ **7. a.** 1   **b.** -1 **c.** 1
d. Does not simplify   **9.** $x, 1, x + 1, x + 3$ **11.** 4   **13.** 0
15. Undefined   **17.** $-\frac{2}{11}$ **19.** Undefined   **21.** $\frac{1}{6}$ **23.** 0
25. 2   **27.** None   **29.** $\frac{1}{2}$ **31.** $-6, 6$ **33.** $-2, 1$ **35.** $\frac{5}{a}$
37. $\frac{3}{2}$ **39.** $\frac{7c^2}{3d^2}$ **41.** $\frac{9a^2}{11b}$ **43.** $\frac{2x + 1}{y}$ **45.** $\frac{1}{3}$ **47.** $\frac{x + 2}{x - 4}$
49. Does not simplify   **51.** $\frac{1}{2b + 1}$ **53.** $\frac{m - n}{7(m + n)}$ or $\frac{m - n}{7m + 7n}$
55. $\frac{10}{3}$ **57.** $\frac{2x}{x - 6}$ **59.** -1 **61.** $-\frac{1}{2}$ **63.** $-\frac{1}{a + 1}$
65. $-\frac{5}{m + 5}$ **67.** $\frac{1}{a}$ **69.** $-\frac{x + 2}{x + 1}$ **71.** -6 **73.** $\frac{x + 1}{x - 1}$ **75.** $\frac{3x}{y}$
77. $\frac{4 - x}{4 + x}$ or $-\frac{x - 4}{x + 4}$ **79.** 4   **81.** $\frac{3(x + 3)}{2x + 1}$ or $\frac{3x + 9}{2x + 1}$ **83.** $-\frac{3x + 11}{x + 3}$
85. Does not simplify   **87.** $\frac{2u - 3}{u^3}$ **89.** $(2x + 3)^2$ **91.** 9
93. $\frac{3x}{5y}$ **95.** $85\frac{1}{3}$ **97.** 2, 1.6, and 1.2 milligrams per liter
103. a. $(a + b) + c = a + (b + c)$ **b.** $ab = ba$

Study Set Section 7.2 (page 569)

1. reciprocal   **3. a.** numerators, denominators, reciprocal
b. AC, BD, D, C **5.** $-\frac{y^2}{y + 1}$ **7.** 1   **9.** ft   **11.** $\frac{3y}{14}$
13. $\frac{3(y + 2)}{y^3}$ or $\frac{3y + 6}{y^3}$ **15.** $\frac{20}{3n}$ **17.** x^2y^2 **19.** $\frac{x}{5}$ **21.** -2

23. $\frac{3}{2x}$ **25.** $x + 1$ **27.** $-(x - 2)$ or $-x + 2$ **29.** $\frac{(x-2)^2}{x}$

31. $\frac{(m-2)(m-3)}{2(m+2)}$ **33.** $\frac{3(a+3)^2}{2a^3}$ **35.** 35   **37.** $3x + 3$

39. $10y - 16$ **41.** $\frac{36a-60}{a}$ **43.** $\frac{3}{2y}$ **45.** $\frac{3a}{5}$ **47.** $\frac{x^2}{3}$ **49.** $\frac{9p^3}{5}$

51. $\frac{5(a-2)}{4a^3}$ **53.** $\frac{-x+2}{3}$ **55.** $-(m+5)$ or $-m-5$ **57.** $t + 7$

59. $\frac{2(x-7)}{x+9}$ **61.** $\frac{d(6c-7d)}{6}$ **63.** $\frac{1}{3}$ **65.** 1   **67.** $\frac{1}{12(2r-3s)}$

69. $\frac{4(n-1)}{3n}$ **71.** 450 ft   **73.** $\frac{3}{4}$ gal   **75.** $\frac{1}{2}$ mi per min

77. 1,800 m per min   **79.** $\frac{b-3}{b}$ **81.** $\frac{1}{(x+1)^2}$ **83.** $25h - 15$

85. $n - 1$ **87.** $\frac{5r^3}{2s^2}$ **89.** $\frac{7(p+2)}{3p^4}$ or $\frac{7p+14}{3p^4}$ **91.** $\frac{x-2}{x-3}$

93. $\frac{2x-3y}{y(2x+3y)}$ **95.** $\frac{x^2}{10}$ ft²   **97.** 4,380,000   **99.** 8 yd²

101. $\frac{1}{2}$ mi per min   **103.** $\frac{1}{4}$ mi²   **109.** $w = 6$ in., $l = 10$ in.

Study Set Section 7.3 (page 579)

1. denominator   **3.** build   **5.** numerators, denominator, $A + B, D, A - B, D$ **7.** $\frac{4}{5}$ **9. a.** Twice   **b.** Once

11. $x - 1, -, +, 6$ **13.** $\frac{11}{x}$ **15.** $\frac{x+5}{18}$ **17.** $\frac{x}{3}$ **19.** $\frac{1}{3a^2}$

21. $\frac{x+4}{y}$ **23.** $\frac{1}{r-5}$ **25.** 9   **27.** $\frac{3}{4}$ **29.** $\frac{x}{25}$ **31.** $\frac{2t}{9}$

33. $\frac{m-6}{6m^2}$ **35.** $\frac{5a}{a+2}$ **37.** $\frac{1}{t+2}$ **39.** $\frac{2}{w(w-9)}$ **41.** $\frac{1}{2}$

43. $2x - 5$ **45.** $\frac{1}{y}$ **47.** 0   **49.** $\frac{2x}{3-x^2}$ **51.** $-\frac{1}{3x-1}$ **53.** $6x$

55. $30a^3$ **57.** $3a^2b^3$ **59.** $c(c+2)$ **61.** $(3x+1)(3x-1)$

63. $12(b+2)$ **65.** $8k(k+2)$ **67.** $(x+1)(x-1)$

69. $(x+1)(x+5)(x-5)$ **71.** $(2n+5)(n+4)^2$ **73.** $\frac{50}{10r}$

75. $\frac{8xy}{x^2y}$ **77.** $\frac{27b}{12b^2}$ **79.** $\frac{3x^2+3x}{(x+1)^2}$ **81.** $\frac{x^2+x-6}{x(x+3)}$ **83.** $\frac{5t+25}{20(t+2)}$

85. $\frac{4y^2+12y}{4y(y-2)(y-3)}$ **87.** $\frac{36-3h}{3(h+9)(h-9)}$ **89.** $\frac{3}{t-7}$ **91.** $\frac{1}{c+d}$

93. $\frac{7n+1}{(n+4)(n-2)}$ **95.** $\frac{5}{9y}$ **97.** $3x - 2$ **99.** $\frac{3}{r}$ **101.** $\frac{2x+6}{x+2}$ ft

109. a. $I = Prt$ **b.** $A = \frac{1}{2}bh$ **c.** $P = 2l + 2w$

Study Set Section 7.4 (page 588)

1. unlike   **3. a.** $2 \cdot 2 \cdot 5 \cdot x \cdot x$ **b.** $(x-2)(x+6)$

5. $(x+6)(x+3)$ **7.** $\frac{5}{5}$ **9.** $3x, 5, 15x, 15x, 35$ **11.** $\frac{13x}{21}$

13. $\frac{5y}{2}$ **15.** $\frac{7t-32}{8t}$ **17.** $\frac{11b}{12}$ **19.** $\frac{7-2m}{m^2}$ **21.** $\frac{17x+3}{x^2}$

23. $\frac{1}{10p}$ **25.** $-\frac{29}{24t}$ **27.** $\frac{41}{30x}$ **29.** $\frac{16c^2+3}{18c^4}$ **31.** $\frac{a+8}{2(a+2)(a-2)}$

33. $\frac{6a+9}{(3a+2)(3a-2)}$ **35.** $\frac{4a+1}{(a+2)^2}$ **37.** $\frac{6-3m}{5m(m-1)}$

39. $\frac{17t+42}{(t+3)(t+2)}$ **41.** $\frac{2x^2+11x}{(2x-1)(2x+3)}$ **43.** $\frac{35x^2+x+5}{5x(x+5)}$

45. $\frac{2x^2-1}{x(x+1)}$ **47.** $\frac{14s+58}{(s+3)(s+7)}$ **49.** $\frac{2m^2+20m-6}{(m-2)(m+5)}$

51. $\frac{s^2+8s+4}{(s+4)(s+1)(s+1)}$ **53.** $\frac{2x+13}{(x-8)(x-1)(x+2)}$ **55.** $\frac{1}{a+1}$

57. $\frac{1}{(y+3)(y+4)}$ **59.** $\frac{6x+8}{x}$ **61.** $\frac{x^2-4x+9}{x-4}$ **63.** $\frac{a^2b-3}{a^2}$

65. $\frac{-4x-3}{x+1}$ or $-\frac{4x+3}{x+1}$ **67.** $\frac{2}{a-4}$ **69.** $\frac{c+d}{7c-d}$ **71.** $\frac{6d-3}{d-9}$

73. $\frac{1}{g+2}$ **75.** $\frac{j-5}{(j+3)(j+5)}$ **77.** $\frac{xy-y+10}{x-1}$ **79.** $\frac{2b+1}{(b+1)(b+2)}$

81. $\frac{y+4}{y-1}$ **83.** $\frac{2n+2}{15}$ **85.** $\frac{y^2+7y+6}{15y^2}$ **87.** $\frac{x+2}{x-2}$

89. $\frac{10a-14}{3a(a-2)}$ **91. a.** $\frac{75+8x^2}{30x}$ **b.** $\frac{2}{3}$ **93. a.** $\frac{t^2+4t}{(t-5)(t+5)}$

b. $t + 5$ **95.** $\frac{20x+9}{6x^2}$ cm   **101.** 8; (0, 2)   **103.** 0

Study Set Section 7.5 (page 597)

1. complex, complex   **3.** simplify, division, reciprocal

5. a. $\frac{x-3}{4}$, yes   **b.** $\frac{1}{12} - \frac{x}{6}$; no   **7.** ÷   **9.** $\frac{8}{9}$ **11.** $\frac{5x}{12}$ **13.** $\frac{x^2}{y}$

15. $\frac{n^3}{8}$ **17.** $\frac{10}{3}$ **19.** $-\frac{x^3}{14}$ **21.** $\frac{5x}{3}$ **23.** $\frac{t^3}{2}$ **25.** $\frac{5}{7}$

27. $\frac{3y+12}{4y^2-6y}$ **29.** $\frac{2-5y}{6}$ **31.** $\frac{24-c^2}{12}$ **33.** $\frac{s^2-s}{2+2s}$ **35.** $\frac{b-5ab}{3ab-7a}$

37. $\frac{5}{4}$ **39.** $\frac{1-3x}{5+2x}$ **41.** $\frac{1+x}{2+x}$ **43.** $\frac{3-x}{x-1}$ **45.** $\frac{32h-1}{96h+6}$

47. $\frac{d-3}{2d}$ **49.** $\frac{x-12}{x+6}$ **51.** $a - 7$ **53.** $\frac{m^2+n^2}{m^2-n^2}$ **55.** $\frac{6d+12}{d}$

57. $\frac{8}{4c+5c^2}$ **59.** $\frac{y}{x-2y}$ **61.** $\frac{1}{x+2}$ **63.** $\frac{1}{x+3}$ **65.** $\frac{x}{x-2}$

67. $2m - 1$ **69.** $\frac{q+p}{q}$ **71.** $18x$ **73.** $\frac{2c}{2-c}$ **75.** $\frac{r-1}{r+1}$

77. $\frac{b+9}{8a}$ **79.** $\frac{t+2}{t-3}$ **81.** $\frac{xy}{y+x}$ **83.** $-\frac{10x^3}{3}$ **85.** $\frac{7}{6}$

87. $\frac{R_1R_2}{R_2+R_1}$ **93.** 1   **95.** $\frac{25}{16x^{12}}$

Study Set Section 7.6 (page 606)

1. rational   **3.** multiply   **5. a.** Yes   **b.** No   **7. a.** 3, 0

b. 3, 0   **c.** 3, 0   **9. a.** y **b.** $(x+2)(x-2)$ **11. a.** 3

b. $3(x+6)$ **13.** $2a, 2a, 2a, 2a, 2a, 4, 7, 4, 3$ **15.** 1   **17.** 0

19. $\frac{3}{5}$ **21.** $-\frac{4}{3}$ **23.** $-\frac{12}{7}$ **25.** -48 **27.** 2, 4   **29.** $-5, 2$

31. $-4, -5$ **33.** $3, -\frac{5}{3}$ **35.** 1   **37.** No solution; 5 is extraneous

39. No solution; -2 is extraneous   **41.** 7   **43.** 1   **45.** $-1, 6$

47. 0, 3   **49.** No solution; 2 is extraneous   **51.** -3 **53.** $\frac{1}{5}$

55. $P = nrt$ **57.** $d = \frac{bc}{a}$ **59.** $A = \frac{h(b+d)}{2}$ **61.** $r = \frac{E-IR}{I}$

63. $a = \frac{b}{b-1}$ **65.** $x = \frac{5yz}{5y+4z}$ **67.** $r = \frac{st}{s-t}$

69. $L^2 = 6dF - 3d^2$ **71.** 6   **73.** 1   **75.** -40 **77.** No

solution; -1 is extraneous   **79.** $-4, 3$ **81.** 3   **83.** $\frac{9}{40}$

85. 0   **87.** 1, 2   **89.** 7   **91. a.** $\frac{2a+3}{5}$ **b.** $-\frac{9}{4}$

93. a. $\frac{x^2-4x+2}{(x-2)(x-3)}$ **b.** 4   **95.** $R = \frac{HB}{B-H}$ **97.** $r = \frac{r_1r_2}{r_2+r_1}$

103. 20

Study Set Section 7.7 (page 616)

1. motion, investment, work   **3.** iii   **5.** $\frac{1}{45}$ of the job per minute

7. a. $t = \frac{d}{r}$ **b.** $P = \frac{I}{rt}$ **9.** $\frac{x}{15}, \frac{x}{8}$ **11.** $6\frac{1}{9}$ days   **13.** 4   **15.** 2

17. 5   **19.** $\frac{2}{3}, \frac{3}{2}$ **21.** 8   **23.** Garin: 16 mph, Armstrong: 26 mph

25. 1st: $1\frac{1}{2}$ ft per sec, 2nd: $\frac{1}{2}$ ft per sec   **27.** Canada goose: 30 mph,

great blue heron: 20 mph   **29.** $\frac{300}{255+x}, \frac{210}{255-x}$; 45 mph

31. $2\frac{6}{11}$ days   **33.** No, after the pipes are opened, the swimming is

scheduled to take place in 6 hours. It takes 7.2 hr (7 hr 12 min) to fill

the pool.   **35.** 8 hr   **37.** 20 min   **39.** $1\frac{4}{5}$ hr = 1.8 hr

41. Credit union: 4%, bonds: 6%   **43.** 7% and 8%   **47.** (1, 3)

49. Yes   **51.** 0

Study Set Section 7.8 (page 628)

1. ratio, rate   **3.** extremes, means   **5.** unit   **7.** equal, ad, bc

9. 25, 2, 1,000, x **11.** 2.19, 1   **13.** x, 288, 18, 18   **15.** as, to

17. $\frac{4}{15}$ **19.** $\frac{3}{4}$ **21.** $\frac{5}{4}$ **23.** $\frac{1}{2}$ **25.** $\frac{1}{2}$ **27.** $\frac{25}{22}$ **29.** Yes

31. No   **33.** 4   **35.** 14   **37.** 0   **39.** $-\frac{3}{2}$ **41.** -2 **43.** -27

45. 2, -2 **47.** 6, -1 **49.** $-1, 16$ **51.** $-\frac{5}{2}, -1$ **53.** 15

55. 8   **57.** $-\frac{1}{3}, 2$ **59.** 2   **61.** $-\frac{27}{2}$ **63.** $-10, 10$ **65.** $-4, 3$

67. $\frac{26}{9}$ **69. a.** $\frac{3}{2}$, 3:2   **b.** $\frac{2}{3}$, 2:3   **71.** $62.50   **73.** 20

75. 84   **77.** 568, 13, 14   **79.** 510   **81.** Not exactly, but close

83. 140   **85.** 522 in.; 43.5 ft   **87.** 10 ft   **89.** 45 min for $25

91. 150 for $12.99   **93.** 6-pack for $1.50   **95.** 24 12-oz bottles
97. 39 ft   **99.** $46\frac{7}{8}$ ft   **105.** 90%   **107.** 480

Chapter 7 Review (page 632)

1. 4, −4   **2.** $-\frac{3}{7}$ **3.** $\frac{1}{2x}$ **4.** $\frac{5}{2x}$ **5.** $\frac{x}{x+1}$ **6.** $a-2$ **7.** −1
8. $-\frac{1}{x+3}$ **9.** $\frac{x}{x-1}$ **10.** Does not simplify   **11.** $\frac{1}{x-y}$ **12.** $\frac{4}{3}$
13. x is not a common factor of the numerator and the denominator; x
is a term of the numerator.   **14.** 150 mg   **15.** $\frac{3x}{y}$ **16.** 96
17. $\frac{x-1}{x+2}$ **18.** $\frac{2x}{x+1}$ **19.** $\frac{3y}{2}$ **20.** $-x-2$ **21. a.** Yes
b. No   **c.** Yes   **d.** Yes   **22.** $\frac{1}{3}$ mi per min   **23.** $\frac{1}{3d}$ **24.** 1
25. $\frac{2x+2}{x-7}$ **26.** $\frac{1}{a-4}$ **27.** $9x$ **28.** $8x^3$ **29.** $m(m-8)$
30. $(5x+1)(5x-1)$ **31.** $(a+5)(a-5)$ **32.** $(2t+7)(t+5)^2$
33. $\frac{63}{7a}$ **34.** $\frac{2xy+x}{x(x-9)}$ **35.** $\frac{2b+14}{6(b-5)}$ **36.** $\frac{9r^2-36r}{(r+1)(r-4)(r+5)}$
37. $\frac{a-7}{7a}$ **38.** $\frac{x^2+x-1}{x(x-1)}$ **39.** $\frac{1}{t+1}$ **40.** $\frac{x^2+4x-4}{2x^2}$
41. $\frac{b+6}{b-1}$ **42.** $\frac{6c+8}{c}$ **43.** $\frac{14n+58}{(n+3)(n+7)}$ **44.** $\frac{4t+1}{(t+2)^2}$
45. $\frac{1}{(a+3)(a+2)}$ **46.** $\frac{17y-2}{12(y-2)(y+2)}$ **47.** Yes
48. $\frac{14x+28}{(x+6)(x-1)}$ units, $\frac{12}{(x+6)(x-1)}$ square units   **49.** $\frac{n^3}{14}$
50. $\frac{r+9}{8s}$ **51.** $\frac{1+y}{1-y}$ **52.** $\frac{21}{3a+10a^2}$ **53.** x^2+3 **54.** $\frac{y-5xy}{3xy-7x}$
55. 3   **56.** No solution; 5 is extraneous   **57.** 3   **58.** 2, 4
59. 0   **60.** −4, 3   **61.** $T_1=\frac{T_2}{1-E}$ **62.** $y=\frac{xz}{z-x}$ **63.** 3
64. 5 mph   **65.** $\frac{1}{4}$ of the job per hr   **66.** $5\frac{5}{6}$ days   **67.** 5%
68. 40 mph   **69.** No   **70.** Yes   **71.** $\frac{9}{2}$ **72.** 0   **73.** 7
74. 4, $-\frac{3}{2}$ **75.** 255   **76.** 20 ft   **77.** 5 ft 6 in.   **78.** 250 for $98
79. $c=kt$ **80.** $f=\frac{k}{L}$ **81.** $2,000   **82.** 600   **83.** 1.25 amps
84. Inverse variation

Chapter 7 Test (page 640)

1. a. rational   **b.** similar   **c.** proportion   **d.** build   **e.** factors
2. 10 words   **3.** 0   **4.** −3, 2   **5.** 3,360,000 or 3,360K bits per
minute   **6.** 5 is not a common factor of the numerator, and therefore
cannot be removed. 5 is a term of the numerator.   **7.** $\frac{8x}{9y}$ **8.** −7
9. $\frac{x+1}{2x+3}$ **10.** 1   **11.** $3c^2d^3$ **12.** $(n+1)(n+5)(n-5)$
13. $\frac{5y^2}{4}$ **14.** $\frac{x+1}{3(x-2)}$ **15.** $-\frac{x^2}{3}$ **16.** $\frac{1}{6}$ **17.** 3   **18.** $\frac{4n-5mn}{10m}$
19. $\frac{2x^2+x+1}{x(x+1)}$ **20.** $\frac{2a+7}{a-1}$ **21.** $\frac{c^2-4c+9}{c-4}$ **22.** $\frac{1}{(t+3)(t+2)}$
23. $\frac{12}{5m}$ **24.** $\frac{a+2s}{2as^2-3a^2}$ **25.** 11   **26.** No solution, 6 is extraneous
27. 1   **28.** 1, 2   **29.** $\frac{2}{3}$ **30.** $B=\frac{HR}{R-H}$ **31.** Yes   **32.** 1,785
33. 171 ft   **34.** 80 sheets for $3.89   **35.** $3\frac{15}{16}$ hr   **36.** 4 mph
37. 2   **38.** We multiply both sides of the equation by the LCD of the
rational expressions appearing in the equation. The resulting equation is
easier to solve.   **39.** 100 lb   **40.** $\frac{80}{3}$ or $26\frac{2}{3}$

Cumulative Review Chapters 1–7 (page 642)

1. a. False   **b.** False   **c.** True   **d.** True   **2.** <   **3.** 36
4. 77   **5.** $6c+62$ **6.** −5   **7.** 3   **8.** About 26%
9. $B=\frac{A-c-r}{2}$ **10.** 104°F   **11.** 240 ft³   **12.** 12 lb of the
$6.40 tea and 8 lb of the $4 tea   **13.** 500 mph

---

14. $[-1, \infty)$ 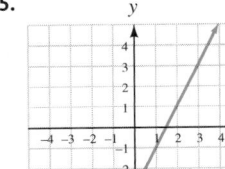   **15.**

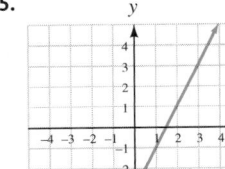

16. −1   **17.** 0.008 mm/m   **18.** $\frac{8}{7}$ **19.** $y=3x+2$
20.   **21.** 0   **22.** domain, range

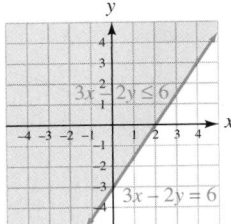

23.

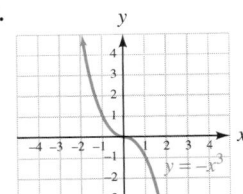

24. (5, 2)   **25.** (1, −1)   **26.** Red:
8, blue: 15   **27.** x^7 **28.** x^{25}
29. $\frac{y^3}{8}$ **30.** $-\frac{32a^5}{b^5}$ **31.** $\frac{a^8}{b^{12}}$
32. $3b^2$ **33.** 2.9×10^5 **34.** 3

35.

36. $A=\pi R^2-\pi r^2$
37. $6x^2+x-5$
38. $-\frac{1}{2}t^3-\frac{7}{4}t^2-\frac{1}{12}$ **39.** $6x^4y^5$
40. $6y^2-y-35$
41. $-12x^4z+4x^2z^2$
42. $9a^2-24a+16$
43. $2x+3$ **44.** x^2+2x-1
45. $k^2t(k-3)$
46. $(b+c)(2a+3)$ **47.** $(u-9)^2$
48. $-(r-2)(r+1)$ **49.** Prime   **50.** $(2x-9)(3x+7)$
51. $2(a+10b)(a-10b)$ **52.** $(b+5)(b^2-5b+25)$ **53.** 0, $-\frac{1}{5}$
54. $\frac{1}{3}$, $\frac{1}{2}$ **55.** 16 in., 10 in.   **56.** 5, −5   **57.** $\frac{2x}{x-2}$ **58.** $-\frac{x^3}{3}$
59. $2m-5$ **60.** $-\frac{1}{x-3}$ **61.** $\frac{m-3}{(m+3)(m+1)}$ **62.** $\frac{x^2}{(x+1)^2}$
63. $\frac{17}{25}$ **64.** 2   **65.** $14\frac{2}{5}$ hr   **66.** 34.8 ft   **67.** 28 in.
68. $\frac{3}{4}=0.75$

Study Set Section 8.1 (page 656)

1. equation   **3.** linear   **5.** identity   **7.** inequality   **9. a.** same
b. same   **11. a.** No   **b.** Yes   **13. a.** iii   **b.** i   **c.** ii   **15.** 3
17. 4   **19.** −2   **21.** 1   **23.** −2   **25.** 24   **27.** 18   **29.** −5
31. No solution, $\varnothing$; contradiction   **33.** All real numbers, $\mathbb{R}$; identity
35. All real numbers, $\mathbb{R}$; identity   **37.** No solution, $\varnothing$; contradiction
39. $w=\frac{P-2l}{2}$ **41.** $B=\frac{3V}{h}$ **43.** $W=T-ma$
45. $x=z\sigma+\mu$ **47.** $l=\frac{2S-na}{n}$ or $l=\frac{2S}{n}-a$ **49.** $d=\frac{l-a}{n-1}$
51. $(2, \infty)$ **53.** $(-\infty, 2]$

55. $(-\infty, 20]$

57. $(0, \infty)$

59. $[-2, \infty)$

61. $\left[-\frac{2}{5}, \infty\right)$

63. $(-\infty, 3)$

65. $(-\infty, 7]$

67. $(-\infty, \infty)$; $\mathbb{R}$

69. No solution; $\varnothing$ **71.** 2   **73.** -11 **75.** $-\frac{5}{2}$

77. $-\frac{1}{2}$ **79.** All real numbers, $\mathbb{R}$; identity   **81.** 2   **83.** $\frac{21}{19}$

85. No solution; $\varnothing$; contradiction   **87.** 1.7   **89.** $\frac{9}{13}$

91. $\left(-\infty, -\frac{8}{5}\right)$

93. $(-\infty, 6]$

95. $(-\infty, \infty)$; $\mathbb{R}$

97. $(-\infty, 1.5]$

99. $[-36, \infty)$

101. $(6, \infty)$

103. 20

105. 310 mi   **107.** 20 ft by 45 ft   **109.** 8 hr   **111.** 10 hr
113. 11   **117.** $\frac{1}{t^{12}}$

Study Set Section 8.2 (page 668)

1. intersection, union   **3.** double   **5. a.** both   **b.** one, both
7. a. intersection   **b.** union   **9. a.** No   **b.** Yes
11. a. $[-2, 1)$ **b.** $[2, 2]$ **c.** $\varnothing$ **13.** union, intersection
15. All real numbers   **17.** $\{4, 6\}$

19. $\{-3, 1, 2\}$ **21.** $\{-3, -1, 0, 1, 2, 4, 6, 8, 10\}$
23. $\{-3, 0, 1, 2, 3, 4, 5, 6, 8\}$ **25.** $(-2, 5]$

27. $(2, 3]$

29. $(-\infty, -6]$

31. No solution; $\varnothing$

33. $[1, 4]$ **35.** $(0.8, 1.1)$

37. $(-\infty, -2] \cup (6, \infty)$

39. $(-\infty, -1) \cup (2, \infty)$

41. $(-\infty, 2) \cup (7, \infty)$

43. $(-\infty, \infty)$ **45.** $(-\infty, 1)$

47. $(-10, -9)$

49. $(-2, 5)$ **51.** $(-\infty, \infty)$

53. $[2, \infty)$ **55.** No solution; $\varnothing$

57. $[-4, 6)$ **59.** $[2, 2]$

61. $(-\infty, 4.8) \cup (6.5, \infty)$

63. $(-0.7, 0.2]$

65. $(-12, -6]$

67. $\left(-\infty, \frac{1}{6}\right] \cup (2, \infty)$

69. $[-2, 4]$ **71.** $[4, 4]$

73. a. 128, 192   **b.** $32 \le s \le 48$, $[32, 48]$ **75. a.** $(67, 77)$
b. $(62, 82)$ **77. a.** 2001   **b.** 2001, 2002, 2003, 2004
c. 2003, 2004   **d.** 2002, 2003, 2004
79. a.   **b.**   **87.** $29,100

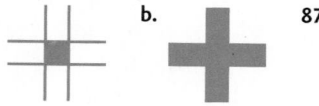

Study Set Section 8.3 (page 682)

1. absolute value   **3.** isolate   **5.** compound   **7.** compound
9. a. $-2, 2$ **b.** $-1.99, -1, 0, 1, 1.99$ **c.** $-3, -2.01, 2.01, 3$
11. a. $8, -8$ **b.** $x - 3, -(x - 3)$ **13. a.** $x = 8$ or $x = -8$
b. $x \le -8$ or $x \ge 8$ **c.** $-8 \le x \le 8$ **d.** $5x - 1 = x + 3$ or
$5x - 1 = -(x + 3)$ **15. a.** ii   **b.** iii   **c.** i   **17.** $23, -23$
19. $\frac{3}{4}, -\frac{3}{4}$ **21.** $13, -3$ **23.** $\frac{14}{3}, -6$ **25.** $50, -50$
27. $3.1, -6.7$ **29.** No solution; $\varnothing$ **31.** No solution; $\varnothing$
33. $25, -19$ **35.** $7, -\frac{7}{3}$ **37.** $2, -\frac{1}{2}$ **39.** $\frac{32}{7}, -16$ **41.** -10
43. -8 **45.** $\frac{9}{2}$ **47.** $-\frac{1}{2}$ **49.** $-4, \frac{28}{9}$ **51.** $-2, \frac{18}{11}$
53. $0, -2$ **55.** $11, \frac{1}{3}$ **57.** $(-4, 4)$

59. $[-21, 3]$

61. $[-24, 0]$

63. $\left(-\frac{8}{3}, 4\right)$ **65.** No solution; $\varnothing$

67. No solution; $\varnothing$ **69.** $(-\infty, -3) \cup (3, \infty)$

71. $(-\infty, -12) \cup (36, \infty)$

73. $\left(-\infty, -\frac{1}{5}\right] \cup \left[\frac{3}{5}, \infty\right)$

75. $(-\infty, \infty)$ **77.** $0, -6$ **79.** $(-3, 1)$

81. $\left(-\infty, -\frac{16}{3}\right) \cup (4, \infty)$

83. $[-10, 14]$ **85.** $40, -20$

87. $(-\infty, \infty)$ **89.** $-3, -\frac{3}{4}$

91. $\left[\frac{2}{3}, \frac{2}{3}\right]$ **93.** No solution; $\varnothing$ **95.** $\frac{12}{11}, -\frac{12}{11}$

97. $(-\infty, -2) \cup (5, \infty)$ **99.** No solution; $\varnothing$

101. $70° \le t \le 86°$ **103. a.** $|c - 0.6°| \le 0.5°$ **b.** $[0.1°, 1.1°]$
105. 26.45%, 24.76%   **111.** $50°, 130°$

Study Set Section 8.4 (page 694)

1. factored   **3.** greatest common factor   **5.** grouping
7. leading, coefficient, 2   **9.** $6xy^2$ **11.** $9, 6, -9, -6$
13. $-1 + (-12) = -13, -2(-6) = 12, -2 + (-6) = -8,$
$-3 + (-4) = -7$ **15.** $5c^2d^4$ **17.** $3m, 3$ **19.** $2x(x - 3)$ **21.**
$5x^2y(3 - 2y)$ **23.** $3z(9z^2 + 4z + 1)$
25. $6s(4s^2 - 2st + t^2)$ **27.** $-8(a + 2)$ **29.** $-3x(2x + y)$ **31.**
$-6ab(3a - 2b)$ **33.** $-4a^2c^8(2a^2 - 7a + 5c)$ **35.** Prime   **37.**
Prime   **39.** $(x + y)(u + v)$ **41.** $(a - b + c)(5 - t)$
43. $(x + y)(a + b)$ **45.** $(x + y)(x - 1)$ **47.** $(t - 3)(t^2 - 7)$
49. $(a + b)(a - 4)$ **51.** $6(x - 1)(x^2 + 2)$
53. $2c(2b^3 + 1)(7a^3 - 1)$ **55.** $h = \frac{2g}{c + d}$ **57.** $r_1 = \frac{rr_2}{r_2 - r}$
59. $a^2 = \frac{b^2x^2}{b^2 - y^2}$ **61.** $n = \frac{360}{180 - S}$ **63.** $(x - 3)(x - 2)$
65. $(x + 6)(x - 5)$ **67.** $3(x + 7y)(x - 3y)$
69. $6(a - 4b)(a - b)$ **71.** $n^2(n - 30t)(n + 2t)$
73. $-3(x - 3y)(x - 2y)$ **75.** $(5x + 3)(x + 2)$
77. $(7a + 5)(a + 1)$ **79.** $(11y - 1)(y + 3)$
81. $(4x - 1)(2x - 5)$ **83.** $(3y - 2)(2y - 3)$
85. $(5b - 2)(3b + 2)$ **87.** $5(3x^2 - 4)(2x^2 + 1)$
89. $8(2x^2 - 3)^2$ **91.** $4h(8h^2 - 1)(2h^2 + 1)$
93. $-(3a^2 + 2b^2)(a^2 + b^2)$ **95.** $(a + b + 4)(a + b - 6)$
97. $(x + a + 1)^2$ **99.** $(m + n + p)(3 + x)$
101. $-7u^2v^3(9uv^3 - 4v^4 + 3u)$ **103.** $x^2(b^2 - 7)(b^2 - 5)$
105. $(1 - m)(1 - n)$ **107.** $-(x + 3y)(x - 7y)$
109. $(a^2 + b)(x - 1)$ **111.** $(2y + 1)^2$ **113.** Prime
115. $(3r - 2)(r + 5)$ **117.** $(y^2 + 3)(y - 4)$
119. $2y(y^2 - 10)(y^2 - 3)$ **121.** $(7q - 7r + 2)(2q - 2r - 3)$
123. $\pi r^2\left(h_1 + \frac{1}{3}h_2\right)$ **125.** $x + 3$ **129.** $4,900

Study Set Section 8.5 (page 703)

1. squares   **3. a.** 1, 4, 9, 16, 25, 36, 49, 64, 81, 100   **b.** 1, 8, 27,
64, 125, 216, 343, 512, 729, 1,000   **5. a.** $F - L$
b. $F^2 - FL + L^2$ **c.** $F^2 + FL + L^2$ **7. a.** $x^2 - 4$; answers
may vary   **b.** $(x - 4)^2$; answers may vary   **c.** $x^2 + 4$; answers
may vary   **d.** $x^3 + 8$; answers may vary   **e.** $(x + 8)^3$; answers
may vary   **9.** $(x + 4)(x - 4)$ **11.** $(3y + 8)(3y - 8)$
13. $(12 + c)(12 - c)$ **15.** $(10m + 1)(10m - 1)$
17. $(9a + 7b)(9a - 7b)$ **19.** Prime
21. $(3r^2 + 11s)(3r^2 - 11s)$ **23.** $(4t + 5w^2)(4t - 5w^2)$
25. $(10rs^2 + t^2)(10rs^2 - t^2)$ **27.** $(6x^2y + 7z^3)(6x^2y - 7z^3)$
29. $(x^2 + y^2)(x + y)(x - y)$ **31.** $(4a^2 + 9b^2)(2a + 3b)(2a - 3b)$
33. $(x + y + z)(x + y - z)$
35. $(r - s + t^2)(r - s - t^2)(r - 2s + t^2)(r - 2s - t^2)$
37. $2(x + 12)(x - 12)$ **39.** $3x(x + 9)(x - 9)$
41. $5a(b^2 + 1)(b + 1)(b - 1)$ **43.** $4b(4 + b^2)(2 + b)(2 - b)$
45. $(c + d)(c - d + 1)$ **47.** $(a - b)(a + b + 2)$
49. $(x + 6 + y)(x + 6 - y)$ **51.** $(x - 1 + 3z)(x - 1 - 3z)$
53. $(a + 5)(a^2 - 5a + 25)$ **55.** $(2r + s)(4r^2 - 2rs + s^2)$
57. $(4t^2 - 3v)(16t^4 + 12t^2v + 9v^2)$
59. $(x - 6y^2)(x^2 + 6xy^2 + 36y^4)$
61. $(a - b + 3)(a^2 - 2ab + b^2 - 3a + 3b + 9)$

63. $(4 - a - b)(16 + 4a + 4b + a^2 + 2ab + b^2)$
65. $(x + 1)(x^2 - x + 1)(x - 1)(x^2 + x + 1)$
67. $(x^2 + y)(x^4 - x^2y + y^2)(x^2 - y)(x^4 + x^2y + y^2)$
69. $5(x + 5)(x^2 - 5x + 25)$ **71.** $4x^2(x - 4)(x^2 + 4x + 16)$
73. $(4a - 5b^2)(16a^2 + 20ab^2 + 25b^4)$
75. $2b^2(12 + b^2)(12 - b^2)$ **77.** $(x + y)(x - y + 8)$
79. $(x + y)(x^2 - xy + y^2)(x^6 - x^3y^3 + y^6)$
81. $(12at + 13b^3)(12at - 13b^3)$ **83.** Prime
85. $(9c^2d^2 + 4t^2)(3cd + 2t)(3cd - 2t)$
87. $2u^2(4v - t)(16v^2 + 4tv + t^2)$ **89.** $(y + 2x - t)(y - 2x + t)$
91. $(x + 10 + 3z)(x + 10 - 3z)$
93. $(c - d + 6)(c^2 - 2cd + d^2 - 6c + 6d + 36)$
95. $\left(\frac{1}{6} + y^2\right)\left(\frac{1}{6} - y^2\right)$
97. $(m + 2)(m^2 - 2m + 4)(m - 2)(m^2 + 2m + 4)$
99. $(a + b)(x + 3)(x^2 - 3x + 9)$
101. $(x^3 - y^4z^5)(x^6 + x^3y^4z^5 + y^8z^{10})$
103. $\frac{4}{3}\pi(r_1 - r_2)(r_1^2 + r_1r_2 + r_2^2)$ **109.** 45 minutes for $25

Study Set Section 8.6 (page 716)

1. rational   **3.** opposites   **5.** common, denominator
7. numerators, denominators, reciprocal   **9.** factor, greatest
11. a. Twice   **b.** Once   **13. a.** $\frac{5x^2 + 35x}{1}$ **b.** $\frac{1}{5x^2 + 35x}$
15. $\frac{3}{5a^6}$ **17.** $\frac{4y}{9x}$ **19.** $\frac{5x}{x - 2}$ **21.** $\frac{3x - 5}{x + 2}$ **23.** $-\frac{x + 2}{x + 1}$
25. $-\frac{p + 2q}{q + 2p}$ **27.** $\frac{8}{b}$ **29.** $\frac{p(p - 4)}{12}$ **31.** $\frac{x + 1}{9}$ **33.** 1
35. $-\frac{a^4}{a^2 + 3}$ **37.** $-\frac{(x + 1)^2(x + 2)}{2c + x}$ **39.** $\frac{3m}{2n^2}$ **41.** $\frac{x - 4}{x + 5}$
43. $\frac{c + 1}{25c^2 - 5c + 1}$ **45.** $\frac{n + 2}{n + 1}$ **47.** $\frac{a + 1}{2(a - 1)}$ **49.** -2
51. $\frac{3 - a}{a + 7}$ **53.** 3   **55.** $\frac{3}{x + 3}$ **57.** $\frac{1}{2}$ **59.** $\frac{17}{12x}$ **61.** $\frac{16y^2 + 3}{18y^4}$
63. $\frac{3a - 10b}{4a^2b^2}$ **65.** $-\frac{y^2 + 5y + 14}{2y^2}$ **67.** $\frac{8x - 2}{(x + 2)(x - 4)}$
69. $\frac{7x + 29}{(x + 5)(x + 7)}$ **71.** $\frac{2x^2 + x}{(x + 3)(x + 2)(x - 2)}$
73. $\frac{-x^2 + 11x + 8}{(3x + 2)(x + 1)(x - 3)}$ **75.** $\frac{6 - 3d}{5d(d - 1)}$ **77.** $\frac{m - 5}{(m + 3)(m + 5)}$
79. $\frac{2y}{3z}$ **81.** $\frac{1}{x - y}$ **83.** $\frac{b + a}{b}$ **85.** $y - x$ **87.** $-\frac{h}{4}$ **89.** $\frac{1}{2y}$
91. $\frac{1}{2}$ **93.** $-\frac{1}{2}$ **95.** 1, 2   **97.** 4, -1 **99.** $-\frac{q}{p}$
101. $\frac{t - 3}{(t + 3)(t + 1)}$ **103.** $\frac{x - 9}{3}$ **105.** 26   **107.** $\frac{-x^2 + 3x + 2}{(x - 1)(x + 1)^2}$
109. $x - 5$ **111.** $\frac{y^2 + xy + x^2}{x + y}$ **113.** $\frac{3n}{2n + 3}$ **115.** $\frac{x^2 - 5x - 5}{x - 5}$
117. $\frac{x + y}{a - b}$ **119.** 1   **121.** 2, -5 **123.** $\frac{1}{c - d}$ **125.** $\frac{x}{x - 3}$
127. $k_1(k_1 + 2), k_2 + 6$ **129. a.** $\frac{3}{r + 5}, \frac{3}{r - 5}$ **b.** $\frac{30}{(r + 5)(r - 5)}$ hr
135. 13, 25

Study Set Section 8.7 (page 732)

1. two   **3.** table   **5.** y-intercept, x-intercept   **7.** vertical,
horizontal   **9.** horizontal, vertical   **11.** Parallel, perpendicular
13. $y = mx + b$ **15. a.** $(-3, 0); (0, 4)$ **b.** False   **17.** sub 1
19.   **21.**

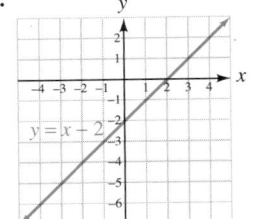

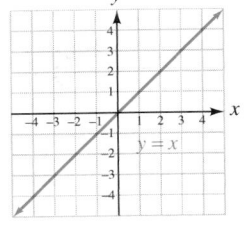

23.
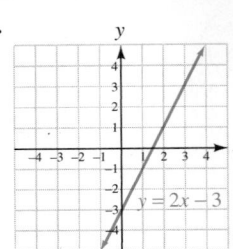
$y = 2x - 3$

25.
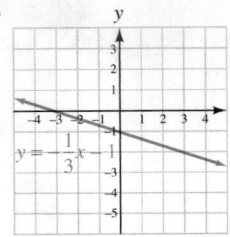
$y = -\frac{1}{3}x - 1$

27.
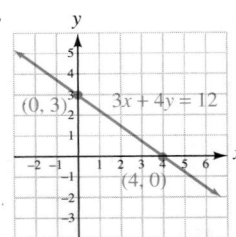
$(0, 3)$ $3x + 4y = 12$ $(4, 0)$

29.
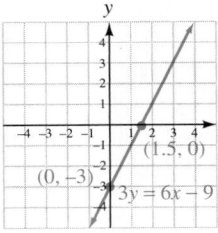
$(1.5, 0)$ $(0, -3)$ $3y = 6x - 9$

31.
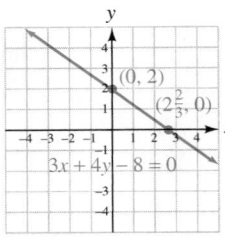
$(0, 2)$ $(2\frac{2}{3}, 0)$ $3x + 4y - 8 = 0$

33.
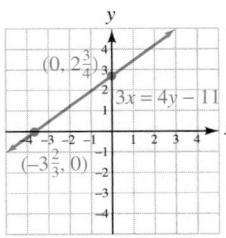
$(0, 2\frac{3}{4})$ $3x = 4y - 11$ $(-3\frac{2}{3}, 0)$

35.
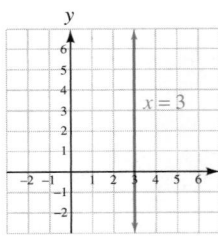
$x = 3$

37.
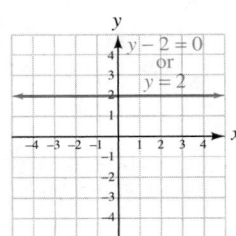
$y - 2 = 0$ or $y = 2$

39. $\frac{6}{7}$ **41.** $-\frac{8}{3}$ **43.** $\frac{5}{2}$ **45.** -1 **47.** $-\frac{1}{3}$ **49.** Undefined
51. $y = -4x + 16$ **53.** $y = -\frac{11}{6}x - \frac{7}{3}$ **55.** $y = 5x - 25$
57. $y = 3x + 17$ **59.** Parallel   **61.** Perpendicular   **63.** Neither
65. Perpendicular   **67.** $y = 4x$ **69.** $y = 4x - 3$ **71.** $y = 3x$
73. $y = -\frac{1}{4}x + \frac{11}{2}$ **75.** $(3, 4)$ **77.** $(9, 12)$ **79.** $\left(\frac{1}{2}, -8\right)$
81. $\left(-\frac{3}{2}, \frac{5}{2}\right)$ **83.** $\frac{1}{25} = 4\%$ **85. a.** An increase of 65.5 million
units/yr   **b.** A decrease of 47 million units/yr
87. $v = -2{,}298x + 19{,}984$ **89. a.** $B = \frac{1}{100}p - 195$ **b.** 905
95. -5 **97.** 5   **99.** $w = \frac{P - 2l}{2}$

Study Set Section 8.8 (page 752)
1. function   **3.** squaring, parabola   **5.** absolute, value
7. a. $\{-5, 0, 1, 9\}$ **b.** $\{-1, 2, 4\}$ **9. a.**
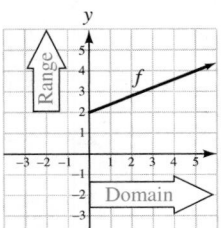

b. D: the set of nonnegative real numbers; R: the set of real numbers
greater than or equal to 2   **11.** of   **13.** 2, 7, 2, 7   **15.** No;
(4, 2), (4, 4), (4, 6)   **17.** Yes   **19.** No; $(-1, 0)$, $(-1, 2)$
21. Yes   **23.** Yes   **25.** No; (1, 1), (1, -1)   **27.** Yes   **29.** Yes
31. No; (1, 1), (1, -1)   **33.** 22, 2   **35.** 3, 11   **37.** $6, 2t^2 - t$
39. $7, t^3 - 1$ **41.** 4, 4   **43.** 2, 2   **45.** $\frac{1}{5}$, 1   **47.** $-2, \frac{2}{5}$
49. $3w - 5, 3w - 2$ **51.** $w^2 + 9, w^2 + 2w + 10$ **53.** 0
55. 1   **57.** D: $\{-2, 4, 6\}$; R: $\{3, 5, 7\}$ **59.** D: the set of real
numbers; R: the set of real numbers   **61.** D: the set of all real
numbers except 4; R: the set of all real numbers except 0   **63.** D: the
set of real numbers; R: the set of nonnegative real numbers

65.
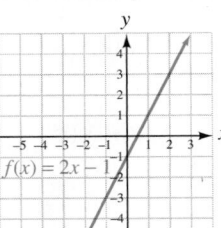
$f(x) = 2x - 1$

67.
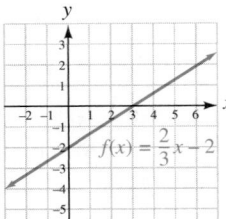
$f(x) = \frac{2}{3}x - 2$

69. Not a function; (0, 2), (0, -2)   **71.** A function   **73.** A function
75. A function   **77. a.** -4 **b.** 0   **c.** 2   **d.** -1 **79. a.** 2
b. 4   **c.** $-1, 1, 5$ **d.** 2, 4
81. D: the set of real numbers;   **83.** D: the set of real numbers;
R: the set of real numbers   R: the set of real numbers
greater than or equal to 2

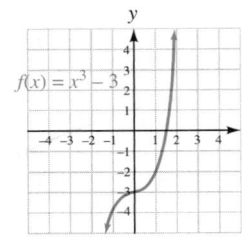
$f(x) = x^2 + 2$
$f(x) = x^3 - 3$

85. D: the set of real numbers;   **87.** D: the set of real numbers;
R: the set of nonnegative real   R: the set of nonnegative real
numbers   numbers

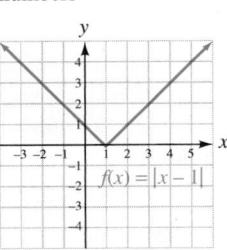
$f(x) = |x - 1|$

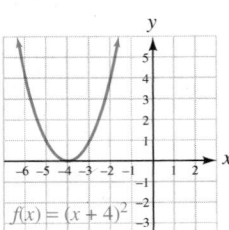
$f(x) = (x + 4)^2$

89. D: the set of real numbers;   **91.** D: the set of real numbers;
R: the set of all real numbers   R: the set of real numbers
greater than or equal to -2

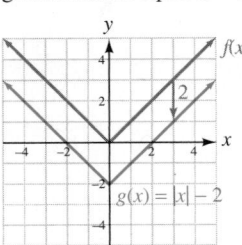
$f(x)$
$g(x) = |x| - 2$

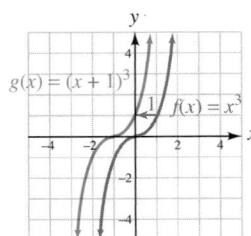
$g(x) = (x + 1)^3$
$f(x) = x^3$

93. D: the set of real numbers; R: the set of real numbers greater than or equal to -3

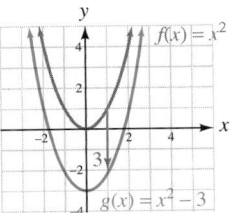

95. D: the set of real numbers; R: the set of all real numbers greater than or equal to 1

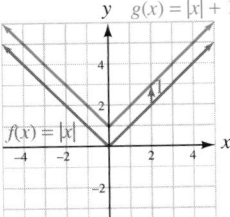

97.

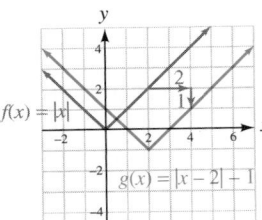

99.

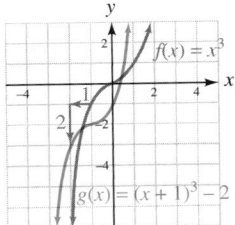

101.

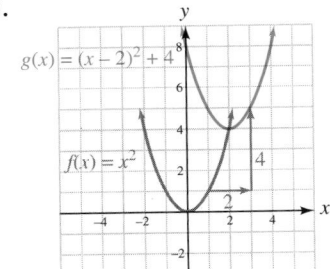

103.

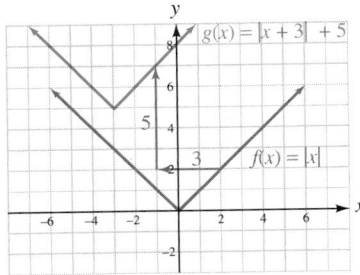

105.

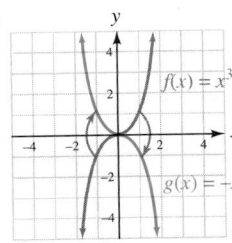

107.

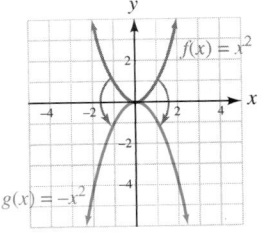

109.

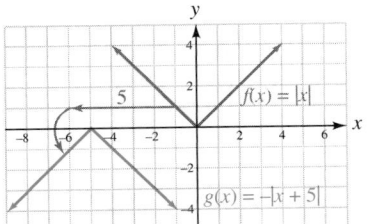

111.

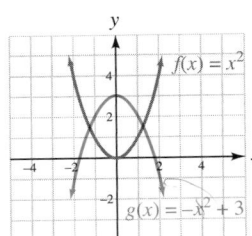

113.

115. 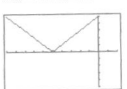   **117.** 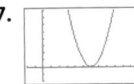   **119.**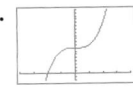

121. Between 20°C and 25°C   **123. a.** $C(f) = 95f + 12,000$
b. \$197,250   **129.** $(0, -6)$

Study Set Section 8.9 (page 761)

1. direct, increases   **3.** number   **5.** directly, inversely   **7.** Inverse
9. Direct   **11. a.** No   **b.** Yes   **c.** No   **d.** Yes   **13.** $t = ks$
15. $v = \frac{k}{r^2}$ **17.** $d = krt$ **19.** $M = \frac{kxz^2}{n^3}$ **21.** A varies directly as
the cube of r.   **23.** d varies inversely as the fourth power of W.
25. L varies jointly as m and n.   **27.** R varies directly as L and
inversely as d^2.   **29.** 117.6 newtons   **31.** 432 mi   **33.** $85\frac{1}{3}$
35. $P = 105f$ **37.** 26,437.5 gal   **39.** 45 w   **41.** 4.4 in.
43. 546 Kelvin   **47.** 3.5×10^{-4} **49.** 25,000

Chapter 8 Review (page 765)

1. $-\frac{12}{5}$ **2.** -9 **3.** 8   **4.** $\frac{11}{7}$ **5.** $\frac{88}{17}$ **6.** 12   **7.** All real
numbers, $\mathbb{R}$; identity   **8.** No solution, $\varnothing$; contradiction   **9.** -8
10. 0   **11.** $h = \frac{V}{\pi r^2}$ **12.** $x = \frac{6v - aby}{ab}$ or $x = \frac{6v}{ab} - y$
13. $[4, \infty)$ **14.** $\left(-\infty, -\frac{51}{11}\right)$
15. $(-\infty, 20)$ **16.** $(-\infty, \infty), \mathbb{R}$
17. 5 ft from one end   **18.** Length: 9 m, width: 5 m   **19.** $\{-3, 3\}$
20. $\{-6, -5, -3, 0, 3, 6, 8\}$ **21.** Yes   **22.** No
23.   **24.**
25. $[-10, -4)$ **26.** $(-\infty, -11)$
27. No solution, $\varnothing$ **28.** $[0, 0]$
29. $\left(-\frac{1}{3}, 2\right)$ **30.** $[1, 9]$
31. Yes   **32.** No   **33.** $(-\infty, -5) \cup (4, \infty)$

34. $(-\infty, \infty)$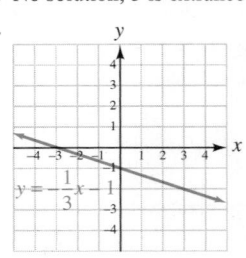

35. $17 \le 4x \le 25$, 4.25 ft $\le x \le 6.25$ ft, $[4.25, 6.25]$

36. a. ii, iv   **b.** i, iii   **37.** $2, -2$ **38.** $3, -\frac{11}{3}$ **39.** $\frac{26}{3}, -\frac{10}{3}$

40. No solution, $\varnothing$ **41.** 3 **42.** $\frac{1}{8}, -\frac{19}{8}$ **43.** $\frac{1}{5}, -5$ **44.** $\frac{13}{12}$

45. $[-3, 3]$

46. $(-5, -2)$

47. $\left[-3, \frac{19}{3}\right]$ **48.** No solution, $\varnothing$

49. $(-\infty, -1) \cup (1, \infty)$

50. $(-\infty, -4] \cup \left[\frac{22}{5}, \infty\right)$

51. $\left(-\infty, \frac{4}{3}\right) \cup (4, \infty)$

52. $(-\infty, \infty)$, $\mathbb{R}$ **53.** Since $|0.04x - 8.8|$ is always greater than or equal to 0 for any real number x, this absolute value inequality has no solution.   **54.** Since $\left|\frac{3x}{50} + \frac{1}{45}\right|$ is always greater than or equal to 0 for any real number x, this absolute value inequality is true for all real numbers.   **55. a.** $8, 2$ **b.** $[6, 10]$

56. $3, -3$ **57.** $(z - 5)(z - 6)$ **58.** $(x^2 + 4)(x^2 + y)$

59. $(4a - 1)(a - 1)$ **60.** $9x^2y^3z^2(3xz + 9x^2y^2 - 10z^5)$

61. $(5b - 2)(3b + 2)$ **62.** $-(x + 7)(x - 4)$

63. $3(5x + y)(x - 4y)$ **64.** $(w^4 - 10)(w^4 + 9)$

65. $(ry - a + 1)(r - 1)$ **66.** $(7a^3 + 6b^2)^2$ **67.** Prime

68. $2a^2(a + 3)(a - 1)$ **69.** $(s + t - 1)^2$ **70.** $m_1 = \frac{mm_2}{m_2 - m}$

71. $(z + 4)(z - 4)$ **72.** $(xy^2 + 8z^3)(xy^2 - 8z^3)$ **73.** Prime

74. $(c + a + b)(c - a - b)$

75. $2c(4a^2 + 9b^2)(2a + 3b)(2a - 3b)$

76. $(k + 1 + 3m)(k + 1 - 3m)$ **77.** $(m + n)(m - n - 1)$

78. $(t + 4)(t^2 - 4t + 16)$ **79.** $(2a - 5b^3)(4a^2 + 10ab^3 + 25b^6)$

80. $\frac{\pi}{2}h(r_1 + r_2)(r_1 - r_2)$ **81.** $\frac{31x}{72y}$ **82.** -2 **83.** $\frac{cd}{7x^2}$

84. $\frac{2a - 1}{a + 2}$ **85.** $\frac{a + b}{(m + p)(m - 3)}$ **86.** $\frac{3x(x - 1)}{(x - 3)(x + 1)}$ **87.** $\frac{1}{c - d}$

88. $-\frac{2}{t - 3}$ **89.** $\frac{40x + 7y^2z}{112z^2}$ **90.** $\frac{12y + 20}{15y(x - 2)}$ **91.** $\frac{14y + 58}{(y + 3)(y + 7)}$

92. $\frac{5x^2 + 11x}{(x + 1)(x + 2)}$ **93.** $\frac{2bc^3}{7}$ **94.** $\frac{p - 3}{2(p + 2)}$ **95.** $\frac{x - 2}{x + 3}$ **96.** -1

97. 5 **98.** $-2, 3$ **99.** $-1, -2$ **100.** $\frac{3}{2}$ **101.** 0

102. No solution; 3 is extraneous

103.

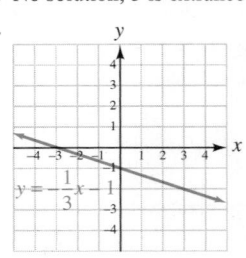

104.

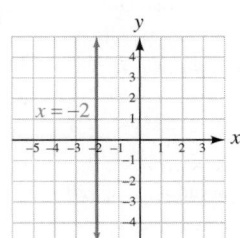

105.

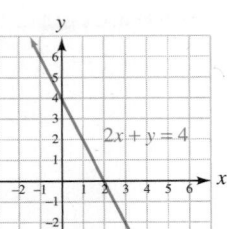

106.

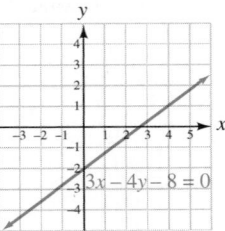

107. $\frac{2}{3}$ **108.** Slope of $l_1 = \frac{4}{5}$; slope of $l_2 = -\frac{8}{5}$ **109.** 1

110. $-\frac{14}{9}$ **111.** 0 **112.** Undefined   **113.** Perpendicular

114. Parallel   **115.** $y = 3x + 29$ **116.** $y = -\frac{13}{8}x + \frac{3}{4}$

117. $y = \frac{3}{2}x - \frac{1}{2}$ **118.** $y = -\frac{2}{3}x - 7$

119. $v = -1,720x + 8,700$ **120.** $\left(\frac{3}{2}, 8\right)$ **121. a.** function

b. function, variable, dependent   **122. a.** Yes   **b.** No;

$(-1, 8), (-1, 9)$ **c.** Yes   **123.** Yes   **124.** No; $(1, 1), (1, -1)$

125. -7 **126.** 18 **127.** 8 **128.** $3t + 2$ **129.** 3 **130.** $\frac{4}{3}$

131. D: the set of real numbers, R: the set of real numbers

132. D: the set of all real numbers except 2, R: the set of all real numbers except 0   **133.** Function   **134.** Not a function; $(0, 2), (0, 3)$

135.

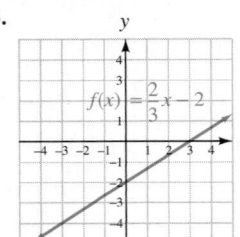

136. a. -4 **b.** 3 **c.** 0

137. a. 6, up   **b.** 6, left

138.

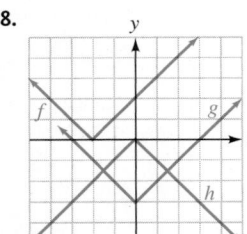

139.

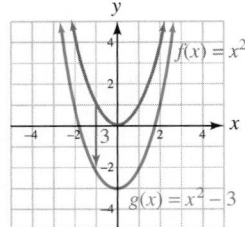

140.

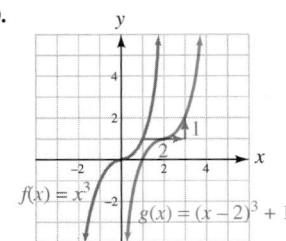

141. D: the set of real numbers; R: the set of real numbers

142. D: the set of real numbers; R: the set of all real numbers greater than or equal to 1   **143.** $\$5,460$ **144.** 1.25 amps   **145.** $\frac{1}{32}$

146. 126.72 lb   **147.** Inverse variation   **148.** 0.2

Chapter 8 Test (page 781)

1. a. solve   **b.** inequality   **c.** binomials   **d.** reciprocal

e. function, variable, dependent   **2.** Yes   **3.** $\frac{21}{5}$ **4.** -1

5. No solution, $\varnothing$; contradiction   **6.** $a = \frac{180n - 360}{n}$

7. 4 cm by 9 cm **8.** More than 78 **9.** $(-\infty, -5]$

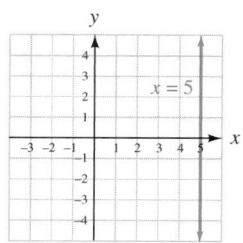

10. $(-2, 16)$

11. $\left[1, \frac{9}{4}\right]$

12. $(-\infty, -3) \cup (8, \infty)$

13. $(-\infty, -9) \cup (13, \infty)$

14. $\left[\frac{4}{3}, \frac{8}{3}\right]$ **15.** 8, −11 **16.** 4, −4

17. $3abc(4a^2b - abc + 2c^2)$ **18.** $4(y^2 + 4)(y + 2)(y - 2)$
19. $(b + 5)(b^2 - 5b + 25)$ **20.** $3(u + 2)(2u - 1)$
21. $(a - y)(x + y)$ **22.** $(5m^4 - 6n)^2$ **23.** Prime
24. $(x + 3 + y)(x + 3 - y)$
25. $(4a - 5b^2)(16a^2 + 20ab^2 + 25b^4)$
26. $(x - y + 5)(x - y - 2)$ **27.** -3 **28.** $\frac{2x + y}{4y}$ **29.** $\frac{(x + y)^2}{2}$
30. $\frac{13}{x + 1}$ **31.** $\frac{2}{t - 4}$ **32.** $\frac{6a - 17}{(a + 1)(a - 2)(a - 3)}$ **33.** $\frac{u^2}{2vw^2}$
34. $\frac{3k^2 + 4k + 4}{3k^2 - 9k - 9}$ **35.** $y = -\frac{3}{2}x - 2$ **36.** $m = -\frac{1}{3}, \left(0, -\frac{3}{2}\right)$
37. $-\frac{11}{7}$ **38. a.** $v = -600x + 4,000$ **b.** $(0, 4,000)$; it gives the
value of the copier when new: $4,000.
39. $(5, 0), (0, -2)$ **40.**

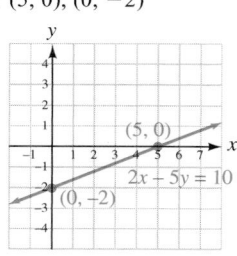

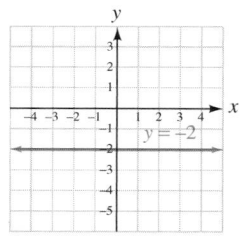

41. Yes **42.** No **43.** -1 **44.** -20
45.

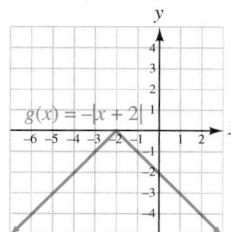

46. D: the set of real numbers; R: the
set of all real numbers greater than or
equal to -5 **47.** -2 **48.** 2
49. $\frac{6}{5}$ **50.** 25 decibels

Cumulative Review Chapters 1–8 (page 784)
1. a. True **b.** False **c.** True **2.** -14 **3.** $-2p - 6z$
4. -2 **5.** 17 lb **6.** 4 in. **7.** 3.5 hr **8.** 80
9. $\left(-\infty, -\frac{3}{4}\right]$

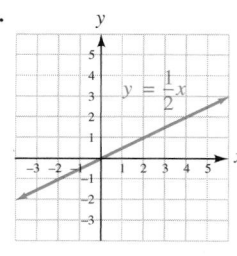

10. No

11.

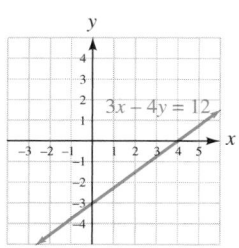

12.

13.

14.

15. $80 billion per yr **16. a.** 3 **b.** $-\frac{2}{3}$ **17.** $y = \frac{1}{2}x + 6$
18. 28 **19.** Yes **20.** $(-1, 5)$

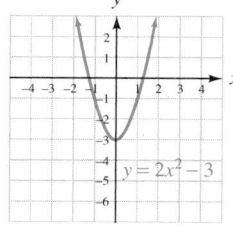

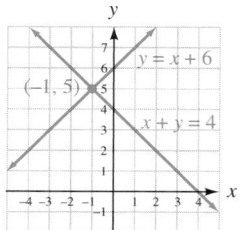

21. $(3, -1)$ **22.** $(-1, 2)$ **23.** 6%: $3,000; 12%: $3,000
24.

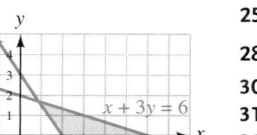

25. x^{31} **26.** $\frac{a^{15}b^5}{c^{20}}$ **27.** 1
28. $\frac{a^8}{16b^{12}}$ **29.** 4.8×10^{18} m
30. 4.3×10^{-9}
31. $2a^2 + a - 3$
32. $0.12p^9 - 1.8p^7$
33. $-6t^2 + 13st - 6s^2$
34. $16b^2 - 64b + 64$
35. $36b^2 - \frac{1}{4}$ **36.** $2x - 1$
37. $3xy(4x - 2y + 3y^2)$ **38.** $(x + y)(2x - 3)$
39. $(x + 5)(x + 2)$ **40.** $(3a + 4)(2a - 5)$ **41.** Prime
42. $(5a - 7b)^2$ **43.** $(a + 2b)(a^2 - 2ab + 4b^2)$
44. $2x(x^2 + 4)(x + 2)(x - 2)$ **45.** $-1, -2$ **46.** 0, 2
47. $\frac{2}{3}, -\frac{1}{2}$ **48.** $-5, 5$ **49.** 5 cm **50.** $\frac{x + 1}{x - 1}$
51. $\frac{(p - 3)(p + 2)}{3(p + 3)}$ **52.** $-\frac{3}{5x}$ **53.** $\frac{1}{3a}$ **54.** $\frac{7x + 29}{(x + 5)(x + 7)}$
55. $\frac{3 - 16b^2}{18b^4}$ **56.** $\frac{y + x}{y - x}$ **57.** 1 **58.** $2\frac{2}{9}$ hr **59.** 6,480
60. 20 **61.** $(-\infty, 2) \cup (7, \infty)$

62. $(-\infty, -2) \cup (2, \infty)$

63. $3, -\frac{3}{2}$ **64.** $5, -\frac{3}{5}$ **65.** $\left[-\frac{2}{3}, 2\right]$

66. $(-\infty, -4) \cup (1, \infty)$

67. $(x + y + 4)(x + y + 3)$
68. $[(a - b)^2 + c^2](a - b + c)(a - b - c)$
69. $(a + b)(a - b + 1)$ **70.** $(a + 2 + b)(a + 2 - b)$
71. $-a + 5$ **72.** $\frac{x - 5}{x - 4}$ **73.** $-\frac{3x}{2}$ **74.** Perpendicular
75. D: $\{-6, 0, 1, 5\}$; R: $\{-12, 0, 4, 8\}$ **76.** -6 **77.** The set of
all real numbers except 2 **78.** D: the set of real numbers; R: the set
of all real numbers greater than or equal to -2 **79.** 9 **80.** 1 day

Study Set Section 9.1 (page 800)
1. square, cube **3.** index, radicand **5.** odd, even **7.** a
9. two, 5 **11.** x **13.** 3, up **15. a.** 3 **b.** 0 **c.** 6

d. D: $[2, \infty)$; R: $[0, \infty)$ **17.** $\sqrt{x^2} = |x|$ **19.** $f(x) = \sqrt{x-5}$
21. 10   **23.** -8 **25.** $\frac{1}{3}$ **27.** 0.5   **29.** Not a real number
31. 11   **33.** 3.4641   **35.** 26.0624   **37.** $2|x|$ **39.** $9h^2$
41. $6|s^3|$ **43.** $12m^4$ **45.** $|y-1|$ **47.** a^2+3
49. a. 2 **b.** 5 **51. a.** 2   **b.** -3
53. $0, -1, -2, -3, -4$; D: $[0, \infty)$;   **55.** D: $[-4, \infty)$; R: $[0, \infty)$
R: $(-\infty, 0]$

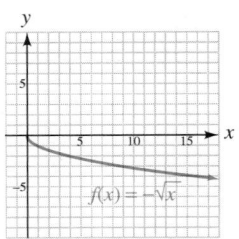

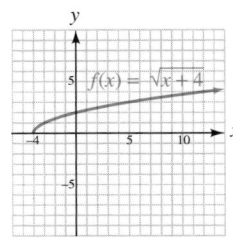

57. 1   **59.** -5
61. $\frac{2}{3}$ **63.** 4   **65.** $-6a$ **67.** $-10p^2q$
69. $-5, -4, -3, -2, -1$;   **71.** D: $(-\infty, \infty)$; R: $(-\infty, \infty)$
D: $(-\infty, \infty)$; R: $(-\infty, \infty)$

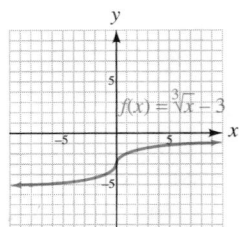

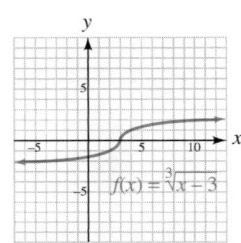

73. 3   **75.** -3 **77.** Not a real number   **79.** $-\frac{1}{2}$ **81.** $2a$
83. $3|a|$ **85.** k^2 **87.** $(m+4)^2$ **89.** $4s^3t^2$ **91.** $-7b^4$
93. $\frac{1}{2}$ **95.** $-5m^2$ **97.** $2|ab|$ **99.** $x+2$ **101.** Not a real
number   **103.** $|n+6|$ **105.** 7.0 in.   **107.** 1,138.5 mm
109. About 58.6 beats/min   **111.** 13.4 ft   **113.** 3.5%   **119.** 1
121. $\frac{3(m^2+2m-1)}{(m+1)(m-1)}$

Study Set Section 9.2 (page 814)

1. rational (or fractional)   **3.** index, radicand
5. $25^{1/5}, \left(\sqrt[3]{-27}\right)^2, 16^{-3/4}, \left(\sqrt{81}\right)^3, -\left(\frac{9}{64}\right)^{1/2}$
7.
$(-125)^{1/3} \quad -(9/100)^{-1/2} \qquad -16^{-1/4} \qquad\qquad 8^{2/3} \qquad\qquad 4^{3/2}$

```
———•—————•———•———•————•———•———•———•———•———•———•———•—→
  -5   -4  -3  -2  -1   0   1   2   3   4   5   6   7   8
```

9. $\sqrt[n]{x}$ **11.** $\frac{1}{x^{m/n}}$ **13.** $100a^4, 10a^2$ **15.** 5   **17.** 3   **19.** 2
21. -6 **23.** -2 **25.** $\frac{1}{2}$ **27.** $2x^2$ **29.** $|x|$ **31.** m^2
33. Not a real number   **35.** $-3n^3$ **37.** $2|x|$ **39.** Not a real number
41. $|x+1|$ **43.** 216   **45.** 8   **47.** $\frac{1}{36}$ **49.** -32 **51.** $125x^6$
53. $4x^4y^2$ **55.** $27x^3y^6$ **57.** $-\frac{x^4}{16}$ **59.** $(8abc)^{1/5}$
61. $(a^2-b^2)^{1/3}$ **63.** $\sqrt[6]{6x^3y}$ **65.** $\sqrt{2s^2-t^2}$ **67.** $\frac{1}{2}$ **69.** $\frac{1}{5}$
71. $\frac{1}{64}$ **73.** $-\frac{1}{100y^2}$ **75.** $\frac{16}{81}$ **77.** $\frac{27y^3}{8}$ **79.** 2   **81.** 243
83. $9^{5/7}$ **85.** $\frac{1}{36}$ **87.** m^6 **89.** $a^{3/4}b^{1/2}$ **91.** 3   **93.** a
95. $y+y^2$ **97.** $x^2-1+x^{3/5}$ **99.** $\sqrt{5}$ **101.** $\sqrt[3]{11}$ **103.** $\sqrt{p}$
105. $\sqrt[5]{xy}$ **107.** $\sqrt[18]{c}$ **109.** $\sqrt[15]{7m}$ **111.** 2.47   **113.** 1.02
115. $5y$ **117.** $-\frac{a^3}{27}$ **119.** p **121.** $-\frac{1}{3x^2}$ **123.** $n^{3/5}-1$

125. 243   **127.** $\sqrt{5b}$ **129.** $-\frac{1}{4a^2b^4}$ **131.** 736 ft/sec
133. 1.96 units   **135.** 145.8 in. or 12.1 ft   **139.** $2\frac{1}{2}$ hr

Study Set Section 9.3 (page 826)

1. like   **3.** factor, perfect   **5.** $\sqrt[n]{a}\sqrt[n]{b}$, product, roots
7. a. $\sqrt{4 \cdot 5}$ **b.** $\sqrt{4}\sqrt{5}$ **c.** $\sqrt{4 \cdot 5} = \sqrt{4}\sqrt{5}$ **9. a.** $\sqrt{5}$,
$\sqrt[3]{5}$ (Answers may vary); no   **b.** $\sqrt{5}, \sqrt{6}$ (Answers may vary); no
11. $8k^3, 8k^3, 4k$ **13.** $5\sqrt{2}$ **15.** $3\sqrt{5}$ **17.** $2\sqrt[3]{4}$ **19.** $2\sqrt[3]{3}$
21. $5a\sqrt{3}$ **23.** $4\sqrt{2b}$ **25.** $8ab^2\sqrt{2ab}$ **27.** $10\sqrt{3xy}$
29. $-3x^2\sqrt[3]{2}$ **31.** $2x^3y\sqrt[4]{2}$ **33.** $11\sqrt{2}$ **35.** $4a\sqrt{7a}$
37. $-2\sqrt[5]{3a^4}$ **39.** $3x^4y\sqrt[3]{15y}$ **41.** $\frac{\sqrt{11}}{3}$ **43.** $\frac{\sqrt[3]{7}}{4}$ **45.** $\frac{\sqrt[4]{3}}{5}$
47. $\frac{x^2\sqrt[5]{3}}{2}$ **49.** $\frac{z}{4x}$ **51.** $\frac{\sqrt[4]{5x}}{2z}$ **53.** 10   **55.** $7x$ **57.** $2x^2$
59. $3a\sqrt[3]{a}$ **61.** $8\sqrt{7}$ **63.** $5\sqrt[3]{4}$ **65.** $3\sqrt{2}$ **67.** $-4\sqrt{2}$
69. $-\sqrt[3]{4}$ **71.** -10 **73.** $13\sqrt[4]{2}$ **75.** $7\sqrt{5}-3\sqrt{3}$
77. $10\sqrt{2x}$ **79.** $\sqrt[5]{7a^2}$ **81.** $3\sqrt{2t}+\sqrt{3t}$ **83.** $-11\sqrt[3]{2}$
85. $12\sqrt[3]{a}$ **87.** $4\sqrt[4]{4}$ **89.** $m\sqrt[6]{m^5}$ **91.** $5y^3\sqrt{2y}$
93. $\frac{5n^2\sqrt{5}}{8}$ **95.** $4x\sqrt[5]{xy^2}$ **97.** $2m\sqrt[4]{13n}$ **99.** $\frac{a^2}{4}$ **101.** $2t^2\sqrt[5]{2t}$
103. $3\sqrt[3]{3x}$ **105.** $8\pi\sqrt{5}$ ft^2; 56.2 ft^2 **107.** $5\sqrt{3}$ amps; 8.7 amps
109. $\left(26\sqrt{5}+10\sqrt{3}\right)$ in.; 75.5 in.   **115.** $-\frac{15x^5}{y}$
117. $3p+4-\frac{5}{2p-5}$

Study Set Section 9.4 (page 838)

1. FOIL   **3.** irrational   **5.** perfect   **7. a.** $6\sqrt{6}$ **b.** 48
c. Can't be simplified   **d.** $-6\sqrt{6}$ **9.** $\sqrt{6}, 48, 16, 4$ **11.** $3\sqrt{5}$
13. $6\sqrt{2}$ **15.** 18   **17.** $2\sqrt[3]{3}$ **19.** $48ab^2$ **21.** $5a\sqrt[4]{a}$
23. $-20r\sqrt[3]{10s}$ **25.** $x^2(x+3)$ **27.** $12\sqrt{5}-15$
29. $8\sqrt{3}+2\sqrt{14}$ **31.** $-8x\sqrt{10}+6\sqrt{15x}$ **33.** $8+2\sqrt[3]{3}$
35. $-1-2\sqrt{2}$ **37.** $3x-2y$
39. $12\sqrt[3]{2}+8\sqrt[3]{5}-18-6\sqrt[3]{20}$ **41.** $\sqrt[3]{25z^2}+3\sqrt[3]{15z}+2\sqrt[3]{9}$
43. 7   **45.** 12   **47.** 18   **49.** $-16x^2$ **51.** $18r-12\sqrt{2r}+4$
53. $3x+6\sqrt{x}+3$ **55.** $\frac{\sqrt{14}}{7}$ **57.** $\frac{2\sqrt{6}}{3}$ **59.** $\frac{2\sqrt{6}}{3}$ **61.** $\frac{\sqrt[3]{4}}{2}$
63. $\sqrt[3]{3}$ **65.** $\frac{\sqrt[4]{4}}{y}$ **67.** $\frac{\sqrt{5y}}{y}$ **69.** $\frac{\sqrt{6y}}{y}$ **71.** $\frac{\sqrt[3]{6t}}{3}$ **73.** $\frac{\sqrt[3]{2ab^2}}{b}$
75. $\frac{23\sqrt{2p}}{10p^3}$ **77.** $\frac{7\sqrt{6b}}{12b^2}$ **79.** $\frac{\sqrt[3]{20}}{4}$ **81.** $\frac{\sqrt[3]{36}}{9}$ **83.** $\frac{19\sqrt[3]{25c}}{5c}$
85. $\frac{\sqrt[3]{12r^2}}{2r}$ **87.** $\frac{\sqrt[4]{54t^2}}{3t}$ **89.** $\frac{25\sqrt[3]{2a^3}}{2a}$ **91.** $\frac{3\sqrt{2}-\sqrt{10}}{4}$
93. $\frac{2(\sqrt{x}-1)}{x-1}$ or $\frac{2\sqrt{x}-2}{x-1}$ **95.** $\frac{9-2\sqrt{14}}{5}$ **97.** $\frac{3\sqrt{6}+4}{2}$
99. $\frac{x-2\sqrt{xy}+y}{x-y}$ **101.** $\sqrt{2z}+1$ **103.** $\frac{x-9}{x(\sqrt{x}-3)}$
105. $\frac{x-y}{\sqrt{x}(\sqrt{x}-\sqrt{y})}$ **107.** $x\sqrt{14}+\sqrt{2x}$ **109.** 2,000x
111. $9p^2+6p\sqrt{5}+5$ **113.** $\frac{6m^2\sqrt{2m}}{5}$ **115.** $3n\sqrt[4]{n}$ **117.** $\frac{\sqrt[3]{3x}}{3}$
119. $\frac{\sqrt{2\pi}}{2\pi\sigma}$ **121.** $\frac{\sqrt{2}}{2}$ **129.** $\frac{1}{3}$

Study Set Section 9.5 (page 851)

1. radical   **3.** extraneous   **5.** $^n, ^n$ **7. a.** Square both sides.
b. Subtract 3 from both sides.   **c.** Add $\sqrt{2x+9}$ to both sides.
9. $x-6\sqrt{x}+9$ **11.** $6, ^2, ^2, 3x+3, 33, 11$, Yes   **13.** 4   **15.** 5
17. 6   **19.** 198   **21.** 7, ⁄   **23.** 14, ∅   **25.** 3, ⁄2
27. 0, ⁄1   **29.** $\frac{4}{5}$, no solution   **31.** ∕4, no solution   **33.** 4

35. $2, -1$ **37.** $1, -3$ **39.** 10   **41.** 85   **43.** -7 **45.** 12
47. 0   **49.** 3   **51.** 3   **53.** 9   **55.** 1   **57.** 4   **59.** 0
61. $\cancel{-3}$, no solution   **63.** 1, 9   **65.** $h = \frac{v^2}{2g}$ **67.** $l = \frac{8T^2}{\pi^2}$
69. $A = P(r + 1)^3$ **71.** $v^2 = c^2\left(1 - \frac{L_A^2}{L_B^2}\right)$ **73.** 16   **75.** $-1, 1$
77. $2, \cancel{1}$ **79.** -16 **81.** 4   **83.** $\frac{5}{2}, \frac{1}{2}$ **85.** $-\frac{1}{2}$ **87.** $\cancel{6}$, no
solution   **89.** 1   **91.** 4, 3   **93.** $\cancel{8}$, no solution   **95.** 1
97. 178 ft   **99.** About 488 watts   **101.** 10 lb   **103.** \$5
113. 2.5 foot-candles   **115.** 0.41511 in.

Study Set Section 9.6 (page 862)

1. hypotenuse   **3.** Pythagorean   **5.** a^2, b^2, c^2 **7.** $\sqrt{2}$ **9.** $\sqrt{3}$
11. $(x_2 - x_1)^2, (y_2 - y_1)^2$ **13.** 6, 52, 4, 2, 7.21   **15.** 10 ft
17. 17 ft   **19.** 40 ft   **21.** 24 cm   **23.** $x = 2, h = 2\sqrt{2} \approx 2.83$
25. $3.2\sqrt{2}$ ft ≈ 4.53 ft   **27.** $x = \frac{3\sqrt{2}}{2} \approx 2.12, y = \frac{3\sqrt{2}}{2} \approx 2.12$
29. $5\sqrt{2}$ in.   **31.** $x = 5\sqrt{3} \approx 8.66, h = 10$
33. $75\sqrt{3}$ cm ≈ 129.90 cm, 150 cm   **35.** $x = \frac{40\sqrt{3}}{3} \approx 23.09$,
$h = \frac{80\sqrt{3}}{3} \approx 46.19$ **37.** $\frac{55\sqrt{3}}{3}$ mm ≈ 31.75 mm,
$\frac{110\sqrt{3}}{3}$ mm ≈ 63.51 mm   **39.** $x = 50, y = 50\sqrt{3} \approx 86.60$
41. 0.75 ft, $0.75\sqrt{3}$ ft ≈ 1.30 ft   **43.** 5   **45.** 13   **47.** 10
49. $\sqrt{34}$ **51.** $2\sqrt{5}$ **53.** $3\sqrt{10}$ **55. a.** 118.73 m
b. 133.14 m   **57.** $7\sqrt{3}$ cm   **59.** $\left(5\sqrt{2}, 0\right), \left(0, 5\sqrt{2}\right), \left(-5\sqrt{2}, 0\right),$
$\left(0, -5\sqrt{2}\right); (7.07, 0), (0, 7.07), (-7.07, 0), (0, -7.07)$
61. $10\sqrt{3}$ mm ≈ 17.32 mm   **63.** $10\sqrt{181}$ ft ≈ 134.54 ft
65. About 0.13 ft   **67. a.** 21.21 units   **b.** 8.25 units
c. 13.00 units   **69.** Yes   **75.** 7   **77.** 9

Study Set Section 9.7 (page 877)

1. imaginary, power   **3.** real, imaginary   **5. a.** $\sqrt{-1}$ **b.** -1
c. $-i$ **d.** 1   **e.** four   **7.** real, imaginary   **9.** $1 + 8i, 1 + 8i$
11.

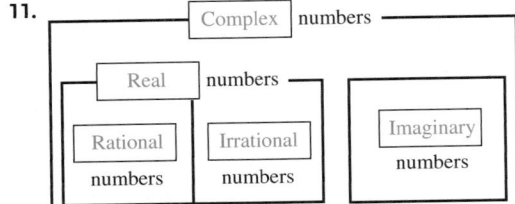

13. $9, 6i, 2, 11$ **15. a.** True   **b.** False   **c.** False   **d.** False
17. $3i$ **19.** $\sqrt{7}i$ or $i\sqrt{7}$ **21.** $2\sqrt{6}i$ or $2i\sqrt{6}$
23. $-6\sqrt{2}i$ or $-6i\sqrt{2}$ **25.** $45i$ **27.** $\frac{5}{3}i$ **29. a.** $5 + 0i$
b. $0 + 7i$ **31. a.** $1 + 5i$ **b.** $-3 + 2i\sqrt{2}$ **33. a.** $76 - 3i\sqrt{6}$
b. $-7 + i\sqrt{19}$ **35. a.** $-6 - 3i$ **b.** $3 + i\sqrt{6}$ **37.** $8 - 2i$
39. $15 + 2i$ **41.** $3 - 5i$ **43.** $1 + 3i$ **45.** -6 **47.** $-2\sqrt{6}$
49. $6 - 27i$ **51.** $35 - 28i$ **53.** $6 + 14i$ **55.** $-25 - 25i$
57. $7 + i$ **59.** $12 + 5i$ **61.** $13 - i$ **63.** $3 + 4i$ **65.** 40
67. 65   **69.** $\frac{45}{26} - \frac{9}{26}i$ **71.** $-\frac{77}{65} + \frac{44}{65}i$ **73.** $\frac{14}{17} - \frac{5}{17}i$
75. $-\frac{6}{29} + \frac{43}{29}i$ **77.** $\frac{11}{10} + \frac{13}{10}i$ **79.** $0 - i$ **81.** $4 + 0i$
83. $-2 + 0i$ **85.** $0 - \frac{5}{3}i$ **87.** $0 + \frac{2}{7}i$ **89.** i **91.** $-i$
93. 1   **95.** -1 **97.** $4 - 11i$ **99.** $14 - 8i$ **101.** $-2 - 6i$

103. $-\frac{4}{13} - \frac{6}{13}i$ **105.** $18 + 12i$ **107.** $0 + \frac{4}{5}i$
109. $8 + \sqrt{2}i$ or $8 + i\sqrt{2}$ **111.** $-2 + 7i$ **113.** $-48 - 64i$
115. $\frac{1}{4} - \frac{\sqrt{15}}{4}i$ **117.** $-1 + i$ **123.** 20 mph

Chapter 9 Review (page 880)

1. 7   **2.** -11 **3.** $\frac{15}{7}$ **4.** Not a real number   **5.** $10a^6$
6. $5|x|$ **7.** x^4 **8.** $|x + 2|$ **9.** -3 **10.** -6 **11.** $4x^2y$
12. $\frac{x^3}{5}$ **13.** 2   **14.** -2 **15.** $4x^2|y|$ **16.** $x + 1$ **17.** $-\frac{1}{2}$
18. Not a real number   **19.** Not a real number   **20.** 0   **21.** 13 ft
22. 24 cm^2
23. D: $[0, \infty)$; R: $[0, \infty)$ **24.** D: $(-\infty, \infty)$; R: $(-\infty, \infty)$

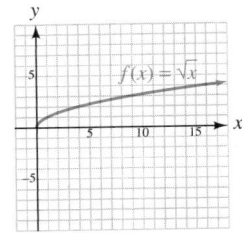
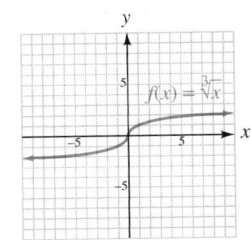

25. D: $[-2, \infty)$; R: $[0, \infty)$ **26.** D: $(-\infty, \infty)$; R: $(-\infty, \infty)$

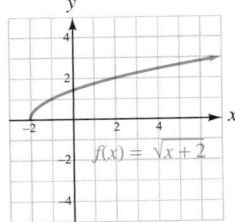
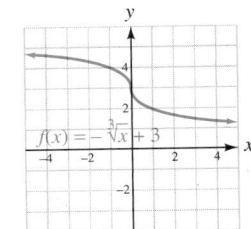

27. $\sqrt{t}$ **28.** $\sqrt[4]{5xy^3}$ **29.** 5   **30.** -6 **31.** Not a real number
32. 1   **33.** $\frac{3}{x}$ **34.** -2 **35.** 5   **36.** $3cd$ **37.** 27   **38.** $\frac{1}{4}$
39. $-16,807$ **40.** 10   **41.** $\frac{27}{8}$ **42.** $\frac{1}{3,125}$ **43.** $125x^3y^6$
44. $\frac{1}{4u^4v^2}$ **45.** $5^{3/4}$ **46.** $a^{1/7}$ **47.** k^8 **48.** $3^{2/3}$ **49.** $u - 1$
50. $v + v^2$ **51.** $\sqrt{a}$ **52.** $\sqrt[6]{c}$ **53.** 183 mi   **54.** Two true
statements result: $32 = 32$.   **55.** $4\sqrt{5}$ **56.** $3\sqrt[3]{2}$ **57.** $2\sqrt[4]{10}$
58. $-2\sqrt[5]{3}$ **59.** $2x^2\sqrt{2x}$ **60.** $r^4\sqrt[4]{r}$ **61.** $-3j^2\sqrt[3]{jk}$
62. $-2xy\sqrt[3]{2x^2y}$ **63.** $\frac{\sqrt{m}}{12n^6}$ **64.** $\frac{\sqrt{17xy}}{8a^2}$ **65.** $2x$ **66.** $3x^3$
67. $3\sqrt{2}$ **68.** $11\sqrt{5}$ **69.** 0   **70.** $-8a\sqrt[4]{2a}$ **71.** $29x\sqrt{2}$
72. $13x\sqrt[3]{2}$ **73.** $9\sqrt[4]{2t^3} - 8\sqrt[4]{6t^3}$ **74.** $12x^2\sqrt[4]{x} + 5x\sqrt[4]{x}$
76. $\left(6\sqrt{2} + 2\sqrt{10}\right)$ in., 14.8 in.   **77.** 7   **78.** $6\sqrt{10}$ **79.** 32
80. $6\sqrt{10}$ **81.** $3x$ **82.** $x + 1$ **83.** $-2x^3\sqrt[3]{x}$ **84.** 3
85. $42t + 9t\sqrt{21t}$ **86.** $-2x^3y^3\sqrt[4]{2x^2y^2}$ **87.** $3b + 6\sqrt{b} + 3$
88. $\sqrt[3]{9p^2} - \sqrt[3]{6p} - 2\sqrt[3]{4}$ **89.** $\frac{10\sqrt{3}}{3}$ **90.** $\frac{\sqrt{15xy}}{5xy}$ **91.** $\frac{\sqrt[3]{6u^2}}{u^2}$
92. $\frac{\sqrt[4]{27ab^2}}{3b}$ **93.** $2\left(\sqrt{2} + 1\right)$ or $2\sqrt{2} + 2$
94. $\frac{12\sqrt{xz} - 16x - 2z}{z - 16x}$ **95.** $\frac{a - b}{a + \sqrt{ab}}$ **96.** $r = \frac{\sqrt[3]{6\pi^2 V}}{2\pi}$
97. 22   **98.** 16, 9   **99.** $\frac{13}{2}$ **100.** $\frac{9}{16}$ **101.** $2, -4$ **102.** 7
103. 1   **104.** $-\frac{3}{2}, 1$ **105.** $\cancel{8}$, no solution   **106.** 6, $\cancel{-2}$
107. $-\frac{1}{2}, 4$ **108.** 2   **109.** $P = \frac{A}{(r + 1)^2}$ **110.** $I = \frac{h^3b}{12}$

111. 17 ft   **112.** 88 yd   **113.** $7\sqrt{2}$ m ≈ 9.90 m

114. $\frac{15\sqrt{2}}{2}$ yd ≈ 10.61 yd   **115.** Shorter leg: 6 cm, longer leg:

$6\sqrt{3}$ cm ≈ 10.39 cm   **116.** $40\sqrt{3}$ ft ≈ 69.28 ft,

$20\sqrt{3}$ ft ≈ 34.64 ft   **117.** $x = 5\sqrt{2} \approx 7.07$, $y = 5$

118. $x = 25\sqrt{3} \approx 43.30$, $y = 25$ **119.** 13   **120.** $2\sqrt{2}$

121. $5i$ **122.** $3i\sqrt{2}$ **123.** $-i\sqrt{6}$ **124.** $\frac{3}{8}i$

125. Real, Imaginary   **126. a.** True   **b.** True   **c.** False

d. False   **127. a.** $3 - 6i$ **b.** $0 - 19i$ **128. a.** $-1 + 7i$

b. $0 + i$ **129.** $8 - 2i$ **130.** $3 - 5i$ **131.** $3 + 6i$

132. $22 + 29i$ **133.** $-3\sqrt{3} + 0i$ **134.** $-81 + 0i$ **135.** $4 + i$

136. $0 - \frac{3}{11}i$ **137.** -1 **138.** i

Chapter 9 Test (page 889)

1. a. radical   **b.** imaginary   **c.** extraneous   **d.** isosceles
e. rationalize   **f.** complex   **2. a.** If $\sqrt[n]{a}$ and $\sqrt[n]{b}$ are real
numbers, then $\sqrt[n]{ab} = \sqrt[n]{a}\sqrt[n]{b}$.   **b.** If $\sqrt[n]{a}$ and $\sqrt[n]{b}$ are real
numbers, then $\sqrt[n]{\frac{a}{b}} = \frac{\sqrt[n]{a}}{\sqrt[n]{b}}$, $(b \neq 0)$.

3. D: $[1, \infty)$; R: $[0, \infty)$

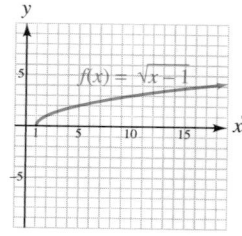

4. 46 ft/sec   **5. a.** -1 **b.** 2   **c.** 1   **d.** D: $(-\infty, \infty)$;
R: $(-\infty, \infty)$ **6.** No real number raised to the fourth power is -16.
7. $7x^2$ **8.** -9 **9.** $\frac{1}{216}$ **10.** $\frac{25n^4}{4}$ **11.** $2^{4/3}$ **12.** $a^{1/9}$
13. $|x|$ **14.** $|y - 5|$ **15.** $-4xy^2$ **16.** $\frac{2}{3}a$ **17.** $t + 8$
18. $6xy^2\sqrt{15xy}$ **19.** $2x^5y\sqrt[3]{3}$ **20.** $2\sqrt[4]{2}$ **21.** $2y^2\sqrt{3y}$
22. $14\sqrt[3]{5}$ **23.** $5z^3\sqrt[4]{3z}$ **24.** $-6x\sqrt{y} - 2xy^2$ **25.** $3 - 7\sqrt{6}$
26. $\sqrt[3]{4a^2} + 18\sqrt[3]{2a} + 81$ **27.** $\frac{4\sqrt{10}}{5}$ **28.** $\sqrt{3t} + 1$
29. $\frac{\sqrt[3]{18a^2}}{2a}$ **30.** $\frac{1}{\sqrt{2}(\sqrt{5} - 3)} = \frac{1}{\sqrt{10} - 3\sqrt{2}}$ **31.** $\frac{1}{15}$ **32.** 10
33. $\varnothing$, no solution   **34.** 3, $\cancel{-3}$ **35.** 2, 3   **36.** $\cancel{108}$, no solution
37. -2 **38.** $G = \frac{4\pi^2r^3}{Mt^2}$ **39.** $x = \frac{8\sqrt{3}}{3}$ cm ≈ 4.62 cm,

$h = \frac{16\sqrt{3}}{3}$ cm ≈ 9.24 cm   **40.** $x = \frac{(12.26)\sqrt{2}}{2}$ in. ≈ 8.67 in.,

$y = \frac{(12.26)\sqrt{2}}{2}$ in. ≈ 8.67 in.   **41.** 25   **42.** 28 in.   **43.** $3i\sqrt{5}$
44. -1 **45.** $-4 + 11i$ **46.** $4 - 7i$ **47.** $75 + 45i$
48. $-46 - 78i$ **49.** $0 - \frac{\sqrt{2}}{2}i$ **50.** $\frac{1}{2} + \frac{1}{2}i$

Study Set Section 10.1 (page 905)

1. quadratic   **3.** square   **5.** $\sqrt{c}, -\sqrt{c}$ **7. a.** 36   **b.** $\frac{25}{4}$
9. a. Subtract 7 from both sides   **b.** Divide both sides by 4
11. a. False   **b.** True   **13.** plus or minus   **15.** $0, -2$
17. $5, -5$ **19.** $-2, -4$ **21.** $2, \frac{1}{2}$ **23.** ± 9 **25.** ± 6
27. $\pm 5\sqrt{2}$ **29.** $\pm\frac{4\sqrt{3}}{3}$ **31.** $-8, -2$ **33.** $0, -8$
35. $-5 \pm \sqrt{3}$ **37.** $2 \pm 2\sqrt{2}$ **39.** $\pm 4i$ **41.** $\pm\frac{9}{2}i$
43. $3 \pm i\sqrt{5}$ **45.** $1 \pm 2i$ **47.** $x^2 + 24x + 144 = (x + 12)^2$

49. $a^2 - 7a + \frac{49}{4} = \left(a - \frac{7}{2}\right)^2$ **51.** $2, -4$ **53.** $2, 6$ **55.** $1, -6$

57. $-1, 4$ **59.** $-\frac{1}{2}, 1$ **61.** $\frac{3}{4}, -\frac{1}{3}$ **63.** $\frac{6 \pm \sqrt{33}}{3}$

65. $\frac{-5 \pm \sqrt{41}}{4}$ **67.** $-1 \pm i$ **69.** $-4 \pm i\sqrt{2}$ **71.** $2, -\frac{4}{3}$

73. $\frac{3 \pm 2\sqrt{3}}{3}$ **75.** $-4 \pm \sqrt{10}$ **77.** $\pm 2i\sqrt{3}$ **79.** $1 \pm 3\sqrt{2}$

81. $\frac{7 \pm \sqrt{37}}{2}$ **83.** $\pm\sqrt{5}$ **85.** $\frac{-7 \pm \sqrt{29}}{10}$ **87.** $-\frac{1}{2} \pm \frac{\sqrt{11}}{2}i$

89. $\frac{-5 \pm 2\sqrt{6}}{8}$ **91.** $-3, 9$ **93.** $-\frac{1}{4} \pm \frac{\sqrt{11}}{4}i$ **95.** 4.4 sec

97. 3,959 mi   **99.** 1.70 in.   **101.** 0.92 in.   **103.** 2.9 ft, 6.9 ft

109. $2ab^2\sqrt[3]{5}$ **111.** $\frac{2}{5}$

Study Set Section 10.2 (page 916)

1. quadratic   **3. a.** $x^2 + 2x + 5 = 0$ **b.** $3x^2 + 2x - 1 = 0$

5. a. True   **b.** True   **7.** $2, -4$ **9. a.** $2; \frac{-2 + \sqrt{3}}{2}, \frac{-2 - \sqrt{3}}{2}$

b.

$$\frac{-2 - \sqrt{3}}{2} \qquad \frac{-2 + \sqrt{3}}{2}$$

number line from -2 to 2 marked at $\frac{-2-\sqrt3}{2}$ and $\frac{-2+\sqrt3}{2}$

11. a. The fraction bar wasn't drawn under both parts of the numerator.
b. A $\pm$ sign wasn't written between $-b$ and the radical.   **13.** 1, 2
15. A repeated solution of -6 **17.** $-\frac{3}{2}, 1$ **19.** $\frac{2}{3}, -\frac{1}{4}$
21. $\frac{1 \pm \sqrt{29}}{2}$; 3.19, -2.19 **23.** $\frac{-5 \pm \sqrt{5}}{10}$; $-0.28, -0.72$
25. $\frac{-3 \pm \sqrt{6}}{3}$; $-0.18, -1.82$ **27.** $\frac{1 \pm 2\sqrt{5}}{2}$; 2.74, -1.74
29. $-\frac{1}{4} \pm \frac{\sqrt{7}}{4}i$ **31.** $\frac{1}{3} \pm \frac{\sqrt{2}}{3}i$ **33.** $1 \pm i$ **35.** $-\frac{1}{2} \pm i$
37. a. $5x^2 - 9x + 2 = 0$ **b.** $16t^2 + 24t - 9 = 0$
39. a. $3x^2 + 2x - 1 = 0$ **b.** $2m^2 - 3m - 2 = 0$ **41.** $\frac{4}{3}, -\frac{2}{5}$
43. $\frac{2}{3} \pm \frac{\sqrt{2}}{3}i$ **45.** $\frac{1}{4}, -\frac{3}{4}$ **47.** $\frac{3 \pm \sqrt{17}}{4}$ **49.** $5 \pm \sqrt{7}$
51. $23, -17$ **53.** $\frac{-5 \pm 3\sqrt{5}}{2}$ **55.** $\frac{1}{3} \pm \frac{\sqrt{6}}{3}i$ **57.** $\frac{-3 \pm \sqrt{29}}{10}$
59. $\frac{10 \pm \sqrt{55}}{30}$ **61.** $2 \pm 2i$ **63.** $3 \pm i\sqrt{5}$ **65.** $\frac{-5 \pm \sqrt{17}}{2}$
67. $\frac{9 \pm \sqrt{89}}{2}$ **69.** $1.22, -0.55$ **71.** $8.98, -3.98$ **73.** 40 ft
75. 0.7, 2.4, 2.5   **77.** 97 ft by 117 ft   **79.** 0.5 mi by 2.5 mi
81. 25 sides   **83.** $4.80 or $5.20   **85.** 4,000   **87.** Late 1976
91. $n^{1/2}$ **93.** $(3b)^{1/4}$ **95.** $\sqrt[3]{t}$ **97.** $\sqrt[4]{3t}$

Study Set Section 10.3 (page 927)

1. discriminant   **3.** conjugates   **5.** rational   **7. a.** x^2 **b.** $\sqrt{x}$
c. $x^{1/3}$ **d.** $\frac{1}{x}$ **e.** $x + 1$ **9.** $4ac$, 5, 6, 24, rational
11. One repeated rational-number solution   **13.** Two imaginary-number solutions (complex conjugates)   **15.** Two different irrational-number solutions   **17.** Two different rational-numbers solutions
19. Two different irrational-number solutions   **21.** Two different rational-numbers solutions   **23.** $-1, 1, -4, 4$ **25.** $2, -2, 3i, -3i$
27. 25, 64   **29.** 1   **31.** $-1, 27$ **33.** $-64, 8$ **35.** 0, 2
37. $\frac{1}{4}, \frac{1}{2}$ **39.** $-\frac{1}{3}, \frac{1}{2}$ **41.** $-4, \frac{2}{3}$ **43.** $\frac{5 \pm \sqrt{65}}{2}$ **45.** $\frac{7 \pm \sqrt{105}}{2}$
47. $\frac{9}{4}$ **49.** $-\frac{1}{3}, 1$ **51.** $-i, i, -3i\sqrt{2}, 3i\sqrt{2}$ **53.** $4 \pm i$
55. $\frac{3 \pm \sqrt{57}}{6}$ **57.** 16, 4   **59.** $\pm i\sqrt{2}, \pm 3\sqrt{2}$
61. Repeated solutions of 1 and -1 **63.** $2, -2, i\sqrt{7}, -i\sqrt{7}$
65. $\frac{243}{32}, 1$ **67.** A repeated solution of $-\frac{3}{2}$ **69.** $3 \pm \sqrt{7}$
71. 49, 225   **73.** $1 \pm i$ **75.** No solution   **77.** $1, -1, \sqrt{5}, -\sqrt{5}$

79. $-\frac{5}{7}, 3$ **81.** $-1, -\frac{27}{13}$ **83.** 10.3 min **85.** 20 ft/sec

87. 32.4 ft **91.** $x = 3$ **93.** $y = \frac{2}{3}x$

Study Set Section 10.4 (page 941)

1. quadratic, parabola **3.** axis of symmetry

5. a. $(1, 0), (3, 0)$ **b.** $(0, -3)$ **c.** $(2, 1)$ **d.** $x = 2$

e. Domain: $(-\infty, \infty)$; range: $(-\infty, 1]$

7.

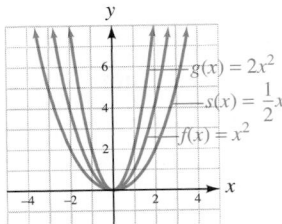

9. a. 2 **b.** 9, 18 **11.** $-3, 5$

13. $h = -1; f(x) = 2[x - (-1)]^2 + 6$

15.

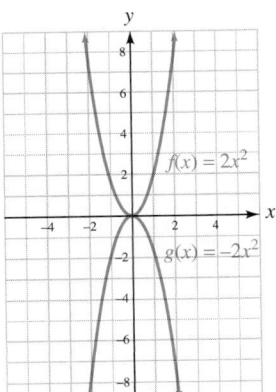

17.

19.

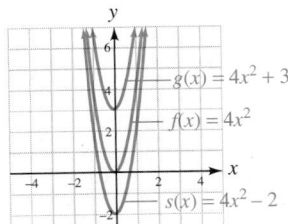

21.

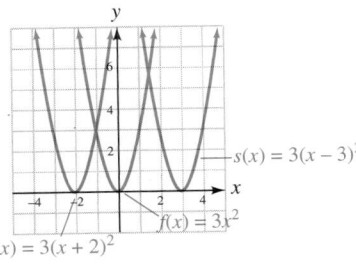

23. $(1, 2); x = 1$; upward **25.** $(-3, -4); x = -3$; downward

27. $(7.5, 8.5); x = 7.5$; downward **29.** $(0, -4); x = 0$; upward

31. $(3, 2)$ **33.** $(2, 0)$

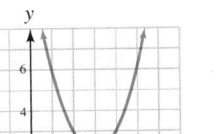
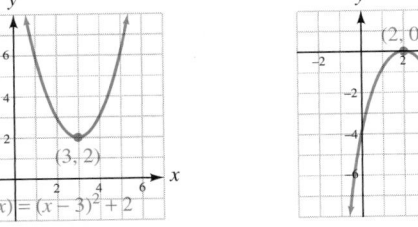

35. $(-3, 4)$

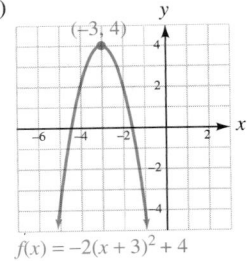

37. $(-1, -3)$

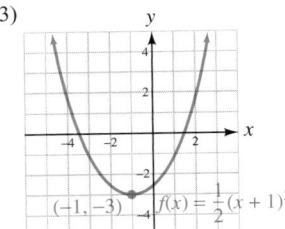

39. $(-2, 1); x = -2$; upward **41.** $(3, -6); x = 3$; downward

43. $f(x) = (x + 1)^2 - 4$;

$(-1, -4), x = -1$

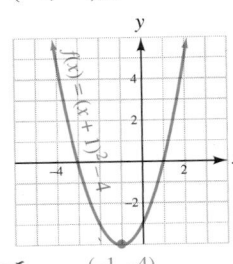

45. $f(x) = 4(x + 3)^2 + 1$;

$(-3, 1), x = -3$

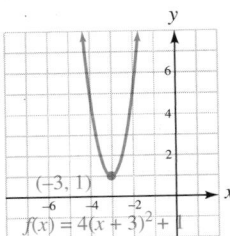

47. $f(x) = \left(x + \frac{1}{2}\right)^2 - \frac{25}{4};$
$\left(-\frac{1}{2}, -\frac{25}{4}\right); x = -\frac{1}{2}$

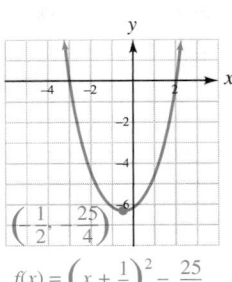

$f(x) = \left(x + \frac{1}{2}\right)^2 - \frac{25}{4}$

49. $f(x) = -4(x - 2)^2 + 6;$
$(2, 6), x = 2$

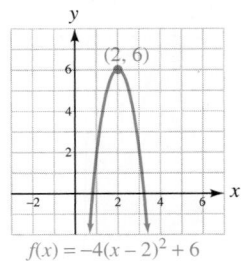

$f(x) = -4(x - 2)^2 + 6$

51. $f(x) = 2(x + 2)^2 - 2;$
$(-2, -2), x = -2$

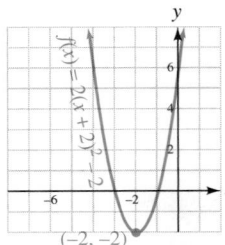

53. $f(x) = -(x + 4)^2 - 1;$
$(-4, -1), x = -4$

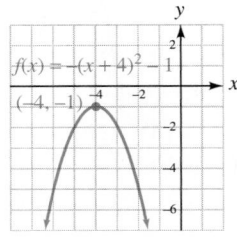

$f(x) = -(x + 4)^2 - 1$

55. $(-1, -6)$ **57.** $\left(\frac{3}{4}, \frac{23}{8}\right)$ **59.** $(-5, 0), (7, 0); (0, -35)$

61. $(0, 0), (2, 0); (0, 0)$

63. $(-2, 0); (-2, 0); (0, 4)$

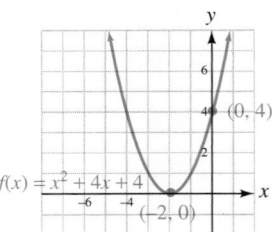

$f(x) = x^2 + 4x + 4$

65. $(1, 0); (1, 0); (0, -1)$

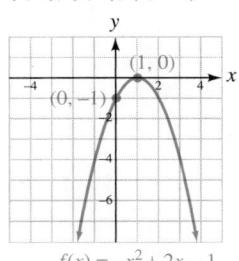

$f(x) = -x^2 + 2x - 1$

67. $(1, -1); (0, 0), (2, 0); (0, 0)$

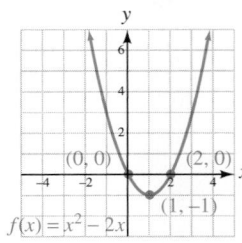

$f(x) = x^2 - 2x$

69. $(2, -2); (1, 0), (3, 0); (0, 6)$

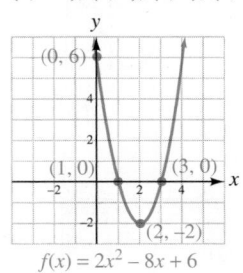

$f(x) = 2x^2 - 8x + 6$

71. $(-1, -2)$; no x-intercept; $(0, -8)$

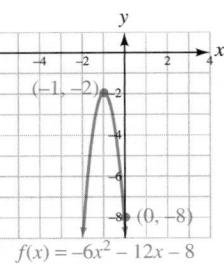

$f(x) = -6x^2 - 12x - 8$

73. $\left(\frac{3}{2}, 0\right); (0, 9)$

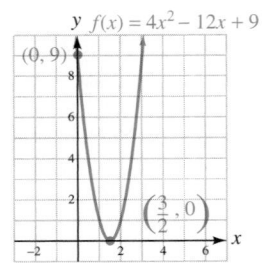

$y\ f(x) = 4x^2 - 12x + 9$

75. $(0.25, 0.88)$ **77.** $(0.50, 7.25)$ **79.** $2, -3$ **81.** $-1.85, 3.25$

83.

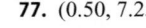

axis of symmetry

85. 15 min, $160
87. 3.75 sec, 225 ft
89. 75 ft by 75 ft, 5,625 ft²
91. 1968, 1.5 million; the U.S. involvement in the war in Vietnam was at its peak
93. 200, $7,000   **101.** $\frac{\sqrt{6}}{10}$
103. $15b - 6\sqrt{15b} + 9$

Study Set Section 10.5 (page 955)

1. quadratic   **3.** two   **5.** $(-\infty, -1), (-1, 4), (4, \infty)$
7. a. Yes   **b.** No   **c.** Yes   **d.** No
9. a. $(-3, 2)$ **b.** $(-\infty, -1] \cup [1, \infty)$ **11. a.** Solid
b. Yes   **13.** $x^2 - 6x - 7 \geq 0$ **15.** $(1, 4)$

17. $(-\infty, 3) \cup (5, \infty)$

19. $(-\infty, -6] \cup [7, \infty)$

21. $[-4, 3]$

23. $(-\infty, 0) \cup \left(\frac{1}{2}, \infty\right)$

25. $\left(-\infty, -\frac{5}{3}\right] \cup (0, \infty)$

27. $(-\infty, -3) \cup (1, 4)$

29. $\left(-\frac{1}{2}, \frac{1}{3}\right] \cup \left[\frac{1}{2}, \infty\right)$

31. $(0, 2) \cup (8, \infty)$

33. $\left[-\frac{34}{5}, -4\right) \cup (3, \infty)$

35.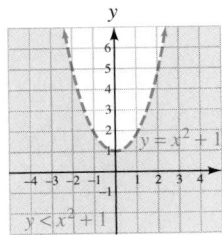

37.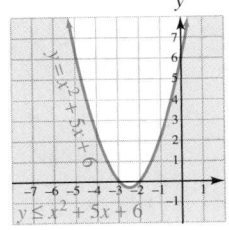

39.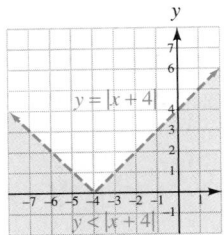

41.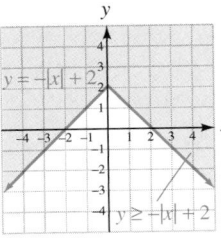

43. $(-1, 3)$ **45.** $(-\infty, -3) \cup (2, \infty)$

47. $(-4, -2] \cup (-1, 2]$

49. $(-\infty, -3] \cup [3, \infty)$

51. $(-\infty, \infty)$

53. $(-2, 1) \cup (3, \infty)$

55. $(-5, 5)$

57. $\left(-\frac{1}{3}, \frac{3}{2}\right)$ **59.** No solutions

61. $(-1, 4) \cup \left(\frac{23}{2}, \infty\right)$

63. $(-2{,}100, -900) \cup (900, 2{,}100)$ **69.** $x = ky$ **71.** $t = kxy$

Chapter 10 Review (page 957)

1. $-5, -4$ **2.** $-\frac{1}{3}, -\frac{5}{2}$ **3.** $\pm 2\sqrt{7}$ **4.** $4, -8$ **5.** $\pm 5i$

6. $\pm \frac{7\sqrt{5}}{5}$ **7.** $r = \frac{\sqrt{\pi A}}{\pi}$ **8.** $x^2 - x + \frac{1}{4} = \left(x - \frac{1}{2}\right)^2$

9. $-4, -2$ **10.** $\frac{3 \pm \sqrt{3}}{2}$ **11.** $\frac{6 \pm \sqrt{30}}{6}$ **12.** $1 \pm 2i\sqrt{3}$

13. Because 7 is an odd number and not divisible by 2, the computations involved in completing the square on $x^2 + 7x$ involve fractions. The computations involved in completing the square on $x^2 + 6x$ do not.   **14.** 2 is not a factor of the numerator—it is a term. Only common factors of the numerator and denominator can be removed.   **15. a.** $(2 + 2x)$ ft   **b.** $(6 + 2x)$ ft

16. 6 seconds before midnight   **17.** $\frac{1}{2}, -7$ **18.** $5 \pm \sqrt{7}$

19. $0, 10$ **20.** $\frac{13 \pm \sqrt{163}}{3}$ **21.** $-\frac{3}{4} \pm \frac{\sqrt{15}}{4}i$ **22.** $\frac{2}{3} \pm \frac{\sqrt{2}}{3}i$

23. $\frac{-3 \pm \sqrt{29}}{10}$ **24.** $\frac{3 \pm 3\sqrt{13}}{2}$ **25.** \$24 or \$26

26. Sides: 1.25 in. wide; top/bottom: 2.5 in. wide

27. 0.7 sec, 1.8 sec   **28.** 33 in., 56 in.

29. Two different irrational-number solutions

30. Two imaginary-number solutions that are complex conjugates

31. One repeated solution, a rational number   **32.** Two different

rational-number solutions   **33.** 1, 144   **34.** $8, -27$

35. $i, -i, \frac{\sqrt{6}}{3}, -\frac{\sqrt{6}}{3}$ **36.** $1, -\frac{8}{5}$ **37.** $4 \pm i$ **38.** Repeated

solutions of -1 and 1 **39.** A repeated solution of $-\frac{2}{5}$ **40.** $\frac{1}{32}, 2$

41. About 81 min   **42.** 30 mph   **43.** 30.8 million   **44.** h, k, x

45.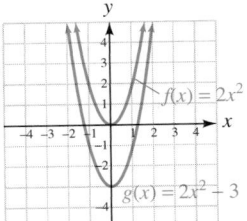

46.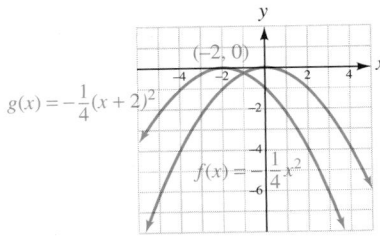

47. $(1, 4), x = 1$ **48.** $f(x) = 4(x + 2)^2 - 7$; $(-2, -7), x = -2$

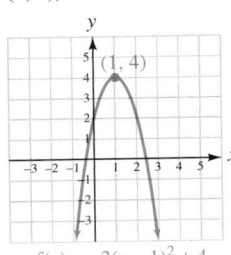 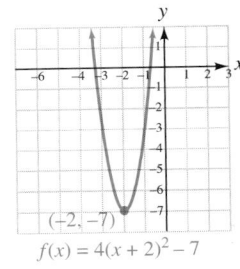

49. $(1, -6)$

50. $\left(-\frac{1}{2}, -\frac{9}{4}\right); x = -\frac{1}{2}; (-2, 0), (1, 0); (0, -2)$

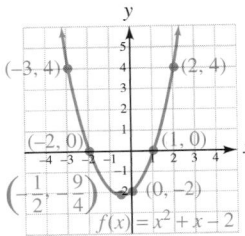

51. 1921; 6,469,326   **52.** $-2, \frac{1}{3}$

53. $(-\infty, -7) \cup (5, \infty)$

54. $[-9, 9]$

55. $(-\infty, 0) \cup \left[\frac{3}{5}, \infty\right)$

56. $\left(-\frac{7}{2}, 1\right) \cup (4, \infty)$

57. $\left[-4, \frac{2}{3}\right]$ **58.** $(-\infty, 0) \cup (1, \infty)$

59. 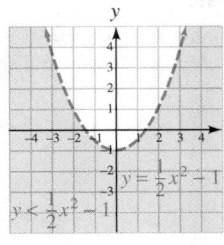   **60.**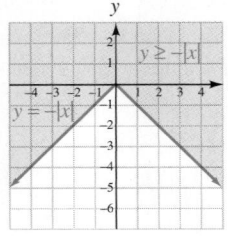

Chapter 10 Test (page 964)

1. a. quadratic   **b.** completed, square   **c.** vertex   **d.** rational
e. nonlinear   **2.** $\pm 3\sqrt{7} \approx 7.94$ **3.** $-7 \pm 5\sqrt{2}$ **4.** $\pm 2i$
5. $x^2 + 11x + \frac{121}{4} = \left(x + \frac{11}{2}\right)^2$ **6.** $\frac{3}{2}, \frac{5}{2}$ **7.** $\frac{-1 \pm \sqrt{2}}{2}$
8. $1 \pm \sqrt{5}$ **9.** $2 \pm 3i$ **10.** $-5, -3$ **11.** $1, \frac{1}{4}$
12. $-1, \frac{1}{3}$ **13.** $2, -2, i\sqrt{3}, -i\sqrt{3}$ **14.** $-\frac{4}{5}, \frac{4}{7}$
15. $2 \pm \sqrt{10}$ **16.** $-8, -\frac{1}{125}$ **17.** $c = \frac{\sqrt{Em}}{m}$
18. a. Two different imaginary-number solutions that are complex
conjugates   **b.** One repeated solution, a rational number
19. 4.5 ft by 1,502 ft   **20.** About 46 min   **21.** 20 in.   **22.** iii
23. $(1, -2), x = 1$

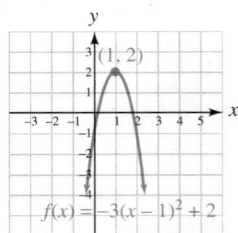

24. $f(x) = 5(x + 1)^2 - 6; (-1, -6), x = -1$

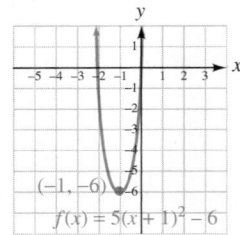

25. $\left(-\frac{1}{4}, -\frac{9}{8}\right), x = -\frac{1}{4}, (-1, 0), \left(\frac{1}{2}, 0\right); (0, -1)$

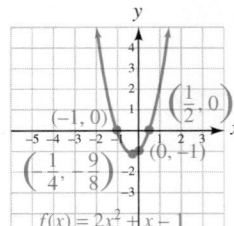

26. 211 ft   **27.** $(-\infty, -2) \cup (4, \infty)$

28. $(-3, 2]$

29. 11,160 gal

30. 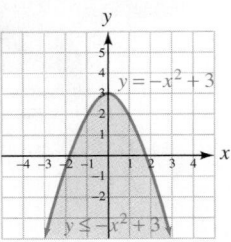   **31.** $-2, 3$ **32.** $[-2, 3]$

Cumulative Review Chapters 1–10 (page 967)

1. -3 **2.** 6 L   **3.** x-intercept: $\left(\frac{6}{5}, 0\right)$; y-intercept: $(0, -2)$
4. Supply: a decrease of 15,000 nurses per year; demand: an increase of
55,000 nurses per year   **5.** $y = 3x + 2$ **6.** $y = -\frac{2}{3}x - 2$
7. $(1, -2)$ **8.** $(3, 8)$

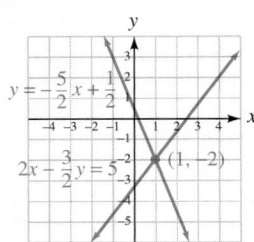

9. 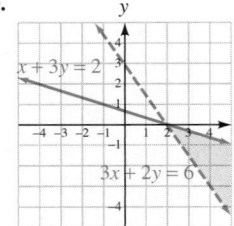   **10.** $\frac{125c^{21}}{8a^{12}b^{12}}$
11. 1.44×10^{10}; 14,400,000,000
12. $\frac{1}{10}s^2 - st - \frac{2}{5}t^2$
13. $-11t^3 + 0.8t^2 - 1.4t$
14. $8a^3 - b^3$ **15.** $6uvw^2(2w - 3v)$
16. $(x - y)(x - 4)$
17. $2a^2(3a + 2)(5a - 4)$
18. $(7s^3 - 6n^2)^2$
19. $(x^2 + 4y^2)(x + 2y)(x - 2y)$
20. $(2x^2 + 5y)(4x^4 - 10x^2y + 25y^2)$ **21.** $2, -\frac{5}{2}$ **22.** $0, \frac{2}{3}, -\frac{1}{2}$
23. $\frac{3x - 5}{x + 2}$ **24.** $\frac{x + y}{x - y}$ **25.** 0   **26.** $\frac{1}{2r(r + 2)}$
27. $\frac{x^4y^4}{2} - \frac{x^3y^9}{4} + \frac{3}{4xy^2}$ **28.** $5a^2 - 3a - 4$ **29.** 5; 3 is extraneous
30. $R = \frac{R_1R_2R_3}{R_2R_3 + R_1R_3 + R_1R_2}$ **31.** 18 sec   **32.** 21 tons
33. $\left(-\infty, -\frac{10}{9}\right)$ **34.** $\left[1, \frac{9}{4}\right]$
35. $(-\infty, -10] \cup [15, \infty)$
36. $(x + 5 + y^4)(x + 5 - y^4)$ **37. a.** Domain: $[0, 24]$
b. 1.5   **c.** At noon, the low tide mark was -2.5 m.   **d.** 0, 2, 9, 17
38. The set of all real numbers except 20   **39.** $9,000
40. D: $[2, \infty)$; R: $[0, \infty)$

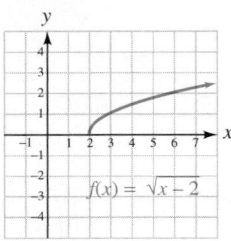

41. $-3x$ **42.** $4t\sqrt{3t}$ **43.** $\frac{1}{16}$ **44.** $x^{17/12}$

45. $-12\sqrt[4]{2} + 10\sqrt[4]{3}$ **46.** $-18\sqrt{6}$ **47.** $\frac{x + 3\sqrt{x} + 2}{x - 1}$

48. $\frac{5\sqrt[3]{x^2}}{x}$ **49.** 2, 7   **50.** $\frac{1}{4}$ **51.** $3\sqrt{2}$ in.   **52.** $2\sqrt{3}$ in.

53. 10   **54.** $-i$ **55.** $-5 + 17i$ **56.** $\frac{3}{2} + \frac{1}{2}i$ **57.** $3 + 4i$

58. $0 - \frac{2}{3}i$ **59.** $\pm 2\sqrt{7}$ **60.** $19 \pm i\sqrt{5}$ **61.** $\frac{3 \pm \sqrt{3}}{2}$

62. $\frac{1}{5} \pm \frac{2}{5}i$ **63.** 10 ft by 18 ft   **64.** 50 m and 120 m

65. $-8, 27$ **66.** Repeated solutions of -1 and 1

67. $(-2, 4), x = -2; (-4, 0), (0, 0); (0, 0)$

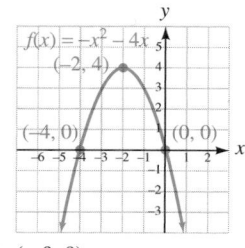

68. $(-9, 9)$

69. $(-4, -2] \cup (-1, 2]$

70. a. $-\frac{3}{4}$ **b.** No solution

Study Set Section 11.1 (page 980)

1. sum, $f(x) + g(x)$, difference, $f(x) - g(x)$ **3.** domain
5. identity   **7. a.** $g(3)$ **b.** $g(3)$
9.

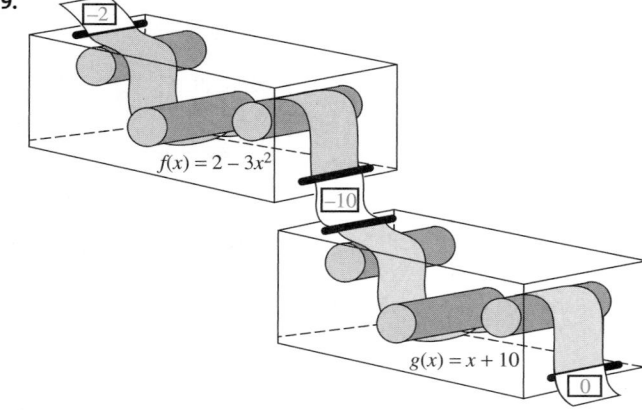

11. $g(x), (3x - 1), 9x, 2x$ **13.** $(f + g)(x) = 7x, (-\infty, \infty)$
15. $(g - f)(x) = x, (-\infty, \infty)$ **17.** $(f \cdot g)(x) = 12x^2, (-\infty, \infty)$
19. $(g/f)(x) = \frac{4}{3}, (-\infty, 0) \cup (0, \infty)$
21. $(f + g)(x) = 3x - 2, (-\infty, \infty)$
23. $(g - f)(x) = -x - 4, (-\infty, \infty)$
25. $(f \cdot g)(x) = 2x^2 - 5x - 3, (-\infty, \infty)$
27. $(g/f)(x) = \frac{x - 3}{2x + 1}, \left(-\infty, -\frac{1}{2}\right) \cup \left(-\frac{1}{2}, \infty\right)$
29. $(f - g)(x) = -2x^2 + 3x - 3, (-\infty, \infty)$
31. $(f/g)(x) = \frac{3x - 2}{2x^2 + 1}, (-\infty, \infty)$ **33.** $(f - g)(x) = 3, (-\infty, \infty)$
35. $(g/f)(x) = \frac{x^2 - 4}{x^2 - 1}, (-\infty, -1) \cup (-1, 1) \cup (1, \infty)$ **37.** 7
39. 24   **41.** -1 **43.** $-\frac{1}{2}$ **45.** $(f \circ g)(x) = 2x^2 - 1$

47. $(g \circ f)(2x) = 16x^2 + 8x$ **49.** 58   **51.** 110   **53.** 2
55. $(g \circ f)(x) = 9x^2 - 9x + 2$ **57.** 16   **59.** $\frac{1}{9}$
61. $(g \circ f)(8x) = 64x^2$ **65. a.** 0   **b.** 6   **c.** -3 **d.** -2
e. 1   **f.** -3 **67. a.** 7   **b.** 8   **c.** 12   **d.** 0   **69.** $3x + 1$
71. $-2x - 5$ **73. a.** 1,003; in 1993, the average combined score on the SAT was 1,003.   **b.** 3; in 1996, the average difference in the math and verbal scores was 3.   **c.** 1028   **d.** 11
75. $C(t) = \frac{5}{9}(2,668 - 200t)$ **77. a.** About $75
b. $C(m) = \frac{3m}{20} = 0.15m$ **83.** $\frac{1}{c - d}$

Study Set Section 11.2 (page 993)

1. one-to-one   **3.** inverses   **5.** one-to-one   **7.** range, domain
9. 1   **11. a.** No   **b.** No   **13.** $-4, 0, 8$ **15.** $y, 2y, 3, y, f^{-1}$
17. inverse, inverse   **19.** Yes   **21.** No   **23.** No   **25.** No
27. One-to-one   **29.** Not one-to-one   **31.** Not one-to-one
33. One-to-one   **35.** $f^{-1}(x) = \frac{x - 4}{2}$ **37.** $f^{-1}(x) = 5x - 4$
39. $f^{-1}(x) = 5x + 4$ **41.** $f^{-1}(x) = \frac{2}{x} + 3$ **43.** $f^{-1}(x) = \frac{4}{x}$
45. $f^{-1}(x) = \sqrt[3]{x - 8}$ **47.** $f^{-1}(x) = x^3$ **49.** $f^{-1}(x) = \sqrt[3]{x} - 10$
51. $f^{-1}(x) = \sqrt[3]{\frac{x + 3}{2}}$ **53.** $f^{-1}(x) = \sqrt[3]{2x}$
59. $f^{-1}(x) = \frac{1}{2}x$ **61.** $f^{-1}(x) = \frac{x - 3}{4}$

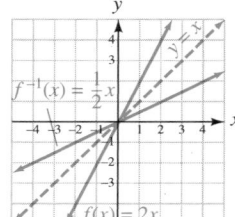 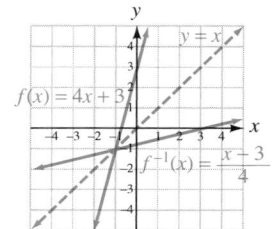

63. $f^{-1}(x) = -\frac{3}{2}x + \frac{9}{2}$ **65.** $f^{-1}(x) = \sqrt[3]{x}$

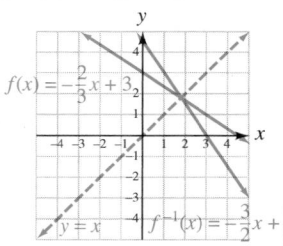 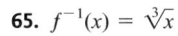 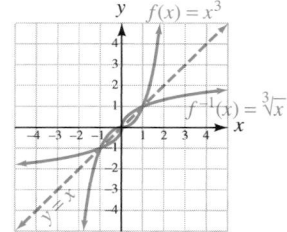

67. $f^{-1}(x) = \sqrt{x + 1}$

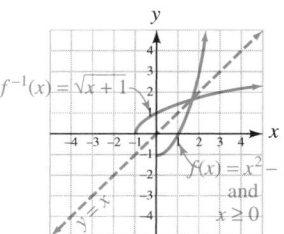

69. a. Yes; no   **b.** No. Twice during this period, the person's anxiety level was at the maximum threshold value.   **77.** $3 - 8i$
79. $18 - i$ **81.** $-28 - 96i$

Study Set Section 11.3 (page 1007)

1. Exponential   **3.** $(0, \infty)$ **5.** Yes   **7.** Increasing   **9. a.** $\frac{1}{9}$

b. $\frac{1}{16}$ **c.** 25   **11. a.** ii   **b.** i   **c.** ii   **d.** ii   **13.** $b > 0, b \neq 1$

15. 8   **17.** $7^{3\sqrt{3}}$ **19.** $3^{\sqrt{7}}$ **21.** $\frac{1}{5\sqrt{5}}$

23.
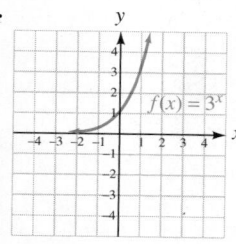

25.
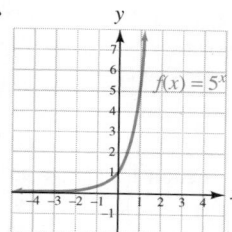

27.
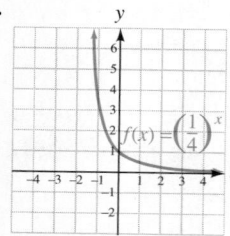

29.
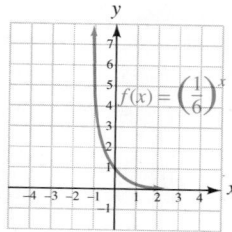

31.
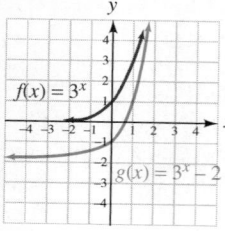

33.
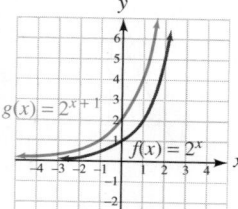

35.
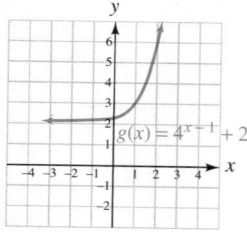

37.
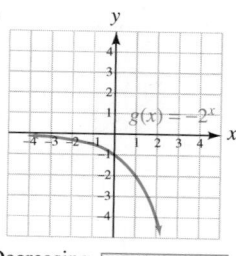

39. Increasing
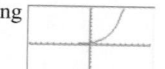

41. Decreasing
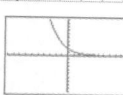

43. a. About 1500, about 1825   **b.** About 6.5 billion
c. Exponential   **45.**

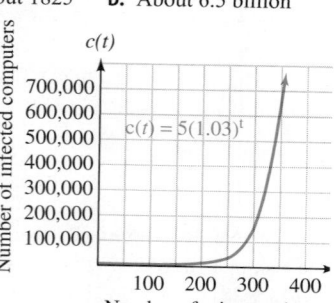

47.
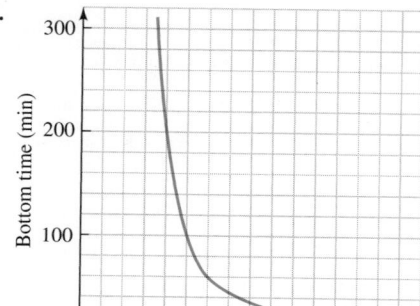

49. $22,080.40
51. $32.03
53. $2,273,996.13
55. About 422 mm
57. 8.7 gm
59. 2,295
67. 40
69. 120°

Study Set Section 11.4 (page 1018)

1. The natural exponential function   **3.** $(0, \infty)$ **5.** Yes
7. Increasing   **9.** continuous   **11.** 2.72
13.
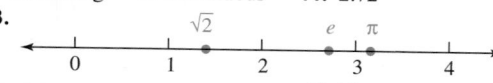

15. 2.7182818...; e **17.** 1,000, ^{10}, $^{0.9}$, 2,459.6   **19.** 24,765.16
21. 3,811,325.37   **23.** 11,542.14   **25.** 542.85

27.
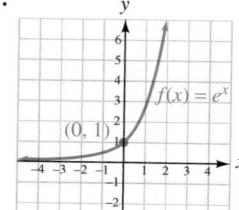

29.
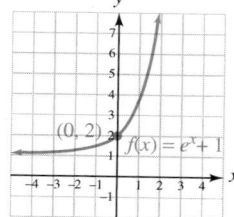

31.
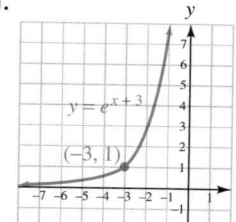

33.
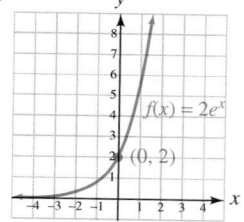

35. About 72 yr   **37.** $13,375.68   **39.** $7,518.28 from annual
compounding, $7,647.95 from continuous compounding
41. $6,849.16

43.

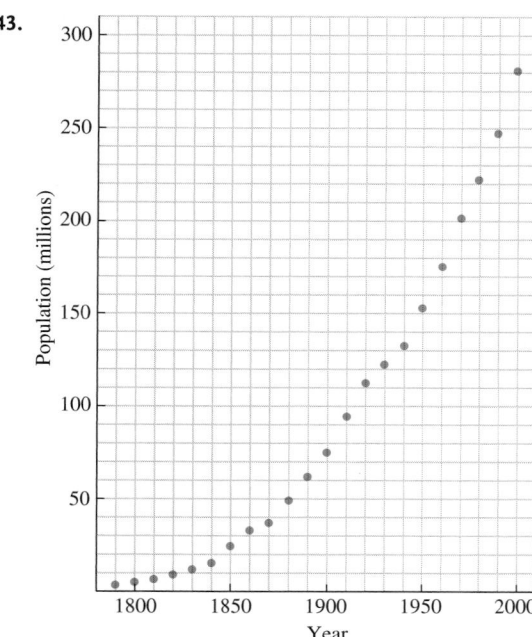

45. 133 million, 140 million, 165 million, 194 million, 206 million, 237 million, 269 million   **47.** 5,995,915; 6,565,316   **49.** 14
51. 51   **53.** 12 hr   **55.** 49 meters per second   **61.** $4x^2\sqrt{15x}$
63. $10y\sqrt{3y}$

Study Set Section 11.5 (page 1032)

1. Logarithmic   **3.** $(-\infty, \infty)$ **5.** Yes   **7.** Increasing   **9.** x, b^y
11. inverse   **13.** $2, -2$ **15.** 1, none, none
17. a. $-0.30, 0, 0.30, 0.60, 0.78, 0.90, 1$
b.

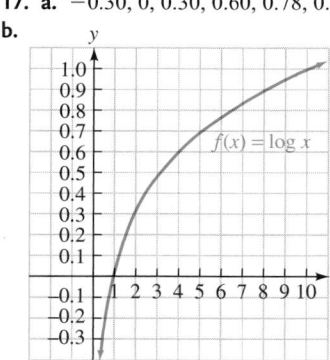

19. a. 10   **b.** x
21. $3^4 = 81$
23. $10^1 = 10$
25. $4^{-3} = \frac{1}{64}$
27. $5^{1/2} = \sqrt{5}$
29. $\log_8 64 = 2$
31. $\log_4 \frac{1}{16} = -2$
33. $\log_{1/2} 32 = -5$
35. $\log_x z = y$ **37.** 9
39. 64   **41.** 3   **43.** $\frac{1}{25}$
45. $\frac{1}{6}$ **47.** 10   **49.** $\frac{2}{3}$
51. 5   **53.** 1,000   **55.** 4   **57.** 1   **59.** $\frac{1}{3}$ **61.** 3   **63.** 2
65. 6   **67.** -1 **69.** 5   **71.** $\frac{1}{2}$ **73.** 0.5119   **75.** -2.3307
77. 6,043.6597   **79.** 0.1726   **81.** 0.3162   **83.** 0.0195
85. Increasing   **87.** Decreasing

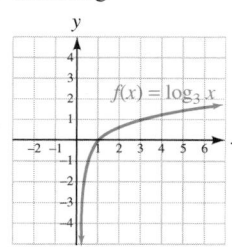   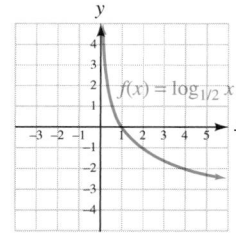

89.

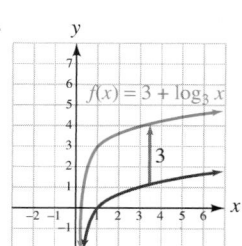

91.

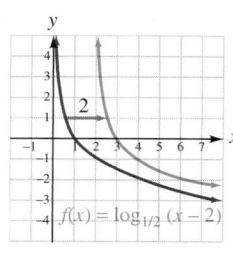

93.

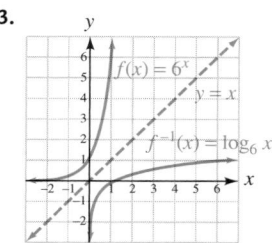

95.

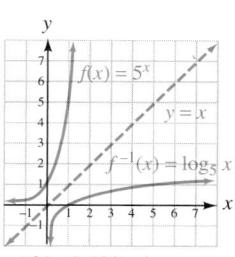

97. 29 db   **99.** 49.5 db   **101.** 4.4   **103.** 2,500 micrometers
105. 77.8%   **107.** 10.8 yr   **113.** 10   **115.** $0; -\frac{5}{9}$ does not check

Study Set Section 11.6 (page 1040)

1. natural, e **3. a.** $-0.69, 0, 0.69, 1.39, 1.79, 2.08, 2.30$
b.

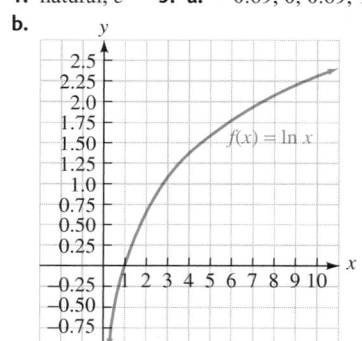

5. $(0, \infty), (-\infty, \infty)$ **7.** $e, 2$
9. $g(x) = 2 + \ln x$ **11.** l, n
13. LOG, LN
15. 5   **17.** 6   **19.** -1
21. $\frac{1}{4}$ **23.** $\frac{2}{3}$ **25.** -7
27. 3.5596   **29.** -5.3709
31. 0.5423   **33.** Undefined
35. 4.0645   **37.** 69.4079
39. 0.0245   **41.** 2.7210

43.   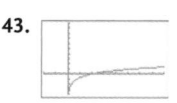   **45.**   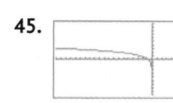   **47.** About $5\frac{1}{4}$ yr

49. About 87 yr   **51.** 9.2 yr   **53.** About 3.5 hr   **61.** $y = 5x$
63. $y = -\frac{3}{2}x + \frac{13}{2}$ **65.** $x = 2$

Study Set Section 11.7 (page 1053)

1. product   **3.** power   **5.** 0   **7.** M, N **9.** $-$ **11.** x
13. $\neq$ **15.** $\log_a b$ **17.** rs, t, r, s **19.** 0   **21.** 7   **23.** 10
25. 2   **27.** 1   **29.** 7   **31.** $2 + \log_2 5$ **33.** $\log 25 + \log y$
35. $2 + \log p + \log q$ **37.** $\log 5 + \log x + \log y + \log z$
39. $2 - \log 9$ **41.** $\log_6 x - 2$ **43.** $\log 7 + \log c - \log 2$
45. $1 + \log x - \log y$ **47.** $1 + \ln x + \ln y - \ln z$
49. $-1 - \log_8 m$ **51.** $7 \ln y$ **53.** $\frac{1}{2}\log 5$ **55.** $-3 \log e$
57. $\frac{3}{5}\log_7 100$ **59.** $\log x + \log y + 2 \log z$
61. $1 + \frac{1}{3}\log_2 x - \log_2 y$ **63.** $3 \log x + 2 \log y$
65. $\frac{1}{2}(\log_b x + \log_b y)$ **67.** $\frac{1}{3}\log_a x - \frac{1}{4}\log_a y - \frac{1}{4}\log_a z$
69. $\ln x + \frac{1}{2}\ln z$ **71.** $\log_2 x^9(x + 1)$ **73.** $\log_3 \frac{x(x + 2)}{8}$
75. $\log_b \frac{\sqrt{z}}{x^3 y^2}$ **77.** $\ln \frac{\frac{x}{z} + x}{\frac{y}{z} + y} = \ln \frac{x}{y}$ **79.** 1.4472   **81.** -1.1972

83. 1.1972   **85.** 1.8063   **87.** 1.7712   **89.** −1.0000
91. 1.8928   **93.** 2.3219   **99.** About 4.8
101. 6.3×10^{-14} gram-ions per liter   **109.** $-\frac{7}{6}$ **111.** $\left(1, -\frac{1}{2}\right)$

Study Set Section 11.8 (page 1065)

1. exponential   **3. a.** equal, $x = y$ **b.** equal, $x = y$ **5.** −2
7. logarithm   **9.** x **11.** It is a solution.   **13.** Yes
15. a. 1.2920   **b.** 2.2619   **17. a.** $A_0 2^{-t/h}$ **b.** $P_0 e^{kt}$
19. log, log 2, 2.8074   **21.** 4   **23.** $-\frac{3}{4} = -0.75$ **25.** 3, −1
27. A repeated solution of 2   **29.** 1.1610   **31.** 1.2702
33. 1.7095   **35.** −8.2144   **37.** 0.5186   **39.** −13.2662
41. 5,000   **43.** 12   **45.** −93   **47.** 0.08   **49.** −7   **51.** 3
53. 50   **55.** $\frac{5}{4} = 1.25$ **57.** 0.5   **59.** 15.75   **61.** 2
63. $e \approx 2.7183$ **65.** ±1.0878   **67.** 10   **69.** 10   **71.** 4
73. 0.2325   **75.** 10, −10   **77.** 10   **79.** 9
81. A repeated solution of 4   **83.** 1, 7   **85.** 0.7324   **87.** 6
89. 1.8   **91.** 2.1   **93.** 20   **95.** 8   **97.** 5.1 yr   **99.** 42.7 days
101. About 4,200 yr   **103.** 5.6 yr   **105.** 5.4 yr
107. Because ln 2 ≈ 0.7   **109.** 25.3 yr   **111.** 2.828 times larger
113. $\frac{1}{3}$ ln 0.75 ≈ −0.0959   **119.** $\sqrt{137}$ in.

Chapter 11 Review (page 1068)

1. $(f + g)(x) = 3x + 1, (-\infty, \infty)$
2. $(f - g)(x) = x - 1, (-\infty, \infty)$
3. $(f \cdot g)(x) = 2x^2 + 2x, (-\infty, \infty)$
4. $(f/g)(x) = \frac{2x}{x + 1}, (-\infty, -1) \cup (-1, \infty)$ **5.** 3   **6.** 5
7. $(f \circ g)(x) = 4x^2 + 4x + 3$ **8.** $(g \circ f)(x) = 2x^2 + 5$
9. a. 0   **b.** −8   **c.** 0   **d.** −1   **10.** $C(m) = \frac{3.25m}{8}$
11. No   **12.** Yes   **13.** Yes   **14.** No   **15.** Yes   **16.** No
17. −6, −1, 7, 20   **18.**

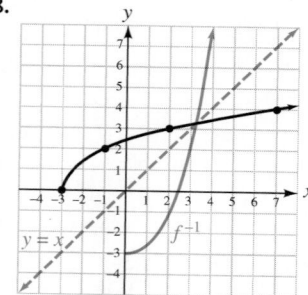

19. $f^{-1}(x) = \frac{x + 3}{6}$ **20.** $f^{-1}(x) = \frac{4}{x} + 1$ **21.** $f^{-1}(x) = \sqrt[3]{x} - 2$
22. $f^{-1}(x) = 6x + 1$
23. $f^{-1}(x) = x^3 + 1$

25. $5^{4\sqrt{6}}$
26. $2^{2\sqrt{7}}$

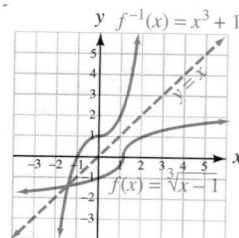

27. D: $(-\infty, \infty)$; R: $(0, \infty)$

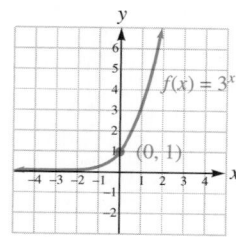

28. D: $(-\infty, \infty)$; R: $(0, \infty)$

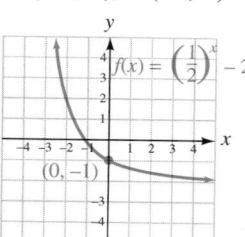

29. D: $(-\infty, \infty)$; R: $(-2, \infty)$

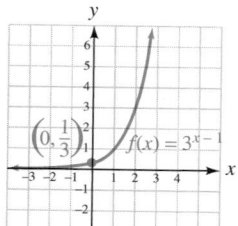

30. D: $(-\infty, \infty)$; R: $(0, \infty)$

31. The x-axis ($y = 0$)
32. An exponential function

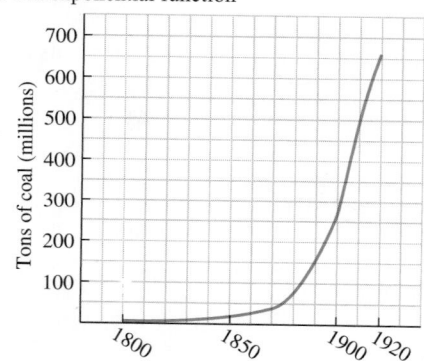

33. $2,189,703.45
34. About $2,015

35. D: $(-\infty, \infty)$; R: $(1, \infty)$

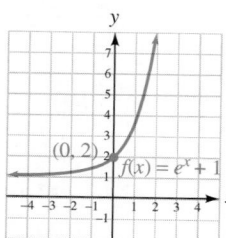

36. D: $(-\infty, \infty)$; R: $(0, \infty)$

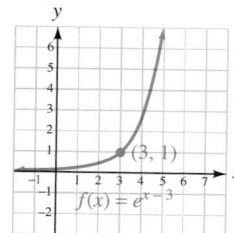

37. $2,324,767.37   **38.** 8,838,365   **39.** 13.9%, 11.67%, 9.80%
40. The exponent on the base e is negative.
41. D: $(0, \infty)$; R: $(-\infty, \infty)$ **42.** Since there is no real number such that $10^? = 0$, log 0 is undefined.   **43.** $4^3 = 64$
44. $\log_7 \frac{1}{7} = -1$ **45.** 2   **46.** −2   **47.** 0   **48.** Undefined
49. $\frac{1}{2}$ **50.** 3   **51.** 32   **52.** $\frac{1}{81}$ **53.** 4   **54.** 10   **55.** $\frac{1}{2}$

56. $\frac{1}{3}$ **57.** 0.6542   **58.** 26.9153   **59.**

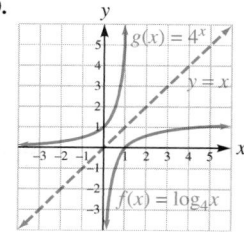

60.

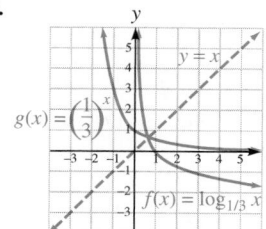

61.

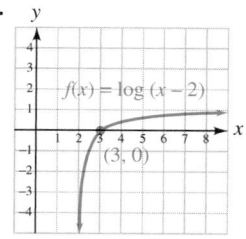

62.

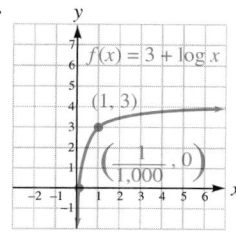

63. About 53   **64.** About 4.4
65. 1   **66.** 2   **67.** -5 **68.** $\frac{1}{2}$
69. Undefined   **70.** Undefined
71. 0   **72.** -7 **73.** 6.1137
74. -0.1625 **75.** 10.3398
76. 0.0002
77. They have different bases:
$\log x = \log_{10} x$ and
$\ln x = \log_e x$.
78. $f^{-1}(x) = e^x$

79.

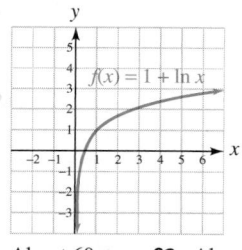

80.

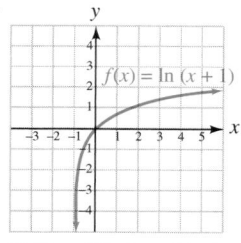

81. About 60 yr   **82.** About 72 in. (6 ft)   **83.** 0   **84.** 1   **85.** 3
86. 4   **87.** $3 + \log_3 x$ **88.** $2 - \log x$ **89.** $\frac{1}{2}\log_5 27$
90. $\log_b 10 + \log_b a + 1$ **91.** $2\log_b x + 3\log_b y - \log_b z$
92. $\frac{1}{2}(\ln x - \ln y - 2\ln z)$ **93.** $\log_2 \frac{x^3 z^7}{y^5}$ **94.** $\log_b \frac{\sqrt{x+2}}{y^3 z^7}$
95. 2.6609   **96.** 3.0000   **97.** 1.7604
98. About 7.9×10^{-4} gram-ions/liter   **99.** -4 **100.** $-3, -1$
101. 1.7712   **102.** 2.7095   **103.** 1.9459   **104.** -8.0472
105. 104   **106.** 9   **107.** 25, 4   **108.** A repeated solution of 4
109. 4, 3   **110.** 2   **111.** 6   **112.** 31
113. $0.76787 \neq -0.27300$ **114.** About 3,300 yr
115. About 91 days   **116.** 2, 5

Chapter 11 Test (page 1079)

1. a. composite   **b.** natural   **c.** continuous   **d.** inverse
e. logarithmic   **2. a.** f composed with g **b.** g of f of eight
3. $(f + g)(x) = 4x^2 - 2x + 11, (-\infty, \infty)$
4. $(g/f)(x) = \frac{4x^2 - 3x + 2}{x + 9}, (-\infty, -9) \cup (-9, \infty)$ **5.** 76
6. $32x^2 - 128x + 131$ **7. a.** -16 **b.** 17   **8. a.** 0   **b.** 3
c. 1   **d.** -2 **e.** -5 **9.** Yes   **10.** No

11. $f^{-1}(x) = -3x$

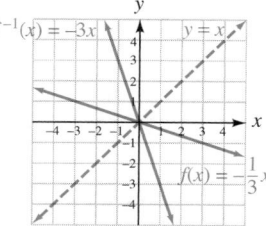

12. $f^{-1}(x) = \sqrt[3]{x} + 15$ **14. a.** Yes   **b.** Yes
c. 80; when the temperature of the tire tread is 260°, the vehicle is
traveling 80 mph
15. D: $(-\infty, \infty)$; R: $(1, \infty)$ **16.** D: $(-\infty, \infty)$; R: $(0, \infty)$

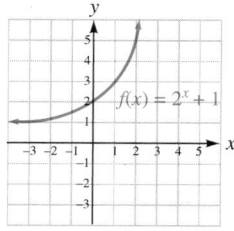   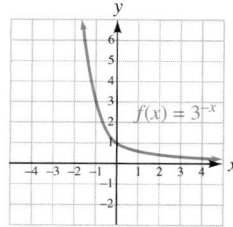

17. $\frac{3}{64}$ g $= 0.046875$ g   **18.** \$1,060.90
19.   **20.** About 1,829,237,435

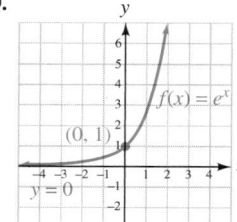

21. $6^{-2} = \frac{1}{36}$
22. a. D: $(0, \infty)$; R: $(-\infty, \infty)$
b. $f^{-1}(x) = 10^x$ **23.** 2   **24.** -2
25. Undefined   **26.** -6 **27.** $\frac{1}{2}$
28. 0   **29.** 2   **30.** 16   **31.** $\frac{1}{27}$
32. e

33.   **34.**

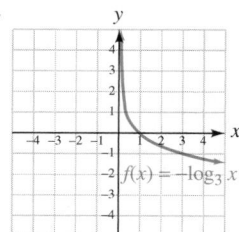   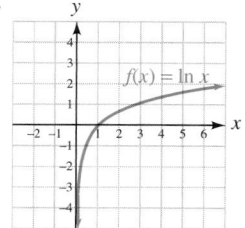

35. 6.4   **36.** About 46   **37.** 0.0871   **38.** 0.5646
39. $2\log_b a + 1 + 3\log_b c$ **40.** $\ln \frac{b\sqrt{a+2}}{c^3}$ **41.** 0.6826
42. 4   **43.** 1   **44.** 10   **45.** 5   **46.** About 20 min

Study Set Section 12.1 (page 1097)

1. system   **3.** infinitely   **5. a.** True   **b.** False   **c.** True
d. True   **7. a.** 3; -4 (Answers may vary)   **b.** 2; -3 (Answers
may vary)   **9.** $3x - 7, 5, 4, 12, 3, 2, 3$ **11.** Yes   **13.** No
15. No   **17.** Yes

19. $(4, 2)$

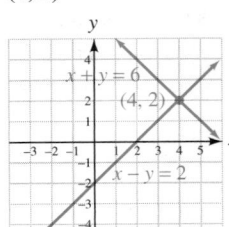

21. $(-1, 3)$

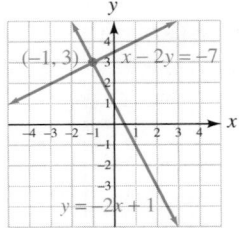

23. No solution, $\varnothing$; inconsistent system

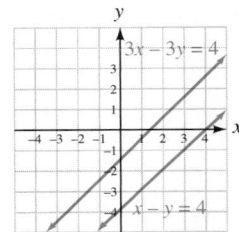

25. Infinitely many solutions, $\{(x, y) \mid x = 3 - 2y\}$ or $\{(x, y) \mid x + 2y = 3\}$; dependent equations

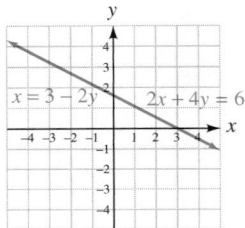

27. $(-3, -3)$

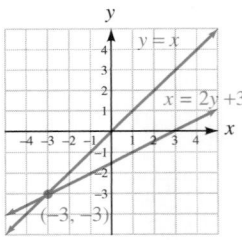

29. $(-2, -1)$

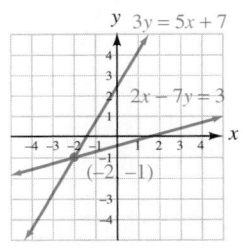

31. $(-0.37, -2.69)$ **33.** $(-7.64, 7.04)$ **35.** $(2, 6)$ **37.** $(5, 3)$
39. $(-2, 4)$ **41.** $(-4, -2)$ **43.** $(9, 2)$ **45.** $(-4, 0)$
47. $(0, -6)$ **49.** $(12, 4)$ **51.** No solution, $\varnothing$; inconsistent system   **53.** Infinitely many solutions, $\{(x, y) \mid 3x - y = 5\}$; dependent equations
55. $\left(\frac{1}{2}, -1\right)$

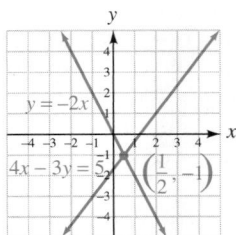

57. $(1, -2)$

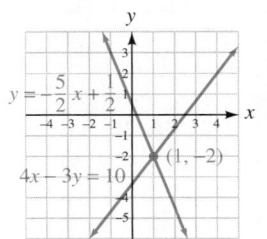

59. No solution, $\varnothing$; inconsistent system

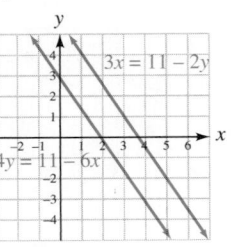

61. $\left(3, \frac{5}{2}\right)$

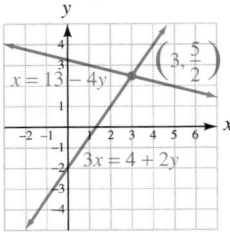

63. $(2, 1)$

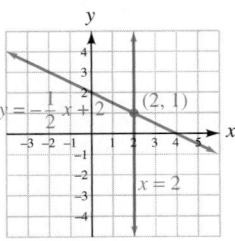

65. Infinitely many solutions, $\{(x, y) \mid x + 3y = 6\}$; dependent equations

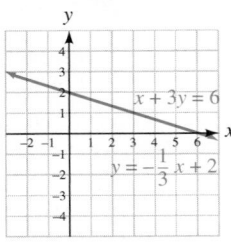

67. $(1, 2)$ **69.** $\left(\frac{1}{2}, \frac{2}{3}\right)$ **71.** $(4, -7)$ **73.** Infinitely many solutions, $\{(x, y) \mid x + y = 12\}$; dependent equations   **75.** $(-2, -3)$
77. $\left(5, \frac{3}{2}\right)$ **79.** $\left(10, \frac{15}{2}\right)$ **81.** $(-6, 16)$ **83.** $(2, -3)$
85. $(9, -1)$ **87.** $(2, 3)$ **89.** $\left(-\frac{1}{3}, 1\right)$ **91.** $(2,000, 50)$
93. a.

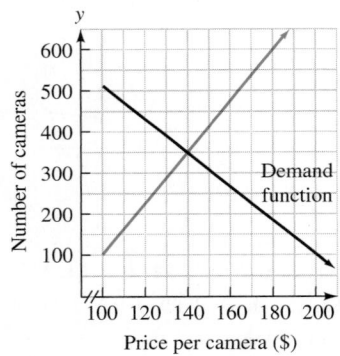

b. $140   **c.** Supply increases, and demand decreases.
95. $(250, 2,500)$; if the company makes 250 widgets, the cost to make them and the revenue obtained from their sale will be equal: $2,500.
97. Adults: 150, children: 420
99. 15 sec: $475, 30 sec: $800
101. $65°, 25°$ **103.** $3,000 at 10%, $5,000 at 12%
105. Snowmobile: 1.5 hr, ski: 4.5 hr   **107.** 85 racing bikes, 120 mountain bikes   **109.** 103   **111.** 148 g of the 0.2%, 37 g of the 0.7%   **117.** $P_1 = \frac{P_2 V_2}{V_1}$ **119.** $r = \frac{S - a}{S - l}$

Study Set Section 12.2 (page 1111)

1. system, standard   **3.** triples   **5.** dependent   **7. a.** No solution
b. No solution   **9.** $x + y + 4z = 3$, $7x - 2y + 8z = 15$, $3x + 2y - z = 4$ **11.** Yes   **13.** No   **15.** $(1, 1, 2)$
17. $(-1, 3, 0)$ **19.** $(3, -1, 3)$ **21.** $(-3, 0, 5)$ **23.** $(7, -6, 3)$
25. $\left(\frac{1}{2}, 2, -1\right)$ **27.** $(40, -20, 5)$ **29.** $\left(\frac{2}{3}, 2, -1\right)$
31. $(9, 0, -4)$ **33.** $(10, -5, 5)$ **35.** No solution, $\varnothing$; inconsistent system   **37.** Infinitely many solutions; dependent equations
39. $(8, 4, 5)$ **41.** $(12, -9, 4)$ **43.** $(3, 2, 1)$ **45.** No solution, $\varnothing$; inconsistent system   **47.** $\left(\frac{3}{4}, \frac{1}{2}, \frac{1}{3}\right)$ **49.** No solution, $\varnothing$;

inconsistent system   **51.** (2, 4, 8)   **53.** Infinitely many solutions; dependent equations   **55.** (2.5, 3, 3.5)   **57.** (2, 6, 9)
59. a. Infinitely many solutions, all lying on the line running down the binding   **b.** 3 parallel planes (shelves); no solution   **c.** Each pair of planes (cards) intersect; no solution   **d.** 3 planes (faces of die) intersect at a corner; 1 solution   **61.** 100, 81, 78

67.

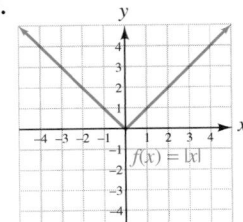

$f(x) = |x|$

69.

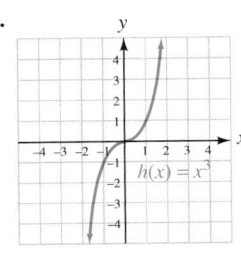

$h(x) = x^3$

Study Set Section 12.3 (page 1119)

1. satisfy   **3.** $\begin{cases} x + y + z = 50 \\ 5x + 6y + 7z = 295 \\ 2x + 3y + 4z = 145 \end{cases}$ **5.** $-3 = 4a + 2b + c$
7. 30 large, 50 medium, 100 small   **9.** Food A: 2, Food B: 3, Food C: 1   **11.** 120 coats, 200 shirts, 150 slacks   **13.** Young: 85, Montana: 55, Gannon: 16   **15.** Nitrogen: 78%, oxygen: 21%, other gasses: 1%   **17.** $\angle A = 40°$, $\angle B = 60°$, $\angle C = 80°$
19. *X-Files:* 201, *Will & Grace:* 194, *Seinfeld:* 180
21. 6 lb rose petals, 3 lb lavender, 1 lb buckwheat hulls
23. Nickels: 20, dimes: 40, quarters: 4   **25.** $y = \frac{1}{2}x^2 - 2x - 1$
27. $x^2 + y^2 - 2x - 2y - 2 = 0$ **31.** Yes   **33.** Yes
35. No; (1, 2), (1, −2)   **37.** No; (4, 2), (4, −2)

Study Set Section 12.4 (page 1131)

1. matrix   **3.** rows, columns, by   **5.** augmented   **7. a.** 2 × 3
b. 3 × 4   **9. a.** $x, y, y, 6, -4$ **b.** $x, z, 2z, z, 4, 0, -20$
11. a. Interchange rows 1 and 2   **b.** Multiply row 1 by $\frac{1}{2}$
c. Add row 3 to 6 times row 2   **13.** $\begin{bmatrix} 1 & 2 & | & 6 \\ 3 & -1 & | & -10 \end{bmatrix}$
15. $\begin{cases} x + 6y = 7 \\ y = 4 \end{cases}$ **17.** $\begin{bmatrix} 1 & -4 & | & 4 \\ -3 & 1 & | & -6 \end{bmatrix}$ **19.** $\begin{bmatrix} 1 & -\frac{1}{3} & | & 2 \\ 1 & -4 & | & 4 \end{bmatrix}$
21. $\begin{bmatrix} 3 & 6 & -9 & | & 0 \\ -2 & 2 & -2 & | & 5 \\ 1 & 5 & -2 & | & 1 \end{bmatrix}$ **23.** $\begin{bmatrix} 3 & 6 & -9 & | & 0 \\ -2 & -1 & 7 & | & 1 \\ -2 & 2 & -2 & | & 5 \end{bmatrix}$ **25.** (1, 1)
27. (2, −3)   **29.** (3, 2, 1)   **31.** (4, 5, 4)   **33.** No solution, $\varnothing$; inconsistent system   **35.** $\{(x, y) \mid x + y = 3\}$, infinitely many solutions; dependent equations   **37.** (−2, −1, 1)   **39.** (−1, −1)
41. (0, −3)   **43.** No solution, $\varnothing$; inconsistent system
45. $\{(x, y) \mid 4x - y = 2\}$, infinitely many solutions; dependent equations   **47.** $\left(\frac{1}{2}, 1, -2\right)$ **49.** (0, 1, 3)   **51.** (−4, 8, 5)
53. 262,144   **55.** 22°, 68°   **57.** 40°, 65°, 75°   **59.** 76°, 104°
65. $m = \frac{y_2 - y_1}{x_2 - x_1}, (x_2 \neq x_1)$ **67.** $y - y_1 = m(x - x_1)$

Study Set Section 12.5 (page 1142)

1. determinant, diagonal   **3.** minor   **5.** *ad, bc*   **7.** The third column   **9.** $\begin{vmatrix} 1 & 2 & 0 \\ 3 & 1 & -1 \\ 8 & 4 & -1 \end{vmatrix}$ **11.** (−2, −1, 1)   **13.** 6, 30

15. 4   **17.** −10   **19.** −170   **21.** 6   **23.** 35   **25.** 0   **27.** 4
29. 26   **31.** −37   **33.** 5   **35.** −79   **37.** 1   **39.** (4, 2)
41. (−5, −8)   **43.** $\left(5, \frac{3}{2}\right)$ **45.** (11, 3)   **47.** No solution, $\varnothing$; inconsistent system   **49.** $\{(x, y) \mid 5x + 6y = 12\}$, infinitely many solutions; dependent equations   **51.** (1, 1, 2)   **53.** (−2, 0, 2)
55. (3, −2, 1)   **57.** (2, −1)   **59.** $\left(-\frac{1}{2}, -1, -\frac{1}{2}\right)$
61. No solution, $\varnothing$; inconsistent system   **63.** No solution, $\varnothing$; inconsistent system   **65.** (−2, 3, 1)   **67.** $\left(-\frac{1}{2}, \frac{1}{3}\right)$
69. Dependent equation; infinitely many solutions   **71.** −46,811
73. −60,527,941   **75.** 50°, 80°   **83.** No   **85.** The graph of function *g* is 2 units below the graph of function *f*.   **87.** *y*-intercept
89. *x*; *y*

Chapter 12 Review (page 1146)

1. a. (1, 3), (2, 1), (4, −3) (Answers may vary)
b. (0, −4), (2, −2), (4, 0) (Answers may vary)   **c.** (3, −1)
2. The point of intersection is (2001, 50%). In 2001, the percent of time spent viewing broadcast networks and cable networks was the same, 50%.   **3.** Yes   **4.** No
5. (3, 5)

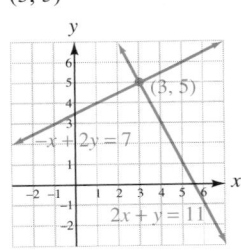

$-x + 2y = 7$
(3, 5)
$2x + y = 11$

6. (−2, 3)

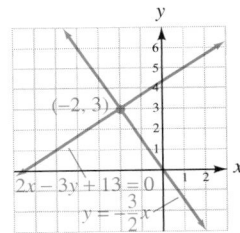

(−2, 3)
$2x - 3y + 13 = 0$
$y = -\frac{3}{2}x$

7. Infinitely many solutions, $\{(x, y) \mid 3x + 2y = 12\}$; dependent equations

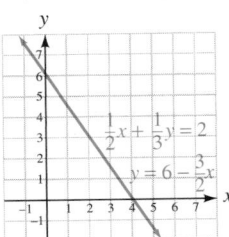

$\frac{1}{2}x + \frac{1}{3}y = 2$
$y = 6 - \frac{3}{2}x$

8. No solution, $\varnothing$; inconsistent system

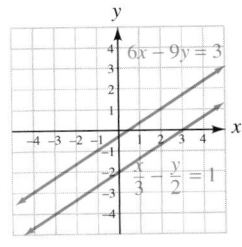

$6x - 9y = 3$
$\frac{x}{3} - \frac{y}{2} = 1$

9. 2   **10.** −1   **11.** (−1, 3)   **12.** (−3, −1)   **13.** (3, 4)
14. Infinitely many solutions, $\{(x, y) \mid x = -2 - 3y\}$; dependent equations   **15.** (−3, 1)   **16.** No solution, $\varnothing$; inconsistent system
17. (9, −4)   **18.** $\left(4, \frac{1}{2}\right)$ **19.** Using the elimination method, the computations are easier.   **20.** (−1, 0.7) (answers may vary); $\left(-1, \frac{2}{3}\right)$
21. 9, 2   **22.** Austin–Houston: 162 mi; Austin–San Antonio: 83 mi
23. 8 mph, 2 mph   **24.** 500 oz of 6%; 250 oz of 18%   **25.** $4,000 at 6%, $6,000 at 12%   **26.** Teaspoon: 5 mL; tablespoon: 15 mL
27. No   **28.** Yes; infinitely many solutions   **29.** (2, 0, −3)
30. (2, −1, 3)   **31.** (−1, 1, 3)   **32.** (−5, −14, −4)
33. (6, −2, 2)   **34.** (1, 4, 1)   **35.** No solution, $\varnothing$; inconsistent system   **36.** Infinitely many solutions; dependent equations
37. Small: 50; medium: 60; large: 40
38. 2 cups mix A, 1 cup mix B, 1 cup mix C
39. $5,000 at 5%, $7,000 at 6%, and $10,000 at 7%

40. $a = -\frac{1}{16}, b = 2, c = 0$ **41.** $\begin{bmatrix} 5 & 4 & | & 3 \\ 1 & -1 & | & -3 \end{bmatrix}$

42. $\begin{bmatrix} 1 & 2 & 3 & | & 6 \\ 1 & -3 & -1 & | & 4 \\ 6 & 1 & -2 & | & -1 \end{bmatrix}$ **43. a.** $\begin{bmatrix} 1 & 3 & | & -2 \\ 6 & 12 & | & -6 \end{bmatrix}$

b. $\begin{bmatrix} 1 & 2 & | & -1 \\ 1 & 3 & | & -2 \end{bmatrix}$ **c.** $\begin{bmatrix} 0 & -6 & | & 6 \\ 1 & 3 & | & -2 \end{bmatrix}$

44. a. $\begin{bmatrix} 1 & 1 & 0 & | & -1 \\ 2 & -1 & 1 & | & 3 \\ 3 & -1 & -2 & | & 7 \end{bmatrix}$ **b.** $\begin{bmatrix} 2 & -1 & 1 & | & 3 \\ 3 & 3 & 0 & | & -3 \\ 3 & -1 & -2 & | & 7 \end{bmatrix}$

c. $\begin{bmatrix} 0 & -3 & 1 & | & 5 \\ 1 & 1 & 0 & | & -1 \\ 3 & -1 & -2 & | & 7 \end{bmatrix}$ **45.** $(1, -3)$ **46.** $(5, -3, -2)$

47. Infinitely many solutions, $\{(x, y) \mid -2x + y = -4\}$; dependent equations   **48.** No solution,$\varnothing$; inconsistent system   **49.** 18
50. 38   **51.** -3 **52.** 28   **53.** $(2, 1)$ **54.** No solution, $\varnothing$;
inconsistent system   **55.** $(1, -2, 3)$ **56.** $(-3, 2, 2)$

Chapter 12 Test (page 1154)

1. a. system   **b.** rows, columns   **c.** triples   **d.** plane   **e.** matrix
2. $(2, 1)$

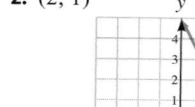

3. It is a solution.   **4.** The point of intersection is (6/06, 44%). That means that Governor Schwarzenegger's job approval and disapproval ratings were the same in June of 2006; approximately 44%.
5. a. Inconsistent system; no solution, $\varnothing$ **b.** Dependent equations; infinitely many solutions   **6.** 3   **7.** $(7, 0)$ **8.** $(2, -3)$
9. $\{(x, y) \mid 2x + 3y = -3\}$, infinitely many solutions; dependent equations   **10.** $\left(\frac{1}{3}, 2\right)$ **11.** 55, 70   **12.** 15 gal of 40%, 5 gal of 80%   **13.** 8×10: $8.75; 5×7: $5.25   **14.** No
15. $(3, 2, -1)$ **16.** $(-6, 2, 5)$ **17.** Children: 60, General
Admission: 30, Seniors: 10   **18.** $\begin{bmatrix} 1 & 7 & | & -3 \\ 0 & -22 & | & 22 \end{bmatrix}$ **19.** $(2, 2)$
20. No solution, $\varnothing$, inconsistent system   **21.** -2 **22.** 4
23. $(-3, 3)$ **24.** -1

Chapters 1–12 Cumulative Review (page 1156)

1. a. True   **b.** True   **c.** True   **2.** 36   **3.** 12   **4.** $-\frac{3}{2}$
5. $r = \frac{T - 2t}{2}$ **6.** 8.9%
7. $(-\infty, -2)$

8.

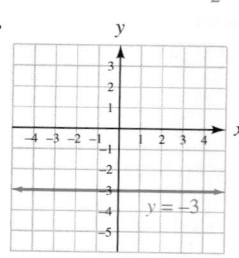

9.

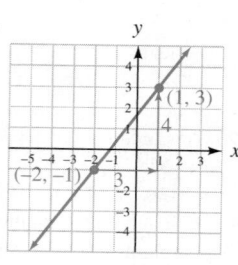

10. A decrease of 750,000 viewers per year   **11.** $y = \frac{1}{4}x - 1$
12. Perpendicular
13. $(-2, 3)$

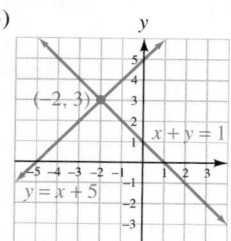

14. $(-3, -1)$
15. $(-1, 2)$
16. 32 lb of Sweet-n-sour; 16 lb Bit-O-Honey
17. 550 mph, 50 mph

18.

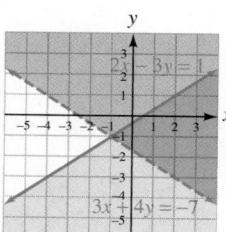

19. y^{21} **20.** $\frac{x^{10}z^4}{9y^6}$
21. 2,600,000 to 1
22. 7.3×10^{-4}
23. $6x^3 + 8x^2 + 3x - 72$
24. $-6a^5$ **25.** $6b^2 + 5b - 4$
26. $6x^3 - 7x^2y + y^3$
27. $4x^2 + 20xy + 25y^2$
28. $81m^4 - 1$ **29.** $2a - \frac{3}{2}b + \frac{1}{2a}$
30. $2x + 1$ **31.** $6a(a - 2a^2b + 6b)$ **32.** $(x + y)(2 + a)$
33. $(5t + 4)(5t - 4)$ **34.** $(b + 5)(b^2 - 5b + 25)$ **35.** $0, -\frac{8}{3}$
36. $\frac{2}{3}, -\frac{1}{5}$ **37.** 5 in.   **38.** $0, 2$ **39.** $\frac{3(x + 3)}{x + 6}$ **40.** -1
41. $\frac{x - 3}{x - 5}$ **42.** $\frac{1}{s - 5}$ **43.** $\frac{x^2 + 4x + 1}{x^2y}$ **44.** $\frac{x^2 + 5x}{x^2 - 4}$ **45.** $\frac{9m^2}{2}$
46. $\frac{9y + 5}{4 - y}$ **47.** 1   **48.** $2\frac{6}{11}$ days   **49.** 875 days   **50.** 39
51. $[-10, 14]$ **52.** $(-\infty, 2) \cup (7, \infty)$

53. $(x + 2 + y)(x + 2 - y)$ **54.** $(b + 1)(b - 1)(b + 4)(b - 4)$
55. -8
56. D: $(-\infty, \infty)$; R: $(-\infty, \infty)$ **57.** Yes   **58.** 1.2   **59.** $2|x|$
60. $\frac{1}{16}$ **61.** $10a^3b^2$

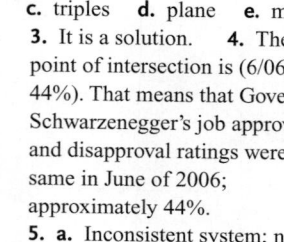

62. $2xy\sqrt[4]{x^3}$ **63.** $9\sqrt{6}$
64. $4x\sqrt[5]{xy^2}$ **65.** $\frac{6x\sqrt{2x}}{y}$
66. $\frac{3m}{2n^2}$ **67.** $\frac{2\sqrt[3]{a^2}}{a}$
68. $\frac{x - 2\sqrt{xy} + y}{x - y}$ **69.** $1, 9$
70. 21.2 in.   **71.** $7i$ **72.** $3i\sqrt{6}$

73. $1 + 5i$ **74.** $16 - 2i$ **75.** $6 - 17i$ **76.** $1 - i$
77. $-2, -6$ **78.** $\frac{1 \pm \sqrt{33}}{8}$; $-0.59, 0.84$ **79.** $0 \pm 4i$
80. $2 \pm i$ **81.** $8, -27$ **82.**

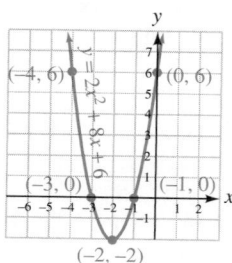

83. 16   **84.** $f^{-1}(x) = -\frac{2}{3}x + 2$

85. D: $(-\infty, \infty)$; R: $(0, \infty)$

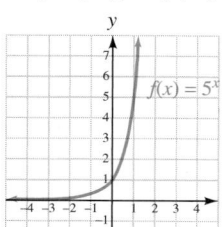

86. D: $(0, \infty)$; R: $(-\infty, \infty)$

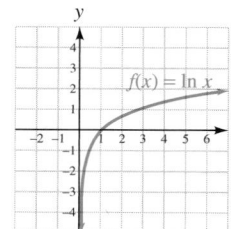

87. 2.7   **88.** About 8.3 billion   **89.** 5   **90.** 64   **91.** -2

92. 1   **93.** $3 \ln y + \frac{1}{2} \ln x - \ln z$ **94.** $\log \frac{x^2 z}{y^3}$ **95.** -8.2144

96. 10   **97.** $(2, -1, 1)$ **98.** $\angle D$: $40°$, $\angle E$: $60°$, $\angle F$: $80°$

99. $(2, -3)$ **100.** $\left(5, \frac{3}{2}\right)$

Study Set Section 13.1 (page 1173)

1. conic sections   **3.** circle, radius

5. a. $(x - h)^2 + (y - k)^2 = r^2$ **b.** $x^2 + y^2 = r^2$

7. a. $(2, -1); r = 4$ **b.** $(x - 2)^2 + (y + 1)^2 = 16$

9. a. $y = a(x - h)^2 + k$ **b.** $x = a(y - k)^2 + h$

11. a. Circle   **b.** Parabola   **c.** Parabola   **d.** Circle

13. $6, -2, 3$

15. $(0, 0), r = 3$

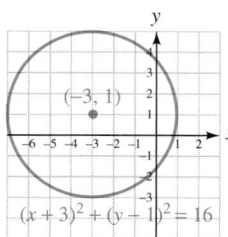

17. $(0, -3), r = 1$

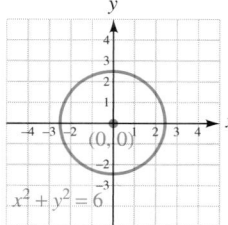

19. $(-3, 1), r = 4$

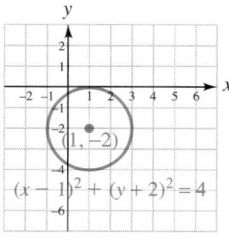

21. $(0, 0), r = \sqrt{6} \approx 2.4$

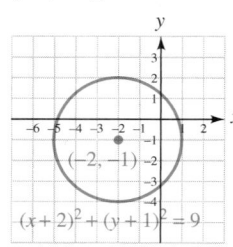

23. $x^2 + y^2 = 1$ **25.** $(x - 6)^2 + (y - 8)^2 = 25$

27. $(x + 2)^2 + (y - 6)^2 = 144$ **29.** $x^2 + y^2 = \frac{1}{16}$

31. $\left(x - \frac{2}{3}\right)^2 + \left(y + \frac{7}{8}\right)^2 = 2$ **33.** $x^2 + y^2 = 8$

35. $(x - 1)^2 + (y + 2)^2 = 4$;   **37.** $(x + 2)^2 + (y + 1)^2 = 9$;

$(1, -2), r = 2$ $(-2, -1), r = 3$

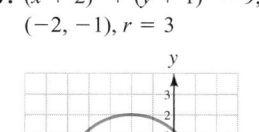

39. $y = 2(x - 1)^2 + 3$;   **41.** $y = -(x + 1)^2 + 4$;

vertex: $(1, 3)$ vertex: $(-1, 4)$

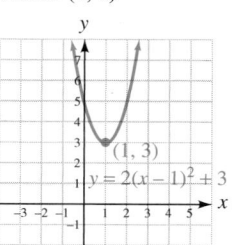   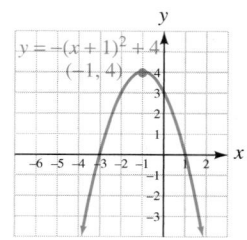

43. Vertex: $(0, 0)$ **45.** Vertex: $(3, -1)$

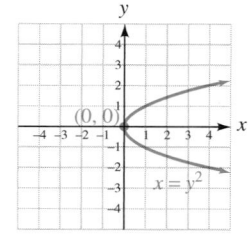   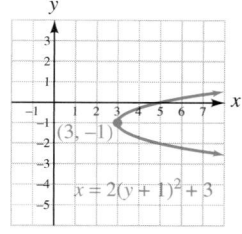

47. $x = (y - 1)^2 + 4$; vertex: $(4, 1)$

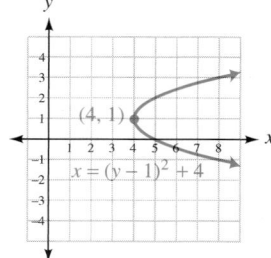

49. $x = -3(y - 3)^2 + 2$; vertex: $(2, 3)$

51.   **53.**

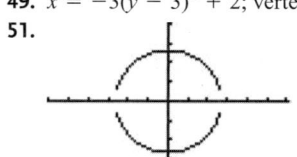   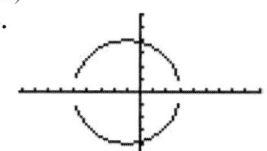

55.   **57.**

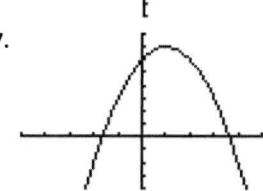

59. $x = (y - 3)^2 - 5$;   **61.** $(2, 0); r = 5$

vertex: $(-5, 3)$

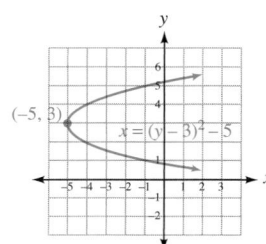   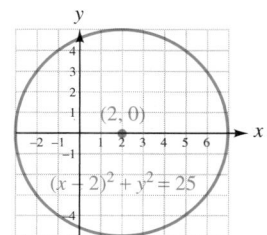

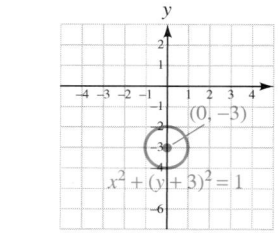

63. $(x - 3)^2 + (y + 4)^2 = 7$;
$(3, -4); r = \sqrt{7} \approx 2.6$

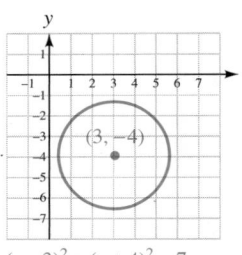

$(x - 3)^2 + (y + 4)^2 = 7$

65. $y = 4(x - 2)^2 + 1$;
vertex: $(2, 1)$

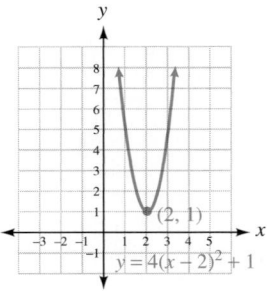

$y = 4(x - 2)^2 + 1$

77. $(x + 1)^2 + y^2 = 9$; $(-1, 0)$;
$r = 3$

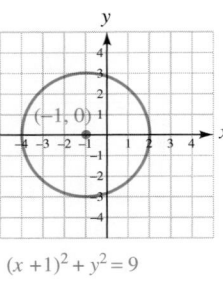

$(x + 1)^2 + y^2 = 9$

79. $x = \frac{1}{2}(y + 2)^2 - 2$;
vertex: $(-2, -2)$

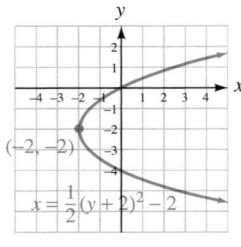

$x = \frac{1}{2}(y + 2)^2 - 2$

67. $(1, 3); r = \sqrt{15} \approx 3.9$

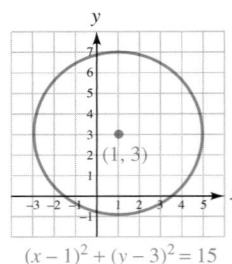

$(x - 1)^2 + (y - 3)^2 = 15$

69. Vertex: $(1, 0)$

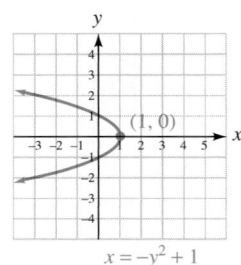

$x = -y^2 + 1$

81. Vertex: $(-5, 5)$

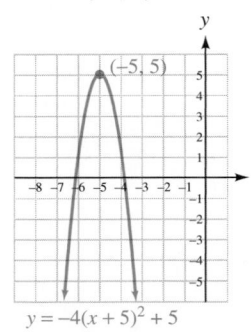

$y = -4(x + 5)^2 + 5$

83. No
85. a. 8 mi
b. 9 mi   **87.** 5 ft   **89.** 2 AU
95. $5, -\dfrac{7}{3}$ **97.** $3, -\dfrac{1}{4}$

71. $(2, 4); r = 6$

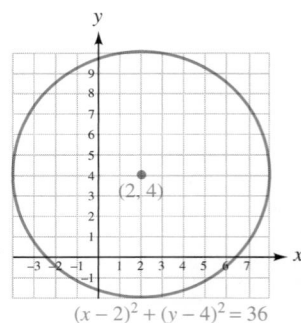

$(x - 2)^2 + (y - 4)^2 = 36$

Study Set Section 13.2 (page 1184)

1. ellipse   **3.** foci, focus   **5.** major   **7.** $\dfrac{x^2}{a^2} + \dfrac{y^2}{b^2} = 1$
9. x-intercepts: $(a, 0), (-a, 0)$; y-intercepts: $(0, b), (0, -b)$
11. a. $(-2, 1); a = 2, b = 5$ **b.** Vertical
c. $\dfrac{(x + 2)^2}{4} + \dfrac{(y - 1)^2}{25} = 1$ **13.** $\dfrac{(x - 1)^2}{16} + \dfrac{(y + 5)^2}{1} = 1$
15. $h = -8, k = 6, a = 10, b = 12$ **17.**

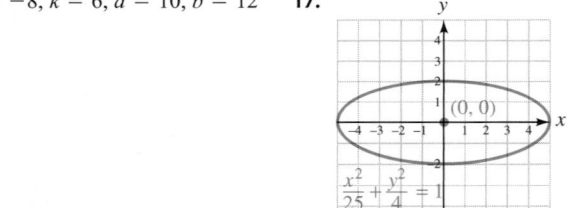

$\dfrac{x^2}{25} + \dfrac{y^2}{4} = 1$

73. Vertex: $(0, 0)$

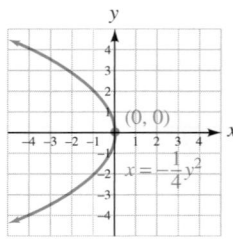

$x = -\dfrac{1}{4}y^2$

75. Vertex: $(3, 1)$

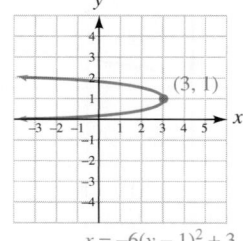

$x = -6(y - 1)^2 + 3$

19.

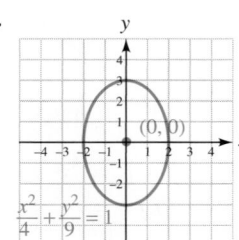

$\dfrac{x^2}{4} + \dfrac{y^2}{9} = 1$

21.

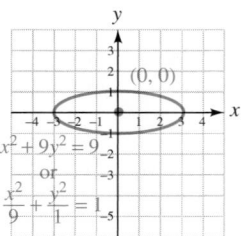

$x^2 + 9y^2 = 9$
or
$\dfrac{x^2}{9} + \dfrac{y^2}{1} = 1$

23.

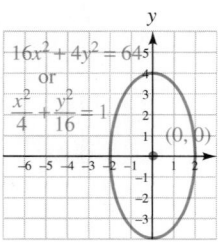

$16x^2 + 4y^2 = 64$
or
$\dfrac{x^2}{4} + \dfrac{y^2}{16} = 1$

25.

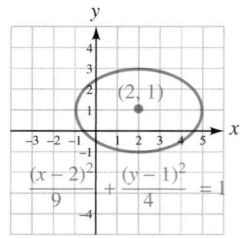

$\dfrac{(x - 2)^2}{9} + \dfrac{(y - 1)^2}{4} = 1$

27.

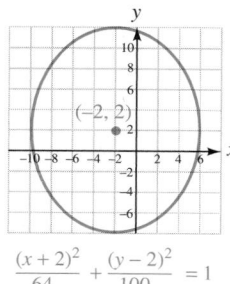

$$\frac{(x+2)^2}{64} + \frac{(y-2)^2}{100} = 1$$

29.

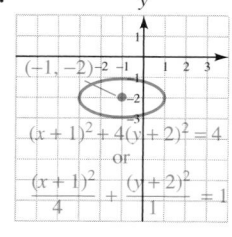

31.

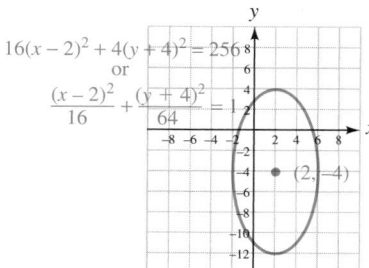

33.

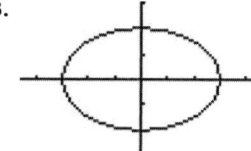

35.

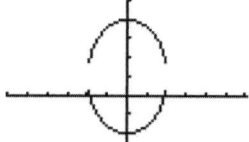

37.

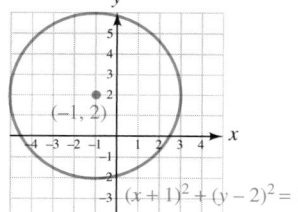

39.

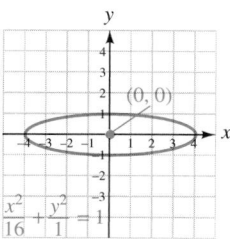

41.

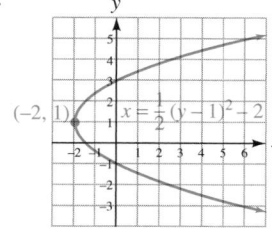

43. $x^2 + y^2 = 25$

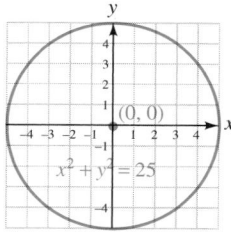

45. $\frac{x^2}{100} + \frac{y^2}{25} = 1$

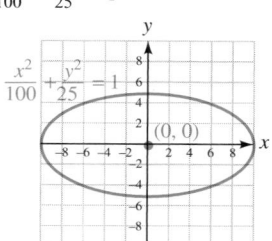

47. $y = -3(x+4)^2 + 5$

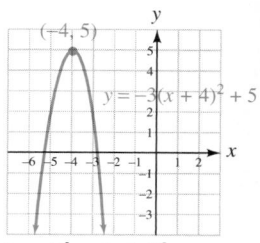

49. $(x-1)^2 + (y+2)^2 = 9$

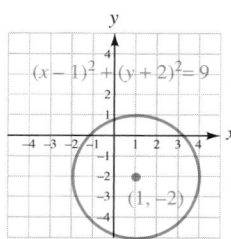

51. $\frac{(x-1)^2}{4} + \frac{(y+2)^2}{9} = 1$

53. $\frac{x^2}{144} + \frac{y^2}{25} = 1$ **55.** $5\sqrt{3}$ ft ≈ 8.7 ft

57. 12π sq. units ≈ 37.7 sq. units   **63.** $12y^2 + \frac{9}{x^2}$ **65.** $\frac{y^2 + x^2}{y^2 - x^2}$

Study Set Section 13.3 (page 1194)

1. hyperbola   **3.** vertices   **5.** diagonals   **7.** $\frac{x^2}{a^2} - \frac{y^2}{b^2} = 1$

9. $\frac{(x-h)^2}{a^2} - \frac{(y-k)^2}{b^2} = 1$ **11. a.** $(-1, -2)$; $a = 3, b = 1$

b. $\frac{(y+2)^2}{9} - \frac{(x+1)^2}{1} = 1$ **13.** $\frac{(x+1)^2}{1} - \frac{(y-5)^2}{4} = 1$

15. $h = 5, k = -11, a = 5, b = 6$

17.

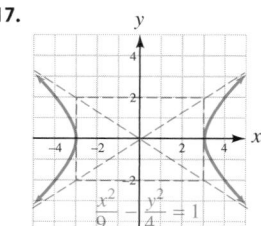

19.

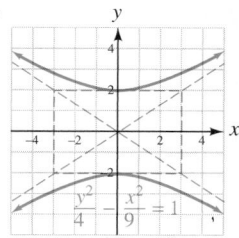

21.

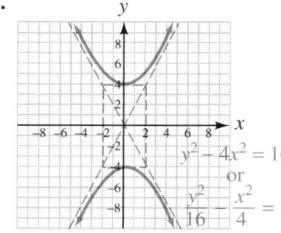

23.

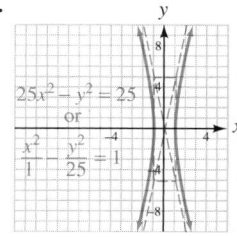

25.

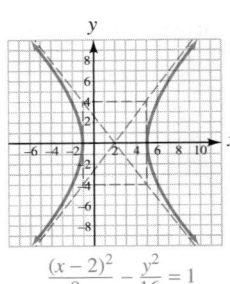

$$\frac{(x-2)^2}{9} - \frac{y^2}{16} = 1$$

27.

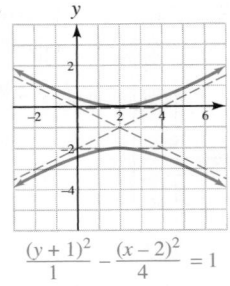

$$\frac{(y+1)^2}{1} - \frac{(x-2)^2}{4} = 1$$

29.

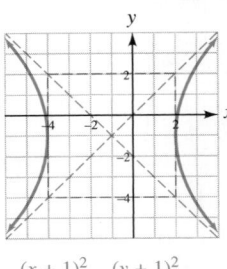

$$\frac{(x+1)^2}{9} - \frac{(y+1)^2}{9} = 1$$

31.

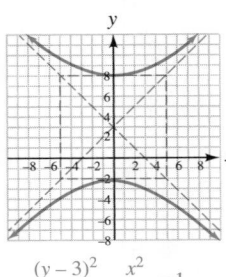

$$\frac{(y-3)^2}{25} - \frac{x^2}{25} = 1$$

33.

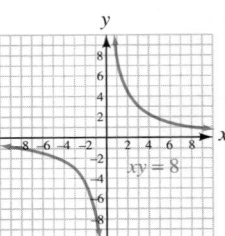

$$xy = 8$$

35.

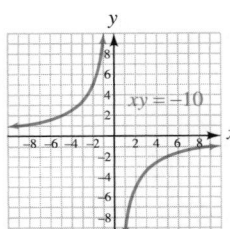

$$xy = -10$$

37.

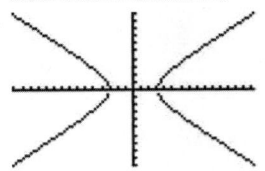

39.

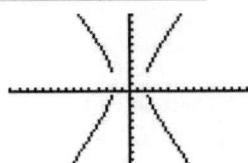

41.

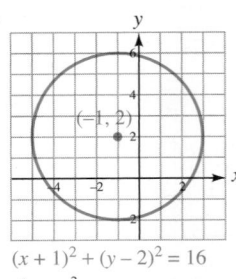

$$(x+1)^2 + (y-2)^2 = 16$$

43. $\frac{x^2}{49} - \frac{y^2}{9} = 1$

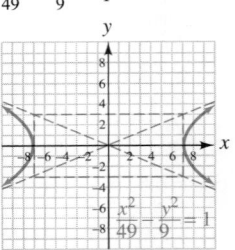

$$\frac{x^2}{49} - \frac{y^2}{9} = 1$$

45. $\frac{(x+1)^2}{9} + \frac{(y+1)^2}{4} = 1$

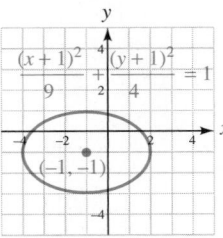

$$\frac{(x+1)^2}{9} + \frac{(y+1)^2}{4} = 1$$

47. $\frac{(x+3)^2}{1} - \frac{(y-1)^2}{4} = 1$

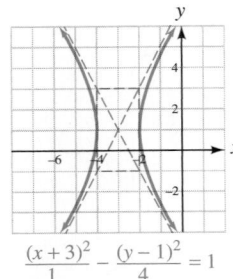

$$\frac{(x+3)^2}{1} - \frac{(y-1)^2}{4} = 1$$

49.

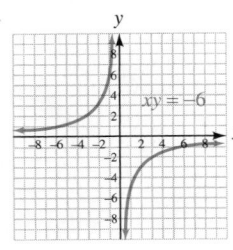

$$xy = -6$$

51.

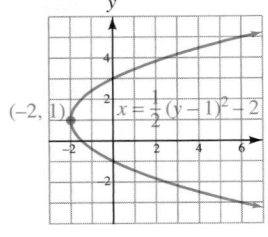

$$x = \frac{1}{2}(y-1)^2 - 2$$

53.

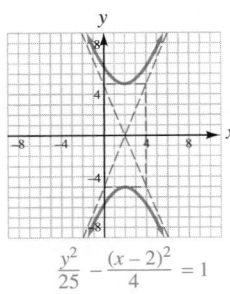

$$\frac{y^2}{25} - \frac{(x-2)^2}{4} = 1$$

55. $y = -(x-3)^2 + 5$

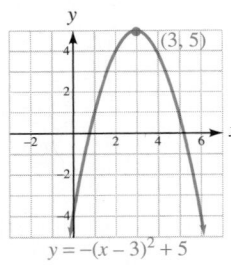

$$y = -(x-3)^2 + 5$$

57.

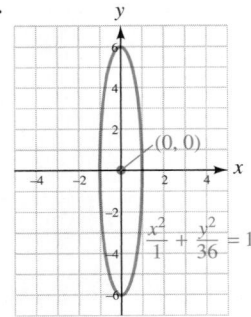

$$\frac{x^2}{1} + \frac{y^2}{36} = 1$$

59. $(x + 2)^2 + (y - 3)^2 = 36$

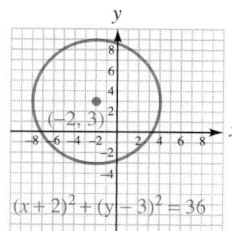

61. 3 units   **63.** $10\sqrt{3}$ miles   **69.** 64   **71.** 3   **73.** $\frac{3}{2}$ **75.** 10

Study Set Section 13.4 (page 1201)

1. system   **3.** intersection   **5.** secant   **7. a.** two   **b.** four
c. four   **d.** four   **9.** $(-3, 2), (3, 2), (-3, -2), (3, -2)$
11. a. -4 **b.** -2 **13.** $2x, 5, 5, 1, 1, -1, -2$
15. $(0, 3), (-3, 0)$ **17.** $(-4, 0), (4, 0)$

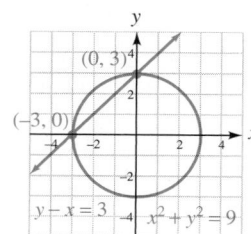   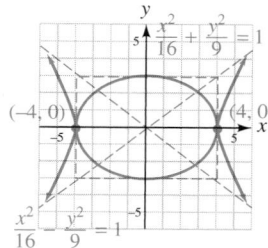

19. $(0, 0), (2, -4)$ **21.** $(-2, 0), (0, -1), (0, 1)$

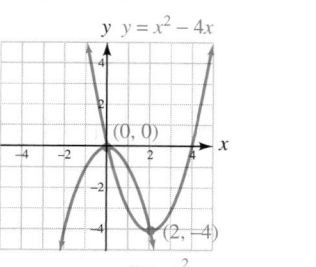   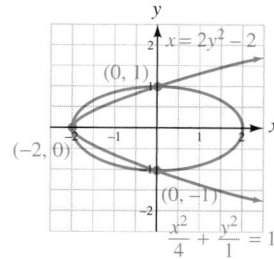

23. $(1, 2), (2, 1)$ **25.** $(-6, 7), (-2, -1)$ **27.** $(-2, 3), (2, 3)$
29. $(\sqrt{5}, 5), (-\sqrt{5}, 5)$ **31.** $(2, 4), (2, -4), (-2, 4), (-2, -4)$
33. $(3, 0), (-3, 0)$ **35.** $(\sqrt{3}, 0), (-\sqrt{3}, 0)$
37. $(-2, 3), (2, 3), (-2, -3), (2, -3)$
39. $(1, 0), (5, 0)$

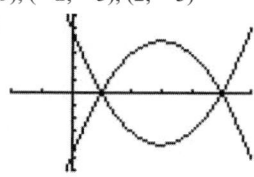

41. $(2, 1), (-2, 1), (2, -1), (-2, -1)$ **43.** $(0, -4), (-3, 5), (3, 5)$
45. $(6, 2), (-6, -2), \left(-\sqrt{42}, 0\right), \left(\sqrt{42}, 0\right)$
47. $\left(-\sqrt{15}, 5\right), \left(\sqrt{15}, 5\right), (-2, -6), (2, -6)$
49. $(0, 3), \left(-\frac{25}{12}, -\frac{13}{4}\right)$ **51.** $(0, 0), (2, 4)$ **53.** No solution, $\varnothing$
55. $(3, 5)$
57. $\left(\frac{\sqrt{10}}{4}, \frac{3\sqrt{6}}{4}\right), \left(\frac{\sqrt{10}}{4}, -\frac{3\sqrt{6}}{4}\right), \left(-\frac{\sqrt{10}}{4}, \frac{3\sqrt{6}}{4}\right), \left(-\frac{\sqrt{10}}{4}, -\frac{3\sqrt{6}}{4}\right)$
59. $\left(\frac{1}{2}, \frac{1}{3}\right), \left(\frac{1}{3}, \frac{1}{2}\right)$ **61.** $(-1, 3), (1, 3)$ **63.** $\left(\frac{5}{2}, \frac{3}{2}\right)$ **65.** 4, 8

67. $\left(10, \frac{10}{3}\right); \frac{10}{3}\sqrt{10}$ m   **69.** 80 ft by 100 ft or 50 ft by 160 ft
71. $2,500 at 9%   **75.** 2,000   **77.** 7

Chapter 13 Review (page 1205)

1.

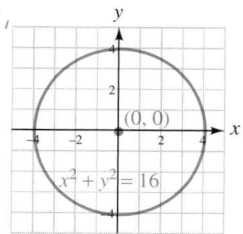

2.

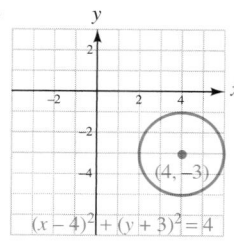

3. $(x + 2)^2 + (y - 1)^2 = 9$

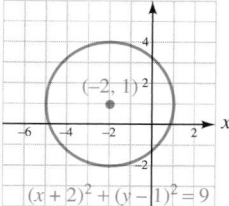

4. $(x - 9)^2 + (y - 9)^2 = 9^2$ or $(x - 9)^2 + (y - 9)^2 = 81$
5. $(-6, 0); r = 2\sqrt{6}$ **6.** center, radius
7. $(0, 0)$ **8.** $(-2, -1)$

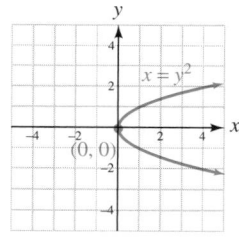   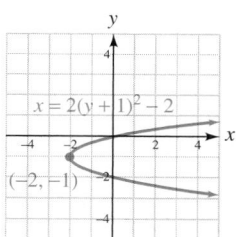

9. $x = -3(y - 2)^2 + 5$ **10.** $y = (x + 4)^2 - 5$

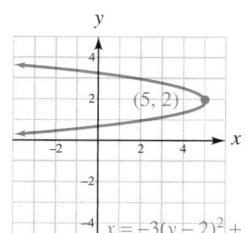   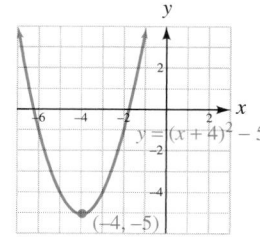

11. $(2, -2), (8, -3)$
12. When $x = 22$, $y = 0$: $-\frac{5}{121}(22 - 11)^2 + 5 = 0$
13.

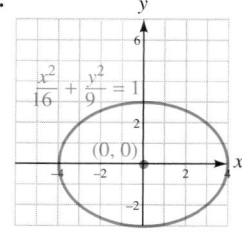

14.

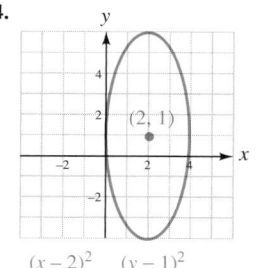

15.

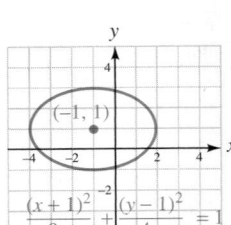

$$\frac{(x+1)^2}{9}+\frac{(y-1)^2}{4}=1$$

16. $\frac{x^2}{12^2}+\frac{y^2}{1^2}=1$ **17. a.** Circle
b. Ellipse   **c.** Parabola
d. Ellipse   **18.** $\frac{x^2}{25}+\frac{y^2}{9}=1$
19. ellipse, focus

20.

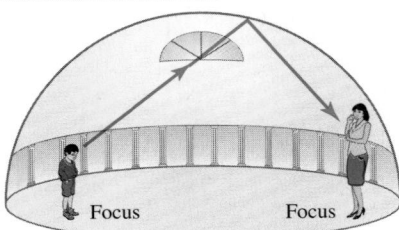

Focus   Focus

21.

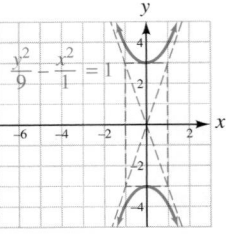

$$\frac{y^2}{9}-\frac{x^2}{1}=1$$

22.

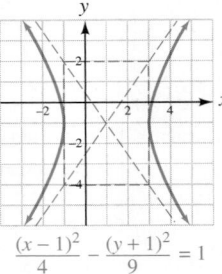

$$\frac{(x-1)^2}{4}-\frac{(y+1)^2}{9}=1$$

23.

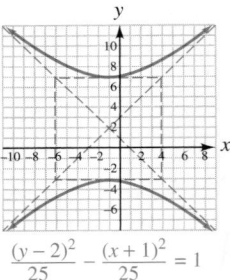

$$\frac{(y-2)^2}{25}-\frac{(x+1)^2}{25}=1$$

24.

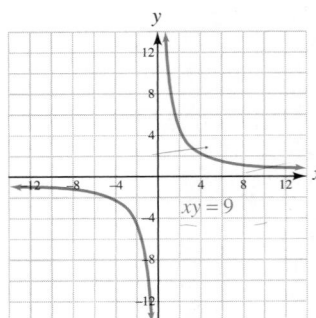

$xy=9$

25. 4 units   **26. a.** Ellipse
b. Hyperbola   **c.** Parabola
d. Circle   **27.** Yes
28. $(0, 3), (0, -3)$

29. $(2, 2), (-1, -4)$

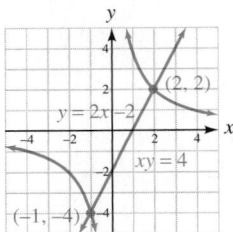

30. a. 2   **b.** 4   **c.** 4   **d.** 4
31. $(0, 1), (0, -1)$
32. $y = 3x - 1$
33. $(0, -4), (-3, 5), (3, 5)$
34. $\left(\sqrt{2}, 0\right), \left(-\sqrt{2}, 0\right)$
35. $(2, 2), \left(-\frac{2}{9}, -\frac{22}{9}\right)$
36. $(4, 2), (4, -2), (-4, 2), (-4, -2)$
37. $(2, 3), (2, -3), (-2, 3), (-2, -3)$
38. $\left(2\sqrt{2}, \sqrt{2}\right), \left(-2\sqrt{2}, -\sqrt{2}\right), (1, 4), (-1, -4)$
39. No solution, $\varnothing$ **40.** $(-2, 1)$

Chapter 13 Test (page 1210)
1. a. conic   **b.** center, radius   **c.** hyperbola   **d.** nonlinear
e. ellipse   **2.** $(0, 0); r = 10$

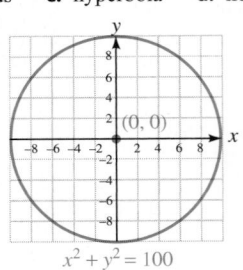

$x^2+y^2=100$

3. $(-2, 3), r = 3\sqrt{2}$ **4.** $(x-4)^2+(y-3)^2=9$

5.

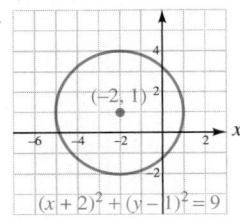

$(x+2)^2+(y-1)^2=9$

6.

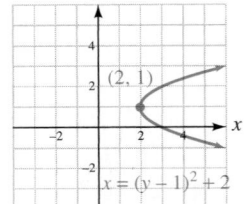

$x=(y-1)^2+2$

7.

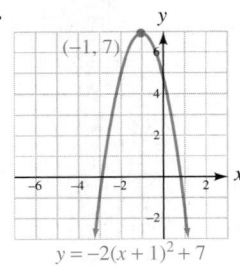

$y=-2(x+1)^2+7$

8.

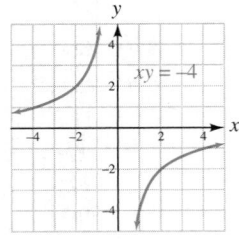

$xy=-4$

9.

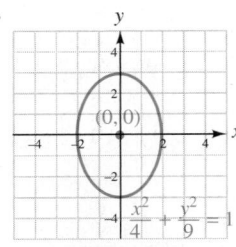

$\frac{x^2}{4}+\frac{y^2}{9}=1$

10.

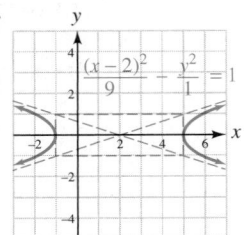

$\frac{(x-2)^2}{9}-\frac{y^2}{1}=1$

11. $\frac{(x-1)^2}{16}+\frac{(y+2)^2}{9}=1$ **12.** 2.5 in.
14. $(-1, 1)$; length: 4 units, width: 4 units   **15.** $\frac{y^2}{16}-\frac{x^2}{36}=1$
16. a. Ellipse   **b.** Hyperbola   **c.** Circle   **d.** Parabola

17. $(-4, -3), (3, 4)$ **18.** $(2, 6), (-2, -2)$

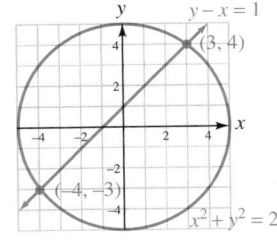

19. $\left(1, \sqrt{2}\right), \left(1, -\sqrt{2}\right), \left(-1, \sqrt{2}\right), \left(-1, -\sqrt{2}\right)$

20. $\left(-1, \frac{9}{2}\right), \left(3, -\frac{3}{2}\right)$ **21.** $(-1, 0)$ **22.** No solution; $\varnothing$

Study Set Section 14.1 (page 1222)

1. binomial   **3.** binomial   **5.** factorial, decreasing   **7.** one
9. $20, 20$ **11.** $1, 1$ **13.** $n, 1$ **15.** 1 **17.** $9, 3$ **19.** $2, 4$
21. $3, 3, y, 2, 2, 3$ **23.** $n, 1$ **25.** $a^3 + 3a^2b + 3ab^2 + b^3$
27. $m^5 - 5m^4p + 10m^3p^2 - 10m^2p^3 + 5mp^4 - p^5$ **29.** 6
31. 120   **33.** 30   **35.** 144   **37.** 40,320   **39.** 2,352   **41.** $\frac{1}{110}$
43. 72   **45.** 5   **47.** 10   **49.** $\frac{1}{168}$ **51.** 21   **53.** 39,916,800
55. $2.432902008 \times 10^{18}$ **57.** $m^4 + 4m^3n + 6m^2n^2 + 4mn^3 + n^4$
59. $c^5 - 5c^4d + 10c^3d^2 - 10c^2d^3 + 5cd^4 - d^5$
61. $a^9 - 9a^8b + 36a^7b^2 - 84a^6b^3 + 126a^5b^4 - 126a^4b^5 + 84a^3b^6 - 36a^2b^7 + 9ab^8 - b^9$
63. $s^6 + 6s^5t + 15s^4t^2 + 20s^3t^3 + 15s^2t^4 + 6st^5 + t^6$
65. $8x^3 + 12x^2y + 6xy^2 + y^3$
67. $32t^5 - 240t^4 + 720t^3 - 1,080t^2 + 810t - 243$
69. $625m^4 - 1,000m^3n + 600m^2n^2 - 160mn^3 + 16n^4$
71. $\frac{x^3}{27} + \frac{x^2y}{6} + \frac{xy^2}{4} + \frac{y^3}{8}$ **73.** $\frac{x^4}{81} - \frac{2x^3y}{27} + \frac{x^2y^2}{6} - \frac{xy^3}{6} + \frac{y^4}{16}$
75. $c^{10} - 5c^8d^2 + 10c^6d^4 - 10c^4d^6 + 5c^2d^8 - d^{10}$ **77.** $28x^6y^2$
79. $15r^2s^4$ **81.** $78x^{10}$ **83.** $-12x^3y$ **85.** $810xy^4$ **87.** $-\frac{1}{6}c^3d$
89. $-70,000t^4$ **91.** $6a^{10}b^2$ **97.** $\log x^2 y^{1/2}$ or $\log x^2\sqrt{y}$ **99.** $\ln y$
105. $3\log x + 2\log y$

Study Set Section 14.2 (page 1232)

1. sequence   **3.** arithmetic   **5.** mean   **7.** $1, 7, 13$
9. a. $a_n = a_1 + (n - 1)d$ **b.** $S_n = \frac{n(a_1 + a_n)}{2}$ **11.** nth
13. sigma   **15.** summation, runs   **17.** $3, 7, 11, 15, 19; 159$
19. $-2, -5, -8, -11, -14; -89$
21. $-1, -4, -9, -16, -25; -400$ **23.** $0, \frac{1}{2}, \frac{2}{3}, \frac{3}{4}, \frac{11}{12}$
25. $-\frac{1}{3}, \frac{1}{9}, -\frac{1}{27}, \frac{1}{81}$ **27.** $-7, 8, -9, 10$ **29.** $3, 5, 7, 9, 11; 21$
31. $-5, -8, -11, -14, -17; -47$ **33.** $7, 19, 31, 43, 53; 355$
35. $-7, -9, -11, -13, -15; -35$ **37.** 88   **39.** 59
41. $5, 11, 17, 23, 29$ **43.** $-4, -11, -18, -25, -32$ **45.** $5, 8$
47. $12, 14, 16, 18$ **49.** $\frac{45}{2}, 25, \frac{55}{2}$ **51.** $\frac{5}{4}$ **53.** 2,555
55. 2,920   **57.** $3 + 6 + 9 + 12$ **59.** $4 + 9 + 16$ **61.** 60
63. 91   **65.** 31   **67.** 12   **69.** 70   **71.** 57   **73.** 12   **75.** 354
77. $-118, -111, -104, -97, -90$ **79.** $34, 31, 28, 25, 22, 19$
81. 1,398   **83.** 1,275   **85.** -179 **87.** -23 **89.** -3
91. 2,500   **93.** $\$60, \$110, \$160, \$210, \$260, \$310; \$6,060$
95. 11,325   **97.** 78   **103.** $1 + \log_2 x - \log_2 y$
105. $3\log x + 2\log y$

Study Set Section 14.3 (page 1245)

1. geometric   **3.** mean   **5.** $16, 4, 1$ **7.** $a_n = a_1 r^{n-1}$
9. a. Yes   **b.** No   **c.** No   **d.** Yes   **11.** $r^2, a_1 r^4$ **13.** n
15. $3, 6, 12, 24, 48; 768$ **17.** $-5, -1, -\frac{1}{5}, -\frac{1}{25}, -\frac{1}{125}; -\frac{1}{15,625}$
19. 1,458   **21.** $-\frac{3,125}{2}$ **23.** $2, 6, 18, 54, 162$
25. $3, 6, 12, 24, 48$ **27.** $6, 18, 54$ **29.** $-20, -100, -500, -2,500$
31. $-16, 16$ **33.** $-10\sqrt{2}, 10\sqrt{2}$ **35.** 728   **37.** 122
39. -255 **41.** 381   **43.** 16   **45.** 81   **47.** $-\frac{81}{2}$ **49.** No sum
51. 8   **53.** $-\frac{135}{4}$ **55.** $\frac{1}{9}$ **57.** $\frac{1}{3}$ **59.** $\frac{4}{33}$ **61.** $\frac{25}{33}$ **63.** 3
65. $-64, 32, -16, 8, -4$ **67.** No geometric mean exists.
69. 3,584   **71.** $-64, -32, -16, -8$ **73.** 4   **75.** $\frac{1}{27}$ **77.** 93
79. $\$1,469.74$ **81.** About $\$539,731$
83. $\left(\frac{1}{2}\right)^{11} \approx 0.0005$ square unit   **85.** 30 m   **87.** 5,000
93. $[-1, 6]$ **95.** $(-\infty, -3) \cup (4, \infty)$

Chapter 14 Review (page 1248)

1. $1, 3, 1, 1, 10, 15; 1, 5, 10, 10, 5, 1$ **2. a.** 13   **b.** 12
c. a^{12}, b^{12} **d.** a: decrease; b: increase   **3.** 144   **4.** 20
5. 15   **6.** 220   **7.** 1   **8.** 8
9. $x^5 + 5x^4y + 10x^3y^2 + 10x^2y^3 + 5xy^4 + y^5$
10. $x^9 - 9x^8y + 36x^7y^2 - 84x^6y^3 + 126x^5y^4 - 126x^4y^5 + 84x^3y^6 - 36x^2y^7 + 9xy^8 - y^9$ **11.** $64x^3 - 48x^2y + 12xy^2 - y^3$
12. $\frac{c^4}{16} + \frac{c^3d}{6} + \frac{c^2d^2}{6} + \frac{2cd^3}{27} + \frac{d^4}{81}$ **13.** $6x^2y^2$ **14.** $-20x^3y^3$
15. $-108x^2y$ **16.** $5u^7v^{12}$ **17.** $-2, 0, 2, 4$
18. $-\frac{1}{2}, \frac{1}{3}, -\frac{1}{4}, \frac{1}{5}, -\frac{1}{6}$ **19.** 75   **20.** 42
21. $122, 137, 152, 167, 182$ **22.** $-1,194$ **23.** -5
24. $\frac{41}{3}, \frac{58}{3}$ **25.** $-\frac{45}{2}$ **26.** 1,568   **27.** $\frac{15}{2}$ **28.** 378
29. 14   **30.** 360   **31.** 20,100   **32.** 1,170   **33.** 4
34. $24, 12, 6, 3, \frac{3}{2}$ **35.** $\frac{1}{27}$ **36.** $24, -96$ **37.** $\frac{2,186}{9}$
38. $-\frac{85}{8}$ **39.** About 1.6 lb   **40.** 125   **41.** $\frac{5}{99}$ **42.** 190 ft

Chapter 14 Test (page 1252)

1. a. Pascal's   **b.** alternate   **c.** arithmetic   **d.** series
e. geometric   **2.** $2, -4, -10, -16$ **3.** $-1, \frac{1}{8}, -\frac{1}{27}, \frac{1}{64}, -\frac{1}{125}$
4. 210   **5.** $a^6 - 6a^5b + 15a^4b^2 - 20a^3b^3 + 15a^2b^4 - 6ab^5 + b^6$
6. $24x^4y^2$ **7.** 66   **8.** 306   **9.** $34, 66$ **10.** $\frac{1}{4}$ **11.** $-1,377$
12. 205   **13.** 1,600 ft   **14.** 3   **15.** -81 **16.** $\frac{364}{27}$ **17.** 6
18. $18, 108$ **19.** $\frac{27}{2}$ **20.** About $\$651,583$ **21.** 2,000 in.
22. $\frac{7}{9}$

Group Project (page 1253)

1. g   **2.** o   **3.** i   **4.** l   **5.** u   **6.** y   **7.** p   **8.** x   **9.** m
10. d   **11.** f   **12.** c   **13.** w   **14.** b   **15.** j   **16.** s   **17.** e
18. z   **19.** n   **20.** k   **21.** h   **22.** v   **23.** q   **24.** a
25. r   **26.** t

Cumulative Review Chapters 1–14 (page 1254)

1. a. 0   **b.** $-\frac{4}{3}, 5.6, 0, -23$ **c.** $\pi, \sqrt{2}, e$
d. $-\frac{4}{3}, \pi, 5.6, \sqrt{2}, 0, -23, e$ **2.** 0
3. $b_2 = \frac{2A - b_1 h}{h}$ or $b_2 = \frac{2A}{h} - b_1$ **4.** $85°, 80°, 15°$ **5.** $\$8,250$

6. $\frac{1}{120}$ decibels/rpm   **7. a.** Parallel   **b.** Perpendicular

8. $y = -3,400x + 28,000$ **9.** $y = -2x + 5$

10. $y = -\frac{9}{13}x + \frac{7}{13}$ **11.** It doesn't pass the vertical line test. The graph passes through $(0, 2)$ and $(0, -2)$.   **12.** $-4; 3a^5 - 2a^2 + 1$

13. $(-7, 7)$ **14.** $\left(\frac{4}{5}, \frac{3}{4}\right)$ **15.** Regular: $29\frac{1}{3}$ lb; Brazillian: $10\frac{2}{3}$ lb

16.

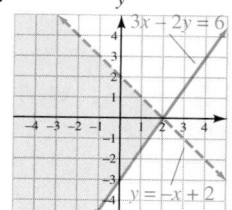

17. $x^8 y^{10}$ **18.** $\frac{16}{y^{16}}$

19. $1.73 \times 10^{14}; 4.6 \times 10^{-8}$ **20.** $1.44 \times 10^{-3}; 0.00144$

21. $\frac{7}{4}rt^2 - \frac{5}{6}rt$ **22.** $-3x^2 + 5x - 7$ **23.** $2x^2y^3 + 13xy + 3y^2$

24. $x^3 - 27y^3$ **25.** $6m^{10} - 23m^5 + 7$ **26.** $81a^2b^4 - 72ab^2 + 16$

27. $xy(3x - 4y)(x - 2)$ **28.** $(b - 4)(b^2 - 3)$

29. $(3y + 2)(4y + 5)$ **30.** $(16x^2y^2 + z^4)(4xy + z^2)(4xy - z^2)$

31. $(3t + u)(9t^2 - 3tu + u^2)$ **32.** $b^2(a + 4)(a - 4)(a + 2)(a - 2)$

33. $-2, \frac{1}{3}$ **34.** A repeated solution of -8 **35.** $0, 1, -9$

36. Length: 21 ft; width: 4 ft   **37.** 6, 8, 10   **38.** $\frac{3x + 2}{3x - 2}$ **39.** $-\frac{q}{p}$

40. $\frac{1}{(x^2 + 9)(x - 3)}$ **41.** $\frac{4a - 1}{(a + 2)(a - 2)}$ **42.** $y - x$ **43.** $-\frac{7}{5}$

44. $R = \frac{R_1R_2R_3}{R_2R_3 + R_1R_3 + R_1R_2}$ **45.** $2\frac{2}{5}$ hr = 2.4 hr   **46.** 93

47. No solution, $\varnothing$; contradiction   **48.** 10 hr

49. $(-\infty, -2)$ 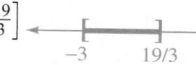   **50.** $\left[-3, \frac{19}{3}\right]$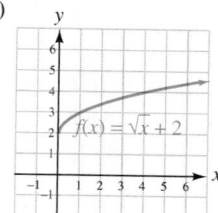

51. $(7x - 7t + 2)(2x - 2t - 3)$ **52.** $\lambda = \frac{4d - 2}{A - 6}$

53. $(x + 5 + 4z)(x + 5 - 4z)$ **54.** No; $(2, 2), (2, -2)$ **55. a.** 4
b. 3   **c.** 0, 2   **d.** 1   **56.** All real numbers except -5 and 5

57. 2 lumens   **58.** D: $[0, \infty)$; R: $[2, \infty)$

59. $\frac{343}{125}$ **60.** $4ab^2\sqrt{7ab}$ **61.** $5\sqrt{2}$ **62.** $81x\sqrt[3]{3x}$

63. $13 - \sqrt{7}$ **64.** $15x - 6\sqrt{15x} + 9$ **65.** $\frac{\sqrt[3]{4b^2}}{b}$

66. $\sqrt{3t} - 1$ **67.** 1, 3   **68.** 5   **69.** 0   **70.** About $21\frac{1}{2}$ in.

71. $5i$ **72.** -1 **73.** $-5 + 17i$ **74.** $-\frac{21}{29} - \frac{20}{29}i$ **75.** $\pm2\sqrt{6}$

76. $-5 \pm 4\sqrt{2}$ **77.** $\frac{-3 \pm \sqrt{5}}{4}$ **78.** $\frac{2}{3} \pm \frac{\sqrt{2}}{3}i$ **79.** $\frac{1}{4}, \frac{1}{2}$

80. $-i, i, -3i\sqrt{2}, 3i\sqrt{2}$ **81. a.** A quadratic function
b. At about 85% and 120% of the suggested inflation

82. $(-1, -2); (0, -8)$; no x-intercepts

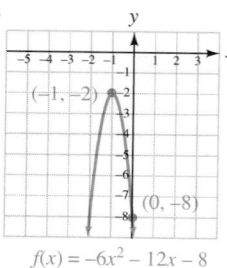

$f(x) = -6x^2 - 12x - 8$

83. $[3, 5]$ 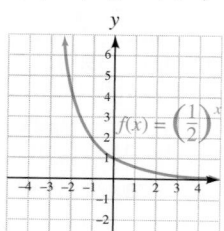   **84.** $4x^2 + 4x - 1$

85. $f^{-1}(x) = \sqrt[3]{\frac{x + 1}{2}}$

86. D: $(-\infty, \infty)$; R: $(0, \infty)$

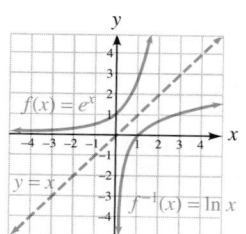

87.

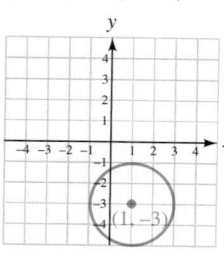

88. About 145 million

89. 3   **90.** 2   **91.** $\frac{1}{27}$ **92.** 5

93. 1   **94.** -1 **95.** Undefined

96. $2 - 3\log_6 x$ **97.** $\ln\frac{y\sqrt{x}}{z}$

98. About 13.4 hr   **99.** $-\frac{3}{4}$

100. 3.4190

101. 1, -10 does not check

102. 9

103. $(3, 2, 1)$ **104.** Small: 150; standard-size: 250; super-size: 100

105. $(2, -3)$ **106.** $\left(-2, \frac{3}{2}\right)$

107. $(x - 1)^2 + (y + 3)^2 = 4$ **108.** $x = -\frac{1}{4}(y - 3)^2 + 2; (2, 3)$

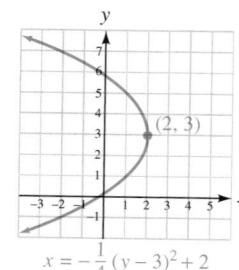

$(x - 1)^2 + (y + 3)^2 = 4$

$x = -\frac{1}{4}(y - 3)^2 + 2$

109.

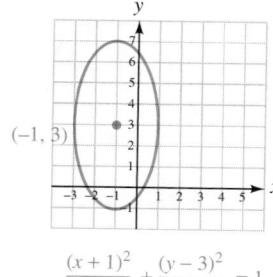

$\frac{(x + 1)^2}{4} + \frac{(y - 3)^2}{16} = 1$

110.

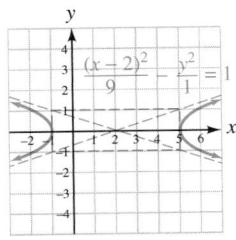

$$\frac{(x-2)^2}{9} - \frac{y^2}{1} = 1$$

111. $81a^4 - 108a^3b + 54a^2b^2 - 12ab^3 + b^4$ **112.** $112x^2y^6$

113. 66   **114.** 103   **115.** 690   **116.** 27   **117.** 27

118. \$2,848.31   **119.** $\frac{1,023}{64}$ **120.** $\frac{27}{2}$

Appendix II (page A-7)

1. synthetic   **3.** divisor   **5.** theorem

7. a. $(5x^3 + x - 3) \div (x + 2)$ **b.** $5x^2 - 10x + 21 - \frac{45}{x+2}$

9. $6x^3 - x^2 - 17x + 9, x - 8$ **11.** 2, 1, 12, 26, 6, 8   **13.** $x + 2$

15. $x - 3$ **17.** $x + 2$ **19.** $x - 7 + \frac{28}{x+2}$ **21.** $3x^2 - x + 2$

23. $2x^2 + 4x + 3$ **25.** $6x^2 - x + 1 + \frac{3}{x+1}$ **27.** $t^2 + 1 + \frac{1}{t+1}$

31. $-5x^4 + 11x^3 - 3x^2 - 7x - 1$ **33.** $8t^2 + 2$

35. $x^3 + 7x^2 - 2$ **37.** $7.2x - 0.66 + \frac{0.368}{x - 0.2}$

39. $9x^2 - 513x + 29,241 + \frac{1,666,762}{x+57}$ **41.** -1 **43.** -37

45. 23   **47.** -1 **49.** 2   **51.** -1 **53.** 18   **55.** 174

57. -8 **59.** 59   **61.** 44   **63.** $\frac{29}{32}$ **65.** yes   **67.** no

73. 0   **75.** 2

INDEX

Enhanced WebAssign

The Start Smart Guide for Students

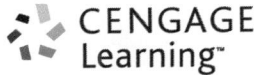

CENGAGE
Learning™

Australia • Brazil • Japan • Korea • Mexico • Singapore • Spain • United Kingdom • United States

Enhanced WebAssign: The Start Smart Guide for Students

For product information and technology assistance, contact us at
Cengage Learning Customer & Sales Support, 1-800-354-9706

For permission to use material from this text or product, submit all requests online at **cengage.com/permissions**
Further permissions questions can be emailed to **permissionrequest@cengage.com**

Executive Editors:
Michele Baird

Maureen Staudt

Michael Stranz

Project Development Manager:
Linda deStefano

Senior Marketing Coordinators:
Sara Mercurio

Lindsay Shapiro

Production/Manufacturing Manager:
Donna M. Brown

PreMedia Services Supervisor:
Rebecca A. Walker

Rights & Permissions Specialist:
Kalina Hintz

Cover Image:
Getty Images*

ISBN-13: 978-0-495-38479-3

ISBN-10: 0-495-38479-8

Cengage Learning
5191 Natorp Boulevard
Mason, Ohio 45040
USA

Cengage Learning is a leading provider of customized learning solutions with office locations around the globe, including Singapore, the United Kingdom, Australia, Mexico, Brazil, and Japan. Locate your local office at: **international.cengage.com/region**

Cengage Learning products are represented in Canada by Nelson Education, Ltd.

For your lifelong learning solutions, visit **custom.cengage.com**

Visit our corporate website at **cengage.com**

* Unless otherwise noted, all cover images used by Custom Solutions, a part of Cengage Learning, have been supplied courtesy of Getty Images with the exception of the Earthview cover image, which has been supplied by the National Aeronautics and Space Administration (NASA).

Printed in Canada
1 2 3 4 5 6 7 12 11 10 09 08

CONTENTS

> **WebAssign works with any recent browser and computer. Some assignments may require plugins like Java, Flash, Shockwave, or Adobe Reader.**
>
> **For technical support go to http://webassign.net/student.html or email support@webassign.net.**

Contents

GETTING STARTED

Welcome to Enhanced WebAssign, the integrated, online learning system that gives you 24/7 access to your math, physics, astronomy, chemistry, biology, and statistics assignments.

Now, you can do homework, take quizzes and exams, and receive your scores and graded assignments from any computer with an Internet connection and web browser, any time of the day or night.

Note: As a live, web-based program, Enhanced WebAssign is updated regularly with new features and improvements. Please refer to WebAssign's online Help for the most current information.

Technical Startup Tips

Before you start, please note the following important points:

○ Most standard web connections should work with WebAssign. We recommend using Firefox 1.0 or later, or Internet Explorer 5.5 or later. *We do not recommend the AOL browser.*

○ You can use a 56 KBPS modem, broadband, or school network connection.

○ Your browser needs to have both JavaScript and Java enabled.

○ *You cannot skip the login page.* WebAssign must know it is you before delivering your assignments.

Note: If you'd like to bookmark WebAssign on your computer, we recommend that you bookmark **https://www.webassign.net/login.html** or the appropriate address for your school.

Login to WebAssign

In order to access WebAssign, you'll need to login with your username, institution and password, which will be provided by your instructor.

➤ **To get started**

1. If you are using a shared computer, completely exit any browsers that are already open.

2. Open a new web browser and go to https://www.webassign.net/login.html, or the web address provided by your instructor.

3. Enter your **Username**, **Institution** (school code), and **Password** *provided by your instructor*.

   Institution

   If you do not know your **Institution,** you can search for it by clicking **(what's this?)** above the **Institution** entry box.

   In the **What's My Institution Code** pop-up window, enter your school name and click **go!**. The **Institution Search Results** table will give you choices of the School Names that most closely match your entry, and the **Institution Code** that you should enter in the **Institution** entry box on the **WebAssign Login** screen.

   Password

   If you have forgotten or do not know your **Password,** click **(Reset Password)** above the **Password** entry box, and follow the directions on the **WebAssign New Password Request** screen.

4. Click **Log In**.

Note: Before starting WebAssign on a shared computer, always exit any browsers and restart your browser application. *If you simply close the browser window or open a new window, login information contained in an encrypted key may not be yours.*

Logout

When you are finished with your work, click the **Logout** link in the upper right corner of your Home page, and *exit the browser completely* to avoid the possibility of someone else accessing your work.

YOUR ENHANCED WEBASSIGN HOME PAGE

Your personalized Home page is your hub for referencing and managing all of your Enhanced WebAssign assignments.

Using Access Codes

Some classes require an **access code** for admission. Please remember:

○ An **access code** is *not* the same as your login password.

○ An **access code** is good for *one class only* unless the textbook includes a two-term **access code**.

○ An **access code** is an alphanumeric code that is *usually* packaged with your textbook. It can begin with 2 or 3 letters, followed by an alphanumeric code, or it can have a longer prefix such as **BCEnhanced-S** followed by four sets of four characters.

○ If your textbook did not include an **access code**, you can buy one at your bookstore, or from your personalized Home page by clicking the **Purchase an access code online** button.

➢ **To enter an Access Code**

1. Under **WebAssign Notices**, select the proper prefix from the **Choose your access code prefix** pull-down menu.

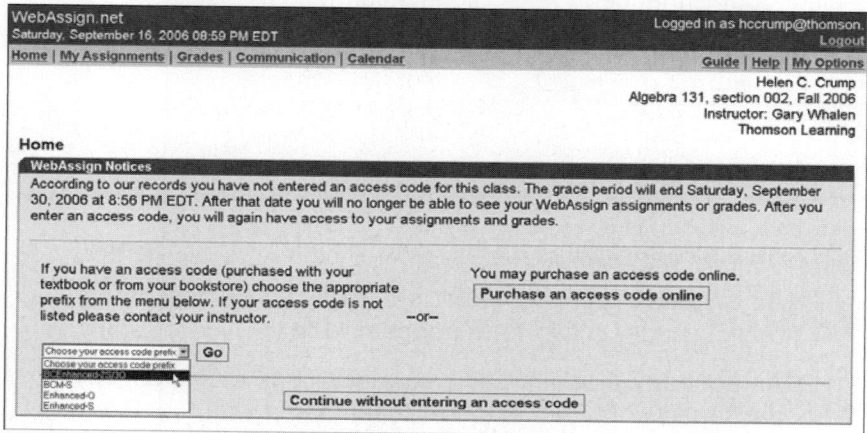

WebAssign notices

2. Click **Go**.

3. In the entry boxes, type in your access code *exactly* as it appears on your card. (When you purchase online, the access code is entered automatically.)

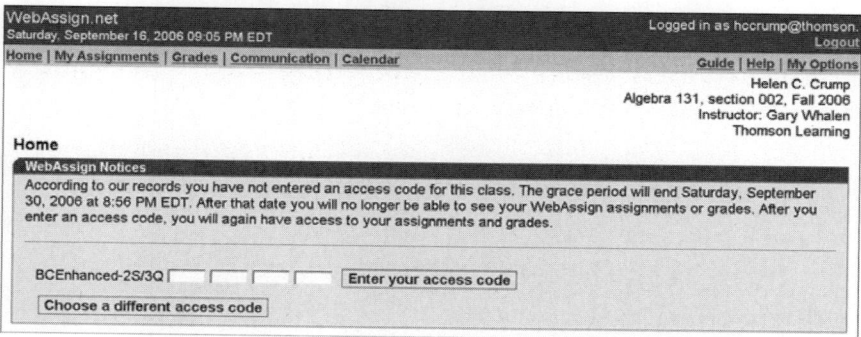

Access code entry

4. Click **Enter your access code.**

   If you have chosen the wrong prefix from the previous screen, you can click the **Choose a different access code** button to try again.

   If your **access code** is a valid unused code, you will receive a message that you have successfully entered the code for the class. Click the **Home** or **My Assignments** button to proceed.

Customizing Your Home Page

Your instructor has initial control over what you see on your Home page to make sure that you have all of the information you need. Your instructor might also set controls so that you can further personalize this page by moving or hiding certain modules.

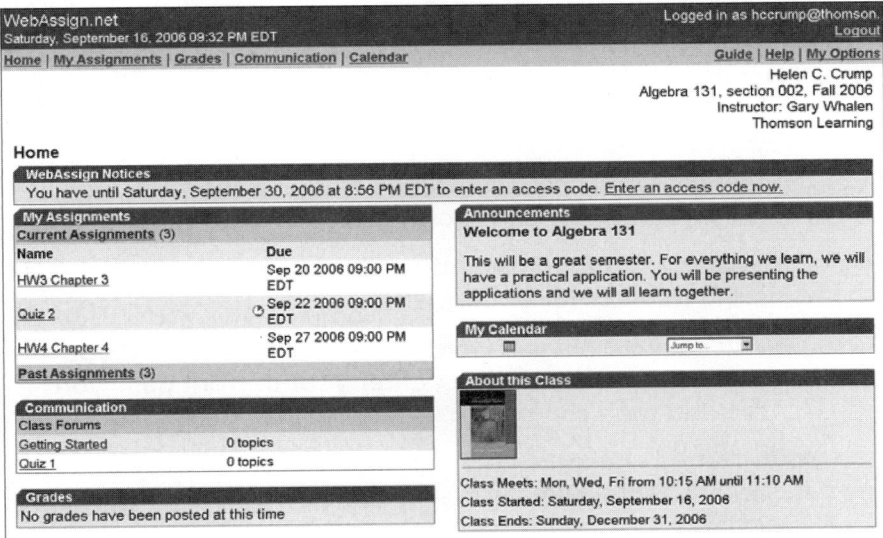

Student Home page

If your instructor has allowed you to personalize your Home page, each module will have markings like this:

Calendar module

To move a module

On the module's heading line, click an up, down, or sideways arrow (indicated by white triangles) until the module is where you'd like it placed on the page.

To minimize a module

On the module's heading line, click the underscore.

To hide a module

On the module's heading line, click the x.

Changing Your Password

For your personal security, it's a good idea to change the initial password provided by your instructor.

➤ To change your password

1. Click the **My Options** link in the upper right of your Home page.

2. In the **My Options** pop-up window, under the **Personal Info** tab:

   Enter your *new* password in the **Change Password** entry box next to **(enter new password)**, then

   Reenter your new password *exactly* the same in the entry box next to **(reenter for confirmation)**.

3. Enter your *current* password in the entry box under **If you made any changes above, enter your current password here and then click save:**, located at the bottom of the pop-up window.

4. Click the **Save** button in the bottom right corner of the pop-up window.

   If the change was successful, you will see the message **Your password has been changed**.

Note: Passwords are case-sensitive. This means that if you capitalize any of the letters, you must remember to capitalize them the same way each time you sign in to Enhanced WebAssign. (Example: "BigKat" and "bigkat" are treated as two different passwords.) Also, be careful with characters that look alike, such as the lowercase letter *l* (as in leopard) the uppercase letter *I* (as in Indiana), and the number *1*.

Changing Your Email Address

If your instructor provided you with an email address, you can easily change it to your own personal email address any time.

➤ To change your email address

1. Click the **My Options** link in the upper right of your Home page.

2. In the **My Options** pop-up window, under the **Personal Info** tab, enter your *valid* email address in the **Email Address** box.

3. Enter your current password in the entry box under **If you made any changes above enter your current password here and then click save:**, located at the bottom of the pop-up screen.

4. Click the **Save** button in the bottom right corner of the pop-up window.

   A confirmation email will be sent to your new email address.

Once you receive the confirmation email, you must click the link in the email to successfully complete and activate this change.

Note: If you do not currently have an email account, you should be able to get one quickly and at little or no expense. Check with your Internet Service Provider, or try free email sites such as Hotmail at http://mail.msn.com, Yahoo Mail at http://mail.yahoo.com, or Gmail at gmail.google.com.

WORKING WITH ASSIGNMENTS

The courses that have been set up for you by your instructor(s) appear on your Enhanced WebAssign personalized Home page. If you have more than one course, simply select the course you want to work with from the pull-down menu.

Assignment Summary

There are two ways to get a quick summary of your assignments. On the Home page:

○ Click the **My Assignments** link in the upper left *menu bar, or*

○ Click the **Current Assignments** link in the **My Assignments** *module* on the Home page.

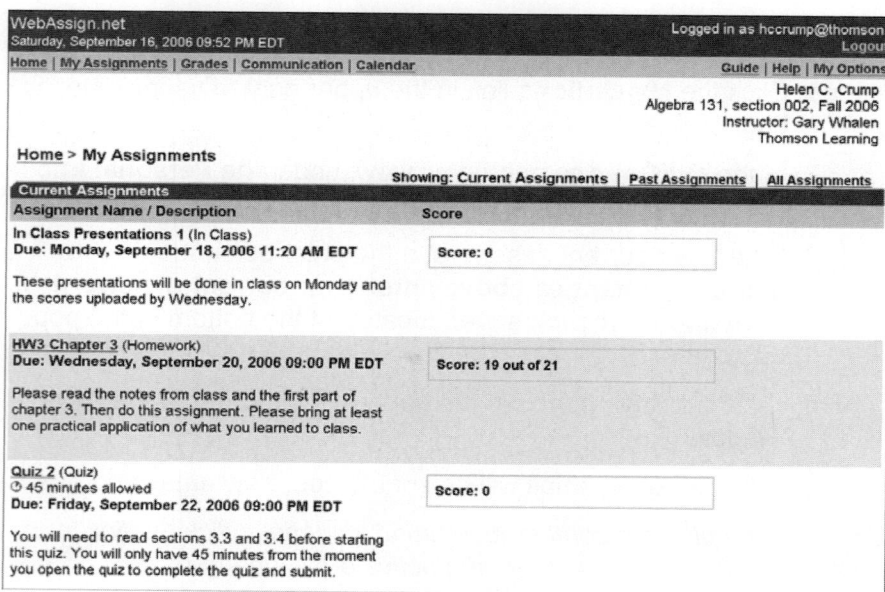

Assignment summary

Accessing an Assignment

Once your assignments are displayed on your Home page, simply click the name of the assignment you'd like to begin.

○ If you have previously submitted an assignment, you will see your most recent responses, if your instructor allows this feature.

○ If you have already submitted the assignment, there will usually be a link to **Review All Submissions** on the page, if your instructor has allowed it.

Using the Assignment Page

When you click on an assignment name, your assignment will load. Within the **About this Assignment** page are links to valuable information about your assignment's score, submission options, and saving your work in progress. Within each question, there might also be "enhanced" action links to useful tutorial material such as book content, videos, animations, active figures, simulations, and practice problems. The links available may vary from one assignment to another.

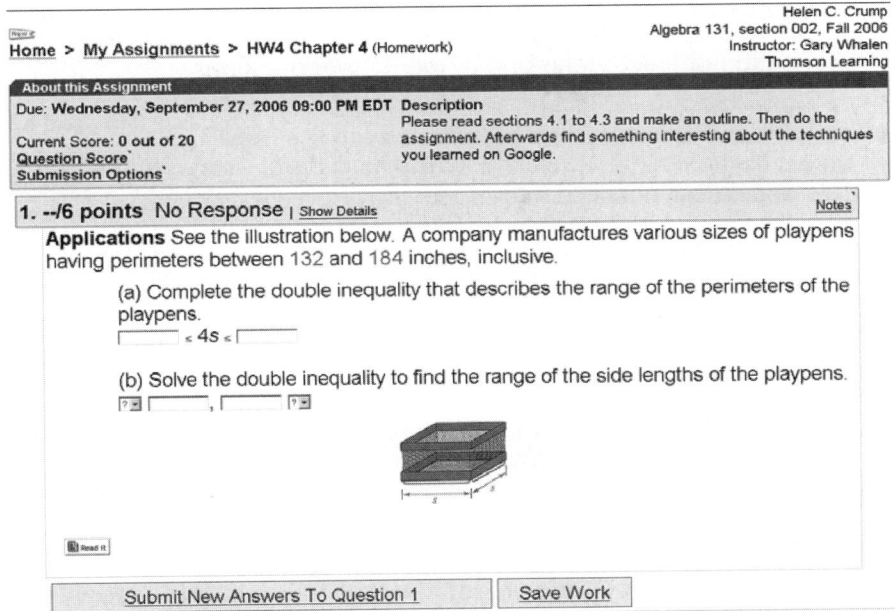

Math assignment

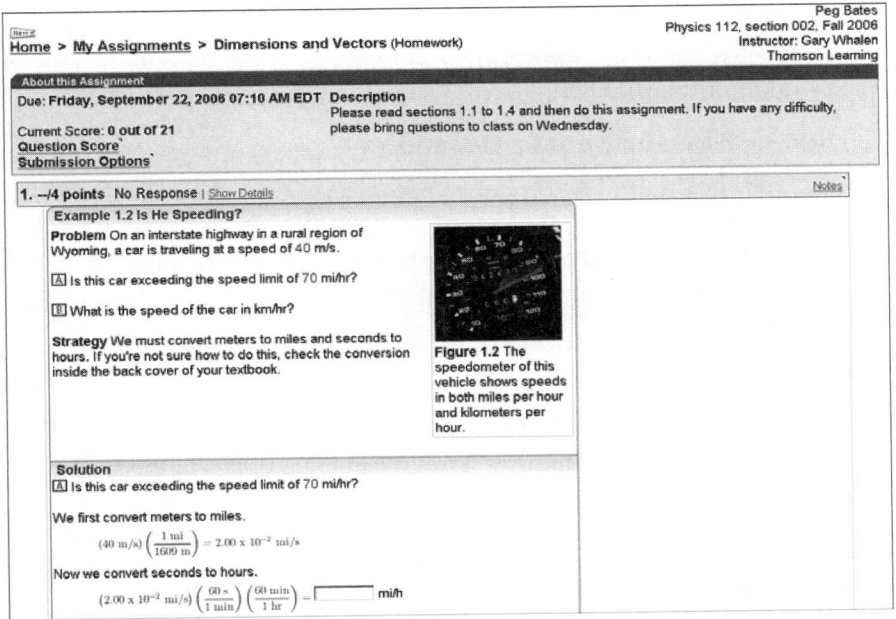

Physics assignment

Actions

Click a button or link to take one of the following actions:

Current Score

This gives you a quick look at your current score versus the maximum possible score.

Question Score

This gives you a pop-up window showing your score for each question.

Submission Options

This gives you a pop-up window explaining how you can submit the assignment and whether it can be submitted by question part, by whole question, or by the whole assignment.

Submissions Made

This shows you the number of submissions you've made. This information is only displayed on assignments that require submission of the entire assignment.

Notes

This feature gives you a pop-up window with a text box in which you can enter and save notes or show your work with a particular question.

Submit New Answers To Question

Use this button when you're ready to submit your answer for the question. This feature allows you to answer just the parts you want scored. If you leave any part of a question unanswered, the submission *will not* be recorded for that part.

Submit Whole Question

Use this button to submit your answer(s) for the entire question. If you leave any part of a question unanswered, the submission *will* be recorded as if the entire question has been answered, and graded as such.

Save Work

This button allows you to save the work you've done so far on a particular question, but does not submit that question for grading.

View Saved Work

Located in the question's header line, this allows you to view work that you previously saved for that question.

Show Details

Located in the question's header line, this link shows your score on each part of the question, how many points each part of the question is worth, and how many submissions are allowed for each part if you can submit each part separately.

Submit All New Answers

This submits all of your new answers for all of the questions in the assignment.

Save All Work

This allows you to save all the work you've done on all of the questions in the assignment, but does not submit your work for grading.

Ask Your Teacher

This feature provides a pop-up window in which you can enter and send a question about the assignment to your instructor.

Extension Request

This link gives you a pop-up window allowing you to enter and submit a request to your instructor for an extension of time on an assignment.

Home

This link takes you to your personalized Enhanced WebAssign Home page.

My Assignments

This link takes you to your assignments page.

Open Math Palette

This opens a tool to use in writing answers that require math notation.

Read it

This links to question-specific textbook material in PDF form.

Practice it

> This links to a practice problem or set of practice problems in a pop-up window. No grade is recorded on the work you do on practice problems.

See it

> This links to a tutorial video.

Hint

> This links to a pop-up window with helpful hints in case you get stuck on a question.

Hint: Active Figure

> This links to an animated simulation to help you better understand the concepts being covered.

Note: Your instructor has the ability to turn on/off many of the options listed above.

ANSWERING QUESTIONS

Enhanced WebAssign uses a variety of question types that you're probably already familiar with using, such as multiple choice, true/false, free response, etc.

Always be sure to pay close attention to any instructions within the question regarding how you are supposed to submit your answers.

Numerical Questions

There are a few key points to keep in mind when working on numerical questions:

- ○ Numbers can be entered in both scientific notation and numerical expressions, such as fractions.

- ○ WebAssign uses the standard scientific notation "E" or "e" for "times 10 raised to the power." (Note: both uppercase E and lowercase e are acceptable in WebAssign.) For example, 1e3 is the scientific notation for 1000.

○ Numerical answers may *not* contain commas (,) or equal signs (=).

○ Numerical answers may only contain:

- Numbers

- E or e for scientific notation

- Mathematical operators +, -, *, /

○ Numerical answers within 1% of the actual answer are counted as correct, unless your instructor chooses a different tolerance. This is to account for rounding errors in calculations. In general, enter three significant figures for numerical answers.

➤ Example: Numerical Question

Let's suppose you're presented a question to which your answer is the fraction "one over sixty-four." Following are examples of Correct and Incorrect answer formats:

Correct Answers

Any of these formats would be correct:

1/64

0.015625

0.0156

.0156

1.5625E-2

Incorrect Answers

These formats would be graded as incorrect:

O.015625	The first character is the letter "O"
0. 015625	There is an improper space in the answer
1.5625 E-2	There is an improper space using E notation
l/64	The first character is lowercase letter "L"
5,400	There is a comma in the answer
1234.5=1230	There is an equal sign in the answer

Numerical Questions with Units

Some Enhanced WebAssign questions require a number and a unit, and this is generally, although not always, indicated in the instructions in the question.

You will know that a unit is expected when there is no unit after the answer box.

When you are expected to enter units and do not, you will get an error message telling you that units are required.

Note: Whether omission of the unit counts as a submission depends on the submission options chosen by the instructor.

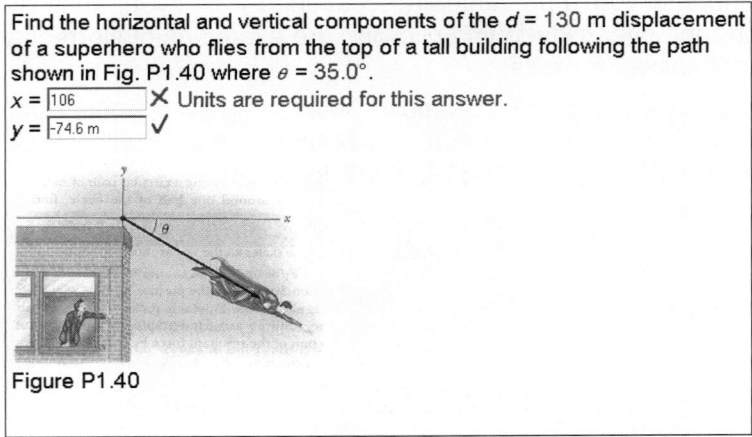

Numerical with units

The easiest units to use in this question are m, but the answer converted to yd would also be scored correct.

Numerical Questions with Significant Figures

Some numerical questions require a specific number of significant figures (sig figs) in your answer. If a question checks sig figs, you will see a sig fig icon next to the answer box.

If you enter the correct value with the wrong number of sig figs, you will not receive credit, but you will receive a hint that your number does not have the correct number of sig figs. The sig fig icon 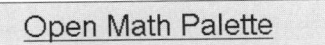 is also a link to the rules used for sig figs in WebAssign.

Carry out the following arithmetic operations. (Use the correct number of significant figures.)

(a) the sum of the measured values 760., 37.2, 0.81, and 2.2

`4.0✓` `8e2`   ✗ Check the number of significant figures.

(b) the product 3.4 × 3.563

`4.0✓` `12`   ✓

(c) the product 5.7 × π

`4.0✓` `18`   ✓

Check for significant figures

Math Notation: Using the Math Palette

In many math questions, Enhanced WebAssign gives you an answer box with a **Math Palette** button. The **Math Palette** provides easy input of math answers, even the more complicated ones.

Open Math Palette

Math Palette button

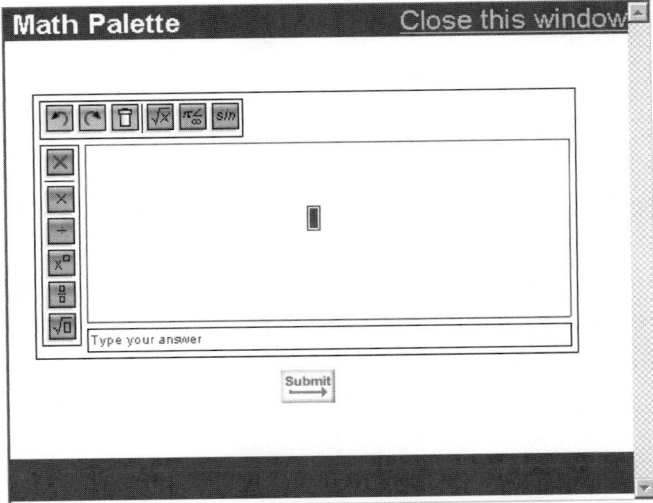

Math Palette tool

Top Symbols

The **Math Palette** has a toolbar of symbols on top that, when selected, give you a drop-down menu with more symbols from which to choose.

Side Symbols

The buttons on the side are single input buttons for frequently used operations.

The yellow buttons are editing buttons:

↺ ↻	Arrow keys are for undo and redo
🗑	Trashcan is to clear your input and start over with an empty field
✕	The red "x" is used as a short cut for the answer "no solution."

After using the **Math Palette** to write your answer, click the **Submit** button. Your answer will appear in the appropriate boxed area with the question. Your answer will be graded once you actually submit your answers for grading.

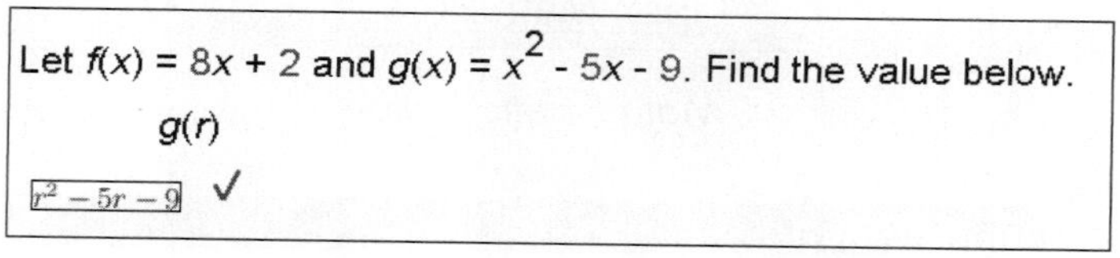

Let $f(x) = 8x + 2$ and $g(x) = x^2 - 5x - 9$. Find the value below.

$g(r)$

$\boxed{r^2 - 5r - 9}$ ✓

After using Math Palette

Math Notation: Using the Keyboard

If you use your keyboard to enter math notation (calculator notation), *you must use the exact variables specified in the questions.*

The order is not important, as long as it is mathematically correct.

➢ Example: Math Notation Using Keyboard

In the example below, the keyboard is used to enter the answer in the answer field.

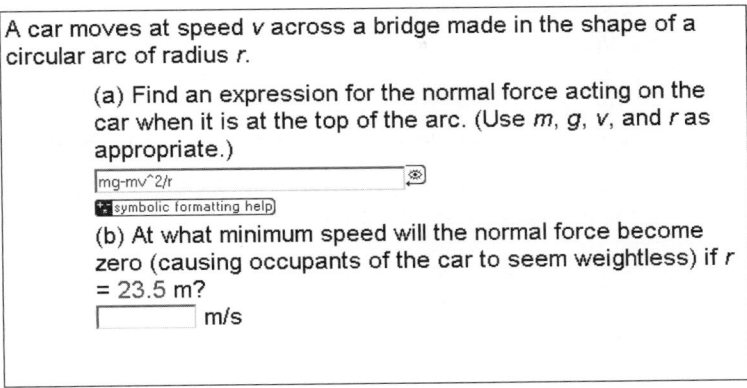

A car moves at speed *v* across a bridge made in the shape of a circular arc of radius *r*.

(a) Find an expression for the normal force acting on the car when it is at the top of the arc. (Use *m*, *g*, *v*, and *r* as appropriate.)

mg-mv^2/r

symbolic formatting help

(b) At what minimum speed will the normal force become zero (causing occupants of the car to seem weightless) if *r* = 23.5 m?

☐ m/s

Symbolic question

Expression Preview

Clicking the eye button ⊛ allows you to preview the expression you've entered in calculator notation.

Use this preview feature to help determine if you have properly placed your parentheses.

Symbolic Formatting Help

If you're unsure about how to symbolically enter your answer properly, use the **symbolic formatting help** button to display allowed notation.

Allowed notation for symbolic formatting

+ for addition	x+1
- for subtraction	x-1, or −x
* or nothing for multiplication	4*x, or 4x
/ for division	x/4
** or ^ for exponential	x**3, or x^3
( ) where necessary to group terms	4/(x+1), or 3(x+1)
abs( ) to take the absolute value of a variable or expression	abs(-5) = 5
sin, cos, tan, sec, csc, cot, asin, acos, atan functions (angle x expressed in radians)	sin(2x)
sqrt( ) for square root of an expression	sqrt(x/5)
x^ (1/n) for the nth root of a number	x^ (1/3), or (x-3)^ (1/5)
pi for 3.14159…	2 pi x
e for scientific notation	1e3 = 1000
ln( ) for natural log	ln(x)
exp( ) for "e to the power of"	exp(x) = e^x

ADDITIONAL FEATURES

Calendar

The Calendar link presents you with a calendar showing all of your assignments on their due dates. You can also click on any date and enter your own personal events.

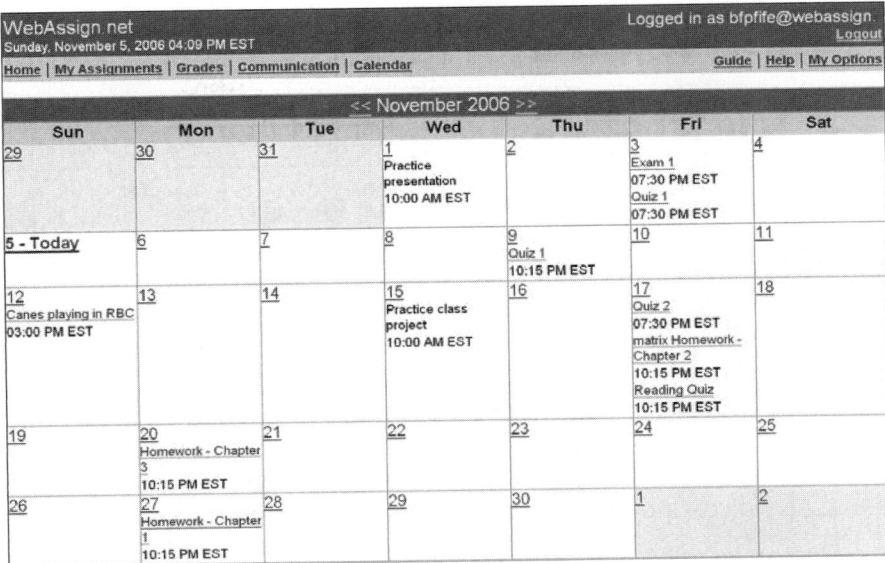

Calendar

Communication

The Communication link gives you access to **Private Messages** and course **Forums**, if your instructor has enabled these features.

Forums

The **Forums** are for discussions with all the members of your class. Your instructor can create forums, and you can create topics within a forum or contribute to a current topic.

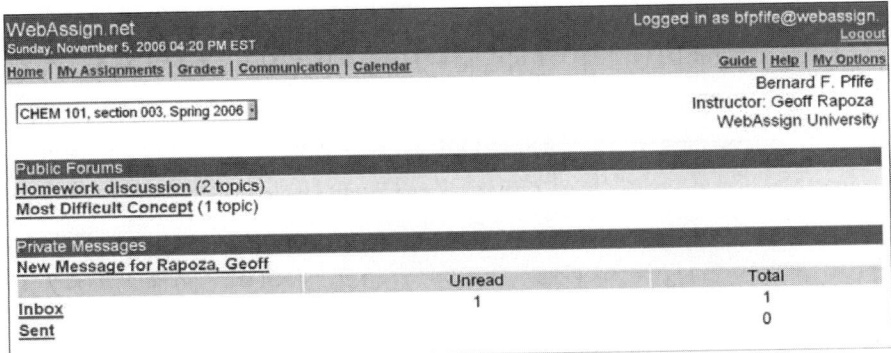

Communication link

Private Messages

Private Messages are for communication between you and your instructor. If your instructor has enabled private messages, click the **New Message** link to send your instructor a message.

GRADES

The **Grades** link at the top of all your WebAssign pages gives you access to the raw scores and grades that your instructor posts. This page may also include statistics on the whole class, and a histogram of scores for each category of assignment and each individual assignment. It may have your individual average for each category of assignment, as well as the score on each of your assignments.

Your instructor will let you know what Scores and Grades will be posted in your course.

If your instructor has enabled all of the options, your display will be similar to the one below.

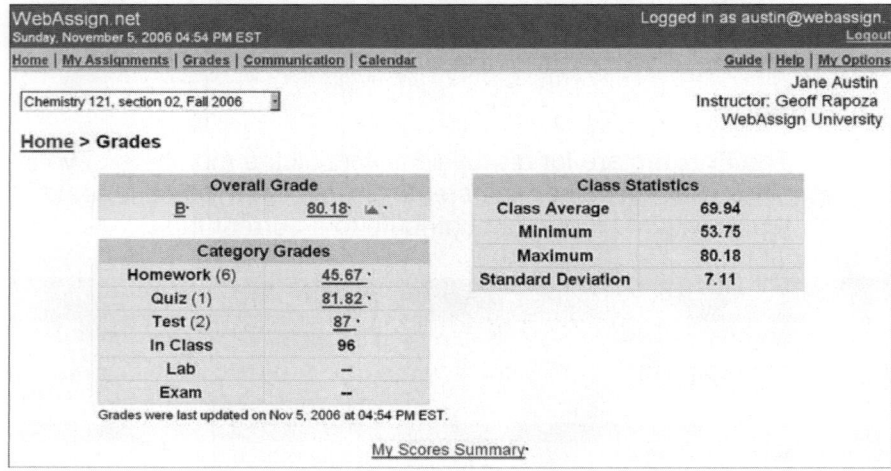

Grades

Overall Grade

This score is calculated from the various categories of assignments, for example, **Homework, Test**, **In Class**, **Quiz**, **Lab**, and **Exam**. Your instructor may have different categories.

Category Grades

The **Category Grades** give the contribution to your overall grade from each of the categories. If you click a grade that is a link, you will get a pop-up window explaining how the number was calculated.

Class Statistics

Class Statistics shows the averages, minimum scores, maximum scores, and standard deviation of the class at large.

My Scores Summary

This link presents a pop-up window with a summary of your raw scores and the class statistics on each assignment, if your teacher has posted these.

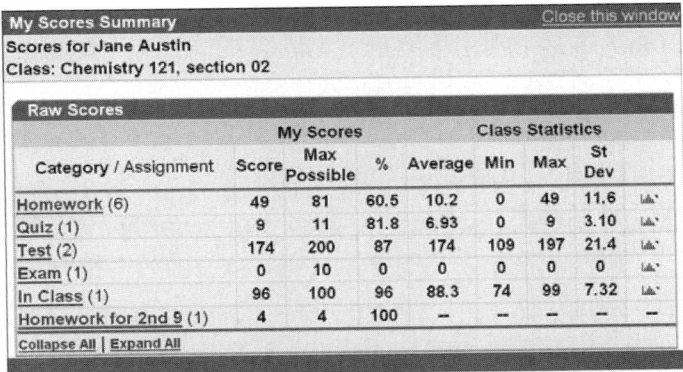

My Scores summary

TECHNICAL TIPS

Enhanced WebAssign relies on web browsers and other related technology that can lead to occasional technical issues. The following technical tips can help you avoid some common problems.

Cookies

Allow your browser to accept cookies.

WebAssign will work if you set your browser to not accept cookies; however, if an encrypted cookie is not saved to your computer during your session, you may be asked to login again more frequently. Once you logout and exit your browser, the cookie is deleted.

Login and Credit

If you see an assignment that does not have your name at the top, you have not logged in properly.

You will not receive credit for any work you do on an assignment if your name is not associated with it. If you find yourself in the midst of this situation, make notes of your solution(s) and start over. Be aware that any randomized values in the questions will probably change.

Logout When You Finish Your Session

If someone wants to use your computer for WebAssign, logout and exit the browser before relinquishing control.

Otherwise, the work you have just completed may be written over by the next user.

**For technical support go to
http://webassign.net/student.html
or email support@webassign.net.**

Server

Although it is very rare, the WebAssign server may occasionally be unavailable.

If the WebAssign server is unavailable, instructors will provide instructions for submitting your assignments—possibly including new due dates. The policy for handling server problems will vary from instructor to instructor.

Use the Latest Browser Software

Use the latest version of Firefox, Mozilla, Netscape, or Internet Explorer browsers.

Older versions of browsers may not be supported by WebAssign.

For technical support go to
http://webassign.net/student.html
or email support@webassign.net.

Notes

Notes

Notes

Notes

Notes

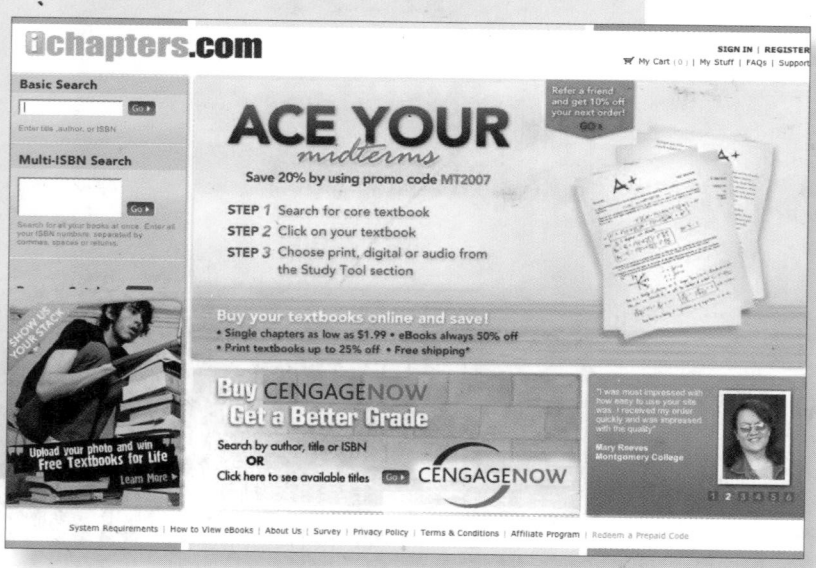